AMPHIBIAN DECLINES

AMPHIBIAN DECLINES

THE CONSERVATION STATUS OF UNITED STATES SPECIES

Edited by

MICHAEL LANNOO

UNIVERSITY OF CALIFORNIA PRESS
Berkeley   Los Angeles   London

University of California Press
Berkeley and Loss Angeles, California

University of California Press, Ltd.
London, England

Library of Congress Cataloging-in-Publication Data

Amphibian declines: The conservation status of United States species

   / edited by Michael Lannoo.
      p. cm.

   Includes bibliographical references and index.
   ISBN 0-520-23592-4 (cloth : alk. paper)
   1. Amphibian declines. 2. Amphibian declines—United States. I.
Lannoo, Michael
   QL644.7.A48 2005
   333.95'78—dc22

2004015272

Printed in Canada

10   09   08   07   06   05
10   9   8   7   6   5   4   3   2   1

We dedicate this effort to David Wake, George Rabb, and Whit Gibbons, around whose energy and insight and influence we in amphibian conservation have gathered. The story of how Dave saw the pattern in the early scattered reports of amphibian declines, organized herpetologists, and helped to create the Declining Amphibian Populations Task Force is part of our folklore. We continue to rely on his wisdom and guidance. George, who furiously works both in front of and behind the scenes, is truly one of the great conservation biologists of our time. Whit's kind and infectious enthusiasm mobilizes both scientists and lay people alike. They are the giants upon whose shoulders we now stand.

We also dedicate this effort to the memories of two great mentors from an earlier generation, who passed when this book was in its early stages: Sherman Minton, Jr. (1920–1999), whose last published paper is included here; and Richard V. Bovbjerg (1919–1999), who in 1964 was one of the first scientists to speak to the issue of amphibian conservation.

CONTENTS

ACKNOWLEDGMENTS

This work truly represents an effort from the community of United States herpetologists, and I thank each and every one of the 215 contributors to this volume for their efforts, for working with me, and for sharing my views on the necessity and urgency of this effort. I am humbled. I also thank a subset of these workers, my Advisory Board, for the guidance and perspective they have offered. Thanks also to Norm Scott, Ronn Altig, and Erin E. Clark for their constructive comments on an earlier draft and to Floyd Scott for tracking down citations.

Many other herpetologists have assisted, including Kraig Adler, James Andrews, Tom Anton, Richard Baker, Val Beasley, Robert Bezy, Sean Bloomquist, James Bogart, Alvin Breisch, Janalee Caldwell, Cynthia Carey, Isaac Chellman, James Collins, Joseph Collins, Jeffrey Davis, Phillip deMaynadier, Katie Distler, James Dixon, Sara E. Faust, John Ferner, Suzanne Fowle, Lindsay D. Funk, Alisa Gallant, Carl Gerhardt, Harry Green, Dave Golden, Robert H. Goodman Jr., Tim Halliday, Dan Helsel, Laura Herbeck, Richard Highton, Julia Holloway, David Hoppe, Arthur C. Hulse, Erin Hyde, John Iverson, Robert Jaeger, Tom R. Johnson, John Jones, Robin E. Jung, Daryl Karns, Robert Klaver, Jon Klingel, Ted Koch, Ken Lang, William Leonard, Lauren J. Livo, Timothy Maret, Bryce A. Maxell, Kelly McAllister, Roy McDiarmid, Sherman A. Minton, David J. Morafka, John Moriarty, Stephen Morreale, David Morris, Michael Mossman, Robert W. Murphy, David Naugle, Holly Niederriter, George Oliver, Deanna Olson, Stanley Orchard, Joseph H.K. Pechmann, Stephen A. Perrill, Charles Peterson, James Petranka, Tom Pluto, Andrew H. Price, John O. Reiss, Alan Resetar, Alan Savitzky, Greg Schneider, A. Floyd Scott, Stan Sessions, Doug Siegel, Melody Stoneham, Bruce Taubert, Howard Whiteman, Jill A. Wicknick, David Withers, and Richard Wassersug. I thank Jim Petranka for his fine book, and the organizers and the numerous contributors of the Catalogue of American Amphibians and Reptiles—you have made a portion of our job much easier.

This project has been supported by funding provided by the following agencies or organizations: National Fish and Wildlife Foundation; Disney Wildlife Conservation Fund; Partners in Amphibian and Reptile Conservation (PARC); USGS National Mapping Division; USGS EROS Data Center; Office of the Provost (Warren Vander Hill) and Faculty Publications Committee (Ione DeOllos), Ball State University; Office of Academic Research and Sponsored Programs (James Pyle, Kristi Koriath), Ball State University; Virginia Herpetological Society; City of Elkhart (Indiana) Environmental Science Center; and Linn County (Iowa) Environmental Council. I am grateful. I am also grateful for the following technical support I have received from the staff at Ball State University: Contracts and Grants Office (Sharon Armbrust, Sharon Harris) and Library Services (Jan Vance).

The research in this volume was supported by an array of federal (e.g., The National Science Foundation and the USGS's Amphibian Research and Monitoring Initiative), state, and private organizations. The authors acknowledge and are grateful for this support. Similarly, the authors acknowledge and thank the many people that contributed their unpublished observations, local knowledge, and critical comments.

This project could not have been done without the talents and dedication of my team: Priya Nanjappa, Laura Blackburn, Laura Guderyahn, Donna Helfst, Molly Schaller, Jason Hall, Marti Tiedeman, Chris Lawhorn, and Susan Johnson Lannoo. Thanks to you all. And finally, a special thanks to Dick and Debby Baker, Gary Casper, Bob Cruden, Bill Cummings, Joe Eastman, Alisa Gallant, Ken Lang, Susie Lannoo, Barb Mendenhall, Hank Miguel, Ole Oldefest, Jane Shuttleworth, Bill Souder, Dan Sutherland, Arnold van der Valk, and Judy and Mark Wehrspann—each a biologist in their own way. They have learned, independent of each other, that the necessity to live life is at least as important as the drive to study life; and that's why we are friends.

ADVISORY BOARD

CONTRIBUTORS

MICHAEL J. ADAMS, USGS Forest and Rangeland Ecosystem Science Center, Corvallis, Oregon

CARL D. ANTHONY, Department of Biology, John Carroll University, University Heights, Ohio

ELIZABETH ARBAUGH, Associate Curator of Education, Detroit Zoological Institute (DZI), Royal Oak, Michigan

SARAH AUCOIN, Urban Park Rangers, New York, New York

DOUGLAS C. BACKLUND, South Dakota Natural Heritage Program, South Dakota Department of Game Fish and Parks, Pierre, South Dakota

MARK A. BAILEY, Conservation Services Southeast, Shorter, Alabama

CHRISTOPHER K. BEACHY, Department of Biology, Minot State University, Minot, North Dakota

DAVID A. BEAMER, Department of Biology, East Carolina University, Greenville, North Carolina

VAL R. BEASLEY, College of Veterinary Medicine, University of Illinois, Urbana, Illinois

LISA K. BELDEN, Department of Zoology, Oregon State University, Corvallis, Oregon

LAURA BLACKBURN, Great Lakes Commission, Ann Arbor, Michigan

ANDREW BLAUSTEIN, Department of Zoology, Oregon State University, Corvallis, Oregon

SEAN M. BLOMQUIST, Nongame Branch, Arizona Game and Fish Department, Phoenix, Arizona

RONALD M. BONETT, Department of Biology, University of Texas at Arlington, Arlington, Texas

JEFF BOUNDY, Louisiana Department of Wildlife and Fisheries, Baton Rouge, Louisiana

DAVID F. BRADFORD, USEPA National Exposure Research Lab, Las Vegas, Nevada

RONALD A. BRANDON, Department of Zoology, Southern Illinois University, Carbondale, Illinois

ALVIN L. BRASWELL, North Carolina State Museum of Natural Sciences, Raleigh, North Carolina

CHRISTINE M. BRIDGES, USGS Columbia Environmental Research Center, Columbia, Missouri

ROBERT BRODMAN, Biology Department, Saint Joseph's College, Rensselaer, Indiana

LAUREN E. BROWN, Department of Biological Sciences, Illinois State University, Normal, Illinois

STEPHEN R. BURTON, Department of Biological Sciences and Idaho Museum of Natural History, Idaho State University, Pocatello, Idaho

R. BRUCE BURY, USGS Forest and Rangeland Ecosystem Science Center, Corvallis, Oregon

BRIAN P. BUTTERFIELD, Department of Biology, Freed–Hardeman University, Henderson, Tennessee

CARLOS D. CAMP, Department of Biology, Piedmont College, Demorest, Georgia

CYNTHIA CAREY, Department of Environmental, Population, and Organismic Biology, University of Colorado, Boulder, Colorado

DEBRA L. CARLSON, Department of Biology, Augustana College, Sioux Falls, South Dakota

GARY S. CASPER, Milwaukee Public Museum, Milwaukee, Wisconsin

PAUL T. CHIPPINDALE, Department of Biology, University of Texas at Arlington, Arlington, Texas

GEORGE R. CLINE, Biology Department, Jacksonville State University, Jacksonville, Alabama

NICHOLAS COHEN, Department of Microbiology and Immunology, University of Rochester School of Medicine and Dentistry, Rochester, New York

REBECCA COLE, USGS National Wildlife Health Center, Madison, Wisconsin

JAMES P. COLLINS, Department of Biology, Arizona State University, Tempe, Arizona

CHRISTOPHER CONNER, Department of Biology, Butler University, Indianapolis, Indiana

PAUL STEPHEN CORN, USGS Aldo Leopold Wildlife Research Institute, Missoula Field Station, Missoula, Montana

JOHN A. CRAWFORD, Division of Biological Sciences, University of Missouri, Columbia, Missouri

JOHN J. CRAYON, USGS Western Ecological Research Center, Department of Biology, University of California, Riverside, California

MARTHA L. CRUMP, Department of Biological Sciences, Northern Arizona University, Flagstaff, Arizona

CARLOS DAVIDSON, Environmental Studies Department, California State University, Sacramento, California

ELIZABETH W. DAVIDSON, Department of Biology, Arizona State University, Tempe, Arizona

JEFFREY G. DAVIS, Geier Collections and Research Center, Museum of Natural History and Science, Cincinnati Museum Center, Cincinnati, Ohio

GAGE H. DAYTON, Department of Wildlife and Fisheries, Texas Cooperative Wildlife Collections, Texas A&M University, College Station, Texas

C. KENNETH DODD, JR., USGS Florida Integrated Science Centers, Gainesville, Florida

MAUREEN A. DONNELLY, College of Arts & Sciences, Florida International University, North Miami, Florida

BROOKE A. DOUTHITT, Department of Biological Sciences, Butler University, Indianapolis, Indiana

DARRIN DOYLE, Department of Biological Sciences, Humboldt State University, Arcata, California

SAM DROEGE, USGS Patuxent Wildlife Research Center, Laurel, Maryland

HAROLD A. DUNDEE, Tulane Museum of Natural History, Tulane University, Belle Chasse, Louisiana

PAIGE EAGLE, USGS Patuxent Wildlife Research Center, Laurel, Maryland

JOYCE EISOLD, Department of Veterinary Biosciences, College of Veterinary Medicine, University of Illinois, Urbana, Illinois

EDWARD L. ERVIN, USGS Western Ecological Research Center, San Diego, California

MICHAEL A. EWERT, Department of Biology, Indiana University, Bloomington, Indiana

SANDRA A. FAEH, Department of Veterinary Biosciences, College of Veterinary Medicine, University of Illinois, Urbana, Illinois

EUGENIA FARRAR, Department of Zoology and Genetics, Iowa State University, Ames, Iowa

ZACHARY I. FELIX, Center for Forestry and Ecology, Alabama A&M University, Normal, Alabama

GARY M. FELLERS, USGS Western Ecological Research Center, Point Reyes National Seashore, Point Reyes, California

DANTE B. FENOLIO, Department of Zoology, University of Oklahoma, Norman, Oklahoma

JOHN W. FERNER, Department of Biology, Thomas More College, Crestview Hills, Kentucky, and Geier Collections and Research Center, Museum of Natural History and Science, Cincinnati Museum Center, Cincinnati, Ohio

KIMBERLEIGH J. FIELD, Arizona Game and Fish Department, Phoenix, Arizona

TATE D. FISCHER, Department of Wildlife and Fisheries Sciences, South Dakota State University, Brookings, South Dakota, and AFMO Fuels, Chippewa National Forest, Walker, Minnesota

M. J. FOUQUETTE, JR., Department of Biology, Arizona State University, Tempe, Arizona

SUZANNE C. FOWLE, Natural Heritage and Endangered Species Program, Massachusetts Division of Fisheries and Wildlife, Westborough, Massachusetts

JOE N. FRIES, USFWS San Marcos National Fish Hatchery and Technology Center, San Marcos, Texas

JULIE A. FRONZUTO, School of Biological Sciences, Washington State University, Pullman, Washington

DARREL R. FROST, Division of Vertebrate Zoology (Herpetology), American Museum of Natural History, New York, New York

ALISA L. GALLANT, USGS EROS Data Center, Sioux Falls, South Dakota

ERIK W.A. GERGUS, Department of Biology, Glendale Community College, Glendale, Arizona

STEVE GIAMBRONE, Department of Philosophy, University of Louisiana, Lafayette, Louisiana

WHIT GIBBONS, University of Georgia, Savannah River Ecology Laboratory, Aiken, South Carolina

ANNA GOEBEL, Department of Environmental, Population, and Organismic Biology, University of Colorado, Boulder, Colorado, and Ponto Gorda, Florida

CAREN S. GOLDBERG, Wildlife and Fisheries Science, University of Arizona, Tucson, Arizona

ROBERT H. GOODMAN, JR., Biological Sciences Department, Citrus College, Glendora, California

BRENT M. GRAVES, Department of Biology, Northern Michigan University, Marquette, Michigan

ROBERT H. GRAY, R.H. Gray and Associates, Richland, Washington

DAVID M. GREEN, Redpath Museum, McGill University, Montreal, Québec, Canada

MARTIN GREENWELL, John G. Shedd Aquarium, Chicago, Illinois

CRAIG GUYER, Department of Biological Sciences, Auburn University, Auburn, Alabama

STEPHEN F. HALE, Tucson, Arizona

TIM HALLIDAY, International Director, Declining Amphibian Populations Task Force, Department of Biological Sciences, The Open University, Milton Keynes, United Kingdom

ROBERT HANSEN, Watershed Protection Department, City of Austin, Texas

ROBERT W. HANSEN, Clovis, California

REID N. HARRIS, Department of Biology, James Madison University, Harrisonburg, Virginia

JULIAN R. HARRISON, Department of Biology, College of Charleston, Charleston, South Carolina

MARC P. HAYES, Washington Department of Fish and Wildlife, Olympia, Washington

RUSS HENDRICKS, Lake County Forest Preserve District, Libertyville, Illinois

K. HENLE, Department of Conservation Biology and Natural Resources, UFZ Leipzig-Halle GmbH, Leipzig, Germany

JEAN-MARC HERO, School of Environmental and Applied Sciences, Griffith University, Queensland, Australia

ROBERT E. HERRINGTON, Department of Biology, Georgia Southwestern State University, Americus, Georgia

JANE HEY, Morningside College, Sioux City, Iowa

W. RONALD HEYER, Department of Vertebrate Zoology, National Museum of Natural History, Smithsonian Institution, Washington, D.C.

KENNETH F. HIGGINS, USGS and South Dakota Cooperative Fish and Wildlife Research Unit, South Dakota State University, Brookings, South Dakota

RICHARD HIGHTON, Department of Biology, University of Maryland, College Park, Maryland

DAVID M. HOPPE, Division of Science and Mathematics, University of Minnesota, Morris, Minnesota

W. JEFFREY HUMPHRIES, Clemson University, Department of Forest Resources, Clemson, South Carolina

TODD W. HUNSINGER, Research and Collections, New York State Museum, Albany, New York

ERIN J. HYDE, Department of Zoology, North Carolina State University, Raleigh, North Carolina, and USGS Forest and Rangeland Ecosystem Science Center Corvallis, Oregon

JASON T. IRWIN, Redpath Museum, McGill University, Montreal, Québec, Canada, and Department of Biology, Bucknell University, Lewisburg, Pennsylvania

KELLY J. IRWIN, Arkansas Game and Fish Commission, Benton, Arkansas

JEF R. JAEGER, Department of Biological Sciences, University of Nevada–Las Vegas Las Vegas, Nevada

ROBERT G. JAEGER, Department of Biology, University of Louisiana, Lafayette, Louisiana

RANDY D. JENNINGS, Department of Natural Sciences, Western New Mexico University, Silver City, New Mexico

JOHN B. JENSEN, Nongame-Endangered Wildlife Program, Georgia Department of Natural Resources, Forsyth, Georgia

PIETER T. J. JOHNSON, Center for Limnology, University of Wisconsin, Madison, Wisconsin

REX R. JOHNSON, USFWS Habitat and Population Evaluation Team, Fergus Falls, Minnesota

STEVE A. JOHNSON, USGS Florida Integrated Science Centers, Gainesville, Florida and Department of Wildlife Ecology and Conservation, University of Florida, Plant City, Florida

LAWRENCE L.C. JONES, Pacific Northwest Research Station, USDA Forest Service, Olympia, Washington, and Coronado National Forest, Tucson, Arizona

MARK S. JONES, Colorado Division of Wildlife, Fort Collins, Colorado

FRANK W. JUDD, University of Texas–Pan American, Edinburg, Texas

J. ERIC JUTERBOCK, Department of Evolution, Ecology, and Organismal Biology, Ohio State University, Lima, Ohio

NANCY E. KARRAKER, USDA Forest Service, Redwood Sciences Laboratory, Arcata, California, and Department of Environmental and Forest Biology, State University of New York, Syracuse, New York

EDWARD D. KOCH, U.S. Fish and Wildlife Service, Pacific Region, Snake River Field Office, Boise, Idaho

KENNETH H. KOZAK, Department of Biology, Washington University, St. Louis, Missouri

FRED KRAUS, Department of Natural Sciences, Bishop Museum, Honolulu, Hawaii

JAMES J. KRUPA, University of Kentucky, School of Biological Sciences, Lexington, Kentucky

PAUL J. KRUSLING, Geier Collections and Research Center, Museum of Natural History and Science, Cincinnati Museum Center, Cincinnati, Ohio

SHAWN R. KUCHTA, Museum of Vertebrate Zoology, Department of Integrative Biology, University of California, Berkeley, California, and Department of Ecology and Evolutionary Biology, University of California at Santa Cruz, Santa Cruz, California

MICHAEL LANNOO, Muncie Center for Medical Education, Indiana University School of Medicine, Ball State University, Muncie, Indiana

JAMES LAZELL, The Conservation Agency, Jamestown, Rhode Island

WILLIAM T. LEJA, Chicago, Illinois

KAREN R. LIPS, Department of Zoology and Center for Systematic Biology, Southern Illinois University, Carbondale, Illinois

LAUREN J. LIVO, Department of Environmental, Population, and Organismic Biology, University of Colorado, Boulder, Colorado, and Colorado Division of Wildlife, Fort Collins, Colorado

CHARLES W. LOEFFLER, Species Conservation Section, Colorado Division of Wildlife, Denver, Colorado

KIRK LOHMAN, Department of Fish and Wildlife, University of Idaho, Moscow, Idaho

JOYCE E. LONGCORE, Department of Biological Sciences, University of Maine, Orono, Maine

KEVIN B. LUNDE, W.M. Keck Science Center, Claremont, California

JOHN H. MALONE, Department of Zoology, University of Oklahoma, Norman, Oklahoma, and Department of Biology, University of Texas-Arlington, Arlington, Texas

SHARYN B. MARKS, Department of Biological Sciences, Humboldt State University, Arcata, California

TIMOTHY O. MATSON, Department of Vertebrate Zoology, Cleveland Museum of Natural History, Cleveland, Ohio

BEVERLY K. MCKINNELL, Department of Genetics, Cell Biology, and Development, University of Minnesota, S. Paul, Minnesota

ROBERT G. MCKINNELL, Department of Genetics, Cell Biology, and Development, University of Minnesota, S. Paul, Minnesota

DAVID S. MCLEOD, School of Biological Sciences, University of Nebraska–Lincoln, Lincoln, Nebraska, and Museum of Natural History, University of Kansas, Lawrence, Kansas

D. BRUCE MEANS, Coastal Plains Institute and Land Conservancy, Tallahassee, Florida

SCOTT M. MELVIN, Natural Heritage and Endangered Species Program, Massachusetts Division of Fisheries and Wildlife, Westborough, Massachusetts

JOSEPH R. MENDELSON III, Department of Biology, Utah State University, Logan, Utah

WALTER E. MESHAKA, JR., Everglades Regional Collections Center, Everglades National Park, Homestead, Florida, and Section of Zoology and Botany, State Museum of Pennsylvania, Harrisburg, Pennsylvania

ANN MESROBIAN, Bastrop County Environmental Network, Bastrop, Texas

SHERMAN A. MINTON, Indiana University School of Medicine, Indianapolis, Indiana

JOSEPH C. MITCHELL, Department of Biology, University of Richmond, Richmond, Virginia

PAUL E. MOLER, Florida Fish and Wildlife Conservation Commission, Gainesville, Florida

DAVID J. MORAFKA, Department of Biology, California State University, Dominguez Hills, Carson, California

STEVEN R. MOREY, U.S. Fish and Wildlife Service, Portland, Oregon

EMILY MORIARTY, University of Texas, Texas Memorial Museum, Austin, Texas

MICHAEL J. MOSSMAN, Wisconsin Department of Natural Resources, Monona, Wisconsin

JENNIFER MUI, Center for Biodiversity, Illinois Natural History Survey, Champaign, Illinois, and Division of Herpetology, Field Museum of Natural History, Chicago, Illinois

JAMES B. MURPHY, Department of Herpetology, National Zoological Park, Smithsonian Institution, Washington, D.C.

ROBERT W. MURPHY, Centre for Biodiversity and Conservation Biology, Royal Ontario Museum, Toronto, Ontario, Canada

TONY P. MURPHY, A Thousand Friends of Frogs, Project Center for Global Environmental Education, Hamline University, St. Paul, Minnesota, and Education Department, The College of St. Catherine, St. Paul, Minnesota

ERIN MUTHS, USGS Biological Resources Division, Fort Collins Science Center, Fort Collins, Colorado

PRIYA NANJAPPA, USGS Patuxent Wildlife Research Center, Laurel, Maryland

DAVID E. NAUGLE, Fish and Wildlife Biology Program, University of Montana, Missoula, Montana

DONALD NICHOLS, Department of Pathology, National Zoological Park, Smithsonian Institution, Washington, D.C.

R. ANDREW ODUM, Department of Herpetology, Toledo Zoological Society, Toledo, Ohio

RICHARD B. OWEN, Florida Department of Environmental Protection, Guana River State Park, Ponte Vedra Beach, Florida, and Wekiva River Basin State Parks, Apopka, Florida

CHARLES W. PAINTER, Endangered Species Program, New Mexico Department of Game and Fish, Santa Fe, New Mexico

JOHN G. PALIS, Jonesboro, Illinois

THEODORE J. PAPENFUSS, Museum of Vertebrate Zoology, University of California, Berkeley, California

DUNCAN PARKS, Museum of Vertebrate Zoology, Department of Integrative Biology, University of California, Berkeley, California, and Portland, Oregon

MATTHEW J. PARRIS, Department of Biology, University of Memphis, Memphis, Tennessee

DEBRA A. PATLA, Department of Biological Sciences and Idaho Museum of Natural History, Idaho State University, Pocatello, Idaho

BETH ANNE PAULEY, Department of Biology, Marshall University, Huntington, West Virginia

THOMAS K. PAULEY, Department of Biological Sciences, Marshall University, Huntington, West Virginia

CHRISTOPHER A. PEARL, USGS Forest and Rangeland Ecosystem Science Center, Corvallis, Oregon

CHARLES R. PETERSON, Department of Biological Sciences, and Idaho Museum of Natural History, Idaho State University, Pocatello, Idaho

CHRISTOPHER A. PHILLIPS, Center for Biodiversity, Illinois Natural History Survey, Champaign, Illinois

DAVID S. PILLIOD, Department of Biological Sciences, Idaho State University, Pocatello, Idaho, and Aldo Leopold Wilderness Research Institute, USDA Forest Service, Missoula, Montana

ANDREW H. PRICE, Texas Parks and Wildlife Department, Austin, Texas

FRED PUNZO, Department of Biology, University of Tampa, Tampa, Florida

FRANK R. QUAMEN, Fish and Wildlife Biology Program, University of Montana, Missoula, Montana

HUGH R. QUINN, Cleveland Metroparks Zoo, Cleveland, Ohio

TIMOTHY QUINN, Habitats Division, Washington Department of Fish and Wildlife, Olympia, Washington

CINDY RAMOTNIK, USGS Fort Collins Science Center and Department of Biology, University of New Mexico, Albuquerque, New Mexico

JAMIE K. REASER, Ecos Systems Institute, Springfield, Virginia

MICHAEL REDMER, U.S. Fish and Wildlife Service, Chicago Field Office, Barrington, Illinois

ROCHELLE B. RENKEN, Missouri Department of Conservation, Columbia, Missouri

STEPHEN C. RICHTER, Department of Zoology and Sam Noble Oklahoma Museum of Natural History, University of Oklahoma, Norman, Oklahoma Department of Biological Sciences, Eastern Kentucky University, Richmond, Kentucky

JAMES C. RORABAUGH, U.S. Fish and Wildlife Service, Phoenix, Arizona

TRAVIS J. RYAN, Department of Biological Sciences, Butler University, Indianapolis, Indiana

WESLEY K. SAVAGE, Section of Evolution and Ecology, University of California, Davis, California

ANNA M. SCHOTTHOEFER, Department of Pathobiology, College of Veterinary Medicine, University of Illinois, Urbana, Illinois

CECIL R. SCHWALBE, USGS Western Ecological Research Center, Sonoran Desert Field Station, University of Arizona, Tucson, Arizona

TERRY D. SCHWANER, Department of Biology, Southern Utah University, Cedar City, Utah, and Department of Biology, North Georgia College & State University, Dahlonega, Georgia

COLLEEN SCOTT, Cleveland Metroparks Zoo, Cleveland, Ohio

DAVID E. SCOTT, Savannah River Ecology Laboratory, Aiken, South Carolina

RAYMOND D. SEMLITSCH, Division of Biological Sciences, University of Missouri, Columbia, Missouri

DAVID M. SEVER, Department of Biology, Saint Mary's College, Notre Dame, Indiana

H. BRADLEY SHAFFER, Section of Evolution and Ecology, University of California, Davis, California

DONALD B. SHEPARD, Center for Biodiversity, Illinois Natural History Survey, Champaign, Illinois, and Department of Zoology, and Sam Noble Oklahoma Museum of Natural History University of Oklahoma Norman, Oklahoma

THEODORE R. SIMONS, Cooperative Fish and Wildlife Research Unit, Department of Zoology, North Carolina State University, Raleigh, North Carolina

HOBART M. SMITH, Department of Environmental, Population, and Organismic Biology, University of Colorado, Boulder, Colorado

ANDREW T. SNIDER, Curator of Herpetology, Detroit Zoological Institute, Royal Oak, Michigan

WILLIAM SOUDER, Stillwater, Minnesota

MICHAEL J. SREDL, Nongame Branch, Arizona Game and Fish Department, Phoenix, Arizona

CRAIG STAEHLE, Department of Veterinary Biosciences, College of Veterinary Medicine, University of Illinois, Urbana, Illinois

NANCY L. STAUB, Biology Department, Gonzaga University, Spokane, Washington, and Museum of Vertebrate Zoology, University of California, Berkeley, California

MARGARET M. STEWART, Department of Biological Sciences, SUNY at Albany, Albany, New York

MELODY STONEHAM, School of Environmental and Applied Sciences, Griffith University, Queensland, Australia

ANDREW STORFER, School of Biological Sciences, Washington State University, Pullman, Washington

BRIAN K. SULLIVAN, Department of Life Sciences, Arizona State University West, Phoenix, Arizona

DANIEL SUTHERLAND, Department of Biology, University of Wisconsin–La Crosse, La Crosse, Wisconsin

SAMUEL S. SWEET, Department of Ecology, Evolution, and Marine Biology, University of California, Santa Barbara, California

STEPHEN G. TILLEY, Department of Biology, Smith College, Northampton, Massachusetts

STANLEY E. TRAUTH, Department of Biological Sciences, Arkansas State University, State University, Arkansas

PETER C. TRENHAM, Section of Evolution and Ecology, University of California, Davis, California

VANCE VREDENBURG, Museum of Vertebrate Zoology, Department of Integrative Biology, University of California, Berkeley, California

DAVID B. WAKE, Museum of Vertebrate Zoology, Department of Integrative Biology, University of California, Berkeley, California

J. ERIC WALLACE, School of Renewable Natural Resources, University of Arizona, Tucson, Arizona

MARK B. WATSON, Allegheny Institute of Natural History, University of Pittsburgh at Bradford, Bradford, Pennsylvania, and Biology Department, University of Charleston, Charleston, West Virginia

LINDA A. WEIR, USGS Patuxent Wildlife Research Center, Laurel, Maryland, and Graduate Program in Sustainable Development and Conservation Biology, Department of Biology, University of Maryland, College Park, Maryland

HARTWELL H. WELSH, JR., Pacific Southwest Research Station, USDA Forest Service, Arcata, California

HOWARD H. WHITEMAN, Department of Biological Sciences, Murray State University, Murray, Kentucky, and Rocky Mountain Biological Laboratory, Crested Butte, Colorado

JILL A. WICKNICK, Department of Biology, John Carroll University, University Heights, Ohio

BRIGIT WIKOFF, Department of Veterinary Biosciences, College of Veterinary Medicine, University of Illinois, Urbana, Illinois

ANTHONY B. WILSON, Law Enforcement Division, Indiana Department of Natural Resources, Taylorsville, Indiana

SCOTT A. WISSINGER, Biology Department, Allegheny College, Meadville, Pennsylvania, and Rocky Mountain Biological Laboratory, Crested Butte, Colorado

KELLY R. ZAMUDIO, Department of Ecology and Evolutionary Biology, Cornell University, Ithaca, New York

PREFACE

MICHAEL LANNOO

For much of the past decade and a half, reports of amphibian declines and developmental malformations have been featured in the scientific and popular press. These reports typically open or close with the question: Are amphibians telling us something about the state of the environment—are they canaries in the coal mine? The answer to this question, we now know, is an unequivocal *yes*.

The first hint that all was not well with amphibians came at the First World Congress of Herpetology, held in Canterbury, England, in 1989. Here, the impression of amphibian losses by many respected herpetologists from around the world led to a 1990 workshop held in Irvine, California, entitled "Declining Amphibian Populations: A Global Phenomenon?" (Wake and Morowitz, 1991). At this meeting it became apparent that there was good reason for concern: amphibian disappearances that biologists had previously felt were simply a local problem were not—the problem was global. And what was most disturbing about these discussions was not that amphibians are disappearing from areas disturbed by humans (this was expected), but rather that they are disappearing in regions of the world once thought to be pristine.

In response, the Species Survival Commission, a division of the World Conservation Union, established the Declining Amphibian Populations Task Force (DAPTF). The DAPTF is organized into six regions—Nearctic, Palearctic, Ethiopian, Oriental, Neotropical, and Australian—encompassing the world. The Nearctic Region includes the United States, Canada, and Mexico. The United States is further divided into twelve divisions—Pacific Northwest, California/Nevada, Rocky Mountains, Southwest, Northern Plains, Southern Plains, Great Lakes, Midwest, Mississippi Delta, Northeast, Appalachia, and Southeast—and includes about 1,500 members. The mission of the DAPTF is *To determine the nature, extent, and causes of declines in amphibians throughout the world, and to promote means by which the declines can be halted or reversed*. Toward these ends, three fundamental questions need to be addressed: (1) Are amphibians declining? (2) If so, why? and (3) What can be done to halt these losses? DAPTF working groups have been meeting and compiling information about their regions. In North America, the Canadian working group has already published two volumes on amphibian declines (Bishop and Pettit, 1992; Green, 1997b). The United States groups have also published several

volumes on sampling procedures for amphibians (Corn and Bury, 1990; Bury and Corn, 1991; Heyer et al., 1994; Olson et al., 1997)—a process necessary to sufficiently document population fluctuations. One working group, the Midwest, has published a book (Lannoo, 1998b) and a symposium proceedings (Kaiser et al., 2000). Corn (1994a) has summarized the state of knowledge for the West.

A second response has come from Partners in Amphibian and Reptile Conservation (PARC). PARC is centered in the United States and in addition to concerns over endangered and threatened amphibian and reptile species has the goal of "keeping common species common." PARC works in several arenas, including a partnership with the National Fish and Wildlife Foundation to fund worthy projects.

Asking the Right Questions

Key questions about amphibian declines are whether known historical declines in amphibian populations have now ceased and populations have leveled off, whether declines are continuing, and whether declines are accelerating. As we will see, these questions must be addressed on a species-by-species and region-by-region basis. For some amphibians, and in some areas, severe declines have occurred and continue to occur.

Documenting Declines

One problem with documenting amphibian declines is that in most regions, and for most species, we have no historical data to compare to the data from our current studies (for a consideration of these issues, see Lannoo, 1996). A second problem is that amphibian populations fluctuate with environmental conditions—wet years favor reproduction, drought years do not. Therefore, to properly document amphibian declines, we tend to focus our studies on areas where historical data are available and sample across many years under a variety of conditions.

A number of such studies have now documented amphibian declines. Drost and Fellers (1996) have described the decline of several frog species in the Yosemite region of the California Sierra Nevada. Also in the Sierra Nevada, Bradford et al. (1993) noted the role that introduced fishes play in reducing anuran populations, a theme that was expanded upon by

Knapp and Matthews (2000). Introduced American bullfrogs are reducing native amphibian numbers in many parts of the West (Kupferberg, 1996b, 1997a; but see Hayes and Jennings, 1986).

In the Midwestern United States, my colleagues and I, working at the long-established Iowa Lakeside Laboratory, have documented changes in the amphibian assemblage over the past 70 years (Lannoo et al., 1994). Interestingly, in the Midwest, fish introductions have played the same role in decimating amphibian populations that they have in California (Lannoo, 1996). Long-term data from the Wisconsin Frog and Toad Survey indicate that numbers of most amphibians in that state are in steady decline, at rates of between 1% and 4% per year (Mossman et al., 1998).

Contradicting reports of declines, other studies designed to look at amphibian abundance indicate that numbers are stable. Long-term studies at the Savannah River Ecology Laboratory, on the Georgia–South Carolina border, show large population fluctuations but little evidence for declines over the past two decades (Pechmann et al., 1991).

From these studies and others, a general picture emerges. Within species, populations may be in decline in some regions yet stable in others. Within regions, some species are in decline, while others are not. For example, in North America, northern populations of Blanchard's cricket frogs *(Acris crepitans blanchardi)* have declined, while southern populations remain robust. These declines in northern populations of cricket frogs have not been mirrored by concomitant declines in other resident species, such as northern leopard frogs *(Rana pipiens)*, American toads *(Bufo americanus)*, western chorus frogs *(Pseudacris triseriata)*, and tiger salamanders *(Ambsytoma tigrinum)*.

Factors Underlying Declines

The reasons underlying amphibian declines are in some cases thought to be known, while in others they are completely unknown. In the United States, and indeed, the world, ultraviolet radiation, acid rain, and pesticide overuse have all been implicated. Global warming also affects amphibians. In England, toads responding to springtime temperatures now breed a full two weeks earlier than they did just two decades ago (Beebee, 1995). As we in North America focus on the benefits of this climate change—shorter, warmer winters—we ignore two negative aspects—fluctuating spring and fall temperatures and drier conditions. The spring and autumn transitions will be characterized by alternating warm and cold periods. Early breeding amphibians responding to prematurely warm weather risk eggs being frozen at a later date. During the summer, global warming will affect summer nighttime temperatures but not summer daytime temperatures—in short we will experience fewer dewy nights and evapotranspiration will take place around the clock. Wetlands will dry faster. Droughts will occur more frequently and be more prolonged, reducing or eliminating amphibian breeding and larval habitats. Indeed, Pounds and Crump (1994) and Pounds et al. (1999) propose that global warming was the factor causing the collapse of the anuran fauna at Monteverde, Costa Rica, in the late 1980s. Global warming will also alter vegetation patterns across much of North America. Under the 2 CO_2 model (doubling the ambient levels of atmospheric carbon dioxide), the lush southeastern forests, currently home to the most stable North American amphibian populations, will convert to a dry chaparral-like ecosystem, conditions inconsistent with the life history requirements of the native amphibian species assemblage (R. Neilson, personal communication).

An additional factor affecting amphibians has been the collecting done for the biological supply companies. This trade provides the dissection specimens for school classrooms. Such collections are not well regulated and have reduced the number of individuals within many populations.

Habitat loss is undoubtedly the largest single factor contributing to amphibian declines, but simple habitat loss alone cannot explain all amphibian disappearances, especially more recent losses. For example, in the Midwestern United States, Frank Blanchard, the University of Michigan herpetologist, reported that leopard frog tadpoles were the "widespread and abundant amphibian of the region" and "at least one specimen was found in nearly every pond seined." In the early 1990s, working at Blanchard's historic field sites, my colleagues and I found leopard frogs in only twenty-four of the thirty-four wetlands sampled, a 70% rate of occurrence (Lannoo et al., 1994; Lannoo, 1996). Over this same time period, it is estimated that this region has lost somewhere between 90% and 98% of its wetlands, a hundred-fold drop. Therefore, as much as a tenfold decline in leopard frog numbers in this region cannot be explained by simple habitat loss. Furthermore, the treefrog that bears Blanchard's name—Blanchard's cricket frog—is in severe decline in the Upper Midwest. Cricket frog declines are occurring in the absence of a similar magnitude of wetland loss (and indeed in the absence of similar declines in other amphibian species) in the northern tier of the midwestern states. While other factors that affect amphibians may be playing a role, in the Midwest, the alteration of remaining wetland habitats by anthropogenic factors (including the growing aquaculture industry) explains in large part the amphibian declines beyond those predicted by simple habitat loss alone.

While the trend in experimental studies has been to focus on single potential causes of amphibian declines, animals in nature rarely face only one threat. Recognizing this, recent studies (e.g., Kiesecker and Blaustein, 1995, 1999; Kiesecker et al., 2001a; Kiesecker and Skelly, 2001; Kiesecker, 2002) have focused on synergistic effects of anthropogenic disturbances on amphibians. A second trend in experimental studies has been to assess the level of insult needed to deliver a knock out punch. But again, recognizing the situation in nature, more recent studies (e.g., Hayes et al., 2002a,c) have focused on the role of sublethal effects.

What We Know

First, in the United States there are several conclusions that can be drawn from recent work:

1. Despite our knowledge, we still do not have a complete picture of the conservation status of amphibians. In essence, by defining what we know we have also defined what we do not know—it is this lack of information, and its magnitude, that now draws our attention.

2. Amphibian species are responding in various ways to the environmental pressures presented by current land management practices and by compromises in air and water quality.

3. We now have enough information on many U.S. amphibian species to begin to make informed decisions about their management.

4. In assessing amphibian declines we must distinguish between naturally rare species and declines for unnatural reasons.

5. Current abundance does not equate with future conservation status.

6. Developmental deformities can cause amphibian declines, but declines are occurring in the absence of high rates of deformities.

7. It is important to realize that while amphibian declines are currently receiving a great degree of publicity, they serve as a proxy for declines that are co-occurring in many other non-game species, including (and especially) wetland species. All species depend on ecosystems, they do not exist in isolation, nor are they alone likely to be affected by detrimental environmental factors. Amphibian declines present no mystery to scientists. Amphibian populations are continually under assault from a variety of anthropogenic factors, and they will continue to be assaulted until they either become extinct or we exhibit the cultural and political will to practice good environmental stewardship.

8. The notion of amphibians as bioindicators is real, but needs some qualification. First, while it is true that amphibians have a number of characteristics that make them potential bioindicators, including unprotected, permeable skin and a lack of long-range dispersal capability, the grunge that appears in water and affects human health also affects exposed amphibians. But realize that our physicians, while sometimes tending to act more like business persons than custodians of public health, can also determine trends. Amphibians cannot tell us any more about human health than a careful epidemiologist could tease apart from an examination of either historical or comparative data.

Second, as environmental indicators, amphibians show the effects of compromised ecosystems. But in certain cases the situation is reversed—rather than amphibians being affected because ecosystems are sick, ecosystems become sick because amphibians are affected. For example, in chytrid fungal outbreaks, amphibians themselves are the targets, and with their loss, ecosystems are damaged. These cause–effect relationships may not be as simple as we imagine they are. Sublethal exposures of pollutants or UV-B may stress amphibians to the point of making them susceptible to pathogens such as chytrid fungi. The factor (or a coincident one) causing malformations at the Crow Wing site in Minnesota is also causing behavioral modifications in northern leopard frogs that is affecting their ability to reproduce. So, amphibians are affected when ecosystems are affected, and amphibians themselves can be targeted even when ecosystems otherwise appear to be healthy.

Approaching Conservation Initiatives

Given this background and the conclusions we have been able to draw from it, the question becomes: How does one take the next step (the subject of this book) and assemble and integrate the information required to address amphibian decline and malformation issues at the national level? We began by establishing an advisory board composed of experienced, respected herpetologists, representing various regions of the country as well as federal and state agencies, academic institutions, and museums. We emphasized the scientific method, which produces a solid, overarching theoretical perspective that is both underpinned by, and provides context to, the facts that comprise the state of our knowledge. The current volume is composed of two parts. The first part, consisting of conservation essays, presents our perspective. These ideas, often theoretical and usually based on the work on one (or a handful of) species, give us our context—how we think about amphibian declines and malformations. To accomplish this, we invited herpetologists and conservation biologists to compose essays about the amphibian conservation issues thought to be most important at this point in time. We divide these essays into several sections, including an Introduction, Declines, Causes, Conservation, Surveys and Monitoring, Education, and a Perspective. The second part of this volume consists of species accounts accompanied by newly developed, digitally based, distribution maps. This is the first time that detailed species accounts have been compiled for all currently recognized U.S. amphibian species. By assembling this basic information on species life history and natural history traits, we gather the range of facts we need to both support and expand our theoretical and practical perspectives. We conclude this volume with a detailed consideration of causes of amphibian declines in the United States.

Beyond the Science

There is consensus among the scientific community that the way we have chosen to address amphibian conservation issues is an effective approach toward making informed management decisions. However, we have learned enough over the past decade to know that this effort, by itself, cannot ensure the conservation of our amphibian species. Even with the best management approaches, amphibians will continue to decline and exhibit gross developmental malformations until we exhibit the social and political will to implement the lifestyle and policy changes necessary to ensure sustainable ecosystems. The consequences of staying the course are obvious, and extend far beyond what we are describing for amphibians. Many fiscal conservatives cite Genesis 1:26 for their biblical justification. I prefer Jeremiah 2:7: "And I brought you into a plentiful country, to eat the fruit thereof and the goodness thereof; but when ye entered, ye defiled my land, and made mine heritage an abomination." We do not live in two worlds—one that we can exploit for profit, another where we raise our children. As we think about conserving amphibians, keep in mind that by doing so, we may be saving ourselves.

PART ONE

CONSERVATION ESSAYS

INTRODUCTION

ONE

Diverse Phenomena Influencing Amphibian Population Declines

TIM HALLIDAY

Twelve years after the Declining Amphibian Populations Task Force (DAPTF) was established, those who have to answer queries from other biologists and the media are still unable to say why amphibians are declining. In this chapter, I briefly discuss a number of issues concerning amphibian declines and, in so doing, will attempt to reveal some of the reasons for these declines.

Declines in Protected Areas and Elsewhere

Scientific and media attention in the last twelve years has been largely focused on the sudden collapses of amphibian faunas in protected areas, notably in Australia, Central America, and the Pacific Northwest of the United States. This attention is appropriate because these changes are deeply disturbing and have profound implications for conservation. In particular, the fact that amphibians can decline catastrophically in protected areas raises serious doubts about the efficacy of protection as a means for conserving biodiversity. However, there is a serious danger that by concentrating on protected areas we are overlooking the fact that amphibian populations in many parts of the world have been declining for a long time outside protected areas—perhaps less dramatically, but equally disturbing.

Protecting an area eliminates only some of the environmental insults that threaten the inhabiting wildlife. Although we do not yet know with certainty the causes of any of the declines that have occurred in protected areas, there is a growing consensus that four causal factors are involved in varying degrees: climate change, pollution, increased ultraviolet B (UV-B) radiation, and disease (Alford and Richards, 1999). None of these causal factors is moderated in the slightest degree by putting a fence around an area and labeling it "protected."

Protection does eliminate one factor that threatens wildlife—habitat destruction. There is little doubt that this has been a major factor in the widespread decline of amphibian populations outside protected areas. However, protected areas are often small in area and isolated from other areas of suitable habitat. The effects of habitat destruction are obvious—amphibians disappear. The effects of habitat fragmentation are less obvious, however, and of greater importance if we are to progress to effective conservation of amphibians. The research

conducted outside protected areas concerning this issue is very important, not only because it tells us a lot about the causes of population declines, but also because it influences the way we create and design protected areas.

The Quest for a "Smoking Gun"

Although the experience of extinction events over many years has shown that the causes for any one species are rarely clear-cut, there was an implicit assumption among many who became concerned about amphibian population declines twelve years ago that there must be a single causal agent—the task was to find out what it was. Because amphibian declines were so sudden and were occurring in protected areas, there was also a presumption that we were looking for something new. As a result, both research and media attention focused on certain possible causes, while others were ignored. This has had both beneficial and detrimental effects on our efforts to understand amphibian declines.

Conservation biologists are not isolated from the sociological pressures that drive science in general. To succeed in science, an ambitious researcher has to carry out groundbreaking work, reveal novel phenomena, and develop new methodologies. The amphibian decline phenomenon has provided the opportunity for some scientists to conduct well-funded, high profile research into topical issues such as ultraviolet radiation, climate change, and emerging infectious disease. Over the last twelve years, very able herpetologists who have made their reputation in other areas of biology have seized these opportunities and turned their attention to the amphibian decline problem. This has been of enormous benefit—bringing powerful minds to bear on the problem and blurring the distinction between "pure biology" and "conservation biology," which has long been a low-prestige activity.

The downside of the quest for a single, novel explanation for amphibian declines is that the familiar causes of extinction, notably habitat loss and pollution, have tended to be overlooked. I suggest that this is slowly beginning to change. Research into the ecology and population dynamics of amphibians has made it clear that their effective conservation requires much more than putting a fence around a little bit of suitable habitat. Work on pollutants, and particularly on their sub-lethal effects, is suggesting that their impact is far more subtle, more varied, and more sinister than had been imagined previously. In particular, there is increasing evidence that pollutants can have a harmful effect on amphibians at great distances from where the pollutants are released and that, although not lethal, these pollutants can act synergistically with other threats to amphibians.

Environmental Pollution

The dramatic amphibian population declines reported by Drost and Fellers (1996) in the Yosemite region of the Sierra Nevada, California, provide a classic example of amphibian declines in a protected area. Recent research (LeNoir et al., 1999) that implicates agricultural pesticides drifting westward from the Central Valley emphasizes that pollution cannot be regarded as a purely local influence on wildlife. This study also makes the point that forty years after the publication of Rachel Carson's *Silent Spring* (1962), pollution is still very high on the causal agenda. Recent work by Tyrone Hayes and his colleagues (2002a,c) shows that a commonly used agricultural pesticide has feminizing effects on several species of frogs.

Recent research in the Pacific Northwest of the United States has revealed that some amphibian species are extremely sensitive to nitrate pollution derived from the widespread use of agricultural fertilizers (Marco et al., 1999). This pollution poses a threat to amphibians throughout the world (Rouse et al., 1999), particularly where more intense forms of agriculture are being introduced.

For those who try to raise public awareness about amphibian population declines and their wider implications, it is frustrating to see how readily reports of deformed amphibians make the headlines. Why is it that people seem more concerned that some frogs lack legs than with the fact that they are headed toward extinction? Although there are a number of reasons for regarding amphibian deformities as somewhat tangential to the amphibian population decline phenomenon (Reaser and Johnson, 1997), it is important that we not totally overlook these deformities. A recent study in New Hampshire (Sower et al., 2000) suggests that both deformities and population declines among anurans may result from hormonal changes caused by endocrine disruptors, a form of pollution that has not yet received enough attention from amphibian biologists. A second study (Kiesecker, 2002) suggests that exposure to agricultural pesticides may make amphibian larvae more susceptible to infection by trematode metacercariae, which in turn can cause malformations.

Amphibians: A Special Case?

It is appropriate that amphibian biologists are intensely interested in amphibian population declines—if we don't care about amphibians, who else will? It is important, however, that we do not close our eyes to what is happening to other components of biodiversity.

It is common in reviews of amphibian declines, and particularly in grant applications, to see the argument that amphibians deserve special attention because they are good indicators of environmental degradation (Halliday, 2000). While there may be some merit in this argument, it is very clear that many other groups are undergoing a similar phenomenon. Corals are only

one example. The World Wide Fund for Nature (WWF) maintains an index of biodiversity, categorized by habitat type (Loh, 2000). The index for freshwater habitats, based on time-series data for 194 species of vertebrates (including amphibians), reveals a decline of 50% between 1970 and 1999. This is a faster rate of decline than that for any other habitat type, including tropical forest. These data suggest that freshwater habitats are under particular threat, and that amphibians may thus be part of a much larger process. Among freshwater species, freshwater bivalves appear to be declining even more dramatically than amphibians and may therefore have an even better claim for special attention.

It is important that amphibian biologists do not become so focused on amphibians that they lose sight of what is happening to other groups. In the context of disease, recent reports of emerging infectious diseases that affect a diversity of organisms suggest that outbreaks of apparently new diseases, like chytridiomycosis among amphibians, are a widespread phenomenon (Carey et al., 1999; Carey, 2000; Dobson and Foufopoulos, 2001). This demonstrates that we should broaden our horizons to look for very general processes that are facilitating either the spread of wildlife diseases or the susceptibility of wild animals. In the context of pollutants, studies of the effects of endocrine disruptors on amphibians lag behind those on fishes or reptiles. Therefore, it is important for amphibian biologists to forge close links with researchers who have studied these effects in other taxa.

Variation in Susceptibility to Extinction

One of the ironies of the amphibian decline phenomenon is that cane toads (*Bufo marinus*) and American bullfrogs (*Rana catesbeiana*) are widespread and serious pests in some parts of the world. Whatever the causes of amphibian declines, it is clear that not all amphibian species are affected in the same way. The variation among species in their susceptibility to decline is also demonstrated through declines that have occurred in protected areas—in all cases, some amphibians have declined while others have not. Until recently, little attention has been devoted to this variation among species. However, Hero's recent (Williams and Hero, 1998) and current (Hero et al, in press) work seeks ecological and life history correlates of population declines and points the way to possibly understanding why some species are more susceptible to the causes of declines than are others.

In all areas of research into amphibian declines, it is important that we consider the variation among species. Variation in susceptibility to the harmful effects of UV-B radiation on amphibians has considerably enhanced our understanding of the mechanisms involved. Most ecotoxicology literature for amphibians is based on laboratory studies; there are indications among these studies that amphibian species vary considerably in their susceptibility to particular pollutants. Our understanding of amphibian diseases, such as chytridiomycosis, would be greatly enhanced if we could find an amphibian species that is both exposed to and immune to specific diseases.

Amphibian Ranges and Population Structure

While much attention has been devoted to amphibian population declines over the last ten years, it is paradoxical that new species of amphibians have been described at an unprecedented rate (Hanken, 1999). Even the comparatively small amphibian fauna of Europe (which has been the subject of study for two centuries) has undergone major taxonomic revision in recent years (Dubois, 1998). One of the many reasons why new amphibian taxa are being described is that many amphibian species have very small geographic ranges, especially in the tropics. Williams and Hero (1998) identified small range size as one of the factors that make some amphibian species more prone to extinction than others.

Before we can devise and implement any global strategy for conserving amphibians, we must complete the mapping of amphibian distributions to determine regions of higher biodiversity and endemism. With the exception of the United States and a few other developed countries, our understanding of geographical patterns of amphibian distribution is woefully deficient.

There is an increasing realization that an understanding of the complex structure of amphibian populations is crucial for effective conservation. Because many amphibians breed at a limited number of suitable breeding sites, their population dynamics probably conform to metapopulation models more closely than other taxa (Alford and Richards, 1999). Detailed studies of amphibians in Europe have looked at the effects of isolation of sub-populations within metapopulations by, for example, road building and changes in land use. There is accumulating evidence that isolation leads to reduced genetic variation, which leads to reduced survival and an increased incidence of developmental abnormalities (e.g., Reh and Seitz, 1990; Hitchings and Beebee, 1998).

It is clearly imperative that such findings influence conservation strategy. Conservation is often focused on protecting individual breeding sites, frequently with good results. Such efforts may prove to be useless, however, if they are not combined with measures that also conserve neighboring sites and ensure that there is suitable dispersal habitat between sites.

Amphibian Reproductive Dynamics

In an important recent review, Alford and Richards (1999) drew attention to a feature of the population dynamics of many amphibians—a feature of which many of us have been aware for many years without realizing its important implications. Some amphibian populations, especially among temperate pond-breeding species, are sustained by successful episodes of reproduction that occur in only a few seasons. In many populations, the occurrence of these years of successful recruitment is sporadic and unpredictable, an effect emphasized by Pechmann et al. (1991). Indeed, in the early days of the DAPTF, the knowledge that this pattern of breeding makes long-term declines very difficult to detect (Pechmann and Wilbur, 1994) prompted some to question whether amphibian population declines were a real phenomenon. For a time, the debate engendered by this skepticism became quite lively and heated (e.g., Blaustein, 1994). I believe, however, that it focused the attention of amphibian ecologists on the aspects of the life history, population structure, and reproductive dynamics of amphibians that we must grasp if we are to (1) immediately understand population declines and (2) ultimately conserve amphibians.

Alford and Richards (1999) emphasize that long-term population declines can only be detected from time-series data on populations that span many years. There is a serious risk that short time-series data (of only a few years) are more likely to suggest declining than stable or increasing populations. This effect was not taken into account by Houlahan et al. (2000), whose analysis of 936 data sets included a very large number of studies that covered time-series data of three or four years. Their conclusion that amphibian declines have been occurring since as early as the 1950s is thus suspect (Alford et al., 2001).

While the variable and unpredictable nature of amphibian breeding dynamics increases the difficulty and cost of setting up and maintaining monitoring programs, it also suggests a way that we can sharpen our focus in looking for the causes of amphibian declines. If the long-term viability of an amphibian population is dependent on infrequent seasons when breeding success is high, we can concentrate our attention on why those seasons are successful and on whether the effect of specific environmental factors is to make them more or less frequent. This approach is likely to be particularly relevant in research into the effects of climate change on amphibian populations.

Lack of Data on Amphibian Habitats

It is widely assumed that the great majority of amphibian population declines throughout the world result from habitat destruction. This assumption is very difficult to test, however, because of the nature of many amphibian habitats. Many amphibians breed in small, often ephemeral pools, ponds, and streams—habitats that are included on only the most detailed maps. For most parts of the world, we have virtually no quantitative data for the extent of amphibian breeding habitats; in contrast, the areas of woodlands, lakes, and rivers can be readily estimated from simple maps.

This problem is exacerbated by our extensive ignorance of the terrestrial habitat requirements of most amphibian species. We know very little about the habitat requirements of terrestrial juveniles and adults, or about how far they disperse from their breeding sites. Moreover, we tend to neglect the many amphibian species that do not breed in water, but which have wholly terrestrial life histories. Knowing so little about the kind of habitat that amphibians require, the task of quantifying the amount of suitable habitat available is a largely futile exercise.

Summary and Concluding Comments

There is no "smoking gun." A single cause of amphibian declines has eluded us because, other than the activities of *Homo sapiens*, no single process is responsible (Pechmann and Wake, 1997). Among those who study amphibian declines, it is widely agreed that there are many causal factors involved—indeed, all the causes that have been suggested are implicated to some degree. This creates problems for us. To conduct effective research into the causes of declines requires specialization in a particular area, such as pollution or UV-B, but we need to carry out focused research while being fully aware that there is a much wider context. For an organization such as the DAPTF, limited human and financial resources make it very difficult to identify priorities. We must try to foster and encourage research over a very wide range.

The bibliography of this chapter is modest, and by no means is it representative of the voluminous literature on amphibian population declines. However, it does serve to illustrate how widely one must read to keep abreast of all aspects of the amphibian decline phenomenon. In just ten years, it is remarkable how the amphibian decline problem has engaged researchers in most disciplines within the biological sciences.

Finally, in a book that is primarily about amphibians in the United States, it is important to remember how different the situation is in other parts of the world. American herpetologists have a pretty good idea what species they have to deal with; there are many parts of the world where amphibian diversity has only scantily been described and is grossly underestimated. While Americans have access to much historical data, there is none for most of the rest of the world. Until recently, the DAPTF awarded most of the Seed Grants for research into amphibian declines to Americans, particularly in the northwest states; virtually no grants went to support research in those parts of the world where amphibian diversity is highest. However, this situation is beginning to change. One of the most important tasks of the DAPTF is to help spread the ideas, skills, and methods that have emerged in the United States and other developed countries into our ever-changing world.

Acknowledgments. I thank David M. Green and Michael Lannoo for their comments on early drafts of this chapter.

TWO

Why Are Some Species in Decline but Others Not?

MARTHA L. CRUMP

Since residents of Monteverde began to informally monitor the golden toads' *(Bufo periglenes)* activity in 1972, these toads have emerged *en masse* from their underground retreats every year in April and May. During 1988, however, something was definitely wrong. I found only one golden toad in the Monteverde Cloud Forest Reserve in northwestern Costa Rica; in the previous year I had seen over 1,500 individuals. During the following month, I could not find harlequin frogs *(Atelopus varius)* at my study site along the Río Lagarto near Monteverde—on a single day just the year before, I had found over 700 individuals. In the literally hundreds of times I had censused the stream in recent years, harlequin frogs were always abundant, no matter the time of year, no matter the weather conditions.

Observations

It has now been more than fifteen years since golden toads last congregated at rain pools in the elfin forest and harlequin frogs sat on moss covered boulders at the Río Lagarto. One-half of the 50 known species of anurans from the vicinity of Monteverde disappeared in the late 1980s (Pounds et al., 1997). A likely culprit was a prolonged drought caused by a severe El Niño weather pattern in 1986–1987. Still, other species persisted. Why did certain species disappear while others remain seemingly unaffected? During the past fifteen years, reports of declines and disappearances of amphibians around the world have revealed a similar pattern: within a given area, only certain species have been strongly affected. If we could identify particular traits of the affected species, perhaps we could better protect those species identified as being the most vulnerable.

By examining studies of declines within anuran assemblages from fairly protected areas around the world in Brazil, Costa Rica, Australia, and California, we can rule out habitat destruction, modification, or degradation as a cause of observed declines. Within these assemblages, some recognizable patterns emerge with regard to which species have declined.

Severe frost likely caused local declines and extinctions of frogs at Boracéia in southeastern Brazil (Heyer et al., 1988). Four of the five species that disappeared have terrestrial modes of reproduction involving either complete independence of water or terrestrial eggs that hatch into tadpoles and then wriggle

their way into water. The fifth species has aquatic eggs and tadpoles requiring an extremely long developmental time. Severe population declines occurred in other species with terrestrial eggs that undergo direct development, as well as still other species that lay their eggs on leaves above water and have aquatic larvae. In contrast, most of the species that lay their eggs in ponds and have aquatic tadpoles were still common during the census periods.

A similar pattern emerges from an assemblage of frogs that breed in mountain streams around Santa Teresa in southeastern Brazil (Weygoldt, 1989). Of 13 species, eight have declined or disappeared. Weygoldt speculated that a likely cause for the declines was extremely dry winter weather. Of the eight species that were strongly affected, four were thought to be particularly susceptible to drought because of long developmental periods of aquatic larvae. The other four species all laid a small number of large eggs and exhibited specialized modes of reproduction that involve at least the egg stage out of water. The five species believed to be holding their own are all treefrogs that have aquatic eggs and larvae.

Of the 25 species of anurans that had disappeared from the vicinity of Monteverde by 1990, only five of these reappeared during 1991–1994. Interestingly, although Pounds et al. (1997) found no significant association between the presence/absence of species and the mode of reproduction, an association between the presence/absence of species and habitat selection was established. Contrary to the pattern found in the assemblages in Brazil, the Costa Rican species completely independent of water were less likely to be affected than those associated with aquatic habitats.

In a recent paper, Williams and Hero (1998) examined patterns of ecological characteristics of various rainforest frogs that are either declining or not declining from the wet tropics of Australia. They identified the combination of low fecundity, high habitat specificity, and stream breeding as characteristics of the declining species. Just as relevant, they found no difference between the declining and nondeclining species in their temporal activity period (diurnal or nocturnal), body size, or use of microhabitat.

High specificity to breeding sites (particularly stream breeding) was likewise associated with anuran declines in the Yosemite area of the California Sierra Nevada mountains (Drost and Fellers, 1996). Three species of *Rana* (red-legged frogs, *R. aurora*; foothill yellow-legged frogs, *R. boylii*; and mountain yellow-legged frogs, *R. muscosa*) closely associated with, and breeding in, lakes and streams have nearly disappeared from the area. Some populations of the two historically present species of *Bufo* (western toads, *B. boreas* and Yosemite toads, *B. canorus*) associated with temporary ponds and wet meadows have declined or disappeared, whereas the populations are persisting in other areas. In contrast, population densities of the one treefrog inhabiting this region (Pacific treefrogs, *Pseudacris regilla*), which breed in a wide variety of aquatic sites, appear to have changed the least.

Clearly, patterns identified within one assemblage in one habitat are not necessarily repeated in other assemblages in other habitats. So, what makes a species vulnerable to population declines? The following attributes are a few that we might consider in our attempt to identify the distinguishing characteristics of declining amphibians.

Phylogeny

Although certain genera seem to be disproportionately affected (e.g., *Atelopus*, *Bufo*, and *Rana*), declines are not restricted to these genera. Furthermore, many other populations of *Atelopus*, *Bufo*, and *Rana* seem to be thriving. No evidence exists to suggest that phylogeny alone determines which species are most likely to be affected.

Distribution and Habitat

Although some declines have occurred in populations existing at the edge of the species' geographical distribution, more examples exist of populations that have declined from well within the species' range. Populations of both widespread and endemic species have declined. Many declines have been reported from high-elevation sites, while many others have occurred at mid and low elevations. Declines have been associated variously with stream, still water, and terrestrial habitats. Species with aquatic, semi-aquatic, terrestrial, fossorial, and arboreal habits have all been affected.

Activity

Populations of both diurnal and nocturnal species have declined. Again, no association seems to occur between the susceptibility of a species to decline and its temporal activity pattern.

Diet

No obvious correlations emerge between declining species and diet, either in the breadth of their diet or in specific prey items. Both specialized species (e.g., those that eat primarily ants) and generalized species (e.g., those that opportunistically feed on anything within an appropriate size range) have declined.

Body Size

Large, medium, and small species have all declined. Thus, body size and associated physiological correlates, such as water balance and metabolism, do not seem to be identifiable traits of declining species.

Skin Characteristics

Declining species do not appear to have thinner, more permeable skin than do nondeclining species. Thin-skinned glass frogs (Centrolenidae) have declined, but so have many thick-skinned toads (Bufonidae).

Life History Characteristics

There is no clear pattern of differences in life history characteristics between declining and nondeclining species. Among rainforest frogs from Australia, low fecundity seems to be a factor. Consider, however, that ranids with large clutch sizes have also declined elsewhere in the world. Species that produce either large or small eggs have declined. Long-lived and short-lived species have declined. Many terrestrially breeding species have declined, but so have many species that deposit eggs in water and have aquatic larvae. The duration of the larval period does not seem to be a distinguishing factor. The length of the breeding season and breeding site specificity, though possibly correlated with declining species in specific assemblages, are not distinguishing correlates on a large scale.

An Investigative Protocol

Do we conclude that just by chance some species are declining and others are not? No. I would argue that the absence of obvious distinguishing characteristics suggests that certain species are simply less able to cope with changes in their environment—changes that are in large part (either directly or indirectly) anthropogenic in nature. It is possible, indeed probable, that declining species are more vulnerable because they exhibit narrow tolerances to moisture, temperature, and other habitat conditions, or they are less able to deal with unpredictability.

For amphibians, three of the major environmental changes that cause stress are habitat modification or destruction, environmental pollution, and drought. How do amphibians respond to these stresses? Basically, they have three options: (1) move to another site, (2) stay and cope with the situation, or (3) die. I suggest that in order to understand the patterns of amphibian declines, we need to examine the responses of individuals of declining (as well as nondeclining) species to these stresses in terms of flexibility of physiology, endocrine function, morphology, ecology, and behavior.

As an example of the sort of questions we could ask, consider the following: when deforestation changes the local weather patterns such that a given area experiences reduced rainfall and the ponds less predictably fill, how do different species cope with the situation? A fundamental difference will be in species-specific physiological responses to the increasingly dry environment. In the case of severe drought, those species most vulnerable to dry conditions may experience sudden catastrophic population crashes. Other species may experience slight declines, and still other species may be able to survive through the crisis with little change in population size.

Those species that are physiologically able to survive the drying conditions may still be affected differentially. For example, some may be more flexible in various life history attributes and thus will be more successful in recruiting, despite the unpredictability of the breeding site. Within a population, a great deal of inter-individual variability exists in the timing of breeding, the clutch size, the number of clutches laid per season, and even the egg size. We do not know, however, how much flexibility an individual has in these characteristics or how much an individual can modify its behavior in response to a deteriorating environment. Despite our lack of understanding, it is reasonable to assume that individuals of different species exhibit varying degrees of flexibility.

Flexibility of reproductive readiness could strongly influence population dynamics in changing environments over time and could explain why populations of some species fluctuate more widely than do others in response to breeding site unpredictability. For example, individuals of some species may be able to remain physiologically ready to breed: females can hold their eggs in a mature, but unovulated state rather than resorbing them, and males can continuously produce active sperm so that they can be ready to fertilize eggs once the pond fills. In contrast, individuals of other species may shut down their reproductive functions during a drought. Once the rains come, these frogs may then require considerable lead time before they can breed.

Another trait that will strongly affect successful recruitment, and therefore population density, is the flexibility that individuals display in the timing of egg deposition relative to rainfall. At one extreme is the mud-nesting frog *(Leptodactylus bufonius)* that lays its eggs suspended in foam inside nests of mud at the edge of small depressions. After the eggs hatch, the larvae can remain inside their protective mud nest for at least 46 days while waiting for rainfall to wash them into the newly formed pond (Philibosian et al., 1974). Because female mud-nesting frogs are not dependent on standing water, they can deposit successive clutches throughout the season. Likewise, a *Physalaemus* species (Leptodactylidae) from Brazil has the flexibility to lay its foam nest full of eggs on the water surface (if available) or to lay its foam nest on the ground or in bromeliads near a depression that will fill with water once it rains (Haddad and Pombal, 1998). Theoretically, the *Leptodactylus* and the *Physalaemus* experience more stable population sizes in unpredictable environments than do species of the other extreme—species that deposit their eggs directly in water and therefore must wait for an appropriate breeding site to form. Within this group of aquatic egg layers (which includes most anurans from the United States), a range of responses exists, from species that breed immediately following light rains to those that must wait for several consecutive heavy rains. The degree of flexibility any given species exhibits in the timing of egg deposition may influence how successfully it can persist through a spectrum of unpredictability in breeding habitat availability, and in turn, how stable it can maintain its population size.

A working hypothesis, then, is that within any one assemblage, the species that decline are those that are the least flexible in dealing with environmental perturbations such as habitat degradation, extreme climatic changes, and contamination. Either the species' tolerances are narrow or their range of responses is limited, or both. They simply cannot cope. The "weed" species—those that are the most tolerant through a range of environmental conditions and the most flexible in their responses to environmental change—will persist.

How can we test this prediction? Answers must come from a combination of field and laboratory studies. We need fieldwork to document how local environments are changing from the perspective of an amphibian: moisture and temperature conditions, levels of contamination, and habitat modification. Laboratory experiments should then be conducted with a wide range of species from the environments studied to determine tolerances to moisture and temperature conditions and pesticide levels. Then, controlled laboratory and field experiments should be carried out on these same species to determine whether some species are inherently more flexible than others in their responses to changing environmental conditions. Perhaps then we will begin to understand why some species are declining while others are not.

THREE

Philosophy, Value Judgments, and Declining Amphibians

SARAH AUCOIN, ROBERT G. JAEGER, AND STEVE GIAMBRONE

> The very idea of the modern natural sciences is bound up with an appreciation that they are objective rather than subjective accounts. They represent *what is* in the natural world, not *what ought to be*, while the possibility of such a radical distinction between scientific 'is-knowledge' and moral 'ought-knowledge' itself depends on separating the objects of natural knowledge from the objects of moral discourse. The objective character of the natural sciences is supposed to be further secured by a method that disciplines practitioners to set aside their passions and interests in the making of scientific knowledge. Science, in this account, fails to report objectively on the world—it fails to *be* science—if it allows considerations of value, morality, or politics to intrude into the processes of making and validating knowledge. When science is being done, society is kept at bay. The broad form of this understanding of science was developed in the seventeenth century, and that is one major reason canonical accounts have identified the Scientific Revolution as the epoch that made the world modern.
>
> STEVEN SHAPIN (1996, P. 162)

> What should be the response to the number of anecdotal accounts and some recent long-term studies suggesting amphibian declines? On the one hand, it is essential that rigorous census studies of a representative sample of amphibian populations be initiated worldwide as a means of assessing the directions, magnitudes, and agents of changes in their numbers. How much information is needed before one can decide whether special efforts should be undertaken to protect or restore populations that are declining? The conservative approach of withholding intervention until extinction rates are conclusively demonstrated to be unusually high might result in an unacceptable loss of populations or entire species. The opposite approach may give mistaken conclusions that a global decline is occurring when populations are simply exhibiting normal ranges of fluctuations. This could waste resources and political capital. Therefore, one must balance the risk of lost credibility, which might seriously compromise future conservation efforts in this and other arenas, against the cost of failing to respond to a potentially serious environmental crisis.
>
> ANDREW BLAUSTEIN (1994, PP. 93–94)

Absolute Rigor versus Moral Concerns

As quoted, Shapin's (1996) view of how the modern natural sciences establish knowledge (based on the historical roots of the Scientific Revolution) provides an interesting framework from which to examine contemporary discourses concerning "declining amphibian populations." If one believes the paradigm advocated by Shapin, then amphibian population and community ecologists should follow a rigorous philosophy of science in their studies of populations and species, in which "society is kept at bay." Being trapped by "the problem of induction" (Hume, 1748; Popper, 1959), these ecologists might emphasize rigor by basing research programs on the hypothetico-deductive model (Lakatos, 1970). Such rigor, then, might allow the ecologists to draw "strong inferences" (Platt, 1964) from the data generated by their research. In the hypothetico-deductive model, conjectures are made (e.g., this amphibian population is declining more than would be expected by chance fluctuations) and refutations of the conjectures are attempted. (Failing to refute a conjecture, however, does not necessarily mean that the conjecture is true). Pechmann and Wilbur (1994) basically advocated this perspective in their essay on declining amphibian populations. Strong inference, as advocated by Platt (1964), is gained by posing alternative hypotheses against which one's data can be compared, which leads to the refutation of one or more of the alternative hypotheses.

On the other hand, ecologists, particularly those interested in conservation biology, may not be able to "set aside their passions and interests in the making of scientific knowledge" (Shapin, 1996). For example, Blaustein (1994) advocated a balance between the scientific rigor in studies of amphibian population dynamics and the concern for "a potentially serious environmental crisis" if, indeed, amphibian populations are declining beyond expected natural fluctuations. Following this view, amphibian ecologists may be influenced by moral considerations, value judgments, and political considerations in shaping their research programs and in shaping the inferences drawn from these programs. Ecologists might believe they have a moral responsibility to do something about declining amphibian populations, even in the absence of strong inferences (Platt, 1964) to support the conclusion that such declines actually exist and are systematically interrelated. Ecologists might also believe that such populations are of value to humans and to the ecosystem per se, and that politicians should be persuaded to join the fight to save declining populations.

Along with McCoy (1994), we agree that debates concerning the philosophy of science versus the ethics of value judgments are inevitable, but we also believe that such debates are extremely useful. Such discourses may lead amphibian ecologists to take different routes in designing research programs, ranging from censuses of amphibian populations (e.g., Corn and Bury, 1989a) to experimental manipulations attempting to discern underlying causes for particular amphibian declines (e.g., Blaustein et al., 1994b,c). Because all inferences ("strong" or otherwise) from studies of "declining amphibian populations" are enslaved by their philosophical underpinnings, we shall examine whether environmental ethics and the philosophy of science can or cannot lead to a unified approach to drawing "strong inferences" from diverse research programs.

Where Deduction Ends and Induction Begins

Most amphibian biologists would agree that some species of amphibians, or at least some populations of some species, are in decline and may be heading for extinction. For example, it is not unusual for one to see a pond destroyed to make a parking lot or shopping center to the detriment of the populations of anurans and caudates that utilized that pond. Such particular cases of amphibian declines are not controversial because they can be mutually deduced from simple observations. What is controversial is whether declines within numerous species are causally interrelated over a global scale. Such issues can be addressed only by using inductive inferences. Pechmann and Wilbur (1994) contrasted these two kinds of conclusions: "Human impacts obviously have reduced or eliminated many populations of amphibians and other organisms. Recent reports, however, have suggested that declines and disappearances of amphibian populations over the last two decades represent a distinct phenomenon that goes beyond this general biodiversity crisis." It is the inductive logic underlying the idea of interrelated amphibian declines that fuels the debate over how to, or even whether to, draw strong inferences from research

data. The idea that interrelated amphibian declines may be global in scope provides even more fuel for this fire.

Two of the authors of this essay have substantially different views concerning desirable approaches to the study of declining amphibian populations. Aucoin, from her studies of anurans, is concerned about how interpretations of data may be used to make management decisions. Jaeger, based on his 35-year study of terrestrial caudates, leans toward the hypothetico-deductive model in drawing inferences. Giambrone's task in this essay was to impose logic on the writings of the other two authors. Our unified purpose is to examine the discrepancy between environmental ethics and the philosophy of science. First, we explore the problem in general. Then, we explore the problem in light of "declining amphibian populations." Finally, we propose an approach for generating testable hypotheses regarding amphibian declines. It is our hope that this approach, or one like it, will lead to strong inferences regarding amphibian declines.

Strong Inference and the Philosophy of Science

It has long been argued that one can remove bias from the scientific pursuit of truth by avoiding the use of value judgments to discern "good" from "bad" scientific theories (Hume, 1748; Chamberlain, 1897; Popper, 1959; Lakatos, 1970). To accomplish this, one can adhere to a strict philosophical approach where the methods reduce the likelihood of incorporating personal bias in the formulation of scientific theories. The discrepancy between adhering to a rigorous scientific methodology and incorporating value judgments (or other scientifically unsupported assumptions) in the pursuit of knowledge is not new to scientific thought. Hume (1748) was the first to suggest that there was an inherent problem with the reasoning employed by most scientists of his day. Hume argued that by using inductive logic, scientists could never claim with certainty that something was true and therefore that the process of induction could not be logically justified as a method of scientific reasoning.

Until the middle of the nineteenth century, however, scientific theories relied on observable examples to predict unobservable, universal phenomena and suffered from the "problem of induction" (Chamberlain, 1897; Popper, 1959). Popper (1959) claimed to have solved this problem and argued that the strongest support for a theory comes from tests that have a good chance of refuting that theory. That is, a theory gains greater credence as it survives repeated attempts to refute it. Drawing from the teachings of Popper, Lakatos (1970) outlined what has come to be called the methodology of "scientific research programmes" for formulating and testing scientific theories. Using this method, one formulates a *hard core*, or a set of claims that cannot be rejected without rejecting the research program altogether (Larvor, 1998). Alternative, testable hypotheses or predictions are drawn from this hard core, subjected to tests, and adjusted or replaced to "defend the thus-hardened core" (Lakatos, 1970, p. 191). According to Lakatos (1970), such an approach is successful if it leads to a progressive problem shift, such as Newton's gravitational theory. A theory is considered progressive if it leads to some new predictions of observable events and at least some of those predictions are confirmed. With regard to the purported amphibian declines, the hard-core claims may be that (1) the declines are occurring on a global scale or (2) there is a global cause responsible for the declines.

Although the idea of alternative or multiple working hypotheses had been proposed earlier (Chamberlain, 1897), the idea was not published in a widely read scientific journal until the middle of the twentieth century (Platt, 1964). Platt (1964) argued that the rapid advancement of certain fields of science (e.g., molecular biology and high-energy physics) was the result of the systematic application of a series of familiar steps to every problem of interest. The process of strong inference begins by formulating alternative hypotheses and carrying out crucial experiments that have a high likelihood of falsifying at least one of the hypotheses. The data collected from the experiments then lead to the rejection or refinement of each hypothesis, and the procedure repeats itself in an open-ended process.

Strong Inference versus Value Judgments

Applying the hypothetico-deductive or strong inference approach to scientific pursuits may reduce the risk of introducing bias into the formulation and testing of scientific theories (Shrader-Frechette and McCoy, 1994). However, many scientific disciplines depend upon value judgments to discern important problems from unimportant problems (Kuhn, 1962; Feyerabend, 1980; Shrader-Frechette and McCoy, 1994). Shrader-Frechette and McCoy (1994) argued that value judgments are inherent in the application of ecological theory and they may render strict hypothesis deduction impossible for scientists attempting to apply ecological theory. Conservation biology relies in part on the application of ecological theories (Simberloff and Abele, 1976), many of which were formulated and tested using the hypothetico-deductive approach (Simberloff and Abele, 1976; Simberloff, 1988). The management decisions that result from these studies, however, are often guided by a set of conservation ethics that is based upon value judgments and, hence, is arguably biased.

The ethical principles that motivate much of the work in conservation biology are that (1) biological diversity is of value to humans and (2) biological diversity has intrinsic value (Primack, 1993). These principles can affect the conduct of biologists in different ways. One obvious way is in the choice of which research projects to pursue at any given point in time—the "type 1 effect." If the scientist and/or society at large believes that biological diversity is substantially valuable then the scientist might well choose now to investigate the purported decline of amphibian populations rather than some other issue that is possibly of greater intellectual significance.

We must make two points about type 1 effects. Science does not independently exist in a vacuum, and people who are scientists have legitimate nonscientific goals—scientists are personal, moral, and aesthetic agents as well as epistemic agents. Hence, inasmuch as one's scientific efforts are publicly supported, the social good has a right to make demands on the current direction of research. Also, to the extent that one's scientific efforts are personally supported, one has a right to indulge one's own interests—intellectual or otherwise. However, one must take into account the fact that a type 1 effect (whether legitimate or not) can have significantly detrimental effects on the development of science. It can lead to the neglect of intellectual problems with solutions that are crucial to the progress of a discipline at a particular stage in its development, and it can affect the type of training that aspiring scientists receive as well as the type of scientists that are hired and, hence, remain in the field. Therefore, type 1 effects can further affect the long-term direction of a field—for good or for ill. However, the effect of non-intellectual values on problem choice does

not in itself compromise the integrity of the scientist in the research that is undertaken.

Ethical principles can also influence a scientist's judgment about what conclusion is supported by the current evidence or at least its degree of support—the "type 2 effect." Such judgments might even lead scientists to make a public, scientific claim that they know is not supported by the evidence. This sort of influence compromises the scientist and undermines the integrity of the scientific enterprise.

Finally, a scientist's ethical commitments might bring one to attempt to influence public action and/or policy decisions—the "type 3 effect." This is not in itself a scientific matter. However, to the extent that one is perceived as acting as a scientist, actions can affect the public perception of, or attitude toward, science. Further, whatever claims one makes about the scientific basis for certain actions or policies are scientific matters. The scientist here must guard against type 2 effects.

In the case of amphibian declines, the relative importance of reported decreases in population densities has been questioned (a type 1 effect; Pechmann and Wilbur, 1994). Data from amphibian and other animal populations suggest that "declines" of some species (e.g., microtine voles) may actually represent natural population fluctuations or cycles (Pechmann and Wilbur, 1994; Sarkar, 1996). However, ecologists concerned with the preservation of biological diversity may be more likely to conclude that population fluctuations represent actual declines (a type 2 effect), erring on the side of protecting a relatively stable population (Pechmann and Wilbur, 1994) rather than on the side of disregarding an actual decline (a type 3 effect). Scientists must publicly emphasize the inconclusiveness of the evidence and the pragmatic nature of their recommendation—prudent action in the face of uncertainty. In this way, both the integrity and reputation of conservation biology can be preserved.

In addition to the concern regarding the use of value judgments to qualify ecological theories (e.g., "amphibians are in decline"), questions about the universality of ecological models (e.g., "amphibians are in decline on a global scale") have been raised by many scientists (e.g., Schoener, 1972; Peters, 1991) and philosophers (e.g., Sagoff, 1985). Pimm (1991) suggested that most exact models in ecology are restricted to scales that are too small to be of use in real life conservation contexts, but that speculation replaces modeling at scales relevant to application. Shrader-Frechette and McCoy (1993) argued that ecology has few general theories that are able to provide prediction for future experiments or environmental applications. In response to these claims and with regard to amphibian declines, Sarkar (1996) advocated shifting the primary goal of ecological research from one of generality to one of "accurately representing and predicting the detailed behavior of individual systems." In other words, Sarkar (1996) suggested that the study of individual systems (single species, populations, or communities) may be more important to conservation efforts than focusing on the generation of global models. While focusing on individual systems may indeed be more helpful with regard to making conservation decisions, this approach also compromises the most basic tenant of ecological research: the idea that ecological processes can be explained by universal patterns (Gurevitch and Hedges, 1993). Furthermore, although shifting the goal of ecological research from one of generality to one of "situation-specific" modeling (Sarkar, 1996) may be an attempt to generate useful information regarding the purported amphibian declines, it does raise the question of how important these declines may be to understanding the general biodiversity crisis—a type 1 effect.

What makes the purported amphibian declines interesting to many ecologists is the idea that there might be an identifiable global cause, one that might pose a risk to other, less sensitive species at some point in the future. For example, Blaustein and Wake (1995) suggested that amphibian declines might be useful as indicators of general environmental degradation. Yet, if each decline or change in population density is the result of a slightly different, perhaps unique, cause, it is unlikely that the information obtained from localized investigations will ever contribute to a global model that explains environmental degradation.

The Rigorous Evaluation of Amphibian Declines

Because not much is known about amphibian population dynamics, we think that the first order of business in the rigorous evaluation of amphibian declines is to determine whether declines are actually occurring at a greater rate than would be expected by random chance. To do this, it is necessary to evaluate exactly what is meant by "declining amphibian populations." As Green (1997c) pointed out, there is a difference between a concern for declining *sizes* of populations and a concern for declining *numbers* of populations. Although some long-term monitoring projects have focused on the demographics of certain populations (Green, 1997c), there is no certainty that changes in population demographics represent anything more than natural population fluctuations. Furthermore, although there is clearly much to be learned about long-term amphibian population demographics, the cause for concern about amphibian declines in general has been about the loss of certain populations or a loss of total population numbers (Barinaga, 1990; Blaustein and Wake, 1990; Wyman, 1990; Wake, 1991). This realization led Green (1997c) to propose a definition for amphibian declines based on the "local loss of populations across the normal range of a species [that] exceeds the rate at which populations may be established, or re-established." Yet, assessing a decline in the number of populations in a certain area is no less difficult than assessing a decline in the size of a certain population, and it requires an understanding of normal extinction and recolonization rates. Metapopulation dynamics may be important, as well, because while an isolated population may not be able to recover from a large decline, recovery may be more likely for a population that has some interaction with other populations. Furthermore, because the likelihood of extinction due to stochastic events depends on the size of the population in question and not on the direction of change of the population size (Blaustein et al., 1994a; Sarkar, 1996), a population may be at risk of extinction regardless of whether its numbers are presently increasing or decreasing.

In response to the suggestion that amphibian declines may be nothing more than natural population fluctuations, Pounds et al. (1997) tested the null hypothesis that the observed number of anuran species that have disappeared (40%) in the Monteverde region of Costa Rica's Cordillera de Tilarán since 1987 was no different than expected for natural fluctuating populations. In order to determine the characteristics of a natural fluctuation, Pounds et al. (1997) used data generated from amphibian and other animal populations. The results of this test indicated that the observed "declines" at Monteverde exceeded natural fluctuations. In addition, Pounds et al. (1997) compared the loss of anuran diversity at Monteverde to the loss of breeding bird diversity in the same area, observing that the relative frequency of absences for frogs and toads was much

greater than that for birds. These data suggest that not only is anuran diversity declining at Monteverde, but that this decline goes beyond the general loss of biodiversity in the area.

To fill in gaps in the knowledge of long-term amphibian population dynamics, some scientists have advocated intense monitoring (Blaustein et al., 1994a; Pechmann and Wilbur, 1994; Green, 1997c). Although monitoring programs can be tailored to address questions about demography or diversity (Green, 1997c), some philosophers have pointed out that such Baconian data collection is unlikely to lead to the formulation of testable hypotheses (Shrader-Frechette and McCoy, 1993; McCoy, 1994; Sarkar, 1996). Furthermore, although establishing monitoring programs may help conservation efforts in the future, it does little to address the issue of current conservation strategies. Ralls (1997) argued that individuals who wish to engage in a debate about a decline in biodiversity while advocating slow or no conservation action are making unintended policy decisions via their inaction. Even those who question the reality behind purported amphibian declines suggest that when in doubt about conservation issues, it is better to err on the side of caution (Pechmann and Wilbur, 1994).

If the general cause for concern about amphibian populations is a loss in population numbers, then monitoring need only consist of collecting current presence or absence data. In the 14 years since amphibian declines were first reported in mass numbers, many studies have compared recent presence or absence data to historical presence or absence data (e.g., Wissinger and Whiteman, 1992; Fellers and Drost, 1993; Hedges, 1993; Drost and Fellers, 1996; Fisher and Shaffer, 1996). Indeed, the data of many of these studies suggest that declines in population numbers are occurring, even in "pristine" areas.

These widespread reports of amphibian declines do little to alleviate concerns about the rigorous evaluation of such declines. Beyond testing the null hypothesis that a certain "decline" does not differ from a natural fluctuation, or testing null hypotheses regarding potential causes for declines, it is easy to understand how studies could become biased by a desire to protect certain populations or species. Instead of advocating the continued examination of individual systems (as Shrader-Frechette and McCoy, 1993, and Sarkar, 1996, suggested) or intense, long-term monitoring (as Pechmann and Wilbur, 1994, suggested), we suggest examining the reported declines and nondeclines in light of information regarding life history parameters and then formulating testable hypotheses based on this information.

Life History Parameters and Amphibian Declines

Perhaps from the general interest in amphibian declines and, hence, an increase in funding sources for studies of amphibian populations, there has been a boom in studies of amphibian populations over the past decade (e.g., Green, 1997b and Lannoo, 1998b, and the studies therein). Although some studies identify potential causes for a global amphibian decline (e.g., Corn and Vertucci, 1992; Blaustein et al., 1994b,c; Carey and Bryant, 1995; Vertucci and Corn, 1996), a large number of studies identify populations that have declined in numbers or have been extirpated locally (e.g., Bradford, 1991; Crump et al., 1992; Fellers and Drost, 1993; Hedges, 1993; Drost and Fellers, 1996; Pounds et al., 1997). In addition, some studies identify populations that have not been extirpated or changed in size (e.g., Harte and Hoffman, 1989; Pechmann et al., 1991; Wissinger and Whiteman, 1992). We suggest that it is now possible to review the body of knowledge regarding individual populations and communities of amphibians to look for interspecific patterns associated with various life history traits. For example, if after reviewing the existing literature one could suggest that amphibians that lay terrestrial eggs appear to be immune to statistically unusual population fluctuations, one could then test this hypothesis experimentally. Ideally, this approach could be used to evaluate amphibian declines on a global scale by taking a comparative approach. Until that time, however, such an approach can be used to generate testable hypotheses on a more local scale.

By way of illustration, one could restrict the focus to the continental United States. After amassing the published data regarding amphibian declines (and nondeclines), one could use a review method to examine general questions about declines in the United States. Examining a body of literature on a specific topic and tallying the number of statistically significant results is a review method that has been applied to ecological data in the past (Connell, 1983; Schoener, 1983, 1985; Simberloff, 1983). Such a method, while quantifying the results of many studies, has been criticized because by focusing on significance levels alone, one does not take into account how the magnitude of the effect and the sample size affect the significance level of a particular study (Guervitch and Hedges, 1993).

Meta-analysis (Gurevitch et al., 1992; Gurevitch and Hedges, 1993) offers a solution to these problems. Meta-analysis is a set of statistical tools that synthesizes the results of separate, independent experiments to estimate the average overall effect of a given phenomenon (Gurevitch et al., 1992; Gurevitch and Hedges, 1993; see Houlahan et al., 2000 and Alford et al., 2001 for specific applications to the amphibian decline problem). By estimating the average overall effect across all studies examined, one can test whether the effect is significantly different from zero. The magnitude of effects is not dependent on sample sizes; for this reason, meta-analysis offers a statistically rigorous approach to the results of independent experiments that have differing sample sizes and experimental designs. Meta-analysis may be especially useful for examining patterns in amphibian declines because it serves as a powerful means of controlling statistical type II errors (the probability of failing to reject a false null hypothesis; Arnqvist and Wooster, 1995; Krzysik, 1998a).

While meta-analysis provides a valuable method for quantitatively assessing the magnitude of a given effect, it can also be used to answer a number of general questions about the given effect. Increasing the number of studies may also raise the number of informative questions that can be asked of the resulting data set (Van Zandt and Mopper, 1998). This may also increase the likelihood of picking out patterns in the data set. For example, using meta-analysis to examine patterns regarding amphibian declines, one could ask if the magnitude of a decline (or change in population size) differs among species with different (a) methods of egg deposition, (b) lengths of larval period, (c) larval feeding habits, or (d) adult feeding habits. By including data from populations that do not appear to be in "decline," one can avoid biasing the results.

As with all reviews of published studies, there are some pitfalls associated with meta-analysis. First, because journals are biased against studies reporting null results, those studies failing to show evidence of a decline may not be as prevalent in the literature. By contacting amphibian population and community ecologists and requesting the use of unpublished data sets (including those in unpublished theses and dissertations), one can attempt to eliminate such a bias. Second, reviewing the existing literature to look for large-scale, interspecific patterns does nothing to evaluate whether reported declines represent real declines or simply natural population fluctuations. Of course, until a population is extirpated from a certain area, there is no way of knowing with certainty that the population is approaching extinction, and this will prove to be a problem for all studies addressing this issue.

Although there are pitfalls associated with the review of any large body of knowledge, there are also important benefits. An essential task in the evaluation of ecological data is assessing the generality of the results of a single experiment (Gurevitch et al., 1992; Gurevitch and Hedges, 1993). Summarizing the results of a number of studies that examine an ecological phenomenon (e.g., amphibian declines) is a valuable way to assess the relative importance of that phenomenon. In addition, summarizing can identify potential large-scale patterns that can then be used to help scientists make predictions on a global scale.

Conclusions and Summary

Although choosing to direct one's research interests toward the examination of the "declining amphibian issue" may be a decision motivated by value judgments, we believe that the formulation of hypotheses and the collection and interpretation of data need not be. Furthermore, as with other contemporary issues that fall within the discipline of conservation biology, we believe that careful consideration of both conservation ethics and philosophy of science is vital if one wishes to draw strong inferences regarding amphibian declines. By rigorously evaluating the evidence regardless of one's own bias, one not only lends credence to the scientific merits of a particular study but to the strength of the subsequent conservation policy and management decisions as well. Policy and management decisions based on strong inferences should have a better chance of protecting species in need of protection and of providing information that may one day be important to human welfare than should decisions based on social values. Researchers concerned about the status of amphibian populations should make it a priority to follow Platt's (1964) advice. We propose the use of a statistical review method to examine interspecific patterns regarding life history parameters and the sensitivity of certain groups of species to decline. By subsequently testing hypotheses generated from the elucidation of these patterns, it is our hope that amphibian biologists will be able to draw strong inferences about the susceptibility of various species of amphibians to decline on a global scale. This information may one day prove useful in the protection of groups of organisms other than amphibians.

Acknowledgments. We thank B. Aucoin, M. Crump, P. Leberg, L. Rania, and the Evolutionary Biology Group, University of Louisiana at Lafayette, for stimulating discussions. P. Leberg and L. Mathews offered valuable editorial advice and J. Gillette provided technical assistance. S. Aucoin was supported by Sigma Xi, a research assistantship at Northern Arizona University (NAU) and a P.E.O. National Scholar Award while writing this manuscript. R.G. Jaeger was supported by National Geographic Society grant 5721-96 during the writing of this manuscript.

FOUR

Embracing Human Diversity in Conservation

WHIT GIBBONS

Amphibians are in decline. It is not important whether decline means the number of species, populations, or individuals, or whether amphibians will always be perceived as declining because of varying numbers of offspring, or whether the natural fluctuations in amphibian communities make quantification difficult. Fewer amphibian species, populations, and individuals are on Earth today than were present in the last century, primarily because most species no longer have as much of their requisite natural habitat as they did a century ago, a decade ago, or even a year ago. The formula is simple, and it does not require intense scientific verification or documentation to be credible.

Disease, pollution, invasive species, overcollecting, global changes, and other causes have been documented or proposed to be responsible for particular or widespread amphibian declines. Yet, finding solutions for any of these causes will not matter for most species in the long term if the basic components of their natural habitats are degraded or eliminated by human use of natural resources. To address and solve the problem of too many people—especially those who inappropriately use too much land, water, and other natural resources—is a daunting task. However, rather than sit, watch, and be discouraged by the global problem of human overpopulation, anyone who desires a world in which native amphibians occupy their natural habitats needs to make a counteractive effort of some sort. I believe the answer lies in PARC.

Partners in Amphibian and Reptile Conservation (PARC) is an initiative designed to find solutions to the loss of habitat and other problems faced by herpetofauna. The PARC mission statement is succinct: *To conserve amphibians, reptiles and their habitats as integral parts of our ecosystem and culture through proactive and coordinated public/private partnerships.* Therefore, I encourage groups or individuals with an agenda involving beleaguered amphibians (i.e., frogs, salamanders, and caecilians) to become involved with PARC.

PARC differs from many other conservation groups. PARC recognizes that reptiles are more globally threatened than are amphibians, and, hence, these two major classes of vertebrates are coupled in conservation efforts. PARC also differs from other groups in that it includes the most diverse group of individuals and organizations ever to work together to address the problems confronting amphibians, as well as reptiles, on a regional, national, and global scale. Diversity is a symbol of strength, health, and well-being in biological communities. Likewise, diversity is essential for societies and organizations that address conservation issues in which habitat is the key—anyone could potentially affect an animal's habitat. Thus, PARC has been organized to involve not only the person or group with a concern for amphibians and their habitats, but also *any* person or group whose actions and attitudes are perceived by some as detrimental to the well-being of amphibians. Representatives of museums, nature centers, state wildlife departments, universities, federal agencies, conservation societies, research laboratories, the forest products industry, the pet trade, the real estate industry, and environmental consultants and contractors are all included. Since the initial organizational meeting in Atlanta in June 1999 and the establishment of the PARC Web site (www.parcplace.org), PARC membership has steadily risen, more than tripling during the first year.

Many of the individuals and groups who have joined PARC are unaccustomed to working together. Although all have an agenda that involves amphibians, the agendas do not all affect amphibians in the same way. Disagreements arise because amphibians and the regulations designed to protect them impede other goals for some groups. They are members of PARC because the achievement of their goals and the protection of amphibians are inextricably intertwined, although not necessarily fully compatible.

The time is upon us to put aside differences of opinion and to hear all sides. Solutions for the conservation of wild populations of amphibians vary among the academic community, government agencies, conservation groups, and private industry. Even within each of these broadly categorized groups, individuals have fervent and divergent views. All groups and individuals must be heard and allowed to participate, because all can contribute to solving the problems and, if not involved, to creating them. A diverse mix of people and organizations can identify the problems confronting native herpetofauna as well as the problems created by protecting them. By involving all groups that have an impact or interest in amphibians, we can identify and act upon the obstacles to solving problems. The involvement of diverse groups and individuals is the most effective means of implementing solutions and providing the support needed to assure effective conservation.

PARC is not looking for scapegoats, but is instead looking for partnerships with people who want to do the right thing—who want to correct the score in conservation efforts toward amphibians while, in some instances, proceeding toward other goals. PARC has a vision of providing remedies by encouraging those people who wish to correct the problems to work cooperatively with those people who are perceived (sometimes wrongly) as causing them. The PARC concept of diversity of participation is the most productive model for any wildlife conservation effort that is to have lasting sustainability. PARC is our best chance to ensure that humans and amphibians can live harmoniously in today's world.

FIVE

Declining Amphibian Populations Task Force

W. RONALD HEYER AND JAMES B. MURPHY

During the 1970s and 1980s, researchers in many parts of the world reported seemingly drastic population declines and disappearances of amphibians. International amphibian and reptile scientific societies held special sessions at annual meetings. In February 1990, the U.S. National Academy of Sciences sponsored an international meeting to determine whether there was cause for alarm.

From these meetings, researchers reached two conclusions: (1) although most of the evidence for amphibian declines was anecdotal, the number and geographically dispersed nature of the informal reports indicated that the situation should be addressed and treated as a possible environmental emergency and (2) an international working group should be established to produce scientifically defensible information to determine the extent of the problem as quickly as possible.

The first sobering lessons from these early consultations related to the state of the science. It was readily apparent that information about amphibian populations was largely based on observations in western North America, Middle and South America, and parts of Australia. No generalized conclusions could be valid without an information base that encompassed other areas—especially the tropics, where most of the world's amphibians are located. There was also a need to improve the reliability and compatibility of data. In the absence of generally accepted standard research protocols, trained investigators were completing work that was not comparable with other studies nor readily replicated by other scientists.

Nevertheless, the anecdotal data, while not conclusive, were sufficiently consistent to require attention. If, as is widely accepted, amphibians are reliable bioindicators of environmental change, then these population declines had to be regarded as early warnings signaling an important biodiversity crisis. With such high stakes and with so many scientific disciplines involved, it was necessary to proceed with caution. The coverage of the data would have to be improved, and the field techniques would have to be standardized to ensure reliability and compatibility before reaching generalized conclusions.

The scientific community responded by forming the IUCN/ SSC Declining Amphibian Populations Task Force (DAPTF) in December 1990. This Task Force is also affiliated with the World Congress of Herpetology. The Task Force operations are linked to the greater conservation community through the IUCN and the international DIVERSITAS program; these links ensure that the broader implications of the amphibian declines will be given appropriate attention. The Task Force has been successively chaired by Dr. David B. Wake, Mr. Robert Johnson, Dr. W. Ronald Heyer, Dr. James Hanken, and currently Dr. James Collins. DAPTF headquarters was initially established at the Center for Analysis of Environmental Change, Oregon State University, with Dr. James L. Vial serving as the coordinator, assisted by Lorelei Saylor. In June 1994, the Task Force transferred its office to Dr. Timothy Halliday's home university, The Open University, in Milton Keynes, United Kingdom. Dr. Halliday assumed the position of International Director and Dr. John M.R. Baker was hired as the first International Coordinator; Mr. John W. Wilkinson currently serves in this latter capacity.

Early in 1992, the Task Force board of directors held its first meeting. At that meeting, the following goals were established:

1. Catalyze, catalogue, and coordinate efforts to gain an understanding of declining amphibian populations.

2. Identify those target populations, species, and regions that merit immediate attention.

3. Gather and critically examine evidence concerning causal factors contributing to amphibian declines and identify remedial action.

4. Promote data collection on amphibian populations on a long-term basis.

5. Enlist the support of appropriate scientific disciplines needed to address the issues.

6. Disseminate information on declines to the scientific community and promote public awareness.

7. Advise the IUCN, other conservation organizations, and appropriate governmental bodies on necessary and immediate action.

Progress Report

Overview

With the formation of the Task Force closely linked to the general scientific community, the work has steadily moved ahead.

The Task Force developed a strategy to maintain its office as the nerve center with a worldwide network of investigators to collect data. The highest priority was given to the preparation of a standard research protocol to ensure compatible and reliable information. A program of seed money grants was set up to fund studies directed to filling important information needs.

Among the first projects the Task Force completed was the preparation and adoption of the standard protocols (Heyer et al., 1994), which are now being used by investigators. A database was then established at Task Force headquarters to receive reports from field investigators. This database is now fully operational. Summaries of the field information are disseminated to Task Force participants through FROGLOG (now a bimonthly newsletter) and other forms of communication as particular situations require. More than 3,000 individuals, including all participating investigators, currently receive FROGLOG.

In developed countries where there is an established base of investigators, the network of regional and sub-regional Working Groups is largely complete and in the field. The Canadian Working Group (known as DAPCAN) has completed its DAPTF charge, and it has now become a more conservation-oriented organization.

The Task Force leadership has been actively involved in media work in order to gain public support for the conservation measures that are needed to address the amphibian decline problem. The International Director, International Coordinator, Chair, Board Members, and Regional Coordinators have published several articles in popular magazines. They have also responded to and participated in many radio, television, and newspaper inquiries and interviews. The Task Force headquarters frequently issues media briefings that are sent to targeted science writers throughout the world and are also posted on the DAPTF Web site.

Specific Accomplishments

While the Task Force headquarters serves as coordinator and nerve center for this worldwide effort, the actual fieldwork is completed by regional and sub-regional Working Groups, which operate under their own leadership and with their own funding. Task Force headquarters provides the database, disseminates progress reports, defines the issues, builds infrastructure, and engages in projects designed to improve research methods and produce reliable results. In other words, the Task Force headquarters empowers local groups so that they can target their work more effectively toward the global effort.

A symposium on recent studies of geography and the causes of amphibian declines was held at the Third World Congress of Herpetology in Prague, Czech Republic, during August 1997. Topics included the role of chemical contaminants, disease processes, declines in specific areas of the world, and controversies related to amphibian declines. The Congress prepared a resolution that commended the DAPTF for its efforts to investigate and understand the decline phenomenon.

On 28–29 May 1998, DAPTF assisted in the development of an amphibian declines workshop, which was organized by the National Science Foundation (NSF) in Washington, D.C., through the efforts of Dr. John A. Phillips. The Chair, Dr. James P. Collins from Arizona State University, invited prominent amphibian biologists and other scientists to participate. The following important topics were covered: the geography of amphibian declines, UV-B radiation, deformities, toxins, viruses and other diseases, climatic change, and immunology. It became clear that an informations system, modified monitoring protocols for specific areas, and a major study of amphibians along the American Cordillera were needed. Several participants mentioned the increased interest by the press (perhaps spurred on by potential human effects) in amphibian deformities (i.e., Kaiser, 1997). The final task was to prepare a list of resolutions that dealt with specific recommendations to address amphibian declines: (1) the need for multi-disciplinary and collaborative studies; (2) increased public and private initiatives to support research, policy, and conservation measures; and (3) the encouragement of researchers to use this broad-based approach as a model for future studies directed toward the larger global biodiversity crisis. Based on information presented at the meeting, NSF funded an additional workshop specifically directed toward disease and immunological issues. Most recently, NSF has funded a multinational, multidisciplinary, multimillion-dollar proposal that addresses the disease aspects of the declining amphibian phenomenon (James Collins, Principal Investigator).

Procedural and Issue-Oriented Working Groups

Standardized Protocols

To assure greater reliability in evaluating amphibian populations, the Monitoring Protocols Working Group was created. This Working Group addressed the fact that reports were often based solely on personal observations or on data that were difficult to evaluate due to dissimilar methodologies. The Monitoring Protocols Working Group produced the book, *Measuring and Monitoring Biological Diversity: Standard Methods for Amphibians*, published by the Smithsonian Institution Press (Heyer et al., 1994). Copies of this book have now been sent to all Regional Working Groups. The Task Force Board set the goal for each Regional Working Group to initiate at least one study in its region utilizing the recommended techniques, if such studies are not already underway. A Spanish translation has also been delivered to a South American press for printing and distribution.

Disease and Pathology Working Group

The Task Force created the Disease and Pathology Working Group as a resource for the Regional Working Groups. It provides expertise for evaluating the impact of diseases in any amphibian decline event. The first Working Group Chair, Dr. D. Earl Green, then of the Maryland Animal Health/Diagnostic Laboratory, assembled 26 individuals with expertise in environmental toxicology, virology, parasitology, radiation, immunology, and pathology. Dr. Green produced a detailed report, "Diagnostic Assistance for Investigating Amphibian Declines and Mortalities," which has been distributed to all interested parties. Andrew Cunningham succeeded as Chair in 1997. He worked with John Wilkinson to produce information for nonpathologists in the DAPTF leaflet, "Amphibian Mortality Information Sheet." This leaflet provides simple guidelines for dealing with dead and diseased amphibians found during a suspected disease outbreak or amphibian mortality event. Recognizing that researchers could inadvertently be spreading disease organisms as they went from site to site, the Disease and Pathology Working Group assisted John Wilkinson in producing a fieldwork protocol leaflet on procedures that fieldworkers should follow to avoid spreading disease. One of the most concrete examples of the importance of this working group occurred recently when researcher Karen Lips discovered many dead and dying frogs in Panama. She was able to interact directly with pathologists through the DAPTF network to obtain advice for the collection of samples. This rapid response resulted in finding that chytridiomycosis, a condition caused by what was thought to be a fungal organism only associated with plants and invertebrates may have been the agent responsible for the mortality witnessed in the field. Since that time, this etiologic agent has been identified in other frog populations in Middle America and Australia (Berger et al., 1998), North America (e.g., Fellers et al., 2001), and in captive groups of frogs in zoos.

Interdisciplinary Working Groups

From the beginning, the Task Force has been aware that amphibians are excellent indicator species for "state of the environment" reporting. The environmental changes affecting their survival may be indicative of underlying threats to life-sustaining ecosystem processes on a global level. The investigation of chemical contaminants, and climatic and atmospheric changes implicated in amphibian population declines requires knowledge beyond the ordinary expertise of many herpetologists. The Task Force has organized two multidisciplinary committees concerned with environmental and climatic changes.

Chemical Contaminants Working Group

This multidisciplinary Working Group was organized by Michael J. Tyler, Professor of Zoology at the University of Adelaide, and is currently chaired by Christine Bishop of the Canadian Wildlife Service. This Working Group addresses the plethora of chemical compounds which, individually or in combination, constitute a danger to life systems.

Climatic and Atmospheric Working Group

This multidisciplinary Working Group was organized and originally chaired by Andrew Blaustein, Professor of Zoology at Oregon State University. He studied a potential link between a global causal factor, increased levels of UV-B radiation, and amphibian population declines (Blaustein et al., 1994c). Cynthia Carey, the current chair, is the principal investigator using a National Aeronautics and Space Administration (NASA) grant to determine whether there are NASA data to either support or reject the hypothesis that climatic change factors are contributing to the amphibian decline phenomenon.

Regional and Sub-Regional Working Groups

A volunteer network of 108 regional and sub-regional Working Groups (with numerous participants from 90 countries) has been organized to secure information from all parts of the world. Each regional and sub-regional group has its own chair. Each group builds its own network of investigators and volunteers, shares research results, pursues answers to the causal questions the Task Force is striving to resolve, and reports its progress to the Task Force headquarters for insertion in the data

bank. Each Working Group determines its own agenda based on guidelines provided by the Task Force office. The progress of each group depends on its ability, enthusiasm, and fundraising efforts.

The organization of Working Groups has rapidly progressed in areas that have traditionally supported amphibian studies, particularly the United States, Canada, and Australia. Efforts to strengthen the Working Groups in Europe and South America advanced notably in 1994, and there is progress toward activating groups to fill the urgent need for data from Africa and Southeast Asia.

The Regional Working Groups' input at an early stage of Task Force activity is summarized in the report prepared by the first Task Force Coordinator, James L. Vial, and Loralei Saylor (1993). Regional information is communicated in FROGLOG and DAPTF reports (Alcala, 1996; Tarkhnishvili, 1997; Wilkinson, 1997). The following are completed and published studies on the status of the amphibians from various regional working groups:

- The Conservation Status of Lesser Antillean Frogs. 1994. Hinrich Kaiser and Robert W. Henderson. *Herpetological Natural History* 2:41–56.
- *Amphibian Populations in the Commonwealth of Independent States: Current Status and Declines.* 1995. Edited by Sergius L. Kuzmin, C. Kenneth Dodd Jr., and Mikhail M. Pikulik. Pensoft Publishers, Moscow, Russia.
- Amphibians in Decline: Canadian Studies of a Global Problem. 1997. Edited by David M. Green. *Herpetological Conservation* 1:1–338.
- *Status and Conservation of Midwestern Amphibians.* 1998. Edited by Michael J. Lannoo. University of Iowa Press, Iowa City, Iowa.
- The present work is the contribution of the U.S. National Working Group.

Seed Grants

Work on amphibian declines prior to the formation of the Task Force reflected the interests of investigators and their ability to secure funding. The Task Force board of directors recognized that the worldwide amphibian decline project required more direction if it was to be expeditiously completed. The Task Force Seed Grant Program addresses this important need by awarding modest grants to assist investigators with closing important gaps in information and in accumulating the necessary preliminary data for obtaining support for large, long-term studies. The seed grant awards are competitive and carefully assessed and monitored.

Specific benefits are inherent in the data collection and research programs supported by Task Force seed grants. These projects deal with the complex biological questions stemming from the impact of environmental change on wildlife populations. Basic research is necessary to understand ecosystem dynamics and to allow development of effective strategies for wildlife management.

Several grants have been made to assist in the initiation of monitoring programs for identified "target" species and critical habitats in developing countries where no other financial support exists. Also, in order to maximize Task Force funds, some grants have funded investigators who are working to secure preliminary data to establish a strong base with which to gain funding for larger, long-term studies. As of 2002, 95 international projects have been awarded almost $210,000 in seed grants, the important results of which are beginning to appear in the scientific literature (e.g., Carcy et al., 1996a,b; Cunningham et al., 1996; Hays et al., 1996; Laurance, 1996; Maniero and Carey, 1997; Berger et al., 1998). Several of the seed grant recipients have successfully obtained further funding based on the results from their seed grant award work.

The State of Current Knowledge

By 1997, we were able to state as a fact that "amphibian populations, in far-flung locations, are indeed disappearing even in seemingly virgin environments" (Halliday and Heyer, 1997, p. 61). The conservation implication of this finding is that saving habitat is not in itself sufficient for saving amphibians. We must find out what is causing the declines in order to stop or reverse the problem. If there were a straightforward single global factor causing amphibian declines, the amount of research that has gone into the effort would have identified it by now. Rather, the most plausible scenario is that there are multiple causes interacting synergistically with one another, which may vary from region to region. The hypothesized causes include the following:

> Local impact causes—these include habitat modification, amphibian collection, and species introduction. Habitat modification is the most important cause of overall amphibian population loss. Collecting amphibians for the pet trade, biological supply houses, and food has also posed problems for a few specific amphibians. The introduction of organisms that prey on native amphibians, such as game fish outside their native range, can introduce diseases as well.

> Regional impact causes—factors such as acid rain and chemical contaminants adversely impact amphibians on a regional scale.

> Global impact causes—these include ozone depletion, environmental estrogens, climatic change factors, and disease. Although the importance of environmental estrogens has been studied for such groups as fishes, this topic has only recently been initiated for amphibians. Early reports suggest that environmental estrogens may play a role in the declining amphibian phenomenon. However, further studies are needed to evaluate the impact of this factor. Climate change factors that amphibians (as well as the rest of the planet) are experiencing require additional study. Climates are changing and we are in a period of human driven global warming. Although the magnitude of these changes seems no different from past historical records (at scales of millions of years), we are in a period when many amphibians in different parts of the world should be experiencing stress due to these changes. While these changes may not be sufficient in themselves to cause total disappearances of amphibian populations, they may be a contributing stress factor that in combination with other stressors is pushing some amphibians over the brink. The recent discoveries of a chytrid fungus as a vector for amphibians indicate that its impact may be global as well. The research community is working hard to understand the implications of the chytrid fungus on the amphibian decline phenomenon.

Whither the Task Force?

The mandate the Task Force took upon itself was to determine the nature, extent, and causes of declines of amphibians throughout the world and to promote means by which those declines could be halted or reversed. The Task Force has recently completed the part of the mandate that could realistically be expected of it. When the Task Force was first formed, the major question was whether the anecdotal observations of amphibians declining and disappearing from relatively undisturbed habitats were scientifically true. We now know that amphibians are declining and disappearing from certain habitats at much greater rates than has occurred throughout the millions of years that amphibians have lived on earth. There remains much research to be done before we can begin to fully understand the causes of amphibian declines from relatively pristine habitats. This research, by its nature, will take several years to complete.

Plans were begun in 1997–1998 to assemble a comprehensive summary of information on the declining and disappearing amphibian population phenomenon. At the June 1999 DAPTF board meeting, the following two actions were endorsed:

> Three major products, targeted for completion in 2004–2005, will comprise the summary: (1) a multi-authored book providing an assessment of knowledge to date, including an analysis of the ecogeography of declines and evaluation of causal factors and case studies of particularly well-studied declines; (2) a compact disc containing the DAPTF data with database software that will allow users to exhaustively query the data; and (3) a compilation of reports from Regional Working Groups that have not published their reports elsewhere.

> The DAPTF Board, after considerable thought and discussion, concluded that it was most unlikely that all the causes of amphibian declines would be resolved by 2001, although substantial progress had been made. Given that situation, the board concluded that the DAPTF should not terminate itself upon completion of the comprehensive summaries described above, but should instead maintain an international presence and focus on the scientific issues involved. The board also enthusiastically agreed that the Seed Grant Program should be continued.

Summary

The Declining Amphibian Populations Task Force (DAPTF) was created in December 1990 under the aegis of the Species Survival Commission (SSC) of IUCN-The World Conservation Union with the support of the international herpetological community. Its purpose is to organize and coordinate a global investigation of unexplained (and sometimes conflicting) data indicating amphibian population and species declines and worldwide disappearances.

In the fourteen years since it was established, the DAPTF has mobilized numerous investigators within 108 regional and sub-regional Working Groups in all parts of the world, and it has begun to develop a database that pinpoints several factors that appear to relate to the phenomenon.

The Task Force's plans specified priority to testing the hypothesis that amphibians are useful biological indicators of important environmental changes. As the work progressed, there was a deepening conviction that some of the factors associated with the declines in amphibian populations were environmental and climatic changes that also influence other life systems. Reflecting this progress, the Task Force priorities were reassessed in 1995. This new assessment called for expanding the participation of specialists from other disciplines who contribute to and review the database information.

Acknowledgments. This manuscript incorporated information from an unpublished report prepared for the Task Force in May 1995. Stanley Cohen, Tim Halliday, and Susan Tressler participated in the preparation of that report. We thank them for allowing us to use the joint effort for this report. Judith A. Block, Tim Halliday, and Miriam Heyer read drafts of this manuscript at various stages of development and made many helpful suggestions for improvement.

Addendum. This article was originally submitted in January 1999 at which time WRH was the DAPTF Chair. The article was first updated in early 2001, at which time WRH was no longer DAPTF Chair. Some updates at the page proof stage in the fall of 2004 have been incorporated into the text, but other items are best incorporated in this addendum. John Wilkinson served as DAPTF International Coordinator from 1996 to 2004. Jeanne McKay is the current International Coordinator. The Climatic and Atmospheric Working Group NASA grant results were reported in Alexander and Eischeid (2001), Carey et al. (2001), Middleton et al. (2001), and Stallard (2001). The current status of the three major products listed in the section "Whither the Task Force" is: (1) the multi-authored book is still in development; (2) the DAPTF database is posted on the DAPTF website, but additional resources are needed to make it readily available to the entire scientific community; and (3) the Regional Working Reports compendium is ready for publication, pending funding.

DECLINES

Meeting the Challenge of Amphibian Declines with an Interdisciplinary Research Program

JAMES P. COLLINS, NICHOLAS COHEN, ELIZABETH W. DAVIDSON, JOYCE E. LONGCORE, AND ANDREW STORFER

Responding to a Global Problem

Amphibian populations fluctuate in size (Pechmann et al., 1991; Alford and Richards, 1999), but around 1989, herpetologists became alarmed by reports that populations and even species were declining—some to extinction (Blaustein and Wake, 1990; Corn, 1994a; Bury et al., 1995; Pounds et al., 1997). By 1997, this problem led three of us (J.P.C., A.S. and E.W.D.) to organize a workshop. This workshop, "Amphibian population dynamics: Is the threat of extinction increasing for amphibians?" was supported by the National Science Foundation (NSF) and held in May 1998 in Washington, D.C. As a practical matter, we could not invite all researchers with an interest in this important problem. Therefore, we focused on investigators with diverse expertise in herpetology, ecology, infectious diseases, ecotoxicology, physiology, climate change, and science policy. We asked participants to address two questions that highlighted the central issues: (1) Is the threat of extinction increasing for amphibians? and (2) What response to evidence of amphibian declines is recommended? The goal of the meeting was to assess the evidence for amphibian declines, and, if warranted, to recommend a strategy for addressing the causes of the declines.

All workshop participants agreed (Wake, 1998) that there is compelling evidence that during the last 20 years unusual and substantial declines have occurred in population sizes and numbers of populations of amphibians (see Appendix 6-A). Population declines were reported worldwide, but especially from a broad region of the Cordilleras of western North America, from southern Saskatchewan south to Costa Rica and western Panama, and at higher elevations from southeastern Australia to north Queensland (Richards et al., 1993; Laurance, 1996; Lips, 1998, 1999); research conducted after the workshop reinforces this conclusion (Houlahan et al., 2000).

Workshop participants also agreed that declines could be traced to four main factors occurring alone, sequentially, or synergistically: habitat destruction, exotic species, disease, and anthropogenic environmental change due to toxic chemicals, ultraviolet (UV-B) radiation, or global climate change. The mechanisms by which habitat destruction or exotic species cause amphibian declines are straightforward—places where frogs and salamanders breed and grow are destroyed, or superior

predators or competitors displace native amphibians. Many areas with declines are reserves that are protected from exotic species and habitat destruction, which suggests that these declines are caused by subtle, complex forces, such as pathogens or environmental alteration due to toxic chemicals, UV-B radiation, or global climate change. The complexity suggested that we needed one or more interdisciplinary research teams to study the interacting factors in multiple locations, which raised questions about how the research should be conducted. What is the best research strategy for understanding and perhaps mitigating amphibian declines? What is the role of individual-investigator research relative to multi-investigator, multi-disciplinary, and interdisciplinary research? How should the research community proceed?

Two steps resulted from the initial workshop. The U.S. Secretary of the Interior, Bruce Babbitt, proposed Step 1 at the Washington workshop: The federal, interagency Task Force on Amphibian Declines and Deformities (TADD, which has since been supplanted by the USGS's Amphibian Research and Monitoring Initiative [ARMI]) was formed because of his concern about declining amphibian populations and the recommendations of workshop participants. TADD was established as a team effort to coordinate the research activities of diverse agencies. Step 2 involved understanding how multiple variables, especially pathogens and environmental change, affect amphibian populations; this would require an interdisciplinary, collaborative research program—"the interdisciplinary perspective that environmental problems axiomatically require" (Metzger and Zare, 1999, p. 642). Talking about interdisciplinary research and actually doing it is something else (Snieder, 2000)—interdisciplinary research is not easy.

Participants in the Washington workshop realized that we needed a better understanding of the dimensions of the declining amphibian problem, and we also needed an opportunity to explore how we could meld the multiple talents of diverse investigators. NSF facilitated both of the objectives by funding two more workshops at the San Diego Zoo. At the first San Diego workshop we tackled the problem of disease; the focus of the second San Diego workshop was environmental change as a cause of amphibian deformities and, perhaps, declines. We sharpened research questions, identified collaborators, built a common vocabulary, learned techniques and concepts from disciplines other than our own, and identified investigators who found the research questions sufficiently interesting to attend meetings and engage in ongoing correspondence. In short, the two workshops were a means for assembling the elements of successful collaborations.

The first San Diego workshop (July 1998) was organized by Cynthia Carey and Louise Rollins-Smith. Ecologists, pathologists, epidemiologists, immunologists, endocrinologists, mycologists, virologists, and toxicologists gathered to review what we knew about amphibian disease and immune function (Carey, 1998). We discussed the fact that amphibians have complex immune systems theoretically capable of defending against all but the most virulent pathogens. Apparent failure of these defenses to prevent infections suggests that there are newly introduced pathogens and/or environmental change(s) altering pathogen virulence or compromising amphibian immune systems. This conclusion is also supported by patterns of losses in the form of *extinction waves* (Laurance et al., 1996), which suggests one or more globally emerging epizootics are placing amphibian populations at risk of extinction. We now have evidence that disease outbreaks in diverse geographic locations are due to viral (Cunningham et al., 1996; Jancovich et al., 1997) and chytrid fungal (Berger et al., 1998; Longcore et al., 1999; Pessier et al., 1999) pathogens.

The first San Diego workshop made it clear that understanding the role of pathogens in amphibian population dynamics would not be easy; so little is known about the basic host-pathogen biology of amphibians, especially pathogenic viruses and chytrid fungi, which emerged as leading candidates causing previously unexplained amphibian declines. How are the pathogens introduced and spread? Are hosts responding immunologically? Are environmental stressors contributing to host susceptibility? Studies of model amphibians should be used to characterize sensitive immunological parameters for identifying environmental stressors that may contribute to immune suppression and increased disease susceptibility. By coupling these studies with field collections and field monitoring, we would enable rapid detection and identification of pathogens, their reservoirs, means of spreading, and host stressors. The effects of stimuli on host resistance to each pathogen following controlled laboratory exposures or field exposures could then demonstrate how suspected stressors affect the survival of selected populations. All this would require considerable basic research since we are so ignorant of the details of host-pathogen biology, even in some of our best-studied models like laboratory rats or mice, let alone a rare Australian frog species. So participants also agreed that a proposal for a vertically integrated research program—from molecular biology to population dynamics—was the best means to understand the causes of amphibian declines. To sharpen the issue further, we needed to understand how long-term environmental change might fit into the picture, and reports of deformed frogs suggested that something was wrong with the environment, especially in the upper midwestern United States and eastern Canada.

"Workshop on Mechanisms of Developmental Disruption in Amphibians," the second San Diego workshop (November 1998), was organized by David Gardiner. We explored developmental and molecular mechanisms underlying environmentally induced, congenital malformations in amphibians—"deformed frogs." The developmental biologists, toxicologists, parasitologists, epidemiologists, immunologists, ecologists, and endocrinologists who met agreed that identifying the causes of amphibian deformities is complicated because some

traumatic injuries resemble developmental abnormalities. New questions and new issues sharpened our focus on the causes of amphibian declines: What causes traumatic injuries in amphibian populations? Do all limb and other developmental abnormalities have different causes? One cause? Or do they result from a complex interaction of multiple factors? Why are some deformities site- or species-specific? What factors and mechanisms cause deformities in amphibian populations where abnormalities are rare or where they are common? This workshop also made it clear that something besides disease, or perhaps in addition to disease, is likely causing increased frequencies of amphibian deformities in the midwestern United States and eastern Canada (Souder, 2000). Further, we do not know how malformations and long-term declines are related, nor of any correlation between malformations and extinctions. After this second San Diego workshop, we had more questions than when we first met in Washington. However, our vision of the dimensions of the declining amphibian problem was becoming sufficiently well circumscribed that a research strategy was visible and a cohesive group of investigators committed to conducting the needed research was identifiable.

The three NSF-sponsored workshops in 1998 made it clear that beyond habitat destruction and exotic species, three research areas offered the best hope of unraveling the complex causes of declining amphibian populations: anthropogenic environmental change due to toxic chemicals; global climate change, including UV-B radiation; and host-pathogen biology, especially disease caused by viruses and chytrid fungi. A comprehensive research program would incorporate all these variables (and perhaps others), and any research program purporting to uncover the causes of amphibian declines had to consider the possibility that one or more of these variables could be a cause. The realities of time, energy, and resources dictate that we cannot simultaneously investigate all areas, but we must divide the problem into solvable units; the complexity of the problem suggested that a team could be even more effective than single investigators. A research team can tackle more dimensions of a problem than can one investigator, and collaboration of participants promises a level of understanding exceeding the sum of the efforts of individual researchers.

Through the workshops, a group of us found that we should forego studying habitat destruction and exotic species as factors; both are important and solving them is complicated, but the concepts and methods needed to understand the mechanisms at work are already being studied by ecologists and conservation biologists. In contrast, the mechanisms underlying the effect of environmental change or host-pathogen ecology on amphibian declines are much less clear. The Environmental Protection Agency (EPA), U.S. Geological Survey (USGS), and various international agencies (especially Canadian) are studying anthropogenic environmental change and the causes of amphibian deformities. Global climate change/UV-B change is complex; as well, university and National Aeronautics and Space Administration (NASA) scientists are working in this area (NASA, 1999). That left disease largely unexplored, and we proposed developing a host-pathogen research program under NSF auspices (Fig. 6-1). The diverse skills of multiple investigators allow planning vertically integrated field and laboratory experiments that will untangle the complex interactions from molecular genetics to population biology, which will enable testing general models of disease ecology. Capitalizing on these integrated approaches, our questions can be richer and our answers more comprehensive.

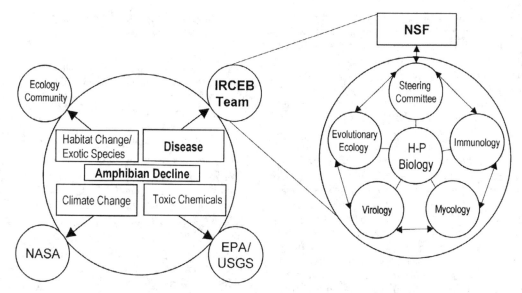

FIGURE 6-1 Illustration of the role of our Integrated Research Challenges in Environmental Biology (IRCEB) team in investigating amphibian declines and how the team is managed around the study of host-pathogen (H-P) biology.

Host-Pathogen Biology: An Integrating Theme

After a year of workshops, discussions, and research, we were convinced that solving the central questions required an integrative research program, and we assembled a diverse team of scientists to pursue this goal. We successfully secured funding through NSF's Integrated Research Challenges in Environmental Biology (IRCEB) program to advance our basic understanding of pathogens as a factor driving population dynamics (Hudson et al., 1998) and as a contributor to amphibian declines (Appendix 6-B).

The virus/fungus-amphibian systems are among the numerous, interrelated host-pathogen (H-P) interactions that alternately fascinate and frustrate biologists. The fascination stems from the opportunities H-P systems offer as models for understanding the exquisite complexity of the mechanisms underlying relationships between organisms. How did these relationships evolve? What makes some pathogens so virulent? How do host and pathogen change over time? What prevents one from driving the other to extinction? How do pathogens affect host population dynamics (Real, 1996; Hudson et al., 1998)? The frustration stems from sufficiently understanding H-P interactions in order to anticipate or react to epidemics or epizootics. What principles are at work in controlling host and pathogen? Why are some hosts in a population susceptible and others are not? How does the environment alter H-P interactions?

Ecologists increasingly acknowledge a role for pathogens in population dynamics (Real, 1996) and in maintaining diverse communities and ecosystems (McCallum and Dobson, 1995), but we generally have a poor understanding of how different pathogens alter host population dynamics, how pathogens infect hosts in a population, and how host and pathogen populations co-evolve. Understanding these details and the way they are connected requires research in molecular genetics, immunology, pathology, and ecology—disciplines from which scientists do not usually work together. Because of these connections, H-P biology is an organizing research theme broader than any one discipline; however, focusing on a problem, like the role of pathogens in amphibian declines, enhances collaboration and provides an incentive for seeking answers beyond the boundaries of traditional disciplines. H-P systems afford excellent opportunities to understand the mechanisms uniting complex biological systems.

Although our project on amphibian disease is "integrative," above all it must be "collaborative." This project has provided us an opportunity to think about the characteristics that make a large, collaborative effort work to the benefit of the overall project and of the contributing scientists. It becomes obvious that true collaboration on a large project requires much more than a collection of persons with the requisite skills. Successful collaboration requires trust, congeniality, respect, and a high level of willingness to work with colleagues who "speak" a different scientific "language." This project has provided the chance for ecologists to discuss molecular biology, modelers to plan projects with pathologists, and evolutionary biologists to learn immunology. Modern communication systems, including e-mail and courier services, make it possible for geographically dispersed researchers to work together. Communication and organization are key features of our program—otherwise, our "integrated" project would become merely a set of individual experiments. As the project has progressed, we find that collaborators tend to cluster into smaller groups working on various aspects of the project, often in ways that we did not originally envision. Reuniting these smaller groups, through the use of annual meetings, e-mail communication, and a Web site, is critical to progress on answering the research questions.

Generalizing from the Particular: An Integrated Research Strategy

Scientists responded decisively to the international recognition that amphibian populations are declining and amphibian species are disappearing—proposals were prepared, agencies funded workshops and research, and government officials formed an interagency task force. The IRCEB host-pathogen research team (Appendix 6-B) proposed a program that will advance our understanding of basic H-P biology in ways that will allow us to answer the central question: Why are pathogens causing some amphibian populations to decline, even to extinction? For our research group in H-P biology, the three workshops in 1998 made two things about this question more obvious than they were when we started.

First, even though we started with a more or less "applied" question (Why are amphibians dying in alarming numbers?) that focused on one group, the answers eluded us because we lacked a basic understanding of many features of H-P biology that extended beyond amphibians. We found ourselves, therefore, facing a problem analogous to Pasteur, who in the course of investigating various "applied" questions developed the discipline of microbiology. Stokes (1997) generalized this lesson, arguing that much progress in science occurs when basic and applied questions draw upon each other. Similarly, while puzzling about the causes of amphibian declines, we found that we had to move beyond a special group of threatened organisms. We realized that although we were ignorant of many of the aspects of basic amphibian biology that we required to save them, amphibians offered a distinctive opportunity and a challenge. Like Pasteur, we first needed to use our amphibian systems to address a series of fundamental research questions in H-P biology before we could ever really understand why amphibian populations are declining.

We also made a second discovery during 1998. As smaller, more species-specific questions metamorphosed into larger, more general questions, we found that the best science emerged when we collaborated, networked, compared results, pooled resources, and integrated our findings. Building on common research questions, methods, and models made it clear that we could do more (even better) research that would help ensure the survival of a distinctive group of animals while serving as a general epistemological model for conserving biodiversity in general. We found that the biocomplexity of the H-P systems we wanted to study required a complex research enterprise as well. We knew, however, that such a research program required funding beyond levels customary for individual-investigator awards. NSF's IRCEB program afforded an opportunity to seek support for a large-scale, interdisciplinary research program. We proposed studies of basic H-P biology from molecular genetics to global climate change with the goal of applying our findings to one of the most important global biodiversity problems facing us today: Why are amphibian populations declining? Frogs, salamanders, and caecilians are clearly not canaries, but they are likely sending us the same message—something has changed, and we must pay attention. The task of studying large-scale questions requires that we reassess how we study Earth's ecosystems; specifically, a fundamental research challenge for the twenty-first century will be better understanding how to conduct team research that involves creating and managing research programs spanning multiple levels of biological organization.

Acknowledgments. Comments by M. Parris improved the paper. The ideas in this essay derive from a research proposal prepared for NSF's Integrated Research Challenges in Environmental Biology program, which we wrote with our collaborators (Appendix 6-B). Preparation of the manuscript was supported by NSF grant #IBN 9977063.

Appendix 6-A: Resolution: Declining Amphibian Populations

Whereas, there is compelling evidence that, over the last 15 years, there have been unusual and substantial declines in abundance and numbers of populations of various species of amphibians in globally distributed geographic regions, and

Whereas, many of the declines are in protected areas or other places not affected by obvious degradation of habitats, and

Whereas, these factors are symptomatic of a general decline in environmental quality, and

Whereas, even where amphibian populations persist, there are factors that may place them at risk, and

Whereas, some patterns of amphibian population decline appear to be linked by causative factors, and

Whereas, declines can occur on multiple scales, in different phases of amphibian life cycles, and can impact species with differing ecology and behavior, and

Whereas, there is no obvious single common cause of these declines, and

Whereas, amphibian declines, including species extinctions, can be caused by multiple environmental factors, including habitat loss and alteration, global change, pathogens, parasites, various chemicals, ultraviolet radiation, invasive species, and stochastic events, and

Whereas, these factors may act alone, sequentially, or synergistically to impact amphibian populations, and

Whereas, to understand, mitigate and preempt the impacts of these factors, a comprehensive, interdisciplinary research program must be undertaken, and

Whereas, this research program must be conducted in several regions around the globe, both in areas of known declines, and in areas where declines have not been documented, and

Whereas, this research must examine issues ranging from environmental quality of landscapes to the condition of individual animals,

Now therefore be it resolved, the signatories hereto call for the establishment of an interdisciplinary and collaborative research program, which will specify and quantify the direct and indirect factors affecting amphibian population dynamics, and

Be it further resolved, that this program will include basic research and monitoring that will test hypotheses of causative factors and examine patterns of change through historical records, field-based correlative data, and controlled, multifactorial experiments, and

Be it further resolved, that interdisciplinary, incident response teams should be assembled in "hot spots" of amphibian decline to identify causative factors to facilitate the mitigation of these sudden declines, and

Be it further resolved that the signatories hereto call upon both public and private agencies and institutions, to promote and support research, policies, and conservation measures that will ameliorate losses and declines of amphibian populations, and

Be it further resolved that this broad-based approach to the study of amphibian population dynamics will serve as a model for study of the global biodiversity crisis.

This resolution was formulated, agreed upon, and signed by the attendees of the May 1998 Conference on Amphibian Declines and Malformations, sponsored by the National Science Foundation. It has been reprinted in *Amphibian and Reptile Conservation* 2:33.

Appendix 6-B: Research Team and Groups for Host-Pathogen Biology and the Global Decline of Amphibians

Project Director

James Collins, Arizona State University, Population Biology (EE)

Project Co-PIs

Nicholas Cohen, University of Rochester, Immunology (I)

Elizabeth Davidson, Arizona State University, Microbial Pathology (V)

Joyce E. Longcore, University of Maine, Mycology (M)

Andrew Storfer, Washington State University, Ecological Genetics (EE)

Senior Researchers

Ross Alford, James Cook University, Evolutionary Ecology (EE)

Seanna Annis, University of Maine, Mycology (M)

Lee Berger, CSIRO, Amphibian Pathology (V, M)

Andrew Blaustein, Oregon State University, Evolutionary Ecology (EE)

Trent Bollinger, University of Saskatchewan, Pathology, Epidemiology (V)

Cynthia Carey, University of Colorado, Physiological Ecology (I)

V. Gregory Chinchar, University of Mississippi, Virology, Molecular Genetics (V)

Peter Daszak, University of Georgia, Electron Microscopy (V, M)

Mark Farmer, University of Georgia, Protistan Systematics (V, M)

David Green, National Wildlife Health Lab, Vertebrate Pathology (V, M)

Laura Hungerford, University of Nebraska, Veterinary Epidemiology (EE)

Alexander Hyatt, CSIRO, Pathology, Diagnostics (V, M)

Bertram Jacobs, Arizona State University, Virology, Molecular Genetics (V)

Sudhir Kumar, Arizona State University, Statistical Genetics (EE)

Karen Lips, Southern Illinois University, Population Ecology (EE)

Allan Pessier, San Diego Zoo, Vertebrate Pathology (V, M)

David Porter, University of Georgia, Mycology, Phylogenetics (M)

Louise Rollins-Smith, Vanderbilt University, Immunology (I)

Richard Speare, James Cook University, Epidemiology (V)

NOTE: EE = evolutionary ecology; V = virology; M = mycology; I = immunology.

SEVEN

Biology of Amphibian Declines

DAVID M. GREEN

The evident decline of amphibian populations today is worrisome and surprising. Infectious disease (Cunningham et al., 1996; Lips, 1999; Morell, 1999), parasitic infection (Sessions and Ruth, 1990; Johnson et al., 1999), ultraviolet radiation (Blaustein et al., 1994c), chemical pollutants (Berrill et al., 1997b; Bonin et al., 1997b; Hayes et al., 2002a,c), introduced predators (Liss and Larson, 1991; Bradford et al., 1993; Morgan and Buttemer, 1996), habitat destruction (Blaustein et al., 1994a; Green, 1997c; Corn, 2000), and climate change (Pounds et al., 1999) are among the many probable causes touted as explanations for present-day declining amphibian populations. Yet, amphibians are abundant, adaptable, and resilient. Many pond-breeding species possess formidable powers of population increase, having clutch sizes in the hundreds or thousands. Wood frog *(Rana sylvatica)* populations, for example, can withstand a year of nil recruitment (Berven, 1995), as long as surviving adults return to breed again the next year—the *storage* effect. Pond-breeding amphibians may also exist as a network of sub-populations interlinked by dispersal (i.e., metapopulations; Hanski and Gilpin, 1991; Sjögren, 1991). In these circumstances, the demise of a local population may be forestalled or rendered temporary by immigration from surrounding sites—the *rescue* effect.

Diamond (1989) considered four categories of agents to be implicated in any population decline, an "Evil Quartet" consisting of (1) overkill, (2) habitat destruction and fragmentation, (3) impact of introduced species, and (4) chains of extinction. As a way to view the proximate causes of population declines (see Caughley, 1994), this is a valid classification of evils, but not necessarily a classification relevant to the reactions of species to the evils imposed upon them. Although both are caused by human activities, habitat destruction and fragmentation are likely to have different effects upon animal populations. On the other hand, overkill, introduced species, or the loss of other species resulting in perturbation of food webs (which leads to chains of extinction) are all aspects of an environment that is in some way disrupted, and is simply no longer hospitable. These disruptive agents are only so many different angels of death. Therefore, I think the environmental stressors acting upon populations, especially those stressors of anthropogenic origin, can be better divided into the following three evils (rather than four).

1. Habitat destruction—the outright loss of all or part of a species' necessary habitat, *ipso facto* resulting in population extinction.

2. Habitat fragmentation—the isolation of all or part of a species' necessary habitat without its alteration or destruction.

3. Habitat degradation—the unnatural alteration of all or part of a species' necessary habitat with neither its destruction nor its isolation.

Different species, with differing susceptibilities to extinction (Slobodkin, 1986), should react to these environmental stressors in various ways (Fahrig, 1997) depending upon their ecological differences, especially in demographic characteristics and population structures. Amphibians exhibit a variety of natural history and life history strategies, and this concept must apply as much to them as to the organisms upon which it was based.

Amphibian populations are dynamic to varying degrees. Stochastic factors (Lande, 1993) push amphibian populations into fluctuations beyond levels predictable from knowledge of deterministic processes. A classic problem in animal demographics and population biology is how populations of small animals (rodents, particularly lemmings, in the classic case) fluctuate in size (Leslie and Ranson, 1940; Chitty, 1967). There has been no shortage of hypotheses. Proximate factors responsible have included population senescence (Boonstra, 1994), Lotka-Volterra style predator/prey interactions (Hanski et al., 1993), and age structure. For some amphibians, the prime factor appears to be variation in recruitment (Banks and Beebee, 1988; Berven, 1995).

As we can see from even the simplest theory, populations will change in size at some intrinsic rate *(r)*, which is ordinarily positive when resources are freely available. Therefore, an explanation for fluctuations in population size is commonly sought among the agents of decline rather than those of increase. If fluctuations in lemming populations have thus far defied explanation (Tkadlec and Zejda, 1998), how may we explain dynamics and declines in amphibian populations?

A truism beloved by paleontologists is that all species eventually disappear (Diamond, 1989) and that when viewed over a long enough time frame, a species is a transient phenomenon. Over ecologically relevant time spans, species do quite handily persist. To understand amphibian populations and their decline, we must seek to understand their persistence as well as their extinction, and their increase as well as their decline; and then place these in the context of the three evils of habitat destruction, habitat fragmentation, and habitat degradation.

Number of Individuals

Population decline, itself, has not proven to be an obvious thing to define. Much conservation biology has, in one way or another, concerned itself with the fate of individual populations either via the largely deterministic factors which may govern diminution of population size or via consideration of stochastic variables affecting population persistence (Caughley, 1994). The initial focus of the Declining Amphibian Populations Task Force (DAPTF) was individual population size decline and a within-population demographic approach to the problem. Thus, Vial and Saylor (1993) defined amphibian decline as a definite downward trend in numbers of individuals exceeding normal population size fluctuations. Yet the immediate deterministic causes of individual population declines are numerous and contingent upon local conditions. There are many published studies of changing population sizes but the proximate reasons for declining populations are not easily generalized. It seems reasonable that the necessary prelude to extinction is consistently fewer individuals. This works passably well with animals that breed slowly and have a high expectation of survival, such as pandas, tigers, and whales, the glamour species on endangered and threatened lists. However, the documentation of true population declines in any species with even moderate population fluctuations is problematic (witness the lemming problem), and changes in population size are not necessarily tightly coupled with the chance of extinction in many of those species (Schoener and Spiller, 1992; Blaustein et al., 1994a). Populations with a high potential for increase, whether lemmings or frogs, will tend to fluctuate naturally.

Rather than deal only with population size and what drives it, considerable theory deals with the time a population takes to become extinct and those inherent forces, largely stochastic, that govern its expected persistence (Lande, 1988; Caughley, 1994). Probabilistic functions based upon the intrinsic rate of population increase and its variance induced by demographic stochasticity (the outrageous fortunes of individuals), environmental stochasticity (the vicissitudes of weather and climate), and catastrophes (earthquakes, volcanoes, shopping plazas, etc.) are beguiling, even in the absence of empirical data (Caughley, 1994). Decades of work on population size fluctuations in small mammals have failed to yield any deterministic theory, perhaps because population size is also contingent upon stochastic events that are therefore not completely predictable.

Classically, average survival times for populations of particular initial sizes can be modeled based upon birth/death processes. The size of an animal population (N) at a time (t) is always a balance between gains of individuals by birth (B) and immigration (I) against losses by death (D) and emigration (E):

$$N_t = N_{t-1} + (B + I) - (D + E)$$

Trivially, extinction is when $n = 0$, but it is the time it takes to get to that point that matters. There are many interacting factors governing a small population's persistence in theory, but the primary determinants are effective population size (N_e), how N_e is affected by environmental and demographic stochasticity, and the rate of increase (r), which is gains $(B + I)$ minus losses $(D + E)$. Also important, but often overlooked, is the magnitude of the parameters, B, I, D, and E. When they are large, a slight change in frequency can translate into a large number of individuals gained or lost—fluctuations. Highly fluctuating populations invite "gambler's ruin" upon themselves. In gambling, no matter how you have been doing, once you have lost your stake $(n = 0)$, you are out of the game (extinct); casinos prefer high rollers, as they always go bust eventually, even if the odds are even. The relevant parameters are turnover (Gains + Losses) or, as an approximation, r_{max} (the maximum possible r) and thus the greatest possible positive difference between gains and losses. I shall argue for the importance of this in amphibian populations.

Number of Populations

However much we may know about a single population, it is the overall loss of populations over landscapes that is the true currency of declines leading to species extinctions. Population losses result in the attenuation and/or contraction of a species' range (Green et al., 1996). At the point where only one population remains, any further local loss obviously equals species extinction. Extreme range contractions to single populations is particularly evident for many restricted species of endemic amphibians in montane tropical regions. By definition, range decline leading to extinction occurs when the loss of populations exceeds the establishment, or re-establishment, of new populations so that there is a definite downward trend in population number (Green, 1997c).

Defining decline in terms of population losses enables us to view the problem as one side of another equilibrium process balancing gain and loss, this time of populations rather than of individuals. The number of populations present in a species' range is determined by the number existing at a previous time plus new ones (gains, G) minus ones that have gone extinct (losses, L). This is also a birth/death process model because new populations come into existence by founders arriving from other, preexisting populations. Stability in numbers of populations is maintained where $G - L = 0$. Declining numbers occur where $G - L < 0$, and increases occur where $G - L > 0$. Regional population decline results not just from increased local extinction rates (larger ΔL) but also from decreased rate of establishment or rescue of populations (smaller ΔG), and is thus a *net* loss of populations.

Decline from Reduced Recolonization and Rescue

Amphibian species with terrestrial adults that breed in ponds tend to have juveniles that disperse widely. Normally, local sub-populations may be connected by the immigration and emigration of these individuals (Gill, 1978; Sjögren, 1991), and the probability of extinction for any one local population necessarily exceeds the probability of extinction of the metapopulation as a whole (Poethke et al., 1996). Loss of a net source population will inevitably invite loss of the surrounding local populations that depend upon its emigrants for continuation (i.e., population sinks; Pulliam, 1988). More often, a metapopulation may consist of two or more local populations in disequilibrium (Hanski and Gilpin, 1991). Dispersal and recruitment have been shown to be important in maintaining populations of northern European toads (*Bufo calamita*; Sinsch, 1997) and northern leopard frogs (*R. pipiens*; Seburn ct al., 1997). Restricting dispersal results in genetic substructuring, as shown in European common frogs (*R. temporaria*) in urbanized landscapes (Hitchings and Beebee, 1997). Stochastic migration maintains genetic variation in population sinks (Gaggiotti and Smouse, 1996).

The extent of dispersal (and the accompanying rescue effect) depends both on the propensity of a particular species to disperse and on the connectivity of the landscape. Among amphibians, different species have different dispersal abilities (Stebbins and Cohen, 1995). Whereas pond-breeding anurans employing a high-risk strategy tend to have high dispersal abilities, salamanders in the genus *Plethodon* may be extremely sedentary and territorial (Mathis, 1989). Any habitat alteration that severs or attenuates dispersal connections between local populations risks altering the metapopulation structure and increases the probability of extinction for erstwhile connected local populations. A local population may be more rapidly driven to extinction if it is deprived of immigrants due to curtailed dispersal or disrupted dispersal routes even though it is in a protected area and experiences no direct impact. In the extreme, if there is no rescue effect the overall rate of recolonizations will decrease. In a species with a high turnover of local populations and with its persistence dependent upon dispersal, this can result in a species-wide decline irrespective and complementary to stochastic local extinctions (direct losses) of populations.

Fluctuating Amphibian Populations

The classic life-style model typical of Temperate Zone pond-breeding amphibians (mainly bufonid, ranid, and hylid frogs, and ambystomatid and salamandrid salamanders) is to lay large clutches of eggs, and to have biphasic (aquatic and terrestrial) life histories and low early survival rates. These demographic traits, characterized by high r_{max}, are often regarded as typical of amphibians. This bias is unconsciously formed by people inhabiting temperate countries, such as Canada, the United States, or Britain, who are most familiar with amphibian faunas consisting most conspicuously of pond-breeding *Bufo, Rana, Hyla, Triturus,* or *Ambystoma.*

A species I have studied extensively, Fowler's toad (*B. fowleri*) lays an average of 3,500 eggs, each with a probability of living to reproduce of close to zero (about 0.0007; Green, 1997a, unpublished data). Breden (1988) found that less than 0.1% of eggs survive to age of first reproduction in northern Indiana, a rate of under 0.001. However, because there are so many eggs and as long as the survival probability does not actually achieve zero, there will be toads. In Fowler's toad populations in southern Ontario, annual survival among adult toads was invariably under 10% over the 11 years of my study, and post-metamorphic toads rarely lived to reproduce even once, much less twice. Using skeletochronology techniques on hundreds of individuals, the oldest toads I found were two individuals that were 5 years old (unpublished data). The great majority of breeding males in each year were 2 years old; in some years a substantial proportion might only have been 1 year old (Kellner and Green, 1995). In Fowler's toads, post-metamorphic juveniles disperse widely (unpublished data; Clarke, 1974a; Breden, 1988), but adults are not territorial and also disperse. Breden (1988) calculated that juveniles may move 174 m median linear distance over 1 year and as far as 2 km. I have captured marked individuals over 4 km away from the site of first capture.

Smooth newts (*Triturus vulgaris*) are longer lived and less fecund (Bell, 1977) than are many pond-breeding toads and frogs such as Fowler's toads. Nevertheless, the survivorship rate from egg to adult in smooth newts is still breathtakingly low. Using mark/recapture and age/size estimation methods, Bell (1977) concluded that the newts attain sexual maturity at between 3–7 years of age, with most individuals reproducing for the first time at age 6 or 7. Annual adult survival was about 50%; annual juvenile survival (although calculated with lesser reliability) was a relatively high 80%. Clutch sizes ranged from an average of 100 eggs among three-year-olds to about 400 eggs at age 12. Significantly, the mean population size among 34 different ponds, calculated using removal sampling, was about 70 adult newts, which is well within the range to be affected by demographic stochasticity for the majority of ponds. Over 4 years, one pond exhibited population size fluctuations in adult newts as follows: 191, 1,298, 315, 62. Other pond-breeding salamanders have similarly high population fluctuations. In a population of tiger salamanders (*Ambystoma tigrinum*) in South Carolina, Semlitsch (1983a) recorded variation in larval production ranging over three orders of magnitude in only a four-year period.

Considering the sizes of many amphibian populations, demographic stochasticity (the variable breeding success of individuals) may be highly important (Pechmann et al., 1991). Among the three factors identified as decreasing time to population extinction (i.e., demographic stochasticity, environmental stochasticity, and catastrophes; Shaffer, 1987), Lande (1993) concluded that demographic stochasticity exerted the least influence, as the other two were independent of population size. True, the influence of demographic stochasticity should wane at suitably large effective population sizes (N_e). Theoretically, beyond $N_e = 100$ (where extinction probability decreases to negligible values), demographic stochasticity can effectively be ignored (Lande, 1988; Burgman et al., 1993). Effective population size (reflecting the true number of breeding individuals in a population over time) is negatively affected by many things (Vucetich et al., 1997) relative to the total number of individuals present. Not all individuals breed, some breed more than once, and clutch sizes vary. Thus, N_e is almost always smaller than the census population size. In amphibians, males usually have greater variance in mating success and so, effectively, fewer males breed than females; N_e is negatively affected by a skewed sex ratio. In fluctuating populations, N_e is progressively smaller compared to mean population size. Vucetich and Waite (1998) show that over short time spans in fluctuating populations, N_e estimates are consistently inflated. Even so, magnitudes of N_e in pond-breeding amphibians have turned out to be even smaller than we might have thought. The effective number of breeding adults in some populations has been shown to be orders of magnitude smaller than the census population size. Scribner et al. (1997) used allelic variance in mini-satellite DNA to estimate the effective number of breeding adults in populations of European toads (*B. bufo*) in England and showed that effective breeding numbers in three populations were but 1.24%, 0.96%, and 0.53% of the number of adults present. Populations of pond-breeding amphibians, such as European toads, would need to be on the order of at least 10,000 individuals in order to have an N_e of at least 100. It may be a rare thing among many, or even most, pond-breeding amphibians for populations to exceed the theoretical threshold level whereby demographic stochasticity is immaterial.

Stable Amphibian Populations

The model of fluctuating populations certainly does not fit all amphibians. Although Hairston (1987) stated that salamanders generally have stable populations, this applies largely to terrestrial breeding, low r_{max} species, especially the various plethodontids that he studied. In stark contrast to the pond-breeding *Triturus*, *Ambystoma*, and *Bufo*, woodland salamanders in the genus *Plethodon* lay small clutches of eggs, care for them and, in return, realize high probabilities of survival to adulthood (Hairston, 1987). As clutch size decreases, demographic and effective population size goes up. Therefore, among species such as these plethodontid salamanders with long lives, small clutches, and mammalian-style demographics, stochasticity will be comparatively low for a given population size (Kokko and Ebenhard, 1996). As Hairston (1987) described, the overall impression of these species, from all studies, is one of great population stability.

In North Carolina, a five-year study showed an annual fall increase in northern slimy salamander *(P. glutinosus)* numbers, associated with recruitment, and a corresponding winter/ spring drop in numbers (Hairston, 1983). Clutch sizes range from 5–18 eggs (Petranka, 1998); Highton (1962b) recorded a mean clutch size of 8 (range = 5–11 eggs) among six nests in northern Florida. In southern populations, both females and males may breed annually; in northern populations, females may breed only every other year and breed first at the age of 5 years (Petranka, 1998). Home ranges for northern slimy salamanders are small. Highton (1956) reported that recaptures of individuals were rarely made more than 60 cm away from the original site of capture. Unlike pond-breeding amphibians, which rely upon the pre-sexual juvenile stage for dispersal, sexual adult northern slimy salamanders move farther afield than do juveniles, even if that "dispersal" is usually only up to about 9 m. Wells and Wells (1976) found that juveniles do not move from their home logs during their first year. Although northern slimy salamanders may reach high densities, as many as 0.52–0.81 salamanders per m² (Semlitsch, 1980b), they are highly territorial and will defend territories aggressively (Thurow, 1976), except during periods of drought. Population densities and age structure in Jordan's salamanders (also termed red-cheeked salamanders; *P. jordani*) and other species of *Plethodon* show similarly minor levels of annual change. Mean generation time in Jordan's salamanders is about 10 years, and adults enjoy about an 81% annual survival rate. Females are not sexually mature until 4–6 years of age (Hairston, 1983).

Many species of *Desmognathus* have aquatic larvae. Yet these species maintain stable populations (Hairston, 1987) because they, like other plethodontids, also follow a low r_{max} life history strategy. Furthermore, they breed in streams rather than ponds. Tilley (1980) estimated stable population sizes over 7 years for Allegheny Mountain dusky salamanders *(D. ochrophaeus)* in North Carolina, with annual survival rates of 0.743 and 0.626 for adults in the two populations studied. Bishop (1941b) recorded 11–14 eggs per nest from among seven nests in a site in New York, whereas Wood and Wood (1955) found a mean clutch size of 20 from among 20 nests located in Virginia. Female Allegheny Mountain dusky salamanders first oviposit at about 3–4 years of age (Keen and Orr,

1980). Allegheny Mountain dusky salamanders are aggressively territorial (Smith and Pough, 1994) and sedentary, and they have high population densities (0.96–1.2 salamanders per m² in Ohio; Orr, 1989) because territories are so small. Holomuzki (1982) found that salamander home ranges in Ohio averaged less than 1 m². In all respects, stream-breeding *Desmognathus* have the same general demographic and reproductive characteristics as terrestrial breeding *Plethodon*, including female egg attendance in the nest. Furthermore, many species of *Desmognathus*, such as pigmy salamanders *(D. wrighti)*, have foregone having stream-dwelling larvae, opting instead for direct terrestrial development.

The low r_{max} strategy adopted by *Plethodon* is seen among other terrestrial-breeding amphibians. These include other plethodontid genera (e.g., *Bolitoglossa*, *Hydromantes*, *Ensatina*, *Pseudoeurycea*), some salamandrid salamanders (e.g., *Salamandra*), and assorted leptodactylid (e.g., *Eleutherodactylus*), dendrobatid (e.g., *Dendrobates*, *Phyllobates*, etc.), leiopelmatid (*Leiopelma*), discoglossid (e.g., *Alytes*), bufonid (e.g., *Nectophynoides*), sooglossid (*Sooglossus*), hylid (e.g., *Gastrotheca*, *Flectonotus*, *Hemiphractus*), myobatrachid (e.g., *Arenophryne*, *Assa*, *Myobatrachus*), and microhylid (e.g., *Sphenophryne*) frogs.

Responses to Stresses

With a firm understanding of the variation in amphibian demographies and recognizing a dichotomy between high r_{max} and low r_{max} species, we can now question how particular species might react to the imposition of habitat destruction, habitat fragmentation, and/or habitat degradation.

Habitat Destruction

Urban or agricultural expansion and development inevitably cause the destruction of habitats and extinctions of populations. For species with complex habitat requirements (such as those using separate breeding, foraging, and/or overwintering sites), it is only necessary for one crucial habitat feature, a breeding pond for example, to be lost to precipitate the eventual, if not the immediate, loss of the population. Habitat destruction may be permanent (urban expansion) or more transient. Clearcut logging is immensely destructive in the short term for both terrestrial and aquatic-breeding amphibians (Dupuis, 1997; Waldick, 1997), but forests regrow and amphibian populations will re-establish themselves in second-growth forests as long as sources of immigrants remain (Ash, 1997).

For these reasons, we should expect any amphibian species to go extinct locally in the face of habitat destruction. For widely dispersing species with high fecundities (and high r_{max}), the effects of habitat destruction may be felt over an extended area if a net source population were eliminated. If only a sink population were destroyed, that local extinction might be inconsequential, and if the habitat loss were not permanent, rapid recolonization should be expected when conditions ameliorate. However, for sedentary and low fecundity species (with low r_{max}), there should be no widespread effect outside of direct elimination of the particular population. If the habitat loss were not permanent, slow recolonization may be expected.

Habitat Fragmentation

The effects of fragmentation will be profoundly different for high r_{max} versus low r_{max} species. By creating barriers to the movement of individuals, it would be expected that widely dispersing species and those with complex habitat requirements would be most affected. Fragmentation turns habitat patches into isolated islands with island populations that have concomitantly high extinction rates (Frankham, 1998). At the microcosmic level, it has been shown that maintenance of dispersal corridors between otherwise isolated habitat patches is sufficient to preserve biodiversity (Gilbert et al., 1998). Without them, local stochastic extinctions in a fragmented landscape should remain permanent, leading to the eventual losses of populations over extended areas. For species with simple habitat requirements, such as terrestrially breeding salamanders, habitat fragmentation may have no effect upon extinction rates other than to promote slow population genetic differentiation and, over the long term, promote the loss of intra-populational genetic diversity due to inbreeding.

Habitat Degradation

Normally, habitats constantly change via both stochastic and deterministic agents such as storms and ecological succession. However, introduced pathogens, increased ultraviolet radiation, chemical pollutants and their degradation products, and alien predators alter, degrade, and render uninhabitable habitats that may appear unchanged outwardly. These agents may directly kill animals, lower their viability, or eliminate essential resources via competition or chains of extinction. However, one species' degraded environment may be another species' enhanced environment. An invasive species that finds the altered habitat favorable becomes itself a degradative agent for the habitat of the native species and replaces it, with local, or α-, diversity retained.

The nature of the degradative effect upon native population survival depends upon the particular species' dispersal ability, the complexity of its habitat requirements, and the permanence of the effect. Local population extinctions in highly dispersing species with high r_{max} may be recovered once the source of mortality is eliminated, as long as colonists are able to arrive (rescue effect) or adults remain (storage effect). Catastrophic local extinctions due to epidemics may be less likely among species with low population densities and more likely in low r_{max} species with low dispersal potential and overlapping generations (i.e., strong storage effect but weak or negligible rescue effect). Although destroyed habitats may regrow and isolating barriers may come down, the agents of habitat degradation, especially chemical contaminants and alien species, may be difficult to eliminate, rendering their effects more likely permanent.

Habitats in which introduced predators severely drive down native prey numbers may also be considered degraded. Overpredation or over-harvest (where it is humankind that is the introduced predator) will differentially affect species depending on their demographic traits. Low r_{max} species cannot easily withstand the loss of reproductive adults. Populations of these species tend to exhibit age structures that are skewed toward older individuals. Because recruitment is low, population size is dictated by survivorship. Although the probability that an individual will breed more than once is high, the numerical success of any one reproductive attempt may be low. Total lifetime reproductive potential of adults is severely compromised if they are harvested. In contrast, high r_{max} species can withstand losses of adults from populations, as long as they occur just after the breeding season. Population size is recruitment driven and skewed toward younger individuals. The probability of an individual breeding twice may be low, and successful lifetime reproductive output might not be jeopardized even if the individual is only allowed to breed once.

Conclusions

Amphibian population declines have been identified mostly among highly fecund, pond-breeding species, at least in North America (Weller and Green, 1997; Corn, 2000). This is partly because pond-breeding anurans produce great numbers of easily caught larvae and juveniles and engage in often noisy, aggregative mating swarms. They are easy to study. If they appear to decline, it will be noticed, whether it is a true decline or only a normal fluctuation. Secretive terrestrial salamanders are more difficult to census and study. Nevertheless, pond-breeding species fluctuate greatly in population size, inviting "gambler's ruin" and local population extinctions (Schoener and Spiller, 1992). Pond breeders counteract the tendency to local extinction by high recolonization potentials and the rescue effect of dispersal. These traits render pond-breeding species more prone to the effects of habitat fragmentation and degradation than are terrestrial species. In the montane tropics, though, relatively few species are widespread (see Lips and Donnelly, this volume) and there are localized endemic species in mountainous regions where there is limited potential for dispersal. For these restricted species, local extinction equals species extinction, whatever their demographics.

Populations naturally change in their distribution and abundance. All populations and species have lifetimes, of course, but some types of populations can be expected to go extinct more quickly than do others. Yet, even though populations of many species normally have short survival times, species themselves do not willy-nilly go extinct. In opposition to population losses are population gains—populations are rescued by immigration. Those species with highly fluctuating population sizes and high risks of local extinction must have high dispersal potentials or else they simply could not persist. Thus, species with different life-history parameters may be expected to react differently to habitat destruction, habitat fragmentation, and habitat degradation in predictive ways.

Summary

Amphibian population declines are surprising because many species have high reproductive potentials. A population decline may be defined either as a downward trend in numbers of individuals within populations (demographic decline) or as a downward trend in numbers of populations across the species range (range decline leading to extinction). Populations decline in size when the number of new recruits and immigrants does not keep pace with the loss of individuals due to death and emigration. In similar fashion, the number of populations of a species is governed by the relative rate of losses of existing populations (local extinctions) versus colonization of new populations. Decline in the total number of populations is due either to an increase in the rate of local extinction or a decrease in the rate of recolonization, or both. Pond-breeding amphibians have high maximum r (the intrinsic rate of increase), large clutch sizes, high mortality, low effective population sizes compared with census population sizes, large fluctuations, and a propensity to local extinctions. Alternatively, direct-developing species have stable population sizes, low maximum r, small clutch sizes, low mortality, high effective population sizes, and persistent populations. These two amphibian demographies portend differing extinction probabilities in the face of the three evils of habitat destruction, habitat fragmentation, and habitat degradation. Many causes have been proposed to explain a global decline of amphibian populations, but pond-breeders, because of their reliance upon dispersal and despite their high clutch sizes, may be expected to be the most severely affected by habitat fragmentation.

Acknowledgments. This work is supported by an NSERC Canada research grant.

EIGHT

Declines of Eastern North American Woodland Salamanders *(Plethodon)*

RICHARD HIGHTON

Recent declines and extinctions of amphibian populations have been reported in many areas of the world. A majority of the documented declines are in easily detectable anuran species. In eastern North America, flatwoods salamanders *(Ambystoma cingulatum)* have declined and have recently been added to the threatened species list, and southern dusky salamanders *(Desmognathus auriculatus)* have disappeared at some sites (Dodd, 1998). A decline has been reported in a small part of the range of green salamanders *(Aneides aeneus;* Snyder, 1991). However, other published reports indicate little change in eastern salamander populations (e.g., Pechmann et al., 1991; Hairston and Wiley, 1993; Pechmann and Wilbur, 1994). In contrast, my data indicate widespread declines in eastern North American populations of the most common woodland salamander genus—*Plethodon,* the largest genus of salamanders in the United States (53 presently recognized species).

Since 1951, I have done a great deal of fieldwork in eastern North America in the course of studies on the life histories, systematics, population genetics, and molecular evolution of the plethodontid genus *Plethodon*. In the late 1980s, I noted a decline in the number of salamanders seen at some sites where several species previously had been common, as well as a reduction in the number of salamanders encountered at new sites in areas where species of *Plethodon* were generally abundant. In the 1990s, I returned to 127 of my sites in order to investigate the apparent declines in salamander populations and to estimate the extent of the declines. Data are available for 205 populations of 38 species.

Background Information and Methods of Analysis

I began my studies of the genus *Plethodon* by taking collections of several species throughout the seasons of their surface activity in order to investigate the life histories of these salamanders (Highton, 1956, 1962b; Sayler, 1966; Angle, 1969). When it was discovered during the 1960s and 1970s that there are far more than the 16 species of *Plethodon* previously recognized (Highton, 1962a), molecular studies were made in order to clarify the taxonomy of the genus before publishing the results of the life history studies on the remaining eastern North American species. There are now 53 species recognized in the genus,

including 45 in eastern North America (Highton and Webster, 1976; Larson and Highton, 1978; Duncan and Highton, 1979; Highton and Larson, 1979; Highton and MacGregor, 1983; Highton, 1984, 1989, 1995b, 1997, 1999a; Wynn et al., 1988; Lazell, 1998; Highton and Peabody, 2000).

I have 109 life history sites in 22 eastern and central states, which include 169 populations of 44 of the 45 eastern species (all but the recently described Bay Springs salamander [also termed Catahoula salamander; Crother et al., 2000], *Plethodon ainsworthi*; Lazell, 1998). Most of the populations were selected for life history analysis because of the abundance of one or more of the species present at the site. Salamanders from these sites were also used in molecular studies. Most of my fieldwork for the life history studies was accomplished before the end of the 1970s, although some populations were sampled into the first half of the 1980s. I became familiar with the patterns of seasonal activity and relative abundance at sites throughout the range of each species. My records from this period indicate consistent patterns of surface activity from year to year. When season, surface moisture, and temperature conditions were taken into account, the abundance of salamanders was usually predictable. When multiple collections were made at many sites during the 1950s through the mid-1980s, I never observed a consistent change in abundance at any of the sites. At one site (Cunningham Falls State Park, Frederick County, Maryland) an exceptionally large number of salamanders was collected from 1956–76 (3,473 eastern red-backed salamanders *[P. cinereus]* and 654 northern slimy salamanders *[P. glutinosus]*); however, both of these species appeared to be just as abundant at the end of the studies (Highton, 1962b, 1972, 1977; Sayler, 1966; Semlitsch, 1980b) as they were initially.

During the 1950s and 1960s, I made extensive field studies on patterns of geographic interactions among, and geographic variation within, each of the *Plethodon* species occurring in five states: Delaware, Maryland, Pennsylvania, Virginia, and West Virginia (Highton, 1972, 1977). Collections were made at over 1,700 sites, with many visited repeatedly in order to obtain adequate sample sizes for statistical analyses. One species, Valley and Ridge salamanders *(P. hoffmani)*, was later found to be composed of two morphologically similar species, valley and ridge salamanders and Shenandoah Mountain salamanders *(P. virginia*; Highton, 1999a). During the 1980s and 1990s,

I returned to almost 100 of my sites in Virginia and West Virginia to obtain material for allozyme studies in order to delineate accurately the ranges of these two species.

During the period from 1961–71, I visited over 1,700 sites in the southern Appalachian Mountains of Georgia, North Carolina, South Carolina, Tennessee, and Virginia to study the ecological and genetic interactions of the species of *Plethodon* of the region, especially the *P. glutinosis* and *P. jordani* complexes (Highton, 1970, 1972). From 1969–73, 70 sites (where members of the *P. jordani* complex and other species of *Plethodon* were abundant) were collected repeatedly for allozyme studies (Highton and Peabody, 2000; unpublished data). Eighteen of these populations were among those selected for life history studies (above).

When it was discovered during the late 1980s that some populations of *Plethodon* were declining, I revisited many of my sites. In the 1990s, in order to determine the frequency that salamanders were encountered compared with earlier visits, I recorded the observed frequencies of salamanders in 205 populations of 38 species at 127 sites in 22 states representing 299 counts of individuals per species per visit.

My records of the number of individuals seen or collected are not sufficient for accurate quantitative estimates of population densities of *Plethodon*. Such a study was made by Merchant (1972), who found high population densities. From mark and recapture data he estimated the number of Jordan's salamanders (*P. jordani*) and southern Appalachian salamanders (*P. teyahalee*) in a plot in the Great Smoky Mountains National Park, North Carolina, as approximately 8,600 and 2,300 individuals per hectare, respectively. In 1961, before Merchant did his fieldwork, two of us collected in an area of about two hectares at his site for about 30 minutes in favorable conditions and found 16 Jordan's salamanders and one southern Appalachian salamander. Thus, by searching for salamanders during the day by turning logs and stones, only a small proportion of each population is observed. Apparently, most salamanders are underground in burrows during the daytime. Studies on populations of eastern red-backed salamanders have also been reported (Test and Bingham, 1948; Klein, 1960; Heatwole, 1962; Burton and Likens, 1975b; Jaeger, 1979, 1980b; Grover, 1998; Maerz and Madison, 2000), with estimates of population densities ranging from 500 to 27,200 per hectare. At one site in New Hampshire, the estimated biomass of eastern red-backed salamanders is twice that of all breeding birds and equal to the biomass of small mammals (Burton and Likens, 1975b). Because I usually selected populations where each species was especially abundant for my life history studies and often collected over several hectares, my collections of each species were probably less than 1% of these estimates per hectare per year. It is unlikely that my pre-1990s collecting appreciably affected the density of any of these populations.

Temperature, moisture conditions, and season are important determinants of the surface activity of these terrestrial salamanders. During dry periods, salamanders are sometimes difficult to find, even during seasons of usual surface activity. During periods of extreme heat and cold, most species also retreat to their underground burrows. These activity patterns vary geographically, and in some species they also are influenced by elevation. For each species in this analysis, I have eliminated observations made during seasons when surface activity is rare, as well as observations made during unfavorable moisture and temperature conditions. I did not include populations in which the mean number of specimens collected per person was less than one in the pre-1990s.

The time of day can influence collecting success. *Plethodon* often forage on the forest floor at night, and we have found significantly different proportions of species in our day and night collections from the same locality on the same day. Success of night collecting varies geographically and fluctuates much more so than day collections, probably due to the sensitivity of salamanders to surface moisture, relative humidity, wind velocity, and light (moonlight) conditions. Except for three sites that were visited either solely at night (populations 171, 196, and 202) or usually at night (populations 45, 76, and 80) in both the pre-1990s and the 1990s, all night searches were eliminated from this analysis.

The time devoted to searching is another factor that obviously influences how many salamanders are seen at a given site. Unfortunately, I usually did not keep records of this variable. The experience and enthusiasm of the collectors also varies. My general impression is that during our earlier studies approximately the same amount of time (about one hour) was spent at most sites, with few exceptions. The exact time spent collecting was recorded for a series of 15 visits made from 1958–69 at Cunningham Falls State Park, Frederick County, Maryland (populations 3 and 125). The mean number of eastern red-backed salamanders observed per person per hour for the 15 visits was 15.7; the mean number of northern slimy salamanders observed was 5.6 for 11 visits (omitting four winter collections when this species is usually not found at the surface; Highton, 1972). These figures are comparable to the data for these populations given in Table 8-1 (29.2 for population 3; 10.9 for population 125); we usually spent twice the amount of time collecting there than we spent collecting at most other sites because more salamanders were required for our studies.

Because there are so many variables that influence the number of salamanders found on a visit, the mean number of salamanders taken per person per visit often has a large standard error. I would be hesitant to reach conclusions on the status of *Plethodon* populations on the basis of my observations at a single site. I consider the data for each population to be much more anecdotal than a quantitative estimate of comparative population densities. However, with data for 205 populations now available, the results of comparisons between the pre-1990s and the 1990s appear to be indicative of widespread declines. The simple non-parametric sign test is used to compare the mean number of salamanders over the 205 populations observed on multiple visits during these two time periods.

Results and Discussion

Variability between the mean number of salamanders found per person was compared for pairs of visits to 166 populations of this study that were collected more than once during the period from 1970–80. This was the period of my most intensive fieldwork for the life history studies. A comparison of the difference between the first and last collections for each of the 166 populations is shown in Figure 8-1A. The mean number taken in the first collection was 8.3 per person (mean year, 1972); for the last collection, the mean was 11.3 per person (mean year, 1978). I have no explanation as to why the mean number was higher late in the decade. Weather conditions or some unknown bias in our collecting may have been responsible or the difference may be due merely to sampling error. The data do not support the premise that any general declines occurred during the 1970s or that my collecting caused a decline in population densities. The increase is primarily due to the fact

TABLE 8-1
The Mean Number of Individuals Per Collector in Visits in the Pre-1990s Compared with Those in the 1990s during Searches of 205 Populations of *Plethodon* at 127 Sites in the Eastern United States

POPULATION NUMBER	STATE	COUNTY	Latitude ° ′ ″	Longitude ° ′ ″	Pre-1990s MEAN	Pre-1990s NUMBER OF VISITS	Pre-1990s YEARS	1990s MEAN	1990s NUMBER OF VISITS	1990s YEARS
Plethodon cinereus Group										
Eastern red-backed salamanders (*P. cinereus*)										
1	IN	Jackson	38 50 44	86 01 25	19.6	3	1974–76	8.0	1	1998
2	IN	Parke	39 53 14	87 11 20	29.9	8	1971–86	14.0	1	1998
3	MD	Frederick	39 37 50	77 28 17	29.2	58	1956–76	14.3	2	1999
4	NY	Tompkins	42 19 55	76 39 34	8.0	6	1974–85	3.0	2	1999
5	NY	Ulster	41 55 44	74 06 10	81.1	10	1961–86	9.0	1	1999
6	NC	Ashe	36 23 52	81 31 43	2.9	5	1970–85	2.0	1	1994
7	NC	Mitchell	36 08 14	82 08 30	1.5	3	1970–72	9.0	1	1994
	TN	Carter								
8	NC	Mitchell	36 06 36	82 21 40	8.7	15	1970–87	0.8	2	1993–95
	TN	Unicoi								
9	PA	Cambria	40 42 07	78 48 08	6.7	8	1970–88	0	1	1996
10	PA	Cameron	41 28 43	78 11 27	4.8	7	1970–79	1.0	1	1996
11	PA	Snyder	40 43 09	76 59 48	11.8	9	1970–85	25.0	1	1996
12	PA	Susquehanna	41 50 06	76 02 15	18.5	10	1970–85	0.3	1	1999
13	VA	Augusta	38 12 47	79 17 47	1.8	4	1968–76	0	1	1998
14	VA	Giles	37 14 56	80 51 48	3.7	8	1969–81	1.5	1	1999
15	VA	Grayson	36 37 50	81 35 22	1.1	5	1969–74	1.0	1	1999
16	VA	Grayson	36 39 30	81 33 25	3.6	12	1957–84	1.0	1	1999
	VA	Smyth								
17	VA	Russell	36 56 55	81 52 33	17.3	7	1969–79	3.0	1	1997
18	VA	Southampton	36 52 26	76 57 28	13.1	10	1969–85	15.0	2	1996–99
19	VA	Washington	36 50 13	82 05 15	4.8	6	1969–82	0	1	1999
20	WV	Hampshire	39 17 08	78 23 58	2.9	19	1964–81	1.8	2	1999
21	WV	Pocahontas	38 25 04	79 46 10	2.7	2	1969	9.5	2	1993–97
22	WV	Pocahontas	38 20 57	80 00 30	5.8	1	1969	17.2	2	1993–94
23	WV	Preston	39 27 24	79 31 08	6.1	4	1967–76	3.1	3	1992–96
24	WV	Summers	37 47 54	80 51 00	3.9	6	1970–86	0.2	3	1990–98
Valley and Ridge salamanders (*P. hoffmani*)										
25	PA	Snyder	40 43 09	76 59 48	1.2	9	1970–85	0	1	1996
26	VA	Alleghany	37 49 44	79 50 08	2.0	1	1956	2.8	2	1993–94
27	VA	Augusta	38 12 47	79 17 47	3.0	4	1968–76	0	1	1999
28	VA	Augusta	38 13 49	79 25 42	1.0	1	1969	1.2	4	1993–98
	VA	Highland								
29	VA	Frederick	39 21 23	78 15 03	2.5	1	1968	4.1	2	1992–93
30	WV	Mineral	39 24 07	78 51 29	1.0	1	1968	1.7	4	1993–95
31	WV	Pendleton	38 47 37	78 24 17	4.5	1	1975	3.8	2	1993
32	WV	Pendleton	38 39 16	79 30 55	4.2	1	1969	6.5	2	1993

TABLE 8-1 (continued)

			Latitude	Longitude	Pre-1990s			1990s		
POPULATION NUMBER	STATE	COUNTY	° ′ ″	° ′ ″	MEAN	NUMBER OF VISITS	YEARS	MEAN	NUMBER OF VISITS	YEARS
33	WV	Pocahontas	38 25 04	79 46 10	1.3	2	1969	2.4	2	1993–97
34	WV	Pocahontas	38 20 57	80 00 30	5.6	1	1969	2.3	2	1993–94
35	WV	Pocahontas	38 08 08	80 10 54	5.3	1	1968	3.7	1	1993
36	WV	Pocahontas	38 06 23	80 07 24	4.0	1	1968	12.0	1	1993

Southern ravine salamanders
(*P. richmondi*)

37	KY	Harlan	36 55 03	82 54 04	1.7	5	1980–85	1.5	1	1996
38	KY	Scott	38 23 23	84 34 02	6.5	4	1974–83	3.0	2	1990–98
39	VA	Grayson	36 37 50	81 35 22	4.7	5	1969–74	0.5	1	1999
40	VA	Grayson	36 45 47	81 13 22	14.4	17	1960–78	5.5	1	1993
	VA	Wythe								
41	VA	Russell	36 56 55	81 52 33	2.6	7	1969–79	0	1	1997
42	VA	Wise	36 53 42	82 37 58	8.2	13	1980–86	7.8	2	1992–95
43	WV	Cabell	38 24 13	82 26 03	4.0	2	1978–80	2.7	1	1991

Southern red-backed salamanders
(*P. serratus*)

44	AR	Montgomery	34 22 35	93 53 45	6.5	3	1975–76	3.5	1	1996
45	AR	Polk	34 39 20	93 57 00	4.9	7	1974–82	0.8	2	1995–96
46	MO	Iron	37 33 37	90 40 16	39.2	12	1971–83	11.5	2	1990–98
47	OK	LeFlore	34 43 10	94 32 33	5.4	3	1975–76	0	1	1996
48	OK	LeFlore	34 36 55	94 37 52	8.6	2	1975	1.5	2	1995–96
49	OK	McCurtain	34 07 29	94 40 15	29.5	12	1974–87	11.3	2	1995–96

Shenendoah Mountain salamanders
(*P. virginia*)

50	VA	Rockingham	38 34 42	79 02 08	7.7	5	1971–85	1.0	2	1994–96
51	WV	Hampshire	39 19 57	78 34 44	1.3	1	1964	1.5	4	1992–94
52	WV	Pendleton	38 41 36	79 05 44	2.5	7	1966–87	0.3	2	1995–96

Plethodon welleri Group
Weller's salamanders
(*P. welleri*)

53	NC	Mitchell	36 06 36	82 21 40	12.7	15	1970–87	0.3	2	1993–95
	TN	Unicoi								
54	VA	Grayson	36 39 30	81 33 25	10.7	12	1957–84	1.0	1	1999
	VA	Smyth								

Plethodon dorsalis Complex
Northern zigzag salamanders
(*P. dorsalis*)

55	IL	Pope	37 22 54	88 40 20	20.2	9	1971–86	12.0	1	1996
56	IN	Parke	39 53 14	87 11 20	40.2	8	1971–86	17.0	1	1998

Southern zigzag salamanders
(*P. ventralis*)

57	AL	Lawrence	34 18 25	87 20 10	4.8	10	1971–86	1.5	1	1996
58	TN	Blount	35 38 20	83 44 52	8.1	8	1962–78	4.0	1	1996

TABLE 8-1 (continued)

			Latitude	Longitude	Pre-1990s			1990s		
POPULATION NUMBER	STATE	COUNTY	° ′ ″	° ′ ″	MEAN	NUMBER OF VISITS	YEARS	MEAN	NUMBER OF VISITS	YEARS
Ozark salamanders (*P. angusticlavius*)										
59	AR	Pope	35 38 28	93 04 03	2.3	6	1973–84	1.5	1	1995
60	AR	Stone	35 59 05	92 16 02	15.9	7	1973–86	5.0	3	1990–96
61	OK	Adair	35 50 13	94 39 20	3.2	5	1974–80	0	1	1996
Webster's salamanders (*P. websteri*)										
62	AL	Etowah	34 04 06	86 18 43	10.6	3	1976–88	7.5	2	1992–94
63	AL	Etowah	34 02 51	86 10 38	13.9	11	1971–83	0	1	1994
64	MS	Winston	33 09 10	89 02 50	10.4	4	1976–84	4.5	1	1996
Plethodon glutinosus Group Yonahlossee salamanders (*P. yonahlossee*)										
65	NC	Mitchell	36 08 14	82 08 30	1.1	3	1970–72	0	1	1994
	TN	Carter								
66	NC	Mitchell	36 06 36	82 21 40	1.2	15	1970–87	0	2	1993–95
	TN	Unicoi								
67	TN	Johnson	36 32 25	81 56 55	4.4	3	1972–86	1.0	1	1999
	TN	Sullivan								
68	VA	Grayson	36 45 47	81 13 22	1.3	17	1960–78	0	1	1993
	VA	Wythe								
Plethodon ouachitae Complex Caddo Mountain salamanders (*P. caddoensis*)										
69	AR	Montgomery	34 22 35	93 52 45	5.3	3	1975–76	2.0	1	1996
Rich Mountain salamanders (*P. ouachitae*)										
70	AR	Polk	34 40 59	94 22 16	6.7	8	1973–78	0.5	1	1991
71	OK	Latimer	34 46 03	95 05 54	9.0	6	1975–83	3.8	2	1995–96
72	OK	LeFlore	34 43 10	94 32 33	11.0	4	1975–78	2.0	1	1996
73	OK	LeFlore	34 40 48	94 36 40	9.0	10	1973–84	1.2	3	1991–96
74	OK	LeFlore	34 36 55	94 37 52	9.4	5	1974–84	3.3	2	1995–96
75	OK	LeFlore	34 36 55	94 29 50	6.0	12	1973–84	4.3	2	1996
Fourche Mountain salamanders (*P. fourchensis*)										
76	AR	Polk	34 39 20	93 57 00	4.1	7	1974–82	1.0	2	1995–96
Plethodon glutinosus Complex Western slimy salamanders (*P. albagula*)										
77	AR	Dallas	33 54 18	92 34 02	2.5	3	1980–83	0	1	1991
78	AR	Garland	34 32 32	93 01 42	9.6	6	1971–86	9.5	1	1996
79	AR	Polk	34 40 59	94 22 16	1.2	8	1973–78	0	1	1991
80	AR	Polk	34 39 20	93 57 00	3.1	7	1974–82	1.3	2	1995–96
81	AR	Pope	35 38 28	93 04 03	4.7	10	1973–84	0	1	1995
82	AR	Stone	35 59 05	92 16 02	10.5	4	1973–79	2.5	3	1990–96

TABLE 8-1 (continued)

POPULATION NUMBER	STATE	COUNTY	Latitude ° ′ ″	Longitude ° ′ ″	Pre-1990s MEAN	Pre-1990s NUMBER OF VISITS	Pre-1990s YEARS	1990s MEAN	1990s NUMBER OF VISITS	1990s YEARS
83	MO	Iron	37 33 37	90 40 16	2.1	10	1971–83	1.5	2	1990–98
84	OK	Adair	35 50 13	94 39 20	6.1	6	1973–80	1.5	1	1996
85	TX	Hays	29 56 27	97 54 14	13.0	4	1973–75	2.7	1	1991
86	TX	Travis	30 18 17	97 46 16	13.0	2	1973–75	16.5	1	1992

Tellico salamanders
(P. aureolus)

| 87 | TN | Monroe | 35 27 45 | 84 01 37 | 7.3 | 6 | 1978–84 | 5.5 | 1 | 1997 |

Chattahoochee slimy salamanders
(P. chattahoochee)

| 88 | GA | Towns | 34 52 21 | 82 48 31 | 11.7 | 9 | 1961–84 | 4.5 | 1 | 1997 |

Atlantic Coast slimy salamanders
(P. chlorobryonis)

89	NC	Columbus	34 19 55	78 52 36	6.4	4	1978–87	0.3	2	1995–99
90	NC	Craven	35 17 19	77 07 39	7.8	10	1953–76	7.0	2	1996–99
91	NC	Halifax	36 12 48	77 34 38	4.9	4	1979–85	1.0	1	1999
92	NC	Pender	34 31 17	77 48 38	5.8	4	1980–85	3.5	1	1999
93	NC	Tyrell	35 48 38	76 05 17	10.0	1	1981	0.5	1	1999
94	SC	Aiken	33 25 32	81 52 48	2.9	4	1977–79	0	1	1996
95	SC	Chesterfield	34 43 55	80 02 57	8.3	7	1973–84	1.0	1	1999
96	SC	Florence	33 54 10	79 26 25	2.7	6	1970–82	0	1	1999
97	SC	McCormick	34 01 47	82 23 55	7.4	8	1971–78	0.5	2	1995–99
98	VA	Dinwiddie	36 56 34	77 29 43	4.3	10	1969–76	0	2	1996–99
	VA	Sussex								
99	VA	Southampton	36 52 26	76 57 28	5.1	12	1969–85	2.5	2	1996–99

White-spotted slimy salamanders
(P. cylindraceus)

100	NC	Ashe	36 23 52	81 31 43	3.2	5	1970–85	1.5	1	1994
101	NC	Burke	35 41 43	81 43 02	6.5	9	1970–79	2.0	1	1997
102	NC	Henderson	35 10 20	82 26 10	2.9	9	1972–87	1.0	1	1999
103	NC	Madison	36 00 30	82 36 32	27.2	5	1974–79	11.0	1	1997
	TN	Unicoi								
104	NC	Madison	35 57 15	82 33 34	3.5	12	1970–88	3.0	1	1994
	TN	Unicoi								
105	NC	Mitchell	36 08 14	82 08 30	2.1	3	1970–72	1.0	1	1994
	TN	Carter								
106	NC	Mitchell	36 06 36	82 21 40	7.1	15	1970–87	1.3	2	1993–95
	TN	Unicoi								
107	NC	Montgomery	35 13 18	79 47 25	5.1	8	1973–84	0	1	1999
108	NC	Yancey	36 01 58	82 25 33	2.1	4	1970–73	0.7	2	1994
109	TN	Johnson	36 32 25	81 56 55	10.6	3	1972–86	3.0	1	1999
	TN	Sullivan								
110	VA	Grayson	36 45 47	81 13 22	6.1	17	1960–78	1.5	1	1993
	VA	Wythe								

TABLE 8-1 (continued)

			Latitude	Longitude	Pre-1990s			1990s		
POPULATION NUMBER	STATE	COUNTY	° ′ ″	° ′ ″	MEAN	NUMBER OF VISITS	YEARS	MEAN	NUMBER OF VISITS	YEARS
111	VA	Pittsylvania	36 34 13	79 26 06	4.6	9	1967–81	0	1	1997
112	VA	Rockingham	38 34 42	79 02 08	1.9	4	1971–85	0	2	1994–96
113	WV	Hampshire	39 17 08	78 23 58	3.9	21	1964–81	1.0	2	1999

Northern slimy salamanders
(*P. glutinosus*)

114	AL	Etowah	34 04 06	86 18 43	2.6	3	1976–88	1.0	2	1992–94
115	AL	Etowah	34 02 51	86 10 38	3.3	11	1971–84	0.5	1	1994
116	AL	Macon	32 29 30	85 36 08	4.2	8	1973–86	2.7	1	1996
117	GA	Fannin	34 52 52	84 33 57	13.9	7	1974–78	0.5	1	1999
118	IL	Pope	37 22 54	88 40 20	11.2	9	1971–86	1.5	1	1996
119	IL	Union	37 32 43	89 26 14	1.2	11	1971–88	2.5	1	1996
120	IN	Crawford	38 16 35	86 32 10	4.2	7	1974–88	1.0	1	1998
121	IN	Jackson	38 50 44	86 01 25	13.6	3	1974–76	1.0	1	1998
122	IN	Parke	39 53 14	87 11 20	35.8	7	1971–77	7.0	1	1998
123	KY	Harlan	36 55 03	82 54 04	2.8	5	1980–85	1.0	1	1996
124	KY	Scott	38 23 23	84 34 02	4.0	7	1956–83	16.5	2	1990–98
125	MD	Frederick	39 37 50	77 28 17	10.9	38	1956–76	4.5	1	1999
126	NJ	Union	40 40 42	74 23 10	5.9	10	1971–86	1.0	1	1999
127	NY	Tompkins	42 19 55	76 39 34	4.8	7	1974–85	1.4	1	1999
128	NY	Ulster	41 55 44	74 06 10	11.6	10	1961–86	2.5	1	1999
129	OH	Meigs	39 02 58	81 58 45	8.4	5	1976–85	0	1	1998
130	PA	Bedford	39 47 24	78 39 40	42.6	8	1957–76	3.1	2	1992–96
131	PA	Cambria	40 42 07	78 48 08	4.2	8	1970–88	0	1	1996
132	PA	Cameron	41 28 43	78 11 27	9.4	7	1970–79	7.5	1	1996
133	PA	Snyder	40 43 09	76 59 48	9.5	9	1970–85	19.0	1	1996
134	PA	Susquehanna	41 50 06	76 02 15	6.3	9	1970–85	1.3	1	1999
135	TN	Blount	35 38 20	83 44 52	4.6	9	1962–78	0.5	1	1996
136	VA	Giles	37 14 56	80 51 48	3.3	8	1969–81	0.5	1	1999
137	VA	Russell	36 56 55	81 52 33	7.0	7	1969–79	1.5	1	1997
138	VA	Washington	36 50 13	82 05 15	4.3	6	1969–82	0.5	2	1999
139	VA	Wise	36 53 42	82 37 58	5.2	11	1980–85	1.3	2	1992–95
140	WV	Clay	38 21 27	81 07 48	4.2	7	1970–85	0	1	1998
141	WV	Pocahontas	38 25 04	79 46 10	1.2	2	1969	2.8	2	1993–97
142	WV	Preston	39 27 24	79 31 08	8.9	4	1967–76	3.1	3	1992–96
143	WV	Summers	37 47 54	80 51 00	3.5	6	1970–86	0.5	3	1990–98

Southeastern slimy salamanders
(*P. grobmani*)

144	AL	Barbour	31 42 17	85 40 13	13.0	1	1980	2.5	1	1999
145	AL	Escambia	31 11 35	87 07 39	2.6	4	1984–86	3.0	1	1996
146	FL	Jackson	30 49 25	85 18 15	18.0	9	1971–76	10.0	1	1999
147	FL	Leon	30 25 48	84 31 45	7.2	6	1979–84	2.5	1	1999
148	GA	Schley	32 10 26	84 22 17	8.2	6	1979–85	7.5	1	1999

TABLE 8-1 (continued)

POPULATION NUMBER	STATE	COUNTY	Latitude ° ′ ″	Longitude ° ′ ″	Pre-1990s MEAN	Pre-1990s NUMBER OF VISITS	Pre-1990s YEARS	1990s MEAN	1990s NUMBER OF VISITS	1990s YEARS
Cumberland Plateau salamanders (*P. kentucki*)										
149	KY	Harlan	36 55 03	82 54 04	7.1	5	1980–85	4.5	1	1996
150	VA	Wise	36 53 42	82 37 58	12.1	11	1980–85	6.9	2	1992–95
151	WV	Cabell	38 24 13	82 26 03	1.3	2	1978–80	0	2	1990–91
152	WV	Mason	38 49 19	82 09 21	9.0	3	1985–87	13.5	2	1990–96
Kiamichi slimy salamanders (*P. kiamichi*)										
153	OK	LeFlore	34 36 55	94 29 50	6.6	12	1973–84	2.5	2	1996
154	OK	LeFlore	34 36 55	94 37 52	1.4	5	1974–84	0.8	2	1995–96
Louisiana slimy salamanders (*P. kisatchie*)										
155	AR	Union	33 17 58	92 31 44	3.8	2	1982–83	0	2	1991–95
156	LA	Grant	31 43 15	92 28 02	1.1	13	1971–88	1.0	2	1992–96
157	LA	Winn	31 52 12	92 33 30	3.1	9	1973–80	0	1	1992
Mississippi slimy salamanders (*P. mississippi*)										
158	AL	Lawrence	34 18 25	87 20 10	9.5	14	1971–86	0.5	1	1996
159	AL	Tuscaloosa	33 03 54	87 43 35	5.4	8	1971–85	4.0	1	1992
160	MS	Forrest	30 55 40	89 10 36	2.3	13	1971–85	1.3	1	1996
161	MS	Lowndes	33 29 25	88 20 52	13.3	6	1973–86	5.0	1	1996
162	MS	Scott	32 24 37	89 29 02	8.5	9	1974–86	2.0	1	1996
163	MS	Winston	33 09 10	89 02 50	3.8	4	1976–84	0	1	1996
164	TN	Henderson	35 48 08	88 15 36	11.5	5	1971–77	1.0	1	1995
165	TN	Shelby	35 15 18	89 45 10	11.2	6	1973–86	2.5	1	1996
Ocmulgee slimy salamanders (*P. ocmulgee*)										
166	GA	Bulloch	32 23 03	81 49 59	4.2	11	1953–88	5.5	1	1995
167	GA	Long	32 42 36	81 45 16	4.1	3	1980–81	0	1	1995
168	GA	Wheeler	32 05 38	82 53 35	7.9	8	1971–80	2.0	1	1999
Savannah slimy salamanders (*P. savannah*)										
169	GA	Richmond	33 19 48	82 03 49	10.1	7	1979–88	0.3	1	1995
Sequoyah slimy salamanders (*P. sequoyah*)										
170	OK	McCurtain	34 07 29	94 40 15	7.5	11	1974–87	5.0	3	1990–96
Southern Appalachian salamanders (*P. teyahalee*)										
171	NC	Graham	35 21 27	83 43 08	5.5	3	1987–88	3.3	6	1992–97
172	NC	Macon	35 06 20	83 17 05	2.9	7	1968–84	0.5	1	1997
173	NC	Madison	35 48 50	82 56 58	11.3	8	1968–79	2.5	1	1997
174	SC	Pickens	34 44 32	82 50 35	1.4	8	1967–77	2.5	1	1997
175	TN	Monroe	35 27 45	84 01 37	4.2	6	1978–84	1.5	1	1997

TABLE 8-1 (continued)

			Latitude	Longitude	Pre-1990s			1990s		
POPULATION NUMBER	STATE	COUNTY	° ′ ″	° ′ ″	MEAN	NUMBER OF VISITS	YEARS	MEAN	NUMBER OF VISITS	YEARS
South Carolina slimy salamanders (*P. variolatus*)										
176	GA	Chatham	32 08 42	81 09 18	6.6	11	1953–79	1.0	1	1995
177	SC	Allendale	33 01 30	81 15 00	14.0	1	1980	1.0	1	1999
178	SC	Berkeley	33 08 00	79 47 06	8.7	15	1970–86	1.5	1	1999
179	SC	Charleston	32 53 44	80 08 12	7.5	8	1971–76	0	1	1999
	SC	Dorchester								
180	SC	Jasper	32 36 14	80 54 08	5.3	11	1971–79	4.3	1	1996
181	SC	Jasper	32 33 10	81 10 10	6.5	12	1953–76	0.7	3	1990–99
Plethodon jordani Complex Northern gray-cheeked salamanders (*P. montanus*)										
182	NC	Ashe	36 23 52	81 31 43	11.9	5	1970–85	12.0	1	1994
183	NC	Madison	35 48 50	82 56 58	27.1	8	1968–79	2.5	1	1997
184	NC	Madison	35 57 15	82 33 34	11.9	12	1970–88	8.3	1	1994
	TN	Unicoi								
185	NC	Mitchell	36 08 14	82 08 30	12.9	3	1970–72	21.0	1	1994
	TN	Carter								
186	NC	Watauga	36 10 15	81 49 38	16.5	7	1970–86	15.5	1	1994
187	NC	Watauga	36 20 53	81 47 23	21.0	9	1970–86	14.0	1	1999
	TN	Johnson								
188	NC	Yancey	36 01 58	82 25 33	11.8	4	1970–73	7.1	2	1994
189	VA	Giles	37 14 56	80 51 48	5.3	8	1969–81	2.0	1	1999
190	VA	Grayson	36 37 50	81 35 22	17.4	5	1969–74	7.0	1	1999
191	VA	Grayson	36 39 30	81 33 25	19.0	9	1957–84	4.0	1	1999
	VA	Smyth								
192	VA	Russell	36 56 55	81 52 33	27.6	7	1969–79	20.5	1	1997
193	VA	Washington	36 50 13	82 05 15	13.6	6	1969–82	2.0	1	1999
Southern gray-cheeked salamanders (*P. metcalfi*)										
194	NC	Henderson	35 10 20	82 26 10	17.5	8	1972–87	0.5	1	1999
195	NC	Macon	35 06 20	83 17 05	15.4	7	1968–84	10.5	1	1997
196	NC	Macon	35 03 36	83 11 28	10.2	4	1971–73	5.0	2	1992–94
197	SC	Pickens	34 44 32	82 50 35	1.7	8	1967–77	2.5	1	1997
South Mountain gray-cheeked salamanders (*P. meridianus*)										
198	NC	Burke	35 35 08	81 41 22	24.4	6	1974–87	0.5	1	1999
	NC	Cleveland								
Jordan's salamanders (red-cheeked salamander) (*P. jordani*)										
199	NC	Swain	35 35 00	83 23 54	14.1	9	1961–75	2.3	1	1994

TABLE 8-1 (continued)

			Latitude	Longitude	Pre-1990s			1990s		
POPULATION NUMBER	STATE	COUNTY	° ′ ″	° ′ ″	MEAN	NUMBER OF VISITS	YEARS	MEAN	NUMBER OF VISITS	YEARS
Red-legged salamanders (*P. shermani*)										
200	NC	Cherokee	35 14 24	84 03 07	12.9	3	1966–73	7.4	2	1994–97
201	NC	Macon	35 09 44	83 35 00	31.8	6	1976–80	6.0	1	1997
Cheoah Bald salamanders (*P. cheoah*)										
202	NC	Graham	35 21 27	83 43 08	8.5	3	1987–88	6.7	6	1992–97
Plethodon wehrlei Group Wehrle's salamanders (*P. wehrlei*)										
203	PA	Cambria	40 42 07	78 48 08	6.1	8	1970–88	0	1	1996
Cow Knob salamanders (*P. punctatus*)										
204	VA	Rockingham	38 34 42	79 02 08	3.7	4	1971–85	0	2	1994–96
205	WV	Pendleton	38 41 36	79 05 44	3.7	7	1966–87	1.3	2	1995–96

that there are 28 populations with a difference of greater than 10 per person per population, but only 12 with a difference of less than 10. Otherwise, the variance of the difference between the earlier and later collections is similar (Fig. 8-1A). Of the 166 populations sampled more than once in the decade of the 1970s, the mean number taken per person in the last collection was lower than that of the first collection of the decade in 69 populations and higher in 93 populations (it was the same for four populations). A sign test indicates that the difference between the first and last collections is not significant at the 0.05 level of significance ($\chi^2 = 3.56$, p > 0.05). Regressions of the mean number of salamanders collected per person plotted against the year of collection do not show slopes significantly different from zero (not shown).

When the entire data set is analyzed in a similar way, except that the mean number of all pre-1990s collections is compared with the mean observed during visits in the 1990s (the latter often is based on a single visit), a trend appears (Fig. 8-1B). The mean number seen at all 205 populations from the pre-1990s is 8.77, compared to 3.65 for the 1990s, which is only 41.6% of that in the pre-1990s. During the 1990s, the mean was higher for only 25 of 205 populations, while it was lower for 180 populations, a startling result of high statistical significance (sign test: $\chi^2 = 117.2$, p < 0.001).

The extent of decline varies among populations, although all species sampled seem to have been affected. Some populations may not have declined at all; in fact, more individuals were recorded in 25 populations in the 1990s than were seen on visits in the pre-1990s. Of populations that have lower counts during the 1990s, 41 had counts of 50% or more of the mean of the earlier collections. However, many populations appear to have crashed—139 of 205 populations sampled in the 1990s have less than 50% of the mean of the pre-1990s' collections, and 50 populations have less than

10%. No salamanders were seen in the 1990s in 32 of the latter 50 populations.

The decline across species appears to be widespread with the exception of Valley and Ridge salamanders and their sibling species, Shenandoah Mountain salamanders. The former is the only species that has more increases (7) than decreases (5) when the 1990s observations are compared with the pre-1990s collections. Only three samples of Shenandoah Mountain salamanders are included in my analysis, and two of these have declined (Table 8-1). During the earlier fieldwork, both Valley and Ridge and Shenandoah Mountain salamanders were rarely as abundant as other small, eastern *Plethodon* species such as eastern red-backed salamanders (Highton, 1972). From our fieldwork in the 1980s and 1990s on almost 100 populations of Valley and Ridge salamanders and Shenandoah Mountain salamanders, I generally found the abundance of these two species to be about the same as it was during the 1950s and 1960s. (I did little fieldwork with these two species during the 1970s.) Most populations of Valley and Ridge salamanders and Shenandoah Mountain salamanders were not included in this study because mean numbers in earlier collections were fewer than one per person per visit. Of almost 100 populations of these two species visited during the 1980s and 1990s, only three populations (27, 50, and 52) appear to have had a noticeable decline in abundance.

Shenandoah salamanders (*P. shenandoah*) are one species not listed in Table 8-1. I conducted class field trips every year throughout the 1990s to Hawksbill Mountain, Page County, Virginia, to observe this species (listed as endangered both by Federal and State agencies); the salamanders appeared to be as abundant on all these visits as they were in the 1960s and 1970s, although no counts were made.

Some species of *Plethodon* are often found in different relative abundance than others (Table 8-1). For example, members

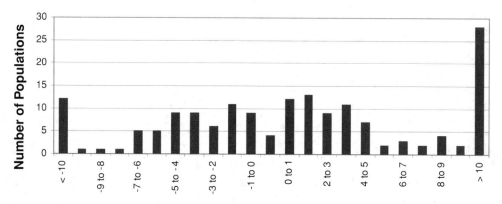

A

Difference in Mean Number of Salamanders
1970 to 1980

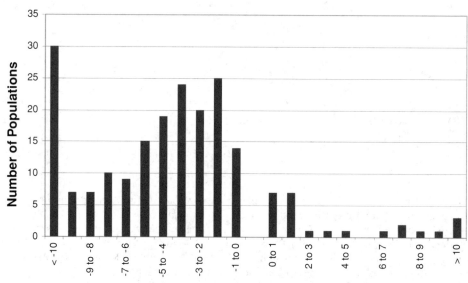

B

Difference in Mean Number of Salamanders
Pre-1990s to 1990s

FIGURE 8-1 **(A)** The difference between the mean number of salamanders found per person in the first and the last collections in 166 populations collected more than once during the period from 1970 through 1980. **(B)** The difference between the mean number of salamanders found per person in 205 populations in the pre-1990s and the mean number per person observed in the 1990s.

of the *P. jordani* and *P. ouachitae* complexes are usually much more abundant that those of the *P. glutinosis* complex where the latter is sympatric with the other two complexes (Highton, 1970, 1995b).

At 70 of the sites, a single species was encountered in sufficient numbers to be included in the analysis. At 57 sites, I have data for more than one species of *Plethodon* (two species at 39 sites, three species at 15 sites, and four species at 3 sites). Sometimes more than that number occur at a single site (up to five species of *Plethodon* sometimes occur syntopically), but some species were so rarely found in the pre-1990s that they were not included in the analysis. The pattern of change in abundance (or lack thereof) is usually concordant among species at the same site. At three sites where four species are included in the analysis, I found that (1) at the site for populations 17, 41, 137, and 192, declines probably have occurred in three species— only northern gray-cheeked salamanders (*P. montanus*) were still abundant there on our 1997 visit; (2) at the site for populations 8, 53, 66, and 106, all four species appear to have declined based on two visits in the 1990s; and (3) at the site for

populations 7, 65, 105, and 185, none of the three species seen on the 1994 visit appear to have declined.

Fifteen sites had three species included in the analysis. At 12 of these sites, the mean of all three species decreased (only two of these sites had been logged), while at one site, the mean of all three species increased slightly. There was a mixture of increases and decreases at only two sites.

Thirty-nine sites had two species included in the analysis. At 35 sites the mean of both species was lower in the 1990s (four sites had been logged), at one site both were higher in the 1990s, and at three sites one was higher and the other was lower. These figures differ from what would be expected by chance, but not much significance can be placed on these data because of the possible lack of independence (e.g., weather conditions might have affected the activity of both species in the same way, so that the observations for different species at the same site might be correlated). However, if there is a tendency for changes in the abundance of all species of *Plethodon* at the same site to be correlated, this might be relevant in reducing the likelihood of some potential causes of declines, such as competition between species (see Jaeger, 1971b; Hairston, 1980a,b).

My overall subjective impression based on 35 years of field experience on eastern North American *Plethodon* (beginning in 1951) was that in populations where salamanders were abundant on early visits they tended to remain common, while in populations where salamanders were initially scarce they continued to be rarely seen. Throughout the 35-year period, when I returned to hundreds of sites more than once to obtain sufficient material for our studies, I noted very few instances in reasonably favorable weather conditions in which no *Plethodon* were found. In the group of 205 populations considered in Table 8-1, there are only three instances of not finding a common species in the pre-1990s (population 87 in 1984, population 95 in 1974, and population 157 in 1980). In the first two of these cases, subsequent visits indicated that salamanders were still there in numbers roughly equivalent to that of earlier visits. In the third case, logging destroyed the forest soon after the 1980 visit, and no salamanders have since been seen. However, beginning in the last half of the 1980s, at several sites that were revisited apparent declines were noted. Three examples of such declines are as follows:

At a site in South Carolina (population 97), Atlantic Coast slimy salamanders (*P. chlorobryonis*) were abundant during all eight visits from 1970–78. On a visit in 1988, none were seen although conditions were dry. Again in 1995, when conditions were ideal, none were found. In 1999, only two were observed by two collectors, again in favorable conditions.

Four species were present at a site in West Virginia, but only two species (eastern red-backed and northern slimy salamanders) were abundant (populations 24 and 143). On six visits in favorable weather conditions from 1970–86, the mean number found per person was 3.9 and 3.5, respectively. On a 1988 visit in favorable conditions no salamanders were seen. Similarly, in three visits during the 1990s, few were observed (mean of 0.2 and 0.5, respectively) and two other species that occur at the site (Valley and Ridge salamanders and Wehrle's salamanders [*P. wehrlei*]) also were much less often encountered.

At a site in South Carolina, South Carolina slimy salamanders (*P. variolatus*; population 181) were abundant from 1953–76. When the site was next visited in early 1990 and again in 1994, no South Carolina slimy salamanders were found. In 1999, only four were seen by two collectors. This population likely crashed in the 1980s. Thus, it is possible that whatever is the cause of the declines may have occurred (or begun) in the mid- or late-1980s. The less frequently observed marbled salamanders (*A. opacum*) probably have not declined at this site during the 1990s; more (4–5) were seen on all three of our visits in the 1990s than on any of our 12 visits from 1953–76 (0–3).

Declines appear to be present in populations of eastern *Plethodon* throughout the range of the genus in the eastern United States. Declines do not appear to be restricted to some areas.

It is well known that destruction of forest habitat is a major cause of population declines in many species of animals including salamanders (Ash, 1988; Petranka et al., 1991). Considerable habitat destruction occurred at 16 of the 127 sites (22 populations) due to logging during the 1980s and 1990s. At four of the sites only a part of the area we collected earlier had been destroyed, while the forest of the entire area has been almost completely logged at the remaining 12 sites. Not surprisingly, most of these populations are among those that have declined. There is evidence that at least one of these

populations declined before the logging took place, and at another two sites there are few salamanders left in adjacent similar habitat that still remains. The following describes our findings from these three populations.

Population 130

This population of northern slimy salamanders from Pennsylvania were sampled seven times (1957–59) for a life history study (Highton, 1962b). Salamanders were found more abundantly at this site than any other site for the 16 species of the slimy salamander complex I have collected (mean = 42.6 per person per visit). Semlitsch (1980b) also made collections at this site during the 1970s. In 1976, I made one collection and northern slimy salamanders were still abundant (57 per person). In 1992, northern slimy salamanders were less commonly observed (4.3 per person). Between 1992 and 1996, the site was logged and only two individuals per person were seen on the 1996 visit. Thus, the decline predated the destruction of the forest.

Population 162

During nine visits to a site in Mississippi from 1974–86, Mississippi slimy salamanders (*P. mississippi*) were always commonly found (8.5 per person per visit). The area had been almost completely logged by 1996, except for a small strip along a highway. Only a few (mean = 2 per person) were found by searching the remaining forest. In an unlogged area (across the highway) of equal size to the destroyed forest, salamanders were also rare in 1996, even though they had been abundant in the 1970s and 1980s.

Population 178

During 15 visits to this site from 1971–86, South Carolina slimy salamanders were always found in abundance (8.7 per person). In 1989, Hurricane Hugo blew down a large number of trees and the U.S. Forest Service cleared the area of downed trees. In the process, most of the forest was clearcut. We found no salamanders in 1990. On a visit in 1999 in an area that apparently was not damaged severely by either the hurricane or the logging operations, two persons found only three individuals.

The only obvious destruction of habitat by human activity (logging) that I observed was at a small proportion of the sites (16 of 127), and in some cases the salamander decline preceded the habitat destruction. The reason(s) for widespread declines in *Plethodon* populations is (are) unknown and in need of investigation. There is no evidence that the observed declines are due to natural fluctuations in population densities (see Pechmann et al., 1991; Pechmann and Wilbur, 1994; Green, this volume). My observations on hundreds of populations do not indicate that the size of any *Plethodon* population fluctuated substantially before the mid-1980s, when declines were first noted.

Although declines in populations of *Plethodon* appear to be threatening, all the species of *Plethodon* reported here are still known to be common at some sites, and most species generally remain the most frequently encountered terrestrial salamanders in eastern North America. I would not recommend placing any of these species on endangered or threatened lists until studies are made to determine their current distribution and abundance. However, in light of the apparent widespread declines in eastern *Plethodon*, it would seem crucial to find explanations for these reductions in this ecologically important component of our eastern forest ecosystems.

Summary

My colleagues and I observed little change in the abundance of 44 species of salamanders of the genus *Plethodon* in eastern North America from 1951–85 at hundreds of sites repeatedly visited. During the late 1980s, we began to observe declines in a few populations. In response, in the 1990s we revisited 127 sites in 22 states and counted the number of salamanders seen in 205 populations of 38 eastern species to compare with those in collections from the same localities in the pre-1990s. The mean number of salamanders observed per person per visit in the 1990s was only 41.6% of that in the pre-1990s. The mean number seen in the 1990s was higher in only 25 populations and it was lower in 180 populations, a statistically significant difference. When compared to the number of salamanders found in the pre-1990s, fewer than 50% of the number of animals were observed in 139 populations and fewer than 10% were seen in 50 populations (including none in 32 populations). The cause(s) of these declines is (are) unknown, except that extensive habitat destruction by logging occurred at 16 sites (22 populations); but this accounts for only a small proportion of the observed declines.

Acknowledgments. I thank all those who have helped me with fieldwork over the years I have been studying *Plethodon*. Special thanks go to D. E. Green, who accompanied me on a majority of my field trips in the 1990s. For making many helpful suggestions on the manuscript, I thank G. M. Fellers, D. E. Gill, S. B. Hedges, J. C. Maerz, S. G. Tilley, and D. B. Wake. Thanks also are due to the National Science Foundation for supporting my fieldwork from 1958–88 and to the various state departments and National Parks and Forests for issuing collecting permits.

NINE

Decline of Northern Cricket Frogs *(Acris crepitans)*

ROBERT H. GRAY AND LAUREN E. BROWN

At the end of the nineteenth century, there were indications that northern cricket frogs *(Acris crepitans)* were numerous in the midwestern United States. Garman (1892) found the species was "one of the most abundant members of the family in all parts of Illinois," and Hay (1892) similarly indicated it was "one of our commonest batrachians" in Indiana. This abundance continued well past the middle of the twentieth century in both of these states (Smith, 1961; Minton, 1972). Campbell (1977) published the first report of a cricket frog decline in the relatively small area occupied by the species in extreme southern Ontario, Canada. This was puzzling because at least part of the area was protected as the Point Pelee National Park. Subsequently, Vogt (1981) reported a sharp cricket frog decline in Wisconsin, and numerous declines have been and continue to be reported throughout the upper Midwest and elsewhere: Illinois (Mierzwa, 1989, 1998a; Ludwig et al., 1992; Greenwell et al., 1996); Indiana (Minton, et al., 1982; Vial and Saylor, 1993; Brodman and Kilmurry, 1998; Minton, 1998); West Virginia (T. Pauley, personal communication); Iowa (Christiansen and Mabry, 1985; Lannoo et al., 1994; Lannoo, 1996, 1998a; Christiansen and Van Gorp, 1998; Hemesath, 1998); Michigan (Harding and Holman, 1992; Harding, 1997; Lee, 1998); Minnesota (Vial and Saylor, 1993; Oldfield and Moriarty, 1994; Moriarty, 1998); Wisconsin (Jung, 1993; Casper, 1996, 1998; Hay, 1998a,b; Mossman et al., 1998); Colorado (H. Smith, personal communication; Hammerson and Livo, 1999); Ohio (Lipps, 2000); and Canada (Oldham, 1990, 1992; Weller and Green, 1997). Despite numerous reports of decline and ample scientific literature on the biology of cricket frogs, there is no clear-cut indication of the cause(s) of this trend toward extinction. However, a number of anthropogenic factors and environmental conditions have been suggested.

Interestingly, as northern cricket frogs have become extinct in a number of peripheral areas, populations remain stable in the more central regions of the species range. Unlike the more celebrated frog declines, this is not a post-mortem case, and cricket frog declines thus present an excellent opportunity to study the process of amphibian extinction. Furthermore, amphibian declines are of obvious significance to the future of humankind (Wake, 1991; K. Phillips, 1994). Consequently, the objective of this contribution is to review the relevant biology and trend toward extinction of this widely distributed species.

Morphological Characteristics

Because northern cricket frogs may become extinct, we provide the following in-depth morphological description. Maximum reported snout-vent length is 38 mm (Conant and Collins, 1991), with males smaller (on average) than females. Skin on the dorsum, including the head, is rough with warts and ridges. The venter is granular. The snout is pointed and the male has a single subgular vocal sac. The hind legs are long, as are the hind toes, which have extensive webbing. There are two metatarsal tubercles and a tarsal fold on each hind limb. Toe pads are small. There are two relatively large palmar tubercles, with one at the base of the thumb and the other being adjacent. A fold occurs across the chest between the axillae. There is a pair of white tubercles (round, elongated, or of irregular shape) below the cloacal opening. The dorsal background coloration is typically black, brown, tan, olive green, or gray, although Gray (1995) reported some individuals to be entirely bright green dorsally. A vertebral stripe shows color polymorphism (red, green, and gray). Small green spots sometimes occur on other areas of the dorsum. Between the eyes is a dark triangular spot with the apex pointing posteriorly. On each side of the body is a dark stripe that is sometimes broken into spots. A dark stripe with irregular edges occurs on the posterior surface of the thigh; this stripe is sometimes broken into elongated spots. A white stripe extends between the base of the forelimb and the eye (this is often not clearly evident in preserved specimens). The limbs have dark bars (or spots) dorsally. The upper jaw has narrow light and wide dark bars. The pattern sometimes appears spotted or reticulated. The venter is white or cream with small dark flecks in the gular region. The male vocal sac is yellow with dark flecks.

A detailed description of cricket frog eggs was given by Livezey (1950). The vitelli average 1.13 mm in diameter and are surrounded by two gelatinous envelopes. Up to 400 eggs are deposited singly or in small clumps of 2–7 eggs. Eggs float on the surface of the water, are scattered on the bottom, or are attached to stems of submergent vegetation.

Northern cricket frog tadpoles are notable for their large size (maximum total length = 47 mm; Wright and Wright, 1949) relative to the snout-vent length of the adult. Typically, the tadpole tail has a black tip, although sometimes this pigmentation is lacking. The free portion of the spiracular tube is one-half (or less) of the tube's length, the throat is light, and the tail musculature is reticulated or mottled (Altig, 1970). Detailed descriptions of the tadpole were provided by Walker (1946), Altig (1970), Minton (1972), Vogt (1981), and Conant and Collins (1991).

Distribution

The historic distribution of northern cricket frogs is primarily across the eastern half of the United States (Conant and Collins, 1991, map 311). The extremities of the historic range extended from southeastern New York, south to the eastern panhandle of Florida, west to western Texas and southeastern New Mexico, north to western Nebraska and southeastern South Dakota, and east across central Wisconsin and central Michigan. They do not occur on the Atlantic Coastal Plain (except for relictual populations) from southeastern Alabama to southeastern Virginia, nor do they occur in the Appalachian Mountains from southwestern North Carolina north or their foothills in eastern Ohio. Cricket frogs also occur in northern Coahuila, Mexico (not far from the U.S./Mexican border), and historically occurred in extreme southern Ontario, Canada (Point Pelee Peninsula and Pelee Island in Lake Erie; Oldham, 1992).

In addition to declines reported in the north, which have produced scattered and isolated populations, northern cricket frog populations are scattered and isolated along the southeastern and western peripheries of their range. For example, there are a number of isolated records along the more arid western edge of the distribution (e.g., Nebraska [Lynch, 1985]; Colorado [Hammerson, 1986; Hammerson and Livo, 1999]; Texas [Milstead, 1960]; New Mexico [Degenhardt et al., 1996]; and Arizona [Frost, 1983]). Furthermore, isolated populations have been reported on the Atlantic Coastal Plain of the southeastern United States in North and South Carolina (Conant, 1975, 1977a).

Habitat

Northern cricket frogs are terrestrial and semiaquatic. Individuals can potentially occur in or near virtually any body of fresh water, although they tend to be rare or absent at large lakes, wide rivers, and polluted sites. Cricket frogs typically occupy a zone along the water's edge sitting on floating vegetation and debris or on exposed beach-like banks. When approached by a potential predator, cricket frogs make several quick zigzag leaps (each often 1 m or more in length) on the bank and/or floating vegetation. They then dive underneath the water or come to rest on the shoreline some distance away. Experimental studies show a progressive decline in locomotory performance over an extended period of consecutive jumps (Zug, 1985). Cricket frogs sometimes occur in cattail (*Typha* sp.) thickets as well as in other terrestrial and/or aquatic shoreline vegetational assemblages. Occasionally, cricket frogs are heard calling from floating vegetation over deeper water a considerable distance from the edge of a pond. In some flood plain forests, cricket frogs are often encountered on damp soil and under logs some distance from aquatic sites. After rains, cricket frogs disperse to more distant aquatic sites (Gray, 1983; Burkett, 1969, 1984) and often pass through drier, suboptimal terrain.

Northern cricket frogs in central Illinois overwinter in cracks in the banks formed by the freezing, drying, and subsequent contracting of the soil (Gray, 1971a). Other workers have found them during the winter under logs or overhanging

banks, under masses of compact dead vegetation or leaves, and buried in gravelly soil (Garman, 1892; Walker, 1946; Pope, 1964a; Irwin, this volume). On warm winter days in southern Illinois, cricket frogs are sometimes active on ice (L.E.B., personal observations). During a period of severe drought in Texas, W.F. Blair (1957) reported that cricket frogs avoided desiccation by taking refuge in the deep cracks of a dry pond bed.

Life History

The timing of various life history events varies throughout the range of northern cricket frogs. Metamorphosis begins in late July or early August, and continues for about three weeks. Population numbers are greatest immediately following metamorphosis. Growth of newly metamorphosed frogs is rapid during the fall. Individuals can grow from 12 to 26 mm in 2–3 months (Gray, 1983). Cricket frogs are opportunistic, and apparently feed both day and night (Johnson and Christiansen, 1976) on whatever small invertebrates are available on the ground or in the air. Food items include aquatic and terrestrial insects and spiders (Garman, 1892; Hartman, 1906; Jameson, 1947; Gehlbach and Collette, 1959; Labanick, 1976b) and occasionally annelids, mollusks, crustaceans, and plant matter (Hartman, 1906; Johnson and Christiansen, 1976). Food consumption is greater in frogs that are larger, female, reproducing, have small, fat bodies, and/or collected in June and July (Johnson and Christiansen, 1976). Growth stops in winter, and individuals may actually be smaller when first observed in the spring, possibly due to overwinter metabolism of (and subsequent reduction in) body lipids and nonlipid material in the liver (Long, 1987a). Growth again resumes in the spring and continues until death (Gray, 1971b, 1983; Fig. 9-1). Even after reproduction, females have significantly more lipid than males (Blem, 1992). Maximum utilization of fat in both sexes occurs during breeding and overwintering (Brenner, 1969; Johnson and Christiansen, 1976).

In Illinois (Gray, 1983) and Kansas (Burkett, 1984), a drastic reduction in the number of northern cricket frogs occurs from metamorphosis until breeding. In Illinois, this reduction ranges from 50% to 97%. Thus, in many populations of cricket frogs, the number of potential breeding individuals is low. The critical period (when cricket frogs in Illinois are most susceptible to predation) is in the fall, when water recedes and pond banks are exposed. Animals known or observed to prey on cricket frogs include American bullfrogs (*Rana catesbeiana*), fishes (e.g., bass, *Micropterus* sp.), snakes (*Thamnophis sirtalis*, *T. radix*, *Nerodia sipedon*, *N. harteri paucimaculata*), turtles (R.H.G., personal observations), birds (*Cassidix mexicanus*, *Falco sparverius*), and mammals (Carpenter, 1952; Lewis, 1962; Fitch, 1965; Wendelken, 1968; Gray, 1978; Fleming et al., 1982; Greene et al., 1994). Adults from the previous year usually have disappeared by the end of October. Interestingly, most cricket frogs that survive to overwintering survive to the spring emergence (Fig. 9-1). Thus, overwintering is *not* a major cause of mortality in Illinois populations (Gray, 1983).

The dispersal of northern cricket frogs from one population to another tends to occur after rains. Gray (1983) observed cricket frogs moving up to 1.3 km between farm ponds in central Illinois. Frogs that dispersed from one location and were later recaptured at another represented 1–7% of the original population. However, many more frogs likely dispersed and either reached other sites or died in transit.

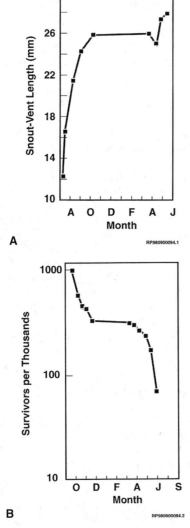

A

RP980900094.1

B

RP980900094.2

FIGURE 9-1 **(A)** Example growth curve for cricket frogs (adapted from Gray, 1983; English Pond A near Bloomington, Illinois, 1969–1970). **(B)** Example survivorship curve for cricket frogs (adapted from Gray, 1983; English Pond B near Bloomington, Illinois, 1968–1969).

As temperatures decline in the fall, northern cricket frogs become less active and usually are last seen in late October in central Iowa (Johnson and Christiansen, 1976) or late November in central Illinois (Gray, 1971a). However, they may be active during the winter on warm days or where there are springs in some areas of Illinois, west-central Indiana, Kansas, Oklahoma, Texas, Louisiana, and elsewhere in the south (Blatchley, 1892; Pope, 1919; Linsdale, 1927; Jameson, 1950a; Blair, 1951b; Pyburn, 1956, 1961a; Gray, 1971a).

Spring emergence for northern cricket frogs typically begins in early March in southern Illinois and in late March or early April in central Illinois (Gray, 1983), and in early April in central Iowa (Johnson and Christiansen, 1976). In some cricket frogs, the male vocal pouch begins to develop in October, while in other individuals, the vocal pouch is not apparent until April or early May. Male cricket frogs are probably the first to arrive at breeding sites (Pyburn, 1958; Gray, 1983). In Illinois, calling males are first heard in late April or early May and last heard in July (Gray, 1983) or early August (L. E. B., personal observations); in central Iowa, calling males are first heard in mid-May and are last heard in late July (Johnson and Christiansen, 1976). The

male mating call is a sharp "glick, glick, glick . . ." and resembles two glass marbles or small stones being struck together. The repetition rate of the "glicks" is faster in the middle of a call sequence than at the beginning or end. Males sit on the bank or on floating aquatic vegetation and call during the day and night.

Distance to the nearest neighbor appears to have the greatest influence on variation in male calling behavior, with the most profound changes occurring during male–male encounters (Wagner, 1989a,b). Calls at the middle and end of a call sequence are modified in response to male–male aggressive interactions. The hormone arginine vasotoxin increases the probability of male calling and noticeably alters the middle and end of a call sequence of more aggressive males from that of less aggressive males (Marler et al., 1995). Males may signal information about their size and other attributes through changes in call dominant frequency (Wagner, 1989a,c, 1992; but see Crawford, 2000). We have not seen any male aggressive behavior over many years of field work in Illinois. Crawford (2000) likewise did not detect such behavior. Mate preference of female northern cricket frogs is influenced by three call characteristics: dominant frequency, number of pulses per call, and number of pulse groups per call (Ryan et al., 1995). Females attracted to calling males are clasped in axillary amplexus.

The number and intensity of male choruses suggests that breeding activity peaks in May, when population sizes are again declining. Females mobilize body lipids during ovarian development (Long, 1987a). Gravid females are identified by developing eggs (visible through skin in the groin) and are present from late April to late August. In Illinois, females mature eggs once annually. The sex ratio of northern cricket frogs in Illinois favors males by about 3:1 and appears even higher at some locations (see "Effective Breeding Size of Populations," and Gray, 1984). Interestingly, some gravid females can be found after the last calling males are heard (Gray, 1983), which suggests that perhaps not all females breed.

Northern cricket frog tadpoles feed on periphyton and phytoplankton (Johnson, 1991). In ponds, tadpoles that co-occur with dragonfly (*Anax* sp.) larvae tend to have black tails, while tadpoles in lakes and creeks where fish predominate are mostly plain tailed (Caldwell, 1982a). This polychromatism may reflect differential selection by predators. In experimental studies, cricket frog tadpole abundance was reduced in acidified macrocosms (as compared to circumneutral ones) and in macrocosms with loam soils (as compared to clay soils; Sparling et al., 1995). Contaminants in the tadpole-inhabited waters may be accumulated in tissues, causing not only acute and chronic effects (Boschulte, 1993, 1995; Greenwell et al., 1996), but also effects that may be passed to higher trophic levels through predation (Hall and Swineford, 1981; Fleming et al., 1982; Sparling and Lowe, 1996).

In several respects, northern cricket frog populations in the south differ from those in the north. Evidence suggests that cricket frogs in Texas and Louisiana are active year round and probably experience two breeding peaks (Pyburn, 1961a; Bayless, 1966); related southern cricket frogs (*A. gryllus*) are also prolonged breeders (Forster and Daniel, 1986). Additionally, sex ratios of northern cricket frogs in the south approximate 1:1 and breeding population sizes are larger (Jameson, 1950a; Blair, 1961a). (See the following sections for additional characteristics of northern populations.)

Color Polymorphism

Juvenile and adult northern cricket frogs exhibit a genetically based color polymorphism (Pyburn, 1961a,b) involving the vertebral stripe. The stripe can be of a varying shade of red (red, brown, reddish brown), green, or gray (matching the non-red or non-green dorsum of the animal). The green stripe is dominant to the non-green dorsum (i.e., red or gray), while the red stripe is dominant to a gray dorsum. With few exceptions, the gray morph is consistently more abundant than the red or green in Kansas (Gorman, 1986), Illinois (Gray, 1983), Indiana (Issacs, 1971), Texas and Louisiana (Pyburn, 1961a,b), and in other parts of the species' range (Nevo, 1973).

Factors that determine color morph frequencies have been the subject of much study and speculation. Possibilities include (1) the founder principle (Mayr, 1942); (2) genetic drift or chance operating in populations with extremely small effective breeding sizes (Gray, 1984); (3) visual selection by predators (Goin, 1947a; Pyburn, 1961a; Mathews, 1971; Nevo, 1973; Stewart, 1974), including apostatic selection (Milstead et al., 1974); (4) selection for a behavioral characteristic associated with a particular morph (Wendelken, 1968); (5) selection for an associated physiological factor (Goin, 1947a; Test, 1955; Main, 1961, 1965; Merrell and Rodell, 1968; Dasgupta and Grewal, 1968; Williams et al., 1968; Nevo, 1973); and (6) a combination of chance, visual selection by predators, and physiological factors (Jameson and Pequegnat, 1971; Nevo, 1973).

Mathews (1971) pointed out that during the tadpole stage, selection would have to act at the physiological or behavioral level, as adult color phases have not yet developed. Thus, the trend has been to attribute the adaptive value of amphibian color morphs to certain cryptic physiological or behavioral differences that may be controlled by pleiotropic (linked) genes. However, there is little conclusive evidence that adaptive differences actually exist, and in populations with small effective breeding sizes (see below), chance may play a major role in determining color morph frequencies (Gray, 1984).

In Texas, Pyburn (1961a) studied northern cricket frogs and found differences in color morph frequencies related to habitat. He suggested that visual selection by predators directly determined color morph proportions. However, from studies of cricket frogs using birds as predators, Wendelken (1968) disagreed with the visual selection hypothesis and suggested that natural selection may favor some behavioral characteristics associated with a particular morph. Gray (1978) conducted behavioral studies and predation experiments with cricket frogs from Illinois using garter snakes, water snakes, and American bullfrogs as predators. He found no evidence for morph-specific behavioral differences or differential predation susceptibility in those populations. Furthermore, he found no differences among morphs related to the timing of various life history events, individual movements, dispersal, growth, or survivorship rates.

Nevo (1973) provided data on color morph proportions at 64 localities throughout the northern cricket frog range and concluded that the polymorphism was stable but varied spatially and temporally. He suggested that color morph frequencies were correlated with local substrate color, thus supporting Pyburn's (1961a) visual selection hypothesis. Although Gray (1983) found some annual and geographic differences among color morph frequencies in Illinois, he found no statistically significant seasonal differences. Nevo (1973) presented evidence that larger gray morphs in Texas were more resistant to drying than red morphs, which may explain the lower frequency of red and green morphs in the arid west and northern parts of the species range. However, Gray (1977) found no differences in desiccation rates and resistance to thermal stress among cricket frog color morphs in Illinois.

The contradictory conclusions of various authors concerning the factors that determine northern cricket frog color morph frequencies suggest that different selective factors may be involved in different geographical regions and/or at different times of the year. For example, selection may depend on physiological (Nevo, 1973) or behavioral (Wendelken, 1968) differences in Texas where breeding sizes are large, but in Illinois (Gray, 1977, 1978), chance may override natural selection when the effective breeding size (see below) of a population is small (Gray, 1984).

Further complicating the issues of color polymorphism are observations that the vertebral stripes of northern cricket frogs are highly variable and may undergo metachrosis (color change). Changes involving background color in cricket frogs are well known, but various authors disagree as to what extent metachrosis involves the vertebral stripe. Le Conte (1855), Abbott (1882), Cope (1889), Hay (1902), and Ditmars (1905) all suggested that cricket frogs could rapidly change vertebral stripe color. Dernell (1902) failed to observe these changes in the laboratory and believed the red and green stripes were permanent. Based on field studies with marked cricket frogs and laboratory observations, Pyburn (1961a) concluded that metachrosis involved the green (but not the red) vertebral stripe. Gray (1972), however, showed that both red and green vertebral stripes can change color and that this change may occur in a large number (>20%) of frogs within a population.

Effective Breeding Size of Populations

The adaptive significance of northern cricket frog color polymorphism has most often been attributed to certain cryptic physiological or behavioral differences among morphs that may be controlled by pleiotropic genes. However, there is little convincing evidence that adaptive differences actually exist, and most authors have paid little attention to Wright's (1931, 1948) concept of effective population breeding size (but see Green this volume). To ascertain the effects of selection versus chance in determining color morph frequencies in a population, effective breeding size (determined by the number of actively reproducing females in the population) must be considered.

There have been few attempts to estimate effective breeding size in natural populations (Dobzhansky and Wright, 1941; Kerster, 1964; Tinkle, 1965; Merrell, 1968; Gray, 1984; Gorman, 1986). When effective breeding populations are as small as those reported by Merrell (1968) for northern leopard frogs (*R. pipiens*), any associated physiological and behavioral differences among morphs may have little meaning. Several studies (see review by Dubois, 1980) indicate that for many anuran populations, different individuals have different chances of breeding due to the temporal and spatial structure of reproductive populations. Behavioral attributes such as territoriality and dominance (as well as other factors) will also reduce effective breeding size.

Northern cricket frogs in Illinois undergo a drastic reduction in numbers during the fall (Gray, 1983). This results in spring breeding populations that are often numerically small. Additionally, other factors (unequal sex ratios and failure of some adults to breed) result in effective breeding populations that are even lower than the actual breeding population size. Based on sex ratios alone, calculated effective breeding sizes of cricket frog populations in Illinois ranged from 27 to 107 (the number of potentially breeding females; Gray, 1983).

Factors that determine sex ratios in northern cricket frog populations have only recently been suggested. Reeder et al. (1998) studied the relationships among intersex gonads, sex ratios, and environmental contaminants in several populations of Illinois cricket frogs. Sex ratios from sites contaminated with polychlorinated biphenyl (PCB) and polychlorinated dibensofuran favored males, suggesting that either females are more susceptible to these contaminants or that these contaminants may influence sexual differentiation, which would exert a male bias, in this species.

In other parts of the species range, factors that reduce effective breeding size may not be as prevalent as in Illinois. For example, in the south (e.g., Texas and Louisiana), several other life history phenomena tend to increase effective breeding size. Larger initial populations, more frequent breeding, and a life span that may exceed one year would reduce the impact on effective breeding size that results from failure of individuals to breed.

Threats

The twentieth century may be best known to future generations as the "Age of Environmental Destruction." Our wetlands and waterways provide poignant examples of this degradation. The plethora of anthropogenic factors that has adversely affected wetlands has potentially contributed to the decline of northern cricket frogs. Natural phenomena, by themselves or in concert with anthropogenic effects, may also be at work. Examples of natural and anthropogenic factors include:

1. biocides (herbicides [terrestrial and aquatic], fungicides, insecticides, rodenticides, lampricides, formalin, etc.) and agricultural fertilizers

2. byproducts of mining (particularly coal and lead); water pollution by railroad byproducts, heavy metals, and PCBs; and other industrial pollutants

3. acid rain; high nitrate accumulation; low dissolved oxygen content of water; human wastewater and sewage; and fecal contamination of water by cattle, hogs, and Canada geese (*Branta canadensis*)

4. predation on metamorphosing stages (large fish, herons, egrets, shorebirds, ducks, turtles, garter snakes, water snakes, American bullfrogs, mink [*Mustela vison*], and raccoon [*Procyon lotor*]); predation on tadpoles (various aquatic insects and fish); and predation on eggs (European carp [*Cyprinus carpio*], other fish, leeches, and bullfrog tadpoles)

5. ultraviolet (UV-B) light

6. parasitism; ingestion of toxic blue-green algae by tadpoles; competition from other frog species; and disease and morphological abnormalities

7. prolonged drought; unseasonable temperature extremes (particularly during chorusing and overwintering); flooding; and vegetational succession and overgrowth of grassy and woody vegetation along the shore, which creates too much shade and eliminates beach-like habitat

8. use of northern cricket frogs for fishing bait

9. spraying of oil to control snakes

10. siltation; erosion and scouring of shores; trampling of shores by cattle; excessive wave production by barges and other boats on rivers and reservoirs; and degradation of breeding areas by European carp

11. highway construction; strip mining; creation of golf courses; use of marshes as landfills; and filling marshes for development

12. agricultural cultivation and associated practices; and aquaculture and fish management

13. recreational development of aquatic sites; mowing around ponds; and burning of vegetation around ponds and along ditches

14. dredging; draw-down of drainage canals and drainage of ponds and other wetlands; excavation of pond edges to increase water depth at the shoreline, thus eliminating habitat; creation of large lakes or impoundments (stocked with fish) in areas that were previously riparian habitat, marshes, or only seasonally inundated; and destruction of ponds during development of residential subdivisions and further urbanization

15. habitat fragmentation

16. chytrid skin fungus

17. riprapping of shores

18. contamination of water by road de-icers; lowering of water table; and lining of irrigation ditches with concrete

19. replacement of prairies by forests

20. ignorance about the species and widespread indifference about its plight among state and federal conservation personnel—as is true for many declining amphibians, this may ultimately prove to be the most serious threat

There are undoubtedly many other factors that could contribute to northern cricket frog declines, however, most localities in the Midwest are probably influenced by one or more of the above-mentioned threats. It is not known which threats are most important in northern cricket frog declines, and their significance has probably varied in time and at different sites.

Listing

Northern cricket frogs are listed as "Endangered" in Minnesota and Wisconsin, "Threatened" in New York, and as a "Species of Special Concern" in Indiana, Michigan, and West Virginia (Brodman and Kilmurry, 1998; Ramus, 1998). In Canada, the species is listed as "Endangered" in Ontario, as well as on the national level (Weller and Green, 1997). The absence of listing in other U.S. states where declines have been noted may reflect the current abundance of cricket frogs in the southern parts of those states, or a lassitude and/or impenetrability of the state's endangered species bureaucracy.

Morphological Abnormalities

Concurrent with the worldwide decline in amphibian populations, reports of malformed individuals have been increasing (Kaiser, 1997; SETAC, 1997; Helgen et al., 1998). The occurrence of malformed northern leopard frogs and other amphibian species is now well documented in the popular press.

Reports include external deformities such as missing and/or extra arms and legs, missing eyes and mandibles, as well as internal abnormalities involving the bladder, digestive system, and testes (Greenwell et al., 1996; Helgen, 1996; SETAC, 1997; Helgen et al., 1998). Recent surveys of anurans in Minnesota (Hoppe and Mottl, 1997) revealed limb and/or eye malformations in six species: northern leopard frogs, mink frogs (R. septentrionalis), wood frogs (R. sylvatica), spring peepers (Pseudacris crucifer), eastern gray treefrogs (Hyla versicolor), and American toads (Bufo americanus). The highest frequency and severity of malformations tend to occur in the most aquatic species (mink frogs). Because adult anurans develop in water from fertilized eggs and their skin is permeable, they are at particular risk from xenobiotic agents in water.

Most past investigations (e.g., Pyburn, 1956, 1958, 1961a; Wendelken, 1968; Gray, 1971a,b, 1972, 1978, 1983, 1984, 1995; Issacs, 1971; Nevo, 1973; Burkett, 1984) and more recent field studies (e.g., Converse et al., 1998; Lee, 1998; Link, 1998; Mierzwa, 1998c) of cricket frogs have not reported on the incidence of malformations. Smith and Powell (1983) reported one specimen collected in 1973 in Missouri that lacked the left eye and orbit, while Johnson (1998a) reported a low incidence of malformed cricket frogs in Missouri. Other historic data on northern cricket frog malformations were recently reported by Gray (2000a,b); from 1968–71, only 0.39% of the frogs collected at 35 sites throughout Illinois were abnormal.

Abnormalities in northern cricket frogs occurred in two categories: missing whole (or parts of) limbs and digits and malformed or extra limbs, digits, and mouth parts. Most abnormalities involved missing arms and legs (0.32%; a subset of which could have been caused by failed predation attempts) rather than extra limbs (0.07%). Only seven confirmed developmental malformities (extra or malformed arms, malformed digits, or underdeveloped mouth) were recorded among almost 10,000 frogs examined.

In Illinois, the incidence of northern cricket frog malformations during the years studied was low and generally less than those reported for other species in different regions during different years. Higher incidences of malformations have been reported by Reynolds and Stevens (1984) for Pacific treefrogs (P. regilla) and by Worthington (1974) for spotted salamanders (Ambystoma maculatum). In Minnesota, Quebec, and Vermont, where abnormalities have recently received widespread attention, the overall frequency of malformations in other amphibian species ranged from 8% to as high as 67% (Kaiser, 1997; Helgen et al., 1998).

Most reported abnormalities are bilaterally asymmetrical and appear to involve missing rather than extra limbs. Ouellet et al. (1997a) suggested that conspicuous abnormalities constitute a survival handicap because they interfere with swimming and hopping. Interestingly, several of the northern cricket frogs marked as abnormal in Illinois were later recaptured and two (of the 39 observed [5.0%]) survived through the winter and following breeding season (Gray, 2000a,b). In these instances, neither the loss of a hand nor the addition of an arm prevented the animal from overwintering and living through the following breeding season. Whether these animals actually bred is unknown, but one female did develop and expel eggs.

In the spring and fall of 1998, students of L.E.B. resampled (using similar methods) the Mackinaw River (which included R.H.G.'s Illinois study area) and found no increase in the incidence of malformations from that observed 30 years ago (Gray, 2000b). Of 140 frogs examined, one had a broken hindlimb and three were each missing a digit. One of those

with a missing digit also had a laceration on the same forelimb. On another frog, the digit was obviously damaged. The broken hindlimb and lacerated forelimb were most likely the result of some type of accident.

Other historical databases need to be evaluated to identify areas where resampling could provide information concerning potential changes in the frequency of abnormalities over time. Resurveying the areas historically studied would indicate if the frequency of malformations in Illinois has changed. A decline in northern cricket frog populations without an observed increase in the frequency of malformations would suggest that these phenomena are caused by different factors.

Extinction Scenarios

Northern cricket frog declines likely reflect a combination of factors: (1) brief adult life span; (2) small effective breeding population sizes; (3) prolonged droughts; and (4) anthropogenic alterations of permanent wetlands. Brief life spans dictate that cricket frogs must breed annually (Gray, 1971b; Burkett, 1984; Lannoo, 1998a). Small effective breeding sizes increase the chance that cataclysmic events, such as environmental contamination, may extirpate a population. If droughts in turn dictate that cricket frogs must breed in permanent wetlands, and if permanent wetlands are rendered unsuitable for successful cricket frog reproduction, population declines would be inevitable (Lannoo, 1998a). Isolation of habitats due to landscape fragmentation would preclude recolonization and further exacerbate these effects.

Habitat loss is undoubtedly the largest single factor contributing to amphibian declines (Wyman, 1990; Wake, 1991; Vial and Saylor, 1993). However, habitat loss alone cannot explain all amphibian losses, even in the highly agricultural Midwest. Most of the prairie in Illinois had been converted to agricultural use by the early 1880s (Ridgway, 1889; Walters, 1997). Surface water in the spring and early summer was a serious problem for tillage in the nineteenth century. In the 1870s in central Illinois, tiling and ditching were initiated to reduce standing water and thus enhance agriculture (Walters, 1997). By the early part of the twentieth century, the majority of these pooled surface waters had disappeared over much of this area due to tiling and ditching. Similar time frames for these changes probably occurred in other areas of the Midwest (Leja, 1998).

Obviously, northern cricket frogs survived this major loss of wetland habitat and retained abundant populations well past the mid-twentieth century (Smith, 1961; Minton, 1972). In many areas, habitat has actually increased in the last 50 years due to construction of rural ponds (L.E.B. personal observations; Leja, 1998). Thus, the cricket frog decline is a phenomenon of the last 20 to 25 years, and it is not associated with the massive earlier loss of wetland habitat. There is considerable wetland habitat presently available in the upper Midwest, but few or no cricket frogs are found in these habitats. Thus, it appears that habitat modification (rather than habitat disappearance) is associated with cricket frog declines (see also Hay, 1998a).

Support for aridity as a factor in northern cricket frog declines can be found in the southwestern part of its range. One of the most severe droughts of the twentieth century in Texas occurred in the early 1950s. Blair (W.F., 1957), using data collected by Pyburn (1956), reported that the "effective breeding population size" of cricket frogs at his study pond dropped from 310 in 1953 to 15–36 individuals in the spring of 1955. However, numbers increased to nearly 600 by August 1955. Recovery was probably due to (1) frogs taking refuge from the drought in deep cracks in the pond bed; (2) a return of water to the pond; (3) a reduction or elimination of predators on tadpoles because of the drying of the pond, which resulted in higher survival to metamorphosis; and (4) a rapid re-establishment of the food supply, which resulted in increased cricket frog growth and early sexual maturity. In a severe drought in the late 1980s in central Illinois, the number of calling male cricket frogs was low and the calling season was truncated (L.E.B., personal observations). After the drought, the number of calling males increased and the calling season lengthened. Thus, cricket frogs seem to be able to recover from a severe drought spanning several years. However, even under non-drought conditions, there is up to 97% mortality of cricket frogs between metamorphosis and breeding, and few individuals survive a second winter (Gray, 1983; Burkett, 1984). Thus, long-term aridity could have a severe effect on the species, and the relictual distribution along the western edge of the range (see "Distribution" above) may reflect increasing aridity during the Holocene (Frost, 1983; Brown and Mesrobian, this volume).

Greenwell et al. (1996) presented another extinction scenario involving herbicides that are often toxic to tadpoles (Tyler, 1989; Boschulte, 1993, 1995). Greenwell et al. studied a number of ponds in Illinois and found that those with substantial plant damage due to herbicide (e.g., atrazine, metolachlor, copper sulfate, and hydrothol) use had the lowest numbers of metamorphosing northern cricket frogs. This was probably a reflection of reduced food (algae) and cover for tadpoles, as well as dissolved oxygen concentrations that approached zero. Juvenile frogs at these locations also suffered stress and had heavy infestations of renal parasites. Recovery occurred in subsequent years after the cessation of herbicide use, and the ponds were recolonized by dispersing frogs from nearby ponds.

A direct kill of any life history stage of northern cricket frogs could also occur, of course, if enough herbicide (or other biocide) enters the frog's microenvironment. Of particular concern is the herbicide atrazine because of its frequent use and slow degradation in aquatic environments. Herbicide use is one of the dominant features of the rural Midwest, and thus, has a high potential to negatively impact cricket frogs. Thirty years ago, a farm pond that was selected for studies of cricket frog polymorphism in Illinois (Gray, 1971b) was later abandoned after the owner applied copper sulfate and extirpated the population (R.H.G., personal observations).

Another extinction scenario involves the contracting range of northern cricket frogs. Peripheral habitats along the edge of a species' range are often inferior to those more centrally located. Thus, populations in peripheral habitats are more often subject to extinction. We previously mentioned cricket frog declines in the upper Midwest and this species' fragmented distribution along the western and southeastern edges of its range. The isolated relicts on the coastal plain in the Carolinas (Conant, 1975, 1977a) also suggest peripheral extinction, although recent colonization cannot be ruled out. Additionally, Conant and Collins (1991) indicated that cricket frogs were extirpated on Long Island, New York, and Vial and Saylor (1993) reported that the species was threatened in the state of New York. Thus, there are suggestions that cricket frogs may be under stress along much of the peripheral portion of their range.

Aquaculture and fish management practices may also be factors in the decline of northern cricket frogs. Raising of game and bait fish have impacted many, if not most, of the permanent wetlands of the upper Midwest and involve both publicly and privately owned basins (Lannoo, 1998a; Leja, 1998). These operations consume considerable potential habitat and influence cricket frogs in a number of ways: (1) aquatic herbicides (aquazine, copper sulfate, etc.), rotenone, and other biocides are often applied to wetlands being managed for fish production and are lethal to tadpoles (Lannoo, 1996); (2) the lampricide TFM (3-trifluoromethyl-4-nitrophenol) is used to kill tadpoles (Kane and Johnson, 1989; see also Matson, 1998), which cause numerous problems in ponds, particularly when fish are being harvested or transferred (Redmer et al., 1999a); (3) bank habitat is destroyed by mowing (L.E. Brown, 1971a); (4) there are frequent draw-downs of water in hatchery ponds (L.E. Brown, 1971a) to facilitate fish management and this can have a detrimental effect on all life history stages of cricket frogs; and (5) fish hatcheries and aquaculture areas frequently have a greater than normal number of frog predators (e.g., large fish, herons, water snakes, and American bullfrogs). However, the actual areas occupied by the primary aquaculture operations are quite small compared to the huge area in which cricket frogs have declined (although there have been marked increases recently in aquaculture in Iowa, Minnesota, and Wisconsin; Lannoo, 1996, personal communication). The important factor is that the byproducts of aquaculture (predatory fishes and bullfrogs) are transported throughout the Midwest. Both Lannoo (1996) and Redmer et al. (1999a) pointed out the significance of the inadvertent transport and subsequent coincidental introduction of tadpoles when fish are being stocked in wetlands.

The introduction of American bullfrogs is particularly harmful because they congregate in and around permanent wetlands and feed heavily on northern cricket frogs (Burkett, 1984). In northwestern Iowa, native amphibians no longer occur in habitats where bullfrogs have been introduced or have invaded. However, other species of native amphibians remain in similar nearby wetlands that do not contain bullfrogs (Lannoo et al., 1994; Lannoo, 1996). Introduced bullfrogs spread during high-water years, and those introduced into northwestern Iowa have now expanded their range into Minnesota (Oldfield and Moriarty, 1994).

To model various extinction scenarios, Veldman (1997) used a VORTEX extinction simulation program (Lacy, 1993; Lacy et al., 1993, 1995), which is a Monte Carlo simulation of stochastic events. Veldman gathered life history and ecological data on northern cricket frogs at eight study sites in central Illinois. Additional information was obtained from the literature. Four different catastrophes considered to be of significance in the area (biocides, drought, habitat destruction, and predation) were programmed at different severities, at different time intervals, and in various combinations. In nearly every simulation, populations declined toward zero in numbers of frogs and in genetic heterozygosity. Biocides and predation had the most adverse effects.

All, some, or none of these extinction scenarios may be operational. Different factors may be operating under different conditions or in different parts of the species range and there may be factors that have not yet been identified. At present there is not enough information to prove or disprove that any of the proposed factors or scenarios have been primarily responsible for the observed declines of northern cricket frog populations.

Summary

Northern cricket frogs *(A. crepitans)* are small terrestrial/semi-aquatic frogs distributed in the eastern half of the United States. They are adapted to many different local conditions of temperature and moisture and potentially can be found in or near any body of fresh water. Cricket frogs exhibit a genetically based color polymorphism in which the vertebral stripe can be various shades of red, green, or gray. Factors that may determine color morph frequencies vary throughout the species range and include the founder principle and chance in populations with small effective breeding sizes, visual selection by predators, pleiotropically linked physiological or behavioral attributes, or a combination of these factors acting in concert. Observations that the vertebral stripes can be highly variable and undergo metachrosis complicate the issue.

Timing of life history events varies throughout the species range. Growth of newly metamorphosed frogs is rapid during the fall and stops in winter, and individuals in some populations may actually be smaller when first observed in spring. Growth resumes again in spring and continues until death. The critical period for northern cricket frogs—when they are most susceptible to predation—is in the late summer and fall immediately after metamorphosis, when water levels recede and the banks of ponds are exposed. Cricket frogs overwinter in cracks on land or buried in gravelly soil, under logs, compact dead vegetation, or overhanging banks, or among leaves. Overwintering is not a major cause of mortality.

From 1968–71, morphological abnormalities in northern cricket frogs occurred in low frequencies (0.39%). Most abnormalities involved missing whole or parts of limbs and digits. Two malformed individuals were observed to survive through the winter and breeding season with no apparent ill effects. Limited resampling 30 years later suggests that the incidence of abnormalities is still low. If the decline in cricket frog populations is occurring in the absence of an increase in the frequency of malformations, these phenomena are likely caused by different factors.

There is currently not enough scientific information to determine which factors are primarily responsible for northern cricket frog declines. However, numerous threats have been identified as potential or actual causes. One or more of these threats probably occur at most localities. A number of extinction scenarios have been suggested. Different scenarios and threats may vary in number and significance by time and site. Unidentified threats may also be at work. Unusual life history characteristics in the upper Midwest (e.g., up to a 97% reduction in population size in the fall; small effective breeding population sizes; brief adult life spans; and females breeding only once or not at all in a given year) may predispose populations to anthropogenic and natural environmental threats that result in population extirpations.

Acknowledgments. We thank C. Buckley for library assistance; J. Brown for critically reviewing the manuscript; and T. Johnson, G. Lipps, T. Pauley, and H.M. Smith for providing information on the status of cricket frogs in their respective states.

TEN

Overwintering in Northern Cricket Frogs *(Acris crepitans)*

JASON T. IRWIN

Although winter weather in north-temperate regions may dominate 6–9 months of the year, this season has received relatively little attention in studies of amphibian life history. While some aspects of behavioral and physiological responses to cold have been elucidated, we generally have not applied this understanding to the management and conservation of amphibian populations. To begin to address this concern, I provide a basic description of the various overwintering methods used by amphibians and of the physiological responses that accompany these methods. Then I describe the unique overwintering method of northern cricket frogs *(Acris crepitans)* and consider how their physiology and winter habitat use may be contributing to their recent severe declines, especially in the northern portions of their range (Gray and Brown, this volume).

Overwintering Methods Employed by Amphibians

Amphibians have colonized temperate regions from tropical centers of diversity (Duellman, 1999), so it is not surprising that the most commonly used method for surviving cold climates is simply to avoid the cold altogether. To do so, many species overwinter in aquatic sites that do not freeze completely. The most familiar examples are species in the family Ranidae, which typically overwinter on the bottom of ponds, lakes, and streams (e.g., Bohnsack, 1951; Bradford, 1983). Even if the surface of the water freezes, liquid water is most dense (and thus heaviest) at 4 °C (Marchand, 1991), so a frog resting on the bottom remains relatively warm and free of ice.

However, hibernating in permanent bodies of water is not without challenges. Predation can be a factor. Fishes are known to eat hibernating frogs (e.g., 20% of the trout captured during winter had frogs in their stomachs; Emery et al., 1972). In response, some ranid tadpoles (e.g., American bullfrogs, *Rana catesbeiana*) are toxic, presumably to protect themselves from predation (Kruse and Francis, 1977). Hypoxia (lack of oxygen) is also a concern, particularly when ponds freeze over and the oxygen exchange between water and air is prevented. The metabolism of living organisms and the organic decomposition in the pond consumes oxygen, eventually lowering oxygen concentrations to levels that are lethal to amphibians (e.g., Barica

and Mathias, 1979; Lannoo, 1998a). Species that typically overwinter in such sites have evolved mechanisms to enhance their tolerance of hypoxic (and even anoxic) conditions (Christiansen and Penney, 1973; Boutilier et al., 1997; Holden and Storey, 1997), but even with such mechanisms, winter kills of amphibians are well known (Barica and Mathias, 1979; Bradford, 1983).

Rather than hibernating in water, other amphibians escape freezing by moving underground, below the frost line. True toads (Bufonidae) and spadefoot toads (Pelobatidae) are actively fossorial, burrowing more than 1.1 m underground as the frost penetrates the soil (Breckenridge and Tester, 1961; Ruibal et al., 1969; Seymour, 1973; Kuyt, 1991). Animals in the mole salamander family (Ambystomatidae) also dig underground or exploit existing burrows to get beneath the frost line. The soft passageways of ant mounds allow underground access for both lungless salamanders (Plethodontidae) and chorus frogs (Hylidae; Caldwell, 1973). Although one might expect that oxygen would be limiting for burrowers, this is not the case, at least not in the desert regions where oxygen levels have been measured (Seymour, 1973). The costs associated with burrowing include morphological adaptations for digging, the energy required to burrow into the soil, and mechanisms to prevent or tolerate water loss to the surrounding soil (Pinder et al., 1992).

Not all amphibians protect themselves from the cold. Some North American anuran species hibernate on the forest floor under only a thin layer of fallen leaves. In such shallow sites, frogs are within the frost zone, when the soil and leaves surrounding the frog freeze, the frog also freezes. These species are said to be "freeze tolerant" and include striped chorus frogs *(Pseudacris triseriata*; Storey and Storey, 1986), spring peepers *(P. crucifer*; Schmid, 1982), gray treefrogs (both eastern, *Hyla versicolor*, and Cope's, *H. chrysoscelis*; Schmid, 1982; Costanzo et al., 1992, respectively), and wood frogs *(R. sylvatica*; Schmid, 1982). Remarkably, these species have evolved physiological mechanisms that enable them to survive freezing of up to 70% of their body water. Adaptations supporting freeze tolerance include glucose production upon freezing (glucose acts as a substrate for anaerobic metabolism and reduces cellular dehydration), tolerance of dehydration (losing water to ice is similar in a physiological sense to being dehydrated), and a suite of other more subtle physiological responses (for a recent review, see Lee and Costanzo, 1998). Other than a few reports of terrestrial hibernation in frogs, the microhabitats chosen by these species and the minimum temperatures and duration of freezing they may experience remain unknown.

Individuals may not follow the pattern of winter habitat use typical for their species. For example, green frogs (R. clamitans) and American bullfrogs are occasionally found overwintering (and surviving) on land (Bohnsack, 1951; personal observations), while aquatic frogs are often found to be active in caves during the winter (e.g., Rand, 1950; Barr, 1953; Campbell, 1970c), as these sites tend to remain wet and at a stable temperature throughout the winter. Substantial proportions of some populations of the European common frog (R. temporaria) overwinter on land rather than in the water, depending on local factors and habitat quality (R. Sinsch, 1991; Pasanen and Sorjonen, 1994).

Overwintering in Northern Cricket Frogs

Northern cricket frogs are unusual among North American treefrogs (Hylidae) in that they are highly aquatic and are most frequently found near permanent water (Wright and Wright, 1995), even on warm days in mid-winter (Gray, 1971a). Cricket frogs do not have the well-developed toe disks typical of the family and, when disturbed, they escape by jumping into the water rather than climbing high into vegetation. The unusual habits of these animals lead one to wonder where they spend the winter. Given their aquatic nature, one could hypothesize that they hibernate underwater, much like the aquatic ranids. However, it is also conceivable that, like the other northern hylids, they hibernate in terrestrial sites and are freeze tolerant. My colleagues and I (Irwin et al., 1999) performed a series of experiments to distinguish between these possibilities. The results provide clues as to why this species is declining in the northern portions of its range.

To begin, we examined whether northern cricket frogs move to land or water as air temperatures fall. Several authors report finding cricket frogs in terrestrial sites (Walker, 1946; Neill, 1948b; Gray, 1971a), so we constructed a laboratory experiment to test whether these were chance occurrences or if cricket frogs habitually hibernate on land. As temperatures dropped in our terrarium with half land and half water, cricket frogs consistently moved to land. Occasionally, individuals moved into natural depressions in the soil, but they did not have the ability to burrow deep into the substrate, even when the upper soil layers were dry and moist soil was only 2 cm below (unpublished data).

Furthermore, northern cricket frogs did not tolerate submergence in water for long periods of time. Cricket frogs die in <24 h when submerged in moderately hypoxic (2 mg·ml^{-1}) or strongly hypoxic (0.5 mg·ml^{-1}) simulated pond water. Even in oxygenated water (8 mg·ml^{-1}), they survived only 7–10 d. This result is consistent with the poor submergence tolerance of other hylid frogs; these species generally are not tolerant of low oxygen conditions and are not well adapted for either cutaneous gas exchange (Czopek, 1962) or maintaining osmotic homeostasis in water (Schmid, 1965b). These physiological limitations prevent aquatic hibernation by cricket frogs and put them at risk should their hibernation sites become flooded with water.

Given the described results of the experiments, it seemed certain that northern cricket frogs hibernate in terrestrial sites.

If this species behaved similar to other northern hylids, we suspected they would be freeze tolerant. But this was not the case; under moderate freezing conditions (>−2.0 °C for 48 h) only 2 of 15 frogs survived. This result is similar to that found in other freeze-intolerant species such as northern leopard frogs (R. pipiens; Layne, 1992). Clearly, cricket frogs must overwinter in terrestrial sites that do not freeze.

At this point, we realized that the overwintering method of northern cricket frogs must be unusual. Cricket frogs cannot survive in aquatic sites, are not fossorial, and do not tolerate freezing. Based on this information and the observations of Gray (1971a), we sought to find and characterize the natural hibernation sites of cricket frogs. Searches of the water, the pond edge, and adjacent forest sites yielded cricket frogs only along the pond edge. These frogs were found >10 cm deep in crayfish burrows and cracks in the pond bank.

The temperature of northern cricket frog hibernation sites is ameliorated by the saturation of the soil by water. Because water has a high specific heat, it cools slowly during cold weather. Also, the latent heat of fusion (produced as the water becomes ice) inhibits further cooling of the soil until the surface layers are thoroughly frozen. The soil of the pond bank required several weeks of subzero air temperatures (as low as −20 °C) before soil temperature at 2 cm depth fell to −0.5 °C, approximately the freezing point of frog tissues (Irwin et al., 1999). Cricket frogs hibernate at even greater depths and are therefore protected from tissue freezing by the wet soil. Thus, cricket frogs require preexisting subterranean sites along the water's edge to successfully overwinter in cold climates.

Relationship Between Hibernation Behavior/ Physiology and Species Decline

Northern cricket frogs have been declining for decades in the northern portions of their range and have disappeared from southern Ontario, Minnesota, most of Wisconsin, and northern Iowa (Green, 1997b; Lannoo, 1998b; and references therein; Gray and Brown, this volume). Because these extensive declines and extirpations generally have been limited to the northern edge of historic distribution, it has been suggested that events during winter play a role (Hay, 1998a; Gray and Brown). No large numbers of dead cricket frogs have been found, which is also consistent with winter mortality—frogs that die in burrows are seldom found by humans. With our new understanding of this species' physiological ecology and winter habitat requirements, I have compared the phenology of cricket frog declines with historical weather patterns, and I propose a mechanism by which these declines could have occurred.

Northern cricket frogs have narrow habitat requirements for successful overwintering owing to their limited tolerances of water loss, desiccation, and freezing. The most important component of an overwintering site is the water content of the soil. In northern areas, cricket frog overwintering sites must be wet enough to provide thermal buffering against freezing. Decreases in the water level would reduce the thermal buffering capacity of the soil, thus increasing frost penetration and the likelihood of freezing. Because cricket frogs are among the least dehydration tolerant of all hylids (Ralin and Rogers, 1972), severe soil drying could kill cricket frogs by desiccation. While the soils must have a high water content, too much water is harmful to cricket frogs. If the soils become so wet that the hibernation sites are flooded, cricket frogs will die because of

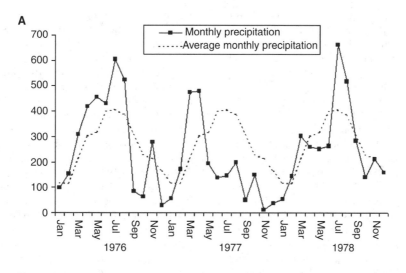

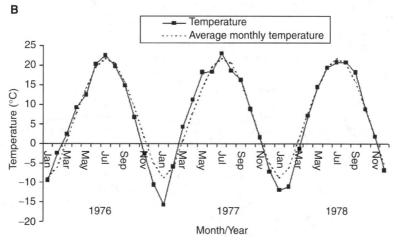

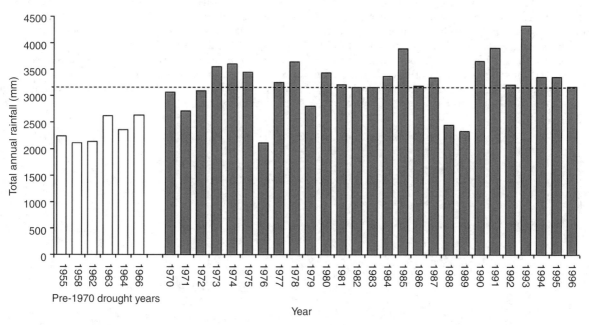

FIGURE 10-2 Annual rainfall for the area around Madison, Wisconsin. Dashed line indicates average annual rainfall for the period 1950–1996.

their inability to tolerate submergence and low oxygen conditions. These narrow requirements limit cricket frogs to overwintering in a relatively small spatial area along a pond or stream edge, and increase the susceptibility of cricket frogs to extremely dry or wet weather conditions.

A historical review of climatic conditions during northern cricket frog declines in Wisconsin supports the hypothesis that winter habitat quality has played a role in these declines. The decline of cricket frogs in Wisconsin was first noted by Vogt (1981) who remarked that, "During the last three years, populations of cricket frogs have diminished rapidly." The time frame of this decline closely matches with a period of severe drought in the autumn of 1976 (Figs. 10-1A, 10-2). (Consider that Vogt's book, although published in 1981, was likely written in 1980.) The winters of 1976 and 1977 were also extremely cold, with average monthly temperatures well below the average (Fig. 10-1B). This represents a worst-case scenario for hibernating cricket frogs—reduced soil water content due to the drought reduces the soil's ability to buffer cricket frog overwintering sites against cold. At the same time, extreme cold would allow deep penetration of frost into the soil. Thus, it is quite reasonable to assume that many cricket frogs died from freezing during these winters. Jung (1993) reported that a 1991 survey of the species indicated that cricket frogs had continued to decline. This follows two years of drought in 1988 and 1989. Temperatures were also below average, particularly in January and, during the following winter, in December of 1989. Earlier declines through the 1960s (Hay, 1998a) correlate with drought years during that decade (Fig. I-10-2), but a lack of historical data prevents a complete analysis.

The correlation between cold, dry winters and the decline of northern cricket frogs can also be made on Pelee Island, Ontario. Weather records from nearby Windsor, Ontario, show that the winters of 1976–1977 and 1977–1978 were both very cold and dry, as they also were in Wisconsin. Following these two severe winters, the northern cricket frogs nearly disappeared from Pelee Island altogether (Oldham and Campbell, 1990), as the number of breeding sites fell from a high of 15 sites in 1976 to only a single site after 1977. Populations of northern cricket frogs have never recovered (Britton, 2000) and it is likely that this species has since been extirpated from Pelee Island and, therefore, the entire province of Ontario.

The scenario I propose best explains long-term extinction of populations, not reduced population size. If only a portion of a population is killed during winter, the population should be able to recover through reproduction. However, if the entire population is eliminated, it will only be re-established through migration into the area. The climate effects I propose would have affected a large geographic area. Re-establishment will require (1) source populations within the area and (2) a long period of suitable climatic conditions.

The impact of cold and dry winters on northern cricket frog declines may have been worsened by anthropogenic factors. Pond dredging, a common technique to deepen ponds, may fill cracks and burrows along the pond edge with bottom mud. Clearing of pond vegetation and shoreline debris may remove an important insulating layer. Finally, the exotic rusty crayfish (Orconectes rusticus), an active predator, has been introduced to most of the northern Midwest and may be consuming cricket frogs (Gunderson, 1995). Rusty crayfish also do not burrow and have displaced native crayfish that do burrow (Gunderson, 1995), possibly reducing the availability of cricket frog hibernation sites. The role of these factors in cricket frog declines has not been studied.

Although these data are suggestive of a correlation between northern cricket frog declines and cold, dry winters, the correlation does not imply direct causation. The scenario I describe is one of several that have been proposed to explain cricket frog declines. Other scenarios include revegetation of mudflats (Hay, 1998a), aquaculture and aquatic habitat modification (Lannoo, 1998a), direct and indirect effects of pesticide use (Beasley et al., this volume), and disease outbreaks (Carey et al., 1999).

The careful monitoring of population size on focal populations (not simply the presence/absence of northern cricket frog populations) is required to fully understand the role of winter weather on cricket frog declines. Although the data presented focus on Wisconsin, similar weather patterns were present in these years over the entire northern Midwest, including Minnesota and Michigan, and also Ontario. These areas, however, lacked the adequate historical data on cricket frog distributions to correlate declines with historical weather patterns.

Summary

Northern cricket frogs are unique among temperate frogs in the mechanism they use to overwinter. Rather than overwintering underwater or overwintering terrestrially by burrowing below the frostline or tolerating freezing, cricket frogs overwinter in preexisting shallow cracks or burrows in a band of moist soil near the edge of ponds and streams. This microclimate insulates against freezing, even during severe cold. The narrow tolerances of this band of moist soil appear to make cricket frogs susceptible to extremely cold and/or dry winters. Support for this view comes from correlations with climate data during the recent periods of decline.

CAUSES

ELEVEN

Repercussions of Global Change

JAMIE K. REASER AND ANDREW BLAUSTEIN

Living organisms must track the climate regimes appropriate for their survival, adapt to new conditions, or go extinct. In the 1970s, climatologists began to warn that Earth would experience rapid changes, induced in part by emissions of "greenhouse" gases resulting from the burning of fossil fuels, intensifying land use, and reduction in forest cover. They projected that global temperatures would rise substantially in the coming decades (e.g., Climate Resources Board, 1979). At approximately the same time, climatologists also became concerned that chloroflourocarborns (CFCs) and other commonly used industrial gases were depleting the earth's protective ozone layer, thereby increasing the amount of cell damaging ultraviolet B (UV-B) radiation that reaches ground level (van der Leun et al., 1998). Scientists projected that species might concurrently respond to some of these global changes; ranges might shift, natural communities might be disrupted, and mass extinctions of some species might occur (e.g., Peters, 1988).

Amphibians warrant substantial conservation attention. They are considered valuable indicators of environmental quality, and they have multiple functional roles in aquatic and terrestrial ecosystems (Blaustein and Wake, 1990; Stebbins and Cohen 1995; Green, 1997b; Lannoo, 1998b). Furthermore, amphibians provide cultural and economic value to human society (Grenard, 1994; Stebbins and Cohen, 1995; Reaser and Galindo-Leal, 1999; Reaser, 2000a).

As part of the overall "biodiversity crisis," many amphibian populations have been declining and undergoing range reductions (reviewed in Blaustein and Wake, 1995; Stebbins and Cohen, 1995; Reaser, 1996a, 2000a). Indeed, during the past decade, the amphibian decline issue has come to be regarded as an ecological emergency in progress. More than a dozen amphibian species are believed to have recently gone extinct, and the population ranges of many species have been dramatically reduced (Stebbins and Cohen, 1995).

Numerous anthropogenic factors have been implicated as causes of amphibian population declines (see Blaustein and Wake, 1995; Stebbins and Cohen, 1995; Reaser, 1996a, 2000a; Blaustein et al., 2001; Hayes et al., 2002a,c; Kiesecker, 2002; Halliday, this volume; Crump, this volume; Blaustein and Belden, this volume; Bridges and Semlitsch a,b, this volume; Beasley et al., this volume). These factors operate across multiple scales, often have synergistic relationships, and can trigger a cascade of impacts on biological communities. For many such reasons, the site-specific causes of amphibian population declines have been difficult to assess. Habitat destruction and the introduction of invasive alien species (e.g., *Tilapia*, trout) are readily apparent causative agents at some sites, and they present obvious resource management and policy options. However, amphibian population declines in areas with little human activity, especially those in protected reserves, invoke particular concern (e.g., Pounds and Crump, 1994; Lips, 1998, 1999; Pounds et al., 1999). Where amphibians are declining without apparent cause, it is difficult to arrest these declines or to identify what the implications are for the rest of the biological community (including humans).

Recent studies investigating site-specific cases of amphibian declines have revealed that global changes may be involved. Regional warming, increases in ultraviolet radiation, and disease epidemics may all be driven by global phenomena. These global changes might be induced, at least in part, by the increasing intensity and extent of the human impact on climatic and ecological systems.

Global Warming

Severe declines in frog populations at Monteverde Cloud Forest Preserve, Costa Rica, were first noted in 1988 when only eleven golden toads (*Bufo periglenes*) of the 1,500 adults noted the previous year showed up to breed. The last of the species, a single adult male, was observed the following year (Pounds and Crump, 1994; Crump, this volume). Over the following decade, 40% of the amphibian species at Monteverde were decimated in a series of synchronous crashes (Pounds et al., 1999).

The rapid declines in Monteverde occurred during peaks of warm and dry conditions, leading scientists to suspect that the frogs had been physiologically stressed through moisture limitation. Pounds et al. (1999) found that the dry season at Monteverde has indeed become warmer and drier. Furthermore, the dry days are now sustained in longer runs. Pounds et al. hypothesized that the cloud bank in this montane cloud forest has lifted, decreasing misting and condensation. A model produced by a separate team of scientists (Still et al., 1999) to simulate the effects of global warming on tropical montane cloud forests lends credence to this hypothesis. In addition to amphibian disappearances, populations of two species of lizards disappeared and the ranges of 15 species of birds shifted upslope. The concurrent changes in frog, lizard, and bird populations are all statistically associated with the same regional patterns of mist frequency and congruent with large-scale climate trends.

Pounds et al. (1999) associate the reduction in moisture with the El Niño Southern Oscillation (El Niño) and longer term increases in sea surface temperatures. Globally, average surface air temperatures are about 0.5 °C (almost 1 °F) higher than the average temperatures in the nineteenth century (NOAA, 1997). Analyses by the National Atmospheric and Space Administration (NASA) indicate that the rate of warming is the most rapid of any previous period of equal length in the time of instrumental records (NASA, 1999).

Over a 17-year period in Britain, Beebee (1995) observed a gradual, significant shift in the timing of amphibian breeding. On average, Natterjack toads (B. calamita) and edible frogs (Rana esculenta) spawned two and three weeks earlier, respectively, in 1990–94, than they did in 1978–82. Three species of European newts (Triturus vulgaris, T. cristatus, and T. heleveticus) showed highly significant tendencies toward early breeding—by 1990–94, the first newts were arriving 5–7 weeks earlier than in 1978–82. All these shifts in timing of breeding correlate with changes in climate over the same period, with winter and spring average temperatures steadily increasing. In a separate 18-year study, Reading (1998) found that the timing of the arrival of the European toads (B. bufo) in south Dorset, United Kingdom, was highly correlated with the mean daily temperatures of the previous 40 days. However, although the five earliest breeding records were within the 10 last years of the study and were associated with particularly mild winters, he did not identify a significant trend toward earlier breeding for this species. In contrast to some of these European data, Blaustein et al. (2001) showed that climate change has not influenced the timing of breeding in at least four species of amphibians in North America for which they had long-term data sets. At one site in Oregon, western toad (B. boreas) breeding has been increasingly early and was associated with increasing temperature. However, at four other sites in Oregon, neither western toads nor Cascades frogs (R. cascadae) showed statistically significant positive trends toward earlier breeding. At three of these four sites, breeding time was associated with warmer temperatures. In Michigan, spring peepers (Pseudacris crucifer) did not show a statistically significant trend to earlier breeding but did show a significant positive relationship between breeding time and temperature. In eastern Canada, Fowler's toads (B. fowleri) did not show a trend for earlier breeding nor was there a positive relationship between breeding time and temperature.

While there will undoubtedly be variation in the type, rate, and degree of response amphibian species make to global warming (Ovaska, 1997), it is important to note that observations of amphibian population declines and range shifts attributed to climate change are being reported in other ecological systems. For example, the range of Edith's checkerspot butterfly in Canada, the western United States, and Mexico changed in accordance with a regional climate shift (Parmesan, 1996). Recent events in the marine environment (where just a slight increase in sea water temperature has lead to massive coral bleaching and mortality [up to 90%] in most tropical oceans) are believed to be a large consequence of a steadily rising baseline of marine temperatures (Pomerance et al., 1999; Wilkinson et al., 1999; Reaser et al., 2000a).

Amphibian populations and species most at risk due to global warming are those that (1) are already at the upper limits of their physiological tolerance to temperature and/or dryness; (2) depend on small, ephemeral wetlands; and/or (3) are bound by barriers to dispersal. Because amphibians are reproductively and physiologically dependent on moisture, moisture uptake is temperature sensitive, and amphibian dispersal capacities are low compared with other groups (Blaustein et al., 1994a; Stebbins and Cohen, 1995), it would not be surprising if they are among the first vertebrates to exhibit broad-scale changes in response to global warming.

Ultraviolet Radiation

Many of the amphibian population declines have taken place in remote regions at relatively high elevations, prompting scientists to consider increased ultraviolet irradiance (especially UV-B) associated with depletion of stratospheric ozone as a probable agent. It has also been noted that certain climate changes such as acidification can increase exposure of aquatic organisms to UV-B via effects on dissolved organic carbon, which normally limits UV-B (Schindler et al., 1996).

Several laboratory studies have shown that ultraviolet-B radiation (UV-B; 280–315 nm) can damage amphibians. These studies showed that slightly enhanced UV-B radiation can cause certain deformities in developing frogs and toads (e.g., Worrest and Kimeldorf, 1976; Blaustein et al., 1994c; Anzalone et al., 1998; Blaustein et al., 1998; Corn, 1998; Lizana and Pedraza, 1998; Broomhall et al., 2000; Blaustein and Belden, this volume). Results of field studies strongly indicate that the hatching success of at least nine species of amphibians (from widely separated locales) is reduced under ambient UV-B radiation (Blaustein et al., 1998). This includes two frog species, one toad species, two salamander species, and a newt from North America; two frog species from Australia; and a species of toad from Europe. These species comprise a taxonomically diverse group that includes two orders, six families, and seven genera of amphibians. Some of these species are found in montane areas, while others are found at sea level. A key behavioral characteristic shared by these species is that they often lay their eggs in shallow water, where they are exposed to solar radiation.

Hatching success of several other species of frogs in North America and Australia and toads in Europe and North America were not affected by UV-B radiation. Thus, there seems to be differential sensitivity of amphibians to UV-B radiation, perhaps even within a species at different locations. Some of the differential sensitivity may reflect differences in the ability to repair UV-B-induced DNA damage. Those species with the highest levels of the photoreactivating enzyme photolyase seem to be the most UV-B resistant species (Blaustein et al., 1998). In addition to studies on the hatching success of embryos, other investigations have shown that ambient levels of UV-B radiation damages eyes in basking frogs (Fite et al., 1998) and causes deformities in developing salamander embryos (Blaustein et al., 1997). Eye damage may impair an individual's ability to avoid predators or find prey, while salamander deformities may affect swimming behavior and cause a number of other problems (Blaustein et al.).

In nature, more than one environmental agent may affect amphibians as they develop. Field experiments have shown that at least three factors may interact synergistically with UV-B: a pathogenic fungus (Saprolegnia ferax), low pH (Long et al., 1995), and fluoranthene, a polycyclic aromatic hydrocarbon that may pollute aquatic environments impacted by petroleum contamination (reviewed in Blaustein et al., 1998). Thus, these agents in combination with UV-B radiation increase the mortality rates of developing embryos.

UV-B radiation obviously is not the only agent that can contribute to an amphibian population decline. It would be an unlikely factor in the declines of species that lay their eggs under logs, in crevices, in deep water, or under dense forest canopy. Nevertheless, the hatching success of many amphibian species is affected by UV-B radiation. Several factors, such as pathogens, low pH, and pesticides may act synergistically with UV-B radiation to enhance mortality in early life stages. These synergistic interactions may eventually contribute to a population decline.

Infectious Disease

Disease is an important indicator of stress. Recent surges in disease outbreaks throughout a diversity of taxonomic groups and ecological systems (e.g., Epstein et al., 1998; Morell, 1999) have scientists posing the following questions:

1. Is a general decline in environmental quality compromising animal immune systems and making them more susceptible to typically benign microbes (Epstein et al., 1998; Carey et al., 1999)?

2. Are climatic shifts in the environment enabling microbes to increase in virulence, range, and/or diversity (Kennedy, 1998; Daszak et al., 2000; Kiesecker et al., 2001a,c)?

3. Are increases in our technological ability to transport people and products further and faster than at any time in the history of the biosphere facilitating the introduction of microbes to novel environments and hosts (Bright, 1998; Morell, 1999)?

4. Are two or more of these processes concurrently operating to lead to population declines in wildlife (Daszak et al., 1999, 2000)?

A wide diversity of microbes are commonly associated with amphibians (e.g., Gibbs et al., 1966; Carr et al., 1976; Brodkin et al., 1992; Blaustein et al., 1994b). However, larval and adult amphibians whose immune systems have been compromised by acute or chronic stressors may be susceptible to infection by a wide variety of pathogens, and amphibian eggs may become diseased if their gelatinous coating is altered in such a way as to permit the entry of microbes. Scientists investigating declines of amphibians in relatively remote, undisturbed regions have frequently pointed to fungi (e.g., Berger et al., 1998), viruses (e.g., Laurance et al., 1996), or bacteria (e.g., Worthylake and Hovingh, 1989) as the proximal cause of death. Few studies, however, have yet to investigate the potential links between disease outbreaks in amphibians and global change.

Observations and field experiments in Oregon showed that, the pathogenic fungus *Saprolegnia ferax* plays an important role in contributing to the mortality of amphibian eggs in Oregon (Blaustein et al., 1994a; Kiesecker and Blaustein, 1995, 1997b, 1999). Moreover, the fungus interacts with UV-B radiation to enhance mortality in the early life stages of frogs and toads (Kiesecker and Blaustein, 1995). Differential susceptibility of amphibian species to fungus and UV-B radiation may lead to profound changes in community structure (Blaustein and

Kiesecker, 1997; Kiesecker and Blaustein, 1999). Recently, Kiesecker et al. (2001a) have shown that there is a complex interaction between climate change, UV radiation, and amphibian susceptibility to *Saprolegnia* infection. Essentially, they suggest that climate-induced reductions in water depth at oviposition sites have caused high mortality of amphibian embryos by increasing their exposure to UV-B radiation and, consequently, their vulnerability to infection.

Fish are intentionally moved from one geographical region to another, even among continents, for a variety of reasons. Fish are also unintentionally relocated. For example, fish are transferred around the world when ships take in ballast water at one location and later release it in another. When fish are introduced to novel environments, their pathogens might also be introduced and thus transmitted to other aquatic species (Bright, 1998), including amphibians. Blaustein et al. (1994c; Kiesecker et al., 2001c) suspect that introducing hatchery-reared fish contributes to the spread of *Saprolegnia ferax* to amphibians. *Saprolegnia*, along with UV-B radiation, seems to be contributing to the decline of the western toad in the Pacific Northwest. Laurance et al. (1996) suggested that a rapidly spreading disease was responsible for the rapid decline of 14 species of endemic, stream-dwelling frogs in the montane rainforests of eastern Australia. They proposed that the pathogen was exotic, brought to the region with aquarium fish, and feared that the thriving international trade in aquarium fish was facilitating the global spread of microbes to which amphibians are susceptible.

We have much to learn about the emergence and spread of amphibian pathogens. Amphibians are themselves traded around the world (Gibbs et al., 1971; Jennings and Hayes, 1985) and are probable vectors when intentionally or unintentionally released into the wild. Because the world trading system is so extensive and complex, controlling trade-mediated epidemics will be a formidable challenge (Bright, 1998). Furthermore, people (including biologists) who come in contact with amphibian pathogens, and thus might transmit them, are increasingly traveling long distances and into remote regions. Every week, about one million people move between the industrial and developing worlds; every day, about two million people cross an international border (Institute of Medicine, 1997).

Conclusion

Because amphibians play multiple functional roles in both aquatic and terrestrial environments, the repercussions of amphibian population declines might be far reaching in time and through space. Profound changes in ecosystems (e.g., Blaustein and Kiesecker, 1997; Kiesecker and Blaustein, 1999) and some socioeconomic systems (e.g., Reaser and Galindo-Leal, 1999; Reaser, 2000a) may occur with a loss of amphibians. Amphibian population declines may become more frequent and severe as temperatures continue to rise, the ozone layer is further depleted, and emerging diseases are rapidly transported around the world.

Clearly, government agencies and other organizations must consider their role in drawing attention to the predicted impacts of global scale perturbations on the resources they manage. Conservation goals can no longer be achieved without taking into account changes in the global system. A comprehensive strategy to maintain amphibian populations must include reducing the emissions of greenhouse and ozone depleting gases, as well as monitoring and managing their infectious diseases—for even those amphibians granted well-enforced legal protection in refuges, sanctuaries, or parks are threatened by global change. While "thinking globally and acting locally" will reduce some stressors on amphibians, we must now think globally and act globally if we are to conserve amphibians and their habitats.

Summary

Recent studies investigating site-specific cases of amphibian declines have revealed that global changes may be involved. Regional warming, increases in ultraviolet radiation, and disease epidemics may all be driven by global phenomena. These global changes might be induced, at least in part, by the increasing intensity and extent of the human impact on climatic and ecological systems. While there will undoubtedly be variation in the type, rate, and degree of response that amphibian species make to global-scale alterations to the environment, it is clear that amphibian conservation can no longer be achieved without taking into account changes in the global system. A comprehensive strategy to maintain amphibian populations must include reducing the emissions of greenhouse and ozone depleting gases, as well as monitoring and managing their infectious diseases—even those amphibians granted well-enforced legal protection in refuges, sanctuaries, or parks are threatened by global-scale change.

TWELVE

Lessons from Europe

K. HENLE

Amphibians and reptiles are coming to be regarded in Europe as indicator groups for a general decline in species diversity (Thielcke et al., 1983; Blab, 1985, 1986). The decline of these groups has been well documented in Europe and on other continents as a result of numerous surveys (e.g., Lemmel, 1977; Feldmann, 1981; Hayes and Jennings, 1986; Hölzinger, 1987; Osborne, 1990; Carey, 1993; Mahony, 1993). Even in areas little affected by human activity, declines seem to have occurred. However, to date, adequate investigations of the declines and their actual causes are lacking (Pechmann et al., 1991; but see Osborne, 1989).

The global decline of amphibians over large land masses is presumed to be attributable to as yet essentially unknown factors (Blaustein and Wake, 1990; Yoffe, 1992), while previously established causes of decline have been neglected. However, the reverse tendency exists for local and regional investigations—in those cases, as a rule, factors are readily identified as causes, but are seldom investigated because the causal relation to observed declines is often difficult to clearly establish (Henle and Streit, 1990). Thus, there is frequently not a clear distinction between potential threats and proven causes, and many opportunities to prove causal connections, or at least to carefully construct foundations for hypotheses, are missed. A scientifically based analysis of causal relations is essential to effective conservation efforts since it leads to the prediction of appropriate countermeasures.

The most important type of database for the documentation of the declines of amphibians and reptiles and their potential causes results from regular surveys of a specific geographic area. Surveys will maintain this important role in the future. Therefore, attempts should be made to fully utilize the potential of these surveys for causal analysis of declines within the bounds of permissible conclusions. That is, the limits of herpetofaunal surveys as a tool for determining the causes of decline must be explicitly defined. The available methods for causal analysis are insufficiently known to many people engaged in surveying projects because of inadequate education in

statistics and research planning. The goal of this work is to demonstrate the available methods of causal analysis, as well as their limitations, by applying the methods in the analysis of a long-term surveying project (Henle and Rimpp, 1994). Further, remarks on the optimization of surveying projects that will facilitate subsequent causal analyses are presented. By suggesting improvements in the planning and assessment of future surveying projects, I hope to contribute to the technical support of herpetological conservation work, and with that, to the protection of our amphibians and reptiles.

Methods for Analysis of the Causes of Species Decline

The principles of planning and analysis of experiments furnish three experimental approaches for evaluating causal relations in ecology: (1) laboratory experiments, (2) field experiments, and (3) unplanned or natural experiments (Diamond, 1986; Henle and Streit, 1990). In all three types of experiments the presence of adequate control populations that remain unaffected by the factor in question (e.g., road traffic) is essential. Likewise, it is important to formulate a precise question in the form of a so-called null hypothesis, for example, "Addition of pollutant chemicals to a water body does not lead to increased mortality in comparison to unpolluted water bodies." The probability that the null hypothesis is correct is then determined statistically. The advantages and disadvantages of the three different experimental approaches are briefly summarized below.

Laboratory Experiments

Laboratory experiments have the advantage that disturbances are largely eliminated; independent variables (in our case, potential threats) are controlled, allowing the effect of a specific threat to be clearly determined. They have a disadvantage in that their relevance to natural systems is highly questionable because of the stark simplicity of the laboratory system. This problem can be minimized, but not fully eliminated, by the use of experimental conditions that simulate natural

* Translation by William T. Leja of Möglichkeiten und Grenzen der Analyse von Ursachen des Artenrückgangs aus herpetofaunistischen Kartierungsdaten am Beispiel einer langjährigen Erfassung. 1996. *Zeitschrift für Feldherpetologie* 3:73–101.

conditions as much as possible. There has been considerable investigation in the laboratory of the threat posed to amphibians by predators (e.g., Glandt, 1984; Kats et al., 1988; Semlitsch, 1993). However, in this case the results are frequently of severely limited applicability or are actually inapplicable to field situations because of the generally simplified and artificial laboratory conditions.

Field Experiments

Field experiments are of far greater relevance. The chief problem is the difficulty of finding a suitable research area for the desired manipulation within a landscape utilized by humans. Further problems are the difficulty in obtaining permission from the responsible government authorities, and the difficulty of holding the effect of disturbance factors (such as the conditions on the shore of amphibian breeding ponds or the vegetation structure of reptile habitats) constant. Because of these difficulties, relatively few field experiments on the influence of threats to amphibians or reptiles exist (for one example, predation by fish, see Breuer and Viertel, 1990; Breuer, 1992).

Unplanned or Natural Experiments

Unplanned or natural experiments offer a better alternative. Two different categories must be distinguished: natural *snapshot* experiments, in which a final steady state (or an intermediate state) resulting from a perturbation is observed by comparison of affected and unaffected systems; and natural *trajectory* experiments, in which the progression of changes in a system caused by a perturbation are observed as they unfold over time. The principal advantage of these experiments lies in their relevance to large areas over long periods of time. For this reason they are more realistic and of greater general validity. Natural experiments "interpreted by experience" generally form the basis of our empirical knowledge. That knowledge is clearly derived from complex nonlinear associations, as very commonly occur in nature conservation, and it can easily be misleading.

A particular disadvantage of natural experiments is the difficulty of eliminating confounding factors. In this regard, it is generally necessary to develop an extensive database from which an adequate partial database can be chosen, or to identify confounding factors and when possible to "remove" their influence with various statistical methods. With snapshot experiments, moreover, two factors can be merely correlated with each other coincidentally or because both depend on a third factor. An example of such a situation is the well-known correlation of the decline of the stork population in East Prussia and the human birth rate (both factors are independent of each other and caused by a third factor, industrial growth; Sachs, 1982). In snapshot experiments, therefore, this possibility must be eliminated as much as possible, or hypotheses need to be supported indirectly (see below). The value of hypotheses resulting from natural trajectory experiments is severely reduced if, as is often the case in nature conservation, a system is first studied during the beginning of a disturbance or even after it is in progress, and if it is not monitored with adequate controls. Natural trajectory experiments can result from fish stocking, entry of pesticides, or enlargement of agricultural fields, among others.

The three types of experiments are mutually complementary. Additional support of an interpretation, particularly for natural snapshot experiments but also for natural trajectory experiments, can be provided through knowledge of the biology of the affected species or the ecology of the affected habitat

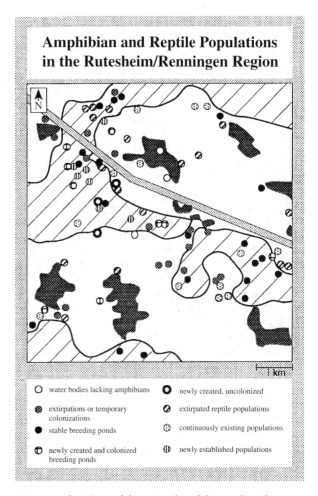

FIGURE 12-1 Location and dynamics of amphibian and reptile populations in the study site. Hatched = forested areas; Dark = villages; White = agricultural areas.

(for example, Schlüpmann, 1982; King, 1984; Dickman et al., 1993). These opportunities are seldom utilized during the analysis of survey data (see below).

Materials and Methods

Research Area and Study Methods

The research area, about 50 square kilometers in size, lies on the western edge of the Stuttgart region (southern Germany). Agricultural fields predominate, occupying 57% of the land surface, followed by forest, occupying 33% of the land. The remaining 10% consists of settlements and small gardens. Aside from expansion of settlements on agricultural land, which is of little value to amphibians and reptiles, land use changed only slightly during the research period. The number of water bodies clearly decreased (see below). The research area is poor in natural water bodies and in thermally favorable sites. Henle and Rimpp (1994) published a more complete description.

Since 1966, herpetofaunal populations, their habitats, and the factors that posed potential threats have been surveyed with varied intensity within the research area (Rimpp and Hermann, 1987; Rimpp, 1992; Henle and Rimpp, 1993, 1994; Fig. 12-1). Breeding waters were characterized by employment of the following parameters, among others: type; surface area and depth (>2 m or <2 m); shading (four categories); shallow water zones—narrow (<1 m wide) or extensive (>1 m wide); shore structure (flat/steep, obstructed/unobstructed); degree of cover of submerged vegetation, floating-leaf vegetation, reeds or shore shrubbery; location (distance to roads; type of surroundings—deciduous forest, wet meadows, settlements, etc.). For reptile habitats the following characteristics were noted: type of habitat; general penetration of sunlight; presence of sunning places and open areas; degree of cover by low vegetation, bush and tree layers (four classes for each parameter); and the location. Further, all potential threats to mapped populations were recorded.

During the annual observations, coarse estimates of relative abundance were made. However, quantitative determinations of population sizes were not attempted. Amphibian populations were considered large if the estimated population at a breeding site exceeded 100 adults or small if the estimate was fewer than 50 adults. Irregular observations of individual reptiles were considered to represent a small population, whereas populations were considered large if there were regular observations of snakes or slow worms (Anguis fragilis) or more than 10 observations of other lizard species.

Observations were made throughout the research area in the course of the 26-year mapping project, with the intention of surveying all potential habitats for amphibians and reptiles at least once. Apart from the smallest habitats and habitats that came into existence after the first traverse, the monitoring of amphibian and reptile habitats is believed to be complete. Numerous habitats were selected for observation after the survey began, as is usually the case for most surveying projects. Annually, the largest and most biologically rich habitats were traversed thoroughly; the remaining habitats were surveyed irregularly (for an overview of the monitored habitats, see Henle and Rimpp, 1994).

Extinction and Colonization Dynamics

A total of 13 amphibian and 7 reptile species were found in the research area (Henle and Rimpp, 1994; Table 12-1), of which palmate newts (Triturus helveticus), European pond turtles (Emys orbicularis), and pond sliders (Trachemys scripta) were definitely released and were not native to the research area. Palmate newts occur in neighboring regions (Rimpp and Hermann, 1987). During the study period, natterjack toads (Bufo calamita) and European treefrogs (Hyla arborea) disappeared, whereas agile frogs (Rana dalmatina) were actively colonizing new areas.

Amphibian populations occurred at 91% of the 54 observed wetland complexes. Twenty (24%) of the 82 water bodies that were investigated within these wetland complexes disappeared: seven disappeared or were rendered unsuitable through human construction activity; five were intentionally filled; four were excavation pits that disappeared through change in use or through abandonment; two disappeared through natural succession to land; one dried out due to a failure of the seal on the bottom; and one disappeared primarily because of a lowering of the water table. Countering these losses were the creation or installation of five ponds (one of them was created in the 1930s) by the forest administration; the installation of six amphibian conservation ponds; the installation of a cattle watering pond as a substitute for a filled sinkhole pond; and the construction of three rainwater interception ponds alongside a highway. However, the rainwater interception ponds were unsuitable amphibian habitats because of their location

and structure. Additionally, one pond was renaturalized to serve as a nature conservation pond.

During the research period a total of 107 colonizations (94 by amphibians and 13 by reptiles) were recorded (Henle and Rimpp, 1993). Of these, however, only 51 amphibian and two reptile populations persisted. The successful colonizations by amphibians occurred in the newly created or renaturalized ponds. Further colonizations occurred at the preexisting ponds, but were as a rule unstable and at most based on few individuals (status as of 1991). On the other hand, 102 amphibian populations and 22 reptile populations (Table 12-1) disappeared.

Data Evaluation

Research aimed at analyzing the degree of isolation of neighboring populations was not conducted. Therefore, the observed colonizations and extirpations cannot be differentiated into populations and subpopulations. Of course there were indirect and accidental observations of dispersions (Henle and Rimpp, 1993, 1994) that suggest that most of the amphibian and some of the reptile species in the research area exist as metapopulations, with limited exchange among subpopulations. However, generally, as expected in larger research areas, a continuum of fully isolated populations to barely separated subpopulations occurred. Arbitrary division of this continuum was not attempted. The term *population* is therefore used independent of the degree of isolation of the individuals in a biotope used for breeding (e.g., a pond) that is spatially isolated by biotopes that are not suitable for breeding. Connected, heterogeneously structured breeding habitats are considered to be one habitat. Because summer habitats of amphibians were not specifically mapped, they cannot be included in the analysis. The database is altogether typical for long-term surveying projects of defined research areas.

Given the time constraints inherent in surveying many sites over large areas, it is difficult to devise a method that distinguishes between areas where a species is simply overlooked, where there are actual disappearances (Den Boer, 1990), or where there are complete shifts to a neighboring breeding site in species that do not exhibit philopatry (Tester, 1990). In keeping with the need for a conservative approach, absences of species that are difficult to find were only judged as losses if no individuals were found in spite of intensive searches for at least 5 years. Five years was chosen because most native amphibian species in climatically comparable regions exhibit a turnover of the breeding population of at most 5 years (Heusser, 1970; Bell, 1977; Ryser, 1986; Tester, 1990; Kuhn, 1994; Wolf, 1994). The following are regarded as difficult to find: newts in large, richly structured ponds; snakes; slow worms; and lizards in small populations within large, potentially suitable habitats. Likewise, losses at ponds that occasionally dry up and are later resettled are only recorded as losses if the interval is at least 5 years—amphibians can also seek breeding ponds in the years following their destruction (Heusser, 1970; Blab, 1986; Kuhn, 1994), and an observed collapse may be the collapse of an age class rather than the loss or emigration of a population.

Detailed reports of the investigated habitats and changes in their species composition were published by Henle and Rimpp (1993, 1994). A few unimportant departures from those publications are based on additional data first available for consideration in the present work. Three previous classifications and one printing error in Henle and Rimpp (1993) were corrected.

TABLE 12-1

Known and Inferred Causes of Decline

Causes of Loss of Populations

Species	Number of Populations	Habitat Loss (L–D)	Chemical Pollutants	Fish	Natural Losses (S–T–E)	Unknown
Salamandra salamandra	11	1–0	–	–	–	–
Triturus alpestris	44	6–1	(1)	–	2–3–3	6
T. cristatus	6	2–0	–	(1)	–	2
T. helveticus	3	–	–	–	0–0–2	–
T. vulgaris	17	2–0	(1)	–	2–2–1	4
Bombina variegata	15	3–0	–	–	2–3–2	1
Bufo bufo	37	4–0	5	–	0–4–1	1
B. calamita	2	2–0	–	–	–	–
B. viridis	5	2–0	–	–	–	–
Hyla arborea	7	1–0 (1)	–	–	0–0–3	2
Rana dalmatina	9	–	–	–	0–0–1	–
R. esculenta/lessonae	12	4–0	2	–	0–0–1	2
R. temporaria	34	3–0	(1)	2	2–1–3	1
(Emys orbicularis)	1	–	–	–	0–0–1	–
(Trachemys scripta)	2	–	–	–	0–0–2	–
Lacerta agilis	19	1–0	–	–	(2)	2
L. vivipara	21	1–8	–	–	–	–
Anguis fragilis	16	1–0 (1)	–	–	–	–
Coronella austriaca	6	–	–	–	–	–
Natrix natrix	10	1–0	–	–	0–0–1	1
Totals	277	43 + (2)	7 + (3)	2 + (1)	42 + (2)	22

NOTE: L = lethal habitat losses; D = habitat losses demonstrated or considered very likely on the basis of dynamic experiments; S = losses due to natural succession; T = losses due to the drying out of spawning sites; E = unsuccessful colonization attempts. The number of observed losses coinciding with assumed causes but with insufficient controls for an analysis of natural experiments are included in parentheses.

The statistical methods of Sachs (1982) were employed. Applications of probability theory were based on Bosch's (1976) textbook.

Results: Analysis of the Causes of Population Extirpations

A limited number of factors causing the extirpation of populations have repeatedly appeared in the literature. These factors have been divided into seven groups (Henle and Streit, 1990). Using these seven groups of factors, the possibilities and limitations of such an analysis of extirpation causes is presented below for the present survey project. The results of the analysis are summarized in Table 12-1.

Destruction and Alteration of Habitats

Sixteen out of 82 breeding ponds studied during the research period were completely destroyed. Along with these ponds, 32 amphibian populations (31% of the losses) and a population of ringed snakes (*Natrix natrix*; 5% of the reptile losses) were extirpated. The time at which a population of European treefrogs disappeared is not precisely known. Although the population may have disappeared before the complete destruction of its breeding pond, that destruction is considered as the presumed cause of the extirpation. However, because the entire surroundings had been developed previously, resulting in decreasing terrestrial habitat, this loss can probably be listed as due to habitat alteration. This extirpation should not be attributed to unknown causes (Table 12-1). Ten large and 10 small populations disappeared; the sizes of the remaining 12 populations are not sufficiently known.

The habitats of four reptile populations were eliminated when they were covered with concrete and sod. Biological knowledge that reptile populations cannot live on concrete and sod substitutes in this case for otherwise indispensable control habitats. Similarly, no control is necessary with the complete destruction of a breeding pond because it is essential to the life of an amphibian.

For two species, common lizards (*Lacerta vivipara*) and alpine newts (*Triturus alpestris*), comprehensive natural trajectory experiments involving habitat changes have been evaluated. During the research period, 21 populations of common lizards were found, with 14 of these inhabiting wooded regions. The habitats of eight of these populations were forested with conifers or dense beech stands; none of these populations has survived. The habitats of the remaining six populations continue to be maintained as open deciduous or mixed deciduous forest with clearings or wide, sunny path edges; all these populations have survived. Under the extremely conservative assumption that exactly eight populations would have to survive, the probability of an accidental extirpation of eight populations exclusively in the changed habitats amounts to:

$$\alpha = \left\lceil \begin{matrix} 8 \\ 14 \end{matrix} \right\rfloor = 0.0003$$

With a probability of error of 0.03%, the null hypothesis that the habitat changes in the forests had no influence on the disappearance of common lizards can be rejected.

Closed forest canopies, which prevent direct sunlight from reaching the ground, offer only infrequent opportunities for common lizards to achieve their preferred temperature range of 29–34 °C (Van Damme et al., 1986). Most essential physiological processes, such as various aspects of feeding, reach 80% of their optimal value at body temperatures of 25–30 °C (Van Damme et al., 1991); common lizards cease all activity in the summer at temperatures below 16 °C (Van Damme et al., 1990). Thermophysiological information, along with appropriate temperature measurements, could also have explained the observed losses without reference to a natural trajectory experiment with control populations. However, this would be uncertain, since the microclimatic conditions severely limit activity but are not directly lethal.

Ten bomb craters or sinkholes of 1–175 m^2 size (depending on annual rainfall) north and west of Rutesheim were surveyed almost every year. Eight of them are situated in beech stands (*Fagus sylvatica*), or in mixed stands primarily of beech with some Norway spruce (*Picea abies*) and a few pines (*Pinus species*); one was located at the transition zone between deciduous forest and old spruce forest. Alpine newts bred annually in nine of these sinkholes. The tenth sinkhole was located in a spruce nursery. As the nursery matured, alpine newts steadily decreased during the 1980s and disappeared in 1986. If one of 10 populations disappears, the chance that this would accidentally occur at the impaired pond is 1 out of 10, or 0.1. Excluding populations that disappeared because of habitat destruction, a total of 15 out of 37 alpine newt populations included in the study disappeared. This indicates an average probability of extirpation of 0.41 (in 26 years). In view of this extinction probability, the chance of an accidental extirpation of the population in the spruce forest is only 0.04. The hypothesis that "the growth of the spruce forest led to the extirpation of the population" is additionally supported by the fact that another population about 1.5 kilometers away disappeared after increased shading from a spruce monoculture. However, the water level of its breeding pond also sharply decreased (Henle and Rimpp, 1994) so that population is not directly applicable as a control.

A parallel snapshot experiment permits confirmation of the conclusion that dense spruce forests (in the research area) led to the disappearance of alpine newt populations. An additional bomb crater, located in a dense spruce forest about 15 m from one of the 10 other bomb craters, was avoided between 1966

and 1989. After intense storm damage in the winter of 1989–90 and subsequent complete deforestation of the surrounding area, alpine newts oviposited in this bomb crater in the following 2 years. Of course, alpine newts breed in other research areas (although infrequently) in shaded ponds within spruce stands (Loske and Rinsche, 1985). For this reason the snapshot experiment by itself must be carefully interpreted. It is conceivable that a lighter stand of spruces or a difference in the buffering capacity of the pond water could have permitted settlement of the ponds observed by Loske and Rinsche (1985).

The snapshot experiment, as well as the result of the above trajectory experiment, is valid in the same way for common toads (*B. bufo*). They were definitely absent in the sinkhole in the matured spruce nursery in which alpine newts disappeared. Therefore, for common toads, a colonization rather than an extirpation was proven to result from habitat change.

The reasons for the unsuitability of the pond and the disappearance of alpine newts cannot be determined with complete certainty. Possibilities other than the cool, shady location are the acidification of the ground or the water. No measurements related to either factor were made at the site. Investigations of the minimal temperature at which alpine newt larvae can still successfully develop or of preferred breeding water temperatures are lacking to my knowledge. In some cases, shady breeding waters in the research area are settled by alpine newts (Henle and Rimpp, 1994). In other places the species lives under substantially more extreme climatic conditions, but breeds, as would be expected, only in thermally favorable ponds (Nöllert and Nöllert, 1992).

In conifer forest monocultures low pH values in the soil (<pH 5) frequently occur (Mückenhausen, 1985). However, this may not cause a problem for terrestrial alpine newts because adults can be encountered in breeding waters of pH less than 5 (Stevens, 1987). In contrast, embryos exhibit increased mortality under pH 4.5, and complete mortality at pH 4 (Böhmer et al., 1990). Acid pulses in spring after snowmelt could therefore be responsible for the avoidance of ponds. Unfortunately no measurements were made, but Loske and Rinsche (1985) measured pH values of less than 4 in pools within spruce stands and observed strings of dead spawn of common toads infected with fungi. Although a slight drop in the water table that occurred at the same time as the increased shading may have contributed to a reduction in the population, it can be excluded as the main cause since three of the remaining sinkholes exhibited considerably lower water levels (of 20 cm maximum) for a period of 1–3 years without abandonment by alpine newts.

In the course of widening a highway along the edge of the habitat of a slow worm population, the reconstructed road shoulder became smaller and much steeper and was fortified with stone blocks. This population disappeared after this qualitative change in its habitat. No other population in the research area was affected by similar habitat changes. In the other 15 slow worm habitats surveyed, severe changes occurred in some cases. In the most extreme case, widely spaced homes were built on the habitat; none of these other populations disappeared. Because a sufficient number of similarly altered habitats (as well as unaltered control habitats) are lacking to allow an experimental evaluation, in this case, habitat alteration is merely presumed as the cause of disappearance of the slow worm population.

At five ponds on fallow land or within meadows under light agricultural use, 3–7 (average = 5.4) amphibian species were found. Common toads and common frogs were always present at the ponds; either smooth newts or alpine newts were also invariably present; water frogs *(R. esculenta/lessonae)*, agile frogs, European treefrogs, or crested newts *(Triturus cristatus)* were sometimes present. Ringed snakes were also present at four of these ponds. In contrast, there were no amphibians at a pond in a meadow in the middle of an intensively farmed area. There were no woods, hedges, field borders, or old fields within 300 m of the pond. This pond is completely encircled by shrubs and lacks a surrounding reed belt, further differentiating it from the others.

Pools with diameters of up to 5 m exhibited similar relationships. Common frogs always oviposited in such water bodies on lightly farmed meadows or fallow land (n = 4), whereas they were lacking at two similar ponds on intensively farmed fields. Essential summer and winter quarters are provided for most amphibians by hedgerows and woodlots (e.g., newts, Beebee, 1985; European treefrogs, Stumpel, 1993), whereas intensively farmed fields and pastures are microclimatically unsuitable (Wolf, 1994) and presumably offer scant protection from predators. Moreover, thermally favorable locations for oviposition by ringed snakes (Zuiderwijk et al., 1993) are lacking in the absence of vegetative cover.

In summary, alteration or destruction of habitats within the research area was definitely responsible for the disappearance of 43 populations or subpopulations and probably responsible for the disappearance of two others. This represents 35–36% of the losses.

Chemical Pollutants

Five water bodies were significantly overloaded with environmental pollutants or fertilizers. From 1972 to the early 1980s near the Rutesheim dog training club, used oil, styrofoam, and other refuse was dumped repeatedly (in violation of the law) into a shallow residual water body of about 200 square meters. This pond is a remnant of a large wetland that was drained in 1928. Animal excrement also entered from a neighboring small-animal rearing facility belonging to a small farm animal fanciers club. Alpine newts and smooth newts disappeared during this period of pollutant loading. The last individuals of both species were found in 1975. A population of common frogs in this water body steadily decreased in the second half of the 1970s, but held out as a small breeding population until 1988 in a bordering depression that was only affected by the pollutant loadings at extraordinarily high water levels.

A quite similar water body about 150 m away, part of the same former wetland complex, serves as a control water body. The two water bodies are separated by an expressway. All three species still occur here. Common frogs also survived at another water body with similar depth, shore structure, size, shading, growth of vegetation, and surroundings (meadows; distance to woods of 100–250 m). Newts were first regularly observed at this water body after 1980, for which reason it cannot serve as a control water body in the present case. Additional control water bodies (with two populations of alpine newts as well as three populations each of smooth newts and common frogs) are of limited applicability here because they are clearly different in size, depth, or shore structure. Another nine potential control populations of alpine newts inhabited water bodies that are quite comparable in size and shore structure, but all of them are situated in deciduous forest and in some cases are significantly deeper. Upon consideration of these populations, the probability of an accidental extirpation

in the polluted water body approaches significance only for alpine newts ($\alpha = 0.09$).

The continued oviposition by common frogs in the unpolluted depression definitely proves that, at least for this species, the reasons for its extirpation must have lain in the breeding habitat rather than the terrestrial habitat. Also, the common disappearance of all three species with the entry of pollutants points to the pollutants as the cause. However, it must be remarked that common toads disappeared before 1970 at the polluted water body, before the occurrence of substantial pollution. The hypothesis that the disappearance of the three species accidentally occurred at the polluted water body cannot be rejected, even though much evidence suggests that it is incorrect. Therefore, these three losses should be designated as "presumably caused by chemical pollutants" (Table 12-1).

The two Längenbühl Ponds, one lying directly behind the other, were polluted with liquid manure from a swine fattening operation. Both ponds were designated as Nature Monuments at the beginning of the 1970s on the basis of formerly large water frog populations (Henle and Rimpp, 1994). During my first year of observations (1973), the water bodies were already severely polluted. At that time, only common toads and common frogs were still breeding at both ponds. The pollution has clearly decreased since the end of the 1980s. Common frogs have continued to breed regularly since the late 1980s, but, as previously, their spawn have always failed to develop. Additionally, every few years there have been attempts at breeding by common toads from a quite large breeding population 250 m away. There were at least five failed attempts. Likewise, the lack of success is probably a result of the severe pollution. Common frogs are not considered to be extirpated at these water bodies, although there definitely has been a lack of successful larval development during the research period. On the basis of the life expectancy of common frogs (Heusser, 1970), the isolated oviposition that has been observed, at least in the 1980s, probably resulted from individuals that annually migrated from a very large breeding population 250 m away. Apart from two populations in dried out water bodies, the only other case of oviposition by common toads without successful embryo development occurred at a water body that was potentially polluted by runoff from a highway drain. With the remaining 15 populations of common toads, which serve adequately as controls, embryo development was successful and there was no indication of pollution. This difference is highly significant ($\chi^2 = 17$; $\alpha = < 0.001$).

Severe pollution due to over-fertilization was observed at two further water bodies: the pond on the southwest edge of the Village of Silberg and the settling pond at the Hardtsee. A buildup of putrid sludge has been observed at both water bodies since the second half of the 1980s. The Silberberg Pond lies in a water protection zone, but that has not prevented a farmer from dumping liquid manure on adjacent pastures and fields. Discoloration of large portions of the bottom and occasional intense releases of sewer gas have occurred at this pond. The Silberberg Pond originally possessed eight amphibian and one reptile species; today only common frogs and common toads remain. The losses (alpine newts, crested newts, smooth newts, yellow-bellied toads *[Bombina variegata]*, European treefrogs, and water frogs) cannot, however, clearly and exclusively be attributed to hypereutrophication. The pond was drained in 1978 and 1979 and was previously reduced in size by partial filling; additionally, a few species had declined or disappeared (crested newts and European treefrogs) before hypereutrophication.

Control water bodies, in which either sludge buildup occurred without habitat changes or corresponding habitat changes occurred without sludge buildup, would be necessary for evaluation of a natural experiment. Intense sludge buildup occurred, however, only at one additional pond—the previously mentioned settling pond at the Hardtsee. It was installed as a settling basin for the removal of sediments in water entering a larger pond, the Hardtsee. Alpine newts, crested newts, smooth newts, common toads, agile frogs, water frogs, and common frogs bred in the settling pond. All the species produced large populations at the beginning. With the exception of common toads, common frogs, and agile frogs, all the populations decreased as the pond filled with sludge, and crested newts completely disappeared. This is consistent with the Silberberg Pond, which was also polluted with putrid sludge. The settling pond at the Hardtsee, however, cannot be employed as a control for this pond since, additionally, fish were released in it. In the Hardtsee, which receives effluent from the settling pond, the same species decreased as in the settling pond, and crested newts likewise disappeared. However, the Hardtsee is unacceptable as a control because it was also stocked with fish (to a considerably higher degree than the settling pond), and nutrient pollution only occurred in some years, in particular as a result of large-scale feeding of mallards (Anas platyrhynchos). Because the population shifts slowly occurred at all three ponds, and other factors (habitat alteration and fish stocking) were probably involved, and further because there were no further potential control water bodies, these losses must be considered to be of unknown (presumably complex) origin rather than due to pollution.

In connection with pollution, it should further be mentioned that common frogs spawned in two of three ponds constructed in the course of the widening of a highway. Runoff from the highway drained into each of the ponds. Common toads also spawned in one of the ponds. All the spawn clumps of common frogs were destroyed by fungi, and larger larvae or juveniles of common toads were never observed. In 1991 and 1992, common toads no longer spawned in these ponds. The lack of successful development suggests that chemical pollution was the cause of the extirpation (see above). Moreover, it was the only case (n = 10) of the colonization of a newly constructed pond by common toads that failed. Chemicals leached from materials used in highway construction can lead to acidification of adjacent water bodies and the loss of amphibian populations (Harte and Hoffman, 1989), and attack of spawn by fungi occurs frequently in acidic waters (Clausnitzer, 1987). Nevertheless, this loss cannot be attributed to pollution, but is classified as "unknown." Chemical pollution can only be the presumed cause because analytical demonstration of the presence of chemicals in toxic concentrations is lacking and other causes cannot be excluded. The three ponds alongside the highway are clearly different from the remaining newly constructed ponds in their location and shore structure. Accidental factors must also be considered in regard to high extirpation rates of small, newly founded populations (Wissel and Stephan, 1994).

Seven populations (6% of the losses of amphibian populations) were extirpated at the three water bodies at which chemical pollution could be proved as the cause.

Predators

Two attempts by common frogs to breed in a trout-rearing pond were unsuccessful because the spawn was consumed. No amphibian spawned on any of the remaining basins used for intensive rearing of trout. However, common frogs regularly spawned on an adjacent basin that was not used for trout rearing.

In one pond, crested newts disappeared after the stocking of fish. However, there were no controls for this observation, so it is not proven that fish were the cause; also, the fish did not reach such an extreme density (as would exist in a trout-rearing installation) that total predation would be certain. However, there are indirect observations that corroborate the hypothesis that fish exert a negative influence. After the disappearance of crested newts, the pond was maintained free of fish for two years. During these years, the survival of common toads and common frogs to metamorphosis increased several-fold (K. Rimpp, personal communication). But since semiquantitative control studies at comparable ponds are lacking in this case, this hypothesis remains speculative—especially since Schlüpmann (1982) observed that tadpoles of common toads remain undisturbed in concrete basins of rainbow trout (Oncorhynchus mykiss).

In order to settle the often hotly debated controversy between amphibian conservationists and anglers, predation experiments were carried out in the laboratory (Bauer, 1983) with various fish species: roach (Scardinius erythrophthalmus), European carp (Cyprinus carpio), eel (Anguilla anguilla), chub (Leuciscus cephalus), tench (Tinca tinca), and rainbow trout. In a 200-L container, 6–8 large larvae of alpine newts or smooth newts (crested newts were not tested) were added in each experiment to each fish species. The investigations indicated that all the fish species tested consumed all the larvae within eight hours to seven days. A detrimental effect of fish on the above-mentioned crested newt population is therefore possible. However, the laboratory investigation is not considered valid for crested newts, nor is it at all likely to have any relevance to field conditions, because amphibians exhibit a variety of defense mechanisms (for example, skin toxins, behavior, and choice of microhabitat) against fish (Duellman and Trueb, 1986). Also, in the above-mentioned laboratory experiments, no cover was offered. Adequate cover substantially alters the survival rate (Kats et al., 1988; Semlitsch, 1993). Laboratory experiments of various authors, for example, illustrate that common toad tadpoles possess better defensive mechanisms against predation from most fish species than other native amphibian species (see literature overview in Breuer, 1992). Moreover, fish species vary in their predation on a given amphibian species under laboratory conditions (Glandt, 1984, 1985; Breuer, 1992).

Seminatural experiments (Semlitsch and Gibbons, 1988) or carefully evaluated natural trajectory experiments are largely lacking so far for European amphibians (Henle and Streit, 1990). However, Breuer and Viertel (1990) and Breuer (1992) have shown in field experiments that European carp and rainbow trout clearly increase the mortality of tadpoles of common toads and common frogs (rainbow trout more considerably affect mortality than do carp). Common frogs have poorer defense mechanisms at their disposal than do common toads, and in contrast to the latter they do not survive to metamorphosis in every experiment.

Clear but unsupported indications of fish as a cause of the loss of breeding populations are available from personal investigations of a pond just outside the research area. After heavy stocking of European carp, grass carp (Ctenopharyngodon idella), rainbow trout, catfish (Silurus silurus), northern pike (Esox lucius), and pike perch (Lucioperca lucioperca), the following amphibian species disappeared from a pond in an abandoned quarry: alpine newts, smooth newts, yellow-bellied toads, green toads *(B. viridis)*, and common frogs. At the same time, green toads and common frogs continued to inhabit four control ponds (likewise outside the research area). There were only two control populations for smooth newts and only one for alpine newts and yellow-bellied toads. For the species with four control populations, the probability of accidental extirpation in the pond where fish had been stocked was 20%. The parallel extirpation of five species would therefore be significant ($\alpha < 0.01$). Because a portion of the terrestrial habitat had been altered at the same time the fish had been stocked, the possibility that the habitat alteration was responsible for the extirpations cannot be excluded. However, a substitute pond in the immediate vicinity that was constructed 2 years after fish stocking as an amphibian breeding pond was utilized successfully for at least 5 years by green toads and smooth newts. Therefore, the changes in the surrounding terrestrial habitat cannot be responsible for the disappearance of these two species in the original breeding pond. Fish stocking remains the sole cause of disappearance of at least these two species.

Isolated observations of predation on spawn by mallards occurred in three ponds, but common frogs did not disappear from any of them. Common frogs are indeed absent from park and farm ponds in Westphalia that are heavily stocked with mallards (Schlüpmann, 1981), but mallards are only seldom likely to present a serious threat to this species (Kwet, 1996).

Predators (fish stocking) were responsible for the loss of 2% of the amphibian populations. Fish predation was the presumptive cause of extirpation for one large population of the crested newt. All other populations involved were small.

Collecting

The influence of the collecting of amphibians and reptiles on the survival of the affected populations is a hotly debated issue for which little supportive information exists (Ehmann and Cogger, 1985; Henle and Streit, 1990). The danger from the collecting of amphibians and reptiles depends, among other factors, on the size of the range and the survival strategy of the affected species (Henle and Streit, 1990), and can only be determined objectively by field experimentation or estimated reliably by extrapolation from thorough investigations of population dynamics (Henle and Streit, 1990).

Animals were taken more or less regularly out of several populations in our research area. These removals were more frequent in the 1960s and 1970s than in the 1980s, when removals were only seldom observed—presumably because of decreasing interest as a consequence of tighter legal restrictions on collecting and rearing. Convincing evidence of endangerment by collecting is unavailable for any of the populations in this study.

The evaluation of a trajectory experiment at nine small ponds in a deciduous forest stand eliminated the possibility of a threat from observed collecting of adult alpine newts as well as from the collecting of spawn and larvae of common toads. In two of the pools, regular collecting was carried out in the 1960s and 1970s by youths. During the breeding period, the removal of perhaps at most 10 newts and fewer than 50 tadpoles per week was observed. In contrast, in the remaining seven pools removals were the exception (fewer than five per year recorded) because of their substantially greater distance from a

woodland path and more difficult accessibility of the shore area. A qualitative decline of the populations in the investigated pools in comparison to the control pools could not be established, and all populations survived the collecting. These results confirm the argument advanced by Henle and Streit (1990) for common toads that endangerment of this species from collection of the larvae can generally be excluded because of its survival strategy—high numbers of offspring, high natural mortality, and, presumably, strong density-dependent regulation of larval mortality (Grossenbacher, 1981; Kuhn, 1994).

Traffic

During the entire research period, with the exception of recently metamorphosed common toads and common frogs and of at least 25 slow worms, only 15 amphibians and reptiles were found dead on the road. These were primarily fire salamanders *(Salamandra salamandra)* on asphalted woodland paths. Although no systematic studies of isolated roadkills were attempted, mass mortality was largely excluded because crossings of considerable numbers occurred only on the highway between Malmsheim and Weil der Stadt, and this highway is closed during the spring migration. No populations have disappeared in the affected breeding waters. Road traffic did not therefore contribute to loss of populations. However, it must be kept in mind that no new highways were constructed or substantially widened within the research area during the period of observation.

Expressways probably indirectly affect populations by isolating them. After its partial cleanup and regeneration, the pond at the Hundesportplatz (mentioned above under "Chemical Pollutants") was not colonized (or recolonized) by common frogs, common toads, alpine newts, smooth newts, or ringed snakes over a period of 3 to more than 12 years, even though populations existed on the other side of the expressway at a distance of about 150 m. In contrast, newly constructed ponds that were separated by a railway track rather than an expressway were settled by these species, and also by crested newts, European treefrogs, and water frogs, within 1–3 years from a distance of at least 750 m.

Underground storm sewers and cellar shafts can lead to substantial losses of amphibians (Bitz and Thiele, 1992; Strotthatte-Moormann and Forman, 1992; Wolf, 1994). The residents of Leonberg-Silberberg gave me an alpine newt and a smooth newt that had fallen into cellar shafts. Systematic controls were of course lacking. Nevertheless, storm sewers and cellar shafts cannot be excluded as causes for the loss of populations within the research area. Amphibian and reptile habitats in the vicinity of villages are lacking, with the exception of the edge of Leonberg-Silberberg and populations of slow worms. The only populations in the vicinity of villages to disappear died out after the destruction of their breeding ponds or after temporary drainage and extreme nutrient pollution.

In summary, highway traffic was not proven to cause the loss of populations, although an expressway has so far prevented colonization of a pond.

Competition with Introduced Species

A few exotic amphibian and reptile species have been released in the research area, but their populations have failed to survive (Table 12-1). They have therefore apparently not exerted a negative influence on the other populations of the research area.

Natural Causes

It is difficult to relate observed species losses in the landscape created by European civilization to natural causes since indirect effects of human activity can seldom be excluded. Drying of ponds or a lowered water level led to the disappearance of four populations of common toads, three populations each of alpine newts and yellow-bellied toads, two populations of smooth newts, and a population of common frogs. Although the proximate cause is natural, human activity may have contributed to the lowered water table and more rapid drying out of the ponds. In one case, the cause was clearly anthropogenic—the plastic sheet that was sealing a pond failed. Only two large populations, one of alpine newts and one of common toads, were affected. Further, an annually decreasing water level in a former stone quarry severely threatened formerly very large populations of alpine newts, crested newts, and fire salamanders; at the time of the last observation in 1991 there was only a little water present under stone rubble. As a result, the likelihood of continued survival is starkly limited or actually nonexistent.

Four road ruts disappeared through natural succession, and with them two small populations each of alpine newts, smooth newts, and yellow-bellied toads. Common frogs spawned at two fens until they became completely overgrown. An attempt by common toads to breed in a complex of road ruts failed to establish a population. Additionally, there were 14 other populations and six releases or human-mediated recolonization attempts that failed to become established—that is, they lasted fewer than 5 years. There was no indication of other possible causes. The number of individuals in all these cases amounted to less than 10.

Two very small populations (<10 individuals) of sand lizards lived for only a few years, each on an area of road shoulder of less than 100 square meters. Populations of this magnitude have only a small chance of survival (see Hildenbrandt et al., 1995, for wall lizards [*Podarcis muralis*]); both sand lizard populations are likely to have disappeared due to stochastic fluctuations.

Natural factors, with the above-mentioned limitations, explain 37% of the losses of amphibian populations and 27% of the losses of reptile populations. These values are respectively reduced to 21% and 9% if unsuccessful colonizations are excluded. Only two large populations were affected.

Unknown Causes

Causes are classified as unknown if there is no indication of a definite cause or if various factors exert an influence, but no factor or complex of factors can be proved to be responsible.

Of a total of 22 populations that disappeared for unknown reasons, 68% were classified as small. Disappearances of large populations for unknown reasons occurred once each for crested newts, smooth newts, and common toads, as well as twice each for European treefrogs and water frogs. Crested newts disappeared at the settling pond at the Hardtsee. Clearly, the addition of fish and increasing buildup of sludge are potential causes, but cannot be proven (see above).

Smooth newts, European treefrogs, and water frogs died out at the Silberberg Pond presumably because of a combination of several factors: the pond became smaller, it was drained for 2 years, and there was a severe buildup of sludge (for details see the section entitled "Chemical Pollutants"). A large European treefrog population and the largest population of water frogs with several hundred adults disappeared at the Renninger See (a pond) without any indication of the cause. Common toads died out at a remnant pond of a large former wetland long after the wetland was drained and after the construction of an expressway about 50 m away. The expressway cut the population off from suitable summer habitat, with the exception of the immediate surroundings and a very small patch of woodland (<1 ha). Possibly, the extirpation was a delayed consequence of these massive landscape alterations, but no evidence of that exists.

Discussion

Causes of Species Decline

Numerous publications indicate that a general decline of amphibians and reptiles has occurred in Central Europe (Grossenbacher, 1974; Brunken and Meineke, 1984; Thurn et al., 1984; Berger, 1987; Löderbusch, 1987; Riffel and Braun, 1987; Rimpp and Hermann, 1987; Stangier, 1988; Rimpp, 1992). For example, in the 26-year period of investigation of Henle and Rimpp (1994), two species were extirpated; on the other hand, one species actively immigrated. Only a few species appear to be unaffected regionally by the general decline. Indeed, green toads (Rimpp, 1992) and agile frogs (Henle and Rimpp, 1994), contrary to the general trend, expanded within the area covered by this study. Water frogs very sharply decreased in the 1980s. With one exception they now exist only in small populations, for which the reasons are only partially known (Table 12-1), while they are increasing in areas of North Rhine-Westphalia because of recolonization attempts and newly created habitats (Klewen, 1988; Kordges, 1988).

Similar relationships (sharp decline, but only of some of the species, and not always the same ones) occur on other continents (Osborne, 1990; Dodd, 1993a; Ingram and McDonald, 1993; Mahony, 1993), for which largely unknown global causes of decline are discussed (Blaustein and Wake, 1990; see also Reaser and Blaustein, this volume). In contrast, unknown factors are only seldom mentioned in Central Europe. The same causes of decline are cited repeatedly here (see Henle and Streit, 1990). Of these, habitat destruction also clearly represents a global, frequently proven cause of the loss of biodiversity (Henle and Streit, 1990; Saunders et al., 1993; McDade et al., 1994).

In order to devise conservation measures that promise success, it is not as important to postulate presumptive global phenomena as it is to provide detailed knowledge of the relative significance of various factors (Henle and Streit, 1990). This knowledge, which can be further developed only by critical analyses, will facilitate the targeting of the actual causes, rather than the symptoms (Henle, 1995).

Because detailed causal analyses of observed changes of amphibian and reptile populations are generally lacking, at this time a sufficiently accurate assessment of the relative significance of the various potential threats is only possible within limits. The causal analysis of the long-term surveying project

(Henle and Rimpp, 1994) discussed here demonstrates that it is generally possible to carry out such analyses in spite of the methodological problems posed by having to repeat surveys over an extended period of time. In this example, the causes remained unknown for only 24% of the extirpations.

Habitat destruction and alteration were proven to represent the leading cause of extirpation within the research area of Henle and Rimpp (1994). Numerically, this is the leading cause for reptiles and the second cause—behind natural losses—for amphibians (Table 12-1). When considering natural losses, however, one must bear in mind that these losses almost exclusively affected the smallest populations. On the other hand, about half of the amphibian populations that disappeared because of habitat destruction were large, in spite of only minor changes in land use. This underscores the greater importance of habitat destruction/alteration as a cause of loss of populations. This conclusion is supported additionally by the fact that habitat alterations are involved in perhaps half of the losses classified as natural. Also, indirect effects of human intervention in some cases cannot be fully excluded.

The sharp decline of ponds and wetlands, essential habitats of amphibians (Honegger, 1981; Stangier, 1988), and of semi-natural ecosystems (Saunders et al., 1993) confirms the essential role that habitat changes play in the global decline of amphibians and reptiles and of biodiversity in general (Settele et al., 1996). Works that analyze population changes of particular Central European species (Comes, 1987; Fritz, 1987; Fritz et al., 1987; Glandt and Podloucky, 1987; Corbett, 1988; Podloucky, 1988) reach similar conclusions; therefore, the first priority of conservation measures must be to deal effectively with habitat destruction and change. It is not enough, however, to place a large number of small areas under protection (see the example above). Rather, strategies must be developed that change land use practices so that habitat destruction and negative habitat alterations are generally counteracted (Saunders, 1996); otherwise symptoms, instead of causes, will be addressed (Henle, 1995).

In other regions, additional factors besides the biotope changes in the present example presumably contribute to local declines, but their relative significance is considerably more difficult to assess at this time. Chemical pollution may be the most important cause after direct habitat alterations, as it was in the research area analyzed here. The situation on islands, where pursuit by man and introduced predators plays a more important role (Henle and Streit, 1990), is apparently an exception. However, losses due to invasive species may frequently be promoted by habitat alterations, and may not occur in the absence of the alterations (King, 1984).

Although a relatively large number of populations in the research area of Henle and Rimpp (1994) died out as a result of stochastic events, this factor cannot be regarded as a fundamental cause of a general decline, since small populations were almost always involved. Generally, clear evidence of a natural factor (such as climate) posing a serious threat to the regional survival of an amphibian or reptile population is rarely encountered (Böhme, 1989; Osborne, 1990), but it must be remembered that it is quite difficult to establish a natural factor as a cause, and in most investigations natural factors are disregarded (Henle and Streit, 1990).

Other factors played no role or only a comparatively unimportant role in our research area. Likewise, Henle and Streit (1990) stress that additional factors contribute importantly to regional or global extinction of amphibians and reptiles only under limited conditions. Among these are, for example, threats to endemic island species from introduced predators and pursuit by humans, as well as commercial trade in species groups with a survival strategy characterized by low reproduction rate, late sexual maturity, and long life expectancy—a combination that is not generally characteristic of native amphibians and reptiles, and generally characteristic of only a few amphibian and reptile groups (Duellman and Treub, 1986; Dunham et al., 1988; Wilbur and Morin, 1988; see also Green, this volume).

Road traffic in heavily populated Central Europe can also contribute substantially to losses of amphibians or other faunal groups (Stubbe et al., 1993) and can lead to a decrease in genetic variability through isolation (Reh and Seitz, 1989). With regard to the influence of traffic on the survival probability of amphibian and reptile populations, the criticism must be clearly stated that, to date, causal analyses are lacking, in spite of a practically overwhelming number of expensive, labor intensive conservation measures and intervention experiments (Henle and Streit, 1990; Wolf, 1994). Field experiments, which are generally best suited for analysis of declines (Mahony, 1993), are required for clarification. Normally, such experiments are rarely justified or performed in order to investigate a threat to a given species, but since new highways are regularly constructed and pose a threat to many species, it would be easy to integrate appropriate field experiments into highway planning.

Detailed Population Vulnerability Analyses (PVAs; Clark and Seebeck, 1990; Hildenbrandt et al., 1995) provide an alternative. These are often quite expensive and restricted to specific, spatially limited problems (but see Henle and Streit, 1990). For these reasons they have been largely reserved for the "flagship species" of nature conservation, for example Leadbeater's possum (*Gymnobelideus leadbeateri*; Lindenmayer and Possingham, 1995). They recently have been developed for routine employment in the practice of nature conservation (Settele et al., 1996). To date, this type of analysis has been applied to only one of our native herpetofauna—wall lizards—after a predictive model was developed and the required knowledge of its population biology was worked out (Hildenbrandt et al., 1995).

Optimization of Surveying Projects

Improvements in contemporary population surveying practices may also furnish better evidence of the relative significance of risk factors. To date, surveying projects primarily have been carried out to obtain an overview of the actual distribution and the specific habitats occupied by the mapped species. Over time there has been increasing focus on the use of extensive survey data to probe the reasons for changes in populations, although most compilations and evaluations have been restricted to the listing of potential risk factors.

Discriminant analysis can be used as a first step in the analysis of snapshot experiments, even to some extent with data that has already been compiled. That is, the parameters investigated in the previously encountered situation are analyzed in order to determine which parameters are particularly useful for distinguishing the occupied habitats from the unoccupied ones (Foeckler, 1990; Ildos and Ancona, 1994). This initial step presupposes, however, that the parameter (for example, measurement of the distance to the nearest road, the concentration of chemical pollutants, or the stocking of fish in a pond) is adequately standardized and quantified. From the results, well-founded hypotheses about the relative significance of various potential risk factors can be derived. However, the disadvantages of snapshot experiments (see above) remain.

Along with discriminant analysis, habitat models can be constructed by means of additional multivariate methods. They can provide a basis for a spatial representation of detailed PVAs (Kuhn and Kleyer, 1996). Systematic recording of critical habitat parameters and potential threats in investigated habitats is essential for discriminant analysis and for construction of habitat models. In particular, biotopes in which amphibians and reptiles are absent must also be included. In this manner, a foundation is created for a better understanding of colonization processes and metapopulation dynamics of the surveyed species (Settele et al., 1996). Repeat surveys are essential for this purpose and additionally serve as an indispensable foundation for the evaluation of trajectory experiments.

The long-term surveying project discussed here indicates that the causes of extirpations as well as the relative significance of the various risk factors can be fairly objectively assessed for a large number of herpetofaunal populations even when the repeat surveys are not systematically executed. However, this requires an extensive data record. It is often difficult to find a sufficient number of affected and unaffected control populations within a restricted area. One must attempt to keep interfering factors to a minimum when choosing control populations. This means that factors other than the risk factor under examination should be as similar as possible among the assessed populations. The fewer comparable populations available, the more important this becomes for reliable interpretation of the results. A unique opportunity for planning of future survey activities in the interval between survey periods (perhaps annual) is available with survey programs in which a number of qualified individuals are providing assistance. Ponds can be selected systematically as investigation or control ponds on the basis of unplanned experiments (for example, habitat changes, fish stocking, etc.) reported in the previous year or years and assigned to particular co-workers for resurvey. Finally, when repeat surveys are planned systematically and standardized survey methods for risk factors based on previous literature are applied, the likelihood of being able to pursue a trajectory experiment at the right point in time is considerably increased, and the number of potential control populations is also increased. Ideally, repeat surveys must be executed at firmly established intervals. By careful planning, the responsibility for conducting repeat surveys can be allocated among the collaborators in keeping with their individual schedules.

With improved methods, as proposed here, efforts can be concentrated on specific causes of decline according to their relative importance. In this manner, the conservation of our native herpetofauna can be more effectively guided.

Summary

Survey data have been major sources for the documentation of amphibian and reptile declines and of potential threats. Often, all potential threats are taken as proven actual causes of decline, although the presented data analysis is seldom sufficient for such claims. The potential for causal inferences is limited because of a general lack of rigorous field designs. Nevertheless, the potential of survey data for such inferences is seldom fully realized. Therefore, comparisons of the relative importance of various potential threats among regions and, thus, the development of effective conservation strategies are hampered.

In this paper, potential methods for causal inferences are outlined briefly. The potential and limits of causal inferences from survey data are illustrated with an example from an approximately 50 square kilometer area west of Stuttgart, Baden-Württemberg, Germany. During a 26-year period of data collecting that lacked a rigorous design, 13 and 7 species of amphibians and reptiles, respectively, were observed. Two amphibian species, European treefrogs (Hyla arborea) and natterjack toads (Bufo calamita) were extirpated, and two species, crested newts (Triturus cristatus) and water frogs (Rana esculenta/lessonae) showed a considerable decline. A total of 124 populations were extirpated, including 54 unsuccessful recolonization attempts out of a total of 107 observed recolonizations. Analyzing the data as natural field experiments, only 24% of the extirpations (22 unknown and 8 uncertain causes) remained unexplained in spite of the lack of a rigorous design. Habitat change was the prime factor responsible for declines (e.g., 24% of the water bodies used for spawning disappeared). Pollution was the second major cause. Many extirpations were due to natural causes; with the exception of two populations, extirpations were limited to small and very small populations. Predation by fish caused two losses. No extirpation could be attributed to other predators, collecting, competition with exotic species, or road traffic. However, a highway built before the start of the survey project acts as a barrier to the reinvasion of a partially restored spawning site at which amphibians became extirpated due to pollution.

Recommendations for planning the collection of survey data are made to improve their potential for causal inferences on declines.

Acknowledgments. My sincerest thanks to Mr. Kurt Rimpp for placing his data at my disposal and for conducting extensive field surveys, without which the present effort would have had to depend on an all too meager database.

THIRTEEN

Risk Factors and Declines in Northern Cricket Frogs *(Acris crepitans)*

VAL R. BEASLEY, SANDRA A. FAEH, BRIGIT WIKOFF,
CRAIG STAEHLE, JOYCE EISOLD, DONALD NICHOLS,
REBECCA COLE, ANNA M. SCHOTTHOEFER,
MARTIN GREENWELL, AND LAUREN E. BROWN

Reports from around the world have indicated declines in numerous amphibian species (e.g., Wake, 1991; Adler, 1992; K. Phillips, 1994; Stebbins and Cohen, 1995; Green, 1997b; Lannoo, 1998b). There have been many proposed causes for these amphibian declines: habitat destruction, acidification of aquatic environments, pesticides, other toxicants, viral, bacterial, and fungal infections, drought, feral pigs, and excessive ultraviolet (UV-B) irradiation linked to ozone depletion from environmental contamination with chloroflourocarbons (Baker, 1985; Harte and Hoffman, 1989; Pechmann et al., 1991; Blaustein et al., 1994c; Kutka, 1994; Stebbins and Cohen, 1995; Berger et al., 1998). One species that has exhibited a marked decline in the midwestern United States is the northern cricket frog *(Acris crepitans)*.

Northern cricket frogs are small (12–30 mm in snout-vent length at maturity), non-arboreal members of the treefrog family Hylidae. Their historic distribution once extended from southern Minnesota and Ontario to northeastern Mexico, along the East Coast and west to Colorado (Conant and Collins, 1998; Gray and Brown, this volume). In the early 1960s, cricket frogs were found in all regions of Illinois and were considered to be the most abundant amphibian in the state (Smith, 1961). By the 1980s, however, the species had experienced a marked decline in the northern third of the state, and it was no longer readily encountered in amphibian surveys (Mierzwa, 1989; Ludwig et al., 1992; chapters in Lannoo, 1998b). A recent survey in northeastern Illinois indicated that cricket frogs were essentially absent at least as far south as the Kankakee River. Only two isolated populations north of the river were found, one in the Midewin National Tallgrass Prairie (formerly the Joliet Army Ammunition Plant) near Lockport, and another near LaSalle (Ludwig et al., 1992; T. Anton and K. Mierzwa, personal communication). Numerous other surveys have also indicated decline or extirpation elsewhere in the northern Midwest (chapters in Lannoo, 1998b; Gray and Brown, this volume), and the species is listed as endangered in Minnesota, Wisconsin, and New York, and as a species of concern in Indiana, Michigan, and West Virginia (chapters in Lannoo, 1998b; Ramus, 1998).

Burkett (1984) reported that the average life expectancy of northern cricket frogs is approximately 4 months, with complete population turnover in 18 months or less. Thus, adults

are usually present for only a single breeding season in the Midwest (Gray, 1983; Burkett, 1984; Lannoo, 1998c). However, skeletochronology (S. Perrill, Butler University, unpublished data) indicates that individual cricket frogs in some wild populations may live as long as 5 years. Like many other amphibians, cricket frog tadpoles are aquatic, whereas the juveniles and adults are terrestrial/semi-aquatic. Cricket frogs have a thin, well-vascularized skin that may render them more vulnerable to environmental contaminants than animals with a thicker, less vascularized integument. Therefore, they may potentially serve as indicators of recent environmental pollution. Moreover, the small effective breeding population size of cricket frogs in the northern part of their range may predispose them to local elimination due to anthropogenic or natural environmental catastrophes (Gray, 1984; Gray and Brown, this volume).

The present study was undertaken to identify risk factors that may be involved in the decline of northern cricket frogs in Illinois. As with all initial studies, while our data set answers some questions, it also raises more questions for further inquiry. In 1993, we located a number of ponds in northern and southern Illinois where cricket frogs were present. We report here a case series of ponds studied in 1994 and 1995; we describe relationships among habitat characteristics, toxicant applications, and concentrations in water and sediment samples, lesions (including those associated with parasites), and relative reproductive success of cricket frogs.

Materials and Methods

In 1994, eight ponds of similar size in Illinois were studied (Fig. 13-1). Five were located in southern Illinois, one was in east-central Illinois (near Monticello), and two others were in the northern areas of the state (one near Peoria and another in the Midewin National Tallgrass Prairie). In 1995, seven of these ponds were re-examined and an additional pond located in an agricultural area in east central Illinois was studied. To gain a thorough history of the sites, the steward(s) of each pond filled out a survey form and was (were) interviewed during both years of the study. Questions addressed age of the pond, current and past uses, recent modifications, applications

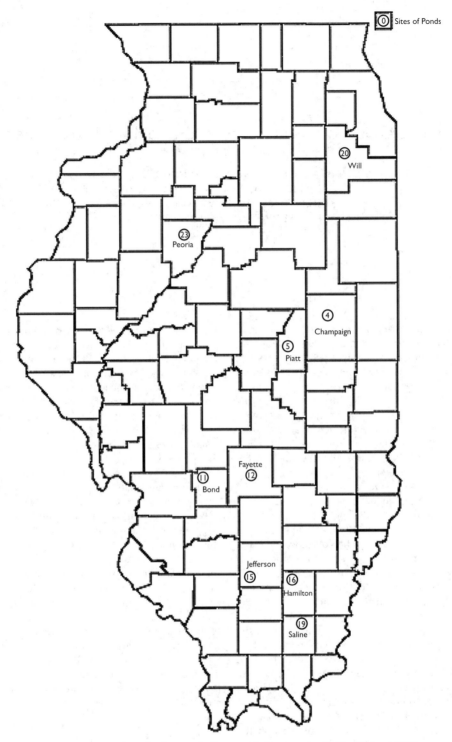

FIGURE 13-1 Locations of ponds where cricket frog populations were studied in 1994 and 1995.

of pesticides on surrounding land or in the pond, and presence of fish and other wildlife observed in and near the pond. During our visits, the types of plants in the water and surrounding shore were identified and any signs of wildlife or human activity were noted. Also, aerial photographs were taken in April 1995 to further characterize nearby land use and to identify the number of water bodies within ca. 2.6 km² surrounding each pond.

Visits to the ponds began with the southern-most site and progressed northward so that variation due to seasonal progression could be minimized. Ponds were visited five times in 1994 and three times in 1995, between late May and early September. Calling surveys were conducted on the first visit of each year (late May/early June). The number of adult male northern cricket frogs calling during eight 15-minute intervals was counted. Surveys at each site began about 2030 CST and ended about 2230 CST. The average number of calling males over this time period was calculated for each site. The total number of calling American bullfrogs (Rana catesbeiana) at each site also was recorded.

At the end of August, counts were made of juvenile northern cricket frogs visually encountered along the edge of the pond where habitat was considered suitable. Suitable habitat was defined as areas with a gently sloping, largely bare bank, that merged with shallow (depth <10 cm at 1 m from shore) water containing abundant aquatic macrophytes. During preliminary surveys in 1993, cricket frogs were never found outside these areas. Furthermore, this observation was consistent with that of other investigations conducted during the breeding season and immediately after metamorphosis (e.g., Burkett, 1984; L. E. Brown, personal observations).

An index of reproductive success (IRS) was obtained by dividing the number of juveniles encountered in late August by an estimate of the breeding population size (calculated as follows). Driscoll (1998) found that counts of calling male Australian myobatrachid frogs (*Geocrinia alba, G. vitellina*) gave good estimates of population sizes, thus we used this value to calculate the relative breeding population size at each pond. A mean male:female ratio was determined from the samples of adults collected from all ponds studied in 1994. This ratio (2.4-1) was similar to that observed by Burkett (1984) for Kansas cricket frog populations (3-1) and, therefore, was used to estimate the number of females present at each pond. Thus, the value used to represent the relative breeding population size at each pond was calculated by adding the mean number of calling males at the site to the estimated number of females present (i.e., the estimated number of females was computed based on the mean number of calling males at the site divided by 2.4). Both the estimates of the adult populations and the numbers of juveniles visually encountered in August were adjusted for the areas of suitable habitat available.

Composite water and sediment samples were collected from multiple locations in each pond at the first and third trips to the ponds. Each sample was collected in acetone-rinsed plastic bottles (Nalgene Products, Nalge Company, Rochester, New York), frozen, and sent to the Centralia Animal Disease Laboratory of Illinois for toxicant analyses (Table 13-1). Water samples (900 ml) were partitioned three times, each time with 50 ml methylene chloride. No pH adjustment was made to the first partition; the second partition was adjusted to pH > 11 with 50% sodium hydroxide; and the third partition was adjusted to pH < 2 with sulfuric acid. The three extracts obtained were combined, mixed with 0.1 ml decane, and evaporated to near dryness at 30–35 °C. Sediment samples were extracted in a 9:1 acetonitrile:water mixture, sonicated for one hour, filtered, and dried. Herbicides were analyzed by resolubilizing residues from the water and sediment extracts with acetonitrile:water (9:1) and performing high performance liquid chromatography using a Shimadzu Series LC-6 (Shimadzu Corporation, Columbia, Maryland) with UV detection at 220 nm. Residues to be assayed for insecticides and polychlorinated biphenyls (PCBs) were resolubilized in iso-octane and analyzed with a Hewlett Packard Series 2 gas chromatograph with an electron capture (for organochlorine insecticides and PCBs) or a nitrogen-phosphorus detector (for organophosphorus and carbamate insecticides; Hewlett Packard, Palo Alto, California). Concentrations of lead were determined by digesting portions of the water and sediment samples in concentrated nitric acid, filtering, and assaying with a Perkin Elmer 4000 atomic absorption spectrophotometer with a graphite furnace (Perkin Elmer Corporation, Norwalk, Connecticut). A cold vapor mercury analysis kit was used with the same spectrophotometer to measure mercury. Aqueous nitrate was measured using a Hach DR/3000 spectrophotometer

and NitraVer V Nitrate Reagent kits (Hach Co., Loveland, Colorado).

Evaluation of the Health of Northern Cricket Frogs

In 1994, northern cricket frogs were collected at each pond and examined for evidence of disease, developmental abnormalities, and/or trauma. Approximately 20 adults from each pond were captured with aquarium dipnets during the first visit (late May/early June). Similarly, tadpoles and juveniles were collected in early and late July, respectively. All animals were killed by immersion in 3-amino benzoic acid ethyl ester methanesulfonate salt (MS-222). Approximately 75% of the collected individuals were then fixed in 10% neutral buffered formalin (NBF), after the body cavity was incised. The remaining frogs were frozen on dry ice.

The heart, lungs, liver, kidneys, stomach, small intestine, gall bladder, pancreas, and gonads of the fixed frog specimens were removed and embedded in paraffin, sectioned at 3 μm, stained with hematoxylin and eosin, and examined for parasites and lesions with a light microscope. The severity of renal infections with encysted larval trematodes (Echinostomatidae) was rated based on the percentage of the tissue section occupied by trematodes as none, light (<25%), moderate (25–50%), or heavy (>50%). The rest of the specimen was decalcified with formic acid, embedded in paraffin, and sectioned and stained as described above. Sections of the brain, spinal cord, sciatic nerve, eyes, skeletal muscle, skin, bone, and bone marrow were examined for lesions using a light microscope.

The frozen frog specimens were examined grossly for internal and external macroparasites. Most protozoans are destroyed in the freeze/thaw process and, therefore, were not considered. After thawing, the carcass was skinned, and the musculature was examined for parasites with the aid of a stereo microscope. Internal organs (heart, liver, kidneys, spleen, urinary, and gall bladder) were removed and examined by compression of the organs between two glass microscope slides. Fat bodies were also examined in this manner. Any helminths seen were removed and placed in 70% ethanol. Processing followed general parasitology protocols (Schmidt, 1992).

The associations between IRS and pond size, pond age, number of surrounding water bodies, number of calling American bullfrogs in late May/early June, and prevalence of renal trematode infections were determined with Spearman rank correlation tests. Correlation analyses were also conducted to evaluate associations among habitat and health variables. An odds ratio analysis was used to determine which factors might be related to high prevalence of renal trematode infections in northern cricket frog populations. Significance in the odds ratio analysis was tested with the Yates' corrected chi-square test (Sokal and Rohlf, 1995). A level of $\alpha = 0.05$ was chosen to detect significant differences.

Results

Habitat Characteristics of Study Ponds

We present information regarding anthropogenic uses and general habitat characteristics for each pond in the order of highest to lowest IRS values from 1994 (Table 13-2), except for site 4, which was only studied in 1995. Sites are summarized as follows.

Table 13-1
Compounds and Elements Analyzed in Water and Sediment Samples

Herbicides		Carbamate	Insecticides		Fungicides and Other Organic Contaminants	Metals
			Organophosphorous	Organochlorine		
Alachlor (5)	Metribuzin (5)	Aldicarb (10)	Chlorpyrifos (1)	Aldrin (0.5)	Benomyl (5)	Lead (5) (100)[a]
Ametryn (5)[b]	Monuron (5)[b]	Aminocarb (5)	Diazinon (1)	Chlordane (0.5)	Hexachlorbenzene (0.5)	Mercury (100)[a]
Atrazine (0.5)	Napropamide (5)	Bendiocarb (5)	Dimethoate (1)	DDD (0.5)	Thiobencarb (10)	
Barban (5)[b]	Naptalam (5)	Carbaryl (5)	Disulfoton (1)	DDE (0.5)	PCBs (0.5)	
Bifenox (5)[b]	Oryzalin (5)	Carbofuran (5)	Ethoprop (1)	DDT (0.5)		
Bromacil (1)	Pebulate (5)	Methiocarb (5)	Ethyl parathion (1)	Dieldrin (0.5)		
Butachlor (5)	Pendimethalin (5)	Lannate (5)	Fenchlorphos (1)	Endosulfan (0.5)		
Butylate (5)	Profluralin (5)	Oxamyl (5)	Fenthion (1)	Endrin (0.5)		
Chloropropham (5)	Prometon (5)		Fonofos (1)	Heptachlor (0.5)		
Chlorothal (5)	Prometryn (5)		Isofenphos (1)	Heptachlor epoxide (0.5)		
Cyanazine (0.5)	Pronamide (5)		Malathion (1)	Lindane & isomers (0.5)		
Diclofop (5)	Propanil (5)		Methidathion (1)	Methoxychlor (0.5)		
Dinitramine (5)	Propazine (5)		Methyl parathion (1)	Mirex (0.5)		
Dipropetryn (5)[b]	Propham (5)		Mevinphos (1)			
Diuron (5)[b]	Simazine (1)		Phorate (1)			
EPTC (5)	Terbacil (5)[c]		Terbufos (1)			
Fluchloralin (5)[b]	Terbuthylazine (5)[b]		Trichlorfon (1)			
Hexazinone (5)	Terbutryn (5)[b]					
Linuron (5)	Trifluralin (5)					
Metolachlor (5)						

[a] Detection level in sediment
[b] Analyzed in 1994 only
[c] Analyzed in 1995 only
NOTE: Detection limits are indicated in parentheses. Concentrations are in μ/L.

TABLE 13-2
Study Sites, Indices of Reproductive Success (IRS), and Pond Characteristics in 1994 and 1995

Site	IRS	Pond Perimeter (m)	Pond Age (years)	Water Bodies	Bullfrogs	Trematode Prevalence (%)	Pesticides Detected or Used	Nitrate Detected	Creeping Water Primrose
1994									
19	33.11	998	5	7	13	7.1	N	N	Y
23	7.0	996	18	1+wlnd	—	6.3	N	N	Y
12	6.96	903	35	0	7	8.0	Y	N	Y
5	6.65	214	19	2	4	36.2	Y	N	N
16	5.21	274	20	0	8	8.7	N	Y	Y
15	1.62	942	24	2	2	3.9	N	Y	Y
20	0.6	176	—	2+river	10	27.3	Y	N	N
11	0.26	628	29	5	7	29.2	Y	N	N
1995									
19	22.46	1239	6	7	6	—	N	Y	Y
11	7.16	740	30	5	5	—	Y	N	Y
16	3.82	431	21	0	6	—	Y	Y	Y
5	2.64	222	20	2	7	—	N	N	N
23	1.83	834	19	1+wlnd	10	—	N	Y	N
12	0.84	644	36	0	10	—	Y	N	Y
20	0.32	289	—	2+river	8	—	N	N	N
4	0.05	512	33	3+river	7	—	Y	Y	N

NOTE: Ponds are listed from highest to lowest IRS. For 1994, the adult population size for site 23 was estimated based on the mean number of calling males per unit of suitable habitat for other sites and adding the expected number of females. Water bodies = number of water bodies in an area ca. 2.6 km^2 surrounding the pond; wlnd = wetland; Y = Yes; N = No.

SITE 19

This manmade pond is located in Saline County near Galatia. During both years of the study, the count of northern cricket frog juveniles relative to the estimate of breeding adults in the area of suitable habitat of the pond was greatest at this site. This pond was the largest surveyed and provided the largest area of suitable habitat. It was also the youngest pond surveyed, having been constructed in 1989 for fishing and recreation. It was completely surrounded by open habitat consisting of grass that was periodically cut to within 1 m of the water's edge. Growth of creeping water primrose (Jussiaea repens), a submergent aquatic plant, was substantial in many shallow areas near the pond margin. Algal mats and a large number of lily pads (Nymphacea odorata) were present as well, and we noticed many male cricket frogs using them for support during our calling survey. We observed many crayfish chimneys, bluegills (Lepomis macrochirus), American bullfrogs, southern leopard frogs (R. sphenocephala), a painted turtle (Chrysemys picta), and green herons (Butorides virescens) at this site. Also, numerous tracks of white-tailed deer (Odocoileus virginianus) and raccoons (Procyon lotor) were evident at this pond. Seven other ponds were located nearby.

SITE 23

This site, located west of Princeville in Peoria County, was constructed in 1976 and used to water and pasture cattle and horses until 1989. Trees grew around most of its edge, shading a substantial portion of the pond's surface. The surrounding land was used primarily to grow corn and soybeans. A small pond, which provided additional suitable habitat for cricket frogs, was located immediately to the southwest. Aquatic vegetation consisted of patches of creeping water primrose, lily pads, and filamentous algae. Unidentified sedges also were present in shallow areas along the shore. Channel catfish (Ictalurus punctatus), red-ear sunfish (Lepomis microlophus), white crappies (Pomoxis annularis), black crappies (P. nigromaculatus), largemouth bass (Micropterus salmoides), and a small number of grass carp (Ctenopharyngodon idella) were reported as being present in the pond by the owner. Coyotes (Canis latrans), several beavers (Castor canadensis), mink (Mustela vison), great egrets (Casmerodius albus) and belted kingfishers (Ceryle alcyon) also frequent this pond. Because this site was added to the study after breeding in 1994, we calculated the IRS value using a breeding population estimate equal to the mean of the other sites.

SITE 12

This pond, southeast of Vandalia in Fayette County, was constructed in 1959 for fishing, swimming, and wildlife. Fields north and west of the pond, occupying approximately 45% of the surrounding land, were leased for corn production in 1994. A stand of deciduous trees was located at the west end of the pond. Prairie grasses grew around the edge of the pond but were regularly mowed to within 1.5 m of the water. The pond possessed a diverse aquatic flora, including creeping water primrose and duckweed *(Lemna sp.)*. No other water bodies were within the ca. 2.6 km^2 surrounding the pond. Great blue herons *(Ardea herodias)*, a belted kingfisher, and southern leopard frogs were noted at this site, as were tadpoles of southern leopard frogs and American bullfrogs. The owner reported a thriving population of bass (not otherwise identified), bluegill, and other sunfish (not identified) in this pond.

SITE 5

This pond, located west of Monticello in Piatt County, was constructed prior to 1975 for recreation and wildlife. Surrounding habitat largely consisted of tall grass and a wooded area between the pond and the Sangamon River. In addition to the river, two other ponds were present within the surrounding ca. 2.6 km^2. A large number of trees, mostly small sandbar willows *(Salix exigua)*, were found along the pond's edge. These trees shaded a substantial portion of the water near the shore. White-tailed deer and raccoon tracks were abundant at the water's edge, and many recently severed willows provided evidence of beaver activity. American bullfrogs were abundant. The pond was shallow and lacked creeping water primrose; the water level varied substantially over the season during the 2 years it was studied.

SITE 16

This pond is north of Dahlgren in Hamilton County; it was created in 1974 for fishing and recreation. The landowner's lawn occupied about 35% of the pond perimeter and the remaining 65% was forested. The lawn was mowed within 2 m of the water's edge and occasionally up to the edge. Creeping water primrose were numerous where northern cricket frogs were seen. To control aquatic plants, the owner applied copper sulfate in July and August in both 1994 and 1995. The chemical was placed in a burlap bag, suspended from a rowboat, and distributed around the pond. Ground limestone was added to the pond to raise the pH. Crayfish were present, and American bullfrogs were common. Southern leopard frogs and channel catfish were present, and the owner reported feeding the latter. The owner also reported having stocked bass and trout (not otherwise identified) as well as green sunfish *(Lepomis cyanellus)* in the pond. No other water bodies were near the pond.

SITE 15

This pond is west of Mount Vernon in Jefferson County and was created in 1970. We studied this pond only in 1994. The pond originally served as a water source for a nearby coal mine, but in recent years it was used to water cattle. At the northeast edge of the pond there was a low-lying area with a cluster of trees that occupied about 10% of the shoreline. Small to large blooms of blue-green algae were consistently recorded, probably due in part to nutrient inputs from excreta of numerous Canada geese *(Branta canadensis)* and domestic geese *(Anser anser)* that were commonly seen, as well as the 25–30 cows that used about 20% of the shoreline for grazing. A small area of the pond supported creeping water primrose. One quarter of the shoreline was covered with riprap, and the owner reported that snakes (species unknown) often used the area for basking. Few American bullfrogs were observed at the pond, and small catfish (not otherwise identified) were noted

in the water. Two other ponds and a small creek were in the vicinity.

SITE 20

This site is located in the Midewin National Tallgrass Prairie in Will County, north of Wilmington. Before establishment of the national reserve, the area belonged to Joliet Army Ammunition Plant. The nearby land was primarily used for hunting. The pond was surrounded by forest, although an agricultural field (used to grow soybeans in 1994 and corn in 1995) was about 300 m from the water. One small river and two ponds were nearby. Duckweed consistently blanketed the pond during the summers, bulrushes *(Juncus effusus)* were present on the southeast edge, and unidentified sedges commonly grew around the pond. Tracks of coyotes or dogs, raccoons, and white-tailed deer were noted. Several potential predators were observed, including great blue herons, great egrets, eastern rat snakes *(Elaphe obsoleta)*, common garter snakes *(Thamnophis sirtalis)*, and possibly northern water snakes *(Nerodia sipedon)*, small American toads *(Bufo americanus)*, northern leopard frogs *(R. pipiens)* and American bullfrogs. Wood ducks *(Aix sponsa)* and signs of beavers (gnawed, fallen trees) also were seen.

SITE 11

This pond is located in Bond County, northeast of Vandalia. It was constructed in 1965 for recreation, but was currently being used to irrigate an apple orchard and berry bushes northeast of the pond. A field south of the pond was used to grow corn in 1994 and soybeans in 1995. All grass around the pond was mowed to within 1 m of the water. Five other ponds were within the ca. 2.6 km^2 area surrounding the pond. Crayfish, grass carp, American bullfrogs, Canada geese and a rail (not otherwise identified) were present. In June 1994, there was a dense growth of creeping water primrose in areas that provided suitable habitat, however, a plant die-off occurred by the beginning of July, which left the water practically free of live macrophytes for the remainder of the summer. At the second visit on 6 July, water had receded and plants in the water and just above the shoreline were brown. The only living plant observed in the water at this time was a black willow tree *(S. nigra)*. By contrast, in 1995, creeping water primrose was again abundant.

SITE 4

This pond, located near Fisher in Champaign County, was studied only in 1995. It was constructed in 1962 for recreational purposes. Small pine trees had been planted recently in an old field to the south, where alfalfa had been grown. Nearby to the east, corn and soybeans were grown in large fields. To the west, there was a large forested bottomland. The landowner's lawn occupied the northern perimeter of the pond. Vegetation immediately surrounding the pond consisted of tall grasses and thorn bushes. Cattails *(Typha latifolia)* were present at the southeast edge, and the only other aquatic plant encountered was water smartweed *(Polygonum amphibium)*. Only 9.75 m^2 (1.9%) of the perimeter provided suitable northern cricket frog habitat. The Sangamon River and three other ponds were nearby. The IRS at this pond was the lowest among all sites in 1995 (Table 13-2). Grass carp, largemouth bass, bluegill, and black crappies were present in the pond. Mallards *(A. platyrhynchos)* and Canada geese were common at the site, as were American bullfrogs and American toads.

TABLE 13-3

Concentrations of Nitrate and Pesticides Detected in Water and Sediment Samples Collected at Each Pond in 1994 and 1995

Site	Nitrate	Atrazine	Cyanazine	Metolachlor	Chlorpyrifos
1994					
5	ND	3	ND	ND	ND
11	ND	6 (70)	ND	2 (40)	ND
12	ND	2	ND	ND	3.1
15	2	ND	ND	ND	ND
16[a]	2	ND	ND	ND	ND
19	ND	ND	ND	ND	ND
20	ND	3	ND	ND	ND
23	ND	ND	ND	ND	ND
1995					
4[a]	19.8	15	17	ND	ND
5	ND	ND	ND	ND	ND
11	ND	2	ND	ND	ND
12	ND	1	ND	ND	ND
16[a]	5.0	ND	ND	ND	ND
19	4.0	ND	ND	ND	ND
20	ND	ND	ND	ND	ND
23[b]	2.2	ND	ND	ND	ND

[a]$CuSO_4$ applied to the water twice during the summer.

[b]Endothall applied to water during the summer.

NOTE: Nitrate concentrations are reported in mg/L and pesticide concentrations are reported in μ/L water or μ/kg (sediment). ND = not detected

Contaminants in Water and Sediment

In Table 13-3, we list nitrate and pesticide concentrations found in water and sediment samples collected from each pond in 1994 and 1995. Contamination with the agricultural herbicide atrazine was most widespread, detected at 4 of the 8 ponds surveyed in 1994, and 3 of the 8 ponds in 1995. The highest aqueous concentration detected in 1994 was 6 μg/L at site 11. Metolachlor, another herbicide, also was detected at site 11 each of the three times that water was collected in 1994. The insecticide chlorpyrifos was detected at site 12 at a concentration that would be lethal to aquatic insects (Biever et al., 1994). At site 4, atrazine, cyanazine, and nitrate were detected in 1995 at comparatively high concentrations. In addition, the landowner reported using copper sulfate to control aquatic vegetation. Copper sulfate was also used at site 16. In June of 1995, the herbicide endothall was applied to the aquatic plants near the shore at site 23. Lead, mercury, and PCB concentrations were considered to be toxicologically insignificant at all sites (results not shown).

Relationships Among Reproductive Success, Habitat Characteristics, and Health of Northern Cricket Frogs

Correlation analyses failed to reveal any habitat characteristics that were significantly associated with IRS, although a positive correlation between IRS values and pond size was suggested in 1994 (n = 8; r_s = 0.64; p = 0.089). IRS values were negatively correlated with the number of calling American bullfrogs in 1995 (n = 8; r_s = −0.73; p = 0.054).

In contrast to the findings with the correlation analyses, the odds ratio analyses indicated associations among trematode infection, the presence of live creeping water primrose (p < 0.001), and higher IRS values. Most notably in 1994, high trematode prevalences were found only at sites that lacked creeping water primrose (i.e., sites 5, 11, and 20); these sites also had low IRS values. Conversely, sites that possessed creeping water primrose had low prevalence of trematode infection and high IRS values (Tables 13-2, 13-4; Fig. 13-2). Risk of trematode infection was also positively associated with herbicide contamination and the percentage of shoreline composed of suitable habitat, but negatively associated with overall pond perimeter (Table 13-4).

The observed scenario at site 11 most strongly suggested a relationship among parasitism, herbicide contamination, absence of creeping water primrose, and a low IRS value. When the pond was first visited in 1994, creeping water primrose was plentiful and adult northern cricket frogs were frequently seen. When the pond was revisited that year, most of the aquatic macrophytes were dead. Analysis of the water sample detected atrazine, as well as metolachlor. Herbicide toxicity to the aquatic vegetation may have been exacerbated by a decrease in dilution due to the removal of water by the landowner for irrigation. Following the plant die-off in 1994, the pond was searched for 2 consecutive days in an attempt to locate cricket frog tadpoles; only seven were found, and these were located among the branches of the one willow tree in the water. Frogs at this site harbored heavy renal trematode infections. Four of the five juveniles examined had heavy infections; similarly, three of the four tadpoles examined had either light or heavy infections. The IRS value was only 0.26. By contrast, in 1995, aquatic vegetation in this pond was abundant, consisting mainly of creeping water primrose. We did not observe a plant die-off. We detected a low concentration of atrazine in the water, but cricket frog juveniles were abundant; and an IRS value of 7.16 was obtained.

Larval trematodes in the kidneys were the only parasites that appeared to cause lesions sufficiently severe as to influence the survival/death of northern cricket frogs (Fig. 13-3). High prevalence and intensity of infections were associated with low IRS values (Fig. 13-2). Infections were encountered in all life stage classes examined. Juveniles tended to be more severely affected than either adults or tadpoles (Table 13-4; Fig. 13-2). Sixty-one percent of the cases of renal trematode infections were found in juveniles, compared to 18.4% and 20.4% in tadpoles and adults, respectively. Subcutaneous and peripharyngeal infections with larval trematodes were also common and showed a similar life stage-related pattern, with prevalences of 55% in juveniles, 33.6% in adults, and 15.5% in tadpoles. These infections, however, were not more prevalent at sites where IRS was low versus sites where IRS was high (Tables 13-5, 13-6).

Other metazoan parasite infections, lesions, and abnormalities were present (Tables 13-5, 13-6). Six out of 107 (5.6%) adults surveyed in 1994 had hyperkeratotic skin lesions that were associated with chytridiomycotic (fungal) infections; one adult male had an extra forelimb; and 4.6% of adults examined had intersex gonads. About 2% of the frogs examined had lesions indicative of injuries, such as remodeled fractures and myositis in the legs.

TABLE 13-4
Odds Ratio Analysis of Risk Factors Possibly Associated with Trematode Infections in Cricket Frog Populations

Risk Factor	Infected Frogs (% of total infected)	Risk[a]	X_c2[b]	P
Age of pond:				
<20 yr	27 (55.1)	1.092	0.0165	0.8996
>20 yr	22 (44.9)	1.00	—	—
Pond perimeter:				
<600 m	37 (75.5)	4.681	20.129	<0.001
>600 m	12 (24.5)	1.00	—	—
% of pond with suitable habitat:				
>10%	34 (69.4)	2.190	5.067	0.0245
<10%	15 (30.6)	1.00	—	—
Isolated from other ponds?				
Yes	8 (16.3)	0.450	3.349	0.0716
No	41 (83.7)	1.00	—	—
Creeping water primrose present?				
Yes	16 (32.7)	0.157	34.44	<0.001
No	33 (67.3)	1.00	—	—
Herbicides detected or used?				
Yes	41 (83.7)	4.255	13.422	<0.001
No	8 (16.3)	1.00	—	—
Frog stage:				
Adult	10 (20.4)	1.143	0.002	0.9656
Juvenile	30 (61.2)	5.196	16.975	<0.001
Tadpole	9 (18.4)	1.00	—	—
Site:				
5	17 (34.7)	14.167	14.613	<0.001
11	7 (14.3)	10.294	7.805	0.0053
12	4 (8.2)	2.174	0.221	0.7129
15	2 (4.1)	1.00	—	—
16	4 (8.2)	2.381	0.333	0.6110
19	4 (8.2)	1.923	0.107	0.8172
20	9 (18.4)	9.38	7.865	0.0051
23	2 (4.1)	1.667	0.001	0.9606

[a]Measured by odds ratio statistic, compares rates of trematode infection in frogs found within different risk factor categories. An odds ratio significantly greater than 1.0 implies an increased rate of infection. An odds ratio significantly less than 1.0 implies a decreased rate.

[b]Significance tested with Yates' corrected chi-square.

Discussion

An important determinant of northern cricket frog breeding was the presence of gently sloping banks, free from tall vegetation and merging with shallow water containing aquatic plants. Ponds completely lacking such microhabitats were devoid of calling cricket frogs. Adults and juveniles were typically found foraging for insects on flat land near the shoreline, and when disturbed, would jump into the water and hide among the aquatic plants. Thus, this type of habitat allows cricket frogs to forage effectively on terrestrial insects, as well as to escape terrestrial predators by retreating to the water (Burkett, 1984). Cricket frogs occupying this microhabitat may also reduce their contact with (and, therefore, avoid) predation by American bullfrogs because bullfrogs typically utilize areas where there is a steep transition between the shoreline and water. Bullfrogs are known to readily consume other frogs (Brooks, 1964; Corse and Metter, 1980), and they have been implicated as a causative agent in the decline of many western North American amphibians (Moyle, 1973; Hayes and Jennings, 1986). In our study, the importance of bullfrog predation on cricket frog success was not clear. Cricket frog IRS values and the number of calling bullfrogs were negatively correlated in 1995, and yet high numbers of calling bullfrogs were

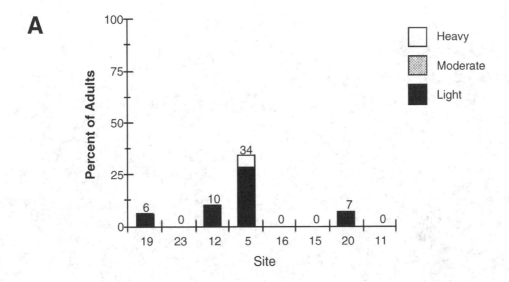

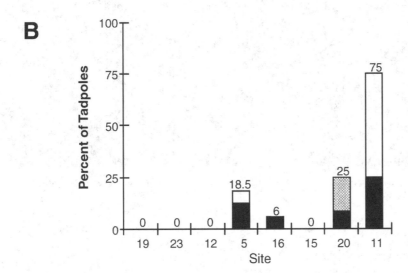

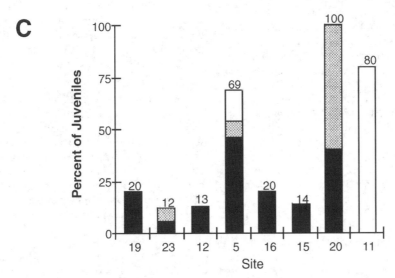

FIGURE 13-2 Distribution of light, moderate, and heavy renal trematode infections in (A) adult, (B) tadpole, and (C) juvenile cricket frogs across all sites surveyed in 1994. Sites are arranged from highest *(left)* to lowest *(right)* index of reproductive success (IRS). Percentages of individuals of each life stage with observed trematode cysts in the kidney sections are shown above the bars. A total of 340 animals were examined; over all sites, 30 of juveniles, 10 of adults, and 9 of the tadpoles were found to be infected.

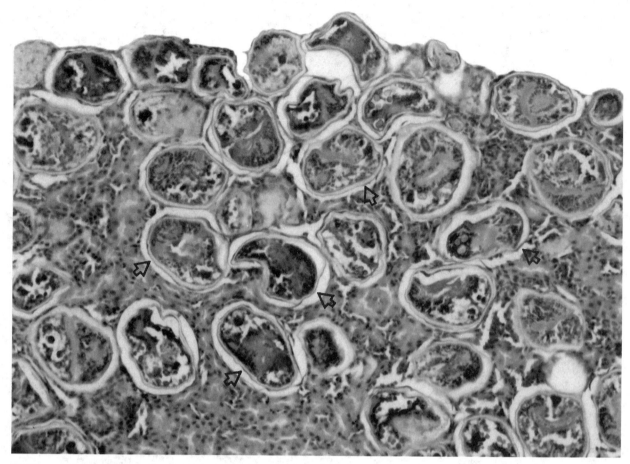

FIGURE 13-3 Photomicrograph of a kidney of a cricket frog with severe infection with echinostomatid trematodes (arrows).

often heard where cricket frogs were abundant (e.g., sites 19 and 12; Table 13-2).

The association between the presence of creeping water primrose and northern cricket frog success suggests that this submerged and emergent plant is beneficial to juvenile survival. We do not know whether this native aquatic macrophyte helps to ensure survival, but our data strongly suggest the relationship should be further investigated. Such an association between aquatic vegetation and cricket frog success seems logical because similar vegetation provides cover to tadpoles, juveniles, and adults from predators such as birds and American bullfrogs. Unfortunately, quantitative assessment of avian predators was not conducted in this study. Aquatic macrophytes also maintain dissolved oxygen concentrations, take up potentially toxic dissolved nutrients such as nitrite and nitrate, and provide an important surface area for growth of periphyton, which tadpoles eat (Johnson, 1991). Vegetation could also be important for cricket frog breeding by providing a substrate for egg attachment (Livezey, 1950) and reducing exposure to UV radiation.

We suggest the following three scenarios to account for the increased prevalence of renal trematodes at sites with decreased aquatic vegetation:

1. The loss of vegetation resulted in a simplified habitat structure and increased the success rate of larval trematodes finding the tadpoles. Impediment of trematode transmission at sites with complex habitat structure was noted by Sousa and Grosholz (1991). Complex habitats

have more refugia in which tadpoles can escape larval trematodes.

2. The loss of vegetation and the resulting decrease in visual obstruction increased bird visitation (either the number of birds or the time birds spent at sites) to sites because of easier predation, which resulted in increased deposition of trematode eggs in the ponds. It is well known that predatory birds are attracted to aquaculture ponds where fish are concentrated and relatively easy prey.

3. A combination of (1) and (2). Increased deposition of parasites into ponds, coupled with a simplified habitat, increased the success rate of parasite transmission. One scenario or all three could explain our observation, including that the intensity of trematodes tended to be lower in ponds with creeping water primrose than in ponds without this plant (Table 13-4). For instance, four of the five juveniles examined from site 11 had more than 50% renal histologic sections occupied by larval trematodes.

The presence and numbers of snails, which act as an intermediate host for this parasite, and how they reacted to increased/decreased vegetation would also play an important role in why sites differed in prevalence and intensity of trematodes.

The mechanisms that accounted for the relationship between low IRS, increased parasites in newly metamorphosed amphibians, and decreased plant cover are not fully revealed

TABLE 13-5

The Number of Cases of Lesions and Parasites Found in the Adult Cricket Frogs Examined in 1994

Site	No.	Intersex Gonads	Epidermal Hyperkeratosis with Fungi[a]	Dermatitis	Rhabdias sp. in Lungs	Subcutaneous and/or Peripharyngeal Parasites	Adult Trematodes (location)	Other	Total
19	16	0	0	0	4	9	0	4[b]	17
23	2	0	0	1	1	1	0	0	3
12	15	0	0	1	0	1	0	1[c]	3
5	17	2	2	0	0	6	0	0	10
16	15	1	3	1	2	6	0	1[d]	14
15	12	0	1	0	1	4	1 (cloaca)	3[e]	10
20	15	1	0	0	3	4	1 (gall bladder)	2[f]	11
11	15	1	0	0	1	5	0	1[g]	8

[a] chytridiomycotic fungi

[b] 1 multifocal myodegeneration and myositis in legs, 2 mild eosinophilic bronchitis, 1 focal ovarian microsporidiosis

[c] microsporidian protozoa

[d] chromomycotic granuloma in liver

[e] 1 focal degeneration and myositis of gluteal muscle, 1 partially remodeled femoral fracture, 1 focal gluteal myofiber regeneration

[f] 2 mild eosinophilic bronchitis

[g] intestinal nematodes

NOTE: Sites are listed from highest to lowest index of reproductive success (IRS).

TABLE 13-6

The Number of Cases of Lesions and Parasites Found in Tadpole and Juvenile Cricket Frogs in 1994

		Tadpoles				Juveniles			
Site	No.	Subcutaneous Parasites	Gill Protozoa	Total	No.	Rhabdias sp. in Lungs	Subcutaneous and/or Peripharyngeal Parasites	Other	Total
19	25	6	0	6	15	1	8	1[b]	10
23	15	3	1	4	15	1	14	1[c]	16
12	15	0	1	2[a]	15	0	12	2[d]	14
5	16	7	0	7	15	0	10	0	10
16	16	0	3	3	15	0	4	0	4
15	25	2	13	15	15	0	1	1[e]	2
20	13	0	6	6	5	0	0	0	0
11	4	2	2	4	5	0	6	1[f]	7

[a] 1 case of mild focal dermatitis

[b] focal ulcerative dermatitis at base of urostyle

[c] partially remodeled fracture of the distal leg

[d] 1 mid-shaft tibial callus, 1 multifocal subacute dermatitis

[e] focal skeletal muscle degeneration and myositis

[f] parasites in nasal mucosa.

NOTE: Sites listed from highest to lowest index of reproductive success (IRS).

by the data set. The increased parasitism at these sites may be a byproduct of increased bird visitation and predation, which could have been the proximate cause for low IRS values. The question of whether low IRS is due to increased predation or increased parasitism cannot be discerned in this study. It is known that *Echinostoma trivolvis* in kidneys of northern leopard frogs *(R. pipiens)* has been linked to mortality, which is parasite intensity dependent (Beaver, 1937; Fried et al., 1997). Moreover, heavy echinostome infections were highly lethal to northern leopard frog tadpoles that were infected early in development when they were reliant on pronephric (larval) kidneys (Schotthoefer et al. 2003). In addition, this parasite can also reduce tadpole growth, which could increase the window of susceptibility of tadpoles to predation (Fried et al., 1997). We need further studies to separate the roles that predation, plant cover, snails, and parasites play in survival, and thus recruitment, of juveniles into the adult population.

In our study, we recognize the adverse effects of accidental or purposeful introduction of herbicides to ponds as a major mode of habitat modification. Agricultural herbicides, probably introduced with runoff, were detected in several of the ponds, and pond owners reported using aquatic herbicides, including endothall and copper sulfate in others. Nonetheless, concentrations of agricultural herbicides (atrazine, metolachlor, and cyanazine) were not high enough to be expected to have acutely toxic effects on northern cricket frogs. For instance, Birge et al. (1980) reported LC50s for atrazine of 0.41 mg/L for post-hatchling American bullfrog embryos and greater than 48 mg/L for American toad tadpoles. However, the concentrations of herbicides detected at some of our study ponds were, or recently had been, high enough to be phytotoxic to some aquatic algae and/or macrophytes (Huber, 1993; Atrazine Ecological Risk Assessment Panel, 1995; Solomon et al., 1996). Atrazine concentrations detected in water samples in this study ranged from 1 to 15 μg/L, and in sediment up to 70 μg/kg. Green algae EC50 values for atrazine range from 4 to 854 μg/L; macrophytes typically exhibit negative effects between 20 and 500 μg/L (Solomon et al., 1996). Because phytoplankton make up a large part of the diet of many tadpoles, it is possible that tadpoles were indirectly affected by atrazine contamination. A reduction in tadpole biomass (primarily American bullfrog tadpoles) was noted by deNoyelles et al. (1989) when atrazine was applied to limnal corrals, and it was suggested that the reduction was related to the observed decline in food availability and cover. The concentrations of cyanazine (observed at site 4 in 1995) and metolachlor (observed at site 11 in 1994) would also be detrimental to algal species (Weed Sciences Society of America, Herbicide Handbook Committee, 1983) and, therefore, could have had an effect on tadpoles by limiting the amount of food and oxygen available to them.

When aquatic macrophytes and/or algae become established in midwestern ponds, owners commonly use copper sulfate as an aquatic herbicide. This was the case at sites 4 and 16, where copper sulfate was applied twice during the summer. The recommended concentrations for controlling aquatic plants and algae range from 0.03 to 1.0 mg/L (Weed Sciences Society of America, Herbicide Handbook Committee, 1983). We know neither how much copper sulfate was applied by landowners nor the concentrations of copper achieved in the water. However, it seems likely that northern cricket frogs were adversely affected because concentrations found to be toxic to amphibians are below those recommended for plant control. A concentration of 0.31 mg/L was lethal to northern leopard frog tadpoles (Lande and Guttman, 1973). Fort and Stover (1997) studied the developmental and lethal toxicity of copper to African clawed frogs (*Xenopus laevis*). They found LC50 values of 1.32 mg/L for embryos and 0.20 mg/L for 12–16-day-old tadpoles, indicating that the susceptibility to copper increased with age. Furthermore, growth was reduced at concentrations as low as 0.048 mg/L and completely inhibited at 1.3 mg/L in embryos. In addition, distal hind limb aplasia, a sensitive indicator of copper toxicosis, occurred in 85% of the larvae exposed to 0.05 mg/L.

Although the concentrations of atrazine, cyanazine, and metolachlor were lower than those expected to have direct effects on developing anurans, it is quite possible that the combinations of herbicides, such as those found at sites 4 and 11, had additive or synergistic toxicity effects. For instance, northern leopard frog and American toad tadpoles were more susceptible to a 50:50 mixture of atrazine and alachlor than either atrazine or alachlor alone (Howe et al., 1998). Complex interactions among atrazine, cyanazine, nitrate, and copper sulfate may explain the low reproductive success of northern cricket frogs observed at site 4.

Summary

We quantified the relative risks associated with a host of factors potentially contributing to the decline of northern cricket frog populations in Illinois, and we identified some factors likely to be important in cricket frog survival. The presence of a gently sloping bank merging with shallow water containing submergent vegetation is an essential requirement for cricket frog breeding and juvenile survival. In the Midwest, landowners often excavate the edges of ponds and lakes to make the water sufficiently deep so that plants cannot become established. Also, ponds are made deeper so that they can sustain carnivorous fish populations. Such anthropogenic practices only aggravate the habitat loss and fragmentation problems that affect cricket frogs. The removal of vegetation by chemical means may similarly limit cricket frog reproduction or juvenile survival. The present study indicates that the removal of aquatic vegetation may increase the risks of severe trematode infection in the kidneys of tadpoles. The frequency of moderate and heavy infections in cricket frogs was high in ponds where IRS values were low. We recommend that additional research be conducted on potential interactions among herbicides, plants, trematodes, top predators, and frogs. Although high contaminant loads may be lethal to cricket frogs and, thus, be of concern for their conservation, the less obvious indirect effects of herbicides via their effects on aquatic plants are much more likely to be important risk factors for amphibian populations. Herbicides, in combination, (e.g., atrazine and metolachlor, or atrazine, cyanazine, and copper sulfate) or alone, (e.g., copper sulfate or endothall) likely affected phytoplankton and macrophyte biomass in the contaminated ponds examined. Ponds that had phytotoxic concentrations of herbicides were associated with lower recruitment of juveniles than ponds in which herbicide impacts were not apparent. We propose that a decline in phytoplankton biomass limits tadpole growth and survival. Most importantly, our study suggests that physically and chemically induced changes in the aquatic plant community, together with the direct toxic effects of contaminants, are likely to be involved in amphibian declines. These findings highlight the need for additional research on ways in which anthropogenic factors interact to cause negative impacts on amphibian species worldwide, and importantly, how human beings can more effectively manage landscapes and water bodies to promote improved amphibian community structure and health.

Acknowledgments. We thank the John G. Shedd Aquarium and Chicago Zoological Society for financial support; S. Kasten, S. Ross, J. Stedelin, and H. Braddock of the Animal Disease Laboratory, Centralia, Illinois, for analytical toxicology expertise; and J. Brown, R. Gray, and K. Beckmen for critically reviewing the manuscript.

FOURTEEN

Ultraviolet Radiation

ANDREW R. BLAUSTEIN AND LISA K. BELDEN

Global climate changes, including changes in atmospheric conditions, may be contributing to amphibian population declines. Thus, a number of recent studies have investigated the effects of the ultraviolet (UV) component of ambient solar radiation on amphibians. Studies of amphibians and UV radiation have concentrated on UV-B (280–315 nm), which is the portion of the spectrum of most biological concern at the earth's surface. Higher wavelengths are less efficiently absorbed by critical biomolecules; lower wavelengths are absorbed by stratospheric ozone (see Blaustein et al., 1994c). UV-B radiation is known to induce the formation of photoproducts that can cause cell death or genetic mutations. Seasonal increases in UV-B irradiance linked to stratospheric ozone depletion are well documented at the poles, and there is evidence that UV-B radiation has increased in temperate latitudes (van der Leun et al., 1998). Laboratory studies show detrimental effects of UV-B radiation on amphibian growth, development, and behavior (e.g., Worrest and Kimeldorf, 1976; Hays et al., 1996; Nagl and Hofer, 1997). Field studies demonstrate that ambient UV-B radiation adversely affects the developing embryos of some, but not all, species (see Blaustein et al., 1998). Moreover, some recent studies have shown that ambient UV-B radiation may cause malformities (Blaustein et al., 1997; Fite et al., 1998). In this chapter, we briefly summarize the methods, evidence, and implications of the effects of UV-B radiation on amphibians based on the results of field experiments.

Experimental Design

Most published amphibian/UV field studies have addressed whether ambient UV-B radiation damages amphibian embryos. Properly designed field experiments are a rigorous method for assessing environmental damage by specific agents. In a properly designed field experiment, all factors naturally and simultaneously vary between experimental and control treatments, except for the variable(s) of interest. Adequate controls are necessary, with controls being treated in the same manner as experimental treatments. Treatment replication ensures that results can be generalized. Employing an adequate number of replicates for each treatment will help ensure that results are not unique to a particular series of treatment.

General Methods of UV Field Studies

In most of the published field studies, investigators have placed eggs or embryos in enclosures that were either (1) shielded or (2) unshielded from UV-B radiation and then compared the hatching success of both environments. Across studies, a variety of UV shields and enclosures, differing in both size and composition, have been used (Blaustein et al., 1998). However, within a single study the enclosures and UV shields have been similar. Some studies have been designed to examine potential synergistic effects of UV-B and other variables.

Results of Field Experiments

The results of field experiments strongly indicate that embryos of some amphibian species are adversely affected by ambient UV-B radiation. Thus, the hatching success of the eggs of some frogs, toads, and salamanders in North America, Europe, and Australia was greater under regimes that filtered out UV-B radiation than those that allowed UV-B penetration (Blaustein et al., 1995a, 1998). Results also show that the hatching success of some species is unaffected by ambient UV-B radiation.

Other detrimental effects of UV-B are observed in amphibians. For example, severe body malformations have been observed in embryonic salamanders exposed to ambient levels of UV-B radiation (Blaustein et al., 1997). These malformations included blistering, edema, and curvature of the body (scoliosis). Moreover, in the mountains of the Cascade Range in Oregon, retinal damage consistent with the adverse affects of UV-B radiation has been observed in frogs that bask (Fite et al., 1998). This retinal damage may cause severe impairment of vision in frogs, and could affect their ability to find food or to detect, and therefore escape from, predators.

Synergistic effects of other factors with UV-B have also been observed. For example, UV-B radiation can interact with low pH, pollutants, and pathogens to adversely affect development and hatching success (e.g., Kiesecker and Blaustein, 1995; Long et al., 1995; Hatch and Burton, 1998).

Factors Affecting Exposure to Ultraviolet Radiation

At the earth's surface, UV-B levels are constantly changing. Levels vary with weather patterns, cloud cover, and water flow, depth, and turbidity. UV levels also change with latitude and altitude. This variability makes it difficult to compare experimental results between different studies, especially those conducted in different regions. However, within a single, well-designed experiment, natural variability is expected. If differences between UV-exposed and non-exposed treatments are apparent, then UV-B is affecting hatching success, regardless of site- and condition-specific variability in exposure levels.

Natural Selection

Why are the eggs and embryos of some species more affected by UV-B than those of other species? One possible reason is that there have been strong selection pressures for certain species to evolve behavioral, ecological, and physiological mechanisms to counteract the harmful effects of UV-B radiation. For example, in some species there may have been selective advantages to laying eggs in areas with less UV-B exposure. Morphological and molecular mechanisms may also help amphibians cope with exposure to UV-B radiation. For example, the composition and thickness of the jelly coat that surrounds amphibian eggs may limit UV-B exposure to developing embryos, and darker egg pigmentation may impede UV-B transmission.

In addition to mechanisms that help limit an amphibian's initial exposure to UV-B radiation, there are several ways UV-induced DNA damage can be repaired once it has occurred. Photoreactivation is one such mechanism. A single enzyme, photolyase, uses visible light energy to remove the most frequent UV-induced lesion in DNA, cyclobutane pyrimidine dimers (Freidberg et al., 1995). Within the limited number of species examined, those species with the highest levels of photolyase are generally the most resistant to UV-B radiation (Blaustein et al., 1994c, 1996, 1999).

Interspecific Differences in Resistance to UV-B Radiation

UV-B does not affect amphibians equally. Some investigators have found that their species of study are resistant to ambient UV-B radiation (e.g., Ovaska et al., 1997; Corn, 1998; Blaustein et al., 1996, 1999). Others (within a single study) using identical methods to examine a number of species, have found that some species are resistant whereas others are highly susceptible to ambient levels of UV-B radiation (e.g., Blaustein et al., 1994c; Anzalone et al., 1998; Lizana and Pedraza, 1998; Broomhall et al., 2000). There is no contradiction in results between studies that do and do not show effects of UV-B radiation on amphibians. Obviously, there are interspecific differences in resistance to UV-B radiation just as there are interspecific differences in responses to other environmental factors such as temperature and pH. Even populations of the same species may show differences in susceptibility to UV-B radiation.

Measuring UV-B in the Field

We must reiterate an important point about measuring UV-B levels in the field. In controlled experiments, measurements of UV-B radiation are not necessary to answer the question posed: Does ambient UV-B radiation damage amphibian embryos? However, measuring UV levels becomes necessary when addressing a further question: What levels of UV radiation cause damage to amphibians? The techniques used in addressing these very different questions must not be confounded.

The Role of UV-B Radiation in Amphibian Population Declines

UV-B radiation is affecting the eggs and embryos of several species of amphibians in widely scattered locales around the world. Continued mortality in early life stages may eventually lead to a population decline. However, the effects of continued mortality at egg and embryo stages may not be observed for many years, especially in long-lived species. Those species that lay their eggs in shallow water exposed to sunlight and with poor abilities to repair UV-induced DNA damage are at most risk, but the extent of this risk is unknown.

UV-B radiation is obviously not the only factor contributing to amphibian population declines. UV-B would be an unlikely factor in the declines of species that lay their eggs under logs, in litter, in cracks in soil and rocks, in deep water, or under dense forest canopy. Moreover, even the eggs of some species that are laid in open sunlight may be unaffected by UV-B radiation if they have the molecular and morphological mechanisms mentioned to counteract the harmful effects of UV-B.

Finally, we emphasize that increases in UV-B radiation are not necessary to demonstrate that UV-B is affecting amphibians in nature. Even if levels of UV-B remain constant or only slightly increase, synergistic interactions with UV and other agents could harm amphibians.

Acknowledgments. We thank Andy Dufresne, Ellis Boyd Redding, Tommy Williams, and Brooks Hatlen for their assistance. Financial support for our research was kindly provided by National Science Foundation grant DEB-9423333, an NSF predoctoral fellowship to L.K. Belden, and the Katherine Bisbee II Fund of the Oregon Community Foundation.

FIFTEEN

Xenobiotics

CHRISTINE M. BRIDGES AND RAYMOND D. SEMLITSCH

Published in 1962, Rachel Carson's *Silent Spring* was an impassioned plea to reduce or eliminate the production and release of xenobiotics into the environment. *Silent Spring* painted a grim portrait of an earth so polluted with these human-produced chemicals that no birds or frogs were left to sing during the spring. Since that time, the federal government has passed many acts and laws in an effort to protect wildlife from the burgeoning number of chemicals finding their way into the environment (e.g., the most recent amendments of the Federal Insecticide, Fungicide and Rodenticide Act [1976], Clean Water Act [1992], Toxic Substances Control Act [1992]). Despite these efforts, our environment is becoming increasingly contaminated and more unfavorable for both plants and wildlife (Connell and Miller, 1984). Indeed, Carson's book now seems rather prophetic, as each passing spring seems to be more silent than the last, and as global reports of annually declining songbird and amphibian populations become more frequent.

While a number of compounds have been reported as toxic to amphibians (see review in Harfenist et al., 1989), until recently, there have been conspicuously few ecotoxicological studies concerning amphibians (Hall and Henry, 1992; Sparling et al., 2000). Studies are now focusing on the effects of xenobiotics on amphibians, an interest likely stimulated by widespread reports of amphibian declines (Barinaga, 1990; Blaustein and Wake, 1990; Wake, 1998; Alford and Richards, 1999; Houlahan et al., 2000). It has been speculated that chemical contamination may be partially to blame for some documented amphibian declines, by disrupting growth, reproduction, and behavior (Bishop, 1992). However, evidence that xenobiotics are directly to blame for population declines is sparse because environmental concentrations are typically not great enough to generate direct mortality (see also Hayes et al., 2002a,c; Kiesecker, 2002). Therefore, it is important to examine the sublethal effects of contaminants that may lead to subtle changes in amphibians, which can ultimately contribute to population declines. For example, xenobiotics that alter behavior can decrease growth and development, increase predation rates, or diminish reproductive success. Such a change among individual amphibians could lead to a concomitant decrease in population numbers.

Classes of xenobiotics having the potential to contaminate aquatic amphibian habitats include inorganics (e.g., metals, ammonia, and nitrates), organics (e.g., polychlorinated biphenyls and polyaromatic hydrocarbons), and radionucleotides. All these classes may either singly enter the environment or in mixtures (Rand et al., 1995; Diana and Beasley, 1998). Aquatic environments are the eventual recipients of most xenobiotics, regardless of their source (e.g., agriculture, industry, or personal use; Anderson and D'Apollonia, 1978). Because most amphibian species are highly associated with freshwater aquatic environments, they may frequently encounter xenobiotics.

Although compounds such as those listed are often considered predominant environmental contaminants, naturally occurring abiotic factors (e.g., temperature, pH, and ultraviolet-B radiation) also pose a threat to amphibians when normal values are exceeded, and they can influence the toxicity of other contaminants (Boone and Bridges, 1999). Gosner and Black (1957) were the first to determine that low pH (acidic condition) is detrimental to amphibians. They demonstrated that within the naturally acidic waters of the New Jersey Pine Barrens the distribution of amphibians was dependent upon pH, and individuals exposed to low pH levels had a reduced survival rate. Since that study, there have been numerous publications on the effects of acidic conditions on amphibian larvae (see reviews in Pierce, 1985; and Freda, 1986). For example, amphibian larvae in acidic waters (pH of lower than approximately 5.0) often properly develop but fail to hatch (Dunson and Connell, 1982; Clark and Hall, 1985). Tadpoles exposed to low pH exhibit diminished swimming behavior (Freda and Taylor, 1992; Kutka, 1994), altered growth and development (Pierce and Montgomery, 1989), and increased mortality prior to metamorphosis, even if exposure is of short duration (e.g., 24 hours; Leftwich and Lilly, 1992). Acidic deposition can also impact the degree to which aquatic amphibian larvae are affected by UV-B radiation. Although dissolved organic carbon (DOC) within water can act as a natural sunscreen and protect organisms from harmful effects of UV-B radiation (Lean, 1998; see also Blaustein and Belden, this volume), DOC breaks down under acidic conditions (Schindler et al., 1996; Yan et al., 1996), thereby increasing the depth at which UV radiation can penetrate the water column. Exposure to UV-B radiation at larval and egg stages has been suspected of negatively impacting some amphibian species directly by reducing hatching success and increasing rates of embryonic deformity (Blaustein et al.,

1998; Lizana and Pedraza, 1998; Broomhall et al., 2000; Starnes et al., 2000) as well as generating adult deformities (Ankley et al., 2000).

Acidic waters often contain toxic levels of metals such as aluminum, cadmium, copper, iron, lead, and zinc, which has led many researchers to investigate the effects that elevated metal concentrations have on amphibian larvae (see review in Freda, 1991; Diana and Beasley, 1998; Linder and Grillitsch, 2000). Aluminum (the metal most commonly associated with decreased pH) reduces tadpole survival (Clark and Hall, 1985; Horne and Dunson, 1995). Other less common metals can also be harmful to amphibian larvae by contributing to decreased survival (lead, Herkovits and Pérez-Coll, 1991; and copper, Horne and Dunson, 1995). Metals are present in the environment due to natural biogeochemical cycles. However, since the industrial revolution, environmental concentrations of metals have increased; these levels now reflect both the natural levels as well as the input from human activity (Goyer, 1996). Although metals are generally not found in nature at concentrations high enough to be harmful to amphibian larvae, metal levels recorded near mines and smelters can reach levels that are toxic to many taxa.

Inorganic pollutants other than metals include nitrates and ammonia, which can enter the environment as runoff from fertilized crops or lawns and feed lots. Concentrations of such compounds can reach high levels in ponds situated within agricultural landscapes, which are common amphibian breeding sites. Although some suggest that nitrate alone may be relatively non-toxic to amphibian larvae (N. Mills, personal communication), Baker and Waights (1994) report that toad tadpoles exposed to nitrates throughout development are smaller upon metamorphosis than control tadpoles. Similarly, high levels of un-ionized ammonia (NH_3) can increase mortality, slow growth and development, and increase the prevalence of tadpole malformations (Jofre and Karasov, 1999; Schuytema and Nebeker, 1999a). Furthermore, both nitrate and nitrite compounds can negatively impact larval behavior (Marco and Blaustein, 1999; Marco et al., 1999), which can ultimately affect growth and development.

Organic pollutants enter ecosystems in the form of pesticides. Many of these compounds tend to persist within the environment because they do not quickly break down, they can be sequestered in sediments, and they can accumulate in animal tissue (i.e., bioconcentration) until the death of the organism, at which time the compound is released back into the environment. Although currently used pesticides (e.g., organophosphates) are less persistent than their historically used forerunners (e.g., chlordane, DDT), residues can commonly be measured in the environment weeks after application. Many pesticides act as neurotoxins, in some fashion affecting the nervous systems of organisms coming in contact with the poison. Amphibian larvae appear to be less acutely sensitive to such compounds than other aquatic organisms (e.g., fish, aquatic insects; Mayer and Ellerseick, 1986), and they have been found to bioconcentrate organophosphate compounds to levels much higher than more sensitive organisms (Hall and Kolbe, 1980; Sparling, 2000). Despite the relatively low sensitivity of amphibians, organophosphates can cause larval mortality as well as embryonic malformations (Pawar et al., 1983; Fulton and Chambers, 1985; Sparling, 2000). Organometals used as fungicides can alter tadpole swimming and feeding behavior (Semlitsch et al., 1995) as well as growth and development (Fioramonti et al., 1997).

The types of organic pollutants that Rachel Carson wrote about (e.g., organochlorines such as DDT) are no longer being released into the environment at the same rate. However, our underlying concerns remain the same: pesticide usage leads to a decline in habitat quality and a gradual reduction in the number of organisms that can tolerate life there, regardless of their taxa. Additionally, the compounds long banned in the United States are inexpensive to produce, and they continue to be used in third world countries (primarily) without stringent laws or guidelines, thereby posing a threat to tropical amphibians.

Because most xenobiotics are a product of human activity, they are likely to continue to be present in the environment. Current concentrations of environmental contaminants are often shown to be detrimental. Therefore, it is necessary to determine ways to limit concentrations in the environment and to protect living organisms, especially those such as amphibians whose numbers are in decline. The increased input of chemicals in the environment requires us to determine the toxicity of these compounds and the impact they have on amphibians. FETAX (Frog Embryo Teratogenesis Assay-*Xenopus*; Bantle et al., 1989; American Society for Testing and Materials, 1991) studies have become a popular component of many toxicity studies but do not utilize native U.S. amphibian species. While these tests are extremely important in comparing the toxicity and teratogenicity of various compounds, the relevance to native frog populations may be limited. Although ecotoxicologists recognize the importance of using amphibians in toxicology tests, many do not have the benefit of training in amphibian biology, and they are unable to connect the important link between laboratory and field. Amphibian biologists face a contrasting predicament: they are well schooled in the biology of amphibians, but they frequently lack the technical knowledge and training to examine the effects of contaminants on this group. The number of individuals with training and expertise in both areas has increased in recent years, but more of these cross-trained individuals are needed.

Chemical hypotheses for amphibian declines are not easy to explain because frequently there is no direct link between contamination and population declines; many declines have occurred in environments having little or no apparent chemical contamination (e.g., Crump et al., 1992). In fact, it has recently been reported that chytrid fungi may be responsible for many mysterious and sporadic amphibian declines worldwide (Carey, 1993; Berger et al., 1998; Lips, 1998; Kaiser, 1999). Although the effects of environmental contaminants on the amphibian immune system are currently unknown (Carey et al., 1999), it is possible that exposure to stressors such as contaminants may depress immune system function, thus allowing greater susceptibility to fungal infections.

Toxicity Testing

Because the release of xenobiotics is regulated, it is necessary for industries to ascertain levels of contaminants that can be released into the environment without harming plants or wildlife. Similarly, it is important to know how much of a pesticide can be applied to a crop to kill weeds or insects, yet remain at a "safe" level for other, non-target organisms. Therefore, ecotoxicologists are enlisted to determine safe and acceptable concentrations of many chemical compounds, and they have created assays routinely used to determine an organism's sensitivity. These tests fall into two broad categories, dependent upon the dose of the chemical used (lethal or sublethal doses) and the length of the exposure (acute or chronic).

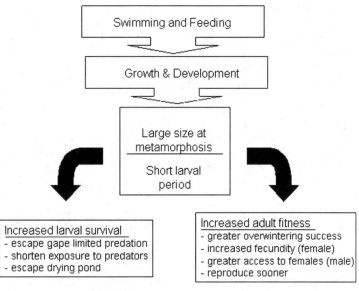

FIGURE 15-1 Illustration of how decreases in swimming and feeding caused by a contaminant can negatively impact growth and development and, ultimately, important life-history characteristics.

Arguably the most widely used assays for determining an organism's sensitivity to a xenobiotic are those generating direct mortality in a short amount of time (i.e., acute lethality test). The most common of these tests is the LC50 (lethal concentration–50%); i.e., the concentration of a chemical required to generate mortality in 50% of a test population within a prescribed amount of time (e.g., 24, 48, or 96 hours). Organisms more sensitive to a contaminant would have a lower LC50 than an organism that is more tolerant. Exact standardization of these tests (e.g., pH, temperature, and water quality) by organizations such as the American Society for Testing and Materials (1989) and the U.S. Environmental Protection Agency (1975) allows for direct comparisons among species and across chemical compounds, regardless of the investigator. Acceptable water quality standards are generally set using data from the most sensitive species. Most toxicity tests are conducted on a few fish species such as trout (Salmonidae) or fathead minnows *(Pimephales promelas)*, or aquatic invertebrates such as water fleas *(Daphnia* sp.) or amphipods *(Hyallela* sp.), whereas studies including native amphibian species remain rare.

While LC50s are widely used, these tests often cannot be conducted with species whose numbers are limited, because they require a large number of animals (~250). Another method of determining the toxicity of a compound is to evaluate the length of time it takes for an organism to die (e.g., time-to-death [TTD]) when exposed to a specific concentration of a chemical (Newman and Dixon, 1996). As with LC50s, higher TTD values indicate more tolerant organisms (i.e., it takes longer for mortality to occur). Time-to-death studies are useful (and perhaps necessary) when testing threatened or endangered species because they can resolve accurately which individuals are most sensitive, while using a minimum number of organisms. Our own research indicates that there is a significant correlation between the 24- and 48-hour LC50s and TTD for tadpoles exposed to the pesticide carbaryl (Bridges, 1999a), indicating that TTD tests yield the same information and may be an efficient substitute for the more traditional LC50s.

In general, concentrations necessary to induce direct mortality are much higher than any expected environmental concentration, and therefore, data generated from LC50 may be irrelevant in a natural setting. While criteria for setting water quality standards commonly include data from sublethal exposures, there are many cases in which legal, sublethal levels have profound and unexpected effects on organisms that may lead indirectly to mortality. Consequently, examining the effects of sublethal concentrations of contaminants is more ecologically relevant. Many acute assays have been developed to determine sublethal concentrations and to incorporate sublethal endpoints, including behavior (e.g., feeding, swimming, mating; see review in Doving, 1991). Chronic sublethal exposures, lasting weeks or months, can determine the effects of a chemical on various life-history characteristics (e.g., growth and development). The benefit of chronic studies is that they demonstrate the effects of long-term exposure, which most closely mimics exposure in the field.

Because of the heightened interest in amphibian species, ecotoxicologists are beginning to include this taxon more frequently when determining the toxicity of a contaminant. The LC50s generated for amphibian species suggest that amphibians are less sensitive than other vertebrate and invertebrate aquatic test species examined (Mayer and Ellerseick, 1986; but see Birge et al., 2000). In contrast, amphibians are often considered more sensitive because of their ability to absorb chemicals through their gills and permeable skin (Boyer and Grue, 1995). Most environmentally safe concentrations of various chemicals are based on values obtained from organisms that are considerably more sensitive. In theory, amphibians will be protected from harm using values generated from these other organisms. However, the growth and development (and ultimately the metamorphosis) of larval amphibians is very sensitive to environmental conditions. For example, tadpoles reduce their activity when exposed to low concentrations of various xenobiotics. A reduction in activity can result in diminished feeding and lead to decreased growth and development (Rist et al., 1997), which can lengthen the larval period and reduce the size at metamorphosis (Fig. 15-1). The length of the larval period and mass at metamorphosis are traits critical in determining an individual's fitness (i.e., survival and future reproductive success). Tadpoles that metamorphose at larger sizes and earlier in the season have a greater chance of surviving over winter and will reproduce at younger ages (Smith, 1987; Semlitsch et al., 1988). Furthermore, a short larval period is especially important to amphibian species breeding in temporary ponds, where any factor that lengthens the time to metamorphosis, such as the presence of an environmental contaminant, can lead to indirect mortality. Consequently, low concentrations of xenobiotics in the environment could disrupt these vital growth and developmental processes and cause a population to slowly decline over time. Where researchers first searched for obvious causes of amphibian population declines, the focus recently has shifted toward examining more of these subtle factors.

A Case Study Using the Insecticide Carbaryl

Much of our research has focused on the subtle effects that sublethal concentrations of the chemical carbaryl have on tadpoles. This insecticide is widely used in agricultural practices throughout the United States and Canada. Additionally, carbaryl can be found on the shelves of gardening centers under the trade name Sevin,® and it is the active ingredient in many common garden insecticides and flea powders. An advantage of using carbaryl for insect control is that it quickly breaks down after application (Liu et al., 1981; Peterson et al., 1994), thus minimizing the exposure risk to non-target organisms such as amphibians. However, aquatic habitats can become contaminated with carbaryl drifting from aerial spray, as well as runoff from gardens or agricultural fields. Tadpoles have 96-hour LC50 values (11.32–22.02 mg/L; Boone and Bridges, 1999) that fall well above maximum concentrations found in the field (<4.8 mg/L; Norris et al., 1983; Peterson et al., 1994). Many other organisms also have LC50 values for carbaryl that are above field concentrations (Mayer and Ellerseick, 1986). This would suggest that carbaryl would be relatively harmless when found in natural environments. However, acute sublethal tests show that tadpole behavior is affected after only 24 hours of exposure at concentrations of as low as 1.25 and 2.50 mg/L. These behavioral modifications could affect interactions between tadpoles and their predators. For example, tadpole swimming (Bridges, 1997) as well as predator avoidance behavior is altered (Bridges, 1999b), potentially increasing susceptibility to predation. However, when tadpoles and predatory newts are dosed simultaneously, predation rates do not increase. Conversely, when tadpoles are exposed and newts are not, or when newts are exposed and tadpoles are not, predation rates increase (Bridges, 1999c). This is important when the breeding season of frogs and the production of tadpoles do not coincide with predator presence in nature. Thus, tadpoles of early breeding species present in the pond and exposed to a contaminant prior to the arrival of predators (e.g., eastern newts [Notophthalmus viridescens]) will experience greater predation rates than tadpoles in uncontaminated ponds. In other cases, larger predators, by nature of their size, may not be as vulnerable to the effects of a contaminant as their prey, thereby increasing predation rates on affected individuals. Because of its rapid evanescence after its release into the environment, several carbaryl applications may be required throughout a season

in agricultural systems. Furthermore, carbaryl can persist in low concentrations for weeks to months once applied (Gibbs et al., 1984; Hastings et al., 1998). Chronic exposure to concentrations as low as 0.16 mg/L carbaryl can cause tadpoles to develop bent tails (Bridges, 2000), which decreases swimming speed and may result in increased predation. Additionally, exposure to this low concentration at any time during development increases the rate of lethal and non-lethal deformities, including visceral malformations and missing or supernumerary limbs. When exposed to carbaryl concentrations between 0.16 and 1.0 mg/L throughout development, there is a significant increase in mortality that is concentration dependent. Additionally, when exposed during the theoretically protected egg stage, these tadpoles have a reduced mass at metamorphosis. Thus, exposure to levels of carbaryl well below standard LC50s, regardless of duration, contributes to negative behavioral, developmental, and morphological effects.

Research designed to determine the effects of environmental contaminants on larval amphibians is necessary, as it may be in the larval phase of the amphibian life cycle that most population regulation occurs (Wilbur, 1980). Additionally, amphibian larvae may be at higher risk of exposure than adults for several reasons. First, a large number of amphibian species, especially frogs, migrate to aquatic habitats to breed. In many cases, breeding seasons coincide with the application of agricultural chemicals and with spring rains or runoff from snowmelt. In cases of acid pollution, spring breeding is often associated with episodes of decreased environmental pH due to snowmelt. Second, amphibian larvae are not only more sensitive to environmental contaminants than adults, they may also be exposed to greater concentrations because they are constrained to the aquatic environment, while the adults are more mobile. Finally, larval amphibians may be exposed to higher concentrations of a contaminant than terrestrial adults because aquatic habitats not only receive chemicals from direct application, but also from the chemical runoff of nearby agricultural lands.

Amphibian ecotoxicological research is especially vital in light of growing concerns of amphibian population declines (Wake, 1998; Houlahan et al., 2000). Xenobiotics have been shown to increase mortality and alter behavior, as well as critical life-history characteristics of larval amphibians. In combination, these alterations can disrupt natural regulatory processes. The impairment or loss of individuals in a single amphibian population can have ramifications at the population (and eventually the community) level through increased larval mortality and reduced juvenile recruitment. Such changes can lead to localized declines over time and ultimately dictate the persistence of a species.

Acknowledgments. We thank M. Boone, V. Burke, J. Dwyer, and W. Gibbons for helpful comments and suggestions on this manuscript. Reported research was supported by a conservation Grant-In-Aid in herpetology from the Society for the Study of Amphibians and Reptiles (CMB), a seed grant from the Declining Amphibian Populations Task Force (CMB), and EPA grant #827095-01 (RDS). Facilities were provided by the USGS Columbia Environmental Research Center. Carbaryl was donated by Rhône-Poulenc.

SIXTEEN

Variation in Pesticide Tolerance

CHRISTINE M. BRIDGES AND RAYMOND D. SEMLITSCH

A growing body of evidence suggests that a number of amphibian populations have declined in recent years (Barinaga, 1990; Blaustein and Wake, 1990; Wake, 1998). The cause of these declines has been difficult to establish because in some instances only a single species is declining while sympatric species are thriving (e.g., K.R. McAllister et al., 1993). Similar variation can be observed within a single species at the population level—there are often instances when some populations of a particular species are declining while others remain unaffected (e.g., northern leopard frogs *[Rana pipiens]*, Corn and Fogleman, 1984; mountain yellow-legged frogs *[R. muscosa]*, Bradford, 1991; western toads *[Bufo boreas]*, Carey, 1993). Because not all populations or species have been affected, it is important for conservation efforts to identify causes for the variability among populations whose environments appear physiographically similar. Research efforts should be focused on addressing what makes some species and populations vulnerable and others resistant to declines. Here we focus on research we have conducted to determine the degree of variation present in amphibians with respect to their response to insecticide exposure. Our goal was to make it possible to establish potential patterns of vulnerability among and within amphibian species.

Chemical contaminants are not homogeneous stressors because their presence in the environment can vary over small temporal and spatial scales. Differences in tolerance to a stressor can thus arise from the differential exposure that results from this variability in the presence of the contaminant. Variation in responses may also be due to genetic differences among or within species. In order for a species to adapt to an environment, a trait must be phenotypically variable, and this variation must have a heritable genetic basis. That is, there must be measurable differences among individuals (which ultimately lead to differential survival), and these differences must be passed on from generation to generation. Populations having little or no heritable genetic variability in tolerance may be more susceptible to decline and local extinction because they cannot adapt to the presence of environmental stressors.

Assessing Variation Among and Within Frog Species

We assessed the degree of variation in response to an anthropogenic stressor among and within species of frogs in the family Ranidae by examining the variation in tolerance of tadpoles to the insecticide carbaryl. Carbaryl is an insecticide that is widely used throughout the United States and Canada, and it can contaminate amphibian habitats via drift from application or in runoff from adjacent agricultural fields or gardens. Carbaryl acts by inhibiting nervous system acetylcholinesterase, which is a common mode of action among insecticides; thus, carbaryl can serve as a model chemical with which to examine amphibian responses.

We examined variation in a hierarchical fashion to identify where variation was the greatest: among nine ranid species (crawfish frogs, *R. areolata*; northern red-legged frogs, *R. aurora*; plains leopard frogs, *R. blairi*; foothill yellow-legged frogs, *R. boylii*; green frogs, *R. clamitans*; pickerel frogs, *R. palustris*; Oregon spotted frogs, *R. pretiosa*; southern leopard frogs, *R. sphenocephala*; wood frogs, *R. sylvatica*), among populations within a single species (southern leopard frogs), and within populations (i.e., among families) of southern leopard frogs (Bridges and Semlitsch, 2000). We used ranid frog species because many declines in the United States have occurred within this family (Hayes and Jennings, 1986). For each group (e.g., family, species) we conducted two assays, both using carbaryl (Bridges, 1999a). The first assay determined sensitivity to a lethal dose of carbaryl, measured by time-to-death. In the second assay, we exposed tadpoles to a sublethal concentration of carbaryl and recorded changes in behavioral activity.

Variation Among Ranid Species

We detected differences in the tolerance of tadpoles of all nine species tested (Fig. 16-1). Localized declines have been noted in all ranid species in the western United States (Corn and Fogleman, 1984; Hayes and Jennings, 1986; Bradford, 1991; Fellers and Drost, 1993; Bradford et al., 1994a,b; Drost and Fellers, 1996), including the three western U.S. species that we examined: Oregon spotted frogs (USFWS, 1993; K.R. McAllister et al., 1993), California red-legged frogs (*R. draytonii*; USFWS, 1996c), and foothill yellow-legged frogs (Corn, 1994a). Therefore, we were able to test an important hypothesis of amphibian conservation efforts: that species in decline are more sensitive to environmental stressors (e.g., chemical contamination), thus partially

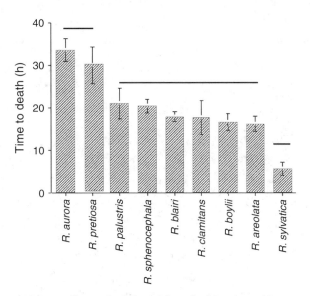

FIGURE 16-1 Time to death (hours) for each of the nine ranid species tested. Horizontal lines group species that do not significantly differ from one another (p = 0.05). Vertical lines on bars represent ± 1 standard error.

explaining observed patterns of decline. An increase in vulnerability could be due to low genetic diversity resulting from small population sizes and restricted ranges or simply a function of evolutionary similarity. Further, the detrimental effects of environmental stressors could be exacerbated in these western species because gene flow among populations may be limited due to fragmented distributions and long distances between populations that make recolonization of declining or extirpated populations more difficult (Blaustein et al., 1994a). If true, increased vulnerability to environmental stressors (e.g., carbaryl or other pesticides) in conjunction with limited dispersal among populations could contribute to observed amphibian declines.

Two of three western U.S. species we examined that have experienced the greatest declines (Oregon spotted frogs and red-legged frogs) are those that were the most tolerant to carbaryl; the third species (foothill yellow-legged frogs) had an average tolerance (Bridges and Semlitsch, 2000). While the theory of general-purpose genotypes (Lynch, 1984) would predict that organisms with broad ranges (e.g., eastern U.S. ranid species) have greater tolerances (which are beneficial for dispersing), this is not always true (Parker and Niklasson, 1995). Oregon spotted frogs and red-legged frogs are more narrowly distributed than the other species we tested and may have become adapted locally to harsh environmental conditions, thus demonstrating higher tolerance to carbaryl in our experiment. Additional sampling of these two species from several populations would be necessary to determine whether the entire species demonstrates the same level of tolerance as the populations we sampled.

Within-Species Variation

Southern leopard frogs are widely distributed in the midwestern and southeastern United States and represent a species having few reports of localized declines. Southern leopard frogs often breed in shallow pools or temporary ponds situated within agricultural areas and, therefore, may encounter

agricultural chemicals. We detected significant differences among eight populations of southern leopard frogs in time-to-death, as well as in the magnitude of activity change, indicating that a large degree of variation for chemical sensitivity exists throughout the range of this species. Furthermore, in four of the eight populations we measured significant differences in carbaryl tolerance among families (i.e., tadpoles from separate egg masses). However, it is unknown whether differences observed at any level are attributable to local adaptation to an environmental stressor rather than random effects or geographic variation associated with other traits. Local adaptation would require that (1) pesticides occur in the environment at low enough concentrations as to not kill all frogs within a population; (2) pesticide tolerance within populations has a heritable genetic basis; (3) stressors persist in the environment between generations; and (4) there are fitness differences between tolerant and sensitive genotypes. Our sampling design precluded us from establishing whether the within-species differences that we observed have a genetic basis or whether collection sites have a history of carbaryl exposure, which might suggest adaptation to environmental stressors. However, we have determined that tolerance to carbaryl is heritable in one of our study populations. Thus, at least this one population likely has the ability to adapt to environmental contaminants (Bridges and Semlitsch, 2001).

Carbaryl decreased the activity of all tadpole groups we tested. The magnitude of the decrease in activity, however, differed among populations of southern leopard frogs, as well as within four of the eight southern leopard frog populations we tested. Tadpole activity is correlated with feeding, which is, in turn, directly related to larval growth and development and ultimately adult fitness (see Bridges and Semlitsch, this volume). Therefore, tadpole groups whose activity is least affected by carbaryl will likely have greater fitness than tadpoles demonstrating a severe decline in activity.

We observed a strong, negative correlation between the magnitude of decreased activity and time-to-death. This suggests that sublethal assays, such as the one used in our experiment, may be as sensitive as more commonly used mortality assays and may be useful for distinguishing relative sensitivities. Sublethal assays would be advantageous when working with threatened or endangered species, where using assays involving the death of individuals may not be acceptable or allowable. Moreover, organisms are likely to encounter sublethal chemical concentrations in their natural habitats.

Conclusions and Conservation Implications

Knowing which species are most sensitive to environmental stressors (e.g., insecticides) is important because it can elucidate which species are in danger of being decimated by such stressors. However, it is often difficult to compare sensitivity of species because of variation at the individual, family, and population levels, as well as uncertainty about which stressors are ecologically relevant. Our study had the advantage of conducting all experiments using the same protocol and chemical stressor, thereby maximizing our ability to directly compare among species. While in some instances the responses of African clawed frog (*Xenopus laevis*) or northern leopard frog tadpoles (both common surrogates) to a toxicant are similar to that of natural populations of native species, there are often discrepancies (Mayer and Ellersieck, 1986). Examining the responses of wild populations of native species, such as southern leopard frogs, is important in order to draw conclusions about variation in the reported declines and to develop management plans to protect native species from future declines.

Conservation efforts for declining species must consider variation (individual, family, population, and geographic) in tolerance to protect them from chemical contamination. It is necessary to obtain this information before populations decline too precipitously, which would result in obliteration of potential adaptive variation. An examination of the geographic distribution of historical chemical contamination in relation to distributions of amphibian population declines would also be worthwhile. This information would allow us to correlate declining populations with the history of exposure to chemical contamination. Although we know that chemical contaminants can affect amphibian population processes (e.g., juvenile recruitment), whether they do in reality is the ultimate, and as yet unanswered, question.

Our results emphasize the importance of broad sampling techniques when determining sensitivity to an environmental stressor, especially when the data will be used for regulatory purposes (e.g., dictating pesticide usage). For instance, if eggs only from a single egg mass or population were used, it is possible for these individuals to be either more or less sensitive than the rest of the individuals or populations. With respect to contaminant exposure and tolerance, each population has a unique environmental and evolutionary history and possesses a unique combination of genes as a result. Therefore, we advocate the collection of eggs from multiple masses, from more than one population if possible, to maximize the chances of including representative genotypes and therefore the full range of responses to chemical stressors. Practicing such guidelines will ensure that when water-quality standards are set using data from amphibian larvae, a maximum number of populations within a species' range will be protected.

Chemical hypotheses for amphibian declines are not easy to test because there is often no direct link between contamination and population declines; in fact, many declines have occurred in environments having little or no apparent chemical contamination (e.g., Crump et al., 1992). Sublethal levels of chemical contaminants and biotic factors may interact synergistically to increase the toxicity of compounds (UV-B, Zaga et al., 1998; temperature, Boone and Bridges, 1999), or contaminants may break down into more toxic compounds, leading to an overestimation of acceptable field concentrations. Thus, multiple factor rather than single factor hypotheses may be necessary to adequately describe the potential effects of chemicals on natural amphibian populations.

Acknowledgments. We thank M. Boone, V. Burke, J. Dwyer, and W. Gibbons for helpful comments and suggestions on this manuscript. Reported research was supported by a conservation Grant-In-Aid in herpetology from the Society for the Study of Amphibians and Reptiles (C.M.B.), a seed grant from the Declining Amphibian Populations Task Force (C.M.B.), and Environmental Protection Agency grant #827095-01 (R.D.S.). Facilities were provided by the U.S. Geological Survey Columbia Environmental Research Center, and carbaryl was donated by Rhône-Poulenc.

SEVENTEEN

Lucké Renal Adenocarcinoma

ROBERT G. MCKINNELL AND DEBRA L. CARLSON

The Lucké renal adenocarcinoma of northern leopard frogs *(Rana pipiens)* was originally described by Balduin Lucké during the 1930s (Lucké, 1934a,b, 1938a). Lucké's contribution to this era in medical history is important for several reasons. At the time, Lucké, a pathologist at the University of Pennsylvania, was treading on dangerous terrain when he suggested that the frog renal adenocarcinoma was caused by a virus—in this case, a herpesvirus (Lucké, 1952). A few years earlier, Rockefeller Institute's Peyton Rous had identified a virus as the etiological agent of a chicken tumor. Rous was castigated by the medical establishment for daring to suggest that a cancer could be caused by a virus. It was believed then that cancers were truly spontaneous. If an agent could be identified as the cause of a malignancy, then the malignancy could not be a "true cancer." It was not until 1966 that Rous received proper recognition of his viral oncogenesis work and was awarded a Nobel Prize (Rous, 1967).

Lucké was a trailblazer in yet another way. Prior to the frog renal adenocarcinoma, no cancer had ever been linked with a herpesvirus. Lucké's herpesvirus, now known as ranid herpesvirus-1 (RaHV-1), was subsequently shown to be the etiological agent of the cancer. It has been well established (two-thirds of a century later) that in addition to several animal cancers, a number of human malignancies are associated with herpesviruses. These human cancers include Burkitt's lymphoma of African children, a nasopharyngeal carcinoma that afflicts males of southeast Asia, and Kaposi's sarcoma (Haverkos, 1996; Howley et al., 1997; Schulz, 1998; McGeoch and Davison, 1999). We should remember that Lucké's studies on a common North American anuran clearly established for the first time the validity of a herpesvirus etiology for a specific cancer.

The depletion of frog populations has been a major concern of herpetologists worldwide (Houlahan et al., 2000). One of the authors of this paper (R.G.M.) has studied the cell biology of the Lucké renal adenocarcinoma since 1958. Early in these studies it became obvious that the best way to obtain frog renal adenocarcinomas for experimental studies was to study frogs in natural populations. Mature frogs were collected and palpated for lumps in the region of the mesonephros. The diagnostic procedure of frog palpation is an imperfect art, but it is sufficiently accurate to permit the identification of many frogs with malignancies. During the early years of this study, northern leopard frogs were adequately abundant to permit the examination of thousands of frogs. Later, it became obvious that frogs were less plentiful. We reported finding reduced frog abundance long before many other scientists recognized the phenomenon (McKinnell et al., 1979). In retrospect, it seems curious that an investigation into the cell biology of a frog neoplasm would lead to early recognition of depleted frog populations.

Differentiation therapy—manipulating a cancer cell genome in such a way that its mitotic progeny are forced to give rise to normal cells—is a relatively new concept in the treatment of cancer (Pierce and Speers, 1988; McKinnell, 1989; Sachs, 1993). Differentiation therapy exploits the origin of cancer from aberrant stem cells. Cloning of vertebrates, which originated with the now classic study of northern leopard frogs (Briggs and King, 1952), involves the transfer of nuclei into enucleated eggs with the intent to characterize the differentiative potential of the genome under consideration. Emerging from the cloning studies of frog renal adenocarcinoma nuclei (and differentiation studies elsewhere) was the notion that the genome of cancer cells could be manipulated to undergo substantial differentiation with the formation of normal, or near normal, nuclear progeny. While this research may seem esoteric to some, it led to the development of a treatment protocol for acute promyelocytic leukemia, which is acknowledged to be the "paradigm of differentiation therapy" (Degos et al., 1995). We can only hope that other human neoplasms will respond to the more gentle differentiation therapy in lieu of highly toxic chemotherapy—if this happens we are at least partly in debt to early studies of a frog renal adenocarcinoma.

The history of the Lucké renal adenocarcinoma of northern leopard frogs is impressive in that these studies led to the notions that herpesviruses can cause cancer, that there was a problem with frog populations in the upper Midwest, and that differentiation therapy may lead to a new form of cancer treatment. It is a pleasure, therefore, to review past and present investigations of this fascinating cancer that afflicts northern leopard frogs.

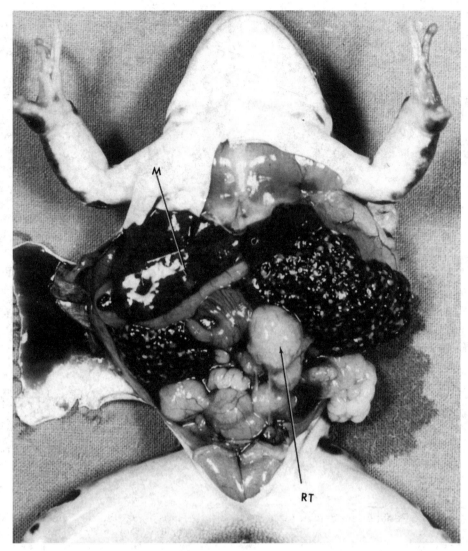

FIGURE 17-1 Renal adenocarcinoma-bearing northern leopard frog. The ventral skin has been cut and folded back to reveal the renal tumor (RT). There is a small metastic mass (M) in the liver. The afflicted frog is a sexually mature female containing many oocytes. From McKinnell et al. (1976).

Early History of the Lucké Renal Adenocarcinoma

Balduin Lucké was a careful pathologist, and he received eponymous recognition for his description of the renal adenocarcinoma of northern leopard frogs (Figs. 17-1 and 17-2; Lucké, 1934a, 1938a,b, 1952). Smallwood (1905) was perhaps less careful (or less experienced) when describing a bilateral (mesonephric) tumor with many large "whitish" lobes. But, Smallwood thought that the tumor was of adrenal origin. Smallwood's slides were examined by Murray (1908) who categorically rejected the adrenal origin of the frog cancer and stated that "histologically, the cells do not present the slightest trace of the characteristic granules of the adrenal tissue." Murray commented on the large size of amphibian cells, which rendered their tumors an "unequalled opportunity" for study in lieu of human and other mammalian cancers with their much smaller cells.

Electron microscopic imaging and molecular analysis are contemporary methods for the identification of herpesviruses. Neither the electron microscope (for the study of cells and tissues) nor modern molecular probes had been developed during the 1930s when Lucké studied the frog cancer. How then was it possible for Lucké to detect the presence of what has become known as RaHV-1? The answer is that Lucké used light microscopy with consummate skill. Of course, herpesviruses are too small to be resolved with glass lenses. However, the effects of their presence can be detected with properly prepared histological slides. Herpesvirus infections often cause the clumping of nucleoprotein ("basophilic chromatin") on the inside surface of the nuclear membrane. This effect of herpesvirus infection has been described as "margination" of the basophilic chromatin and is categorized as a Type A nuclear inclusion typical of herpes infections (Cowdry, 1934; see fig. 7 of McKinnell, 1965). Clearly, Lucké believed the cytological stigmata indicated a herpesvirus infection. To be certain, Lucké sent slides to E. V. Cowdry at Washington University in Saint Louis, Missouri. Cowdry agreed with the diagnosis made by Lucké and listed the frog tumor cells with other cells containing the Type A inclusions typical of herpes infections (Cowdry, 1934). From that point on, there was no doubt concerning an association of the herpesvirus with the frog cancer.

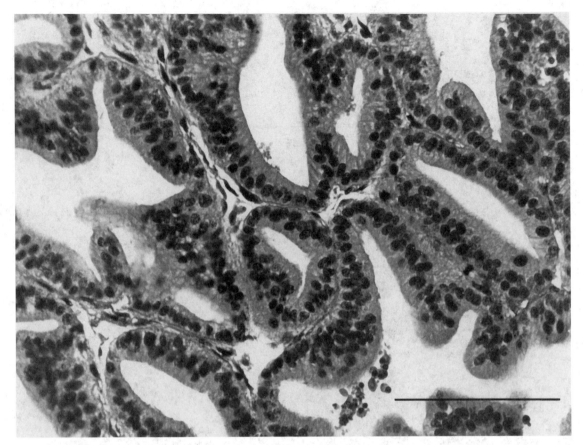

FIGURE 17-2 Renal adenocarcinoma histology. Malignant epithelial cells are surrounded by minimal connective tissue stroma. Length of index line is 100 micrometers. From McKinnell and Duplantier (1970).

Pathology of the Lucké Renal Adenocarcinoma

Malignancy is defined by the capacity of a neoplasm to transgress the basement membrane, invade contiguous normal tissue, intravasate (gain access to the vasculature), disseminate, and set up metastatic colonies distant from the primary tumor. These several steps constitute the metastatic cascade, and metastases absolutely define malignancy. Metastases of the frog tumors occur spontaneously (Lucké, 1934a, 1938b), but perhaps of much greater importance is the temperature dependence of metastasis in the frog renal adenocarcinoma. Metastatic colonies are found in fewer than 6% of tumor-bearing frogs maintained at 7 °C for 98–140 days. This may be contrasted with a metastatic rate of 76% of tumor-bearing frogs kept at 28 °C for 50 days (Lucké and Schlumberger, 1949).

The phenomenon of temperature-dependent metastasis (identified by Lucké and Schlumberger, 1949), provides for an analysis of metastasis mechanisms. Any variable that reproducibly associates with metastasis-permissive conditions (when the ambient temperature is changed between permissive and inhibitory conditions) is a valid candidate as a causal element in the metastatic cascade. Like metastasis invivo, invasion in vitro, either by Lucké tumor fragments or by cells harvested from a Lucké tumor cell line, was shown to be strictly temperature-dependent (Mareel et al., 1985; McKinnell et al., 1986; Anver, 1992). Enzymes from the frog cancer not only degrade basement membrane collagen (Shields et al., 1984), but a metalloprotease that digests collagen type I is elaborated in a temperature-dependent manner (Ogilvie et al., 1984). An

intact cytoplasmic microtubule complex (CMTC) is a requisite for directional cell motility. The CMTC of the Lucké renal adenocarcinoma is intact at invasion-permissive temperature but depolymerizes at invasion-restrictive temperatures (McKinnell et al., 1984). The temperature-dependent pathology of the Lucké cancer is more completely reviewed in McKinnell and Tarin (1984).

Epidemiology

Geographic Pathology

Northern leopard frogs are distributed in the United States from New England into parts of New York, Pennsylvania, Indiana, and northern Kentucky, covering all of Ohio and the upper Midwest to the Rocky Mountains, and also into much of the Southwest. They are found in Canada in areas contiguous to their distribution in the United States, from the Canadian provinces of New Brunswick, Nova Scotia, and Prince Edward Island in the east to portions of Alberta and British Columbia provinces in the west—a distance of over 4,000 km (Pace, 1974; Cook, 1984; Conant and Collins, 1991; Behler and King, 1995; Harding, 1997).

The extent to which the Lucké renal adenocarcinoma afflicts widely distributed northern leopard frogs is not now known and perhaps will never be known. The only certain way to know the limits of distribution is to autopsy frogs. Autopsies were performed in the days when frogs were thought to be plentiful. However, it has become unacceptable to kill large numbers of frogs because of the worldwide decline in amphibian populations. Abdominal palpation was earlier described in this paper, but digital palpation is less precise than autopsy because only mid- to large-sized tumors are discovered by palpation. Some human cancers can be detected with x-ray analysis. Obviously frogs can be x-rayed, but tumors of soft tissues such as the mesonephros are difficult to detect without inflation of the body with a gas. This has been done, but it has the disadvantages of high cost and of unknown and possibly harmful effects on the frogs. We are thus left with historic reports on the geographic distribution of mesonephric cancer in northern leopard frogs.

Frogs with spontaneous tumors (i.e., tumors that occur in untreated frogs taken from the wild) have been reported from Vermont (Auclair, 1961), Minnesota (see below), Wisconsin (Rose and Rose, 1952), and Michigan (Nace and Richards, 1969). Spontaneous renal tumors probably occur in populations of frogs at other sites between South Dakota and Vermont. This judgment is based on observations that frogs held under laboratory conditions for several months develop cancer (Rafferty and Rafferty, 1961; Roberts, 1963; Di Berardino and King, 1965; McKinnell and John, 1995). Rafferty and one of the authors of this paper (R.G.M.) reported that tumors were detected in frogs collected from South Dakota, Minnesota, Wisconsin, New York, Vermont, and Quebec, Canada, after laboratory maintenance (Rafferty, 1967). Inasmuch as early exposure to RaHV-1 infection is a prerequisite for oncogenesis, the frogs that developed tumors after laboratory treatment must have already been exposed to the virus in the wild. Because the only source of RaHV-1 is tumor-bearing frogs, tumors must occur at those sites.

Tumor Prevalence

The frog renal adenocarcinoma has varied in prevalence over time. In November 1963, an autopsy study of 884 untreated frogs from Minnesota and Wisconsin showed a tumor prevalence of just under 9% (McKinnell, 1965). Frogs collected during different seasons also reveal variation in tumor prevalence. Tumors are more common in overwintering frogs than in frogs collected during the summer (McKinnell, 1967; McKinnell and McKinnell, 1968). It is doubtful that tumor regression occurs during the warm months. We believe that the low frequency of tumors during the summer is due to rapid tumor growth in response to warm temperature, metastasis, and subsequent loss of vitality due to tumor burden. Frogs bereft of stamina due to tumor burden become especially vulnerable to predation by frog-eating birds, snakes, and small mammals. Subsequent studies reported a decrease in tumor prevalence (McKinnell et al., 1979) with one notable exception (Hunter et al., 1989). We support the notion that the observed reduction in tumor frequency is simply the result of diminished frog populations. Exposure of eggs and young embryos to the oncogenic RaHV-1 occurs in breeding ponds. We believe that when renal adenocarcinomas are common, the likelihood of one or more tumor-bearing frogs occupying any particular breeding pond is high. With a reduced population of frogs, there is a decreased chance that an infected frog will occupy the pond. No tumors would be expected from a breeding pond lacking a tumor-bearing frog. A similar density-dependent transmission of an etiological agent has been described for bovine leukemia (R.K. Anderson et al., 1971).

The Lucké Tumor Herpesvirus—RaHV-1

Herpesviruses

The etiological agent of the Lucké renal adenocarcinoma is a herpesvirus known as RaHV-1 (Fig. 17-3). All herpesviruses contain a single, linear, double-stranded DNA molecule wrapped on a protein spool enclosed within the nucleocapsid. The nucleocapsid, composed of 162 capsomeres, possesses icosahedral symmetry and is contained within a lipid bilayer envelope. Replication of herpesviruses occurs within the nucleus of the infected cell. All these characteristics are found in RaHV-1.

A brief note on etymology is of interest. The term *herpes* is derived from the Greek *herpein*, which means "to creep" or "creeping." Shingles (produced by the same herpesvirus that causes chickenpox) causes an unpleasant superficial tingling feeling in the affected skin. That symptom has also been described as a "creepy" or "crawling" feeling, hence the name "herpes." Herpetology is the study of creeping or crawling animals and their kin, such as frogs. This paper thus concerns a herpetological herpesvirus.

Viral Etiology of the Lucké Renal Adenocarcinoma

The evidence that a herpesvirus causes the Lucké renal adenocarcinoma has accumulated over many years. Cytological characteristics (marginated chromatin) led Balduin Lucké (in the 1930s) to suspect that a virus was associated with the frog tumor. Virus particles were reported by Fawcett and all others who examined a tumor with the electron microscope (Fawcett, 1956; Zambernard et al., 1966; McKinnell and Zambernard, 1968; McKinnell and Ellis, 1972a,b). Furthermore, herpesviruses were described in metastatic colonies of the neoplasm (McKinnell and Cunningham, 1982).

Koch's first postulate requires that a putative etiological agent be present in all cases of a disease. Early on, there were difficulties with the electron microscope studies. Viruses could be found in only some tumors. Herpesviruses have a propensity for becoming latent from time to time. It was thought that RaHV-1 was probably present but not in a replicative state that could be visualized with electron microscopy. We made the judgment that if RaHV-1 was indeed the etiological agent of the frog tumor, it would be unlikely to have a capricious or erratic relationship with the host. We already knew that cold temperature favored virus replication. Thus, instead of studying random tumors of frogs purchased from dealers (who had little idea of how the frogs were previously treated or from where the frogs were obtained) we began a study of frogs that we collected from known sites in Minnesota during the cold season.

Eleven renal tumors were obtained from wild Minnesota frogs just prior to emergence from overwintering lakes or soon thereafter in frog breeding ponds. Lake water temperature varied from 2–7 °C at the time. Herpesviruses were detected in each of the 11 tumors (McKinnell and Zambernard, 1968). Subsequent to that study, we collected frogs directly from overwintering lakes prior to the spring melt of ice cover. We were successful in obtaining 18 additional renal tumors from hibernating conditions and 6 shortly thereafter. Electron microscopic examination revealed virus particles in each of the tumors (McKinnell and Ellis, 1972b). Thus, 35 consecutive cancers obtained during or just after overwintering were characterized by the presence of RaHV-1. These are the only electron microscope

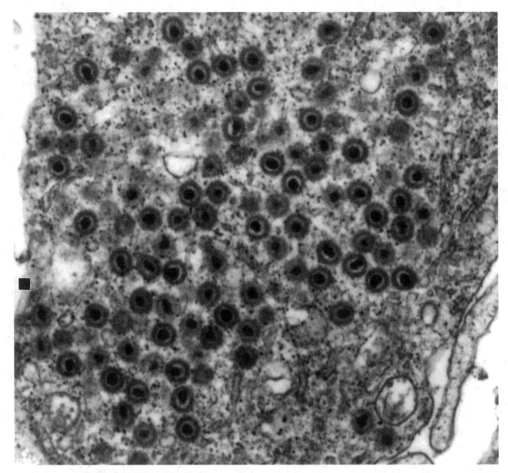

FIGURE 17-3 Electron micrograph of renal adenocarcinoma cell fragment containing many RaHV-1 nucleocapsids (which appear as dark circles) with enclosed double stranded DNA (the dense material within the nucleocapsids). From Sauerbier et al. (1995).

studies of RaHV-1 in frog tumors directly taken from the wild, and there is a remarkable consistency in the change to virus' presence. The chance of our selecting only virus-containing tumors from a population containing equal numbers of virus-laden and "virus-free" tumors is 2^{35}—an obviously implausible event.

In contrast, electron microscope examination of 11 tumors from late summer or early autumn frogs revealed no viruses (Zambernard and McKinnell, 1969). There was, of course, evidence of a viral genome in the "virus-free" tumors; viruses appeared after low temperature treatment of fragments of these tumors (Stackpole, 1969; Morek, 1972). This indirect indication for viral genome presence was confirmed with direct evidence many years later. The polymerase chain reaction (PCR) was used to amplify a restriction enzyme fragment of RaHV-1 DNA, and the identity of the amplified viral DNA was confirmed by Southern hybridization. All the warm (virus-free) spontaneous tumors that were taken from the wild contained the RaHV-1 restriction enzyme fragment (Carlson et al., 1994a; McKinnell et al., 1995). When combined, this direct evidence of the presence of RaHV-1 in warm tumors and the electron microscope evidence of viral presence in cold tumors makes it manifestly clear that all Lucké renal adenocarcinomas harbor a RaHV-1 infection. Thus, after many years, Koch's first postulate was confirmed.

At about the time of the electron microscopy studies, Kenyon Tweedell performed a critical experiment. He injected viruses obtained from frog tumor homogenates into pre-feeding embryos of northern leopard frogs. Most of the injected animals developed renal tumors at or near the time of metamorphosis (Tweedell, 1967). Tumor homogenates lacking viruses did not induce renal carcinomas in these animals. This notable study clearly indicated the oncogenic potential of RaHV-1, as well as fulfilling Koch's second postulate. Later, viruses were obtained from experimental tumors and were also shown to induce cancer. All of Koch's postulates were satisfied, and the tumor became a model for herpesvirus cancer etiology (Naegele et al., 1974).

Genomic Studies of RaHV-1

Lucké tumor herpesvirus DNA, prepared in Minnesota from viral nucleocapsids, was shown by field inversion gel electrophoresis to have an estimated genomic size of 220 kbp (Sauerbier et al., 1995). The preparation of viral DNA permitted further studies by Andrew Davison in Glasgow. A BamHI map for the viral genome was derived from cosmid libraries, and fragment summation yielded a genomic size of 217 kbp, a figure in close harmony with the previous estimate from field inversion gel electrophoresis. The nucleotide sequence of a 39,757-bp insert in one cosmid was established, and it was predicted to contain 21 complete and three partial genes. Twelve of those genes are similar to counterparts in the fish herpesvirus (ictalurid herpesvirus 1). These studies suggest that RaHV-1 is associated with the fish herpesvirus lineage rather than with mammalian or avian herpesviruses. A gene novel to herpesviruses, a putative DNA (cytosine-5) methyltransferase, was described, which raises the possibility that RaHV-1 may be involved in host methylation (Davison et al., 1999). The genome sequence of RaHV-1 is nearly completed (Andrew Davison, personal communication).

Differentiative Potential of Mitotic Progeny of the Lucké Renal Adenocarcinoma

Cloning by Nuclear Transplantation

It has been approximately 50 years since the momentous pioneering study of frog nuclear transplantation (Briggs and King, 1952) in which it was shown that embryonic nuclei retain the same genomic potential as the zygote nucleus. Perhaps more importantly, it was recognized that the cytoplasm of oocytes had the potential to direct the introduced embryonic nucleus to act as if it were a zygote nucleus. It was soon deemed important to ascertain if the controlling elements in oocyte cytoplasm could redirect the aberrant expression of the genes in a cancer cell (McKinnell and Di Berardino, 1999). The first cloning experiments were performed with eggs and embryonic nuclei of northern leopard frogs. The best known malignancy of frogs (then and now) is the Lucké renal adenocarcinoma. Thus, it was both appropriate and convenient to study the differentiative potential of the Lucké tumor.

Normal embryonic differentiation requires a normal set of chromosomes. In order to characterize the differentiation potential of cancer cells by the cloning procedure, the cancer under consideration must have a normal or near-normal set of chromosomes. It was fortunate for the purposes of cloning to learn that most cells of the Lucké renal adenocarcinoma have a normal karyotype (Di Berardino et al., 1963). Moreover, experimentally produced triploid tumors were also shown to have three normal-appearing haploid sets of chromosomes (J.W. Williams et al., 1993).

Frog tumor nuclei were inserted into enucleated host oocytes. Abnormal embryos (some of which looked remarkably normal) developed from the operated oocytes (King and McKinnell, 1960; King and Di Berardino, 1965; McKinnell et al., 1969). It is obvious that embryos reveal limited differentiative potential, but it has been known for many years that the differentiative potential of embryos that are fated to die can be enhanced by grafting (Hadorn, 1961). We exploited allografting to explore the differentiative potential of cancer nuclei. Small fragments of the embryos cloned from renal tumor nuclei were grafted to the tails of normally developing embryos produced by ordinary fertilization. We reported that successful grafts grew for about 40 days. The tumor genome grafts differentiated in such a way that they could not be distinguished from grafts of normal tissue. The 40-day limit was not related to the malignant origin of the donor tissue fragment but was due to the immune response of the host (Volpe and McKinnell, 1966; Rollins and McKinnell, 1980). Obviously, the cloned tumor embryos were not genetically identical to the normally fertilized hosts, and the lack of histocompatibility resulted in their rejection. Tissue fragments of normally fertilized embryos were rejected at the same time as the experimental allografts (Lust et al., 1991). The important result was that the mitotic progeny of a cancer genome, the Lucké renal adenocarcinoma, gave rise to normally differentiated tissue.

The differentiation study was not designed to cure cancer in frogs. Rather, it was devised to ascertain if a cancer cell genome could be manipulated in such a way that its mitotic progeny are forced to give rise to normal cells. That rationale has now awakened an area of human cancer treatment known as *differentiation therapy* (Pierce and Speers, 1988; McKinnell, 1989; Sachs, 1993; Degos et al., 1995).

Fate of RaHV-1 in Embryos Cloned from Tumor Nuclei

Herpesviruses replicate in the nuclei of infected cells. It seemed reasonable to seek viral DNA in normal or near normal embryos cloned from tumor nuclei because retention of viral DNA could result in the eventual reversion of the normally differentiated cells to the malignant phenotype. For this reason, normal control embryos and embryos cloned from tumor nuclei were assayed for the presence of viral DNA by amplification of a RaHV-1 restriction fragment. No control embryo contained these DNA sequences. Only three of 34 embryos cloned from Lucké tumor nuclei were positive for the sequences (Carlson et al., 1994b).

It was postulated that the rate of cell division of the cloned embryo was faster than the replicative process in RaHV-1 and, thus, the etiological agent was diluted and lost in the rapid sequence of cell divisions. The absence of viral DNA could be due to the selection of tumor cells bereft of a herpesvirus infection (this possible reason assumes that not all nuclei of a Lucké renal adenocarcinoma contain the etiological agent). Another study was unable to detect evidence of RaHV-1 in a frog tumor-derived cell line (J.W. Williams et al., 1996). We mention the issue of lost herpesvirus infections because of their truculent and intractable nature. In two different situations, the herpesvirus was "lost" from infected cells. Might not this be considered a "cure" of the infection? If so, were the phenomenon more thoroughly studied, would it be possible to use the knowledge gained from frog tumors to develop a cure for otherwise incurable herpesvirus infections?

The Lucké Renal Adenocarcinoma and Frog Declines

The extent to which the Lucké renal adenocarcinoma has contributed to frog population declines is not known. While tumors in juvenile frogs can be induced under laboratory conditions, in natural populations the cancer afflicts only sexually mature frogs. The reproductive potential of individual frogs is so great that there is little or no effect of cancer on frog abundance. In a stable population, only two frogs will survive on average from each pair of adult frogs (throughout their entire reproductive life). If the female has 4 years of sexual maturity, only two of her 12,000 potential progeny (four times 3,000 oocytes per year) need survive. The number of frogs that live at a particular site is *not* due to an enormous reproductive potential but *is* related to the carrying capacity of the habitat. Thus, loss of one or more mature females (or males) with their huge reproductive potential would be compensated by the reproduction of non–tumor-bearing frogs.

What is the effect of the renal cancer in a declining population? As long as only a low percentage of adults manifests the malignancy, there probably is sufficient reproductive potential of unafflicted frogs to fill in for the loss of frogs with cancer. Thus, at the present time, there appears to be little about the Lucké renal adenocarcinoma to concern population biologists.

However, it should be stressed that nothing is known about why only a low percentage of frogs in nature manifest the malignancy. Probably many (most?) embryos and larvae in a breeding pond are exposed to the virus. There must be genetic and ecological factors that make certain frogs less susceptible to the oncogenic RaVH-1. If one or several of the environmental insults escalate (e.g., an increase in pesticide runoff or elevated UV-B radiation), and if these insults are involved in cancer susceptibility, then tumor prevalence may increase dramatically. Increased tumor prevalence at a time of population crisis might thus bode a catastrophe for frog populations. This has the potential for disaster because, as we stated earlier, frogs manifest the Lucké renal adenocarcinoma during their peak reproductive years. Until we learn more about what factors other than the etiological herpesvirus affect tumor vulnerability, we cannot become too sanguine concerning the potential danger of the Lucké cancer to frog populations.

Summary

The Lucké renal adenocarcinoma was the first cancer known to have a herpesvirus etiology. The herpesvirus has been detected both by electron microscopy and by PCR amplification of a restriction enzyme fragment of the herpesviral DNA. The herpesviral DNA has been sized by field inversion gel electrophoresis, and a large fraction of this DNA has been sequenced. While cold temperature inhibits metastasis, warm temperatures favor metastasis. The tumor occurs in wild populations of northern leopard frogs in Minnesota and Vermont and probably many intermediate sites. The prevalence of the malignancy of the frog varies with season and between years. The competence for normal cell differentiation has been studied in mitotic progeny of the Lucké renal adenocarcinoma genome. The extensive normal reprogramming of the malignant genome is a model for differentiation therapy of other cancers. There is no evidence that the Lucké renal adenocarcinoma is a factor in frog population declines, but a change in ecological conditions could cause the tumor to exacerbate the decline in frog abundance.

Acknowledgments. The research of R.G.M. reported in this paper was supported in part by Grant Number 2675AR1 from the Council for Tobacco Research, U.S.A., Inc. We thank David Gartner, University of Minnesota Imaging Center, for his help in preparing illustrations.

EIGHTEEN

Malformed Frogs in Minnesota: History and Interspecific Differences

DAVID M. HOPPE

Sporadic reports of malformed amphibians are abundant in the literature, and these reports have been thoroughly reviewed prior to the recent "outbreak" of malformations (Van Valen, 1974) as well as in papers related to the current malformation phenomenon (Ouellet, 2000; Lannoo et al., 2003). Merrell (1969) reported finding limb malformations in northern leopard frogs *(Rana pipiens)* from a Minnesota site at a frequency of 14.8% during late July 1965. There were no subsequent Minnesota reports until 1993, when residents near Granite Falls, Minnesota, reported "large numbers" of abnormal leopard frogs exhibiting extra limbs, missing limbs, and a missing eye (Helgen, 1996). Investigators did not find malformed frogs in that area during the summer of 1994, nor did the residents report additional abnormal animals. Major media attention first focused on malformed Minnesota frogs in 1995 after a group of students reported abnormal leopard frogs near Henderson, Minnesota (Helgen et al., 1998). The frequency of malformations was about 33%, with a more serious array of defects reported. These defects were described at that time as "webbed" leg skin, "bony protrusions," "contorted legs," missing legs, extra legs, and missing eyes, as illustrated in Helgen et al. (1998). Newspaper coverage of these frogs led to more reports by citizens in 1995, including a site near Litchfield where more than 90% of the leopard frogs were missing limbs or parts of limbs. One of those citizen reports came to me from a landowner near Brainerd; that site subsequently became an important study site coded "CWB," to which I refer extensively in following sections.

I joined the Minnesota investigations in 1993 as a herpetological consultant, and I began field surveys in 1994 that continue today. Here I report findings relative to two questions about frog malformations: (1) Is this anything new? Because deformed frogs have both an anecdotal and a literary history, are the types of malformations or their frequencies any different in the 1990s? (2) Which anuran species are affected and are they affected differently?

Historical Comparisons

Literature Comparisons

Interviews with David Merrell of the University of Minnesota revealed that his 1965 findings were quite different from what we were seeing in the 1990s. Merrell saw only two types of malformation, ectromelia and "deformed toes" (Merrell, 1969); I have seen up to 12 different types of malformation in single collections from my intensive sites. Other differences also exist. The 1965 malformation frequency decreased as the season progressed, from 14.8% in July to 3.6% in September; at sites where I am able to do multiple surveys throughout the season, the malformation frequency increases later in the season. Merrell did not see a recurrence of malformations in subsequent years, while my intensive sites have had numerous malformations during 5 consecutive years.

Museum Comparisons

I also made historical comparisons in two other ways (reported in Hoppe, 2000). One was to examine northern leopard frogs in the collections of the Bell Museum of Natural History, University of Minnesota, particularly a large set of frogs collected from 1958–63 by David Merrell and his collaborators. Only 18 abnormal frogs were found among 2,433 juvenile museum specimens examined in collections representing 43 sites in 26 counties, a "background frequency" of 0.7% abnormalities. These consisted of 12 predator amputations (0.5%) based on scar tissue and 6 malformations (0.2%). From anecdotal information and a few published findings, investigators have been using "about 1%" as a background level of frog abnormalities. Read and Tyler (1994) reported "natural levels of abnormalities" in *Neobatrachus centralis* to be about 1.6%. Ouellet et al. (1997a) reported 0.7% malformations among four ranid species sampled at pesticide-free sites in Canada. Gray (2000b) reported 0.39% abnormalities among nearly 10,000 northern cricket frogs *(Acris crepitans)* in Illinois surveys from 1968–71. The 0.2% historical frequency of malformations in northern leopard frogs based on the Bell Museum collections is lower than other published estimates, and it is considerably lower than the arbitrary "about 1%" value. Literature reports often do not distinguish age categories and, hence, are hard to compare. Because malformations are scarce among adult frogs regardless of the frequency among juveniles (Helgen et al., 2000), it is best to exclude adults when reporting malformation frequencies or to report them separately.

TABLE 18-1

Malformed *Rana pipiens* from 1958–63 Museum Collections Compared to 1996–99 Field Surveys of Juvenile Frogs at the Same Sites

County	Site	Museum (1958–63)			Field (1996–99)		
		N	# Malformations	% Malformations	N	# Malformations	% Malformations
Big Stone	1	85	0	0.0	129	1	0.8
Big Stone	2	35	0	0.0	120	4	3.3
Douglas	3	40	0	0.0	85	1	1.2
Grant	4	50	0	0.0	24	0	0.0
Traverse	5	66	1	1.5	244	9	3.7
Crow Wing	6	161	1	0.6	47	2	4.3
Rice	7	55	0	0.0	37	0	0.0
Becker	8	32	0	0.0	35	0	0.0
Pope	9	109	0	0.0	75	2	2.7
Total		633	2	0.3	796	19	2.4

From 1996–99, I surveyed many of the same sites represented by the Merrell collections. I was able to find frogs at only 5 of the 14 sites visited in 1996–97. Data from these sites are presented and discussed in Hoppe (2000). Visits to nine more sites during 1998–99 yielded frogs at four sites. It is noteworthy that only 39% of Merrell's original sites surveyed from 1996–99 still support northern leopard frog populations. The combined survey data from 1996–99 are presented in Table 18-1. The 0.3% malformation frequency among museum frogs from these sites is representative of the overall museum malformation frequency of 0.2%. However, the follow-up survey malformation frequency of 2.4% is significantly higher. To account for both variation between sites and through time, homogeneity of association across sites was tested by the Breslow-Day procedure, and differences across time were tested using the Cochran-Mantel-Haenszel procedure to combine evidence of association across sites (Agresti, 1996). The Breslow-Day test did not reveal evidence of different risks across sites (p = 0.883), while the Cochran-Mantel-Haenszel procedure revealed a significant difference in malformation frequencies in recent years compared to museum collections (p <0.01). This difference does not come from finding any particular "hot spots" of malformation comparable to the original Henderson site or CWB, but from finding a few malformations at most sites, at frequencies ranging from 0.8–4.3%.

Qualitatively, the array of malformation types observed in museum frogs is a small and unrepresentative sample of the northern leopard frog malformations as seen in recent years. The specific malformations seen in museum frogs are listed in Table 18-2. Only four broad categories of malformations were seen in museum frogs: ectrodactyly, syndactyly, polydactyly, and ectromelia. A broader array, including more severe malformations, has been observed in recent years (Hoppe, 1996; Helgen et al., 1998). The two most common malformations in my 1996–99 leopard frog surveys are hindlimb cutaneous fusions and hindlimb anteversions, neither of which was found in museum frogs. Other malformation categories found in significant frequencies in 1996–99 (but never in museum frogs) include split limbs, missing eyes, malformed jaws, forelimb ectromelia, and hip displasia. Such observations may be the strongest evidence that malformations are a different phenomenon now than in the past.

TABLE 18-2

Types of *Rana pipiens* Anomalies Among Abnormal Museum Frogs (Including Likely Predator Amputations and Malformations)

Category of Anomaly	# Abnormal Frogs	# Malformed Frogs
Hindlimb ectrodactyly	7	3
Hindlimb syndactyly	1	1
Hindlimb ectromelia	8	3
Forelimb ectrodactyly	3	0
Forelimb syndactyly	1	1
Forelimb polydactyly	1	1
Forelimb ectromelia	1	1

NOTE: Four frogs had two malformations each (from Hoppe, 2000).

West-Central Minnesota Field Comparisons

The other type of historical comparison reported in Hoppe (2000) compared historical data to recent northern leopard frog survey data. These data were from counties I had surveyed during my northern leopard frog pattern polymorphism research (see McKinnell et al., this volume). Historical data are those collected prior to 1993; recent data are post-1995 surveys. Nine sites having sample sizes of n >50 in both pre-1993 and post-1995 surveys were compared in that study. In the historical polymorphism surveys, I recorded obvious abnormalities such as hemi- and ectromelia, but I did not examine for scar tissue or count digits as in the recent malformation research surveys. Hence, rather than malformations, I compared total abnormalities in these data and excluded digital abnormalities. Historical surveys noted only three abnormal frogs among 1,772 juveniles examined (0.2%), while recent surveys noted 59 abnormal frogs among 2,548 examined at those same sites (2.3%). The Breslow-Day test for site differences was not significant (p = 0.104), while the Cochran-Mantel-Haenszel test confirms significant temporal increases in these west-central Minnesota counties (p <0.0001). Again, no Henderson- or

TABLE 18-3
Malformation Frequencies Among Anuran Species at Site CWB, 1996–99

Species	N	# Malformed	% Malformed	Degree aquatic[a]
Mink frog *(Rana septentrionalis)*	695	473	68.1	++++++++++
Green frog *(R. clamitans)*	219	80	36.5	++++++++++
Northern leopard frog *(R. pipiens)*	1657	191	11.5	+++++++[b]
American toad *(Bufo americanus)*	119	8	6.7	+
Eastern gray treefrog *(Hyla versicolor)*	19	1	5.3	+
Wood frog *(R. sylvatica)*	24	1	4.2	+
Spring peeper *(Pseudacris crucifer)*	95	2	2.1	+

[a]Number of + is a semi-quantification of aquatic behavior in the species' life styles, including duration of embryonic and larval stages, feeding habitat, and overwintering habitat.

[b]This species often breeds and overwinters in two different bodies of water; in this population both activities are in the same water as with *R. septentrionalis* and *R. clamitans*.

CWB-like "hotspots" were uncovered, but more sites had at least a few abnormal frogs (at frequencies of 0.8–9.4%) to reveal a much higher risk of abnormality compared with historical data.

Qualitatively, the types of abnormalities historically occurring versus currently occurring are similar to the museum findings described above. The only types of abnormality recorded before 1993 were hindlimb hemimelia and ectromelia (although I had recorded one frog pre-1993 with hindlimb polymelia at a site that no longer had frogs in post-1995 surveys). Again, the abnormality categories of hindlimb cutaneous fusions, hindlimb anteversions, split limbs, missing eyes, malformed jaws, and forelimb ectromelia were seen after 1995 but not before 1993.

I concluded from these described museum and field findings that "recent findings of anuran abnormalities in Minnesota do represent a new phenomenon. Frog abnormalities were more frequent, more varied, more severe, and more widely distributed in 1996–1999 than in 1958–1992" (Hoppe, 2000).

Interspecific Differences

Malformation Differences

All the confirmed reports of malformed anurans in Minnesota in 1965 (Merrell, 1969) and 1995 (Helgen et al., 1998) were northern leopard frogs. I immediately became interested in whether other species were also affected here, and if so, to what comparable degrees. The Minnesota surveys in 1995 had already revealed considerable frequency variation between sites for malformed leopard frogs; a high degree of site variation for many species is also currently being shown on the NARCAM (1997) reports. With so much site variation for individual species, the only valid way to compare different species may be where they occur in the same anuran community, sharing resources. Minnesota does not have a diverse anuran fauna— only 14 species (Oldfield and Moriarty, 1994). It does, however, have habitat diversity, as the state includes tallgrass prairie, deciduous forest, coniferous forest, and their respective ecotones. With so few species spread among many habitat types, it is unusual for me to encounter a breeding pond having as many as three species. Within a month of surveys at the CWB site in

central Minnesota, I realized that as many as nine species may be breeding there in the same small lake, presenting a good opportunity for interspecific comparisons. By the end of the 1996 season I had data on seven species.

The CWB anuran species and their frequencies of malformation from 1996–99 are summarized in Table 18-3. What became obvious was a correlation between the occurrence of malformations and the aquatic habits of various species (as shown in the last two columns of Table 18-3). Two highly aquatic species, mink frogs *(R. septentrionalis)* and green frogs *(R. clamitans)* had the highest malformation frequencies (68.1% and 36.5%, respectively). Chi square tests reveal these frequencies to be significantly higher than all other species, as well as different from each other (p <0.01). These two ranid species spend much of the summer in the water, venturing only a jump or two onto land in feeding activities, and they then overwinter in the same water. Both species breed late (June and July), spend the first winter as larvae, and do not metamorphose until the following summer. Northern leopard frogs are less aquatic than mink frogs and green frogs in that they become highly terrestrial while feeding during the summer and early autumn. They also overwinter in the water. They breed early in the spring and their tadpoles usually complete metamorphosis in mid-summer of the same year. Their malformation frequency (11.5%) was significantly lower than mink frogs or green frogs. By comparison, American toads *(B. americanus)*, eastern gray treefrogs *(Hyla versicolor)*, wood frogs *(R. sylvatica)*, and spring peepers *(Pseudacris crucifer)* are the least aquatic. They feed and overwinter terrestrially. Adults come to the water for a short time to breed, and the tadpoles reach metamorphic climax in a relatively short time during the same season. The collective malformation frequency among those four species was 4.7%, again significantly lower than in the highly aquatic mink and green frogs and semi-aquatic northern leopard frogs (p <0.01). Interestingly, although they were not among the severely malformed species, wood frogs were the first to show a drastic decline at CWB, disappearing altogether after the 1996 season.

The types of malformations found in different species at the same site are not strikingly different. Table 18-4 shows the eight most common types of malformations found in the three species having the most malformations (mink frogs, green frogs, and northern leopard frogs) for one year (1997) when capture success

TABLE 18-4

The Eight Most Common Malformations Found Among Three Species at CWB in 1997

	Mink Frog	Green Frog	Northern Leopard Frog	Total
Number with malformations	259	73	70	402
Percent with multiple malformations	44%	63%	23%	45%
	Number of Frogs Showing Indicated Malformation			
Cutaneous fusion (hindlimb[s] webbed)	160	51	53	264
Polymelia (extra hindlimbs)	98	39	7	145
Anteversions (bent/twisted hindlimbs)[a]	41	24	9	77
Split hindlimbs or feet[b]	25	5	7	37
Hindlimb truncations and hypoplasia[c]	20	8	2	30
Ectromelia (missing hindlimbs)	7	7	0	14
Polydactyly (extra digits)	8	3	3	14
Ectrodactyly (missing digits)	6	1	4	14

[a]Most of these are seen after clearing and staining as bony triangle growth (Gardiner and Hoppe, 1999).

[b]Clearing and staining reveals most of these to be extra limbs with proximal elements contained within same skin and (usually) muscle tissue, "splitting" distally (Gardiner and Hoppe, 1999).

[c]Pooling of truncated and vestigial hindlimbs and limbs missing elements proximal to digits (e.g. missing tibia-fibula segment).

was best for all three species. Cutaneous fusion (a webbing of skin behind the knee joint) was by far the most common malformation among all three species. Beyond those skin fusions, no particular pattern of malformation types is seen for leopard frogs at CWB, although pooling with other sites (Hoppe, unpublished data) reveals anteversions to be clearly the second most common in that species, and polymelia to be relatively uncommon. Interestingly, ectromelia is uncommon in CWB northern leopard frogs, but is the most common malformation at many other sites in that species (Helgen et al., 1998). Beyond skin fusions among mink and green frogs, polymelia and anteversions are clearly the next most common malformations, followed by progressively fewer of the other types.

It is difficult to grade the various malformation categories as to severity. The scarcity of most of these malformation types in adult frogs suggests high juvenile mortality, labeling all of those types as "severe." I have seen no cutaneous fusions, polymelia, anteversions, split hindlimbs, or underdeveloped hindlimbs among 128 adults of mink frogs, green frogs, and northern leopard frogs captured at CWB. As one attempt to quantify severity, I have looked at the number of frogs showing two or more different malformations in the 1997 data (Table 18-4). With that criterion, the aquatic mink frogs and green frogs are more "severely" affected than the semi-aquatic leopard frogs (p <0.05), showing increased risks of multiple malformation of 1.9% and 2.7%, respectively. None of the four "terrestrial" species described showed multiple malformations.

Interspecific differences in malformation frequencies and/or malformation types and the relationships to the species' life styles may provide clues to investigators seeking the causes of these malformations. Granted, all these species are 100% aquatic during the embryonic and larval stages during which the teratogenic effects must be expressed. But they differ in the dates when various species reach certain stages, so species may be affected differently because the teratogens are not present

continuously. The rates of embryonic and larval development differ between species, based both on water temperature and their own programmed timing of development. Thus, some may develop more slowly than others through critical stages of teratogenic exposure. Mink and green frog larvae typically spend five to six times longer in pre- and prometamorphosis than northern leopard frogs, hence their larvae have more time to accumulate potential teratogens. Adult frogs overwintering in the water have a long period of time to absorb and accumulate aquatic teratogens, and maternal transfer of causative agents (e.g., in the egg yolk) has not yet been ruled out.

Species Declines

In addition to the wood frog's disappearance, three other species (mink frogs, northern leopard frogs, and American toads) appear to have declined drastically at CWB. For example, newly metamorphosed northern leopard frogs became progressively more scarce in 1997–99, and increasing numbers of dead larvae and metamorphs were found in the lake (Souder, 2000). Studies are in progress to quantify those declines through nightly recordings of breeding choruses, direct counts of breeding adults, egg mass surveys, and catch-per-unit-effort estimators.

Notable Malformed Specimens

A feature of the CWB site that becomes interesting, yet quite discouraging, is the appearance of gross malformations and combinations of malformations that I have not seen at any other sites. Most of these notable oddities belong to mink frogs, samples of which are described and illustrated in Gardiner and Hoppe (1999). One mink frog showed eight different malformation types (displaced hip, extra pelvic bone, three extra legs,

A

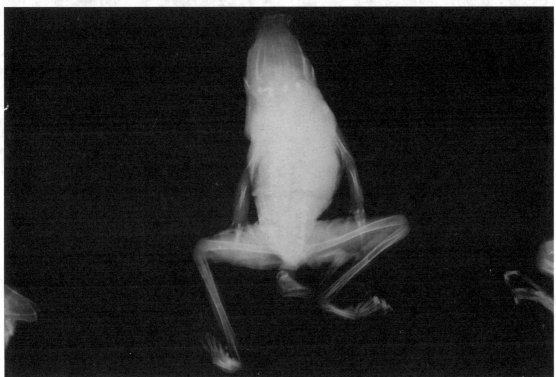

B

FIGURE 18-1 Juvenile northern leopard frog with supernumerary hindlimb fusing distally with the torso. (A) Live specimen, ventral view. (B) Radiograph of same specimen showing "split" at the ankle (courtesy of Michael Lannoo).

severe cutaneous fusions, one primary leg split into two feet, the other primary leg having an underdeveloped tibia-fibula, missing metatarsals, and only two digits (see cover photos, Journal of Experimental Zoology, 1 January 2000). A northern leopard frog had an extra hindlimb "splitting" from the primary limb at the ankle, then fusing with the torso in the groin area (Fig. 18-1). Another mink frog had an extra "lower unit" consisting of two limbs joined at an underdeveloped pelvis, with the whole structure dragging behind the frog while attached to the lower abdomen by a thin string of skin. One other notable mink frog had a mass of scrambled hindlimb elements and extra feet medial and posterior to the urostyle, but with

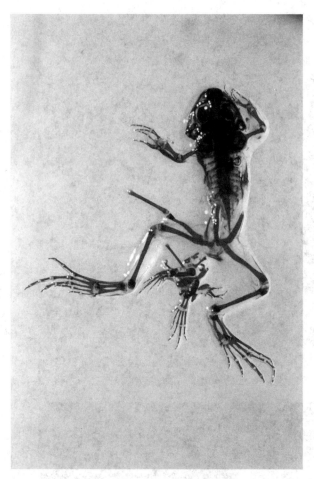

FIGURE 18-2 Juvenile mink frog having supernumerary pelvis with truncated hindlimb, and medial mass of hindlimb elements with extra feet, bony triangles, and amorphous bony growths; cleared and stained specimen.

I once attended a symposium on the behavioral significance of color where Jack Hailman was the keynote speaker. He closed his presentation with a slide of a drake wood duck *(Aix sponsa)* and remarked that if we ever understood the significance of all that brilliant plumage we might have some understanding of the significance of color in general. Perhaps if we ever understand what causes a frog like that in Fig. 18-2, we will have some understanding of malformed frogs in general.

Summary

Older literature, interviews, comparisons of museum specimens of northern leopard frogs with recent collections at the same sites, and comparisons of older and more recent northern leopard frog surveys in west-central Minnesota were used to determine whether the malformed frog phenomenon in the last 5 years is different from historical reports. These comparisons reveal that recent findings of anuran abnormalities in Minnesota do represent a new phenomenon. Frog abnormalities were more frequent, more varied, more severe, and more widely distributed in 1996–99 than in 1958–92. Four years of surveys at one central Minnesota site were used to reveal interspecific differences in malformation frequencies and types. Among seven anuran species breeding in the same pond, malformations were more frequent and varied among species with more aquatic lifestyles and became less frequent and varied as species' lifestyles were less aquatic. Mink frogs were the most severely affected using a combination of measures including the highest malformation frequency, the widest array of malformation types, a high percentage having multiple malformations, and the presence of particularly unique and gross malformations. Three of the species at this pond have noticeably declined and one species has disappeared.

Acknowledgments. Thanks to Erik and Larissa Mottl for field assistance and for helping to score thousands of pickled museum frogs. Able field assistance at CWB was provided by Dennis, Rhonda, Troy, and Brandon Bock, Matt Hoppe, Carlene Ness, and Bill Souder. Sean Menke cleared and stained the specimen in Fig. 18-2.

no articulation to the frog's primary skeleton (i.e., "satellite limbs"). Clearing and staining this specimen revealed an extra pelvic bone associated with a truncated supernumerary limb, as well as bony triangles and amorphous bony growths within the medial mass (Fig. 18-2).

NINETEEN

Parasites of North American Frogs

DANIEL SUTHERLAND

Every species of vertebrate has co-evolved with its own diverse parasite fauna. Harboring huge numbers of parasitic worms of several different species might seem to imply that the host's health is severely compromised and may even be fatal. After all, malaria and hookworms annually kill millions of people worldwide. Fortunately for most wild animal populations, especially those inhabiting relatively natural environments that have not been severely altered by humans, heavy parasite burdens are usually not detrimental to an individual host or host population. During co-evolution, parasite and host have "learned," through natural selection, to tolerate each other relatively well. Many host-parasite relationships are so innocuous as to probably best be classified as commensalistic. In my 25 years of experience, when an aquatic wild animal population harbors few parasites it is often an indicator of habitat degradation or alteration, where intermediate host populations (usually free-living aquatic invertebrates) have been reduced and life cycles of parasites are not completed.

I have two reasons to be interested in the parasites of anurans. First, as a source of instruction; frogs are often utilized in introductory parasitology courses because of the healthy diversity of parasites and the large numbers of worms and protozoans that an individual frog harbors. (Helminths, including trematodes, cestodes, and nematodes, are metazoan [multicellular] worm parasites; protozoans usually refer to unicellular organisms familiar to most introductory biology students as flagellates, ciliates, sporozoans, and amoebae.) Secondly, I am interested in the role trematode parasites may be playing in amphibian malformations, and, therefore, in amphibian declines. In this chapter, I will review the more frequently encountered parasites found in North American frogs.

Trematodes

The most visible and abundant parasites inhabiting North American anurans are trematodes. Both adult and larval trematodes routinely occur in frogs. Trematodes are often called *flukes* and are characterized by a complex indirect life cycle involving at least one (often more) intermediate hosts. The first intermediate host for a frog trematode is a mollusk (usually a snail, but less frequently a freshwater mussel). Many trematode species include a second, or even a third, intermediate host in their life cycle. Depending upon the species of trematode, frogs may serve as intermediate hosts harboring larval trematodes or definitive (or final) hosts, where adult flukes develop. When serving as an intermediate host, infected frogs transmit parasites to the definitive host when the frogs are eaten.

The typical life cycle of a trematode is as follows:

1. Because most adult trematodes inhabit the gastrointestinal tract of the definitive host, eggs typically exit in the host's feces.

2. A ciliated, free-swimming larva—the miracidium—hatches from the egg and penetrates a molluscan first intermediate host.

3. The miracidum develops into a simple, saclike larva—the sporocyst.

4. Within the sporocyst, a number of "germ balls" asexually produce multiple larvae—the rediae.

5. Rediae asexually produce hundreds to thousands of tailed cercariae that escape from the mollusk.

6. These short-lived cercariae must find the next intermediate host, which they actively penetrate and encyst as metacercariae within various tissues of the second intermediate host.

7. The life cycle is typically completed when second intermediate hosts are eaten by definitive hosts, where metacercariae are excysted and migrate to their final sites within the host to develop as egg-producing adult worms.

Adult trematodes are monecious (complete male and female reproductive systems develop in an individual worm). Being members of the phylum Platyhelminthes ("flatworms"), most flukes are dorsoventrally flattened and oval in shape. The typical adult trematode has two muscular suckers used for attachment. An anterior oral sucker surrounds the mouth and a powerful ventral acetabulum is located somewhere behind the oral sucker. The digestive system consists of a mouth, a muscular pharynx used for ingesting food, and a short esophagus that branches into two intestinal cecae. Trematodes do not have an anus and all undigested and unabsorbed materials within the ceceae must exit by way of the mouth. Adult trematodes feed on various host tissues and fluids that are ingested through the mouth. The remainder of the structures visible within the adult fluke are reproductive organs of which the most conspicuous are a single ovary, two testes, lateral fields of follicular yolk glands, and a highly convoluted uterus, which may contain hundreds to thousands of golden brown eggs. The average adult trematode found in frogs is between one and five millimeters long.

Adult Trematodes in Frogs

Haematoloechus are one of the most commonly encountered frog trematodes (Kennedy, 1981). Adult haematoloechids (Fig. 19-1F,G,H) are large inactive flukes that inhabit the lungs (Fig. 19-2A); various authors have described as many as 30–75 haematoloechids in one frog (Ulmer, 1970). The uterus is voluminous and obscures other internal organs. Placement of gravid flukes into tap water results in expulsion of thousands of eggs. Second intermediate hosts are dragonflies and damselflies (Odonata), which become infected when cercariae are swept into the branchial baskets of naiads (Krull, 1931, 1932; Schell, 1965). Metacercariae encyst on lamellae of the rectal gills and remain viable when naiads metamorphose to adults. Frogs and toads become infected when they ingest an infected adult. Numerous investigators have noted that lung flukes regularly ingest blood. Because frogs also respire through their skin and oral mucosa, the presence of large numbers of haematoloechids in their lungs appears to cause minimal harm.

Halipegus (Fig. 19-1I) reside in the eustachian tubes, pharynx, and occasionally the stomach of frogs. Several workers have reported that *Halipegus* occur infrequently, and rarely in large numbers. In Wisconsin, we find *Halipegus* more often in green frogs *(Rana clamitans)* than in other ranids. Their life cycle consists of embryonated eggs being eaten by a suitable snail host (usually *Physa* sp. and *Helisoma* sp.) in which sporocyst and redial generations develop. Cercariae are eaten by copepods or ostracods. Metacercariae develop in the hemocoel of these crustaceans but do not encyst. Infected crustaceans may be eaten by either tadpoles or dragonfly naiads, and adult flukes develop when infected tadpoles metamorphose or when frogs ingest infected dragonflies (Thomas, 1939).

Two genera of gorgoderid flukes (Fig. 19-1A,B,C) are routinely recovered from urinary bladders of frogs and newts (Salamandridae). Gorgoderids are active muscular worms. *Gorgodera* possess 9–11 testes arranged in two rows; *Gorgoderina* possess only two testes. Fingernail clams (Pelycypoda) inhale miracidia through incurrent siphons. Cercariae (produced by daughter sporocysts in the gills) escape by way of the excurrent siphon and are ingested by tadpoles and crayfish (Crustacea). Adult frogs acquire infections when they ingest infected second intermediate hosts (Rankin, 1939; Goodchild, 1948).

Individuals of *Megalodiscus temperatus* are common inhabitants of the rectums of various frogs and toads. This thick, inactive fluke (Fig. 19-1D) is characterized by its ventral sucker, which is located at the posterior end of the worm and is larger in diameter than any other portion of the worm. Their life cycle includes a miracidium penetrating a suitable snail host, development of a sporocyst generation, and redial generation. Cercariae encyst on the epidermis of tadpoles and frogs. Frogs become infected when they eat tadpoles or ingest bits of their sloughed epidermis (Krull and Price, 1932; Smith, 1967).

Glypthelmins quieta (Fig. 19-1J) occur in the intestine of various ranids. Physid snails are first intermediate hosts for mother and daughter sporocysts. Cercariae encyst in the skin of frogs and enter the frog intestine when bits of epidermis are shed and ingested (Rankin, 1944; Leigh, 1946; Schell, 1962). Other species of *Glypthelmins* (and the closely related *Hylotrema*) infect treefrogs (Hylidae) and apparently have life cycles where cercariae penetrate to the coelom of tadpoles and become unencysted metacercariae (e.g., Brooks, 1976a). They remain in the coelom until metamorphosis, at which time metacercariae migrate to the adult's intestine (Schell, 1985).

Cephalogonimus also inhabit the intestines of frogs of the genus *Rana*. Embryonated eggs are ingested by snails (*Helisoma* sp). Cercariae then develop in daughter sporocysts in the hemocoel. After leaving the snails, cercariae penetrate and encyst in tissues of tadpoles. Adult flukes develop in the intestine when infected tadpoles are eaten (Lang, 1968).

Brachycoelium salamandrae have been frequently reported to be in the intestines of salamanders and frogs. The unique life cycle of these worms involve eggs being eaten by terrestrial snails. As tailless cercariae leave the snail, they secrete a gelatinous capsule around themselves. Metacercariae develop in land snails that eat these encapsulated cercariae (Jordan and Byrd, 1967).

Bunoderella metteri was described from larvae and adults of tailed frogs, *Ascaphus truei*, in the Pacific Northwest (Schell, 1964). *Bunoderella* belongs to a group of flukes called the papillose allocreadids, which possess muscular papillae lateral and dorsal to the oral sucker. Cercariae develop in rediae in the fingernail clam *Pisidium idahoensis*. Caddisfly larvae and pupae and chironomid larvae serve as hosts for metacercariae (Anderson et al., 1965).

Two genera of lecithodendriids occupy unique locations within frog hosts. Adult *Loxogenoides* occur in the lumen of the bile duct, and adult *Loxogenes* are usually encysted in the intestinal or gastric wall in the region of the pyloric sphincter. Metacercariae of *Loxogenes* encyst in abdominal muscles of numerous species of dragonfly naiads, which are then eaten by frogs acting as definitive hosts (Schell, 1985).

Metacercariae in Frogs

The presence of trematode metacercarial stages in frogs has received increased attention during the past decade with the advancement of a hypothesis that metacercariae of one or several flukes may be responsible in part for occurrence of malformed frogs in North America. In ponds in northern California, Sessions and Ruth (1990) found a positive correlation between the presence of supernumerary limbs in Pacific treefrogs *(Pseudacris regilla)* and long-toed salamanders *(Ambystoma macrodactylum)* and metacercariae encysted in or near affected limbs. The working hypothesis is that encystment of metacercariae within the limb bud might mechanically (or alternatively by secretion of some biochemical product) partition the limb bud into extra limbs or disrupt the limb bud sufficiently to prevent or halt limb formation (see Johnson and Lunde, this volume). In laboratory studies, Sessions and Ruth (1990) were able to elicit malformations in tadpoles of Pacific treefrogs and African clawed frogs *(Xenopus laevis)* by placing resin beads the size of metacercariae into limb buds. One species of trematode metacercaria found by Sessions and Ruth (1990) in malformed treefrogs from northern California was identified as *Manodistomum*. A second type could not be identified. *Manodistomum* mature in

Trematodes of Frogs

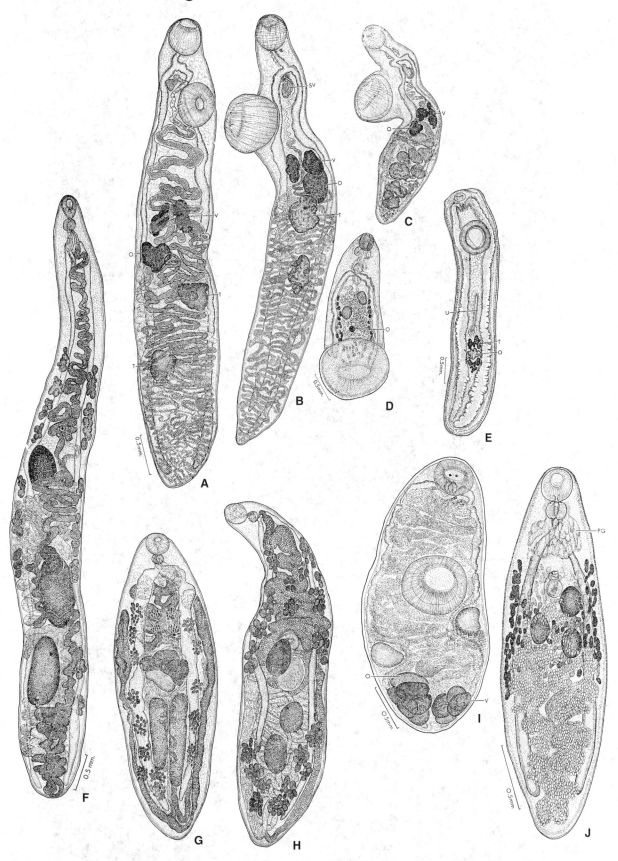

FIGURE 19-1 Trematodes found in North American frogs. **(A)** *Gorgoderina simplex*, **(B)** *Gorgoderina attenuata*, **(C)** *Gorgodera amplicava*,
(D) *Megalodiscus temperatus*, **(E)** *Clinostomum attenuatum* (metacercaria), **(F)** *Haematoloechus medioplexus*, **(G)** *Haematoloechus longiplexus*,
(H) *Haematoloechus similiplexus*, **(I)** *Halipegus* sp., **(J)** *Glypthelmins quieta*. (From Ulmer, Martin J. (1970). Studies on the Helminth Fauna of Iowa.
I. Trematodes of Amphibians. The American Midland Naturalist 83(1):38–64, with permission from University of Notre Dame).

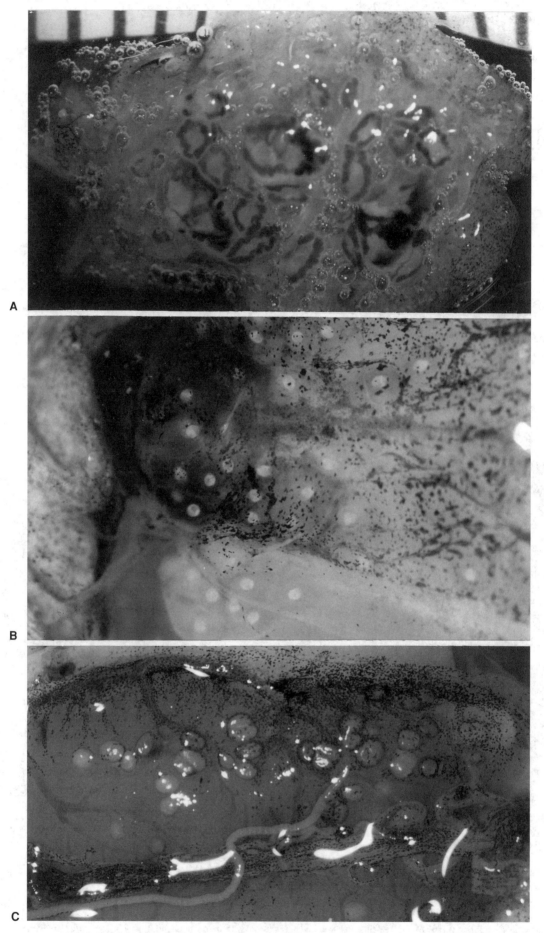

FIGURE 19-2 **(A)** Thirty-six *Haematoloechus* in one lung of a Wisconsin leopard frog. **(B)** *Fibricola cratera* metacercariae in posterodorsal trunk area of Minnesota leopard frog. **(C)** Tetrathyridial metacestodes of *Mesocestoides* in kidney of Wisconsin green frog.

the roof of the mouth and anterior portion of the esophagus of garter snakes (*Thamnophis* sp.) and use frogs as second intermediate hosts. Snakes become infected by eating tadpoles and frogs harboring metacercariae.

Recently, Johnson et al. (1999) reported the presence of *Ribeiroia* metacercariae in malformed Pacific treefrogs in northern California ponds, near where Sessions and Ruth (1990) sampled. To test the hypothesis that trematode infections induce abnormalities, Pacific treefrog tadpoles were exposed to *Ribeiroia* cercariae at varying concentrations. Metamorphic animals exhibited severe limb malformations, including extra and missing limbs. Increased dosages of cercariae resulted in an increased frequency of abnormalities and increased mortality of metamorphic frogs. To establish controls, Johnson and his team exposed Pacific treefrog tadpoles to cercariae of *Alaria* (another trematode found in malformed frogs) from their ponds. No malformations occurred even at cercarial dosages considerably heavier than those used with *Ribeiroia*. Johnson recently elicited similar deformities in western toads (*B. boreas)* when they were exposed to *Ribeiroia* using a similar experimental protocol. During the summer of 1999, Johnson and Lunde visited 103 ponds in six northwestern states (California, Oregon, Washington, Idaho, Colorado, and Minnesota) including 42 ponds where deformities ranged from 5% to 90% in six anuran species. Only two of the 42 ponds where deformities were observed did not harbor *Ribeiroia*; ponds where no deformities were observed almost never had *Ribeiroia* (Johnson and Lunde, this volume).

In nature, *Ribeiroia* occur as adults in the intestines of fish-eating and frog-eating birds and mammals, including common loons, bald eagles, ospreys, brown pelicans, hawks, herons, badgers, and rats (Table 19-1). In our laboratory, we have successfully fed *Ribeiroia* metacercariae from Minnesota, Montana, California, and Washington frogs to surrogate hosts (chicks, ducklings, pigeons, and rats) and obtained what appear to be the type species for the genus *Ribeiroia ondatrae* (Fig. 19-3A). We are currently evaluating the taxonomic validity of *Ribeiroia marini, R. insignis,* and *R. thomasi* (Fig. 19-3B). *Ribeiroia ondatrae* was described as *Psilostomum ondatrae* by Price (1931, 1942) from the liver of muskrat (*Ondatra zibethicus)* and proventriculus of California gulls (*Larus californicus*; Table 19-1). Even Price seemed to have reservations about the location of the type material in muskrat liver since he felt obliged to add the notation "according to the label" after the type host. The most distinguishing morphological feature of *Ribeiroia* is a pair of lateral esophageal diverticulae. These diverticulae are present in cercariae, metacercariae (Fig. 19-4B), and adult worms (Fig. 19-3B). The genus *Trifolium* is closely related to *Ribeiroia* in that it also possesses esophageal diverticulae. *Trifolium* contains two species both from anhingas (*Anhinga anhinga)* in Cuba and Brazil. *Trifolium* and *Ribeiroia* differ only in location of the testes and anterior distribution of the vitellaria (Yamaguti, 1971).

Ribeioria seem to exhibit low host specificity throughout their life cycle. In addition to being found in a wide range of avian and mammalian definitive hosts (Table 19-1), *Ribeioria* seem to occur in virtually all amphibians that inhabit ponds where *Ribeioria* occur (Johnson and Lunde, this volume). According to Beaver (1939), metacercariae of *R. ondatrae* are normally found in the lateral line canal, and to a lesser extent under the scales, of freshwater fish. In addition, Beaver found *R. ondatrae* metacercariae in the nostrils of tadpoles and fish and occasionally in the cloaca and "associated ducts after prolonged exposure to heavy suspensions of cercariae." Huizinga and Nadakavukaren (1997) demonstrated that fish mobilize an active inflammatory response that

seems to eliminate *R. marini* metacercariae from the lateral line canals of laboratory exposed goldfish. Snail hosts for *Ribeiroia* include helisomes in the United States and *Biomphalaria glabrata* in the Caribbean (Nassi, 1978).

In 1997, I became interested in frog metacercariae from the Upper Mississippi River Valley after reading about Sessions and Ruth's (1990) parasite hypothesis in the popular press. Following three field seasons during which we examined frogs from a large number of sites throughout Minnesota and Wisconsin, we have identified an exceptionally diverse fauna of trematode metacercariae consisting of at least ten species. The presence of so many trematodes was not unexpected because the Mississippi River Flyway provides for optimal dispersal of flukes into and out of western Wisconsin and eastern Minnesota.

Metacercariae were frequently encountered in the viscera, mesenteries, muscles, coelom, and beneath the skin of local frog hosts (Figs. 19-5, 19-6). Northern leopard frogs *(R. pipiens)* were the anuran host most commonly found to harbor metacercariae.

Ribeiroia (Figs. 19-5C, 19-7) are found sporadically throughout Minnesota and Wisconsin, and in our surveys they occur in both malformed and normal animals. At one Minnesota pond (CWB) that was notorious for frog malformations and population declines, we observed massive infections of *Ribeiroia* metacercariae in association with satellite limbs (Fig. 19-7A, B) and along the mandibles (Fig. 19-7C, D) of malformed mink frogs *(R. septentrionalis)*. *Ribeiroia* metacercariae were so abundant that packets of encysted metacercariae tumbled from the inguinal region as the skin was removed (Fig. 19-4A). At another Minnesota pond (TRD), which has recently been identified as harboring frogs with severe limb malformations (including both supernumerary limbs and limb truncations), no *Ribeiroia* were recovered. Toxicologists from the National Institute of Environmental Health Sciences have recently isolated compounds from CWB water and sediments, which is suggestive that toxic and/or bioactive agents are responsible for frog malformations (Fort et al., 1999a,b; Lannoo et al., 2003).

Clinostomum attenuatum (Fig. 19-1E) are frequently called "yellow grubs." Because of their large size (cysts up to several millimeters in diameter) and bright yellowish color, they are conspicuous parasites when located just under a frog's skin. They also encyst at deeper locations, including the fascia of the cervical region, pharyngeal areas, mesenteries and viscera (Ulmer, 1970). These large metacercariae are progenetic in that most reproductive structures are well developed. Progenesis in migratory bird helminths is a common phenomenon and may represent an evolutionary trend that shortens the time between metacercarial recruitment into definitive hosts and worm egg production. This would allow reintroduction of the parasite back into the pond from which the parasite was recruited, prior to the bird host leaving the area. Adult *C. attenuatum* usually develop in the roof of the mouth of bitterns (Hunter and Hunter, 1934). *Clinostomum marginatum* metacercariae are morphologically similar to those of *C. attenuatum* (Hopkins, 1933; Price, 1938; Hoffman, 1999). The two species can be partially distinguished by the fact that metacercariae of *C. marginatum* apparently use fish as second intermediate hosts (where they cause "yellow grub disease") and also exhibit a wider definitive host specificity by infecting the oral cavities of herons, gulls, and bitterns.

Among the numerically most abundant metacercariae to occur in local frogs are those of *Fibricola cratera*. These parasites usually are located throughout the gastrocnemius and other muscles of the pelvic appendages and less frequently in the coelomic cavity and beneath the skin (Fig. 19-5B). We have

TABLE 19-1
Definitive Hosts for *Ribeiroia* and the Closely Related (Synonymous?) *Trifolium*

Parasite Species	Host Species	Locality	Citation
Ribeiroia ondatrae	Muskrat	Ontario	Price, 1931
	Ondatra zibethicus		
	California gull	Oregon	Price, 1931
	Larus californicus		
	Domestic chicken	Colorado	Newsom and Stout, 1933
	Gallus domesticus		
	Osprey	Pennsylvania	Beaver, 1939
	Pandion haliaetus		
		Virginia	Cole, personal communication
		Massachusetts	Cole, personal communication
	Cooper's hawk	Michigan	Beaver, 1939
	Accipter cooperii		
	Louisiana heron	Louisiana	Lumsden and Zischke, 1963
	Hydranassa tricolor		
	Common loon	Florida	Kinsella and Forrester, 1999
	Gavia immer		
	Little blue heron	Puerto Rico	Cable et al., 1960
	Egretta caerulea		
	Green heron	Puerto Rico	Riggin, 1956
	Butorides virescens		
	American egret	Brazil	Travassos, 1939
	Casmerodius egretta		
	Domestic chicken, duck, pigeon, canary	Experimental	Beaver, 1939
Ribeioria marini	Finches, canary, pigeon	Experimental, West Indies	Basch and Sturrock, 1969
Ribeioria "congolensis"	Goliath heron	Central Africa	Dollfus, 1950
	Ardea goliath		
	Common egret	Rhodesia	Mettrick, 1963
	Casmerodius albus		
	Hammerkop	Rhodesia	Mettrick, 1963
	Scopus umbretta		
	Giant kingfisher	Rhodesia	Mettrick, 1963
	Megaceryle maxima		
Ribeioria sp.	Osprey	Illinois	Cole, personal communication
	P. haliaetus		
Trifolium trifolium	Anhinga	Brazil	Travassos, 1922
	Anhinga anhinga		
T. travassosi	Anhinga	Cuba	Perez Vigueras, 1940
	A. anhinga		

found several metamorphic northern leopard frogs that each harbored in excess of 1,000 *F. cratera* metacercariae. Hoffman (1955) used a pepsin digest technique and recovered 3,550 metacercariae from one northern leopard frog. Locally, adult *F. cratera* have been recovered from intestines of mink, river otter, and raccoon. Cercariae develop in physid snails. Cercariae penetrate into tadpoles and migrate to the coelom where they develop into diplostomula-type metacercariae that actively crawl over the viscera. After frog metamorphosis, diplostomula migrate to pelvic muscles and encapsulate (Chandler, 1942). At

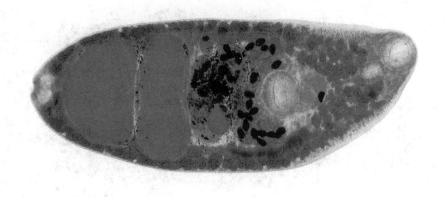

A

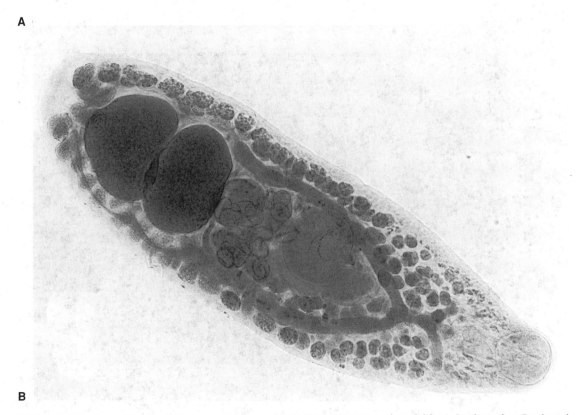

B

FIGURE 19-3 **(A)** Gravid *Ribeiroia ondatrae* from pigeon surrogate host fed metacercariae from California Pacific treefrog. Esophageal diverticulae obscured by abundant yolk glands. **(B)** Gravid *Ribeiroia thomasi* from Wisconsin great-horned owl. Note esophageal diverticulae. Specimen courtesy of Dr. Stephen Taft, University of Wisconsin-Stevens Point.

La Crosse area sites, northern leopard frogs also routinely harbor an unidentified diplostomulum (Fig. 19-8B) in subdermal yellowish packets on the basal elements of the middle toes of the hind feet. It is not unusual to find over a hundred of these diplostomula in a communal connective tissue sheath on the toe dorsum. When the sheath is ruptured, these active metacercariae spill out and writhe about repeatedly elongating and contracting their bodies.

Another strigeid that infects frogs from the Upper Mississippi River Valley is *Apharyngostrigea pipientis*. Tetracotyle metacercariae of this fluke occur in mesenteries supporting the liver, intestine, urinary bladder, and just under the hepatic visceral peritoneum (Hughes, 1928; Olivier, 1940). Ulmer (1970) found that Iowa northern leopard frogs frequently harbor grape-like clusters of *A. pipientis* metacercariae near the urinary bladder. In Minnesota frogs we found *A. pipientis* in viscera and muscles.

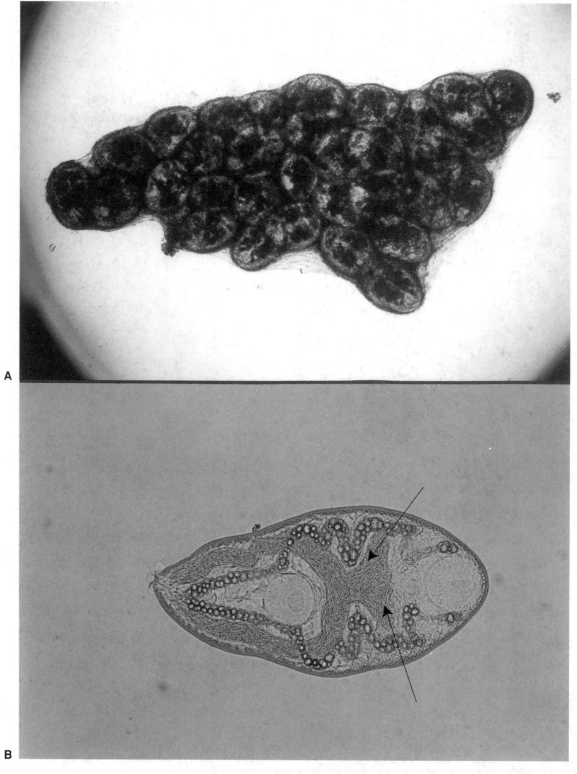

FIGURE 19-4 **(A)** Packet of *Ribeiroia* metacercariae from inguinal region of a mink frog from CWB. **(B)** Excysted *Ribeiroia* metacercaria from a Montana Pacific treefrog *(Pseudacris regilla)*. Note esophageal diverticulae *(arrows)*.

One mink frog harbored 163 of these extremely thick-walled metacercariae (Fig. 19-2B) distributed throughout the body. Natural definitive hosts appear to be herons and ibises.

Alaria is a genus of strigeid trematode that utilizes felids, canids, and mustelids as definitive hosts (Schell, 1985). We have obtained gravid adults from the intestines of red fox, mink, raccoon, and skunk. Planorbid snails serve as first intermediate hosts. Fork-tailed cercariae penetrate and develop as unencysted mesocercariae in muscles and fascial connective tissues of tadpoles and frogs. In local northern leopard frogs, we almost always find *Alaria* (Fig. 19-8A) mesocercariae located in intramuscular fascia at the distal end of the femur. *Alaria* frequently makes use

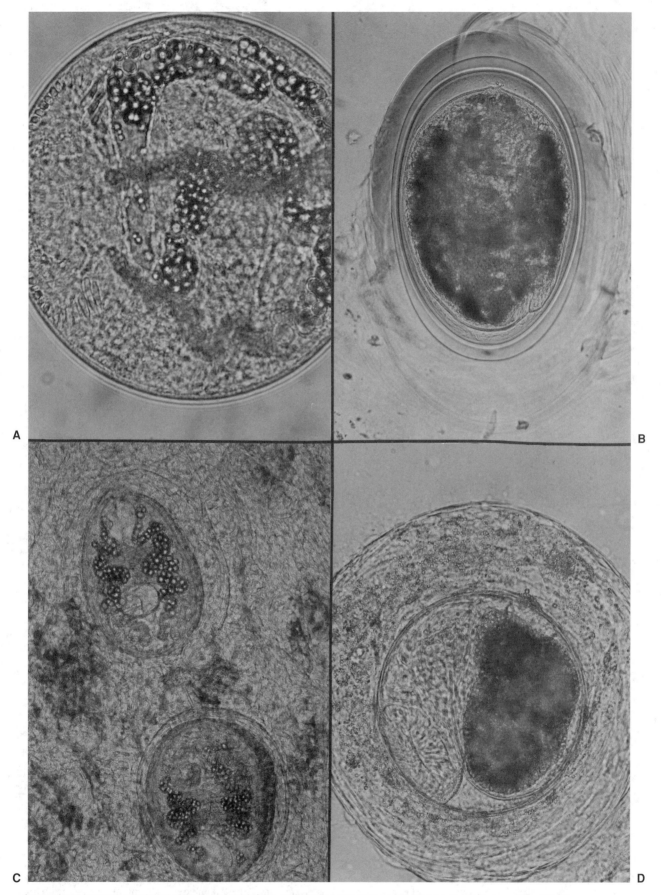

FIGURE 19-5 Metacercariae found in North American frogs. (A) Encysted echinostome, (B) Tetracotyle of *Apharyngostrigea pipientis*, (C) Encysted *Ribeioria* sp., (D) Encysted Meta A.

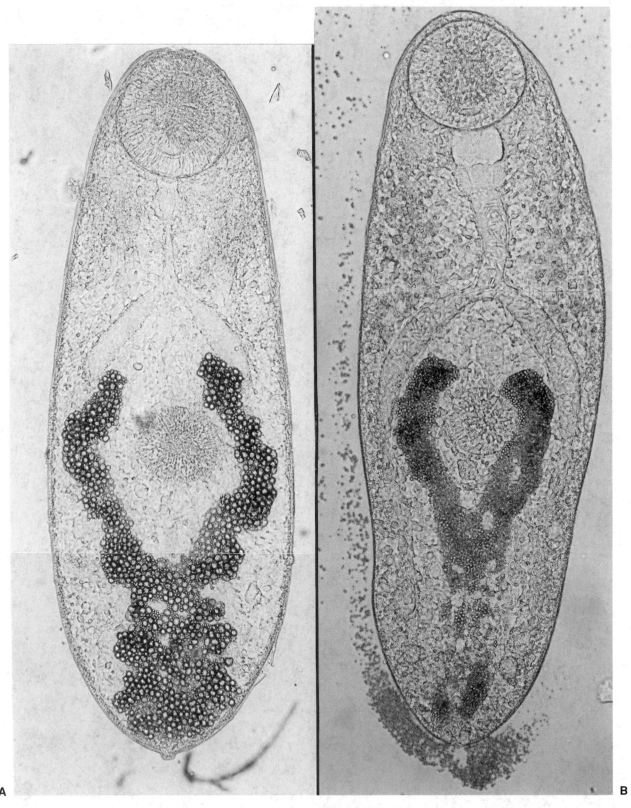

FIGURE 19-6 Metacercariae found in North American frogs. (A) Ochetosomatid. (B) Meta B.

of small rodent and snake paratenic (carrier) hosts, in which no further development of the parasite occurs. Paratenic hosts can recruit massive numbers of mesocercariae and fat bodies from many local snakes can be riddled with thousands of *Alaria* mesocercariae. Other common paratenic hosts for mesocercariae include various rodents. When suitable definitive hosts ingest mesocercariae, flukes migrate to the lungs and develop into metacercariae. After a suitable encystment, flukes migrate up the respiratory tree and are swallowed (Pearson, 1956). Shoop and Corkum (1987) demonstrated that transmammary infection can

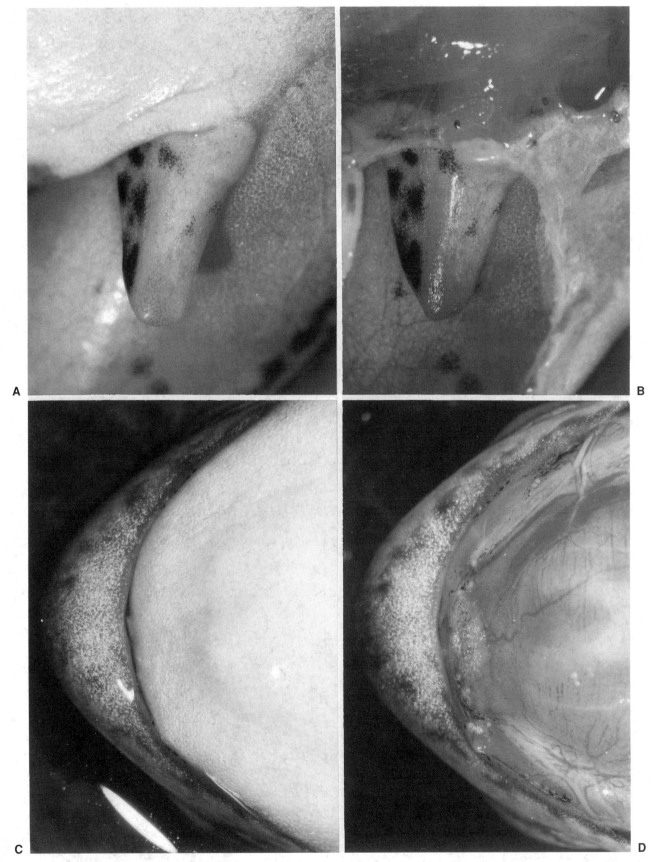

FIGURE 19-7 *Ribeiroia* metacercariae associated with satellite appendage of mink frog from CWB. **(A)** Before removal of skin. **(B)** After removal of skin. *Ribeiroia* metacercariae associated with abbreviated mandible of mink frog from CWB. **(C)** Before removal of skin. **(D)** After removal of skin.

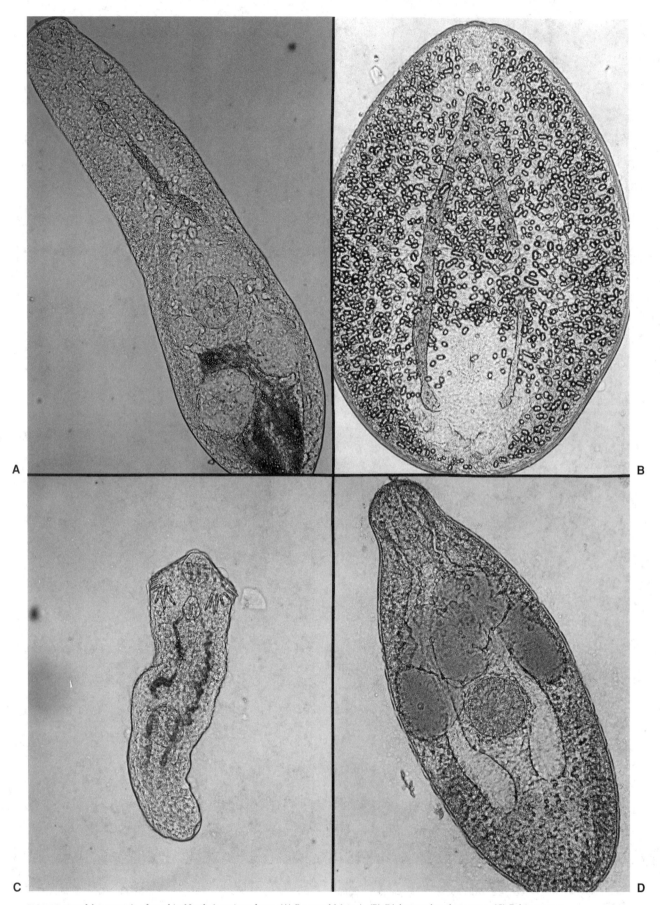

FIGURE 19-8 Metacercariae found in North American frogs. (A) Excysted Meta A. (B) *Diplostomulum* from toes. (C) Echinostome. (D) *Alaria* mesocercaria.

occur when kittens suckle on females that harbor mesocercariae in their mammary glands. Successful transfer of mesocercariae from a lactating marmoset (primate) to her offspring has also been demonstrated (Shoop et al., 1990).

While not yet recovered from local frogs, *Diplostomum micradenum* metacercariae were reported by Olivier (1940) to be common parasites of northern leopard frog and American toad (*B. americanus*) tadpoles and northern leopard frog adults from Michigan. The metacercariae are unencysted and occur within cavities of the brain and spinal cord or under the meninges.

Metacercariae of a heterophyid trematode in the genus *Euryhelmis* encyst in the skin and subcutaneous connective tissue of tadpoles and frogs of the genus *Rana* and *Ascaphus*. Natural definitive hosts for these uniquely rectangular-shaped flukes are raccoons and mink (Ameel, 1938; Anderson and Pratt, 1965). *Metagonimoides oregonensis* is another heterophyid of raccoons that uses frogs as an intermediate host. Cercariae develop in operculate snails. Cercariae can either remain in rediae and develop into infective metacercariae or they can leave the rediae and snail and encyst in frogs (Ingles, 1935; Burns and Pratt, 1953; Lang and Gleason, 1967).

Ochetosomatids (Fig. 19-6A) are common metacercariae in frogs from our study sites where they infect primarily fascia binding skin to underlying musculature (Talbot, 1933; Sogandares-Bernal and Grenier, 1971). Ochetosomatids are a family of flukes that are found as adults in the mouth, esophagus, and lungs of snakes. Sessions and Ruth (1990) found ochetosomatid metacercariae associated with leg deformities in Pacific treefrogs and long-toed salamanders from northern California. These metacercariae were identified as belonging to the genus *Manodistomum* (now considered a synonym of *Paralechriorchis*; Schell, 1985). The life cycle of this fluke involves physid snail first intermediate hosts and garter snake definitive hosts (Ingles, 1933b). Garter snakes from our study sites routinely harbor gravid *Paralechriorchis* and *Ochetosoma* attached to the mouth and anterior reaches of the esophagus and hundreds of *Lechriorchis* in the lungs.

Green frogs from North-central Wisconsin harbor metacercariae of *Auridistomum chelydrae*, which are extremely common flukes in the intestines of snapping turtles (*Chelydra serpentina*). First intermediate hosts are *Helisoma trivolvis* (Sizemore, 1936; Ralph, 1938). In green frogs, these metacercariae are found in the gastrocnemius, foot, tail resorption site, and around the orbits of the eye.

Other metacercariae from local frogs have yet to be identified to genus and species. These include echinostomes that infect the kidneys of ranids. Frogs from several sites in Minnesota also harbored echinostomes in the tail resorption site. Echinostomes (Fig. 19-5A, 19-8D) are characterized by a circlet or two of peg-like spines mounted on an oral hood surrounding the oral sucker. Echinostomes are a frequent component of the trematode faunas of waterfowl and aquatic mammals (muskrats, otters, mink, etc.). Echinostomes, strigeids, and several other metacercariae from local frogs (for example, two metacercariae preliminarily identified only as Meta A [Fig. 19-5D] and Meta B [Fig. 19-6B]) will need to be fed to surrogate definitive hosts to obtain gravid worms in order to determine their identity. Finding adult worms in naturally infected definitive hosts is usually precluded by the fact that final hosts are often federally protected species such as bald eagles, loons, hawks, and pelicans.

Cestodes

Cestodes (or tapeworms) represent another group of flatworms that parasitize frogs. Cestodes from frogs are characterized by an elongated body (strobila) consisting of numerous attached segments (proglottids), with each proglottid containing both male and female reproductive systems. A holdfast structure (scolex) at the anterior end of the strobila bears four muscular suckers (acetabula). A neck region immediately posterior to the scolex buds off immature proglottids, which rapidly develop mature male and female reproductive systems. As eggs are produced, the uterus within each proglottid swells with eggs. By the time proglottids are located at the posterior end of the strobila they are completely filled with hundreds to thousands of eggs and are called gravid proglottids. Gravid proglottids either rupture by way of a medial suture to release eggs or are shed from the strobila to decompose and release eggs. All known life cycles for frog tapeworms require intermediate hosts for completion.

Cestodes of North American frogs are singularly few when compared with trematodes. Surveys of amphibian parasites often refer to a paucity of cestodes (Ulmer and James, 1976). When tapeworms are found in abundance in frogs, larvae are usually far more abundant than are adults. One of the most commonly encountered adult tapeworms from frogs in our locality is *Ophiotaenia saphena* from northern leopard frogs and green frogs. Less frequently, local frogs harbor larval tetrathyridia of *Mesocestoides* in the mesentaries, kidneys (Fig. 19-2C), and outside of the parietal peritoneum. *Mesocestoides* are common cestodes in raccoons, opossums, skunks, and several other carnivorous mammals (McAllister and Conn, 1990). Local garter snakes often harbor thousands of *Mesocestoides* tetrathyridia in mesenteric fat deposits. A number of Wisconsin northern leopard frogs were infected with plerocercoids, characterized by a well-developed apical organ on the scolex. Ulmer and James (1976) suggested that these plerocercoids may be larvae of *Ophiotaenia perspicua*, cestodes of garter snakes and water snakes (*Neroidea* sp.). The apical organ is hypothesized to be responsible for secreting lytic enzymes that allow plerocercoids to undergo parenteral migration within the second intermediate host and definitive host. Several nematotaeniid cestodes have been reported as adult parasites from North American anurans. Nematotaeniids are characterized by gravid proglottids with conspicuous parauterine organs.

Nematodes

Nematodes are commonly called *roundworms* (in contrast to trematodes and cestodes, which are *flatworms*). Nematodes are circular in cross section and are usually pointed at both ends. A mouth and muscular pharynx are found at the anterior end and an anus occurs at or near the posterior end. Nematodes are typically dioecious (having separate sexes in comparison to monecious flatworms). Nematodes are usually sexually dimorphic, with males often possessing copulatory spicules that can be everted into the female's vulva during copulation. Males of some nematodes have evolved a copulatory bursa (which resembles a baseball glove) at the posterior end, which allows the male to caress the female during copulation. Sensory papillae arranged in taxonomically important patterns often adorn the posterior end of male nematodes.

Rhabdias ranae and *R. bufonis* are ubiquitous nematodes in the lungs of Wisconsin frogs and toads, respectively. These stout roundworms have an interesting life cycle (Walton, 1929). The parasitic adult is a protandrous hermaphrodite (i.e., an individual is a functional male prior to becoming an adult female). Males produce and store sperm, and they then develop into females that produce ova that are fertilized with the stored sperm. Eggs embryonate as they are swept up the respiratory tree and swallowed. Eggs hatch in the intestine, and the juveniles accumulate in the frog rectum before being voided in feces. Away from host frogs, juveniles molt several times to produce a dioecious generation of free-living males and females that feed on soil bacteria. Ova produced by the free-living female hatch in utero, and the juveniles then consume their mother from within. Juvenile worms escaping from the mother's body frequently undergo developmental arrest (hypobiosis) until they come into contact with a frog or toad, which they then penetrate. Inside frogs, juveniles lodge in various tissues. Juveniles that reach the lungs develop into hermaphroditic adults, whereas the rest apparently die.

Several species of *Oswaldocruzia* taken from the intestines of various North American frogs and toads have been described (Walton, 1929, 1935). Midwestern frogs are frequently infected with *Oswaldocruzia*. These long (6-7 mm), slender worms are easily identified when a male is present, because of his conspicuous copulatory bursa and robust copulatory spicules, which characteristically branch distally into a number of stubby processes.

Cosmocercoides are common oxyuroid nematodes found in the rectums and intestines of frogs, other amphibians, and even reptiles (Walton, 1929; Harwood, 1930, 1932; Ogren, 1953). Oxyuroids are pinworms, and they all possess a bulbous expansion of the posterior end of the pharynx (esophagus). Other oxyuroids that routinely infect frogs in North America include *Cosmocerella*, *Aplectana*, and *Oxysomatium* (Skrjabin et al., 1961).

Gyrinicola batrachiensis are unique pinworms found in most tadpoles examined from eastern and central Canada (Adamson, 1980). Metamorphic frogs and toads do not appear to retain infections. The genus is distinguished by the presence of two egg types in females. Thick-shelled eggs containing embryos in an early stage of cleavage are produced by one ovary; thin-shelled eggs containing juveniles are produced by the other ovary. Thick-shelled eggs transmit these parasites to other frogs when these eggs leave in feces and are ingested by tadpoles. Thin-shelled eggs are autoinfective, in that they hatch within the gastrointestinal tract of the tadpole in which they were produced and substantially increase worm numbers in the tadpole (Adamson, 1981a).

Numerous authors have reported physalopteroid juveniles (usually identified as *Abbreviata*) from North American frogs. Juveniles of *Abbreviata* are usually found encysted in mesenteries and walls of the gastrointestinal tract. Juvenile *Spiroxys* are also reported to be encysted in frog viscera and mesenteries.

Foleylla are filarial worms that have been described from the body cavities and mesenteries of North American ranids (Sonin, 1968). Filariae are elongated, thread-like (Latin *filum* = "thread") nematodes. Female *Foleylla* reach 7.0 cm in length and give birth to juveniles called microfilariae. Microfilariae find their way into the peripheral circulation and are ingested by *Culex* and *Aedes* mosquitoes. Infective stage juveniles develop in thoracic muscles and then disperse throughout the mosquito body. The life cycle is completed when infected mosquitoes feed again on frogs (Causey, 1939a,b,c; Kotcher, 1941).

Protozoans

Because of their small size, many protozoans are easily overlooked unless fecal smears, blood smears, or wet mounts from scrapings of various tissues are prepared and examined with a compound microscope at high magnifications. The rectums of frogs are choice locations for a number of protozoans. Using the hand polished lenses in his simple microscopes, Leeuwenhoek found *Opalina* and *Nyctotherus* in the rectums of frogs in the seventeenth century (Roberts and Janovy, 2000). Opalinids are among the most interesting protozoan inhabitants of the frog rectum. Four genera of opalinids have been described from frogs (Affa'a and Lynn, 1994; Delvinquier and Desser, 1996). Opalinids are multinucleate (*Opalina* and *Cepedea*) or binucleate (*Protoopalina* and *Zelleriella*) and, depending on the genus, are either flattened or circular in cross section. These large protozoans are covered with rows of elongated flagella. Taxonomists originally grouped opalinids with ciliates but more recent revisions align opalinids with flagellates. Opalinids may number in the thousands in an individual frog. When rectum and posterior reaches of the intestine are opened in saline, opalinids swim slowly, spiralling around their longitudinal axes in a fashion that my parasitologist wife describes as "swimming potato chips." The study of opalinids may contribute to the understanding of the evolution and zoogeography of their anuran hosts (Earl, 1979). El Mofty and Sadek (1973) have demonstrated that the reproductive cycle of opalinids is controlled by levels of host hormones.

Nyctotherus cordiformis are frequently encountered ciliates in the rectums of frogs. A conspicuous macronucleus is present in these organisms. Food material is swept into a longitudinal cytopharynx located along the lateral margin of the cell. Several species of *Balantidium* have also been reported from frog rectums. Similar to *Nyctotherus*, *Balantidium* have a shorter, anteroterminally located cytopharynx. *Tritrichomonas augusta* occur in the rectums of anurans and are characterized by a conspicuous undulating membrane along the margin of the cell, three anteriorly projecting flagella, and a stiff axostyle projecting posteriorly. *Hexamita intestinalis* are flagellates of the anuran intestine that possess two nuclei and four pairs of flagella. *Entamoeba ranarum* are amoebae, occasionally isolated from the rectums of North American frogs. Stabler and Chen (1936) described a symbiotic relationship between *Entamoeba* sp. and *Zelleriella opisthocarya*, an opalinid of toads, in which more than 200 cysts of the amoeba were present in one opalinid.

Blood smears from frogs may reveal extracellular *Trypanosoma rotatorium*. Trypanosomes are elongated cells with a conspicuous undulating membrane originating at the anterior end of the cell, continuing along the length of the cell and extending off the posterior extremity of the cell as a free flagellum. A single, large nucleus occurs in the center of the cell. *Hepatozoon catesbianae* are malaria-like parasites that enter erythrocytes of American bullfrogs (*R. catesbeiana*) and asexually reproduce large numbers of parasites that escape from erythrocytes and invade uninfected cells. Trypanosomes are transmitted to frogs when infected leeches take a bloodmeal. *Hepatozoon* use the mosquito *Culex territans* as their vector (Desser et al., 1995). Trypanosomes are readily apparent, thrashing about in fresh blood applied to a slide and coverslipped. Haemogregarines are best seen in Giemsa-stained blood films.

Eimeria fitchi have recently been described from wood frogs (*R. sylvatica*) in Arkansas (McAllister et al., 1995e). *Eimeria* are most easily diagnosed by examining frog feces for oocysts. Sporulated oocysts contain four ovoid sporocysts. The life cycle is direct, and parasites reside and reproduce within columnar epithelial cells of the intestinal mucosa.

Myxosporans are spore-forming metazoan parasites that are primarily parasites of fish. (Trout fishers may be familiar with *Myxobolus cerebralis*, the causative agent of salmonid whirling disease.) *Myxidium serotinum* are found in the gall bladder of several species of frogs (Kudo, 1943; McAllister, 1991). Myxosporans have indirect life cycles that utilize aquatic oligochaetes as intermediate hosts.

Miscellaneous Groups

The glossiphoniid leeches *Desserobdella* (syn. *Batrachobdella*) *picta* are frequently encountered on mating frogs, toads, and salamanders (Sawyer, 1972). These leeches frequent small mud- and leaf-bottomed woodland ponds. *D. picta* can be extremely abundant locally and can be found attached to most, if not all, species of amphibians that cohabit ponds. These leeches seem to feed exclusively on amphibians. Brockelman (1969) found *D. picta* at densities as high as 66 per square meter and determined them to be the largest single source of mortality of American toad tadpoles. Young leeches are found on tadpoles of moderate to large size, usually at the base of the tail.

Larval trombiculid mites *(Hannemania)* are often found encapsulated on undersides of the legs, on venters, and near cloacal openings of frogs (Murphy, 1965; McAllister, 1991).

Trombiculids include the annoying chigger mites, which cause cutaneous irritations of humans in southern states.

Summary

North American frogs harbor extremely diverse communities of parasites. While I had previously considered harm caused by these parasites to be minimal, recent studies by Johnson and his colleagues show that *Ribeiroia* can elicit substantial deformities and mortalities under laboratory conditions. Our studies in Minnesota and Wisconsin provide preliminary evidence that *Ribeiroia* are not present at all sites where deformities occur. This means that frog deformities at sites where *Ribeiroia* appear to be absent are caused by other (as yet undetermined) means (such as toxicants or bioactive agents; see Souder, 2000; Lannoo et al., 2003). Yet, we have documented the presence of *Ribeiroia* in tremendous numbers at other sites in Minnesota, and as being associated with satellite limb formation and mandibular abbreviation. Almost certainly, frog deformities in Minnesota and Wisconsin are the result of a suite of causes working alone in certain situations or in concert in others. Multiple causes of deformities suggest that stresses predisposing frog populations to malformations are involved and that continued research into causes and stressors is justified. At any rate, frogs living in relatively healthy environments will continue to harbor parasites in profusion.

TWENTY

Parasite Infection and Limb Malformations: A Growing Problem in Amphibian Conservation

PIETER T. J. JOHNSON AND KEVIN B. LUNDE

Over the last two decades, scientists have become increasingly concerned about ongoing trends of amphibian population decline and extinction (Blaustein and Wake, 1990, 1995; Phillips, 1990; Pechmann et al., 1991; Wake, 1998). Parasitic pathogens, including certain bacteria, fungi, viruses, and helminths (see also Sutherland, this volume), have frequently been implicated as causes of gross pathology and mass die-offs, often in synergism with environmental stressors (Dusi, 1949; Elkan, 1960; Elkan and Reichenbach-Klinke, 1974; Nyman, 1986; Worthylake and Hovingh, 1989; Carey, 1993; Blaustein et al., 1994b; Faeh et al., 1998; Berger et al., 1998; Morell, 1999; Kiesecker, 2002; Lannoo et al., 2003). The effects of human activity on the epidemiology of endemic and emerging diseases remain largely unknown (Kiesecker and Blaustein, 1995; Laurance et al., 1996).

More recently, malformed amphibians have been reported with increasing frequency in several parts of North America. Since 1992, severe limb abnormalities, including extra, missing, and malformed limbs, have appeared in dozens of species from diverse habitats, including several species in marked decline (Ouellet et al., 1997a; Helgen et al., 1998; Burkhart et al., 1998; NARCAM, 1999). While it is unlikely that these abnormalities have been a major source of historical amphibian population declines, they may represent an emerging threat, particularly if they are increasing in frequency (Wake, 1998; Hoppe, 2000). A generally low (0–5%) baseline rate of abnormalities, due to mutation, trauma, and developmental errors, may be expected in most amphibian populations (Martof, 1956b; Meyer-Rochow and Asashima, 1988; Zaffaroni et al., 1992; Tyler, 1998), but many recent accounts consistently document frequencies of 15% or greater (e.g., Hoppe and Mottl, 1997; Johnson et al., 2001b). Limb malformations are widely suspected to impair the survival of affected individuals, and some researchers have recorded corresponding mass die-offs (Gardiner and Hoppe, 1999).

The potential danger of these malformations to amphibians and possibly other vertebrate species depends, in part, on the agent(s) responsible. Among those currently under investigation are ultraviolet radiation (Blaustein et al., 1997; Ankley et al., 1998), pesticides (Ouellet et al., 1997a; Burkhart et al., 1998; Fort et al., 1999a; Kiesecker, 2002), retinoids (Gardiner

and Hoppe, 1999), and parasites (Sessions and Ruth, 1990; Johnson et al., 1999; Sessions et al., 1999). The recent and widespread nature of the malformations has focused the investigation on the direct and indirect impacts of human activity. One of the prime suspects in the East and Midwest has been water contaminants. Ouellet et al. (1997a) documented high rates of limb malformations in amphibians from agricultural ponds in southeastern Canada; laboratory researchers in Minnesota have found that water from affected ponds causes some malformations in African-clawed frog (*Xenopus laevis*) embryos and larvae. Identification of the active compounds and their role in amphibian limb malformations in the field remains under investigation (Fort et al., 1999b). The wide geographic range in North America with reports of abnormal amphibians suggests multiple causes are responsible, either interactively within ponds, independently among ponds/regions, or both.

In the western United States many of the reported sites have been linked directly to intense infections of a little-studied trematode (*Ribeiroia ondatrae*; Johnson et al., 1999, 2001b, 2002; Figs. 19-3, 19-4, and 19-7 in Sutherland, this volume). In laboratory infections, *Ribeiroia* parasites target the developing limb tissue of anuran larvae, inducing high rates of limb malformations strikingly similar to those recorded in field reports (Johnson et al., 1999; 2001a). Laboratory-infected larvae also suffer greater mortality than uninfected conspecifics. At this point, the impacts of *Ribeiroia* infection and the resulting limb abnormalities on natural amphibian populations are still poorly understood. Concerns about the potential role of human activity in increasing the range or abundance of *Ribeiroia* have yet to be addressed. In other parts of the world, however, human impacts have dramatically influenced the prevalence and infection intensity of a variety of trematodes, often with devastating consequences on related host species.

Here we evaluate two issues concerning *Ribeiroia* infection and amphibian limb malformations. First we ask if parasite-induced malformations in amphibians have increased in frequency, severity, or geographic prevalence. To this end, we examine (1) published accounts of malformed amphibians in the western United States; (2) the frequency of abnormalities in vouchered museum specimens; and (3) the historic range and host records of *Ribeiroia*, as recorded in the parasitology literature. In the second section, we assume that an increase in the above-stated attributes has occurred, and we investigate the most likely causal factors. Of particular concern are human impacts on trematodes, and we review the various mechanisms of influence and evaluate their relevance to *Ribeiroia* epidemiology.

Trematode Infections and Amphibian Malformations

Digenetic trematodes are a diverse group of exclusively parasitic flatworms common within most vertebrate groups. Of great interest to many parasitologists are trematodes' complex life cycles, which typically involve two or more hosts (see Sutherland, this volume). Sessions and Ruth (1990) suggested a connection between trematode infection and abnormal amphibians after observing a close association between metacercariae *(Manodistomum syntomentera)* and limb malformations in a California pond. The abnormalities included extra and deformed hindlimbs and affected two amphibian species: Pacific treefrogs *(Pseudacris regilla)* and Santa Cruz long-toed salamanders *(Ambystoma macrodactylum croceum)*. As an indirect test of their hypothesis, Sessions and Ruth (1990) implanted inert resin beads of a similar size to metacercariae in the developing limbs of African-clawed frog and spotted salamander *(Ambystoma maculatum)* larvae. Twenty percent of the treated animals six percent of the controls exhibited minor limb anomalies. This result suggests that mechanical disruption of limb development caused by the parasitic cysts might be contributing to the limb abnormalities observed in the field (Sessions and Ruth, 1990).

In an independent study, we reported a field correlation between infection with a different trematode species and sites with high rates (10–45%) of amphibian limb malformations (Johnson et al., 1999; Fig. 20-1). Over a two-year period, we inspected 15,000 amphibians from 35 ponds in northern California. Malformations were recorded only from sites that supported the trematode *Ribeiroia* and its snail host, *Planorbella tenuis*. To test the hypothesis that *Ribeiroia* infection caused the observed malformations, we exposed laboratory-raised Pacific treefrog larvae to specific numbers of *Ribeiroia* cercariae isolated from infected snails. The infections had dramatic effects, and severe hindlimb deformities similar to those observed in the field were recorded in all *Ribeiroia* treatments. The frequency of abnormalities ranged from 70% (light infection) to 100% (heavy infection), whereas the number of hindlimbs ranged from zero to as many as eight. Larvae either exposed to a second species of trematode *(Alaria mustelae*; Fig. 19-8A in Sutherland, this volume) or not exposed to parasites did not exhibit abnormalities and had high survivorship. Subsequent experiments with western toads *(Bufo boreas)* have replicated and expanded these results. Using the same protocol, we again induced high rates of limb deformities, which in toads included extra, missing, and malformed forelimbs as well as hindlimbs (Johnson et al., 2001a).

Among trematodes with complex life cycles, species that reduce intermediate host fitness are reported with increasing frequency (reviewed by Moore and Gotelli, 1990; Kuris, 1997; Poulin, 1998). As the completion of these worms' life cycles depends on predation of the intermediate host by the definitive

FIGURE 20-1 Pacific treefrog *(Pseudacris regilla)* with extra hind limbs.

host (trophic transmission), selection favors traits impairing the intermediate host's ability to evade predation. In a classic example, Bethel and Holmes (1977) documented an amphipod that, when infected with a species of acanthocephalan, swims erratically near the water's surface and is more frequently consumed by foraging ducks, which are the parasite's definitive host. In a similar fashion, we suspect malformed limbs increase the likelihood that an infected frog will be eaten by a definitive host of *Ribeiroia*, which include muskrats, waterbirds, and raptors (Fig. 20-2). As with many other impairing parasites, *Ribeiroia* exhibits intensity-dependent pathogenicity—greater infections cause higher rates of limb malformations (Kuris, 1997; Johnson et al., 1999). Unlike most other parasites, *Ribeiroia* induces profound structural changes in its host's body plan. Changes in a host's morphological development (such as the number of limbs) are unusual among the many other accounts of parasite manipulation, which often report changes in the intermediate host's behavior (Moore, 1983, 1984a,b; Lafferty and Morris, 1996) or physical conspicuousness (Camp and Huizinga, 1979; Bakker et al., 1997). The mechanism through which *Ribeiroia* affects limb development is not known, but probably involves mechanical disturbance of the developing limb field (Sessions et al., 1999), secretion by the parasite of a limb growth factor (e.g., retinoic acid, sonic hedgehog, FGF-8), or a combination of the two (Johnson et al., 1999).

Recently, limb deformities associated with *Ribeiroia* infection were recorded in frogs, toads, and salamanders from across the western United States (Johnson et al., 2002; Fig. 20-3; Table 20-1). In a survey of more than a hundred ponds throughout California, Oregon, Washington, Idaho, and Montana, Johnson et al. (2002) reported a clear association between *Ribeiroia* infection and amphibian deformities; approximately 90% of the sites where limb malformations were observed supported *Ribeiroia*. No association between pesticide presence and amphibian malformations was detected. The abnormalities, which ranged in frequency from 1–90%, included missing limbs, extra limbs, skin webbings, and bony triangles. Affected sites were diverse, encompassing eutrophic stock ponds, mitigation ponds, agricultural run-off

FIGURE 20-2 The life cycle of *Ribeiroia ondatrae*. Depicted hosts are as follows: first intermediate host, planorbid snail (*Planorbella* sp.); second intermediate host, Pacific treefrog *(Pseudacris regilla)*; final host, great blue heron *(Ardea herodias)*.

canals, and montane lakes, but almost all shared a common feature: the presence of *Ribeiroia* and one of its requisite snail hosts.

Are Parasite-Induced Deformities on the Rise?

Since 1995, it has been suggested frequently that amphibian malformations are more severe, more widespread, and affect-

ing a greater percentage of a given population than in the historical record. Hoppe (2000; this volume), after examining museum vouchers and resurveying historic field sites, effectively concluded that malformations in Minnesota northern leopard frogs (*Rana pipiens*) have become more common and more severe since 1993. Drawing upon recent studies of *Ribeiroia* infection and amphibian deformities, we examine the uniqueness of the current malformation phenomenon in the western United States, narrowing our focus to a single cause (parasite

A

B

FIGURE 20-3 A bullfrog *(Rana catesbeiana)* from California with two extra forelimbs (above), and a western toad *(Bufo boreas)* from Oregon with a partially missing hindlimb (below).

TABLE 20-1

Conservation Status of Amphibian Species with Limb Abnormalities in the Western United States

Species	Malformation Type	Reported States	Species of Concern Listing[a]
Ambystoma macrodactylum	EX	MT	
A. macrodactylum croceum	EX	CA	CA[b]
Bufo boreas	MIS, MAL, EX	CA, OR, WA	OR, WA
Pseudacris regilla	EX, MAL, MIS	CA, OR, WA, MT	
Rana aurora aurora	MAL	OR	CA, OR, WA
R. cascadae	MAL, EX	OR	CA, OR, WA
R. catesbeiana	MIS, EX, MAL	CA	
R. luteiventris	MAL	WA, MT	WA
R. pipiens	EX	AZ	CA, AZ, OR, WA, MT

[a] Federal or state listing of Species of Concern as offered by the California Department of Fish and Game, the Oregon Department of Fish and Wildlife, the Washington Department of Fish and Wildlife, the Idaho Department of Fish and Game, Arizona Game and Fish, and Montana Fish, Wildlife, and Parks.

[b] Federally listed as Endangered.

NOTE: The types of observed limb abnormalities are abbreviated in the second column and are presented in the approximate order of observed occurrence for each species: EX = extra limb(s); MAL = malformed limb(s); MIS = missing limb(s). The western states in which these species have been recorded with limb abnormalities are also presented (California, Oregon, Washington, Montana, and Arizona). Data adapted from Johnson et al. (1999, 2001b, 2002), Sessions et al. (1999), and Sessions and Ruth (1990).

infection) and to a confined array of abnormalities (limb malformations). This approach offers several advantages over attempts to evaluate the national issue at large. First, it limits the geographic area to a particular region, and the western United States has been an active area of amphibian research for nearly a century. Second, *Ribeiroia* infection is correlated strongly with the presence of amphibian abnormalities in the West, potentially reducing the number of causes under consideration. Finally, one of the most common malformations associated with *Ribeiroia* infection—supernumerary limbs—is highly noticeable and has a rich historical literature. Operating with the null hypothesis that limb deformities in western amphibians (and the presence of *Ribeiroia* metacercariae) have not changed in frequency or prevalence, we use three lines of historical information to examine this notion: (1) historical reports of amphibian limb deformities, (2) abnormality rates in museum voucher specimens, and (3) previous host records and documented range information for *Ribeiroia*. While we recognize the substantial limitations of each data set and their inability to provide a rigorous test of the hypothesis, they may be suggestive of general trends.

Historical Reports

Setting 1990 as our dividing line, we classify reports published prior to 1990 as "historical" and all subsequent publications as "recent." Among the recent reports are five studies offering information on malformed amphibians in the West. Taken together, these reports document severe limb abnormalities in eight species from 47 sites across California, Arizona, Washington, Montana, and especially Oregon (Table 20-1). The majority of the limb abnormalities involve supernumerary limbs (1–12 extra), but also include large numbers of missing limbs, skin webbings, and malformed limbs. The frequency of affected amphibians at a given site varied dramatically, ranging from 1–90%.

Although not new in the western United States, amphibian limb malformations are uncommonly reported. Reports date back as far as 1899 and have steadily increased through the current decade. This increase may be due to increases in parasite abundance, the amount of amphibian research, or the number of available scientific journals. In total, our survey yielded ten independent publications on amphibian limb abnormalities. The reports offer accounts from four western states, encompassing only ten sites and five species (Table 20-2). With two exceptions, all describe extra-legged frogs (1–3 extra). Cunningham (1955)

TABLE 20-2

Historic Reports of Amphibian Limb Abnormalities in the Western United States

Species	Year	State	Abnormality[a]	Frequency[b]	Citation
Bufo boreas	1896	OR	Extra fore	1	Washburn (1899)
B. boreas	1920	CA	Extra hind	1	Crosswhite and Wyman (1920)
B. boreas	1923	CA	Missing digits, feet, limbs	50%	Storer (1925)
Rana aurora draytonii	1935	CA	Extra hind; missing fore	2	Cunningham (1955)
R. catesbeiana	1948	CA	3 extra fore	1	Pelgen (1951)
R. catesbeiana	1960	CA	3 extra hind	1	Ruth (1961)
Pseudacris regilla	1958–1961	MT	1–2 extra hind	21% (86)	Hebard and Brunson (1963)
R. boylii	1962	CA	2 extra hind	1	Banta (1966)
P. regilla	1967	WA	1–2 extra hind	some	Miller (1968)
P. regilla	1981	ID	1–2 extra hind	22% (54)	Reynolds and Stephens (1984)

[a] All abnormalities involve the limbs, either forelimbs or hindlimbs.

[b] The number of described individuals or the frequency and (sample size).

notes a California red-legged frog (*R. draytonii*) with a complete missing limb, and Storer (1925) describes a series of adult western toads (*B. boreas*) with partially missing limbs and feet. Storer (1925) estimated that approximately 50% of the toads were affected and, based on the recent nature of their abnormalities, suspected injury from a lawn mower was responsible.

Storer's report excluded, six of the nine remaining historic accounts document only one or two abnormal specimens and provide no mention of frequency. Three sites however, offer more interesting and reliable comparisons to recent material. All discuss Pacific treefrogs with supernumerary hindlimbs, which also accounts for most of the recent abnormality reports linked to *Ribeiroia* infection in the West (Johnson et al., 1999, 2002). Miller (1968), unfortunately, provided no quantitative frequency data for the "several" five-legged treefrogs and the "some" that had six limbs. Reynolds and Stephens (1984) reported an abnormality rate of 22% in metamorphic treefrogs from an Idaho pond; all frogs examined had one or more extra limbs (n = 54). The most interesting data, however, come from Jette Pond in western Montana (Fig. 20-4A). Between 1958–61, Hebard and Brunson (1963) consistently found extra-legged treefrogs at a rate of 20–25%. In a recent re-survey of Jette Pond, Johnson et al. (2002) found that approximately 50% of the treefrog larvae and metamorphic frogs exhibited severe malformations (n = 215). Correspondingly, high intensities of *Ribeiroia* infection were universal within affected frogs. Although the frequency of abnormalites is known to vary across a season, recent collections were made within one week of historic sampling dates. Undoubtedly, more data are needed from the Jette site, but these preliminary results suggest a local increase in the frequency of limb malformations. Additionally, two of the four species breeding at this site in the 1950s, Columbia spotted frogs (*R. luteiventris*) and northern leopard frogs are no longer present (Hebard and Brunson, 1963; Anderson, 1977). The fourth species, long-toed salamanders (*A. macrodactylum*), which Hebard and Brunson (1963) reported as unaffected by limb anomalies, recently has been recorded at Jette Pond with supernumerary limbs at frequencies of 10% (J. Werner, personal communications).

Museum Specimens

Within the collections of the San Diego Natural History Museum (SDNHM) and the Los Angeles County Museum of Natural History (LACMNH), limb abnormalities in Pacific treefrogs are minor and of low frequency. These collections include specimens from 36 counties in California, Oregon, and Washington, as well as vouchers from Canada and Mexico. Of the 1,328 frogs examined at LACMNH and the 658 at SDNHM, 1.74% and 0.91%, respectively, suffered limb abnormalities. Digit amputations, which are confounded by the common marking practice of toe-clipping, were not included in this calculation. Sixty percent of the abnormalities involved missing or partially missing limbs. Three frogs (10% of the abnormalities) with a split digit (minor polydactyly) were also noted. We found no supernumerary limbs, skin webbings, or other serious limb malformations. While the inclusion of collections from other museums would greatly supplement this line of study, we noticed collection biases with respect to life history stage (adult over larval) and habitat (lotic over lentic waters), which questions the utility of this type of evidence. In both historic and recent literature, abnormalities are most commonly recorded in metamorphosing frogs from pond habitats.

Parasitology Literature

Although Johnson et al. (2002) documented *Ribeiroia* in nearly a dozen species of frogs, toads, and salamanders, we found no previous amphibian host records for the western United States (e.g., Ingles, 1933a,b, 1936; Turner, 1958b; Frandsen and Grundmann, 1960; Lehmann, 1960; Pratt and McCauley, 1961; Waitz, 1961; Koller and Gaudin, 1977; Moravec, 1984; Goodman, 1989; Goldberg et al., 1996c, 1998b). In fact, as far as we can determine, *Ribeiroia* has never previously been recorded in a wild amphibian, despite extensive volumes dedicated to amphibian helminths (Brandt, 1936b; Walton, 1947; Smyth and Smyth, 1980; Prudhoe and Bray, 1982; Brooks, 1984; Aho, 1990; Andrews et al., 1992).

FIGURE 20-4 **(A)** Jette Pond in western Montana exemplifies a culturally eutrophic stock pond in which dense populations of helisome snails are found. **(B)** In a naturally eutrophic pond in Lassen National Park, California, helisome snail numbers are low.

However, Beaver (1939) and Riggin (1956) did conduct experimental infections with several anurans and salamanders in the Midwest and Puerto Rico, respectively, suggesting amphibians historically have served as suitable hosts for *Ribeiroia*. Unfortunately, most helminth surveys focus on the adult parasites of frogs, decreasing the likelihood *Ribeiroia* metacercariae would be noticed. Additionally, trematode taxonomy depends primarily on adult morphology, and larval stages (meta- and mesocercariae), when recorded, are rarely identified.

Caveats and Conclusions

Based on our review of the historic literature on amphibian abnormalities, limb malformations are not new to western amphibians. They are, however, uncommonly reported, especially in light of the amount of amphibian research in the West and on Pacific treefrogs—the most commonly affected species (e.g., Brattstrom and Warren, 1955; Jameson 1956b, 1957; Johnson and Bury, 1965; Altig and Brodie, 1968; White and Kolb, 1974; Whitney and Krebs, 1975; Schaub and Larsen, 1978; Fellers, 1979a; Kupferberg, 1998). The main question then, becomes whether the frequency or severity of limb malformations has increased. In a comparison of recent (1990–2000) and historical (1899–1989) publications on the subject, we found that recent reports document (1) a wider range of severe malformations; (2) a greater number of affected amphibian species; (3) a larger number of affected sites; and (4) a higher frequency range of affected individuals at reported sites. These conclusions parallel those of Hoppe (2000; this volume), whose Minnesota study was conducted with a unique synthesis of museum voucher inspections and re-survey data.

What we cannot know for certain is the degree to which historical publications and museum specimens are representative of the historical condition. Substantial biases with respect to sampling methods, numbers of researchers, and the focus of historic and recent studies preclude a rigorous testing of our null hypothesis. Of interest, however, are the historic reports documenting large numbers of malformed amphibians. While similar to recent reports in many respects, it is notable that these accounts are few in number (n = 4), never describe more than one affected species from a site, and discuss no abnormalities more severe than two extra limbs. A re-survey of one of these sites recorded higher rates of abnormalities and more affected species than did the survey 40 years prior. Qualitatively then, sufficient evidence exists to suggest that *Ribeiroia*-induced deformities may have increased in the West, either in scale (number of affected sites), intensity (frequency and severity of malformations), or both.

Human Impacts on Trematodes

Throughout the recent investigation of amphibian malformations, the potential causes have been divided artificially into "natural" (e.g., parasites and predators) and "anthropogenic" (e.g., pesticides and elevated UV-B radiation) factors. Use of this grossly over-simplified classification scheme leads to two perspectives: (1) assuming amphibian deformities recently have become more common or severe, "natural" agents, such as parasitism, are unlikely to explain such changes; and (2) sites for which *Ribeiroia* infection has been identified as the likely cause represent no concern from a conservation standpoint. Both perspectives are false. In truth, parasite infection is a *biotic* agent, which may be influenced substantially by interactions with other factors, whether biotic or abiotic. Human activity, for example, has frequently had a tremendous impact on the virulence of trematode-induced disease. We review the understudied realm of human impacts on the distribution and abundance of trematode populations. Each of the discussed mechanisms is then evaluated with respect to *Ribeiroia* infection in western North America.

Non-Indigenous Parasites

As global transportation achieves greater speed and efficiency, the introductions of non-indigenous animals, plants, and their respective parasites have become increasingly problematic (Elton, 1958; Vitousek et al., 1996; Lodge et al., 1998). Introduced parasites that successfully establish may have devastating effects on ecosystems as well as the economy (Stewart, 1991; Amin and Minckley, 1996; Barton, 1997). Native fauna, having no evolutionary resistance to the invading parasite, frequently exhibit heightened pathology and mortality following infection relative to hosts that have co-evolved with the parasite (Dobson and May, 1986). The colonization of a trematode to a new region requires the presence of suitable host species to support the life cycle, and records of digenetic trematode introductions are less common than many other direct life cycle parasites (Table 20-3; Kennedy, 1993; Dubois et al., 1996; Font, 1997). However, considering our limited knowledge on the range of most trematodes as well as their complicated taxonomy, it is likely the frequency of intercontinental (and especially intracontinental) trematode introductions remain underestimated.

Two of the most well-known examples of introduced trematodes are *Schistosoma mansoni* and *Fasciola hepatica*, both of which may have serious consequences on society. *Schistosoma mansoni*, the human blood fluke, traveled from Africa to South America, Central America, and the Caribbean Islands via the slave trade (D.S. Brown, 1978). *Fasciola hepatica*, the ruminant liver fluke of Europe, has colonized North America, Australia, New Zealand, Hawaii, and Papua New Guinea with the transport of infected stock (Boray, 1978). Both parasites remain important problems, with cases of human schistosomiasis estimated at 200 million and livestock industry losses due to liver flukes at over U.S. $2 billion (Stauffer et al., 1997; Boray and Munro, 1998).

Considering that several of the earliest records of *R. ondatrae* are in birds from Oregon (Price, 1931), Washington (McNeil, 1948), and California (Dubois and Mahon, 1959), we think it unlikely that *Ribeiroia* represents a recently introduced trematode in the western United States. Moreover, *Ribeiroia* has probably caused amphibian limb deformities for at least 40 years and probably much longer. Johnson et al. (2002) recently found dense infections of *Ribeiroia* in severely deformed Pacific treefrogs from a pond in western Montana which, during the late 1950s, supported similarly abnormal treefrogs (Hebard and Brunson, 1963). We suspect that *R. ondatrae* infection explains most of the limb malformations in both accounts.

Although not exotic in western North America, *R. ondatrae* may have been aided greatly in its dispersal by the intentional and unintentional introductions of non-indigenous amphibians and game fish. Both groups have functioned as important vectors for trematode introduction in other parts of the world (Table 20-3) and have been disseminated widely within the western states. Introduced frog species currently established in western regions include Rio Grande leopard frogs (*R. berlandieri*), American bullfrogs (*R. catesbeiana*), green frogs (*R. clamitans*), southern leopard frogs (*R. sphenocephala*), and African-clawed frogs. Bullfrogs, in particular, have spread widely following their introduction at the turn of the century (Hayes and Jennings, 1986). The stocking of game fish, particularly salmonids and centrarchids, remains a widespread practice impacting a diverse array of lentic and lotic habitats. Greater than 50% of the freshwater fish in California are non-indigenous (Moyle, 1976; Vitousek et al., 1996).

TABLE 20-3
Introductions of Non-indigenous Trematodes

Species	Subclass	Family	Native	Recently Colonized	Vector
Amurotrema dombrowskaje	Digenea	Diplodiscidae	East Asia	Central Asia	exotic fish[a]
Diplodiscus megalodiscus	Digenea	Diplodiscidae	Australia	New Zealand	frog[f]
Ascocotyle tenuicollis	Digenea	Heterphyiidae	North America	Hawaii	exotic fish[b]
Brachylaima apoplania	Digenea	Brachylaimidae	Southeast Asia	New Zealand	polynesian rat[c]
Calicophoron calicophorum	Digenea	Paramphistomidae	Europe	Australia	ruminants[d]
Orthocoelium streptocoelium	Digenea	Paramphistomidae	Europe	Australia	ruminants[d]
Paramphistomum ichikawai	Digenea	Paramphistomidae	Europe	Australia	ruminants[d]
Cyathocotyle bushiensis	Digenea	Cyathocotylidae	North America	Canada	infected snail[e]
Fasciola hepatica	Digenea	Fasciolidae	Europe	Australia, North America, New Zealand, Hawaii, Papua New Guinea	ruminants[d,g]
Gorgodera australiensis	Digenea	Gorgoderidae	Australia	New Zealand	frog[f]
Lissorchis attenuatum	Digenea	Lissorchidae	Canada	Canada	exotic fish[h]
Mesocoelium monas	Digenea	Mesocoeliidae	South America	New Guinea	toad[f]
Notocotylus gippyensis	Digenea	Notocotylidae	New Zealand	Europe	infected snail[d]
Ornithodiplostomum ptychocheilus	Digenea	Diplostomatidae	North America	Arizona Reservoir	exotic fish[i]
Posthodiplostomum minimum	Digenea	Diplostomatidae	North America	Arizona Reservoir	exotic fish[i]
Sanguinicola inermis	Digenea	Sanguincolidae	Europe	Europe	[j]
Schistosoma mansoni	Digenea	Schistomatidae	Africa	South America, Caribbean	slave trade[k]
Cleidodiscus pricei	Monogenea	Dactylogyridae	North America	Europe	exotic fish[a]
Dactylogyrus extensus	Monogenea	Dactylogyridae	Europe	North America	exotic fish[a,j]
Dactylogyrus vastator	Monogenea	Dactylogyridae	Japan	North America, Europe	exotic fish[a,j]
Dactylogyrus anchoratus	Monogenea	Dactylogyridae	Japan	North America, Europe	exotic fish[a,j]
Dactylogyrus aristichthys	Monogenea	Dactylogyridae	East Asia	Central Asia, North America, Europe	exotic fish[l]
Dactylogyrus baueri	Monogenea	Dactylogyridae	Japan	North America	exotic fish[l]
Dactylogyrus chenshuchenae	Monogenea	Dactylogyridae	East Asia	Central Asia, North America, Europe	exotic fish[l]
Dactylogyrus ctenopharyngodonis	Monogenea	Dactylogyridae	East Asia	Central Asia, North America, Europe	exotic fish[l]
Dactylogyrus formosus	Monogenea	Dactylogyridae	Japan	North America	exotic fish[l]
Dactylogyrus hypothalmichthys	Monogenea	Dactylogyridae	East Asia	Central Asia, North America, Europe	exotic fish[l]
Dactylogyrus lamellatus	Monogenea	Dactylogyridae	East Asia	Central Asia, North America, Europe	exotic fish[l]
Dactylogyrus minutus	Monogenea	Dactylogyridae	Europe	North America	exotic fish[a]

TABLE 20-3 (continued)

Species	Subclass	Family	Native	Recently Colonized	Vector
Dactylogyrus skrjabini	Monogenea	Dactylogyridae	East Asia	Central Asia, North America, Europe	exotic fish[l]
Dactylogyrus suchengtai	Monogenea	Dactylogyridae	East Asia	Central Asia, North America, Europe	exotic fish[l]
Pseudodactylogyrus anguillae	Monogenea	Dactylogyridae		Europe	j
Pseudodactylogyrus bini	Monogenea	Dactylogyridae	East Asia	Europe	j
Gyrodactylus ctenopharyngodonis	Monogenea	Gyrodactylidae		Central Asia, North America, Europe	exotic fish[l]
Gyrodactylus cyprini	Monogenea	Gyrodactylidae	Europe	North America	exotic fish[a]
Gyrodactylus salaris	Monogenea	Gyrodactylidae	Sweden	Norway	exotic fish[a]
Nitzschia sturionis	Monogenea	Capsalidae	Caspian Sea	Aral Sea	exotic fish[a]
Urocleidus principalis	Monogenea	Ancyrocephalidae	North America	Europe	j
Urocleidus dispar	Monogenea	Ancyrocephalidae	North America	Europe	exotic fish[a]
Urocleidus furcattus	Monogenea	Ancyrocephalidae	North America	Europe	exotic fish[a]
Urocleidus similis	Monogenea	Ancyrocephalidae	North America	Europe	exotic fish[a]

[a]Bauer (1991), [b]Font (1997), [c]Wheeler et al. (1989), [d]Cribb (1990), [e]Dobson and May (1986), [f]Prudhoe and Bray (1982), [g]Boray (1978), [h]Dubois et al. (1996), [i]Amin and Minckley (1996), [j]Kennedy (1993), [k]Brown (1978), and [l]Bauer and Hoffman (1976).

NOTE: When available, the suspected transport vector is provided in the final column.

Non-indigenous Host Species

Many trematodes exhibit considerable specificity in the use of a first intermediate host, and it is often the distribution of the snail host that limits a parasite's range. Mammalian and avian definitive hosts are highly vagile and facilitate trematode dispersal, but establishment will only occur if the proper intermediate hosts are available (Esch et al., 1988). Unfortunately, human-mediated introductions and the subsequent establishment of freshwater mollusks, including many hosts of waterborne diseases (e.g., schistosomiasis, fascioliasis, and paragonimiasis), are frequent and may result in substantial financial (e.g., zebra mussel, *Dreissena polymorpha*) or ecological (e.g., Asian clam, *Corbicula fluminea*) repercussions (Table 20-4). As with trematodes, many more introductions of non-indigenous snails undoubtedly go unreported.

Under natural conditions, aquatic snails and their eggs are distributed passively on the plumage and feet of birds and, to a lesser extent, mammals and insects (Rees, 1965; Prentice, 1983; Boag, 1986). Human activity considerably increases the range and likelihood of such passive transport and acts primarily through the following four pathways:

1. Aquatic plant trade (Boray, 1978; Pointier et al., 1993; Madsen and Frandsen, 1989)

2. Aquarium fish (Woodruff et al., 1985; Bowler and Frest, 1992; Warren, 1997)

3. Aquaculture and fish farms (D.S. Brown, 1980)

4. Ballast water (Stewart, 1991; Carlton, 1992)

Many of these activities are only loosely regulated and, while snails are routinely intercepted by national inspectors (Hanna, 1966), it is clear that many escape detection.

Although less likely to be infected at the time of introduction, the establishment of non-native snails may drastically alter the epidemiology of endemic pathogens. The most notable examples come from the fasciolid trematodes or liver flukes. *Pseudocussinea columella*, a North American snail, has been translocated across the globe (Table 20-4) with important consequences on the transmission of *F. hepatica*. In both New Zealand and Australia, the semi-amphibious *P. columella* expanded into habitats historically uncolonized by native lymnaeids, leading to fascioliasis in previously unaffected regions (Boray, 1978; D.S. Brown, 1978). Similarly, the rapid colonization of a series of dams in South Africa by *P. columella* has increased the prevalence and intensity of infection in local ruminants, resulting in substantial financial loss (D.S. Brown, 1980).

The influence of introduced snails on the epidemiology of *Ribeiroia* is difficult to determine. All available data on *Ribeiroia* indicate that species in the genus require snails in the family Planorbidae as their first intermediate hosts (Beaver, 1939; Basch and Sturrock, 1969; Schmidt and Fried, 1997; Johnson et al., 2002). Planorbid snails, including a number of documented *Ribeiroia* host species, have been widely introduced in North America (Table 20-3) but their influence on the parasite's range has not been studied. Malek (1977) recorded *Ribeiroia* in *Biomphalaria obstructa*, a South American planorbid, at a site in Louisiana. Intracontinental transfers of snails may also be important, and Johnson et al. (2002) collected two planorbid species in the western United States that were beyond their known range. Populations of each species hosted *Ribeiroia* infections at several wetlands (Johnson et al., 2002). Another

source of intracontinental transfers is aquarium stores. *Planorbella tenuis*, a known *Ribeiroia* host, and several other planorbid species have been routinely observed at southern California pet stores (Johnson and Lunde, personal observation). If released in the wild, these snails could establish at previously uncolonized wetlands or increase the local planorbid density at established sites, as has been recorded in other parts of the United States (Bowler and Frest, 1992) and South America (Correa et al., 1980). Finally, aquatic researchers themselves may act as unwitting vectors, transporting snails in their muddy boots and nets as they move among wetlands. This emphasizes the importance of rigorously sterilizing sampling equipment between sampling sites.

Habitat Modification—Snail Hosts

Human modification of aquatic systems can lead to increases of human and bovine diseases resulting from trematode infection, such as schistosomiasis, paragonimiasis, clonorchiasis, and fascioliasis (Madsen and Frandsen, 1989). Water impoundments (e.g., reservoirs and hydroelectric dams), often improve snail habitat and concentrate human populations, both of which facilitate schistosome transmission and have historically been a source of great misery in the tropics and neotropics (Jordan et al., 1980; Roberts and Janovy, 1996; Sutherland, this volume). Intermediate hosts of mesenteric and urinary schistosomiasis, planorbid snails of the genus *Biomphalaria* and *Bulinus*, respectively, rapidly colonize and establish dense populations in such habitats, leading to disease ouvreaks as documented from Lakes Volta, Kariba, and Nasser in Africa (D.S. Brown, 1980; Jordan et al., 1980). More recently, Southgate (1997) reported a severe increase in both schistosome species following the construction of two dams in the Senegal River Basin. Prior to 1985, the diseases were virtually foreign to the populace, but within 5 years of the dams' completion many of the surrounding villages suffered human infection prevalences of 80% or greater.

This problem is not confined only to large dams. Irrigation schemes, borrow pits, quarries, drainage ditches, small impoundments, and aquaculture ponds have all been associated with elevated snail populations and trematode infection intensities, sometimes with serious economic impacts (Correa et al., 1980; D.S. Brown, 1980; Jordan et al., 1980; Woodruff et al., 1985; Ripert and Raccurt, 1987; Madsen and Frandsen, 1989; Bowler and Frest, 1992). Coates and Redding-Coates (1981), examining agricultural practices in Sudan, identified the increased snail densities associated with irrigation canals as responsible for the high prevalence of schistosomes and the resultant financial loss. Similar irrigaton systems outside of western Australia have recently been colonized by the lymnaeid snail host of *F. hepatica;* in Rhodesia the construction of numerous small dams was blamed for the subsequent increase in *Fasciola* infection and the consequent condemnation of bovine liver meat (D.S. Brown, 1980; Boray and Munro, 1998).

In the western United States, as in other parts of the world, aquatic systems have been altered extensively by human activity. In geologically active California, where natural lakes and ponds are rare, nearly every river and stream with a large and constant flow has been dammed, frequently facilitating the establishment of exotic species (Moyle, 1976; McGinnis, 1984; Schoenherr, 1992). One of the most common impoundments is the farm pond, approximately 80% of which were built

Table 20-4

The Global Translocations of Freshwater Snails (Pulmonates and Prosobranchs)

Species	Family	Native	Recently Colonized
Biomphalaria straminea	Planorbidae	South America	China, Australia, Caribbean, Central America[a,b,c]
Biomphalaria glabrata	Planorbidae	South America, West Indies	North America[d,e]
Biomphalaria havanesis	Planorbidae	Cuba, West Indies	North America[e]
Biomphalaria obstructa	Planorbidae	Mexico, South America	North America[f,g,h]
Planorbella duryi	Planorbidae	North America (Florida)	Europe, Africa, Middle East, South America, (Western) North America[a,f,i]
Helisoma anceps	Planorbidae	North America	Europe[j]
Planorbis corneus	Planorbidae	Europe	Australia[k]
Menetus dilatatus	Planorbidae	North America	Europe[l]
Indoplanorbis exustus	Planorbidae	India, Southeast Asia	Africa, Australia[a]
Drepanotrema aeruginosus	Planorbidae	Central America	North America[d]
Drepanotrema cimex	Planorbidae	Central America	North America[d]
Drepanotrema kermatoides	Planorbidae	Central America	North America[d]
Pseudosuccinea columella	Lymnaeidae	(Eastern) North America	Australia, New Zealand, Africa, South America, Caribbean, Europe, (Western) North America, Hawaii[a,f,k,m,n]
Lymnaea viridis	Lymnaeidae	Papua New Guinea, Guam	Australia, China, Japan, Philippines[k,m]
Lymnaea auricularia rubiginosa	Lymnaeidae	India	Australia[a,k,m]
Lymnaea peregra	Lymnaeidae	Europe	Australia[m]
Lymnaea stagnalis	Lymnaeidae	Europe	New Zealand, Australia[k]
Fossaria truncatula	Lymnaeidae	Europe	New Zealand[m]
Radix auricularia	Lymnaeidae	Europe, Asia	North America[d,e,f,o,p]
Lymnaea natalensis	Lymnaeidae	South Africa	Denmark[a]
Physella acuta	Physidae	North America	Europe, Africa, Australia, Middle East, Far East[a,d,n]
Physa waterloti	Physidae		Denmark[a]
Stenophysa marmorata	Physidae	South America, West Indies	North America[d]
Stenophysa maugeriae	Physidae	Mexico	North America[d]
Thiara granifera	Thiaridae	Far East	North America, South America, Caribbean[a,d,q,r,s]
Thiara granifera mauiensis	Thiaridae	Hawaii	California[f]
Tarebia tuberculata	Thiaridae	South America, Asia, Africa, India, East Indies	North America, Europe, Caribbean[a,d,f,s]
Melanoides turriculus	Thiaridae	Asia, Africa	North America[d,s]
Bithynia tentaculata	Bithyniidae	Europe	North America[d,f,q]
Marisa cornuarietis	Ampullaridae	South America	North America, Caribbean[d,s,t,u]
Pomacea bridgesi	Ampullaridae	South America	North America[d,s]
Pomacea canaliculata	Ampullaridae	South America	North America, Far East, Southeast Asia, Philippines[d,v]
Pomacea haustrum	Ampullaridae	South America	North America[d]
Potamopyrgus antipodarum	Hydrobiidae	New Zealand	Europe, North America, Australia, Corsica[d,e,k,t,w]
Cipangopaludina chinensis malleata	Viviparidae	Japan	North America[d,f]

TABLE 20-4 (continued)

Species	Family	Native	Recently Colonized
Cipangopaludina japonica	Viviparidae	Japan	North America[d]
Elimia livescens livescens	Pleuroceridae	Great Lakes to S. Quebec, Indiana, Illinois, Ohio	Hudson River Drainage[e]
Elimia virginica	Pleuroceridae	Massachusetts to Virginia	Great Lakes Basin[e]
Valvata piscinalis	Valvatidae	Europe	North America[q]

[a]Madsen and Frandsen (1989), [b]Pointier et al. (1993), [c]Woodruff et al. (1985), [d]Turgeon et al. (1998), [e]Bowler and Frest (1992), [f]Taylor (1981), [g]U.S. Congress (1993), [h]Malek (1977), [i]Roushdy and El-Eniani (1981), [j]Henrard (1968), [k]Walker (1998), [l]Boycott (1936), [m]Boray (1978), [n]D.S. Brown (1967), [o]Metcalf and Smartt (1972), [p]Hanna (1966), [q]Abbott (1950), [r]Prentice (1983), [s]Warren (1997), [t]Brown (1980), [u]Burch (1989), [v]Vitousek et al. (1996), and [w]Cribb (1990).

to support stock grazing (Bennett, 1971). Since the 1930s, extensive pond building by organizations such as the Soil Conservation Service has resulted in an increase of artificial wetlands in the United States of 2.9 million acres (Bennett, 1971; Tiner, 1984; Dahl and Johnson, 1991; Leja, 1998). California's Central Valley has gained 27.9 thousand acres since 1939—a 300% increase (Frayer et al., 1989). Unfortunately, little is known about the ecological importance of such habitats. Some native amphibians clearly utilize farm ponds if the non-indigenous species load (i.e., introduced fishes and American bullfrogs) is not extreme (Fisher and Shaffer, 1996; Hayes and Jennings, 1989; Stebbins, 1985). Similar ponds in England provide important snail habitats (Boycott, 1936). Considering the abiotic and biotic profiles of such ponds, we suspect these anthropogenic habitats, in contrast to larger reservoirs, are particularly conducive to *Ribeiroia* parasites and amphibian limb malformations, primarily through their effects on snail hosts.

One important feature of farm ponds is their productivity. Frequently a source of water for cattle and agricultural irrigation, these ponds are often eutrophic. Productivity is considered an important determinant of snail distributions, and mildly eutrophic water systems are associated with greater snail diversity and total gastropod biomass (Carr and Hiltunen, 1965; Harman and Forney, 1970; Russell-Hunter, 1978). What little experimental work has been done suggests that, in more productive environments, aquatic snails (1) grow faster, (2) reach a larger maximum size, (3) are more fecund, and (4) achieve greater densities (Eversole, 1978; Brown and DeVries, 1985; K.M. Brown et al., 1988). Helisome snails, the first intermediate hosts for trematodes in the genus *Ribeiroia*, may be particularly favored in nutrient-enriched environments (but see Bovbjerg and Ulmer, 1960; Bovbjerg, 1980). These snails are commonly found in permanent, eutrophic, often manmade bodies of water (Boycott, 1936; Boerger, 1975; Eversole, 1978; Taylor, 1981; K.M. Brown, 1982; Fernandez and Esch, 1991). Chase (1998) found a positive relationship between productivity and the relative biomass of the snail *H. trivolvis* among 29 temperate ponds in Michigan, suggesting that these snails are more successful in productive environments. According to Chase (1998), elevated productivity can have important mediating effects on the predation of helisomes in permanent ponds, where predation is an important regulator of snail species composition and abundance (Brown and DeVries, 1985; Lodge et al., 1987; K.M. Brown, 1991). In a series of mesocosm and *in situ* field experiments, Chase (1998) determined that more productive environments favored helisome

snails by facilitating faster growth, thus allowing them to achieve size refugia from predators and a greater total biomass in the long term.

Taken together, these studies suggest that eutrophic ponds will support greater densities of helisome snails which, depending on definitive host activity, could lead to an increased number of *Ribeiroia*-infected snails. More infected snails, in turn, would lead to higher intensities of infection in larval amphibians. According to the results of Johnson et al. (1999, 2001a), heavier *Ribeiroia* infections cause higher rates of more severe limb deformities in metamorphic anurans, suggesting such ponds might have important consequences on the local frequency of malformations.

Although eutrophication is a natural event, humans frequently accelerate the process via increased nitrogen and phosphorus inputs from fertilizer run-off, cattle defecation, and domestic sewage. This "cultural" eutrophication is an increasing problem in aquatic systems, where it often occurs more rapidly than the ecosystem can adapt (Valtonen et al., 1997). If extreme, the result is a community with low species diversity and high densities of a few, over-represented species (Dobson and May, 1986). In the course of our own fieldwork, we observed substantial differences in the *Ribeiroia* host-parasite complex between culturally eutrophied ponds (e.g., Jette Pond in Montana; Fig. 20-4A) and naturally eutrophic ponds (e.g., Lassen Volcanic National Park in California; Fig. 20-4B). With a 100-year intensive grazing history, Jette Pond is highly eutrophic with an abnormality frequency of 50% and a dense helisome population. Several ponds at Lassen Park, in contrast, exhibit low abnormality frequencies (4–8%) and support small snail populations (Johnson et al., 2002). Although more field and experimental research is necessary to test a hypothesis causally connecting increasing productivity and amphibian abnormalities, eutrophication is already an implicated cause in helminth outbreaks from other parts of the world (D.S. Brown, 1980; Northcote, 1992; de Gentile et al., 1996; Bohl, 1997). This relationship should not, however, be expected in all eutrophic habitats. Hypereutrophic ponds, in which the oxygen levels become depleted from the excess primary productivity, or ponds contaminated by pesticides associated with agricultural habitats, are likely to prove detrimental to snails and amphibians alike (Harman, 1974; Leja, 1998).

Habitat Modification—Avian Hosts

Human modifications to the aquatic environment have also changed the activity and population sizes of many waterbirds, some of which serve as definitive hosts for *Ribeiroia*. By extension, these alterations affect the distribution and abundance of avian trematode communities. Aquatic habitat modifications could contribute to the apparent increase of amphibian malformations by either (1) increasing the number of water bodies supporting the *Ribeiroia* life cycle or (2) increasing the intensity of *Ribeiroia* at sites already supporting the parasite. Through their impacts on bird populations, the following trends support one or both of the above mechanisms: disappearing natural wetlands, growing numbers of artificial aquatic habitats, and increasing bird populations.

Since colonial times, the United States has developed an estimated 117 million acres of natural wetlands, with more than 4.5 million acres lost in California alone—a 91% decrease (Dahl, 1990). The destruction of natural wetlands displaces definitive hosts into alternate wetlands, causing birds to concentrate in remaining wetlands (Banks and Springer, 1994; Krapu, 1996). Higher densities of birds at these sites can lead to higher trematode infection rates of intermediate snail hosts (Bustnes and Galaktionov, 1999) and thus higher infection intensities in second intermediate hosts. In the case of *Ribeiroia*, this could directly increase both the frequency and severity of amphibian limb malformations (Johnson et al., 1999). Indeed, increased densities of waterfowl at several European lakes have been associated with outbreaks of swimmers itch, a skin condition caused by avian trematodes (de Gentile et al., 1996).

The colonization of manmade wetlands by definitive hosts can broaden the distribution of the host's trematode community. As birds are perpetually forced out of disappearing natural habitats, they are becoming increasingly dependent on the growing number of artificial wetlands (Ruwaldt et al., 1979; Hudson, 1983). These alternative sites are not always suboptimal; adaptable species may even thrive in such conditions. Destruction of coastal wetlands have displaced lesser snow geese (form of *Chen caerulescens*) to inland rice fields, where overwintering populations in Texas have nearly doubled in 30 years (Robertson and Slack, 1995). Cattle and farm ponds built from the 1950s to the 1980s comprised the majority of the 2.7-million-acre increase in palustrine, non-vegetated wetlands in the United States (Tiner, 1984; Dahl and Johnson, 1991). These abundant aquatic habitats provide important colonization opportunities for wildlife. They may also be eutrophic, and, as previously discussed, are therefore more likely to support dense helisome populations—the first intermediate hosts of *Ribeiroia*. Artificial ponds, reservoirs, and fish farms also attract definitive hosts of *Ribeiroia* such as herons, egrets, and mallards (Ruwaldt et al., 1979; Bildstein et al., 1994; Glahn et al., 1999). By increasing the likelihood that appropriate hosts will converge in aquatic habitats, these anthropogenic water bodies provide opportunities for the *Ribeiroia* life cycle to establish, possibly contributing to the apparent increase in the number of wetlands with deformed amphibians. However, by its very dependence on artifical (as opposed to natural) wetlands, this pattern also brings malformations in closer contact with human populations where they are more likely to be observed and reported.

Birds that can adapt to, or even take advantage of, massive alterations in their environment may expand their populations, as reported for certain *Ribeiroia* hosts (Table 20-5). Great blue heron populations have increased nationwide by 47% in the last 30 years and are growing substantially in California, Oregon, and Washington (p < 0.05; Price et al., 1995; Sauer et al., 1996). As a group, wading birds are highly adaptable to human-modified environments such as rice fields, fish farms, reservoirs, farm ponds, and wastewater treatment ponds (Hoy,

1994; Stickley et al., 1995; Day and Colwell, 1998; Elphick and Oring, 1998). The flexible foraging habitats of great blue herons, along with their broad diet (e.g., fish, frogs, insects, and small mammals) are likely contributors to their population expansions (Butler, 1992). Increases in the population size of *Ribeiroia* definitive hosts will either concentrate birds at existing wetlands or encourage the colonization of new wetlands. Either of these outcomes could effectively increase the prevalence and/or severity of amphibian malformations, depending on the abundance of *Ribeiroia* and its intermediate hosts (helisome snails and amphibians). Unfortunately, we know little about the definitive hosts of *Ribeiroia* in the western United States, and the importance of different waterbird species to the transmission of this parasite requires further investigation.

Conclusions and Continued Study

Increasing reports of malformed amphibians from across North America have generated substantial concern over the implications of the phenomenon for environmental health. In the western United States, amphibian malformations are frequently associated with infection by a trematode in the genus *Ribeiroia*. This paper evaluates the importance of *Ribeiroia* infection and the resulting pathology of limb malformations from a conservation standpoint—do they represent a concern?

Trematode parasites, of which more than 150 species utilize amphibians as intermediate hosts, are infrequently studied with respect to their impacts on amphibian distributions and populations dynamics (but see Kiesecker and Skelly, 2000, 2001). Field and laboratory studies of *Ribeiroia* infection suggest this trematode may substantially reduce amphibian survivorship through two mechanisms: direct mortality due to infection and indirect mortality resulting from impaired fitness associated with limb malformations. Experimental exposures of larval Pacific treefrogs to cercariae resulted in nearly 50% direct mortality, and malformed metamorphic frogs in the wild are not expected to survive to sexual maturity (Johnson et al., 1999). The long-term effects of infection on a population are not yet known, but *Ribeiroia* has recently been recorded in a number of amphibian species of concern status in the West (Table 20-1). While it is unlikely the parasite is the major factor behind these declines, amphibian deformities may represent an added insult to already precarious populations, particularly if such deformities are increasing in frequency.

The conservation importance of *Ribeiroia* infection and limb malformations largely depends on whether the rate or prevalence of the induced abnormalities has increased in recent years. If the current rate of parasite-induced deformities is equivalent to the historical baseline, we would expect that native amphibians have adapted to accommodate them or previously gone extinct. If however, *Ribeiroia* is a recent invader or has increased in intensity or range, these deformities may be cause for concern. On the basis of comparisons between recent and historic literature on amphibian abnormalities, there is evidence suggestive of an increase in the severity and distribution of limb malformations. At this time, questions regarding the representative nature of the historic data prevent a rigorous, quantitative test of this hypothesis. However, reports over the last decade do document higher rates of more severe malformations from a larger number of sites and species than noted in the historical precedent. Within a limited examination of amphibian vouchers from two museum collections, abnormalities were infrequent and typically minor, and records

TABLE 20-5
Population Trends in Avian *Ribeiroia* Hosts in North America

Species	United States Population Change[a]	United States Population Change[b]	States with Substantial Increase in the West[c]	Source
Common loon (*Gavia immer*)	+	N/A	N/A	Kinsella and Forrester (1999); (1994 Conboy USNPC)[d]
Brown pelican (*Pelecanus occidentalis*)	0	+		(1994 Roderick USNPC)[d]
Olivaceus cormorant (*Phalacrocorax olivaceus*)	0	+		Ramos Ramos (1995)
Great blue heron (*Ardea herodias*)	+	+	CA, OR, WA	M. Kinsella, unpublished data
Great egret (*Ardea albus*)	0	+	CA	Travassos (1939) as cited by Cable et al. (1960); Soledad Sepulveda et al. (1999); HWML[e]
Little blue heron (*Egretta caerulea*)	0	0		Cable et al. (1960); Basch and Sturrock (1969); Soledad Sepulveda et al. (1996)
Tricolored heron (*Egretta tricolor*)	0	0		Lumsden and Zishke (1963)
Green heron (*Butorides virescens*)	0	+		Riggin (1956)
Black-crowned night-heron (*Nycticorax nyticorax*)	0	+	CA, OR	Pineda et al. (1985)
Reddish egret (*Egretta rufescens*)	0	0		(1977 Paul USNPC)[d]
Wood duck (*Aix sponsa*)	+	0	OR, WA	Thul et al. (1985)
Mallard (*Anas platyrhynchos*)	0	−	CA	(1993 Fedynich USNPC)[d]; HWML[e]
Common merganser (*Mergus merganser*)	+	0	CA, OR	(1947 Zimmerman USNPC)[d]
Red-breasted merganser (*Mergus serrator*)	0	0		(1932 Woodbury USNPC)[d]
Osprey (*Pandion haliaetus*)	+	+	CA	Beaver (1939); Taft et al. (1993); Kinsella et al. (1996)
Bald eagle (*Haliaeetus leucocephalus*)	+	0	OR, WA, ID, MT	Tuggle and Schmeling (1982)
Cooper's hawk (*Accipiter cooperii*)	+	0	CA, WA, ID	Beaver (1939)
Broad-winged hawk (*Buteo platypterus*)	0	0		Taft et al. (1993)
Red-tailed hawk (*Buteo jamaicensis*)	+	+	CA, OR, WA, ID	Taft et al. (1993)
California gull (*Larus californicus*)	0	N/A	N/A	Price (1931)
Great horned owl (*Bubo virginianus*)	0	+	CA, OR, WA	Taft et al. (1993); HWML[e]

[a] Population trends (1966–93) from Price et al. (1995). Significant change when $p < 0.05$; 0 = no significant change.

[b] Population trends (1959–88) from Sauer et al. (1996). Significant change when $p < 0.05$; 0 = no significant change.

[c] Western States include California, Oregon, Washington, Idaho, and Montana. Population trends from Sauer et al. (1996). Significant change when $p < 0.05$.

[d] United States National Parasite Collection at http://www.lpsi.barc.usda.gov/BNPCU/parasrch.htm

[e] Harold W. Manter Laboratory, Division of Parasitology, University of Nebraska State Museum, Lincoln, Nebraska

NOTE: N/A designates when data were not available.

of *Ribeiroia* infection from wild amphibians are non-existent prior to 1999.

Over the last 300 years, humans have altered dramatically the distribution of plants, animals, and their respective parasites. In reviewing some of this literature, it is immediately apparent that human activity frequently favors trematode populations, sometimes with costly repercussions. The interactions between human disturbance and wildlife diseases are exceedingly complex and may assume direct, indirect, or synergistic mechanisms (Möller, 1987; Khan and Thulin, 1991). Based on existing evidence, *Ribeiroia* is not a recently introduced species nor has its distribution been dramatically affected by the colonization of a non-indigenous host species. An interaction between pesticide contamination and elevated parasite infection is also not suspected, given the rarity of pesticide presence at amphibian malformation field sites in the western United States. If *Ribeiroia* has increased or shifted its range within the western United States, we suspect it is a result of extensive human alterations to aquatic systems. The life cycle of *Ribeiroia* (Fig. 20-2), which depends on helisome snails, amphibian larvae, and waterbirds, appears to thrive in artificial impoundments. More specifically, highly productive farm ponds can support dense helisome snail populations. Widespread construction of these understudied habitats may have facilitated large range shifts of helisome snails and, consequently, of their trematodes. The accelerated, cultural eutrophication of farm ponds, due to cattle and agricultural fertilizers, create the appropriate conditions for elevated snail densities, possibly increasing *Ribeiroia* infection intensities in amphibians from mild, non-pathological levels to extremely heavy infections, with the resulting malformations threatening the longevity of amphibian host populations. There is also evidence that, as natural wetlands are continually altered or destroyed, these anthropogenic habitats become increasingly important foraging areas for waterbirds, including several known *Ribeiroia* definitive hosts.

At this point, our hypotheses, while well founded in theory and precedents, remain speculative. The study of *Ribeiroia*, its ecology, and unique pathology in amphibians is still relatively new. In the continued research of amphibian limb malformations in the West and the *Ribeiroia* host-parasite complex, we emphasize the following areas of study: (1) broad-based field surveys comparing parasite dynamics (e.g., host densities, infection intensities, and abnormality rates) between natural and human-modified systems; (2) *in situ* experimental studies on the effects of increasing productivity on helisome density and *Ribeiroia* infection prevalence; (3) waterbird helminth surveys to determine the definitive hosts of *Ribeiroia* in the West; and (4) long-term monitoring on the impacts of *Ribeiroia* infection and limb deformities on amphibians, particularly for those species and populations currently in decline.

Summary

Numerous reports of malformed amphibians from across North America have prompted investigations into the cause(s) of the abnormalities and their implications for affected populations. While many agents may be responsible for the national phenomenon, reports of amphibian limb malformations in the western United States correlate strongly with infection by a parasitic trematode *(Ribeiroia ondatrae)*. In laboratory experiments *Ribeiroia* causes high frequencies of limb malformations and reduced survivorship in metamorphosing anurans. Although similar malformations have been recorded in the West over the last century, reports from the last decade detail a greater frequency range of more severe malformations from a larger number of sites and species. The intensity of trematode infections in other parts of the world has been dramatically influenced by human activity, particularly via the transport of non-indigenous host species and the modification of wetland habitats. These changes have sometimes caused outbreaks of trematode-related diseases in intermediate host populations. Although not an exotic genus in the western United States, *Ribeiroia* may have increased recently in range or density in response to substantial alterations of western aquatic systems. Planorbid snails, the first intermediate hosts in the *Ribeiroia* life cycle, thrive in the permanent, highly productive ponds that have become an increasingly common habitat over the last 70 years. Cultural eutrophication, a frequent consequence of human activity around wetlands, can facilitate elevated densities of these snails and their resident parasites. Higher intensities of *Ribeiroia* infection translate directly to higher rates of more severe malformations in amphibian hosts. Additionally, human-modified wetlands serve as important foraging grounds for waterfowl, including many known *Ribeiroia* hosts. Increases in some of these species' populations have recently been recorded in the western United States.

Considering the debilitating nature of the observed limb malformations, their widespread occurrence in the West, and the number of amphibian species affected, we believe parasite-induced malformations may represent a concern deserving further investigation. In the continued study of this issue we advocate (1) an evaluation of the long-term, population-level threat to amphibians presented by *Ribeiroia* infection and the resulting malformations, and (2) field and laboratory studies on the effects of human activity, particularly elevated nutrient levels, on the *Ribeiroia* host-parasite complex and its pathology in amphibians.

TWENTY-ONE

Pine Silviculture

D. BRUCE MEANS

The Coastal Plain of the southeastern United States is a vast natural area that has gone largely unheralded. Geographically, it fringes the southeastern corner of the continent, stretching 3,200 km from Long Island, New York, to the Mexican border, and including all of Florida. Geologically, it is a region composed entirely of sedimentary deposits of limestone, clay, sand, a small amount of gravel, and peat. Biologically, it is one of the country's richest centers of biodiversity and endemism, yet basic studies and even surveys of its biota are often wanting (Dodd, 1997; Means, 2000).

The Coastal Plain contains the highest species density of amphibians and reptiles in the United States and Canada (Kiester, 1971; Duellman and Sweet, 1999). Among the reptiles, both turtles and snakes reach their highest species densities in the Gulf Coastal Plain (Kiester, 1971; Iverson and Etchberger, 1989; Iverson, 1992). Among amphibians, species richness of frogs peaks in the Coastal Plain (Kiester, 1971) and salamanders are not far behind (Duellman and Sweet, 1999; Means, 2000). Only lizards are depauperate (Kiester, 1971), probably because the climate is not arid.

A little over a decade ago, biologists alerted the public to declining amphibian populations around the world (Barinaga, 1990; Blaustein and Wake, 1990; Phillips, 1990; Wyman, 1990), and a great deal of attention has since been focused on amphibians of the American West (Bradford, 1991; Carey, 1993; Fellers and Drost, 1993), Midwest (Lannoo, 1998b), and Canada (Green, 1997b). Much less attention has been given to the status of amphibians in the southeastern United States (Hairston and Wiley, 1993), but a review by Dodd (1997) has shown that many Coastal Plain amphibians are in trouble.

Of the 77 species of salamanders and frogs native to the Coastal Plain (42 and 35, respectively), 21 (27%) are dependent upon longleaf pine (Pinus palustris) forests as habitat in their adult or juvenile stages (Table 21-1). Among the 77 species, 15 (19%) have been ranked by The Nature Conservancy from G-1 (critically imperiled) to G-4 (apparently secure, but not demonstrably widespread, abundant, and secure; Table 21-1). Among the 21 longleaf pine forest-dependent species, however, 7 (33.3%) species are ranked G-1 to G-4. In one study of the amphibians and reptiles of longleaf pine communities, the authors found that over half of the specialist taxa dependent upon longleaf pine savanna were rare enough to be listed by state heritage programs, and they felt that this reflected a severe loss of the ancestral habitat (Guyer and Bailey, 1993). Considering that fully one-third of the amphibians that are dependent upon longleaf pine habitats may be declining, it would seem that preservation and proper management of the native longleaf pine upland habitats of these species should be a high conservation priority throughout the Coastal Plain.

The uplands of the largest portion of the Coastal Plain, from the Pine Barrens of New Jersey to Florida and east Texas, were once continuously more or less vegetated by longleaf pine forests (Wahlenberg, 1946; Means, 1996c; Platt, 1999). (Hereafter I follow Platt [1999] in referring to all longleaf pine forests as longleaf pine savanna.) Longleaf pine grows in many different soil types and therefore occurs in several facies, including sandhills, clayhills, and flatwoods. Over the longleaf pine belt, it has been estimated that more than 60% of the landscape, about 82.5 million acres, was in longleaf pine savanna (Wahlenberg, 1946; Ware et al., 1993). This is an exceptionally large percentage of the relatively low-lying Coastal Plain, which was otherwise full of swamps, marshes, lakes, and other types of wetlands.

By the end of the twentieth century, longleaf pine savanna had been reduced to less than 2% of its original coverage (Ware et al., 1993; Means, 1996c; Platt, 1999) and old-growth stands with virgin trees amounted to less than 10,000 acres (~0.01%; Means, 1996c). In Florida alone, for example, longleaf pine forests declined 88% in the 51-year period from 1936–87 (Kautz, 1993). Kautz predicted that if the same rate of attrition of longleaf pine forests continued in Florida, longleaf pine forests would disappear on all but public lands by 1995.

Today, traveling through the South (including the Coastal Plain) can be misleading to the uninformed because pine forests or pine-hardwood mixtures are the dominant vegetation. A large percentage of these are old-field successional forests or mixed pine-hardwood forests that grew up on cutover or abandoned lands (Ware et al., 1993); the rest are pine plantations.

Among the principal reasons for the extensive loss of longleaf pine forests are (1) large-scale industrial logging of the original forests from the 1880s through the 1930s; (2) interruption of natural fire cycles that are necessary for the long-term recycling of longleaf pine vegetation; (3) subsistence and industrial agriculture; (4) urbanization; and (5) silviculture (pine tree farming).

TABLE 21-1

Amphibians in the Coastal Plain of the Southeastern United States (from Conant and Collins, 1998)

Species	Longleaf Pine Dependent	Declining
Genus: Anura		
Family: Bufonidae		
Bufo americanus		
Bufo houstonensis	X	G1
Bufo quercicus	X	
Bufo terrestris	X	
Bufo nebulifer		
Bufo woodhousii		
Family: Hylidae		
Acris crepitans		
Acris gryllus		
Hyla andersonii	X	G4
Hyla avivoca		
Hyla chrysoscelis		
Hyla cinerea		
Hyla femoralis	X	
Hyla gratiosa	X	
Hyla squirella	X	
Pseudacris brimleyi		
Pseudacris crucifer		
Pseudacris nigrita	X	
Pseudacris ocularis	X	
Pseudacris ornata	X	
Pseudacris streckeri		
Pseudacris triseriata		
Family: Microhylidae		
Gastrophryne carolinensis	X	
Gastrophryne olivacea		
Family: Ranidae		
Rana areolata	X	G4
Rana capito	X	G3
Rana catesbeiana		
Rana clamitans		
Rana grylio		
Rana heckscheri		
Rana okaloosae	?	G2G3
Rana palustris		
Rana sphenocephala		
Rana virgatipes		
Family: Pelobatidae		
Scaphiopus holbrookii	X	

TABLE 21-1 (continued)

Species	Longleaf Pine Dependent	Declining
Genus: Caudata		
Family: Proteidae		
Necturus beyeri		G4
Necturus lewisi		G3
Necturus maculosus		
Necturus punctatus		G5
Family: Amphiumidae		
Amphiuma means		
Amphiuma pholeter		G3
Amphiuma tridactylum		
Family: Sirenidae		
Siren intermedia		
Siren lacertina		
Pseudobranchus axanthus		
Pseudobranchus striatus		
Family: Ambystomatidae		
Ambystoma cingulatum	X	G2G3
Ambystoma mabeei	X	G4
Ambystoma maculatum		
Ambystoma opacum		
Ambystoma talpoideum	X	
Ambystoma texanum		
Ambystoma tigrinum	X	
Family: Salamandridae		
Notophthalmus perstriatus	X	G2G3
Notophthalmus viridescens	X	
Family: Plethodontidae		
Desmognathus apalachicolae		
Desmognathus auriculatus		K
Desmognathus conanti		
Desmognathus monticola		
Eurycea cirrigera		
Eurycea guttolineata		
Eurycea quadridigitata (4)	X	
Haideotriton wallacei		G2
Hemidactylium scutatum		
Phaeognathus hubrichti	?	G2
Plethodon albagula		
Plethodon grobmani	?	
Plethodon kisatchie		
Plethodon mississippi		
Plethodon ocmulgee		

TABLE 21-1 (continued)

Species	Longleaf Pine Dependent	Declining
Plethodon sequoyah		
Plethodon serratus		
Plethodon variolatus		
Plethodon websteri		
Pseudotriton montanus		
Pseudotriton ruber		
Stereochilus marginatus		

NOTE: Many more species may be present when the systematics of groups such as *Desmognathus fuscus*, *Eurycea quadridigitata*, *Pseudobranchus*, and *Siren* are further researched. Those species marked under "Longleaf Pine Dependent" spend part of their life cycles in longleaf pine forests. Species whose habitats are surrounded by longleaf pine forest are marked "?" because it is not known whether fire or some other aspect of longleaf pine forests are important in their ecology. Those species marked under "Declining" are listed by their global ranking used by The Nature Conservancy. K = known to be declining but not yet globally ranked as such (personal observation).

While considerable attention has been devoted to the history of the logging of old-growth longleaf forests (Wahlenberg, 1946; Ware et al., 1993) and fire ecology (Means, 1996c; Platt, 1999), very little study of the effects of pine silviculture on native biota have been undertaken. Pine plantations have probably been as responsible for replacing native longleaf pine forests as has agriculture. The U.S. Forest Service estimates that pine plantations now make up 36% of all pine stands in the South, and it projects that within 20 years they will account for 70% (McWilliams et al., 1993). Pine plantations are very different habitats from the native longleaf pine savanna in many ways. They are so different, in fact, that one should expect that native animals and even plants might have difficulty surviving in them.

In this paper, I discuss the limited scientific literature dealing with the effects of pine silviculture on amphibians in the longleaf pine forests of the Coastal Plain. I also offer observations and opinions about the import of such effects on amphibians based upon these limited data and my 40 years of experience in the longleaf pine savanna.

Case Studies

Incredibly, I found only five research papers that have attempted to assess the effects of pine silviculture on Coastal Plain amphibians.

Bennett et al. (1980) compared diversity indices and relative abundance values for amphibians (5 salamanders and 11 frogs) inhabiting three different sites contiguous around a small lake in the upper South Carolina Coastal Plain—a 24-year-old planted slash pine (P. elliottii) stand; a 25-year-old planted loblolly pine (P. taeda) stand; and a second-growth (no age given) predominantly oak-hickory hardwood forest. The three sites were similar in species diversity and evenness, while the hardwood forest yielded approximately 50% more individual amphibians than either pine forest. It was not specified what the original natural vegetation might have been.

Enge and Marion (1986) compared herpetofaunas among three sites in a north Florida flatwoods—a 40-year-old, naturally regenerated slash pine forest; and two adjacent three- to four-year-old clearcuts subjected to minimum and maximum site preparation practices. They reported that clearcutting and site preparation treatments did not affect amphibian species richness (4 salamanders and 14 frogs), but that clearcutting reduced amphibian abundance ten-fold.

Raymond and Hardy (1991) studied the use of a breeding pond by mole salamanders (Ambystoma talpoideum) in northwestern Louisiana. They compared the survival and displacement of adults moving from terrestrial habitat and adjacent to a recent clearcut that was being converted to a pine plantation on one side of the pond versus the uncut pine-hardwoods habitat on the far side of the pond. The clearcut near the study pond appeared to affect the mole salamander population by (1) lowering the survival of adults immigrating from the clearcut side of the pond, and (2) displacing adults to the terrestrial habitat on the east side of the pond, which appeared to be a less suitable habitat.

Grant et al. (1994) studied the effects on amphibians (7 salamanders and 14 frogs) among replicated stands of planted loblolly pine either 1, 3, 8, or 26 years old in 1979. They found that intermediate stages of pine regeneration had higher amphibian species diversity than did either the youngest or oldest stands. They attributed this result to the fact that the intermediate-aged stands had the most dense foliage and the highest forest structural diversity as opposed to those stands following clearcutting and replanting or in the older stages following canopy closure and management activities (periodic burning and thinning) that eliminated understory vegetation. This was not a study of the effects of pine monoculture versus native vegetation, and the authors did not specify what the original native forest vegetation was prior to loblolly pine silviculture.

Means et al. (1996) observed, during a 22-year period, a dramatic decline of adult flatwoods salamanders (Ambystoma cingulatum) migrating to breeding ponds across a 4.3 km stretch of paved road in Liberty County, Florida. They believed that the decline was caused by the conversion of the native longleaf pine savanna surrounding breeding ponds to bedded slash pine plantations on one side of the road. They postulated that the raised beds might have interfered with migration, successful hatching, larval life, and finding suitable cover post-metamorphosis.

Discussion

Basically, although these five studies give the impression that pine silviculture can have some negative effects on amphibian populations, they collectively say very little about the regional magnitude of pine silviculture on amphibians. It seems incredible that in the last half of the twentieth century, no one has designed a study to examine what happens to any of the 21 species of native amphibians during pine silviculture. Such studies could be conducted by (1) monitoring amphibian population levels on longleaf pine savanna before silvicultural activities on replicated sites, and then monitoring amphibians on the same sites through time after site-preparation silviculture treatments; and/or (2) comparing replicated sites among undisturbed longleaf savanna and silvicultural treatments. This simply has not been done.

There are ecological reasons why one should expect that intensive pine silviculture would have negative effects on amphibian populations. Most importantly, the physiognomy of longleaf pine savanna is dramatically different from that of pine plantations. Longleaf pine savanna is an open-canopied pine forest (Fig. 21-1A), which allows large amounts of sunlight to reach the forest floor (Noel et al., 1998). Because of the high light levels, the groundcover contains a diverse mixture of hundreds of species of grasses and forbs (Christensen, 1988; Platt et al., 1988; Hardin and White, 1989; Walker, 1993), as well as a relatively large number (60+) of woody species (Peet and Allard, 1993). Because of its high plant species richness, including ground-level structural diversity, animal species diversity is also high (Means, 1996d). Four to five thousand estimated arthropod species have an important and complex role in the functioning of longleaf pine savanna (Folkerts et al., 1993), especially as food for amphibians. In addition to the fact that the highest U.S. reptile and amphibian species richness occurs on longleaf range, the number of breeding bird species was higher in old-growth longleaf pine stands than in other forest types in Florida, for example (Engstrom et al., 1984).

In pine plantations, the main objective is to maximize pine cellulose production. This means that a site is managed so that as much primary productivity takes place in the pine trees (not the groundcover or midstory) as is economically feasible to accomplish. Longleaf pine is not replanted, therefore, because (among several reasons) it is much more sensitive to competition than are other pines (Wahlenberg, 1946; Boyer, 1993). Prior to planting, a pine plantation is often "site prepared" to eliminate or severely reduce competition from the other native woody plants that were part of the longleaf pine savanna or that seed in from adjacent southern temperate hardwood forests (Batista and Platt, 1997). Under normal circumstances in longleaf pine savanna, these woody plants are kept suppressed by naturally occurring, periodic fires (Stoddard, 1935; Komarek, 1974; Christensen, 1981; Means, 1996c; Gilliam and Platt, 1999; Platt, 1999), but site preparation techniques suppress potentially competitive plants through mechanical actions such as scraping, roller chopping, harrowing, windrowing, and bedding. Ultimately, canopy closure in pine plantations ensures that the rich groundcover is severely suppressed or eliminated by shading (Halls, 1955; Hebb, 1971; Means, 1997).

A pine plantation, therefore, is a forest ecosystem dominated by a single densely stocked species in which most of the primary productivity takes place in the pine needle canopy. After conversion to pine plantation, the vigorous ground-level primary productivity of longleaf pine savanna is replaced by detrital inputs, which support wholly different arthropod detritivore faunas. Finally, the rich microhabitat structural diversity of hundreds of species of grasses, forbs, and woody groundcover and understory plants is replaced in older pine plantations by dense carpets of pine needles, twigs, and bark (Fig. 21-1B). These physiognomic differences in comparison with longleaf pine savanna are equally true for plantations of slash and loblolly pine, and especially sand pine (P. clausa).

Throughout the Coastal Plain there is no natural ecological analog of the modern pine plantation, except possibly in Florida's natural stands of sand pine. Native closed canopy hardwood forests exist in which the primary productivity is mostly in the canopy, but most longleaf pine amphibians do not inhabit these forest types. The closed canopy conditions of natural sand pine forests do not generally support robust populations of longleaf pine specialists, but sand pine stands naturally succeed to longleaf pine savanna under certain fire regimes anyway (Myers, 1990). In short, longleaf pine amphibians are generally poorly adapted to living under the ecologically different conditions in pine plantations.

Similarly, pine plantations are unsuitable habitat for other Coastal Plain vertebrates. White et al. (1975) compared habitat differences 9 years after the planting of slash pine on replicated longleaf pine flatwoods sites among three levels of site preparation ranging from clearcut, burn, and plant (low intensity), to clearcut, burn, KG, single harrow, bed and plant (high intensity). They found that where grasses and forbs were least productive (high site preparation intensity) ground arthropods, small mammals, birds, and other herbivores and insectivores were least abundant. They concluded that growth and development of slash pine overstories was favored by intensive site preparation at the expense of understory wildlife habitat.

Breeding and wintering bird populations were compared among natural longleaf pine forests and four age classes (1, 10, 14, and 40 years) of slash pine plantations in Florida (Repenning and Labisky, 1985). These authors found little similarity in the composition of the breeding bird communities between longleaf pine and any age-class of slash pine plantation, although older pine plantations and natural longleaf pine forest had relatively similar wintering bird communities. Likewise, Means and Means (unpublished data) found a much reduced breeding bird community in sand pine plantations compared to those adjacent to longleaf pine stands.

Generally, other studies reported increased or similar levels of use of pine plantations and longleaf pine savanna by some vertebrates (white-tailed deer, cottontail rabbit, and gopher tortoise) in the first 10–15 years after initial planting, but declines thereafter (e.g., Umber and Harris, 1974; Aresco and Guyer, 1999). I have found no published studies, however, that compare the population status or use by vertebrates of older pine plantations (25–40+years) with longleaf pine savanna, nor that compare the vegetative differences between them. Maybe the reason is that it is patently obvious that old pine plantations differ in plant and vertebrate species abundance and composition.

From the scant data presented here, it seems reasonable to conclude that pine plantations in the Coastal Plain, at least in their older stages, are not suitable habitat for native biota. The argument might be made, however, that the native biota can survive at the landscape level in patches of native habitat

A

B

FIGURE 21-1 **(A)** Longleaf pine savanna is an open-canopied pine forest that allows large amounts of sunlight to reach the forest floor. **(B)** The rich microhabitat structural diversity of hundreds of species of grasses, forbs, and woody groundcover and understory plants is replaced in older pine plantations by dense carpets of pine needles, twigs, and bark.

interspersed among pine plantations on different logging rotations—that native biota can recolonize newly recycled pine plantations from native habitat and younger plantation patches. This idea has never been tested experimentally in the field and for the reasons I mention below, I do not believe it has merit.

Throughout the Coastal Plain on privately owned industrial silvicultural land, few (if any) patches of native vegetation are left as recolonizing sources for native biota such as arthropods, rodents, amphibians, or reptiles. There simply has been no economic incentive for leaving patches of native habitat and no governmental regulations requiring it. Moreover, pine plantation patches are often huge, involving tens to hundreds of acres, so that large blocks of uninhabitable pine monoculture dominate the landscape. Additionally, since it is on the pine plantations that are 1–10 years old that native biota stand a chance of establishing a population for a short while, most of the landscape is in the 10–40+ year age classes, and is unsuitable habitat. There is also a larger problem that has become apparent in the past two decades. Many regions of the Coastal Plain, indeed of the entire Southeast, are now in their second or third planting since the advent of extensive pine silviculture following World War II. We simply do not know what effect repeated pine planting and harvest has on populations of any native plant or animal. However, highly specialized species (such as wiregrasses *[Aristida beyrichiana, A. stricta]*), which might have survived an initial plantation, are most certainly ripe for local extinction when faced with repeated cycles of site preparation followed by intense shading and competition from "off-site" pines (as species of pines are called that did not occur initially on a site).

Conclusions and Summary

In conclusion, I believe that pine tree farms are false forests (Williams, 2000) and that pine silviculture has played as large a role as any other factor in the decline of Coastal Plain amphibians such as flatwoods salamanders, striped newts (*Notophthalmus perstriatus*), and gopher frogs (*Rana capito*). Before these animals become extinct, and before others such as tiger salamanders (*Ambystoma tigrinum*), mole salamanders (*A. talpoideum*), Mabee's salamanders (*A. mabeei*), ornate chorus frogs (*Pseudacris ornata*), eastern spadefoot toads (*Scaphiopus holbrookii*), oak toads (*Bufo quercicus*), and barking treefrogs (*Hyla gratiosa*) are discovered to have also declined to threatened levels, I urge students of amphibian ecology to study amphibian response to pine silviculture.

TWENTY-TWO

Commercial Trade

ANTHONY B. WILSON

The international and interstate trade in amphibians is enormous and legally complex. There are also ramifications to this trade. In addition to posing a threat to native populations from overcollecting, the herpetofauna trade (including the bait industry) imports native animals that may be diseased or of different genetic stock and exotic species that may be invasive.

The Legal Amphibian Trade

The U.S. Federal Animal Welfare Act of 1966 regulates the use of animals for research and exhibition, as well as for the pet trade. Many state regulators misunderstand this law, believing that it totally regulates the pet industry. In fact, this law defines animals as "warm blooded" (7 U.S. Code § 2131-2159 and 9 CFR, Chapter 1, Parts 1–4), and by definition leaves no provisions for the regulation of "cold blooded" animals, including amphibians. Individual states often take their lead from the federal government in proposing regulations. As a result, few laws have been passed to regulate the harvest and importation of amphibians (see Appendix 22-A).

The Convention on International Trade in Endangered Species (CITES; e.g., 2000) of Wild Fauna and Flora became law in 1973. CITES prohibits the importing, exporting, or re-exporting of wildlife or plants (or their parts or derivatives of certain species) unless permitted to do so by both the importing and exporting country. The CITES treaty has been signed by 145 member nations. This treaty is a step in the right direction, but it has not stopped the illegal trade in amphibians and reptiles.

For example, the exportation of Australian amphibians and reptiles is strictly regulated by the Australian Wildlife Protection Act. The only wildlife that legally leave Australia must have permits and can go only to zoos or research institutions. Yet, an Australian wildlife law enforcement officer revealed that at this time some Australian species are more readily available in the United States than in Australia. How can this be true? It turns out that Australian amphibians and reptiles are being smuggled illegally out of the country via ships or aircraft, as well as simply being mailed out of the country. Some Australian amphibians and reptiles come from zoos and research institutions that had obtained them legally.

As another example, protected South American species and products readily find their way into European land-holding countries where they are traded freely and legally. Eventually, some of these animals and products make their way to the lucrative United States market.

Almost all countries (including the United States) permit an extremely large, legal trade in captive-bred amphibians and reptiles. Because there is no way (short of DNA fingerprinting) to tell if an amphibian or reptile is captive raised, commercial dealers have been known to supplement their legal captive-born stock with legal and/or illegal wild-caught specimens.

Some regulators are too cautious and would like to see population studies before they take action. It is true that no one is absolutely sure what will happen when amphibians and reptiles are harvested. This is simply because there has not been enough research. Usually one would expect to see a decrease if harvesting were allowed, especially in specific isolated populations. American bullfrogs (*Rana catesbeiana*) are a good example. I have been checking frog hunters in the same area for twenty-six years and the populations have decreased substantially. The habitat is still there, with no apparent pollution, so over-harvesting must be suspected. Waiting for a population problem to exist or to be confirmed before protecting a species is reactively regulating, and it may result in action being taken too late to save a population.

Another problem is a lack of communication between states. An embarrassing example is an individual who was made a member of the Indiana Department of Natural Resource's Non-Game Amphibian and Reptile Technical Advisory Committee. Two years previously he had been arrested in West Virginia for violation of amphibian and reptile laws.

The commercial interest in amphibians and reptiles has grown rapidly (Jones, 1994), driven in part by the general interest in amphibian declines and the popularity of amphibians and reptiles in the entertainment and advertisement industries. Today, millions of amphibians and reptiles are being kept as pets in the United States alone.

There are relatively few laws across the United States that deal with harvesting amphibians and reptiles for the pet industry or the release of non-native pet amphibians and reptiles to the wild (see Levell, 1995). Where such laws exist, the effectiveness of state wildlife enforcement officers depends on staffing and funding, as well as the attitudes of the enforcing agency. Most states have no proactive enforcement policy that deals with amphibians or reptiles. Normally, the officers respond to complaints from the public rather than patrolling for collectors.

I believe that the over-harvesting by some amphibian and reptile dealers is far more prevalent than anyone expects; however, I can only offer one case as an example. During my investigation into the commercialization of amphibians and reptiles, I learned from commercial collectors that several of them had been legally collecting salamanders from breeding ponds in Indiana's Yellowwood State Forest. Dr. Spencer Courtwright, who had been conducting a long-term study of those breeding ponds, confirmed my suspicions. His data show a population decline during the period of the commercial dealers' activity. After the Indiana amphibian and reptile law prohibiting such collections went into effect, the study showed population increases.

Regional amphibian and reptile shows have increased since the first All-Ohio Reptile Shows started in 1988, and many shows are now held across the country. The largest amphibian and reptile show is probably the Orlando Reptile Breeders Expo. The Midwest Reptile Show in Indianapolis is also one of the largest in the United States, and it meets every month.

Biological supply companies also contribute to the herpetofauna trade problem. One can purchase African clawed frogs (Xenopus laevis) through an "Educational Frog Hatching Kit" for $18.99, packaged by "Nasco, Fort Atkinson, Wisconsin." The kit comes with a small aquarium, supplies, and a coupon that is sent to Nasco and is redeemable for "6–10 early stage live frog embryos." Clawed frogs are now established in California and Arizona (see Crayon, this volume). In areas where they have been established, African clawed frogs prey upon native amphibians. The kit coupon is not redeemable May through October in Arizona, California, Nevada, Utah, Washington, and all provinces of Canada, where the import and release of clawed frogs is illegal.

In general, there is a need for non-game animal laws that deal with the over-harvesting of wildlife or the spread of non-native species. Although some states have proactive statutes (Levell, 1995), many are concerned with only a certain species of wildlife or simply react to an individual incident. Inclusive, thoughtful, wildlife laws are more likely to safeguard a variety of species in most situations.

Examples of Illegal Trade Cases

Many years ago, before I became a conservation officer, an internationally known herpetologist told me about smuggling amphibians and reptiles into the United States. There seems to be a general misconception among those who certainly should know better that the laws do not apply to them.

In the late 1980s, the director of zoological and botanical collections of the Indianapolis Zoo violated both state and federal laws. While in Peru to direct a tour sponsored by the Indianapolis Zoo, he personally collected red and blue poison-arrow frogs (Dendrobates reticulatus), Amazonian poison-arrow frogs (D. quinquevittatus), and other amphibians. He returned to the United States without any export documents from Peru for the animals. This person was obviously in a position to know the laws about wildlife but chose to ignore them. He was subsequently charged by the U.S. Fish and Wildlife Service and pled guilty.

In October 1999, a well-known British author and amphibian expert, Marc Staniszewski, was apprehended by Special Agents of the United States Fish and Wildlife Service and the United States Customs Service while trying to smuggle sixty salamanders out of the United States. He was charged with violations of both the Lacey Act and the Endangered Species Act. He had illegally collected the amphibians, including twenty Pacific giant salamanders (Dicamptodon tenebrosus) in California. During an interview, Staniszewski admitted that he had smuggled amphibians both in and out of the United States several times, and he recently had sold smuggled salamanders at a Baltimore, Maryland, amphibian show. Faced with these charges, Staniszewski pled guilty.

Following a tip, I conducted a covert investigation into the exportation of Indiana amphibians and reptiles to Florida. The dealers I spoke with claimed to import and export amphibians and reptiles internationally and were willing to buy all the native Indiana amphibians and reptiles I could supply. They reported buying amphibians and reptiles from Indiana every year, and they were especially interested in hellbenders (Cryptobranchus alleganiensis), four-toed salamanders (Hemidactylium scutatum), green salamanders (Aneides aeneus), and northern red salamanders (Pseudotriton ruber ruber), all of which are rare in Indiana.

Summary: What Can be Done?

The answer to the problem of commercial exploitation of amphibians and reptiles in general, and amphibians in particular, lies in communication, education, legislation, and enforcement. States should agree to compacts or reciprocal agreements. At a minimum, these agreements should be among contiguous states. Through these agreements, agencies could keep track of violators, the type of amphibians and reptiles of current interest, and what species other states are planning to list as endangered.

Education is critical. Herpetologists should volunteer to help enforcement branches of the regulatory agencies with such tasks as training officers, talking with administrators, and working to change the attitudes of agencies that are resistant. Herpetologists should also be talking with members of the commercial amphibian and reptile trade, most of whom are responsible people who truly care about their animals.

Where there are loopholes with wild populations being affected, we must make our laws tougher. When the director of the Indiana Department of Natural Resources was made aware of the information the Law Enforcement Division had gathered on the commercialization of our native amphibians and reptiles, he did not hesitate to sign an emergency order to halt the sale of our native (as well as dangerous) amphibians and reptiles until a permanent law could be put in place.

Finally, it is not good enough simply to pass regulations and then not fund the enforcement effort, including the education of the conservation officers. Funding to educate the public must not be overlooked. This can be easily achieved through brochures, Internet Web sites, and videos. Public awareness can lead to change.

I believe that next to the loss of habitat, the over-collecting of herpetofauna and the release of diseased, hybrid, and non-native amphibians and reptiles are the main problems facing us today. Everyone involved—scientists, representatives of commercial concerns, and regulators—must shoulder some responsibility. We have a simple choice: we can develop the social and political will to conserve amphibian populations, or we can watch numbers dwindle and species become extinct. Sustainable and wise commercial use of amphibians must be a component of the social change we seek to effect.

Appendix 22: Federal and International Laws Governing the Amphibian and Reptile Trade

(16 U.S. Code § 1531) Endangered Species Act

http://endangered.fws.gov/esa.html

http://endangered.fws.gov/

(7 U.S. Code § 2131-2159) Animal Welfare Act

http://www.api4animals.org/doc.asp?ID = 851

(9 CFR Ch. 1, Parts 1–4) Animal Welfare Act

http://www.access.gpo.gov/cgi-bin/cfrassemble.cgi?title = 199909

(16 U.S. Code § 3371) Lacey Act

http://iris.biosci.ohio-state.edu/regs/lacey/laceyamn.html

The Convention on International Trade in Endangered Species (CITES) of Wild Fauna and Flora

http://international.fws.gov/global/citestxt.html

http://international.fws.gov/cites/cites.html

NOTE : Although Internet addresses are included, some sites may move to a new address or become obsolete. We suggest using search engines to search by the name of the law.

CONSERVATION

TWENTY-THREE

Houston Toads and Texas Politics

LAUREN E. BROWN AND ANN MESROBIAN

Media hype has inaccurately portrayed amphibian declines as cataclysmic events of recent origin. It is more probable that declines in the United States began in earnest at least as far back as the mid-nineteenth century, shortly after the invention of the steel moldboard plow. Repeated use of this agricultural implement results in a 5–8 cm thick compacted layer of hardpan 5–30 cm beneath the surface, which is probably impenetrable to many animals including amphibians (Bromfield, 1955; Brown and Morris, 1990). The moldboard plow continues to be widely used in agricultural areas across the United States to the present day. In the latter part of the nineteenth century, the laying of drainage tiles in tilled fields began and continues to the present day (Walters, 1997). This eliminates or reduces springtime pooled surface waters, which are used as amphibian breeding sites. Thus, these two factors (moldboard plow and drainage tile) made much of the vast agricultural land across the United States unavailable as habitat for amphibians long before the post-World War II initiation of heavy biocide use and other recent causes of decline. Furthermore, industrialization of the country starting in the nineteenth century polluted many wetlands and undoubtedly further contributed to the amphibian decline. Numerous other factors were also likely of significance in the early decline of amphibians in the United States.

A number of species of amphibians have long been recognized as sliding toward extinction. For example, Houston toads (*Bufo houstonensis* Sanders, 1953) were first thought to be nearing extinction forty years ago, in 1962 (L.E. Brown, 1995). There is evidence that humans began to effect their decline over a half century ago (in the late 1940s; L.E. Brown et al., 1984), and research began on the trend toward extinction in 1965 (L.E. Brown, 1967). These events occurred long before the recent media hype about amphibian declines. The objective of this chapter is to review the relevant biology of Houston toads and to discuss some of the major political interactions concerning the species. While most amphibians recently recognized as declining have not been involved in negative politics, this chapter suggests what may lie ahead for those species.

Biology

Unlike so many of the media-publicized declining species, Houston toads can hardly be considered spectacular in appearance.

Nor are the toads known to be unusual in any aspect of their life cycle. Nonetheless, other attributes of Houston toads (e.g., mating call and natural hybridization) are unusual and of considerable interest. A review of the more important aspects of the biology of the species follows.

Characteristics and Life Cycle

Houston toads are moderately small, with snout-vent lengths for sexually mature preserved individuals ranging from 45–70 mm for males and 52–80 mm for females (L.E. Brown, 1973a; Brown et al., 1984; Shepard and Brown, this volume, Part Two). The parotoid glands are elongated but otherwise variable in configuration with a heavily pitted surface and one or more dark spots (L.E. Brown, 1973a). Parotoid glands contact the posterior ends of the preparotoid cranial crests or sometimes the postorbital cranial crests. The parietal cranial crests are reduced and not enlarged as bosses. The postorbital crests are only moderately thickened, usually not excessively (Brown and Thomas, 1982). The dorsal surface of the body is warty with 1–5 larger warts per spot, most frequently 1–3 warts per spot (L.E. Brown, 1973a). The dark brown or black spots on the dorsum lack accentuated borders, and the medial external surface of the upper jaw has dark spots. The cream-colored venter has a pectoral spot, with other small dark spots in the pectoral area and occasionally on the throat (L.E. Brown, 1973a; L.E. Brown et al., 1984). The dorsums of alcohol-stored (and many living) specimens are various shades of brown between the dark spots. Some living specimens are reddish between dorsal spots and warts. For additional morphological detail see Sanders (1953) and Shepard and Brown (this volume, Part Two). Tadpoles were briefly described by L.E. Brown et al. (1984).

The mating call of Houston toads is a long, high-pitched trill (Blair, 1956a; L.E. Brown, 1967, 1971b). Brown (L.E., 1975) indicated, "It resembles the tinkling of a small bell and is esthetically the most pleasing vocalization of all the toad species I have heard." The release call was described by Brown and Littlejohn (1972).

Mating typically takes place in early spring, although it may occur later in the spring and into early summer (Kennedy, 1961; L.E. Brown, 1971b; Hillis et al., 1984; Jacobson, 1989).

Amplexus is axillary and eggs are laid in strings in the water. Eggs hatch into tadpoles that metamorphose later in the spring or summer. Sexual maturity takes 1 year for males and apparently 2 years for females (Quinn and Mengden, 1984).

Distribution

Distribution records for Houston toads have been reported from the following counties in central and southeastern Texas: Austin, Bastrop, Burleson, Colorado, Fort Bend, Harris, Lavaca, Lee, Leon, Liberty, Milam, and Robertson (Sanders, 1953; Blair, 1956a; L. E. Brown, 1971b; Seal, 1994; Kuhl, 1997). The localities are isolated and scattered except in Bastrop County where they cluster in the Lost Pines.

Taxonomic Status

Houston toads are members of the *Bufo americanus* species group (Blair, 1963a). This assemblage contains five allopatric species plus one species (Woodhouse's toad, *Bufo woodhousii*) that is partially sympatric with the others. Artificial laboratory hybridizations indicate considerable genetic compatibility among the six species (Blair, 1963a). Furthermore, limited natural hybridization has occurred between Woodhouse's toads and four of the species with which they are sympatric (including Houston toads). The taxa within the species group that are most closely related to Houston toads are eastern American toads *(B. americanus americanus)* and dwarf American toads *(B. americanus charlesmithi)*. *Bufo americanus* are distributed in the eastern half of North America (with the exception of the southeastern coastal plain); dwarf American toads occur in the southwestern part of the range from southern Indiana to northeastern Texas; eastern American toads occupy the northern, eastern, and southeastern portions of the species range (Conant and Collins, 1998). There appears to be considerable overlap in morphological variation among Houston toads, eastern American toads, and dwarf American toads (as is also true for other members of the *B. americanus* species group). Thus, A. P. Blair (1957a) considered Houston toads as a subspecies of *B. americanus*, but this taxonomic arrangement was not followed by subsequent workers. Blair's (W. F., 1963a,b) suggestion that Houston toads are closely related to American toads has been accepted by many workers.

The anuran mating call has long been recognized (Blair, 1958a) as an important reproductive isolating mechanism, and it has played a prominent role in many speciation studies. Brown (L. E., 1973a) compared mating call characteristics of Houston toads recorded by L. E. Brown (1967) with characteristics of mating calls of eastern American toads (from New Jersey) recorded by Zweifel (1968a). Use of predictive equations allowed comparison at similar temperatures. The two taxa showed considerable difference in pulse rates, call duration, and dominant frequencies of their mating calls. Brown and Littlejohn (1972) also found release call differences between Houston toads and eastern American toads. Unfortunately, similar comparisons are not available for Houston toads and dwarf American toads.

The ultimate proof of speciation is maintenance of sympatry without extensive breakdown of reproductive isolation. However, this is a moot test for Houston toads and American toads because the taxa are not presently known to be sympatric. In light of the foregoing, it seems best to consider Houston toads as an incipient species unless new evidence proves differently.

Habitat

An essential component of the environment of Houston toads is sandy soil (Kennedy, 1961; L. E. Brown, 1971b). Bragg (1960a) found Houston toads to be weak burrowers that had difficulty digging in hard packed soil. L. E. B. (personal observations) found that six Houston toads spent most daylight hours buried under sand in the laboratory. Recently, the natural sand substrates were further characterized (Seal, 1994). They are friable, fairly deep, have subsurface moisture, and occur in five geologic formations that run in a northeast-southwest direction.

Pristine breeding sites are primarily ephemeral rain pools (Kennedy, 1961; L. E. Brown, 1971b), but a variety of other natural and human-made aquatic sites are utilized. Pristine vegetational associations inhabited include loblolly pine *(Pinus taeda)* forest, mixed deciduous forest, post oak *(Quercus stellata)* savannah, and coastal prairie (Kennedy, 1961; L. E. Brown, 1971b; L. E. Brown et al., 1984; Seal, 1994). The largest number of Houston toads occurs in the 70-square mile loblolly pine forest known as the Lost Pines of Bastrop Country, and Brown and Thomas (1982) pointed out that this supports the common misconception that Houston toads occur only in areas where there are pines. In fact, the species is not reliant on the presence of pines, and Houston toads are known to occur on sandy soils where there are no pines. By coincidence, sand is conducive to both pine growth and habitation by Houston toads.

A plethora of types of environmental destruction may actually or potentially be detrimental to Houston toads (L. E. Brown, 1971b; L. E. Brown et al., 1984; Seal, 1994; and others). Major threats include urbanization, recreational over-development, agriculture, and deforestation. Most Houston toad localities suffer from one or more types of habitat degradation.

Zoogeography

Houston toads are probably a Pleistocene relic left behind in central and southeastern Texas after the widely distributed parental species (precursor of the present *B. americanus*) underwent a range contraction to the north some 10,000 years ago after the retreat of the Wisconsin glaciation (Blair, 1958a; L. E. Brown, 1971b). The increasing aridity that characterized the Holocene in the southwestern United States was probably instrumental in isolating Houston toads into restricted favorable habitats (e.g., friable sandy soils into which the species could easily burrow to escape desiccation).

Natural Hybridization

Houston toads occasionally form natural hybrids with Woodhouse's toads and Gulf Coast toads *(B. nebulifer* [formerly *valliceps]*) in the Lost Pines area of Bastrop County (L. E. Brown, 1967, 1971b; Hillis et al., 1984). A number of species of *Bufo* have been found to form natural hybrids periodically (L. E. Brown, 1967a), but it is unusual to find two types of hybrids in the same area. The hybrid Houston toad × Woodhouse's toad is difficult to visually distinguish from both parental species because of overlapping variation. However, the pulse rate, duration, and dominant frequency of the hybrid mating calls are intermediate between those characteristics of the parental mating calls (L. E. Brown, 1971b). Furthermore, the pulse rate of the release call vibration of the hybrid is intermediate between those of the parental species (Brown and Littlejohn, 1972). These hybrids are fertile (L. E. Brown, 1971b).

The Houston toad × Gulf Coast toad natural hybrids are easily identifiable by the shape of the parotoid glands and intermediacy between the parental species in other characteristics (L. E. Brown, 1971b). The hybrid mating call is abnormal, poorly developed, and not always intermediate between those of the parental species. These hybrids are sterile (L. E. Brown, 1971b).

In many cases of natural hybridization, habitat destruction has been implicated as the cause of breakdown in reproductive isolating mechanisms, and this may be true in the Lost Pines (L. E. Brown, 1971b). However, a low number of Houston toads in interspecies choruses involving a larger number of the other species of *Bufo* may also be important. Under such conditions, accidental contact and hybridization are much more likely (L. E. Brown, 1971b).

Brown and Thomas (1982) pointed out the widely held misconception that natural hybridization is a cause of the trend toward extinction of Houston toads. Brown (L. E., 1971b) only mentioned the potential of natural hybridization to contribute to the decline. There is no evidence of widespread natural hybridization influencing the decline of Houston toads. The opposite effect is actually more likely (e.g., a low number of Houston toads being a contributing cause of the natural hybridization).

Rarity

Houston toads have been extirpated in the city of Houston and in Harris County (including the type locality). No specimens have been encountered there since 1976 (L. E. Brown et al., 1984). Extirpation is also suspected in Burleson, Fort Bend, and Liberty counties, but monitoring has been intermittent or scanty. The localities in Austin, Colorado, Lavaca, Leon, Milam, and Robertson counties are characterized by small or moderate numbers of Houston toads that easily could be extirpated. Recent surveys in Lee County confirmed the presence of a healthy chorus near the Bastrop County line (Kuhl, 1997, personal communication). The largest number of Houston toads is found in the Lost Pines of Bastrop County, but numbers seem to vary in time and space. Attempts to estimate the number of Houston toads using Lincoln-Petersen type indices are unrealistic and unreliable because of the considerable problems in fulfilling the restrictive conditions of the mathematical model and the difficulties inherent to the study of a fossorial animal (Poole, 1974; Brown and Moll, 1979; Brown and Cima, 1998).

Causes of the Trend Toward Extinction

Scientists, environmentalists, the press, and laypersons often look for a single cause of the decline of a given (or number of) species. Such is not the case for Houston toads. Several causes of the trend toward extinction seem evident, and their effects varied in time and place. Lundelius (1967) concluded that a drier, warmer climate with greater seasonal contrast was responsible for the disappearance of a number of extant northern-adapted species of mammals from central

Texas during the late Pleistocene and Holocene. Brown (L. E., 1971b) suggested that because Houston toads are closely related to the northern, more mesic-adapted American toads, the inability of the former to adapt to the increasingly arid and warmer climate of central Texas during the Holocene may be a cause of the reduction in range and numbers.

There was considerable over-collecting of Houston toads by both amateurs and professional scientists in the Houston area during the late 1940s and early 1950s (L. E. Brown, 1994). For example, the field notes of John C. Wottring (who first discovered the toads with his daughter Margaret in southern Houston) indicate he collected 66 specimens at a single chorus in 1949 (L. E. Brown et al., 1984). Shortly after the toads were described as new (Sanders, 1953), the species was considered a prize that many museums and individuals desired for their collections (L. E. Brown, 1994).

One of the most severe droughts of the twentieth century in central Texas occurred from 1951–56, and it had a marked effect on the abundance of amphibians and reptiles (W. F. Blair, 1957). This and other droughts undoubtedly affected Houston toads. The city of Houston experienced a tremendous and long-lasting post-World War II housing boom (L. E. Brown et al., 1984; L. E. Brown, 1994), which was the cause of the extirpation of Houston toads in Harris County by the late 1970s. The last half of the twentieth century has seen profound environmental destruction throughout much of central and southeastern Texas, including the former and present range of Houston toads. Undoubtedly, a number of types of habitat destruction affected Houston toads, and the impacts of these factors can be expected to increase in the future as human population growth continues.

Texas Politics

It seems almost axiomatic that endangered species become involved in political entanglements, and Houston toads are no exception. Political adversaries of Houston toads have included the U.S. Fish and Wildlife Service (USFWS), Texas Parks and Wildlife Department (TPWD), politicians in Texas and Washington, D.C., real estate developers and other moneyed interests, and human population growth. Following are accounts of the major battles in which Houston toads played a central role.

Listing and Potential De-Listing

The federal listing of a species is a lengthy, bureaucratic process that is often less concerned with the endangerment of the taxon than the politics of the moment (e.g., Who will be offended by this listing?). In marked contrast, the first federal listing of Houston toads was a simple, noncontroversial event that rapidly transpired. Surprisingly, one individual, Clark Hubbs, was primarily responsible for the listing (L. E. Brown, 1994). Hubbs is a well-known ichthyologist and Professor of Zoology at the University of Texas at Austin. While serving as a member of L. E. B.'s dissertation committee, Hubbs became a strong advocate for the conservation of Houston toads. After reading L. E. B.'s dissertation in June 1967, Hubbs left for the annual meeting of the American Society of Ichthyologists and Herpetologists, where he conferred with James A. Peters, Curator of Reptiles and Amphibians at the National Museum of Natural History, Washington, D.C. Peters was in charge of the "Redbook" of "Rare and Endangered Fish and Wildlife of the United States" published by the federal Bureau of Sport Fisheries and Wildlife. Hubbs was highly respected by ichthyologists and herpetologists, and his effusive enthusiasm convinced Peters of the grave problems faced by Houston toads. Thus, Houston toads received their first federal listing as endangered in the next edition of the Redbook (Peters, 1968).

The species was then automatically indicated as the only endangered anuran in the next federal list (Gottschalk, 1970) and was so designated in all subsequent official listings. In the 1970s, the concept of "endangered species" was new and exciting to many scientists, environmentalists, and the media. States developed their own lists of endangered species, and other independent compilers included Houston toads as endangered in their lists (e.g., Honegger, 1970; Gehlbach et al., 1975; Ashton, 1976). The Endangered Species Act (ESA) of 1973 (U.S. Senate and House of Representatives, 1973), the Lacey Act, other federal and state laws, and international agreements also potentially protected (at least on paper) Houston toads as an endangered species (L. E. Brown et al., 1984).

The listing and potential protection of Houston toads as an endangered species came quite easily, primarily because it was one of the first anurans to be nationally and internationally recognized as declining toward extinction. In subsequent years this process became much more difficult because of the rise of the endangered species bureaucracy in the federal government, the strong opposition by much of corporate America, and the realization that the listing of a species implied a major commitment of money to conserve the species (which might well prove to be an unsuccessful endeavor). Thus, instead of recognizing evidence in *support* of listing, the prevailing attitude has often seemed to involve developing any excuse possible to *prevent* listing.

In recent years, there have been suggestions bandied about by government officials concerning the "advantage" of de-listing Houston toads in spite of an absence of evidence for, or attempts at, biological recovery of the species. De-listing is a forced objective by the USFWS administrators even though it is not realistic in many cases (C. K. Dodd, personal communication). Furthermore, a petition to de-list Houston toads was sent to Bruce Babbitt, former Secretary of the U.S. Department of the Interior, from the National Wildlife Institute (Gordon, 1997). The justification was "data error" and waste of taxpayer dollars. We hope that the process of de-listing involves a bureaucratic quagmire fully as daunting as the present process of listing. The potential for de-listing before recovery may prove to be the greatest threat faced by Houston toads.

Proposed Critical Habitat Causes an Uproar in Houston

After the passage of the ESA in 1973, USFWS gave notification that it intended to start proposing "critical habitat" for endangered and threatened species (Greenwalt and Gehringer, 1975). Critical habitat was therein defined as "the entire habitat or any portion thereof, if, and only if, any constituent element is necessary to the normal needs or survival of that species." Shortly, another notice (Greenwalt, 1975) specified a list of species for which critical habitat would be determined. C. K. Dodd (Staff Herpetologist, Office of Endangered Species [OES], USFWS, Washington, D.C., personal communication) successfully pushed to get Houston toads (the only anuran) on the list in spite of a lack of interest in the species by anyone else in USFWS. In 1976, Robert A. Thomas, a herpetologist at Texas

A&M University, was asked by USFWS to recommend critical habitat for Houston toads (Thomas, 1997). He selected areas in Bastrop, Burleson, and Harris counties as critical habitat where Houston toads were then known to be extant and sent his recommendation to the Albuquerque Regional Office of USFWS on 3 November 1976 (Thomas, 1976a; USFWS, 1977). It was then forwarded to Washington, D.C. On 7 December 1976, Harold O'Conner (acting Associate Director of the USFWS' Washington, D.C., office) responded to the regional office that there were "serious difficulties" with Thomas' recommendations in Harris County because too much of the proposed critical habitat was valuable urban land (O'Conner, 1976). This complaint was sent to Thomas by the regional office with a memo (Johnson, 1976) stating, "Shoot-em down and send the critique to us for forwarding." In a letter to the regional office Thomas (1976b) defended his recommendation by stating (among other things), "The localities submitted on the referenced map are genuine localities from which Houston toads have been observed during the past two years by competent scientists."

Nothing was heard until five months later when USFWS (1977) published its version of proposed critical habitat for Houston toads. The Office of Endangered Species had consulted a map entitled "Known Location of Populations of Houston Toads in Harris County, Texas" dated February 1975 (Thomas, 1997). This map delineated 14 sand outcroppings within the zone suggested by Thomas in south Harris County, but contained no plotted locations for Houston toads (Thomas, 1997). OES chose seven of these sand outcroppings for inclusion as proposed critical habitat (Thomas, 1997), presumably assuming that because they had a sand substrate they must be suitable habitat for Houston toads. Unfortunately, three of these areas were not known to harbor Houston toads, and one site was 100% occupied by the Sharpstown Shopping Center (Thomas, 1997). It was an innocent but profound error due in part to lack of communication between Thomas and OES. Later, OES admitted they "goofed in listing the busy Sharpstown Shopping Center" (Scarlett, 1977b). However, the damage had already been done.

The publication of the proposed critical habitat created an unbelievable uproar in Houston, particularly among land developers and building contractors (L. E. Brown, 1994). This was because the proposed critical habitats were soon to be developed into commercial and residential areas. Federal agencies such as the Federal Housing Administration would not be able to insure housing developments in critical habitat and federally financed roads could not be built because the U.S. government cannot finance any program in critical habitat that might harm an endangered species. Thus, the expansion of Houston would receive a critical blow. Some land developers had spent large sums of money to purchase land in what was proposed as critical habitat. They envisioned considerable losses in profit if these areas were so designated. Inclusion of the Sharpstown Shopping Center in proposed critical habitat made the problem particularly ludicrous. Thus, letters were written to and from the White House (Yaffee, 1982). David Wolff, a Houston developer, was particularly active: "He wrote the White House repeatedly. The White House staff responded by pressuring the Interior Department. Ken Dodd, the OES staffer in charge of the toads, received phone calls from Senator Proxmire's office inquiring as to the status of the designation. In Dodd's telephone record, he wrote, 'I smell a Wolff here.'" (Yaffee, 1982). C. K. Dodd (personal communication) indicated, "In mid-late 1977, the White House called the Washington, D.C., USFWS and demanded that all critical habitat proposals for Houston toads should be withdrawn. President Carter was going to Houston to give a speech, during which he was to announce this favor to Houston politicians. I immediately refused—if someone wrote such a document, it damn well was not going to be me, even if the demand came from the White House. The matter was then dropped."

The critical habitat controversy received considerable coverage by the press, both in Texas and nationally (e.g., Anonymous, 1977a,b; Scarlett, 1977a,b,c). Newspaper cartoons ridiculed Houston toads (e.g., Houston Chronicle, 25 June 1977; Houston Post, 19 June 1977). An article in Fortune magazine (Anonymous, 1978b) made fun of Texans and demeaned Houston toads by publishing a drawing of an ugly toad wearing a large cowboy hat. At the high point of political activity, the widely watched Weekend television show (NBC) was devoted to the controversy. The Washington office of USFWS received written letters from 26 individuals and organizations commenting on proposed critical habitat. Sixteen of these were in favor of the proposal, seven were opposed, and three had no opinion (Dodd, 1978).

The regional office of USFWS had defended Thomas' original critical habitat proposal against the complaints of the Washington office (Yaffee, 1982). However, after proposed critical habitat was published, the regional office reversed its position, and office managers decided to recommend only the Bastrop County and Burleson County sites for critical habitat (Yaffee, 1982). The Washington office followed this advice and published the final determination on critical habitat on 31 January 1978 (Dodd, 1978). Even though Houston toads had been collected in Houston (Harris County) as recently as 1976 (L.E. Brown et al., 1984) and were not considered to be extirpated there in 1978, USFWS stated, "There is [sic] insufficient data at present on which to base a Critical Habitat designation for those remaining areas in Harris County. Therefore, these areas will not be acted on in this final rulemaking" (Dodd, 1978). Dodd (1977c,d; personal communication) was vehemently against this decision but was ordered to write the determination by his superior. At that time, Dodd was a temporary employee and his appointment was up for renewal (Dodd, 1977c). USFWS had clearly demonstrated that political pressure took priority over scientific data in making decisions on critical habitat. Unfortunately, no Houston toads were ever documented in Harris County after the final rulemaking.

Thus, an unglamorous little toad with a beautiful mating call struck fear in the hearts of Houston developers and politicians and shook the hallowed governmental halls of Washington, D.C. The toads may have lost the Critical Habitat War and become extinct in Houston, but their notoriety made them a symbol of declining species that have suffered from uncontrolled urban expansion. Furthermore, Houston toads had now won the support of environmentalists, and they still survived in the Lost Pines, over 100 miles to the northwest.

Development of the Recovery Plan and Subsequent Failure to Implement Conservation Efforts

An important provision of the ESA of 1973 is that the Secretary of the Interior is required to establish and implement a conservation program for each federally listed threatened and endangered species. In January 1975, L.E.B. wrote to Ronald Skoog (Chief, Office of Endangered Species, USFWS, Washington, D.C.) and asked what conservation plan they had in mind for

Houston toads. Skoog indicated that they had no plan but hoped to develop one. In October 1976, C.K. Dodd sent L.E.B. a copy of a short recovery plan authored by another USFWS employee (Anonymous, 1973) from the Washington office. This document had so many serious defects that L.E.B. wrote a highly critical, three-page response (L.E. Brown, 1976). Yaffee (1982) obtained access to L.E.B.'s letter and quoted its essence, "If this recovery plan is put into action, its main effect will be to hasten the trend toward extinction of B. houstonensis. I don't know who the Yahoos were that wrote up this plan, but they didn't know anything about anuran ecology." L.E.B. went on, "I think we can feel happy that the U.S. Congress has not adequately funded the Endangered Species Act, for if it had, the recovery plan for B. houstonensis would probably have been put into effect and as a result the species would probably now be extinct."

Perhaps hoping to recoup from the sting of the critical habitat debacle, or perhaps reacting to L.E.B.'s letters, USFWS decided to form a recovery team for Houston toads. It is the responsibility of the recovery team to formulate a recovery plan and attempt to resolve problems that arise that may adversely affect the listed species. In May 1978, Lynn A. Greenwalt (Director, USFWS, Washington, D.C.) appointed the first Houston Toad Recovery Team: Lauren E. Brown, William L. McClure (Texas Department of Highways and Public Transportation), Floyd E. Potter, Jr. (Leader; Texas Parks and Wildlife Department), Norman J. Scott, Jr. (USFWS), and Robert A. Thomas (Louisiana Nature Center).

Meetings of the Houston Toad Recovery Team were usually fiery affairs, with arguing not only about scientific matters (e.g., the pros and cons of using fire as a management tool in the Lost Pines), but also about bureaucratic pressures to compromise on political issues. The latter is exemplified by the replacement of the forthright terminology "environmental destruction" with the more palatable "habitat modification" in the Recovery Plan (L.E. Brown et al., 1984). The resulting compromise guaranteed that corporate America and governmental bureaucracy would not be offended. Arguing was so extensive that progress was slowed in writing the Recovery Plan. Eventually, USFWS became so exasperated they insisted that the recovery team concentrate on completing the plan. The team was told that the regional office needed the Recovery Plan so that Washington would know that progress was being made toward the conservation of Houston toads. Thus, USFWS sent in one of their representatives (D.B. Bowman) to oversee writing of the plan. L.E.B. and R. Thomas made progress by publishing a paper (Brown and Thomas, 1982) that was also incorporated into the Recovery Plan. In 1984, the Recovery Plan was finally completed and published (L.E. Brown et al., 1984).

USFWS then had in hand a recovery plan for a well-known endangered species that they could wave around Washington, D.C., and proclaim that they were finally doing something substantial to save a listed species and, thus, comply with the requirements of the ESA. This was of special importance because complaints were beginning to be voiced by some environmentalists that conservation efforts were mainly being focused on mammals and birds. Houston toads obviously were not cuddly or beautifully plumed, and moreover, as a veteran of the Critical Habitat War, the toads had attained considerable notoriety across the nation.

After 1984, no meetings of the Recovery Team were scheduled. USFWS gave no notice of termination, but after a few years it became evident that the team had ceased to exist. As the years passed, we hopefully looked for direct action from

USFWS in conserving Houston toads by implementing the Recovery Plan. There was none. Clearly, the only value of the developed Recovery Plan to USFWS was publicity.

In May 1994, USFWS and 14 other agencies sponsored a "Population and Habitat Viability Assessment Workshop" on Houston toads in Austin, Texas, that was attended by about 50 persons (including L.E.B. and A.M.). At that meeting, two employees of USFWS indicated to L.E.B. that a new recovery plan was needed. L.E.B. asked why, because there was nothing wrong with the first Recovery Plan and it had never been implemented. That question went unanswered.

In February 1998, the second Houston Toad Recovery Team was appointed: Lauren E. Brown, James R. Dixon (Texas A&M University at College Station), Jeff Hatfield (Patuxent Wildlife Research Center), David M. Hillis (University of Texas at Austin), Michael W. Klemens (Wildlife Conservation Society), Lisa O'Donnell (Leader, USFWS), Andrew H. Price (TPWD), Cecil R. Schwalbe (University of Arizona), Robert A. Thomas (Loyola University–New Orleans), and C. Richard Tracy (University of Nevada). Kathy Nemec (USFWS) and Eric Simandle (University of Nevada) were later added as team members. The new team met the following month and began work on the revised Recovery Plan.

We are confident that the new Recovery Team can produce a quality product. However, The question remains . . . will it be implemented by USFWS? Members of both Recovery Teams invested vast amounts of time and energy in this project, serving without salary. Their consulting services are normally $800–$1,000 per day (1998 value), plus expenses. Travel, lodging, sustenance, and secretarial help all involved substantial expenditures. Thus, total monetary value of the Recovery Plan(s) is considerable. If both plans end up collecting dust on the shelf because USFWS has not the will or the resources to implement them, the investment in this effort made by the scientific community will have been for naught.

Tunnels of Love

In the hiatus between Houston Toad Recovery Teams, the toads made national news again. In 1989, the Texas Department of Transportation (TXDOT) announced plans to improve traffic safety on a scenic 5.7-mile stretch of State Highway 21 in the Lost Pines, northeast of Bastrop. Citing an above-average incidence of accidents and fatalities along the forested route, TXDOT proposed utilizing $700,000 of federal funds to modify the highway. The agency's decision to remove over 1,000 of the majestic pine trees (about 30%) to improve visibility along the route ignited panic and outrage from local residents (Anonymous, 1989). A "Save the Pines Committee" was formed and, after energetic lobbying, a compromise plan was reached whereby only 160 trees were removed (Anonymous, 1990a).

The project area runs alongside Bastrop State Park and lies entirely within critical habitat for Houston toads. The Save the Pines Committee delayed the chainsaw massacre by citing the ESA and pulling USFWS into the fray (Yom, 1990). An $18,500 toad survey (Price, 1990) discovered (to no one's surprise) that many Houston toads were being run over on the highway during breeding season; of the 18 found, 12 were already flattened. TXDOT considered installing amphibian tunnels to channel the toads under the road at a cost of $628,000 (Anonymous, 1990c). While tunnels have been used successfully in Europe, Bastrop would have been the first place in the United States to build them (Matustik, 1990). The trees, tree huggers, lusty

toads, and their pricey tunnels of love made such a cute story begging for language play amongst copy-weary journalists that it was picked up by the *Wall Street Journal*, the *Chicago Tribune*, *Cox News Service* and the *Houston Chronicle* (Anonymous, 1990a McAuley, 1990; Yom, 1990).

In the end, the tunnels were nixed and fences were constructed to channel toads through existing drainage culverts, at a cost of $51,250 (Anonymous, 1990b,c). The fences were poorly designed, and, admittedly, a bad idea (A. Price, personal communication). The "toad tunnels," which were never built, continued to provide ample fodder for toad jokes and ridicule, usually in urban settings far from the source of amusement (McAuley, 1990; Kelso, 1995).

Golf Course Expansion: A Seething Cauldron of Science, Politics, and Houston Toads

The most widely publicized "taking" of Houston toads occurred during the 1995–97 expansion of the Lost Pines Golf Course inside Bastrop State Park. As this project unfolded, politics undermined science, popular opinion, and common sense, made a mockery of the ESA, and allowed federal funds to finance the destruction of the habitat of Houston toads.

Despite widespread grassroots opposition, Texas Parks and Wildlife Department (TPWD) transformed 35 acres of Houston toad habitat into nine golf course holes; degraded an additional 200 acres due to edge effect and runoff (Hamilton, 1995); eroded the reasonable and prudent measures intended to offset take (Grote, 1995a,b; Royder, 1995e); and failed to address mitigation promptly (Hess, 1998). While many state parks in Texas languished in disrepair on threadbare budgets, this pork barrel project siphoned off $500,000 from the Land and Water Conservation Fund (LWCF) of the National Park Service (NPS) and flaunted published guidelines for the use of these funds (Araujo et al., 1990; Anonymous, 1994a) to pamper a local golf club that had been operating rent-free on state land since 1936 (Anonymous, 1960, 1976, 1995j). The project traded a highly rated, historically preserved 9-hole bargain for a mediocre 18-hole course that cannot compete with much better ones a short distance away. While TPWD held their official position that the new nine holes honored a 30-year-old promise to compensate the city of Bastrop for donating 1,000+ acres to Bastrop State Park, insiders admitted that the agency was responding to intense political pressure that amounted to budgetary blackmail (Dawson, 1996). A full understanding of how this golf course expansion came about requires a brief history of Bastrop State Park, located in the Lost Pines east of Bastrop.

In 1933, the city of Bastrop deeded 435 acres of pine forest to the state of Texas to be developed into a park (Anonymous, 1933). Several months later, the city acquired an adjacent, logged-over, 1,145-acre tract, stipulating that "said land is intended to be used by said city for State Park Purposes . . ." (Anonymous, 1934). Additional acreage acquired over the years brought the park to its present 3,504 acres.

From 1933–39, the Works Progress Administration (WPA) and Civilian Conservation Corps (CCC) constructed park facilities on the original 435 acres using native red sandstone, local eastern red cedar and pine, and a rustic architectural style similar to that used in national parks. Improvements included a swimming pool, bathhouse, refectory, 12 cabins, roads and trails, camping and picnic areas, and a 9-hole golf course (Newlan et al., 1997). In 1936, the brand new 9-hole golf course was turned over to the Lost Pines Golf Club, a group of local golfers organized to manage and maintain the course (Anonymous, 1936).

Unfortunately, careless paperwork tainted the 1,145-acre acquisition—the deed was never transferred to the state. This error was caught in 1946 by an alert State Parks Board clerk (Camiade, 1946), and again in 1963 by a State Highway Department surveyor (Anonymous, 1963), but neither the city of Bastrop nor the state of Texas took action to resolve it (Anonymous, 1964).

In 1965, the same year L.E.B. discovered Houston toads in Bastrop State Park (L.E. Brown, 1995), the Texas Legislature appropriated $160,900 for park improvements at Bastrop (Gosdin, 1975). While trying to augment this budget with a NPS–LWCF match, the State Parks Board, now renamed Texas Parks and Wildlife Department (TPWD), stumbled upon the delinquent deed during the grant application process. With matching funds at stake, TPWD approached the city of Bastrop, hat in hand, to request the deed. Eleven days after a special meeting between city leaders and TPWD's Executive Director, J. Weldon Watson, and Park Services Director, William Gosdin, the city signed over the land to the state (Anonymous, 1966a,b). With the deed question resolved, TPWD began an upgrade of the park's roads and electrical and water systems and renovated the swimming pool, cabins, and other buildings, completing the work in 1968 (Anonymous, 1966–68).

Bastrop Mayor Jack Griesenbeck and city council members came away from the meeting with Gosdin and Watson believing that the city had been promised a number of concessions in return for the deed (Griesenbeck and Patton, 1966; Griesenbeck, undated, 1984; Kragh, 1984; Long, 1984; Simpson, 1984). As dedicated members of the Lost Pines Golf Club, the favor they wanted most was another nine holes. Because TPWD gave no indication that the golf course expansion was anywhere on the horizon, the Bastrop officials sought help from State Representative John Wilson, who nudged the department into producing a "feasibility study" for the golf course expansion (Wilson, 1975a,b). This document was presented to TPWD Commissioners in October 1975 (Anonymous, 1975).

Bastrop officials were devastated by the study, which showed no regional need for additional golf holes. The project would benefit a small fraction of park users at the expense of higher priority recreational facilities. "Additional golf course development [was] considered incompatible with the character of the park since it would severely impact a portion of the park's natural resources" (Anonymous, 1975). The study denounced the adverse effects on biotic communities within the 18-hole footprint and the secondary impacts that would degrade a much larger area. The "Environmental Assessment of the Proposed Golf Course Addition," written in 1974 by Larry Lodwick (TPWD Biologist for Resource Management), identified two at-risk species threatened by the expansion: the endangered Houston toad and "a species of ladies tresses orchids (Spiranthes tuberosa), which is known from only eight counties in Texas and is on the periphery of its Texas range in the uplands of the Bastrop County Pineywoods" (Lodwick, 1974). Citing the ESA, the study concluded that expansion would trigger a federal Environmental Assessment (EA) and Environmental Impact Study (EIS). Because so little was known about Houston toads, it would be impossible to design an adequate mitigation plan; the best advice staff could offer was to avoid any manner of habitat alteration.

A stern rebuttal from former Mayor Greisenbeck (who was Bastrop County Judge [county administrator] in 1976) accused Watson and Gosdin of misrepresentation, demanded to know where the federal money had gone, and belittled the importance of toad habitat (Greisenbeck, undated). A 1978 resolution passed by the Bastrop City Council requested that TPWD either build the additional nine holes or return the land to the city (Anonymous, 1978a).

No credible evidence of the alleged "promise" ever surfaced, and Gosdin (1975) even drafted a memorandum denying it, which TPWD added to the list of reasons they should not pursue the project. In 1975, the ESA was a new and intimidating law, and TPWD was unwilling to test it for nine new holes. The notion of expanding the course appeared dead.

Thirteen years later, a disgruntled Bastrop resident contacted State Representative Robert Saunders, briefed him about the 1966 deed transfer (McClure, 1988), and provided him with written statements from the former mayor and three city council members claiming that a promise to expand the golf course had been violated by the state of Texas (Griesenbeck, 1984; Kragh, 1984; Long, 1984; Simpson, 1984). As Chair of the House Environmental Affairs Committee, Saunders held the purse strings for TPWD and many other state agencies. He took a personal interest in the golf course expansion, and over the next eight years he found ways to remove every obstacle standing in its way.

When the issue of the golf course expansion splashed across the headlines of the Bastrop Advertiser in September 1990, it was painted as a conflict between city-backed local golfers who felt that they were promised the expansion in 1966 and outside environmentalist meddlers (two Audubon Society chapters) defending the Lost Pines and its endangered resident toad (Anonymous, 1990b).

Three months later, a fledgling, all-volunteer watchdog group, Bastrop County Environmental Network (BCEN), met with members of the Lost Pines Golf Club and TPWD zoologist Andy Price to explore ways the expansion could occur without destroying toad habitat. BCEN's preferred option put the new nine holes across Business Loop 150 from the park, in a heavily modified area devoid of suitable habitat for Houston toads. TPWD asserted that an expensive overpass or tunnel would be needed to safely transport golfers across the roadway. Price suggested as an alternative that, were the expansion to occur inside the park, a "toad fee" tacked onto the greens fee could fund a regional Habitat Conservation Plan that would help preserve the species while Bastrop boomed from nearby Austin's sprawl. A workable solution seemed at hand, and BCEN published an optimistic article about this meeting in its journal, Beacon (Rassner, 1991).

To implement this compromise, BCEN wrote to TPWD Executive Director Andrew Sansom in January 1991, offering to work with the department and golf club toward expanding the course and protecting the toads (Miller, 1991). BCEN suggested using interpretive exhibits in the park and golf course to educate park users about Houston toads. The letter recommended siting the additional nine holes outside toad habitat or mitigating in-park expansion with the purchase of additional habitat funded through a greens fee surcharge. Sansom did not respond.

After hearing of progress in the golf course expansion, BCEN wrote to Sansom again in February 1992, reiterating points made in the January 1991 letter (Miller, 1992a). Again, there was no answer. Frustrated by the lack of response from the department, BCEN published an article that detailed the history of recovery efforts for Houston toads and urged that expansion occur outside toad habitat (Miller, 1992b). The piece included a cartoon of a Houston toad choking on a golf ball (Fig. 23-1).

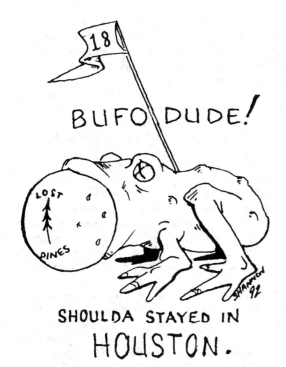

BUFO DUDE!

SHOULDA STAYED IN
HOUSTON.

FIGURE 23-1 Political cartoon by Shannon Cassel satirizing the effect on Houston toads *(Bufo houstonensis)* of the proposed golf course expansion in Bastrop State Park, Bastrop County, Texas (Beacon, Volume 2, Number 3, p. 11, Summer 1992).

In August 1992, BCEN distributed a position paper on the proposed golf course expansion (Miller et al., 1992) to Texas Governor Ann Richards, U.S. Representative Greg Laughlin, State Representative Robert Saunders, Andrew Sansom, Lost Pines Golf Club President Gilbert Cervantes, Bastrop Mayor David Lock, Bastrop County Judge Randy Fritz, the Bastrop County Audubon Society, and others. Key elements of the paper (as published in the *Beacon*; Miller, 1992c) included forming a Mitigation Advisory Panel; building outside of toad habitat or purchasing additional habitat to offset take; minimizing chemical and water use on the golf course; implementing a five-year chemical impact study; and launching an area-wide education program and habitat management plan.

In mid-August, USFWS visited Bastrop State Park to inspect the proposed expansion site (Hamilton, 1995). TPWD delivered a draft Habitat Conservation Plan (HCP) to USFWS in September and requested authorization for incidental take of toads under section 10(a)(1)(B) of the ESA. BCEN, the Lone Star Chapter of the Sierra Club, the Southwest Regional Office of the National Audubon Society, and others who opposed the destruction of the Lost Pines and habitat of Houston toads were unhappy with TPWD's intention to build inside the park. However, the Section 10(a) process called for public comment, and an EIS would likely be part of the HCP, so opponents believed due process would provide ample opportunity to defeat the in-park proposal. Besides, both the Audubon Society and the Sierra Club had opposed the project on record and intended to argue against its funding in TPWD's budget.

However, in January 1993, TPWD approved funding the golf course expansion in their 1993 capital budget without allowing public comment as had been promised (Anonymous, 1993a). Scott Royder (Sierra Club) and Dede Armentrout (Audubon) protested the decision. TPWD assured them that public comment would be allowed before the project could receive final approval.

In the summer of 1993, TPWD applied for a $500,000 Land and Water Conservation Fund grant from the National Park Service for golf course expansion (Hamilton, 1995). The money was to match $250,000 from the Lost Pines Golf Club and $250,000 from TPWD (Anonymous, 1994b). By introducing federal involvement to the project, TPWS traded the cumbersome Section 10(a) permitting process for the streamlined Section 7 consultation, skirting public comment and an EIS. To access federal funds, TPWD needed only a favorable biological opinion from USFWS, and a Finding of No Significant Impact (FONSI) from NPS.

BCEN declared a red alert (Miller, 1993) and rushed a letter to Executive Director Sansom, accusing TPWD of fast-tracking the expansion to avoid public debate. The *Bastrop Advertiser* featured BCEN's letter (Dureka, 1993) and Sansom's reply (Sansom, 1993) on its editorial page (McAuley, 1993). Sansom vehemently denied that the mitigation plan had been compromised, saying, "We intend this consultation to be a model, not a blight on the ESA and TPWD . . . the golf course expansion will take Houston toad habitat, but it will also be the vehicle to fund a habitat plan for the entire area and thus make a significant contribution to recovery of the species." In an effort to quell opposition, TPWD invited BCEN (including A.M.), Bastrop County Audubon members, and others to walk through the areas chosen for the new nine holes, assuring them that the largest trees would be spared. The visit had the opposite effect—toad or no toad, invitees found the idea of bulldozing in the park obscene.

In the *Houston Chronicle*, an article covering the controversy muddled the issue (Harper, 1994), quoting Representative Robert Saunders, "We're not going to go out there and bulldoze frog (sic) habitat. As I understand, we're not going to hurt the habitat of the frog . . . actually, we have a lot of support from environmental groups . . . the only opposition . . . is coming from the Audubon group, which is worried about losing trees." In the same article, Andy Price of TPWD is quoted, "Habitat will be cleared if we expand in that area. We need to ensure that enough habitat is left to keep a robust population."

In the wake of the golf course publicity, USFWS logged a dramatic increase in phone inquiries (Dureka, 1994) about Houston toads. Unprecedented growth was anticipated in Bastrop County due to construction of the new Austin-Bergstrom International Airport on Austin's southeast side. Furthermore, the habitat of Houston toads occurs in the Lost Pines, an area favored for residential development. In early 1994, USFWS developed guidelines for building in Houston toad habitat (Anonymous, 1994c). This seven-page document was available from USFWS on request; it contained an unreadable map and a cumbersome procedure intended to help residents determine if a 10(a) permit would be necessary for proposed construction projects.

Perhaps realizing that the user-unfriendly guidelines would lead to haphazard conservation measures at best, USFWS offered to help Bastrop County adopt a regional HCP for Houston toads before urbanization got out of hand. Although Section 7 of the ESA did not require it, TPWD still claimed commitment to an HCP and offered the greens "toad fee" as partial funding (Sansom, 1993). Bastrop County Judge Randy Fritz, a golf course supporter, believed that the best (and possibly only) way to ensure the new holes were built was for the county to take a leadership role in designing a regional HCP (R. Fritz, personal communication). He organized steering committee meetings to study its feasibility (Vinklarek, 1993, 1994; Zappler, 1994) and invited participants from the environmental community (including A.M.), landowners, realtors, bankers, utility providers, city and county officials, and biologists from USFWS, TPWD, and the Lower Colorado River Authority.

At the first of these meetings, a handful of local pro-business types predicted dire economic consequences for Bastrop County due to growth restrictions (Rassner, 1994), but community leaders eventually agreed that an HCP could help guide development. The Texas Constitution contains no provision for planning, zoning, or ordinance-making authority in unincorporated areas outside city limits. However, if an HCP could empower Bastrop County to impose restrictions on development within the habitat of Houston toads, a portion of the county could be protected from high density, low-end development.

Three months into the HCP meetings, USFWS released its draft biological opinion for the golf course expansion (Rogers, 1994), which vetoed the in-park expansion. It recommended "an alternative site that does not support Houston toad habitat," such as the Loop 150 location, an option TPWD had already abandoned. Bastrop officials felt betrayed (R. Fritz, personal communication); after months of friendly HCP meetings, USFWS had pulled the plug on the only funding source identified for the plan. Within days of the document's release, the city of Bastrop, spurred by County Commissioner Johnny Sanders, threatened to sue TPWD for the return of the (now resurveyed) 1,134 acres if it did not expand the golf course (McAuley, 1994a). Representative Saunders stepped in to mediate the conflict, and the city backed off.

At TPWD headquarters in Austin, USFWS Texas State Administrator Sam Hamilton explained to TPWD officials, Representative Saunders, and representatives from BCEN (including A.M.), the Audubon Society, and the Lost Pines Golf Club that TPWD could achieve a more favorable biological opinion by modifying its proposal or by discovering new scientific data about Houston toads that proved no adverse affect from the proposed action. He warned, however, that because Bastrop State Park contains the world's largest population of Houston toads and some of the finest critical habitat, this opinion would be difficult to overturn. Hamilton further indicated that to adequately fund a mitigation plan, the "toad fee" would reach $5–$10 per round, making the course unaffordable to its users. Saunders may have felt that Bastrop County was encouraging a hard-line position from USFWS, because shortly after this meeting, he exerted intense pressure on Judge Fritz to abandon the "unnecessary and burdensome" regional HCP (R. Fritz, personal communication).

Three days after the *Bastrop Advertiser* headlined the city's potential lawsuit, USFWS sponsored a Houston Toad Population and Habitat Viability Assessment (PHVA) workshop in Austin (attended by A.M. and L.E.B.). The workshop fulfilled one of TPWD's mitigation measures for golf course expansion, and it was a necessary step to set up the HCP for Houston toads. However, the pending suit cast a heavy pall over the event, and HCP stakeholders found it hard to get past a sense of distrust to achieve cooperation. Throughout the workshop, attendees identified threats to Houston toads and their habitat and presented a generous dose of research results and population modeling. Yet on the last day, an astonished facilitator, Ulysses S. Seal, had no answer when BCEN member Robert Schmitt abruptly asked where people (i.e., local residents) fit into the equations. The workshop lacked what Bastrop County participants needed most—a concrete strategy they could take home and implement to preserve Houston toads and their habitat.

A month later, the HCP steering committee (including A.M.) met for the last time. Release of the negative draft biological opinion had been a tragic accident of timing and circumstance, but it made USFWS appear untrustworthy. The city's suit and golf course expansion were in gridlock, and Judge Fritz, a relative newcomer to politics, saw no reason to challenge the experienced and powerful Saunders (R. Fritz, personal communication). The HCP committee agreed that politics had killed the plan (Mesrobian, 1994b), and TPWD's Price opined that the golf course project had been set back two years.

Under USFWS guidance, two groups emerged from the HCP committee: public outreach and biological research. The public outreach group (including A.M.) met only twice. While Bastrop County residents sought updates on the golf course expansion and straight answers about land use restrictions within the habitat of Houston toads, USFWS personnel wanted to discuss bumper stickers and T-shirt designs (Anonymous, 1994d). By September 1994, USFWS had abandoned public outreach meetings.

The biological research group conducted surveys of potential habitat for Houston toads during the 1994–95 breeding season (see Price, 2003). Led by professional biologists, this volunteer effort sparked considerable participation among local residents, and it inspired BCEN to seek a grant from World Wildlife Fund for a survey the following year. A drought in 1996 made an accurate census of the toads unfeasible and the effort was rescheduled for the next year. Over 80 local volunteers joined ToadWatch '97, driving county roads to listen for Houston toads, helping to define privately owned habitat, and raising public awareness about this endangered mascot of the Lost Pines (Mesrobian, 1996, 1997).

Six weeks after the Population and Habitat Viability Assessment Workshop, A.M. attended a second meeting called by TPWD to discuss the draft biological opinion with Sam Hamilton from USFWS and other interested parties. Paying lip service to the USFWS recommendation, TPWD rejected two options that would not take Houston toad habitat. The plan placing all new holes within the footprint of the existing nine holes would yield a cramped, unsatisfactory course. Overpass and land acquisition costs ruled out the location across Loop 150, plus the site lacked aesthetic appeal. TPWD then promoted the third option—taking 56 acres inside the park. Yet, the golfers claimed they could afford a toad fee of only 25 cents per round.

Hamilton countered with five crucial mitigation elements: (1) design a chemical-free golf course; (2) charge an adequate toad fee to build a "Toad Fund;" (3) dedicate the remainder of Bastrop State Park as a natural area in perpetuity; (4) acquire and protect additional habitat; and (5) launch a public education and outreach program and continue to monitor and perform research on Houston toads in Bastrop State Park.

Realizing that TPWD intended to bulldoze nine holes in the park at all costs, Lost Pines Golf Course superintendent Mike McCracken stunned the meeting attendees (including A.M.) by suggesting that if the Lost Pines Golf Club and TPWD needed an 18-hole course they should relocate entirely and start from scratch. He further asserted that the in-park option was so fraught with restrictions and controversy that by the time it was built, nobody would be happy. These prophetic words went unheeded.

A third mid-August meeting to discuss the biological opinion never occurred. Instead, at TPWD's annual commission meeting on 25 August, the agency approved $500,000 of LWCF monies for the expansion, pending NPS approval (Mesrobian, 1994a). The decision had clearly been made in advance, so the heated discussion and public comment from Sierra Club's Scott Royder and BCEN's A.M. were mere formalities. They were barely noticed amidst the pleas of a dozen Lost Pines golfers, coached by Saunders to beg for the expanded course.

Two commissioners refused to support the action—Mickey Burleson abstained and Terry Hershey vehemently opposed tearing down the Lost Pines for golf (Leggett, 1994). Sansom and the others almost apologized for the project, referring to the mythical 1966 "gentlemen's agreement" as a 28-year-old obligation to the city of Bastrop. Chairman Ygnacio Garza admitted, "That's not what we should be doing with park land," and that he would like TPWD to consider a policy change so they would "never build another golf course" (McAuley, 1994b). Reluctance aside, the vote was cast under the watchful eye of Representative Saunders.

BCEN immediately circulated a petition demanding a full Environmental Impact Statement for the project and sent over 2,000 signatures to NPS, TPWD, the city of Bastrop, USFWS, and Representative Saunders. Saunders fired back a mass mailing, claiming the holes had been reworked to satisfy the environmentalists and that everything was on track. He proclaimed, "The Bastrop economy will greatly benefit, the golfers will have the best public walk course in the state, the great loblolly pines will be enhanced along the fairways, and resource protectionists will receive a perpetual fund for the Houston Toad and a large amount of mitigated land to improve the existing population of the Houston Toad" (Saunders, 1994). Unfortunately, most of this did not occur.

In January 1995, buckling to intense political pressure, USFWS released the final biological opinion of "no adverse effect" (Hamilton, 1995). The proposed action directly impacted 35 acres, and it included a mitigation plan so grandiose and noncommittal that NPS balked at having sole responsibility for its implementation. An amendment to the opinion requiring NPS and TPWD to share that burden was added before NPS would consider the LWCF grant application (Grote, 1995a). Within months, USFWS Texas State Administrator Sam Hamilton, who had defended the habitats of many endangered species during his tenure in Texas, was transferred out of state. His position was subsequently eliminated when the agency reorganized, and for the next three years, the Austin office of USFWS floundered without strong leadership.

It would take an obedient and trustworthy Boy Scout to design an effective mitigation plan based on the final opinion. The "Toad Fund" so crucial to John Roger's original plan, and defended vigorously by Sam Hamilton just months earlier, was demoted to a mere "conservation recommendation." Other measures such as imposing a moratorium on all other construction inside Bastrop State Park and restricting (not forbidding) chemical use on the golf course are inherently unenforceable; no language requires TPWD to adopt rules to ensure that the measures are followed.

While the opinion required TPWD to "secure and maintain, in perpetuity, from 1,000 to 2,000" acres of additional toad habitat through conservation easements with private landowners, it did not require that the easements themselves be perpetual. Nor did the plan explain who would relocate the resident toads when each temporary easement expired. Further, it did not address a likely result of such piecemeal, short-term easements—that all habitat would one day be developed, leaving no willing landowners remaining to provide safe harbor.

Shortly after USFWS issued the biological opinion, TPWD gave a two-week notice of a public meeting at Bastrop State Park to discuss plans for the golf course expansion. Construction was to begin two days after the meeting, on 15 February. The Sierra Club immediately faxed an urgent letter to NPS, requesting that the project be halted until an EIS was initiated (Royder, 1995a). Sierra (1) criticized TPWD for shutting the public out of the planning process and denying the project's adverse impacts to non-golfing park users; (2) challenged the need for another nine holes in the park, given the 50 golf courses within 50 miles of Bastrop, including an 18-hole course just two miles away; (3) decried TPWD's woefully inadequate Environmental Assessment (EA; Anonymous, 1994e), which did not address the adverse impacts to the park's cultural resources, ground or surface water, highly erodible soils, or federally protected bird species; and (4) questioned how USFWS could so handily reverse its opinion without substantial new biological information about Houston toads. Sierra also pointed out that the biological opinion prohibited construction during the breeding season, which includes February.

While the Sierra Club feverishly tried to prevent the project from moving forward, TPWD furiously greased the skids. Within days of announcing the public meeting, TPWD sent a letter to USFWS requesting a revision to the biological opinion so that dirt work could commence during breeding season (Sansom, 1995). USFWS immediately responded (Grote, 1995a) with an amendment stating that construction could occur during breeding season if "a herpetologist-led crew" was on site to "relocate any toads encountered in the construction area." This absurd amendment was protested by the Sierra Club (Royder, 1995e; Anonymous, 1995d) and ridiculed in the *Smithville Times* by a cartoon of a golfer in a TPWD hat and a "Toad Relocation Team" shirt, shown at the top of his backswing, teeing off at a Houston toad (Fig. 23-2).

Opponents of the project used the two weeks to organize, and they turned the public meeting (attended by A.M.) into a protest rally against the waste of public funds on gratuitous golf. While freezing rain dribbled on the overflow crowd outside, Representative Saunders and TPWD brass preached their "win-win" proposition to 300 people, many of whom held anti-golf course signs and wore buttons proclaiming "No Tax Dollars For Golf" and "No Pork In My Park."

TPWD hogged the only microphone, forcing participants to shout. Furious park users demanded to know why TPWD could afford $250,000 for the golf course when it could not keep the swimming pool open or provide a playground for youngsters. Many participants demanded proof of the 1966 "promise," and a vote on the issue since they had been offered no opportunity for meaningful public comment. Bird-watchers, hikers, campers, and picnickers asked why Representative Saunders and TPWD would rather serve the 2% of park users who golf than the 98% who do not, referring to the mitigation plan's park-wide moratorium on all other construction. Saunders defended the golfers as his "constituents," disdaining all others as "a special interest group."

At times, the meeting resembled a farce. Larry McKinney, TPWD's Director of Resource Protection, was asked to comment on one ludicrous excerpt from the EA: "Long-term air quality will be improved as a result of oxygen being produced in the expanded turf areas." McKinney sputtered, "If it's good enough for the U.S. Fish and Wildlife Service, it's good enough for me."

Shady details soon surfaced about funding. When asked if the Lost Pines Golf Club had its $250,000 share of the project in hand, TPWD admitted that the club could pay it back later out of revenue generated on the expanded course. When asked if TPWD had misrepresented local funds available in the NPS

FIGURE 23-2 Political cartoon by Shawn Janak satirizing the relocation team for Houston toads *(Bufo houstonensis)* during construction of the golf course expansion in Bastrop State Park, Bastrop County, Texas (Smithville Times, 25 January 1996).

grant application in order to receive the match, Sansom had no answer. A. M. reminded Sansom that NPS had not approved the grant application and that without the federal match, TPWD could not skirt the 10(a) permit, biological opinion notwithstanding.

After three hours of heated discussion, the meeting broke up. Fearing for his own safety, Representative Saunders was escorted to his car by four police officers. A *Smithville Times* editorial raked him over the coals for disrespectful behavior unbefitting of a public servant (Laird, 1995).

On 13 February 1995 (the day of the public meeting), the Sierra Club alerted the Advisory Council on Historic Preservation that the Texas Historical Commission had not cleared TPWD to begin work on this federally funded project, and it requested Interested Party Status in the Section 106 review process (Royder, 1995b). On 15 February, BCEN members videotaped a van full of prisoners arriving at the park to begin clearing. The supervisor chose to avoid controversy and retreated.

On 16 February, three days after the public meeting, NPS denied TPWD's $500,000 grant application (Rogers, 1995a), citing three reasons. First, Bastrop State Park had been proposed for nomination as a National Historic Landmark District, and it would be inappropriate to use federal money for the project. Second, NPS had received approximately one hundred letters, 2,000 petition signatures, and numerous phone calls opposing the grant request. This degree of opposition constituted sufficient controversy to require an EIS for the project. Third, the grant application was missing the comments of the State Historic Preservation Officer, a justification of need, and a more thorough EA.

Encouraged by the denial of federal funds, the Bastrop County Concerned Taxpayers Association (BCCTA) denounced the expansion as a waste of tax dollars and launched a campaign to block the expenditure of any public funds benefiting the project. BCCTA utilized the Texas Open Records Act to obtain evidence that the non-profit Lost Pines Golf Club actually made a tidy profit off greens fees despite the rent-free use of state-owned land, and it suggested that this park parasite pay for the expansion out of its own pocket (Anonymous, 1995c). BCCTA lobbied Bastrop County to withdraw in-kind support for the project, but only succeeded in a reduction of the amount pledged.

Loss of the federal grant seemed to nail the coffin lid on the golf course expansion; the historic value of Bastrop State Park and the authority wielded by the National Historic Preservation Act blindsided TPWD (Cromeens, 1995a). However, the well-oiled political machine had gathered too much momentum to stop. During a phone conversation, Representative Saunders badgered Jerry Rogers to track down the source of this unforeseen obstacle (Rogers, 1995b).

Through sheer coincidence, historical researchers at NPS inventorying CCC-built parks throughout the country had recommended Bastrop State Park be nominated to the National Register as one of the best-preserved examples. Under these circumstances, adding nine new holes to the quaint, charming layout of the CCC-era golf course would detract from its historic value. For the project to proceed, it would need clearance from the Texas Historical Commission and its federal equivalent, the Advisory Council of Historic Preservation.

State approval came easily. The Texas Legislature was in session and Saunders chaired the House Environmental Affairs Committee, which approves the budget for the Texas Historical Commission. The Commission signed off on the project

immediately (Bruseth and Perttula, 1995a,b). Federal approval took longer.

The Sierra Club continued to pressure the Advisory Council on Historic Preservation about the Section 106 Review, which was still not initiated. During confirmation hearings for new TPWD commissioners, State Senator Gonzalo Barrientos (who grew up in Bastrop) made an offhand remark supporting the golf course expansion (Anonymous, 1995a). Lone Star Sierrans jammed the Senator's phone lines for three days in protest.

By now the project had become TPWD's number one priority (Cromeens, 1995b). Andy Sansom met with his friend Bob Armstrong, Assistant Secretary for Land and Mineral Management at the U.S. Department of the Interior (and formerly a Texas Land Commissioner and TPWD Commissioner; MacCormack, 1995). Ten days later, NPS Director Roger Kennedy sent Destry Jarvis (Assistant Director for External Affairs) and Historian Linda McClelland (National Register of Historic Places) to Bastrop State Park to find a way for TPWD to build the nine holes yet comply with historic concerns (Anonymous, 1995b). Revised plans squeezed the new holes onto smaller acreage and deleted the driving range.

A Memorandum of Agreement incorporating the changes was drafted between TPWD, NPS, the Texas Historical Commission, and the Advisory Council of Historic Preservation (Anonymous, 1995f). TPWD agreed to remove a trashy row of golf cart sheds and to restore two historic buildings used by the golf club. One building was to become an interpretive center for the CCC/WPA, the Lost Pines, and Houston toads. Less than three months after funding was denied, NPS reversed itself, issued a FONSI, and approved the LWCF grant (Anonymous, 1995e).

Through the Freedom of Information Act, BCEN acquired federal documents and reconstructed the paper trail leading to approval of this misguided project. Documents evaluating the need for the golf course expansion based on LWCF guidelines revealed that it had earned the lowest score of any project funded in Texas between 1991 and 1995 (Anonymous, 1994a). High-level soul searching at the Department of the Interior allowed NPS to back-pedal on the need for a more thorough EA and an EIS. Though it would be "generous to call the EA weak" (Rogers, 1995c), there was "major political pressure to proceed with a FONSI" (Sovik, 1995). Bob Baum, Associate Solicitor, Conservation and Wildlife, ultimately proclaimed the EA to be adequate and denied the need for an EIS based solely on public controversy (Rogers, 1995d). BCEN (Mesrobian, 1995) and the Sierra Club (Royder, 1995b,c,d) tried to appeal the new NPS decision, arguing that TPWD had been allowed an appeal, but NPS closed the case. NPS admitted in the FONSI that "local government constituents and a small number of supporting public" had pushed the project through (Anonymous, 1995e).

Several TPWD employees, angered by the project early on, expressed their opinions in "Tricks & Trials," an underground spoof of the employee newsletter "Tracks & Trails." They accused Sansom of promising the golf course in Bastrop in order to get his budget passed in the Legislature (Anonymous, 1993b). The following year, Andy Price acknowledged "significant political pressure" to complete the project (Harper, 1994). In 1996, TPWD finally admitted to the media that Representative Saunders, "a leading golf course supporter, had been strategically positioned in the Legislature to affect the outcome of a crucial agency funding initiative. Under those circumstances, the agency's leadership did not believe it had any alternative to building the golf course . . . 'There was a political gun to our temple,' [one high-ranking TPWD] official said. 'We

FIGURE 23-3 Political cartoon by Shawn Janak satirizing the attitude of the Texas Parks and Wildlife Department and the effects of its expanded golf course on Houston toads *(Bufo houstonensis)* in Bastrop State Park, Bastrop County, Texas (Smithville Times, 2 November 1995).

had to do it.'" (Dawson, 1996.) After 16 years in office, Representative Robert Saunders chose not to seek reelection in 1996.

By late June, TPWD had laboriously flagged the areas to be cleared for the new nine holes. Days before land clearing was to begin, every scrap of flagging tape that defined the proposed fairways and greens mysteriously vanished in the dark of night, presumably removed by phantom tree huggers (Anonymous, 1995g). Weeks later, on the first day of heavy machinery work, EarthFirst! member Eric Daniel Dollinger secured himself—using a bicycle lock around his neck—to a bulldozer poised to level the forest and was arrested. Thirty people joined in the protest, representing BCEN, BCCTA, Sierra Club, and Audubon. Dollinger received front page headlines in Bastrop and Smithville, mention in the Austin and Elgin papers, a $3,000 fine, and hundreds of Brazilian fire ant *(Solenopsis invicta)* bites for his efforts (Anonymous, 1995h,i; Kaye, 1995; Spencer, 1995; Todd, 1995).

Construction proceeded fitfully, and two years elapsed before the course was ready for play (Todd, 1996). TPWD endured heavy criticism for destroying thousands of Bastrop's fabled Lost Pines in a park where they should have been preserved—and resentment toward the Lost Pines Golf Club festered

(Anonymous, 1995d; Dawson, 1996; Dureka, 1996). The *Smithville Times* ran a cartoon of a wealthy couple dining at the "TPWD Supper Club" hosted by Andy Sansom—while the man orders the "Houston toad plate" his wife reminds him he has a golf game that afternoon (Fig. 23-3). Sansom found this cartoon so amusing he purchased the original from the *Times* (L. Zappler, personal communication).

Doubts about TPWD's intention to implement the mitigation plan mounted (Anonymous, 1995k; Royder, 1995e). In April 1995, TPWD and USFWS co-sponsored a workshop in Bastrop to define agricultural land use activities compatible with the conservation of Houston toads (Linam and Grote, 1995). Workshop participants concluded that prescribed burning could occur, provided a 10(a) permit was first acquired. Just eight months later, TPWD conducted an unpermitted burn in an area inhabited by the toad in Buescher State Park, adjacent to Bastrop State Park. USFWS sent a letter of reprimand (Grote, 1996), but took no action.

The same meeting also addressed pond construction. Participants determined that to prevent problems caused by hybridization with other *Bufo* species, any new ponds should be constructed at least 300 feet downstream from existing ponds.

USFWS had always opposed construction of new ponds on the golf course because of the hybridization problem, but six months into the project, TPWD informed USFWS that two new ponds upstream from the existing pond were necessary to irrigate the enlarged greens. TPWD suggested that they could police the ponds in order to remove potentially hybrid eggs, or, better yet, stock the ponds with predatory fish (Williford, 1996). USFWS refused. TPWD attempted to piecemeal its way around several other mitigation measures, saying they would be implemented "pending identification of funding" (Williford, 1995a,b). USFWS issued stern warnings (Grote, 1995b), but did nothing as the violations mounted.

The expanded course opened with much fanfare in August 1997 (Harvey, 1997), but even the golfers were less than happy with the "improvements." Last-minute modifications of the already crowded 35-acre plan yielded a 6,152 yard course, too short by today's standards to be a challenging layout (Widener, 1998). Coupled with irksome pesticide restrictions, the course quickly become a maintenance nightmare (Kramer, 1998). Expensive new greens fees drove away many local golfers, including a large number of senior golfers who had been using the course for decades (P. Cook, personal communication).

TPWD's commitment to protect Houston toad habitat seemed to end when the new nine holes opened. The mitigation plan that TPWD designed, and of which NPS tried to wash its hands, contains nine non-discretionary "reasonable and prudent measures" (RPMs) that TPWD was required to implement by July 1997—two years after construction began. On 23 January 1998, exactly three years to the day after the biological opinion had been signed for the golf course expansion, the Lone Star Chapter of the Sierra Club and the Southwest Center for Biodiversity filed a 60-day letter of intent to sue TPWD and NPS for violations of the ESA (Hess, 1998). The letter cited TPWD's failure to comply with at least six (and possibly eight) of the RPMs. That same day, the Southern Plains Regional Working Group, Declining Amphibian Populations Task Force (DAPTF) of the World Conservation Union's Species Survival Commission met at Bastrop State Park (McAuley, 1998a). TPWD hosted this prestigious conference at the very site of its own violations and invited A.M. to present results of BCEN's World Wildlife Fund-assisted ToadWatch '97 project. After her presentation, A.M. informed the conference hosts and attendees of the lawsuit and distributed copies of the 60-day letter and Sierra Club's press release.

Though no lawsuit was actually filed, the threat of one apparently spurred TPWD to implement long-overdue provisions of the mitigation plan (Haurwitz, 2000). Within a year, the department secured a conservation easement on 1,000 acres of prime habitat in the 5,000-acre Griffith League Ranch, a property that for years had been tied up in probate but was finally awarded to the Boy Scouts of America. In early 2000, this one-year easement was replaced by acquisition of 1,000 acres adjacent to the park, a $1.7 million purchase facilitated by The Nature Conservancy of Texas (Anonymous, 2000). The size and location of the new acreage made up for the fact that it was not all optimum habitat. TPWD has indicated a commitment to rehabilitate its overgrazed pastures (about 25% of the property); with proper management, the property could potentially serve as an experimental laboratory for habitat restoration.

Perhaps because no pending lawsuit forced the issue, TPWD was less responsive to concerns of historical nature. Despite non-compliance, Bastrop State Park was dedicated as a National Historic Landmark in October 1998 (Anonymous, 1998a). By early 2001, many violations of the historic preservation MOA

still remained. No interpretive center had moved into the CCC structure because it was still in use as a clubhouse. The new clubhouse was not yet started and the unsightly tin golf cart sheds were still standing.

Why did TPWD renege on its part of the deal? The money ran out. The $250,000 local "match" from the Lost Pines Golf Club was never there. TPWD struck an agreement with the golf club (signed days before the new nine holes opened) that allowed the club to raise its 25% of the construction costs through a $1 surcharge on every non-member round of golf (Anonymous, 1995j). Interest free and open-ended, this unsecured loan from the state of Texas could take a decade or more to repay, if repaid at all. TPWD suffered through a humiliating fiscal audit in 1998, which revealed a shocking, systemic lack of accountability throughout the agency (Leggett, 1998).

When the new nine holes and the clubhouse parking lot were completed, some $200,000 of the original grant remained to build the clubhouse. In 1997, the lowest bid for clubhouse construction came in at $450,000, and the golf club balked at the additional expense. Two years later, the lowest bid of $287,000 was accepted.

In October 2000, TPWD bailed out the Lost Pines Golf Club yet again. Their 1995 agreement stipulated that the club would pay TPWD $1 per non-member round to pay off the golf course improvements, and $1.25 per round for park entry fee (historically, golfers were excused from entry fees, which had climbed to $3 per person by that time). Perhaps realizing that the debt would never be repaid on the original schedule, TPWD modified the agreement such that both fees would count toward construction expense, effectively excusing golfers once again from park entry fees (DeKunder, 2000).

Sansom's proclamations that the expanded course would provide a "vehicle to fund a habitat plan" or make "a significant contribution to the recovery of the species," have not played out. When USFWS demoted the Toad Fund to a discretionary measure in the biological opinion, it allowed TPWD the option of dropping the "toad fee" altogether from its agreement with the golf club. Though TPWD still collects this fee— a surcharge of $25 per membership and $1 per non-member round (Anonymous, 1995j)—it amounts to less than $30,000 annually (Anonymous, undated), a paltry mitigation fund.

In 1966, TPWD *may* have promised an additional nine holes to Bastrop city officials in order to access federal funds for park improvements. In 1996, TPWD *certainly* promised that the mitigation plan for the golf course expansion would benefit Houston toads in the long run. It took a 30-year wait, eight years of political power play, and $500,000 of federal money to get the 1966 "promise" fulfilled. If it takes that much time and arm twisting to convince TPWD to fully implement this mitigation plan, Houston toads may join the extinct dodo bird before an interpretive center in their honor is built at Bastrop State Park.

Lost Pines Logging Crisis Catalyzes Conservation of Houston Toads

Far from the sawmills of east Texas, Bastrop County's 70-square-mile pocket of loblolly pines is attractive to loggers only from December through March, when the east Texas Pineywoods are too wet to log. Out-of-town timber companies post classified ads in Bastrop County's three newspapers and send mass mailings to find potential sellers. However, logging has become unpopular among the forest-dwelling humans, many of whom are urban refugees who moved to Bastrop County because of the Lost Pines. Those who share habitat with Houston toads once believed that the ESA provided protection from logging and other environmental degradation in their area. In 1998, they learned they were wrong.

The first week of January 1998, BCEN learned that industrial logging machinery, rarely seen in the area, was parked on the right-of-way to a 256-acre tract in the Lost Pines—within critical habitat for Houston toads. The headwaters of Alum Creek run through the property, which includes a five-acre bog, a federally designated wetland (National Wetlands Inventory, 1993). The property had recently been sold to a logger from the east Texas town of Chester, whose two house-sized hydro-axes and six skidders stood ready. Cutting seemed imminent (Dureka, 1998).

BCEN immediately alerted USFWS, certain that this clear violation of the ESA during the breeding season of Houston toads would spur the Service into action. Several days of calls were needed to prompt USFWS personnel to drive 45 miles from their Austin office to inspect the site. By that time, many panic-stricken neighbors had also contacted USFWS about the industrial logger, whose clearcutting had commenced with ruthless efficiency. The machinery flattened smaller trees to get at the 70-year old pines, rapidly erasing the forest and leaving behind a mangled wasteland.

Ten days after BCEN's first alert (just before the three-day Martin Luther King weekend), USFWS asked N.D. Hooks, owner of the tract, to stop cutting and meet with them the following Tuesday to set up a mitigation plan for Houston toads. Instead, Hooks brought in additional crews at dawn on Sunday and logged almost nonstop for the two days before the meeting (Suchomel, 1998a). Two Austin television stations rushed to film the frenetic deforestation and to interview the outraged neighbors, many of whom now looked out on a clearcut instead of the pine forest that had attracted them to the area. USFWS had taken the federal holiday off, and its one law enforcement officer had to be called back from a hunting trip (R. Nino, personal communication). When the cutting finally stopped on Tuesday morning, some 80 acres had been scalped, and Alum Creek oozed with spilled oil and choked on sediments eroding from the denuded slopes (Dureka, 1998).

After meeting with USFWS, Hooks hauled out logs for a few days, tearing up more ground, and then abandoned the tract, leaving dozens of huge logs to rot in place (A.M., L.E.B., personal observations). BCEN pressured USFWS to fine Hooks for violating the ESA, and it informed the U.S. Army Corps of Engineers (COE) about the despoiling of Alum Creek. Despite BCEN's petition and phone campaign to U.S. Representative Ron Paul urging that USFWS prosecute Hooks, the Service waffled on enforcement. At one point, USFWS did contact the biologist who surveyed the transmission line right-of-way bisecting the tract. Although he was ready to testify that Houston toads had been found on the property, USFWS abandoned the issue (G. Galbraith, personal communication).

USFWS looked worse yet when Acting Field Supervisor Bill Seawell gave conflicting interpretations of the ESA to two local newspapers. The *Smithville Times* first reported Seawell as saying, "logging . . . or other activity that could result in habitat disruption requires a 10-A [sic] permit" (Neilson, 1998a). Later reports softened the tone: "since there is no way to guarantee that logging operations will have no impact on the toad, USFWS recommends that a permit be acquired" (Neilson, 1998c). Meanwhile, the *Bastrop Advertiser* quoted him saying, "There are proper seasons of the year when they don't need a permit" [to log inside

critical habitat for Houston toads]. The article went on to say that "normally, January is one of those periods" (Suchomel, 1998b). Yet, the Houston Toad Habitat Evaluation Procedure document produced by USFWS (Anonymous, 1994c) contains no provision for seasonal exemption from Section 10(a)(1)(B) of the ESA, and it specifically recommends avoiding habitat disturbance during the breeding season (January through June). Alarmed that this published quote would lead to more incidents of unmitigated "take," the Lone Star Chapter of the Sierra Club requested that Seawell immediately issue a correction to the media (Royder, 1998). Instead, Seawell (1998) wrote to Sierra claiming that USFWS was "pursuing several ways to correct this misinformation" and was "preparing a comprehensive story for publication in area newspapers and meeting with logging companies, citizens' groups and elected officials to provide information on what constitutes take." Though USFWS did organize a forestry issues workshop in Bastrop one month after the Hooks incident (Neilson, 1998c), no comprehensive story produced by USFWS ever appeared in local papers to clarify the definition of "take." By recommending—but not requiring—a permit, USFWS invited potential violators to skirt the law, it being easier to beg forgiveness than to ask permission. USFWS eventually maintained that to prove "take" they would need a dead Houston toad. Given the voracious and ubiquitous fire ant, the chances of recovering any dead toads were slim indeed. Thus, under this definition of "take," no violation of the ESA in the habitat of Houston toads would ever be punishable.

Intense negative publicity did plague Hooks (Neilson, 1998a; Suchomel, 1998a,b). Aggressive Austin-based TV reporting and front-page coverage in local papers so infuriated the logger that he invited BCEN (including A.M.) to his easement in February to meet with him and 11 government officials representing USFWS, COE, the Environmental Protection Agency, and the Texas Forest Service to "talk things out" (Neilson, 1998b). The two-hour inspection prompted a cease-and-desist order three weeks later from COE, citing Hooks for violating Section 301(a) of the Clean Water Act and directing him to restore Alum Creek to its original configuration at his own expense (Lea, 1998). The logger has yet to comply with this COE directive. He sold the land, which was subdivided and sold again. His behavior left many Bastrop County residents bitter about the logging industry and agog at the government's unwillingness to take decisive action against flagrant violators (Dureka, 1998).

The reluctance of USFWS to prevent or punish this transgression seems misplaced, particularly since many in the Bastrop community felt violated by this outsider who extracted a precious resource and hauled away the profits. At the February meeting, Hooks threatened to sue BCEN (Suchomel, 1998c), and his aggressive attitude suggested that he would fight back if USFWS tried to prosecute. He also threatened to file complaints against every other ESA and Clean Water Act violation he could find in Bastrop County, which could potentially keep the USFWS and COE law enforcement divisions occupied for many years. Fortunately, Hooks' threats were more smoke than fire.

Habitat Conservation Planning Rises as a Phoenix from the Ashes

While the logging crisis smoldered, a new supervisor took the helm at the Austin Ecological Services Office of USFWS. A veteran of Pacific Northwest spotted owl (*Strix occidentalis*) battles, David C. Frederick brought a new management style and strength of character that had been lacking since Sam Hamilton's 1995 departure. Frederick wisely chose to avoid a potentially litigious situation in Bastrop County, and he instead sought a way to involve the local community in a conservation plan that would address the needs of the toads and the people.

In late 1998, USFWS entered into a cost-share agreement with the Lower Colorado River Authority (LCRA), Bastrop County, the city of Bastrop, Bastrop Economic Development Corporation, Bastrop County Board of Realtors, Champion International Inc., and BCEN to develop and implement a conservation plan for Houston toads in the Lost Pines (Anonymous, 1998c; McAuley, 1998b, Neilson, 1998e). Project funding came from the USFWS regional office in Albuquerque ($50,000) and from the cooperators (some of whom pledged cash and others in-kind services), yielding a total budget of $100,000. The project was to support a two-year LCRA staff position (filled by a Bastrop County resident) who would work with USFWS, the grant cooperators, and the local community to undertake regional conservation planning, fund-raising, and public outreach programs for Houston toads and the Lost Pines (Frederick, 1998).

This challenge grant took many months to develop and exemplifies the spirit of cooperation that USFWS far prefers over enforcement of the ESA in response to violations (Frederick, 1998). Litigation can polarize communities with long-lasting effects. The 1994 effort to develop a habitat conservation plan for Houston toads had dissolved overnight when a lawsuit almost erupted between two of the cooperators (McAuley, 1994a) and political repercussions came down on a third (R. Fritz, personal communication). In late 1998, given the Hooks incident, anticipation of more logging crises, and intense development pressures, Bastrop County was ready once again to spearhead a regional HCP (Anonymous, 1998b). County Judge Peggy Walicek appointed A.M. to the task force overseeing the project; at the first meeting, A.M. was elected Vice Chairman.

Despite an inspiring presentation early in the process by Rick Alexander, a San Diego consultant who specializes in environmental planning and public policy development, USFWS had trouble spurring local leaders to take charge of the project and move forward. Meanwhile, the stakeholders expected more direction from USFWS. Realtors, Bastrop County officials, and the Boy Scouts of America (key cooperators by virtue of owning the 5,000-acre Griffith League ranch in prime toad habitat) were frustrated by the agency's reluctance to name actual acreage figures for the amount of preserve lands that would be necessary for a successful HCP.

BCEN helped to identify likely allies in the community who would support conservation measures, and it acted as a repository and distributor of GIS maps and data. This all-volunteer group had received two basic GIS grant packages from the Conservation Technology Support Program (a non-profit partnership between the Smithsonian Institution, Hewlett-Packard, and software giant ESRI), and it had spent hundreds of hours gathering data from various sources in the region. BCEN provided USFWS and LCRA with the parcel layer from the Bastrop Central Appraisal District, which showed property ownership in Bastrop County. When overlaid on the already-digitized deep sands favored by Houston toads and the roughly delineated "known or potential habitat" that extends beyond the legally defined critical habitat, it became clear that the toads were in trouble.

"The habitat is mincemeat," proclaimed Lisa O'Donnell of USFWS when LCRA first unveiled the large-scale GIS maps it had produced (Dureka and Mesrobian, 1999). Though much of the habitat appeared intact and forested in the field, huge areas actually consisted of approved-but-not-built subdivisions, tracts of which were selling like hotcakes. Pervasive fragmentation of privately owned habitat coupled with the declining populations at the protected breeding ponds monitored by TPWD's Andy Price made it clear that the toads were in a tailspin. Faced with an obvious crisis, USFWS was forced to quantify the acreage needed for a preserve, based on the recommendations made by the Houston Toad Recovery Team.

In April 1999, USFWS met with Bastrop County leaders to announce that in order for Houston toads to escape extinction, 15,000 acres of habitat needed to be managed for the toads—roughly 2.5% of the land in Bastrop County (Gee, 1999). Of that amount, some 5,000 acres were already protected within Bastrop and Buescher State Parks, under conservation easement through TPWD, or located in undeveloped buffer land surrounding an LCRA power plant. Given the likely purchase price of some $15 million (which would devour most of the county's annual budget), Bastrop County balked at acquiring the 10,000 "new" acres of habitat. Three months passed, and the haphazard development in toad habitat continued.

In early July, USFWS met with Bastrop County officials to inform them that they might be breaking the law by continuing to issue septic permits for new construction within toad habitat. The county immediately stopped issuing septic permits but then realized that such action ran counter to a brand new law, Senate Bill 1272, which made it illegal for a county to refuse construction permits solely because the platted land lay within endangered species habitat (Gee, 1999). Two weeks later, the county sidestepped the federal threat and resumed issuing septic permits, attaching a letter to each application informing landowners that they might also need a Section 10(a) incidental take permit from USFWS prior to construction (McAuley, 1999a).

In August 1999, USFWS held a public meeting in Bastrop to explain why the agency had finally "dropped the hammer" on the county and to explain what the HCP process entailed. Several townspeople who attended the meeting expected to deliver public comments, but the forum did not allow it. Rather, following a lengthy presentation, USFWS personnel circulated throughout the crowd to answer questions one on one. A frustrated resident with strong family ties to LCRA wrote a vitriolic editorial letter to the local weekly papers, accusing the federal government of plotting to confiscate lands and violate private property rights (Long, 1999). Ann M. responded with an explanation of how the conservation effort could be good for the whole community (Mesrobian, 1999), and USFWS followed with an open letter outlining the agency's policy and specifically disclaiming any intent to condemn or confiscate lands (Frederick, 1999).

Despite efforts to contain the damage, this attempt to develop a regional HCP collapsed. In early September, LCRA, the agency charged with hiring a two-year employee to oversee the plan, withdrew its cash and administrative support for the cost–share agreement, citing the failure of Bastrop County to take a leadership role in the effort (McAuley, 1999b).

Only temporarily daunted by this major upheaval, USFWS forged ahead with efforts to preserve toad habitat. Two "mini HCPs" were drafted for the most rapidly growing subdivisions in the county. A habitat-wide incidental take fee of $1,500 per lot, earmarked for land acquisition or conservation easements, was imposed on all new construction. In October, Bastrop County Commissioners Court passed a resolution that endorsed a toad conservation effort while also maintaining the local quality of life (McAuley, 1999c).

By November, USFWS had recruited the Texas Agricultural Extension Service of Texas A&M University to act as administrative coordinator for a revised cost–share agreement (Anonymous, 1999). A newly revitalized Houston Toad Working Group came to understand that while USFWS could not force Bastrop County to preserve Houston toads, the agency could definitely make the county regret its failure to do so after the fact.

In January, Texas Parks and Wildlife acquired 1,000 acres of land adjacent to Bastrop State Park to fulfill part of its mitigation plan for the golf course expansion (Anonymous, 2000). On the heels of that announcement, Bastrop County applied for a $1.5 million grant from USFWS to acquire additional lands for the toads. The funds were awarded in August 2000, shortly after Texas A&M hired a full-time Houston Toad Conservation Plan administrator.

The outright purchase for set-aside land provides one avenue for Houston toad habitat conservation, but other options also exist. In 1995, Texas voters approved a state constitution amendment that allows property taken out of agricultural use to be actively managed for wildlife without forfeiting the low agricultural tax rate (Anonymous, 1996; Neilson, 1997b). This new provision in the tax code was originally intended to allow overgrazed ranch lands adequate time to recover without losing their "ag value." However, in rapidly developing areas such as Bastrop County, it provides compelling economic benefits to landowners that adopt wildlife management plans when they subdivide formerly agricultural lands. Rather than being charged five years of roll-back taxes on the entire parcel, the landowner can designate portions of the property to be managed for wildlife, thereby avoiding roll-back taxes and maintaining the low ag value on the designated acreage. A dramatic example of this has already occurred on a 70-acre tract subdivided out of logger Hooks' 256 acres. The new owner has set up a wildlife management plan to rehabilitate the land and preserve habitat for Houston toads (Gelhausen et al., 1998).

In the spring of 2000, BCEN launched a new non-profit organization, the Pines and Prairies Land Trust, which can hold conservation easements and property and offer tax incentives to private landowners who wish to practice conservation in perpetuity. The land trust provides another vehicle to integrate private lands into an effective habitat plan for Houston toads. By late 2000, conservation efforts in Bastrop County had a great deal of momentum and strong support from many local leaders who realized that the quality of life in this largely rural county would be well served by a planning tool that preserves green space.

When USFWS included the Lost Pines in its conservation plan, not only did the agency widen its scope to include ecosystem preservation, but it also capitalized on the relationship between the trees and the toads. The Lost Pines have no legal protection, but these charismatic trees give Bastrop County marketable charm. Houston toads may lack popular appeal, but their endangered status has on occasion preserved some loblolly pines (see "Tunnels of Love," above). For years, toad advocates have pondered how changing the common name of Houston toads to "Lost Pines toads" would better reflect their present habitat location and likely boost their image (given Houston's reputation as an environmental disaster).

Within the core habitat, the political climate continues to evolve as the human residents discover they cannot take the Lost Pines for granted. In early 1998, TPWD encouraged the formation of the Friends of the Lost Pines State Parks. This local, non-profit group can direct funds generated by programs at Bastrop and Buescher State Parks toward special projects at

those parks, rather than surrendering them to TPWD's general fund (as would otherwise occur). In its first three years, Friends refurbished the historic refectory at Bastrop State Park, purchased and constructed playgrounds at both parks, and built a rustic barn at Buescher State Park to house rental canoes, which in turn generate more funds for Friends projects. While the Friends group involves the community in hands-on stewardship projects with "warm fuzzy" appeal, it also lobbies TPWD to be vigilant in protecting the Lost Pines and its endangered residents.

In October 2000, TPWD sponsored the first annual Endowment Weekend, a statewide celebration featuring free park entry and special fund-raising events in partnership with local organizations. This outreach effort attempts to build community with civic groups who share proceeds from the events with the brand new Lone Star Legacy Endowment Fund, which is dedicated to addressing funding shortfalls identified in TPWD's 1998 audit. When the Kiwanis Club of Bastrop sponsored a 5K "Toad Trot," complete with a charming promotional graphic of a sprinting toad in jogging shoes, it was clear that Houston toads had finally achieved their rightful place as the Lost Pines' beloved mascot.

Crossfire at the Crosswaters

Biologists now understand that both Houston toads and the Lost Pines rely on deep sands for survival, but distribution of the two species is restricted to very limited areas within the geological formation containing those sands. In fact, the Lost Pines are located at the crosswaters of the Colorado River and the Carrizo-Wilcox aquifer, Bastrop County's sole source of drinking water. Quite possibly, survival of the Lost Pines hinges on the proximity of adequate groundwater to moisten those well-drained sands during long, hot Texas summers and drought cycles.

Responding to serious drought conditions in 1996, Bastrop County launched efforts to address future water needs (Neilson, 1997a, 1998d; Stolarek, 1997). At the turn of the millennium, Texas must now address a statewide problem—a rapidly growing population that faces diminishing water supplies. Though surface waters are owned by the state, groundwater is privately owned and governed by the archaic "rule of capture" law, also known as "rule of the biggest pump" (Cook, 1997). In 1999, the Texas Supreme Court upheld the rule of capture in a widely publicized case involving the Ozarka bottled water company, and it punted the issue to the Legislature, where it has yet to be addressed. Texas law presently offers one remedy for the rule of capture—the formation of groundwater conservation districts that can regulate withdrawals to preserve aquifers.

Bastrop County soon became acutely aware of threats to the Carrizo-Wilcox aquifer. In early 1999, the county's efforts to establish the Lost Pines Groundwater Conservation District (LPGCD) collided with contracts between the Alcoa Corporation and the San Antonio Water System (SAWS; Blakeslee, 1999). Through these contracts, Alcoa promised to provide at least 40,000 acre feet of groundwater per year to SAWS, pumping water from strip mining operations in Bastrop, Lee, and Milam counties and transporting it over one hundred miles south to the Alamo city. Though subsequent hydrological reports indicated that such quantities could not be sustainably recharged by the aquifer, Alcoa and SAWS would not back down. BCEN helped organize a new citizens group, Neighbors For Neighbors (NFN), to fight both the water grab and a 16,000-acre lignite stripmine straddling the Bastrop/Lee County line. The lignite was to fuel Alcoa's 50-year-old smelter, the #1 grandfathered air polluter in Texas. Because stripmining operations fall under the Texas Railroad Commission (an industry-friendly agency with a rubber stamp reputation), groundwater withdrawals would be exempt from district regulation.

One of us (A.M.) recalled that during the World Wildlife Fund-assisted ToadWatch '97 project a handful of biologists had discovered Houston toads in the vicinity of the proposed new stripmine (Kuhl, 1997). She produced a GIS map showing the 1997 toad sightings barely four miles from the new mine and within the deepest projected groundwater drawdowns (Dureka and Mesrobian, 2000). A drawdown of at least 50 feet extends south throughout the entire Lost Pines and Houston toad habitat to the Colorado River.

Widely distributed among interested parties and agencies, the "toads and drawdowns" map transformed Houston toads from a villain who violated private property rights to a hero who might save the lands of many Lee County residents threatened by the Alcoa/SAWS deal. Hydrologists admit that the Carrizo-Wilcox aquifer is poorly understood, and nobody can predict with certainty if large withdrawals from deep water-bearing sands will dry up the surface seeps and springs that keep toad habitat inhabitable, but USFWS indicated that desiccation of habitat could be considered "take." As of late 2000, the agency opted to monitor NFN's legal battles from a distance. Nevertheless, Houston toads will be remembered as an important player in the as-yet-unfolding Alcoa/SAWS/NFN drama (Dureka, 2001).

Summary

Federally endangered Houston toads (B. houstonensis) have been found at various isolated locations with sandy soil in central and southeastern Texas. Toads are most numerous in the Lost Pines of Bastrop County, but they have been extirpated or are nearing extirpation at several other localities. Holocene aridity, historical droughts, over-collecting, agriculture, recreational over-development, urbanization, and deforestation have been major causes of the decline.

Houston toads have been at the center of numerous political debacles in the last 35 years. Like many endangered species that seem to lack intrinsic value, the toads are often perceived to be an obstacle to development and economic growth. For this reason, their main adversaries have been those entities that feel threatened or frustrated by their presence. Politicians in Texas and Washington, D.C., who wield influence over the law, and the moneyed interests that fund those politicians, have a financial stake in the issue. Depending on the political climate of the time, the two agencies charged with implementation of the law—U.S. Fish and Wildlife Service and Texas Parks and Wildlife Department—can either champion or betray endangered species. Without a doubt, politics will always play a role in conservation—sometimes to the benefit and often to the detriment of the species or ecosystem at risk. Bastrop County's achievements in preserving habitat for Houston toads illustrate how such efforts advance more readily when tied to human quality-of-life issues, and they serve as a model for similar efforts in other communities harboring species in decline.

Acknowledgments. We thank C.K. Dodd and R. Thomas for providing supplemental information on the determination of critical habitat; MIT Press for giving us permission to quote passages from Yaffee (1982); the *Beacon* for granting permission to reproduce the "BUFO DUDE" cartoon (Fig. 23-1); the *Smithville Times* and Shawn Janak for granting permission to reproduce the two political cartoons (Figs. 23-2 and 23-3); C. Ruyle for copyright information; and J. Brown, T. Dureka, P. Cook, C.K. Dodd, L. Goldstein, R. Gray, and R. Thomas for critically reading the manuscript.

TWENTY-FOUR

Amphibian Conservation Needs

EDWARD D. KOCH AND CHARLES R. PETERSON

We have experienced many scientific and management challenges and opportunities through our 15 years working together to understand and conserve amphibians in the northern Rocky Mountain and Pacific Northwest regions of the United States. We have observed that many herpetologists are unaware of or poorly informed on management needs and opportunities for conserving amphibian species. Because of this lack of awareness and a relative lack of attention paid by herpetologists to serving specific research and management needs, many natural resource managers lack the sound scientific information and experience needed to conserve amphibians and their habitats.

Citizens have created many tools for conserving species that are often unfamiliar to research scientists, many conservationists, and even some managers. In this essay we present some ideas and examples of opportunities to promote amphibian species conservation. We have developed these ideas and experienced these examples as of 2001, in part through our jobs as an endangered species biologist for the U.S. Fish and Wildlife Service in the Pacific Region (Koch) and a professor of herpetology at Idaho State University (Peterson).

The Framework for Legal Protection and Conservation of Species

Throughout their history, citizens in the United States have expressed an interest in conserving native species of fish, wildlife, and plants through support of legislation, policy initiatives, and other avenues. Some initiatives represent voluntary opportunities for conserving species or habitats (e.g., making yards "wildlife friendly"), whereas others represent sweeping legislation authorizing regulatory protection of endangered species, for example.

In order for governments and other organizations to serve this interest in conserving species, we believe they must successfully complete the following four general management tasks:

1. Determine what group of organisms constitutes a "species." Scientists must help species and habitat managers to clearly identify and categorize the biological entity they are trying to manage. Without some type of discrete identification of a "species" for management purposes, there will be little likelihood of a discrete management emphasis or focusing of management dollars

2. Classify that species' management status (e.g., abundance and trend). Managers must be able to prioritize their expenditure of management dollars. One of the most useful methods for prioritization includes evaluating a species' management status. Species that are relatively more rare or in greater decline would likely receive more immediate attention than more common species.

3. Identify future threats. Combined with rarity, the type and degree of threat is perhaps one of the most relevant factors to a species or habitat manager in considering when, where, and how to spend limited management dollars. Scientists can help managers identify the type and degree of threat that amphibians face. Managers should focus conservation resources on the most important threats.

4. Develop and implement programs to remove those threats where necessary. If managers can identify the biological entity on which they are focusing attention and successfully characterize its abundance and trend and how immediate and important the threats to the species are, then they are faced with the need or opportunity to prescribe specific management actions to remove those threats. The most relevant information would include describing specific threat factors and methods for minimizing or mitigating the effects of those threats. In our experience, most fish and wildlife species are threatened with adverse impacts to and/or loss of habitat. Identification of specific habitat needs, specific measures of adequate habitat protection, and ideas for effective conservation actions to protect the species would all be helpful.

Tools for completing the first three species conservation tasks—species taxonomic definition, management classification, and threat identification—have been available for a few to several decades within federal and state governments and within private organizations in the northwestern United States. However, use of these tools is still relatively new and re-markably imprecise when applied to amphibians, and their ul-timate effectiveness remains largely unseen. Tools for completing the fourth task—developing and implementing programs for conserving species—are only beginning to be developed.

Currently in the northwest, conservation programs protect-ing species are or may be available through legal mandates that already exist at both the federal and state levels (in addition to private means such as land conservation purchases or volun-tary management actions). The Tenth Amendment to the United States Constitution delegates all authorities not other-wise reserved by Congress to the states. This generally includes the authority to manage wildlife. Notable exceptions to this federal delegation of species protection authority to states in-clude the Migratory Bird Treaty Act, which authorizes the fed-eral government to negotiate international treaties on behalf of the states to conserve migratory birds; the Lacey Act, which bars interstate transport of illegally harvested wildlife; and (po-tentially the most relevant to amphibian conservation) the En-dangered Species Act (ESA), which is intended to prevent species from becoming extinct.

Historically in the United States, states have developed laws regulating the harvest of fish and wildlife of interest to hunters and fishermen. In Idaho, the Idaho Department of Fish and Game (IDFG) retains primary management responsibility for almost all fish and wildlife species in the state and is charged under Idaho Code Section 36-103 to "preserve, protect, perpet-uate, and manage" all wildlife. However, many western states, including Idaho, have substantially inadequate state-funded programs for conserving non-game fish and wildlife species.

Because almost all funding for the IDFG comes from hunter and angler license dollars and related revenues, the vast major-ity of IDFG's species and habitat management efforts are fo-cused on managing harvestable game fish and wildlife species. There is no state-level "Endangered Species Act" type of fund-ing program or legislative protection for on-the-ground con-servation of fish, wildlife, or plant species in Idaho, except for nominal protection of listed "Species of Concern." The state's existing non-game program is only a small fraction of the IDFG's total budget, and it is widely considered to be inade-quate for addressing statewide non-game species conservation needs, of which amphibians are only a small portion. Game species habitat management programs may not be comprehen-sive enough to benefit many non-game species such as native amphibians. Consequently, almost all fish, wildlife, and plant species in Idaho—all those not of interest to hunters and fish-ermen, including all native amphibians—receive little or no management attention from the state, despite the state's "right" to manage them.

A result of the state's inattention is that few data are avail-able at the state level to inform the first three steps necessary for identifying and conserving non-game species, including amphibians. In general, non-game species in the region only receive serious management consideration when species de-clines and threats become so egregious that consideration for federal protection under the ESA may be warranted. State-listed "Species of Special Concern" are legally protected from "take," and this seemingly strict prohibition could potentially provide important state-level protection to listed amphibians. How-ever, such protection has rarely (if ever) been exercised on be-half of the state's four native amphibians currently identified as Species of Special Concern or for any other state-listed taxa. States like Idaho often decry the intrusion of the federal gov-ernment (under laws like the ESA) on their "state's rights." However, when states fail to accept the responsibility that comes with their rights, federal laws are often triggered to pro-vide a "safety net," including the ESA protection of species.

Until recently, of the 37 species of amphibians and reptiles known to occur in Idaho, only one species received specific, proactive management protection—the introduced American bullfrog *(Rana catesbiana)*. The IDFG implemented harvest re-strictions for bullfrogs as a game animal, the purpose for which the bullfrog was introduced in the state. In 1996, the conserva-tion organization Idaho Herpetological Society (IHS), with support from the Idaho Wildlife Federation and others, suc-cessfully convinced the Idaho Fish and Game Commission to adopt a regulation to protect native amphibians and reptiles. This regulation requires persons to obtain a collecting permit from the IDFG if they possess more than four live individuals of any one native amphibian or reptile species.

The purpose for IHS to secure this regulation was to provide data to IDFG to monitor the extent of animal collection to supply the pet trade. In the event that over-collection of a species becomes a problem, the IDFG now has the authority to restrict legally permitted collecting activities. As of 2000, the IDFG has successfully prosecuted at least one individual for failure to comply with this regulation, and that was for posses-sion of a reptile species. There is no "bag limit" on the number of dead native amphibians or reptiles one may possess, but so far this has not been a practical concern.

An alternate strategy for directly conserving species is to conserve the ecosystems upon which they depend. However, scientists' ability to define and identify sufficiently discrete "ecosystems" for management purposes is limited, and an ecosystem management or regulatory framework for imple-menting protections in the United States is even less well de-veloped. In the northwestern United States, efforts to conserve ecosystems are often limited by land ownership status (e.g., federal land strategies such as the Northwest Forest Plan). Even management approaches under such overarching laws as the ESA tend to be on a "species-by-species" basis for listing and re-covery, despite the ESA's first-stated purpose to "conserve ecosystems."

General Observations Regarding Classification and Conservation

In general, we feel that few (if any) species' management needs have been identified adequately and met in the region under the four steps we propose, be they widespread and lo-cally common or restricted in distribution. Most amphibian species that occur in Idaho—and much of the northwest—are widely distributed and often locally common. However, our greatest conservation concerns involve some of these wide-spread and locally common species, such as boreal toads *(Bufo b. boreas)* and northern leopard frogs *(R. pipiens)*. In our expe-rience, these species (and possibly other relatively widespread species) are declining in or disappearing from substantial parts of their range in the northwest. So far, no management tool or opportunity has become available to evaluate and ad-dress these declines adequately and comprehensively. Surpris-ingly, one of the greatest challenges so far has been completing the first step of simply identifying a "species" to evaluate and conserve.

Past taxonomic classification suggests that the boreal toad and leopard frog "species" are widely distributed throughout either western North America or the entire continent (Stebbins, 1985). In fact, however, identifying what a "species" of toad or frog is for taxonomic or management purposes is becoming increasingly uncertain, and definitions of what a species is for management purposes may not be consistent with definitions used by taxonomists. Recent advances in biochemical systematics have created a world where managers are increasingly unclear which taxonomic unit may warrant management attention for purposes of achieving conservation.

Because of this taxonomic uncertainty for many amphibian species in Idaho and the Northwest, successfully completing the first management task—identifying the "species" that may be subject to the other three management tasks—is difficult. Obviously, then, completing the other three tasks, including classifying management status and identifying threats to a "species" of toad or frog, is also extremely difficult. (See the discussion below regarding boreal toad taxonomic status for an example of the difficulties that managers face in dealing with uncertain taxonomy.)

Interestingly, of those endemic or indigenous amphibian species that have a restricted distribution in and adjacent to Idaho, we generally believe that management status is not rare and threats are not large or immediate, at least when compared to species such as boreal toads or northern leopard frogs (Koch and Peterson, unpublished data). Also notable is the fact that these species with restricted distributions are relatively easier to identify as a species for management purposes. This fact may bias our view of their management status because we may be more confident that a species is adequately conserved if we are more confident that we can identify that species. However, our opinions of the status of all amphibian species are based on precious few data. Therefore, our sense that indigenous species in the region are not rare is tempered by the fact that virtually all native Idaho amphibians have likely declined to some degree from pre-European settlement.

Management Classification Status for Species with an Evaluation

Maintaining lists of species identifying management status can be important for several reasons. First, the fact that a species can be put on a list suggests that its taxonomic status is sufficiently clear to call it a species or other taxonomic entity that should be recognized by natural resource managers. Second, understanding which species may be rare and which are common is relevant to identifying which species may warrant further consideration for conservation. Third, identifying species as rare, or potentially rare, often elicits a variety of interests, independent of the goals of the organization maintaining the list. These interests may include making and reporting observations and focusing independent research efforts, which often results in more information becoming known about that species over time. Fourth, if species are present on a management agency's list, for example, then that agency will be more likely to consider potential adverse impacts to that species from other management actions.

In Idaho, five different management classification lists are currently in use. The IDFG, under Idaho Code 36-201, maintains a list of Species of Special Concern. The U.S. Fish and Wildlife Service, under the authority of the federal ESA, maintains a list of endangered and threatened species and species

that are candidates for listing under the ESA. The two major federal land management agencies—the U.S. Forest Service, under the authority of the federal National Forest Management Act, and the U.S. Bureau of Land Management, under the authority of the Federal Land Policy Management Act—also maintain lists of sensitive species. Finally, the private Idaho Conservation Data Center (CDC, formerly The Nature Conservancy Natural Heritage Program) maintains a list of globally ranked and state-ranked rare species. In general, we believe the four government management agency lists, though different in structure, are relatively comparable (Table 24-1). In our view, the CDC program list is the least similar among the five classification lists, in part because it focuses on the degree of rarity and does not emphasize the threats to species to the extent that the other lists do. For example, one CDC program list category includes species that are "Demonstrably widespread, abundant, and secure" (a "5" level ranking), whereas those species would not occur on any agency's list. We feel this diversity among lists is useful in facilitating the completion of the four management tasks.

In our experience, species must be officially present on an organization's list in order to receive management attention. For example, IDFG identified Coeur d'Alene salamanders *(Plethodon idahoensis)* as a Species of Special Concern in the late 1980s, soon after they were identified as a species separate from Van Dyke's salamanders *(P. vandykei)* in western Washington and Oregon. IDFG's non-game biologist at that time led a field survey program and status review of the species, with support from the U.S. Forest Service, and determined that Coeur d'Alene salamanders occurred at many sites (Groves et al., 1996). If not for the fact that this species was identified on the state list as "sensitive," the research likely would not have occurred. This is one species of amphibian for which the state has taken a lead role in gathering scientific information and developing conservation ideas. Because the species is actually not rare, management actions to conserve the species were not warranted; this is fortunate because as we discussed above, despite the authority for the state to prohibit "take" of Species of Special Concern, the state has few resources or directives to provide such conservation.

The U.S. Forest Service and the U.S. Bureau of Land Management in Idaho maintain lists of sensitive species. These lists recognize the federal ESA list of protected species, as well as state lists and state Natural Heritage Program lists, among others. These agencies, by law, use their lists to identify species that may warrant special management consideration before implementing other land management actions, such as timber harvest, road building, or livestock grazing. These agencies also use their lists to direct limited funds for research and conservation, primarily through challenge-cost-share funding programs.

Such agency funds have been invaluable in helping us to conduct fieldwork and develop a more informed opinion of the status and threats of the region's amphibians. However, having research funding sources tied to "listed" species creates a tricky situation for scientists who might advocate for certain species to be on these lists. On one hand, having a species on a list can be beneficial because it provides the opportunity for funding for research and management. On the other hand, superfluously putting species on these lists can dilute limited funding resources and degrade the agencies' confidence that all the species on the list truly warrant special management consideration. We have worked closely with these agencies to carefully inform them of our opinions regarding species that are appropriate to include on their lists.

TABLE 24-1
Categories of Rare Species Status Listed by Agency Ranking System, with Relative Comparisons Among Ranking Systems Suggested Across Rows.

Biological Factors	U.S. Fish and Wildlife Service	U.S. Forest Service	U.S. Bureau of Land Management	Idaho Dept. of Fish and Game	Idaho Conservation Data Center[a]
One or more of "five factors" under ESA = significant threats DO exist	Listed threatened and endangered species	Recognize FWS list, must "promote conservation"	Recognize FWS list, must "promote conservation"	Species of Special Concern— Priority A	Global 1 State 1
One or more of "five factors" under ESA = significant threats MAY exist	Proposed and candidate species	Sensitive species (recognize proposed/ candidate status)	Sensitive species (recognize proposed/ candidate status)	Species of Special Concern— Priority A	Global 2 State 2
Some declines documented, and/ or recognized by other agencies	Species of concern	May be a sensitive species	May be a sensitive species	Possible Species of Special Concern— Priority A	Global 3 State 3
Small population, limited quantity of habitat, or lack of information	Watch species	*No Equivalent*	Watch list	Species of Special Concern—Priority B and/or C	Global 4 State 4

[a] The Idaho Conservation Data Center categorization scheme is sufficiently different that we believe each global and state rank could match up very differently depending on what species is being classified. For example, a candidate species under the Endangered Species Act could be a G4 and an S1, depending on species distribution, status, and threats relative to political boundaries (state and national borders). The other categorization schemes (notably, all designed by government agencies) are relatively more comparable across categories.

NOTE: The Internet Web sites for each organization provide detailed definitions of each of the management categories. Because URLs frequently change, please execute a keyword search to find the current Web site for each organization.

Placing species on the list maintained under the authority of the federal ESA requires completion of a more difficult process than that for any other list, but these species then benefit from the most substantial conservation opportunities of any of the five lists. Placing a species on the ESA list requires a formal, federal rule-making process that takes at least two years. Notably, under the ESA, a species may be a species, subspecies, or "distinct population segment," making identification of a "species" for management purposes under the ESA particularly difficult. However, once on the list, any unauthorized "take" of that species (including harm or harassment) is prohibited by federal law. No amphibian species are included on the federal list in virtually all the Pacific Northwest region of the United States. Consequently, most amphibian species currently receive very little attention from state management agencies and no regulatory protection under the ESA. However, there is an increasing number of opportunities to gain voluntary, proactive conservation under the ESA for amphibians through habitat conservation planning (described in more detail below in the similarly titled section of this chapter).

The other four management classification lists can change much more rapidly than the federal ESA list, and there are often fewer conservation opportunities and requirements for species on these lists. However, in our experience, these lists can sometimes be slow to change. For example, we have recommended since 1991 that the U.S. Forest Service in two different regions (Region 1, which includes northern Idaho, and Region 4, which includes southern Idaho) place boreal toads and northern leopard frogs on their species lists. However, these species did not get onto the Region 1 list until 1998, and they still are not on the Region 4 list. We believe that at least

part of the reason for this inattention to adding appropriate species to these lists is because of a general lack of diligent attention paid to maintaining these lists. Also, the U.S. Forest Service may be poised to soon do away with its current list and begin using the Natural Heritage Program list, creating further disinterest in maintaining its current list.

We evaluated all five management list classification criteria and the current classification status assigned for five amphibian species in Idaho. We then re-evaluated our view of the status of all amphibian species in Idaho, and we made recommendations for changes in status and additions to each of the five lists (Table 24-2). We found that in addition to the five species already occurring on one or more management classification lists, three more species warranted being added to two or more agency lists, with some type of management classification status designation. Of the original five species, four warranted at least some changes to their identified status. Columbia spotted frogs (*R. luteiventris*), which did not warrant any changes in their designation, had recently been split taxonomically from Oregon spotted frogs (*R. pretiosa*; Green et al., 1996). Concurrent with this taxonomic split, all five management organizations updated their status designations to reflect current thinking for spotted frogs, which included no status designation north of the Snake River in Idaho and some type of designation for all five organizations south of the Snake River in Owyhee County.

In general, we felt data were lacking for a confident status designation for all eight species, as well as for some other species that we did not recommend appear on any of the lists. We have almost no population or abundance trend data for any of the species that can be used effectively for this management

TABLE 24-2
Past Status of Five Amphibian Species Systems in Idaho

Species	U.S. Fish and Wildlife Service	U.S. Forest Service	U.S. Bureau of Land Management	Idaho Dept. of Fish and Game	Idaho Conservation Data Center
Coeur d'Alene salamander (*Plethodon idahoensis*)	Species of concern	Sensitive	Sensitive	Species of special concern— Priority A	G3, S3 **G4, S4**
Boreal toad (*Bufo b. boreas*)	Species of concern **Candidate Species in S.E. Idaho**	**Sensitive**	Sensitive	Species of special concern— Priority C Priority A	G4, S4, G4, **S2/S4**
Woodhouse's toad (*Bufo woodhousii*)	Species of concern	*Not ranked -*	*Not ranked -* Watch	*Not ranked -* **Species of Special Concern— Priority C**	G5, S3?
Northern leopard frog (*Rana pipiens*)	Species of concern	Sensitive in Region 1, not 4; Sensitive in both Regions	Sensitive	Species of special concern— Priority A	G5, S3, G5, **S2/S4**
Columbia spotted frog (*Rana luteiventris*) south of Snake River	Candidate species	Sensitive—USFS Region 4 only	Sensitive	Species of special concern— Priority A	G4, S2/S3
Rough-skinned newt (*Taricha granulosa*)	**Species of Concern**	***Not ranked***	***Not ranked***	**Species of Special Concern— Priority C**	G5, **S3Q**
Idaho giant salamander (*Dicamptodon aterrimus*)	**Species of Concern**	***Not ranked***	**Watch species**	**Species of Special Concern— Priority A**	G3/G4, **S3/S4**
Wood frog (*Rana sylvatica*)	**Species of Concern**	**Sensitive, Region 1**	**Sensitive**	**Species of Special Concern— Priority B, C**	G5, **S1**

NOTE: [1]Boldface marks recommended changes in status as well as the suggested addition of three species.

classification exercise. Also, because we have a poor understanding of the current and historical occurrence and abundance of these species, identifying important risks to species was particularly difficult. For example, we have a poor understanding of the proximal reasons why boreal toads and northern leopard frogs are disappearing from many parts of their range in and around Idaho. In some cases, boreal toads are gone from otherwise seemingly intact habitat, so it is particularly difficult to specify threat factors.

We generally used comparisons between our observations of where species occur today and their relative abundance, and where they were reported to occur previous to 1990 (either in the literature or from museum collection records). We also relied on our own field observations, on anecdotal information obtained from others, and on an opinion survey of Idaho biologists (Groves and Peterson, 1992). For at least one of the species (boreal toads), taxonomic and/or management classification status is so uncertain that we are not confident we have the most appropriate "species" identified in the first column of the table. Roughskin newts (*Taricha granulosa*) may be an introduced species in Idaho, in which case they should not occur on any of these lists, which are generally intended to identify and conserve native species.

We believe that each list tends to reflect each organization's management interests, and that no one list is the best or the only

"right" way to list species. For example, the state list includes species that are rare in the state, even though they may be common elsewhere in the western United States; the U.S. Fish and Wildlife Service is interested in listing only those species that are truly threatened with extinction; the U.S. Forest Service and U.S. Bureau of Land Management lists reflect species relevant to their management considerations; and the CDC program list focuses more on unique or imperiled species, with less emphasis on threats than the other agency lists. What we have found through our compilation of species status under all five lists is that using the five lists in combination provides the most powerful approach for understanding both the species management classification status and which management organizations may be best positioned to remove threats.

The Boreal Toad Taxonomic and Management Challenge

The historic range of the *Bufo boreas* species complex covers the western half of North America, from southeast Alaska to central New Mexico and west to the Pacific Ocean. Recent analyses of biochemical characteristics of these animals suggest there may be evolutionary relationships among groups of toads that would allow taxonomists and managers to classify different species, subspecies, or distinct population segments within the boreal toad species complex (Goebel, this volume). However, what would constitute the correct grouping of toads (e.g., lump more or split more) and what those groupings would be called (e.g., new species or populations) is unclear, from both a taxonomic perspective and a management perspective, particularly under the ESA. These uncertainties pose important challenges to addressing species-management needs for toads.

As we mentioned above, there is strong evidence that members of the boreal toad species complex are disappearing from substantial portions of their range in the central and southern Rocky Mountains (Corn and Fogleman, 1984; USFWS, 1995a). However, relatively little attention has been paid to identifying and addressing the conservation needs for boreal toads. The need for protection under the ESA could be particularly difficult to identify and address if a "listable entity" (a species, subspecies, or distinct population segment under the ESA) cannot be clearly identified. Based on population and geographic data available prior to the recently available biochemical data, the Service identified a group of toads in the southern Rocky Mountains (the "southern Rocky Mountain population," or SRMP, of toads) as a candidate species for listing under the ESA (USFWS, 1995a). This group of toads is geographically discrete, and it has suffered disappearances or declines in many areas.

Recent biochemical data suggest that, while this SRMP of toads is somewhat unique and discrete, there are three other groups of toads in the southern and central Rocky Mountains that are closely related to the SRMP group, as well as to each other (Goebel, this volume; personal communication). These other three groups of toads occur in southern Utah, northern Utah, and southeastern Idaho. The two questions that arise are (1) should a sub-group (or groups) of boreal toads in the Rocky Mountains be identified as separate species of toad from the rest of the boreal toad complex throughout western North America? (2) what is the appropriate level or grouping at which management and "species" conservation efforts should be aimed (particularly under the ESA, which has a potentially broad regulatory definition of a "species")?

If the Service were to apply the same species-definition standard (based on biochemical data) to boreal toads as it currently applies to Wyoming toads *(B. baxteri)* and Houston toads *(B. houstonensis)*, which are both listed as "species" under the ESA, then it would view all four groups of boreal toads in the southern and central Rocky Mountains as a different "species" (Goebel, this volume; personal communication). In fact, the current, official "candidate" species status of the SRMP group of toads by itself is consistent with this management classification approach, where the four groups of toads would each be "split off" into different species under the ESA.

With this approach, the Service would then be faced with evaluating four separate, narrowly restricted, probably critically imperiled "species" of toad in the southern and central Rocky Mountains, and it would be more likely that all four "species" would warrant protection under the ESA. Specifically, the southeast Idaho population of toads, which has been found at only two different sites in just one watershed, would become perhaps the most narrowly distributed, highly imperiled species of amphibian on the continent. Also, this rare, highly imperiled "species" of toad in southeastern Idaho would occur in the middle of the historic range of the species complex from which it was split.

At some point in our efforts to correctly identify and conserve "species," scientific credibility and management practicality are called into question. We do not know how to address the issue of correctly addressing scientific taxonomy in this case. If taxonomists believe that each of the four groups of toads are clearly unique evolutionary lineages and should be identified as separate species, then what constitutes a "species" for management purposes under the ESA would likely be consistent with this taxonomic assessment. However, in the face of taxonomic uncertainty and a lack of confirmation of clearly unique evolutionary lineages among the four groups of toads, a better management approach may be to deal with the four groups of toads as one "species."

The purpose of the ESA is to protect the ecosystems upon which "species" depend, and to recover those species. From an ESA management perspective, splitting toads into numerous, small groups or "species" in the face of scientific taxonomic uncertainty may be unnecessary and possibly counter to the purpose of the ESA. Perhaps a more plausible classification and management scenario would be that the four groups of toads in the southern Rocky Mountains—the SRMP group, the southern Utah group, the northern Utah group, and the southeast Idaho group—could be viewed as one "species" with four separate populations worth conserving. If listing of this combined central and southern Rocky Mountain toad "species" were warranted under the ESA, all four expressions of evolutionary uniqueness could still be conserved by regulation, if warranted.

The advantages of a unified management approach to all four groups of toads as one "species" under the ESA include (1) completing only one multi-year, high-cost species listing status review rather than four separate status reviews; (2) having a unified approach to recovery planning; (3) developing and implementing specific conservation measures that would likely be applicable throughout the range of the combined "species," rather than four separate, duplicative efforts; (4) maintaining a management focus on conserving a larger expression of evolutionary uniqueness within the boreal toad complex; (5) heeding Congress' specific guidance that the Service subdivide species for management purposes only "sparingly;" and (6) avoiding having to defend the scientific position that the southeast Idaho group of toads is one of the most narrowly distributed, highly imperiled species of amphibian on the continent. It is theoretically possible to achieve the first four advantages even with listing four separate toad species, rather than one, combined toad species. However, it is more likely to achieve all these advantages by managing these four groups together, and there are few discernable advantages to separately managing them.

By way of this example, we urge scientists to be sensitive to the challenges management biologists face. The ESA requires the Service to use the best scientific information available when making species listing and management decisions. Obviously, the best course of action for boreal toads in this case is to gather more data. However, the cry for "more data" forever burdens scientifically based management decisions. If the Service were forced through court action, for example, to make a decision today on whether to list toads under the ESA based on current information, the challenges would be large.

Endangered Species Act Conservation Planning

Historically, even when species or populations of amphibians were discernable and identifiable, there has been little effort made to conserve them, and most species conservation efforts in the United States have been focused on "charismatic megafauna," including animals like bald eagles, grizzly bears, and wolves. Other species have been important to conservationists because of their commercial or recreational value, including elk, deer, turkeys, salmon, and other game fish and wildlife species. Management and conservation of amphibians have been largely ignored, unless those amphibians provided an opportunity for commercial gain, such as with harvest of northern leopard frogs or American bullfrogs in some parts of the country.

With the passage of the ESA in 1973, many "charismatic megafauna" were listed, and conservation measures were developed. In 1990, spotted owls (Strix occidentalis) in the Pacific Northwest were listed, and some conservation measures included restrictions or prohibitions on commercial timber harvest on private land. At the same time, many salmon and steelhead runs in rivers in the northwestern United States were listed under the ESA. Private landowners began approaching the federal government asking for endangered species permits with long-term regulatory assurances for owls, salmon, and other species. These permits would allow landowners to commercially harvest timber in exchange for agreeing to implement species conservation measures. Along with these conservation-planning efforts for species listed under the ESA, landowners requested regulatory assurances for other, unlisted species (including some amphibian species) in exchange for conservation measures specific to those species.

On one hand, this request for regulatory assurances for amphibians provided the Service with the opportunity to gain specific commitments for conservation of these species that are not currently listed under the ESA. Without other conservation programs directed specifically at amphibians, this represented the first and only opportunity to gain such conservation for this group of animals, especially on private land, absent new regulatory actions (e.g., ESA listing of amphibian species). On the other hand, knowing so little about the conservation needs of many amphibian species posed some risk to the Service in offering long-term regulatory assurances to private landowners in the form of an endangered species permit. If amphibian species conservation needs are uncertain, then how could the Service offer a permit with long-term regulatory assurance and remain confident that the original conservation measures would adequately protect species?

The most reasonable approach to the dilemma of whether and how to offer regulatory assurances for conservation measures in the face of scientific uncertainty is to first obtain the best possible up-front commitments to conservation from the landowner, and then to gain assurance that management can be adapted in the future, when and where necessary to adequately conserve species. It is this dual approach that the Service seeks to pursue in most recent conservation planning efforts on private lands in the northwestern United States, including one we worked on specifically—the Pacific Lumber Company Headwaters Habitat Conservation Plan (HCP) in northern California (USFWS, 1998b).

The Headwaters HCP includes regulatory assurances for three amphibian species: tailed frogs (Ascaphus truei), foothill yellow-legged frogs (R. boylii), and southern torrent salamanders (Rhyacotriton variegatus). The Service had little or no experience managing or conserving any of these four amphibian species. To try to gain adequate species conservation, we first went to the literature to locate information on species habitat needs and for ideas on how to conserve these species. From information in the literature and consultation with local species experts, we developed an idea of habitat needs that then influenced our negotiating position with the private landowner to gain specific commitments to conservation.

For torrent salamanders, we were fortunate to find an excellent source of data and analysis for amphibian species managers seeking to identify habitat characteristics and specific measures for the purpose of conserving torrent salamanders in Welsh and Lind (1996). In our experience, this Welsh and Lind paper—and specifically its table 5—provided the most directed, specific characterization of species habitat needs that we have found for any amphibian species management effort. We used the data in Welsh and Lind, combined with data from other sources, to construct species habitat matrices for salamanders and other species to use as guides for requesting species-specific conservation commitments from the landowner (Appendix 24-A; USFWS, 1998b). We shared these matrices with Pacific Lumber to facilitate its understanding of our perception of species' habitat needs.

Some of the specific commitments from the landowner to conserve amphibian species included (1) leaving forested riparian buffers of specific widths and with specific densities of trees of certain sizes; (2) minimizing clearcutting of timber; (3) excluding machinery from riparian areas; (4) minimizing and repairing road stream crossings; and (5) removing or repairing old roads throughout the property. The landowner had agreed to implement some form of each of these types of commitments to protect salmon spawning and rearing habitat. In addition, an increment of protection was added (especially in non-fish bearing, headwater-stream areas) for amphibian species conservation.

Finally, the Service gained a commitment that the landowner would agree to work with management agencies to monitor the effectiveness of the conservation commitments for the three amphibian species. Furthermore, if monitoring data suggested that the existing commitments were inadequate to conserve species, then the Service could request that the landowner adapt their management and species conservation practices to ensure adequate conservation was provided. All this was accomplished for species not listed under the ESA and with few other opportunities available to gain conservation commitments (for example, through prohibitions of "take" of these species).

There have been past scientific criticisms of ESA conservation planning efforts. These criticisms include not gaining enough conservation commitments from a landowner "up front" by, for example, not having wide enough riparian buffers or aggressive enough road treatments on commercial forest lands. Also, many scientists have questioned providing regulatory assurances in the form of an endangered species permit for long time periods (e.g., 50 years) when our current scientific knowledge of species conservation needs is in many cases so inadequate.

These criticisms reflect very well the most basic uncertainties with which endangered species biologists constantly struggle. In its most basic form, the question we ask ourselves is if the level of conservation assurance in a plan from the landowner is commensurate with the level of regulatory assurance offered by the agency via a permit? Striking the correct balance lies in the degree of conservation assurance received up-front in on-the-ground commitments, and the ability to revisit those commitments and adapt management in the future, when necessary. For example, if a landowner commits up front to avoid all disturbance of a species' habitat, then there is little or no need for adaptive management. Conversely, the more a landowner wishes to disturb a species' habitat, and the lower the assurance of adequate up-front conservation, then the less regulatory assurance should be offered in return. Regulatory assurance can be reduced by offering a shorter permit period

Appendix 24: Draft Habitat Needs Matrix for the Southern Torrent Salamander *(Rhyacotriton variegatus)* for the Pacific Lumber Company Habitat Conservation Plan, March 17, 1997.

BIOLOGICAL VARIABLE INFLUENCED	PARAMETER	NUMERIC/NARRATIVE TARGET	REFERENCES	SUGGESTED METHOD
Maintain **water temperature** necessary to support life history traits	Water temperature range and maximum	6.5–15EC (immediate behavioral stress level 17.2EC), optimum, 8–13E	Welsh and Lind (1996); Diller and Wallace (1996)	Provide stream shading, riparian buffers to reduce conductive heat gain to stream from warm air
Provide specific **microhabitats associated** with mature forest types	Age of timber stand, size and number of trees per unit area, disturbance history	Old growth, undisturbed forest, forests >100y old, forests with 22–38 conifers >53cm DBH/ha, except coast	Diller and Wallace (1996); Welsh and Lind (1996); Bury and Corn (1989)	Maximize late seral forest types in riparian areas, increase particulate organic matter transport
Provide **tree canopy closure** sufficient to protect **stream and forest temperature and humidity levels**	% canopy closure, air and soil temperature and relative humidity, short-wave radiation	>80% canopy closure; OR <22EC air temp, <14EC soil temp, >40% relative humidity, low radiation on hottest days	Welsh and Lind (1996); Welsh, (unpublished data); Bury and Corn (1989); Chen et al. (1993)	Allow riparian vegetation to provide maximum canopy closure (as soon as 10y post-logging)
Provide the necessary **riparian buffer width** to protect **stream and forest temperature and humidity levels**	Width of forest habitat adjacent to stream, air and soil temperature and relative humidity, short-wave radiation	>30m, <125m, <>60m avg. buffer width; OR <22EC air temp, <14EC soil temp, >40% relative humidity, low radiation on hottest days	Welsh and Lind (1996); Ledwith (1996); Chen et al. (1993, 1995)	Avoid tree removal nearest stream; minimize removal beyond 30m of stream
Reduce sediment impacts and provide the necessary **substrate size** to maintain interstitial spaces	Proportions of size classes of sediment, degree of cementedness	>=68% gravel, boulder, bedrock, <50% cobble with gravel (2-16mm), <18-33% cementedness w/low % sand	Diller and Wallace (1996); Welsh and Lind (1996)	Reduce sediment delivery to stream, reduce cementedness, allow for fine particular organic matter transport (<0.063mm)
Provide adequate **woody debris** to maintain microhabitats on land and in water	Number of downed trees, mass of downed wood, % cover provided by downed wood, downed wood/stream length	Increase no. of downed logs per unit area over what would occur following intensive, even-aged management including logs in streams	Bury and Corn (1988b); Welsh and Lind (1996)	Leave trees to recruit to downed wood in stream and on land

NOTE: Developed by the U.S. Fish and Wildlife Service, Arcata, California, with California Department of Fish and Game and others. Prepared by Ted Koch and Amidee Brickee. March 1997. U.S. Fish and Wildlife Service, Arcata, California.

This habitat matrix has been developed using the most current peer-reviewed, published data, and other data available, in consultation with local experts. The matrix is intended to be used to help guide identification of conditions which may be necessary to support the southern torrent salamander *(Rhyacotriton variegatus)* within its range in northwestern California. From this matrix, land management prescriptions may be developed that will allow some certainty for long-term protection of habitat for this species while allowing for timber harvest or other proposed actions to occur. Not all variables identified in the matrix may be required in all locations, and some of the targets may not have to be met completely in every location in order for the species to persist over time.

and/or greater flexibility for the Service to revisit the commitments and require adaptations in management to retain the permit into the future. Ultimately, conservation planning under the ESA is a tool—representing one of the only proactive amphibian conservation opportunities currently available—and it is only as good or effective as people make it.

Summary

United States citizens have supported species and habitat conservation through a variety of initiatives. For herpetologists and other scientists to be successful in promoting or facilitating adequate conservation of amphibians, they must help ensure successful completion of the four general management tasks: (1) determine what group of organisms constitutes a "species;" (2) classify that species' management status (e.g., abundance and trend); (3) identify future threats; and (4) develop and implement programs to remove those threats where necessary.

A variety of tools have been and are now available for completing each of these four tasks. Tools for completing the fourth task are the least well developed. Also, though conservation tools and opportunities have been present for a while, managers have little experience using these tools to conserve amphibians. Our greatest conservation concerns generally involve widespread and locally common species that appear to be declining and disappearing in many parts of their range.

In our experience, successfully completing even the first task can be difficult for species and habitat managers. For example, boreal toad taxonomy in the central and southern Rocky Mountains is sufficiently unclear to us that making informed decisions on whether and how to protect toads there is very challenging.

Data quality is generally very poor for completing the second task of identifying management status. For most species in the Pacific Northwest, species status and trends are poorly known. There are few historic data to which we can compare current species status data, and current data usually do not include population or trend estimates. In general, we suspect most species have suffered declines since European settlement, and boreal toads and northern leopard frogs have suffered the most severe declines in the northern Rockies.

It is difficult for most managers to complete the third task of identifying future threats to species because of unclear "species" taxonomy, incomplete information on species' abundance and trends, and incomplete information on species' habitat needs. Then, even if these first three tasks of identifying a "species," its management status, and future threats can be successfully completed, completing the fourth task of developing and implementing programs to remove those threats is fraught with uncertainty and can often only be accomplished opportunistically under existing regulatory and management circumstances. Scientists should help species and habitat managers conserve amphibians by resolving as much uncertainty as possible throughout all four management steps.

TWENTY-FIVE

Amphibian Population Cycles and Long-Term Data Sets

HOWARD H. WHITEMAN AND SCOTT A. WISSINGER

The loss of biodiversity throughout the world is increasing at an alarming rate, with habitat destruction and fragmentation as the leading causes for extinction rates that are 100 to 1,000 times greater than pre-human levels (Pimm et al., 1995; Global Biodiversity Assessment, 1996; Chapin et al., 1998). Such losses are particularly evident among amphibian populations, which have disappeared or are declining in a wide range of environments (Wyman, 1990; Carey, 1993; Pounds and Crump, 1994; Blaustein and Wake, 1995; Vertucci and Corn, 1996; Lannoo, 1998b; Wake, 1998; Houlahan et al., 2000). Although most researchers agree that many amphibian populations are declining, there is debate about how to distinguish human-induced declines from natural population fluctuations (Pechmann and Wilbur, 1994; Sarkar, 1996). As with many species, amphibian populations are regulated by a variety of intrinsic and extrinsic factors that can create cyclic population fluctuations (see below). However, when compared to organisms such as insects (Hassell, 1986; Cappuccino and Price, 1995; Turchin et al., 1999) and small mammals (Krebs and Meyers, 1974; Krebs, 1992; Stenseth, 1993), there is a dearth of basic ecological information about the factors that underlie amphibian population cycles (see Green, this volume). This information is critical for determining whether the cause is manmade or natural and whether the effects are short-term or permanent (Pechmann et al., 1991; Pechmann and Wilbur, 1994; Sarkar, 1996).

Here we argue that systematic, long-term research on amphibian populations is necessary to provide basic information about the amplitude and frequency of natural fluctuations. Such baseline information is essential for posing and testing alternative hypotheses to explain declines in amphibian populations (e.g., Pounds et al., 1997). We review 20 years of research on a population of Arizona tiger salamanders (*Ambystoma tigrinum nebulosum*) at the Mexican Cut Nature Preserve (MCNP) in south-central Colorado. This review includes data that we have collected from 1989–99 (Wissinger and Whiteman, 1992; Whiteman et al., 1994, 1996; Whiteman, 1997; Wissinger et al., 1999a), and the work of other researchers from the previous 10 years (Dodson, 1982; Harte and Hoffman, 1989). The salamander population that we have been studying has fluctuated dramatically, but as with many studies (e.g., Crump et al., 1992; Laurance et al., 1996, 1997), fluctuations cannot be clearly linked to human impacts. We then consider which of the many possible hypotheses for population fluctuations are the most likely to explain the observed cycle of decline and recovery. Finally, we discuss the ability of long-term demographic studies to provide the background information necessary to distinguish natural fluctuations from human-induced declines.

Natural History of Mexican Cut Salamanders

Understanding the natural history of a species and the community within which it is embedded is fundamental for posing hypotheses about the factors that regulate population size. The population we study is isolated on a subalpine (3,500 m elevation) shelf that contains numerous adjacent open wetland basins. MCNP is owned by The Nature Conservancy and managed for low-impact ecological research by the Rocky Mountain Biological Laboratory (RMBL). MCNP has been the site of numerous ecological studies over the past three decades (e.g., Dodson, 1970, 1974, 1982; Sprules, 1972, 1974; Maly and Maly, 1974; Sexton and Bizer, 1978; Maly et al., 1980; Harte and Hoffman, 1989; Wissinger and Whiteman, 1992; Whiteman et al., 1994, 1996, 1999; Wissinger et al., 1996, 1999a,b; Whiteman, 1997; Bohonak, 1999; Bohonak and Whiteman, 1999).

The following summary of the natural history of the MCNP Arizona tiger salamanders is based on our long-term data set. In early summer (from late June to early July, depending on the year; Whiteman et al., 1999) metamorphic (terrestrial morphology, sexually mature) adults emigrate from terrestrial overwintering sites to breed in the largest permanent and semipermanent wetland basins. After breeding, these adults then migrate to semipermanent and seasonal basins, where they feed for 6–8 weeks before returning to the surrounding forest to overwinter (Whiteman et al., 1994). Eggs typically hatch in mid-July. Due to cold water temperatures and the short summer at this altitude, larval development is prolonged compared with lower-elevation populations, and metamorphosis occurs during the second or third summer. Thus, only larvae in permanent habitats or semipermanent habitats that do not dry can survive to successive summers (Wissinger and Whiteman, 1992). Many individuals forego metamorphosis and become sexually mature as larvae—i.e., become paedomorphic (larval morphology, sexually mature) adults (Whiteman, 1994a). Salamander life history stages are tied to wetland type. Permanent wetlands typically contain paedomorphic adults and several year classes of larvae that act as top predators within these ponds. Semipermanent and temporary basins typically contain metamorphic adults and hatchling larvae (Wissinger et al., 1999a). Semipermanent wetlands may have two cohorts during the summer after a year in which they retained water.

Salamanders at MCNP are keystone predators (Payne, 1966; Power et al., 1996) that have important impacts on the distribution and abundance of aquatic invertebrates (Dodson, 1970, 1974; Bohonak and Whiteman, 1999; Sprules, 1972; Wissinger et al., 1999a,b). Dietary analyses from stomach pumping indicate that most of the more than 100 species of aquatic invertebrates in these wetlands are consumed by one or more life history stages of salamanders (Whiteman et al., 1994, 1996; Wissinger et al., 1999a). Metamorphic adults specialize on fairy shrimp (*Branchinecta coloradensis*), which comprise up to 99% of their diet. Larval stages exhibit an ontogenetic dietary niche shift in prey size and type, from small planktonic to larger benthic invertebrates. Large larvae and paedomorphic adults often cannibalize small larvae.

Fluctuations in Salamander Population Size at Mexican Cut

Over the past 20 years, estimates of the total size of the Arizona tiger salamander population (all life stages combined) at Mexican Cut have fluctuated from fewer than 200 to over 3,000 individuals (Fig. 25-1A). Although there are no quantitative census data from the 1970s, researchers working at that time report that salamanders were abundant and likely numbered in the thousands (S. Willey, S.I. Dodson, personal communication). We began marking individuals in 1989, including adults and larvae from the 1988 and subsequent cohorts. Through recapture censuses we have been able to estimate the population size, document the basic demography of the population, and detail the behavior of individuals (see Whiteman, 1997; Wissinger et al., 1999a).

During the early 1980s a population decline was documented by visual censuses (Harte and Hoffman, 1989). We now know that visual censuses considerably underestimate actual population sizes (unpublished data). However, given that the same methodology was used by Harte and Hoffman each year during the 1980s and based on the number of adults we found during the early 1990s, the general trend of decline was undoubtedly real. The decline abruptly ended in 1988 with the recruitment of over 3,000 juveniles into the population (Wissinger and Whiteman, 1992; Fig. 25-1A). From 1989–91 total population size totaled around 3,200 individuals, declined during 1992–93, and slowly increased from 1993–97 (Fig. 25-1). The number of adults also declined throughout the early 1980s, remained relatively constant during the latter part of that decade, then increased during the 1990s with the maturation of the 1988 and subsequent cohorts (Fig. 25-1). There is anecdotal evidence that similar fluctuations in population size have occurred before at MCNP. Long-term researchers at the RMBL who worked at MCNP recall similar boom and bust cycles in this population dating back to the 1940s (J. Cairns, S.I. Dodson, S. Willey, personal communication).

Hypotheses for Cyclic Population Fluctuations

Population fluctuations are common in nature (Varley et al., 1973; Myers, 1988; Royama, 1992; see examples in Hanski et al., 1993) and the causes of cyclic fluctuations have been the subject of considerable debate among ecologists (e.g., Murdoch, 1994; Turchin, 1995 and references therein). Underlying mechanisms for population cycles can be divided broadly into three categories: (1) fluctuations in the abiotic environment; (2) coupled oscillations associated with interspecific interactions; and (3) density-dependent regulation mechanisms within populations (Table 25-1). Historically, ecologists have most often associated population fluctuations with variation in abiotic factors such as climate (e.g., rainfall and temperature; Andrewartha and Birch, 1954; Kingsland, 1985). Variation in climate affects survival and reproduction (Stearns, 1992) and can be linked to both an initial fluctuation in population size and subsequent cycles related to demographic effects (e.g., Kalela, 1962; Stafford, 1971; Stacey and Taper, 1992; Woiwood and Hanski, 1992). Examples of amphibian fluctuations associated with natural climatic variation include the effects of drought (e.g., Pechmann et al., 1991; Semlitsch et al., 1996) and winterkill (typically due to oxygen depletion; e.g., Bradford, 1983) on recruitment.

Perhaps the best examples of regular cycles of population fluctuation are associated with interspecific interactions, especially predator-prey and parasite- or pathogen-host dynamics (e.g., Krebs and Myers, 1974; Anderson and May, 1980; Myers, 1988; Crawley, 1989; Hanski et al., 1993; Hudson et al., 1998). In some cases, prey and predator oscillations are truly coupled and mutually density-dependent. In other cases, predator numbers are determined by prey numbers that result from fluctuations in abiotic conditions (e.g., Dempster and Lakhani, 1979). Thus, it is useful to consider separately the degree to which a population is regulated by top-down (predators, parasites, pathogens), or bottom-up (prey, hosts, or their resources) processes (Harrison and Cappuccino, 1995). Populations can also be regulated by lateral or horizontal interactions such as interspecific competition and mutualism (Auerbach et al., 1995; Harrison and Cappuccino, 1995). Coupled oscillations among competitors ultimately can be driven by shifts in the abiotic environment that alternatively favor different competitors. Cyclic oscillations can also be facilitated by spatial dynamics (Kareiva, 1989) including metapopulation dynamics (Taylor, 1990, 1998; Hanski and Gilpin, 1997).

Although there are numerous studies that have shown the importance of predator-prey and competitive interactions among amphibians (see review by Wilbur, 1997), such interactions rarely have been linked to population fluctuations in nature. One exception is population fluctuations that apparently are associated with pathogen outbreaks (Kiesecker and Blaustein, 1997b; Laurance et al., 1996, 1997; Lips, 1999).

Fluctuations in populations that result from temporal or spatial variation in the abiotic environment and/or from interspecific interactions (described above) are considered to be extrinsic regulators of population size. In contrast, fluctuations related to some form of within-population, density-dependent mechanism (such as cohort effects, time lags associated with energy storage, intraspecific competition, cannibalism, and density-dependent dispersal; Denno and Peterson, 1995; Turchin, 1995) are considered to be intrinsic regulators. For species with isolated or fragmented populations, intrinsic regulators have become increasingly framed in the context of metapopulation dynamics (Levins, 1969, 1970; Hanski and Gilpin,

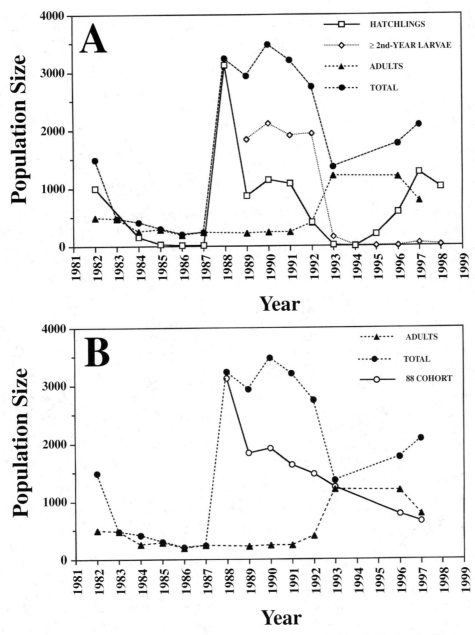

FIGURE 25-1 Population fluctuations of Arizona tiger salamanders from 1982–98 at the Mexican Cut Nature Preserve, based on life-history stage or cohort. Data from 1982–87 are from Harte and Hoffman (1989), which are based on visual counts and are therefore, likely underestimates of the true population size. (A) Population estimates of hatchlings, other larvae (2nd year and older), adults (paedomorphic and metamorphic adults combined), and total population size. (B) Impact of the 1988 larval cohort relative to the adult and total population size.

1997; Diffendorfer, 1998). Numerous studies on amphibians, particularly salamanders, provide evidence for the potential for intrinsic population regulation (e.g., Petranka and Sih, 1986; Van Buskirk and Smith, 1991; Scott, 1994), but there are few data to document the population outcomes of such regulation.

Evaluation of Alternate Hypotheses

Here we consider the degree to which these various potential causes fit the population fluctuations of salamanders at the MCNP study site. Due to the nature of these fluctuations, we are especially interested in the degree to which different mechanisms

could be related to the genesis of large cohorts. We have the following three goals in mind for future study: (1) synthesize current observational and experimental data on population fluctuations to develop a predictive model; (2) use the model to identify the types of data that should be gathered during the course of a continued monitoring program; and (3) experimentally test hypotheses for the causes of these fluctuations.

Salamander Decline from Acidification

The 1980s decline in size of the MCNP salamander population was hypothesized to be related to acid precipitation (Harte and

TABLE 25-1

Summary of Hypotheses and Evidence for Population Fluctuations in Tiger Salamanders at the Mexican Cut Nature Preserve

Hypothesis	Evidence	References
Human Effects		
Episodic acidification	indirect and weak inference for decline in 1980s; not observed in 1990s	Harte and Hoffman, 1989; Wissinger and Whiteman, 1992; Vertucci and Corn, 1996
Climatic Fluctuations		
Winter conditions	minimal winter mortality in some years apparently due to oxygen depletion; not clearly correlated with boom and bust cycles of recruitment	
Terrestrial conditions	inter-annual variation in migration of breeding adults; not clearly correlated with observed population fluctuation	Whiteman, 1997; Whiteman et al., 1999
Drought cycles and survival	annual variation in drying affects hatchling survival; not clearly correlated with observed population fluctuation	Wissinger and Whiteman, 1992; this chapter
Interspecific Interactions		
Predators	no evidence for linked fluctuations in egg predators (caddisflies and leeches); no evidence for linked fluctuations in larval predators (odonates and beetles); no known aquatic predators on paedomorphic or metamorphic adults; no information for predation on metamorphic adults in terrestrial environment	Wissinger et al., 1999a
Parasites and pathogens	no evidence for or against in this population; not correlated with population fluctuations	
Prey	over 100 species exploited; no clear temporal links with prey cycles	Wissinger et al., 1999a,b
Competitors	no evidence for interspecific competition; top predator in the system	Wissinger et al., 1999a
Intraspecific Effects		
Life history and cohort dynamics	lag-time associated with maturation and breeding frequency; life-table recruitment effects of large cohorts	Whiteman, 1994b, 1997
Density-dependent dispersal	isolated population—immigration/emigration unlikely	
Density-dependent growth	salamander growth is density-dependent potential effects on reproduction and mortality	Whiteman, 1994b; Whiteman et al., 1996 this chapter
Density-dependent resource abundance and time lags	benthic invertebrates have declined with rise of 1988 cohort; time lags may influence effects of resources on fluctuations	unpublished data; this chapter
Cannibalism	adults in large cohorts reduce survival of subsequent cohorts; could lead to long-term population cycles	this chapter

NOTE: See text for explanation and discussion of potential interactions among hypotheses.

Hoffman, 1989, 1994). Harte and Hoffman (1989) conducted field experiments using salamander eggs from a nearby habitat and found that pH levels similar to those in the MCNP ponds during snowmelt resulted in reduced egg survival. Although Harte and Hoffman (1989) never monitored the survival of eggs in the ponds themselves, they speculated that episodic acidification during snowmelt in spring was a likely cause of the decline in this population. The recruitment boom in 1988 and subsequent years led us to question the acid pulse-egg mortality hypothesis. For this hypothesis to be correct, acid pulses must occur when eggs are present. During our study, acid pulses always occurred between late May and early June, but always rebounded to circumneutral levels by late June to early July when female salamanders deposited their eggs. Eggs

monitored in the field had low mortality that did not vary with pH levels (Wissinger and Whiteman, 1992; unpublished data). Furthermore, A. tigrinum are known to be relatively acid-tolerant (Whiteman et al., 1995; Kiesecker, 1996), although geographic variation in pH tolerance can occur (see review by Rowe and Freda, 2000). In short, because of the temporal disparity between acid pulses and egg deposition, we rejected the acid pulse hypothesis as a cause of MCNP population declines (Wissinger and Whiteman, 1992; Vertucci and Corn, 1994, 1996).

Climate and Population Fluctuations

The hypotheses suggesting that climatic fluctuations affect reproduction and/or survival of one or more life history stages in

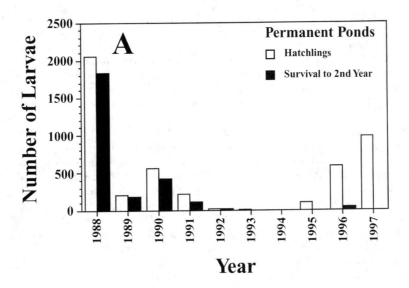

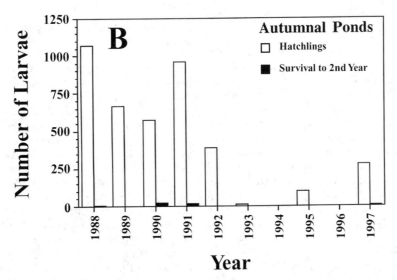

FIGURE 25-2 Larval recruitment at the Mexican Cut Nature Preserve, 1988–98 in permanent (**A**) and semipermanent (**B**) ponds. Open bars represent estimated hatchling production, and dark bars represent estimates of hatchlings surviving to their second year.

this population are supported by data showing that population variation is related to pulses in juvenile recruitment or mortality. Several abiotic factors could be simultaneously affecting different life stages of this species and subsequently leading to such fluctuations. First, as has been shown in many ambystomatids (Douglas, 1979; Semlitsch and Pechmann, 1985; Phillips and Sexton, 1989; Semlitsch et al., 1996), there is considerable year-to-year variation in the number of metamorphic adults that enter MCNP ponds during breeding migrations. Our data show that not all metamorphic females breed every year and that there is considerable variation in the proportion of females that breed in a particular year (Whiteman, 1997). We have not yet determined the causes or even correlates (factors could include variation in snowpack, timing of snowmelt, and invertebrate abundance during the previous year) of this variation, but it is likely that it is driven to some degree by environmental variability.

Second, as in other populations (Pechmann et al., 1991; Rowe and Dunson, 1995), at MCNP there is variation in the

survival of larvae depending on whether or not semipermanent wetlands dry (Fig. 25-2B; Wissinger and Whiteman, 1992). Why salamanders continue to deposit eggs in semipermanent habitats in the face of this mortality is probably related to the food-resource benefits of exploiting these habitats in wet years. Larvae that metamorphose from semipermanent habitats are often much larger and in better condition than those from permanent basins (Whiteman et al., unpublished data). However, our data do not show a correlation between recruitment over the entire population and basin drying; wetlands dried in several of the largest recruitment years (e.g., 1988 and 1990) and remained wet in some years with low recruitment (e.g., 1992, 1993, and 1995). Thus, while drying has a strong effect on recruitment in individual semipermanent wetlands (Fig. 25-2B), drying alone does not provide a sufficient explanation for the observed patterns of fluctuation in the overall population.

Third, there is considerable variation across years in mortality among larvae and paedomorphic adults in permanent wetlands. In some years, large numbers of dead animals are observed in the ponds in early spring after snowmelt. Measurements taken through the ice suggest that this mortality is related to low oxygen levels during winter (a classic winterkill scenario; see Bradford, 1983; Lannoo, 1998c; Larson, 1998). Such winter mortality could kill large numbers of paedomorphic adults generated by boom cohorts (see below), but is not likely to drive the variability that produces boom cohorts (also see Vertucci and Corn, 1996). This is true in part because paedomorphic adults are male-biased in this population and paedomorphic females breed less frequently than metamorphic females (Whiteman, 1997), so paedomorphic egg production is not as important to cohort production as metamorphic reproduction (see below). Thus, winter mortality might contribute to the decline phase of salamander population cycles, but in and of itself is not sufficient to initiate these cycles. Because there is some evidence that smaller larvae are more susceptible to winter mortality than larger ones (Whiteman et al., unpublished data), it is also possible that winter conditions could play a role in lengthening population cycles by reducing the size of larval cohorts, and thus their impact on population dynamics.

Interspecific Interactions

Cyclic fluctuations in population size can be the result of interactions with competitors, predators, parasites, pathogens, prey, or hosts (Table 25-1). Several of these population interactions are unlikely explanations for the fluctuations that we have observed in this population. As described above, Arizona tiger salamanders are the top predators in this system and large larvae and adults are unlikely to be regulated by top-down processes related to predation. Salamander eggs are eaten by caddisfly larvae and leeches, while odonate and dytiscid beetle larvae consume hatchling larvae. However, the overall effect of these predators on salamander mortality appears to be minimal (Wissinger et al., 1999a). Thus, we currently have no evidence that the fluctuations in this population are predator induced (Table 25-1).

We currently have no evidence that parasites and pathogens are important in this system. However, because parasites often underlie cyclic population fluctuations in vertebrate hosts (e.g., Anderson and May, 1980; Dobson and Hudson, 1992; Ranta, 1992; Sait et al., 1994; Poulin, 1995; Hudson et al., 1998), we recognize the importance of testing this hypothesis. We also have no evidence for competitor-induced cycles. None of the top invertebrate predators (beetles, odonates) that are likely to compete with larvae have fluctuated during the ten years of our study (Wissinger, unpublished data). Finally, while there have been changes in the overall abundance of prey taxa during our study (see below), there have been no obvious patterns of fluctuation in species that dominate the diets of metamorphic adults (fairy shrimp), paedomorphic adults and older larvae (immature stages of dipteran flies, caddisflies, and beetles), or hatchlings (cladocerans and copepods).

Life History and Cohort Dynamics

The observed population fluctuations at MCNP could be related to the effects of large cohorts on population demography. Several types of demographic mechanisms are known to lead to cyclic population fluctuations. First, because female sexual maturation in this population requires a minimum of three years, and usually four to six years (Whiteman, 1994b; unpublished data), there is a time-lag in reproduction. Thus the contribution of a large or small cohort to annual recruitment will lag behind the appearance of that cohort. Maturation varies by morph with metamorphic females breeding approximately one to two years earlier than paedomorphic females. Furthermore, metamorphic females breed every 1.8 years on average, whereas paedomorphic females breed every 2.4 years (Whiteman, 1994b, 1997). When metamorphic and paedomorphic females from the same cohort converge in reproduction (as they might at ages 6, 11, and 13), one could expect recruitment booms followed by a slow decline in population size on a twelve to nineteen-year cycle. Second, this is a long-lived species, and a single cohort can have long-term demographic impacts. The large 1988 cohort has dominated the Mexican Cut population over the past 12 years (Fig. 25-1B). Major changes in population size due to mortality have all been associated with the events that affect the 1988 cohort (e.g., winter mortality). However, no reproductive pulses associated with this cohort have been documented.

Density-Dependence, Resource Abundance, Time Lags, and Cannibalism

Birth rate, death rate, emigration, and immigration vary with population density, and time lags in the responses of these variables with respect to density can lead to cyclic fluctuations in population size. We do not suspect that density-dependent emigration or immigration is important in this extremely isolated population. There is experimental evidence for density-dependent reproduction and survival in amphibians (e.g., Wilbur, 1977a,b; Petranka, 1989c; Scott, 1990, 1994), but there are relatively few data from multiple cycles of fluctuation in natural populations. At MCNP, we have documented density-dependent growth both in larvae (Whiteman, 1994b) and adults (Whiteman et al., 1996). We have also found that the overall abundance of prey resources has decreased as the total biomass of the 1988 cohort has increased (Whiteman et al., unpublished data). Population fluctuations associated with density-dependent effects are most likely when there is a lag time between resource depression and a decline in reproduction and survival (May, 1976). For example, stored energy can sustain reproduction and survival in populations that leads to cycles of carrying capacity overshoot and subsequent declines (e.g., Goulden and Hornig, 1980). The potential for such time lags could be inferred by assessing changes in body fat or body condition with respect to the timing of food availability. For example, we found that metamorphic adults increase their body condition (mass per SVL) by feeding on fairy shrimp in semipermanent wetlands (Whiteman et al., 1996). Although the effect of these food resources on subsequent reproduction is unclear, fairy shrimp consumption may account for the reduced interval between breeding attempts in metamorphic females (Whiteman, 1997). This, in turn, should affect the overlap in breeding between the two morphs and thus produce a potential mechanism for boom cohorts (see above).

Cannibalism can give rise to population fluctuation cycles (Fox, 1975; Polis, 1981, 1988; reviews by Elgar and Crespi, 1992). At MCNP, large larvae and paedomorphic adults cannibalize small larvae (Wissinger et al., 1999a; Whiteman, unpublished data). Hatchlings avoid larger conspecifics by foraging at different times and in different microhabitats (Marcus and Whiteman, in preparation; Wissinger et al., unpublished data), and hatchling survival is greater in ponds without larger larvae (Whiteman et al., unpublished data). If cannibalism is a major source of hatchling mortality, then large cohorts such as the 1988 generation could reduce or eliminate recruitment in subsequent years. For example, second-year recruitment was high between 1988 and 1991 in permanent ponds when adult densities were low (Figs. 25-1A, 25-2A). As the number of paedomorphic adults increased from the maturation of the 1988 cohort, hatchling survival declined dramatically (1995–97). We cannot rule out that reduced resource abundance (see above) was responsible for this lack of recruitment success, but the ontogenetic diet shift that occurs in these salamanders (Wissinger et al., 1999a) suggests that competition between paedomorphic adults and hatchlings is unlikely to substantially reduce hatchling survival.

Cannibalism has been linked to cyclic fluctuations in population size in a variety of other stage-structured populations (e.g., Alm, 1952; Brinkhurst, 1966; Mertz and Robertson, 1970; Mertz, 1972; King and Dawson, 1973; Orr et al., 1990; Ruxton et al., 1992; Van Buskirk, 1992). The role of cannibalism in generating cyclic fluctuations in fish populations provides an intriguing model for the MCNP salamander population. Cannibalism by adult perch in large cohorts can reduce or completely eliminate subsequent recruitment for years. Cohort suppression is most likely to lead to long-term (10–15 years) population fluctuations in low nutrient habitats and in the absence of potential competitors (Le Cren, 1955, 1965; McCormack, 1965). Interestingly, MCNP is ultraoligotrophic and Arizona tiger salamanders are the lone vertebrate predator at the top of the food web (Wissinger et al., 1999a).

Lessons from MCNP for Population Regulation and Amphibian Conservation

In some cases of amphibian decline (e.g., habitat destruction), human causes are so obvious that alternative hypotheses only distract (e.g., Blaustein and Wake, 1995). In other cases, declines are not as easily linked to a specific causal agent and it is important to consider alternative explanations (Pechmann and Wilbur, 1994; Sarkar, 1996). For the latter situation, we must understand how to distinguish between true declines and population fluctuations. It is equally important, however, to determine the demographic warning signs of amphibian populations that are susceptible to decline.

Multiple Hypotheses and Key-Factor Analysis

The approach that we have taken loosely resembles a key factor analysis, a method that is often used for insect populations. A key factor analysis attempts to identify which of the many sources of mortality at different life history stages are the density-dependent (and therefore key) regulators of population size (Varley and Gradwell, 1960, 1970; Varley et al., 1973; Manly, 1977; Stiling, 1988; Yamamura, 1999).

A similar approach to population regulation in amphibians should be useful for distinguishing between anthropogenic declines and natural population fluctuations (c.f. Murdoch, 1994). At MCNP, we are in the process of evaluating the most likely hypotheses of population regulation by documenting a variety of key demographic variables at different population sizes. One result of this research will be an understanding of

how salamander demography changes during a cycle, which will be used to pinpoint variables that can predict future cycles. For example, three variables appear to have a strong impact on MCNP Arizona tiger salamanders: age structure, age at maturity, and frequency of breeding. The interaction of these three variables (and the forces that influence them, such as variation in climate and cannibalism) might explain the boom and bust pattern we have observed. If similar basic demographic data were available for a presumed declining species, or a closely related congener, we would have valuable insight into whether true declines have occurred or if the species has an equilibrial life history, such as the Arizona tiger salamanders at MCNP. Unfortunately, such basic demographic data are not available for most amphibians (Duellman and Trueb, 1986; see also Pechmann and Wilbur, 1994).

We propose that similar multiple-hypothesis, key-factor studies be used across several representative taxa to identify alternative potential causes (and their interactions) for population fluctuations. Organisms should be sampled in areas with minimal human impacts so that baseline information can be collected from presumed healthy populations. Such analyses may provide insights into the mechanisms of population regulation, identify the variables most closely associated with fluctuations in each group or geographic domain, and potentially help predict the early warning signs of declining populations. We do not suggest that such data sets replace current conservation efforts. Instead, we are promoting the idea of sound scientific knowledge of amphibian population regulation to better understand and predict population fluctuations and declines.

There are several caveats to this approach. First, for organisms with complex life cycles, identifying a key factor in one life stage does not necessarily mean that factor regulates the overall population (Turchin, 1995). Second, population regulation can often be the result of several mechanisms that are simultaneously operating or are of varying importance at different stages of population cycles (Myers and Rothman, 1995). Third, factors can differ between locations within species (e.g., Berven, 1995), so intraspecific comparisons must be conducted with caution.

Long-Term Monitoring

Many studies have shown that understanding population fluctuations and their causes requires long-term population monitoring (e.g., Blaustein et al., 1994a; Semlitsch et al., 1996). Turchin (1995) suggests that the absence of evidence for population cycles is usually the result of a short-term data set—the longer we look, the more likely it is that we find evidence for fluctuations and their causes. The inherent cost to this approach is that if populations are truly declining, by the time it is clearly recognized it may be too late.

Another problem associated with long-term population monitoring is that causes will likely differ among species and habitats. For example, density-dependent dispersal is a likely hypothesis for fluctuations in many situations (reviewed by Denno and Peterson, 1995), but it is probably not important for an isolated population such as the one at MCNP. Similarly, although interspecific interactions are important in some amphibian communities (Wilbur, 1982; Stenhouse et al., 1983; Morin, 1986; Cortwright, 1988; Semlitsch et al., 1996; reviewed by Wilbur, 1997), they are unlikely to be important at our study site (one species, top generalist predator). However, the true utility of multiple, long-term studies is an understanding of how geographic and/or species-specific differences affect population fluctuations.

Computational Models and Ecological Experiments

One way to shorten the minimum time needed to understand the dynamics of population fluctuations is to develop predictive/computational models to transcend the long-term, real-time nature of the observed population fluctuations (see Sarkar, 1996). In collaboration with Dr. Ian Billick, we are using our basic demographic and ecological information to parameterize such models and determine what additional data will be necessary to test our hypotheses. For example, knowing the per capita effects of cannibalism on juvenile survival is important for testing the hypothesis that large cohorts can suppress recruitment and cause cyclic fluctuations. Similarly, the experimental manipulation of a particularly important food resource (e.g., fairy shrimp) would allow the inclusion of prey resources into a computational model predicting reproductive output. Incorporating a variety of such data into computational models will allow for the simultaneous testing of a subset of hypotheses that might operate alone or jointly to produce observed fluctuations. We agree with Sarkar (1996) that long-term monitoring in combination with the interim use of explicit and predictive models will help direct monitoring efforts. This, in turn, will allow researchers to maximize the collection of data most likely to be useful for the conservation and restoration of amphibian populations.

Summary and Conclusions

Our work with the MCNP Arizona tiger salamanders suggests a fluctuating population that is regulated by abiotic (pond hydroperiod) and biotic (resource abundance, life history, and cannibalism) factors. Although there is still much to learn about the interaction of these variables, a multiple-hypothesis, long-term demographic monitoring approach, when combined with planned models and experiments, should help predict future fluctuations and, perhaps, help understand cyclic phenomena in other species. Our results thus far suggest that similar approaches on relevant taxa throughout the globe will provide important and useful information for understanding amphibian population changes and the associated natural or anthropogenic factors that influence them.

Acknowledgments. This long-term research could not have been completed without numerous field assistants and collaborators, especially Wendy Brown, John Gutrich, Andy Bohonak, and Steve Horn. We are grateful to the Rocky Mountain Biological Laboratory and The Nature Conservancy for access to the Mexican Cut Nature Preserve, and particularly to B. Barr from the RMBL for facilitating our research. We thank the following funding sources for continued support throughout this study: the NSF (DEB-9122981 and DEB-0109436 to HHW; BSR-8958253, DEB-9407856, and DEB-0108931 to SAW), the Colorado Division of Wildlife, The Nature Conservancy, the American Museum of Natural History, the Purdue Research Foundation, the Allegheny College Faculty Development Committee, and the Committee on Institutional Studies and Research at Murray State University. Finally, we thank Ian Billick for critical review of the manuscript.

TWENTY-SIX

Landscape Ecology

DAVID E. NAUGLE, KENNETH F. HIGGINS, REX R. JOHNSON,
TATE D. FISCHER, AND FRANK R. QUAMEN

The increased public interest in amphibian conservation and the growing evidence of detrimental effects of habitat fragmentation on biological diversity (see Saunders et al., 1991 for a review) has prompted land managers to seek ways of managing amphibian populations at landscape scales. For example, principles of landscape ecology are now being used by avian ecologists to direct conservation efforts and design nature reserve systems (e.g., Robbins et al., 1989; Pearson, 1993; Flather and Sauer, 1996).

Landscape ecology emphasizes landscape patterning, species interactions across landscape mosaics, and the change in these patterns and interactions over time (e.g., Gardner et al., 1987; Turner, 1989; Turner and Gardner, 1991; Forman, 1995). Central to landscape-scale studies of amphibians is an assessment of relevant spatial and temporal scales. Wetlands in the prairie pothole region (PPR) of eastern South Dakota provide breeding habitat for 11 frog and toad species (Naugle et al., this volume). Because wetlands are discontinuous habitats embedded within an upland matrix (Winter, 1988; Johnson and Higgins, 1997), amphibians here have clumped distributions with interactions that may produce metapopulations (Sjögren, 1991; Hanski and Gilpin, 1997). Metapopulations are susceptible to wetland losses that fragment habitat because subpopulations that go extinct must be recolonized by dispersing amphibians from nearby wetlands.

Despite a 45% loss in wetland numbers (Johnson and Higgins, 1997), glaciated landscapes in eastern South Dakota have been less impacted by habitat fragmentation than elsewhere in the Midwest, where wetland losses exceed 90% (Tiner, 1984; Dahl, 1990; Leja, 1998). Wetland size and permanence in the PPR vary along a continuum from small, temporary and seasonal wetlands to large, semipermanent and permanent wetlands. Variable hydrologic cycles cause seasonal and semipermanent wetlands to go completely dry for 2 or more consecutive years. The abundance of wetland complexes and dynamic nature of the hydrologic cycle make eastern South Dakota an ideal landscape in which to study amphibian wetland use and to examine how use might change over time with fluctuating water conditions (following Lannoo, 1998a). Here, we introduce the potential role of landscape ecology in amphibian conservation, provide an interpretation of landscape analyses using real data on amphibian populations, and outline insights gained from this study that might enhance future landscape-level management decisions.

Methods

Survey Methodology

We used call survey data collected in 1997–98 in eastern South Dakota (Naugle et al., this volume) to evaluate anuran habitat use in temporary, seasonal, and semipermanent (Stewart and Kantrud, 1971) wetlands. Male mating calls were recorded during nighttime auditory surveys at randomly selected wetlands positioned along roadside transect routes. The timing of surveys coincided with peak calling periods for amphibians in the northern Great Plains (see Fig. 47-2 in Naugle et al., this volume).

We initially conducted call surveys at 1,496 wetland sites to obtain information on distributions of anurans over a large geographic region (Naugle et al., this volume). Survey methodology followed an established sampling protocol that has been extensively used to monitor amphibians (Hemesath, 1998; Johnson, 1998b; Mossman et al., 1998). Although call surveys enable researchers to evaluate large regional databases (Mossman et al., 1998; Knutson et al., 1999), some investigators (e.g., Kline, 1998) have indicated that call survey data may be a poor substitute for intensive survey work in a small number of wetlands. To minimize the probability of misclassifying occupied wetlands using call surveys, we considered the influences of species occurrence rates and distributions, survey timing, and weather conditions. In these analyses, we used six species—Woodhouse's toads (*Bufo woodhousii*), Canadian toads (*B. hemiophrys*), American toads (*B. americanus*), Great Plains toads (*B. cognatus*), northern leopard frogs (*Rana pipiens*), and striped chorus frogs (*Pseudacris triseriata triseriata* or *maculata*)—that occurred in at least 90 of the original wetlands surveyed. Ranges for these six species were sympatric (see Fig. 47-1 in Naugle et al., this volume) in 13 eastern South Dakota counties (Fig. 26-1). Calling in all six species was recorded in the highest proportion of wetlands between 2100–0059 hr (Fig. 26-2). The wetlands included in analyses met the above criteria, were surveyed during the best weather

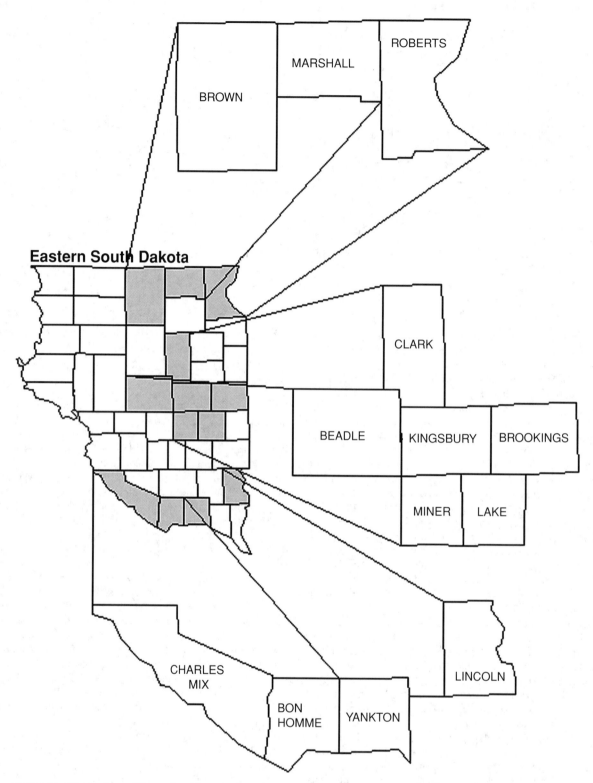

FIGURE 26-1 Distribution of 13 counties with wetlands used to evaluate amphibian habitat use in eastern South Dakota.

conditions (low winds [<5 km/h] and water temperatures >10 °C), and were located more than 3,000 m apart.

Local Habitat Measures

The area (ha) of surveyed wetlands was estimated using a wetland geographic information system (GIS) that was constructed from National Wetland Inventory (NWI) data for eastern South Dakota (Johnson and Higgins, 1997). The percent of the wetland area containing emergent cover was visually estimated into class intervals as follows: (1) <1%; (2) 1–5%; (3) 6–25%; (4) 26–50%; (5) 51–75%; (6) 76–95%; and (7) >95%. The proportion of the wetland perimeter occupied by grassland habitat was also estimated and categorized into one of the seven class intervals defined above. Class interval midpoints were used to

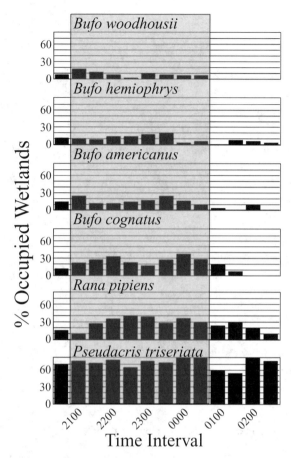

FIGURE 26-2 Proportion of wetlands where calling anurans were recorded during 13 time intervals (2030–0230 hrs.). Across six species surveyed, calling anurans were recorded in the highest proportion of wetlands between 2100–0059 hrs.

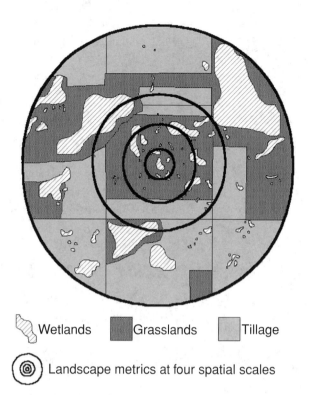

Wetlands   Grasslands   Tillage

Landscape metrics at four spatial scales

FIGURE 26-3 Land-use coverage from the GIS. Four scales are 250; 500; 1,000; and 2,500 m.

analyze categorical data. We recorded the number of NWI wetlands (Cowardin et al., 1979) within surveyed wetland basins as an index to wetland basin heterogeneity.

Landscape Habitat Measures

We quantified landscape pattern and land cover using the wetland GIS and Landsat satellite imagery at four spatial scales: 250 m (20 ha); 500 m (79 ha); 1,000 m (314 ha); and 2,500 m (1,963 ha) (Fig. 26-3). The total areas of temporary, seasonal, semipermanent, and permanent wetlands at each scale were calculated to characterize the wetland complex surrounding each surveyed site (Fig. 26-3). We also calculated the proportion of grassland area as the sum of untilled area divided by the sum of non-wetland area at each scale to assess whether grassland abundance was related to anuran wetland use (Fig. 26-3). Grain crops (e.g., corn [*Zea mays*] and soybeans [*Glycine max*]), small grains (e.g., wheat [*Triticum aestivum*]), and annually fallowed lands were considered tilled. Untilled lands were permanent pastures, Conservation Reserve Program grasslands, and alfalfa (*Medicago sativa*). Trees, which constitute <1% of the area of eastern South Dakota, were not distinguished from untilled lands. Overall, per-pixel classification accuracy for this coverage was 97%. (Detailed classification methods for the land-use coverage used in this study are available in Naugle et al. [1999].)

We calculated the distance (m) from surveyed wetlands to the nearest neighbor wetland and the average distance to the nearest five and 10 wetland neighbors as a measure of wetland isolation. We also tallied the number of times that eight equally spaced azimuths radiating out 1,000 m from the center of the surveyed wetlands intersected another wetland or grassland patch. The number (0–8 possible) of intersects was used to index the abundance of wetland and grassland habitat available for dispersing anurans.

Statistical Analyses

We used forward stepwise logistic regression (Hosmer and Lemeshow, 1989) to evaluate the relationships between the occurrence of the six species and local- and landscape-level habitat variables at four spatial scales. Correlation coefficients were calculated for pair-wise combinations of independent variables (Table 26-1). The least significant variable in a pair of correlated attributes ($r > 0.5$) was removed. This process continued until a set of uncorrelated variables was found at each spatial scale. All uncorrelated variables began in the logistic model. Threshold p values for entering and removing variables from a model were 0.15 and 0.05, respectively. The significance of each regression was tested using the Wald chi-square statistic (maximum likelihood estimate). McFadden's Rho^2 statistic, which was used to evaluate model fit, is a transformation of the likelihood ratio statistic, intended to mimic R^2 values in logistic regression. Rho^2 values are lower than R^2 values. We considered Rho^2 values between 0.20 and 0.42 as satisfactory (Hensher and Johnson, 1981). Natural history attributes of individual species were the principle criteria used in the selection of a final model from multiple competing models. Following final model selection, we assessed the relative importance of each

TABLE 26-1
Description of Variables Measured to Evaluate Anuran Use of Wetland Habitats in Eastern South Dakota, 1997–98

Variable Name	Description
AREA	Natural logarithm of the area (ha) of the wetland
COVER	Percent vegetated area of the wetland surveyed
PERIM	Proportion of the wetland perimeter surrounded by grassland
HETERO	Index to wetland basin complexity measured as the number of National Wetland Inventory wetlands (Cowardin et al., 1979) within the wetland basin surveyed
TEMP	Sum of total temporary wetland area within 250, 500, 1,000, and 2,500 m radius from surveyed wetland
SEAS	Sum of total seasonal wetland area within 250, 500, 1,000, and 2,500 m radius from surveyed wetland
SEMI	Sum of total semipermanent wetland area within 250, 500, 1,000, and 2,500 m radius from surveyed wetland
PERM	Sum of total permanent wetland area within 250, 500, 1,000, and 2,500 m radius from surveyed wetland
GRASS	Proportion of untilled upland habitat within 250, 500, 1,000, and 2,500 m radius from surveyed wetland
DIST	Distance (m) from the surveyed wetland to the nearest wetland and the mean distance to the nearest 5 and 10 wetlands
DISPERSW	Number of times that eight equally spaced azimuths radiating out from surveyed wetlands intersected another wetland; used to index abundance of potential habitat available to dispersing anurans
DISPERSG	Number of times that eight equally spaced azimuths radiating out from surveyed wetlands intersected a patch of grass; used to index abundance of grassland habitat available to dispersing anurans

variable by partitioning the total explained variation among the individual variables within models.

Results

Analyses were conducted using 84 temporary, 96 seasonal, and 56 semipermanent wetlands located within species ranges and surveyed during peak calling periods under optimal weather conditions (see "Methods" for data screening process). The results focus on competing models for bufonids, because habitat relationships for toad species were stronger (Rho^2 = 0.20–0.42) than those for frogs (Rho^2 < 0.14). Variation explained in logistic equations was generally higher at larger spatial scales (for an example of a set of models; Table 26-2). The most parsimonious and biologically meaningful relationships for three of four bufonids were estimated at the largest spatial scale studied (2,500 m radius; the exception was Woodhouse's toads at 500 m). The relationships between the presence of toad species and habitat attributes were weak in temporary and seasonal wetlands. In contrast, habitat relationships were strongest in semipermanent wetlands (the wetland type among those considered that holds water for the longest time; Table 26-2). Bufonid habitat use was related to the structure of the wetland landscape surrounding semipermanent wetlands (Table 26-3). Canadian and American toads were more likely to occur in semipermanent and permanent wetland landscapes, whereas Woodhouse's toads were found in semipermanent wetlands embedded within high-density temporary wetland landscapes. Great Plains toads occurred in landscapes largely devoid of other seasonal and semipermanent wetlands (Table 26-3). Woodhouse's toads were more likely to occur in less heterogeneous semipermanent wetlands, while the occurrence of American toads was associated with agricultural landscapes. The area of wetlands surveyed did not enter into logistic models, which indicates that small wetlands provided breeding habitat comparable to that of larger wetlands.

Discussion

The Importance of Wetland Complexes

Landscapes differ in their potential to provide habitat for prairie bufonids because species have contrasting life histories that influence the way that each interacts with the landscape. Results from this study indicate that Canadian and American toads, the most aquatic bufonids in this region (Breckenridge and Tester, 1961; Oldfield and Moriarty, 1994), inhabit landscapes with an abundance of semipermanent and permanent wetland types (Table 26-3). In contrast, Woodhouse's toads are more likely to occupy semipermanent wetlands embedded within temporary wetland landscapes. Great Plains toads, the most terrestrial species studied (Wright and Wright, 1949; Oldfield and Moriarty, 1994), are commonly found in landscapes largely devoid of other seasonal and semipermanent wetlands. A holistic wetland conservation strategy that incorporates the characteristics of entire landscapes is more likely to provide suitable habitat for the complete assemblage of prairie anurans.

A View of Amphibian Conservation Under Fluctuating Hydrologic Conditions

Wetland permanence and predation are important factors shaping amphibian use of wetlands in the PPR, an ecosystem with highly variable hydrologic cycles (see Skelly, 1997; Lannoo, 1998a, for similar views of factors influencing amphibian wetland use). Wetland types vary along a continuum from small temporary and seasonal wetlands to large semipermanent and permanent wetlands. Seasonal and semipermanent wetlands that pond water year-round in a protracted wet cycle may go dry for prolonged periods (e.g., 2–3 years) during drought (Fig. 26-4). Prairie bufonids either forego reproduction or restrict their breeding activities to permanent wetlands that still pond water during drought (the number and area of

TABLE 26-2

An Example of Competing Habitat Models for the Canadian Toad at Four Spatial Scales in
Three Wetland Types

Wetland Type	Scale	Habitat Model	McFadden's Rho^2
Temporary	250	+DIST + GRASS (0.05) (0.05)	0.10
	500	+DIST5 + SEAS (0.07) (0.05)	0.12
	1,000	+DIST10 + AREA (0.06) (0.06)	0.12
	2,500	+DIST10 + PERIM (0.11) (0.07)	0.18
Seasonal	500	−PERCO−TEMP (0.05) (0.05)	0.10
	1,000	−PERCO + PERM (0.06) (0.05)	0.11
	2,500	−PERCO − SEAS + PERM (0.07) (0.05) (0.05)	0.17
Semipermanent	1,000	−SEAS + PERM (0.10) (0.06)	0.16
	2,500	−SEAS + PERM + SEMI (0.24) (0.13) (0.05)	0.42

NOTE: McFadden's Rho^2 is a transformation of the likelihood ratio statistic intended to mimic R^2 values in logistic regression. Rho^2 values between 0.20 and 0.40 are considered satisfactory (Hensher and Johnson, 1981), because these values are usually much lower than R^2 values in linear multiple regression. The proportion of explained variance attributed to individual variables is in parentheses. The variation explained in logistic equations was generally higher at larger spatial scales. Only the strongest models that characterize toad use of semipermanent wetlands are presented because temporary and seasonal wetland models were weak (Rho^2 < 0.20). Habitat models for four toad species in semipermanent wetlands are presented in Table 26-3.

TABLE 26-3

Habitat Models Generated Using Stepwise Multiple Logistic Regression for Four Bufonid
Species in Semipermanent Wetlands

Species	Habitat Model	McFadden's Rho^2	% Occupied Wetlands
Bufo hemiophrys	− SEAS + PERM + SEMI (0.24) (0.13) (0.05)	0.42	30.4
Bufo woodhousii	+ TEMP − HETERO (0.19) (0.14)	0.32	10.7
Bufo americanus	− GRASS + PERM (0.14) (0.12)	0.26	25.0
Bufo cognatus	− SEAS − SEMI (0.12) (0.08)	0.20	21.4

NOTE: All habitat variables in models have P < 0.05. Wald chi-square statistics for complete models were each P < 0.004. Variables (see Table 26-1) are listed in the order they entered into equations. McFadden's Rho^2 is a transformation of the likelihood ratio statistic intended to mimic R^2 values in logistic regression, and it is, therefore, an index to explained variation for the entire equation. Rho^2 values between 0.20 and 0.40 are considered satisfactory (Hensher and Johnson, 1981) because these values are usually lower than R^2 values in linear multiple regression. The proportion of explained variance attributed to the individual variables is in parentheses.

remaining wetlands depends on drought severity). Aquaculture initiatives are a constant threat to maintenance of amphibian populations because permanent wetlands no longer function as refugia for breeding amphibians when predatory fishes are present (Bradford et al., 1993; Bronmark and Edenhamn, 1994; Lannoo, 1998a).

We hypothesize that pulses (i.e., rapid population increases) in amphibian productivity typically occur during wet hydrologic conditions when ephemeral wetland types remain wet throughout spring (temporary wetlands) and summer (seasonal and semipermanent wetlands) (bottom of Fig. 26-4; line A). Adult toads that survive a drought readily colonize nearby wetlands

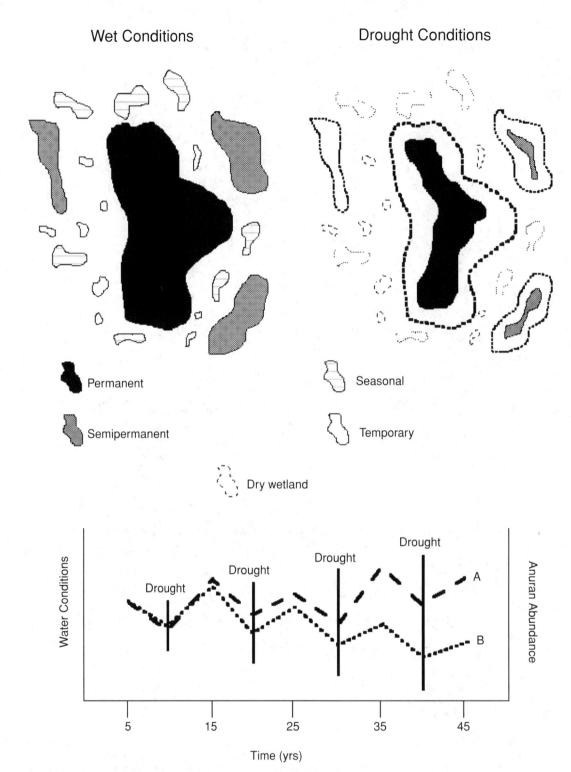

Wet Conditions   Drought Conditions

Permanent

Semipermanent

Seasonal

Temporary

Dry wetland

Drought

Drought

Drought

Drought

A

B

Water Conditions

Anuran Abundance

5   15   25   35   45

Time (yrs)

FIGURE 26-4 Changes in wetland conditions coincident with hydrologic cycles in the prairie pothole region of eastern South Dakota *(top)*. Wetlands that pond water year-round during wet conditions but go dry in a drought may cause amphibians to either forego reproduction or restrict their breeding activities to permanent wetlands. We hypothesize that periodic pulses in amphibian productivity resulting from colonization of ephemeral wetland types likely supplement amphibian populations that otherwise decline during drought *(bottom; line A)*. Long-term viability of amphibian populations may be jeopardized if small wetlands that maintain the pulse potential of prairie landscapes are lost *(bottom; line B)*.

that pond water during favorable hydrologic conditions (Lannoo, 1996). The pulse in productivity resulting from periodic colonization of ephemeral wetland types likely supplements toad populations that decline during drought (Fig. 26-4, line *A*). Ephemeral wetland types, when viewed within the hydrologic context under which these species evolved, are a vital link in maintaining the pulse potential of amphibian populations.

The Value of Small Wetlands. Resource managers acknowledge the current lack of empirical evidence needed to rebut opponents who question the value of small wetlands (Gibbs,

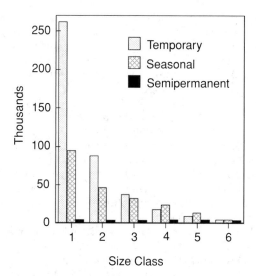

FIGURE 26-5 Number of unprotected temporary, seasonal, and semipermanent wetlands in eastern South Dakota. Size classes (ha) are (1) 0–0.2; (2) 0.3–0.5; (3) 0.6–1.0; (4) 1.1–2.0; (5) 2.1–5.0; and (6) > 5.0.

1993; Haig et al., 1998; Semlitsch and Bodie, 1998; Lehtinen et al., 1999). Opponents that view small or isolated wetlands as individual, disjunct sites quickly devalue their importance as potential amphibian habitat when other taxa (e.g., wetland bird species) are abundant in large semipermanent wetlands (Brown and Dinsmore, 1986; Naugle et al., 1999). The area of wetlands surveyed in this study did not enter into logistic models for any species in any water regime, indicating that small wetlands provide habitat comparable to that of large wetlands for breeding amphibians. A heightened awareness of the value of small wetlands to long-term persistence of amphibian populations is critical because the median size of all wetlands in eastern South Dakota is only 0.16 ha (Johnson and Higgins, 1997). Johnson et al. (1996) estimates that the deregulation of small wetlands (<0.40 ha) in agricultural fields could result in drainage of up to 49% of the remaining wetlands in eastern South Dakota. We hypothesize that the loss of small wetlands that maintain the pulse potential of prairie landscapes could jeopardize the long-term viability of amphibian populations (bottom of Fig. 26-4, line B). From our most recent research, we suggest that we risk losing periodic pulses in amphibian productivity if wetland drainage laws change because most unprotected wetlands are small (<0.50 ha) ephemeral wetland types (Fig. 26-5; Naugle et al., this volume).

When wetlands are viewed as components of a larger landscape, our results indicate that small wetlands are critical landscape elements that influence amphibian species composition. Our habitat model for Great Plains toads indicates that this species is found in wetlands (regardless of their size) isolated from other wetlands. At 12 separate survey sites visited during the course of this study, Great Plains toads were heard calling from small, isolated "wet spots" that were not delineated as wetlands by NWI personnel. In contrast, Woodhouse's toads were more likely to occur in wetland complexes such as those in the Central Lowlands of eastern South Dakota, where temporary and seasonal wetland densities commonly exceed 180 wetlands/25.9 km² (Johnson and Higgins, 1997). Studies involving metapopulation dynamics have shown that small temporary wetlands increase connectivity for amphibians dispersing between wetlands (Sjögren, 1991; Vos and Stumpel,

1995) and that the loss of temporary wetlands may increase inter-wetland distances, which subjects amphibian populations to greater risks of local extinction (Richter and Azous, 1995; Lannoo, 1998a; Semlitsch and Bodie, 1998; Lehtinen et al., 1999). The most effective strategies for amphibian conservation seem to be those that consider the characteristics of entire wetland complexes in addition to the attributes of individual wetlands. Long-term data are much needed in the PPR to evaluate temporal patch dynamics and to determine threshold levels at which populations become vulnerable to habitat fragmentation. Studies assessing dispersal patterns, extinction and recolonization rates, and population dynamics during each phase of the hydrologic cycle would enhance our current understanding of amphibian habitat needs.

Suggestions for Future Landscape Amphibian Studies

Scale Selection

A landscape may be defined as a "heterogeneous area composed of a cluster of interacting ecosystems that is repeated in similar form throughout" (Forman and Godron, 1986). In application to this study, our landscapes consist of a mosaic of upland grasslands and agricultural lands in which habitat patches of four wetland types are embedded. The selection of an appropriate scale at which to study species interactions among habitat patches and how patterns of use change over time is challenging because each organism scales the environment differently. For example, the scale at which a vagile marsh bird species such as the black tern (Chlidonias niger) interacts with the landscape (Naugle et al., 2000, 2001) is likely much larger than that of a less mobile amphibian. In general, landscapes occur at a spatial scale intermediate between an organism's home range and its regional distribution. Researchers usually either conduct studies at arbitrarily defined scales or use natural history attributes of organisms (e.g., behavioral traits, movement patterns, or dispersal distances) to investigate relationships at multiple scales. The selection of appropriate scales at which to study amphibian habitat use is somewhat difficult because little landscape-level work has been done and only limited natural history information is available for these species. In this study, we found that habitat relationships for toads were generally better at larger spatial scales. We recommend that future studies investigating the influence of scale on amphibian habitat use include a spatial scale that exceeds the largest used in this investigation.

Avoid a Shotgun Approach

Landscape ecologists have developed a tremendous variety of spatial metrics (i.e., measurements and indices) to investigate relationships between landscape patterns and ecological processes (e.g., O'Neill et al., 1988; Milne, 1991; Gustafson and Parker, 1992; Plotnick et al., 1993) and free software for quantifying metrics is available (e.g., APACK, FRAGSTATS, or RULE). Although the use of metrics enables ecologists to quantify landscape structure, it is often applied without a clear understanding of the strengths and limitations of each measure. As a result, some investigators will take a "shotgun approach" to quantifying landscape structure by calculating every available metric. Such an approach commonly leads to findings that are either erroneous or difficult to interpret. We recommend that readers wishing to conduct landscape analyses first read Hargis et al. (1998) and McGarigal and Marks (1995) to develop a clear understanding of the landscape metrics they intend to use.

Further Consideration of Sampling Protocol

Herpetologists are polarized in their views of the usefulness of call surveys. Although call surveys have been successful in monitoring regional distributions and trends in amphibian populations (Mossman et al., 1998; and others), other investigators present compelling arguments for the use of larval surveys to confirm a species presence/absence. Kline (1998) has indicated that call survey data are a poor substitute for larval surveys because negative call survey results may not confirm the absence of some species. Resolution of this topic is necessary because replication (i.e., adequate sample sizes in multiple landscape types) is of paramount importance in landscape studies. Misclassification of wetlands occupied by anurans whose calling activity did not coincide with observer presence would obscure habitat relationships for individual species. In contrast, findings from a recent study in Minnesota (Lehtinen et al., 1999) were restricted to species richness analyses because intensive fieldwork (larval sampling, chorusing, and visual encounter surveys) in a small number of wetlands decreased sample sizes. We recommend that a combination of call surveys and larval sampling be used in future studies, with adjusted sample sizes that allow landscape analyses to be conducted on individual species.

Summary

The emerging field of landscape ecology provides a theoretical framework for investigating spatial and temporal relationships in amphibian wetland use. We evaluated wetland habitat use for six anuran species in the glaciated landscapes of eastern South Dakota. We identified occupied wetlands using existing call survey data (Naugle et al., this volume) and quantified landscape attributes at four scales: 250, 500, 1,000, and 2,500 m. Our findings focus on competing models for four bufonids because habitat relationships for toads were stronger than those relationships for frogs. The most parsimonious and biologically meaningful logistic models were estimated at larger spatial scales. Our findings indicate that wetland complexes differ in their potential to provide suitable habitat because species have contrasting life histories influencing how each interacts with the landscape. We hypothesize that pulses in amphibian productivity typically occur during protracted wet cycles when amphibians recolonize ephemeral wetland types that pond water. When wetlands are viewed as components of a larger landscape, results from this study indicate that small wetlands are critical landscape elements that influence the types of amphibians inhabiting a landscape. A heightened awareness of the value of small wetlands is essential in our region of study where median wetland size is only 0.16 ha. We risk losing periodic pulses in amphibian productivity if wetland drainage laws change because most unprotected wetlands in eastern South Dakota are small (< 0.50 ha) ephemeral wetland types. The most effective strategies for amphibian conservation are likely those that consider characteristics of entire wetland landscapes in addition to attributes of individual wetlands.

Acknowledgments. We thank Scott Stolz, Chad and Brenda Kopplin, Jim Bauer, Kris Skare, and David Giese for assisting with data collection. We also thank J. A. Jenks and J. E. Vogelmann for providing technical support and computer facilities during production of the digital land use cover. Land cover was produced as part of the South Dakota GAP Analysis Project funded by the U.S. Geological Survey. Funding for this project was provided by the U.S. Fish and Wildlife Service (RWO-60), Division of Refuges, Denver, Colorado, through the South Dakota State Cooperative Fish and Wildlife Research Unit in cooperation with the U.S. Geological Survey; the South Dakota Department of Game, Fish and Parks; South Dakota State University; and the Wildlife Management Institute.

TWENTY-SEVEN

Conservation of Texas Spring and Cave Salamanders *(Eurycea)*

PAUL T. CHIPPINDALE AND ANDREW H. PRICE

Many species of endemic aquatic organisms inhabit the springs and water-filled caves of the Edwards Plateau region of central Texas. Most have limited distributions, and their existence is dependent upon the availability of clean water from subterranean sources (the Edwards Aquifer and associated aquifers). The Edwards Plateau is composed of uplifted karst limestone; water percolates through the limestone, recharges the underground reservoirs, and re-emerges from a large number of springs. The biggest and most well known of these springs are located along the southern and eastern margins of the Edwards Plateau (the Balcones Escarpment) and include Barton Springs in the city of Austin (Travis County); San Marcos Springs in the city of San Marcos, southwest of Austin in Hays County; and Comal Springs in the city of New Braunfels, northeast of San Antonio in Comal County. North of the Colorado River (which flows through the city of Austin), a smaller aquifer system supplies water to the springs, creeks, and caves of Travis, Williamson, and Bell counties at the northeastern edge of the plateau. Sweet (1982) provides a useful overview of the hydrogeology of the region; also see Abbott (1975), various authors in Abbott and Woodruff (1986), and Veni (1988).

Throughout the southern and eastern portions of the Edwards Plateau, numerous populations of salamanders of the genus *Eurycea* are known, and all are restricted to caves with water and/or the vicinity of spring outflows (Hamilton, 1973; Sweet, 1982, 1984; Chippindale, 2000; Chippindale et al., 2000). Here we include in the genus *Eurycea* two species formerly assigned to the genus *Typhlomolge*, based on the phylogenetic work conducted by Chippindale et al. (2000) and discussed below. Nearly all populations of *Eurycea* in central Texas are paedomorphic (i.e., attain reproductive maturity without undergoing metamorphosis). The only known exceptions are a few transforming populations in mesic canyons of the Hill Country, in the southwestern Edwards Plateau (Bogart, 1967; Sweet, 1977b). All members of the group are aquatic, although transformed individuals may venture short distances onto land (Sweet, 1977b, 1978b).

Taxonomic History

The taxonomic history of the Texas *Eurycea* is somewhat complicated. Chippindale et al. (2000) provide a systematic revision that includes a detailed history and list of synonymies as well as the description of three new species, and Chippindale (2000) outlines past and current perspectives on species diversity in the group. Hillis et al. (2001) describe another new species (Austin blind salamanders *[Eurycea waterlooensis]*) and provide further taxonomic revisions. Here we provide a brief summary of species that were recognized prior to molecular-based systematic studies of the group.

In 1895, large aquatic plethodontid salamanders that exhibited highly cave-associated morphological features (e.g., lack of pigmentation, wide, flattened skulls, vestigial eyes, elongated limbs, and reduced numbers of vertebrae) were discovered in the outflows of a 58-meter-deep artesian well drilled at San Marcos, Hays County. Stejneger (1896) described this species as *Typhlomolge rathbuni* (Texas blind salamanders). Decades later, the status of the genus *Typhlomolge* became controversial when Mitchell and Reddell (1965) and Mitchell and Smith (1972) provided evidence that *T. rathbuni* should be considered a member of the genus *Eurycea*. This view was contradicted by Potter and Sweet (1981), who argued for continued recognition of the genus *Typhlomolge* as distinct from *Eurycea*. However, molecular phylogenetic evidence supports synonymization of *Typhlomolge* under *Eurycea* (Chippindale, 1995, 2000; Chippindale et al., 2000; Hillis et al., 2001; see discussion below).

Bishop and Wright (1937) described the next member of the group, *Eurycea neotenes* (Texas salamanders; see also Bruce, 1976), from a spring at Helotes, Bexar County, north of San Antonio. Until recently, nearly all populations of spring and cave *Eurycea* in the Edwards Plateau region were assigned to this species; now (as discussed below) Texas salamanders are restricted in distribution to springs in the general area of the type locality (Chippindale et al., 2000). *Eurycea nana* (San Marcos salamanders) were described by Bishop (1941a; see also B.C. Brown, 1967b) from San Marcos Springs, Hays County. Despite confusion in the literature regarding the distribution of this species, it is clear that these animals occur only at the type locality (Chippindale et al., 1998). Smith and Potter (1946; see also B.C. Brown, 1967a) described *E. latitans* (Cascade Caverns salamanders) from Cascade Caverns, Kendall County (northeast of San Antonio), and Burger et al. (1950) described *E. pterophila* (Fern Bank salamanders) from

Fern Bank Springs, Hays County (north of San Marcos). Baker (1957) described *E. troglodytes* (Valdina Farms salamanders) from Valdina Farms Sinkhole, a deep and extensive cave in Medina County in the southwestern Edwards Plateau region. Mitchell and Reddell (1965) described *E. tridentifera* (Comal blind salamanders, which exhibit a strongly cave-associated morphology second only to that of the members of the formerly recognized genus *Typhlomolge*) from Honey Creek Cave, Comal County. Sweet (1977a, 1978b, 1984) later extended the distribution of this species to include additional caves in Comal and Bexar counties, a move supported by the work of Chippindale et al. (2000).

Potter (1963), in a master's thesis, described a second species of *Typhlomolge, T. robusta* (Blanco blind salamanders), based on a single specimen collected in 1951 from a hole drilled in the dry bed of the Blanco River just east of San Marcos. However, this description cannot be considered valid under current rules of zoological nomenclature. Technically, Longley (1978) must be credited with an accidental description of the species. Potter and Sweet (1981) redescribed *T. robusta* and argued for continued recognition of the genus *Typhlomolge*. A description recently was published of a third species, Austin blind salamanders *(E. waterlooensis)*, most closely related to *E. rathbuni* and *E. robusta* (Hillis et al., 2001).

Wake (1966) assigned plethodontids of the genera *Eurycea* and *Typhlomolge* (plus the genera *Gyrinophilus, Haideotriton, Hemidactylium, Pseudotriton, Stereochilus,* and *Typhlotriton)* to the tribe Hemidactyliini, within the subfamily Plethodontinae. However, relationships among and within the genera and species boundaries in the Hemidactyliini remain uncertain. With respect to the central Texas hemidactyliines, most authors prior to the work of Sweet (1977a, 1978a,b, 1982, 1984) recognized at most six species of central Texas *Eurycea (E. latitans, E. nana, E. neotenes, E. pterophila, E. tridentifera,* and *E. troglodytes)* and one species of *Typhlomolge (T. rathbuni*; the recognition of *T. robusta* did not become widespread until Potter and Sweet published their 1981 redescription). Until recently, nearly all workers have regarded *E. neotenes* (Texas salamanders) as widely distributed in the Edwards Plateau region and assigned the majority of known populations to this species (e.g., Bishop, 1943; B. C. Brown, 1950, 1967c; Schmidt, 1953; Conant, 1958a, 1975; Baker, 1961; Mitchell and Smith, 1972; Sweet, 1977b, 1978a,b, 1982, 1984; Behler and King, 1979; Dixon, 1987; Conant and Collins, 1991; Petranka, 1998). However, several authors suggested that additional species remained to be discovered in the group (e.g., B.C. Brown, 1950, 1967c; Baker, 1961; Bogart, 1967; Mitchell and Smith, 1972).

Sweet (1977a,b, 1978a,b, 1982, 1984) conducted comprehensive studies of geographic distribution and morphological variation in the group. He reduced the number of recognized species of central Texas *Eurycea* to three: *E. neotenes* (which he viewed as widespread in springs and caves throughout the Edwards Plateau); *E. nana* (which he restricted to San Marcos and perhaps Comal Springs); and *E. tridentifera* (from caves of the Cibolo Sinkhole Plain of Comal and Bexar counties; Sweet also believed that this species might range underground into the Southwestern Plateau region). Sweet (1978a, 1984) also suggested that the population from Barton Springs, Travis County, was a distinct species, but did not formally describe it (this population was described as *E. sosorum* [Barton Springs salamanders] by Chippindale et al., 1993). Sweet (1978b) considered *E. pterophila* conspecific with *E. neotenes,* and Sweet (1978a, 1984) viewed *E. latitans* and *E. troglodytes* as hybrids between *E. neotenes* and a subterranean taxon, most likely *E.*

tridentifera. As described above, Potter and Sweet (1981) also recognized *T. rathbuni* and *T. robusta.*

Current Views of Species Diversity in the Central Texas *Eurycea*

Assessments of species boundaries in the central Texas *Eurycea* based on morphology alone have been complicated by three factors. First (as described above), nearly all members of the group are paedomorphic, retaining larval morphologies throughout their lives. Thus, adult characters that typically are used to differentiate and diagnose species of plethodontids are not available for examination (although larval morphological characters have been used successfully to distinguish some species in the group; e.g., Chippindale et al., 1993, 2000). Second, plethodontid salamanders often exhibit morphological evolutionary stasis despite long periods of isolation and evolutionary divergence. This has resulted in the existence of numerous morphologically cryptic species in the family (e.g., Wake et al., 1983; Highton et al., 1989, 1990; Larson and Chippindale, 1993; Highton, 2000), and the Texas *Eurycea* contain several noteworthy examples (Chippindale et al., 2000). Third, the history of the central Texas hemidactyliines is likely to have involved repeated instances of isolation in islands of aquatic habitat followed by convergent or parallel morphological evolution, especially in caves (Chippindale et al., 2000; Wiens et al., unpublished data). As has been the case for other groups of plethodontids (reviewed by Larson and Chippindale, 1993; Chippindale, 2000; Highton, 2000), reliable assessments of species boundaries in the central Texas *Eurycea* have required application of molecular techniques.

Chippindale (1995, 2000) and Chippindale et al. (1993, 1998, 2000) used allozyme electrophoresis, sequencing of mitochondrial DNA, analysis of other molecular data, and, in some cases, external morphology and osteology to reconstruct the phylogeny of the central Texas hemidactyliines and reassess species boundaries in the group. Chippindale et al. (2000) provide a systematic revision of the central Texas *Eurycea* and conclude the following:

1. The central Texas hemidactyliines *(Eurycea* and *Typhlomolge)* represent a monophyletic group within the genus *Eurycea.* Therefore, the genus *Typhlomolge* must be placed in the synonymy of *Eurycea* because it is inappropriate for one genus to be nested within another under the Linnean system of classification.

2. Levels and patterns of divergence and phylogenetic relationships in the group are inconsistent with recognition of *E. neotenes* as widespread within the Edwards Plateau, and indicate the presence of numerous evolutionarily distinct lineages. Sweet's (1978a, 1984) hypothesis of a hybrid origin for *E. latitans and E. troglodytes* was rejected. Therefore, Chippindale et al. (2000) make the following taxonomic recommendations with respect to previously recognized species:

    a. *Eurycea neotenes* is restricted in distribution to springs in the vicinity of the type locality in Bexar and Kendall counties.

    b. *Eurycea nana, E. sosorum, E. tridentifera, E.* (formerly *Typhlomolge*) *rathbuni* and *E.* (formerly *Typhlomolge*) *robusta* are valid species and should continue to be recognized.

c. The names *E. latitans*, *E. troglodytes*, and *E. pterophila* should be resurrected and applied to populations at the type localities. However, the range of each probably extends beyond the type locality. *Eurycea pterophila* occurs in springs and caves of the Blanco River drainage of Blanco, Hays, and Kendall counties, although evidence for its status as a distinct species should be considered relatively weak. Chippindale et al. (2000) assigned numerous populations from the southeastern Edwards Plateau to the *E. latitans* complex, but emphasized that species boundaries in this group remain to be investigated more thoroughly. Chippindale et al. (2000) also recognized the *E. troglodytes* complex encompassing all populations from the southwestern portion of the Edwards Plateau. This group is well supported as monophyletic and exhibits deep genetic divergences among populations.

3. There remain additional, previously unrecognized species in the group. Populations from northeast of the Colorado River (the "northern" group) are extremely divergent from all others based on a variety of nuclear and mitochondrial molecular data sets, consistent with the presumed great age of the Colorado's entrenchment. Isolation of the northern populations probably occurred millions of years ago. Within the northern group, there are at least three distinct species, which Chippindale et al. (2000) formally described as *E. tonkawae* (Jollyville Plateau salamanders; Travis and Williamson counties), *E. naufragia* (Georgetown salamanders; Williamson County), and *E. chisholmensis* (Salado salamanders; Bell County). Jollyville Plateau salamanders occur within the rapidly growing Austin metropolitan area; the areas northeast of Austin that are inhabited by Jollyville Plateau salamanders and Georgetown salamanders are undergoing rapid urbanization. Very recently, a blind subterranean species, *E. waterlooensis* (Austin blind salamanders), was discovered in the Barton Springs Aquifer beneath the southern portion of the city of Austin, and a formal description was published (Hillis et al., 2001). Based on morphological and DNA evidence, this new species is closely related to *E. rathbuni* (and presumably *E. robusta*) and is partially sympatric with *E. sosorum*, to which it is much more distantly related. Many undescribed species probably exist within the southwestern group (*E. troglodytes* complex), and Chippindale et al. (2000) identified two potential candidates within the southeastern group that probably are distinct species from Pedernales Springs (Travis County) and Comal Springs (Comal County). Molecular work on these species groups is in progress and it is likely that additional species will be described in the near future.

Conservation Status of the Central Texas *Eurycea*

Several Texas hemidactyliines currently enjoy protection at the state and/or federal level. *Eurycea* (formerly *Typhlomolge*) *rathbuni* was federally listed as endangered in 1967, prior to the final version of the Endangered Species Act (ESA) in 1973, followed by state listing in 1974. San Marcos salamanders (*E. nana*) were listed as a threatened species by the state of Texas in 1977, followed by federal listing under the ESA in 1980. The history of the conservation status of Barton Springs salamanders

(*E. sosorum*) is especially interesting, if not sobering; one popular account can be found in Stearns and Stearns (1999). Barton Springs salamanders are known only from Barton Springs, located in the Zilker Park recreation area of downtown Austin. Most of the population inhabits the outflows of Parthenia Springs that fill Barton Springs pool, an extremely popular semi-natural community swimming area for over a century (Brune, 1981). The propriety of including the Barton Springs salamander in the nascent Balcones Canyons Conservation Plan was brought to the attention of the U.S. Fish and Wildlife Service (USFWS) as early as 1990. Concerns over increased development within the Barton Springs watershed and the concomitant decline in water quality led to the establishment of a coalition of local citizens known as the Save Our Springs Alliance (S.O.S.). This group succeeded in placing a restrictive water quality ordinance on the ballot in 1992, and following a postponement of the referendum scheduled for the preceding May by the Austin City Council, the measure passed in August by a 2:1 margin of those citizens voting (20% of those eligible). A petition was filed in January 1992 to list Barton Springs salamanders as an endangered species under the ESA, and this generated considerable controversy at both local and national levels. The species was described by Chippindale et al. (1993), and although they saw no conflict between human use of the pool and continued survival of the salamander, swimmers feared that listing might jeopardize the recreational use of Barton Springs. This fear was fueled by those in favor of development in the Barton Springs watershed; listing of the salamander posed a major threat to planned growth in the region because development held the potential for increased siltation and impervious cover that could threaten recharge of the Barton Springs segment of the Edwards Aquifer. While the petition for listing was being considered by the USFWS, considerable local support arose for the salamander as a totem for the collective environmental protection efforts in the Austin region. Posters, t-shirts, and frisbees appeared celebrating the salamander. At the same time, many of those interested in the protection of the salamander joined forces with city of Austin personnel to formulate salamander-friendly methods for pool maintenance (for many years prior to the recognition of Barton Springs salamanders as a distinct species the city had drained the pool and used chlorine and high-pressure hoses to clean algae from pool surfaces).

A proposal to list Barton Springs salamanders as a federally listed endangered species was published in February 1994 (O'Donnell, 1994). A storm of controversy ensued, leading the business community and development interests to file a lawsuit in Hays County District Court successfully challenging the legality of the S.O.S. ordinance and prompting the Texas Legislature to enact laws exempting several large developments within the Austin extraterritorial jurisdiction and Barton Springs watershed from water quality regulations. At the same time, the USFWS extended the deadline for final action on the proposed rule (normally a year following a proposal to list) to August 1995. Congress, prompted in part by the Texas delegation, enacted a moratorium on listing actions in April 1995 and eliminated funding to conduct final listings. A Federal Court upheld a lawsuit by the S.O.S. Legal Defense Fund charging the USFWS with failure to enforce the ESA in this case and issued an order at the end of November 1995 requiring that a final determination of listing status be made within 14 days. A stay was granted pending an appeal by the USFWS on the grounds that Congressional actions prevented the agency from complying with the order. The listing moratorium was lifted at the end of

April 1995 by Presidential decree, and a new public comment period was closed in mid-July by U.S. District Court order.

At the local level, the city of Austin and the Texas Parks and Wildlife Department (TPWD) formed the Aquatic Biological Advisory Team (ABAT) in 1994, a group of five experts charged with independently reviewing the listing proposal and salamander issues. ABAT issued a final report (Bowles, 1995) recommending a regional approach to salamander conservation in the Austin area. As a result of this report, a "Barton Springs Salamander Conservation Agreement and Strategy" was drafted and signed in August 1996 by representatives of three state agencies (TPWD, Texas Natural Resource Conservation Commission, and Texas Department of Transportation) and the USFWS. The Barton Springs Salamander Conservation Team (BSSCT), consisting of staff from these agencies and other individuals with knowledge and expertise on relevant issues, was formed to draft a plan to implement the recommendations of the ABAT report, and the proposal to list Barton Springs salamanders as a federally endangered species was withdrawn by the USFWS in September (Helfert, 1996). The S.O.S. ordinance was reinstated by the Third District Court of Appeals in August 1996 as well. The activities of BSSCT were abruptly terminated by the state agencies involved when the U.S. Tenth District Court ruled that the withdrawal of the listing proposal violated the ESA and ordered a listing decision within 30 days. The Barton Springs salamander was listed as federally endangered at the end of April 1997 (O'Donnell, 1997).

In recent years, the city of Austin has exhibited great dedication to the conservation of both the Barton Springs salamanders and the newly described Jollyville Plateau salamander, and its efforts represent a model for conservation of endangered species in urban areas. Swimmers continue to coexist with Barton Springs salamanders at Barton Springs, and available evidence indicates that protection of water resources for salamanders is consistent with protection of water resources for human use (USFWS, 1998a). The city is currently engaged in a regional planning effort, including the acquisition of sensitive tracts of land within the recharge and contributing zones for Barton Springs (Barton Springs Salamander Recovery Plan, in preparation). The city recently completed a comprehensive two-year study of water quality parameters and populations of Jollyville Plateau salamanders (Davis et al., 2001). To address threats to Jollyville Plateau salamanders, the USFWS recently established a task force involving the city of Austin, several state and county agencies, and private groups (R. Hansen, personal communication). The recent, surprising discovery of another endemic species of *Eurycea* from the aquifer beneath southern Austin (Hillis et al., 2001) adds a new dimension to salamander conservation efforts in the region, but the city of Austin is already working with this species and establishing a group in captivity (Hillis et al., 2001; R. Hansen, D. A. Chamberlain personal communication).

One other species of central Texas hemidactyliine has been the subject of a petition for federal listing as an endangered species—Blanco blind salamanders (*E. robusta*); these cave dwellers have been found only in the San Marcos pool of the Edwards Aquifer (Russell, 1976; Chippindale et al., 1990). However, the petition, filed in 1995, was rejected in 1998 due to insufficient information indicating that listing of this species is warranted (O'Donnell, 1998).

Several species of central Texas *Eurycea* (Cascade Caverns salamanders, San Marcos salamanders, Comal blind salamanders, Barton Springs salamanders, Texas blind salamanders, and Blanco blind salamanders) are protected by the state of Texas. A permit is required to possess or collect individuals of these species, but importantly, these measures (as do most state regulations) specifically omit any form of habitat protection. Valdina Farms salamanders may already be extirpated at the type locality, Valdina Farms Sinkhole, due to human-induced flooding of the cave and introduction of surface predators (Veni and Associates, 1987; Chippindale et al., 2000; G. Veni, personal communication).

Future Prospects

The Edwards Plateau region of central Texas is inhabited by an ancient and diverse assemblage of hemidactyliine plethodontid salamanders of the genus *Eurycea*. Given the recent taxonomic revisions by Chippindale et al. (2000) and Hillis et al. (2001), re-evaluation of their conservation status is in order. Most of the recognized species have extremely restricted distributions, and many distinct species remain to be described formally. All species in the group are dependent on the maintenance of water quality and adequate water levels in the Edwards and associated aquifers, and many (Cascade Caverns salamanders, Georgetown salamanders, Fern Bank salamanders, Texas blind salamanders, Blanco blind salamanders, Austin blind salamanders, Barton Springs salamanders, Jollyville Plateau salamanders, Comal blind salamanders, and several putative, undescribed species) occur in or near the metropolitan areas of Austin and San Antonio. This part of Texas is undergoing tremendous development and urban growth (USFWS, 1995b; Rowell, 1999; Barton Springs Salamander Recovery Plan, in preparation) and the water supply for the entire city of San Antonio comes exclusively from the Edwards Aquifer. Based on recent projections for use of waters of the southern Edwards Aquifer, Comal Springs (inhabited by a probable new species of *Eurycea* as well as federally endangered species of fishes and invertebrates) was expected to cease flowing by 2000 if pumping continued to increase at historical rates and a drought of record were to occur (Technical Advisory Panel, 1990; Comal Springs ceased flowing from 13 June to 4 November 1956, the sixth year of a seven-year drought). San Marcos Springs (home to San Marcos salamanders, Texas blind salamanders, and numerous endangered species of invertebrates, fishes, and plants) is expected to go dry by 2010 under similar scenarios (Klemt et al., 1979). Water quality and quantity within the Barton Springs segment of the Edwards Aquifer continues to decline (Barton Springs/Edwards Aquifer Conservation District, 1997; Mahler and Lynch, 1999; Mahler et al., 1999; Rowell, 1999), and much of the proximate recharge zone for Barton Springs is already developed (Hauwert et al., 1998).

North of the Colorado River, water supplies for springs and caves inhabited by the newly described species Jollyville Plateau salamanders, Georgetown salamanders, and Salado salamanders are derived from small, localized aquifers that are highly subject to pollution and depletion (see references in Rowell, 1999). Not surprisingly, Davis et al. (2001) found an inverse relationship between the physicochemical integrity of spring sites and abundance of Jollyville Plateau salamanders. Large quantities of foam of unknown chemical composition have been observed flowing from springs at the type locality of Jollyville Plateau salamanders, and some individuals recently collected at this locality exhibit spinal deformities. As Rowell (1999) and Chippindale et al. (2000) emphasized, development in the northern region can be expected only to increase. The human population of Travis, Williamson, and Bell counties (encompassing the entire range of the five species of central Texas *Eurycea* described since 1990) was stable at about 650,000 people for the first seven decades of the twentieth century; that number has more than doubled in the last two decades. Groundwater levels in artesian zones in this region declined by more than 30 m between 1975 and 1985 (see references in Rowell, 1999), and total water consumption in these three counties has doubled in these two decades and has been projected to increase 6.5 times by the year 2050. Strategies for the protection of the northern species must be implemented quickly to ensure their survival. The city of Austin has been proactive in this respect, and has instituted an intensive monitoring program for Jollyville Plateau salamanders, with the goal of developing a protection plan whether or not federal or state listing of this species occurs.

If the diversity of central Texas *Eurycea* is to be preserved, it is essential that the springs and caves inhabited by these species and the aquifers that supply water to these habitats be protected. Doing so represents a major challenge. In the face of unprecedented urban growth along the southern and eastern margins of the Balcones Escarpment during the last decade and the unmitigated demand for human water consumption throughout the central Texas region, we are not optimistic. We can find no compelling reason to obviate the words of Gunnar Brune, who said, " The story of Texas' springs is largely a story of the past. In the not very distant future most of Texas' springs will exist only in a legend of a glorious past . . . destroyed by pollution and overpopulation" (Brune, 1981). Our view and challenge to our fellow Texans remains closer to the final paragraph in Morowitz (1991), " As human population goes up, biological species diversity goes down. We might be able to moderate the rate of decline, but we cannot fend off the inevitable. As species number goes down, we might, of course, change our valuation system and subsequent responses; they are, after all, cultural, not metaphysical. The answer to 'How much is a species worth?' is 'What kind of a world do you want to live in?'" The experiment in central Texas is well underway.

Summary

Recent studies of central Texas salamanders of the genus *Eurycea* have revealed extensive geographic and genetic fragmentation and have greatly increased the number of recognized species. Most species have very restricted distributions in caves and spring waters associated with the aquifers of the Edwards Plateau region, and the health of the many aquifer-dependent ecosystems is threatened by human activities. Several species of central Texas *Eurycea* are the focus of conservation plans at the local, state, and/or federal levels, and intensive efforts will be necessary to preserve the diversity in the group.

TWENTY-EIGHT

Lessons from the Tropics

KAREN R. LIPS AND MAUREEN A. DONNELLY

Abrupt declines in amphibian populations have been reported in the media and the scientific literature for more than a decade. Scientists have detected declines in amphibian populations in North America (Corn and Fogleman, 1984; Blaustein and Wake, 1990; Bishop and Pettit, 1992; Carey, 1993; Kagarise-Sherman and Morton, 1993; Scott, 1993; Drost and Fellers, 1996; Green, 1997b; Lannoo, 1998b), Central America (Crump et al., 1992; Pounds et al., 1997; Lips, 1998, 1999; Wilson and McCranie, 1998), and South America (Heyer et al., 1988; Weygoldt, 1989; LaMarca and Reinthaler, 1991; Lynch and Grant, 1998), as well as in Australia (Richards et al., 1993). Declines may have resulted in the extirpation of up to 50% of the anuran species (Pounds et al., 1997) at a site and reduced the abundance of surviving species. There are two major types of decline—those in which obvious causes such as habitat destruction or environmental contamination can be identified and those "mysterious" declines that have no obvious cause. Most scientists generally agree that the majority of historical and current amphibian declines are due to habitat destruction or habitat alteration.

The causes of mysterious declines are more intractable. These declines have tended to occur at mid- to high-elevation sites and in protected regions; they happen rapidly and are selective (i.e., not all species at a given locality are affected). Declines with no apparent cause generate the greatest concern, as do declines occurring in protected reserves, such as those from Costa Rica (Crump et al., 1992; Lips, 1998), Panama (Berger et al., 1998; Lips, 1999), Colombia (Lynch and Grant, 1998), Brazil (Heyer et al., 1988; Weygoldt, 1989), Venezuela (LaMarca and Reinthaler, 1991), and Australia (Richards et al., 1993). These declines suggest global factors that could threaten human welfare.

There are several differences between temperate and tropical amphibian declines. While temperate declines generally occur more slowly, affect mostly pond-breeding species, and include salamanders as well as anurans (Stebbins and Cohen, 1995; Jancovich et al., 1997), tropical declines have tended to involve entire anuran faunas that abruptly crash (the "faunal collapse" of Drost and Fellers, 1996). One of the biggest differences between declines in the two regions is the occurrence of malformities, which have been found in many temperate areas (California, Oregon, all states bordering the Great Lakes, the

St. Lawrence River, and Lake Champlain), but have not been reported from tropical areas. Nor have deformities been associated with declines at any site. Hypothesized causes for these recurring malformations in temperate frogs include parasites (Sessions and Ruth, 1990; Johnson et al., 1999; Sessions et al., 1999), chemical contaminants (Ouellet et al., 1997a), and ultraviolet B (UV-B) radiation (Ankely et al., 1998; summarized in Souder, 2000). Most mysterious North American declines occurred in the 1970s–1980s, primarily in the montane areas of the western United States and Canada (Carey, 1993; Stebbins and Cohen, 1995), although more recent declines in upper midwestern northern cricket frog (*Acris crepitans*) populations have raised serious concerns. Similarities between historic disappearances in North America (Corn and Fogleman, 1984; Carey, 1993; Kagarise-Sherman and Morton, 1993; Scott, 1993; Blaustein, 1994; Drost and Fellers, 1996) and those now happening in Central America suggest that similar agents might have affected amphibians in both sites (Carey et al., 2003).

At least 13 countries in Latin America and the Carribbean have experienced declines or extinctions of protected anuran populations over the past ten years (Young et al., 2001). Each of these unexplained tropical declines has occurred in upland areas, suggesting a possible synergistic interaction of the causative agent with some environmental condition that varies with elevation (e.g., temperature, UV-B radiation, precipitation, and wind patterns). Tropical declines especially involve species associated with streams during some portion of their life cycle (Pounds et al., 1997; Lips, 1998, 1999; Lynch and Grant, 1998), although some tropical declines have been reported in terrestrial salamander populations (Parra-Olea et al., 1999; see below), pond- or pool-breeding frogs (Pounds et al., 1997), and terrestrial frogs (Stewart, 1995; Lips, 1999).

Few tropical declines have actually been observed as they occurred, so it is not always clear how or in which life history stage the amphibians are affected. The rapidity of most declines suggests that adults are being killed. In some tropical declines, life stages other than adults are known to be affected. Additionally, deformities of tadpole mouth parts and dying tadpoles or newly metamorphosed animals have been reported from some tropical sites (J. Campbell, personal communication; J. Mendelson, personal communication; Weygoldt, 1989; Wilson and McCranie, 1998; Lips, 1999) and may contribute to declines. In 1999, reports of tadpoles of temperate anurans lacking mouth parts were announced on the amphibian decline list server from California (G.M. Fellers, personal communication) and Washington State (K. McAllister, personal communication), suggesting that tropical and temperate declines might not be so different after all.

The failure to recover following the sudden disappearance of adults at many tropical sites suggests that the causative agent is either affecting several life history stages at once or that it persists at these sites. Congdon et al. (1994) found that long-lived species whose reproductive effort carries across many years have certain traits that constrain their ability to respond to chronic disturbances. Therefore, protection of all life stages is necessary to ensure the survival of these species.

Possible Causes of Declines

Blaustein et al. (1998) reviewed recent studies of the possible effects of UV-B radiation on amphibian population declines. From their review it is clear that UV-B radiation can kill amphibian embryos, and that these lethal levels exist at several sites in the western United States and Canada and in southeastern Australia. Blaustein and his colleagues have also shown synergistic effects of UV-B radiation with a species of alga, low pH, and fluoranthene (an environmental contaminant). Amphibians affected by these agents included salamanders, treefrogs, toads, and ranid frogs. The effects of UV-B radiation on tropical amphibians have not yet been investigated. Because all unexplained declines described from the tropics have occurred in upland areas and because high elevation equatorial areas receive high levels of solar radiation (Kricher, 1997; Fite et al., 1998), we believe that there is a serious need to extend UV-B and UV-B synergistic studies to Neotropical sites.

Donnelly and Crump (1998) discussed possible effects of how predicted levels of climate change could affect leaf-litter and pond-breeding assemblages of tropical amphibians. They concluded that increased temperature, increased length of dry season, decreased soil moisture, and increased inter-annual rainfall variability would likely affect amphibians and would do so at individual, population, and community levels. They suggested that species with restricted ranges would be most affected. Few sites that have experienced amphibian declines have long-term weather data, but Pounds and Crump (1994) and Stewart (1995) suggested that changes in rainfall might contribute to amphibian declines (but see Laurance, 1996). Pounds et al. (1999) have correlated both the loss of amphibian species and the uphill migration of birds and lizards with a rise in the cloud cover elevation at the Monteverde Cloud Forest Preserve in Costa Rica.

Introduced predators are common to many areas where amphibian declines have been reported, including gamefish and American bullfrogs (Rana catesbeiana) in the western and midwestern United States (Hayes and Jennings, 1986; Lannoo, 1996) and tilapia, European carp, and rainbow trout throughout Latin America. These exotics have been implicated in amphibian declines either by preying on them, competing with them, or by transporting some detrimental pathogen. As in the Sierra Nevada mountains of California (Bradford, 1989; Fellers and Drost, 1993), trout may be especially disruptive to the montane amphibians of Central America because there are no native fish above 1,500 m elevation (Hildebrand, 1938; Bussing, 1987; J.M. Savage, personal communication). Despite the potential impact of introduced fishes on native Neotropical amphibians, we are not aware of studies comparing reproductive success or species richness of tropical amphibian faunas in areas with introduced trout to those faunas in areas without trout.

Pathologists have isolated several pathogens from dead Australian frogs (Berger et al., 1998) and have posited infectious disease as a cause of mortality. A chytrid fungus (Berger et al., 1998) has been associated with die-offs in Central America (D.E. Green, personal communication; Lips, 1998, 1999) and Australia (Laurance et al., 1996), and these fungi have been shown to kill healthy frogs (Longcore et al., 1999). These fungi are typically decomposers of plant and animal material in aquatic and moist, terrestrial habitats and were not previously known as vertebrate pathogens (Berger et al., 1998). This raises the possibility that geographically distinct declines have been caused by the same agent (Wake, 1998). Based on the similarities of declines, habitats, and species affected in both Central America and Australia, Lips (1999) hypothesized that declines in Costa Rica and Panama were caused by this or a similar infectious agent moving southward throughout the isthmus, affecting Monteverde in 1987–88 (Crump et al., 1992), Las Tablas in 1993–94 (Lips, 1998), and Fortuna, Panama, in 1996 (Lips, 1999).

Similar symptoms of decline have been reported from other Central American countries (Wilson and McCranie, 1998). Anecdotal reports suggest that population declines may be widespread in the region, however, detailed surveys and analyses are lacking. Based on the patterns of declines similar to those of Costa Rica and Panama, the riparian amphibians of Atlantic coastal Brazil (Heyer et al., 1988; Weygoldt, 1989), and the Venezuelan (LaMarca and Reinthaler, 1991) and Colombian (Lynch and Grant, 1998) Andes may also have experienced declines resulting from infectious agents.

Other diseases implicated in amphibian declines include ranavirus in Arizona salamanders (Jancovich et al., 1997) and British frogs (Cunningham et al., 1996; summarized in Faeh et al., 1998). These two viral outbreaks occurred in human-induced habitats, and these outbreaks may have been facilitated by human-modified changes to these environments. Alternatively, pathogens might consist of non-native species introduced to a novel environment where they take advantage of abundant hosts that lack effective defenses. Because these animals would not have evolved immunological resistance to foreign diseases (Pimm, 1991; Leighton, 1995), this may explain the widespread impact on amphibian populations (Pounds et al., 1997; Lips, 1999).

Many agrichemicals (including many banned from use in the United States and Canada) are used extensively in the tropics (Colborn et al., 1993). Many of these chemicals readily vaporize and can be transported long distances by normal atmospheric movements (Eisenreich et al., 1981; Rapaport et al., 1985; Blais et al., 1998). It is now recognized that "pristine" or otherwise legally preserved or isolated habitats are not protected from airborne contaminants. Throughout most of the Neotropics, expansive lowland agricultural regions receive high levels of chemical applications (Abdullah et al., 1997; Castillo et al., 1997). In Central America, prevailing winds blowing from the Caribbean pick up moist air loaded with particulate matter, including agrichemical fertilizers and pesticides (Eklund et al., 1997). As the air then moves over mountains, it rises and cools, releasing its moisture and trapped particulate matter as rain, fog, or cloud. Analyses have shown that snowpack from remote areas of the Canadian Rockies has high levels of organochlorine contaminants from cold condensation. Blais et al. (1998) predicted that upland tropical areas such as those in the vicinity of Mexico City might receive even greater accumulations of toxic compounds by this process.

Environmental contamination cannot be the sole cause of amphibian declines, however, because there have been no reported population declines in the lowlands where these chemicals are directly applied. Many agrichemicals can poison and kill a variety of organisms, including amphibians, but may also indirectly act through effects from bioaccumulation within the food chain. Lacher and Goldstein (1997) described how tropical environments differ from temperate environments and why these differences dictate the necessity to develop new methods and techniques for ecotoxicological studies in the tropics. For example, the high biodiversity of the tropics produces more interactions among species than occur in the temperate zone. Thus, the indirect effects of a contaminant are likely to be more complex.

The geographically and taxonomically widespread patterns of declines in the tropics might also indicate that anurans have become immunosuppressed and susceptible to previously nonlethal organisms. If 20–50% of the species at a site disappear, one might imagine a widespread agent (e.g., UV-B radiation or acid rain) was at work. Where only one or a few species disappear, a variety of agents acting on varied spatiotemporal scales might be operating (e.g., predation, competition [Griffis and Jaeger, 1998], unusual weather, disease, or poor recruitment). Synergistic effects between two or more agents are also possible, with environmental stress weakening amphibians so that they become susceptible to disease (Carey, 1993). Because amphibian life history stages may occur in different habitats, amphibians may be vulnerable to stressors present in any of these environments. Multiple El Niño events in recent years (Weylan et al., 1996; Pounds et al., 1999), combined with deforestation-induced changes in precipitation patterns (Pounds and Crump, 1994; Laurance, 1996), could have compromised the immune responses of tropical amphibians (Pounds and Crump, 1994). While this scenario is possible, given that it requires an event that would have simultaneously weakened the immune systems of numerous species representing a wide range of taxa and ecological conditions, it seems less likely than other extinction scenarios.

What Can We Learn From the Tropics?

Tropical areas are those areas within approximately 30° latitude of the equator. Because of the diverse array of threats to the great diversity of tropical ecosystems, it has been said that tropical biology is essentially the study of rare species and their conservation. We believe that the study of the declining amphibian phenomenon is similarly defined and that certain conditions of tropical ecosystems might make amphibians either more susceptible to population declines or less likely to rebound following a decline. We will use amphibian declines in Central America as an example of the kind of data that could be collected from the Old World Tropics, where the status of amphibian populations is unknown. We will then describe how tropical ecology might help us decipher certain aspects of temperate amphibian declines. Lastly, we present a case study of a tropical salamander that exemplifies current reports of declining amphibian populations and we provide suggestions for future research.

Species Richness and Endemism

We suggest that the high species richness of tropical anuran communities makes patterns of tropical amphibian disappearance

more obvious than in temperate zones. For example, Santa Cecilia in lowland Ecuador has about 80 amphibian species (Crump, 1974; Donnelly and Guyer, 1994), while the Savannah River Reserve in South Carolina has only 41 species (W. Gibbons, personal communication). Imagine both sites have the same number of individuals in total, but the temperate site (with half the species) has twice as many individuals within a species. Declines of a similar magnitude in both sites would tend to produce more easily detected extirpations in the tropical site and less easily detected declines in the temperate site.

Many hypotheses have been proposed to explain the high species richness of the tropics (Kricher, 1997), including historical (long "stable" climate), abiotic (spatial heterogeneity), and biotic (competition, predation, and productivity) factors. No hypothesis, however, has been able to sufficiently explain the greater tropical species richness, in part because we do not fully understand the factors that influence speciation rates or those that regulate populations and whether they differ in either quality or quantity between tropical and temperate regions.

The generally lower abundance of most tropical species suggests that essential differences in the population and community dynamics should exist between temperate and tropical organisms. We do not yet understand the relative contributions of density-dependent and density-independent processes to demography in the tropics. Schoener (1986) elucidated several features of organisms and their environments that affect survival and reproductive success. These features include body size, diet, mobility, generation time, number of life stages, recruitment (organismal), severity of physical environment, spatial fragmentation, long-term climatic variation, resource availability, and partitioning of environmental resources. While our attention has been focused on the environmental differences between temperate and tropical areas, works emphasizing differences in population dynamics between temperate and tropical amphibians are lacking. For example, some long-term studies of temperate amphibians have shown population size to fluctuate annually (e.g., Bragg, 1960b; Turner, 1960b; Berven, 1990; Pechmann et al., 1991; Hairston and Wiley, 1993; Cortwright, 1998), but few "long-term" (>five years) studies are available for tropical amphibian populations (but see Woolbright, 1991; Stewart, 1995; Voris and Inger, 1995). Regardless, high tropical biodiversity will result in complex interactions among species, and extreme changes in amphibian assemblages are sure to have serious ramifications throughout the ecosystem, including both top-down and bottom-up effects.

Anuran species richness decreases with altitude; montane sites have fewer species of frogs and toads than lowland areas. In contrast, tropical salamander richness increases with elevation in Mexico and Central America. Tropical frog species are more numerous than salamanders, so montane amphibian assemblages are not as species rich as lowland assemblages, but they often have numerous endemic species considered regionally rare (Hoffman and Blows, 1993; Donnelly and Crump, 1998). For example, in northern Oaxaca, Mexico, two salamander species occur in the lowland rainforest, with one (50%) species being endemic; nine species occur in the cloud forest between 1,000 and 2,800 m elevation, with seven (78%) species being endemic; and eleven species occur above 2,800 m, with eight (73%) species being endemic (Wake, 1987). Widespread population declines in upland areas would produce extinctions and severely reduce the richness of the amphibian fauna, especially salamanders. It follows that conservation of Central American endemics will require adequate protection of upland areas.

Many tropical endemics are "island" endemics, occurring on the tops of mountains. These species are thought to be physically or physiologically incapable of occupying the intervening lowland areas, virtually eliminating any chance of future immigration. The high elevation tepuis of South America are almost exclusively populated by endemic amphibian species (Myers and Donnelly, 1996), and many cordilleras and peaks in the Andes are home to endemics (Frost, 1985). The ancient uplands of Guatemala (Campbell and Vannini, 1989; Campbell and Frost, 1993), Mexico (Flores-Villela, 1993; Ceballos et al., 1998), and Honduras (Wilson et al., 2001) are dominated by endemics. In Honduras, montane moist forests (1,500–2,700 m elevation) have the greatest herpetological endemism (Wilson et al., 2001). Of the described Honduran herpetofauna, 15 species (20.5%) of anurans and three species (15%) of salamanders are restricted to this elevational band (Wilson et al., 2001). However, even the relatively young mountains of southern Nicaragua, Costa Rica, and Panama have a high number of endemics (Savage, 1982). For example, 32 (27%) of the 120 amphibian species known from Costa Rica (Savage and Villa, 1986) are only found in the three Costa Rican mountain chains, which include upland areas over 1,200 m elevation.

Vertebrate distribution patterns may also suggest potential causes of amphibian declines. Some general rules are that geographic ranges in vertebrates typically consist of many species with a small or restricted distribution, and a few species with large or widespread distributions. Further, species distribution ranges are generally smaller as they approach the equator (Arita et al., 1997). It is also well established that taxa with small distributions are more prone to extinction than are taxa with large distributions (MacArthur and Wilson, 1967; Meffe and Carroll, 1997; Ceballos et al., 1998). In mammals, extinction risks correlate with large body mass, increased specialization, and lower population density (Ceballos et al., 1998); in reptiles, extinctions correlate with population abundance and habitat specialization (Foufopoulos and Ives, 1999). Interestingly, recent amphibian declines seem to contradict these patterns—most amphibians are small (Pough et al., 1998), and have generalized insectivorous diets. Some widespread tropical species (Donnelly and Crump, 1998) have declined in certain parts of their ranges, as have some abundant tropical species (Lips, 1999), but there have been no obvious patterns of larger, more specialized or rarer amphibian species disappearing more often than common taxa (Richards et al., 1993; Pounds et al., 1997; Lips, 1999; but see Highton, this volume, for plethodontid salamanders in the eastern United States). Williams and Hero (1998) found that the declining species of Australian rainforest frogs were characterized by the combination of low fecundity, habitat specialization, and reproduction in streams. Perhaps declines have more to do with behavior, microhabitat use, or breeding modes (see below).

Differences in tropical and temperate patterns of population distribution and density also may contribute to differences in amphibian declines. For instance, the floodplain of the southeastern United States is a large tract of fairly homogeneous habitat inhabited throughout by a similar assemblage of species. Amphibian populations fluctuate here, but faunal collapses have not been reported (Pechmann et al., 1991; but see Lannoo, 1998b). Is this because the agents causing faunal collapses in the western United States (Drost and Fellers, 1996) and Central America (Pounds et al., 1997; Lips, 1998, 1999; Wilson and McCranie, 1998) are not present? Or are they present and causing periodic declines within a population but are then becoming obscured by immigrating individuals allowing the population to recover? Or is the seasonality of the temperate zone so different that it minimizes the interaction between the amphibians and agents causing declines? We encourage monitoring of diverse upland amphibian assemblages in the temperate zone for signs of decline and causative agents. Salamanders in the Appalachian Mountains would provide an excellent comparative model (indeed, see Highton, this volume).

Diversity of Reproductive Modes

In addition to having amphibian faunas characterized by high species richness, the tropics tend to have a greater ecological diversity. This is illustrated by a consideration of anuran reproductive modes. Reproductive modes (*sensu* Crump, 1974) describe the variety of ways in which amphibians can reproduce and include characteristics such as the location of eggs and larvae; presence and kind of larvae; kind and source of nutrition; and morphological and behavioral adaptations of parents and offspring. Amphibians from tropical areas have many more types of reproductive modes than do temperate amphibians, especially Neotropical amphibians (Duellman and Trueb, 1986). This diversity is a combination of a wide taxonomic representation and a large number of uniquely tropical reproductive modes. Families of primarily tropical anurans that show either a variety of reproductive modes or unique reproductive modes include the Leptodactylidae, Centrolenidae, Dendrobatidae, Microhylidae, Pipidae, and Brachycephalidae. The widespread and species-rich families of Bufonidae and Hylidae are better represented and have a greater diversity of reproductive modes in tropical regions than in the temperate zones.

Neotropical amphibians have the greatest number of reproductive modes (21/39, 54%) with eight unique modes (21%). The next most diverse regions are the Ethiopian and Australo-Papuan with 12 (31%) different modes of reproduction. Combined, the three tropical regions account for 33 reproductive modes, of which 11 (33% of total) are unique to that zone (Duellman and Trueb, 1986). Even within a geographical area, tropical areas have a greater diversity of reproductive modes, with between 8–14 reproductive modes recorded from five well-studied lowland sites (Duellman, 1990) compared to only two to three reproductive modes at a given temperate site (Duellman and Trueb, 1986). If there was a microhabitat-specific agent that was responsible for extirpating amphibians, one could compare survival and mortality among the different reproductive modes present at a site. This is the case in Central America where primarily stream-breeding amphibians have disappeared, while exclusively forest-dwelling and bromeliad-dwelling species still exist. This pattern of disappearance suggests that the causative agent is in the water (Lips, 1998). Investigators examining potential causes of future declines may be wise to focus their efforts in riparian habitats (Lips, 1999).

Compared to temperate amphibians, tropical anurans show a trend towards reproduction away from standing water and the utilization of many microhabitats that either do not exist for (phytotelmata) or are rarely utilized by (egg and larval transport) temperate species. In the temperate zone, many amphibians reproduce by laying relatively large clutches of eggs directly into water. Reduction in clutch size, increase in yolk stores, increase in egg diameter, and increase in length of development are representative changes in life history traits of frogs with non-aquatic reproductive modes. These traits can make a population more prone to extinction because they all reduce the number of propagules produced at one time and may allow for potentially catastrophic mortality in, and slow recovery of, a population (Williams and Hero, 1998). Perhaps the specialized reproductive modes of tropical species limit recovery from population declines. These factors should be investigated to understand patterns of survival and mortality among species at sites of decline.

The reduced dependence on standing water, through such means as direct development and terrestrial or aerial oviposition sites, might make some tropical anurans more resistant to changes in climate, especially changes in patterns of precipitation. Incorporating the evolution of reproductive modes into amphibian conservation questions might provide insight into the environmental forces affecting survival and ecology of early life stages. Heyer (1969) described progressive evolution towards terrestrial reproduction within a group of *Leptodactylus* species. He described how behavior, morphology, and egg material evolved to be more resistant to desiccation. Duellman and Trueb (1986) provided a diagrammatic version (their fig. 2-5) of how the evolution of diverse amphibian reproduction modes might have occurred. Duellman and Trueb do not identify the group (or groups) to which these hypotheses apply, nor do they provide any supporting data for their predictions. Additional studies based on observations and experiments are needed if we are to understand the evolution of reproductive modes among amphibian taxa and the environmental factors important in shaping amphibian life histories.

Biotic and abiotic factors can have a variety of effects on each amphibian life history stage. A reproductively diverse anuran fauna can be examined for evidence of declines by comparing species' abundance among microhabitats or life stages. For example, if all stream breeders were disappearing yet the direct developing plethodontid salamanders and *Eleutherodactylus* sp. were not (e.g., Pounds et al., 1997; Lips, 1999), one would first look in the streams for a causative agent. Likewise, if declines appeared to be independent of reproductive mode, one should examine environment-wide agents such as UV-B radiation, acid rain, or chemical contamination. Examination of numerous life stages also has been useful in localizing the source of infection. Where the unusual loss of mouth parts was seen in riparian tadpoles (Lips, 1999), examination of preserved material revealed chytrid fungal spores on the remaining mouth parts. Three sick tadpoles collected during a Panamanian die-off also had lesions on their mouth parts formed by a ranavirus (D. E. Green, personal communication). This is the only life stage and the only site at which both a fungus and a virus have been found. We encourage researchers to examine larvae for the loss of mouth parts and to have their specimens tested for both kinds of pathogen.

Climate Change and Amphibian Physiology

Unique combinations of temperature and precipitation occur in the tropics that do not exist in temperate regions (Mabberley, 1992; Pounds et al., 1999; Pringle, 2000), and climate may be an important regulating agent in amphibian population dynamics (Andrewartha and Birch, 1954). If abiotic factors control rates of birth, death, immigration, and emigration, then abiotic factors may regulate population density. Some differences in population biology probably exist between temperate and tropical amphibians, especially in traits (such as length of reproductive season) that are shaped by climatic factors. The long rainy season characteristic of wet tropical forests allows

for prolonged reproduction and the potential for continuous production of offspring. This type of asynchronous breeding could limit the impact of catastrophic events because some adults would be likely to avoid mortality, and thus allow the population to recover.

Tropical seasonality is driven by the variation in precipitation creating distinct wet and dry seasons, rather than a variation in temperature. This does not mean that the tropics are aseasonal or stable, only that the type of seasonality differs. For tropical amphibian populations, water availability may be more of a regulating factor than temperature, because the permeable skin of amphibians facilitates water and gas exchange. Plethodontid salamanders, a widespread group that ranges from North to South America (see Highton, this volume), would serve as good indicators for changes in moisture for both temperate and tropical areas because all species lack lungs and depend exclusively on their skin for respiration. Some salamander populations have disappeared (see case study below) and some species have become increasingly rare (Parra-Olea et al., 1999), but some populations survive (Hairston and Wiley, 1993), even in areas where anuran populations have crashed (Lips, 1998, 1999).

Changes in microclimates will potentially have complex and interacting physiological effects; unfortunately, we know very little about physiology of tropical amphibians (Donnelly, 1994). Tropical amphibians have lower resting metabolic rates than temperate species (Feder, 1978; Duellman and Trueb, 1986), resulting in a reduced capacity for thermal acclimation (Duellman and Trueb, 1986). For this reason, tropical amphibians might be more susceptible to slight environmental changes than temperate taxa (Feder, 1978). High latitude species are more cold tolerant than lowland tropical populations, and some tropical species prefer higher rather than cooler temperatures (Brattstrom, 1968). Environmental temperature shifts could trigger increases in amphibian body temperatures that could suppress the immune system (Carey, 1993) and increase metabolism (Dunham, 1993). Increased metabolism requires increased amounts of oxygen, water, and food, which could change the amount of energy allocated to reproduction and growth, and therefore population dynamics. Carey (1993) predicted that cold temperatures would limit the ability of the amphibian immune system to fight infections.

Central American amphibians historically have experienced a much drier environment (Colinvaux, 1997). Species in the Talamancan uplands of Costa Rica and Panama may even have experienced glacial environments. What we do not know is how those frogs responded to past temperature changes—did they move downslope and subsequently reinvade the uplands once the climate warmed? Or did they go extinct, with these habitats being repopulated by other or newly evolved species? (See Holman, 1995, for a discussion of postglacial migrations in North America.) Insight into selectivity among species following population extinctions could be obtained through molecular studies that examine relationships among extant populations and species.

In response to global warming, species will have to adapt to higher temperatures, migrate to cooler sites, or die (Miles, 1993). It is generally assumed (Travis and Futuyma, 1993) that global climatic change is too rapid for organisms to evolve or migrate (Miles, 1993) and that peripheral populations will be most susceptible to global temperature changes (Hoffman and Blows, 1993). Miles (1993) found that this was the case for populations of the lizard *Urosaurus ornatus*. He predicted that high-elevation populations would be more susceptible to stress associated with climate change and that they would not have the option of dispersing into other habitats. Similarly, amphibian species inhabiting cooler montane areas may be adversely affected by slight increases in temperature because most high-elevation populations are close to the highest elevation, and have few places to go (i.e., they would run out of mountain). Certain aspects of global climate change favor species with particular traits (Dukes and Mooney, 1999). A herpetological example might include marine toads (*Bufo marinus*), a generalist species with large clutch size. Slight increases in temperature could allow this species to invade high elevations. Global climate change could operate indirectly on amphibian assemblages because of associated changes in environmental factors that promote the range extension by pests, pathogens, and predators (Dukes and Mooney, 1999).

Conservation

Conservation issues in the tropics fundamentally differ from those in the temperate zone (Janzen, 1994). Tropical countries contain a greater number of species, endemic species, and rare species compared to temperate countries. Tropical countries are also faced with a greater number and variety of threats to their biological diversity (Meffe and Carroll, 1997). These threats include recent rapid population growth, and economic, social, and political issues that are not easily solved by biologists. Most temperate zone countries are relatively wealthy and can afford to invest in well-trained staff and comprehensive protection for biological preserves. Most tropical countries, however, are relatively poor and do not have the means to effectively protect biodiversity (Meffe and Carroll, 1997). Campbell and Frost (1993) spoke eloquently for many in their plea for herpetologists to get involved in issues of conservation in those countries where they work.

Conservation and management plans in tropical countries are often based on little data and supported by little capital or infrastructure (Janzen, 1994; McDiarmid, 1994). Despite these considerable barriers, most Neotropical countries have a well-trained corps of biologists that are limited by lack of long-term funding and innovative policy (Janzen, 1994). To understand patterns of amphibian population declines and the possibility of recovery, data on population and community dynamics of tropical amphibians are needed. This will require long-term collaborations between tropical biologists from temperate and tropical countries, increased information and technology transfer between countries, and continued promotion of international graduate education (Young et al., 2001).

Cerro Salamanders: A Case Study

The Organization for Tropical Studies (OTS) is one institution committed to the goals of international cooperation for tropical conservation. OTS was established in 1963 to train graduate students in tropical field biology. One of the classic field exercises conducted by OTS students is to survey populations of the plethodontid salamander, *Bolitoglossa subpalmata*, on the Cerro de la Muerte, Costa Rica. This exercise has been conducted for the past 25 years; numerous OTS courses have revisited the same sites along the Interamerican Highway to describe various features of this terrestrial salamander, including habitat preference, color patterns, growth rates, and population parameters. Surprisingly, other than the original population

study by Vial (1968), no other field studies have been published on this abundant salamander. Thus, the OTS data set, stored as field reports in course books, is one of the few long-term population studies of a tropical amphibian, and the only one of a Neotropical salamander. While this series of field projects was meant to investigate other issues, we use it as an example of how other amphibian populations could be studied over the long term. We will describe the study, summarize results from a 15-year mark–recapture study, and describe the problems and benefits of this approach.

For population studies of Cerro salamanders, we reviewed the course books (1970–96) from the three annual courses in *Tropical Biology: An Ecological Approach.* Using personal observations and data from final reports, we summarized population parameters from two populations: one located on a rocky slope at the 89.3 km marker along the InterAmerican Highway (previously referred to as the 90.7 km site) and another from a rocky site at 84 km. We usually calculated the captures per 100 m^2 by using descriptions in the reports; when these values were not available, we relied on site measurements described by Maple (OTS 94-3).

Following conflicting reports about field growth rates of *B. subpalmata* (Vial, 1968; Houck, 1977a, b), David Wake and the OTS 84-1 course initiated a mark–recapture study at the 89.3 km marker to resolve this question. At least six subsequent OTS courses have resurveyed this site by turning over the rocky rubble, collecting, anesthetizing, measuring, toe clipping, and releasing all captured animals. Subsequent OTS courses found numerous salamanders along these transects but with limited (<10%) recapture success. This site was abandoned when salamander captures were too low for population studies (Table 28-1). Donnelly initiated a similar study at the 84 km marker using the same methodology.

Salamanders initially were abundant at both sites, with 100+ captures for the first three surveys. High abundance was followed by a few years with 30–50% reductions in captures, and surveys during the last three years found a 90% reduction in capture rates (Fig. 28-1). Both sites differed in initial density, but within seven surveys the sites had declined to less than one salamander per 100 m^2. Perhaps even more interesting are identical changes in both the adult sex ratio and the juvenile to adult ratio (Table 28-1). Initially, populations at both sites were dominated by males and had almost as many juveniles as adults. When total abundance dropped to less than half of the initial levels, the relative juvenile abundance increased from one juvenile per every adult to twice that (or more) in both sites. Declines are not an artifact of low sample size, because as salamanders became more rare, students searched more intensively (Table 28-1).

Neither of these areas currently supports the large numbers of salamanders seen in previous years. In both sites, declines became noticeable about two years after studies at each site began (Fig. 28-1). In both cases, the density also declined substantially, with a noticeable drop in the number of adults found, especially males. No dead animals were ever found. Given the results of earlier OTS studies (see below) it is possible that marked animals burrowed into the soil and/or moved out of the study area.

Alternatively, adult mortality could have resulted from increased exposure to predation, environmental contaminants, or handling during the sampling. Students have proposed all these explanations in field reports, and all hypotheses require experimental testing. Toe-clipping of other plethodontids may cause them to migrate (Nishikawa and Service, 1988), and early

TABLE 28-1

Population Parameters of *Bolitoglossa subpalmata* at Two Paramo Sites Along the Interamerican Highway in the Cerro de la Muerte, Costa Rica. Columns Indicate the Location of the Study, the OTS Course Responsible for the Data, Number of Rocks Turned Over, the Number of Salamanders Found, Salamander Density, Juvenile to Adult Ratio, and the Sex Ratio.

Site	Course	Season	Rocks	Individuals	Individuals/ 100 m^2	Juveniles: Adults	Males: Females
89.3 km	84-1	dry	31	124	104.00	1.70	1.20
	84-3	wet	—	132	54.00	1.17	1.58
	85-3	wet	—	105	61.00	1.16	0.96
	86-3	wet	—	42	7.00	0.91	0.69
	87-3	wet	—	4	7.80	2.50	0.50
	94-3	wet	695	2	0.30	2.00	—
	95-1	dry	485	0	0.00	0.00	0.00
84.0 km	89-3	wet	—	107	14.08	1.04 west	1.17
	90-1	dry	120	108	6.05	0.80 west	1.20
					4.73	1.60 east	
	90-3	wet	811	117	7.60	0.68 west	1.20 west
					4.50	0.61 east	1.10 east
	91-1	dry	—	23	1.11	0.60	1.10
	94-3	wet	1510	14	0.51	13.30	0.30
	95-1	dry	669	0	0.00	0.00	0.00
	95-3	wet	500+	2	0.07	2.00 west	0.00
	96-1	dry	1120	1	0.05	—	0.00

studies of *B. subpalmata* (OTS 76-2) showed that when released, recently toe-clipped adults were found only by digging into the substrate, rather than just turning over surface cover objects. Vial (1968) found that an adult plethodontid salamander had an average home range of 3.4 m^2. With the expectation of finding higher densities, several field exercises involved searching areas surrounding the plots, but these searches were unproductive. In 1994, the initial study site (89.3 km marker) was resampled and only two juveniles were found. If the adult decline indicated emigration, then hatchlings and juveniles from 1986–87 should have been present as adults in 1994.

The cause(s) of Cerro salamander declines, as with most other declines, are unknown. This project was designed to follow a plot with marked individuals over multiple years to document population dynamics. Access to baseline data and project continuity has allowed this project to evolve from a study of growth rates and color patterns to one that attempts to understand the patterns of population change. By studying this species in the field and developing hypotheses to explain the patterns, the students involved are better prepared to participate in the discussions of amphibian declines and the loss of biodiversity. We also use the Cerro salamander study as an example of how experimental studies are necessary to understand the mechanisms of amphibian declines. The lack of sufficient controls and replicates in these exercises limits our conclusions regarding the cause of these salamander declines, and we hope future studies establish a series of permanently marked, replicated plots to be sampled at different frequencies. We also suggest that future studies quantify the time, area, and number of rocks sampled in a manner consistent with previous

survey methods, and perform experimental manipulations to identify individual responses to pollution, disturbance, and microhabitat variation in moisture and temperature.

Future Prospects

We believe that the Cerro salamander project is a good example of how various institutions (e.g., field stations, universities, and field courses) could collect long-term data for studies of declining amphibian populations while simultaneously training new investigators in basic sampling techniques. This approach has been used successfully at locations such as the Savannah River site, where some amphibian breeding sites have been monitored for almost 20 continuous years (Semlitsch et al., 1996), and at some sites that have been established in the midwestern United States (Lannoo, 1998b). A commitment to fund and staff these studies for their duration is needed, as in the Long Term Ecological Research program of the National Science Foundation.

Geographically extensive surveys using the same methodology from both decline and control sites are needed, so that researchers can document and compare declines. Plethodontid salamanders, bufonid, hylid, and ranid frogs might all be good representative taxa. These animals could provide a phylogenetically balanced comparison between tropical and temperate areas for two reasons. First, species of each group occur in temperate and tropical regions. Second, species of each group occur in areas where declines have or have not occurred. Follow-up studies are needed at sites of historical declines in the western

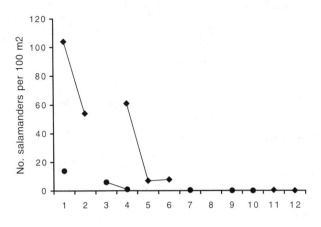

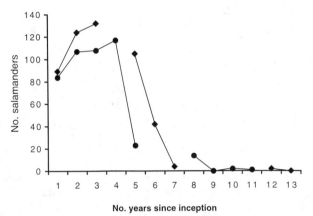

No. years since inception

FIGURE 28-1 Summary data from two populations: one (upper) located on a rocky slope at the 89.3 km marker along the Interamerican Highway (previously referred to as the 90.7 km site), and another (lower) from a rocky site at 84 km. Note that high abundance was followed by a few years with 30–50% reductions in captures, and surveys during the last 3 years found a 90% reduction in capture rates. Neither of these areas currently supports the large numbers of salamanders seen in previous years. In both sites, declines became noticeable about 2 years after studies at each site began. Note also that in both cases, the density also declined substantially, with a noticeable drop in the number of adults found, especially males.

United States and in Central America to document the current status of amphibian populations and the potential causes for observed declines.

While additional ecological studies of tropical amphibians are being published, few represent montane sites and even fewer report on salamander population dynamics. For example,

Blaustein et al.'s (1994a) review of amphibian population studies lasting over 4 years included only one tropical reference (Woolbright, 1991), and five of the six recent "long-term" studies published since that time are sites of amphibian decline (Richards et al., 1993; Stewart, 1995; Voris and Inger, 1995; Pounds et al., 1997; Lips, 1998, 1999). The Cerro salamander study is critical because it offers one of the longest running studies of various aspects of the biology of tropical amphibians. In conclusion, relevant empirical data are especially scarce for upland tropical amphibians, for tropical vertebrate physiology, and for field-based immunology, but we believe long-term research into any area of tropical amphibian biology has the potential to contribute to understanding population declines.

Summary

Herpetologists are alarmed because a number of sites around the world have experienced declines or extinctions of amphibian species. Tropical patterns of decline differ from declines in some, but not all, temperate regions. We describe the nature, extent, and possible causes of amphibian declines in the Neotropics. We argue that reported declines from a few Neotropical sites have played a key role in our understanding of the global amphibian crisis and that collection of comparative data at temperate and tropical sites will be necessary to understand the extent of this phenomenon. We conclude by noting ecological differences between temperate and tropical areas, and examine whether these differences might contribute to observed differences in patterns of declines. Throughout, we highlight areas of research that might contribute to understanding aspects of this phenomenon but which have received little attention to date.

Acknowledgments. We thank the Organization for Tropical Studies for use of the CRO library and the hundreds of students and faculty who have contributed to the Cerro salamander study. We thank A. Blaustein, W. Gibbons, D. Wake, and L. D. Wilson for access to unpublished data or manuscripts, and M. Crump, C. Guyer, and R. McDiarmid for comments on previous drafts. We especially thank numerous participants from various DAPTF-, NASA-, and NSF-sponsored meetings for ideas and discussions. We thank the National Science Foundation (Grant IBN 9807583 to K.R.L.), the Chicago Zoological Society, J. Holden, and the Bay and Paul Foundation for their generous support. This is contribution 82 to the Program in Tropical Biology at Florida International University.

TWENTY-NINE

Taxonomy and Amphibian Declines

SHERMAN A. MINTON

Any attempt to describe the extent and significance of biodiversity requires a clear and workable system of classification. This is especially true of conservation efforts with a goal of the recognition and protection of threatened populations. However, the past decade has seen a major revision in systematics with some authorities going so far as to state that the Linneaen system of classification and nomenclature has outlived its usefulness.

The species concept is a particular point of controversy in the taxonomy of higher organisms. For much of this century, the genetic or biological species concept using the touchstone of reproductive isolation has been widely accepted by zoologists. However, there is growing support for a new idea in which a species is defined as the largest lineage on a single phylogenetic trajectory—the evolutionary species concept introduced around 1980 (Wiley, 1978). This phylogenetic species concept defines the species as "the smallest diagnosable cluster of individual organisms within which there is a parental pattern of ancestry and descent" (Cracraft, 1983a). This field, termed *cladistics,* is a product of the computer age, and it has greatly facilitated the definition of evolutionary and phylogenetic species, but not all taxonomists find it acceptable. Those interested in the species controversy as it relates to herpetology are referred to papers of Frost and Hillis (1990), Echelle (1990), Frost et al. (1992), Pritchard (1994), De Queiroz (1995), and Meylan (1995).

The subspecies concept is even more controversial, although it has been used effectively in herpetology for many years. However, Collins (1991, 1992) contended that the subspecies category camouflages valid species, and he cited 55 examples in the North American herpetofauna. His position was contested by Van Devender et al. (1992) and Dowling (1993). Smith et al. (1997) present the case for retaining the subspecies category.

In this paper, I provide a history and call attention to some taxonomic problems involving amphibians of the midwestern United States and southern Canada, a region defined as the Great Lakes states including Ohio, the Dakotas, Nebraska, Kansas, Iowa, and Missouri, and southeastern Manitoba and southern Ontario. I do not offer solutions, but I do point out potential topics for research.

Chromosome Numbers in the Salamander Genus *Ambystoma*

Perhaps the most interesting problem involves salamanders of the *Ambystoma jeffersonianum* complex. Originally described in 1827, Jefferson salamanders (*A. jeffersonianum*) were recognized for more than a century as rather nondescript urodeles found from Labrador and Hudson Bay south through the northeastern United States. Clanton (1934) called attention to a peculiar situation involving this species near Ann Arbor, Michigan. Here they exist as a small, dark form (average body length about 55 mm) with a normal sex ratio, and a larger, lighter form (average body length about 70 mm) composed almost entirely of females. There are also differences in the number and appearance of eggs laid by the two types of females. Clanton reported populations with a similarly abnormal sex ratio from western New York.

In 1954, I reported that Jefferson salamander populations in Indiana were composed of a small, dark, blue-spotted form in the northern counties and a larger, lighter, unspotted form in the southern hills. Both populations seemed to have a normal ratio of males to females. In the territory between there were several populations that appeared intermediate and seemed to be composed entirely of females. I revived the name *A. laterale* (blue-spotted salamanders) for the northern population and assumed it was identical with Clanton's (1934) small, dark form; I retained *A. jeffersonianum* for the southern Indiana populations (Minton, 1954).

About a decade later, studies by Thomas Uzzell did much to clarify the Jefferson salamander problem. Animals from the all-female populations were found to have a triploid (3N—organisms with an extra chromosome) number of chromosomes (Uzzell, 1963). In contrast, the northern *A. laterale* and the southern *A. jeffersonianum* were typically diploid (2N) salamander species. Uzzell (1964) divided the triploids into two species, *A. tremblayi* (morphologically similar to and geographically associated with *A. laterale*) and *A. platineum* (similarly related to *A. jeffersonianum*). Triploids apparently arose by hybridization when *A. jeffersonianum* and *A. laterale* were united after separation by the Pleistocene glaciation.

Eggs of triploid females develop by gynogenesis, requiring sperm of a related salamander species only to initiate development while the sperm contributes nothing genetically. Sperm incorporated at temperatures about 6 °C usually produce triploids by gynogenesis; at 15 °C (warmer than most breeding ponds), hybrids are usually produced (Bogart et al., 1989). Diploid species of the *A. jeffersonianum* complex were assumed to be the usual sperm donors (Uzzell, 1964, 1969).

Additional observations have revealed further complexity. Uzzell and I set up drift fence and traps at a vernal pond (now destroyed) near Whitestown, a few miles northwest of Indianapolis. In two seasons, we trapped dozens of *A. platineum* triploids and dozens of small-mouthed salamanders (*A. texanum*) but no diploids of the *A. jeffersonianum* complex. Either the triploids were reproducing by parthenogenesis (the production of all females in the absence of sperm) or they were using *A. texanum* sperm (Uzzell and Goldblatt, 1967). Some years later, Morris and Brandon (1984) reported both gynogenesis and hybridization between *A. platineum* triploids and *A. texanum* at Kickapoo State Park in eastern Illinois. Continued studies at this site show tetraploid (4N) animals make up 7–28% of polyploid animals each year. Pentaploid (5N) animals are produced occasionally but have low viability (Phillips et al., 1997). Triploid populations resulting from gynogenesis using *A. texanum* sperm have been reported from four central Indiana sites (Spolsky et al., 1992).

Kraus (1985a) described *A. nothagenes* from Kelley's Island in Lake Erie. This population incorporates genes from *A. laterale, A texanum,* and *A. tigrinum.* Kraus et al. (1991) described additional triparental *Ambystoma* from Michigan and Ohio, but did not apply names to them. Morris (1985) reported a *platineum-tigrinum* hybrid from St. Joseph County, Indiana, and I obtained a similar hybrid from Steuben County, Indiana.

The triploid females evidently will utilize sperm from any northeastern *Ambystoma* species, but sperm of *maculatum* rarely result in viable eggs. Most current workers reject a species ranking for triploid populations in this group of salamanders, although a nomenclature for the various hybrids has been proposed (Lowcock et al., 1987). Identification of *A. jeffersonianum* complex females can be difficult. Males can be assigned to *A. jeffersonianum* or *A. laterale,* although males have been reported a few times in triploid populations.

Cell size is influenced by chromosome number. Identification of polyploids by erythrocyte size is quick and fairly easy if known reference cells are available. Fresh cells from shed skin can also be used. Genomic identification of individuals requires electrophoretic analysis and chromosome counts. A recent and detailed study of the *A. jeffersonianum* complex in New York and New England revealed triploid females made up 70% of a sample of 1,002 animals representing six hybrid combinations (Bogart and Klemens, 1997).

Stream Breeding Small-mouthed Salamanders

A less complicated situation in the genus *Ambystoma* involves the salamanders formerly known as "stream-breeding *A. texanum* (small-mouthed salamanders)" but now known as *A. barbouri,* streamside salamanders. The range of this species was described by Kraus and Petranka (1989c) as from southwestern Ohio and southeastern Indiana to central Kentucky, with an isolated population in western Kentucky. Field identification of *A. barbouri* is difficult, although finding animals in typical breeding sites is helpful. Positive identification entails sending the specimens to experts. Externally, *A. barbouri* is identical with *A. texanum,* differing only in features of the teeth and premaxillary bone. However, the breeding biology of the two is quite different: *A. texanum,* as with most *Ambystoma,* lay eggs in vernal ponds or other shallow (often temporary) collections of standing water and then attach them to twigs and plant stems. In contrast, *A. barbouri* breeds in shallow rocky creeks and then attaches eggs to the underside of flat stones. Their eggs are fewer and larger than those of *A. texanum,* and *A. barbouri* larvae are larger on hatching.

Kraus and Petranka (1989) found that *A. barbouri* and *A. texanum* coexist in Scott County, Indiana, and two Kentucky counties, with little evidence of genetic introgression. The range of *A. barbouri* has recently been extended in the Ohio Valley to Hancock County, Kentucky, and Perry County, Indiana. In this region, it is probable the species are parapatric (occur separately), with *A. barbouri* breeding in rocky creeks and *A. texanum* in floodplain ponds, but with adults of both species sharing woodland habitat. It has been suggested that selective destruction of breeding sites could break down the isolating mechanisms separating these species. (Although there are no data, the *A. barbouri* and *A. jeffersonianum* complex triploids in southwestern Ohio present another interesting relationship that merits further study.)

Species Identification in the Toad Genus *Bufo*

Toads are familiar creatures to most Midwestern residents, although they are no longer as plentiful as they once were. East of the Mississippi, only two toad species occur in the area considered in this article, American toads (*Bufo americanus*) and Fowler's toads (*B. fowleri*). Both these species are also found west of the Mississippi where additional species occur. Woodhouse's toads (*B. woodhousii*) and Great Plains toads (*B. cognatus*) have extensive distributions. Canadian toads (*B. hemiophrys*), red-spotted toads (*B. punctatus*), and green toads (*B. debilis*) just enter the Midwest as defined here. Most of the taxonomic problems with the genus *Bufo* in the United States involve American toads, Fowler's toads, Woodhouse's toads, and Canadian toads. Great Plains toads, red-spotted toads, and green toads are distinct and readily identified species.

The name *B. americanus* originated with Holbrook (1836), although his description could fit any of the three larger toads found in the Carolinas. American toads are characterized by large tibial warts, moderate to very dark ventral pigmentation, dorsal dark spots that enclose only one or two warts or may be absent, and cranial crests with transverse ridges that do not contact the parotoid glands. *Bufo americanus* has a long taxonomic and nomenclatural history. Around the turn of the century, the name *B. lentiginosus* was widely used for all toads of Ohio, Michigan, Illinois, and Indiana. In the early years of this century, *B. americanus* began to be used in its present sense. In 1946, it was proposed that *B. americanus* be considered a subspecies of the southern *B. terrestris,* but this was a short-lived arrangement.

Fowler's toads (*B. fowleri*) were described in 1882 from specimens collected in Massachusetts. This species is characterized by tibial (lower leg) warts no larger than those on the back, a white belly usually with a dark pectoral (on the belly between the forelimbs) spot, dark dorsal spots that enclose several warts, and cranial crests whose transverse ridges are in contact with the parotid gland. For about 20 years, they were either ignored or considered a variety of some other toad species. *Bufo fowleri* achieved gradual recognition as a distinct form, but from 1934 until recently (Sullivan et al., 1996a) it has been considered a subspecies of *B. woodhousii* (*B. woodhousii fowleri*). Recently, Sullivan et al. (1996a) presented evidence for regarding *B. fowleri* as a valid species (see also Crother et al., 2000). This had been proposed previously by Sanders (1987), who described *B. hobarti* (type locality Shades State Park, Montgomery County, Indiana) as a sibling species of *B. fowleri*. Sanders' (1987) arrangement has not been widely accepted.

Canadian toads enter the Midwest in northwestern Minnesota and adjacent North and South Dakota. Originally described as *B. hemiophrys* in 1886, these animals have been considered a subspecies of *B. americanus*, a subspecies of *B. woodhousii*, and a distinct species, which is its present status with most herpetologists. *Bufo hemiophrys* are characterized by a prominent boss (bony protuberance) between the eyes, which may fuse with the cranial crests. Spotting is variable, warts are rather small; there is usually some ventral pigmentation.

There are several problems concerning the status and identification of toad species in the Midwest. One is the status of the form *B. charlesmithi*, characterized by small size (body length less than 65 mm), reddish color, and reduced ventral pigment. These animals are generally considered a subspecies of *B. americanus* (*B.a. charlesmithi*). Their range is usually given as from northeast Texas and eastern Oklahoma to southern Illinois and Indiana. However, many *B. americanus* populations from central Indiana (and some from northern Indiana) fit this description. Interspersed with them are populations of large *B. americanus* (average body length 71 mm) that are not reddish and have heavy ventral pigmentation. This larger form, presumably typical *B. americanus*, occurs in northern Illinois, most of Ohio and Michigan, and probably all of Wisconsin and Iowa.

Hybridization between American and Fowler's toads has been investigated most extensively in Indiana. Blair (1941) considered 82 of 813 adult toads (about 10%) from Monroe and Brown counties as hybrids. However, Jones (1973) collected in essentially the same areas 30 years later, identified only 7 of 713 toads as hybrids morphologically, and stated that even those could be allocated to one species or the other by call. In the South Bend area, Cory and Manion (1955) found predominantly hybrid toads in one breeding pond, while two other ponds had breeding aggregations of almost pure *B. americanus* in one case and *B. fowleri* in another. Blair (1941) reported five hybrids in a sample of 39 toads from Vincennes, a curious finding because *B. americanus* are extremely rare in the lower Wabash Valley, although *B. fowleri* are plentiful. In the course of basically random collecting in Indiana over many years, I have collected nine adult toads that I consider *B. americanus-fowleri* hybrids. Most were from the central till plains.

There is little information from other states. Blair (1941) mentions hybrids from Olney, Illinois, but this is a lower Wabash Valley locality where *B. americanus* have never been reported. Smith (1961) does not mention hybrids in Illinois. In Missouri, hybridization is said to occur in many areas. Additionally, in a zone between central and southwestern Missouri, there is genetic interchange between *B. fowleri and B. woodhousii*. Long interpreted as intergradation between subspecies, this zone may reflect hybridization between species. In western Iowa, *B. americanus-woodhousii* hybridization may be increasing (Christiansen and Bailey, 1991). Walker (1946) described *B. americanus-fowleri* hybrids from Ohio but gave no indication of

their prevalence. Because hybrids occur at South Bend, Indiana, they almost certainly occur in southern Michigan.

Several mechanisms discourage hybridization between American and Fowler's toads. Although varying to some extent with ambient temperature, advertisement calls of the male toads are distinctly different, even to the unpracticed ear. The peak of the breeding season for Fowler's toads is about 3 weeks later than that of American toads, although there is always overlap that varies with spring weather conditions. Most observers note a difference in habitat between the species, with Fowler's toads preferring loose or sandy soil and open terrain, while American toads prefer denser soils and woodlands. There are some exceptions to this generalization about habitat, however. In the black soil prairies of northern Illinois and Indiana, a large race of *B. americanus* are the only toads found, while in the wooded hills of southern Illinois and Indiana, *B. fowleri* usually outnumber *B. americanus*. Some have suggested that human modification of the environment has increased hybridization between the species.

Frogs in the *Rana pipiens* Complex

For generations of American zoologists and biology students, "the frog" has been the leopard frog, usually designated *Rana pipiens*. Until about 1970, this was considered a single polytypic species ranging from northern Canada to Costa Rica. Current authorities recognize up to 20 species in the *R. pipiens* complex, with about 11 found in the United States. Five species are found in the Midwest; however two, pickerel frogs (*R. palustris*) and crawfish frogs (*R. areolata*) have long been recognized as distinct and are easily recognized. For over a century, it was generally believed that at least two other varieties of leopard frogs occurred in parts of the Midwest. These were identified by the names *R. pipiens, R. virescens, R. halecina, R. sphenocephala, R. brachycephala*, and *R. utricularia* used in a variety of combinations. Smith (1961) applied the name *R. pipiens pipiens* to the leopard frogs of northern Illinois and *R. p. sphenocephala* to southern Illinois populations with a wide area of intergradation in the central part of the state. I used the same nomenclature in Indiana (Minton, 1972) but commented that west of Indianapolis intergrades were uncommon and there seemed to be little gene interchange between *R. sphenocephala* and *R. pipiens*. Mecham et al. (1973) described a common and widely distributed leopard frog of the central United States as *R. blairi* (plains leopard frogs). Pace (1974) confirmed the validity of three leopard frog species in the Midwest but used the name *R. utricularia* instead of *R. sphenocephala* for the southern species. This taxonomy (i.e., *R. pipiens, R. utricularia*, and *R. blairi*) is the generally accepted arrangement today (but see Crother et al., 2000). Species are identified by dorsal pattern and color, presence or absence of a light tympanic spot, head shape, posterior interruption of the dorsolateral folds, and position of the male vocal sacs.

Northern leopard frogs (*R. pipiens*) are the most widely distributed member of the complex and the only one found in Ontario, Minnesota, Michigan, Wisconsin, and in all but the southernmost counties of Ohio and Iowa. At one time they were extremely abundant in the northern part of this range. Literally millions were shipped annually from the Okoboji wetlands of northern Iowa in the early years of this century. Interestingly enough, they remain the most numerous amphibian in the area today, although their numbers have been drastically reduced (Lannoo, 1996). The frogs were just as heavily exploited in Minnesota (Wright and Wright, 1949). Most of the catch went to restaurants and biological supply houses. In the 1960s, *R. pipiens* populations began to decline sharply, even in states such as Indiana where there had never been much commercial collecting. The reasons are unclear, although habitat destruction, chemical pollution, disease, and predation all have been blamed.

Southern leopard frogs *(R. utricularia)* are found throughout most of Missouri, the southern half of Illinois, southeastern Kansas, southern Indiana, extreme southern Ohio, and extreme southeastern Iowa. At least in Indiana, they have not undergone a serious population decrease.

The Midwest range of *R. blairi* includes all of Kansas, most of Nebraska, northern Missouri, and southern and western Iowa with small, presumably relict populations in Indiana, southern Illinois, and central Iowa.

All three of these frog species adapt to a variety of wetland habitats, but they prefer those in open country with shallow water and abundant aquatic vegetation. They avoid densely forested habitat. In my experience, it is unusual for two of these species to occur together.

Hybridization between the species in the *R. pipiens* complex occurs in nature to a varying degree. Hybrids can usually be detected by external examination. Prior to about 1974, they were usually identified as intergrades between subspecies. Most reports of hybridization are from western and southwestern states; there is little information from the Midwest. Breeding season, male advertisement calls, and habitat all may act as reproductive isolating mechanisms. Throughout their range, northern leopard frogs appear to have a rather short breeding season, only in spring. Plains leopard frogs may breed any time from February to October (L.E. Brown, 1992); late March to early May is the season in Missouri (Johnson, 1977). Southern leopard frogs are also spring breeders in the Midwest, but autumn breeding has been reported (Wright and Wright, 1949; Minton, 1972). Calls of males can be distinguished, but it takes an experienced ear or an audiospectrogram. There is evidence that plains leopard frogs occupy a greater variety of habitats than do northern leopard frogs in the western portions of their ranges. In eastern Colorado, habitat differences kept the two apart until the 1960s, when construction of ponds brought *R. pipiens* close to the more riparian habitat of *R. blairi*. The result was the creation of a large hybrid population by 1975. By 1988, most of the hybrid and all the *R. pipiens* populations had disappeared (Cosineau and Rogers, 1991). In central Indiana, southern leopard frogs tend to live in floodplain swamps and sluggish streams, while northern leopard frogs inhabit marshes and peat bogs that do not have substantial stream connections.

Treefrogs in the *Hyla versicolor* Complex

Another curious taxonomic problem involves the gray treefrogs known as *Hyla versicolor* and *H. chrysoscelis*. These species cannot reliably be distinguished by morphology, however the advertisement calls of the males are distinctly different, that of *H. versicolor* being a mellow, musical trill lasting up to three seconds while that of *H. chrysoscelis* is harsh, nasally, and short. Moreover, *H. chrysoscelis* has a diploid number of 24 chromosomes, while *H. versicolor* has a tetraploid number of 48. The combined distribution of the two species is from Maine to southeastern Manitoba and south to the Gulf of Mexico. Generally speaking, *H. versicolor* has a northern and eastern distribution and *H. chrysoscelis* southern and western. Apparently these species widely overlap over much of the Midwest, however much of the distributional data are fragmentary. Based on call surveys in Wisconsin dating back to 1981, *H. versicolor* occurs virtually statewide with stable populations; *H. chrysoscelis* occurs largely in the northwest and central parts of the state, and their populations seem to be declining (Mossman et al., 1998). In southeast Manitoba, *H. versicolor* is found in more humid parts of the region than is *H. chrysoscelis* (Preston, 1982). In Indiana, records of *H. versicolor* are from Indianapolis northward, while most *H. chrysoscelis* records are south of Indianapolis. The picture is similar in Ohio except that *H. versicolor* records are almost statewide (Pfingsten, 1998).

Over most of the range, these two species are found in similar habitat, breed at the same season, and may use the same ponds. In spite of this, hybridization has not been reported, and no striking decline in numbers has been reported for either species in the Midwest.

Summary

The examples given here illustrate some taxonomic problems that may confront those concerned with the status and conservation of amphibians, an intrinsically difficult group for taxonomists. Nevertheless, a uniform and stable nomenclature is needed for the preparation of regional faunal lists and particularly for legislation regulating the hunting, sale, and possession of amphibians and their protection in natural habitats. However, such a nomenclature is not likely to evolve in the foreseeable future. For a long time, we will have to be constantly making judgments. Just remember that a new taxonomic arrangement does not mean compliance. Consider the source of the opinions and review the facts. In making legislative regulations and local species lists, it is wise to cite an authority for the nomenclature used.

THIRTY

Conservation Systematics: The *Bufo boreas* Species Group

ANNA M. GOEBEL

Systematics and taxonomy play critical roles in conservation (May, 1990; Eldredge, 1992; Systematics Agenda, 2000, 1994a,b; Wheeler, 1995; Koch and Peterson, this volume; Minton, this volume). Taxonomic names are important for recognition and clear communication about the units to be conserved; conservation efforts have been compromised when taxonomy did not accurately reflect systematic relationships (e.g., Greig, 1979; Avise and Nelson, 1989; Daugherty et al., 1990; O'Brien and Mayr, 1991; Mishler, 1995; but see Zink and Kale, 1995). However, new developments in systematics and taxonomy that recognize, describe, and quantify organismic diversity have not been adequately incorporated into conservation programs. Extinction of the divergent island populations of tuataras (Daugherty et al., 1990; Finch and Lambert, 1996) is exemplary. Despite the description of subspecies occupying different islands, this diversity was ignored and several subspecies were allowed to go extinct because divergent tuatara lineages did not have the Linnaean rank of species. Furthermore, the remaining evolutionary diversity within the few extant tuatara subspecies is valuable for the conservation of organismic diversity because tuataras are not simply a distinct species group. In fact, the phylogenetic lineage, of which they are the sole representatives, is sister to a lineage that is represented by about 6,000 species of snakes, lizards, and amphisbaenians (May, 1990).

The major tasks of systematics (which includes taxonomy; Quicke, 1993) are to (1) classify organisms into species; (2) provide species names that are explicit, universal, and stable; and (3) combine species into the more inclusive categories of the Linnaean hierarchy (Futuyma, 1986). The general focus of systematics is to discover the genealogical relationships among the inclusive categories and describe patterns of evolutionary change (Futuyma, 1986). As a conservation biologist interested in systematics, my goals are to identify organismic units for conservation (whether they are species, Evolutionarily Significant Units [Ryder, 1986], or other units), describe and name these units, and quantify the divergence among them to assist in conservation efforts. My focus is similar to other systematists, but differs in the specific intent for conservation. Discovering relationships is critical because quantification of diversity depends on the pattern of evolutionary relationships. Deducing evolutionary processes from patterns of change is critical because it is the processes, as well as the end products (populations and individual organisms),

that must be conserved. Biologists who focus on conservation have coined the terms "conservation biology" (Soulé and Wilcox, 1980) and "conservation genetics" (Shonewald-Cox et al., 1983; Avise and Hamrick, 1996). In that sense, my focus is on "conservation systematics" and "conservation taxonomy."

The purpose of this essay is to describe how systematics and taxonomy can better address conservation issues in both theoretical and utilitarian ways. I begin with a discussion of organismic diversity and how systematics and Linnaean taxonomy have failed to meet the needed description and quantification of diversity for conservation purposes. I then argue that recognizing diversity is more critical than recognizing species, and I suggest how diversity can be incorporated into systematics using measures of phylogenetic diversity and phylogenetic taxonomy. In the final section, I suggest three utilitarian ways conservation systematics can incorporate diversity into management and politics: (1) set priorities for conservation; (2) reconstruct the Endangered Species Act (ESA); and (3) mitigate loss of total diversity by a procedure that identifies acceptable losses. To illustrate problems and solutions, I use examples from North American bufonids, especially the western toad *(Bufo boreas)* species group (Examples 1–3, below).

The Critical Role of Diversity

Diversity

The intent of both systematics and taxonomy is to describe organismic diversity in a general way ("systematics is the study of organismic diversity," Wiley, 1981). However, conservation biologists need to be more specific when quantifying diversity. Conservation biologists frequently are asked: (1) How divergent are two units/taxa from one another—is a subspecies or population of special interest really a different species? (2) Is a particular species/taxon made up of many diverse lineages that should have independent conservation programs or is it a single lineage that lacks diversity and can be managed as a single unit? (3) Are there genetic or taxonomic restrictions to translocating organisms? (4) Are the organisms in a particular U.S. state the same taxon/population as those that have been listed as endangered or threatened by another state or the Federal Government? (5) Is a species/population divergent enough for an expensive conservation program to be biologically and politically defensible or should the money, time, and credibility of the conservation program be spent elsewhere? (6) Which species/taxa should have the highest conservation priority and how should those priorities be set? The answers to these questions require not only a delineation of the organismic units in question and an assessment of speciation, but also a quantification of diversity within and/or among such units.

Conservation depends on understanding many kinds of diversity, including organismic, ecological, climatic, and landscape diversity (Moss, 2000). However, the purpose of systematics and taxonomy is to describe organismic diversity, which will remain the focus of this essay. Because organismic diversity is a broad term, let me define my use in this essay. Organismic diversity is comprised of the different attributes or traits (e.g., molecular, biochemical, physiological [Spicer and Gaston, 1999], morphological, behavioral, etc.) that are passed down through evolutionary lineages and within individuals and populations. Because such traits are inherited, organisms that are closely related have a high probability of sharing many traits; those that are more distant will share few. Traits evolve through time; novel traits arise through random mutations and rearrangements of existing traits. Diversity is critical because it is the fuel for evolutionary change and essential for adaptation to changing environments. The extinction of any lineage represents not only the loss of novel traits but also the loss of knowledge about what kinds of traits, trait combinations, and evolutionary pathways are possible.

The ability to describe organismic diversity has increased dramatically in the last 10 years with advances in molecular systematics (e.g., Hillis et al., 1996; Smith and Wayne, 1996; Karp et al., 1998; Goebel et al., 1999; Hall, 2001). Unlike frequently used morphological characters, molecular characters not only recognize patterns of inheritance, but also recognize inherited diversity on a continuum from parent-offspring relationships to diversity within and among lineages as well as among higher taxa. Although the connection between specific molecular changes and the presence of a particular physiological, behavioral, or morphological trait is rarely known, the greater the degree of molecular divergence, the higher the probability that unique physiological, behavioral, or morphological traits have evolved by random chance alone. Thus, a calculus of divergence based on independently evolving molecular characters may provide an estimate of the probability of other divergent traits (e.g., physiological) that are not measurable at this time (Faith, 1992a,b, 2002; Crozier, 1997; but see Pearman, 2001). Another advantage of molecular phylogenetic analyses is that they identify genetic diversity even if they cannot unambiguously identify clades as specific taxonomic units in the Linnaean hierarchy (e.g., classes, genera, and species). But molecular data are not a panacea for describing diversity (Pritchard, 1999). For example, discriminating between gene and organismic lineages may be difficult (Neigel and Avise, 1986; Pamilo and Nei, 1988; Quinn et al., 1991; but see also Moore, 1995, 1997; Nichols, 2001) and rates of change may vary among lineages and genes (Wu and Li, 1985; Martin et al., 1992; Zhang and Ryder, 1995). These difficulties result in a discouraging sense that more data will always be needed to correctly identify phylogenetic relationships. However, molecular data continue to provide valuable insights even while better methods to collect, interpret, and analyze data are being developed to circumvent these problems. In spite of an explosion in the ability to identify and describe diversity, the incorporation of measures of diversity into systematics and taxonomy has been slow (Soltis and Gitzendanner, 1999).

Inadequacies in Systematics and Linnaean Taxonomy for Conservation

The Linnaean system (Linnaeus, 1737, and described in the codes of nomenclature: International Commission on Zoological Nomenclature, 1999; International Botanical Congress, 2000; International Association of Microbiological Societies, 1992) applies specific ranks to all lineages (e.g., class, order, family, genus, species, and variety) independent of the diversity within or among them. Ranks are an imprecise measure of diversity because they identify only a few categories in a world that can have a near infinite number of hierarchical levels. Even so, many conservation efforts are based on the species rank—it is seen as the fundamental unit of evolution and therefore as the fundamental unit for conservation (e.g., Wilson, 1992; Caughley and Gunn, 1996). However, conservation efforts based on any rank have resulted (and continue to result) in a critical loss of diversity. For example, conservation efforts have been inhibited by the lack of recognition of species (Greig, 1979; Daugherty et al., 1990), disagreement over the recognition of species (Daugherty et al., 1990; Sangster, 2000; see also Hille and Thiollay, 2000), arguments over the importance of hybridization (O'Brien et al., 1990; Lehman et al., 1991; Wayne and Jenks, 1991; Nowak, 1992; Roy et al., 1994, 1996), and a lack of understanding of the phylogenetic (genealogical) relationships among species (Avise and Nelson, 1989). Conservation efforts may also have been misplaced with programs for poorly defined taxa (e.g., Bowen and Karl, 1999; Karl and Bowen, 1999; Zink et al., 2000; but see also Pritchard, 1999; Grady and Quattro, 1999).

Conservation of diversity grounded on a species-based system is inadequate. Species definitions continue to abound (Mayden, 1997; Soltis and Gitzendanner, 1999; Wheeler and Meier, 2000; Hey, 2001) yet there is little consensus on what a species is (Cantino and de Queiroz, 2000; Barton, 2001). The species category is not defined on the basis of divergence, but frequently on the basis of qualities such as reproductive isolation (Biological Species Concept; Mayr, 1942, 1982, 1996), ability of mates to recognize one other (Recognition Species Concept; Paterson, 1985; Lambert and Spencer, 1995), occupation of an adaptive zone (Ecological Species Concept; Van Valen, 1976), and potential for phenotypic cohesion (Cohesion Species Concept; Templeton, 1989). Other species definitions are based primarily on the historical pattern of evolution including Evolutionary Species (lineages have their own evolutionary tendencies and historical fate; Simpson, 1961; Wiley, 1978), Plesiomorphic Species (lineages identified by a unique combination of characters even if uniquely derived characters have not evolved or are as yet undetected, Olmstead, 1995), and Phylogenetic Species (species are monophyletic groups, Mishler and Donoghue, 1982; or species are the smallest diagnosable clusters of individual organisms within which there is a parental pattern of ancestry and descent; Cracraft, 1983b; see also Goldstein et al., 2000, and Soltis and Gitzendanner, 1999, and references therein). Contention over which qualities or phylogenetic patterns are most appropriate for the delineation of species may further confound species-based conservation programs if accepted definitions change in time or vary among conservation agencies or legislative decisions. However, even if criteria for identifying species were not contentious and were uniformly applied, species-based conservation programs are problematic because the pattern and quantity of diversity are unique to each lineage and are not defined by rank (e.g., Karl and Bowen, 1999).

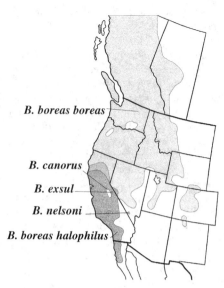

FIGURE 30-1 Distribution of four species of the *B. boreas* group (Blair, 1964b; Blair, 1972a; Schmidt, 1953; Feder, 1973; Stebbins, 1985). Intermediate shading represents range overlap between the subspecies *B. b. boreas* and *B. b. halophilus*.

This is especially apparent for paraphyletic species (e.g., Shaffer et al., 2000).

In real lineages, the amount of diversity within and divergence among species varies widely (Examples 1 and 3). Attempts to describe diversity within ranking systems include the incorporation of the many super-, sub-, and infra-categories (e.g., subspecies, subgenus) to the Linnaean system (Simpson, 1961; Mayr, 1969), as well as definitions for non-Linnaean categories such as Evolutionarily Significant Units (ESU; Ryder, 1986; Waples, 1991, 1995, 1998; Dizon et al., 1992; Rojas, 1992; Vogler and DeSalle, 1994; Karl and Bowen, 1999; Paetkau, 1999; but see Cracraft, 1997; Crandall et al., 2000), Management Units (MU; Moritz, 1994, 1995; Paetkau, 1999), and Evolutionary Units (EU; Clegg et al., 1995). Populations have also been considered the appropriate unit for conservation (Crozier, 1992; Crozier and Kusmierski, 1994; Pennock and Dimmick, 2000). However, these all suffer from the same difficulties as the species rank: (1) criteria can be applied to ranks that have widely differing levels of diversity within and among them and (2) naming specific ranks across lineages will falsely imply similar levels of divergence.

One approach to conservation is a rush to write new species descriptions. Only a small fraction of all species (about 1.75 million of an estimated 13.6 million or more total) have been taxonomically described (Hammond, 1992, 1995). Amphibian species descriptions have increased an estimated 20% in 15 years (4,103 species in 1985 to an estimated 5,000 in 2000 A.D.; Frost, 1985; Duellman, 1993; Glaw and Kohler, 1998). Even this rate, however, may not be enough to document the world's diversity due to increasing rates of species extinctions (Pimm and Brooks, 2000), especially in poorly studied taxa (McKinney, 1999). Furthermore, species descriptions may result in increased conservation efforts for the single species of interest but could ignore conservation of equally divergent and threatened lineages (Paetkau, 1999) because the descriptions rarely identify the diversity of the whole group to which the species of interest belongs. Similar difficulties are seen with bird species (Peterson, 1998) to the point that systematics is seen as a threat to conservation (Sangster, 2000) due to the lack of taxonomic stability.

An emphasis on conserving diversity rather than a taxonomic rank (e.g., species) will allow the design of conservation programs appropriate for the unique pattern of diversity found in each lineage, at all levels of the evolutionary hierarchy, rather than the sole preservation of the characters or criteria on which a species was determined. Suggestions for such diversity-based conservation programs for bufonids in North America and for the western toad species group are described in Examples 1 and 2.

Using Linnaean taxonomy for conservation is inadequate because the identification of Linnaean species is a slow process and providing a Linnaean species name does not mean that the species is a legal biological entity. An estimated 6 months to several years is needed for documentation of characteristics defining species and for the publication of formal species descriptions in peer-reviewed journals. Even with formal descriptions, only the name is a legal entity and it belongs to a single (or few) type specimen(s). The recognition of species status need not be accepted (Lazell, 1992) and may be contentious for years, due to the many criteria for recognizing species (described above; Goebel et al., this volume, Part Two). A system where diversity is formally recognized from phylogenetic analyses, rather than by recognizing only species from descriptions, may provide a faster and more stable base for conservation programs. Arguments will surely continue about which lineages constitute species (e.g., the disagreements among "splitters" and "lumpers") and which criteria should be used to recognize species, but there will be much less contention about the presence of lineages or clades once they are discovered. Due to the current extinction crisis and limited time and funding, inclusive descriptions of diversity may be more useful than single species descriptions.

Conserving species (or any unit) alone is inadequate because it protects only organisms. In contrast, an emphasis on diversity requires conservation of the evolutionary processes that maintain diversity (e.g., natural selection, mutation, adaptation, gene flow, random drift, vicariance, polyploidy, population and community dynamics, ecological shifts, geological changes, etc.; Dimmick et al., 1999; Crandall et al., 2000; Owens and Bennett, 2000) although the importance of specific processes is debated (e.g., adaptation [Storfer, 1996; Crandall et al., 2000; Young, 2001] versus vicariance [Dimmick et al, 1999; Dimmick et al., 2001]). Conserving diversity requires an understanding of the critical components of diversity, how diversity is apportioned, and how it evolves.

Finally, populations may be the units within which evolutionary processes are most critical. Extinction rates for populations are staggering; if population extinction is a linear function of habitat loss, then about 1,800 populations are being lost per hour in tropical forests alone (Hughes et al., 1997). If populations are the units to be conserved (Crozier, 1992; Crozier and Kusmierski, 1994; Pennock and Dimmick, 2000) then formal recognition of diversity within and among populations, identification of population lineages, and formal names for populations may provide much assistance to the conservation of evolutionary processes.

EXAMPLE 1: PHYLOGENY OF THE WESTERN TOAD (*BUFO BOREAS*) SPECIES GROUP AND MEASURES OF PHYLOGENETIC DIVERSITY.

The *Bufo boreas* species group contains four species distributed across western North America (Fig. 30-1). Three (Yosemite toads *[B. canorus]*, black toads *[B. exsul]*, and Amargosa toads *[B. nelsoni]*) are thought to be localized relictual isolates from Pleistocene glaciations. (Although the species status of *B. nelsoni* is not recognized by all [Crother et al., 2000], here it is treated as a species.) The fourth and nominal form, *B. boreas*, comprises two subspecies (boreal toads *[B. b. boreas]* and California toads *[B. b. halophilus]* occurring over the rest of the range.

Molecular diversity is not distributed evenly among species (Fig. 30-2). For example, *B. exsul* and *B. nelsoni* compose clades (nodes 5, 7) of closely related individuals. In contrast, specimens of *B. boreas* are found in multiple divergent clades (e.g., 10, 9, 3, 2), some of which may comprise previously unrecognized species (e.g., nodes 2,3). A conservation program based solely on species could result in much loss of diversity within *B. boreas*. Conservation programs based on clades, whether they are currently recognized as species or not, would preserve much more diversity.

The geographic distribution of clades (Figs. 30-2 and 30-3) can identify units for conservation in a hierarchical manner. For example, independent conservation programs could be identified for the species *B. nelsoni* (node 7), the southern clade (node 6), and the species *B. exsul* (node 5). These could work in cooperation with national conservation programs for the entire southwest clade (node 4) and international programs for the entire *B. boreas* species group (node 1).

Priorities for conservation could be identified from phylogenetic diversity calculated from branch lengths (Faith, 1992a,b). Setting priorities based on diversity would provide protection for some populations of *B. boreas*. For example, the southern Utah population, node 2, comprises 8.71% of the total diversity of the species group (Fig. 30-2). Similarly, the previously unrecognized Southern Rocky Mountain clade (node 3) comprises more diversity than the species *B. exsul* or *B. nelsoni*.

Incorporating Diversity into Systematics and Taxonomy

Diversity can be incorporated into systematics and taxonomy by combining phylogenetic taxonomy (de Queiroz and Gauthier, 1992, 1994; Cantino and de Queiroz, 2000) above and below the species level with quantitative measures of phylogenetic diversity (PD) calculated from branch lengths of molecular analyses (Faith, 1992a,b). Phylogenetic taxonomy can identify an infinite number of names based on the hierarchical pattern of relationships, while measures of phylogenetic diversity provide a quantitative measure of the probability that any organism will have unique or divergent traits. By recognizing phylogenetic species (Cracraft, 1983b), more diversity will be identified than species definitions based on other criteria (e.g., ability to interbreed), but phylogenetic species will not identify all of the diversity useful for conservation purposes.

Phylogenetic Taxonomy

Organismic units need name recognition for conservation. As described above, Linnaean taxonomy will always be inadequate because it identifies in a few categories a world that can have a near infinite number of hierarchical levels. Linnaean taxonomy also insists that the same ranks be applied across all lineages, which can be misinterpreted as similar levels of divergence for each rank. For conservation purposes, a taxonomy that provides names for clades that are unique to each lineage and names on a hierarchical scale from groups of higher taxa down to populations is needed. Because conservation systematics includes conservation of evolutionary processes, a taxonomy that is based on historical evolutionary clades is also desirable.

Phylogenetic taxonomy (de Queiroz and Gauthier, 1992, 1994; Cantino et al., 1999; Pleijel, 1999; Cantino, 2000; Cantino and de Queiroz, 2000, and references therein) differs from Linnaean taxonomy in that it provides names unique to each hierarchical division in a lineage (clade). Names are unique to each clade, are not assumed to be equivalent ranks across lineages, and are based on the pattern of evolutionary descent. Phylogenetic taxonomy was originally described for higher taxa, not for subspecific categories (de Queiroz and Gauthier, 1992). However, the basic tenets can be applied to any clade. Names for some clades may be identical in phylogenetic and Linnaean taxonomy; however, they may differ, because in phylogenetic taxonomy: (1) taxon names always identify clades (Linnaean names are not necessarily clades) and (2) names for all Linnaean ranks are not necessary or sufficient (e.g., names for clades other than order, genus, and species can be included and names for all Linnaean ranks need not be included). Lastly, phylogenetic taxonomy will allow a much more rapid translation of phylogenetic information into taxonomy reflecting the rapid accumulation of phylogenetic information from advances in molecular biology and computer technology (Cantino and de Queiroz, 2000).

Here, phylogenetic taxonomy is applied to the western toad species group with an attempt to identify names for clades, without attempting to identify these clades as species or other Linnaean ranks (Example 2). Applying names to divergent lineages rather than to species alone will have two effects: the burden of identifying which lineages have the characteristics of species is somewhat alleviated, but the burden of identifying diversity is increased dramatically. Both effects will assist conservation efforts by promoting conservation programs that are based on evolutionary hierarchies, are lineage-specific, and incorporate varying levels of diversity.

EXAMPLE 2: PHYLOGENETIC TAXONOMY OF THE WESTERN TOAD *(BUFO BOREAS)* SPECIES GROUP

Linnaean taxonomy currently recognizes four species (Table 30-1). This taxonomic classification does not recognize all the diversity identified in *B. boreas* (Example 1). Even if some of the clades (e.g., nodes 2, 3; Example 1, Fig. 30-2) are eventually identified as species or species groups, the current taxonomy does not recognize clades that comprise multiple taxa (e.g., nodes 8, Northwest clade, and node 4, Southwest clade).

Phylogenetic taxonomy (Table 30-2) can provide names for all the hierarchical clades identified with DNA data (Example 1, Fig. 30-2), whether they are species or multiple supra- or sub-specific categories. Phylogenetic names are identical to Linnaean names when they identify the same clades (e.g., the species Yosemite toads *[B. canorus]*, black toads *[B. exsul]*, and Amargosa toads *[B. nelsoni]* except that in phylogenetic taxonomy the binomial may be combined with a hyphen [Cantino and de Queiroz, 2000]). Phylogenetic names follow the Linnaean rule of first use, but recognize first use of a clade, not a Linnaean category. For example, clade 3 is identified as Clade Pictus because the name *B. pictus* was first used for a toad collected near Provo, Utah (Cope, 1875c). Names of clades identified with the phylogenetic system that are not recognized as species do not follow all rules of Linnaean nomenclature. They do not have to be binomials and they do not have to be differ-

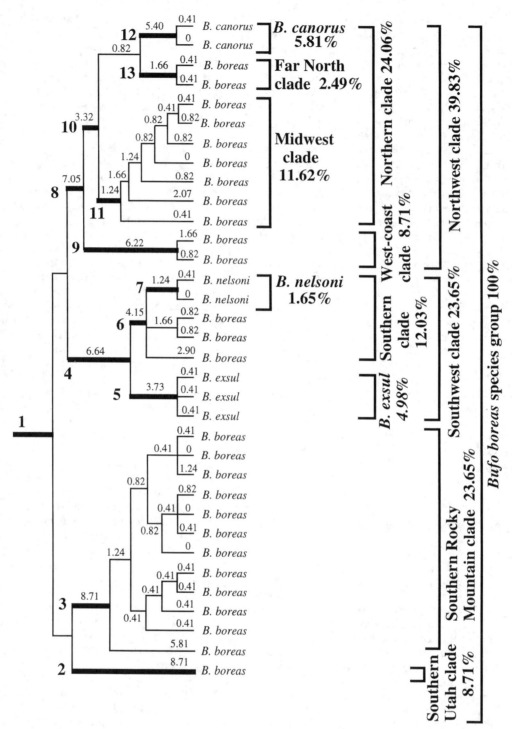

FIGURE 30-2 Phylogenetic analysis of the *B. boreas* species group hypothesized from parsimony analyses (Swofford, 1993) of three mitochondrial DNA regions (12S ribosomal DNA, the control region, and cytochrome oxidase I). Detailed methods, results, and discussion are provided elsewhere (Goebel, 1996a; Goebel, T. Ranker, S. Corn, R. Olmstead and T. Bergren, in preparation). The branching pattern of the phylogenetic tree provides a hypothesis of the genealogical relationships of individual organisms that are represented at the tips (or terminals) of the branches. A clade includes all individuals from the specified node to the tips of the tree (e.g., clade 12 includes the first two terminals only and represents the species *B. canorus*). The branch that identifies each strongly supported clade is indicated by a bold line and names for these clades (based on geographic regions) are identified with brackets to the right of the tree. Strongly supported clades are identified with bold numbers (1–13) whether they are species (e.g., *B. exsul*, node 5, and *B. nelsoni*, node 7) or other units (e.g., 10, 9, 3). Small numbers on the branches are branch lengths, calculated as a percent of the total length of the tree. Phylogenetic diversity (Faith, 1992a,b) is calculated from branch lengths. Measures for whole clades can be identified by summing all the branch lengths within the clade (e.g., phylogenetic diversity of *B. exsul* would be 0.41 + 0.41 + 0.41 + 3.73 = 4.98% of the total value of the species group). Diversity measures for the strongly supported clades are identified to the right of the tree below the names of clades.

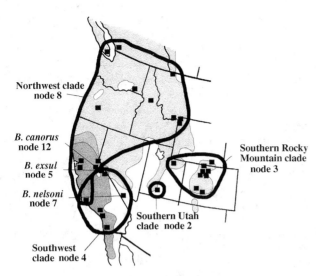

FIGURE 30-3 Distribution of several clades in the *B. boreas* species group. Species identification and shading are as in Figure 30-1. Squares indicate locations of populations sampled in the analysis. Names for clades and node numbers are as in Figure 30-2.

entiated by type font (Linnaean species typically are italicized within non-italicized text). No claim is made that all phylogenetic names refer to species, rather, names are provided for strongly supported clades because the names provide a basis for communication.

Phylogenetic nomenclature will assist conservation efforts by providing names for hierarchical units whether they are species or higher or lower categories. Names can be provided for the full range of diversity; they do not imply equivalent ranks across lineages and they are based on the pattern of evolution of the group. For example, the diversity of Clade Pictus and Clade Bufo-sevieri can be recognized even though one (Clade Bufo-sevieri) is likely a species and the other (Clade Pictus) is likely made of multiple species. In addition, naming "groups" where evolutionary independence is unclear at this time will assist in recognizing diversity before species delimitations are discovered.

Measures of Phylogenetic Diversity

The ability to identify and conserve organismic diversity depends on quantifying the probability that any taxon will have unique traits (also called features or attributes; Faith, 1992a,b). All traits cannot possibly be measured or conserved, nor is it possible to predict which traits should be given the highest value. However, preservation of the greatest number of traits (options) for the future is desirable. IUCN (1980) defined Option Value as "a safety net of biological diversity for future generations," and the preservation of Option Value (Weisbrod, 1964; Hanemann, 1989; Weitzman, 1992a,b) is seen here as the preservation of the greatest array of diversity.

Measures of organismic diversity have been defined based on species counts (Pielou, 1967; May, 1981; Smith and van Belle, 1984) or on higher taxonomic categories (Gaston and Williams, 1993), where species (or higher categories) are given equal weight. However, these measures are still based on ranks (species, genera, etc.) that may not reflect diversity accurately. Measures based on cladistic analyses, rather than taxonomy alone, include node counts within diversity measures (May,

1990; Vane-Wright et al., 1991; Williams et al., 1991; Nixon and Wheeler, 1992), but these still do not attempt to quantify diversity within and among nodes any more than do species counts. Finally, branch lengths (Altschul and Lipman, 1990) were incorporated into diversity measures based on both genetic distances (Crozier, 1992; Crozier and Kusmierski, 1994) and phylogenetic diversity (Faith, 1992a,b, 1993, 1994a,b). Some methods incorporate both phylogenetic pattern and branch length information (Faith and Walker, 1993; Faith, 1994c) in an attempt to describe phylogenetic diversity within and among clades.

Molecular phylogenetic analyses are especially amenable to quantifying diversity in the form of branch lengths (Faith, 1992a,b). Phylogenetic analyses of gene sequences are used as surrogate data for the evolutionary diversity of many features (e.g., morphological, chemical, or developmental attributes) that cannot be measured independently at this time (Faith, 2002). Because different character suites are likely to follow different evolutionary patterns (Eldredge and Gould, 1972; but see Omland, 1997), DNA analyses of potentially neutral characters on multiple unlinked genes may provide an estimate of the probability of underlying feature diversity. Measures of molecular phylogenetic diversity can be seen within the western toad species group (Example 1, Fig. 30-2) and among North American bufonids (Example 3, Fig. 30-4).

Although it seems intuitive that making decisions about loss of diversity is better than allowing it to proceed at random, Nee and May (1997) suggest that this may not be the case. They suggest that algorithms designed to maximize the amount of diversity (described above) are not much better than choosing the survivors at random, and that 80% of the underlying tree of life can survive even when 95% of the species are lost. However, historical and current extinctions are not randomly distributed among species (Bennett and Owens, 1997; Russell et al., 1998; Purvis et al., 2000, and references therein), which may cause a greater loss of diversity. In addition, it is not clear what level of diversity can be lost before human life cannot be sustained. It may be that the additional 10% of diversity that might be saved with careful planning (Nee and May, 1997) is critical.

EXAMPLE 3: PHYLOGENETIC DIVERSITY AMONG BUFONID TAXA PROTECTED BY THE U.S. ENDANGERED SPECIES ACT

Four bufonid taxa are recognized as endangered in the United States (Wyoming toads *[B. baxteri]*, Houston toads *[B. houstonensis]*, arroyo toads *[B. californicus]*, and Puerto Rican crested toads *[Peltophryne lemur]*; Fig. 30-4). Eight taxa (Colorado River toads *[B. alvarius]*, western toads *[B. boreas]*, Yosemite toads *[B. canorus]*, black toads *[B. exsul]*, Amargosa toads *[B. nelsoni]*, Arizona toads *[B. microscaphus]*, green toads *[B. debilis]*, and red spotted toads *[B. punctatus]*; for alternate nomenclature see Crother et al., 2000) are provided some form of protection within individual U.S. states (Frank and Ramus, 1994; Levell, 1995).

Species-based conservation programs for U.S. taxa are inappropriate because phylogenetic diversity is not distributed evenly among species. Twelve of 22 bufonid species in the United States are found in two clades of closely related species (*B. boreas* and *B. americanus* species groups; Fig. 30-4). The threatened taxon *(P. lemur)* is quite divergent and comprises the most basal lineage among U.S. taxa. The remaining three endangered taxa are closely related within the *B. americanus* species group. This pattern of low taxonomic divergence of endangered species

Table 30-1

(Example 2) Taxonomy of the *Bufo boreas* Species Group Based on the Linnaean Taxonomic System

Taxon (Common name)	Type Locality and First Description	Distribution
Taxa Currently Recognized		
Bufo boreas	Type locality: Columbia River and Puget Sound, WA; restricted to vicinity of Puget Sound First described by Baird and Girard (1852b)	See subspecies below
B. boreas boreas	Type locality: Mouth of the Columbia River, WA; First described as *B. columbiensis* (Baird and Girard, 1853a)	Coastal Alaska south to northern CA, east into MT, WY, CO and extreme northern NM
B. boreas halophilus	Type locality: Benicia, Solano Co., CA; First described as *Bufo halophila* (Baird and Girard, 1853b)	Extreme western NV, central valleys of CA, and mid-coastal CA south into Baha California of Mexico
B. exsul	Type locality: Deep Springs, Deep Springs Valley, Inyo Co., CA; First described by Myers (1942a)	Deep Springs Valley of eastern CA
B. nelsoni	Type locality: Oasis Valley, Nye Co., NV; First described as *B. boreas nelsoni* (Stejneger, 1893)	Amargosa River drainage of southwestern NV
B. canorus	Type locality: Porcupine Flat, Yosemite National Park, CA; First described by Camp (1916a)	High Sierra Nevada of CA
Taxa Not Currently Recognized		
B. pictus	No locality given; designated as Provo, UT (Schmidt, 1953); First described by Cope (1875c); later determined to be *B. boreas* (Schmidt, 1953)	
B. nestor	Fossil specimens from La Brea deposits, Los Angeles Co., CA; First described by Camp (1917a); later changed to *B. b. halophilus* (Tihen, 1962)	
B. politus	Greytown, Nicaragua; First described by Cope (1862); later determined to be *B. boreas* (Savage, 1967) and locality presumed to be in error	Location assigned to be in the Pacific coast region (Savage, 1967)

NOTE: From Schmidt, 1953; Blair, 1964, 1972; Feder, 1977; Stebbins, 1985. Distributions identified in Figure 30-1.

has also been seen in birds where many endangered taxa are actually peripheral populations when the entire range of the species is considered (Godown and Peterson, 2000).

For comparative purposes, phylogenetic diversity (PD) values were estimated from U.S. taxa alone (Figs. 30-4, 30-5). The United States was chosen as a geo-political unit for evaluation of diversity because laws and funding sources for conservation efforts frequently are geo-politically based. Bufonid taxa nationally listed as endangered include two with low PD values *(B. baxteri* and *B. houstonensis)*, one with an intermediate value *(B. californicus)*, and the taxon with the highest U.S. PD value *(P. lemur)*. On diversity criteria alone, these data provide support for a strong national conservation effort for *P. lemur*, and comparatively less support for *B. baxteri* and *B. houstonensis*.

Phylogenetic diversity values provide baseline measures with which to evaluate conservation needs of North American bufonids. For example, *B. alvarius* has a high PD value, a limited distribution, and may be a species at risk due to recent over collection. Preservation of this species before it declines would be prudent because it would conserve a high level of diversity both within the United States and globally. In contrast, giant toads *(B. marinus)* have the second highest U.S. PD value but are a common species with near worldwide distribution (due to human translocations) and have closely related taxa outside the United States. Conservation of *B. marinus* within the United States would not preserve a high level of diversity on a global scale.

Individual PD values are very low (0.05–0.74%) among the five taxa of the *B. boreas* group. However, when any one taxon within the *B. boreas* group (labeled *B. boreas* [1 taxon] in Fig. 30-5) is compared to other taxa in the United States outside of the *B. boreas* group, PD values substantially rises to the high category. These results suggest a high conservation value for the species group, but comparatively low values for any particular taxon

TABLE 30-2

(Example 2) Suggested Taxonomy of the *Bufo boreas* Species Group Based on Principles of Phylogenetic Taxonomy

Node Number (Clade Name) (Fig. 30-2, Example 1)	Current Taxon Name	Revised Linnaean (Based on potential species revisions)	Phylogenetic Taxonomy
1. (*B. boreas* species group)	No name	No name, multiple species	Clade Boreas
2. (Southern Utah)	*Bufo boreas boreas*	*Bufo sevieri**	Clade *Bufo-sevieri* [a]
3. (Southern Rocky Mountain)	*Bufo boreas boreas*	No name, multiple species	Clade Pictus
4. (Southwest)	No name	No name, multiple species	Group Nestor
5. (*B. exsul*)	*Bufo exsul*	*Bufo exsul*	Clade *Bufo-exsul*
6. (Southern)	*Bufo boreas halophilus*	No name, multiple species	Group Halophilus
7. (*B. nelsoni*)	*Bufo nelsoni*	*Bufo nelsoni*	Clade *Bufo-nelsoni*
8. (Northwest)	No name	*Bufo boreas*	Group Boreas
9. (West Coast)	*Bufo boreas halophilus* and *Bufo boreas boreas*	No name, not a species	Group Orarius
10. (Northern)	No name	No name, not a species	Group Politus
12. (*B. canorus*)	*Bufo canorus*	*Bufo canorus*	*Bufo-canorus*

[a] This clade may be a species. MtDNA data are consistent with nuclear data (Goebel, unpublished data) and with comments concerning morphological variation within the group. However, the name *Bufo sevieri* is provided for discussion purposes only. I do not provide a formal description, which is needed for recognition within Linnaean and Phylogenetic taxonomy.

NOTE: From de Queiroz and Gauthier, 1992; Cantino and de Queiroz, 2000. Clade names and node numbers are identified in Figure 30-2. Within the phylogenetic taxonomy, names of subordinate clades are indented to indicate hierarchical relationships. For each pair of sister clades, the first listed has fewer subordinate clades than the second. This phylogenetic taxonomy differs from de Queiroz and Gauthier (1992) and Cantino and de Queiroz (2000). First, "Groups" are identified by prefix in the way "Clades" are identified. The designation "clade" infers that the organisms are independently evolving units (e.g., *B. exsul*) or have a historical monophyletic evolutionary pattern (e.g., Clade Boreas). In contrast, "groups" are made up of organisms with divergent but sympatric mtDNAs. Toads in sympatry probably interbreed, although the significance of the interbreeding is not clear at this time. The phylogenetic taxonomy is similar to some suggestions in de Queiroz and Gauthier (1992) and Cantino and de Queiroz (2000) in that clades are identified and named; species names remain identical except that they are joined by a hyphen; not all possible clades are named (clades 11 and 13 are not named); and the taxonomy is presented in a hierarchical fashion. At this time, *Bufo canorus* is not presented as a clade in the phylogenetic taxonomy due to conflicting data presented here and in Shaffer et al., 2000.

within the group. A similar increase is also seen for single taxa in the Sonoran green toad *(B. retiformis)* group.

Examination of comparative levels of phylogenetic diversity can redirect conservation priorities. For example, managers in captive breeding programs in U.S. zoos want to know whether two populations of *P. lemur* should be managed as independent evolutionary units. Phylogenetic analyses identified the two populations as divergent, but more importantly, analyses identified these toads as the most divergent toads covered by the U.S. ESA. Therefore, both populations have extremely high value—not only because they are divergent from one another but primarily because both will be needed to increase the survival probability of the lineage endemic to Puerto Rico.

Ideally, cooperative global conservation strategies, in addition to national and regional priorities examined here, could be established. Phylogenetic diversity values based on taxa worldwide would vary slightly for a few taxa. For example, taxa with the three highest PD values (Gulf Coast toads [*B. nebulifer* (formerly *valliceps*)], *B. marinus,* and *P. lemur*) have sister taxa outside the United States. When these sister taxa are included in PD estimates, PD values decrease from those presented here.

Finally, much of the diversity in U.S. bufonid taxa (47–50%) is not within terminal lineages, but is shared among taxa within the deeper branches of the tree (Fig. 30-4) as may be the case for much of the tree of life (Nee and May, 1997). These data suggest that conservation programs should be coordinated among phylogenetic clusters rather than based solely on the terminal lineages of individual species.

Phylogenetic Species

If phylogenetic species concepts (Cracraft, 1983b) were applied, then many genetically divergent or geographically distinct groups that are currently considered to be subspecies or populations would be elevated to species level (e.g., McKitrick and Zink, 1988; Cracraft, 1992; Peterson and Navarro-Sigüenza, 1999). Further definitions of phylogenetic species (Cracraft, 1987, 1997; Nixon and Wheeler, 1990; Davis and Nixon, 1992; Vogler and DeSalle, 1994; Mayden and Wood, 1995), in which they are diagnosed by a unique combination of traits, would identify even finer units (ESUs).

Adoption of phylogenetic species concepts still may not preserve diversity because the diversity of species varies widely. For example, the western toad species is composed of two subspecies (boreal toads [*B. b. boreas*] and California toads [*B. b. halophilus*]). These two subspecies are not likely to be independent phylogenetic species because of the high probability of interbreeding where they are sympatric. However, recognizing a single species (e.g., western toads) would fail to recognize the unique divergence contributed by the southern populations of California toads. Alternate conservation strategies are appropriate in the regions where diversity is dramatically different. If systematic considerations are fully incorporated into recovery programs, species would neither be considered in isolation nor without information regarding such intraspecific variation.

Phylogenetic species concepts can provide a unique perspective for conservation. For example, 20 years ago, western toads

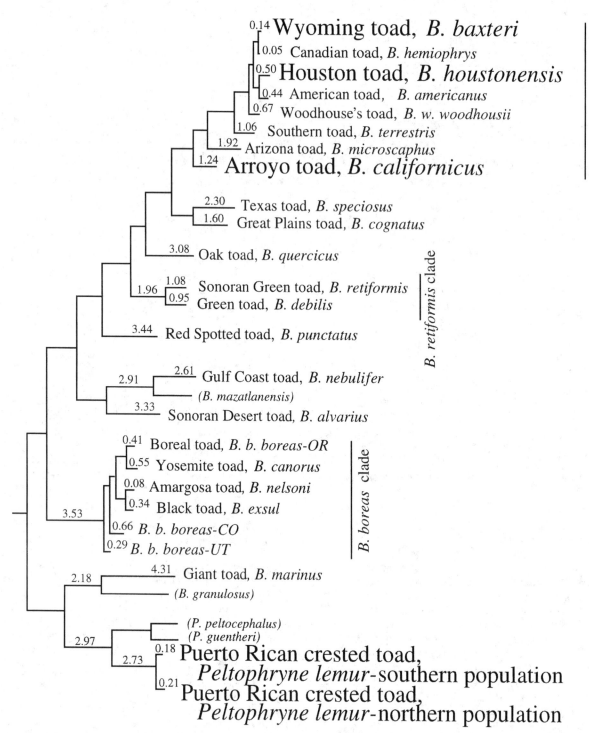

FIGURE 30-4 Phylogenetic relationships among bufonid taxa protected by the U.S. Endangered Species Act. Relationships are based on parsimony analyses (Swofford, 1993) of mitochondrial DNA gene sequences of the 12S ribosomal DNA, control region, cytochrome oxidase I, and cytochrome b regions (details including site-specific and transition/transversion weights are in Goebel, 1996a,b). Taxa listed as endangered or threatened by the ESA are in large print. Taxa that are in parentheses are not protected by the ESA but were included in the analysis because they are closely related (putative sister taxa) to taxa that are protected by the ESA. In the phylogenetic tree, branches are drawn in proportion to their lengths. Numbers on the branches are the percent of the length of a single branch based on the total length of the tree. The tree is rooted; outgroups and taxa not relevant to this analysis were excluded from the figure (details in Goebel, 1996a).

had a near continuous distribution across the high elevations in central Colorado. Recent declines (Goettl and BTRT, 1997) left three viable "populations" in Rocky Mountain National Park, Chaffee and Clear Creek counties, and only scattered individuals across the rest of the state (see Carey et al., this volume). Are these three populations now species? Classifying them as species under the Phylogenetic Species Concept would provide a better evolutionary perspective to their conservation. If the three populations are considered to be a single species, translocations among the three, as well as translocations from all three (either singly or in combination) to regions where the toads have been extirpated would

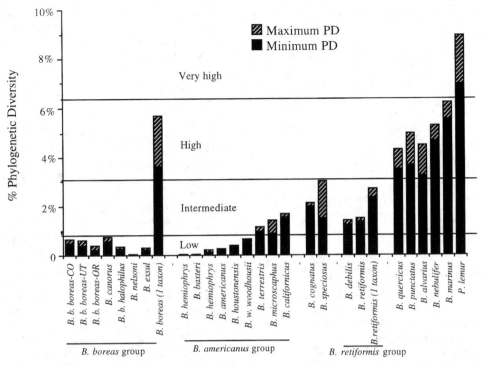

FIGURE 30-5 Comparative phylogenetic diversity values (PD) of taxa protected by the U.S. Endangered Species Act. Phylogenetic diversity values for individual taxa are branch lengths (Faith, 1992a), from the terminal to the first node shared with any other taxon also protected by the U.S. ESA. For example, the PD value for *B. nebulifer* is 5.52 (2.61 + 2.91) where a node is shared with *B. alvarius* (Fig. 30-4). Phylogenetic diversity values are presented as a range (maximum and minimum), based on multiple most-parsimonious trees and alternate weighting techniques in the phylogenetic analysis (Goebel, 1996a,b). Phylogenetic diversity values are classified as low (0–0.074), intermediate (0.88–3.04), high (3.49–6.24), and very high (above 6.24).

be considered to be reasonable conservation alternatives (see Dodd, this volume). However, while gene flow might have occurred through multiple generations along great distances, it is unlikely that gene flow occurred directly among those populations as would be proposed with translocation programs. The distinct evolutionary histories of the three populations may well have allowed unique adaptations to different environmental factors in each region (Crandall et al., 2000). Therefore, a recovery program based on the evolutionary tenets of phylogenetic species would only encourage translocation to expand each population into habitat that previously was accessible to each population. By multiple expansions of each population (until they are once again contiguous across the state) the number of animals can be increased without interfering with evolutionary processes occurring within each population. In addition, by emphasizing a program of "assisted dispersal" rather than long-distance translocations, corridor habitats would be protected, allowing natural levels of gene flow rather than by relying on continued translocations by humans.

Conservation Systematics in Management and Politics

Setting Priorities with Measures of Phylogenetic Diversity

Not all species or lineages have equal value to humans or to ecological and evolutionary processes. Rather than allowing the high rates of human-induced extinctions to proceed haphazardly, humans might be wise to make choices and set priorities

locally and globally. A variety of criteria for setting priorities for conservation have been considered, including ecosystem attributes (Scott et al., 1993), ecological or functional diversity (Williams et al., 1994), the ability to evolve (e.g., giving high priority to species rich groups; Erwin, 1991; Brooks et al., 1992; Linder, 1995), rarity (Gaston, 1994), morphology or phenotype (Owens and Bennett, 2000; but see Crozier, 1992; Williams and Humphries, 1994; Faith, 2002), allelic richness (Petit et al., 1998), and priorities for basal taxa (Stiassny, 1992; Stiassny and de Pinna, 1994). No single criterion should be considered in isolation (Pennock and Dimmick, 1997; Parker et al., 1999; Taylor and Dizon, 1999; Asquith, 2001), but incorporating phylogenetic diversity into all decisions will provide a needed perspective.

While it is critical to conserve lineages with traits of known high value (e.g., human food crops, species with valuable ecosystem functions), humans cannot predict which traits are going to have a high value in the future. However, each lineage has some value representing a unique set of traits. Setting priorities for conservation based on phylogenetic diversity is an attempt to preserve the highest degree of organismic diversity while recognizing that more closely related taxa have a higher chance of sharing traits. Methods for setting conservation priorities based on systematic principles are being developed (Krajewski, 1991, 1994; Vane-Wright et al., 1991; Faith, 1992a,b, 1993, 1994a,b, 1996; Nixon and Wheeler, 1992; Gaston and Williams, 1993; Weitzman, 1993; P. H. Williams et al., 1993; Crozier and Kusmierski, 1994; Forey et al., 1994; Vane-Wright et al., 1994; Williams and Gaston, 1994; Williams and Humphries, 1994; Humphries et al., 1995; Lande et al., 1995; Walker and Faith, 1995). Crozier (1997) reviews methods and

Bininda-Edmonds et al. (2000) suggest directions for future research.

To date, the most common use for systematic diversity measures has been to identify geographic regions that contain the greatest organismic diversity (e.g., methods of reserve design, Humphries et al., 1991; Vane-Wright et al., 1991; Williams, 1996; Moritz and Faith, 1998). The same principles can be used to set priorities for scarce conservation funds, to maximize representation of other features such as environmental diversity (Faith, 1994c, 1966; Faith and Walker, 1995), to assess environmental impacts (Clarke and Warwick, 1998; Warwick and Clarke, 1998), to provide some protection of highly divergent lineages before they become rare (Witting and Loeschcke, 1995), and to identify hierarchical management regions based on hierarchical levels of diversity rather than political boundaries alone (Examples 1 and 2).

Reconstruction of the Endangered Species Act (ESA)

The U.S. Endangered Species Act has come under increasing criticism. For example, political and economic forces have fought against the ESA because it costs too much, prevents development, infringes on private property rights, and preserves "useless" plants and animals (Sward, 1990; Rohlf, 1994). Criticisms have also come from those that perceive the ESA as not doing enough to preserve diversity, especially through ecosystem approaches (Losos, 1993; Scheuer, 1993; Murphy et al., 1994; Clegg et al., 1995; Sampson and Knopf, 1996; Yaffee et al., 1996). However, other than attempts to define "distinct populations" (Waples, 1991, 1998; Pennock and Dimmick, 1997), there is little discussion of ESA modifications to protect organismic diversity within and among a variety of hierarchical lineages.

The mandate of the ESA to preserve unique or divergent populations and subspecies, in addition to species (U.S. Forest Service, 1988), provides evidence that diversity, rather than a particular taxonomic rank (e.g., species), was the intended basis for conservation in the ESA (Waples, 1991, 1998; Gleaves et al., 1992; Pennock and Dimmick, 1997, 2000; but see also Rohlf, 1994). Problems have arisen because (1) there is no biological definition of the species category within the ESA and (2) authors of the ESA could not have foreseen the recent usefulness of molecular data in describing and quantifying diversity. In order to utilize new knowledge, measures of phylogenetic diversity should now be included in conservation legislation in the following ways:

> Priorities for conservation should include discussions of diversity. Currently, taxa are evaluated as endangered or threatened, with the highest priority given to endangered taxa. New categories that incorporate diversity should be developed. For example, taxa that are listed as threatened but also have a very high phylogenetic diversity (e.g., Puerto Rican crested toads, *P. lemur*; Example 3) could be given priority equal to taxa that are listed as endangered but have less phylogenetic diversity (e.g., Wyoming toads [*B. baxteri*] and Houston toads [*B. houstonensis*]).

> Conservation efforts should be based on hierarchical lineages, not primarily on the species category, in order to speed conservation programs and tailor them to each unique lineage. For example, newly discovered but highly divergent lineages (e.g., western toads in southern Utah, Example 1) could be given a high priority long before species descriptions

are published. In addition, clades that contain multiple endangered species could be listed as endangered rather than having to list multiple species independently (e.g., conservation for Clade Pictus [Fig. 30-3; Table 30-2] rather than conservation of *B. boreas boreas* in Colorado only).

Some priority should be given to highly divergent lineages that are not yet listed in order to ensure their survival before a crisis occurs.

Broader systematic analyses (not just species descriptions) should be encouraged for listing on the ESA. Estimates of intraspecific diversity across the range of the group would encourage the discovery of cryptic species/lineages and their inclusion in the listing process. Identification of supraspecific diversity would assist in providing priorities for the most divergent lineages. The ability to list supraspecific categories may be most valuable if extinctions are clustered within certain amphibian genera or families as they are in mammals and birds (Russell et al., 1997).

A national and international effort to identify organismic diversity among all taxa should begin with the cooperation of agencies regulating the ESA. Much like the human genome project, a database for Earth's diversity would provide both a better understanding of life itself and worldwide priorities for diversity. Such programs are beginning with the "Tree of Life" project (Maddison, 1998) and TreeBASE archive at Harvard, but these are not receiving serious levels of federal attention for funding (like the human genome project).

Defining Acceptable Loss

Not all individuals or populations need to be conserved. Knowledge of phylogenetic diversity can be used to mitigate the loss of diversity if levels of acceptable loss can be identified. While in some cases there may be no acceptable loss (e.g., preservation of rare species may dictate that all lineages and organisms be preserved in order to ensure long-term survival), many species or clades can tolerate some loss. Identification of acceptable loss based on phylogenetic diversity might assist in mitigating losses until the high rates of loss due to the extinction crisis can be stabilized.

Within molecular measures of phylogenetic diversity, there are no inherent criteria for identifying an "acceptable loss" or a value for survival. When faced with calculating a measure of acceptable loss in the absence of biological data, a procedure based on phylogenetic diversity might be used. For purposes of this example, an acceptable loss of 5% phylogenetic diversity will be used; greater loss will be seen as statistically significant and biologically unacceptable. I suggest the following procedure:

1. A clade-specific measure of acceptable loss is defined (either based on 5% phylogenetic diversity or other biological criteria).

2. The value of any clade will be maintained at 100% of the total known diversity. If additional diversity is discovered, the value of less inclusive units will be reduced proportionally but the total value will remain at 100%.

3. If diversity is lost due to extinction, diversity will be subtracted from the whole rather than resetting the total value to 100% (e.g., total known diversity will remain at 100% but total extant diversity may be less).

4. A charge will be levied for loss of diversity and collected funds applied to conservation efforts for the rest of the

clade. An exponential cost scale will be applied, such that low levels of loss (e.g., 0.001–0.10%) have a proportionately low charge but would increase exponentially with higher levels of loss.

Implementing the above produces several desirable effects. First, decisions can be made to allow loss with the least impact on diversity. Pruning short branches of the phylogenetic tree and pruning tips evenly throughout the tree would preserve the greatest dispersion of traits (Williams et al., 1991). Second, an exponential charge scale would discourage the loss of whole clades. In the western toads example, a high value would be placed on the single population comprising the southern Utah clade (8.71%; Example 1, Fig. 30-5), but much lower values would be placed on most single populations throughout the rest of the range. Third, placing a 100% value on the whole group provides an incentive to identify additional diversity. If additional diversity were found, the comparative value for individual clades throughout the tree is decreased relative to the whole. Fourth, priorities could be used by regional planning agencies to determine areas most appropriate for development. A high financial cost of development would prohibit high losses of diversity. Previously developed land without biological diversity might be less costly to develop than pristine habitat. Finally, if an environmental disaster occurred due to human actions, responsible parties could be charged in proportion to the lost diversity. The remaining diversity would automatically increase in value, reducing the risk of future loss that business or managing agencies might be willing to take. For example, in a historical context, tuataras have already lost a high level of phylogenetic diversity (Daugherty et al., 1990); the cost of reducing diversity further should be prohibitively high. Another example is the American bald eagle, which may have lost similarly high levels of phylogenetic diversity when populations declined. Under the current ESA, bald eagles have recently been down-listed due to an increase in the number of individuals. With diversity-based legislation, any agent that might cause a repeated decline would be monitored carefully. This is because a large proportion of the phylogenetic diversity within these species may have already been lost, reducing their evolutionary potential to adapt to future change.

Update

Many of the ideas presented in this chapter were developed almost a decade ago and changed through time. Most were first initiated as a graduate student in the early 1990s and were part of a dissertation (1996). I am encouraged to see some ideas closer to reality such as the implementation of phylogenetic taxonomy (Cantino and de Queiroz, 2000), large phylogenetic projects (Tree of Life [Maddison, 1989]; TreeBASE), and coordinated methods for identifying conservation priorities for geographic regions (WORLDMAP [Williams, 1996]). Other ideas, such as directly modifying the ESA to formally recognize diversity in all its forms, identifying acceptable levels of loss, or initiating financial charges for loss of diversity do not seem close to implementation. A few early ideas, such as an exchange program for "diversity credits" (paralleled after the exchange program for "pollution credits") now seem so out of favor that they are not included here. Research on biodiversity is being outlined by working groups (Bininda-Edmonds et al., 2000; Wall et al., 2001) as the final version of this chapter is submitted. Throughout the decade, and now, I feel sadness for lost lineages (*B. boreas* in Colorado may be lost in the next few years), and I hope that we can conserve more lineages than we can discover relationships for and name.

Acknowledgments. Thanks to Hobart Smith, Brian Miller, Sharon Collinge, and John Wortman for providing comments on the manuscript. Some of these ideas were first developed within a Ph. D. dissertation supervised by Richard G. Olmstead. Continued support and funding for much of the work on toads was provided by Bob Johnson (Metropolitan Toronto Zoo), P. Stephen Corn (USGS Midcontinent Ecological Sciences Center and the Aldo Leopold Wilderness Research Institute), and Thomas Ranker (University of Colorado at Boulder and the University of Colorado Museum). Hobart Smith suggested species *(B. sevieri)* and clade names (Orarius) in the phylogenetic taxonomy. As always, Hobart Smith provided personal encouragement and gummi bears without limits even while disagreeing with ideas presented. I am grateful to Dan Faith and Kevin de Queiroz, who provided manuscripts in press and references for current work.

THIRTY-ONE

Factors Limiting the Recovery of Boreal Toads *(Bufo b. boreas)*

CYNTHIA CAREY, PAUL STEPHEN CORN, MARK S. JONES,
LAUREN J. LIVO, ERIN MUTHS, AND CHARLES W. LOEFFLER

Boreal toads *(Bufo b. boreas)* are widely distributed over much of the mountainous western United States. Populations in the Southern Rocky Mountains suffered extensive declines in the late 1970s through early 1980s (Carey, 1993). At the time, these mass mortalities were thought to be associated with a bacterial infection (Carey, 1993). Although the few populations that survived the mass die-offs were not systematically monitored until at least 1993, no mass mortalities had been observed until 1996 when die-offs were observed. A mycotic skin infection associated with a chytrid fungus is now causing mortality of toads in at least two of the populations (M.S. Jones and D.E. Green, unpublished data; Muths et al., 2003). Boreal toads are now absent throughout large areas of their former distribution in Colorado and southern Wyoming and may be extinct in New Mexico (Corn et al., 1989; Carey, 1993; Stuart and Painter, 1994). These toads are classified as "endangered" by Colorado and New Mexico and are designated as a protected non-game species in Wyoming. The U.S. Fish and Wildlife Service has categorized the Southern Rocky Mountain populations for federal listing and is currently reviewing their designation as a "warranted but precluded" species for possible listing in the next few years. For the management of boreal toads and their habitats, a Boreal Toad Recovery Team was formed by the Colorado Division of Wildlife in 1995 as part of a collaborative effort with federal agencies within the United States' departments of the Interior and Agriculture and with agencies in two adjoining states. To date, conservation agreements have been signed by eight state and federal agencies and by the Colorado Natural Heritage Program.

Although boreal toads were considered common throughout their range in Colorado, no comprehensive surveys of the numbers and sizes of their populations were conducted prior to mass die-offs in the 1970s. Surveys completed in the late 1980s to early 1990s, however, indicated that the toads were still present in only one of 377 historical sites in parts of western Colorado (Hammerson, 1992) and in about 17% of previously known sites in the Colorado Front Range (Corn et al., 1989). Once common in Rocky Mountain National Park, boreal toads are now found at only seven localities, with just two or three of these populations likely to be reproducing successfully each year (Corn et al., 1997). Intensive surveys by the Boreal Toad Recovery Team since 1995 have found about 50 breeding sites

comprising 25 distinct populations within Colorado (Loeffler, unpublished data). Of these populations, most are small (fewer than five egg clutches laid per year) and may not survive over the long term.

Efforts to restore population sizes and expand the geographical distribution of boreal toads in the southern Rocky Mountains have involved considerable person-hours and financial commitments. Special care has been taken to protect habitats and, when feasible, to improve sites where breeding populations currently exist. However, initial attempts to repatriate these toads in historic habitats in which boreal toads were present before 1975 have generally proven unsuccessful (Carey, unpublished data; Muths et al., 2003). It is too early to determine if recent repatriations will establish breeding populations (Scherff-Norris, 1999), but these efforts will likely continue. Despite the best human intentions and efforts, the recovery of former population sizes and the historical distribution of boreal toads will greatly depend on its own life history characteristics. However, as we will review in this paper, environmental factors affect many life history attributes in a manner that poses serious obstacles for recovery.

Geographical Locations and Time Periods of Observations

Data presented in this study were gathered at several localities in Colorado (Fig. 31-1), both before and after the mass mortalities of the 1970s–80s. Boreal toads were studied prior to their population crash, which started in 1973 or 1974 in the East River Valley of the West Elk Mountains, near Crested Butte, Colorado (106° 59' W, 38° 58' N; Carey, 1993). A map of the breeding sites, located at altitudes between 2990 and 3550 m within the East River Valley, has been published in Carey (1993). Data collected at these sites are designated as "West Elk Mountains."

Two sites in the North Fork of the Big Thompson River within Rocky Mountain National Park have been monitored intensively since 1991 (Corn et al., 1997; Corn, 1998). Kettle Tarn (40° 30' N, 105° 31' W, 2810 m) is a shallow glacial kettle pond. Lost Lake (3,266 m) is a drainage lake. Data from these sites are identified as "RMNP."

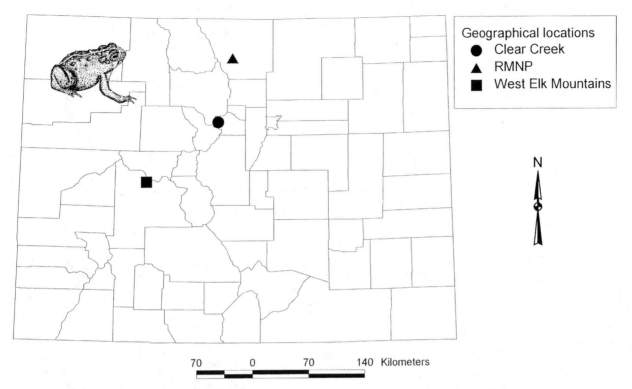

FIGURE 31-1 Colorado sites at which boreal toad life history characteristics reported in this study were observed. Observations were conducted at West Elk Mountains (1971–76), Rocky Mountain National Park (RMNP; 1991–present), and Clear Creek (1995–present).

Several sites in the vicinity of Woods Mountain (39° 43′ N, 105° 50′ W), Clear Creek County, have been monitored since 1995. The most intensively studied sites are in the Urad Valley (Woods Creek drainage); including sites informally known as Hesbo, Lower Urad, Treatment, Donut, Anne's Pond, and Upper Urad. In West Fork, the adjacent drainage, we collected information from sites informally known as Power Alley, JSP, and 1 Pond. The Herman Gulch and Mt. Bethel sites are located in the Clear Creek drainage. All locations are at elevations between 3,048 and 3,170 m. Data from these sites are designated as "Clear Creek."

Environmental Characteristics

Boreal toad populations in the Southern Rocky Mountains historically were found between approximately 2,500 and 3,600 m (Campbell, 1970c,d; Livo and Yackley, 1997). Breeding sites range from large lakes, beaver ponds, and glacial kettle ponds to temporary pools formed by snowmelt in depressions such as tire ruts. The high montane environment where the toads live is dominated by cold temperatures; snow generally covers the ground for six to eight months. In response, boreal toads hibernate in locations such as rodent burrows, cavities in dams of beaver ponds, and under the overhanging edges of stream banks (Campbell, 1970c; Jones and Goettl, 1998). Emergence is governed by snowmelt, which can occur anytime between mid-May through late June. During the short summer growing season, boreal toads must accumulate sufficient energy to sustain basic metabolism, growth, and reproduction. At this time, the daily body temperatures of boreal toads fluctuate from near freezing to around 30 °C, depending on cloud conditions and behavioral

tendencies (Carey, 1978). High body temperatures, which foster high rates of passage and absorption of food through the digestive tract, circulation of blood, immune function, and overall metabolism (Carey, 1976, 1979; Maniero and Carey, 1997), are achieved by basking in direct sunlight. However, frequent cloud cover during the day reduces body temperatures to 10–20 °C, and body temperatures fall to near freezing during summer nights (Carey, 1978). In addition to support of basic metabolic and reproductive functions, boreal toads must also store energy in the form of fat and glycogen, some of which is used to survive the subsequent winter. Toads enter hibernation sometime between late August and early October, depending upon weather conditions. Therefore, the combination of short growing seasons and cold summer temperatures lowers cumulative metabolic rate and restricts the rate of accumulation of energy by the toads.

We believe these environmental factors are major determinants of the life history characteristics of boreal toads. These factors differ from those encountered by lowland *Bufo* populations, and they likely limit the ability of Southern Rocky Mountain populations to recover from mass mortalities.

Life History Attributes

Timing of Breeding

At all three locations (Fig. 31-1), breeding has almost always occurred in the spring or early summer when the snow melted from most, if not all, of the body of water. However, in one unusual case, a clutch was laid in August (Fetkavitch and Livo, 1998). Because snow usually melts earlier at lower altitudes, breeding commences earlier at lower locations. We observed

TABLE 31-1

Clutch Sizes and Sizes of *Bufo boreas* Females at Three Localities in Colorado

Locality	Mean + SE Clutch Size	n	Range	Female Snout-Vent Length (mm)
West Elk Mountains	5897 ± 363	21	3,239–8,663	72.5–87.2
RMNP	8155 ± 954	4	6,655–10,872	83.0–84.0
Clear Creek	7440 ± 435	13	4887–9333	71.7–83.95
Overall mean	6661 ± 294	38		

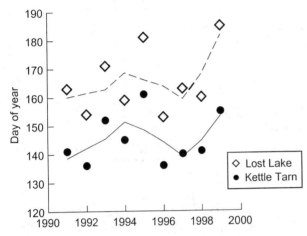

FIGURE 31-2 Dates of peak breeding activity by boreal toads at two locations in Rocky Mountain National Park, Colorado. Trend lines were drawn using LOWESS regression (Cleveland, 1979).

considerable variation in the timing of breeding by toads at two sites along the North Fork of the Big Thompson River in Rocky Mountain National Park (Fig. 31-2). From 1991–99, the date of peak breeding activity at Kettle Tarn averaged 24 May, but varied from 15 May 1992 and 1996 to 10 June 1995. At Lost Lake, breeding peaked, on average, around 13 June. The onset of breeding at Lost Lake varied from 1 June 1996, to 4 July 1996, to 10 June 1995 (Fig. 31-2).

Between 1971–75 in the West Elk Mountains, the earliest date on which egg laying began was May 25. Frequently, no breeding occurred at the highest elevations (3,550 m) until the end of June. Breeding was initiated at lower elevation sites in Clear Creek County (such as Hesbo and Herman Gulch) from mid- to late May and at higher elevations from late May to early June.

Reproductive Size

The minimum snout-vent length recorded for a breeding female was 71.7 at the Urad Valley site of the Clear Creek population and 72.5 mm at the West Elk Mountains populations. In the Clear Creek population, the mean snout-vent length of breeding females and males was 77.6 ± 1.1 and 67.0 ± 1.1 mm, respectively, but a 55-mm male was caught in amplexus in the Clear Creek population (Livo, 1999). In the West Elk

Mountains, mass and snout-vent length of breeding females captured immediately after egg laying averaged 46.1 ± 1.4 g (range 37.9–52.7 g) and 78.6 ± 3.3 mm (range 72.5–87.2 mm), respectively.

Clutch Size

Three different methods were used to estimate clutch size. To investigate clutch size in the West Elk Mountains during the 1973–74 breeding season, males and females in amplexus were located just prior to egg laying. A thread was tied around the hind foot of each female to prevent her from leaving the breeding site before her snout-vent length and body mass could be determined. Snout-vent length and body mass were measured with vernier calipers and an Ohaus balance. After egg laying was completed, eggs were gathered in a plastic container and counted. Because this method disrupted the protective properties of the egg jelly, egg masses treated in this manner usually developed fungal infections and few larvae hatched. Therefore, although this procedure provides an accurate count, we do not recommended it for future studies of clutch size, especially on endangered or threatened species.

Clutch sizes at RMNP were recorded *in situ* by gently pulling the egg string through the observer's hands and counting the eggs. The jelly was largely undisturbed by this process and hatching success appeared normal.

Prior to egg laying, pairs of boreal toads found in amplexus during the 1999 breeding season in the Urad Valley of Clear Creek County were placed in a 30 × 44 cm plastic container. The eggs of each clutch were arranged in a single layer and photographed. (For details on the method used to estimate egg number from the photographs see Livo [1999].) Diameters of 20 eggs from two positions in each egg mass were measured using a dissecting microscope with an ocular micrometer. All toads and resulting eggs were released in the ponds at which the toads were captured. Using this procedure, no unusual fungal infections were noted and hatching rates of these clutches approached those of undisturbed clutches.

An overall mean clutch size (number of eggs per clutch) calculated for the three boreal toad populations was 6,661 ± 294 eggs (Table 31-1). The number of eggs in 21 clutches laid between 1973–74 in the West Elk Mountains was significantly lower (ANOVA, F = 5.46, p = 0.009) than the average clutch sizes in Clear Creek and RMNP (Table 31-2).

TABLE 31-2

Known Sources of Complete Boreal Toad *(Bufo boreas)* Embryo and Tadpole Mortality at Seven Breeding Sites in Clear Creek County in 1999

	Hesbo	Donut	Lower Urad	Power Alley	Anne's Pond	Mt. Bethel	Herman Gulch	Total
Freezing	4							4
Desiccation				1	21		6[a]	28
Sterility	2							2
Water quality							4[a]	4
Fungus		3						3
Temperatures			2					2
Unknown						1		1
Total egg masses	23	17	2	1	21	2	10	76

[a] Egg masses in pools were moved to more permanent pools to prevent desiccation; poor water quality was probably the cause of mortality of eggs found covered with fungus.

NOTE: Numbers are egg masses.

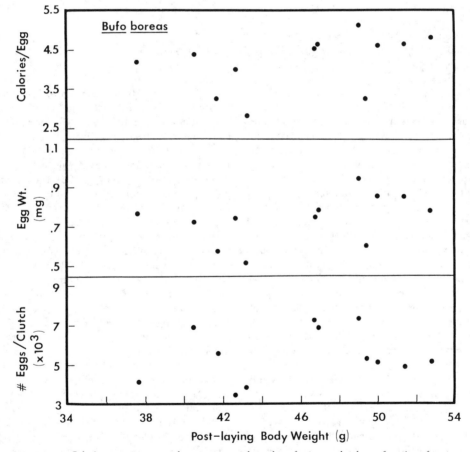

FIGURE 31-3 Caloric content per egg, dry egg mass, and number of eggs per clutch as a function of post-reproductive body mass of females in the West Elk Mountains. Each female provided one point for each variable.

The mean initial mass of 70 ± 3 g of 12 females in the Clear Creek population decreased an average of 17 ± 1 g following egg laying, representing an average loss of 24 ± 5 percent of the initial body mass. A significant positive relationship existed between the initial mass of 12 females and the number of eggs they produced ($r = 0.851$, df = 10, $p < 0.01$).

Within individual populations, such as the West Elk population, no significant relation existed between clutch size and either female post-breeding body mass ($p = 0.22$; Fig. 31-3) or

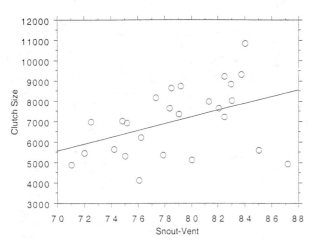

FIGURE 31-4 Clutch size of individual females as a function of snout-vent length (mm). Data were combined from populations in the West Elk Mountains, RMNP, and Clear Creek. The least-square regression equation best describing the relation between clutch size and snout-vent length (SVL) is "Clutch Size = −6046.2 + 166.2 Snout-vent Length;" it is shown by the thin line.

snout-vent length (p = 0.70). However, when data were combined for the three populations, a significant (r = 0.43, df = 25, p = 0.03) relation existed between clutch size and female snout-vent length (Fig. 31-4).

Egg Characteristics

Ten eggs from each of 18 clutches gathered from the West Elk Range were freeze-dried for 4 days until they reached constant mass. These were then weighed on a Mettler balance, accurate to 0.001 g, to determine dry mass. The values from the ten eggs were averaged for an estimate of the dry egg mass for a given clutch. Dry eggs from each clutch were combined and homogenized into pellets for analysis of caloric content. The caloric contents of three pellets per clutch were analyzed with a Phillips microbomb calorimeter, calibrated with benzoic acid. The average of the results from the three pellets was used as the estimate of caloric content per g egg. The caloric content per clutch was obtained by multiplying the clutch size by the dry egg mass to obtain dry clutch mass, and then multiplying that value by the caloric content per g egg mass. (Average values will be presented throughout this paper as mean ± SE. Least squares regression lines were calculated to define the relation of various types of fecundity data to body mass or snout-vent length.)

Mean egg diameter at the Urad site in Clear Creek County was 1.6 ± 0.07 mm. A negative correlation between egg diameter and number of eggs per clutch existed at that location (r = −0.524, df = 11, p < 0.05). Total volume for an egg mass can be estimated as $(3.14 \times n \times d^3) \times 6^{-1}$, where n is the egg number and d is the mean egg diameter in mm. A significant, positive relationship existed between total egg mass volume and female snout-vent length (r = 0.604, df = 11, p < 0.05). Mean egg diameter and female snout-vent length were not significantly related (r = −0.505, df = 11, p > 0.05).

Dry mass of eggs laid in the West Elk Mountains averaged 0.74 ± 0.02 mg (range 0.51–0.94 mg). The dry mass of a clutch averaged 4.40 ± 0.21 g (range 1.96–6.94 g). Mean caloric content per egg and per clutch were 4.18 ± 0.03 cal (range 2.80–5.12) and 24,786 ± 1,818 cal (range 10,782–43,108),

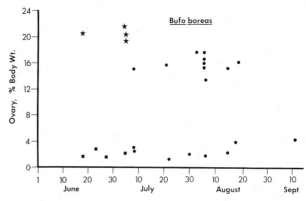

FIGURE 31-5 Ovarian masses of female boreal toads (% body weight) captured between 1972–74 in the West Elk Mountains. Ovary masses of females captured in amplexus before ovulation and those that just finished laying eggs are represented by stars and squares, respectively. Ovarian masses of those that were captured after breeding are shown by circles. Each symbol represents the value for a single female.

respectively. Calories per egg, dry egg mass, or calories per clutch were not significantly related to either female post-breeding body mass (Fig. 31-3) or to snout-vent length.

Frequency of Breeding

Several lines of evidence indicate that boreal toad females rarely, if ever, breed two years in a row. First, morphological evidence indicates that the ovaries of some females of breeding size contain only immature, pre-vitellogenic oocytes during a given summer, and that these females could not lay eggs the subsequent summer. In the West Elk Mountains, ovarian masses of females in amplexus prior to ovulation ranged between 19–22% body mass (Fig. 31-5). These ovaries were dominated by large, highly pigmented oocytes. Ovaries collected from females just after laying a clutch averaged 1–3% of body mass and contained non-pigmented, immature oocytes, and atretic oocytes (those that began vitellogenesis, but then stopped development and were not ovulated; Fig. 31-5). Ovaries of females collected after the breeding season fell into two categories: (1) ovaries weighing between 1–3% of body mass and containing non-pigmented oocytes and atretic follicles and (2) ovaries weighing between 13–17% of body mass and containing pigmented oocytes that were undergoing vitellogenesis. These data suggest that the ovaries of females collected in July and August and weighing 1–3% of body mass are in a "resting" condition, in which no oocytes are being prepared for breeding the following year. Ovaries of other females are enlarged with oocytes undergoing vitellogenesis in preparation for laying the following summer.

Mark–recapture studies provide a second line of evidence supporting the idea that female boreal toads do not usually breed every year. Adult toads inhabiting the North Fork of the Big Thompson River in RMNP have been marked using PIT tags since 1991 (Corn et al., 1997). Far more males than females are actually captured. In the North Fork (Kettle Tarn, Lost Lake, and other localities) from 1991–98, 874 males (691 recaptures) and 246 females (31 recaptures) were marked, but the mean estimated population sizes (412 ± 17.5 males and 591 ± 314.4 females) were not significantly different. Females had lower capture probabilities (0.098 ± 0.054, hence the poor precision of the estimate) than did males (0.49 ± 0.02), and

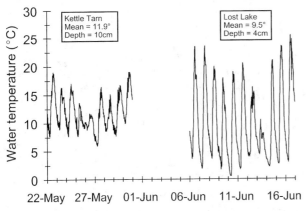

FIGURE 31-6 1994 water temperatures at two egg deposition sites in RMNP. Temperatures were recorded at 10–12 min intervals using thermometers and single-channel data loggers (Corn, 1998).

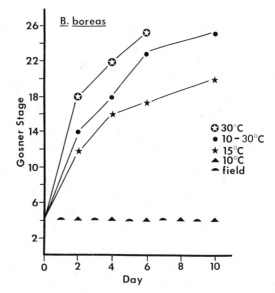

FIGURE 31-7 Effect of temperature on median Gosner (1960) stage of boreal toad embryos held at a variety of temperatures (in the field under a natural thermal regime which involved two snowstorms and eventual freezing of the pond); constant temperatures of 10 °C, 15 °C, 30 °C and a 12-hour alternating cycle of 10 °C to 30 °C for 10 days.

most recaptures of females occurred after breeding. From 1991–98, only 6 of the 31 recaptures were of female toads marked during one breeding season and recaptured in a later breeding season.

Finally, the ratio of effective to total population size supports the idea that boreal toad females rarely, if ever, breed two years in a row. The number of egg masses averages 12% of the estimated number of males at Kettle Tarn and Lost Lake. Given that the sex ratio for a population is probably close to 1:1, the low number of females breeding in a given year indicates that most do not breed in consecutive years.

Thermal Influences on Egg and Tadpole Development

Boreal toads typically lay eggs on the north, northwest, or northeastern shores of large ponds or lakes—areas that receive the most daily sunlight. In most cases, egg laying occurs in shallow water (< 10 cm) on gently sloping pond edges. Solar radiation warms shallow water to temperatures as high as 30 °C during the day, but water temperatures often cool to near freezing at night. Therefore, eggs of boreal toads are exposed to considerable fluctuations in daily temperatures (Fig. 31-6).

The importance of egg laying in sunlit, shallow water is highlighted by the fact that tadpole development does not occur at constant, cold temperatures and requires high water temperatures for at least a portion of the day. As an experiment, we removed four groups of about 50 eggs each from a clutch of eggs laid on 2 June 1973 in the West Elk Mountains. Remaining eggs were left in the field to develop. It snowed on day 2 and day 5, and the pond froze on day 7. The eggs left in these conditions did not show any development past Gosner stage 4 (Gosner, 1960), and all these eggs died when the pond froze (Fig. 31-7). The four groups that were taken to a laboratory at the Rocky Mountain Biological Laboratory were acclimated to one of four temperatures (10 °C, 15 °C, or 30 °C) and a cyclical regime of 12 hours at 10 °C and 12 hours at 30 °C. Embryos showed no development past stage 4 at a constant temperature of 10 °C (Fig. 31-7). Development was most rapid at 30 °C and was slowest at 15 °C. Development of embryos/larvae exposed to the fluctuating thermal regime (which most closely represented typical daily changes in temperature) was intermediate between the rate at 30 °C and 15 °C. Therefore, exposure to 10 °C was not harmful to the embryos as long as they experienced warmer temperatures

for a portion of the day, but it did retard development to some degree. These data indicate that cooler water temperatures, like those associated with cloud cover and/or late season snowstorms, can delay development of larvae to metamorphosis. Therefore, failure to undergo metamorphosis before freezing temperatures occur in the fall can result not only from heavy spring snowpacks that delay egg laying, but also from unusually cold summer temperatures that retard egg and embryo development.

In Clear Creek County, the Lower Urad site is across a dirt road from the Hesbo site. Although the two sites are located at the same elevation and experience the same atmospheric conditions, thermal regimes differ substantially. The Lower Urad site is a cold, stream-fed reservoir; the Hesbo site is a temporary pool that warms considerably during the summer. In 1999, females deposited two egg masses at the Lower Urad site during the same time that breeding was underway at the Hesbo site. Metamorphosis at the Hesbo site began on August 13, 81 days after the first eggs had been deposited. In contrast, no tadpoles developed to metamorphosis at the colder Lower Urad site (Table 31-3).

Breeding Success

Boreal toads suffer high mortality at both the egg and larval stages. Desiccation of egg masses appears to be the single largest source of egg mortality. For instance, 15 clutches laid in 1973 at the top of Schofield Pass in the West Elk Mountains were destroyed when a June drought caused the pond to dry. In 44 out of 76 clutches that suffered complete mortality at the Clear Creek County sites, pond desiccation caused over half of the cases (Table 31-2). Corn (1998) reported a loss of 5 of 35 egg masses laid at Lost Lake in 1994 due to dropping water levels.

Egg masses can also be lost when water levels rise rapidly. In 1995 at Kettle Tarn, an abnormally high amount of snow fell during May and delayed breeding. However, some pairs bred

TABLE 31-3

Dates of Egg Deposition and Onset of Metamorphosis of Boreal Toads at Clear Creek County
Breeding Sites

Site	Year	Egg Deposition Date	Date of Metamorphosis	Approximate Days Elapsed
Hesbo	1999	24 May	13 August	81
Hesbo	1998	20 May	5 August	77
Donut	1999	3 June	20 August	78
Donut	1998	2 June	12 August	71
Donut	1997	4 June	26 August	83
Mt. Bethel	1998	2 June	20 August	79
Mt. Bethel	1997	2 June	2 September	92
Mt. Bethel	1996	29 May	21 August	84
Herman Gulch	1998	16 May	6 August	82
Herman Gulch	1997	22 May	18 August	88
Anne's Pond	1998	2 June	5 August	64
Anne's Pond	1997	4 June	26 August	83

early; the resulting massive snowmelt raised the level of the pond by at least 50 cm and, subsequently, no trace of the early egg masses was observed. Other factors, such as gamete sterility, water quality, and fungal infections also contribute to high egg mortality at boreal toad breeding sites (Table 31-2).

Some mortality of boreal toad larvae occurs because they are unable to metamorphose before their breeding pond freezes in the fall (Fetkavitch and Livo, 1998). If egg laying is delayed by a heavy winter snowpack, or if larval development is retarded by unusually cold summer days, larvae may not have enough time to complete metamorphosis. In general, at lower elevations in RMNP, variation in breeding dates does not affect the eventual metamorphosis by tadpoles. Breeding occurs an average of 20 days earlier at the lower elevation Kettle Tarn site in RMNP than the higher Lost Lake locality. Additionally, water temperatures are warmer and tadpoles develop faster at the lower altitude site (Corn and Muths, unpublished data). Breeding was delayed by heavy spring snowpack at Lost Lake in 1995 and 1999, and no tadpoles were observed to have metamorphosed in either year. In most years at Lost Lake, at least some tadpoles metamorphose; however, almost every year at this site we observe small tadpoles in September and October that would be unable to reach metamorphosis prior to freeze-up. Many of the shallower pools at this site held tadpoles and metamorphic animals killed by freezing or drying.

Predation is another major source of mortality of boreal toad larvae. Major predators of boreal toad tadpoles and toadlets include several species of birds (Campbell, 1970c; Beiswenger, 1981; Jones et al., 1999; Livo, 1999). Spotted sandpipers (Actitis macularia) consumed numerous metamorphosing toadlets in Clear Creek County and virtually eliminated all the toadlets at one site (Jones et al., 1999). Mallard ducks (Anas platyrhynchos) consumed all the tadpoles at several sites in the Denny Creek drainage of Chaffee County (C. Fetkavitch, unpublished data). Other predators include terrestrial garter snakes (Thamnophis elegans), tiger salamanders (Ambystoma tigrinum), and predaceous diving beetle larvae (Dytiscus sp.; Reese, 1969; Campbell, 1970c; Livo, 1999).

Metamorphosis

In the Clear Creek County sites, metamorphosis begins within 64–92 days after egg laying (Table 31-3). Typical body masses and snout-vent lengths of newly metamorphosed boreal toads at these sites were about 0.30–0.35 g and 15 mm, respectively. Comparable values for toadlets in the West Elk Mountains averaged 0.25 g and 13.1 mm, respectively. These differences probably represent normal variability in boreal toad metamorphosis because considerable site-to-site and year-to-year variation in size at metamorphosis exists within Clear Creek populations (Fig. 31-8). When toadlet sizes at two Clear Creek sites were compared between years, the production of smaller toadlets occurred in the comparatively hotter, drier summer. No significant relation ($r = 0.58$, df = 9, $p < 0.05$) existed between toadlet mass at metamorphosis and the minimum length of the larval period at five sites in the Clear Creek area.

Some variation in size at metamorphosis is due to genetic differences in growth rates, even among siblings. About 200 eggs from a single clutch laid at one of the Clear Creek sites were raised in the laboratory at 25 °C with food for larvae provided ad libitum. An opportunity for raising body temperature above 25 °C was provided in one corner of the cage by a heat lamp. Body masses of the newly metamorphosed toadlets varied from 0.42–1.16 g.

Due to colder mean temperatures, tadpoles at higher elevations frequently have difficulty completing development before winter. Although we do not have quantitative data addressing this issue, we believe that survival through metamorphosis is only one of a series of hurdles that are encountered by young boreal toads. The fact that many more toadlets are usually observed soon after metamorphosis in the late summer or fall than are found the following spring suggests that considerable mortality occurs during their first winter.

Growth of Metamorphosed Individuals

Temperature is the primary determinant of growth rate in metamorphosed boreal toads. Forty-five sub-adult boreal toads

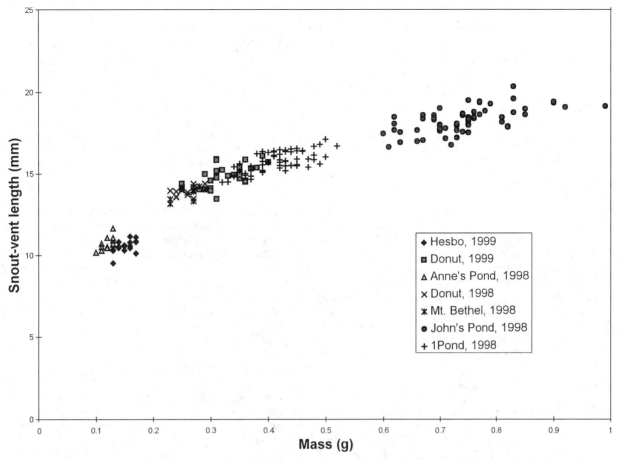

FIGURE 31-8 Variation in mass of newly metamorphosed boreal toad toadlets at various sites in Clear Creek County.

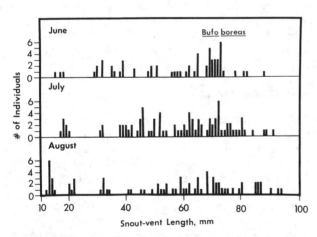

FIGURE 31-9 Number of boreal toads of various snout-vent lengths that were captured from June through August, 1971–74. These individuals were captured at the Trail 401 breeding area near the Rocky Mountain Biological Laboratory in the West Elk Mountains (for a map, see Carey, 1993).

regime most closely approximates the change in body temperatures that would be experienced by boreal toads in their montane habitat during the summer (Carey, 1978). For the 4-week duration of the experiment, food (*Tenebrio* larvae) and access to water were provided *ad libitum*. Although the toads held at 7 °C consumed food, they lost an average of 3% body mass, and snout-vent length did not increase during the 4-week trial. Body mass and snout-vent length of the toads held on the cyclical 7–25 °C regime increased an average of 29% and 4%, respectively, over the values at the beginning of the experiment. Toads held at a constant 25 °C for the four-week period gained an average of 40% of initial body mass. Snout-vent lengths of these animals increased an average of 7%. Therefore, growth of metamorphosed boreal toads does not occur at the cold temperatures that prevail at night during the summer in their montane habitat. The high body temperatures achieved during daytime basking contribute importantly to the rate of growth of these toads in the field. Growth rates of the toads exposed to the cyclical regime in this experiment are more rapid than those observed in the field (see below) because cloud cover precludes basking on some summer days and the amount of food supplied in this experiment likely exceeded that available in the field.

A continuum of post-metamorphic sizes exists in boreal toad populations that produce some metamorphic young each year or nearly each year (Fig. 31-9). A total of 102 metamorphosed toads from the Trail 401 population in the West Elk Mountains were toe-clipped over three summers between 1971–73. Of that number, 25 marked individuals were recaptured at approximately the

collected in the West Elk Mountains in 1972 were divided randomly into three groups. Initial mass and snout-vent lengths of all 45 toads averaged 10.82 ± 0.4 g and 47.71 ± 1.4 mm, respectively. Fifteen toads were held at 7 °C, another 15 at 25 °C, and a third 15 were exposed to temperatures that cycled between 7 °C and 25 °C every 12 hours. The cyclical thermal

TABLE 31-4

Average (± SE) Increase in Snout-vent Length (mm) Per Year of Various Sizes of *B. boreas* from the West Elk Mountains

Snout-vent length (mm) at first capture:

10–20 mm	30–40 mm	40–50 mm	50–60 mm	60–70 mm	70+ mm
8.0 (1)	8.0 ± 1.1 (5)	6.86 ± 1.5 (7)	7.38 ± 0.9 (8)	5.67 ± 0.9 (3)	0.17 + 0.4 (5)

NOTE: A value for a particular animal was placed in the size category in which snout-vent length fell the year the animal was first captured. No data were available for the 20–30 mm size range. Sexes of individuals could not be determined until they were about 50 mm. All individuals within the 70 + category were females. Sample size is given in parentheses.

same time each summer at least one year after they were marked. Snout-vent lengths of these individuals were measured (Table 31-4). As judged by a one-way ANOVA, the average annual increase in length does not significantly change until the toads reach about 70 mm, when growth nearly ceases. Based on our values (Table 31-4), it would take an average of 4–5 years for a male, and about 6–7 years for a female, to reach the minimum breeding size. The differences in growth rates among siblings described above, however, suggest that some individuals may reach sexual maturity more rapidly than most of their age cohort.

Movements of Metamorphosed Individuals

Because newly metamorphosed and juvenile toads cannot be marked permanently and individually marked, no information exists concerning the movements of these toads in the years before they breed. When individuals in these age groups have been repatriated into historical habitats in which no known breeding populations have existed for a number of years, they have gradually disappeared from the site. We do not know whether they have died or dispersed to other areas, or if they will ever return to breed.

Because adult toads that were PIT-tagged at Lost Lake in RMNP have been caught a year or two later 8 km downstream at Kettle Tarn, we know that some movement of older toads occurs. However, data from the Clear Creek sites indicate that most breeding adults tend to remain (at least for several summers) in the same valley in which they participated in breeding. At these sites, 26 toads (9 males and 17 females) were radio-tagged between May and June 1998, with Holohil BD-2G radio transmitters that weighed 2 g and had a projected battery life of 6 months. A harness composed of fishing leader in vinyl tubing was used to fasten a radio around the waist of each toad. Toads were monitored once per week from May until hibernation. Locations were recorded in Universal Transverse Mercator (UTM) coordinates using a Trimble Pathfinder Basic Plus global positions system (GPS). Locations were imported into the ARC/INFO program for analysis.

Movements after breeding depended in part upon the type of habitats surrounding the breeding pond. If wetlands existed next to the pond, toads remained closer to the breeding site; if no wetlands existed, toads tended to move uphill to relatively moist areas that provided protection in the form of rodent burrows, overhanging shrubs, and other habitat features.

Movements varied considerably among individuals (Figs. 31-10, 31-11). Females left breeding sites immediately after breeding and, compared to males, they moved longer distances and more quickly from the breeding site. Females also had a greater tendency to move upland than males. After breeding

terminated, males tended to move a greater distance per day (14.1 m per day) than females (6.3 m per day), but the averages do not differ statistically due to the effect of large individual variability. The maximum (summed between location points) distances traveled by radio-tracked toads during the summer of 1999 was 5,756 m in 70 days (82.2 m/d) by one female, and 6,485 m in 106 days (61.2 m/d) for another female. None of the adults monitored with radio telemetry left the valley in which breeding occurred (Jones, 1999).

Interpopulation and Interspecific Comparisons

The three investigations that contributed data to this paper approached the study of boreal toad life histories from different directions. Clutch size was the only common variable that was examined in all three studies. Therefore, it is difficult to determine whether the RMNP and Clear Creek populations possessed variations in life history attributes that may have contributed to their ability to survive the mass die-offs that occurred throughout most of the Colorado Rockies during the 1970s. Mean clutch size of the West Elk Mountain populations was significantly lower than those found in RMNP and Clear Creek. The meaning of this difference cannot be determined at this time.

Annual Period of Activity

The Southern Rocky Mountain environment in which boreal toads breed is characterized by short, snow-free growing seasons that last from approximately mid-May through late September. The near freezing temperatures during summer nights restrict prey activity and the rate of digestion and absorption. In contrast, in the lowland temperate habitats experienced by many other *Bufo*, moderate temperatures foster nocturnal feeding and digestion (Carey, 1976). Therefore, the amount of time in which boreal toad females can accumulate the energy and nutrients necessary for egg production is relatively limited. Lowland Great Plains toads (*B. cognatus*) breed as early as March (Bragg, 1937b, 1940a; Krupa, 1994), and red-spotted toads (*B. punctatus*) breed in late April through early May in Kansas (Smith, 1934). Neither of these species enters hibernation until late October, at the earliest. Low altitude, high latitude European toads (*B. bufo*) are active in Denmark from April through early November (Jørgensen et al., 1979).

The seasonality and thermal environment of boreal toads in the southern Rockies is most similar to that of Yosemite toads (*B. canorus*) in the Sierra Nevada of California. Yosemite toads are primarily found between 2,500–3,000 m, with an upper altitudinal

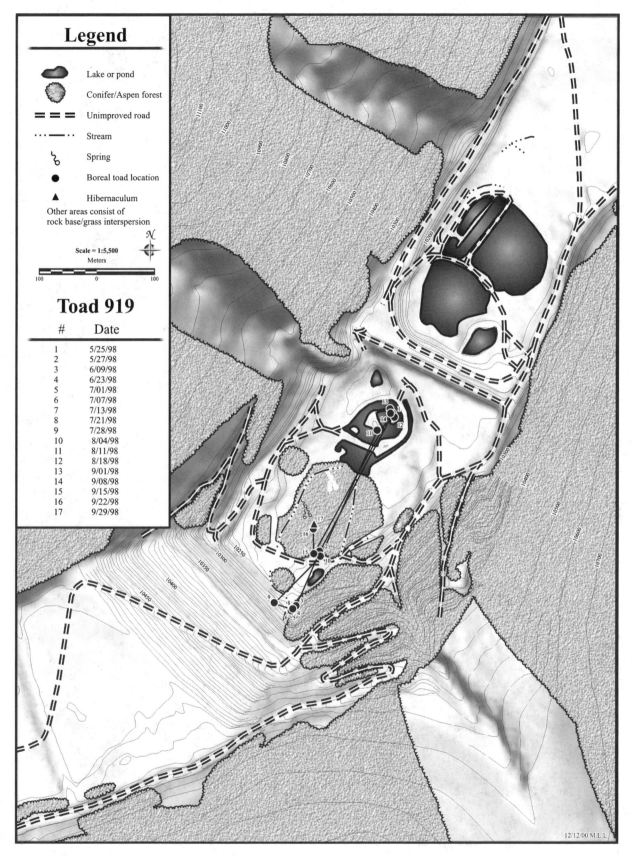

FIGURE 31-10 1998 movements, measured by radio telemetry, of Toad 919 (adult male) at the Henderson/Urad study area in Clear Creek County.

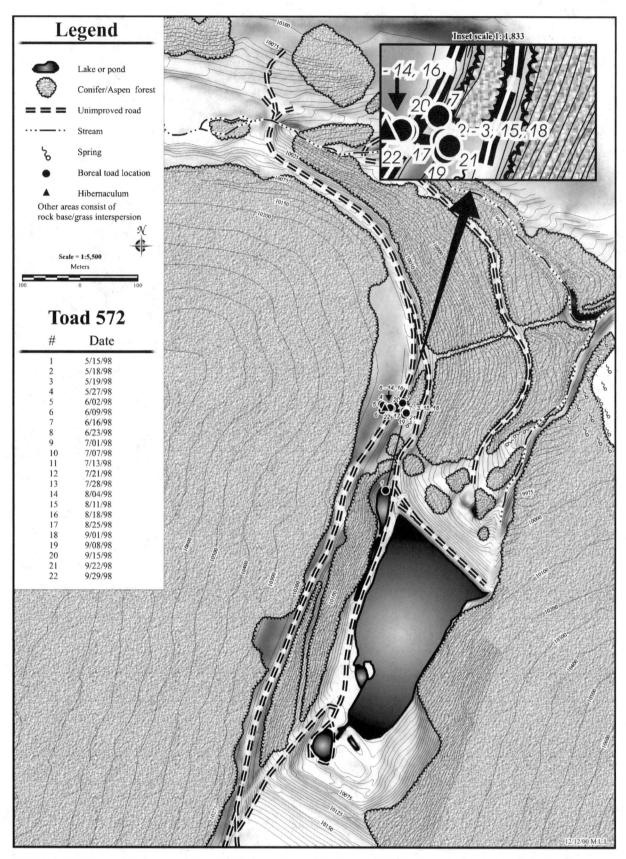

Legend

- **Lake or pond**
- **Conifer/Aspen forest**
- = = = **Unimproved road**
- ·—··—·· **Stream**
- **Spring**
- ● **Boreal toad location**
- ▲ **Hibernaculum**

Other areas consist of rock base/grass interspersion

N

Scale = 1:5,500
Meters
100   0   100

Inset scale 1: 1,833

Toad 572

#	Date
1	5/15/98
2	5/18/98
3	5/19/98
4	5/27/98
5	6/02/98
6	6/09/98
7	6/16/98
8	6/23/98
9	7/01/98
10	7/07/98
11	7/13/98
12	7/21/98
13	7/28/98
14	8/04/98
15	8/11/98
16	8/18/98
17	8/25/98
18	9/01/98
19	9/08/98
20	9/15/98
21	9/22/98
22	9/29/98

FIGURE 31-11 1998 movements, measured by radio telemetry, of Toad 572 (adult female) at the Henderson/Urad study area in Clear Creek County.

limit of about 3,450 m (Karlstrom, 1962; Fellers and Davidson, this volume, part II). Heavy snow accumulations (300–400 cm depth on 1 April; Morton and Allan, 1990) are not unusual in the habitats of Yosemite toads and can delay emergence from hibernation into July. Populations of Yosemite toads in the lower portion of their altitudinal range are active typically from late May through early October. In contrast, the activity periods of high altitude populations may be restricted by one to two months (Karlstrom, 1962).

Clutch and Egg Characteristics

Females can be expected to experience natural selection both for number of offspring and fitness of offspring (Smith and Fretwell, 1974; Wilbur, 1977c). Consequently, in many anuran species there are positive correlations between female size and egg number and size (Crump, 1974; Ryser, 1988; Corn and Livo, 1990; Tejedo, 1992). Although clutch size is significantly related to body size in Great Plains toads (Krupa, 1994), common toads (B. bufo) and natterjack toads (B. calamita) in Britain (Banks and Beebee, 1986), and a number of other temperate anurans (Jørgensen et al., 1979), interestingly, there is no correlation in several tropical anurans (Lang, 1995). In our data set, the association between clutch size and female body size in boreal toads was significant only when data from all three study sites were combined; within each population, no significant relation existed. Furthermore, caloric content per egg and egg weight were not significantly related to female size (Fig. 31-3). Although larger boreal toad females may be physiologically capable of laying larger clutches or increasing egg mass or caloric content per egg in some years, snowpack, cold summer temperatures, and/or other environmental factors that affect food availability may prevent them from producing additional eggs.

Many low altitude, temperate Bufo females appear to lay eggs every year, or in some species, such as Great Plains toads, even occasionally produce more than one clutch per season. (Krupa, 1994). Most of our data support the hypothesis that boreal toad females do not breed every year. A similar finding was reached by Olson (1992), who studied three populations of boreal toads in Oregon at altitudes ranging from 1,200–2,000 m. She marked over 1,700 female toads during a five-year period, between 1982–86. Only 88 of these females were recaptured in a subsequent year. Of those that returned to breed, 80% skipped at least one year between breeding events. In contrast, males in Oregon and in our Clear Creek populations bred each year. The males tended to return to the same breeding pond each year; only 6% of the males switched breeding sites at least once during a five-year period at the Urad Valley. Biennial breeding is common in other montane amphibians. European toads in Switzerland typically lay eggs every other year (Heusser, 1968), while they undergo vitellogenesis each summer in preparation for egg laying in the subsequent spring in lowland Denmark (Jørgensen et al., 1979). Female northwestern salamanders (Ambystoma gracile) breed annually in low altitude (100 m) populations in British Columbia, whereas those at 1,200 m spawn biennially (Eagleson, 1976).

Montane Bufo theoretically might respond to the restriction in the yearly accumulation of energy and materials for egg synthesis with a reduction in clutch size or alteration in the energy and material content of the egg, but their egg and clutch sizes are not abnormal compared to other Bufo. The mean egg diameter of boreal toads (1.6 mm at Clear Creek

TABLE 31-5

Clutch Sizes and Snout-vent Lengths of Female *Bufo* of Various Species

Species	Snout-Vent Length (mm)	Clutch Size (reference)
B. alvarius	87–178[b]	7,500–8,000[b]
B. americanus	56–100[b]	4,000–8,000[b]
		20,603[c]
	56–76[d]	1,700–5,600[d]
B. arenarum		23,000[e]
B. boreas	72–87[a]	6,661[a]
B. boreas halophilus	60–120[b]	16,500[f]
B. cognatus	60–99[b]	19,617[g]
	80	9,376[k]
	60–115	1,342–45,054[k]
B. canorus	45–75[b]	1,500–2,000[h]
B. punctatus	42–64[b]	30–5,000[i]
B. quercicus	20–32[b]	600–800[j]
B. terrestris	44–92[b]	2,500–3,000[b]
B. fowleri	56–82[b]	8,000[b]
		2,500–10,300[j]
		5,000–10,000[f]
B. woodhousii	58–118[b]	25,644[c]

[a] This study; [b] Wright and Wright, 1949; [c] Smith, 1934; [d] Collins, 1975; [e] de Allende and Orias, 1955; [f] Livezey and Wright, 1947; [g] Bragg, 1937; [h] Karlstrom and Livezey, 1955; [i] Tevis, 1966; [j] Clarke, 1973; [k] Krupka, 1994.

sites) falls within the range of egg sizes of various lowland *Bufo*: 1.0 mm in oak toads (B. quercicus) to 2.0 mm in California toads (B. boreas halophilus; Karlstrom, 1962; see other references in Jørgensen et al., 1979). Examples of egg diameters of other lowland North American bufonids are 1.2 mm in Great Plains toads (Bragg, 1940a) and 1.43 mm in American toads (B. americanus; Smith, 1934).

The range of clutch sizes (number of eggs laid) of boreal toads generally matches those of other, similarly sized bufonids (Table 31-5). Another indication that the clutch size of boreal toads is similar to those of other bufonids is ovarian mass relative to body size (Fig. 31-5). The pre-laying and post-laying values for ovarian mass as a proportion of body mass are similar to those of European toads (Jørgensen et al., 1979) and Fowler's toads (B. fowleri in Georgia; Bush, 1963). The ovaries of prebreeding European toads in Denmark weigh 18–19% of body mass, and after breeding about 2.5%. However, the ovarian cycles of European and boreal toads diverge following egg laying. While the ovaries of boreal toads contain only undeveloped and atretic oocytes for at least a year after breeding, vitellogenesis of oocytes of European toads begins after a short resting period of a month or two, culminating with ovaries weighing about 14% of body mass prior to hibernation. Ovarian mass remains constant during hibernation. After emergence from hibernation, final maturation of the eggs increases ovary mass to the pre-laying value (Jørgensen et al., 1979).

Although egg and clutch sizes of boreal toads do not differ appreciably from those of low altitude bufonids, Yosemite toads apparently have reduced clutch size in favor of larger eggs. Yosemite toads lay the largest eggs (2.1 mm) and among the smallest clutch sizes of any North American bufonid (Karlstrom, 1962; Table 31-5). Other montane species appear to lay larger eggs at higher altitudes, but data on clutch size are lacking. For example, European common frogs *(Rana temporaria)* lay larger eggs at higher latitudes and altitudes (Koszlowska, 1971; Koskela and Pasanen, 1975), and western chorus frogs *(Pseudacris triseriata)* in the Colorado Rockies show the same pattern (Pettus and Angleton, 1967). These authors speculate that a relatively large egg size might contribute to more rapid development and faster metamorphosis, a characteristic that could prove advantageous for tadpoles that must complete metamorphosis in one, short season. We hypothesize that this reasoning holds for boreal toads.

Egg diameter is inversely correlated with clutch size in both European (Jørgensen et al., 1979) and boreal toads. Coefficients of variation for intraclutch egg diameters in boreal toads range from 4.1–12.9%. Jørgensen et al. (1979) noted considerable interclutch variation in egg size within a population of European toads. Crump (1984) suggested that intraclutch variation in egg size may be a form of "bet-hedging" as a means of dealing with environmental uncertainty.

If boreal toads were able to lay eggs every year, and if egg diameter and caloric content were maintained, the limitations on energy acquisition forced by the short growing seasons and cold night temperatures would require a reduction in clutch size. Some of the clutches in the West Elk population were small enough that they could potentially represent reproductive output by females selected to lay smaller clutches more frequently. Until more data become available from PIT-tagged individuals, however, this possibility will remain untested. However, the tendency of most females to skip years between breeding results in the possibility that a female may successfully produce young only rarely in her lifetime because egg and larval mortality are so high. This tendency contributes to the difficulty in recovery of boreal toad populations.

Survival to Metamorphosis

The tendency of boreal toads to lay eggs in the shallow areas of large bodies of water or in small puddles that warm rapidly on sunny days is related to the fact that developing larvae require temperatures above 10 °C (Fig. 31-7). Comparative data are lacking on thermal minima for development in other *Bufo*. Similarly, no data on comparative size at metamorphosis exist in other *Bufo*. Several factors (including food availability, tadpole density, site permanence, degree of predation pressure, and temperature) interact in a complex manner to influence anuran size at metamorphosis (Wilbur and Collins, 1973; Brockelman, 1975; Smith-Gill and Berven, 1979; Werner, 1986; Pfennig et al., 1991; Tejedo and Reques, 1994; Reques and Tejedo, 1995; Newman, 1998).

Many causes of egg and larval mortality in boreal toad populations in the southern Rockies are identical to those of lowland species of *Bufo* and include desiccation, predation, infertility, and disease. For example, a population of Great Plains toads suffered high larval mortality due to desiccation (Krupa, 1994). As noted by Burger and Bragg (1947, p. 62), boreal toads "exercise little discrimination in the selection of breeding sites." While smaller puddles provide thermal advantages by

warming rapidly during the day (Young and Zimmerman, 1956), they commonly dry before the larvae can metamorphose unless summer rains are unusually frequent.

One source of larval mortality of boreal toads in the southern Rockies appears to differ from lowland *Bufo*. If egg laying is delayed by spring snowmelt and/or if growth is retarded due to unusually cold summer temperatures, some larvae do not have time to metamorphose before the pond freezes in the fall. In common with larvae of Yosemite toads (Karlstrom, 1962), boreal toad larvae do not survive the winter under ice. The minimal larval period of boreal toads in the Clear Creek population varied from 64–92 days (Table 31-3). The average length of time to metamorphosis (71 days) for Yosemite toads in the Sierra Nevada falls within this range (Karlstrom, 1962). Larval periods of lowland *Bufo*, however, are considerably shorter, due in part to relatively high average temperatures, particularly at night (e.g., 49 days for Great Plains toads; Bragg, 1940a; Krupa, 1994; and 45 days for Woodhouse's toads *[B. woodhousii]*; Carey, unpublished data). Metamorphosis in these species can occur as early as mid-June through July, long before ponds freeze. One unusual case of mortality recorded elsewhere (Oregon) in the range of boreal toads has not been observed (but is plausible) for boreal toad populations in the Southern Rocky Mountains: boreal toad clutches in shallow water were washed out of the water onto the shore by a seiche (wave) generated when a massive snow bank fell into the water on the opposite side of a pond (Ferguson, 1954a).

Because of the different causes of mortality and the difficulty of tracking the number of eggs that develop through metamorphosis, we cannot estimate the average survival of boreal toad clutches that do not suffer complete mortality. Estimates for survival through metamorphosis for various species of *Bufo*, *Rana*, and *Ambystoma* range from 1–13% (Miller, 1909, cited in Bragg, 1940a; J. D. Anderson et al., 1971; Shoop, 1974).

Growth

Metamorphosed boreal toads of all ages, including newly metamorphosed toads, bask for considerable amounts of time each day (Lillywhite, 1974a; Carey, 1978; Muths and Corn, 1997). In newly metamorphic toadlets, this behavior differs from that of larger, older toads in that toadlets will aggregate in piles. These aggregations maximize body temperatures while minimizing water loss (Livo, 1998). Furthermore, these aggregations may be particularly important for protecting toadlets from the damaging effects of UV-B light. Newly metamorphosed Woodhouse's toads will bask, but do not appear to form aggregations (Carey, unpublished data).

In North America, post-metamorphic growth is rapid in lowland anuran populations. Canadian toads *(B. hemiophrys)* in Minnesota metamorphose in late June to early July. Before hibernation in mid-September, they grow to an average of 31 mm SVL. In their second year, they reach an average length of 51 mm; they are able to breed in their third summer (second year) of life (Breckenridge and Tester, 1961). Great Plains toads reach their minimal breeding size (60 mm for females, somewhat smaller for males) by their second full year (Krupa, 1994), as do Wyoming toads *(B. baxteri;* Corn, unpublished data). Breeding desert bufonids, including Great Plains toads, red-spotted toads, and Sonoran Desert toads *(B. alvarius)* are between 2–4 years old (Sullivan and Fernandez, 1999). In species in which the females become sexually mature at larger sizes than males, the size differential may result from faster growth rates in females (Krupa, 1994) or extra seasons of growth.

Boreal toads in the Southern Rocky Mountains, however, need considerably more than 2–4 years to reach sexual maturity. The minimal size at breeding for males and females in the West Elk Mountains and Clear Creek populations was about 55 and 70 mm, respectively. Our estimates that males and females take at least 4–5 years and 6–7 years, respectively, to become sexually mature is difficult to verify with PIT tags, because these devices cannot be applied until an individual reaches a critical size, usually over 10 g (about 40 mm). Olson (1992) estimated that the age at maturity of male and female boreal toads in a population at 2,000 m in the Oregon Cascades was 3 years and 4–5 years, respectively, but it is unclear on what measurements this estimate is based. Male Yosemite toads in the Sierra Nevada were thought to need 3 years to grow to sexual maturity at 53–68 mm (Karlstrom, 1962). The slow rate of growth of boreal toads to sexual maturity is, however, correlated with a long life span. Data appear to be unavailable on the longevity of other toad populations, but boreal toads in Oregon are thought to live at least 10–11 years (Olson, 1992). Measurement of growth rings in toes of museum specimens by D. M. Green (personal communication) indicates that one boreal toad female was at least 13 years at death.

Growth rates of boreal toads above 70 mm are markedly lower than rates of smaller individuals (Table 31-4). A similar drop in growth rates of larger, sexually mature individuals occurs in wood frogs (*Rana sylvatica*; Bellis, 1961a). Depending on the rate of growth during those later years, the largest individuals caught in this study could be considerably older than 10 years. We predict that boreal toads grow only in the summer. Even if low temperatures did not retard or prevent growth, other *Bufo* appear to cease growth in the fall, even if kept at warm temperatures (Jørgensen, 1983). We also predict that growth rates are not homogeneous throughout the active season. Food supplies may be lacking early in the season after emergence from hibernation and exponential rates of liver glycogen deposition in preparation for hibernation would also limit energy availability for growth in late summer (Carey, unpublished data).

Movements of Metamorphosed Individuals

We are unsure about the movements of pre-adult boreal toads from the time they metamorphose until they breed. A large number of toadlets were reintroduced to Lost Lake, in Boulder County, Colorado, in 1996 and 1997. A few were recaptured in 1998, having survived at least one winter (Scherff-Norris, 1999). At least some toadlets from this reintroduction experiment were observed as far away as 600 m (straight-line distance) from Lost Lake, indicating that some individuals may have dispersed (Scherff-Norris, 1999). However, no information is available concerning the long-term geographical distribution or survival of introduced toadlets.

The tendencies of radio-telemetered boreal toad adults to remain in the vicinity of breeding localities, and movements within those localities during the summer after breeding, are similar to that of Canadian toads (Breckenridge and Tester, 1961). Adult male boreal toads may exhibit more breeding site fidelity than females (this study; Olson, 1992).

Boreal toad habitat occurs in the valleys of all mountain ranges in Colorado. Historically, at least several breeding areas typically occurred within a valley or just over an adjacent ridge.

Therefore, an individual dispersing from one breeding area would likely encounter another breeding site within a short distance. Prior to the mass mortalities in the 1970s, movement of individuals from one breeding area to another would have fostered genetic exchange and establishment of new populations. Currently, many of the remaining sites at which breeding still occurs are isolated from others by long distances or by unsuitable habitat between mountain ranges. Individuals dispersing away from these isolated sites may not find another breeding site in their lifetime and could therefore represent a genetic loss to the population.

The loss of individuals by dispersal is particularly critical for boreal toad recovery because many of the remaining populations are very small (one to five clutches laid per year). The Allee effect may increase the likelihood of extinctions for small and low-density populations through the reduced ability of dispersing individuals to find mates (Courchamp et al., 1999; Stephens and Sutherland, 1999). Surveys at historical boreal toad breeding sites sometimes find small numbers of individuals but no evidence of current breeding. In these situations, the individuals tend to be females rather than males, an observation that may reflect a longer lifespan of females or their possible tendency to disperse greater distances than males.

Final Comments

The sizes and numbers of boreal toad populations in the southern Rockies dramatically dropped in the late 1970s through early 1980s. The few populations that remain began to be monitored in 1993, and further mass mortalities were not observed until the last few years. A mycotic skin infection associated with a chytrid fungus is now causing mortality of toads in at least two metapopulations (Jones and Green, unpublished data). The etiology of these fungal infections matches the pattern of the previous die-offs; in particular, metamorphosed individuals, including breeding adults, succumb. This epidemic, combined with the environment and life history of this species, challenge the recovery of these toads.

One aspect of the current recovery effort involves the repatriation of eggs, larvae, or metamorphosed toadlets from existing breeding areas into historical habitat. Translocation efforts for anurans have generally proved unsuccessful (Dodd and Seigel, 1991). For example, no new populations of Houston toads (*B. houstonensis*) became established as of 1991 despite the translocation of 500,000 individuals of various life history stages (Dodd and Seigel, 1991). In some cases, insufficient time has elapsed to evaluate the success of translocation efforts. Since 1996, approximately 10,000 Wyoming toads have been released in southern Wyoming in an effort to restore this endangered species (Spencer, 1999). However, survival rates of these repatriated individuals typically have been low (Jennings and Anderson, 1997; Spencer, 1999). Some reports of successful translocations of *Bufo* exist (Beebee, 1996; Denton et al., 1997). Unfortunately, species of *Bufo* and other anurans that have been involved in successful translocations typically require less time to reach sexual maturity and have higher reproductive outputs than boreal toads. Therefore, the life history characteristics of boreal toads may restrict the success of this technique to help these toads recover their former geographical distribution.

Global climate change may also affect the long-term prospects for recovery of boreal toad populations in the southern Rocky Mountains. Global mean temperatures are predicted to increase 1–5 °C during the next 100 years, primarily due to the greenhouse effect of increasing atmospheric CO_2 concentrations, and distributions of vertebrates are predicted to move north or to higher elevation (Schneider and Root, 1998). For a variety of reasons, global change is expected to decrease the survival of species with small populations and limited dispersal ability (e.g., boreal toads), but warming temperatures could possibly benefit high-elevation populations of boreal toads in two ways. First, earlier breeding due to warmer spring temperatures would provide more time to complete metamorphosis. Indeed, some British amphibians, including natterjack toads, are breeding significantly earlier now than they did 20 years ago (Beebee, 1995). However, there has yet been no trend toward earlier breeding during the 1990s among boreal toads in Rocky Mountain National Park (Fig. 31-2). Second, warmer summer temperatures would allow faster rates of development by tadpoles, again resulting in greater chances to escape aquatic environments that are lethal during winter. However, summer temperatures in the Front Range of Colorado, including Rocky Mountain National Park, are actually decreasing, possibly due to the influence of increased agriculture and urbanization on the plains to the east (Stohlgren et al., 1998). Thus, it is not apparent that a more favorable thermal environment yet exists for high-elevation populations of boreal toads, and we do not expect that hypothesized benefits will measurably increase the probability of survival for boreal toads in the Southern Rockies.

Summary

Boreal toad (*B. b. boreas*) populations in the southern Rocky Mountains suffered dramatic declines in the late 1970s through early 1980s. Recovery efforts to protect the few remaining populations began in 1995. Many aspects of the life history of these toads, such as clutch size, size at maturity, and wet egg mass, mirror those of low altitude, temperate *Bufo*. However, environmental factors in the montane environment of these toads, such as short growing seasons and cold nighttime summer temperatures, force changes in other life history characteristics that restrict the ability of boreal toads to recover their original geographic distribution or population sizes. Breeding starts one to three months later than in lowland *Bufo* in temperate climates, and boreal toads are forced into hibernation one to three months sooner than lowland counterparts. Population recruitment is limited by the prolonged length of the larval period, mortality of larvae that fail to complete metamorphosis before onset of freezing temperatures in fall, slow growth rates of metamorphosed toads to breeding size, and the tendencies of females to skip one or more years between breeding. Although many adult toads of both sexes appear to remain near breeding sites, dispersal of metamorphosed individuals from isolated populations will additionally limit population size. These factors, plus the recent resurgence of outbreaks of infectious disease, create enormous challenges for the recovery of boreal toads in the Southern Rocky Mountains.

Acknowledgments

The research presented in this paper was supported by grants from Great Outdoors Colorado, the Colorado Division of Wildlife, the National Science Foundation, and the Rackham Graduate School of the University of Michigan. Bud Stiles assisted with the fieldwork. We thank the personnel of the Rocky Mountain Biological Laboratory for their assistance throughout the study on boreal toads from 1971–76. We thank David Armstrong, David Chiszar, Gregory Carey, Ruth Bernstein, and Alan de Quieroz for their assistance to Lauren Livo.

THIRTY-TWO

Southwestern Desert Bufonids

BRIAN K. SULLIVAN

The anuran family Bufonidae is a large, cosmopolitan group comprising of almost 400 species that inhabit a great variety of environments. Three bufonids with relatively limited distribution in the United States are federally listed as "endangered:" Wyoming toads *(Bufo baxteri)*, arroyo toads *(B. californicus)*, and Houston toads *(B. houstonensis)*; golden toads *(B. periglenes)* from Central America are perhaps the best known example of an anuran that has recently declined. By contrast, marine toads *(B. marinus)* are one of the most successful anuran introductions, having spread throughout much of the Southern Hemisphere. In spite of the attention these forms have received, many bufonids remain relatively "understudied," and their conservation status is unknown.

Twenty-one bufonid species are recognized in the United States, and eleven of these occur in the arid Southwest (Collins, 1997; see also Crother et al., 2000). Although most of the desert dwelling forms have not been surveyed in any detail, a number have been superficially inventoried (Table 32-1). My purpose here is to review what is known about the status of bufonids of the arid southwestern United States and, in so doing, to draw attention to emerging general patterns regarding their status. I first review case histories of some bufonids that have been surveyed over the past decade.

Spring Breeding, Perennial Streams, and Arizona Toads

Arizona toads *(B. microscaphus)* occur in extreme southern Nevada and southwestern Utah, northwestern, central, and eastern Arizona, and far west-central New Mexico (Stebbins, 1985; Sullivan, 1993). Throughout this region they are associated with riparian systems from desert scrub at low elevations to pine forest habitats over 2,500 m. They typically breed in running water during the early spring, although the breeding period varies depending on rainfall, temperature, and elevation (Stebbins, 1985; Schwaner et al., this volume, Part Two).

Arizona toads are members of the *Bufo microscaphus* complex, which includes three species: arroyo toads *(B. californicus)* of southwestern California and extreme northwestern Mexico; Arizona toads of central Arizona; and Mexican toads *(B. mexicanus)* of north-central Mexico. Regarded as subspecies (of *B.*

microscaphus) by many workers, recent analysis indicates substantial genetic divergence among the three lineages; given their allopatric distribution and diagnosability, they should be regarded as full species (Gergus, 1998; Crother et al., 2000). Interestingly, the genetic divergence among these lineages occurred without divergence in mating calls used by males and females during courtship (Gergus et al., 1997). Although speculative, this may be related to the similarity in habitats occupied by the toads in these three geographic areas or to the absence of other syntopically breeding bufonids.

Arizona toads have been documented at a number of new sites over the past 10 years. Of the 98 element occurrences for the Arizona toad listed in the Arizona Game and Fish Department Heritage Data Management System, 55 (56%) are new localities recently (relatively) documented (1986–96; Anonymous, 1996a). All these localities are within the historic range of the species and suggest that additional search efforts will document these anurans in other appropriate habitats throughout their suspected range in the Southwest. Long-term viability of any of the recent or most historic populations is unknown, but, given their persistence at many historic localities and the documentation of occurrence at many new sites, it is reasonable to conclude that substantial declines have not occurred.

Three historic sites are no longer occupied by Arizona toads in central Arizona: Alamo Lake on the Bill Williams River, Lake Pleasant on the lower Agua Fria River, and the Verde Valley. All three sites are similar in having experienced dramatic alteration of the native riparian communities due to construction of impoundments. Each area is now occupied by Woodhouse's toads *(B. woodhousii)*, a close relative that hybridizes with Arizona toads in all areas of distributional overlap. Interestingly, in southwestern Utah and northwestern Arizona spatial dynamics of hybrid zones between these toads have been stable over the past 50 years (Sullivan, 1995). However, in central Arizona the hybrid zone between these two toads along the lower Agua Fria River (near Phoenix) appears to have expanded over the past 40 years from the vicinity of Phoenix north to Black Canyon City (about 60 km; see Lamb et al., 2000; Malmos et al., 2001). It remains for future study to determine if Woodhouse's toads can supplant Arizona toads in the absence of major habitat alterations, such as the development of impoundments.

TABLE 32-1
Bufonid Status in Relation to Breeding Habitat and Breeding Period

	Breeding Habitat	Breeding Period	Status
Chihuahuan Desert Scrub Biome			
Colorado River toads *Bufo alvarius*	temporary	summer	localized
Great Plains toads *Bufo cognatus*	temporary	summer	stable
Green toads *Bufo debilis*	temporary	summer	stable(?)
Red-spotted toads *Bufo punctatus*	mixed	spring, summer	stable
Texas toads *Bufo speciosus*	temporary	summer	stable(?)
Woodhouse's toads *Bufo woodhousii*	mixed	spring, summer	stable(?)
Great Basin Desert Scrub Biome			
Western toads *Bufo boreas*	mixed	spring	local declines
Black toads *Bufo exsul*	permanent	spring	localized
Amargosa toads *Bufo nelsoni*	permanent	spring	localized
Red-spotted toads *Bufo punctatus*	mixed	spring, summer	stable
Woodhouse's toads *Bufo woodhousii*	mixed	spring, summer	expanding
Mojave Desert Scrub Biome			
Western toads *Bufo boreas*	mixed	spring	localized
Arroyo toads *Bufo californicus*	permanent	spring	localized, declining
Red-spotted toads *Bufo punctatus*	mixed	spring, summer (?)	stable
Woodhouse's toads *Bufo woodhousii*	mixed	spring, summer (?)	expanding(?)
Sonoran Desert Scrub Biome			
Colorado River toads *Bufo alvarius*	temporary	summer	stable
Great Plains toads *Bufo cognatus*	temporary	summer	stable
Arizona toads *Bufo microscaphus*	permanent	spring	local declines
Red-spotted toads *Bufo punctatus*	mixed	spring, summer	stable
Sonoran green toads *Bufo retiformis*	temporary	summer	stable
Woodhouse's toads *Bufo woodhousii*	mixed	spring, summer	expanding

NOTE: Breeding habitats of these desert scrub bufonid species range from permanent to temporary water bodies. Status assessments are subjective "best guesses" based on published reports and personal communications from workers familiar with the taxon. "Localized" refers to historically restricted distribution.

An analysis and a more complete understanding of ecological factors influencing population dynamics and hybridization in Arizona toads will be vital to the conservation of these amphibians. We know surprisingly little about the behavior and ecology of juvenile and adult toads—we know even less about the ecology of larvae. Differences in larval ecology will no doubt provide important clues as to why the Arizona toad declines when aquatic habitats are altered even though such environments appear well-suited to Woodhouse's toads.

Summer Breeding, Rain-Formed Pools, and Sonoran Green Toads

Sonoran green toads (*B. retiformis*) are a poorly known anuran species endemic to the Sonoran Desert in south-central Arizona and north-central Sonora, Mexico (Sullivan et al., 2000). Like many explosive-breeding, desert anurans, Sonoran green toads are primarily active during July and August following rainfall and will take advantage of a variety of water sources for reproduction.

Based upon overall similarity, Sanders and Smith (1951) initially described Sonoran green toads as a subspecies of green toads (*B. debilis*), which occur to the east in the Chihuahuan Desert and southern Great Plains (Stebbins, 1985). Given an absence of intergradation with a second putative subspecies, little Mexican toads (*B. debilis kelloggi*) found in Sonora, Mexico, and differences in behavior (e.g., calls), Bogert (1962) recognized Sonoran green toads as a species. Based on morphology, advertisement calls, and laboratory crosses, Ferguson and Lowe (1969) placed Sonoran green toads in the *B. punctatus* (red-spotted toad) species group, along with green toads and their close relatives, little Mexican toads. More recently, on the basis of mtDNA sequence divergence and morphology, Graybeal (1997) concluded that green toads, red-spotted toads, and Sonoran green toads do not form a monophyletic group (her analysis did not include little Mexican toads). Sullivan et al. (2000) analyzed advertisement calls and suggested that green toads and Sonoran green toads are more closely related to one another than either species is to red-spotted toads.

The distribution of green toads, little Mexican toads, and Sonoran green toads is mostly allopatric, but red-spotted toads are sympatric with all three taxa (Hulse, 1978; Stebbins, 1985). Red-spotted toads typically occur in upland habitats, whereas green toads and Sonoran green toads generally inhabit open creosote or mesquite flats and grasslands. Although rare, hybridization between red-spotted toads and Sonoran green toads has been documented where they co-occur in Arizona, possibly due to the widespread use of cattle tanks for breeding by both forms (Sullivan et al., 1996b). Bowker and Sullivan (1991) documented a naturally occurring hybrid between Sonoran green toads and red-spotted toads, but only three additional hybrids were observed subsequently in the same region during an extensive survey (Sullivan et al., 1996b); hybrids between Arizona toads and Woodhouse's toads are typically common in virtually all areas of sympatry. Given the apparent rareness of hybrids, it is unlikely that hybridization represents a concern for the population status of either Sonoran green toads or red-spotted toads.

Surveys were conducted from 1993–94 to determine the presence or absence of Sonoran green toads across most historic localities in southern Arizona (Sullivan et al., 1996b). They were found at all historical sites that were visited when conditions were appropriate (i.e., during summer rainstorms) and were especially abundant 0–40 km north of Quijotoa, an area associated with the flood plain of Santa Rosa Wash. Chorusing activity was observed in cattle tanks and roadside pools associated with washes.

Hulse (1978) indicated that Sonoran green toads might be expanding their range northward from the Santa Rosa floodplain into areas of agricultural activity (e.g., southern Pinal County), although Sullivan et al. (1996b) could find no evidence in support of this claim. During 1999, additional sightings of Sonoran green toads were recorded in west-central Pinal County, indicating that Sonoran green toads might indeed be more common in this area than previously assumed (R. Bowker and K. Malmos, personal communication).

Overall, the presence of Sonoran green toads at most historic localities, the rareness of hybrids with other taxa, and the documentation of their presence at new sites on the periphery of its historic range suggest that there has been no general decline for this species. Additional study is necessary to assess the relationship between habitat alterations (such as the construction of cattle tanks) and agricultural practices and the population responses of this taxon.

Summer Breeding in an Urbanized Environment

Three bufonids that breed primarily during the summer rainy season are found in the Phoenix Metropolitan area: Colorado River toads (*B. alvarius*), Great Plains toads (*B. cognatus*), and red-spotted toads. These three bufonids co-occur throughout the Sonoran Desert of southern Arizona, breeding in ephemeral waterbodies after summer rainstorms (much like Sonoran green toads). In order to document the anuran community and assess responses to increasing urbanization, Adobe Dam and Cave Buttes (two Maricopa County Flood Control properties on the northern boundary of Phoenix) were surveyed for anuran activity over six seasons (Sullivan and Fernandez, 1999).

The three bufonids at each of these sites were active only a limited number of nights during the six (1990–95) summer rainy seasons of the survey: Colorado River toads chorused on 12 nights; Great Plains toads chorused on 17 nights; and red-spotted toads chorused on 20 nights. All three species often exhibited breeding behavior for one or two nights following a large rainfall event at a particular site. On some occasions, some taxa (e.g., Great Plains toads) chorused on the first night following rain, whereas others (e.g., Colorado River toads and red-spotted toads) initiated chorus activity on the second or even third night following rain (Sullivan and Fernandez, 1999).

During 1990–95, chorus size (number of males) for Colorado River toads at the two flood control sites averaged from 13–21 toads/chorus (Sullivan and Fernandez, 1999). On average, choruses of Great Plains toads were larger, ranging from 11–38 toads/chorus during 1990–93, when estimates were available for breeding choruses. Chorus sizes for the red-spotted toad ranged from 2–65 during 1990–93.

Hybridization among the three bufonids in the north Phoenix area has been rare. A single hybrid male Great Plains toad × red-spotted toad was observed at Cave Buttes in 1990 (Sullivan, 1990); three Colorado River toad × Woodhouse's toad hybrids were observed immediately south of Adobe Dam site in 1995–97; one Great Plains toad × Woodhouse's toad hybrid was observed on the southwest edge of the Phoenix Metropolitan region in 1994 (Gergus et al., 1999); and a single Colorado River toad × Great Plains toad hybrid was found in 1996 south of the Gila River, 75 km south of the flood control sites (L. Thirkill, personal communication). Given that hundreds of individuals of each of these taxa have been observed over the past 20 years in central Arizona, these rare hybrids indicate that hybridization is not a serious threat to the population status of any of these bufonids.

It is unclear what impact the construction of dikes across Skunk Creek and Cave Creek has had on the natural breeding habitat of anurans that use the Adobe Dam and Cave Buttes sites. The anurans present at these study sites typically use rain-formed pools in low-lying areas or along low gradient flood plains. Hence, habitat created by dikes may be similar to that historically used by these species in this region. Given that some habitat was retained with the preservation of these county flood control properties, it remains for future monitoring to determine if these three bufonids can continue to exist in the face of increasing urbanization surrounding these sites. Presently, however, it does not appear that any of these taxa have experienced recent, dramatic declines.

Comparison with other Western Bufonids

Two bufonids occurring in more mesic environments and typically higher elevations in the western United States, western toads *(B. boreas)* of the Rocky Mountains and Pacific Northwest and Yosemite toads *(B. canorus)* of the Sierra Nevada in California, have experienced declines over the past three decades (Carey, 1993; Corn, 1994a, 1998; Drost and Fellers, 1996; Carey et al., this volume). Multiple causes have been implicated in the decline of both forms, including natural cycles of drought and predation and possible influences of pathogens (fungal and bacterial origins). Populations of western toads at both low and high elevations have declined in the Rocky Mountains and Pacific Northwest (Muths and Nanjappa, this volume, Part Two). Although few data are available, it appears that populations of western toads (and close relatives) occurring at lower-elevation arid sites (e.g., in the Great Basin Desert of Nevada) have either not declined or declined only in areas impacted by direct habitat alteration (E. Simandle, personal communication).

As suggested in the case studies reviewed above, for bufonids occurring in the deserts of the southwestern United States, declines have been noted for only three: arroyo toads (Schwaner et al., this volume, Part two), and to a lesser extent, Arizona toads (described above) and western toads (Carey et al., this volume). These declining populations are similar in that they use sources of permanent water for reproduction during the spring—a habitat that is often dramatically altered by humankind. Arroyo toads are similar to Arizona toads in their preference for small streams and rivers. They are restricted to southern, coastal California and northern coastal Baja California, Mexico, and have declined across much of their former range in southern California, primarily as a result of direct habitat loss (Sweet, 1992). Only a few populations historically inhabited the western fringe of the Mojave Desert, and it is unclear to what extent these have declined; they have been documented along the upper Mojave and Whitewater Rivers over the past decade. The declines observed for Arizona toads in southern Nevada and central Arizona can be similarly traced primarily to local effects of habitat alteration and possibly to hybridization with Woodhouse's toads (Sullivan, 1993; Bradford et al., personal communication). Apparently, pathogens, global climate change, and introduced species have not impacted arroyo toads and Arizona toads to the extent that they have other bufonids, such as western toads (Carey et al., this volume) or Yosemite toads (Fellers and Davidson, this volume, Part Two) in the higher elevations of the western United States. It is critical to note that some recent, preliminary observations suggest that stream or spring breeding bufonids can succumb to pathogenic forces—a small number of Arizona toads and Woodhouse's toads at two Arizona sites have

been found dead in stream habitats, suggesting death by a pathogenic agent (M. Sredl and V. Meira, personal communication).

Implications

Only those southwestern bufonids using perennial sources of water for breeding have experienced declines, primarily as a result of habitat loss. It may be that we have not noted declines for summer breeding bufonids of the southwestern United States because those species use temporary aquatic habitats that may be less conducive to transmission of pathogenic vectors (fungal, bacterial, or viral) or less susceptible to anthropogenic habitat alteration such as groundwater pumping. It is also reasonable to assume that larvae (and adults) of these desert forms might tolerate high ambient levels of UV radiation and high temperatures and are less likely to be affected negatively by any widespread, global changes in temperature or ozone level, which some hypothesize to have negatively impacted certain anurans.

Temporary habitats used for breeding are also typically avoided by introduced predators (e.g., American bullfrogs *[Rana catesbeiana]* and salmonid fish) that have been implicated in the declines of other anurans. Additionally, many bufonid larvae are unpalatable and, perhaps, less susceptible to declines due to introduced predators. It should be noted that tiger salamanders *(Ambystoma tigrinum)* experiencing declines in southern Arizona use similar aquatic environments for breeding (cattle tanks) and have been affected by both bacterial and fungal pathogens (Collins et al., this volume). A more complete examination of the hypothesis that bufonids of arid environments are less susceptible to factors causing declines awaits long-term monitoring and continued assessment of populations in the deserts of the world.

Summary

Bufonids inhabiting the harsh deserts of the southwestern United States, especially those breeding during summer rainy seasons, seem to have largely escaped the declines recently experienced by many anurans. Some spring-breeding bufonids (those making use of permanent streams or other water bodies for reproduction) have been eliminated from some areas, primarily through habitat alteration. Hybridization with closely related taxa, especially in areas that have been modified by construction of impoundments, may represent a substantial local threat to some populations. However, it appears that widespread die-offs attributable to pathogens or other inexplicable causes have not affected toads of the southwestern United States.

Acknowledgments. Randy Babb, Rob Bowker, Mike Demlong, Matt Flowers, Erik Gergus, Paul Hamilton, Tom Jones, Keith Malmos, and Mike Sredl assisted with field observations. Dave Bradford, Eric Simandle, and Luke Thirkill shared their experiences with bufonids. Erik Gergus and J. Dale Roberts provided helpful comments on an earlier version of the manuscript. Funding was provided in part by the Arizona Game and Fish Department Heritage Fund and the Office of Sponsored Programs, Arizona State University, Tempe, Arizona.

THIRTY-THREE

Amphibian Ecotoxicology

RAYMOND D. SEMLITSCH AND CHRISTINE M. BRIDGES

The imperiled status of numerous amphibian species worldwide suggests that current research efforts need to take new, biologically relevant directions in order to understand the influence of chemical contamination. A recent summary of the current state of understanding concerning declining amphibians indicates that airborne contaminants are important but that "existing test protocols might be inappropriate" to evaluate their influence (Wake, 1998). We advocate adopting several approaches, experimental designs, and analyses that will promote a better understanding of the effects that chemicals can have on individuals, populations, and communities. We also provide examples of these approaches and the types of hypotheses that could be tested. Further, we attempt to show how the results of these approaches pertain to population and community regulation, and thus, how these results may be linked to conservation efforts.

Incorporating Realism into Experiments

Future research should incorporate greater diversity into studies (e.g., more species, genetic variation within and among geographic populations, and ontogenetic effects) and should consider the effects on multiple life history stages. One of the primary weaknesses in past studies has been the use of non-native amphibian species (e.g., African clawed frogs [*Xenopus laevis*]) or broadly distributed species (e.g., northern leopard frogs [*Rana pipiens*]) with little consideration to local chemical contamination or specific environmental problems. Although we understand why such studies were initially useful and efficient, targeting ecologically relevant species is now more valuable for solving conservation problems. For example, direct comparisons of larvae of species with different feeding modes (e.g., planktivores versus detritivores) could increase our understanding of how morphology and behavior influence susceptibility. Chemicals that quickly adhere to sediments, for instance, may affect only detritivores whereas chemicals taken up by algae would primarily affect planktivores.

Future research should also focus on the diversity of life modes and the rates of development, such as comparing species with direct versus indirect development or short versus long larval periods. A short period of larval development in a contaminated environment would reduce exposure. Species with a high degree of plasticity in length of the larval period may be better able to adapt to contaminated environments by minimizing their exposure as larvae (but see Larson, 1998). Further, we know little about the chemical susceptibility of species with narrow versus wide distributions, which may reflect evolutionary history or other ecological adaptations. Such a comparison might yield insight concerning declining species or endemic species that have small or restricted ranges that may make them particularly vulnerable to environmental stressors. A study comparing the susceptibility of multiple species tested against a well-known phylogeny (e.g., Ranidae) would aid our understanding of the evolution of chemical tolerance. We could evaluate important evolutionary considerations such as if declining species share common ancestors or whether unaffected species share common features enabling them to persist in the presence of higher levels of contamination. Thus, by incorporating the diverse features that taxa possess into toxicology studies, we can achieve a better understanding of how chemical exposure affects or is affected by morphology, physiology, behavior, ecology, and evolutionary history.

One of the most interesting features of many amphibian species is their complex life cycle. Unfortunately, this added complexity necessitates understanding the chemical effects at several stages (e.g., egg, larva, and metamorphic individuals) and usually in completely different environments (e.g., aquatic versus terrestrial). It is likely that for some contaminants only one environment, and hence only one stage, is affected, thereby making single-stage studies of less practical use. Studies that directly compare the susceptibility of aquatic larvae to post-metamorphic terrestrial juveniles or adults have the benefit of elucidating the relative importance of aquatic versus terrestrial pollution. Further, such studies may uncover correlations in responses (phenotypic or genetic) between stages that can enhance our understanding of the differential sensitivity of species or if tolerance to chemicals is constrained or promoted. In addition, studying the ontogenetic changes in susceptibility may increase our understanding of the effects that the timing of chemical application has on a species. For example, pulsed applications of herbicides and insecticides in an agricultural landscape might coincide with particular amphibian developmental stages. Finally, examining carryover effects of chemical exposure from one life stage to another (e.g., tadpole to adult) across seasons can help reveal detrimental effects that are not apparent initially but are expressed in a later stage or at a later time. For instance, there is little known about how chemical exposure in the larval stage might affect the development of reproductive organs or mating behavior many months or even years after exposure. Likewise, could exposure during the summer months affect overwintering survival? These questions reveal the need for longer-term studies that monitor individuals across seasons and throughout their life cycle.

Explicit incorporation of genetic variation into ecotoxicology studies is perhaps the most ignored yet critical factor for understanding the differential susceptibility of populations (Forbes, 1999). Genetic variation, as measured by life history traits or molecular markers, within and among amphibian populations is well established (e.g., Travis, 1980; Berven and Gill, 1983; Blouin, 1992a; C. Phillips, 1994). However, standard procedures for using multiple populations, parents, or full-sibship families of eggs in toxicological tests are seldom utilized. We propose a hierarchical approach to understanding genetic variation that incorporates individual, population, and geographic variation into experiments. This can be achieved with nested factors at each level of genetic variation (e.g., individual, population, and region) that is analyzed statistically. An alternative and simpler technique can be used that pools or mixes levels of variation (i.e., pooling multiple full-sibship clutches of eggs thereby using "sacrificial" replication) and allows genetic variation to be evaluated in the experiment but not analyzed explicitly. The hierarchical design yields information on the relative amount of genetic variation and the level at which it resides and thus indicates how the scale of contamination might affect the persistence of a species. The pooled design yields results that are consistent with the full range of genetic variation present but fails to partition relative contributions at each level. Either approach can take advantage of several tractable features of amphibians (at least for most anurans). For example, experimental crosses can create full- and half-sibship clutches of eggs via external fertilization. These eggs can then be tested to examine genetic variation within populations. Another feature, female oviposition of large egg masses in the shallows of ponds, can facilitate the collection of full-sibship family groups. The ability to generate eggs in the laboratory from field-collected paired adults makes species recognition unambiguous when the identification of egg masses in the field is suspect. Finally, the sheer number of eggs generated by most amphibians (e.g., most female frogs of the family Ranidae produce >1,000 eggs; extremes such as American bullfrogs (*R. catesbeiana*) can produce up to 20,000 eggs; Mount, 1975; Smith, 1961) makes them ideal subjects for the large hierarchical experiments necessary to demonstrate genetic variation.

Level or Scale of Experimentation

For amphibians, we usually aim traditional experimental studies in toxicology at understanding the physiological basis for the mode of action of a chemical and its effect on responses such as mortality, growth, development, and morphological deformities (e.g., LC50s or Frog Embryo Teratogenesis Assay-*Xenopus*; American Society for Testing and Materials, 1991) of individuals. Although this approach is still necessary and yields a clear understanding of the direct effects chemicals can have on individual traits, we assert that these individual responses extend to populations and even communities and, thus, have far reaching effects that are seldom tested. There are two directions that future research should take in order to understand the role chemicals play in population and regional declines of species, as well as their impact on the community in which they are embedded.

First, spatial scale must be increased to accommodate whole populations or communities of a realistic size. For amphibians, the use of field cages or artificial ponds (e.g., cattle watering tanks or wading pools) has proven to be useful in experimental studies that necessitate whole population or community manipulations (Rowe and Dunson, 1994; Britson and Threlkeld, 1998a). More recently, enclosures have been used to examine the effects of abiotic factors such as pH (Warner et al., 1993) and insecticides (Boone and Semlitsch, 2001) on species interactions in anurans. This approach represents an excellent compromise between unrealistic laboratory studies and unreplicated field studies. In addition, the use of enclosures can accommodate multiple predator and prey species that are realistic for natural aquatic habitats in many regions (e.g., Morin, 1983b; Wilbur and Fauth, 1990). Further, dividing natural or manmade ponds into replicate quadrants is also feasible for incorporating greater scale into experiments (Scott, 1990). However, this approach can suffer from pseudoreplication at the level of pond habitats (Hurlbert, 1984). The use of replicate ponds embedded in natural landscapes is also possible and can produce maximum realism. This approach would yield a wealth of information on population and community-level responses if followed for a number of years. Yet, we are not aware of any attempts to examine amphibian communities in such a manner. Such studies necessitate long-term monitoring and, thus, an increase in the temporal scale (i.e., longer than an individual's life span). Because of the scale of destruction and mortality, ethical arguments might be developed for not contaminating natural communities; however, watching species decline to the point of extinction without taking action is more repugnant.

The second direction we need to take is toward understanding the relative role of direct versus indirect effects of chemicals, especially at the community or ecosystem level. Indirect effects are responses of species or components of a community or ecosystem that are not a result of direct exposure to chemical contamination. They may be obvious or be so far removed as to affect species many links further up or down the food chain. A simple example would be when an herbicide reduces algal and periphyton productivity, which, in turn, reduces the available food, thereby slowing the growth rate of tadpoles that are present well after the chemical degrades or dissipates. Summer breeding anurans might experience these indirect effects after an herbicide is applied in early spring or when chemicals enter the aquatic environment during snowmelt. Another more complex example might be when an insecticide is present that reduces zooplankton numbers but has no effect on the algal or periphyton community (e.g., carbaryl; Fairchild et al., 1992). Predators, such as newts (*Notophthalmus* sp.), entering the aquatic environment after the chemical dissipates may be negatively affected because of the reduced food supply. This can cause two indirect effects: tadpoles within gape limitations of newts experience increased predation pressure; or tadpoles exceeding gape limitations of newts benefit (e.g., increased growth rate) from enhanced phytoplankton food supply through reduced competition with zooplankton (Boone and Semlitsch, 2001). The methods for studying the indirect effects necessitate experimental manipulation of each community component in all single- and multi-factor combinations and thus have a large number of treatments. Simplified communities of one or two predators and just a few prey species are useful to initially understand the response pathways following chemical exposure (e.g., Wilbur and Fauth, 1990); however, as more data accumulate, more complex communities could be examined.

Testing Interaction Effects

The traditional experimental approach in biology, as well as in toxicology, has been to test single factors in isolation of all other confounding variables. Unfortunately, such simple designs yield simple results that lack realism and are inefficient in terms of test subjects, time, and other resources (Cochran and Cox, 1957). There are numerous reasons for using multi-factor designs in ecotoxicology. The major biological reason is that environmental problems associated with contaminants are complex and involve multiple factors, both abiotic and biotic. Understanding the effects on communities or ecosystems and reaching solutions, such as how to reverse the trend of declining amphibian populations or species extinction, requires knowledge of both antagonistic and synergistic interactions. Multi-factor designs provide a rigorous experimental approach that yields direct statistical tests of interaction effects, thereby providing essential information not possible with single factor approaches. An excellent recent example involves the photo-enhanced toxicity of a chemical resulting from its interaction with UV-B radiation (Zaga et al., 1998). The results of this two-factor design indicate UV-B and the insecticide carbaryl each affected tadpole responses separately, but the combination of factors (i.e., photoactivated) caused carbaryl toxicity to increase ten-fold above either factor alone. Another good example using a two-factor design is when water temperature interacts with the concentration of a chemical. The increase in water temperature (17 °C, 22 °C, 27 °C) increased the mortality of green frog (*R. clamitans*) tadpoles at four of the seven highest concentrations of the chemical carbaryl (Boone and Bridges, 1999). Natural stressors such as crowding or larval density can also interact in complex ways with chemical contaminants. Boone and Semlitsch (2001) have found that sublethal concentrations of carbaryl enhanced the survival of gray treefrog (*Hyla versicolor*) tadpoles at high density compared to those at low density or in controls without chemicals. They determined that the chemical differentially eliminated the zooplankton, thereby facilitating algal productivity and increasing the growth and development of tadpoles reared at high density (i.e., with low per capita resources).

In the cases of declining, threatened, or endangered species where the availability of specimens for experimentation is limited, multi-factor designs are more efficient. Multi-factor designs require significantly fewer test subjects to yield the same information on each factor and to provide tests of the same power as single-factor designs, but additionally, to provide tests of interaction effects. With the convenience of software programs for analyzing or even designing multi-factor experiments (e.g., SAS, SYSTAT; see Krzysik, 1998a), there is little reason for limiting testing to single-factor approaches.

Linking Ecotoxicology Studies to Conservation

The biological link between ecotoxicological studies and conservation lies in factors that disrupt the natural regulation of species populations and community structure. A reasonable hypothesis is that variation in natural environmental factors can interact with anthropogenic stressors (such as sublethal concentrations of chemicals) to cause the local or regional declines of amphibian populations that have been reported in the literature. When considering how chemical contamination can disrupt natural processes, it is important to recognize that pond-breeding amphibians exhibit complex life cycles with aquatic larval stages for growth and development and terrestrial adult stages for reproduction and dispersal (Wilbur, 1980). Regulation of populations most likely occurs in the aquatic larval stage. Hence, chemicals that directly or indirectly enter aquatic habitats are potential stressors on critical regulatory processes of amphibians.

Metamorphosis from the aquatic larval habitat to the terrestrial environment is the critical step for individuals to be recruited into the breeding adult population. Larval growth and development—upon which metamorphosis depends—is extremely sensitive to environmental factors. Any factor, biotic or abiotic, that impedes the process of metamorphosis (such as low food, low temperature, high larval density, rapid pond drying, or the presence of chemical contaminants) reduces juvenile recruitment and the probability of a population's persistence. Further knowledge of how chemicals affect species attributes, including competitive or predator-prey interactions, is critical to understanding how community processes can be disrupted. It is also important to understand that in natural populations of amphibians only 3–5% of all offspring produced annually reach metamorphosis, and the production of any metamorphic individuals from year to year is episodic (Semlitsch et al., 1996). Thus, anthropogenic stress from chemical contamination may reduce recruitment further in years when metamorphic individuals are produced and increase the time interval between bouts of successful recruitment. If the adult reproductive life span is short or chemical contamination is long-lived, either effect could lead to species declines and local extinction.

Summary and Conclusions

Studies using realistic and expanded designs will help increase our understanding of species differences in life histories, the genetic basis for population differences, interactions among factors (natural or anthropogenic), direct and indirect effects, and the role of chemical contamination in disrupting population and community processes. If we are to preserve the remaining biodiversity in our amphibian fauna, novel and pluralistic approaches are necessary to develop a clear understanding of the processes leading to species declines.

Acknowledgments. We thank M. Boone, V. Burke, J. Dwyer, and W. Gibbons for helpful comments and suggestions on this manuscript. Reported research was supported by EPA grant #827095-01 (RDS).

THIRTY-FOUR

Museum Collections

JOHN W. FERNER, JEFFREY G. DAVIS, AND PAUL J. KRUSLING

Most professional herpetologists have had a course in college or graduate school that required some form of field collection. Perhaps it was a class in vertebrate natural history, where we were expected to collect, preserve, document, and catalog one or more specimens from each of the major vertebrate classes. These specimens were often expected to be of "museum quality" and handed in to the professor as part of our evaluation for the course. The resulting collections were then routinely retained at the institution (or in some cases in the personal collection of the professor) as a teaching resource and record of the vertebrate fauna of the region. This tradition of the past century has resulted in hundreds of smaller collections at institutions of higher education and regional museums around the United States. When combined with the vertebrate collections at our major research museums, the specimens held in these repositories provide an extremely important resource for teaching and research in vertebrate and conservation biology. Here we review the relationship of museum collections (and those of amphibians in particular) to the study of current conservation issues; we look at both the benefits and costs of collections to scientific institutions and society, and then attempt to suggest future directions for these efforts.

Benefits of Collections

While the value of scientific collections seems obvious to those using specimens on a daily basis, it is important that all members of the scientific community, as well as those in disciplines such as land management and conservation biology, understand and respect the benefits of this natural history resource. The museum literature provides good examples of why we maintain collections. Barr's (1974) early review of the importance of systematic collections to environmental assessment remains relevant today. Reasons for maintaining collections include documenting the past and biogeography, supporting the study of the relationships among organisms (systematics), and hands-on education (Allmon, 1994). In the following sections, we elaborate on some of the most compelling arguments for preserving, maintaining, and continuing to build museum collections of amphibians.

Voucher and Research Specimens

Many specimens are collected in order to describe variation within a population, reproductive condition, a new geographic location, or a new species. These voucher specimens are deposited in museums to provide verification of the initial observation and to be available for further research. It is important that these specimens have accurate and sufficient associated data to make them useful as vouchers, as well as to provide information that will assist future research projects. For example, if a frog is collected and the date is simply given as "summer 1945," this information would be too imprecise for someone studying the details of seasonal changes in the reproductive anatomy of that species. The value of the specimen is directly proportional to the precision of the data provided in the field notes. Although specimens with no data may still be used in teaching collections, for skeletal preparations, or for more general studies, by no means should anyone be encouraged to collect a specimen without gathering complete data.

In some limited cases, photographic records are now considered to be an acceptable means to document geographic distributions; these photographs are also deposited in museum collections and catalogued as vouchers. Photographic records need to be of high quality, and they should often provide more than one view. An ideal series of photographs would record dorsal, lateral, and ventral views, and also include diagnostic features such as those noted in field guides (Conant and Collins, 1991). If a specimen is released after being photographed, there is always the risk that the images will not be adequate for later identification. Therefore, when photographing specimens of significance, we recommend holding the animal in captivity (if allowable under the collecting permit) until the film can be developed and the quality of the documentation verified.

Professional herpetologists and naturalists are familiar with the distribution maps published in field guides, such as Conant and Collins (1991). These maps are constructed from data for museum voucher specimens. On a more restricted (but more precise) scale, books about the statewide distribution of particular taxonomic groups, such as Casper's (1996) distribution maps of Wisconsin herpetofauna, are of equal importance. Every dot, triangle, or other figure on these maps represents not just a locality, but a voucher specimen as well. To see an extensive museum collection makes one appreciate the collective efforts of the people whose data make field guides and similar publications possible. When conservation efforts focus on a particular taxon, a first step is typically to delineate the range of that organism. Therefore, while it may seem hypocritical to euthanize and preserve animals for conservation biology, it must be emphasized that museum collections are critical resources for amphibian conservation projects.

Historical Records

Over the past 10 years, amphibian collections throughout the world have been used to study historical conditions related to the observed decline in many populations. Resetar (1998) has amply reviewed the methods to obtain historical information on amphibians from collections. As we replace natural ecosystems with urbanization and agricultural systems, we can determine the impact by studying museum vouchers to ascertain historical distributions. In considering restoration of wetlands, for example, it should be routine practice to determine what historical records tell us about pre-disturbance species assemblages. These collections could also be used to examine stomach contents to see what impact historical land use changes have had on the ecology of a particular species' habitat as reflected by diet.

When the recent alarming malformations in frogs became front-page news, we all referred to our preserved specimens to determine if historical records of abnormalities were present. These searches indicated that while malformations had occurred in the past, they were probably not as widespread as the more recent reports.

There is no "industry" standard for how often a voucher specimen should be re-collected from a site to document a species' continued existence. This is a relatively new concern in terms of collection strategy for conservation biologists. With the current rate of environmental change, to collect a voucher every 50 years or so would be quite conservative. We would suggest that collecting vouchers from known populations every 5 years would provide much more data useful for long-term monitoring.

Education

Teaching collections are usually separated from those with research potential so that students may benefit from hands-on experience working with specimens. Many of us learned our initial facts about species from studying them as preserved specimens in teaching collections. Scientific collections are valuable sources of data and should not be handled by students in general education programs; specimens are damaged when students are learning to locate and document key characteristics. Therefore, any specimens used for teaching should be duplicated in the scientific collection of the museum. In instances when scientific specimens must be used in an educational setting, these specimens should not be handled by students, but rather viewed in their container.

Large institutions have collections that help educate the public about biodiversity. Museum collections have been instrumental in environmental education issues, such as elucidating the declining amphibian population problem. Examples from frog population studies of the past 10 years have been broadly integrated into new museum exhibits and educational programs across the United States. The attention given to amphibians by the media has certainly been a success story in environmental education and awareness.

Cultural Heritage

As we look back on the building of our collections at the Cincinnati Museum of Natural History and Science, we can see that they not only help to document much of our local natural history, but have themselves become a part of our cultural heritage. A favorite pastime of volunteers working in our museum is to browse through the catalog and shelves looking not only at the specimens, but who collected it and where and when it was collected. These collections reflect the work of scores of scientists, students, and volunteers who over the years have had a curiosity about the natural world. Allmon (1994) stresses the importance of collections beyond simply being a resource for scientists to being a source of inspiration and insight. As in a library, collections of biological specimens can inspire us in unexpected ways and provide us with a new sense of meaning about life. This notion is difficult to articulate, as Allmon admits, and may only be understood by those who have had this vicarious field experience as they immerse themselves in the collections.

This experiential aspect of museum work may be demonstrated most poignantly by extending the library analogy. To see specimens of amphibians drying up in their jars in neglected, orphaned collections is no less an atrocity than book burnings. Museum collections have become a part of our cultural heritage, particularly since the environmental revolution of the past three decades. To handle and study specimens of an extinct species is not simply a scientific experience—it inspires awe and can serve as a grim reminder of humanity's lack of respect for biodiversity.

Another example of unexpected inspiration sparked by collections comes from the story of a Cincinnati high school senior. Almost 70 years ago, Worth "Buzz" Weller was a major contributor to the growing amphibian and reptile collection at what was then known as the Cincinnati Society of Natural History. Some of the salamanders he collected in North Carolina became the type specimens for a new species. While looking for more of these salamanders shortly after his graduation, Weller died in a tragic fall on Grandfather Mountain in North Carolina. The salamander species that Buzz helped discover was later named for him, *Plethodon welleri* (Weller's salamander). Stories such as these are reflected in the collections on our shelves, and they should serve as an inspiration to current and future generations.

Costs of Supporting Collections

Collections obviously need resources in order to be obtained and maintained (Simmons, 1987). Each museum or university must have a clear mission statement as to the scope of its collections. We look upon collections management as a challenge to be met for each individual collection and not as a reason to stop collecting or to downsize.

Space, Materials, and Supplies

Each institution has its own limitations on the maintenance and growth of its collections, and there must be a commitment to the mission of the collection. Van Devender (1998) has started a dialogue concerning setting limits on vertebrate collections at smaller schools and managing collections at an optimum size. Research grant proposals that rely on any aspect of a collection should include funding for long-term maintenance costs. Many museums are now charging fees for providing information from their collections in order to raise operating funds. Consolidation of some collections within a region is a concept that seems to be gaining support. Recently, the invertebrate paleontology collections at the University of Cincinnati and the Cincinnati Museum of Natural History and Science were combined with the help of funding from the National Science Foundation. Duplication of efforts, especially within institutions of such close proximity, should be eliminated whenever possible.

Curators, Staff, and Volunteers

For most institutions, the expense of personnel to maintain and use collections is more costly than the needs for space and supplies. How much release time will a department give a faculty member to curate a collection? Most small academic institutions would not consider this expense; the faculty does most collections management in support of its teaching and research efforts. "Orphaned collections" are becoming a problem throughout the country as the more traditional vertebrate biologists retire and are not replaced. Unfortunately, if there is no institutional commitment to the collections, they may deteriorate rapidly. New faculty may not have the training to curate a collection and specimens may be lost. We recently visited an orphaned collection where catalogs had been discarded and hundreds of amphibians and reptiles were drying up in their jars. If an institution can no longer care for its collections (e.g., due to lack of personnel or space), it is imperative that it find a new home for the specimens. All too frequently, "territorial" issues occur. An institution unwilling to relinquish ownership of its collection may work against stated interests and only contribute to its further deterioration. Procedures for adopting orphaned collections are reviewed in detail by Lane (2001).

Some museum management and curation must be part of the curriculum in our traditional courses in vertebrate zoology and herpetology. Museums have a growing reliance on volunteers, and this needs to be supported by college faculty through recruitment of volunteers for both their own and other regional collections. We received a collection from a university a few years ago that would have been discarded if not for a new faculty member's volunteer experience during high school at the Field Museum of Natural History in Chicago—his appreciation for the value of museum specimens runs deep. In addition, these specimens were particularly important, as some were voucher specimens of Roger Conant's (1938) work on Ohio reptiles.

Permits and Restrictions on Collecting

With increasing legislation to regulate collecting, museums are faced with the escalating cost of obtaining permits. Not only have collecting fees and travel costs increased, but more importantly, the paperwork involved with filing reports and tracking specimens has become burdensome. In the past, over-collecting may have occurred occasionally by scientists or their paid field assistants, but this rarely occurs today. The threats to

our biodiversity do not include scientific collecting (Goodman and Lanyon, 1994).

In some cases, it may become difficult to obtain a needed voucher specimen from a particular site. At many museums, photographs may now be deposited as vouchers when specimens are not available. A general benefit from the increased regulation of collecting may be the reduction of personal collections, which, because they are not available for general use, have little recognition in the scientific community. We emphasize one caution related to free access to museum information: collection sites for some amphibians should not be made available for fear of people (even well-intentioned herpetologists) over-visiting the habitat and causing damage. We also recognize that government agencies must regulate collecting in order to protect species from commercial collectors in particular.

Summary: The Future of Collections

With the rapid change in the populations of amphibians around the world, it is essential that our current collections be maintained and that the acquisitions be continued (Goodman and Lanyon, 1994). In essence, we may be documenting a portion of the largest mass extinction in the history of life on earth. Whatever their use, these collections will be valuable to future generations of scientists. Some or all of the following suggestions should be considered for college, university, and museum collections:

- Incorporate modern technologies such as maintaining frozen tissue (see Jacobs and Heyer, 1994) and photographic collections.
- Coordinate collection specialties with other museums as part of a regional consolidation.
- Expand the curatorial support base through more museum training in the college curriculum and build a strong volunteer system.
- Seek funding in support of regional surveys that encourage the collection of appropriate voucher specimens.
- Seek out and acquire orphaned collections that are within the realm of your mission.
- Relocate specimens that are not within your mission.
- Proactively seek specimens from regions not adequately documented.
- Encourage local naturalists to deposit their field notes and photographs in your museum.

Collections are the ultimate library for historic information about our natural heritage. They provide a resource that will continue to have value in the conservation of amphibians for generations to come.

Acknowledgments. We thank R.S. Kennedy, Curator of Vertebrate Zoology at the Cincinnati Museum Center, for his never-ending support of our work and the importance of collections for conservation biology. The Hamilton County (Ohio) Parks District and Cinergy Corporation have provided long-term support for the amphibian surveys we have conducted. J.P. Ferner provided helpful comments on an early draft of this manuscript.

THIRTY-FIVE

Critical Areas

HUGH R. QUINN AND COLLEEN SCOTT

The global decline of amphibians has received a great deal of attention (Wake, 1991; Wake and Morwitz, 1991; K. Phillips, 1994) and serves as an indicator of a larger problem involving the decline of overall biodiversity associated with uncontrolled human population growth. In North America, amphibians have declined due to environmental alteration associated with timber harvesting, agriculture, wetland drainage, urbanization, stream pollution and siltation, and the introduction of exotic predators (Orser and Shure, 1972; Bury, 1983; Gore, 1983; Hayes and Jennings, 1986; Pierce and Harvey, 1987; Ash, 1988; Welsh and Lind, 1988; Blaustein and Wake, 1990, 1995; Pechmann et al., 1991; Petranka et al., 1993, 1994; Vial and Saylor, 1993; Fisher and Shaffer, 1996; Gamradt and Kats, 1996). In the midwestern United States, which includes Ohio, severe declines of certain amphibian species have occurred and are continuing to occur (Lannoo, 1998b). Disturbingly, broad-ranging, common species such as Fowler's toads (Bufo fowleri) and Blanchard's cricket frogs (Acris crepitans blanchardi) have recently disappeared from certain portions of their ranges in Ohio and other states (J. Davis, J. Harding, and S. Moody, personal communication). This alarming trend in population declines of an array of amphibian species in and around Ohio indicates that additional conservation measures are needed within the state, not only for those species considered endangered, but for all species. Adding to this concern is the destruction of 90% of Ohio's wetlands within historic times, which represents a decline in habitat critical for many amphibian species (Sibbing, 1995).

Ohio is home to 14 species and subspecies of frogs and toads, and 26 species of salamanders. Nine families are represented among these two groups, demonstrating the high level of amphibian diversity within the state. Of the 40 species, five are listed as endangered and one of special interest by the Ohio Department of Natural Resources, Division of Wildlife. Endangered species include eastern hellbenders (Cryptobranchus a. alleganiensis), blue-spotted salamanders (Ambystoma laterale), green salamanders (Aneides aeneus), cave salamanders (Eurycea lucifuga), and eastern spadefoot toads (Scaphiopus h. holbrookii). Four-toed salamanders (Hemidactylium scutatum) are listed as special interest. Thus, about 19% of Ohio's salamanders are listed as endangered or special interest, as are about 7% of the frogs and toads. Overall, 15% of the state's amphibians are contained within these protected categories.

Due to these alarming trends of amphibian population declines, decisive conservation measures must be taken. It is the purpose of this paper to present practical methods, thought processes, and other considerations to stimulate establishment of a reserve network for all Ohio amphibian species based on distribution. Specifically, areas critical for amphibian conservation are evaluated by (1) defining biological "hotspots;" (2) defining a minimum reserve network (minimum number of reserves) to conserve all Ohio amphibian species; and (3) evaluating the use of existing protected land in forming an amphibian reserve network. We emphasize that the information we present is only a first stage or coarse analysis in defining a reserve network and that further refinement is needed to define precise locations of reserves. Further, the presence of viable amphibian populations, habitat quality, land availability, current land use practices, land cost, and other factors important in reserve formation and function at the sites indicated are unknown. Such information can be obtained through field surveys and consultation with individuals familiar with the chosen areas.

Materials and Methods

The Algorithms

SOURCES OF DATA

Ohio amphibian distribution data were taken from Pfingsten and Downs (1989), Pfingsten (1998), and Davis and Menze (2000). Tremblay's salamanders (A. tremblayi) and silvery salamanders (A. platineum) reported from Ohio (Pfingsten and Downs, 1989) are considered hybrids, and thus are not included in this analysis (for summary of hybrid discussion see Klemens, 1993; for an alternative viewpoint see Phillips and Mui, this volume, Part Two). Locality records were reported by township, with these areas used as the distribution units. For subsequent analyses an absence/presence matrix was formed comparing all species to all townships. This resulted in 8,066 records in the 1,343 Ohio townships. Of these townships, 86 had no amphibian records. Townships average 30 mi^2 within the 40,740 mi^2 political area of Ohio, and most townships are roughly square. Size range is approximately 5–65 mi^2, however, in the south-central portion of the state, their shape and size are more irregular.

CHOOSING ALGORITHMS

There are 19 algorithms that have been utilized in selecting nature reserve networks, all of which were reviewed by Csuti et al. (1997). From these we selected the (1) richness-based heuristic or greedy, (2) progressive rarity-based heuristic, and (3) linear programming-based branch-and-bound algorithms for our analyses in defining minimum reserve networks that represent all Ohio amphibian species at least once. The simple greedy method was utilized due to its computational ease compared to the other methods. The progressive rarity method was chosen because it performed well in selecting the minimum number of sites to represent all species at least once when compared to the other 18 algorithms mentioned above (Csuti et al., 1997). The branch-and-bound method was chosen because, by definition, it identifies the optimal solution or solutions of representing each species a given number of times among a minimal number of sites. Therefore, it can be used as a standard against which the performance of all other algorithms can be compared (Csuti et al., 1997). However, the disadvantage of this method is that it is computationally intense. In addition to these three algorithms, we utilized a site selection method commonly employed by several conservation organizations that focuses on species rich areas, or "hotspots."

SIMPLE GREEDY ALGORITHM

With the simple greedy algorithm, the first township chosen is the one that has maximal species richness. Subsequent sites are selected so that at each step the inclusion of the next site adds the most additional (new and/or different) species to those already represented. This process is repeated until all species are represented in the analysis. When two or more sites add the same number of additional species, we break the tie by choosing the township that appears first alphabetically. However, the identity of all tied sites is retained so the various pathways that may develop from these sites can be followed separately to provide alternative site choices from the analysis. This flexibility provides realistic alternatives during subsequent fine-grained evaluations of precise reserve site locations. The greedy algorithm was again applied to the data in examining the number of townships required to include all species at least twice, and then three times.

PROGRESSIVE RARITY ALGORITHM

In the rarity-based algorithm, the first site selected is the one that contains species unique to that site (i.e., occur in no other sites). Next, a site containing the least frequent, unrepresented species (i.e., the species occurring in just two sites, if any, if not, in three sites, etc.) is chosen. This process is repeated until all species are represented in the analysis. Ties are handled in the same manner as that described for the greedy algorithm. Selected sites are then arranged from most species-rich to least so that species accumulation rates can be compared to the other selection methods utilized. Such comparisons demonstrate the number of new species added by each method with each township selected throughout the site selection process.

BRANCH-AND-BOUND ALGORITHM

With the linear programming-based branch-and-bound algorithm, as described by Csuti et al. (1997), the linear programming (LP) relaxation of the integer programming model (that is, the binary restrictions replaced by lower and upper bounds of 0 and 1) is solved. If the LP relaxation is an integer, the optimal solution has been found. If one (or more) of the variables is fractional, branching takes place. A branch is the creation of two new problems (nodes), one with the fractional variable set to 0 and the other with the fractional variable set to 1. The LP relaxation for each of the new problems is solved and the process is repeated until all nodes under consideration have been fathomed. A node is fathomed if its solution is infeasible, is an integer, or has a value worse than the current incumbent integer solution.

The branch-and-bound analysis represents the optimal solution or solutions in finding the minimum number of sites to represent all species at least once. The analysis is generally reported as a series of steps, beginning with the assumption that only one site (township, in our case) can be chosen, with a solution then generated under the constraints of that assumption. This solution is the site demonstrating the highest species richness. For the next step, an optimal solution is then reported if only two sites are available, and so on until the step is reached at which all species are included. Sites included in any one step may or may not be included in the other steps, as the process examines all possible combinations of sites in finding the optimal solution for the number of sites indicated at a given step. Therefore, the result of each step is independent from those of all other steps. This step-wise information is useful in finding optimal solutions given less than the minimum number of sites to represent all species, as well as in illustrating the proportion of total diversity represented as sites are accumulated.

SPECIES RICHNESS METHOD

The richness method chooses the most species rich area (township) as the first site, the second richest next, and so on until all species are represented. If two or more sites share the same number of species, the sites are arranged alphabetically.

Representation of Rare Species

To compare the ability of each reserve selection method to capture single or multiple representations of rare species, a rarity score was derived for each. First, such a score was calculated for each species as the inverse of the number of townships in which it occurs throughout the state (Usher, 1986; Avery and Leslie, 1990; Howard, 1991; Williams, 1993; P.H. Williams et al., 1996). This value was then multiplied by the number of times the given species was represented in the analysis being evaluated (i.e., the number of townships chosen by the analysis containing the species). All such products for all species were then summed to give an overall rarity score for the analysis. Thus, the rarity score for the suite of sites selected by each reserve selection method is represented as:

$$\text{Rarity score} = \sum (1/T_i)(t_i)$$
$$\{i{:}T \neq 0,\ 1 \leq i \leq n\}$$

where T_i is the number of statewide townships occupied by species i, and t_i is the number of townships chosen by the reserve selection method containing the species. Thus, higher scores indicate greater representation of rare species.

Habitat Information

Ohio's five physiographic sections were used as crude indicators of habitat variation between sites (Anderson, 1983). For a description of the influence of these sections on amphibian distribution, see Pfingsten and Downs (1989).

Existing Reserve System

Existing reserves were identified from the Ohio Natural Heritage Database of the Department of Natural Resources, Division of Natural Areas and Preserves. Additional information about the reserves was obtained from Anonymous (1995l, 1996b), Weber et al. (1993), and through phone conversations with numerous agencies and institutions. Of the 600 reserves listed, only the 519 for which size and location (by township) could be determined were used in the analyses. Of these, 99 are classified numerically by the Ohio Division of Natural Areas and Preserves as protection status code 1 (*preserved:* an area which is dedicated under the Ohio Natural Areas law); 286 as status code 2 (*protected:* an area whose management policies include some protection for the species and features on the property, although the area is not legally protected); 127 as status code 3 (*unprotected:* an area which may have important rare species or features, but the protection of these species or features is not part of the purpose of the area); and seven were unclassified. To indicate the amphibian species potentially found within each reserve, it was assumed that if a species was reported from the same township within which a reserve was located, then it also occurred in the reserve.

Results

Algorithm Analyses

SPECIES ACCUMULATIONS

The simple greedy, progressive rarity, and branch-and-bound analyses each embraced all 40 Ohio amphibian species in six townships or steps (Table 35-1), but the species richness analysis required 119.

In terms of quickly and efficiently representing all species during the township selection process, the branch-and-bound method (as described in the step-wise process in the Methods section, not as represented in the final solution presented in Table 35-1) performed better than the other methods. Species accumulation at each step for the branch-and-bound step-wise process was 29, 35, 37, 38, 39, and 40 for steps 1–6, respectively. The simple greedy and progressive rarity methods performed equally, except in the initial step, where the greedy method captured three more species. The species richness method accumulated species as quickly as the greedy method during the first two steps, but thereafter did not, and it failed to represent all species in as few townships as did the other three methods. Because species accumulation was slow after the initial two steps of the species richness method, and because species richness remained high in the townships chosen, these amphibian "hotspots" contain similar species suites. Interestingly, the progressive rarity and the final solution of the branch-and-bound

methods indicated the same townships (Table 35-1). Therefore, in comparing species accumulations of all methods to the optimal solution (branch-and-bound), the simple greedy method was superior, but the progressive rarity method also performed well.

ALTERNATIVE SOLUTIONS

Having more than one optimal solution from the algorithms presented here is advantageous in that choices exist during the actual site selection process. For example, if some of the alternative sites are superior to others in such attributes as habitat quality, target species population size, or land affordability, then the one with the highest advantage can be chosen. The species richness method, by definition, does not have alternative solutions, as it embraces all sites based on species richness. If two sites contain the same number of species, both are utilized and neither is considered as alternative. There were no alternative townships chosen during the first three steps of the simple greedy analysis. Step 4 (representing an increase of only one species, blue-spotted salamanders [*A. laterale*]) had two townships containing the targeted species (two alternative solutions); Step 5 (representing an increase of only one species, Allegheny Mountain dusky salamanders [*Desmognathus ochrophaeus*]) revealed 25 alternative solutions; and Step 6 (representing an increase of only one species, Wehrle's salamanders [*Plethodon wehrlei*]) demonstrated two alternative solutions. The progressive rarity method had no alternative solutions until Step 6, which contained 25 solutions to select a single species, Allegheny Mountain dusky salamanders. The branch-and-bound analysis demonstrated no alternative solutions (Table 35-1). Therefore, although all solutions were fairly rigid and did not present alternative sites until late in the analysis, the simple greedy method provided alternative choices at more steps than did the other two algorithms used.

SITE HABITAT VARIABILITY

Ohio contains five physiographic regions. Only one of these regions was represented in the first six townships selected by the species richness method, whereas three regions were represented in the six sites selected by each of the other analysis methods (Table 35-1). The first six townships selected by the species richness method were located in the south-central portion of the state, whereas the sites selected by the other methods represented a more disperse pattern (Fig. 35-1). Therefore, the algorithms represented greater habitat diversity among sites than did the first six steps of the species richness method. This is probably a result of varying habitat requirements among species, with the algorithms capturing all species (and their respective habitats) and the species richness method only capturing a subset of species and their habitats.

MULTIPLE REPRESENTATION OF SPECIES

Multiple representations of each species in a reserve network are important for conservation purposes as a hedge against extinction. This holds particularly true for rare species, where each population represents a much larger portion of the entire species than it does for more common species. The number of times each species was represented in the first six sites selected by each method ranged from 0–6 with the species richness method and 1–5 with each of the other methods. The mean number of times each species was represented utilizing the species richness, simple greedy, progressive rarity, and branch-and-bound analyses was 4.2 ± 2.1; 2.6 ± 1.2; 2.1 ± 1.2; and 2.1 ± 1.2, respectively.

TABLE 35-1
Comparison of Ohio Amphibian Reserve Selection Methods

Step	Site Description	Species Richness	Simple Greedy	Progressive Rarity	Branch-and-Bound[a]
1	C/T	Athens/Athens	Athens/Athens	Adams/Green	Adams/Green
	SR	29	29	26	26
	CS (%)	29 (73%)	29 (73%)	26 (65%)	26 (65%)
	PS	4	4	4	4
2	C/T	Adams/Green	Adams/Green	Hamilton/Mill Creek	Hamilton/Mill Creek
	SR	26	26	17	17
	CS (%)	34 (85%)	34 (85%)	34 (85%)	34 (85%)
	PS	4	4	2	20
3	C/T	Hocking/Good Hope	Hamilton/Mill Creek	Henry/Washington	Henry/Washington
	SR	26	17	13	13
	CS (%)	35 (88%)	37 (93%)	37 (93%)	37 (93%)
	PS	4	2	1	1
4	C/T	Scioto/Nile	Henry/Washington[b]	Monroe/Center	Monroe/Center
	SR	25	13	13	13
	CS (%)	35 (88%)	38 (95%)	38 (95%)	38 (95%)
	PS	4	1	4	4
5	C/T	Vinton/Eagle	Ashtabula/Austinburg[c]	Lawrence/Fayette	Lawrence/Fayette
	SR	25	6	9	9
	CS (%)	35 (88%)	39 (98%)	39 (98%)	39 (98%)
	PS	4	1	4	4
6	C/T	Adams/Meigs	Monroe/Center[b]	Ashtabula/Austinburg[c]	Ashtabula/Austinburg
	SR	24	13	6	6
	CS (%)	36 (90%)	40 (100%)	40 (100%)	40 (100%)
	PS	4	4	1	1

[a] The information in this column is the final solution of representing all species at least once in the minimum number of townships.

[b] One alternative township choice exists.

[c] Twenty-four alternative township choices exist.

NOTE: C/T = county/township, SR = species richness, CS (%) = the cumulative number of species and percent of total species represented through the designated step, and PS = physiographic section. Physiographic section 1 is the Great Lake Section, 2 Till Plains, 3 Glaciated Allegheny Plateau, 4 Unglaciated Allegheny Plateau, and 5 Bluegrass Section.

Thus, on average, the species that were represented by the species richness method occurred in nearly twice the number of chosen townships as did those of the other methods (Table 35-1).

Twelve species were represented in all of the first six sites selected by the species richness method. These were American toads (*B. americanus*), northern two-lined salamanders (*E. bislineata*), southern two-lined salamanders (*E. cirrigera*), long-tailed salamanders (*E. longicauda*), eastern gray treefrogs (*Hyla versicolor*), eastern newts (*Notophthalmus viridescens*), northern slimy salamanders (*P. glutinosus*), mountain chorus frogs (*Pseudacris brachyphona*), spring peepers (*P. crucifer*), American bullfrogs (*Rana catesbeiana*), green frogs (*R. clamitans*), and wood frogs (*R. sylvatica*). With the exception of mountain chorus frogs and eastern newts, the above species represent 10 of the 15 most common amphibian species in Ohio (in terms of the number of townships in which they occur). The species not represented by the species richness method in the first six steps of the analysis were blue-spotted salamanders, small-mouthed

salamanders (*A. texanum*), Allegheny Mountain dusky salamanders, and Wehrle's salamanders; it took 113 additional sites utilizing this method to capture these four remaining species. Excluding small-mouthed salamanders, the other three are among the nine most rare Ohio amphibian species (in terms of the number of townships in which they occur). Therefore, the species richness method readily captured common species at multiple sites, but had difficulty representing the rare species.

To compare the ability of each of our four reserve selection methods to represent rare species, rarity scores were calculated for each. The species richness (first six sites chosen), simple greedy, progressive rarity, and branch-and-bound methods had scores of 1.82; 2.18; 2.16; and 2.16, respectively. Thus, the three algorithms better represent rare species than does the species richness method, and the simple greedy algorithm represents these species only slightly better than do the other two algorithms.

FIGURE 35-1 Locations of townships chosen to form an Ohio amphibian reserve network by the species richness (first six chosen sites), simple greedy, progressive rarity, and branch-and-bound methods.

If it is desirable to ensure that all species are represented in the reserve system multiple times, site selection can continue utilizing each method until each species is represented the desired number of times. For example, the number of townships required to represent all species at least twice and also three times by utilizing the simple greedy algorithm was 14 and 19, respectively. However, two species, blue-spotted salamanders and Wehrle's salamanders, could not be represented three times, as each is recorded from only two townships in the state. Obviously, the higher the number of times each species is represented, the higher the number of reserves required. Such species representation may not be practical due to limited resources in establishing a large number of reserves.

SUMMARY OF ALGORITHM ANALYSES

Compared to the other methods utilized here, the simple greedy algorithm performed well in (1) rapidly accumulating all species, (2) providing alternative solutions, (3) providing multiple representations of species, and (4) representing rare species. This, coupled with its ease of use, makes it a desirable method to analyze our data. This method selected six key areas for amphibian conservation in Ohio that collectively encompass all species within the state (Table 35-1). However, other data sets may not produce the same comparative results among methods (Csuti et al., 1997).

Existing Nature Reserves

SPECIES REPRESENTED

All 40 Ohio amphibian species are represented in existing Ohio nature reserves. Additionally, all species except blue-spotted salamanders, Wehrle's salamanders, and eastern spadefoot toads are located in status code 1 reserves (Table 35-2).

SELECTING TARGET AMPHIBIAN AREAS FROM ALL RESERVES

In conducting a simple greedy analysis on all Ohio reserves we found that seven reserves captured all 40 species (Table 35-3). These reserves varied in size from 14–63,380 acres and contained status code rankings of 1, 2, and 3. Therefore, concern may exist as to the feasibility of long-term population maintenance due to the size of some reserves and the protection some may receive from disturbance.

To address the size issue, we performed another simple greedy analysis on the maximum-sized reserves recorded for each species, which represented 15 reserves varying in size from 518–99,122 acres (Table 35-2). We found that all 40 species are represented in seven of these reserves and that the species are a mix of status codes 1–3 (Table 35-3).

To address the protection concern, we performed a simple greedy analysis on all status code 1 reserves, and found that the 37 species represented could be captured within six reserves (Table 35-3). However, five of the six reserves selected by the algorithm were comparatively small, ranging in size from 9–277 acres; this raises concerns about long-term population maintenance due to reserve size constraints. Status code 1 reserves are generally small, with a mean size of 222 acres and a size range from 0.5–2,234 acres.

To address both the protection and size concerns, we examined the largest status code 1 reserves and found that there are only five with over 1,000 acres; Blackhand Gorge State Nature Preserve (1,109 acres), Clear Creek Nature Preserve–Beck Compon (2,234 acres), Crane Hollow Nature Preserve (1,155 acres), Lake Katherine State Nature Preserve (1,850 acres), and Lawrence Woods State Nature Preserve (1,059 acres). Collectively, these preserves contained 31 species. Species not contained in these reserves include streamside salamanders (*A. barbouri*), blue-spotted salamanders, green salamanders, hellbenders, Allegheny Mountain dusky salamanders, cave salamanders, Kentucky spring salamanders (*Gyrinophilus porphyriticus duryi*), Wehrle's salamanders, and eastern spadefoot toads. We further found that there are only nine reserves over 500 acres, and these added no new species to those already contained in the five larger reserves (listed above). By performing a simple greedy analysis on the four largest reserves, we found that Blackhand Gorge State Nature Preserve is not needed to represent all 31 species.

SUMMARY OF EXISTING RESERVES

All of Ohio's 40 amphibian species are potentially represented in existing nature reserves. Concerns examined here regarding long-term population sustainability are reserve size and protection status. In addressing both of these concerns, we found that status code 1 reserves over 1,000 acres embrace 31 species.

Discussion

Choosing High Priority Amphibian Conservation Areas

UTILIZING THE SITE SELECTION METHODS

If our objective is to create a new reserve network for all Ohio amphibian species, then the set of sites chosen by the simple greedy algorithm should be utilized. The progressive rarity and branch-and-bound algorithms also performed well in selecting sites efficiently, but they did not perform as well as the simple greedy method when collectively considering the rapid accumulation of species, the provision of alternative solutions, the multiple representation of species, and the representation of rare species. The species richness method required substantially more reserves to incorporate all species than the other methods.

While we have identified key conservation areas for Ohio amphibians through the greedy analysis (Table 35-1), we do not mean to imply that natural and semi-natural lands outside these areas do not play an important role in maintaining regional species and ecosystem diversity (Scott et al., 1990). Additionally, even though these key conservation areas are reported as townships, it is not our suggestion to set aside these entire civil districts as reserves, but to allocate appropriate sites within these areas for reserve formation.

Several conservation organizations designate "hotspots" as key conservation areas. Such "hotspots" were located by the species richness method (Table 35-1) and contain an impressive number of amphibian species; however, recent studies (Pendergast et al., 1993; Saetersdal et al., 1993; Lawton et al., 1994; Williams and Gaston, 1994) have pointed out that areas of high species richness for different major taxa may not coincide. This raises questions about the appropriateness of selecting a reserve network to conserve maximal biodiversity utilizing this method on only one or a few major taxa. The concern of not representing different major taxa in chosen sites is emolliated when the algorithms described here are used in the selection process. For example, consider that in selecting a set

TABLE 35-2
Potential Amphibian Species Represented Within All Existing Ohio Reserves

	Reserves	Size (Acres)			Reserves in each Classification			
		Min	Max	Mean ± Standard Deviation	1	2	3	U[1]
Blanchard's cricket frog (Acris crepitans blanchardi)	142	1	99122	2776.7 ± 10445.5	22	84	35	1
Streamside salamander (Ambystoma barbouri)	44	14	10223	933.5 ± 2155.7	9	22	13	0
Jefferson salamander (Ambystoma jeffersonianum)	106	14	26377	1881.0 ± 4350.2	20	64	19	3
Blue-spotted salamander (Ambystoma laterale)	1	3103	3103	—	0	0	1	0
Spotted salamander (Ambystoma maculatum)	161	8	99122	2362.5 ± 8634.4	28	85	46	2
Marbled salamander (Ambystoma opacum)	66	1	67246	2318.6 ± 8890.1	14	38	14	0
Eastern tiger salamander (Ambystoma t. tigrinum)	73	2	5794	578.4 ± 1086.2	14	40	18	1
Small-mouthed salamander (Ambystoma texanum)	120	2	9532	1049.6 ± 2013.9	30	60	30	
Green salamander (Aneides aeneus)	8	22	518	209.8 ± 160.5	2	6	0	0
Eastern American toad (Bufo a. americanus)	324	1	99122	2067.6 ± 7978.0	65	180	75	4
Fowler's toad (Bufo fowleri)	192	1	99122	2103.3 ± 8981.7	39	99	49	5
Eastern hellbender (Cryptobranchus a. alleganiensis)	49	9	16119	1378.0 ± 2839.0	4	21	21	3
Northern dusky salamander (Desmognathus fuscus)	216	3	99122	2770.7 ± 9676.3	37	115	60	4
Allegheny Mountain dusky salamander (Desmognathus ochrophaeus)	55	14	22000	1541.2 ± 3954.5	6	37	12	0
Northern two-lined salamander (Eurycea bislineata)	248	2	99122	2607.4 ± 9170.9	37	139	67	5
Southern two-lined salamander (Eurycea cirrigera)	165	2	99122	3050.5 ± 10872.5	32	85	44	4
Long-tailed salamander (Eurycea l. longicauda)	153	3	99122	3234.5 ± 11179.0	30	74	46	3
Cave salamander (Eurycea lucifuga)	8	23	3907	1057.2 ± 1494.3	1	7	0	0
Kentucky spring salamander (Gyrinophilus p. duryi)	15	14	11980	2403.5 ± 4461.7	5	6	4	0
Northern spring salamander (Gyrinophilus p. porphyriticus)	63	14	26377	1616.4 ± 4109.4	12	30	19	2
Four-toed salamander (Hemidactylium scutatum)	73	11	26377	1990.3 ± 4503.7	23	31	18	1
Cope's gray treefrog (Hyla chrysoscelis)	48	2	99122	6964.4 ± 19162.3	8	29	10	1
Gray treefrog (Hyla versicolor)	178	2	63380	2381.7 ± 7409.7	29	106	38	5
Common mudpuppy (Necturus m. maculosus)	113	1	99122	3476.8 ± 12668.2	7	61	43	2

TABLE 35-2 (continued)

Reserves		Size (Acres)			Reserves in each Classification			
		Min	Max	Mean ± Standard Deviation	1	2	3	U [1]
Red-spotted newts (*Notophthalmus v. viridescens*)	169	9	99122	3446.1 ± 11766.5	32	88	47	2
Eastern red-backed salamander (*Plethodon cinereus*)	308	1	67246	2133.0 ± 8149.0	56	170	78	
Northern slimy salamander (*Plethodon glutinosus*)	193	4	99122	3003.2 ± 10002.1	28	107	54	4
Ravine salamander (*Plethodon richmondi*)	135	4	99122	3376.4 ± 11490.0	24	73	34	4
Wehrle's salamander (*Plethodon wehrlei*)	1	63380	63380	—	0	0	1	0
Mountain chorus frog (*Pseudacris brachyphona*)	44	12	99122	8473.3 ± 21033.9	11	17	16	0
Spring peeper (*Pseudacris crucifer*)	262	1	99122	2683.2 ± 9597.7	47	140	72	3
Western chorus frog (*Pseudacris triseriata*)	174	1	99122	2401.6 ± 9189.2	35	98	39	2
Midland mud salamander (*Pseudotriton montanus diastictus*)	15	88	61724	7810.7 ± 16634.8	3	6	6	0
Red salamander (*Pseudotriton ruber*)	94	5	61724	2393.2 ± 7287.0	15	52	26	1
American bullfrog (*Rana catesbeiana*)	279	0.5	99122	2635.1 ± 9285.2	44	165	67	3
Northern green frog (*Rana clamitans melanota*)	334	2	99122	2366.0 ± 8523.7	61	196	71	6
Pickerel frog (*Rana palustris*)	134	3	99122	4007.2 ± 13054.6	21	73	37	3
Northern leopard frog (*Rana pipiens*)	249	0.5	99122	2852.9 ± 9818.5	40	147	59	3
Wood frog (*Rana sylvatica*)	203	2	99122	2677.9 ± 9776.7	39	116	46	2
Eastern spadefoot toad (*Scaphiopus holbrookii*)	2	3	11961	4188.0 ± 6738.2	0	1	1	0

NOTE: Occurrence of species within reserves is not confirmed, but is suggested from amphibian distribution data, which is reported by township. See text for reserve classification definitions.

[1] Unclassified reserves.

of areas in which all species of one major taxon are represented, the chosen areas reflect dissimilar species lists, often the result of dissimilar environments. Such dissimilar environments may then also reflect differing species lists within other major taxa. With Ohio amphibians, the species richness method selected townships contained solely within one physiographic section of the state, and these sites were confined to Ohio's south-central portion. However, the other methods utilized selected sites within several physiographic sections and were more scattered around the state, indicating more variation in habitat among sites. The algorithms used here select complementary areas (those containing different suites of species), which may contain relatively few species, but are included as reserve sites because of maximally different species content. As a result, a set of areas in which one major taxon is completely represented

may also fairly represent the diversity of unrelated taxa, a phenomenon known as the "sweep effect" (Kiester et al., 1996). Therefore, the amphibian sites selected by the complimentary analyses we utilized (simple greedy, progressive rarity, and branch-and-bound methods) may also represent diversity of other major taxa.

UTILIZING EXISTING RESERVES

If our objective is to utilize existing reserves to conserve Ohio's amphibians, it seems reasonable to utilize four of the five largest (over 1,000 acres) status code 1 reserves to represent 31 of the state's 40 amphibian species (78%). These reserves may be large enough to sustain long-term viable populations (although no data are available to support this), and they are

TABLE 35-3

Simple Greedy Analysis of Status Code 1 Reserves, All Ohio Existing Reserves, and Maximum Size Reserve for Each Species

Step Number	Reserve Description	Status Code 1 Reserves	All Reserves	Maximum Size Reserve/Species
1	Name	Clear Creek Nature Preserve-Beck Compone	Tar Hollow State Forest	Tar Hollow State Forest
	C/T	Hocking/Good Hope	Hocking/Salt Creek; Vinton/Eagle; Ross/ Colerain, Harrison	Hocking/Salt Creek; Vinton/Eagle; Ross/ Colerain, Harrison
	Size	2234	16,119	16,119
	SR	26	26	26
	CS (%)	26 (70%)	26 (65%)	26 (65%)
	C	1	3	3
2	Name	Newberry Wildlife Sanctuary	Edge of Appalachia-Buzzardroost Rock[a]	Edge of Appalachia-Buzzardroost Rock
	C/T	Hamilton/Colerain	Adams/Brush Creek	Adams/Brush Creek
	Size	100	518.5	518.5
	SR	15	21	21
	CS (%)	30 (81%)	32 (80%)	32 (80%)
	C	1	2	2
3	Name	Davis Memorial State Nature Preserve	Killbuck Wildlife Area[a]	Ravenna Arsenal[e]
	C/T	Adams/Meigs	Holmes/Holmes; Wayne/Clinton, Franklin, Wooster	Portage/Charlestown, Freedom, Paris, Windham; Trumbull/Braceville
	Size	88	5,484	22,000
	SR	24	22	20
	CS (%)	33 (89%)	36 (90%)	35 (88%)
	C	1	2	3
4	Name	Liberty Fen Nature Preserve[a]	Wayne National Forest-Marietta Unit[c]	Wayne National Forest-Marietta Unit[f]
	C/T	Logan/Liberty	Monroe/Bethel, Benton, Center, Green, Jackson, Perry, Washington, Wayne	Monroe/Bethel, Benton, Center, Green, Jackson, Perry, Washington, Wayne
	Size	9	63,380	63,380
	SR	3	19	19
	CS (%)	35 (95%)	37 (93%)	37 (93%)
	C	1	3	3
5	Name	Adams Lake Prairie Nature Preserve[a]	Madison Township Park[d]	Pike State Forest[e]
	C/T	Adams/Tiffin	Lake/Madison	Pike/Benton, Mifflin, Perry
	Size	22	14	11,961
	SR	9	20	23
	CS (%)	36 (97%)	38 (95%)	38 (95%)
	C	1	3	3
6	Name	Burton Wetlands Nature Preserve[b]	Pike State Forest[e]	Ottawa National Wildlife Reserve[f]
	C/T	Geauga/Burton, Newberry	Pike/Benton, Mifflin, Perry	Lucas/Jerusalem, Oregon; Ottawa/Benton, Carroll
	Size	277	11,961	5,794
	SR	17	23	8
	CS (%)	37 (100%)	39 (98%)	39 (98%)
	C	1	3	2

TABLE 35-3 (continued)

Step Number	Reserve Description	Status Code 1 Reserves	All Reserves	Maximum Size Reserve/Species
7	Name		Maumee State Forest	Maumee State Forest
	C/T		Fulton/Swan Creek; Lucas/Providence; Henry/Washington	Fulton/Swan Creek; Lucas/Providence; Henry/Washington
	Size		3,102	3,102
	SR		16	16
	CS (%)		40 (100%)	40 (100%)
	C		3	3

[a] One alternative reserve choice exists

[b] Five alternative reserve choices exist

[c] Fifty-nine alternative reserve choices exist

[d] Fifty-eight alternative choices exist

[e] Three alternative choices exist

[f] Two alternative choices exist

NOTE: There are 40 species found in all reserves, and 37 in status code 1 reserves. Reserve size is in acres, C/T = county and township in which the reserve is located, SR = species richness, CS (%) = the cumulative number of species and percent of total species represented through the designated step, and C = the reserve status code (see text for definition).

provided protection from disturbance under Ohio law. The remaining nine species occur in one or more status code 2 reserves and could be designated to their protection.

Combining Existing and New Reserves into a Network System

Many possible combinations of existing and new reserves are available to represent all Ohio amphibian species. Factors presented here influencing these combinations include species content, size, status code, and whether the reserve currently exists. If our objective is to contain all species in highly protected areas, then we would want to either form a novel reserve network with status code 1 designation or utilize the existing status code 1 reserves in some combination with novel reserves. For example, we identified that 31 species are contained in a minimum network of an existing four, large (> 1,000 acres) status code 1 reserves. By establishing additional status code 1 reserves to protect the remaining nine species, all 40 species would be represented. Identification of the additional reserve locations could be aided by using a simple greedy analysis. If we feel comfortable in utilizing status code 2 reserves, then we could also include these in our combination of reserves. Doing so would enhance the size of reserves chosen for many species.

Considerations in Forming Ohio Amphibian Reserves

OVERALL CONSIDERATIONS

While we propose to form a reserve network for all Ohio amphibian species, it is beyond the scope of this paper to specifically identify the size, exact location, and configuration of each of these reserves necessary to maintain viable amphibian populations or functional ecosystems. However, to begin the examination of necessary Ohio reserve characteristics, the preliminary considerations regarding reserve attributes for conservation of amphibians are presented below.

RESERVE SIZE

Concern regarding reserve size has mainly focused on mammals and birds (Schoenwald-Cox, 1983; Shaffer, 1983; Newmark, 1987; Salwasser et al., 1987; Grumbine, 1990; Beier, 1993). From these and other studies, larger reserves are generally considered better for maintenance of individual species, biodiversity, and ecological function. For example, larger reserves capture a greater number of species than do smaller reserves, and continued subsistence of these species over long periods of time may depend on reserve size. Persistence of species in reserves varies greatly by taxon and life history characteristics. In general, large-bodied, low-population density, upper-trophic-level species with large individual ranges need a greater area to maintain long-term viable populations than do smaller-bodied, higher-density, lower trophic-level species with smaller ranges. Thus, when compared to such animals as bears and cougars, amphibians require smaller reserves to maintain long-term viable populations. However, due to the extensive areas needed to conserve animals like bears (600–1,600 km^2; 148,262–395,364 acres; Shaffer, 1983) and cougars (1,000–2,200 km^2; 247,103–543,626 acres; Beier, 1993), addressing the spatial needs of these and similar species can also accomplish conserving the viability of the entire ecosystem in which they live (Soule, 1987; Noss, 1991). There are currently no methods to determine the minimum areas of reserves with reference only to ecosystem properties, so these "indicator" or "umbrella" species represent an efficient way to address ecosystem conservation. Establishing amphibian reserves of this magnitude in appropriate amphibian habitat would clearly be large enough to conserve populations well into the future, and it would have the added enhancement of conserving entire ecosystems. However, forming such expansive reserves for amphibians in Ohio is not realistic.

MINIMUM VIABLE POPULATIONS

To get an idea about how large an amphibian reserve must be in order to sustain viable populations of targeted species for an extended period of time, several factors need to be considered. We must know how many individuals of a given amphibian species are necessary to sustain the population for the time interval indicated, and then make sure the reserve is large enough to accommodate that population. A minimum viable population (MVP; a population size below which extinction is quite likely) for any given amphibian species can be estimated from a population viability analysis (a structured, systematic, and comprehensive examination of the interacting factors that place a population or species at risk) using such variables as demographic, genetic, and environmental stochasticities, habitat fragmentation, and suitability (Shaffer, 1990). The accuracy of the estimations is dependent on the amount and quality of the data available for analysis. Unfortunately, few such data sets are available for amphibians. This problem is compounded by the possibility that the variables analyzed are not constant across the range of any given species or through time. In other words, life history parameters like age specific fertility, age specific mortality, and average generation time can vary across the range of a species and even from year to year within a given area. For example, annual oogenic cycles may be normal for some populations of eastern red-backed salamanders (*P. cinereus*), northern slimy salamanders, and Wehrle's salamanders, while other populations of these species apparently have biennial cycles (Houck, 1977b; Tilley, 1977).

Thinking in more simplistic terms, there has been discussion in the conservation biology literature on the number 500 as an effective population size for long-term survival (Dawson et al., 1987; Lande and Barrowclough, 1987). This number originated with Franklin (1980), and additional support was provided by others (see Soule, 1987). However, Lande and Barrowclough (1987) point out that 500 is only about the right order of magnitude (i.e., as low as 50 and as high as 5,000 individuals). Because amphibians are comparatively short-lived vertebrates, viable populations are probably at the higher end of this scale.

Assuming that we could estimate minimum viable population sizes for the 40 Ohio amphibian species, it would seem that we could then combine this figure with known densities or territory sizes to estimate the needed reserve size. However, there are several problems with this method. Population densities and territory sizes can vary from population to population within a given species, so what is identified as a suitably sized area for population maintenance in one location may not be sufficient in another. For example, eastern red-backed salamanders demonstrate great variation in the densities of local populations (see Burton and Likens, 1975b, for references). To further confuse the issue, methods of censusing can yield differing results. A surface census in Michigan provided an estimate of 0.0496/m^2, whereas digging plots of dry and wet litter resulted in estimates of 0.0900 and 0.8900/m^2, respectively. Additionally, published densities normally are from plots of appropriate habitat where the species is known to occur and do not take into consideration areas of inappropriate habitat (even those near the sample areas) where specimens of the target species are not found. Further, densities can change within a given area depending on the time of year. For example, densities of tiger salamanders (*A. tigrinum*) may be quite high in plots near breeding pools in the spring, yet much lower in those same plots during the summer months.

Generally, reserves will contain patches of appropriate habitat for targeted species, so simple division of density estimates into minimum viable population size would likely underestimate the required reserve size for long-term maintenance of the targeted species. Additionally, if reserve size is estimated by summing the population sizes among patches until this figure reaches the calculated minimum viable population size, then gene flow must be assured between patches if a viable metapopulation (network of semi-isolated populations) is to be maintained. Therefore, amphibian reserves must not only be of sufficient size to accommodate a MVP, but also be of appropriate configuration to assure gene flow among patches of suitable habitat (Semlitsch and Bodie, 1998; Semlitsch 2000a, 2000b). Adding to this complexity are migration or annual movement distances and patterns. Several species such as spotted salamanders (*A. maculatum*) and northern leopard frogs (*R. pipiens*) can move substantial distances during the course of a year.

HABITAT HETEROGENEITY

Certain species utilize different habitats during different life stages or during the course of a year, so all habitat types utilized by the target species must be represented (Semlitsch, 1999; Semlitsch and Jensen, 2001). For example, northern leopard frogs overwinter in permanent waters (i.e., large ponds, lakes, and streams). In early spring they seek shallower water for breeding (i.e., temporary ponds, stream backwaters, and marsh pools), sometimes migrating over one kilometer (0.6 miles). In summer they often disperse well away from the water into meadows and other grassy places, and more rarely into open woodlands. In the fall they migrate back to their hibernation sites (Harding, 1997). So in designing a reserve to include leopard frogs, all habitat types mentioned above must be represented and the reserve must be large enough to accommodate the frog migrations.

If the goal of an amphibian reserve is to maintain several species that have different habitat needs, then an area that is spatially and temporally heterogeneous is superior to one that is not. For example, tiger salamanders and southern ravine salamanders (*P. richmondi*) are both recorded from Athens County/Athens Township. The former species lives not only in woodlands, but also in more open habitats such as marshes, grasslands, farmlands, and even suburbs; the latter species inhabits the slopes of woodland ravines and valleys. If a reserve was to be established for these two species in Athens Township, appropriate habitat for both would have to be included. Encompassing this heterogeneity within the reserve may require a large and appropriately configured area. Heterogeneity of habitat types promotes heterogeneity in species composition due to the array of habitats available. Such habitat heterogeneity also better accommodates disturbance than does a homogeneous reserve by offering species a diversity of habitat types at any given time. If a particular habitat patch is altered or destroyed by disturbance, or if it undergoes succession and becomes unsuitable for a given species, then other appropriate patches may be available for colonization within the reserve. Thus, patches come and go, and the populations of specialists living within them must migrate to suitable patches or become extinct. Pickett and Thompson (1978) described this changing system as patch dynamics, and they emphasize its importance in maintaining overall diversity.

EDGE EFFECTS

Reserve size and shape have implications for the expected degree of edge effects. In the negative sense, edge effects are the detrimental influences of habitat edges on the interior conditions of that habitat or on the species that use the interior of the habitat (Meffe and Carroll, 1997). In conservation biology, focus has been given to the edge effects resulting from habitat destruction—where once there was a continuous area of habitat supporting a variety of species, there is now an area of remaining habitat patches fragmented by such activities as forestry, agriculture, or housing development. From the more positive perspective, edge effects can also be considered as the effect of adjoining natural habitat types on populations in the edge ecotone, with some of these species requiring this ecotone in order to survive.

Small or narrow reserves have a greater edge-to-interior ratio, thereby increasing edge effects and decreasing the amount of true interior habitat. In fact, edge effects may invade throughout small or narrow patches. This shift can make the reserve more vulnerable to invasion by exotic, domestic, and edge species, and subject it to edge environmental influences such as temperature extremes and increased winds (Saunders et al., 1991). Therefore, amphibian species that are specialists for a given habitat, like Allegheny Mountain dusky salamanders, may be negatively impacted by edge effects. This species is dependent on the existence of shady woodlands with cool, flowing waters such as small streams and springs. The potential impact of edge effects raising ground and stream temperatures could prove detrimental. Therefore reserves need to be of appropriate size and shape to contain core areas free of these effects. However, species that are generalists, such as American toads, may benefit from edge effects, and such ecotones should be components of reserve selection for them.

HUMAN ACTIVITIES

Human activities adjacent to or within defined amphibian reserves can have an impact on reserves. Because amphibians are very moisture dependent and their skins absorbent of chemicals contained in water, water contaminated by certain farming and industrial practices which then flows into the reserve from surrounding areas can negatively impact amphibian populations.

Alteration of drainage patterns through channeling and damming, or heavy water consumption for irrigation or drinking outside the reserve may alter the amount of water entering a reserve through seeps, streams, springs, and rivers. Amphibian populations such as those of northern dusky salamanders (*D. fuscus*) that are most common in shaded streamside habitats could be harmed by stream pollution from agricultural runoff or siltation. Northern spring salamanders (*G. p. porphyriticus*) require clear, cool waters free of predatory fish. Their relatively large size compared to other lungless salamanders (with less oxygen-absorbing surfaces compared to body mass) suggests that they would be very sensitive to changes in dissolved oxygen levels (Harding, 1997). Stream pollution or siltation from activities outside the reserve and even the introduction of predatory fish could adversely affect this species. Therefore, in selecting an amphibian reserve site, it is desirable to find a location where surrounding activities have little potential of negatively impacting amphibian populations within the reserve.

BUFFER ZONES

Some reserves establish buffer zones, often a mix of public and private lands, to dilute the impacts of detrimental activities on their borders from its core (Meffe and Carroll, 1997). Buffer zones often allow activities not permitted in the core of the reserve, such as manipulative research, habitat rehabilitation, and harvest of natural crops (i.e., nuts and mushrooms), yet do not allow activities that will damage the core. Activities in the core are permitted, but they are of a non-destructive nature

such as ecological monitoring, photography, hiking, and bird watching. Some reserves even specify an area outside the buffer zone, referred to as the transition area, which typically includes human settlements and associated activities such as fishing, forestry, agriculture, and other sustainable economic pursuits consistent within the protection of the core area.

CORRIDORS

The importance of facilitating gene flow within reserves was emphasized in the "Minimum Viable Population" section above. In the broader context, it can also be important to provide gene flow corridors between reserves, especially if reserves are small. Because the townships within the sets chosen by each algorithm used here are not adjacent to one another (Fig. 35-1), establishing corridors between these sites may not be possible. However, if currently protected land is near the chosen sites, then establishment of corridors linking these should be explored.

SUMMARY OF AMPHIBIAN RESERVE CONSIDERATIONS

An Ohio amphibian reserve network encompassing all the state's 40 species can be formed through either a new set of reserves (as outlined in Table 35-1), existing reserves, or a combination of existing and new reserves. However the network is formed, reserve size, minimum viable populations, habitat heterogeneity, edge effects, human activities, buffer zones, and corridors must be considered. It is important that the attributes of each reserve are compatible with the long-term population survival of each of its targeted species; and these attributes vary from species to species. Additionally, within a species, reserve design needs to be based on the population characteristics at the targeted reserve site, as such characteristics can vary from area to area. Therefore, each reserve must be individually designed.

Conclusions and Recommendations

The algorithms we utilized selected efficient combinations of new reserves to incorporate all amphibian species in Ohio (Table 35-1). Reserves currently exist that potentially incorporate all 40 amphibian species. Existing reserves vary in size from 0.5–99,122 acres, and most are classified as status code 1, 2, or 3 (with 1 receiving the highest protection). All species occur in reserves greater than 3,000 acres, except green salamanders, which are found in reserves up to 518 acres. These reserves may be large enough to sustain long-term viable amphibian populations, although no data are available to support this. Existing status code 1 reserves contain only 37 (93%) of the state's species, so if it is desirable to contain all species in code 1 reserves, new such reserves must be established. However, status code 1 reserves are on average smaller than the other reserves, and only 31 species (78% of total) occur in reserves of 1,000 acres or larger. Reserves of this size may be large enough to sustain long-term viable amphibian populations, although no data are available to support this. Therefore, if only status code 1 reserves of 1,000 acres or larger are utilized, new such reserves must be established if all 40 Ohio amphibian species are to be represented. However, efficient combinations of existing, large status code 1 and status code 2 reserves can also be utilized to represent all species.

These analyses, however, do not address the issues of shape, quality, or minimal size of the natural areas needed to support amphibian populations. Neither do they indicate that viable populations or even extant populations of species indicated from the distribution data still exist in the targeted areas. Field surveys and interactions with those familiar with the areas selected are needed to confirm the conservation value of potential reserve sites. Further, landscape context should be considered during natural area design (Noss, 1987; Semlitsch, 2000b). Additional factors, such as acquisition and management costs, political constraints, and proximity to other reserves may also need to be addressed when recommending a realistic natural area network (Pressey et al., 1996).

It is not the purpose of this paper to emphatically determine which areas should be components of an Ohio amphibian network, but rather to present practical methods, thought processes, and other considerations to begin implementation of such an initiative. We propose that the next phase of this reserve network formation is to conduct a multidisciplinary workshop of participants knowledgeable in the above factors to refine the results of our analyses into a real-world context. Analyses can be re-run as new information is provided by the experts at this gathering, creating a realistic amphibian reserve network.

Accompanying the establishment of a reserve network should be educational programming and marketing that defines the purpose, uniqueness, and identity of each component of the network. Such a reserve network will be unique among all U.S. states, one that targets conservation of all amphibian species within a state's political boundaries.

Summary

Amphibians are declining globally. To address this issue locally, we present practical methods, thought processes, and other considerations to stimulate establishment of a reserve network for all of Ohio's amphibian species. Potential reserve locations were determined through amphibian distribution data as analyzed utilizing four techniques: (1) simple greedy algorithm, (2) progressive rarity algorithm, (3) branch-and-bound algorithm, and (4) species richness method. The three algorithms each represented all 40 species of Ohio amphibians within six sites, whereas the species richness method required 119 sites. Alternative sites were available for many of the sites selected by the algorithms, adding flexibility to the site selection process. On average, the species that were represented by the species richness method occurred in nearly twice the number of chosen sites as did those of the other methods; however, the algorithms demonstrated better representation of rare species. The simple greedy method was utilized to determine that the number of sites required to represent all species at least two and three times was 14 and 19, respectively. It is suggested that the three algorithms not only selected sites for amphibians efficiently, but that these same sites are also important conservation areas for other groups of animals and plants. The simple greedy method was utilized on large, existing Ohio reserves, and it was found that only seven such reserves were required to embrace all 40 species. Most of Ohio's existing reserves are classified status code 1, 2, or 3, with code 1 reserves considered the most highly protected. Large code 1 reserves represent habitats for all but nine species. We discuss utilizing combinations of existing and new reserves in forming an Ohio amphibian reserve network. It is beyond the scope of this paper to specifically identify the size, exact location, and configuration of each of these reserves necessary to maintain viable amphibian populations or functional ecosystems. However, we do present preliminary considerations regarding amphibian reserve attributes.

Acknowledgments. We are grateful to Jeffrey Camm for conducting the branch-and-bound analysis of data provided and we thank Blair Csuti for guidance with current nature reserve formation literature and for reviewing the manuscript. Jeff Davis generously provided recent anuran distribution records. We extend gratitude to Chris Kuhar, Meghan McNamara, Linda Quast, Gloria Rivera, Kisha Roach, and Jordan Schaul for assisting with data entry and analysis and to Holly Quinn for reviewing the manuscript. For providing valuable information about existing Ohio nature reserves we thank Nancy Strader, Ashtabula County Parks District, Big Brothers–Big Sisters Association, Boy Scouts of America, Cincinnati Museum of Natural History, Cleveland Audubon Society, Cleveland Museum of Natural History, Cleveland Metroparks, Columbus and Franklin County Metro Park District, Erie Metro Parks, Geauga Park District, Hamilton County Park District, Kent State University, Lake Metroparks, Lorain County Metroparks, Montgomery County Parks, NASA, Oberlin College, Ohio Department of Natural Resources Division of Natural Areas and Preserves, Ohio University, Ohio Wesleyan University, Summit County Park District, The Nature Conservancy, and Toledo Metropark District.

THIRTY-SIX

Creating Habitat Reserves for Migratory Salamanders

SUZANNE C. FOWLE AND SCOTT M. MELVIN

Habitat loss and fragmentation results in the reduction and isolation of amphibian populations (Reh and Seitz, 1990; Gulve, 1994) and the subsequent increased risk of local extinction (Saccheri et al., 1998). While local extinctions are often part of amphibian population dynamics, amphibian populations persist because such extirpations are compensated for by recolonization and the resulting rescue effect (Skelly et al., 1999). However, the fragmentation of amphibian habitats inhibits dispersal and thereby hinders or prevents the rescue effect (Reh and Seitz, 1990; Gulve, 1994).

Massachusetts provides a prime example of this conservation challenge and of the need to protect connected habitat complexes for amphibians, especially for migratory salamanders. Massachusetts is the third most densely populated state in the union, with an average of 765 residents per square mile, ten times the national average of 74 residents per square mile (MEOTC, 1998). Massachusetts contains 68,642 lane miles of roads. The largest roadless area is the 3,400 acres (0.07% of the state's total land area) that surrounds the Quabbin Reservoir in central Massachusetts. This "roadless" area does include limited-access dirt roads.

Despite relatively strong state laws protecting wildlife and habitats, populations of ambystomatid salamanders are threatened by the densities of development and roads in Massachusetts. Of the four native ambystomatids in Massachusetts, three are listed by the Massachusetts Division of Fisheries and Wildlife (DFW) as *Threatened* (marbled salamanders, *Ambystoma opacum*) or *Species of Special Concern* (Jefferson salamanders, *A. jeffersonianum*, and blue-spotted salamanders, *A. laterale*). Only spotted salamanders *(A. maculatum)* are considered common.

Regulatory measures alone are unlikely to protect rare salamanders from further decline in Massachusetts. Regulations do not recognize the full extent of the habitats on which ambystomatid salamanders depend. Even when breeding habitats are protected by the Massachusetts Wetlands Protection Act and its implementing regulations, these regulations do not apply to upland habitat, nor do they require that upland and wetland habitats remain connected and free of barriers to salamander movement. Although the Massachusetts Endangered Species Act can protect both wetland and upland habitats of rare salamanders, upland habitats between breeding sites are

difficult to delineate, and, subsequently, to regulate (Melvin and Roble, 1990). In addition, because state environmental regulations are applied to individual projects as they are proposed, the regulations fail to prevent the cumulative loss of habitat brought on by the gradual encroachment of urban and suburban sprawl on remaining wildlands.

An effective strategy for statewide conservation of ambystomatid salamanders involves proactive protection of habitats, rather than solely relying on regulations. The Natural Heritage and Endangered Species Program of the DFW is currently working to identify and map conservation areas for rare herpetofauna, including marbled, Jefferson, and blue-spotted salamanders. To design effective reserves, we need to know the minimum areas necessary to sustain viable populations, as well as the landscape and habitat features that these areas should contain (see Quinn and Scott, this volume). To address these questions, we need information on (1) the minimum viable population size for each species; (2) population density; (3) dispersal distances; and (4) habitat requirements of each species, including the entire array of habitat types used during an annual cycle as well as during a drought cycle (see Lannoo, 1998a).

Our project illustrates both a conservation strategy and the information needs of managers attempting to protect amphibian habitats. Initially, the effectiveness of habitat reserves and the process of designing them may be limited by a lack of empirical data to guide our efforts. We have limited information on upland habitat use and dispersal distances of these animals, so designing reserves for ambystomatid salamanders involves working with some degree of uncertainty about the types and extents of critical habitats. However, waiting until we answer all the unknowns will certainly result in continued loss and fragmentation of habitats.

Current Legal Protection for Rare Herpetofauna in Massachusetts

Conservation laws in Massachusetts have not succeeded in comprehensive protection of connected upland and wetland complexes on which populations of ambystomatid salamanders depend (Melvin and Roble, 1990). Wetland habitats are

TABLE 36-1
Maximum Dispersal Distances of *Ambystoma maculatum* and *A. jeffersonianum* from Breeding Sites.

Location	Individuals Studied	Maximum distance moved (m)	Source
Ambystoma maculatum:			
Michigan	2	108	Wacasey, 1961
Michigan	6	249	Kleeberger and Werner, 1983
Indiana	7	125	Williams, 1973
New York	8	210	Madison, 1997
Kentucky	8	220	Douglas and Monroe, 1981
Michigan	14	200	Wacasey, 1961
Ambystoma jeffersonianum:			
Michigan	6	108	Wacasey, 1961
Michigan	45	231	Wacasey, 1961
Indiana	86	625[1]	Williams, 1973

[1] We are basing reserve boundaries for blue-spotted salamanders on the known migration distance for Jefferson salamanders; blue-spotted dispersal and migration distances have not been recorded and hybrids of these species are often not distinguishable from pure genotypes.

NOTE: From Semlitsch, 1998.

protected from direct impact by the Massachusetts Wetlands Protection Act, which prohibits "short or long-term adverse impacts" to habitats of state-listed, wetland-dependent wildlife (MGL c. 131 s. 40). This protection extends to a 100-foot-wide buffer zone around most jurisdictional wetlands, but it otherwise fails to protect the full extent of uplands used by migratory salamanders (Semlitsch, 1998). In addition, only a subset of seasonal pools, the primary breeding habitat of all state-listed ambystomatids in Massachusetts, is protected.

The Massachusetts Endangered Species Act (MESA) (MGL c. 131A) is a more effective regulatory tool for protecting upland habitats of state-listed species. MESA prohibits the "taking" of any species of animal or plant listed as *Endangered*, *Threatened*, or *of Special Concern*. For animals, "taking" is defined as "to harass, harm, pursue, hunt, shoot, hound, kill, trap, capture, collect, process, disrupt the nesting, breeding, feeding, or migratory activity or attempt to engage in any such conduct, or to assist in any such conduct." Although upland habitats can therefore be protected under MESA, these habitats are difficult to delineate, especially given the dearth of site-specific information on upland habitat use by salamanders. As a result, the determination that a proposed activity will result in a "taking" is often difficult to defend.

Without strategic conservation planning and proactive identification of priority habitat areas, both the Wetlands Protection Act and MESA may fail to adequately protect salamander habitat from long-term cumulative effects of multiple development projects. Substantial, gradual erosion of habitat quantity and quality may occur over the period of several decades. Three types of development projects can lead to this erosion: (1) multiple small projects with individual adverse effects that are relatively small or difficult to document; (2) larger projects that are permitted with compensatory mitigation or that are modified to minimize adverse effects; or (3) projects permitted in areas where regulators are unaware of the occurrence and distribution of rare species. Regulatory

protection will be most effective for sites that have been proactively identified, described, and mapped as priorities for conservation.

In response to the need for more effective, proactive habitat protection, the Massachusetts DFW has begun a program to identify and protect habitat reserves for rare herpetofauna. We are starting by identifying and mapping priority habitats for three species of migratory salamanders, all of which are state-listed: marbled salamanders, Jefferson salamanders, and blue-spotted salamanders.

Selecting Sites for Potential Reserves

The first stage of our conservation strategy is to identify potential reserve sites. Because the available landscape data (e.g., GIS [Geographic Information System] data layers) for Massachusetts do not include most vernal pools and other small wetlands, we are using a qualitative approach to initial site selection. We are using the following criteria to develop a preliminary list of sites: (1) known occurence of target species (from records contained in our Natural Heritage database); (2) relative lack of habitat fragmentation; (3) diversity of wetland habitat types; (4) presence of clusters of three or more vernal pools within 625 m of each other (based on the maximum overland migration or dispersal distance that has been published for any of our target species, in this case, adult Jefferson salamanders; Williams, 1973; Table 36-1); and (5) proximity to existing conservation land.

Research Needs

As more detailed landscape data are compiled for Massachusetts, we would like to systematically search (e.g., using GIS) for areas of sufficient size and habitat type to enable the persistence of our target species. To successfully conduct such a query, we need answers to the following questions:

1. What is the minimum area that could support a viable ambystomatid population?

2. If we assume the minimum area is a function of the number and density of breeding sites and other wetlands (see Semlitsch et al., 1996; Skelly et al., 1999), how many breeding sites are necessary and at what density?

3. Assuming the minimum area is a function not only of the number and density of breeding sites but also of the diversity of wetlands available, which enables populations to persist through drought cycles (see Lannoo, 1998a) and wetland succession, what types of wetlands are necessary and at what density?

Site Prioritization

Our next step is to rank the sites according to their importance to the persistence of rare salamanders. Our data needs include estimates of relative abundance, local distribution patterns, local habitat conditions, and threats. In the field, we have focused our surveys on breeding pools, tallying numbers of egg masses of Jefferson and blue-spotted salamanders and larvae of marbled salamanders. However, these rapid assessment techniques have not been tested for their accuracy and efficiency in determining relative or absolute abundance of marbled, Jefferson, or blue-spotted salamanders. Several important questions remain about the value of one or two seasons' data in indicating breeding population size (Pechmann et al., 1991; Skelly et al., 1999). In addition, eggs and larvae of rare ambystomatid salamanders are difficult to locate. As a result, we have low sample sizes of egg masses and larvae, compounding the problems associated with short-term surveys.

Research Needs

We must be able to conduct more effective and efficient rapid assessments in the field that provide us with data for identifying and prioritizing potential reserve sites. For example, a model that allows us to predict population size from egg mass abundance would help direct our conservation planning. Important questions include the following:

1. Is there a strong relationship between numbers of eggs or egg masses and breeding population sizes of marbled, Jefferson, and blue-spotted salamanders?

2. If this relationship exists, will counts over multiple years significantly improve estimates?

3. If this relationship does not exist, are there ways to rapidly assess other age classes (e.g., survey for larvae or breeding adults) and gain insight into breeding population size, actual population size, or relative densities of breeding or entire populations?

4. Is there a strong relationship between breeding population size and entire population size?

A key but poorly understood element of our conservation planning is the minimum size of local salamander populations that should be targeted for protection. Information on sensitivity of persistence probabilities to initial population size would improve our ability to rank sites from our initial list and ensure the effectiveness of reserves, but this information is not currently available for these species. With growing evidence that populations of ambystomatids persist in a constant local extinction-recolonization dynamic (Semlitsch et al., 1996; Skelly et al., 1999), population viability analyses are further complicated. For example, viability will likely vary according to the distance between local populations and the habitat types connecting those populations.

Reserve Design

A logical approach to reserve design for migratory salamanders is to draw radii out from the edges of known breeding pools, encircling the pools with protected upland. Based on available information, an effective reserve for our target species would encompass critical upland habitat (defined as the maximum dispersal or migration distance for the species to be protected) surrounded by a 35 m buffer zone to protect against edge effects (deMaynadier and Hunter, 1998). The upland within a reserve (including the buffer) should be free of barriers and unsuitable habitats, such as roads and clearcuts (Gibbs, 1998; deMaynadier and Hunter, 1999), so that dispersal and the extinction–recolonization dynamic can continue.

The maximum distances away from breeding pools that adult Jefferson and marbled salamanders have been recorded and published are 625 m and 450 m, respectively (Williams, 1973). Based on these data, we recommend that a reserve designed to protect marbled salamanders alone encompass 450 m of suitable upland around each breeding pool; a reserve for Jefferson and blue-spotted salamanders should encompass 625 m of suitable upland around each breeding pool (Table 36-1). These boundaries should then be extended by a 35 m buffer zone (deMaynadier and Hunter, 1998).

In the absence of population viability analyses for our target species, maximum dispersal distances must suffice as our criteria for delineating reserve boundaries. However, as estimates of minimum viable population sizes and extinction probabilities are developed, a more effective strategy might be to determine the area and habitat features necessary to support viable populations. In other words, if we can develop a better understanding of how population size is related to long-term population persistence, and if we have estimates of actual population densities in representative habitats, we can then begin to estimate the minimum areas necessary to support viable populations.

Research Needs

The current lack of information on salamander dispersal and habitat requirements impedes our efforts to confidently delineate reserve boundaries. Based on published studies of salamander movements, Semlitsch (1998) recommends a 164 m-wide "life zone" around salamander breeding pools. However, examination of the literature on dispersal of ambystomatid salamanders indicates that the best available information may be underestimates. Among the spotted salamander and Jefferson salamander studies that Semlitsch (1998) summarized, the maximum dispersal distance recorded appears to be positively correlated with sample size (Table 36-1). Williams (1973) noted this as well. When he increased his sample size of radio-tagged marbled salamanders, from six in one breeding season to twelve in the next breeding season, the average dispersal distance increased. Increased sample sizes and study duration may indicate that ambystomatids travel farther than is currently reported in the literature.

Our project has led us to the following dispersal and migration-related questions for researchers:

1. How far do ambystomatid salamanders migrate and disperse?

2. What is suitable migratory and dispersal habitat? Will ambystomatid salamanders disperse via a relatively narrow "corridor" or do they require a wider connecting zone?

As previously noted, we need information on population viability in ambystomatid salamanders. What are the minimum population sizes (or metapopulation sizes and configurations) necessary to ensure high probabilities of long-term persistence? How strongly are extinction probabilities related to initial population size, changes in survival rates or reproductive success, and variability in demographic parameters? For example, do we gain substantially, in terms of probability of population persistence, by protecting a population of 1,000 salamanders versus 100?

Reserve Management

When we designate critical upland habitat and buffer zones and draw boundaries around proposed reserves, we will inevitably capture residences and roads within those boundaries. Because this is unavoidable in Massachusetts, we are developing conservation guidelines for landowners in and near rare salamander habitat. Fortunately, the implementing regulations of MESA and the Wetlands Protection Act provide legal grounds for these guidelines, although the guidelines themselves are not regulations. Given that the Massachusetts Wetlands Protection Act does not allow any "short or long-term adverse impacts" to wetland habitats of rare wildlife, and that MESA prohibits "takings" of rare species, our guidelines will outline activities that may constitute violations of these regulations. Examples of such activities that may adversely affect salamanders and/or their habitats include the following:

- Destroying breeding pools or any portion of them by filling or draining.

- Degrading breeding pools by increasing erosion and sedimentation, clearing trees in and around pools, or discharging runoff and contaminants into pools.

- Altering the hydrology of breeding pools by adding impermeable surfaces nearby (such as pavement and buildings) or by adding storm-water detention systems (see Skelly et al., 1999). Changing the elevation or grade of land adjacent to pools may also alter the amount of runoff.

- Destroying or degrading upland habitats by clearcutting (see deMaynadier and Hunter, 1995, 1999) or removing substrates such as logs, rocks, and leaf litter used by salamanders as well as their prey.

- Adding sources of direct mortality in upland or breeding habitats by removing or disturbing burrowing substrates,

adding or increasing vehicular traffic, or introducing or increasing environmental contaminants such as pesticides, fertilizers, and runoff from roadways (see Berrill et al., 1997a; Bonin et al., 1997a; Diana and Beasley, 1998).

- Impeding connectivity between upland, breeding, or dispersal habitats (see Reh and Seitz, 1990; Gulve, 1994; Semlitsch et al., 1996; Skelly et al., 1999). Roads, walls, and curbs are examples of impediments to salamander movement.

Research Needs

Although the scientific literature on amphibians has provided us with evidence of adverse impacts from deforestation (deMaynadier and Hunter, 1995, 1999), roads, and other impediments to amphibian movements (Laan and Verboom, 1990; Reh and Seitz, 1990; Gulve, 1994; Gibbs, 1998), as well as adding pesticides and other contaminants (Berrill et al., 1997a; Bonin et al., 1997a; Diana and Beasley, 1998), critical, unanswered questions remain:

1. What is the spatial extent within which we can defensibly apply these guidelines and regulatory protection?

2. Assuming we adopt Semlitsch's (1998) suggestion of protecting 95% of a population, how much suitable upland habitat is necessary to protect 95% of a population of marbled, Jefferson, or blue-spotted salamanders?

3. How is population persistence influenced by protecting habitat that is only sufficient to support ≤ 95% of the local population? Can losing a portion of the habitat base and population be compensated for by a density-dependent population response (see Scott, 1990; Taylor and Scott, 1997)?

The issue of road mortality and the subsequent adverse impacts to the population raises questions about underpass design and effectiveness as a means of maintaining connectivity and reducing mortality. Jackson (1996) has recommended an underpass and retaining wall scheme based on his experience with salamander tunnels in Amherst, Massachusetts. However, monitoring post-installation is necessary to determine the success of this and other designs (Jackson, 1999). Wildlife underpasses and overpasses are still in an experimental phase, and the management of herpetofaunal reserves illustrates the need for further investigation into the following questions:

1. Are underpasses effective in re-connecting salamander habitat fragmented by roads?

2. Are underpasses effective at preventing road mortality?

3. Can underpasses help maintain metapopulation dynamics across roads?

4. What design(s) of underpasses is most successful? Considerations include: underpass length (e.g., road width); underpass width; type of barrier that keeps salamanders off the road and directs them into underpasses; amount of ambient light and moisture necessary; frequency of underpasses along a roadway; and potential for predators to take advantage of the funnel effect of underpasses.

5. Do breeding pools adjacent to roads act as population sinks?

6. Once we identify sites in need of underpasses, how do we prioritize their installation? Should roads with a higher traffic volume be a priority, or should we focus on roads that are not busy enough to have previously depleted the population?

Discussion and Conclusions

With only limited information on minimum viable population size, metapopulation dynamics, and upland habitat use and requirements, we are taking a cautious approach to reserve design, and we recognize that reserves may have to be modified as new information becomes available. We also recognize that our reserves could fail to encompass enough individuals (or enough suitable habitat) to protect a given population. However, waiting until we answer all the unknowns will certainly result in a greater loss and fragmentation of amphibian habitats and will likely result in lost opportunities to protect habitat areas that are of sufficient size and connectivity to support viable populations. Ironically, the severe habitat alteration and fragmentation that has already occurred in Massachusetts helps to simplify our conservation strategy. That is, we can simply seek to identify and protect the largest and least fragmented areas of suitable habitat that remains.

While Massachusetts' conservation laws alone do not achieve comprehensive protection of salamander habitats, they can play an important role during the process of designing reserves. While site selection and reserve design are in progress, and until priority habitats are protected through acquisition or conservation easements, the Massachusetts Endangered Species Act and the Wetlands Protection Act can be used to protect areas within potential reserves before they receive this designation.

In identifying potential reserves, delineating their boundaries, and developing management guidelines, we have identified many research needs. Our project illuminates the gaps that exist between theoretical conservation biology and applied, on-the-ground efforts to manage and protect amphibian populations. Radio-tracking technology has already greatly increased our knowledge of habitat requirements, and it will undoubtedly continue to do so. In addition, continued studies on dispersal, habitat use, and metapopulation dynamics (see Semlitsch et al., 1996; Windmiller, 1996; Skelly et al., 1999) are essential to our management efforts. As researchers address these questions and convey their results, managers will be able to more efficiently use their resources in conserving amphibian habitats. Environmental regulators will also benefit from increased confidence and effectiveness in preventing habitat loss and alteration.

Reserves designed for ambystomatid salamanders will benefit other species of amphibians as well. Wood frogs also depend on seasonal pools for breeding. Other northeastern amphibians that commonly use these wetlands include American bullfrogs (*Rana catesbeiana*), green frogs (*R. clamitans*), eastern newts (*Notophthalmus viridescens*), four-toed salamanders (*Hemidactylium scutatum*), spring peepers (*Pseudacris crucifer*), and American toads (*Bufo americanus*). These species are all threatened by the advanced stage of habitat loss and fragmentation in Massachusetts, and all are vulnerable to road mortality. Therefore, they all stand to benefit from proactive efforts to protect connected habitat complexes.

Despite our unanswered questions, we believe that a proactive strategy of identifying and protecting high-quality salamander habitats is necessary to effectively protect rare salamanders and their habitats in Massachusetts. Waiting for the answers in a region where human pressures are extreme will likely result in further depletion and fragmentation of remaining habitats and more difficulty in protecting them later. We are willing to start conserving habitats now, based on the best available information, and work with the current uncertainties. We also challenge researchers to conduct studies that will improve the efficiency and success of conservation efforts for migratory salamanders and other amphibians.

Summary

Massachusetts is the third most densely populated state in the country, and the resulting high degree of habitat fragmentation makes Massachusetts a prime example of the need to protect connected habitat complexes for amphibians. Migratory salamanders are especially threatened by fragmentation due to roads and residential and commercial development, and three of four ambystomatid species native to Massachusetts are currently state-listed. Because regulatory measures neither succeed in protecting both wetlands and uplands nor in preventing the cumulative impact of multiple development projects, an effective strategy for statewide conservation of migratory salamanders is the proactive identification and protection of habitats.

Establishing reserves for ambystomatids involves four general stages: (1) selecting sites, (2) prioritizing sites, (3) delineating reserve boundaries, and (4) developing reserve management guidelines. Each of these tasks raises several research questions concerning population viability, extinction probabilities, dispersal distances, habitat requirements, and metapopulation dynamics. We challenge researchers to address these questions, thereby increasing the effectiveness and efficiency of habitat conservation efforts for migratory salamanders.

THIRTY-SEVEN

Population Manipulations

C. KENNETH DODD, JR.

In recent years in North America and in other locales, there has been a surge of interest in the status and conservation of amphibian populations. Concern centers on the disappearance or decline of individual populations, species, and even geographic assemblages of amphibians, particularly anurans. The declines are real, despite much initial skepticism that downward trends in numbers might result from natural population fluctuations. That skepticism has been valuable, however, in that it forced researchers to double efforts to ensure that they were not crying wolf over a phenomenon that could be of natural occurrence. Although there is likely no one cause for declines in many scattered regions or for the deformities reported in midwestern North America, researchers are now feverishly developing monitoring and research programs that can only aid in our understanding of amphibian population dynamics and the importance of amphibians to ecosystem function.

Amphibian declines did not begin with the currently observed population crashes. Human beings have been modifying habitats throughout recorded time, and habitat alteration may be a trait associated with human cultural and social evolution. In North America, Native Americans extensively modified habitats to clear land for agriculture, increase hunting success, and for building materials (Hammett, 1992). It seems unlikely, however, that serious impacts on amphibians occurred prior to settlement by Europeans. In the 1800s and 1900s, habitat modification, especially draining wetlands, clearing forests, plowing grasslands, and mining desert water, increased to the point of frenzy as human population exploded, technology became more advanced and less expensive, human lifestyles became more affluent (the term "affluenza" has been applied to the materialistic culture of the late Twentieth Century; O'Neill, 1996), and chemicals pervaded every aspect of the environment. Although sometimes lost in the hubris of the press and in the halls of academia and government, the piecemeal destruction of habitats required during various phases of an amphibian's life cycle has been, still is, and will continue to be the greatest threat to the long-term survival of most amphibian species (e.g., Delis et al., 1996).

Lost in the current press is the fact that concern about declining amphibians did not begin with the First World Congress of Herpetology in Canterbury, England, in 1989. Prior to this important meeting, field biologists in North America had expressed concern about the status of species such as Houston toads *(Bufo houstonensis)*, Texas blind salamanders *(Eurycea rathbuni)*, and others as early as the mid-1960s. In the U.S. Department of the Interior Red Data Book of 1973, 11 amphibians were included (BSFW, 1973). In a later review, Bury et al. (1980) listed 39 amphibians as needing conservation-related research and management. Summaries of amphibian status were published in many states (e.g., Virginia, North Carolina, and California) prior to 1980. In most cases, habitat destruction or alteration was identified as the cause for concern, although restricted range, apparent rarity, or simply "unknown" factors influenced the inclusion of species.

Serious declines of ranid frogs were observed in the Upper Midwest and Great Lakes Region at least as far back as the 1970s, although the cause was partly attributed to overcollection for the biological supply trade. "Mysterious" declines were noted for the Tarahumara frog *(Rana tarahumara)* in the southwestern United States (Hale et al., 1995) and for several species of *Eleutherodactylus* in Puerto Rico, especially webbed coquís *(E. karlschmidti*; last seen in 1974) and golden coquís *(E. jasperi*; last seen in 1981; summarized by Joglar and Burrowes, 1996). Recognition of the precarious status of certain amphibian species, and even the disappearance of some for unknown reasons, did not generate much public, scientific, or governmental awareness until widespread and sudden declines and disappearances were "discovered" in Canterbury and reported in the popular press. Since then, amphibians have received much publicity and, I hope, will now receive the research, management, and habitat acquisition funding that their conservation requires.

Before the current interest, few amphibians were considered in conservation planning documents. At the federal level, the U.S. Fish and Wildlife Service (FWS) was required to develop Recovery Plans that might lead to eventual recovery and delisting for amphibians protected under provisions of the U.S. Endangered Species Act of 1973 (e.g., Houston toads, Wyoming toads *(B. baxteri)*, Texas blind salamanders, desert slender salamanders *(Batrachoseps aridus)*, and golden coquís. Recovery Plans were prepared for all listed amphibians either by Recovery Teams or by biologists within FWS regional and field offices. Unfortunately, some of the plans were never implemented, some were clearly inadequate, and one (that for golden coquís) was completed after the species was likely extinct. Most plans called for the requisite statutory protection and habitat acquisition necessary to protect a seriously declining or vulnerable species, although funding was rarely available to implement the plans.

As a herpetologist in the Endangered Species Program (1976–1984), I was often asked to review Recovery Plans for amphibians and reptiles. Even as far back as the early 1980s, I noticed a trend in Recovery Plans to advocate elaborate efforts based on captive breeding, head-starting, reintroduction, or simply relocating existing populations of animals; in a few cases, it seemed as though the manipulative approach was included because the obvious approach (protect the integrity of natural habitats) was too obvious and simple. The underlying philosophy seemed to be that amphibian conservation demanded a technologically intensive approach, just as it often did with large mammals and other charismatic endangered species. In this respect, the conservation of imperiled species of amphibians followed a traditional resource management approach, that is, one that relies substantially upon manipulation of a species and its habitat (see Seigel and Dodd, 2000, for a review of this approach). To paraphrase a popular television program, "this bothered me."

I discussed the apparent favor of manipulative conservation approaches among natural resource agencies with Richard Seigel (Southeastern Louisiana University), curiously enough, at the First World Congress of Herpetology. Our discussion eventually led to a review of repatriation, relocation, and translocation (RRT) as strategies in herpetofaunal conservation; this review was later published in *Herpetologica* in the "Points of View" series (Dodd and Seigel, 1991). The article generated a great deal of interest. Most biologists agreed with our common-sense approach, but we did manage to step on a few toes. When "sacred cows" are disturbed, one cannot expect calm to reign. Still, the article fulfilled its purpose—to get conservation biologists and natural resource managers working in herpetology to question the validity of assumptions, to think about the limitations of their techniques, and, perhaps, to better plan.

Head-Starting and RRT for Amphibians

Head-Starting

Head-starting (HS) is a management technique whereby small animals or early life-history stages (eggs, larvae, and juveniles) are raised in captivity to a larger size (subadults and adults), then released into native habitats to fend for themselves. Programs employing head-starting assume that there is a very high level of mortality in the small size classes. In theory, overall mortality should be decreased, survivorship should be enhanced, and the population size would likely increase by raising an animal to a size larger than most predators can eat. Of course, this management option assumes that the solution to the perceived problem (decline or disappearance) is to "make more," which may or may not be the case. Conservation problems involving amphibians are rarely that simple.

Implicitly, head-starting programs assume that appropriate habitats are available into which the animals can be released, that released animals will behave exactly as those reared in natural habitats (e.g., in food choice or in movements back to breeding ponds), and that the problem causing low recruitment or decline can be identified and corrected, or at least satiated at

the location in which the animals are to be released. Somewhat surprisingly, these assumptions are often not addressed. It is simply taken as faith that by putting more and larger animals into the environment, the problem will somehow correct itself. For example, Bloxam and Tonge (1995) report that "In the late 1980s, biologists identified gorges in which Mallorcan midwife toads [*Alytes muletensis*] had recently become extinct and, within them, *identified sites suitable for release of captive-bred specimens*" (emphasis added). One might question the propriety of releasing captive-bred specimens into a habitat where natural populations "had recently become extinct" without knowing why the natural populations had disappeared. Fortunately, a great deal is now known about the life history of the ferreret (e.g., Hemmer and Alcover, 1984).

Captive breeding is always a manipulative option, but one not often employed in amphibian conservation despite the success of herpetoculturists in keeping many amphibians. Head-starting may or may not be used in conjunction with captive breeding. For aquatic-breeding amphibians, egg masses can be moved to a secure location, the tadpoles raised to metamorphosis, and the juveniles or subadults released back into the wild. It may not be necessary or advisable to have the parents actually breed in captivity. However, captive breeding is always followed by release into the wild, at least if carried out for conservation purposes. Then, the assumptions are similar to those for head-starting—the solution to a problem is simply "more."

There are very few captive breeding programs that have operated solely for amphibian conservation. These include programs for Mallorcan midwife toads at the Jersey Wildlife Preservation Trust and other European institutions (Bloxam and Tonge, 1995), Puerto Rican crested toads (*Peltophryne lemur*) at the Buffalo and Metro Toronto Zoos (Miller, 1985; Paine et al., 1989; Bloxam and Tonge, 1995), Houston toads at the Houston Zoo (references in Dodd and Seigel, 1991), and Wyoming toads at a state hatchery and other cooperating facilities (Swaringen, 1996). Head-starting is an integral part of these programs inasmuch as the breeding facilities are located some distance from the natural habitat. As such, juveniles and adults, rather than freshly deposited eggs, are released into the wild. Head-starting, without captive breeding, has been used by researchers in Arizona to augment populations of Ramsey Canyon leopard frogs (*R. subaquavocalis*). In this case, eggs are taken to the Phoenix Zoo, hatched, and the larvae are raised through metamorphosis (Demlong, 1997). Both metamorphic animals and large larvae are released into natural habitats.

Relocation

As defined by Dodd and Seigel (1991), a relocation "involves the moving of animals or populations of animals away from an area where they are immediately threatened to an area where they would be less prone to habitat loss; ideally, relocated animals should be moved to habitats where they historically occurred, but this is not always the case." Relocations have been employed mostly in reptile, particularly tortoise, conservation projects. In Florida, individual Florida gopher frogs (*R. capito aesopus*) sometimes may be relocated in conjunction with gopher tortoise relocation, but no consideration is given to their biological requirements.

To my knowledge, only one amphibian breeding population has been deliberately relocated (Schlupp and Podloucky, 1994). In this case, the breeding pond of a population of common toads (*B. bufo*) in Germany, whose habitat was bisected by a road thus separating the breeding pond from the non-breeding terrestrial habitat, was relocated to the same side of the road as the terrestrial habitat. Relocation thus eliminated previously high mortality as the toads crossed a busy highway on their way from terrestrial habitat to their breeding pond. Fortunately, common toads readily accepted a newly constructed breeding pond on the same side of the road as their terrestrial habitat, but this might have been anticipated by knowing the biology of common toads.

The assumptions of relocation are that the animals can be moved successfully; habitat is available on which to relocate them (and which is presumably protected from further disruption); relocated animals will assume "normal" behavior; local populations of the same species (if any) will not be harmed (demographically, socially, or by disease); relocated animals will not attempt to return home; and the new habitat will provide all requirements necessary to satisfy the biological constraints (see Dodd and Seigel, 1991) of the species. Relocation projects that do not understand or meet these assumptions will usually fail.

Repatriation

Repatriation is defined as releasing animals into an area formerly or currently occupied by that species (Dodd and Seigel, 1991). Repatriations are carried out to augment (*sensu* Reinert, 1991) existing populations or to re-establish populations in areas where the species was formerly extant. In this regard, most head-starting projects involve repatriation, such as those for Ramsey Canyon leopard frogs, Mallorcan midwife toads (in part), Puerto Rican crested toads, Houston toads, Wyoming toads, and natterjack toads (*B. calamita*) in Britain (Denton et al., 1997). In addition to the assumptions listed for relocations, knowledge of habitat suitability in relation to what caused the species or population to be extirpated in the first place is absolutely essential for the success of a repatriation. It is completely senseless to put animals raised or moved at great cost into habitats that are still unsuitable for the species to survive. Yet, this apparently has occurred (see Bloxam and Tonge, 1995).

The experience of British researchers with the repatriation of natterjack toads is most enlightening. For the last 25 years, various researchers have conducted extensive ecological investigations of this narrowly distributed (in Britain) species. Conservation activities were carried out at 29 of 39 sites, and 20 repatriations were attempted. The species was known to have declined because of habitat loss, habitat alteration, and interference by common toads; management was directed at eliminating these problems. Still, most of the early repatriations failed, in part because of a lack of knowledge about the natterjack's breeding requirements—ponds in repatriated sites were often too acidic. Repatriation success improved once ponds were restored to circumneutrality (Denton et al., 1997). Now, six repatriations have resulted in successful establishment of populations and eight others have shown signs of initial success (Denton et al., 1997).

Repatriations of natterjack toads in Britain occurred with a great amount of concurrent ecological research and prior knowledge about the species' biology and habitat requirements. Still, initial repatriations failed because of a lack of habitat suitability. While it is impossible to know everything there is about a species prior to undertaking a repatriation program,

researchers must plan for comprehensive research concurrently with repatriation attempts. If the natterjack recovery program is successful, it can only be attributed to the dedication and tenacity of the researchers involved, coupled with favorable biological characteristics of toads for repatriation. Conservation biologists contemplating repatriations of other amphibians should carefully review the natterjack experience.

Translocation

A translocation is defined as the release of animals into areas where they are not historically known to occur. Usually, the reason given for a translocation is that it might be necessary to establish auxiliary populations in areas away from the core population as insurance against a catastrophic occurrence (e.g., habitat loss or disease). At least this is the reason given for actual or proposed translocations of Mallorcan midwife toads, Ramsey Canyon leopard frogs, desert slender salamanders (USFWS, 1982), Hamilton's frogs (*Leiopelma hamiltoni*; D. Brown, 1994), and western toads (*B. boreas*; E. Muths et al., 2001). This certainly is a laudable goal, but one that should be carried out only after extremely careful analysis of the biological and habitat requirements of the species.

In at least one other series of translocations, species not known to occur in the area within immediate historic times were translocated from the east to west-central Illinois by a single researcher. Three species (eastern red-backed salamanders [*Plethodon cinereus*], southern two-lined salamanders [*E. cirrigera*], and wood frogs [*R. sylvatica*]) were translocated to Brown and McDonough counties, Illinois, beginning in the early 1970s and extending through the early 1990s (Thurow, 1994, 1997, unpublished data). Adults and eggs were used, depending on species, and literally hundreds of individuals of each species were moved. The rationale presented was that Thurow wanted to determine the best methods to translocate individuals by choosing common species, improving habitats, and creating shelters and other structures to facilitate the success of the projects. Translocations were carried out somewhat randomly without clear schedules or protocols. Some of the translocations appear to have been successful at certain sites, whereas others apparently failed.

Amphibians as Candidates for HS/RRT Projects

Bloxam and Tonge (1995) suggested that amphibians might be good candidates for captive breeding and HS/RRT (head-starting/relocation, repatriation, and translocation) programs because of their reproductive strategies and their "low levels of behavioral complexity." They examined two repatriation projects in support for their position (for Mallorcan midwife toads and Puerto Rican crested toads), curiously one of which has not been demonstrated to be successful (although it may well be in time) and the other of which seems not to have worked. Amphibian rearing was said to be cost effective (only US $150 for the entire Mallorcan midwife toad-breeding program). It was also reported that amphibians retain natural behavior after many generations in captivity, and that, in general, amphibians produce large numbers of "behaviorally normal offspring" within one or two generations.

Unfortunately, this type of information is highly misleading. Amphibians may or may not be candidates for captive breeding and HS/RRT projects, but such determination must be made on a case-by-case basis. Contrary to the generalizations in Bloxam and Tonge (1995), many amphibians do not rely on water for reproduction (many terrestrial salamanders and direct-developing frogs whose reproductive behavior is often quite complex and poorly understood), many amphibians lay small numbers of eggs that produce small numbers of offspring (many salamanders and frogs only deposit from one to a few eggs), and I would posit that it has not been demonstrated anywhere that amphibians raised in captivity through several generations demonstrate the behavioral characteristics and complexity that they would need to survive in natural habitats. In addition, Bloxam and Tonge's (1995) cost analysis does not consider researcher/technician salaries, travel expenses, field time, long-term monitoring, habitat protection, nor many other factors. Most of these considerations are necessary in any conservation program, but to suggest that captive breeding is inexpensive is gravely misleading. Specific recommendations are needed instead of blanket generalizations.

Those HS/RRT projects that have considered the five criteria elaborated upon by Dodd and Seigel (1991) are, not surprisingly, those projects with the best records of success, at least for species whose survival is in doubt. These include the natterjack toad project in Britain and, potentially, the Ramsey Canyon leopard frog project in Arizona. In general, the other projects have not succeeded (e.g., Houston toads and Puerto Rican crested toads), or it is still too early to tell whether they will be successful (Mallorcan midwife toads and Wyoming toads). Successful projects have a long-term commitment, a multifaceted approach (involving genetic, life history, and environmental physiology research and monitoring), and a reasonable understanding of the factors that caused the original decline or disappearance. Those projects that have not been successful did not address all the biological constraints affecting the species, or do (did) not understand or address corrective management to eliminate threats.

I suggest that those amphibians that might benefit from HS/RRT projects are those that have intact habitat available free from threats, a large reproductive output (eggs appear to be the most effective life stage with which to start new populations), long life spans (at least several years), and those that are generalists rather than specialists. Unfortunately, most amphibians under threat do not have these characteristics. Still, if a habitat remained intact after a predator was removed, repatriation would be appropriate; relocating a toad population might be appropriate because the life history characters (large reproductive output, and metapopulation demography) suggest possible success. Amphibians with low reproductive output, short life spans, low natural population numbers or density, and specialized behaviors or habitat requirements would seem to be poor candidates for HS/RRT projects. Other amphibians may benefit from habitat restoration or the construction of artificial shelters (Thurow, 1994, 1997; Denton et al., 1997). Thus, some amphibians may be good candidates for manipulative conservation-related technology, but most species that need attention probably are not. HS/RRT programs should only be considered for amphibian conservation after careful planning, and then only if a long-term commitment is present and more cost effective and in situ options have been exhausted. To reiterate (from Dodd and Seigel, 1991), one must know:

Causes of decline: If the cause or causes of declines are not known, it is possible that they will still affect amphibians placed back into the habitat. If this happens, HS/RRT programs

waste time, money, and precious individuals. At the very least, intensive research should accompany the HS/RRT project to monitor the success of individuals. Intensive concurrent research ensured that a serious environmental problem was detected and corrected in the natterjack toad project, even though most initial repatriations were unsuccessful. If research had not occurred, it is likely that few repatriations would have been successful despite seemingly favorable intensive habitat restoration.

Biological constraints: If you do not know the biological (habitat, demographic, and biophysical) constraints of the life history of the amphibian to be released, you cannot plan for its conservation. The first step of any conservation program is to know the life history of the animal. If that is known, it will help determine whether it is a suitable candidate for HS/RRT. Life history data tell you where, when, and how to carry out HS/RRT projects. For example, repatriating animals into habitats too small to meet minimum population sizes for a self-sustaining population would certainly be self defeating and wasteful, regardless of intent.

Population genetics and social structure: It is essential to know something about the social structure of animals released in a repatriation program (Dodd and Seigel, 1991). For example, in some toads the males are philopatric to a particular breeding site but the females are not—they will go to the nearest calling male, regardless of where they bred the previous year. Hence, if a species with this type of behavior was to be repatriated, it might be best to repatriate adult males during the breeding season to get them established. Females could be released when males were calling from established territories. Otherwise, females might wander away from an area and the repatriation would not succeed.

Knowledge of population genetics and social structure also will give clues as to how many animals need to be released and ensure that populations of varying genetic composition (or even cryptic species) are not intermingled during HS/RRT programs. For example, populations of the Australian endangered frog *Geocrinia alba* are highly subdivided genetically (i.e., they are very differentiated). This suggests that although they are reproductively compatible, populations of this frog species are following different evolutionary trajectories. Thus, repatriation or population augmentation, if desirable, would have to avoid mixing genomes since that could actually decrease genetic diversity (Driscoll et al., 1994; Driscoll, 1997). Fortunately, several HS/RRT projects (e.g., Puerto Rican crested toads and Ramsey Canyon leopard frogs) have incorporated genetic analysis into their conservation plan.

Disease transmission: In 1991, Rich Seigel and I were primarily concerned with disease transmission among reptile species and populations when we cautioned about the threat of disease in HS/RRT programs. Today, amphibians involved in HS/RRT projects have been moved across continents (e.g., Jersey Islands to Mallorca; Canada and the United States to Puerto Rico) seemingly with little attention to the threat posed by introduced pathogens. It is now known that a lethal fungus (phylum Chytridiomycota) is responsible for mass amphibian mortality in Australia and Central America (Berger et al., 1998). How an identical pathogen affected amphibian assemblages on two widely separated continents remains unknown, but it is certainly possible that it was introduced by exotic animals. With a virtual panmixis of animal faunas due to international trade, especially in exotic pets (including amphibians), it is likely that disease prevention will play an increasing role in amphibian conservation. HS/RRT projects should take every precaution to avoid introducing exotic disease pathogens either to con- or heterospecifics in local amphibian assemblages.

Need for long-term monitoring: Without doubt, HS/RRT projects must be monitored for a long period after introducing animals into new habitats. For amphibians, I suggest a period covering several generations (at minimum) to ensure that animals are breeding and maintaining population numbers. As such, monitoring repatriated amphibians often will require 10–15 years or longer before researchers can be reasonably assured that the release has been effective at establishing or re-establishing populations. Simply recording reproduction in a few individuals shortly after repatriation is not sufficient to claim success. As in any worthy endeavor, success is earned only after a long period of hard work and continuous vigilance. If anything can be learned from the lessons of the natterjack toad recovery program (Denton et al., 1997) and the strange translocations of Thurow (1994, 1997, unpublished data), it is that persistence often pays off and may be absolutely necessary for amphibians to successfully colonize new habitats or to recolonize old habitats.

Halfway Technology and Amphibian Conservation

In a series of thoughtful essays on turtle conservation, Nat Frazer noted that the way a problem is defined often determines the methodologies that are considered as solutions to the problem (Frazer, 1992, 1997). In turtle conservation (and for many other species), the problem is often defined in a numbers context, that is, there are too few individuals present of species X in natural habitats. If the problem is "too few," then the solution is to increase numbers, hence the proliferation of intensive conservation programs involving captive breeding, head-starting, and relocation, repatriation, and translocation. Often, but not always, such programs involve treating the symptoms of the problem rather than the actual cause. As such, they become halfway solutions to the real causes of population decline or disappearance.

Recently, Rich Seigel and I have noted the connection between technology-intensive applications historically prevalent in natural resource agencies (e.g., controlling predators, re-introductions, and increasing numbers in existing populations of game animals) with the tendency of these agencies or organizations to advocate similar "high-tech" solutions to conservation problems (Seigel and Dodd, 2000). Certain conservation organizations with a historic tradition of captive breeding also fall into the trap of advocating such without a careful examination of other options or alternatives. It is our experience that most HS/RRT projects are set forth as solutions to herpetofaunal conservation with little consideration of the assumptions underlying the proposed solution, the limitations surrounding the techniques that are advocated, or an adequate consideration of what the problems are in relation to possible solutions. If the problem is not "too few animals," why should technology-intensive, labor-intensive, and potentially capital expensive programs be advocated? We suggest that the answer to this riddle involves outdated historical biases, a lack of clear understanding of the problems faced by many species, a lack of will to address the real causes of problems, and a number of other reasons (Dodd and Seigel, 1991; Seigel and Dodd, 2000).

I am particularly concerned that the reason some HS/RRT programs are advocated has nothing to do with biology, but with publicity for the sponsoring organization (often under the guise of concern for the species). For example, Bloxam and Tonge (1995) state that "Regardless of whether or not the translocation or reintroduction of amphibians can be shown to have resulted in stable populations in the long term, there is little doubt that, in the short term, such efforts generate excellent publicity and have a role to play in raising the awareness of local people about the problems of their local amphibian fauna." On the contrary, ill-conceived publicity about biologically unsound solutions gives the public the false impression that conservation concerns are being addressed effectively, when in fact programs may be having no impact, or at worst may actually be contributing to the continued drain on species and habitats.

Rather than advocate halfway, intensive technology to generate publicity and drain finances from habitat conservation or research on actual causes of problems, I suggest that conservation biologists focus on the biology of the animal, the nature of perceived threats, and a careful examination of all underlying assumptions surrounding the techniques offered as solutions. HS/RRT may indeed be appropriate in certain cases, but only after other cost-effective and parsimonious options have been carefully considered. Conservation programs involving amphibians must (1) focus on the causes of problems, (2) adopt solutions within the biological constraints imposed by their complex life cycles, (3) devise solutions that treat causes, not symptoms, and (4) not mislead either the public or sponsoring agencies about the difficulties inherent in carrying out effective and biologically meaningful actions. They should be considered experiments, not proven management practices. To do anything less or for any other underlying motives is unconscionable.

Addenda

In our 1991 paper, Rich Seigel and I could not say for certain that any HS/RRT program involving amphibians (n = 5) had proven successful. We overlooked a one-time repatriation of tiger salamander *(Ambystoma tigrinum)* eggs in Maryland (Enge and Stine, 1987), but the fate of the repatriation is unknown. The program for natterjack toads has now had partial success, although a number of repatriations proved unsuccessful (Denton et al., 1997). Unfortunately, the programs involving the other species we noted in 1991 cannot be evaluated since nothing has ever been published on them. Projects involving Houston toads, western toads in the Rocky Mountains, and Puerto Rican crested toads appear to have failed; projects involving Mallorcan midwife toads, Ramsey Canyon leopard frogs, and Wyoming toads have not been in place long enough to say for certain whether they will succeed.

The outlook appears favorable for Mallorcan midwife toads and Ramsey Canyon leopard frogs, but grim for Wyoming toads. To date, more than 7,000 tadpoles and toadlets of the Wyoming toad have been released, mostly at Mortenson Lake in the Laramie Basin. In 1997, toads were heard calling for the first time after being declared extinct in the wild in 1994; in 1998, four egg masses were found, all of which resulted from toads raised in captivity and released into native habitats. Currently, 12 rearing institutions hold approximately 250 adults (AZA, 1998). Project sponsors are hopeful that the wild populations will increase.

On other fronts, it appears as though a new population of Hamilton's frogs has been successfully established near the only previously known location (D. Brown, 1994). Eastern tiger salamanders *(A. t. tigrinum)* may have been successfully established at a new pond mitigation site in New Jersey. From 1982–85, about 1,000 salamander eggs were transplanted from a donor pond (20 km away) to the newly created pond; egg masses were found at the site at least through 1990, suggesting the establishment of a reproductive population (Zappalorti, 1998). Sexton et al. (1998) report that repatriation (or augmentation, it is unclear from the paper) of spotted salamanders *(A. maculatum)* and wood frogs to both natural and manmade ponds in Missouri has been successful. These repatriations appear to have been carried out on only one occasion, and population status and dispersal have been monitored since the mid-1960s. However, much crucial information on this project (e.g., the project rationale, the number of wood frog eggs moved, and the status of these species within both the immediate ponds and the nearby areas prior to repatriation) remains to be published. The translocations carried out by Gordon Thurow for eastern red-backed salamanders, southern two-lined salamanders, and wood frogs over a period of 20 years or more appear to have resulted in the establishment of new populations. In Alabama, Braid et al. (1994) conducted laboratory observations on the best methods of captive-rearing larval dusky gopher frogs *(R. c. sevosa)* preliminary to head-starting. Some larvae were returned to ponds, but the success of the effort or whether it was continued is unknown. At Jamaica Bay, New York, five amphibians (both adults and larvae, depending on the species) were reintroduced to the Gateway National Recreation Area (spotted salamanders, eastern red-backed salamanders, eastern gray treefrogs *[Hyla versicolor]*, spring peepers *[Pseudacris crucifer]*, and green frogs *[R. clamitans]*) between 1980 and 1987. Although the fate of these reintroductions is presently unknown, initial results with regard to spring peepers and eastern gray treefrogs were promising (Cook, 1989).

It is likely that I have missed other HS/RRT programs that have been undertaken during the last nine years. I reiterate, however, that it is absolutely essential that the rationale, techniques, and results of HS/RRT projects, both positive and negative, be published in the peer-reviewed literature if past mistakes are to be avoided and future successes enhanced.

Summary

Head-starting, relocation, repatriation, and translocation (HS/RRT), often in conjunction with captive breeding, have frequently been suggested as viable options in the conservation of amphibians. In this essay, I review recent projects employing HS/RRT solutions to problems facing imperiled amphibians. If the sole conservation concern is that there are "too few" individuals in a population, high-technology solutions may be appropriate. However, if the real causes of a species' decline and disappearance are not addressed, HS/RRT projects become halfway solutions that will not result in conservation, and may even harm more effective programs. HS/RRT may be effective in some situations, but they should only be considered as a last resort. Long-term research and monitoring, absolutely essential in any conservation program, are doubly important to ensure the success of HS/RRT projects.

Acknowledgments. Preparation of this paper was supported by the Inventory and Monitoring Program of the U.S. Geological Survey, Great Smoky Mountains National Park Project. Andrew Odum directed me to information on the Wyoming toad project. I thank Russell Hall, Joseph Mitchell, and Marian L. Griffey for reviewing the manuscript.

THIRTY-EIGHT

Exotic Species

WALTER E. MESHAKA, JR.

Within the framework of species conservation resides the increasingly relevant and disconcerting topic of exotic species. A conservative definition of exotic amphibian species includes only those species that have colonized the United States by human-mediated dispersal. This definition covers six anuran species (Table 38-1): green and black dart-poison frogs *(Dendrobates auratus)*; Cuban treefrogs *(Osteopilus septentrionalis)*; coquis *(Eleutherodactylus coqui)*, greenhouse frogs *(E. planirostris)*; African clawed frogs *(Xenopus laevis)*; and wrinkled frogs *(Rana rugosa)*. By including any extralimital populations of species otherwise native within the continental United States and whose dispersal to extralimital sites was human-mediated, I expand this list to include American bullfrogs *(R. catesbeiana)* and marine toads *(Bufo marinus)*. Two of these exotic species, green and black dart-poison frogs and wrinkled frogs are found only in Hawaii. Three other species are found only in the continental United States: coquis, greenhouse frogs, and African clawed frogs. Three species, Cuban treefrogs, marine toads, and American bullfrogs, are found in both Hawaii and the continental United States (Table 38-1).

The role of humans in the geographic expansion of species has a long and checkered past (di Castri, 1989). As humanity builds upon an age of unprecedented strides in technology that enables humans and their commensals to travel long distances more frequently, more quickly, and more safely than ever before, the stage is set for both disturbed habitats and the dispersal of species over distances that are unlikely to have ever occurred naturally. Unwitting at best and indifferent at worst, humanity's role in biogeography brings with it the potential for negative ecological impacts on indigenous species by exotic species. This has been noted with Cuban treefrogs (Meshaka, 1994, 2001), marine toads (Covacevich and Archer, 1975), and African clawed frogs (Lenaker, 1972). Although evidence of American bullfrogs as a cause of anuran declines in extralimital U.S. populations is equivocal (Hayes and Jennings, 1986; but see Dumas, 1966; Moyle, 1973; Lannoo et al., 1994), their broad diet (summary by Bury and Whelan, 1984), and large body size predispose them to be a likely threat to a wide range of indigenous species, including anurans. In that regard, frogs can constitute a major portion of the diet of wild American bullfrogs (Stewart and Sandison, 1972).

A framework to interpret colonization dynamics is found in a suite of ecological correlates that provide a test to explain colonization success in species (J. H. Brown, 1989; Ehrlich, 1989; Gaston and Kunin, 1997). Among these are synanthropy (the ability to coexist with humans), similarity between introduced and native habitats, high vagility, and high fecundity. Although meeting these correlates does not guarantee colonization success (and indeed some species succeed having only met one or two of them), meeting these correlates most certainly increases the likelihood of colonization success with respect to dispersal and establishment. Having a thorough understanding of the life history of a species not only predisposes the researcher to test these correlates for the sake of explaining colonization success, but also to predict colonization success and impacts in nearby regions.

The literature provides sufficient ecological and historical information to examine modes and rates of dispersal of exotic anurans in the United States and to test some of the ecological correlates associated with colonization success. For the eight listed species, I compare and contrast the circumstances associated with colonization and test the four aforementioned correlates of colonization that relate to both the structure of the introduced habitat and the biology of the invading species.

Nature of Initial Introduction

Human agency was responsible for the initial extralimital introductions of all eight exotic anuran species in the United States. For two species, Cuban treefrogs and greenhouse frogs, dispersal by humans from Cuba to the Florida Keys was almost certainly incidental with trade during the turn of the nineteenth century (Meshaka, 1994, 1996a, 2001; Meshaka et al., 2004). However, introductions were equally likely by the highly mobile Spaniards, and perhaps even earlier by aboriginal Indians in extreme southern Florida, who were known to trade between Key West and Cuba (Griffin, 1988). The completion of a highway connecting the Florida Keys with the Florida mainland coincides with the early records of Cuban treefrogs in Miami and of greenhouse frogs in southern and central Florida (Meshaka et al., 2004). Commerce may have reinforced

TABLE 38-1

List of Exotic Amphibian Species of which (1) Natural Geographic Distribution does not Include the United States and (2) Exotic Status is Found Extralimitally within its Native U.S. Range

	Non-native to United States	Extralimital within native United States	Found in Continental United States	Found in Hawaii	Native Range
Bufonidae					
Marine toad *Bufo marinus*		X	X	X	Neotropics
Dendrobatidae					
Green and black dart-poison frog *Dendrobates auratus*	X			X	Neotropics
Hylidae					
Cuban treefrog *Osteopilus septentrionalis*	X		X	X	West Indies
Leptodactylidae					
Coqui *Eleutherodactylus coqui*	X		X		West Indies
Greenhouse frog *E. planirostris*	X		X		West Indies
Pipidae					
African clawed frog *Xenopus laevis*	X		X		Africa
Ranidae					
American bullfrog *Rana catesbeiana*		X	X	X	North America
Wrinkled frog *Rana rugosa*	X			X	Asia

NOTE: Primary literature and review articles relating to their ecology and/or colonization success accompany the species list.

or introduced populations in the West Indies and Florida as it continues to do today (Meshaka, 1996a). Extant populations of coquis, restricted to nurseries in extreme southern Florida, are derived and probably maintained incidentally through an active bromeliad trade (Meshaka et al., 2004).

For three species, initial introductions were associated with pest control. Marine toads were introduced to Florida and Hawaii (McKeown, 1996; Meshaka et al., 2004) in the early 1900s and, as elsewhere, were associated with agricultural pest control. However, in Florida these early introductions failed (Meshaka et al., 2004). Wrinkled frogs and green and black dart-poison frogs were introduced to Hawaii in the late 1800s for insect control (McKeown, 1996). In Hawaii, American bullfrogs were introduced in the late 1800s for aquatic invertebrate control as well as human consumption (McKeown, 1996). In a curious turn of events, Hawaii not only received American bullfrogs from California but also served as the source for subsequent introductions back to California (McKeown, 1996). Incidental release associated with fish stocking is also responsible for the introductions of some populations of American bullfrogs (McAuliffe, 1978; Lannoo et al., 1994).

The pet trade is associated with the successful introductions of three species. In Hawaii, Cuban treefrogs colonized one island in the 1980s after the release of pets by a hobbyist (McKeown, 1996); the accidental release of individuals destined for the pet trade is credited for the establishment of marine toads in

southern Florida in the mid-1900s (King and Krakauer, 1966). Some of the initial introductions of African clawed frogs were the result of escapes or releases associated with the pet trade (Crayon, this volume, Part Two). African clawed frogs associated with biomedical research also escaped and/or were released into the wild (Crayon, this volume, Part Two).

These sources of initial introductions establish three trends. First, anuran species that colonized the United States were generally species that humans valued. Second, the earliest introductions of these species varied in time, space, and sometimes in mode. Third, and especially noteworthy, is that as with other exotic species, some anurans, such as Cuban treefrogs and greenhouse frogs, were successful despite (or because of) our disinterest in them.

Secondary Dispersal Modes and Dispersal Rate

Once in the United States, anuran species dispersal occurs in various ways. Cuban treefrogs and greenhouse frogs are highly mobile species, particularly in the agency of humans, readily dispersing incidentally with ornamental plants (Meshaka, 1996a; Meshaka et al., 2004). Greenhouse frogs have the advantage of living in small colonies and completing their life cycle in potted plants. Cuban treefrogs can also disperse directly on vehicles (Meshaka, 1996a). The occurrence of marine toads in ornamental nurseries, along with their burrowing habits and small juveniles, predispose them to dispersal in nursery shipments. Furthermore, marine toads are a popular pet trade species because of their large body size, docile habits, and voracious appetites. Organized and private releases and releases incidental with fish stocking further the geographic expansion of American bullfrogs. African clawed frogs passively disperse downstream and actively disperse upstream; although usually thought to be aquatic, this species will migrate overland across short distances (Crayon, this volume, Part Two). The presence of African clawed frogs in widely distributed impoundments that are too far apart for natural overland dispersal (Crayon, this volume, Part Two) suggests continued human mediation in the dispersal process.

Among the species for which dispersal data are available, the degree of natural mobility responsible for their geographic expansion within the United States has been overshadowed by human-mediated dispersal. In connection with this dispersal, geographic range expansions for species introduced onto the U.S. mainland have been scattershot in pattern; peripheral range records pre-date interior range records (Bury and Whelan, 1984; Meshaka et al., 2004; Crayon, this volume, Part Two). Both active and incidental human-mediated dispersal best explains the large discontinuous ranges of recent introductions and the speed at which these sites interconnect. Indeed, few gaps now remain within the geographic ranges of exotic Florida anurans since the half century (marine toads and Cuban treefrogs) and century (greenhouse frogs) when they were first recorded.

The rapid dispersal rate of these species reduces the time to fully realize their final extralimital geographic ranges. For all eight species, some introduced sites are geographically distant from the centers of their native distribution. For all but green and black dart-poison frogs, wrinkled frogs, and coquis, introduced ranges are also large and represent substantial additions to their historic native range. In this regard, the colonization success of green and black dart-poison frogs and wrinkled frogs is the weakest and is in sharp contrast to the highly successful marine toads. The same is true when one compares the instability of the few small colonies of coquis with the phenomenal success of its congener, marine toads, and Cuban treefrogs in Florida.

Correlates of Colonization Success

Among the ecological correlates associated with colonization success, four (synanthropy, similarity of introduced and native habitats, vagility [mobility], and high fecundity) are relatively easy to measure in the field and with the use of museum specimens.

Across taxa, a close association with humans is a strong correlate of colonization success (J. H. Brown, 1989). In light of humanity's increasing presence worldwide, association with humans confers an enormous advantage to colonization. For example, marine toads, ubiquitous in southern Florida, are unknown in natural systems in the state (Meshaka et al., 2004). Cuban treefrogs are most numerous in the presence of humans (Meshaka, 1994, 2001), and greenhouse frogs in Florida are nearly always found around dwellings (Meshaka et al., 2004). Precarious though it is, colonization of coquis within the confines of a few southern Florida bromeliad nurseries is probably entirely explained by this correlate. American bullfrogs thrive in a wide range of artificial aquatic impoundments (Bury and Whelan, 1984), and African clawed frogs use a variety of human-created and human-disturbed water courses (Crayon, this volume, Part Two).

In this connection, manmade structures provide many of these species with not only breeding sites, an abundance of vertebrate prey, and the opportunity to disperse, but also with protection from predators, temperature extremes, and potential competitors (Meshaka, 1994, 2001; Meshaka et al., 2004). For example, in the southern Everglades, Cuban treefrogs are most abundant on buildings, a habitat that has fewer predators and more refuges than even tropical hardwood hammocks, their natural habitat (Meshaka, 1994, 2001). A thriving, disjunct population of marine toads exists in a severely modified habitat at the southern end of the Lake Wales Ridge in south-central Florida—a housing development with a far more tropical and mesic environment than the surrounding sand pine (Pinus clausa) scrub that is considered uninhabitable to this species (unpublished data).

Habitat similarity between native and introduced sites is also correlated with successful colonization (J. H. Brown, 1989). For example, the westernmost habitats of American bullfrogs in their native range in the United States are similar to those of extralimital populations in the western United States. Lentic impoundments, natural or otherwise, are familiar to this species in introduced sites, including those in southern Florida. Marine toads are found in seasonally xeric farmlands and open disturbed sites in Florida that mimic neotropical open savannah habitats (Zug et al., 1975). Greenhouse frogs inhabit xeric pineland, tropical hardwood hammocks, and mangrove forests in mainland Florida and the southern Everglades that are similar to their native habitats in Cuba (Barbour and Ramsden, 1919; Dalrymple, 1988; Schwartz and Henderson, 1991; Meshaka, 1994, 2001; L. Moreno, personal communication). Structurally similar, even if floristically different, hammocks in central and northern Florida are also inhabited by this species (Meshaka et al., 2004). The forest, streamside-dwelling green and black dart-poison frogs occupy structurally similar habitats in Hawaii. Permanent ponds and rivers are used by African clawed frogs both in their native Africa and in the United States (Crayon, this volume, Part Two). Wrinkled frogs, known from mountain streams in Japan (Stejneger, 1907), are likewise found in forested mountains in Hawaii (McKeown, 1996). In contrast to these aforementioned species, coquis are not successful colonizers in Florida. Their native ecological systems markedly differ from Florida, where harsh dry seasons and periodic frosts restrict them to greenhouses (Loftus and Herndon, 1984).

Highly vagile species are often successful (Ehrlich, 1989). The ability of a species to disperse increases its opportunities to colonize or recolonize a site, and under the best of circumstances can result in a rapid expansion over a broad geographic range. Cuban treefrogs seek cavities and folds for refuge in vehicles and in nursery shipments of palm trees (Meshaka, 1996a,b). Greenhouse frogs and, to a lesser degree, juvenile marine toads likewise disperse in nursery shipments.

High fecundity is also often associated with successful colonization (Gaston and Kunin, 1997). Marine toads, Cuban treefrogs, African clawed frogs, and American bullfrogs are highly fecund species (Krakauer, 1968; McAuliffe, 1978; Zug and Zug, 1979; Lampo and De Leo, 1998; Meshaka, 1994, 2001; Crayon, this volume, Part Two). For these four species, breeding seasons in some introduced sites exceed those in parts of their native ranges (e.g., Carr, 1940a; Bury and Whelan, 1984). Little information is available on the reproductive biology of green and black dart-poison frogs or wrinkled frogs in Hawaii. However, neither species has a body size that would favor large clutch production even if the breeding season were extended and multiple clutches produced. Furthermore, both species require more specialized conditions for reproduction—the former species is limited by the availability of both moist cover for egg deposition and nearby pools for larval development, and the latter species is limited by the availability of lentic habitats for egg deposition and larval development.

Greenhouse frogs provide an exception to test the rule of high fecundity as a correlate of colonization success. This species, probably the least fecund of the North American exotic anurans, is also among the most successful. Although clutch sizes are small (Goin, 1947a), the breeding season is extended and development is direct and rapid (about 3 weeks). Additionally, eggs require only a moist cover for development. Perhaps low fecundity is offset by an abundance of suitable breeding sites throughout a long breeding season and by rapid and high recruitment.

Conclusions and Summary

Eight anuran species, African clawed frogs, marine toads, green and black dart-poison frogs, Cuban treefrogs, coquis, greenhouse frogs, wrinkled frogs, and American bullfrogs, are considered exotic species in the United States. For these species, I compare and contrast the circumstances associated with colonization and test four correlates of colonization that relate to both the structure of the introduced habitat and the biology of the invading species. Using synanthropy, similarity of introduced and native habitats, vagility, and high fecundity as tools, I evaluate the reasons for exotic species colonization success and their impacts on components of natural systems. Clearly, we have learned nothing if we do not yet understand that the combination of fractured systems and human-mediated dispersal is the greatest magnet for exotic species of all taxa. For this reason, our management goal should be to keep natural areas natural, large, numerous, and intact. Finally, the United States, with Florida, California, and Hawaii as prime lessons, should be tightly regulated with respect to transport of any exotic plant or animal. To be clear, this approach will not eliminate the exotic species problem. However, if made a priority, it will most certainly minimize the numbers of invasions and the subsequent ecological disasters that can accompany them.

Acknowledgments. Thanks are due to Betty Ferster and Samuel D. Marshall for reviewing an earlier version of the manuscript.

THIRTY-NINE

Protecting Amphibians While Restoring Fish Populations

DEBRA PATLA

Park and wildlife managers are facing an ironic dilemma as they work to restore and protect aquatic ecosystems—must amphibians be sacrificed if native fish are to return?

Over the past century throughout the United States, resource managers sought to enhance the recreational value of lakes and streams by stocking non-native (exotic) game fish. The "success" of this effort is now recognized as a serious impediment to conserving natural aquatic biodiversity. Introduced fish endanger and replace native fish species through predation, competition, hybridization, and disease transmission. In many cases, bringing back the natives is doomed unless the introduced fish are eradicated.

In Yellowstone National Park, for example, over 300 million fish were planted between the years 1889 and 1955. Four species of non-native trout (the ancestors of transplants from elsewhere in the United States and Europe) have become established in the park: brown, rainbow, eastern brook, and lake trout. They inhabit historically fishless waters as well as the habitat of indigenous fish, such as the famous Yellowstone cutthroat trout *(Oncorhynchus clarkii bouvieri)*. Native westslope cutthroat trout *(O. clarkii lewisi)* and arctic grayling *(Thymallus arcticus)* have almost disappeared, and they can be restored only if the non-native trout are removed. The park's recent decision to attempt restoration is supported by its mandate to protect indigenous wildlife, as well as by science-based principles of ecosystem management. There are, however, two large obstacles: politics, in the form of opposition from a segment of the sport-fishing public; and biology, in the formidable difficulty of removing established exotic fish populations.

The method most likely to effectively remove unwanted fish is the application of chemical piscicides, such as antimycin or rotenone. These toxins can kill all or most fish in the waters of streams and small lakes when dissolved at sufficient concentration. When the toxins clear and barriers are put in place to prevent re-invasions by unwanted fish, the native fish can be reintroduced. By working segment by segment along drainages, fishery managers in Yellowstone and other areas hope that they can eventually reclaim large portions of watersheds for populations of native fish.

The catch in terms of native biodiversity is that amphibians may suffer high levels of mortality from exposure to chemical piscicides, which kill amphibian larvae and may also affect or kill adult amphibians. Amphibian populations, particularly if small and spatially isolated, may be at risk. Caught between the well-justified need to restore native fish species and the potential for inadvertently killing native amphibians, what is the proper course of action? Considering that the success of fish removal projects is not certain, can amphibian conservationists accept a potential sacrifice of the animals we strive to protect?

This dilemma may be eased if amphibians are considered throughout all phases of the proposed fish removal projects. As a first step, the presence of amphibians and patterns of habitat use in areas targeted for chemical application must be studied. How and when do amphibians use these areas? How might the toxins affect breeding success or other aspects of life history? Are some areas so important for amphibian diversity and abundance that they should be protected from all human-caused risks? Are genetically unique populations of amphibians present in the project area? Are other populations of the same species close enough to allow recolonization of the treated area? Considering the water temperatures and chemistry of the targeted areas, how toxic is the piscicide to amphibian larvae and adults? If risks to amphibians appear to be high, can non-chemical methods of fish removal (such as electroshocking) be used effectively?

With the knowledge that piscicides may affect resident amphibians, measures can be taken to avoid or limit mortality. For example, treatments may be scheduled to avoid periods when amphibians would be most vulnerable. It may be possible to construct barriers that prevent the spread of toxins into breeding sites or other important habitat areas. Amphibian larvae and adults can be live-captured and temporarily maintained in safety until the waters are cleared of the toxin. Nevertheless, planning for piscicide use should include consideration of the worst case scenario, where amphibian populations are diminished or extirpated despite precautions and restoration is required. Ensuring protection of amphibians requires sufficient financial resources for monitoring, conservation actions, and potential restoration.

Through this effort, which will be unique for each project and region, we may find that removing non-native fish and conserving native amphibians need not be contradictory. Indeed, amphibians stand to benefit from the removal of fish species that prey on amphibians or compete for resources—and in some areas amphibian species conservation depends on non-native fish eradication efforts. Sharing information about how amphibians are affected by fishery projects and whether conservation efforts fail or succeed is now crucially important. This underscores the need for amphibian survey and monitoring before, during, and for several years after fish removal and restoration projects.

Summary

In Yellowstone National Park, four species of non-native trout were introduced, have become established, and threaten the survival of indigenous fish. A program to remove non-native trout has begun, but techniques used to remove these fish also threaten amphibians. Recommendations are made to eliminate or reduce the threat to amphibians, including making managers aware of the presence of amphibians, and offering strategies for reducing amphibian vulnerability to fish removal techniques.

FORTY

Reflections Upon Amphibian Conservation

THOMAS K. PAULEY

I began my fieldwork with amphibians in West Virginia in 1963 at the age of 23, and I have spent most of my time (when not teaching) walking the mountains of West Virginia searching for amphibians and reptiles. My major study sites include the New River Gorge National River, the Bluestone National Scenic River, the Gauley River National Recreational Area, and the high Alleghenies. Animals of particular interest include Cheat Mountain salamanders *(Plethodon nettingi)* and Cow Knob salamanders *(P. punctatus)*. I draw from these years of experience to present the following information on amphibian conservation.

Habitat Disturbances Can Create Amphibian Habitats: West Virginia Coal Mines and Logging Roads

The New River Gorge National River (NRGNR) includes an approximate 85 km (53 mi) portion of the New River with a steep gorge that passes through the Appalachian Plateau in Summers, Raleigh, and Fayette counties in southeastern West Virginia. The plateau overlies rich coal seams that were deep mined from the turn of the century to the mid-1900s. Today, numerous abandoned deep coal mines line the walls of the gorge. Many of these are wet with running water that seeps from the overlying forests. These abandoned mine portals are known to support bats, wood rats, and other mammal species. I thought that cave salamanders *(Eurycea lucifuga)* might also inhabit these mines, but all known populations of cave salamanders in West Virginia were from natural limestone caves in the eastern counties of the state.

One night in August 1991, while examining the entrance to an abandoned deep coal mine, my graduate students and I were indeed surprised to find several cave salamanders. Over the next year, we examined 40 mine portals and found cave salamanders in nine portals (22.5%). We found all size classes, indicating that cave salamanders were successfully reproducing. These mines had water pools that provided reproductive habitat. Ensuing fieldwork at rock cliffs and overhangs (both without caves) in the NRGNR and the Bluestone National Scenic River (a tributary of the New River) resulted in the observation of cave salamanders in large fissures. Obviously, cave salamanders had been in rock formations in the NRGNR for

many years. Because of the lack of pre-mining data on cave salamanders, it is impossible to assess whether or not the mines enhanced the population sizes of cave salamanders. However, there are no data (or good reasons) to suggest that the mines harmed populations of cave salamanders.

For approximately 10 years prior to my studies in the NRGNR, I had been monitoring several populations of mountain chorus frogs *(Pseudacris brachyphona)*, and I had noticed declines. However, as I walked the old mining and timbering haul roads and skid roads in the NRGNR, I found mountain chorus frogs to be abundant in road rut pools and roadside ditches. In 1992, with the cooperation and support of the U.S. Forest Service Northeastern Research Station in Parsons, West Virginia, I began a two-year study of 19 created road rut pools in skid roads in a newly clearcut area. The study area was on McGowan Mountain in Tucker County, West Virginia. The clearcut had a northwest aspect and an elevational range from 890–987 m. The forest had been clearcut in 1992, and skid road pools were constructed in October of that year. My study commenced in April 1993 and ended in October 1994. In 1993, 11 pools were used as breeding sites by wood frogs *(Rana sylvatica)*, mountain chorus frogs, Cope's gray treefrogs *(Hyla chrysoscelis)*, and eastern American toads *(Bufo americanus americanus)*. In 1994 in an effort to determine the reproductive success of these species, we constructed drift fences with pitfall traps to capture newly metamorphosed frogs and toads as they exited these breeding pools. Individuals from all four species were trapped leaving the pools. Two additional species, red-spotted newts *(Notophthalmus v. viridescens)* and northern water snakes *(Nerodia s. sipedon)* entered the pools in 1994. Red-spotted newt larvae were found in one pond, indicating that this species had also reproduced successfully in this habitat. The water snakes (two juveniles) were foraging on the juvenile frogs.

We searched the area for established breeding pools that might explain where the toads and frogs that used these created pools originated, but we could not locate even one pool. The nearest stream that could support water snakes was 700 m downslope. It is apparent that these species of anurans will colonize newly created habitats that were obviously not their natal wetlands, and they will travel great distances to do so. It is also apparent that these species spend the nonbreeding portion of the warm months moving through the forest foraging, otherwise they would not have found these created pools as they migrated to established breeding pools in the spring.

Several studies have demonstrated that clearcutting practices kill forest salamanders. Nevertheless, silvicultural practices do not eliminate all salamanders in a region, because nearly all of the eastern forests were clearcut by the early 1900s and salamanders persist. (However, historical data are lacking and we cannot know the true extent of either losses or rebounds.) This shows that clearcuts likely (at least in some cases) shift the environment to favor a different type of amphibian assemblage. With proper management during the regrowth period of a forest, created pools can provide excellent reproductive habitats for pond-breeding species of amphibians.

Species Considered Uncommon May in Fact be Common

Much discussion has transpired over the last few years about the declines of amphibians on a local scale as well as worldwide. Many field herpetologists have reported some species to be in decline and consider other species to be uncommon or rare. In fact, it is important to realize that because amphibians are ectothermic and can respire through their integument, many species are attracted to cooler temperatures and more moist conditions—factors that regulate their surface activity and therefore their ability to be detected by humans.

I have been working in the mountains of West Virginia since 1966, and I have conducted inventories for amphibians at thousands of sites. In the early 1980s, I was concerned that species such as mountain chorus frogs (*P. brachyphona*), black-bellied salamanders (*Desmognathus quadramaculatus*), and green salamanders (*Aneides aeneus*) were uncommon in the state, were in decline, or were possibly even gone. In examining my previous inventory sites, I discovered that usually I had conducted extensive searches in places easily accessible by vehicles. Therefore, during the late 1980s and throughout the 1990s, I began to examine (on foot) some of the more rugged areas of the state, including NRGNR and Bluestone National Scenic River in West Virginia. New River and Bluestone gorges are both national parks in West Virginia; as with nearly all of West Virginia, the forests in these gorges have been clearcut at least twice since the late 1880s. The New River area is "riddled" with strip mines and "honey-combed" with deep coal mines. There are miles of abandoned haul roads. Both gorges have potentially interesting zoogeographical significance because their rivers originate in the southeast United States and flow north through West Virginia.

Prior to my work in NRGNR (1989–91), I was concerned that mountain chorus frogs were showing signs of decline in the state. Several populations I had monitored for 9–10 years in the north-central sections of the state had declined and many had disappeared. Although herpetologists from surrounding states (e.g., Ohio and Maryland) reported similar concerns, as I began to walk the old haul roads in the New River and Bluestone gorges, I soon learned that these animals were abundant in the ditches, along the abandoned haul roads, and in strip mines. These old roads now provided excellent breeding and foraging habitats for mountain chorus frogs and other species of amphibians, such as northern spring peepers (*P. c. crucifer*), green frogs (*R. clamitans*), red-spotted newts, four-toed salamanders (*Hemidactylium scutatum*), spotted salamanders (*Ambystoma maculatum*), and Jefferson salamanders (*A. jeffersonianum*).

In our book, *Amphibians and Reptiles in West Virginia*, N. B. Green and I reported that we feared black-bellied salamanders may have been extirpated in the state because of the combined effects of water pollution from sewage runoff, acidic mine drainage, and pressure from fisherman and commercial bait collectors. Black-bellied salamanders reach the most northern extent of their distribution in West Virginia, near the point where Gauley River merges with New River to form the Kanawha River. We surveyed 103 tributaries (first and second order) in the NRGNR for black-bellied salamanders, and what we found was surprising. Of the 103 streams sampled, we found black-bellied salamanders in 54 (52.4%), and in many of these streams they were quite abundant. Since this study, we have found black-bellied salamanders several miles upstream in major tributaries of the New River (Greenbrier, Bluestone, and Gauley rivers). We suspected that fishing pressure was having the biggest impact on these salamanders. In nearly every stream examined, black-bellied salamanders were absent near the confluence of rivers where they could be conveniently found by fishermen. As we examined these streams, it was evident that rocks near the mouths of the tributaries had been turned many times by fishermen in search of black-bellied salamanders for fishing bait. We found these salamanders in good numbers in surveys farther up the tributaries where bait collecting is infrequent.

We were also surprised by the apparent lack of impact of water pollution on black-bellied salamanders. We found black-bellied salamanders in streams laden with sewage and in streams where water flowed out of deep coal mines. Although when tested, the pH levels in these streams were near neutral, ranging from 6.0–8.0, as the rock strata here has a great buffering capacity.

It has been known for many years that the range of green salamanders extends through most of the central and western portions of West Virginia, presumably at lower elevations. Reports of historic surveys conducted in the New River and Bluestone Gorges suggested that green salamanders were absent. However, in our surveys of these river gorges and the Gauley River Gorge, we found green salamanders to be abundant—even on rocks that one might suppose were too dry to support this species and at elevations from a high of 914 m in the eastern mountains down to a few hundred meters along the Ohio River.

In searching for many species, including green salamanders, an investigator should be cognizant of habitat and life history. It is particularly important to know the times (seasonally or daily) when species are active on the surface of their habitats. Green salamanders become less active on the soil surface from mid-June to late-August in central West Virginia, and after July at higher elevations. During these months, it can be difficult to find them on rock outcrops, even where you know they are abundant.

The requirement to know species habits also holds true for other species of salamanders, such as Wehrle's salamanders (*Plethodon wehrlei*) that span a large elevational gradient. Wehrle's salamanders appear on the surface of the forest floor in the Allegheny Plateau in West Virginia in March, April, and May. At lower elevations, Wehrle's salamanders move to cooler and moister underground habitats from June through September, and emerge again in October. In the Allegheny Mountains, Wehrle's salamanders are active on the surface of the forest floor throughout the summer, from April to mid-October. Searches for Wehrle's salamanders at times inappropriate for the locations would provide erroneous data on population status and conservation priority.

To achieve a true measure of the presence of amphibians, it is important to know their habits and search in optimal seasons, in optimal weather conditions, and throughout optimal habitat types.

Habitat Restoration is Not Always Sufficient: Historical Contingencies Mark the Current Distribution of Cheat Mountain Salamanders (Plethodon nettingi)

Cheat Mountain salamanders (*P. nettingi*) were first found on White Top Mountain, Randolph County, West Virginia, in 1935, and were described by the late N.B. Green (co-author of *Amphibians and Reptiles in West Virginia*) in 1938 from specimens taken from Barton Knob, Randolph County, approximately 3 miles west of White Top. Cheat Mountain salamanders were originally thought to only occur on Cheat Mountain.

Since 1976, I have surveyed 1,300 potential Cheat Mountain salamander sites and found the species in 135 (10.9%). After determining the areal extent of these sites, I now believe they make up 60 disjunct populations. In addition to the 1,300 sites surveyed, I have made habitat assessments of an additional 275 sites within the elevational and total range of which I did not believe the habitat was appropriate to support this species. These populations extend to five counties: Pendleton, Pocahontas, Randolph, Tucker, and Grant. The total range extends from Blackwater Falls State Park in the north to a point southwest of Bald Knob in the south, approximately 42 × 121 km (19 × 55 mi). The vertical range extends from 1,482 m (4,862 ft; the highest point in West Virginia) down to 804.7 m (2,640 ft) at Blackwater Falls.

Based on their limited distribution and the habitat disturbances that have been occurring throughout their range, Cheat Mountain salamanders were listed as a threatened species by the U.S. Fish and Wildlife Service on 28 September 1989 (Federal Register, Volume 53, Number 188, 37814-37818).

Ecology

The original Cheat Mountain salamander habitat consisted of red spruce/yellow birch forest with *Bazzania* sp. on the forest floor. I have found a few populations in deciduous forest (once a red spruce forest) and in hemlock stands. I hypothesize that when the original Appalachian Forest was cut (and many areas burned), the only Cheat Mountain salamander populations that survived were those that gained refuge around large rocks or in narrow ravines protected by thick growths of *Rhododendron*. It is important to realize that most of the original forest within the range of Cheat Mountain salamanders has been completely cut within the last 100 years. In fact, nearly all red spruce had been taken by 1920 (Clarkson, 1964). After examining each site where Cheat Mountain salamanders are known to occur, I realized that current habitat sites are all associated with either larger, emergent rocks, boulder fields, or steep, narrow ravines with *Rhododendron* thickets.

The distribution of Cheat Mountain salamanders overlaps with two other plethodontids, eastern red-backed salamanders (*P. cinereus*) and Allegheny Mountain dusky salamanders (*D. ochrophaeus*), that compete with Cheat Mountain salamanders for resources. I have found that Cheat Mountain salamanders occur in microsites that are more moist than those where eastern red-backed salamanders are found, but no more moist than

where Allegheny Mountain dusky salamanders are found. Critical thermal maximum and dehydration tests conducted in the laboratory concurred with the moisture conditions observed in the field (i.e., Cheat Mountain salamanders and Allegheny Mountain dusky salamanders lose body moisture faster than eastern red-backed salamanders). Consistent with these data, eastern red-backed salamanders can tolerate a greater percentage of lost body moisture than can Cheat Mountain salamanders and Allegheny Mountain dusky salamanders. Further, the many nests of Cheat Mountain salamanders and eastern red-backed salamanders that I have found during the past 24 years have been located in the same habitat—under rocks, logs, and bark. (Eggs of both species are deposited in May or June and hatch in August or September.) Based on these data, I have concluded that there is a three-way competition that prevents Cheat Mountain salamanders from invading newly mature forest from their refugia of boulder fields and narrow, heavily vegetated ravines. In these new habitats, Cheat Mountain salamanders simply cannot compete with eastern red-backed salamanders for the nesting sites nor with Allegheny Mountain dusky salamanders for the moist sites.

A Lesson for Conservationists

While many species of amphibians respond well to habitat restoration efforts (e.g., amphibians in the Upper Midwest quickly recolonize restored prairie potholes; Galatowitsch and van der Valk, 1994), the maturation of forests in the original range of Cheat Mountain salamanders has not resulted in this species establishing its presumed contiguous historical distribution. From this, I conclude that the disturbance brought about by clearcutting was more stressful on Cheat Mountain salamanders than on eastern red-backed salamanders and Allegheny Mountain dusky salamanders. These latter two species are now established, and they compete with and restrict the current distribution of Cheat Mountain salamanders. It appears that restoring habitat is not, by itself, sufficient to restore Cheat Mountain salamanders to their historical distribution. Restoration would require eliminating the competitive pressures by removing eastern red-backed salamanders and Allegheny Mountain dusky salamanders from historic Cheat Mountain salamander habitats—a nearly impossible task.

The Role of Subtle Forest Fragmentation in Amphibian Declines

Cheat Mountain salamanders are known to occur in only 60 locations in five counties in the eastern high mountains of West Virginia. Each of these 60 populations experiences some type of habitat disturbance, ranging from hiking trails to ski slopes. I examined the potential effects of habitat fragmentation in two perturbations in the Monongahela National Forest in West Virginia. Both of these studies are ongoing; I have studied the trails for 4 years (1991–1992, 1999–2000) and the ski slope for 14 years.

My study of the hiking trail includes one trail with heavy foot traffic, one with moderate foot traffic, and a third with light foot traffic. The heavily traveled trail is located in the most visited State Park in West Virginia; the moderately traveled trail is in a popular tourist area of the Monongahela National Forest; and the infrequently traveled trail is in a less known area of the National Forest. The heavily traveled trail has significantly less leaf litter during the summer months than do the moderately and lightly traveled trails. Salamanders captured in the treads of each trail or within one meter of the edge of each trail are measured for SVL, sexed, and marked (toe-clipped). The three most common species of salamanders observed are Cheat Mountain salamanders (n = 167), eastern red-backed salamanders (n = 24), and Wehrle's salamanders (n = 82).

To date, Cheat Mountain salamanders and eastern red-backed salamanders have not been observed on the heavily traveled trail, and none have been found to cross this trail. Wehrle's salamanders have not been observed on the trail, however, four individuals have crossed the trail, as they have been recaptured on the opposite side of the trail from where they were found initially. I have not observed any of the three species on the moderately traveled trail or crossing the trail. I have found all three species on the lightly traveled trail throughout the summer and autumn months.

In my research, I have found that Cheat Mountain salamanders become more diurnal in the autumn when leaf fall greatly increases the leaf litter on the forest floor. Adults of both sexes can be found readily between the surface of the soil and the leaf litter. They are active and display obvious sexual dimorphic characteristics—males with swollen cloacas and squared snouts and females with well-developed follicles that are visible through the body wall. I have observed courtship display in Cheat Mountain salamanders in late summer, so some mating probably occurs before winter. Heavy leaf litter in the autumn may allow some individuals to cross trails (and mate) that they would not otherwise traverse. However, most mating occurs in May and June, and all nesting occurs from May to August. By this time, the previous autumn leaves have decayed and hiking activity has increased substantially.

Bare trail treads may prevent gravid females and males from meeting, thus interfering with gene flow and decreasing genetic diversity in these populations. Given the small size of some of the Cheat Mountain salamander populations, further division into subpopulations by hiking trails may result in the loss of these population fragments due to competition with other salamanders, predation, or disease. In fact, since 1992 there has been a decrease in the number of salamanders observed on the heavily traveled trail.

The effects of larger disturbances such as graveled roads, skid and haul roads, ski slopes, and utility rights-of-way that fragment populations are more obvious. I have been monitoring potential effects of a ski slope on a population of Cheat Mountain salamanders since 1986. The ski slope originates on private land but makes a large curve around a Cheat Mountain salamander population on Monongahela National Forest land. This curve positions a portion of the slope above the population, a portion around one end, and a portion below the population. The construction of the ski slope invaded the margins of this population, pushing the edge of the population back at least 10 meters. The ski slope now comes within a few meters of the population on two sides as it turns and goes down hill. I established four transects between the upper and lower slopes of the curve, two near the upper slope and two near the lower slope. There are 10 monitoring sites (impact sites) adjacent to the slope (within 20 m) and 33 monitoring sites (non-impact) over 20 m from the slope. I conduct inventories of the salamanders in all 43 monitoring sites two times each year, once in the spring and once in early autumn. All salamanders are measured (SVL) for size classes, sexed, and the reproductive status of females determined (gravid or non-

gravid). I measure air temperature and relative humidity (at ground level), litter weight and moisture, and soil temperature and moisture. Soil temperature and soil moisture are taken from the first 3–5 cm. (Deeper readings give erroneous data because the trees have been removed from the ski slope thus limiting the uptake of water.) Over these 14 years, I have found substantially lower numbers of salamanders in the affected sites. The major differences in the population structure are the lack of gravid females and juveniles. In the impact sites, relative humidity, soil moisture, litter weight, and litter moisture are substantially lower, while air and soil temperatures are substantially higher. Drier and warmer leaf litter and soil reduce potential nesting sites, thus excluding gravid females and consequently, young of the year and second- and third-year juveniles. Because of the critical importance of moisture and temperature for the nesting success of terrestrial plethodontids, the impact from the ski slope is causing a Cheat Mountain salamander decline.

Possible Synergistic Effects of Multiple Sublethal Stressors

Research has demonstrated that alterations in habitats can result in threats to amphibian biodiversity. Generally speaking, biodiversity is lost through habitat destruction, fragmentation, and degradation. For amphibians, habitat destruction varies from clearcutting a forest to draining a wetland. Habitat fragmentation results from many influences by people such as roads, rights-of-way, ski slopes, and parking lots. Habitat degradation comes from factors such as pesticides, species introductions, and atmospheric anomalies such as ultraviolet (UV-B) radiation and acid deposition. Scientists realize that threats to some amphibians are synergistic, that is that independent factors may combine to have a negative impact on species.

Many upland fens in West Virginia have low pH values because of natural tannic acids that leach into streams from dominant vegetation types (red spruce, hemlock, mountain laurel, and *Rhododendron*) in these higher elevations. Tannic acids cause water in some of these streams to have the color of coffee or tea, and streams in this area of the state have names like Blackwater River and Yellow Creek. Most of these fens support species of amphibians that include spotted salamanders, red-spotted newts, four-toed salamanders, northern spring peepers, green frogs, and wood frogs. In one such fen supplied by Yellow Creek, hundreds of wood frogs migrate each year to mate and deposit eggs. In a two-year study (1995–96) of amphibians at Yellow Creek, we did not find one wood frog egg that successfully developed from the thousands of eggs that were deposited. All embryos died within a few days after deposition. In three other fens we studied at the same time, wood frog eggs successfully hatched and larvae transformed.

Something was clearly different about Yellow Creek. To determine what this might be, we measured levels of potential threats, including UV-B radiation, pH, and dissolved aluminum in the water in all four fens. Data were collected twice per month from April–October. To examine the potential effects of UV-B radiation on the amphibian development at each site, newly laid wood frog eggs were placed in three plastic boxes: one covered with Mylar, one with acetate, and one not covered (method following Blaustein et al., 1994c). Water pH values were taken on site. Water samples were delivered to a water analysis laboratory to measure dissolved aluminum.

UV-B wavelengths from 290–320 nm are considered to be detrimental to organisms (Worrest and Kimeldorf, 1975, 1976; Blaustein et al., 1994c; Grant and Licht, 1995). UV-B wavelengths at all four fens were below 290 nm, except for one reading at Yellow Creek that reached 323 nm. Over the two-year period, mean UV-B wavelengths were significantly lower at Yellow Creek than at one other study site, but significantly higher than at another study site. Most UV studies on amphibians reported in the literature have been conducted in the western United States, where the elevations are higher and the skies clearer than in West Virginia. Elevations at the four study sites in West Virginia ranged from 914–1,158 m. The Allegheny Mountain Physiographic Province of West Virginia receives more precipitation in the form of rain and fog during the spring, summer, and autumn seasons than any other physiographic province in the state. These climatic conditions probably reduce the effects of UV-B on amphibians in this section of the Allegheny Mountains.

Yellow Creek also had the highest dissolved aluminum values (0.52–0.58 mg/L), and significantly lower pH values (3.81–4.13). The West Virginia Division of Natural Resources has been sporadically collecting pH data from Yellow Creek since 1946. The average pH values were 4.75 from 1946–52 and 4.12 from 1957–85. These data illustrate a decline in water pH of Yellow Creek between 1946 and 1996, a decline that amphibians may not be able to tolerate.

Amphibian breeding habitats abound at Yellow Creek. In 2 years of searching, we did not find evidence (eggs or larvae) of successful reproduction in wood frogs, spotted salamanders, red-spotted newts, northern spring peepers, or green frogs. However, we did find nests of four-toed salamanders with viable eggs. Acidic fens are common breeding habitats for four-toed salamanders in the higher elevations in West Virginia. Kilpatrick (1997) examined median tolerance limits (Tlm) of eggs and larvae of four-toed salamanders and wood frogs from these same fens and found that four-toed salamander embryos were more tolerant of acid conditions than were wood frog embryos, thus allowing them to survive here.

A Lesson for Conservationists

It is unfortunate that historical data for all these values are not available. The West Virginia Division of Natural Resources water pH data is invaluable. It provides us with data that shows that the decline in pH values is from a source(s) other than tannic acids, which have been present in this area for thousands of years. We can speculate that the cause of change in pH is atmospheric. The probable cause of the demise of wood frogs at Yellow Creek is the combined influence of high aluminum and low pH values. It is doubtful that one UV-B reading above 290 nm out of over 70 readings recorded at Yellow Creek would be detrimental to the development of wood frog eggs. Perhaps field herpetologists should make an effort to record as much environmental data as possible and develop multiple databases throughout the United States. This effort will allow the next generation of herpetologists to make inferences about the causes of changes in species diversity. Today's data create tomorrow's historical baseline.

Conclusion

While natural history data are often dismissed as being old school or old fashioned, there can be no substitute for such knowledge when it comes to making intelligent management decisions. In these modern days, when cutting edge science requires big grants and expensive equipment, to conserve amphibians it may be that all you need are a sturdy pair of legs and an inquisitive mind.

Acknowledgments. The fieldwork at New River Gorge National River and Bluestone Scenic River was funded by the United States National Park Service and by the West Virginia Division of Natural Resources to develop an amphibian and reptile atlas. The U.S. Forest Service funded the fieldwork on Yellow Creek. I acknowledge the support of biologists and staff members of Monongahela National Forest, the West Virginia Field Office of the United States Fish and Wildlife Service, and the West Virginia Division of Natural Resources. I also thank numerous students who have worked with me in the field these many years.

SURVEYS AND MONITORING

FORTY-ONE

Distribution of South Dakota Anurans

DAVID E. NAUGLE, TATE D. FISCHER, KENNETH F. HIGGINS,
AND DOUGLAS C. BACKLUND

In a little over a century, the prairie pothole region of the northern Great Plains has been transformed from a contiguous expanse of wetlands and grasslands into a highly fragmented agricultural landscape. Regional wetland losses due to agricultural activities and urbanization have been extensive and widespread, exceeding 90% in northwestern Iowa and western Minnesota (Tiner, 1984; Dahl, 1990; Leja, 1998). Declines in amphibian numbers coinciding with the habitat loss have heightened concerns over the future of amphibian populations (e.g., Barinaga, 1990; Blaustein and Wake, 1990, 1995; Wake, 1991; Lannoo, 1998b).

South Dakota (199,500 km^2 in area), a state that lies in the prairie pothole region, is bisected into two nearly equal eastern and western segments by the upper Missouri River. The eastern portion is a poorly drained, glaciated landscape with an abundance of wetlands covering approximately 10% of the total land area (Johnson and Higgins, 1997). In contrast, the western portion is a well-drained, non-glaciated landscape composed of ancient sediments. Despite a 45% decline in wetland numbers in eastern South Dakota, more than 930,000 wetland basins remain (Johnson and Higgins, 1997). While concern about amphibian declines have produced conservation initiatives in other Midwestern states (Lannoo, 1998b), in South Dakota such programs have not been implemented and there remains a lack of basic knowledge concerning the distribution and habitat use of South Dakota frogs, toads, and salamanders (Fischer, 1998; Fischer et al., 1999).

To address this lack of data on South Dakota anurans, we present distribution maps for 12 frog and toad species. These maps include new records from fieldwork conducted in 1997–98, where we surveyed 1,496 wetland sites throughout all 44 counties in eastern South Dakota (Fischer, 1998). Because no one had conducted extensive survey work in western South Dakota, we construct distribution maps using historical data from the South Dakota Natural Heritage Database. We also assess how breeding anurans use habitats throughout eastern South Dakota by using occurrence rates of 11 species in seven different wetland types. Finally, to assist future monitoring programs, we follow Mossman et al. (1998) in using field data to recommend ranges of dates for sampling each species.

Methods

We conducted anuran auditory surveys throughout eastern South Dakota. In 1997, we selected 14 transects by partitioning eastern South Dakota into 13 km-wide strata. Within each stratum, we randomly selected one Public Land Survey section line as a transect origin. We then established transects on east-west roads closest to the latitude of origin; transects were separated by at least 3.2 km. We eliminated major roadways (e.g., Interstate 90) from transect selection to reduce auditory disturbances. In 1998, we positioned eight new transects along latitudes bisecting counties in which few anurans were recorded during the 1997 surveys. To supplement these transect surveys, we conducted auditory surveys at wetlands along every available stretch of roadway in areas where historical records have indicated the presence of isolated anuran populations.

We overlaid transects onto a wetland geographic information system developed for eastern South Dakota (Johnson and Higgins, 1997). We classified wetlands according to Stewart and Kantrud (1971) into seven classes: temporary, seasonal, semipermanent, permanent, manmade (e.g., stock dams), undifferentiated tillage ponds, and riverine. We randomly selected roadside wetlands to survey sites that were within 100 m of the transect line. In counties with high wetland densities (> 75 wetlands/km^2), we surveyed at least three wetlands in each class along each transect. In counties with low wetland densities, we surveyed all wetlands within 100 m of the transect.

We followed an established protocol that has been extensively used to monitor anurans in the Midwest (Hemesath, 1998; Johnson, 1998b; Mossman et al., 1998). We surveyed wetlands in the spring when water temperatures reached 10 °C (Johnson and Batie, 1996; Mossman et al., 1998). We conducted surveys from 22 April–19 May and 2 June–30 June in 1997, and from 25 April–28 May and 28 May–29 June in 1998. We began surveys in the south and proceeded northward to account for latitudinal differences in breeding phenology.

We recorded male mating calls during three-minute survey periods at each wetland (after Zimmerman, 1994; Shirose et al., 1995; see also Mossman et al., 1998). We identified calling anurans to species and assigned the relative abundance of each species to one of the following classes: 1 = individuals could be counted with space between calls; 2 = individuals were distinguishable with some overlap of calls; and 3 = a constant, continuous, and overlapping full chorus of calls. We conducted surveys during peak calling periods (one-half hour after sunset to 0200 h). We traversed transects before nighttime surveys to evaluate road accessibility and identify target wetlands. We recorded Universal Transverse Mercator coordinates at each wetland edge using a global positioning system (GPS) to ensure that the same sites were surveyed on subsequent visits. We also referenced wetland locations on county maps using section lines, adjacent wetlands, and landmarks. We did not conduct surveys during unseasonably cool or windy nights.

We supplemented call survey records with anuran capture records because some species call sporadically (e.g., northern leopard frogs [Rana pipiens] or following heavy rains (plains spadefoot toads [Spea bombifrons]). We captured anurans on public lands using straight-line drift fences with pitfall traps (after Dalrymple, 1988; Corn, 1994b). We placed traps in areas of expected high amphibian density and checked them daily. We shaded pitfall traps, which contained soil for use by burrowing anurans. We also systematically searched surrounding public lands for anurans (after Crump and Scott, 1994). We searched blacktop and secondary gravel roads at night as we traveled along call survey routes (after Simmons, 1987; Shaffer and Juterbock, 1994).

Our methods of euthanasia and preservation of voucher specimens follow Pisani (1973) and Simmons (1987). Vouchers are archived at the Museum of Natural History, Division of Herpetology, University of Kansas, Lawrence, Kansas.

We produced distribution maps by shading counties in eastern South Dakota where anuran species were found. Current and historical anuran distributions were compared using data compiled from college and museum collections for South Dakota (O'Roke, 1926; Over, 1943; Underhill, 1958; Fishbeck and Underhill, 1959, 1960; Dunlap, 1963, 1967; Smith, 1963; Thompson, 1976). We produced distribution maps for western South Dakota anurans using existing information compiled from the South Dakota Natural Heritage Database in Pierre, South Dakota.

In eastern South Dakota, we calculated occurrence rates of breeding anurans in seven different wetland classes (i.e., temporary, seasonal, semipermanent, permanent, manmade, tillage, and riverine) by dividing the number of wetlands used by a species by the number of surveyed wetlands. The denominator was derived by counting the number of surveyed wetlands from counties within a particular species distribution. We determined calling phenologies of breeding anurans in eastern South Dakota on the basis of their peak calling periods in both 1997 and 1998.

Results

Statewide Patterns in Anuran Distribution and Abundance

We conducted auditory surveys in 1,496 wetlands (122 temporary, 243 seasonal, 319 semipermanent, 301 permanent, 115 tilled, 163 manmade, and 233 riverine) along 7,712 linear km of roadway in eastern South Dakota. Spatial distribution patterns and wetland occupancy rates varied among species in eastern and western South Dakota (Fig. 41-1; Table 41-1).

Western chorus frogs (Pseudacris triseriata), northern leopard frogs, and Great Plains toads (Bufo cognatus) occupied the greatest proportion of wetlands (Table 41-1) and are the most widespread anurans in eastern South Dakota (Fig. 41-1). Historical data from the Natural Heritage Database indicate that these species are also the most widespread species in western South Dakota (Fig. 41-1). American toads (B. americanus) and Canadian toads (B. hemiophrys) were most common in the eastern half of eastern South Dakota and occurred in fewer wetlands than the most common species (Fig. 41-1; Table 41-1). Woodhouse's toads (B. woodhousii) are the least common bufonid in eastern South Dakota (Table 41-1) that also occur in western South Dakota (Fig. 41-1). Plains spadefoot toads also occur throughout the state but may be more common in the west (Fig. 41-1). American bullfrogs (R. catesbeiana) are limited to southern counties in South Dakota (Fig. 41-1). Eastern gray treefrogs (Hyla versicolor) were only recorded in the extreme northeastern and southeastern corners of the state. Cope's gray treefrogs (H. chrysoscelis) were only detected in the extreme southeastern corner of the state (Fig. 41-1). The authors caution that historical records for eastern gray treefrogs could be Cope's gray treefrogs (Christiansen and Bailey, 1991; Gerhardt et al., 1994). Although present historically in southeastern South Dakota (Fig. 41-1), we did not detect Blanchard's cricket frogs (Acris crepitans blanchardi) or plains leopard frogs (R. blairi) in this survey. Two observations of wood frogs (R. sylvatica) were made in northern Roberts County, the most northeastern county in the state.

Anuran Habitat Use in Eastern South Dakota

The proportion of wetlands occupied by breeding anurans in eastern South Dakota varied (Fig. 41-1; Table 41-1). Species with statewide distributions (i.e., western chorus frogs, northern leopard frogs, and Great Plains toads; Fig. 41-1) were those most likely to occur in each of the seven wetland classes (Table 41-1).

Anuran Calling Phenology

Peak calling activity for wood frogs was in April (Fig. 41-2). Northern leopard frogs and chorus frogs began calling in April and continued through May (Fig. 41-2). A pulse in calling activity was apparent for three Bufo species (American, Canadian, and Great Plains toads) from mid-May through June (Fig. 41-2). From early June to mid-July, eastern and Cope's gray treefrogs, plains spadefoot toads, Woodhouse's toads, and American bullfrogs intensely called (Fig. 41-2).

Discussion

Distribution maps resulting from this study may serve as a benchmark on which to assess the future status of anurans in eastern South Dakota. In western South Dakota, current anuran distributions remain speculative because extensive survey work has yet to be conducted. New distribution records for some species (e.g., chorus frogs, northern leopard frogs, and American toads) indicate that anuran species diversity may be greater than previously known. Current distributions indicate that anuran diversity is highest in the south and east (Fig. 41-1). Although results from this study indicate that a greater number of species occur in glaciated wetland habitats, we clearly need more data from western South

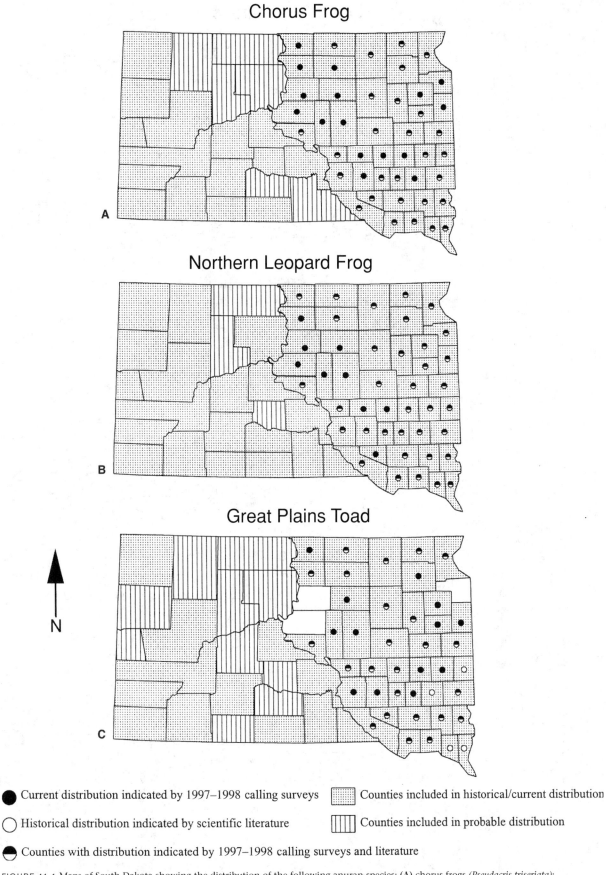

Chorus Frog

A

Northern Leopard Frog

B

N

Great Plains Toad

C

● Current distribution indicated by 1997–1998 calling surveys · · · · Counties included in historical/current distribution

○ Historical distribution indicated by scientific literature ||||| Counties included in probable distribution

◖ Counties with distribution indicated by 1997–1998 calling surveys and literature

FIGURE 41-1 Maps of South Dakota showing the distribution of the following anuran species: (A) chorus frogs *(Pseudacris triseriata)*;
(B) northern leopard frogs *(Rana pipiens)*; (C) Great Plains toads *(Bufo cognatus)*; (D) Woodhouse's toads *(B. woodhousii)*; (E) American toads
(B. americanus); (F) Canadian toads *(B. hemiophrys)*; (G) plains spadefoot toads *(Spea bombifrons)*; (H) American bullfrogs *(Rana catesbeiana)*;
(I) eastern gray treefrogs *(Hyla versicolor)*; (J) Cope's gray treefrogs *(H. chrysoscelis)*; (K) Blanchard's cricket frogs *(Acris crepitans blanchardi)*; and
(L) plains leopard frogs *(R. blairi)*.

Woodhouse's Toad

American Toad

Canadian Toad

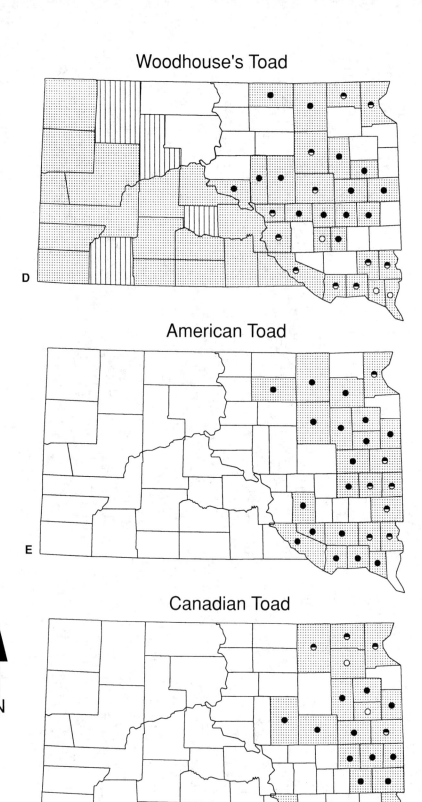

FIGURE 41-1 (continued)

Plains Spadefoot Toad

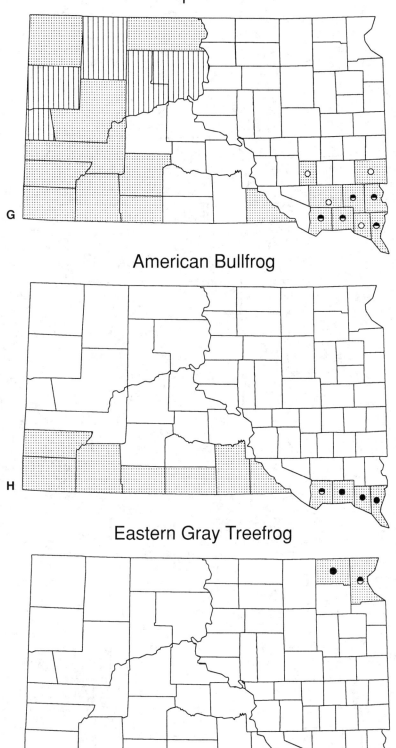

American Bullfrog

Eastern Gray Treefrog

N

FIGURE 41-1 (continued)

Cope's Gray Treefrog

Blanchard's Cricket Frog

Plains Leopard Frog

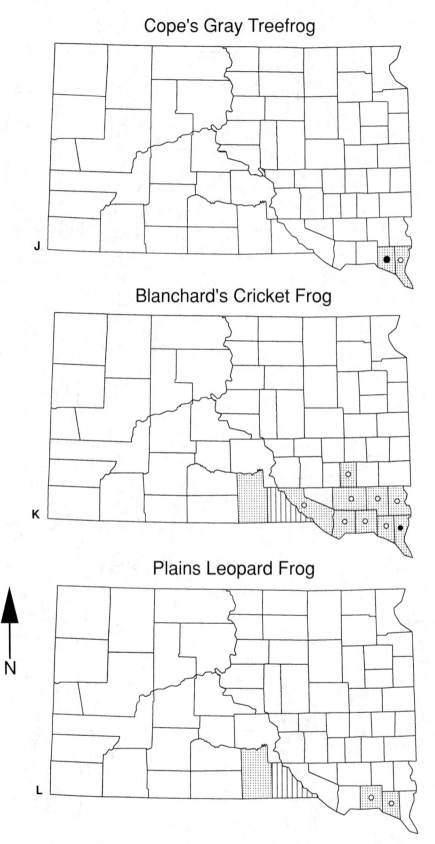

FIGURE 41-1 (continued)

TABLE 41-1

Occurrence Rates of 11 Breeding Anurans in Seven Different Wetland Classes in Eastern South Dakota, 1997–98

	Number Occupied	Temporary	Seasonal	Semi-Permanent	Permanent	Tillage	Manmade	Riverine
Wetlands Surveyed	1496	122	243	319	301	115	163	233
Species:								
Pseudacris triseriata[a]	1102	70	85	84	82	75	53	52
Rana pipiens	438	13	21	39	49	31	17	17
Bufo cognatus	249	29	17	17	20	13	11	11
Bufo americanus	107	5	4	6	13	11	5	5
Bufo hemiophrys	100	tr[b]	4	5	11	19	6	3
Bufo woodhousii	90	3	4	6	8	9	7	5
Rana sylvatica	2	0	0	0	2	0	0	0
Spea bombifrons	12	0	6	2	0	0	4	0
Rana catesbeiana	4	0	5	4	0	0	0	0
Hyla versicolor	3	0	0	2	0	0	0	0
Hyla chrysoscelis	1	0	0	1	0	0	0	0

[a] *Pseudacris triseriata triseriata* and *P. t. maculata* were grouped for this study because we were unable to differentiate between their breeding calls in the field.

[b] tr < 1%.

NOTE: Occurrence rates are calculated as number of wetlands used divided by number of wetlands surveyed. Species are listed from the greatest to the least abundant. The denominator used to calculate occurrence rates reflects the number of wetlands surveyed within a species distribution (Fig. 41-1).

Dakota to make legitimate comparisons. Plains leopard frogs and Blanchard's cricket frogs are two species known to occur historically in South Dakota (Over, 1943; Fishbeck and Underhill, 1959, 1960; Fig. 41-1), but were not recorded by us. Since completion of this study, Blanchard's cricket frogs were collected in summer 1999 (by author D.C.B.) at a site along the Big Sioux River in eastern South Dakota (Union County; updated in current distribution map, Fig. 41-1). Blanchard's cricket frogs are known to be in severe decline, especially along the margins of their distribution (Gray and Brown, this volume).

Anuran occurrence rates were highly variable across the seven wetland classes surveyed, indicating that no single wetland type can fulfill all the habitat requirements of the diverse assemblage of anurans found in South Dakota (see also Lannoo, 1996, 1998a). Overall, the highest occurrence rates were in semipermanent and permanent wetlands (Table 41-1). In contrast, the lowest occurrence rates were in manmade (e.g., stock dams) and riverine wetlands (Table 41-1). In the northern Great Plains, temporary, seasonal, and tillage wetlands located within agricultural fields are at the greatest risk of drainage (Johnson et al., 1996; Leja, 1998). The importance of these wetland classes as foraging and pairing habitats for early nesting waterfowl (e.g., Drewien, 1968; Dzubin, 1969; Krapu, 1974; LaGrange and Dinsmore, 1989) and shorebirds (Skagen and Knopf, 1994; Farmer and Parent, 1997) is well known. The high proportion of temporary, seasonal and tillage wetlands that were occupied (13–85%) by species with statewide distributions (Table 41-1) attests to the importance of these wetlands to amphibians (see also Lannoo, 1998a). Although research designed to assess the importance of temporary and seasonal wetlands to anurans in the northern Great Plains is yet to be conducted

(see Lannoo, 1996, 1998a), recent European studies have shown that temporary wetlands increase connectivity for amphibians dispersing between breeding and overwintering ponds (Sjögren, 1991; Vos and Stumpel, 1995). Loss of temporary and seasonal wetlands may increase inter-wetland distances and subject amphibian populations to greater risks of local extinction (Richter and Azous, 1995; Semlitsch and Bodie, 1998; Lehtinen et al., 1999).

Wetland protection efforts in the northern Great Plains are usually based around habitat requirements of waterfowl communities. Results from this study indicate that amphibian communities in the northern Great Plains may also have diverse habitat requirements, making them potential indicators of ecosystem health. Statewide conservation initiatives and large-scale monitoring programs could be implemented using the basic biological knowledge of anuran distributions and habitat use obtained from this study. We suggest that anurans in South Dakota be monitored using three survey periods that coincide with the peak calling periods identified in our study (i.e., approximately 1 April–31 May, 15 May–15 June, 20 June–20 July; Fig. 41-2). A statewide monitoring initiative would fill the current gap in our understanding of anuran distributions in western South Dakota. We further recommend that studies be initiated in the northern Great Plains to identify local- and landscape-level attributes influencing habitat suitability for both native anurans and salamanders.

Summary

We present distribution maps for 12 anurans occupying wetland habitats in South Dakota. These maps include new records from auditory surveys conducted in 1997–98 at 1,496 wetland

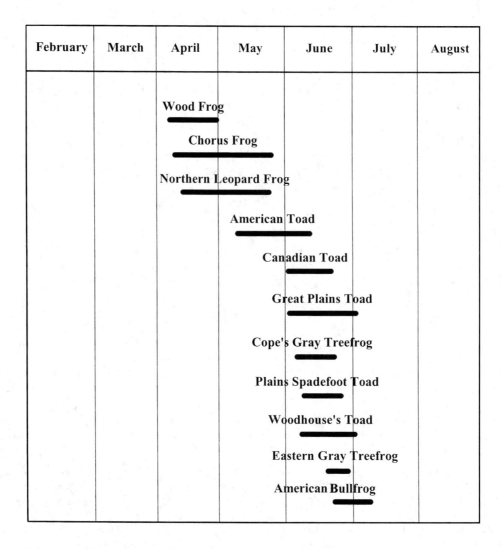

February	March	April	May	June	July	August

Wood Frog

Chorus Frog

Northern Leopard Frog

American Toad

Canadian Toad

Great Plains Toad

Cope's Gray Treefrog

Plains Spadefoot Toad

Woodhouse's Toad

Eastern Gray Treefrog

American Bullfrog

Dates which hosted the greatest occurrence of calling frogs and toads in eastern South Dakota during 1997 and 1998.

FIGURE 41-2 Peak calling periods for 11 anuran species in eastern South Dakota during 1997 and 1998.

sites in eastern South Dakota. Historical records compiled from the South Dakota Natural Heritage Database were used to construct distribution maps for eight anurans in western South Dakota, where extensive survey work is lacking. In eastern South Dakota, chorus frogs, northern leopard frogs, and Great Plains toads are the most widespread anurans and occupy the greatest proportion of wetlands. Historical data indicate that these three species are also the most widespread species in western South Dakota.

American toads and Canadian toads were most common in the eastern half of eastern South Dakota and occurred in fewer wetlands than the most common species. Woodhouse's toads are the least common bufonid in eastern South Dakota that also occur in western South Dakota. Plains spadefoot toads may be more widespread in western South Dakota. American bullfrogs are limited to southern counties along the Missouri River. Eastern gray treefrogs were only recorded in

the extreme northeast and/or southeast corners of eastern South Dakota, while Cope's gray treefrogs were detected in the extreme southeastern corner. Although Blanchard's cricket frogs and plains leopard frogs were not recorded in this study, Blanchard's cricket frogs were found (by D.C.B.) in 1999 at one site along the Big Sioux River in eastern South Dakota. Occurrence rates across the seven wetland classes were highly variable, indicating that no single wetland type can fulfill the habitat requirements of all anurans. Highest occurrence rates were in semipermanent and permanent wetlands, while the lowest were in manmade and riverine wetlands. High proportions (13–85%) of temporary, seasonal, and tillage wetlands were occupied by species with statewide distributions, attesting to the importance of these wetlands. We suggest that anurans be monitored in South Dakota using three survey periods that coincide with the peak calling times (1 April–31 May, 15 May–15 June, 20 June–20 July) identified in this study. We further recommend that studies be initiated in northern prairie ecosystems to identify local- and landscape-level attributes influencing habitat suitability for native anurans and salamanders.

Acknowledgments. We thank Scott Stolz, Chad and Brenda Kopplin, Jim Bauer, Kris Skare, and David Giese for assisting with data collection. Funding for this project was provided by the U.S. Fish and Wildlife Service (RWO-60), Division of Refuges, Denver, Colorado, through the South Dakota State Cooperative Fish and Wildlife Research Unit in cooperation with the U.S. Geological Survey, the South Dakota Department of Game, Fish and Parks, South Dakota State University, and the Wildlife Management Institute.

FORTY-TWO

Nebraska's Declining Amphibians

DAVID S. MCLEOD

During the 1970s, John D. Lynch and his students conducted extensive fieldwork in Nebraska to complete a herpetological survey begun by George E. Hudson nearly three decades earlier (Hudson, 1942; Lynch, 1985). In the wake of alarm calls concerning the status of amphibians around the globe since the late 1980s and early 1990s numerous studies have been conducted to determine whether or not amphibians are in fact declining and if changes are the results of local or global phenomena. During the late 1990s, I revisited the work done by Lynch to address questions of amphibian population declines at the state level in Nebraska since the 1970s.

During his survey, Lynch employed larval sampling techniques as a means of rapidly acquiring distributional data on amphibian populations in the state. The absence of terrestrial-breeding amphibians made larval sampling an effective methodology for sampling all 13 species of amphibians found within the state. In Nebraska, the amphibian breeding season begins in April and typically extends through August (Hudson, 1942; Billings, 1973; Lynch, 1985). A window of opportunity exists for approximately 6 weeks during the early summer in which all species that have bred will have larvae in ponds across the state; thus, those species can be sampled simultaneously.

In 1997 and 1998 I resurveyed Nebraska amphibian populations and compared these data to Lynch's 1970s' data (presumably pre-decline) to determine the status of this state's amphibians (McLeod, 1999). I revisited Lynch's original sites, repeated his methodologies, and added new and replacement sites as needed to determine whether changes had occurred during the intervening 20+ years. Because changes had occurred, I addressed the following questions: (1) Had all species of amphibians been affected? (2) Were changes regional or statewide? (3) What possible causes may explain these changes?

Over a two-year period (3 June to 9 July 1997 and 1 June to 10 July 1998) I visited 267 sites, of which 181 were original sites sampled by Lynch. Ninety-six of Lynch's original sites were located and sampled. Eighty-five of the original sites were unavailable for sampling due to land-use changes or disappearance; these were replaced with nearby ponds selected on the basis of ecological characteristics given in Lynch's original field notes and distance from the original site. Thirty-nine original sites visited were abandoned because they were neither available for

sampling nor able to be replaced appropriately. Forty-six new sites were added to the study. I replicated Lynch's methodology of collecting presence/absence data by sampling larvae with a one-man push seine and/or a fine mesh dip net. Larvae were identified, recorded, and released. I made general notes on amphibian relative abundance and appearance, as well as on pond characteristics. Current land owners were interviewed and pond history obtained whenever possible.

Eleven species of amphibians commonly occur in the state of Nebraska (Lynch, 1985): northern cricket frogs (*Acris crepitans*), small-mouthed salamanders (*Ambystoma texanum*), eastern tiger salamanders (*A. t. tigrinum*), Great Plains toads (*Bufo cognatus*), Woodhouse's toads (*B. woodhousii*), Cope's gray treefrogs (*Hyla chrysoscelis*), western chorus frogs (*Pseudacris triseriata*), plains leopard frogs (*Rana blairi*), American bullfrogs (*R. catesbeiana*), northern leopard frogs (*R. pipiens*), and plains spadefoot toads (*Spea bombifrons*). Two other historically present species, American toads (*B. americanus*) and western narrow-mouthed toads (*Gastrophryne olivacea*) were not collected during either study and are consequently not included in this study. Lynch found larvae at all (100%) of his 269 sites. During my survey, I found 40.9% of 227 ponds sampled to be unoccupied (McLeod, 1999). Seven of 11 species exhibited significant declines in occurrence at the state level (Table 42-1). Species that declined included northern cricket frogs, eastern tiger salamanders, Woodhouse's toads, western chorus frogs, plains leopard frogs, northern leopard frogs, and plains spadefoot toads. Of the four remaining species, Cope's gray treefrogs increased significantly in occurrence, but they were collected only in the 1998 field season (McLeod, 1999).

To determine if amphibian declines represented regional or statewide phenomena, presence/absence data were sorted into four, independently designated ecological areas based on zones of native vegetation (Kaul and Rolfsmeier, 1993). Collection values for each species were compared between decades and across pond types within each zone. The number of amphibian species known to occur in each zone are: (1) eleven species occur in upland tall grass prairie (Zone 1), which comprises the eastern third of the state; (2) seven species in loess and mixed grass prairie (Zone 2), which is found primarily in south-central Nebraska; (3) six species are known from the sand hills mixed grass prairie (Zone 3),

TABLE 42-1
Representation of Species Among Sites Sampled

	1970s (n=253) %	1990s (n=218) %	Habitat
Acris crepitans	4.0	0.45	PS
Ambystoma texanum	0.4	0.45	TS
Ambystoma tigrinum	26.0	6.4	IS
Bufo cognatus	1.0	0.0	TS
Bufo woodhousii	23.0	15.5	IS
Hyla chrysoscelis	3.0	6.0	G
Pseudacris triseriata	45.0	25.6	TS
Rana blairi	39.0	18.3	G
Rana catesbeiana	7.0	9.6	PS
Rana pipiens	13.0	7.3	G
Spea bombifrons	25.0	9.6	TS

NOTE: Total number of species collected was converted to a percentage of the total number of sites for both decades. Pond types are abbreviated as Ephemeral (E), Temporary (T), Semi-permanent (S), and Permanent (P).

located in northwestern Nebraska; (4) seven species occur in the lowland tall grass prairie (Zone 4), including the areas surrounding the Platte and Loup (main branch) River basins. In Zone 1, four species (eastern tiger salamanders, western chorus frogs, plains leopard frogs, and plains spadefoot toads) showed declines, while Cope's gray treefrogs demonstrated an increase (McLeod, 1999). Zones 3 and 4 each had one species (western chorus frogs and Woodhouse's toads, respectively) that exhibited a decline. No significant changes in the frequency of occurrence were detected for any species in Zone 2 (McLeod, 1999).

Certain amphibian species have specific reproductive habitat requirements. In Nebraska, for example, plains spadefoot toads (an explosive breeder) most frequently uses ephemeral ponds, and American bullfrogs, which overwinters one to two years in its larval form, is most often found in permanent ponds. Based on Billings (1973) and Lynch (1985) each species was identified by its breeding habitat requirements as follows: temporary specialist (TS), permanent specialist (PS), intermediate specialist (IS), and generalist (G) (Table 42-1). Of those species exhibiting declines, examples of each of the reproductive habitat requirements were observed.

The declines observed in Nebraska have a number of possible explanations. The most evident influences on amphibian populations are human impacts, such as the loss or alteration of habitats, change in land-use practices, and agrichemical pollution. These factors may be working independently or synergistically to cause Nebraska's amphibian declines. During 1998, I observed ponds in Grant County (which I had sampled the previous year) being drained as a new railway bed was being laid. In 1997, Lynch and I revisited a site in Lancaster County, only to discover that the series of three ponds, which had been sampled frequently in the 1970s, had been removed from the property to make room for row crop agriculture. In some cases in the 1990s, temporary ponds were found to be considerably deeper and larger due to excavation by landowners and were frequently stocked with fish.

Compared with Lynch's 1970s field notes on the southeast portion of the state, I observed a reduction in the number of ephemeral and temporary ponds but an increase in the number of semi-permanent and permanent ponds. One explanation is that land formerly used as ranch and range land is now being used for row crop agriculture. A small ephemeral or temporary pond would likely be ignored, if not preferred, in a ranching situation, but it may be seen as problematic in a row crop agriculture setting. The loss of ephemeral and temporary ponds or a change to more permanent water bodies alters the availability and suitability of breeding sites for certain amphibian species.

The application of herbicides and pesticides has long been involved in agricultural practices. Pesticides have been implicated as a possible cause for amphibian declines elsewhere (Boschulte, 1993, 1995; Stebbins and Cohen, 1995; Diana and Beasley, 1998; Howe et al., 1998; Larson, 1998). While different agricultural chemicals have varied effects on amphibians, there is little doubt that chemical contamination of aquatic habitats is a factor in the decline of amphibians (Diana and Beasley, 1998). Amphibians are especially sensitive to aquatic chemical pollutants because of their highly permeable skin and aquatic development stages (Howe et al., 1998). Further, the uptake of some chemical compounds may be biomagnified in their tissues to levels greater than 10 times the concentrations found in the same environments (Clark et al., 1998).

In Nebraska, the highest levels of herbicides in surface water occur during spring and early summer after application of pre-plant chemicals (Spalding and Snow, 1989; Spalding et al., 1994). This coincides with the timing of egg deposition and development for all species of amphibians in Nebraska. It has also been determined that the half-life of chemicals, such as atrazine, is considerably longer in aqueous environments than in soil. Spalding et al. (1994) found that atrazine had an estimated half-life of 124–193 days versus the published soil half-life of 60 days. This may have serious implications for those species that have longer developmental periods such as eastern tiger salamanders and American bullfrogs, which overwinter in their larval form (Duellman and Trueb, 1986). Even with species of relatively short developmental periods such as plains spadefoot toads (normally fewer than 30 days in Nebraska [Lynch, 1985]), sensitivity to chemicals may vary. In addition, spikes in concentration levels after rainstorms may impact development, survivorship, or reproductive potential.

During 1997 and 1998, species that were previously either abundant or relatively rare at historical sites were found less frequently, if at all. No collections of Cope's gray treefrogs, northern cricket frogs, small-mouthed salamanders, or Great Plains toads were made in 1997. Northern cricket frogs and small-mouthed salamanders were found only once in 1998. Great Plains toad larvae, rarely collected in the 1970s' survey, were absent during my study. Cope's gray treefrogs were collected frequently (14 sites) during 1998. One explanation for these observations could be inadequate rainfall affecting the reproductive timing of these amphibians. However, this does not appear to be the case, as mean monthly precipitation values (March–August) were not significantly different between the two studies (McLeod, 1999). Therefore, natural fluctuations can be seen in populations of amphibians, even over a short period of time, and caution must be exercised when concluding that these results represent real declines. Pechmann et al. (1991) argued that what biologists termed "declines" might in fact be a time-limited perspective of normal stochastic events for a given population.

Summary and Conclusions

Four important conclusions can be drawn from this study:

The majority of commonly occurring amphibian species have declined at the state level.

Declining populations of amphibians occurred in at least three of four ecological regions, with the eastern third of the state (Zone 1) experiencing the greatest declines.

Occurrences of declines are not limited to those species with specific breeding habitat requirements, but are seen even in breeding site generalists.

There is an alarming decrease in the proportion of ponds that were historically occupied by amphibian larvae.

Further investigation into the effects of land use practices on amphibians, including the use of agricultural chemicals, should be pursued. The Nebraska Game and Parks Commission has recently conducted calling surveys (1997 and 1998), and I would urge the continued monitoring of Nebraska's amphibian populations. However, I would encourage larval sampling, as this is a rapid and less time-sensitive technique for acquiring population presence/absence data, and it documents breeding.

Acknowledgments.   I thank Drs. J. D. Lynch and R. E. Ballinger for historical data, support, and supervision of this project. Funding for this study was provided by Nebraska's Lower Loup Natural Resource District and the USGS Biological Resources Division, Northern Prairie Science Center.

FORTY-THREE

Museum Collections Can Assess Population Trends

JEFF BOUNDY

Early warnings of amphibian declines have been realized as species have vanished, or disappeared from large portions of their ranges (Blaustein et al., 1994a; Pechmann and Wilbur, 1994; Lannoo, 1998b,c). These declines have alerted biologists and conservation agencies to the need to combine proactive evaluation of status and trends with retroactive research on causes of disappearances (Heyer et al., 1994; Fellers and Freel, 1995; Green, 1997b; Lannoo, 1998a). An initial step in detecting population declines in seemingly stable species can be accomplished by comparing current population levels with historical data. However, detecting shallow declines is difficult in the absence of replicable baseline data for most amphibian species (see also Green, this volume). A number of programs have been initiated to gather baseline data on amphibian status (e.g., Wisconsin's Frog and Toad Survey [Mossman et al., 1998]), with various state and federal monitoring programs based on this approach (Johnson, 1998b; Hemesath, 1998; Droege and Eagle, this volume; Weir and Mossman, this volume). However, trends in the data generated by these programs may not be discernible for a decade or more (Mossman et al., 1998; Droege and Eagle, this volume). Because of this absence of baseline data, researchers must often rely on anecdotal data—as one venerable Louisiana herpetologist said, "You don't see them like you used to."

An abundance of baseline data exists in the form of museum voucher specimens. Far from being just specimen "libraries" for systematists, museum vouchers offer a wide range of natural history data for a number of biological disciplines (Greene and Losos, 1988; Winker, 1996; Resetar, 1998; Ferner et al., this volume). Under some circumstances (long-term, generalized acquisition of specimens from a specific region), museum specimens offer verifiable historical records that can be used in trend analyses (Jennings and Hayes, 1994; Burgman et al., 1995; Fisher and Shaffer, 1996). Under the assumption of equal collection effort per species per time period, collection trends for amphibians can be assessed by comparing percentage of recent captures to historical captures. To determine the utility of this method in identifying amphibian population trends, I analyzed data from four Louisiana museums that focused collection efforts on Louisiana's Florida parishes. Because actual collecting activity may vary over time, and because interspecific differences exist in the ability of animals to be captured (Hyde and Simons, this volume), data are evaluated by interspecific comparisons of trends in capture percentages. An assumption of this analysis is that the variability in collection efforts by different individuals and institutions is reduced by combining results over periods of one or two decades.

Materials and Methods

I recorded the locality, year of capture, and number of specimens per accession for each of the 44 amphibian species occurring in the Florida parishes (Table 43-1) from the following collections: Louisiana State University (LSUMZ; n = 9,170), Southeastern Louisiana University (SLU; n = 1,088), Tulane University (TU; n = 11,504), and the University of Southwestern Louisiana (USL; n = 1,540). Each locality, when possible, was given latitude-longitude coordinates to the nearest minute, which assigned localities to approximate one-square mile blocks (hereafter locality). Localities, all species collected at a locality, and the year of most recent collection were assembled in a LOTUS® spreadsheet. Numbers of specimens for each species were summed in five-year periods beginning with 1945–49, with an additional block for pre-1945 counts. Trend data for each species were determined by (1) comparing percentage of specimens collected in recent years (1980–99) to previous years, and (2) by comparing percentage of localities from which each was collected in recent years.

I examined possible causes for observed trends by regressing recent specimen and locality percentages against the following parameters and their respective coded values:

Duration of larval stage: (1) <60 days, (2) 60–200 days, (3) >200 days

Breeding site: (1) terrestrial, (2) permanent aquatic, (3) ephemeral aquatic

Non-breeding site: (1) terrestrial, (2) facultative aquatic, (3) obligate lacustrine, (4) obligate riverine

Geographic distribution: (1) <5 localities, (2) <50% of total land area of Florida parishes (except for species coded 1), (3) >50% of total area

Habitat: (1) lowland, (2) upland, (3) both

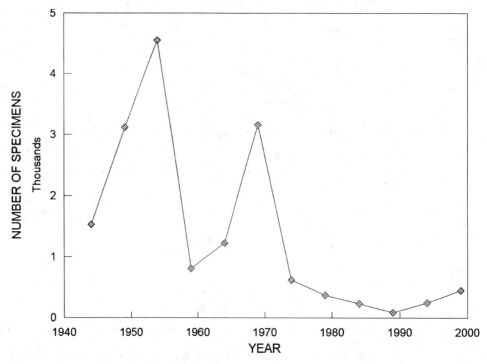

FIGURE 43-1 Trends in numbers of amphibian specimens added to four Louisiana collections between 1945 and 1999. The numbers are grouped in five-year periods starting with pre-1945–49.

Trophic level: (1) microinvertebrates, (2) macroinvertebrates, (3) vertebrates and macroinvertebrates

Movements: (1) small home territory including breeding site, (2) some seasonal movements, or relatively large territory, (3) mass migrations to and from breeding sites

Commercial use: (1) none, (2) some, (3) extensive

Clutch size: (1) <10 ova, (2) 11–100 ova, (3) >100 ova

Breeding season: (1) winter, (2) intermediate or unconcentrated, (3) late spring-summer

TABLE 43-1

Number of Amphibian Specimens Collected from Each Parish (1918–99)

Parish	Number	Percentage[a]
Ascension	289	1.8
East Baton Rouge	2,650	16.4
East Feliciana	317	2.0
Iberville	89	0.6
Livingston	699	4.2
St. Helena	978	6.1
St. Tammany	6,688	41.5
Tangipahoa	1,313	8.1
Washington	2,478	15.4
West Feliciana	627	3.9
*Total	16,128	100.0

[a]May vary due to rounding

Results

Collection Effort

The ten Florida parishes were unevenly surveyed, with 73% of specimens coming from East Baton Rouge, St. Tammany, and Washington Parishes (Table 43-1). However, these three parishes contain 43 of the 44 species of Florida parish amphibians (except Webster's salamanders [Plethodon websteri]), so that the collection effort could have theoretically been equal across species, except for Webster's salamanders.

The collection effort diminished overall with time, with some fluctuations that appeared to be related to changes in museum personnel and activity (Fig. 43-1). The bulk of the Tulane collection was accumulated from 1946–54. SLU was active from 1964–82, and USL from 1964–74. LSUMZ had several activity periods: 1950–52, 1964–69, and 1992–present. Thus, the late 1950s and 1980s are poorly represented by museum vouchers. Temporal acquisition of specimens was as follows: 9,997 (61%) specimens through 1959; 5,372 (33%) from 1960–79; and 1,018 (6%) since 1980. Temporal acquisition of anurans differed from that of salamanders, with trends of 58, 37, 5% for anurans, and 65, 27, 8% for salamanders. For this reason, trends are examined separately for the two groups.

Species Trends

Percentage of specimens collected by decade varied between species (Table 43-2). The mean percentage of recent (1980–99) collections of salamanders and anurans were 0.075 and 0.052, respectively. Assuming that these values represent collection effort, five species of salamanders and nine species of anurans are significantly under-represented ($c^2 < 0.5$ level) in recent

TABLE 43-2
Trends in Numbers of Specimens and Localities Collected of Amphibians in the Florida Parishes of Louisiana

	Pre-1950s	1950s	1960s	1970s	1980s	1990s	N	LOC	%R
Siren intermedia	0.20	0.21	0.35	0.06	0.02	0.17	66	42	0.19
Necturus beyeri	0.36	0.05	0.25	0.14	0.17	0.04	587	42	0.19
Amphiuma means	0.66	0.03	0.08	0.13	0.03	0.07	92	19	0.32
Amphiuma tridactylum	0.16	0.29	0.24	0.05	0.03	0.23	146	53	0.36
Notophthalmus viridescens	0.51	0.25	0.17	0.03	0.01	0.04	404	93	0.13
Ambystoma maculatum	0.46	0.13	0.26	0.04	0.00	0.11	46	25	0.16
Ambystoma opacum	0.49	0.22	0.20	0.02	0.03	0.05	737	136	0.08
Ambystoma talpoideum	0.18	0.61	0.16	0.01	**	0.05	391	47	0.19
Ambystoma texanum	0.16	0.69	0.05	0.01	0.00	0.09	178	29	0.10
Ambystoma tigrinum	0.00	0.00	1.00	0.00	0.00	0.00	1	1	0.00
Desmognatus auriculatus	0.62	0.27	0.10	**	0.00	0.01	1,794	110	0.06
Desmognathus conanti	0.11	0.19	0.48	0.08	0.03	0.12	341	93	0.20
Eurycea cirrigera	0.38	0.14	0.25	0.14	0.04	0.05	304	83	0.16
Eurycea guttolineata	0.19	0.20	0.24	0.23	0.04	0.09	345	111	0.21
Eurycea quadridigitata	0.32	0.16	0.40	0.05	0.00	0.07	480	134	0.15
Hemidactylium scutatum	0.50	0.00	0.00	0.00	0.00	0.50	4	3	0.67
Plethodon mississippi	0.37	0.22	0.31	0.06	0.02	0.03	1,354	203	0.16
Plethodon websteri	0.00	0.00	0.00	0.50	0.50	0.00	4	1	1.00
Pseudotriton montanus	0.07	0.29	0.57	0.00	0.00	0.07	14	8	0.13
Pseudotriton ruber	0.17	0.38	0.08	0.00	0.08	0.17	24	9	0.44
Scaphiopus holbrookii	0.23	0.43	0.29	0.03	0.01	**	361	54	0.06
Bufo americanus	0.00	0.01	0.15	0.59	0.17	0.08	95	29	0.41
Bufo terrestris	0.13	0.40	0.40	0.04	0.01	0.02	349	70	0.14
Bufo quercicus	0.55	0.34	0.08	0.01	0.00	0.02	216	26	0.12
Bufo nebulifer	0.11	0.43	0.29	0.09	0.02	0.08	344	117	0.23
Bufo fowleri	0.08	0.43	0.39	0.05	0.01	0.03	752	218	0.12
Acris crepitans	0.08	0.34	0.51	0.03	0.00	0.05	419	119	0.07
Acris gryllus	0.29	0.33	0.36	0.05	0.01	0.05	952	226	0.11
Hyla chrysoscelis	0.14	0.46	0.30	0.07	0.02	0.02	679	136	0.09
Hyla avivoca	0.37	0.29	0.33	**	0.00	**	440	63	0.03
Hyla femoralis	0.42	0.29	0.14	0.05	0.02	0.08	167	33	0.36
Hyla squirella	0.19	0.44	0.24	0.07	0.03	0.03	495	108	0.10
Hyla cinerea	0.04	0.48	0.31	0.12	0.01	0.04	393	120	0.08
Hyla gratiosa	0.30	0.24	0.42	0.02	0.01	0.01	161	36	0.06
Pseudacris crucifer	0.16	0.34	0.38	0.08	0.01	0.03	344	104	0.09
Pseudacris feriarum	0.17	0.39	0.30	0.12	0.01	0.01	567	110	0.05
Pseudacris ornata	0.74	0.26	0.00	0.00	0.00	0.00	70	8	0.00
Rana capito	0.56	0.18	0.26	0.00	0.00	0.00	45	12	0.00
Rana clamitans	0.19	0.36	0.26	0.07	0.02	0.09	705	235	0.19
Rana sphenocephala	0.10	0.43	0.26	0.17	0.01	0.03	140	50	0.06
Rana catesbeiana	0.07	0.37	0.41	0.09	0.02	0.04	241	112	0.13
Rana palustris	0.22	0.11	0.56	0.00	0.00	0.11	9	6	0.17
Rana sphenocephala	0.26	0.38	0.25	0.03	0.02	0.06	481	156	0.16
Gastrophryne carolinensis	0.11	0.65	0.16	0.05	0.01	0.03	635	148	0.14

NOTE: The number of specimens per decade are expressed as a percent of the total (N). LOC = total number of localities for each species, %R = percentage of localities from which each species has been collected since 1980; ** indicates < 0.01.

TABLE 43-3

Correlation of Recent Amphibian Collection Numbers and Localities with Life History Parameters

| | Salamanders | | | | Anurans | | | |
| | Specimens | | Localities | | Specimens | | Localities | |
PARAMETER	R	P	R	P	R	P	R	P
Larval duration	0.65	0.01	ns		ns		ns	
Breeding site	ns		0.57	0.03	ns		ns	
Non-breeding site	0.67	0.01	ns		0.49	0.05	ns	
Distribution	0.71	0.01	ns		ns		ns	
Habitat	ns		ns		0.53	0.03	ns	
Trophic level	ns		ns		ns		ns	
Movements	ns		0.44	0.05	ns		ns	
Commerce	ns		ns		ns		ns	
Clutch size	ns		ns		ns		ns	
Breeding season	ns		0.58	0.03	ns		ns	

NOTE: Abbreviations are as follows: ns = not significant, r = correlation coefficient, p = significance value.

collections: eastern newts *(Notophthalmus viridescens)*, mole salamanders *(Ambystoma talpoideum)*, tiger salamanders *(A. tigrinum)*, southern dusky salamanders *(Desmognathus auriculatus)*, Mississippi slimy salamanders *(P. mississippi)*, eastern spadefoot toads *(Scaphiopus holbrookii)*, southern toads *(Bufo terrestris)*, oak toads *(B. quercicus)*, Cope's gray treefrogs *(Hyla chrysoscelis)*, bird-voiced treefrogs *(H. avivoca)*, barking treefrogs *(H. gratiosa)*, ornate chorus frogs *(Pseudacris ornata)*, upland chorus frogs *(P. feriarum)*, and gopher frogs *(Rana capito)*.

The relative number of recent versus historical (total) localities for each species also varies (Table 43-2). The mean percentage of recently collected localities for salamanders and anurans was 0.166 and 0.124, respectively. Ten salamander and 15 anuran species were collected at fewer than the mean number of localities. The following species were significantly undercollected ($c^2 < 0.5$ level): eastern newts, marbled salamanders *(A. opacum)*, small-mouthed salamanders *(A. texanum)*, tiger salamanders, southern dusky salamanders, mud salamanders *(Pseudotriton montanus)*, eastern spadefoot toads, northern cricket frogs *(Acris crepitans)*, Cope's gray treefrogs, bird-voiced treefrogs, green treefrogs *(H. cinerea)*, barking treefrogs, ornate chorus frogs, spring peepers *(P. crucifer)*, upland chorus frogs, gopher frogs and pig frogs *(R. grylio)*. All the species listed here and in the preceding paragraph are hereinafter referred to as "declining." Species that have values near the mean are termed "stable," and those with values above the mean are termed "highly stable."

Correlations with Life History Parameters

Trends in the numbers of specimens and localities of amphibians collected are correlated in different ways with potential causative factors (Table 43-3). Commercial use, trophic level, and clutch size are not correlated with any of the trends in percentages. Trends in anuran numbers are correlated with non-breeding site and habitat; most of the stable species are facultatively or obligately aquatic, and all the lowland species are declining. Locality trends are not correlated with any parameters.

The trends in salamander numbers are directly correlated with the duration of larval period and aquatic non-breeding sites, and inversely correlated with the size of geographic distribution. The correlation with larval period is based on a tendency for stable species to have extended larval periods (> 200 days). The correlation with geographic distribution is based on several highly stable species (four-toed salamanders *[Hemidactylium scutatum]*, Webster's salamanders, and red salamanders *[P. ruber]*) having relatively small ranges. The correlation with non-breeding site is based on a tendency toward obligately aquatic species to be highly stable. None of the declining species were obligately aquatic. Locality trends were correlated with type of breeding site, movements, and breeding season. Declining species exhibited a greater tendency to use ephemeral breeding sites and to exhibit mass movements. Highly stable species tended to be summer (rather than winter) breeders.

Discussion

The chief assumption of the present analysis is the temporal uniformity of collection effort for each species. If this assumption is met, then trends in species abundance can be measured in relation to those of the other species. Many of the curators and principle collectors at the four collections have provided ample evidence that the great majority of collecting activity upon which this study is based has been of a general nature: field crews or individuals sought and collected samples of any species encountered. However, there has been a tendency to collect smaller samples since the 1960s, with a focus on obtaining locality vouchers rather than large series for life history and variation studies. The change in collection purpose has probably accounted for the marked drop in total specimens collected in recent years. Recent fieldwork (by myself and others) indicates that large series of many species are still potentially obtainable from the wild.

Based on the trend analysis, eight salamander and 13 anuran species are declining in relative numbers, localities, or both. Percentage values for rarely collected or localized species are sensitive to collection; until a few years ago, four-toed salamanders, mud salamanders, and pickerel frogs (R. palustris) had not been found in eastern Louisiana since the mid-1960s. The collection of a single tiger salamander would change its recent percentages from 0 to 0.5%, and would shift it from a declining to a highly stable species under the present scheme. However, the effect that rare/localized species have on the overall analysis is negligible; unlike the results of Brodman and Kilmurry (1998), no correlation exists between total specimen number and percentage value, or locality number and percentage value, per species ($F = 0.12–4.19$, $p = 0.06–0.88$).

Life history correlations indicate a general tendency for stability in upland, riverine species, perhaps due to the fact that no Florida-parish streams (except the Pearl River) are dammed or bordered by industrial areas. Except for the effects from headcutting in watersheds and sand/gravel operations on some streams, riverine taxa enjoy non-polluted, permanent water sources. Declining species have a general tendency to occupy lowlands and to migrate to ephemeral breeding sites, often during winter. Thus, locality-based decreases appear to be based on the loss of small lacustrine sites versus permanence of riverine habitats. Species that breed in ephemeral water usually exhibit mass migrations to breeding sites, and the correlation of declines with migratory species is probably a result of use of ephemeral sites rather than a cause of declines in itself. Lowland, lacustrine sites may serve as sinks for biotoxins (either through drainage or ground water), which may explain the disappearance of southern dusky salamanders from most of their historic range (D. Means, personal communication), whereas its upland, riverine sister taxon, spotted dusky salamanders (D. conanti), remain highly stable. However, different species have exhibited varied responses to changes at low, ephemeral sites. Twenty years after the region around an ephemeral flatwoods pond in St. Tammany Parish was clearcut, tiger salamanders, southern dusky salamanders, and gopher frogs appear to have been extirpated; however, mole salamanders, dwarf salamanders (Eurycea quadridigitata) and barking treefrogs remain common (A. Ballew, personal communication). Aside from some species-specific variability, the overall spectrum from declining to highly stable species appears to follow a lowland, ephemeral pond-breeding to upland, riverine species continuum. Thus,

the identification and conservation of ephemeral lowland sites should be prioritized in stabilizing amphibian populations in eastern Louisiana.

Summary

The preceding analysis indicates that museum-based data have promise in detecting amphibian declines in historically well-collected areas. The analysis has detected several potentially imperiled species whose declines are corroborated by population studies in other states (e.g., southern dusky salamanders and gopher frogs), and it has flagged others that are currently not recognized as Special Concern Species by the Louisiana Natural Heritage Program (LNHP; e.g., eastern newts and eastern spadefoot toads). On the other hand, museum-based analyses may flag stable trends in species that presently are considered threatened. For example, four-toed salamanders are listed as special concern or threatened in Louisiana, Missouri, and elsewhere, but the historical scarcity of this species in margins of its range has created a false concern for its security (Lannoo, 1998c; T. Johnson, D. Rossman, personal communication). Of the 21 species identified here as declining, only four are currently listed as Special Concern species by the LNHP, one had been deleted from the list, and a study is underway to assess the status of a sixth. Most of the 17 non-listed declining species are widespread in the Florida parishes and are still observed with regularity. The first step in conserving populations of the latter group is to forecast potential problems and to take proactive measures to insure stability of their populations.

Acknowledgments. I thank Douglas Rossman, David Good, and Frank Burbrink of LSUMZ, Richard Seigel and Brian Crother of SLU, and Harold Dundee of TU for allowing me access to collections and records in their care and for assisting in the recovery of historical data. Thank you to James Jackson and Robert Jaeger for donating the USL collection to LSUMZ. I thank the following individuals for providing data relevant to this project: Amelia Ballew, Paul Conzelmann, Jim Delahoussaye, Harold Dundee, Danny Edler, Donald Hahn, Tom Johnson, Steve Karsen, Ed Keiser, Curtis Kennedy, Ernie Liner, Frank Marabella, Bruce Means, Tim Michel, Paul Moler, Buck Prima, Douglas Rossman, Rich Seigel, Steve Shively, Johnnie Tarver, and Avery Williams. I also thank Harold Dundee for reviewing the manuscript.

FORTY-FOUR

Monitoring Salamander Populations in Great Smoky Mountains National Park

ERIN J. HYDE AND THEODORE R. SIMONS

Recent evidence of worldwide amphibian population declines has highlighted the need for a better understanding of both species-specific habitat associations and methodologies for monitoring long-term population trends (Barinaga, 1990; Blaustein and Wake, 1990; Wake, 1991; Lannoo, 1998b). For decades, studies have relied on relative abundance indices to evaluate salamander populations across space and time. However, little effort has been made to evaluate the underlying assumptions of these indices or their relationship to the true population. Heatwole (1962) has shown that eastern red-backed salamanders (Plethodon cinereus) change their micro-habitat use in response to precipitation events, differentially using cover objects and leaf litter refugia in response to changing humidity levels. This behavior may suggest that eastern red-backed salamanders are not equally detectable at different moisture levels. Because salamander surface activity is likely to vary with topography, season, humidity, climate, or other landscape variables, the ability to detect animals may also vary across space or time.

Comparing two populations over time or space requires that $n1/n2 = [C_1/b_1]/[C_2/b_2] = C_1/C_2$, where n = population size, C = number of individuals counted, and b = the ability of individuals to be detected. To compare two count indices we assume that $b_1 = b_2$. Finally, we must assume a linear relationship between salamander counts (C) and population size (n), $E(C) = bN$ (Lancia et al., 1994).

Great Smoky Mountains National Park (GRSM) is committed to incorporating salamander population monitoring into the park's long-term inventory and monitoring program because of the large number of unique species in the park, as well as evidence that salamanders are finely tuned indicators of environmental quality (Duellman and Trueb, 1986; Corn and Bury, 1989a; Dodd, 2003). Data from ongoing research in GRSM designed to assess spatial and temporal patterns in salamander diversity and abundance are being used to evaluate sampling effectiveness and bias across a variety of habitat types. Here we present evidence that some common salamander sampling techniques may not be appropriate indices for salamander abundance and, therefore, may not be suitable methodologies for use in long-term monitoring programs in the southern Appalachians and perhaps elsewhere.

Study Area and Methods

Great Smoky Mountains National Park is an internationally recognized refugia of temperate forest biodiversity. Geography and geology, combined with steep, complex topography, promote extreme gradients of temperature and moisture across the park's environments. In many groups, including salamanders (Jackson, 1989), these gradients produce levels of species diversity that are unmatched elsewhere in North America. Perhaps as many as 10% of the world's salamander species are found in the region (Petranka, 1998).

As part of a larger study designed to assess the distribution, abundance, and habitat associations of salamanders in GRSM (Simons and Johnson, 1999; Hyde, 2000), 111 terrestrial sampling sites were established within the Mount LeConte USGS quadrangle. Sampling sites were stratified by elevation, land use history, and plant community type. Because salamander life histories are variable, we employed four different sampling methods: searches of natural cover objects along transects (Jaeger, 1970, 1994), opportunistic nighttime surveys along transects (Ash and Bruce, 1994), artificial cover boards (Fellers and Drost, 1994; Jung et al., 1997), and leaf litter searches (Pauley, 1995c). Prior to this study, we did not know the relationship between species abundance estimates obtained by these sampling methodologies.

The sampling framework at each site was comprised of up to four (approximately parallel) 50 m transects, one transect for each sampling method (Fig. 44-1). Where possible, each site included all three diurnal transects. However, impassable terrain and safety concerns prohibited the inclusion of all transects at some sites. Transects for nighttime surface counts were included at a subset of sites and were opportunistically sampled on warm rainy nights when temperature and humidity conditions favored surface activity by terrestrial salamanders. For a detailed description of our site design or sampling techniques see Hyde (2000).

Results

All sites were sampled a minimum of three times between 27 May and 5 August 1998, and five times between 5 April and 27

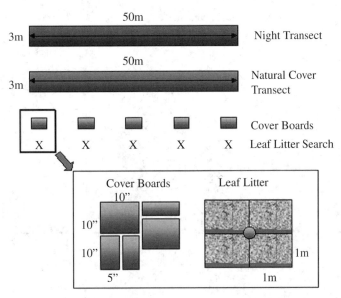

FIGURE 44-1 The sampling framework consisted of four transects: night, natural cover, artificial cover board, and leaf litter. The cover board transect contained five cover board stations, each 10 m apart. Each station contained two large boards and three small boards. Leaf litter stations were 1–2 m from each cover board station.

TABLE 44-1

Relative Efficiency of Salamander Sampling by Four Methods

	Number of:			Salamanders per Transect:	
	Transects	Samples	Salamanders	Undisturbed	Disturbed
Cover boards	101	837	1,224	1.9	1.1
Natural cover	92	700	2,651	7.0	2.0
Leaf litter	97	731	566	1.4	0.4
Night	26	97	1,669	25.2	8.5

NOTE: Salamander captures were higher on undisturbed sites than on sites with a history of disturbance prior to the formation of the park 65 years ago.

June 1999. During the 1998 and 1999 field seasons, 98 opportunistic night transect samples were recorded at 39 different sites. In addition to regular sampling, three sites were sampled every 7 to 10 days during the 1998 and 1999 field seasons as part of a separate study and to determine within site variation in salamander abundance.

The distribution of salamander catches in 1998 and 1999 was non-normal for all sampling methods. Because we were unable to satisfy the requirements of parametric statistics even after transforming the data, hypothesis testing was conducted by non-parametric statistics.

An analysis of salamanders captured per unit effort (as measured by the number of transect samples) suggests that night surveys and natural cover transects are more efficient for capturing salamanders than are artificial cover boards or leaf litter counts (Table 44-1). However, this measure of catch per unit effort does not account for the variability of counts, expense, effort required per sample, practicality, or disturbance caused by sampling, all of which must be considered when choosing a sampling method.

Species Specific Bias

Our data indicate that sampling techniques may have species-specific biases (Fig. 44-2; Table 44-2). For example, Jordan's salamanders (also termed red-cheeked salamanders, *P. jordani*) were 20–30% more likely to be captured on night transects than with any other method. In contrast, pigmy salamanders (*Desmognathus wrighti*) were more likely to be detected on leaf litter searches (43% of all salamander captures) than under cover boards (15% of captures) or on natural cover transects (25% of captures). These variations in measures of species composition probably reflect differences in the microhabitat used by different species. Although Huhey and Stupka (1967) state that pigmy salamanders are relatively rare in GRSM, we found pigmy salamanders to be the second most abundant species on our undisturbed sites. Because this species spends most of its time in leaf litter, it may be underrepresented by traditional search techniques that target natural cover. Therefore, methods that target specific microhabitats may produce misleading

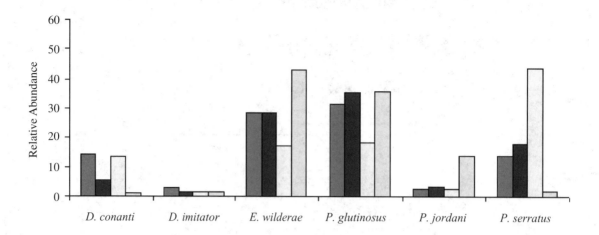

Relative Species Abundance on Disturbed Sites

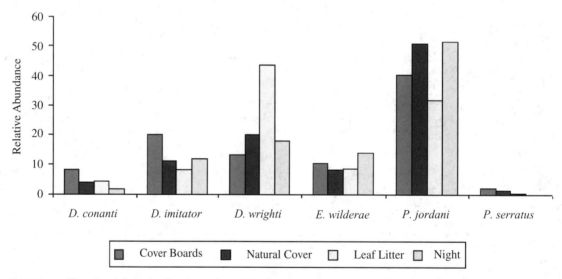

Relative Species Abundance on Undisturbed Sites

Cover Boards   Natural Cover   Leaf Litter   Night

FIGURE 44-2 The estimated relative abundance of species in the salamander community depends on the sampling method employed. If species were represented equally by all sampling methods, relative abundance (% of total captures) would be equal across methods.

conclusions about the distribution and abundance of salamanders across larger areas. We do not currently know the relationship between relative abundance measures obtained by these methods and the true relative abundance of salamanders.

Lack of Conformity Among Methods

Relatively large sample sizes (2,400 samples of 320 transects) permit us to explore relationships among several sampling methods. Because measures of relative abundance should reflect characteristics of the true population, we expect a positive relationship between any two methods that sample the same population. In other words, an increase in one method

(theoretically reflecting a larger population) should correspond to an increase in another method. Our data show a poor correlation between cover board samples and samples on adjacent natural cover transects ($r^2 = 0.11$; Fig. 44-3). Equally weak relationships exist between abundance indices obtained with other sampling methods (Fig. 44-3).

Effect of Natural Ground Cover on Salamander Abundance Indices

Because natural ground cover provides habitat for many salamanders, the quantity of natural cover on a study site may bias sampling results. Differences in natural ground cover

TABLE 44-2

Salamander Composition at Disturbed and Undisturbed Sites Differs According to the Method Used for Sampling

| Disturbed Sites | | | |
Cover Boards	Natural Cover	Leaf Litter	Night
Northern slimy salamanders (*Plethodon glutinosus*)	Northern slimy salamanders (*Plethodon glutinosus*)	Southern red-backed salamanders (*Plethodon serratus*)	Blue Ridge two-lined salamanders (*Eurycea wilderae*)
Southern red-backed salamanders (*Plethodon serratus*)	Southern red-backed salamanders (*Plethodon serratus*)	Northern slimy salamanders (*Plethodon glutinosus*)	Northern slimy salamanders (*Plethodon glutinosus*)
Blue Ridge two-lined salamanders (*Eurycea wilderae*)	Blue Ridge two-lined salamanders (*Eurycea wilderae*)	Blue Ridge two-lined salamanders (*Eurycea wilderae*)	Southern red-backed salamanders (*Plethodon serratus*)
Dusky salamander complex (*Desmognathus* sp.)	Dusky salamander complex (*Desmognathus* sp.)	Dusky salamander complex (*Desmognathus* sp.)	Red-cheeked salamanders (*Plethodon jordani*)

Undisturbed Sites

Cover Boards	Natural Cover	Leaf Litter	Night
Red-cheeked salamanders (*Plethodon jordani*)	Red-cheeked salamanders (*Plethodon jordani*)	Pigmy salamanders (*Desmognathus wrighti*)	Red-cheeked salamanders (*Plethodon jordani*)
Imitator salamanders (*Desmognathus imitator*)	Blue Ridge two-lined salamanders (*Eurycea wilderae*)	Red-cheeked salamanders (*Plethodon jordani*)	Pigmy salamanders (*Desmognathus wrighti*)
Pigmy salamanders (*Desmognathus wrighti*)	Southern red-backed salamanders (*Plethodon serratus*)	Blue Ridge two-lined salamanders (*Eurycea wilderae*)	Blue Ridge two-lined salamanders (*Eurycea wilderae*)
Southern red-backed salamanders (*Plethodon serratus*)	Pigmy salamanders (*Desmognathus wrighti*)	Imitator salamanders (*Desmognathus imitator*)	Imitator salamanders (*Desmognathus imitator*)

NOTE: Species listed in order of decreasing abundance.

densities across time or space may alter the effectiveness of sampling methods, thereby violating the assumption of equal detectability required for comparing indices. We found a poor relationship between the quantity of natural cover and the number of salamanders under cover boards ($r^2 = 0.11$; Fig. 44-4). Only 40% of the variation in natural cover transect counts could be attributed to variations in the density of natural cover (Fig. 44-5). Additional research is needed to clarify the relationship between natural ground cover and relative abundance indices.

Effect of Land Use History on Salamander Sampling Success

Salamander diversity and abundance in GRSM were significantly higher in undisturbed areas than in areas with a history of disturbance prior to the establishment of the park 65 years ago. The relationship between abundance and disturbance history was unaffected by the sampling method used (Fig. 44-6).

Measures of salamander abundance were more variable on disturbed sites for all three diurnal sampling methods (Fig. 44-7). This result may reflect differences in soil and leaf litter moisture at disturbed and undisturbed sites. We currently are analyzing soil moisture data to test this hypothesis. Because increased sampling variability decreases our ability to detect population trends, salamander population declines may be most difficult to detect on disturbed habitats, the very regions that are most likely to suffer declines in salamander diversity or abundance.

All sampling methods show high within and across site variability. Total variation (coefficient of variation [CV] across time and space) for salamander samples ranged from 90–205% depending on the sampling method used (Fig. 44-8). Total variation for individual species counts were much higher. Locally rare species or species that are poorly sampled by a given sampling method regularly experienced CVs as high as 300–900%. Such high variability reduces the power to detect population trends without very large sample sizes. Decreasing the inherent variability of salamander sampling techniques will be critical to developing efficient and reliable methods for monitoring terrestrial salamander populations. Although all our sampling techniques showed high variation, we found lower variability with natural cover transects (n = 89) and cover board transects (n = 88) than with leaf litter transects (n = 80; 8 samples per transect; Fig. 44-8). Frequent sampling (n = 14) on three sites shows that natural cover transects have lower variability than other methods.

Summary

Data collected over two years in GRSM suggest that salamander sampling techniques may violate assumptions required

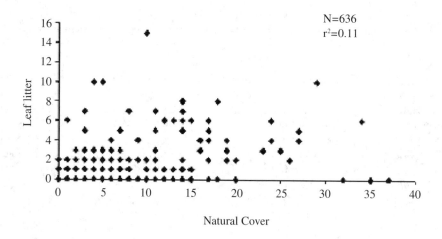

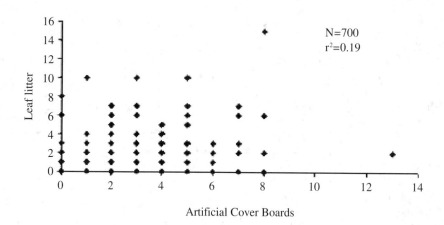

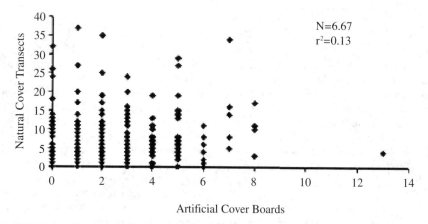

FIGURE 44-3 Abundance indices by various methods are poorly related.

to make valid comparisons of population indices over space or time. Land use history and temporal changes in habitat characteristics may affect the abundance or surface activity of salamanders. During drought years, salamanders may spend less time actively feeding at the surface and more time in underground refugia where they would not be detected by our sampling methods. Such a change in detection probability (b) violates one assumption required for comparing relative abundance indices. Changes in microhabitat use may also affect the probability of detecting a salamander with a given sampling method. Additional research is necessary to

clarify how salamander detectability changes over space and time.

We have shown that detectability (b) varies across species according to the sampling method used. Differences in the ability to detect salamanders may result from species-specific microhabitat use that is unequally represented by sampling methods. Therefore, a single method may not be reliable for comparing relative abundance.

We have shown that salamander counts in GRSM are not normally distributed, exhibit enormous variation, and display unequal variances across habitat types. Our findings suggest

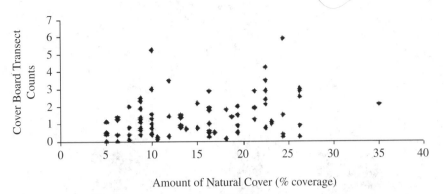

FIGURE 44-4 Salamander abundance under cover boards is poorly related to the abundance of natural cover ($r^2 = 0.11$).

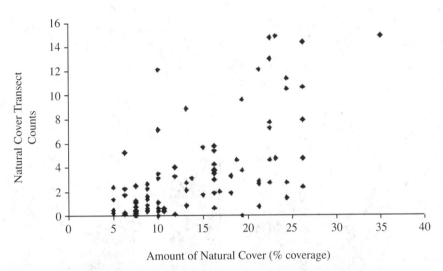

FIGURE 44-5 Only 40% of the variation in salamander abundance on natural cover transects is explained by the density of natural cover ($r^2 = 0.40$).

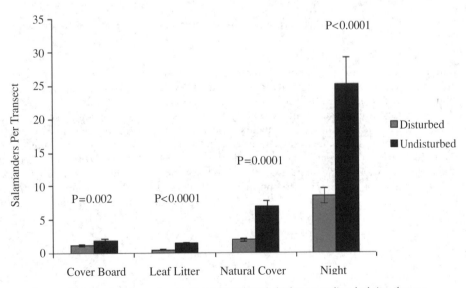

FIGURE 44-6 Variation in salamander counts was significantly higher on undisturbed sites than on disturbed sites. This relationship was consistent for all four sampling methods.

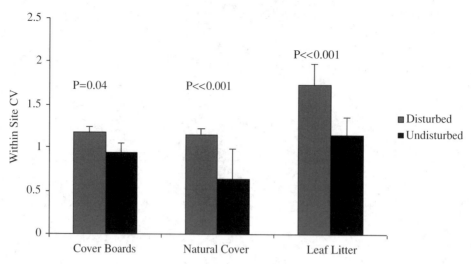

FIGURE 44-7 Within-site variation in salamander counts was significantly higher on disturbed sites than on undisturbed sites. This relationship was consistent for all sampling methods.

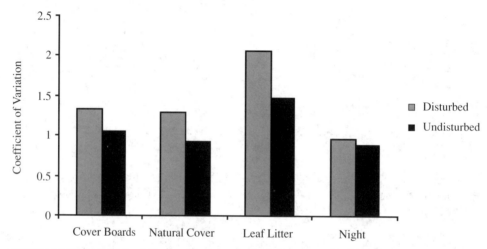

FIGURE 44-8 The high total variation in salamander samples (CV across time) will limit power for detecting population trends.

that different sampling methods are sampling different "populations." These characteristics reduce our ability to detect long-term population trends and limit the use of parametric statistics. It is important to understand the biases of the sampling methods used before substantial investments are made to inventory or monitor salamander populations. Additional research is necessary to design salamander monitoring methods that reduce variation, adequately sample important microhabitats, and meet necessary assumptions about detectability.

Acknowledgments. Funding for this research was provided by Friends of Great Smoky Mountains National Park and the United States Geological Survey. We thank the staff of GRSM for their assistance. Keith Langdon and Dana Soehn were particularly helpful with logistic and administrative assistance. Ken Pollock and Larissa Bailey have provided valuable suggestions and insight into issues of salamander sampling and variability. Finally we thank our dedicated field crew: R. Bain, M. Beall, J. Csakany, M. Hoffheimer, D. Jackan, P. Larson, T. Lossen, E. Raulerson, H. Stephenson, S. Marcum, and W. Ward.

FORTY-FIVE

North American Amphibian Monitoring Program (NAAMP)

LINDA A. WEIR AND MICHAEL J. MOSSMAN

Declines in amphibian populations have been noted since at least the 1970s (Gibbs et al., 1971; Hayes and Jennings, 1986; Tyler, 1991; Pounds and Crump, 1994; Bradford et al., 1994a; Drost and Fellers, 1996; Green, 1997b; Lannoo, 1998b; Bury, 1999; Campbell, 1999). In 1989 at the World Congress of Herpetology, informal conversations among scientists led to a concern that amphibian declines were more than local phenomena and may be a global issue (Wake and Morowitz, 1991; see also K. Phillips, 1994). In 1991, out of this concern, scientists and resource managers established the Declining Amphibian Populations Task Force (DAPTF) and formed regional working groups. In 1994, the Midwestern Working Group of the DAPTF met with biologists from around the United States at the Indiana Dunes National Lakeshore; the group discussed monitoring needs and set a goal of establishing a continent-wide monitoring program. Specifically, the group wished to provide a statistically defensible program to monitor the distributions and relative abundance of amphibian populations in North America, with applicability at state, provincial, eco-regional, and continental scales.

The Indiana Dunes meeting led to the formation of the North American Amphibian Monitoring Program (NAAMP), a partnership among state, provincial, academic, and nonprofit groups working regionally to gather monitoring data on amphibian populations. The NAAMP framework and protocols were developed during subsequent meetings held at the Canada Centre for Inland Waters in 1995, over the Internet in 1996, and in subsequent discussions. Personnel of the U.S. Geological Survey at the USGS Patuxent Wildlife Research Center in Laurel, Maryland, agreed to administer and coordinate the central database and Internet Web site.

Across North America, amphibians have variable life history and natural history features. Therefore, the original goal of the NAAMP effort was to employ several survey approaches and protocols to ensure that all species could be monitored. The first NAAMP protocol to be implemented was a roadside calling survey, with the goal of monitoring anuran (frog and toad) species that make distinctive vocalizations during courtship. This roadside calling survey is now the sole focus of NAAMP.

Roadside calling surveys are most useful in regions of North America where all (or most) of the anuran species in an

assemblage vocalize in a relatively predictable manner. This protocol can be used in the eastern half of North America (eastward from the Great Plains states and provinces) and to a lesser extent in the remaining Canadian provinces and Pacific Coast region. Due to its usefulness over a large geographic scale, its non-invasive nature, and the success of auditory surveys such as the Wisconsin Frog and Toad Survey and the Breeding Bird Survey, a volunteer-based program for this protocol appeared feasible.

Some regions of North America began to monitor anurans with calling surveys prior to the creation of NAAMP. In 1981, the Wisconsin Frog and Toad Survey was the first to begin (Mossman and Hine, 1984; Mossman et al., 1998). Modeled after the successful long-term North American Breeding Bird Survey (BBS; e.g., Peterjohn et al., 1995), the Wisconsin survey is based on permanent routes, each consisting of 10 roadside listening stations. Routes and listening stations were subjectively selected. At each station, observers listen for 5 minutes and record the species heard (within an unlimited detection radius) and a rough index of abundance (calling index) for each species. Each route is completed once during three prescribed seasonal periods calculated to cover the breeding periods of all Wisconsin anurans. The Wisconsin program has served as the model for programs initiated in other states and provinces, such as Iowa in 1991 (Hemesath, 1998); Ontario in 1992 (Bishop et al., 1997); Minnesota, which started a pilot program in 1993 and expanded in 1996 (Moriarty, 1998); and Michigan in 1996 (L. G. Sargent, personal communication). Some of these programs modified the Wisconsin protocols (e.g., using a three-minute listening period or selecting routes in a more objective manner). Provinces and states where roadside calling surveys have been attempted for amphibian monitoring are listed in Appendix 45-A.

The first NAAMP calling survey field season was held in 1997, and it included both new and previously established regional programs. The 1997 NAAMP protocol was based on the experience gained by the regional surveys, the results of a technique-validation study (Shirose et al., 1997), preliminary power and trend analyses of the Wisconsin data set (e.g., Mossman et al., 1998), and experience from the BBS, which provided a useful framework for constructing a large-scale monitoring program. The protocol resembled that of the Wisconsin program

Appendix 45-A: States and Provinces with Anuran Roadside Calling Surveys as of August 2000

| | | *Roadside Surveys* | | |
| | | Active | | |
State or Province	INACTIVE	LIMITED[a]	COMPLETE*	NAAMP[b]
Alaska		X		
Alberta		X		
Arkansas		X		X
California		X		X
Connecticut			X	X
Delaware			X	X
Florida	X			
Georgia	X			
Illinois	X			
Indiana			X	X
Iowa			X	
Kansas			X	X
Kentucky			X	X
Louisiana			X	X
Maine			X	X
Manitoba		X		
Maryland			X	X
Massachusetts			X	X
Michigan			X	X
Minnesota			X	X
Montana			X	X
Nebraska	X			
New Hampshire			X	X
New Jersey			X	X
New York			X	X
Ohio		X		X
Ontario			X	
Pennsylvania			X	X
Quebec			X	
Rhode Island			X	X
Saskatchewan			X	
Tennessee			X	X
Texas		X		X
Vermont			X	X
Virginia			X	X
West Virginia			X	X
Wisconsin			X	X

[a]Refers to whether all or part of a state/province is being managed for roadside surveys; complete does not imply that every route is assigned.
[b]Indicates which active programs have agreed to the NAAMP unified protocols.

except for the use of a three-minute listening period and the selection of routes according to a geographically stratified random design. Regional Coordinators, each responsible for a portion of (or entire) a state or province, were recruited from current programs and from among other interested individuals and organizations. NAAMP route maps were generated at Patuxent and distributed to Regional Coordinators, along with the 1997 NAAMP protocol. However, the NAAMP protocol was not adopted universally because some aspects of the protocol were loosely defined, some programs were invested in protocols initiated before the NAAMP existed, and some new programs created protocols based on an amalgamation of protocols from existing programs. This diversity of regional protocols and experiences, along with additional analyses of the Wisconsin data set, indicated a need for further discussion to develop a truly unified protocol so that data collected by regional programs could be compared and collectively used for population trend analyses in accordance with the goals of NAAMP.

To move towards a goal of a unified protocol, a meeting was held at Patuxent on 12–14 January 2000. Because population trend analyses should be based upon random sampling design, only regional programs that were interested or invested in stratified-random routes were invited to the January meeting (Appendix 45-A, invited NAAMP members). The reason for this limited invitation was utilitarian; discussions of unification needed to move beyond random versus nonrandom sampling design. If this smaller group could not agree to pursue a unified protocol, then further discussion would not have been necessary and USGS involvement in NAAMP would have been reevaluated.

The participants of the January meeting agreed that the benefits to a unified protocol outweighed the difficulties involved in altering their current respective protocols. To be successful, the unified protocol would need to balance the biological, statistical, and logistical requirements involved. A timeline was developed to implement the unified protocol for the 2001 field season (recognizing the need for approval by Regional Coordinators who were unable to attend the January meeting). A second meeting took place on 12–14 April 2000 to resolve remaining protocol decisions. Of the 26 states within the NAAMP, 14 were represented at one or both meetings. An Internet discussion list (listserve) of Regional Coordinators facilitated discussion among the NAAMP partners. Regional Coordinators commented upon a unified protocol draft; a revised unified protocol was then adopted.

NAAMP Calling Survey: Unified Protocol

An overview of the unified NAAMP Calling Survey Protocol is summarized in Appendix 45-B and detailed here.

Route Creation Protocol

The NAAMP calling survey uses randomly placed roadside routes, stratified by degree block of latitude and longitude; five random points per block are computer-generated and plotted using mapping software (initially Delorme's MapExpert version 2, later Delorme's StreetAtlas USA version 5.0). The nearest road intersection to the random point is chosen as the starting point for the calling survey route. From this intersection a 15-mile route is drawn along secondary or

Appendix 45-B Overview of Unified Protocol and Previous Protocols of NAAMP Partners

PROTOCOL CATEGORY	UNIFIED PROTOCOL	PREVIOUS VARIATIONS IN PROTOCOLS
Listening duration	5 min	3 min; 4 min; 5 min; 3–5 min
Initial waiting period (before listening duration)	No waiting period	No waiting; 30 s wait; 1 min wait
Significant noise disturbance during sampling	If disturbance is 1 min or longer and is not a constant background noise, then a "time out" is taken. The remainder of the 5-min sampling period is completed after the disturbance. Document when "time out" is used.	No change to listening duration; add 1 min; add 2 min. No documentation of when listening duration extended due to noise disturbance
Start and end time	Start survey 30 min after sunset or later	Start 30 min before sunset; start 30 min after sunset; start 30–60 min after sunset; start 60 min after sunset; start anytime after sunset. No end time specified
Survey completion	Finish by 1 a.m. Route completed in one night	Route completed in one night; allowed to complete over two nights
Order of listening	Listening stations completed in assigned numerical order	Numerical order; reverse numerical order allowed; skip and return to station allowed
Number of listening stations per route	10 listening stations	10 listening stations; 10–12 stations; 10–15 stations
Number of observers per datasheet	One observer per datasheet	One observer per datasheet; individuals or pairs allowed; unspecified number of observers per datasheet
Minimum distance between listening stations	0.5 mi minimum distance (though some routes already established with 0.3 mi minimum distance will continue)	0.25 mi; 0.3 mi; 0.5 mi; 1 mi
Calling index definitions	Adopt original Wisconsin Index (Table 39–3)	Most variation is in definitions of codes one and two. Examples of code one variation: Individuals can be counted, there is space between calls. One or two individuals calling or calls not overlapping. You can clearly hear all individuals of a species and can easily count them. There is ample space (time) between calling individuals.
Number of sampling occasions per year	Three survey periods, plus one optional period to target early explosive breeders	Three survey periods; four survey periods; five survey periods

smaller roads, adhering as closely as possible to a randomly selected direction (north, northeast, east, southeast, south, southwest, west, and northwest). If appropriate roads do not exist in the selected direction, another direction is determined randomly. Routes cannot cross state, provincial, or international boundaries.

Additional stratified-random routes may be generated for regional programs where Regional Coordinators want larger samples. Additional routes are also created for an entire state or province to maintain consistent route density within a political boundary. If the degree block is shared by two or more states/provinces or by a state/province and large body of water, routes are generated in proportion to the land area of each region within the block. Previous to this protocol, New Jersey and Tennessee had requested and received additional routes stratified by physiographic region. Some regional programs

also collect data on nonrandom or site-selected routes; data from these routes may be included in the NAAMP database but will probably not be used in most trend analyses.

Routes are then checked by ground truthing and rerouted if the designated road does not exist, it is a private road, it does not meet safety standards, or it has excessive ambient noise. When rerouting, the original route's starting point and random direction are maintained where possible. In a case where the whole route needs to be adjusted, a neighboring, parallel route may be substituted if it meets the requirements. When no substitution is available, the route is documented as inaccessible and a replacement random route is newly generated to maintain consistent route density within the region.

During the ground truthing phase, listening stations are also determined. Each route has 10 listening stations spaced 0.5 miles or more apart (although a few routes established in

1996 have some listening stations positioned 0.3 miles or more apart). Listening stations are assigned by one of two methods: equidistant placement or stratified by habitat. For equidistant placement, listening stations are placed 0.5 miles apart regardless of habitat. When listening stations are stratified by habitat, each is placed within hearing distance of a wetland that provides potential breeding habitat (e.g., pond, marsh, vernal pool, or roadside ditch), but without regard to the presence or absence of calling anurans. In this approach, if the starting point of the roadside route has appropriate wetland habitat within hearing distance, it becomes the first listening station; if it does not, then the nearest appropriate habitat along the direction of travel becomes the first listening station. The second station becomes the next appropriate habitat that is at least 0.5 miles from the first listening station. This process continues until all ten listening stations are designated. For both methods, the 0.5 mile separation among listening stations should be a measured distance "as the crow flies" rather than measured by an odometer across roads.

NAAMP Route Creation

To make valid conclusions from data sets, statistical analyses generally require data to be collected in a random manner (Zar, 1996). With nonrandom samples, there is risk of developing biases that do not represent the entire population. For this reason, NAAMP routes are generated from a random starting location and random direction. The 10 listening stations are then assigned to the route by one of two methods: equidistant sampling or stratified by habitat. Equidistant listening stations are preferable statistically because they presumably sample different habitat types in proportion to relative abundance. Also, equidistant stations should monitor equally well those population increases or decreases due to habitat creation or destruction. However, it also reduces the survey's ability to detect a trend (due to the large number of zero counts expected in most landscapes) and, therefore, requires a greater number of routes to monitor populations. Drawbacks to this approach are that more observers are needed and that observers are reluctant to conduct this type of survey. For example, Maryland attempted equidistant sampling and was unable to retain its observers for the second year.

The stratified by habitat method locates listening stations near potential breeding habitats, which increases the probability of hearing breeding anurans but may also introduce a bias. At existing wetlands, habitat destruction may be more likely to occur than habitat creation, whereas at sites without existing wetlands, only habitat creation could occur. Thus, if the initial selection of listening stations requires wetlands to be present, the chance to detect population trends may be biased towards detecting declines. However, NAAMP listening stations are established at wetland habitats without regard to the presence of amphibians, and this reduces the potential sampling bias associated with local extinction events. Another factor mitigating against this bias is that an observer records all anurans heard in any direction from the listening station, not just those at the wetland for which the station was selected. Observers prefer these routes and retention of volunteers is higher. As a result of these logistical considerations, most participating regional programs chose the stratified random sampling technique, although it is acceptable for regions to adopt either method; Massachusetts and Wisconsin have adopted equidistant sampling.

The use of roadside routes allows for efficient coverage of a large geographic area. It is not feasible at this geographic scale to conduct off-road sampling or to receive permission to survey on private property. If roadside habitat is changing in a manner different from landscape changes as a whole, then population trends based upon roadside sampling may be biased. This issue has already been examined for the BBS. A study using aerial photographs from 1963 and 1988 of areas along Ohio BBS routes examined the change in forest cover compared to the larger landscape and found that trends in habitat change were within 1% of being congruent (Bart et al., 1995). A second study found that habitat changes along roadsides reflect landscape changes, though change may be greater along the road (Keller and Scallan, 1999). As Keller and Scallan point out, if roadsides and the larger landscape agree in the direction of habitat change but differ in the extent of change, then the over- or under-estimation can be corrected. However, if the direction of habitat change differs between roads and the landscape, trend estimates from roadside surveys will not be representative of regional trends. Similar research examining wetland habitat change along roads and the larger landscape should be conducted to better understand the applicability of NAAMP population trend estimates to regional and continental populations.

Because route density may differ among regional programs, when the data are analyzed for population trends, the routes can be weighted to accommodate the different densities of random routes among regions (as is done with the BBS routes for bird populations; Peterjohn et al., 1995). New routes are identified in the database, which allows data analysis using only the original routes, if desired. This practice allows regions some flexibility in the number of routes administered.

Seasonal Sampling Periods

There are three seasonal sampling periods (separated by intervals of at least two weeks) selected by each region to cover the calling periods of its local species. Regions may also elect to add a fourth earlier period targeted towards wood frogs *(Rana sylvatica)* or other early explosive breeding species that breed during a brief, relatively unpredictable period in early spring when vernal ponds first melt and fill. There is no required time interval between this optional sampling period and the first sampling period, and in some years these may overlap.

The total number of potential sampling days varies regionally in relation to the length of the calling season. The maximum number of potential sampling days is 60% of the calling season duration (i.e., from the beginning of the peak of the species with the earliest calling phenology through the peak of the last species). Thus, the number of potential sampling days is greatest in southern states. Each sampling period is no longer than 6 weeks; for many regional programs the sampling periods are 2–3 weeks. The regional program may divide the available number of sampling days equally or unequally among the sampling periods.

Regional Coordinators set the sampling periods based upon experience and the available data on breeding phenology. Different sampling periods may be set within a given regional program (e.g., in two or three bands within a given state) to accommodate phenological differences due to elevation or latitude. Neighboring regional programs coordinate sampling periods as much as possible to encourage consistency across political boundaries. More data on breeding phenology are needed to confirm or refine the seasonal sampling periods in many regions.

Nightly Sampling Conditions

Observers begin surveys 30 minutes after sunset or later, but all routes must be completed by 0100 hours. Appropriate sampling conditions are based upon wind, sky, and air temperature conditions. The Beaufort Wind Codes are as follows:

0: Calm (<1 mph); smoke rises vertically

1: Light air (1–3 mph); smoke drifts, weather vane inactive

2: Light breeze (4–7 mph); leaves rustle, can feel wind on face

3: Gentle breeze (8–12 mph); leaves and twigs move around, small flag extends

4: Moderate breeze (13–18 mph) moves thin branches, raises loose paper; *do not conduct survey, unless in Great Plains region.*

5: Fresh breeze (19–24 mph); trees sway; *do not conduct survey in any regions.*

Sky Codes are as follows:

0: Few clouds

1: Partly cloudy (scattered) or variable sky

2: Cloudy (broken) or overcast

4: Fog or smoke

5: Drizzle or light rain (not affecting hearing ability)

7: Snow

8: Showers (affecting hearing ability); *do not conduct survey in any regin.*

In most regions, to conduct the survey the Beaufort Wind Code should be category 3 or less, although in the Great Plains category 4 is acceptable. The survey is not conducted during rain showers that are heavy enough to affect hearing ability, but light rain is acceptable. The air temperature criteria are the minimum allowable temperatures, which vary for each sampling period. For the optional survey period and the first survey period, the minimum temperature is 5.6 °C (42 °F). For the second survey period, the minimum temperature is 10 °C (50 °F); the third survey period minimum temperature is 12.8 °C (55 °F). A regional program may choose to set higher minimum temperatures based upon regional phenology information. For example, the New Jersey program raised the third survey period temperature minimum to 15.6 °C (60 °F), as research indicates that Pine Barrens treefrogs *(Hyla andersonii)* are more likely to vocalize when temperatures are above this minimum (John Bunnell, Pinelands Commission, unpublished data). In addition to the wind, sky, and air temperature conditions, the survey night should be selected based upon the likelihood that amphibians would vocalize, such that relative humidity is high and/or there has been a recent rainfall event.

The calling survey temporal period (30 minutes past sunset through 0100 hours) is based upon biological and practical reasons. Because almost all observers are volunteers, the survey is designed to accommodate the times when most volunteers are available and most anurans are calling. It is not feasible to sample all species at their peak calling periods, and

the recommended period is designed to sample when most species are actively calling. Temporal phenology data would confirm which species are most likely to vocalize during the nighttime period. In North Dakota during 5 May–8 July, Bowers et al. (1998) found that the six most common species called within the nightly time period, and the calls of one rarely heard taxa *(H. versicolor/chrysoscelis)* appeared to peak later. In a one-month study conducted from mid-June to mid-July in South Carolina, eight of the nine species examined called during the nightly time period of the unified protocols (Bridges and Dorcas, 2000). The ninth species, southern leopard frogs *(R. sphenocephala)*, was found to vocalize in the early morning hours. Bridges and Dorcas (2000) found that several species end peak calling activity around midnight. If similar patterns are confirmed in other locations and/or species, the NAAMP survey period may need to be shortened. More data are needed to confirm which species are most likely to vocalize during the nightly time period and to interpret population monitoring data for individual species across various geographic areas.

NAAMP protocol accommodates biotic and abiotic differences among regions. In southern states, the NAAMP calling survey protocol recommends surveys occur after recent rainfall (within 3 days) because southern amphibian vocalizations appear to be associated with rainfall (L. A. Linam, personal communication; S. Shively, personal communication). Amphibian breeding phenology at more northern locations indicates that rainfall is less important, although sampling should occur during humid conditions. Appropriate humidities appear to vary regionally, however, research is needed to verify these assumptions. Observations are not permitted when there are strong winds or heavy rains, as both may affect the observer's ability to hear (Robbins, 1981). However, the Great Plains area is allowed to sample during category 4 wind speed (see above) to make sampling possible in this wind-prone area; the BBS makes a similar allowance for the Great Plains region (Robbins et al., 1986).

Minimum temperature thresholds were established based upon the knowledge of Regional Coordinators. While amphibians may call at temperatures below these thresholds, the minimum temperatures were established to increase the likelihood that amphibians, if present, would be calling. Research examining temperature and calling activity should be conducted to confirm or refine these standards.

Data Collection

Listening stations are conducted in numerical order on one night by one observer. We encourage (but do not require) one observer to conduct all surveys of a route in a given year. Because some observers have assistants who may also wish to collect data, multiple observers are instructed to fill out their own datasheet separately and independently. One observer is the official recorder of the route whose data will be entered into the NAAMP database. All datasheets are returned to the Regional Coordinator for archival purposes.

Each datasheet includes the observer's name and contact information, route number, route name, seasonal sampling period, and survey date. At the beginning and end of each survey, observers record time, sky code, and wind code. Observers listen for 5 minutes per listening station and assign an amphibian calling index value for each amphibian species heard. The calling codes for the Amphibian Calling Index are as follows:

1: Individuals can be counted; space between calls

2: Call of individuals can be distinguished; some overlapping of calls

3: Full chorus, calls are constant, continuous, and overlapping

There is no initial waiting period before beginning the listening period. At each listening station, observers also record air temperature, presence of ambient noise, and whether an interruption in the listening period occurs. Ambient noise disturbance is indicated as follows:

0: No appreciable effect (e.g., owl calling)

1: Slightly affecting sampling; (e.g., distant traffic, dog barking, one car passing)

2: Moderately affecting sampling; (e.g., nearby traffic, 2–5 cars passing)

3: Seriously affecting sampling; (e.g., continuous traffic nearby, 6–10 cars passing)

4: Profoundly affecting sampling; (e.g., continuous traffic passing, construction noise)

Only the last two levels signal ambient noise disturbance. A regional program may require observers to answer whether an ambient noise affected the observer's listening abilities in either a yes/no format or by using the noise index definitions.

In Gulf Coast states where air temperature varies little over the course of an evening, observers are required to record air temperature only at the start and end of their survey. The ambient noise data indicate whether the observer's ability to hear calling amphibians was impaired. If a disturbance "event" of one-minute or longer occurs, then the observer interrupts the five-minute listening period in order to eliminate the disturbance from the sampling time; the occurrence of an interruption within the listening period is noted on the data sheet. To qualify as a disturbance event, the noise should be sudden or intermittent in nature (such as a train) and not a constant background noise. In addition to the required information listed above, regional programs may ask or require observers to collect additional data considered optional by the NAAMP protocol. Examples of optional data include the number of days since last rainfall, water temperature, presence of snow cover, the number of vehicles that pass during the listening period, and identification of night-calling birds.

Once the route has been ground truthed and listening stations are established, these locations are permanent and the locations may not be changed unless a safety issue arises. If habitat destruction occurs at a listening station and a local extinction of amphibians occurs, it is important information. To document habitat destruction, the location should be surveyed for three seasons beyond the destruction date. After three seasons of non-activity, the listening station may be retired and null data will be assumed for this site. A listening station cannot be retired merely because the wetlands are uninhabited by anurans. Retired stops should be visited periodically to verify that no suitable habitat exists, but the five-minute listening period is no longer required.

Observers record the time, sky code, and wind code at the beginning and end of each survey to verify that the sampling conditions were met on the evening of the survey. Gulf Coast states also request documentation of the last rainfall event. The Louisiana program suggests that a local contact in the immediate vicinity of the route should be established, instead of relying on rainfall information from local news media or the observer's residence. Observers are not asked to record humidity levels, as humidity reports from local news media may be based upon a regional weather station miles from a route and measuring local humidity along a route is not feasible for most observers. At each listening station, air temperature is recorded to verify that sampling conditions were met on the sampling night; at least 8 of the 10 listening stations must meet the temperature guidelines. For southern states that record air temperature only at the beginning and end of a survey, both temperature readings must meet these guidelines.

The relationship between the amphibian calling index and the actual population for each amphibian species monitored should be investigated. For example, to reach a calling index value of 3 requires that calls are constant, continuous, and overlapping. It may take only a few individuals to reach this definition for a species with extended calls, such as the long trill of American toads (B. americanus). However, another species with a short call, such as green frogs (R. clamitans), may require more individuals to reach the same index value. Similar issues exist with the other index values, except zero. Until such index-to-species population relationships are established, the calling index values may be translated into presence/absence data that can be used for population trend analysis.

A limitation of the calling index is that level 3 is a threshold above which population fluctuations are not detected. If one hundred individuals are enough to reach a level 3 calling index, then hundreds (or thousands) of individuals would still register as the same level. Thus, population fluctuations above this threshold are not detected. Also, when a population is at level 3, only stability or decline are possible to record and may result in a bias toward decline. Using the data as presence/absence data in population trend analyses would avoid this potential bias.

Observer Training

Regional Coordinators recruit and train volunteer observers to conduct the amphibian calling survey routes and have discretion to determine who will be accepted as an observer. Most programs allow any interested adult to join the survey, although the West Virginia program requires previous experience or study. When programs accept inexperienced observers, the Regional Coordinator provides training on protocols, calling phenology, call identification, and completion of datasheets. Regional Coordinators decide how observer training is conducted for their program. Some Regional Coordinators have annual training meetings, while others have meetings only when training is required for new observers. The Wisconsin program encourages new observers to accompany an experienced observer at least once before adopting their own route. In the first year of sampling, the New Hampshire program had an initial training meeting and small group follow-up meetings where calls recorded by observers were reviewed. In all programs, observers are provided with (or asked to purchase) a training audiocassette. The Virginia program has a calling quiz on the training tape. Observers are also instructed in the importance of gathering all required data. For training purposes, each Regional Coordinator is provided with a slide presentation of the unified protocol, which discusses NAAMP history, goals, and unified protocols.

In a study examining observer agreement during amphibian calling surveys, Shirose et al. (1997) compared experienced observers to intermediate and novice observers and found that all groups performed well in amphibian species identification. When calling index values were combined to form presence/absence data, agreement among the observer levels was good: 98.0% between experts, 96.9% between experts and intermediates, and 96.8% between experts and novice observers. Agreement on calling index values (when a species was present) ranged from 46.7% to 83.3%; thus, inter-observer results more closely agreed at species identification than at assigning an index value. Two other studies have found similar results, with good species identification agreement but variable calling index assignment (Bishop et al., 1997; Hemesath, 1998).

The use of volunteer observers is necessary in large-scale monitoring efforts such as the NAAMP. State agencies and other organizations that support NAAMP amphibian calling surveys would not be able to afford such large-scale programs if observers were paid technicians or biologists. While further research on inter-observer agreement is needed, particularly for species not represented in the Midwest or Canada, these studies indicate that calling survey data can be used in presence/absence analyses. Indeed presence/absence data may be sufficient for population trend analysis (Green, 1997c). However, improving index determination by volunteer observers should be the emphasis of future training efforts. If research can determine an index-to-species population relationship and inter-observer agreement on index assignment is improved, then analyses using index values may be possible. It should be noted that increasing observer agreement on index determination does not reduce the bias associated with the threshold.

Summary

The NAAMP calling survey is currently active in 17 states where the unified protocols have been adopted. Protocols will be modified as necessary when (and if) new information suggests more appropriate methods for sampling calling amphibian populations. A stratified random route design is used to create roadside routes, each of which has ten listening stations placed 0.5 miles or more apart. Listening stations are determined by one of two methods: equidistant placement or stratified by habitat. Data are collected three to four times per year during the designated seasonal sampling periods in order to reach as many amphibian species as possible. All surveys are initiated at least 30 minutes after sunset and are completed by 0100 hours during appropriate sampling conditions for the region. Observers record amphibian and environmental data during a five-minute listening period at each listening station.

Further research is needed to confirm or refine the seasonal sampling periods and nightly sampling conditions on a regional basis. The NAAMP data may be used for population trend analyses in presence/absence format, similar to that developed for the Wisconsin Frog and Toad Survey (Mossman et al., 1998). If a relationship between the amphibian calling index and actual population for amphibian species can be determined, then analyses beyond a presence/absence format may be developed. Data collected under the unified protocol may be combined to provide trend analyses at multiple geographic scales, including sub-state, state, multi-state, physiographic regions, and species range. Provinces and other states are encouraged to join the unified protocols, which would allow for a more comprehensive coverage of the status and trends of the continent's anuran populations.

Acknowledgments. Special thanks to all the people who have donated their time to making the NAAMP calling survey a reality; without the dedication of our volunteers such a large-scale monitoring effort would not be possible. To all the Regional Coordinators, thank you for the hard work that goes into making the calling survey a reality each season. Special appreciation and acknowledgment is owed to those Regional Coordinators (or their representatives) who participated in one or both protocol meetings or provided important commentary: Aram Calhoun, Joe Collins, Jim Delahoussaye, Alice Doolittle, Carol Drummond, Steve Faccio, Zach Felix, Mark Gumbert, Lisa Hartman, Wayne Hildebrand, Susan Hitchcox, Scott Jackson, Glenn Johnson, Lee Ann Linam, Sue Marden, Tom Pauley, Lori Sargent, Steve Shively, and Eric Stiles. We thank John Bunnell of the Pinelands Commission for permission to use his unpublished data. From L. W., thanks also to Sam Droege, Jude Griffin, Robin Jung, and Bruce Peterjohn.

Addendum. In 2004 the seasonal sampling periods of unified protocol were modified. The minimum interval between sampling periods changed from two weeks to one week, in order to better accommodate fluctuations in peak calling activity for northern states. States initiating programs and involvement in NAAMP have changed over time. As of November 2004, the following states are active in NAAMP: Delaware, Florida, Indiana, Kentucky, Louisiana, Maine, Maryland, Massachusetts, Michigan, Minnesota, Mississippi, New Hampshire, New Jersey, Tennessee, Texas, Vermont, and West Virginia. Montana is exploring the use of NAAMP protocols in a western state. Arkansas, California, and Connecticut are inactive, Kansas and Ohio have programs separate from NAAMP, and Florida and Mississippi have joined NAAMP. All other states listed as active NAAMP states in Table 45-A (New York, Pennsylvania, Rhode Island, Virginia, and Wisconsin) have not yet submitted data to the NAAMP database or their status within NAAMP is uncertain. North Carolina and Missouri are planning to start NAAMP in 2005.

FORTY-SIX

Evaluating Calling Surveys

SAM DROEGE AND PAIGE EAGLE

In North America, approximately 55 of the 103 species of anurans can be surveyed readily by using counts of vocalizing males as an index to their presence or population size. Such surveys are most applicable to eastern and northern parts of the continent, where almost all species regularly vocalize and breeding seasons extend over several weeks. In the West, members of the chorus frog *(Pseudacris)* complex and American bullfrogs *(Rana catesbeiana)* consistently call during their breeding seasons, but many of the other species call either infrequently, quietly, or sporadically only following heavy rains and are thus unsuited to be surveyed using calling count techniques.

Calling surveys of amphibians are designed to provide an index to changes in amphibian populations. The technique can be applied in monitoring sampling designs to provide estimates of abundance and change at scales ranging from continental down to single wetlands. A powerful attribute of a uniform, randomly allocated system of calling surveys is that they can be repartitioned in many meaningful ways. The estimates of trends from the same system of surveys can be developed for watershed, political, physiological, or survey-wide units of scale (depending upon research and management requirements) in a way similar to the North American Breeding Bird Survey (Droege, 1990). By correlating observed trends and distribution patterns with ancillary variables, it is also possible to gain insight into the factors affecting population changes over time (e.g., declines may be associated with agricultural regions, while increases may occur in regions where wetlands have increased).

Calling Survey Protocol

Most calling survey protocols in North America use a simple index to population size based on the one developed for the Wisconsin Frog and Toad Survey (for details see Mossman et al., 1998; Weir and Mossman, this volume). Surveys are usually run three to four times per year in synchrony with the mating and peak calling seasons of the local species. Surveys are often strung together as a "route" (a series of stops) at wetlands along rural roads. Starting time is 30–60 minutes after sunset, and a route typically takes about 2 hours to complete. To reduce disturbance effects, an observer usually waits one minute at each

wetland stop, then starts a three- to five-minute listening period. The observer records which species were heard along with a calling index value.

Variation

Land managers, researchers, states, and provinces initiate calling survey programs with the idea that these counts act as an index to the true number of frogs and toads present in the area being sampled. The assumption is that this index is both an accurate (unbiased) and precise (little variation) reflection of what we really want to track—the rate of change of the amphibians in the study areas. In this section, we explore the known and suspected factors that affect calling surveys and reflect on what impact they may have on the interpretation of the resulting trends.

Any monitoring program's ability (power) to estimate a change (within a given level of precision) is determined by the amount of year-to-year variation in the counts, with high variation in those counts tending to diminish the ability to estimate trends. The intrinsic variability of a monitoring program's counts also influences the sampling effort (e.g., the number of routes, the number of stops, the time spent at a stop, and the number of replicates of stops) needed to detect trends (if any) of the magnitude set by the organizers of the count; more routes are required when there is high variability.

To investigate the variability in calling survey counts, we used analyses by John R. Sauer examining 14 years of data provided by Mike Mossman and Lisa Hartman from the Wisconsin Frog and Toad Survey. These analyses used the number of stops (out of 10) at which a species occurred on each route. The full 0–3 Wisconsin Index was not used because further work is needed in the analysis and interpretation of the Index in calculating trends. Consequently, this simplified index represents changes in the frequency of occurrence of amphibians at wetlands. The estimates of variation in counts calculated using the annual indices produced by Sauer's analysis were then used to determine whether calling surveys have the ability to detect changes in populations of frogs and toads in Wisconsin.

The currency we used for evaluating the variation in counts is a simple proportion called the coefficient of variation (CV).

Appendix 46: The Relationship Between the Coefficient of Variation of Yearly Counts on a Survey and an Estimate of the Power of a Monitoring Program to Detect Changes

	Power over 10 y		Power over 20 y	
	−2% per y	−3% per y	−2% per y	−3% per y
CV (%)				
5	100	100	100	100
10	86	98	100	100
15	63	76	97	100
20	53	79	92	99
25	46	65	81	96
30	38	54	76	91
35	36	50	73	88
40	37	48	68	82

NOTE: Power values were calculated using Monitor4.exe (Gibbs et al., 1998). The parameters that entered into power calculations were alpha = 0.20; iterations = 500; whole number rounding, one-sided test, exponential trends, and 0.1 constant added.

The CV is the ratio of the standard deviation (a measure of the variability of the counts) to the mean (average) of the counts. With respect to detecting ongoing trends, a small CV is considered good, while a large CV is considered bad. As a rule of thumb, anytime you are presented with a set of counts where the standard deviation is 50% or more of the mean, then there will be substantial costs (more sampling effort) to overcome that variability.

Appendix 46-A lists the power to detect trends for a range of CVs. Under the given assumptions, and for the indicated CVs, we list the estimated probability of detecting a decline (or its power) for rates of decline of 2% and 3% per year over 10- and 20-year time periods. We will use the convention that any power over 90% is adequate, anything over 80% is still not unreasonable, and anything below 80% is generally inadequate. There are four columns showing our ability to detect trends at various CVs at time periods of 10 and 20 years with actual decline rates of 2% and 3% per year. All estimates of power were made with the MONITOR software package (Gibbs et al., 1998).

The CVs calculated for each species in the Wisconsin Frog and Toad Survey data varied from 4% to 26% (with an average of 15%) when calculated at the statewide monitoring scale (Table 46-1). This relatively low variability indicates that all species pass the minimum power requirements to detect statewide population declines over a 20-year time interval as presented in Appendix 46-A.

To examine the ability of calling surveys to detect trends at the regional level, we looked at the species' CVs for each of the six physiographic subregions of Wisconsin. There were a total of 55 species-region estimates. Of the CVs calculated for the species-region units, 52 out of 55 (95%) met the detection criteria for the 20 years/3% decline situation. For the three that failed, increasing the number of survey routes might improve their detection ability, as each of these units had only three routes. Note that while many of these regions were composed of fewer than 10 routes, most could pass our criteria for low CVs and high power. Species differences do exist; notably, spring peepers (*P. crucifer*) exhibit extremely low variation in their counts, while toads (*Bufo* sp.) and wood frogs (*R. sylvatica*) have higher than average amounts of variation. This is most likely a reflection of the relatively greater consistency in detection of spring peepers during the sampling period as compared to the more pulsed seasons for toads and wood frogs.

Trends for fewer than 10 years were not listed in Appendix 46-A because (1) trends in counts over fewer than 10 years are probably reflections of yearly rainfall patterns and winter temperatures rather than real long-term trends; and (2) detecting trends with fewer than 10 years of data is difficult due to the magnitude of variability that counts of animals exhibit, autocorrelation in their counts, and the difficulties of estimating the variance of counts. Consequently, detection of anything other than large population changes over even a 10-year period should not be expected from a monitoring program.

In this analysis, each estimated CV for a region is based on a mix of species, routes, observers, and a pattern of complete and incomplete survey routes unique to that region and this data set. Thus, these data are useful for evaluating the general utility of calling surveys in detecting trends; the individual CVs, however, are difficult to compare to one another, as each CV was created from a different underlying pattern of completed surveys over the years of the program along with a different number of routes.

Bias

While variation in counts has the effect of hiding long-term trends, bias in population indices can cause monitoring programs to reach false conclusions about the trajectory of those

TABLE 46-1
Means, Detrended Standard Errors from a Linear Regression of the Indices, and Coefficient of Variation from Regional and Statewide Annual Indices of Frogs and Toads Calculated Using Estimating Equations with 14 Years of Wisconsin Frog and Toad Survey Data

Latin Name	Region	N	x	SE	CV(%)
Acris crepitans	Driftless area	2	0.293	0.069	23.7
	Wisconsin (all)`	2	0.292	0.068	23.5
Bufo americanus	Eastern forest	12	4.288	1.313	30.6
	Northcentral forest	35	4.066	0.918	22.6
	Northwest forest	3	5.219	2.017	38.7
	Central sands	3	4.630	1.985	42.9
	Southeast	38	4.174	1.016	24.3
	Driftless area	26	5.996	1.312	21.9
	Wisconsin (all)	117	4.514	0.873	19.3
Hyla chrysoscelis	Northcentral forest	6	0.276	0.084	30.5
	Northwest forest	3	4.582	1.238	27.0
	Central sands	2	3.509	1.020	29.1
	Southeast	24	1.884	0.319	16.9
	Driftless area	19	2.140	0.289	13.5
	Wisconsin (all)	56	1.295	0.177	13.7
Hyla versicolor	Eastern forest	13	5.522	1.616	29.3
	Northcentral forest	35	6.496	1.139	17.5
	Northwest forest	3	7.467	2.373	31.8
	Central sands	3	6.732	0.326	4.8
	Southeast	30	2.823	0.514	18.2
	Driftless area	25	7.369	1.303	17.7
	Wisconsin (all)	109	5.846	0.493	8.4
Pseudacris crucifer	Eastern forest	14	7.092	1.488	21.0
	Northcentral forest	36	8.319	0.352	4.2
	Northwest forest	3	9.430	0.576	6.1
	Central sands	3	7.315	0.253	3.5
	Southeast	34	5.115	0.381	7.5
	Driftless area	25	9.941	0.289	2.9
	Wisconsin (all)	115	7.922	0.325	4.1
Pseudacris triseriata	Eastern forest	12	2.093	0.367	17.5
	Northcentral forest	33	3.259	0.865	26.5
	Northwest forest	3	3.741	1.567	41.9
	Central sands	3	3.747	0.514	14.4
	Southeast	39	6.612	0.415	6.3
	Driftless area	24	5.305	0.937	17.7
	Wisconsin (all)	114	4.203	0.537	12.8
Rana catesbeiana	Eastern forest	2	0.266	0.098	36.7
	Northcentral forest	17	1.279	0.268	21.0
	Southeast	13	0.640	0.126	19.7
	Driftless area	6	0.470	0.107	22.8
	Wisconsin (all)	40	0.784	0.097	12.4

TABLE 46-1 (continued)

Latin Name	Region	N	x	SE	CV(%)
Rana clamitans	Eastern forest	10	3.460	0.751	21.7
	Northcentral forest	36	6.078	0.685	11.3
	Northwest forest	3	6.522	1.077	16.5
	Central sands	3	5.773	0.825	14.3
	Southeast	38	5.870	1.058	18.0
	Driftless area	25	7.141	0.732	10.2
	Wisconsin (all)	115	5.974	0.597	10.0
Rana palustris	Driftless area	9	0.980	0.175	17.9
	Wisconsin (all)	20	0.299	0.049	16.2
Rana pipiens	Eastern forest	12	3.748	0.912	24.3
	Northcentral forest	23	1.434	0.503	35.1
	Northwest forest	3	2.849	0.970	34.1
	Central sands	3	3.596	1.473	41.0
	Southeast	38	3.428	0.495	14.5
	Driftless area	22	3.647	0.498	13.7
	Wisconsin (all)	101	2.616	0.449	17.1
Rana septentrionalis	Northcentral forest	11	0.594	0.164	27.6
	Northwest forest	2	2.530	0.673	26.6
	Wisconsin (all)	13	0.776	0.199	25.7
Rana sylvatica	Eastern forest	13	5.783	1.251	21.6
	Northcentral forest	36	6.343	1.341	21.1
	Northwest forest	3	6.563	2.150	32.8
	Central sands	3	4.766	1.703	35.7
	Southeast southeast	24	1.761	0.209	11.9
	Driftless area	16	2.351	0.850	36.2
	Wisconsin (all)	95	4.414	0.809	18.3

population changes. In this paper, we characterize the relationship of the "true" population of amphibians (i.e., the real number of frogs and toads out there) with that of the Wisconsin Index.

Traditionally, index validation requires the simultaneous sampling of the index protocol and an accepted method of determining "true" population (such as total counts, capture-recapture, removal techniques, and/or distance methods) at the same site, and then comparing the two. Because estimating true population size is time and cost intensive, index validation is typically conducted at a small number of sites and the relationships found are assumed to hold for larger areas. In the literature, there are only a handful of examples where researchers make an attempt to validate animal indices (Conroy, 1996). Realistically, there is no reasonable way to completely validate large-scale surveys, such as a statewide amphibian calling survey. However, it would be possible to validate calling indices at individual wetlands, though that still would be a major undertaking.

Another approach to evaluating bias (other than simply ignoring the problem) is to list known or suspected factors that affect the detection rate of calling surveys (i.e., the ratio of animals detected to the entire population). This list will help to establish a general notion of whether an index will (or is likely to) yield interpretable information, and in turn, what types of conclusions may be reasonably drawn from this information.

Biasing Agents

The following factors could act as potential biasing agents in a system of amphibian calling surveys: call saturation or obfuscation, weather effects, behavioral effects, landscape changes, and observer effects. We will discuss the potential implications of each of these factors on a monitoring program based on calling surveys below.

Calling Saturation

Call saturation is a bias linked to changing detection rates with different densities of calling amphibians. As the number of individual frogs or toads calling increases at a site, the ability of the human observer to differentiate or count individuals rapidly declines (to which anyone who has experienced the painful impact of a large chorus of spring peepers can attest). The calling index scale acknowledges this human limitation and uses only three intensity levels. For example, a true population could range broadly, yet the index could peak at intensity level 3 and stay constant from then on. Thus, the population could fluctuate widely over time, but the index has no ability to detect those fluctuations as long as most counts remain above the threshold where a level 2 becomes a level 3. At lower population levels, the index tracks the population change, but only as far as the scale of 0–3 permits.

The reality of most calling surveys is that a series of survey locations never has all points saturated with detections of a species at the highest calling level (level 3). Rarely is a species even detected at all the points on a night's round of surveys. Therefore, a conservative approach to the analysis of calling frog and toad data is to collapse all the calling categories into a simple detected/not detected category. This does not eliminate bias in detections associated with population size (i.e., large populations are likely to be more detectable on an average night than small populations), but it makes the analysis more understandable. For example, it is easier to understand the changes in the frequency of points with green frogs (*R. clamitans*) over time than it is to understand the changes to a set of index values for green frogs taken at the same point. The frequency of points is clearly related to the number of wetlands occupied by green frogs, while the changes in the index are harder to interpret. This is because it is not exactly clear how an index of 1, 2, or 3 relates to the real number of green frogs, nor what would be the best means of combining index scores across several wetlands.

The practical consequences of using a simple 1 or 0 index (and acknowledging that large populations are more detectable than small ones) will be an over-estimation of the rate of decline. That is, the slope will be correctly determined to be negative, but it will be steeper than what is actually occurring. Thus, a monitoring program based on calling surveys is likely (analyzed either with the index or as presence/absence) to prematurely sound an alert to declines. In some ways this is not bad, as it gives a longer period of time to react to a potential problem by more closely examining the data and initiating additional studies and surveys to address the potential problem.

Obfuscation

A less clear problem is call obfuscation, which occurs when one species' call interferes with the detection of another, quieter species. Bias could potentially be severe enough that a population increase in the quiet species could be misidentified as a decline if the louder species (population also increasing) were to overwhelm the calls. It is unclear how many stops in a set of surveys would ever be affected by such a situation. An evaluation of regional chorus timing and loudness would be simple to conduct.

Weather

Most calling survey protocols minimize the biasing effects of weather on sampling by providing guidelines for appropriate sampling conditions, with prohibitions on sampling in nights with strong wind or heavy rainfall. Despite the guidelines,

weather induced variation does exist, as droughts and wetter-than-normal years affect the number and rate of frogs calling. Consequently, short-term comparisons (2–5 years) of population counts are difficult to interpret because weather effects on detectability and on population fluctuations are intertwined. Over longer periods of time, the effects of weather diminish and simply become "extra noise" in the system, not a bias.

Behavioral Effects

Another source of variation and potential bias in calling surveys is the onset of calling behavior. Advent of the calling season for a frog or toad begins earlier in warm years (often associated with large rainfalls), at times by several weeks (Mossman et al., 1998). To mitigate that effect, the sampling windows of regional programs are often based on temperature cues. In addition, monitoring programs in the United States and Canada have begun to track the amphibian calling phenology. The U.S. program, which is based on three successful programs in Canada, is called Frogwatch USA (www.frogwatch.org). The goals of Frogwatch USA are to document the yearly phenological pattern of calls and regional detection rates, determine amphibian population changes at individual wetland sites, and educate the public about amphibians.

Landscape Changes

Biasing factors can also creep in over extended periods of time if landscape or other initial conditions change and affect the detection of amphibians on calling surveys. Over time, forests, ponds, fields, human development, and wetlands change in their predominance, succession, and interspersion. As these habitats change, so does the mix of amphibian populations—some species are favored while others are not. These represent the types of changes we would expect a monitoring program to track. However, another consequence of these changes may be shifts in the acoustical aspects of the landscape. With changes in vegetation, calling populations in some areas may become more easily detectable, while the sounds of others may be impeded (Varhegyi et al., 1998). Any large and uncorrected shift in the detectability of calls will affect (i.e., bias) the resulting calculated trends. It may be possible to correct landscape bias by documenting habitat changes along routes or using groups of years as co-variables (to diminish the effects of long-term habitat changes, which will present more realistic data). Fortunately, such landscape changes occur over decades, providing some lead time in determining and developing a correction when warranted.

Observer Effects

Observer effects include bias and errors caused by the survey participants themselves, factors that may be caused by hearing losses or identification errors. One could argue that if you have many observers participating, the relative hearing levels would be randomly distributed throughout that population yielding extra noise in the system but not bias. However, many researchers now statistically factor out the differences among observers in their analyses of trend (e.g., they control for the folks who are over-estimators and those who are under-estimators) by creating observer co-variables. This reduces the noise in the data caused by differences in how people count, but it is based on the assumption that an observer's hearing does not change over time (which it can and ultimately does). Thus, a negative bias is built into the system (because the observers' hearing deteriorates as they age, possibly resulting in their recording relatively fewer frogs and toads over time). Hearing loss in observers can be managed in several ways. The best way would be to test observers' hearing and to impose strict guidelines for participation based on the results of those tests.

Most existing amphibian calling surveys use volunteer observers to collect survey data. This use of non-biologists to gather scientific data has been criticized. It should be noted that all observers (whether volunteers or not) have the potential for adding both bias and unwanted variance to counts of calling frogs. At issue here, however, is what additional bias or variance volunteers and laypeople bring to a system over and above the paid technician.

Based on a small sample survey of participants in the Wisconsin Frog and Toad Survey, most observers reported that they had no or minimal prior experiences with amphibian identification (Mossman et al., 1998). While demographics in most other regions are unstudied, there is reason to suspect that calling surveys for amphibians will be more attractive to non-professional biologists than are bird surveys—amphibian vocalizations are simpler and there are fewer species in any given region (roughly 7–30 for amphibians versus 100–225 for birds).

Three independent studies (Shirose et al., 1997; Hemeseth, 1998; and Kline, 1998) have investigated the ability of previously untrained lay-observers to learn and identify calling frogs. In Kline's (1998) study, students in grades 5–8 (note that almost no students in this age group are currently participants in calling surveys in most states and/or provinces) listened to a training tape of calling amphibians and went out on one practice run prior to being tested. Groups of students were sent to a set of ponds during a five-day interval and then asked to record species and calling intensity level. The results demonstrated that variation among observers was less than the variation that occurred from day to day. The ability to detect and distinguish species was not reported. Hemeseth's (1998) study demonstrated that variance in species identification among observers was low, but that assignment of calling intensity showed high

variance for some species. Shirose et al. (1997) demonstrated similar results in Ontario, with no significant differences between raw recruits and professionals in species identification, but variation in assignment of calling intensity between both participant groups.

We can expect volunteers to have longer tenure on routes than technicians, which decreases observer turnover and consequently improves the statistical adjustment of counts for observers (Sauer et al., 1994; Mossman et al., 1998). Note too that many technicians start on the job at fundamentally the same place that the volunteer does—that is, with no prior knowledge of the calls of amphibians. Finally, it is unlikely that most of the existing monitoring programs could afford to pay technicians to run surveys. With training, tests, and the vetting of the resulting data, the quality assurance of volunteer-collected data can be safeguarded. An Internet-based amphibian call identification quiz has been developed by the USGS (http://www.pwre.usgs.gov/frogquiz/), which will help volunteers to improve their identification skills and help us to assess volunteer identification abilities.

Conclusions and Summary

While calling surveys permit a very coarse estimation of population increases and decreases at landscape levels, they can also be used to assess whether the number of sites occupied by amphibians are increasing or decreasing. However, once validation studies are completed, calling surveys likely will be found to overestimate rates of declines. Calling survey programs have sufficient power to detect changes in amphibian species and sample sizes do not need to be extremely high in order to detect trends for some species. Trained volunteers and laypeople are appropriate and economical choices for running calling surveys. Additional research on calculation of detection rates for these surveys will help diminish bias and increase the uses of these data.

Acknowledgments. We thank Mike Mossman and Lisa Hartman for so readily sharing their data from the Wisconsin Frog and Toad Survey, Linda Weir for her extensive help structuring and editing the manuscript, John Sauer for providing analyses of population indices of Wisconsin frogs and toads, and Jim Nichols for his review and comments of this manuscript.

FORTY-SEVEN

Geographical Information Systems and Survey Designs

CHARLES R. PETERSON, STEPHEN R. BURTON,
AND DEBRA A. PATLA

The availability and utility of Geographical Information Systems (GIS) has increased greatly within the past 20 years. For general descriptions of GIS, see Clarke (1997), Heywood et al. (1998), and Krzysik (1998a). In the past, GIS required expensive workstations, software that was difficult to use, expert technicians, and considerable resources for acquiring spatial data in a digital format. Now, systems using relatively inexpensive desktop computers and programs (e.g., ArcView) have the ability to perform many workstation GIS functions. The software is easier to use, training is widely available, and large amounts of spatial data in digital format can be obtained on CD ROM and over the Internet. For many survey projects, it is now expected that the results will be provided in a format that can be readily brought into the sponsoring organization's GIS. The role of GIS in designing animal surveys, analyzing survey results, and sharing information will undoubtedly continue to increase in the future.

The objective of this paper is to introduce (rather than comprehensively describe) how to use a GIS for designing amphibian surveys. Although several amphibian studies have used GIS for measuring landscape variables (e.g., Richter and Azous, 1995; Bosakowski, 1999; Diller and Wallace, 1999; and Knutson et al., 1999), relatively little has been written about how to use GIS to plan amphibian surveys (but see Hayek and McDiarmid, 1994; Fellers, 1997). Here we indicate what is possible, how to get started, some limitations, and where to go for further information. Examples are taken primarily from our experiences using GIS to design amphibian surveys in the Northern Intermountain West for a variety of state and federal agencies, private corporations, and conservation organizations.

GIS can aid in designing amphibian surveys in several ways. A GIS can give biologists a better understanding of the landscapes in which they will be working (e.g., topography, geology, hydrology, climate, vegetation, and roads). It facilitates the organization and visualization of pre-existing amphibian data and, thus, helps to identify data gaps. It can incorporate a variety of constraints (e.g., access) into the site selection process. Probably most importantly, a GIS can help identify areas of suitable amphibian habitat for sampling. If the appropriate data are available, a GIS can be used to select sampling sites in an efficient and less biased way (e.g., implementation of systematic, random, and/or stratified sampling approaches). In addition to helping with the design of surveys, GIS also can play an important role after the surveys are completed, through visualizing and analyzing results and modeling the effects of habitat change or management actions.

The task of developing an appreciation for the characteristics of a study area has been greatly facilitated by the increasing availability of GIS data (see below and Appendix 47-A). Much of this information, such as USGS topographic maps, aerial photographs, and National Wetland Inventory (NWI) maps (Cowardin et al., 1979), has been available in non-digital format for a number of years. After being digitized, this information can be layered in a GIS with other data, which provides a better understanding of the landscape in a study area. For instance, aerial photos in the form of DOQQs (digital ortho quarter quads) are useful in identifying topographic features. Color DOQQs are available in many areas of the country making it easier to identify potential wetalnds. The availability of other types of GIS data, such as vegetation, soil type, precipitation, and temperature, makes it more feasible to incorporate these variables into a survey design. Until recently, detailed GIS data availability for regions was limited, but now it is common for an organization such as a national forest or timber company to have an extensive GIS database available. Existing data can be used to derive data layers (e.g., slope and aspect from Digital Elevation Models [DEMs]). Spatial relationships, such as the isolation of wetlands from other wetland habitat or the distance of sites to roads, can also be determined through the use of GIS. Available GIS data are listed as follows (many sources are available via the Internet; see Tabel 47-1 for URLs of selected sources):

Digital Line Graphics (DLG)

Digital Elevation Models (DEM)

Digital Raster Graphics of USGS Topographic Maps (DRG)

Digital Orthophoto Quadrangles (DOQ) or Quarter Quadrangles (DOQQ)

National Wetland Inventory (NWI) classification maps

Satellite Imagery (Landsat, SPOT, etc.)

Temperature

Precipitation

Selected Internet GIS Data Sources and the Data They Provide

Data Source	Available Data	URL
U.S. Geological Survey (USGS) http://www.usgs.gov/	DRG[a] Topos: 1:250K, 1:100K, 1:63K, 1:20–30K	http://mcmcweb.er.usgs.gov/drg/
	DLG[b]: 1:2000K, 1:100K, 1:20–25K	http://edcwww.cr.usgs.gov/nsdi/gendlg.htm
	DOQ[c]: Variable availability (50,000 available), remaining available by 2004 1–2 resolution	http://mapping.usgs.gov/www/ndop/
	DEM[d]: 2-Arc seconds, 1-degree, 1:20–25k (10 or 30 m resolution)	http://mcmcweb.cr.usgs.gov/status/dlgstst.html
	Geology and mineral resources	http://mrdata.usgs.gov/index.html
	Hydrography: Hydrological unit codes, rivers, streams, lakes, springs, wells	http://water.usgs.gov/GIS/
	Land Cover Resolution: 30 m	http://edcwww.cr.usgs.gov/programs//ccp/nationalland cover.html
U.S. Environmental Protection Agency	Various data at the state and county levels: census, TIGER[e] roads, TIGER railways, TIGER political boundaries, TIGER hydrography	Enviromapper for Watersheds http://www.epa.gov/iwi/iwimapper/
National Wetland Inventory Maps	Digital data available for 39% of conterminous U.S. Resolution: 1/4–5 acres	http://wetlands.fws.gov/
National Land Cover Data http://edcwww.cr.usgs.gov/programs /lccp/index.html	Land cover: conterminous U.S. Resolution: 30 m	http://edcwww.cr.usgs.gov/programs/lccp/mrlcreg.html
National Gap Analysis Program	State land cover, predicted species distribution	http://www.gap.uidaho.edu/
National Geophysical Data Center	Various: Satellite imagery, glaciology, climate, topography, geothermal, habitat, and ecosystems	http://www.ngdc.noaa.gov
PRISM Climate Mapping	Precipitation, humidity, temperature	http://www.ocs.orst.edu/prism/prism_new.html
GIS Data Depot	DRG Topos, DOQQ[f], DEM, hydrography, NWI[f], land use/land cover, political boundaries	http://www.gisdatadepot.com/
ESRI ArcData Online	DRG Topos: Aerial photos (limited availability)	http://www.esri.com/data/online/index.html
Sure!Maps Raster	DRG Topos, DEMs	http://www.suremaps.com/suremaps
All Topo Maps	DRG Topos: Limited (15 states)	http://www.igage.com/
National Spatial Data Infrastructure	Geospatial data clearinghouse	http://www.fgdc.gov/nsdi/nsdi.html
Rocky Mountain Mapping Center	More links	http://rockyweb.cr.usgs.gov/outreach/rockylinkdata.html

[a]Digital Raster Graphics, [b]Digital Line Graphics, [c]Digital Orthophoto Quadrangle, [d]Digital Elevation Model, [e]Topologically Integrated Geographic Encoding and Referencing, [f]Digital Orthophoto Quarter Quadrangle

NOTE: Although Internet addresses are included, some sites may move to a new address or become obsolete. We suggest using search engines to search on "GIS Data" "Geographic."

Climate

Soil

Vegetation or Land Cover

A GIS facilitates the organization and visualization of pre-existing data and, thus, helps to identify data gaps. One of the first steps in planning an amphibian survey usually is to determine which species of amphibians possibly occur within a study area. This is necessary to ensure that the appropriate sampling techniques are selected so that some species are not missed. Typically, this has involved examining field guides,

published literature, and museum collection records. Other sources of information include The Association for Biodiversity Information (http://www.abi.org), state Gap Analysis projects (http://www.gap.uidaho.edu), and state or regional herpetological databases. Incorporating these data into a GIS requires that they be organized into a tabular format and that locality descriptions be converted into digital coordinates (e.g., decimal degrees and Universal Transverse Mercator coordinates; see Krzysik, 1998a). In the past we have used written descriptions from observations and museum specimen records to find locations on paper topographic maps. We then determined the coordinates using a digitizing table. Each record location was

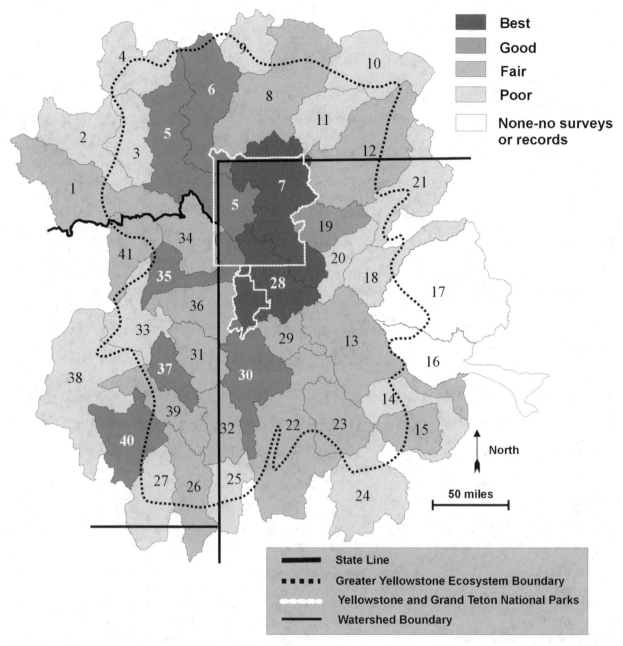

Best
Good
Fair
Poor
None-no surveys
or records

North

50 miles

State Line
Greater Yellowstone Ecosystem Boundary
Yellowstone and Grand Teton National Parks
Watershed Boundary

FIGURE 47-1 Relative information availability for amphibians in the Greater Yellowstone Ecosystem (GYE). Numbered watersheds are USGS 4th-level hydrological units. Information availability was based on the number and intensity of surveys within a watershed. Our ratings system found that 17 of 41 watersheds (41%) have little or no information on amphibian occurrence. The best information exists for watersheds in the core of the GYE. (Figure from Van Kirk et al., 2000)

assigned an accuracy rating (in meters) that represented the radius of a circle around the digitized point in which we were highly confident that the actual specimen location occurred. We now determine coordinates on screen, directly from digital topographic maps (Digital Raster Graphics [DRGs]), which is easier, less expensive, and more accurate. These locality data can be used to generate point distribution maps for individual species. We have found that summarizing previous survey efforts by watershed can provide a way to access information availability across a large region (Fig. 47-1). This helps determine if a lack of locality data for a species is due to species rarity or possibly to insufficient survey effort. Pre-existing data can be especially valuable for determining changes in occurrence and distribution. However, they are usually less useful for

determining abundance and habitat relationships because of differences in the way the data were obtained in comparison with current surveys and the lack of negative data (i.e., sites without observations or specimens not described).

Identifying areas of suitable amphibian habitat for sampling is potentially the most important use of a GIS for designing amphibian surveys. Because amphibians generally use their environments at a relatively fine spatial scale (Inger, 1994; Bartelt, 2000), data that can distinguish among various habitat features (e.g., water, soil, and vegetation) are required to accurately indicate appropriate habitat. Distinguishing among small habitat features (such as ephemeral wetlands) often requires data with both high spatial and high spectral resolution (many narrow radiation bands versus a few broad bands). High spatial resolution is required to delineate the features; high spectral resolution is required to distinguish among different wetland types. Unfortunately, data with fine enough resolution are rarely available. *The lack of high resolution data is probably the single most important limiting factor in using a GIS to design, analyze, and apply the results of amphibian surveys.* For example, we discovered that by using a GIS to design an amphibian survey for Potlatch Corporation in northern Idaho, we missed many sites that were important to species such as long-toed salamanders *(Ambystoma macrodactylum)* and Pacific treefrogs *(Pseudacris regilla).* This occurred because many of the small, temporary wetlands simply were not represented in the GIS (G. B. Hamilton, C. R. Peterson and W. A. Wall, unpublished data).

The lack of sufficient resolution data can sometimes be partially addressed by combining different types of spatial data or by modeling. For example, we evaluated the ability of different GIS data sources to predict ponds within portions of Caribou National Forest. The digital data included USGS DRGs, aerial photos (DOQQ), NWI maps, hydrology (streams), and DEMs (used to model low slope areas where water might accumulate). We found that individually these data sources would either underpredict the number of ponds (DRG, DOQQ, NWI) or overpredict, indicating areas too extensive to efficiently survey (DEM). By combining these data sets, we were able to predict correctly most of the wetland sites (> 94%) while minimizing the total area to be searched. Other modeling approaches also exist for predicting the occurrence of wetlands. For example, in the Spatial Analyst module of ArcView, DEMs can be used to model water flow and accumulation within a study area.

New types of remotely sensed data such as high spatial resolution, hyperspectral imagery (Jia and Richards, 1999) may prove useful for generating detailed wetland habitat maps in areas with low canopy cover. Results from a preliminary study using high spatial resolution, hyperspectral imagery to map amphibian habitat in Yellowstone National Park have been promising. Using 5 m-resolution hyperspectral images obtained with a PROBE 1 sensor (128 spectral bands) on a fixed-wing aircraft, we were able to generate predicted maps of amphibian habitat (sedge wetlands and pools with algae) that were relatively accurate (about 85% correct classification of 32 sites; C. R. Peterson and W. A. Marcus, unpublished data). We found two new western toad *(Bufo boreas)* sites and five new Columbia spotted frog *(Rana luteiventris)* sites. We probably would not have identified several of these sites if we had only been using conventional techniques (e.g., topographic and NWI maps).

GIS can be very useful for incorporating a variety of constraints into the selection of sampling sites. If a study area is small enough or if sufficient resources are available, constraints are less of an issue and a complete survey (census) of all wetland habitats is possible. Unfortunately, most studies are faced with numerous constraints making it necessary to subsample within a study area. GIS can assist by incorporating many of these constraints into a design a priori. These constraints can arise from several sources, including limited resources, the topography of the study area, road access, the presence of other wildlife, and restrictions due to other uses (e.g., tourism). For example, to avoid spending excessive time traveling to sites, researchers may limit sampling to areas within 5 km of a road or trail. In Yellowstone National Park, we are not allowed to conduct surveys in grizzly bear management areas during certain times of the year. National Parks may also restrict sampling to areas out of the view of the public. Determining viewsheds (polygon maps

resulting from a visibility analysis showing all the locations visible from a specified viewpoint [Heywood et al., 1998]) in a GIS can be used to help find such areas. Viewshed analysis could also be used to select trapping sites that would be less likely to be disturbed by vandals.

If the appropriate data are available, a GIS can be used to select sampling sites in a more efficient and less biased way. Using systematic, random, and/or stratified sampling approaches to select amphibian sampling sites is described in Hayek (1994), Fellers (1997), and Thompson et al. (1998). Systematic schemes such as a grid can be imposed on a study area (see Yellowstone case study below). While random selection of sampling areas or sites does not require previous knowledge of amphibian habitat relationships, it may require many sites to adequately describe the distribution of available areas or sites. Options for randomly selecting sites within GIS thematic layers exist in programs such as ArcInfo and the Animal Movement extension (Hooge and Eichenlaub, 1997) for ArcView. However, even with random schemes, sampling areas will typically be constrained to aquatic and wetland areas to avoid inefficiently spending large amounts of effort to determine where amphibians do not occur. GIS layers and their attributes can be used to efficiently stratify site selection but require considerable previous knowledge of amphibian habitat relationships to avoid biasing the results. If the required habitat information is already in the GIS, the sampling scheme needs to incorporate the range of environmental variation, but it does not need to determine the actual distribution of those variables. Gradsect analysis provides a way to sample the range of environmental variation within the study area using two key habitat variables (such as temperature and precipitation) to define a two-dimensional environmental space (Austin and Heyligers, 1991).

All these approaches may be combined and applied at multiple spatial scales. For example, a systematic grid may be used to spatially distribute sampling effort at a broad scale. At an intermediate spatial scale, watersheds within a specific grid cell may be selected randomly. Within a watershed, sampling areas could be selected randomly from stratification categories (e.g., ponds and stream reaches) or, if the area is small enough, complete surveys could be conducted.

How a GIS is used to select sampling sites is likely to vary with the objectives of the survey (i.e., determining presence, distribution, abundance, habitat relationships, or status/trends). To simply determine occurrence within a study area (e.g., a national park), using a GIS to focus first on those habitats most likely to be occupied by amphibians would be the most appropriate approach. Selecting sampling sites with a systematic scheme might be the best approach to determining the spatial distribution of amphibians within the study area. A study of habitat relationships should minimize bias and include sites where the species does and does not occur, requiring some random component to the site selection. Including sites with historical data would be appropriate for a survey attempting to determine species status and trends in the study area. Complete surveys will probably need to be conducted to evaluate the probability of amphibians persisting within a study area (Pilliod and Peterson, 2000).

Spatially defining a site using a GIS may vary in difficulty depending on the landscape in which it occurs. Isolated, individual wetland sites are relatively easy to delineate; wetland complexes of ponds, streams, and wet meadows are more difficult. However, several studies indicate that amphibians may move considerable distances (e.g., kilometers; Berven and Grudzien, 1990; Reaser, 1996b; Seburn et al., 1997; Bartelt, 2000;

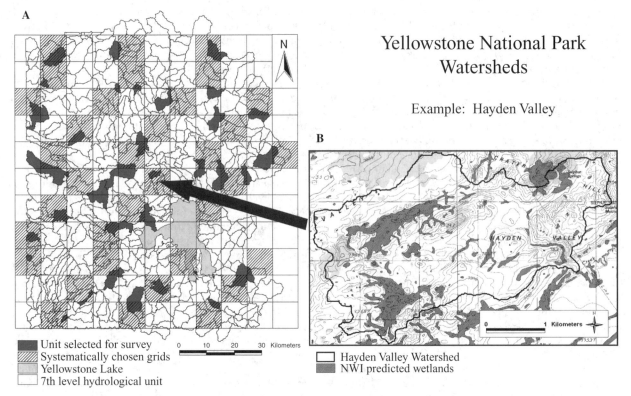

A

Yellowstone National Park Watersheds

Example: Hayden Valley

B

Unit selected for survey
Systematically chosen grids
Yellowstone Lake
7th level hydrological unit

0   10   20   30 Kilometers

Hayden Valley Watershed
NWI predicted wetlands

0   1 Kilometers

FIGURE 47-2 Methods for survey site selection within Yellowstone National Park. **(A)** A 10 × 10 km grid was overlain on the 7th-level hydrological units and numbered. A starting grid was randomly selected and every third grid was systematically chosen thereafter. A subwatershed within each selected grid was randomly chosen for surveying. **(B)** Within a selected subwatershed, all possible wetlands were identified for survey by using 1:24,000 USGS topographic maps and National Wetlands Inventory (NWI) classifications.

Pilliod, 2001), suggesting that sites should be more widely delineated to be biologically meaningful. Selecting sampling sites in extensive habitats such as large marshes is especially difficult and probably will require subsampling using a grid system or transects rather than a complete survey.

In addition to helping with the design of surveys, GIS can play important roles after the survey data are gathered. A GIS can be used for visualizing and analyzing the results of the survey (e.g., viewing distributional patterns, determining distance between survey sites to a nearest feature, calculating landscape variables, and assessing habitat associations), and it is useful for modeling the effects of habitat change and management activities. A GIS can be used for effectively communicating information to managers and policy makers (e.g., maps to supplement oral presentations or as graphics in a report [Peterson, 1997]). GIS data can also be integrated into an organization's information system so that data will be more readily available when it is needed for planning purposes. Spatial data and maps can also be distributed to the public over the Internet with programs such as ArcIMS (ESRI, 1999).

Yellowstone Case Study

We provide here an example of using GIS in a sampling scheme with multiple spatial scales. In this case, the objective was to describe the status and trends of amphibians in Yellowstone National Park (YNP) for the U.S. Department of the Interior's national Amphibian Research and Monitoring Initiative (ARMI). One of the objectives of the ARMI project is to document the *presence/absence* (or *detected/not detected*) status of local or breed-

ing populations so that the net gain or loss of such populations over time can be used to discern declines or other trends. One approach is to conduct surveys of all habitats within randomly selected drainages or geographically delimited areas in large parks or tracts of public land. GIS provided us with tools to select sampling areas spatially distributed across YNP's vast area (2.2 million acres) while incorporating random selection to ensure that survey results could be extrapolated to the entire park. GIS also allowed the identification of potential amphibian habitat within the selected units.

To first select sampling areas with geographic representation of the park's area, we stratified the park into sampling blocks by placing a 10 × 10 km grid over YNP using an ArcView layer constructed for this purpose. YNP has approximately 107 blocks in this grid. We consecutively numbered the blocks, starting from the upper left-hand corner. We then randomly selected a number from 1–3 and, starting from the block with the selected number, took every third square as a sampling block.

Second, to randomly select study areas, we acquired a GIS layer depicting hydrological units (HU), also known as watersheds. The 7th-level HUs (provided by YNP's GIS department) are sub-watersheds nested within higher-level HUs defined and used by national and state agencies. The 7th-level HUs range in size from about 1–60 km^2, covering stream or river segments and the topographic area drained by these segments. We overlayed these watershed boundaries on the grid of 10 × 10 km squares. Within each square selected as a sampling block, we randomly selected one watershed unit from the list of units (listed by identification numbers) falling within the sampling block (Fig. 47-2a).

Finally, to target areas within the units for field surveys, we imported digital 1:24,000 topographic maps into our GIS and superimposed layers showing the selected 7th-level watershed unit boundaries and NWI maps (Fig. 47-2b). This made it possible to pre-determine the presence and location of potential amphibian breeding habitat (ponds, lakes, and wetlands) in the selected units. Field crews were then instructed to visit all these potential habitat areas. Recent advances in digital mapping and GPS (global positioning system) technology make it easy to upload the coordinates of targeted areas from a GIS into a GPS unit for navigation to an identified location in the field. Furthermore, GPS can provide relatively precise locality information and a means of recording the locations of breeding sites and apparently suitable but unoccupied habitat.

Our initial tests of this method of area and site selection for field surveys revealed the need for better tools to identify potential amphibian habitat. Wetlands depicted by NWI and 1:24,000 topographic maps do not always correspond to existing wetland features, due in part to errors in mapping or changes that have occurred since mapping was completed. In addition, many areas defined as wetlands lack features suitable for the pond-breeding amphibians we are seeking, and some areas where amphibians breed are too small for NWI to recognize. With development of higher resolution and more accurate GIS layers, complete surveys of watersheds (or other geographically delineated areas) will become more feasible and practical. In the meantime, a combination of existing GIS mapping tools with on-the-ground searches for additional suitable habitat provides a means of documenting amphibian population occurrence and establishing baselines.

Data from the YNP watershed surveys are entered in a database that keeps records of sites surveyed (whether or not amphibians were present) and sightings of amphibians, including data on life stages. Brought in as a GIS layer, these data depict the distribution and abundance of breeding populations within the watershed units. When surveys of all the selected watershed units are completed, we will be able to describe the status of amphibian species in YNP in quantitative, distributional, and relative terms. In 5, 10, or 20 years, when similar surveys are repeated, it should be possible to distinguish whether or not significant changes have occurred in amphibian species richness and distribution.

Summary

A GIS facilitates the organization and visualization of data, and can give biologists a better understanding of their landscapes, as well as identify data gaps. The objective of this paper was to introduce how to use a GIS to design and undertake amphibian surveys. Using Yellowstone National Park as an example, we used GIS to select sampling areas using randomized techniques, and to select watersheds within these areas. Within these watersheds, we then used GIS to visualize potential amphibian pond breeding habitats for field sampling. Results from this sampling effort were then organized as a GIS layer depicting the distribution and abundance of amphibian breeding populations. The GIS layer created from these data has become the baseline for comparisons with future surveys.

Acknowledgments. Stephen Corn, Ted Koch, John Lee, and David Pilliod helped develop the ideas presented here. Paul Bartelt, Alisa Gallant, Jason Jolley, Robert Klaver, Bryce Maxell and Brian Smith reviewed an early version of the manuscript. Ann Rodman and Shannon Savage provided the 7th-level Hydrological Unit GIS layer in draft form for our use. Yellowstone Ecosystems Studies generously supported the hyperspectral analysis study.

FORTY-EIGHT

Impacts of Forest Management on Amphibians

ROCHELLE B. RENKEN

Researchers with the Missouri Department of Conservation, the Missouri Department of Natural Resources, the U.S. Forest Service, and cooperating universities (University of Missouri–Columbia, University of Missouri–St. Louis, Michigan Technological University, and University of Tennessee–Chattanooga) are conducting a bold experiment to examine the long-term, large-scale impacts of forest management practices on the biotic and abiotic components of the oak-hickory forests of southern Missouri. Entitled the Missouri Ozark Forest Ecosystem Project (MOFEP), the project's purpose is to conduct a controlled experiment to document the effects of typical forest management practices on the numbers and types of forest plants and animals, including amphibians (see Renken et al., 2004). In this experiment, we examine the impact of the forest management practices of clearcutting (technically called even-aged management) and select tree cutting (called uneven-aged management). MOFEP is unique both in its duration and its scale—data will be collected for the next 100+ years and on the landscape scale of 1,000-acre forests.

The basic design of MOFEP is to use nine forests scattered in Reynolds, Carter, and Shannon counties in southern Missouri. Each of these nine forests is about 1,000 acres in size. When the MOFEP project began in 1990, each forest was covered by 60–80-year-old oak and hickory trees, as well as other tree species typical of Missouri's hardwood forest. In 1990, each forest was randomly assigned a management fate. Three of the nine forests would never have trees cut during the next 105 years. For three others, the forests would be managed through the use of clearcuts. The remaining three forests would be managed through the use of select tree cutting methods during the duration of the experiment.

For several reasons, drift fence arrays were chosen as the MOFEP technique for trapping amphibians. First, arrays would not continually disturb the forest and compromise the sampling protocols of other MOFEP researchers. Second, because array trapping is a passive sampling technique, trapping efforts would not be biased by the abilities of the hundreds of technicians who will be working on MOFEP during its lifetime. Third, once arrays are installed they allow sampling at that same spot throughout MOFEP's duration. While arrays are not the best technique for catching some amphibians (such as treefrogs), no

other single trapping technique is as effective when considering the entire amphibian assemblage.

Each array consists of three 7.5-meter-long (25-foot-long), 61-centimeter-high (24-inch-high) aluminum fences, buried 4–6 inches in the ground and placed to form a Y-shaped design (Fig. 48-1). Pitfall traps (five-gallon buckets) are positioned at the junction of the three fences. Funnel traps (constructed from aluminum window screen wire) are placed along the sides and at the ends of the fences. As the animals crawl along the forest floor and bump into the fence, they can not climb over nor go through the fence, so they crawl along the fence until they fall into the pitfall trap or crawl into the funnel trap. Once caught, an animal is identified to species, measured, and marked. After processing, the animal is released outside the perimeter of the array. An array is located at 12 permanent sample points within each of the nine forests.

To collect pre-treatment data, amphibians were trapped from March–June and September–October from 1992–95. Because of logging activity, amphibians were not collected in 1996 and early 1997. Following logging, post-treatment data were collected when trapping resumed in the fall of 1997. Trapping continued every year through 2001; from that point forward, trapping will then be conducted periodically during the next 100 years.

At this writing, it is still too early to determine what effect the cutting of trees in both the clearcut and select tree cutting forests has had on amphibians. During the pre-treatment period (1992–95), we captured at least one individual from all but two species of amphibians that might inhabit that region of Missouri. We probably did not capture ringed salamanders (*Ambystoma annulatum*) or wood frogs *(Rana sylvatica)* because the MOFEP forests are at the edge of those species' range within Missouri. The following 19 species made up the approximately 3,500 amphibians that were caught each year (Table 48-1).

During 1992–95, two species of amphibians were widespread across the nine forests. Spotted salamanders *(A. maculatum)* and American toads *(B. americanus)* were captured at all the arrays. These species were also some of the most abundant species captured. American toads were captured at a rate of at least one animal for every 100 days the arrays were open. Spotted salamanders were a slightly less abundant species. They were captured at a rate of at least one animal for every 200 days

TABLE 48-1

Amphibians captured on the MOFEP sites from 1992–97

Salamanders

Central newts	*Notophthalmus viridescens louisianensis*
Spotted salamanders	*Ambystoma maculatum*
Marbled salamanders	*Ambystoma opacum*
Eastern tiger salamanders	*Ambystoma tigrinum tigrinum*
Long-tailed salamanders	*Eurycea longicauda*
Cave salamanders	*Eurycea lucifuga*
Four-toed salamanders	*Hemidactylium scutatum*
Western slimy salamanders	*Plethodon albagula*
Southern red-backed salamanders	*Plethodon serratus*

Frogs/Toads

Eastern narrow-mouthed toads	*Gastrophryne carolinensis*
Eastern American toads	*Bufo americanus americanus*
Fowler's toads	*Bufo fowleri*
Blanchard's cricket frogs	*Acris crepitans blanchardi*
Gray treefrogs	*Hyla chrysoscelis-Hyla versicolor complex*
American bullfrogs	*Rana catesbeiana*
Northern green frogs	*Rana clamitans melanota*
Pickerel frogs	*Rana palustris*
Southern leopard frogs	*Rana sphenocephala*
Northern spring peeper	*Pseudacris crucifer crucifer*

the arrays were open. Salamanders, as a group, were our most abundant captures. On average, we captured 5–15 salamanders for every 100 days the arrays were open in each forest.

Summary and Additional Information

The Missouri Ozark Forest Ecosystem Project (MOFEP) purpose is to conduct a controlled experiment to document the effects of typical forest management practices on the numbers and types of forest plants and animals, including amphibians. At this writing, it is still too early to determine what effect the cutting of trees in both the clearcut and select tree cutting forests has had on amphibians. If you desire further information about the MOFEP project, consult "Proceedings of the Missouri Ozark Forest Ecosystem Project: an experimental approach to landscape research" (General Technical Report NC-193), 1997. This publication can be obtained through the U.S. Forest Service, Department of Agriculture, 1992 Folwell Avenue, St. Paul, Minnesota, 55108, or visit the MOFEP Web site at http://www.snr.missouri.edu/mofep/.

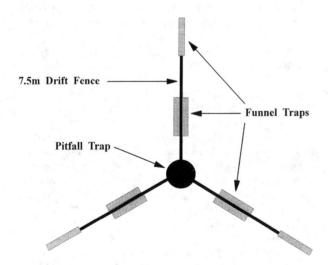

7.5m Drift Fence

Funnel Traps

Pitfall Trap

FIGURE 48-1 The design of a MOFEP amphibian sampling array with three drift fences, a pitfall trap, and nine funnel traps.

FORTY-NINE

Monitoring Pigment Pattern Morphs of Northern Leopard Frogs

ROBERT G. MCKINNELL, DAVID M. HOPPE,
AND BEVERLY K. MCKINNELL

This study has emerged from the coupling of exceedingly rare events. Here we report on the progeny of northern leopard frogs *(Rana pipiens)* that have undergone mutations affecting their pigment patterns. These extraordinarily uncommon pattern mutations have increased to polymorphic frequencies in populations of frogs in the upper Midwest. The vast majority of mutations are lost, and the success of mutant genes (i.e., their increase and retention in a population) is in itself a rare event (Crow and Kimura, 1970). However, the rare burnsi and kandiyohi mutations have indeed become polymorphic in Minnesota and its contiguous states. We cannot address the cause of these unlikely mutational events nor their success, but we can and do record the results. It is fair to ask if the frequency and distribution of pigment pattern morphs resulting from past mutation have any relevance to contemporary problems in amphibian population declines. We believe that they do in that the presence and stability of these morphs become a marker for the health of the much more widely distributed wild-type frogs, and a model system for studying the effects of genetic bottlenecks that occur in declining species.

Northern leopard frogs (Fig. 49-1A) are distributed in the United States from New England into parts of New York, Pennsylvania, Indiana, and northern Kentucky, covering all of Ohio and the upper Midwest to the Rocky Mountains, and also into much of the Southwest. They are found in Canada in areas contiguous to their distribution in the United States, from the Canadian provinces of New Brunswick, Nova Scotia, and Prince Edward Island in the east to portions of Alberta and British Columbia in the west, a distance of over 4,000 km (Pace, 1974; Cook, 1984; Conant and Collins, 1991; Behler and King, 1995; Harding, 1997).

Northern leopard frogs with the burnsi allele are characterized by a severe reduction or complete absence of spotting on the dorsal body surfaces (Fig. 49-1B). This "spotless" frog was originally described as a new species (*R. burnsi*; Weed, 1922), but was subsequently shown to differ from the wild-type by a single allele dominant to the wild-type spotting (Moore, 1942). Kandiyohi frogs have vermiculated mottling between the dorsal spots on their bodies and limbs (Fig. 49-1C). They were also described as a new species (*R. kandiyohi*; Weed, 1922), but were later shown to differ from wild-type frogs by a dominant allele

(Volpe, 1955a). The fact that these mutant alleles are dominant rather than recessive is of genetic interest in itself.

Variants of northern leopard frogs lacking dorsal spots have been reported from many locations within the range of the species, but they are known to occur in polymorphic frequencies only in the northern United States in an area surrounding central Minnesota (Breckenridge, 1944; Volpe, 1961; Merrell, 1965, 1970; Oldfield and Moriarty, 1994). (The limited area of Minnesota and its contiguous states occupied by burnsi and kandiyohi frogs is larger than England, yet represents a greatly restricted range within widely distributed, monomorphic northern leopard frogs.) *Polymorphism* was defined by Ford (1945) as the "occurrence together in the same habitat of two or more distinct forms of a species in such proportions that the rarest of them cannot be maintained by recurrent mutation." Clearly, both burnsi and kandiyohi may be considered as polymorphic variants of northern leopard frogs in parts of Minnesota and contiguous states, typically occurring at frequencies of 1% to more than 20%. Burnsi and kandiyohi may be examples of either transient or balanced polymorphism. Volpe (1961) defines the transient form as "a temporary state occasioned by the displacement of a gene by a more favorable allele." Balanced polymorphism is, according to Volpe (1961), a "condition when a heterozygote is superior to both of its corresponding homozygotes." Dobzhansky (1951), however, more broadly considers balanced polymorphism as maintenance of relatively unchanged frequencies, with heterozygote superiority as only one mechanism for such maintenance.

Merrell (1965) reported that the highest frequencies of burnsi were found in east-central Minnesota, in an area known as the Anoka Sand Plain. Even there, frequencies of burnsi frogs were low (4.0–7.1%) and appeared to decrease in all directions from that area, with burnsi polymorphism disappearing in eastern South Dakota, northern Iowa, western Wisconsin, and northern Minnesota. Exceptions to this generalization include populations in Stevens County, Minnesota (9.1%), and Kingsbury County, South Dakota (10.3%). However, these frequencies were calculated from collections of fewer than 30 frogs each. McKinnell and McKinnell (1967), McKinnell and Dapkus (1973), and Hoppe and McKinnell (1991) updated the distribution records for burnsi and kandiyohi, and in the present report we extend distribution and prevalence records farther

FIGURE 49-1 (A) Northern leopard frog *(Rana pipiens)* with typical spots on the dorsum of the legs and body. The frog was collected in Otter Tail County, Minnesota, on 12 August 1996. (B) The burnsi morph of northern leopard frogs *(Rana pipiens)* showing a lack of dorsal spotting on the body and legs. The frog was collected in Beadle County, South Dakota, on 25 July 1996. (C) The kandiyohi morph of northern leopard frogs *(Rana pipiens)* collected in McIntosh County, North Dakota, on 29 September 1999. The frog has mottled patches of pigment between the spots on the dorsal aspects of the body and legs.

C

FIGURE 49-1 Continued

west into western Minnesota and eastern North Dakota and South Dakota. Of particular interest in our study is a site in Otter Tail County, Minnesota, with a burnsi prevalence far in excess of any site previously reported. This site is 175 km from the approximate center of the Anoka Sand Plain.

In examining data from 1967–99, including the phenotypic classification of 3,643 frogs, we observed that the burnsi frequency in one population (Block Lake, Otter Tail County, Minnesota) has undergone a statistically significant increase and subsequently remained elevated for a long time. Furthermore, we see neither uniformly low frequencies nor gradually lower frequencies at greater distances from previously reported sites of high burnsi frequency (Anoka Sand Plain and southeastern Otter Tail County) of the burnsi range. We report both temporal and geographic differences in the frequencies of the burnsi morph and add 17 counties to the previously known distribution of burnsi frogs and seven counties to the distribution of kandiyohi frogs.

It is not known what restricts the burnsi and kandiyohi pigment pattern polymorphism to only a small fraction of the range of northern leopard frogs. However, the documentation of geographic and temporal differences in their prevalence may ultimately permit the identification of factors permissive to the reproduction, distribution, and survival of these interesting morphs in a circumscribed and limited area in the range of northern leopard frogs. Despite their limited distribution, these morphs, especially the more common burnsi form, have become important research subjects in a variety of investigations.

Following the genetic studies of Moore (1942) and Volpe (1955a), much burnsi and kandiyohi research dealt with maintenance of the polymorphism. Merrell (1970) concluded that frequency differences due to genetic drift, expected in populations of low effective breeding size, are not seen for burnsi. He

reported the gene to be relatively uniformly distributed within its range at low frequencies, with the exception of some Anoka Sand Plain populations. One possible explanation in support of Merrell's view concerning genetic drift is that natural selection affects burnsi frequencies, perhaps maintaining the polymorphisms through seasonal selection. Merrell (1965) found that the number of dorsal spots on wild-type northern leopard frogs was lower within the range of the burnsi morph than either to the north or south of the range. He suggested that selection may favor less spotting in that area through reduced spotting in wild-type frogs and increased burnsi prevalence. Low heritability estimates for dorsal spotting ($h^2 = 0.14$) would allow for only very small or slow selection effects through the first mechanism (Underhill, 1968a). Dapkus (1976) found that burnsi frogs survived stressful captive conditions better than wild-type frogs. Merrell and Rodell (1968) reported an apparent winter survival advantage of burnsi morphs. However, a balancing superiority of wild-type frogs over burnsi during the summer has not been demonstrated.

Alternatively, uniformly low frequencies could be maintained by migration of frogs throughout the burnsi range. Merrell (1970) described the movement patterns of northern leopard frogs that lead to a wide dispersal of individual frogs and may thus prevent local differences in burnsi frequency from being established. He suggested that the low frequencies of burnsi made migration a more reasonable hypothesis than selection for counteracting drift and producing uniformly low frequencies throughout the burnsi range.

Both burnsi and kandiyohi have been used in the genetic analysis of pigment patterns in frogs (Baker, 1951; Volpe, 1961; Browder and Davison, 1964; Davison, 1964; Smith-Gill, 1974; Mangano et al., 1992), and they have been used in studies of transplantation immunity (Volpe, 1963, 1980; Davison, 1966;

Volpe and Gebhardt, 1966). Animal cloning, also known as *nuclear transplantation*, was first developed in northern leopard frogs (Briggs and King, 1952; McKinnell and Di Berardino, 1999). From the onset of cloning to the present time, genetic markers ("genetic tags") associated with donor nuclei have been essential to provide unequivocal evidence that the nucleus inserted into the wild-type recipient oocyte is responsible for development of the clone. The burnsi and kandiyohi alleles served as donor genetic markers in the early days of amphibian cloning (McKinnell, 1962, 1964; Simpson and McKinnell, 1964).

Northern leopard frogs are vulnerable to a herpesvirus-induced cancer of the mesonephros (Lucké 1934b, 1938a; McKinnell, 1973; McKinnell and Carlson, 1997, this volume) and burnsi differ little in susceptibility to the cancer compared with wild-type frogs (McKinnell, 1965). Ranid herpesvirus 1 (RaHV-1; Davison et al., 1999) is the etiological agent of the mesonephric cancer, and the virus is readily detectable in renal malignancies of burnsi frogs (Zambernard et al., 1966).

The present study reports the presence and distribution of the two pigment pattern variants of northern leopard frogs, burnsi and kandiyohi. We describe the capture and examination of 18,887 frogs.

Materials and Methods

Sites were selected in Minnesota, North Dakota, and South Dakota counties that were contiguous to or near the previously published range of the burnsi and kandiyohi morphs. Frogs were individually captured by hand or by net, and data were recorded as to size, sex, color, and pigment pattern. Frogs representing new county records were photographed. After data acquisition and photography, the frogs were released on site (with the exception of frogs in early years which were used for cancer prevalence studies; McKinnell and Duplantier, 1970; McKinnell et al., 1972). Most collections were made in late summer or early autumn when the frogs congregated on the margins of over-wintering lakes or streams. Collection procedures were designed to minimize stress to the animals under study.

We do not believe that our sampling procedure introduces a burnsi or kandiyohi bias in our data. Frequently, a frog thought to be burnsi was seen and collected, but after evaluation in the hand it was shown to be wild-type. Similarly, frogs thought to be wild-type were occasionally shown to be burnsi after capture. Years of collecting experience have taught us to identify only captured animals. (We do not subscribe to the phenotypic scoring of leaping batrachians.)

All procedures were performed with the approval of the Animal Care Committee of the University of Minnesota (Protocol Number 9206006). Frogs are currently collected under Special Permit Number 7998 issued by the Minnesota Department of Natural Resources, Division of Fish and Wildlife, Saint Paul, Minnesota; Scientific Collecting Permit number 000102038938 issued by the North Dakota Game and Fish Department, Bismarck, North Dakota; and License Number 25 issued by the Department of Game, Fish, and Parks, State of South Dakota, Pierre, South Dakota.

Results

Temporal Fluctuations

Data for populations in which substantial numbers of frogs were captured at individual sites over a period of years are summarized in Tables 49-1–49-4. The Block Lake population in Otter Tail County, Minnesota, has been successfully sampled for the longest time (1967–99), and year-to-year comparisons reveal a lack of temporal stability with regard to burnsi abundance (Table 49-1). Data from these collections of frogs (n = 3,643) reveal a major increase at some time between 1967 and 1989 and subsequent bi-directional fluctuations in burnsi frequency. The prevalence of burnsi in 1967 was 3.1%, and it reached a high of 24.7% in 1996 with an average of 18.5% during the years 1985–99. The burnsi frequency at Block Lake in 1996 was, to our knowledge, the highest ever reported in a large collection at any site within the range of the morph. Year-to-year fluctuations in burnsi abundance are statistically detectable in these data (p < 0.01 using a chi-square test for equal proportions).

The increase in (and extraordinarily high frequencies of) burnsi morphs at Block Lake stimulated our interest in the prevalence of the morph at nearby lakes. Lake Vermont in Douglas County, Minnesota (approximately 13 kilometers southwest of Block Lake), had a burnsi prevalence of 4.7% in 1997 (Table 49-2). Eagle Lake in Otter Tail County, 16 kilometers west of Block Lake, had a burnsi prevalence of 4.5% in both 1990 and 1997 (Table 49-2). Burnsi prevalence at Fish Lake in Otter Tail County, 8 kilometers east of Block Lake, was 5.8% in 1997 (Table 49-2). We collected only 28 frogs at Ellingson Lake in Otter Tail County in 1997. Two of the 28 were burnsi, giving a prevalence of 7.1%. In contrast to these relatively lower frequencies of neighboring lakes, an unnamed wetland 7 kilometers north of Block Lake had a burnsi prevalence of 22.6%; Long Lake, 23 kilometers east of Block Lake, had a burnsi prevalence of 16% (Table 49-2). With the exception of Long Lake, the 1997 Block Lake prevalence (21.0%) is significantly higher than the prevalence of burnsi at neighboring lakes (p < 0.01), but it does not significantly vary from the prevalence of burnsi in the unnamed wetland north of the lake.

Two other populations having substantial sample sizes (n = 501 and 374) and high burnsi frequencies suggest temporal stability through a number of years of sampling (Table 49-3). The Pomme de Terre Park population in Stevens County, Minnesota, fluctuated between 4.2% and 9.3% between 1975–96. The Lake Hansen population, also in Stevens County, fluctuated between 0% and 5.1% from 1978–91. However, these year-to-year fluctuations in low percentages are not distinguishable from sampling variation (p > 0.05). In contrast to Block Lake, Otter Tail County collections, the stability of the Stevens County collections suggest that there was no area-wide change in burnsi prevalence.

Geographic Variation

Within the range of burnsi distribution, there is significant geographic variation of burnsi abundance. Table 49-2 records diversity in burnsi abundance in a number of burnsi-containing collections. It may be noted that collections vary from 0.3% to a high of 22.6% (Table 49-2) or even higher at Block Lake, 24.7% in 1996 (Table 49-1).

Additional collections of 6,290 frogs in 37 counties in or near the known distribution of burnsi and kandiyohi contained neither of the variants. These collections were designed to ascertain the western limits of the burnsi variant. For example, extensive sampling was done in Stutsman County, North Dakota, during 1968–69 and again in 1996. This site is 175 kilometers northwest of the burnsi site in LaMoure

TABLE 49-1
Frequencies of Burnsi Frogs in Block Lake, Otter Tail County, Minnesota,
from 1967–99

Year	Frogs	Burnsi	% Burnsi
1967	192	6	3.1
1971	40	0	0.0
1985	546	76	13.9
1986	136	21	15.4
1989	134	26	19.4
1990	614	98	16.0
1991	95	20	21.0
1992	405	64	15.8
1993	190	31	16.3
1994	243	42	17.3
1995	130	22	16.9
1996	223	55	24.7
1997	214	45	21.0
1998	188	45	23.9
1999	293	58	19.8

	Total Frogs	Total Burnsi	% Burnsi
1967–1986	914	103	11.3
1985–1999	2,729	506	18.5
Total	3,643	609	16.6

County and 110 kilometers from the Barnes County burnsi site. No burnsi frogs were found among 1,030 northern leopard frogs captured. No burnsi frogs were taken in a number of other collections from counties in North Dakota and South Dakota west of the reported burnsi range. These burnsi negative collections were made in Cass, McIntosh, Steele, Logan, Emmons, Burleigh, and Trail counties of North Dakota. Similarly, Faulk, Jerauld, Sanborn, and Brown counties in South Dakota were burnsi negative. Although the counties cited here may in fact be outside the true range of burnsi, caution is expressed with this statement because additional collections at different sites within the counties could prove to be positive.

Distribution Maps

The known geographic distributions of burnsi and kandiyohi frogs are given in Figs. 49-2 and 49-3. These maps are drawn from historic sources (Breckenridge, 1944; Merrell, 1965; McKinnell and McKinnell, 1967; McKinnell and Dapkus, 1973; Hoppe and McKinnell, 1991) plus the recent collections reported herein. There are 17 new county records for burnsi and seven new county records for kandiyohi.

Burnsi frogs are reported for the first time in seven western counties of Minnesota: Lac Qui Parle, Chippewa, Lincoln, Lyon, Pipestone, Traverse, and Murray. We have also added

six southeastern North Dakota counties to the known burnsi distribution: Richland, LaMoure, Dickey, Sargent, Barnes, and Ransom. Burnsi frogs were also found for the first time in Roberts, Spink, Beadle, and Hand counties, South Dakota (Fig. 49-2; Table 49-4). We report kandiyohi frogs from seven counties where they were not previously known: Lincoln and Traverse counties of Minnesota; Richland, Sargent, and McIntosh counties of North Dakota; and Coddington and Aurora counties of South Dakota (Fig. 49-3; Table 49-4).

Discussion

Polymorphism, Distribution, and Prevalence

Whether burnsi polymorphism is transient or balanced cannot be answered with the relatively few years' data available on frequencies. In either case, we do not know what factors are changing or stabilizing the polymorphism, nor why polymorphic populations are found only in the limited area of the immense range of the parent species.

As stated above and in contrast to Merrell's studies, we report that extensive variation in frequency of burnsi occurs both in time and in space (Tables 49-1–49-4). Our Block Lake population (Table 49-1) shows temporal variation in burnsi frequency; our other collections in the same county (Otter Tail) show significant geographic variation (Table 49-2). Further, the

TABLE 49-2

Collections Showing Variations in Frequencies of Burnsi Frogs in Minnesota, North Dakota,
and South Dakota

County	Population	Frogs	Burnsi	% Burnsi	Year
Minnesota					
Anoka	Lake Martin	17	1	5.9	1970
Douglas	Lake Mary	483	9	1.9	1990–92
	Vermont Lake	127	6	4.7	1997
Isanti	Lake Francis	21	2	9.5	1970
Kandiyohi	Diamond Lake	978	3	0.3	1967–71
	Lake Calhoun	556	4	0.7	1970
Otter Tail	Lake Lida	98	9	9.2	1966
	Western sites	458	21	4.6	1967–71
	Pelican River	60	4	6.7	1971
	Unnamed wetland	62	14	22.6	1997
	Unnamed wetland	27	3	11.1	1998
	Ellingson Lake	28	2	7.1	1997
	Fish Lake	121	7	5.8	1997
	Eagle Lake	88	4	4.5	1990
	Eagle Lake	110	5	4.5	1997
	Long Lake	50	8	16.0	1998
Lincoln	Steep Bank Lake	148	4	2.7	1990
Lyon	Cottonwood Lake	44	1	2.3	1990
Stevens	Gausman farm	118	3	2.5	1987
Swift	Lake Oliver	65	1	1.5	1990
North Dakota					
Richland	Mud Lake	54	7	13.0	1990
LaMoure	Lake LaMoure	126	13	10.3	1993
	Lake LaMoure	65	3	4.6	1998
Dickey	James River	30	2	6.6	1993
Sargent	Silver Lake	26	4	15.3	1993
	Silver Lake	197	10	5.1	1998
South Dakota					
Roberts	Big Stone Lake	46	4	8.7	1991
Hand	Rose Hill Lake	75	3	4.0	1996
Spink	Timber Creek	2	1	50.0	1997
	Cottonwood Lake	182	4	2.1	1995–96
Beadle	Lake Cavour	74	1	1.4	1996
	Lake Byron	46	1	2.2	1996

average frequency of burnsi from 1985–99 in the Block Lake site (18.5%) is considerably higher than the burnsi prevalence was reported to be in the Anoka Sand Plain to the east (see discussion in the next paragraph). We do not have a hypothesis as to why our data differ from that published earlier.

It was previously believed that the highest frequency of burnsi frogs was found in east central Minnesota in an area known as the Anoka Sand Plain (Merrell, 1965). We report here a burnsi prevalence of 10.3% (1993) and 4.6% (1998; Table 49-2) from near Lake LaMoure in LaMoure County, North Dakota, a site approximately 20 km west of 98° longitude west and 400 km from the Anoka Sand Plain. The 1993 Lake LaMoure, North Dakota, burnsi frequency is as high as any of the collections previously reported (Merrell, 1965) and the average prevalence in the combined 1993 and 1998 collections (8.3%) exceeds any site in the Anoka Sand Plain. Minnesota burnsi frogs are reported to occur at somewhat higher frequencies in the spring than in the fall (Merrell and Rodell, 1968). If seasonal differences in prevalence also hold for North Dakota frogs, then the frequency of burnsi in the spring at the LaMoure site would be

TABLE 49-3

Frequencies of Burnsi Frogs in Two Populations from Stevens County, Minnesota, from 1975–91

Population	Year	Frogs	Burnsi	% Burnsi
Pomme de Terre Park	1975	86	8	9.3
	1978	69	3	4.3
	1979	170	10	5.9
	1980	106	7	6.6
	1996	70	3	4.2
Hansen Lake	1978	43	2	4.7
	1985	91	4	4.4
	1987	39	2	5.1
	1990	100	0	0.0
	1991	101	2	2.0

TABLE 49-4

New County Records in Minnesota, North Dakota, and South Dakota for Burnsi and Kandiyohi Variants of the Northern Leopard Frog (*Rana pipiens*)

County	Date Da/Mo/Yr	Number of Frogs	Number of Mutants	% Mutants
Burnsi				
Chippewa, MN	23/9/93	237	5	2.1
Lac Qui Parle, MN	16/9/93	120	3	2.5
Lincoln, MN	23/9/90	148	4	2.7
Lyon, MN	6/10/90	48	1	2.0
Pipestone, MN	4/10/01	12	1	8.3
Traverse, MN	14/9/88	31	1	3.2
Murray, MN	8/8/97	3	1	33.3
Dickey, ND	28/8/93	30	2	6.6
La Moure, ND	27/9/93	126	13	10.3
Richland, ND	14/9/92	54	7	13.0
Sargent, ND	29/8/93	26	4	15.3
Ransom, ND	26/9/98	135	1	0.7
Ransom, ND	13/9/99	201	2	1.0
Barnes, ND	14/9/99	79	1	1.3
Roberts, SD	14/9/90	23	2	8.7
Spink, SD	12/10/95	89	2	2.5
Beadle, SD	15/6/96	120	2	1.6
Hand, SD	10/9/96	75	3	4.0
Kandiyohi				
Lincoln, MN	23/9/90	148	3	2.0
Traverse, MN	19/9/90	95	2	2.1
Richland, ND	14/9/92	54	3	5.6
McIntosh, ND	29/9/99	320	8	2.5
Sargent, ND	28/9/01	370	1	0.3
Codington, SD	11/10/95	150	1	0.7
Aurora, SD	25/9/97	16	1	6.3

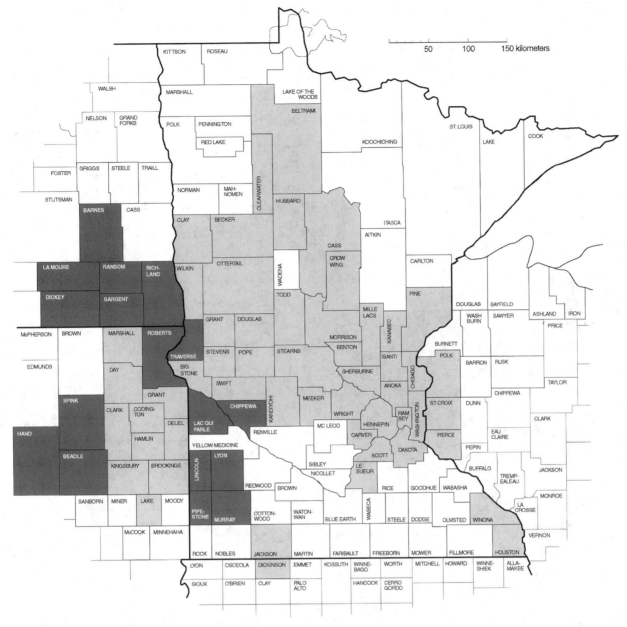

FIGURE 49-2 Distribution of the burnsi mutant of northern leopard frogs *(Rana pipiens)* in Minnesota and contiguous states. Lightly shaded counties were previously described. Darkly shaded areas are new county records of burnsi mutants described in this Chapter (see page 332, Distribution Maps).

expected to be even greater than the percentages reported here for early fall collections.

We do not know how far west in North Dakota burnsi frogs occur. We sought burnsi in a total of 1,720 frogs collected in Barnes, Cass, Logan, Emmons, Trail, Burleigh, Stutsman, and McIntosh counties, but we found no burnsi. To date, the Lake LaMoure site is the western limit of the burnsi range in North Dakota (Fig. 49-2). The Hand County, South Dakota, burnsi frogs represent the western limit of known burnsi in that state (Fig. 49-2). Obviously, additional collections may extend the western limits of the burnsi range in both North Dakota and South Dakota.

We believe that our sites in North Dakota with burnsi frogs are the first from that state to be reported in the scientific literature. Wheeler and Wheeler (1966) stated that northern leopard frogs are found throughout North Dakota and vary from pond to pond, but they make no mention of spotless leopard frogs. There is, however, an intriguing mention of leopard frogs "without spots" that occur in North Dakota, but with no reference as to where, when, or at what prevalence (Hoberg and Gause, 1992). That report finds support in the assertion of a commercial frog dealer located in Oshkosh, Wisconsin, who wrote that burnsi extend from northeastern South Dakota "into North Dakota a short distance" (see p. 76, Merrell, 1965).

All the new counties identified herein with the burnsi morph (Table 49-4) are contiguous to counties previously reported to contain these frogs (Fig. 49-2). However, this statement is not applicable to the new sites for kandiyohi frogs.

The most western collections of North Dakota kandiyohi frogs yet reported are those of the McIntosh collection (Table 49-4; Fig. 49-3). McIntosh County, North Dakota, is not contiguous to any other county where kandiyohi frogs have been found, although it is near Brown County, South Dakota. Similarly,

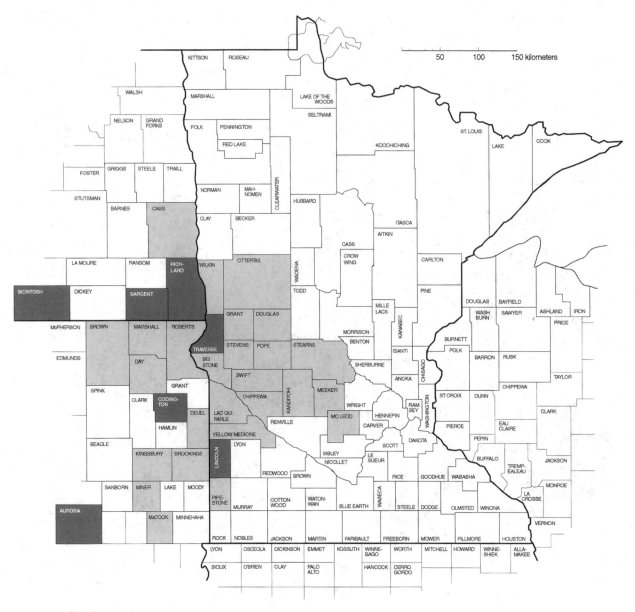

FIGURE 49-3 Distribution of the kandiyohi mutant of northern leopard frogs *(Rana pipiens)* in Minnesota and contiguous states. Lightly shaded counties were previously described. Darkly shaded areas are new county records of kandiyohi frogs described in this Chapter (see page 332, Distribution Maps).

Aurora County, South Dakota, the most western county in South Dakota with reported kandiyohi (Table 49-4; Fig. 49-3), is not contiguous with other known kandiyohi-containing counties.

An examination of a distribution map often raises the question of whether it more accurately reflects the animal's presence or the investigators' selective effort. Indeed, we more intensively sampled in the western portions of the burnsi and kandiyohi ranges, hence, we cannot say with confidence that burnsi morphs have spread westward. This morph may have been present for decades in some or all of the counties we list as new records and simply not recorded due to lack of sampling there. However, our similar surveys to the north and south in 1996–98 suggest that there may have been a true expansion to the west. Samples including 421 frogs from three counties contiguous to the burnsi range in northern Minnesota (Mahnomen, Lake of the Woods, and Aitkin) contained no burnsi frogs. Samples including 748 frogs from five counties contiguous to the burnsi range in southern Minnesota

(Yellow Medicine, McLeod, Sibley, Rice, and Wabasha) also contained no burnsi frogs. More surveys in other directions (especially to the east) may further reflect on whether the burnsi range can be extended only to the west.

If one assumes a continuous distribution of both mutant forms, then candidate counties become obvious upon scrutiny of the maps (Figs. 49-2, 49-3). For example, Grant County, South Dakota, is bounded by kandiyohi-containing counties in every direction. We examined 121 frogs at one site in this county in 1993 without finding kandiyohi, but collections of a larger size or at different sites within the county could well fill this gap in the known kandiyohi distribution. Similarly, there are several counties in the burnsi range which seem likely candidates for burnsi-positive populations. An example is Cass County, North Dakota, which is surrounded to the west, south, and east by burnsi-containing populations. We collected 246 wild-type frogs in Cass County in 1999 but, as stated above, additional collections might prove fruitful. Wadena County,

Minnesota, merits special mention. It is completely surrounded by burnsi-containing populations (Fig. 49-2). Although we collected 151 wild-type frogs in that county with no burnsi, we did encounter a burnsi morph in a Wadena County collection at the University of Minnesota Bell Museum of Natural History. We elected not to shade in the county (Fig. 49-2) because we had not personally collected the frog and there are no literature reports of finding burnsi there.

We do not know why both burnsi- and kandiyohi-containing populations are primarily found in Minnesota and its contiguous states. Spotless frogs are reported to occur in other sites within the range of northern leopard frogs, but these occasional sightings are at a frequency that precludes the use of the term "polymorphism," which is appropriate for the frogs mapped in the present study.

Summary

Two pigment pattern morphs of northern leopard frogs (*R. pipiens*) are found in polymorphic frequencies in Minnesota and contiguous states in North America. Burnsi frogs have dorsal spots that are either greatly reduced in number or completely lacking. Kandiyohi morphs are characterized by a mottling of the pigment pattern between the dorsal spots. We report that the frequency of the burnsi morph is now significantly greater at sites in western Minnesota and North Dakota than at the Anoka Sand Plain. Hence, the center of burnsi prevalence does not appear to be stable and may have moved as much as 400 kilometers to the west.

Burnsi frequencies were recorded and examined for differences between populations and for temporal stability. Burnsi frequency in one population increased from 3.1% in 1967 to a high of 24.7% in 1996. From 1985–99, the frequency of burnsi in that population remained elevated on an average of 18.5%, but fluctuated from year to year. The burnsi frequency varied from 0.3% to 22.6% among other frog populations sampled through 1999. Several populations of unusually high burnsi frequency were found in North Dakota, both far from the area where burnsi previously had been thought to be most prevalent. Genetic drift may explain some of the temporal and geographic population differences reported here, but there is also the possibility of selection for the burnsi morph.

We sought to ascertain the current boundaries of the burnsi and kandiyohi morphs and construct updated distribution maps. We report collecting burnsi frogs in 17 counties of Minnesota, North Dakota, and South Dakota and kandiyohi frogs in three North Dakota, two South Dakota, and two Minnesota counties in which these variants were not previously reported.

Acknowledgments. This study was supported in part by funds from a Morse Minnesota Alumni Association Award and by Grant Number 2675AR1 from the Council for Tobacco Research, U.S.A., Incorporated. The Nongame Wildlife Research Program of the Department of Natural Resources, State of Minnesota, provided support at the beginning of the study. We thank Rolando Mazzoni, Debra L. Carlson, Kristine S. Klos, Erik and Larissa Mottl, Jeanine Refsnider, and Matt and Rachel Hoppe for their help in collections. David Gartner, University of Minnesota Imaging Center, assisted with map production. Daniel J. Meinhardt critically read an early version of this paper.

EDUCATION

FIFTY

The National Amphibian Conservation Center

ANDREW T. SNIDER AND ELIZABETH ARBAUGH

Zoos are evolving, becoming more than just a place to take children on a warm summer day. The best zoos provide not only the recreational opportunities but the educational and science-based conservation opportunities as well. Most zoos must rely upon the charismatic megafauna (elephants, rhinos, lions, etc.) to bring the public through the gates—it is a simple matter of economics. This does not mean, however, that the smaller, lesser-known species, such as amphibians, should be excluded. In fact, amphibians have become big business. With the overwhelming number of frog-related items currently in the marketplace (including Kermit© and ads with frogs croaking out beer brand names), amphibians are in the public's consciousness as never before.

The history of amphibians at the Detroit Zoo extends back to 1960, when The Holden Museum of Living Reptiles opened to the public. At that time, the facility was considered state-of-the-art, and a smattering of small amphibian enclosures were included among the larger reptile habitats. Amphibian breeding efforts first occurred in 1970, when we began producing and rearing clutches of axolotls (*Ambystoma mexicanum*). In 1990, to acknowledge that amphibians were also included in the building, the facility's name was officially changed to The Holden Museum of Living Reptiles and Amphibians.

Since 1994, the Detroit Zoological Institute has intensified its commitment to amphibian husbandry and conservation. Our successes, which include several programs that have won national awards for long-term propagation and conservation of amphibians, have encouraged an even greater role with these animals. In light of the global decline in amphibian populations, the need for a national conservation center for amphibians became more urgent and an idea was born: The National Amphibian Conservation Center (NACC).

The NACC is the first major conservation facility dedicated entirely to conserving and exhibiting amphibians. In fact, only two other zoos in the United States (the Columbus Zoo and the San Antonio Zoo) have ever had separate public exhibits for amphibians, and both of these exhibits were converted from small, other-use buildings. The San Antonio Zoo is the only other U.S. zoo that currently exhibits amphibians separate from reptiles and/or fish.

The goals of the NACC can be summarized as follows:

To create a sense of wonder and excitement about these fascinating creatures

To show the diversity and complexity of the amphibian world through the use of living exhibits and interpretive messages

To create a sense of public stewardship for these animals and their environments

To serve as a model for future facilities around the country and the world

To serve as a resource for scientists and researchers.

One critical element of the NACC is its placement near the front of the zoo. The NACC is located along the southeast corner of Island Lake and is "immersed" in a created wetland habitat. A free-flowing contour has been constructed with lush plantings of Great Lakes plants representing bog, sedge, and wet meadow habitats. Native turtles, fish, birds, and amphibians are exhibited in this wetland area. The building itself appears slightly sunken into the habitat, allowing for a more immersed and naturalistic feel. The nearby river otter exhibit and wetland-themed children's playscape area enhance the wetlands emphasis in this part of the zoo.

Although the NACC is best experienced rather than described, we will attempt to provide an overview. The NACC is designed to be both intrinsically beautiful and functional. It occupies over 12,000 square feet, of which approximately half is dedicated to public-access space; the other half contains off-exhibit research, propagation, and holding space. Exhibits define and describe amphibians, metamorphosis, amphibian evolution and diversity, and aspects of amphibian ecology, as well as conservation biology. The Orientation Theater, a circular room with multimedia capabilities, is open to the public, school groups, and other organizations. The outside wall of this room is glass, which allows visitors to observe the outdoor wetland. A demonstration pond appears to come inside from the outdoors. This glass-sided pond features aquatic plants, tadpoles, and the occasional aquatic salamander. The pond allows visitors an opportunity to get up close and personal with an aquatic ecosystem. Docents (zoo volunteers) have an opportunity to use this area as a living classroom, where they can discuss amphibians with the general public and school groups. A new short film produced for the Detroit Zoological Institute by Sue Marx, an Academy Award-winning producer, tells the amphibian story. Other features of the NACC include a wildlife viewing area, where visitors can spot frogs, turtles, butterflies, dragonflies, and birds through the glass wall; the Immersion Gallery, where visitors walk through a re-created ecosystem with free-ranging amphibians, reptiles, birds, and fish; and the Michigan Gallery, which highlights species in our own backyard and the threats that they encounter. Here, a large vanishing-habitat mural shows the ever-continuing loss of amphibian habitat.

Off-exhibit facilities are designed for the proper husbandry and propagation of a wide variety of species. Because most zoo propagation efforts with amphibians have been based on anurans, the NACC will concentrate heavily on caudates and caecilians. Over the years, the Detroit Zoological Institute has been successful in propagating seven species of caudates and one species of caecilian, as well as several anurans. We will try to expand on these programs by providing the proper conditions and the expertise necessary for continued and enhanced success. Cave-dwelling salamanders remain a high priority in captive husbandry due to their precarious habitat in various parts of the country. Caecilians and certain anurans, such as the mantellas of Madagascar, will be a priority. Although very few concerted reintroduction efforts have taken place with amphibians, some have been accomplished with success, such as those for Wyoming toads *(Bufo baxteri)*, Puerto Rican crested toads *(Peltophryne lemur)*, and Houston toads *(B. houstonensis)*. We hope to continue to contribute to these and other programs in the future.

An endowed chair of amphibian conservation has been established by the Detroit Zoological Institute. A research room, library, and office are provided for visiting scientists who wish to use our facilities to conduct non-invasive research on amphibians. Many conservation organizations, including members of the Declining Amphibian Populations Task Force, have expressed interest in planning collaborative projects with Detroit Zoological Institute. We expect that dedicating this resource to amphibians will contribute substantially to successful conservation, education, and propagation programs on regional, national, and international levels.

Summary

The Detroit Zoological Institution has recently established the National Amphibian Conservation Center (NACC), the first major conservation facility dedicated entirely to conserving and exhibiting amphibians. The goals of the NACC are to create a sense of wonder and excitement about these fascinating creatures; show the diversity and complexity of the amphibian world; create a sense of public stewardship for these animals and their environments; serve as a model for future facilities around the country and the world; and serve as a resource for scientists and researchers.

FIFTY-ONE

A Thousand Friends of Frogs: Its Origins

TONY P. MURPHY

What do malformed frogs in Minnesota and Japan have in common? Students on field trips discovered them. The problem of malformed frogs drew national attention in 1995 when Cindy Reinitz and her students at the Minnesota New Country School in Henderson, Minnesota, found large numbers of malformed frogs. Images of frogs with gross abnormalities were broadcast on national news programs and printed in newspapers across the country. Minnesota scientists were already studying the phenomenon, but the large number of malformed frogs found by these students created intense interest, both in the public and in politicians.

Following this, Cindy Reinitz, Judy Helgen (Minnesota Pollution Control Agency [MPCA] scientist), and Tracy Fredin (Director of the Center for Global Environmental Education at Hamline University) met to discuss opportunities for scientists, educators, and students to work together to find the causes of this phenomenon. With this, *A Thousand Friends of Frogs* was born. In 1995, the Minnesota State Legislature provided funding to begin the initiative. A reporting structure was established with a survey protocol and reporting form, which was distributed throughout the state. A toll-free number was also dedicated as a *Frog Hotline* for malformed reports.

In 1996, a group of partners under the lead of *A Thousand Friends of Frogs* proposed to the state legislature to begin an amphibian population monitoring project in addition to continuing the malformation reporting effort. Part of this funding would provide a set of educational deliverables that could be used by educators to teach about amphibian conservation and environmental awareness. This state funding was granted after the students who had originally found the malformed frogs testified in front of the Minnesota Legislature. The monitoring program that was established as Minnesota Frog Watch is coordinated with the Minnesota Department of Natural Resources (MDNR). The staff has trained over 100 volunteers for a statewide anuran call survey that will eventually become a part of the wildlife monitoring programs administered by the MDNR.

The mission of *A Thousand Friends of Frogs* is twofold: to educate citizens about the environment through the issues facing amphibians, and in particular frogs; and to involve the public in local environment issues, including ones that affect amphibians. To accomplish this mission the program has the following goals:

1. Create, disseminate, and then encourage teachers to use an education program that is based on amphibian conservation but also teaches about broader environmental issues.

2. Work at local, state, regional, national, and international levels with other groups concerned about amphibians and the environment.

3. Use the best possible aspects of current and future technologies to educate citizens about issues facing amphibians.

4. Encourage citizens to become involved in environmental issues and with monitoring their own local environment. The Web site for A Thousand Friends of Frogs (http://cgee.hamline.edu/frogs) details these objectives and the current programs.

Since the inception of *A Thousand Friends of Frogs*, the staff has received thousands of malformity reports, worked with MPCA scientists, and created kits with a complete array of teaching resources for educators (available through the Minnesota Science Museum). A play called *Frog Call*, based on the challenges facing amphibians, has toured schools in the Minneapolis/St. Paul area. The project Web site has won numerous awards for its content. The staff is also working with national and international organizations, such as the Declining Amphibian Populations Task Force (DAPTF), Partners for Amphibian and Reptile Conservation (PARC), and Frog Force, to educate the public about the plight of amphibians.

In 1999, through a grant from the National Fish and Wildlife Foundation, U.S. Fish and Wildlife Service, and Best Buy Children's Foundation, staff developed an online course for educators. The course, "Helping Your Local Amphibians" (or HYLA project), was made available to 20 educators nationwide. During the course, participants received background information on amphibian biology and the global decline of amphibians. They were then asked to discover their local amphibians and explore the status of populations in their state. As part of the final component, educators had to devise a project involving their students to promote amphibian awareness and conservation in their area. HYLA is scheduled to be offered on an annual basis.

Summary

A Thousand Friends of Frogs offers a successful example of constructive student and community involvement in an environmental issue. Students helped to make the public and government more aware of the amphibian malformity issue. They showed that there are certain areas within the environmental arena where students can be just as effective as scientists with their data. In this case, perhaps *more* effective in highlighting this phenomenon.

A PERSPECTIVE

FIFTY-TWO

Of Men and Deformed Frogs: A Journalist's Lament

WILLIAM SOUDER

In August 1995, a group of middle school students on a field trip in south-central Minnesota made an odd and, to their minds, alarming discovery while exploring a bean field surrounding a large wetland. A year later I started writing about what they had found, and once I started writing I learned that it was hard to stop. A mystery in biology is a mystery without end, for inside every question are the seeds of a thousand more questions. The book of life is a narrative running in two directions—into the past and toward the future. For a biologist, the challenge is to delimit a problem to a set of testable inferences. For a journalist, the upshot is that stories in biology have murky beginnings and no endings at all, happy or otherwise.

Science and journalism share an objective—the rational portrayal of the true nature of existence—but not a common language. Naturally, this leads to a recurrent lack of agreement between what scientists actually mean and what journalists *think* they mean—and vice versa. Something always gets lost in translation. But what those kids in Minnesota stumbled onto that warm, drizzly summer morning seemed obviously important to scientists and journalists alike. Teeming in the tall grasses at the water's edge and distributed in robust numbers uphill and among the soybean rows were hundreds—probably thousands—of young-of-the-year leopard frogs. Northern leopard frogs *(Rana pipiens)* are ubiquitous in Minnesota, where they are found in two common wild types (brown and green) as well as two single-gene variants, the burnsi and kandiyohi mutants (see McKinnell et al., this volume). The frogs caught by the students were all ordinary wild types. But fully half of them were anything but normal. One out of every two frogs caught that day exhibited some kind of hindlimb abnormality.

The assortment of limb deformities discovered over the course of this and many subsequent visits to the pond included a now familiar litany of extensive and grotesque dysmorphologies. There were missing legs. There were also extra legs—often thin, sometimes translucent, parodies of normal limbs. In certain cases they were "complete." In others, the extra legs tapered at the ends. Some of the supernumeraries bore an orientation to the primary legs, as if paired with the normal limbs. Others seemed to be mirror image duplicates of each other. Some appeared singly or unrelated to any other structure on the frog. Some primary legs were "withered" in appearance

or were in fixed, rigid positions. Or both. Some limbs were proportional miniatures. Others were reduced segmentally, usually at the distal end. Some were twisted like corkscrews. A great many frogs were found to have webbings of skin spanning joints—typically from the back of the thigh to the back of the calf, with the knee bent so that normal extension of the limb was constrained and, in extreme cases, impossible. Most of the deformities were either unilateral or bilaterally asymmetrical. Over time, other sorts of malformations were found in more limited frequencies: missing or misplaced eyes, abnormal forelimbs, incomplete jaws, and gut and gonad anomalies.

The story goes on from there, as I've already said, but let's stop a moment. Let's look at the moment of discovery and examine the data as we knew it then.

On the ground in Minnesota, simultaneously with their find, the schoolchildren drew several inferences. The first of these was that deformed frogs are not a routine feature of nature. In fact, this is debatable—reviews of the literature in the days and months that followed turned up many reports of abnormal limb development in frogs, some dating back to the eighteenth century. One researcher had even noted that people who happen onto deformed frogs are typically unaware that such malformations are already known to science. What at first appeared to be unprecedented turned out to have precedent after all.

But the fact that limb malformations may occur naturally in some frog populations from time to time is not proof that all such outbreaks are "normal" or that they are common. The students, who were immediately concerned that the deformities signaled something toxic in the local environment, posted their discovery on the Internet. Presently, other reports began streaming in from around Minnesota. Within a year, more than 200 outbreaks of deformed frogs had been recorded, from one end of the state to the other. It then emerged that similar deformities had also been found at multiple sites in Canada, some three years earlier. The Canadian Wildlife Service and researchers from McGill University in Montreal were actively investigating the situation, which

William Souder, a freelance writer, has written extensively about frog deformities and population declines for The Washington Post. *His book,* A Plague of Frogs, *was published in spring 2000 by Hyperion.*

presented a notable concordance with the reports in the States: a majority of the malformities consisted of missing or reduced legs—phenotypes almost entirely absent from the literature.

The second inference made by the students in Minnesota was that something in the water where the frogs lived must have caused the deformities. At that moment, this connection seemed quite apparent. The frogs—juvenile, newly transformed frogs—were coming from the water. The pond and its fringes were where the frogs were still concentrated, owing in part, the students assumed, to the animals' inhibited locomotion. Eventually, all the researchers who came to work on the deformities problem would share this belief—with some refinements. For example, the water in question might not be the water in the breeding pond, because northern leopard frogs typically overwinter elsewhere at deep-water sites. The possibility of adult frogs accumulating teratogenic pollutants during hibernation and then passing them on to their offspring also had to be considered. And the phrase "something in the water" clearly had to include both biological and xenobiotic agents.

Lastly, the students worried that whatever was causing the deformities in the frogs could do the same—or something of equal concern—to people. The idea that limb deformities in frogs indicated a risk to human health would become the most controversial, yet essential, aspect of the unfolding story.

The course of the investigation into the deformities that has since taken place has been well documented elsewhere—in newspaper articles, on television, and in my own book, not to mention other sections of the work you hold in your hands—and will not be revisited in detail here. Suffice it to say this: during a few hours in the field on a summer morning, eight teenagers identified a problem and correctly outlined the key issues that have since framed a massive scientific inquiry costing millions of dollars and involving researchers from academia as well as a number of state and federal agencies. And the best evidence that they got it right is that, as this is being written, not one of those issues has been resolved conclusively.

From the beginning, scientists looking into the deformities took up positions on either side of a conceptual fault line. One group saw only nature at work in the deformities and believed that the alarm over them was misplaced. This small but vocal minority viewed the outbreaks as routine ecological events—not manmade disasters. The main (and much larger) group believed that the incidence of limb abnormalities in frogs had grown explosively in recent years, either because of a sudden rearrangement of the ecological factors that produced such abnormalities in nature or because anthropogenic agents were creating new outbreaks of deformities that were now superimposed over the natural frequency. Apart from the debate over causation, these divergent views led to a protracted (and at times contentious) discussion about the "baseline rate" of deformities in wild amphibian populations.

The data are inconclusive, and the question is really two questions confused as one. Two important studies, one by David Hoppe of the University of Minnesota Morris and the other by Martin Ouellet of McGill University, put the incidence of developmental malformations in normal frog populations at well under one percent. Hoppe examined several thousand museum specimens collected in Minnesota in the 1950s and 1960s. Ouellet captured an even bigger sample from numerous populations of frogs living in a large, well-isolated research preserve atop Mt. St. Hilaire in the St. Lawrence River valley of Quebec. Both found minor deformities in only an occasional individual. This gave credence to claims that frog populations

showing rates of malformation as low as a few percent were out of the ordinary. It also raised doubts about claims that the many missing limbs being reported might be the result of failed predation and other ordinary traumas. And it also reinforced the judgment of most field herpetologists (who had never seen a deformed frog in the wild and doubted that anybody did very often) that something unusual was going on.

But there was more than one baseline rate to consider. None of these findings shed any light on the claim that the frequency of frog populations exhibiting higher-than-normal rates of deformity was on the rise. The literature showed that such outbreaks turn up now and then—but how often was anybody's guess. It still is, and my hunch is it always will be because the idea of sampling enough frog populations in enough places to make even a reliable estimate is mind-boggling. The North American Reporting Center for Amphibian Malformations (NARCAM), a government-run Web site that has compiled reports of deformities for several years, shows hundreds of outbreaks of malformations widely distributed across the northern tier of the United States. But there is no way to know whether this is typical or not. Thus, an essential premise of the deformities investigation—that numerous new outbreaks have been superimposed over the ones that naturally occur—is without empirical foundation.

Anecdotally, it's a different story. There is wide agreement among herpetologists and frog dealers—people who have "sampled" frog populations for many years—that the apparent explosion of malformation episodes is well outside the norm. In football parlance, it's gut-check time—in more ways than one. If, as most researchers still suspect, there is a human factor behind at least some of the deformities, then it will not matter that the work of teasing out the details got started on a hunch. Ultimately, the rate at which deformity outbreaks occur is irrelevant except in quantifying the extent of a phenomenon that may or may not be of concern. In the end, it is the cause or causes of the malformations that count. Based on the data that we have or that we are ever likely to obtain, there is no way anyone can argue that everything going on out there is "natural."

Early in the investigation, efforts focused on the search for some common factor among the sites where the deformities had been found. This was a mistake—and also the point at which people began talking past one another. *The premise, remember, was that more than one thing was happening.* The assumption that some outbreaks were natural and some were not implied at least two causes—or at least similar proximate causes mediated by more than one factor. But not everyone involved seemed to get this, and the journalists who covered the investigation soon lost sight of its starting point. It had to be expected, indeed it was necessary, that an ecological explanation (or even more than one) would be found. But when a group of Stanford undergraduate students reported results of an experiment that linked many of the observed abnormal phenotypes to infection by an aquatic trematode of the genus *Ribeiroia*, all but a handful of the media outlets that covered the news took this as the last word on the deformities problem. A number of major newspapers and television networks reported—with a sense of both relief and irony—that the mystery of the deformed frogs had been "solved" and that it had been only a quirk of Mother Nature after all.

The media got it wrong.

Parasites may explain some, even many, of the deformities outbreaks. Something in the normal ecology of frog populations *must.* But there is clear and compelling evidence that this is not the whole story. The whole story remains a work in progress.

Careful necropsies and morphological analyses by several techniques (e.g., external examination, dissection, soft-tissue histology, radiography, and clearing-and-staining) have demonstrated that (1) not all deformed frogs are infected by parasites; (2) not all parasite infections cause deformities; (3) not all types of deformities seen in the wild have been observed in experimental treatments with parasites; and (4) not all species react the same way to trematode infections if, indeed, they react at all. The mechanism by which an encysted parasite may cause limb dysmorphogenesis is hotly disputed by developmental biologists who have long studied limb induction and patterning. Even where *Ribeiroia* has been found in abnormal frogs it has not always been clear in every specimen that the infection was heavy enough or in close enough physical association with the deformity to be the cause. Most importantly, there is ambiguous evidence that *Ribeiroia* infection causes the most widely seen patterns of deformity—those involving missing or reduced legs only.

Stan Sessions, of Hartwick College in Oneonta, New York, who first proposed a parasite link to amphibian malformations, has long cautioned that missing limbs are "difficult to interpret," and that various traumas—predation, cannibalism, or accidents—may account for some or even most of such abnormalities in the wild. Difficult is not the same as impossible, however, and other researchers have come to quite a different conclusion for some missing-leg phenotypes that show no obvious sign of injury such as scarring, bleeding, or inflammation. Specifically, some partial legs show skin pattern disruptions proximal to the truncation. Others are associated with a corresponding absence of internal skeletal structures—like half a pelvis, for instance—that cannot be plausibly explained by a trauma, even one occurring at a very early stage of limb development. Meanwhile, Martin Ouellet has established a strong correlation between outbreaks predominated by missing limbs and the application of agrichemical pesticides local to the affected habitats, just as researchers in the United States continue to observe deformities in animals exposed to hot-spot site water or its extracts. These findings have, at this writing, not been published and are thus all but invisible to the popular media.

Why? Because journalists do not cover science in the same way that they cover politics or government or business or sports. Science coverage is discontinuous and, as a result, misleading as often as it is informative. There are, I believe, three broad shortcomings in the way science is communicated to the lay public by the media.

First, reporters—and more importantly their editors—tend not to see science as a developing story, but rather as a perplexing and boring process that produces "news" only intermittently, usually in the form of a readily digestible "discovery." This, for the most part, eliminates from science reporting what is elsewhere the gold standard in journalism—enterprise. A reporter trying to cover an ongoing story in science is likely to find room for only fragments of it in the paper or on the evening news. Very small fragments. Imagine that newspaper and TV journalists reported the results of elections, but said not a word about the campaigns leading up to voting day, and you begin to get an idea of the disparity.

Second, journalists are overly reliant on findings published in the scientific literature. In most forms of journalism getting "scooped" is a disaster. In science reporting, it's almost a requirement. The surest way to convince an editor to go with

a science story is to show him or her that it has already been published in a scientific journal—or, preferably, that it will appear in one on the very same day you are proposing you run with your version. Here, I think, journalism and science must shoulder the blame equally. Journalists, in choosing only to cover periodic developments, give a false picture of the nature of scientific progress. A paper in a journal reporting a set of findings rarely represents a comprehensive view of the whole field of knowledge about a particular issue; rather it is a snapshot of one facet of that knowledge, incomplete and lacking context. No wonder the public often sees scientific discoveries as contradictory of one another. The public—that is to say, you and I—may feel a little like it's listening to a radio broadcast of a football game in which the plays aren't reported, but only the score is given every few minutes. In a seesaw game (and science *is* very much a seesaw game) you would never know when one reality might supplant another. At the same time, the scientific community makes better, more continuing coverage of science difficult when most journals require researchers to embargo their findings as a condition of publication. Why don't reporters do a better job of keeping track of what scientists are up to? Much of the time it's because scientists keep it a secret. Embargoing scientific findings that are in press enhances the status and confirms the supreme power of scientific journals—but it inhibits a full public understanding of what science does or does not know about many subjects of vital importance.

Finally, journalists, and (to a lesser but still substantial degree) scientists as well, place an inordinate significance on human health concerns with respect to ecological problems. Human health, of course, is a paramount consideration, but you should not have to have evidence of people keeling over or growing extra legs to sell an editor on a story about deformed frogs (or ozone depletion, global warming, endocrine disruption, water scarcity, shrinking biodiversity, etc., etc., unto oblivion). From the beginning of the frog story, there were worries over whether frogs are "sentinel species" for human environmental welfare. There is one thing I'm sure of. As we enter a century when environmental problems are likely to surpass traditional sources of human misery such as war, racism, and fanatical nationalism, if we continue to see wildlife ecology as separate from human ecology, we are going to be (however well-intentioned) dead meat. If we get to the point where the only sentinels left are a handful of opportunistic species—rodents, white-tailed deer, cockroaches, or the weeds in your lawn—it's time to turn out the lights. If you are in the water with a great white shark, you can think of your feet as "sentinel organs" if you want to. But if you get a message from them, it is going to be too little, too late. Deformed frogs are nothing if not proof that environmental degradation is circling us with its fin up, coming in ever and ever closer.

So we've made a bit of a mess of this story, as both scientists and journalists. We pretend to see beginnings and endings where they do not exist, in the interim sounding and then squelching the alarm by turns. Inference, which should be the tool that tightens the bolts on well-built experiments, has instead become a kind of blunt instrument with which researchers with competing views beat one another about the head and shoulders. It is no surprise that deformed frogs have been treated in the press as either a very big deal or no big deal at all.

Here's a thought, then—it's a big deal to the frogs. Among the vertebrates, amphibian limb induction is the only major form of organogenesis that takes place in direct contact with the surrounding environment. At best, limb development is a plastic, potentially divertible process. At worst—well, the worst is just what we've been seeing, isn't it? One issue on which there has been general agreement all along is that limb deformities are maladaptive in frogs, which, after all, are little more than mouths and legs on opposite ends of a digestive and reproductive system. Take away a frog's legs and you put it out of business, for good. For the past several seasons here in Minnesota, as biologists have sparred with one another and the media have looked on with occasional passing interest, frogs have been growing scarce. At deformities hot-spots that have been under study for years now, it's getting harder with the passage of time to find frogs, period. Could it be a natural downward cycle? Sure. Wanna bet on it? Not me. I don't like the odds.

As I write this, the weather here in Minnesota remains unusually balmy for October. From the open window in my office I can hear a smattering of chorus frogs calling a bit erratically many months after the breeding season. These are young of the year, trying out their vocal equipment for the first time, getting ready to hunker down for a long winter and an uncertain future. Come spring, they will be part of the music of the night, embarked on yet another chapter in a story hundreds of millions of years in the making. Now that their narrative has taken a chilling turn, it's more important than ever for us to pay attention, to try to get it right. We owe it to ourselves. We owe it to the frogs. Who knows? One day we might begin to understand how little difference there is between those two obligations.

PART TWO

SPECIES ACCOUNTS

Introduction

MICHAEL LANNOO, ALISA L. GALLANT, PRIYA NANJAPPA,
LAURA BLACKBURN, AND RUSSELL HENDRICKS

Worldwide reports of amphibian population declines and malformations prompt concern about species protection. At this point in time in the United States, we recognize 289 extant amphibian species: 103 species of frogs and 186 species of salamanders (Appendix IN-A), although the identity and relationships of species in several genera remain unresolved. Amphibians occur in nearly all habitats, and they exhibit variations in life history and natural history features that enable them to accommodate a wide range of environmental conditions (Duellman, 1999; Zug et al., 2001; Pough et al., 2004). Some species are entirely aquatic, while others spend their complete life cycles in trees or in caves. For most species, however, very little is known, and much of what we do know comes from breeding season observations. Even so, there are still species for which eggs have never been seen, so we have no idea of typical clutch size or location, and for species observed only during the breeding season, their habitat requirements at other times of the year remain unknown. With so little information available about basic species life history and natural history traits, and with some species being inaccessible for monitoring, how do we begin to manage for amphibians? And, if we cannot address important management issues, how can we hope to address conservation issues?

Our objective was to develop a toolbox for managers. This toolbox consists of species accounts (the work that follows and forms the bulk of this volume) and a new digital database that links each species with geospatially referenced units. With these tools, it is possible to map distribution patterns for individual species as well as for groups of species sharing common phylogenetic, life history, ecological, or behavioral traits as revealed in the following species accounts. This digital database can be easily imported into a geographic information system (GIS) to generate such maps, as we have done here. This technology provides the means for readily viewing patterns that once were prohibitively labor intensive. The species accounts were assembled by experts and compiled in a standardized, outline format. Our rationale for compiling accounts in this manner is as follows:

1. No detailed accounts for all U.S. species currently exist. Current accounts are either unreferenced or lightly referenced summaries (e.g., Stebbins, 1985; Behler and King, 1998; Conant and Collins, 1998), based on state or regional analyses (many works), taxonomically specialized (e.g., Bishop, 1943; Wright and Wright, 1949; Petranka, 1998), or incomplete (Catalogue of American Amphibians and Reptiles, Society for the Study of Amphibians and Reptiles, St. Louis, Missouri). Stebbins and Cohen (1995) do a masterful job of summarizing the highlights of amphibian life history and natural history features, but they illustrate their points using examples (rather than lists) of species that exhibit particular traits (as they must in this successful effort to provide a popular account).

2. The detailed species accounts that do exist tend to be written as narratives. Our goal was to make this information easily accessible; therefore, we organized life history, natural history, and status data into a standardized outline format.

3. Our species accounts were written or reviewed by experts on the species.

4. Species account entries are backed by references to the scientific literature, references to unpublished data, or personal communications.

5. Species accounts document not only what is known, but also indicate what we do *not* know about a species. By noting these information gaps, we hope to encourage future research in these areas. Collectively, our goal from using this approach coupled with our maps is to establish a foundation of facts.

In the tradition of the classic North American amphibian species accounts (Bishop, 1943, for salamanders; Wright and Wright, 1949, for frogs and toads), we tie these accounts together with a considered introduction. In particular, we present here a preliminary comparison of amphibian distributions with distributions of pertinent environmental features. We also consider the distribution patterns of a number of important life history and natural history features, such as terrestrial versus aquatic development and types of aquatic development, and then supplement this analysis with data presented in tabular form. We employ an ecoregion framework to help us interpret these distribution patterns, relying not only on the mapped regional boundaries but also on an understanding of the environmental characteristics that define each ecoregion. We conclude by demonstrating how to use the toolbox with an example of an application focused on conservation.

Background and Methods

This concept of comparing available information about species distributions with distributions of environmental features is not new. For amphibians, such relationships were first determined on a broad scale (with an emphasis on temperature and precipitation patterns) by Kiester (1971). More recently, general linkages between environmental attributes and fauna were articulated by Krzysik (1998b), who constructed a classification (based primarily on characteristics of hydrologic regime and setting) of ecosystems relevant to amphibians at the genus level. Duellman and Sweet (1999) compared amphibian species distributions with regions derived from information on physiography, climate, and geology, though they did not detail the linkages between these characteristics and amphibian life history traits. Cumulatively, these authors provide an exhaustive description of amphibian distributions with reference to physiographic divisions, natural regions, temperature and precipitation patterns, and geological history. Duellman and Sweet (1999) also considered species shared across physiographic regions and areas of high diversity and endemism.

It is not difficult to surmise that environmental features that provide hospitable conditions for amphibian survival can be related to broader, regional patterns of past and present climate, terrain, hydrology, vegetation, and substrate. Natural and human-caused disturbances that affect amphibian habitats are also related to environmental characteristics at the regional level. By understanding patterns of regional ecological characteristics that relate to amphibian species distributions, we can gain insight into some of the biotic and abiotic processes that are affecting habitat conditions.

As were our predecessors, we, too, are interested in the correlation between environment and species distributions. We build upon the past broad-scale comparative analyses by using a regional framework derived from a larger suite of environmental characteristics than those employed by Duellman and Sweet (1999), and we suggest associations between amphibian life history traits and regional ecological characteristics. Additionally, we have two advantages over all previous authors (Kiester, 1971; Krzysik, 1998b; Duellman and Sweet, 1999): first, in the availability of a new, high quality, geospatially referenced, digital, national database (Blackburn et al., 2000) of species distributions that facilitates national analyses at the species level; and second, new species descriptions standardized across all U.S. amphibian taxa (this volume).

Our interest in determining which species patterns can be explained by ecoregions derives from a region-level approach to conservation. Instead of focusing on how to protect individual species, we draw attention to environmental characteristics that are important for a diverse set of amphibian species and are pertinent to broader land management and policy-making issues.

Ecoregions

Ecoregions have been defined by Wiken (1986) as ecologically distinct areas resulting from "the mesh and interplay of the geologic, landform, soil, vegetative, climatic, wildlife, water and human factors" where "the dominance of any one or a number of these factors varies with the given ecological land

unit." This latter phrase is key because while a number of ecological frameworks have been derived from information on multiple environmental variables, very few frameworks have been developed in a manner that recognizes that environmental driving forces vary in type and importance across space and time. There has been contention among geographers as to whether a consideration of human factors is appropriate in delineating ecoregions (e.g., see discussion in Omernik, 1994). For biological assessments, it is critical to consider anthropogenic effects, because humans have imposed wholesale changes in region-wide vegetation cover and, correspondingly, in soil and hydrologic characteristics, disturbance regimes, fauna, and even local to regional climate. In areas where vegetation cover has not been so drastically altered, consideration of the human context is still regionally relevant because of the strong relationship between natural resources and the capacities of an area to support or withstand different types of human use. For example, widespread water withdrawal for irrigation in arid and semiarid areas of the United States has lowered water tables, reducing the availability of surface waters and concentrating naturally occurring and human-introduced chemical compounds in the water—often to toxic levels. Extensive timber harvesting has reduced the availability of old growth stands in the western and eastern United States and, likewise, the biotic species that rely on ecosystem components that exist only in old stands. So, while regional ecological characteristics dictate what humans can do with the landscape, human use results in broad-scale feedbacks that affect and reshape regions.

The ecoregion framework that we selected for our analysis was derived from environmental information on (but not limited to) climate, surficial and bedrock geology, Pleistocene glaciation, soils, landforms, hydrology, vegetation (current, potential natural, and, where available, past), and land use. The resultant regions represent areas within which there is relative homogeneity with respect to these characteristics and to the types, spatial patterns, and processes of the ecological systems created by the combinations of these characteristics. These ecoregions were originally defined by Omernik (1987), but have since been revised (see U.S. Environmental Protection Agency, 1999). Our analysis is limited to the conterminous United States. Though a comparable ecoregion map exists for Alaska (Gallant et al., 1995), the information available on amphibians there needs additional work, and geospatial map units for Alaska are too large for assessing ecoregional relationships. Hawaii is also excluded from our analysis; a comparable ecoregion map has not been developed for the Hawaiian Islands, and all Hawaiian amphibian species have been introduced.

Ecoregions can be defined over a range of scales from global to local, depending upon the objectives of the user and the availability of sufficient environmental information. The framework that we used provided three levels of hierarchy for the conterminous United States (Gallant et al., 1995; Commission for Environmental Cooperation, 1997; U.S. Environmental Protection Agency, 2000a), with a fourth, more detailed level existing for some states (e.g., Woods et al., 1996, 1998; Bryce et al., 1998; Griffith et al., 1998; Pater et al., 1998; U.S. Environmental Protection Agency, 2000b).

Database Design

To compile amphibian distribution data, we first consulted the most recent field guides or herpetological atlases for each state. For those states without such works, we contacted museums (such as the Field Museum of Natural History) to obtain distribution information from catalogue data on voucher specimens. We also contacted state and regional experts, from whom we were able to obtain additional county records, updates to less current field guides, the U.S. Geological Survey's Gap Analysis Program data (GAP; Scott et al., 1996), and unpublished data. We then filled any remaining distributional gaps using regional data (e.g., DeGraaf and Rudis, 1981; Nussbaum et al., 1983) and national data/field guides (Stebbins, 1985; Conant and Collins, 1998; Petranka, 1998).

There are three potential sources of error in our database. The first is the difference between the existing knowledge of the community of herpetologists and the actual distribution of animals. In some cases, especially where there is taxonomic flux (e.g., the *Pseudacris triseriata* complex) or difficulty with identification (e.g., between eastern gray treefrogs *[Hyla chrysoscelis]* and Cope's gray treefrogs *[H. versicolor]*), this error can be large. The second source of error is between our database and the existing knowledge of the community of herpetologists. To minimize this error, we posted our maps to a secure Web site for 2 years, from 2000 to early 2002, and invited approximately 150 experienced herpetologists from around the United States to critique the maps. The third source of error is temporal—where changes in species distributions have occurred rapidly. This causes a disconnection between, for example, an older dataset from one state abutting a newer dataset from another state. We have tried to incorporate all documentation of a species presence, regardless of date; we do not attempt to document declines. Further, we do not tend to believe the species distributions that follow state boundaries where there are no corresponding geographical or biological boundaries, but to the extent that they reflect what we know, we must tolerate these results until more work can be done.

The database encompasses information for all amphibians occurring in the United States (Appendix IN-A for a complete listing of species). We define species according to the standardized list compiled by Crother et al. (2000, 2003) with the following exceptions: (1) we recognize the existence of silvery salamanders *(Ambystoma platineum)* and Tremblay's salamanders *(A. tremblayi)*, and we recognize the existence of unisexual *Ambystoma* hybrids; (2) following Gergus (1998), we recognize Arizona toads *(Bufo microscaphus)* and California toads *(B. californicus)* as separate species; (3) following the opinion of several workers including A. Goebel (this volume), we consider *B. nelsoni* and *B. boreas* to be separate species; (4) following the work of G. Fellers and B. Shaffer (this volume), we consider *Rana draytonii* and *R. aurora* to be separate species; (5) in accordance with the state of Georgia, we acknowledge the existence of dark-sided salamanders *(E. aquatica)*; (6) we recognize that green and black dart-poison frogs *(Dendrobates auratus)* and Martinique greenhouse frogs *(Eleutherodactylus martinicensis)* have been introduced to Hawaii; (7) we include *Ascaphus montanus, Batrachoseps aridus, B. gavilanensis, B. incognitus, B luciae, B. minor, B. robustus, Desmognathus folkertsi, E. chamberlaini,* and *E. waterlooensis,* recently described species not recognized when Crother et al. (2000), was assembled (but see Crother et al., 2003 for a partial update); and (8) we include two species of *Eurycea* and one species of *Necturus* that have been recognized as distinct, but not yet described.

Information on species occurrence was interpreted to the county level; however, for five western states having exceptionally large counties (Arizona, California, Nevada, Oregon and Washington), information was interpreted for minor civil divisions-subcounty units defined by the U.S. Census Bureau. For each county (or minor civil division), we included fields in the database indicating the presence or absence of each species, source of information, and status of voucher specimens. For species "absence," we were unable to consistently distinguish between documented absence versus lack of survey effort. Therefore, species "absence" connotes some ambiguity. Regarding voucher specimens, the presence of a voucher represents either recent or historical documentation; the lack of a voucher represents either no voucher specimen collected or unavailable voucher information.

Database Refinement

Following this initial assembly of the available distribution data, we created a set of distribution maps (see below for the process of map creation) and posted them to the previously mentioned protected Web site for peer-review. We asked that all suggestions for adjustments to the maps be supported by documentation in the literature or by a voucher specimen. For a period of about 2 years, we continuously adjusted and updated the maps on the Web site based on the comments we received. Adjustments included incorporating new literature sources, new county records, new voucher data (to support existing [historical] voucher data and [assumed] presence data), changes to taxonomic organization of species, and deleting invalid county records. In 2001, copies of our dataset were transferred to the USGS Patuxent Wildlife Research Center and AmphibiaWeb (AmphibiaWeb.com). The dataset will be managed at USGS Patuxent Wildlife Research Center (http://www.pwrc.usgs.gov/armiatlas).

Map Development

We used GIS software to generate maps of species distributions and frequency of occurrence by county/subcounty. Digital files of county and minor civil division boundaries can be obtained from the U.S. Census Bureau (see TIGER/Line* files at www.census.gov). We merged the subcounty units of the five western states (listed above) with county units for the remaining states to create a single set of map units (Fig. IN-1). These map units were then linked with our amphibian species database through a common data field that provided a unique identifier for each unit based on FIPS (Federal Information Processing Standards) reference codes for states, counties, and minor civil divisions. Thus, by knowing the list of FIPS codes indicating where a species occurs, a map can be generated with those units shaded (Fig. IN-2).

To map national patterns of species richness, we summed the number of species occurring in each mapping unit and depicted the resultant range of values as a monochromatic gradient from 0 to the maximum number of species encountered. To map national patterns of richness at genus or family levels, or by various behavioral traits or status conditions, we created a second data file that cross-tabulated each species with these attributes. The list of species that occur in a given map unit (our original amphibian database) can be crosslinked with information on genus, family, or other attribute (our second data file), in order to depict frequency of occurrence per map unit of species sharing certain attributes.

---

* TIGER® stands for Topologically Integrated Geographic Encoding and Referencing, a system and digital database developed at the Census Bureau. The TIGER/Line files are a digital database of geographic features, such as roads, railroads, rivers, lakes, political boundaries, census statistical boundaries, etc., covering the United States.

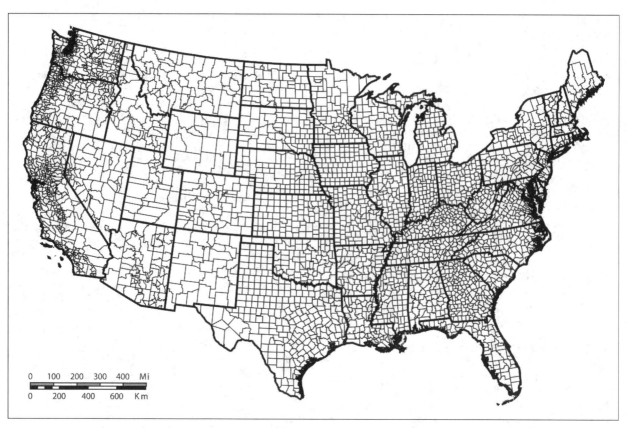

FIGURE IN-1 Political (county and minor civil division [for California, Oregon, Washington, Nevada, and Arizona]) boundaries in the conterminous United States that are the minimum mapping units for our analysis (U.S. Census Bureau).

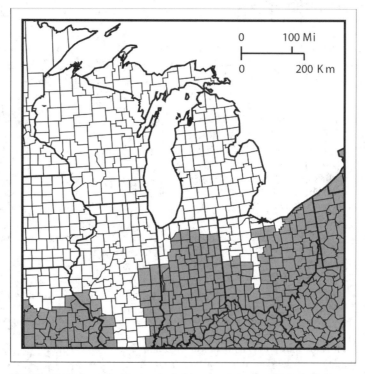

FIGURE IN-2 An example showing shaded (species present) and non-shaded (species absent) counties. Note that once a species has been documented in a county, the whole county is shaded. This method allows for a high resolution in situations where counties are small relative to the range of a species. However, where counties are large, this technique can produce a distorted range. For this reason, when determining distributions in five western states, we have used subcounty, minor civil divisions (townships) as our political mapping unit (see text for complete explanation).

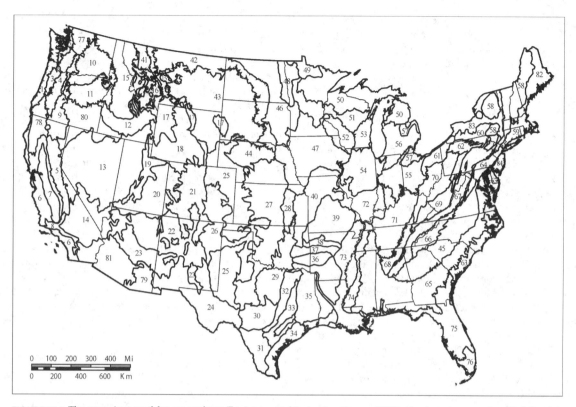

FIGURE IN-3 The ecoregions used for our analyses (Environmental Protection Agency, 2000). Numbers correspond to the following ecoregions: 1 = Coast Range; 2 = Puget Lowland; 3 = Willamette Valley; 4 = Cascades; 5 = Sierra Nevada; 6 = Southern and Central California Chaparral and Oak Woodlands; 7 = Central California Valley; 8 = Southern California Mountains; 9 = Eastern Cascades Slopes and Foothills; 10 = Columbia Plateau; 11 = Blue Mountains; 12 = Snake River Basin; 13 = Central Basin and Range; 14 = Mojave Basin and Range; 15 = Northern Rockies; 16 = Montana Valley and Foothill Prairies; 17 = Middle Rockies; 18 = Wyoming Basin; 19 = Wasatch and Uinta Mountains; 20 = Colorado Plateaus; 21 = Southern Rockies; 22 = Arizona/New Mexico Plateau; 23 = Arizona/New Mexico Mountains; 24 = Chihuahuan Desert; 25 = Western High Plains; 26 = Southwestern Tablelands; 27 = Central Great Plains; 28 = Flint Hills; 29 = Central Oklahoma/Texas Plains; 30 = Edwards Plateau; 31 = Southern Texas Plains; 32 = Texas Blackland Prairies; 33 = East Central Texas Plains; 34 = Western Gulf Coastal Plain; 35 = South Central Plains; 36 = Ouachita Mountains; 37 = Arkansas Valley; 38 = Boston Mountains; 39 = Ozark Highlands; 40 = Central Irregular Plains; 41 = Canadian Rockies; 42 = Northwestern Glaciated Plains; 43 = Northwestern Great Plains; 44 = Nebraska Sand Hills; 45 = Piedmont; 46 = Northern Glaciated Plains; 47 = Western Corn Belt Plains; 48 = Lake Agassiz Plain; 49 = Northern Minnesota Wetlands; 50 = Northern Lakes and Forests; 51 = North Central Hardwood Forests; 52 = Driftless Area; 53 = Southeastern Wisconsin Till Plains; 54 = Central Corn Belt Plains; 55 = Eastern Corn Belt Plains; 56 = S. Michigan/N. Indiana Drift Plains; 57 = Huron/Erie Lake Plains; 58 = Northeastern Highlands; 59 = Northeastern Coastal Zone; 60 = Northern Appalachian Plateau and Uplands; 61 = Erie Drift Plains; 62 = North Central Appalachians; 63 = Middle Atlantic Coastal Plain; 64 = Northern Piedmont; 65 = Southeastern Plains; 66 = Blue Ridge Mountains; 67 = Ridge and Valley; 68 = Southwestern Appalachians; 69 = Central Appalachians; 70 = Western Allegheny Plateau; 71 = Interior Plateau; 72 = Interior River Lowland; 73 = Mississippi Alluvial Plain; 74 = Mississippi Valley Loess Plains; 75 = Southern Coastal Plain; 76 = Southern Florida Coastal Plain; 77 = North Cascades; 78 = Klamath Mountains; 79 = Madrean Archipelago; 80 = Northern Basin and Range; 81 = Sonoran Basin and Range; 82 = Laurentian Plains and Hills; 83 = Eastern Great Lakes and Hudson Lowlands; 84 = Atlantic Coastal Pine Barrens.

Comparison Analysis and Interpretation

As discussed earlier, we used the set of ecoregions originally defined by Omernik (1987), but since refined. In fact, these regional boundaries are continually being refined by Omernik and others (e.g., Bryce et al., 1998; Griffith et al., 1998; Woods et al., 1999) as more detailed information is learned. The map version used for our analyses is M-1 (Fig. IN-3), which was provided in digital format by the U.S. Environmental Protection Agency (2000).

We generated a series of maps concurrently depicting ecoregion boundaries with amphibian richness patterns. Interpretation of these maps was conducted visually so we could mentally adjust for the fact that politically defined county units do not hierarchically nest within ecologically defined ecoregions. Our approach was to determine the general environmental characteristics held in common by ecoregions associated with distributions of amphibians.

Results and Discussion

Amphibian Distributions

In the United States, amphibian taxa are not distributed equally, but, rather, exhibit regional patterns. While the genera *Ambystoma*, *Aneides*, *Plethodon*, *Bufo*, *Hyla*, *Pseudacris*, *Rana*, and *Scaphiopus* are distributed throughout much of the continental United States, most taxa are not. For example, the genera *Acris*, *Amphiuma*, *Cryptobranchus*, *Desmognathus*, *Eurycea*, *Gyrinophilus*, *Haideotriton*, *Hemidactylium*, *Necturus*, *Notophthalmus*, *Phaeognathus*, *Pseudobranchus*, *Pseudotriton*, *Siren*, *Stereochilus*, and *Typhlotriton* are eastern taxa. The genera *Ascaphus*, *Batrachoseps*, *Dendrobates* (introduced to Hawaii), *Dicamptodon*, *Ensatina*, *Hydromantes*, *Pternohyla*, *Rhyacotriton*, and *Spea* are western taxa. *Ascaphus* and *Rhyacotriton*, in the Pacific Northwest, are our only northern genera. *Amphiuma*,

Eleutherodactylus (two species introduced), *Gastrophryne*, *Hypopachus*, *Leptodactylus*, *Osteopilus* (introduced), *Phaeognathus*, *Pseudobranchus*, *Pternohyla*, *Rhynophrynus*, *Smilisca*, *Stereochilus*, and *Xenopus* (introduced) are southern taxa.

Amphibian Richness

Amphibian richness, the total number of species occurring anywhere within a given area ("density" in the terminology of Kiester, 1971; and Duellman and Sweet, 1999), varies by taxonomic level and is uneven across the United States (Fig. IN-4). At the family level, amphibian richness is highest in the South and Southeast (Southeastern Plains, Middle Atlantic Coastal Plain, Southern Coastal Plains, Mississippi Valley Loess Plains, and South Central Plains Ecoregions) and the Pacific Northwest (Coast Range, Cascades, and Klamath Mountains Ecoregions [Fig. IN-4A]). From the South and Southeast, amphibian family richness tapers off to the north and west, as conditions become cooler and drier. From the Pacific Northwest, amphibian family richness tapers off in association with drier ecoregions. In these family-rich portions of the country, certain ecoregions stand out as having lower richness than their surrounding regions (e.g., the Piedmont Region in the Southeast, the Mississippi Alluvial Plain Ecoregion in the South, and the Puget Lowland and Williamette Valley Ecoregions in the Pacific Northwest). What these four ecoregions share is a history of widespread removal of forests and replacement with agriculture and urban development. Nationally, the lowest levels of amphibian family richness occur in the Mojave and Sonoran Basin and Range Ecoregions, the Snake River Basin Ecoregion, and the Montana Valley and Foothill Prairies.

At the genus level, amphibian richness exhibits the same basic pattern as the richness of families, but it is more exaggerated (Fig. IN-4B). At this taxonomic level, the highest richness values in the Southeast and Northwest stand in contrast to the low values in the Great Plains, Rockies, and desert ecoregions.

At the species level, the trend in contrast between high and low richness continues, with the highest richness occurring in the Southeast (Fig. IN-4C). Interestingly, the eastern portion of the Mississippi Alluvial Plains Ecoregion has notably greater species richness than the western portion.

Anuran Richness

Anuran family richness is centered in the south-central and southeastern ecoregions, and extends east along the Atlantic Coastal Plain and west throughout much of the West (Fig. IN-5A). Areas of highest richness in the Southeast correspond with ecoregions having high humidity throughout the year, hot summers, warm winters, and more than 130 cm of annual precipitation. The regions of highest family richness in the south-central United States are also associated with warm climates, but are much drier. These ecoregions may, however, offer suitable habitats on a finer ecological scale. For example, the Edwards Plateau Ecoregion in central Texas offers moist caves and a network of perennial streams that, though sparse in distribution, are cooler and clearer than those of surrounding ecoregions. Areas of lowest family richness occur in the Mojave and Sonoran Basin and Range Ecoregions, in scattered areas throughout the West (including the Sierra Nevada, the Snake River Basin, the Northern Rockies, and the Montana Valley and Foothill Prairies), and in the

Upper Midwest, extending northeast into the northern Appalachians and northern New England (Fig. IN-5A). This latter observation is interesting, as this drop in anuran family richness in the Upper Midwest and New England corresponds with the limits of the Pleistocene glaciation. At the genus level, anuran richness is also centered in the south-central, southeastern, and Atlantic Coastal regions, and associated with ecoregions that tend to be warm and have moderate to high relative humidity throughout the year. Genus richness is lower in the West, Upper Great Plains, Upper Midwest, Appalachians, and northern New England (Fig. IN-5B). Compared with family richness, differences at the genus level between high richness regions and low richness regions are more pronounced. At the species level, anuran richness peaks in the Southeastern Coastal and Middle Atlantic Coastal Plains, and in the southern Great Plains (Fig. IN-5C). At the species level, the differences between high richness regions and low richness regions are even more pronounced than at the genus level, with the high richness regions representing conditions of hot summers, warm winters, moderate to high annual precipitation, and high relative humidity throughout the year.

Salamander Richness

At the level of families, salamander richness is discontinuous, with the highest values east of the eastern Great Plains, in the northern Rockies, in the Sierra Nevada, and along the West Coast (Fig. IN-6A). In general, the regions of highest family richness are those that provide closed canopy forests with fairly mild winter climates and moderate to high annual precipitation. Salamander family richness is lowest in the semiarid and arid ecoregions of the Great Plains, Rockies, and deserts of the West and Southwest. In the Southeast, richness values are lower in the Piedmont than in the surrounding Appalachian highlands and Atlantic Coastal Plain. Originally forested, the Piedmont was extensively cleared for agriculture, and later heavily developed for urban expansion. Surface waters in the Piedmont Ecoregion have high sediment loads, compared with those of surrounding regions. In many ways, salamander genus richness mirrors family richness (Fig. IN-6B). Again, high values occur in the East and the Pacific Northwest. In the East, the Piedmont shows reduced salamander richness compared with surrounding ecoregions. Salamander species richness peaks along the spine of the southern Appalachians, along the North Carolina—Tennessee border (Fig. IN-6C), where endemic species are often associated with individual mountaintops. Richness is also high in the Southeastern Plains Ecoregion. Compared with family and genus richness, salamander species richness is relatively low in the Pacific Northwest.

Contributions of Anuran and Salamander Patterns to the Overall Pattern Observed in Amphibians

The patterns presented above for amphibians (Fig. IN-4) represent a composite of different patterns exhibited by anurans (Fig. IN-5) and salamanders (Fig. IN-6), with a larger influence imposed by the 186 species of salamanders than the 102 species of anurans (this number does not include *Rana rugosa*, which only occurs in Hawaii). At the family level, frogs are much more evenly distributed than are salamanders (compare Fig. IN-5A to Fig. IN-6A). Indeed, tiger salamanders (*A. tigrinum*)

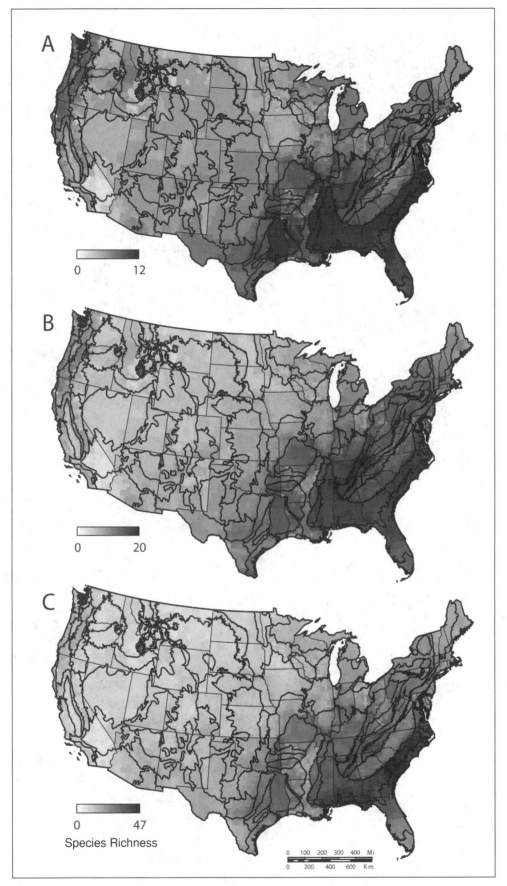

FIGURE IN-4 Amphibian richness illustrated at the family level (A), genus level (B), and species level (C).

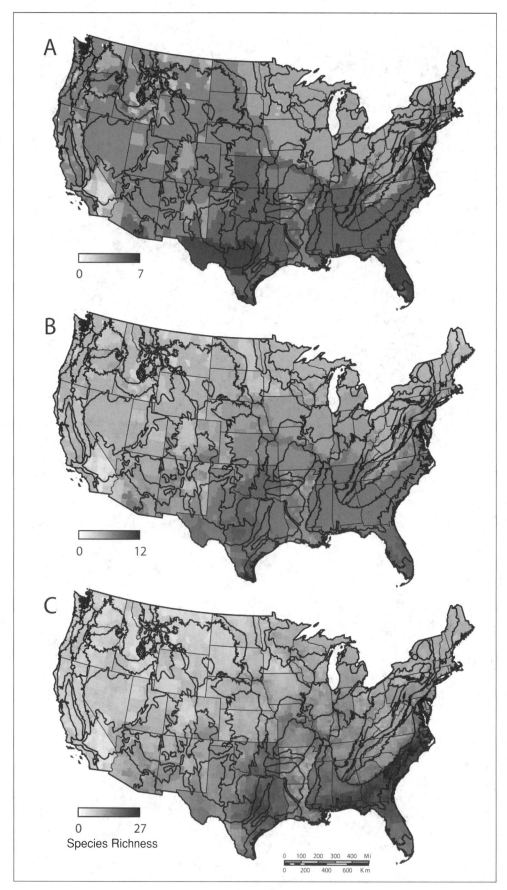

FIGURE IN-5 Anuran richness illustrated at the family level (A), genus level (B), and species level (C).

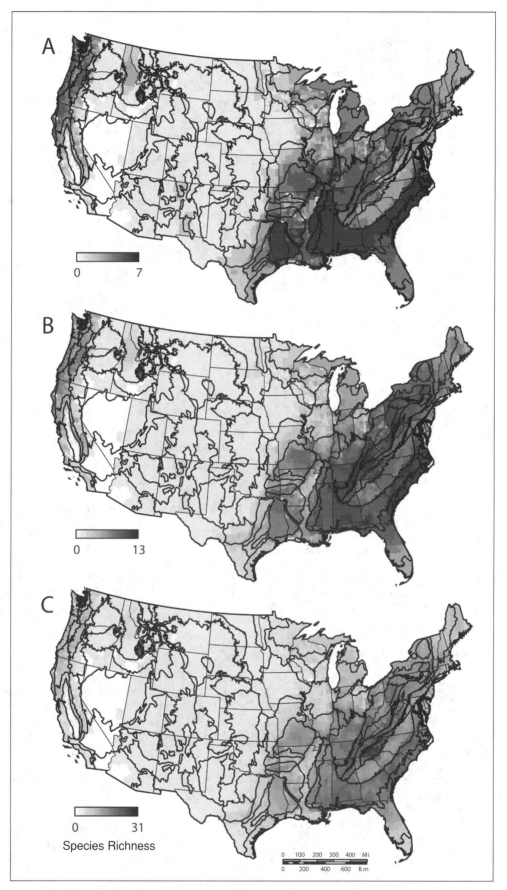

FIGURE IN-6 Salamander richness illustrated at the family level (**A**), genus level (**B**), and species level (**C**).

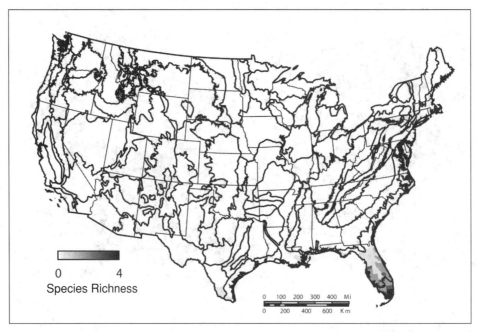

FIGURE IN-7 The richness of introduced amphibians in the U.S. (exclusive of Hawaii).

are the only salamander species present through much of the Great Plains and Rockies. Furthermore, the tendency for amphibian richness to become increasingly biased towards the Southeast, with a ratcheting down from family through genus to species levels, is largely due to this same tendency in anurans. Increased amphibian richness in the Appalachians is almost exclusively due to salamander richness. Similarly, in the Pacific Northwest, increased amphibian richness (especially at the family and genus level) is largely due to salamander richness at these taxonomic levels.

From these maps we hypothesize that (1) desiccation appears to limit salamander distributions far more than it limits anuran distributions, and (2) cold appears to limit frog distributions much more than it limits salamander distributions (although wood frogs [R. sylvatica] are the only North American amphibians to live above the Arctic Circle).

The Distribution of Introduced Amphibians

With the exclusion of native species that have expanded their range, introduced amphibians, including Cuban treefrogs (Osteopilus septentrionalis), coquis (E. coqui), greenhouse frogs (E. planirostris), African clawed frogs (Xenopus laevis), wrinkled frogs (R. rugosa [in Hawaii, not shown]), and green and black dart-poison frogs (D. auratus [in Hawaii, not shown]) are shown in Fig. IN-7. All these introductions are restricted to ecoregions with warm tropical or subtropical habitats (see Meshaka, 2001).

In addition to these introduced species, at least six and possibly seven (there is debate about whether marine toads [B. marinus] are native to the United States) native species are known to have expanded their range within the United States, and, in at least one case (American bullfrogs; R. catesbeiana), have become a nuisance:

Woodhouse's toads, B. woodhousii

Rio Grande chirping frogs, E. cystignathoides

Rio Grande leopard frogs, R. berlandieri

American bullfrogs, R. catesbeiana

Tiger salamanders, A. tigrinum

Shovel-nosed salamanders, D. marmoratus

Marine toads (?), B. marinus

The Distribution of Extinctions and Extirpations

In the United States, Vegas Valley leopard frogs (R. fisheri) and Bay Springs salamanders (Plethodon ainsworthi) have not been observed for more than 60 and almost 40 years, respectively. Vegas Valley leopard frogs are thought to be extinct. The outlook for Bay Springs salamanders is better. Reasonable habitat remains at the one site where these burrowing animals have been collected, and it is thought that with more effort, and perhaps different searching techniques, Bay Springs salamanders will be re-discovered. Tarahumara frogs (R. tarahumarae) are thought to be extirpated in the United States, but survive in Mexico. Wyoming toads (B. baxteri) have been nearly extirpated in the wild and captive breeding colonies have been established. Captive bred Wyoming toads have been repatriated into the wild in the hopes of re-establishing wild populations.

The Role of Phylogenetic Features in Determining Amphibian Distributions

The phylogenetic patterns underlying amphibian distributions in North America have been elucidated by Duellman and Sweet (1999). Here, we extend Duellman and Sweet's analysis for the United States by using the digital maps (described above) in combination with ecoregion maps as the basis for our analysis, and include all 289 known U.S. species. Additionally, rather than describing amphibian distributions with reference to physiographic and historical features, we attempt to explain

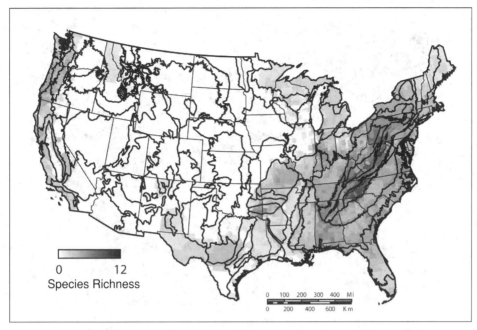

FIGURE IN-8 The richness of U.S. amphibian species with terrestrial egg, larval, and adult stages.

these distributional patterns in terms of amphibian life history and natural history features. Both approaches are justified, and in expanding Duellman and Sweet's (1999) analysis, we feel our approach complements theirs. Additionally, by using GIS technology, we provide the basis for the next generation of computer-based amphibian ecoregional analyses.

Many life history features can be explained by phylogenetic factors. For example, all species in the genus *Plethodon* have terrestrial development and all *Rana* have aquatic development. However, many life history features cannot be explained by phylogenetic factors. Species in the genus *Desmognathus* will breed in water or on land. And while most mole salamanders are characterized by a terrestrial adult stage, adults in some tiger salamander populations (especially in the Great Plains, Rocky Mountains, and desert regions) exhibit neoteny and trend from facultatively to obligately aquatic—indeed the pattern in several amphibian lineages is towards variability. Some of these variations on phylogenetic themes must have great ecological consequences. Many speciose higher-level taxa contain species with widespread distribution as well as species with restricted distributions.

FERTILIZATION

All frogs, except the two tailed frog species *(A. montanus and A. truei)* and coquis *(E. coqui)*, exhibit external fertilization. All salamanders, except hellbenders *(Cryptobranchus alleganiensis)*, lesser sirens *(Siren intermedia)*, greater sirens *(S. lacertina)*, southern dwarf sirens *(Pseudobranchus axanthus)*, and northern dwarf sirens *(P. striatus)*, exhibit internal fertilization.

TERRESTRIAL DEVELOPMENT

Taxa with direct development include all members of the salamander genera *Plethodon*, *Aneides*, and *Batrachoseps*, all members of the frog genus *Eleutherodactylus*, and pigmy salamanders *(D. wrighti*; Appendix IN-B). The distribution of amphibians with terrestrial development is scattered but biased towards the Appalachian uplands (Fig. IN-8).

PARENTAL CARE

While several groups of salamanders exhibit parental care (Appendix IN-C), among frogs, only leptodactylids in the genera *Eleutherodactylus* and *Dendrobates* exhibit parental care.

AQUATIC DEVELOPMENT

Most U.S. amphibians have aquatic development, but only a few species have an aquatic adult stage (Appendix IN-D). The distribution of species with aquatic egg, larval, and adult stages is shown in Fig. IN-9. Note that these species are concentrated in the Southeastern Plains Ecoregion, but salamandrids (which may have a terrestrial eft stage) extend throughout the East and the Northwest, and lesser sirens range up the Mississippi River drainage into Michigan.

Most U.S. amphibian species have aquatic egg and larval stages. With the notable exception of the Appalachian ecoregions, the richness of species with aquatic development that breed in standing water (seasonal, semipermanent, and permanent wetlands, or lakes; Fig. IN-10) mirrors the richness of U.S. amphibians (Fig. IN-4C).

A relatively small subset of aquatic breeding amphibians reproduce in association with flowing water (small streams, springs, and seeps; Table Appendix IN-E; Fig. IN-11). These species are found in the East, in association with the Appalachian and Coastal Plain regions, in the Ozarks and south through Louisiana and eastern to central Texas, and in the Pacific Northwest and northern Rockies. The similarity in species richness between flowing water breeding species and terrestrial breeding species (compare Fig. IN-11 with Fig. IN-8) is interesting in light of the recently controversial hypothesis that terrestrial-breeding plethodons evolved from stream-breeding species (Wilder and Dunn, 1920; Dunn, 1926; Ruben and Boucot, 1989; Beachy and Bruce, 1992; Ruben et al., 1993).

CLUTCH SIZE

Clutch sizes vary across family, across genera within families, within genera, and within species according to size of the

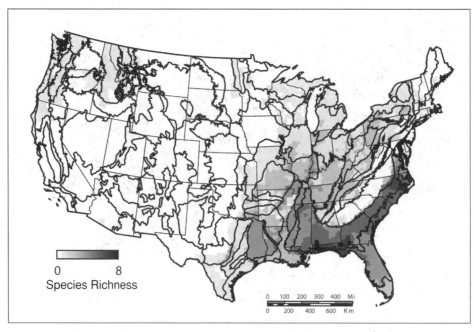

FIGURE IN-9 The richness of U.S. amphibian species with aquatic egg, larval, and adult stages.

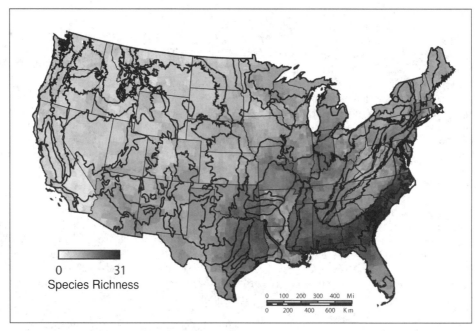

FIGURE IN-10 The richness of species with aquatic development that breed in standing water (seasonal, semipermanent, and permanent wetlands or lakes).

female and geographic distribution. Data on clutch size (see Appendix IN-A) are summarized here:

Ambystomatidae   3 (*Ambystoma opacum*)–7,631 (*Ambystoma tigrinum*)

Amphiumidae   98 (*Amphiuma tridactylum*)–354 (*Amphiuma means*)

Ascaphidae   28 (*Ascaphus truei*) –97 (*Ascaphus montanus*)

Bufonidae   300 (*Bufo quercicus*)–45,054 (*Bufo cognatus*)

Cryptobranchidae   200–400 (*Cryptobranchus alleganiensis*)

Dendrobatidae   5–7 (*Dendrobates auratus*)

Dicamptodontidae   50 (*Dicamptodon copei*)–200 (*Dicamptodon aterrimus*)

Hylidae   63 (*Pseudacris regilla*)–4,000 (*Hyla gratiosa*)

Leptodactylidae   2 (*Eleutherodactylus planirostris*)– 250 (*Leptodactylus fragilis*)

Microhylidae   152–2,100 (*Gastrophryne carolinensis*)

Pelobatidae   300 (*Spea intermontana*)–5,468 (*Scaphiopus holbrookii*)

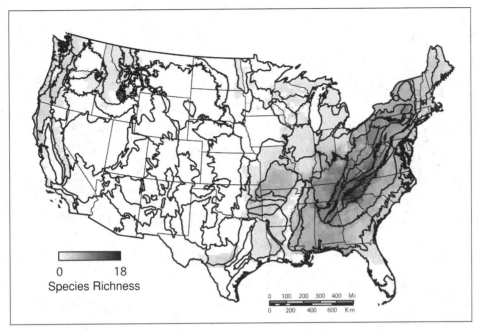

FIGURE IN-11 A relatively small subset of aquatic breeding amphibians reproduce in association with flowing water (small streams, springs, and seeps).

Pipidae	< 17,000 (*Xenopus laevis*)
Plethodontidae	1 (e.g. *Aneides hardii*)–192 (*Pseudotriton montanus*)
Proteidae	15 (*Necturus punctatus*)–83 (*Necturus maculosus*)
Ranidae	100 (*Rana muscosa*)–20,000 (*Rana catesbeiana*)
Rhinophrynidae	?
Rhyacotritonidae	2 (*Rhyacotriton cascadae, R. kezeri*)–16 (*Rhyacotriton kezeri*)
Salamandridae	6 (*Taricha rivularis*)–375 (*Notophthalmus viridescens*)
Sirenidae	200–500 (*Siren intermedia*)

Considered by family, clutch size ranges from 1 (Plethodontidae) to 40,000 (Bufonidae). Salamanders in the genus *Plethodon* tend to have the smallest clutch sizes; ranids and bufonids tend to have the largest. Data on clutch sizes for pond breeding animals tend to come from egg counts, while data for terrestrially breeding salamanders tend to come from follicle counts, which likely overestimate the actual number of eggs that are laid.

The Role of Ecology in Determining Amphibian Distributions

CAVE DWELLING AMPHIBIANS

The distribution of cave salamanders is largely (but not solely) determined by the distribution of solution caves (Appendix IN-F). For example, the extensive karst cave systems of the Appalachian regions and the Texas Edwards Plateau Ecoregion are populated, some with highly derived (sightless, pigmentless, neotenic) troglodytic species. However, not all cave systems (e.g., in the Driftless Area of Southeastern Wisconsin) are occupied. A specialized population of tiger salamanders occupies a cave system in Arizona.

While not obligate cave dwellers with derived morphologies, cave salamanders (*Eurycea lucifuga*) and long-tailed salamanders (*E. longicauda*) are often found living in or around the mouths of caves. Both Shasta salamanders (*Hydromantes shastae*) and Kentucky spring salamanders (*Gyrinophilus porphyriticus duryi*) are found both in caves and in the open. Further, while larvae of grotto salamanders (*Typhlotriton [Eurycea] spelaeus*) are present in greatest abundance in streams in the open, adults are partial to underground waters and caves.

While there are cave-associated populations of anurans, there are no sightless, albino populations or species of frogs. Two features of anurans probably resist permanent troglobitic habitation. First, most adult anurans rely on visual cues to detect prey. Second, even if other sensory cues were enhanced (e.g., olfaction or tactile cues) many species of frogs assume a feeding posture that positions the head above the substrate, a location that de-emphasizes olfactory and tactile cues. If frogs were obligately troglobitic, a shift to nonvisual cues and an associated shift in their neurosensory apparatus would likely have to be accompanied by a neuromotor postural shift.

BREEDING AND DISPERSAL MIGRATIONS

Members of the following genera have breeding and dispersal migrations: *Rana, Bufo, Scaphiopus, Spea, Hyla, Pseudacris, Acris, Ambystoma, Notophthalmus,* and *Taricha*.

ARBOREAL HABITATION

Climbing salamanders (*Aneides lugubris*) and wandering salamanders (*A. vagrans*) can be arboreal at all stages of their life cycle (eggs, juveniles, and adults). Hylid anurans in the genus *Hyla* and Cuban treefrogs (*O. septentrionalis*) are arboreal as adults (although canyon treefrogs [*H. arenicolor*] are not arboreal). Alone among North American bufonids, Coastal-Plain toads (*B. nebulifer*) can be arboreal. Many species of frogs call from perches above the ground, including members of the hylid genus *Pseudacris* and the leptodactylid genus *Eleutherodactylus*.

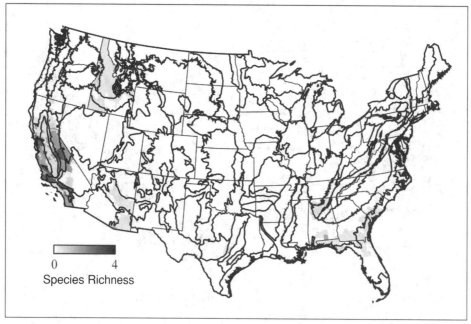

FIGURE IN-12 The richness patterns of U.S. endangered species.

NEOTENY

Neoteny could be included under either phylogenetic or ecological factors—for while neoteny is a response to specific sets of environmental factors and is exhibited to a greater or lesser extent in all U.S. salamander families, there are no neotenic anurans. Permanently aquatic *Ascaphus* and *Xenopus* (introduced) have adult morphologies.

TROPHIC SPECIALISTS

In anurans, narrow-mouthed toads (Microhylidae) and burrowing toads *(Rhinophrynus dorsalis)* specialize on hymenopterans (ants and termites). Carnivorous morph spadefoot toad tadpoles (genus *Spea*) specialize on fairy shrimp. There are few trophic specialists among salamanders, but tiger salamanders in western populations and long-toed salamander *(A. macrodactylum)* larvae in some populations develop cannibal morphs, which may be either facultatively or obligately cannibalistic, depending on the region and the subspecies.

UNDERGROUND TAXA

The following taxa live underground and emerge to breed: *Rhinophrynus* (feed underground), *Scaphiopus* (do not feed underground), *Spea* (do not feed underground), and Strecker's chorus frogs *(Pseudacris streckeri;* feed underground). Amphibians in many other genera, including *Ambystoma, Batrachoseps, Gastrophryne, Hypopachus, Phaeognathus, Pternohyla,* and *Plethodon* may spend a considerable amount of time underground, avoiding desiccating and cold conditions or tending eggs.

OVERWINTERING

Genera that overwinter by burrowing include *Ambystoma, Bufo, Plethodon,* terrestrial plethodontids, and *Notophthalmus viridescens* (under certain situations). Species that overwinter by utilizing glucose or glycerol antifreeze include wood frogs *(R. sylvatica),* eastern gray treefrogs *(H. versicolor),* Cope's gray treefrogs *(H. chrysoscelis),* spring peepers *(P. crucifer),* western chorus frogs *(P. triseriata),* and boreal chorus frogs *(P. maculata).*

Species that overwinter using aquatic habitats include American bullfrogs, green frogs *(R. clamitans),* pickerel frogs *(R. palustris),* northern leopard frogs *(R. pipiens),* mink frogs *(R. septentrionalis),* lesser sirens *(Siren intermedia),* eastern newts *(Notophthalmus viridescens;* under certain situations), and aquatic plethodontids, including Allegheny Mountain dusky salamanders *(D. ochrophaeus).* Species that overwinter using microhabitat selection include northern cricket frogs *(Acris crepitans).*

AESTIVATION

The following species aestivate by forming cocoons: lesser sirens, greater sirens, southern dwarf sirens, northern dwarf sirens, lowland burrowing treefrogs *(Pternohyla fodiens)* and Mexican smiliscas *(Smilisca baudinii).* Bishop (1943) notes that in exceptionally dry seasons in the North, some salamanders burrow and remain hidden until the return of the rains. Northern slimy salamanders *(Plethodon glutinosus)* are particularly susceptible to drying and will burrow deeply during a drought. Eastern red-backed salamanders *(P. cinereus),* while less sensitive than northern slimy salamanders, will sometimes disappear for weeks in the absence of rainfall.

NOXIOUS SKIN SECRETIONS

Species that produce noxious skin secretions as a component of predator defense include *Notophthalmus* sp., *Taricha* sp., *Ambystoma,* spring salamanders *(G. porphyriticus),* red salamanders *(Pseudotriton ruber),* mud salamanders *(P. montanus),* members of the *Plethodon jordani* complex, and *Bufo* sp.

ACTIVITY PATTERNS

North American amphibians tend to be active at night. For example, most anurans form evening or nighttime breeding choruses; this in fact may be the best time for researchers to sample populations. At night, woodland salamanders emerge from daytime retreats to forage; again, this is the time of day when most researchers choose to sample populations. There are exceptions to this nocturnal tendency, large ranids, newts,

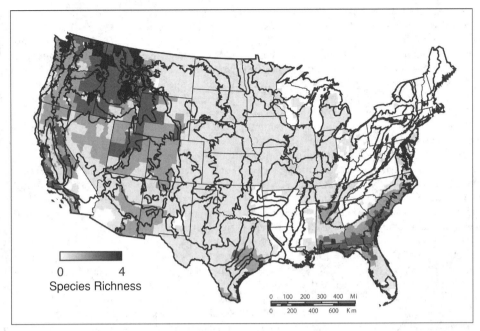

FIGURE IN-13 The richness patterns of U.S. species of concern, including species nationally endangered or threatened, species that have been extirpated, and species considered extinct.

some species of *Desmognathus*, and sirens may be active during the day. Activity patterns can vary by life history stage: larvae may be active while adults are in hiding. Activity patterns can be seasonal: normally nocturnal species that migrate may move during the day, especially following heavy rains.

FOOD

In general, anuran tadpoles are herbivorous, feeding on filamentous algae, other plant matter, and the plants and animals that grow on aquatic plant and rock surfaces. Some tadpoles, for example in the pelobatid genus *Spea*, are carnivorous, feeding on the fairy shrimp (Anostraca) that also occupy their ephemeral wetland habitats (see "*Trophic Specialists*," above). Adult frogs are carnivorous. Most species take invertebrates, but larger ranids, such as American bullfrogs, will also ingest any vertebrates they can manage to swallow.

As with adult frogs, salamander diets at all life history stages tend to consist of living animals. There are exceptions: *Cryptobranchus* will gorge on animal remains; sirens, which usually feed on crayfish, shrimp, and other animal types, will occasionally fill themselves with filamentous algae. *Gyrinophilus*, *Amphiuma*, larval *Ambystoma*, and (to a lesser extent) larvae and adults of many other species are cannibalistic (see "*Trophic Specialists*," above). Victims of cannibalism are typically younger, smaller animals.

Carnivorous amphibians tend to swallow their prey whole—they do not rip prey apart. Therefore their diet is restricted to animals small enough to swallow.

Case Study: Using the Toolbox to Address Conservation Issues

How can these species accounts, digital distribution maps, and the perspectives gained from the accompanying volume be applied to address concerns about land management and

species conservation? Here, we offer an example of how we envision this toolbox operating by considering the U.S. species that have gone extinct, been extirpated (but persist elsewhere), are recognized as threatened or endangered, or have been petitioned for threatened or endangered status.

First, we compile a list of threatened or endangered species (Appendix IN-G) and ask whether there are phylogenetic patterns to these species in severe decline—indeed, there are. Of the 19 families of amphibians found in the United States (Appendix IN-A), 4 (26%) are represented in Appendix IN-G (Ambystomatidae, Plethodontidae, Bufonidae, and Ranidae). Of the 42 genera found in the United States (Appendix IN-A), 7 (*Ambystoma, Batrachoseps, Eurycea, Phaeognathus, Plethodon, Bufo,* and *Rana*) representing 17%, are listed in Appendix IN-G. Species in the genera *Rana, Ambystoma, Bufo, Eurycea,* and *Phaeognathus* are over-represented compared to their national representation. Species in the genera *Batrachoseps* and *Plethodon* are under-represented.*

Second, we use the digital database to examine the distributional patterns of endangered species (Fig. IN-12) or species of concern (Fig. IN-13). With this map we can then ask questions about regional and ecoregional biases in the distribution of these species. We note that when compared with the distributional

---

* Twenty taxa are listed in Appendix IN-G, representing 6.9% of the U.S. total number of species. The genus with the highest number of listed species is *Rana* (with 6), representing 20% of all listed species. In the United States, species in the genus *Rana* constitute only 10.4% of the U.S. total (30/289). Four taxa in the genus *Ambystoma* are listed, representing 16% of all taxa listed, while species in the genus *Ambystoma* constitute only 5.9% of all U.S. species (17/289). Three taxa of *Bufo* represent 12% of the listed species, while *Bufo* constitute 8.0% of all U.S. species (23/289). Likewise, 3 species of *Eurycea* are listed (12%), while *Eurycea* compose only 9.0% of the U.S. total (26/289). The only known species of *Phaegnathus* is listed. In contrast, only two species of *Plethodon* (representing 8% of listed species) are listed, while *Plethodon* constitute 18.4% of all U.S. species (53/289). Finally, among *Batrachoseps,* one (4% of listed species) of the 20 known species (6.9% of the U.S. total) is listed.

maps presented in Figs. IN-4, IN-5, and IN-6, there is a bias in listed species towards the Western and Southern portions of the United States. There may be several reasons for this bias, including restricted distributions of species in these areas (compared to congeners in the East), increased environmental stressors, and regional tendencies by humans to pursue litigation. In particular the combination of many endemic species with restricted distributions and a large growing human population have produced a hot spot of species in trouble in California (Fig. IN-12). Seven of the 20 species listed in Appendix IN-G (35%) are endemic to California and northern Baja California. Of the 59 native species in California, at least two and probably several more, have been extirpated in the state, and 12% are federally listed (N. Scott, personal communication).

Third, we use the species account information to examine shared life history and natural history features of threatened species. This analysis reveals that aquatic species are disproportionately threatened, for while aquatic breeding species represent 69% of the total U.S. species (191/276), 80% (16/20) of the species listed in Appendix IN-G live and/or breed in aquatic habitats. Many of these aquatic habitats are small—seasonal or semipermanent—and, therefore, are not afforded environmental protection, are associated with agriculture, and are susceptible to aquacultural modifications. Furthermore, these habitats must be entered (by breeding adults) and abandoned (by newly metamorphosed juveniles); these seasonal migrations entail risk, and migration routes can be blocked or compromised by habitat fragmentation (e.g., road building).

Summary

To address amphibian conservation issues in the United States, we have assembled the available information on basic species life history and natural history traits and developed a toolbox for managers. This toolbox consists of species accounts (the work that follows) and a new digital database that links each species with geospatially referenced units. With these tools, we map distribution patterns for individual species, as well as for groups of species sharing common phylogenetic, life history, ecological, or behavioral traits as revealed in the following species accounts. We present here a preliminary comparison of amphibian distributions with distributions of pertinent environmental features. We also consider distribution patterns of a number of important life history and natural history features, such as terrestrial versus aquatic development and types of aquatic development, and then supplement this analysis with data presented in tabular form. We conclude by demonstrating how to use the toolbox with an example of an application focused on threatened species.

Acknowledgments. We thank the species accounts authors (this volume) for providing us with the life history and natural history information necessary to conduct our analyses. We also thank the many people who critiqued our maps (see book acknowledgments). We thank Doug Brown and Bob Klaver for consultations as this manuscript was being assembled. We also thank David Bradford, Donna Helfst, and Susan Lannoo for commenting on previous drafts of this manuscript.

Addendum. While this book was in final proof. Highton (2004) described *Plethodon sherando* (Big Levels salamander). At this time, Big Levels salamanders are known only from Augusta County, Virginia.

Further, Bonett and Chippindale (2004) have recently shown that based on molecular data, the distribution of *Eurycea multiplicata* may be much more limited than is shown here. Readers are referred to this ppapter for details.

Species List of U.S. Amphibians, in Alphabetical Order by Family

Ambystomatidae

Ambystoma annulatum, Ringed salamander
(eggs: 205–390 average)

Ambystoma barbouri, Streamside salamander
(follicles: 260 average)

Ambystoma californiense, California tiger salamander
(eggs: 814 average, range 413–1,340)

Ambystoma cingulatum, Flatwoods salamander
(follicles: 166–225 average)

Ambystoma gracile, Northwestern salamander
(egg masses: range 30–270)

Ambystoma jeffersonianum, Jefferson salamander
(follicles: range 140–280)

Ambystoma laterale, Blue-spotted salamander
(follicles: 196–250 average)

Ambystoma mabeei, Mabee's salamander (?)

Ambystoma macrodactylum, Long-toed salamander
(eggs: 90–411)

Ambystoma maculatum, Spotted salamander
(eggs: 1–250 in each of 2–4 masses)

Ambystoma opacum, Marbled salamander
(eggs: 3– > 200)

Ambystoma platineum, Silvery salamander
(eggs: 130 average, range 124–207)

Ambystoma talpoideum, Mole salamander
(eggs: 300–590 average)

Ambystoma texanum, Small-mouthed salamander
(follicles: range 550–700)

Ambystoma tigrinum, Tiger salamander (eggs 421–7,631)

Ambystoma tremblayi, Tremblay's salamander
(follicles: 136–156 average)

Ambystoma sp., Unisexual members of the *Ambystoma jeffersonianum* complex (?, but likely to average ≤ 200 eggs)

Amphiumidae

Amphiuma means, Two-toed amphiuma
(eggs: 201 average, range 106–354)

Amphiuma pholeter, One-toed amphiuma (?)

Amphiuma tridactylum, Three-toed amphiuma
(follicles: 98–201 average)

Ascaphidae

Ascaphus montanus, Montana tailed frog
(eggs: range 33–97)

Ascaphus truei, Tailed frog
(eggs: range 28–96)

Bufonidae

Bufo alvarius, Colorado River toad
(eggs: 7,500–8,000 average)

Bufo americanus, American toad
(eggs: range 2,000–20,000)

Bufo baxteri, Wyoming toad
(eggs: range 1,000–8,000)

Bufo boreas, Western toad
(eggs: 5,200 average, < 12,000)

Bufo californicus, Arroyo toad
(eggs: 4,714 average; range 2,013–10,368)

Bufo canorus, Yosemite toad
(eggs: 1,500–2,000 estimate)

Bufo cognatus, Great Plains toad
(eggs: 11,074 average, range 1,342–45,054)

Bufo debilis, Green toad (eggs: 1,287)

Bufo exsul, Black toad (?)

Bufo fowleri, Fowler's toad
(eggs:3,700 average, range 2,000– > 6,300)

Bufo hemiophrys, Canadian toad (eggs: 3,354–5,842)

Bufo houstonensis, Houston toad
(eggs: 728 average, range 513–6,199)

Bufo marinus, Marine toad
(eggs: 10,000 average, range 6,970–36,100)

Bufo microscaphus, Arizona toad
(eggs: 4,500 average)

Bufo nebulifer, Coastal Plain toad
(eggs: 20,000 estimate)

Bufo nelsoni, Amargosa toad (?)

Bufo punctatus, Red-spotted toad
(eggs: 1,500 average, <5,000)

Bufo quercicus, Oak toad
(eggs: range 300–500)

Bufo retiformis, Sonoran green toad
(follicles: 50–200)

Bufo speciosus, Texas toad
(eggs: 11,000 estimate)

Bufo terrestris, Southern toad
(eggs: range 2,500–4,000)

Bufo woodhousii, Woodhouse's toad
(eggs: > 28,000)

Cryptobranchidae

Cryptobranchus alleganiensis, Hellbender
(eggs: range 200 – >400)

Dendrobatidae

Dendrobates auratus, Green and black dart-poison frog
(eggs: range 5–7)

Dicamptodontidae

Dicamptodon aterrimus, Idaho giant salamander
(eggs: range 135–200)

Dicamptodon copei, Cope's giant salamander
(eggs: 50 average, range 25–115)

Dicamptodon ensatus, California giant salamander
(eggs: range 70–100)

Dicamptodon tenebrosus, Coastal giant salamander
(eggs: range 83–146)

Hylidae

Acris crepitans, Northern cricket frog (Blanchard's, Eastern and Coastal) (eggs: <400)

Acris gryllus, Southern cricket frog (Florida and Coastal)
(eggs: <250)

Hyla andersonii, Pine Barrens treefrog (eggs: range 800–1,000)

Hyla arenicolor, Canyon treefrog (?)

Hyla avivoca, Bird-voiced treefrog
(eggs: 632 average, range 409–811)

Hyla chrysoscelis, Cope's gray treefrog (?)

Hyla cinerea, Green treefrog
(eggs: 700–2,152 average, range 478–3,946)

Hyla femoralis, Pine woods treefrog
(eggs: range 800–2,000)

Hyla gratiosa, Barking treefrog
(eggs: range 1,500–4,000)

Hyla squirella, Squirrel treefrog
(eggs: range 900–1,059)

Hyla versicolor, Gray treefrog
(eggs: 1,800 estimate)

Hyla wrightorum (eximia), Arizona treefrog (?)

Osteopilus septentrionalis, Cuban treefrog
(eggs: 3,961 average)

Pseudacris brachyphona, Mountain chorus frog
(eggs: range 300–1,500)

Pseudacris brimleyi, Brimley's chorus frog
(eggs: <300)

Pseudacris cadaverina, California treefrog (?)

Pseudacris clarkii, Spotted chorus frog
(eggs: 1,000 estimate)

Pseudacris crucifer, Spring peeper
(eggs: range 700–1,000)

Pseudacris feriarum, Southeastern chorus frog
(eggs: 500–1,459 range; geographic variation noted)

Pseudacris maculata, Boreal chorus frog (?)

Pseudacris nigrita, Southern chorus frog
(eggs: range 78–157)

Pseudacris ocularis, Little grass frog
(eggs: 100 – >200)

Pseudacris ornata, Ornate chorus frog
(eggs: >100)

Pseudacris regilla, Pacific treefrog
(eggs: 268 average, range 63–1,250)

Pseudacris streckeri, Strecker's chorus frog
(eggs: 469–608 average, range 148–1,012)

Pseudacris triseriata, Western chorus frog
(eggs: range 500–1,459)

Pternohyla fodiens, Lowland burrowing treefrog (?)

Smilisca baudinii, Mexican smilisca
(eggs: range 480–560; follicles: range 2,620–3,320)

Leptodactylidae

Eleutherodactylus augusti, Barking frog
(eggs: range 50–76)

Eleutherodactylus coqui, Coqui
(eggs: range 12–24)

Eleutherodactylus cystignathoides, Rio Grande chirping frog
(eggs: range 5–13)

Eleutherodactylus guttilatus, Spotted chirping frog
(eggs: range 5 – <15)

Eleutherodactylus marnockii, Cliff chirping frog
(eggs: range 8–20)

Eleutherodactylus martinicensis, Martinique greenhouse frog (?)

Eleutherodactylus planirostris, Greenhouse frog
(eggs: average 11–16; range 2–26)

Leptodactylus fragilis, White-lipped frog
(eggs: range 25–250, foam nest)

Microhylidae

Gastrophryne carolinensis, Eastern narrow-mouthed toad
(eggs: range 152–1,600)

Gastrophryne olivacea, Western narrow-mouthed toad
(eggs: range 650–2,100)

Hypopachus variolosus, Sheep Frog
(eggs: 700 estimate)

Pelobatidae

Scaphiopus couchii, Couch's spadefoot toad
(eggs: average >3,000)

Scaphiopus holbrookii, Eastern spadefoot toad (average 3,872,
range 2,332–5,468)

Scaphiopus hurterii, Hurter's spadefoot toad (?)

Spea bombifrons, Plains spadefoot toad
(eggs: <2,000)

Spea hammondii, Western spadefoot toad
(eggs: range 300–500)

Spea intermontana, Great Basin spadefoot toad
(eggs: range 300–1,000)

Spea multiplicata, Mexican spadefoot toad
(eggs: >1,000)

Pipidae

Xenopus laevis, African clawed frog (follicles: <17,000)

Plethodontidae

Aneides aeneus, Green salamander
(eggs: range 10–32)

Aneides ferreus, Clouded salamander
(eggs: range 9–17)

Aneides flavipunctatus, Black salamander
(follicles: range 8–25)

Aneides hardii, Sacramento Mountains salamander
(eggs: range 1–10)

Aneides lugubris, Arboreal salamander
(eggs: range 5–24; follicles: 5–26)

Aneides vagrans, Wandering salamander
(eggs: range 6–13; follicles: average 18, average 14–26)

Batrachoseps aridus, Desert slender salamander (?)

Batrachoseps attenuatus, California slender salamander
(eggs: average 7−>13)

Batrachoseps campi, Inyo Mountains salamander (?)

Batrachoseps diabolicus, Hell Hollow slender salamander (?)

Batrachoseps gabrieli, San Gabriel Mountains slender
salamander (?)

Batrachoseps gavilanensis, Gabilan Mountains slender
salamander (?)

Batrachoseps gregarius, Gregarious slender salamander
(eggs: average 7.3–15.3 average)

Batrachoseps incognitus, San Simeon slender salamander (?)

Batrachoseps kawia, Sequoia slender salamander (?)

Batrachoseps luciae, Santa Lucia Mountains slender
salamander (5.1–10.6)

Batrachoseps major, Garden slender salamander
(follicles: range 15–20)

Batrachoseps minor, Lesser slender salamander (?)

Batrachoseps nigriventris, Black-bellied slender salamander (?)

Batrachoseps pacificus, Channel Islands slender salamander
(follicles: 20 average)

Batrachoseps regius, Kings River slender salamander (?)

Batrachoseps relictus, Relictual slender salamander (?)

Batrachoseps robustus, Kern Plateau slender salamander
(eggs and follicles: range 4–6)

Batrachoseps simatus, Kern Canyon slender salamander (?)

Batrachoseps stebbinsi, Tehachapi slender salamander (?)

Batrachoseps wrighti, Oregon slender salamander
(eggs: 5.1 average; follicles: 6.3 average, range 3–11)

Desmognathus abditus, Cumberland dusky salamander (?)

Desmognathus aeneus, Seepage salamander
(eggs: range 3–19)

Desmognathus apalachicolae, Apalachicola dusky
salamander (?)

Desmognathus auriculatus, Southern dusky salamander
(eggs: range 9–26)

Desmognathus brimleyorum, Ouachita dusky salamander
(eggs: range 20–29)

Desmognathus carolinensis, Carolina Mountain dusky
salamander (follicles: range 12–40)

Desmognathus conanti, Spotted dusky salamander
(eggs: range 13–24)

Desmognathus folkertsi, Dwarf black-bellied salamander
(follicles: average 39, range 2–62)

Desmognathus fuscus, Northern dusky salamander
(eggs: range 13–34)

Desmognathus imitator, Imitator salamander
(eggs: 19 average, range 13–30)

Desmognathus marmoratus, Shovel-nosed salamander
(eggs: range 20–65)

Desmognathus monticola, Seal salamander
(eggs: 22–27 average, range 13–39)

Desmognathus ochrophaeus, Allegheny Mountain dusky
salamander (eggs: range 3–37)

Desmognathus ocoee, Ocoee salamander
(eggs and follicles: range 5–29)

Desmognathus orestes, Blue Ridge dusky salamander
(eggs: 11 average, range 14–16)

Desmognathus quadramaculatus, Black-bellied salamander
(eggs: 45 average, range 38–55)

Desmognathus santeetlah, Santeetlah dusky salamander
(eggs: 17–20 average)

Desmognathus welteri, Black Mountain salamander
(eggs: 26 average, range 18–33)

Desmognathus wrighti, Pigmy salamander
(eggs: range 3–8)

Ensatina eschscholtzii, Ensatina
(eggs: range 3–25, but usually 9–16)

Eurycea aquatica, Dark-sided salamander
(eggs: 80 average, range 60–96)

Eurycea bislineata, Northern two-lined salamander
(eggs: 18–30 average, range 12–43; follicles: 46 average,
range 19–86)

Eurycea chamberlaini, Chamberlain's dwarf salamander
(follicles: average 45, range 35–64)

Eurycea chisholmensis, Salado salamander (?)

Eurycea cirrigera, Southern two-lined salamander
(eggs: 18–50 average, range 12–110)

Eurycea guttolineata, Three-lined salamander
(eggs: range 8–14)

Eurycea junaluska, Junaluska salamander
(eggs: 38 average, range 30–49; follicles: 51 average,
range 41–68)

Eurycea latitans, Cascade Caverns salamander (?)

Eurycea longicauda, Long-tailed salamander
(eggs: range 61–106)

Eurycea lucifuga, Cave salamander
(follicles: 68–78 average, range 49–120)

Eurycea multiplicata, Many-ribbed salamander
(eggs: 10–13 average, range 2–21)

Eurycea nana, San Marcos salamander
(eggs: 33 average)

Eurycea naufragia, Georgetown salamander (?)

Eurycea neotenes, Texas salamander (eggs: 12)

Eurycea pterophila, Fern Bank salamander (?)

Eurycea quadridigitata, Dwarf salamander
(eggs: range 7–48; follicles: range 14–59)

Eurycea rathbuni, Texas blind salamander (?)

Eurycea robusta, Blanco blind salamander (?)

Eurycea sosorum, Barton Springs salamander
(eggs: 18–29)

Eurycea tonkawae, Jollyville Plateau salamander (?)

Eurycea tridentifera, Comal blind salamander
(eggs: range 7–18)

Eurycea troglodytes, Valdina Farms salamander (?)

Eurycea tynerensis, Oklahoma salamander
(follicles: range 1–11)

Eurycea waterlooensis, Austin blind salamander (?)

Eurycea wilderae, Blue Ridge two-lined salamander
(eggs: 8–56)

Eurycea sp. 1, Comal Springs salamander
(eggs: range 19–>50)

Eurycea sp. 2, Pedernales Springs salamander (?)

Gyrinophilus gulolineatus, Berry Cave salamander (?)

Gyrinophilus palleucus, Tennessee cave salamander (?)

Gyrinophilus porphyriticus, Spring salamander
(follicles: range 16–106)

Gyrinophilus subterraneus, West Virginia spring salamander (?)

Haideotriton wallacei, Georgia blind salamander (?)

Hemidactylium scutatum, Four-toed salamander
(follicles: range 4–80)

Hydromantes brunus, Limestone salamander (?)

Hydromantes platycephalus, Mt. Lyell salamander (?)

Hydromantes shastae, Shasta salamander
(eggs: 9 average)

Phaeognathus hubrichti, Red Hills salamander
(eggs: 16)

Plethodon ainsworthi, Bay Springs salamander (?)

Plethodon albagula, Western slimy salamander
(eggs: range 6–15)

Plethodon amplus, Blue Ridge gray-cheeked salamander (?)

Plethodon angusticlavius, Ozark salamander
(eggs: 5.3 average, range 3–9)

Plethodon aureolus, Tellico salamander (?)

Plethodon caddoensis, Caddo Mountain salamander
(eggs: 7.8 average, range 4–11; follicles: 11.3 average)

Plethodon chattahoochee, Chattahoochee slimy salamander (?)

Plethodon cheoah, Cheoah Bald salamander (?)

Plethodon chlorobryonis, Atlantic Coast slimy salamander
(follicles: 16, 19)

Plethodon cinereus, Eastern red-backed salamander
(eggs: 6–9 average, range 1–14)

Plethodon cylindraceus, White-spotted slimy salamander (?)

Plethodon dorsalis, Northern zigzag salamander
(eggs: 5.3 average, range 3–9)

Plethodon dunni, Dunn's salamander
(eggs: 9.4 average, range 4–15)

Plethodon electromorphus, Northern ravine salamander
(eggs: 12; follicles: 9–15)

Plethodon elongatus, Del Norte salamander
(eggs: 7 average, range 3–11)

Plethodon fourchensis, Fourche Mountain salamander
(follicles: 14 average)

Plethodon glutinosus, Northern slimy salamander
(eggs: range 13–38)

Plethodon grobmani, Southeastern slimy salamander
(eggs: range 5–11; follicles range 10–22)

Plethodon hoffmani, Valley and Ridge salamander
(follicles: 4.6 average, range 3–8)

Plethodon hubrichti, Peaks of Otter salamander
(eggs: range 5–15)

Plethodon idahoensis, Coeur d' Alene salamander
(eggs: range 1–13)

Plethodon jordani, Jordan's salamander (?)

Plethodon kentucki, Cumberland Plateau salamander
(eggs: range 9–12)

Plethodon kiamichi, Kiamichi slimy salamander (?)

Plethodon kisatchie, Louisiana slimy salamander (?)

Plethodon larselli, Larch Mountain salamander
(eggs: 7.3 average, range 2–12)

Plethodon meridianus, South Mountain gray-cheeked
salamander (?)

Plethodon metcalfi, Southern gray-cheeked salamander (?)

Plethodon mississippi, Mississippi slimy salamander
(eggs: 17)

Plethodon montanus, Northern gray-cheeked salamander (?)

Plethodon neomexicanus, Jemez Mountains salamander
(eggs: 7.7 average, range 5–12)

Plethodon nettingi, Cheat Mountain salamander
(eggs: range 4–17)

Plethodon ocmulgee, Ocmulgee slimy salamander (?)

Plethodon ouachitae, Rich Mountain salamander
(follicles: 15.4–16.7 average)

Plethodon petraeus, Pigeon Mountain salamander
(eggs: 19.3 average)

Plethodon punctatus, Cow Knob salamander
(eggs: range 7–16)

Plethodon richmondi, Southern ravine salamander
(follicles: 8.3–8.5 average)

Plethodon savannah, Savannah slimy salamander (?)

Plethodon sequoyah, Sequoyah slimy salamander (?)

Plethodon serratus, Southern red-backed salamander
(eggs: 5; follicles: 5.9–7.0 average)

Plethodon shenandoah, Shenandoah salamander
(eggs: 13 average, range 4–19)

Plethodon shermani, Red-legged salamander (?)

Plethodon stormi, Siskiyou Mountains salamander
(follicles: 9 average, range 2–18)

Plethodon teyahalee, Southern Appalachian salamander (?)

Plethodon vandykei, Van Dyke's salamander
(follicles: 11, 14)

Plethodon variolatus, South Carolina slimy salamander (?)

Plethodon vehiculum, Western red-backed salamander
(eggs: range 4–19)

Plethodon ventralis, Southern zigzag salamander (?)

Plethodon virginia, Shenandoah Mountain salamander
(eggs: 4.6 average)

Plethodon websteri, Webster's salamander
(eggs: <30; follicles: 5.8 average, range 3–8)

Plethodon wehrlei, Wehrle's salamander
(follicles: range 7–24)

Plethodon welleri, Weller's salamander
(eggs: range 4–11)

Plethodon yonahlossee, Yonahlossee salamander
(follicles: range 19–27)

Pseudotriton montanus, Mud salamander
(follicles: range 77–192)

Pseudotriton ruber, Red salamander
(eggs: 80 average, range 29–130)

Stereochilus marginatus, Many-lined salamander
(eggs: 57 average; range 16–121)

Typhlotriton (Eurycea) spelaeus, Grotto salamander
(eggs: 13)

Proteidae

Necturus alabamensis, Blackwarrior waterdog (?)

Necturus beyeri, Gulf Coast waterdog
(eggs: range 4–40 in untended nests,
range 26–37 in tended nests; follicles: 47.5
average, range 28–76)

Necturus lewisi, Neuse River waterdog
(eggs: range 19–35)

Necturus maculosus, Mudpuppy
(eggs: range 36–83)

Necturus punctatus, Dwarf waterdog
(follicles: range 15–55)

Necturus cf. beyeri, Loding's waterdog
(eggs: range 20–40 estimate)

Ranidae

Rana areolata, Crawfish frog
(eggs: range 2,000–7,000)

Rana aurora, Northern red-legged frog
(eggs: 530–830 average, range 200–1,100)

Rana berlandieri, Rio Grande leopard frog (?)

Rana blairi, Plains leopard frog
(eggs: range 4,000–6,500)

Rana boylii, Foothill yellow-legged frog
(eggs: 900 average, range 300–2,000)

Rana capito, Gopher frog
(eggs: 1,244 average, range 1,500–2,000)

Rana cascadae, Cascades frog
(eggs: range 300–800)

Rana catesbeiana, American bullfrog
(eggs: 11,149 average, range ≤ 20,000)

Rana chiricahuensis, Chiricahua leopard frog
(eggs: range 300–1,485)

Rana clamitans, Green frog (eggs: first clutches range
1,000–7,000; second clutches range 1,000–1,500)

Rana draytonii, California red-legged frog
(eggs: 2,000 average, range 300–4,000)

Rana fisheri, Vegas Valley leopard frog (?)

Rana grylio, Pig frog
(eggs: ≤15,000)

Rana heckscheri, River frog
(eggs: 5,000–8,000 estimate)

Rana luteiventris, Columbia spotted frog
(eggs: 600 average, range 150–2,400)

Rana muscosa, Mountain yellow-legged frog
(eggs: 233 average, range 100–350)

Rana okaloosae, Florida bog frog (?)

Rana onca, Relict leopard frog ("several hundred eggs")

Rana palustris, Pickerel frog (?)

Rana pipiens, Northern leopard frog
(eggs: range 645–7,648)

Rana pretiosa, Oregon spotted frog
(eggs: 598–643 average)

Rana rugosa, Wrinkled frog
(eggs: range 400–1,350)

Rana septentrionalis, Mink frog
(eggs: range 500–4,000)

Rana sevosa, Dusky gopher frog
(egg: range 500–2,800)

Rana sphenocephala, Southern leopard frog
(eggs: range 1,200 –<5,000)

Rana subaquavocalis, Ramsey Canyon leopard frog
(eggs: 1,518 average, range 1,200–2,040)

Rana sylvatica, Wood frog
(eggs: range 300–1,500)

Rana tarahumarae, Tarahumara frog
(eggs: 1,083 average, range 527–1,635)

Rana virgatipes, Carpenter frog
(eggs: range 200–600)

Rana yavapaiensis, Lowland leopard frog (?)

Rhinophrynidae

Rhinophrynus dorsalis, Burrowing toad (?)

Rhyacotritonidae

Rhyacotriton cascadae, Cascade torrent salamander
(eggs: 8 average, range 2–14)

Rhyacotriton kezeri, Columbia torrent salamander
(eggs: range 2–16 estimate)

Rhyacotriton olympicus, Olympic torrent salamander
(follicles: 8 average)

Rhyacotriton variegatus, Southern torrent salamander
(eggs: 8, 11)

Salamandridae

Notophthalmus meridionalis, Black-spotted newt (?)

Notophthalmus perstriatus, Striped newt (?)

Notophthalmus viridescens, Eastern newt
(eggs: range 200–375)

Taricha granulosa, Rough-skinned newt (?)

Taricha rivularis, Red-bellied newt
(eggs: 10 average, range 6–16)

Taricha torosa, California newt
(follicles: range 130–160)

Sirenidae

Pseudobranchus axanthus, Southern dwarf siren
(Everglades and Narrow-striped) (?)

Pseudobranchus striatus, Northern dwarf siren
(Broad-striped, Gulf hammock and Slender) (?)

Siren intermedia, Lesser siren,
(eggs: 200–500 average)

Siren lacertina, Greater siren
(eggs: 500 estimate)

NOTE: Egg counts in parentheses. As modified (see text) from Crother et al. (2000, 2003).

Species with Terrestrial Eggs

Salamanders

Amphiuma means, Two-toed amphiuma (nest in water or on land, aquatic larvae)

Ambystoma cingulatum, Flatwoods salamander (nest on land, aquatic larvae)

Ambystoma opacum, Marbled salamander (nest on land, aquatic larvae)

Aneides aeneus, Green salamander

Aneides ferreus, Clouded salamander

Aneides flavipunctatus, Black salamander

Aneides hardii, Sacramento Mountain salamander

Aneides lugubris, Arboreal salamander

Aneides vagrans, Wandering salamander

Ensatina eschscholtzii, Ensatina

Batrachoseps aridus, Desert slender salamander

Batrachoseps attenuatus, California slender salamander (communal nesting)

Batrachoseps campi, Inyo Mountains salamander

Batrachoseps diabolicus, Hell Hollow slender salamander

Batrachoseps gabrieli, San Gabriel slender salamander

Batrachoseps gavilanensis, Gabilan Mountains slender salamander

Batrachoseps gregarius, Gregarious slender salamander

Batrachoseps incognitus, San Simeon slender salamander

Batrachoseps kawia, Sequoia slender salamander

Batrachoseps luciae, Santa Lucia Mountains slender salamander

Batrachoseps major, Garden slender salamander

Batrachoseps minor, Lesser slender salamander

Batrachoseps nigriventris, Black-bellied slender salamander

Batrachoseps pacificus, Pacific slender salamander

Batrachoseps regius, Kings River slender salamander

Batrachoseps relictus, Relictual slender salamander

Batrachoseps robustus, Kern Plateau salamander

Batrachoseps simatus, Kern Canyon slender salamander

Batrachoseps stebbensi, Tehachapi slender salamander

Batrachoseps wrighti, Oregon slender salamander

Desmognathus aeneus, Seepage salamander (non-feeding larvae)

Desmognathus auriculatus, Southern dusky salamander (nest on land, aquatic larvae, larvae migrate into water)

Desmognathus ochrophaeus, Allegheny Mountain dusky salamander (nest on land, aquatic larvae, larvae migrate into water [Virginia])

Desmognathus wrighti, Pigmy salamander (no larvae, non-feeding larvae)

Hemidactylium scutatum, Four-toed salamander (nest on land, aquatic larvae, larvae drop into water)

Hydromantes shastae, Shasta salamander

Phaeognathus hubrichti, Red Hills salamander

Plethodon ainsworthi, Catahoula salamander

Plethodon albagula, Western slimy salamander

Plethodon amplus, Blue Ridge gray-cheeked salamander

Plethodon angusticlavius, Ozark salamander

Plethodon aureolus, Tellico salamander

Plethodon caddoensis, Caddo Mountain salamander

Plethodon chattahoochee, Chattahoochee slimy salamander

Plethodon cheoah, Cheoah Bald salamander

Plethodon chlorobryonis, Atlantic Coast slimy salamander

Plethodon cinereus, Eastern red-backed salamander

Plethodon cylindraceus, White-spotted slimy salamander

Plethodon dorsalis, Northern zigzag salamander

Plethodon dunni, Dunn's salamander

Plethodon electromorphus, Northern ravine salamander

Plethodon elongatus, Del Norte salamander

Plethodon fourchensis, Fourche Mountain salamander

Plethodon glutinosus, Northern slimy salamander

Plethodon grobmani, Southeastern slimy salamander

Plethodon hoffmani, Valley and Ridge salamander

Plethodon hubrichti, Peaks of Otter salamander

Plethodon idahoensis, Cour d'Alene salamander

Plethodon jordani, Jordan's salamander

Plethodon kentucki, Cumberland Plateau salamander

Plethodon kiamichi, Kiamichi slimy salamander

Plethodon kisatchie, Louisiana slimy salamander

Plethodon larselli, Larch Mountain salamander

Plethodon meridianus, South Mountain gray-cheeked salamander

Plethodon metcalfi, Southern gray-cheeked salamander

Plethodon mississippi, Mississippi slimy salamander

Plethodon montanus, Northern gray-cheeked salamander

Plethodon neomexicanus, Jemez Mountains salamander

Plethodon nettingi, Cheat Mountain salamander

Plethodon ocmulgee, Ocmulgee slimy salamander

Plethodon ouachitae, Rich Mountain salamander

Plethodon petraeus, Pigeon Mountain salamander

Plethodon punctatus, Cow Knob salamander

Plethodon richmondi, Southern ravine salamander

Plethodon savannah, Savannah slimy salamander

Plethodon sequoyah, Sequoyah slimy salamander

Plethodon serratus, Southern red-backed salamander

Plethodon shenandoah, Shenandoah salamander

Plethodon shermani, Red-legged salamander

Plethodon stormi, Siskiyou Mountains salamander

Plethodon teyahalee, Southern Appalachian salamander

Plethodon vandykei, Van Dyke's salamander

Plethodon variolatus, South Carolina slimy salamander

Plethodon vehiculum, Western red-backed salamander

Plethodon ventralis, Southern zigzag salamander

Plethodon virginia, Shenandoah Mountain salamander

Plethodon websteri, Webster's salamander

Plethodon wehrlei, Wehrle's salamander

Plethodon welleri, Weller's salamander

Plethodon yonahlossee, Yonahlossee salamander

Frogs

Dendrobates auratus, Green and black dart-poison frog

Eleutherodactylus augusti, Barking frog

Eleutherodactylus coqui, Coqui

Eleutherodactylus cystignathoides, Rio Grande chirping frog

Eleutherodactylus guttilatus, Spotted chirping frog

Eleutherodactylus marnockii, Cliff chirping frog

Eleutherodactylus planirostris, Greenhouse frog

Leptodactylus fragilis, White-lipped frog

Species of U.S. Amphibians that Exhibit Some Form of Parental Care or Communal Nesting

Ambystoma opacum, Marbled salamander (F)

Amphiuma means, Two-toed amphiuma (F)

Amphiuma pholeter, One-toed amphiuma (?)

Amphiuma tridactylum, Three-toed amphiuma (F)

Aneides aeneus, Green salamander (F)

Aneides ferreus, Clouded salamander (F)

Aneides flavipunctatus, Black salamander (F)

Aneides hardii, Sacramento Mountain salamander (F)

Aneides lugubris, Arboreal salamander (F; M often present)

Cryptobranchus alleganiesis, Hellbender (M)

Desmognathus abditus, Cumberland Dusky salamander (F?)

Desmognathus aeneus, Seepage salamander (F)

Desmognathus apalachicolae, Apalachicola dusky salamander (F)

Desmognathus auriculatus, Southern dusky salamander (F)

Desmognathus brimleyorum, Ouachita dusky salamander (F)

Desmognathus carolinensis, Carolina Mountain dusky salamander (F)

Desmognathus conanti, Spotted dusky salamander (F) (communal)

Desmognathus folkertsi, Dwarf black-bellied salamander (F) (presumed)

Desmognathus fuscus, Dusky salamander (F) (communal)

Desmognathus imitator, Imitator salamander (F)

Desmognathus marmoratus, Shovel-nosed salamander (F)

Desmognathus monticola, Seal salamander (F)

Desmognathus ochrophaeus, Allegheny Mountain dusky salamander (F)

Desmognathus ocoee, Ocoee salamander (F) (communal)

Desmognathus orestes, Blue Ridge dusky salamander (F)

Desmognathus quadramaculatus, Black-bellied salamander (F)

Desmognathus santeetlah, Santeetlah dusky salamander (F) (communal)

Desmognathus welteri, Black Mountain dusky salamander (F)

Desmognathus wrighti, Pigmy salamander (F)

Dicamptodon aterrimus, Idaho giant salamander (?)

Dicamptodon copei, Cope's giant salamander (F)

Dicamptodon ensatus, California giant salamander (?)

Dicamptodon tenebrosus, Pacific giant salamander (F)

Eleutherodactylus augusti, Barking frog (F)

Eleutherodactylus coqui, Coqui (M)

Ensatina eschscholtzii, Ensatina (F)

Hydromantes shastae, Limestone salamander (F)

Hemidactylium scutatum, Four-toed salamander (F) (communal)

Necturus alabamensis, Alabama waterdog (?)

Necturus beyeri, Gulf Coast waterdog (F)

Necturus lewisi, Neuse River waterdog (M?)

Necturus maculosus, Mudpuppy

Necturus punctatus, Dwarf waterdog (F)

Necturus cf. *beyeri*, Loding's waterdog (F?)

Plethodon ainsworthi, Catahoula salamander (?)

Plethodon albagula, Western slimy salamander (?)

Plethodon amplus, Blue Ridge gray-cheeked salamander (?)

Plethodon angusticlavius, Ozark zigzag salamander (F) (see *P. dunni*)

Plethodon aureolus, Tellico salamander (F) (presumably brood)

Plethodon caddoensis, Caddo Mountain salamander (?)

Plethodon chattahoochee, Chattahoochee slimy salamander (?)

Plethodon cheoah, Cheoah Bald salamander (?)

Plethodon chlorobryonis, Atlantic Coast slimy salamander (?)

Plethodon cinereus, Eastern red-backed salamander (F)

Plethodon cylindraceus, White-spotted slimy salamander (?)

Plethodon dorsalis, Zigzag salamander (?)

Plethodon dunni, Dunn's salamander (F)

Plethodon electromorphus, Northern ravine salamander (?)

Plethodon elongatus, Del Norte salamander (F)

Plethodon fourchensis, Fourche Mountain salamander (?)

Plethodon glutinosus, Northern slimy salamander (F or M/F)

Plethodon grobmani, Southeastern slimy salamander (?)

Plethodon hoffmani, Valley and Ridge salamander (?)

Plethodon hubrichti, Peaks of Otter salamander (?)

Plethodon idahoensis, Coeur d'Alene salamander (?)

Plethodon jordani, Jordan's salamander (?)

Plethodon kentucki, Cumberland Plateau salamander (F)

Plethodon kiamichi, Kiamichi slimy salamander (?)

Plethodon kisatchie, Louisiana slimy salamander (?)

Plethodon larselli, Larch Mountain salamander (?)

Plethodon meridianus, South Mountain gray-cheeked salamander (?)

Plethodon metcalfi, Southern gray-cheeked salamander (?)

Plethodon mississippi, Mississippi slimy salamander (?)

Plethodon montanus, Northern gray-cheeked salamander (?)

Plethodon neomexicanus, Jemez Mountains salamander (?)

Plethodon nettingi, Cheat Mountain salamander (F)

Plethodon ocmulgee, Ocmulgee slimy salamander (?)

Plethodon ouachitae, Rich Mountain salamander (?)

Plethodon petraeus, Pigeon Mountain salamander (?)

Plethodon punctatus, White-spotted salamander (?)

Plethodon richmondi, Ravine salamander (F) (presumably brood)

Plethodon savannah, Savannah slimy salamander (?)

Plethodon sequoyah, Sequoyah slimy salamander (?)

Plethodon serratus, Southern red-backed salamander (F) (presumably brood)

Plethodon shenandoah, Shenandoah salamander (F)

Plethodon shermani, Red-legged salamander (?)

Plethodon stormi, Siskiyou Mountains salamander (F)

Plethodon teyahalee, Southern Appalachian salamander (?)

Plethodon vandykei, Van Dyke's salamander (F)

Plethodon variolatus, South Carolina slimy salamander (?)

Plethodon vehiculum, Western red-backed salamander (?)

Plethodon ventralis, Southern zigzag salamander (?)

Plethodon virginia, Shenandoah Mountain salamander (?)

Plethodon websteri, Webster's salamander (?)

Plethodon wehrlei, Wehrle's salamander (F)

Plethodon welleri, Weller's salamander (F)

Plethodon yonahlossee, Yonahlossee salamander (?)

Pseudotriton montanus, Red salamander (F) (communal)

Pseudotriton ruber, Mud salamander (F) (communal)

Rana sylvatica, Wood frog (Eggs laid communally,) (females do not brood)

Rhyacotriton kezeri, Columbia torrent salamander (Eggs laid communally, (females do not brood)

Rhyacotriton variegatus, Southern torrent salamander (Eggs laid communally, (females do not brood)

Siren intermedia, Lesser siren (F)

NOTE: F = female parental care, M = male parental care.

Genera or Species with Aquatic Egg, Larval, and Adult Life History Stages

Ascaphus

Xenopus

Siren

Amphiuma

Cryptobranchus

Necturus

Notophthalmus (except terrestrial eft stage)

Taricha

Haideotriton

Eurycea chisholmensis, Salado salamander

Eurycea latitans, Cascade Caverns salamander

Eurycea multiplicata, Many-ribbed salamander

Eurycea nana, San Marcos salamander

Eurycea naufragia, Georgetown salamander

Eurycea neotenes, Texas salamander

Eurycea pterophila, Fern Bank salamander

Eurycea rathbuni, Texas blind salamander

Eurycea robusta, Blanco blind salamander

Eurycea tonkawae, Jollyville Plateau salamander

Eurycea troglodytes, Valdina Farms salamander

Eurycea tynerensis, Oklahoma salamander

Eurycea waterlooensis, Austin blind salamander

Eurycea sp. 1, Comal Springs salamander

Eurycea sp. 2, Pedernales Springs salamander

U.S. Amphibians that Breed in Seeps, Springs, or Small Streams

Ascaphus montanus, Montana tailed frog

Ascaphus truei, Tailed frog

Ambystoma barbouri, Streamside salamander

Ambystoma texanum, Small-mouthed salamander (rarely)

Cryptobranchus alleganiensis, Hellbender

Desmognathus abditus, Cumberland Dusky salamander (?)

Desmognathus aeneus, Seepage salamander

Desmognathus apalachicolae, Apalachicola dusky salamander

Desmognathus brimleyorum, Ouachita Mountain dusky salamander

Desmognathus carolinensis, Carolina dusky salamander

Desmognathus conanti, Spotted dusky salamander

Desmognathus folkertsi, Dwarf black-bellied salamander

Desmognathus fuscus, Dusky salamander

Desmognathus imitator, Imitator salamander,

Desmognathus marmoratus, Shovel-nosed salamander

Desmognathus monticola, Seal salamander

Desmognathus ochrophaeus, Allegheny Mountain dusky salamander

Desmognathus ocoee, Ocoee salamander

Desmognathus orestes, Blue Ridge dusky salamander

Desmognathus quadramaculatus, Black-bellied salamander

Desmognathus santeetlah, Santeetlah dusky salamander

Desmognathus welteri, Black Mountain dusky salamander

Desmognathus wrighti, Pigmy salamander

Dicamptodon aterrimus, Idaho giant salamander

Dicamptodon copei, Cope's giant salamander

Dicamptodon ensatus, California giant salamander

Dicamptodon tenebrosus, Pacific giant salamander

Eurycea bislineata, Two-lined salamander

Eurycea chamberlani, Chamberlain's dwarf salamander

Eurycea cirrigera, Southern two-lined salamander

Eurycea guttolineata, Three-lined salamander

Eurycea junaluska, Junaluska salamander

Eurycea longicauda, Long-tailed salamander

Eurycea lucifuga, Cave salamander

Eurycea multiplicata, Many-ribbed salamander

Eurycea rathbuni, Texas blind salamander

Eurycea robusta, Blanco blind salamander

Eurycea sosorum, Barton Springs salamander

Eurycea tynerensis, Oklahoma salamander

Eurycea wilderae, Blue Ridge two-lined salamander

Gyrinophilus porphyriticus, Spring salamander

Gyrinophilus subterraneus, West Virginia spring salamander

Haideotriton wallacei, Georgia blind salamander

Necturus alabamensis, Alabama waterdog

Necturus beyeri, Gulf Coast waterdog

Necturus lewisi, Neuse River waterdog

Necturus maculosus, Mudpuppy

Necturus cf. *beyeri*, Loding's waterdog

Pseudotriton montanus, Mud salamander

Pseudotriton ruber, Red salamander

Rhyacotriton cascadae, Cascade torrent salamander

Rhyacotriton kezeri, Columbia torrent salamander

Rhyacotriton olympicus, Olympic torrent salamander

Rhyacotriton variegatus, Southern torrent salamander

Taricha rivularis, Red-bellied newt

Taricha torosa, California newt

Typhlotriton (Eurycea) spelaeus, Grotto salamander

Species of Amphibians Found in Caves

Ambystoma jeffersonianum (Jefferson salamander; in part)

Ambystoma maculatum (Spotted salamander; in part)

Ambystoma tigrinum (Tiger salamander; Chaves County, Arizona, in part)

Bufo nebulifer (Coastal-Plain toad; in part)

Desmognathus monticola (Seal salamander; in part)

Eleutherodactylus augusti (Barking frog; in part)

Eleutherodactylus marnockii (Cliff chirping frog; in part)

Eurycea latitans (Cascade Caverns salamander; in part)

Eurycea longicauda (Long-tailed salamander; in part)

Eurycea lucifuga (Cave salamander; in part)

Eurycea multiplicata (Many-ribbed salamander; in part)

Eurycea naufragia (Georgetown salamander; in part)

Eurycea neotenes (Texas salamander; in part)

Eurycea pterophila (Fern Bank salamander; in part)

Eurycea rathbuni (Texas blind salamander; in part)

Eurycea robusta (Blanco blind salamander; in part)

Eurycea tonkawae (Jollyville Plateau salamander; in part)

Eurycea tridentifera (Comal blind salamander; in part)

Eurycea troglodytes (Valdina Farms salamander; in part)

Eurycea waterlooensis (Austin blind salamander; in part)

Gyrinophilus gulolineatus (Berry Cave salamander; in part)

Gyrinophilus palleucus (Tennessee cave salamander; in part)

Gyrinophilus porphyriticus (Kentucky spring salamander; in part)

Gyrinophilus subterraneus (West Virginia spring salamander; in part)

Hydromantes shastae (Shasta salamander; in part)

Plethodon albagula (Western slimy salamander; in part)

Plethodon glutinosus (Northern slimy salamander; in part)

Plethodon hoffmani (Valley and Ridge salamander; in part)

Plethodon petraeus (Pigeon Mountain salamander ; in part)

Plethodon wehrlei (Weller's salamander; in part)

Rana catesbeiana (American bullfrog; in part)

Rana clamitans (Green frog; in part)

Rana palustris (Pickerel frog; in part)

Typhlotriton (Eurycea) spelaeus (Grotto salamander; in part)

Notophthalmus viridescens (Eastern newt; in part)

NOTE: See text for full explanation.

List of Federally Listed Species

Salamanders

Ambystoma californiense, California tiger salamander, E

Ambystoma cingulatum, Flatwoods salamander, T

Ambystoma macrodactylum croceum, Santa Cruz long-toed salamander, E

Ambystoma tigrinum stebbinsi, Sonoran tiger salamander, E

Batrachoseps aridus, Desert slender salamander, E

Eurycea nana, San Marcos salamander, T

Eurycea rathbuni, Texas blind salamander, E

Eurycea sosorum, Barton Springs salamander, E

Phaeognathus hubrichti, Red Hills salamander, T

Plethodon nettingi, Cheat Mountain salamander, T

Plethodon shenandoah, Shenandoah salamander, E

Frogs

Bufo baxteri, Wyoming toad, E

Bufo californicus, Arroyo toad, E

Bufo houstonensis, Houston toad, E

Rana capito sevosa, Mississippi gopher frog, E

Rana chircahuensis, Chiricahua leopard frog, T

Rana draytonii, California red-legged frog, T

Rana fisheri Vegas, Valley leopard frog, Extn

Rana muscosa, Mountain yellow-legged frog, E*

Rana tarahumarae, Tarahumara frog, Extr

NOTE: This list excludes Puerto Rico and includes Tarahumara frogs, which are thought to be extirpated from the United States and Vegas Valley leopard frogs, which are presumed extinct. E = endangered; Extn = extinct; Extr = extirpated; T = threatened; * = southern populations. Data obtained from http://ecos.fws.gov/tess_public/ TESSWebpageVipListed?code+V&listings = 0#D, on 20 August 2003.

ANURA

Family Ascaphidae

Ascaphus montanus Nielson, Lohman, and Sullivan, 2001

MONTANA (ROCKY MOUNTAIN) TAILED FROG

Michael J. Adams

Nielson et al. (2001) recommended that the genus *Ascaphus* be split into two species: tailed frogs *(A. truei)* and Montana tailed frogs *(A. montanus)*. Their analysis was based on divergence of mitochondrial DNA and was consistent with previous allozyme work (Daugherty, 1979). The following account highlights references that are specific to the new species. Refer to the *Ascaphus truei* account for a complete review of the genus *Ascaphus*.

few comparisons have been made. Montana tailed frogs have a 3-yr larval period (Metter, 1967; Daugherty and Sheldon, 1982a) and appear highly philopatric (Daugherty and Sheldon, 1982b). Clutch sizes reported from natural nests (n = 2 nests; Franz, 1970) and from dissected females (Metter, 1964a, 1967) range from 33–97, which appears somewhat higher than tailed frog clutches from the Cascade Mountains and Coastal Ranges (reviewed by Bury et al., 2001). Montana tailed frogs may only oviposit every other year (Metter, 1964a). Adams and Frissell (2001) reported seasonal movements of adults consistent with an avoidance of warm water temperatures.

lions of years (since the late Cretaceous; Metter and Pauken, 1969). They may have once had a nearly continuous distribution, but since the late Miocene or early Pliocene their range has been shrinking due to geological events including lava flows, mountain uplifting, and glaciation (Metter and Pauken, 1969; Nielson et al., 2001). Tailed frogs, including the recently recognized *A. montanus* (see Nielson et al., 2001 and account above) currently inhabit the Cascade Mountains of British Columbia (Ricker and Logier, 1935; Dupuis et al., 2000), Washington (Van Winkle, 1922; Svihla, 1933; Svihla and Svihla, 1933; Visalli and Leonard, 1994), and Oregon (Fitch, 1936; Metter, 1967; Smith, 1997); the Coast Range of British Columbia north to the Noss River (Carl, 1945; Dupuis et al., 2000); the Olympic Mountains (Gaige, 1920; Noble and Putnam, 1931), the Willipa Hills (Adams and Wilson, 1993; Manlow, 1994), the Oregon Coast Range (Metter, 1967; Bury et al., 1991a); and northern California south along the coast to Mendocino County (Myers, 1931a, 1943; Shapovalov, 1937; Salt, 1952; Bury, 1968; Welsh, 1985) and east to Shasta County (Simons and Simons, 1998). Inland, tailed frogs (now recognized as *A. montanus*) occur in the Blue Mountains of southeast Washington (Metter, 1964a); the Wallowa Mountains of northeast Oregon (Ferguson, 1952; Bull and Carter, 1996), central Idaho and the panhandle (Linsdale, 1933a; Corbit, 1960; Maughan et al., 1980); and western Montana (Smith, 1932; Franz and Lee, 1970), including at least one population on the eastern slope of the Rocky Mountains (Donaldson, 1934). Tailed frogs occur from sea level to near timberline: 1,600 m on Mount Rainier; 2,100 m in the Wallowa Mountains (Nussbaum et al., 1983; Leonard et al., 1993). The distribution of tailed frogs does not appear to have changed drastically in the past 100 yr. They are recovering rapidly from extirpations that occurred in

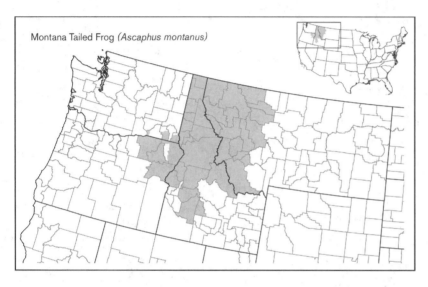

Montana Tailed Frog *(Ascaphus montanus)*

Montana tailed frogs occur in the Blue Mountains of southeast Washington (Metter, 1964a), central Idaho and the panhandle (Linsdale, 1933a; Corbit, 1960; Maughan et al., 1980), and western Montana (Smith, 1932; Franz and Lee, 1970) including at least one population on the eastern slope of the Rocky Mountains (Donaldson, 1934). Populations in the Wallowa Mountains of northeastern Oregon and the Seven Devils Mountains of western Idaho were not examined by Nielson et al. (2001) but were most similar to the Montana populations in Daugherty's (1979) allozyme analysis. Ritland et al. (2000) found that interior British Columbia populations of *Ascaphus* were genetically distinct from coastal populations suggesting that the interior populations in Canada may also be Montana tailed frogs.

Mittleman and Myers (1949) reported that Montana tailed frogs have larger eyes and greater head width relative to body size than tailed frogs. However, Metter's (1967) range-wide analysis failed to find geographic patterns in morphology for *Ascaphus*.

The life history of Montana tailed frogs appears similar to tailed frogs although

Ascaphus truei Stejneger, 1899

(COASTAL) TAILED FROG

Michael J. Adams, Christopher A. Pearl

1. Historical versus Current Distribution.

Tailed frogs *(Ascaphus truei)* have been present in the Pacific Northwest for mil-

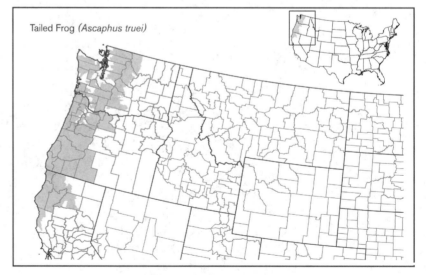

Tailed Frog *(Ascaphus truei)*

the Mt. St. Helens' blast zone (Hawkins et al., 1988).

2. Historical versus Current Abundance.
No change documented. In general, sedimentation and warm water temperatures are associated with lower abundances (see "Conservation" below), but there is no published evidence of broad scale decline in this species.

3. Life History Features.
A. Breeding. Reproduction is aquatic.

i. Breeding migrations. None reported. Adults breed in streams they inhabit. Amplexus has been reported from May–October (Gaige, 1920; Noble and Putnam, 1931; Slater, 1931; Metter, 1964b; Wernz, 1969), but Nussbaum et al. (1983) suggest that most breeding occurs in the fall. Females retain sperm in oviductal sperm storage tubules (Metter, 1964b; Sever et al., 2001), and oviposition does not occur until the following spring or summer after spring runoff (Gaige, 1920; Franz, 1970; H.A. Brown, 1975a; Daugherty and Sheldon, 1982a; Adams, 1993; Karraker and Beyersdorf, 1997). Females appear to breed every other year in inland populations; however, smaller clutch sizes suggest that coastal females may breed every year (Metter, 1964b).

ii. Breeding habitat. Breeding and oviposition occur in the streams occupied by adults.

B. Eggs.

i. Egg deposition sites. Eggs are deposited in strings under large rocks (Gaige, 1920; Metter, 1964a; Franz, 1970; H.A. Brown, 1975a; Adams, 1993; Karraker and Beyersdorf, 1997).

ii. Clutch size. Egg numbers are low (28–96), with coastal populations having fewer (Noble and Putnam, 1931; Metter, 1964a; Franz, 1970; Bury et al., 2001) but larger (3.7–4.5 mm diameter; Metter, 1964a; H.A. Brown, 1989a) eggs. Eggs take about 3–6 wk to hatch (Franz, 1970; H.A. Brown, 1989a). Embryos from the North Cascades of Washington hatched after 28 d in the laboratory at 11 °C (H.A. Brown, 1989a). Belton and Owczarzak (1968) describe the pre-ovulatory deposition and storage of hepatic lipids. Wernz and Storm (1969) provide a staging table for embryos.

C. Larvae/Metamorphosis.

i. Length of larval stage. Tadpoles hatch from August–September (Metter, 1964a; Franz, 1970; Adams, 1993) and always overwinter at least once. Hatchlings are unpigmented, 13–15 mm total length, and do not fully absorb yolk until they reach 20–21 mm TL (Metter, 1964a; H.A. Brown, 1989a). They reach metamorphosis after 1–2 yr in coastal areas (Wallace and Diller, 1998; Bury and Adams, 1999), usually after 2–3 yr in the Cascades (Metter, 1967) and after 3 yr in inland regions (Metter, 1967; Daugherty and Sheldon,

1982a; Lohman, 2002). A 4-yr larval period was reported near Mount Baker, Washington (H.A. Brown, 1990). Laboratory behavior of tadpoles was described by Altig and Brodie (1972). Cranial osteology were described by Altig (1969), and variations in cranial ossification were described by Moore and Townsend (2003).

ii. Larval requirements. Tadpoles occupy swift mountain streams with cobble substrates (Gaige, 1920; Metter, 1964a; Bury, 1968; Nussbaum et al., 1983; Leonard et al., 1993; Bull and Carter, 1996; Diller and Wallace, 1999; Adams and Bury, 2002). They adhere to smooth rocks using their large, suctorial mouths (Gradwell, 1971). Tadpoles prefer cold water temperatures and are seldom found in streams >16 °C (de Vlaming and Bury, 1970; Claussen, 1973; Welsh, 1990). A strong association with consolidated surface lithologies has been documented in harvested forests (Diller and Wallace, 1999; Dupuis et al., 2000; Wilkins and Peterson, 2000), which may be due partially to a greater sensitivity of streams with sedimentary substrates to logging (Adams and Bury, 2002; Welsh and Lind, 2002). Because of their long larval period, tadpoles typically require permanent water, but some populations with 1-yr larval periods may persist in streams that occasionally dry (Wallace and Diller, 1998; Waters et al., 2001). Sedimentation appears to degrade habitat quality (Noble and Putnam, 1931; Bury and Corn, 1988b; Corn and Bury, 1989a; Welsh and Ollivier, 1998; Dupuis and Steventon, 1999). Severe floods that scour stream bottoms can remove entire tadpole populations (Metter, 1968; Lohman, 2002).

a. Food. Mostly diatoms (Metter, 1964a); some filamentous algae and desmids found in guts. Conifer pollen is seasonally abundant in guts. Fine sand grains are also ingested (Metter, 1964a). Noble (1927) felt that tadpoles ingest most of their food through the external nares. However, Altig and Brodie (1972) found that more food is scraped from rocks and orally ingested. Tadpoles will sometimes crawl up rocks out of the water, possibly to feed (Noble and Putnam, 1931). Altig and Kelly (1974) suggest that the relatively short gut length of tailed frogs may be due to cold water temperatures and slow passage times rather than a carnivorous feeding habit. Addition of phosphate to streams can increase periphyton abundance and coastal tailed frog tadpole growth rate, suggesting nutrient limitation (Kiffney and Richardson, 2001).

b. Cover. Tadpoles use interstices for cover during daylight and emerge to the surface of stones at night (Metter, 1964a; Feminella and Hawkins, 1994). Tadpoles will increase cover use in the presence of non-visual cues from predators (Feminella and Hawkins, 1994).

iii. Larval polymorphisms. Albinism was reported by Pearl et al. (2002). The white on the tip of the tail is absent in Blue Mountain populations (Metter, 1967).

iv. Features of metamorphosis. Metamorphosis (from front-limb emergence to tail resorption) occurs in late summer and lasts approximately 1 mo, although the tail may not be completely resorbed for months (H.A. Brown, 1990; Bury and Adams, 1999).

v. Post-metamorphic migrations. Adults and juveniles are highly philopatric in Montana (Daugherty and Sheldon, 1982b), but movements directly after metamorphosis have not been well documented. Bury and Corn (1987, 1988b) captured numerous recently transformed animals in pitfall traps set in forested stands. These captures suggest a pattern of fall dispersal by recently transformed animals.

D. Juvenile Habitat. Same as adults.

E. Adult Habitat. Adults are aquatic, occupying the streams needed by their eggs and tadpoles. Adults may use thermal microhabitats to avoid warm water temperatures (Adams and Frissell, 2001). After heavy rains or dews, adults may be found in moist woods (Nussbaum et al., 1983; Welsh and Reynolds, 1986). Upland captures of tailed frogs can be common during the autumn and spring (Bury and Corn, 1988a; Gomez and Anthony, 1996).

F. Home Range Size. Marked adults in Montana seldom moved >10 m upstream or downstream (Daugherty and Sheldon, 1982b).

G. Territories. Not reported.

H. Aestivation/Avoiding Desiccation. These behaviors have not been reported.

I. Seasonal Migrations. Daugherty and Sheldon (1982b) report none. Metter (1964a) suggests adults may move up into smaller, more shaded streams during the summer, although Landreth and Ferguson (1967) question this conclusion. Adams and Frissell (2001) reported seasonal movements of adults consistent with an avoidance of warm water temperatures. Wahbe and Bunnell (2001) describe movement of tadpoles in streams flowing through harvested and natural forests in southwestern British Columbia.

J. Torpor (Hibernation). Not reported.

K. Interspecific Associations/Exclusions. Frequently occupy small streams with giant salamanders (*Dicamptodon* sp.) and torrent salamanders (*Rhyacotriton* sp.), although each species is associated with somewhat different microhabitats (Metter, 1964a; Bury et al., 1991b; Welsh and Ollivier, 1998; Adams and Bury, 2002). Fish can be predators (Feminella and Hawkins, 1994); Noble and Putnam (1931) suggest that salmonids may exclude tailed frogs, but tailed frogs have been found coexisting with a variety of fish including salmonids (Metter, 1964a; Feminella and Hawkins, 1994).

L. Age/Size at Reproductive Maturity.
Tailed frogs are remarkable among anurans in their delayed maturity. In a Montana study, reproductive maturity was reached in yr 7 (4 yr after metamorphosis), but most males did not breed until yr 8 and most females did not breed until yr 9 (Daugherty and Sheldon, 1982a). Mature males were >34 mm (snout to base of "tail") and mature females were >44 mm.

M. Longevity. Long lived. Daugherty and Sheldon (1982a) recovered a wild female with a known minimum age of 14 yr. They speculate that females may live 15–20 yr and males somewhat less.

N. Feeding Behavior. Adult tailed frogs appear to be generalist predators, feeding on insects and other invertebrates as they are available (Fitch, 1936; Metter, 1964a; Bury, 1970). They feed along stream banks and adjacent forest at night (Metter, 1964a; Daugherty and Sheldon, 1982b). Nishikawa and Cannatella (1991) describe the kinematics of prey capture. Abourachid and Green (1999) describe swimming behavior.

O. Predators. Known predators include western terrestrial garter snakes (*Thamnophis elegans*), common garter snakes (*T. sirtalis*), giant salamanders (*Dicamptodon* sp.), trout (Salmonidae), sculpins (*Cottus confusus*), dippers (*Cinclus mexicanus*), and hellgrammites (Megaloptera; Metter, 1963; Metter, 1964a; Daugherty and Sheldon, 1982a; Feminella and Hawkins, 1994; Jones and Raphael, 1998; Karraker, 2001). Blair and Wassersug (2000) describe predator-induced tail damage.

P. Anti-Predator Mechanisms. Nocturnally active, seek cover under rocks during daylight or in presence of non-visual cues from predators (Feminella and Hawkins, 1994). Adults have been observed to fold their limbs against their body and let the current carry them away when disturbed (Metter, 1964a).

Q. Diseases. None reported.

R. Parasites. The gut ciliate, *Protoopalina*, is found in tadpoles (Metcalf, 1928; Metter, 1964a). Both larvae and adults can host subcutaneous, encysted parasitic flukes (Anderson, 1964; Metter, 1964a).

4. Conservation.
Population densities vary considerably (Lohman, 2002), but lower abundances have been documented following timber harvest (Gaige, 1920; Noble and Putnam, 1931; Metter, 1964a; Bury and Corn, 1988b; Corn and Bury, 1989a; Bury et al., 1991b; Bull and Carter, 1996; Aubry, 2000) and road construction (Welsh and Ollivier, 1998). Tailed frogs have been characterized as both environmentally sensitive and resilient to large scale disturbance. They were one of the first vertebrates to recover following the 1980 eruption of Mt. St. Helens (Hawkins et al., 1988; Crisafulli and Hawkins, 1998).

While Metter (1964a) reported long-term absence of a tailed frog population following timber harvest, some populations persist in forests with extensive timber extraction (Diller and Wallace, 1999; Dupuis et al., 2000; Wilkins and Peterson, 2000). Riparian buffers may help protect populations from timber harvest impacts (Dupuis and Steventon, 1999). In general, sedimentation and warm water temperatures are associated with lower abundances (Bury and Corn, 1988b; Hawkins et al., 1988; Corn and Bury, 1989a; Welsh and Ollivier, 1998; Diller and Wallace, 1999), but there is no published evidence of broad scale decline in this species.

Family Bufonidae

Bufo alvarius Girard, 1859
COLORADO RIVER TOAD

M. J. Fouquette, Jr., Charles W. Painter, Priya Nanjappa

1. Historical versus Current Distribution.
Colorado River toads (*Bufo alvarius*) are one of the least known toads in the United States (Fouquette, 1970; Degenhardt et al., 1996). The type locality is Fort Yuma, Imperial County, California (Fouquette, 1968, 1970). In the United States, Colorado River toads range from extreme southwestern New Mexico (southwestern Hidalgo County only) throughout southern Arizona to southeastern California (Wright and Wright, 1949; Fouquette, 1970; Peters and McCoy, 1978), essentially restricted to the Sonoran Biotic Province (Dice, 1939). Stebbins (1951) suggested a possible extension north along the Colorado River bottomlands to southern Nevada and Utah. No records of this species are known for Nevada or Utah or for California north of the area of the type locality. Colorado River toads range from near sea level to 1,600 m (Cole, 1962; Fouquette, 1970).

2. Historical versus Current Abundance.
Both King (1932) and Cole (1962) reported Colorado River toads as common near Tucson, Arizona, and west to the Colorado River. Stebbins (1985) indicated these toads are widespread throughout the desert. Today, while Colorado River toads seem to be abundant at many desert localities in Arizona, they appear to have declined in New Mexico (Degenhardt et al., 1996) and California (Jennings and Hayes, 1994a).

3. Life History Features.

A. Breeding. Reproduction is aquatic. Males sometimes give advertisement calls, especially when few other males are present, but may rely more on mate searching (Sullivan and Malmos, 1994).

i. Breeding migrations. During dry, pre-monsoon periods, Colorado River toads seek shelter, often in rodent burrows. With the onset of summer monsoon rains, adults move overland to breeding sites. The extent of these overland migrations has not been investigated. Breeding usually occurs on one night, 2 or 3 d following a major rainfall event (Sullivan and Malmos, 1994).

ii. Breeding habitat. Seasonal and permanent pools (Fouquette, 1970); Musgrave and Cochran (1930) noted that irrigation ditches provide ideal breeding sites, and stock tanks are commonly utilized (Blair and Pettus, 1954; Degenhardt et al., 1996). According to the field notes of J. J. Thornber, adults appear before summer showers and congregate in seasonal pools once the rains begin (Ruthven, 1907; Fouquette, 1970). Stebbins (1985) notes that while breeding activity is stimulated by rainfall, Colorado River toads are not dependent on rainfall for breeding (supported by Arnold, 1943). Sullivan and Malmos (1994) found that all breeding activity occurred on one or two nights following rainfalls of over 25 mm; if rains exceed 75 mm, causing flooding, breeding

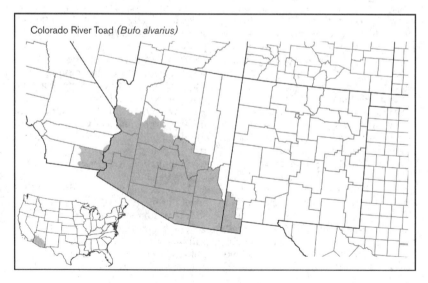

Colorado River Toad (*Bufo alvarius*)

activity may be delayed for two or three nights. Males exhibit one of two behaviors: they either call for females from shallow water (larger males) or they actively search for females. Sullivan and Malmos (1994) suggest that the number of males in a breeding chorus is inversely related to the likelihood that any single male will call. The breeding season itself can be long (D. Pettus, in Degenhardt et al., 1996).

B. Eggs.

i. Egg deposition sites. Shallow waters of seasonal and permanent pools.

ii. Clutch size. Average 7,500–8,000 eggs/female encased in a long, single tube of jelly, with a loose but distinct outline. Egg capsule gelatin is clear and not very adhesive; eggs are not partitioned, average 1.6 mm in diameter, and are packed between about 5–7/cm (12–18/in; Wright and Wright, 1949; see also Savage and Schuierer, 1961).

C. Larvae/Metamorphosis.

i. Length of larval stage. Not longer than 1 mo (Ruthven, 1907; Musgrave and Cochran, 1930). Tadpoles may reach 57 mm TL (Degenhardt et al., 1996).

ii. Larval requirements.

a. Food. Undescribed; presumably tadpoles are algivorous/omnivorous.

b. Cover. Undescribed.

iii. Larval polymorphisms. Unknown and unlikely.

iv. Features of metamorphosis. Undescribed.

v. Post-metamorphic migrations. Unknown.

D. Juvenile Habitat. Similar to adults.

E. Adult Habitat. Colorado River toads occur primarily in deserts, including mesquite-creosote bush lowlands, but are also found in arid grasslands, rocky riparian zones with sycamore and cottonwoods, and oak-walnut woodlands in mountain canyons (Schmidt, 1953; Fouquette, 1970; Stebbins, 1985; Holycross et al., 1999).

Largely nocturnal, they will take refuge in rodent burrows (Lowe, 1964), rocky outcrops (D. Beck, unpublished data), or in hollows under watering troughs (Wright and Wright, 1949); they breed in seasonal pools (Fouquette, 1970). Wright and Wright (1949) describe Colorado River toads as semiaquatic, found only in wet places around cattle watering troughs and in seasonal wetlands where Couch's spadefoot toads (*Scaphiopus couchii*) and western spadefoot toads (*Spea multiplicata*) breed. Wright and Wright (1949) cite E. A. Mearns' 1907 description: "Nothing was seen nor heard of them until the advent of early summer rains, which formed a large shallow lake . . . These large toads then filled the air with their loud cries until a deafening roar was produced." Likely this description actually was of a chorus of either Great Plains toads (*B. cognatus*) or spadefoot toads (*Scaphiopus* or *Spea*), as the calls of Colorado River toads are not very loud and choruses are small (Blair and Pettus, 1954; Sullivan and Malmos, 1994); while the call may carry >100 m (Degenhardt et al., 1996; Sullivan and Malmos, 1994), their choruses would hardly be considered deafening.

F. Home Range Size. Dan Beck (unpublished data) radio-tracked an adult Colorado River toad for a period of 390 d in the Tucson Mountains, Pima County, Arizona. During that time activity centered around the release site, although movements of >400 m were noted in a single day.

G. Territories. Undescribed.

H. Aestivation/Avoiding Desiccation. Degenhardt et al. (1996) suggest that large size and smooth skin may predispose Colorado River toads to desiccation, but there is no experimental evidence for this. Indeed, larger size also means lower surface to volume ratio, thus relatively lower expectation of evaporation. There is no direct evidence for aestivation by the species, and it is not likely that they utilize any form of torpor to any greater extent than other species of *Bufo*. Dan Beck (unpublished data) radio-tracked an adult Colorado River toad that remained in the same burrow under a railroad tie from 26 September 1988–17 June 1989. Body temperature during that time ranged from 11.7 °C–29.7 °C. It is possible that during part of that period below the surface the toad may have been in a state of torpor or aestivation.

I. Seasonal Migrations. May migrate several hundred meters from permanent to seasonal pools (Wright and Wright, 1949) following heavy rains.

J. Torpor (Hibernation). Unknown, but see "Aestivation/Avoiding Desiccation" above.

K. Interspecific Associations/Exclusions. Will associate with allotopic spadefoot toads *Scaphiopus* or *Spea* sp. in seasonal wetlands (Wright and Wright, 1949). Found with Great Plains toads in association with watering troughs (Wright and Wright, 1949), and at stock tanks with Great Plains toads, red-spotted toads (*B. punctatus*), and Couch's spadefoot toads (Blair and Pettus, 1954; Sullivan and Malmos, 1994). Gergus et al. (1999) reported hybridization between *B. alvarius* and *B. woodhousii* (Woodhouse's toads) in central Arizona, and Painter (2001, personal observations) documented the first known calling *B. woodhousii* males within the habitat of *B. alvarius* in New Mexico, thus increasing the possibility of a breakdown in reproductive isolation.

L. Age/Size at Reproductive Maturity. Males, 80–156 mm; females, 87–178 mm (Wright and Wright, 1949). Adults may be as large as 190 mm (Heringhi, 1969, in Fouquette, 1970; Degenhardt et al., 1996). Sullivan and Fernandez (1999) studied populations in which adult males were mostly 2–4 yr of age.

M. Longevity. Conant and Hudson (1949) cited a record of 2 yr (and 0 mo) in captive toads, while Snider and Bowler (1992) reported a wild-caught adult lived 15 yr, 5 mo, 16 d in captivity; however, based on data from other bufonids, longevity in nature is likely to be at least 4–5 yr.

N. Feeding Behavior. Colorado River toads are active foragers. Prey includes beetles, grasshoppers, wasps, centipedes, millipedes, ants, termites, solpugids, spiders, snails, scorpions, Great Plains toads, Couch's spadefoot toads, small lizards, and mice (King, 1932; Arnold, 1943; Gates, 1957; Cole, 1962; Degenhardt et al., 1996). Colorado River toads will eat almost any prey they can subdue and ingest, including those with defensive stinging capabilities (Cole, 1962; Degenhardt et al., 1996).

O. Predators. Raccoons (*Procyon lotor*; Wright, 1966) and probably birds, other mammals, and reptiles; adults may be safe from most predators other than raccoons due to the toxicity of their parotoid secretions (see "Anti-Predator Mechanisms" below). The secretory product from the parotoid glands is strongly hallucinogenic (Weil and Davis, 1994), making this species attractive to exploitation by members of certain drug subcultures (Most, 1984).

P. Anti-Predator Mechanisms. Colorado River toads are poisonous (Musgrave and Cochran, 1930; Hanson and Vial, 1956) and exhibit defensive behaviors that include assuming a butting pose with their parotoid glands directed toward the intruder (Hanson and Vial, 1956; Stebbins, 1985). In addition to having parotoid glands, Colorado River toads possess glands resembling parotoid glands on the dorsal surfaces of both fore- and hindlimbs (Fouquette, 1970). Musgrave (in Musgrave and Cochran, 1930) witnessed a fox terrier become paralyzed and die after picking up one of these toads in its mouth; a "police dog" (= German shepherd) became paralyzed and distressed for most of an hour after mere facial contact (see also Noble, 1931; Hanson and Vial, 1956; Wright, 1966). Erspamer et al. (1967) and Cei et al. (1968) report the presence of presumably toxic indolealklamines from the skin and parotoid glands of Colorado River toads. Cannon and Hostetler (1976) and McGill and Brindley (1978) report that the secretory product from the parotoid glands of these toads contains steroid-derived bufogenins.

Wright (1966) reported a raccoon slitting the bellies of Colorado River toads, avoiding the glandular skin and eating the entrails; other predators may not have learned this means of feeding on these toads. Musgrave and Cochran (1930) report that Colorado River toads will "inflate to a remarkable degree and remain in this condition for some time" when disturbed (see photo in Hanson and Vial, 1956).

Q. Diseases. Unknown.

R. Parasites. Goldberg and Bursey (1991a) reported four nematodes from the gastrointestinal tract of the Colorado River toad (*Aplectana itzocamensis, Physaloptera* sp., *Physocephalus* sp., and *Oswaldocruzia pipiens*). They also found one species of cestode (*Nematotaenia dispar*) as well as a species of nematode from the lungs *(Rhabdias americanus).*

4. Conservation.

Colorado River toads are listed as Endangered by the New Mexico Department of Fish and Game (Degenhardt et al., 1996). Jennings and Hayes (1994a) commented on the status of Colorado River toads and noted that they appear to be extirpated from most or all sites in California and " . . . some investigators have suggested that Colorado River toads are imperiled throughout much of [their] range. . . ." In southern Arizona, these toads seem to be abundant at many desert localities.

Acknowledgments. Thanks to David Bradford, who commented on an earlier draft of this manuscript. Dan Beck provided unpublished field notes from a radio telemetry study.

Bufo americanus Holbrook, 1836
AMERICAN TOAD

David M. Green

1. Historical versus Current Distribution.

In the United States, American toads *(Bufo americanus)* are found throughout the eastern states from the Canadian border south to the edge of the Coastal Plain (Conant and Collins, 1998). The western edge is a line running south from western Minnesota and extreme east North Dakota, through South Dakota, Nebraska, Kansas, and Oklahoma to northeast Texas (Henrich, 1968; Collins, 1974, 1982; Seifert, 1978; Cochran, 1986a; Olson, 1987; Oldfield and Moriarty, 1994; Whiting and Price, 1994). In the south, the edge of the range runs from northeast Texas across northern Louisiana, except for an extension south along the Mississippi River to about Baton Rouge and across the middle of Mississippi (Dundee and Rossman, 1989; Lazell and Mann, 1991; Whiting and Price, 1994; Himes and Bryan, 1998). There is a gap in northwest Alabama, and the range bulges into northeast Alabama and northern Georgia (Mount, 1975). Toward the Atlantic Coast, the range encompasses only extreme northwest South Carolina, misses southeast North Carolina, and approaches the coast from Virginia and Maryland (Miller, 1979) north through New England (Hunter et al., 1992; Klemens, 1993) except for Delaware, coastal New Jersey, and extreme southeast New York including Long Island. American toads are found throughout the Great Lakes region and northeastern states to

Canada (Corin, 1976; Kraus and Schuett, 1982; Cochran, 1986b; Harding, 1997). A gap in the range occurs in southern Illinois and southwest Indiana (Garman, 1892; Smith, 1947, 1961; Minton, 1972). American toads are found ≤1,200 m in the peaks of the Allegheny Mountains (Green and Pauley, 1987) and to >1,524 m (5,000 ft) in the Great Smoky Mountains (Huheey and Stupka, 1967).

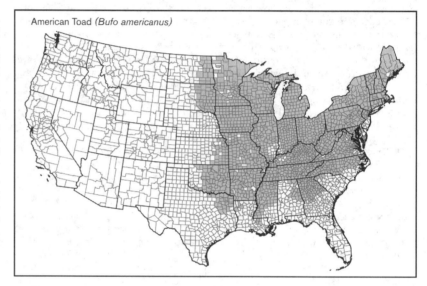

American Toad *(Bufo americanus)*

Comparisons of maps in Conant and Collins (1998), Conant (1958a, 1975), and Wright and Wright (1949) reveal no fundamental changes in the species' range during the past 50 yr. The current distribution probably reflects the distribution prior to Euro-American settlement; however, many populations within this range have likely been extirpated, leaving many more populations isolated.

Two subspecies are currently recognized: eastern American toads *(B. a. americanus)* and dwarf American toads *(B. a. charlesmithi).* Dwarf American toads are smaller and redder and have an unspotted belly, unlike the often profusely spotted underside of eastern American toads (Blair, 1943; Smith, 1961). Dwarf American toads occupy the southwest of the species' range, and there is a wide band of intergradation between these two subspecies; in Missouri it lies between St. Louis to an area south of Kansas City (Johnson, 1987), in Kentucky in the Jackson Purchase area (Barbour, 1971), and in Illinois through their range in the southern third of the state (Smith, 1961). A northern and colorful subspecies in Canada, *B. a. copei* (Ashton et al., 1973) is no longer recognized as it cannot be distinguished genetically from other populations (Guttman, 1975) despite Gaige's (1932) morphological diagnosis. Cook (1983) proposed the synonymy of *B. americanus* and Canadian toads (*B. hemiophrys*) based upon morphological variation

across a cline between the two forms in eastern Manitoba and recognized *B. a. hemiophrys* as a subspecies of *B. americanus.* Although this system had been followed by several publications originating in Canada (e.g., Cook and Cook, 1981; Preston, 1982; Kuyt, 1991), the nomenclature advocated by Cook generally has not been accepted (Green, 1983; Frost, 1985; Collins, 1997; Conant and Collins, 1998;

Green and Pustowka, 1998). American toads were held for some time to be in taxonomic synonymy with southern toads *(B. terrestris)* as *B.t. americanus* (Netting and Goin, 1946), but have been considered distinct from that species since the publication of Schmidt's (1953) checklist. Based on sequences of mitochondrial genes, Masta et al. (2002) found that American toads are more closely related to Woodhouse's toads *(B. woodhousii)* than to southern toads or any other species.

2. Historical versus Current Abundance.

Always noted as common, even "ubiquitous" (Barbour, 1971; Kolozsvary and Swihart, 1999), American toads are frequently abundant throughout their range (Wright and Wright, 1949). Conant and Collins (1998) call American toads the "common and abundant 'hoptoad' of the Northeast." Nowhere is this species considered to be under any threat of extirpation (Lannoo, 1998b). Nevertheless, this species is subject to fluctuations in population size (Casebere and Taylor, 1976; Christein and Taylor, 1978; Heyer, 1979; Hecnar and M'Closkey, 1996).

American toads do not appear to be as sensitive to habitat fragmentation as are many other species of co-occurring amphibians (Hager, 1998; Kolozsvary and Swihart, 1999; Lehtinen et al., 1999). They are the most readily able amphibian to reinvade clearcut and burned areas of forest (Blymyer and McGinnes, 1977; Kirkland et al., 1996). However, American toad larvae are susceptible to low pH due to acid precipitation which sublethally affects their nitrogen balance (Clark and LaZerte, 1985, 1987; Dale et al., 1985a,b; Tattersall and Wright, 1996) and thus, ultimately their survival (Clark and Hall, 1985; Freda and Dunson, 1985; Leftwich and Lilly, 1992). In consequence, acid precipitation has affected the distribution of American toads in affected regions of their range (Clark, 1986a; Wyman, 1988a). Enhanced ultraviolet radiation, particularly UV-B, however, has not been demonstrated to have a substantially detrimental effect upon American toad eggs or larvae under natural conditions, largely due to the rapid attenuation of UV-B penetration into the water (Grant and Licht, 1995; Crump et al., 1999). Chemical contaminants, including pesticides such as benzopyrene, methoxychlor, toxaphene, and endrin, as well as lead and ammonium nitrate, have both lethal and sublethal effects upon American toads, particularly at the larval stage (Clark and Diamond, 1971; Hall and Swineford, 1979, 1981; Bracher and Bider, 1982; Steele et al., 1991, 1999; Hecnar, 1995; Jofre and Karasov, 1999; Harris et al., 2000) against which they appear to have no defense. Hindlimb deformities have been noted among juvenile toads in agricultural habitats fraught with pesticides, although to a more limited extent than among some other co-occurring species, especially ranids (Ouellet et al., 1997a). Metamorphosing individuals may be particularly susceptible to certain forms of environmental stress, as they have reduced high temperature tolerance compared to both larvae and adults (Hathaway, 1928; Cupp, 1980).

3. Life History Features.
A. Breeding. Reproduction is aquatic.
i. Breeding migrations. American toad males move to breeding sites, to which they have some measure of site fidelity (Ewert, 1969), and begin to call almost immediately after emergence from winter hibernation. Males call in large choruses usually beginning in early evening (Aronson, 1944a; Bogert, 1960; Brown and Littlejohn, 1972; Gerhardt, 1975; Tyning, 1990) but will sing during the daytime on particularly warm and humid days at the height of the breeding season (Dirig, 1978). In Alabama, movement to breeding sites may commence as early as mid-January or February, although the peak of breeding activity is usually mid-March (Mount, 1975). Toads begin to breed in late February in Louisiana (Dundee and Rossman, 1989); February to March in the Carolinas (Martof et al., 1980); early March in West Virginia (Green and Pauley, 1987); March with a peak in mid-April in Kentucky (Barbour, 1971); mid-March to early April in Connecticut (Klemens, 1993);

late March to early April in Massachusetts (Miller, 1909b); March to early May in Kansas (Collins, 1982); mid-April to mid-May in Maine (Hunter et al., 1992); late April to early May in Wisconsin (Vogt, 1981) and the Great Lakes region (Werner and McCune, 1979; Harding, 1997); and early May to mid-June in Minnesota (Oldfield and Moriarty, 1994). The peak of breeding activity usually lasts just under 2 wk (Hunter et al., 1992; Harding, 1997), although even after the peak, there are persistent toads that will sing even into the first days of June in more northern localities (Aronson, 1944a; Klemens, 1993).

Females appear at the breeding sites from a few days to one week after the males (Miller, 1909b; Aronson, 1944a; Collins and Wilbur, 1979) and do not persist there for as long a time. Some authors have noted an apparently non-random bias in mating based on size (Licht, 1976; Gatz, 1981a; Fairchild, 1984), but females appear to choose their mates depending upon individual characteristics of the males' calls (Wilbur et al., 1978; Howard, 1988; Sullivan, 1992a; Howard and Palmer, 1995; Howard and Young, 1998) and orient towards the calls of particular males (Schmidt, 1971). This results in competition among males in breeding choruses for mates (Kruse, 1982; Wells and Taigen, 1984). Depending upon the density of the breeding congregation, a proportion of the males present will not sing but nevertheless attempt to intercept females as they arrive (Forester and Thompson, 1998). Males recognize toads of their own sex, regardless of species, with the aid of a well-developed and stereotypic release call (Leary, 2001b).

ii. Breeding habitat. American toads congregate to breed in shallow, often grassy, areas within lakes, ponds, streams, ephemeral wetlands, prairie potholes, farm ponds, floodplain pools, ditches, or streamside pools (Miller, 1909a; Wright and Wright, 1949; Mount, 1975; Dundee and Rossman, 1989; Holomuzki, 1995).

B. Eggs.
i. Egg deposition sites. Eggs are 1.0–2.0 mm in diameter (Wright and Wright, 1949; Smith, 1961; Dundee and Rossman, 1989) and are laid in two long strings, usually entwined around vegetation or resting on the pond bottom (Miller, 1909a; Wright and Wright, 1949; Tyning, 1990; Holomuzki, 1997; Behler and King, 1998) in shallow water 5–10 cm deep (Miller, 1909a).

ii. Clutch size. Clutch size has been reported to range from 2,000–12,000 eggs (Gilhen, 1984); 4,000–12,000 eggs (Livezey and Wright, 1947); 4,000–8,000 eggs (Wright and Wright, 1949); 4,000–15,000 eggs (Miller, 1909a); or 2,000–20,000 eggs (Johnson, 1987). Numbers of eggs/in (2.5 mm) vary from 15–17 in eastern American toads to a mean of about 7.5/in dwarf American toads (Livezey and Wright, 1947; Dundee and Rossman, 1989).

C. Larvae/Metamorphosis.
i. Length of larval stage. Eggs hatch in 3–12 d (Smith, 1961; Martof et al., 1980; Vogt, 1981). The larval period takes 50–60 d (Wright and Wright, 1949), and tadpoles transform at 7–12 mm SVL (Wright and Wright, 1949) or 6–8 mm SVL in dwarf American toads (Dundee and Rossman, 1989). High tadpole density negatively affects growth and timing of metamorphosis (Brockelman, 1969; Wilbur, 1977a; Breden and Kelly, 1982).

ii. Larval requirements. American toad tadpoles are black and reach a length of about 2.5 cm (Altig, 1970; Dundee and Rossman, 1989; Oldfield and Moriarty, 1994). They evidently require small, open, stillwater ponds and pools devoid of fishes (Wright and Wright, 1949), including ephemeral ponds (Wilbur, 1990), preferentially not overly shaded by trees (Werner and Glennemeier, 1999).

a. Food. Tadpoles are omnivorous and will consume vegetable matter, suspended organic matter and algae, detritus, and dead fish or tadpoles (Test and McCann, 1976; Seale and Beckvar, 1980; Bowen, 1984; Ahlgren and Bowen, 1991). American toad tadpoles compete for food with the tadpoles of other anuran species to variable extents (Alford and Wilbur, 1985; Wilbur and Alford, 1985; Alford, 1989a,b).

b. Cover. Not needed.

iii. Larval polymorphisms. None.

iv. Features of metamorphosis. Metamorphosis occurs in mid-May in Louisiana (Dundee and Rossman, 1989) and West Virginia (Green and Pauley, 1987); early June in Illinois (Smith, 1961); mid-June in Massachusetts (Miller, 1909a); mid-June to early August in Connecticut (Klemens, 1993); late June to early July in Wisconsin (Vogt, 1981); and late July in Minnesota (Oldfield and Moriarty, 1994). Metamorphosis is rapid (Miller, 1909a); newly metamorphosed animals will quit the water within 1 d of acquiring their front legs and will have completely resorbed the tail a mere 2 d later. A whole cohort will transform within 6 d.

v. Post-metamorphic migrations. Newly metamorphosed animals and juveniles disperse widely but not aimlessly, evidently orienting themselves in part through use of celestial cues (Dole, 1972c, 1973).

D. Juvenile Habitat.
Similar to adults, although large aggregations of post-metamorphic toads can often be seen on mudflats surrounding breeding wetlands. American toads grow rapidly following transformation, doubling in length and more than quadrupling their weight from June–August (Hamilton, 1934).

E. Adult Habitat.
American toads evidently tolerate humans well. They can be common in gardens, fields, lawns, and barnyards (Dickerson, 1906; Smith, 1947, 1961; Klemens, 1993). Mainly nocturnal, during the day toads seek cover

under stones, boards, woodpiles, walkways, porches, or other cover (Wright and Wright, 1949). American toads are tolerant of brackish estuarine waters (Kiviat and Stapleton, 1983). Mount (1975), in Alabama, and Johnson (1987), in Missouri, note that American toads are common in and along the edges of forested areas. Oldfield and Moriarty (1994) note that American toads will thrive in prairie parkland habitats, although they seem to be predominantly animals of the forest, a point contradicted by Guerry and Hunter (2002), who found a negative association of American toads in relation to forest area. Green and Pauley (1987) observed that in West Virginia individuals may be found in dense woods, but are more frequently found in more open pastures, agricultural areas, and gardens. Fitch (1958) considered the optimum habitat for American toads in Kansas to be rocky habitats in open woods or wooded edges, where toads were found under large flat rocks covering loose, damp soil. Nevertheless, due to their predilection for orchards, gardens, and lawns in Illinois, Smith (1947) considered American toads to be essentially a prairie species.

F. Home Range Size. Fitch (1958) and Collins (1974, 1982) report that American toads in Kansas will establish home ranges of 0.16 ac or more in size. Toads will repeatedly use particular hiding places (Dole, 1972c) and will return daily to "forms," a small depression in a lawn or under stones (M. Stewart, personal observations), but during wet periods will travel great distances ≤1 km (Ewert, 1969).

G. Territories. Not territorial (Fairchild, 1984; Tyning, 1990).

H. Aestivation/Avoiding Desiccation. In summer, American toads may be little in evidence during extended periods of hot, dry weather. Toads likely aestivate during the summer in Alabama (Mount, 1975).

I. Seasonal Migrations. Toads move in spring to breeding sites from wherever they had been hibernating during the winter (Miller, 1909a; Oldham, 1969), both over land and along streams (Maynard, 1934). Weather, particularly humidity and rainfall, influences movement (Fitzgerald and Bider, 1974a,b).

Nocturnal post-breeding movements have been recorded ≤1,000 m, but most excursions are considerably shorter (Ewert, 1969). Dole (1972c) found that toads displaced ≤235 m from point of capture will move and orient themselves but not necessarily toward the point of capture. Both vision and olfaction are used by adult toads to orient (Dole, 1972c; Adler and Taylor, 1981).

J. Torpor (Hibernation). During cold winter weather, American toads hibernate terrestrially (Miller, 1909b). They dig backwards into the soil where they reside or find another hibernation site that permits them to burrow below the frost line

(Wright and Wright, 1949; Tester et al., 1965; Ewert, 1969). American toads are not freeze tolerant (Miller, 1909b; Storey and Storey, 1986) and evidently have no mechanism for freeze tolerance (Holzwart and Hall, 1984). Hibernation begins as the temperature falls below their normal activity minimum of about 9 °C, which is usually October in northern U.S. populations (Oldfield and Moriarty, 1994).

K. Interspecific Associations/Exclusions. The members of the *B. americanus* group of toads (W.F. Blair, 1963a, 1972a; Guttman, 1969; Martin, 1973) are notorious hybridizers (A.P. Blair, 1941; Sanders, 1961, 1987; W.F. Blair, 1964a, 1972b; Green, 1983, 1984, 1996; Green and Parent, 2003). American toads form a hybrid zone in the northwest with Canadian toads (Henrich, 1968, Cook, 1983, Green, 1983, Green and Pustowka, 1997) and another in the southeast with southern toads (Neill, 1949a; Volpe, 1955b; Mount, 1975; Weatherby, 1982). In Louisiana and Alabama, American toads are not usually found breeding in sympatry with southern toads, due to some separations in range and because American toads breed earlier (Mount, 1975; Dundee and Rossman, 1989) by 4–6 wk. The same difference in timing of breeding is also true for American toads relative to Canadian toads and Fowler's toads *(B. fowleri)*; American toads breed earlier and at colder temperatures (Henrich, 1968; Barbour, 1971; Green, 1982). Nevertheless, overlaps in breeding seasons do occur (Green, 1984).

American toads are broadly sympatric with Fowler's toads, with which they are known to hybridize in numerous scattered localities (Allard, 1908; Miller and Chaplin, 1910; Hubbs, 1918; Myers, 1927; Pickens, 1927a; Blair, 1941; Volpe, 1952, 1955b; Cory and Manion, 1955; Zweifel, 1968a; L.E. Brown, 1970; Jones, 1973; Green, 1982, 1984; Green and Parent, 2003). The two species tend to have different temperature tolerances (Frost and Martin, 1971) and habitat preferences: American toads in forests, Fowler's toads in more open sandy areas and savannas (Smith, 1961; Mount, 1975; Green and Pauley, 1987; Johnson, 1987). There had been long-standing confusion between the two species in the literature until their identities were clearly defined by Myers (1927, 1931b) and explained by Wright and Wright (1949). Dickerson (1906), for instance, includes a photograph labeled American toads although the animals portrayed are clearly Fowler's toads.

L. Age/Size at Reproductive Maturity. American toads are strongly sexually dimorphic in body size (Licht, 1976; Gatz, 1981a; Kruse, 1981) with an apparent south to north gradient of increasing average size. Throughout most of their range, male eastern American toads reach 54–85 mm, but females are 56–111 mm (Wright and Wright, 1949; Conant and Collins,

1998). In Connecticut, Klemens (1993) found 33 males measuring from 51–72 mm, averaging 60.9 mm, whereas 18 females measured from 68–85 mm, averaging 75.1 mm. Particularly large individuals, ≤155 mm, have been reported on islands in northern Lake Michigan (Long, 1982). Dwarf American toads are comparatively smaller, reported to be 44–70 mm in Louisiana (Dundee and Rossman, 1989). In Illinois, Smith (1961) reports that 47 eastern American toads ranged from 47.0–90.0 mm SVL, whereas 19 dwarf American toads ranged from 42.8–69.5 mm, stating also that dwarf American toads are usually under 60 mm.

American toads reach sexual maturity in 2–4 yr (Dickerson, 1906; Hamilton, 1934), but the sexes differ: males are mature by the end of their second summer and breed the following spring. Females mature at the age of 3 (Acker et al., 1986; Green and Pauley, 1987) or 4 (Kalb and Zug, 1990).

M. Longevity. Dickerson (1906) mentions, as "authentic record," the story of an American toad living to be 36 yr of age, but this is undoubtedly an exaggeration. Both Acker et al. (1986), in Illinois, and Kalb and Zug (1990), in Virginia, found breeding males up to 4 yr old and breeding females up to 5 yr old using skeletochronology. An American toad of unrecorded sex survived 4 yr, 8 mo, and 25 d at the Philadelphia Zoo (Bowler, 1977), and a captive male from Illinois lived 5 yr, 6 mo, and 21 d (Morris and Meyer, 1980).

N. Feeding Behavior. American toads opportunistically eat large numbers of insects and other invertebrates (Miller, 1909a; Hamilton, 1934; Wright and Wright, 1949; Oliver, 1955a; Leclair and Vallieres, 1981; Gilhen, 1984; Garrett and Barker, 1987; Green and Pauley, 1987; Jennings et al., 1991; Lannoo, 1996) and their methods are lovingly described by Dickerson (1906). They eat beetles in large numbers (Dickerson, 1906; Miller, 1909a; Wright and Wright, 1949; Oliver, 1955a; Gould and Massey, 1984), with the consumption of ground beetles (Carabidae) especially noted (Larochelle, 1974, 1975a,b, 1976, 1977a,b,c; Charlebois, 1977), although they are evidently stymied by bombardier beetles (Dean, 1980a,b). They also consume large numbers of ants (Oliver, 1955a; Bellocq et al., 2000) and moths (Babbitt, 1937). Primarily crepuscular, American toads will feed at night on insects attracted to lawn and garden lights (Klemens, 1993; Lannoo, 1996), although they will also search for food during late afternoons.

O. Predators. American toads are eaten readily by eastern hog-nosed snakes *(Heterodon platirhinos*; Barbour, 1973), water snakes *(Nerodia sp.)*, garter snakes *(Thamnophis* sp.), and occasionally by ducks (Mueller, 1980), crows, and screech owls (Miller, 1909b). Raccoons *(Procyon lotor)* will consume toads, generally by eating the belly and leaving the dorsal skin behind (Schaaf and Garton, 1970), as will striped skunks *(Mephitis mephitis;* Groves, 1980). Huheey and Stupka (1967) mention the remains of an American toad in the stomach of a road-killed opossum. American bullfrogs *(Rana catesbeiana)* try but fail to eat toads (Tucker and Sullivan, 1975). According to Miller (1909b), "Boys are very destructive to toads in the spawning season."

American toad tadpoles are consumed, in general order of importance (Miller, 1909b), by larvae of predaceous diving beetles (Dytiscidae; Kruse, 1983; Leclair et al., 1986), newts *(Notophthalmus* sp.; Miller, 1909b), dragonfly naiads (Van Buskirk, 1988), giant water bugs (Belostomatidae; Miller, 1909b), and crayfish (Miller, 1909b). Least sandpipers have been known to consume them as well (Stangel, 1983).

P. Anti-Predator Mechanisms. When confronted by a potential predator, a toad will inflate its lungs, lower its head, and lift up its body, while remaining immobile (Brown and Thrall, 1974; Dodd and Cupp, 1978; Marchisin and Anderson, 1978; Vogt, 1981). Otherwise a toad relies upon immobility and its camouflaging resemblance to a clod of dirt for protection (Heinen, 1985, 1994). The skin secretions of the warts are harmless to human skin but will irritate mucous membranes thus making the toads distasteful to many animals, including shrews (Formanowicz and Brodie, 1982) and domestic dogs (Dickerson, 1906; Green and Pauley, 1987). Toads have little defense, though, against boys (Miller, 1909b).

As in other species of *Bufo,* toxins contained in the skin secretions are also secondarily deposited in the ova (Licht, 1968) and are thought thereby to afford some protection to newly laid, fertilized eggs, which then decreases as the zygotes develop (Phisalix, 1922).

Many vertebrate predators find the tadpoles distasteful (Brodie et al., 1978; Brodie and Formanowicz, 1987). However, invertebrate predators, as they suck body contents and do not consume the skin (Wassersug, 1973), generally are reported to find bufonid larvae palatable (Kruse, 1983; Leclair et al., 1986; Van Buskirk, 1988), even though they are repulsed by the skin secretions of newly metamorphosed animals (Brodie et al., 1978; Formanowicz and Brodie, 1982). Tadpoles will adjust their behavior depending upon the particular predator encountered (Reylea, 2001). Tadpoles respond to the presence of predatory odonate larvae, which they detect by olfaction, by decreasing feeding rate and exhibiting immobility and avoidance (Anholt et al., 1996; Petranka and Hayes, 1998). They also transform earlier at a

smaller size (Skelly and Werner, 1990). Density of predators and food availability do not appear to appreciably affect these responses (Petranka, 1989a; Pearman, 1995; Anholt et al., 1996), nor does learning play a role (Gallie et al., 2001). Tadpoles commonly accrue injuries to the tail tip due to attempted predation (Blair and Wassersug, 2000).

American toad tadpoles form schools (Black, 1971a, 1975; Beiswenger, 1975, 1977, 1978; Feder, 1984; Dupre and Petranka, 1985) abetted by kin recognition (Waldman and Adler, 1979; Waldman, 1981, 1982a, 1985a,b, 1986; Dawson, 1982). Kruse and Stone (1984) found that largemouth bass *(Micropterus salmoides)* learned to distinguish American toad tadpoles from spring peeper *(Pseudacris crucifer)* tadpoles, but surmised that this was likely due to the schooling behavior of the tadpoles that may thus also serve as a deterrent to predation.

Newly metamorphosed toadlets also form aggregations that may be a part of an anti-predation defense against snakes (Hayes, 1989; Heinen, 1993a,b, 1994, 1995). Juveniles may avoid snakes by relying upon chemical clues of the snakes' presence (Flowers and Graves, 1997). As with adults, newly metamorphosed toadlets respond to the touch of garter snakes by crouching and remaining immobile, which affords them some success in thwarting predation; unlike adults, juveniles do not noticeably inflate themselves (Hayes, 1989).

Q. Diseases. An incidence of a xanthoma cancer has been noted by Counts and Taylor (1977), and the incidence of assorted hindlimb deformities was noted by Ouellet et al. (1997a).

R. Parasites. American toads harbor a host of internal and external parasites, including protozoans, helminths of various sorts, and other organisms. Protozoan parasites include the sarcomastigote cloacal parasite *Opalina* (Delvinquier and Desser, 1996), the myxosporean *Myxidium serotinum* (McAllister and Trauth, 1995), and the protozoan pathogen *Toxoplasma* (Stone and Manwell, 1969). Trypanosomiasis blood parasitism is widespread (Werner and Walewski, 1976; Barta and Desser, 1984; Werner et al., 1988; Lun and Desser, 1996), particularly involving the species *Trypanosoma fallisi* (Martin and Desser, 1990, 1991a,b; Martin et al., 1992, 2002). Nematode parasites include *Gyrinicola batrachiensis* of tadpoles (Adamson, 1981a,b) and *Oswaldocruzia pipiens* (Baker, 1977), *Rhabdias* sp. (Baker, 1978a, 1979), *Cosmocercoides dukae* (Baker, 1978b), and *C. variabilis* (Vanderburgh and Anderson, 1987a,b; Joy and Bunten, 1997) of adults. The apicomplexan *Hepatozoon clamatae* (Kim et al., 1998), trematodes of the genus *Allassostomoides* (Brooks, 1975), cestodes of the genus *Mesocestoides,* assorted other helminths (James, 1969; Ulmer, 1970;

Ulmer and James, 1976; Ashton and Rabalais, 1978; D. D. Williams, 1978; Williams and Taft, 1980; Coggins and Sajdak, 1982; Bolek and Coggins, 2000), dipteran flies *(Bufolucilia elongata;* Briggs, 1975), the leech *Desserobdella picta* (Briggler et al., 2001), and the parasitic fungus *Dermosporidium penneri* (Jay and Pohley, 1981) are also known to parasitize American toads.

4. Conservation. Conant and Collins (1998) call American toads the "common and abundant 'hoptoad' of the Northeast." Nowhere is this species considered to be under any threat of extirpation (Lannoo, 1998b). Nevertheless, this species is subject to fluctuations in population size (Casebere and Taylor, 1976; Christein and Taylor, 1978; Heyer, 1979; Hecnar and M'Closkey, 1996).

American toads do not appear to be as sensitive to habitat fragmentation as are many other species of co-occurring amphibians (Hager, 1998; Kolozsvary and Swihart, 1999; Lehtinen et al., 1999). They are the most readily able amphibian to reinvade clearcut and burned areas of forest (Blymyer and McGinnes, 1977; Kirkland et al., 1996). However, American toad larvae are susceptible to low pH due to acid precipitation, which sublethally affects their nitrogen balance (Clark and LaZerte, 1985, 1987; Dale et al., 1985a,b; Tattersall and Wright, 1996) and thus, ultimately, their survival (Clark and Hall, 1985; Freda and Dunson, 1985; Leftwich and Lilly, 1992). In consequence, acid precipitation has reduced the distribution of American toads in affected regions of their range (Clark, 1986; Wyman, 1988a). Enhanced ultraviolet radiation, particularly UV-B, has however not been demonstrated to have a substantial detrimental effect upon American toad eggs or larvae under natural conditions, largely due to the rapid attenuation of UV-B penetration into the water (Grant and Licht, 1995; Crump et al., 1999). Chemical contaminants, including pesticides such as benzopyrene, methoxychlor, toxaphene and endrin, as well as lead and ammonium nitrate, have both lethal and sublethal effects upon American toads, particularly at the larval stage (Clark and Diamond, 1971; Hall and Swineford, 1979, 1981; Bracher and Bider, 1982; Steele et al., 1991, 1999; Hecnar, 1995; Jofre and Karasov, 1999; Harris et al., 2000), against which they appear to have no defense. However, Allran and Karasov (2001) indicate that atrazine concentrations in the field may not be high enough to cause direct effects. Hindlimb deformities have been noted among juvenile toads in agricultural habitats fraught with pesticides, although to a more limited extent than among some other co-occurring species, especially ranids (Ouellet et al., 1997). Metamorphosing individuals may be particularly susceptible to certain forms of environmental stress, as they have reduced high temperature tolerance compared to both larvae and adults (Hathaway, 1928; Cupp, 1980).

Bufo baxteri Porter, 1968
WYOMING TOAD

R. Andrew Odum, Paul Stephen Corn

1. Historical versus Current Distribution.
Historically, Wyoming toads *(Bufo baxteri)* were abundant in the vicinity of Laramie, Albany County, Wyoming (Baxter, 1952; Corn, 1991), where they were found in the flood plains of the Big and Little Laramie rivers (Stebbins, 1985), an area of only 2,330 km^2 (Lewis et al., 1985). A rapid decline was observed in the mid-1970s (Lewis et al., 1985) when they disappeared from most of their range. Surveys in the early 1980s yielded few animals, and Lewis et al. (1985) reported their possible extinction in 1983. Wyoming toads were listed as Endangered under the Endangered Species Act in February 1984.

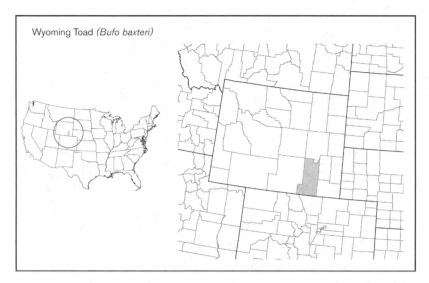

Wyoming Toad *(Bufo baxteri)*

Currently, Wyoming toads have been reintroduced under a U.S. Fish and Wildlife Service recovery plan (Stone, 1991) to Mortenson Lake, ~23 km southwest of Laramie and the site of the last known population of toads. This population has yet to return to its 1988 levels and may not be sustainable (see "Historical versus Current Abundance" below). Attempts were also made to reintroduce Wyoming toads at two locations on the Hutton Lake National Wildlife Refuge, 10 km southeast of Mortenson Lake. These animals did not establish breeding populations and all released toads disappeared. The Mortenson Lake population is considered the only known current distribution in nature for Wyoming toads (Jennings et al., 2001).

Captive populations are being maintained at seven American Zoo and Aquarium Association (AZA) institutions (Central

Park Zoo, New York; Cheyenne Mountain Zoo, Colorado Springs, Colorado; Detroit Zoo, Detroit, Michigan; Houston Zoo, Houston, Texas; Sedgwick County Zoo, Wichita, Kansas; St. Louis Zoo, St. Louis, Missouri; and Toledo Zoo, Toledo, Ohio); at the Saratoga National Fish Hatchery, Saratoga, Wyoming, and the Wyoming Game and Fish Sybille Wildlife Research Unit, Sybille, Wyoming. These animals are managed cooperatively in an AZA Species Survival Plan (SSP®) to maximize genetic diversity in captivity and produce offspring for reintroduction (Spencer, 1999).

Wyoming toads reflect relict populations of Canadian toads *(B. hemiophrys)* isolated at higher altitudes (2,164 m) of the Laramie Basin at the end of the Pleistocene when glaciers receded and climatic conditions warmed (Blair, 1972a). Currently, the closest population of Canadian toads is ~750 km from the Laramie Basin. Canadian toads are more aquatic than most toads and are frequently found on the shores of small lakes (Conant and Collins, 1991; see the Canadian toad account). This aquatic tendency is shared by Wyoming toads.

The presence of Canadian toads in the Laramie Basin was first recorded by Baxter (1947, 1952) and included on the map accompanying the species account in Stebbins (1954a, p. 145). The Laramie population, with its long-term isolation and distinctive characteristics, was later elevated to subspecies status (*B. hemiophrys baxteri*; Porter, 1968) and then to full species status (*B. baxteri*; Smith, 1998). Debate continues about whether Wyoming toads deserve separate species status (Packard, 1971; Smith et al., 1998). We follow Crother et al. (2000) in designating Wyoming toads as a valid species.

2. Historical versus Current Abundance.
The occurrence of natural populations has dropped precipitously in the past 30 yr.

Lewis et al. (1985), Parker (1998), and others (see "Historical versus Current Distribution" above) report that Wyoming toads were abundant in the 1950s and 1960s, but by the mid-1970s experienced a drastic decline (Baxter and Stone, 1985). A 1980 survey revealed Wyoming toads in a 5 km^2 area (Van Kirk, 1980). In 1981, only one male and one female were observed during the breeding season (Baxter and Meyer, 1982). No toads were observed in 1982; the discovery of two juveniles in 1983 represented the first evidence that reproduction had occurred since 1975 (Lewis et al., 1985). An intensive survey in 1984 located ~30 males, but similar surveys in 1985 and 1986 found no animals (Stone, 1991). In 1987, a single population of Wyoming toads was located at Mortenson Lake (Anonymous, 1987). This is the last natural population of toads that has been located.

There is no evidence for what the size was of a "typical" population before the decline of this species, but the size of the Mortenson Lake population has been monitored with varying degrees of rigor since it was discovered. Based on single surveys conducted each year about 1 September, this population was thought to include about 100–150 adults in 1987–89 (Withers, 1992; Corn, 1993a; Parker, 2000). Beginning in 1990, a capture-recapture study was conducted using photography to identify individual toads. In 1990, there were 120 (s.e. = 22) adult-sized toads (Corn, 1991). The population estimate in 1991 was 415 (s.e. = 99), but the majority of these were subadult toads that had metamorphosed the previous summer (Corn, 1992). The population estimate in 1992 was 155 (s.e. = 21), and a size-frequency analysis estimated that 36% of the population was composed of toads that were 2 yr old (Corn, 1993a). Few toads emerged in 1993, and only two adults were located during the annual September survey; these were taken into captivity in an effort to salvage the species (Parker, 2000).

Reproduction was recorded at Mortenson Lake beginning in 1988 when at least nine egg masses were found (Freda et al., 1988). Numbers of egg masses declined in subsequent years, with two found in 1989 (McLeary, 1989), three in 1990 (Chamberlin, 1990), and three in 1991 (Withers, 1992). No reproduction was observed in 1992 or 1993 (Corn, 1993a; Parker, 2000).

Corn (1993a) observed that with a mortality rate of about 80% for adults, the population was unsustainable. This was supported by population modeling, which resulted in a negative population growth rate and certain extinction within 10 yr (Jennings et al., 2001). When the last naturally bred wild toad was collected in 1994 at Mortenson Lake, Wyoming toads probably became extinct in the wild (M. Jennings, unpublished data).

Efforts to establish a second population were initiated in 1992 at Lake George, Albany County, Wyoming. Four "managed breedings" in protective enclosures were successful. The breeding stock came from animals collected at Mortenson Lake in 1990 and 1991 and held in captivity until the spring of 1992. These breedings resulted in large numbers of newly metamorphosed animals in July (Withers, 1992), but a survey performed the following spring found only 22 individuals. This number decreased to two in the fall of 1993. Subsequent efforts to release animals at Lake George have not been successful (total releases = 3,963 larvae or metamorphosed animals, 56 juveniles [1 yr olds]; M. Jennings, unpublished data).

In 1994, 77 Wyoming toads were produced in the first captive breeding of this species at the Sybille Wildlife Research Unit, Sybille, Wyoming. These animals were held in captivity for further breeding stock. In 1995, a concentrated effort to reintroduce toads back to Mortenson Lake commenced from captive produced offspring. In 1995, 1,300 Wyoming toad young (< 4 mo) were released. This was followed by 2,638 young in 1996 and 213 in 1997 (M. Jennings, unpublished data). In 1998, free-ranging toads from the 1995 releases at Mortenson Lake produced four egg masses, the first known wild breedings since 1991 (Parker, 2000). The same year, these breedings were augmented with the release of 3,428 captive-born young. Three wild breedings were observed at Mortenson Lake in 1999 and the population was estimated to be 159 adults. Two thousand captive bred toads were also released in 1999. It appeared that the population was on its way to recovery; however, the following year reproduction was limited to two egg masses (Parker, 2000). In 2000, no reproduction was observed, census numbers were very low, and sick and dead toads were observed. The outcome for this reintroduction effort is still guarded.

3. Life History Features.

A. Breeding. Reproduction is aquatic.

i. Breeding migrations. Adult toads first emerge from hibernation when air temperatures reach 21 °C (usually in early to mid-May; Stone, 1991; Withers, 1992; Parker, 2000). Males usually start calling a week later. The call of Wyoming toads is a harsh trill. Calling peaks when air and water temperatures are between 21.1–26.7 °C, decreases as temperatures fall, and ceases at 10 °C (Withers, 1992).

ii. Breeding habitat. Withers (1992) reported that vegetation at sites selected by calling males was a sedge/rush (Carex sp./Scirpus americanus) mix. These plants were also associated with egg masses.

B. Eggs.

i. Egg deposition sites. Wyoming toads deposit their eggs in shallow areas of ponds and small lakes (Baxter, 1952). Withers (1992) reported egg masses in water 3.5–6.3 cm deep where mean daytime water temperatures ranged from 20.6–21.8 °C.

ii. Clutch size. Egg masses may contain 1,000–6,000 individual 2–3 mm ova in strings (Freda et al., 1988, Withers, 1992). One captive produced egg mass was estimated at 8,000 eggs (M. Bock, personal communication). The female involved in this captive breeding was especially robust and this number may exceed the largest clutch of eggs produced in the wild. Eggs may hatch in 4–6 d (Freda et al., 1988; Withers, 1992) in the wild (2 d in a controlled captive situation; R. A. O., personal observations).

C. Larvae/Metamorphosis.

i. Length of larval stage. Wyoming toad tadpoles are darkly pigmented and 5–7 mm in length at hatching. Length of larval stage is approximately 1 mo, with metamorphosis in early July (Withers, 1992). Larval periods are reduced under more thermally homogeneous conditions in captivity (R. A. O., personal observations).

ii. Larval requirements.

a. Food. Although not well studied, Wyoming toad tadpoles are generalized suspension feeders, grazing on the organic and inorganic matter associated with rock, plant, and log surfaces.

b. Cover. Larvae select areas of higher temperature as high as 31 °C (Withers, 1992). Beiswenger (1978) collected tadpoles from shallow water at 30.5 °C and noted that Wyoming toad tadpoles seek out warm regions in thermally stratified environments. In one case, shallow depressions described as "water cow paths" were areas of congregation for tadpoles from three separate clutches (Withers, 1992). In the laboratory, Beiswenger (1978) observed that Wyoming toad tadpoles aggregate in temperatures between 28–34 °C (preferred temperature was ~31 °C) when placed in a stratified thermal environment.

iii. Larval polymorphisms. Do not occur.

iv. Features of metamorphosis. Metamorphosis usually takes place in early July, which corresponds to the annual bloom of small black flies (Diptera), possibly a food resource for young toads (Withers, 1992; A. Anderson, personal communication).

v. Post-metamorphic migrations. Following metamorphosis, young toads disperse along or near the shoreline. By late summer, many migrated from the north side of Mortenson Lake to the east and southeast sides (Withers, 1992).

D. Juvenile Habitat. Similar to those of adults.

E. Adult Habitat. Wyoming toads are found close to water in short grass prairie (Stebbins, 1985; Luce et al., 1997; see Withers, 1992 for a more detailed analysis of habitat). Adults almost always restrict their habitat use to the shoreline within 10 m of water (Withers, 1992; Parker, 2000). This may be due to the low humidity associated with the Laramie Basin uplands and the need for toads to have access to moisture for rehydration. Historically, Wyoming toads frequented a variety of habitats including lakes, ponds, streams, marshes, roadside ditches, and floodplains (Stebbins, 1985; Luce et al., 1997). At Mortenson Lake, toads use areas along the shoreline as well as adjacent marshes and ditches (Corn, 1991; Withers, 1992). When disturbed, Wyoming toads frequently seek escape by diving into open water and submerging (Parker, 2000). Adults tend to be nocturnal (Luce et al., 1997) or diurnal (Parker, 2000), with activity levels largely dependent on temperature (Parker, 2000).

F. Home Range Size. Has not been clearly delineated. At Mortenson Lake, Withers (1992) suggested a post-breeding migration from the north shore to the south shore. This migration would require the toads to move several hundred meters. Parker (2000) observed a maximum 1 d movement of 151.8 m and a mean daily movement of 5.05 m.

G. Territories. Unknown, but unlikely.

H. Aestivation/Avoiding Desiccation. Parker (2000) observed that adult toads active during the day often sought "night refuge" under dense grass on moist substrate. While in these nocturnal refugia, toads were observed in a state of torpor. Wyoming toads also use rodent burrows for refuge (Withers, 1992; Parker, 2000).

I. Seasonal Migrations. Seasonal migrations involve breeding (see "Breeding migrations" above), post-metamorphic (see "Post-metamorphic migrations" above), and overwintering (see "Torpor [Hibernation]" below) movements.

J. Torpor (Hibernation). Only anecdotal data exist on the hibernation sites of Wyoming toads. Withers (1992) observed toads near pocket gopher or ground squirrel burrows in the spring and fall and assumed these structures were used for hibernation. One captive born and released toad with an implanted radio transmitter was found in the fall burrowed in a badger excavation. The toad was at a depth of 35 cm and a temperature of 13 °C. The badger had broken through the hard soil crust, allowing the toad to dig into the softer substrata (Parker, 2000).

K. Interspecific Associations/Exclusions. Wyoming toads may hybridize with American toads (B. americanus; Henrich, 1968).

L. Age/Size at Reproductive Maturity. Wyoming toads are smaller than Canadian toads. Smith et al. (1998) recorded a maximum length of only 59.5 mm, with only 3 of 23 preserved specimens measuring over 56 mm. However, Corn (1993a) measured live toads and recorded a maximum length of 68 mm. Sexually mature males and females are dimorphic for size.

In 1992 at Mortenson Lake, the average size of 1-yr-old males was estimated to be 48 mm and the average size of 2-yr-old males was 52 mm. One-yr-old females were the same size (47 mm) as males of the same age, but the average size of 2-yr-old females was estimated to be 61 mm. Growth after metamorphosis is rapid, and males reach adult size by August of their second summer and breed the following spring (2 yr old, similar to Canadian toads; Breckenridge and Tester, 1961). Females most likely require an additional year to reach maturity (Baxter, 1952; Corn, 1991; Withers, 1992). In the artificial environment of captivity, 1-yr-old toads have produced offspring (D. Roberts, personal communication).

M. Longevity. No long-term studies have been performed to determine the maximum longevity of Wyoming toads in the wild. Corn (1993a) observed that few adult toads lived >2 yr at Mortenson Lake, but that was in a population afflicted with chytrid fungus. In captivity, one female toad with an estimated birth date in 1989 (studbook #5) lived in captivity from 1994 until its death in 1997 (Callaway, 1998). This animal produced large numbers of healthy young from 1994–96.

N. Feeding Behavior. Adults feed on ants, beetles, and other small insects and invertebrates (Luce et al., 1997). Stomach contents and fecal pellets of adult toads contained substantial numbers of ants, which appear to be the main component of the adult diet (G. Lipps, personal communication; R. A. O., personal observations).

O. Predators. Predation may be an important cause of mortality in Wyoming toads (Parker, 2000; see "Conservation" below).

P. Anti-Predator Mechanisms. Nocturnal activity patterns. Wyoming toads will swim out into the water away from shore when frightened (Parker, 2000).

Q. Diseases. An overview of diseases of Wyoming toads is presented in Taylor et al., 1999. A substantial number of deaths (71% in cases where cause was known) resulted from mycotic dermatitis with secondary bacterial infections. The authors misidentified the primary pathogen as *Basidiobolus ranarum*, which was later determined to be *Batrachochytrium dendrobatidis* (chytrid fungus; A. Pessier, personal communication; E. Williams, personal communication). The majority (90%) of these deaths in free-ranging toads occur in September–October when temperatures start to fall. Chytrid fungal infections are also the most frequent cause of mortality in captive populations and again are most prevalent in autumn. A treatment for this disease has been developed for captive animals (A. Pessier, personal communication).

R. Parasites. Unreported in wild populations. In captivity, a wide variety of protozoan and helminth parasites have been isolated (R. A. O., personal observations).

4. Conservation.
Wyoming toads were extirpated in the wild in 1994 and have been reintroduced. Chytrid fungus appears to have been the proximate cause of the decline of these toads (Taylor et al., 1999; A. Pessier, personal communication). Disease still remains important and impacts Wyoming toad recovery efforts (Jennings et al., 2001).

Other factors have been hypothesized as responsible for the precipitous Wyoming toad decline, including the aerial application of baytex (fenthion), an organophosphate pesticide used for mosquito control; predation by California gulls (*Larus californicus*), American white pelicans (*Pelecanus erythrorhynchos*), and raccoons (*Procyon lotor*), which have been increasing in numbers in this region; and habitat modification due to the irrigation regions of the native hay meadows (see Jennings and Anderson, 1997). However, there is little evidence that any of these factors are the primary cause of toad declines. Parker (2000) found radio-tagged toads preyed upon in his study and suggested additional predators include skunks, weasels, badgers, coyotes, mink, domestic cats, and herons.

Another factor impacting Wyoming toad recovery is the small amount of genetic diversity remaining in the species. The entire captive, and thus the reintroduced wild, population contains the equivalent genetic diversity of ~2.4 unrelated animals (Founder Genome Equivalents). This is equal to 79.4% of the expected gene diversity that should have been contained in the original wild population. The mean inbreeding coefficient (F) of the population is 0.155, which is greater than a first cousin cross. Inbreeding depression has not been conclusively documented in Wyoming toads, although there is some correlation between inbreeding and lower fecundity (Lipps and Odum, 2001).

Attempts have been made to manage the captive and reintroduced populations to maximize the retention of gene diversity. Using the studbook data, mean kinship is calculated to determine which pairs will maximize the amount of gene diversity retained in the next generation. Unfortunately due to husbandry difficulties, this method has not been entirely successful. Many scheduled priority breedings have not produced offspring and other less important pairings have. Analysis of the husbandry data (G. Lipps and R. A. O., unpublished 2000 data) show that there is a slightly less than significant negative correlation between the inbreeding coefficient of the offspring produced by a pairing and breeding success (as number of offspring produced from the breeding). These data are still preliminary, and results from 2001 appear to be contradictory.

Captive Wyoming toads currently are being managed as two separate populations. The first group (Group A toads) is composed of animals that have fully

known pedigrees that can be traced back to the original collection of animals in the early 1990s. All other animals are derived from these earlier collections, and Group A represents all of the known diversity for the species. The second group (Group B) is composed of animals that were collected as wild-born offspring of reintroduced Wyoming toads from Group A. Although Group B represents less gene diversity than Group A, they are the offspring of successful survivors in nature, which may indicate increased fitness.

Wyoming toad recovery is still largely dependent upon an influx of individuals from the captive populations. Large numbers of animals have been released at Mortenson and two other sites (Lake George and Rush Lake at Hutton Lake NWR). Only the Mortenson Lake site has had any recent breeding activity and continues to have animals present. The recovery of Wyoming toads is still uncertain. Considering the limited numbers, low genetic diversity, reliance on captive breeding, and continuing problems with disease, Wyoming toads are the most endangered anuran in the United States, if not the world.

Bufo boreas Baird and Girard, 1852(b)
WESTERN TOAD

Erin Muths, Priya Nanjappa

1. Historical versus Current Distribution.
Western toads (*Bufo boreas*) occur throughout much of western North America. They are found throughout the Pacific Northwest, north through western Canada and into the southeastern portion of Alaska (Wright and Wright, 1949; Herreid, 1963; Stebbins, 1985; Norman, 1988a).

In the United States, western toads occur from north-central New Mexico (Campbell and Degenhardt, 1971; Stuart and Painter, 1994; Degenhardt et al., 1996) into western Colorado, where they have been found in most of the mountain ranges at elevations between 2,615 and 3,557 m (except in the Sangre de Cristo Range, Wet Mountains, and Pikes Peak; Hammerson, 1986, 1999; Livo et al., 1999). Their distribution extends north into the higher elevations (between 2,461 and 3,385 m) of western Wyoming (Luce et al., 1997) and into Montana (Maxell, 1999); northwest into Oregon and Washington, where they are found nearly statewide (except for the low-lying areas of the Columbia Basin [Washington], the Willamette Valley and the northern Coast Range [Oregon]; Leonard et al., 1993); west into the High Plateaus and the Wasatch Mountains of Utah (Ross et al., 1995; Oliver, 1997); across areas of high elevation on the Utah–Nevada border and throughout Nevada (Nevada Natural Heritage Program, 1999); and into northern California (Stebbins, 1985). Western toads

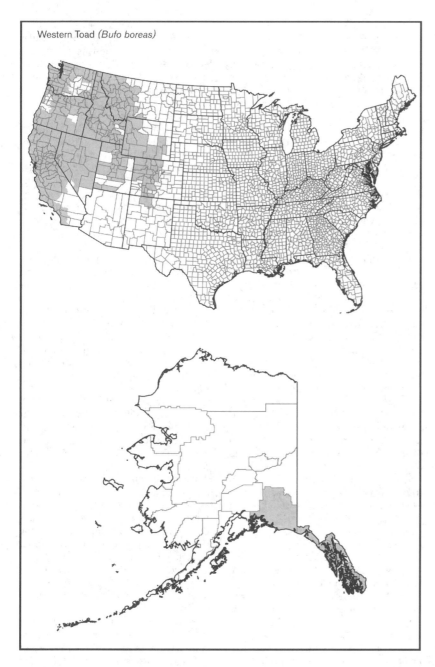

Western Toad *(Bufo boreas)*

are also found in the coastal forests of southeastern and south-central Alaska from the British Columbia border north to Prince William Sound (Hodge, 1976).

According to Crother et al. (2000), three subspecies of *B. boreas* are recognized: boreal toads *(B. b. boreas)*, California toads *(B. b. halophilus)*, and Amargosa toads *(B. b. nelsoni)*. The systematics of this group is provided in Goebel (this volume, Part One). However, this view is controversial (see Crother et al., 2003). In particular, most, perhaps all, workers familiar with Amargosa toads agree that they deserve species status. The arguments for this position and what is known about the conservation status of Amargosa toads are presented in this volume by Anna Goebel in the separate *Bufo nelsoni* account.

Severe declines and extirpations of many populations have occurred in areas where western toads were once abundant (Leonard et al., 1993; Carey et al., 2003; this volume, Part One). Between 1986 and 1988, Corn et al. (1989) searched 59 known western toad localities in Colorado and Wyoming and found them at only 17% of the localities. According to Loeffler (1999), western toads are documented to occur in only two counties in Wyoming. Western toads were found in 6 of 13 historical localities and in eight sites without previous records during surveys between 1987 and 1994 in Rocky Mountain National Park (RMNP; Corn et al., 1997). Surveys in 2000 and 2002 found toads in only 3 of 22 historical localities in RMNP and in one new locality (non-breeding;

E.M., unpublished data). In New Mexico, a small number of western toad localities were documented in Rio Arriba County between the elevations of 2,775 and 3,200 m, however these populations have declined rapidly and are now thought to be extirpated (Stuart and Painter, 1994; Degenhardt et al., 1996; Loeffler, 1999).

There is no published evidence that western toads have declined in the mountains of the Pacific Northwest to the same extent as have populations in the Rocky Mountains (Corn, 2000). According to Crisafulli and Hawkins (1998), toads currently are abundant within the blast zone of the Mount St. Helens 1980 eruption. However, Blaustein et al. (1994b) described mortality in large numbers of eggs, and Olson (1992) described mortality in adults in Oregon. In the Puget Sound lowlands of Washington and other lower elevations in the Pacific Northwest, toads are rare. Recent surveys have found western toads present at only 0–22% of the sites surveyed (Richter and Azous, 1995; Adams et al., 1998, 1999).

Factors suspected to be contributing to the decline of western toads include habitat degradation and destruction (Hammerson, 1999), fungal infections and other pathogens (Carey, 1993; Hammerson, 1999; Muths et al., 2003), acid and mineral pollution from mine water drainage (Porter and Hakanson, 1976), and increased ultraviolet radiation (Kiesecker and Blaustein, 1995; Blaustein et al., 1998; but see Corn, 1998).

2. Historical versus Current Abundance.

Western toads are considered Endangered in several states, and their status is under review in other states. The U.S. Fish and Wildlife Service lists Rocky Mountain populations of western toads, specifically in Colorado, New Mexico, and Wyoming as a candidate species (U.S. Department of the Interior, 1991). The Colorado Division of Wildlife and the New Mexico Department of Game and Fish lists western toads as Endangered, and western toads are a Protected Species in Wyoming. According to Loeffler (1999), they are believed to occur in 15 counties in Colorado and possibly one county in New Mexico. The Utah Division of Wildlife Resources as well as the U.S. Bureau of Land Management has listed this species as Sensitive in Utah, which is defined as "declining population, distribution and/or habitat" (Oliver, 1997).

3. Life History Features.
A. Breeding. Reproduction is aquatic.
i. Breeding migrations. Adults emerge from hibernation sites and migrate to breeding wetlands. In Washington and Oregon, breeding occurs as early as February–April in the lower elevations west of the Cascade Mountains, and from May–July in the Cascades (Leonard et al.,

1993). Males can sometimes exceed females by 20:1 at breeding sites and actively search for gravid females during the night.

Male western toads lack vocal sacs (Hammerson, 1999), although they will emit a chirp when grasped either by other toads or by researchers (Hammerson, 1999; E.M. and P.S. Corn, personal observations). According to Blair (1972a), western toads have lost their mating call. Western toad vocalizations are not typical anuran advertisement calls. There is some controversy about calling in western toads; Awbrey (1972) recorded a putative advertisement call produced by an individual of the western toad subspecies in California (California toad; *B. boreas halophilus*). The calling male was atypical of western toads because it had a vocal pouch, and the call produced was audible at greater distances than the release call (Awbrey, 1972). Whether the call of western toads is simply a release call or functions in the formation of breeding aggregations of males and the attraction of females is unresolved.

In Colorado, breeding depends on the timing of snowmelt, occurring at higher elevations from mid-May to mid-July (Hammerson, 1999; E.M., personal observations). In the vicinity of Juneau, Alaska, breeding occurs between May and July (J. Van Denburgh, in Wright and Wright, 1949; Hodge, 1976). Although female western toads have been found in Colorado at elevations from 3,462–3,557 m, there is little evidence of successful breeding at these elevations (Campbell, 1970d). In Rocky Mountain National Park, however, western toad eggs and tadpoles have been found in a pond at an elevation of 3,380 m (Corn et al., 1997).

ii. Breeding habitat. Breeding sites are in still or barely moving water, typically ponds and small lakes, streams, rain pools, and ditches (Hodge, 1976; Luce et al., 1997; Hammerson, 1999).

B. Eggs.

i. Egg deposition sites. In many western toad populations, egg deposition occurs after snowmelt when breeding ponds are refilled (Leonard et al., 1993; Degenhardt et al., 1996). In the Blue and Wallowa Mountains of northeastern Oregon and southeastern Washington, egg deposition can sometimes occur as late as June–July (Leonard et al., 1993). In New Mexican montane populations of western toads, eggs are laid in mid to late July (Degenhardt et al., 1996). Clutches have been laid as late as August in Colorado (Fetkavich and Livo, 1998). Eggs are deposited in shallow water, generally not > 15 cm deep (Hammerson, 1999).

ii. Clutch size. Eggs are laid in strings of double rows of up to 12,000 eggs/clutch (Wright and Wright, 1949; Samollow, 1980; Leonard et al., 1993) with an average of 5,200 eggs/clutch (Carey, 1976). Eggs are large; the vitellus has a diameter of 1.5–1.75 mm (Wright and Wright, 1949).

At an elevation of about 2,440 m (8,000 ft), Ferguson (1954a) observed an instance of western toad egg masses being washed ashore due to large chunks of snow-pack and ice falling into the breeding lake. He suspected that this phenomenon may be common at these and higher elevations and may affect reproductive success.

C. Larvae/Metamorphosis.

i. Length of larval stage. Egg and larval development is dependent on the temperature of the water (Smith-Gill and Berven, 1979; Ultsch et al., 1999). Western toad eggs hatch in 3–10 d (Leonard et al., 1993) but often take longer at higher elevations. The duration of the larval stage is 30–45 d, with metamorphosis occurring between May and September (Wright and Wright, 1949; G. Fellers, personal communication). Forelimbs appear at Gosner (1960) stage 42, which generally indicates the onset of metamorphosis. In California, metamorphosis occurs from mid-May to late September (G. Fellers, personal communication). In Alaska, metamorphosis occurs between July and August (Hodge, 1976) and has been observed to occur as late as October in Colorado (E.M., personal observations). In Wyoming, tadpoles will metamorphose before the end of summer at elevations below 3,050 m (10,000 ft; Luce et al., 1997). Juvenile toads emerge at 1.0–1.6 cm in length (Leonard et al., 1993; Hammerson, 1999). At higher elevations, tadpoles may not metamorphose before early autumn freezes occur, and overwinter survival of larvae has not been observed during extensive monitoring in Colorado (Fetkavich and Livo, 1998). In Washington, metamorphosis at high elevations may occur 1–2 wk prior to snow accumulation covering the breeding pools (Leonard et al., 1993).

ii. Larval requirements.

a. Food. Larvae will feed upon filamentous algae, detritus, and may even scavenge carrion (Leonard et al., 1993). Western toad tadpoles have been observed to feed on a fish carcasses and on the bodies of conspecifics (E.M., personal observations).

b. Cover. Larvae generally seek out the warmer, shallower portions of their habitat during the day and retreat to deeper waters as temperatures drop. They will swim to cover, rocks, vegetation, or shadows when disturbed.

iii. Larval polymorphisms. Unknown and unlikely.

iv. Features of metamorphosis. Large metamorphic aggregations will form at the edge of the pond, sometimes two or more individuals deep (Lillywhite and Wassersug, 1974; L.J. Livo, personal communication) probably to conserve moisture (Livo, 1998).

v. Post-metamorphic migrations. Upon completion of metamorphosis, juveniles move from natal wetlands to nearby terrestrial sites or to other nearby wetlands (Hammerson, 1999; E.M., personal observations). Hammerson (1999) notes that juveniles may stay to overwinter along the border of their natal wetland.

D. Juvenile Habitat. The habitat characteristics of juvenile western toads are unknown. Presumably, they use the same habitats as adults, but probably use wetland habitat more than terrestrial adults because they are more susceptible to desiccation (Livo, 1998). Because of their small size, newly metamorphosed individuals can use small cracks in the substrate and other very tiny refugia to overwinter, in contrast to adults that typically use larger ground squirrel burrows or other underground hibernacula (see Mullally, 1952).

E. Adult Habitat. Western toads are often found at the water's edge or basking on partially submerged logs in the spring and early summer (Muths and Corn, 1997; Hammerson, 1999). Later in the year, they are often found in more terrestrial habitats, although often in damp areas or near water (Hammerson, 1999). Historically, western toads were thought to be more terrestrial except when breeding, and tolerant of dry habitats (Ruthven and Gaige, 1915; Wright and Wright, 1949).

Bartelt (2000) documented extensive (>75% of the time) use of terrestrial habitats by western toads in Idaho. He found that conditions at microsites with toads were more humid but approximately the same temperature as randomly chosen microsites without toads (Bartelt, 2000).

In California, western toads are found from valley floors (sea level) to high mountains (>2,740 m [9,000 ft]) in grassy tussocks near lakeshores, streams, and in mountain meadows (Grinnell and Camp, in Wright and Wright, 1949). In Wyoming, western toads are found in the wet areas of higher elevations between 2,461 and 3,385 m in the foothills, subalpine zones, and mountainous regions (Luce et al., 1997).

During cold weather (temperatures below 3 °C), western toads will use gopher and ground squirrel holes as retreats, where temperatures remain between 4.8 °C and 7 °C at varying depths even when freezing temperatures occurred above ground (Mullally, 1952). At higher elevations, western toads hibernate in rocklined chambers near creeks (Campbell, 1970b), in ground squirrel (*Spermophilus lateralis*) burrows (Jones and Goettl, 1998), in and under root systems of evergreen trees (E.M. and P.S. Corn, personal observations), and possibly in beaver dams (Goettl et al., 1997). Smits (1984) reported that western toads were diurnal following emergence and just prior to hibernation, with an abrupt switch to nocturnality in June. Smits (1984) also observed that western toads avoided extreme temperatures (hot or cold) by retreating to burrows. Experimental results from Smits and Crawford (1984) suggest that thermal cues in western toads direct their daily emergence from subterranean refugia. Lillywhite et al. (1973) found that the preferred body temperature among western toads in the laboratory is approximately 26–27 °C. Their results also demonstrated that basking versus seeking cooler temperatures is dependent on both food and moisture availability (Lillywhite et al., 1973).

F. Home Range Size. Western toads with embedded radioactive tags moved 900 m from their summer habitat to their hibernacula, then emerged in the spring to return to their summer habitats (Campbell, 1970b). Western toads were studied at ponds and wetlands at a heavily disturbed site (due to molybdenum mining) in Colorado. Data from this location suggest that toads use a variety of habitat types from upland aspen/conifer stands to rocky areas (Jones, 2000). Bartelt (2000) reported daily movements up to 439 m and found males to travel shorter distances from breeding ponds than did females. Home range estimates using a fixed kernel method (Worton, 1989) for males (n = 11) and females (n = 12) are 7.09 ha and 16.96 ha respectively (Jones, 2000). In Rocky Mountain National Park, toads have moved over 4 km between breeding locations in 1 yr (E.M. and P.S. Corn, unpublished data). Female toads move farther from breeding sites and use larger home ranges than males (Muths, 2003) and tend to be found more often at drier sites (E.M., unpublished data).

G. Territories. Although there is no evidence of territoriality in western toads, amplexed males will defend females from approaching males (Black and Brunson, 1971). Black and Brunson (1971) observed a male of an amplexing pair kick away six males that approached. During the breeding season, Black and Brunson (1971) also observed males to actively approach and briefly wrestle with newly arriving males.

Male western toads displaced from their breeding sites returned rapidly and directly to the site of capture, under both clear and overcast skies (Tracy and Dole, 1969). Tracy and Dole (1969) found under test conditions that olfactory cues appear to play a large role in orienting displaced animals towards their breeding sites, and immature toads are reported to use celestial cues in dispersal (Tracy, 1971).

H. Aestivation/Avoiding Desiccation. Not documented.

I. Seasonal Migrations. Western toads emerge from hibernation and travel to breeding pools in the spring, move to terrestrial sites during the summer, and return to hibernacula in the autumn (Hammerson, 1999). In Colorado, western toads generally emerge from hibernacula in May (Hammerson, 1999) and begin to move toward hibernacula in late August to early September (Campbell, 1970b; Hammerson, 1999). Activity varies among locations and seasons. Western toads are active in California from February (in some years) to late October (Mullally, 1952; G. Fellers, personal communication). Toads in a controlled environment were observed to stay close to their burrows during seasonal transitions (immediately pre-hibernation/post-emergence or between diurnal and nocturnal activity; Smits, 1984).

J. Torpor (Hibernation). According to Pinder et al. (1992), in amphibians that are not freeze tolerant, the metabolism of hibernating animals does not appear to decrease more than is expected from temperature effects alone. Campbell (1970b) found western toads hibernating beneath or near large boulders along a spring-fed brook supplemented by extensive snow accumulation to maintain the flow of groundwater. The toads were found with feet and venters in contact with moist substrate, not in physical contact with other toads.

K. Interspecific Associations/Exclusions. Western toads will breed in association with spadefoot toads (*Spea* sp.; Cope, in Wright and Wright, 1949). They will also breed in association with Pacific treefrogs (*Pseudacris regilla*), Yosemite toads (*B. canorus*), and several species of *Rana* (G. Fellers, personal communication). Western toads will naturally hybridize with red-spotted toads (*B. punctatus*) in the southwestern United States and in Baja, Mexico (Feder, 1979).

L. Age/Size at Reproductive Maturity. Sexually mature adult males range from 56–108 mm SVL; females from 60–125 mm (Wright and Wright, 1949). Western toads become sexually mature at 4–6 yr old (Carey, 1976; Carey et al., 2001).

M. Longevity. Western toads that survive to adulthood have a long life expectancy (Campbell, 1970c, in Samollow, 1980; Hammerson, 1999). Snider and Bowler (1992) report a wild-caught western toad lived almost 6 yr, 3 mo. Western toads in Colorado live at least 9 yr (Campbell, 1976) and probably much longer (Muths and Corn, 2000).

N. Feeding Behavior. Prey items are ingested by a quick extension of the tongue, termed a "zot," to snap up a prey item (Tracy, 1973). Primary food sources include spiders, worms, ants, moths, beetles, and other arthropods (Campbell, 1970a,c; Barrentine, 1991b; Leonard et al., 1993; Luce et al., 1997). Billbug weevils (*Sphenophorus* sp.) are also ingested frequently. However, Barrentine (1991a) found that approximately 68% of ingested billbug weevils survived through the digestive tract and emerged from fecal pellets; thus, little nutrition is derived from ingested billbug weevils. Cunningham (1954) documented an instance of adult cannibalism in western toads.

O. Predators. Western toads are preyed upon by garter snakes (*Thamnophis* sp.), coyotes (*Canis latrans*), raccoons (*Procyon lotor*), and some birds (Leonard et al., 1993). Badgers will also eat western toads (Long, 1964). Ravens (*Corvus corax*) eviscerate toads and leave them partially eaten, presumably to avoid the toxins in the skin (Olson, 1989; Corn, 1993b). Beiswenger (1981) reported tadpole predation by gray jays (*Perisoreus canadensis*), with no indication that the birds found the tadpoles to be distasteful (see "Anti-Predator Mechanisms" below). Predaceous diving beetle larvae (*Dytiscus* sp.) and garter snakes were observed to prey upon western toad larvae (Livo, 1998), as were spotted sandpipers (*Actitis macularia*) and mallard ducks (*Anas platyrhynchos*; Jones et al., 1999).

P. Anti-Predator Mechanisms. When threatened by predators, adult western toads secrete a mild toxin from the parotoid glands (Cei et al., 1968; Stebbins, 1985). In *Bufo* sp., this toxin is secondarily deposited in the ova and thought to afford some protection to newly laid, fertilized eggs, although the toxicity of the eggs is thought to decrease as the zygotes develop (Brodie et al., 1978). Generally, reports on bufonid larvae show that invertebrate predators find them palatable, while many vertebrate predators find them unpalatable (Kruse and Stone, 1984). Hews and Blaustein (1985) demonstrated that chemoreception can elicit an alarm response in western toad tadpoles and thus may contribute to predator avoidance.

Q. Diseases. Amphibian chytridiomycosis (chytrid fungus) is a recently described disease caused by infection by the fungus *Batrachochytrium dendrobatidis*. Chytrids, fungi in the phylum Chytridiomycota, live in water and soil and are considered to be ubiquitous. There are many genera and species of chytrid fungus, most of which are beneficial saprobes. *Batrachochytrium dendrobatidis*, cultured, identified, and named by Longcore et al. (1999), is the first member of this phylum to be described as a parasite or pathogen of a vertebrate species. *Batrachochytrium dendrobatidis* has been implicated as a contributor to amphibian declines in Australia, Central America (Berger et al., 1998), and the United States (Nichols et al., 1998; Daszak et al., 1999). This chytrid fungus has been identified in declining wild populations of western toads in Colorado and Wyoming (Muths et al., 2003; D.E. Green, personal communication).

Upon infection, *B. dendrobatidis* lives inside cells in the superficial layer of amphibian skin (Longcore et al., 1999; Pessier et al., 1999). Amphibians rely on their permeable skin for a variety of uses. Severe infection by this chytrid fungus causes a thickening, or *hyperkeratosis*, of the skin. Because amphibians obtain moisture by absorbing it through their skin, infection may compromise water absorption and/or osmoregulation enough to cause death (Nichols et al., 1998). Alternately, *B. dendrobatidis* infection may affect immune systems and may damage the skin sufficiently to allow micro-organisms such as bacteria to cause lethal secondary infections (D. E. Green, personal communication). Information on sublethal infections of or spontaneous recovery from chytridiomycosis in western toads is lacking.

Olson (1989) found that disease caused by the bacteria *Aeromonas hydrophila* (often called "red-leg") was fairly common among western toads. Pathologists suspect that bacterial infections in the 1970s and 1980s may have been secondary to chytrid fungus infections (D. E. Green, personal communication). Carey (1993) proposed that suppression of the immune system leaves toads vulnerable to bacteria such as *Aeromonas*, but there has been no convincing evidence of what might be causing immunosuppression.

R. Parasites. James and Maslin (1947) report parasitization of western toads by fly larvae and Ingles (1936) discusses "worm parasites" of California amphibians including western toad species. More recently, Paperna and Lainson (1995) reported parasites in marine toads (*B. marinus*), but specifics about western toads are not known.

4. Conservation.
Western toad populations were last documented in New Mexico in 1986. Causes of this extirpation may include increased competition due to the presence of introduced fathead minnows (*Pimephales promelas*; Carey, 1987), mortality due to natural predators (Corn, 1993b), mortality due to trampling from domestic livestock usage of wetlands (Bartelt, 1998), and decreased hatching success due to UV-B radiation exposure (Blaustein et al., 1994c, see also Degenhardt et al., 1996; Corn, 1998). High altitude populations in Colorado and elsewhere in the southern Rocky Mountains are in severe decline (for a discussion of reasons, see Carey et al., 2003; this volume, Part One).

Bufo californicus Camp, 1915
ARROYO TOAD

Samuel S. Sweet, Brian K. Sullivan

We follow Gergus et al. (1997), Gergus (1998), and Crother et al. (2000) in separating arroyo toads (*B. californicus*) from Arizona toads (*B. microscaphus*); most recent literature refers to arroyo toads as *B. microscaphus californicus*. Arroyo toads were described originally by Camp (1915)

as *B. cognatus californicus*, type locality: "Santa Paula, 800 feet altitude, Ventura County, California." Myers (1930a) was the first to recommend species status, but in the interim, *B. californicus* was regarded as a subspecies of *B. compactilis* (Linsdale, 1940), *B. woodhousii* (Shannon, 1949), and most recently *B. microscaphus* (Stebbins, 1951). A complete synonymy is presented in Price and Sullivan (1988).

I. Historical versus Current Distribution.
Arroyo toads (*B.californicus*) occur in southwestern California and adjacent Baja California del Norte, Mexico, mostly on the coastal slopes, from the San Antonio River of southern Monterey County southward through the Transverse and Peninsular ranges to the Rio Santo Domingo. They also occupy a few drainages on the desert slopes of the San Gabriel and San Bernardino ranges (Jennings and Hayes, 1994a), though reported localities on the desert slopes of the Peninsular Ranges recently have been shown to involve misidentifications (E. Ervin, personal communication, 2001). The mapping format used in this volume substantially overestimates both the areal distribution and the continuity of arroyo toad populations, such that this map should not be used for any but the most general purposes. See U.S. Fish and Wildlife Service (1999a) for accurate point maps detailing the historical and current distributions of this species.

The current distribution of arroyo toads is highly fragmented, largely through the alienation of coastal lowlands by development and the widespread alteration of the middle reaches of larger drainages by dams and flood control projects (Sweet, 1992;

USFWS., 1994a; Campbell, 1996). Jennings and Hayes (1994a) estimated that arroyo toads had been eliminated from 76% of their historical range, though subsequent discoveries of new localities and remnant populations reduce this figure to about 65%. Many populations are now isolated, either (1) restricted to small headwater drainages above impoundments where conditions are marginal or (2) confined to narrow riparian corridors along larger drainages that are subject to extensive disturbance from water flow management practices, gravel mining, urbanization, and military training (USFWS, 1999a).

2. Historical versus Current Abundance.
Arroyo toads have declined in abundance (often to extirpation) at most sites where historical records exist. High density populations described by Sanders (1950), Stebbins (1951), and Cunningham (1962), for example, are now extirpated, and the large series of museum specimens collected by L. M. Klauber and others in the 1930s can no longer be duplicated at the great majority of their historical sites (S.S.S., personal observation). In the early 1990s, surviving populations reached a low ebb following an extended drought and several decades of adverse land-use and water management practices (Sweet, 1992). Most populations on public lands have recovered substantially following the Federal listing of arroyo toads in late 1994 (USFWS, 1994a). However, many

populations remain well below apparent carrying capacity despite high recruitment, probably as a result of as-yet-unexplained high mortality during aestivation in autumn and winter (Sweet, 1992, 1993, and unpublished data). Six of eight populations

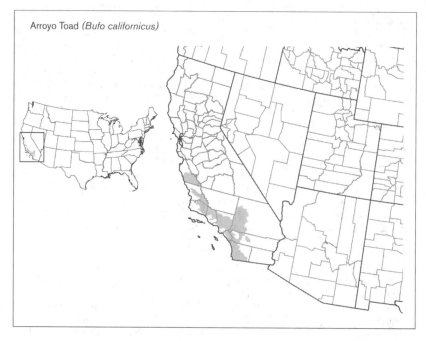

Arroyo Toad (*Bufo californicus*)

being monitored on MCB (Marine Corps Base) Camp Pendleton (San Diego County) from 1996–2000 showed continuous declines of ≤90% (D.C. Holland, personal communication).

In general terms, population densities of arroyo toads are relatively low (about 12 adults/ha) along second- to fourth-order streams in montane and foothill areas (Sweet, 1993), where it is rare to find >5 calling males/100 m of suitable habitat (Sweet, 1992, 1993; Ramirez, 2000). Densities are often higher along coastal streams, where a density of ten calling males/100 m is not unusual, and densities may reach 100/km locally, though there is considerable heterogeneity in density (D.C. Holland, personal communication, 2001). Still, these figures are well below reported densities for ecologically similar Arizona toads (*B. microscaphus*; Sullivan, 1993; Schwaner et al., 1998).

3. Life History Features.

A. Breeding. Reproduction is aquatic. Arroyo toads breed in shallow streams; their reproductive biology is generally similar to that of other north-temperate species of *Bufo*. The advertisement call of arroyo toads is a whistling trill, averaging 8 s (4–9 s) in duration, with a dominant frequency of 1.46 kHz and pulse rate of 44 pulses/s at 15 °C (Sullivan, 1992b; Gergus et al., 1997). Individual calling males are audible to humans at ≤300 m under ideal conditions, but the sound is easily masked by noise resulting from wind or stream riffles.

i. Breeding migrations. Sweet (1992, 1993) and Holland and Goodman (1998) report a prolonged breeding cycle for arroyo toads, beginning in February (rarely January) in coastal areas and late March or early April at montane sites and extending through July. Males begin calling about 10 d before any females respond; individual males may call almost nightly for the entire duration of the breeding cycle (Sweet, 1993). Breeding is not triggered by rain, but seems to require air and water temperatures above 11–13 °C (Myers, 1930a; Sweet, 1992; Holland and Goodman, 1998); breeding activities are suspended during floods and resume when stream flow rates decline sufficiently to provide shallow edges with minimal flow rates.

Male arroyo toads do not form calling aggregations, and satellite behavior has not been observed. Females generally avoid streamside areas until they are ready to breed; they appear to select a male from a distance and approach him directly. Egg deposition occurs at the male's calling site and typically requires <1 hr (Sweet, 1992).

ii. Breeding habitat. Arroyo toads breed in the quiet margins of open streams and avoid sites with deep or swift water, tree canopy cover, or steeply incised banks. Males typically call in water 2–4 cm deep 1–2 m from shore, facing a low shoreline with a horizon unobstructed by nearby vegetation (Sweet, 1992). Substrates are most often gravel and sand, less frequently silt or cobble, and rarely bare bedrock or boulders. The toads do not breed in riffle areas and almost never use pools that are isolated from the flowing channel; side channels and washouts may be utilized as long as there is some flow through them, but they are abandoned as soon as this flow ceases (S.S.S., personal observations).

B. Eggs.

i. Egg deposition sites. Female arroyo toads do not transport males but instead deposit eggs at the male's calling site. Intertwined masses of 3–5 clutches of different developmental ages are sometimes encountered, reflecting high site fidelity by males. Eggs are laid in water averaging 9 cm deep (range of 1.3–31.8 cm, n = 65; Sweet, 1992) on fine sediments where there is no appreciable current; they are virtually never entangled in twigs, roots, or other submerged debris. Deposition sites are fully exposed to the sky. Detailed site data on 135 clutches are presented in Sweet (1992).

ii. Clutch size. Eggs of arroyo toads are deposited in two strands simultaneously. A sample of 40 clutches averaged 6.8 m (range 3–10.6 m) in length and contained an average of 4,714 eggs (range, 2,013–10,368 eggs). Individual fertilized ova average 1.7 mm in diameter and have a pale vegetal pole; the fully hydrated envelope is 3.3–4.2 mm in width and is relatively inelastic (data from Sweet, 1992). There is no evidence that individual females can lay a second clutch.

C. Larvae/Metamorphosis.

i. Length of larval stage. Eggs require 4–6 d to hatch at water temperatures of 12–16 °C. Larvae remain associated with the degenerating envelopes for 5–6 d and do not begin active dispersal from the clutch deposition site until 15–18 d post-hatching (Sweet, 1992). Arroyo toad eggs and larvae are described and illustrated in Stebbins (1951, 1985) and Sweet (1992; reprinted in USFWS, 1999a). From hatching (Gosner stage 18, 4.0 mm TL) to stage 26 (10–12 mm TL) larvae are colored black. After stage 26, tan crossbars appear on the tailbase. By stage 30 (18–20 mm TL, 24–27 d post-hatching) the dorsum becomes tan, leaving dark crossbars on the tail base, an irregular black lateral stripe on the tail, and an opaque white venter. Larvae reach an average maximum of 34 mm TL (exceptionally to 40 mm) and require a minimum of 65 d post-hatching (most often 72–80 d) to metamorphose (all data from Sweet, 1992).

ii. Larval requirements.

a. Food. Larval arroyo toads do not aggregate once they have become free swimming. They are substrate gleaners and feed by processing detritus and microbial mats from just beneath the surface layer of fine sediments or within the interstices of gravel deposits. They do not consume macroscopic algae or other aquatic vegetation and will not take plant material in the lab (Sweet, 1992). Larvae remain stationary for long periods and generally travel <5 cm/movement.

b. Cover. Not needed. Larvae occupy shallow areas of open streambeds on substrates ranging from silt to cobble, with preferences for sand or gravel. Areas under tree canopies are avoided, as are portions of the streambed with submerged or emergent vegetation. Larvae are cryptic, but are strong swimmers, darting 2–4 m toward deeper water when disturbed.

iii. Larval polymorphisms. None.

iv. Features of metamorphosis. Newly metamorphosed arroyo toads may occur as early as late April or as late as early October, but metamorphosis is concentrated in the interval from late May to early July in most years. There is a positive correlation between metamorphic date and elevation in most years, but the high variability in rainfall patterns among years results in a broad range of breeding phenology. Larvae from a single cohort may metamorphose over a period of 10 d or more (data from Sweet, 1992). Forelimbs emerge 2–3 d before the tail is fully resorbed; during metamorphosis, larvae seek shallow water on exposed shorelines and become terrestrial when the tail is reduced to a stub. Newly metamorphosed juveniles are usually 12–15 mm SVL, but may be as small as 9 mm or exceptionally to 22 mm (Holland and Goodman, 1998).

v. Post-metamorphic migrations. Newly metamorphosed animals and juveniles remain on sparsely vegetated sand and gravel bars bordering the natal pool for 3–5 wk (Sweet, 1992).

D. Juvenile Habitat. Juvenile arroyo toads are extremely cryptic and remain within a few meters of the margin of their breeding pool for an extended period, where they forage, chiefly diurnally, on sparsely vegetated sand or gravel bars in full sun at substrate temperatures of 30–43 °C. During windy or cool weather and at night, juveniles shelter in damp depressions in the gravel. Juveniles begin to actively burrow in dry sand at 20–25 mm SVL; they switch to nocturnal activity at this stage, sheltering by day in shallow burrows in stream-edge vegetation and foraging widely in the lower riparian zone after dark. Dispersal from the immediate vicinity of the stream is a prominent behavioral transition that is mediated by drying conditions and/or attainment of around 30 mm SVL, usually occurring within 3–5 wk following metamorphosis (data from Sweet, 1992). Juvenile toads >30 mm SVL disperse throughout habitats used by adults and remain active well into late summer or early fall, long after adults have stopped foraging. Many juveniles reach 45–50 mm SVL by late September to early October, and some males may attempt to breed in the following spring (Sweet, 1993).

E. Adult Habitat. Arroyo toads are closely associated with low gradient drainages from near sea level to about 1,400 m elevation (to a maximum of 2,440 m in Baja California del Norte), with most remaining populations residing in the 300–1,000 m range (USFWS, 1999a). They may locally occupy first-order drainages, but are usually associated with second- to sixth-order streams that have extensive terrace systems, braided channels, and large areas of fine sediment deposits that are episodically reworked by flooding. Vegetation reflects the frequency and intensity of flood events, with live oak, sycamore, and cottonwood groves interspersed with grasslands and sage scrub on high terraces, and patches of willows, alder, and mulefat on temporary alluvial benches adjoining the active channels. Streams may be either permanent or seasonal; seasonal streams must flow for at least 4–5 mo in spring and summer in most years to support breeding populations. Barto (1999) provided a habitat suitability model for arroyo toads that does not consider areas beyond the riparian border.

Mark–recapture (Sweet, 1992, 1993) and radiotracking studies (Griffin, 1999; Ramirez, 2000) have demonstrated that in many areas arroyo toads rarely disperse beyond the upland margins of the stream terraces, where dense hard chaparral, steep slopes, and stony soils replace fine alluvium and patchy vegetation. However, pitfall trapping work by Holland and Sisk (2000) has documented extensive use of upland grasslands and sage scrub on compacted soils in near-coastal localities, with animals being found as far as 1.2 km away from the riparian/upland ecotone. This clear regional difference in habitat use may indicate that much larger areas of coastal southern California that are now extensively urbanized were formerly inhabited by arroyo toads, and that the ecological scope of the species is underestimated by reliance on data from remnant populations.

F. Home Range Size. In addition to extensive upland movements in near-coastal sites, arroyo toads show several patterns of seasonal dispersal (Sweet, 1993). The largest adult male toads may remain at a single breeding pool for an entire season, while smaller males typically move consistently upstream or downstream, calling for a few nights at each pool encountered. Between years, these males tend to travel within a 2–3 km zone along the stream and often become sedentary as they reach large size. Adult female toads are highly sedentary, with a continuously used activity area usually <100 m in diameter; such animals nearly always breed in the same pool in successive years. Older juvenile and

subadult (yearling) toads make extensive along-stream movements during their extended activity season. Radiotracking studies at both near-coastal (Griffin, 1999) and montane sites (Ramirez, 2000) document qualitatively similar patterns among individual toads, although Griffin (1999) noted upstream dispersal by adult females prior to breeding.

Sweet (1992, 1993), Griffin (1999), and Ramirez (2000) showed that arroyo toads construct shallow burrows within the riparian zone where they shelter by day during the active season. Toads tend to prefer sand over finer or coarser substrates and most often burrow in relatively open microsites, but otherwise show no strong preference for available vegetation types. Ramirez (2000) found that toads construct burrows closer to stream channels as the dry season progresses, presumably to maintain water balance. Griffin (1999), Holland and Sisk (2000), and Sweet (personal observations) have noted occasional use of small mammal burrows as refugia.

G. Territories. Territorial or activity area fidelity may occur among adult female arroyo toads, and at least some male toads return to the same calling site for many nights in succession (Sweet, 1992, 1993); changes in stream level result in local adjustments in calling sites over the course of a breeding season. Other males, particularly smaller individuals, tend to "drift" upstream or downstream, calling for a few nights at each pool encountered (Sweet, 1993). Calling males display a minimum spacing of 1–8 m, with no evidence of agonistic or territorial behavior. Satellite behavior by males has not been observed (Sweet, 1992). While newly metamorphosed and juvenile arroyo toads often display patchy distributions involving some clustering, this appears to be a response to physical habitat features rather than aggregation per se (S.S.S., personal observations).

H. Aestivation/Avoiding Desiccation. Adult arroyo toads do not appear to aestivate regularly during the active season (February–July), though they may remain inactive for several days during cold or windy weather. Calling by males is greatly reduced on two or three nights preceding and following a full moon (Sweet, 1992). Subadult and adult toads, especially, become progressively less active after early July, and few adults can be found by August; juvenile and subadult toads may remain active at reduced levels into October and occasionally later in the fall following rains (Sweet, 1993; Holland and Sisk, 2000; Ramirez, 2000).

I. Seasonal Migrations. No patterns are evident beyond age- and sex-related movements described above.

J. Torpor (Hibernation). Most adult arroyo toads are inactive from August or September to February or March; subadults may remain active into early November

(Sweet, 1992, 1993; Holland and Sisk, 2000). The locations and characteristics of hibernation sites are poorly known; toads are presumed to select higher stream terraces where the likelihood of severe flooding is reduced. However, Ramirez (2000) documented two individuals that ceased surface activity in early August by burrowing into the stream channel; one of these, when excavated, had produced a thin "cocoon" of shed epidermis. The extent to which toads might use upland areas for hibernation is presently unknown.

K. Interspecific Associations/Exclusions. Arroyo toads are usually microsympatric with western toads (*B. boreas*) and California and Pacific treefrogs (*Pseudacris cadaverina* and *P. regilla*, respectively); they also occur frequently with northern red-legged frogs (*Rana aurora*) and American bullfrogs (*R. catesbeiana*), and occasionally with western spadefoot toads (*Spea hammondii*; Sweet, 1992; Holland and Goodman, 1998). In the recent past they co-occurred with foothill yellow-legged frogs (*R. boylii*) as well, but foothill yellow-legged frogs have now been extirpated within the range of arroyo toads. Arroyo toads are eaten regularly by American bullfrogs (Holland and Goodman, 1998; Griffin, 1999; S.S.S., personal observations).

Male arroyo toads have been observed in amplexus with small, late-breeding female western toads on three occasions; hybrid larvae hatch, but have massive deformities; most die by Gosner stage 25 (Sweet, 1992). Dan C. Holland (personal communication) has also observed male arroyo toads in amplexus with large female western toads. Small male western toads are sometimes closely associated with calling male arroyo toads and will repeatedly attempt amplexus with them; male arroyo toads with an attending western toad seldom complete a call without being grasped, and this interference probably reduces mating success (Sweet, 1992).

Differences in peak breeding season, oviposition sites, and larval behavior serve to minimize interactions between arroyo toad and western toad larvae (Sweet, 1992); larval habitat overlap is greatest between arroyo toads and California treefrogs.

Some potential for competitive interactions between arroyo toads and western toads exists during the juvenile stage. While western toads usually metamorphose earlier in the year than do arroyo toads and are predominantly nocturnal, normal variation in breeding phenology can result in temporal and spatial overlap with arroyo toads. In laboratory tests, groups of size-matched individuals of both species grew at similar rates when reared separately, but arroyo toads grew at only half the rate of western toads in trials where the two were housed together. Juvenile western toads consumed prey (crickets) at three times the foraging rate of arroyo toads (Sweet, 1992). There is, however, no evidence for substantial interference competition in the field.

L. Age/Size at Reproductive Maturity. In the population studied intensively by Sweet (1992, 1993), calling male arroyo toads ranged from 51–67 mm SVL, while gravid or breeding females ranged from 66–78 mm SVL. A few males marked in the late summer of their natal year were mature and calling the following spring, but none of these individuals was observed to mate. Most males (and a small number of females) mature in their second year, with most females breeding for the first time as they reach age 3. Calling males do not grow during the breeding season (many, in fact, shrink), and thus males that forgo breeding until age 2 are substantially larger. Because females strongly prefer the largest males in a local chorus and satellite behavior is unknown, there is no evident advantage in the early maturation displayed by some male arroyo toads (Sweet, 1993).

M. Longevity. Mark–recapture studies suggest that few arroyo toads survive into their fifth year, and that these are predominantly females (Sweet, 1993). In the absence of American bullfrogs, adult arroyo toads have a high survivorship during the active season, but suffer 55–80% mortality as they overwinter (Sweet, 1993; Holland and Sisk, unpublished data). American bullfrogs target calling male arroyo toads and are associated with sex ratio biases of up to 1:19, leading to local extirpations (Sweet, 1992).

N. Feeding Behavior. Newly metamorphosed and juvenile arroyo toads feed mostly on ants and small flies, while larger individuals eat a wider range of invertebrates (Cunningham, 1962; Sweet, 1992). Adult toads feed predominantly on ants, especially nocturnal, trail-forming tree ants *(Liometopum occidentale)*. Ant foraging columns follow the same trails for many weeks, and most individual arroyo toads will return to the same site each night, consuming up to 25% of their body mass in 1–1.5 hr of feeding (Sweet, 1992, 1993). Feces of adult arroyo toads typically contain >95% *Liometopum* exoskeletons by mass and superficially resemble feces of horned lizards *(Phrynosoma* sp.), which contain only diurnal ant taxa (Sweet, 1992).

O. Predators. Arroyo toad eggs and young larvae have few or no predators (Sweet, 1992), though mosquitofish *(Gambusia affinis)* and crayfish *(Procambarus clarkii)* have been observed to remove and consume individual eggs (D.C. Holland, personal communication). Two-striped garter snakes *(T. hammondii)* have been observed to repeatedly ingest then reject egg strands (Sweet, 1992). Free-swimming larvae are actively hunted by garter snakes and are subject to high predation by introduced fishes (especially green sunfish *(Lepomis cyanellus)*

and prickly sculpins *(Cottus asper;* Sweet, 1992).

Juvenile arroyo toads may suffer heavy mortality locally from killdeer *(Charadrius vociferus)* and are often essentially eliminated by trampling by humans on streamside flats at popular recreational sites (Sweet, 1992). Trampling may also occur where cattle frequent the riparian zone, but there is a larger negative effect from enhanced evaporation because the microrelief of the surfaces of sand and gravel bars is greatly increased by trampling (Sweet, 1992).

Predation intensity declines as juvenile toads become nocturnal and disperse, and subadult and adult toads show very high survivorship during the active season at sites lacking American bullfrogs; two-striped garter snakes and common garter snakes *(T. sirtalis)* occasionally attempt to consume mature arroyo toads with variable success (though some toads that escape die later of injuries sustained; Sweet, 1993, Griffin, 1999). Other than killdeer, no predation by birds or mammals has been documented, though adult toads occasionally display injuries consistent with attempted predation by shrews.

Where the two species co-occur, American bullfrogs are major predators on arroyo toads. Focusing mainly on calling males (but also consuming pairs in amplexus), individual bullfrogs can essentially eliminate local populations of arroyo toads (USFWS, 1994a, 1999a).

P. Anti-Predator Mechanisms. Eggs and small (black) larvae appear to be distasteful, but larger larvae are readily eaten by various predators without evident discomfort (S.S.S., personal observations). The cryptic appearance, behavior, and flight responses of larger larvae are effective against foraging garter snakes but are ill-adapted to attacks by introduced centrarchid fishes (Sweet, 1992). When disturbed, larger larvae usually dart towards deeper water, making an arc that often contacts the surface. Green sunfish are proficient at intercepting fleeing larvae and have been seen to roll laterally to take them even in shallow water (S.S.S., D.C. Holland, personal observations).

Juvenile arroyo toads are extremely cryptic, with pattern elements that closely match the color frequencies and patch sizes of the damp gravel with evaporite deposits where they forage, and they remain motionless when approached. On land at night, subadult and adult toads will move rapidly into dense vegetation, and calling males will retreat underwater in response to movements by a human observer ≤ about 25 m away. No specialized postures or visible release of parotoid gland secretions in response to capture or handling are known (S.S.S., personal observations).

Q. Diseases. No instances of disease are known among wild populations, but a

chytrid fungal epidemic killed all juvenile arroyo toads being reared in lab in 1991. Symptoms appeared too soon after collection for a lab-acquired origin, but no unexplained mortality was observed in the wild source populations during the remainder of the season. This fungus could not be maintained for study in a lab colony of boreal toads (Sweet, 1992).

Egg clutches swept into deeper and cooler water by changes in water level are usually attacked by fungus (Sweet, 1992; see "Conservation" below).

R. Parasites. Larval arroyo toads are often heavily infected with encysted metacercariae of an unidentified trematode whose definitive hosts are birds. At metamorphosis, the cysts become concentrated in the groin and around the urostyle as the tail is resorbed. A small, unidentified cestode is regularly present in the body wall musculature. No adverse effects attributable to parasitism have been noted (Sweet, 1992).

4. Conservation.
Egg clutches are subject to high mortality from changes in water level; losses either by stranding as stream flow declines or displacement by even minor flooding events are commonplace. In particular, late rains and associated minor flooding or water management via dam operations can minimize survival of clutches and small larvae across entire drainage systems in some years (Sweet, 1992). Most clutches that survive changes in water level hatch, except that those portions swept into deeper and cooler water are usually attacked by fungus (Sweet, 1992).

Arroyo toads have declined in abundance (often to extirpation) at most sites where historical records exist, and were federally listed as Endangered in late 1994 (USFWS, 1994a). Jennings and Hayes (1994a) estimated that arroyo toads had been eliminated from 76% of their historical range, though subsequent discoveries of new localities and remnant populations reduce this figure to about 65% (see "Historical versus Current Distribution" above). Surviving populations reached a low ebb in the early 1990s following an extended drought and several decades of adverse land-use and water management practices (Sweet, 1992). Many of these populations are now isolated. Most populations on public lands recovered substantially following the federal listing; however, the high-density populations described by Sanders (1950), Stebbins (1951), and Cunningham (1962) are now extirpated, and many populations remain well below apparent carrying capacity (Sweet, 1992, 1993, unpublished data). These populations continue to face threats due to disturbance from water flow management practices, gravel mining, urbanization, and military training (USFWS, 1999a).

Acknowledgments. S.S.S. acknowledges contract support and logistical assistance from the U.S. Forest Service, Los Padres National Forest, invaluable field assistance by former USFS biologist Nancy Sandberg, and permits and encouragement from the U.S. Fish and Wildlife Service. B.K.S. acknowledges support of the Arizona Game and Fish Department and the assistance of Rob Bowker, Mike Demlong, Matt Kwiatkowski, Erik Gergus, and Keith Malmos. We thank Dan C. Holland for comments on the manuscript and permission to cite unpublished data and observations.

Bufo canorus Camp, 1916(a)
YOSEMITE TOAD

Carlos Davidson, Gary M. Fellers

This species account is dedicated to the memory of Cynthia Kagarise Sherman.

1. Historical versus Current Distribution.

Yosemite toads *(Bufo canorus)* are endemic to the Sierra Nevada, California, from Ebbetts Pass, Alpine County to the Spanish Mountain area, Fresno County (Karlstrom, 1962, 1973; Stebbins 1966; unpublished Sierra National Forest survey data, 1995, 2002). Sites occur from 1,950–3,444 m elevation, with the majority of sites between 2,590–3,048 m (Karlstrom, 1962). Jennings and Hayes (1994a) estimate that populations have disappeared from 50% of historically reported sites, although the overall range of the species may have only contracted in the far north and in western Fresno County. Disappearances have been concentrated at lower elevation sites on the western edge of the range, with greater persistence at higher elevation sites (Davidson et al., 2002).

2. Historical versus Current Abundance.

Trends in Yosemite toad population size have not been evaluated at most sites. Kagarise Sherman and Morton (1993) report sharp population declines at seven sites in the eastern Sierra Nevada from 1971–91. At one well-studied site at Tioga Pass with a 20-yr history of counts, the number of marked males entering breeding pools had declined nine-fold (Kagarise Sherman and Morton, 1993).

3. Life History Features.

A. Breeding. Reproduction is aquatic.

i. *Breeding migrations.* Males arrive at breeding pools several days before females (Kagarise Sherman, 1980; Kagarise Sherman and Morton, 1984). Individual males stay at breeding ponds for 1–2 wk, and females for only a few days (Kagarise Sherman, 1980; Kagarise Sherman and Morton, 1984). Breeding takes place from mid-May to mid-August (Kagarise Sherman, 1980; G.M.F., unpublished data). Both sexes are primarily active during the day (Kagarise Sherman, 1980; Kagarise Sherman and Morton, 1984). There may be 10 times as many males as females at a breeding site (Karlstrom, 1962; Kagarise Sherman, 1980). Males call diurnally and can be heard from a distance of over 100 m (Kagarise Sherman, 1980). Grinnell and Storer (1924) remarked on the call, "Its mellow notes are pleasing additions to the chorus of bird songs just after the snow leaves." Camp (1916a) gave the species the name *canorus,* which in Latin means "tuneful." After breeding, both sexes move into meadow areas to feed for 2–3 mo before winter snows arrive (Kagarise Sherman, 1980; Kagarise Sherman and Morton, 1984). At Tioga Pass, adults traveled 150–230 m between sites where they spent the winter at breeding ponds, with females tending to move farther than males (Kagarise Sherman, 1980).

ii. *Breeding habitat.* Breeding sites are typically meadow edges without deep water or adjacent steep terrain (Karlstrom, 1962). Yosemite toads will occasionally breed in the shallows of lakes (G.M.F., personal observations).

B. Eggs.

i. *Egg deposition sites.* Typical egg deposition sites include shallow (<7.5 cm) pools and small, slow moving, shallow streams usually in meadows, with short emergent vegetation and loose silt substrate (Karlstrom, 1962).

ii. *Clutch size.* Females lay an estimated 1,500–2,000 eggs (Karlstrom, 1962) in single or double strands or in a radiating network 4–5 eggs deep (Karlstrom and Livezey, 1955). In the Tioga Pass area, eggs hatched in about 10–12 d (Kagarise Sherman, 1980; Kagarise Sherman and Morton, 1984). Karlstrom (1962) estimates critical thermal maximum of 36–38 °C for larvae and 31 °C as upper limiting temperature for egg development.

C. Larvae/Metamorphosis.

i. *Length of larval stage.* In the Tioga Pass area, tadpoles metamorphosed 52–63 d after the eggs were laid (Kagarise Sherman, 1980; Kagarise Sherman and Morton, 1984).

ii. *Larval requirements.*

a. Food. Unknown, but presumably tadpoles are grazing feeders that ingest algae and other material suspended when they scrape rocks, plants, and other submerged substrates.

b. Cover. Unknown.

iii. *Larval polymorphisms.* Unknown.

iv. *Features of metamorphosis.* Unknown.

v. *Post-metamorphic migrations.* Unknown.

D. Juvenile Habitat. Believed to be similar to adults.

E. Adult Habitat. High elevation, open, montane meadows, willow thickets, and adjoining forests. Although adult toads spend little time actually in water, they are seldom found more than about 100 m from permanent water (Karlstrom, 1962). Adults take cover in rodent burrows, under surface objects, and in willow thickets (Karlstrom, 1962). In the Tioga Pass area, animals primarily utilized burrows of meadow voles *(Microtus montanus)* and pocket gophers *(Thomomys monticola;* Karlstrom, 1962).

Critical thermal maximum for adults was estimated at 38–40 °C and a lower thermal limit at −1 °C (Karlstrom, 1962).

Yosemite Toad *(Bufo canorus)*

Adults seem to exhibit little temperature preference in the field, where temperatures between 2–30 °C were reported, with no sign of temperature stress (Karlstrom, 1962).

F. Home Range Size. Unknown.

G. Territories. Similar to most toads, calling males defend space around themselves from intrusion by other males. Males are more likely to defend calling areas when male density at breeding pools is low (Kagarise Sherman and Morton, 1984).

H. Aestivation/Avoiding Desiccation. Does not occur.

I. Seasonal Migrations. See "Breeding migrations" above.

J. Torpor (Hibernation). Animals become inactive for the winter in late September or early October and become active again from April–July, as soon as the snow melts from breeding pools (Karlstrom, 1962; Kagarise Sherman, 1980). During the winter, toads utilize rodent burrows, crevices under rocks, or the root tangle at the base of willows (Kagarise Sherman, 1980). In the Tioga Pass area, toads used rodent burrows, including those of meadow voles, pocket gophers, Belding ground squirrels (*Spermophilus beldingi*), and yellow-bellied marmots (*Marmota flaviventris*; Kagarise Sherman, 1980). First-year juveniles spend the winter in similar habitats near the pools from which they emerged (Kagarise Sherman, 1980).

K. Interspecific Associations/Exclusions. See "Torpor (Hibernation)" above.

L. Age/Size at Reproductive Maturity. At Tioga Pass Meadows, minimum reproductive size was 48 mm SVL for males and 60 mm SVL for females (Kagarise Sherman, 1980). Females first breed at 4–6 yr, and not every year thereafter. Males first breed at 3–5 yr of age (Kagarise Sherman, 1980).

M. Longevity. Kagarise Sherman and Morton (1984) estimate that some females may live at least 15 yr and males at least 12 yr.

N. Feeding Behavior. Adults and juveniles eat tenebrionid beetles, weevils, large ants, centipedes, spiders, ladybird beetles, dragonfly naiads, bees, wasps, millipedes, flies, mosquitoes, and lepidopteran larvae (Grinnell and Storer, 1924; Mullally, 1953; Kagarise Sherman and Morton, 1984). Wood (1977) reported that hymenopterans composed almost 80% of the summer diet.

O. Predators. Predation on tadpoles has been reported for mountain yellow-legged frogs (*Rana muscosa*), dragonfly naiads (Mullally, 1953), robins (*Turdus migratorius*), and diving beetles (*Dytiscus* sp.; Kagarise Sherman, 1980; Kagarise Sherman and Morton, 1984, 1993; G.M.F., personal observations). Garter snakes, especially western terrestrial garter snakes (*Thamnophis elegans*), probably eat large numbers of larvae and juveniles (Karlstrom, 1962;

Kagarise Sherman, 1980). Karlstrom (1962) reported that Brewer's black birds (*Euphagus cyanocephalus*) and California gulls (*Larus californicus*) have been observed eating other species of tadpoles and are present in the range of Yosemite toads, but there are no records of them eating Yosemite toad tadpoles. In most years, desiccation of breeding pools, not predation, is the single major cause of larval mortality (Kagarise Sherman, 1980). Kagarise Sherman (1980) and Kagarise Sherman and Morton (1984, 1993) report on adults being killed by Clark's nutcrackers (*Nucifraga columbiana*) and California gulls. Common ravens (*Corvus corax*) may also eat adults (Kagarise Sherman and Morton, 1993).

P. Anti-Predator Mechanisms. Adults and juveniles can release toxic secretions from the parotoid glands. Frightened adults and juveniles may leap into the water or retreat into rodent burrows (Mullally, 1953).

Q. Diseases. Green and Kagarise Sherman (2001) examined preserved specimens from a mass die-off of Yosemite toads in the 1970s at Tioga Pass and found indications of numerous diseases. Chytridiomycosis (chytrid fungus infection) and bacillary bacterial septicemia (redleg disease) were considered the cause of death for four specimens. In addition they found *Dermosporidium* sp. (a fungal infection), myxozoan infection (*Leptotheca ohlmacheri*, a cnidarian), and a roundworm infection (larval *Rhabdias* sp.).

R. Parasites. Helminth worms were observed by Wolton (1941). Green and Kagarise Sherman (2001) found helminths in the gut, lungs, and bladders, as well as trematodes in the lungs and bladder.

4. Conservation.
The State of California lists Yosemite toads as a Species of Special Concern. Environmental groups petitioned the U.S. Fish and Wildlife Service to list the toad under the Endangered Species Act. In November 2002, the U.S. Fish and Wildlife Service decided that the listing was "warranted, but precluded," meaning that listing was warranted based on the status and threats facing the species, but that the agency had higher priorities. This leaves the toad as a Candidate species that may be reconsidered for listing in the future.

The causes of declines for Yosemite toads are unclear. Leading hypotheses for the declines are disease, airborne contaminants, and livestock grazing. In an examination of preserved specimens from a 1970 die-off, Green and Kagarise Sherman (2001) found multiple pathogens, but no single pathogen was present in more than 25% of the specimens, suggesting that the animals suffered from suppressed immune systems possibly due to a virus or chemical contaminants. Davidson et al. (2002) found that historic sites where

Yosemite toads are absent had twice as much agricultural land upwind, compared to historic sites that still have toads (suggesting that windborne agrichemicals may have contributed to declines), but these differences were not statistically significant. Livestock grazing may have detrimental impacts on Yosemite toads through trampling, alteration of meadow habitat, and possible lowered water quality (D. Martin, personal communication). Other factors that may have contributed to declines are the 1980s' California drought, fish predation, and increased predation by common ravens (Kagarise Sherman and Morton, 1993). According to the Breeding Bird Survey (USFWS, unpublished data) the number of common ravens has increased in the Sierra Nevada by 9.5% annually over the period from 1966–89. Because ravens may feed on Yosemite toads, an increase in the raven population could contribute to a decline of these toads. Increases in ravens may be related to human activities (Kagarise Sherman and Morton, 1993).

The U.S. Forest Service, which manages most of the land within the range of Yosemite toads, has developed plans for new management practices to help conserve Yosemite toads. Under the plans, grazing during the Yosemite toads' breeding season would be excluded from wet meadows with known populations. In addition, the Forest Service would conduct surveys for toads, monitor a sampling of known populations, study the impact of grazing, and try to avoid application of pesticides within 152 m (500 ft) of known toad sites.

Bufo cognatus Say, 1823
GREAT PLAINS TOAD

Brent M. Graves, James J. Krupa

1. Historical versus Current Distribution.
Great Plains toads (*Bufo cognatus*) occur throughout the U.S. Great Plains, from western Minnesota and Iowa, northwestern Missouri and the western 2/3 of Kansas, Oklahoma, and Texas, west to the Imperial Valley of California and up the Colorado River though eastern Nevada, Colorado, and eastern Wyoming and Montana. Their range extends south to central Mexico and north into the southwestern corner of Manitoba, across southern Saskatchewan and the southeastern corner of Alberta (Wright and Wright, 1949; Dixon, 1987, 2000; Johnson, 1987; Krupa, 1990; Collins and Collins, 1993; Oldfield and Moriarty, 1994). Great Plains toads generally are found at elevations <1,900 m, but are found from 2,286–2,438 m in the San Luis Valley of Colorado (Hammerson, 1986, 1999; Degenhardt et al., 1996; Luce et al., 1997). Lannoo et al. (1994) compared their results to Bailey and Bailey (1941) and documented an apparent eastward

range expansion of Great Plains toads in northwestern Iowa. King (1932) anecdotally described Great Plains toads as the most common toad of the southwest, while more recent accounts describe them as less common and localized (Stebbins, 1985; J.J.K., personal observations). Similarly, Great Plains toads were described as having a widespread distribution in eastern Montana (Black, 1970b, 1971b), but were later described as locally distributed in that state with large gaps in the known range (Reichel and Flath, 1995, cited in James, 1998). Busby and Parmelee (1996) did not find Great Plains toads in their survey of the Fort Riley Military Reservation in Kansas in 1993, although they were reported to be present in 1927. It is not known whether these patterns reflect long-term changes in species distribution, short-term population fluctuations, or inconsistent and superficial population monitoring (James, 1998).

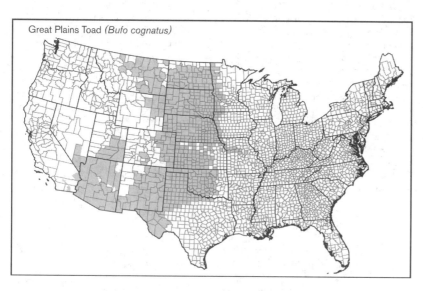

Great Plains Toad (Bufo cognatus)

2. Historical versus Current Abundance.
Unknown, but Great Plains toads are common in many portions of their range (J.J.K., personal observations). For example, they were one of the three most common anurans in playa wetlands in the high plains of Texas (Anderson et al., 1999a). In contrast, Luce et al. (1997) suggest they are uncommon in Wyoming. They were described as common in Alberta in 1984, but were recommended for Potentially Threatened status in 1987 and Endangered status in 1988 (James, 1998). They are considered rare in Saskatchewan (Seburn, 1992) and in southwestern Minnesota (Lehtinen et al., 1999). Population densities are known to fluctuate widely in association with periodic droughts. Reproduction may not occur during years with insufficient rainfall (Sullivan and Fernandez, 1999). Furthermore, Great Plains toads are largely fossorial during times of inactivity, making them difficult to detect except during breeding periods (James, 1998).

3. Life History Features.
A. Breeding. Reproduction is aquatic.

i. Breeding migrations. Great Plains toads emerge from burrows below the frostline following heavy spring rains and move to wetland breeding areas (Ewert, 1969). Individuals may move >1 km from breeding ponds (James, 1998), and mass, unidirectional migrations have been reported (Smith and Bragg, 1949; Bragg and Brooks, 1958), although it is not known if such movements are associated with feeding or reproduction. They exhibit an explosive breeding pattern (Wells, 1977; Sullivan, 1989a), but females have large follicles and reproduction may occur throughout most of the active season (Bragg and Smith, 1942; Long, 1987b; Krupa, 1989). Additionally, individual females may produce multiple clutches in a single year (Krupa, 1986a), suggesting that reproduction may be prolonged if conditions are appropriate. Reproductive activity is initiated by rainfall. The amount of rainfall necessary to trigger breeding averaged 4.4 cm in Oklahoma (Krupa, 1994), although chorus activation can be independent of rainfall (Brown and Pierce, 1967). Large choruses of 200–500 individuals have been recorded (Brown and Pierce, 1967; Degenhardt et al., 1996). Males call from vegetated shorelines to attract females or position themselves near other calling males to intercept females (satellite behavior; Sullivan, 1983; Krupa, 1989).

ii. Breeding habitat. Great Plains toads breed from March–September throughout their range and May–July in the north (Bragg 1937b, 1940a; Wright and Wright, 1949; Krupa, 1986b, 1994). Timing of reproduction is associated with the availability of temporary aquatic breeding habitat (Krupa, 1994), although breeding occurs in both temporary and permanent ponds (Woodward, 1987a). Brown and Pierce (1967) reported the highest intensity choruses in temporary floodings, with lower intensity choruses in apparently permanent ponds and the lowest intensity choruses in temporary lotic habitats. Other researchers have reported that Great Plains toads breed primarily in shallow temporary pools (Mackay et al., 1990; Krupa, 1994; Luce et al., 1997; Fischer et al., 1999). Some authors have emphasized that breeding will not occur in muddy, turbid habitats (Bragg and Smith, 1942, 1943; James, 1998). In Alberta, floodings produced by irrigation, as well as natural floodings, were used as breeding sites, but all were in regions of native vegetation rather than cultivated areas (Wershler and Smith, 1992, cited in James, 1998).

B. Eggs.
i. Egg deposition sites. Egg laying usually begins at dawn and is completed before noon, although oviposition can occur at other times (Krupa, 1994). Eggs are usually deposited in temporary floodings having clear water with little current (see "Breeding habitat" above). Breeding females may lay eggs in close proximity to each other (Krupa, 1994). Eggs in natural conditions normally hatch in 2–7 d (Hammerson, 1986; Johnson, 1987; Collins and Collins, 1993; Krupa, 1994; Luce et al., 1997). Eggs from a population in Maricopa County, Arizona, maintained under laboratory conditions hatched in 6–13 d at temperatures from 15.9–20.8 °C, in 4–5 d from 21.6–25.7 °C, and in 3 d at temperatures between 29.4–34.5 °C. Development was incomplete or did not occur at all at temperatures <13.6 °C and >39.1 °C. Water temperatures at natural breeding sites were from 15–28 °C (Ballinger and McKinney, 1966). Embryos at the mid-gastrulation stage survived 6 h at 40.5 °C.

ii. Clutch size. Mean clutch size for 27 females in Oklahoma was 11,074, with a range of 1,342–45,054. Clutch size was positively and exponentially related to female size. Females are capable of laying multiple clutches over the course of a breeding season (Krupa, 1986a). Krupa (1994) reported that only 6.8% of males and 7.5% of females were captured at breeding ponds in multiple years of a 3-yr study. This suggests that either mortality is high, individuals move to new breeding habitats frequently, or reproduction occurs less frequently than annually. In Krupa's (1988) study, fertilization rate in nature was 89%. Mean egg diameter is 1.18 mm (Bragg, in Wright and Wright, 1949). Egg strings are usually single, sometimes double.

C. Larvae/Metamorphosis.
i. Length of larval stage. Tadpoles begin to metamorphose from 17–45 d following hatching (Bragg, 1937a; Hammerson, 1986; Degenhardt et al., 1996; Luce et al., 1997), at about 10 mm SVL (Johnson, 1987; Degenhardt et al., 1996). The larval period is shorter for clutches laid later in

the spring or summer, probably as a result of higher water temperatures. However, higher temperatures increase evaporation and desiccation of temporary ponds reduces the area of available habitat. This increases larval density, which slows developmental rates (Krupa, 1994). Density dependent, chemically mediated growth suppression (Semlitsch and Caldwell, 1982) may also occur.

ii. Larval requirements. Temporary ponds are preferred because of the absence of many aquatic predators such as fish, and lotic habitats are extremely ephemeral in arid habitats. Desiccation of aquatic habitats either kills tadpoles outright or makes them more vulnerable to predation and may be the primary cause of larval mortality. Warmer water temperatures speed larval development, but increase the rate that ephemeral aquatic habitats evaporate. Low larval density enhances growth and development (Krupa, 1994).

a. Food. Great Plains toad tadpoles are similar to other bufonids in being suspension feeders that graze on organic and inorganic material associated with the submerged surfaces of plants, rocks, and other substrates. They are reported to feed on algae and decomposing invertebrate remains (Bragg, 1940a).

b. Cover. Great Plains toad tadpoles do not appear to be influenced by the use of cover. In wetlands, they tend to space themselves evenly or haphazardly, except to aggregate in warm, shallow waters in full sun (J.J.K., personal observations).

iii. Larval polymorphisms. Do not occur.

iv. Features of metamorphosis. Krupa (1986b, 1994) notes that the survival rate of tadpoles in natural ponds is so low that the emergence of newly metamorphosed animals is a rare event. However, large numbers of newly metamorphosed Great Plains toads have been reported (Bragg and Brooks, 1958; Graves et al., 1993), supporting the hypothesis of great year-to-year and geographic variation in reproductive success. Metamorphosis may be synchronous (Krupa, 1986b, 1994).

v. Post-metamorphic migrations. Newly metamorphosed toads remain near the natal pond for about 1 mo or until it dries. They commonly move in large numbers into agricultural fields where they find it easier to burrow into the soil (Smith and Bragg, 1949). Post-metamorphic Great Plains toads will form aggregations (Bragg and Brooks, 1958; Graves et al., 1993). Individuals are apparently attracted to each other, and this attraction is chemically mediated (Graves et al., 1993). Ewert (1969) found two first-year toads 1 km from the nearest oviposition site.

D. Juvenile Habitat. While adults are largely nocturnal, juveniles are active diurnally and remain near natal ponds. If pools dry while juveniles are small (30–35 mm), toads may move to agricultural fields.

E. Adult Habitat. Great Plains toads are found in habitats including short grass to tallgrass prairies, sandhills, desert mesquite, and desert scrub (Wright and Wright, 1949; Ewert, 1969; Stebbins, 1985; Lehtinen et al., 1999). Degenhardt et al. (1996) report that Great Plains toads are generally found in warmer grassland areas, rarely in upland woodlands. They are associated with temporary ponds, irrigation ditches, and bottom lands (Wright and Wright, 1949; Stebbins, 1985; Hammerson, 1986; Krupa, 1990; Fischer et al., 1999). Great Plains toads tolerate agriculture and drier conditions better than most bufonids (Degenhardt et al., 1996). They are also tolerant of urban conditions (Krupa, 1994). They are nocturnal, proficient burrowers (Ewert, 1969; Lannoo, 1996; James, 1998) and commonly form shallow superficial burrows that they occupy for 1–6 d (Ewert, 1969). They also form deeper burrows in the shape of an inverted question mark, with the toad positioned at the upper, terminal end (Tihen, 1937; in Collins and Collins, 1993). Ewert (1969) suggested that woods and cattail marshes are avoided. Mean preferred body temperature is 27.7 °C (Sievert, 1991), and ambient temperatures of 40 °C are lethal in <40 min (Schmid, 1965a).

F. Home Range Size. May not establish home ranges, but some individuals return to the same breeding pond for many years (J.J.K., personal observations), and individuals may return to specific overwintering sites (Ewert, 1969). Great Plains toads in Minnesota moved about 462 m from overwintering sites to breeding sites and about 308 m from breeding sites to foraging areas (Ewert, 1969). Early summer daily movement distances were 0–3 m for 43% of observations, 3–31 m for 29% of observations, and >31 m for 29% of observations. One animal moved 815 m in 1 d. During late summer and fall, the movements in these three distance categories were 69%, 7%, and 24%, respectively. Individuals tend to move gradually but directly from summer foraging areas toward overwintering sites, stopping to forage and remaining in shallow burrows (i.e., "forms") for several days at a time. Hence, home ranges were described as distances (averaging at least 615 m with few < 308 m) rather than areas (Ewert, 1969).

G. Territories. Unknown.

H. Aestivation/Avoiding Desiccation. Commonly use forms <5 cm deep during short periods of inactivity (1–6 d; Ewert, 1969). Will aestivate deeper and for longer periods of time during hot, dry weather (Fischer et al., 1999). Ewert (1969) observed two toads dig to depths of 15 and 55 cm in July and August, respectively. The first remained burrowed for 14 d, while the second remained burrowed through the winter. Great Plains toads survive the loss of 42.9% ± 1.0% (mean ± SE) of initial body weight due to dehydration

(Hillman, 1980). They absorb water from the soil at a soil moisture tension of 2.5 atmospheres (Walker and Whitford, 1970). They survive temperatures of 40 °C for 40 min, 43.5 °C for only 5 min (Schmid, 1965a).

I. Seasonal Migrations. Movements from overwintering sites to breeding sites were at least 462 m (Ewert, 1969). Following breeding, Great Plains toads moved to feeding sites from 300–1,300 m away from wetlands (Ewert, 1969; Fischer et al., 1999). Toads moved from 100–1,100 m between breeding sites and overwintering sites (Ewert, 1969).

J. Torpor (Hibernation). As with all northern bufonids, Great Plains toads are not freeze tolerant (Swanson et al., 1996) and overwinter by burrowing below the frostline (Collins and Collins, 1993; Irwin, this volume, Part One). Toads entered overwintering burrows between 10 August–16 September in Minnesota (Ewert, 1969). Sixteen toads overwintered in roadbanks and six survived the winter, while two overwintered in grasslands and both died. Those that overwintered in the road bank did so near the crest. A preference for elevated overwintering sites may be general, but these habitats are limited. Toads that survived burrowed to maximum depths of 74–104 cm. Toads move up in the soil to avoid rising groundwater (Ewert, 1969), but have physiological responses to aid survival during short periods of anoxia (Armentrout and Rose, 1971). Individual toads spend 63–77% of the year in dormancy (Ewert, 1969). Spring emergence and breeding are triggered by heavy rains (Fischer et al., 1999), although Brown and Pierce (1967) found breeding to occur independent of rainfall, and Ewert (1969) observed emergence from hibernation to occur over at least a 5-wk period.

K. Interspecific Associations/Exclusions. In Arizona, Great Plains toads are found in watering troughs with Colorado River toads (*B. alvarius*) and are found calling on pond banks in association with spadefoot toads (*Scaphiopus* sp.; A.I. and R.D. Ortenburger, in Wright and Wright, 1949). In an Arizona playa, they breed concurrently with Mexican spadefoot toads (*S. multiplicata*; MacKay et al., 1990). In Utah, Great Plains toads breed in association with Woodhouse's toads (*B. woodhousii*; V.M. Tanner, in Wright and Wright, 1949). In Colorado and Iowa, they breed in association with plains spadefoot toads (*Spea bombifrons*), plains leopard frogs (*Rana blairi*), Woodhouse's toads (*B. woodhousii*), and striped chorus frogs (*Pseudacris triseriata*). In Oklahoma, Great Plains toads are found in association with tiger salamanders (*Ambystoma tigrinum*), western narrow-mouthed toads (*Gastrophryne olivacea*), spotted chorus frogs (*P. clarkii*), and Strecker's chorus frogs (*P. streckeri*) in early spring (J.J.K., personal observations). In South Dakota,

Great Plains toads, Woodhouse's toads, chorus frogs, and plains spadefoot toads will breed concurrently in the same ponds (Flowers and Graves, 1995; B.M.G., personal observations). Great Plains toads hybridize with American toads (*B. americanus*), Woodhouse's toads (Collins and Collins, 1993; Gergus et al., 1999), red-spotted toads (*B. punctatus*; Sullivan, 1990; Degenhardt et al., 1996), Texas toads (*B. speciosus*; Rogers, 1973), Canadian toads (*B. hemiophrys*; Brown and Ewert, 1971), and Colorado River toads (Gergus et al., 1999).

L. Age/Size at Reproductive Maturity. Males range from 47–103 mm SVL, females from 49–115 mm (Wright and Wright, 1949; Krupa, 1990, 1994; Collins and Collins, 1993). Hammerson (1986) reported that Great Plains toads begin to breed when 2–5 yr old. In Arizona, reproductive maturity for both sexes is achieved at 2 yr (Sullivan and Fernandez, 1999). Both Krupa (1994) and Degenhardt et al. (1996) note that males first breed at 56 mm SVL, females at about 60 mm. Great Plains toads in the higher altitudes of the San Luis Valley, Colorado, breed at smaller sizes than lowland animals (Hammerson, 1986).

M. Longevity. Great Plains toads commonly live 10 yr, perhaps 20 (James, 1998; J.J.K., personal observations). The captive longevity record is 10 yr, 8 mo, 10 d (Snider and Bowler, 1992; Collins and Collins, 1993). The oldest individuals in a Sonoran Desert population were 6 yr, as determined by skeletochronology (Sullivan and Fernandez, 1999).

N. Feeding Behavior. Adults feed primarily at night, feeding on arthropods almost exclusively (Smith and Bragg, 1949). Primary prey include insects such as lepidopterans, dipterans, hymenopterans, and coleopterans, as well as other small invertebrates including centipedes and mites (Smith and Bragg, 1949; Dimmitt and Ruibal, 1980b; Hammerson, 1986; Luce et al., 1997). Ants and termites (Hymenoptera) predominated in the diets of animals examined by Dimmitt and Ruibal (1980b), whereas coleopterans were the most common prey of toads in New Mexico (Anderson et al., 1999b) and Oklahoma (Smith and Bragg, 1949). Great Plains toads forage daily in surrounding habitat, then return to agricultural fields to burrow (Smith and Bragg, 1949). Postmetamorphic toads feed day and night during the first month after metamorphosis, and feed almost exclusively on arthropods (only 1 out of 1,412 stomach content items was an annelid; Smith and Bragg, 1949). In Oklahoma, juveniles fed on acarinans (mites), hymenopterans, and coleopterans (beetles), with a decrease in mites and an increase in hymenopterans in the first month of growth (Smith and Bragg, 1949). In South Dakota, collembolans (springtails) and mites were eaten

after metamorphosis, with a shift in preference to coleopterans and hymenopterans within 1 mo (Smith and Bragg, 1949; Flowers and Graves, 1995).

O. Predators. Adults are eaten by larger mammals, birds, and snakes, including badgers, skunks (Mustelidae), opossums (*Didelphis marsupialis*; Jense and Linder, 1970; J.J.K., personal observations), western hog-nosed snakes (*Heterodon nasicus*; Wershler and Smith, 1992, cited in James, 1998; J.J.K., personal observations), plains garter snakes (*Thamnophis radix;* Flowers and Graves, 1997), and crows (Bragg, 1940a). Larvae are eaten by birds, insect larvae, and spadefoot toad (*Scaphiopus* sp.) tadpoles (Bragg, 1940a). Woodward (1983) found that tadpoles in permanent ponds suffered higher predation rates than those in temporary ponds. Ewert (1969) reported that 3 of 21 road crossings by Great Plains toads resulted in fatal interactions with automobiles.

P. Anti-Predator Mechanisms. Nocturnal activity patterns, cryptic coloration, and parotoid gland secretions reduce predation. Great Plains toads inflate when attacked by western hog-nosed snakes (J.J.K., personal observations). Juveniles avoid areas containing chemical cues from plains garter snakes (Flowers and Graves, 1997).

Q. Diseases. Mortality from "red-leg disease" (i.e., *Pseudomonas* bacterial infection) is known (Ewert, 1969). Cutaneous, hepatic, respiratory, and intestinal infections with *Mycobacterium marinum* can be fatal (Shively et al., 1981).

While reports of malformations have appeared with some regularity in recent years among many anuran species (Lannoo, 2000), malformations were not reported for Great Plains toads in four recent surveys conducted within the species' range (Converse et al., 2000; Helgen et al., 2000; Johnson et al., 2002; Lannoo et al., 2003).

R. Parasites. Nematodes and cestodes may occur in the gastrointestinal tract and

lungs (Goldberg and Bursey, 1991a; Goldberg et al., 1995). Ulmer (1970) examined two Great Plains toads from Iowa, and found no trematodes; Ulmer and James (1976) examined four Great Plains toads from Iowa and found no cestodes.

4. Conservation.
Great Plains toads are fossorial and difficult to monitor except during their explosive breeding bouts. In some portions of their range, populations may be scattered and isolated compared to the historical condition. There is at least one report of a range expansion to the east (see "Historical versus Current Distribution" above). It is not known whether these observations reflect long-term changes in species status, short-term population fluctuations, or inconsistent and superficial population monitoring. They currently receive no federal or state protection in the United States.

Bufo debilis Girard, 1854
GREEN TOAD

Charles W. Painter

1. Historical versus Current Distribution.
Green toads (*Bufo debilis*) were originally described by Girard (1854) with the type locality presented as "found in the lower part of the Rio Bravo (Rio Grande del Norte), and in the Province of Tamaulipas." Kellogg (1932) later changed this designation to Matamoros, Tamaulipas. Flores-Villela (1993) pointed out that both Schmidt (1953) and Frost (1985) erroneously stated that the type locality was restricted to Brownsville, Texas, by Sanders and Smith (1951), although there is no such restriction in that publication. Green toads consist of two recognized subspecies, eastern green toads (*B. d. debilis*) and western green toads (*B. d. insidor*; Crother et al., 2000), and range from southeastern Colorado and adjacent southwestern Kansas to Tamaulipas, San

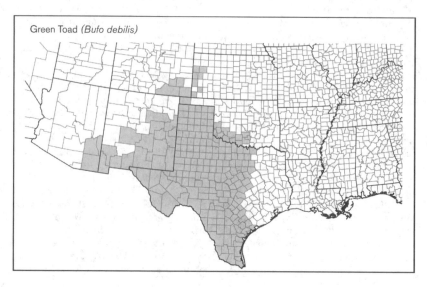

Green Toad (*Bufo debilis*)

Luis Potosí, and Zacatecas (Mexico), and from southeastern Arizona to eastern Texas (Frost, 1985; Conant and Collins, 1991). Western green toads were described as green toads by Girard (1854). Knowledge of the current distribution has not been greatly expanded since early descriptions (Sanders and Smith, 1951). The current distribution is expected to be similar to the historical distribution detailed as follows: Oklahoma (Bragg, 1950a,b; Black and Sievert, 1989); Texas (Lowe, 1964; Morafka, 1977; Dixon, 1987, 2000; Garrett and Barker, 1987); Kansas (Rundquist, 1979; Collins, 1982; Taggart, 1997a); Colorado (Hammerson, 1982a, 1999); New Mexico (Degenhardt et al., 1996); Arizona (Rosen et al., 1996); and Mexico (Flores-Villela, 1993).

2. Historical versus Current Abundance. Historical abundance is unknown, however localized populations have likely declined due to the conversion and disappearance of wetlands and low-lying areas used in reproduction. Green toads are localized but common in remaining areas of suitable habitat.

3. Life History Features.

A. Breeding. Reproduction is aquatic.

i. Breeding migrations. Stimulated by summer rains, males move from drier, terrestrial habitat to aquatic breeding sites, followed by females who are attracted by chorusing males.

Breeding occurs from late March to mid-June (Wright and Wright, 1949) and into late July (Sullivan, 1984; Degenhardt et al., 1996). Taylor (1929) reported chorusing on 8 August in extreme southwestern Kansas, and Bogert (1962) reported males calling during the first 2 wk of August in southwestern New Mexico. Taggart (1997a) observed breeding behavior from 12 June–2 September in Kansas. Breeding aggregations are usually of short duration, lasting only 1–3 d in southeastern Arizona (Sullivan, 1984).

ii. Breeding habitat. Breeding occurs during or after summer rains. Green toads breed in temporary water sources including stock tanks, temporary rain pools, roadside ditches, or shallow pools in streams of intermittent flow.

B. Eggs.

i. Egg deposition sites. The eggs of green toads are poorly known and are not produced in the long strings typical of most *Bufo* (Zweifel, 1968b). Sites for egg attachment may be limited in the ephemeral pools generally used as breeding sites by green toads. Strecker (1926) indicated the eggs are laid in small strings that are attached to grass and weed stems.

ii. Clutch size. Taggart (1997a) reported a single clutch of 1,287 non-adhesive eggs laid by a female while in amplexus. The eggs were laid singly and no egg-strings were observed in the clutch.

Twenty-five ova averaged 0.98 mm and another sample of 32 (measured before preservation) averaged 1.15 mm. Working in southeast Arizona, Zweifel (1968b) reported embryos hatched within 24 hr at 33.1 °C water temperature, while at 18.2 °C, hatching required 140 hr. Length at hatching (generally at Gosner stage 18–20) is about 3.1–3.4 mm.

C. Larvae/Metamorphosis. The labial tooth row formula of 2/2 is unique and diagnostic of the *B. debilis* group (*B. debilis*, the Mexican *B. kelloggi*, and Sonoran green toads [*B. retiformis*]) in North America (Zweifel, 1968b, 1970).

i. Length of larval stage. Uncertain and is dependent upon numerous environmental factors. Strecker (1926; Texas) suggested metamorphosis took fewer than 3 wk. Burkhart (1984) estimated larval life exceeded 25 d in Kansas, while Taggart (1997a) observed development from zygote to juvenile in 8 d.

ii. Larval Requirements.

a. Food. Tadpoles of green toads are a morphological type IV tadpole (Orton, 1953). They are a generalized pond-type tadpole that likely feeds on a variety of algae, detritus, and periphyton on or near the bottom of ephemeral pools.

b. Cover. Cover is generally unavailable in the ephemeral pools that support green toad tadpoles, although increased water turbidity, wind-blown debris, or vegetation, especially clumps of grass that has been inundated from flood-waters, may provide cover.

iii. Larval polymorphisms. Carnivorous or cannibal morphs are unknown in green toads.

iv. Features of metamorphosis. Metamorphosis is complete at about 19–20 mm SVL (Savage, 1954). Newly transformed toadlets have been found in southwestern New Mexico as late as 27 September (Seymour, 1972), although in New Mexico most metamorphosis takes place during late July to early August (Degenhardt et al., 1996).

v. Post-metamorphic migrations. Newly metamorphosed toadlets remain near the parent pool and disperse only when the pool evaporates or when they reach an age of about 2–3 mo (Bragg, 1950c). Creusere and Whitford (1976) reported juvenile green toads that remained at the hatching site for 55 d, using vegetative cover and fissures in the soil for shelter and for protection from desiccation.

D. Juvenile Habitat. Believed to be the same as adult habitat characteristics. Seymour (1972) reported green toad juveniles are active by day at 15 °C, often basking on mud banks in direct sunlight and increasing body temperature to 30 °C. Recently transformed toadlets often hide in the fissures formed in the mud of drying pools (Seymour, 1972; Creusere and Whitford, 1976; Taggart, 1997a).

E. Adult Habitat. Generally arid regions below 1,524 m (5,000 ft). Specific habitat

characteristics include grasslands in Arizona (Lowe, 1964); plains grasslands in Colorado (Hammerson, 1982a); open grass plains and native prairie vegetation with considerable topographic relief in Kansas (Collins, 1982; Taggart, 1997a); lower desert grasslands; areas grown to mesquite and creosotebush; playa bottom grasslands to open bajada creosote community in New Mexico (Creusere and Whitford, 1976; Degenhardt et al., 1996); mesquite-shortgrass prairies, especially along valleys of small creeks; short-grass plains, mesquite savannas, and gypsum-hill regions; ecotone between short-grass plains and mixed prairie in Oklahoma (Bragg, 1950a,c; Black and Sievert, 1989); arid and semiarid plains and grasslands in Texas (Garrett and Barker, 1987); and xerophilic brushland and grasslands in Mexico (Flores-Villela, 1993). Green toads often take refuge under rocks or in existing rodent or other burrows and may occur in grasslands that have been converted to agriculture where herbicide and/or pesticide levels do not exceed lethal limits.

F. Home Range Size. Unknown.

G. Territories. Territorial behavior other than during breeding aggregations is unknown. Generally only single individuals are encountered under rocks during the dry season. During breeding aggregations, calling males are usually separated by 0.5–3.0 m, however on occasion may call within 0.1 m of one another (Sullivan, 1984). Bogert (1962) and Sullivan (1984) provide information on advertisement call characteristics and variation.

H. Aestivation/Avoiding Desiccation. Unknown. Green toads are not surface active unless sufficient moisture is available in the form of summer rainfall or unusually humid nights.

I. Seasonal Migrations. None reported.

J. Topor (Hibernation). Generally below ground and inactive September–May in southern New Mexico (personal observations). The period of seasonal inactivity is likely longer at higher elevations and shorter at southerly latitudes.

K. Interspecific Associations/Exclusions. Breeding associates in New Mexico are Great Plains toads (*B. cognatus*), western narrow-mouthed toads (*Gastrophryne olivacea*), Couch's spadefoot toads (*Scaphiopus couchii*), plains spadefoot toads (*Spea bombifrons*), Mexican spadefoot toads (*S. multiplicata*), and, marginally, American bullfrogs (*Rana catesbeiana*; Creusere and Whitford, 1976; Stuart and Painter, 1996). Bragg (1950a,b) reported Texas toads (*B. speciosus*), Woodhouse's toads (*B. woodhousii*), Couch's spadefoot toads, plains spadefoot toads, western narrow-mouthed toads, and tiger salamanders (*Ambystoma tigrinum*) as breeding associates in Oklahoma. Additional breeding associates include plains leopard frogs (*R. blairi*), northern cricket frogs (*Acris crepitans*;

Kansas; Taggart, 1997a), and Colorado River toads (*B. alvarius*; Arizona; B.K. Sullivan, personal communication). Although predation by American bullfrogs has been reported (Stuart, 1995), it is unlikely that American bullfrogs would exclude green toads from an area.

L. Age/Size at Reproductive Maturity. Sullivan (B.K., personal communication) reported males in southwest New Mexico calling at 37 mm SVL (range 37–46 mm; n = 22); females found in amplexus ranged from 44–54 mm SVL (n = 6).

M. Longevity. Unknown. Fewer than 1% of the juveniles survive to adult size (Creusere and Whitford, 1976).

N. Feeding Behavior. No comprehensive study of the food habits of green toads has been published, although individuals likely feed on a variety of small arthropods and other invertebrates. Taggart (1997a) reported observing green toads feeding on small insects, especially Hymenoptera (ants) and small Lepidoptera, Coleoptera, and Orthoptera.

O. Predators. Little information is available on the predators of green toads. Known predators on adults and juveniles include American bullfrogs (Stuart, 1995), checkered garter snakes (*Thamnophis marcianus*; Stuart and Painter, 1996), plains garter snakes (*T. radix*), and tiger salamanders (Taggart, 1997a). Likely predators include western hog-nosed snakes (*Heterodon nasicus*), coachwhips (*Masticophis flagellum*), and gopher snakes (*Pituophis catenifer*; Creusere and Whitford, 1976). Eggs or tadpoles may be attacked by the fungus *Saprolegnia* sp. and by a host of invertebrate predators including Odonata (dragonfly larvae), Hemiptera (Notonectidae—backswimmers; Belostomatidae–giant water bugs), Coleoptera (Dytiscidae—predaceous diving beetles; Hydrophilidae—water scavenger beetles). Tadpoles and adults may also be preyed upon by black-necked garter snakes (*T. cyrtopsis*), ravens, raccoons, and skunks. Cei et al. (1968) report the presence of presumably toxic indolealkalamines from the skin and parotoid glands of green toads. The high mortality rate of juveniles is likely accounted for by poor microhabitat selection (Creusere and Whitford, 1976).

P. Anti-Predator Mechanisms. Nocturnal and secretive, usually seeking shelter in underground burrows or beneath surface objects. Adults rarely are seen except during the breeding season. Adults in breeding choruses are quick to retreat underwater or into surrounding vegetation at slight disturbance. Parotoid glands produce cardiotoxic steroids, possibly rendering the species unpalatable to some predators (Duellman and Trueb, 1986).

Q. Diseases. Unknown. Malformations in green toads have not been reported by the North American Reporting Center for Amphibian Malformations (Northern Prairie Wildlife Research Center, 1997).

R. Parasites. Metcalf (1923) reported an unknown species of Opalinidae (Protoza) from 11% of the green toads examined from Texas. McAllister et al. (1989) reported the following endoparasites from a series of 27 green toads collected in north-central Texas: *Nyctotherus cordiformis* (Protoza); *Myxidium serotinum* (Myxozoa); *Istoichometra bufonis* (Cestoidea); and *Cosmoceroides variabilis* (Nematoda). All blood samples were negative for Apicomplexa or trypanosomes, and no Protoza were found. Goldberg et al. (1995) found 98% of 49 green toads examined from Doña Ana County, New Mexico, harbored helminths including the cestoid *Distoichometra bufonis* and the nematodes *Aplecctana incerta*, *A. itzocanenis*, *Rhabdias americanus*, and *Physaloptera* sp.

4. Conservation.
The current distribution of green toads is thought to be similar to the historical distribution. Their historical abundance is unknown, however localized populations have likely declined due to the disappearance of habitat, especially wetlands. Green toads are localized, but common in remaining areas of suitable habitat. They are listed as Protected by Kansas.

Bufo exsul Myers, 1942(a)
BLACK TOAD

Gary M. Fellers

1. Historical versus Current Distribution.
Black toads (*Bufo exsul*) are known historically from four different spring systems in Deep Springs Valley, Inyo County, California. Buckhorn, Corral, and Bog Mound springs are in the immediate vicinity of Deep Springs Lake, while Antelope Spring is 7 km to the NW. Schuierer (1962) suggested that the population at Antelope Spring might have been introduced there in the early 1900s, but recent genetic research could not corroborate that the Antelope Springs toads were introduced (E. Simandle, personal communication).

A small, introduced population of black toads occupies a flowing well near Salt Lake (Saline Valley, 65 km SWS of Buckhorn Springs) in Death Valley National Park. In 1961, black toads were released at Cottonwood Springs in the Owens Valley (30 km to the west of Deep Springs; Schuierer, 1962), but there are no subsequent reports of this species in the valley.

The entire natural range of this species encompasses approximately 15 ha, one of the smallest ranges for any North American amphibian.

2. Historical versus Current Abundance.
Myers (1942) estimated the population at 600–700 individuals based on one visit to the site in 1940. In 1954, Schuierer (1961) noted that this estimate was much too low and stated that there was "an excess of 10,000 adults." Schuierer also noted that population numbers appeared to fluctuate with fewer toads seen in 1958 compared with 1960. In 1971, Schuierer (1972) enlisted the members of a biology class to systematically count the toads at Deep

Black Toad (*Bufo exsul*)

Springs. They counted 3,967 toads in the sloughs, ponds, and bog. Difficulty of counting toads in the boggy areas likely resulted in a tally that was too low, but he did not try to resolve the difference between his earlier estimate of 10,000 toads and the count of approximately 4,000 toads. As part of the same census, Schuierer (1972) reported large numbers of tadpoles: "three cubic meters of suitable habitat may contain as many as five thousand larvae."

In 1971, the population appeared to be "stable and maintaining itself" (Schuierer, 1972), but he found the most accessible and frequently visited part of the toad habitat had more smaller, presumably younger, toads. He attributed this to "collecting pressure" from herpetologists and indicated that the population could "not withstand the collectors, amateurs or professionals, who harvest numbers of adult toads."

In 1999 Murphy et al. (in preparation) evaluated population size in one area (Corral Springs) using PIT tags and mark–recapture. They estimated the population to be >8,400 toads, a number that is similar to that of Kagarise Sherman (1980) for the same spring. The population at Antelope Springs has always been reported as small (Schuierer, 1961, 1962; Stebbins, 1985), but the actual size is difficult to estimate due to the difficulty of seeing toads in the dense vegetation. There are no population estimates for Bog Mound or Buckhorn Springs, but the habitat at Buckhorn is more extensive than at Corral Springs. It is likely that the total number of black toads in Deep Springs Valley is about 24,000 individuals (J. Murphy and E. Simandle, personal communication). Though geographically restricted, the population appears to have been stable for the last 15–20 yr.

3. Life History Features.

A. Breeding. Reproduction is aquatic.

i. Breeding migrations. There is no migration, but Kagarise Sherman (1980) noted that males tend to stay near breeding sites while females move away. Black toads lack vocal sacs; they have a release call that is somewhat higher in pitch than that of western toads (B. boreas; Stebbins, 1985), but males apparently do not have an advertisement call (Schuierer, 1961).

ii. Breeding habitat. The breeding (and non-breeding) habitat in Deep Springs Valley includes a series of about 10 springs and the associated marshes, ponds, bogs, and sloughs. The springs are located at an elevation of 1,450–1,700 m (4,900–5,600 ft). The valley is in an isolated, desert basin located between the Inyo and White Mountains of northeastern Inyo County, California. Breeding begins in late March and continues into April (Schuierer, 1961).

B. Eggs.

i. Egg deposition sites. Shallow marshes or quiet pools bordering the larger wetlands (Schuierer, 1972). In the laboratory, eggs hatched in 4–5 d at 20 °C (Schuierer, 1972). Egg laying takes place from late March to late May (Stebbins, 1985).

ii. Clutch size. There are no published estimates of clutch size. Livezey (1960) described the eggs and oviposition sites. Eggs are deposited in strings, generally within a 30–35 cm area. The strings are entwined among grasses and watercress. Eggs masses are either at the water surface or up to 10 cm deep. Eggs generally are laid in the shallower parts of the marsh (Schuierer, 1961) where the water depth ranges from 15–20 cm deep (Livezey, 1960).

C. Larvae/Metamorphosis.

i. Length of larval stage. Schuierer (1961) states that the larval period is from 3–5 wk. Wright and Wright (1949) mention that two or three size-classes of tadpoles were present on 12 May 1942, but there seems to be no evidence that tadpoles overwinter because Schuierer (1972) reported seeing tadpoles of all sizes in May 1971.

ii. Larval requirements. Tadpoles are commonly found in shallow water of marshes or pools bordering the sloughs (Schuierer, 1961, 1972). These areas are subject to wide fluctuations in temperature (Schuierer, 1961).

a. Food. Undescribed; tadpoles probably feed on both organic and inorganic matter suspended in the water and on the surface of plants, rocks, and other substrates.

b. Cover. Schuierer (1962) describes breeding sites as "principally in unshaded marshes that border the sloughs." In the water, there are "dense mats of vegetation interlaced with swamp-like sedge areas. These boggy sites overlay a semi-soft black anaerobic mud, 1–3 feet thick" (Schuierer, 1972).

iii. Larval polymorphisms. Unknown and unlikely.

iv. Features of metamorphosis. Tadpoles reach a total length of 35 mm (Wright and Wright, 1949). Schuierer (1961) reported the size of newly transformed juveniles as 14–19 mm. Juveniles will grow to 22–33 mm SVL by November of their first year (Schuierer, 1961).

v. Post-metamorphic migrations. None.

D. Juvenile Habitat. Similar to adults, juveniles generally remain in shallow marshy areas.

E. Adult Habitat. "Like other desert valleys to the east of the Sierra Nevada, Deep Springs is exceedingly dry, and on its floor the vegetation consists of sparse low desert brush (Chrysothamnus)" (Myers, 1942). Schuierer (1961) provides a figure that details the plant associations for the alkali area, marsh, slough, stream margins, and head of the springs. The vegetation includes Polypogon, Juncus, Ranunculus,

Carex, Lemna, Muhlenbergia, Plantago, Scirpus, Zennichella, Roippa (Nasturtium), Mimulus, Sisyrinchium, Epipectis, and Salix. Toads do not frequent rocky portions of creeks that have no vegetation (Wright and Wright, 1949). The mountains that form Deep Springs Valley support juniper (Juniperus californica) and pinyon pine (Pinus monophylla).

Adults are the most aquatic members of the Bufo boreas group (Schuierer, 1961). Black toads have never been reported >12m from water (Schuierer, 1961).

Black toads are active primarily from late May to mid-September (Schuierer, 1962) in conditions that range from snowfall to hot summer days (Schuierer, 1972). Adults generally are diurnal, but they are active at night during the late spring and early summer. Schuierer (1961) reported toads active during the morning and early evening hours when air temperatures ranged from 17–22 °C. Toads were less active when temperatures were in excess of 25 °C. At temperatures above 30 °C, toads were sluggish even when prodded.

"Just below the first spring, in the water runways between tussocks, [we] began to see Bufo exsul. The bottoms are mud, water 2–4 inches deep. These areas more or less shaded by tussocks. Soil dark and mucky. Depressions between tussocks 8 inches to 1.5 feet deep. The toads often dashed into holes in tussocks or into shaded runways" (Wright and Wright, 1949).

F. Home Range Size. Unknown, but probably small.

G. Territories. Unknown.

H. Aestivation/Avoiding Desiccation. Black toads shift their activity from diurnal to nocturnal during the warmer months, but aestivation has not been reported.

I. Seasonal Migrations. None.

J. Torpor (Hibernation). Black toads use rodent burrows located 25–50 cm above the water level of the streams and sloughs. Toads found in burrows during November 1958 were huddled together and torpid (air temperature 12 °C), but became active when handled (Schuierer, 1961). At that same time, some toads were active near springs where water temperatures were about 20 °C.

K. Interspecific Associations/Exclusions. The only other amphibian reported within the range of black toads (Deep Springs Valley) is the western spadefoot toad (Spea hammondii; Schuierer, 1961).

L. Age/Size at Reproductive Maturity. The average SVL of males is 50 mm (range 44–59) and females is 52 mm (range 46–69; Schuierer, 1961). There are no published data on age at reproduction or on size versus age. Murphy et al. (in preparation) note that toads at Corral Springs tend to be small, especially compared with those at Antelope Spring (51.1 mm versus 74.7 mm SVL). The reason for this is unclear, but they note that European carp (Cyprinus carpio) at Antelope Spring are also small.

Murphy et al. (in preparation) found the sex ratio skewed significantly in favor of females (62.7% of population). This was unexpected since Kagarise Sherman (1980) found fewer females than males. Murphy et al. believe that the shift in sex ratio may be due to changes in habitat.

M. Longevity. Unknown.

N. Feeding Behavior. Stebbins (1951) examined the stomachs of several dozen black toads and found the remains of beetles, ants, and lepidopteran larvae. Livezey (1961) examined the food habits of 23 adult and 24 juvenile toads and provided a detailed table listing number and percent frequency of each food type. He noted that ". . . juvenile toads eat more small food materials than the adults, as well as more larval stages of dipterans and coleopterans . . . juvenile toads consumed considerably greater absolute numbers of food items." The primary food items were Diptera, Coleoptera, Hymenoptera, Homoptera, with lesser numbers of Hemiptera, Odonata, Collembola, Corrodentia, fairy shrimp (Anostraca), spiders, mites, and mollusks.

Schuierer (1961) examined the stomachs of 26 toads and found Hymenoptera, Diptera, Coleoptera, and small numbers of Hemiptera, Gastropoda, and Arachnida.

O. Predators. None reported, but eviscerated toads have been observed suggesting predation from common ravens (*Corvus corax*; E. Simandle, personal communication).

P. Anti-Predator Mechanisms. Black toads are "more aquatic than other members of the *Bufo boreas* group. Adults are often found resting on the surface of the water near water cress clumps or along the margins of the sloughs and ditches in tussocks of the dwarf bulrush. When disturbed, they swim to the bottom where their coloration blends with that of the dark substratum. Their swimming is more frog-like than toad-like. Locomotion on land is by walking, with the body raised high off the ground" (Schuierer, 1961). "When prodded, they progress by short, ineffective hops" (Schuierer, 1961). "Unless pressed, the toads seemed to prefer to walk rather than to hop" Stebbins (1951).

"The brownish black ground color with yellow-white mottling blends in with the substratum, both on land and in the dark-colored water courses . . ." (Schuierer, 1961).

As with other *Bufo*, black toads produce skin secretions that are toxic and function as a deterrent to predators.

Q. Diseases. None reported.

R. Parasites. "Some toads were parasitized by an undetermined intestinal nematode. Occasionally leeches were found on toads" (Schuierer, 1961).

4. Conservation.
Schuierer (1961) noted that "the annual recanalization of the water courses for irrigation has notable effect upon the population. When stream modification occurred after oviposition, the marsh area dried before metamorphosis was completed." He also noted that collecting had affected the population. Subsequently, Schuierer (1972) reviewed the status of black toads in 1971 and concluded that the population appeared to be in good condition (compared with 1962). Schuierer indicated that the primary threat to the population was from "collectors, amateurs or professionals" who were removing toads from the population. His conclusions were based on the relative lack of large adult toads from the most accessible areas. Schuierer (1972) also noted that the toads seem to have "adjusted to various climatic conditions and periodic droughts, as well as changes in the irrigation pattern of its habitat."

Busack and Bury (1975) reported that livestock grazing at the springs and canals was a potential problem for the toads. Bury et al. (1980) made five specific conservation recommendations that addressed issues of spring management, introduced predators, livestock grazing, marsh burning, channel modifications, and management of the area by both state and federal government and Deep Springs College.

All native habitat for the black toad is owned by Deep Springs College. Cattle graze throughout Deep Springs Valley, but the college fenced the springs in the early 1970s and currently allows only brief, seasonal grazing. This has resulted in an increase in vegetation, especially sedges (*Scirpus* sp.) and cattails (*Typha* sp.; Murphy et al., in preparation). The college has also eliminated the practice of diverting water from the springs for irrigation and no longer allows the raking and burning of vegetation. All these management practices are in keeping with the recommendations of Bury et al. (1980).

As with any species with a highly restricted distribution, there is concern that a catastrophic event or introduction of a disease or predator could eliminate the entire species.

Black toads were listed as Rare by the California Fish and Game Commission in 1971, but that status was changed to Threatened in 1984.

Acknowledgments. I thank John Murphy and Eric Simandle for valuable comments based on their extensive experience with black toads.

Bufo fowleri Hinckley, 1882
FOWLER'S TOAD

David M. Green

1. Historical versus Current Distribution.
Fowler's toads (*Bufo fowleri*) occur throughout the eastern United States, exclusive of the southern Atlantic Coastal Plain and the Florida Peninsula (Burt, 1932; Wright and Wright, 1949; Conant and Collins, 1998). Their northern range edge is along an irregular line reaching from southeastern Iowa through northern Illinois and turning north to the south shore of Lake Michigan (Bailey, 1944; Smith, 1947, 1961; Minton, 1972; Harding, 1997) to Chicago (Higginbotham, 1939; Edgren and Stille, 1948). They reach their northern limit on the west side of Michigan's Lower Peninsula, reaching almost to the Mackinac Strait in the north (Smith, 1961; Harding, 1997). The northern limit also encompasses Lake Erie in Ohio (Walker, 1946; Kraus and Schuett, 1982) and extends across northern Pennsylvania, New Jersey, southeast New York and southern Vermont, through southern New Hampshire to reach the Atlantic Coast almost to Maine (Stewart and Rossi, 1981; Kraus and Schuett, 1982; Shaffer, 1991; Harding, 1997).

At their eastern limit, Fowler's toads occur along the Atlantic coast from New

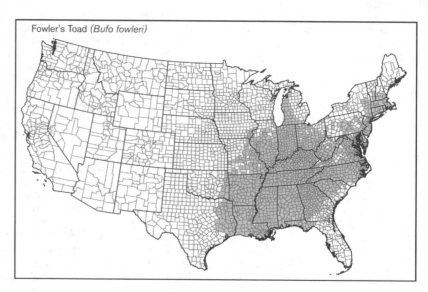

Fowler's Toad (*Bufo fowleri*)

Hampshire to southern North Carolina, including Massachusetts, Connecticut, New York, Delaware, Maryland, New Jersey, and Virginia (Burt, 1932; Babbitt, 1937; Latham, 1971a,b; Lazell, 1976; Miller, 1979; Klemens, 1993; Given, 1999). Fowler's toads occur on many nearshore islands along this stretch of the Atlantic coast including Cape Cod and its adjacent islands, Long Island, and islands off the Virginia coast (Harper, 1935; Latham, 1971a,b; Lazell, 1976; Hranitz et al., 1993). The range skirts the Coastal Plain from southern North Carolina across the middle of South Carolina and Georgia to the panhandle of Florida at the Gulf of Mexico (Miller, 1979; Martof et al., 1980; Ashton and Ashton, 1988; Hunter et al., 1992). Fowler's toads range across the Gulf Coast to the Florida Parishes of southeast Louisiana (Anderson et al., 1952; Liner, 1954; Mount, 1975; Ashton and Ashton, 1988; Dundee and Rossman, 1989).

The western edge of the range of Fowler's toads is ill defined, according to most sources, due to their intergradation with the more westerly Woodhouse's toads (*B. woodhousii*; Meacham, 1962; Johnson, 1987; Black and Sievert, 1989; Dundee and Rossman, 1989). Conant and Collins (1998) indicate a wide overlap of the two forms encompassing eastern Oklahoma, eastern Texas, western Arkansas, and much of Louisiana. The two toads' interrelationship is especially complicated in the region of east Texas and western Louisiana where there exists a confusion of difficult-to-identify toads with intermediate morphologies. Bragg and Sanders (1951) named these animals *B. woodhousii velatus* (or simply *B. velatus* according to Sanders [1987]; East Texas toads). The status of *B. w. velatus* is uncertain (see *B. woodhousii* account, this volume). Conant and Collins (1991), Sullivan et al. (1996a), and Dundee and Rossman (1989) view them as a variant of Woodhouse's toads. Dundee and Rossman (1989) also question the existence of this hybridization zone in Missouri and Oklahoma. The western edge of the range of unequivocal Fowler's toads runs from southeast Louisiana (Dundee and Rossman, 1989) through most of Arkansas almost to the Oklahoma border, and north across central Missouri to southeast Iowa, avoiding Kansas (Conant and Collins, 1998; Smith and Johnson, 1999). Delimiting the distribution of Fowler's toads in the southeastern portion of their range may also be complicated due to their tendency to hybridize with southern toads (*B. terrestris*; L.E. Brown, 1970) and Gulf Coast toads (*B. valliceps* [now considered to be Coastal-Plain toads, *B. nebulifer*; see Mulcahy and Mendelson, 2000; Mendelson, this volume]; Fox et al., 1961; Wittliff, 1964).

The large range of Fowler's toads was neither fully appreciated nor well mapped until about 50 yr ago (Wright and Wright,

1949). In the late 1800s and for many years subsequent to the description of the species, the range was considered to be confined merely to the vicinity of the type locality in Massachusetts, as reported by Dickerson (1906). This is truly remarkable, as Fowler's toads are exceedingly common all through the Atlantic states and across their presently known range. At last, Allard (1907, 1908) reported what is now obvious: Fowler's toads occur commonly along the eastern seaboard as far south as at least northern Georgia. Ruthven (1917) and then Hubbs (1918) reported them in Michigan, Indiana, and Illinois, and then the true extent of their range rapidly began to be filled in. Compared to more recent maps in Conant and Collins (1998) and Conant (1958a, 1975), Wright and Wright (1949) did not register Fowler's toads as occurring along the Gulf of Mexico coast except in Louisiana, considered the range to encompass Louisiana and east Texas, and missed much of their occurrence in Arkansas, Pennsylvania, and along the coast of North Carolina.

The early nomenclatural history of Fowler's toads, reviewed by Kluge (1983), Sanders (1987), and Green (1989), is confused. According to Dexter (1973), naturalist S.P. Fowler (1862), at a meeting of the Essex Institute on 30 June 1858, reported the existence of a new and "hitherto undescribed" species of toad collected from Danvers, Massachusetts, and F.W. Putnam dubbed it "*Bufo fowleri*," a name printed in the Proceedings of the Essex Institute without description. Authorship of the species from that time until 1921 was generally ascribed to Putnam (e.g., Allen, 1868; Garman, 1884; Jordan, 1888; Cope, 1889; Allard, 1907, 1908), yet there existed no published description by Putnam, only his 1863 intent to publish (Dexter, 1966), which finally saw print in an article by Dexter (1973). Garman (1884) gave a brief description of *B. fowleri* Putnam and, with the realization that Putnam's intended description had not been published, *Bufo fowleri* Garman was recognized. However, Myers (1931b), in reviewing the problem, concluded that Hinckley's (1882) description of the tadpole as distinct from the tadpole of the American toad (*B. americanus*) was the first actual description of the species and therefore had taxonomic priority over Garman's (1884) account according to the rules of zoological nomenclature. The attribution to Hinckley has been accepted ever since.

Following Smith (1934), Woodhouse's toads and Fowler's toads had been considered conspecific as subspecies of *B. woodhousii* (also spelled *woodhousei*) by most, but not all, authorities. Frost (1975), too, listed *fowleri* as a junior synonym of *B. woodhousii* but acknowledged that this taxonomy was controversial. Smith (1934), in his consideration of the amphibians of

Kansas, including the toads, concluded that both Woodhouse's toads and Fowler's toads were equally different, and different in equal ways from American toads. He stated that their similarities were striking, citing egg membranes, song, cranial crests, and the spinyness of the skin. Differences he noted were size, clutch size, and color pattern. It is now evident, however, that Fowler's toads do not occur in Kansas (Collins, 1974, 1982). Burt (1935) considered *woodhousii* and *fowleri* to be conspecific, as he could distinguish no significant diagnostic differences between toads from the southeast and toads from the middle west, both of which may have been *fowleri*. Meacham (1962) added apparent weight to Smith's (1934) taxonomy with an analysis of morphological characters of toads in east Texas, where Fowler's toads likewise do not occur. Meacham concluded that there was a zone of intergradation between Woodhouse's toads on the west and Fowler's toads on the east and therefore, under a biological species concept, they were one species. Meacham (1962) used six head characters, largely from the cranial crests, in his analysis, but the characters he used are not independently variable (Green, 1989), invalidating many of the results. Sullivan et al. (1996a) summarized this and other evidence from behavior, morphology, and genetics with a conclusion (followed by Collins [1997]) that supported the recognition of these taxa as separate species.

Sanders (1987), in an idiosyncratic analysis of skull characters principally concerning the cranial crests, concluded that only Fowler's toads from Danvers, Massachusetts, and vicinity were truly *B. fowleri*. According to Sanders, these toads had more closely spaced interorbital cranial crests than did Fowler's toads from elsewhere in the range, which he dubbed *B. hobarti*. This taxonomic arrangement has resisted general acceptance.

2. Historical versus Current Abundance.
Fowler's toads have been noted by numerous authors to be extremely abundant throughout much of their range (Allard, 1908; Martof, 1962a; Lazell, 1976; Klemens, 1993; Zampella and Bunnell, 2000), but especially in the northeast of their distribution. In the extreme south of their range, however, they are replaced by other species as the most common toad (Mount, 1975; Dundee and Rossman, 1989). Fowler's toads are more tolerant and dependent upon higher temperatures than are American toads, coincident with their generally more southerly distribution (Frost and Martin, 1971).

Historical abundances are unknown, but populations can vary widely in size in different years and at different places and, at times, may consist of large numbers of individuals (Breden, 1988; Green, 1992, 1997a; Hranitz et al., 1993). Lazell (1976)

notes that Fowler's toads, with their penchant for dry scrub and open country, probably benefited from land clearing during the early days of European settlement in eastern North America. Fowler's toads once occurred virtually everywhere on Cape Cod and adjacent islands. Recently, however, Lazell (1976) reports that they were extirpated from Nantucket, Muskeget, Cuttyhunk, and, probably, Tuckernut islands between 1940 and 1960, which Lazell attributes to indiscriminate use of pesticides, notably DDT (See "Conservation" below).

3. Life History Features.

A. Breeding. Reproduction is aquatic.

i. Breeding migrations. Fowler's toads converge at breeding ponds in late spring (Minton, 1972; Grogan and D. Bystrak, 1973a; Laurin and Green, 1990), usually from one to a few weeks after sympatric American toads (Hoopes, 1930; Babbitt, 1937). In Florida, Fowler's toads breed from February–April (Ashton and Ashton, 1988); in Alabama, breeding may commence in March or April, although the peak of breeding activity is usually mid-May (Mount, 1975). Toads may begin to breed in March in Louisiana but more commonly not until April (Dundee and Rossman, 1989). Fowler's toads breed in March–May in South Carolina and April–July in Virginia (Martof et al., 1980); April in West Virginia (Green and Pauley, 1987) and Tennessee (Huheey and Stupka, 1967); late April to July with its peak in mid-May in Kentucky (Barbour, 1971); mid-May to mid-June in Michigan (Harding and Holman, 1992); mid-May to June in Connecticut (Clarke, 1974a); late April to late June in Illinois (Smith, 1961); and late April to early July in the Great Lakes region (Harding, 1997). Breden (1988) recorded choruses beginning from 4–11 May at a minimum air temperature of 10 °C at Indiana Dunes on Lake Michigan. Also, at the toads' northern range limit on Lake Erie, Green (1997a; unpublished data) has recorded Fowler's toads beginning to sing any time from the last days of April–24 May, at a minimum body temperature of 14 °C (Green, 1997; Blaustein et al., 2001; unpublished data). In Georgia, however, Fowler's toads do not call at air temperatures below 19 °C (Martof, 1962a).

Many authors have documented the male-biased sex ratios of breeding aggregations of Fowler's toads (Aronson, 1944a; Fairchild, 1981; Breden, 1988; Hranitz et al., 1989; Laurin and Green, 1990). Breden (1988) found the ratio of females to total number of toads present to range from 0–0.6 but averaging 0.13 and 0.10 in two different years. However, this represents differences in behavior of the two sexes, as the population sex ratio did not differ significantly from 1:1, a finding echoed by Green (1997a).

Descriptions of the call of the male Fowler's toad have been given by many authors (e.g., Harper, 1928; Wright and Wright, 1949; Bogert, 1960; Zweifel, 1968a; Green, 1982; Given, 1996; Sullivan et al., 1996a). Garman (1892) described it as a "prolonged and rather shrill scream," although both he and Allen (1899) erroneously attributed the source to American toads late in their breeding season without suspecting that the sound came from a different kind of toad. At last tracing it to its true originators, H. A. Allard variously likened the call of Fowler's toads to an "almost agonized wail" (Allard, 1907), a "penetrating droning scream" (Allard, 1908), or a "weird, wailing scream" (Allard, 1916). It usually sounds rather like a sort of muffled scream from a small and distressed sheep. Structurally, the call has the same characteristics as other *Bufo*, including American toads, which make trilled calls, but the pulse frequency is so high that individual pulses cannot be discerned (Zweifel, 1968a; Green, 1982). The vocal sac of the male radiates the sounds more to the front of the animal than behind it (Gerhardt, 1975). The males in a chorus respond to each others' vocalizations and females orient to particular males' calls based on persistence and intensity (Given, 1993, 1996). Not all males in a breeding aggregation will call, however, thus the intensity of chorus is not a good indicator of the actual number of animals present (Shirose et al., 1997). Males issue a grumbling, vibrating release call when handled, whether by humans or by other toads (Brown and Littlejohn, 1972; Leary, 2001a,b).

ii. Breeding habitat. Fowler's toads breed in the shallow water of permanent ponds, flooded low ground, temporary pools, farm ponds, roadside ditches, quiet streams, lake shores, or along the shallows of rivers (Wright and Wright, 1949; Smith, 1961; Mount, 1975; Collins and Wilbur, 1979; Green and Pauley, 1987; Breden, 1988; Dundee and Rossman, 1989). Breden (1988) described breeding ponds used by the toads as shallow with sandy bottoms and gradually sloping banks, vegetated primarily with sedges and bulrushes.

B. Eggs.

i. Egg deposition sites. Same as breeding habitat. Eggs are laid in long twin strands and are 3–4 mm in diameter (Wright and Wright, 1949).

ii. Clutch size. Ovarian eggs are reported as 2,000–4,000 eggs by Birge et al. (2000), but clutch sizes have been reported as up to 7,000 (Martof et al., 1980), 8,000 (Wright and Wright, 1949; Barbour, 1971; Green and Pauley, 1987; Ashton and Ashton, 1988), or even 10,000 (Mount, 1975) eggs. The average clutch size, based on data from Clarke (1974b) from Connecticut, was 5,221 eggs/clutch. However, based on direct counts of eggs laid, clutch sizes ranging from 2,000–6,300,

an average around 3,700, were found (unpublished data). Eggs hatch about 1 wk after laying (Smith, 1961; Martof et al., 1980).

C. Larvae/Metamorphosis. Larvae are black overall, except for the pale venter of the tail (Hinckley, 1882; Altig, 1970; Altig et al., 1998). Larvae are often observed in large aggregations (Breden et al., 1982) in stream and pond habitats, typically in standing water and resting on a muddy or sandy substrate. Details on larval development can be found in Bragg (1940b). Gosner and Black (1957b) found Fowler's toads to use ponds with pH between 5.5 and 6.6 in the New Jersey Pine Barrens, avoiding sphagnaceous ponds with lower pHs.

i. Length of larval stage. The larval period takes 40–60 d (Wright and Wright, 1949; Ashton and Ashton, 1988), and tadpoles transform at about 8–12 mm SVL (Wright and Wright, 1949; Martof et al., 1980; Breden, 1988). Fowler's toad larvae tend not to do well in artificial ponds with higher pH and floating vegetation (Bunnell and Zampella, 1999).

ii. Larval requirements.

a. Food. Larvae are suspension feeders and will take a variety of organic and inorganic material.

b. Cover. Fowler's toad tadpoles are active and benthic in behavior compared to co-occuring other species of tadpoles in the New Jersey Pine Barrens, and they decrease their activity in the presence of eastern newts (*Notophthalmus viridescens*) and black-banded sunfish (*Enneacanthus obesus*) even though the fish find them unpalatable (Lawler, 1989).

iii. Larval polymorphisms. Do not occur.

iv. Features of metamorphosis. Metamorphosis takes place from late June to July in Illinois (Smith, 1961), mid-June to August in New York (Wright and Wright, 1949), and late July to August in Kentucky (Barbour, 1971). Newly metamorphosed animals tend to average smaller in size under increasingly crowded conditions, with concomitant reduction in their initial stamina and jumping ability (John-Alder and Morn, 1990).

v. Post-metamorphic migrations. From wetland breeding sites to upland feeding sites.

D. Juvenile Habitat. Similar to habitat characteristics of adults.

E. Adult Habitat. Fowler's toads occur in areas with loose, well-drained gravelly or sandy soils, including sand dunes, sandy deciduous woodland, and rocky, poorly vegetated areas (Hubbs, 1918; Smith, 1961; Minton, 1972; R.L. Brown, 1974; Klemens, 1993). Wright and Wright (1949) note that Fowler's toads can be common along roadsides, near homes, and in fields, pastures, gardens, and sand dunes. They are a typical species of the New Jersey Pine Barrens (Zampella and Bunnell, 2000). Lazell states that whereas American toads on Cape Cod occur in wet deciduous woodlands and uplands, Fowler's toads prefer dry scrub, sand dunes, and open country. Klemens (1993) likewise observes in Connecticut that American toads are located in moist shady woodland but Fowler's toads are to be found on dry sunny rock ledges. In Louisiana, where Fowler's toads co-occur with southern toads (*B. terrestris*), Fowler's toads are in the bottomlands whereas the southern toads occupy higher ground (Dundee and Rossman, 1989). Huheey and Stupka (1967) observe that Fowler's toads occur up to 1,200 m elevation in the Great Smoky Mountains of Tennessee but are much more common at lower elevations. Bossert et al. (2003) report that Fowler's toads will use northern diamond-backed terrapin *(Malaclemys terrapin terrapin)* borrows as refugia.

F. Home Range Size. Fowler's toads will establish small home ranges, along the shorelines of lakes or large ponds. On Lake Michigan, Stille (1952) recorded them emerging from under the sand about 60–210 m from the water's edge and moving to the beach to rehydrate and forage over about 8 m of shorefront. Tracking three different individuals, Clarke (1974a) calculated their home range minimum polygons to be 2,742 m^2, 2,398 m^2, and 526 m^2, respectively. Adult Fowler's toads will return to their home ranges when displaced up to 1.28 km (R. J. Nichol, quoted in Oliver, 1955a) and will occupy the same home range year after year (Clarke, 1974a). They apparently can orient themselves using olfactory cues and a sun compass (Landreth and Ferguson, 1968; Grubb, 1973a).

G. Territories. Fowler's toads are not territorial in any way.

H. Aestivation/Avoiding Desiccation. Fowler's toads escape hot and dry conditions by burrowing into the ground (Harding and Holman, 1992) or finding burrows (Bossert et al., 2003) and become less active at temperatures above about 25 °C (Hadfield, 1966).

I. Seasonal Migrations. After the breeding season, adult Fowler's toads tend to move to beaches alongside larger water bodies to take up their "summer quarters" (Breden, 1988). This movement pattern has been observed at Indiana Dunes on Lake Michigan by Shelford (1913), Hubbs (1918), Stille (1952), and Breden (1982), and at Long Point on Lake Erie, Canada, by Green (1997a). They are later joined on the beach by the young of the year after metamorphosis (Breden, 1988). Non-breeding juveniles tend to remain in "summer quarters" throughout their active season. Juveniles tend to disperse more widely than adults (Clarke, 1974b; Breden, 1988). Fowler's toads are generally nocturnal (Higginbotham, 1939; Clarke, 1974a), but can be seen during the day in humid, overcast weather (Smith, 1961; Green and Pauley, 1987).

J. Torpor (Hibernation). In Connecticut, Fowler's toads are dormant for 7 mo of the year (Clarke, 1974a). They burrow into the sand or soil during the winter (Harding and Holman, 1992) to a depth of about 15–30 cm by late winter (R. Latham quoted in Oliver, 1955a). They may also overwinter in burrows (Bossert et al., 2003).

K. Interspecific Associations/Exclusions. Wright and Wright (1949) note that wherever Fowler's toads are sympatric with American toads, Fowler's toads occur in rivers, streams, or lake beaches. The two species tend to have different temperature tolerances (Frost and Martin, 1971) and habitat preferences: American toads in forests, Fowler's toads in more open sandy areas and savannas (Smith, 1961; Mount, 1975; Green and Pauley, 1987; Johnson, 1987). Although both Fowler's toads and American toads are found throughout the Midwest, where one species commonly is found the other is typically rare (Minton, 1972). Compared to American toads, Fowler's toads breed later and at warmer temperatures (Barbour, 1971; Green, 1982). Nevertheless, overlaps in breeding season do occur (Green, 1984). There had been long-standing confusion between the two species in the literature until their identities were clearly defined by Myers (1927, 1931b) and Netting (1930) and explained by Wright and Wright (1949). Dickerson (1906), for instance, includes a photograph labeled American toads although the animals portrayed are clearly Fowler's toads. Usually unmentioned by accounts of characters that differentiate Fowler's toads from American toads is the distinctive smell of Fowler's toads, which is reminiscent of the smell of unroasted peanuts or, according to Miller and Chapin (1910), *Ailanthus* sp. (Tree of Heaven) wood.

Fowler's toads, as with all the members of the *B. americanus* group of toads (Blair, 1959, 1963a, 1972a; Guttman, 1969; Martin, 1973) are notorious hybridizers (A.P. Blair, 1941; W.F. Blair, 1964a, 1972b; Green, 1984, 1996; Sanders, 1987; Green and Parent, 2003). Fowler's toads are broadly sympatric with American toads and hybridize with them in numerous scattered localities (Allard, 1908; Miller and Chapin, 1910; Hubbs, 1918; Myers, 1927; Pickens, 1927a; Blair, 1941; Volpe, 1952, 1955b; Cory and Manion, 1955; Zweifel, 1968a; L.E. Brown, 1970; Jones, 1973; Green, 1982, 1984; Green and Parent, 2003). They also have been known to hybridize with Coastal-Plain toads (Fox et al., 1961; Wittliff, 1964) and southern toads (L.E. Brown, 1970). Furthermore, diagnostic characters, such as the extent of ventral spotting, are not necessarily consistent from place to place (Blair, 1943), and newly metamorphosed animals may not have their distinguishing features fully developed, which continues to engender

confusion and misidentification. On the basis of mitochondrial DNA sequences, Masta et al. (2002) discerned three distinct mitochrondrial phylogroups within what is currently known as Fowler's toad.

L. Age/Size at Reproductive Maturity. Post-metamorphic growth rate is rapid (Labanick, 1976a; Claussen and Layne, 1983). In Connecticut, Clarke (1974b) re-counts an average 6.58-fold increase in length during the first year, whereas Labanick (1976a) calculated an even faster rate of growth, averaging 0.36 mm/d among post-metamorphic toads in Indiana from mid-June to mid-August. Females grow faster than males to reach larger size (Clarke, 1974b), and both males and females usually reach reproductive maturity at 2 yr of age (Breden, 1987). Particularly rapidly growing individuals may reach maturity within 1 yr of metamorphosis (Kellner and Green, 1994). Breeding males range from 42–74.5 mm, females from 55–82 mm SVL (Wright and Wright, 1949). Martof et al. (1980) say that adult Fowler's toads range in size from 50–82 mm in the Carolinas. Mount (1975) recounts a maximum SVL of 85 mm in Alabama. Dundee and Rossman (1989) say that the toads range from 51–76 mm in Louisiana. Green and Pauley (1987) give a size range of 51–84 mm for toads in West Virginia. Breden (1988) found adult males to average 58.8 mm ± 0.09 (s.e.) and adult females to average 62.2 mm ± 0.27 (s.e.) in northern Indiana. The average size of Fowler's toads tends to be negatively correlated with their abundance (Hranitz et al., 1993; Green, 1997a).

M. Longevity. Fowler's toads live a maximum 5 yr in the wild (Kellner and Green, 1995). An adult Fowler's toad survived at the Philadelphia Zoo for 2 yr, 5 mo, and 3 d (Bowler, 1977). Survivorship is low. Clarke (1972) concluded that toe-clipping of toads in order to study them lowered their survivorship, and Parris and McCarthy (2001) found that the effect was correlated with the number of toes removed. In any case, the survivorship rate is very low. The curve is Type III and decidedly J-shaped (Breden, 1988). Clarke (1977) calculated about 22.5% annual survival among post-metamorphic toads in Connecticut.

N. Feeding Behavior. Fowler's toads feed on a variety of invertebrates, especially beetles and ants (Metcalf, 1921; Bush, 1959; Bush and Melnick, 1962; Klimstra and Myers, 1965; Latham, 1968; R.L. Brown, 1974; Clarke, 1974c; Labanick and Schleuter, 1976; Gould and Massey, 1984). Cope (1889) states that they feed readily on flies but will not eat earthworms. Smith (1961) reiterates this aversion to earthworms, mentioning that captives are reluctant to eat them. Post-metamorphic toadlets unsurprisingly consume smaller prey than adults, largely collembolans, aphids, and fly larvae (Clarke, 1974c). Although Fowler's toads routinely hop, especially when disturbed, they approach prey by walking (Heatwole and Heatwole, 1968).

O. Predators. Eastern hog-nosed snakes *(Heterodon platyrhinos)* appear to be immune to Fowler's toad toxins and thus can and will feed on them (Surface, 1906; Edgren, 1955; Lazell, 1976; Shaffer, 1991). Some birds, such as shrikes (Laniidae) and bitterns (Ardeidae), have been known to take Fowler's toads (Latham, 1970, 1971c). Clarke (1977) notes that American bullfrogs *(Rana catesbeiana)* and raccoons *(Procyon lotor)* will eat Fowler's toads.

P. Anti-Predatory Mechanisms. A toad's noxious skin secretions afford protection against some predators (Kruse and Stone, 1984; Harding and Holman, 1992). Clarke (1977) recounts that sometimes Fowler's toads may be found with the scars of mammal bites or bird pecks from unsuccessful attempts to make them food. Fowler's toads also rely upon their general resemblance to sand and dirt, coupled with immobility (Dodd, 1977a; Harding and Holman, 1992) to escape detection. A Fowler's toad's normal locomotor gait, which consists of both walking and hopping, switches entirely to a more energetically efficient hopping when the toad seeks to escape after detection (Walton, 1988; Walton and Anderson, 1988; Anderson et al., 1991). Under duress, Fowler's toads can cover a distance of up to 37 cm/hop on sand, but generally manage about 13–15 cm/hop (Rand, 1952). Fowler's toad tadpoles are active and benthic in behavior compared to other co-occuring species of tadpoles in the New Jersey Pine Barrens, and they decrease their activity in the presence of eastern newts and black-banded sunfish even though the fish find them unpalatable (Lawler, 1989).

Q. Diseases. Infection by *Mycobacterium* is known (Shiveley et al., 1981).

R. Parasites. The nematode *Spinitectus gracilis* (Jilek and Wolff, 1978), among other endoparasitic helminths (Ashton and Rabalais, 1978; McAllister et al., 1989), plagues Fowler's toads. Unidentified cilate parasites have recently been described from Fowler's toad tadpoles (Vences et al., 2003).

4. Conservation.
Lazell (1976) attributes the extirpation of Fowler's toads from Nantucket, Muskeget, Cuttyhunk, and, probably, Tuckernut islands between 1940 and 1960 to indiscriminate use of pesticides, notably DDT, which is one among many environmental contaminants to which Fowler's toads have been or continue to be exposed. Birge et al. (2000) consider Fowler's toads, compared to other amphibians, to be tolerant of organic contaminants, including, for example, carbon tetrachloride and chloroform, with a mean LC50 of 13.3 mg/L, but Fowler's toads are less tolerant of atrazine

than expected. However, the organophosphate insecticide azinphos-methyl (Guthion®) is the most toxic of many such chemicals for Fowler's toads, with LC50 measured at 0.13 mg/L (Sanders, 1970) or 0.109 mg/L (Mayer and Ellersieck, 1986). The organochlorides endrin, toxaphene, dieldrin, aldrin, DDT, and lindane, in decreasing order, are also highly toxic to larval Fowler's toads (Sanders, 1970). The LC50 for hatchling Fowler's toad tadpoles exposed to the organic contaminant arochlor 1254 was 38.2 Fg/L (Birge et al., 1978). Four days post-hatching, the tadpoles' sensitivity increased to 3.7 Fg/L. Effects of the contaminant include lordosis, scoliosis, and abdominal edema. Adult Fowler's toads are also susceptible to organochloride poisoning (Ferguson and Gilbert, 1968) and exhibited accumulated residues of fenvalerate, a pyrethroid insecticide, after aerial spraying of cotton fields (Bennett et al., 1983).

In the presence of contaminating metals, according to Birge et al. (2000), Fowler's toads' highest tolerances are for cesium (LC50 = 1,076 mg/L) and magnesium (LC50 = 807 mg/L) but they are least tolerant of chromium (LC50 = 0.11 mg/L), gallium (LC50 = 0.13 mg/L), titanium (LC50 = 0.24 mg/L), or aluminum (LC50 = 0.28 mg/L). Fowler's toads are considered more tolerant of copper (LC50 = 25 mg/L; Linder and Grillitsch, 2000) and zinc (LC50 = 87 mg/L; Birge et al., 2000) than are most other amphibians.

Fowler's toads also show effects of lowered pH due to acid precipitation. Juveniles suffer decreased growth rate when raised as larvae at their lowest tolerable pH (Freda and Dunson, 1986).

Bufo hemiophrys Cope, 1886
CANADIAN TOAD

Michael A. Ewert, Michael J. Lannoo

1. Historical versus Current Distribution.
Canadian toads *(Bufo hemiophrys)* have their widest distribution across the prairies

and aspen parklands of Canada from south of Great Slave Lake in the Northwest Territories, across eastern Alberta, central Saskatchewan, and southern Manitoba (Hamilton et al., 1998). In the United States, their historical range includes an isolated site in northeastern Montana and extends across northern North Dakota, then southeastward into northeastern South Dakota, and also into the Red River Valley and adjacent areas of western Minnesota. Reports during the 1970s extend their known distribution into southwestern North Dakota (e.g., Seabloom et al., 1978). This distribution is unique among amphibians lying completely within regions of North America covered by the late Wisconsinan ice sheets (Flint, 1947; Underhill, 1961). Further, the range of Canadian toads tends to mirror the former extent of glacial Lake Agassiz (Underhill, 1961). Historically, this glaciation may have allowed Canadian toads (or their precursor) to range south and west of their current range. This notion is supported by the relict population of closely related Wyoming toads *(B. baxteri,* formerly designated as a subspecies, *B. h. baxteri)* in southeastern Wyoming (Henrich, 1968). Canadian toads range in altitude from 300–2,130 m (Stebbins, 1985).

2. Historical versus Current Abundance.
In the recent past, Canadian toads have been considered "very abundant" in northeastern North Dakota (Fishbeck and Underhill, 1960). In northwestern Minnesota, at the 640-acre Waubun Prairie Research Area, Mahnomen County, the population size exceeded 5,100 individuals in 1961; numbers fluctuated at lower levels in other years (Tester and Breckenridge, 1964b). In any given year, juveniles (second-year animals) outnumber adults with juvenile/adult ratios varying from 1.5–14.1 (Kelleher and Tester, 1969). Two amphibian calling surveys were conducted in northern and eastern North Dakota during 1995 (Johnson and Batie,

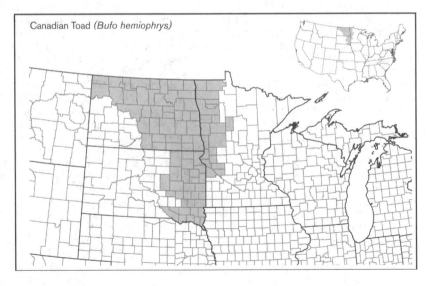

Canadian Toad *(Bufo hemiophrys)*

1996; Bowers et al., 1998). Both surveys sampled according to a predetermined grid of focal areas and found that the apparent distribution of Canadian toads was much smaller than the historical one given by Wheeler and Wheeler (1966). The more extensive of the two surveys (Johnson and Batie, 1996) found Canadian toads at only 4 northeastern focal areas (11 call listening posts) among about 30 sampled focal areas within the range known in 1966. However, such surveys may underestimate species with special habitat needs and spotty distributions. The historical status of Canadian toads in Alberta is well documented (Hamilton et al., 1998) and provides a perspective on the near present. At very least, a range contraction appears to have occurred in south-central Alberta since the mid-1980s (Roberts, 1992). Declines pertain especially to populations along the lower Medicine River and adjacent portions of the Red Deer River in Alberta. Sufficient concern for this apparent decline has resulted in yellow-listing ("may be at risk") in Alberta. Saskatchewan is without a comprehensive survey for Canadian toads, however the province also lacks obvious signs of population declines (Didiuk, 1997). In the United States, only Montana, which is peripheral in the species range, protects Canadian toads and lists them as Endangered (review in Hamilton et al., 1998).

3. Life History Features.

A. Breeding. Reproduction is aquatic.

i. Breeding migrations. In northwestern Minnesota, adults generally move from overwintering sites to breeding sites soon after emergence during late April to May (Tester and Breckenridge, 1964b). Atypically warm early spring weather fostered emergence as early as 28 March (1958) at Delta Marsh (southern Manitoba); however, return to freezing weather for an additional month stalled breeding and may have killed the toads that emerged early (Tamsitt, 1962). Chorusing normally commences during May. In flooded ditches in Minnesota, breeding congresses tend to include <20 individuals. At Waubun Prairie, however, breeding congresses composed mostly of males varied from 43–112 individuals and totaled 207 individuals on 21 May 1966 (Tester et al., 1970; M.A.E., personal observations). Not all males in a population appear to breed in any given year. Females approach the chorus singly, lay eggs, then leave (Tester and Breckenridge (1964b). Chorusing activities tend to decline during June; but at Delta Marsh, Manitoba, calling has continued sporadically into July and early August and even to 12 September 1958 (Tamsitt, 1962), which seems non-adaptive. The call is a series of low-pitched, soft trills of short duration, suggesting slightly the call of American toads (*B. americanus*), but much softer (Breckenridge, 1944; Conant and Collins, 1998).

ii. Breeding habitat. Breeding habitat includes the shallows of lakes, in ponds, and other bodies of water (Wright and Wright, 1949; Roberts and Lewin, 1979; Oldfield and Moriarty, 1994). In one instance, 14 sexually mature adults occurred in a lake in mid-May in Day County, South Dakota. Sampling at this same lake in early July yielded seven newly metamorphosed animals (Underhill, 1958). Toads also use flooded roadside ditches, especially in the flat agricultural country that currently occupies much of the species' range in Minnesota (Breckenridge, 1944; Brown and Ewert, 1971; M.A.E., personal observations). Breeding habitat may include flowing water, such as the Red Deer River of Alberta (Roberts and Lewin, 1979). Water associated with breeding habitat may be in permanent, semi-permanent (Tamsitt, 1962; Tester et al., 1970), or temporary basins (roadside ditches). At Delta Marsh, Manitoba, Canadian toads favored interdunal ponds near Lake Manitoba as opposed to more distant marshes (Tamsitt, 1962). Canadian toads can be syntopic with Great Plains toads (*B. cognatus*) in temporary basins in the Red River Valley but rarely hybridize with them (Brown and Ewert, 1971). The presence of fishes has not been noted and seems unlikely at places where Canadian toads breed in northwestern Minnesota (M.A.E., personal observation).

B. Eggs.

i. Egg deposition sites. Chorusing and egg deposition sites are often in 10–50 cm deep water in areas with a mix of open water and vascular plant debris, new blades of grass or sedges, and/or sparse cattail. Canadian toads avoid calling and egg deposition in dense beds of cattail or rushes (M.A.E., personal observations, Mahnomen and Norman counties, Minnesota). In Alberta, some oviposition sites lack any vegetation and may render the eggs and early tadpoles at risk from water movement (Roberts and Lewin, 1979).

ii. Clutch and egg size. Egg counts from three individuals in northeastern Alberta ranged from 3,354–5,842 eggs (Roberts and Lewin, 1979). These eggs are joined in long strings (Breckenridge, 1944; Roberts and Lewin, 1979). The early embryo is about 1 mm in diameter and positioned about 1 mm from the next embryo in the string (Breckenridge, 1944).

C. Larvae/Metamorphosis.

i. Length of larval stage. The exact duration for individual freely ranging tadpoles is unknown; metamorphosing toadlets appear at Waubun Prairie from late June to mid-July. Thus, if eggs are laid from 15–20 May and hatch within 4–5 d, the tadpole stage lasts approximately 6–8 wk (see Breckenridge and Tester, 1961).

ii. Larval requirements. Canadian toad tadpoles spend daylight hours in warm areas of wetlands that are thermally stratified. Further, in response to these thermal preferences, several bufonids will aggregate, including closely related American toad and Wyoming toad tadpoles (Beiswenger, 1978).

a. Food. Tadpoles of Canadian toads, as with all North American bufonids, are generalized suspension feeders, ingesting a range of organic and inorganic material associated with mud, plant, and other surfaces.

b. Cover. The need for cover, such as to seek shielding from UV-B radiation, remains undocumented and seems unlikely.

iii. Larval polymorphisms. No larval polymorphisms occur.

iv. Features of metamorphosis. Canadian toads metamorphose between 9.0–15 mm SVL (Wright and Wright, 1949; Breckenridge and Tester, 1961).

v. Post-metamorphic migrations. At Waubun Prairie, Minnesota, large numbers of juveniles remain near their natal wetland to feed along open mudflats (Breckenridge and Tester, 1961). At Delta Marsh, Manitoba, newly metamorphosed toads move to the edges of sand dune ridges and gradually move out onto them (Tamsitt, 1962).

D. Juvenile Habitat. Juveniles (defined as animals <45 mm; Breckenridge and Tester, 1961) between 18–36 mm were found near the remaining pools of a drying stream in northeastern North Dakota (Wright and Wright, 1949). Young of the year often occur in high numbers along the margins of spawning areas in Alberta (Roberts and Lewin, 1979). At Delta Marsh, Manitoba, second-year juveniles were common on the dune ridges and by midsummer frequented the open beaches of Lake Manitoba at night and regularly moved across beaches, between interior marshes and the lake side—75–120 m (Tamsitt, 1962). Age-specific growth rates are described in Breckenridge and Tester (1961); first-year growth rates are detailed in Roberts and Lewin (1979).

E. Adult Habitat. Canadian toads occur in mesic (but not arid) prairies and savannas, usually near streams, lakes, pothole wetlands, irrigation ditches, and flooded fields (Wright and Wright, 1949; Roberts and Lewin, 1979; Stebbins, 1985). They appear to have evolved to associate with wet prairie or pothole country in places with enough water in their hydrocycle to not render them saline (M.A.E., personal observations). They favor the aspen parkland country (Hamilton et al., 1998), which has groups of trees interspersed with grassland. In northeastern Alberta, Canadian toads are more abundant in areas of grass or willow bog than in areas of aspen or spruce (Roberts and Lewin, 1979). In a pilot study of choice for seminatural cover in Minnesota, adult male Canadian, American, and Great Plains toads were released within a fenced area (19 × 6 m) that ranged from mostly white spruce woods at one end to short grass at the other. Whereas Great Plains toads clearly favored the short grass, Canadian toads were more moderate in their preference for grassy areas, and American toads favored tall grass and woods (M.A.E. and J.R. Tester, unpublished data). During droughts, Canadian toads will enter gutters and sewers in small towns, such as Walhalla, North Dakota (Wright and Wright, 1949). Canadian toads are good swimmers (Stebbins, 1985), more aquatic than either American toads or Woodhouse's toads (*B. woodhousii*; Underhill, 1961), and the most aquatic of Minnesota's toads (Oldfield and Moriarty, 1994). Canadian toads are active around the clock during the breeding season (Stebbins, 1985; Fischer et al., 1999; M.A.E., personal observations). On cool nights, Canadian toads tend to burrow into sandy or loamy soils, but can be active on warm nights (Tamsitt, 1962; Stebbins, 1985).

F. Home Range Size. In a traditional sense, home ranges probably do not exist. Canadian toads have local focal areas for breeding, summer feeding, and overwintering. Individual toads often remain in one place for several days and then suddenly move some distance to another. The extent of these daily movements varies from almost no movement to >225 m (Breckenridge and Tester, 1961, table 2). The day-to-day movements of Canadian and American toads are similar in distance but differ from the longer movements of Great Plains toads (Ewert, 1969).

G. Territories. As with other species of North American *Bufo*, Canadian toads move among focal areas but do not defend territories (e.g., Breckenridge and Tester, 1961; Ewert, 1969).

H. Aestivation/Avoiding Desiccation. During the numerous observations of Breckenridge and Tester (1961), they "... found only one animal that exhibited anything suggesting aestivation." This animal remained dormant, buried in grass and surface litter for periods of 5, 7, and 8 d, from late June to early August. Upper lethal temperatures are about 40 °C (Schmid, 1965a, fig. 2). In a laboratory comparison of deep body temperatures with adjacent environmental surface temperatures in American, Canadian, and Great Plains toads, deep body temperatures of American and Canadian toads were cooler than the substrate, especially at high temperatures. Deep body temperatures within Great Plains toads remained similar to substrate surface temperatures (Tester et al., 1965). These differences seem unlikely to reflect simple evaporative cooling and instead could represent behavioral thermoregulation (Schmid, 1965b).

I. Seasonal Migrations. The annual period of activity for Canadian toads at Waubun Prairie, Minnesota, begins in May with their emergence from hibernation and their congregating along the pond margins. These events are dependent on weather conditions (Breckenridge and Tester, 1961). Findings of post-breeding adults most frequently occur along the ponds margins, with no differences between sexes. The cessation of activity in autumn is gradual, with animals burrowing to hibernate over the course of several weeks. The distances of "long movements" of toads are given in Breckenridge and Tester (1961, table 3). These movements are not correlated with rainfall events.

J. Torpor (Hibernation). Retreating to overwintering sites at Waubun Prairie takes place over a period of several weeks, with many of the larger toads burrowing in late August to early September (Breckenridge and Tester, 1961). Here, Canadian toads overwinter communally in small earthen (mima) mounds (Breckenridge and Tester, 1961; Tester and Breckenridge, 1964a; Ross et al., 1968; Oldfield and Moriarty, 1994; Fischer et al., 1999). The mounds, apparently derived through the burrowing activities of the toads and several mammals, are ~0.5 m high, 3–12 m in diameter, and located about 25 m from wetlands (Breckenridge and Tester, 1961; Tester and Breckenridge, 1964a; Ross et al., 1968). Individual toads repeatedly overwinter in the same mima mounds and show 87–95% homing fidelity to these areas (Kelleher and Tester, 1969).

Within the mima-type mounds, Canadian toads move vertically in the soil horizon in response to temperatures. By mid-October, marked toads in one study had burrowed to a depth of 46–66 cm and tended to stay there until January–February. Then, when soil temperatures dropped dramatically, toads burrowed deeper to a maximum depth of 117 cm (Breckenridge and Tester, 1961; Tester and Breckenridge, 1964a). Emergence appears to be stimulated by thawing of the soil. Spring soils thaw from both above and below the frozen layer, and toads begin to emerge once the thaw is complete. Breckenridge and Tester (1961; Tester and Breckenridge, 1964a) note that emergence occurs over the course of several weeks. The earliest toads to emerge from hibernation are adult males, with females and juveniles following (see also Kelleher and Tester, 1969). The final stage in emergence is often triggered by rains.

Canadian toads occur, at least in small populations, in many places in the United States where mima mounds are absent or not evident. One of us (M.A.E.) suspects that, as with Great Plains toads (Ewert, 1969), Canadian toads select roadside berms and other spots with good drainage as places for overwintering. In the Northwest Territories, Canada, Canadian toads overwinter communally (apparently >500 individuals) but in a sandy hillside with a southern exposure, somewhat reformed by a road cut (Kuyt, 1991). The sandy nature of this location contrasts sharply with the structure of the mima-type mounds. The mounds have a thick layer of black silt loam overlying dense yellow clay (Ross et al., 1968). A comparison of soil characteristics with depths attained by burrowing toads (Tester and Breckenridge, 1964b) suggests that toads normally enter the dense clay layer.

K. Interspecific Associations/Exclusions. In North Dakota, Canadian toads, American toads, and Woodhouse's toads have complementary distributions (Fishbeck and Underhill, 1960). Canadian toads will hybridize with Great Plains toads in northwestern Minnesota (Brown and Ewert, 1971). Canadian and American toads have high survival in reciprocal hybridization crosses in the laboratory (Blair, 1972b). Natural hybridization with American toads occurs in eastern South Dakota (Henrich, 1968) and has been studied extensively in southeastern Manitoba (Cook, 1983; Green, 1983; Stebbins, 1985; Green and Pustowka, 1997). Based on morphology and allozyme data, the steepest gradient in the transition from *B. hemiophrys* (west) to *B. americanus* (east) is about 20 km across. This zone appears to have drifted nearly 10 km westward (toward Winnipeg) between 1968–69 and 1977–79. Although there is some evidence for hybridization on either side of the transition, the width and character of the overall zone seems to be stable; the introgression is not expanding. Green (1983) and Green and Pustowka (1997) conclude that the zone has natural causes dating back to the retreat of glacial Lake Agassiz during the early Holocene epoch. Hybridization with American toads seldom occurs, if at all, across a sharp forest-prairie transitional zone in northwestern Minnesota, directly west of Itasca State Park (J.R. Tester, unpublished data; M.A.E., personal observations).

L. Age/Size at Reproductive Maturity. Males grow to 56–68 mm and females tend to reach 56–80 mm; hence, females tend to be larger (Wright and Wright, 1949; Breckenridge and Tester, 1961; but see Underhill, 1961). Individuals over 45 mm SVL enter breeding choruses and have thus attained adult size (Tester and Breckenridge, 1964b). This length is achieved during the second summer, and the first opportunity to breed follows winter when the toads are nearly 3 yr old.

M. Longevity. Populations that emerged from overwintering sites (mima-type mounds) at Waubun Prairie fluctuated between approximately 1,000–2,500 toads during the 5-yr period, 1962–66 (Kelleher and Tester, 1969). Post-metamorphic mortality tends to be high, such that 1–2-yr-old juveniles constitute 93% of a spring emergence—only 7% are adults (Tester and Breckenridge, 1964b). Fewer than 1% of yearlings survive to be 6 yr old (Kelleher and Tester, 1969).

N. Feeding Behavior. The bulk of the diet of 16 adults from southern Alberta consisted mainly of ants and secondarily of ground-dwelling beetles, but included flies and insects from six additional orders and some small spiders. Small ground beetles dominated the diet of 19 juveniles from the same region. Small flies, mites, and springtails followed in abundance, and ants were not a large dietary component of these juvenile toads (Moore and Strickland, 1954). Moore and Strickland (1955) concluded that the diet of adult Canadian toads did not vary appreciably from that of adults and subadults of western toads (*B. boreas*), which they had studied personally.

O. Predators. Documented predators of Canadian toads are plains garter snakes (*Thamnophis radix*; Breckenridge and Tester, 1961; Tester and Breckenridge, 1964b), badgers (*Taxidea taxus*; field notes of M.K. Nelson, cited in Breckenridge and Tester, 1961), and red-tailed hawks (*Buteo jamacensis*; Tester and Breckenridge, 1964b). Additional predators likely include raccoons (*Procyon lotor*; Tester and Breckenridge, 1964b) and other mammals, birds, and snakes.

P. Anti-Predator Mechanisms. Canadian toads avoid some kinds of predators by being mostly nocturnal during warm weather. They are also cryptically colored and produce toxic secretions from their parotoid glands. When they are disturbed near water, Canadian toads often move toward the water for escape, where they may swim over 30 m and either float at the surface or dive to the bottom (Breckenridge and Tester, 1961; Underhill, 1961). However, fleeing toads have sought other alternatives on the beach near Lake Manitoba and moved toward upland vegetation rather than toward the open lake (Tamsitt, 1962).

Q. Diseases. Field-collected but laboratory-maintained Canadian toads died from mycotic dermatitis caused by the fungus *Basidiobolus ranarum* (Taylor et al., 1999). Two malformed Canadian toads, one from Minnesota and one from North Dakota, have been reported (NARCAM, 1997).

R. Parasites. In a sample of 40 adult Canadian toads from Alberta, Canada, there was one species of trematode (*Gordoderina simplex*) and three species of nematodes (*Cosmocercoides variabilis, Oswalocruzia pipiens,* and *Rhabdias americanus*). The nematode *R. americanus* had the highest prevalence (73% of the toads sampled), the nematode *C. variabilis* had the highest intensity (average 26 individuals/toad; Bursey and Goldberg, 1998).

4. Conservation.
Despite evidence for species declines in Alberta (Hamilton et al., 1998), Canadian toads retain an overall large geographic distribution. This large distribution clearly distinguishes Canadian toads from their closest relative, critically endangered Wyoming toads. Even though the most proximate threats to Wyoming toads are not clear (see *B. baxteri* account, this volume), Canadian toads seem unlikely to become endangered globally. Montana, which represents the northwestern extreme of the U.S. distribution, protects Canadian toads and lists them as Endangered (reviewed in Hamilton et al., 1998).

Canadian toad populations in the southern portion of their range (i.e., in the United States) experience a periodic (10–11 yr) natural stress from severe droughts. Population contractions and expansions due to the availability of surface water are no doubt a natural occurrence, but droughts likely have a more profound effect on amphibian populations in today's fragmented landscapes than they have had in the past (Lannoo, 1998a).

In the United States, Canadian toads range mainly within the prairie pothole breeding grounds of waterfowl and other wetland birds. This area remains rich in wetlands, as federal and state governments and NGOs have protected habitat for waterfowl. Within the historical range of Canadian toads in northern and eastern North Dakota, there are at least ten separately named National Wildlife Refuges. Minnesota has at least three named National Wildlife Refuges along the eastern edge of Canadian toad range, and South Dakota includes at least two National Wildlife Refuges. It is beyond the present scope to assess how much management within these refuges is optimally compatible with the needs of toads. However, a cursory review of internet information suggests that some wetland-upland combinations (e.g., Type III wetlands with adjacent prairie grass uplands and permeable soils) are maintained. These refuges should buffer Canadian toads against extirpation during extreme droughts.

Information on whether Canadian toads or other amphibians and reptiles actually occur on the refuges is less readily available than the internet postings of resident bird and mammal checklists. Sand Lake National Wildlife Refuge (in South Dakota) lists Canadian toads as present, but Audubon National Wildlife Refuge (North Dakota) does not include them among their four listed amphibian species. It is unclear whether amphibian and reptile surveys have occurred at many of the refuges, although any environmental impact assessment would require an evaluation. In general, keeping with state non-game programs, lists of all vertebrate wildlife should be created and maintained. While amphibians and reptiles did not generate the funding to create these refuges or to maintain them, the problem of declining amphibian populations has generated wide public interest, including visiting refuges, and non-game funding for many species is increasing.

Canadian toads occur along the northern edge of agricultural lands with extensive pesticide applications for row crops. Overlap between toads and pesticide applications seems likely in the Red River Valley of Minnesota and North Dakota. Some pesticides formerly assumed to be harmless are now known to act as endocrine disrupters on amphibians (e.g., atrazine acts as a testosterone suppressor, which in very low concentrations leads to feminized male amphibians; Hayes et al., 2002a). Actual effects of pesticides on Canadian toads remain unknown (Sparling et al., 2000), and this gap in the toxicological literature must be filled.

Acknowledgments. M.A.E. offers warm thanks to John R. Tester for providing guidance, encouragement, and financial support during the author's studies on toads in northwestern Minnesota during the 1960s.

Bufo houstonensis Sanders, 1953
HOUSTON TOAD

Donald B. Shepard, Lauren E. Brown

1. Historical versus Current Distribution.
Houston toads (*Bufo houstonensis*) are considered a Pleistocene relict left behind in Texas after the widely distributed parental species (precursor to the present American toad *[B. americanus]*) withdrew to the north following the retreat of the Wisconsin glaciation 10,000 yr ago (Blair, 1958a, 1965; L.E. Brown, 1971b). Increasing temperatures and aridity during the Holocene were probably instrumental in isolating Houston toads into areas of friable sandy soil where they could burrow to avoid desiccation (Brown and Mesrobian, this volume, Part One).

Only a few scattered populations of Houston toads are known to be extant in central and southeastern Texas (Seal, 1994). Houston toads previously have been reported from the following counties: Austin, Bastrop, Burleson, Colorado, Fort Bend, Harris, Lavaca, Lee, Leon, Liberty, Milam, and Roberston (Sanders, 1953; Blair, 1956a: L.E. Brown, 1971b; Seal, 1994; Brown and Mesrobian, this volume, Part One; Price, 2003). However, they are now extirpated from Harris County (no encounters since 1976) and believed extirpated from Burleson, Fort Bend, and Liberty counties (L.E. Brown et al., 1984; Brown and Mesrobian, this volume, Part One).

2. Historical versus Current Abundance.
Since the 1940s, Houston toad populations have declined drastically. For example, in Harris County, John Wottring, the collector of the holotype, collected 66 individuals from a single chorus in 1949 and indicated large numbers were still present in 1953 (L.E. Brown et al., 1984). In the same county, L.E. Brown (1967, 1971b)

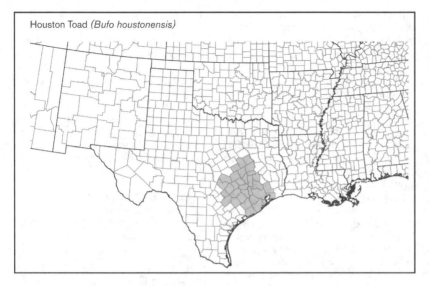

Houston Toad *(Bufo houstonensis)*

found only three individuals from 1965–67; three individuals were found from 1974–76 (L.E. Brown et al., 1984) and no Houston toads have been found since (L.E. Brown et al., 1984). The largest number of toads occurs in Bastrop County, but population estimates have varied considerably among authors and over time (see Seal, 1994; Price, 2003).

3. Life History Features.

A. Breeding. Reproduction is aquatic.

i. *Breeding migrations.* Breeding may occur from late January to late June, but usually earlier than May (Kennedy, 1961; L.E. Brown, 1971b; Hillis et al., 1984; Jacobson, 1989). Kennedy (1961) observed ephemeral calling and spawning triggered by warm temperatures and heavy rainfall. Hillis et al. (1984) found initiation of breeding was not associated with rainfall and began when minimum air temperatures for the preceding 24 hr did not fall below 14 °C. However, males will call below 14 °C (L.E. Brown, 1967). Lack of calling when temperature is suitable and breeding asynchrony among populations experiencing similar environmental conditions suggest other factors besides temperature may influence the timing and intensity of breeding (Hillis et al., 1984).

Around sunset, toads move to breeding sites from underground burrows ≤40 m away. Males will call from burrows prior to migrating to the breeding site and may return to the same burrow the next day (Hillis et al., 1984).

ii. *Breeding habitat.* Houston toads use rain pools, flooded fields, roadside ditches, and natural or manmade ponds for breeding (L.E. Brown et al., 1984). Optimum habitats are non-flowing, fishless pools that persist for at least 60 d (L.E. Brown et al., 1984; Hillis et al., 1984; Price, 2003).

B. Eggs.

i. *Egg deposition sites.* Eggs are deposited in strings in water (Kennedy, 1961; L.E. Brown, 1971b). Within the jelly tube, eggs are separated from one another in compartments (Sanders, 1953; L.E. Brown et al., 1984).

ii. *Clutch size.* Kennedy (1961) reported a female laying 728 eggs while Quinn and Mengden (1984) found clutch size of seven females ranged from 513–6,199. On average, eggs are 1.9 mm in diameter (Quinn and Mengden, 1984). Eggs hatch in about 7 d, but can hatch in as fast as 2 d at warm temperatures (Hillis et al., 1984; Quinn and Mengden, 1984). Hatchlings measure 6.1–6.7 mm TL (Hillis et al., 1984).

C. Larvae/Metamorphosis.

i. *Length of larval stage.* About 53–58 d, depending on water temperature (Hillis et al., 1984).

ii. *Larval requirements.*

a. *Food.* Houston toads can use algae for food and will also feed on loblolly pine *(Pinus taeda)* pollen (Hillis et al., 1984). Where two cohorts of tadpoles co-occur, older tadpoles will eat the jelly envelopes of the discarded egg capsules of the younger cohort (Hillis et al., 1984).

b. *Cover.* Tadpoles form aggregations during the day but disperse at night (Hillis et al., 1984).

iii. *Larval polymorphisms.* Not known to occur.

iv. *Features of metamorphosis.* Tadpoles metamorphose at 20–22 mm TL; newly metamorphosed juveniles are 7–9 mm TL (Hillis et al., 1984).

v. *Post-metamorphic migrations.* Hillis et al. (1984) reported post-metamorphic Houston toads left ponds and moved in large numbers up gullies that fed ponds. These movements occurred during day and night with individuals found ≤100 m from the breeding pond.

D. Juvenile Habitat. Same as adult habitat. Juvenile growth has recently been described by Greuter and Forstner (2003).

E. Adult Habitat. Houston toads are poor burrowers in compact soil (Bragg, 1960a) and are restricted to areas with sandy, friable soil (L.E. Brown, 1971b). These sands are deep, have subsurface moisture, and occur in five geological formations that run in a northeast-southwest direction (Seal, 1994; Brown and Mesrobian, this volume, Part One). Houston toads are nocturnal, spending daylight hours burrowed in sand (Brown et al., 1984).

Many Houston toad populations are located in or near loblolly pine forests (L.E. Brown, 1973a); however, this is due more to both Houston toads and loblolly pines being dependent on sandy soils than the dependence of Houston toads on pine woods (Brown and Thomas, 1982; Brown and Mesrobian, this volume, Part One). Other pristine vegetational associations include mixed deciduous forest, post oak *(Quercus stellata)* savanna, and coastal prairie (Kennedy, 1961; L.E. Brown, 1971b; L.E. Brown et al., 1984; Seal, 1994; Brown and Mesrobian, this volume, Part One; Price, 2003).

F. Home Range Size. Individual males have been observed making round-trip movements greater than 1,400 m between breeding sites (Price, 1992, 2003).

G. Territories. Houston toads are explosive breeders and not known to defend territories (Jacobson, 1989).

H. Aestivation/Avoiding Desiccation. Occurs underground in sandy soils (Seal, 1994).

I. Seasonal Migrations. Unknown.

J. Torpor (Hibernation). Occurs underground in sandy soils (Seal, 1994).

K. Interspecific Associations/Exclusions. Licht (1967a) reported that Houston toad larval growth was inhibited when tadpoles were raised in water conditioned by Woodhouse's toad *(B. woodhousii)* tadpoles, an effect not seen when raised in water conditioned by Texas toad *(B. speciosus)* tadpoles. Houston toads naturally hybridize with Woodhouse's toads and Gulf Coast toads *(B. valliceps* [now considered to be Coastal-Plain toads, *B. nebulifer;* see Mulcahy and Mendelson, 2000; Mendelson, this volume) in the Lost Pines area of Bastrop County (L.E. Brown, 1967, 1971b; Hillis et al., 1984).

L. Age/Size at Reproductive Maturity. Males are sexually mature at 45–70 mm SVL and females at 52–80 mm (L.E. Brown, 1973a; L.E. Brown et al., 1984). Sexual maturity takes 1 yr for males and apparently 2 yr for females (Quinn and Mengden, 1984). Most males breed during their first and second breeding season and rarely breed more than twice (Hillis et al., 1984).

M. Longevity. Snider and Bowler (1992) report a maximum longevity of 4 yr, 3 mo, and 10 d for a wild-caught individual.

N. Feeding Behavior. Houston toads utilize active searching and sit-and-wait methods to acquire prey (Thomas and Allen, 1997). They have been observed to eat beetles (Coleoptera), flies (Diptera), lacewings (Neuroptera), moths (Lepidoptera), ants (Hymenoptera: Formicidae), and other anurans (Bragg, 1960a; Thomas and Allen, 1997).

O. Predators. Freed and Neitman (1988) reported tadpole predation by blotched water snakes (*Nerodia erythrogaster transversa*) and Gulf Coast ribbon snakes (*Thamnophis proximus orarius*), and toadlet predation by fire ants (*Solenopsis invicta*). Brown et al. (L.E., 1984) reported predation on adults by blotched water snakes and discussed other potential vertebrate predators.

P. Anti-Predator Mechanisms. Tadpoles form aggregations during daytime hours (Hillis et al., 1984). The eggs and tadpoles of many toad species are unpalatable to predators (Voris and Bacon, 1966; Licht, 1968), and adults have enlarged parotoid glands that produce toxins (L.E. Brown, 1973a).

Q. Diseases. Unknown.

R. Parasites. *Cosmocercoides dukae, Oswaldocruzia pipiens, Physaloptera ranae, Rhabdias ranae,* and *Brachycoelium storeriae* (Harwood, 1932; Thomas et al., 1984).

4. Conservation.
Houston toads are federally Endangered (Peters, 1968; Gottschalk, 1970). Since the 1940s, Houston toad populations have declined drastically (L.E. Brown et al., 1984), and only a few scattered populations are known to be extant in central and southeastern Texas (Seal, 1994; Price, 2003). Populations are considered relictual and population sizes appear to be small (L.E. Brown, 1973a). Primary causes of the decline have been urbanization, prolonged droughts during the 1950s, and other types of anthropogenic habitat modification. Major threats to the Houston toad include continued urbanization, recreational over-development, road mortality, agriculture, and deforestation. Introduced red fire ants, known predators of recently metamorphosed toads (Freed and Neitman, 1988), and increasing temperatures and aridity associated with recent warming trends may also threaten the future survival of the Houston toad.

Past conservation efforts have included 1970s land acquisitions by the State of Texas within designated critical habitat in Bastrop County adjacent to Buescher and Bastrop state parks (L.E. Brown et al., 1984). Additionally, an effort was started in 1978 by the Houston Zoo to identify remaining Houston toad populations and supplement them or establish new populations in protected areas using wild caught adults, naturally deposited eggs, or captive-reared juveniles and adults. However, in spite of introducing 500,000 individuals (adults, juveniles, tadpoles) since 1982 into sites at the Attwater Prairie Chicken National Wildlife Refuge, new populations have not been established (Dodd and Seigel, 1991). Conflicting interests have plagued the conservation of Houston toads and created frequent problematic political issues (reviewed by Brown and Mesrobian, this volume, Part One).

Bufo marinus (Linnaeus, 1758)
MARINE TOAD, CANE TOAD

Jean-Marc Hero, Melody Stoneham

1. Historical versus Current Distribution.
Marine toads (*Bufo marinus*) are native from Sorona to Tamaulipas in Mexico in an area forming a continuous arc into the Orinoco and Amazon River Basins in South America (Tyler, 1975; Easteal, 1986); they are also native to extreme southern Texas (Easteal, 1986). They have been introduced to most tropical regions as a control for agricultural pests (Von Volkenberg, 1935; Oliver, 1949, 1955b; Mead, 1961; Krakauer, 1968; Easteal, 1981, 1986; Evans et al., 1996), including Jamaica and the Philippines in the late 1800s to control rats, and Puerto Rico, Fiji, New Guinea, America, and Australia in the early 1900s to control sugar cane pests (Freeland, 1985). In 1919, marine toads were introduced to Puerto Rico. By the early 1930s their numbers had grown and they were distributed both inland and on the coast (Grant, 1931). The demise of white-grub (*Phyllophaga* sp.) in Puerto Rico's cane fields was attributed to the introduction of marine toads in the 1920s (Freeland, 1985). However, it is unknown whether this decline was a result of toad predation or due to unusual weather conditions that prevented the emergence of *Phyllophaga* pupae (Freeland, 1985). It was the assumption that marine toads were a successful biological control for cane pests that led to their introduction throughout the Pacific Basin (Freeland, 1985).

While native to extreme southern Texas, marine toads were introduced elsewhere in the United States (Florida and Hawaii) to control insect pests. In Florida, specimens from Puerto Rico were released in 1936 at Canal Point and Belle Glade, Palm Beach County, as a control for sugar cane pests (Lobdell, 1936; Krakauer, 1968). These introductions, as well as introductions into the Florida interior were unsuccessful (Reimer, 1958; Krakauer, 1968). Present populations in Florida were established from accidental releases at Miami International Airport, where they remained until the completion of a canal dug in 1958 that linked the airport rock pits to the extensive south Florida canal system (Krakauer, 1968). They are also sold as pets, and releases or escapes facilitate range expansion (Bartlett and Bartlett, 1999a). Altitudinal limits vary from sea level to 1,600 m (in Venezuela; Zug and Zug, 1979; see also Easteal, 1986).

In 1935 in Australia, marine toads (known in Australia as cane toads) were introduced to Gordonvale, just south of

Marine Toad (*Bufo marinus*)

Hawaiian Islands

Cairns on the east coast of Queensland, to control cane beetles (grey-backed *[Dermolepida albohirtum]* and Frenchi beetles *[Lepidiota frenchi*; Straughan, 1966; Freeland, 1985; Alford et al., 1995a,b]). Their failure to control cane beetles and their subsequent spread north, south, and west at rates of 25–30 km/yr quickly led to their current status as a pest species (Grigg, 2000). By the 1950s, marine toads had spread throughout most of the eastern seaboard of Queensland and northern New South Wales; in 1986, they had reached Calvert Hills Station in the Northern Territory (Alford et al., 1995a). To date, marine toads have colonized 500,000–785,000 km^2 of eastern Australia, including 50% of Queensland, and continue their northwesterly advance at 27–40 km/yr and their southerly advance at 1.07–5 km/yr (Beurden and Grigg, 1980; Sabath et al., 1981; Freeland and Martin, 1985; Sutherst et al., 1995; Caneris and Oliver, 1999). In northern Australia, marine toads reached Mataranka (40 km south of Darwin, Northern Territory) in 1999 (Caneris and Oliver, 1999) and entered Kakadu National Park (a World Heritage area) in the summer of 2000/2001.

The potential geographical distribution of marine toads in Australia has been predicted based on climatic conditions around the country and tolerance limits of the toad (Sutherst et al., 1995). These workers predicted that marine toads have the potential to expand their range as far south as Bega, near the New South Wales and Victorian border, and across the top of Australia to Broome in Western Australia (Sutherst et al., 1995). Temperature and rainfall extremes are foreshadowed to prevent marine toad occupation of other regions (Sutherst et al., 1995).

2. Historical versus Current Abundance. Because marine toads have been introduced to parts of the United States, Australia, and throughout the Indo-Pacific region, they are more abundant now than they have been historically. In Oahu, Hawaii, the number of individuals increased from 148 to ≥100,000 in 2 yr (Oliver, 1949).

In Australia, 101 adult marine toads were introduced in 1935 (Freeland, 1985, 1986). The number of toads currently in Australia is unknown, but populations have been observed to increase dramatically within a short period, and the density of adult toads in Australia is now greater than that in their native homeland, South America (Alford et al., 1995b). Long-term studies on toad populations in Townsville, Queensland, show that populations fluctuate greatly between seasons with males reaching higher densities in the early to mid-wet season, females in early dry season, and juveniles late in the wet season (Alford et al., 1995a). Compar-

isons of old and newly established toad populations indicate that toads exhibit a "boom and bust" population growth pattern common to many pest species (Olding, 1994). For example, immediately after colonization the population increases exponentially (boom) until it reaches a peak, and then it declines (bust) and finds a "natural" level (Olding, 1994). Populations also contain a higher proportion of juveniles than adults and more males than females (Alford et al., 1995a); despite being introduced in relatively small numbers, the populations of marine toads in Australia and the United States have been shown to be genetically divergent (Easteal, 1985).

3. Life History Features.
   A. Breeding. Reproduction is aquatic.
   i. *Breeding migrations.* In Australia, male toads begin to move toward breeding sites from aestivation sites when the temperature begins to rise after winter—usually about August–September (Tyler, 1975). It is not known when the females arrive at the breeding sites, but they only appear at these sites when their oocytes have matured and they are ready to mate (Floyd and Bendow, 1984).
   ii. *Breeding habitat.* Includes brackish water: ". . . low concentrations of sea water constitute a favorable environment for the development of *B. marinus* larvae" (Ely, in Wright and Wright, 1949). Within their natural range in Venezuela, breeding habitat is determined by the transparency and pH of the water (with preferences for clearer water with a higher pH), the density of the vegetation surrounding the water body (with preferences for less dense vegetation), and permanence of the waterbody (with preferences for temporary, shallow, and warm water; Hero, 1992; Evans et al., 1996). The size and abundance of fish is apparently not important in breeding site choice (Hero, 1992; Evans et al., 1996). In central Amazonas, Brazil, marine toads breed in ephemeral ponds, permanent lakes, or large temporary ponds and can breed in sites with fish (Hero, 1990, 1992; Azevedo-Ramos, 1992). In the United States., marine toads are associated with and will breed in a variety of wetland sites, both natural and man made (Meshaka, this volume, Part One).

The breeding season varies between seasons and geographic location. In Florida, females have mature eggs throughout the year and have been observed breeding all through the year within their natural range—with a tendency towards spring and summer (Krakauer, 1968). Chorusing is sporadic until late March, when it becomes nightly through early September (Krakauer, 1968). Choruses can occasionally be heard during the day, in bright sunshine. Wright and Wright (1949) note that there is controversy about whether a female can breed once or twice a year.

In Australia, toads breed from September–March (peaking in January) in static or slow-flowing water (Tyler, 1975). Breeding sites are usually relatively free from dense ground vegetation, and while only large toads will crawl through dense ground vegetation to get to ponds, they generally avoid these habitats (Hero, 1994; Olding, 1994). Marine toads prefer sparse, patchy fringing vegetation (Dickman, 1991). Adult toads have also been shown to prefer breeding habitats where there are trees within 5 m of the pond for shelter (Hero, 1994; Olding, 1994). Other studies have noted that there are substantially more toads in areas cleared of ground vegetation and canopy cover (such as paddocks) than in naturally vegetated areas (Dickman, 1991). Their presence in natural areas is usually facilitated by tracks that cut into the vegetation (Dickman, 1991).

   B. Eggs.
   i. *Egg deposition sites.* Temporary to permanent, fresh to brackish waters (Wright and Wright, 1949). Evans et al. (1996) report that sites with high water transparency, high density of macrophytic vegetation, and neutral pHs are preferred. Females select a mate and may travel for days with him in amplexus before depositing eggs (Reed and Borowsky, 1967). One pair was recorded to have been in amplexus for 2 wk prior to breeding and several days after breeding (Reed and Borowsky, 1967). In Guyana, females (carrying males) have been observed constructing shallow nests filled with water, usually in sandy areas at the edges of pools (Reed and Borowsky, 1967).
   ii. *Clutch size.* Bartlett and Bartlett (1999a) state females can lay up to 20,000 eggs in long jelly strings, which are usually attached to submergent or emergent vegetation. Oliver (1949) gives a value of 10,000 eggs/female; Crump (1974) found a range between 4,240–12,700 eggs in Ecuador. Zug and Zug (1979) found a range between 6,050–23,000 eggs in Panama; Hearnden (1991) found a range between 6,970–36,100 eggs in northern Australia. The number of eggs/female increases with body size (Alford et al., 1995b).
   C. Larvae/Metamorphosis. Tadpoles emerge from the jelly surrounding the eggs approximately 48–72 hr after they are laid (Tyler, 1975).
   i. *Length of larval stage.* Tadpoles metamorphosed about 45 or 46 d following egg laying (Ruthven, 1919, in Wright and Wright, 1949), or 27 d post-hatching (Wright and Wright, 1949). In Australia, marine toad cohorts were observed to take between 37–40 d to reach metamorphosis from eggs (Alford et al., 1995b). The length of the tadpole stage has been known to vary considerably, between 10 d (J.-M.H., unpublished data) and 6 mo (Tyler, 1975). This may reflect differences

between climatic zones, environmental factors, competition, or a lack of food available to tadpoles, therefore delaying metamorphosis (Tyler, 1975). A correlation between density and growth rates exists, such that tadpoles at lower densities tend to metamorphose more quickly than tadpoles in higher densities (Alford et al., 1995a). In Australia, marine toad tadpoles have been observed at densities of between 15–61/m^2 (Alford et al., 1995a). The pressure from competition may be higher in this species because of their tendency to travel in large aggregations.

ii. Larval requirements. Descriptions of tadpoles and geographic variation in tadpole morphology, are given by Savage (1960a). Tadpoles will develop in brackish water at about 5% to <10‰ sea water (similar concentrations are also given in Ely, 1944). In 1971, experiments were carried out on the resistance of marine toad tadpoles to desiccation. These experiments determined that pre-metamorphic toad tadpoles can survive without water for 10 hr, provided the substrate is damp (Valerio, 1971). This indicates a limited ability to survive through pond drying or drought.

Marine toads can tolerate temperatures between 16.8 °C and 42 °C (Krakauer, 1970). Although in the lab tadpoles recovered after 24 hr at 0.4 °C, below 16.8 °C they became movement impaired and would not survive prolonged periods in this state due to the cessation of feeding (Krakauer, 1970). Tadpoles have highest survival rates at temperatures around 29 °C (Floyd, 1983), and larger tadpoles tend to be more tolerant of changes in temperature (Floyd, 1984). Earlier studies recorded tadpoles surviving prolonged periods in temperatures between 8 °C and 43.7 °C (Heatwole et al., 1968). The turbidity of the water also appears to influence the distribution of toad tadpoles, with tadpole abundance decreasing as turbidity increases (Olding, 1994).

a. Food. Toad tadpoles eat algae (Hinckley, 1962). The availability of food is known to affect the time required for tadpoles to become toadlets (Tyler, 1975). Competition for food becomes an issue at higher densities (Alford et al., 1995a).

b. Cover. As they prefer warmer water and are adapted to high water temperatures, marine toad tadpoles are often exposed in pools with little to no vegetative cover beside or in the water body (Krakauer, 1970). Tadpoles remain active throughout their development and swim in large aggregations mid-water column (Straughan, 1966).

iii. Larval polymorphisms. Under natural conditions, studies have revealed that the main source of predation on hatchlings appeared to be marine toad tadpoles from earlier cohorts (Alford et al., 1995a). For example, hatchling survival in the presence of older tadpoles was reduced to 1.7%

(from 88% without predators). Therefore, it would appear that cannibalistic behavior develops at mid to late tadpole stages. Interestingly, marine toad tadpoles do not prey heavily upon tadpoles of other species (Crossland, 1998a).

iv. Features of metamorphosis. In the United States, metamorphosis takes place from late spring to midsummer, although time to metamorphosis is known to vary considerably (Tyler, 1975). This variation is due to factors such as differences between climatic zones, competition, tadpole food availability (Tyler, 1975), and tadpole density (Alford et al., 1995a).

In northern Australia, tadpoles metamorphose in October to early April. The size at metamorphosis is relatively small (11 mm SVL) and toadlets are quite underdeveloped upon emergence (Cohen and Alford, 1993). Their weight represents only 0.01% of the eventual adult mass (Cohen and Alford, 1993). Emergence at such a small size is thought to reduce the risk of larval mortality through desiccation, however it may lead to high mortality of newly metamorphosed animals (Cohen and Alford, 1993).

v. Post-metamorphic migrations. Newly metamorphosed marine toads require easy access to water to facilitate gas exchange and to prevent desiccation, therefore they often stay within 1–5 m of the water source (Straughan, 1966; Alford et al., 1995a). Within 3–4 d following metamorphosis, juveniles (<30 mm) disperse from the banks of their breeding canals and lakes and do not return until they are about 90 mm (Alford et al., 1995a). As toads get older and larger they are found at greater distances from water (Alford et al., 1995a). Only 10–47% of metamorphosed toads will survive through their first dry season (Alford et al., 1995a).

Studies have indicated that juvenile toads act as dispersalists and colonizers within the toad life cycle (Freeland and Martin, 1985). For example, surveys in northern Australia found two immature toads <90 mm) approximately 12 mo before the establishment of substantial populations, and similar occurrences have been recorded in western Queensland (Freeland and Martin, 1985). Other studies have suggested that as shelter positions are taken up by adults at the onset of the dry season, juvenile toads are forced to move away; when this happens at the edge of the toads distribution, new breeding sites are established (Straughan, 1966).

D. Juvenile Habitat. Newly metamorphosed marine toads are primarily diurnal until they are about 3–4 d old, at which time they establish nocturnal activity patterns (Krakauer, 1968). Post-metamorphic toads (<30 mm) generally do not travel far from a water source (remaining within 0–5 m of the water's edge) because their heart, lungs, and aerobic capacity are poorly developed and most gas exchange

occurs across the skin (which must be moist for respiration to occur; Cohen and Alford, 1993). The necessity to be in such close proximity to a water source means that the density of newly metamorphosed toads in this area is high and detracts from survival and growth rates of toads (Cohen and Alford, 1993). As the toads develop they are generally able to move further away from the water source, but will return for the duration of the wet season (Cohen and Alford, 1993).

Juveniles (30–70 mm) have different habitat preferences and activity patterns than do newly metamorphosed animals and adults (Krakauer, 1968). Juveniles are found in lawns or associated with buildings, where they emerge at dusk and are active at night—they are frequently found under lights, feeding on the insects attracted to the light (Krakauer, 1968).

E. Adult Habitat. Adults are tolerant of humans and found in gardens, around houses, and in water tanks (Wright and Wright, 1949). Krakauer (1968, 1970) notes that marine toads are frequently found in disturbed areas and rarely encountered in undisturbed habitats. Marine toads are nocturnal and attracted to house and patio lights that also attract the insects on which toads feed (Wright and Wright, 1949). Toads are only active 1 out of every 3, 4, or 5 nights (Brattstrom, 1962a; Zug and Zug, 1979; Floyd and Benbow, 1984), and their activity tends to be correlated with rain (Floyd and Benbow, 1984). During the day, the toads are secretive, hiding under rocks and boards, in burrows (Wright and Wright, 1949), and under long grass clumps out of direct sunlight (Cohen and Williams, 1992).

As their name suggests, marine toads are generally found along rivers and coasts in association with fresh and/or brackish water, including mangrove swamps. In a study by Krakauer (1970), adult toads were found to survive in 10‰ sea water, but quickly died in 15‰ sea water.

Johnson (1972) found that marine toads also have a broad temperature tolerance and could survive at temperatures from 5–41.8 °C. These temperature tolerance limits influence the altitudes and latitudes where the toad is found (Brattstrom, 1968). The immediate response of a toad to heat stress is to escape; failing this, they are often found floating in water with their lungs inflated and their heart rates raised (Stuart, 1951; Novotney, 1976; Sherman, 1980). As temperatures reach the lower limits of tolerance levels, marine toads become less active and lose reflexes (Stuart, 1951; Novotney, 1976; Sherman, 1980). Temperature also has an influence on the respiration of marine toads. Experiments reveal that toads generally rely on pulmonary (lung) respiration (as opposed to cutaneous respiration) more than co-occurring tropical frogs (Hutchison et al.,

1968). However, as the temperature increases, marine toads rely increasingly on cutaneous respiration for gas exchange (Hutchison et al., 1968).

F. Home Range Size. Variable, dependent to an extent on the size of their water bodies and feeding sites (Brattstrom, 1962a; Carpenter and Gillingham, 1987). Displaced animals will home (return to their capture site), with local landscape and visual cues providing the key inputs for orientation (Brattstrom, 1962a). In Queensland, Australia, mark–recapture studies were done using several toads over a period of ten nights. The average minimum home range was calculated at 340 m^2 (Pearse, 1979). A similar study by Zug and Zug (1979) found that at least some toads were familiar to an area of 2,812 m^2. Spooling studies on marine toads have shown that they rarely move >25 m away from the water's edge, but some adult toads were spooled a distance of 200 m (Cohen and Williams, 1992). The greater distance traveled by adult marine toads may be related to their ability to jump further than juveniles. For example, as the body size of marine toads increases, the distance they can jump also increases by an equivalent amount, such that a toad twice as large as another can jump twice as far (Rand and Rand, 1966).

G. Territories. Marine toads do not establish defended territories during the reproductive season at the breeding site or in the terrestrial foraging zone (Sabath, 1980). Adult toads display some fidelity to shelter sites and prefer shelters with high soil moisture (they often increase soil moisture by urinating on the soil; Alford et al., 1995a). Adult toads seem to establish long-term foraging territory associations, and therefore it is more likely that newly metamorphosed and juvenile toads act as the dispersalists in the life cycle of the toad (Sabath et al., 1981). Juvenile toads are often excluded from breeding sites at the onset of the dry season, as all appropriate shelter sites are taken up by adult toads (Straughan, 1966). The juveniles then move away to establish new breeding colonies (Straughan, 1966).

H. Aestivation/Avoiding Desiccation. Marine toads aestivate during the dry season under boulders along rivers, under leaf litter, in old burrows of other animals, under long grass, and in hollow logs (Straughan, 1966). Captive specimens from Guyana kept in aquarium conditions were observed to dig nests in moist soil and bury themselves so only the eyes and top of the head were visible (Reed and Borowsky, 1967). Marine toads can lose 52.5% of body water before desiccation and will store water in their bladder. They therefore have the ability to survive for long periods without water (Krakauer, 1970).

I. Seasonal Migrations. The only migrations made by marine toads are to and from breeding sites at the onset and the closure of the wet/breeding season.

J. Torpor (Hibernation). Marine toads are intolerant of freezing conditions, and low temperatures appear to be restricting their spread northward and inland in Florida (Krakauer, 1968). Juveniles living near buildings and suburban lawns become inactive for long periods during cold weather.

K. Interspecific Associations/Exclusions. In their natural habitat in Venezuela, toads are known to co-occur with 21 other species, but usually occur by themselves or with one other species at any one time or location (Hero, 1992). This suggests that marine toads select waterbodies at times or locations when other species are absent, or that other species are avoiding marine toads (Hero, 1992). In 12% of cases, marine toads co-occur with a South American treefrog species (*Hyla crepitans*), whose larvae are known to consume anuran eggs and therefore may be a potential predator of marine toads (Hero, 1992).

In the United States, marine toads occur in regions with southern toads (*B. terrestris*), but unlike marine toads, southern toads are found in drier pine lands and on drier ground within the Everglades.

L. Age/Size at Reproductive Maturity. Toads are able to reproduce from 66–220 mm, with males averaging about 13 mm shorter than females (Wright and Wright, 1949; Easteal, 1986). Bartlett and Bartlett (1999a) note that animals in northern populations are smaller and speculate that cooler winter temperatures may inhibit growth. It takes 1 yr for toads to reach reproductive maturity in tropical regions and 2 yr in temperate zones (Easteal, 1982). In northern Australia, toads must be from 65–90 mm and usually in their second wet season before they are capable of reproduction (Cohen and Alford, 1993). Rapid growth follows emergence and lasts through the wet season but slows at the approach of the dry season, probably reflecting food availability (Zug and Zug, 1979). Once the toads reach adult size, little growth is experienced (Zug and Zug, 1979).

M. Longevity. The lifespan of toads under natural conditions is not known (Tyler, 1975). An age of 40 yr has been attributed to a similar species of *Bufo* in captivity (Tyler, 1975). There are two records of captive marine toads surviving beyond 15 yr (Tyler, 1975). A female marine toad held in captivity to determine longevity lived for 15 yr, 10 mo, and 13 d (Pemberton, 1949).

N. Feeding Behavior. Marine toads feed on a wide variety of prey, especially terrestrial arthropods, including tenebrionid and carabid beetles (Taylor and Wright, in Wright and Wright, 1949; Krakauer, 1968), crabs, spiders, centipedes, millipedes, scorpions (Easteal, 1982), and cockroaches (Easteal, 1986; see list in Krakauer,

1968; see also Rabor, 1952 [Phillipines] and Strüssmann et al., 1984 [Brazil]). In one study, the most popular prey items in the stomach contents of 100 toads were ants, bees, caterpillars, millipedes, beetles, snails, bugs, slugs, and leafhoppers (in that order; Hinckley, 1962). The proportions of prey consumed largely reflect the availability of the prey at that time (Hinckley, 1962). A study on stomach contents in northern Australia showed ants and beetles to be the most popular prey items (Cohen and Williams, 1992). Krakauer (1968) noted a low percentage of empty stomachs.

Marine toads are considered to be nonspecific and aggressive predators and will occasionally consume native frogs and toads, even dog food and feces (Alexander, 1964; Tyler, 1975; Rossi, 1983; Bartlett and Bartlett, 1999a). Other prey include snakes (Rabor, 1952), birds (Krakauer, 1968), and mammals (Oliver, 1949). Alexander (1964) notes that individuals ingest plant material and repeats Oliver's (1955b) observation of toads being killed by strychnine after ingesting fallen blossoms of the strychnine trees. A study by Ingle and McKinley (1978) on the effects of stimulus on prey-catching behavior found that striking behavior in marine toads was more commonly elicited by dark objects rather than lightly colored objects and that the toads struck at the leading edge of moving objects representing prey. While many bufonids appear to rely on visual cues to detect and capture prey, marine toads differ in also using strictly olfactory cues (Tyler, 1975; Rossi, 1983). The prey that a toad will eat is largely limited by the gape of its jaws and the distention of its stomach (Tyler, 1975).

O. Predators. Due to the presence of noxious chemicals in all stages of the toad's life cycle, marine toads have few predators. At the tadpole stage, Australian studies have revealed that several species of native dragonfly naiads will readily consume marine toad tadpoles and eggs, as will dytiscid beetles, water scorpions (*Lethocerus sp.*), notonectids (*Anisops* sp.), leeches, tortoises, *Macrobrachium* sp., and crayfish (*Cherax quadricarinatus*; Crossland, 1992, 1993; Alford et al., 1995a). Native fishes have been found to ignore or taste and reject toad tadpoles unharmed (Alford et al., 1995a; Lawler and Hero, 1997). The most frequent predators of toad eggs and tadpoles, however, are older cohorts of marine toad tadpoles (Alford et al., 1995b).

Toads may be most vulnerable to predation immediately following metamorphosis, whilst the development of terrestrial skin glands is occurring (Cohen and Alford, 1993). Although there are no studies on predators of newly metamorphosed toads, several animals have been observed to eat them, including adult marine toads, ants, centipedes, wolf spiders, small mammals, and some birds (e.g., *Ibis* sp.; Cohen and Alford, 1993).

Predators of adult toads include small mammals (Krakauer, 1968; Cintra, 1988; Garrett and Boyer, 1993), snakes, including common garter snakes (*Thamnophis sirtalis*; Licht and Low, 1968, and references therein), and birds (Krakauer, 1968). Automobiles are likely the major source of mortality in Florida (Krakauer, 1968).

In Australia, several vertebrate species have been observed to eat juvenile and adult marine toads, including fork-tailed kites (Lavery, 1969; Mitchell et al., 1995), ibises (Goodacre, 1947), koels (Cassels, 1970), tawny frogmouth owls (Freeland, 1985), crows, common rats (Adams, 1967), and white-tailed water rats (St. Cloud, 1966). These animals have apparently learned to flip the toad on its back, slit its belly open and eat its insides, therefore avoiding the toxic skin (Freeland, 1985). In northern Australia, keelback snakes (*Amphiesma mairii*) are unaffected by marine toad poison (Freeland, 1985) and readily consume juvenile toads in preference to native frog species (J.-M.H., unpublished data). The mortality of juvenile and adult toads in South America (87%/yr) is much greater than that in Australia (30–70%/yr) due to the larger number of co-evolved aquatic and terrestrial predators (Alford et al., 1995a).

P. Anti-Predator Mechanisms. Marine toads are highly poisonous and secrete a whitish, viscous compound from their parotoid glands (in Wright and Wright, 1949; Allen and Neill, 1956; Licht, 1967b; Easteal, 1986). The parotoid glands produce and store a mixture of bufotenine and epinephrine—steroid-like substances that are toxic to most animals (Chen and Osuch, 1969; Freeland, 1986). Bartlett and Bartlett (1999a) describe the head-down defensive position marine toads assume to present their parotoid glands to potential predators. These toads are known to approach potential predators and attempt to force contact with their parotoid glands. Toad-eating snakes have apparently evolved tolerances to bufonid parotoid gland venom (Licht and Low, 1968).

Toad eggs and tadpoles are also known to be toxic, although there are ontogenic shifts in palatability and toxicity such that older tadpoles are less palatable and more noxious than younger ones (Azevedo-Ramos, 1992; Lawler and Hero, 1997; Crossland, 1998b). This shift in palatability coincides with the development of poison-producing glands in the skin (Crossland, 1998). Unpalatability offers marine toad tadpoles protection from most aquatic vertebrate predators, especially fish. The conspicuous dark color of toad tadpoles makes them easy to recognize, and it is thought that fish learn to avoid them (Lawler and Hero, 1997).

Experiments on the effects of *Bufo* toxins on Australian and Brazilian tadpoles reveal that tadpoles native to Brazil (where marine toads are endemic) readily consume marine toad tadpoles without any ill effect (Crossland and Azevedo-Ramos, 1999). However, native Australian tadpoles were found to display varied behaviors towards marine toad tadpoles and consequently had varied mortality rates (Crossland and Azevedo-Ramos, 1999). For example, marine toad tadpoles were avoided by most *Litoria alboguttata, L. gracilenta,* and *L. rubella* tadpoles. These tadpoles had a high rate of survival in the presence of toxic toad tadpoles. Of the native Australian tadpoles that consumed marine toad tadpoles, *Bufo* toxins were only fatal to half the tadpoles of *L. alboguttata and Cyclorana brevipes and always toxic to Limnodynastes ornatus and Litoria gracilenta.* Crossland and Azevedo-Ramos (1999) suggested that the differences in the responses of Brazilian and Australian tadpoles to toxic marine toads may result from differences in their evolutionary histories of exposure to *Bufo* toxins.

Tadpoles of native species that prey on the noxious toad eggs have been found to suffer high mortality rates, between 60% (*Litoria nigrofrenata*) and 100% (*L. bicolor and L. infrafrenata*) within a span of 24 hr (Crossland, 1992; Crossland and Azevedo-Ramos, 1999). Small tadpoles have higher survival rates than large tadpoles in the presence of marine toad eggs, as small tadpoles cannot effectively penetrate the jelly surrounding the egg strings to graze on the toxic eggs (Crossland, 1998c). There is also some suggestion that the toxins within toad eggs can leach out into the water body and poison potential predators, although this theory remains to be verified (Crossland, 1992). Experimentation on native Australian aquatic predators have found that toad eggs are always lethal to snails and fish, but that notonectids and leeches experienced differential mortality, and nepids, dytiscid larvae, belostomatids, and crustaceans were unaffected (Crossland and Alford, 1998). Some invertebrate species, including dytiscid beetles, dragonfly naiads, and crayfishes, also seem to be unaffected by the unpalatability and toxicity of toad tadpoles (Crossland, 1992, 1998; Alford et al., 1995a). These predators have piercing and sucking mouthparts, and either avoid the glands in the skin that produce the toxins or simply lack the ability to taste (Crossland, 1998).

Q. Diseases. Speare (1990) lists the diseases that have been found in marine toads. A fatal disease of unknown etiology arose in a Philippine population (Alcala, 1957). This disease was also observed by Tyler (1975) in New Britain in 1967. The individuals suffering from this disease appeared emaciated and ultimately died (Tyler, 1975). Upon dissection they were found to have food in their stomachs, however the liver and some muscles were atrophied (Tyler, 1975). Some heritable diseases occur in toads that cause myotonia (Bretag et al., 1980). This disease affects muscle membranes and often results in muscle spasms and stiffness (Bretag et al., 1980). Animals in this condition suffer reduced movement and may starve to death. Research has also discovered six iridoviruses in toads from Venezuela (Hyatt et al., 1995). In experiments that involved bathing native Australian frog spawn and toad spawn in inoculum, substantial mortality was observed in toad spawn (Hyatt et al., 1995). The reason for the survival of the Australian frog spawn is unknown (Hyatt et al., 1995). Marine toads have been shown to act as a host for ranaviruses that infect native fish and amphibians (Hyatt et al., 1995). Marine toads are also known to be a disease vector in areas with poor hygiene standards, where some toads have been found to harbor *Salmonella* (Tyler, 1975).

Chytrid fungus is also lethal to marine toads and is reported to have had an important role in the disappearance of many native frogs in the Australian Tropics and Panama (Hyatt et al., 1995).

R. Parasites. Lehmann (1967; see also Easteal, 1986) found two blood parasites in marine toads. Kloss (1974, in Easteal, 1986) reported on the nematodes of marine toads, including the round worm *Ascaris lumbricoides*. Marinkelle and Willems (1964) commented on the toad's potential to act as a vector of the eggs of *A. lumbricoides*, which may then infect small mammals. In laboratory experiments, *Rhabdias sphaerocephala*, a parasitic worm found in the lungs of *Bufo*, was found to be both highly infectious and fatal to toads (Williams, 1960). Interestingly, further experimentation showed that treefrogs were more resistant to the worm (suffering smaller parasitic loads) and were not a natural host to *R. sphaerocephala* (Williams, 1960). Brooks (1976a) reported five species of platyhelminths from marine toads. The trematode *Mesocoelium danforthi* is believed to have reached the West Indies through its marine toad host (Tyler, 1975). Levels of endoparasite infection rates are greater in South America than in Australia (Alford et al., 1995b). Marine toads have also been shown to act as a host for endoparasites that infect native fish and amphibians (Barton, 1995).

In 1995, a microsporidian was discovered in tadpoles and post-metamorphic toads. It was previously thought to be a hyperparasite (occurring in trematodes within toads; Paperna and Lainson, 1995). This parasite forms cysts within the gut walls, spleen, and kidney and is passed on via cannibalism of dead tadpoles (Paperna and Lainson, 1995). It is not known to be fatal, as infected tadpoles survive and successfully metamorphose and the infection soon disappears as the toad develops (Paperna and Lainson, 1995).

In South America, the presence of debilitating ticks (*Amblyomma dissimile* and *A. rotundatum*) reduces survival and fecundity of adult marine toads, having a large impact on toad numbers (Lampo, 1995).

4. Conservation.
In the United States, marine toads are apparently native to southern Texas but were introduced to Florida and Hawaii to control insect pests. They continue to be sold as pets, and releases and escapes facilitate range expansion. Outside of their native range, marine toad populations should be considered introduced or invasive, and attempts should be made to control them.

Krakauer (1968) suggests that competition between marine toads and other anurans may be minimal. Krakauer (1968) also notes the rapid expansion of the Miami metropolitan area is destroying habitat for southern toads while creating habitat for marine toads.

In Australia, marine toads are known to co-occur with several other species of anurans, and their effects on these native species is widely varied. For example, when raised with tadpoles of *Limnodynastes ornatus*, marine toad tadpoles suffer greatly reduced growth and fail to survive to metamorphosis (Alford et al., 1995a). However, when raised with *Litoria rubella*, *Limnodynastes terareginae*, *Limnodynastes tasmaniensis*, and *Notaden bennetti*, this situation is reversed and the native tadpoles fail to reach metamorphosis (Alford et al., 1995a). Marine toad tadpoles have not been found to exert predation pressure on native populations, and experiments have shown them to eat very few native Australian frog eggs, hatchlings, or tadpoles in comparison to native species of anuran larvae (Crossland, 1998a). However, many species of native tadpoles are known to suffer high mortality via direct consumption of marine toad tadpoles and eggs and show varying ability to detect and avoid marine toad toxins (Crossland, 1992; Alford et al., 1995a; Crossland, 1995). Some frogs may avoid using breeding sites used by marine toads (Williamson, 1995). For example, pond experiments have found the presence of toad tadpoles significantly reduces populations of predatory *Limnodynastes ornatus* tadpoles (Crossland, 2000). This in turn has an effect on other species of native tadpoles that co-occur with *Limnodynastes ornatus*, because it reduces the predation pressure they would normally suffer, and species such as *L. rubella* have been shown to have increased survivorship (Crossland, 2000). In northern Australia, native fish tend to learn to avoid marine toad eggs and tadpoles (Crossland, 1995; Lawler and Hero, 1997).

There is no conclusive evidence to support the theory that in Australia adult marine toads have had a negative impact on native adult frog populations, although

predation and competition for food, shelter, and breeding sites is a probable result of marine toad introductions (Freeland, 1985; Williamson, 1995; Grigg, 2000). Some frogs may avoid using breeding sites used by marine toads (Williamson, 1995). In the Northern Territory, Australia, research is currently underway to assess the impact of toads on native frog populations. This research involves monitoring breeding activity of native frogs before and after the arrival of marine toads (Grigg, 2000).

Marine toads have also had an impact on a number of endemic Australian predators, including goannas/ monitors (*Varanus* sp.), native 'cats' or quolls (Dasyuridae), several snakes (brown snakes, death adders, and tiger snakes; Covacevich and Archer, 1975; Burnett, 1996), and dingoes (*Canis lupus dingo*; Catling, 1995). As toads colonize new areas, these animals show a substantial drop in numbers, presumably a result of being poisoned after attempting to eat the toads (Burnett, 1996). Three quoll species and 8 of 20 monitor species are now considered to be at risk because they include amphibians in their diet and their distributions overlap with current and potential marine toad distributions (Burnett, 1996). Studies in the Northern Territory comparing the fauna at sites before and after toad invasion have also indicated the presence of a long-term effect on dingoes, coleopterans, and reptiles, particularly small reptiles (Catling et al., 1999).

Bufo microscaphus Cope, 1867 "1866"
ARIZONA TOAD

Terry D. Schwaner, Brian K. Sullivan

We follow Gergus et al. (1997), Gergus (1998), and Crother et al. (2000) in recognizing Arizona toads as a distinct species.

1. Historical versus Current Distribution.
Arizona toads (*Bufo microscaphus*) were originally described by Cope (1867), type locality: "Arizona . . . near the parallel of 35° and along the valley of the Colorado from Fort Mojave to Fort Yuma." Shannon (1949) restricted the type locality to Fort Mohave, Mohave County, Arizona. Based on allozyme evidence, morphological diagnosability, and allopatry, Gergus (1998) argued that each of the three members of the *B. microscaphus* complex previously recognized as subspecies (*B. m. microscaphus*, *B. m. californicus*, and *B. m. mexicanus*; Price and Sullivan, 1988) should be recognized as full species. Arizona toads range from the Mogollon plateau of southwestern New Mexico westward to the Colorado and Virgin River basins of northwestern Arizona, southern Nevada, and southwestern Utah.

The current distribution of Arizona toads is probably largely similar to their historical distribution (e.g., A.P. Blair, 1955; Stebbins, 1985; Price and Sullivan, 1988; Sullivan, 1993, 1995; Hovingh, 1997). It appears that Arizona toads have declined at some sites following habitat disturbance; they are apparently being replaced by Woodhouse's toads (*B. woodhousii*) at some localities in Arizona (Sullivan, 1993) and southern Nevada (D. Bradford and colleagues, unpublished data).

2. Historical versus Current Abundance.
Historical abundance is unknown; current abundance is presumably higher in areas not disturbed by damming, introduced

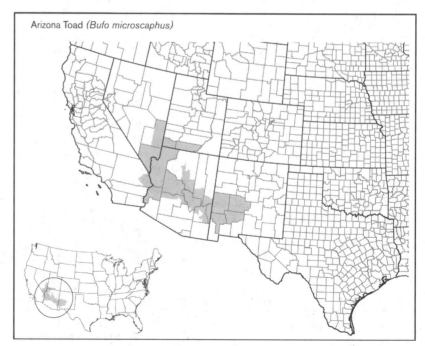

Arizona Toad (*Bufo microscaphus*)

predators, degrading land-use practices, and human use of recreational areas. In Arizona, 55 new localities have been documented for Arizona toads over the past 15 yr (1980–95); given their presence at many historical sites, it appears a general decline has not occurred in this region (Sullivan, 1993; B. K. S., unpublished data).

Mark–recapture studies conducted at Lytle Preserve, Beaver Dam Wash, a tributary of the Virgin River in extreme southwestern Utah and southern Nevada, estimated 95% confidence limits of 436–972 adult toads in a 4-m wide, 1,500-m long road and stream transect in July–August 1996 (Schwaner et al., 1998). At Birch Creek, Zion National Park, in a 1,000 m long stream transect of similar width, 357–507 (95% confidence limits) adults were estimated in April–August 2000 (T. D. S., unpublished data). High densities at Lytle Preserve may be atypical compared to most other areas due to availability of irrigated fields for foraging following the breeding period. Schwaner (personal observations) found average monthly captures along the road and stream transect at Lytle Preserve were 167 (range = 31 in October and 330 in August). In a stream transect of similar proportions at Oak Creek, a tributary to the Virgin River in Zion National Park, southwestern Utah, average monthly adult captures were only 60 (range = 22 in May and 113 in June).

3. Life History Features.

A. Breeding. Reproduction is aquatic. Amplectic position is axillary; oviposition is aided by the "basket method" (A. Brown et al., 2000). The advertisement call of Arizona toads is a trill, averaging 5.7 s in duration, with a dominant frequency of about 1.380 kHz and a pulse rate of roughly 46 pulses/s at 15 °C (Gergus et al., 1997).

i. Breeding migrations. In west-central Arizona (Sullivan, 1992b), and at Lytle Preserve in Utah (T. D. S., personal observations), calling males have been heard as early as February following warm days. Breeding begins in early to late February in Arizona, but in early to late March to early April in Utah or at higher elevations in Arizona (A. P. Blair, 1955; Sullivan, 1992b, 1995; T. D. S., personal observations). Breeding is not triggered by rain (B. K. S., personal observations), but more likely by warming nocturnal, ambient temperatures (air, 8–18 °C, water 12–18 °C; Sullivan, 1992b). Spring flooding delays breeding (Sullivan, 1992b; T. D. S., personal observations). Breeding choruses may be of short duration, lasting 10–12 d for less dense populations, and may be interrupted by flooding, only to resume following warmer, drier weather (Sullivan, 1992b).

Sullivan (1992b) observed female selection of calling males in Arizona toads.

Working at Lytle Preserve, T. D. S. (personal observations) found only a few calling male Arizona toads among dense aggregations of non-calling and searching satellite males who invariably mated with approaching females. This behavior was much like Sullivan's (1982a) description of mating in Great Plains toads (*B. cognatus*), except that females were either already clasped by a male before reaching the water or immediately clasped by any male sensing her presence in the water.

ii. Breeding habitat. Breeding sites are the edges of streams or shallows, backwashes, and side-pools where flow is minimal. In central Arizona, males often call from rocks along the edge of flowing streams and shallow rivers if pools or backwaters are unavailable (B. K. S., personal observations). Cottonwoods *(Populus fremontii)*, willows (*Salix* sp.), and seep willows (*Baccharis* sp.) are common plants associated with lower elevation riparian areas used for breeding in Arizona and New Mexico. Factors explaining site selection at Oak Creek, Utah, in 1997 included areas of the stream with numerous side-pools and warmer diurnal water temperatures, perhaps facilitated by a reduction in tall trees and their canopy cover (Dahl et al., 2000).

B. Eggs.

i. Egg deposition sites. Riparian areas of streams, shallows, backwashes, and side-pools.

ii. Clutch size. Blair (A.P., 1955) indicated an average of 4,500 eggs/clutch. Schwaner et al. (1998) observed 227 egg masses laid in a 2-km section of stream at Lytle Preserve over a period of 16 d in March 1997. Although breeding ceased, some males continued to call until early June of that year (see Sullivan, 1992b). Eggs hatch in 3–6 d, depending on water temperature (Schwaner et al., 1998).

C. Larvae/Metamorphosis. Larvae and eggs are described in Stebbins (1951, 1985). Altig et al. (1998) provide a key to larvae of Arizona toads and their relatives. Sweet (1992) reported larval development and metamorphosis of arroyo toads from April–August (even September at higher elevation sites); Schwaner et al. (1998) indicated a shorter cycle, March–June or mid-July, at Lytle Preserve, but longer cycles exist for populations in cooler water at higher elevations in the headwaters of the Beaver Dam Wash and Birch Creek, Zion National Park (T. D. S., unpublished data). Clearly, mating systems and developmental cycles vary according to environmental variables related to seasonal temperature and water cycles, which in turn are dependent in part on latitude and elevation. Most clutches of Arizona toads have a high survival rate, with close to 100% of the embryos hatching and reaching the dispersal phase. Through the course of larval life, metamorphosis, and early juvenile stages, predation appears to

be the chief cause of mortality (Sweet, 1992; T. D. S., unpublished data), although flooding in late spring (April–June) can also cause high mortality (T. D. S., B. K. S., personal observations).

D. Juvenile Habitat. Detailed observations are available only for the closely related arroyo toads, and so at this time the reader is referred to that account (Sweet and Sullivan, 2001).

E. Adult Habitat. Arizona toad populations in Arizona and Utah utilize sandy marginal zones or terraces with a mixture of dense willow clumps and open flats or flood channels situated within 100 m or so of the stream, plus adjacent terraces with cottonwoods or live oaks (see Sweet, 1992). At higher elevations (2,000 m) in Arizona and New Mexico, Arizona toads are still associated with riparian communities but may move more widely in associated forest biotic communities (e.g., Ponderosa pine) during summer rains (B. K. S., personal observations).

F. Home Range Size. Juvenile (young of the year, >30 mm SVL) Arizona toads migrated from streamside to irrigated fields (at distances of 50–200 m) at Lytle Preserve, Utah, in August 1996–98; adult females migrated to the fields shortly after breeding in March–April 1997–98; and adult males appeared in the fields from May–June 1997–98 (T. D. S., personal observations).

G. Territories. At sites in central Arizona in 1990 and 1991, male Arizona toads were grouped in choruses of 2–15 individuals (but usually groups of 2–4) along a 50–100 m stretch of river with interindividual spacing of about 1 m (Sullivan, 1992b). Subsequently, following a series of wet years, males still maintained interindividual distances of about 1 m, but densities were considerably higher (1997) with as many as 40 males observed in a chorus along an approximately 80 m stretch of the Agua Fria River (B. K. S., personal observations). Sweet (1992) reported a minimum distance between adult, calling male arroyo toads of 1–8 m, with no evidence of agonistic or territorial behavior, although the densities of males/pool were typically low (1–3 males). Neither author reported active searching or satellite male behavior. However, T. D. S. (personal observations) saw several groups of up to 15 adult male Arizona toads along stretches of the Beaver Dam Wash, 1997–99, at average distances between males of 0.5–1.5 m. Only a few males/group called regularly, but movement of a male caused immediate clasping from an adjacent male, whereupon the clasped male gave a release call. The clasping male then released its hold whereupon the released male called; the clasping male then moved 0.5–1 m away. Subsequent movements often prompted the released male to approach and clasp the previously clasping male resulting in the same stereotyped behavior. This behavior often brought other adjacent males into the area so that male–male interactions occurred frequently and repeatedly among the same group of males.

H. Aestivation/Avoiding Desiccation. Unknown. Adults are nocturnal, emerging from sandy burrows at dusk (Sweet, 1992; T.D.S., B.K.S., personal observations).

I. Seasonal Migrations. None observed or reported.

J. Topor (Hibernation). Presumably September–February; activity for Arizona toads at Lytle Preserve in Utah can begin in February and last to September (T.D.S., personal observations).

K. Interspecific Associations/Exclusions. Woodhouse's toads hybridize with Arizona toads at a number of sites in Arizona and at the junction of tributaries to the main course of the Virgin River in southwestern Utah, southern Nevada, and northwestern Arizona (Shannon, 1949; Stebbins, 1951; A.P. Blair, 1955; Sullivan, 1986b, 1995; Sullivan and Lamb, 1988). Blair (A.P., 1955) reported clasped Arizona toads and red-spotted toads *(B. punctatus)* without hybridization. Arizona toads have been observed breeding with Great Basin spadefoot toads *(Spea intermontana)*, red-spotted toads, canyon treefrogs *(Hyla arenicolor)*, and American bullfrogs *(Rana catesbeiana)* in southwestern Utah (T.D.S., personal observations), and with Great Plains toads, red-spotted toads, Woodhouse's toads, canyon treefrogs, and lowland leopard frogs *(R. yavapaiensis)* in central Arizona (B.K.S., personal observations).

L. Age/Size at Reproductive Maturity. In central Arizona, males at a breeding aggregation along the Hassayampa River (n = 26 over 2 yr) ranged from 53–79 mm SVL (B.K.S., personal observations). The smallest male Arizona toads in 30 male-female clasping pairs observed at Lytle Preserve in Utah was 53 mm SVL; the smallest female was 56 mm SVL (T.D.S., unpublished data).

M. Longevity. Size frequency distributions, growth rates, and skeletochronology all indicated only four generations of Arizona toads at Lytle Preserve (Schwaner et al., 1998). However, these measures suggest five generations at Birch Creek and Zion National Park (T.D.S., unpublished data).

N. Feeding Behavior. No comprehensive analysis of feeding behavior has been conducted. Larvae presumably feed on algae and small, unicellular organisms attached to local substrates.

O. Predators. Killdeer *(Charadrius vociferus)* and intermountain wandering garter snakes *(Thamnophis elegans vagrans)* preyed on Arizona toad larvae in southwestern Utah, and raccoons killed and consumed many adult male toads during the breeding season (T.D.S., personal observations). Predation on breeding males

and females by small mammals has also been observed for populations in central Arizona (B.K.S., personal observations).

P. Anti-Predator Mechanisms. Eggs are presumably distasteful to snakes (see Sweet, 1992). Adults in breeding choruses are quick to retreat underwater at slight disturbances (T.D.S, personal observations). Parotoid glands produce steroids, possibly rendering adults unpalatable to some predators (Duellman and Trueb, 1986).

Q. Diseases. There is no evidence of disease among field populations.

R. Parasites. Goldberg et al. (1996c) described the helminths of Arizona toads, Woodhouse's toads, and their hybrids from a site in central Arizona.

4. Conservation.

Populations of Arizona toads in central Arizona are threatened by habitat destruction and interspecific hybridization (Sullivan, 1986, 1993). Damming along the Agua Fria River has reduced lotic habitats favored by Arizona toads for breeding and provided lentic environments favored by Woodhouse's toads. Interspecific hybridization is occurring in these areas and appears to be unidirectional (Malmos et al., 2001), with Woodhouse's toad females mated by Arizona toad males. Historically, these species hybridized at the junctions of several tributaries of the Virgin River, where branch streams and washes entered the main course of the river, and along stretches between these locations (Blair, 1955; Sullivan, 1995). Most recently, T.D.S. and B.K.S. (unpublished data) found genetic evidence for hybrid swarms at these locations, and introgression of *B. woodhousii* (16S rRNA) genes into putatively "pure" *B. microscaphus* populations 100 km upstream from the junction of the Beaver Dam Wash and the Virgin River. They also found genetic evidence for hybridization (marker alleles for aspartate amino transferase; Sullivan and Lamb, 1988) in populations near the junction of Ash-Laverkin Creek and the Virgin River, near Zion National Park. Although anecdotal, based on observations in a single season, morphotypes representing the range of phenotypes of the two species and their hybrids bred throughout March 2001 (T.D.S., unpublished data) at the junction of the Beaver Dam Wash and the Virgin River. Populations of putatively "pure" *B. woodhousii* did not begin to breed until the beginning of April 2001 at Mesquite, only 20 km away and at the same elevation. This evidence suggests that effects of hybridization are not localized and perhaps behavioral as well as ecological, that genetic barriers to introgression have broken down, and most, if not all, hybrid zones are in areas of human disturbance (dams and golf courses along tributaries and rivers). Southwestern Utah, northwestern Arizona, and southeastern

Nevada contain large and continuous populations of *B. microscaphus*. However, the integrity of these populations is challenged by hybridization, at least in part due to human activities. As reported in the local media, human population growth in this area is expected to increase from tens of thousands to hundreds of thousands of people (many retirees) in the next 30 yr.

Acknowledgments. T.D.S. thanks the Utah Division of Wildlife Resources for permits, Brigham Young University for use of the facilities at Lytle Preserve, Zion National Park for access to Oak Creek populations of toads, and Wes Adams, Anna Brown, Brooke Christensen, D. Rijeana Hadley, and Kim Jenkins for undergraduate research assistance. B.K.S. acknowledges support of the Arizona Game and Fish Department, and the assistance of Rob Bowker, Mike Demlong, Matt Kwiatkowski, Erik Gergus, and Keith Malmos.

Bufo nebulifer Girard, 1854
COASTAL-PLAIN TOAD

Joseph R. Mendelson III

I follow Mulcahy and Mendelson (2000) in recognizing Coastal-Plain toads *(Bufo nebulifer)* as a species distinct from Gulf Coast toads *(B. valliceps)*. This study was based on an analysis of mtDNA sequences and corroborated comments by Mendelson (1998) that showed the northern and southern populations of the widespread taxon *B. valliceps* were likely not conspecific. The taxon *B. nebulifer* was proposed by Girard (1854) as a replacement name for the preoccupied taxon *B. granulosus* Baird and Girard, 1852a, type locality: "between Indianola and San Antonio, Texas." Mulcahy and Mendelson (2000) resurrected the taxon *B. nebulifer* from the synonymy of *B. valliceps* Wiegmann, 1833, and applied it to populations generally north of the coastal village of Palma Sola, Veracruz, Mexico. Further discussions of variation in *B. valliceps* with respect to *B. nebulifer* appear in Mendelson (1998) and McCranie and Köhler (2000). Note that the vast majority of literature regarding *B. valliceps*—especially that relevant to natural history and behavior—actually considers northerly populations now referred to as Coastal-Plain toads *(B. nebulifer)*.

1. Historical versus Current Distribution.

Coastal-Plain toads *(B. nebulifer)* are found in a wide variety of habitats along the Gulf Coastal Plain from central Veracruz, Mexico, northward to north-central Texas and extending eastward into southern Louisiana and extreme southwestern Mississippi. The species ranges westward in Texas, occurring in the Hill Country of the Edwards Plateau region and along the Rio Grande, Devil's, and Pecos River systems. A record from further up the Rio Grande system, near El Paso, Texas, represents an accidental introduction (Dixon, 2000). Apparently disjunct populations in southern Arkansas and northern Louisiana may also represent dispersal along river systems (Dundee and Rossman, 1989). The range map published by Porter (1970) is misleading because it shows the combined ranges of *B. valliceps* and *B. nebulifer*, includes localities for numerous misidentified specimens (see Mendelson, 1997a,b; Frost, 2000c; Mendelson, 2001) and lacks many records reported since 1970 (e.g., additional records from Texas reported by Dixon, 2000). The map presented by Mulcahy and Mendelson (2000, fig. 5) and represented here for U.S. populations more accurately illustrates the distribution of both *B. nebulifer* and *B. valliceps*.

Coastal-Plain Toad *(Bufo nebulifer)*

2. Historical versus Current Abundance.
Reports concerning the abundance of Coastal-Plain toads typically state that they are common or abundant wherever they occur (e.g., Penn, 1943; Raun, 1959). Size of local populations appears to vary with respect to annual rainfall patterns; in some cases appearing to increase or decrease by 50% between years. Despite this variation, the sex ratio of the population remained male biased by about 2:1 over a 5-yr study (Blair, 1960a). Recent reports from the United States (e.g., Foley, 1994; Dixon, 2000) indicate that they remain common or abundant in all areas of their distribution. Less information is available from the area of the range occurring in Mexico, but Martin (1958) reported this species to be widespread and common in every variety of habitat ≤1,550 m in the Gómez Farías region of the Sierra Madre Oriental, in southern Tamaulipas, Mexico. Large series of museum specimens from other localities in northeastern Mexico suggest these toads have historically been abundant in these areas and continue to be so today. They were observed to be common in several localities in northern

Veracruz between 1996–2000 (personal observations). With regard to Mexico, virtually all published reports and most series of museum specimens correspond to populations referable to *B. valliceps* in the southern areas of the country.

3. Life History Features.
A. Breeding. Reproduction is aquatic. Coastal-Plain toads breed in many varieties of still water, and their reproductive biology is generally similar to that of other north-temperate species of *Bufo*.

i. Breeding migrations. Spring or summer rains in excess of about 3 cm usually stimulate formation of choruses of calling males (Blair, 1960a). Individuals congregate at suitable breeding sites throughout the summer rainy season; reproductive behavior has been observed from March–August (Martin, 1958; Thornton, 1960; Wiest, 1982; Foley, 1994); Thornton (1960) observed a few calling males, but no females, as early as February. Calling males and breeding activity were observed at temperatures from 15–30 °C (Wiest, 1982). Sullivan and Wagner (1988) reported variation in the frequency and duration of advertisement calls with respect to size of the individual; call rate and effort also varied with respect to density of calling males and their proximity to one another. Females do not remain at the breeding site after eggs are deposited (Thornton, 1960). Larvae have been observed in ponds from April–August (Wiest, 1982). Orientation to breeding sites by Coastal-Plain toads apparently is at least partially based on olfactory cues (Grubb, 1973a).

ii. Breeding habitat. Coastal-Plain toads breed in a wide variety of still-water habitats, including ponds, ephemeral wetlands, roadside ditches, and artificial impoundments (Wright and Wright, 1949; Wiest, 1982; Foley, 1994; personal

observations). Toads may breed occasionally along the margins of rivers (e.g., Awbrey, 1963). Dundee and Rossman (1989) detail and provisionally accept the arguments that Coastal-Plain toads breed in brackish waters. Adults began breeding in a series of artificial pools within 1 yr of their installation at a field site in Texas (Hubbs and Martin, 1967).

B. Eggs.
i. Egg deposition sites. Eggs are typically deposited in still, shallow water, with or without emergent vegetation.

ii. Clutch size. The eggs of Coastal-Plain toads are usually deposited in double rows that are strung out and pelagic; egg density in the rows ranges from 7–10 to 25–27 eggs/3-cm section of an egg string (Wright and Wright, 1949; Dundee and Rossman, 1989). A single clutch may contain approximately 20,000 eggs (Blair, 1960a). A few females have been observed to produce two clutches of eggs during a single, extended reproductive season (Blair, 1960a).

C. Larvae/Metamorphosis.
i. Length of larval stage. At a laboratory-controlled temperature of 25 °C, >50% of a group of tadpoles hatched 24 hr after egg deposition and began metamorphosis 19 d after egg deposition (Limbaugh and Volpe, 1957). In the wild, eggs hatch 1–2 d after deposition and, after a 20–30 d larval period, tadpoles metamorphose at 7.5–12 mm (Wright and Wright, 1949).

ii. Larval requirements. Eggs of Coastal-Plain toads exposed to temperatures <18 °C or >30 °C did not hatch (Hubbs et al., 1963). Eggs exposed to temperatures >30 °C also produced tadpoles that developed a variety of deformities (Volpe, 1957; Hubbs et al., 1963). Similar experiments by Ballinger and McKinney (1966) found the low lethal temperature also to be 18 °C, but the high lethal temperature to be 38 °C. Short-term exposure (3 d) to acidic (pH 4) conditions reduced body weights of tadpoles, but by 7 d after return to neutral (pH 7.2–7.6) conditions, tadpoles were at normal weight and metamorphosed normally (Pierce and Montgomery, 1989). Exposure to acidic conditions did not result in mortality of tadpoles (Rosenberg and Pierce, 1995).

a. Food. Tadpoles are presumed to be generally algivorous, scraping plant and animal material off submerged substrates.

b. Cover. Cover objects are not specifically sought by tadpoles. Tadpoles usually occur in groups and are conspicuous in the water.

iii. Larval polymorphisms. Not known to occur.

iv. Features of metamorphosis. In a laboratory study with temperatures held at 25 °C, most Coastal-Plain toad tadpoles hatched 24 hr after egg deposition (Limbaugh and Volpe, 1957). Hindlimb buds appeared after 5 d, forelimbs protruded around 21 d; metamorphosis began around 19 d and was complete by around 28 d.

v. Post-metamorphic migrations. Newly metamorphosed Coastal-Plain toads exhibit a short period of residency at their natal pond, then migrate away in a direction perpendicular to the shoreline (Grubb, 1973b). However, under dry conditions, recently metamorphosed animals may remain by the pond margin for extended periods of time until favorably wet conditions allow dispersal; mortality is high during this time (Blair, 1953).

D. Juvenile Habitat. Similar to adults. Juvenile Coastal-Plain toads may remain concentrated along margins of their natal pond for some time before dispersing (Blair, 1953; personal observations).

E. Adult Habitat. Adult Coastal-Plain toads are associated with almost every variety of habitat, including suburban and urban areas, and are frequently found in agricultural and wet hardwood areas, although they are relatively uncommon in pinelands (Dundee and Rossman, 1989; personal observations). They are commonly seen in suburban and some urban areas, foraging for prey on lawns and under streetlights (personal observations). Coastal-Plain toads are frequently found in railroad ditches or roadside pools, garbage dumps, and storm sewers (McAllister et al., 1989; personal observations). In urban environments, they are found frequently under concrete slabs and gathered in groups in cracks and holes under sidewalks (Awbrey, 1963; Moore, 1976). During the day, Coastal-Plain toads seek cover and can be found under logs, other cover objects (personal observations), or in rodent burrows (Wilks, 1963). Neill and Grubb (1971) found Coastal-Plain toads from 2–5 m above the ground in oak trees; they can be considered arboreal in that individual toads will find tree holes and may use them repeatedly for periods of weeks. In Texas, individuals have been found at the mouth of caves and, in one case, well down in a cave system (see Reddell, 1970). This species does not appear to be adversely affected by alteration of native vegetation or by invasion of non-native vegetation. Adults may be found in vegetated areas of coastal barrier islands within 30 m of saltwater (personal observations). Timber harvesting practices at a site in southeast Texas did not affect overall abundance of toads (Foley, 1994). Body temperatures of active toads ranged from 22.3–27 °C (Brattstrom, 1963).

F. Home Range Size. A male Coastal-Plain toad moved approximately 800 m in about 1 wk (Thornton, 1960). Another individual marked by Thornton moved approximately 1,200 m in 2 d and was located approximately 1,600 m from its original point of capture 1 yr later. Thornton also reported many toads regularly moving between two ponds about 100 m apart. Although toads clearly do wander relatively great distances, a mark–recapture study (Awbrey, 1963) found that the majority of his marked toads (n = 49) remained within a "home range" area within approximately 46 m from their original site of capture. Homing experiments (Awbrey, 1963) indicated that individual toads could relocate their site of capture after being removed distances up to approximately 220 m.

G. Territories. Individuals do not appear to defend territories. At breeding aggregations, male combat and satellite behavior have not been observed (Wagner and Sullivan, 1992). Toads appear to exhibit homing behavior toward breeding sites, based on visual and olfactory cues (Grubb, 1970).

H. Aestivation/Avoiding Desiccation. Toads in breeding season have been known to take refuge in rodent burrows during the day (Grubb, 1970) and under debris piles, logs, and stones during drier parts of the summer (personal observations).

I. Seasonal Migrations. Distinct migrations among habitats do not occur, but aggregations of adults appear suddenly at temporary and permanent bodies of water during the prolonged summer breeding season.

J. Torpor (Hibernation). Coastal-Plain toads are typically inactive during winter and/or dry seasons. Presumably they take refuge in burrows of other animals or under larger debris piles.

K. Interspecific Associations/Exclusions. Coastal-Plain and Fowler's toads (*B. fowleri*) are broadly sympatric in Texas and Louisiana and natural hybrids are relatively frequent (Blair, 1956b; Gosner and Black, 1958a; Dundee and Rossman, 1989). At ponds shared by both species in Texas, 7–9% of amplectant pairs were interspecific (Thornton, 1955, 1960; L.E. Brown, 1971a). Volpe (1960) reported *B. nebulifer-woodhousii* (Woodhouse's toad) hybrids to be viable and vigorous, but sterile. Hybridization between *B. nebulifer* and *B. houstonensis* (Houston toads) has been implicated as one of several factors resulting in the Endangered status in the latter species (L.E. Brown, 1971b; Brown and Mesrobian, this volume, Part One). Axtell (1958) described a pond harboring a mixed chorus of Coastal-Plain toads, Texas toads (*B. speciosus*), plains spadefoot toads (*Spea bombifrons*), Couch's spadefoot toads (*Scaphiopus couchii*), and spotted chorus frogs (*Pseudacris clarkii*). Moore (1976) described mixed choruses of Texas toads and Coastal-Plain toads, in which the former species was more abundant. In these choruses, Coastal-Plain toads called from the margins of ponds rather than the knolls and ledges around the pond seemingly favored by Texas toads. Kelly Irwin (personal communication), working in Starr County, Texas, observed a mixed chorus of Coastal-Plain toads, Texas toads, Couch's spadefoot toads, western narrow-mouthed toads (*Gastrophryne olivacea*), sheep frogs (*Hypopachus variolosus*), and Rio Grande leopard frogs (*Rana berlandieri*). Coastal-Plain toads use pocket gopher burrows as refuges (Wilks, 1963). Where Coastal-Plain toads are sympatric with marine toads (*B. marinus*; extreme southern Texas and northeastern Mexico), adults of these two species frequently are found together in a variety of habitats. I have not observed these two species to form mixed breeding choruses or to share breeding sites.

L. Age/Size at Reproductive Maturity. Males 53–98 mm; females 54–125 mm (Wright and Wright, 1949; Blair, 1956b, 1963b; Dundee and Rossman, 1989). Sexual maturity appears to occur during the second year (Blair, 1953). Blair (1953) notes that 25 of 357 marked juvenile toads were recovered the following summer when they were apparently sexually mature. Allowing for differences between sexes, Blair (1953) suggests about 11% of juveniles reach sexual maturity. Growth rates are fastest during the first year; one male grew 39 mm between mid-August and late September.

M. Longevity. Individuals in the wild are known to live 8 yr (Blair, 1960a).

N. Feeding Behavior. Campbell and Davis (1968) examined the stomachs of 21 Coastal-Plain toads and found the bulk of their diet to be composed of invertebrates such as isopods and coleopterans, but also included were an eastern fence lizard (*Sceloporus undulatus*) and a juvenile Fowler's toad. McGehee et al. (2001) also reported a preponderance of isopods, along with numerous scorpions and relatively few coleopterans. Most likely, Coastal-Plain toads are opportunistic predators taking a wide variety of small animals based on their availability.

O. Predators. Western ribbon snakes (*Thamnophis proximus*) and diamond-backed water snakes (*Nerodia rhombifera*) will feed on Coastal-Plain toads (Wright and Wright, 1949). Martin (1958) reported predation by cat-eyed snakes (*Leptodeira septentrionalis*). In captivity, tiger salamanders (*Ambystoma tigrinum*) will eat juveniles (Strecker, 1927). Predation on tadpoles by diving beetles (*Acilius semisulcatus*) has been reported (Neill, 1968). Brown (L.E., 1974) described an unsuccessful predation attempt by an American bullfrog (*Rana catesbeiana*). Kelly Irwin (personal communication) observed predation by an indigo snake (*Drymarchon corais*) in Cameron County, Texas.

P. Anti-Predator Mechanisms. Coastal-Plain toad eggs are toxic and/or distasteful to a wide variety of potential predators (Licht, 1968). Their parotoid gland secretions are presumably distasteful to some predators. Brown (L.E., 1974) described an unsuccessful predation attempt by an American bullfrog in which the parotoid secretions by the toad apparently caused the bullfrog to release it, unharmed, after an approximately 5-min attempt to swallow it.

Q. Diseases. No instances of diseases are known from wild populations.

R. Parasites. Walton (1946) reported *Nyctotherus cordiformis* (Protozoa: Ciliophora) from Coastal-Plain toads. Hoffpauir and Morrison (1966) reported *Rhabdias ranae* (Nematoda) from the lungs of a specimen from eastern Texas. McAllister et al. (1989) describe the following endoparasites from Coastal-Plain toads: *Opalina* sp. (Protozoa: Sarcomastigophra); *Adelina* sp. and *Eimeria* sp. (Apicomplexa: Eucoccidiorida); *Myxidium serotinum* (Myxozoa: Bivalvulida); *Mesocestoides* sp. (Cestoidea: Cyclophyllidea); and *Cosmocercoides variabilis* (Nematoda: Ascaidida).

4. Conservation.

Coastal-Plain toads are widespread and common throughout their range. They are adaptable and seem to tolerate habitat alterations caused by humans. In Mexico, they are more common in secondary habitats rather than primary forest habitats, suggesting that this species may expand its range or increase in local density following disturbance. Coastal-Plain toads are found in a wide variety of habitats, including suburban and urban areas, and are commonly seen in suburban areas foraging for prey on lawns and under streetlights. They are also found in railroad ditches or roadside pools, garbage dumps, and storm sewers (McAllister et al., 1989; personal observations). All of these observations suggest that Coastal-Plain toads are in no need of special protection.

Acknowledgments. I was ably assisted in compiling literature relevant to this account by J. Minnick, and the original manuscript was greatly improved by the critical comments of the students in the U.S.U. Herpetology Group. I offer my thanks to all these persons.

Bufo nelsoni Stejneger, 1893
AMARGOSA TOAD

Anna Goebel , Hobart M. Smith, Robert W. Murphy, David J. Morafka

Editor's note: Following Frost (2000a), we have covered Amargosa toads *(Bufo nelsoni)* as *Bufo boreas nelsoni* in the boreal toad (*B. boreas*) species account (Muths and Nanjappa, this volume). However, most workers familiar with these animals agree that they deserve species status. The arguments for this position (further arguments can be found on the discussion board of AmphibiaWeb [2002]) and what is known about the conservation status of Amargosa toads are presented as follows.

Taxa in the *B. boreas* group have had varied histories, being considered species and subspecies at different times. For example, Stejneger (1893) considered *B. halophilus* a species, but it is now com-

monly treated as a subspecies of *B. boreas*. *Bufo exsul* (black toads) has been particularly volatile, switching rank several times; they were described as a species (Myers, 1942a), treated as a subspecies of *B. boreas* (Tihen, 1962; Schmidt, 1953), and again as a species (Schuierer, 1962, 1963; Feder, 1977; Stebbins, 1985). Some nominal taxa such as *B. boreas columbiensis*, *B. nestor*, and *B. pictus* are no longer considered by current authorities. Similar taxonomic instability is seen in other North American bufonids, such as the *B. americanus* group (Blair, 1941; Frost, 2000a).

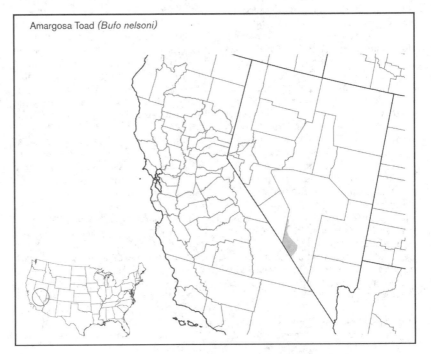

Amargosa Toad *(Bufo nelsoni)*

Nomenclatural History of *B. nelsoni*.
Bufo nelsoni was described as a subspecies of *Bufo boreas* (Stejneger, 1893). Camp (1917b) stated that the two samples described by Stejneger as *nelsoni* represented *B. halophilus*; subsequently, Stejneger did not include *nelsoni* as a subspecies or species in later checklists (Stejneger and Barbour, 1917, 1923, 1933, 1939). Despite this history of rejection, many species lists, accounts, and field guides included *nelsoni* and treated it as a subspecies of *B. boreas* (Grinnell and Camp, 1917; Slevin, 1928; Linsdale, 1940; Myers, 1942a; Stejneger and Barbour, 1943; Pickwell, 1948; Wright and Wright, 1949; Stebbins, 1951, 1954a, 1972; Schmidt, 1953; Conant, 1956; Cochran, 1961; Tihen 1962; Stebbins, 1966; Banta, 1965; Collins et al., 1978, 1982; Collins, 1990, 1997; Stebbins and Cohen, 1995; Frost, 2000a,b; Center for North American Herpetology, 2002). A species list by Frank and Ramus (1995) treated *nelsoni* as a subspecies of *B. halophilus*, perhaps based on Camp (1917b).

Savage (1960b) was the first to refer to *nelsoni* as a species, and this stands as a proper nomenclatural change according to the rules of the International Code of Zoological Nomenclature (hereinafter the "Code," International Commission on Zoological Nomenclature [ICZN], 1999). After Savage (1960b), many guides and papers treated *nelsoni* as a species (Savage and Schuierer, 1961; Karlstrom, 1962; Schuierer, 1963; Altig, 1970; Blair, 1972a; Bury et al., 1980; Maxson et al., 1981; Stebbins, 1985; Altig and Johnson, 1986; Altig and Dodd, 1987; Graybeal, 1993; Goebel,

1996a; Altig et al., 1998; Powell et al., 1998; Burroughs, 1999; Duellman and Sweet, 1999) as did many unpublished documents and dissertations (Feder, 1977; Altig, 1981; Maciolek, 1983a,b; Hoff, 1994, 1996; Clemmer, 1995; Goebel, 1996a; Heinrich, 1996; Stein, 1996, 1999; Stein et al., 2000; Sjoberg, 2000, 2001). *Bufo nelsoni* was listed as a full species on the Integrated Taxonomic Information System (2002) database compiled by the Smithsonian Institution with the highest credibility rating by their Taxonomic Working Group. *Bufo nelsoni* was consistently treated as a species within the federal listing process (Federal Register, 1977, 1982, 1985, 1989, 1991, 1994, 1995, 1996) and in a petition to list *Bufo nelsoni* as an Endangered Species (Biodiversity Legal Foundation, 1994). Conservation programs treated *Bufo nelsoni* as a species (Burroughs, 1999; Nevada Division of Wildlife, 2000; Bureau of Land Management, 2002; Nature Conservancy, 2002). Thus, since 1960, *nelsoni* has been treated both as a species and a subspecies, although there has been considerable consensus by authors of original research and conservation measures in treating *B. nelsoni* as a species.

Frost (2000a,b) and Collins (Collins et al., 1978, 1982; Collins, 1990, 1997) provided critically important lists and commentaries on the scientific and standard English names for amphibians. These have been invaluable and influential to the scientific community (e.g., Muths and Nanjappa, this volume). But the variable treatment of *nelsoni* as a subspecies by some and a species by others has led to some unfortunate consequences. For example, species accounts on AmphibiaWeb (2002) provided no mention of *nelsoni* at all. *Bufo nelsoni* was not included as a subspecies of *B. boreas* (Garcia, 2002), perhaps because she believed it was a species. However, *nelsoni* was not given an independent species account because AmphibiaWeb provided accounts only for species recognized by the American Museum of Natural History (Frost, 2000b).

Similarly, by initially following Crother et al. (2000; but see Crother et al., 2003), *B. nelsoni* was not given a separate treatment in this volume (Muths and Nanjappa, this volume), which we now try to provide. First we review why *nelsoni* may not have been treated as a species. We then discuss morphological and genetic evidence and implications from the biogeographical history of the Great Basin. We conclude that *B. nelsoni* is a species and review conservation measures.

Is *Bufo nelsoni* a subspecies?
Justification and publication.
Both Frost (2000a,b) and Collins (Collins et al., 1978, 1982; Collins, 1990, 1997) treated *nelsoni* as a subspecies of *B. boreas* within their influential species lists. While neither provided a reason, it was probably due to a lack of any published, taxonomically sound justification for treating *nelsoni* as a species.

Frost (2000b) recognized Savage (1960b) as the first to treat *nelsoni* as a species, and that stands nomenclaturally. However, Savage did not provide any justification for that change. While justification is not required by the Code, acceptance by others generally requires some justification. Perhaps none of the many authors in the published literature cited above supplied taxonomically sound evidence for elevation, but certainly Feder (1977) did. She clearly stated that *B. nelsoni* was a species and provided allozymic data sufficient for taxonomic justification. Unfortunately, Feder's work was not published in a format that is commonly accepted. Again, the rules of the Code do not apply to non-nomenclatural matters, such as publication issues discussed here. But the taxonomic literature generally follows guidance by the Code on a variety of matters, and the Code does not consider master's and doctoral dissertations as acceptable publications (Article 8). Thus, *nelsoni* may have been treated as a subspecies because Feder's data were poorly distributed (due to lack of publication), not because her conclusions were discredited.

If unpublished sources are available, exceptions that do not involve new scientific names have been incorporated into the taxonomic literature, especially when they provide credit to authors (e.g., Myers, 1931). Feder's dissertation is available on microfilm from both the University of California at Berkeley and the University of Montana. On these grounds, we regard Feder (1977) as a legitimate part of the citable (not necessarily published) literature on *B. nelsoni*. In fact, her dissertation has been cited in the published literature (Graybeal, 1993). Below, we provide some details from Feder (1977). We encourage readers interested in further documentation to examine her dissertation in its entirety.

Recently, Frost (2002) treated *nelsoni* as a species and included a reference to Feder's (1977) work. Similarly, Collins and Taggart (2002) treated *nelsoni* as a species. However, we provide a more thorough account of *nelsoni* here.

Morphological evidence. Taxonomic instability among North American bufonids, including *B. nelsoni*, may be due to limited morphological change and a lack of understanding of the evolutionary history of diagnosable morphological characters (Wright and Wright, 1949; Baldauf, 1959; Karlstrom, 1962; Tihen, 1962; Martin, 1972; Winokur and Hillyard, 1992). Many bufonid species groups have been identified based on morphological similarity, but the relationships within and among those groups have been difficult to determine (Tihen, 1962; W. F. Blair, 1964b, 1972a).

The original species account for *nelsoni* (Stejneger, 1893) did not give many diagnostic characters: "Diagnosis – Similar to *B. boreas*. Skin between warts smooth; snout protracted, pointed in profile; webs of hind legs very large; soles rather smooth; limbs shorter, elbows and knees not meeting when adpressed to the sides of the body; inner metacarpal tubercle usually very large." Both Linsdale (1940) and Wright and Wright (1949) noted the toad had small feet and reduced webbing, which was just the reverse of Stejneger's (1893) account and may have been Stejneger's original intention.

Stejneger concluded *nelsoni* was a subspecies: "This seems to be the southern form of *Bufo boreas*, distinguished from the latter as above. Extreme examples of both forms are very different and would readily pass for distinct species, but specimens occur in which one or the other of the characters are less developed, making it expedient to use a trinomial appellation." Linsdale (1940) concluded *nelsoni* was a valid subspecies of *B. boreas* based on its narrow wedge-shaped head, protracted snout, short limbs, and reduced webbing of the hind feet. Myers (1942a) distinguished *B. exsul* from *nelsoni* largely on the dark color of the former. Morphological characters of *nelsoni* are discussed by Wright and Wright (1949) and Karlstrom (1962), but both suggested that more comprehensive analyses were needed. Morphological analyses of tadpoles by Altig (1970), Altig and Johnson (1986), and Altig et al. (1998) treated *nelsoni* as a species based on allopatry, not on unique morphological characters.

Confusion about morphological characters may have led to specimen misidentification. *Bufo nelsoni* was reported from Pahranagat Valley, Lincoln County, Nevada and Hot Creek Valley, Nye County, Nevada (Linsdale, 1940), the north slope of Bullfrog Hills, Nye County, Nevada (Maciolek, 1983b), and Resting Springs, Morans, and Lone Pine, Owens Valley, California (Wright and Wright, 1949). Schuierer (1963) and Maciolek (1983b) concluded these sightings were not *B. nelsoni*, which is believed to occur only in Oasis Valley, Nye County, Nevada (Savage, 1960; Schuierer, 1963, Nevada Division of Wildlife, 2000). *Bufo nelsoni* may be a cryptic species better identified by collection site rather than specific morphological characters. It has been >100 yr since the original description appeared, yet an understanding of the evolution of the North American bufonid groups based on morphology eludes us.

Genetic evidence. Feder (1977) analyzed allozymes in all taxa of the *B. boreas* group and found that *nelsoni* was distinct from all other members of the boreas group. Feder (1977, p. 37) concluded: "The designation of the two desert isolates as subspecies of *B. boreas* must now be reconsidered in light of this new genetic evidence. Specific status best describes the position of *B. exsul* and *B. nelsoni* within the Boreas group." Her conclusion was based on analyses of 360 metamorphosed individuals (302 from the *B. boreas* group, including 13 *nelsoni* and 58 outgroup individuals). Genetic distances were calculated from allele frequencies of 21 loci, 11 of which were variable within the *B. boreas* group. All samples of *nelsoni* had a unique allele for PGM, and *nelsoni* had unique allele frequencies at other loci. One allele at the ME locus (ME88) was fixed in both *B. exsul* and *nelsoni* but was found in low frequencies in populations of *B. boreas* nearest *B. exsul* and *nelsoni*. Phenetic dendrograms based on UPGMA and distance-Wagner cluster analyses both identified *B. nelsoni* as a unique lineage similar to *B. exsul*.

Goebel's (1996a, this volume, Part One) mitochondrial DNA analysis examined

194 samples from the *B. boreas* group including two from *nelsoni*. Phylogenetic analyses of restriction site data (378 base pairs) and control region sequences (834 base pairs) identified *nelsoni* as a unique lineage closely related to both *B. exsul* and some specimens of *B. b. halophilus* (*B. b. halophilus* was considered paraphyletic, closely related to both *B. b. boreas* and *B. nelsoni*).

Eric Simandle is analyzing microsatellite data from Great Basin boreal toads including *B. nelsoni* (E. Simandle, personal communication). More data from Simandle and others should give new perspectives on the evolution of *B. nelsoni* and the *B. boreas* group.

Implications from Biogeography.

Isolated populations in the B. boreas group. Bufo nelsoni and *B. exsul* may be two of many species that persist in small, isolated populations surrounding the Great Basin. Myers (1942a) and Karlstrom (1962) hypothesized that *B. boreas* was widespread in the Pliocene, covering much of the coastal, Sierra, and Great Basin regions, and speciation in the group resulted from isolation in the Pleistocene. Climatic and geologic fluctuations in the Great Basin region during the Pleistocene resulted in repeated wet and dry cycles with an increasing trend toward dryness that resulted in increased fragmentation of drainages (Mifflin and Wheat, 1979; Axelrod and Raven, 1985; Stokes, 1986; Hovingh, 1997). Present distributions of northern leopard frogs (*Rana pipiens*), Oregon spotted frogs (*R. pretiosa*), western toads (*B. boreas*), Woodhouse's toads (*B. woodhousii*), red-spotted toads (*B. punctatus*), and tiger salamanders (*Ambystoma tigrinum*) are all consistent with fragmentation of populations in the Pliocene and Pleistocene within the Great Basin region (Green et al., 1996; Hovingh, 1997). Immunological data (Maxson et al., 1981) suggest that lineages in the *B. boreas* group separated in the Pleistocene. Schuierer (1963) suggested that toads in the lower Owens Valley were distinct and should receive subspecific status based on morphological variation. The pattern of mtDNA evolution is consistent with fragmentation of the *B. boreas* group in and surrounding the Great Basin as well. Isolated and genetically differentiated populations include boreal toads in Kane County, Utah (Goebel, this volume, Part One), and in Haiwee Springs, California (described below). Frost (2002) felt that geographic variation within *B. boreas* was poorly studied and might mask a number of cryptic species.

Murphy and Morafka (unpublished data) investigated an isolated population of "*B. boreas*" from Haiwee Springs (about 1,500 m elevation) in the Coso Range, China Lake Naval Weapons Center Station, Inyo County, California, using 834 base pairs of homologous control region nucleotides. The sequences from this population in the Mojave Desert were compared with those obtained from a number of *B. b. halophilus* from coastal areas of California and with *nelsoni*. Toads from Haiwee Springs clustered with *nelsoni*; only a single base pair separated the two. However, the population from Haiwee Springs also differed from *B. b. halophilus* by only one base pair. These small differences allowed genetic identification of toads from the Mojave Desert, and this population is certainly a distinct population segment (DPS) deserving of additional protection. Importantly, if *nelsoni* retains recognition as a species, then the Mojave Desert population may also merit species status (i.e. description as a new species). We do not regard it at the present time as representative of *B. nelsoni*. Numerous other springs in the Coso Range provide potential habitats for toads. Additional sampling could reveal further variation and evidence for either gene flow or isolation. The disjunct mesic riparian and montane habitats of the northwestern Mojave Desert are rich in vertebrate, arthropod, and vascular plant endemics, some of which are very locally restricted and isolated from one another.

Specific Status of *Bufo nelsoni*.

We recognize *B. nelsoni* as a species. *B. nelsoni* was diagnosed by allozymic data (Feder, 1977). Therefore, the first justification of species rank for *B. nelsoni* should be credited to Feder (1977), validating Savage's (1960b) strictly nomenclatural elevation. Mitochondrial DNA data by Goebel (1996a, this volume, Part One) and Murphy and Morafka (above) are consistent with species status for *B. nelsoni*. *Bufo nelsoni* is allopatric to all other members of the *B. boreas* group (Nevada Division of Wildlife, 2000), and both Altig et al. (1998) and Powell et al. (1998) recognize *nelsoni* as a species based on allopatry. We have not identified any morphological characters that distinguish *B. nelsoni*; it may be a cryptic species. The Taxonomic Working Group for the Integrated Taxonomic Information System database (ITIS, 2002) makes judgments on species status; they independently concluded that *nelsoni* was a species, although criteria were not provided. Recently, Frost (2002) treated *nelsoni* as a species, also. Since 1960, most authors of original data on *B. nelsoni* and the *B. boreas* group concur.

Conservation.

Amargosa toads are endemic to Oasis Valley in Nye County, Nevada. Their historical range may be limited to a 16-km stretch of the Amargosa River and nearby spring systems between Springdale and Beatty. They are isolated; there are no known or probable connections between them and other members of the *B. boreas* species group (Nevada Division of Wildlife, 2000). Early anecdotal population counts were highly variable. Linsdale (1940) saw "many small young toads" in 1931; Wright and Wright (1949) found 5 and 4 on consecutive days in 1942; Savage (1960) found 5, "a few," "thousands," and 40, during March, May, June, and November 1958. More recent opportunistic counts varied between 20 adults (Altig and Dodd, 1987) and larger numbers (e.g., 98 adults, Altig, 1981; 49 adults, Maciolek, 1983a; 15 breeding pairs, Hoff, 1994; 40 adults, Heinrich, 1996).

Population estimates based on mark–recapture data (Stein, 1996, 1999; Sjoberg, 2000, 2001; Stein et al., 2000) led to higher numbers. A population size of 536 adults was estimated in 1997 (Stein, 1999). More accurate measures based on animals tagged with passive internal transponders (PIT tags) were made in 1998–2001. Between 13 and 15 localities were surveyed three times each year and 2,895 adults were tagged by 2001. Data for 1999 were thoroughly analyzed and the population size was estimated at 4,697 ± 715 adults at surveyed localities (Stein, 1999; Stein et al., 2000). A potential population of 25,000 toads throughout Oasis Valley was estimated based on the number of toads at surveyed sites and potential habitat in the Valley as estimated from satellite imagery (Stein et al., 2000; Sjoberg, 2002; J. Sjoberg, personal communication). Preliminary analyses from surveys in 2000 (1,415 adults captured, 875 new captures) and 2001 (1,185 adults captured, 667 new captures) suggest similar population sizes and that the population may be stable in the short term (Sjoberg, 2000, 2001; J. Sjoberg, personal communication). Stein (2000) concluded that there was no evidence of a long-term population decline of Amargosa toads in the Oasis Valley, but fluctuating populations might account for variability in past population estimates. Perhaps the greatest concern is the limited distribution of a species that numbers 25,000 at best (and may be much smaller) and the potential susceptibility of its riparian habitat in a desert environment.

Federal designations for the Amargosa toad varied similarly. *Bufo nelsoni* was a candidate species (Category 2) for listing under the Endangered Species Act as early as 1977 (Federal Register, 1977). They were elevated to a Category 1 candidate (Federal Register, 1982), but reduced to Category 2 (Federal Register, 1985). In 1994, the U.S. Fish and Wildlife Service was petitioned to emergency list *B. nelsoni* as Endangered (Biodiversity Legal Foundation, 1994; petition date 19 September 1994). *Bufo nelsoni* was again elevated to a Category 1 species (Federal Register, 1994 [15 November]) and in a 90-d finding the U.S. Fish and Wildlife Service found that the petition action might be warranted

(Federal Register, 1995c). However, after finding that toads were more abundant and widespread than stated in the petition (Heinrich, 1996) and habitat was being managed by the Bureau of Land Management, The Nature Conservancy, and private land owners (Stein, 1999), the U.S. Fish and Wildlife Service recommended removal of the Amargosa toad from Category 1 (USFWS, 1995) and determined listing was not warranted (Federal Register, 1996a). Amargosa toads are currently classified as a Species of Concern (a new category similar to Category 2, Federal Register, 1996b), which does not provide any federal protection to the species or its habitat.

Amargosa toads were classified as a Protected amphibian by the State of Nevada in 1998, which does provide some protection. The Nevada Natural Heritage Program gave Amargosa toads a designation of highest global and state concern (G1S1; Clemmer, 1995). An Amargosa Toad Working Group was formed in 1994 and meets biannually to implement and provide oversight for conservation actions (Burroughs, 1999; Nevada Division of Wildlife, 2000). Threats to these toads include direct kill and habitat disturbance by wild burros, livestock, and off-road vehicles; predation and competition by nonnative aquatic species; and alteration of riparian plant communities by invasion of tamarisk (*Tamarix* sp.; Burroughs, 1999; Nevada Division of Wildlife, 2000). Habitats have been affected by lowered water tables due to increased groundwater use for municipal, agricultural, and industrial purposes. Habitats also have been changed or eliminated by human activities including the alteration of springs or direct diversion of water for agricultural use and urban development. The Amargosa River channel and its riparian corridor were altered for flood control and vehicular traffic. The river channel is assumed to have served as the primary route for toad movements between populations and disruption may have resulted in fragmentation of toad habitats.

A formal multiparty conservation agreement was completed (Nevada Division of Wildlife, 2000), and a long-term species management plan is being developed (J. Sjoberg, personal communication). Some conservation measures have been implemented. Protective fencing was provided for several localities (Nevada Division of Wildlife, 2000; Burroughs, 2001), including Indian Springs in 1993 and Crystal Springs in 1995. The Bureau of Land Management (BLM) removed 900 feral burros in 1995 and 1996 (Nevada Division of Wildlife, 2000; Bureau of Land Management, 2002). The BLM and cooperators also initiated a program to remove invasive and non-native tamarisk trees (Nevada Division of Wildlife, 2000; Burroughs, 2001; Bureau of Land Management, 2002). Measures were initiated

to control nonnative crayfish (*Procambarus* sp.), American bullfrogs (*Rana catesbeiana*), and black bullheads (*Ameiurus melas*, Nevada Division of Wildlife, 2000; Burroughs, 2001), which may prey on or compete with *B. nelsoni* (Hoff, 1994; Nevada Division of Wildlife, 2000). Local volunteer landowners are creating and maintaining Amargosa toad habitat and participating in long-term planning for conservation and habitat protection (Burroughs, 1999). Beginning in 1998, surveys were conducted at least three times annually to identify population size, habitat use, movement and distribution of individuals, and life history parameters (Nevada Division of Wildlife, 2000; Sjoberg, 2000, 2001; Stein et al., 2000; Burroughs, 2001). Much of the potential and historical range of Amargosa toads is now protected either formally (land acquisitions and agreements) or by informal agreements with landowners (J. Sjoberg, personal communication). The Nature Conservancy purchased over 263 ha of land in the Oasis Valley (Rogers, 1999; Nature Conservancy, 2002). Habitat has been restored or modified at several sites to control vegetation and manage water in riparian habitats (Burroughs, 2001). A nature trail and educational visitor center is planned near Beatty (Burroughs, 2001). Databases for *B. nelsoni* and all sensitive species in the Oasis Valley are being collected and managed by the Nevada National Heritage Program (G. Clemmer, personal communication). A Web site (Simandle, 2002) has been developed from which information on *B. nelsoni* can be obtained.

Information on the ecology and life history of Amargosa toads is fragmentary, but aspects of habitat and habitat preference, life history and population biology parameters, behavior, and causes of mortality including cannibalism and captive rearing have been described (Wright and Wright, 1949; Savage, 1960b; Savage and Schuierer, 1961; Altig, 1981; Maciolek,

1983a; Hoff, 1994, 1996; Clemmer, 1995; Heinrich, 1996; Stein et al., 2000) and are reviewed elsewhere (Nevada Division of Wildlife, 2000). Researchers at the University of Nevada, Reno, are currently investigating metapopulation structure and its implications for conservation and habitat restoration in Amargosa toads and quantifying characteristics of oviposition sites and habitats of adult toads (Burroughs, 2001; D. Jones and E. Simandle, personal communication). Recently, Jones et al. (2003) report predation on Amargosa toads by introduced American bullfrogs.

Acknowledgments. Thanks to Steve Corn, Darrel Frost, David Wake, Eric Simandle, Glenn Clemmer, and Jon Sjoberg who all provided comments that contributed to this paper. Thanks to Glenn Clemmer (Nevada Natural Heritage Program) for providing many unpublished reports and correct citations from the Federal Register. Thanks to Jon Sjoberg (Nevada Division of Wildlife), who provided unpublished reports from 2000 and 2001. Research on toad genetics of the population from Haiwee Springs was supported by a contract issued to California State University, Dominguez Hills by the Directorate of Public Works, National Training Center, Fort Irwin, California, to D.J.M., and by the Natural Sciences and Engineering Research Council of Canada grant A3148 to R.W.M.

Bufo punctatus Baird and Girard, 1852(a)
RED-SPOTTED TOAD

Brian K. Sullivan

1. Historical versus Current Distribution.
The type locality of red-spotted toads (*Bufo punctatus*) is the Rio San Pedro, a tributary of the Rio Grande, in Val Verde County, Texas (Baird and Girard, 1852a). Their range extends from southwestern Kansas, western and southern Oklahoma, central

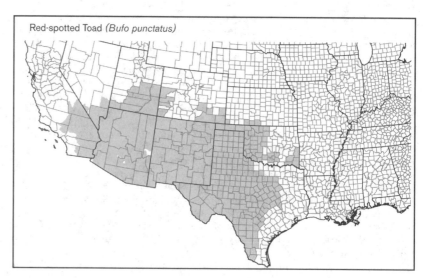

Red-spotted Toad (*Bufo punctatus*)

and western Texas, New Mexico, south-western Colorado, Arizona, southern Utah, southern Nevada, and southeastern California, south into Mexico. Red-spotted toads are found at elevations from below sea level to almost 2,000 m (Stebbins, 1951). Additional comments on their distribution can be found in Collins (1982), Creusere and Whitford (1976), Dixon (1987), Hammerson (1982a), Lowe (1964), and Turner and Wauer (1963).

2. Historical versus Current Abundance. Historical abundance is unknown. In central Arizona, localized populations can be abundant. For example, in early April 1991, 20 males were observed chorusing in a 30 m stretch of the New River, Maricopa County, Arizona; 65–75 d later (June), hundreds of toadlets metamorphosed at this site. Another population breeding during the summer (August 1992) in a cattle tank in north-central Maricopa County contained a minimum of 50 calling males and 20 amplectant pairs (personal observations).

David Bradford (personal communication) and colleagues surveyed sites in southern Nevada and found red-spotted toads at every (n = 16) historical collecting locality examined; they noted that red-spotted toads appeared to be thriving at many disturbed sites (e.g., sites altered for livestock use). Similarly, central Arizona populations occurring in rocky upland sites often use disturbed sites such as cattle tanks for breeding following summer rainstorms (personal observations).

3. Life History Features.
A. Breeding. Reproduction is aquatic.
i. Breeding migrations. Breeding occurs from March–September (Stebbins, 1951; personal observations). In the Sonoran Desert, populations along small streams breed in the spring (March–June), whereas populations occurring in rocky desert uplands breed in rain-formed pools during the summer rainy season (June–September). In the Sonoran Desert of southern California (Palm Springs), stream-dwelling populations breed from April–June, depending on the availability of water (Tevis, 1966). Long-distance migrations of large numbers of individuals have not been reported. Adults in relatively arid regions probably reside throughout the year in the vicinity of streams used for breeding. Breeding periods span 2–4 wk for stream-breeding populations and 1–5 nights for rain-pool breeding populations (personal observations). Males call in shallow water, completely exposed on rocks in or near the water, or on land away from water including from burrows and under rocks (Stebbins, 1951; Turner, 1959; Tevis, 1966; Ferguson and Lowe, 1969; personal observations).
ii. Breeding habitat. Small, rocky streams and springs with shallow pools are used for breeding in the spring; rain-formed

pools in rocky desert upland areas are used during the summer (Stebbins, 1951; Turner, 1959; Tevis, 1966; Creusere and Whitford, 1976; Sullivan, 1984; personal observations).
B. Eggs.
i. Egg deposition sites. Eggs are usually laid singly, rather than in gelatinous strands, in rocky streams, springs, and shallow pools; egg capsules are small, 3–4 mm in diameter.
ii. Clutch size. According to Tevis (1966), the "number of eggs per mass varied from 30–5,000 (average 1,500)." Stebbins (1951, 1985) has also described eggs and larvae.
C. Larvae/Metamorphosis. Larvae are generally black with metallic bronze flecks and are often observed clustered in large aggregations in stream habitats, resting on muddy substrates. In central Arizona, the larval period lasts about 8 wk in stream-breeding populations (31 March–1 June). Altig et al. (1998) provided a key to larvae of red-spotted toads and other sympatric anurans in the Southwest. Tevis (1966) described larval development and behavior in a stream-dwelling population in southern California. Luepschen (1981) described albino larvae.
D. Juvenile Habitat. Unknown; recently metamorphosed young have been observed near breeding habitats in Arizona and California (Stebbins, 1951; personal observations).
E. Adult Habitat. Adults are typically associated with rocky streams in the arid West, although may occur away from drainages in well-drained, rocky soils of the Arizona Upland division of the Sonoran Desert Biotic Community (e.g., south-central Arizona). Bradford et al. (2003) propose that red-spotted toads exist as "patchy populations," implying frequent dispersal among patches and virtually no local extinctions. Aspects of water balance are discussed by Brekke et al. (1991) and Propper et al. (1995).
F. Home Range Size. Along a stream in southern California, males moved ≤185 m and females ≤460 m over a season (February–June; Tevis, 1966). Additional data on movements for a population in Death Valley, California, are presented in Turner (1959).
G. Territories. Unknown. Males defend calling territories at breeding sites; interindividual distances typically range from 1–3 m (Sullivan, 1984; personal observations). Males engage in amplexus-like wrestling bouts during territorial disputes (Sullivan, 1984).
H. Aestivation/Avoiding Desiccation. Unknown. Adults presumably aestivate in areas lacking summer rainfall.
I. Seasonal Migrations. Unknown. Tevis (1966) considered relatively large (between 450–850 m) movements of a small number of individuals as migrations (see "Home Range" above).

J. Torpor (Hibernation). Unknown, although adults presumably hibernate during cooler and/or drier months (October–February; see Johnson et al., 1948) over most of the range.
K. Interspecific Associations/Exclusions. Hybridization documented between red-spotted toads and western toads (B. boreas) in California (Feder, 1979), Great Plains toads (B. cognatus) in central Arizona (Sullivan, 1990), Sonoran green toads (B. retiformis) in southern Arizona (Bowker and Sullivan, 1991), and Woodhouse's toads (B. woodhousii) in northern Arizona and southern Colorado (Malmos et al., 1995). Hybrids are generally easily identified by their morphology (intermediate to parentals) and aberrant calls.

Red-spotted toads breed in association with Arizona toads (B. microscaphus), Woodhouse's toads, canyon treefrogs (Hyla arenicolor), and Great Basin spadefoot toads (Spea intermontana) in southern Utah. In south-central Arizona, red-spotted toads breed in association with Arizona toads, Woodhouse's toads, canyon treefrogs, and lowland leopard frogs (Rana yavapaiensis) in the spring, and with Colorado River toads (B. alvarius), Great Plains toads, Sonoran green toads, and Couch's spadefoot toads (Scaphiopus couchii) in the summer (personal observations). In southern Nevada, red-spotted toads breed with Woodhouse's toads, Pacific treefrogs (Pseudacris regilla), and relict leopard frogs (R. onca; D. Bradford, personal communication). Creusere and Whitford (1976) suggested that red-spotted toad larvae could not develop to metamorphosis in the presence of larvae of Spea sp. or Couch's spadefoot toads due to heavy predation by their carnivorous larvae.
L. Age/Size at Reproductive Maturity. In central Arizona, calling males at breeding aggregations of five populations (n = 71 males sampled) ranged in size from 47–63 mm SVL. Females in amplexus (n = 8) in a single population in central Arizona ranged from 52–66 mm SVL. For a population in central Arizona, skeletochronology indicated that males matured in their first or second full season following metamorphosis (Sullivan and Fernandez, 1999).
M. Longevity. In a population from central Arizona, average age was 2 yr for both males (n = 22) and females (n = 8); no individuals were >6 yr of age (Sullivan and Fernandez, 1999). Individually marked males have been captured over four seasons in a population in southern California (Tevis, 1966). Additional mark–recapture data are presented by Turner (1959) for a population from Death Valley, California.
N. Feeding Behavior. Red-spotted toads feed on a variety of invertebrates (Stebbins, 1951); reports on diet include Little and Keller (1937), Smith (1950), and Tanner (1931).

O. Predators. Blazquez (1996) reported predation by a water snake (*Nerodia valida*) in Baja California, Mexico; red-spotted toads may be preyed upon by a variety of snakes, birds, and small mammals.

P. Anti-Predatory Mechanisms. Cei et al. (1968) report the presence of presumably toxic indolealkamines from the skin and paratoid glands of red-spotted toads.

Q. Diseases. Unknown.

R. Parasites. Goldberg and Bursey (1991b) reported on the helminth fauna (cestodes, nematodes) for populations in southern Arizona.

4. Conservation.

Widespread declines have not been noted for red-spotted toads, but they are listed by the State of Kansas (in the extreme northeastern portion of their range) as a species in need of conservation (Levell, 1997). David Bradford and his colleagues (personal communication; see also Bradford et al., 2003) documented the continued presence of red-spotted toads at all historical localities they surveyed in southern Nevada. One local decline is suspected and requires continued monitoring: red-spotted toads appear absent from the vicinity of Austin, Texas (D. Hillis, personal communication).

Acknowledgments. David Bradford and Jef Jaeger shared their observations and provided helpful comments on the account. Rob Bowker, Mike Demlong, Matt Kwiatkowski, Erik Gergus, and Keith Malmos assisted with some observations. Funding was provided in part by the Heritage Fund, Arizona Game and Fish Department.

Bufo quercicus Holbrook, 1840
OAK TOAD
Fred Punzo

1. Historical versus Current Distribution.

Oak toads (*Bufo quercicus*) range from the Coastal Plain of southeast Virginia, west to Louisiana. They are found throughout Florida and on some of the lower Florida Keys (Conant, 1975; Dalrymple, 1990; Dodd, 1994; Behler and King, 1998). Their current range has been extended since the early description by Holbrook (1842; see Duellman and Schwartz, 1958; Ashton and Ashton, 1988).

2. Historical versus Current Abundance.

Oak toads are abundant in southern pine wood habitats. Wright (1932; Wright and Wright, 1949) reported as early as 1921 that they were locally abundant in the Okefenokee Swamp (Georgia). Their local abundance was also reported in several locations in Florida (in 1895) and Louisiana (in 1923; Wright and Wright, 1949; Hamilton, 1955). In subtropical Florida, these toads are most common in low elevation pinelands (Dodd, 1994). They remain abundant in undisturbed preferred habitats throughout their range. However, due to rapid human development throughout much of the south and concomitant habitat destruction, as well as the introduction of exotic species via the pet trade (Meshaka, this volume, Part One), their numbers can be expected to steadily decrease (Wilson and Porras, 1983).

3. Life History Features.

A. Breeding. Reproduction is aquatic.

i. Breeding migrations. Mating occurs from April–October, with peak activity in the early spring (Harper, 1931; Einem and Ober, 1956). Heavy, warm spring rains stimulate mating behavior (Wright and Wright, 1949). Males typically call from flooded fields, the edges of ponds and puddles, and from clumps of understory vegetation (Dalrymple, 1990). Oak toads are distinctive in that they are primarily active during the day (Dalrymple, 1990), although both daytime and nighttime breeding has been observed (Punzo, 1992a). Most adults are collected during the summer throughout their range, whereas juveniles are most active during late summer and early autumn (Dodd, 1994).

ii. Breeding habitat. Oak toads prefer shallow pools, cypress and flatwood ponds, and ditches. Eggs are usually attached to grass blades 4–12 cm beneath the surface of the water (Hamilton, 1955; Ashton and Ashton, 1988). Following breeding, adults exhibit a y-axis orientation to exit ponds, suggesting animals use celestial cues to navigate (Goodyear, 1971).

B. Eggs.

i. Egg deposition sites. The eggs are laid in strings or bars that are typically attached to vegetation; each string contains 3–8 eggs (Wright, 1932).

ii. Clutch size. Females typically lay 300–500 eggs, each approximately 1 mm in diameter (Hamilton, 1955; Dodd, 1994). Embryos usually hatch from 24–36 hr after fertilization.

C. Larvae/Metamorphosis. Oak toad tadpoles reach a maximum length of 18–19.4 mm at stage 41 in 4–5 wk, and toadlets range from 7.2–8.9 mm SVL (Volpe and Dobie, 1959). Tadpoles are non-selective filter feeders and ingest a diverse array of algae as well as decaying animal material (Dalrymple, 1990).

D. Juvenile Habitat. Same as adult habitat.

E. Adult Habitat. Oak toads are most often associated with open canopied oak and pine forests containing shallow temporary ponds and ditches (Duellman and Schwartz, 1958; Dodd, 1994) and wet prairies, characterized by short hydroperiods, of the southeastern coastal plain (Hamilton, 1955; Pechmann et al., 1989). Oak toads prefer areas without permanent water and well- to poorly drained soils. They commonly seek refuge under boards and logs or in shallow depressions or burrows surrounded by vegetation, including cabbage palms (*Sabal palmetto*) and saw palmettos (*Serenoa repens*; Hamilton, 1955; Duellman and Schwartz, 1958). The minimum habitat area has not been delineated for oak toads (Dalrymple, 1990), however, an earlier study by Hamilton (1954) suggested that an area as small as 1 ac could sustain a breeding population.

F. Home Range Size. Unknown.

G. Territories. Unknown.

H. Aestivation/Avoiding Desiccation. During the winter in the northern part of their range, oak toads remain underground in burrows and shallow depressions for intermittent intervals depending on ambient temperature conditions (Harper, 1931; Wright and Wright, 1949). During cold weather they have also been found in rotten oak logs and under pine bark (Hamilton, 1955).

I. Seasonal Migrations. During heavy rains, adults will move up to 80 m from refuge sites to breeding ponds (Wright, 1932; Dodd, 1994).

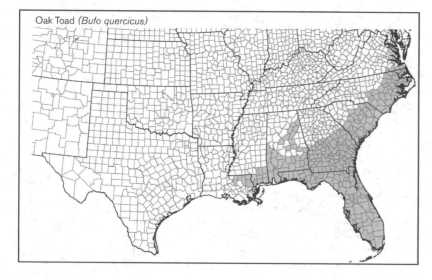
Oak Toad (*Bufo quercicus*)

J. Torpor (Hibernation). May hibernate from early December to early March (Harper, 1931).

K. Interspecific Associations/Exclusions. Unknown.

L. Age/Size at Reproductive Maturity. Oak toads are the smallest toads in North America, with an adult body size ranging from 20–26 mm SVL for breeding males and 24–30 mm for females (Hamilton, 1955). Age at first reproduction is 1.5–2.3 yr (Ashton and Ashton, 1988; Dalrymple, 1989).

M. Longevity. Oak toads are known to live 4 yr (Wright, 1932; Ashton and Ashton, 1988).

N. Feeding Behavior. Adults are insectivorous with a strong preference for ants (Punzo, 1995). Adults also feed on beetles, lepidopterans, aphids, dipterans, orthopterans, spiders, pseudoscorpions, centipedes, and mollusks (Duellman and Schwartz, 1958; Punzo, 1995). Juvenile diets include a high percentage of collembolans, ants, small spiders, and mites (Crosby and Bishop, 1925; Punzo, 1995).

O. Predators. A number of animals have been observed attacking (and in some cases killing and ingesting) oak toads, including raccoons, crows, hog-nosed snakes (*Heterodon* sp.), garter snakes (*Thamnophis* sp.), gopher frogs (*Rana capito*), and marine toads (*B. marinus*; Harper, 1931; Wright, 1932; Hamilton, 1954).

P. Anti-Predator Mechanisms. As with many other bufonids, oak toads will inflate their bodies (unken reflex) when confronted by potential snake predators (Duellman and Trueb, 1986; Heinen, 1995). They also are capable of secreting toxins from the parotoid glands (Licht, 1967c). Their eggs appear to possess some toxic properties (Licht, 1968).

Q. Diseases. Unknown.

R. Parasites. Nematodes of the genera *Oxysomatium* and *Oswaldocruzia* are reported to parasitize oak toads (Hamilton, 1955).

4. Conservation.
In Florida, progressive development, drainage of wetlands, and urbanization always pose a potential threat to amphibian populations. Such anthropogenic activities have substantially reduced natural areas where oak toads breed, including wet prairies, freshwater marshes, margins surrounding cypress swamps, hardwood swamps, and upland areas. Sensitive wetland habitats tend to be small and localized and are therefore more readily threatened by habitat destruction and fragmentation. Sedimentation and runoff of silt during highway and home construction projects, as well as pollution from agricultural pesticides and heavy metals, pose a serious threat to the quality of aquatic breeding sites and survival of tadpoles (Punzo, 1997). State and federal agencies should carefully monitor oak toad populations throughout the state and implement specific habitat management programs to ensure the long-term survival of this species. Further studies should focus on a more detailed analysis of post-breeding movements and home ranges of juveniles and adults.

Bufo retiformis Sanders and Smith, 1951
SONORAN GREEN TOAD

Sean M. Blomquist

1. Historical versus Current Distribution.
Sonoran green toads (*Bufo retiformis*) are known only from Pima and Pinal counties in south-central Arizona and extend south through west-central Sonora to north of Guaymas, Mexico (Hulse, 1978; Stebbins, 1985). The range of Sonoran green toads in the United States extends from San Cristobal Wash and Organ Pipe Cactus National Monument, east to San Xavier Mission and the Altar Valley, and north to Waterman Wash near Mobile, Arizona (Nickerson and Mays, 1968; Hulse, 1978; Stebbins, 1985; Rosen and Lowe, 1996; Sullivan et al., 1996b). Sonoran green toads are found at elevations from 150–900 m (Bogert, 1962; Stebbins, 1985; Sullivan et al., 1996b).

The range of Sonoran green toads is thought to be limited to semi-arid habitats and may be expanding due to irrigation associated with increasing agricultural activity (Bogert, 1962). Hulse (1978) suggests that if this trend continues, Sonoran green toads will expand northward into the irrigated lands of Santa Rosa and Gila Valleys in Arizona. Sullivan et al. (1996b) did not find evidence to support this trend, but recent sightings near Mobile, Arizona, possibly support the northward expansion of Sonoran green toads (Sullivan, this volume, Part One; B. K. Sullivan, personal communication).

2. Historical versus Current Abundance.
No estimates of abundance or censuses have been published. Large breeding aggregations (e.g., 30–200 individuals) have been reported by Bogert (1962) and Sullivan et al. (1996b, 2000), but the current status of the species remains unknown (Bury et al., 1980; USFWS, 1989).

3. Life History Features.
A. Breeding. Reproduction is aquatic.

i. Breeding migrations. Sonoran green toads are explosive breeders (Sullivan et al., 1996b). Breeding occurs opportunistically in July–August with the onset of the summer rains (Savage, 1954; Bogert, 1962; Sullivan et al., 1996b; Behler and King, 1998). Males arrive at temporary pools 1–2 d after rains begin (Bogert, 1962), and it is speculated that Sonoran green toads may delay breeding at temporary pools until water levels have stabilized (Bogert, 1962). Breeding has been recorded at air temperatures from 22.5–32.6 °C (Bogert, 1962; Ferguson and Lowe, 1969; Sullivan et al., 2000).

In low-density breeding aggregations, all males call actively (Sullivan et al., 1996b). Alternative mating approaches (e.g., satellite males and actively searching

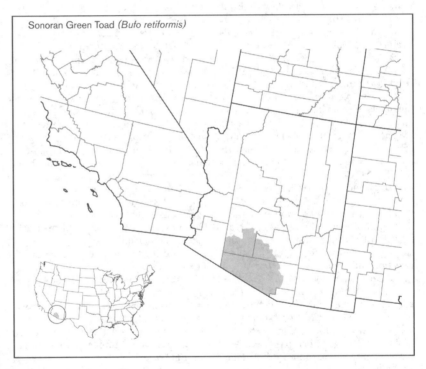

Sonoran Green Toad (*Bufo retiformis*)

males) have been observed at large breeding aggregations (Sullivan et al., 1996b). Their call (Bogert, 1998) sounds to the human ear like "the buzzer on an electric alarm clock" (Savage, 1954) or "a rapid cricket-like trill, on one pitch … a mixture of a buzz and a whistle" (Stebbins, 1954a). Pulse rate and dominant frequency of the call varies with temperature and are negatively correlated with male size (Sullivan et al., 2000).

ii. Breeding habitat. Breeding has been observed in temporary pools formed in roadside ditches, rainwater sumps, cattle tanks, and wash bottoms (Hulse, 1978; Stebbins, 1985; Sullivan et al., 1996b). Males usually call from clumps of vegetation (e.g., grasses or shrubs) on dry, damp, or wet ground within 1–5 m of water (Ferguson and Lowe, 1969; Sullivan et al., 1996b). Males have been observed calling up to 18 m from water (Ferguson and Lowe, 1969). Females approach a calling male on land, and the male continues to call until the female elicits amplexus by touching the male (Bogert, 1962; Sullivan et al., 1996b). The amplexing pair moves to the water to oviposit (Bogert, 1962; Sullivan et al., 1996b).

B. Eggs.

i. Egg deposition sites. Zweifel (1970) suggests that unlike most other true toads, eggs are not laid in strands, but rather individually or in small clumps (see also Ferguson and Lowe, 1969; Hulse, 1978). Eggs are about 1.2 mm in diameter and hatch at a more advanced stage (i.e., Limbaugh and Volpe stage 19) than other true toads (i.e., Limbaugh and Volpe stage 16 or 17; Limbaugh and Volpe, 1957; Zweifel, 1970; Hulse, 1978).

ii. Clutch size. Clutch size is unknown in the wild. Ferguson and Lowe (1969) artificially extracted 50–200 eggs from Sonoran green toad females by hormone injection and stripping.

C. Larvae/Metamorphosis.

i. Length of larval stage. Duration of larval stage in the wild has not been reported, but is expected to be short (e.g., 2–3 wk) based on the duration of temporary pools. Savage (1954) reports collecting a recently metamorphosed individual 13 d after observing breeding in the area. Tadpoles hatch after 2–3 d at 3.1–3.4 mm TL, a size similar to other members of the green toad group (Zweifel, 1970).

ii. Larval requirements.

a. Food. Feeding habits of larvae have not been observed in the wild. Larvae have been raised with limited success in the laboratory on boiled lettuce (Ferguson and Lowe, 1969). Unlike most other true toads, larvae in the green toad group have only two rows of labial teeth posterior to the beak (Zweifel, 1970).

b. Cover. Unknown.

iii. Larval polymorphisms. Unknown.

iv. Features of metamorphosis. Knowledge of newly metamorphosed animals is limited.

Savage (1954) reports collecting one recently metamorphosed individual (20 mm SVL) moving diurnally across hot sand at temperatures of 38 °C.

v. Post-metamorphic migrations. See "Features of metamorphosis" above.

D. Juvenile Habitat. Little is known of juvenile habitat characteristics. Juveniles have been collected in localities with adults and are easily confused with adults of other species in the green toad group (Savage, 1954; Riemer, 1955; Bogert, 1962; Jones et al., 1983; Sullivan et al., 1996b).

E. Adult Habitat. Adult Sonoran green toads have been observed in creosote flats, upland saguaro-palo verde associations, mesquite-grasslands, and arid and semi-arid grasslands; they also extend into the Pacific Coastal Plain near Hermosillo, Mexico (Bogert, 1962; Hulse, 1978; Stebbins, 1985; Sullivan et al., 1996b; Behler and King, 1998). Adults are nocturnal and rarely seen except at breeding aggregations (Stebbins, 1985; Behler and King, 1998). The habits of Sonoran green toads away from breeding aggregations have not been studied.

F. Home Range Size. Unknown.

G. Territories. Unknown.

H. Aestivation/Avoiding Desiccation. Based on the morphology of the head and palmar and metatarsal tubercles, Sonoran green toads are speculated to be fossorial for most of the year, emerging only to breed during summer rains (Savage, 1954). However, whether Sonoran green toads enter a physiological state such as aestivation or torpor has not been studied.

I. Seasonal Migrations. Adult frogs migrate to breeding pools opportunistically at the onset of rains in July–August (Savage, 1954; Bogert, 1962; Sullivan et al., 1996b; Behler and King, 1998). Movements during the nonbreeding season are unknown.

J. Torpor (Hibernation). Unknown. See "Aestivation/Avoiding Desiccation" above.

K. Interspecific Associations/Exclusions. Sonoran green toads will hybridize with little Mexican toads (*B. kelloggi*; Riemer, 1955; Smith and Chrapliwy, 1958; Ferguson and Lowe, 1969; but see Bogert, 1962), green toads (*B. debilis*; Ferguson and Lowe, 1969; Hulse, 1978), and red-spotted toads (*B. punctatus*; Ferguson and Lowe, 1969; Bowker and Sullivan, 1991; Sullivan et al., 1996b; for a discussion of hybridization see Blair, 1972b).

Sonoran green toads have been observed at breeding pools with lowland burrowing treefrogs (*Pternohyla fodiens*), western narrow-mouthed toads (*Gastrophyne olivacea*), red-spotted toads, little Mexican toads, Great Plains toads (*B. cognatus*), Colorado River toads (*B. alvarius*), Couch's spadefoot toads (*Scaphiopus couchii*), and Mexican spadefoot toads (*Spea multiplicata*; Bogert, 1962; Sullivan et al., 1996b).

Bufo debilis, *B. kelloggi*, and *B. retiformis* form a species group referred to as the *B. debilis* group based on morphology and behavior (Bogert, 1962). While the *B. debilis* group was placed in the *B. punctatus* species group (Ferguson and Lowe, 1969), recent analysis of morphology, mtDNA sequence divergence, and behavior indicates the *B. punctatus* group is not monophyletic, and *B. debilis* and *B. retiformis* are closely related (Graybeal, 1997; Sullivan et al., 2000).

L. Age/Size at Reproductive Maturity. Based on 12 specimens from across the range, breeding males are 40–47 mm SVL and breeding females are 45–49 mm (Savage, 1954; Hulse, 1978). Bogert (1962) reports size range of 39–47 mm SVL for males and 46–57 mm SVL for females based on 42 specimens from Sonora, Mexico. Stebbins (1985) reports a size range of 28–56 mm SVL. Behler and King (1998) report a size range of 38–57 mm. The largest individual recorded is a 60 mm SVL female collected near Why, Arizona (Boundy and Balgooyen, 1988).

M. Longevity. The longevity of Sonoran green toads is unknown in the wild. A wild-caught adult lived 3 yr, 4 mo in captivity (Bowler, 1977). As of 29 March 2001, the Arizona Sonora Desert Museum housed a living individual that was collected 17 November 1983, and the museum has records of individuals that lived 15 yr, 3 mo and 14 yr, 7 mo in captivity (C.S. Ivanyi, personal communication).

N. Feeding Behavior. Unknown.

O. Predators. Little is known about predators of Sonoran green toads. American bullfrogs (*Rana catesbeiana*) have been reported to prey on adults toads of the *B. debilis* group (Stuart, 1995).

P. Anti-Predator Mechanisms. Sonoran green toads have large parotoid glands that secrete toxins that are harmful if ingested by a predator (Sanders and Smith, 1951; Lutz, 1971; Hulse, 1978). Release calls are easily elicited by males when handled (Sullivan et al., 1996b). While many authors have commented on the striking green coloration of Sonoran green toads (e.g., Stebbins, 1985; Behler and King, 1998), any function of this coloration is unknown.

Q. Diseases. Unknown.

R. Parasites. The helminths *Distoichometra bufonis* (Cestoda), *Aplectana incerta* (Nematoda), *Aplectana itzocanensis* (Nematoda), *Oswaldocruzia pipiens* (Nematoda), *Psyaloptera* sp. (Nematoda), and *Rhabdias americanus* (Nematoda) have been found infecting Sonoran green toads (Goldberg et al., 1996a).

4. Conservation.
The status of Sonoran green toad populations is unknown, and there is no monitoring program for this species. Scientists currently working in the Sonoran Desert in Arizona observe large breeding aggregations

of toads during and after summer monsoon storms (S.M.B., C.R. Schwalbe, B. K. Sullivan, personal observations).

Sonoran green toads were removed from C.I.T.E.S. Appendix II because the species is thought to have no major international trade or human threats and is protected by state and federal laws (C.I.T.E.S., 2000; USFWS, 2001b). Sonoran green toads are protected as Sujeta a Protección Especial (Determined Subject to Special Protection) in Mexico, meaning utilization is limited due to reduced populations, restricted distribution, or to favor recovery and conservation of the taxon or associated taxa. Sonoran green toads are ranked at G3G4 and S4 by the State of Arizona, meaning the species is apparently secure but uncommon in parts of its range. Collection of Sonoran green toads in Arizona is limited to 10 toads per year with a fishing license.

Bufo speciosus Girard, 1854
TEXAS TOAD

Gage H. Dayton, Charles W. Painter

1. Historical versus Current Distribution.
In the United States, Texas toads (*Bufo speciosus*) are distributed throughout the

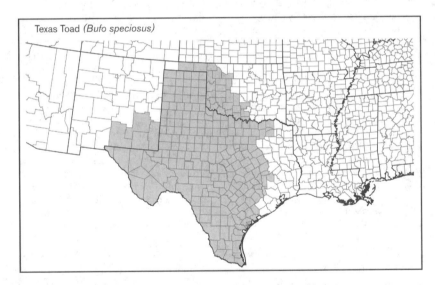

Texas Toad (*Bufo speciosus*)

western 2/3 of Texas, west into southeastern New Mexico, and north into western Oklahoma, including the panhandle (McAllister and Trauth, 1993; Killebrew et al., 1995; Dixon, 2000). The distribution of Texas toads has not changed as dramatically as other *Bufo* species throughout the United States (J. Dixon, personal communication).

2. Historical versus Current Abundance.
Brown (B.C., 1950) reported that Texas toads are one of Texas' most abundant toads. However, data on changes in abundance of Texas toads over their entire range do not exist. Dixon (2000) reports that populations of Texas toads in

the Rio Grande Valley, Texas, are declining due to the heavy use of pesticides and herbicides on agricultural lands. The decline of Texas toads in the Rio Grande Valley is somewhat alarming, as Texas toads represent one of the most abundant toads in the region (Thorton, 1977). Population trends of Texas toads in southeast New Mexico appear stable (C.W.P., personal observations).

3. Life History Features.
A. Breeding. Reproduction is aquatic.
i. Breeding migrations. Adults aggregate at temporary pools within a few hours after a rain event. Breeding males call any time of day for up to 4 d after a rain event, calling most intensely at night (Moore, 1976; G.H.D., unpublished data). Males call while clinging to partially submerged vegetation or in the shallow regions of the breeding site. Their call consists of a continuous series of explosive trills, each lasting approximately 1.5 s, given at intervals of about 1 s (Stebbins, 1985). Axtell (1958) reported that calling males were "passive in their relationship with the females until they approached and made body contact." Axtell (1958) also reported that females preferred the loudest calling male

(also the largest) over smaller individuals in the vicinity.
ii. Breeding habitat. Texas toads breed in temporary rain pools in open areas near streams, in pools in creek valleys, and in irrigation ditches, cattle tanks, and temporary desert pools (Bragg and Smith, 1943; Wright and Wright, 1949; Moore, 1976; Dayton and Fitzgerald, 2001). Breeding season is from April–September following heavy rains (Bragg and Smith, 1943; Moore, 1976).
B. Eggs.
i. Egg deposition sites. Eggs are typically deposited around the base of submerged vegetation. Multiple pairs will deposit eggs around the same clump of grass.

ii. Clutch size. Eggs are brown and yellow, crowded (14–20 eggs/30 mm), about 1.4 mm in diameter, and hatch in 2 d (Wright and Wright, 1949). There are no data on clutch size for Texas toads. However, the clutch of an average sized Great Plains toad (*B. cognatus*), closely related to Texas toads, is about 11,000 eggs (Krupa, 1994); this number could be used as an estimate until data on Texas toads are available.
C. Larvae/Metamorphosis.
i. Length of larval stage. The tadpole stage lasts from as few as 18 d (Moore, 1976) to approximately 60 d (Wright and Wright, 1949).
ii. Larval requirements.
a. Food. Tadpoles scrape algae off of submerged vegetation and substrate. There are no data on the specific dietary preference of Texas toad tadpoles.
b. Cover. Cover may be aquatic vegetation or rocky substrate, although this may vary from site to site. When startled, tadpoles will often retreat to deeper areas within the pool.
iii. Larval polymorphisms. Have not been documented.
iv. Features of metamorphosis. Studies have reported time to metamorphosis from 18 d–60 d (Wright and Wright, 1949; Moore, 1976). Thus, timing and duration of metamorphosis varies depending upon site characteristics (i.e., water temperature, presence of predators, densities, and resource availability).
v. Post-metamorphic migrations. Recently metamorphosed toadlets usually remain near the natal pool until the pool has dried (Degenhardt et al., 1996). No data are available on migration distance; however, dispersal is likely correlated to the proximity of suitable adult habitat. Moore (1976) monitored the growth of recently metamorphosed toads for 7 mo in southern Texas. He calculated a growth rate of approximately 10 mm/mo, which indicates that toads born early in the breeding season would reach sexual maturity in time to breed the following year. Toadlets will burrow into the soil, take refuge in mud cracks, or retreat beneath vegetation and other cover to prevent desiccation.
D. Juvenile Habitat. Although Texas toads burrow, they also utilize habitat beneath rocks and in mud cracks and gopher burrows (Wilks, 1963; G.H.D., personal observation). The critical habitat characteristics for juvenile Texas toads are likely similar to those of the adults, that is, high water retention in soils and/or relatively high humidity in refuge sites.
E. Adult Habitat. Texas toads are a desert species found in association with permanent streams, irrigation ditches, watering tanks, and ephemeral pools (Jameson and Flury, 1949; Wright and Wright, 1949; Dayton, 2000a). They are found in grassland, open woodland, and mesquite-savanna habitats. Texas toads prefer sandy soils, where they burrow. However, they

are associated with soils that are frequently inundated and have a relatively high percentage of clay (Dayton, 2000a). Habitat characteristics for adults are likely similar to those of juveniles. It is crucial for adults to take refuge in sites with relatively high humidity and/or water holding capacity to prevent desiccation during dry periods that can last for several months.

F. Home Range Size. Unknown.

G. Territories. Unknown.

H. Aestivation/Avoiding Desiccation. Texas toads burrow in mud, inhabit gopher burrows, and take refuge beneath rocks and in mud cracks to avoid desiccation (Wright and Wright, 1949; Wilks, 1963; G.H.D., personal observation).

I. Seasonal Migrations. During the breeding season, migrations to breeding sites occur. Distance traveled by the toads is unknown and undoubtedly varies depending upon suitable habitat surrounding each breeding site.

J. Torpor (Hibernation). During dry periods, toads remain dormant in refuge sites.

K. Interspecific Associations/Exclusions. Texas toads breed in sites also occupied by Coastal-Plain toads (*B. nebulifer*), green toads (*B. debilis*), western narrow-mouthed toads (*Gastrophryne olivacea*) and Couch's spadefoot toads (*Scaphiopus couchii*, Wright and Wright, 1949; Bragg, 1955a; G.H.D., personal observation). They will also breed and hybridize with Woodhouse's toads (*B. woodhousii*) and Great Plains toads (Blair, 1961b; Ballinger, 1966; Rogers, 1973a,b).

L. Age/Size at Reproductive Maturity. Males 52–78 mm; females 54–91 mm (Wright and Wright, 1949).

M. Longevity. The life span of Texas toads in the wild is unknown. However, a wild-caught specimen was kept in the Philadelphia Zoo for 4 yr, 3 mo. Sex and age at the time of capture of this individual was unknown (Bowler, 1977).

N. Feeding Behavior. Texas toads are opportunistic feeders that take a diverse array of terrestrial and flying arthropods (Degenhardt et al., 1996). The stomach contents from one individual included hemipterans, coleopterans, and hymenopterans (Malone, 1999).

O. Predators. Grackles (*Quiscalus quiscala*), beetle larvae (Dytiscidae, Hydrophilidae), and yellow mud turtles (*Kinosternon flavescens*) will feed on tadpoles in drying pools (Wright and Wright, 1949; Dayton and Fitzgerald, 2001). Garter snakes (*Thamnophis* sp.) also likely prey upon all life stages.

P. Anti-Predator Mechanisms. Texas toads secrete toxic substances to deter predators. Tissue samples from 13 specimens from two localities in Texas revealed the presence of 5-hydroxytryptamine, N-methyl-5-hydroxytryptamine, and dehydrobufotenine (Cei et al., 1968).

Q. Diseases. No disease outbreaks have been reported in Texas toad populations.

R. Parasites. The gall bladder myxosporean *Myxidium serotinum* has been found in Texas toads (McAllister and Trauth, 1995; McAllister et al., 1995a). Kuntz (1941) reported the nematotaeniid tapeworm, *Distoichometra bufonis*, from Texas toads from Oklahoma.

4. Conservation.
Although historical data on the abundance and distribution of Texas toads over their entire range do not exist, population and abundance trends throughout the majority of their range appear to be stable (J. Dixon, personal communication; C.W.P. personal observations). Declines have been observed in agricultural areas in southern Texas where pesticide and herbicide use is heavy (Dixon, 2000). Texas toads can be locally abundant and do not seem to be undergoing the dramatic declines that are being observed for some of the other amphibian species in North America.

Bufo terrestris (Bonnaterre, 1789)
SOUTHERN TOAD

John B. Jensen

1. Historical versus Current Distribution.
Southern toads (*Bufo terrestris*) primarily occur in the Atlantic Coastal Plain from southeastern Virginia to the Florida Keys, and the Gulf Coastal Plain west through the Florida Parishes of Louisiana (Blem, 1979). Gergus (1993) reported disjunct occurrences of southern toads from western Louisiana, however, skepticism exists over the origin of these animals. Although records of southern toads outside of the Coastal Plain in Georgia were misidentified (Laerm and Hopkins, 1997), this species does occur in disjunct portions of the Blue Ridge and Piedmont of neighboring South Carolina (Blem, 1979; Conant and Collins, 1998). Southern

toads are also known from the southern portion of the Ridge and Valley region of Alabama (Mount, 1975). No substantial changes in this species' distribution have been noted.

2. Historical versus Current Abundance. Southern toads are common throughout most of their range (Wright and Wright, 1949; Mount, 1975; Blem, 1979), although they have become increasingly uncommon in areas of Florida where non-native marine toads (*B. marinus*) have become established (Bartlett and Bartlett, 1999a). Krakauer (1968) suggested that the destruction of preferred habitats of southern toads in southern Florida has created better marine toad habitats and that direct competition between the two species is unlikely responsible for the population changes. Bartlett and Bartlett (1999a) also note that despite massive habitat destruction, southern toads remain commonly seen, especially during the breeding season.

3. Life History Features.
A. Breeding. Reproduction is aquatic.

i. Breeding migrations. Southern toads migrate from their sandy upland habitats to the wetland sites in which they breed from February (perhaps earlier) to October (Wright and Wright, 1949; Duellman and Schwartz, 1958; Dundee and Rossman, 1989), although most breeding activity ceases by early summer (personal observations). Breeding aggregations can be enormous (Bartlett and Bartlett, 1999a). Heavy rains at anytime of the year may stimulate chorusing (Krakauer, 1968), although actual breeding may not coincide.

ii. Breeding habitat. Shallow waters, from lake margins to seasonal pools, including cypress ponds and wooded (Carolina) bays (Wright and Wright, 1949), ditches and canals (Dundee and Rossman, 1989; Bartlett and Bartlett, 1999a). Southern toads breed in both temporary and

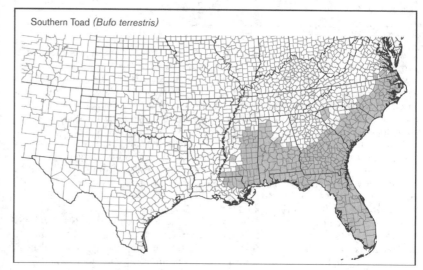

Southern Toad *(Bufo terrestris)*

permanent aquatic habitats (Gibbons and Semlitsch, 1991). Wright and Wright (1949) commented: "On one occasion their eggs were so plentiful in impermanent pools that we wrote: What a frightful waste of frog life in transient pools." Mount (1975) notes that, as opposed to Fowler's toads, he has never seen southern toads breeding in streams or rivers.

B. Eggs.

i. Egg deposition sites. Shallow waters from the littoral regions of lakes to seasonal wetlands, usually amongst aquatic vegetation.

ii. Clutch size. Southern toads lay 2,500–4,000 eggs, 1–1.4 mm in diameter, in long coils that hatch in 2–4 d, depending on water temperature (Wright and Wright, 1949; Ashton and Ashton, 1988; Bartlett and Bartlett, 1999a).

C. Larvae/Metamorphosis.

i. Length of larval stage. Southern toad tadpoles hatch and develop in 30–55 d; tadpoles transform at 6.5–11 mm (Wright and Wright, 1949; Ashton and Ashton, 1988).

ii. Larval requirements.

a. Food. Algae scraped from aquatic vegetation (Ashton and Ashton, 1988). Tadpoles have been observed opportunistically feeding on the eggs of previously killed gravid female southern toads (Babbitt, 1995).

b. Cover. As with most *Bufo* tadpoles, those of southern toads are unpalatable or toxic to many potential predators (Michl and Kaiser, 1963; Lefcort, 1998), therefore cover may not be critically important. However, Babbitt and Jordan (1996) found that southern toads had increased survival in dense plant refugia when pursued by odonates. Southern toad tadpoles frequently school together (Lefcort, 1998) and are often quite conspicuous in open areas of water despite having nearby cover of emergent and submerged plants available (personal observations). Further, southern toad tadpoles select areas of higher temperature (i.e., exposed areas) within their aquatic habitats that favor a higher metabolic rate and faster development (Noland and Ultsch, 1981).

iii. Larval polymorphisms. Unknown for this species.

iv. Features of metamorphosis. The age at metamorphosis for toadlets studied in South Carolina was 54–63 d at a mean SVL of 7.46 mm (Beck and Congdon, 1999).

v. Post-metamorphic migrations. Following transformation and prior to emigration, juvenile southern toads forage for several weeks around the edge of the pond from which they emerged (Beck and Congdon, 1999).

D. Juvenile Habitat. Unknown, though likely similar to adults.

E. Adult Habitat. Southern toads can be common in a variety of terrestrial habitats including agricultural fields, pine woodlands, hammocks, and maritime forests (Wright and Wright, 1949; Kraukauer, 1968; Wilson, 1995). Sandy soils are preferred (Blem, 1979; Martof et al., 1980) to accommodate their burrowing needs. Adults may also take refuge under logs or other debris during the day (Ashton and Ashton, 1998). Southern toads tolerate humans well, and Ashton and Ashton (1988) note that they are a common yard and garden toad, frequently found under lights at night feeding on attracted insects (see also Neill, 1950a; Behler and King, 1998). It is not known whether the habitat characteristics of males differ substantially from that of females.

F. Home Range Size. "May encompass an area a mile [1.6 km] wide" (Bogert, 1947).

G. Territories. Unknown.

H. Aestivation/Avoiding Desiccation. Southern toads remain nocturnally active throughout the summer.

I. Seasonal Migrations. See "Breeding migrations" above.

J. Torpor (Hibernation). During the coldest months, southern toads are much less conspicuous and may remain inactive in their burrows or under cover items. Neill (1950a) reported that southern toads dig into the ground to a depth of a foot (30 cm) or more during this time. However, Einem and Ober (1956) reported feeding activity of southern toads throughout the winter in central Florida.

K. Interspecific Associations/Exclusions. Southern toads will hybridize with Fowler's toads (*B. fowleri*; Neill, 1949a; L. E. Brown, 1969; Blem, 1979) and American toads (*B. americanus*; Mount, 1975; Blem, 1979). In Georgia and perhaps elsewhere, southern toads and American toads appear strongly allopatric along the Fall Line (Neill, 1949a; personal observations). Many other anurans share breeding sites with southern toads including, but not limited to, oak toads (*B. quercicus*; Bogert, 1947), southern cricket frogs (*Acris gryllus*), barking treefrogs (*Hyla gratiosa*), pine woods treefrogs (*H. femoralis*), squirrel treefrogs (*H. squirella*), and southern leopard frogs (*Rana sphenocephala*; personal observations).

L. Age/Size at Reproductive Maturity. Males 42–82 mm SVL; females 44–92 mm SVL (Wright and Wright, 1949). Dundee and Rossmann (1989) cite a maximum of 113 mm, although note that the largest animal known from Louisiana is 64 mm. Adults found on many of the coastal islands of South Carolina, Georgia, Mississippi, and Florida can reach exceptionally large sizes (Smith and List, 1955; Sanders, 1961; Mount, 1975; personal observations); however, island populations of dwarfed southern toads have also been reported (Duellman and Schwartz, 1958).

M. Longevity. Perhaps at least 10 yr (Ashton and Ashton, 1988).

N. Feeding Behavior. Typically will eat small invertebrates including beetles, earwigs, ants, cockroaches, mole crickets, and snails (Duellman and Schwartz, 1958), but are known to eat anything they can swallow (Ashton and Ashton, 1988). Beetles were especially abundant in the diet of southern toads from Georgia and southern Florida (Neill and Allen, 1956; Duellman and Schwartz, 1958). Huheey (1980) found that southern toads in captivity continued to eat honeybees despite being stung during previous consumptions. Wilson (1995) suggested that this species is an insectivore that captures its prey with its sticky tongue. Southern toads accumulate large fat reserves to sustain them through a continuous period without food, from late fall until breeding is completed (Smith, 1976).

O. Predators. Hog-nosed snakes (both *Heterodon platyrhinos* and *H. simus*) feed heavily on adult southern toads (Bogert, 1947; Gibbons and Semlitsch, 1991; Palmer and Braswell, 1995). Adult southern toads have also been found in the diets of water snakes (both *Nerodia erythrogaster* and *N. fasciata*; Neill and Allen, 1956; Palmer and Braswell, 1995) and eastern indigo snakes (*Drymarchon couperi*; Bogert, 1947). Two-toed amphiumas (*Amphiuma means*), lesser sirens (*Siren intermedia*), and odonates are known predators of southern toad tadpoles (Babbit and Jordan, 1996; Lefcort, 1998). Additionally, giant water bugs (Hemiptera, Belcstomatidae) will eat breeding adults (McCoy, 2003).

P. Anti-Predator Mechanisms. As with other *Bufo* sp., southern toads will inflate their lungs, thereby puffing up their body in an effort to appear much larger to a potential predator (Duellman and Trueb, 1986; personal observations). Bufotoxin, a cardiotoxic steroid (Duellman and Trueb, 1986) released from the parotoid glands of *Bufo* sp. is toxic and/or unpalatable to many would-be predators. In fact, when triggered to inflate their bodies, *Bufo* sp. flex their head downward to present the parotoid glands to potential predators. Truitt (1964) observed defensive behavior of a southern toad towards an eastern hog-nosed snake. These behaviors included crab-like movements, charges, and sand-kicking directed at the snake, but were ultimately unsuccessful in deterring the snake and the toad was consumed. Jensen (1996) found that eggs were fatally toxic, by contact, to southern cricket frogs that fell into pitfall buckets where captured female southern toads previously oviposited in the few inches of water present in the bottom of the buckets. *Bufo* eggs are also unpalatable or toxic to many animals that eat them (Licht, 1968), and the affected water surrounding the eggs may provide a chemical barrier that repels potential amphibian egg predators such as newts (*Notophthalmus* sp.; Jensen, 1996).

Q. Diseases. Unknown.

R. Parasites. The nematode *Cosmocercoides dukae* has been found in southern toads from south Florida (Walton, 1940).

4. Conservation.
Southern toads are common throughout most of their range, although they have become increasingly uncommon in areas of Florida where non-native marine toads have become established. Southern toads remain commonly seen, especially during the breeding season, even in areas with extensive habitat destruction. They are not protected by either state or federal regulations.

Bufo valliceps Wiegmann, 1833
GULF COAST TOAD

See Joseph Mendelson's account for *Bufo nebulifer.*

Bufo velatus Bragg and Sanders, 1951
EAST TEXAS TOAD

Editor's note: While considered to be a subspecies of Woodhouse's toads *(Bufo woodhousii)* by Crother et al. (2000) the arguments for considering East Texas toads *(Bufo velatus)* as distinct species are presented by Dixon (2000). Those interested in the biology of "*Bufo velatus*" are referred to Bragg and Sanders (1951) and Sanderss (1953, 1978, 1986, 1987).

Bufo woodhousii Girard, 1854
WOODHOUSE'S TOAD
Brian K. Sullivan

1. Historical versus Current Distribution.
The type locality of Woodhouse's toads *(Bufo woodhousii)* is from San Francisco Peaks, New Mexico (=Arizona; Girard, 1854). Woodhouse's toads occur from south-central Washington, far eastern Oregon and California, Nevada, south-central Idaho, southern and eastern Montana, and southern North Dakota, south through Utah, Colorado, South Dakota, Nebraska, Kansas, Oklahoma, Texas, New Mexico, and Arizona to Mexico. They range from near sea level (Salton Sea, California) to almost 2,500 m (Stebbins, 1951; personal observations). Additional comments on distribution can be found in Tanner (1931), Bragg (1940b), Smith (1950), Lowe (1964), Collins (1982), Hammerson (1982a), Dixon (1987), Conant and Collins (1991), and Degenhardt et al. (1996).

Although in the past some have considered Woodhouse's toads and Fowler's toads *(B. fowleri)* to be conspecific (e.g., Conant and Collins, 1991), Sullivan et al. (1996a) summarized evidence (behavior, morphology, genetics) supporting the recognition of these taxa as separate, full species. One western subspecies is recognized *(B. w. australis;* Shannon and Lowe, 1955); the status of a second subspecies *(B. w. velatus),* reported to occupy the zone of intergradation between Woodhouse's and Fowler's toads, is uncertain (Conant and Collins, 1991; Sullivan et al., 1996a).

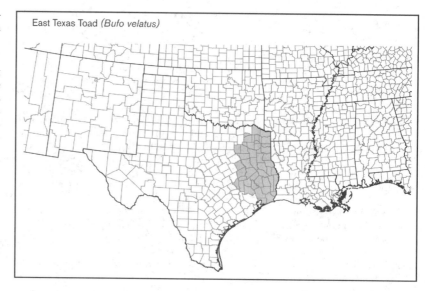

East Texas Toad *(Bufo velatus)*

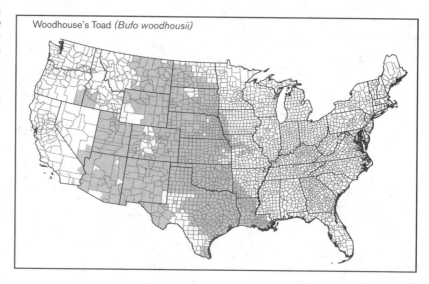

Woodhouse's Toad *(Bufo woodhousii)*

2. Historical versus Current Abundance.
Historical abundance is unknown; localized populations can be abundant, even in urbanized and other disturbed habitats (e.g., Hammerson, 1982a). David Bradford and colleagues (personal communication) surveyed sites in southern Nevada and found Woodhouse's toads at all (n=4) historical collecting localities they examined. They noted that Woodhouse's toads are now present at sites near Las Vegas that were not occupied historically. Similarly, Sullivan (1986b, 1993) noted the apparent range expansion of Woodhouse's toads and coincident decline of Arizona toads *(B. microscaphus)* at degraded riparian sites in Arizona. Hammerson (1982a) noted that Woodhouse's toads appear to have expanded their range in agricultural areas formerly occupied by red-spotted toads *(B. punctatus)* in Colorado. Woodhouse's toads also appear to be expanding their range in southeastern California in the vicinity of Palm Springs.

3. Life History Features.
   A. Breeding. Reproduction is aquatic.
   i. Breeding migrations. Across their range, breeding occurs from February–September (Stebbins, 1951; personal observations). In the Sonoran Desert, populations along larger streams and rivers breed on warm (air temperature >15 °C), rainless nights in the spring (February–June), whereas populations occurring in open desert flats breed immediately after summer rainstorms (June–September). In the Great Plains, populations typically breed in ponds, lakes, and rain-formed pools from February–July. Breeding periods span 2–4 mo for riparian-breeding populations and 1–5 nights for pool-breeding populations. For example, in central Arizona, chorus activity occurred on a total of 25–56 nights over the course of three spring seasons, whereas in central Texas, chorus activity occurred on only a total of 15–17 nights through spring and summer in each of two seasons (Sullivan, 1989b).

Males have been observed calling in shallow water, completely exposed on shore at the water's edge, and on land near water (Stebbins, 1951; personal observation). The advertisement call is a nasal "waaah," from 1–4 s in duration, 1.4–2.0 kHz in frequency, and rapidly pulsed (80–200 p/s; see Stebbins, 1951; Sullivan et al., 1996a).

Long-distance migrations of large numbers of individuals have not been reported. Adults in populations from relatively arid regions may reside throughout the year in the vicinity of streams used for breeding.

ii. Breeding habitat. Standing water is apparently preferred for breeding—either pools in river channels following spring run-off, artificial ponds and reservoirs, or rain-formed pools and cattle tanks in open desert flats (Stebbins, 1951; Sullivan, 1982b; personal observation). A great variety of breeding habitats are used (Bragg, 1940b; Stebbins, 1951; Hammerson, 1982a; Sullivan, 1982b, 1989b).

B. Eggs.

i. Egg deposition sites. Descriptions of eggs and larvae can be found in Stebbins (1951, 1985). Altig et al. (1998) provide a key to larvae of Woodhouse's toads and their relatives. Eggs are laid in long gelatinous strands and are small, 3–4 mm in diameter.

ii. Clutch size. Krupa (1995) provided data on maximal clutch size (>28,000 eggs) for midwestern Woodhouse's toads.

C. Larvae/Metamorphosis. Larvae are generally black, although the venter of the tail is pale. Larvae are often observed in large aggregations in stream and pond habitats, typically in standing water, resting on muddy substrate. In central Arizona, the length of the larval period was 8 wk for pond breeding populations (31 March–1 June; personal observations), whereas development lasted 5–7 wk in Oklahoma (Bragg, 1940b). Woodward (1987) found that larvae grew more slowly when housed with larger conspecific larvae. Metamorphosis begins at a tadpole length of about 30 mm total length (Bragg, 1940b; personal observation). Recently metamorphosed toadlets average 12–15 mm SVL (Bragg, 1940b; personal observation). Details on larval development can be found in Bragg (1940b). Verma and Pierce (1994) described acid tolerance of larvae. Cupp (1980) analyzed thermal preferences of larvae.

D. Juvenile Habitat. Juveniles may be largely restricted to relatively mesic environments associated with breeding habitats of adults. Recently metamorphosed young have been observed near breeding pools in Arizona in April and May (Stebbins, 1951; personal observation). Bragg (1940b) documented rapid growth of almost 10 mm/mo in recently metamorphosed toadlets. Labanick (1976a) and Clarke (1977) described post-metamorphic growth in the closely related Fowler's toad.

E. Adult Habitat. Adults are found in a variety of habitats (Stebbins, 1951). In the western states, Woodhouse's toads are often associated with larger riparian corridors at lower elevations, and moist meadows, ponds, lakes, and reservoirs at higher elevations. They can be common in disturbed habitats such as canals and irrigated fields, golf courses, and urban parks with ponds. O'Connor and Tracy (1992) described thermoregulation and thermal preferences in juveniles, and Swanson et al. (1996) described freeze-tolerance of adult Woodhouse's toads. Additional comments on habitat use can be found in Bragg (1940b).

F. Home Range Size. Unknown.

G. Territories. Unknown. Males defend a calling territory at breeding sites (Sullivan, 1982b).

H. Aestivation/Avoiding Desiccation. Populations breeding along streams may aestivate during the dry summer.

I. Seasonal Migrations. Unknown.

J. Torpor (Hibernation). Unknown, although populations are presumed to hibernate during cooler and/or dryer months (about October–February) over most of the range.

K. Interspecific Associations/Exclusions. Hybridization has been documented between Woodhouse's toads and both Colorado River toads (*B. alvarius*) and Great Plains toads (*B. cognatus*) in Arizona (Gergus et al., 1999), American toads (*B. americanus*) in the Midwest (Malmos, 1992), Fowler's toads in the Midwest (Meacham, 1962; Sullivan et al., 1996a), Houston toads (*B. houstonensis*) in Texas (Hillis et al., 1984), Arizona toads in Arizona and Utah (Sullivan and Lamb, 1988; Sullivan, 1995), red-spotted toads in northern Arizona and southern Colorado (Malmos et al., 1995), and Coastal-Plain toads (*B. nebulifer*) in Texas (Thornton, 1955). Hybrids are generally easily identified by their morphology (intermediate to parentals) and aberrant calls.

In southern Arizona, Woodhouse's toads have been observed breeding with Sonoran Desert toads, Great Plains toads, green toads (*B. debilis*), Arizona toads, red-spotted toads, canyon treefrogs (*Hyla arenicolor*), Couch's spadefoot toads (*Scaphiopus couchii*), American bullfrogs (*Rana catesbeiana*), and lowland leopard frogs (*R. yavapaiensis*). In central Texas, Woodhouse's toads breed sympatrically with northern cricket frogs (*Acris crepitans*), green toads, Coastal-Plain toads, western narrow-mouthed toads (*Gastrophryne olivacea*), spotted chorus frogs (*Pseudacris clarkii*), Strecker's chorus frogs (*P. streckeri*), and Rio Grande leopard frogs (*R. berlandieri*).

L. Age/Size at Reproductive Maturity. In central Arizona, calling males at breeding aggregations (n=61 males sampled) ranged in size from 69–98 mm SVL (Sullivan, 1982b). Reproductive females (n=38)

in this same population ranged from 84–109 mm SVL (Sullivan, 1982b). In central Texas, calling males at breeding aggregations (n=98 males sampled) ranged in size from 73–104 mm SVL (Sullivan, 1989b). Reproductive females (n=22) in this same population ranged from 79–110 mm SVL (Sullivan, 1989b).

M. Longevity. Given observations on close relatives (e.g., Fowler's toads and Arizona toads), it is presumed that males typically reach reproductive maturity within 1 yr of metamorphosis, females within 2 yr.

N. Feeding Behavior. Presumably feed on a variety of invertebrates; descriptions of their diet can be found in Bragg (1940b) and Stebbins (1951). Flowers and Graves (1995) described the diet of juvenile toads immediately following metamorphosis and reviewed previous reports on diet in Woodhouse's toads. Benally et al. (1996) discussed the feeding mechanics of adults.

O. Predators. Bragg (1940b) listed predators of adults, including bullsnakes (*Pituophis melanoleucus*), rat snakes (*Elaphe* sp.), American bullfrogs, and a hawk. Woodward and Mitchell (1990) documented evidence of predation on male (15 of 120) Woodhouse's toads in a breeding aggregation in New Mexico and suggested that a mammalian predator, possibly a skunk, was responsible. I observed similar disemboweled carcasses in central Texas at a breeding aggregation in 1987, and I agree with Woodward and Mitchell (1990) that a small mammalian predator was likely responsible. Parmley (1982) documented predation on Woodhouse's toads by a roadrunner in Texas. Morin (1995) documented predation (in artificial enclosures) by eastern newts (*Notophthalmus viridescens*) and marbled salamanders (*Ambystoma opacum*) on larvae of the closely related Fowler's toads. Similarly, Lawler (1989) documented larval behavior in response to the presence of potential predators, eastern newts and black-banded sunfish (*Enneacanthus obesus*), in Fowler's toads. Kruse and Stone (1984) noted that largemouth bass (*Micropterus salmoides*) find Woodhouse's toad larvae distasteful and learn to avoid them as prey.

P. Anti-Predatory Mechanisms. Skin secretions presumably afford some protection against some predators for larvae (Kruse and Stone, 1984) and adults (Cei et al., 1968).

Q. Diseases. Shiveley et al. (1981) reported on bacterial infections.

R. Parasites. Goldberg et al. (1996c) reported on the helminth fauna (cestodes, nematodes) for populations sympatric with Arizona toads in southern Arizona; Hardin and Janovy (1988) described cestode infections. McAllister et al. (1989) presented a comparative analysis of endoparasites in a community of three bufonids in Texas.

4. Conservation.
Widespread declines have not been noted for Woodhouse's toads, although they are listed as a Protected species by the State of Oregon (Levell, 1997). While Hovingh (1997) indicated that Woodhouse's toads might be absent from southeastern Idaho and adjacent, historically occupied, parts of northern Utah, Mulcahy et al. (2002) confirmed the presence of Woodhouse's toads in both regions. Similarly, D. Bradford and colleagues (personal communication) documented the continued presence of Woodhouse's toads at historical localities in southern Nevada. The expansion of Woodhouse's toads into areas formerly occupied by Arizona toads in central Arizona (Sullivan, 1986, 1993) stands in contrast to the declines noted for other anurans in the past decade. However, some local declines are suspected and require continued monitoring: Woodhouse's toads appear absent from the Santa Cruz River floodplain in the vicinity of Tucson, Arizona (P. Rosen, personal communication), and from the vicinity of Austin, Texas (D. Hillis, personal communication).

Acknowledgments. Rob Bowker, Mike Demlong, Erik Gergus, Tom Jones, Keith Malmos, Kit Murphy, Dave Pfennig, Elizabeth Sullivan and Bill Wagner assisted with some observations. Fieldwork was supported in part by the Heritage Fund, Arizona Game and Fish Department.

Family Dendrobatidae

Dendrobates auratus Girard, 1855
GREEN AND BLACK DART-POISON FROG

Michael J. Lannoo, Priya Nanjappa

1. Historical versus Current Distribution.
Native to Central America, green and black dart-poison frogs (*Dendrobates auratus*) were purposely introduced onto Hawaii by the entomologist David T. Fullway as a part of a larger program by the Territory (at that time) of Hawaii to introduce beneficial animals to assist in controlling non-native insects (McKeown, 1996). According to McKeown (1996), the source of these frogs was from populations on Tobaga or Tobagilla Island off the Pacific Coast of Panama. In Hawaii, green and black dart-poison frogs currently reside only on Oahu, in a few well-vegetated, moist valleys.

2. Historical versus Current Abundance.
Because green and black dart-poison frogs were introduced to Hawaii, their U.S. abundance is obviously higher now than before their introduction. However, these frogs have a restricted range and, according to McKeown (1996), are sensitive to habitat destruction.

3. Life History Features.
A. Breeding. Reproduction is complex. Eggs are laid and tended in leaf litter, and males transport hatchlings to water bodies (see "Breeding migrations" below).

i. Breeding migrations. Unknown per se. Males call to attract females, then lead them to oviposition sites in the leaf litter.

ii. Breeding habitat. Leaf litter. Males care for eggs and transport tadpoles to water. Males can and will tend >1 clutch simultaneously (Wells, 1978).

B. Eggs.

i. Egg deposition sites. Leaf litter (Wells, 1978, and references therein).

ii. Clutch size. Wells (1978) reports clutches of 5, 6, and 7 eggs; Dunn (1941) notes 6 eggs. Eggs hatch in 10–13 d (Wells, 1978).

C. Larvae/Metamorphosis. Newly hatched tadpoles are deposited in water-filled tree cavities, bromeliad tanks, and stagnant pools (Eaton, 1941; McDiarmid and Foster, 1975; Silverstone, 1975; see also Lannoo et al., 1987). Some of these cavities are small and support notably small volumes of water (from 13 ml to about 7,600 ml; Eaton, 1941). Different males will deposit tadpoles in the same tree cavity (Eaton, 1941).

i. Length of larval stage. Eaton (1941) suggests 43 d of larval life (Dunn, 1941, cites 42–60 d), although Eaton (1941) notes that tadpoles vary in their rates of development and therefore, time to metamorphosis is also likely to vary.

ii. Larval requirements.

a. Food. Tadpoles are carnivorous, and their diets include protozoans, rotifers, wood fragments, and perhaps diatoms (Eaton, 1941; summarized in Lannoo et al., 1987).

b. Cover. Will hide in the "frass from dead leaves" (Eaton, 1941).

iii. Larval polymorphisms. No formal polymorphisms, but as with many cavity-dwelling species, tadpole size varies considerably within sibships (Eaton, 1941; Lannoo et al., 1987). Some features of tadpoles may be neotenic (see discussion in Lannoo et al., 1987).

iv. Features of metamorphosis. Metamorphosis in an individual proceeds rapidly, but the timing of metamorphosis within a cohort varies considerably (see Eaton, 1941).

v. Post-metamorphic migrations. Presumably from tanks to terrestrial sites to feed and grow.

D. Juvenile Habitat. Unknown, but probably similar to adults.

E. Adult Habitat. Green and black dart-poison frogs are found on the forest floor as well as in trees. They are diurnal, as are most members of the genus *Dendrobates* (Dunn, 1941; Eaton, 1941; Wells, 1978), although McKeown (1996) notes that in Hawaii these dart-poison frogs are less active on sunny afternoons. Dunn (1941) notes the greatest activity is on mornings after rain.

F. Home Range Size. Unknown, but Wells (1978) cites an example of a male tending two clutches located about 5 m apart.

G. Territories. During courtship, males are nonterritorial but will occasionally engage in aggressive competitive behavior (Wells, 1978). Summers (1989) notes territorial behavior in males. Females, while not territorial, will use aggression to compete for males (Wells, 1978; Summers, 1989).

H. Aestivation/Avoiding Desiccation. Undescribed.

I. Seasonal Migrations. Undescribed, but other than to disperse from breeding tanks and for males to deposit tadpoles in tanks, probably do not occur.

Green and Black Dart-poison Frog *(Dendrobates auratus)*

Hawaiian Islands

J. Torpor (Hibernation). Does not occur.

K. Interspecific Associations/Exclusions. Have not been described.

L. Age/Size at Reproductive Maturity. Unknown in Hawaii (McKeown, 1996), but in captivity, *Dendobates* sp. will breed in their first or second year (Heselhaus, 1994; C. LeBlanc, personal communication).

M. Longevity. Unknown in Hawaii (McKeown, 1996) but in general *Dendrobates* sp. in captivity live 8–10 years, and often longer (Heselhaus, 1994; C. LeBlanc, personal communication). Lifespans in the wild are probably shorter.

N. Feeding Behavior. In general, members of the genus *Dendrobates* feed during the daytime and are active foragers (Toft, 1995). They exhibit a wide range of total prey/stomach but have narrow feeding niches. Ants form a staple of their diets (Toft, 1981, 1995; McKeown, 1996).

O. Predators. Females will eat the eggs of other females (Wells, 1978; Summers, 1989). Adults will avoid large ants and will flee from small land crabs (Dunn, 1941).

P. Anti-Predator Mechanisms. Dendrobatids contain histrionicotoxins and 3,5 disubstituted indolizidine alkaloid skin toxins.

Q. Diseases. Juvenile captive green and black dart-poison frogs displayed at the National Zoological Park in Washington, D.C., were diagnosed with chytridiomycosis (*Batrachochytrium dendrobatidis*; Longcore et al., 1999; Pessier et al., 1999). There is no evidence of this chytrid fungus affecting wild native or exotic populations.

R. Parasites. Unknown.

4. Conservation.

Green and black dart-poison frogs have been introduced to Oahu, Hawaii, and as such are considered an introduced or invasive species. Steps to eradicate these animals on this island are recommended.

Family Hylidae

Acris crepitans Baird, 1854(b)
NORTHERN CRICKET FROG

Robert H. Gray, Lauren E. Brown, Laura Blackburn

1. Historical versus Current Distribution.

The historical distribution of northern cricket frogs (*Acris crepitans*) is primarily across the eastern half of the United States. The extremities of the historical range extended from southeastern New York state, south to the eastern panhandle of Florida, west to western Texas and southeastern New Mexico, north to western Nebraska and southeastern South Dakota, and east across central Wisconsin and central Michigan. The species does not occur on the Atlantic Coastal Plain (except for relictual populations) from southeastern Alabama to southeastern Virginia, nor do they occur in the Appalachian Mountains from southwestern North Carolina northward, nor their foothills in eastern Ohio. Cricket frogs also occur in northern Coahuila, Mexico (not far from the U.S./Mexican border), and historically occurred in extreme southern Ontario, Canada (Point Pelee Peninsula and Pelee Island in Lake Erie; Oldham, 1992).

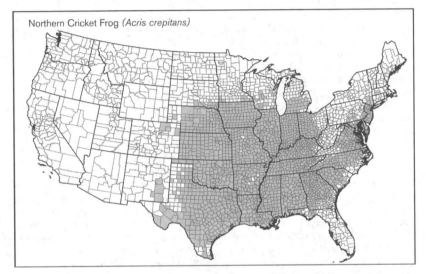

Northern Cricket Frog *(Acris crepitans)*

Three subspecies of northern cricket frogs are recognized: Blanchard's (*A. c. blanchardi*), eastern (*A. c. crepitans*), and coastal (*A. c. paludicola*). Blanchard's cricket frogs are represented in the West and Midwest; eastern cricket frogs are found in east Texas, the Gulf Coast and southern forest regions, and the Piedmont north to New York; coastal cricket frogs are found in the Gulf Coast region of western Louisiana and eastern Texas.

Severe declines of northern cricket frogs have been reported throughout the northern extent of their range (primarily Blanchard's cricket frogs), which have produced scattered and isolated populations. In addition to these declines, northern cricket frog populations are also scattered and isolated along the western and southeastern peripheries of their range. For example, there are a number of isolated records along the more arid western edge of their distribution (e.g., Nebraska [Lynch, 1985]; Colorado [Hammerson, 1986; Hammerson and Livo, 1999]; Texas [Milstead, 1960]; New Mexico [Degenhardt et al., 1996]; and Arizona [Frost, 1983]. Furthermore, isolated populations have been also reported on the Atlantic Coastal Plain of the southeastern United States in North Carolina and South Carolina (Conant, 1975, 1977a).

2. Historical versus Current Abundance.

At the end of the nineteenth century, there were indications that northern cricket frogs were numerous in the midwestern United States. Garman (1892) found that they were "one of the most abundant members of the family in all parts of Illinois," and Hay (1892) similarly indicated they were "one of our commonest batrachians" in Indiana. This abundance continued well past the mid-twentieth century in both of these states (Smith, 1961; Minton, 1972). Campbell (1977) published the first report of a northern cricket frog decline in extreme southern Ontario, Canada—puzzling because at least part of the area was protected as the Point Pelee National Park. Vogt (1981) next reported a sharp northern cricket frog decline in Wisconsin, and subsequently numerous declines have been, and continue to be, reported throughout the upper midwest and elsewhere: Illinois (Mierzwa, 1989, 1998a; Ludwig et al., 1992; Greenwell et al., 1996); Indiana (Minton et al., 1982; Vial and Saylor, 1993; Brodman and Kilmurry, 1998; Minton, 1998); West Virginia (T. Pauley, personal communication); Iowa (Christiansen and Mabry, 1985; Lannoo et al., 1994; Lannoo, 1996, 1998a; Christiansen and Van Gorp, 1998; Hemesath, 1998); Michigan (Harding and Holman, 1992; Harding, 1997; Lee, 1998); Minnesota (Vial and Saylor, 1993; Oldfield and Moriarty, 1994; Moriarity, 1994); Wisconsin (Jung, 1993; Casper, 1996, 1998; Hay, 1998a,b; Mossman et al., 1998); Colorado (H. Smith, personal communication, 1998; Hammerson and Livo, 1999); Ohio (Lipps, 2000); and Canada (Oldham, 1990, 1992; Weller and Green, 1997). Despite numerous reports of declines, and ample scientific literature on the biology of northern cricket frogs, there is no clearcut indication of the cause(s) of this trend toward extinction, although a number of anthropogenic factors and environmental conditions have been suggested.

3. Life History Features.

A. Breeding. Reproduction is aquatic.

i. Breeding migrations. Spring emergence for northern cricket frogs typically begins in early March in southern Illinois, late March or early April in central Illinois (Gray, 1983), and early April in central Iowa (Johnson and Christiansen, 1976). The male vocal pouch begins to develop in some northern cricket frogs in October, while in other individuals the vocal pouch is not apparent until April or early May. Male northern cricket frogs probably are the first to arrive at breeding sites (Pyburn, 1958; Gray, 1983). In Illinois, calling males are first heard in late April or early May and last heard in July (Gray, 1983) or early August (L. E. B., personal observations). In central Iowa, calling males are first heard in mid-May and are last heard in late July (Johnson and Christiansen, 1976). The male mating call is a sharp "glick, glick, glick . . ." and resembles two glass marbles or small stones being struck together. The repetition rate of the "glicks" is faster in the middle of a call sequence than at the beginning or end. Males sit on the bank or on floating aquatic vegetation and call during the day and night.

The number and intensity of male choruses suggests that breeding activity peaks in May, when population sizes are declining. Females mobilize body lipids during ovarian development (Long, 1987a). Gravid females are identified by developing eggs—visible through the skin in the groin—and are present from late April to late August. In Illinois, females mature eggs once annually. The sex ratio of northern cricket frogs in Illinois favors males by about 3:1 and appears even higher at some locations (see Gray, 1984). Mate preference of female northern cricket frogs is influenced by three call characteristics: dominant frequency, number of pulses/call, and number of pulse groups/call (Ryan et al., 1995). Females attracted to calling males are clasped in axillary amplexus. Interestingly, some gravid females can be found after the last calling males are heard (Gray, 1983), suggesting that all females may not breed.

Northern cricket frog populations in the south differ from those in the north in several respects. Evidence suggests that northern cricket frogs in Texas and Louisiana are active year-round and probably experience two breeding peaks (Pyburn, 1961a; Bayless, 1966); southern cricket frogs (*A. gryllus*) are also prolonged breeders (Forester and Daniel, 1986). Additionally, sex ratios of northern cricket frogs in the south approximate 1:1, and breeding population sizes are larger (Jameson, 1950a; Blair, 1961a).

ii. Breeding habitat. Semi-permanent and permanent wetlands and river backwaters.

B. Eggs.

i. Egg deposition sites. Eggs either float on the surface of the water, are scattered on the bottom, or are attached to stems of submergent vegetation.

ii. Clutch size. Up to 400 eggs are deposited singly or in small clumps of 2–7 eggs. A detailed description of northern cricket frog eggs is given in Livezey (1950). The vitelli average 1.13 mm in diameter and are surrounded by two gelatinous envelopes.

C. Larvae/Metamorphosis.

i. Length of larval stage. From 50–90 d or longer (Wright and Wright, 1949). Burkett (1984) noted the tadpole stage lasts 35–70 d (5–10 wk). Working in the laboratory, Pyburn (1956) noted a larval period from 29–90 d.

ii. Larval requirements.

a. Food. Cricket frog tadpoles feed on periphyton and phytoplankton (Johnson, 1991).

b. Cover. Tadpoles are secretive and seek shelter in aquatic vegetation when startled (L. E. B., personal observations).

iii. Larval polymorphisms. Tadpoles that co-occur with dragonfly (*Anax* sp.) larvae in ponds tend to have black tails, while tadpoles in lakes and creeks where fish predominate are mostly plain tailed (Caldwell, 1982a). This polychromatism may reflect differential selection by these predators.

iv. Features of metamorphosis. Metamorphosis begins in late July or early August and continues for about 3 wk.

v. Post-metamorphic migrations. Juvenile northern cricket frogs can disperse from breeding wetlands 1.3 km to other aquatic sites (Gray, 1983).

D. Juvenile Habitat. Same as adults. Growth of newly metamorphosed frogs is rapid during the fall. Individuals can grow from 12–26 mm in 2–3 mo (Gray, 1983).

E. Adult Habitat. Northern cricket frogs are terrestrial and semiaquatic. Individuals potentially can occur in or near virtually any body of fresh water, although they tend to be rare or absent at large lakes, wide rivers, and polluted sites. Northern cricket frogs typically occupy a zone along the water's edge, sitting on floating vegetation and debris or on exposed beach-like banks (e.g., Smith et al., 2003). When approached by a potential predator, northern cricket frogs make several quick zigzag leaps (each often 1 m or more in length) on the bank and/or floating vegetation. They then dive underneath the water or come to rest on the shoreline some distance away. Experimental studies show a progressive decline in locomotory performance over an extended period of consecutive jumps (Zug, 1985). Northern cricket frogs sometimes occur in cattail (*Typha* sp.) thickets as well as in other terrestrial and/or aquatic shoreline vegetational assemblages. Northern cricket frogs occasionally are heard calling on floating vegetation over deeper water a considerable distance from the

edge of a pond. In some flood plain forests, northern cricket frogs are often encountered on damp soil and under logs some distance from aquatic sites. After rains, northern cricket frogs disperse to more distant aquatic sites (Gray, 1983; Burkett, 1984) and often pass through drier, suboptimal terrain.

F. Home Range Size. Adults generally remain near wetlands, but migrate between them. The subject of home ranges has not been studied.

G. Territories. Unknown but unlikely. During breeding, distance to the nearest neighbor appears to have the greatest influence on variation in male calling behavior, with the most profound changes occurring during male-male encounters (Wagner, 1989a,b). Calls at the middle and end of a call sequence are modified in response to male–male aggressive interactions. The hormone arginine vasotoxin increases the probability of male calling and alters the middle and end of a call sequence of more aggressive males to that of less aggressive males (Marler et al., 1995). Males may signal information about their size and fighting ability through changes in call dominant frequency (Wagner, 1989a,c, 1992).

H. Aestivation/Avoiding Desiccation. During a period of severe drought in Texas, W.F. Blair (1957) reported that northern cricket frogs avoided desiccation by taking refuge in deep cracks in the bed of a dry pond. Similar behavior has been observed in Indiana (L.B., personal observations).

I. Seasonal Migrations. Dispersal of northern cricket frogs from one population to another tends to occur after rains. Gray (1983) observed northern cricket frogs moving up to 1.3 km between farm ponds in central Illinois. Frogs that dispersed from one location and were later recaptured at another represented 1–7% of the original population. However, many more frogs likely dispersed and either reached other sites or died in transit.

J. Torpor (Hibernation). As temperatures decline in the fall, northern cricket frogs become less active and are usually last seen in late October in central Iowa (Johnson and Christiansen, 1976) or late November in central Illinois (Gray, 1971a). However, they may be active during the winter in some areas of Illinois, west-central Indiana, Kansas, Oklahoma, Texas, Louisiana, and elsewhere in the south on warm days or where there are springs (Blatchley, 1892; Pope, 1919; Linsdale, 1927; Jameson, 1950a; Blair, 1951b; Pyburn, 1956, 1961a; Gray, 1971a). Northern cricket frogs in central Illinois overwinter in cracks in the banks formed by freezing, drying, and subsequent contraction of the soil (Gray, 1971a; see Irwin et al., 1999; Irwin, this volume, Part One). Other workers have found them during the winter under logs, under a mass of compact dead vegetation, and buried in gravelly soil (Garman, 1892; Walker, 1946;

Pope, 1964a; Irwin et al., 1999; Irwin, this volume, Part One). On warm winter days in southern Illinois, northern cricket frogs are sometimes active on ice (L.E.B., personal observations). Growth stops in winter, and individuals may actually be smaller when first observed in spring (Gray, 1983), possibly due to the overwinter metabolism of and subsequent reduction in body lipids and nonlipid material in the liver (Long, 1987a). Interestingly, most northern cricket frogs that survive to overwintering also survive to the spring emergence. Thus, overwintering is not a major cause of mortality in Illinois populations (Gray, 1983). McCallum and Trauth (2003) report communal hibernation on the Ozark Plateau.

K. Interspecific Associations/Exclusions. Northern cricket frogs can be associated with green frogs (*Rana clamitans*; Jung, 1993; L.E.B., personal observations), northern leopard frogs (*R. pipiens*), wood frogs (*R. sylvatica*), American bullfrogs (*R. catesbeiana*), Fowler's toads (*B. fowleri*), American toads (*B. americanus*), eastern gray treefrogs (*Hyla versicolor*), western chorus frogs (*Pseudacris triseriata*), spring peepers (*P. crucifer*; L.B. personal observations), and a great many other anuran species.

L. Age/Size at Reproductive Maturity. Maximum reported length is 38 mm SVL (Conant and Collins, 1991), with males smaller on average than females.

M. Longevity. Burkett (1984) states that northern cricket frogs have an average life expectancy of about 4 mo and a complete population turnover in 16 mo. Gray (1983) found marked animals that survived two winters. Using skeletochronology, S. Perrill (personal communication) has found 3-yr-old northern cricket frogs.

N. Feeding Behavior. Northern cricket frogs are opportunistic and apparently feed both day and night (Johnson and Christiansen, 1976) on whatever small invertebrates are available on the ground or in the air. Food items include aquatic and terrestrial insects and spiders (Garman, 1892; Hartman, 1906; Jameson, 1947; Gehlbach and Collette, 1959; Labanick, 1976b), and occasionally annelids, mollusks, crustaceans, and plant matter (Hartman, 1906; Johnson and Christiansen, 1976). Food consumption is greater in larger frogs, female frogs, frogs that are reproducing, small frogs with fat bodies, and frogs collected in June and July (Johnson and Christiansen, 1976).

O. Predators. A drastic reduction in number of northern cricket frogs occurs from metamorphosis until breeding in Illinois (Gray, 1983) and Kansas (Burkett, 1984). In Illinois, this reduction ranges from 50–97%. Thus, in many populations of northern cricket frogs, numbers of potential breeding individuals are low. The critical period—when northern cricket frogs in Illinois are most susceptible to predation—is in the fall when water re-

cedes and pond banks are exposed. Animals known or observed to prey on northern cricket frogs include American bullfrogs, fishes (e.g., bass, *Micropterus* sp.), common and plains garter snakes (*Thamnophis sirtalis* and *T. radix*), northern and Concho water snakes (*Nerodia sipedon* and *N. paucimaculata*), turtles (R.H.G., personal observations), kestrels (*Falco sparverius*), great-tailed grackles (*Cassidix mexicanus*), other birds, and mammals (Carpenter, 1952; Lewis, 1962; Fitch, 1965; Wendelken, 1968; Gray, 1978; Fleming et al., 1982; Greene et al., 1994). Adults from the previous year have usually disappeared by the end of October.

P. Anti-Predator Mechanisms. Animals are cryptic, exhibit color polymorphism (Pyburn, 1956, 1961a,b; Wendelken, 1968; Gray, 1971b, 1972, 1977, 1995; Issacs, 1971; Jameson and Pequegnat, 1971; Nevo, 1973; Milstead et al., 1974), and have a remarkable ability to leap (to 1 m). Details of escape behavior have recently been described by Johnson (2003).

Q. Diseases. Gray (1995, 2000 a,b) documented morphological abnormalities and frequencies of abnormalities from northern cricket frog populations in Illinois (see also Beasley et al., this volume, Part One; L.B., personal observations).

R. Parasites. Greenwell et al. (1996) found a massive amount of kidney parasitism in juvenile northern cricket frogs from Illinois (see also Beasley et al., this volume, Part One).

4. Conservation.
Northern cricket frogs are listed as Endangered in Minnesota and Wisconsin, Threatened in New York, and as a Species of Special Concern in Indiana, Michigan, and West Virginia (Brodman and Kilmurry, 1998; Ramus, 1998). In Canada, they are listed as Endangered in Ontario and on the national level (Weller and Green, 1997). Interestingly, as northern cricket frogs are becoming extirpated in a number of

peripheral areas, populations remain stable in the more central regions of their range. Absence of listings in other U.S. states where declines have been noted may reflect the current abundance of northern cricket frogs in the southern parts of those states or a lassitude and/or impenetrability of the state's endangered species bureaucracy.

While causes of northern cricket frog declines remain elusive, in experimental studies, northern cricket frog tadpole abundance was reduced in acidified macrocosms compared with circumneutral ones and reduced in macrocosms with loam soils compared with clay soils (Sparling et al., 1995). Contaminants in waters inhabited by northern cricket frog tadpoles may be accumulated in tissues, causing not only acute and chronic effects (Boschulte, 1995; Greenwell et al., 1996), but also effects that may also be passed to higher trophic levels through predation (Hall and Swineford, 1981; Fleming et al., 1982; Sparling and Lowe, 1996).

Acris gryllus (LeConte, 1825)
SOUTHERN CRICKET FROG

John B. Jensen

1. Historical versus Current Distribution. Southern cricket frogs (*Acris gryllus*) are found primarily below the Fall Line, in both the Atlantic and Gulf Coastal Plains, from southeastern Virginia south and west to eastern Louisiana, including all of Florida (Mecham, 1964). Populations are also known above the Fall Line in the Cumberland Plateau and Ridge and Valley of Alabama (Mount, 1975), as well as the Piedmont of Georgia (Williamson and Moulis, 1994). Two subspecies are recognized: Florida cricket frogs (*A. g. dorsalis*), found in Florida and adjacent portions of Alabama and Georgia, and the larger Coastal-Plain cricket frogs (*A. g. gryllus*), found throughout the remainder of the range (Conant and Collins, 1991). No changes in the distribution of either subspecies have been noted.

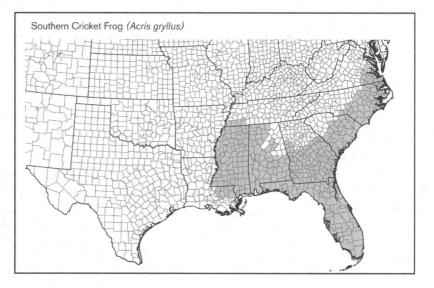

Southern Cricket Frog (*Acris gryllus*)

2. Historical versus Current Abundance.

The naturalist J. D. Corrington (1929) noted that southern cricket frogs are "the most abundant amphibian in the [southeastern United States]. Each roadside pool or ditch contained numerous individuals and the swamps and marshes literally swarmed with them." Deckert (1915) commented that they are "one of the commonest frogs." Many regional guides have also indicated that this species is abundant (Martof et al., 1980; Dundee and Rossman, 1989; Gibbons and Semlitsch, 1991; Wilson, 1995). More recently, however, there have been indications that southern cricket frogs may be declining locally. Means and Means (2000) found that the number of breeding populations of southern cricket frogs in the Munson Sand Hills of panhandle Florida occur at a much lower percentage on silviculture lands than in nearby native habitat. They hypothesize that elimination or severe alteration of the upland habitat, resulting from intensive soil disturbance, is the principal reason. This is corroborated by a 1996–98 rare amphibian survey conducted at 444 sites on industrial forest lands in south Georgia, south Alabama, and north Florida (Wigley et al., 1999). This study revealed that ponds where southern cricket frogs were found had a substantially lower frequency of intensive site preparation and lower density of planted pines in pond edges and surrounding upland habitats than those ponds where this species was not detected.

3. Life History Features.

A. Breeding. Reproduction is aquatic.

i. Breeding migrations. The adult breeding and non-breeding habitats do not differ substantially, therefore true breeding migrations may not exist for this species. Southern cricket frogs typically breed February–October (Wright and Wright, 1949). Although southern cricket frogs may breed throughout the year in Florida, the peak of their breeding activity occurs April to autumn (Carr, 1940b; Einem and Ober, 1956). Mecham (1964) noted an April to early June breeding peak in Alabama. Surges in breeding activity are strongly correlated with periodic rain events (Turnipseed and Altig, 1975). Males call both day and night (Deckert, 1915; Mount, 1975).

ii. Breeding habitat. Nearly every type of freshwater habitat, both temporary and permanent, found within the range of southern cricket frogs has been indicated by one authority or more as suitable breeding habitat, including lake margins, rivers, creeks, sinkhole ponds, cypress ponds, open grassy ponds, bogs, marshes, Carolina bays, bottomland swamps, shallow pools, and roadside ditches (Wright and Wright, 1949; Mount, 1975; Martof et al., 1980; Ashton and Ashton, 1988; Gibbons and Semlitsch, 1991). Southern cricket frogs have also been found breeding in interdunal pools within 18.3 m (20 yd) of the ocean (Neill, 1958a). Males call from mats of floating vegetation in the water or from protected areas along the shore (Mount, 1975). Breeding sites may or may not contain predatory fish (Gibbons and Semlitsch, 1991). In Louisiana and perhaps elsewhere, southern cricket frogs are typically associated with more acidic waters than sympatric northern cricket frogs (*A. crepitans*; Viosca, 1944).

B. Eggs.

i. Egg deposition sites. Eggs are attached to the stems of vegetation or stones or spread on the pond bottom (Wright and Wright, 1949; Mount, 1975).

ii. Clutch size. Eggs are laid singly (Wright, 1923) or in small clusters of 7–10, and each female may lay up to 250 eggs (Ashton and Ashton, 1988). Hatching occurs $\geq$ 4 d following oviposition, depending on the water temperature (Ashton and Ashton, 1988).

C. Larvae/Metamorphosis.

i. Length of larval stage. Typically 50–90 d (Wright and Wright, 1949), though possibly as quickly as 41 d (Ashton and Ashton, 1988). Southern cricket frogs transform after tadpoles reach 9–15 mm SVL (Wright and Wright, 1949).

ii. Larval requirements.

a. *Food.* Unknown, though tadpoles likely graze on algae.

b. *Cover.* Tadpoles are most readily captured in submerged and emergent vegetation, suggesting they use this microhabitat for cover and/or feeding.

iii. Larval polymorphisms. Unknown for this species.

iv. Features of metamorphosis. Metamorphosis occurs from April–October and newly metamorphosed animals are 9–15 mm SVL (Wright and Wright, 1949).

v. Post-metamorphic migrations. Unknown.

D. Juvenile Habitat. Unknown, but thought to be similar to adults.

E. Adult Habitat. Not known to differ substantially from breeding habitats (see "Breeding habitat" above). However, southern cricket frogs are occasionally observed in uplands quite distant from the nearest aquatic habitat (personal observations). In fact, Wright and Wright (1949) indicated that they are often terrestrial in meadows or wooded edges. Whether or not upland habitats are used substantially by this species other than as corridors between aquatic sites is unknown. The relatively few southern cricket frogs found at wetlands surrounded by intensive silviculture versus those surrounded by natural habitats or lower intensity silviculture (Wigley et al., 1999; Means and Means, 2000) may indicate that terrestrial habitats are important to some aspect of this species' life history. Terrestrial foraging, noted by Ashton and Ashton (1988), may be one such aspect. Mount (1975) noted that above the Fall Line, southern cricket frogs are more often found associated with sandy soils. There is no information to indicate a difference in habitat characteristics between the sexes.

F. Home Range Size. Unknown.

G. Territories. Unknown.

H. Aestivation/Avoiding Desiccation. Southern cricket frogs may be found year-round (Corrington, 1929; Mount, 1975) and breed during summer (Wright and Wright, 1949), suggesting that aestivation is unlikely.

I. Seasonal Migrations. Unknown.

J. Torpor (Hibernation). Fewer southern cricket frogs are seen during mid-winter than at other times of the year, which may indicate that the majority of individuals hibernate (Corrington, 1939). However, animals in southern populations remain active, and perhaps breed, throughout the year (Wright and Wright, 1949; Mount, 1975).

K. Interspecific Associations/Exclusions. Southern cricket frogs will occasionally hybridize with northern cricket frogs (Neill, 1954; Mount, 1975). Southern cricket frogs call and breed in association with many other anurans, including northern cricket frogs, American bullfrogs (*Rana catesbeiana*), southern leopard frogs (*R. sphenocephala*), spring peepers (*Pseudacris crucifer*), barking treefrogs (*Hyla gratiosa*), Cope's gray treefrogs (*H. chrysoscelis*), and Fowler's toads (*B. fowleri*; Cahn, 1939), as well as oak toads (*B. quercicus*), southern toads (*B. terrestris*), pine woods treefrogs (*H. femoralis*), southern chorus frogs (*P. nigrita*), ornate chorus frogs (*P. ornata*), gopher frogs (*R. capito*), pig frogs (*R. grylio*), and carpenter frogs (*R. virgatipes*; personal observations).

L. Age/Size at Reproductive Maturity. Males 15–29 mm SVL; females 16–33 mm (Wright and Wright, 1949).

M. Longevity. Unknown.

N. Feeding Behavior. Southern cricket frogs may forage a good distance from the water's edge, especially during the day (Ashton and Ashton, 1988). Bayless (1969) examined the stomachs of southern cricket frogs and found a variety of arthropods, especially springtails, hymenopterans, spiders, dipterans, beetles, and homopterans.

O. Predators. Southern cricket frogs are commonly preyed upon by water and garter snakes, as well as by a large variety of aquatic predators such as other frogs, fishes, and birds (Ashton and Ashton, 1988). Fish predators include redfin pickerel (*Esox americanus*), largemouth bass (*Micropterus salmoides*), and bluegill (*Lepomis macrochirus*; Ferguson et al., 1965). Pine woods litter snakes (*Rhadinaea flavilata*) have been reported to "readily" eat southern cricket frogs (Allen, 1939).

P. Anti-Predator Mechanisms. Southern cricket frogs will make one or more long, erratic leaps to escape predation (Conant and Collins, 1991). They shun open water; following leaps into water they will immediately and quickly return back to shore (Mount, 1975). However, southern cricket frogs may attempt to elude detection by hiding in the debris on the pond bottom

(Blem et al., 1978). They are also known to hop from the ground into bushes and back down again (Wright and Wright, 1949). Additionally, astronomical orientation may be used to avoid predation by fish (Ferguson et al., 1965).

Q. Diseases. Unknown.

R. Parasites. Unknown.

4. Conservation.
Southern cricket frogs remain common throughout their range, although they may be declining locally. In the Munson Sand Hills region of panhandle Florida, there are lower numbers of breeding populations of southern cricket frogs on silviculture lands than in nearby native habitat; Means and Means (2000) hypothesize that elimination or severe alteration of the upland habitat, resulting from intensive soil disturbance, is the principal reason. A 1996–98 amphibian survey conducted on industrial forest lands in south Georgia, south Alabama, and north Florida (Wigley et al., 1999) corroborates Means and Means (2000) interpretation. Wigley et al. (1999) found that ponds with substantially higher frequencies of intensive site preparation and higher densities of planted pines along pond edges and in surrounding upland habitats had reduced numbers of southern cricket frogs.

Hyla andersonii Baird, 1854(b)
PINE BARRENS TREEFROG

D. Bruce Means

1. Historical versus Current Distribution.
The type locality of Pine Barrens treefrogs (*Hyla andersonii*) is Anderson, South Carolina, but most authorities believe that neither the holotype specimen nor the species ever occurred there (Noble and Noble, 1923; Neill, 1948a, 1957a; Wright and Wright, 1949; Gosner and Black, 1967; E.E. Brown, 1980; Karlin et al., 1982). Today, populations are known from three widely disjunct areas of the southeastern United States (New Jersey Pine Barrens, the Fall Line sandhills of North and South Carolina, and the Florida Panhandle and adjacent south Alabama), but the knowledge of this species' distribution has come slowly. The second record and first New Jersey specimen was reported by Cope (1862). Then, the first North Carolina record was reported 60 yr later (Davis, 1922), and it was 86 yr after its formal description before anyone verified that the species really did occur in South Carolina (Brimley, 1940). Amazingly, Pine Barrens treefrogs were not discovered in Florida and southern Alabama until 116 yr after they were named (Christman, 1970), but even the determination of its Florida/Alabama distribution required time and was fraught with difficulty (Means and Longden, 1976; Means, 1978, 1983; Means and Moler, 1979; Moler, 1981; Printiss and Hipes, 1999). Because the geographical distribution of Pine Barrens treefrogs is just now becoming understood, we have no way of comparing historical versus present distributions.

2. Historical versus Current Abundance.
The fact that local populations call erratically is probably the most important reason why knowledge about the biology and distribution of Pine Barrens treefrogs have been slow to accumulate. Means and Longden (1976) and Means and Moler (1979) noted that known breeding sites in Florida were sometimes silent on nights when frogs were calling elsewhere. Moler (1981) in Florida and Cely and Sorrow (1983) in South Carolina verified this and demonstrated that the seepage status of breeding sites may be a more important variable in predicting calling behavior than temperature, humidity, and thundershower activity. At one time it was believed that the New Jersey population was the largest of the three isolates (Wright and Wright, 1949; Means and Longden, 1976). Fieldwork in the Carolinas (Montanucci and Wilson, 1980; Tardell et al., 1981; Cely and Sorrow, 1983), Florida (Means and Moler, 1979; Moler, 1981; Printiss and Hipes, 1999), and Alabama (Mount, 1980; Jensen, 1991), however, has shown in the past two decades that Pine Barrens treefrogs are more abundant in these enclaves than was once believed. Populations in all three regions have dwindled from pre-settlement levels because of agriculture, development, and plant succession following the interruption of the natural fire cycles in Pine Barrens treefrog habitats (Means and Longden, 1976; Means and Moler, 1979; Ehrenfeld, 1983; Ehrenfeld and Schneider, 1983; Freda and Morin, 1984; Cely and Sorrow, 1986).

3. Life History Features.
A. Breeding.
i. Breeding migrations. None known, but adult frogs disperse ≤105 m away from breeding ponds within 30 d after breeding (Freda and Gonzalez, 1986), and therefore must return to breed again.

ii. Breeding habitat. An important characteristic of most breeding sites is low pH (Wright and Wright, 1949; Gosner and Black, 1957b; Means and Longden, 1976; Means and Moler, 1979; Freda and Morin, 1984; Cely and Sorrow, 1986). In New Jersey, Pine Barrens treefrogs breed in small shallow ponds, acid pools, seepage streams, and bogs usually surrounded by heavy shrub growth (Gosner and Black, 1957b; Freda and Morin, 1984). The largest breeding colonies are found in open canopy, early successional sites dominated by shrubs and herbs; the water is always acidic (< pH 4.5) with dense mats of sphagnum or other aquatic vegetation (Freda and Morin, 1984). Breeding habitat in the Carolinas is evergreen shrub-herb bog along small blackwater tributaries in the sandhills. Florida and Alabama breeding sites are similar to those in New Jersey and the Carolinas: acid, hillside seepage bogs with small pools of clear, acid seepage water and a diverse wetland flora of *Sphagnum* sp., sedges, rushes, bladderwort, pipeworts, sundews (*Drosera* sp.), pitcher plants (*Sarracenia purpurea*), clubmoss, and filamentous algae. These sites are usually at the interface between heliophilic herb bogs upslope from evergreen shrub bogs that are dominated by black titi (*Cliftonia monophylla*), swamp titi (*Cyrilla racemiflora*), tall gallberry (*Ilex*

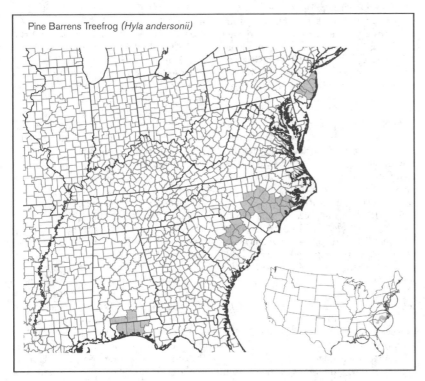

Pine Barrens Treefrog *(Hyla andersonii)*

coriacea), and sweet bay (*Magnolia virginiana*; Means and Longden, 1976; Means and Moler, 1979; Moler, 1980; Mount, 1980). Generally, herb bogs are characterized by sundews, pitcher plants, orchids, sedges, and grasses, especially beakrush (*Rhynchospora* sp.), broomsedge (*Andropogon* sp.), and needlerush (*Juncus* sp.), with sphagnum moss present at most sites (Gosner and Black, 1956; Montanucci and Wilson, 1980; Tardell et al., 1981; Cely and Sorrow, 1983, 1986). In all three enclaves, researchers noted that breeding habitat is often associated with disturbances such as logging clearcuts, or gas, powerline, or railroad rights-of-way (Means and Moler, 1979; Tardell et al., 1981). Means and Moler (1979) believe this is because the specialized breeding sites of Pine Barrens treefrogs depend upon adequate seepage water that is reduced through evapotranspiration when herb bogs become encroached by woody plants. They argued that fire was the dominant force in pre-settlement times maintaining herb bogs in early successional stages, which is mimicked by logging, bush-hogging, and mowing (Means and Moler, 1979). Researchers in the Carolinas (Cely and Sorrow, 1986) and New Jersey (Freda and Morin, 1984) concurred.

B. Eggs.

i. Egg deposition sites. Noble and Noble (1923) and Gosner and Black (1957a) described and figured the diagnostic stages and characters of developing eggs, larvae, and recently metamorphosed froglets in New Jersey. Oviposition occurs in New Jersey from about 20 May–8 July (Wright, 1932), but published information is not available for the Carolinas, Florida, or Alabama.

ii. Clutch size. From 800–1,000 (Wright and Wright, 1949). Hatching occurred in 4 d in the laboratory (Noble and Noble, 1923) and has been estimated to take place in 3 or 4 d in nature (Wright, 1932).

C. Larvae/Metamorphosis. Larvae were collected in Florida from the last week in May to the third week in August (Means, 1992a). The earliest date for transformation in New Jersey is 25 June, and the average length of the larval period in laboratory-raised groups reared at room temperature was 48 d (Wright and Wright, 1949). Wright (1932) estimated length of larval period in the field to be 50–75 d, with transformation occurring from 20 June–31 August. It is estimated that Florida larvae metamorphose by the end of September (Means, 1992a).

D. Juvenile Habitat. Post-metamorphic juvenile habitat is not known to differ from that of adults.

E. Adult Habitat. The preferred habitat of adult Pine Barrens treefrogs when not engaged in breeding is unknown. Most observations of adult Pine Barrens treefrogs have been made on calling males whose habitat preferences were likely to have been biased by their drive to breed. In spite of this, researchers report the habitat of adults to be shrub bogs dominated by

mixtures of the evergreen woody plants where calling males and amplexed females have been found. Shrub bogs in South Carolina (Cely and Sorrow, 1986) are characterized by a sparse overstory of yellow poplar (*Liriodendron tulipifera*), red maple (*Acer rubrum*), blackgum (*Nyssa sylvatica*), and pond pine (*Pinus serotina*); an understory of red maple, sweet bay, yellow poplar, swamp titi, pine, blackgum, and black alder (*Alnus serrulata*); and a shrub layer of sweet pepper bush (*Clethra alnifolia*), cinnamon fern (*Osmunda cinnamomea*), greenbriar (*Smilax* sp.), red maple, switch cane (*Arundinaria gigantea*), wax myrtle (*Myrica cerifera*), fetterbush (*Lyonia lucida*), and tall gallberry (*Ilex coriacea*). In addition to most of these species, in New Jersey shrub bogs there was a greater representation of Atlantic white cedar (*Chamaecyparis thyoides*) and *Vaccinium* sp. (Wright, 1932); in Florida black titi and swamp titi were usually the most common shrubs (Means and Longden, 1976).

F. Home Range Size. Has not been studied, except that Freda and Gonzalez (1986) generated data on movement away from breeding ponds by eight adults (seven males, one female) over periods of 4–30 d.

G. Territories. Unknown.

H. Aestivation/Avoiding Desiccation. Pine barrens treefrogs are active March–November and have been observed calling and breeding throughout these months during rainy periods (Means, 1992a). Whether individuals seek certain microhabitats as refuges or become inactive during droughts and between rainy periods is unknown.

I. Seasonal Migrations. Not known to occur, except adults may move at least 105 m through shrub bogs to breeding ponds (Freda and Gonzalez, 1986).

J. Torpor (Hibernation). Almost nothing is known about the circumstances of overwintering in Pine Barrens treefrogs, but Neill (1948b) reported a Richmond County, Georgia, specimen that was "found beneath loose bark of a standing tree on a cool day." Other than this casual mention, the species has not been reported from Georgia despite intense searching by herpetologists over the past half century.

K. Interspecific Associations/Exclusions. Few other frogs have been found associated with Pine Barrens treefrogs in larval habitats because of the high acidity of the water (Gosner and Black, 1957a,b; Pehek, 1995). However, tadpoles of pine woods treefrogs (*H. femoralis*), squirrel treefrogs (*H. squirella*), and southern cricket frogs (*Acris gryllus*) were reported in larval habitats with Pine Barrens treefrog tadpoles from Florida (Means and Longden, 1976). Adult green frogs (*Rana clamitans*) and southern leopard frogs (*R. sphenocephala*) were associated with Pine Barrens treefrogs in New Jersey breeding ponds (Freda and Morin, 1984).

L. Age/Size at Reproductive Maturity. In aquaria and terraria, Means and Longden

(1976) raised tadpoles to adult size in 1 yr, but did not know if they were sexually mature.

M. Longevity. Adult frogs have been kept in captivity for periods of up to 20 mo (personal observations), 2 yr (Wright, 1932), and 7 yr (Moler, 1980).

N. Feeding Behavior. In aquaria, tadpoles readily eat boiled lettuce, filamentous algae brought in from larval habitats, and commercial pet fish food (Means and Longden, 1976). Newly metamorphosed animals were successfully raised to adult size on tiny insects that accumulated at night lights (Means and Longden, 1976). Ten males examined from New Jersey had grasshoppers, beetles, ants, dipterans, and unidentified insect remains in their stomachs (Noble and Noble, 1923). Captive specimens fed on crickets and flies (Bullard, 1965) and moths and mealworms (Means and Longden, 1976).

O. Predators. Potential tadpole predators found in Pine Barrens treefrog larval habitats included adult bronze frogs (*R. c. clamitans*), two-toed amphiumas (*Amphiuma means*), red salamanders (*Pseudotriton ruber*), and banded pigmy sunfish (*Elassoma zonatum*; Means and Longden, 1976). Also, painted turtles (*Chrysemys picta*), spotted turtles (*Clemmys guttata*), and common snapping turtles (*Chelydra serpentina*) may be tadpole predators wherever their geographic distributions overlap with that of Pine Barrens treefrogs (Wright, 1923). Aquatic predaceous insects, eastern mudminnows (*Umbra pygmea*), banded sunfish (*Enneacanthus* sp.), and pickerels (*Esox* sp.) were presumed to be major predators in New Jersey (Wright, 1923; Freda and Morin, 1984). The only documented predators of adult Pine Barrens treefrogs are banded water snakes (*Nerodia sipedon*; Kauffeld, 1957) and ribbon snakes (*Thamnophis sauritus*; Freda and Gonzalez, 1986).

P. Anti-Predator Mechanisms. When frightened, adult frogs "invariably leap to the ground and, with a series of short jumps, disappear among the grass and sphagnum of the bog" (Noble and Noble, 1923). No other anti-predator mechanisms are known.

Q. Diseases. Unknown.

R. Parasites. Unknown.

4. Conservation.

Pine barrens treefrogs are considered Endangered by the State of New Jersey (New Jersey Natural Heritage Program, 1995), Significantly Rare in North Carolina (LeGrand and Hall, 1999), Threatened in South Carolina (Garton and Sill, 1979), and have no status in Georgia. Based upon existing information in the early 1970s, Florida populations were listed as Endangered by the U.S. Fish and Wildlife Service (Federal Register, 1977, 42(218):58754–58756), but over time, knowledge about the Florida/Alabama distribution of Pine Barrens treefrogs has steadily increased (Means and

Longden, 1976; Means and Moler, 1979; Mount, 1980; Moler, 1981; Jensen, 1991; Printiss and Hipes, 1999). This has led to delisting the species at the federal level, so that Pine Barrens treefrogs are now considered Rare by the State of Florida (Means, 1992a), where >135 localities are known, with many on publicly owned lands (Moler, 1981; Printiss and Hipes, 1999). About 20 localities are known from Alabama, where the species is considered Threatened (Means, 1986a). Conservation of Pine Barrens treefrogs depends upon the long-term existence of suitable breeding and adult habitat in all three enclaves.

Freda and Morin (1984) pointed out that the seepage waters where Pine Barrens treefrogs breed and develop are some of the most acidic, dilute, and nutrient-deprived freshwater ecosystems in North America. Pine barrens treefrogs have specific physiological, behavioral, and biochemical adaptations to survive in such demanding aquatic environments. Any perturbations of the water chemistry of herb bogs and shrub bogs, caused for example by agricultural and residential development upstream, upslope, or in the preferred habitats, would probably have dramatic negative impacts on Pine Barrens treefrogs. Also, changes in groundwater chemistry or activities that lower the water table would be highly detrimental

water supply in breeding habitats, the most important management actions are to periodically set back plant succession in herb bogs. Lightning-ignited fires that ran downslope from longleaf pine forests through herb bogs and then into shrub bogs were once the natural agent that kept streambottom hardwood trees out of shrub bogs and evergreen shrubs out of herb bogs (Means and Moler, 1979). Plant succession was thus set back by frequent fires. Because natural wildfires are no longer a feature of southeastern U.S. landscapes, management of Pine Barrens treefrog habitat must be accomplished by controlled burns or prescribed fires.

Hyla arenicolor Cope, 1866(a)
CANYON TREEFROG
Charles W. Painter

1. Historical versus Current Distribution.
In the United States, canyon treefrogs *(Hyla arenicolor)* occur from southern Utah and southern Colorado, south through Arizona and New Mexico, and into western Texas, where disjunct populations occur in the Chisos and Davis Mountains and canyons of Trans-Pecos, Texas, and in New Mexico (Degenhardt et al., 1996; Hammerson, 1999). Historical versus current distribution is unknown.

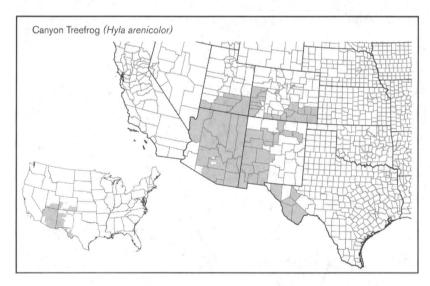

Canyon Treefrog *(Hyla arenicolor)*

because they would pollute the breeding habitat or cause the drying of bogs. Liming of ponds and food plots within wildlife management areas, for example, should be prohibited.

Having populations on publicly owned lands, however, is in itself not sufficient to ensure the survival of the species. Active management must be undertaken to maintain a certain level of quality of the preferred habitat of Pine Barrens treefrogs. Next to ensuring an adequate high-quality

2. Historical versus Current Abundance. Unknown. Breeding success varies with rainfall and populations fluctuate. Chytrid fungus has been reported in Arizona populations (Sredl and Caldwell, 2000a), which may be leading to a decline in animals in these populations.

3. Life History Features.
  A. Breeding. Reproduction is aquatic.
  i. Breeding migrations. Do not occur. Adults reproduce in their natal pools,

which may be bounded by solid rock. Calling takes place in the early evening and at night, from vertical, smooth walls above the waterline. Breeding peaks after rains in May–June. Some breeding may occur in April and as late as July if the breeding season has been dry. Gelbach (1965) studied a high-altitude (2,380 m) population and found metamorphosing frogs in early July and freshly laid eggs 2 wk later.

  ii. Breeding habitat. Adult populations may be found in pools and rocky streams, which may be in the solid rock of canyon bottoms or in rain pools on top of rock cliffs (Degenhardt et al., 1996).
  B. Eggs.
  i. Egg deposition sites. Eggs are generally attached to objects at the bottom of pools.
  ii. Clutch size. Has not been reported. Length of incubation has not been reported.
  C. Larvae/Metamorphosis.
  i. Length of larval stage. From 40–75 d after oviposition (Wright and Wright, 1949; Zweifel, 1961).
  ii. Larval requirements.
  a. Food. Tadpoles ingest organic matter and filamentous algae scraped from rock and plant surfaces, as well as bottom debris, and will sometimes feed on floating organic material at the water surface (Hammerson, 1999).
  b. Cover. Does not appear to be necessary. Pools generally have clear water; tadpoles and adults are visible on the bottom.
  iii. Larval polymorphisms. Unknown and unlikely.
  iv. Features of metamorphosis. Newly metamorphosed animals range from 15–28 mm SVL (Zweifel, 1961; Gehlbach, 1965).
  v. Post-metamorphic migrations. Unlikely. Newly metamorphosed animals move from breeding pools to rocks surrounding breeding pools.
  D. Juvenile Habitat. Similar to adult habitat.
  E. Adult Habitat. Canyon treefrogs are found in deep, rocky canyons along intermittent streams in permanent, canyon-bottom pools. They are often found associated with areas of large boulders and rock outcrops. Vegetation in these areas includes cottonwood trees *(Populus deltoides)* along the stream courses and pinyon-juniper woodlands along the slopes (Hammerson, 1999).

Canyon treefrogs generally do not climb trees. During warm rains, canyon treefrogs are active and will feed by perching on rocks. They will roam when feeding but typically stay close to—generally within a single leap of—their pool (Hammerson, 1999).

Most activity takes place at night from May–December (Hammerson, 1999). Canyon treefrogs are unusual in that, especially during the spring and early summer, they will spend their entire day at one site, on steep-sloped rocks, in full sunlight within about 1 m from their breeding site, where they maintain a body temperature of between 29 and 31 °C (Snyder and

Hammerson, 1993; Hammerson, 1999). As Hammerson (1999) points out: "If undisturbed, they do not visit the water at all during the day and tolerate conditions that certainly would be fatal to a typical aquatic frog." A perched frog will keep its head down and tuck its limbs under its body. It will typically return to the same site in successive days (Hammerson, 1999).

F. Home Range Size. Small and restricted to the area around their breeding pools and streams.

G. Territories. Unlikely, groups of >10 adults have been found

H. Aestivation/Avoiding Desiccation. During dry conditions, canyon treefrogs retreat to rock crevices near their breeding sites (Hammerson, 1999).

I. Seasonal Migrations. Unlikely.

J. Torpor (Hibernation). During cold conditions, canyon treefrogs retreat to rock crevices near their breeding sites (Hammerson, 1999).

K. Interspecific Associations/Exclusions. Canyon treefrogs will breed in association with members in the chorus frog (*Pseudacris triseriata*) complex (Degenhardt et al., 1996).

L. Age/Size at Reproductive Maturity. Adult males range from 42–52 mm SVL, females from 45–55 mm (Hammerson, 1999). Hammerson (1999) argues that the presence of subadults suggests breeding does not occur before animals are in their third year.

M. Longevity. Unknown.

N. Feeding Behavior. Feeding typically takes place at night; canyon treefrogs may feed from perches or may roam and search (Hammerson, 1999). Prey items include annelids, small scorpions, arachnids, centipedes, and insects such as coleopterans, hymenopterans (ants), lepidopterans, hemipterans, and ephemeropterans (Painter, 1985; Degenhardt et al., 1996; Hammerson, 1999).

O. Predators. Poorly understood, but possible predators include carnivorous mammals, birds, and snakes (Tanner, 1929; Hammerson, 1999). Adults are attacked by mosquitoes (Hammerson, 1999). Tadpoles are susceptible to invertebrate predators.

P. Anti-Predator Mechanisms. Proximity to pools is important. Individuals flushed from perching sites will jump into the water and dive to the bottom. When handled, canyon treefrogs release fluid from their vent; they may also produce a sticky skin secretion that deters predators (Hammerson, 1999). This secretion is also toxic (Powell and Lieb, 2003). Adults have bright yellow-orange flash colors in their inguinal region that may confuse or startle predators (Hammerson, 1999).

Q. Diseases. Chytrid fungus has been discovered in canyon treefrog populations in Arizona (Sredl and Caldwell, 2000a).

R. Parasites. Unknown.

4. Conservation.
Hammerson (1999) notes that most canyon treefrog populations occur on public lands; direct impacts from humans may constitute the biggest threat to canyon treefrogs. This danger comes from both collectors and the increasing recreational use of public lands. Canyon treefrogs are listed as a Species of Special Concern in Colorado, but are not otherwise offered protection by state governments or the U.S. Federal Government.

Hyla avivoca Viosca, 1928
BIRD-VOICED TREEFROG
Michael Redmer

1. Historical versus Current Distribution.
Bird-voiced treefrogs (*Hyla avivoca*) range from extreme southwestern South Carolina, southwest across Georgia to the Florida Panhandle, west across the Gulf Coast (roughly including the southern half of Alabama and statewide in Mississippi) to the east side of the Mississippi River drainage, and north through western Kentucky and Tennessee to extreme southern Illinois (Drury and Gessing, 1940; Gentry, 1955; Smith, 1961, 1966a; Barbour, 1971; Mount, 1975; Martof et al., 1980; Ashton and Ashton, 1988; Conant and Collins, 1991; Redmond and Scott, 1996). Bird-voiced treefrogs also occur west of the Mississippi River, in isolated populations in central and northwestern Louisiana, the Red River Drainage of extreme southeastern Oklahoma, and in eastern, central, and southern Arkansas (Blair and Lindsay, 1961; Krupa, 1986c; Dundee and Rossman, 1989; Trauth, 1992).

2. Historical versus Current Abundance.
Historical abundance is unknown and therefore cannot be compared with current abundance. Bird-voiced treefrogs are currently listed as Threatened in Illinois (Redmer and Kruse, 1998), where many remaining colonies are isolated due to past drainage of hardwood swamps.

3. Life History Features.
A. Breeding. Reproduction is aquatic.
i. Breeding migrations. Males are stimulated by warm temperatures and will chorus by day in treetops in the same swamps used for breeding. Chorusing often begins ≥ 1 mo before breeding (Dundee and Rossman, 1989; Redmer et al., 1999a), an indication that males spend the entire year in the vicinity of the breeding site. The prolonged breeding season begins in late spring (April in the southern part of the range; May in the north) and lasts throughout much of the summer (Fortman and Altig, 1974; Mount, 1975; Krupa, 1986c; Ashton and Ashton, 1988; Trauth, 1992; Redmer et al., 1999a). In Illinois, chorusing commences at dusk and lasts until temperatures drop below about 16 °C or until most males are paired; few individual males chorus after midnight. Individuals are occasionally found alive on roads adjacent to swamps in which they breed (Redmer et al., 1999a,b).

ii. Breeding habitat. Hardwood swamps and forested flood-plains, especially those consisting of cypress (*Taxodium distichum*) and tupelo gum (esp. *Nyssa aquatica*) trees (Viosca, 1928; Fouquette and Dalahoussaye, 1966; Trauth and Robinette, 1990; Redmer et al., 1999a). Most males call from elevated positions in trees and other woody vegetation over water (Parker, 1951; Secor, 1988; Trauth and Robinette, 1990; Redmer et al., 1999a). Females approach and touch calling males to stimulate amplexus (Redmer, 1998), or males

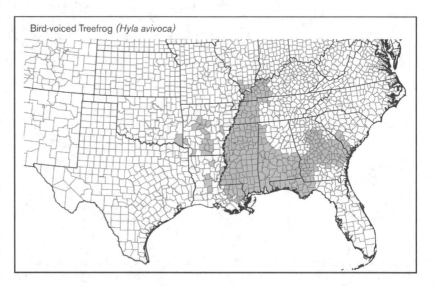

Bird-voiced Treefrog *(Hyla avivoca)*

may initiate amplexus with approaching females (Trauth and Robinette, 1990).

B. Eggs.
i. Egg deposition sites. Amplexed pairs have been observed on the branches of trees and shrubs, in reeds (*Phragmites* sp.), willow thickets, and a shrubby farm pond (Parker, 1951; Turnipseed and Altig, 1975;

Trauth and Robinette, 1990; Redmer, 1998). After amplexus begins, the female descends (carrying the male) head first to the water surface to oviposit. When the female reaches the water she rotates and backs her posterior into the water. Oviposition takes place with the female clinging to the vegetation, and eggs sink to the bottom substrate (Redmer, 1998).

ii. Clutch size. I have examined clutch size in 12 females and found a mean of 632 eggs (range 409–811; unpublished data). Because of the prolonged breeding season and presence of yolked ovarian eggs in individuals that have oviposited, it is possible that females of this species lay multiple clutches (as do females of some other North American *Hyla*), though direct evidence of this is lacking (Trauth and Robinette, 1990; Redmer, 1998a).

C. Larvae/Metamorphosis. Tadpoles have been collected in the water under the perches where amplexus occurs.

i. Larval period. Lasts approximately 30 d (Hellman, 1953; Volpe et al., 1961; Trauth and Robinette, 1990).

D. Juvenile Habitat. Juveniles are sometimes found perched on low vegetation or on the ground adjacent to swamps used for breeding (Redmer et al., 1999a).

E. Adult Habitat. Prior to and following the chorusing season, adults are sometimes found perched on low vegetation, on the ground, under logs, in shrub thickets, or in tree crevices in or adjacent to swamps used for breeding (Parker, 1951; Dundee and Rossman, 1989; Redmer et al., 1999a).

F. Home Range Size. Unknown.

G. Territories. Males combat to defend chorusing sites from other males (Altig, 1974).

H. Aestivation/Avoiding Desiccation. Unknown.

I. Seasonal Migrations. Unknown. Daytime chorusing in treetops over the same swamps used for breeding might indicate that males spend the entire year in the vicinity of the breeding site.

J. Torpor (Hibernation). Conditions under which torpor occurs in wild are unknown. Captive individuals refuse food and become sluggish; their movement is noticeably uncoordinated at air temperatures below 15 °C (unpublished data).

K. Interspecific Associations/Exclusions. Bird-voiced treefrogs comprised 0.8% of the total population of three species (also including northern cricket frogs [*Acris crepitans*] and green treefrogs [*H. cinerea*]) of hylid tadpoles in an upland pond in Mississippi (Turnipseed and Altig, 1975). In Illinois, large choruses of bird-voiced treefrogs are most often associated with choruses of cricket frogs, American bullfrogs (*Rana catesbeiana*), and green frogs (*R. clamitans*; Redmer et al., 1999a). Males may chorus with or near choruses of green treefrogs, though male perch sites rarely overlap (Turnipseed and Altig, 1975; Secor, 1988; Redmer et al., 1999a). While it has

been reported that green treefrogs often predominate in more shallow water areas with emergent herbaceous vegetation, whereas bird-voiced treefrogs prefer deeper areas with woody structure (Turnipseed and Altig, 1975; Redmer et al., 1999a), the opposite also has been reported (Secor, 1988). Differences in chorus sites and breeding call may usually act as isolating mechanisms, but hybridization with other *Hyla* is known to occur naturally and has been demonstrated in the lab (Mecham, 1960a, 1965; Fortman and Altig, 1974).

L. Age/Size at Reproductive Maturity. Mature females from Illinois were found to be 2–4 yr old (unpublished skeletochronology data). Male ages at sexual maturity are unknown. Females (32–52 mm SVL) are larger than males (28–39 mm SVL; Neill, 1948a; Wright and Wright, 1949; Smith, 1961; Trauth and Robinette, 1990; Redmer et al., 1999a; unpublished data).

M. Longevity. Maximum known age is 4 yr for females (unpublished data). Male longevity is unknown.

N. Feeding Behavior. Individuals feed primarily (if not entirely) on arboreal arthropods (Jamieson et al., 1993; Redmer et al., 1999b).

O. Predators. Observations of predation in the wild are unknown, but birds, water snakes, and other vertebrates probably prey on juveniles and adults. Tadpoles are probably consumed by aquatic invertebrates and vertebrates.

P. Anti-Predator Mechanisms. Unknown.

Q. Diseases. Unknown.

R. Parasites. Protozoans, trematodes, and cestodes have been reported (Reiber, 1941; C.T. McAllister et al., 1993a). A report of a trypanosome (Woo and Bogart, 1984) is questionable because the host reportedly was collected in Ohio, a state from which this species has not been documented.

4. Conservation.
Bird-voiced treefrogs are currently listed as Threatened in Illinois, where many re-

maining colonies are isolated. They have fairly specific habitat characteristics, including bottomland hardwood swamps and forested flood-plains, especially those consisting of cypress and tupelo gum. This habitat specificity may have had, and may continue to have, consequences for the conservation of this species.

Hyla chrysoscelis Cope, 1880
COPE'S GRAY TREEFROG
George R. Cline

Cope's gray treefrogs (*Hyla chrysoscelis*) and eastern gray treefrogs (*H. versicolor*) are members of a cryptic, diploid-tetraploid species complex. This has resulted in considerable taxonomic confusion, especially in early reports. As a result of this confusion, many authors have chosen to combine the accounts and distributions of these species. In this account, I have attempted to separate, as much as possible, the literature related to *H. chrysoscelis* and *H. versicolor*. For categories where data are lacking or where there is a question regarding the species identification, data from the sister species are reported. This approach is at least partially validated by the genetic analysis of this complex by Ptacek et al. (1994) and the growth and development studies of Ptacek (1996).

1. Historical versus Current Distribution.
Because of the cryptic nature of Cope's gray treefrogs and eastern gray treefrogs, discussions of these species are nearly always intertwined. Cope's gray treefrogs were originally designated as a subspecies of pine woods treefrogs (*H. femoralis chrysoscelis*; Cope, 1880). Wright and Wright (1949) listed Cope's gray treefrogs as subspecies of eastern gray treefrogs (*H. versicolor chrysoscelis*). The distribution of Cope's gray treefrogs has long been associated, and usually combined, with that of eastern gray treefrogs. Noble and Hassler (1936) reported two call types along the

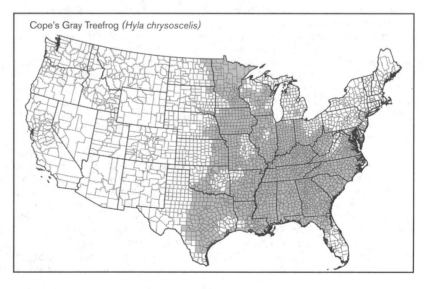

Cope's Gray Treefrog (*Hyla chrysoscelis*)

East Coast of the United States. A fast-trilling, harsh call type was found in the southern United States, while a slower trilling, more mellow-sounding call type was found at higher latitudes. The two call types coexisted in a railroad yard in Baltimore, Maryland. Wright and Wright (1949), apparently not realizing the importance of the call differences, restricted the distribution of *H. v. chrysoscelis* to east-central Texas, east into southern Arkansas and northwestern Louisiana. Blair (1958a) reported distinct geographic distributions for two call types of what was then considered *H. versicolor*. Recent depictions show the distribution of the gray treefrog complex to encompass much of the eastern United States, including the eastern Great Plains (eastern portions of Nebraska, Kansas, Oklahoma, and eastern Texas), Minnesota south to Louisiana, and east of the Mississippi River to the Atlantic coastline, excluding northern Maine and peninsular Florida, south of the panhandle (Conant and Collins, 1998).

Species identification has long been problematic in this group. Noble and Hassler (1936) first reported discrete call types along a latitudinal cline. Blair (1958a) plotted distinct geographic distributions of two call types across the range of what was then considered *H. versicolor*. Johnson (1959, 1961, 1963, 1966) reported genetic incompatibility between the call types, prompting him to designate the faster trilling call type as *H. chrysoscelis* and the slower trilling call type as *H. versicolor*. The diploid-tetraploid nature of this complex was first suggested when Wasserman (1970) reported that *H. versicolor* was a tetraploid (4N = 48). Later, Bogart and Wasserman (1972) confirmed the diploid (2N = 24) nature of *H. chrysoscelis*. Ralin (1977a) discussed call variation between *H. chrysoscelis* and *H. versicolor* along an east-west transect from central Texas to Louisiana. Gerhardt (1974a), Ralin (1977a), and Cline (1990) quantified call differences between eastern and western races of *H. chrysoscelis*. Bachmann and Bogart (1975), Cash and Bogart (1978), and Green (1980) demonstrated partial discrimination between the two species by measuring cell diameter. Ralin and Rogers (1979) reported that *H. versicolor* were morphologically and genetically (Ralin and Selander, 1979) intermediate between eastern and western populations of *H. chrysoscelis*. Cline (1990, using Ralin 1977a) combined acoustical, morphological, and genetic analyses in central Oklahoma, Kansas, Missouri, and Arkansas to show Cope's gray treefrogs range from east-central Texas (excluding the eastern 1/5 of the state), north into central Oklahoma (excluding the eastern 1/4 of the state), eastern Kansas, extreme eastern Nebraska, extreme eastern South Dakota, extreme eastern North Dakota, western and southern Minnesota, Iowa, extreme southwest Wisconsin, most of Missouri, east in a band across south-central Illinois, much

of central Indiana, and much of western and central Ohio, south through most of Arkansas and Louisiana, Mississippi, southwestern Tennessee, Alabama, the Florida Panhandle, Georgia, South Carolina, eastern North Carolina, southeastern Virginia, and southeastern Maryland.

Numerous attempts have been made to identify and determine the distributions of these cryptic species at the state level. Hoffman (1946, Virginia), Walker (1946, Ohio), Brown and Brown (1972, Illinois), Bogart and Jaslow (1979, Michigan), Johnson (1987, Missouri), Matson (1988, Ohio), and Little (1983, Ohio/West Virginia) used call characteristics. Olson (1984) used calls and karyotypes to verify *H. chrysoscelis* in Minnesota. Jaslow and Vogt (1977, Wisconsin), Chapin and Trauth (1987, Arkansas), and Hillis et al. (1987, Kansas) used histological techniques. Recent research at the state level has begun to elucidate the separate distributions of Cope's and eastern gray treefrogs.

2. Historical versus Current Abundance. Little data comparing historical and current abundance are available. With the advent of the North American Amphibian Monitoring Program (NAAMP) and its standardized protocols, this paucity of data should be resolved. Most of the early reports suggest, at least qualitatively, that members of the gray treefrog complex were common historically. Personal experience in the Southwest suggests that both species remain common. They occur densely around ponds in Alabama, but diffuse "populations" of calling Cope's gray treefrogs are found in suburban habitats.

3. Life History Features. Ritke et al. (1991a) discussed the life history of a western Tennessee population of Cope's gray treefrogs.

A. Breeding. Reproduction is aquatic.

i. Breeding migrations. Calling is probably stimulated by a combination of day length and temperature. Calling typically begins earlier in the southern portion of their range. Wright (1932) reports *H. versicolor* (almost certainly *H. chrysoscelis*) from the Okefenokee Swamp calling from 10 June–13 August. In northeastern Alabama, calling can begin in late March and continue through July, but calling is most intense in April–May. In north-central Oklahoma, calling usually begins in April, with a sharp peak in May to early June. Calling usually does not extend very far into July in Oklahoma. Fellers (1997b) reports calling activity in Maryland from April–July or August.

Cope's gray treefrogs breed from mid-March to July (Wright and Wright, 1949). It seems reasonable to suggest that the peak of the breeding season for both species is late spring (May–June).

ii. Breeding Habitat. During the breeding season, Cope's gray treefrogs are found calling near the edges of ponds, ephemeral

wetlands, and ditches, and from floating algae and emergent vegetation (Fellers, 1979b; Godwin and Roble, 1983; Conant and Collins, 1998; personal observations). At dusk, Cope's gray treefrogs begin calling from high in the trees surrounding a pond. As the evening progresses, individuals move down the trees (sometimes calling along the way) until they reach lower branches or shrubs, or they continue until they reach the ground and move to a point usually within 1.5 m of the water's edge (personal observations). Fellers (1979b) detailed calling site characteristics. Ptacek (1992) compared calling sites of *H. chrysoscelis* and *H. versicolor*. Godwin and Roble (1983) suggest that females may mate on the first day they arrive at the breeding pond. Males and females may mate ≤3 times/breeding season (Godwin and Roble, 1983). Ritke and Lessman (1994) present biopsy data indicating that females develop follicles throughout the breeding season, and production of later follicles relate to foraging success. Ritke et al. (1991b) report strong breeding pond philopatry.

B. Eggs.

i. Egg deposition sites. Egg deposition sites are similar for both Cope's gray treefrogs and eastern gray treefrogs. Eggs are loosely attached to emergent vegetation at the surface of shallow ponds and pools (permanent or temporary). These wetlands may be natural or created and can be highly disturbed. Hillis et al. (1987) reports Cope's gray treefrogs breeding in the rain-filled furrows of cornfields in Kansas.

ii. Clutch size. Eggs are laid in packets (10 × 12.5 cm, 30–40 eggs/packet) as a surface film loosely attached to emergent vegetation (Wright, 1932).

C. Larvae/Metamorphosis. Development is aquatic. Thibaudeaux and Altig (1998) documented the ontogenic development of the oral apparatus of Cope's gray treefrogs.

i. Length of larval stage. Dickerson (1906) reported a larval period of 3 wk from eggs to hatching and 4 wk to metamorphosis. Wright (1932) reported 45–65 d from eggs to metamorphosis. Because developmental rates are generally positively linked to temperature, one would expect that the average developmental times are shorter in the South.

ii. Larval requirements. McDiarmid and Altig (1999) have summarized most of what has been published on tadpoles. The sources cited herein are largely from this text.

a. Food. No data are available on the food habits of Cope's gray treefrogs or eastern gray treefrogs. Tadpoles generally feed by filtering food from the water column or by scraping periphyton from submerged substrates (Hoff et al., 1999). Steinwascher and Travis (1983) reported that Cope's gray treefrog tadpoles grew faster on diets with high protein-to-carbohydrate ratios.

b. Cover. There are no published reports of cover requirements for Cope's gray treefrog tadpoles. During the day, tadpoles can be seen resting on a variety of substrates including exposed sediment, leaf litter, and fallen tree limbs. I have observed hylid tadpoles resting on top of these same substrates at night.

iii. Larval polymorphisms. McCollum and Van Buskirk (1996) reported that reddish and yellowish tail pigments in Cope's gray treefrogs were induced by the presence of odonate naiad predators.

iv. Features of metamorphosis. Larval Cope's gray treefrogs are approximately 16 mm TL at metamorphosis (Wright, 1932).

v. Post-metamorphic migrations. Roble (1979) describes post-metamorphic migrations of eastern gray treefrogs from central Wisconsin. Within a week, juveniles dispersed from their natal ponds. Juveniles moved an average of 1.58 m/d, with maximum dispersal distances approaching 125 m (Roble, 1979). Juveniles were active throughout the day from July–September.

D. Juvenile Habitat. Roble (1979) reported that juvenile eastern gray treefrogs from Wisconsin were captured on sedges (*Carex* sp.) about 1/3 of the time, with false nettle (*Boehmeria cylindrica*), reed grass (*Phalaris arundinacea*), and swamp white oak (*Quercus bicolor*) saplings as preferred habitat. Nearly 3/4 (70.3%) of all captures were below 50 cm above the forest floor, while >2% were found above 1.2 m. Roble (1979) surmised that young frogs did not ascend into trees during their first season.

E. Adult Habitat. Outside of the breeding season, Cope's gray treefrogs are found on trees or on mossy or lichen-covered fences, usually above ground (Conant and Collins, 1998), and will utilize knothole cavities (Ritke and Babb, 1991) and bluebird nesting boxes (personal observations).

F. Home Range Size. Little information is known regarding movements outside the breeding season. Adults are thought to spend the remaining part of the activity season high in trees where they forage on insects and insect larvae. Short-term movements are probably limited, but during dry seasons, low relative humidity may drive treefrogs to seek out high relative humidity microhabitats.

G. Territories. Little is known about territoriality in these frogs. Gray treefrogs are known to produce specialized calls (called "turkey roots" by Wright, 1932) when approached while calling. Fellers (1979a) describes territorial behavior in treefrogs. Wells and Taigen (1986) note that call duration increases in high density (mean distance between individuals ~1 m) populations, presumably a response to intrusion in a territory.

H. Aestivation/Avoiding Desiccation. Not documented.

I. Seasonal Migrations. The only migrations reported for gray treefrogs are those to the breeding ponds beginning in March and continuing through June. In July, individuals may call during periods of high humidity or after rains, but populations tend to be diffuse. Newly metamorphosed eastern gray treefrogs move ≤1.1 km (0.7 mi) from the breeding ponds (Roble, 1979). Dispersal to winter hibernacula have not been described, but such movements are probably short and asynchronous.

J. Torpor (Hibernation). Wright (1932) speculated that gray treefrogs remained active in Georgia "until November at least." Harlan (1835, in Wright, 1932) relates collecting a specimen several feet below the surface of the ground from under a root of an apple tree in winter. Burkholder (1998) discovered hibernacula of Cope's gray treefrogs near the bases of sugar maple trees (*Acer saccharum*), 2.5–5.0 cm below the soil surface, as well as within leaf litter of varying depths; all individuals were found <1 m from the base of each tree investigated.

Freeze tolerance has been described for eastern gray treefrogs. Schmid (1982) reported that glycerol production in eastern gray treefrogs allowed individuals to survive −6 °C for 5–7 d. Storey and Storey (1985) reported additional production of glucose as a cryoprotectant. Gray treefrogs were able to survive –2 °C for 5 d and were also capable of surviving repeated freezing and thawing events. Storey and Storey (1986) found eastern gray treefrogs could survive moderate freezing temperatures (−2 to −4 °C) for ≤2 wk. Cope's gray treefrogs exhibit natural freeze tolerance via production of elevated levels of glucose in their blood plasma (Costanzo et al., 1992).

K. Interspecific Associations/Exclusions. Maxson et al. (1977) reported immunological hybrids between *H. chrysoscelis* and *H. versicolor* in southeastern Oklahoma. Ralin et al. (1983) reported on electrophoretic hybrids from Illinois. Gerhardt et al. (1994) cytologically confirmed the natural occurrence of a triploid hybrid between Cope's gray treefrogs (diploid) and eastern gray treefrogs (tetraploid).

Petranka (1989b) reported possible chemical inhibition of southern leopard frog (*R. sphenocephala*) tadpoles by Cope's gray treefrog tadpoles. Alford and Wilbur (1985) and Morin (1987) reported that Cope's gray treefrog tadpoles were smaller in the presence of other tadpoles, and the order in which species appeared in a pond influenced the community composition at that pond.

Ralin (1981) demonstrated that Cope's gray treefrogs and eastern gray treefrogs from Texas had similar abilities to cope with desiccation in the same habitats. He further noted greater variability among populations of the same species than he saw between the species.

L. Age/Size at Reproductive Maturity. Cope's gray treefrogs range in size from 32–60 mm (Wright and Wright, 1949; Conant and Collins, 1998). Wright (1932) suggests that gray treefrogs from the Okefenokee Swamp begin breeding at 2 yr of age.

M. Longevity. There are no published reports on longevity of Cope's gray treefrogs.

N. Feeding Behavior. General reports list insects as the prey of gray treefrogs (Holbrook, 1842, in Wright, 1932). Dickerson (1906) and Ritke and Babb (1991) found grey treefrogs to be "sit-and-wait" predators, consuming caterpillars, beetles, flies, wood roaches (*Parcoblatta* sp., *Ischnoptera deropeltiformis*), and camel crickets (*Ceuthophilus* sp.).

O. Predators. Numerous potential predators exist for hylids, ranging from invertebrates through vertebrates. Dickerson (1906) indicated that diving beetles preyed upon Cope's gray treefrog tadpoles. Resetarits (1998) reported that odonate naiads and larval dytiscids (diving beetles) both preyed on Cope's gray treefrog tadpoles, but larval dytiscids were major egg predators as well. Dragonfly naiads were also used in experimental studies of tadpole response to predators (McCollum and Van Buskirk, 1996). A wide variety of fish could prey upon all life stages of Cope's gray treefrogs. Petranka et al. (1987) noted the ability of Cope's gray treefrog tadpoles to detect chemicals associated with fish predators (green sunfish, *Lepomis cyanellus*). Smith et al. (1999) reported that bluegill sunfish (*L. macrochirus*) substantially reduced eastern gray treefrog tadpole abundance in field experiments. Potential amphibian predators include salamanders and salamander larvae (chiefly newts [*Notophthalmus* sp.] and ambystomatids [*Ambystoma* sp.]) and some adult frogs (i.e., American bullfrogs [*Rana catesbeiana*]). Skelly (1992) used tiger salamanders (*A. tigrinum*) in field experiments of antipredator costs. Numerous turtles and snakes represent potential predators on the various stages of the life cycle of frogs. Wading birds, especially herons (Ardeidae), prey upon tadpoles and frogs of many species. Additionally, raccoons (*Procyon lotor*) and striped skunks (*Mephitis mephitis*) are potential mammalian predators.

P. Anti-Predator Mechanisms. Cope's gray treefrogs produce mucous secretions that are foul tasting and cause burning sensation and inflammation in mucous membranes of eyes (personal observations). While these secretions have antipredator functions, it is possible that they also function as antimicrobial agents. Cline (1986) reported death feigning (thanatosis) in Cope's gray treefrogs from northeastern Oklahoma.

Tadpoles may not use chemical defense compounds. Kats et al. (1988) categorized Cope's gray treefrogs tadpoles as "palatable" in their predation studies. McCollum and Van Buskirk (1996) reported that predators (odonate naiads) induced production of reddish or yellowish tail pigments in

Cope's gray treefrogs. Subsequently, two papers (Semlitsch, 1990; Figiel and Semlitsch, 1991) noted that substantial tail damage (>75%) must occur before tadpoles suffer increased mortality (using odonate and crayfish predators). Thus, it appears that tadpoles distribute these pigments in the tail to misdirect predator attacks. Several studies have also reported decreased tadpole activity (Fauth, 1990) and a shift in habitat usage (Petranka et al., 1987; Kats et al., 1988) in the presence of predators. Resetarits and Wilbur (1989) concluded that Cope's gray treefrog tadpoles were capable of detecting chemical odors of potential predators in water conditioned by these predators.

Q. Diseases. So far, no known reports of hylid frog declines have been related to diseases such as those caused by chytrid fungi or ranaviruses.

R. Parasites. A variety of parasite hosts have been suggested for gray treefrogs. Armstrong et al. (1997) studied reproduction of a monogenean flatworm parasite in Cope's gray treefrogs. Delvinquier and Dresser (1996) reported an opalinid (Sarcomastigophora) from eastern gray treefrogs. Hausfater et al. (1990) chose eastern gray treefrogs as a model organism for studying the effect of parasitism on mate choice, in part because males harbored a "wide range of helminth parasites."

4. Conservation.

Cope's gray treefrogs are moderately tolerant to the pollutants tested so far, and they are tolerant to human habitat disturbance.

Confusion between this species and eastern gray treefrogs continues to confound conservation efforts. Cope's gray treefrogs are listed as Endangered in New Jersey, where a permit issued by the New Jersey Division of Fish, Game, and Wildlife is required for all activities involving this species.

Hyla cinerea (Schneider, 1799)
GREEN TREEFROG

Michael Redmer, Ronald A. Brandon

1. Historical versus Current Distribution.

The range of green treefrogs (*Hyla cinerea*) extends southward from the Chesapeake Bay region of Delaware, Maryland, and Virginia, through the Carolinas, Georgia, and Florida, then west through most of Alabama and Mississippi, statewide in Louisiana, the eastern half of Texas, and the Red River drainage in southeast Oklahoma. In the Mississippi River drainage, their range extends north from Louisiana and Mississippi through floodplains in southern and eastern Arkansas, western Tennessee, and Kentucky, to extreme southeastern Missouri and southern Illinois (Conant and Collins, 1998; Redmer and Brandon, 2003). Introduced populations occur in Puerto Rico (Schwartz and

Thomas, 1975; Rivero, 1978; Schwartz and Henderson, 1985, 1991; Hedges, 1996), Brownsville, Texas (Conant, 1977b; Smith and Kohler, 1977), in central Missouri (Johnson, 1987), and perhaps a coastal island in Florida (Smith et al., 1993). A population that was introduced in east-central Kansas is now probably extirpated (Collins, 1993). There is evidence of recent peripheral range expansion in Illinois (Redmer and Ballard, 1995; Redmer et al., 1999), Missouri (Powell et al., 1995, 1996), and South Carolina (Snyder and Platt, 1997; Platt et al., 1999).

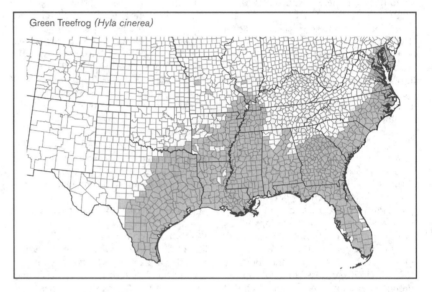

Green Treefrog (*Hyla cinerea*)

2. Historical versus Current Abundance.

There are few data on green treefrog abundance, although there have been few localized reviews of population status. Most authors have commented that they are generally common or locally abundant throughout most of their range (Anderson et al., 1952; Liner, 1954, 1955; O.B. Goin, 1958; Smith, 1961; Bider, 1962; Dyrkacz, 1974; Ackerman, 1975; Mount, 1975; Ashton, 1976; Martof et al., 1980; Bancroft et al., 1983; Gibbons, 1983; Johnson, 1987; Ashton and Ashton, 1988; Dundee and Rossman, 1989; Gibbons and Semlitsch, 1991; Redmond and Scott, 1996; Lannoo, 1998d; Bartlett and Bartlett, 1999a; Mitchell and Reay, 1999; Redmer et al., 1999). There is one report of a localized population decline in Florida (Delis et al., 1996; Dodd and Griffey, 2002).

3. Life History Features.

A. Breeding. Reproduction is aquatic.

i. Breeding migrations. There are few data on this topic. Movements from upland habitats into adjacent wetlands have been reported (O.B. Goin, 1958). Roberts and Page (2003) report an instance of a male trying to dislodge an amplectant male.

ii. Breeding habitats. Most data on habitats are based upon adults captured within,

or adjacent to, breeding habitat. During the breeding season, adult habitats are variously described as floating or emergent herbaceous vegetation, shrubs, or trees around the margins of aquatic habitats including swamps, sloughs, marshes, ponds, and lakes (Hurter, 1911; Wright, 1932; Carr, 1940a; Jobson, 1940; Cagle, 1942; Goin, 1943; Bartsch, 1944; Neill, 1951a; Werler and McCallion, 1951; Anderson et al., 1952; Moulton, 1954a; Carr and Goin, 1955; Tinkle, 1959; Rossman, 1960; Smith, 1961; Lee, 1969a; Mount, 1975; Garton and Brandon, 1975; McDi-

armid et al., 1983; Dundee and Rossman, 1989; Mitchell and Miller, 1991; Scott and Koons, 1993a,b; Redmer et al., 1999).

B. Eggs.

i. Egg deposition sites. Oviposition takes place in association with floating mats of vegetation, such as duckweed (Garton and Brandon, 1975; Mount, 1975; Turnipseed and Altig, 1975).

ii. Clutch size. Clutch sizes vary considerably, ranging from 478–1,061 (mean = 700) in southern Illinois (Garton and Brandon, 1975); from 1,348–3,946 (mean = 2,152) in Arkansas (Trauth et al., 1990); and 275–1,160 (mean = 814) in Georgia (Perrill and Daniel, 1983). Green treefrogs may spawn several times over the course of a single prolonged breeding season (Perrill and Daniel, 1983), with minimum fecundities (based on eggs counted from mutiple spawns of single females) calculated to be from 1,088–3,272 (mean = 1,688). However, these authors may not have counted all clutches produced by a single female, so actual fecundity may be substantially higher.

C. Larvae/Metamorphosis.

i. Length of larval stage. The larval period ranges from 24–45 d (Orton, 1947; Wright and Wright, 1949; Garton and Brandon, 1975; Turnipseed and Altig, 1975; Blouin, 1991, 1992a,b; Blouin and Loeb, 1991).

ii. Larval requirements.

a. Food. Unknown.

b. Cover. Larval habitat has been reported by Garton and Brandon (1975), Redmer et al. (1999), and Turnipseed and Altig (1975) as ponds with large areas of emergent vegetation. Roth and Jackson (1987) conducted experiments in pools of three sizes and found that tadpole survival was greatest in the smallest pools, which contained the lowest natural densities of predatory insects.

iii. Larval polymorphisms. Tadpoles have been described by or included in keys devised by several authors (Dickerson, 1906; Wright, 1929, 1932; Brimley, 1944; Morris, 1944; Altig, 1970; Cochran and Goin, 1970; Travis, 1981; Ashton and Ashton, 1988; Redmer et al., 1999). Tadpoles are 4.5–5.5 mm TL at hatching and grow to approximately 60 mm before metamorphosis. Ontogenetic color change is continuous until tadpoles reach Gosner (1960) stages 25 or 26, when light blotches fuse to form interorbital and transverse body bands. After this, the tadpoles have a green body, a yellow to buff venter, a yellow tail with dark mottling or reticulations, and distinct yellow orbitonasal stripes. The yellow interorbital stripe present earlier in development is sometimes retained, thus forming a triangle on the head. The tail is long; the distinctly arched dorsal tail fin originates on the back of the body is proportional in height to the ventral fin. Other external characteristics include laterally bulging eyes, a dextral anus, and a sinistral spiracle.

iv. Features of metamorphosis. Tadpoles metamorphose within 28–44 d of hatching (Wright and Wright, 1949; Garton and Brandon, 1975; Turnipseed and Altig, 1975).

v. Post-metamorphic migrations. At one site in southern Illinois, post-metamorphic juveniles moved to upland habitats <60 m from the breeding pond (Garton and Brandon, 1975). Elsewhere in southern Illinois, dispersing juveniles were found ≤0.5 km from breeding ponds (Redmer et al., 1999). Some of those juveniles were found ≤3 mo of the initiation of the breeding season (M.R., unpublished data).

D. Juvenile Habitat. Juveniles are often found in emergent vegetation in and around breeding ponds, though they may also migrate to adjacent uplands (Garton and Brandon, 1975; Redmer et al., 1999).

E. Adult Habitat. Green treefrogs are adaptable to a number of habitats, although sites are typically associated with permanent bodies of water containing abundant emergent vegetation. Reported aquatic habitats include swamps, sloughs, marshes, lakes, farm ponds, sewage ponds, fish-farm ponds, flooded borrow pits, flooded sink-holes, and ditches (Hurter, 1911; Wright, 1932; Carr, 1940a; Jobson, 1940; Cagle, 1942; Goin, 1943; Bartsch, 1944; Babbit and Babbit, 1951; Neill, 1951a; Werler and McCallion, 1951;

Anderson et al., 1952; Carr and Goin, 1955; Tinkle, 1959; Rossman, 1960; Smith, 1961; Lee, 1969a; R.L. Brown, 1974; Garton and Brandon, 1975; Mount, 1975; Turnipseed and Altig, 1975; Dundee and Rossman, 1989; Scott and Koons, 1993a,b; Phelps and Lancia, 1995; Redmer et al., 1999). Green treefrogs are commonly reported from barrier islands and other coastal areas where they apparently are tolerant of brackish water (Allen, 1932; Dunn, 1937; Oliver, 1955a; Neill, 1958a; Martof, 1963; Diener, 1965; Moore, 1976; Mueller, 1985; Smith et al., 1993; Mitchell and Anderson, 1994). There are a number of reports of refugia or hibernacula, including rock crevices, bird houses, and human litter such as tin cans; green treefrogs are frequently found around human dwellings (O.B. Goin, 1958; Tinkle, 1959; Grzimek, 1974; Garton and Brandon, 1975; Delnicki and Bolen, 1977; McComb and Noble, 1981). PVC pipes placed as artificial refugia have been used in efforts to sample this and other species of *Hyla* (Mouton et al., 1997).

F. Home Range Size. Unknown.

G. Territories. Unknown, but adult males are known to defend calling sites (Garton and Brandon, 1975).

H. Aestivation/Avoiding Desiccation. Unknown, but studies of temperature and water balance have emphasized thermoregulation (O.B. Goin, 1958; Freed, 1980b), body temperatures (Brattstrom, 1963, 1968), cooling (Wygoda, 1988b), factors affecting evaporative water loss (Wygoda, 1984, 1988a,b, 1989a,b; Wygoda and Williams, 1991; Wygoda and Garman, 1993), fever (Kluger, 1977; Muchlinsky, 1985), thermal acclimation/tolerances (Brattstrom, 1963; Layne and Romano, 1985; Blem et al., 1986; Layne et al., 1989), tolerance of desiccation (Layne et al., 1989), and water absorption (Walker and Whitford, 1970). Ballinger and McKinney (1966) reported a lower lethal developmental temperature of 20 °C.

I. Seasonal Migrations. Unknown, but juveniles will disperse onto wooded hillsides and into open fields (Bartsch, 1944; Garton and Brandon, 1975; Redmer et al., 1999).

J. Torpor (Hibernation). Captive specimens become lethargic, and movements are uncoordinated at temperatures below 16 °C (M.R., unpublished data; for further information see "Aestivation/Avoiding Desiccation" above).

K. Interspecific Associations/Exclusions. A number of authors have reported associations with other species of anurans that reproduce in similar habitats—for example, northern cricket frogs (*Acris crepitans*), southern cricket frogs (*A. gryllus*), bird-voiced treefrogs (*H. avivoca*), American bullfrogs (*Rana catesbeiana*), and green frogs (*R. clamitans*; Wright, 1932; Cagle, 1942; Livezey and Johnson, 1948; Wright and Wright, 1949; Peterson et al., 1952; O.B. Goin, 1958; Brown and Pierce, 1965; Hardy, 1972; Garton and Brandon, 1975;

Turnipseed and Altig, 1975; Moore, 1976; Trauth, 1992; Grimké and Jaeger, 1998; Redmer et al., 1999).

Numerous studies report on interspecific isolating mechanisms or their failure, producing hybridization, between this and other hylids (Blair, 1958b; Mecham, 1960b, 1965; Littlejohn, 1961; Pyburn and Kennedy, 1961; Kennedy, 1964; Lee, 1968a; Fortman and Altig, 1974; Gerhardt, 1974b; Oldham and Gerhardt, 1975; Pierce, 1975; Ralin, 1977b; Anderson and Moler, 1986; Schlefer et al., 1986; Lamb and Avise, 1986, 1987; Lamb, 1987; Maxon et al., 1987; Lamb et al., 1990; and Mable and Rye, 1992).

L. Age/Size at Reproductive Maturity. Uncertain. Based on growth rates, several authors have suggested that sexual maturity is reached in the second year of life (O.B. Goin, 1958; Garton and Brandon, 1975). Other demographic information is based primarily upon studies of growth (O.B. Goin, 1958; Garton and Brandon, 1975; Blouin, 1991, 1992a,b).

M. Longevity. Unknown.

N. Feeding Behavior. Their postmetamorphic diet includes a variety of arthropods and other small invertebrates (Haber, 1926; Kilby, 1945; Oliver, 1955a; R.L. Brown, 1974; Freed, 1982a; Ritchie, 1982). Green treefrogs are visually-oriented predators that respond to prey size, shape, and speed (Deban and Nishikawa, 1992; Freed, 1980a,b, 1982a,b, 1988; and Hueey, 1980). Leips and Travis (1994) studied the effects of food availability on newly metamorphosed animals.

O. Predators. Reported by a number of authors to consist of a variety of vertebrates, including several species of snakes and wading birds, and invertebrates, including spiders (Wright, 1932; Wright and Wright, 1949; Bowers, 1966; Jenni, 1969; Garton and Brandon, 1975; Schardien and Jackson, 1982; Lockley, 1990; Bishop and Farrell, 1994; Mitchell, 1994a; Palmer and Braswell, 1995).

P. Anti-Predator mechanisms. Behavioral defense by adults was reported by Marchison and Anderson (1976). There is some evidence that green treefrog tadpoles are more unpalatable to predatory fishes than are tadpoles of related barking treefrogs (*H. gratiosa*; Blouin, 1990).

Q. Diseases. Unknown, but the effects of environmental contaminants on larval development and behavior have been studied (Webber and Cochran, 1984; Mahaney, 1994). For example, Jung and Jagoe (1995) demonstrated that aluminum, often present in elevated concentrations in acidfied waters, can have lethal or non-lethal effects. Among the non-lethal effects, exposed tadpoles demonstrate slower swimming speeds, which increases their vulnerability to predators. The distribution of radiocesium in a contaminated population has also been studied (Dapson and Kaplan, 1975).

R. Parasites. Endoparasites reported from this species include several helminths

(Steiner, 1924; Brooks, 1979), and a *Basidiobolus* sp. (Okafor et al., 1984). Transmission of a trypanosome to this species (by a dipteran) has been described (Johnson et al., 1993). McKeever (1977) has observed mosquitoes preying on adults. Green treefrogs may be beneficial to some agricultural crops because they are a known vector of a blight *(Colletotrichtum gloeosporioides)* known to affect certain vetch plants considered to be weeds (Yang and TeBeest, 1992; Yang et al., 1992).

4. Conservation.
Though there have been few localized reviews of population status, it is generally agreed that green treefrogs are common or locally abundant throughout most of their range. There is one report of a localized population decline in Florida, and reports of recent range expansion in Illinois, Missouri, and South Carolina (see "Historical versus Current Distribution" and "Historical versus Current Abundance," above). They are not protected under any state or federal laws.

Hyla femoralis Bosc, 1800
PINE WOODS TREEFROG

Joseph C. Mitchell

1. Historical versus Current Distribution.
The historical distribution of pine woods treefrogs *(Hyla femoralis)* is unknown but may have been larger than today due to the presence of isolated disjunct populations in Alabama and Mississippi (Hoffman, 1988). There is a questionable record from southern Maryland. Pine woods treefrogs occur from the middle Coastal Plain of Virginia, south through most of Florida, except the Everglades, and west to eastern Louisiana.

2. Historical versus Current Abundance.
Historical abundance is unknown. Pine woods treefrogs are locally common in most of their range.

3. Life History Features.
Aspects of the life history of pine woods treefrogs are in Wright (1932), Harper (1932), and Wright and Wright (1949), but a complete modern synthesis is not available.

A. Breeding. Reproduction is aquatic.

i. Breeding migrations. Adults migrate from arboreal retreats to breeding sites, but mass migrations have not been reported.

ii. Breeding habitat. Breeding sites include depressional wetlands in pine flatwoods, Carolina bays, vernal pools, ditches and other shallow water-filled depressions, and swamps. Pine woods treefrogs are early colonizers and often call from pools in clearcuts and fields.

Calling sites are usually on vegetation in and around breeding sites, as high as several meters above the substrate. Large choruses are heard after heavy rains. Calling periods are early April to October in central Florida (Carr, 1940a; Einem and Ober, 1956); March to mid-September in Georgia (Harper, 1932; Brandt, 1953); April–August in Alabama (Mount, 1975); early April to mid-August in Louisiana (Dundee and Rossman, 1989); April–August in North Carolina (Brandt, 1936a; personal observations); and May–August in Virginia (Mitchell, 1986). Einem and Ober (1956) noted that males stopped calling when a hurricane passed through Florida in 1953. Egg-laying occurs in May–July in North Carolina (Brandt, 1936a; personal observations) and late May early August in Virginia (Mitchell, 1986).

B. Eggs.

i. Egg deposition sites. Females lay their eggs attached to vegetation or debris no more than 2–3 cm below shallow water (Wright, 1932; Livezey and Wright, 1947; Mount, 1975).

ii. Clutch size. Females lay about 800–2,000 eggs in clusters of about 100 eggs (unpublished data).

C. Larvae/Metamorphosis. Larvae may be found from May–September. Wright (1932) reported a larval period of 50–75

d. Metamorphic animals were found early July to mid-August in North Carolina (Travis, 1980b) and from 16 June–16 September in Virginia (Mitchell, 1986). SVL at metamorphosis is 11–15 mm (Wright, 1932). Tadpole survivorship in the laboratory was 65–100% (Travis, 1980b).

D. Juvenile Habitat. Similar to adult habitat.

E. Adult Habitat. Pine woods treefrogs are strongly associated with pine forests, but also may be found in hammocks, swamps, cypress ponds, vernal pools, Carolina bays, mixed hardwood/pine forests, and brackish marshes (Harper, 1932; Duellman and Schwartz, 1958; Neill, 1958a). Pine woods treefrogs climb to the tops of longleaf pines (Carr, 1940a), and seek shelter in cabbage palms, bromeliads, and pitcher plants (Smith and List, 1955; Einem and Ober, 1956).

F. Home Range Size. Unknown.

G. Territories. Territorial defense is unknown.

H. Aestivation/Avoiding Desiccation. Unknown.

I. Seasonal Migrations. Pine woods treefrogs do not migrate.

J. Torpor (Hibernation). Overwintering sites include inside decomposing pine logs and underground about 0.6 m (Carr, 1940a). Pine woods treefrogs are active on warm days in winter, emerging from shelters under pine bark and inside the axils of cabbage palms and bromeliads (Einem and Ober, 1956).

K. Interspecific Associations/Exclusions. Pine woods treefrogs are syntopic with oak toads *(Bufo quercicus)*, green treefrogs *(H. cinerea)*, barking treefrogs *(H. gratiosa)*, bird-voiced treefrogs *(H. avivoca)*, little grass frogs *(Pseudacris ocularis)*, American bullfrogs *(Rana catesbeiana)*, Cope's gray treefrogs *(H. chrysoscelis)*, Fowler's toads *(B. fowleri)*, green frogs *(R. clamitans)*, southern cricket frogs *(Acris gryllus)*, southern leopard frogs *(R. sphenocephala)*, eastern spadefoot toads *(Scaphiopus holbrookii)*, spring peepers *(P. crucifer)*, squirrel treefrogs *(H. squirella)*, and eastern narrow-mouthed toads *(Gastrophryne carolinensis*; Harper, 1932; Wright, 1932; Einem and Ober, 1956; personal observations).

Natural and experimental hybrids between pine woods treefrogs and Cope's gray treefrogs were reported from Alabama, Mississippi, and Texas by Blair (1958a), Pyburn (1960), Pyburn and Kennedy (1961), Mecham (1965), Fortman and Altig (1973, 1974), and Gerhardt (1974c).

L. Age/Size at Reproductive Maturity. These life history traits are unknown for any population.

M. Longevity. Known longevity in captivity is 4 yr, 5 mo (Snider and Bowler, 1992).

N. Feeding Behavior. Pine woods treefrogs prey on grasshoppers, crickets, beetles, caddisflies, ants, wasps, craneflies, moths, and jumping spiders (Carr, 1940a; Duellman and Schwartz, 1958).

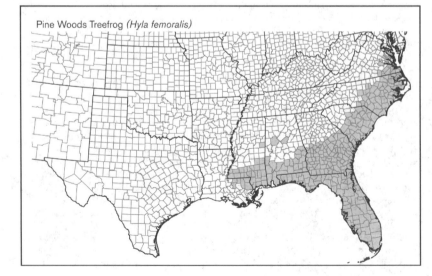

Pine Woods Treefrog *(Hyla femoralis)*

O. Predators. Known predators include banded water snakes (*Nerodia fasciata*), black racers (*Coluber constrictor*), black rat snakes (*Elaphe obsoleta*), common garter snakes (*Thamnophis sirtalis*), and ribbon snakes (*Thamnophis sauritus*; Wright, 1932; E.E. Brown, 1979; Palmer and Braswell, 1995).

P. Anti-Predator Mechanisms. Adult pine wood treefrogs used the following escape behaviors when placed in enclosures with potential snake predators: flight, remaining motionless, crouching, climbing away, body inflation, walking, and hiding (Marchisin and Anderson, 1978). Tadpoles possess bright coloration, usually red, on the posterior margins of the tail fins and a flagellum, presumably to distract invertebrate predators away from the body (Altig, 1972b; personal observations).

Q. Diseases. None reported.

R. Parasites. None reported.

4. Conservation.
Pine woods treefrogs are locally common in most of their range and are not listed in any state in which they occur. However, several states (Florida, Louisiana, Mississippi, and Virginia) have permit regulations on this and other species regulating the number collected and sold commercially (Levell, 1997).

Hyla gratiosa LeConte, 1857 "1856"
BARKING TREEFROG

Joseph C. Mitchell

1. Historical versus Current Distribution.
The historical distribution of barking treefrogs (*Hyla gratiosa*) was probably more widespread than is currently known. Disjunct populations in Kentucky and Tennessee, and in Maryland, Delaware, and southern New Jersey in the Atlantic Coastal Plain (Black and Gosner, 1958; Monroe and Giannini, 1977; Anderson and Dowling, 1982; Caldwell, 1982b; Arndt and White,

1988; Redmond and Scott, 1996) suggest that this species was once more widespread than it is today. Barking treefrogs currently occur in the Coastal Plain of the southeastern United States from southern New Jersey and Delaware, through southeastern Virginia, south through most of the Florida Peninsula, and westward to southeastern Louisiana (Caldwell, 1982b). Populations occur in the Piedmont of Alabama and parts of Georgia and South Carolina. Barking treefrogs are absent from most of the Atlantic Coastal Barrier Islands (Caldwell, 1982b).

2. Historical versus Current Abundance.
Historical abundance is unknown. Estimates of current abundance range from common in North Carolina, Florida, and Louisiana, to uncommon to rare in Kentucky, Tennessee, Virginia, and northern states. Barking treefrogs are not often encountered except after hard rains (Neill, 1958b). VanNorman and Scott (1987) reported an average of 3.3 calling males/site in Kentucky and Tennessee, but numbers of 130 males or more have been observed at breeding sites in other parts of the range of this species (Goin, 1938; personal observations). Murphy et al. (1993) reported a census population size of 1,082 adults for a site in Florida.

3. Life History Features.
Aspects of the life history and biology of these treefrogs have been summarized by Wright (1932), Wright and Wright (1949), Murphy et al. (1993), and Murphy (1994a,b).

A. Breeding. Reproduction is aquatic.

i. Breeding migrations. Adults migrate from arboreal and terrestrial retreats to breeding sites. Mass migrations have not been reported. Neill (1958b) and Murphy (1994a) reported males entering ponds along the ground and by jumping into the water from tree branches overhanging the water.

ii. Breeding habitat. Barking treefrogs breed in a wide variety of shallow wetlands, including ephemeral pools, semipermanent ponds, and permanent ponds. Observed habitats include corn fields, cypress ponds, acid bogs, flooded ditches, swamps, flooded sand pits, woodland ponds, sinkhole ponds, Carolina bays, live oak hammock wetlands, human impoundments surrounded by fields, and backwaters of streams (see summaries in Wright and Wright, 1949; Mount, 1975; VanNorman and Scott, 1987). Most populations occur in wetlands with aquatic vegetation. Wetlands with fish are seldom used for breeding.

The breeding period occurs from mid-June to late July in Kentucky (VanNorman and Scott, 1987); late April to July in Alabama and North Carolina (Mount, 1975; personal observations); mid-March to mid-August in Louisiana (Dundee and Rossman, 1989); and March–August in Florida (Murphy, 1994a). Initial dates of calling vary due to temperature and dry versus wet conditions; calling dates were 25 May–18 June in one North Carolina population (Travis, 1983b). Males call from perches on vegetation in and out of the water, but primarily while floating on the water's surface. Neill (1958b) described the barking call given from arboreal perches during the day and the breeding call produced when calling from water. Carr (1940a) noted that the barking call was heard in Florida from 1 March–23 October. In Florida, breeding is sporadic until mid-April after which it is nearly continuous with females mating on 72–95% of the nights during this period (Murphy, 1994a). Most males are present for ≤8 nights at a breeding site, but some are present for as many as 26–46 nights (Murphy, 1994b). However, Travis (1983b) stated that nearly all eggs are laid on two nights in North Carolina. Brandt (1936a) noted egg laying as late as 7 August in North Carolina. Females select males in vegetation and in water, and ovulation occurs only after amplexus is initiated (Fortman and Altig, 1973). Amplexus lasts from 5 hr to all night and females move around the pond while depositing eggs (Travis, 1983b; Murphy, 1994b). Both sexes move to the surrounding vegetation after mating. Females are not known to produce >1 clutch/yr. Amplexus, or a cue associated with amplexus, triggers oviposition (Scarlata and Murphy, 2003).

B. Eggs.

i. Egg deposition sites. Females deposit eggs singly on the substrate of the breeding pond (Livezey and Wright, 1947; Travis, 1983b).

ii. Clutch size. Females lay about 1,500–4,000 eggs (Livezey and Wright, 1947; Travis, 1983b). Number of eggs laid is positively correlated with time spent in amplexus (Scarlata and Murphy, 2003).

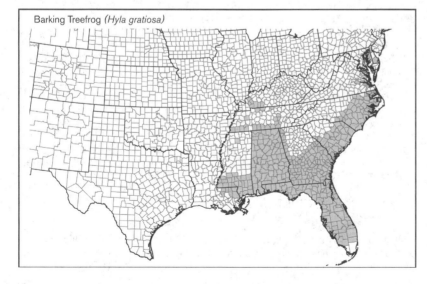

Barking Treefrog (*Hyla gratiosa*)

C. Larvae/Metamorphosis. The large and uniquely shaped tadpoles develop in 41–160 d (Wright, 1932; Travis, 1984; Dundee and Rossman, 1989) and metamorphose in June–August, depending on latitude, pond hydrology, and temperature. Tadpole densities reach 4–8/m² in some ponds (Caldwell et al., 1980). Survivorship in field enclosures was 0.1–0.22 and varied due to invertebrate densities (Travis, 1983a). Size at metamorphosis is 18–28 mm SVL and is affected by food availability, genetic relationships, and temperature (Travis, 1980a,b; Dundee and Rossman, 1989; Leips and Travis, 1994).

D. Juvenile Habitat. Juveniles remain in low vegetation following metamorphosis, sometimes dispersing widely from natal ponds.

E. Adult Habitat. Adult barking treefrogs remain in trees and shrubs or burrow into damp sand under logs or grass tussocks around the pond border when they are not engaged in calling or reproduction in water (Neill, 1952, 1958b). Neill (1952) found them no more than 1.5 m above ground on vegetation, but I have seen them about 2.5 above ground on tree limbs. Murphy (1994a) reported movements of 100 m between breeding ponds by several males in Florida.

F. Home Range Size. Home range size is unknown.

G. Territories. Barking treefrogs do not maintain or defend territories.

H. Aestivation/Avoiding Desiccation. Aestivation occurs during dry periods in summer. Neill (1952) noted that barking treefrogs aestivate in sandy soil beneath grass tussocks.

I. Seasonal Migrations. Barking treefrogs do not migrate seasonally, but remain in the vicinity of breeding sites.

J. Torpor (Hibernation). Barking treefrogs burrow into sandy substrates in Georgia and Florida and use gopher tortoise (*Gopherus polyphemus*) burrows and other burrows for overwintering, although Neill (1952) suggested that they may dig backwards similar to toads to create their own subterranean retreats. Carr (1940a) found adults 1.2 m below the surface. Lee (1968b) found adults in Florida gopher mouse (*Peromyscus floridanus*) burrows.

K. Interspecific Associations/Exclusions. This species occurs with other frogs that use shallow wetlands for breeding, including northern cricket frogs (*Acris crepitans*), southern cricket frogs (*A. gryllus*), Fowler's toads (*Bufo fowleri*), southern toads (*B. terrestris*), eastern narrow-mouthed toads (*Gastrophryne carolinensis*), Cope's gray treefrogs (*H. chrysoscelis*), eastern gray treefrogs (*H. versicolor*), green treefrogs (*H. cinerea*), pine woods treefrogs (*H. femoralis*), squirrel treefrogs (*H. squirella*), Brimley's chorus frogs (*Pseudacris brimleyi*), spring peepers (*P. crucifer*), upland chorus frogs (*P. feriarum*), American bullfrogs (*Rana catesbeiana*), green frogs (*R.

clamitans), and southern leopard frogs (*R. sphenocephala*; de Rageot, 1969; VanNorman and Scott, 1987; personal observations). Syntopic salamanders include Mabee's salamanders (*Ambystoma mabeei*), mole salamanders (*A. talpoideum*), tiger salamanders (*A. tigrinum*), and eastern newts (*Notophthalmus viridescens*).

Mecham (1960b), Gerhardt et al. (1980), and Lamb and Avise (1986, 1987) evaluated hybridization between barking treefrogs and green treefrogs in Alabama and Georgia.

L. Age/Size at Reproductive Maturity. Age at sexual maturity is unknown for barking treefrogs. VanNorman and Scott (1987) reported that the smallest adult male was 45 mm SVL; the smallest adult female 57 mm.

M. Longevity. Maximum known longevity for a captive adult was 10 yr, 3 mo (Snider and Bowler, 1992).

N. Feeding Behavior. These gape-limited predators eat anything that they can swallow. Murphy et al. (1993) found that prey in 30 adult males represented nine orders from three classes of arthropods, with the majority being insects and beetles (Coleoptera) being the most common insect taxa. Barking treefrogs are opportunistic foragers, consuming arboreal and terrestrial prey.

O. Predators. Known predators of barking treefrogs are dragonfly naiads, mole salamanders, banded water snakes (*Nerodia fasciata*), cottonmouths (*Agkistrodon piscivorus*), and southern hog-nosed snakes (*Heterodon simus*; Neill, 1952; E.E. Brown, 1979; Caldwell et al., 1980).

P. Anti-Predator Mechanisms. Adults inflate with air when disturbed but are not known to be toxic. Small tadpoles have a black saddle in the middle of the tail musculature (Altig, 1972b) that may direct predators away from the body. Caldwell et al. (1980) found that tadpoles showed an average immobile response time of >4 min when predation was simulated. When frightened or prodded, tadpoles initially dart off, then sink slowly a short distance as they become immobile (Caldwell et al., 1980).

Q. Diseases. None reported, although Mitchell and Green (2003) recently documented an intestinal hernia representing a developmental malformation.

R. Parasites. None reported.

4. Conservation.
Barking treefrogs are locally common throughout much of their range from North Carolina southward. However, this species is listed as state Endangered in Delaware, Threatened in Virginia, and legally Protected in Maryland and Tennessee (Levell, 1997). Disjunct populations in Kentucky are not protected. Habitat loss and the small number of occurrences contribute to their protected status in the northern portions of their range.

Murphy et al. (1993) made three management recommendations for a location in central Florida: (1) decisions on this species should be based on estimates of effective population size and not short-term census results; (2) multiple ponds in the landscape should be protected because barking treefrogs migrate among breeding sites; and (3) habitats between breeding ponds should be protected to allow dispersal. Similar recommendations would be appropriate for other areas where barking treefrogs breed in complexes of ponds.

Hyla squirella Bosc, 1800
SQUIRREL TREEFROG

Joseph C. Mitchell, Michael J. Lannoo

1. Historical versus Current Distribution.
Squirrel treefrogs (*Hyla squirella*) are found along the Atlantic Coastal Plain from Virginia to the Florida Keys, and along the Gulf Coastal Plain from south Florida to eastern Texas (Wright, 1932; Burt, 1938b; Wright and Wright, 1949; Hoffman, 1955; Duellman and Schwartz, 1958; Hendrickson, 1974; Martof, 1975a; Mount, 1975; Gotte and Ernst, 1987; Conant and Collins, 1991; Petzing and Phillips, 1998a; Mitchell and Reay, 1999; Dixon, 2000). They occur on numerous barrier islands off the southeastern Atlantic coast and Florida Gulf Coast (Martof, 1963; Blaney, 1971; Gibbons and Coker, 1978; Braswell, 1988). Squirrel treefrogs have been introduced into the Bahamas (Crombie, 1972).

2. Historical versus Current Abundance.
Squirrel treefrogs are one of the most common species in Florida (Carr, 1940a; Bartlett and Bartlett, 1999a). Deckert (1914, cited in Wright, 1932) noted squirrel treefrogs in southeastern Georgia are "... the commonest of the southern tree toads ... found everywhere, in corn fields, sugar cane, about wells and under eaves of stable roofs, barns, outhouses, etc." Large populations of squirrel treefrogs were observed in the low country of southern Alabama by Mount (1975), the Mississippi Gulf Coast by Smith and List (1955), and in eastern North Carolina by Robertson and Tyson (1950) and Palmer and Whitehead (1961). In Princess Anne County, Virginia, Hoffman (1955) observed numerous treefrogs that "... came to the bright light of our Coleman lantern." Werler and McCallion (1951) did not encounter this species in their surveys of the early 1940s in the same area. Although squirrel treefrogs can be locally abundant (J.C.M., personal observations), historical and current estimates of population sizes are lacking.

3. Life History Features.
  A. Breeding. Reproduction is aquatic.
    i. Breeding migrations. Adults migrate from upland sites to breeding pools during

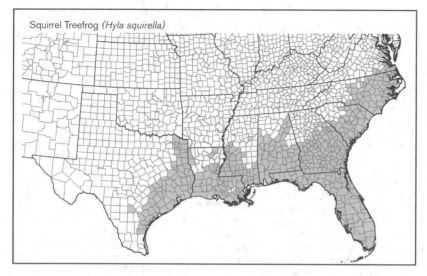

Squirrel Treefrog (Hyla squirella)

rains (Wright, 1932; Wright and Wright, 1949; J.C.M., personal observations).

ii. Breeding habitat. Woodland or pasture wetlands, flooded roadside ditches, stock ponds, or other shallow bodies of water (Wright, 1932; Wright and Wright, 1949; Gosner and Black, 1956; Mount, 1975). Breeding males call while sitting on shorelines, perched on debris or vegetation, hidden in clumps of grass, and while sitting in 1–2 cm of water (Carr, 1940a; Wright and Wright, 1949; Gosner and Black, 1956; Duellman and Schwartz, 1958). Mount (1975) reported that males call from a perch 30–60 cm above the water or from a bank near the water's edge.

iii. Breeding periods. Seasonal breeding times vary with latitude but usually coincide with spring and summer rains (Wright, 1932; Wright and Wright, 1949). The common name of "rain frog" reflects the tendency for this species to call during the day when rain is approaching, although males produce a raspy, squirrel-like call dissimilar from the breeding call (Wright and Wright, 1949; Duellman and Schwartz, 1958; Mount, 1975; Garrett and Barker, 1987). Squirrel treefrogs breed from mid-April to mid-August in Alabama (Mount, 1975); early April to August in northern Florida (Carr, 1940a; Einem and Ober, 1956); late March to early August in southern Florida (Duellman and Schwartz, 1958); May–August in North Carolina (Brandt, 1936a); and March–October in Texas (Garrett and Barker, 1987). In Florida, "Full choruses come with July electrical storms" (Carr, 1940a).

Diel activity and breeding cycles have been understudied. Squirrel treefrogs are nocturnal and seldom seen abroad during the day. Buchanan (1992) observed treefrogs in Louisiana leave and re-enter diurnal retreats at dusk and dawn, respectively, and experience peak periods of activity at 2000–2300 and 0300–0500 h.

B. Eggs.

i. Egg deposition sites. Wright (1932) stated that eggs are laid in shallow pools and that "no end of tadpoles must be lost from drying of breeding places," indicating the ephemeral nature of the larval habitat. Eggs may be deposited singly or in clusters on the bottom or attached to vegetation (Wright, 1932; Wright and Wright, 1949; Livezey and Wright, 1947; Brugger, 1984; Bartlett and Bartlett, 1999a). Orton (1947) noted that eggs "were scattered singly or in pairs on debris in water about 3–6 inches [7.5–15 cm] deep."

ii. Clutch size. Females lay >950 eggs (Livezey and Wright, 1947). Wright (1932) cited egg counts of 942 and 972, and Martof et al. (1980) noted 1,000 eggs/clutch. Clutch size from five females in northern Florida averaged 1,059 (Brugger, 1984). Production of >1 clutch/year has not been reported.

C. Larvae/Metamorphosis.

i. Length of larval stage. Time from deposition to hatching is 24–43 hr following ovulation (Wright, 1932; Beck, 1997). Tadpoles metamorphose after 40–50 d (Wright and Wright, 1949; Garrett and Barker, 1987) at 11–13 mm (Wright, 1932; Martof et al., 1980). Beck (1997) reported experimental larval periods of 54–77 d at 22 °C with low to constant feeding rates and body mass at metamorphosis from 0.25–0.52 g.

ii. Larval requirements.

a. Food. Tadpoles are suspension feeders that eat organic and inorganic food particles they scrape from rock, plant, and log substrates.

b. Cover. Tadpoles use high habitat complexity to escape predation by dragonfly naiads (*Anax junius*) and giant water bugs (*Lethocerus americanus*; Babbitt and Tanner, 1997). Some ephemeral pools, however, lack cover (J.C.M., personal observation), and tadpoles probably experience variable rates of mortality.

iii. Larval polymorphisms. Not known to occur.

iv. Features of metamorphosis. Metamorphosis appears to occur over extended periods. Wright (1932) reported a 2-wk duration.

v. Post-metamorphic migrations. Recently metamorphosed animals emerge from breeding pools and migrate to upland feeding sites and shelters. Goin and Goin (1957) observed numerous squirrel treefrog juveniles entering their yard in northern Florida from October–December, but none of the migrations were en masse. Pool emergence and post-metamorphic migrations have not been quantified.

D. Juvenile Habitat. Juveniles occupy habitats similar to adults, although Neill (1951a) noted that many squirrel treefrogs found in bromeliad plants are recently metamorphosed animals. Goin and Goin (1957) found that newly arriving juveniles initially used palmettos for cover, but subsequently moved to more permanent shelters including oaks, holly trees, and magnolias. They noted that suitability of holly leaves as overwintering sites reflects the structure of the leaves, as they are leathery and tend to curl toward the underside. Juvenile squirrel treefrogs frequently select a position between two leaves that lie one over the other. Juveniles also remained under inverted tin cans atop stakes in their yard during the winter. Site shifts occurred from natural to unnatural cover from autumn to spring.

E. Adult Habitat. Squirrel treefrogs show little discrimination in their selection of major habitat types (Carr, 1940a). They occur in and around buildings, in gardens, grasslands, weed or brush tangles, bottomland hardwoods, riparian zones, open woodlands, pinelands, trees, vines, cypress stands, longleaf pine/turkey oak/wiregrass associations, and longleaf pine/slash pine flatwoods—in almost any place associated with moisture, food, and cover (Wright, 1932; Carr, 1940a; Wright and Wright, 1949; Anderson et al., 1952; Funderburg, 1955; Goin and Goin, 1957; Goin, 1958; Duellman and Schwartz, 1958; McComb and Noble, 1981; Enge and Marion, 1986; Garrett and Barker, 1987; Conant and Collins, 1991; Dodd, 1992; Lamb et al., 1998). Squirrel treefrogs exhibit some preference for open woodlands, such as mature pine and mixed hammock forests and open woody areas (Wright, 1932; Carr, 1940a; Wright and Wright, 1949; Delzell, 1979). They occur in trash piles (Ashton and Ashton, 1988) and bromeliad plants (Neill, 1951a), and often select narrow spaces in human structures (Goin and Goin, 1957). Squirrel treefrogs are known for their ability to change colors to match their backgrounds (Wright, 1932; Wright and Wright, 1949).

Populations of squirrel treefrogs occupy coastal estuarine and harsh barrier island habitats. Engles (1952) found them on, in, and under logs surrounded by meters of bare, dry sand in the wrack zone, and in all vegetated habitats on the Outer Banks of North Carolina. Neill (1958a) noted that this treefrog breeds in rainwater pools affected by saltwater spray in the Florida Keys. Webb (1965) determined that the salinity of a pool with squirrel treefrog tadpoles on Bogue Bank (a barrier island off North Carolina) was 47% of full-strength sea water.

F. Home Range Size. Home ranges have not been reported, but Neill (1957b) noted that squirrel treefrogs apparently establish temporary home ranges, which they occupy for days to weeks. Adults in northern Florida returned to the same resting places day after day, even after retreating to other sites during inclement weather (Goin and Goin, 1957).

G. Territories. Analyses of territorial behavior have not been published.

H. Aestivation/Avoiding Desiccation. Squirrel treefrogs will often aestivate communally. Carr (1940a) wrote, "Several are found in the same hollow tree or in the axils of the same royal or coco-palm petiole when numerous available and apparently identical retreats are unoccupied." Of the eight species of hylid frogs studied by Farrell and MacMahon (1969), squirrel treefrogs had the lowest water content, and females were composed of less water than males. Squirrel treefrogs will tolerate an average of 34% of total weight loss in desiccating conditions before succumbing—about average for the species studied by Farrell and MacMahon (1969).

I. Seasonal Migrations. Migrations are tied to breeding and metamorphic events; most are apparently tied to periods of rain.

J. Torpor (Hibernation). Carr (1940a) writes: "They are gregarious hibernators, thirty or forty sometimes congregating under a loose slab of bark." Squirrel treefrogs hibernate beneath bark of pine stumps and logs and decaying bark of trees (Allen, 1932; Neill, 1948b). Adults may be seen on warm days in northern Florida during the winter; the lowest air temperature at which a juvenile squirrel treefrog was active measured 6.7 °C (Goin and Goin, 1957; Goin, 1958). They also noted that upon the approach of a cold front, frogs sought more secure shelters, such as under leaf mold and rotten bark.

K. Interspecific Associations/Exclusions. Wright and Wright (1949) noted that squirrel treefrogs breed in association with eastern spadefoot toads (*Scaphiopus holbrookii*) and gopher frogs (*Rana capito*). In New Orleans, squirrel treefrogs called with Coastal-Plain toads (*Bufo nebulifer*), Fowler's toads (*B. fowleri*), and eastern narrow-mouth toads (*Gastrophryne carolinensis*; Volpe, 1956). One of us (J.C.M.) observed breeding in a shallow wetland in eastern North Carolina along with southern toads (*B. terrestris*), barking treefrogs (*H. gratiosa*), eastern spadefoot toads, and southern leopard frogs (*R. sphenocephala*), and in Virginia with Fowler's toads, Cope's gray treefrogs

(*H. chrysoscelis*), green treefrogs (*H. cinerea*), eastern narrow-mouthed toads, and eastern spadefoot toads.

Goin and Goin (1953) and Goin (1958) noted that where squirrel treefrogs co-occur with green treefrogs in Alachua County, Florida, the latter prefer low shrubbery and more permanent breeding sites and the former prefer the canopy and ephemeral breeding sites. Squirrel treefrogs were found trapped in a ditch dug to accommodate a pipeline along with seven species of salamanders, eight other species of hylids, five species of ranids, five species of bufonids, and eastern narrow-mouthed toads Anderson et al., 1952). Volpe (1956) found amplexed pairs of squirrel treefrogs and eastern narrow-mouthed toads (two pairs of male squirrel treefrogs with female eastern narrow-mouthed toads, three pairs of male eastern narrow-mouthed toads with female squirrel treefrogs). None of the pairs produced viable eggs; the sperm did not penetrate the eggs.

L. Age/Size at Reproductive Maturity. Wright (1932) noted 13 mm body length as size at metamorphosis and estimated that juveniles grow to 19–23 mm during their first year and that 2-yr-olds were ≥26 mm in size. Adult body length is 22–41 mm (Wright and Wright, 1949; Martof, 1975a; Martof et al., 1980; Conant and Collins, 1991) and maximum size will reach 45 mm (Mount, 1975). Males and females are similar in size (Wright and Wright, 1949; Duellman and Schwartz, 1958).

M. Longevity. A wild-caught adult lived 8.5 yr in captivity (Snider and Bowler, 1992).

N. Feeding Behavior. Squirrel treefrogs are aggressive predators that feed on insects and other invertebrates (Wright, 1932; Garrett and Barker, 1987). Carr (1940a) noted that they are "Often present in enormous numbers along lake-shores when chironomids (Diptera) are emerging; they also collect around lamp posts and lighted windows at night. I once saw nine young in a circle around a pile of newly deposited cow-dung, awaiting and devouring the midges attracted thereto." Duellman and Schwartz (1958) examined the stomachs of 20 individuals and found that 9 were empty, 2 contained only plant debris, and 4 contained beetles (Coleoptera); other stomachs contained crustacean (crayfish?) remains, a spider (Arachnida), a cricket (Orthoptera), and an ant (Hymenoptera). Brugger (1984) found that two groups of arthropods (Arachnida, Isoptera) and three orders of insects (Coleoptera, Orthoptera, Hymenoptera) dominated the diet of frogs from northern Florida. Goin and Goin (1957) noted that differences in habitats selected by adults and juveniles may result in dietary differences.

O. Predators. Duellman and Schwartz (1958) observed ribbon snakes (*Thamnophis sauritus*) feeding on squirrel treefrogs. Other predators undoubtedly include

small mammals, birds, other frogs, other snakes, and aquatic invertebrates. Dragonfly naiads and giant water bugs eat tadpoles (Babbitt and Tanner, 1997).

P. Anti-Predator Mechanisms. Their small size, propensity for seeking small hiding places, and ability to change colors to match their background probably aid in avoiding visual predators.

Q. Diseases. None reported.

R. Parasites. None described.

4. Conservation.
Pague and Mitchell (1987) noted that some populations in urbanized areas of southeastern Virginia had become extirpated. However, Neill (1950a) and Delis et al. (1996) found squirrel treefrogs relatively abundant in urban areas of Augusta, Georgia, and Tampa, Florida, respectively. Although squirrel treefrogs are well known to cross roads at night in rains (e.g., Mount, 1975), the effect of road mortality on size and structure of urban populations is unknown.

Hyla versicolor LeConte, 1825
EASTERN GRAY TREEFROG
George R. Cline

Eastern gray treefrogs (*Hyla versicolor*) and Cope's gray treefrogs (*H. chrysoscelis*) are members of a cryptic, diploid-tetraploid species complex. This has resulted in considerable taxonomic confusion, especially in early reports. As a result of this confusion, many authors have chosen to combine the descriptions and distributions of these species. In this account, I have attempted to separate, as much as possible, the literature related to *H. chrysoscelis* and *H. versicolor*. For categories where data are lacking or where there is a question regarding the species identification, data from the sister species are reported. This approach is at least partially validated by the genetic analysis of this complex by Ptacek et al. (1994) and the growth and development studies of Ptacek (1996).

1. Historical versus Current Distribution.
Because of the cryptic nature of Cope's gray treefrogs and eastern gray treefrogs, discussions of these species are nearly always intertwined. *Hyla versicolor* was originally described by LeConte (1825). Wright and Wright (1949) listed Cope's gray treefrogs as subspecies of eastern gray treefrogs (*H. versicolor chrysoscelis*). Fitzgerald et al. (1981) reviewed the early taxonomy of this complex. The distribution of Cope's gray treefrog has long been associated, and usually combined, with that of eastern gray treefrogs. Noble and Hassler (1936) reported two call types along the East Coast of the United States. A fast-trilling, harsh call type was found in the southern United States, while a slower trilling, more mellow-sounding call type was found at higher latitudes. The two call

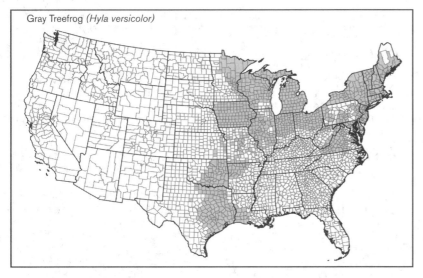

Gray Treefrog (*Hyla versicolor*)

types coexisted in a railroad yard in Baltimore. Wright and Wright (1949), apparently not realizing the importance of the call differences, restricted the distribution of *H. chrysoscelis* to east-central Texas, east into southern Arkansas and northwestern Louisiana. Blair (1958a) reported distinct geographic distributions for two call types of what was then considered *H. versicolor*. Fouquette and Johnson (1960) demonstrated call discrimination by females of both call types. Recent depictions show the distribution of the *H. chrysoscelis/versicolor* complex to encompass much of the eastern United States, including the eastern Great Plains (eastern portions of Nebraska, Kansas, Oklahoma, and eastern Texas), Minnesota south to Louisiana, and east of the Mississippi River to the Atlantic coastline, excluding northern Maine and peninsular Florida, south of the panhandle (Conant and Collins, 1998).

Species identification has long been problematic in this group. Noble and Hassler (1936) first reported discrete call types along a latitudinal cline. Blair (1958a) plotted distinct geographic distributions of two call types across the range of what was then considered *H. versicolor*. Johnson (1959, 1961, 1963, 1966) reported genetic incompatibility between the call types, prompting him to designate the faster trilling call type as *H. chrysoscelis* and the slower trilling call type as *H. versicolor*. The diploid–tetraploid nature of this complex was first suggested when Wasserman (1970) reported that *H. versicolor* was a tetraploid (4N = 48). Later, Bogart and Wasserman (1972) confirmed the diploid (2N = 24) nature of *H. chrysoscelis*. Ralin (1977a) discussed call variation between *H. chrysoscelis* and *H. versicolor* along an east–west transect from central Texas to Louisiana. Gerhardt (1974a), Ralin (1977a), and Cline (1990) quantified call differences between eastern and western races of *H. chrysoscelis*. Bachmann and Bog-

art (1975), Cash and Bogart (1978), and Green (1980) demonstrated partial discrimination between the two species measuring cell diameter. Later, Ralin and Rogers (1979) reported that *H. versicolor* were morphologically and genetically (Ralin and Selander, 1979) intermediate between eastern and western populations of *H. chrysoscelis*. Cline (1990) combined acoustical, morphological, and genetic analyses in central Oklahoma, Kansas, Missouri, and Arkansas. Cline (1990, using Ralin, 1977a) shows Cope's gray treefrogs in east-central Texas (excluding the eastern 1/5 of the state), north into central Oklahoma (excluding the eastern 1/4 of the state), eastern Kansas, extreme eastern Nebraska, extreme eastern South Dakota, extreme eastern North Dakota, western and southern Minnesota, Iowa, extreme southwest Wisconsin, most of Missouri, east in a band across south-central Illinois, much of central Indiana, and much of western and central Ohio, south through most of Arkansas and Louisiana, Mississippi, southwestern Tennessee, Alabama, the Florida Panhandle, Georgia, South Carolina, eastern North Carolina, southeastern Virginia, and southeastern Maryland. *Hyla versicolor* is divided into two groups. A southwestern group extends from east-central Texas north through central and eastern Oklahoma and western Arkansas into eastern Kansas and southwestern Missouri. A northern and eastern group extends north from roughly southern New Jersey through central Missouri, and south along the Appalachian Mountains, possibly as far south as Alabama and Georgia. Ralin and Selander (1979) reported isozyme differences between eastern and western forms of *H. chrysoscelis*. Gerhardt (personal communication) has narrowed the gap between northern and southern *H. versicolor* groups.

Numerous attempts have been made to identify species and determine distribution

of these cryptic species at the state level. Hoffman (1946, Virginia), Walker (1946, Ohio), Brown and Brown (1972, Illinois), Bogart and Jaslow (1979, Michigan), Matson (1988, Ohio), Little (1983, and Little et al., 1989, Ohio/West Virginia), Johnson (1987, Missouri) used call characteristics. Jaslow and Vogt (1977, Wisconsin), Hillis et al. (1987, Kansas), Chapin and Trauth (1987, Arkansas) used histological techniques. Recent research at the state level has begun to elucidate the separate distributions of Cope's and eastern gray treefrogs. The abstract of McAlpine et al. (1991) makes no reference to the method used to identify gray treefrogs, although it is nearly certain that the New Brunswick and Maine gray treefrogs are eastern gray treefrogs.

2. Historical versus Current Abundance.
Little data comparing historical and current abundance are available. With the advent of the North American Amphibian Monitoring Program [NAAMP]) and its standardized protocols, this paucity of data should be resolved. Most of the early reports of members of this complex suggest, at least qualitatively, that gray treefrogs were common historically. Dickerson (1906) claims that gray treefrogs are second only to American toads (*Bufo americanus*) in local abundance. Cope (1889, probably *H. versicolor* in Pennsylvania), LeConte (1825; in Wright, 1932; probably *H. versicolor* in New York), and Holbrook (1842; in Wright, 1932) refer to gray treefrogs as common. Personal experience in the south suggests that both species are still common. Dense populations occur around ponds in northern and eastern Oklahoma.

3. Life History Features.
Ritke et al. (1990, 1991a) discussed the life history of a western Tennessee population of Cope's gray treefrogs.

A. Breeding. Reproduction is aquatic.
i. Breeding migrations. Calling is probably stimulated by a combination of day length and temperature. Calling typically begins earlier in the southern than in the northern portions of the species range. Wright (1932) reports gray treefrogs (almost certainly Cope's gray treefrogs) from the Okefenokee Swamp calling from 10 June–13 August. In northeastern Alabama, calling can begin in late March and continue through July, but calling is most intense in April–May. In north-central Oklahoma, calling usually begins in April, with a sharp peak in May and early June. Calling usually does not extend very far into July in Oklahoma. Fellers (1997b) reports calling activity in Maryland from April–July or August.

Cope's gray treefrogs breed from mid-March to July (Wright and Wright, 1949). Wright and Wright (1942) report breeding in Ithaca, New York, (most probably eastern gray treefrogs) from April–July. It seems reasonable to suggest that the peak of the breeding season for both species is late spring (May–June).

ii. Breeding habitat. During the breeding season, eastern gray treefrogs are found calling near the edges of ponds, ephemeral wetlands and ditches, and from floating algae and emergent vegetation (Fellers, 1979b; Godwin and Roble, 1983; Conant and Collins, 1998; personal observations). At dusk, gray treefrogs may begin calling from high in the trees surrounding a pond. As the evening progresses, individuals move down the trees (sometimes calling along the way) until they reach lower branches or shrubs, or they continue until they reach the ground and move to a point usually within 1.5 m of the water's edge (personal observations). Godwin and Roble (1983) suggest that females may mate on the first day they arrive at the breeding pond. Both males and females have been observed to mate ≤3 times/breeding season (Godwin and Roble, 1983). Ritke and Lessman (1994) present biopsy data that indicate that females have developing follicles throughout the breeding season, and production of later follicles is a function of foraging success. Ritke et al (1991b) report strong breeding pond philopatry.

B. Eggs.

i. Egg deposition sites. Eggs are deposited in shallow ponds and pools (permanent or temporary), which may be natural or artificial, pristine or disturbed. Eggs are loosely attached to emergent vegetation at the surface. Runoff and splash at the base of a waterslide enticed calling males, while a nearby dilapidated swimming pool supported breeding in eastern gray treefrogs near Kimerling City, Missouri (personal observations).

ii. Clutch size. Eggs are laid in packets (10 x 12.5 cm, 30–40 eggs/packet) as a surface film loosely attached to emergent vegetation (Wright, 1932). Egg number ~1,800 (Wright, 1932).

C. Larvae/Metamorphosis. Larval length near metamorphosis is approximately 50 mm (Wright and Wright, 1949). Cope (1889) presented drawings of developing larvae of eastern gray treefrogs (Plate 78, Figs. 23–26). Thibaudeaux and Altig (1998) documented the ontogenic development of the oral apparatus of Cope's gray treefrogs.

i. Length of larval stage. Dickerson (1906) reported a larval period of 3 wk from eggs to hatching and 4 wk to metamorphosis. Wright (1932) reported 45–65 d for eggs to metamorphosis. Because developmental rates are generally positively linked to temperature, average developmental times are shorter in the South.

ii. Larval requirements. McDiarmid and Altig (1999) have summarized most of what has been published on tadpoles.

a. Food. Little data are available on the food habits of Cope's gray treefrogs or eastern gray treefrogs. Tadpoles generally feed by filtering food from the water column or by scraping periphyton from submerged substrates (Hoff et al., 1999). Steinwascher and Travis (1983) reported that Cope's gray treefrog tadpoles grew faster on diets with high protein-to-carbohydrate ratios.

b. Cover. There are no published reports of cover requirements for gray treefrog tadpoles. During the day, tadpoles can be seen resting on a variety of substrates including exposed sediment, leaf litter, and fallen tree limbs. I have observed hylid tadpoles resting on top of these same substrates at night.

iii. Larval polymorphisms. McCollum and Van Buskirk (1996) reported that reddish and yellowish tail pigments are induced by the presence of odonate naiad predators.

iv. Features of metamorphosis. Transformation of tadpoles occurs from late June to August, between 45–65 d post-hatching (Wright and Wright, 1949). Larval Cope's gray treefrogs are approximately 16 mm TL at metamorphosis (Wright, 1932; Wright and Wright, 1932, 1949).

v. Post-metamorphic migrations. Roble (1979) describes post-metamorphic migrations of eastern gray treefrogs from central Wisconsin. Within a week, juveniles dispersed from their natal ponds. Juveniles moved an average of 1.58 m/d, with maximum dispersal distances approaching 125 m (Roble, 1979). Juveniles were active throughout the day from July–September.

D. Juvenile Habitat. Roble (1979) reported that juvenile eastern gray treefrogs from Wisconsin were captured on sedges (*Carex* sp.) about 1/3 of the time, with false nettle (*Boehmeria cylindrica*), reed grass (*Phalaris arundinacea*), and swamp white oak (*Quercus bicolor*) saplings as preferred habitat. Nearly 3/4 (70.3%) of all captures were below 50 cm above the forest floor, while >2% were found above 1.2 m. Roble (1979) surmised that young frogs did not ascend into trees during their first season.

E. Adult Habitat. Outside of the breeding season, eastern gray treefrogs are found in trees or on mossy or lichen-covered fences, usually above ground (Wright and Wright, 1949; Conant and Collins, 1998), and have been found using abandoned bird houses for shelter (Murphy, 1968). Eastern gray treefrogs will use introduced tree species such as Chinese tallowtrees (*Triadica sebifera*) as microhabitat (Fontenot, 2003).

F. Home Range Size. Little information is known regarding movements outside the breeding season. Adults are thought to spend the remaining part of the activity season high in trees where they forage on insects and insect larvae. Short term movements are probably limited, but during dry seasons, low relative humidity may drive treefrogs to seek out high relative humidity microhabitats.

G. Territories. Little is known about territoriality in frogs. Eastern gray treefrogs are known to produce specialized calls (called "turkey roots" by Wright, 1932) when approached while calling. Fellers (1979a) describes territorial behavior in treefrogs. Wells and Taigen (1986) note that call duration increases in high density (mean distance between individuals ~1 m) populations of eastern gray treefrogs. Presumably this is a response to intrusion in a territory. Eastern gray treefrogs will also call prior to thunderstorms. These "rain calls" are common and produced by many hylids (C. M. Bogert, 1960; M. Stewart, personal communication).

H. Aestivation/Avoiding Desiccation. There are no published reports of aestivation in eastern gray treefrogs.

I. Seasonal Migrations. The only migrations reported for eastern gray treefrogs are those to the breeding ponds from March–June. In July, individuals may call during periods of high humidity or after rains, but populations tend to be diffuse. Newly metamorphosed eastern gray treefrogs move up to 800 m (0.5 mi) from the breeding ponds (Roble, 1979). Dispersal to winter hibernacula have not been described, but such movements are probably short and asynchronous.

J. Torpor (Hibernation). Wright (1932) speculated that gray treefrogs remained active in Georgia "until November at least." Harlan (1835, in Wright, 1932) relates collecting a specimen several feet below the surface of the ground from under a root of an apple tree in winter.

Freeze tolerance has been described for eastern gray treefrogs. Schmid (1982) reported that glycerol production in eastern gray treefrogs allowed individuals to survive −6 °C for 5–7 d. Storey and Storey (1985, 1986) reported additional production of glucose as a cryoprotectant. Eastern gray treefrogs were able to survive −2 °C for 5 d and were also capable of surviving repeated freezing and thawing events. Storey and Storey (1986) found eastern gray treefrogs could survive moderate freezing temperatures (−2 to −4 °C) for ≤2 wk. In a laboratory setting, Layne (1991) found that contact with external ice formation triggers freezing in eastern gray treefrogs.

K. Interspecific Associations/Exclusions. While all authors agree that eastern gray treefrogs arose from Cope's gray treefrogs, there has been considerable disagreement over the possibility of multiple origins for the tetraploid. Maxson et al. (1977) reported immunological hybrids between *H. chrysoscelis* and *H. versicolor* in southeastern Oklahoma. Ralin et al. (1983) reported on electrophoretic hybrids from Illinois and argued for a single origin of the tetraploid. However, Ptacek et al. (1994) identified three different *H. versicolor* lineages based on cytochrome b sequencing. Later, Gerhardt et al. (1994) cytologically confirmed the natural occurrence of a natural triploid hybrids and argued for strong selection against these hybrids in the field. Most recently, Wiley and Little (2000) demonstrated multiple

H. versicolor lineages based on the position of the nuclear organizing region and replication banding patterns.

Petranka (1989b) reported possible chemical inhibition of southern leopard frog *(R. sphenocephala)* tadpoles by Cope's gray treefrog tadpoles. Alford and Wilbur (1985) and Morin (1987) reported that both eastern and Cope's gray treefrog tadpoles inhibit the presence of other tadpoles, and the order in which species appeared in a pond influenced the community composition at that pond. Ralin (1981) demonstrated that both Cope's and eastern gray treefrogs from Texas had similar abilities to respond to desiccation in the same habitats. He further noted greater variability among populations of the same species than he saw between the species.

L. Age/Size at Reproductive Maturity. Eastern gray treefrogs range in size from 32–60 mm (Wright and Wright, 1949; Conant and Collins, 1998). Wright (1932) suggests that gray treefrogs from the Okefenokee Swamp began breeding at 2 yr of age.

M. Longevity. Snider and Bowler (1992) report a wild-caught animal lived 7 yr, 9 mo, and 20 d in captivity.

N. Feeding Behavior. General reports list insects as the prey of gray treefrogs (Holbrook, 1842, in Wright, 1932). Puckette (1962) listed roaches, earthworms, and other invertebrates as the major diet items for gray treefrogs, but also noted the presence of small snakes. Ralin (1968) reported diet partitioning between diploids and tetraploids: eastern gray treefrogs consumed more terrestrial insects, while Cope's gray treefrogs ate more arboreal insects. Dickerson (1906) and Ritke and Babb (1991) found gray treefrogs to be "sit-and-wait" predators, consuming caterpillars, beetles, flies, wood roaches (*Parcoblatta* sp., *Ischnoptera deropeltiformis*), and camel crickets (*Ceuthophilus* sp.).

O. Predators. Numerous potential predators exist for hylids, ranging from invertebrates through vertebrates. Dickerson (1906) indicated that diving beetles preyed upon eastern gray treefrog tadpoles. Resetarits (1998) reported that odonate naiads and larval dytiscids (diving beetles) both preyed on Cope's gray treefrog tadpoles, but larval dytiscids were major egg predators as well. Dragonfly naiads were also used in experimental studies of tadpole response to predators (McCollum and van Buskirk, 1996). A wide variety of fish could prey upon all life stages of eastern gray treefrogs. Smith et al. (1999) reported that bluegill sunfish (*Lepomis macrochirus*) significantly reduced eastern gray treefrog tadpole abundance in field experiments. Potential amphibian predators include salamanders and salamander larvae (chiefly *Notophthalmus* sp. and *Ambystoma* sp.) and some adult frogs (i.e., American bullfrogs, *Rana catesbeiana*). Skelly (1992) used tiger salamanders *(A. tigrinum)* in field experiments of

antipredator costs. Numerous turtles and snakes represent potential predators on the various stages of the life cycle of frogs. Wading birds, especially herons (Ardeidae), prey upon tadpoles and frogs of many species. Additionally, raccoons *(Procyon lotor)* and striped skunks *(Mephitis mephitis)* are potential mammalian predators.

P. Anti-Predator Mechanisms. Banta and Carl (1967) noted the occurrence of "death-feigning" behaviors in male frogs following handling or capture. Brodie and Formanowicz (1981b) noted that eastern gray treefrogs are apparently unpalatable to short-tailed shrews (*Blarina brevicauda*), which may be due to noxious skin secretions. Mucus secretions are foul tasting and cause burning sensation and inflammation in the mucus membranes of the eyes (personal observation). While these secretions have antipredator functions, it is also possible that they also function as antimicrobial agents.

Tadpoles may not use chemical defense compounds. Kats et al. (1988) categorized Cope's gray treefrog tadpoles as "palatable" in their predation studies. Tadpole morphology has a significant influence on predator escape. Van Buskirk and McCollum (1999) observed that eastern gray treefrog tadpoles from predator-rich ponds tended towards having longer, shallower bodies, and shallow, brightly pigmented tails. Following up that observation, van Buskirk and McCollum (2000a) found that these tadpoles accelerated faster than other shapes. Further, van Buskirk and McCollum (2000b) demonstrated that 30% of the tail must be removed before swimming performance is affected. These studies suggest that tadpoles distribute pigments in the tail to misdirect predator attacks. Several studies have also reported decreased tadpole activity (Lawler, 1989) and a shift in habitat usage (Formanowicz and Bobka, 1989) in the presence of predators. Resetarits and Wilbur (1989) concluded that Cope's gray treefrog tadpoles were capable of detecting chemical odors of potential predators in predator conditioned water.

Q. Diseases. Emerging infectious diseases have been increasingly implicated in the local extinction of many amphibian populations worldwide. Chytrid fungi and ranaviruses have been implicated in local extinctions in ranid, bufonid, pelodryadid treefrogs, newt, and ambystomatid populations worldwide (Daszak et al., 1999). So far, no known reports of hylid frog declines are related to chytrid fungi or ranaviruses.

R. Parasites. A variety of parasite hosts have been suggested for Cope's and eastern gray treefrogs. Woo and Bogart (1984) reported two species of trypanosomes from Cope's and eastern gray treefrogs. Delvinquier and Dresser (1996) reported an opalinid (Sarcomastigophora) from

eastern gray treefrogs. Hausfater et al. (1990) chose eastern gray treefrogs as a model organism for studying the effect of parasitism on mate choice, in part because males harbored a "wide range of helminth parasites."

4. Conservation.

McAlpine et al. (1991) suggested that the range of eastern gray treefrogs is expanding in Maine and New Brunswick as the result of human activity. Bunnell and Zampella (1999) report similar conclusions for eastern gray treefrogs from the border of the New Jersey Pine Barrens. Lannoo et al. (1994) report that gray treefrogs have persisted in an Iowa county 70 yr since an earlier report. McAlpine et al. (1991) indicated that eastern gray treefrogs from New Brunswick and Maine were neither rare nor endangered. Kolozsvary and Swihart (1999) consider eastern gray treefrogs ubiquitous in Indiana and not affected by habitat patch size. This disagrees with Hager (1998), who concludes that eastern gray treefrogs are sensitive to patch size because they were excluded from smaller islands in Lake Erie, Georgian Bay, and the St. Lawrence River.

Several studies (Karns, 1992; Grant and Licht, 1993; Pehek, 1995) indicate that eastern gray treefrogs are at least moderately tolerant of pH levels as low as 3.5. Ambient levels of UV-B radiation had no negative effects on development or survival of eggs and larvae of eastern gray treefrogs from Ontario (Grant and Licht, 1995). Zaga et al. (1998) reported a negative effect of UV-B radiation on swimming activity of eastern gray treefrogs. Additionally, Zaga et al. (1998) reported no significant effect of the carbamate pesticide, carbaryl, but they did find that UV-B radiation photoenhanced the toxicity of carbaryl as measured by swimming activity. The pesticide carbaryl had no significant effect on activity in eastern gray treefrogs tadpoles (Bridges, 1999b; Zaga et al., 1998).

The overall impression is that eastern gray treefrogs are moderately tolerant to the pollutants tested so far, and that they are tolerant to human habitat disturbance.

Hyla wrightorum: (eximia) Taylor, 1938(a)
ARIZONA TREEFROG

Erik W.A. Gergus, J. Eric Wallace, Brian K. Sullivan

1. Historical versus Current Distribution.

The type locality of Arizona treefrogs *(Hyla wrightorum)* is 17.7 km (11 mi) south of Springerville, Apache County, Arizona (Taylor, 1938a). Their range includes forested uplands of central Arizona from the vicinity of Williams, south and east to west-central New Mexico; isolated populations occur in the Sierra Anchas Mountains in central Arizona, the Huachuca

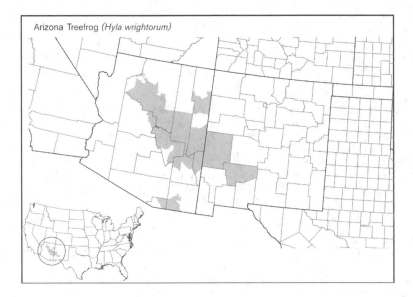

Arizona Treefrog (*Hyla wrightorum*)

Mountains and adjacent Canelo Hills of southeastern Arizona, and the Sierra Madre Occidental of northern Mexico, south to near Michoacan. Arizona treefrogs occur from 900–2,900 m (Stebbins, 1985). Additional comments on their distribution (and to some extent, natural history) can be found in Chapel (1939), Smith and Taylor (1948), Martin (1958), Lowe (1964), Duellman (1970), Morafka (1977), Van Devender and Lowe (1977), Painter (1986), McCranie and Wilson (1987), Tanner (1989a), Korky and Webb (1991), Degenhardt et al. (1996), and Gergus (1999).

Considerable controversy has persisted as to whether Arizona treefrogs should be recognized or synonymized with mountain treefrogs (*H. eximia*; e.g., Duellman, 1970; Renaud, 1977; Sullivan, 1986a). Mountain treefrogs were first described by Baird (1854b) with a type locality from "Valley of Mexico" (Districto Federal), Mexico. Arizona treefrogs were diagnosed by Taylor (1938a) as a species separate from mountain treefrogs based on the presence in the former of larger size, anterior edge of tibia with heavy brown spots and lacking a white line, and proportionately longer legs. The contact zone between Arizona and mountain treefrogs, according to Taylor (1938a), is somewhere in Chihuahua and Sonora, Mexico, with Arizona treefrogs being distributed north into Arizona and New Mexico. Schmidt (1953) arbitrarily listed Arizona treefrogs as a subspecies of mountain treefrogs, although Blair (1960b) provided evidence from mating calls indicating subspecies designation was premature and apparently incorrect. Duellman (1970) synonymized Arizona and mountain treefrogs based upon similarity in tadpole morphology, adult morphology, and mating calls. Although Duellman (1970) contended that a

mosaic pattern of variation exists in advertisement calls of mountain treefrogs, he failed to take into account the size and temperature of recorded individuals when analyzing geographic variation in calls. In light of Blair's (1960b) identification of "fast" and "slow" calls in mountain treefrogs (pulse rates of advertisement calls at similar recording temperatures were dramatically different between some samples in southern Mexico), Duellman's taxonomic conclusions are suspect. Maxson and Wilson (1974) compared serum albumins of mountain treefrogs, Pacific treefrogs (*Pseudacris regilla*), and Arizona treefrogs and supported Duellman's (1970) contention that mountain and Arizona treefrogs are closely related and together relatively divergent from Pacific treefrogs. Renaud (1977) subsequently compared morphometric, allozyme, and advertisement call variation of Mogollon Rim and mainland Mexico populations and concluded that the Arizona populations could be diagnosed from those in Mexico based on differences in size (SVL), shape, and dominant frequency of male advertisement calls. Based on these differences, Renaud (1977) referred the treefrogs of central Arizona and western New Mexico to Arizona treefrogs and those of mainland Mexico to mountain treefrogs. Gergus (1999) documented that mountain treefrogs are restricted to southern Mexico and may be diagnosed from populations to the north (e.g., Arizona and Sonora) by unique allozyme character states for at least four independent loci, greater than 7% divergence in a 575 base-pair sequence of the cytochrome b gene of mtDNA, and advertisement call pulse rates approximately half that of populations to the north. The contact zone between Arizona and mountain treefrogs is unknown (Ger-

gus, 1999), but based upon advertisement call differences, may occur in the vicinity of Michoacan (Blair, 1960b). The Huachucan Mountain population of Arizona treefrogs may be recognized as a separate species following additional study (Gergus and Reeder, in preparation).

2. Historical versus Current Abundance. Historical abundance in the United States and Mexico is unknown. In 1937, during a summer with "sufficient rainfall," Chapel (1939) found Arizona treefrogs to be relatively scarce at the western extent of their range (vicinity of Williams, Arizona), with abundance increasing to the east along the Mogollon Rim. He noted: "their numbers in summer ponds were very great" in the vicinity of Hart Canyon, and Pinetop and McNary in east-central Arizona. Collins (1996) found Arizona treefrogs in 17 of 96 localities sampled in central Arizona during 1993–94, and similar to Chapel (1939), reported the highest abundance in east-central Arizona (vicinity of Heber). Degenhardt et al. (1996) stated that Arizona treefrogs were common in appropriate habitat in west-central New Mexico. In the Huachuca Mountains, Arizona treefrogs were observed to have relatively low abundance (between 2 and 30 adults observed at any one breeding locality) and may be susceptible to extirpation by virtue of their small population sizes (Gergus, 1999).

3. Life History Features.
A. Breeding. Reproduction is aquatic.
  i. Breeding migrations. Breeding occurs primarily in June, July, and August during the summer rains (Chapel, 1939; Stebbins, 1951; Degenhardt et al., 1996; B. K. S., personal observations). Chorusing is generally restricted to a few nights following rainfall events (Sullivan, 1986a, personal observations; E. W. A. G., personal observations). During the day, Arizona treefrogs in the Huachuca Mountains take refuge under rocks and logs in nearby mesic oak groves near breeding sites (Holm and Lowe, 1995). Arizona treefrogs congregate at breeding ponds beginning around dusk and often remain there until dawn of the next day (Holm and Lowe, 1995).
  ii. Breeding habitat. "Large, grassy, shallow ponds appear to be favored, but the animals also may be found at permanent lakes, brooks, wells, and in nearly any place where rain water collects in sufficient quantities" (Stebbins, 1951; B. K. S., personal observations). In the Huachuca Mountains, Arizona treefrogs were observed only to breed in shallow, rain-filled pools with abundant aquatic vegetation available for male calling sites. In contrast to Stebbins' (1951) observations, Gergus (1999) never found Arizona treefrogs associated with permanent bodies of water, such as cattle tanks, or those lacking aquatic vegetation.

B. Eggs.

i. Egg deposition sites. Include shallow ponds, but also in permanent lakes, brooks, wells, and in nearly any place where water collects and there is adequate aquatic vegetation (see "Breeding habitat" above).

ii. Clutch size. Unknown; neither Wright and Wright (1949) nor Degenhardt et al. (1996) mention clutch size for this species.

C. Larvae/Metamorphosis. Measurements of eggs are given by Livezey and Wright (1947) and egg masses were briefly described in Wright and Wright (1949) and Zweifel (1961). Zweifel (1961) provided a detailed account of larval development, from hatching to metamorphosis, for eggs collected in east-central Arizona. Zweifel (1961) also showed that the description of larval Arizona treefrogs provided by Stebbins (1951) actually represented canyon treefrogs (*H. arenicolor*) and suggested the same for the larval measurements provided by Wright and Wright (1949). Korky and Webb (1991) also provide detailed measurements for Arizona treefrogs from three sites in Mexico and compared them with other members of the species group. Altig et al. (1998) provides a key to the larvae.

In central Arizona, larvae have been collected as early as 21 June and as late as 6 November. Post-metamorphic froglets have been collected as early as 15 June and as late as 6 November in central and east-central Arizona. Larvae have been observed in July, August, and September, and newly metamorphosed froglets in mid-September to early October (Sredl and Collins, 1992; Degenhardt et al., 1996).

D. Juvenile Habitat. Largely unknown; anecdotal observations are available for young of the year for the Huachuca Mountains population. Holm and Lowe (1995) observed metamorphic animals to be most abundant in saturated soils with emergent vegetation near the breeding ponds during the fall. Sheridan Stone (personal communication) observed a post-metamorphic Arizona treefrog approximately 250 m downstream from a breeding pond in September, possibly suggesting an emigration event away from its natal site.

E. Adult Habitat. Arizona treefrogs inhabit meadows or areas near slow-moving streams in pine-oak or pine-fir forests, generally at elevations above 550 m (Duellman, 1970; Stebbins, 1985). Adults can be found in trees at some elevation off the ground (Chapel, 1939).

F. Home Range Size. Unknown.

G. Territories. Unknown.

H. Aestivation/Avoiding Desiccation. Unknown.

I. Seasonal Migrations. Unknown; most individuals reported have been found during the summer.

J. Torpor (Hibernation). Largely unknown. A single record for an overwin-tering individual is represented by a frog collected from a debris pile in January just below the Mogollon Rim in the vicinity of Christopher Creek, Arizona. A series of Arizona treefrogs, both adults and juveniles, was collected in March from beneath boulders surrounding a small pond in Durango, Mexico (Holman, 1965a).

K. Interspecific Associations/Exclusions. Arizona treefrogs have been observed breeding in association with Arizona toads (*Bufo microscaphus*), canyon treefrogs, western chorus frogs (*Pseudacris triseriata*), various species of true frogs (*Rana* sp.), Mexican spadefoot toads (*Spea multiplicata*), and tiger salamanders (*Ambystoma tigrinum*). Sredl and Collins (1992) found that survival of larvae of Arizona treefrogs was greatly reduced in the presence of larval tiger salamanders.

L. Age/Size at Reproductive Maturity. Calling males in a breeding aggregation near Baker Lake, Coconino County, Arizona, ranged in size from 37–47 mm SVL (Sullivan, 1986a). Mean SVL for seven calling males near Hannagan Meadow, Greenlee County, Arizona, was 41.1 mm (Gergus, 1999). Mass/length relationships for adults (and other life stages) were provided by Holm and Lowe (1995).

M. Longevity. Unknown.

N. Feeding Behavior. Dietary observations for stomachs of seven individuals were provided in Chapel (1939).

O. Predators. Larval tiger salamanders prey upon larval Arizona treefrogs (Sredl and Collins, 1992). Holm and Lowe (1995) observed Mexican garter snakes (*Thamnophis eques*) preying upon larval and adult Arizona treefrogs. Other garter snakes (e.g., black-necked garter snakes [*T. cyrtopsis*] and western terrestrial garter snakes [*T. elegans*]) co-occur with Arizona treefrogs and presumably prey on them as well. Holm and Lowe (1995) observed giant water bugs (*Lethocerus* sp.) feeding on eggs, larvae, and adult frogs.

Introduced crayfish (*Orconectes* sp.), centrarchid fish (e.g., *Lepomis* sp.), and American bullfrogs (*R. catesbeiana*) are presumed predators of larval and adult Arizona treefrogs, although actual predation has not been observed (Holm and Lowe, 1995; Fernandez and Rosen, 1996). Collins (1996) included dragonflies and water beetle larvae as predators of Arizona treefrogs, but provided no detailed observations.

P. Anti-Predator Mechanisms. Degenhardt et al. (1996) described the allergic reaction of humans following the handling of adult Arizona treefrogs.

Q. Diseases. Unknown. Malformed individuals of this species have been noted from southern Apache County, Arizona (NARCAM, 1997), but to date no causal agent has been implicated.

R. Parasites. Goldberg et al. (1996b) reported on the helminth fauna (cestodes, nematodes) for populations in Arizona.

4. Conservation.
Degenhardt et al. (1996) stated that Arizona treefrogs were common in appropriate habitat in west-central New Mexico, but they may be susceptible to extirpation by virtue of their small population sizes (Gergus, 1999). Arizona treefrogs are not protected by any states or by the U.S. Federal Government.

Acknowledgments. Robert Bowker, Ryan Sawby, Sheridan Stone, and Elizabeth Sullivan assisted with some field observations. This work was supported in part by the Heritage Fund, Arizona Game and Fish Department.

Osteopilus septentrionalis (Duméril and Bibron, 1841)
CUBAN TREEFROG
Walter E. Meshaka Jr.

1. Historical versus Current Distribution.
The centers of distribution for Cuban treefrogs (*Osteopilus septentrionalis*) are Cuba and the Isle of Pines. They are also native to the Cayman Islands and the Bahamas (Schwartz and Henderson, 1991). Human-mediated dispersal is responsible for introductions in Puerto Rico, St. Croix, St. Thomas, Necker Island, and mainland Florida (Schwartz and Henderson, 1991; Meshaka, 1996a; this volume, Part One), where Cuban treefrogs are established as far north as north-central Florida. Their status as native or nonindigenous on Key West is unresolved.

2. Historical versus Current Abundance.
On mainland Florida, Cuban treefrogs were first recorded in Miami in the early 1950s (Schwartz, 1952). Since then, dispersal has been rapid, with most of their current geographic range being reached by the mid-1970s (Wilson and Porras, 1983; Meshaka, 1996a). Newly and greatly expanded geographic distributions and broad habitat distributions have resulted in greater overall abundances now than in the past.

3. Life History Features.
A. Breeding. Reproduction is aquatic.
i. Breeding migrations. Breeding is year-round in Cuba (Ruiz Garcia, 1987) and southern Florida (Meshaka, 1994, 2001). Reproductive activity in the Everglades is most common during the wet season of May–October (Meshaka, 1994, 2001). Optimal conditions associated with reproductive activity are on average wet (1.3 cm rain), warm (25.7 °C), humid (97.8%) nights (Meshaka, 1994, 2001). Tropical storms incite a strong reproductive response (Meshaka, 1993, 2001).

ii. Breeding habitat. Undescribed in Florida, but likely to be fishless freshwater

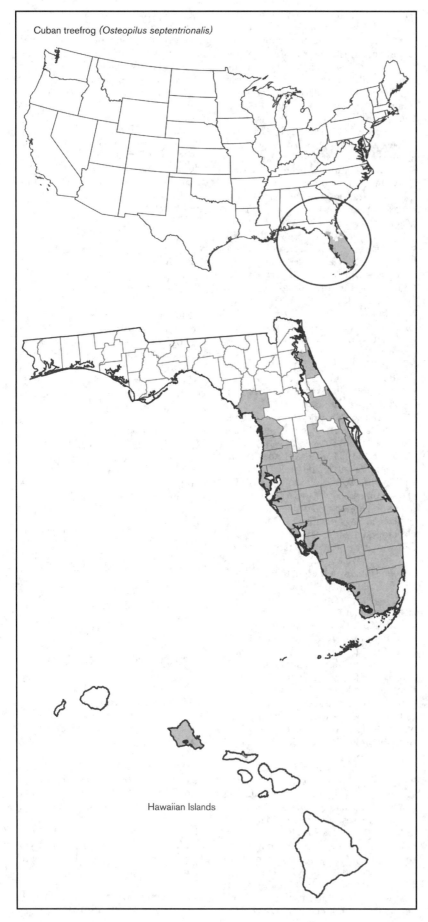

Cuban treefrog (*Osteopilus septentrionalis*)

Hawaiian Islands

wetlands, similar to their breeding habitat in Cuba (Duellman and Schwartz, 1958; Lannoo et al., 1987).

B. Eggs.

i. Egg deposition sites. Eggs are laid in a surface film (Duellman and Schwartz, 1958; Meshaka, 1994, 2001).

ii. Clutch size. Eggs are laid in partial clutches of 100–1,000 (Duellman and Schwartz, 1958; Meshaka, 1994, 2001). In the Everglades, clutch size averages 3,961 eggs and varies in proportion to female body size (Meshaka, 2001).

C. Larvae/Metamorphosis.

i. Length of larval stage. Eggs hatch in <30 hr. Development can occur in <1 mo (Meshaka, 1994, 2001).

ii. Larval requirements. Tadpoles are most often found in shallow, hot, ephemeral pools devoid of predatory fish or in similar microhabitat along the edges of permanent water (Meshaka, 1994, 2001). Tadpoles eat algae and will facultatively eat other tadpoles, including conspecifics.

iii. Larval polymorphisms. None reported, although Cuban treefrog tadpoles are highly variable in color pattern.

iv. Features of metamorphosis. In the Everglades, newly metamorphosed Cuban treefrogs appear in any month but are most common at the beginning and end of the wet season (Meshaka, 1994, 2001).

v. Post-metamorphic migrations. Newly metamorphosed animals move from breeding sites but not with any directionality other than to available cover.

D. Juvenile Habitat. In Florida, vertical vegetative structure, water, refuges, and an invertebrate prey base will sustain populations (Meshaka, 1994, 2001).

E. Adult Habitat. Similar to that of juveniles but adults are better able to withstand desiccation than smaller individuals (Meshaka, 1996b).

F. Home Range Size. Unknown.

G. Territories. Unknown, although large individuals are philopatric to refuges (personal observations).

H. Aestivation/Avoiding Desiccation. Unknown.

I. Seasonal Migrations. Following heavy initial rainstorms of the wet season in the Everglades, large numbers of adults will move *en masse* to breeding sites (Meshaka, 1994, 2001).

J. Torpor (Hibernation). Activity ceases in temperatures below 9.5 °C (Meshaka, 1994, 2001).

K. Interspecific Associations/Exclusions. In the Everglades, Cuban treefrogs occupy an otherwise underexploited niche with respect to habitat and diet and are more fecund than their potential anuran competitors, green treefrogs (*Hyla cinerea*) and squirrel treefrogs (*H. squirella*). Both of these native species, as well as other vertebrates including brown anoles (*Anolis sagrei*), Indo-Pacific geckos (*Hemidactylus garnotii*), tropical house geckos (*H. mabouia*), southern toads (*Bufo terrestris*), eastern narrow-mouthed toads *(Gastrophryne carolinensis)*, and southern leopard frogs *(Rana sphenocephala)*, are eaten by large Cuban treefrogs, primarily female (Meshaka, 1994, 2001).

L. Age/Size at Reproductive Maturity. Both sexes are reproductively mature ≤1 yr; males (27.0 mm SVL) within about 3 mo, females (45.0 mm SVL) within 7–8 mo following metamorphosis (Meshaka, 1994, 2001). Mittleman (1950) reports size ranges of 46–73 mm for males; 52.5–130 mm for females.

M. Longevity. Captives have lived ≤13 yr (Bowler, 1977). In the Everglades, individuals are much shorter lived, with males living about 1 yr and females 2–3 yr (Meshaka, 2001).

N. Feeding Behavior. Post-metamorphic individuals are "sit-and-wait" foragers perched with front feet somewhat folded over each other. Individuals in this hunting posture generally stare at surfaces about 0.5 m below them, but also stare at vertical structures in front of them or on a trail before them while waiting for moving prey. Prey include roaches and beetles, but they will also eat small vertebrates (Meshaka, 1994, 2001). Maskell et al. (2003) report a Cuban treefrog preying on a Florida brownsnake *(Storeria victa)*.

O. Predators. Primarily avian and ophidian snakes (Meshaka, 1994, 2001).

P. Anti-Predator Mechanisms. Large body size, toxic skin, and use of confining refuges (Meshaka, 1996b) serve as mechanisms to deter predators. Once caught, individuals will scream, and inflate their bodies with air to hinder ingestion by the predator (Meshaka, 1994, 2001).

Q. Diseases. Unknown.

R. Parasites. The nematode *Skrjabinoptera scelopori* is found in post-metamorphic individuals (Meshaka, 1996c) throughout their range, including native and introduced populations.

4. Conservation.
Cuban treefrogs were introduced onto mainland Florida, where they are considered an invasive species. Their rapid expansion into a variety of both natural and disturbed habitats, potentially high densities, and demonstrable negative impacts—both predatory and competitive—on native amphibians makes controlling Cuban treefrog populations an essential component of native amphibian conservation efforts in Florida.

Pseudacris brachyphona (Cope, 1889)
MOUNTAIN CHORUS FROG
Joseph C. Mitchell, Thomas K. Pauley

1. Historical versus Current Distribution.
The historical range of mountain chorus frogs *(Pseudacris brachyphona)* was undoubtedly wider than is currently known due to the occurrence of disjunct populations in

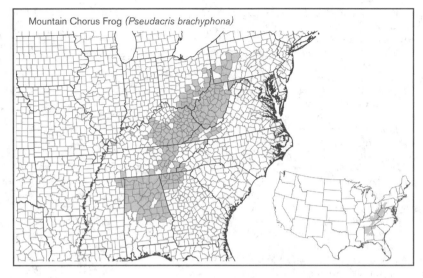

Mountain Chorus Frog *(Pseudacris brachyphona)*

several states (Hoffman, 1980). Current distribution is primarily in the Appalachian Plateau Physiographic Province from central western Pennsylvania southwestward through the Appalachians to southwestern Georgia, central Alabama, and northeastern Mississippi (Hoffman, 1980).

2. Historical versus Current Abundance.
Green (1952) estimated 1,069–1,815 individuals in 10 pools over a 4-yr period in West Virginia. McClure (1996) examined three of Green's pools in 1993 (the other sites had been destroyed) and estimated population sizes of 6–116.

3. Life History Features.
The life history and ecology of mountain chorus frogs has been summarized by Green (1938, 1964), Wright and Wright (1949), Barbour (1957, 1971), and Green and Pauley (1987).

A. Breeding. Reproduction is aquatic.

i. Breeding migrations. Green (1952) observed mass movements of breeding frogs at the beginning of the breeding season. Initiation of the calling period depends on temperature, but usually begins in late February to early March after males emerge from terrestrial hibernacula (Green, 1964). Males call from the edges of pools and do not conceal themselves as do most other species in this genus. Calling periods are sporadic in winter, occurring during warm spells in January. Breeding activity occurs late February to early June in West Virginia (Green, 1964) and mid-February to May in Kentucky (Barbour, 1971). Males will also call periodically throughout the summer, returning to breeding sites after each rainfall (Green and Pauley, 1987).

ii. Breeding habitat. Mountain chorus frogs breed in shallow wetlands at elevations from about 365–1,341 m (Green, 1938; Hoffman, 1981; Pauley, 1993a). Habitats include temporary pools, roadside ditches, pools in road-ruts, mountaintop bogs, furrows in plowed fields, seepages, and woodland springs.

B. Eggs.

i. Egg deposition sites. Females arrive at breeding sites after males, select a mate, and lay eggs attached to vegetation in several eggs masses containing 10–50 eggs or on the bottom of the pool unattached to vegetation. More have been laid unattached than attached (T. K. P., unpublished data).

ii. Clutch size. Total clutch size is 300–1,500 (Green, 1938; Barbour, 1971). Dates of egg laying are 20 March–2 July (Green, 1938). Hoffman (1955, 1981) found egg masses in southwestern Virginia on 30 March and 1 July. Barbour (1971) found egg masses at a high elevation site in August. Females leave breeding sites the same night of egg laying (Green, 1964).

C. Larvae/Metamorphosis. Eggs hatch in about 7–10 d; the larval period is short, lasting about 30–64 d, depending on temperature (Green, 1938; Barbour, 1971). Newly metamorphosed animals were found on 2 June in West Virginia, and size at metamorphosis was 11–13 mm SVL (Green, 1964; Green and Pauley, 1987).

D. Juvenile Habitat. Similar to adult habitats.

E. Adult Habitat. Green (1952) determined that mountain chorus frogs move downhill to breeding pools. He found that males move between pools during the breeding season. Maximum distance traveled was 610 m within the same season, and 1,219 m between breeding seasons. Mountain chorus frogs presumably move upslope after the breeding season to forage and seek shelter in surrounding terrestrial habitats.

F. Home Range Size. Unknown.

G. Territories. Unknown.

H. Aestivation/Avoiding Desiccation. Unknown.

I. Seasonal Migrations. Mountain chorus frogs migrate from breeding sites to foraging sites and shelters on hillsides in the forest after the breeding season.

J. Torpor (Hibernation). Overwintering sites are underground (Green, 1964).

K. Interspecific Associations/Exclusions.
Mountain chorus frogs are sympatric with American toads *(Bufo americanus)*, Fowler's toads *(B. fowleri)*, northern cricket frogs *(Acris crepitans)*, southeastern chorus frogs *(P. feriarum)*, spring peepers *(P. crucifer)*, southern chorus frogs *(P. nigrita)*, Cope's gray treefrog *(Hyla chrysoscelis)*, eastern gray treefrogs *(H. versicolor)*, American bullfrogs *(Rana catesbeiana)*, green frogs *(R. clamitans)*, pickerel frogs *(R. palustris)*, southern leopard frogs *(R. sphenocephala)*, and wood frogs *(R. sylvatica*; Barbour, 1957; Hoffman, 1981).

Experimental hybrids between mountain chorus frogs and Brimley's chorus frogs *(P. brimleyi)*, southern cricket frogs *(A. gryllus)*, Strecker's chorus frogs *(P. streckeri)*, and southeastern chorus frogs were obtained by Mecham (1965). A natural hybrid between mountain chorus frogs and southeastern chorus frogs was collected by Mecham in Alabama.

L. Age/Size at Reproductive Maturity.
The smallest male measured 22 mm; the smallest female, 28 mm (Green, 1964). Age at first reproduction is unknown.

M. Longevity. Unknown.

N. Feeding Behavior. The diet of mountain chorus frogs consists of ants, beetles, true bugs, leaf hoppers, flies, lepidopteran larvae, earthworms, centipedes, and spiders (Green and Pauley, 1987). Most prey were terrestrial species.

O. Predators. Barbour (1957) reported a mountain chorus frog in the stomach of an American bullfrog from Kentucky.

P. Anti-Predator Mechanisms. Unknown.

Q. Diseases. Unknown.

R. Parasites. Unknown.

4. Conservation.
There is a single record for this species in North Carolina (Schwartz, 1955). That state lists mountain chorus frogs as a Species of Special Concern and historical, given that it has not been seen since the original observation. Deforestation, urbanization, and loss of floodplain pools have apparently caused a decline in appropriate habitat (Murdock, 1994). Mountain chorus frogs are not listed in any other state where they occur, but regulations in Mississippi, Pennsylvania, Tennessee, and Virginia restrict commercialization (Levell, 1997).

Pseudacris brimleyi Brandt and Walker, 1933
BRIMLEY'S CHORUS FROG
Joseph C. Mitchell

1. Historical versus Current Distribution.
The historical range of Brimley's chorus frogs *(Pseudacris brimleyi)* is unknown. Their current distribution is in the Atlantic Coastal Plain from northeastern Georgia to southern Caroline County, Virginia (Hoffman, 1983; Mitchell and Reay, 1999).

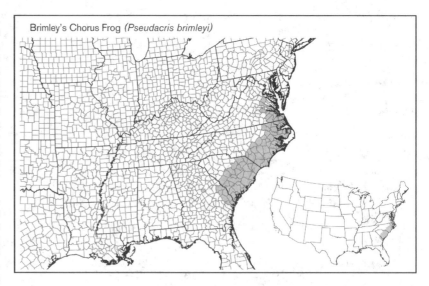

Brimley's Chorus Frog *(Pseudacris brimleyi)*

2. Historical versus Current Abundance.
There are no estimates of population size or densities. Brimley's chorus frogs may be locally abundant.

3. Life History Features.
The life history and ecology of Brimley's chorus frogs were summarized by Wright and Wright (1949) and Gosner and Black (1958b).

A. Breeding. Reproduction is aquatic.

i. Breeding migrations. Brimley's chorus frogs apparently do not migrate. Initiation of the calling period depends on temperature but usually begins in February or March. Males call from perches on or beneath vegetation and usually partially submerged in water. Extreme dates of calling are 7 March–22 April in Virginia (Mitchell, 1986) and 19 February–17 April in North Carolina (Brandt and Walker, 1933). Brandt (1936a) heard calls in November. Mitchell (1986) noted egg laying from 10 March–22 April in Virginia.

ii. Breeding habitat. This species breeds in shallow flooded fields, red maple *(Acer rubrum)* swamps, ditches, floodplain pools, woodland pools with sphagnum, ditches or pools in logged areas, and shrub thickets (Gosner and Black, 1958b; Mitchell, 1986; personal observations).

B. Eggs.

i. Egg deposition sites. Eggs are laid in small gelatinous masses attached to vegetation, usually grass stems, in shallow water.

ii. Clutch size. Maximum known clutch size is 300 (Gosner and Black, 1958b).

C. Larvae/Metamorphosis. The larval period is short, lasting about 30–35 d and as long as 60 d (Gosner and Black, 1958b). Mitchell (1986) reported finding newly metamorphosed animals from 12 May–17 April. Size at metamorphosis is 8.6–11.0 mm (Gosner and Black, 1958b).

D. Juvenile Habitat. Apparently similar to that of adults.

E. Adult Habitat. Adults have been found well away from water in mixed pine

and hardwood forests, pine forests, secondary dune scrub forest, forested wetlands dominated by red maple, loblolly pine *(Pinus taeda)*, and sweetgum *(Liquidambar styraciflua)*, and cultivated fields (Gosner and Black, 1958b; Buhlmann et al., 1994; personal observations).

F. Home Range Size. Unknown.

G. Territories. Unknown.

H. Aestivation/Avoiding Desiccation. Unknown.

I. Seasonal Migrations. Brimley's chorus frogs apparently do not migrate.

J. Torpor (Hibernation). Overwintering sites are apparently underground. Brandt (1936a) noted that this frog appears to remain active throughout the winter except during freezing weather conditions.

K. Interspecific Associations/Exclusions.
Brimley's chorus frogs are sympatric with American toads *(Bufo americanus)*, southeastern chorus frogs *(P. feriarum)*, spring peepers *(P. crucifer)*, southern chorus frogs *(P. nigrita)*, and southern leopard frogs *(Rana sphenocephala)*.

Mecham (1965) obtained experimental hybrids between Brimley's chorus frog and mountain *(P. brachyphona)*, ornate *(P. ornata)*, southern, and southeastern chorus frogs.

L. Age/Size at Reproductive Maturity.
Brandt and Walker (1933) noted that the smallest male measured 24 mm and the smallest female was 27 mm. Age at first reproduction is unknown.

M. Longevity. Unknown.

N. Feeding Behavior. The diet of this frog has not been studied, but Brandt (1936b) noted that specimens he examined contained insects, spiders, and debris.

O. Predators. Brown (E.E., 1979) reported a Brimley's chorus frog in an eastern ribbon snake *(Thamnophis sauritus)* from North Carolina.

P. Anti-Predator Mechanisms. Unknown.

Q. Diseases. Unknown.

R. Parasites. Brandt (1936b) found ten species of protozoans (mostly opalinids), two species of trematodes, seven nematodes, and one acanthocephalan in 55 individuals from North Carolina.

4. Conservation.
Brimley's chorus frogs are not listed in any state in which they occur, but regulations in Virginia do not allow commercialization (Levell, 1997).

Pseudacris cadaverina (Cope, 1866[a])
CALIFORNIA TREEFROG

Edward L. Ervin

This account is dedicated to the memory of the late Dr. Boris I. Kuperman, my friend and mentor.

1. Historical versus Current Distribution.
The geographic distribution of California treefrogs *(Pseudacris cadaverina)* extends from coastal southern California, to Baja Norte, Baja California, Mexico. Within the United States, California treefrogs are restricted to California, ranging southward along the Coast Ranges from San Luis Obispo County, to and across the Transverse Ranges, extending east to Joshua Tree National Park, and south along the Peninsular Ranges to the Mexican border. California treefrogs occur from near sea level to around 2,290 m (7,500 ft). They have a discontinuous distribution within their range but are often locally abundant (Gaudin, 1979; Stebbins, 1985). A previous report of an isolated population of California treefrogs occurring in the Granite Mountains, Granite Mountains Preserve in the East Mojave Desert (Greene and Luke, 1996), is unconfirmed and considered questionable (G. Stewart, personal communication).

2. Historical versus Current Abundance.
Jennings and Hayes (1994a) reviewed the data on 80 amphibian and reptile species native to California to assess the possible need for special listing and/or protection. Data were assembled from individuals having experience with each species, the scientific literature, museum collections, unpublished field notes, field reconnaissance, and archival records. Jennings and Hayes concluded that California treefrogs did not warrant any state-level status or legal protection. Since that time, no information or findings have become available that would suggest the overall status of this species has changed. However, California treefrogs are difficult to find in presumably high-quality habitat along stream segments where populations of non-native predatory fish (i.e., green sunfish *[Lepomis cyanellus]*) have become established, suggesting that some populations may be experiencing declines (R. Fisher, unpublished data; personal observations).

3. Life History Features.
A. Breeding. Reproduction is aquatic.
i. Breeding migrations. Unknown.
ii. Breeding habitat. Oviposition takes place in pools of still or slow moving water usually surrounded by large waterworn rocks and boulders (Storer, 1925; Stebbins, 1951). California treefrogs and Pacific treefrogs *(P. regilla)* are often syntopic, with California treefrogs generally calling from banks and islands and rarely from the water; Pacific treefrogs call from shallow water, often in contact with emergent vegetation (Littlejohn, 1971; personal observations).

B. Eggs. The mean measurement of 15 eggs preserved on 5% formalin was 1.95 mm (vitellus; range 1.83–2.10) and 4.39 (envelope; range 4.14–4.68; Storer, 1925).

i. Egg deposition sites. Eggs are deposited in the quiet pools of intermittent and perennial streams. Egg capsules are surrounded by a colorless gelatinous envelope with adhesive properties that secures most to stationary debris on or near the bottom of the pool (Storer, 1925; Stebbins, 1951; Gaudin, 1965). Eggs are exuded singly and have a tendency to adhere together. Breeding and eggs deposition occur from early February to early October (Stebbins, 1985), after high flows from seasonal storms have begun to subside. During rainfall events and the subsequent increase in stream flow, eggs that dropped into interstitial pockets of the substrate are less susceptible to displacement. A photograph showing eggs attached singly to a sycamore leaf appears in Storer (1925, plate 13, fig. 39).

Anzalone et al. (1998) conducted experiments investigating the effects of solar UV-B on the survivorship and hatching success of California treefrog embryos. The study demonstrated that the groups of embryos shielded from UV-B displayed a significant increased survival rate, while embryos directly exposed to solar UV-B had a decreased survival rate. Laying eggs individually lower in the water column, as opposed to egg masses or strings close to the surface, provides protection from the deleterious effects of direct solar radiation and the unpredictable hydrologic conditions of lotic environments of the southwest.

ii. Clutch size. Unknown.
C. Larvae/Metamorphosis.
i. Length of larval stage. The larval period ranges from 40–75 d (Stebbins, 1951).

ii. Larval requirements. Larvae are found in pools of still or slow-moving water usually surrounded by large water-worn rocks and boulders (Storer, 1925; Stebbins, 1951). As a result of rainfall and the subsequent increased velocity of water, tadpoles are occasionally redistributed downstream. Tadpoles are most abundant in fishless pools and stream reaches (Hemphill and Cooper, 1984; personal observations).

a. Food. California treefrog larvae are classified as generalists and typically feed on detritus, periphyton from algal crusts and mats, and from the surface of submerged objects such as leaves, sticks, and rocks (Duellman and Trueb, 1986; Stebbins and Cohen, 1995).

b. Cover. When not actively foraging, tadpoles often seek out the warmer water found along shallow pool margins. When disturbed or threatened, they quickly retreat to deeper waters and seek cover among algae mats, submergent vegetation, leaf litter, or gaps between cobble stones. Tadpoles possess robust tail musculature that reaches almost to the tail tip and are consequently strong swimmers

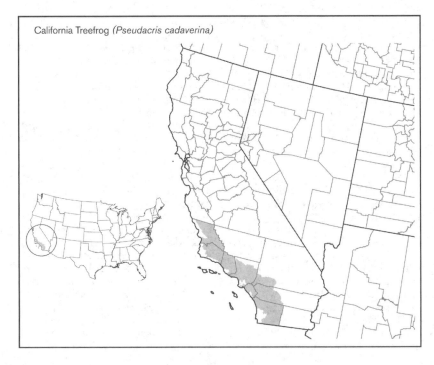
California Treefrog *(Pseudacris cadaverina)*

(Cunningham, 1964; Gaudin, 1964; personal observations).

    iii. Larval polymorphisms. None reported.

    iv. Features of metamorphosis. Metamorphosis has been observed in California from June–August (Stebbins, 1951).

    v. Post-metamorphic migrations. Cunningham (1964) wrote: "Immediately after metamorphosing, juveniles are extremely common, sometimes numbering several individuals per square yard for distances of 6.1 m (20 ft) or more surrounding the pond from which they emerged. In mid-August, however, in such sites as along the Mojave River, juveniles suddenly become quite uncommon. Mortality may be high or the toads [sic] may simply change their habitats and become difficult to find."

    D. Juvenile Habitat. Recently metamorphosed individuals are often found aggregated under and between small rocks and cobble and at the base of annual plants growing along the margins of breeding pools. Newly metamorphosed animals show average daily movements of 1 m (range 0.5–2 m) and move from refugia in pursuit of prey (Harris, 1975). One exceptional juvenile was discovered beneath damp leaves in a small depression 46 m (150 ft) from water (Cunningham, 1964).

    E. Adult Habitat. During active periods, adults most commonly are found in close proximity to and along stream channels. During the daytime, individuals seek refuge in cavities or small depressions on the surfaces of the boulders lining streams, often fully exposed to direct sunlight. These perches are usually within a few jumps from the nearest pool. Stebbins (1951) writes: "Typical habitat includes clean rock surfaces, crevices, shade, and during the breeding season, quiet, clean water." Lillywhite and Light (1975) discovered that while California treefrogs bask in direct sunlight, they discharge a clear, non-viscous fluid onto their integument to prevent the underlying epithelium from drying. Minimal loss of water from the whole animal nevertheless occurs by evaporation and renewal of the mucous film. Outside the breeding season, these treefrogs spend little time in the water.

    Individuals are discovered occasionally in upland habitats far from the drainage during the autumn and winter (S. Sweet, personal communication). In April, two individuals were observed approximately 46 m (50 yd) almost vertically from a small stream; in mid-June, an individual was discovered in a rodent burrow located in dry soil approximately 46 m (50 yd) from a small stream (Cunningham, 1964). As cited in Stebbins (1951), "Klauber has found it on granite boulders to about 100 feet horizontally and 50 feet vertically from the nearest stream (Storer ms.)." No male/female differences in habitat characteristics or utilization have been reported.

    F. Home Range Size. California treefrogs occupy relatively small portions of the streamside habitat available to them during the spring and summer (Dole, 1974; Harris, 1975). Adult treefrogs move an average distance of 3 m (1–5 m)/d while juveniles move less, an average of 1 m (0.5–2 m)/d. Adult movements are associated with foraging and breeding, while juvenile activities are primarily in pursuit of prey (Harris, 1975). Dole (1974) showed that adults rarely moved >3–4 m in higher quality habitat.

    Five females that Dole (1974) studied were recaptured in the same area that each had occupied the previous year. Two of them were recaptured in the same location while three others were within 2 m of their original locations. One treefrog was captured six times over a 593-d period spanning two winters. This individual was always found to be within 2 m of her original capture location. Movements of the four other females captured in both years were greater. One of these females, originally captured in July, was recaptured 80 m downstream the following March, but by June she had returned to her previous spot. The three others had moved downstream. By the following summer all three had returned 91–218 m upstream to where they had been observed originally.

    Of the male treefrogs recaptured during both years of the study, the first moved 13 m, the second moved 6 m in 8 mo, and the third moved 75 m upstream. Diminishing levels of pooled water and available moisture were suspected to be the reason some adults made extensive downstream movements to an area containing deeper pools and a greater tree canopy.

    California treefrogs will move long distances (Kay, 1989). These movements tend to occur between areas of favorable habitat characteristics such as boulders and pooled water and are not thought to be related to population density (Kay, 1989).

    G. Territories. The first report of a California treefrog encounter call (sensu McDiarmid and Adler, 1974) appeared in Littlejohn (1971), although no details were provided. Male California treefrogs have been reported to warn male intruders and maintain territories with encounter calls, and if necessary, defend their calling sites with male–male aggressive encounters (F.T. Awbrey, cited in Wells, 1977a; Fellers, 1979a).

    The following observations are noteworthy in that they serve as examples of site-specific territoriality for California treefrogs. Male aggressiveness has been observed in the field on two separate occasions with both encounters being similar in nature (personal observations). On the evening of 18 May 1999 (1725 hr) in Hot Springs Canyon (Orange County, California), a vocal response was elicited from a vocalizing male California treefrog by imitating his advertisement call. While this in itself is not unique, this individual also reoriented toward the call and hopped up onto a partially buried, 50 cm high, granite boulder. While slowly being approached from 3 m away, imitation calls were again presented to him. He quickly responded by hoping about 20 cm in a single leap to the highest point of the rock, remained trained on the "intruder," and began to return calls with greater intensity. The second encounter occurred on the afternoon of 21 March 2000 (1310 hr) in Tenaja Canyon (Riverside County, California). A male California treefrog responded to the imitation calls in much the same manner. However, this individual approached from the water's edge with a series of hops across the bedrock towards the observer. This treefrog traveled approximately 2 m to within 30 cm of the observer while continuing to respond to the imitation calls, also with increased intensity. The vocalizations reported here have been interpreted as encounter calls because they were accompanied by territorial behavior (i.e., approaching the intruder) and were acoustically distinct from the typical advertisement call (i.e., greater intensity).

    These observations suggest that the drive to defend a preferred site or territory can be strong. Most likely, this aggressive behavior is restricted to the breeding season. At this time, no one has analyzed or described the entire call repertoire of California treefrogs.

    H. Aestivation/Avoiding Desiccation. The distribution of California treefrogs suggests that they have adapted to a variety of weather conditions. Desert populations must aestivate in late summer and early autumn to avoid the hot dry conditions, and high elevation populations must hibernate in the winter months to avoid freezing temperatures (Miller and Stebbins, 1964; Ball and Jameson, 1970; also see "Torpor [Hibernation]" below).

    I. Seasonal Migrations. Seasonal habitat usage patterns can be summarized from Harris (1975) as follows. During spring (mid-March to mid-June) and summer (mid-June to mid-September), individuals aggregate on granitic boulders along stream and river courses associated with pools of water. In the fall (mid-September to November) and winter (December to mid-March), individuals move to crevices located on higher ground bordering the drainages. The migration from summer habitats to winter habitats appears to function in avoiding desiccation, predation, and high water from unpredictable and occasional heavy rains during the winter season (Dole, 1974). During the transition from late winter to early spring (i.e., March–April), individuals begin to reappear along the stream and river courses in greater numbers.

    J. Torpor (Hibernation). By late fall (mid-November) the great majority of California treefrogs have moved to higher ground. From December to mid-March, California treefrogs are seldom encountered (Harris, 1975; unpublished data). Adults seek deep moist crevice microhabitats located on hillsides (Cunningham, 1964; Harris, 1975) and occasionally in damp portions of mine adits (R. Fisher, D. Stokes, personal communication). High concentrations of urea accumulate in the body fluids (140 mM) during dehydration or periods of reduced water turnover. This increases the body water potential to levels where net cutaneous water uptake is possible and reduces the gradient for the net loss of water to the environment (Jones, 1982).

K. Interspecific Associations/Exclusions. California treefrogs occur across a wide elevational range and therefore are associated with a diverse herpetofauna. For example, riparian woodlands along lower gradient stream segments, whether in valley bottoms or at the foot of mountainous terrain, support species such as Coast Range newts *(Taricha t. torosa)*, California toads *(Bufo boreas halophilus)*, arroyo toads *(B. californicus)*, Pacific treefrogs *(P. regilla)*, California red-legged frogs *(Rana draytonii)*, foothill yellow-legged frogs *(R. boylii)*, two-striped gartersnakes *(Thamnophis hammondii)*, and southern Pacific pond turtles *(Clemmys marmorata pallida)*. Co-existence with these species often occurs in ecotones, or transitional zone habitats that include stream reaches bordered by rocks and boulders that are favored by California treefrogs (Schoenherr, 1976; DeLisle et al., 1986; personal observations).

The channel of steeper gradient streams in the mountains and foothills are characterized by a stair-step watercourse forming a series of riffles, runs, and still to slow-flowing pools, often confined to rocky canyons. Coast Range newts, mountain yellow-legged frogs *(R. muscosa)*, two-striped garter snakes, and southern Pacific pond turtles are found in these habitat conditions. Where the distributional ranges of these species overlap with California treefrogs, they often co-occur (Schoenherr, 1976; DeLisle, 1985; Anzalone et al., 1998). Red-spotted toads *(B. punctatus)* and California treefrogs co-occur in wetlands, including canyons, springs and oases, of desert regions (Miller and Stebbins, 1964; Glaser, 1970).

California treefrogs are known to share the same stream reaches and macrohabitats with several native fish species including rainbow trout (*Oncorhynchus mykiss*, freshwater form), southern steelhead (*O. mykiss*, sea-run form), threespine stickleback (*Gasterosteus aculeatus*), Santa Ana sucker (*Catostomus santaanae*), arroyo chub (*Gila orcutti*), and speckled dace (*Rhinichthys osculus*; R. Fisher, unpublished data; personal observations). These associations are possible because these fish are thought to feed primarily on invertebrates and aquatic vegetation and are presumably less reliant on amphibian eggs or larva (Moyle, 1976, 2000; McGinnis, 1984). However, California treefrogs are often most abundant in streams lacking fish fauna (Hemphill and Cooper, 1984; Cooper et al., 1986; personal observations). Strategic placement of eggs and the availability of refugia and structurally complex tadpole foraging areas may be important factors determining these interspecific associations with California treefrogs.

A variety of introduced species are well established in the aquatic habitats utilized by California treefrogs. The most widespread of these are red swamp crayfish *(Procambarus clarkii)*, African clawed-frogs *(Xenopus laevis)*, American bullfrogs *(R. catesbeiana)*, black bullheads *(Ameiurus melas)*, mosquitofish *(Gambusia affinis)*, hatchery stock rainbow trout, largemouth bass *(Micropterus salmoides)*, bluegill *(Lepomis macrochirus)*, green sunfish *(L. cyanellus)*, and European carp *(Cyprinus carpio;* Gamradt and Kats, 1996; Stephenson and Calcarone, 1999; R. Fisher, unpublished data; personal observations). The relative impact of introduced aquatic species on California treefrog populations has not been determined. Interspecific exclusions are currently unknown.

Probable natural adult hybrids between California treefrogs and Pacific treefrogs have been reported (Brattstrom and Warren, 1955; Gorman, 1960). However, attempts to hybridize these species in the laboratory have resulted in the production of inviable crosses that failed at the earliest stages of development (Maxon and Jameson, 1968; Ball and Jameson, 1970; Gaudin, 1979).

L. Age/Size at Reproductive Maturity. Storer (1925) concluded that California treefrogs reach their adult size and breed when 2 yr old. This interpretation is based on finding no more than two size classes among metamorphosed individuals.

M. Longevity. Unknown.

N. Feeding. A stomach content analysis of 15 adult California treefrogs recovered the following prey types: grasshoppers (Orthoptera), spiders (Arachnida), ants (Hymenoptera), beetles (Coleoptera), moths (Lepidoptera), sowbugs (Isopoda), true bugs (Hemiptera), and lacewings (Neuroptera; Cunningham, 1964).

O. Predators. Cooper et al (1986) conducted field experiments to investigate predation by rainbow trout and demonstrated that they have the capacity to completely eliminate California treefrog larvae from stream pools. Introduced non-native green sunfish have been shown to prey on adult California treefrogs (Ervin et al., 2001a). Established populations of this predatory fish occur in many coastal southern California drainages that currently support California treefrog populations (Stephenson and Calcarone, 1999; personal observations). Highly aquatic two-striped garter snakes, which have a geographic distribution similar to California treefrogs (Rossman et al., 1996), prey on larvae and metamorphosed individuals (Cunningham, 1959; Schoenherr, 1976; personal observations).

P. Anti-Predator Mechanisms. The pigmentation patterns of the California treefrog larva, ranging from light to dark brown and with varying amounts of gold flecking (Gaudin, 1965), closely resembles the appearance of the sand and rocky stream substrates. This cryptic coloration, in combination with the tendency of larvae to seek refuge in the presence of trout, would likely reduce or delay predation (Cooper, 1988).

Juvenile and adult California treefrogs possess coloration and marking patterns on the dorsum that resemble those of the rocks and boulders they inhabit (Storer, 1925; Stebbins, 1951). Their ability to remain undetected is enhanced by the tendency to remain virtually still when approached (Storer, 1925; Stebbins, 1951; Cunningham, 1964). Most diurnal perches are within 1 m from water's edge, enabling a rapid escape requiring one or two jumps to the nearest still pool or, occasionally, a swift current (Storer, 1925; Stebbins, 1951; Cunningham, 1964). Noxious and or toxic properties in eggs, larvae, and adults are currently unknown.

Q. Diseases. Unknown. There is a single record of a California treefrog exhibiting gross morphological abnormalities associated with limbs, and it is one of the few reports to document an extra-legged frog from lotic habitat (E. L. E. and P. T. J. Johnson, unpublished data). This specimen possessed three normal extremities with two malformed hindlimbs on the right side. The primary limb, the femur and surrounding musculature were greatly reduced and the tibiofibula folded back upon itself to form a distinct bony triangle (taumelia). The foot extended anteriorly and exhibited only one clearly defined digit (ectrodactyly). The supernumerary right limb (polymelia), which was independent of and ventral to the primary limb, was also poorly developed with a truncated femur and only three digits. Trematode metacercariae, which have been shown to cause developmental abnormalities in the rear limbs of other species of frogs, were not found in this specimen.

R. Parasites. The larval stage of chiggers (*Hannemania hylae*; Acarina: Trombiculidae) has been shown to embed in the skin of adult California treefrogs (Welbourn and Loomis, 1975). Approximately 98.9% of the treefrogs in this study had a mean of 21.4 chiggers/treefrog. While Welbourn and Loomis (1975) determined that one life cycle is completed each year, unengorged larvae were found on treefrogs throughout the summer.

Goldberg and Bursey (2001) examined the helminth communities in California treefrogs and found the trematode *Langeronia burseyi* and metacercariae of *Alaria* sp., *Fibricola* sp., and *Gorgoderina* sp.; the cestode *Distoichometra bufonis*; and two species of nematodes (*Rhabdias ranae* and larvae of *Physaloptera* sp.). They also provided a breakdown of the infection site, number of helminths, prevalence, and mean intensity for helminths from California treefrogs from three counties in southern California.

Fifteen adult California treefrogs from Cedar Creek (San Diego County, California) were examined and found to be infected by three groups of parasites. *Hannemania hylae* were located in the abdominal skin and the bottom surface of the front and rear feet. Prevalence was 100%, while mean intensity was 28 (range 4–57). *Ribeiroia* sp. (Trematoda) metacercariae were found encysted and excysted in the musculature of the rear feet and pelvic area with 87% prevalence and a mean intensity of 34 (range 7–92). Finally, two protozoans were found in the intestines, the flagellate *Opalina* sp. and the ciliate *Balantidium* sp., with a prevalence of 60% and 37%, respectively (B. Kuperman and V. Matey, unpublished data).

4. Conservation.
While Jennings and Hayes (1994a) concluded that California treefrogs did not warrant any state-level status and/or legal protection, California treefrogs are difficult to find in presumably high-quality habitat where populations of non-native predatory fish have become established, suggesting that some populations may be experiencing declines (R. Fisher, personal communication; personal observations).

Pseudacris clarkii (Baird, 1854[b])
SPOTTED CHORUS FROG

Michael J. Sredl

1. Historical versus Current Distribution.
Spotted chorus frogs (*Pseudacris clarkii*) are distributed from extreme northeast Tamaulipas through central Texas and north through central Oklahoma and into central Kansas (Wright and Wright, 1949; Conant, 1975; Pierce and Whitehurst, 1990). A single adult male spotted chorus frog collected in McLean County, Illinois, was likely introduced as a tadpole during fish stocking (Shepard and Burdett, 2000). A population in Chouteau County, Montana, is also likely introduced (Conant, 1975). The report of spotted chorus frogs from New Mexico (Painter and Burkett, 1991) is based on a misidentification of a western chorus frog (*P. triseriata*; Stuart, 1992), although spotted chorus frogs have been reported within 32.2 km of New Mexico (Tinkle and Knopf, 1964). Spotted chorus frogs may readily expand their range in dry years (Bragg, 1960b). No assessment of current range of this species has been done.

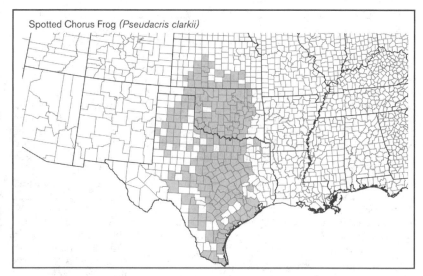

Spotted Chorus Frog (*Pseudacris clarkii*)

2. Historical versus Current Abundance.
Little is known of historical abundance of spotted chorus frogs, and comments on historical abundance are few and largely anecdotal. Strecker (1902, in Wright and Wright, 1949) characterizes them as "an abundant species, especially so in the vicinity of marshes." Similarly, Cope (1894, in Wright and Wright, 1949) noted them to be "abundant and noisy in pools near the Cimmaron River, at Tucker, Oklahoma." No comprehensive studies of recent abundance have been conducted.

3. Life History Features.
A. Breeding. Reproduction is aquatic.
i. Breeding migrations. Spotted chorus frogs breed from January to early June, occasionally in summer to early October; breeding usually follows rains (Bragg, 1943a; Livezey and Wright, 1947; Wright and Wright, 1949; Smith, 1950; Kennedy, 1958; Blair, 1961a; Wiest, 1982). Olfactory cues may be important in locating natal ponds. Grubb (1973c) found that male and female spotted chorus frogs moved toward the odor of water from their home ponds in preference to that from foreign ponds that were sometimes only a few meters distant from the home pond. As male spotted chorus frogs approach the breeding pond, they will produce a call that differs from their typical breeding call (Bragg, 1943a). This call has been characterized as having a "pensive" or "pleading" quality to it. Bragg (1943a) speculated that this call may function to attract other males to appropriate breeding ponds. Large choruses quickly form after heavy rains in late winter, spring, late summer, and early fall (Blair, 1961a). Winter rains may induce a few males to call briefly. Spotted chorus frogs may delay breeding if spring is dry (Smith, 1950). Late breeding congresses may compensate partially for lack of breeding following spring droughts (Kennedy, 1958).

ii. Breeding habitat. Spotted chorus frogs breed in temporary and semi-permanent ponds, and in marshes, shallow water-lily ponds, roadside ditches, grassy ponds, mesquite ponds, buffalo wallows, flooded fields, and other transient pools (Burt, 1936; Bragg, 1943a; Wright and Wright, 1949; Jameson, 1956a; Kennedy, 1958). The water level of these ponds is usually high during the winter, so the fringe of grassland is included in the area covered by water (Burt, 1936). Jameson (1950a) noted a depth of 0.6–1.2 m in breeding ponds in Texas. Spotted frogs never breed in permanent water in Texas (Lord and Davis, 1956). In Oklahoma, they were observed to avoid deep water (Bragg, 1943a). Similarly, Smith (1950) stated that in Kansas, breeding "occurs in temporary, shallow pools, never in deep pools." They also seem to avoid muddier, less protected livestock ponds where spadefoot toads (*Scaphiopus* sp.) and true toads (*Bufo* sp.) breed (Smith, 1934). Site fidelity may be high. In central Texas, spotted chorus frogs returned to the same breeding pond 2 yr in a row (Jameson, 1956a).

After breeding congresses form, males call from the grassy edges of the pond, while hidden (Wright and Wright, 1949). Males call while immersed in water to the level of the throat, with only the head exposed to the air (Bellis, 1957). Most calling is done at night (Wright and Wright, 1949; Blair, 1961a), but at the peak of breeding, males call day and night (Bragg, 1943a; Blair, 1961a). Air temperatures of a pond in Texas where males were observed calling ranged between 3.0 and 23.2 °C, averaging 14.5 °C in 1973 (Wiest, 1982). There may be "a considerable influence of one individual on the other" in the induction of vocalization or "leader effect" (Bragg, 1950d). The call has been described as "a loud, medium pitched, and rapidly repeated 'whank, whank, whank'" (Bragg, 1943a). At temperatures below 12 °C, the call is a slow grinding note. Bellis (1957) found a positive correlation between water temperature and call frequency. Amplexus is axillary, and may last as long as 24 h (Smith, 1950d). The pair move about, chiefly through the efforts of the female.

B. Eggs.

i. Egg deposition sites. Eggs are deposited here and there in loose, irregular masses attached to plant stems (Livezey and Wright, 1947; Eaton and Imagawa, 1948; Wright and Wright, 1949), never >75 mm from the surface of the water (Bragg, 1943a). In Oklahoma, typical oviposition sites were upright sedges or grasses, never twigs. Oviposition sites were never on slanting vegetation, as is the habit of chorus frogs (P. "triseriata," Bragg, 1943b).

ii. Clutch size. Approximately 1,000 eggs are laid in masses ranging from 3–50 (Bragg, 1943a,b; Livezey and Wright, 1947; Wright and Wright, 1949; Smith, 1950d). Females likely lay >1 clutch of eggs/year, but this has not been verified (Blair, 1961a).

Blair (1961a) reported an air temperature range of 8.3–27.2 °C, and a median air temperature of 17.2 °C during breeding. In 1973, air temperatures of a pond in Texas where spotted chorus frogs were ovipositing ranged from 11.8–21.3 °C, averaging 17.8 °C (Wiest, 1982). These frogs have been reported to breed at air temperatures as low as 5.5 °C (Bragg, 1943a). Laboratory experiments indicate that the lower lethal temperature limit of developing larvae is between 5.5–11.3 °C, while the upper lethal temperature limit is between 35.0–39.0 °C (Ballinger and McKinney, 1966). Eggs hatch in 2.5–3 d (Bragg, 1943a).

C. Larvae/Metamorphosis. Altig et al. (1998) describe the tadpoles of spotted chorus frogs. Bragg (1957) noted that color of larvae varies greatly, depending on the environment. Tadpoles collected from clear, stained waters were darker than tadpoles collected from a muddy ditch. Those collected from the muddy ditch were characteristic of the original description (Bragg, 1957). Tadpoles reach a maximum body length of 30 mm (Wright and Wright, 1949).

i. Length of larval stage. Spotted chorus frog larvae metamorphose in 30–45 d at a SUL between 8.0–13.0 mm in Texas (Wright and Wright, 1949; Blair, 1961a), while those in Kansas have been reported to reach between 13.5–17.0 mm (Smith, 1934). In laboratory experiments, length of larval period had a weak inverse correlation with multilocus heterozygosity (Whitehurst and Pierce, 1991).

ii. Larval requirements. Larval requirements are poorly understood. Rosenberg and Pierce (1995) found that growth of spotted chorus frog tadpoles was depressed in acidic conditions. Although a prairie species, they occur in east Texas, a region where acidic conditions potentially can be found (Pierce and Whitehurst, 1995).

a. Food. No detailed study of tadpole diet has been conducted.

b. Cover. No detailed study of tadpole habitat needs has been conducted.

iii. Larval polymorphisms. None reported.

iv. Features of metamorphosis. Recently metamorphosed spotted chorus frogs are diurnal (except when it is extremely hot and dry), agile, and difficult to catch. Wright and Wright (1949) noted that they have very delicate skin. They remain at the pond for a few weeks, then scatter widely (Bragg, 1943b).

v. Post-metamorphic migrations. Except that they disperse widely, little is known of post-metamorphic migrations.

D. Juvenile Habitat. No detailed study of the habitat characteristics of juvenile spotted chorus frogs has been conducted.

E. Adult Habitat. Very little is known about adult habitat characteristics. Spotted chorus frogs principally are found in prairies and prairie islands in savannas (Bragg, 1943b; Conant, 1975; Pierce and Whitehurst, 1990). After emerging from hibernacula, adults are thought to range through pastures and fields while foraging; they do not frequent stream margins or pools except to breed (Bragg, 1943a).

F. Home Range Size. Unknown.

G. Territories. Unknown.

H. Aestivation/Avoiding Desiccation. Aestivation is not reported, but spotted chorus frogs have been noted to have periods of inactivity during hot, dry weather (Bragg, 1943a).

I. Seasonal Migrations. Unknown.

J. Torpor (Hibernation). Burt (1936) observed that an adult spotted chorus frog found in mid-March under a rock may have hollowed out earthen tunnels where it may have undergone at least temporary hibernation. Little else is known about hibernation.

K. Interspecific Associations/Exclusions. Spotted chorus frogs have been reported to occur with numerous temporary and semi-permanent pond breeding amphibians, including plains spadefoot toads (Spea bombifrons), Couch's spadefoot toads (S. couchii), eastern spadefoot toads (Scaphiopus holbrookii), Great Plains toads (B. cognatus), Texas toads (B. speciosus), southern toads (B. terrestris), Coastal-Plain toads (B. nebulifer), Woodhouse's toads (B. woodhousii), northern cricket frogs (Acris crepitans), eastern gray treefrogs (Hyla versicolor), spring peepers (P. crucifer), Strecker's chorus frogs (P. streckeri), southeastern chorus frogs (P. feriarum), Rio Grande leopard frogs (Rana berlandieri), American bullfrogs (R. catesbeiana), southern leopard frogs (R. sphenocephala), and western narrow-mouthed toads (Gastrophryne olivacea; Burt, 1936; Bragg, 1943a; Wright and Wright, 1949; Jameson, 1950a, 1956a; Blair, 1961a; Wiest, 1982; Altig et al., 1998).

Interactions of spotted chorus frogs and the Pseudacris nigrita complex (P. nigrita, P. clarkii, P. triseriata, P. brachyphona,

P. brimleyi) have been of great interest to biologists studying speciation (Mecham, 1958; W.F. Blair, 1964a). Call, habitat differences, and time of breeding are thought to be important isolating mechanisms. Michaud (1962) demonstrated that female spotted and southeastern chorus frogs can resolve calls of their own species. Time of breeding in the overlap zone of these two species in Texas and Oklahoma is less clear. In some areas, the breeding peak of spotted chorus frogs comes later than that of southeastern chorus frogs (Lindsay, 1958, in Mecham, 1958), while in other areas breeding of these two species peaks about the same time (Lord and Davis, 1956). Lord and Davis (1956) observed that in pools where spotted chorus frogs abounded, southeastern chorus frogs were not common. However, in pools where southeastern chorus frogs were abundant, spotted chorus frogs were just as abundant. Bragg (1960b) studied populations of spotted and southeastern chorus frogs in a prairie-savanna ecotone between 1935–59. In wet years, southeastern chorus frogs (savanna dwelling) expanded their range westward into more prairie-like habitat, while in dry years spotted chorus frogs (prairie dwelling) expanded their range eastward into more savanna-like habitat (Bragg, 1960b). Although these two species often co-occur in the same breeding ponds, there have been few reports of natural hybrids or observation of interamplexion (Lord and Davis, 1956; W.F. Blair, 1964a).

The range of spotted chorus frogs also overlaps with Strecker's chorus frogs. Although spotted chorus frogs breed a little later than Strecker's chorus frogs, there is overlap, and time of breeding does not appear to be as important an isolating mechanism as differences in calls (Blair, 1961a).

L. Age/Size at Reproductive Maturity. Wright and Wright (1949) report breeding sizes of males to be between 20–29 mm; females, 21–31 mm SUL. Age at reproductive maturity and growth rates have not been investigated.

M. Longevity. In the wild, spotted chorus frogs live at least 2 yr (Jameson, 1956a).

N. Feeding Behavior. Adults are thought to forage in pastures and fields, but little else on feeding behavior or diet has been reported.

O. Predators. Predators of spotted chorus frogs are not well known. Western ribbon snakes (Thamnophis proximus; Clark, 1974) are known predators of spotted chorus frogs. Other garter snakes (Thamnophis sp.) and water snakes (Nerodia sp.) are likely important predators.

P. Anti-Predator Mechanisms. Spotted chorus frogs are wary: upon slight disturbance spotted chorus frogs duck beneath the water's surface (Burt, 1936). Other anti-predator mechanisms have not been reported.

Q. Diseases. None reported.

R. Parasites. Spotted chorus frogs have been found to be infected with protozoans (*Hexamita intestinalis, Tritrichomonas augusta, Opalina* sp., *Nyctotherus cordiformis,* and *Myxidium serotinum*), nematodes *(Cosmocercoides variabilis)*, cestodes *(Cylindrotaenia americana)*, and larval intradermal mites (*Hannemania* sp.; McAllister and Upton, 1987a; McAllister, 1991).

4. Conservation.

Spotted chorus frogs are not protected by any state or by the U.S. Federal Government. Little is known of their historical abundance, and what is known is largely anecdotal. Similarly, no assessment of their current range has been done. Spotted chorus frogs are known to naturally expand their range in dry years (Bragg, 1960b). At least one, human-mediated, introduced population has been established in Chouteau County, Montana. According to Levell (1997), in Kansas, spotted chorus frogs can be commercially collected for sale as bait.

Pseudacris crucifer (Wied-Neuwid, 1838)
SPRING PEEPER

Brian P. Butterfield, Michael J. Lannoo, Priya Nanjappa

1. Historical versus Current Distribution.

In the United States, spring peepers *(Pseudacris crucifer)* are distributed throughout much of the east, reflecting the historical extent of the eastern deciduous and mixed forests. They occur east of a line from eastern Texas, north through extreme eastern Oklahoma and Kansas, though northwestern Missouri, eastern Iowa, and eastern and northern Minnesota (e.g., Wright and Wright, 1949; Conant and Collins, 1991). Spring peepers generally are found in most eastern wooded habitats, but are not found in the southern half of the Florida Peninsula. Two subspecies are recognized: northern spring peepers *(P. c. crucifer)*, distributed throughout much of the species' range, and southern spring peepers *(P. c. bartramiana)*, present in southeastern Georgia and northern Florida (Conant and Collins, 1991). The current distribution of spring peepers likely resembles their historical distribution, however populations have undoubtedly been lost with logging, the conversion from forest to agricultural land use, mining activities, road building, urbanization, and suburbanization.

2. Historical versus Current Abundance.

Spring peepers are moderately common and occasionally abundant during the breeding season; they are difficult to collect in numbers during summer and fall (Wright and Wright, 1949). Spring peepers are one of the most ubiquitous and abundant frogs of the northern 40% of the Florida Peninsula and the eastern half

Spring Peeper *(Pseudacris crucifer)*

of the panhandle (Bartlett and Bartlett, 1999a). At Portage Lake, Washtenaw County, Michigan, surveys in March, April, and May found that spring peepers were the most abundant animals (Carpenter and Delzell, 1951). These authors also note that 75% of the animals observed were road killed, illustrating the great potential automobiles have in causing mortality.

Blanchard (1928b) notes that spring peepers are found occasionally in bog forests in places that provide protection and breeding sites and are probably more common than they appear to be. In Lebanon County, Pennsylvania, Burger (1933) notes: "myriads" were heard and were taken along every stream and water body in the early spring, after which they are difficult to find. Welter and Carr (1939) note that spring peepers are "Very common throughout eastern Kentucky." For his studies on life history and food habits, Oplinger (1966, 1967) collected and sacrificed 1,322 spring peepers found in the marshes adjacent to Tomkins County Airport, near Ithaca, New York. Minton (2001) suggests that spring peeper numbers have been increasing in southwestern Indiana, but also notes that they are local and uncommon in former prairie areas and usually disappear from intensely cultivated areas.

3. Life History Features.

A. Breeding. Reproduction is aquatic. Spring peepers typically are among the first amphibians to emerge from hibernation, and for many people in the eastern and upper midwestern portions of the United States, their calling signals the arrival of spring.

i. Breeding migrations. From overwintering sites underground and in leaf litter to breeding ponds. Near Athens, Georgia, spring peepers breed in smaller, more open field ponds and pools rather than those on flood plains (Martof, 1955). In

the southern United States, spring peepers begin to call on warm nights in December and January (Van Hymning, 1933; Voice, 1938; Wright and Wright, 1949). The breeding season occurs much earlier in Florida than in the north (Carr, 1940b; Wright and Wright, 1949). "Will breed when the temperature is [2 °C]. Ornate chorus frogs (*P. ornata*) are the only other Florida frog which breeds in such cold weather" (A. Carr, cited in Wright and Wright, 1949). Brandt (1936a) reported amplexed pairs in January in Beaufort County, North Carolina. Spring peepers are active on rainy winter nights when temperatures are above 10 °C (Ashton and Ashton, 1988; Bartlett and Bartlett, 1999a). In Charlton County, Georgia, the breeding season, as far as may be judged by vigorous choruses, extends from late November more or less continuously to early March (F. Harper, cited in Wright and Wright, 1949). Near Athens, Georgia, spring peepers begin calling in the second week of January (Martof, 1955). In eastern Kentucky, they are first heard in the second week in February (Welter and Carr, 1939). Near Indianapolis, Minton (2001) notes choruses during the first week in March during mild, rainy nights and that chorusing begins about 2 wk later in northern Indiana. He further notes that the breeding season extends into early May. In Kalamazoo County, Michigan, spring peepers were heard in full chorus in early April, choruses continued until early June (Allen, 1937). Allen observed the highest abundances " . . . in the shallow swales, among the dead grass and sedges. A few were heard in the marshy borders of the lake but they were not numerous here in comparison." Martof (1960) reported the occurrence of autumn (27 September) breeding in a spring peeper population near Athens, Georgia.

Call characteristics of spring peepers have been documented by Rosen and Lemon (1974), Forester and Lykens (1986),

Etges (1987), Lykens and Forester (1987), Schwartz (1989), Sullivan and Hinshaw (1990), and Lance and Wells (1993). Spring peepers exhibit size-selective mating with larger males breeding more frequently (Gatz, 1981b).

ii. Breeding habitat. Males sing in buttonbrush, briars, willows, *Decodon*, etc. and emergent herbaceous vegetation at the water's edge in small wooded ponds, larger swamps, vernal pools, flooded ditches, wet meadows, Carolina Bays, cypress heads, and sandy coastal and pine barrens habitats (Wright, 1914; A. Carr, cited in Wright and Wright, 1949; Smith, 1961; Mount, 1975; Vogt, 1981; Green and Pauley, 1987; Dundee and Rossman, 1989; Klemens, 1993; Oldfield and Moriarty, 1994; Hunter et al., 1999). Minton (2001) observes that suitable breeding ponds are near woods and are large and permanent enough to support emergent vegetation, but too shallow and temporary to contain fishes.

B. Eggs.

i. Egg deposition sites. Eggs are attached singly or in clumps to submerged vegetation in seasonal and semipermanent wetlands (e.g., Olson, 1956; Minton, 2001). Ovulation precedes amplexus (Oplinger, 1966).

ii. Clutch size. Loraine (1984) collected a female that had laid 702 eggs. Oplinger (1966) gives an average of >700 eggs from females >30 mm, with a maximum estimate of 1,000 eggs (Wright and Wright, 1949). Eggs hatch in 1–2 wk (Ashton and Ashton, 1988; 6–15 d, Minton, 2001). In southern spring peepers, eggs are laid singly, mean diameter is 1.1 mm (vitellus), 2.6 mm (jelly envelope). Eggs hatch in 5.5–6.5 d "at room temperature" (Gosner and Rossman, 1960).

C. Larvae/Metamorphosis. Southern spring peeper tadpoles average 8–10% larger than those of northern spring peepers (Gosner and Rossman, 1960). As is true for many hylids, spring peepers possess a melanophore covering within their peritoneum that forms a complete black lining of the coelom.

i. Length of larval stage. Tadpoles metamorphose in approximately 3 mo (Wright and Wright, 1949; Minton [2001] states 90–100 d). In the laboratory, the minimum size to metamorphosis in southern spring peepers was 45 d, at a mean size of 10.3 mm SVL (Gosner and Rossman, 1960). In New Jersey, Gosner and Black (1957a) note the earliest they observed newly metamorphosed spring peepers was 30 May.

ii. Larval requirements.

a. Food. As is true of most North American anurans, spring peeper tadpoles are suspension feeders that graze on organic and inorganic material typically associated with submerged surfaces.

b. Cover. Tadpoles will aggregate. Brattstrom (1962b) documented several aggregations of spring peeper tadpoles, including one composed of about 75 individuals, in early May in Lake Panamoka, New York. However, laboratory observations using spatial affinity as a recognition assay did not demonstrate sibling recognition (Fishwald et al., 1990).

iii. Larval polymorphisms. Do not occur.

iv. Features of metamorphosis. At metamorphosis, spring peepers are from 12–14 mm TL. Mass migrations of post-metamorphic animals have not been recorded.

v. Post-metamorphic migrations. From wetlands to the forest floor and into low brush.

D. Juvenile Habitat. Similar to adults, although juveniles can be diel and will feed in adjacent grasslands or on the vegetation supported by sphagnum bogs (M. J. L., personal observations).

E. Adult Habitat. Adults are found in lowland marshes, wetlands at the sources of streams whether wooded or open, associated with sphagnum bogs and cattail wetlands, ponds, pools, and ditches in and near woods (Wright and Wright, 1949). Mesophytic and low hammock, swamp borders, the more open bay-heads, and tangles along the smaller streams. In Beaufort County, North Carolina, non-breeding adults retreat to woodlands to feed and for cover (Brandt, 1936a). In Indiana, Minton (2001) notes that the optimal habitat for spring peepers is moist, upland woods with shallow ponds. They avoid floodplain forest. In Connecticut, Gibbs (1998b) notes that populations of spring peepers are resistant to habitat fragmentation.

F. Home Range Size. Unknown.

G. Territories. Unknown but unlikely, except for calling males during the breeding season.

H. Aestivation/Avoiding Desiccation. Spring peepers avoid dry and hot conditions by retreating under logs and bark and perhaps in knot-holes (Wright and Wright, 1949; see also Hudson, 1950). These frogs reach their critical activity point (cannot perform buccal movements in addition to other criteria) when they lose about 33% of their water in desiccating environments (Farrell and MacMahon, 1969). Farrell (1971) demonstrated that desiccation tolerance varies, perhaps due to seasonal physiological adjustments. Spring peepers exhibit a water absorption response (Stille, 1958).

I. Seasonal Migrations. Mass migrations do not occur. Instead, individuals are generally able to feed, breed, and overwinter within the vicinity of forested wetlands.

J. Torpor (Hibernation). Spring peepers hibernate under logs and bark and in knotholes—sites that are the same or similar to retreat sites used to avoid hot, dry conditions (Wright and Wright, 1949; see also Hudson, 1950; and "Aestivation/ Avoiding Desiccation" above). They produce an antifreeze consisting of low molecular weight cryoprotectants (glucose, with a rise to 150–300 mM) that assist in limiting cell volume reduction of up to 65% of total body water as extracellular ice (Churchill and Storey, 1996). Spring peepers typically are among the first amphibians to emerge from hibernation, and animals that form early choruses show good freezing survival (Storey and Storey, 1987; Churchill and Storey, 1996).

In reporting on a population from Beaufort County, North Carolina, Brandt (1936a) notes that there is no evidence of a prolonged hibernation, and that animals are observed throughout the winter "at all temperatures above freezing."

K. Interspecific Associations/Exclusions. In Alachua County, Florida, spring peepers occupy essentially the same habitat as the much rarer eastern gray treefrogs (*Hyla versicolor*; A. Carr, cited in Wright and Wright, 1949). At a historical site near Bloomington, Indiana, spring peeper egg masses were found in association with egg masses of green frogs (*Rana clamitans*), leopard frogs (*R. pipiens* complex), and crawfish frogs (*R. areolata*; Wright and Myers, 1927). In North Carolina, Schwartz (1955) noted spring peepers chorusing in association with mountain chorus frogs (*P. brachyphona*; see also Barbour and Walters, 1941), wood frogs (*R. sylvatica*), and spotted salamanders (*Ambystoma maculatum*). In April in Morgan County, Alabama, Cahn (1939) described a breeding wetland with spring peepers, American bullfrogs (*R. catesbeiana*), southern leopard frogs (*R. sphenocephala*), northern cricket frogs (*Acris crepitans*), southern cricket frogs (*A. gryllus*), eastern gray treefrogs, barking treefrogs (*H. gratiosa*), and Fowler's toads (*Bufo fowleri*). In west-central Wisconsin, spring peepers breed in association with wood frogs, western chorus frogs (*P. triseriata*), eastern gray treefrogs, and American toads (*B. americanus*; M. J. L., personal observations).

Spring peeper and western chorus frog tadpoles often co-occur in ponds across the Upper Midwest (Skelly, 1996).

L. Age/Size at Reproductive Maturity. At 28 mm (Wright and Wright, 1949); 10–31 cm (Bartlett and Bartlett, 1999a). Oplinger (1966) noted that the smallest males with mature spermatozoa were 18 mm; the smallest females with eggs were 23 mm.

M. Longevity. One animal lived 2 yr, 2 mo, and 9 d following capture in the wild (Snider and Bowler, 1992).

N. Feeding Behavior. McAlister (1963) noted a post-breeding, feeding assemblage of nearly adult spring peepers (numbering 37; 17.8–26.7 mm SVL) associated with two juvenile eastern gray treefrogs and two juvenile northern leopard frogs (*R. pipiens*) on the evening of 2 September 1962. All but one leopard frog were found on the upper branches of goldenrods (*Solidago* sp.), elderberries (*Sambucus* sp.), boneset (*Eupatorium perfoliatum*), and Joe-pye weed *(E. purpureum)*. The frogs were feeding on insects attracted to the flowers. Gut contents of 25 spring peepers showed that they were feeding on small arthropods, spiders, phalangids, and mites (Tetranychidae).

Oplinger (1967) gives a detailed list of the food habits of 545 young-of-the-year spring peepers and makes several observations. He notes that food habits are based on prey availability rather than preference and that prey such as arachnids, ants (Hymenoptera), and beetles (Coleoptera) that are generally found throughout the year are eaten throughout the year. Slow-moving, crawling animals were preyed upon more often than active flying animals. No aquatic prey items were eaten. Some seeds, which when windborn could be mistaken for prey, were ingested. Oplinger (1967) did not find a correlation between the size of the spring peeper and the number of prey items in their stomach. Small peepers tended to feed on smaller prey. Feeding activity for younger animals peaked in the early morning and late afternoon; adults tended to feed during the day, from late afternoon to early evening.

O. Predators. Known predators include Butler's garter snakes *(Thamnophis butleri*; Test, 1958), giant water bugs (belostomatids; Hinshaw and Sullivan, 1990), predaceous diving beetles (Coleoptera; Formanowicz and Brodie, 1982), and other pond invertebrates such as odonates (Skelly, 1996). Cochran and Cochran (2003) report spring peepers eaten by brown trout *(Salmo trutta)*.

P. Anti-Predator Mechanisms. Spring peepers are small, inconspicuous, and considered to be strong jumpers (Zug, 1985), and as with most frogs, will use their jumping ability to escape prey (Inger, 1962; Gans and Parsons, 1966). Among the species studied by Zug (1985; including members in the following genera: *Bufo*, *Acris*, *Pseudacris*, *Hyla*, and *Rana*), spring peepers had the least degradation of performance in repeated jumps.

Q. Diseases. Unknown.

R. Parasites. The trematode *Glypthelmins pennsylvaniensis* has been reported in spring peepers from Pennsylvania (Cheng, 1961), Wisconsin (Coggins and Sajdak, 1982), Michigan (Muzzall and Peebles, 1991), and West Virginia (Joy and Dowell, 1994). Muzzall and Peebles (1991) reported the nematodes *Cosmocercoides* sp., *Oswaldocruzia pipiens*, *Rhabdias ranae*, and *Spiroxys* sp. in spring peepers from southern Michigan.

4. Conservation.
Spring peepers are listed as Threatened in Kansas and Protected in New Jersey (Levell, 1997). These designations offer legal protection and require that permits be obtained before undertaking any activities involving this species.

Spring peepers are difficult to collect in numbers during summer and fall, making population assessments difficult. Minton (2001) suggests that spring peeper numbers have been increasing in southwestern Indiana, but also notes that in other parts of the state they usually disappear from areas of intense human activity.

Pseudacris feriarum (Baird, 1854[b])
SOUTHEASTERN CHORUS FROG

See the *Pseudaeris triseriata* complex account.

Pseudacris maculata (Agassiz, 1850)
BOREAL CHORUS FROG

See the *Pseudaeris triseriata* complex account.

Pseudacris nigrita (LeConte, 1825)
SOUTHERN CHORUS FROG

William T. Leja

1. Historical versus Current Distribution.
Southern chorus frogs *(Pseudacris nigrita)* inhabit the Gulf Coastal Plain from just north of the Tar and Pamlico rivers in North Carolina to the Pearl River in southern Mississippi (Schwartz, 1957a; Gates, 1988). They are absent from the Everglades and the Florida Keys (Stevenson, 1976). There is no evidence to indicate any change in distribution over the historical period.

flatwoods, habitat in which southeastern chorus frogs are a characteristic species (Carr, 1940a), have been reduced drastically by conversion to rapidly cropped slash pine plantations throughout the Gulf Coastal Plain (Means et al., 1996; Means, this volume, Part One). Ponds within these flatwoods have been connected by ditches in order to drain them more rapidly after rains (Vickers et al., 1985), and fire has been suppressed in the plantations (Enge and Marion, 1986). It is likely that the abundance and distribution of amphibians inhabiting these flatwoods generally have been affected by these changes. Enge and Marion (1986) compared amphibian species richness and abundance in a naturally regenerated 40-yr-old slash pine forest to recent clearcuts within the forest. They concluded that clearcutting reduced amphibian abundance ten-fold by affecting reproductive success but did not affect species richness. However, a comparison has not been made of current amphibian species abundance and distribution in these plantations to a pine flatwoods forest where fire has not been suppressed and that has not been cut and ditched. The time for such a comparison may have passed, as only about 0.01% of old-growth longleaf pine forest remains (Means et al., 1996; Means, this volume, Part One).

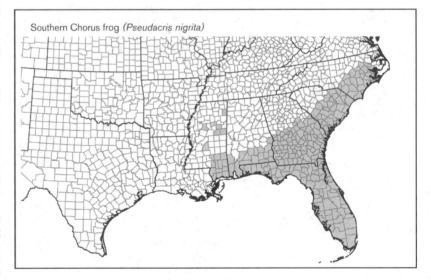

Southern Chorus frog *(Pseudacris nigrita)*

2. Historical versus Current Abundance.
Common to abundant in Georgia (Martof, 1956a). Abundant in Harrison County, Mississippi, through the winter and early spring (Allen, 1932). Not common in Florida (Carr, 1940a; Wright and Wright, 1949). Moderately common in Alachua County, Florida (Van Hyning, 1933). There is no evidence in the literature to indicate a change in abundance of this species over time. However, longleaf pine

3. Life History Features.

A. Breeding. Reproduction is aquatic.

i. Breeding migrations. Breeding migrations occur almost exclusively at night (Pechmann and Semlitsch, 1986).

ii. Breeding habitat. Southern chorus frogs breed in temporary pools, roadside ditches, woodland ponds, and Carolina bays (Caldwell, 1987); flooded fields, roadside ditches, the weedy margins of shallow flatwoods ponds, and temporary woodland pools (Mount, 1975); ditches, bogs, and shallow ponds (Martof et al., 1980); flooded pine-wood ditches, borrow pits, gum ponds in pine woods, and flooded fields surrounded by pine woods (Schwartz, 1957a). Southern chorus frogs breed in wet seasons in otherwise xeric habitats (Schwartz, 1957a; Duellman and Schwartz, 1958).

Southern chorus frog males are secretive and generally call from the bases of grass tussocks and under overhanging grass on the edges of ponds, with only their heads protruding above the water (Einem and Ober, 1956; Mount, 1975; Gartside, 1980). Males call from locations where the vegetation is most dense. Ornate chorus frogs (*P. ornata*) call from open, exposed situations at the same localities (Schwartz, 1957a). In southern Florida, males call from within clumps of grasses, from the ground, or from holes and cracks in the limestone borders of sinkholes (Duellman and Schwartz, 1958). All members of the *nigrita* species group are basically similar in that the males call just out of the water or (most often) partially immersed, or even floating in the water with the forelimbs supported by emergent stems or twigs (Mecham, 1965).

In southwestern Georgia, breeding sites of southern chorus frogs are widely scattered in the uplands, whereas southeastern chorus frogs (*P. feriarum*) breed in floodplain pools in terrain more similar to that of the Piedmont to the north (Crenshaw and Blair, 1959).

In South Carolina and Alabama, the breeding season begins as early as December and extends well into April and possibly into May (Mount, 1975; Gibbons and Semlitsch, 1991). Heavy thundershower activity may initiate vigorous calling during summer or even fall, but breeding activity at this time is unlikely (Mount, 1975). In contrast, southern chorus frogs in southern Florida breed from January–September, with peak activity in June and July (Duellman and Schwartz, 1958). Southern chorus frogs are considered prolonged breeders (Caldwell, 1987).

B. Eggs.

i. Egg deposition sites. Females deposit small, irregular egg clusters of about 15 eggs each on stems, leaves, or other objects in shallow water (Martof and Thompson, 1958; Martof et al., 1980), but sometimes deposit on the bottom of shallow pools (Brandt, 1953).

ii. Clutch size. About 2 hr are required to lay a total of 78–157 eggs (Martof and Thompson, 1958). In southern Florida, a female laid 160 eggs, a few separately but the majority in a single loose mass (Brady and Harper, 1935). Hatching took place within 60 hr in January in southern Florida (Brady and Harper, 1935).

C. Larvae/Metamorphosis. The larval period lasts from about 40–60 d (Wright and Wright, 1949; Martof et al., 1980) to

about 4 mo (Caldwell, 1987). Juveniles metamorphose at 9–15 mm SVL (Wright and Wright, 1949).

D. Juvenile Habitat. Juveniles often remain for several weeks near the ponds from which they emerged (Carr, 1940a).

E. Adult Habitat. Southern chorus frogs are considered pine savanna (Martof et al., 1980) or pine flatwoods species (Carr and Goin, 1959). In Alabama, where they are sympatric with southeastern chorus frogs, southern chorus frogs are likely to be found where the soil is sandy and friable; southeastern chorus frogs occur mostly on places where the soil is heavier (Mount, 1975). In southern Florida, southern chorus frogs are always associated with limestone sinkholes, especially bordering wet prairies, but are absent from wet prairies proper and from sandy country (Duellman and Schwartz, 1958). Adults move to drier hammocks and ridges of the Pine Barrens (Wright and Wright, 1949).

F. Home Range Size. Unknown.

G. Territories. Unknown.

H. Aestivation/Avoiding Desiccation. Carr (1940a) found that captive individuals almost invariably burrow deep in the soil of their cages after a few days of confinement and suggested that the adults may lead an almost wholly subterranean existence. Because these frogs prefer xeric habitats that are only occasionally inundated (Schwartz, 1957a; Duellman and Schwartz, 1958), they may spend time underground in aestivation.

I. Seasonal Migrations. Unknown.

J. Torpor (Hibernation). Unknown.

K. Interspecific Associations/Exclusions. In southwestern Georgia, southern chorus frogs and southeastern chorus frogs inhabit the Upper Coastal Plain of North Carolina, but no hybridization occurs, even when they breed in the same locality (Batts, 1960). However, Gartside and Dessauer (1976) found a narrow (10–20 km wide) zone of hybridization along the Pearl River, which forms the southern border between Louisiana and Mississippi. In an area of sympatry along the Georgia–Alabama border, Fouquette (1975) observed character displacement in the breeding calls (both pulse rate and number of pulses) that resulted in reproductive isolation. To the north and south of this area of sympatry, the difference in pulse rate and number was reduced to a degree probably not sufficient for effective female discrimination of male calls.

L. Age/Size at Reproductive Maturity. Age at first reproduction 12–14 mo, including the approximately 4-mo-long tadpole stage (Caldwell, 1987; Gibbons and Semlitsch, 1991). Adult size is 25–33 mm SVL (Caldwell, 1987).

M. Longevity. The population turnover is nearly annual. A few individuals may live for 2, rarely 3 yr (Caldwell, 1987). Survival of males at a breeding site ranged from 15–34%, while survival of females

ranged from 27–55%. The average time spent at the breeding site varied from 4–25 d, depending on the year. The highest survival rates occurred in years when less time was spent at the breeding site (Caldwell, 1987).

N. Feeding Behavior. Examination of the stomach contents of 10 individuals revealed the remains of ants and small beetles (Duellman and Schwartz, 1958). Carr (1940a) watched adults and recently emerged young catching grasshopper nymphs in short grass at the edge of ponds. He commented that their feeding behavior is similar to that of cricket frogs (*Acris* sp.).

O. Predators. Salamanders, particularly tiger salamander (*Ambystoma tigrinum*) larvae, and aquatic insects are likely predators of southern chorus frog tadpoles (Caldwell, 1987).

P. Anti-Predator Mechanisms. Breeding migrations of southern chorus frogs occur almost exclusively at night, probably to reduce predation risk from visual predators, particularly from diurnal birds such as crows and shrikes (Pechmann and Semlitsch, 1986). The presence of grasses in a shallow water body indicates that it has been dry in recent months and is likely to be relatively free of predators. Grassy calling sites, preferred by southern chorus frog males, may thus constitute a proximate cue indicating a relatively predator-free larval environment (Caldwell, 1987).

Q. Diseases. Unknown.

R. Parasites. Unknown.

4. Conservation.
Southern chorus frogs are not protected either by state or federal laws. There are no data to indicate any change in distribution, however, their native longleaf pine flatwood habitat has been drastically reduced (only 0.01% of old-growth longleaf pine forest remains) by (1) conversion to slash pine plantations, (2) connecting ponds for drainage, and (3) fire suppression. In the face of these activities, it is likely that southern chorus frogs have experienced population reductions and extirpations.

Pseudacris ocularis, (Bosc and Daudin, 1801)
LITTLE GRASS FROG
John B. Jensen

1. Historical versus Current Distribution.
Little grass frogs (*Pseudacris ocularis*) are found in the southeastern Coastal Plain (Harper, 1939) from southeastern Virginia to the southern tip of Florida, inland to the Fall Line and west to Choctawhatchee Bay in the Florida Panhandle (Moler, 1982; Conant and Collins, 1991; Jensen, 1994). Harper (1935) reported little grass frogs from Key West, Florida, however this record was challenged by Duellman and

Schwartz (1958). A record of little grass frogs from Texas (Burt, 1936) was later determined to be a misidentified western chorus frog (*P. triseriata*; Franz and Chantell, 1978). Brandt (1936a) reported little grass frogs from Mississippi that were almost certainly misidentifications. A report of little grass frogs from the upper Gulf Coastal Plain of Georgia, specifically Fort Benning Military Reservation (Goodman, 1958), is quite distant from the nearest confirmed record. Many colleagues and I have spent considerable time in this area without encountering little grass frogs, therefore this record should be considered suspect as well. Misidentifications are not surprising, especially when one considers that several authorities have addressed possibilities that the original description of little grass frogs was based on a specimen of *Acris* (Harper, 1939; Mittleman, 1946). Indeed, some field guides state possible confusion between juvenile *Acris*, other *Pseudacris* sp., and adult little grass frogs (including Martof et al., 1980; Ashton and Ashton, 1988). Unless suspected misidentifications were in fact little grass frogs, there is no information available to indicate a change in this species' distribution.

ii. Breeding habitat. Little grass frogs breed in shallow, grassy, rain-filled depressional wetlands, including roadside ditches and semi-permanent ponds (Harper, 1939; Mount, 1975; Gibbons and Semlitsch, 1991).

B. Eggs.

i. Egg deposition sites. Eggs are deposited on the pond bottom or on submerged vegetation (Wright, 1923).

ii. Clutch size. Wright and Wright (1949) reported that little grass frogs deposit about 100 eggs, although Bartlett and Bartlett (1999a) suggested that >200 eggs may be laid by a single female. Eggs are laid singly (Wright, 1923) or in several clusters of 25 or more (Bartlett and Bartlett, 1999a). Hatching occurs in 1–2 d (Ashton and Ashton, 1988).

C. Larvae/Metamorphosis.

i. Length of larval stage. From 45–70 d (Wright and Wright, 1949).

ii. Larval requirements.

a. Food. Unknown, but tadpoles likely graze on algae.

b. Cover. Tadpoles are most readily captured in relatively dense emergent and submerged vegetation.

iii. Larval polymorphisms. Unknown and unlikely for this species.

or successful in such waters. Adults are capable of climbing vines, tree trunks, and bushes to a height of 1.5 m high or more (Harper, 1939; Wright and Wright, 1949). Harper (1939) found an inactive individual beneath a log in a dried-up cypress pond. Little grass frogs are rarely found away from water and are often active both day and night. There is no information to indicate differing habitat characteristics between the sexes.

F. Home Range Size. Unknown.

G. Territories. Unknown.

H. Aestivation/Avoiding Desiccation. In some areas, little grass frogs are active year-round (Carr, 1940a; Einem and Ober, 1956).

I. Seasonal Migrations. Related to breeding (see "Breeding migrations" above).

J. Torpor (Hibernation). In some areas, little grass frogs are active throughout the year (Carr, 1940a; Einem and Ober, 1956). In fact, Harper (1939) stated that except "possible suspension of activity during cold spells of a few days' duration," little grass frogs doubtfully hibernate in the Okefenokee (southeastern Georgia) region.

K. Interspecific Associations/Exclusions. Little grass frogs breed along with other chorus frogs (Ashton and Ashton, 1988) including ornate chorus frogs (*P. ornata*; Mount, 1975) and southern chorus frogs (*P. nigrita*), as well with other frogs such as southern leopard frogs (*Rana sphenocephala*), Cope's gray treefrogs (*Hyla chrysoscelis*), and southern cricket frogs (*Acris gryllus*; personal observations).

L. Age/Size at Reproductive Maturity. Little grass frogs are the smallest North American frog (Conant and Collins, 1991) with males ranging in size from 11.5–15.5 mm and females 12.0–17.5 mm SVL (Wright and Wright, 1949), although the maximum reported size is 20 mm (Franz and Chantell, 1978).

M. Longevity. Unknown.

N. Feeding Behavior. Marshall and Camp (1995) found that little grass frogs eat a wide variety of arthropods, especially insects. Springtails, hymenopterans (mainly ants and parasitic wasps), rove beetles, and homopterans, in that order, were the most abundant prey items consumed. Arachnids (primarily mites) were the only non-insect prey eaten. Most of the prey items are associated with leaf litter and/or soil, suggesting that little grass frogs frequently forage on the ground.

O. Predators. Owen and Johnson (1997) reported predation on a little grass frog by a wolf spider (*Lycosa* sp.) and suggested that little grass frogs may be important prey items for many species of vertebrates and invertebrates.

P. Anti-Predator Mechanisms. Despite their tiny size, little grass frogs can leap 15–22 cm (1–1.5 ft; Wright and Wright, 1949) to avoid predation. Little grass frogs are often cryptically colored similar to the dead grass and sedge blades in which they

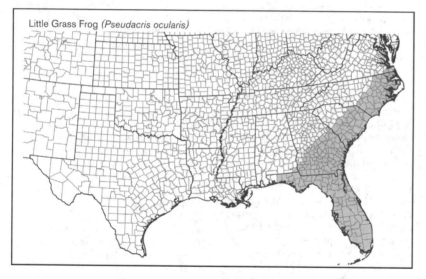

Little Grass Frog (*Pseudacris ocularis*)

2. Historical versus Current Abundance. These are common frogs throughout much of their range, and no significant change in their abundance has been noted.

3. Life History Features.

A. Breeding. Reproduction is aquatic.

i. Breeding migrations. Throughout most of their range, little grass frogs breed from January–September (Harper, 1939) with a peak in Florida during March–April (Ashton and Ashton, 1988). Little grass frogs breed throughout the year in Florida (Carr, 1940a; Einem and Ober, 1956).

iv. Features of metamorphosis. Newly transformed froglets are 7–9 mm SVL (Wright, 1932; Gosner and Rossman, 1960).

v. Post-metamorphic migrations. Unknown.

D. Juvenile Habitat. Unknown, but likely similar to adults.

E. Adult Habitat. Little grass frogs use grass, sedge, and/or sphagnum habitats in or near cypress ponds, bogs, pine flatwoods and savannas, river swamps, and ditches (Harper, 1939; Wright and Wright, 1949; Mount, 1975; Gibbons and Semlitsch, 1991). They have been found calling from vegetation within brackish ditches (Neill, 1958a), though it is unknown whether reproduction is attempted

inhabit. The stripes through their eyes and along the sides may help to break-up their outline to visually oriented predators.

Q. Diseases. Unknown.

R. Parasites. The nematode *Spironoura catesbianae* has been found in little grass frogs (Yamaguti, 1961).

4. Conservation.
Little grass frogs are not protected by either state or federal laws. They remain common throughout much of their range, and no substantial changes in their abundance have been noted.

Pseudacris ornata (Holbrook, 1836)
ORNATE CHORUS FROG

John B. Jensen

1. Historical versus Current Distribution.
Ornate chorus frogs (*Pseudacris ornata*) are restricted to the southeastern Coastal Plain (Mount, 1975) from extreme eastern Louisiana to North Carolina (Conant and Collins, 1991). Brown and Means (1984) identified the southernmost locality known as Lake County, Florida, approximately 230 km north of the incorrect southern limits illustrated on maps in Conant and Collins (1991) and Behler and King (1998). The availability of sandy soils influences the distribution of this species (Brown and Means, 1984). No substantial changes in their distribution have been noted.

that ornate chorus frogs may be declining locally. Means and Means (2000) found that the number of breeding populations of ornate chorus frogs in the Munson Sand Hills of panhandle Florida occur in much lower densities on silvicultural lands than in nearby native habitat. They hypothesized that elimination or severe alteration of the upland habitat, resulting from intensive soil disturbance, is the principal reason. This is corroborated by a 1996–98 rare amphibian survey conducted at 444 sites on industrial forest lands in south Georgia, south Alabama, and north Florida (Wigley et al., 1999). This study revealed that ponds where ornate chorus frogs were found had a substantially lower frequency of intensive site preparation in pond edges and surrounding upland habitats than those ponds where this species was not detected. Further, ornate chorus frogs were substantially less likely to be present if bedding had been used in primary upland stands. Therefore, it is likely that the accelerating conversion of natural pine habitats to industrial pine plantations throughout this species' range (Dodd, 1995a) is reducing their abundance and will continue to do so without major changes in management practices. However, ornate chorus frogs are capable of reestablishing populations on abandoned agricultural lands (Brown and Means, 1984; Caldwell, 1987).

do not occur until January–February (Gibbons and Semlitsch, 1991). Calling males, although not indicative of breeding activity, may be heard as early as late October (Carr, 1940b; Brown and Means, 1984) and at temperatures as low as −2.8 °C (27 °F; Harper, 1937). Males arrive prior to females and begin calling, presumably to attract females to the site (Caldwell, 1987), and remain in the ponds longer than females (Caldwell, 1987). Emigration of adults from the breeding sites occurs from January–March (Gibbons and Semlitsch, 1991).

ii. Breeding habitat. Ornate chorus frogs breed in temporary wetland pools and ponds, including cypress ponds and rain-filled meadows (Harper, 1937), flooded fields and ditches (Martof et al., 1980; Caldwell, 1987), Carolina bays (Caldwell, 1987; Gibbons and Semlitsch, 1991), sink-hole ponds, and borrow pits (personal observations). Mount (1975) suggested that ornate chorus frogs are less likely to breed in shallow roadside ditches and seepage areas than other congeners, and I concur. Males usually call while sitting 2.5–25 cm (1–10 in) above the water in clumps of grass or on floating debris such as logs (Harper, 1937; Mount, 1975) or call while floating on the surface of open water (personal observations). Males often use several breeding ponds over the season (Ashton and Ashton, 1988).

B. Eggs.

i. Egg deposition sites. Egg masses are attached to submerged grass and sedge stems in shallow water areas open to full sunlight (Seyle and Trauth, 1982). Eggs hatch within 1 wk (Ashton and Ashton, 1988).

ii. Clutch size. Typically 10–100 eggs (Wright and Wright, 1949) are laid in small, loose clusters (Mount, 1975). Most of the egg masses Seyle and Trauth (1982) examined from Marion County, Georgia, contained 20–40 eggs, with a maximum of 106 in a single mass.

C. Larvae/Metamorphosis.

i. Length of larval stage. Approximately 90 d (Dundee and Rossman, 1989) in Louisiana, but ≤4 mo in South Carolina (Caldwell, 1987). Tadpoles may reach 43 mm TL before transformation (Dundee and Rossman, 1989).

ii. Larval requirements.

a. Food. Unknown, but tadpoles likely graze on algae.

b. Cover. Larvae are most readily found by dip-netting in submerged and emergent vegetation, suggesting they seek shelter in such cover (personal observations).

iii. Larval polymorphisms. Unknown and unlikely for this species.

iv. Features of metamorphosis. Larvae transform at 14–16 mm (Wright and Wright, 1949).

v. Post-metamorphic migrations. Newly transformed young may remain around the breeding pond to forage (Carr, 1940b)

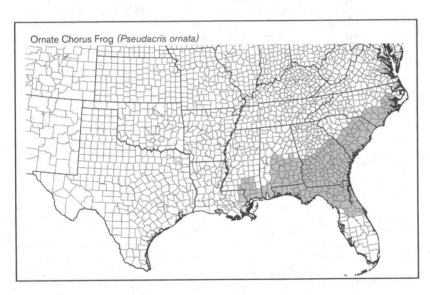

Ornate Chorus Frog (*Pseudacris ornata*)

2. Historical versus Current Abundance.
Some authors have regarded ornate chorus frogs as common throughout their range (Wilson, 1995) or even abundant (Martof et al., 1980). Gibbons and Semlitsch (1991) reported that ornate chorus frogs are 2–3 times more abundant than southern chorus frogs (*P. nigrita*) at South Carolina sites studied in detail. Recently, however, there have been indications

3. Life History Features.
A. Breeding. Reproduction is aquatic.

i. Breeding migrations. Ornate chorus frogs migrate from their upland retreats to aquatic sites in late fall and early winter, where they breed from November–March, depending on rains (Wright and Wright, 1949; Neill, 1957c). Behler and King (1998) report breeding activity into April. In South Carolina, peaks of immigration

prior to emigration. Brown and Means (1984) reported the capture of 42 recently metamorphosed young 20 m from the edge of a breeding pond in late April and several young 200–370 m upslope from the pond in mid-May of the same year, indicating that emigration may commence shortly after transformation. Caldwell (1987) also found that newly metamorphosed animals emigrate to the surrounding forest during late April to May.

D. Juvenile Habitat. Brown and Means (1984) found newly metamorphosed young in turkey oak (*Quercus laevis*) habitat on a ridge crest. Recently metamorphosed animals have been found using the refugia beneath logs and loose pieces of bark on the ground (Harper, 1937).

E. Adult Habitat. Outside of the breeding season, ornate chorus frogs are fossorial (Deckert, 1915; Carr, 1940b; Neill, 1952; Brown and Means, 1984), often burrowed among the roots of herbaceous vegetation (Deckert, 1915; Carr, 1940b; Neill, 1952). Terrestrial adult habitats include pine woodlands (Harper, 1937; Gerhardt, 1973; Martof et al., 1980), pine-oak forest (Dundee and Rossman, 1989), and fallow fields (Harper, 1937; Brown and Means, 1984; Caldwell, 1987). Habitats with sandy substrates are needed to accommodate their burrowing needs (Brown and Means, 1984). Upland retreats can be at least 425 m from the closest available breeding pond (Brown and Means, 1984). There are no data available to indicate a difference in habitats used by males and females.

F. Home Range Size. About 100 m² (Ashton and Ashton, 1988).

G. Territories. Unknown.

H. Aestivation/Avoiding Desiccation. This species remains burrowed and inactive during the summer and early fall. However, heavy rains will bring them to the surface at any time, despite the temperature (Carr, 1940b).

I. Seasonal Migrations. See "Breeding migrations" above.

J. Torpor (Hibernation). Ornate chorus frogs breed during winter (Wright and Wright, 1949; Neill, 1957c), therefore hibernation does not occur. In fact, they may bask in the sun even when snow is on the ground (Ashton and Ashton, 1988).

K. Interspecific Associations/Exclusions. Ornate chorus frogs often call in association with southern chorus frogs, little grass frogs (*P. ocularis*), spring peepers (*P. crucifer*), and southern leopard frogs (*Rana sphenocephala*; Harper, 1937). Ornate chorus frogs may also be found in breeding aggregations of eastern spadefoot toads (*Scaphiopus holbrookii*; Harper, 1937; Neill, 1957c) and gopher frogs (*Rana capito*; personal observations).

L. Age/Size at Reproductive Maturity. Males range from 25–35 mm SVL and females from 28–36 mm (Wright and

Wright, 1949), although working in South Carolina, Gibbons and Semlitsch (1991) report males reach 39 mm; females, 40 mm.

M. Longevity. Most adults in natural habitats rarely live beyond their second breeding season due to predation and intraspecific fighting, although they may live 3.5 yr in captivity (Caldwell, 1987).

N. Feeding Behavior. Adults feed primarily upon small insects (Wilson, 1995). Brown and Means (1984) suggested that earthworms, nematodes, and certain insect larvae may be attracted to the root masses in which ornate chorus frogs often burrow, providing a potential food source. Further, the above-ground portions of the plants attract other insects that may be consumed by ornate chorus frogs positioned at the mouth of their burrows. Newly transformed ornate chorus frogs feed on nymphal orthopterans around the breeding ponds (Carr, 1940b).

O. Predators. Salamander larvae, including tiger salamanders (*Ambystoma tigrinum*), dragonfly naiads, and other aquatic invertebrates may be important predators of ornate chorus frog tadpoles (Caldwell, 1987). Neill (1952) reported the predation of an adult ornate chorus frog by a southern hog-nosed snake (*Heterodon simus*).

P. Anti-Predator Mechanisms. Blouin (1989) suggested that the extreme color polymorphism displayed by ornate chorus frogs is maintained through direct selection by visually oriented predators. One author reported that ornate chorus frogs make immense leaps when pursued (Holbrook, cited in Wright and Wright, 1949), while another observed that this species lies flat against the ground when surprised (Ashton and Ashton, 1988). Their habit of burrowing among the roots of herbaceous vegetation (Deckert, 1915; Carr, 1940b; Neill, 1952) may provide protection from certain predators (Brown and Means, 1984). Harper (1937) suggested that a mid-winter breeding season may avoid predation by water snakes.

Q. Diseases. Unknown.

R. Parasites. Unknown.

4. Conservation.
No substantial change in the distribution of ornate chorus frogs has been noted, however there are indications that populations may be locally declining (see "Historical versus Current Abundance" above). Means and Means (2000) hypothesized that elimination or severe alteration of the upland habitat, resulting from intensive soil disturbance, is the principal reason. It is likely that the accelerating conversion of natural pine habitats to industrial pine plantations throughout this species' range (Dodd, 1995a; Wigley et al., 1999) is reducing their abundance and will continue to do so without major changes in management practices. Ornate chorus frogs are capable of reestablishing

populations on abandoned agricultural lands (see "Historical versus Current Abundance" above). This species is not protected by either state or federal regulations.

Pseudacris regilla (Baird and Girard, 1852[b])
PACIFIC TREEFROG

James C. Rorabaugh, Michael J. Lannoo

1. Historical versus Current Distribution.
The distribution of Pacific treefrogs (*Pseudacris regilla*) includes southeastern Alaska (Waters, 1992), southern British Columbia (including Vancouver Island; Cook, 1980; Weller and Green, 1997), most of California, Oregon, Washington, and Nevada, the western half of Idaho (Nussbaum et al., 1983; Stebbins, 1985; Leonard et al., 1993), the western extremes of Montana (Marnell, 1997) and Arizona, southwestern Utah, and most of Baja California (Stebbins, 1985; Hollingsworth and Roberts, 2001; Grismer, 2002). Pacific treefrogs have been introduced to the Queen Charlotte Islands, British Columbia (Reimchen, 1991), and probably to California City and Soda Springs in southern California (Stebbins, 1985), and in Arizona to plant nurseries near Phoenix and livestock waters in the Virgin Mountains, Mohave County (Rorabaugh et al., 2004). Pacific treefrogs occur from below sea level to an altitude of 3,536 m (11,600 ft) in California (Brattstrom and Warren, 1955; Bezy and Goldberg, 1997). They occur to at least 1,585 m (5,200 ft) in Washington and to 2,247 m (7,370 ft) in Oregon (Leonard et al., 1993). In the Sierra Nevada, California, their distribution extends above timberline, indeed to all zones below Alpine–Arctic regions (J. Grinnell, C.L. Camp and J.R. Slevin, in Wright and Wright, 1949; Cochran and Goin, 1970; Stebbins, 1985; Leonard et al., 1993; Behler and King, 1998; G. Fellers, personal communication). In the Sonoran and Mojave deserts, including the arid regions of Baja California, Pacific treefrogs are generally restricted to springs, oases, rivers, and agricultural areas (Stebbins, 1985; Bezy and Goldberg, 1997; Hollingsworth and Roberts, 2001; Grismer, 2002; Rorabaugh et al., 2004). Reports of Pacific treefrogs from the mountains of central Arizona (Hobbs, 1932) are based on misidentified specimens of mountain treefrogs (*Hyla wrightorum*; Stebbins, 1951; see *H. wrightorum* account, this volume).

The current range of Pacific treefrogs is similar to their historical range. They are not considered to be declining in Canada (Weller and Green, 1997). Their range has not changed in 50–75 yr in Glacier National Park, Montana (Marnell, 1997). In the high Sierra Nevada of California, Pacific treefrogs are not declining, or they are declining much less dramatically than other species, such as mountain yellow-legged

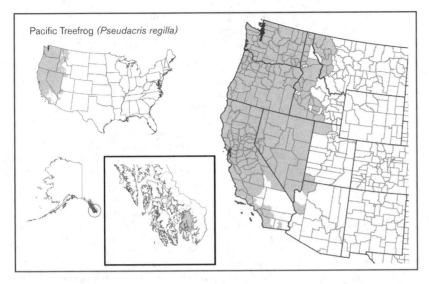

Pacific Treefrog (*Pseudacris regilla*)

frogs (*Rana muscosa*), and Yosemite toads (*Bufo canorus*; Bradford et al., 1994a; see also accounts, this volume). Surveys in Yosemite National Park suggest declines (G.M. Fellers and C.A. Drost, personal communication in Bradford et al., 1994a). In the Central Valley of California during 1990–92, Pacific treefrogs were present in 27 of 28 counties in which they had been found historically (Fisher and Shaffer, 1996). At U.S. Naval holdings on Kitsap and Toandos peninsulas in western Washington, Pacific treefrogs are the most widespread lentic-breeding amphibian (Adams et al., 1999). Banta (1961) suggested that Pacific treefrogs were eliminated from a reach of the lower Colorado River, Nevada and Arizona, when the waters rose behind Davis Dam to form Lake Mohave.

2. Historical versus Current Abundance.

Brattstrom and Warren (1955) comment that Pacific treefrogs are one of the most abundant amphibians in western North America. Slevin (in Wright and Wright, 1949) notes that Pacific treefrogs are probably the most abundant anuran in California. Klauber (in Wright and Wright, 1949) referred to the Pacific treefrog as "the common frog of the coastal areas." Wright and Wright (1949) found one pond in Olympic Park, Washington, that contained between 500–800 egg masses. Working in Wallowa County, Oregon, Ferguson (1954b) observed that Pacific treefrogs were recorded from ten widespread locations and commented that they were common over most of the county. In the Willamette Valley, Oregon, Jameson (1957) suggests that the distribution of breeding sites serves as a limiting factor to both the occurrence and abundance of Pacific treefrog populations. He further notes that "Most of the semipermanent ponds and roadside ditches in the level flood plain of the Willamette Valley are utilized by these frogs." Jewett (1936)

notes that around the Portland, Oregon, area in Multnomah County, Pacific treefrogs are "one of our commonest amphibians . . . found in numbers during March and April along the borders of sloughs and marshes." In a survey of Glacier National Park, Manville (1957) noted that while Pacific treefrogs were considered common, the Park collection contained only one specimen. Similarly, in Stevens County, Washington, Blanchard (1921) considered Pacific treefrogs to be undoubtedly common, although he collected only three specimens. Svihla and Svihla (1933b) working in Whitman County near Pullman, Washington, noted, "In the springtime, practically every roadside pool around Pullman resounds with the calls of these tree-toads. Egg masses are abundant in these pools by the middle of April." In the Willamette Valley, Oregon, Jameson (1957) observed that some breeding choruses occasionally contain over 500 males. Jameson (1956b, 1957) documented isolation among Pacific treefrog populations. His interpretation was that rapid selection could occur, but that isolation puts populations at risk.

At most localities, Pacific treefrogs are probably as common today as they were historically; indeed, Pacific treefrogs are usually the most abundant amphibians where they occur (G. Fellers, personal communication). Leonard et al. (1993) describe them as the most common frog in the Pacific Northwest. In the Sierra Nevada above 2,440 m, they are, by far, the most commonly encountered amphibians (Jennings et al., 1992). Pacific treefrogs are also common in artificial ponds in northwestern Idaho (Monello and Wright, 1999). Fellers (personal communication) has observed 1,500–2,000 egg masses at individual sites.

Pacific treefrog population sizes vary due to climatic conditions. Near Reno, Nevada, treefrog populations were elimi-

nated at a breeding pond 10 times from 1905–89 due to flooding or drought, but were recolonized by individuals that avoided climatic extremes by taking refuge in moist, cool retreats (Weitzel and Panik, 1993). Fewer male treefrogs call during cold, windy nights, as compared to warm, calm nights (Brenowitz and Rose, 1999); thus, any assessments of relative abundance based on call counts must take into account weather conditions.

South of Reno, Nevada, breeding Pacific treefrogs at a semipermanent pond, 21 m in length, peaked at a mean of 60 frogs (53–66; Weitzel and Panik, 1993). In northern Idaho, breeding males at five ponds numbered 360 during 1 yr, 160 the next (Schaub and Larsen, 1978). Calling male treefrogs space themselves in wetland habitats about every 75 cm, limiting the number of calling males in an area (Whitney and Krebs, 1975). In June 1992 at a site in northern California, Pacific treefrog tadpole density was $16.25 \pm 7/m^2$ (mean $\pm$ 1 SD) and ranged from 0–350. One month later, mean density had dropped to 8.5 tadpoles/m2. Tadpole density was judged to be relatively high in 1992 (Kupferberg, 1997a).

3. Life History Features.

A. Breeding. Reproduction is aquatic.

i. Breeding migrations. From November–July, Pacific treefrogs move from the cool, moist terrestrial retreats they use as overwintering sites to aquatic breeding sites (Weitzel and Panik, 1993). In western Oregon and Washington, breeding migrations are triggered by warm (5–10 °C) winter rains (Nussbaum et al., 1983; Weitzel and Panik, 1993). In northern Idaho, no adult frogs were found in breeding sites from mid-June to early April (Schaub and Larsen, 1978). In Los Angeles County, California, Pacific treefrog females were present in large numbers and actively approached males only when wet bulb air temperatures ranged from 12.9–14.7 °C (Straughan, 1975). Individuals probably home to the same general area every year and then migrate to the nearest pond or to a pond with the loudest chorusing (Schaub and Larsen, 1978). A few male Pacific treefrogs enter breeding ponds first and begin calling, which attracts other males and females (Weitzel and Panik, 1993).

Gary Fellers (personal communication) observes that the peak of egg laying is from late winter to late spring, but that egg laying lasts well into the summer in the mountains. Leonard et al. (1993) indicate a breeding season from January–July, with this latter date observed in a population at 2,200–2,620 m in the Sierra Nevada (Livezey, 1953). Pacific treefrogs breed from November–July in southern California (Brattstrom and Warren, 1955; Perrill, 1984; Hollingsworth and Roberts,

2001) and were heard calling from November–June on the Colorado River, Arizona and California (Rorabaugh et al., 2004). South of Reno, Nevada, Pacific treefrogs breed from February–May, with breeding peaking in March (Weitzel and Panik, 1993). In Glacier National Park, Montana, frogs breed in April–May, when water temperatures are 10–12 °C (Marnell, 1997). Jameson (1957; see also Schaub and Larsen, 1978) notes that males will call as late as August at elevations above 610 m, and that during the height of the breeding season, males will call 24 hr/d, with the greatest intensity during the hours following sunset. In the Willamette Valley, Washington, females stay at the breeding sites an average of 9.6 d (1–27) and males stay on average 33 d (1–90). In northern Idaho, males spent 10–13 d at breeding ponds (Schaub and Larsen, 1978). However, in Nevada, males stayed at the breeding ponds for the entire breeding season (February–May; Weitzel and Panik, 1993).

After breeding, frogs leave the pond and move about on the ground or in low shrubbery (Leonard et al., 1993), or move back to cool, moist retreats used for overwintering or aestivation; however, a few individuals remain at the breeding pond (Weitzel and Panik, 1993). Breeding males may switch wetlands during the breeding season and can move at least 400 m to adjacent ponds (Schaub and Larsen, 1978).

There is considerable geographic variation in call patterns (Snyder and Jameson, 1965; see also Foster, 1967; Allan, 1973; Straughan, 1975; Whitney, 1980, 1981). Calls are given infrequently during the day and reach a maximum about 1 h after sundown (Foster, 1967). Male breeding behavior is detailed by Perrill (1984). Breeding animals displaced 274 m returned home to their breeding wetland (Jameson, 1957). Svihla and Svihla (1933b) noted a post-breeding adult in an alfalfa field moving back to the woods.

ii. Breeding habitat. Includes most aquatic habitats, including lakes, ponds, slow-moving streams, backwaters of large rivers, wet meadows, emergent marshes, forested swamps, reservoirs, muskegs, supratidal pools, golf course ponds, and irrigation ditches (Wright and Wright, 1949; Brattstrom and Warren, 1955; Stebbins, 1985; Waters, 1992; Leonard et al., 1993; Gardner, 1995; Rorabaugh et al., 2004). Pacific treefrogs breed in both temporary and permanent waters. In the Pacific Northwest, they are often found breeding in fishless, ephemeral wetlands that dry up before mid summer (Leonard et al., 1993). They are most likely to use shallow, quiet waters for breeding, especially waters with submerged and/or emergent vegetation (Nussbaum et al., 1983). Breeding may occur in weakly brackish waters (Stebbins, 1951; Gardner,

1995). At the Angelo Coast Range Reserve in northern California, Pacific treefrogs breed in wet meadows from early spring to late summer and then switch to breeding in a river as the dry season progresses (Kupferberg, 1997a). In northeastern California, Pacific treefrogs breed in isolated temporary ponds surrounded by semiarid habitats (Watkins, 1996). Within wetlands, breeding populations appear to be clumped, and the location of these aggregations may be the result of frogs being attracted to calling individuals and choruses rather than characteristics of the habitat (Whitney and Krebs, 1975).

In southern California, frogs call at water temperatures of 10–12 °C and avoid large lakes or cold, spring-fed streams in which water temperatures remain low. Frogs seldom, if ever, call at water temperatures above 20 °C (Brattstrom and Warren, 1955). In northern Idaho, frogs chorused in water temperatures as low as 2.0 °C and air temperatures of 0.5 °C. However, they preferred warmer, more open ponds (Schaub and Larsen, 1978).

B. Eggs.

i. Egg deposition sites. In Glacier National Park, Montana, Pacific treefrogs breed in April–May when water temperatures are 10–12 °C (Marnell, 1997). Jameson (1957) observed that within ponds, shallower, more heavily vegetated portions are used for breeding. Leonard et al. (1993) concurred, noting that eggs are laid on submerged aquatic vegetation, including grass, stems, and sticks; however, in shallow water eggs may be deposited on the bottom (Hollingsworth and Roberts, 2001). Egg masses are commonly found at depths ≤10 cm and may be found floating, attached to objects at the surface (Stebbins, 1951). Egg masses are observed in waters with temperatures from 3–30 °C (average about 11 °C; G. Fellers, personal communication).

On Vancouver Island, British Columbia, ambient levels of UV-B radiation, and levels 15–30% above ambient, did not affect the hatching success of Pacific treefrog eggs (Ovaska et al., 1977). Hatching success under ambient levels of UV-B radiation was also unaffected in the Oregon Cascade Mountains (Blaustein et al., 1994c) and in the Santa Monica Mountains, California (Anzalone et al., 1998). Compared to some other anurans, Pacific treefrog eggs are resistant to the effects of UV-B radiation due to differential activity of a photolyase that functions to repair UV-B-induced DNA damage through its action on cyclobutane pyrimidine dimmers (Blaustein et al., 1994c).

Acidified habitats may not be affecting egg survival of Pacific treefrogs in the Sierra Nevada. The estimated extreme pH in Sierra Nevada waters is 5.0; in the laboratory, sensitivity to acidic waters (LC_{50}) for Pacific treefrog eggs averages a pH of 4.3 (Bradford et al., 1994b,c).

Eggs can survive temperatures as low as −5 to −7 °C for up to 2 hr, and as high as 34 °C (Brattstrom and Warren, 1955). Embryos develop and hatch in 1–5 wk (Nussbaum et al., 1983; Leonard et al., 1993; Hollingsworth and Roberts, 2001). In the laboratory, larvae may hatch in 6 d (Pickwell, 1947).

ii. Clutch size. Females lay between 400–750 eggs (maximum 1,250), 1.3 mm in diameter, in clusters composed of 9–80 eggs (average 18; see also Stebbins, 1951; Gaudin, 1965; Nussbaum et al., 1983; Perrill and Daniel, 1983; Leonard et al., 1993; and Marnell, 1997). There may be geographic variation in clutch size. In San Diego, California, mean clutch size was 267.7 (range 62–633; Perrill and Daniel, 1983). Further, Storer (in Wright and Wright, 1949) speculates that size of egg clutches may reflect degree of solitude in mating pairs. Where few pairs mate, egg cluches are larger. Stebbins (1951) notes that towards the close of egg laying, cluster size tapers off to 3 or 4 or even single eggs. Females may produce ≥3 clutches/season. Average time between clutch deposition is 37.3 d (range of 13–69; Perrill and Daniel, 1983).

C. Larvae/Metamorphosis.

i. Length of larval stage. In western Oregon, 2 mo; 2.5 mo in northern Idaho (Nussbaum et al., 1983). Metamorphosis occurs from June–October, depending on altitude and latitude. In the Pacific Northwest, tadpoles metamorphose in June at low elevations and late August at higher elevations (Leonard et al., 1993). In northwestern Idaho, tadpoles metamorphose from mid-July to mid-September (Schaub and Larsen, 1978). Dill (1977) reported tadpoles in British Columbia as late as August.

ii. Larval requirements. Pacific treefrog tadpoles typically are found in quiet waters or sluggish, slow streams. At a pond near Reno, Nevada, tadpoles tended to aggregate in deeper (≤ 1.2 m) water. In enclosure experiments, larval survival was lower in permanent ponds as compared to ephemeral wetlands, however, no cause of this difference was evident (Adams, 2000). Compared with other anuran species, including two species of spadefoot toads (Pelobatidae), Pacific treefrog tadpoles are the least heat resistant (H.A. Brown, 1969). Pacific treefrog tadpoles apparently prefer temperatures around 19–20 °C, but can tolerate low temperatures of 0–2 °C and high temperatures to 33 °C, with lethal temperatures about 5 °C higher (Brattstrom and Warren, 1955). Weitzel and Panik (1993) note that tadpoles grew rapidly at water temperatures of 12–18 °C. Tadpoles are tolerant of weakly saline conditions (Stebbins, 1951; Gardner, 1995).

Tadpoles are highly sensitive to nitrites, which cause reduced feeding, less vigorous activity, disequilibrium, other abnormalities, and increased mortality.

Substantial mortality occurred at Environmental Protection Agency recommended nitrite concentrations (1 mg N–NO2/L) for drinking water (Marco et al., 1999). Nitrogen compounds from agricultural runoff or drainage into Pacific treefrog habitats could harm tadpoles (Schuytema and Nebeker, 1999a,b). In the laboratory, the herbicide diuron causes limb deformities and reduced growth in tadpoles, but experimental concentrations were much higher than those found in the field (Schuytema and Nebeker, 1998). Compared to other vertebrates, Pacific treefrog tadpoles are relatively tolerant of the pesticides guthion and guthion 2S (Nebeker et al., 1998).

The LC_{50} pH value for Pacific treefrog tadpoles is 4.23 (Bradford et al., 1994b). The lowest pH likely to occur in surface waters of the Sierra Nevada is 5.0 (Bradford et al., 1992), thus acidification is not likely to cause acute toxicity to Pacific treefrog tadpoles in the Sierra Nevada. This supposition was borne out by data from Sequoia National Park (Soiseth, 1992) and elsewhere in the Sierra Nevada (Bradford et al., 1994c). Ambient levels of UV-B radiation near Victoria, British Columbia, did not affect tadpole survivorship; however, survival in the first 2 mo of development was substantially lower under enhanced (15–30% over ambient) levels of UV-B (Ovaska et al., 1997).

a. Food. Wassersug (1976) considered Pacific treefrog tadpoles to be typical pond tadpoles, possessing a generalized oral morphology. Typical pond tadpoles are generally considered to be nondiscriminatory suspension feeders that ingest a variety of prey including green algae, blue-green algae, bacteria, diatoms, protozoa, and a wide variety of organic and inorganic debris (Wassersug, 1975; Wagner, 1986). However, Kupferberg (1997b) demonstrates that Pacific treefrog tadpoles can select those foods that favor optimal growth. In feeding trials, tadpoles selected algae with relatively high protein content. Tadpole diets may be low in protein, and addition of protein to their diets increases size at metamorphosis (Kupferberg, 1997b). Wagner (1986) describes Pacific treefrog tadpoles—ingesting pine (*Pinus* sp.) and fir (*Abies* sp.) pollen, when it is seasonally available on the water surface. In the presence of pollen, tadpoles altered their behavior to feed at the surface. Tadpoles that consumed diets rich in diatoms had enhanced growth, development, and survival to metamorphosis (Kupferberg et al., 1994). Pacific treefrog tadpoles may be good at extracting low quality or low availability resources. Kupferberg (1997a) observed tadpoles swimming on their backs at the surface, grazing on epineustic films of diatoms.

b. Cover. Brattstrom (1962b) noted that tadpoles will aggregate and speculated that by doing so they were thermoregulating. Aggregation is also an anti-predator mechanism (De Vito et al., 1999). Pearl et al. (2003) found that Pacific treefrog tadpoles will not seek cover in the presence of either fish or invertebrate predators.

iii. Larval polymorphisms. Do not occur.

iv. Features of metamorphosis. Tadpoles attain a total length of 45–55 mm prior to metamorphosis (Nussbaum et al., 1983). Newly metamorphosed frogs are <10–17.0 mm SVL (Wright and Wright, 1949; G. Fellers, personal communication). In one western Oregon study, recently metamorphosed frogs averaged 13.8 mm (Nussbaum et al., 1983). Metamorphosis is prolonged, for example from mid-July to mid-September in northern Idaho (Schaub and Larsen, 1978), and May–October in California (G. Fellers, personal communication).

v. Post-metamorphic migrations. Metamorphosed frogs leave the natal ponds soon after transformation and move to upland habitats or overwintering sites in midsummer to early fall (Schaub and Larsen, 1978), similar to the adults (see "Breeding migrations" above and "Hibernation" below). In Nevada, all frogs had exited the natal ponds by early October (Weitzel and Panik, 1993). In Montana, metamorphosed frogs disperse by late August and return in the spring as adults. After leaving the ponds, dispersing individuals were found perched 1–1.5 m above the ground in broadleaf shrubs (Marnell, 1997). Newly metamorphosed animals disperse, but apparently not far enough in some populations to form strong metapopulation links (Jameson, 1956b, 1957; Schaub and Larsen, 1978). Juveniles have been recaptured as far as 238 m from their home pond (Jameson, 1956b).

D. Juvenile Habitat. Similar to adult habitats. In Glacier National Park, Montana, juveniles were found in shallow waters among shoreline vegetation at the edges of ponds (Marnell, 1997).

E. Adult Habitat. Adults use a variety of aquatic habitats for breeding, then move upland where they move about in low shrubbery during moist weather. In dry periods or habitats, Pacific treefrogs tend to be more nocturnal (Leonard et al., 1993), and will seek moist, cool retreats for aestivation and for hibernation in the fall (Leonard et al., 1993; Weitzel and Panik, 1993). No differences between male and female habitat characteristics are known.

The name "treefrog" is a misnomer because, although Pacific treefrogs sometimes climb short distances into shrubs or trees, they are usually found on the ground (Dickerson, 1906; Stebbins, 1951; Badaracco, 1962). They are found especially near streams, springs, ponds, wetlands, irrigation ditches, and other moist places (Wright and Wright, 1949; Stebbins, 1985). In these habitats, Pacific treefrogs are found in low plant growth, damp recesses among rocks and logs, under tree bark, in trees in damp forests, and in animal burrows in open country (Wright and Wright, 1949). They can be found far from water outside of the spring–summer breeding season (Stebbins, 1951, Badaracco, 1962; Nussbaum et al., 1983; Leonard et al., 1993). Svihla and Svihla (1933b) collected adults in irrigation ditches in "sagebrush country along the Snake River Canyon." Schaub and Larsen (1978) note that few North American anurans exploit such a variety of habitats, including deserts, grasslands, mountains, and the rain forests of the Pacific Northwest. In western Washington, Pacific treefrogs are associated with open wetland habitats in forests of saplings and in clearcuts (Bosakowski, 1999). Raphael (1988) found Pacific treefrogs to be associated with early successional Douglas-fir forests of northwestern California. However, Welsh and Lind (1988, 1991) found that captures increased from young to older growth forests and were higher in mesic versus wet forests in northwestern California and southwestern Oregon.

Cunningham and Mullally (1956) note that Pacific treefrog adults are active to 4 °C and detail other thermal relations using a combination of field observations and experimental trials. Mean critical thermal maxima for Pacific treefrogs from near Clinton, Montana, ranged from 34.8–35.2 °C. Pacific treefrogs prefer lower temperatures at night than during the day (Claussen, 1973). Brattstrom (1963) reports 3.8 and 24.0 °C as the minimum and maximum temperatures voluntarily tolerated by Pacific treefrogs. Croes and Thomas (2000) demonstrated freeze tolerance in Pacific treefrogs from northern California. In response to freezing, plasma glucose increased 5–14-fold. The liver is the organ responsible for cryoprotectant synthesis.

Adults are polymorphic in their dorsal body color. Individual frogs become lighter or darker and can lose their spots in response to environmental conditions, but the green and brown color phases and the black eye stripe are genetically determined and do not change (Brattstrom and Warren, 1955; Weitzel and Panik, 1993). The green color phase is more absorptive of solar radiation and may be favored in aquatic habitats. Green may, however, be a disadvantage in hot, dry, terrestrial conditions, where the brown phase may be favored (Jameson and Pequegnat, 1971; Stebbins and Cohen, 1995).

F. Home Range Size. Home ranges include upland activity, overwintering and aestivation sites, breeding ponds, and migratory corridors among these habitats. However, activities and locations of Pacific treefrogs when not at the breeding ponds are poorly documented. Apparent overwintering sites were about 60 m above and 150–300 m from breeding habitats in southern California (Brattstrom and Warren, 1955). Leonard et al. (1993) report active frogs on the ground and in low shrubbery during moist weather outside of the breeding season. During the breeding season, adults usually remain at the same breeding pond; however, Schaub and Larsen (1978) documented movements by breeding males of up to 400 m.

G. Territories. Male Pacific treefrogs produce two types of advertisement calls (monophasic and diphasic) and an aggressive encounter or staccato/trill call (Allan, 1973). The encounter call is given at the beginning of chorusing each evening, possibly to establish spacing between males (Allan, 1973; but see Whitney, 1981). Encounter calls may also be given through the night if another male approaches too closely (Awbry, 1978). Brenowitz and Rose (1999) found that the amplitude of either the encounter or advertisement call perceived by a male is an important trigger in eliciting the encounter call. Switching to the encounter call also occurs in response to movement by an intruder (Snyder and Jameson, 1965; Allan, 1973; Whitney, 1981). Typical spacing among males is about 75 cm (Whitney and Krebs, 1975; Awbrey, 1978). Males switch from the advertisement call to the encounter call when another frog approaches within about 20–50 cm (Awbrey, 1978; Whitney, 1980) or the amplitude of the call exceeds 87 dB (Brenowitz, 1989). Some authors have found that calling males have a strong connection to a particular location, as evidenced by displacement experiments (Perrill, 1984) and mark–recapture data (Jameson, 1957). However, Whitney and Krebs (1975) found that calling males usually occupied calling sites for only ≤1 night. As a result, they concluded that the frogs established spacing or individual distances, rather than territories. However, the frogs defend these "spaces" with territorial behavior (Fellers, 1979a).

Whitney (1980) suggests calling males have three lines of defense: (1) the advertisement call that establishes the male's presence, (2) the encounter call when another male approaches, and (3) fighting. In response to the encounter call, males may submit by retreating or ceasing to call, they may perform a "bouncing" behavior or may resort to physical encounters that involve butting and wrestling (Fellers, 1979a; Whitney, 1980; Perrill, 1984). Fights typically end with one frog clasping the other anterior to the front limbs, causing the clasped frog's vocal sacs to deflate. The deflated frog then moves away (Whitney and Krebs, 1975) or becomes subordinate and may sit silently close by the victorious, calling male (Fellers, 1979a). Some males are silent (satellite males) and intercept and mate with females that are attracted to calling, territorial males (Perrill, 1984). Territorial behavior of Pacific treefrogs is similar to several other North American Hylidae (Fellers, 1979a).

H. Aestivation/Avoiding Desiccation. Habitat use of Pacific treefrogs is poorly known outside of the breeding season. However, when away from breeding areas, particularly during dry periods or in dry areas, the species may be found in cool, moist retreats such as piles of debris, dense vegetation, rock or log crevices, mammal burrows, artificial drains, basements of homes and buildings, spring boxes, housing units for sprinkler system valves, and other protected places (Brattstrom and Warren, 1955; Nussbaum et al., 1983; Leonard et al., 1993; Weitzel and Panik, 1993). Brattstrom and Warren (1955) found that in California, Pacific treefrogs must seek hiding places during the hot, dry months of July–October. Pacific treefrogs were one of the few vertebrates to survive in the 150,000 acre-blast zone during the eruption of Mount St. Helens. Apparently, individuals that were underground were spared (Weyerhaeuser, 1999).

I. Seasonal Migrations. Most Pacific treefrogs move from overwintering sites to breeding sites in winter or spring and then leave the breeding sites by early fall. Habitat use and movements outside of the breeding season are poorly known. See "Breeding migrations" and "Aestivation/ Avoiding Desiccation" above and "Hibernation" below for further details.

J. Torpor (Hibernation). In some localities at lower elevations, Pacific treefrogs are active throughout the year (Stebbins, 1951); elsewhere, Pacific treefrogs must avoid cold temperatures. In western Montana, Pacific treefrogs hibernate in subterranean shelters (Cunningham and Mullally, 1956). At elevations above 2,500 m in the Sierra Nevada, California, frogs overwinter in terrestrial shelters (Bradford, 1989). A drought in the winter of 1976–77 in northern Idaho severely reduced snowpacks, which likely resulted in ground freeze at greater depths than normal. Ground freeze probably increased overwintering mortality of Pacific treefrogs, which was reflected in reduced populations in the spring and summer of 1977 (Schaub and Larsen, 1978). Near Gorman, California, Pacific treefrogs were found calling in early January from 2.5 cm diameter holes about 60 m above and 150–300 m from breeding habitat. Holes contained 1–5 frogs each (Brattstrom and Warren, 1955).

K. Interspecific Associations/Exclusions. Pacific treefrogs will breed in association with many species of western amphibians. At Fort Lewis Military Reservation, Washington, they are found in the same wetlands as northwestern salamanders (*Ambystoma gracile*), long-toed salamanders (*A. macrodactylum*), rough-skinned newts (*Taricha granulosa*), western toads (*Bufo boreas*), northern red-legged frogs (*R. aurora*), and American bullfrogs (*R. catesbeiana*; Adams et al., 1998). In the Sierra Nevada, Pacific treefrogs occur with mountain yellow-legged frogs (Pope, 1999). On the lower Colorado River, Arizona–Nevada–California, Pacific treefrogs breed in the same habitats as Great Plains toads (*B. cognatus*), Woodhouse's toad (*B. woodhousii*), and bullfrogs (J.C.R., personal observations). In August 1925 near Las Vegas, Nevada, Wright and Wright (1949) found Pacific treefrogs in association with Vegas Valley leopard frogs (*R. fisheri*, now extinct) and Great Plains toads. Pacific treefrogs occur, or occurred until recently, with relict leopard frogs (*R. onca*) and red-spotted toads (*B. punctatus*) near Hoover Dam, Nevada–Utah (R.D. Jennings, 1995b), and occurred historically with lowland leopard frogs (*R. yavapaiensis*) at San Felipe Creek, California (Ruibal, 1959), before lowland leopard frogs were extirpated from that site. Brattstrom and Warren (1955) observed breeding in association with California toads (*B. boreas halophilus*) and western spadefoot toads (*Spea hammondi*). In southern Nevada, Pacific treefrogs occur with Amargosa toads (*B. nelsoni*; Wright and Wright, 1949). Wright and Wright (1949) note an observation of a male western spadefoot toad in amplexus with a female Pacific treefrog. The latter died soon after, apparently from a rupture of the abdominal wall. In the Central Valley and in southern California, Pacific treefrogs show less variation in recruitment across years, and presence of tadpoles is less reliant on rainfall compared to California toads or western spadefoot toads (Fisher and Shafer, 1996).

Brattstrom and Warren (1955) suggest that Pacific treefrogs compete with California treefrogs (*P. cadaverina*), noting that Pacific treefrogs are less abundant, even absent, where California treefrogs occur. These authors collected an apparent hybrid of these two species in San Diego County, California. Littlejohn (1971) observed mixed choruses of Pacific treefrogs and California treefrogs in Whitewater Canyon, southern California.

Tadpoles of Cascade frogs (*R. cascadae*) and Pacific treefrogs have similar diets and larval periods and frequently breed in the same ponds in the Oregon Cascades (Nussbaum et al., 1983). Keisecker and Blaustein (1999) demonstrated that Cascade frog tadpoles had strong negative effects on the growth, development, and survival of Pacific treefrog tadpoles. However, in the presence of the water mold (*Saprolegnia ferax*), competitive interactions were reversed. Pacific treefrog tadpoles had higher survival, faster development, and were larger at metamorphosis when exposed to both *Saprolegnia* and Cascade frog tadpoles, as compared to exposure to Cascade frog tadpoles alone.

Pacific treefrogs are also found in ponds with introduced American bullfrogs, although Brattstrom and Warren (1955) never found them closer than 1.2 m to the bullfrogs. Average daily survival rate for Pacific treefrog tadpoles did not differ between ponds that contained or did not contain bullfrog tadpoles, although in the presence of bullfrogs, later development was slowed (Govindarajulu, 2000). In enclosure experiments, Pacific treefrog larval survival was not affected by the presence of American bullfrog larvae (Adams, 2000). However, in northern California, exploitative competition from large overwintering bullfrog tadpoles reduced survivorship and growth of Pacific treefrog tadpoles. Competition from recently hatched bullfrog tadpoles also reduced survivorship of Pacific treefrog tadpoles. Nevertheless, Pacific treefrogs are tolerant of bullfrog invasions because their tadpoles are good competitors, and they have a broad range of physical tolerances and can reproduce in shallow, ephemeral habitats unsuitable for bullfrogs (Kupferberg, 1997a).

In the presence of Oregon garter snakes (*Thamnophis atratus hydrophilus*), Pacific treefrog tadpoles moved less and spent more time in lower quality food patches. At these sites, reduced activity resulted in the tadpoles sinking away from floating *Cladophora* algal mats, which are high quality foods for tadpoles (Kupferberg, 1997b). In the Sierra Nevada above 2,440 m, the presence of amphibians is a prerequisite for the presence of western terrestrial garter snakes (*T. elegans*). As the most common amphibian in the region, if Pacific treefrogs declined in the Sierra Nevada, sympatric populations of garter snakes could also disappear (Jennings et al., 1992; Matthews et al., 2002). Jameson (1956b) attributed a rapid increase in a Pacific treefrog population to removal of garter snakes and American bullfrogs.

In high elevation lakes of the Sierra Nevada, Pacific treefrog tadpoles are not found in lakes that support salmonid fish (rainbow trout [*Oncorhynchus mykiss*] and/or brook char [*Salvelinus fontinalis*]). However, Pacific treefrogs often breed in lakes that are shallow and ephemeral, which are not suitable for fishes (Bradford, 1989). At Glacier National Park and ponds in northwestern Idaho, Pacific treefrogs were only found in waters that lacked fish (Marnell, 1977; Monello and Wright, 1999). Monello and Wright (1999) suggested egg predation by goldfish (*Carassius auratus*) eliminated Pacific treefrogs and other amphibians from northwestern Idaho ponds. In enclosure experiments, presence of sunfish (Centrarchidae) reduced survival of Pacific treefrog tadpoles to near zero (Adams, 2000).

Overwintering Pacific treefrogs were found in holes near Gorman, California,

with side-blotched lizards (*Uta stansburiana*; Brattstrom and Warren, 1955).

L. Age/Size at Reproductive Maturity. Males at 25.5–48.0 mm SUL; females at 25.0–47.0 mm SUL. Schaub and Larsen (1978) and Leonard et al. (1993) note that females are larger than males. Sexual maturity probably occurs in <1 yr (Jameson, 1956b, 1957; Nussbaum et al., 1983; Weitzel and Panik, 1993), although Cochran and Goin (1970) and Pickwell (1947) state that Pacific treefrogs take at least 2 yr to reach sexual maturity.

M. Longevity. Unknown. Of 65 Pacific treefrogs captured in 1977 at a series of wetlands in northern Idaho, 9 (13.8%) had been marked the year before (Schaub and Larsen, 1978).

N. Feeding Behavior. Pacific treefrogs are primarily nocturnal, terrestrial foragers (Brattstrom and Warren, 1955; Johnson and Bury, 1965). However, during the breeding season, male Pacific treefrogs apparently feed during the day (Whitney and Krebs, 1975). Food habits of Pacific treefrogs include isopods, spiders, snails, and a variety of insects first reported by Needham (1924, in Brattstrom and Warren, 1955), summarized in Brattstrom and Warren (1955), and listed in Johnson and Bury (1965). Insects constituted 73.5% of the winter diet of 135 Pacific treefrogs collected from northern California. More adult insects were eaten than larvae, indicating frogs primarily catch flying insects (Johnson and Bury, 1965). Pacific treefrogs typically feed above water, either at the surface or in vegetation above the water surface (Brattstrom and Warren, 1955).

O. Predators. Predators on Pacific treefrogs include mountain garter snakes (*T. e. elegans*), common garter snakes (*T. sirtalis*), Oregon garter snakes (Livezey, 1953; Schaub and Larsen, 1978; Kupferberg, 1998; Mathews et al., 2002), northern red-legged frogs (Arnold and Halliday, 1986), mountain yellow-legged frogs (Pope, 1999), American bullfrogs (Cook and Jennings, 2001), northwestern salamanders, egrets and herons (Ardeidae), belted kingfishers (*Megaceryle alcyon*), various species of fish (Bradford, 1989; Goodsell and Kats, 1999; Monello and Wright, 1999), and mammals such as raccoons (*Procyon lotor*), skunks (Mustelidae), feral cats (*Felix domestica*), and opossums (*Didelphis marsupialis*; Storer, 1925; Brattstrom and Warren, 1955; Peterson and Blaustein, 1991; Weitzel and Panik, 1993). In the Oregon Cascade Mountains, some populations of Pacific treefrogs undergo intense predation of eggs by predatory leeches (Glossiphonidae and Erpobdellidae; Chivers et al., 2001). Pacific treefrog populations remain robust in the Santa Monica Mountains, California, despite heavy predation by mosquito fish (*Gambusia affinis*; Goodsell and Kats, 1999).

P. Anti-Predator Mechanisms. Tevis (in Wright and Wright, 1949) observed that in response to disturbance, Pacific treefrogs swim to and hide in masses of filamentous algae or hop upslope to dry brush cover, rather than into water. The latter observation is considered unreliable by G. Fellers (personal communication). Brattstrom and Warren (1955) found that if flushed, frogs jump into water. In a laboratory experiment, frogs presented with a model predator typically jumped at a mean angle of 70° from the frog's initial bearing. Most frogs have longer right limbs and generally jump to the right, suggesting "handedness" (Dill, 1977).

Movement is not always an advantageous predator defense, however. As mentioned above (see "Adult Habitat"), Pacific treefrogs are polymorphic for dorsal body color, being either green or brown. When stationary on a matching substrate, laboratory studies have shown they are able to avoid predation by western terrestrial garter snakes by remaining motionless (Morey, 1990).

In the laboratory, Pacific treefrog tadpoles aggregate as an anti-predator mechanism (De Vito et al., 1999). Tadpoles also exhibit burst swimming in response to predators, and the faster tadpoles are more likely to avoid predation from common garter snakes (Watkins, 1996). However, on the Eel River in northern California, tadpoles reduced their activity in the presence of Oregon garter snakes and tended to sink toward the bottom, away from high quality food resources—floating algal mats. Garter snake predation did not substantially reduce the number of tadpoles on the Eel River, but reduced tadpole growth by 28%, because tadpoles spent less time feeding in floating algal mats (Kupferberg, 1998).

In experimental trials, as the relative size of tadpoles increased, the anti-predator response decreased in the presence of larval northwestern salamanders. The results provide evidence that Pacific treefrog tadpoles are able to assess their individual vulnerability to northwestern salamander larvae and adjust their anti-predator response according to their level of risk (Puttlitz et al., 1999).

Pacific treefrog eggs show plasticity in timing of hatching in response to the threat of predation. Eggs hatch sooner and at an earlier developmental stage when eggs come in contact with predatory leeches, chemical cues of leeches, or chemicals released from injured eggs (Chivers et al., 2001).

Dickerson (1906) found that when Pacific treefrogs were "greatly annoyed" or injured, they would exude a milky secretion from the skin of the dorsum. However, this behavior has not been noted by other authors.

Q. Diseases. A pathogenic fungus (*Saprolegnia ferax*) infects egg masses of Pacific treefrogs. Eggs in communal masses and eggs laid later in the season are most susceptible to infection (Kiesecker and Blaustein, 1997b). Chytridiomycosis, an amphibian fungal disease of global distribution, has been found in a California population of Pacific treefrogs (Fellers et al., 2001). Brattstrom and Warren (1955) mention a haemorrhagic condition of the gut that develops in captive tadpoles reared without access to natural (mud and sand) substrates.

Chlorpyrifos and diazinon, potent organophosphorus cholinesterase inhibitors, and several other pesticides are carried via prevailing summer winds from the agricultural areas of the Central Valley to the Sierra Nevada, California. Chlorpyrifos and diazinon bind with cholinesterase and disrupt neural function. Cholinesterase activity in Pacific treefrog tadpoles from the Sierra Nevada downwind of the Central Valley was lower than at sites on the coast or to the north, and was also lower in areas where ranid population status was poor or moderate compared to sites with good ranid populations. In affected areas, up to 50% of sampled tadpoles had detectable levels of organophosphorus residues. Endosulfan, 4,4'-dichlorodiphenyldichloroethylene, 4,4'-DDT, and 2,4'-DDT residues were also commonly found (Sparling et al., 2001). Presence of polychlorinated biphenyls (PCBs) and toxaphene in tadpoles from the Sierra Nevada showed a trend of increasing concentrations from high to low elevation and from east to west; the latter suggests a rain shadow effect (Angermann et al., 2002). Greater survivorship of Pacific treefrog tadpoles was noted at Lassen National Park than in the affected areas of Yosemite and Sequoia National Parks in the Sierra Nevada. Relatively high (25%) deformity rates were also detected at Yosemite (Cowman et al., 2002). Pacific treefrogs may be less affected by pesticide drift than ranid frogs because they are less dependent on aquatic sites (e.g. the tadpoles metamorphose in the same year they are laid and adults often spend considerable time in the uplands; Sparling et al., 2001).

R. Parasites. Lehman (1964) reported nematodes and trematodes in Pacific treefrogs from eastern Washington and Oregon. In central California, Lehman (1960) reported only *Opalina* sp., a protozoan. Waitz (1961) found no parasitic worms in Pacific treefrogs in Idaho. Frogs from two localities in southern California were infected with the helminths *Rhabdias*, *Oswaldocruzia*, *Cosmocercoides*, and *Distoichometra*. Mean intensity of infection was 10.3 helminths at Malibu and 8.4 helminths at Big Tujunga (Koller and Gaudin, 1977).

Pacific treefrogs with a variety of morphological abnormalities, particularly malformed hindlimbs, have been found in western Montana (Hebard and Brunson,

1963; Van Valen, 1974), Spokane, Washington (Miller, 1968, in Reynolds and Stephens, 1984), Boise, Idaho (Reynolds and Stephens, 1984), and northern California (Johnson et al., 1999, 2001b). At two ponds in northern California, 10–25% of larval and post-metamorphic Pacific treefrogs exhibited abnormalities; of those, >60% were severe malformations involving extra hindlimbs, femoral projections, and skin webbings that probably reduced survivorship (Johnson et al., 2001b). Potential causes include UV-B radiation, retinoid exposure, genetic mutation, pesticide contamination, predation, microbes, and trematode parasites (see Van Valen, 1974; Sessions et al., 1999; Johnson et al., 2001b). Abnormalities observed in northern California are likely caused by a cathaemasiid trematode, *Ribeiroia* sp. Refer to Sessions and Ruth (1990), Johnson et al. (1999, 2001b), Sessions et al. (1999), and Souder (2000) for discussions of trematode parasites and their role in producing limb malformations.

4. Conservation.
Pacific treefrogs have no status under C.I.T.E.S., the U.S. Endangered Species Act, with the Canadian government (COSEWIC, Committee on the Status of Endangered Wildlife in Canada, 2002), or with the government of Mexico (Secretaria de Desarrollo Social, 1994). Where they occur, Pacific treefrogs are typically one of the most common amphibians, often exhibiting robust populations. With a few exceptions, noted in "Historical versus Current Distribution" and "Historical versus Current Abundance," the species is not declining and has not been targeted for conservation or recovery actions.

Acknowledgments. We thank Gary Fellers for valuable comments reflecting his experience with Pacific treefrogs.

Pseudacris streckeri Wright and Wright, 1933
STRECKER'S CHORUS FROG

Donald B. Shepard, Lauren E. Brown, Brian P. Butterfield

1. Historical versus Current Distribution.
Strecker's chorus frogs (*Pseudacris streckeri*) are found from extreme south-central Kansas, south through Oklahoma, northwestern Louisiana, and Texas to the Gulf of Mexico, with isolated localities in southern Texas, western Oklahoma, west-central Illinois, and the Mississippi River Valley of northeastern Arkansas, southeastern Missouri, and southwestern Illinois (Conant and Collins, 1991). Two subspecies are recognized: Strecker's chorus frogs (*P.s. streckeri*) and Illinois chorus frogs (*P.s. illinoensis*).

The occurrence of *P. streckeri* in highly disjunct localities to the northeast of their main range is thought to be the result of a dispersal route across Arkansas, along the Arkansas River floodplain, then north along the Mississippi and Illinois rivers during the Xerothermic period (~6,000–4,000 yr ago; Axtell and Haskell, 1977). Illinois chorus frogs were probably more widespread before the extensive habitat modification along the Illinois and Mississippi rivers during the late 1800s and early 1900s.

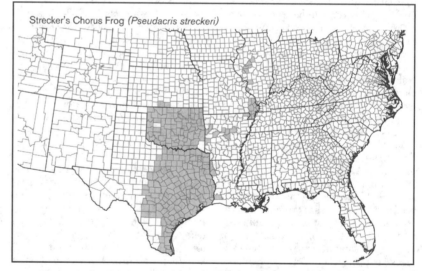

Strecker's Chorus Frog (*Pseudacris streckeri*)

2. Historical versus Current Abundance.
Pseudacris s. streckeri are apparently abundant but rarely observed outside their breeding season (Smith, 1966b). *Pseudacris s. illinoensis* are uncommon in Illinois (Herkert, 1992), with choruses usually comprised of <10 males (Brown and Rose, 1988; Tucker and Philipp, 1993). Their historical abundance is unknown.

3. Life History Features.
A. Breeding. Reproduction is aquatic.
i. Breeding migrations. Mating can occur from late January to mid-April in northeast

Arkansas (Butterfield, 1988); from January to mid-May in Oklahoma (Bragg, 1942); and from late February to mid-May in Illinois (Taubert et al., 1982; Brandon and Ballard, 1998). Breeding activity is higher following heavy rains (Bragg, 1942).

ii. Breeding habitat. In Oklahoma, breeding sites include sloughs, muddy cattle tanks, flooded ditches, flooded fields, and temporary pools in prairies (Bragg, 1942). In Illinois, the most common breeding sites are flooded depressions in fields, although a variety of sites are utilized (Brown and Rose, 1988).

B. Eggs.

i. Egg deposition sites. Eggs are attached to sticks, twigs, or vegetation just under the water surface (Bragg, 1942; Taubert et al., 1982; Butterfield et al., 1989). The egg complement is deposited in 10–100 egg masses (Bragg, 1942). Egg masses become obscured with a coating of silt and debris shortly after oviposition (Bragg, 1942; Tucker, 1997a).

ii. Clutch size. Females produced an average of 469 eggs (range 148–1,012) in Arkansas (Butterfield et al., 1989) and 608 eggs (range 411–783) in Illinois (Tucker, 1997a). The average number of eggs/mass was 41 (range 8–79) in Arkansas (Butterfield et al., 1989) and 22 (range 8–42) in Illinois (Tucker, 1997a). Average egg diameter was 1.24 mm (range 1.17–1.29 mm) in Oklahoma (Bragg, 1942) and 2.3 mm (range 1.9–2.6 mm) in Arkansas (Butterfield et al., 1989).

C. Larvae/Metamorphosis.

i. Length of larval stage. About 2 mo in Oklahoma (Bragg, 1942), about 35–50 d in Illinois (Taubert et al., 1982), and up to 60 d in Missouri (Johnson, 1987).

ii. Larval requirements.

a. Food. Algae (Bragg, 1942; see also McCallum and Trauth, 2001).

b. Cover. Vegetation and debris (Bragg, 1942). The formation of "tadpole nests" due to feeding activities may help prevent desiccation (Black, 1975).

iii. Larval polymorphisms. None.

iv. Features of metamorphosis. In Oklahoma, 18 February is the earliest date that embryos have been observed (Bragg, 1942). Transformation occurs at 12–13.5 mm SVL in Texas (Wright and Wright, 1949) and a mean of 19 mm SVL (range 16–21 mm) in Illinois (Tucker, 1997b).

v. Post-metamorphic migrations. Individuals migrating from a natal pond traveled 0.1 km in a mean time of 46.4 hr (Tucker, 1995). Mean distance moved between years for newly transformed frogs was 0.52 km (range 0–0.9 km) in Illinois (Tucker, 1998).

D. Juvenile Habitat. Same as adult habitat.

E. Adult Habitat. Found in prairie areas, cultivated fields, marshy vacant lots, wooded floodplains, and flatwoods (Smith, 1966b). Illinois chorus frogs are closely associated with sand prairies (Smith, 1966b) or restricted to other areas with a

sand substrate (Brown and Brown, 1973; Axtell and Haskell, 1977). Much of the year is spent underground in burrows excavated using the forelimbs (Brown et al., 1972), in areas devoid of or lacking heavy vegetation (Axtell and Haskell, 1977; Tucker et al., 1995).

F. Home Range Size. Unknown.

G. Territories. Unknown.

H. Aestivation/Avoiding Desiccation. Unknown.

I. Seasonal Migrations. Unknown.

J. Torpor (Hibernation). Occur underground in burrows, presumably at depths below the frost line (Packard et al., 1998), although frost injuries have been reported (Tucker, 2000a).

K. Interspecific Associations/Exclusions. Unknown.

L. Age/Size at Reproductive Maturity. On average, females mature at 39 mm SVL (range 37–40 mm) and males at 38 mm SVL (range 35–41 mm) in Arkansas (Butterfield, 1988; Butterfield et al., 1989). Females mature at an average of 39.8 mm SVL (range 37–43 mm) and males at 36.4 mm SVL (range 32–39 mm) in Illinois (Tucker, 1995). Individuals of both sexes are able to reach sexual maturity in one growing season (Tucker, 1995).

M. Longevity. A minimum of 2.5 yr (Jameson, 1956a) in Texas and 4 yr in Illinois (Tucker, 2000b). Annual survivorship in Illinois was estimated to be 26% for adults (Tucker, 2000b) and 2.8% for juveniles (Tucker, 1998).

N. Feeding Behavior. Prey consists primarily of invertebrates including Araneae, Hemiptera, Coleoptera, Diptera, Lepidoptera, and Hymenoptera (Tucker, 1997c). Brown (L.E., 1978) showed that Illinois chorus frogs are capable of feeding underground.

O. Predators. Larvae: predatory fishes (Tucker and Philipp, 1993) and aquatic insects including dragonfly larvae (D.B.S., personal observations). Adults: possibly hogs (Brown and Rose, 1988), hog-nosed snakes (*Heterodon* sp.), and short-tailed shrews (*Blarina brevicauda*; Murphy, 1979).

P. Anti-Predator Mechanisms. In general, the subterranean lifestyle and early breeding season limit exposure to predators. Tiny particles of soil and algae stick to the jelly of the egg mass protecting it from detection (Bragg, 1942). Tadpoles swim to the bottom and hide under debris when approached, while metamorphosing and recently metamorphosed young escape to the water (Bragg, 1942).

Q. Diseases. Unknown.

R. Parasites. Include the protozoans *Opalina* sp., *Nyctotherus cordiformis*, and *Myxidium serotinum* in tadpoles and adults (McAllister, 1987), and *Eimeria flexuosa*, *Eimeria streckeri*, and *Isospora delicatus* in adults (Upton and McAllister, 1988); also the cestodes *Mesocestoides* sp. and the

nemotodes *Oswaldocruzia* sp. in adults (McAllister, 1987).

4. Conservation.
Strecker's chorus frogs (*P.s. streckeri*) are listed as Threatened in Kansas and a Species of Special Concern in Arkansas and Louisiana. Illinois chorus frogs (*P.s. illinoensis*) are listed as Threatened in Illinois, Rare in Missouri, and a Species of Special Concern in Arkansas. Additionally, Illinois chorus frogs have been designated as a Category 2 taxon by the U.S. Fish and Wildlife Service (Dodd et al., 1985).

The primary cause of the decline of Illinois chorus frogs has been drainage and conversion of wetlands to agriculture. Besides habitat loss, practices associated with agriculture such as tilling, soil compaction from heavy machinery, and use of herbicides and pesticides may also have a negative impact (Brown and Rose, 1988). Many populations are apparently able to persist in disturbed habitats, but others are continuing to decline. In Madison County, Illinois, the range of Illinois chorus frogs has contracted from that of the 1970s (Tucker, 1998). Because most choruses consist of <10 males, it appears that population sizes are typically small and therefore may be prone to local extirpation. Because of low survivorship and extensive post-metamorphic dispersal (Tucker, 2000b), conservation efforts should focus on increasing the number of metamorphosing frogs and establishing buffer zones of at least 1 km around breeding sites while managing the vegetation within these areas to prevent thickening or sod formation (Tucker et al., 1995; Brandon and Ballard, 1998; Tucker, 1998, 2000b).

Pseudacris triseriata (Wied-Neuwied, 1838)
WESTERN CHORUS FROG

See the *Pseudacris triseriata* complex account.

Pseudacris triseriata complex (including *feriarum*, *kalmi*, *triseriata*, and *maculata*)
STRIPED (UPLAND, NEW JERSEY, WESTERN, BOREAL) CHORUS FROG
Emily Moriarty, Michael J. Lannoo

1. Historical versus Current Distribution.
Striped chorus frogs (*Pseudacris triseriata* complex) are among the most broadly distributed anurans in North America. These frogs range from central Arizona to northern Canada, to northern New York, and to the Florida Panhandle (Conant and Collins, 1998). The four "subspecies" of the *triseriata* complex that are currently recognized by Conant and Collins (1998) were initially characterized based on color pattern and morphological variation across their geographic distribution (Smith and Smith, 1952; Harper, 1955; Smith, 1956). Platz and Forester (1988) and Platz (1989)

elevated the "subspecies" to specific status using morphometric data and advertisement call characters. However, recent phylogenetic analyses based on DNA sequence data indicate that the range borders of subspecies in the *triseriata* complex, as currently drawn, do not reflect the distributions of true evolutionary lineages (Moriarty and Cannatella, 2004). Preliminary results suggest that the Mississippi River forms a barrier to gene flow between western and eastern *triseriata* populations. Additionally, southeastern *triseriata* populations are more closely related to *P. nigrita* (southern chorus frogs) than to northeastern *triseriata* populations. The geographic limits of these lineages must be elucidated by broader population sampling.

Striped Chorus Frog Complex *(Pseudacris triseriata Complex)*

2. Historical versus Current Abundance.
Generally unknown. Estimates of population size can only be made during late winter and early spring, when frogs gather in breeding pools. In 1948, Jacobs (1950) found striped chorus frogs to be "breeding abundantly along the [north] shore of Lake Superior wherever I made collections." Carpenter (1953) noted that striped chorus frogs were "more abundant in this [Grand Teton-Jackson Hole, Wyoming] area than I had suspected." Hudson (1954) writes: "The Unami Valley [Pennsylvania] is one of the few scattered localities near Philadelphia where the [striped] chorus frog has not been exterminated." Corn et al. (1989) reported a substantial decline in a Colorado population monitored from 1961–88, but found that in a broader region, including this population, which was monitored from 1986–88, striped chorus frogs did not appear to be in decline. There is evidence that striped chorus frogs are declining in the northeastern United States and southeastern Canada (Schueler, 1999; Weeber and Vallianatos, 2000).

3. Life History Features.
A. Breeding. Reproduction is aquatic.

i. Breeding migrations. From shallow underground overwintering sites to wetlands where mating and egg laying takes place (Whitaker, 1971; E.C.M., personal observations). During times of peak breeding activity, males call vigorously both day and night (Landreth and Ferguson, 1966). Breeding may occur in partially frozen ponds, for example in April along the north shore of Lake Superior (Jacobs, 1950). Timing of breeding varies by latitude, but spring breeding begins in the southern part of the range in January and is mostly finished by May–June in the northernmost populations (Bragg, 1948; Livezey, 1952; Whitaker, 1971; E.C.M., personal observations). Kramer (1973) observed Indiana populations of striped chorus frogs at breeding pools from late March to early August. Matthews and Pettus (1966) report that striped chorus frogs at higher elevations in Colorado breed following the spring thaw beginning in late May to early June. Migrations to breeding sites are typically during or following torrential rains (Landreth and Ferguson, 1966). Males can be found calling from the middle of clumps of emergent vegetation or from shallow water (Lord and Davis, 1956; Landreth and Ferguson, 1966). Landreth and Ferguson (1966) suggest that striped chorus frogs use both auditory and visual cues to locate potential breeding sites. Following the mating season, Kramer (1973) recaptured marked adults over 200 m from breeding pools, with the majority of recaptures occurring within 100 m of the pools.

ii. Breeding habitat. From seasonal to permanent bodies of water. During the 1–3 mo of the spring when striped chorus frogs are active, they are usually ubiquitous, calling from nearly any available aquatic habitat. Typical breeding sites are shallow temporary or semipermanent ditches or pools with few or no fish predators (Bragg, 1948; Livezey, 1952; Lord and Davis, 1956; Whitaker, 1971; Skelly, 1996; E.C.M., personal observations). Jacobs (1950) describes the breeding habitat along the north shore of Lake Superior as water-filled depressions in the "rugged Archean rocks" from 1 m to >6 m in diameter, and from a few to about 60 cm in depth, only 1–15 m above Lake Superior. These pools are unusual in that they did not contain macrophytic vegetation. On Isle Royale, striped chorus frogs breed on rocky shorelines in small pools exposed to storm waves from Lake Superior (Smith, 1983). Striped chorus frogs breed throughout the Prairie Pothole Region of the northern Great Plains, primarily in fishless areas (M.J.L., personal observations). For Ontario striped chorus frog populations, Hecnar and Hecnar (1999) observed egg masses in temporary ponds in savanna habitat.

B. Eggs.

i. Egg deposition sites. Most spawning occurs at night on grassy stems, twigs, and leaves in ephemeral wetlands. Hatching occurs after approximately 1–2 wk (Gosner and Rossman, 1959; Whitaker, 1971; Collins, 1982; Heinrich, 1985; Hecnar and Hecnar, 1999). Eggs discovered in late April by Hudson (1954), working in Pennsylvania, were in the process of hatching. In Jackson County, Illinois, Gosner and Rossman (1959) collected breeding animals in February and March. These wetlands were small (15–20 m), shallow (<60 cm), and partly bordered by pin oak (*Quercus palustris*), river birch (*Betulus nigra*), sycamore (*Plantanus occidentalis*), cottonwoods (*Populus deltoides*), and willows (*Salix,* sp.).

ii. Clutch size. Eggs numbers/female vary geographically, from 500–800 in western New York to 1,081–1,459 in Oklahoma (Wright, 1914; Gosner and Rossman, 1959; Whitaker, 1971). In Illinois populations, egg size averages 3.4 mm (Gosner and Rossman, 1959). Eggs typically are laid 5–10 cm below the surface in water 15–50 cm deep. Following breeding, immature eggs (unpigmented oocytes to pigmented ova) remain in the ovaries. Eggs hatch in 4–46 d, depending on water temperature, and newly hatched tadpoles are 4.8–6.1 mm SVL (Whitaker, 1971).

C. Larvae/Metamorphosis.

i. Length of larval stage. From 6–13 wk after hatching (Jacobs, 1950; Whitaker, 1971; Smith, 1983). In a controlled setting, both Smith (1983) and Heinrich (1985) observed the first stages of metamorphosis (appearance of appendages) within 7–8 wk of hatching. Within 11 wk post-hatching, Heinrich (1985) observed completely metamorphosed frogs.

ii. Larval requirements.

a. Food. Tadpoles are suspension feeders, ingesting a variety of organic and inorganic material associated with substrates such as rocks, plant leaves, and wetland bottoms (Whitaker, 1971; Altig and Kelly, 1974). Britson and Kissell (1996) noted that in Tennessee, striped chorus frog tadpoles co-occur with shortleaf pine (*Pinus echinata*) and southern red oak (*Quercus falcata*) pollen, a potential food source. However, their experiments demonstrate that these pollens had a detrimental effect on normal larval development.

b. Cover. Smith (1983) observed that tadpoles tend to concentrate in warm, shallow water. Larvae may escape to leaf litter or vegetation in deeper water when startled by predators (E.C.M., personal observations).

iii. Larval polymorphisms. Do not occur.

iv. Features of metamorphosis. Metamorphosis takes place over the course of at least 2–3 wk and perhaps longer (Whitaker, 1971). Newly metamorphosed animals in the Elk Mountains of Colorado ranged from 7–8 mm SVL in one population, 10–12 mm in another (Blair, 1951a). Metamorphosing individuals may be more susceptible to garter snake (*Thamnophis* sp.) predation than tadpoles or fully metamorphosed juveniles (see also "Predators" below; Wassersug and Sperry, 1977).

v. Post-metamorphic migrations. Mass migrations do not occur. Newly metamorphosed animals will migrate away from the immediate proximity of the wetland, presumably to upland feeding areas (E. C. M., personal observations).

D. Juvenile Habitat. Juveniles feed on soil algae, springtails (collembolans), mites (arachnids), ants (hymenopterans), and other insects (see Whitaker, 1971, table 3). Whitaker (1971) notes growth rates for newly metamorphosed animals. High tadpole densities increase food competition, thereby limiting growth rates and increasing the length of larval development. This may result in decreased survivorship to metamorphosis (Smith, 1983).

E. Adult Habitat. Smith (1961) described striped chorus frogs in Illinois as "essentially a forest animal" found in flood plains and mesic woods—a bottomland woods species (see also Alexander, 1965; Whitaker, 1971). Outside of the breeding season, striped chorus frogs may be found hidden in leaf litter, among dead vegetation, in cracks in the ground, under logs, in crayfish burrows, or under woody debris (Kramer, 1973). During the day, these frogs are often difficult to locate because they are well camouflaged (Lord and Davis, 1956). Striped chorus frogs exhibit nocturnal activity during the non-breeding season. However, in higher, cooler environments these frogs may be more diurnally active (Matthews and Pettus, 1966; Kramer, 1973). In November 1950, Blair (1951b) observed terrestrial activity of frogs and found individuals in water during the day (temperatures ~10 °C) as well as individuals calling at dusk (temperatures near 0 °C) in Oklahoma populations.

F. Home Range Size. Kramer (1973) recaptured marked individuals over 200 m from their breeding sites with the majority of recaptures occurring within 100 m of these pools. The minimum area home range of nine male frogs tracked for at least 2.5 mo was estimated at 641–6,024 m² (Kramer, 1974).

G. Territories. Generally unknown. During the breeding season, Roble (1985) observed several instances of conspecific, satellite males associated with calling males; no agonistic behavior was noted.

H. Aestivation/Avoiding Desiccation. Adults are found during the daytime under terrestrial cover objects in areas that are moister than ambient conditions (Kramer, 1973).

I. Seasonal Migrations. Aside from migrations to and from breeding wetlands, other migrations (e.g., to hibernation sites) are unknown and unlikely.

J. Torpor (Hibernation). Striped chorus frogs overwinter under leaves, logs, tree roots, or rocks, usually also under a layer of snow (Whitaker, 1971; Froom, 1982; Storey and Storey, 1987). Storey and Storey (1987) have shown that frogs from northern populations produce and accumulate the cryoprotectant glucose within their cells as a result of catabolism of liver glycogen reserves. The cryoprotectant stabilizes cellular structure and function and limits dehydration while the animals are in a frozen state. Respiration, blood flow, and heartbeat are temporarily suspended while frozen. Following emergence in spring, striped chorus frogs are able to withstand subsequent, short bouts of sub-zero temperatures, but freeze tolerance and the amount of cryoprotectant produced appears to decrease rapidly as spring proceeds (Storey and Storey, 1986, 1987).

K. Interspecific Associations/Exclusions. Striped chorus frogs will breed in association with eastern newts (*Notophthalmus viridescens*), blue-spotted salamanders (*Ambystoma laterale*), Arizona tiger salamanders (*A. tigrinum nebulosum*), American toads (*B. americanus*), western toads (*B. boreas*), Great Plains toads (*B. cognatus*), Woodhouse's toads (*B. woodhousii*), plains spadefoot toads (*Spea bombifrons*), western narrow-mouthed toads (*Gastrophryne olivacea*), wood frogs (*Rana sylvatica*), northern leopard frogs (*R. pipiens*), American bullfrogs (*R. catesbeiana*), crawfish frogs (*R. areolata*), green frogs (*R. clamitans*), mink frogs (*R. septentrionalis*), pickerel frogs (*R. palustris*), and spring peepers (*P. crucifer*), as well as most other chorus frog species (Jacobs, 1950; Blair, 1951a; Whitaker, 1971; Smith, 1983; Mitchell, 1990; Skelly, 1996; E.C.M., personal observations). In breeding ponds, striped chorus frogs and spring peepers exhibit spatial segregation (Whitaker, 1971).

L. Age/Size at Reproductive Maturity. First or second year following metamorphosis (Smith, 1987).

M. Longevity. Unknown. However, for southern chorus frogs (*P. nigrita*), a closely related species, Caldwell (1987) found that lifespan is approximately 1–3 yr.

N. Feeding Behavior. Food items vary seasonally and include snails, arachnids, and a wide variety of insects (see lists in Whitaker, 1971, tables 4, 6; Christian, 1982). In a laboratory setting, chorus frogs ate wax worms (*Galeria* sp.) and meal worms (*Tenebrio* sp.; Matthews and Pettus, 1966).

O. Predators. Documented predators on adult frogs include fishing spiders (*Dolomedes* sp.), robins (*Turdus migratorius*), gray jays (*Perisoreus canadensis*), and garter snakes (*Thamnophis* sp.; Matthews and Pettus, 1966; Wassersug and Sperry, 1977; Smith, 1983; Mitchell, 1990). Adult dytiscid beetles (*Rhantus binotatus, Dytiscus,* sp.) and dragonfly naiads (*Anax junius*) prey upon chorus frog tadpoles (Smith, 1983; Skelly, 1996). Other probable predators include various fishes and water snakes (*Neroidia,* sp.; Whitaker, 1971). Metamorphosing individuals may be more susceptible to garter snake predation than tadpoles or fully metamorphosed juveniles due to hindrance of locomotion by their intermediate body form (Wassersug and Sperry, 1977).

P. Anti-Predator Mechanisms. Animals are cryptic and small, which reduces their visual conspicuousness to predators. Metamorphosing striped chorus frogs tend to emerge at night, allowing them to avoid predation by diurnal predators (Wassersug and Sperry, 1977). Striped chorus frogs may swim into deeper water to evade land predators or escape to shore to avoid aquatic predators (Landreth and Ferguson, 1966). Although not strictly an anti-predator mechanism, striped chorus frogs tend not to breed in permanent bodies of water (see Lord and Davis, 1956; Skelly, 1996), which may help them avoid potential predators more frequently found in permanent water, such as fish, salamander larvae, and dragonfly larvae (Skelly, 1996).

Q. Diseases. Unknown.

R. Parasites. Whitaker (1971) examined the stomachs and intestines of 712 male chorus frogs and found trematodes (*Glypthelmins* sp.) and nematodes (*Aplectana* sp. and *Oswaldocruzia* sp.) in almost 63% (Whitaker, 1971). Of 120 female frogs, 60% also had these parasites. Ubelaker et al. (1967) also reported the trematode *Glypthelmins pennsylvaniensis* in Colorado populations. Snail hosts for these parasites include members of the genus *Helisoma* (Whitaker, 1971).

4. Conservation.

Striped chorus frogs are afforded legal protection as follows. They are listed as Protected in New Jersey (as *P. t. feriarum* and *P. t. kalmi*) and Endangered in Pennsylvania (as *P. t. kalmi*) and Vermont (as *P. t. triseriata*; Levell, 1997). There is evidence that declines are occurring in portions of their range, especially in the northeastern and western United States, but that other populations are stable. Striped chorus frogs appear relatively tolerant of human activities, although they may be susceptible to certain agricultural chemicals. Sanders (1970) found that the pesticide endrin was highly toxic to striped chorus frog tadpoles and may affect populations in agricultural areas.

Pternohyla fodiens Boulenger, 1882
LOWLAND BURROWING TREEFROG

Michael J. Sredl

1. Historical versus Current Distribution.
Lowland burrowing treefrogs (*Pternohyla fodiens*) are found from the extreme southwestern United States southward through western Sonora to Michoacán, at sites ranging in elevation from sea level to 1,500 m (Trueb, 1969; Duellman, 1970; Stebbins, 1985). Distribution may be restricted to regions where soil temperatures do not drop below freezing (Ruibal and Hillman, 1981).

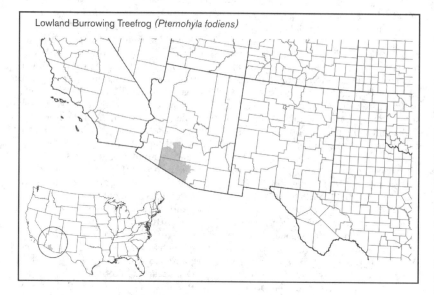

Lowland Burrowing Treefrog (*Pternohyla fodiens*)

The distribution of lowland burrowing treefrogs in the United States is restricted; they are only known to occur in washes associated with the Quijota, Gu Oidak, Santa Rosa, and Vekol Valleys in Pima, Pinal, and Maricopa counties, Arizona (Sullivan et al., 1996b; Enderson and Bezy, 2000).

Because lowland burrowing treefrogs were not discovered in the United States until 1957 (Chrapliwy and Williams, 1957), very little is known of their historical distribution. Sullivan et al. (1996b) surveyed Arizona populations and found them to be present at most historical localities.

2. Historical versus Current Abundance.
No studies of direct counts of breeding aggregations or population estimates have

been conducted, therefore, little is known of historical or current abundance. Sullivan and Bowker (unpublished, in Sullivan et al., 1996b) have observed large aggregations at many Arizona localities over a 30-yr period.

3. Life History Features.
A. Breeding. Reproduction is aquatic.

i. Breeding migrations. Lowland burrowing treefrogs breed June–September, coincident with the summer rainy season (Webb, 1963; Duellman, 1970; Stebbins, 1985). They are extremely explosive breeders (Sullivan et al., 1996b). Male lowland burrowing treefrogs have been observed to call while fully exposed on mud flats, from sparse cover such as the base of bushes or from secluded places such as under rocks or in clumps of grass (Webb, 1963; Hardy and McDiarmid, 1969; Duellman, 1970; Sullivan et al., 1996b; personal observation). The call of lowland burrowing treefrogs has been described as consisting of a series of low-pitched notes, resembling the quacking of a duck (Duellman, 1970). Male lowland burrowing

treefrogs have been noted to call in duets (Bogert, 1958). Choruses form and disband quickly and usually do not last longer than 36 hr after rainfall (Sullivan et al., 1996b).

ii. Breeding habitat. Lowland burrowing treefrogs breed in rain-filled temporary pools that form in washes, depressions, and along roads and in earthen cattle tanks (Chrapliwy and Williams, 1957; Webb, 1963; Sullivan et al., 1996b; personal observations). Vegetation in the vicinity of breeding ponds can be dense (Allen, 1933; personal observations). Air and water temperatures at a temporary pond in Nayarit, Mexico, where lowland leopard frogs *(Rana yavapaiensis)* were breeding, were 24 °C (Porter, 1962).

B. Eggs.

i. Egg deposition sites. Few details on oviposition sites have been reported. Hardy and McDiarmid (1969) observed that jelly envelopes and eggs were scattered about the bottom of the pond.

ii. Clutch size. Has not been reported.

C. Larvae/Metamorphosis.

i. Length of larval stage. Altig et al. (1998) described the tadpoles of lowland burrowing treefrogs. Larvae hatch at about 10 mm TL (Hardy and McDiarmid, 1969) and reach a maximum size of between 45–50 mm TL (Webb, 1963).

ii. Larval requirements. Little is known of general larval habitat characteristics of lowland burrowing treefrogs.

a. Food. Observations on tadpole diet or feeding behavior are few. Hardy and McDiarmid (1969) observed tadpoles clustered at the surface that may have been feeding on surface scum.

b. Cover. Larvae have been observed floating in clusters at the surface of the water, with their tails directed downward. When a cluster was disturbed, larvae would disperse, some swimming away and others sinking to the bottom (Hardy and McDiarmid, 1969).

iii. Larval polymorphisms. None reported.

iv. Features of metamorphosis. Lowland burrowing treefrogs metamorphose at 18–24 mm SUL (Webb, 1963).

v. Post-metamorphic migrations. Webb (1963) noted that post-metamorphic individuals were abundant and moving over land day and night near a pond in Sinaloa.

D. Juvenile Habitat. Little is known of juvenile habitat characteristics of lowland burrowing treefrogs. At metamorphosis, lowland burrowing treefrogs are greenish and resemble Arizona treefrogs (*Hyla wrightorum*; Sullivan et al., 1994).

E. Adult Habitat. Little is known of habitat characteristics and ecology of wild adults. Trueb (1969) and Duellman (1970) reviewed the morphological adaptations to a fossorial existence. Green (1979) examined the functional significance of toes and toe pads of lowland burrowing treefrogs and other treefrogs. Brattstrom (1968) reviewed the thermal ecology of a variety of anurans, including lowland burrowing treefrogs. Taigen et al. (1982) and Taigen (1983) reviewed the comparative physiology of lowland burrowing treefrogs and other anurans.

F. Home Range Size. Unknown.

G. Territories. Males produce a call that sounds similar to the territorial calls of other hylids (Sullivan et al., 1996b). However, it is unknown whether males establish territories.

H. Aestivation/Avoiding Desiccation. The fossorial habits of lowland burrowing treefrogs, among other things, aid in reducing evaporative water loss both in the short and long term. Excavation of temporary, shallow burrows is one method they use to reduce water loss in the short term. Captive lowland burrowing treefrogs have been observed to dig shallow burrows with their metatarsal tubercles (Firschein, 1951a).

Another method lowland burrowing treefrogs use to reduce water loss in the short term is by sealing themselves in an existing burrow or crevice. This species has been observed to assume an immobile position in which they lower their head, arch their back, and raise their limbs (Firschein, 1951a). This reaction and subsequent immobile posture has been called an "unken reflex" in other amphibians (Noble, 1931) and has been considered an anti-predator mechanism that exposes warning colors to potential predators. Lowland burrowing treefrogs do not possess warning colors, and they likely use this posture to seal themselves in a preexisting burrow or crevice using their skull. In addition to preventing intruders from entering (Firschein, 1951a), this entombment may also reduce evaporative water loss.

It is unknown whether lowland burrowing treefrogs primarily dig their own burrows or use preexisting ones to reduce evaporative water loss or avoid predation. Some think the metatarsal tubercles make effective digging tools (Ruibal and Hillman, 1981), while others do not (Firschein, 1951a).

An adaptation to reduce evaporative water loss during long dry periods is the ability to excavate shallow burrows and form an epidermal cocoon (Ruibal and Hillman, 1981). This cocoon consists of multiple shedding of the upper layers of the skin interspersed with mucus and has been measured to reduce the rate of water loss to 0.6 mg g^{-1} hr^{-1} (Ruibal and Hillman, 1981). During cocoon formation, lowland burrowing treefrogs remain motionless, with their limbs folded tightly against their body. When cocoon formation is complete, the frogs do not move and remain at the depth to which they excavated. Because the burrows they dig are relatively shallow, lowland burrowing treefrogs may be restricted to subtropical and tropical habitats, where soil temperatures do not drop below freezing (Ruibal and Hillman, 1981).

I. Seasonal Migrations. Unknown.

J. Torpor (Hibernation). See "Aestivation/Avoiding Desiccation" above.

K. Interspecific Associations/Exclusions. Lowland burrowing treefrogs have been reported to occur with numerous temporary pond breeding amphibians, including Couch's spadefoot toads (*Scaphiopus couchii*), Colorado River toads (*Bufo alvarius*), Great Plains toads (*B. cognatus*), Sonoran green toads (*B. retiformis*), redspotted toads (*B. punctatus*), and western narrow-mouthed toads (*Gastrophryne olivacea*; Chrapliwy and Williams, 1957; Sullivan et al., 1996b; personal observations).

L. Age/Size at Reproductive Maturity. Males look similar to females but possess a dark patch on each side of the throat. The maximum SUL to be recorded for males is 62.6 mm, while that of females is 63.7 mm (Duellman, 1970).

M. Longevity. Nothing is known of longevity in the wild. A captive lowland burrowing treefrog lived 5 yr, 1 mo (Bowler, 1977).

N. Feeding Behavior. In Jalisco, Mexico, Firschein (1951a) noted that lowland burrowing treefrogs wander in fields in nightly forays for food. Little else has been noted of their diet or feeding behavior.

O. Predators. Little is known of predators of lowland burrowing treefrogs in the United States.

P. Anti-Predator Mechanisms. On open ground, adult lowland burrowing treefrogs are easily captured (Allen, 1933; Chrapliwy and Williams, 1957; Webb, 1963; personal observations), although this experience is not universal (Taylor, 1936). Davis and Dixon (1957) speculate that these animals are unwary at the onset of breeding season, but become exceedingly difficult to catch later in the breeding season. Their habit of sealing themselves in a burrow using their skull and "unken reflex" prevents intruders from entering a burrow, but it is ineffective outside its burrow in deterring attack or preventing predation (Firschein, 1951a).

Q. Diseases. Unknown.

R. Parasites. Lowland burrowing treefrogs have been found to be infected with five species of nematodes (*Aplectana itzocanensis, Cosmocercella haberi, Rhabdias americanus, Physaloptera* sp., *Skrjabinoptera* sp.) and one species of cestode (*Distoichometra bufonis*; Goldberg et al., 1999).

4. Conservation.
Lowland burrowing treefrogs are not offered legal protection in the United States. Because they were not discovered in the United States until 1957, very little is known of their historical distribution. Lowland burrowing treefrogs seem to be present at most known historical localities (Sullivan et al., 1996b), suggesting widespread declines have not occurred. However, their U.S. distribution is restricted, making them vulnerable.

Smilisca baudinii (Duméril and Bibron, 1841)
MEXICAN TREEFROG

John H. Malone

1. Historical versus Current Distribution.
In the United States, Mexican treefrogs (*Smilisca baudinii*) are mainly found in Cameron and Hidalgo counties in the extreme southern tip of Texas (Dixon, 1987, 2000; Conant and Collins, 1998). A specimen exists from Bexar County (Murphy and Drewes, 1976), and Mexican treefrogs have been reported from Refugio County (Strecker, 1908b). Both these extraneous locales are widely separated from extreme southern Texas. These disparate records may be the result of accidental introductions due to human transport of potted plants from southern Texas (Dixon, 2000). However, Wright and Wright (1949) report that Mexican treefrogs have "been found by Mr. Marnock in the low country southwest of San Antonio, commencing with San Miguel Creek, a tributary of the Medina." This report is confirmed by a specimen record (Murphy and Drewes, 1976). These reports differ from Raun and Gelbauch (1972) who state that both the Bexar (but see Murphy and Drewes, 1976) and Refugio records are erroneous. Garrett and Barker (1987) believe that the Bexar County population may no longer exist, and no additional data exist for the Refugio County record. Texas represents the northern extreme of the Mexican treefrog distribution that extends south to Costa Rica. Mexican treefrogs inhabit lowlands and foothills below 2,000 m, but most localities are below 1,000 m (Duellman and Trueb, 1966; Duellman, 1968, 2001).

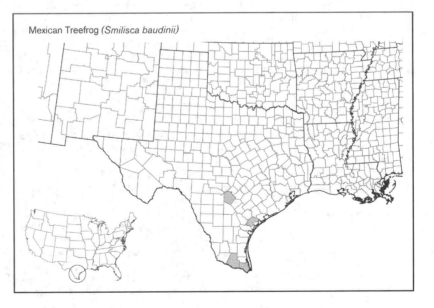

Mexican Treefrog (*Smilisca baudinii*)

2. Historical versus Current Abundance.

Mexican treefrogs are listed as Threatened in Texas but have no federal status. In Texas, Mexican treefrogs are thought to consist of small, patchy populations (G. Stolz, Santa Ana Wildlife Refuge, personal communication), but to date no formal census has been conducted to assess population status or viability. In the early 1990s, Kelly Irwin (Arkansas Game and Fish Commission, personal communication) performed an amphibian and reptile inventory of the Lower Rio Grande Valley and noted the presence of Mexican treefrogs at sites near Santa Ana National Wildlife Refuge and the Sabal Palm Grove Sanctuary. Following a 75 mm rain on 17 June 1991, Kelly Irwin (personal communication) found approximately 10 males calling from a flooded ditch on the Santa Ana National Wildlife Refuge. Gadow (1908) estimated 45,000 Mexican treefrogs at a breeding site near Cordoba, Veracruz, Mexico. Lee (2000) calls Mexican treefrogs "the most abundant and ubiquitous amphibian in the Yucatán Peninsula." Stuart (1934) found a Mexican treefrog chorus so large that the weight of individuals bent the branches. Duellman (2001) reports observing several hundred to thousands of individuals at breeding sites in Mexico, Guatemala, and Costa Rica. Campbell (1998) calls Mexican treefrogs the most common treefrog species in the Petén region of Guatemala. Based on these reports, it appears that Mexican treefrogs are more abundant in parts of Central America than they are in Texas.

3. Life History Features.

A. Breeding. Reproduction is aquatic.

i. Breeding migrations. Mexican treefrogs breed any time of the year following adequate rainfall (Webb, 1971; Duellman, 2001). In 1992, Kelly Irwin (personal communication) noted Mexican treefrogs calling at Santa Ana National Wildlife Refuge from June–August. Mexican treefrogs typically call in duets, and the initial calls of a duet cause other duets to respond (Duellman, 1967). Males will call during the day (Webb, 1971). Amplexus is axillary.

ii. Breeding habitat. Males will vocalize from nearly any body of water, but usual breeding sites are small, temporary pools (Duellman, 2001).

B. Eggs. Eggs have a diameter of 1.3 mm and a vitelline membrane of 1.5 mm (Duellman and Trueb, 1966).

i. Egg deposition sites. Eggs are initially deposited as clusters and then disperse into a surface film. Webb (1971; see also his fig. 1) describes breeding sites as shallow pools that may or may not have vegetation along the shoreline. In Yucatán, Mexico, Maslin (1963a,b) observed tadpoles in small (60 cm in diameter, 25 cm deep) limestone basins, where eggs must have been deposited.

ii. Clutch size. Estimates ranged from 480–560 (Webb, 1971). Duellman and Trueb (1966) removed and counted ovulated eggs from three female Mexican treefrogs. They reported 2,620; 2,940; and 3,320 eggs/female; these clutch sizes are considerably larger than those reported by Webb (1971). Webb (1971) did not observe the total number of eggs deposited by each female and thus his clutch size observations may be conservative.

C. Larvae/Metamorphosis. Series of tadpoles were described by Duellman and Trueb (1966). Mexican treefrogs have a generalized hylid type tadpole. They are exotrophic with an anteroventral oral apparatus and a dextral anus. The color is a uniform brown with a pale dorsolateral stripe on the body and light banding on the dorsal tail musculature (Altig and McDiarmid, 1999).

i. Length of larval stage. Larval stage duration is apparently unknown, but probably is rapid. On 4 July 1959, Maslin (1963a,b) collected a series of 12 specimens in Gosner (1960) stages 29–36 ranging in size from 19–32 mm TL. Duellman and Trueb (1966) showed that Mexican treefrog tadpoles increase from 5–35 mm TL (from Gosner stages 21–38, respectively). As metamorphosis began and tails resorbed, total length decreased to 13 mm at Gosner stage 46.

ii. Larval requirements.

a. Food. Larvae use suspension feeding to ingest small organic and inorganic particles.

b. Cover. Larval habitat characteristics are unknown. Maslin (1963a,b) observed clouds of tadpoles in the small limestone basins he studied.

iii. Larval polymorphisms. Polymorphisms have not been reported, although Maslin (1963a,b) noted morphological differences between the tadpoles he studied and those studied by Stuart (1948). Duellman and Trueb (1966) noted that degree of pigmentation is variable and likely correlated with light levels. Additionally, length and depth of the tail are variable (Duellman and Trueb, 1966).

iv. Features of metamorphosis. Tadpoles metamorphose when total length (newly metamorphosed animals) is 13 mm (Duellman, 2001).

v. Post-metamorphic migrations. Migrations are probably from shallow, drying breeding wetlands to upland cover sites; however, data do not exist for juvenile dispersal.

D. Juvenile Habitat. Unknown.

E. Adult Habitat. Mexican treefrogs are nocturnal and most active following rains. In the arid and semiarid places where they are found, Mexican treefrogs inhabit forested and brushy areas around streams, resacas, and roadside ditches. Mexican treefrogs have been observed living in the tops of palm trees. Individuals seek shelter from heat and dry conditions under loose tree bark, in tree holes, in damp soil, and in the leaves of banana plants, bromeliads, and heliconias (Meyer and Wilson, 1971; Duellman, 2001).

F. Home Range Size. Unknown.

G. Territories. Unknown.

H. Aestivation/Avoiding Desiccation. Mexican treefrogs and lowland burrowing treefrogs *(Pternohyla fodiens)* are the only frogs found in the United States known to form cocoons (Ruibal and Hillman, 1981; McDiarmid and Foster, 1987).

I. Seasonal Migrations. Following rains, Mexican treefrogs move to and from wetlands to breed. These migrations can occur at any time of the year.

J. Torpor (Hibernation). Does not occur.

K. Interspecific Associations/Exclusions. Unknown.

L. Age/Size at Reproductive Maturity. Body size dimorphism exists between males (51–76 mm SVL) and females (≤90 mm; Duellman, 1968; Conant and Collins, 1998).

M. Longevity. Unknown.

N. Feeding Behavior. Mexican treefrogs feed on invertebrates, especially insects and spiders (Lee, 2000).

O. Predators. In Mexico, Duellman and Trueb (1966) observed the xenodontin snake *Leptodeira maculata* engulfing a male Mexican treefrog. Frog-eating snakes are likely to be important predators on both adults and juveniles. Webb (1971) suggested that predaceous aquatic insects feed on tadpoles, but he did not observe this.

P. Anti-Predator Mechanisms. Both males and females emit a distress vocalization that may serve to attract secondary predators (Duellman, 2001).

Q. Diseases. Unknown.

R. Parasites. The trematode *Mesocoelium monas* was found in the intestines of Mexican treefrogs from Los Tuxtlas, Veracruz, Mexico (Guillén-Hernández et al., 2000; Pérez-Ponce de Leon et al., 2000). Goldberg and Bursey (2002) examined 19 specimens (5 females and 14 males) of Mexican treefrogs from southern Sonora, Mexico, for helminth parasites. Small and large intestines of Mexican treefrogs were infected with the nematodes *Aplectana incerta* and *A. itzocamensis* (Goldberg and Bursey, 2002). The nematode *Rhabdias americanus* was found in the lungs, and cysts of *Physaloptera* sp. were found in the stomach (Goldberg and Bursey, 2002). The presence of *Physaloptera* sp. suggests that Mexican treefrogs may serve as a paratenic host for *Physaloptera* sp. (Goldberg and Bursey, 2002). Four individuals of Mexican treefrogs from the Guanacaste Conservation Area, Costa Rica were examined for blood parasites but none were found (Dresser, 2001).

4. Conservation.

Mexican treefrogs are a widely distributed species in Central America and are considered a common species. In the United

States, Mexican treefrogs are at the most northern limit of their distribution; specifically, they occur in three counties in southern Texas, but populations in Texas are considered to be small and patchy. Mexican treefrogs are listed as Threatened by the State of Texas (Levell, 1997) but no federal protection exists or has been proposed. Due to the paucity of data for U.S. populations of Mexican treefrogs, efforts should be made to identify remaining populations and remaining habitats.

Acknowledgments. Thanks to David Bradford and Janalee Caldwell, who commented on this manuscript.

Family Leptodactylidae

Eleutherodactylus augusti (Dugés, 1879)
BARKING FROG

Cecil R. Schwalbe, Caren S. Goldberg

1. Historical versus Current Distribution.

Barking frogs *(Eleutherodactylus augusti)* were first described from Guanajuato, Mexico, by Dugés in 1868 (field notes can be found in Brocchi, 1882). In the United States they were first described from southwestern Texas (Cope, 1878b). Most of the range of barking frogs is in Mexico, from Oaxaca up through the Sierra Madre Occidental and Oriental into the United States (Zweifel, 1967). Barking frogs are now known to range into southeastern Arizona, southeastern and south-central New Mexico, and western and central Texas (Bezy et al., 1966; Degenhardt et al., 1996; Dixon, 2000; Murray and Painter, 2003). In Arizona, barking frogs have been documented in the Santa Rita (Slevin, 1931), Pajarito (Bezy et al., 1966), Huachuca (Schwalbe et al., 1997), and Quinlan (Enderson, 2002) Mountains at elevations of 1,280–1,890 m. There is also report of a barking frog caught in the Sierra Ancha of central Arizona (Wright and Wright, 1949). In New Mexico, barking frogs are documented in Doña Ana, Chaves, Eddy, and Otero counties between elevations of 900–1,200 m (Degenhardt et al., 1996; Murray and Painter, 2003). In Texas, barking frogs are distributed along the Balcones Escarpment (Smith and Buechner, 1947) and are found in isolated populations throughout the western panhandle (Dixon, 2000).

Currently, barking frogs in Texas and New Mexico are members of the subspecies *E. a. latrans* and barking frogs in Arizona belong to the subspecies *E. a. cactorum* (Zweifel, 1956a), but there has been taxonomic confusion with this group. These subspecies were originally considered separate species (Taylor, 1938b), but Zweifel (1956a) combined them into a single species due to their lack of morphological distinctiveness. Before this analysis, barking frogs from Texas, New Mexico,

and Arizona were all considered to be members of the species *E. latrans* (Wright and Wright, 1949; Stebbins, 1951). Differences in call structure between barking frogs in Arizona and those in New Mexico and Texas suggest that *Eleutherodactylus augusti* may currently represent more than one species (C.G., B. Sullivan, J. Malone, and Goldberg et al., 2004a).

Barking frog fossils have been found in late Pleistocene deposits in a cave in Bexar County, Texas (Mecham, 1958), and in middle Holocene deposits in a limestone cave in Kerr County, Texas (Parmley, 1988b). A comparison of historical versus current distributions is not possible due to lack of information, but we are not aware of any reports of extirpations from former localities. For barking frog distribution in Mexico, see Hardy and McDiarmid (1969), Webb (1984), and references in Zweifel (1967).

2. Historical versus Current Abundance.
Abundance of barking frogs in a canyon in the Huachuca Mountains of southern Arizona has remained steady since 1993, based on call counts (1993–2002) and mark–recapture analysis (1996–2000; Goldberg and Schwalbe, 2000; C.G., unpublished data). No other information is available.

3. Life History Features.
A. Breeding. Reproduction is terrestrial. Barking frogs typically call from February–August during periods of rain in central Texas, with the breeding peak apparently taking place in April–May (Jameson, 1954). In New Mexico, barking frogs call from May–July (Radke, 1998); one individual, estimated to be a few days old, was found on 16 July 1991 (Degenhardt et al., 1996). In Arizona, barking frogs call for 2–4 wk on rainy nights after the start of the summer rains in June–July (Goldberg and Schwalbe, 2004a). A sonogram of a

Texas barking frog call can be found in Fouquette (1960). In Texas, gravid females were found in May (McAlister, 1954) and an egg clutch was found in April (Jameson, 1950b). In Arizona, gravid females have been found in June (Goldberg and Schwalbe, 2000; C.G., unpublished data).

i. Breeding migrations. In Arizona, barking frogs moved up to 50 m from overwintering to calling sites at the beginning of the breeding season (Goldberg and Schwalbe, 2000). There is some speculation that males may lead females to a predetermined nest site (Arizona Game and Fish Department, 1999c).

ii. Breeding habitat. Barking frogs usually call from rock fissures and crevices in the rock outcrops they occupy (Jameson, 1954; Schwalbe et al., 1997; Goldberg and Schwalbe, 2000). They have also been observed calling from under vegetation and in the open (B. Alberti, C.G., C.R.S., unpublished data).

B. Eggs.
i. Egg deposition sites. Females likely deposit eggs in moist or rain-filled cracks, fissures, and caves (Wright and Wright, 1949). Eggs may also be deposited in moist earth under rocks (Jameson, 1950b).

ii. Clutch size. Clutches contain from 50–76 eggs (Wright and Wright, 1949; McAlister, 1954; Degenhardt et al., 1996). One clutch contained eggs with diameters of 8–8.5 mm (Valett and Jameson, 1961).

C. Direct Development. Complete metamorphosis occurs within the egg; young hatch fully developed (Jameson, 1950b). Hatching is estimated to occur after 25–35 d of development in Texas (Jameson, 1950b); anecdotal evidence from Arizona suggests that one clutch may have hatched in 21 d (Goldberg and Schwalbe, 2000).

i. Brood sites. Unknown.

ii. Parental care. Jameson (1950b) hypothesized that male barking frogs guard

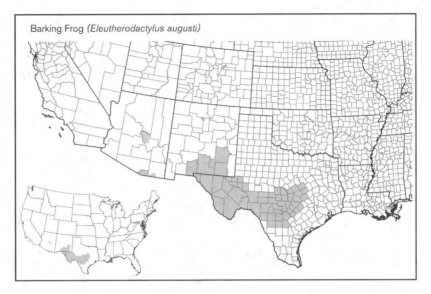

Barking Frog *(Eleutherodactylus augusti)*

the egg clutch and maintain moisture levels by excretion. However, radio-tracking data from Arizona suggest that males move too frequently to guard eggs and that females may stay with the clutch (Goldberg and Schwalbe, 2000).

D. Juvenile Habitat. Juveniles are found in the same general areas as adults (Strecker, 1933; Radke, 1998; C.R.S., unpublished data). However, immediately before and during onset of a 6 hr rainstorm, five hatchlings were found along a footpath at least 30 m from the nearest substantial rock outcrop in the Sierra de Alamos (Sonora, Mexico; C.R.S., personal observations).

E. Adult Habitat. Barking frogs are terrestrial and commonly found in or near cliffs, caves, and limestone or other rock outcrops in a variety of biotic communities (Smith and Buechner, 1947; Wright and Wright, 1949; Bezy et al., 1966; Reddell, 1971). In Arizona, barking frogs have been found on limestone, rhyolite, and other rock outcrops in Madrean evergreen woodland (Bezy et al., 1966; Goldberg and Schwalbe, 2004a). In New Mexico and west Texas, barking frogs are found in creosote bush flats near rodent burrows on gypsum soils (Degenhardt et al., 1996; Bartlett and Bartlett, 1999b). In central Texas, barking frogs are found in juniper-oak scrub forest (Blair, 1950) on or near limestone outcrops (Conant and Collins, 1998). In Mexico, barking frogs are found in a variety of biotic communities, from xeric, yucca-covered hills (Martin, 1958) to open pine forests (Duellman, 1961). Barking frogs have been found in caves and abandoned mines throughout their range (Reddell, 1971; Hubbard et al., 1979; Goldberg and Schwalbe, 2000; C.R.S., personal observations).

F. Home Range Size. In Arizona, the average 95% kernel home range size of 10 barking frogs was 1,087 m^2 (95% C.I. 518–2,327 m^2; range 207–5,498 m^2; Goldberg and Schwalbe, 2004b).

G. Territories. There is no indication that barking frogs are territorial in Arizona. We have observed two males calling from the same rock crevice simultaneously on at least two occasions (C.G., personal observations). Frogs with overlapping home ranges use the same crevices as daytime refugia, but not simultaneously (Goldberg and Schwalbe, 2000). At least four overwintering (aestivating) frogs have been found piled into a single crevice (C.G. and co-workers, unpublished data; see "Aestivation/Avoiding Desiccation" below).

H. Aestivation/Avoiding Desiccation. In Arizona, barking frogs have been observed to be active in an abandoned mine throughout the winter (dry season), except for the warm, wet summer, when they are surface active. Barking frogs in the Huachuca Mountains of southern Arizona leave winter (dry season) refugia just before the start of the summer rains and return from August–October (Goldberg

and Schwalbe, 2000). In Texas, barking frogs are active at any time of year when there is sufficient ground moisture (Jameson, 1954).

I. Seasonal Migrations. In Arizona, barking frogs moved from 12–50 m between summer activity areas and overwintering sites (Goldberg and Schwalbe, 2000).

J. Torpor (Hibernation). Unknown. Likely in more exposed microsites in northern portions of their range.

K. Interspecific Associations/Exclusions. In Texas, barking frogs have sympatric distributions and shared habitat characteristics with cliff chirping frogs (*E. marnockii*; Jameson, 1954). Barking frogs have also been found near Gulf Coast toads (*Bufo valliceps* [now considered to be Coastal-Plain toads, *B. nebulifer*; see Mulcahy and Mendelson, 2000; Mendelson, this volume]), cliff chirping frogs, and within the same cave as a western slimy salamander (*Plethodon albagula*; McAlister, 1954). In Mexico, barking frogs have been found near *Eleutherodactylus* (= *Syrrhophus*) *dennisi* (Martin, 1958), *E. vocalis* (Webb, 1960), and red-spotted toads (*Bufo punctatus*; C.R.S., unpublished data). Barking frogs and *E. hidalgoensis* may exclude each other from areas in Tamaulipas, Mexico (Martin, 1958).

L. Age/Size at Reproductive Maturity. Adult size ranges from 47–94 mm (Wright and Wright, 1949; Zweifel, 1956a; Anderson and Lidicker, 1963). Adult females in the Huachuca Mountains of Arizona were larger than males by an average of 7.1 mm (Goldberg and Schwalbe, 2000). Males in Arizona have dark tympana and adult males have dark throats during the calling season; females have pink tympana and white throats year-round (Goldberg and Schwalbe, 2000).

M. Longevity. Unknown, but the same adults in Arizona have been caught yearly for 7 yr (Goldberg and Schwalbe, 2004a; unpublished data). An adult from Sonora, Mexico, lived in captivity for 11 yr at the Arizona-Sonora Desert Museum.

N. Feeding Behavior. In the wild, barking frogs have been known to eat camel crickets (*Ceuthophilus* sp.), field crickets (*Acheta assimilis*), Gladston grasshoppers (*Melanoplus gladstoni*), longhorned katydids (Tettiganiidae), short-horned grasshoppers (probably Acrididae), land snails (*Bulimului* sp. and *Succinea* sp.), silverfish (*Lepisma saccharina*), centipedes (*Scolopendra* sp.), scorpions (*Vaejovis* sp.), kissing bugs (*Triatoma* sp.), spiders, and adult ant lions (*Hesperoleon niger*; McAlister, 1954; Olson, 1959; Schwalbe et al., 1997; Radke, 1998; Goldberg and Schwalbe, 2000). In captivity, barking frogs have eaten cave crickets (*Pholeogryllus geertsi*; Olson, 1959) and cliff chirping frogs (Jameson, 1955).

O. Predators. Unknown.

P. Anti-Predator Mechanisms. Upon capture, barking frogs will swell to in-

crease their size and avoid predation (Wright and Wright, 1949; McAlister, 1954; Martin, 1958). Barking frogs in Texas have skin secretions that are irritating to the eyes and to open cuts (Wright and Wright, 1949; McAlister, 1954), but this is apparently not the case with Arizona frogs (C.G., personal observations). Barking frog females from Texas and Puebla, Mexico, will screech when hand-captured (Taylor, 1938b; Jameson, 1954); females from Arizona do not display this behavior (unpublished data).

Q. Diseases. Unknown.
R. Parasites. Unknown.

4. Conservation.

Barking frogs have no special protection at the state or federal level. They are considered a Species of Special Concern in Arizona (Arizona Game and Fish Department, 1996a) which does not provide any legal protection. There is no evidence of decline in the only Arizona population studied (Goldberg and Schwalbe, 2000), but data are unavailable to assess status on a larger scale. The secretive habits of barking frogs make detection difficult; their distribution in Arizona is still largely unknown. Known barking frog localities in Arizona are on Forest Service and National Park Service properties. In New Mexico and Texas, barking frog populations are mostly located on private lands, but they are also found on state, Fish and Wildlife Service, and National Park Service properties (Degenhardt et al., 1996; Radke, 1998; Murray and Painter, 2003).

Eleutherodactylus coqui Thomas, 1966
COQUI

Margaret M. Stewart, Michael J. Lannoo

1. Historical versus Current Distribution.
Coquis (*Eleutherodactylus coqui*) are native to Puerto Rico where they are widespread (Thomas, 1966). They were introduced into southeastern Florida and southeastern Louisiana (Schwartz and Henderson, 1991; Conant and Collins, 1998). In Florida, coquis have been documented only from Dade County (Bartlett and Bartlett, 1999a). In the late 1990s, coquis and their congener, greenhouse frogs (*E. planirostris*), native to Cuba but successfully introduced to south Florida, were introduced into Hawaii. Coquis were likely initially transported as egg masses in the nursery trade. They are now found on the islands of Hawaii, Maui, Oahu, and Kauai and continue to expand their range. In the town of Hilo, they entered via the Wal-Mart nursery (W. Mautz, R. Niino-DuPonte, personal communications). Populations there are expanding rapidly.

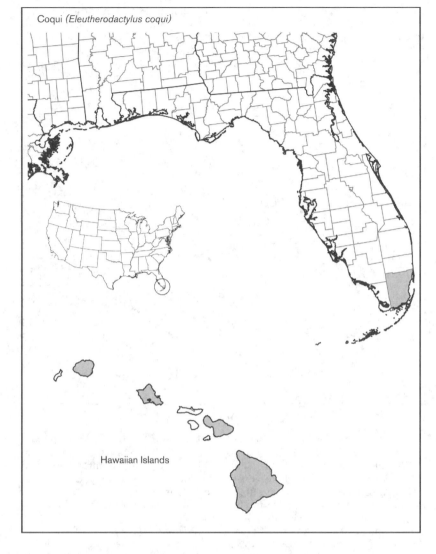

Coqui (*Eleutherodactylus coqui*)

Hawaiian Islands

2. Historical versus Current Abundance.
Coquis have been introduced to the United States and are limited to areas near bromeliad nurseries or greenhouses. Populations do not appear to be self-sustaining and instead may be replenished by new bromeliad shipments from Puerto Rico (Bartlett and Bartlett, 1999a).

3. Life History Features.
(Information comes primarily from studies at 350 m in Puerto Rico.)

A. Breeding. Reproduction is terrestrial.

i. Breeding. Coquis breed year-round, but breeding is more common during times of heat and high humidity, from late spring to early fall (Conant and Collins, 1998; Bartlett and Bartlett, 1999a). Females approach calling males, who then lead the females to potential nest sites where amplexus occurs (Townsend and Stewart, 1986). Fertilization is internal (Townsend et al., 1981).

ii. Breeding habitat. Males will call from open, elevated areas as well as from leaf surfaces, axils of palms, and tree trunks (Townsend, 1989).

B. Eggs.

i. Egg deposition sites. Coquis deposit eggs in partially enclosed spaces such as rolled leaves, palm fronds, and bromeliad axils, where they are guarded by the male parent (Stewart and Pough, 1983; Bartlett and Bartlett, 1999a).

ii. Clutch size. Clutch size varies with size of female, which varies with locality. Clutch size varies from 12–24 eggs in Florida (Bartlett and Bartlett, 1999a) and averages 26 eggs at 350 m in Puerto Rico (Townsend et al., 1984).

C. Direct Development. Complete metamorphosis occurs within the egg. Young hatch as tiny froglets, 6 mm long (Townsend and Stewart, 1985).

i. Brood sites. Coquis brood in enclosed areas such as in curled, fallen leaves and rolled palm fronds (Stewart and Pough, 1983; Townsend, 1989; Bartlett and Bartlett, 1999a).

ii. Parental care. Following deposition, clutches are tended and protected by the male parent until hatching occurs, often extending care to several days post-hatching (Townsend et al., 1984; Bartlett and Bartlett, 1999a). Males will defend the eggs against conspecifics by biting, wrestling, aggressive calling, and preventing access to the nest (Townsend et al., 1984).

D. Juvenile Habitat. In their native range, juveniles tend to remain on or near litter and ground vegetation, moving higher when foraging as they age (Stewart, 1995; Beard et al., 2003).

E. Adult Habitat. In their native range of Puerto Rico, coquis are a widespread generalist species. Adults spend the day in the litter and are arboreal at night (Stewart, 1985). Those observed in Florida tend to remain from 15–220 cm (6 in–7 ft) off the ground (Bartlett and Bartlett, 1999a). In Florida, coquis are limited to areas around greenhouses (Conant and Collins, 1998). During the day, coquis are found in secluded areas such as rock piles, axils of bromeliad fronds, tree-holes, and under rocks or logs (Stewart, 1985). However, on rainy or overcast days coquis are sometimes active (Bartlett and Bartlett, 1999a). Males are most vocal after dark on moist nights when they give their advertisement calls (Woolbright and Stewart, 1987).

F. Home Range Size. In Puerto Rico, the horizontal home range is <5 m; the vertical range is the entire vertical spectrum of the forest; the average calling height of males is approximately 1 m off ground (Woolbright, 1985; Townsend, 1989).

G. Territories. Male coquis give warning calls to defend their calling sites or diurnal retreats, while females produce soft, rasping vocalizations to defend their feeding sites (Stewart and Rand, 1991). Males tending nests will defend the eggs against conspecifics (see "Parental care" above; Townsend et al., 1984).

H. Aestivation/Avoiding Desiccation. Aestivation does not occur. Other than nocturnality, behaviors associated with avoiding desiccation have been unreported.

I. Seasonal Migrations. None.

J. Torpor (Hibernation). None.

K. Interspecific Associations/Exclusions. Personal space increases with size. Juveniles may sit next to each other, but adults do not allow another adult close by if visible (Stewart, 1995).

L. Age/Size at Reproductive Maturity. Size is dependent on locality, with larger individuals at higher elevations. Size ranges from 2.5–5.8 cm (1–2.25 in; Schwartz and Henderson, 1991; Conant and Collins, 1998). Females are larger than males (Woolbright, 1989), and frogs may reach reproductive size in 1 yr (Stewart and Woolbright, 1996). Individuals observed in Florida tend to be small (Bartlett and Bartlett, 1999a), as are lowland frogs in Puerto Rico.

M. Longevity. In nature, most do not live to the second year; the oldest individuals on record are 6 yr (Stewart and Woolbright, 1996).

N. Feeding Behavior. Coquis are dietary generalists eating a wide range of primarily invertebrates in addition to conspecific eggs (Stewart and Woolbright, 1996).

O. Predators. Eggs may be preyed upon by conspecifics, ants, snails, centipedes, and phorid flies; frogs are eaten by birds, lizards, snakes, spiders, scorpions, mongoose, and feral cats (Stewart and Woolbright, 1996).

P. Anti-Predator Mechanisms. Protection is provided primarily by cryptic colors and patterns, jumping and hiding under objects, and urination when disturbed. When attacked by crab spiders, larger frogs can kick themselves free (Formanowicz et al., 1981).

Q. Diseases. Unknown.

R. Parasites. Nematodes in lungs, intestines, and body cavity or under the skin (M.M.S., personal observations).

4. Conservation.

In Puerto Rico, coquis are widespread and abundant (Rivero, 1998). Although other species of *Eleutherodactylus* have declined and disappeared, we have no evidence that coquis have declined in recent years except where habitats have been destroyed (Joglar, 1998). Most population estimates are from montane rather than lowland habitats (Fogarty and Vilella, 2001). Because the coqui is an introduced species in the United States, concerns about its conservation have not been expressed. In Hawaii, where populations are expanding rapidly, the main concerns are not conservation but how to eradicate the frogs (L. Woolbright, personal communication). It is assumed, without evidence, that the frogs compete for food with native birds.

Eleutherodactylus (= *Syrrhophus*) *cystignathoides* (Cope, 1878a "1877")
RIO GRANDE CHIRPING FROG

J. Eric Wallace

The genus *Syrrhophus* was synonymized with the genus *Eleutherodactylus* and considered a subgenus (including the genus *Tomodactylus*) by Hedges (1989). But Dixon (2000) notes that this synonymy does not consider behavior, morphology, or genetic characters, and so this author maintains *Syrrhophus* at the generic level.

1. Historical versus Current Distribution.

The type locality of Rio Grande chirping frogs (*Eleutherodactylus cystignathoides*) is Potrero, near Cordoba, Veracruz, Mexico (Cope, 1877). The type locality of the subspecies found in the United States (*E. c. campi*) is Brownsville, Cameron County, Texas (Stejneger, 1915). In the United States, Rio Grande chirping frogs are native to extreme south Texas along the lower Rio Grande Valley in Cameron and Hildago counties. They have been introduced, probably by way of the potted plant trade, into Corpus Christi, Dallas, Houston, Kingsville, Tyler, San Antonio,

Huntsville, and the vicinity of LaGrange, Texas (Dixon, 2000). Some of these introduced populations have become established in the natural environment (McGown et al., 1994; Lutterschmidt and Thies, 1999).

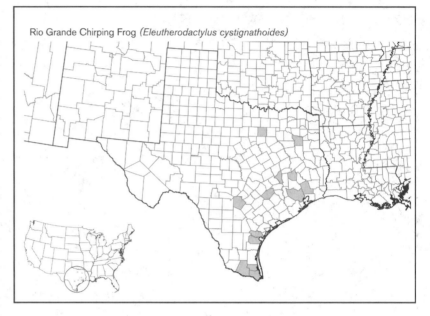

Rio Grande Chirping Frog *(Eleutherodactylus cystignathoides)*

For information regarding this species in Mexico, see Martin (1958), Lynch (1970), and Flores-Villela (1993). Dixon (2000) provides a current, comprehensive bibliography of Rio Grande chirping frogs. Popular accounts in field guide format may be found in Cochran and Goin (1970), Smith and Barlowe (1978), Garrett and Barker (1987), Behler and King (1998), Conant and Collins (1998), and Bartlett and Bartlett (1999a).

For discussions of the historical and current biogeography of the subgenus *Syrrhophus* (Hedges, 1989) and the species *cystignathoides*, see Blair (1950), Morafka (1977), and Duellman and Sweet (1999).

2. Historical versus Current Abundance.

Rio Grande chirping frogs are abundant and thrive in the presence of humans (Wright and Wright, 1949; Conant and Collins, 1998). Brach (1992) suggests that these semi-tropical frogs are able to persist north of their native range by seeking cover under ornamental rockwork that retains heat and moisture.

3. Life History Features.

A. Breeding. Reproduction is terrestrial.

i. Breeding migrations. Rio Grande chirping frogs do not form breeding aggregations and, as with cliff chirping frogs *(E. marnockii)*, may maintain territories (Hayes-Odum, 1990). Breeding occurs from April–May, when males may be heard both night and day (Wright and Wright, 1949). Hayes-Odum (1990) found

that adults called actively when the relative humidity was above 82% and when air and substrate temperatures were 16–29 °C. In an introduced population in Houston, Texas, males called on rainy nights from April–July (Hayes-Odum, 1990). Courtship behavior was observed in captivity and amplexus was axillary (Hayes-Odum, 1990). Wright and Wright (1949) noted inguinal amplexus with captive cliff chirping frogs.

ii. Breeding habitat. See "Adult Habitat" below.

B. Eggs.

i. Egg deposition sites. Eggs were laid just under the soil surface in the laboratory (Hayes-Odum, 1990).

ii. Clutch size. Eggs are laid in clutches consisting of 5–13 large eggs (3–3.5 mm in diameter; Wright and Wright, 1949; Hayes-Odum, 1990).

C. Direct Development. Complete metamorphosis occurs within the egg; young hatch as froglets, ranging in size from approximately 5–8.5 mm (Wright and Wright, 1949; Hayes-Odum, 1990). In a laboratory setting, eggs hatched after 14–16 d of artificial incubation (Hayes-Odum, 1990).

i. Brood sites. Brooding has not been observed (see "Parental care" below).

ii. Parental care. In a laboratory setting, Hayes-Odum (1990) observed one female leave her eggs shortly after laying them.

D. Juvenile Habitat. Hayes-Odum (1990) reported that hatchlings were found on wet nights. General juvenile habitat characteristics are likely similar to adult characteristics.

E. Adult Habitat. Rio Grande chirping frogs inhabit low elevation coastal plains in the Tamaulipan Province (Blair, 1950). Most published accounts of these frogs in the United States (including introduced populations) consist of observations in urban settings. Rio Grande chirping frogs are often associated with debris piles and watered lawns and gardens (Stejneger, 1915; Wright and Wright, 1949; Hayes-Odum, 1990; Brach, 1992). More natural environments include dense vegetation along riparian areas and the edges of lotic, semi-permanent waters (Martin, 1958; Conant and Collins, 1998). Rio Grande chirping frogs can be found under cover objects during the day (Wright and Wright, 1949). Brach (1992) suggests that these semi-tropical frogs are able to persist well north of their native range (e.g., Tyler, Texas) by seeking cover under ornamental rockwork that retains heat and moisture. They are known to utilize arboreal perches 0.2–1.5 m above the ground (Martin, 1958; Hayes-Odum, 1990; Brach, 1992).

F. Home Range Size. Unknown.

G. Territories. Hayes-Odum (1990) observed several instances of one frog moving toward another stationary frog, resulting in the displacement of the stationary frog. From these observations, she suggested that Rio Grande chirping frogs may be territorial.

H. Aestivation/Avoiding Desiccation. As with other members of the genus, activity diminishes with dry conditions.

I. Seasonal Migrations. Unknown.

J. Torpor (Hibernation). Unknown.

K. Interspecific Associations/Exclusions. Unknown, but may include (as inferred from microhabitat similarities) white-lipped frogs (*Leptodactylus fragilis*), marine toads (*Bufo marinus*), Coastal-Plain toads (*B. nebulifer*), sheep frogs (*Hypopachus variolosus*), and western narrow-mouthed toads (*Gastrophryne olivacea*; Blair, 1950; Conant and Collins, 1998).

L. Age/Size at Reproductive Maturity. Rio Grande chirping frogs range in size from 15–25.5 mm SVL (Wright and Wright, 1949). Measurements of preserved material from animals throughout their range (both subspecies) were 16.0–23.5 mm for males and 16.0–25.8 mm for females (Lynch, 1970). The largest individual observed by Wright and Wright (1949) was a female; females tend to be larger (Hayes-Odum, 1990). In reproductively mature females, eggs may be visible through the thin skin of their abdomen (Hayes-Odum, 1990).

M. Longevity. Unknown.

N. Feeding Behavior. Generally unknown, but one individual regurgitated cockroach eggs during capture (Wright and Wright, 1949).

O. Predators. Unknown.

P. Anti-Predator Mechanisms. To avoid capture, Rio Grande chirping frogs are quick to leap to shelter in crevices or under cover objects (Wright and Wright, 1949).

Q. Diseases. Unknown.

R. Parasites. Larval forms of *Abbreviata* sp. have been found in Houston populations (introduced) of Rio Grande chirping frogs (McAllister and Freed, 1992).

4. Conservation. Rio Grande chirping frogs have no federal or state conservation status. They are common and relatively abundant in a number of different urban settings and have been since their early discovery (e.g., Wright and Wright, 1949). They obviously can cope with at least certain types of human-mediated habitat disturbance. This is further evidenced by their ability to acclimate to novel habitats into which they are introduced, some very different from their native environment (e.g., Brach, 1992).

The recent introduction of Rio Grande chirping frogs through the potted plant trade into some urban areas (e.g., San Antonio, Texas; Dixon, 2000) may place introduced frogs in direct contact with resident populations of cliff chirping frogs and other amphibians. The repercussions of this interaction are unknown, but detrimental effects could include introduction of novel disease pathogens, competition, and/or hybridization. Measures should be instituted to decrease the likelihood of further introductions.

Eleutherodactylus (= *Syrrhophus*) *guttilatus* (Cope, 1879)
SPOTTED CHIRPING FROG

J. Eric Wallace

The genus *Syrrhophus* was synonymized with the genus *Eleutherodactylus* and considered a subgenus (including the genus *Tomodactylus*) by Hedges (1989). But Dixon (2000) notes that this synonymy does not consider behavior, morphology, or genetic characters and so this author maintains *Syrrhophus* at the generic level.

1. Historical versus Current Distribution. The type locality of spotted chirping frogs (*Eleutherodactylus guttilatus*) is Guanajuato, Mexico (Cope, 1879). In the United States, spotted chirping frogs are known from the Big Bend Region of West Texas in Brewster, Presidio, and Pecos counties (Dixon, 2000). Most survey work has focused on the southwestern portion of the region. More survey work is needed in northern Brewster, Jeff Davis, and Terrell counties, the possible contact zone with cliff chirping frogs (*E. marnockii*). The taxonomy of chirping frogs in this region has had a dynamic and controversial history that is necessary to discuss in order to interpret the available literature.

Chirping frogs from the Chisos Mountains, Brewster County, Texas, were described as a distinct species (*Syrrhophus gaigeae*) based on slight differences in color pattern and morphometrics, but primarily on the supposed geographic separation of approximately 400 km (250 mi) from cliff chirping frog populations to the east (Schmidt and Smith, 1944). Subsequently, Milstead et al. (1950) looked at frogs of both regions and compared them with animals from an area that fell between the two ranges. They were unable to discern the differences described by the previous authors and considered the two species conspecific. These results were in turn supported by the detailed morphological analysis of *S. marnockii* conducted by Jameson (1955). Lynch (1970), in the most recent taxonomic revision of the genus, recognized Big Bend frogs as unique from central Texas frogs based on color pattern and their allopatric distribution, but synonymized *S. gaigeae* with *S. guttilatus* of Mexico. This is the current taxonomy, yet others contend that most chirping frogs from the Edwards Plateau to the Trans-Pecos region of Texas are of the same species (Morafka, 1977; Dixon, 2000).

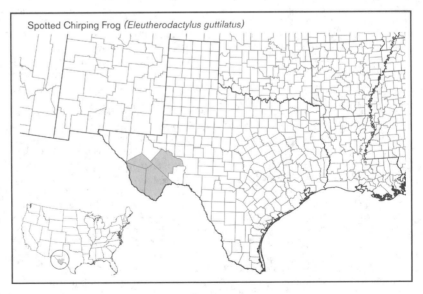

Spotted Chirping Frog (*Eleutherodactylus guttilatus*)

References to the distribution of spotted chirping frogs in Mexico can be found in Lynch (1970), Conant (1983), and Flores-Villela (1993). For discussions of the historical and current biogeography of the subgenus *Syrrhophus* and the species *guttilatus*, see Blair (1950), Milstead et al. (1950), Morafka (1977), Campbell (1999), and Duellman and Sweet (1999).

Little research, beyond field surveys, has been conducted on this frog in this region, and the reader is largely referred to the information in the cliff chirping frog *(E. marnockii)* account. Dixon (2000) provides a current and comprehensive bibliography of this species with notes on taxonomy (see his *Syrrhophus marnockii* account). Popular accounts in field guide format may be found in Smith and Barlowe (1978), Garrett and Barker (1987), Behler and King (1998), Conant and Collins (1998), and Bartlett and Bartlett (1999a).

2. Historical versus Current Abundance.
Based on collecting trips in 1957–58, Milstead (1960) states "that the population [of chirping frogs] is still quite strong" in Big Bend National Park. During surveys in the Park a few years earlier (i.e., 1955), Minton (1959) and his field party were unable to locate individuals of this species and attributed this to drought conditions. Surveys during the wet summer of 1999 confirmed this frog as common and widespread throughout the Park (unpublished data).

3. Life History Features.
   A. Breeding. Reproduction is terrestrial. Little is known on the breeding behavior of spotted chirping frogs. Breeding activity (i.e., calling males) begins in earnest with the onset of summer (June–July) rains continuing for about 2 wk and tapering off over the next about 1.5 mo (S. Droege, unpublished data; unpublished data). Males tend to vocalize from rock cracks and crevices that are 0.2–1.3 m above the ground (unpublished data). An audio-spectrogram of a call from an individual from Durango, Mexico, is provided in Conant (1983).
      i. Breeding migrations. This highly terrestrial frog does not undertake breeding migrations or form breeding aggregations. Similar to cliff chirping frogs, they probably maintain a home range.
      ii. Breeding habitat. Unknown.
   B. Eggs.
      i. Egg deposition sites. Unknown.
      ii. Clutch size. One female collected in July contained five unpigmented eggs, each about 4 mm in diameter (Gaige, 1931). Garrett and Barker (1987) state that females deposit fewer than 15 eggs and Bartlett and Bartlett (1999a) note clutches of 5–12 eggs, yet no reference as to the origin of these data are given.

   C. Direct Development. Development in spotted chirping frogs occurs within the egg; young hatch as froglets (Conant and Collins, 1998).
      i. Brood sites. Unknown.
      ii. Parental care. Unknown.
   D. Juvenile Habitat. Unknown, but probably similar to adult habitat characteristics.
   E. Adult Habitat. Spotted chirping frogs are found in a variety of habitats in the Chihuahuan Biotic Province (Blair, 1950). In the mountains, they inhabit rocky outcrops in ravines, along bluffs, and manmade rock walls in the oak-juniper woodland. At lower elevations they have been found in mines, along road cuts, and along limestone bluffs on the Rio Grande (Scudday, 1996; J. F. Scudday, personal communication; unpublished data). Spotted chirping frogs are nocturnal,and may be found by day under rocks, leaf litter, and debris (Bartlett and Bartlett, 1999a).
   F. Home Range Size. Unknown.
   G. Territories. Unknown.
   H. Aestivation/Avoiding Desiccation. Periods of drought may severely inhibit the activity of this frog (Minton, 1959; J. F. Scudday, personal communication).
   I. Seasonal Migrations. Spotted chirping frogs are not known to make seasonal migrations and probably maintain a home range similar to cliff chirping frogs.
   J. Torpor (Hibernation). Unknown.
   K. Interspecific Associations/Exclusions. No other saxicolous amphibians are known to co-occur with spotted chirping frogs. Red-spotted toads (*Bufo punctatus*) and canyon treefrogs (*Hyla arenicolor*) have been observed in the same canyons as spotted chirping frogs (unpublished data).
   L. Age/Size at Reproductive Maturity. In a study of preserved material from the United States and Mexico, adult male and female spotted chirping frogs ranged from 20.6–29.0 mm SVL and 25.7–31.0 mm, respectively (Lynch, 1970). Adult, calling males from the Chisos Mountains, Texas, were 30–33 mm (n = 6) and weighed 1.8–2 g (n = 3; unpublished data).
   M. Longevity. Unknown.
   N. Feeding Behavior. Gaige (1931) found ants, beetles, a termite, and an isopod in the stomach of one female from the Chisos Mountains. While releasing an individual during daylight I observed it capture and eat a winged termite. The peak of activity in summer 1999 coincided with a large flight of termites (unpublished data), suggesting that they constitute an important and abundant prey source during this period. It is suspected that spotted chirping frogs feed on other small insects and invertebrates, including small spiders and small crustaceans (Garrett and Barker, 1987).
   O. Predators. Unknown.
   P. Anti-Predator Mechanisms. Spotted chirping frogs are quick to jump to cover when disturbed (unpublished data).

Although Conant and Collins (1998) state that spotted chirping frogs will run when disturbed, rather than leaping or hopping.
   Q. Diseases. Unknown.
   R. Parasites. Larval mites (*Hannemania hylae*) are known parasites of spotted chirping frogs (Gaige, 1931; Lynch, 1970; Jung et al., 2001).

4. Conservation.
Spotted chirping frogs have no federal or state conservation status. They are common and relatively abundant in several areas of Big Bend National Park (unpublished data). There is little large-scale development in the Park and in those places where there is (e.g., the Basin), spotted chirping frogs are still found. In fact, on several occasions, attempts to locate random individuals in remote areas resulted in observing frogs in manmade rock walls (trail retaining walls, erosion control barriers) despite the availability of more natural habitat nearby. Like their other generic counterparts in the United States, spotted chirping frogs seem to thrive with certain types of human-mediated habitat disturbance. Due to the remote nature of much of the spotted chirping frogs' native habitat, they are most likely under no immediate population threats.

Acknowledgments. Funding for work in Big Bend National Park was provided by PRIMENet (Park Research and Intensive Monitoring of Ecosystems Network), an interagency effort of the Environmental Protection Agency and National Park Service. I thank Raymond Skiles, wildlife biologist at Big Bend National Park, for making available his vast knowledge of the area, and James Dixon and James F. Scudday, both long-time Texas herpetologists, who provided me with their expertise regarding this species in the region. I also thank Robin Jung and Sam Droege of the USGS Patuxent Wildlife Research Center, who provided support during my time of study.

Eleutherodactylus (= Syrrhophus) marnockii (Cope, 1878[a])
CLIFF CHIRPING FROG

J. Eric Wallace

The genus *Syrrhophus* was synonymized with the genus *Eleutherodactylus* and considered a subgenus (including the genus *Tomodactylus*) by Hedges (1989). But Dixon (2000) notes that this synonymy does not consider behavior, morphology, or genetic characters and so maintains *Syrrhophus* at the generic level.

1. Historical versus Current Distribution.
The type locality of cliff chirping frogs *(Eleutherodactylus marnockii)* is "near San Antonio" (Cope, 1878a). This is further restricted to Helotes, Bexar County, Texas, by Strecker (1933). Cliff chirping frogs are common in rocky areas of central Texas on

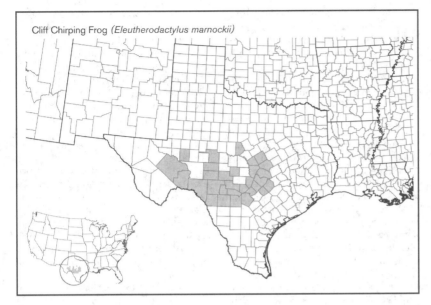

Cliff Chirping Frog (*Eleutherodactylus marnockii*)

the Edwards Plateau west to the Stockton Plateau (Jameson, 1955; Dixon, 2000). Currently, most gaps in the distribution of cliff chirping frogs are likely from lack of survey effort. Searches in appropriate habitats throughout their range will likely continue to increase our knowledge of their distribution (Husak, 1998; Malone, 1998) and are especially needed along the western edge of their known range—the possible contact zone with spotted chirping frogs (*E. guttilatus*; see taxonomic discussion under *E. guttilatus* account). Fossil records exist from Edwards, Foard, Knox, and Kerr counties, Texas (Parmley, 1988a,b; Dixon, 2000) suggesting a Pleistocene distribution approximately 320 km to the north of present day populations.

Jameson (1955) presents the only comprehensive study on the ecology of cliff chirping frogs. His work on the population dynamics of cliff chirping frogs combines extensive field observations, intensive experimental field manipulations, and laboratory work. Much of the information in this species account is extracted directly from there. Dixon (2000) provides a current, comprehensive bibliography of cliff chirping frogs with notes on taxonomy. Popular accounts in field guide format may be found in Cochran and Goin (1970), Smith and Barlow (1978), Garrett and Barker (1987), Behler and King (1989), Conant and Collins (1998), and Bartlett and Bartlett (1999a).

For discussions of the historical and current biogeography of the subgenus (= *Syrrhophus*; Hedges, 1989) and for the species *marnockii*, see Smith and Buechner (1947), Blair (1950), Milstead et al. (1950), Morafka (1977), Duellman and Sweet (1999), and Hedges (1999).

2. Historical versus Current Abundance. Jameson (1955) conducted population studies at 12 sites in the vicinity of Austin, Texas, and estimated population densities of cliff chirping frogs from 1.2–8.9 frogs/ac. He attributed variation in local densities to availability and structure of rocky substrate and moisture (as evidenced by amount of vegetation). Areas with more moisture and vegetation and greater rock structure (e.g., talus slopes and rock piles) supported denser populations. During his research, he noted reductions in frog densities at two of his study sites and suggested disturbance by fire to be the cause in one case and road construction in the other.

3. Life History Features.

A. Breeding. Reproduction is terrestrial. Breeding occurs primarily from April–May with a second pulse during September–October; both peaks are coincident with rain events (Jameson, 1955). Based on observations of gravid females, he suggests that breeding may take place from late February to early December. Calling is most pronounced during the early evening following dusk and tends to diminish throughout the night; during the height of the breeding season(s), calls may be heard until dawn. Cliff chirping frogs exhibit complex mating behaviors including preamplectic courtship and mating-specific vocalizations (Jameson, 1954). Amplexus is axillary. He noted what appeared to be a call hierarchy in which groups of 3–4 individuals followed the lead of a dominant male. Fouquette (1960) provides descriptions of two types of calls produced by cliff chirping frogs. Jameson (1955) comments on the extreme difficulty of sexing frogs in the field. The only means was the presence of eggs observable through the stomach wall in gravid females and the vocalizations of males. He notes that while males may mate several times/year, females may mate from 1–3 times/year. Jameson (1955) provides additional, detailed information on breeding behaviors and general activity patterns under various environmental conditions.

i. Breeding migrations. This highly terrestrial frog does not form breeding migrations or aggregations.

ii. Breeding habitat. Little is known regarding cliff chirping frog breeding microhabitats. Frogs have been observed in amplexus twice in the wild, once on an exposed rock substrate and once in a rock crack above dry soil. In neither case was there any moist soil nearby. Chirping frogs in the wild continue to vocalize while in amplexus. Jameson (1955) suggests this may be an attempt to lead the female to moist soil for egg deposition or may be necessary to instigate egg laying and/or to ensure proper fertilization. In captivity, mating occurred only when males called females to a site with moist soil after which males did not continue to call while in amplexus. All eggs laid in captivity were sterile (Jameson, 1955).

B. Eggs.

i. Egg deposition sites. Egg deposition sites in nature are unknown despite considerable effort to locate them (Jameson, 1955).

ii. Clutch size. In captivity, 8–20 sterile eggs were laid in trenches dug in moist soil by the female and covered by both frogs with their legs as they moved forward.

C. Direct Development. Larvae undergo direct development within the egg capsule and hatch as froglets into the terrestrial environment (Jameson, 1955; Lynch, 1971). Specific information on timing and duration of development is lacking. The incubation period for the closely related Rio Grand chirping frog (*E. cystignathoides campi*) under captive and experimental conditions was 14–16 d (Hayes-Odum, 1990).

i. Brood sites. Unknown.

ii. Parental care. Parental care occurs in some *Eleutherodactylus* and has been suggested to occur in barking frogs (*E. augusti*; Jameson, 1950b). Because barking frogs co-occur with cliff chirping frogs in the same habitats and environmental conditions, parental care in chirping frogs requires further investigation.

D. Juvenile Habitat. Juvenile habitat is likely similar to adults, that is they require a similar physiography in the form of rocky and vegetative structure providing for cover and moist refugia. In Jameson's (1955) intensive study, juveniles were observed in all but the driest and coldest months (December, January) of the year, with peak activity similar to that noted for adults and coincident with moist periods.

During a field manipulation study on dispersing juveniles, Jameson (1955) recorded an average dispersal distance of 211.4 ± 25.6 m (231 ± 28 yd) for those individuals that apparently recruited into the resident population. He found that most juveniles dispersed parallel to available habitat (in these cases, rocky exposures) and considered time of dispersal dependent on availability of moisture and time of breeding and hatching.

E. Adult Habitat. Cliff chirping frogs are saxicolous and to some extent cavernicolous in nature and are primarily found in the juniper-oak association of the Balconian Province of Blair (1950). They are always found in association with rocks and may be found underneath and in crevices in rocks occurring along limestone ledges and bluffs, on talus slides, in ravines, and near streams (Smith and Buechner, 1947; Wright and Wright, 1949; Jameson, 1955). Caves often support large populations (Jameson, 1955). These frogs are primarily nocturnal with most activity recorded in the hours following dusk, especially following spring and fall rains (Wright and Wright, 1949; Jameson, 1955). During the height of breeding season(s), individuals remain active throughout the night and males may continue to vocalize during daylight hours (Jameson, 1955). During the day, these frogs can be found under cover objects (Wright and Wright, 1949; Jameson, 1955). On a few occasions, Jameson (1955) observed individuals on arboreal perches from 1.2–2.4 m (4–8 ft) above the ground.

Jameson (1955) showed that these highly terrestrial frogs are relatively incapable of swimming, as demonstrated by individuals that drowned within 2 hr of being placed in a jar of water, and attributes their inactivity during intense rainfall events to this inability.

F. Home Range Size. Jameson (1955) estimated average home ranges of cliff chirping frogs during the breeding season from four distinct types of habitat. Average home ranges by type ranged between 0.052–0.139 ac. Size of home ranges were significantly correlated with the unique physiography of each type of site (i.e., the more structurally diverse the habitat, the smaller the home range). Home ranges tended to be occupied throughout their lifespan (Jameson, 1955).

G. Territories. Unknown.

H. Aestivation/Avoiding Desiccation. Cliff chirping frogs may be found at nearly any time during the year if environmental conditions are appropriate. Activity during hot, dry summer periods is generally minimal. Jameson (1955) observed frogs at temperatures as high as about 32 °C (90 °F) but does not elaborate on their activities.

I. Seasonal Migrations. Chirping frogs do not undertake seasonal migrations per se, however, through experimental field manipulations, Jameson (1955) showed that cliff chirping frogs will move into areas of decreased population density from areas of high population density.

J. Torpor (Hibernation). Activity during cold wet winter periods is generally minimal. Jameson (1955) found that during cold, moist periods individuals could be observed in the same crevice over the course of several days. He has observed frogs at temperatures as low as 1 °C (34 °F), but does not elaborate on their activities.

K. Interspecific Associations/Exclusions. Cliff chirping frogs are sympatric with barking frogs throughout much of their range in Texas and often share the same rocky microhabitats (Jameson, 1954; Dixon, 2000). Cliff chirping frogs have also been found associated with Gulf Coast toads (*Bufo valliceps* [now considered to be Coastal-Plain toads, *B. nebulifer*], see Mulcahy and Mendelson, 2000; Mendelson, this volume), red-spotted toads (*B. punctatus*), and western slimy salamanders (*Plethodon albagula*; McAlister, 1954; Jameson, 1955; Hubbs and Martin, 1967). Jameson (1955) lists several vertebrates and invertebrates that co-occur with cliff chirping frogs and might compete for prey.

L. Age/Size at Reproductive Maturity. Based on measurements of gravid females and calling males, frogs reach sexual maturity at 19–22 mm and 18–22 mm, respectively (Jameson, 1955). Lynch (1970) measured 103 museum specimens and found adult males and females ranging between 18.4–28.9 and 20.4–35.4 mm, respectively. They attain a maximum length of 38 mm (Cope, 1889). The period of fastest growth occurs during the spring and summer (Jameson, 1952, fig. 4), and growth rates decrease as frogs become larger.

M. Longevity. Jameson (1955) suggests cliff chirping frogs may live ≤3 yr in the wild.

N. Feeding Behavior. Documented prey of wild cliff chirping frogs include ants, small beetles, camel crickets, termites, and small spiders (see Jameson, 1952, for more detailed information). Captive animals also feed on a variety of insects (Jameson, 1955). Jameson (1952) noted the interesting behavior of chirping frogs "robbing" insects entrapped in the webs of orbit web spiders (Argiopidae).

O. Predators. Known predators of cliff chirping frogs include western diamond-backed rattlesnakes (*Crotalus atrox*), mottled rock rattlesnakes (*C. lepidus*), copperheads (*Agkistrodon contortrix*), black-necked garter snakes (*Thamnophis cyrtopsis*), and large wolf spiders (Milstead et al., 1950; Fouquette, 1954; Jameson, 1955). Jameson (1955) observed predation in the laboratory by barking frogs, Coastal-Plain toads, and tarantulas.

P. Anti-Predator Mechanisms. To avoid capture, cliff chirping frogs are capable of quick movements and will hop or run for the nearest cover, usually in the form of rock cracks or crevices (Jameson, 1955).

Q. Diseases. Unknown.

R. Parasites. Mites of the genus *Hannemania* are known ectoparasites of cliff chirping frogs (Jameson, 1952; Lynch, 1970).

4. Conservation.
Cliff chirping frogs have no federal or state conservation status. Based on Jameson's (1955) findings, cliff chirping frogs were common and relatively abundant in several urban areas in Austin, Texas. This suggests an ability to cope with at least certain types of human-mediated habitat disturbance. They are still found in city parks today (Bartlett and Bartlett, 1999a).

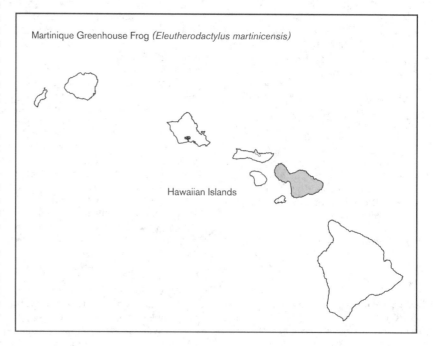

Martinique Greenhouse Frog (*Eleutherodactylus martinicensis*)

Hawaiian Islands

Resurveys at Jameson's urban sites would provide insight into the long-term ability of this frog to persist in urban settings.

The recent introduction of Rio Grande chirping frogs through the potted plant trade into some urban areas (e.g., San Antonio, Texas; Dixon, 2000) may place introduced frogs in direct contact with resident populations of cliff chirping frogs. The repercussions of this interaction are unknown, but detrimental effects could include introduction of novel disease pathogens, competition, and/or hybridization. Measures should be instituted to decrease the likelihood of further introductions.

Eleutherodactylus martinicensis Tschudi, 1838
MARTINIQUE GREENHOUSE FROG

Editor's note: Introduced to Maui in the Hawaiian Islands (Kraus et al., 1999). According to these authors, Martinique greenhouse frogs are known from two sites, at 430 m and about 670 m in elevation. The former site is apparently an established population. Martinique greenhouse frogs co-occur with introduced coquis (*E. coqui*; Kraus et al., 1999). Attempts are being made by the State of Hawaii to eradicate these *Eleutherodactylus*, although these efforts are controversial (F. Kraus, personal communication).

Eleutherodactylus planirostris (Cope, 1862)
GREENHOUSE FROG

Walter E. Meshaka Jr.

1. Historical versus Current Distribution.
Cuba is the center of distribution for greenhouse frogs (*Eleutherodactylus planirostris*; Schwartz and Henderson, 1991); they are also found on other islands in the West Indies (Carr and Goin, 1955) and from southern Florida and Key West (Cope, 1875a, 1889; Carr, 1940a; Goin, 1944, 1947a). Their present distribution in the continental United States is throughout peninsular Florida, the Florida Keys, and a few isolated populations in the panhandle (Carr and Goin, 1955; Duellman and Schwartz, 1958; Wilson and Porras, 1983; Lazell, 1989; Bartlett and Bartlett, 1999a). Greenhouse frogs have also been documented in extreme eastern Louisiana (Schwartz, 1974; Plotkin and Atkinson, 1979; Conant and Collins, 1998), and in isolated counties in both southern Alabama (Baldwin County; Carey, 1982) and southern Georgia (Chatham County; Winn et al., 1999). They have also been introduced in Hawaii on Hawaii Island and Oahu (Kraus et al., 1999; M. Stewart, personal communication). Their natural patterns of dispersal have doubtless been overshadowed by human-mediated dispersal through the potted plant trade (Goin, 1944). Butterfield et al. (1997) have documented the occurrence of these frogs in suitable natural habitats away from human influence.

2. Historical versus Current Abundance.
Carr (1940a) notes that greenhouse frogs are locally abundant in Gainesville, Florida, and are the most numerous anuran on Key West. Rapid dispersal through Florida has been a boon to greenhouse frogs; however, data that measure and assess trends in abundance within populations are lacking. Populations of greenhouse frogs are successfully established in Florida, with their distribution continuing to extend northward and encompassing a wide range of mesic and upland habitats (Bartlett and Bartlett, 1999a; Meshaka et al., 2004).

3. Life History Features.
A. Breeding. Reproduction is terrestrial.
i. Breeding migrations. None known. In the Everglades, calling is heard from April–September. The peak months of calling are May–June (Meshaka et al., 2004). Calling is heard most frequently on wet (≥ 2 cm rain), warm (25 °C), humid (96%) nights. In Homestead, Florida, greenhouse frogs are all but silent from mid-November to mid-February, after which time choruses, which sound like soft chirping, are almost nightly events that intensify with rain (Meshaka et al., 2004). Homestead choruses have been heard in ambient temperatures of at least 20 °C. Rain, watering the garden, and sultry days initiate diurnal choruses. In Gainesville, males call from April–September, and breeding occurs from late May to late September with a peak in July (Goin, 1947a). Amplexus is axillary (Goin, 1947). Hatchlings appear in June in Gainesville (Goin, 1947a) and from late May to early June in Key West (Lazell, 1989).

ii. Breeding habitat. Breeding occurs under moist cover (Carr, 1940a; Goin, 1947a).

B. Eggs.
i. Egg deposition sites. Goin (1944) reports an incident of greenhouse frog eggs found in a backyard flowerpot in Jacksonville, Florida. In general, eggs are laid in moist depressions in the earth or in moist debris (Lazell, 1989; Bartlett and Bartlett, 1999a).

ii. Clutch size. From 3–26 eggs with an average of 16 (Goin, 1947a; Lazell, 1989; Bartlett and Bartlett, 1999a). In Jamaica, the range is 2–22 yolked eggs with an average of 11 (Stewart, 1979).

C. Direct Development. Terrestrial. Larval development occurs within the egg;

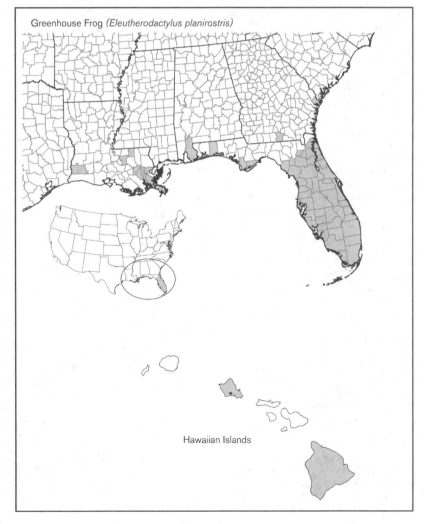

Greenhouse Frog *(Eleutherodactylus planirostris)*

Hawaiian Islands

young hatch as miniature froglets (Lazell, 1989; Bartlett and Bartlett, 1999a). Development may be accelerated by warmer temperatures, with hatching occurring as early as 13 d post-deposition (Lazell, 1989). Lazell (1989) notes that hatching appears to be most successful with 100% humidity.

　　i. Brood sites. Eggs are laid on the ground under moist cover (Lazell, 1989), but females are not known to brood (Goin, 1947a).

　　ii. Parental care. None (Goin, 1947a).

　D. Juvenile Habitat. Humid areas that provide cover, such as leaf mold, flower beds, and moist litter (Goin, 1947a; Carr and Goin, 1955).

　E. Adult Habitat. Same as juveniles; adults are found in humid habitats that provide cover (Carr, 1940a). Greenhouse frogs are particularly common in gardens, greenhouses, nurseries, and other moist substrates (Goin, 1944; Bartlett and Bartlett, 1999a). These frogs will hide beneath leaf litter, mulch, boards, or stepping stones (Carr, 1940a; Bartlett and Bartlett, 1999a). Greenhouse frogs are also found outside of human influence in suitable natural habitats (Butterfield et al., 1997). They are secretive and nocturnal except on warm, overcast, or rainy days (Carr, 1940a; Bartlett and Bartlett, 1999a).

　F. Home Range Size. Unknown.

　G. Territories. Unknown.

　H. Aestivation/Avoiding Desiccation. Unknown.

　I. Seasonal Migrations. None.

　J. Torpor (Hibernation). In late March in the southern Everglades of Florida, greenhouse frogs have been found hibernating underneath the loose bark of wild tamarind (*Lysiloma* sp.), a small tree common in the hammocks (Harper, 1935).

　K. Interspecific Associations/Exclusions. Greenhouse frogs are found under the same cover objects as eastern narrow-mouthed toads (*Gastrophryne carolinensis*; Carr, 1940a; personal observations) and in the same habitats as many small fossorial and semi-fossorial amphibians and reptiles (Dalrymple, 1988; Conant and Collins, 1998; Meshaka et al., 2000; Meshaka and Layne, 2002, 2005), with which they may or may not compete.

　L. Age/Size at Reproductive Maturity. Sexually mature male greenhouse frogs range in size from 15.0–17.5 mm SVL; females range from 19.5–25.0 mm SVL (Bartlett and Bartlett, 1999a). Lazell (1989) records a large adult measuring 32 mm. Sexual maturity is reached in 1 yr (Goin, 1947a; Duellman and Schwartz, 1958).

　M. Longevity. Unknown.

　N. Feeding Behavior. In order of occurrence, greenhouse frogs eat ants, beetles, and roaches, but include other types of small invertebrates (Goin, 1947a; Duellman and Schwartz, 1958; Lazell, 1989). In Jamaica, diet did not include roaches, but

animals ate numerous ants, mites, spiders, and longlegs (Stewart, 1979).

　O. Predators. In the Everglades, greenhouse frogs are eaten by Cuban treefrogs *(Osteopilus septentrionalis)* and ring-necked snakes (*Diadophis punctatus*; Wilson and Porras, 1983; Meshaka, 1994, 2001; Meshaka et al., 2004).

　P. Anti-Predator Mechanisms. Unknown.

　Q. Diseases. Unknown.

　R. Parasites. None reported in the United States.

4. Conservation.

Greenhouse frog populations appear to be stable across much of their native range. They have been rapidly expanding their distribution up peninsular Florida and along the Gulf Coast and have been introduced in Hawaii on Hawaii Island and Oahu, where they are considered an invasive species (Bartlett and Bartlett, 1999a; Kraus et al., 1999; Meshaka et al., 2004; M. Stewart, personal communication). Greenhouse frogs are locally abundant and occur across a wide range of habitats and regions in Florida (Carr, 1940a; Dalrymple, 1988; Lips, 1991; Meshaka et al., 2000). In this connection, greenhouse frogs have the potential to compete for food with lizards such as reef geckos *(Sphaerodactylus notatus)* and mole skinks *(Eumeces egregius)*, which are already threatened because they are habitat specialists and their habitat is rapidly disappearing.

Leptodactylus fragilis Brocchi, 1877
WHITE-LIPPED FROG

W. Ronald Heyer

Author's note: *Leptodactylus fragilis* has been cited historically as either *L. albilabris* (e.g., Metcalf, 1923, in part) or *L. labialis* (e.g., Mulaik, 1937; Maslin, 1963a,b; Dixon and Heyer, 1968; Heyer, 1971;

Villa, 1972; Meyer and Foster, 1996; Levell, 1997; McCranie and Wilson, 2002).

1. Historical versus Current Distribution.

White-lipped frogs (*Leptodactylus fragilis*) are known throughout lowland Middle America to the north coast of South America as far as Venezuela (Heyer, 1978, 2002). In the United States, these frogs marginally occur in southernmost Texas, specifically in the extreme southern edge of the Lower Rio Grande Valley (Garrett and Barker, 1987). They are known historically from one locality in Cameron County, two localities in Hidalgo County, and one locality from Starr County (Heyer, 1978). Dixon (1987, 1996) reported no new localities.

2. Historical versus Current Abundance.

Dixon (1987, p. 66) commented: "This frog may be extirpated from Texas through the continuous dispersal of organophosphate chemicals in the Rio Grande Valley."

3. Life History Features.

　A. Breeding.

　　i. Breeding migrations. Males have been found calling under clumps of grass, dirt clods, and from small depressions (Garrett and Barker, 1987).

　　ii. Breeding habitat. Brooding chambers, excavated by males, are found under rocks, logs, or debris in clay soil (Dixon and Heyer, 1968). Dixon and Heyer (1968) note that these chambers may contain calling males or foam nests.

　B. Eggs.

　　i. Egg deposition sites. Eggs are laid in foam nests, created by body secretions, in small brooding chambers excavated by adults (Maslin, 1963a,b; Garrett and Barker, 1987). The foam aids in preventing desiccation of the eggs during development (Garrett and Barker, 1987).

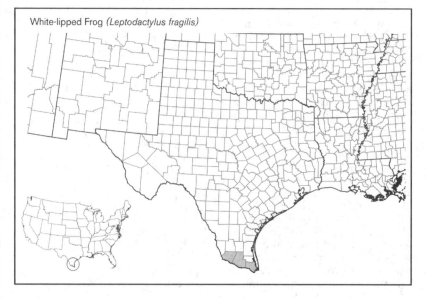

White-lipped Frog (*Leptodactylus fragilis*)

ii. Clutch size. A foam nest from Texas contained 86 yellow eggs (Mulaik, 1937). Villa (1972, p. 122) reported 25–250 eggs/nest.

C. Larvae/Metamorphosis. Early development takes place in the foam nests within the brooding chamber; larvae are released to ponds during heavy rains (Dixon and Heyer, 1968).

i. Length of larval stage. Mulaik (1937) reported a larval duration of 30–35 d in a Texas population. Meyer and Foster (1996) state that in Belize the tadpoles usually metamorphose in <2 wk.

ii. Larval requirements. Unknown.

iii. Larval polymorphisms. None.

iv. Features of metamorphosis. See "Length of larval stage" above.

v. Post-metamorphic migrations. Unknown.

D. Juvenile Habitat. Likely the same as adults.

E. Adult Habitat. Adult white-lipped frogs have been found in a variety of habitats where moisture is sufficient (Garrett and Barker, 1987). These frogs can be found in semi-permanent water bodies such as prairie potholes, oxbow lakes, and resacas (Edwards et al., 1989). Garrett and Barker (1987) note that white-lipped frogs may be encountered in irrigated agricultural fields, irrigation ditches, low grasslands, and runoff areas. These frogs are nocturnally active and may hide in burrows during the day (Garrett and Barker, 1987).

F. Home Range Size. Unknown.

G. Territories. Males call in association with incubating chambers (Dixon and Heyer, 1968) and would be expected to defend these sites from other males, although this has yet to be documented.

H. Aestivation/Avoiding Desiccation. Unknown.

I. Seasonal Migrations. Unknown.

J. Torpor (Hibernation). Unknown and unlikely.

K. Interspecific Associations/Exclusions. Unknown.

L. Age/Size at Reproductive Maturity. Males reach a maximum length of 36 mm SVL, females 40 mm SVL (Heyer, 1971).

M. Longevity. Unknown.

N. Feeding Behavior. White-lipped frogs forage nocturnally in open areas (Garrett and Barker, 1987).

O. Predators. Unknown.

P. Anti-Predator Mechanisms. Unknown.

Q. Diseases. Unknown.

R. Parasites. Metcalf (1923) reported the ciliate opalinid *Zelleriella leptodactyli* from specimens from Tehuantepec, Mexico.

4. Conservation.
Listed as Threatened, and therefore protected, by the State of Texas (Levell, 1997; www.tpwd.state.tx.us). Historically, white-lipped frogs are known from one locality in Cameron County, two localities in Hidalgo County, and one locality from Starr

County (Heyer, 1978; see above). Dixon (1987, 1996) reported no new localities and comments (1987, p. 66) "This frog may be extirpated from Texas through the continuous dispersal of organophosphate chemicals in the Rio Grande Valley."

McCranie and Wilson (2002, p. 533) state that white-lipped frogs are of low vulnerability in at least parts of their distribution.

Family Microhylidae

Gastrophryne carolinensis (Holbrook, 1836)
EASTERN NARROW-MOUTHED TOAD

Joseph C. Mitchell, Michael J. Lannoo

1. Historical versus Current Distribution.
Eastern narrow-mouthed toads (*Gastrophryne carolinensis*) occur in the southeastern and lower midwestern United States, from Maryland south to the Florida Keys, west to central Texas, and north to Kentucky, southern Illinois, southern Missouri, and extreme southeastern Nebraska (Carr, 1940a; Wright and Wright, 1949; Duellman and Schwartz, 1958; Nelson, 1972c; Harris, 1975; Martof et al., 1980; Ashton and Ashton, 1988; Dundee and Rossman, 1989; Conant and Collins, 1991; Redmond and Scott, 1996; Bartlett and Bartlett, 1999a; Mitchell and Reay, 1999; Phillips et al., 1999; Dixon, 2000; Johnson, 2000). Minton (2001) speculated that this anuran may occur in Indiana. They reach elevations to 549 m in the Smoky Mountains National Park (Huheey and Stupka, 1967), although they are absent from most of the Blue Ridge Mountains and the Appalachian region north of Tennessee (Redmond and Scott, 1996). Nelson (1972c) noted that they occur as high as 732 m in Oklahoma. Eastern narrow-mouthed toads also occur on numerous barrier islands in the Gulf Coast and off the southeastern Atlantic Coast (En-

gles, 1952; Blaney, 1971; Gibbons and Coker, 1978; Braswell, 1988; Learm et al., 1999). The disjunct population in southeastern Iowa may be extirpated (Klimstra, 1950; Nelson, 1972c).

2. Historical versus Current Abundance.
Eastern narrow-mouthed toads are solitary, secretive, and common throughout Alabama (Mount, 1975), Florida (Carr, 1940a), and Louisiana (Penn, 1943; Anderson, 1954), and locally common in Illinois (Phillips et al., 2000) and Virginia (Hoffman and Mitchell, 1996). Eastern narrow-mouthed toads are commonly captured in drift fences with pitfall traps (Gibbons and Bennett, 1974; Buhlmann et al., 1994; Enge, 1997). Some studies have revealed large numbers in local populations (e.g., Bennett et al., 1980; Gibbons and Semlitsch, 1981; Dodd, 1992, 1995b), although annual variation in numbers encountered varies among sites and years (Dodd, 1995b; Semlitsch et al., 1996; Enge, 1997).

Abundance estimates or counts are highly dependent on the sampling technique used. Funnel traps along drift fences revealed few individuals in a study in northern Florida (Vickers et al., 1985). Lamb et al. (1998) reported only 19 eastern narrow-mouthed toads among 1,070 frogs encountered under coverboards in a North Carolina floodplain. Coverboards in South Carolina produced few captures, whereas drift fences with pitfall traps yielded thousands of captures over the same time period (Grant et al., 1992). However, Grant et al. (1994) recorded high numbers of narrow-mouthed toads under coverboards in different aged pine stands. Other studies reporting high numbers of captures using drift fences with pitfall traps include Bennett et al. (1980), Gibbons and Semlitsch (1981), Dodd (1992), Enge (1997), and Hanlin et al.

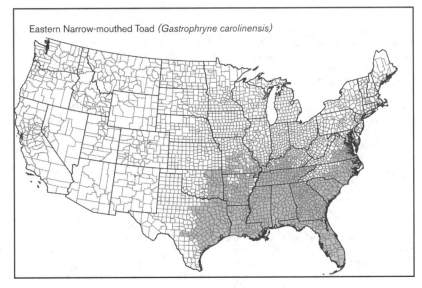

Eastern Narrow-mouthed Toad (*Gastrophryne carolinensis*)

(2000). More individuals of this species were trapped in a deep ditch in Louisiana than any other amphibian (Anderson, 1952).

Changes in local abundance have been little documented. Fowler and Stine (1953) noted the reduction of suitable habitat for this species in southern Maryland due to ditching, agriculture, and logging. Forestry operations may affect abundance in some areas, as more individuals were captured in a naturally regenerated slash pine forest than in two sites intensively prepared for pine plantations in Florida (Enge and Marion, 1986). There was no successful reproduction in a small pond created for mitigation in South Carolina despite the abundance of adults that used this habitat for several years following construction (Pechmann et al., 2001). Delis et al. (1996) found that narrow-mouthed toads were less abundant after urbanization occurred at a Tampa, Florida, area, but were the most abundant species in a nearby, relatively undisturbed, pine flatwoods habitat.

3. Life History Features.

A. Breeding. Reproduction is aquatic.

i. Breeding migrations. Breeding occurs from March–October in southern populations but is more restricted in the northern portion of their range (e.g., Barbour, 1941; Wright and Wright, 1949; Brandt, 1953; Hoffman, 1955; Martof, 1955; Duellman and Schwartz, 1958; Nelson, 1972c). Calling is usually initiated by heavy rains. Eastern narrow-mouthed toads call from April–September in Alabama (Mount, 1975); 1 April–3 September in Florida (Carr, 1940a; Einem and Ober, 1956); May–August in Georgia (Wright, 1932); late March to early October in Louisiana (Anderson, 1954; Dundee and Rossman, 1989); 30 March–September in eastern North Carolina (Funderburg, 1955); late spring to September in South Carolina (Gibbons and Semlitsch, 1991; Hall, 1994); and May–August in Virginia (J. C. M., personal observations). Terrestrial activity in Virginia extends from 28 April–6 October (Hoffman and Mitchell, 1996). Mittleman (1950) notes that breeding males exhibit the secondary sexual characters of enlarged tubercles on the chin and anterior edge of the lower jaw.

ii. Breeding habitat. Males call from temporary ponds, flooded pastures, shallow depressions in open fields, rain-filled ditches, edges of permanent ponds, and open grassy habitats, usually while partially submerged or hidden under grasses and other vegetation or sitting in the mouths of their burrows (Allen, 1932; Brandt, 1936a; Gosner and Black, 1956; Gibbons and Semlitsch, 1991). Males conceal themselves in grass and under vegetation and debris while calling (Carr, 1940a). They often call with just their noses protruding above the water line and

are very difficult to observe. Wright (1932) observed males wiggle their bodies into the wet sand and call with only their snouts protruding.

Male mating calls and calling behavior have been described by Wright (1932), Wright and Wright (1949), Anderson (1954), Nelson (1973), and Conant and Collins (1991), among others. The breeding call sounds like the strong bleat of a lamb. Males apparently do not grip females during amplexus, although Anderson (1954) noted strong axillary amplexus in laboratory matings. Special glands in the sternal region of males secrete a substance that allows adhesion of breeding pairs (Conaway and Metter, 1967; Holloway and Dapson, 1971).

B. Eggs.

i. Egg deposition sites. Females deposit a small sheet of eggs on the water's surface in thick, grassy vegetation or in smaller packets of 10–90 eggs in highly ephemeral pools of water (Wright, 1932; Wright and Wright, 1949; Gibbons and Semlitsch, 1991). Wright (1932) found some attached to vegetation 2.5–5 cm (1–2 in) under water.

ii. Clutch size. Wright and Wright (1949) reported a total of 850 eggs in southeastern Georgia, and Mitchell (1986) noted a total of 1,600 eggs in Virginia. Counts of ovarian follicles ranged from 152–1,089 for a sample of 66 females from Louisiana (Anderson, 1954) and from 186–1,459 for 24 females in Arkansas (Trauth et al., 1999).

C. Larvae/Metamorphosis.

i. Length of larval stage. Eggs hatch in 1–1.5 d (Wright, 1932; Orton, 1946). Tadpoles metamorphose 20–70 d after egg deposition (Wright, 1932; Martof et al., 1980). Larval development occurs quickly, from having "hindlimbs well developed" to complete metamorphosis in 6–10 d (Anderson, 1951). Wright (1932) pointed out the ephemeral nature of the breeding pools and the high rates of mortality due to rapid pool drying.

ii. Larval requirements.

a. Food. Eastern narrow-mouthed toad tadpoles are planktonic feeders but lack the keratinized mouthparts characteristic of other tadpoles in eastern North America (Altig and Kelly, 1974).

b. Cover. Tadpoles will hide under bottom debris and leaves if they are present, but these larvae often occur in pools without any cover on mud to hard clay bottoms (Anderson, 1954; J.C.M., personal observations).

iii. Larval polymorphisms. Unknown and unlikely.

iv. Features of metamorphosis. Size at metamorphosis was 11–13 mm SVL in southeastern Georgia (Wright, 1932), 10–12.5 mm in Louisiana (Orton, 1946; Anderson, 1954), 11 mm in northern Florida (Dodd, 1995b), and 8 mm in southeastern Virginia (Mitchell et al.,

1998). Timing of metamorphosis was mid-June to September in Georgia (Wright, 1932), May–October in Louisiana (Anderson, 1954), and 26 August–9 September in Virginia (Mitchell, 1986).

v. Post-metamorphic migrations. Dodd and Cade (1998) determined that movement patterns of eastern narrow-mouthed toads into and out of a breeding pond in Florida were nonrandom in orientation. Juvenile frogs emigrated toward high pine habitats rather than to xeric oak hammocks and grass meadows.

D. Juvenile Habitat. Juveniles have been found in the same habitats as adults, although Dodd and Cade (1998) found differences in habitats selected by emergent juveniles and older juveniles and adults immigrating to the breeding pond in northern Florida.

E. Adult Habitat. Eastern narrow-mouthed toads are secretive and highly terrestrial. The related aspects of cover and moisture are the two most important environmental variables influencing their presence (Carter, 1934; Robeson and Tyson, 1950; Anderson, 1954; Martof, 1955). General habitat types include cypress-gum swamps, bottomland hardwoods, live-oak ridges, pine-oak uplands, sandy woodlands and hillsides, open woods, prairies, mixed hardwoods, pine forests, longleaf pine sandhills, riparian floodplains, brackish marshes, coastal secondary dune scrub forest, and maritime forests (Blanchard, 1922; Wright, 1932; Brandt, 1936a; Wood, 1948; Blair, 1950; Werler and McCallion, 1951; Anderson, 1954; Tinkle, 1959; Dodd, 1992; Buhlmann et al., 1994; Learm et al., 1999).

Eastern narrow-mouthed toads appear to remain in the vicinity of breeding pools awaiting high humidity and heavy rains (Brandt, 1936a). Cover objects such as rocks, decaying logs, mats of vegetation, bark of logs and stumps, and boards along the edges of ponds and streams are often used for shelter (Holbrook, 1842; Wright, 1932; J. C. M., personal observations). Eastern narrow-mouthed toads may also use crayfish burrows, loose leaf mold, and other vegetation for shelter (Carr, 1940a). Goin (1943) found them under mats of dead water hyacinths that had been tossed on the shore. They have been found around human dwellings, often being taken under trash and boards (Duellman and Schwartz, 1958). Duellman and Schwartz (1958) noted a single individual taken from a wood rat (*Neotoma floridana*) nest on Key Largo.

Adults are tolerant of brackish water (Noble and Hassler, 1936; Hardy, 1953; Conant, 1958b; Neill, 1958a). Their ability to tolerate and disperse over saltwater has allowed them to colonize barrier islands and live in brackish marsh habitats.

F. Home Range Size. Tinkle (1959) found that eastern narrow-mouthed toads marked over a 5-mo period in Louisiana moved no more than 3 m and stayed within the same 5-m quadrant. Dodd (1996) reported movement distances away from a breeding pond in Florida of 42–914 m. Males move more than females (Anderson, 1954).

G. Territories. Eastern narrow-mouthed toads are not known to establish territories. Anderson (1954) noted that calling positions are as close as 2 cm or as distant as several meters apart.

H. Aestivation/Avoiding Desiccation. Werler and McCallion (1951) noted a "spittle-like" substance on a narrow-mouthed toad found in summer under a piece of cardboard in sand and suggested that it may have been secreted by the individual frog. However, true aestivation has not been documented. Individuals occasionally seek refuge behind and in between dead fronds of cabbage palms during droughts (Lee, 1969a).

I. Seasonal Migrations. Eastern narrow-mouthed toads are active during most warm months and may be found in all months of the year in the southern portions of its range. Dodd (1995b) found them in all months, although adult activity was reduced October–April. Migrations to breeding sites are less influenced by seasonal conditions than by rainfall.

J. Torpor (Hibernation). Eastern narrow-mouthed toads overwinter in rotten logs, beneath bark of pine stumps as high as 1 m above ground, buried in sand at the base of small, fallen trees, and under leaf litter (Neill, 1948b; Engles, 1952). Individuals are found occasionally behind and between overlapping dead fronds at the base of cabbage palms during cold periods in northern Florida (Lee, 1969a).

K. Interspecific Associations/Exclusions. Eastern narrow-mouthed toads breed in association with northern cricket frogs (*Acris crepitans*), green treefrogs (*Hyla cinerea*), Cope's gray treefrogs (*H chrysoscelis*), pine woods treefrogs (*H. femoralis*), squirrel treefrogs (*H. squirella*), eastern gray treefrogs (*H. versicolor*), Fowler's toads (*Bufo fowleri*), oak toads (*B. quercicus*), southern toads (*B. terrestris*), Coastal-Plain toads (*B. nebulifer*), and eastern spadefoot toads (*Scaphiopus holbrookii*), as well as eastern newts (*Notophthalmus viridescens*), marbled salamanders (*Ambystoma opacum*), mole salamanders (*A. talpoideum*), tiger salamanders (*A. tigrinum*), and dwarf salamanders (*Eurycea quadridigitata*; Barbour, 1941; Wright and Wright, 1949; Gordon, 1955; Hoffman, 1955; Volpe, 1956; Mount, 1975; Bennett et al., 1980; Pechmann et al., 2000; J.C.M., personal observation). Pechmann et al. (2001) found the following species of anurans in the same ponds with eastern narrow-mouthed toads in South Carolina: northern cricket frogs, oak toads, southern toads, Cope's gray treefrogs, pine woods treefrogs, barking treefrogs (*H. gratiosa*), spring peepers (*Pseudacris crucifer*), southern chorus frogs (*P. nigrita*), ornate chorus frogs (*P. ornata*), American bullfrogs (*Rana*

catesbeiana), green frogs (*R. clamitans*), southern leopard frogs (*R. sphenocephala*), and eastern spadefoot toads. Carr (1940a) found adults with introduced geckos and native skinks under debris in Key West.

Eastern narrow-mouth toads occasionally hybridize with western narrow-mouthed toads (*G. olivacea*) along their zone of sympatry in eastern Oklahoma and Texas (Hecht and Matalas, 1946; Nelson, 1972c). Studies of call patterns, larval development, and body size variation in the overlap zone by A. P. Blair (1950, 1952) and W. F. Blair (1955a,b) indicated that differences are more pronounced in sympatry than in allopatry and that interspecific hybridization, when it occurs, is reinforced by isolating mechanisms (Nelson, 1972c). Volpe (1956) reported finding two mating pairs of male squirrel treefrogs with female eastern narrow-mouth toads and three pairs of male eastern narrow-mouth toads with female squirrel treefrogs in New Orleans, although none of the eggs were fertile.

L. Age/Size at Reproductive Maturity. Adult body sizes range from 20–36 mm SVL (Wright, 1932; Wright and Wright, 1949). Females usually attain larger sizes than males. Wright (1932) noted that males matured in the Okefenokee Swamp in Georgia at 21 mm and females at 22 mm. Maximum body size was 29 mm for males, 30 mm for females. Anderson (1954) noted that males in Louisiana had mature spermatozoa at 18.4 mm, females had pigmented eggs at 23 mm. In southern Florida, males measured 18.8–30.5 mm and females were 22.4–32.5 mm (Duellman and Schwartz, 1958). In northern Florida, males measured 22–34 mm and females 21–35 mm (Dodd, 1995b). Maximum body sizes for narrow-mouthed toads in Louisiana was 32.2 mm for males and 34.3 for females (Anderson, 1954), and 29 mm for males and 32 mm for females in a central Virginia population (Mitchell, 1986). Trauth et al. (1999) reported size at maturity for males in Arkansas at 21.5 mm and females at 26.7 mm, both in their second year of life. He also noted maximum body sizes of 33.6 mm and 36.5 mm, respectively. In Louisiana, males first reproduce after a year of post-metamorphic growth; females may reach maturity at either 1–2 yr (Anderson, 1954).

M. Longevity. Conant and Hudson (1949) reported 6 yr, 9.5 mo for a wild-caught adult. Dodd (1995b) documented a wild-caught individual in its fourth year in northern Florida and noted that average life spans under natural conditions remain unknown.

N. Feeding Behavior. Although this small frog will consume a wide variety of prey, ants, termites, and small beetles are the primary taxa in most stomachs (Wood, 1948; Anderson, 1954; Martof, 1955; Du-

ellman and Schwartz, 1958). Other prey include small (maximum length 6.3 mm) snails, isopods, spiders, mites, collembolans, and lepidopterans, many of which are secretive (Anderson, 1954).

O. Predators. Known predators of adults are glossy water snakes (*Regina rigida*), common garter snakes (*Thamnophis sirtalis*), copperheads (*Agkistrodon contortrix*), eastern cottonmouths (*A. piscivorus*), and cattle egrets (*Bubulcus ibis*; Wright, 1932; Anderson, 1942; Hamilton and Pollack, 1955, 1956; Jenni, 1969; Trauth and McAllister, 1995).

P. Anti-Predator Mechanisms. Eastern narrow-mouthed toads avoid predators by burrowing and seeking cover, and by nocturnal activity patterns. Carr (1940a) described them as "nimble and active when frightened." He noted that they dive into the mouths of crayfish burrows after exposure of their hiding places under logs, and that they rapidly burrow out of sight in loose leaf mold. Mucous secretions produce a violent burning sensation to one's eyes, irritate membranes in the mouth and throat, and may be toxic to other amphibians (Anderson, 1954). He described how one frog's secretions caused attacking ants (*Iridomyrmex* sp.) to become entangled in the thick layer on the skin and that the ants were brushed off as the frog dove under a lump of dirt. Garton and Mushinsky (1979) demonstrated that the skin secretions were unpalatable and deterred predators.

Q. Diseases. Unknown.

R. Parasites. Unknown.

4. Conservation.

The northernmost populations of eastern narrow-mouthed toads known are in Maryland. Their listing as a state Endangered species here reflects loss of habitat due to urban sprawl and other ways the landscape has been affected by human activities (Fowler and Stine, 1953; Harris, 1975; Levell, 1997).

Gastrophryne olivacea Hallowell, 1857 "1856"

WESTERN NARROW-MOUTHED TOAD

Michael J. Sredl, Kimberleigh J. Field

1. Historical versus Current Distribution.

Western narrow-mouthed toads (*Gastrophryne olivacea*) are found in the north from central Missouri along the Missouri River valley and extreme southern Nebraska, south through most of Kansas, Oklahoma, Texas (with the exception of the northern portion of the panhandle and western extremes), and the Mexican Plateau. Disjunct populations occur in the Oklahoma Panhandle, southeastern Colorado, western Kansas, central Arkansas, southwestern New Mexico, northeastern New Mexico, and south-central Arizona (Wright and Wright, 1949; Metter et al.,

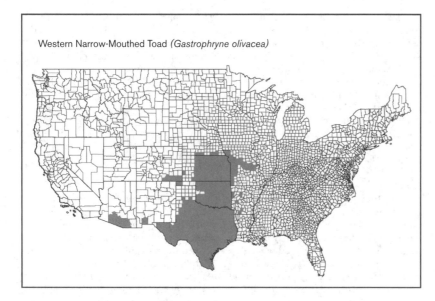

Western Narrow-Mouthed Toad *(Gastrophryne olivacea)*

1970; Collins, 1979; Hammerson, 1980; Neck, 1980a; Price and Price, 1991; Moriarty et al., 2000). Arizona animals are sometimes considered to be a separate subspecies (Stebbins, 1985). Bogert and Oliver (1945) suggest that western narrow-mouthed toads are excluded from California by desert habitats.

Metter et al. (1970) noted that the range of western narrow-mouthed toads in Kansas had expanded into central Kansas along the Missouri River. Blair (W. F., 1955b) observed that as more land in Texas is converted to agriculture, the range of western narrow-mouthed toads should increase.

Sullivan et al. (1996b) surveyed Arizona populations of western narrow-mouthed toads and found them to be present at most historical localities except those on the eastern margin of their Arizona range.

2. Historical versus Current Abundance.
Little is known of historical abundance of western narrow-mouthed toads, and comments on historical abundance are largely anecdotal. Bragg (1941) noted that western narrow-mouthed toads were historically "common and numerous in all parts of Oklahoma. [They are] especially abundant in the prairies and in the oak-hickory savanna areas." Data collected by Bragg (1960b) showed that between 1935–59, populations of western narrow-mouthed toads near Norman, Oklahoma, appeared to be stable. Similarly, Fitch (1956a), using data collected in Kansas, noted that they were one of the most numerous vertebrates and had a biomass that often exceeded that of larger vertebrates. No studies of current abundance for this species have been published.

3. Life History Features.
    A. Breeding. Reproduction is aquatic.
    i. Breeding migrations. Migration to breeding ponds is stimulated by rains >5 cm (Bragg, 1943b; Stebbins, 1954a; Fitch, 1956a), which fill temporary ponds and moisten the earth for overland migration (Fitch, 1956a). Choruses build up rapidly, followed by dwindling activity (Blair, 1961a). Frogs appear to travel a different route when returning to their home range after breeding, perhaps because conditions are drier than those that prevailed during the pondward trip (Fitch, 1956a). Warm rains appear key in initiating reproductive activity, and air and water temperatures below 18.0 °C inhibit breeding (Wiest, 1982).

Breeding takes place from mid-March to September in Texas (Wright and Wright, 1949; Blair, 1961a), but is more restricted farther north (Smith, 1934; Bragg, 1943b; Fitch, 1956a; see also Nelson, 1972a). Rainfall patterns in the southwestern United States may restrict breeding in this area.

    ii. Breeding habitat. Western narrow-mouthed toads utilize permanent and temporary aquatic systems for breeding and have been observed to breed in springs, temporary rain pools, stock ponds, flooded fields, and drainage and roadside ditches (Wright and Wright, 1949; Stebbins, 1954a; Fitch, 1956a; Jameson, 1956a; Metter et al., 1970). Site fidelity may be high. In central Texas, western narrow-mouthed toads returned to the same breeding pond 2 yr in a row (Jameson, 1956a).

When males arrive at the breeding pond, they call from the margins of the pond while hidden in grass hillocks, next to roots of trees, or near bases of large rocks. If they are submerged, only their head protrudes above the water surface (Bragg, 1943b; Stebbins, 1951; Stuart and Painter, 1996). Some males have been observed to call while floating (similar to spadefoot toads), but most maintain an upright posture while calling in shallow water (Stebbins, 1951; Nelson, 1973; Stuart and Painter, 1996). Bragg (1943b) observed that males often clumped at the breeding pond and suggested they were attracted to the calls of one another. Most calling is done at night and usually ceases or is greatly reduced toward midnight or slightly later (Freiburg, 1951), but males may call during the day when breeding is at its peak (Fitch, 1956a).

The call of western narrow-mouthed toads has been described as "a very short peep followed by a buzz like that of an angry bee," a "high, shrill buzz," or "a short whit followed by a low nasal buzz" (Smith, 1934; Conant, 1975; Stebbins, 1985). A single call cannot be heard from >30 m distance (Smith, 1934). A full chorus sounds like a band saw, a swarm of angry bees, or, from a distance, a flock of sheep (Smith, 1934; Stebbins, 1985). Amplexus is axillary and the male "glues" himself to the female using an adhesive produced by specialized secretory cells in the dermis of the venter (Fitch, 1956a). Clasping pairs float in water deeper than where males call, sometimes clinging to vegetation (Henderson, 1961). Males are known to breed at least twice in 1 yr; it is unknown if females breed more than once per year (Fitch, 1956a).

    B. Eggs.
    i. Egg deposition sites. Egg masses of western narrow-mouthed toads have been most frequently described as floating films (Livezey and Wright, 1947; Wright and Wright, 1949; Stebbins, 1951; Salthe, 1963), but occasionally masses may become attached to submerged leaves and stems of grasses and weeds or sink to the bottom of the pond (Livezey and Wright, 1947; Stebbins, 1951; Fitch, 1956a).

    ii. Clutch size. Females can lay at least 650 eggs, and perhaps as many as 2,100 eggs (Livezey and Wright, 1947; Freiburg, 1951; Henderson, 1961), that form clusters of 100–200 (Stebbins, 1951). Water temperatures of oviposition sites in Texas ranged between 19.0–32.0 °C and averaged 24.9 °C (W. F. Blair, 1955b). Water temperature of an oviposition site in Arizona was 28.6 °C (Stebbins, 1951).

Laboratory experiments indicate that the lower lethal temperature limit of developing larvae is between 17.0–18.0 °C, while the upper lethal temperature limit is between 27.0–32.6 °C (Hubbs and Armstrong, 1961; Hubbs et al., 1963; Ballinger and McKinney, 1966) and may be as high as high as 38.9 °C (Ballinger and McKinney, 1966).

    C. Larvae/Metamorphosis. Larvae hatch in 2 d (Wright and Wright, 1949; Fitch, 1956a).

    i. Length of larval stage. Length of larval period reported is generally between 28–50 d (Bragg, 1947; Wright and Wright, 1949; Duellman and Trueb, 1986; see also Nelson, 1972a), although Fitch (1956a) reported larvae that transformed in 24 d in a Kansas pond. Tadpoles reach a maximum body length of 25 mm (Wright and Wright, 1949).

ii. *Larval requirements*. Western narrow-mouthed toad larvae can tolerate temperatures over 35 °C (Bragg, 1950e; Ballinger and McKinney, 1966).

a. Food. Because tadpoles lack keratinized mouthparts, they cannot graze on anything larger than can fit in their mouth (Hoff et al., 1999) and are restricted to filter feeding (Nelson, 1972b). No detailed study of tadpole diet has been conducted, but likely food items are small plants and animals that accumulate at the surface of the water (Stebbins, 1954a; Altig and Kelly, 1974).

b. Cover. Tadpoles have been observed to float motionless near the surface of the water (Bragg, 1947b; Stuart and Painter, 1996). The significance of this behavior, if any, is unknown.

iii. *Larval polymorphisms*. None reported.

iv. *Features of metamorphosis*. Western narrow-mouthed toads metamorphose from mid-April to October (Wright and Wright, 1949; Fitch, 1956a; Duellman and Trueb, 1986). Size at metamorphosis and average adult body size vary geographically (Nelson, 1972b). Wright and Wright (1949) report size at metamorphosis of 10–12 mm for the species. Fitch (1956a) reports a size at metamorphosis of 15–16 mm, rarely as small as 14.5 mm for Kansas frogs.

v. *Post-metamorphic migrations*. Fitch (1956a) observed that juvenile western narrow-mouthed toads disperse widely. In years when juveniles were unable to disperse from breeding ponds because of drought, predation took a heavy toll (Fitch, 1956a).

D. Juvenile Habitat. Little is known of juvenile habitat characteristics, but they are likely similar to those of adults. Following post-metamorphic migrations in Kansas, juveniles were found in a variety of habitat types (Fitch, 1956a).

E. Adult Habitat. Western narrow-mouthed toads are terrestrial, nocturnal, and secretive. They are usually found in burrows and in retreats such as rocks, mud cracks, tree bark and roots, logs, and litter, usually near the vicinity of water (Dickerson, 1906; Smith, 1934; Wright and Wright, 1949; Stebbins, 1954a; Fitch, 1956a). Conant (1975) succinctly described their habitat needs as anywhere where there is "shelter and moisture."

Freiburg (1951) found that western narrow-mouthed toads used rocks of varying sizes as cover, with the usual choice being a flat rock approximately 8 cm thick and 50 cm in diameter. Rocks that were in groups were more often frequented than isolated rocks. Average soil moisture measured between early June and early August at five rock shelters varied between 9–29% (Freiburg, 1951).

While not usually active during the day, Fitch (1956a) recorded a body temperature of 37.6 °C for one western narrow-mouthed toad. This temperature was taken from a frog that "froze" and remained motionless in the sunshine for 30 s after the rock sheltering it was overturned. It is among the highest body temperatures recorded for any anuran (Hutchison and Dupré, 1992).

Throughout their range, western narrow-mouthed toads are found in varied habitats. In general, these habitats are considered prairies and open woodlands (Nelson, 1972b). In Arizona, they are found from desert grasslands to oak woodland habitats (Jones et al., 1983; Stebbins, 1985). Oklahoma and Kansas habitats are described as flood plains, prairies, and deciduous forests (Bragg, 1941; Blair, 1950e; Fitch, 1956a). Elevational ranges of western narrow-mouthed toads is reported from sea level to 1,250 m (Stebbins, 1985).

F. Home Range Size. Some have suggested that western narrow-mouthed toads do not have well-defined home ranges and seem to wander in any direction where suitable habitat is encountered (Freiburg, 1951). Fitch (1956a) suggested that toads become familiar with an area and the available shelters within it, yet may shift their home ranges throughout their lives. In a Kansas population, the typical home range size was approximated to be 22 m in radius, and males tended to travel farther than did females (Fitch, 1956a).

G. Territories. Males will produce a call that differs from the breeding call. This call appears to be agonistic because it is used in establishing territories (Dayton, 2000b).

H. Aestivation/Avoiding Desiccation. Aestivation has not been reported, but western narrow-mouthed toads are known to spend long periods of inactivity in subterranean retreats during droughts (Dickerson, 1906; Wright and Wright, 1949; Freiburg, 1951; Fitch, 1956a; Blair, 1961a; Bragg, 1965; Sullivan, 1996).

I. Seasonal Migrations. Seasonal movements by western narrow-mouthed toads occur during the breeding season. In Kansas, toads moved to breeding ponds during rains. Interestingly, they may take different routes when moving toward and away from breeding ponds. The change in route is likely necessitated by the drier conditions when leaving the pond as compared to the rainstorms when arriving (Fitch, 1956a). In winter, western narrow-mouthed toads are not likely to be found in their usual spots after the first frost, but may have moved to deeper subterranean sites (Fitch, 1956a) rather than migrating out of their active season home range areas.

J. Torpor (Hibernation). While many have commented on the subterranean habits of western narrow-mouthed toads, hibernation has not been studied in detail. Fitch (1956a) noted that western nar-

row-mouthed toads hibernated between mid-October and early May in Kansas. Freiburg (1951) found that in March and April the few western narrow-mouthed toads that could be found were always discovered under the most massive rocks in the area, which may possibly serve as hibernacula.

K. Interspecific Associations/Exclusions. Western narrow-mouthed toads have been reported to occur with numerous temporary and permanent pond breeding amphibians, including burrowing toads (*Rhinophrynus dorsalis*), plains spadefoot toads (*Spea bombifrons*), Mexican spadefoot toads (*S. multiplicata*), Couch's spadefoot toads (*Scaphiopus couchii*), eastern spadefoot toads (*S. holbrookii*), Colorado River toads (*Bufo alvarius*), Great Plains toads (*B. cognatus*), green toads (*B. debilis*), Sonoran green toads (*B. retiformis*), Texas toads (*B. speciosus*), southern toads (*B. terrestris*), Coastal-Plain toads (*B. nebulifer*), Woodhouse's toads (*B. woodhousii*), northern cricket frogs (*Acris crepitans*), canyon treefrogs (*Hyla arenicolor*), eastern gray treefrogs (*H. versicolor*), spotted chorus frogs (*Pseudacris clarkii*), spring peepers (*P. crucifer*), Strecker's chorus frogs (*P. streckeri*), upland chorus frogs (*P. T. feriarum*), lowland burrowing frogs (*Pternohyla fodiens*), plains leopard frogs (*Rana blairi*), Rio Grande leopard frogs (*R. berlandieri*), American bullfrogs (*R. catesbeiana*), green frogs (*R. clamitans*), southern leopard frogs (*R. sphenocephala*), lowland leopard frogs (*R. yavapaiensis*), eastern narrow-mouthed toads (*Gastrophryne carolinensis*), and sheep frogs (*Hypopachus variolosus*; Dickerson, 1906; Campbell, 1934; Wright and Wright, 1949; Jameson, 1950a, 1956a; Freiburg, 1951; Lindsay, 1954; Minton, 1959; Blair, 1961a; Wake, 1961; Bragg, 1965; Wiest, 1982; Jones et al., 1983; Stuart, 1995; Stuart and Painter, 1996; Altig et al., 1998).

Western narrow-mouthed toads hybridize with eastern narrow-mouthed toads and sheep frogs (Nelson, 1972a). Of particular interest have been zones of sympatry between western narrow-mouthed toads and eastern narrow-mouthed toads. The principal isolating mechanisms between these taxa appear to be (1) habitat differences and (2) call differences (W. F. Blair, 1955b). In the zone of sympatry, eastern narrow-mouthed toads are usually found in the more mesic, forested habitats. Blair (W. F., 1955b) studied calls from localities throughout the range of these species and found that calls from the overlap zone differed to the greatest degree. Blair (W. F., 1955b) noted that clearing land for agriculture in the overlap zone may create more xeric habitat, thereby allowing western narrow-mouthed toads to expand their range and replace eastern narrow-mouthed toads.

Interactions between western narrow-mouthed toads and other anuran species have been less studied. Most attention has been paid to breeding phenology (Blair, 1961a; Wiest, 1982). Licht (1967a) describes inhibition of western narrow-mouthed toad tadpoles' growth and even death due to the presence of tadpoles of other genera in laboratory experiments.

Western narrow-mouthed toads have been found in burrows with eastern moles *(Scalopus aquaticus)*, collared lizards *(Crotaphytus collaris)*, common five-lined skinks *(Eumeces fasciatus)*, Great Plains skinks *(E. obsoletus)*, and tarantulas (Blair, 1936; Freiburg, 1951; Hunt, 1980; McAllister and Tabor, 1985). Of the tarantula burrows checked (n = 100), 75% had from 1–3 western narrow-mouthed toads in them (Blair, 1936). Western narrow-mouthed toads may derive protection from tarantulas, while tarantulas may benefit from having their nests kept free of ants (Hunt, 1980).

L. Age/Size at Reproductive Maturity. Western narrow-mouthed toads reach sexual maturity at 1–2 yr of age; adult sizes range between 19–42 mm (Wright and Wright, 1949; Fitch, 1956b; Stebbins, 1985; Duellman and Trueb, 1986). Size varies geographically (W. F. Blair, 1955a; Nelson, 1972a). The best indicator of a sexually mature male is the presence of a dark, distensible throat pouch (Fitch, 1956a; Degenhardt et al., 1996).

M. Longevity. Western narrow-mouthed toads have been documented to live as long as 7–8 yr in the wild, although only an extremely small percentage do so (Fitch, 1956b).

N. Feeding Behavior. Western narrow-mouthed toads, once referred to as "ant-eating frogs," are indeed ant specialists (Nelson, 1972b; Behler and King, 1998); adults will live next to ant colonies (Tanner, 1950; Fitch, 1956a; see also Nelson, 1972a). In addition to ants, beetle remains have been found in their feces (Freiburg, 1951). The pointed head and leathery skin that is characteristic of western narrow-mouthed toads is typical of other fossorial, ant specialists in the family Microhylidae (Nelson, 1972b).

O. Predators. Anuran predators of western narrow-mouthed toads are American bullfrogs and leopard frogs *(R. pipiens* complex). Increased reproductive success has been linked to decreased numbers of bullfrogs and leopard frogs in breeding ponds (Fitch, 1956a). Documented snake predators of the western narrow-mouthed toad include northern water snakes *(Nerodia sipedon)*, common garter snakes *(Thamnophis sirtalis)*, western ribbon snakes *(Thamnophis proximus)*, checkered garter snakes *(Thamnophis marcianus)*, and copperheads *(Agkistrodon contortrix;* Freiburg, 1951; Clark, 1974; Stuart and Painter, 1996). Short-tailed shrews *(Blarina brevicauda)* are also thought to prey on western narrow-mouthed toads and may be capable of eating a dozen western narrow-mouthed toads in one night (Freiburg, 1951).

P. Anti-Predator Mechanisms. Crypticity, movements, and skin secretions are all used by western narrow-mouthed toads to avoid predators. If hiding cover is removed, western narrow-mouthed toads frequently will remain motionless until further disturbed. Escape movements are sporadic, yet not always swift. The western narrow-mouthed toad's "gait is a combination of running and short hops that are usually only an inch or two in length" (Fitch, 1956a). By changing directions many times and easily finding cover, the toad may be able to escape from a predator. At other times, western narrow-mouthed toads have been seen to crawl through grasses in a fast, mouse-like fashion (Wright and Wright, 1949). Once captured, the toad's slippery (Fitch, 1956a), irritating, and distasteful (Conant, 1975) dermal secretions give it yet another chance for escape. When in water, they float with only the tips of their snouts above water and slip beneath the water without leaving a ripple (Cope, 1889; Dickerson, 1906; Stebbins, 1951).

Q. Diseases. None reported.

R. Parasites. Western narrow-mouthed toads have been found to be infected with myxosporeans *(Myxidium serotinum)*, coccidians *(Isospora fragusum)*, nematodes *(Cosmocercoides dukae, Aplectana itzocanensis)*, cestodes *(Cylindrotaenia americana)*, and chiggers (Freiburg, 1951; Nelson, 1972a; McAllister and Upton, 1987a; Upton and McAllister, 1988; McAllister and Trauth, 1995; Goldberg et al., 1998a). McAllister and Upton (1987a) checked western narrow-mouthed toads for hematozoans, but found none (n = 63). They appear to have a resistance to *Saprolegnia* sp. fungus (Bragg, 1962a).

4. Conservation.
Sullivan et al. (1996b) surveyed Arizona populations of western narrow-mouthed toads and found them to be present at most historical localities except those on the eastern margin of their Arizona range. Nevertheless, western narrow-mouthed toads are included on the Arizona Department of Game and Fish's list of Wildlife of Special Concern as a Threatened Species (Arizona Game and Fish Department, 1996). Western marrow-mouthed toads are also listed as a Species of Special Concern in Colorado (http://wildlife.state.co.us) and are given Endangered status in New Mexico (www.nmcpr.state.nm.us).

More optimistically, Metter et al. (1970) noted that the range of western narrow-mouthed toads has expanded into central Kansas along the Missouri River, and Blair (W. F., 1955b) hypothesized that as more land in Texas is converted to agriculture, the range of western narrow-mouthed toads should increase.

Hypopachus variolosus (Cope, 1866[b])
SHEEP FROG

Frank W. Judd, Kelly J. Irwin

1. Historical versus Current Distribution.
Sheep frogs *(Hypopachus variolosus)* range from Costa Rica northward through Mexico and reach the northern limit of their distribution in southern Texas (Smith and Barlowe, 1978; Conant and Collins, 1991). In Texas, B.C. Brown (1950) gave the range as extending from Cameron and Hidalgo counties along the Rio Grande, and north to Duval County; subsequent maps have extended the range farther north. The distribution as mapped by Raun and Gelbach (1972) showed sheep frog records from 11 counties. Dixon (2000) mapped additional records filling in the gap along those counties bordering the coast, but provided no additional records to the north or west. Based on county records, Dixon (2000) provides the most current distribution in Texas. Sheep frogs are currently known from 15 coun-

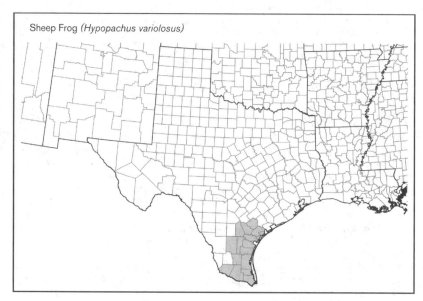

Sheep Frog *(Hypopachus variolosus)*

ties ranging from Cameron, Hidalgo, and Starr along the Rio Grande, north to Aransas and Refugio in the Coastal Bend. It is not known if these range extensions reflect more extensive collecting or a northward movement of the species. Sheep frogs are found where the relative humidity is high (typically 80% or greater) and therefore do not extend inland from the Gulf of Mexico >160 km. Conant and Collins (1991) show a hiatus in the geographic range from central Tamaulipas, Mexico, to the Rio Grande, thus making the Texas population disjunct.

Sheep frogs are currently listed as Threatened by the Texas Parks and Wildlife Department and are protected from collection.

2. Historical versus Current Abundance.

Sheep frogs have never been considered abundant in the United States. Publications from early in the nineteenth century regard them as uncommon and rarely seen (Wright and Wright, 1933; Mulaik and Sollberger, 1938). However, in areas with intact native brushland, after sufficient rainfall, I have found them to be common in Hidalgo, Starr, and Willacy counties, Texas (K.J.I., personal observations).

3. Life History Features.

A. Breeding. Reproduction is aquatic. Wright and Wright (1949) state that breeding takes place from March–September following heavy rains; Mulaik and Sollberger (1938) report that eggs are laid from April–October. Breeding also may be stimulated by irrigation of fields (Wright and Wright, 1949; Conant and Collins, 1991). Wiewandt et al. (1972) found a breeding chorus in southern Sonora, Mexico, at an air temperature of 24.5 °C and a water temperature of 27.7 °C. Nelson (1973) recorded choruses from 26 locations in Texas, Mexico, and Central America at temperatures ranging from 18–30 °C. The combined information from these reports suggests that breeding may occur at any time from March–October. This is in keeping with the temperature and rainfall regimes in southern Texas. Mean air temperature for March at Brownsville, Cameron County, Texas, is 20.2 °C (Lonard et al., 1991) and 69% of the annual rainfall of 68.2 cm occurs from May–October (Lonard et al., 1991). One of us (K.J.I., personal observations) recorded breeding from mid-July to mid-September in Hidalgo, Starr, and Willacy counties in southern Texas. Recently metamorphosed juveniles were observed on 15 July in Willacy County, Texas, indicating that breeding had occurred earlier in the season (K.J.I., personal observations).

i. Breeding migrations. Migrations occur from underground sites to ephemeral wetlands formed during and after rains (Wright and Wright, 1933). In Willacy County, Texas, on 18 July, I found adults active on the surface at 26 °C from 2100–2400 h after rainfall earlier in the day. Although calling males were not heard, available breeding sites were observed nearby (K.J.I., personal observations).

The call of sheep frogs is a clear, resonant bleat, resembling a sheep, from which the species derives its common name. Wiewandt et al. (1972) characterize the call as sounding like a deep, droning, nasal "aaaaaa" (as in bat). The call duration is 1.8–2.7 s and is emitted at 10–20 s intervals. The dominant frequency is in the range of 960–1,272 Hz (Wiewandt et al., 1972). Nelson (1973) reports the harmonic interval increases at about 5 Hz/°C. His analyses of calls from 26 localities throughout the range of the species showed no evidence of geographic variation.

ii. Breeding habitat. Mulaik and Sollberger (1938) report that breeding typically occurs in temporary pools following heavy rains. Wiewandt et al. (1972) found a breeding chorus in a temporary pond in a deciduous subtropical thorn forest in Sonora, Mexico, and McDiarmid and Foster (1975) found a tadpole in the hollow of a tree 87 cm above the ground in Costa Rica. One of us (K.J.I., personal observations) found breeding choruses in areas of intact Tamaulipan brushland of Hidalgo, Starr, and Willacy counties in southern Texas. Sheep frogs have been observed calling in temporary wetlands ranging from roadside ditches, railroad right-of-way ditches, and natural pothole basins. Thus, it appears that breeding may occur in a variety of temporary to permanent aquatic habitats.

Mulaik and Sollberger (1938) report males call while floating free at the water's surface. Conant and Collins (1991) add that floating males at times hold on to stems as they call. In Sonora, Mexico, calling males were at the water's edge, by sticks, or in shallow depressions in mud (Wiewandt et al., 1972).

B. Eggs. Laid within 24 hr following heavy rains.

i. Egg deposition sites. Eggs are laid in rafts loosely held together (Mulaik and Sollberger, 1938) and usually in temporary pools, but see "Breeding habitat" (above) for departures.

ii. Clutch size. Mulaik and Sollberger (1938) report collecting about 700 eggs from a breeding pair. Ova with one envelope range from 1.5–2.0 mm in diameter (Mulaik and Sollberger, 1938).

C. Larvae/Metamorphosis.

i. Length of larval stage. Eggs hatch in approximately 12 hr and larvae transform in 30 d (Mulaik and Sollberger, 1938).

ii. Larval requirements.

a. Food. Unknown, but the mouth parts suggest tadpoles are generalized feeders that take a variety of organic and inorganic material.

b. Cover. Unknown.

iii. Larval polymorphisms. Unknown and unlikely.

iv. Features of metamorphosis. Tadpoles are small (27–30 mm TL). Newly metamorphosed animals measure from 11.0–16.0 mm (Wright and Wright, 1949). Gills are resorbed about 30 hr after hatching (Wright and Wright, 1949).

v. Post-metamorphic migrations. Newly metamorphosed sheep frogs migrate from wetlands to underground upland sites. During these migrations they seek cover under surface objects, including cow dung, old shoes, and litter (H.C. Blanchard, in Wright and Wright, 1933). Recently transformed individuals were found in wet sedges of a water lily pond in San Benito, Cameron County, Texas (Wright and Wright, 1949). On 15 July, five recently metamorphosed juveniles were found active on the surface at 1200 h around the margins of a temporary pond in Willacy County, Texas (K.J.I., personal observations). The weather was overcast, with light rain and 27–29 °C. Although the sample size is small, this observation indicates that newly metamorphosed animals make dispersal movements from breeding sites when humidity or moisture conditions are favorable.

D. Juvenile Habitat. Unknown, presumably similar to adults.

E. Adult Habitat. Sheep frogs occur in warm temperate and tropical thorn scrub and savanna habitats (Nelson, 1974). Throughout their geographic range, sheep frogs are most frequently encountered in open woodlands or pasturelands with abundant short-grass cover (Nelson, 1974). In these habitats, they are found in moist subterranean burrows. Mulaik and Sollberger (1938) state that they do not come out into the open except when driven out by excessive rains, and that they can be found in burrows among the trash (perhaps they mean dens) of packrats and in hollows under trees (presumably fallen trees). Sheep frogs are also commonly found in vegetative debris near ponds and irrigation ditches (Garrett and Barker, 1987). Nelson (1974) suggests that unsuitability of mesic areas may partly account for the northern limit of the species in Texas as "adjacent regions to the north and east (but not the west) are predominantly tall grass or forest." One of us (K.J.I., personal observations) once found four adults under the same log, adjacent to a breeding pond on 15 July in Willacy County, Texas. The surrounding habitat was mature coastal brushland (see Jahrsdoerfer and Leslie, 1988).

F. Home Range Size. Unknown.

G. Territories. Unknown.

H. Aestivation/Avoiding Desiccation. The moist, underground or covered sites used by adults prevent desiccation. Mulaik and Sollberger (1938) note that as humidity decreases, sheep frogs seek out deeper and moister situations or they burrow backwards into the soil where they are not subjected to rapid drying. Brown (B.C., 1950) reported that specimens were found at a depth of 76 cm at the bottom of post-holes during the dry season.

I. Seasonal Migrations. Unknown, other than breeding migrations.

J. Torpor (Hibernation). Unknown.

K. Interspecific Associations/Exclusions. Wright and Wright (1949) note co-occurrences with Coastal-Plain toads (*Bufo nebulifer*) and chorus frogs (*Pseudacris* sp.). One of us (K.J.I., personal observations) found sheep frogs calling syntopically with western narrow-mouthed toads (*Gastrophryne olivacea*) and Couch's spadefoot toads (*Scaphiopus couchii*) in Starr County, Texas. Sympatric amphibians observed in or near sheep frog breeding sites in Willacy County, Texas, include Texas toads (*B. speciosus*), spotted chorus frogs (*P. clarkii*), western narrow-mouthed toads, Rio Grande leopard frogs (*R. berlandieri*), plains spadefoot toads (*Spea bombifrons*), Coastal-Plain toads, and Couch's spadefoot toads. Reptiles observed include yellow mud turtles (*Kinosternon flavescens*), Texas tortoises (*Gopherus berlandieri*), Texas spiny lizards (*Sceloporus olivaceus*), Texas patch-nosed snakes (*Salvadora grahamiae*), western ribbon snakes (*Thamnophis proximus*), checkered garter snakes (*T. marcianus*), Mexican milksnakes (*Lampropeltis triangulum*), coachwhips (*Masticophis flagellum*), Schott's whipsnakes (*M. schotti*), and Texas indigo snakes (*Drymarchon corais*; K.J.I., personal observations).

L. Age/Size at Reproductive Maturity. Males 25.0–37.5 mm SVL, females 29.0–41.0 mm (Wright and Wright, 1949).

M. Longevity. Unknown.

N. Feeding Behavior. Sheep frogs have a specialized diet, taking predominantly ants and termites (Hymenoptera). Mulaik and Sollberger (1938) comment: "Adults driven from under mesquite trees during irrigation were found to have been feeding on termites and minute dipterous insects. Specimens in captivity likewise fed readily upon termites. Even though buried several inches underground, while in captivity, the placing of termites upon the surface readily brought them into the open." Dundee and Liner (1985) recorded the presence of ants in the gut of a sheep frog taken as a prey item of the hylid *Phrynohyas venulosa* in Yucatán, Mexico. Two adults were found in drift fence funnel traps in Hidalgo County, Texas, on the morning of 20 August 1992 (K.J.I., personal observations). This surface activity was prompted by rainfall the previous night and considered to be foraging-induced, as no breeding sites were found nearby nor were choruses heard.

O. Predators. Single ribbon snakes may take up to 9–10 newly metamorphosed sheep frogs (Wright and Wright, 1949). These authorities also suggest that cooters (*Pseudemys* sp.) might prey on newly metamorphosed sheep frogs.

P. Anti-Predator Mechanisms. Nocturnal, secretive.

Q. Diseases. Unknown.

R. Parasites. Mulaik (1945) described a new species of mite *Caeculus hypopachus* (Caeculidae) taken from the venter of a specimen from Hidalgo County, Texas.

4. Conservation.

Sheep frogs are currently listed as Threatened by the Texas Parks and Wildlife Department and are therefore protected from collection. They are not listed by the U.S. Fish and Wildlife Service, nor are they proposed for listing. There are no data on population density anywhere within the species' geographic range, thus statements about their conservation status are based on the general impressions of scientists and resource managers familiar with the species. Certainly, there are no quantitative data to determine if numbers are increasing or decreasing. These data are sorely needed, but are unlikely to be produced unless the Texas Parks and Wildlife Department or the U.S. Fish and Wildlife Service initiate a comprehensive study of amphibian abundance in southernmost Texas. A 3–5 yr mark–recapture study of the population density of all the amphibians at ≥5 sites in the Lower Rio Grande Valley of Texas (LRGV) would yield valuable information of the abundance of 23 amphibian species, five of which are listed as Threatened in Texas.

Proposed efforts by the U.S. Fish and Wildlife Service, Texas Parks and Wildlife Department, and Texas Nature Conservancy to re-vegetate the LRGV with late successional species promise to restore habitat for sheep frogs and other species. However, if we are to know how effective these re-vegetation programs are, we need baseline data on species richness and abundance at the time of the plantings.

Family Pelobatidae

Scaphiopus couchii Baird, 1854(b)
COUCH'S SPADEFOOT

Steven R. Morey

1. Historical versus Current Distribution.

Couch's spadefoot toads (*Scaphiopus couchii*) range from central Texas and southwestern Oklahoma, south and east from central New Mexico and Arizona, into Nayarit, Zacatecas, and Queretaro, Mexico, throughout much of Baja California, Mexico, and into extreme southeastern California (Stebbins, 1985; Conant and Collins, 1991; Degenhardt et al., 1996). A disjunct population occurs in southeastern Colorado (Livo, 1977). Couch's spadefoot toads have been eliminated wherever urban development and irrigated agriculture have destroyed areas where they once lived. However, as with other desert spadefoot toads, Couch's spadefoot toads readily breed in ephemeral artificial impoundments such as stock tanks and pools that form at the base of road and railroad grades. They have colonized many areas where natural pools are rare or nonexistent. Thus, the distribution differs somewhat from the historical

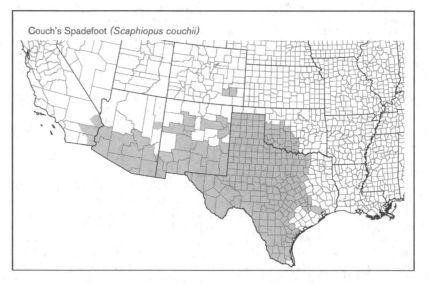

Couch's Spadefoot (*Scaphiopus couchii*)

pattern because it reflects the effects of habitat destruction and colonization of new areas.

2. Historical versus Current Abundance.

Couch's spadefoot toads are probably more abundant than in the past wherever open country still exists and human activities have created ephemeral impoundments (Dimmitt, 1977). Examples of this can be found in southeastern California where, in some places, road and railroad construction has inadvertently increased the number of ephemeral pools, many of which have been colonized by Couch's spadefoot toads. The predominant change in historical versus current abundance, however, is that Couch's spadefoot toads are now absent wherever urban development and irrigated agriculture have destroyed places where they were once abundant. Dimmitt and Ruibal (1980a) encountered Couch's spadefoot toads at a frequency of 0.5/100 km on dry nights and 22/100 km on rainy nights in the San Simon Valley, southeastern Arizona and southwestern New Mexico.

3. Life History Features.
A. Breeding. Reproduction is aquatic.

i. Breeding migrations. Adults are terrestrial and must move from underground refuges to reach breeding sites. Breeding is triggered when summer rainfall fills temporary pools. Breeding is markedly synchronous (Woodward, 1984; Sullivan, 1989a). Breeding usually occurs following heavy rains, from April–August (Wright and Wright, 1949; Stebbins, 1951), even into September in the western portions of the range (Mayhew, 1965). After the first bout of breeding, if showers refill a pool, a second bout of breeding can occur, but this usually involves a smaller number of breeding adults. Little is known about what portion of the population moves to breeding sites each year or how far individuals move to reach the breeding sites.

ii. Breeding habitat. Most breeding occurs in temporary pools that form following intense summer showers. Common breeding sites include ephemeral pools and playas, tanks in rocky streambeds, isolated pools in arroyos, stock tanks, and pools that form at the base of road and railroad grades. In order to support metamorphosis, the breeding pools must remain filled long enough to accommodate the period between egg deposition and hatching, usually about 1 d, and at least the minimum larval period, which is about 7–8 d in the wild (Newman, 1987, 1988; Morey, 1994). Couch's spadefoot toads breed in such ephemeral settings that their larvae seem constantly at risk of desiccating before the aquatic phase is complete. For example, in Big Bend National Park, Newman observed that desiccation was the primary cause of mortality among larvae in 49 of 81 pools surveyed. Likewise, in southeastern California, I observed that 8 of 13 pools surveyed dried completely on or before the day that the first larvae transformed (Morey, 1994).

B. Eggs.

i. Egg deposition sites. Eggs, described in Stebbins (1954a, 1985), are usually deposited on plant stems.

ii. Clutch size. Average clutch size >3,000 eggs (Woodward, 1987a,b).

C. Larvae/Metamorphosis.
i. Length of larval stage. The larval period in Big Bend National Park, Texas, is 8–16 d and is positively correlated with pool duration (Newman, 1987, 1989). In southeastern California, the larval period is 7.5 d (range 7–8.5) and is not distinctly correlated with the duration of the natal pool (Morey, 1994). Newman (1994) showed with experiments that Couch's spadefoot toad larvae respond to low or decreasing food supplies by transforming earlier than larvae reared at higher food levels. Morey and Reznick (2000) demonstrated that the apparent acceleration of development among slow-growing larvae occurs because at low food levels, larvae transform as soon as they reach the minimum size that will support metamorphosis. Fast-growing larvae, on the other hand, delay metamorphosis beyond the minimum, presumably to capitalize on growth in the larval environment.

ii. Larval requirements.

a. Food. Diet of larvae is not well known. Larvae of other spadefoot toad species eat animals, plants, and organic detritus (Pomeroy, 1981; Pfennig, 1990).

b. Cover. Larvae most often occur in turbid pools with little or no other cover. Dried vegetation from the previous wet season is sometimes present.

iii. Larval polymorphisms. None reported.

iv. Features of metamorphosis. As with other spadefoot toad species, body size at metamorphosis is variable. For example, in Big Bend National Park, Texas, the average body length at metamorphosis ranges from 9.5–12.9 mm (Newman, 1989). In southeastern California, body mass at metamorphosis is 0.5g, range 0.2–0.8 (Morey, 1994). A vivid example of the way environment can influence size at metamorphosis was described by Morey and Janes (1994), who compared size and body fat reserves in newly metamorphosed Couch's spadefoot toads that developed in adjacent pools. Both pools filled following a summer shower. One of the pools dried 9 d later, the other remained full much longer. There was no mortality due to drying, even in the short-lived pool, but toadlets from the longer-lived pool were almost four times more massive and had nearly three times more stored body fat than toadlets from the shorter-lived pool. Morey and Reznick (2001) showed that the effects of this type of size variation on characteristics such as survival, growth, and behavior, persist at least several months after metamorphosis. Newman and Dunham (1994) showed that larger toadlets have lower mass-specific rates of water loss. They postulated an advantage for larger toadlets after they leave the natal pond because they can survive longer in dry areas and may have more time to locate suitable refuges. Newman (1999) also showed experimentally that larger toadlets

were better at capturing prey (pinhead crickets) than smaller toadlets. In 1992, I observed in several pools that larvae from bouts of breeding other than the first one of the season have slower growth rates and transform at only about half the size of the larvae from parents that bred with the first pond filling (Morey, 1994). There is some uncertainty about what factors account for the size difference. The first larval cohort may deplete the food resources available to later larval cohorts (Seale, 1980; Loring et al., 1988). This is certainly possible because Newman (1987) demonstrated that food is limiting by manipulating the food supply available to larvae in the wild. Woodward (1987b) suggested that some of the size differences in toadlets from the first versus later cohorts is attributable to quality differences between adults participating in early versus later breeding bouts. In pools where second bouts of breeding occur, there are generally large numbers of tadpole shrimp *Triops* sp. about the same size as Couch's spadefoot toad larvae. Tadpole shrimp are active compared to Couch's spadefoot toad larvae; under these conditions, almost all the Couch's spadefoot toad larvae have damaged tails that appear to be nipped severely, probably by the tadpole shrimp. The accumulation of predators and competitors in desert pools that remain full a long time, or refill, is a well-known phenomenon (Pomeroy, 1981). Harassment and possibly competition for food by tadpole shrimp, which usually does not occur significantly in the first larval cohort, could easily account for the size differences I observed.

v. Post-metamorphic migrations. Post-metamorphic juveniles remain for a few days in the vicinity of the natal pool and are active on the surface as long as the soil remains moist. Eventually, juveniles emigrate from the natal pool and unless showers moisten the soil, the search for suitable refuges must take place over dry soil. Little is known about how far they travel or how they survive the harsh conditions that are typical in the desert summer when these movements take place. Mayhew (1965) observed newly transformed juveniles dispersing across sand dunes 400 m from and 30 m in elevation above a breeding pool. Newman and Dunham (1994) suggest that small juveniles may be incapable of burrowing and rely instead on finding refuge in cracks and holes, such as mammal burrows.

D. Juvenile Habitat. Once they leave the margin of the natal pool, the habitat characteristics of juveniles are probably similar to adults. On rainy nights, adults and juveniles can be encountered together on roads. An exception to the similarity of juvenile and adult habitat characteristics might be in the selection of refuge sites. If, as Newman and Dunham (1994) suggest, small juveniles are unable to burrow, then they must rely on cracks and mammal burrows for refuges, whereas adults almost always dig their own burrows (Ruibal et al., 1969).

E. Adult Habitat. Mesquite and mesquite-yucca, short-grass plains, and creosote desert (Bragg, 1944; Stebbins, 1951, 1985; Mayhew, 1965), as long as temporary rain-filled pools exist. Degenhardt et al. (1996) consider sandy, well-drained soils an important habitat element in New Mexico.

F. Home Range Size. Unknown.

G. Territories. Unknown. There is little evidence of agonistic or territorial behavior. Males seem to maintain individual space while chorusing. Spadefoot toads are solitary during periods of inactivity in burrows (Ruibal et al., 1969). Woodward (1982a) found positive size-assortive mating with larger males mating most often. The pattern was not attributable to female choice, so the result may best be explained by male–male interactions in the form of scramble competition, as postulated by Wells (1977).

H. Aestivation/Avoiding Desiccation. Surface activity is restricted to short periods following summer showers (Ruibal et al., 1969; Dimmitt and Ruibal, 1980a). Thus, much time is spent in underground retreats. In the desert southwest, spadefoot toads spend 8–10 mo in soil-filled "winter" burrows (20 – 90 cm in depth), which they dig themselves (Ruibal et al., 1969). Couch's spadefoot toad burrows sometimes coincide with mammal burrows (McClanahan, 1967). Spadefoot toads survive periods of osmotic stress during long periods of dormancy in burrows by accumulating urea in their body fluids. This allows them to absorb water from the surrounding soil, as long as it has a water potential higher than that of the body fluids (Shoemaker et al., 1969; Jones, 1980). By flooding and excavating pools in Arizona, Ruibal et al. (1969) demonstrated that spadefoot toads do not burrow into the drying mud of a breeding site.

I. Seasonal Migrations. Not known for subadults. Adults make seasonal movements to and from breeding pools. These movements are nocturnal, but little is known about the distance between breeding pools and the winter burrow or about what proportion of the adult population moves to and from the breeding site each year.

J. Torpor (Hibernation). During the period of summer showers, quiet periods are spent in shallow soil-filled summer burrows 1.3–10 cm deep (McClanahan, 1967; Ruibal et al., 1969), often under dense vegetation (Mayhew, 1965).

K. Interspecific Associations/Exclusions. Amphibian communities in the desert southwest tend to be simple compared to those in the eastern United States (Woodward and Mitchell, 1991; Dayton and Fitzgerald, 2001). In southeastern Califor-

nia, for example, Couch's spadefoot toads usually breed alone. Only rarely in this part of the range do red-spotted toads (*Bufo punctatus*) breed in the same pools as Couch's spadefoot toads. Even in more complex desert anuran communities, only one to three species usually make up the breeding assemblage in any one pool (Woodward and Mitchell, 1991, and references therein). The other amphibians that may be encountered at Couch's spadefoot toad breeding sites are usually other spadefoot toads and toads in the genus *Bufo*. Even predators tend to be scarce (Newman, 1987; Woodward and Mitchell, 1991). Among temporary pool breeders in the desert southwest (e.g., *Scaphiopus*, *Spea*, *Bufo*), community structure is probably influenced by intraspecific interactions, such as competition for food, which in turn can influence the outcome of interspecific interactions such as competition and, occasionally, predation (Woodward, 1982b, 1983b, 1987a; Dayton and Fitzgerald, 2001).

L. Age/Size at Reproductive Maturity. Sullivan and Fernandez (1999) used skeletochronology to estimate that most breeding Couch's spadefoot toads are 2–3 yr old. Tinsley and Tocque (1995) also used skeletochronology and estimated that the majority of the breeding population is 5–10 yr old. Stebbins (1985) reports adult body lengths of 56–87 mm. The largest individuals are usually females. Wright and Wright (1949) give the length of adult males as 48–70 mm, adult females as 50–80 mm. A series of 13 calling males measured by Sullivan and Sullivan (1985) ranged from 62–84 mm (mean = 70 mm).

M. Longevity. Using skeletochronology, Tinsley and Tocque (1995) estimated that 5% of the breeding population was >10 yr old. They estimated a maximum longevity of about 13 yr for females and 11 yr for males. A wild-caught adult survived almost 7 yr in captivity (Snider and Bowler, 1992).

N. Feeding Behavior. Adults are nocturnal, but newly transformed juveniles will feed out in the open during the day. Transforming juveniles capture prey with great difficulty until the tail is almost completely resorbed. Morey and Janes (1994) illustrated this when they compared stomach contents of same-aged juveniles from two adjacent pools. The population from one pool was just completing metamorphosis, and individuals still retained remnants of the larval tail (mean = 1.35 mm, n = 50). Eighty percent of the toadlets from this pool had empty stomachs or only a small amount of food in the stomach, and the intestines still contained contents from the larval gut. Fifty-one same-age toadlets from an adjacent pool developed slightly faster and retained no sign of the larval tail. Seventy percent of these had full stomachs, and the intestines had been cleared of larval food.

Newly transformed toadlets eat arachnids and insects, mainly Coleoptera, Collembola, Diptera, and Hymenoptera (Newman, 1999). Adults eat winged and nymphal termites in great numbers when they are available. Other common prey are beetles, especially Carabidae, lygaeid bugs, ants, grasshoppers, crickets, and spiders (Whitaker et al., 1977; Dimmitt and Ruibal, 1980b; Punzo, 1991a). Arthropods with well-known chemical defenses, such as blister beetles, velvet ants, stink bugs, and millipedes, seem to be avoided. Adult Couch's spadefoot toads have an impressive stomach capacity and in experiments voluntarily eat up to 55% of their body weight. Calculations of assimilation efficiency suggest that this is enough food to provide energy reserves sufficient to last ≥1 yr (Dimmitt and Ruibal, 1980b).

O. Predators. Predators on larvae include larval water scavenging beetles (*Hydrophilus* sp.), larval tiger salamanders (*Ambystoma tigrinum*), carnivorous Mexican spadefoot toad (*Spea multiplicata*), larvae yellow mud turtles (*Kinosternon flavescens*), grackles (*Quiscalus* sp.), and skunks (*Spilogale putorius*; Wright and Wright, 1949; Woodward, 1983a,b; Newman, 1987). In the summer of 2000, Mike Westphal discovered a juvenile western diamondback rattlesnake (*Crotalus atrox*) in the process of swallowing a subadult Couch's spadefoot toad on a road in the San Simon Valley, Arizona. Woodward and Mitchell (1991) suggest that, because desert predators are well established in permanent water, proximity to permanent water may increase the level of predation in nearby temporary pools. They also suggest that irrigation practices in the desert may increase the influence of predation in structuring desert anuran communities.

P. Anti-Predator Mechanisms. As with other spadefoot toads, Couch's spadefoot toads can produce volatile skin secretions that cause sneezing and a runny nose in some humans. These secretions are most noticeable after an injury or struggle. Smaller amounts can be produced during handling. Several authors (e.g., Stebbins and Cohen, 1995; Waye and Shewchuk, 1995; Degenhardt et al., 1996) have commented on the irritating quality of the secretions if they come into contact with the eyes, nose, or broken skin. It is probable that the secretions are noxious and repulse some predators.

Q. Diseases. Unknown.

R. Parasites. Couch's spadefoot toads are host to polystomatid monogenean trematode parasites (Tinsley and Earle, 1983). These flukes invade adult toads at breeding pools where they enter through the nostrils and migrate into the lungs and then into the urinary bladder, where they mature and reproduce while the toad is aestivating. When adult toads return to breed the following year, encapsulated fluke larvae are released. Tocque (1993) showed that at the beginning of the active period, infected adults of both sexes had slightly smaller fat bodies than uninfected adults, but the difference disappeared after 2 wk of foraging. Infections from these trematodes apparently do not lead to substantial disease outbreaks in the wild (Tinsley, 1995), and females do not avoid parasitized mates (Pfennig and Tinsley, 2002).

4. Conservation.
Couch's spadefoot toads are now absent wherever urban development and irrigated agriculture have destroyed historical habitats. They are listed as a Species of Special Concern in California (Jennings and Hayes, 1994a) and Colorado (http://wildlife.state.co.us).

In contrast, Couch's spadefoot toads will readily breed in ephemeral artificial impoundments such as stock tanks and pools that form at the base of road and railroad grades. They have colonized many areas where natural pools are rare or nonexistent, and therefore they are probably more abundant than in the past wherever open country exists and humans have created ephemeral impoundments (Dimmitt, 1977).

Acknowledgments. Thanks to Sean Barry and Brian Sullivan for comments on an earlier version of this account.

Scaphiopus holbrookii Harlan, 1935
EASTERN SPADEFOOT

John G. Palis

1. Historical versus Current Distribution.
Eastern spadefoot toads *(Scaphiopus holbrookii)* range from Massachusetts and southeastern New York, south through the Atlantic Coastal Plain to the Florida Keys, west to southeastern Louisiana, southeastern and northeastern Arkansas, and southeastern Missouri. They range northward through western and eastern Kentucky into southwestern Illinois, southern Indiana, southeastern Ohio, and northwestern West Virginia (Conant and Collins, 1991; see also Giovannoli, 1936; Moody, 1986; Tugwell and Schwartz, 1991). Due to their cryptic habits and brief, irregular breeding bouts, the presence of eastern spadefoot toads is difficult to ascertain. The peripheral extent of eastern spadefoot toads is still being determined (e.g., Brandon and Austin, 1966; Klemens, 1993; Redmer and Ballard, 1995). Eastern spadefoot toads no longer inhabit portions of their range due to habitat destruction (McCoy, 1982; Klemens, 1993).

2. Historical versus Current Abundance.
Eastern spadefoot toads can be locally abundant. For example, Carr (1940a) collected over 100 non-breeding individuals during one January night near Gainesville, Florida, and Pearson (1955) estimated populations of 196–621 spadefoots/ac (79–251/ha) near Gainesville. Natural fluctuations in population size associated with annual variations in weather and reproductive success render status assessments of spadefoot populations difficult (Klemens, 1993; Semlitsch et al., 1996). However, historical and recent accounts of the same Connecticut population provide examples of the magnitude of reduction that can occur as a result of habitat loss. Ball (1936) estimated the breeding population of spadefoots at one pond to be 1,000 individuals in 1933 and 800 in 1935. Klemens (1993) reports that this population is now extirpated.

3. Life History Features.
  A. Breeding. Reproduction is aquatic.
  i. Breeding migrations. Eastern spadefoot toads are explosive breeders (Wells, 1977). Breeding activity is stimulated by sufficient rain at minimum temperatures of 7.2–10 °C (45–50 °F; Hansen, 1958; see also Green, 1948; Judy, 1969). Low temperatures can delay emergence for breeding by as much as 7 d following rains (Gosner and Black, 1955). Capable of breeding during any month in the south, eastern spadefoots breed from March–August in the north (Hansen, 1958; Mount, 1975; Ashton and Ashton, 1988; Klemens, 1993). Eastern spadefoots have been captured up to 914 m (1,000 yd) from the nearest body of water (Dodd, 1996). Pearson (1955) determined that eastern spadefoot toads returned to their home sites, sometimes even to the same burrow, after breeding in a pond 0.4 km (0.25 mi) away.

  ii. Breeding habitat. Eastern spadefoot toads breed in a variety of temporary water bodies, including temporary ponds in uplands and bottomlands, flooded fields and roads, roadside ditches and borrow pits (Carr, 1940a; Smith, 1961; Minton, 1972; Mount, 1975; Gibbons and Semlitsch, 1991). Rarely, they breed in permanent ponds (Hansen, 1958), including backyard goldfish ponds (Neill, 1950a). Neill (1957c) observed spadefoots breeding in a freshly dug ditch.

  B. Eggs.
  i. Egg deposition sites. Eggs are attached to submerged vegetation, including the leaves and stems of grasses and forbs, stems and branches of shrubs and small trees, and twigs (Overton, 1915; Wright, 1932; Ball, 1936; Driver, 1936; personal observation). Egg masses are often deposited in select portions of breeding sites (Ball, 1936; Richmond, 1947; personal observation).

  ii. Clutch size. In Illinois, eastern spadefoot toads lay 3,078–5,468 eggs (mean = 3,872, n = 11; unpublished data). Wright (1932) reported a single clutch of 2,332 eggs from Georgia.

  C. Larvae/Metamorphosis.
  i. Length of larval stage. Eggs develop in 1–15 d, and the larval period ranges from 14–60 d depending upon water temperature (Wright and Wright, 1933; Richmond, 1947; Gosner and Black, 1955; Oliver, 1955a, cited in DeGraaf and Rudis, 1983; unpublished data).
  ii. Larval requirements.
  a. Food. Larvae feed on phytoplankton, zooplankton, periphyton, dead plants and animals (e.g., earthworms, tadpoles), and anuran eggs, including their own (Driver, 1936; Richmond, 1947).
  b. Cover. Eastern spadefoot toad tadpoles do not use cover; depending on their age, they swim about individually in the water column or near the bottom or in large schools consisting of thousands of individuals. Compact schools of crowded tadpoles can be 1.2 × 2.4 m (4 × 8 ft), whereas less dense concentrations can be 0.6 × 9.1 m (2 × 30 ft; Richmond, 1947). Ball (1936) estimated that one school consisted of 12,000 tadpoles.

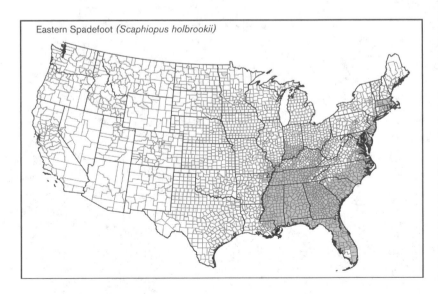

Eastern Spadefoot *(Scaphiopus holbrookii)*

iii. Larval polymorphisms. Not known to occur (Whiteman and Howard, 1998).

iv. Features of metamorphosis. At metamorphosis, spadefoots are 8.5–12.0 mm SVL (Wright, 1932). In Illinois, metamorphic animals average 14.8–15.3 mm and 0.3–0.5g (unpublished data).

v. Post-metamorphic migrations. Eastern spadefoot toads metamorphose and emigrate from breeding sites en masse (Neill, 1957c, and references cited within; personal observation). Ball (1936) found a juvenile 200 m from the breeding site 35 d after metamorphosis.

D. Juvenile Habitat. Assumed to be similar to that of adults. The disappearance of juveniles from areas where they were previously observed suggests that they become fossorial 3.4–6.2 wk following metamorphosis in Illinois (personal observation) and about 6 wk post-metamorphosis in Florida (Pearson, 1955).

E. Adult Habitat. Eastern spadefoot toads occur in open and forested uplands and bottomlands, including ruderal habitats, that have friable, sandy to loamy soils (Stone, 1932; Driver, 1936; Pearson, 1955; Ashton and Ashton, 1988; Dundee and Rossman, 1989). Individuals can sometimes be found at the surface under logs (Smith, 1961). Pearson (1955) determined that spadefoots use the same burrow for 1–713 d (0–24 mo) and emerge about 29 nights annually. Returning to his study area 2 yr after completing his investigation, Pearson (1957) found some individuals within the same home range and using the same burrows as when he last observed them. He concluded that these individuals had used the same burrow for 51 mo and inhabited the same home range for 59.5 mo.

F. Home Range Size. Eastern spadefoot toads have an average home range of 10.1 m^2 (108.4 ft^2; Pearson, 1955).

G. Territories. Territorial behavior has not been observed in eastern spadefoots. However, Pearson (1955) noted that individuals do not share burrows and that home ranges rarely overlap, suggesting that individuals avoid each other.

H. Aestivation/Avoiding Desiccation. In Florida, Pearson (1955) determined that reduced surface activity of eastern spadefoots in summer and winter is not attributable to aestivation or hibernation, but rather to a reduction in the number of nights having conditions suitable for spadefoot surface activity during those seasons.

I. Seasonal Migrations. The only migrations known are those associated with reproduction.

J. Torpor (Hibernation). Eastern spadefoot toads hibernate at the northern end of their range (Klemens, 1993) but remain active throughout the year in the south (Pearson, 1955; Mount, 1975; Ashton and Ashton, 1988; Dundee and Rossman, 1989). Hibernation is sometimes interrupted by surface activity during unusually

warm weather (Redmer and Karsen, 1990; personal observation).

K. Interspecific Associations/Exclusions. Eastern spadefoots breed alone or in association with a variety of temporary-pond breeding anurans including southern cricket frogs (*Acris gryllus*), American toads (*Bufo americanus*), Fowler's toads (*B. fowleri*), eastern narrow-mouthed toads (*Gastrophryne carolinensis*), Cope's gray treefrogs (*Hyla chrysoscelis*), pine woods treefrogs (*H. femoralis*), barking treefrogs (*H. gratiosa*), spring peepers (*Pseudacris crucifer*), ornate chorus frogs (*P. ornata*), Strecker's chorus frogs (*P. streckeri*), western chorus frogs (*P. triseriata*), gopher frogs (*Rana capito*), and southern leopard frogs (*R. sphenocephala*; Brimley, 1896, in Wright, 1932; Wright, 1932; Driver, 1936; Minton, 1972; Brown and Brown, 1973; personal observation). In the absence of predators, eastern spadefoot toad tadpoles out compete hylid tadpoles in artificial ponds (Morin, 1981, 1983b; Wilbur, 1987).

L. Age/Size at Reproductive Maturity. Eastern spadefoots attain sexual maturity between 15–19 mo after metamorphosis (Pearson, 1955), at about 44 mm SVL (Conant and Collins, 1991).

M. Longevity. Captives have lived for 12 yr (Duellman and Trueb, 1986), whereas wild individuals at least 9 yr old have been reported (Pearson, 1955).

N. Feeding Behavior. Eastern spadefoots feed on a variety of terrestrial arthropods (Carr, 1940a; Pearson, 1955; Punzo, 1992a; Jamieson and Trauth, 1996) during nocturnal forays away from or at the mouth of their burrows (Pearson, 1955; Punzo, 1992a).

O. Predators. Confirmed predators of juvenile and/or adult eastern spadefoots include southern toads (*B. terrestris*; Goin, 1955), American bullfrogs (*R. catesbeiana*; Holman, 1957), cottonmouths (*Agkistrodon piscivorus*; Gloyd and Conant, 1990), eastern hog-nosed snakes (*Heterodon platyrhinos*; Hamilton and Pollack, 1956), southern

hog-nosed snakes (*H. simus*; Goin, 1947a), banded water snakes (*Nerodia fasciata*; Palis, 2000a), northern water snakes (*Nerodia sipedon*; Palmer and Braswell, 1995), southern black racers (*Coluber constrictr priapus*; Lynch, 1964), gulls (*Larus* sp.; Carr, 1940a), cattle egrets (*Bulbulcus ibis*; Jenni, 1969), starlings (*Sturnus vulgaris*; Palis, 2000a), opossums (*Didelphis virginiana*; Lynch, 1964), and raccoons (*Procyon lotor*; Lynch, 1964).

P. Anti-Predator Mechanisms. In the presence of predators, eastern spadefoots inflate their lungs to enlarge their bodies, crouch, or attempt to escape by burrowing backwards (Marchisin and Anderson, 1978; Duellman and Trueb, 1986). The fossorial habits of eastern spadefoots probably provide protection from visually oriented predators. Tadpoles appear to lack escape behaviors (Morin, 1985).

Q. Diseases. Unknown.

R. Parasites. Unknown.

4. Conservation.
Due to their secretive habits, the presence of eastern spadefoot toads is difficult to determine, and the peripheral extent of their populations is still being defined. McCoy (1982) and Klemens (1993) state that eastern spadefoot toads no longer inhabit portions of their range due to habitat destruction. Eastern spadefoot toads are considered Rare in Missouri, a Species of Special Concern in Indiana, and Endangered in Ohio (Lannoo, 1998d).

Scaphiopus hurterii Strecker, 1910
HURTER'S SPADEFOOT

Editor's note: While we follow Crother et al. (2000) and consider Hurter's spadefoot toads (*S. hurterii*) to be a distinct species, they historically have been considered a subspecies of eastern spadefoot toads (*S. holbrookii*). The range of Hurter's spadefoot toads is from eastern extensions in central Louisiana west to the Balcones Escarpment of the Edwards Plateau, and

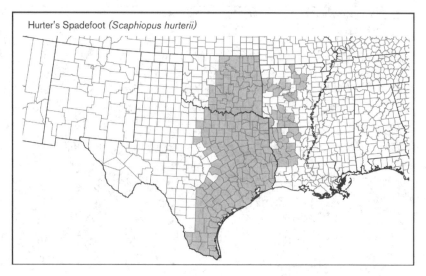

Hurter's Spadefoot (*Scaphiopus hurterii*)

from northern populations in eastern Oklahoma and western Arkansas south to the Rio Grande (Wasserman, 1968). The Mississippi River generally separates Hurter's spadefoot toads from eastern spadefoot toads, however, populations of eastern spadefoot toads occur west of the Mississippi River in Arkansas and Missouri. Hurter's spadefoot toads and eastern spadefoot toads will interbreed (Wasserman, 1957, 1958), and because most aspects of their ecology are similar, features of their life history and natural history are covered under the *Scaphiopus holbrookii* account (Palis, this volume). There is at least one noted difference between these species—Hurter's spadefoot toad tadpoles readily take live invertebrate prey, including mosquito (dipteran) larvae and fairy shrimp (Anostraca; Bragg, 1962b).

Spea bombifrons Cope, 1863
PLAINS SPADEFOOT

Eugenia Farrar, Jane Hey

1. Historical versus Current Distribution.

Plains spadefoot toads *(Spea bombifrons)* were originally described by Cope (1863) with type localities near Fort Union on the Missouri River (latitude 48 °N, on the Platte River 322 km [200 mi] west of Fort Kearney) and Llano Estacado, Texas (Degenhardt et al., 1996). Their range has been described as southern Alberta to northern Mexico, west into southeast Arizona (Shannon, 1953, 1957) or questionably to near Gila Bend (Walters, 1955), east to Nebraska and western Missouri, including central and western Oklahoma and southern Texas (Conant, 1958a). More recently, their range has been extended to include southwestern Manitoba (Preston and Hatch, 1986), new localities in southern Alberta (Lauzon and Balaus, 1998), Saskatchewan (Morlan and Matthews, 1992), the Missouri River

Floodplain and Loess Hills in western Iowa (Huggins, 1971; Christiansen and Mabry, 1985; Farrar and Hey, 1997), and extending across Missouri along the Missouri River (Metter et al., 1970; Easterla, 1972; Femmer and Metter, 1979). Disjunct populations occur in extreme southern Texas and northeast Mexico (Conant and Collins, 1991), along the Arkansas River in Arkansas (Plummer and Turnipseed, 1982; Trauth et al., 1989a), and in southern Colorado (Conant and Collins, 1991).

2. Historical versus Current Abundance.

Relative abundance has not been well documented. The local distribution and secretive nature of plains spadefoot toads make them difficult to census (Farrar and Hey, 1997). Plains spadefoot toads appear to be stable or perhaps increasing in abundance in certain regions, if conclusions are drawn based solely on range expansion records.

3. Life History Features.

A. Breeding. Reproduction is aquatic.

i. Breeding migrations. Migration to breeding sites over distances of at least 1 km/night (Landreth and Christensen, 1971) occur during or following heavy rains.

ii. Breeding habitat. Ephemeral pools such as cattle tanks, flooded farm fields, or playa lakes.

B. Eggs.

i. Egg deposition sites. Eggs are deposited on the bottom or attached to partly submerged vegetation or other objects (Collins and Collins, 1993).

ii. Clutch size. Females lay up to 2,000 eggs in masses of 10–250 each (Collins and Collins, 1993). Hatching takes place in 20 hr at 30 °C (Justus et al., 1977).

C. Larvae/Metamorphosis.

i. Length of larval stage. Time to metamorphosis takes from 13–20 d (King, 1960; Voss, 1961), depending on tempera-

ture and other environmental factors. Larvae tolerate high (39–40 °C) and greatly fluctuating temperatures (Bragg, 1945a).

ii. Larval requirements.

a. Food. Some tadpoles are cannibalistic (see below) or carnivorous, taking small invertebrates; others eat algae and detritus and form aggregations for feeding (Bragg, 1965, 1967; Pomeroy, 1981; Black and Sievert, 1989; Farrar and Hey, 1997). Plains spadefoot toads are stronger competitors for fairy shrimp (Anostraca) than are Mexican spadefoot toads *(S. multiplicata)* when living in sympatry (Pfennig and Murphy, 2002).

b. Cover. Unknown, but larvae live in pools that are frequently murky with little aquatic vegetation (personal observations). Some larvae form aggregations for protection (Bragg, 1965a, 1967; Pomeroy, 1981; Black and Sievert, 1989; Farrar and Hey, 1997).

iii. Larval polymorphisms. Tadpoles occur as two trophic morphs (Bragg, 1956; Bragg and Bragg, 1958; Farrar and Hey, 1997) that appear to be similar in morphology and behavior to those described for Mexican spadefoot toads (Pomeroy, 1981; Pfennig, 1990). Carnivorous morphs are large, flat-headed tadpoles with enlarged jaw muscles, a long snout, and beak with upper cusp and lower notch. Omnivores have smaller jaw muscles, flat beaks, and round bodies. Carnivores develop more rapidly, metamorphose at a larger size (Pfennig et al., 1991; Farrar and Hey, 1997; Hey and Farrar, 1997), and metamorphose sooner than omnivores (unpublished data). Cannibalism is common, and the frequency of cannibalistic morphs may depend on kinship patterns (Pfennig and Frankino, 1997).

D. Juvenile Habitat. Newly metamorphosed animals burrow in mud along the edge of their natal ponds (Bragg, 1945a), hide in cracks in the hard earth, or seek cover in litter near the breeding site (Cornejo, 1982).

E. Adult Habitat. Nocturnal, feed on the surface under humid conditions. When not feeding they occupy shallow summer burrows or deeper winter burrows (Bragg, 1944). Plains spadefoot toads require loose, well-drained soils such as those found in floodplains, prairies, or loess hills in the northeastern part of their range (Christiansen and Bailey, 1991), and grasslands, sandhills, semi-desert shrub, and desert scrub in the southwest (Bragg, 1965; Hammerson, 1986).

F. Home Range Size. Unknown.

G. Territories. Unknown.

H. Aestivation/Avoiding Desiccation. Burrow to depths necessary to remain moist (Hammerson, 1986). Summer burrows are shallow; winter burrows may be soil-filled and as deep as 4.6 m under dry conditions (Bragg, 1965). Rodent burrows may be used for overwintering

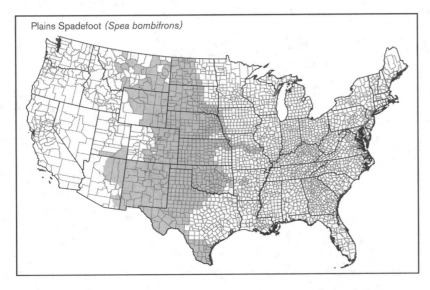

Plains Spadefoot *(Spea bombifrons)*

(Hammerson, 1986). Emergence occurs during the monsoon season in the southwest and may be cued by low frequency sound, as is the case for Couch's spadefoot toads (*Scaphiopus couchii*; Dimmitt and Ruibal, 1980a). In more temperate regions such as Iowa, emergence can occur with spring rains as early as mid-April when temperatures rise above 10 °C (unpublished data). Animals continue feeding throughout the summer on warm humid nights or after rainfall until mid-September (Mabry and Christiansen, 1991).

I. Seasonal Migrations. Unknown.

J. Torpor (Hibernation). See also "Aestivation/Avoiding Desiccation" above. Juveniles are thought to be intolerant of freezing and avoid freezing through burrowing and by having low supercooling temperatures (Swanson and Graves, 1995). These authors estimate that to avoid freezing during mild South Dakota winters, juveniles must burrow to depths of 20–50 cm.

K. Interspecific Associations/Exclusions. Three spadefoot toad species often use the same breeding pond in New Mexico (Cornejo, 1982; Woodward, 1984). Zones of interbreeding between Mexican and plains spadefoot toads have been described in Texas, New Mexico, and Arizona (Sattler, 1985; Simovich et al., 1991; Simovich, 1994). Hybridization has been induced in the laboratory between Mexican spadefoot toads and plains spadefoot toads (Forester, 1975) and may occur in nature (Hughes, 1965). Mating calls serve as reproductive isolating mechanisms (Forester, 1973). Two mating call types have been described for plains spadefoot toads (Pierce, 1976).

Throughout the Great Plains, plains spadefoot toads and Great Plains toads (*Bufo cognatus*) commonly use the same ephemeral breeding sites. Here, plains spadefoot toad tadpoles feed on the smaller Great Plains toad tadpoles (Bragg, 1940a). Plains spadefoot toads also have been reported to share breeding ponds with Woodhouse's toads (*B. woodhousii*; Collins and Collins, 1993).

L. Age/Size at Reproductive Maturity. Reproductive maturity is reached during the second year (Trowbridge and Trowbridge, 1937; Mabry and Christiansen, 1991). Iowa males range from 31–38 mm SVL, females 32–40 mm (Mabry and Christiansen, 1991). Females are also heavier than males (Collins and Collins, 1993). Mature males have keratinized nuptial pads on their thumbs.

M. Longevity. As long as 13 yr based on skeletochronology (Tinsley, 1997).

N. Feeding Behavior. Nocturnal, feeding on beetles, crickets, grasshoppers, ants, spiders, termites, moths, caterpillars, and other small arthropods (Bragg, 1944; Whitaker et al., 1977; Collins and Collins, 1993; Anderson et al., 1999b).

O. Predators. *Hydrophilus triangularis* (hydrophilid beetle larva), *Apus* sp. (a crustacean), western terrestrial garter snakes (*Thamnophis elegans*), and cannibalistic congeners feed on plains spadefoot toad tadpoles (Black, 1968). Predators of adults include prairie rattlesnakes (*Crotalus v. viridis*), Swainson's hawks (*Buteo swainsoni*), and burrowing rodents (Hammerson, 1986).

P. Anti-Predator Mechanisms. Skin gland secretions may cause sneezing, mucous discharge, and watering eyes in humans (Conant, 1975) and presumably could affect predators. Tadpole feeding aggregations only occur in the presence of predators and may deter them (Black, 1968).

Q. Diseases. Unknown.

R. Parasites. Tadpoles have been observed with *Saprolegnia* sp. fungus (Bragg and Bragg, 1957) and adults with intradermal mites (Duszynski and Jones, 1973). They may also serve as hosts for the monogean parasite, *Pseudodiplorchis americanus*, which is known to infect Couch's spadefoot toads.

4. Conservation.
Plains spadefoot toads tend to be locally abundant and currently are not listed by any state or by the U.S. Federal Government. They are listed as Protected in Manitoba (Weller and Green, 1997); their status in Alberta is uncertain (Klassen, 1998). Plains spadefoot toads require loose, well-drained soils for burrowing and aestivating; grasslands (Anderson et al., 1999a) or grassy buffer zones for foraging; and shallow, sparsely vegetated, fish-free, ephemeral wetlands for breeding. In parts of their range, natural breeding habitat has been severely reduced by agriculture, industrial, and other types of floodplain development.

Spea hammondii (Baird, 1859, "1857")
WESTERN SPADEFOOT

Steven R. Morey

There are two recognizable groups of North American spadefoot toads, *Scaphiopus* (Holbrook, 1836) and *Spea* (Cope, 1863). With respect to those species that are referable to *Spea*, the literature is divided, with some authors following Bragg (1944, 1945b), Stebbins (1951, 1985), Blair (W.F., 1955c), Zweifel (1956b), and Hall (1998), who treat the two groups as subgenera. We follow B.C. Brown (1950), Smith (1950), Tanner (1989b), Wiens and Titus (1991), Maglia (1998, 1999), and Crother et al. (2000), who recognize the generic distinctness of *Spea*.

1. Historical versus Current Distribution. Western spadefoot toads (*Spea hammondii*) previously were considered to have a wide range in the arid regions of the southwestern United States and northern Mexico, possibly even central and southern Mexico, with a disjunct portion of the range in cismontane California (Tanner, 1939; Bragg, 1944; Smith and Taylor, 1948; Stebbins, 1954a, 1966; Zweifel, 1956b; Tanner, 1989b). Brown (H.A., 1976a) proposed that the populations east of California be

Western Spadefoot (*Spea hammondii*)

recognized as *Spea multiplicata* Cope, 1863 (Mexican spadefoot toads), citing marked differences in morphology, mating calls, and ecology. Patterns of allozyme variation (Sattler, 1980; Wiens and Titus, 1991) subsequently have supported the elevation of Mexican toads to species status.

In the United States, western spadefoot toads are restricted to California, where their range includes the Great Valley and associated foothills and the Coast Ranges east and south of San Francisco Bay (Stebbins, 1985; Jennings and Hayes, 1994a). Western spadefoot toads also occur in northwestern Baja California, Mexico, at lower elevations west of the Sierra San Pedro Martir Crest, south to Mesa de San Carlos (Stebbins, 1985). In California, their historical and current range are nearly identical (Jennings and Hayes, 1994a, see fig. 26). Some gaps in the distribution are probably artifacts of uneven surveys, particularly in the northern portion of their range. Western spadefoot toads have been eliminated wherever urban development and irrigated agriculture have destroyed areas where they once lived (e.g., Fisher and Shaffer, 1996). On the other hand, as with all North American spadefoot toads, western spadefoot toads readily breed in ephemeral artificial impoundments such as stock tanks and pools that form at the base of road and railroad grades, and they have colonized many areas where natural pools are rare or nonexistent. Thus, the distribution of western spadefoot toads probably differs from historical conditions.

2. Historical versus Current Abundance.
Patterns of abundance have been influenced substantially by human activities. Wherever human activities have created ephemeral impoundments where natural pools are rare or absent, western spadefoot abundance is probably higher than in the past. Examples can be found in the Coast Ranges south of San Francisco Bay, which have an abundant supply of manmade stock tanks, and in places along the eastern edge of the Great Valley and the nearby foothills, where a variety of human activities on the grazing lands have created breeding pools where suitable natural pools are rare or absent. The predominant change in historical versus current abundance, however, is that western spadefoot toads are now absent wherever urban development and irrigated agriculture have destroyed places where they were once abundant. For example, western spadefoot toads formerly were widespread but apparently are now absent on the Los Angeles coastal plain (Stebbins, 1972; Jennings and Hayes, 1994a) and much of lowland southern California. Habitat conversion has also caused losses in the Great Valley and its associated foothills (Jennings and Hayes, 1994a; Fisher and Shaffer, 1996). Historical abundance is unknown, but on

rainy nights along a road intersecting a relatively undisturbed vernal pool complex, Morey and Guinn (1992) found an average of 1.16 individuals/km of roadway during the fairly wet winter of 1982–83, and 0.68 individuals/km during the dry winter of 1984–85.

3. Life History Features.
A. Breeding. Reproduction is aquatic.

i. Breeding migrations. Adults are terrestrial and must move from summer refuges to reach breeding sites. Breeding appears to be triggered by rainfall and is generally synchronous, usually occurring 1 or 2 d after late winter or spring rains. Additional bouts of breeding, usually by smaller numbers of individuals, can occur and pools can contain larval cohorts of different ages. Little is known about what portion of the population moves to breeding sites each year, how far individuals move to reach the breeding sites, or how long they spend at breeding sites. Throughout their range, breeding is most likely during February and March (Storer, 1925; Burgess, 1950; Stebbins, 1954a; Feaver, 1971; H.A. Brown, 1976a; Morey, 1998). Males can be heard chorusing intermittently at breeding pools for up to several weeks following the rains that precede the main bout of breeding (Stebbins, 1954a; H.A. Brown, 1976a).

ii. Breeding habitat. Western spadefoot toads breed most frequently in temporary pools such as vernal playas, vernal pools, stock tanks, and pools that form at the base of road and railroad grades, but they occasionally breed in intermittent streams where larvae develop in more or less isolated pools as the streams dry. In order to support metamorphosis, breeding pools must remain filled long enough to accommodate at least the minimum larval period—in nature, about 30 d. There is almost always substantial mortality due to desiccation among larvae born in pools lasting fewer than 35 d after the eggs are laid (Feaver, 1971; Morey, 1998). Nevertheless, it is not uncommon for western spadefoot toads to breed in pools that remain filled for only 3–4 wk, usually with unfortunate consequences for larvae. Feaver (1971), for example, observed the desiccation of entire cohorts of larvae in 17 of 23 vernal pools in the Central Valley (Fresno County).

B. Eggs.

i. Egg deposition sites. Eggs are attached to plant stems or other objects (Stebbins, 1985).

ii. Clutch size. Females lay 300–500 eggs (Stebbins, 1951) in small clusters (10–42 eggs/cluster; Stebbins, 1985).

C. Larvae/Metamorphosis.

i. Length of larval stage. Larvae and eggs are described in Stebbins (1985). In nature, eggs usually hatch in 3–4 d. Larval development is completed in about 58 d (range 30–79) and is positively correlated

with pool duration (Morey, 1998). As with other *Spea*, age at metamorphosis is flexible. Denver et al. (1998) demonstrated that larvae in drying environments transformed earlier than control larvae in constant volume environments. The cessation of feeding in drying environments was suggested as a partial explanation for the apparent acceleration of development. Morey and Reznick (2000) observed a similar effect and demonstrated that slow-growing larvae transform at near the minimum possible size, while fast-growing larvae delay metamorphosis, presumably to capitalize on growth in the larval environment.

ii. Larval requirements.

a. Food. The food of larvae has not been reported, but larvae of other spadefoot toad species eat animals, plants, and organic detritus (Pomeroy, 1981; Pfennig, 1990).

b. Cover. Larvae most often occur in turbid pools with little or no other cover.

iii. Larval polymorphisms. A characteristic carnivorous morph with a broad head, large jaw muscles, a short gut, and rapid development (Pomeroy, 1981) is uncommon but has been observed in San Luis Obispo and Riverside counties, California.

iv. Features of metamorphosis. Once the first front forelimb has emerged (Gosner stage 42; see Gosner, 1960), transforming larvae begin to make short, nocturnal terrestrial excursions—even though at this stage they still possess a long tail—returning by day to the pool if it retains water. Transforming larvae in drying pools often take refuge in moist cracks around the drying edge. The time between emergence of the front forelimbs and the complete resorption of the tail is 2–6 d. During this time, transforming individuals do not eat and can lose 30% or more of their body mass and a comparable amount of stored body fat. Juveniles are unable to capture prey until tail resorption is complete. In nature, body mass at metamorphosis (Gosner stage 42) averages 3.7 g (range 1.5–10.4 g; Morey, 1998).

v. Post-metamorphic migrations. Juveniles migrate from their natal pool a few days after metamorphosis, but little is known about how far they travel or how they survive the dry conditions that are typical in April–June when these movements usually take place. Migrations by juveniles away from the natal pool sometimes coincide with rainfall, but rainfall during this season is rare. Most of these movements take place on calm, humid nights.

D. Juvenile Habitat. Once they leave the margin of the natal pool, the habitat characteristics of juveniles are probably similar to adults. Morey and Guinn (1992) found juveniles and adults together on rainy nights on roads intersecting a vernal pool complex.

E. Adult Habitat. Grasslands, oak woodlands, occasionally coastal sage scrub, or even chaparral in the vicinity of pools suitable for breeding. Most western spadefoot toads are found at elevations below 365 m (1,000 ft). Stebbins (1985) gives an upper elevation extreme of 910 m (3,000 ft), but Ervin et al. (2001b) observed western spadefoot toads as high as 1,365 m (4,500 ft) in the mountains of San Diego County.

F. Home Range Size. Unknown.

G. Territories. Unknown. Little evidence of agonistic or territorial behavior. Males seem to maintain individual space while chorusing. Other species of *Spea* are solitary during periods of inactivity in burrows (Ruibal et al., 1969).

H. Aestivation/Avoiding Desiccation. Surface activity declines during the unbroken hot, dry periods of late spring, summer, and fall. By late summer, adults and juveniles are quiescent, usually in earth-filled burrows they construct themselves. During dry periods, western spadefoot toads are similar to other spadefoot toad species that burrow ≤1 m (Ruibal et al., 1969) and survive periods of osmotic stress during long periods of dormancy by accumulating urea in their body fluids. This allows them to absorb water from the surrounding soil, as long as the soil has a higher water potential than that of the body fluids (Shoemaker et al., 1969; Jones, 1980). Morey and Reznick (2001) found that even juvenile western spadefoot toads can construct burrows 10–20 cm deep in hard dry soil. Juveniles and subadults sometimes share burrows, but most are solitary. Stebbins (1951) reports an observation of western spadefoot toads emerging from kangaroo rat (*Dipodomys* sp.) burrows.

I. Seasonal Migrations. Not known for juveniles and subadults. Adults make seasonal movements to and from breeding pools. These movements are nocturnal and often coincide with rainfall (Morey and Guinn, 1992), but little is known about the distance between breeding pools and the site of the summer burrow. As with other spadefoot toads, western spadefoot toads probably do not spend the summer at the dried breeding pool. By excavating a pool in Arizona, Ruibal et al. (1969) demonstrated that spadefoot toads do not burrow into the drying mud of a breeding site.

J. Torpor (Hibernation). For several months following the first rains of autumn, periods of inactivity are spent in shallow winter burrows, where, if it is not too cold or too dry, individuals can be encountered just after sunset at their burrow entrance with only their eyes protruding from the soil.

K. Interspecific Associations/Exclusions. Western spadefoot toads tend to co-occur with other obligate ephemeral pool breeders including California tiger salamanders (*Ambystoma californiense*), Pacific treefrogs (*Pseudacris regilla*), western toads (*Bufo boreas*), and an introduced

species, American bullfrogs (*Rana catesbeiana*). Introduced fishes, typically mosquitofish (*Gambusia affinis*), green sunfish (*Lepomis cyanellus*), and various bullhead species (*Ameiurus* sp.) occasionally make their way into the ephemeral pools where western spadefoot toads breed. This happens when they swim or are washed into pools during flooding rains, or when they are intentionally released. In the past, large numbers of mosquitofish were broadcast into vernal pool complexes to control mosquitoes. Even though fishes can survive only as long as pools remain filled, I have observed that they sometimes eliminate the larvae of native amphibians, including western spadefoot toads.

Wherever western spadefoot toad larvae occur in pools occupied by larval California tiger salamanders, which are carnivorous, very few toad larvae escape predation. Under these conditions, competition for food, which must be intense elsewhere, is greatly reduced; though low in number, record-sized spadefoot toad larvae are produced (over 10 g; Morey, 1998). The largest transforming western spadefoot toads consistently come from pools inhabited by California tiger salamanders. Morey and Reznick (2001) showed that the effects of size at metamorphosis on terrestrial characteristics of western spadefoot toads, such as survival, growth, and behavior, persist at least several months after metamorphosis.

L. Age/Size at Reproductive Maturity. Age at maturity is unknown. Morey and Reznick (2001) reared western spadefoot toads under a variety of conditions in the laboratory and in outdoor enclosures and found that under high-food conditions, most males developed secondary sexual characteristics by the beginning of the first breeding season following metamorphosis. Females reared under similar conditions made the transition from juvenile to adult dorsal coloration, but had small ovaries that had not reached the vitellogenic stage of the first ovarian cycle. Thus, it seems reasonable that males mature 1–2 yr after metamorphosis, while females probably are not sexually mature until at least the second breeding season following metamorphosis. Stebbins (1985) reports adult body lengths of 37–62 mm. Most individuals mature at 40–45 mm SVL, and adults <40 mm or >60 mm SVL are uncommon (Storer, 1925; Morey and Guinn, 1992). The largest individuals are usually females, but adult size differences between the sexes are not great.

M. Longevity. Unknown. Other North American spadefoot toads live several years (Snider and Bowler, 1992; Tinsley and Tocque, 1995).

N. Feeding Behavior. Adults feed mainly on insects. Morey and Guinn (1992) found that the stomach contents of 14 adult west-

ern spadefoot toads contained predominantly adult beetles (Coleoptera, mostly Carabidae) and larval and adult moths (Lepidoptera, mostly Noctuidae); crickets (Orthoptera), true bugs (Hemiptera), flies (Diptera), ants (Hymenoptera), and earthworms (Annelida) were also present.

O. Predators. Reports of predators include California tiger salamander larvae (Feaver, 1971), adult American bullfrogs (Hayes and Warner, 1985, Morey and Guinn, 1992), garter snakes (*Thamnophis*, sp.; Feaver, 1971; Ervin and Fisher, 2001), and raccoons (*Procyon lotor*; Childs, 1953). Large larvae (roughly Gosner stage 38–39) seem to be particularly attractive to vertebrate predators. At this time, for example, dabbling ducks, especially mallards (*Anas platyrhynchos*), sometimes take up short-term residence at a pool, where they consume all or almost all of the western spadefoot tadpoles. This happens most often in pools with clear water where, presumably, the larvae are easier to detect. Larvae also are vulnerable to predation by vertebrate predators as pools dry. I have found that numerous bird and mammal tracks in the drying mud at the pool's edge usually indicate that the larvae within have been eaten.

P. Anti-Predator Mechanisms. Postmetamorphic juveniles and adults produce skin secretions that may make them unpalatable to some predators. Some support for the theory that the skin is unpalatable is found in the observation that, on rainy nights, small mammals sometimes eat the carcasses of western spadefoot toads killed on roadways, leaving the skin behind. When injured or sometimes when handled, adults produce volatile skin secretions that cause sneezing and a runny nose in some humans. Stebbins (1951) describes the smell as being similar to popcorn or roasted peanuts. To the taste, the sticky skin secretions of an injured western spadefoot toad are strongly suggestive of a pharmacologically active substance; in the eyes or nose, the secretions cause a burning sensation.

Q. Diseases. Unknown.

R. Parasites. Unknown. Other spadefoot toads are host to polystomatid monogenean trematode parasites (Tinsley and Earle, 1983). In the wild, infections from these trematodes apparently do not lead to significant disease (Tinsley, 1995).

4. Conservation.

Western spadefoot toads are listed as a Species of Special Concern in California (Jennings and Hayes, 1994a) and Federally as a Species of Concern (http://sacramento.fws.gov). The habitats of the western spadefoot toad are coincident with some large areas of urban and agricultural development. Western spadefoot toads have been eliminated from some parts of their range by these land uses and more habitat losses are expected. Fortunately, the ephemeral breeding pools of this and most other spadefoot toads do not support many introduced predators and competitors (e.g., the southeastern U.S. fish fauna, bullfrogs). Because of this, western spadefoot toads present a simpler conservation problem than some other species, such as the western ranid frogs. At present, the prospects for the western spadefoot toad seem to be dependent primarily on the equilibrium between urban and agricultural development and undeveloped places where they can live. Introduced tiger salamanders (*Ambystoma tigrinum*) are established in some California localities; it will be worth following any dispersion from these sites, because tiger salamanders can breed in ephemeral pools and their larvae are potent predators of larval western spadefoot toads.

Acknowledgments. Thanks to Sean Barry, Robert W. Hansen, and Michael Westphal for constructive comments on an earlier version of this manuscript.

Spea intermontana (Cope, 1883)
GREAT BASIN SPADEFOOT

Steven R. Morey

There are two recognizable groups of North American spadefoot toads, *Scaphiopus* (Holbrook, 1836) and *Spea* (Cope, 1863). With respect to those species that are referable to *Spea*, the literature is divided, with some authors following Bragg (1944, 1945b), Stebbins (1951, 1985), Blair (W.F., 1955c), Zweifel (1956b), and Hall (1998), who treat the two groups as subgenera. We follow B.C. Brown (1950), Smith (1950), Tanner (1989b), Wiens and Titus (1991), Maglia (1998, 1999), and Crother et al. (2000), who recognize the generic distinctness of *Spea*.

1. Historical versus Current Distribution.
Great Basin spadefoot toads (*Spea intermontana*) occur from south-central British Columbia, Canada, south into the United States where they range from eastern Washington, Oregon, and California through Nevada and Utah, into southern Idaho, northwestern Colorado, and southwestern Wyoming (Stebbins, 1985; Leonard et al., 1993; Hall, 1998). Hall (1998) gives a detailed review of the distribution, which can be confusing because of the complicated taxonomic history of this species and the North American pelobatids in general. The historical and current ranges are similar. The distribution within some parts of the range differs from the historical pattern due to human activities. Great Basin spadefoot toads no longer live in areas where urbanization and other land uses have destroyed habitat (Orchard, 1992; Leonard et al., 1993). On the other hand, they have colonized some new areas

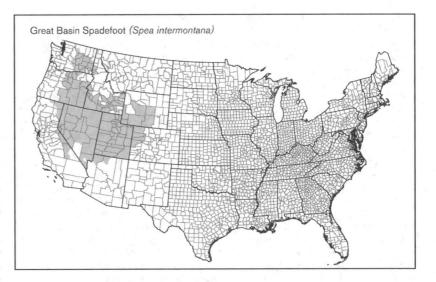

Great Basin Spadefoot (*Spea intermontana*)

where human land uses, such as the construction of reservoirs, have inadvertently created artificial breeding sites where none previously existed. Hovingh et al. (1985), for example, found that 57% of the Great Basin spadefoot toad breeding sites in the Bonneville Basin, Utah, were manmade water sources.

2. Historical versus Current Abundance.
In earlier reports, Great Basin spadefoot toads were usually considered to be common in suitable habitats (Grinnell and Storer, 1924; Tanner, 1939; Wright and Wright, 1949), and they continue to be today. There can be little doubt, however, that patterns of abundance have been influenced by human activities. In grazing country, which includes most of their range, springs and streams have been dammed, diverted into ditches and impoundments, or otherwise altered, and reservoirs have created artificial water sources where natural water sources do not exist. In some cases, Great Basin spadefoot toads have been able to capitalize on these changes, becoming more abundant than under pristine conditions; but where urbanization, agriculture other than grazing, and other land conversions have destroyed or harmed habitats, Great Basin spadefoot toads are absent or less abundant than under pristine conditions.

3. Life History Features.
  A. Breeding. Reproduction is aquatic.
  i. Breeding migrations. Adults are terrestrial and must move from winter refuges to reach breeding sites. Adults become active on the surface during the first warm evenings of spring. Activity is greatest during or following evening rainfall, but daytime activity is not extraordinary. The first such evenings sufficiently warm to promote activity usually occur in April, but newly emerged adults do not necessarily move immediately to breeding sites. The factors that stimulate breeding are not

well known. Linsdale (1938), Hovingh et al. (1985), and others have noted that breeding need not be stimulated by rainfall, as it often is in other North American spadefoot toads. In some areas, breeding occurs in playas or pools that form following spring or summer showers, and the observation by Leonard et al. (1993) that water diversions for irrigation can stimulate breeding is a common one, particularly in pastures and on the margins of agricultural fields. Breeding has been observed in April–July (Wright and Wright, 1949; Nussbaum et al., 1983). Not only is there a great deal of year-to-year variation in the timing of breeding, it is also asynchronous at a site in the same year. For example, in 1989 at Mono Lake, California, the first clutches of eggs were laid on 2–5 June, with other clutches appearing in the same pool on 23 and 25 June. At the same site in 1990, the first clutch appeared on 20 April, and new clutches were laid on 28 April. Males can be expected to chorus intermittently at breeding sites any time from April–June, occasionally as late as July. Linsdale's (1938) estimate of adult migrations of ≤0.8 km (0.5 mi) seems reasonable, but most adults are encountered much closer to breeding sites. Hovingh et al. (1985) seem to suggest that migrations to breeding sites of up to 5 km are not out of the question.
  ii. Breeding habitat. Great Basin spadefoot toads breed in springs, sluggish streams, and other permanent or ephemeral water sources (Wright and Wright, 1949; Nussbaum et al., 1983; Stebbins, 1985; Hall, 1998; and references therein). In the Bonneville Basin, Utah, over half of the breeding sites are manmade reservoirs, the remainder being permanent or temporary springs (Hovingh et al., 1985). In the extreme western edge of the range, just east of the central Sierra Nevada, Morey (1994) found that breeding was restricted to permanent streams and springs. This is because little rainfall occurs in the California

portion of the Great Basin, and snowmelt is insufficient to fill pools. Moving eastward, spring rainfall is more common in the central Great Basin, until, on the Colorado Plateau, most rainfall occurs in the form of summer showers (Kay, 1982) that can be torrential, easily filling playas and other temporary pools. Thus, the reliance on temporary rain-filled pools for breeding increases from west to east across the range. In order to support metamorphosis, the breeding site must remain filled long enough to accommodate the period between egg deposition and hatching (2–4 d) and the minimum larval period, which in the wild is about 36 d (Morey, 1994). In the western portion of the range, larval mortality due to drying is uncommon except when humans divert flows before larval development is complete.

B. Eggs.

i. Egg deposition sites. In the water of temporary rain-filled pools.

ii. Clutch size. Stebbins (1985) reports that females lay 300–500 eggs in packets of 20–40. Leonard et al. (1993) report that females may lay as many as 800 eggs. I have estimated or counted 300; 400; 855; 980; and 1,000 fertilized eggs in clutches produced by captive pairs.

C. Larvae/Metamorphosis.

i. Length of larval stage. The eggs and larvae are described by Stebbins (1985). In the wild, eggs usually hatch in 2–4 d. In the laboratory at 25 °C, embryos hatch in 2 d (Hall, 1998). In the wild, larval development (hatching to the emergence of the first forelimb) is completed in about 47 d, with a range of 36–60 d (Morey, 1994). Hall (1993) and Hall et al. (1997) describe larval development and report a larval period of about 31 d at 25 °C in the laboratory. Brown (H.A., 1989b) reports a larval period of 36 d at 23 °C in the laboratory. As with other *Spea*, the larval period is flexible. Morey (1994, Chapter 2) reared tadpoles at 27 °C and was able to increase the larval period from 16–26 d by manipulating the food supply. Morey and Reznick (2000) demonstrated that slow-growing larvae transform near the minimum size possible, while fast-growing larvae delay metamorphosis beyond the minimum, presumably to capitalize on growth in the larval environment. In the wild, metamorphosis usually occurs from late May to September (Wright and Wright, 1949).

ii. Larval requirements.

a. Food. Specifics of the larval diet have not been reported. The larvae of other spadefoot toads eat animal and plant foods and organic detritus (Pomeroy, 1981; Pfennig, 1990). Great Basin spadefoot toad larvae are routinely reported to feed on conspecific carcasses (Linsdale, 1938) and carrion (Nussbaum et al., 1983).

b. Cover. Amount of cover varies. Unlike other desert spadefoot toads that usually breed in turbid pools, Great Basin

spadefoot toads often breed in clear springs and streams. The amount of emergent vegetation varies from rain-filled pools that have been scoured free of vegetation, to playas and alkaline streams that are ringed with emergent vegetation but otherwise bare, to perennial springs that are choked with aquatic vegetation. Hovingh et al. (1985) reported that successful breeding sites were characterized by being partially free of aquatic vegetation.

iii. Larval polymorphisms. Cannibalism has been reported (Bragg, 1946, 1950f; Durham, 1956). The carnivorous larval morph characteristic of some other spadefoot toad species (Pomeroy, 1981) does not seem prevalent in the wild. Hall and Larsen (1998) and Hall et al. (2002) mention a carnivorous morph, but it has not been described in detail.

iv. Features of metamorphosis. In nature, body mass at metamorphosis (Gosner stage 42; see Gosner, 1960) averages 3.6 g (range 1.8–6.5 g; Morey, 1994). As with other *Spea*, once the front forelimbs emerge (Gosner stage 42), transforming larvae begin to make short, temporary excursions onto land even while still possessing a long tail. The nature of these excursions, which in Great Basin spadefoot toads can occur by day or night, is not known, but may have something to do with avoidance of aquatic predators. The time between emergence of the front forelimbs and the complete resorption of the tail is 2–6 d. During this time, transforming individuals do not eat, losing 30% or more of their body mass and about 16% of their total fat reserves. Cope's (1889) much repeated observation of transforming juveniles, some with complete tails, hopping about on land and gorging on grasshoppers, stretches the imagination, because juveniles have great difficulty feeding on even small, slow-moving prey until tail resorption is complete or nearly so.

v. Post-metamorphic migrations. Juveniles emigrate from their natal site a few days to several weeks after metamorphosis. Little is known about how far they travel or how they survive the harsh, dry conditions that are typical in the Great Basin when these movements usually take place. Migrations by juveniles away from the natal site sometimes coincide with rainfall, but summer rains are unpredictable over much of the range.

D. Juvenile Habitat. Once they leave the margin of the natal site, the habitat characteristics of juveniles are probably similar to adults. Juveniles and adults can be found together on roads on warm or rainy nights.

E. Adult Habitat. Found primarily in sagebrush country. Also found in bunchgrass prairie, alkali flats, semi-desert shrublands, pinyon-juniper woodland to open ponderosa pine communities, and

high elevation (2,800 m) spruce-fir forests (Nussbaum et al., 1983; Stebbins, 1985; Leonard et al., 1993). Hall (1998) cites several other authorities on the habitat associations of Great Basin spadefoot toads.

F. Home Range Size. Unknown.

G. Territories. There is little evidence of agonistic or territorial behavior in Great Basin spadefoot toads. Males seem to maintain individual space while chorusing. Other *Spea* are solitary during periods of inactivity in burrows (Ruibal et al., 1969).

H. Aestivation/Avoiding Desiccation. Great Basin spadefoot toads spend long periods of cold weather, generally October–March, in self-constructed, earth-filled burrows. Nussbaum et al. (1983) indicate that mammal burrows may be used instead of self-made burrows, but no details are provided. Great Basin spadefoot toads are similar to other spadefoot toads, which burrow as deep as ≤1 m (Ruibal et al., 1969) and survive osmotic stress during long periods of dormancy by accumulating urea in their body fluids. This allows them to absorb water from the surrounding soil as long as the soil has a higher water potential than that of the body fluids (Shoemaker et al., 1969; Jones, 1980).

I. Seasonal Migrations. Not known for juveniles and subadults. Adults make seasonal movements to and from breeding sites. These movements are usually nocturnal and do not necessarily coincide with rainfall. Little is known about what proportion of the adult population moves to breeding sites each year or how far individuals move between the winter burrow and the breeding site.

J. Torpor (Hibernation). From April–September, periods of inactivity are spent in shallow burrows; if it is not too cold, individuals can be encountered just after sunset with only their eyes protruding above the surface. Svihla (1953) describes adults retreating during the day beneath rocks near a breeding site in Washington.

K. Interspecific Associations/Exclusions. One notable feature of the breeding sites used by Great Basin spadefoot toads is the absence of other amphibians. In the western part of the range in California, the author has observed no other amphibians breeding at spadefoot sites, and sites occupied by breeding populations of either western toads (*Bufo boreas*) or introduced tiger salamanders (*Ambystoma tigrinum*) seem to be avoided by Great Basin spadefoot toads. Of 151 sites inventoried by Hovingh et al. (1985), only one site contained another amphibian species. An exception to this generality occurs in Deep Springs Valley, California, where Great Basin spadefoot toads and black toads (*B. exsul*) use the same breeding sites. A fascinating experience in the Great Basin is the predictable appearance of intermountain wandering garter snakes (*Thamnophis* *elegans vagrans)* at breeding sites just as Great Basin spadefoot toad larvae reach their maximum size and approach metamorphosis. Over about 1 wk, these predators eat large numbers of larvae (between Gosner stages 38 and 42). Garter snakes usually ignore or are unable to capture smaller, less developed larvae, if any are present.

L. Age/Size at Reproductive Maturity. Unknown. Morey and Reznick (2001) reared closely related western spadefoot toads under a variety of conditions in the laboratory and in outdoor enclosures and found that under high-food conditions, most males developed secondary sexual characteristics by the beginning of the first breeding season following metamorphosis. Females reared under similar conditions made the transition from juvenile to adult dorsal coloration, but had small ovaries that had not reached the vitellogenic stage of the first ovarian cycle. Thus, it seems reasonable that males mature in the first 1–2 yr after metamorphosis, while females probably are not sexually mature until at least the second breeding season after metamorphosis. Nussbaum et al. (1983) speculated that individuals could achieve adult size by their third summer. Nussbaum et al. (1983) report that males mature at a body length of about 40 mm; females mature at about 45 mm. Stebbins (1985) reports adult body lengths of 37–62 mm. Wright and Wright (1949) report adult males as 40–59 mm and adult females as 45–63 mm. In California, males average 57 mm (range 51–65 mm, n = 18) and females average 57 mm (range 51–66 mm, n = 33; unpublished data). Females under 51 mm were not reliably gravid.

M. Longevity. Unknown. Hall (1998) describes a male that must have been at least 6–7 yr old. Tinsley and Tocque (1995) analyzed skeletal growth rings to estimate age structure in a population of Couch's spadefoot toads *(Scaphiopus couchii)*. They estimated that females live approximately 13 yr and males live approximately 11 yr in the wild.

N. Feeding Behavior. Tanner (1931, summarized in Whitaker et al., 1977) found the diet of Great Basin spadefoot toads consisted mostly of ants, with smaller proportions of tenebrionid beetles, adult and larval carabid beetles, larval dytiscid beetles (Coleoptera), Gryllidae, and Ichneumonidae. Adults are generally nocturnal, but a number of authors (e.g., Linsdale, 1938) have noted that juveniles will feed in the open during the day.

O. Predators. Reports of predators on adult Great Basin spadefoot toads include western rattlesnakes *(Crotalus viridis)*, coyotes *(Canis latrans;* Wright and Wright, 1949), and burrowing owls *(Athene cunicularia;* Gleason and Craig, 1979; Green et al., 1993). Larvae are preyed upon by American crows *(Corvus brachyrhynchos;*

Harestad, 1985) and intermountain wandering garter snakes (Wood, 1935). In the eastern Sierra Nevada, California, rainbow trout *(Oncorhynchus mykiss)* and brown trout *(Salmo trutta)* sometimes gorge on larvae or transforming juveniles when rising stream waters flood quiet overflow pools where adults breed. In 1989 near Mono Lake, California, I startled four snowy egrets *(Egretta thula)* as I approached a stream occupied by large larvae (Gosner stage 38–40). All that remained of the entire cohort of larvae when I arrived were thousands of coiled intestines resting in the stream bottom and a few large surviving larvae that were injured and trailing long lengths of intestine. Apparently the intestines were distasteful, and the egrets "popped" the larvae and flicked away the intestines, swallowing the empty carcass. Only smaller, less developed larvae from a younger cohort were uninjured.

P. Anti-Predator Mechanisms. As with other *Spea*, injured or handled adults produce volatile skin secretions that cause an allergic reaction (sneezing and a runny nose) in some humans. Stebbins (1951) and Waye and Shewchuk (1995) describe the smell as being similar to popcorn or roasted peanuts. In the eyes or nose, the sticky skin secretions of an injured Great Basin spadefoot toad can cause a burning sensation. Nussbaum et al. (1983) believe the skin secretions are noxious and probably repulse predators.

Q. Diseases. Unknown.

R. Parasites. Unknown. Other spadefoot toads are host to polystomatid monogenean trematode parasites (Tinsley and Earle, 1983). In nature, infections from these trematodes apparently do not lead to major disease outbreaks (Tinsley, 1995).

4. Conservation.
Great Basin spadefoot toads continue to be common in suitable habitats, however,

patterns of abundance have been influenced by human activities. Throughout most of their range, springs and streams have been dammed or diverted into ditches and impoundments, and reservoirs have created artificial water sources where natural water sources did not exist. In some cases, Great Basin spadefoot toads have been able to capitalize on these changes, becoming more abundant than under pristine conditions; where urbanization, agriculture other than grazing, and other land conversions have destroyed or harmed habitats, Great Basin spadefoot toad populations have been extirpated or have declined. In Colorado, Great Basin spadefoot toads are listed as a Species of Special Concern.

Acknowledgments. Thanks to Sean Barry, Robert W. Hansen, and Michael Westphal for constructive comments on an earlier version of this account.

Spea multiplicata (Cope, 1863)
MEXICAN SPADEFOOT

Steven R. Morey

There are two recognizable groups of North American spadefoot toads, *Scaphiopus* (Holbrook, 1836) and *Spea* (Cope, 1866[a]). With respect to those species that are referable to *Spea*, the literature is divided, with some authors following Bragg (1944, 1945b), Stebbins (1951, 1985), Blair (W.F., 1955c), Zweifel (1956b), and Hall (1998), who treat the two groups as subgenera. We follow B.C. Brown (1950), Smith (1950), Tanner (1989b), Wiens and Titus (1991), Maglia (1998, 1999), and Crother et al. (2000), who recognize the generic distinctness of *Spea*.

1. Historical versus Current Distribution.
Prior to the mid-1970s, Mexican spadefoot toads *(Spea multiplicata)* were widely recognized as a subspecies of western spadefoot toads *(S. hammondii)*. Brown

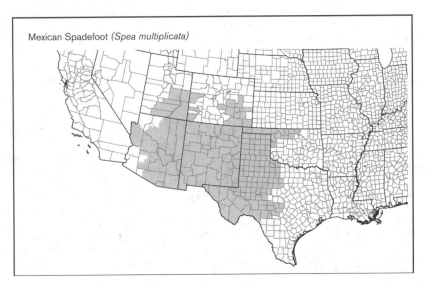

Mexican Spadefoot *(Spea multiplicata)*

(H.A., 1976a) proposed that the populations east of California be recognized as a separate species, *Scaphiopus multiplicatus* Cope, 1863, citing marked differences in morphology, mating calls, and ecology. Patterns of allozyme variation (Sattler, 1980; Wiens and Titus, 1991) have subsequently supported the elevation of *S. multiplicata* to species status. Much of the literature on this species is under the name *Scaphiopus hammondii*. Mexican spadefoot toads are found from southwestern Kansas, western Oklahoma, and central Texas, through New Mexico, southern Colorado, southeastern Utah, Arizona, and in many Mexican states (Stebbins, 1951, 1985; Tanner, 1989b; Conant and Collins, 1991). Unexplained declines have not been observed. The current and historical distributions are similar, but there are minor differences because the current distribution reflects the effects of human activities. For example, Mexican spadefoot toads are now absent where urbanization and water development have destroyed places they once occupied. As with other western spadefoot toads, however, Mexican spadefoot toads readily breed in ephemeral artificial impoundments such as stock tanks and pools that form at the base of road and railroad grades. This has resulted in their colonization of some areas where suitable natural pools are rare or nonexistent.

2. Historical versus Current Abundance.
Human activities have influenced abundance. Mexican spadefoot toads are probably more abundant than in the past wherever open country still exists and artificial ephemeral impoundments have been created where natural pools were rare or absent. Urbanization, water projects, and irrigated agriculture have, in contrast, altered some natural habitats so severely that Mexican spadefoot toads are less abundant than in the past or have been eliminated. Dimmitt and Ruibal (1980a) surveyed roads in the San Simon Valley, southeastern Arizona and southwestern New Mexico, on both dry and rainy nights and encountered Mexican spadefoot toads at a frequency of 0.5/100 km on dry nights and 17/100 km on rainy nights.

3. Life History Features.
A. Breeding. Reproduction is aquatic.

i. Breeding migrations. Adults are terrestrial and must move from underground refuges to reach breeding sites. Migrations are triggered when monsoon rainfall fills temporary pools, usually in July but sometimes later (Wright and Wright, 1949; Stebbins, 1954a; Degenhardt et al., 1996). Breeding is synchronous (Woodward, 1984; Sullivan, 1989a). Little is known about what portion of the population moves to breeding sites each year or how far individuals move to reach breeding sites.

ii. Breeding habitat. Most breeding occurs in temporary pools that form following monsoon rains. Breeding sites include ephemeral pools and playas, cienegas, tanks in rocky streambeds, isolated pools in temporary streams and arroyos, stock tanks, and pools that form at the base of road and railroad grades. Breeding can occur in pools that remain filled from 5 d to several months, and pools can dry and refill in the same season (Pfennig, 1990). In order to support metamorphosis, breeding pools must remain filled at least long enough for eggs to hatch and for the developing larvae to complete metamorphosis, which in nature takes about 3 wk. In the vicinity of Portal, Arizona, Pfennig (1990) observed that pools where eggs were deposited ranged in depth from 5–200 cm (mean = 46 cm) and had a surface-to-volume ratio of 1.35–55 (mean = 11.5). There is a risk of larval mortality due to pool drying. Pfennig (1990) observed that among 37 pools where breeding occurred, only 4 dried completely before any larvae transformed. The food supply available for larvae in desert pools can decline rapidly after filling (Loring et al., 1988).

B. Eggs.

i. Egg deposition sites. Clutches are deposited in clusters on plant stems or rocks (Stebbins, 1985). Eggs hatch in 0.6–6.0 d depending on temperature (H.A. Brown, 1967a,b; Zweifel, 1968b), but usually in < 48 hr in nature.

ii. Clutch size. Clutches average just over 1,000 eggs (Woodward, 1987; Simovitch et al., 1991).

C. Larvae/Metamorphosis.

i. Length of larval stage. Larval period is flexible, but in the wild is usually 12–19 d (Pomeroy, 1981; Pfennig, 1990) and can be as long as 44 d (Pfennig et al., 1991). Under low-food experimental conditions, larval period can exceed 50 d (Pfennig et al., 1991; Simovich et al., 1991).

ii. Larval requirements.

a. Food. Larvae eat organic detritus, algae, crustaceans, and tadpoles (Pomeroy, 1981; MacKay et al., 1990; Pfennig, 1990). There are distinct carnivorous and omnivorous larval forms. Carnivores eat mainly fairy shrimp (*Thamnocephalus* sp. and *Streptocephalus* sp.). Omnivores consume more detritus and algae and fewer fairy shrimp than carnivores (Pomeroy, 1981).

b. Cover. Larvae most often are found in turbid pools with little or no other cover. In some pools, areas with dried vegetation from the previous wet season are present. Some breeding sites are marshy.

iii. Larval polymorphisms. An inducible carnivorous morph, characterized by a broad head, large jaw muscles, a short gut, and rapid development can develop when larvae ingest fairy shrimp (Pomeroy, 1981). Carnivores are most likely to be

found in rapidly shrinking pools with abundant fairy shrimp and low levels of organic debris (Pfennig, 1990). Larval condition, such as body size, plays a role in determining whether the carnivore morphology is expressed (Frankino and Pfennig, 2001). In the San Simon Valley, Arizona, and the Animas Valley, New Mexico, the proportion of carnivores ranges from 0–100% of the population (Pfennig, 1990).

iv. Features of metamorphosis. In nature, carnivorous larvae have more rapid development and transform at smaller sizes and with lower body fat reserves. Because of their more rapid rate of development, carnivores have a better chance than omnivores of completing the larval stage before the pool dries. In longer-lived pools, omnivores, with comparatively slow development, transform at larger sizes and with higher amounts of stored body fat. Pfennig et al. (1991) demonstrated that larger size at metamorphosis was positively correlated with higher post-metamorphic survival in the laboratory, and there is abundant evidence for a terrestrial stage advantage conferred by large size at metamorphosis in spadefoot toads (Newman and Dunham, 1994; Newman, 1999; Morey and Reznick, 2001). Thus, there is a tradeoff between the larval carnivore's advantage in rapidly shrinking pools and the omnivore's post-metamorphic advantage in longer duration pools.

v. Post-metamorphic migrations. Post-metamorphic juveniles remain for a few days in the vicinity of the natal pool and are active on the surface as long as the soil remains moist. Eventually juveniles emigrate from the natal pool and unless showers moisten the soil, the search for suitable refuges must take place over dry soil. Little is known about how far post-metamorphic juveniles travel or how they survive the harsh conditions that are typical in the desert summer when these movements take place. Creusere and Whitford (1976) estimated that as few as 1% of juveniles survive to 6 wk post-metamorphosis. They cite insufficient nutrition, predation, and desiccation among the important causes of mortality. Weintraub (1980) confirmed Creusere and Whitford's (1976) findings, and found that newly metamorphosed Mexican spadefoots will use piles of cow dung as daytime retreats.

D. Juvenile Habitat. Once they leave the margin of the natal pool, the habitats of juveniles are probably similar to adults. On rainy nights, adults and juveniles can be encountered together on roads. One aspect of juveniles that is not well known is their selection of refuge sites. In excavations, juveniles have not been encountered in burrows as commonly as adults. Newman and Dunham (1994) suggested that tiny Couch's spadefoot (*Scaphiopus* *couchii*) toad juveniles are unable to burrow on their own. If this applies to Mexican spadefoot toads, then small juveniles may rely more on cracks and mammal burrows for refuges than adults, which almost always dig their own burrows (Ruibal et al., 1969).

E. Adult Habitat. Mexican spadefoot toads occur in a wide range of arid and semi-arid habitat types, as long as breeding pools exist. They are often found where the soil is sandy or gravelly (Stebbins, 1985). In the vicinity of Portal, Arizona, Pfennig (1990) found Mexican spadefoot toads breeding in pools in semidesert grasslands or Chihuahuan desert scrub. In New Mexico, Mexican spadefoot toads are found in grasslands, sagebrush flats, semi-arid shrublands, river valleys, and agricultural lands (Degenhardt et al., 1996).

F. Home Range Size. Unknown.

G. Territories. There is little evidence of agonistic or territorial behavior. Males seem to maintain an individual space while chorusing. Adults are solitary during periods of inactivity in burrows (Ruibal et al., 1969).

H. Aestivation/Avoiding Desiccation. Surface activity is restricted to short periods following summer showers (Ruibal et al., 1969; Dimmitt and Ruibal, 1980a). Thus, much time is spent in underground retreats. In the desert southwest, spadefoot toads spend 8–10 mo in soil-filled "winter" burrows (20–90 cm in depth) that they dig themselves (Ruibal et al., 1969). Spadefoot toads survive periods of osmotic stress during long periods of dormancy in burrows by accumulating urea in their body fluids. This allows them to absorb water from the surrounding soil, as long as the soil has a water potential higher than that of the body fluids (Shoemaker et al., 1969; Jones, 1980). By flooding and excavating pools in Arizona, Ruibal et al. (1969) demonstrated experimentally that spadefoot toads do not burrow into the drying mud of breeding sites.

I. Seasonal Migrations. Not known for subadults. Adults make seasonal movements to and from breeding pools. These movements are nocturnal, but little is known about the distance between breeding pools and the winter burrow or about what proportion of the adult population moves to and from the breeding site each year.

J. Torpor (Hibernation). During the season of summer showers, periods of activity are spent in shallow, soil-filled summer burrows, 1.3–10 cm deep (Ruibal et al., 1969).

K. Interspecific Associations/Exclusions. In areas of sympatry, matings between Mexican spadefoot toads and other spadefoot toads—usually plains spadefoot toads (*Spea bombifrons*) or Couch's spadefoot toads—can occur. Matings between Mexican spadefoot toads and Couch's spadefoot toads are uncommon and believed to result in inviable eggs. However, hybridization between Mexican spadefoot toads and plains spadefoot toads is well known (H. A. Brown, 1976a; Sattler, 1985); in some areas, such as the San Simon Valley, Arizona, high frequencies of F[1] hybrids and backcross progeny occur, with hybrid adults representing up to 31% of the breeding congress at some pools (Simovich et al., 1991; Simovich, 1994). Hybrid males are sterile; fertile hybrid females produce only about half as many eggs as females of either parental species. Simovich et al. (1991) showed that under controlled conditions, hybrid larvae had better survivorship and developed faster than larvae of either parental species, so there may be a survival advantage for hybrids under some conditions in the wild. Pfennig (2000) conducted phonotaxis experiments on female Mexican spadefoot toads to assess mate choice and found that females from pools with other spadefoot toad species made choices that ensured conspecific matings.

L. Age/Size at Reproductive Maturity. Age at sexual maturity is probably 2–3 yr for males. Woodward used skeletochronology and estimated the average age of breeding males was 2.8 yr (Woodward, 1982). I reared newly transformed juveniles of closely related western spadefoot toads and found that under high-food conditions, most males developed secondary sexual characteristics by the beginning of the first breeding season following metamorphosis. At the same age, females reared under similar conditions had small ovaries that had not reached the vitellogenic stage of the first ovarian cycle. Thus, it seems reasonable that female Mexican spadefoot toads probably mature at a slightly older average age than males. Adult body size is 37–64 mm (Stebbins, 1985). Females may be a bit larger, on average, than males. For example, a series of adult males described by Woodward (1982) ranged from 38–56 mm. During breeding aggregations, Sullivan and Sullivan (1985) collected males that ranged from 41–59 mm. A series of adult females ranged from 42–60 mm (Long, 1989).

M. Longevity. Unknown. Captive wild-caught adult spadefoot toads can live several years in captivity (Snider and Bowler, 1992). In studies on Couch's spadefoot toads, Tinsley and Tocque (1995) used skeletochronology to estimate that 5% of the breeding population was >10 yr old. They made estimates of a maximum longevity of about 13 yr for females and 11 yr for males.

N. Feeding Behavior. Adults are mainly nocturnal, but newly transformed juveniles will feed in the open during the day. Transforming juveniles capture prey with great difficulty until the tail is almost completely resorbed. Adults eat winged and nymphal termites in great numbers when they are available. Other common prey are beetles (Coleoptera, especially Carabidae and Curculionidae), lygaeid bugs (Hemiptera), ants (Formicidae), grasshoppers and crickets (Orthoptera), and spiders (Arachnida; Whitaker et al., 1977; Dimmitt and Ruibal, 1980b; Punzo, 1991a; Anderson et al., 1999b).

O. Predators. Predators on the larvae of desert spadefoot toads include larval water scavenging beetles (*Hydrophilus* sp.), larval tiger salamanders (*Ambystoma tigrinum*), carnivorous conspecific larvae, yellow mud turtles (*Kinosternon flavescens*), grackles (*Quiscalus* sp.), and skunks (*Spilogale putorius*; Wright and Wright, 1949; Woodward, 1983; Newman, 1987). In the summer of 1981, Marie Simovich and Clay Sassaman found spadefoot toad larvae among the stomach contents of 25 of the 35 juvenile American bullfrogs (*Rana catesbeiana*) they examined from a single pool near Portal, Arizona. Woodward and Mitchell (1991) suggest that, because desert predators are well established in permanent water, proximity to permanent water may increase the level of predation in nearby temporary pools. They also suggest that irrigation practices in the desert may increase the influence of predation in structuring desert anuran communities.

P. Anti-Predator Mechanisms. As with other spadefoot toads, Mexican spadefoot toads can produce volatile skin secretions that cause sneezing and a runny nose in some humans; while these secretions are most noticeable after an injury or struggle, a smaller amount can be produced during handling. Several authors (e.g., Stebbins, 1951; Stebbins and Cohen, 1995; Waye and Shewchuk, 1995; Degenhardt et al., 1996) have commented on the irritating quality of the secretions if they come into contact with the eyes, nose, or broken skin. It is probable that the secretions are noxious and repulse some predators.

Q. Diseases. Unknown.

R. Parasites. Spadefoot toads are host to polystomatid monogenean trematode parasites (Tinsley and Earle, 1983). In the wild, infections from these trematodes apparently do not lead to substantial health problems (Tinsley, 1995).

4. Conservation.
Human activities have influenced abundance. Urbanization, water projects, and irrigated agriculture have altered some natural habitats so severely that Mexican spadefoot toads are less abundant than in the past or have been eliminated. Unexplained declines have not been observed. In Colorado, Mexican spadefoot toads are listed as a Species of Special Concern. As with other western spadefoot toads, however, Mexican spadefoot toads readily breed in ephemeral artificial impoundments such as stock tanks and pools that form at the base of road and railroad grades. This has resulted in their colonization of some

areas where suitable natural pools are rare or nonexistent.

Acknowledgments. Thanks to Sean Barry and Brian Sullivan for comments on an earlier version of this account.

Family Pipidae

Xenopus laevis (Daudin, 1802)
AFRICAN CLAWED FROG

John J. Crayon

1. Historical versus Current Distribution.
Introduced in the United States. African clawed frogs *(Xenopus laevis)* were used widely during the 1940s and 1950s as laboratory animals for human pregnancy testing. Animals that were released or had escaped from laboratory stocks, and from the pet trade are the sources of introduced populations. Clawed frogs are still popular in the pet trade in many states. However, concern over the potential impacts of introductions has led to banning their possession in Arizona, California, Florida, Louisiana, Nevada, and Utah (St. Amant et al., 1973; Badger and Netherton, 1995; Tinsley and McCoid, 1996).

Feral clawed frogs have been found in Colorado, Florida, Nevada, New Mexico, North Carolina, Texas, Utah, Virginia, Wisconsin, and Wyoming (Mardht and Knefler, 1973; Bacchus et al., 1993; Tinsley and McCoid, 1996; Blair et al., 1997), but have not established reproducing populations that have persisted over time. Arizona and California are the only states in which apparently permanent populations are known. Other extralimital populations have been documented in Great Britain, the Netherlands, Chile, and Ascension Island (Loveridge, 1959; Veloso and Navarro, 1988; Tinsley and McCoid, 1996).

Most authorities believe that the introduced populations of African clawed frogs found around the world are the subspecies *Xenopus l. laevis*, descendants of animals originally collected from native populations in South Africa (Carr et al., 1987; Tinsley and McCoid, 1996). There has been speculation that >1 taxon may be established in California (Stebbins, 1985; M.R. Jennings, 1987a), which has not been confirmed by molecular-level investigations (Carr et al., 1987). All specimens examined from California populations have been morphologically identical to *X. l. laevis*. In addition, only females of *X. l. laevis* attain sizes greater than 100 mm in Africa (Kobel et al., 1996), and all long-established populations in California produce individuals larger than this.

Reproduction by feral clawed frogs has been documented in five states: Arizona, California, North Carolina, Virginia, and Wisconsin. The North Carolina and Wisconsin frogs have not persisted, largely due to winter temperature extremes (Tinsley and McCoid, 1996). The Virginia

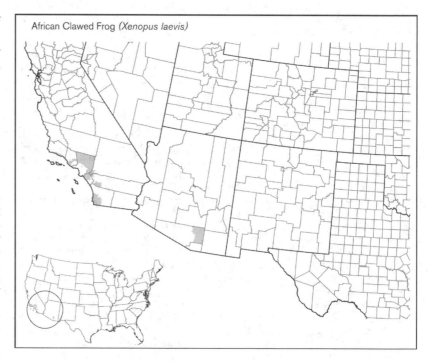

African Clawed Frog *(Xenopus laevis)*

population, first noted in 1982 at the Gulf Branch Nature Center, Arlington, (Zell, 1986), was extirpated by the late 1980s (Tinsley and McCoid, 1996). The Arizona population was introduced in the 1960s to artificial bodies of water on the Arthur Pack Golf Course in Tucson and has remained confined to these sites (Tinsley and McCoid, 1996).

First found as a feral animal in California in 1968 (St. Amant and Hoover, 1969), clawed frogs have since become established in many Southern California drainages, including sites in Los Angeles, Orange, Riverside, San Bernardino, San Diego, Santa Barbara, and Ventura counties (McCoid and Fritts, 1980b; Lafferty and Page, 1997; S. Sweet, personal communication; D. Holland, personal communication). Citations listing clawed frogs in Imperial County (Mardht and Knefler, 1972; Stebbins, 1985) have never been substantiated. The discovery of disjunct populations in California is summarized in Table SP-1. The details surrounding the collection and reporting of almost all of these populations indicate that frogs were present for some period of time prior to their discovery, in some cases probably as long as several years.

The initial establishment of these widely separated, discontinuous populations was clearly the result of separate introduction events. Today, 25–30 yr after the initial discovery of these populations, clawed frogs have now spread throughout most of the drainages that contained the original release sites. For example, populations in Orange County have spread from the coastal plain in the north to the San Diego Creek and Upper Newport Bay drainages and the foothills of the Santa

Ana Mountains as far south as Aliso Creek (B. Goodman, personal communication). In Orange County, clawed frogs have also spread throughout the Santa Ana River drainage into western Riverside and San Bernardino counties (B. Goodman, personal communication).

Clawed frogs originally discovered in tributaries of the Santa Clara River (Zacuto, 1975) have invaded the river and colonized both upstream (J. Dole, personal communication) and downstream to the river's mouth (Lafferty and Page, 1997), now inhabiting the tributaries in Agua Dulce, Soledad, Placerita, and San Francisquito canyons (Tinsley and McCoid, 1996; S. Bautista, personal communication). The clawed frogs first noted in Lake Munz have now colonized neighboring Hughes Lake and Elizabeth Lake and are found throughout the Leona Valley in Amargosa Creek and its tributaries (personal observations). In San Diego County, the Sweetwater, Otay, and Tijuana rivers are now colonized by clawed frogs (R. Fisher, personal communication). These patterns of dispersal mimic patterns of clawed frog dispersal in Africa, where frogs in rivers are carried downstream from breeding habitats, actively move upstream toward headwaters, and utilize human-created bodies of water as "stepping stones" to invade new habitats (van Dijk, 1977).

Some clawed frog populations have stayed confined to relatively discrete locations. The Goleta Slough and Edwards Air Force Base populations have not expanded (the latter primarily due to lack of suitable nearby habitat), and >20 yr after their discovery, clawed frogs from Vail Lake have not moved downstream to invade the main part of the Santa Margarita

First Occurrences of African Clawed Frogs *(Xenopus laevis)* in California Counties

Date	Site	County	Source
1968	Westminster flood control channel	Orange	St. Amant and Hoover (1969)
1969	La Mesa	San Diego	Mardht and Knefler (1973)
1972	Munz Lake	Los Angeles	St. Amant et al. (1973)
1974	U.C. Davis Campus (extirpated)	Yolo	Tinsley and McCoid (1996)
1974	Upper Santa Clara River Drainage	Los Angeles	Zacuto (1975)
1975	Vail Lake	Riverside	McCoid and Fritts (1980b)
1992	Goleta Slough	Santa Barbara	S. Sweet, personal commun.
1992	Chino Hills State Park	San Bernardino	B. Goodman, personal commun.
1995	Santa Clara River Estuary	Ventura	Lafferty and Page (1997)
1996	Piute Ponds, Edwards Air Force Base	Los Angeles	Crayon and Hothem (1998)

River drainage (R. Fisher, personal communication).

Clawed frogs are intolerant of water loss (Hillman, 1980) and not capable of sustained overland travel through the xeric habitats of Southern California. In some areas, the sheer distance between suitable aquatic habitats may thus present an insurmountable barrier to invasion. Most of the spread of the species between drainages has likely been the result of human intervention.

Thirty years after their introduction to California, some patterns of distribution and dispersal of clawed frogs emerge: (1) populations are derived from independent introduction events in five of the counties they now inhabit (San Bernardino and Ventura counties were colonized from neighboring counties); (2) lotic (flowing water) systems are susceptible to complete colonization, including into their brackish interface with tidal waters (e.g., Santa Clara and Sweetwater rivers); (3) some climatic and biological barriers seem to prevent or retard their spread (discussed below); clawed frogs are not present in all apparently suitable habitats (e.g., Santa Barbara and Southern Riverside counties); (4) desert wetlands can sustain clawed frog populations (e.g., Piute Ponds on Edwards Air Force Base); (5) if human-aided introductions continue, there are few freshwater aquatic habitats in California that are not at risk for colonization. Waters that flow either rapidly all year or freeze over completely are among the few systems likely to remain free from invasion.

2. Historical versus Current Abundance.

Although clawed frogs can avoid and survive dry conditions, drought can inhibit both breeding and recruitment, and drought has probably retarded its spread in parts of Southern California. McCoid et al. (1993) recorded declines in populations at some California sites after the extended drought from 1987–91.

The introduction of the largemouth bass *(Micropterus salmoides)* has been implicated in the decline of clawed frogs in Africa, where these frogs are a popular fish bait (Hey, 1949; Mardht and Knefler, 1973). In Southern California, observers believe that the presence of fish has limited the expansion and/or population levels of clawed frogs (Zacuto, 1975; McCoid and Fritts, 1980a).

Introduced animals, including anurans, when released from the limiting effects of predation and interspecific competition, may reach higher densities at colonized sites than where they are indigenous (for anurans see Cohen and Howard, 1958; Lampo and Bayliss, 1996; Lampo and De Leo, 1998). At some sites clawed frogs exist in truly remarkable densities, for example in African fish-free lakes (Tinsley et al., 1996) as aquaculture pests (Schoonbee et al., 1979; Hepher and Pruginin, 1981; Prinsloo et al., 1981; Schramm, 1986), and in California sites free from predatory fish, where their numbers expand to fill the trophic levels fish would normally occupy (Crayon and Hothem, 1998). For example, visual surveys of surfacing frogs, combined with sampling by seining, produced a population estimate of at least 150,000 frogs at Piute Ponds on Edwards Air Force Base, Los Angeles County (unpublished data).

Efforts to eradicate clawed frogs in California are not usually successful (St. Amant, 1975; Zacuto, 1975). There is one known case in which a population in California was permanently eradicated by human efforts—the population on the University of California, Davis campus, Yolo County, cited above. The colonized habitat was a constructed, discrete body of water, which limited frogs to an area where they could be poisoned effectively.

Adult clawed frogs left the water during unsuccessful attempts to poison using rotenone at Vasquez Rock, the site of one of the first populations detected in the upper Santa Clara River drainage (St. Amant, 1975; Zacuto, 1975). Subsequent attempts to develop effective protocols for poisoning in the Santa Clara River also were unsuccessful (J. Dole, personal communication).

Other strategies for removal of clawed frogs include draining all the water from ponds, collecting frogs by seine or trap, electroshocking (Zacuto, 1975; Branning, 1979), and introducing predatory fish (Prinsloo et al., 1981). Early suggestions of biological control (i.e., introducing spiny softshell turtles [*Trionyx spiniferus*; Zacuto, 1975] and American alligators [*Alligator mississippiensis*; Mardht and Knefler, 1973]) wisely were never implemented.

None of these eradication methods have been proven capable of removing all frogs from a habitat or preventing their reintroduction from nearby unaffected areas. The extirpations of clawed frogs from sites in North Carolina, Wisconsin, and Virginia were apparently aided by sub-freezing winter temperatures (Tinsley and McCoid, 1996).

Larvicides are often used to remove anuran larvae when they become pests to aquaculture operations (Helms, 1967; Carmichael, 1983; Kane and Johnson, 1989; Theron et al., 1992; Gabbadon and Chapman, 1996), however, they have not been used against clawed frog larvae in nature.

3. Life History Features.

A. Breeding. Reproduction is aquatic.

i. Breeding migrations. In Africa, clawed frogs are known to migrate to newly filled seasonal rainpools for breeding (Hey, 1949; Thurston, 1967; Balinsky, 1969; Mardht and Knefler, 1973). Feral clawed frogs in Wales migrated 0.2 km in late spring to a spawning site (Tinsley and McCoid, 1996). Such migrations have not been observed in U.S. populations.

Reproduction occurs any time from January–November in California, but is usually confined to late spring, between March and June (McCoid and Fritts, 1989). Male calling intensity peaks in April–May (McCoid, 1985). Patterns of reproduction in Arizona populations have not been examined.

ii. Breeding habitat. Same as adult habitat.

B. Eggs.

i. Eggs deposition sites. Eggs are deposited singly or a few at a time on aquatic plants, rocks, and other benthic structures.

ii. Clutch size. In California, females exhibit asynchronous breeding periods and multiple ovipositions in a single season, depositing hundreds to several thousand eggs at a time; a large female may contain up to 17,000 ova (McCoid and Fritts, 1995). Larvae hatch within 2–3 d (Bles, 1905). Eggs and newly hatched larvae of clawed frogs and western toads (*Bufo boreas*) can often co-occur at sites (personal observations).

C. Larvae/Metamorphosis.

i. Length of larval stage. Unknown under natural conditions; 10–12 wk under laboratory conditions (Bles, 1905; Parker et al., 1947).

ii. Larval requirements.

a. Food. Larvae filter feed while suspended in open water. Food items include phytoplankton, especially unicellular algae and diatoms, protozoans, and bacteria (Bles, 1905; Deuchar, 1975; Schoonbee et al., 1992). Larvae are capable of filtering virus-size particles from the water (Wassersug, 1996).

b. Cover. Larvae are free-swimming within 1–2 d after hatching (stage 47; McCoid and Fritts, 1980a). Because they are weak swimmers (Hoff and Wassersug, 1986), larvae are especially vulnerable to fish predation, and they school in the middle of deeper water to feed, rather than hiding in shallows (Wassersug and Hessler, 1971).

iii. Larval polymorphisms. None reported.

iv. Features of metamorphosis. Developmental stages were described in detail by Nieuwkoop and Faber (1994).

v. Post-metamorphic migrations. Recently metamorphosed clawed frogs migrated overland to colonize a site 0.8 km from their natal pond in Riverside County. Metamorphic animals were also observed migrating overland during a rain in San Diego County, using sheet flooding to facilitate movement (McCoid and Fritts, 1980b).

D. Juvenile Habitat. Not known to be different from adult habitat.

E. Adult Habitat. In extralimital populations, clawed frogs have repeatedly shown plasticity in habitat characteristics such as food availability, vegetation, substrate, turbidity, salinity, water temperature, and hydrology. This makes a precise characterization of habitat characteristics difficult. Highest densities of frogs are reached in permanent, eutrophic, fish-free waters that have soft substrates and submerged vegetation, and do not freeze over but remain above 20 °C for most of the year. Southern California's milder climate (relative to South Africa) seems to accelerate larval development, expand the breeding period, and result in greater adult growth and fat deposition (McCoid and Fritts, 1993, 1995).

Many introduced clawed frog populations are in disturbed or human-made bodies of water (McCoid and Fritts, 1993), such as drainage ditches, flood control channels, golf course ponds, manmade lakes, irrigation canals, cattle tanks, and sewage plant effluent ponds. This affinity for opportunistic colonizing of disturbed habitats is also seen in the parts of Africa where the species' range is expanding. Human-made irrigation canals, lakes, and ponds are especially favored habitats there (Picker, 1985; van Dijk, 1997; Curtis et al., 1998).

F. Home Range Size. Mark–recapture data for two populations in South Wales revealed that distances individuals traveled between captures were under 100 m at one location and 0.2–2 km at another (Measey and Tinsley, 1998).

G. Territories. Group feeding and the swarming of individuals to baited traps suggests that territories are not strictly maintained, if present. Densities can be extremely variable in response to seasonal changes in the extent of the aquatic habitat. Densities greater than $1/m^2$ can be achieved when populations are compressed in drying habitats in California (unpublished data).

H. Aestivation/Avoiding Desiccation. Clawed frogs address the problems of increasing temperatures and decreasing water depth in summer in a number of different ways. In Southern California, individuals construct pits 30–40 cm deep in the mud of evaporating ponds, wherein water remains around 10 °C below the surface temperature of the water (McCoid and Fritts, 1980b). In arid parts of South Africa where ponds dry up seasonally, clawed frogs often burrow deep into the mud to outlast a dry period (Wager, 1965; Balinsky et al., 1967). They are capable of surviving at least 8 mo of starvation in this state (Hewitt and Power, 1913). Clawed frogs also alter the osmotic concentration of body fluids by the retention of urea and, in this hypertonic state, minimize water loss to the surrounding substrate (Balinsky et al., 1967; Stebbins and Cohen, 1995). This ability, coincidentally, makes them one of the most tolerant frogs to saltwater, and predisposes them to invading brackish habitats (Munsey, 1972; Romspert, 1976).

I. Seasonal Migrations. *Xenopus* species in Africa occasionally undertake mass migrations to other water sources when ponds dry up (Hewitt and Power, 1913; Hey, 1949; Loveridge, 1953). Similar behavior was observed during the draining of the San Joachin Reservoir in Newport Beach, California, in 1984. At some critically low water level, a resident population of clawed frogs migrated en masse from the reservoir in a single night and were seen traveling over nearby roads (B. Taylor and D. Otsuka, personal communication). This reservoir had a solid asphalt bottom that precluded frogs from digging down to avoid desiccation and probably precipitated their exodus.

J. Torpor (Hibernation). Poorly documented. Surfacing activity is greatly diminished during colder months in California populations. Nevertheless, frogs in water bodies that ice up at the edges during the winter remain active enough to come to baited traps (unpublished data). Frogs in the extirpated Virginia populations were also sampled by trapping when the pond's surface was frozen (McCoid and Fritts, 1995).

K. Interspecific Associations/Exclusions. In California, clawed frogs are sympatric with western toads, red-spotted toads (*B. punctatus*), California red-legged frogs (*Rana draytonii*), introduced American bullfrogs (*R. catesbeiana*), Pacific treefrogs (*Pseudacris regilla*), and western spadefoot toads (*Spea hammondii*). Concern over the impacts on these native anurans is not unreasonable; clawed frogs are known to eat *Rana* and *Bufo* larvae in Africa (Schramm, 1986). Under confined conditions, such as in aquaria, clawed frogs readily consume larvae of California's anurans (*Bufo, Rana,* and *Hyla*; Lenaker, 1972).

The strongest evidence of negative interactions is for western toads. Clawed frogs have been documented eating both larval and recent metamorphic western toads (Lenaker, 1972). In a sample of 39 adult clawed frogs collected from the Amargosa Creek drainage by David Muth in June 1998, the stomachs of 4 frogs contained 10 small (1.5–2.0 cm SVL) western toads. Because clawed frogs have also invaded the habitats of Pacific treefrogs (Lenaker, 1972; Mardht and Knefler, 1972), red-spotted toads (Crayon, 1997), and western spadefoot toads (E. Ervin, personal communication), there is no reason to expect these species would be exempt from predation.

Early accounts of clawed frogs in California claimed that they were eating bullfrog tadpoles and were responsible for eliminating California red-legged frogs from certain sites (Branson, 1975; Zacuto, 1975). These claims have not been substantiated. California red-legged frogs are virtually extirpated from Southern California south of the Santa Clara River, including from areas that clawed frogs have not invaded. The causes of California red-legged frog declines are the subject of some debate, and are likely symptomatic of a suite of problems (Moyle, 1973; Hayes and Jennings, 1986). Clawed frogs are sympatric with California red-legged frogs at a site in northern Los Angeles County (K. Swaim, unpublished data), but their interactions remain unexamined.

Where thorough systematic surveys were conducted for herpetofauna in Southern California, clawed frogs and American bullfrogs were not found to co-occur frequently (R. Fisher, unpublished data). It is unknown whether this is due to exclusion by competition or predation or due to divergent habitat preferences.

L. Age/Size at Reproductive Maturity.
Females in California mature at approximately 6 mo post-metamorphosis and 65 mm SVL (McCoid and Fritts, 1989, 1995).

In African populations, the average size of adult females is 110 mm SVL (maximum 130 mm SVL); males are 3/4 as large (Kobel et al., 1996). Adults attain a larger maximum size in California populations. Two females collected on Edwards Air Force Base exceeded 140 mm SVL, which is larger than published records for the species in Africa. Early size records in California were likely from populations that were not yet fully mature. If extralimital clawed frogs reach a maximum age ≥ 16 yr, recorded for frogs in South Wales (Measey and Tinsley, 1998), then populations in California may not have contained the oldest and largest size classes before the mid-1980s.

M. Longevity. May attain 15 yr in captivity (Flower, 1936); at least 16 yr in feral populations in Wales (Measey and Tinsley, 1998).

N. Feeding Behavior. Adults are primarily aquatic consumers of slow-moving invertebrates; they are often characterized as rather inept at capturing actively swimming prey (Avila and Frye, 1978; however see Lafferty and Page, 1997). They rely upon olfaction and the lateral line system retained after metamorphosis to detect waterborne scents and the movements of aquatic prey; they can even find food and feed when blinded (Elepfandt, 1996).

A passage from Tinsley et al. (1996, p. 44) concerning feeding ability is illuminating: *Xenopus* represents a much more formidable predator than most anurans, which rely on the tongue for selective capture of rather small prey items. In *Xenopus*, prey capture employs a combination of toothed jaws that improve the grip on the prey, forelimbs that are used to fork the prey into the mouth, and the strong hindlimbs that can be used to rake the prey with the sharp claws. This shredding action enables *Xenopus* to tackle larger food items than could otherwise be ingested whole; indeed, groups of *Xenopus* may attack the same prey and can tear the body into fragments that can then be ingested. This method of feeding is particularly useful for scavenging.

Few studies have been undertaken of diets in the wild. Studies of clawed frogs and their congeners in Africa show a strong reliance upon aquatic invertebrates, especially zooplankton and benthic forms. Terrestrial invertebrates, aquatic vertebrates, and conspecifics occasionally are important components of diet as well (Noble, 1924; Buxton, 1936; Loveridge, 1936; Aronson, 1944b; Rose, 1950; Inger and Marx, 1961; Tinsley, 1973; Tinsley et al., 1979; Schramm, 1986; De Bruyn et al., 1996). Stomach contents analysis of populations in South Wales (Measey, 1998a; Measey and Tinsley, 1998) and California (Lenaker, 1972; McCoid and Fritts, 1980a) revealed similar dietary tendencies.

Accounts of terrestrial prey in the stomachs of *Xenopus* species from Africa were often characterized as ambiguous, since their ability to forage out of water was undocumented (Noble, 1924; Inger and Marx, 1961; Tinsley, 1973; Tinsley et al., 1979; De Bruyn et al., 1996). Measey (1998b) recently has demonstrated that clawed frogs have the ability to capture prey out of the water and that feeding on terrestrial prey may indeed be a normal component of their foraging behavior. Nevertheless, aquatic prey predominate in all other diet studies.

In bodies of water where there are limited prey, adults will cannibalize young (Buxton, 1936; McCoid and Fritts, 1993; Picker, 1994). Larvae may act as collectors of nutrients such as seasonal single-celled algal blooms, which are unavailable to adults. Adults that cannibalize these larvae can thus rely indirectly on this phytoplankton food base (Savage, 1963; Picker, 1994; Tinsley et al., 1996). Cannibalism allows clawed frogs to colonize a body of water that does not offer a large prey base for the adults or to stay in a body of water that has been depleted of prey.

Although there has been much speculation about the possibility of predation on fishes, to date there have been no documented cases where clawed frogs have caused fish declines. Although not particularly adept at capturing actively swimming prey such as fish (Avila and Frye, 1978), clawed frogs have the capacity to find and eat fish eggs. Under confined conditions with unnaturally high densities of fish, such as occur in aquaculture, clawed frogs are indeed capable of consuming high numbers of fish (Branson, 1975; Hepher and Pruginin, 1981; Prinsloo et al., 1981; Schramm, 1986), but observations of fish predation have rarely been documented in nature in California (Lenaker, 1972; Zacuto, 1975; Lafferty and Page, 1997). Early accounts of fish in clawed frog stomachs may have been an artifact of collection methods (McCoid and Fritts, 1980b).

Lafferty and Page (1997) reported clawed frog predation on tidewater gobies (*Eucyclogobius newberryi*), a Federally Endangered species; tidewater gobies are sympatric with clawed frogs in Goleta Slough and Orange County estuaries (Swift et al., 1989). The Federally Endangered, unarmored threespine stickleback (*Gasterosteus aculeatus williamsoni*) is also sympatric with clawed frogs in the Santa Clara River (St. Amant, 1975). California's indigenous pupfish (*Cyprinodon* sp.) may be at particular risk from clawed frog introductions because their populations are confined, and clawed frogs can tolerate the chemical conditions and temperatures of the pupfish habitats.

The presence of high numbers of clawed frog larvae has been shown to directly impact Chinese silver carp (*Hypothalmichthys molitrix*), which feed on phytoplankton and suspended detritus (Schramm, 1986). Based on food preferences, clawed frogs will likely compete with phytophagous and benthic feeding fish (Schoonbee et al., 1992).

O. Predators. Known predators of clawed frogs in Africa include members of many of the same taxa that consume native frogs in North America: invertebrates (Buxton, 1936), fish (Prinsloo et al., 1981), snakes (Sweeney, 1961), turtles (Rose, 1950), birds, and mammals (Worthington and Worthington, 1933; Loveridge, 1942; Rowe-Rowe, 1977). Even barnyard animals such as ducks and pigs are known to eat clawed frogs (Dreyer, 1913; Hey, 1949).

Birds are especially well documented as predators of clawed frogs. Predation has been noted in Africa by egrets (Bates et al., 1992), herons (O'Connor, 1984), storks (Loveridge, 1953; Kahl, 1966, 1967; Siegfried, 1975), cormorants (Kopij, 1996, 1998), owls (Gichuki, 1987), and shrikes (Ryan, 1992).

In California, avian predators on clawed frogs include black-crowned night-herons (*Nycticorax nycticorax*), great blue herons (*Ardea herodias*), great egrets (*Casmerodius albus*), green herons (*Butorides striatus*), common ravens (*Corvus corax*) and western gulls (*Larus occidentalis*; Lafferty and Page, 1997; Crayon and Hothem, 1998; unpublished data).

Clawed frogs are heavily preyed upon by centrarchid fishes, such as bass, crappie, and sunfish (McCoid and Fritts, 1980a; Prinsloo et al., 1981). Two-striped garter snakes (*Thamnophis hammondi*) apparently will also feed on clawed frogs (Mardht and Kneffler, 1972; Ervin and Fisher, 2001).

P. Anti-Predator Mechanisms. Harassed clawed frogs produce large amounts of an extremely slippery mucus from skin glands. Clawed frogs are purported to make dogs that attempt to eat them foam at the mouth in response to these skin secretions (Hey, 1949). Oral dyskinesia is produced in some North American and African water snakes when attempting to ingest clawed frogs (Barthalmus and Zielinski, 1988; Barthalmus, 1989; Zielinski and Barthalmus, 1989).

Surfacing behavior can become synchronized within a group of clawed frogs, apparently in response to the presence of potential predators (Baird, 1983).

Q. Diseases. Unknown in wild populations, but outbreaks of red-leg disease commonly occur in laboratory stocks (A.L. Brown, 1970; Deuchar, 1975).

R. Parasites. Clawed frogs are infected by a remarkably rich parasite fauna that includes over 25 genera from 7 major invertebrate groups (Tinsley, 1996). Most of the parasites probably do not persist in extralimital populations, due to a lack of suitable intermediate hosts. However, Lafferty and Page (1997) found African tapeworms *(Cephalochlamys namaquensis)*, ciliates *(Nyctotherus* sp.), and encysted larval nematodes in California populations. Other clawed frog populations sampled at four Southern California sites (n = 8, 39, 23, 30) exhibited tapeworm infestation levels of 100%, 92%, 91%, and 27%, respectively, with individual infestation levels ranging from 0–50 tapeworms (unpublished data). This tapeworm species has demonstrated an ability to infect other amphibians *(Rana angolensis, Dicroglossus occipitalis, Pleurodeles poireti;* Tinsley, 1996), which raises the possibility of parasite transmission to native fauna. An examination of a sample of clawed frogs from San Diego County (n = 21) revealed infections by an additional two species of protozoan ciliates *(Balantidium xenopodis* and *Protoopalina xenopodus),* a flagellate *(Cryptobia* sp.), and two monogeneans *(Protopolystoma xenopodis* and *Gyrdicotylus gallieni),* all of which are endemic in clawed frog populations in Africa (Kuperman et al., 2000).

4. Conservation.
African clawed frogs are considered an invasive species and every reasonable attempt should be made to eradicate populations.

Acknowledgments. Many individuals provided intellectual stimulation, distribution information, field assistance, specimens, and data for this account, and I offer them my sincerest thanks: Patrick Davis, David Muth, Shawna Bautista, Jim Dole, Bill Taylor, Dennis Otsuka, Bobby Goodman, Mark Jennings, Ed Ervin, Robert Fisher, Karen Swaim, Dan Holland, Michael McCoid, Kevin Lafferty, Sam Sweet, and especially, Roger Hothem. For financial support while working on Edwards A.F.B., I am indebted to Wanda Deal and Mark Hagan of the Environmental Management Division of Edwards A.F.B., and the Davis Field Station of the Biological Resources Division, USGS.

Family Ranidae

Rana areolata Baird and Girard, 1852(a)
CRAWFISH FROG

Matthew J. Parris, Michael Redmer

1. Historical versus Current Distribution.
Crawfish frogs *(Rana areolata)* have a disjunct distribution, with populations localized in areas of suitable habitat. The distribution forms an arc that encircles the eastern, northern, and western boundaries of the Ozark Plateau, from western Indiana and southern Illinois, west through south-central Iowa, central

and southwestern Missouri, southeastern Kansas, eastern Oklahoma, and eastern Texas, and from extreme western Kentucky south along the Mississippi Drainage to central Mississippi, and across southern Arkansas and northwestern Louisiana (Wright and Myers, 1927; Goin and Netting, 1940; Bailey, 1943; Bragg, 1953; Smith, 1961; Altig and Lohoefener, 1983; Garrett and Barker, 1987; Johnson, 1987; Dundee and Rossman, 1989; Conant and Collins, 1998; Young and Crother, 2001). They are also present in a narrow band along the Arkansas River Valley in central Arkansas (S. Trauth, Arkansas State University, personal communication). Crawfish frogs are absent from the Ozark Plateau in Missouri and Arkansas and the Mississippi River Delta in Arkansas and Mississippi (R. Altig, Mississippi State University, personal communication).

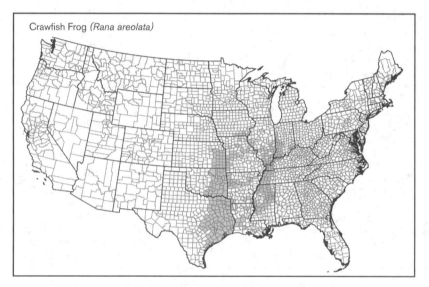

Crawfish Frog *(Rana areolata)*

Two subspecies are recognized, based on morphological differences among larvae and adults (Goin and Netting, 1940; Bragg, 1953). Southern crawfish frogs *(R. a. areolata)* occur from east-central Texas east through central and northern Louisiana and southern Arkansas, and north into southeastern Oklahoma. Northern crawfish frogs *(R. a. circulosa)* occur in the northern and eastern portions of the species distribution. Sympatry between the subspecies occurs in the tallgrass prairie region between the Canadian and Arkansas rivers in Oklahoma, and in southern Arkansas (Bragg, 1953; Conant and Collins, 1998). The similarity of *R. areolata* to the related Florida gopher frog–dusky gopher frog *(R. capito–R. sevosa)* complex may contribute to inaccurate distribution records (Mount, 1975; Martof et al., 1980; Altig and Lohoefener, 1983; Dundee and Rossman, 1989; Young and Crother, 2001). *Rana areolata, R. capito,* and *R. sevosa* are

distinct species (Goin and Netting, 1940; Hillis et al., 1983; Hillis, 1988; Young and Crother, 2001) and allopatrically distributed, although they are only separated by a narrow region in south-central Mississippi and central Louisiana (Dundee and Rossman, 1989; Conant and Collins, 1998), where morphologically intermediate animals were noted by Neill (1957d).

2. Historical versus Current Abundance.
Listed as Endangered or of Conservation Concern in several states (Johnson, 1987; Dundee and Rossman, 1989; Christiansen and Bailey, 1991; Simon et al., 1992; Busby and Brecheisen, 1997). Probably extinct near the type locality for *R. a. circulosa* (Minton, 1998; M. R., unpublished data). Northern crawfish frogs have been reported from mountain valleys, oak-hickory-pine forest, and along the Gulf Coastal Plain in Texas and Louisiana, although population status is unknown in many areas (Bragg, 1953; Garrett and Barker, 1987; Dundee and Rossman, 1989). Most extant midwestern populations of northern crawfish frogs are located in remnant prairie habitats, where they may be considered a locally common species (Bragg, 1953; Busby and Brecheisen, 1997; M. R., unpublished data). Others note that populations have declined or been extirpated with the conversion of prairie to agriculture (Platt et al., 1973; Johnson, 1987; Busby and Brecheisen, 1997). Because of the secretive nature of both subspecies, population status and successful reproduction are often difficult to detect. Long-term monitoring programs of breeding adults and metamorphosing juveniles would benefit from surveys at known and potential sites (Semlitsch et al., 1995; Busby and Brecheisen, 1997; Parris and Semlitsch, 1998).

3. Life History Features.

A. Breeding. Reproduction is aquatic.

i. Breeding migrations. Breeding occurs in late winter to early spring: late February to early May in northern regions, January to early April in southern regions (Bragg, 1953; Smith, 1961; Garrett and Barker, 1987; Johnson, 1987; Dundee and Rossman, 1989; Busby and Brecheisen, 1997). Immigration begins after rainfall has filled temporary ponds and a saturation has occurred (Smith et al., 1947; Busby and Brecheisen, 1997). Ambient air temperatures of 10–12 °C (minimum 8 °C) are critical to initiate and maintain breeding activity (Smith et al., 1948; Bragg, 1953; Busby and Brecheisen, 1997; M. R., unpublished data). Some claim that for breeding, crawfish frogs require the most restrictive moisture and temperature conditions of any North American *Rana* (Smith et al., 1947), but others cite flexibility in reproductive behavior (Bragg, 1953; Busby and Brecheisen, 1997; M. R., unpublished data). Males migrate to temporary ponds 5–6 d prior to the arrival of females (Smith et al., 1948). Breeding occurs explosively, with most reproduction occurring in a short period of intense chorusing at the beginning of the breeding season, although calling and periodic reproductive activity may persist for 22–55 d (Smith et al., 1948; Busby and Brecheisen, 1997). In marginal habitats or in smaller populations, reproduction may not occur every year (Thompson, 1915; M. J. P., personal observations).

ii. Breeding habitat. Northern crawfish frogs utilize a variety of breeding habitats, including shallow ditches, temporary ponds, flooded overflows from small streams, pasture ponds, and prairie wetlands (Bragg, 1953; Dundee and Rossman, 1989). Southern crawfish frogs are largely limited to tallgrass or outlier prairie, degraded grasslands, wooded regions along streams associated with grasslands, and pastures, although smaller breeding choruses may be found in areas where prairie has been converted for agriculture (Bragg, 1953; Smith, 1961; Johnson, 1987; Busby and Brecheisen, 1997).

B. Eggs.

i. Egg deposition sites. Oviposition occurs in shallow water. Emergent aquatic vegetation often supports eggs masses, although not invariably so (Smith, 1934; Bragg, 1953; Smith, 1961; Dundee and Rossman, 1989; Busby and Brecheisen, 1997). Egg masses are often clustered within a pond (Bragg, 1953; Busby and Brecheisen, 1997).

ii. Clutch size. Eggs are deposited at or slightly beneath the water surface in plinth, cylindrical, or disc-shaped masses of 2,000–7,000 eggs (Wright and Myers, 1927; Smith et al., 1948; Bragg, 1953; Smith, 1961; Johnson, 1987; Dundee and Rossman, 1989; Trauth et al., 1990; Redmer, 2000). Eggs are large, with a vitellus

of 2–3 mm (Bragg, 1953; Smith, 1961; Redmer, 2000).

C. Larvae/Metamorphosis. Larvae hatch within 7–15 d after oviposition and metamorphose in 63–75 d in the field (Bragg, 1953; Johnson, 1987; Busby and Brecheisen, 1997). Hatching occurs in 7–10 d and metamorphosis in 80–95 d in the laboratory (M. Redmer, personal observations). Parris and Semlitsch (1998) demonstrated that crawfish frogs have an average larval period of 65 d when reared in artificial ponds, and that this increases to an upper limit of 90 d when reared with larvae of other *Rana* species. Larvae feed primarily on phytoplankton and algae (Bragg, 1953; Parris and Semlitsch, 1998). Larvae reach up to 63 mm in length before metamorphosing and are distinguishable from other *Rana* larvae by having a relatively narrow interruption between the two halves of the inner row of upper labial teeth (Smith et al., 1948; Bragg, 1953) and by a series of dermal pits along the lateral line (Bragg, 1953; Dundee and Rossman, 1989). Newly metamorphosed individuals are 30 mm in SVL and weigh 2–3 g (Wright and Myers, 1927; Smith, 1961; Parris and Semlitsch, 1998).

D. Juvenile Habitat. Unknown, presumably similar to that of adults. Newly metamorphosed crawfish frogs use existing holes, cracks, and other substrate irregularities as sites to initiate burrowing under experimental xeric conditions (Parris, 1998).

E. Adult Habitat. Northern crawfish frogs occupy several different habitats, including open wet woodlands, wooded valleys, prairies, river floodplains, pine forests, and meadows (Bragg, 1953; Garrett and Barker, 1987; Dundee and Rossman, 1989). Southern crawfish frogs are largely limited to prairie, wet pastures, or grassland habitats, although populations may also persist in low-lying hay fields, and occasionally in woodland stream watersheds and river floodplains (Bragg, 1953; Johnson, 1987; Busby and Brecheisen, 1997). Crawfish frogs are also common in the hardpan, clay soil region of southern Illinois (Smith, 1961). Adults of both subspecies are extremely fossorial, seldom emerging from the abandoned crayfish or other small animal burrows they use as shelter. Adults can be found under logs, in road-side banks, and in sewers (Goin and Netting, 1940; Dundee and Rossman, 1989). Burrows may exceed 1–1.5 m in depth, often have flattened platforms at the entrance, and may be located several meters away from breeding ponds (Thompson, 1915; Smith, 1934; Johnson, 1987). Agricultural practices (i.e., cultivation and mowing) may reduce or limit crawfish frog abundance by eliminating suitable burrowing habitat (Thompson, 1915).

F. Home Range Size. Unknown.

G. Territories. Unknown.

H. Aestivation/Avoiding Desiccation. Unknown.

I. Seasonal Migrations. Individuals migrate from overwintering sites to breeding ponds in early spring (Smith et al., 1948; Johnson, 1987; Dundee and Rossman, 1989; Busby and Brecheisen, 1997). Females leave ponds shortly after oviposition, and males occasionally call from burrows after breeding (Smith et al., 1948). They are seldom encountered above ground outside of their breeding season.

J. Torpor (Hibernation). Assumed to use burrows as hibernacula (Thompson, 1915; Smith et al., 1948).

K. Interspecific Associations/Exclusions. Crawfish frogs may be excluded from larval communities containing plains leopard frogs *(Rana blairi)* or southern leopard frogs *(R. sphenocephala)* because of weak interspecific competitive ability (Parris and Semlitsch, 1998). They may breed syntopically with northern cricket frogs *(Acris crepitans)*, American toads *(Bufo americanus)*, Cope's gray treefrogs *(Hyla chrysoscelis)*, eastern gray treefrogs *(H. versicolor)*, spring peepers *(Pseudacris crucifer)*, western chorus frogs *(P. triseriata)*, plains leopard frogs, green frogs *(R. clamitans)*, and southern leopard frogs (Wright and Myers, 1927; Busby and Brecheisen, 1997; M. J. P., personal observation).

L. Age/Size at Reproductive Maturity. Females are larger than males at maturity. Estimated age (using skeletochronology) of mature adults was 2–5 yr (mean = 3.5 yr) for males, and 3–5 yr (mean = 3.8 yr) for females in a southern Illinois population (Redmer, 2000). SVLs for those specimens were 71–90 mm (mean = 82.8 mm) for males and 79–102 mm (mean = 89.6 mm) for females. Other reported sizes include modal SVLs of 100 mm (males) and 95 mm (females), and maximum SVLs of 117 mm (males) and 118 mm (females; Smith et al., 1948). Bragg (1953) reported mean SVLs of adults from three Oklahoma populations of 80.6 mm (range 67–89 mm), 81.8 mm (61–95 mm), and 92.3 mm (72–108 mm). Reported size maxima are 99 mm in Missouri (Powell et al., 1982); 104 mm for males and 108 mm for females in Indiana (Minton, 1972); and 110 mm in Kansas (Collins, 1993). The presence of small (61–71 mm) but sexually mature adults at breeding choruses has been noted (Smith et al., 1948; Bragg, 1953; Trauth et al., 1990; Redmer, 2000). No male <77 mm SVL has been found in amplexus in a southern Illinois population (Redmer, 2000).

M. Longevity. Five years for both males and females (Redmer, 2000).

N. Feeding Behavior. Larvae forage nocturnally in benthic regions of shallow water when young, in deeper water when older (Bragg, 1953). Adults feed from the platforms of their burrows at dawn and dusk (Thompson, 1915). Stomach contents of adults include ants, beetles (Carabidae, Scarabidae), centipedes, crayfish, crickets, millipedes, and spiders (Thompson, 1915; Goin and Netting, 1940; Smith et al., 1948). Captive adults accept crickets, earthworms, small crayfish, and small mice (M.R., unpublished data).

O. Predators. Natural predators are unknown. Other vertebrates may be important predators on adults (e.g., snakes, birds, raccoons). Larval experiments identified backswimmers (*Notonecta* sp.) and dragonfly naiads (*Anax* sp.) as effective predators (Travis et al., 1984; Cronin and Travis, 1986). Adults frequently are killed by automobiles during breeding migrations (M.R., unpublished data).

P. Anti-Predator Mechanisms. Larvae are nocturnally active and swim away from any disturbance (Bragg, 1953). Adults are extremely wary. Breeding activity may cease following slight disturbances (Bragg, 1953; Busby and Brecheisen, 1997; M.R., unpublished data). When alarmed, adults quickly retreat to their burrows and remain there for long periods of time (Thompson, 1915; Busby and Brecheisen, 1997). Juveniles and adults may deter predators with defensive postures and by occasionally snapping their jaws when cornered (Thompson, 1915; Altig, 1974; Marchisin and Anderson, 1978). Adults may produce a foul-smelling skin secretion (Dundee and Rossman, 1989).

Q. Diseases. Unknown.

R. Parasites. Trematodes, nematodes, and mites (Kuntz, 1941; Walton, 1949).

4. Conservation.
Crawfish frogs are of considerable conservation concern and are listed as Endangered or In Decline in several states including Iowa, Indiana, and Kansas (Levell, 1997). Crawfish frogs may be susceptible to regional extinction because they are often associated with tallgrass prairies, outlier prairies, or other native grasslands—habitats that increasingly are being fragmented or converted to cropland. Crawfish frogs may be strongly influenced by larval interactions with other ranid species, potentially leading to reduced recruitment or competitive exclusion in sympatric areas. Furthermore, because of their fossorial nature, accurate accounts of population viability and recruitment are difficult to develop.

Rana aurora (Baird and Girard, 1852[b])
NORTHERN RED-LEGGED FROG

Christopher A. Pearl

1. Historical versus Current Distribution.
Biochemical, ecological, and life-history differences between northern red-legged frogs (historically known as *Rana aurora aurora*) and California red-legged frogs (historically known as *Rana aurora draytonii*) support recognition of these taxa as

distinct species (Hayes and Miyamoto, 1984; H. B. Shaffer et al., 2004). Northern red-legged frogs occur along the California coast north of Elk Creek (Mendocino County) into southwestern Oregon, where their historical range broadens eastward through the Rogue River drainage into the lower elevations of the Cascade Range (Fitch, 1936; Dunlap, 1955; Dumas, 1966; Nussbaum et al., 1983; Jennings and Hayes, 1994a; G.M. Fellers, personal communication). Northern red-legged frogs breed in wetlands between sea level and 1,200 m in elevation, west of the Cascade crest through northwestern Oregon, western Washington, and southwestern British Columbia. Northernmost populations occur near the northern end of Vancouver Island and Sullivan Bay, British Columbia (Green and Campbell, 1984; Stebbins, 1985). Oregon's Willamette Valley (and potentially the Rogue Valley) appear to be the most reduced and fragmented portions of the range, potentially the result of intensive land use and establishment of a variety of non-native predators (Nussbaum et al., 1983; St. John, 1987; Blaustein and Wake, 1990).

2. Historical versus Current Abundance.
Several herpetologists have suggested that the abundance of northern red-legged frogs in Oregon's Willamette Valley has declined (Nussbaum et al., 1983; St. John, 1987; Blaustein and Wake, 1990). Recent surveys suggest reproductive populations remain on much of the valley floor (Pearl et al., 2005). Northern red-legged

frogs are relatively widespread in portions of western Washington (K.R. McAllister et al., 1993; Richter and Azous, 1995; Adams et al., 1999), although analysis of present occurrence at historical sites has not been conducted. They also were widespread historically in southwestern British Columbia and remain common in at least segments of that range (R. Haycock, S. Orchard, personal communications).

3. Life History Features.
A. Breeding. Reproduction is aquatic.
i. Breeding migrations. Northern red-legged frogs often make extensive movements to breeding wetlands from summer habitats (Nussbaum et al., 1983; Hayes et al., 2001). Males generally reach breeding sites before females, sometimes as early as October, but usually arrive in larger numbers in November–December in Oregon and northern California (Storm and Pimentel, 1954; Storm, 1960; Twedt, 1993).
ii. Breeding habitat. Oviposition generally occurs in vegetated shallows of wetlands with little flow (Storm, 1960; Licht, 1971), but egg masses can be deposited in water up to 5 m (Calef, 1973b). Breeding

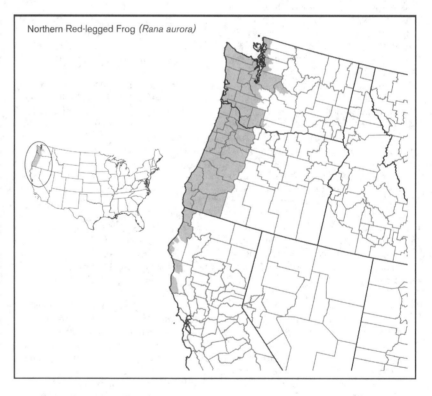

Northern Red-legged Frog *(Rana aurora)*

sites can be permanent or temporary, with inundation usually necessary into June for successful metamorphosis in the Willamette Valley (Storm, 1960; Nussbaum et al., 1983). Breeding is initiated when water temperatures exceed 6–7 °C (usually in January), and can extend through March (Storm, 1960; personal observations).

B. Eggs.

i. Egg deposition sites. Egg masses are usually attached to herbaceous vegetation in areas with little or no flow (Storm, 1960; Calef, 1973b).

ii. Clutch size. Females deposit an average of 530–830 eggs/mass, with a range between 200–1,100 (Storm, 1960; Calef, 1973a; Licht, 1974).

C. Larvae/Metamorphosis.

i. Length of larval stage. In Oregon and Washington, eggs generally hatch after about 10–30 d (often in March–April) and reach metamorphosis 11–14 wk later in June–July (Storm, 1960; Calef, 1973a; Licht, 1974; H. A. Brown, 1975b).

ii. Larval requirements.

a. Food. Larvae consume a variety of epiphytic algae and can alter species composition and standing crop in laboratory conditions (Dickman, 1968).

b. Cover. Larvae appear to utilize relatively dense vegetation as cover (Nussbaum et al., 1983; personal observations). Weins (1970) found that larvae became conditioned to prefer microhabitats possessing complex background patterns in laboratory trials, potentially suggesting an affinity for vegetated areas.

iii. Larval polymorphisms. Not reported.

iv. Features of metamorphosis. Approximately 5% of embryos survived to metamorphosis at two British Columbia breeding sites (Calef, 1973a; Licht, 1974). Northern red-legged frogs generally transform at 20–25 mm SVL (Storm, 1960; Calef, 1973a). At a northwestern Washington site, developmental periods for larvae averaged 110 d; mean size at metamorphosis was 28.7 mm (H. A. Brown, 1975b).

v. Post-metamorphic migrations. Juveniles often remain around edges of breeding ponds for short periods (days to weeks) before dispersing, but cues for emigration are not well known (Licht, 1986; Twedt, 1993). Dispersal distances of newly transformed juveniles appear to be related to size at metamorphosis (N. Chelgren, Oregon State University, unpublished data). In the fall, juveniles have been observed in riparian areas >0.5 km from nearest known breeding site (C. A. P. and M. Hayes, personal observations).

D. Juvenile Habitat. After dispersing from breeding habitats, juveniles tend to occupy relatively moist, densely vegetated riparian microhabitats during the summer (Twedt, 1993; personal observations). Movements away from these microenvironments may be related to elevated moisture levels (Licht, 1986). Hayes and Hayes (2003) have recently documented juvenile growth rates.

E. Adult Habitat. Most northern red-legged frog adults appear to leave breeding sites relatively soon after the breeding period and may move substantial distances (>300 m) from breeding pools in mesic forests and riparian areas (Nussbaum et al., 1983; Licht, 1986; Gomez and An-

thony, 1996; Hayes et al., 2001). Summer habitats of adults in the mid-elevations of the Oregon Cascade range include streambanks and moist riparian areas (Hayes et al., 2001; personal observations). At one northern California breeding lagoon, adults tended to use microhabitats adjacent to standing water rather than remaining in standing water (Twedt, 1993).

Greater numbers of adult northern red-legged frogs have been trapped in coniferous stands of moderate moisture than in drier stands in the Oregon and Washington Cascades (Aubry and Hall, 1991; Bury et al., 1991). One study found red-legged frog adults more frequently in older managed forest stands (Aubry, 2000), but other terrestrial studies have not documented clear preferences for any stand age in managed and unmanaged forests (Aubry and Hall, 1991; Bury et al., 1991; Bosakowski, 1999). Conclusions of the aforementioned terrestrial studies are limited by low captures in pitfall traps and variable juxtaposition of sampled stands relative to breeding sites.

F. Home Range Size. Unknown, but adults are wide ranging (see "Adult Habitat" and "Breeding migrations" above).

G. Territories. Northern red-legged frogs generally are not considered territorial, but males can act aggressively toward one another at breeding sites, and egg masses are often deposited in a dispersed fashion, in contrast to other northwestern lentic ranid frogs (Calef, 1973b; Nussbaum et al., 1983; personal observation).

H. Aestivation/Avoiding Desiccation. Not documented.

I. Seasonal Migrations. See "Breeding migrations" and "Post-metamorphic migrations" above.

J. Torpor (Hibernation). Unknown, but likely in northern ranges and higher elevations. Adults in southern and coastal ranges can remain active through winter (Nussbaum et al., 1983; Twedt, 1993).

K. Interspecific Associations/Exclusions. Northern red-legged frogs often share breeding sites with northwestern salamanders (Ambystoma gracile), long-toed salamanders (A. macrodactylum), Pacific chorus frogs (Pseudacris regilla), rough-skinned newts (Taricha granulosa), and introduced American bullfrogs (Rana catesbeiana). In moderate elevations of the Cascade Range (about 600–1,000 m), northern red-legged frogs may share breeding ponds and co-occur along streams with Cascade frogs (R. cascadae; Nussbaum et al., 1983; Hayes, 1996; B. Bury and D. Major, unpublished data). Historically, northern red-legged frogs occurred syntopically with Oregon spotted frogs (R. pretiosa) in lowland western Oregon, Washington, and southwestern British Columbia (Licht, 1971, 1974, 1986; Nussbaum et al., 1983; K. R. McAllister et al., 1993). Negative interactions

have been suggested between northern red-legged frogs and introduced bullfrogs and sport-fish in Oregon (Hayes and Jennings, 1986; Kiesecker and Blaustein, 1997a, 1998).

L. Age/Size at Reproductive Maturity. Males can become sexually mature the breeding season following metamorphosis, but the majority breed only after 2 yr of age (Licht, 1974; see also Hayes and Hayes, 2003). Females usually reproduce after they reach 3 yr old, although a small portion may be able to breed in the second season after transforming (Licht, 1974).

M. Longevity. Poorly known in field situations, but reportedly can exceed 10 yr in captivity (Cowan, 1941).

N. Feeding Behavior. Northern red-legged frogs consume a variety of small insects, arachnids, and mollusks (Fitch, 1936; Licht, 1986). Larger adults are able to take larger food items, including juvenile conspecifics and salamanders (Licht, 1986; Rabinowe et al., 2002). Young frogs are believed to forage in moist areas close to water, pursuing food farther away only during wet periods (Licht, 1986).

O. Predators. Larval northern red-legged frogs are eaten by fish, rough-skinned newts, northwestern salamanders, giant water bugs (Belostomatidae), larval diving beetles (Dytiscidae), and anisopteran odonates (Calef, 1973a; Licht, 1974). Garter snakes (particularly common garter snakes [Thamnophis sirtalis]) and adult northern red-legged frogs consume juvenile frogs, and herons and raccoons also prey on adults (Licht, 1974, 1986; Gregory, 1979; personal observations). Predation by introduced game-fish and American bullfrogs may represent important threats to northern red-legged frogs throughout their range (Hayes and Jennings, 1986; Twedt, 1993; Kiesecker and Blaustein, 1997a, 1998).

P. Anti-Predator Mechanisms. Tadpoles reduce activity when exposed to chemical cues of injured conspecifics in lab trials (Wilson and Lefcort, 1993), and ammonium (NH_4^+) may be a component of exudates from disturbed larvae (Kiesecker et al., 1999).

Larval northern red-legged frogs reduced activity and distanced themselves from native odonate predators in lab trials and were larger at metamorphosis when the predator was present (Barnett and Richardson, 2002). A laboratory study found that northern red-legged frog tadpoles reared in the presence of predators fed conspecifics or cues of injured conspecifics transformed earlier and smaller than controls (Kiesecker et al., 2002). One field study did not detect northern red-legged frog tadpole avoidance of injured conspecifics (Adams and Claeson, 1998).

Adults possess long rear legs and exceptional leaping ability, which, in tandem with cryptic coloration and stationary evasion behavior, can make juveniles and adults difficult for terrestrial predators to apprehend (Gregory, 1979, Pearl et al., 2004).

Q. Diseases. An iridovirus has been reported from northern California (Mao et al., 1999). Lefcort and Blaustein (1995) reported altered behavior when larvae were exposed to the yeast *Candida humicola*, which may be transmitted through water and fecal material (Richards, 1958).

R. Parasites. The trematodes *Megalodiscus microphagus* and *Prosthopycoides lynchi* have been documented in the digestive tracts of Oregon northern red-legged frogs (Macy, 1960; Martin, 1966a). Johnson et al. (2002) report moderate levels of morphological abnormalities in juvenile northern red-legged frogs relative to other Pacific northwestern amphibians. These malformations potentially are associated with infection by the trematode *Ribeiroia ondatrae* (Johnson et al., 2002).

4. Conservation.
Thorough field studies documenting declines in northern red-legged frogs are lacking, but a broad array of potential stressors may affect this species. Non-native American bullfrogs are established throughout much of the lowland range of northern red-legged frogs west of the Cascades, and bullfrogs have been hypothesized as displacing northern red-legged frogs here (Nussbaum et al., 1983; St. John, 1987; Kiesecker and Blaustein, 1997a, 1998). In Oregon, northern red-legged frog larvae have been found to compete poorly with bullfrog larvae when food resources are concentrated (Kiesecker and Blaustein, 1998; Kiesecker et al., 2001b). However, other studies conducted in Washington have failed to identify direct effects of competition on northern red-legged frog larvae or exclusion from wetlands supporting bullfrogs (Richter and Azous, 1995; Adams, 2000). Negative associations of nonnative fish appear to be of greater importance (Adams, 1999, 2000), and the interactive effects of fishes and bullfrogs may be greater than either separately (Kiesecker and Blaustein, 1998).

Water quality degradation and altered hydrological regimes associated with agricultural and urban land uses are also of concern in the Puget Trough and Willamette Valley (Nussbaum et al., 1983; Platin, 1994; Richter and Azous, 1995; De Solla et al., 2002b). However, northern red-legged frogs persist in some urbanized habitats in the region (Richter and Azous, 1995, 2000). Laboratory investigations of nitrogenous by-products from agricultural fertilizers suggest northern red-legged frog larvae are intermediate among Willamette Valley amphibians in their sensitivity to ammonium sulfate, but may be susceptible to ammonium ions derived from related compounds (Schuytema and Nebeker, 1999b; Nebeker and Schuytema, 2000). Larval northern red-legged frogs are relatively susceptible to nitrite, a form that is

usually short-lived in aerobic field conditions (Marco et al., 1999). De Solla et al (2002a) found organochlorine pesticides and PCB residues in eggs of northern red-legged frogs, but levels were not high relative to reference sites. Hatching rates of northern red-legged frogs in their lowland range do not appear to be reduced by ambient UV-B radiation, although embryos may be susceptible to future increases in UV-B irradiance (Blaustein et al., 1996; Ovaska et al., 1997; Belden and Blaustein, 2002a). Modeling studies suggest stressors that impact juvenile red-legged frogs have the greatest potential to influence population fluctuations (Biek et al., 2002).

Northern red-legged frogs are considered a Species of Special Concern in California (California Department of Fish and Game, 1999), Sensitive-Vulnerable in Oregon's Willamette Valley, and Sensitive-Unknown elsewhere in Oregon (Oregon Natural Heritage Program, 1995).

Rana berlandieri Baird, 1854(a)
RIO GRANDE LEOPARD FROG

James C. Rorabaugh

1. Historical versus Current Distribution.
Rio Grande leopard frogs *(R. berlandieri)* occur from central and western Texas and the Pecos River drainage in Eddy County, southeastern New Mexico, south along the Atlantic slope through southeastern Mexico (Platz, 1991; Degenhardt et al., 1996; Conant and Collins, 1998; Dixon, 2000). Hillis et al. (1983) considered the southern limit of the distribution to be near Veracruz and frogs from farther south in the Campeche region to be *R. brownorum* (= *R. berlandieri brownorum*, Sanders, 1973). Frost (1982) noted populations of "*berlandieri*-like frogs of uncertain taxonomic affiliation" from farther south in Mexico, Guatemala, Honduras, and Costa Rica (but see Zaldivar-Riveron et al., 2004). Platz (1991) considered specimens from as

far south as northeastern Nicaragua to be Rio Grande leopard frogs, but recognized no subspecies. Lee (1996) considered frogs in southwestern Campeche, Tabasco, and southern Veracruz to be *R. b. brownorum*, and frogs from farther south on the Yucatán Peninsula to be *R. b. berlandieri*. Rio Grande leopard frogs are currently extant throughout their historical range and are well established as an introduced species in Arizona on the Gila River drainage and associated croplands from Phoenix to the Colorado River confluence; on the Colorado River, Arizona–California near Yuma, Arizona; in the Imperial Valley of southeastern California; and on the Rio Colorado, Sonora, and Baja California Norte, Mexico (Platz et al., 1990; Jennings and Hayes, 1994b; Miera and Sredl, 2000; Rorabaugh et al., 2002).

2. Historical versus Current Abundance.
No historical information. However, in New Mexico, R.D. Jennings (1987) found Rio Grande leopard frogs to be most abundant at sites with large, flowing springs that formed pools along the spring run. On the Yucatán Peninsula, they are "common" in virtually all freshwater habitats, but seem to reach especially high densities in open, disturbed situations (Lee, 1996). Numbers of frogs at some sites apparently vary greatly among years and seasons (Jung et al., 1999; personal observations; C. Painter, personal communication).

3. Life History Features.
    A. Breeding. Reproduction is aquatic.
    i. Breeding migrations. Breeding migrations are unknown in Rio Grande leopard frogs. In warm climates, the species may breed year-round (Garrett and Barker, 1987; Davidson, 1996). However, in New Mexico, breeding probably peaks early in the spring, and calling has been heard from April–August (R.D. Jennings, 1987; Degenhardt et al., 1996). In central

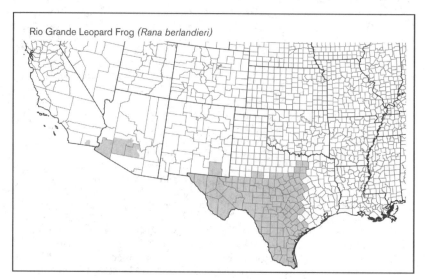

Rio Grande Leopard Frog *(Rana berlandieri)*

Texas, the species breeds in spring and fall, but in areas of sympatry with other leopard frog species, Rio Grande leopard frogs breed in fall and early winter (Hillis, 1981; but see Platz, 1972). Ideuer (1979) reported that the Rio Grande leopard frog breeding season covers at least 11 mo in central Texas. On the Yucatán Peninsula, breeding is associated with the rainy season (Lee, 1996); tadpoles have been observed from February–September, and newly metamorphosed frogs have been found in March–April (Campbell, 1998). In California and Arizona, introduced Rio Grande leopard frogs probably breed nearly year-round. Individuals have been heard calling as early as 10 February and as late as 29 October. A pair in amplexus was observed on 28 October near Yuma, Arizona (personal observation). In an experimental setting, Oldham (1974) found no evidence that female Rio Grande leopard frogs were attracted to playbacks of recorded mating calls. Blair (1961a) described the phenology and other aspects of the breeding ecology of "*Rana pipiens*" near Austin, Texas. However, it is unclear as to whether he was working with Rio Grande leopard frogs or southern leopard frogs (*R. sphenocephala*).

ii. Breeding habitat. In Texas, Rio Grande leopard frogs are found in arid regions along streams or rivers and near cattle tanks, ponds, or ditches (Garrett and Barker, 1987). Hillis (1981) found that in central Texas the species breeds almost always in pools along flowing streams or rivers; but breeding also occurs in artificial ponds and tanks. In New Mexico, Rio Grande leopard frogs generally are found in clear, flowing streams or permanent pools in intermittent stream drainages that originate from springs. Animals are occasionally found in cattle tanks, as well (Fritts et al., 1984; R.D. Jennings, 1987). At Black Gap Wildlife Management Area, Brewster County, Texas, Rio Grande leopard frogs were found only in earthen cattle tanks (Axtell, 1959). In northern Guatemala and the Yucatán Peninsula, including Belize, Rio Grande leopard frogs usually occur at permanent water in wet, moist, or dry tropical forest (Campbell, 1998). They are a common inhabitant of a cienega (wetland) on the northern coast of the Yucatán Peninsula, where the water is possibly brackish, and in cenotes (natural wells resulting from collapse of the limestone roofing of subterranean chambers) in the northern peninsula (Lee, 1996). In Arizona and California, Rio Grande leopard frogs are typically found on the edges of large slow-moving rivers, in agricultural ditches, drains, canals, and sumps; and in one case, they were found in a fish hatchery (Platz et al., 1990; Jennings and Hayes, 1994b; Rorabaugh et al., 2002). Hutchison and Hazard (1984) examined diel and seasonal cycles of erythrocytic organic phosphates in frogs they

called Rio Grande leopard frogs, which may have some relevance to use of, or behavior in, hypoxic environments. However, the frogs used in their experiments were from Sinaloa, Mexico, outside of the range of Rio Grande leopard frogs (the frogs used were *R. forreri*).

B. Eggs.

i. Egg deposition sites. Egg masses in Eddy County, New Mexico, were 7–9 cm across and attached to emergent vegetation in 9–15 cm deep, quiet water along a stream (Degenhardt et al., 1996).

ii. Clutch size. Unknown; however, clutches of other leopard frog species typically contain from hundreds to several thousand eggs.

C. Larvae/Metamorphosis.

i. Length of larval stage. Tadpoles are known to overwinter in Texas (Hillis, 1982) and Arizona (personal observations). Bragg (1961) reported on comparative behavior and development of "*Rana pipiens berlandieri*" and two other anurans. However, the study was conducted near Norman, Oklahoma, where Rio Grande leopard frogs do not occur (he was probably working with the plains leopard frog, *R. blairi*).

ii. Larval requirements.

a. Food. Tadpole stomachs in Texas contained blue green algae, green algae, inorganic particles, and many diatoms. The predominance of diatoms is probably due to indiscriminate feeding of bottom detritus and the indigestibility of diatoms. In the laboratory, larvae are cannibalistic (Hillis, 1982).

b. Cover. Probably necessary to avoid currents. Tadpoles are probably transported passively downstream in drainages. Tadpoles are adapted to life in streams (Hillis, 1982) and will swim upstream against a current (Wassersug and Feder, 1983). Rio Grande leopard frog tadpoles will supplement cutaneous gas exchange by gulping air at the water surface. Air breathing contributes substantially to total oxygen uptake, even in normoxia, and increases stamina of tadpoles under hypoxic conditions (Feder, 1983a; Wassersug and Feder, 1983). Maness and Hutchison (1980) reported on heat "hardening" or acute adjustment of thermal tolerance in Rio Grande leopard frogs; however, the animals used in their experiment were from Sinaloa, Mexico, and thus were probably *R. forreri*.

iii. Larval polymorphisms. Unknown.

iv. Features of metamorphosis. Although variable across their range, breeding occurs over a period of a few to many months each year (see "Breeding Migrations" above). It is not unusual to find tadpoles of different size classes in the same body of water (Jennings, 1985), and newly metamorphic frogs can be found over an extended time period during warm seasons. In areas of hybridization between Rio Grande and southern leopard

frogs in Texas, the former metamorphose earlier than either hybrids or southern leopard frogs (Kocher and Sage, 1986).

v. Post-metamorphic migrations. Unknown.

D. Juvenile Habitat. No differences have been described between adult and juvenile habitat use.

E. Adult Habitat. See "Breeding habitat" above. The presence of holes or burrows where frogs can take refuge may be an important habitat feature (Jennings and Hayes, 1994b).

F. Home Range Size. Unknown.

G. Territories. Playback of mating trills of Rio Grande leopard frogs typically elicits chuckle calling, which is probably associated with male territorial behavior (Mecham, 1971; Gambs and Littlejohn, 1979).

H. Aestivation/Avoiding Desiccation. Rio Grande leopard frogs are apparently active throughout warm periods and seasons. No periods of aestivation have been described.

I. Seasonal Migrations. No seasonal migrations have been described; however, individuals have been observed to disperse at least 1.6 km from any known water source during the summer rainy season in Arizona (personal observations), and after the first rains in the Yucatán Peninsula, frogs have been collected several km from water (Campbell, 1998). In New Mexico, R.D. Jennings (1987) noted collections of Rio Grande leopard frogs from intermittent water sources and suggested these were frogs that had dispersed from permanent water during wet periods.

J. Torpor (Hibernation). Rio Grande leopard frogs are inactive during the cold winter; although in warmer areas they may be active year-round. In Arizona, active Rio Grande leopard frogs have been observed as early as 9 January and as late as 29 October.

K. Interspecific Associations/Exclusions. In New Mexico, Rio Grande leopard frogs were not found in association with other anurans, although several anuran species were found nearby (R.D. Jennings, 1987). At Big Bend National Park, Texas, Rio Grande leopard frogs occurred with low-density American bullfrog (*R. catesbeiana*) populations. In Texas, Hillis (1981) found areas of sympatry between Rio Grande leopard frogs and southern leopard frogs in McLennan and Falls counties, and between Rio Grande leopard frogs and plains leopard frogs in Brown, Coleman, and Comanche counties. In areas of sympatry, habitat, temporal, and behavioral differences among ranid species may act as premating isolating mechanisms. However, hybridization between Rio Grande leopard frogs and southern leopard frogs occurs in central Texas (Sage and Selander, 1979; Kocher and Sage, 1986) and between Rio Grande leopard frogs and plains leopard frogs in Coke and Mitchell counties, Texas (Platz,

1972). At 13 sites in central Texas where Rio Grande leopard frogs hybridized with southern leopard frogs, the hybrid zone appeared to be stable over a 3–5-yr period (Kochler and Sage, 1986). Platz (1981) found Rio Grande leopard frogs replacing plains leopard frogs in Coke and Mitchell counties, Texas, during 1969–75. Rio Grande leopard frogs are often found in association with numerous predatory fish species (R.D. Jennings, 1987; Rorabaugh et al., 2002). In California and Arizona, Rio Grande leopard frogs commonly occur with Woodhouse's toads (*Bufo woodhousii*) and American bullfrogs; however, Rio Grande leopard frogs appear to thrive in habitats that support limited numbers of bullfrogs (Jennings and Hayes, 1994b; personal observations). Rio Grande leopard frogs are found occasionally with Great Plains toads (*B. cognatus*), Colorado River toads (*B. alvarius*), and Couch's spadefoot toads (*Scaphiopus couchii*) in Arizona (personal observations).

L. Age/Size at Reproductive Maturity. Unknown.

M. Longevity. Unknown.

N. Feeding. Degenhardt et al. (1996) note that Rio Grande leopard frogs probably feed on a variety of insects and invertebrates. In Texas, frog stomachs often contained small leopard frogs (Platz et al., 1990; see also Parker and Goldstein, 2004).

O. Predators. In New Mexico, checkered garter snakes (*Thamnophis marcianus*) prey on Rio Grande leopard frog tadpoles and metamorphic animals, and predation by American bullfrogs and fishes was identified as a potential threat to the species (R.D. Jennings, 1987b). Sanders and Smith (1971) suggest crayfish, turtles, fishes, birds, small mammals, and man prey on Rio Grande leopard frogs. An adult aquatic beetle (*Cybister fimbriolatus*) preyed upon a Rio Grande leopard frog tadpole near Austin, Texas, and larvae and adults of this beetle, and early instar larvae of another beetle (*Hydrophilus triangularis*), preyed upon tadpoles in the laboratory (Ideker, 1979). Aggregations of tadpoles near shorelines may increase predation rates by grackles (*Quiscalus mexicanus*; Ideker, 1976). Swimming and air breathing increased the rate at which Rio Grande leopard frog tadpoles were attacked by painted turtles (*Chrysemys picta*; Feder, 1983). I watched a hunter gig a large Rio Grande leopard frog near Yuma, Arizona.

P. Anti-Predator Mechanisms. Frogs seek shelter under rocks and in streamside vegetation during the day. If startled while active, individuals quickly hop into the water or thick vegetation (Degenhardt et al., 1996; personal observations). Where Rio Grande leopard frogs are found in association with predatory fishes in New Mexico, dense aquatic vegetation that could serve as cover for tadpoles and egg

masses was typically present (R.D. Jennings, 1987). Burst swimming is used by tadpoles to avoid predation by turtles (Feder, 1983b).

Q. Disease. In 1992–93, M. Sredl (personal communication) noted a die-off of Rio Grande leopard frogs at a pond near Phoenix, Arizona. Dead and moribund frogs were symptomatic for "red-leg," a bacterial infection. Histological analysis of dermal tissues demonstrated the presence of chytridiomycosis (Sredl and Caldwell, 2000b), a fungal disease implicated in declines of anurans in Australia, Central America, and elsewhere (Berger et al., 1998).

R. Parasites. Two species of mites, *Hannemania hylae* and *H.* sp. were found on Rio Grande leopard frogs in Big Bend National Park, Texas (Jung et al., 2001). Guillen-Hernandez et al. (2000) describe trematodes (digeneans) in Rio Grande leopard frogs collected from Los Tuxtlas, Veracruz, Mexico. A tetrathyridia (*Mesocestoides* sp.) was found in the liver and mesenteries of a single specimen of Rio Grande leopard frog from Texas (McAllister and Conn, 1990).

4. Conservation.
Rio Grande leopard frogs have no status under the Federal Endangered Species Act or C.I.T.E.S., but are considered a Species of Special Protection by the Government of Mexico (Secretaria de Desarrollo Social, 1994). The species is not considered a Species of Special Concern, Threatened, or Endangered by the states of Texas, New Mexico, Arizona, or California. Rorabaugh and Sredl (2002) recommend development of strategies to curb the ongoing invasion of this species into Arizona and California as a means to protect native anurans and other species that may be vulnerable to predation, competition, or diseases carried by Rio Grande leopard frogs.

Rana blairi Mecham, Littlejohn, Oldham, Brown and Brown, 1973
PLAINS LEOPARD FROG

John A. Crawford, Lauren E. Brown, Charles W. Painter

1. Historical versus Current Distribution.
Plains leopard frogs (*Rana blairi*) occur throughout much of the Great Plains and into the central Midwest (prairie peninsula). Their range extends eastward from central New Mexico, central Colorado, and Nebraska (excluding the panhandle) to central Indiana. The northern limit of this range occurs in southern South Dakota and central Iowa and extends south into northern Texas, excluding most of southern Missouri, Arkansas, and southeastern Oklahoma. Isolated population clusters occur in southern Illinois, New Mexico, and southeastern Arizona (L.E. Brown, 1992; L.E. Brown et al., 1993; Conant and Collins, 1998). In New Mexico, plains leopard frogs are found up to 2,130 m in altitude near Sierra Blanca, Lincoln County (Stebbins, 1985; Degenhardt et al., 1996). In Arizona, plains leopard frogs are isolated on the western side of the Chiricahua Mountains and adjoining Sulfer Springs Valley, where they have a range in altitude of 110–2,590 m (Stebbins, 1985). However, in Colorado, plains leopard frogs are found only below 1,828 m (Hammerson, 1982a, 1999). In Missouri, plains leopard frogs occur everywhere except the Ozark Plateau and in the extreme southeast (L.E. Brown, 1992; L.E. Brown et al., 1993; Johnson, 1997). In Indiana, Minton (2001) notes that their range is poorly known, but plains leopard frogs occur primarily along the western border of the state. Plains leopard frogs were first described in 1973, so earlier studies of "leopard frogs" (typically *R. pipiens*) from sites within the range of plains leopard frogs (L.E. Brown, 1973b, 1992) may instead refer to *R. blairi*.

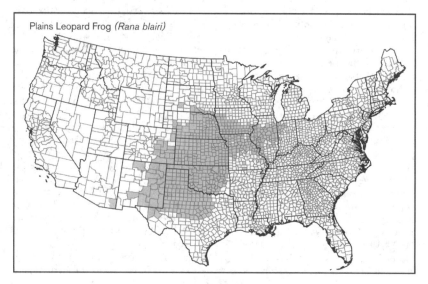

Plains Leopard Frog (*Rana blairi*)

2. Historical versus Current Abundance.
The decline and extirpation of populations (particularly in the west) has been documented by Christiansen and Bailey (1991), Frost and Bagnara (1977), Platz (1981), Hammerson (1982a,b), Frost (1983), Hayes and Jennings (1986), Clarkson and Rorabaugh (1989), Cousineau and Rogers (1991), and L. E. Brown (1992). Hammerson (1999) notes that plains leopard frogs remain widely distributed within their historical range in eastern Colorado, but in areas now occupied by American bullfrogs (*R. catesbeiana*), plains leopard frogs have become scarce.

3. Life History Features.
A. Breeding. Amplexus and reproduction are aquatic (Johnson, 1997).

i. Breeding migrations. Likely from upland (southern distribution) or overwintering (northern distribution) sites to seasonal, semi-permanent, or fishless permanent wetlands.

ii. Breeding habitat. A variety of aquatic habitats are used, including water-filled ditches, ponds, river sloughs, streams, temporary pools, marshes, wetlands, canyon pools, etc. (Stebbins, 1985; L. E. Brown, 1992; Degenhardt et al., 1996; Johnson, 1997). Breeding has been reported from February–October (L. E. Brown, 1992; Degenhardt et al., 1996).

B. Eggs.

i. Egg deposition sites. Eggs are usually attached to vegetation in shallow water (Frost and Bagnara, 1977; Hammerson, 1982a; Lynch, 1985; Degenhardt et al., 1996).

ii. Clutch size. Each female may produce a clutch of 4,000–6,500 eggs (Frost and Bagnara, 1977; Hammerson, 1982a; Lynch, 1985; Degenhardt et al., 1996; Johnson, 1997). Eggs are reported to hatch in "a few days" (Minton, 2001) to 3 wk (Collins, 1993).

C. Larvae/Metamorphosis.

i. Length of larval stage. Tadpoles usually metamorphose during midsummer; however, those tadpoles that hatch late in the breeding season may overwinter and transform the following spring (Gillis, 1975; Scott and Jennings, 1985; L. E. Brown, 1992; Degenhardt et al., 1996; Johnson, 1997). Late stage tadpoles were described by Korky (1978; see also Scott and Jennings, 1985).

ii. Larval requirements.

a. Food. Likely algivorous–omnivorous suspension feeders, similar to other species of tadpoles within the *R. pipiens* complex.

b. Cover. Aquatic macrophytes in seasonal and semi-permanent wetlands provide cover.

iii. Larval polymorphisms. Unknown and unlikely.

iv. Features of metamorphosis. Tadpoles metamorphose at 27–30 mm SVL (Degenhardt et al., 1996; Johnson, 1997).

v. Post-metamorphic migrations. From breeding wetlands to upland feeding sites. Fitch (1958; see also Collins, 1993) recorded successive mass migrations of newly metamorphosed juveniles and adults leaving an upland pond in northeastern Kansas. He speculated that many die when they fail to find aquatic habitat during their movements.

D. Juvenile Habitat. Similar to adults. Lynch (1985; see also L. E. Brown, 1992) found juveniles and adults along streams during autumn.

E. Adult Habitat. Originally a prairie species, plains leopard frogs are now found primarily on grasslands, but can also be found in a variety of other habitats (e.g., oak, oak savanna, oak–pine forests; Stebbins, 1985; L. E. Brown, 1992). Plains leopard frogs are more drought and heat resistant than northern leopard frogs (*R. pipiens*; Gillis, 1979; Stebbins, 1985; Minton, 2001). Adults are often associated with prairie potholes, pools in rocky canyons, livestock tanks, streams, and irrigation ditches (Stebbins, 1985; Degenhardt et al., 1996). Although considered by some parasitologists to be aquatic (Brooks, 1976b; see also L. E. Brown, 1992), during the summer, plains leopard frogs can be found some distance from water (Johnson, 1997). Brown and Morris (1990; see also L. E. Brown, 1992) noted that plains leopard frogs were never found in cultivated fields or in mature upland forests.

F. Home Range Size. Unknown.

G. Territories. Unknown.

H. Aestivation/Avoiding Desiccation. Unknown. However, when subjected to desiccating conditions, plains leopard frogs will assume a water conservation position of legs tucked near the body and ventral surface pressed against the substrate (Gillis, 1979).

I. Seasonal Migrations. May be more pronounced in the northeastern portion of their range, where breeding ponds may differ and be some distance from overwintering ponds, and moist conditions allow foraging away from open water (Collins, 1993). Collins (1993) also noted that after summer rains, both adults and juveniles traveled long distances from the breeding site.

J. Torpor (Hibernation). Plains leopard frogs overwinter in the mud and dead leaves in ponds and streams (L. E. Brown, 1992; Collins, 1993; Johnson, 1997). They are active under the ice (Black et al., 1976; see also L. E. Brown, 1992) and on land during the winter (Collins, 1993).

K. Interspecific Associations/Exclusions. Plains leopard frogs coexist with Chiricahua leopard frogs (*R. chiricahuensis*) and their range overlaps with Rio Grande leopard frogs (*R. berlandieri*; Stebbins, 1985). In central Texas, plains leopard frogs hybridize with Rio Grande leopard frogs (Platz, 1981; Stebbins, 1985). Plains leopard frogs can overlap with northern

leopard frogs in the northern part of their range and southern leopard frogs (*R. sphenocephala*) in the southern part of their range (Conant and Collins, 1998). Hybridization with both species may occur, but this normally is caused by habitat disturbance and alteration due to human activity (Hammerson, 1982a; L. E. Brown, 1992; Collins, 1993; Johnson, 1997; Minton, 2001).

L. Age/Size at Reproductive Maturity. Adults range from 51–111 mm SVL, with western animals being larger (Minton, 2001). In Indiana, males range from 47.0–61.7 mm, and females from 50.5–59.3 mm (Minton, 2001).

M. Longevity. Unknown.

N. Feeding Behavior. Feed on a variety of insects, spiders, annelids, snails, and other invertebrates (Hammerson, 1982a; L. E. Brown, 1992; Degenhardt et al., 1996; Johnson, 1997). Collins (1993) states that plains leopard frogs feed primarily on nonaquatic insects. Examination of stomach contents of Kansas specimens revealed beetles, grasshoppers, crickets, worms, and aquatic snails (Collins, 1993). Plains leopard frogs have also eaten bats (Creel, 1963; Degenhardt et al., 1996).

O. Predators. Predatory fishes (Kruse and Francis, 1977), American bullfrogs (Kruse and Francis, 1977; Smith, 1977; Ehrlich, 1979; Hammerson, 1982b; Hammerson, 1999), and snakes including western terrestrial garter snakes (*Thamnophis elegans*) and black-necked garter snakes (*T. cyrtopsis*; Hammerson, 1982a, 1999). Avian predators include Mississippi kites (*Ictina mississippiensis*; Robinson, 1957) and perhaps burrowing owls (*Athene cunicularia*; Parmley and Tyler, 1978). Mammalian predators include blacktail prairie dogs (*Cyonomys ludovicianus*; Parmley and Tyler, 1978; L. E. Brown, 1992), raccoons, opossums, and skunks (Collins, 1993).

P. Anti-Predator Mechanisms. When frightened, plains leopard frogs will leap away from, rather than towards, water (Frost and Bagnara, 1977). When captured by a predator, plains leopard frogs often emit a loud distress call (Degenhardt et al., 1996). Breeding in seasonal and semi-permanent wetlands reduces the risk of American bullfrog predation (Hammerson, 1999).

Q. Diseases. Unreported.

R. Parasites. Platyhelminths (Brooks, 1976b) and cephalogonimid trematodes (Brooks and Welch, 1976).

4. Conservation.
Plains leopard frogs are generally widespread and abundant throughout most of their range (Conant and Collins, 1998). Isolated populations occur in southern Illinois, New Mexico, and southeastern Arizona (L. E. Brown, 1992; L. E. Brown et al., 1993; Conant and Collins, 1998). Arizona has legally protected plains leopard

frogs, which means a permit is required to collect and/or possess these animals (Levell, 1997).

The decline and extirpation of populations has been documented by a number of researchers (e.g., Frost and Bagnara, 1977; Platz, 1981; Hammerson, 1982a,b; Frost, 1983; Hayes and Jennings, 1986; Clarkson and Rorabaugh, 1989; Christiansen and Bailey, 1991; Cousineau and Rogers, 1991; L.E. Brown, 1992). Hammerson (1999) noted that in areas in Colorado now occupied by American bullfrogs, plains leopard frogs have become scarce. Brown and Morris (1990; see also L.E. Brown, 1992) never found plains leopard frogs in cultivated fields, and the conversion of the Midwest prairies to agriculture has most likely led to a decline in plains leopard frog populations.

Rana boylii Baird, 1854(b)
FOOTHILL YELLOW-LEGGED FROG

Gary M. Fellers

1. Historical versus Current Distribution.
Historically, foothill yellow-legged frogs *(Rana boylii)* ranged throughout much of southwestern Oregon (west of the crest of the Cascade Mountains). The northern-most records are from the Santiam River system in Marion County, Oregon. In California, foothill yellow-legged frogs were found in most of the northwest and south throughout the foothill regions of the coast range (south to the San Gabriel River system, Los Angeles County) and along the western slopes of the Sierra Nevada south to Kern County, and through the Tehachapis and San Gabriel Mountains in southern California. An isolated population has been reported from the Sierra San Pedro Martir of Baja California (Loomis, 1965). A live animal from there was examined and confirmed by R.C. Stebbins, but no specimen exists today (R.C. Stebbins, personal communication). Zweifel (1955) provided a detailed map of the historical range. Foothill yellow-legged frogs range from near sea level to 1,800 m in Oregon (Leonard et al., 1993) and to 1940 m in California (Hemphill, 1952).

Since 1993, my field crews and I have conducted extensive surveys for foothill yellow-legged frogs in California, visiting 804 sites (in 40 counties) that had suitable habitat within the historical range. We found at least one foothill yellow-legged frog at 213 of these sites (26.5% of sites), representing 28 counties.

Extant populations of foothill yellow-legged frogs are not evenly distributed in California. In the Pacific northwest, 40% of the streams support populations of foothill yellow-legged frogs, while that number drops to 30% in the Cascade Mountains (north of the Sierra Nevada), 30% in the south coast range (south of San Francisco), and 12% in the Sierra Nevada foothills.

Foothill Yellow-legged Frog *(Rana boylii)*

2. Historical versus Current Abundance.
While the number of populations is important, population size is also critical. Only 30 of the 213 sites in California with foothill yellow-legged frogs have populations estimated to be 20 or more adult frogs.

The situation for foothill yellow-legged frogs in the Sierra Nevada is bleak; there are no populations in the southern Sierra Nevada foothills that are likely to remain viable for more than a decade. Populations in the northern Sierra foothills are more numerous and generally larger, but they may be in decline as well. Additionally, many of the foothill streams in the northern Sierra Nevada have recreational gold mining activities, which alter the streambed and are likely having a serious, negative impact on the frog fauna.

In the south coast range, several populations of foothill yellow-legged frogs along streams draining into the Central Valley appear to be doing well, in spite of heavy livestock grazing. There are almost certainly other good foothill yellow-legged frog populations in this region, but they are on lands that are privately owned and are thus inaccessible.

The largest populations in California are in the north coast range where the estimated number of adult frogs exceeds 100 at six sites, and an additional nine populations have >50 adult frogs. The Pacific Northwest is clearly the stronghold

for foothill yellow-legged frogs in California, with healthy populations scattered throughout the region.

In Oregon, foothill yellow-legged frogs were once one of the most abundant amphibians in the Rogue River area of southwestern Oregon (Fitch, 1936). Now they are rare or absent throughout the entire western half of their range. There is only one known population in the Cascade foothills on the east side of the Willamette Valley (C. Pearl and D. Olson, personal communications). Farther south, foothill yellow-legged frogs are rare in the Klamath Basin. In the western half of the range, there are moderately good populations in the Umpqua River drainage, and frogs become more common farther south toward California.

3. Life History Features.
A. Breeding. Reproduction is aquatic. Oviposition behavior was recently described by Wheeler et al., 2003.

i. Breeding migrations. Adult migrations appear to be limited to modest movements along stream corridors (Ashton et al., 1998), but the magnitude of such movements, any seasonal component, and differences between sexes remains largely unknown.

ii. Breeding habitat. Unlike other ranid frogs in California, Oregon, and Washington, foothill yellow-legged frogs mate and lays eggs exclusively in streams and rivers. Males typically vocalize underwater (MacTeague and Northern, 1993), but frogs occasionally call above water (Ashton et al., 1998). Their calls are rarely heard.

Timing and duration of breeding activity vary geographically and across populations, but generally occurs during the spring. In California, we have found egg masses between 22 April–6 July, with an average of 3 May. In some areas such as the Trinity River (Trinity County, California), foothill yellow-legged frogs lay eggs throughout the 3 mo period of April–June (Ashton et al., 1998). Other authors cite shorter periods of breeding, that is, within a 2-wk window that occurs between late March and May (Storer, 1925; Grinnell et al., 1930; Wright and Wright, 1949; Zweifel, 1955). Kupferberg (1996a,b) reports an approximate breeding period of 1 mo beginning late April to late May. A Marin County, California population generally lays eggs within a much smaller window of a few weeks around late April. Rainfall during the breeding season can delay oviposition (Kupferberg, 1996a,b). Lind et al. (1996) found that water releases from a dam on the Trinity River washed away most foothill yellow-legged frog egg masses in the main stream of the river.

B. Eggs.

i. Egg deposition sites. Oviposition sites are generally shallow, slow-moving water with a cobble or pebble substrate that is used to anchor each egg mass. On occasion, egg masses may be attached to aquatic vegetation, woody debris, and gravel. Masses are usually attached to the downstream side of rocks, at the stream margin, and at depths of <0.5 m (Stebbins, 1985; Fuller and Lind, 1992; Ashton et al., 1998).

ii. Clutch size. Egg masses vary in size and in the number of eggs/mass. The size of an egg mass after it has absorbed water (usually a few hours after oviposition) is 5–10 cm in diameter and "resembles a cluster of grapes" (Stebbins, 1985). The number of eggs/mass can range from 300–2,000 (Storer, 1925; Fitch, 1936; Zweifel, 1955), with an average of about 900 eggs (Ashton et al., 1998).

Egg masses observed in the field frequently have silt accumulation on the outer surface (Stebbins, 1985). It is not known if silt accumulation affects egg development, but the silt makes the masses less conspicuous and may reduce predation by visual predators.

Eggs generally hatch within 5–37 d (Zweifel, 1955; Ashton et al., 1998). Hatching rates are influenced by temperature, with faster developmental times in warmer waters, up to the critical thermal maximum temperature of about 26 °C (Zweifel, 1955; Duellman and Trueb, 1986). Tadpoles move away from their egg mass after hatching (Ashton et al., 1998).

C. Larvae/Metamorphosis.

i. Length of larval stage. Larval development is, in part, temperature dependent. Typically, tadpoles metamorphose 3–4 mo after hatching.

ii. Larval requirements.

a. Food. Tadpoles feed on algae, diatoms, and detritus by grazing the surface of rocks and vegetation. Diatom rich diets, particularly epiphytic diatoms that contain protein and fat, enhance growth, development, and survival to metamorphosis (Kupferberg, 1996a,b). Tadpoles have also been observed in the field feeding on necrotic tissue of other tadpoles and bivalves (Ashton et al., 1998).

b. Cover. Cover is essential for tadpoles. During the first week of life, tadpoles can often be found in the vicinity of the hatched egg mass. They then move to nearby areas, between and beneath cobble and gravel. When fleeing from threats, their swimming pattern is described as frantic (Ashton et al., 1998).

iii. Larval polymorphisms. None.

iv. Features of metamorphosis. As with most other frog species, major events of metamorphosis include reorganization of the digestive tract, absorption of the tail, and the emergence of front limbs (Duellman and Trueb, 1986). Foothill yellow-legged frogs metamorphose at a size of 1.4–1.7 cm SUL.

v. Post-metamorphic migrations. Young, post-metamorphic frogs tend to migrate upstream from their hatching site (Twitty et al., 1967).

D. Juvenile Habitat. Believed to be similar to adults.

E. Adult Habitat. Foothill yellow-legged frogs are primarily stream dwelling. Stebbins (1985) describes foothill yellow-legged frogs as stream or river frogs found mostly near water with rocky substrate, as found in riffles, and on open, sunny banks. Other authors have expanded this description, and/or offer variations. Critical habitat (i.e., habitat suitable for egg laying) is defined by Jennings and Hayes (1994a) as a stream with riffles containing cobble-sized (7.5 cm diameter) or larger rocks as substrate, which can be used as egg laying sites. These streams are generally small to mid-sized with some shallow, flowing water (Jennings, 1988). Fuller and Lind (1992) observed subadults on partly shaded (20%) pebble/cobble river bars near riffles and pools.

Less typical streams lack a rocky, cobble substrate (Fitch, 1938). Other types of riparian habitats include isolated pools and vegetated backwaters (Hayes and Jennings, 1988). Adult frogs have been observed in deep, shady, spring-fed pools (personal communication).

F. Home Range Size. Unknown.

G. Territories. Unknown, but other ranid frogs are well known to defend breeding areas (Wells, 1977).

H. Aestivation/Avoiding Desiccation. Unknown.

I. Seasonal Migrations. See "Breeding migrations" above.

J. Torpor (Hibernation). None reported.

K. Interspecific Associations/Exclusions. Foothill yellow-legged frogs are frequently found in association with Pacific treefrogs (*Pseudacris regilla*), western toads (*Bufo boreas*), Sierra garter snakes (*Thamnophis couchii*), and Pacific pond turtles (*Clemmys marmorata*). Less frequent associates include coastal giant salamanders (*Dicamptodon tenebrosus*), California newts (*Taricha torosa*), American bullfrogs (*R. catesbeiana*), northern red-legged frogs (*R. aurora*), western terrestrial garter snakes (*Thamnophis elegans*), and common garter snakes (*T. sirtalis*). There are records of foothill yellow-legged frogs co-occurring with California giant salamanders (*Dicamptodon ensatus*), southern torrent salamanders (*Rhyacotriton variegatus*), rough-skinned newts (*T. granulosa*), tailed frogs (*Ascaphus truei*), and northwestern salamanders (*Ambystoma gracile*; personal observations). Lind et al. (2003) recently found a male foothill yellow-legged frog amplexing a female American bullfrog.

L. Age/Size at Reproductive Maturity. It is generally thought that individuals reach reproductive maturity in the second year after metamorphosis (Storer, 1925; Zweifel, 1955), but Jennings (1988) reports that individuals may reproduce as early as 6 mo after metamorphosis. Also, there may be differences by sex. Additional work in this area is needed.

M. Longevity. The life span of foothill yellow-legged frogs is not known, and comparisons with the closely related mountain yellow-legged frogs (*R. muscosa*) may not be appropriate because these two species live under such different environmental regimes.

N. Feeding Behavior. Most of the literature regarding the diet of foothill yellow-legged frogs is rather general. Nussbaum et al. (1983) reports that the diet includes flies, moths, hornets, ants, beetles, grasshoppers, water striders, and snails. Terrestrial arthropods (87.5% insects, 12.6% arachnids) were the primary prey items of recently metamorphosed foothill yellow-legged frogs at a single site studied by Van Wagner (1996). Storer (1925) and Fitch (1936) note that terrestrial and aquatic insects are probable food for post-metamorphic frogs.

O. Predators. A host of vertebrates and perhaps some aquatic invertebrates feed on foothill yellow-legged frogs. Most species of garter snakes (*Thamnophis* sp.), which co-exist with foothill yellow-legged frogs, prey upon both tadpoles and juvenile frogs. This includes common garter snakes, terrestrial garter snakes, and Sierra garter snakes (*T. couchii*). All but Oregon garter snakes (*T. atratus hydrophilus*), which prefer tadpoles (Jennings and Hayes, 1994a), are reported to primarily eat young, post-metamorphic individuals (Fitch, 1941; Zweifel, 1955; Lind, 1990).

Several species of amphibians prey on foothill yellow-legged frogs. Rough-skinned newts eat foothill yellow-legged frog egg masses (Evenden, 1948). Non-native American bullfrogs prey on foothill yellow-legged frogs (Crayon, 1998). Bullfrog larvae apparently do not feed on foothill yellow-legged frogs, but through competitive interactions for algal resources, they can cause a substantial reduction of survivorship and a decreased mass in post-metamorphic individuals (Kupferberg, 1997).

There are no reports of native salmonids preying on foothill yellow-legged frogs, though a variety of introduced trout and warm water fishes eat both the eggs and tadpoles. Green sunfish (*Lepomis cyanellus*, Centrarchidae) are especially pernicious and will systematically eat eggs and larvae (Werschkul and Christensen, 1977). Native Sacramento squawfish (*Ptychocheilus grandis*) feed on adult frogs and eggs (Brown and Moyle, 1997; D. Ashton and R. Nakamoto, personal communication).

P. Anti-Predator Mechanisms. None reported.

Q. Diseases. None reported.

R. Parasites. None reported.

4. Conservation.
Foothill yellow-legged frogs are susceptible to a wide range of environmental impacts including loss of habitat, pesticides, competition/predation from nonnative species (e.g. warm-water fish, bullfrogs, crayfish), disease, water impoundments, logging, mining, and grazing in riparian zones. In the Sierra Nevada foothills of California, air-borne pesticides (that move east on the prevailing winds blowing across the highly agriculturalized Central Valley) are likely to be the primary threat to foothill yellow-legged frogs (LeNoir et al., 1999; Sparling et al., 2001; Hayes et al., 2002b). It is unknown whether pesticides are contributing to the decline of foothill yellow-legged frogs in Oregon (especially east of the agricultural parts of the Willamette Valley), but it should be examined. The populations of foothill yellow-legged frogs in greatest decline are all downwind of highly impacted (mostly agriculturalized) areas, while the largest, most robust frog populations are along the Pacific coast.

Many non-native species are likely to be competitors and/or predators of foothill yellow-legged frogs, but few studies have examined the impacts of these nonnatives. Chytrid fungus has been found in foothill yellow-legged frogs, but it is not known what the effect on foothill yellow-legged frogs might be, or even whether the fungus is a native pathogen.

In some areas, non-native American bullfrogs co-occur with foothill yellow-legged frogs and are known to have a negative impact (Kupferberg, 1996b, 1997a), but it is unclear whether this is sufficient to cause population-level declines. The role of other non-natives needs to be investigated.

There is concern that dams along many river drainages negatively impact foothill yellow-legged frogs. Dams not only interfere with normal dispersal and movements, but also provide refugia for non-native species that are likely affecting foothill yellow-legged frogs. Unfortunately, there is little research on the role of dams and how they relate to native amphibians.

Rana capito Le Conte, 1855
GOPHER FROG

John B. Jensen, Stephen C. Richter

A recent genetic study of gopher frog (*Rana capito*) populations across the current geographic distribution (Mississippi–North Carolina) by Young and Crother (2001) indicated that the Mississippi population was genetically distinct. Young and Crother (2001) therefore elevated the Mississippi population to specific status by resurrecting *R. sevosa* Goin and Netting (1940; dusky gopher frog), because this population is the only one remaining in the historical geographic range of *R. sevosa* (Louisiana to Mobile County, Alabama) as described by Goin and Netting (1940). Assigning the population discovered in Baldwin County, Alabama, after the publication of Goin and Netting (1940) to *R. sevosa* or *R. capito* is difficult because this population was not included in their study and has since gone extinct. Netting and Goin (1942a) assigned this population to *R. sevosa* based on the then-known distribution of *R. sevosa* (Louisiana to Mobile County, Alabama) and *R. capito* (Florida to North Carolina). However, populations later

discovered in counties proximate to Baldwin County were included in the genetic study of Young and Crother (2001) and were not genetically distinct from all other populations sampled east of the Mobile Bay (thus remaining *R. capito*). Based on this evidence, we assign the Baldwin County population to *R. capito* (as indicated by the geographic range maps) and note the inherent uncertainty. It currently is not possible to determine the location of the contact zone between *R. sevosa* and *R. capito*, but the extensive Mobile Basin creates a logical barrier to dispersal and thus probably separates the two species.

1. Historical versus Current Distribution.
Historically, gopher frogs (*R. capito*) were distributed throughout the Gulf and Atlantic Coastal Plain from southeastern Alabama to North Carolina (Bailey, 1991), with one isolated population known from the Ridge and Valley Province of Alabama (Mount, 1975). Though the species continues to be documented from many new localities within its range (for example, Palis and Jensen, 1995; Stevenson and Davis, 1995), this is undoubtedly due to the increased survey efforts of remaining suitable lands rather than recent colonization. The recent discovery of two *Rana* cf. *capito* individuals from a site in the Cumberland Plateau of Tennessee (Miller and Campbell, 1996; B.T. Miller, personal communication) is of great interest and may represent a substantial range extension. Young and Crother (2001) recently provided allozyme data supporting the separation of *R. capito* from both *R. areolata* (crawfish frogs) and *R. sevosa*.

2. Historical versus Current Abundance.
Gopher frogs are considered Endangered, Threatened, or of Special Concern in all of the states within their range (Mount, 1975; Martof et al., 1980; Moler, 1992a; Levell, 1997). Though little data are available

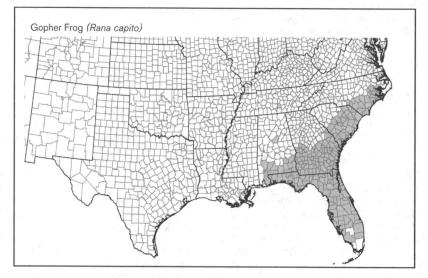

Gopher Frog *(Rana capito)*

concerning changes in the species' abundance, much information exists on the reduction of their habitat and breeding sites. The longleaf pine *(Pinus palustris)* community, the ecosystem primarily inhabited by gopher frogs, has been reduced to < 5% of its historical range (Frost, 1993; Outcalt and Sheffield, 1996; Means, this volume, Part One), and gopher frogs are just one of many component endemic or nearly endemic species declining as a result of this loss. Means and Means (2000) found that the number of breeding populations of gopher frogs in the Munson Sand Hills of panhandle Florida occur at a much lower percentage on sand pine silviculture lands than in nearby native longleaf pine habitat, and that some historical breeding populations have been extirpated. They hypothesized that the principal reason is intensive soil disturbance resulting in the elimination or severe alteration of the upland habitat. Bailey (M.A., 1994) reported that 8 of 14 known breeding sites in Alabama are considered historical, and of the six extant ponds, only three supported substantial populations. However, since that study, one "historical" site has been reconfirmed as extant, two additional breeding ponds have been discovered, and one previously known breeding site has been extirpated due to the introduction of predatory fish (M.A. Bailey, personal communication; J.B.J., personal observations). Although a few new North Carolina sites have been documented in recent years (Beane and Hoffman, 1995, 1997), Braswell (1993) found only 11 of 32 previously known breeding sites to be active in North Carolina. Twelve sites are known from South Carolina, though only four remain extant (S. Bennett, personal communication). Of 23 historical Georgia gopher frog breeding sites investigated by Seyle (1994), 12 were judged suitable, 8 were considered degraded but marginally suitable, and 3 were judged unsuitable. Only one site was found to contain gopher frogs during two extensive surveys of Georgia's Coastal Plain in late winter and early spring of 1995; however, heavy rains during the autumn of the previous year may have contributed to early breeding, and therefore, low detectability during the survey period (Moulis, 1995a,b). Extensive surveys are lacking for Florida.

3. Life History Features.

A. Breeding. Reproduction is aquatic.

i. Breeding migrations. Throughout most of their range, gopher frogs migrate from terrestrial habitats to breeding ponds in winter and early spring (primarily January–April). However, breeding migrations may occur at other times of the year following exceptionally heavy rains, especially those associated with tropical disturbances or hurricanes. Fall breeding (typically October–November) has been documented in Alabama (Bailey, 1991), Florida (Palis, 1998;

J.B.J., personal observations) and Georgia (J.B.J., personal observations). Gopher frogs often breed during summer in central and south Florida (Godley, 1992). Julian Harrison (personal communication to S. Bennett) has heard calling males, not necessarily indicative of breeding, in every month of the year in South Carolina (S. Bennett, personal communication).

Several migrations may occur throughout the breeding season, with males arriving at reproductive sites prior to females and remaining there longer (Bailey, 1991). Local populations increase during the breeding season as frogs from distant populations take up temporary residence in terrestrial retreats along the way (Franz, 1986). Movements are positively correlated with rainfall and warm temperatures and take place between sunset and sunrise (Bailey, 1991). Individuals enter and exit breeding ponds by approximately the same route (Bailey, 1991; Palis, 1998). Franz et al. (1988) recorded the longest known migration distance (2 km) between breeding site and terrestrial retreat.

ii. Breeding habitat. Typically, gopher frogs breed in either temporary or semi-permanent ponds that are shallow, have an open canopy and emergent herbaceous vegetation, and lack large, predatory fish (Moler and Franz, 1987; Bailey, 1991). Cypress *(Taxodium ascendens)* ponds are often utilized in Florida and southeast Georgia (Godley, 1992; Stevenson and Davis, 1995; J.B.J., personal observations). Anthropogenic habitats, such as ditches and borrow pits, are occasionally used (Means, 1986b; Jensen and LaClaire, 1995; J.B.J., personal observations).

B. Eggs.

i. Egg deposition sites. Within breeding wetlands, egg masses are typically attached to vertical stems of upright emergent or submergent herbaceous vegetation (Bailey, 1990; J.B.J., personal observations). The stems of inundated shrubs such as St. John's wort *(Hypericum* sp.) and myrtle-leaf holly *(Ilex myrtifolia)* also serve as egg deposition substrates (Bailey and Jensen, 1993; Palis, 1998).

ii. Clutch size. Egg masses are fist-sized clusters (Bailey, 1990), and the mean number of eggs/mass have been reported as 1,244 in Georgia (Phillips, 1995); 2,210 in Florida (Palis, 1998); and 1,500–2,000 in North Carolina (Braswell, 1993). Bailey (1989) reported 1,709 eggs for a single mass in Alabama. Apparently, each female deposits only a single egg mass (Palis, 1998).

C. Larvae/Metamorphosis. Studies have shown the length of larval period in the lab ranges from 113–225 days (Phillips, 1995). Palis (1998) documented a larval period of 7 mo at a breeding site in western Florida. Cooler water temperatures may inhibit growth (Phillips, 1995). Larvae transform at 28–37 mm SVL (Franz, 1986; Palis and Jensen, 1995; Semlitsch et al., 1995). Larvae are grazing herbivores

and utilize dense emergent and submergent vegetation for cover (J.B.J., personal observations).

D. Juvenile Habitat. Recently metamorphosed and immature gopher frogs are thought to require the same terrestrial habitats as adults.

E. Adult Habitat. Xeric, fire enhanced habitats, especially longleaf pine–turkey oak *(Quercus laevis)* sandhill (Palis, 1995a). Other habitats used include pine flatwoods, sand pine *(Pinus clausa)* scrub, and xeric hammocks (Godley, 1992). Adults seek refuge in the burrows of gopher tortoises *(Gopherus polyphemus;* Franz, 1986; Jackson and Milstrey, 1989), oldfield mice *(Peromyscus polionotus;* Gentry and Smith, 1968; Lee, 1968b), and crayfish (Godley, 1992; Phillips, 1995), as well as within stump holes (Wright and Wright, 1949). Nickerson and Celino (2003) report that gopher frogs use the hollow interiors of previously submerged and partially decomposed willow *(Salix* sp.) tree branches as shelters during droughts.

F. Home Range Size. Franz et al. (1988) documented a 2-km movement between an upland retreat and a breeding site in Florida. Phillips (1995) followed two adults for 43 d at a Georgia site and found that both remained within a 10-m radius of the specific burrow they selected. Blihovde (1999) also found strong burrow fidelity, especially among females, at sites monitored in central Florida.

G. Territories. In terrestrial habitats, rarely is >1 individual in occupancy of a single burrow (Wright and Wright, 1949; R. Franz, personal communication). Male–male combat was observed in dusky gopher frogs at a breeding pond in Mississippi (Doody et al., 1995), which may suggest that calling territories are established by these closely related species. Additionally, Jensen et al. (1995) suggested that the unusual submerged calling behavior often exhibited by male gopher frogs may be associated with territory establishment.

H. Aestivation/Avoiding Desiccation. Unknown. On warm evenings, gopher frogs are known to emerge from burrows to feed (Wright and Wright, 1949; Means, 1986b; Blihovde, 1999). Nickerson and Celino (2003) observed gopher frogs using partially decomposed willow branches as shelters during droughts.

I. Seasonal Migrations. See "Breeding migrations" above.

J. Torpor (Hibernation). Because gopher frogs breed during the winter (Mount, 1975), hibernation or torpor is thought to be nonexistent.

K. Interspecific Associations/Exclusions. Gopher frogs are one of at least 57 vertebrates known to use the burrows of gopher tortoises (Jackson and Milstrey, 1989) for fire avoidance, predator protection, feeding, and/or escape from excessively cold or warm temperatures. Within, and especially outside of, the range of gopher tortoises, gopher frogs may also seek shelter in burrows created by oldfield mice (Gentry and Smith, 1968; Lee, 1968b) and crayfish (Godley, 1992; Phillips, 1995). Blihovde (2000b) found that in addition to gopher tortoise burrows, which were common at his study site in peninsular Florida, gopher frogs used the burrows of southeastern pocket gophers (*Geomys pinetis*) as non-breeding seasonal retreats. However, gopher frogs are not wholly dependent on the burrows made by other animals since they also retreat into stump holes (Wright and Wright, 1949). Other frog species found in breeding aggregations with gopher frogs include southern leopard frogs (*R. sphenocephala*), ornate chorus frogs (*Pseudacris ornata*), southern chorus frogs (*P. nigrita*), southern cricket frogs (*Acris gryllus*), and southern toads (*Bufo terrestris*).

L. Age/Size at Reproductive Maturity. Females are 75–76 mm SVL upon reaching their first breeding season (Bailey, 1991) and are presumed to be 2 yr old (Bailey, 1991; Phillips, 1995). A study conducted in the Florida Panhandle suggests that males reach reproductive maturity at 1.5 yr (Palis, 1998).

M. Longevity. Bailey (1991) indicated that females have a life span of perhaps 6 yr, although that figure was based on a statistically insignificant cluster of size classes. In captivity, Braswell (1993) maintained a gopher frog to 7 yr old (sex unnoted).

N. Feeding Behavior. Adult gopher frogs feed on invertebrates, such as beetles, hemipterans, orthopterans, arachnids, and annelids (Deckert, 1920; Carr, 1940a; Wright and Wright, 1949), as well as on other anurans (Godley, 1992), especially toads (Dickerson, 1906). Feeding is primarily nocturnal (Means, 1986b) and is thought to occur in close vicinity to, and possibly within, utilized burrows. Captives feed readily on crickets, earthworms, and young mice (Braswell, 1993).

O. Predators. On two occasions, Jensen (2000) observed the predation of adult gopher frogs by banded water snakes (*Nerodia fasciata fasciata*) in a western Florida breeding pond. At another pond in western Florida, a Florida softshell turtle (*Apalone ferox*) was observed feeding on an adult gopher frog (J.B.J, unpublished data). However, it was not known whether the turtle had killed the frog or the frog had first died and was being scavenged.

Banded water snakes will also eat gopher frog tadpoles (Aresco and Reed, 1998). Additionally, under laboratory conditions, tadpoles have been consumed by common garter snakes (*Thamnophis sirtalis*), black swamp snakes (*Seminatrix pygea*), snapping turtles (*Chelydra serpentina*), chicken turtles (*Deirochelys reticularia*), common musk turtles (*Sternotherus odoratus*), dragonfly naiads, and backswimmers (Travis et al., 1985; Cronin and Travis, 1986; Phillips, 1995; M.A. Bailey, personal observations).

Tadpoles are palatable to predatory fishes (LaClaire and Franz, 1991) that occasionally colonize or become unnaturally introduced into the isolated wetlands used by gopher frogs. The introduction of bluegill (*Lepomis macrochirus*) and mosquitofish (*Gambusia affinis*) into a pond in south-central Alabama is strongly suspected for the lack of continued breeding by gopher frogs there (Jensen, 1995).

Eastern newts (*Notophthalmus viridescens*) and trichopteran larvae (Insecta) have been observed feeding on the eggs of gopher frogs (Bailey, 1989).

P. Anti-Predator Mechanisms. As with many other anurans, adult gopher frogs will inflate their bodies when grasped by potential predators (J.B.J., personal observations). Additionally, adults often slightly curl up and cover their eyes with their forefeet when threatened (J.B.J., personal observations). However, this behavior may have evolved as a defense from potential eye injury as a result of trampling by gopher tortoises in co-occupied burrows (D.B. Means, personal communication).

Q. Diseases. Unknown.

R. Parasites. Ticks (*Ornithorus turcicata*) have been found on gopher frogs and other vertebrates that share gopher tortoise burrows, including the tortoises themselves. However, these ticks apparently cause no harm to gopher frogs (Milstrey, 1984; Blihovde, 2000a). This tick species is capable of transmitting relapsing fever to humans and other animals and is also a potential vector of African swine fever (Milstrey, 1984).

4. Conservation.
There is concern about the status of gopher frogs in all the states within their range (Mount, 1975; Martof et al., 1980; Moler, 1992a; Levell, 1997), but they are afforded legal protection only in North Carolina (Special Concern), Florida (Special Concern), and Alabama (Protected). The U.S. Fish and Wildlife Service is currently evaluating their range-wide status to determine if they warrant listing as an Endangered species (L. LaClaire, personal communication).

The greatest threat to gopher frogs is the loss and alteration of both upland and wetland habitats resulting from commercial, residential, silvicultural, and agricultural development, as well as from fire suppression. Exclusion and suppression of fire from wetlands and the concomitant build-up of peat may also threaten gopher frogs by increasing water acidity past tolerance levels (Smith and Braswell, 1994). The introduction of predacious fish into gopher frog breeding ponds may render these ponds useless for successful reproduction. In areas where gopher frogs rely heavily on the burrows of gopher tortoises for refuge, tortoise declines may reduce gopher frog populations as well. The prac-

tice of removing tree stumps ("stumping") in silvicultural areas further reduces the availability of subterranean retreats.

Acknowledgments. We thank Carlos Camp, Julie Chastain, and Robert Moulis for their assistance in assembling this manuscript.

Rana cascadae Slater, 1939
CASCADE FROG

Christopher A. Pearl, Michael J. Adams

1. Historical versus Current Distribution.
Cascade frogs (*Rana cascadae*) historically occupied moderate and high elevation (about 400–2,500 m) lentic habitats throughout the Cascade Range, from the very northern edge of California's Sierra Nevada to within 25 km of the British Columbia border (Dunlap and Storm, 1951; Dunlap, 1955; Dumas, 1966; Bury, 1973a; Hayes and Cliff, 1982; Nussbaum et al., 1983; Fellers and Drost, 1993; Jennings and Hayes, 1994a; K.R. McAllister, 1995). Population systems occurring in the Olympic Mountains of Washington and the Trinity Alps, Mt. Shasta, and Mt. Lassen areas of California are notably disjunct from the main Cascade axis and may warrant investigation for cryptic taxa (Jennings and Hayes, 1994a; Monson and Blouin, 2003). Range contractions have been documented in the southern end of their range (Fellers and Drost, 1993; Jennings and Hayes, 1994a). Jennings and Hayes (1994a) and Fellers and Drost (1993) estimate that Cascade frogs are extirpated from about 99% of their southernmost population clusters (Mt. Lassen and surroundings), and 50% of their total historical distribution in California.

2. Historical versus Current Abundance.
Cascade frogs are among the most commonly encountered lentic breeders in Olympic and Mount Rainier national parks, where extensive surveys suggest no evidence of declines (Adams et al., 2001; Tyler et al., 2002). Cascade frogs were among the first amphibians to recolonize sites post-eruption and are now commonly encountered in Mount St. Helens Volcanic Monument in the southern Washington Cascades (Karlstrom, 1986; Crisafulli and Hawkins, 1998; C. Crisafulli, personal communication). Some concerns for declines in the central Oregon Cascade Range have been reported (Nussbaum et al., 1983; Blaustein and Wake, 1990), but other surveys do not suggest exceptionally low site occupancy rates (C. Brown, 1997; C.A.P., B. Bury, and M.J.A., unpublished data). Recent surveys suggest that Cascade frogs remain present in portions of the Trinity Alps and Marble Mountains, but are rare to nonexistent in other Californian portions of their historical range (G. Fellers, H. Welsh, personal communications). Declines have been documented around Lassen Volcanic National Park in northeastern California, where Cascade frogs were historically

Cascade Frog (*Rana cascadae*)

widespread and abundant between 1,450 and 2,480 m (Grinnell et al., 1930; Badaracco, 1962). In 1991, Cascade frogs were detected at only 1 of 50 sites in the vicinity of Lassen Volcanic National Park, 16 of which historically supported this species (Fellers and Drost, 1993). More recent surveys (1992–2002) have detected Cascade frogs at only 4 of 400 sites (G. Fellers, personal communication). Population sizes were small at all four sites, and as of 2002, frogs had disappeared from two of these four sites (G. Fellers, personal communication).

3. Life History Features.

A. Breeding. Reproduction is aquatic.

i. Breeding migrations. Available information suggests that Cascade frog adults overwinter near breeding localities and, because they often become active during ice thaw, are thought not to make extensive breeding migrations (Briggs, 1976, 1987; O'Hara, 1981).

ii. Breeding habitat. Cascade frogs breed in temporary and permanent lentic waters, often in smaller bodies of water nearer larger lakes (Nussbaum et al., 1983; Briggs, 1987; Bury and Major, 1997). In Olympic National Park, Washington, Cascade frog breeding sites generally have silt/mud substrates, lack fish, and have low UV-B transmission (Adams et al., 2001).

B. Eggs.

i. Egg deposition sites. First egg masses are deposited in comparatively warm water along gradually sloping shorelines, often over soft substrates protected from severe wave action (Sype, 1975; O'Hara, 1981). Breeding is explosive (often completed in < 1 wk), and egg masses are commonly deposited in aggregations (Sype, 1975; Briggs, 1987). The placement of clusters of egg masses in shallow water soon after first thaw can make them susceptible to freezing mortality, pathogen transmission between adjacent masses, or desiccation associated with receding water levels (Blaustein and Olson, 1991; Kiesecker and Blaustein, 1997b; C.A.P., personal observations).

ii. Clutch size. Egg masses contain between 300–800 ova (O'Hara, 1981; Nussbaum et al., 1983). Adults appear to be strongly philopatric, using the same breeding sites for several years (O'Hara, 1981; Olson, 1992). Females breed no more than once/year, but whether they skip years remains unknown (Sype, 1975; Nussbaum et al., 1983).

C. Larvae/Metamorphosis.

i. Length of larval stage. Water temperatures strongly influence rates of development, hatching, and time to metamorphosis (Sype, 1975). Metamorphosis is generally achieved about 2 mo after hatching (Nussbaum et al., 1983; Briggs, 1987). There is no documentation of overwintering larvae, but the topic warrants further investigation.

ii. Larval requirements.

a. Food. Larvae are thought to be primarily benthic feeders, but specific preferences are not well known (Nussbaum et al., 1983).

b. Cover. Larvae often form loose aggregations near their oviposition sites. Aggregations usually include kin (O'Hara and Blaustein, 1985) and are generally associated with flooded vegetation in warm microhabitats (O'Hara, 1981; Wollmuth et al., 1987).

iii. Larval polymorphisms. Not reported. Albino larvae have been observed three times in the central Cascade Range of Oregon (Altig and Brodie, 1968; B. McCreary, personal communication).

vi. Features of metamorphosis. In laboratory trials, size at metamorphosis varies from 15–30 mm SVL and depends on larval density, food availability, and water temperature (Blaustein et al., 1984). Larvae reared in 23 °C lab water averaged 20–21 mm SVL at metamorphosis after 37.5 d (Nussbaum et al., 1983). Larval growth rate can affect shape (e.g., leg length) at metamorphosis (Blouin and Brown, 2000).

v. Post-metamorphic migrations. Specific information is lacking.

D. Juvenile Habitat. Similar to adults, especially marshy fringes of lentic environments, but detailed information is not yet available (C.A.P, personal observation).

E. Adult Habitat. Cascade frog adults utilize an array of habitat types, but are generally associated with open wetland habitats at higher elevations (C. Brown, 1997; Bury and Major, 1997; Bosakowski, 1999; Olson, 2001). Cascade frog adults commonly occupy moist meadows and can be found in relatively small permanent and temporary ponds (Sype, 1975; O'Hara, 1981; Olson, 1992). Adults are also found along streams in summer, especially at lower elevations where lentic habitats are less common (Dunlap, 1955; C.A.P, personal observations). Adults generally stay close to water, particularly along sunny shores, under dry summer conditions, but can be found traversing uplands during high humidity (Nussbaum et al., 1983). Adults and breeding can occur in anthropogenic wetland habitats such as pump chances (Quinn et al., 2001).

F. Home Range Size. Unknown.

G. Territories. While adults are not considered territorial (Nussbaum et al., 1983), males will behave aggressively toward one another during breeding (C.A.P, personal observation).

H. Aestivation/Avoiding Desiccation. Not reported; summer is their active season.

I. Seasonal Migrations. See "Breeding migrations" and "Juvenile Habitat" above.

J. Torpor (Hibernation). Cascade frogs hibernate through the long, snowy winters typical of most of their range (Nussbaum et al., 1983). Frogs have been recovered from mud beneath 0.3–1.0 m of water and from spring-saturated areas around ponds (Briggs, 1987).

K. Interspecific Associations/Exclusions.
Cascade frogs may share breeding sites with western toads *(Bufo boreas)*, northwestern salamanders *(Ambystoma gracile)*, long-toed salamanders *(A. macrodactylum)*, rough-skinned newts *(Taricha granulosa)*, and Pacific treefrogs *(Pseudacris regilla)*. They also occur syntopically with Oregon spotted frogs *(R. pretiosa)* at <10 sites along the Oregon Cascades and can co-occur with northern red-legged frogs *(R. aurora)* at sites in lower elevations (C.A.P, personal observations). Several researchers have suggested that Cascade frogs may be reduced in number by introduced sportfish (Fellers and Drost, 1993; Jennings and Hayes, 1994a).

L. Age/Size at Reproductive Maturity. In the Oregon Cascades, males and females are thought to reach sexual maturity at 2–3 yr (about 35–40 mm) and 4 yr (about 50–55 mm) post-metamorphosis, respectively (Briggs and Storm, 1970; Olson, 1992; Jennings and Hayes, 1994a).

M. Longevity. In the Oregon Cascades, both males and females are thought to live >5 yr, sometimes reaching 7 yr (Briggs and Storm, 1970; Olson, 1992).

N. Feeding Behavior. Diet composition of Cascade frogs is poorly known, but adults are thought to consume a variety of invertebrate prey and will occasionally consume conspecifics (Rombough et al., 2003).

O. Predators. Native predators of adult frogs include water bugs (Belostomatidae), garter snakes (especially common garter snakes, *Thamnophis sirtalis*), mustelid mammals, raccoons, and several bird species (Briggs and Storm, 1970; Nussbaum et al., 1983; Nauman and Dettlaff, 1999). Larvae and newly metamorphosed individuals are taken by invertebrates (Belostomatidae, Dytiscidae, Odonata), and aquatic salamanders (rough-skinned newts, northwestern salamanders, and long-toed salamanders; Briggs and Storm, 1970; Peterson and Blaustein, 1991, 1992). Juvenile and adult Cascade frogs are known to cannibalize larvae and newly metamorphosed animals (Rombough et al., 2003). Introduced salmonids are now widespread in high lakes throughout the range of Cascade frogs and may represent a common predator of larvae and small adults (Hayes and Jennings, 1986; Fellers and Drost, 1993; Jennings and Hayes, 1994a; Simons, 1998).

P. Anti-Predator Mechanisms. In laboratory trials, larvae increased activity when exposed to waterborne chemical cues of injured conspecifics (Hews and Blaustein, 1985). Chemical cues of predatory leeches can elicit early hatching of eggs (Chivers et al., 2001). Adult escape behavior often includes a series of rapid bounds into water, followed by burrowing head-first into unconsolidated substrate (C.A.P, personal observations).

Q. Diseases. Field experiments suggest that the oomycete fungus, *Saprolegnia ferax*, is related to embryonic mortality in Cascade frogs and may be enhanced by other stressors such as UV-B radiation (Kiesecker and Blaustein, 1995, 1997b).

R. Parasites. The blood parasite *Lankesterella* sp. was reported in Cascade frogs by Clark et al. (1969). Cascade frogs exhibited lower malformation frequencies attributable to metacercariae of the trematode *Ribeiroia ondatrae* relative to other Pacific Northwestern amphibians (Johnson et al., 2002).

4. Conservation.
A latitudinal gradient of conservation status apparently exists for Cascade frogs, with declines documented in the southern portion of their range but not in study areas to the north (see "Historical versus Current Distribution" and "Historical versus Current Abundance" above). Recent surveys in northern California suggest that populations in the region are often small, and that their abundance and distribution are negatively associated with introduced salmonids (Fellers and Drost, 1993; H. Welsh, G. Fellers, personal communications; but see also Davidson et al., 2002). Declines have been referenced in Oregon (Blaustein and Wake, 1990), but field data from the northern and central Oregon Cascades suggest the species remains widespread in some areas and has the capacity to rebound from short-term declines (Olson, 1992; C. Brown, 1997; C.A.P and colleagues, personal observations). Cascade frogs remain widespread and common in Olympic and Mount Rainier national parks, and in the Mt. St. Helens Volcanic Monument in Washington (Karlstrom, 1986; Crisafulli and Hawkins, 1998; Adams et al., 2001; Tyler et al., 2002). Cascade frogs are considered a Species of Special Concern in California (California Department of Fish and Game, 1999), and Sensitive-Vulnerable in Oregon (Oregon Natural Heritage Program, 1995).

Causes of Cascade frog declines are not fully known, but introduced trout, UV-B radiation, fungal pathogens, and loss of open meadow habitat due to fire suppression have been suggested (Hayes and Jennings, 1986; Fellers and Drost, 1993; Blaustein et al., 1994b,c; Kiesecker and Blaustein, 1995; Fite et al., 1998; Adams et al., 2001). In Oregon, Cascde frog embryos have low photolyase levels and therefore low capacity to repair UV-B damage; they demonstrate reduced hatching in unshielded relative to shielded *in situ* enclosures (Blaustein et al., 1994c). However, breeding sites in the Oregon and Washington portions of the species' range may be afforded some protection from lethal UV-B doses by dissolved organic matter (Palen et al., 2002). One Oregon study failed to detect short-term changes in breeding phenology that might be attributable to climate change

(Blaustein et al., 2001). Fertilizers such as urea may pose a threat to Cascade frogs, as juveniles do not appear capable of sensing and avoiding toxic levels in laboratory studies (Hatch et al., 2001). Nitrites may affect behavior and metamorphosis of Cascade frog larvae (Marco and Blaustein, 1999). An improved understanding of microhabitat associations and interactions with introduced fish is needed to assist with conservation measures for the species (Olson, 2001).

Rana catesbeiana Shaw, 1802
AMERICAN BULLFROG
Gary S. Casper, Russ Hendricks

1. Historical versus Current Distribution.
Historically, American bullfrogs *(Rana catesbeiana;* hereafter in this account referred to as bullfrogs) had one of the largest ranges of any North American amphibian (Bury and Whelan, 1984), occurring from Nova Scotia south to central Florida and west across the Great Plains (Conant, 1975), and probably including Tamaulipas and northern Vera Cruz in northern Mexico (Kellog, 1932). Although the historical range was large, bullfrogs were naturally limited in distribution (Bleakney, 1958). Accurate determination of the original distribution will be forever confused as a result of introductions outside the natural range in the western United States (Bury and Whelan, 1984; see "Interspecific Association/Exclusions" below), as well as within the natural range in the Midwest (Lannoo et al., 1994; Casper, 1996; Christiansen, 1998; Hemesath, 1998; Moriarty, 1998). Bullfrogs were introduced into Oregon as early as the 1920s or 1930s (Nussbaum et al., 1983). Bullfrogs have also been introduced into Europe (Stumpel, 1992; Thiesmeier et al., 1994), South America, Asia (Kupferberg, 1997), British Columbia, Mexico, the Caribbean Islands, Brazil, and Hawaii (Bury and Whelan, 1984), due to their culinary appeal. Local extirpations have been reported in southwestern Ontario (Hecnar, 1997).

2. Historical versus Current Abundance.
In general, frog populations fluctuate widely in response to seasonal precipitation, disease, predation, and other factors, making the discernment of overall trends problematic. Shirose and Brooks (1997) estimated that bullfrog populations in Algonquin Park, Ontario, likely fluctuate by as much as 50–80%, even in the absence of long-term trends in population size. Nevertheless, bullfrogs remain common throughout most of their natural range, despite general depression of frog numbers in many regions.

Historically, bullfrogs were common in the United States and abundant despite harvesting (Baker, 1942). Today many

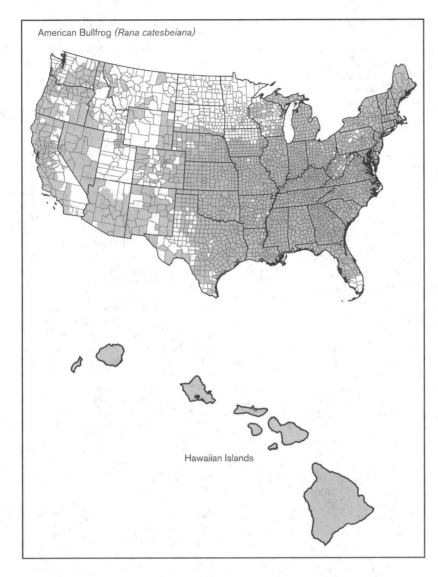

American Bullfrog *(Rana catesbeiana)*

Hawaiian Islands

native bullfrog populations appear to be declining, with habitat loss and degradation, water pollution, and pesticide contamination (see "Conservation" below) commonly invoked as causal factors (Texas, Baker, 1942; Louisiana, Walker, 1963; United States, Bury and Whelan, 1984; Great Lakes, Harding, 1997; Wisconsin, Mossman et al., 1998). Wetland drainage, shoreline development, and damage to the wetland fringes of lakes from home building and recreation have greatly decreased bullfrog habitat quality and availability in many areas (Bury and Whelan, 1984; Hunter et al., 1992; Casper, 1998). Over-harvesting has also been implicated in bullfrog decline (Baker, 1942; Gibbs et al., 1971; Treanor and Nichola, 1972; Treanor, 1975; Vogt, 1981).

The creation of farm and golf course ponds, where high bullfrog densities can develop in altered wetland habitats, has increased bullfrog abundance in some areas (Willis et al., 1956; G. S. C. personal observation, 1990s), and widespread introductions have greatly expanded the range. Introduced populations appear to be experiencing range expansions in Iowa (+ 279% increase; Christiansen, 1998), Minnesota (Moriarty, 1998), and California (Bury and Luckenbach, 1976), although Treanor and Nichola (1972) report declines in the late 1960s in California.

3. Life History Features.

A. Breeding. Reproduction is aquatic.

i. *Breeding migrations.* Breeding takes place in spring and early summer and is earliest in southern latitudes (Bury and Whelan, 1984). In Texas, bullfrogs breed from March–October (Blair, 1961a); in Québec, breeding occurs from late May to mid-July (Bruneau and Magnin, 1980); and in the Great Lakes in June–July (Pentecost and Vogt, 1976; Harding, 1997; Mossman et al., 1998; Varhegyi et al., 1998). Lepage et al. (1997) provides phenology data for Québec, with the peak calling period occurring from mid-June to July. However, Bishop et al. (1997) give 17 May to 8 June as peak calling dates in southern Ontario. In Kansas, calling

begins when air temperatures are above 21 °C (Fitch, 1956c). In Michigan (and probably elsewhere, especially in southern latitudes), double clutching occurs, with some females active early and then again 3 wk later (Emlen, 1977). Older females may produce two clutches/year, with second clutches containing substantially smaller eggs (Howard, 1978b).

ii. *Breeding habitat.* Bullfrogs breed in the vegetation-choked shallows (Pope, 1964a) of permanent bodies of water. Male bullfrogs will aggressively defend oviposition sites (Ryan, 1980) by pushing, shoving, and biting. Females select a mate by entering his territory (Ryan, 1980). Female mate choice becomes more discriminating with age, with older females consistently selecting the oldest, largest males as mates (Howard, 1978a). Older females sometimes vocalize within male choruses, which may elicit higher levels of male–male competition and assist females in selecting high-quality males (Judge et al., 2000). Embryo mortality depends on female choice of the oviposition site, the best of which are controlled by the largest males and where water temperatures do not exceed 32 °C, over which developmental abnormalities occur (Howard, 1978b). Howard (1979) estimated male reproductive success in a Michigan population. The most successful males fertilized three females, the least successful had no matings, and the average was 0.71 matings/male. Males defending territories in the middle of ponds are susceptible to predation by snapping turtles (*Chelydra serpentina;* Emlen, 1976; Howard, 1978a).

B. Eggs.

i. *Egg deposition sites.* Eggs are laid in thin sheets on the water surface, covering 0.5–1 m², and hatching in 3–5 d (Bury and Whelan, 1984).

ii. *Clutch size.* Bullfrogs are extremely prolific, producing up to 20,000 eggs/clutch (Schwalbe and Rosen, 1999). The number of zygotes sired by successfully mating males ranges from 4,928–59,035 (mean 11,149), with the number of resultant hatchlings ranging from 299–29,377 (mean 5,582; Howard, 1978b). Females may lose up to 27% of their body mass during oviposition (Judge et al., 2000).

C. Larvae/Metamorphosis. Tadpoles favor warm water environments (24–30 °C; Brattstrom, 1962b) and have preferred body temperatures ranging from 15–26.7°C, which changes with acclimation (exposure to new temperature ranges), developmental stage, and season (Ultsch et al., 1999). Oxygen consumption increases about exponentially as a function of temperature (Burggren et al., 1983; Feder, 1985).

i. *Length of larval stage.* The time to metamorphosis varies from a few months in the south to 3 yr in Michigan and Nova Scotia (Collins, 1979; Bury and Whelan, 1984). The length of the larval period is negatively correlated with mean length of the frost-free period, and tadpoles in a Michigan study transformed in either the first or second season following the hatching year (Collins, 1979; see also Crawshaw et al., 1992).

ii. Larval requirements.

a. Food. The diet of larvae consists mainly of algae, aquatic plant material, and some invertebrates (Treanor and Nichola, 1972; Bury and Whelan, 1984). Growth rates and digestive abilities of tadpoles fed differing algal diets was studied by Pryor (2003). Conflicting reports exist as to whether larvae will scavenge dead animal material (Dickerson, 1906; Funkhouser, 1976). Coprophagy was demonstrated by Steinwascher (1978a), who suggested that this increased the amount of time food was resident in the digestive tract and may also allow some microbial digestion.

b. Cover. Tadpoles are solitary (Punzo, 1992b) and move into deeper water until just prior to metamorphosis, at which time their depth preference is reversed (Goodyear and Altig, 1971; Smith, 1999). Bullfrog tadpoles are bimodal breathers (Crowder et al., 1998; Ultsch et al., 1999) and their distribution within ponds is likely influenced by both temperature and oxygen availability in addition to predator avoidance (Nie et al., 1999). Bullfrog tadpoles avoid predators by seeking cover (e.g., Pearl et al., 2003).

iii. Larval polymorphisms. Not reported.

vi. Features of metamorphosis. Metampophosis is asynchronous. Newly metamorphosed animals generally frequent shorelines where vegetative cover is thick enough to afford protection from predators (G.S.C., personal observation). Thrall (1971) found newly metamorphosed bullfrogs occupying small circular pits along bare mud flats of a farm pond in Illinois.

v. Post-metamorphic migrations. Specific information is lacking.

D. Juvenile Habitat. Bullfrogs are adept at colonizing newly created ponds (Merovich and Howard, 2000) and the juvenile stage is likely where most dispersal occurs.

E. Adult Habitat. Adult bullfrogs are mostly aquatic and prefer warmer, lentic habitats (George, 1940) such as vegetated shoals, sluggish backwaters and oxbows, farm ponds, reservoirs, marshes, and still waters with dead woody debris (Holbrook, 1842; Storer, 1922) and dense and often emergent vegetation (Bury and Whelan, 1984). They also occupy shorelines of lakes and streams. In a long-term study of ponds in southern Michigan, Skelly et al. (1999) found bullfrogs restricted to open canopy ponds with permanent water. Bullfrogs prefer relatively high body temperatures (26–33 °C; Lillywhite, 1970) and actively thermoregulate, basking and posturing to control body temperatures (Lillywhite, 1970, 1971, 1974b). Adults generally remain inactive until water temperatures approach 15 °C (Harding, 1997). Unlike many other frogs, bullfrogs can co-exist with predatory fishes (Hecnar, 1997).

F. Home Range Size. Currie and Bellis (1969) found bullfrogs to have a mean activity radius of 2.6 meters in an Ontario pond. Their data suggest that home range size is reduced at higher densities and increases as the size of an individual increases, and that males have a larger home range than females. Raney (1940) found no evidence of homing in a New York population. Dispersals of at least 3.2 km from the home pond have been recorded, with dispersal distances of 7–8 km likely (Schwalbe and Rosen, 1999). Maximum movement distances of up to 1,600 m (mean 402), as well as homing, were reported by Ingram and Raney (1943) in New York. McAtee (1921) noted a bullfrog that returned to its lakeshore territory within 2 d after being moved ≤0.4 km away. Schwalbe and Rosen (1999) reported an adult bullfrog every 1.8 m along some shorelines in Arizona ponds. In Kentucky, Cecil and Just (1979) calculated densities of 0.9–13.2 tadpoles/m² of pond, biomass of tadpoles as 11–103 g/m² of pond, and survival rates of 11.8%, 13.1%, and 17.6%. Treanor (1975) reports densities in California canals from 6.6–119 frogs/km, with average numbers reduced by 20–25% in areas where harvest occurred. Bullfrog density in an Ontario pond was 0.9 and 1.3/m² (Currie and Bellis, 1969) and 8.8–45.8 frogs/ha in a 7 ha Illinois lake (Durham and Bennett, 1963).

G. Territories. Bullfrogs are highly territorial and have a polygynous mating system, with the largest males in a population controlling the highest quality oviposition sites preferred by females (Howard, 1978a,b). Males defend roughly circular territories 2–5 m in diameter (Harding, 1997). Bullfrogs recognize the calls of neighbors and respond more strongly to the calls of strangers (Davis, 1987). Most physical encounters are won by larger males by engaging in wrestling, shoving, and pouncing (A.P. Blair, 1963; Durham and Bennet, 1963). Younger males unable to hold territories against larger, older males, employ male parasitism (intercepting females attracted to large males) and opportunism (temporarily holding a vacated territory until threatened by a larger male) as mating strategies (Howard, 1978a). In an Illinois population, Mauger (1988) observed medium-sized males engage in opportunism, where they vocalized from the fringes of a large male territory until forced to flee or they inhabited recently vacated territories. The smallest males practiced parasitism, maintaining a low posture without vocalizing in an attempt to intercept females en route to a large-male territory. Males near a large-male territory also invade such territory when the large male is amplexed, without challenging or displacing the resident male.

H. Aestivation/Avoiding Desiccation. No information regarding aestivation has been reported.

I. Seasonal Migrations. Bullfrogs are the last North American ranid to emerge in spring (Voice, 1931; Smith, 1934; Ryan, 1953), following a general northward progression (Bury and Whelan, 1984). However, bullfrogs in Louisiana may be active throughout the year (George, 1940). Overland movements, such as the dispersal of newly metamorphosed animals and adult movements, occur on warm, rainy nights (Raney, 1940; Gibbons and Bennett, 1974; Johnson, 1987; G.S.C., personal observation). Summer movements appear to be limited to adverse or drought conditions (Jameson, 1956a). In Missouri, only a small percentage of frogs migrated between ponds, and the few that did traveled 0.16–2.8 km (Willis et al., 1956).

J. Torpor (Hibernation). Bullfrogs are not freeze tolerant. In autumn, adults become torpid before frost occurs, while juvenile activity continues until freezing weather (Willis et al., 1956). Winter hibernation usually takes place under water (Treanor and Nichola, 1972), where individuals bury themselves in surface mud or construct protective pits (Emery et al., 1972; Johnson, 1987; Harding, 1997) or cave-like holes (Ruffner, 1933). Bohnsack (1952) reported a torpid bullfrog from a terrestrial hibernaculum, buried under leaf litter in a soil pocket of an oak-hickory woods in Michigan. Radio-tagged bullfrogs in Nova Scotia hibernated in a pond and tolerated prolonged hypoxia (sustained PO2 below 20 Torr for >8 wk; Friet and Pinder, 1990). In a radio telemetry study in Ohio, bullfrogs overwintered in relatively shallow areas (< 1 m) near small inlet streams of ponds, as well as 1–2 m off-shore of ponds, and remained active throughout the hibernation period (Stinner et al., 1994). These bullfrogs laid on the pond bottom and were not covered by silt. Juvenile bullfrogs cease motor activity between 0–1 °C (Lotshaw, 1977).

K. Interspecific Associations/Exclusions. Bullfrogs are often the predominant frog species in permanent aquatic habitats, naturally occurring with species such as green frogs (*R. clamitans*; Werner, 1991). Introduced bullfrogs have been implicated in the decline or displacement of many amphibians and a few reptiles, including Mexican garter snakes (*Thamnophis eques*), lowland leopard frogs (*R. yavapaiensis*), and Chiricahua leopard frogs (*R. chiricahuensis*; Arizona, Schwalbe and Rosen, 1988); foothill yellow-legged frogs (*R. boylii*; California, Kupferberg, 1997); northern red-legged frogs (*R. aurora*; Oregon, Kiesecker and Blaustein, 1997a, 1998); the apparently now-extinct Vegas Valley leopard frog (*R. fisheri*; Nevada, Moyle, 1973; Cohen, 1975); northern leopard frogs and plains leopard frogs (*R. blairi*; Colorado, Hammerson, 1982a,b); northern leopard frogs and Oregon spotted frogs (*R. pretiosa*; Montana, Black, 1969a); Oregon spotted frogs (Dumas, 1966); Blanchard's cricket frogs (*Acris crepitans blanchardi*; Ontario, Oldham, 1992); Pacific treefrogs (*Pseudacris regilla*; Oregon, Jameson, 1956b; California, Kupferberg, 1997); and native amphibians in Iowa (Lannoo et al., 1994). Schwalbe and Rosen (1988, 1999) believe bullfrogs to be responsible for declines in Mexican garter snakes from the San Bernardino National Wildlife Refuge and present supporting evidence.

An analysis of multiple factors correlated with the decline of northern red-legged frogs in western Washington did not support the hypothesis that bullfrogs were excluding this species, implicating instead the presence of exotic fishes (Adams, 1999). In controlled enclosure experiments testing survival of northern red-legged frog and Pacific treefrog larvae, the presence of sunfish (Centrarchidae) reduced survivorship of both species to near zero, while the presence of bullfrog tadpoles had a weak survival effect only in northern red-legged frogs (Adams, 2000). In another enclosure experiment, testing the effects of fishes and bullfrogs on northern red-legged frogs, Kiesecker and Blaustein (1998) found that time to metamorphosis increased and mass at metamorphosis decreased when northern red-legged frog tadpoles were exposed to either larval or adult bullfrogs.

In Kupferberg's (1997) enclosure experiments, the presence of bullfrog tadpoles resulted in a 48% reduction of survivorship of foothill yellow-legged frog tadpoles, and a 24% decline in mass at metamorphosis. Kupferberg (1997) also found a 16% reduction in the size of newly metamorphosed Pacific treefrogs, but no significant effect on survivorship.

Fisher and Shaffer (1996) found a negative correlation between the presence of introduced exotics (bullfrogs and fishes) and native amphibians in California. However, they did not discriminate between fishes and bullfrogs in their analyses. Hayes and Jennings (1986) argued that while bullfrogs are often invoked as being responsible for declines in native ranid frogs in western North America, existing data cannot distinguish adequately among three major causal hypotheses for this decline: bullfrogs, habitat alteration, and introduced predatory fish. They believed that fish predation was the most compelling hypothesis.

In Ontario, Hecnar and M'Closkey (1997) observed the relative abundance of green frogs increase four-fold after bullfrog extirpation, where competition between these species and predation of bullfrogs upon green frogs is postulated. In a study of bullfrog/green frog interactions in Michigan, Werner et al. (1995)

concluded that bullfrogs relied more heavily on aquatic prey than did green frogs, that overlap between the species in diet resources declined with increasing body size disparity, and that juvenile frogs were a major component of the diet of bullfrogs but not green frogs. Stewart and Sandison (1972) found both dietary and habitat differences, with some overlap, in sympatric populations of bullfrogs, mink frogs (*R. septentrionalis*), and green frogs in New York. Werner (1991) demonstrated that small larvae of bullfrogs and green frogs reduced activity in the presence of predatory *Anax* dragonfly naiads, but bullfrogs did less so, thereby gaining a competitive advantage over green frogs when naiads were present. This was a size-dependent effect, as larger bullfrog tadpoles largely ignored the presence of naiads.

In pair-wise competition experiments with eight species of desert anurans, Woodward (1982) found that bullfrog tadpoles were the only permanent pond species able to persist in the presence of Couch's spadefoot (*Scaphiopus couchii*) tadpoles, and that temporary pond species were competitively superior to species breeding in permanent waters.

In laboratory experiments, Alford (1989c) studied competition between Cope's gray treefrog (*H. chrysoscelis*) and bullfrog tadpoles and found that the presence of bullfrogs substantially reduced the mass of Cope's gray treefrogs, but that reproductive phenology influenced this effect. Peacor and Werner (1997) conducted experiments in cattle watering tanks with bullfrog tadpoles, green frog tadpoles, adult bullfrogs, the lethal presence of the larval odonate predator *Tramea lacerata*, and the nonlethal (caged) presence of the larval odonate predators *Anax junius* and *A. longipes*. They demonstrated that the presence of large bullfrog competitors increased the rate of *Tramea* predation on small green frogs and small bullfrogs, and that the presence of nonlethal *Anax* increased the competitive advantage of bullfrogs over green frogs. The proposed mechanisms were changes in small tadpole activity.

Moore (1952) attributed the southern distributional limit of mink frogs to competitive exclusion by bullfrogs. In sum, interactions among bullfrogs, other amphibians, and the environment is complex and is influenced by both the predators and resources present in the system, as well as the phenology of life history events.

L. Age/Size at Reproductive Maturity. Sexual maturity in bullfrogs usually occurs in 1–2 yr in males, and 2–3 yr in females (Howard, 1981), with males maturing at an earlier age and/or smaller size than females. Turner (1960) reported maturity in 1 yr; Raney and Ingram (1941) reported 2–3 yr in a New York population; Minton (1972) gives 1 yr after metamor-

TABLE SP-2

Mean Body Length of American Bullfrogs (mm) in Relationship to Age After Transformation

Age	Size
1	103
2	124
3	133
4	145
5	152
6	162

NOTE: From Schroeder and Baskett, 1968.

phosis; and Carr et al. (1984) recorded ovarian maturity in females 12 mo after metamorphosis under lab conditions. Harding (1997) reports bullfrogs reaching breeding size in 2–4 yr when they are 10–12 cm.

The size reached at sexual maturity is influenced by geographic location (Bury and Whelan, 1984) and occurs at 95–110 mm in Québec (Bruneau and Magnin, 1980) and 125 mm or greater for females in Missouri (Willis et al., 1956). Growth rates of lab-raised juveniles slow from 27 to 14 to 8 mm/yr in the first, second, and third years, respectively (Jorgensen, 1992). The growth data of Schroeder and Baskett (1968) are summarized in Table SP-2. Minimum snout-ischium length of reproductive individuals in Michigan was 95 mm for males, and 108 mm for females (Howard, 1981).

M. Longevity. Longevity for wild bullfrogs is estimated to be 8–10 yr, although a captive specimen survived for nearly 16 yr (Oliver, 1955a; Goin and Goin, 1962). A maximum age of >6 yr was revealed by osteology in bullfrogs from Missouri (Schroeder and Baskett, 1968). Survival rates of larvae are largely unknown but were reported as 11.8%, 13.1%, and 17.6% in Kentucky, where predation was the controlling factor (Cecil and Just, 1979). Larval survival rates are subject to wide yearly fluctuations in north temperate climates (Turner, 1962).

N. Feeding Behavior. Adult bullfrogs are voracious, opportunistic predators (Schwalbe and Rosen, 1988) that employ a sit-and-wait approach and will readily attack any live animal smaller than themselves, including conspecifics and other frogs (Bury and Whelan, 1984). Bullfrogs may locate and eat smaller frogs by orienting to breeding (Green and Pauley, 1987; Harding, 1997) or distress (Collins and Collins, 1991) calls. Cannibalism is well documented (Bury and Whelan, 1984), and conspecifics can comprise up to 80% of the diet (Stuart and Painter, 1993). An instance of scavenging on a road-killed

Prey Items Reported for Adult American Bullfrogs

Prey Item	References
vegetation	Bury and Whelan (1984)
earthworms	Bury and Whelan (1984)
leeches	Mahon and Aikens (1977)
insects	beetles, moths, dragonflies and damselflies, wasps (Schwalbe and Rosen, 1999); beetles (Stuart, 1995); Araneae, Collembola, Odonata, Coleoptera, Hemiptera, Neuroptera, larval Trichoptera, larval Lepidoptera, Diptera, Hymenoptera (Stewart and Sandison, 1972); scarab beetles, ground beetles, caterpillars, moths, cicadas, dragonflies (Korschgen and Baskett, 1963); Bury and Whelan (1984)
centipedes and millipedes	centipedes (Stuart, 1995; Schwalbe and Rosen, 1999); Bury and Whelan (1984)
spiders	Korschgen and Baskett (1963); tarantulas (Schwalbe and Rosen, 1999); Bury and Whelan (1984)
scorpions	Schwalbe and Rosen (1999)
crayfish	Korschgen and Baskett (1963); Beringer and Johnson (1995); Bury and Whelan (1984)
snails	Bury and Whelan (1984); Schwalbe and Rosen (1988)
salamanders	long-toed salamanders (*Ambystoma macrodactylum*; Nussbaum et al., 1983; Bury and Whelan, 1984)
tadpoles	American bullfrogs and green frogs (Beard and Baillie, 1998; Bury and Whelan, 1984)
frogs and toads	lowland leopard frogs (Schwalbe and Rosen, 1999); plains leopard frogs (Guarisco, 1985); eastern gray treefrogs (*Hyla versicolor*; Schwartz et al., 2000); Great Plains toads *(Bufo cognatus)* and green toads (*Bufo debilis*; Stuart, 1995); Gulf Coast toads (*Bufo valliceps* [now considered to be Coastal-Plain toads, *B. nebulifer*; see Mulcahy and Mendelson, 2000; Mendelson, this volume]; Platt and Fontenot, 1993); foothill yellow-legged frogs (Crayon, 1998); spring peepers (*Pseudacris crucifer*) and toads (*Bufo* sp.; Green and Pauley, 1987; Bury and Whelan, 1984)
fish	Bury and Whelan (1984); Schwalbe and Rosen (1999)
small alligator	Bury and Whelan (1984)
turtles	Sonoran mud turtles (*Kinosternon sonoriense*; Schwalbe and Rosen, 1999); common musk turtles (*Sternotherus odoratus*; Ernst, 1986; Bury and Whelan, 1984)
lizards	"alligator lizards" and lesser earless lizards (*Holbrookia maculata*; Korschgen and Baskett, 1963; Schwalbe and Rosen, 1999); western fence lizards *(Sceloporus occidentalis)* and southern alligator lizards (*Elgaria multicarinata*; Crayon, 1998)
snakes	western diamondback rattlesnakes (*Crotalus atrox*; Clarkson and de Vos, 1986); Mexican garter snakes, checkered garter snakes *(Thamnophis marcianus)*, western patch-nosed snakes (*Salvadora hexalepis*; Schwalbe and Rosen, 1999); ring-necked snakes (*Diadophis punctatus*; Fitch, 1975); glossy crayfish snakes (*Regina rigida*; Ernst and Barbour, 1989); common garter snakes (*Thamnophis sirtalis*; Brooks, 1964); Oregon garter snakes (*Thamnophis atratus hydrophilus*; Crayon, 1998; Bury and Whelan, 1984)
small mammals	cotton rats (Schwalbe and Rosen, 1999); shrews (Brooks, 1964); mice (Stuart, 1995); voles (Beringer and Johnson, 1995; Bury and Whelan, 1984)
mink	young mink (Beringer and Johnson, 1995)
bats	Schwalbe and Rosen (1999); Bury and Whelan (1984)
birds	cedar waxwings (Gollob, 1978); red-winged blackbirds and common yellowthroats (Schwalbe and Rosen, 1999); golden-crowned sparrows (M. R. Jennings, 1987b); ducklings (McAtee, 1921; Hewitt, 1950). Also see Frost (1935), Howard (1950), Cochran and Goolish (1980), and Bury and Whelan (1984)

Note: Modified from Bury and Whelan, 1984.

plains leopard frog was reported by Guarisco (1985). Frost (1935) performed stomach dissections and found that smaller bullfrogs eat mostly insects, while larger bullfrogs typically eat frogs, crayfish, and mice. In Arizona, the majority of stomach contents of adults consisted of invertebrates such as snails and insects (Schwalbe and Rosen, 1988), while two other studies showed 80% and 87% of stomach contents by volume to be conspecifics, mainly larvae and newly metamorphosed animals (Kansas, Smith, 1977; New Mexico, Stuart and Painter, 1993). In a New York bullfrog population sympatric with mink frogs and green frogs, the diet of bullfrogs differed from the other species by having fewer food types and by small ranids comprising the largest part of the diet (Stewart and Sandison, 1972). While bullfrogs are known to consume toads (*Bufo* sp.), the extent of their immunity to the toxic parotoid gland secretions is debatable (Tucker, 1994). Items included in the diet of adult bullfrogs are summarized in Table SP-3.

O. Predators. Bullfrogs are an important food source for other wildlife, often providing an abundant and localized source of protein. In many states and provinces, bullfrogs are considered a game species and hunted for their legs. In Kentucky, Cecil and Just (1979) believed predation was the major factor controlling bullfrog population size in permanent ponds. Bullfrog tadpoles are relatively

Predators of American Bullfrogs

Predator	Life Stages Preyed Upon	References
leeches	eggs	Licht (1969a)
insects	tadpoles, juveniles	general (Werner and McPeek, 1994); predaceous diving beetles and dragonfly naiads (Hunter et al., 1992)
spiders	tadpoles, juveniles	six-spotted fishing spiders (*Dolomedes triton*; Rogers, 1996)
fish	eggs, tadpoles, juveniles	smallmouth bass (*Micropterus dolomieu*; Saumure, 1993); largemouth bass *(Micropterus salmoides)* and green sunfish (*Lepomis cyanellus*; Kruse and Francis, 1977); see also Licht (1969a)
salamanders	eggs, tadpoles, juveniles	tiger salamanders (*Ambystoma tigrinum*; Werner and McPeek, 1994); see also Licht (1969a)
frogs	tadpoles, juveniles	Bury and Whelan (1984)
turtles	tadpoles, juveniles	snapping turtles (Emlen, 1976); general (Howard, 1978a); see also Bury and Whelan (1984)
alligators	tadpoles, juveniles, adults	Bury and Whelan (1984)
snakes	tadpoles, juveniles, adults	cottonmouths *(Agkistrodon piscivorus piscivorus)* and copperheads (*Agkistrodon contortrix contortrix*; Heatwole et al., 1999); northern water snakes (*Nerodia sipedon*; Uhler et al., 1939); northern ribbon snakes (*Thamnophis sauritus*; Mitchell, 1994a); western aquatic garter snakes (*Thamnophis atratus*; Kupferberg, 1994)
birds	tadpoles, juveniles, adults	birds (Hayes and Schaffner, 1986; Applegate, 1990); herons (Hunter et al., 1992)
mammals	tadpoles, juveniles, adults	raccoons, skunks, minks, opossums, coyotes, otters (Bury and Whelan, 1984); humans (bullfrogs are a game species in many states)

immune to fish predation because of unpalatability (Walters, 1975; Kruse and Francis, 1977; Nelson, 1980; Kats et al., 1988; Werner and McPeek, 1994) and are one of only a few species likely to persist after fish invasion (Seale, 1980). Typically, higher rates of activity make bullfrog tadpoles more vulnerable to invertebrate and salamander predation in fishless habitats (Werner and McPeek, 1994; Skelly et al., 1999). Table SP-4 lists known predators of bullfrogs. Large males suffer higher mortality than large females (Howard, 1981). A list of prey items taken by introduced bullfrogs in southern California has recently been published by Hovey and Bergen (2003).

P. Anti-Predator Mechanisms. Eggs and larvae avoid predation by fish and some salamanders by being unpalatable (see "Predators" above) or reduced larval activity (Woodward, 1983). Hobson et al. (1967) reported that adults have a state of rest characterized by alertness without a loss of reactivity, aiding in predator avoidance. Upon disturbance, adults retreat to deeper water with a series of long leaps with a great deal of splashing (Smith, 1961). Bullfrogs usually squawk when fleeing, often setting off a mass bullfrog exodus from a shoreline into deeper waters (Schwalbe and Rosen, 1999). Bullfrogs may emit a piercing scream when seized, which may startle a

predator enough to allow escape (Harding, 1997). Partial resistance to the venoms of cottonmouths *(Agkistrodon piscivorus piscivorus)* and copperheads (*A. contortrix contortrix*) has also been reported (Heatwole et al., 1999). Bullfrog tadpoles avoid novel predators by seeking shelter, but ignore predators from their native range (Pearl et al., 2003).

Q. Diseases. Bullfrogs are known hosts to many bacteria and viruses, but only a few of these pathogens appear to be of major importance in nature (Reichenbach-Klinke and Elkan, 1965; Carr et al., 1976; Gruia-Gray et al., 1989; Gruia-Gray and Desser, 1992; Faeh et al., 1998). An intraerythrocytic virus outbreak was reported in Canada (Crawshaw, 1997). Fungi are reported as well (Goodchild, 1953; Hill and Parnell, 1996), including a recent outbreak of chytrid fungus in Arizona (Sredl et al., 2000), which has been implicated as the cause of amphibian declines in Australia and Panama (Berger et al., 1998). In captivity, infections of *Trypanosoma pipientis* (Siddal and Desser, 1992), frog septicemia (Glorioso et al., 1974), Lucké tumor herpes virus (National Academy of Sciences, 1974), and polyhedral cytoplasmic amphibian virus (Faeh et al., 1998) are reported. Lefcort and Eiger (1993) demonstrated how physiological responses to pathogen infections in tadpoles (i.e., fever, malaise) can lead to increased predation.

In Kentucky, Cecil and Just (1979) considered microbial infections as a secondary factor controlling bullfrog population size in permanent ponds. Disease can be an important factor in tadpole survival (Gibbs et al., 1971).

R. Parasites. Bullfrogs host many parasites including helminths (Bursey and DeWolf, 1998; McAlpine and Burt, 1998), platyhelminths such as trematodes (Kennedy, 1980; Shields, 1987; Crawshaw, 1997) and nematodes (Modzelewski and Culley, 1974), protozoans and mesozoans (Bury and Whelan, 1984; Desser et al., 1986, 1990; Desser, 1987; Chen and Desser, 1989; Desser et al., 1995; Smith et al., 1996, 2000), and leeches (Barta and Desser, 1984; Siddall and Desser, 1992). Trematode metacercariae have been implicated in bullfrog limb abnormalities (Christiansen and Feltman, 2000).

4. Conservation.

Historically, American bullfrogs were common in the United States. Today many native bullfrog populations appear to be declining, with habitat loss and degradation, water pollution, pesticide contamination, and over-harvesting commonly invoked as causal factors (Texas, Baker, 1942; Louisiana, Walker, 1963; United States, Bury and Whelan, 1984; Great Lakes, Harding, 1997; Wisconsin, Mossman et al., 1998). Recent increases in the incidence of malformations (or the causal factors of same) may also be contributing to declines (Lopez and Maxson, 1990; Ouellet et al., 1997a; Rowe et al., 1998a). Exposure to pollutants (including the trace elements arsenic, cadmium, selenium, copper, chromium, and vanadium) in early life stages has been shown to increase the incidence of malformations, compromise swimming performance, and elevate maintenance costs (Raimondo et al., 1998; Rowe et al., 1998b; Hopkins et al., 2000). Berrill et al. (1997b) studied the effects of a variety of pesticides and herbicides on the embryos and larvae of five anuran and one salamander species. They found that bullfrogs were the most sensitive species to fenitrothion, but that bullfrogs in general exhibited moderate sensitivity to most contaminants, with wood frogs *(R. sylvatica)* and American toads *(B. americanus)* most tolerant, and spotted salamanders *(Ambystoma maculatum)* least tolerant. Lethal pH (100% mortality) for both embryos and larvae is reported as 3.5–3.8, with a critical pH (where survivorship declines below levels found for neutral water) of 4.0–4.5 for embryos and 4.0 for larvae (Grant and Licht, 1993). Nixdorf et al. (1997) studied the toxic effects of lead on the central nervous system of bullfrog tadpoles, demonstrating learning impairment. In experiments exposing bullfrog and northern leopard frog tadpoles to the organochlorine insecticide DDT, tail regeneration and mortality rates were less in bullfrog tadpoles (Weis, 1975). Bullfrogs are also surprisingly resistant to the anticholinesterase drug, Sarin (isopropyl methyl phosphonofluoridate), with an LD50 of 6 mg/kg body weight (Wilber, 1954). Diana and Beasley (1998) review effects of various toxins on bullfrogs and other amphibians. Because bullfrog eggs are laid in thin sheets at the water surface, they are more exposed to potentially damaging solar UV-B radiation (Ovaska, 1997), which may result in embryonic mortality and abnormal development (Zimskind and Schisgall, 1955).

Widespread introductions have greatly expanded the range. Introduced populations appear to be experiencing range expansions in Iowa (Christiansen, 1998), Minnesota (Moriarty, 1998), and California (Bury and Luckenbach, 1976), although Treanor and Nichola (1972) report declines in the late 1960s in California.

American bullfrog conservation issues vary by region. In the eastern United States, where bullfrogs are native, attempts should be made to conserve populations by halting the causes of their declines (see above). In the western United States, efforts should be made to reduce or eliminate introduced/invasive populations.

Rana chiricahuensis Platz and Mecham, 1979
CHIRICAHUA LEOPARD FROGS
Michael J. Sredl, Randy D. Jennings

1. Historical versus Current Distribution.
Chiricahua leopard frogs *(Rana chiricahuensis)* are found in Arizona, New Mexico, and Mexico (Platz and Mecham, 1979; see also Goldberg et al., 2004b). The range of this species is divided into two areas. The first includes northern montane populations along the southern edge of the Colorado Plateau (= Mogollon Rim) in central and eastern Arizona and west-central New Mexico. The second includes southern populations located in the mountains and valleys south of the Gila River in southeastern Arizona and southwestern New Mexico, and extends into Mexico along the eastern slopes of the Sierra Madre Occidental (Platz and Mecham, 1979). The distribution of this species within its range is fragmented due to the aridity of this region (Mecham, 1968c). Populations in the northern portion of the range may soon be described as a new species (J. Platz, personal communication).

Historical records for Chiricahua leopard frogs are known from Coconino, Yavapai, Gila, Navajo, Apache, Greenlee, Pima, Santa Cruz, Graham, and Cochise counties, Arizona; and Catron, Soccoro, Sierra, Grant, Hidalgo, and Luna counties, New Mexico (Jennings and Scott, 1991; Sredl et al., 1997). A single specimen collected from Rio Arriba County, New Mexico, is sufficiently distant from the established range to suggest that it was likely mismapped (Fritts et al., 1984). Historical records for Chiricahua leopard frogs also exist from Chihuahua, extreme northern

Durango, and northern Sonora, Mexico (Platz and Mecham, 1979). Elevations of Chiricahua leopard frog localities range from 1,000–2,710 m (Platz and Mecham, 1979; Sredl et al., 1997).

During surveys conducted in Arizona between 1983 and 1987, Clarkson and Rorabaugh (1989) found Chiricahua leopard frogs at only 2 of 36 historical sites and at 2 new sites. Sredl et al. (1997) conducted 871 surveys within the northern portion of the Chiricahua leopard frog range and confirmed their presence at 4 of 25 historical sites and at 11 new sites. They also conducted 656 surveys within the range for southern populations and found frogs at 17 of 84 historical sites and at 44 new sites. In New Mexico, R.D. Jennings (1995a) found Chiricahua leopard frogs at 6 of 50 historical sites and 5 of 22 new sites. The status of populations in Mexico is unknown.

2. Historical versus Current Abundance.
Little is known of historical abundance. Zweifel (1968b) characterized leopard frogs (cf. *R. chiricahuensis*) as "abundant" in well-watered sites in valleys such as at San Bernardino Ranch on the Mexican

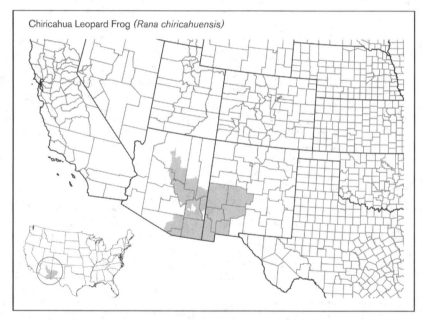

Chiricahua Leopard Frog *(Rana chiricahuensis)*

border in southeastern Arizona. In other parts of southeastern and east-central Arizona, frogs that were likely Chiricahua leopard frogs were more abundant and widespread than they are at present (Wright and Wright, 1949; J.T. Bagnara, personal communication). These frogs can be locally abundant (10–20 frogs jump with each step along the shoreline) or can be scarce. This variation seems to be associated with environmental conditions found in different types of habitats. It is likely that similar patterns of abundance occurred historically.

3. Life History Features.

A. Breeding. Reproduction is aquatic.

i. Breeding migrations. Breeding migrations described for some amphibians have not been noted in Chiricahua leopard frogs. In most populations, breeding usually follows springtime emergence after a period of winter inactivity and may continue through the summer and into the fall.

Male Chiricahua leopard frogs typically call above water, but may also advertise underwater (Degenhardt et al., 1996). Proximate cues that stimulate mating have not been well studied. Using data collected from a long-term captive colony, Fernandez (1996) states that oviposition may be stimulated by rainstorms. Platz (1997), studying wild populations of the closely related Ramsey Canyon leopard frog (R. subaquavocalis), noted that oviposition in that species does not appear to be correlated with rain, but instead may be correlated with changes in water temperature. Oviposition occurred on 10 of 11 nights shortly before or slightly after a decrease in water temperature. Altig et al. (1998) describe the tadpoles of Chiricahua leopard frogs.

Egg masses of Chiricahua leopard frogs have been reported in all months, but reports of oviposition in June are uncommon (Zweifel, 1968b; Frost and Bagnara, 1977; Frost and Platz, 1983; Scott and Jennings, 1985; M. J.S., R.D.J. unpublished data). Zweifel (1968b) noted that breeding in the early part of the year appeared to be limited to sites where the water temperatures do not get too low, such as spring-fed sites. Frogs at some of these sites may oviposit year-round (Scott and Jennings, 1985).

Frost and Platz (1983) studied populations of Chiricahua leopard frogs in Arizona and New Mexico and noted egg masses in March–August. They divided egg-laying activity into two distinct periods with respect to elevation. Populations at elevations below 1,800 m tended to oviposit from spring through late summer, with most activity taking place before June. Populations above 1,800 m bred in June–August.

Scott and Jennings (1985) found a similar seasonal pattern of reproductive activity (February–September) as Frost and Platz (1983), although they did not note elevational differences. Additionally, they noted reduced oviposition in May–June.

ii. Breeding habitat. Chiricahua leopard frogs are habitat generalists and breed in slack waters in a variety of natural and manmade aquatic systems (Mecham, 1968c; Zweifel, 1968b; Frost and Bagnara, 1977; Scott and Jennings, 1985; Sredl and Saylor, 1998). Natural systems include rivers, permanent streams, permanent pools in intermittent streams, beaver ponds, cienegas (= wetlands), and springs.

Manmade systems in which they have been recorded include earthen cattle tanks, livestock drinkers, irrigation sloughs, wells, abandoned swimming pools, ornamental backyard ponds, and mine adits. The year-round flow, constant water temperature that permits year-round adult activity and winter breeding, and depauperate fish community of thermal springs make these sites particularly important breeding habitat for Chiricahua leopard frogs in New Mexico (Scott and Jennings, 1985).

In the Sulfur Springs Valley of southeastern Arizona, egg masses were found most frequently between late March and late May, although occasional egg masses were found in the summer and early fall (Frost and Bagnara, 1977).

Jennings (1988, 1990) studied five populations of Chiricahua leopard frogs in New Mexico from 1987–89. Populations in warm springs have a reproductive period more than twice as long as a population in a cold stream and likely reproduce year-round (Scott and Jennings, 1985). Jennings (1990) also found a great deal of annual and site-specific variation in all breeding activities.

B. Eggs.

i. Egg deposition sites. Females deposit spherical masses attached to submerged vegetation. Jennings and Scott (1991) found egg masses to be suspended within 5 cm of the surface, attached to vegetation. Zweifel (1968b) found the minimum–maximum temperatures for Chiricahua leopard frog embryos to be 12.0–31.5 °C. Zweifel (1968b) reported the highest temperature at which an egg mass was found in the wild was 27.8 °C. In New Mexico, egg mass temperatures ranged from 12.6 °C, recorded from a stock tank at 2,385 m elevation, to 29.5 °C, recorded at a warm spring at 1,885 m (R. D. J., personal observations).

ii. Clutch size. Estimates of the number of eggs per egg mass ranged from 300–1,485 (Jennings and Scott, 1991). Vegetation associated with egg masses included Potamogeton sp., Rorippa sp., Echinochloa sp., and Leersia sp.

C. Larvae/Metamorphosis. Hatching time of egg masses in the wild has not been studied in detail. Eggs of the closely related Ramsey Canyon leopard frog hatch in approximately 14 d depending on temperature (Platz, 1997); hatching time may be as short as 8 d in geothermally influenced springs (R.D.J., unpublished data).

Tadpoles are known to overwinter (Frost and Platz, 1983). Length of larval period may be as short as 3 mo or as long as 9 mo (Jennings, 1988, 1990). Jennings (1990) found that tadpoles in warm springs appear to grow continuously, while growth of those in cold water sites appeared to be arrested or retarded during the winter. Tadpole activity has been observed under ice in water at 5.0 °C

(R. D. J., personal observations). Tadpoles from stream habitats have more contrasting melanic patterns on the tail, thicker dorsal fins, and somewhat larger tail muscles than tadpoles from ponds (Jennings and Scott, 1993).

D. Juvenile Habitat. Not well studied, but some spatial and temporal separation of Chiricahua leopard frog adults and juveniles may enhance survivorship. Seim and Sredl (1994) studied the association of juvenile–adult stages and pool size in the closely related lowland leopard frog (R. yavapaiensis) and found that juveniles were more frequently associated with small pools and marshy areas while adults were associated with large pools. Fernandez (1996) speculated that lack of cover and cannibalism was the reason for low juvenile survival in a captive colony of Chiricahua leopard frogs. Jennings (1988) noted that juveniles were more active during the day, adults more active at night.

E. Adult Habitat. Habitats in which Chiricahua leopard frogs are found range from perennial to near perennial. Mechanisms by which Chiricahua leopard frogs survive the loss of surface water are unknown. However, other species of leopard frogs in the southwestern United States have been observed to survive drought by burrowing into mud cracks (Howland et al., 1997).

Although no studies examining habitat use by adult Chiricahua leopard frogs have been conducted, these frogs are known to be habitat generalists and occupy a variety of natural and manmade aquatic systems within their range. The role of habitat heterogeneity within the aquatic and terrestrial environment is unknown, but is likely to be important. Shallow water with emergent and perimeter vegetation provide tadpole and adult basking habitats, while deeper water, root masses, and undercut banks provide refuge from predators and potential hibernacula (M. J.S., unpublished data). Most perennial water supporting Chiricahua leopard frogs possess fractured rock substrata, emergent or submergent vegetation, deep water, root masses, undercut banks, or some combination of these features that frogs may use as refugia from predators and extreme climatic conditions (R. D. J., unpublished data).

Occupation of natural and artificial aquatic systems presents interesting opportunities and dilemmas for conservation of native leopard frogs. Sredl and Saylor (1998) found artificial aquatic systems such as earthen cattle tanks to be important for the continued viability of populations of Chiricahua leopard frogs and other members of the leopard frog complex in Arizona. Of the nine extant populations of Chiricahua leopard frogs on the Mogollon Rim in Arizona, one is a natural aquatic system and the remainder are artificial or highly modified aquatic systems (Sredl et al., 1997). In New Mexico, only 5 of 33 known, extant populations persist in stock tanks (three earthen tanks and two concrete storage tanks). Streams and springs appear to be the most important habitats for Chiricahua leopard frogs in New Mexico (R. D. J., personal observations).

Severe fragmentation and alteration of aquatic habitats in the southwestern United States has likely constricted many wide ranging aquatic species into isolated pockets, and maintenance of aquatic corridors may be critical in preserving organisms in the arid Southwest (Jennings and Scott, 1991). Sredl and Howland (1995) speculated that distribution of extant Chiricahua leopard frog populations in Arizona may be reflective of habitat fragmentation and extinction without recolonization, as well as habitat quality.

Effects of livestock grazing on amphibian populations may be positive or negative (Jennings, 1988; Rosen and Schwalbe, 1998; Sredl and Saylor, 1998). In the late 1800s and early 1900s, construction of earthen cattle tanks in upland drainages became a common range management practice (U.S. General Accounting Office, 1991), one that continues to this day. Because these tanks were primarily built to water livestock, a positive secondary benefit of these systems is the water and aquatic habitat they provide to many species of wildlife, including amphibians. Overgrazing negatively impacts amphibian habitat by removing bankside cover, increasing ambient ground and water temperatures, destroying bank structure (e.g., eliminating undercut banks), trampling egg masses, and adding high levels of organic wastes (Jennings, 1988). Overgrazing in upland habitats may degrade amphibian habitat by increasing runoff and sedimentation rates (Jennings, 1988; Belsky and Blumenthal, 1997).

Although the relationship between the Chiricahua leopard frog and non-native predators (e.g., American bullfrogs [R. catesbeiana], crayfish, and predatory fishes) has not been studied in detail, there is a negative co-occurrence between them (Rosen et al., 1995; Fernandez and Rosen, 1996). Jennings and Scott (1991) questioned the importance of American bullfrogs as a factor leading to regional declines of native amphibians in the southwestern United States, arguing that many populations that have declined in New Mexico were not impacted by bullfrogs or were impacted by factors other than bullfrogs.

F. Home Range Size. Male home range sizes (dry season mean = 161.0 m^2; wet season mean = 375.7 m^2) tended to be larger than those of females (dry season mean = 57.1 m^2; wet season mean = 92.2 m^2). The largest home range size documented for the species was that of a male who used approximately 23,390 m^2 (2,339 m × 10 m) of an intermittent, low elevation canyon (1,775 m) in New Mexico during July–August 1999. Another male moved 3.5 km (> 2 mi) in one direction during that same time period. The largest home range size documented for a female frog was about 9,500 m^2 (950 m × 10 m). Male frogs tended to expand home range size to a greater degree than females when ranges during the dry season (early July) were compared to wet season (late July to August; R.D.J., unpublished data).

G. Territories. Calling male Chiricahua leopard frogs will engage in fisticuffs with other males, presumably defending calling sites. This site defense appears to be transient, however. Other forms of territorial defense are not known (R.D.J., unpublished data).

H. Aestivation/Avoiding Desiccation. Unknown.

I. Seasonal Migrations. Jennings and Scott (1991) noted that maintenance of corridors used by dispersing juveniles and adults that connect disjunct populations may be critical to preserve populations of frogs and other aquatic organisms.

J. Torpor (Hibernation). Although postmetamorphic Chiricahua leopard frogs are generally inactive from November–February, a detailed study of wintertime activity or habitat use has not been done. Jennings (1988, 1990) studied five populations of Chiricahua leopard frogs in New Mexico from 1987–89. Among sites, the number of frogs observed during diurnal surveys was best predicted by month of the year, diurnal air temperature, and time of day. Time of day was negatively associated with frog numbers, indicating frogs were more numerous early in the day before temperatures elevated. Number of frogs observed during nocturnal surveys among sites was best predicted by nocturnal water temperature and amount of wind. Frogs were most abundant when water temperatures were warmer and when winds were calmer. The number of egg masses observed during diurnal surveys of all sites was best predicted by the number of frogs observed during diurnal surveys. Only diurnal water temperature provided predictive power of number of egg masses at any single site included in the study.

Jennings (1990) found that populations in warm springs exhibited prolonged activity (year-round at one site) and reproductive patterns, while ponds without geothermal input exhibited relatively brief periods of activity and reproduction (also in Jennings, 1988). Chiricahua leopard frogs exhibit greater variation in activity than has been reported for any other species of leopard frog (Jennings, 1990). Chiricahua leopard frogs likely overwinter near breeding sites, although microsites for these "hibernacula" have not been studied.

K. Interspecific Associations/Exclusions. Throughout their range, Chiricahua leopard frogs occur with tiger salamanders (*Ambystoma tigrinum*), Tarahumara salamanders (*A. rosaceum*), Arizona toads (*Bufo microscaphus*), red-spotted toads (*B. punctatus*), Woodhouse's toads (*B. woodhousii*), canyon treefrogs (*Hyla arenicolor*), Arizona treefrogs (*H. wrightorum*), western chorus frogs (*Pseudacris triseriata*), plains leopard frogs (*R. blairi*), northern leopard frogs (*R. pipiens*), Tarahumara frogs (*R. tarahumarae*), and lowland leopard frogs (Frost and Bagnara, 1977; Hale and May, 1983; Platz and Frost, 1984; Clarkson and Rorabaugh, 1989; Jennings, 1990; M. J. S., unpublished data).

Of interest has been sympatry between Chiricahua leopard frogs and four members of the *R. pipiens* complex: northern, lowland, and plains, and one undescribed species of leopard frog (Platz and Mecham, 1979). In east-central Arizona, Mecham (1968c) found northern leopard frogs to predominate in meadow-like habitats and Chiricahua leopard frogs to predominate in rocky streams. In the Sulfur Springs Valley of southeastern Arizona, Frost and Bagnara (1977) found plains leopard frogs to predominate in non-permanent and most semi-permanent tanks and sloughs, while Chiricahua leopard frogs predominate in permanent tanks and streams. In New Mexico along Cuchillo Negro Creek, plains leopard frogs and Chiricahua leopard frogs were observed along the same stretch of stream, in similar microhabitats and in similar numbers (R. D. J., personal observations).

Rosen et al. (1995) noted a strong negative co-occurrence between Chiricahua leopard frogs and introduced American bullfrogs. Jennings and Scott (1991) questioned the role that American bullfrogs have played in declines of native amphibians in New Mexico, where several local extinctions did not involve the presence of American bullfrogs.

L. Age/Size at Reproductive Maturity. Age and size at reproductive maturity are poorly known. In southeastern Arizona, juvenile frogs and late-stage tadpoles introduced to an outdoor enclosure in May–June 1994 reproduced in September 1994 (Rosen and Schwalbe, 1998). The smallest males to exhibit secondary sexual characteristics from study sites in Socorro and Catron counties, New Mexico, were 53.5 mm and 56.2 mm SUL, respectively (R.D.J., unpublished data). Size at which females reach sexual maturity is not known.

M. Longevity. Although scoring of annuli in Chiricahua leopard frogs is more difficult than in lowland leopard frogs (Collins et al., 1996), preliminarily, skeletochronology of Chiricahua leopard frogs indicate that they can live ≤6 yr (Durkin, 1995).

N. Feeding Behavior. No comprehensive studies of the feeding behavior or diet of Chiricahua leopard frog larvae or adults have been conducted. Larval Chiricahua leopard frogs are herbivorous. Available food items at one site examined within the range of this species include bacteria, diatoms, phytoplankton, filamentous green algae, water milfoil (*Myriophyllum* sp.), duckweed (*Lemna minor*), and detritus (Marti and Fisher, 1998). Captive larvae eat spinach, romaine lettuce, cucumber slices, frozen trout, duckweed, spirulina-type fish foods, and rabbit pellets. Captive juvenile frogs will eat crickets (Demlong, 1997).

The diet of Chiricahua leopard frog adults likely contains a wide variety of insects and other arthropods (Degenhardt et al., 1996). Stomach analyses of other members of the leopard frog complex from the western United States show a wide variety of prey items including many types of aquatic and terrestrial invertebrates (e.g., snails, spiders, and insects) and vertebrates (e.g., fish, other anurans [including conspecifics], and small birds; Stebbins, 1951).

O. Predators. Detailed studies of predators of Chiricahua leopard frogs have not been conducted. Tadpoles are likely preyed upon by aquatic insects, including belostomatids, notonectids, dytiscids, and anisopterans, and vertebrates including native and non-native fishes, garter snakes (*Thamnophis* sp.), great blue herons (*Ardea herodias*), and other birds. Predators of juvenile and adult frogs likely include native and non-native fishes, American bullfrogs, garter snakes, and great blue herons, and mammals including rats, coyotes, gray foxes, raccoons, ringtail cats, coatis, black bears, badgers, skunks, bobcats, and mountain lions.

P. Anti-Predator Mechanisms. Adult and juvenile Chiricahua leopard frogs avoid predation by hopping to water (Frost and Bagnara, 1977). Among members of the *R. pipiens* complex, Chiricahua leopard frogs possess the unusual ability to profoundly darken their ventral skin under conditions of low albedo (reflectance) and low temperature (Fernandez and Bagnara, 1991; Fernandez and Bagnara, 1993). In the clear, swiftly moving streams they inhabit (low albedo environments), this trait is thought to aid in escape from predators by reducing the amount of attention that bright flashes of white ventral skin would bring. At low temperatures, poikilotherms (cold-blooded animals) are unable to flee swiftly. Under these conditions, crypsis may be the most effective form of predator avoidance. Other anti-predator mechanisms have not been identified, but deep water, vegetation, undercut banks, root masses, and other cover sites have been mentioned as being important retreats.

Q. Diseases. Post-metamorphic death syndrome was implicated in the extirpation of Chiricahua leopard frog populations in New Mexico in the late 1980s (Anonymous, 1993). This syndrome affected earthen cattle tank populations, which became extirpated over a 3-yr period. The syndrome was characterized by the death of all post-metamorphic or adult-form frogs over winter. Dead or moribund frogs were often found during, or immediately following, winter dormancy or unusually cold periods. The syndrome appeared to spread among adjacent populations causing regional loss of populations or metapopulations.

In 1998, chytrid fungus was implicated in declines of amphibians in Australia and Panama (Berger et al., 1998). That same year, it was first identified in Arizona (Milius, 1998). Presently in Arizona, one salamander species, Sonoran tiger salamanders (*A. tigrinum stebbinsi*), seven species of ranid frogs (Rio Grande leopard frogs [*R. berlandieri*], plains leopard frogs, American bullfrogs, Chiricahua leopard frogs, Ramsey Canyon leopard frogs, Tarahumara frogs, and lowland leopard frogs), and one treefrog, Canyon treefrogs, have been affected by this fungus. All outbreaks have been a cool season phenomena, and the pathogen is well distributed in central and southeastern Arizona (Sredl et al., 2000). In southwestern New Mexico in 2000, the presence of chytrid fungus was confirmed in a population of Chiricahua leopard frogs exhibiting declines. It is likely that chytrid fungus was responsible for population declines observed in the late 1980s in New Mexico, including those characterized as post-metamorphic death syndrome (R.D.J., personal observations).

R. Parasites. Goldberg et al. (1998b) examined parasites of Chiricahua and lowland leopard frogs and American bullfrogs collected in Arizona. Chiricahua leopard frogs were found to be infected with six species of trematode (*Cephalogonimus brevicirrus*, *Glypthelmins quieta*, *Gorgoderina attenuata*, *Haematoloechus complexus*, *Megalodiscus temperatus*, and *Clinostomum* sp.) and one species of nematode (*Physaloptera* sp.). None of the helminths identified from the two native species were found in American bullfrogs.

4. Conservation.
Because of documented declines and extirpations at historical localities, Chiricahua leopard frogs were added to the list of Category 2 candidate species (USFWS, 1991). Beginning with the 28 February 1996, candidate notice of review, the U.S. Fish and Wildlife Service discontinued the designation of multiple categories of candidates, and only those taxa meeting the definition for the former Category 1 candidate were considered candidates (USFWS, 1996a). In that 28 February 1996 notice, Chiricahua leopard frogs were listed as a candidate species (USFWS, 1996a). In 2000, they were proposed for Federal listing as Threatened (USFWS., 2000a), and Chiricahua leopard frogs are now federally listed as Threatened (http://ecos.fws.gov.tess_public/TESSSpeciesReport).

Rana clamitans Latreille, 1801
GREEN FROG

Thomas K. Pauley, Michael J. Lannoo

1. Historical versus Current Distribution.
The range of green frogs (*Rana clamitans*) encompasses most of the eastern United States, from the Canadian border south to the Gulf of Mexico. Green frogs occur almost everywhere east of a line drawn from central Minnesota south through central Iowa, southeast through Missouri (excluding the northwestern corner), and south through central Oklahoma and eastern Texas (Conant and Collins, 1998). Green frogs are absent from the southern half of Florida, from much of central Illinois (a distribution Smith [1961] termed "puzzling"), and from half a dozen scattered counties in north-central Arkansas. Geographic isolates occur in western Iowa, northern Utah, and in several locations in the state of Washington (Stebbins, 1985; Leonard et al., 1993). Two subspecies are recognized, northern green frogs (*R.c. melanota*) and bronze frogs (*R.c. clamitans*). They have distributions that are roughly separated along a line from southeastern Oklahoma arcing northeast through Missouri to southern Illinois, then arcing south through western Tennessee, and east through northeastern Mississippi, northern Alabama, and central Georgia, and northeast through central South Carolina and extreme southeastern North Carolina (Conant and Collins, 1998). Mecham (1954) discusses the geographic variations of the morphologic features throughout the range, and Arndt (1977) reports a blue variant from Delaware. Green frogs have been introduced in the states of Washington and Utah, and in western Iowa.

2. Historical versus Current Abundance.
Little data are available, although green frogs are generally considered common or abundant (Smith, 1961; Mount, 1975; Vogt, 1981; Dundee and Rossman, 1989; Klemens, 1993; Hunter et al., 1999; Mitchell and Reay, 1999.) Populations have undoubtedly been affected by shoreline development throughout much of the range. Lannoo et al. (1998) documented a mass mortality event apparently associated with a cold snap in northern Wisconsin. Ouellet et al. (1997a) documented green frog malformations from agriculturally impacted sites in Québec.

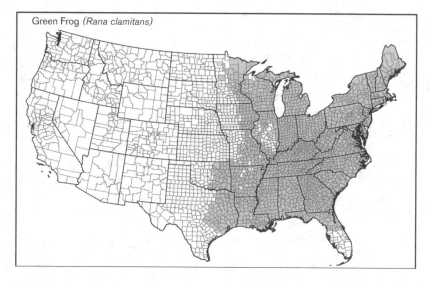

Green Frog (*Rana clamitans*)

3. Life History Features.

A. Breeding. Reproduction is aquatic.

i. Breeding migrations. Adults generally breed in the same lakes and permanent wetlands they inhabit during the non-breeding season, so migrations to breeding sites are rare. Males remain at the breeding sites for ≤2 mo, but females usually only stay for about 1 wk when they mate and deposit eggs (Martof, 1953).

The breeding period is extended, from April through the summer, depending on latitude. Examples of breeding times include early May to early July in Ohio (Walker, 1946), Indiana (Minton, 1972), Connecticut (Klemens, 1993), Maryland, and Delaware (Lee, 1973); May–August in Kentucky (Barbour, 1971), Maine (Hunter et al., 1992; Hunter et al., 1999), and Minnesota (Breckenridge, 1944); May–September in Illinois (Phillips et al., 1999); April to late June in Kansas (Collins, 1993); April–August or September in Alabama (Mount, 1975) and Missouri (Johnson, 1987); and March–September in Louisiana (Dundee and Rossman, 1989). In Michigan, green frogs breed from mid-May to well into the summer (Harding and Holman, 1999). Wright and Wright (1949) note that green frogs breed late in the South. In West Virginia, green frogs breed from mid-April to July in the South and June–August in the North (Pauley and Barron, 1995; Rogers, 1999).

ii. Breeding habitat. Adult green frogs inhabit shorelines of lakes and permanent wetlands such as ponds, bogs, fens, marshes, swamps, and streams.

B. Eggs.

i. Egg deposition sites. Eggs are deposited in shallow water among emergent vegetation (sedges, cattails, rushes) along the shores of lakes and permanent wetlands. Eggs are in a foamy surface film (plinth) that is usually <30 cm in diameter (Walker, 1946; Green and Pauley, 1987).

ii. Clutch size. Clutch sizes vary with the size of females, from 1,000 to nearly 7,000 eggs (Pope, 1944; Wright and Wright, 1949; Minton, 1972; Wells, 1976; Barbour, 1971; Dundee and Rossman, 1989; Oldfield and Moriarty, 1994; Hunter et al., 1999). Eggs masses of females breeding simultaneously can coalesce into a single mass (Wright and Wright, 1949). Some females are known to breed twice annually in the area around Ithaca, New York (Wells, 1976; Martof, 1956b). Second clutches number between 1,000–1,500 eggs.

Eggs are black above and white below and measure about 1.5 mm in diameter (Wright and Wright, 1949; Green and Pauley, 1987; Rogers, 1999). Hatching occurs a few days after deposition (Walker, 1946; Smith, 1961; Green and Pauley, 1987; Johnson, 1987; Harding and Holman, 1999; Hunter et al., 1999). Hatchlings range in size from 10–12 mm TL in Indiana (Minton, 1972) and 9.8–16 mm in West Virginia (Rogers, 1999).

C. Larvae/Metamorphosis.

i. Length of larval stage. Ryan (1953), Whitaker (1961), and Wells (1976) report metamorphosing young from the end of June to mid-July in the area around Ithaca, New York. Larvae metamorphose from April–September in Kentucky (Barbour, 1971), and from early June to late September in Connecticut (Klemens, 1993). In northern West Virginia, Rogers (1999) found newly transformed young from June–October, with a peak in numbers in August. In Kansas, where tadpoles do not typically overwinter, the larval period is about 3 mo (Collins, 1993).

Tadpoles overwinter in many areas throughout their range (Babbitt, 1937; Breckenridge, 1944; Walker, 1946; Wright and Wright, 1949; Smith, 1961; Minton, 1972; Mitchell and Anderson, 1994; Phillips et al., 1999; Rogers, 1999). In a laboratory experiment, Ting (1951) determined that tadpoles from early breeders should be able to metamorphose during

the same year, but those from late broods may not transform until the next year (see also Vogt, 1981).

ii. Larval requirements.

a. Food. Tadpoles feed during the day on a variety of organic debris (Jenssen, 1967; Warkentin, 1992a,b). Werner and McPeek (1994) reported that they are mostly benthic feeders and feed primarily on algae, especially diatoms, but will take entomostracans and fungi.

b. Cover. Both large and small green frog tadpoles seek cover in vegetation during the day (Warkentin, 1992a,b). They also have been found under stones, logs, and bark near water (Minton, 1972).

iii. Larval polymorphisms. Larvae that have overwintered will prey on wood frog (*R. sylvatica*) eggs and larvae (Wegmann, 1997; T.K.P., personal observations).

iv. Features of metamorphosis. Tadpoles are 60–80 mm TL at metamorphosis (Minton, 1972; Johnson, 1977; Dundee and Rossman, 1989; Rogers, 1999); following tail resorption, their sizes range from 21–38 mm. Smaller sizes have been reported in the southern part of the range (Wright and Wright, 1949; Martof, 1956b; Minton, 1972). Metamorphosing frogs are found throughout the summer and early fall in Indiana (Minton, 1972) and West Virginia (Rogers, 1999). First-year frogs will grow about 34 mm (Minton, 1972).

v. Post-metamorphic migrations. Juveniles will disperse into woods and meadows during rainy weather (Harding and Holman, 1999). Subadults may move from 200 m–4.8 km (3 mi) during their first season following transformation (Schroeder, 1968b).

D. Juvenile Habitat. Smaller green frogs are generally found in less dense vegetation than are adults (Martof, 1953). They hide and feed along the margins of ponds and streams (Hunter et al., 1999) and are commonly found in road rut ponds (T.K.P., personal observation). Vogt (1981) reported that froglets are terrestrial and spend their first summer on land in Wisconsin.

E. Adult Habitat. Green frogs are found in most permanent aquatic habitats. Habitats include shorelines of lakes and permanent wetlands, in areas where emergent vegetation such as sedges, cattails, and rushes predominate (Whitaker, 1961; Collins, 1993), and in marshes, swamps, streams, springs, and quarry and farm ponds (Walker, 1946; Minton, 1972; Mount, 1975; Green and Pauley, 1987; Johnson, 1987; Dundee and Rossman, 1989; Collins, 1993; Klemens, 1993; Harding and Holman, 1999; Hunter et al., 1999; Phillips et al., 1999). Smith (1961) observed that in northern Illinois, green frogs occur in a variety of habitats, while in southern Illinois, they are restricted to clear streams associated with rock outcrops. Minton (1972) found that adults rarely go more than 1 m from water except on rainy nights.

Green frogs have been found associated with the mouths of caves (Black, 1973; Trauth and McAllister, 1983; Garton et al., 1993).

F. Home Range Size. Adults have an average home range of 62 m2 (Hamilton, 1948). In Michigan, Martof (1953) found that green frogs have a terrestrial home range of about 20–200 m^2 (average of about 60 m2) that they leave during the breeding season and return to after breeding.

G. Territories. During their prolonged breeding season, males are territorial, with an average residency of about 1 wk (with a range of <1–7 wk). Males inhabit their territories throughout most of the evening during breeding and defend these territories using vocalizations, posturing, and physical combat (Jenssen and Preston, 1968; Wells, 1977, 1978; Ramer et al., 1983; Given, 1990). Schroeder (1968a) observed two males in combat over an apparent breeding site. Green frogs have been shown to use advertisement calls to assess the size of opponents during aggressive encounters (Bee et al., 1999).

Territories are usually centered around clumps of rushes or sedges, artificial shelters, and occasionally abandoned muskrat tunnels (Wells, 1977b). Territory size is dependent on cover density. In dense vegetation, males may be only 1.0 to 1.5 m apart. In more open shorelines, males may be spaced 4.0 to 6.0 m apart (Wells, 1977b).

During non-breeding, Wright and Wright (1949) note that green frogs are solitary. Further, Martof (1956b) showed that most males are solitary during the breeding season, but may later congregate in large groups (called congresses).

H. Aestivation/Avoiding Desiccation. Minton (1972) notes that under dry conditions adults will aggregate around springs or watering holes.

I. Seasonal Migrations. Green frogs probably do not migrate en masse to breeding sites. Lamoureux and Madison (1999) tracked 23 green frogs and found that they made extensive movements in late fall away from summer breeding ponds to areas of flowing water in streams and seeps. They suggested that flowing water is used because it remains unfrozen and provides adequate oxygen.

J. Torpor (Hibernation). Green frog tadpoles will typically, but not always, overwinter for 1 yr prior to metamorphosing the following spring (Martof, 1952; Richmond, 1964; Vogt, 1981). While overwintering, and despite cold temperatures, tadpoles remain active and likely feed (Getz, 1958). Tadpoles probably remain active and grow during the winter in Louisiana but do not transform until warm weather (Dundee and Rossman, 1989). In Maine, they remain under silt and dead vegetation during the winter (Hunter et al., 1999). Both tadpoles and

adults require aquatic sites that do not completely freeze and that also maintain enough oxygen during the winter (Oldfield and Moriarty, 1994).

Green frog adults typically overwinter in water (Dickerson, 1906; Walker, 1946; Pope, 1947; Wright and Wright, 1949; Harding and Holman, 1999) but will occasionally overwinter on land (Bohnsack, 1951). The frog followed by Bohnsack (1951) was found beneath 5 cm (2 in) of compact leaf litter in an oak-hickory forest near Pickney, Michigan. Lannoo et al. (1998) cites an instance where a large number (25, including 4 gravid females) of green frogs were caught on land near a wetland during a cold snap in northern Wisconsin. Minton (1972) notes that adults near Indianapolis, Indiana, seek cover from early December to early March and that juveniles emerge earlier in the season than do adults. In Maine, green frogs hibernate either underwater or underground from October–March (Hunter et al., 1999). In Ohio, several hibernating individuals have been found in springs and in masses of leaves and aquatic vegetation on the bottom of small ponds (Walker, 1946).

K. Interspecific Associations/Exclusions. In New Jersey, green frogs and carpenter frogs (*R. virgatipes*) breed in the same lakes and have overlapping breeding seasons, calling site preferences, and vocal repertoires; they also exhibit generally similar territorial behaviors (Wells, 1977, 1978; Given, 1987, 1990). Nevertheless, Given (1990) shows that carpenter frogs are less territorial but more aggressive than are green frogs in these breeding aggregations. Working in West Virginia, Pauley and Barron (1995) observed that green frogs breed in the same ponds as leopard frogs, pickerel frogs, and bullfrogs, but only American bullfrogs (*R. catesbeiana*) have an overlapping breeding season with green frogs. In northern Minnesota, green frogs commonly occur with mink frogs (*R. septentrionalis*; Oldfield and Moriarty, 1994). However, mink frogs usually inhabit floating vegetation in deeper water while green frogs are found along the edges of the water. This habitat partitioning reduces interspecific competition for food (Fleming, 1976). Minton (1972) reports two cases of male green frogs in amplexus with northern leopard frogs (*R. pipiens*), once with a male and once with a female.

Trauth and McAllister (1983) reported a male using a cave associated with American bullfrogs, long-tailed salamanders (*Eurycea longicauda*), cave salamanders (*E. lucifuga*), gray-bellied salamanders (*E. multiplicata griseogaster*), and northern slimy salamanders (*Plethodon glutinosus*).

L. Age/Size at Reproductive Maturity. Females are slightly larger than males (Smith, 1961). In Indiana, males range from 60–84.5 mm TL, females from 64–88

mm (Minton, 1972); in Ohio, females range from 72–92 mm, males from 69–90 mm (Walker, 1946); and in Connecticut, females range from 52–84 mm, males from 52–94 mm (Klemens, 1993). Males in the South are smaller than in the North (Wright and Wright, 1949). Minton (1972) and Hunter et al. (1999) note that green frogs become sexually mature about 1 yr after metamorphosis. In Maine, females reach sexual maturity at 65–75 mm and males at 60–65 mm (Hunter et al., 1999). Four to five years are required to reach adult size in Michigan (Minton, 1972; Harding and Holman, 1999).

M. Longevity. Cortwright (1998) estimated that at least two population turnovers occurred during a 10–11-yr study in south-central Indiana.

N. Feeding Behavior. As pointed out by Forstner et al. (1998), frogs, including green frogs, are opportunistic feeders, choosing and eating prey from an assortment of moving animals that are large enough to detect and small enough to swallow. Thus, prey choice reflects both habitat and availability (Jenssen and Klimstra, 1966; Hedeen, 1972b; Kramek, 1972; Stewart and Sandison, 1972), although both species and individuals within species can specialize (Sweetman, 1944; Forstner et al., 1998).

Green frog adults are "sit-and-wait" predators (Hamilton, 1948) and will feed both day and night (Minton, 1972). Several authors, including Hamilton (1948), Whitaker (1961), Stewart and Sandison (1972), and Forstner et al. (1998) provide a list of green frog prey that includes invertebrates such as annelids, mollusks, millipedes, centipedes, crustaceans, and arachnids; insects such as coleopterans, dipterans, ephemeropterans, hemipterans, lepidopterans, odonates, orthopterans, and trichopterans; and vertebrates such as fishes and other frogs; vegetable matter; and shed skins. Further, Forstner et al. (1998) and other authors have shown that food habits vary over the summer as prey availability changes.

O. Predators. Martof (1956b) reports that all stages of the life history of green frogs have several predators. Eggs are eaten by turtles. Tadpoles are eaten by larvae of diving beetles and whirligig beetles; dragonfly naiads; nymphs and adults of giant water bugs; water scorpions; and back swimmers. Green frogs are preyed upon by ducks, herons, bitterns, rails, northern harriers, and crows. Large green frogs will take smaller ones and American bullfrogs take almost all sizes of green frogs. Green frogs showed a four-fold increase in 4 yr after bullfrogs disappeared in Point Pelee National Park in Ontario (Hecnar and M'Closkey, 1997). Klemens (1993) reports green frogs in stomachs of water snakes and garter snakes. People use green frogs for food and for sport (Hamilton, 1948).

P. Anti-Predator Mechanisms. Startled green frogs will emit a squawk when they jump (Smith, 1961).

Q. Diseases. Mikaelian et al. (2000) document myositis associated with an infection from *Ichthyophonus*-like protists from animals collected in Québec, Canada. Other reported diseases include frog erythrocytic virus (Faeh et al., 1998).

R. Parasites. Trematodes reported include *Halipegus occidentalis* (Goater et al., 1990; Wetzel and Esch, 1996; Zelmer et al., 1999) and *H. eccentricus* (Goater et al., 1990; Wetzel and Esch, 1996). Other parasites reported include digeneans, cestodes, acanthocephalans, and *Glypthelmins quieta* (McAlpine, 1997c).

4. Conservation.
Green frogs are relatively common throughout most of their range. They are classified as a Game Species in some states (e.g., Missouri, Mississippi, Massachusetts, New York, New Jersey, and Pennsylvania), thereby providing them with protection in terms of hunting season and bag limits (Levell, 1997). In Kansas, green frogs are protected by state law (Collins, 1993) and are listed as a Threatened species (Levell, 1997). Populations have undoubtedly been lost as shorelines have been developed for recreational, business, and domestic uses. Green frogs fall victim to vehicular traffic (Ashley and Robinson, 1996).

Rana draytonii Baird and Girard, 1852(b)
CALIFORNIA RED-LEGGED FROG

Gary M. Fellers

1. Historical versus Current Distribution.
California red-legged frogs *(Rana draytonii)* once ranged throughout the Sierra Nevada foothills and Coast Range Mountains (south of Elk Creek, Mendocino County) in California (M.R. Jennings, 1995; Shaffer et al., 2004). These frogs also were found in northwestern Baja California, Mexico, south to the San Domingo River drainage (Linsdale, 1932). It is unlikely that populations of California red-legged frogs persisted on the floor of the Central Valley due to extensive flooding that occurred during heavy winter storms or during spring snowmelt.

The current distribution is considerably smaller. There are only six known populations in the Sierra foothills, ranging from Butte to El Dorado County. All of the Sierra populations were discovered since 1997 (Barry, 1999; personal observations). A small population found near Coulterville in 1992 (Drost and Fellers, 1996) was apparently extirpated by 1993, but it is always difficult to confirm the absence of a species. In the central California coast range, California red-legged frogs are still present throughout much of their former range, though the number of ex-

California Red-legged Frog *(Rana draytonii)*

tant populations has been reduced substantially by the loss of suitable habitat. In southern California (south of Santa Barbara) there are only two extant populations.

California red-legged frogs have been introduced in Ely, Nye, and White Pine counties in Nevada (Green, 1985a; Reaser, 2003), but the populations are largely inaccessible, and their current status is uncertain (A. Cook, personal communication). There is a 1919 record of an introduced population near Pelican Bay on Santa Cruz Island off Southern California (Jennings, 1988c), but that population has not existed for more than 50 yrs and probably died out shortly after 1919 (P. Collins, personal communication).

2. Historical versus Current Abundance.
California red-legged frogs were once abundant throughout much of California (Jennings and Hayes, 1985). Now the species is nearly extirpated in both the Sierra Nevada foothills and in the southern quarter of its range. The Sierra Nevada populations consist of one site in El Dorado County with <10 frogs (and no reproduction in 1999), one site in Yuba County with <5 adult frogs, and another site in Butte County with <25 adult frogs (Barry, 1999; personal observations). The two southern California populations consist of one site in Riverside County with <10 frogs (USFWS, 1999b) and one site in Ventura County with <25 adult frogs. The status of California red-legged frogs in Mexico is uncertain, but single frogs were caught in the Sierra San Pedro Martir of Baja, California, in both 1997 and 1998

(B. Christman, personal communication). It is unlikely that populations south of Santa Barbara County will survive without intervention, and perhaps not even then.

In a few parts of the central coast range, there are still large, vigorous populations of California red-legged frogs, some populations probably rival what was present 200 yr ago. The largest populations are in Marin County (north of San Francisco), where there are >120 breeding sites with a total adult population of several thousand frogs. Most of these sites are artificial stock ponds constructed on lands that have been grazed by cattle for 150 yr.

Though not as concentrated, there are good populations elsewhere in the San Francisco Bay area (especially Alameda and Contra Costa counties) and in the coastal drainages from San Mateo County (just south of San Francisco) south to Santa Barbara County. One of the largest single populations consists of an estimated 350 adult frogs at Pescadero Marsh (San Mateo County).

3. Life History Features.
  A. Breeding. Reproduction is aquatic.
    i. *Breeding migrations.* California red-legged frogs do not have a distinct breeding migration. Adult frogs are often associated with permanent bodies of water. Some frogs remain at breeding sites all year while others disperse. Dispersal distances are typically <0.5 km, with a few individuals moving up to 2–3 km (G.M.F., in preparation; G. Rathbun, personal communication). Movements are typically along riparian corridors, but

some individuals (especially on rainy nights) move directly from one site to another through normally inhospitable habitats (e.g., heavily grazed pastures or oak-grassland savannas).

In the central coast range, California red-legged frogs spend a highly variable amount of time at breeding sites (radiotelemetry data; G.M.F., in preparation). Some individuals remain at breeding ponds throughout the year, moving only if the pond dries up. Other individuals spend only a few weeks at breeding sites before moving back to non-breeding habitat where they spend up to 11 mo.

ii. Breeding habitat. Historically, California red-legged frogs were found from sea level to about 1,525 m in the Sierra Nevada (Swamp Lake, Yosemite National Park). The highest known extant population is at 975 m in El Dorado County.

California red-legged frogs breed primarily in ponds (Stebbins, 1985), though individuals also breed in slow-moving, pond-like parts of streams, marshes, and lagoons. There is usually some emergent vegetation, most often cattails (*Typha* sp.), rushes (*Scirpus* sp.), or willows (*Salix* sp.). Water depth is generally >0.5 m, but California red-legged frogs occasionally reproduce successfully in ponds with a maximum depth of only 0.25 m. California red-legged frogs breed in both ephemeral and permanent bodies of water. Ponds and streams that dry up in the fall at least every few years are ideal since fish and introduced American bullfrogs (*R. catesbeiana*; which have tadpoles that require >1 yr to metamorphose) do not survive periodic drying.

B. Eggs.

i. Egg deposition sites. Egg masses are typically attached to emergent vegetation near the water surface (Storer, 1925), unlike the closely related northern red-legged frogs *(R. aurora)*, which are known to oviposit at depths exceeding 3 m (C. Pearl, personal communication).

ii. Clutch size. California red-legged frogs lay eggs in clusters. The total number of eggs laid/female ranges between 300–4,000 (Storer, 1925; personal observation) with an average of about 2,000.

As with other species of *Rana*, the timing of breeding for California red-legged frogs varies geographically. Across their range, breeding takes place from late November to late April (Jennings and Hayes, 1994a). In Marin County, the range is from 9 December–14 March, with an average of 12 January.

C. Larvae/Metamorphosis.

i. Length of larval stage. Larval development is variable and probably temperature dependent. Typically, tadpoles metamorphose from May–September, 3.5–7 mo after hatching (Storer, 1925; Wright and Wright, 1949; Jennings and Hayes, 1989), but there are several sites where tadpoles overwinter and metamor-

phose the following summer (Fellers et al., 2004).

ii. Larval requirements.

a. Food. The diet of tadpoles has apparently not been studied, but their diet is probably similar to other ranid frogs that feed on algae, diatoms, and detritus by grazing the surface of rocks and vegetation (Kupferberg, 1996a,b).

b. Cover. California red-legged frog tadpoles are often less conspicuous than other anuran larvae, but the role and importance of cover has not been investigated.

iii. Larval polymorphisms. None.

iv. Features of metamorphosis. California red-legged frogs metamorphose at a size of 1.6–2.0 cm SUL and a weight of 0.3–0.9 g (personal observations).

v. Post-metamorphic migrations. The movements of post-metamorphic frogs are similar to that of adults (see "Breeding Migrations" above).

D. Juvenile Habitat. Similar to adults, but there is some spatial segregation of adult and juveniles in riparian areas during the non-breeding part of the year. It is not known whether this represents active exclusion from preferred areas by adult frogs or whether there are subtle differences in habitat preference.

E. Adult Habitat. California red-legged frogs are primarily pond frogs (Stebbins, 1985), but they also inhabit marshes, streams, and lagoons during the breeding season. During other parts of the year, some frogs remain at breeding sites while others disperse to other areas. Non-breeding habitat includes nearly any area within 2–3 km of a breeding site that stays fresh and cool through the summer. This includes coyote bush (*Baccharis pilularis*) and California blackberry (*Rubus ursinus*) thickets, and root masses associated with willow (*Salix* sp.) and California bay trees (*Umbellularia californica*). Sometimes the non-breeding habitat used by California red-legged frogs is extremely limited in size, for example., a 1–2 m wide *Baccharis* thicket growing along a tiny, intermittent creek surrounded by heavily grazed grassland.

F. Home Range Size. Unknown, but Calef (1973b) reported males of the closely related northern red-legged frogs calling from within a few cm of each other without showing signs of aggression.

G. Territories. Unknown, but other ranid frogs are well known to defend breeding areas (Wells, 1977).

H. Aestivation/Avoiding Desiccation. Does not occur.

I. Seasonal Migrations. See "Breeding migrations" above.

J. Torpor (Hibernation). None reported.

K. Interspecific Associations/Exclusions. California red-legged frogs frequently are associated with Pacific treefrogs (*Pseudacris regilla*), rough-skinned newts (*Taricha granulosa*), and American bullfrogs. Less frequent associates include western toads

(*Bufo boreas*), western pond turtles (*Emys marmorata*), California newts (*T. torosa*), Sierra garter snakes (*Thamnophis couchii*), terrestrial garter snakes (*T. elegans*), and common garter snakes (*T. sirtalis*). There are records of California red-legged frogs co-occurring with arroyo toads (*B. californicus*), California giant salamanders (*Dicamptodon ensatus*), and California treefrogs (*P. cadaverina*; personal observations).

Hayes and Jennings (1986) argue that predation by introduced fish is an important factor contributing to the decline of western ranids, including California red-legged frogs. Though American bullfrogs have been widely implicated as being responsible for declines of California red-legged frogs, their relationship with California red-legged frogs is largely unknown.

L. Age/Size at Reproductive Maturity. Males 2 yr, females 3 yr (Jennings and Hayes, 1985).

M. Longevity. Unknown.

N. Feeding Behavior. Baldwin and Stanford (1987) report a large (about 11 cm SVL) California red-legged frog feeding on a California tiger salamander (*Ambystoma californiense*) larva, and Arnold and Halliday (1986) observed a California red-legged frog (8 cm SVL) with an adult male Pacific treefrog in its jaws. Hayes and Tennant (1985) found that California red-legged frogs feed on 42 different taxa. Invertebrates make up the majority of the diet, including Arachnida, Amphipoda, Isopoda, Insecta (including nine orders), and Mollusca. California red-legged frogs also feed to a limited extent on three-spined stickleback fish (*Gasterosteus aculeatus*), Pacific treefrogs, and California mice (*Peromyscus californicus*).

O. Predators. Rathbun (1998) reported newts (*Taricha* sp.) feeding on California red-legged frog eggs. Fellers and Freel (unpublished data) have regularly observed rough-skinned newts feeding on California red-legged frog eggs. Recently metamorphosed individuals are particularly vulnerable to fish predation (USFWS, 1999b). Adult frogs and tadpoles have been preyed upon by opossums (*Didelphis virginiana*), raccoons (*Procyon lotor*), striped skunks (*Mephitis mephitis*), spotted skunks (*Spilogale putorius*), great blue herons (*Ardea herodias*), American bitterns (*Botaurus lentiginosus*), black-crowned night herons (*Nycticorax nycticorax*), red-shouldered hawks (*Buteo lineatus*), garter snakes (*Thamnophis* sp.), American bullfrogs, various native and non-native species of fish, red swamp crayfish (*Procambarus clarkii*), and signal crayfish (*Pacifastacus leniusculus*; Fitch, 1940; Fox, 1952; Lowery, 1966; Calef, 1973a; Jennings and Hayes, 1989; Rathbun and Murphey, 1996; Lawler et al., 1999; USFWS, 1999b; Fellers and Wood, 2004).

P. Anti-Predator Mechanisms. Gregory (1979) examined responses of the closely related northern red-legged frog to humans and common garter snakes and found that frogs mostly relied on being immobile, but would jump into the water with close approach.

Q. Diseases. Mao et al. (1999) report northern red-legged frogs infected with an iridovirus, which was also present in sympatric three-spined stickleback fish in northwestern California. The virus had caused some mortality in the frog population.

R. Parasites. Ingles reported four species of trematodes from California red-legged frogs, but later synonomized two of the species he described (Ingles, 1932a,b, 1933c).

4. Conservation.

Rana draytonii was federally listed as a Threatened species in June 1996 (USFWS, 1996c). The most serious threats to this frog are loss of habitat from urbanization and agriculture and exposure to pesticides. Historically, much of the prime habitat for California red-legged frogs occurred in the grasslands and rolling hills of the Coast Range and the lower elevations of the Sierra Nevada. Substantial portions of the original range have been converted to other land uses, especially in the Los Angeles and San Francisco Bay areas, and in the Sierra foothills.

In California, pesticides are contributing to amphibian declines, especially in the Sierra Nevada, an area immediately downwind from the highly agriculturalized Central Valley in California (LeNoir et al., 1999; Davidson et al., 2001; Sparling et al., 2001). As a result of habitat loss and possibly pesticides, California red-legged frog populations are almost entirely gone from the foothills east of the Central Valley.

Many non-native species are likely to be California red-legged frog competitors and/or predators (USFWS, 2002c), but the population-level impact of most non-natives is unknown. For example, American bullfrogs are widespread within the range of California red-legged frogs, and predation and competition would seem to be inevitable. But there are almost no studies that evaluate this relationship, even though bullfrogs are routinely cited as a serious problem. There is little doubt that warm-water fish and mosquitofish have an impact. Chytrid fungus has been detected in California red-legged frogs, but the role of this disease in population declines is entirely unknown. The *R. draytonii* recovery plan lists many other potential causes for observed declines (USFWS, 2002c).

Rana fisheri Stejneger, 1893
VEGAS VALLEY LEOPARD FROG

Randy D. Jennings

1. Historical versus Current Distribution.

Vegas Valley leopard frogs *(Rana fisheri)* consist only of populations of leopard

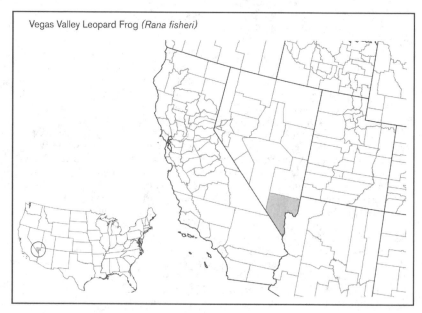

Vegas Valley Leopard Frog *(Rana fisheri)*

frogs found in the Las Vegas Valley, Clark County, Nevada (see Linsdale, 1940; Jennings et al., 1995), that have been often ascribed to relict leopard frogs *(R. onca;* Jennings, 1988). Relict leopard frogs (populations near Black Canyon and Overton, Nevada; Littlefield, Arizona; and St. George, Utah) are treated separately following Jaeger et al. (this volume).

The historical distribution of Vegas Valley leopard frogs included, and was probably restricted to, springs, seeps, and streams of the Vegas Valley, Clark County, Nevada (Frost, 1985; Jennings and Hayes, 1994b), at about 610 m (2,000 ft) in altitude (Linsdale, 1940). Historically, the main drainage through the Valley, Las Vegas Creek, exhibited riparian vegetation only to an area about 13 km (8 mi) north of present day Henderson and, typically, did not flow into the Colorado River (Malmberg, 1965). This suggests that aquatic habitats used by Vegas Valley leopard frogs were isolated from habitats used by other leopard frog species further east. As the population of Las Vegas grew, downstream flows in Las Vegas Creek were augmented by secondary sewage effluent and irrigation run-off; upstream, headwater springs were used progressively to supply water for the various needs of a burgeoning town (Malmberg, 1965; URS, 1977). Because the habitat of Vegas Valley leopard frogs was restricted, the status of these animals has always been considered to be precarious (Wright and Wright, 1949). Indeed, no documented sightings of Vegas Valley leopard frogs have been reported since 1942. Vegas Valley leopard frogs currently hold the dubious distinction of being the only species of North American anuran considered extinct (Stebbins, 1962; Frost, 1985; Jennings et al., 1995).

2. Historical versus Current Abundance.

Within their restricted distribution, Vegas Valley leopard frogs were abundant enough that J. R. Slevin collected 20 animals on 1 May 1913 at "Las Vegas Ranch," and from 10–13 August 1913, 79 specimens spanning age classes and including both sexes were collected (Vandenburgh and Slevin, 1921; Slevin, 1928). Further, A. Vanderhorst collected at least 13 frogs (including several adults) from Tule Springs in January 1942. But Wright and Wright (1949) write:

> May 16 [1942] . . . What frog hunters we are! I thought I was good at it . . . Las Vegas has grown . . . Took us most of the day to locate where the old artesian well and the springs were. At the U.S. Fish Hatchery found bullfrogs. The municipal golf course and possibly the hatcheries are where the springs were. Looked these over but no *R. fisheri*. Tried Las Vegas Creek upper stretches. Found a minnow and plenty of crayfish but no frogs. . . . May 17 . . . Went to Fifth St. crossing of Las Vegas Creek. Looked it over. West of this crossing in tules heard one splash of frogs. Never saw them. . . . went to Main Street crossing and walked up to old artesian well, a mile or so. Some minnows in stream, lots of crayfish—heard four splashes in tules but never saw frogs. . . . Our *R. fisheri* may go with the old springs gone, the creek a mess.

These may have been the last Vegas Valley leopard frogs to be observed (Behler and King, 1998). Stebbins (1962) looked for this species on 5 July 1949 in Tule Springs, near the site of the U.S. Fish Hatchery, and in upper Las Vegas Creek, but found capped artesian wells, covered reservoirs, a golf course, dead tules, restricted aquatic habitat, and American bullfrogs *(R. catesbeiana).* The disappearance of Vegas Valley leopard frogs seems correlated with heavy groundwater pumping by the City of Las Vegas that caused the cessation of flow from artesian springs at the headwaters of Las Vegas Creek (see URS, 1977).

3. Life History Features.
Because Vegas Valley leopard frogs have not been observed since 1942, little is known of their life history. Information presented here is based on limited published observations and inference drawn from those observations and museum specimens.

A. Breeding. Reproduction was aquatic.
i. Breeding migrations. Unlikely. Vegas Valley leopard frogs probably bred in the springs, streams, and pools inhabited by the adults (Wright and Wright, 1949), because little other aquatic habitat was available. Early investigators suggested that Vegas Valley leopard frogs probably bred in the spring (Wright and Wright, 1949). No egg masses were ever collected, but adults have been collected in January–August (see Stebbins, 1962; unpublished data). The lack of specimens from September–December likely represents a lack of activity by investigators rather than by frogs. Collections of frogs spanning much of the year indicate that this species had a prolonged period of activity. Springs along Las Vegas Creek exhibited water temperatures ranging from 21.6–25.5 °C (Carpenter, 1915; Jones and Cahlan, 1975, in Jennings and Hayes, 1994b). The warm climate of Las Vegas coupled with warm water temperatures could make it possible for Vegas Valley leopard frogs to possess a prolonged reproductive period. The presence of recently metamorphosed frogs (SVL 28–33 mm) in collections from April–August supports this contention.

Vegas Valley leopard frogs were reported to "have vocal sacs like *Rana pipiens*" (Wright and Wright, 1949). Vocalizations characterized as "very low croak" were heard on 20 August 1925.

ii. Breeding habitat. Wright and Wright (1949) noted that eggs were unknown. Vegas Valley leopard frogs probably used lentic microhabitats within the springs, seeps, and streams where they were found.

B. Eggs.
i. Egg deposition sites. The importance of vegetation in egg deposition is unknown.
ii. Clutch size. Unknown.
C. Larvae/Metamorphosis.
i. Length of larval stage. Wright and Wright (1949) noted that a tadpole measuring 85 mm TL was collected on 15 July 1938. Size at metamorphosis ranged from 28–33.5 mm SVL (Wright and Wright, 1949). The presence of metamorphic animals in April suggests that tadpoles may have overwintered.
ii. Larval requirements. Unknown.
a. Food. Unknown, but diet was probably similar to other ranid tadpoles.

b. Cover. Unknown.
iii. Larval polymorphisms. Unlikely.
iv. Features of metamorphosis. That recently metamorphosed frogs were collected from April–August and that habitats formerly used by Vegas Valley leopard frogs were largely spring fed with relatively stable water temperatures (Jones and Cahlan, 1975, in Jennings and Hayes, 1994b) suggests metamorphosis could have occurred throughout much of the year.
v. Post-metamorphic migrations. As with other members of the *R. pipiens* complex, newly metamorphosed individuals likely dispersed from natal aquatic habitats, but the distance of dispersal would have been restricted by the aridity of surrounding desert habitats.
D. Juvenile Habitat. Probably similar to adults (see Wright and Wright, 1949).
E. Adult Habitat. Isolated cool spring basins or trickling streams in saturated soils (Wright and Wright, 1949; Jennings and Hayes, 1994b). Wright and Wright (1949) comment that where they collected Vegas Valley leopard frogs the springs looked "marly or alkali." Historically, artesian springs and Las Vegas Creek, the habitat of Vegas Valley leopard frogs, were bordered by cottonwoods, willows, bulrushes, sedges, and cattails (Jennings and Hayes, 1994b). This vegetation would have constituted the tules that Wright and Wright (1949) indicated frogs used as cover to elude capture on 17 May 1942 and that gave Tule Spring its name. Most of this cover was lost, especially willows and cottonwoods, by 1938 (Jennings and Hayes, 1994b). Stebbins (1962) noted the near elimination of aquatic habitats and riparian vegetation by 1949.
F. Home Range Size. Because of the surrounding desert, Vegas Valley leopard frogs probably remained close to water and riparian habitats.
G. Territories. None reported.
H. Aestivation/Avoiding Desiccation. Linsdale (1940) commented that Vegas Valley leopard frogs were most active during the spring and early summer, but dates of capture include January–August. The implication is that animals were capable of year-round activity.
I. Seasonal Migrations. None reported, nor were they likely.
J. Torpor (Hibernation). None reported, nor was hibernation likely.
K. Interspecific Associations/Exclusions. Vegas Valley leopard frogs were found associated with Pacific treefrogs *(Pseudacris regilla)* and Great Plains toads *(Bufo cognatus*; Wright and Wright, 1949). Linsdale (1940) indicated Woodhouse's toads *(B. woodhousii)* were present regionally, and that species is currently found within the historical range of Vegas Valley leopard frogs. American bullfrogs were introduced into the Vegas Valley leopard frog habitats prior to their extinction (Wright and Wright, 1949;

Stebbins, 1962; Jennings and Hayes, 1994b).
L. Age/Size at Reproductive Maturity. Breeding sizes were reported in Wright and Wright (1949) as 44–64 mm for males, 46–74 mm for females. Wright and Wright (1949) cited the size of Vegas Valley leopard frogs as "Adults, 1 3/4–3 inches" (44–76 mm), but the largest frog reported individually was 74 mm. Others have reported the maximum size as 76 mm (Behler and King, 1998). This discrepancy likely represents conversion rounding error.
M. Longevity. Unknown.
N. Feeding Behavior. The diet and feeding behavior of Vegas Valley leopard frogs was probably similar to that of other species of leopard frogs.
O. Predators. Unknown.
P. Anti-Predator Mechanisms. Crypsis and catalepsis were probably effective anti-predator mechanisms for Vegas Valley leopard frogs within dense tule habitats where they occurred. Because of stocky body proportions, upon detection by predators this frog likely sought refuge by jumping short distances into the water and submergent vegetation (see Wright and Wright, 1949).
Q. Diseases. Unknown.
R. Parasites. Unknown.

4. Conservation.
Vegas Valley leopard frogs have not been seen since 1942 and are considered extinct.

Rana grylio Stejneger, 1901
PIG FROG

Stephen C. Richter

1. Historical versus Current Distribution.
Pig frogs *(Rana grylio)* are endemic to the southeastern coastal plain of the United States, along the Atlantic Coast from southern South Carolina to the Everglades of Florida, and west along the Gulf Coast to extreme eastern Texas (Altig and Lohoefener, 1982; Conant and Collins, 1991). Though not native, pig frogs have been found on barrier islands off the Atlantic Coast (Martof, 1963; Kiviat, 1982). They have been introduced on Andros and New Providence islands in the Bahamas (Neill, 1964a; Conant and Collins, 1991) and are well established in northern Puerto Rico (Rios-López and Joglar, 1999). Fossils are known from the Pleistocene (Illinoinan–Sangamonian glacial age; 100,000–1,300,000 ybp) of north-central Florida (Tihen, 1952; Lynch, 1965).

2. Historical versus Current Abundance.
No records that quantify historical and current abundance exist, but pig frogs have been and still are considered common throughout their range (Deckert, 1914a; Voice, 1923; Wright, 1932; Dixon, 1987). Populations do not appear to have

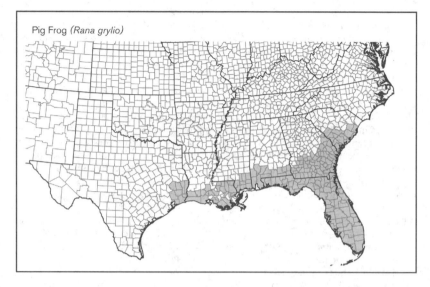

Pig Frog (*Rana grylio*)

been diminished; in fact, their geographic distribution has expanded (Neill, 1964b; Rios-López and Joglar, 1999). Pig frogs, unlike most other anuran species, are positively affected by residential development. For example, Delis et al. (1996) found higher abundances of pig frogs in a developed area that was once pine flatwoods than an adjacent pine flatwoods habitat. No long-term studies exist that quantify changes in population-level abundance over time. Wood et al. (1998) found that females were typically more abundant than males at most sampling locations and that juvenile abundance varied widely among sites, but because this study occurred over a 3-yr period, it could not address long-term variation in abundance.

3. Life History Features.

A. Breeding. Reproduction is aquatic.

i. Breeding migrations. In South Carolina, Lamb (1984) found that aquatic pig frogs were confined to the cypress-hardwood zone during the non-breeding season but moved into the grass-shrub zone to breed. Males made this movement earlier and remained within the grass-shrub zone longer than females. Males typically call at night but are often heard during the day (Deckert, 1914b; Barbour, 1920; Mount, 1975; Lamb, 1984; personal observations). Pig frogs are reproductively active throughout most of the year in Puerto Rico, where males have been heard calling from March–November (Rios-López and Joglar, 1999). Lamb (1984) observed calling males for 6–7 mo and gravid females for 4 mo. Carr (1940a) and Duellman and Schwartz (1958) reported males calling in every month of the year (rarely below 21 °C), however, calling is typically most intense from February–September in the continental United States (Wright and Wright, 1949; Lamb, 1984; Dundee and Rossman, 1989). Males have mature spermatozoa

in their testes year-round; females have mature ova April–July, with development beginning in August (Lamb, 1984). The next clutch of eggs begins to develop soon after the first clutch is deposited, and females overwinter in this condition (Lamb, 1984).

ii. Breeding habitat. Pig frogs typically breed in open, permanent freshwater lakes, cypress ponds, marshes, brushy swamps, roadside ditches, and overflowed river banks containing emergent vegetation (Wright, 1932; Smith and List, 1955; Mount, 1975; Ashton and Ashton, 1988; Dundee and Rossman, 1989), though Voice (1923) found them in wetlands with moderate salinity. Pig frogs have also been seen in ephemeral ponds, streams, and spring runs (Carr, 1940a). Neill (1947b) found them breeding in old, flooded rice fields, and according to Uzzell (1952), their distribution may be limited in South Carolina by these old rice reserves and similar habitats where they are commonly found in the lower coastal area.

B. Eggs.

i. Egg deposition sites. Eggs are deposited as a surface film, typically attached to vegetation, from March–September (Wright, 1932; Carr, 1940a; Wright and Wright, 1949). Wells (1976) suggested that pig frogs might deposit >1 clutch/yr, but data are lacking.

ii. Clutch size. Females deposit 8–15,000 eggs in a surface film (typically attached to vegetation) that ranges in diameter from 305–762 mm (Wright, 1932; Wright and Wright, 1949; Dundee and Rossman, 1989). Eggs are surrounded by two gelatinous capsules and take 2–3.5 d to develop and hatch (Wright and Wright, 1924, 1949; Wright, 1932; Ashton and Ashton, 1988).

C. Larvae/Metamorphosis. Larvae are large and known to reach 110 mm total length and 35.6 mm body length (Wright, 1932; Wright and Wright, 1949).

The larval period ranges from 365–730 d—thus, tadpoles overwinter twice in some populations (Wright, 1932; Wright and Wright, 1949; Dundee and Rossman, 1989); transformation typically occurs from late April to mid-July (Wright, 1932). Gosner (1959) described tooth morphology of larval pig frogs, and Altig and Pace (1974) provided scanning electron photomicrographs and measurements of individual labial teeth (length, width = 0.55, 0.36 mm). Wright (1932) found average size at metamorphosis to be 44 mm (range = 32–49 mm).

D. Juvenile Habitat. No studies have addressed ontogenic shifts in habitat preference/characteristics, but juveniles probably utilize habitats similar to adults.

E. Adult Habitat. Pig frogs are largely aquatic, typically remaining within permanent habitats (as described in "Breeding habitat" above) throughout the year (Wright, 1932; Wright and Wright, 1949; Lamb, 1984). Other than Lamb (1984), no data exist regarding non-breeding season activity, so it is possible that periodic land migrations occur.

F. Home Range Size. Unknown.

G. Territories. Agonistic behavior was observed by Lamb (1984), who observed males wrestling for >3 min and noted that grappling terminated when one individual retreated. Lamb (1984) also observed calling males with "high posture," as described for other territorial ranids by Wells (1977). These observations suggest that male pig frogs are territorial and that the mating system is complex.

H. Aestivation/Avoiding Desiccation. As drought conditions approach, pig frogs will burrow in mud and peat (Ligas, 1960; Wood et al., 1998).

I. Seasonal Migrations. Wood et al. (1998) found that pig frogs tend to remain in one location when food and water conditions are suitable, but that substantial movement is possible when water conditions change. Short, breeding movements occur; see "Breeding migrations" above.

J. Torpor (Hibernation). No data exist addressing slowed metabolism or inactivity in the winter. Based on lack of sightings, Wright (1932) estimated a hibernation period of 4–5 mo, typically occurring from November–March, though geographic variation is probable.

K. Interspecific Associations/ Exclusions. Birkenholz (1963) observed and trapped pig frogs in the "houses" of round-tailed muskrats (*Neofiber alleni*) in Florida.

L. Age/Size at Reproductive Maturity. Post-metamorphic pig frogs range in size from 32–161 mm SVL (Wright, 1932). Growth rates of males and females are similar until about 100 mm SVL, when females begin growing faster and reach larger maximum sizes (Wright, 1932; Ligas, 1960). Wright (1932) estimated size at maturity at 91–102 mm; however, Wright and Wright (1949) reported a minimum size at maturity as 82 mm for males and 85 for females. Mount (1975) observed two calling males 52 mm and 59 mm SVL, one of which was dissected and found to be sexually mature. Maximum SVL records vary: 150 mm in Florida (Ashton and Ashton, 1988); 152 mm (male) and 161 mm (female) in Georgia (Wright, 1932); 162 mm in Louisiana (Dundee and Rossman, 1989); and 165 mm in Alabama (Mount, 1975).

Wright (1932) estimated age at maturity (based on body size) as 3 yr with no intersexual differences. Controversy exists regarding the correlation between size and age (Halliday and Verrell, 1988; Platz and Lathrop, 1993), so caution should be taken in applying this estimate. A skeletochronology and/or multiyear mark–recapture study is necessary to obtain accurate data on population age structure.

Wood et al. (1998) found the mean monthly survival estimate for adult females (0.83 ± 0.058) was higher than for adult males (0.75 ± 0.088) and juveniles (0.72 ± 0.095; all estimates reported as mean ± SE).

M. Longevity. Wright (1932) estimates maximum age based on body size as 6 yr. Again, controversy exists in predicting age from body size (see "Age/Size at Reproductive Maturity" above).

N. Feeding Behavior. Lamb (1984) found the diet of pig frogs consisted primarily (95%) of arthropods (Coleoptera 24.3%, Decopoda 19.8%, and Odonata 11.9%). Crayfish are typically the most commonly taken single food item: 20% of diet (Lamb, 1984), 40% of diet (Carr, 1940a), and 75% of diet (Ligas, 1960). Carr (1940a) found sailfin mollies (*Poecilia latipinna*), smaller frogs, and aquatic insects in stomach contents. Other known food items include leeches (*Placobdella rugosa*), cyprinodont fishes, green treefrogs (*Hyla cinerea*), southern leopard frogs (*R. sphenocephala*), and northern water snakes (*Nerodia sipedon*); adult pig frogs will also cannibalize juveniles (Duellman and Schwartz, 1958). Lamb (1984) found that males fed infrequently during the breeding season and attributed intersexual differences in diet during the breeding season primarily to intersexual differences in behavior and habitat.

O. Predators. Wright (1932) reported predation by water snakes (*Nerodia* sp.), cottonmouths (*Agkistrodon piscivorus*), herons (Ardeidae), and ibises (Threskiornithidae). Coward (1984) reported predation by an osprey (*Pandion haliaetus*).

P. Anti-Predator Mechanisms. When captured or handled, pig frogs occasionally emit what Allen (1932) described as "an unmistakable dank, musty odor and a slime that is bitter to the taste." The biochemical components of this "slime" are unknown, but the secretion of closely related dusky gopher frogs (*R. sevosa*), which

also have a musty smell and bitter taste, contains complex, irritant compounds (C. Graham, S.C.R., P. Flatt, and C. Shaw, unpublished data).

Q. Diseases. Unknown.

R. Parasites. Christian and White (1973) reported 56 specimens of *Allassostomoides louisianaensis* (Trematoda: Paramphistomidae) from the large intestines of 10 pig frogs in Louisiana. Boyce (1985) dosed a single pig frog with 20 nymphs of *Sebekia mississippiensis* (a blood-sucking, endoparasitic arthropod and natural parasite of *Alligator mississippiensis*); soon after infection, the frog became lethargic and died within 24 hr.

4. Conservation.
Pig frogs are not considered Threatened, Endangered, or of Concern by any state or federal agency. In Florida, pig frogs are the second most abundant frog, and the frog most commonly gigged by humans (Carr, 1940a; Ashton and Ashton, 1988). The only legal "protection" given pig frogs is a state fishing license, which is required for Louisiana's limited hunting season (Dundee and Rossman, 1989). Populations do not appear diminished; in fact, their geographic distribution has expanded (Neill, 1964b; Rios-López and Joglar, 1999). Pig frogs, unlike most other anuran species, are positively affected by residential development (see "Historical versus Current Abundance" above). Although pig frogs appear to be doing well, more data are needed that address populations' status across their range.

Rana heckscheri Wright, 1924
RIVER FROG

Brian P. Butterfield, Michael J. Lannoo

1. Historical versus Current Distribution.
River frogs (*Rana heckscheri*) were described by Wright (1924) and are found

along the Atlantic and Gulf Coastal Plains from North Carolina south to the Oklawaha River, Florida, and west to the Mississippi River (Wright, 1932; Simmons and Hardy, 1959; Lodato, 1974; Martof et al., 1980; Conant and Collins, 1991; Bartlett and Bartlett, 1999a). Mount (1975) notes that river frogs are locally distributed. Hansen (1957) describes river frogs as "the least known ranid of the southeast." Beane (1998) failed to find any evidence of the existence of river frogs in North Carolina and stated that river frogs were last documented from the state in 1975. Additional range expansions or contractions have not been described.

2. Historical versus Current Abundance.
Unknown. Both tadpoles and adults can be abundant. Wright and Wright (1949), quote Allen (1938): "*Rana heckscheri* is very common in the vicinity of Silver Springs, Florida, in the streams and rivers and in the lakes connected by streams which flow into the Oklawaha River." Can be common in appropriate habitats (Bartlett and Bartlett, 1999a). "In Alachua County, Florida, [tadpoles] transform consistently in April and May. H. K. Wallace and I found them emerging in tremendous numbers in a backwater of the Santa Fe River 1 May 1933; we probably could have collected a thousand with little difficulty. This disparity in relative abundance between the adults and the young of this species indicates an extremely low survival potential" (A.F. Carr, in Wright and Wright, 1949).

3. Life History Features.
  A. Breeding. Reproduction is aquatic.
    i. Breeding migrations. Adults breed in ponds with emergent vegetation (Martof et al., 1980; Bartlett and Bartlett, 1999a). Breeding begins in April and will continue, when conditions are favorable, into August (Wright and Wright, 1949;

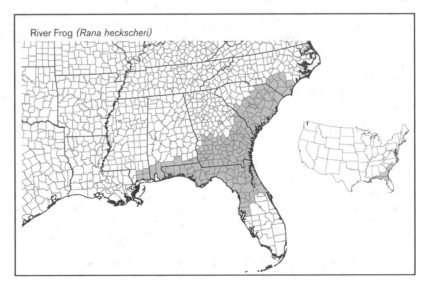

River Frog (*Rana heckscheri*)

Mount, 1975; Behler and King, 1998; see also Wright, 1932). In Florida, Carr (1940a) notes that river frogs begin breeding in April and will breed sporadically throughout the summer. Males can be heard calling from April–July (Ashton and Ashton, 1988). Separate tadpole size classes suggest multiple breeding events in particular wetlands (Wright and Wright, 1949).

ii. Breeding habitat. In habitats ranging from river edges to adjacent, upland ponds.

B. Eggs.

i. Egg deposition sites. Eggs are laid in a surface film in ponds near rivers and streams (Wright and Wright, 1949; Martof et al., 1980; Bartlett and Bartlett, 1999a).

ii. Clutch size. Three amplectic pairs laid an estimated 5,000 eggs (Allen, 1938). Based on counts of ova, Wright (1932) estimates clutch sizes between 6,000 and 8,000. Egg diameter 1.5–2.0 mm (vitellus; Wright, 1932). Ashton and Ashton (1988) report eggs hatch about 3 d after being laid, while Allen (1938) reports eggs hatching in 10–15 d.

C. Larvae/Metamorphosis.

i. Length of larval stage. Tadpoles will overwinter for 1 yr, perhaps 2 yr (Allen, 1938; Wright and Wright, 1949; Mount, 1975; Ashton and Ashton, 1988). Tadpoles reach a length (SVL) of 97 mm (Wright, 1932; Wright and Wright, 1949; Martof et al., 1980), 160 mm TL (Wright and Wright, 1949; Mount, 1975).

ii. Larval requirements.

a. Food. Wright and Wright (1949) note that tadpoles not only fed on meat thrown into a lake to feed the alligators, but also on plant material. Feeding activity of schooling tadpoles can be stimulated by a single tadpole, even in the absence of olfactory stimuli (Altig and Christensen, 1981; see also Punzo, 1991b).

b. Cover. Tadpoles tend to aggregate in shallows during the day but return to deeper water at night (Wright and Wright, 1949). Tadpoles school in masses ("as no other big tadpoles do;" Wright and Wright, 1949) that can contain thousands of individuals (Altig and Christiansen, 1981; Ashton and Ashton, 1988). Wright (1932) notes "seething masses" of river frog tadpoles. Archie Carr, in Wright and Wright (1949), notes: "The number of tadpoles produced in a given breeding site is astounding." Schooling is unusual among ranids (Altig and Christiansen, 1981). It remains unclear whether schooling serves as an anti-predator device or serves some other, as yet unknown, function (Altig and Christiansen, 1981).

iii. Larval polymorphisms. Larvae change color from small, dark tadpoles to large, light tadpoles with a dark-edged tailfin (Allen, 1938; Wright and Wright, 1949; Bartlett and Bartlett, 1999a). Wright (1932) notes: "No other tadpole is so distinctive in the eastern United States."

iv. Features of metamorphosis. Newly metamorphosed animals are 30–52 mm

long (Simmons and Hardy, 1959; Martof et al., 1980). In Alachua County, Florida, they transform consistently in April–May. River frogs will sometimes transform *en masse,* with metamorphosis at a single site lasting 2–3 wk. But Ashton and Ashton (1988) note that due to prolonged breeding and the requirement for a 1-yr developmental period, newly metamorphosed frogs can be observed throughout much of the year. Allen (1938) notes a 20% mortality at metamorphosis.

v. Post-metamorphic migrations. Long migrations are unlikely. Allen (1938) reports a juvenile found about 100 m from its water body, 1–2 mo after metamorphosing.

D. Juvenile Habitat. Appears to be more variable than adult habitats. Hansen (1957) notes that juveniles move greater distances than adults, in large part because adults confine their movements to a home range. Allen (1938) found that for several months after metamorphosis, no juveniles were found around the margins of their water body.

E. Adult Habitat. "Swampy edges of rivers and streams, a truly fluvatile species" (Wright, 1932). Nocturnal, terrestrial, associated with emergent vegetation (Wright, 1932; Wright and Wright, 1949; Behler and King, 1998). Can be found at night in shrubbery and the bases of trees along banks (Wright and Wright, 1949); Hansen (1957) notes an association with *Sphagnum* sp. In Alabama, Mount (1975) notes that river frogs also occur along the edges of shallow impoundments, such as beaver ponds, associated with vegetation such as titi, bay, and cypress. Nocturnal. Inhabit river swamps and swampy shores of ponds. Bottom land forests (Martof, 1980). Hansen (1957) notes that river frogs actively seek temperatures around 25 °C. Adults are not wary (Conant and Collins, 1991).

F. Home Range Size. Hansen (1957) calculated home range sizes for nine animals, a mean of 16 m^2, and notes that this estimate is similar to that calculated for green frogs (*R. clamitans;* 18.6 m^2) by Martof (1953).

G. Territories. Defended territories have not been reported.

H. Aestivation/Avoiding Desiccation. Hansen (1957) notes that river frogs are usually found around water bodies and noted that the fact that animals are "consistently found seated upon moist or wet substrates suggests that this type of substrate is actively sought to prevent desiccation."

I. Seasonal Migrations. Undocumented.

J. Torpor (Hibernation). When temperatures fall below 17 °C, Hansen (1957) found that river frogs are never observed and suggested, by comparison with green frogs, that river frogs avoid cold by moving into bodies of water.

K. Interspecific Associations/Exclusions. Hansen (1957) notes the following species

at his study ponds: bronze frogs (*R.c. clamitans*), southern leopard frogs (*R. sphenocephala*), green treefrogs (*Hyla cinerea*), and Florida cricket frogs (*Acris gryllus dorsalis*). Will breed in same wetlands as pig frogs (*R. grylio*; Wright, 1932).

L. Age/Size at Reproductive Maturity. Males 82–131 mm; females 102–155 mm (Wright, 1932; Wright and Wright, 1949; Conant and Collins, 1991).

M. Longevity. Unknown.

N. Feeding Behavior. River frogs feed largely on invertebrates, especially insects. They will also take small vertebrates, including other ranid frogs (Hill, 2000).

O. Predators. Southern water snakes (*Nerodea fasciata*) feed on tadpoles (Wright, 1932; Allen, 1938; Wright and Wright, 1949). Hansen (1957) also noted an association between river frogs and water snakes (*Nerodia* sp.). Wright (1932) suggests largemouth bass (*Micropterus salmoides*) prey on tadpoles, and grackles (*Quiscalus* sp.) feed on newly transformed animals.

P. Anti-Predator Mechanisms. Passive, not easily alarmed nor evasive; clumsy in their escape (Wright, 1932; Carr, 1940a; Wright and Wright, 1949; Hansen, 1957; Conant and Collins, 1991). Can be caught by humans using just their bare hands. Behler and King (1998) note that adults may have toxic skin secretions because water snakes and indigo snakes (*Drymarchon* sp.) become violently ill after ingesting river frogs; after regurgitating the animal they continue to wipe their mouths on the ground. Schooling in tadpoles is associated with predator avoidance and location of food (Wassersug et al., 1981; see also Punzo, 1991b; Bartlett and Bartlett, 1999a). Large tadpoles and adults will produce noxious and odorous skin secretions (Bartlett and Bartlett, 1999a). Large size at metamorphosis and in adults must deter gape-limited predators.

Q. Diseases. Largely unknown. Punzo (1993) has, however, observed the effects sublethal concentration of mercuric chloride have on the ovaries of river frog females.

R. Parasites. Not well documented. A parasitic copepod (*Argulus americanus*) was found on a river frog tadpole (Goin and Ogren, 1956).

4. Conservation.

River frogs are locally distributed and perhaps the least known ranid of the southeast. River frogs were last documented in North Carolina in 1975 (Beane, 1998), where they are considered a Species of Special Concern (Levell, 1997). Additional range expansions or contractions have not been described. According to Archie Carr (in Wright and Wright, 1949), the disparity in relative abundance between the adults and the young of this species indicates an extremely low survival potential.

Rana luteiventris Thompson, 1913
COLUMBIA SPOTTED FROG
Jamie K. Reaser, David S. Pilliod

1. Historical versus Current Distribution.

Columbia spotted frogs *(Rana luteiventris)* were initially recognized as *R. pretiosa* (Baird and Girard, 1853c) and considered widespread throughout much of western North America, ranging in elevation from sea level to nearly 3,040 m in the Rocky Mountains. Taxonomic subdivision into *R. pretiosa luteiventris* for the Great Basin region and *R. pretiosa pretiosa* for the rest of the species range was proposed (Thompson, 1913), and these subspecies names occur in the scientific literature until the early 1970s. However, the morphological variations (e.g., ventral coloring) were not consistent, and the subspecies designation was contested and eventually abandoned (Turner and Dumas, 1972). In 1996, studies of allozyme variation discerned two spotted frog species whose distributions were not concordant with the original subspecies boundaries (Green et al., 1996, 1997). Oregon spotted frogs *(R. pretiosa)* occur in south-central Washington, the Cascade Mountains of Oregon, and extreme southwestern British Columbia. Historically, Oregon spotted frogs occurred in the Pit River system in California (Hayes, 1997), but have not been seen there since 1911 (Stebbins and Cohen, 1995). Columbia spotted frogs, considered in this account, range from extreme southeast Alaska south through British Columbia and Alberta, western Montana and Wyoming, northern and central Idaho, northeastern Oregon, and eastern Washington. Isolated relict populations of Columbia spotted frogs persist in the Great Basin paleodrainages of the Humboldt River (eastern Lahontan Basin), Bonneville Basin, southeastern Oregon, southwestern Idaho, and northern Nevada drainages of the Snake River, and the Big Horn Mountains east of the continental divide of Wyoming (Dunlap, 1977; Bos and Sites, 2001). Columbia spotted frogs might persist in low numbers on the east side of Warner Mountain Chain in California (Jennings and Hayes, 1994a). This species has been reported at elevations up to 1,951 m in Washington and 2,247 m in Oregon (Leonard et al., 1993); 3,036 m in Montana (Maxell et al., 2003); 2,725 m in Idaho (Munger et al., 1997a,b; Pilliod, 2001); 2,890 m in Wyoming (Turner, 1960; Dunlap, 1977; Patla, 1997; D. Patla and C. Peterson, unpublished data); and 2,650 m in the Toiyabe Range of central Nevada (Reaser, 1997a).

No subspecies are currently recognized. However, allozyme data suggest delineation of distinct populations (Green et al., 1996, 1997): Northern, Great Basin, West Bonneville, and Wasatch Front. The recent molecular study by Bos and Sites

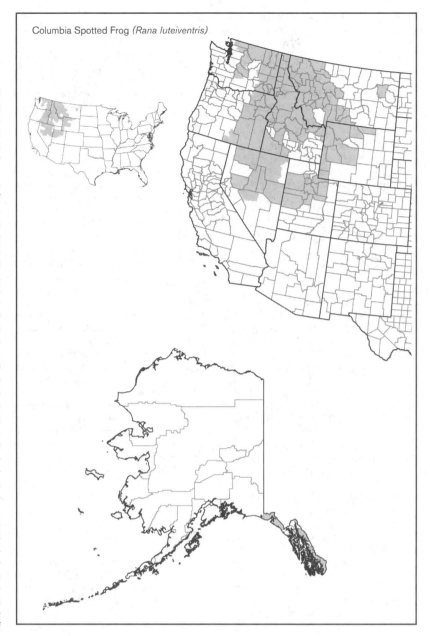

Columbia Spotted Frog *(Rana luteiventris)*

(2001) revealed four well-supported clades defined on the basis of monophyly of a mtDNA sequence; these included Northern Rocky Mountain, Lahonton, Bonneville, and Deep Creek clades that were recognized by those authors as "candidate Evolutionary Significant Units." Other studies are in progress and are also expected to reveal divisions within the species currently recognized as *R. luteiventris* (J. H. Howard and J. C. Munger, in progress; D. M. Green, personal communication). These divisions may prove to be distinguishable on both a molecular and morphological basis (Hayes, 1997; J. K. R., personal observation).

The similarity of native ranids of western North America has led to many misidentifications of other species as *R. pretiosa* (= *R. luteiventris* [Turner and Dumas,

1972]). Reaser (2003) evaluated 57 voucher collections (1912–73) with specimens identified as *R. pretiosa* (= *R. luteiventris*) from Nevada. As a result, vouchers collected from 11 surveys, representing at least five independent sites, were reclassified as *R. draytonii* (California red-legged frogs).

2. Historical versus Current Abundance.

Only one historical population data set exists for which to make direct comparisons. In Wyoming, Patla and Peterson (1994, 1999) documented an 80% decline in a population of Columbia spotted frogs near Lodge Creek in Yellowstone National Park that went from 1,000–1,900 frogs in the 1950s (Turner, 1960) to 200–400 frogs in the 1990s. This decline is likely the result of construction of a road bisecting habitats used by frogs and of the development of the key spring head where the frogs spend the winter (Patla and Peterson, 1999).

Recent surveys provide anecdotal evidence of changes in abundance and provide baseline data for future comparisons. Across much of the range of the northern population (Idaho, western Montana, eastern Oregon, and northwestern Wyoming), Columbia spotted frogs are common and usually abundant in many areas. In central and northern Idaho and western Montana, Columbia spotted frogs are the most commonly encountered frog species with locally abundant populations, especially in fishless lakes and ponds (O'Siggins, 1995; Munger et al., 1997a,b; Beck et al., 1998; Llewellyn and Peterson, 1998; Yeo and Peterson, 1998; Pilliod, 2001; Maxell et al., 2003). Columbia spotted frogs are often locally abundant across the northern half of eastern Oregon, especially in the Blue and Wallowa Mountain system (M. Hayes and E.L. Bull, personal communication). More than a decade (1989–2002) of amphibian surveys in the Greater Yellowstone Ecosystem of Wyoming indicate that Columbia spotted frogs are geographically widespread and the second most abundant amphibian in the region (Koch and Peterson, 1995; Patla and Peterson, 1995; Patla, 1997; Van Kirk et al., 2000).

Isolated distinct populations along the southern boundary of the species' range have patchy distributions—while some populations are high density, others are not. In the Bighorn Mountains in Wyoming, the population is restricted to a single watershed and only three breeding sites have been identified (Dunlap, 1977; Bos and Sites, 2001). Surveys conducted in southwestern Idaho determined that Columbia spotted frogs are sometimes locally abundant, although limited in distribution (Munger et al., 1996, 1997a,b; Engle and Munger, 1998). In eastern Oregon, anecdotal data suggest that Columbia spotted frogs have disappeared from some lower elevation areas (e.g., some of the Grande Ronde Valley, in the vicinity of Pendleton and Prineville). In this region, populations on private land may be critical for the persistence of the species. Recent surveys of potential breeding habitats in the southern half of eastern Oregon and northern Nevada suggest that Columbia spotted frogs are rare, populations are small, and some declines may have occurred (Wente and Adams, 2002). For example, Wente and Adams (2002) found frogs in 13 of 20 (65%) historical sites visited.

In the Toiyabe Range (central) and the Ruby Mountains (northeast) of Nevada, Reaser (1997a) found Columbia spotted frogs to be uncommon but at high densities (10–20/150 m²) at some sites. These sites may have been acting as refugia during especially dry summer seasons. In

northern Nevada, in the vicinity of the Jarbidge and Independence ranges, Reaser (1997a) found Columbia spotted frogs to be more common, but typically at low densities. Columbia spotted frogs often were encountered at beaver ponds. Private land management does not appear to be favoring spotted frogs in Nevada; frogs could not be detected at 14 of the 15 historical sites on private land at which spotted frogs were known to occur in Nevada prior to 1993. Half of historical sites owned by the U.S. Forest Service and 1/3 of the historical Bureau of Land Management sites were also without frogs. However, these surveys also documented 78 sites at which spotted frogs had not been recorded previously.

In Utah, Columbia spotted frogs were found in 2 historical sites out of 25 locations in the Wasatch Front populations (Utah Division of Wildlife Resources Progress Report, 1991). Seven populations are currently known to exist in the Wasatch Front, and although some threats to these populations still exist, ongoing conservation actions have improved the long-term viability of the species (USFWS, 2002d). Molecular evidence suggests that Columbia spotted frogs within the Bonneville basin occur as 13 genetically distinct populations that currently have restricted gene flow (Bos and Sites, 2001). The West Bonneville populations have been inventoried thoroughly, and the many disjunct populations appear to be large enough to be considered secure at this time (Ross et al., 1994).

3. Life History Features.

A. Breeding. Reproduction is aquatic.

i. Breeding migrations. Adult Columbia spotted frogs emerge in late February to early July depending on the latitude, elevation, and seasonal weather patterns (generally later at higher latitudes and elevations). Males typically are reported to emerge before females (3–4 d), but in southwest Idaho populations in 1999, male and female breeders emerged simultaneously in mid-April. There were, however, many more males than females (J.C. Engle, personal communication). If wintering habitat does not meet criteria for reproduction, spotted frogs are required to migrate to breeding sites. These movements often occur along wetland (ephemeral or permanent) corridors and sometimes overland (see Turner, 1960; Morris and Tanner, 1969; Patla, 1997; Pilliod et al., 2002). The arid landscape and present climatic conditions in Tule Valley of Utah likely prevent frequent movement of frogs between the six spring complexes (Hovingh, 1993), except perhaps in wet years.

ii. Breeding habitat. Population-specific reports are highly variable, reflecting the diversity of habitats available within the

range of the species. In general, Columbia spotted frogs use the shallows (only a few cm deep) of lentic habitats (lakes, ponds, marshes, and small springs) for breeding and egg deposition. These habitats are usually permanent (or were so prior to recent degradation), although naturally ephemeral pools are used successfully by some populations. Springs are often nearby. Aquatic substrates may be fine textured and unconsolidated. Floating and/or emergent vegetation are usually present. Percent sun exposure is typically high. Spotted frogs usually breed in fish-free habitats, or in isolated pools adjacent to lakes with fish, when available (Munger et al., 1997a,b; Reaser, 1997a; Hoffman and Pilliod, 1999; Pilliod and Peterson, 2001). However, Columbia spotted frogs can breed successfully in some water bodies with fish (i.e., sites with dense emergent vegetation in the littoral zone). Columbia spotted frogs will use human-created wetlands (Monello and Wright, 1999). Surrounding habitat is quite variable, from arid desert shrub to montane forest. Water temperatures are often well below 18 °C (65 °F; see Turner, 1960, 1962; Stebbins, 1962, 1985; Morris and Tanner, 1969; Hovingh, 1993; Reaser, 1997a; Behler and King, 1998).

B. Eggs.

i. Egg deposition sites. Egg laying occurs in the spring, often under conditions where water freezes at night. This occasionally results in high egg mortality (Licht, 1971; D.S.P., unpublished data). Oviposition sites are normally in the part of the wetland that has the warmest water temperatures, typically associated with shallow water (10–20 cm deep) and high solar radiation (often the north side). Egg masses sometimes become attached to vegetation and/or one another. Egg masses usually become colonized by green algae (possibly *Oophila amblystomatis*), which may increase egg mass temperatures and increase developmental rates (Gilbert, 1944; D.S.P, unpublished data).

ii. Clutch size. Reports of clutch size vary greatly, even within the same population (ranges of 150–2,400, average about 600; Turner, 1957, 1958; Morris and Tanner, 1969; Cuellar, 1994; Werner et al., 1999; Maxell, 2000). Time to hatching ranges from 8–21 d and is influenced by temperature.

C. Larvae/Metamorphosis. Tadpoles are often found in lentic habitats, but may also occupy lotic (flowing) systems where they move up and down stream, grazing on vegetation (especially green algae such as *Spirogyra*) and detritus. Tadpoles typically metamorphose mid to late summer, although they have been observed in Nevada as late as October (Morris and Tanner, 1969; J.K.R., personal observations) and in southwestern Idaho in mid November (J.C. Engle, personal communication). Despite suggestions that spotted frog tadpoles may overwinter at some sites (Logier, 1932; Turner, 1958a), there is no evidence for this in any part of their range.

D. Juvenile Habitat. Juveniles are found in habitats similar to adults (see "Adult Habitat" below; Pilliod et al., 2002), however they are more often found associated with dense emergent and/or floating vegetation (which likely provide cover; Reaser, 1997a). In drying streambeds, juveniles have been found in pooled water under rocks (K. Hatch, personal communication).

E. Adult Habitat. Post- and non-breeding individuals may continuously occupy breeding sites or move into substantially different habitats (e.g., Pilliod et al., 2002). Radio telemetry data suggest that after breeding, adults may move away from breeding sites in order to reduce their risk of predation by garter snakes (*Thamnophis* sp.) that gather to feed on tadpoles and newly metamorphosed animals (M. Hayes and E.L. Bull, personal communication). Columbia spotted frogs frequent pooled or flowing wetlands (ranging in size from lakes to ditches), moist meadows, and forests. Wetlands may be permanent or ephemeral. Floating and/or emergent vegetation is usually present. In some populations, Columbia spotted frogs are found within dense willow clumps or along talus. Basking sites (exposed banks or vegetation mats) are important. This species will use human-created wetlands. Surrounding habitat ranges from mixed coniferous and subalpine forests to arid grass and brushlands (see Turner, 1960; Morris and Tanner, 1969; Stebbins, 1985; Reaser, 1997a; Pilliod, 2001).

F. Home Range Size. Movement is likely greatest in places where predator risk is high, food is scarce (or more unpredictable), or wetlands are ephemeral; these factors likely vary from site to site. Carpenter (1954) and Turner (1960) reported that individually marked Columbia spotted frogs moved within a region of tens to thousands of m², generally returning to the vicinity of initial capture. However, spotted frogs within a study conducted by Hollenbeck (1976) did not exhibit behavior indicative of a restricted home range. Reaser (1997a; unpublished data) found that site fidelity typically was high, however movement patterns varied with extent of rainfall, with animals dispersing more widely (as much as 5 km/yr; Reaser, 1996b) in "wet years." In some populations, individuals annually migrate between breeding, summer foraging, and overwintering sites, covering areas of >20 ha (Pilliod, 2001). In a high elevation basin of Idaho, Pilliod et al. (2002) found that females frequently make longer migrations than males; males appear to remain closer to breeding sites. However, in southwestern Idaho, J.C. Engle (personal communication) observed that males and subadults tend to move greater distances than females.

G. Territories. Columbia spotted frogs are not known to defend occupied sites.

H. Aestivation/Avoiding Desiccation. Skeletochronology (counting of growth rings in bone) reveals that some individuals within Nevada's populations may undergo brief periods of arrested growth during the summer (Reaser, 2000b). However, in this same region, frogs were observed to die under severe drying conditions rather than aestivate (Ross et al., 1999). One radio-tagged frog in southwestern Idaho remained buried in mud for several weeks during mid summer (T. Carrigan, personal communication).

I. Seasonal Migrations. When individual wetlands do not meet all life history requirements, frogs are required to move multiple times within a season in order to reach suitable breeding, foraging, and wintering sites. These movements often occur along wetland (ephemeral or permanent) corridors and sometimes overland (Turner, 1960; Patla and Peterson, 1994; Patla, 1997; Bull and Hayes, 2001; Pilliod et al., 2002). Annual migrations of ≥ 2 km between wintering, breeding, and foraging sites (total round-trip distance) have been observed in some high elevation populations (Pilliod et al., 2002). In Oregon, telemetered spotted frogs migrated up to 560 m from breeding ponds to occupy other ponds and river stretches during the summer (Bull and Hayes, 2001).

J. Torpor (Hibernation). Overwintering occurs in springs, spring-fed water holes, beaver dams, on pond bottoms, and on the bottoms and beneath the undercut banks of permanent streams (Turner, 1960; T. Carrigan, personal communication). During the autumn and early spring, frogs have been observed moving into and out of spring mouths and openings in stream banks. In higher latitudes and upper elevation areas, wintering sites probably are characterized by water that does not freeze due to renewal by flow from underground sources or the inlets and outlets of large lakes (Pilliod et al., 2002). Skeletochronology reveals that Columbia spotted frogs experience an extended period of arrested growth corresponding with the winter season (Reaser, 2000b). However, Columbia spotted frogs are not necessarily overwintering in a torpid state. Radiotelemetry data suggest that Columbia spotted frogs can and do move, even under cap ice during the winter season (Bull and Hayes, 2002).

K. Interspecific Associations/Exclusions. "Healthy" populations are often found in habitats supporting a wide variety of other native aquatic species, including dace and suckers (*Catostomus* sp.). However, introduced char and trout species, such as brook trout (*Salvelinus fontinalis*), cutthroat trout (*Oncorhynchus clarki*), and rainbow trout (*O. mykiss*), have negatively affected the distribution and abundance Columbia spotted frogs (Munger et al., 1997a,b; Pilliod and Peterson, 2000, 2001). Cutthroat trout prey on spotted frog tadpoles and juveniles (Pilliod, 2001), reducing recruitment in lakes with fish (Hoffman and Pilliod, 1999). In southwestern Idaho, spotted frogs occur in streams supporting native redband trout (J.C. Munger, personal communication). There is little evidence that other species of native amphibians compete for habitat with Columbia spotted frogs (Reaser, 1997a). However, in central Idaho and Montana, long-toed salamanders (*Ambystoma macrodactylum*) were found at >80% of wetlands containing Columbia spotted frogs (Werner and Reichel, 1994; O'Siggins, 1995; Munger et al., 1997b; Llewelyn and Peterson, 1998; Pilliod, 2001).

L. Age/Size at Reproductive Maturity. Based on mark-release-recapture studies, Turner (1960) reports that first breeding occurs at 5–6 yr (60 mm TL) for females and 4 yr for males (45 mm TL) in Wyoming. Skeletochronology reveals that males in central Nevada reach reproductive maturity after 1–2 winters (at 35 mm minimum length) and females 1–2 yr later (Reaser, 2000b).

M. Longevity. Turner (1960) estimated (from growth rates) that frogs at a high elevation site in Wyoming reached 12–13 yr for females and 10 yr for males. These estimates are consistent with skeletochronology data from Columbia spotted frog populations at similar elevations in central Idaho (D.S.P, unpublished data). In central Nevada, the oldest female observed via skeletochronology was 7 yr old, while males did not exceed 3 yr (Reaser, 2000b). In southwestern Idaho, the oldest female via skeletochronology was 9 yr (Engle and Munger, 1998).

N. Feeding Behavior. Feeding takes place both day and night (J.K.R. and R.E. Dexter, unpublished data). Spotted frogs appear to be generalists and opportunistic, even capturing and consuming juvenile conspecifics (Pilliod, 1999). Their primary food includes insects, arachnids, and mollusks. For lists of food items see Thompson (1913), Tanner (1931), Schonberger (1945), Turner (1959), Miller (1978), and Whitaker et al. (1983).

O. Predators. A wide variety of birds (especially herons [Ardeidae] and ravens [*Corvus corax*; Turner, 1960], American kestrels [*Falco sparvericus*; Smith et al., 1972], and sandhill cranes [*Grus canadensis*; J.C. Engle, personal communication]) will feed on Columbia spotted frogs. Other predators include fishes (cutthroat trout [Pilliod, 2001]) and snakes (especially garter snakes [*Thamnophis* sp.; Reaser and Dexter, 1996a; J.C. Munger, personal communication]). Mammals, including weasels (*Mustela* sp.), river otters (*Lutra canadensis* [Pilliod, 2001]), and coyotes (*Canis latrans* [Turner, 1960; Hovingh, 1993]) also prey upon spotted frogs.

Tadpoles are particularly susceptible to predation by birds, including robins *(Turdus migratorius)*, Brewer's blackbirds *(Euphagus cyanocephalus)*, and ravens (D.A. Patla, personal communication); Clark's nutcrackers (*Nucifraga columbiana* [Pilliod, 2002]); and belted kingfishers (*Megaceryle alcyon* [Licht, 1974]). Other tadpole predators include aquatic insects (e.g., predacious diving beetles [Coleoptera] and water bugs [Hemiptera]), macrophagous leeches (Hovingh, 1993), and snakes (J.K.R., personal observation). Non-native fishes, including trout, char, and goldfish *(Carasius auratus)* are thought to be potentially important predators of spotted frogs (Ross et al., 1993; Munger et al., 1997a,b; Reaser, 1997a; Hoffman and Pilliod, 1999; Monello and Wright, 1999; Pilliod and Peterson, 2000, 2001; Bull and Marx, 2002). Invasive crayfish and American bullfrogs *(R. catesbeiana)* might also be aggressive predators on various life history stages of spotted frogs (Reaser, 1997a).

P. Anti-Predator Mechanisms. Spotted frogs are sluggish relative to other ranids, and their best defense against sight-cued predators may be to remain still or descend below the water's surface (J.K.R., unpublished data). The color of spotted frogs may vary with habitat type to reduce visibility. For example, darker brown frogs are found along muddy seeps and lighter green frogs are found in algal mats (J.C. Engle, personal communication). In lakes with predacious fish, adults that are startled by researchers are more likely to return immediately to shore as compared with frogs in fishless lakes (D.S.P, personal communication). When captured by snakes, spotted frogs will often twist and turn while thrashing wildly (Reaser and Dexter, 1996a). Occasionally, captured adults utter a "deep clicking" release call. A similar behavior is often observed when they are placed in zip-lock bags. Subadult frogs captured by predacious water bugs emit a piercing, high-pitched scream (J.C. Engle, personal communication); juveniles and adults captured by garter snakes have also been observed to scream (M. Hayes, personal communication; J. K. R, personal observation). Remarkably, few frogs scream when handled by researchers (M. Hayes, personal communication; J.K.R., personal observation). Columbia spotted frogs may release a mild skin toxin—the water in zip-lock bags becomes opaque and foamy, and stings mildly when in contact with open cuts. Some people experience skin that is dried, cracked, and irritated after prolonged handling of the species (J.K.R., personal observation). Heavy metals may decrease the fright response of spotted frog tadpoles to predators in aquatic habitats contaminated by mine tailings (Lefcort et al., 1998).

Q. Diseases. Sickly Columbia spotted frogs have been documented in central Nevada. Observations include emaciation, skin and cornea lesions, ulcerated digits and tarsus, and prolapse of urinary bladder (J.K.R. and D. E. Green, personal observations). Similar symptoms (e.g., emaciation, cornea lesions and blindness, prolapsed bladder) were observed in several individuals in central Idaho and were termed the "wasting disease" by researchers (D.S.P., personal observation). During ice melt-off at a high elevation pond (ca. 2,000 m [6,100 ft]) in the Wallowa Mountains of northeastern Oregon, >20% of frogs in a 150+ frog sample had a fungal brush-border on the eyelids and/or loss of integrity of the skin of the feet that resulted in spontaneous bleeding with minor handling (M. Hayes and E.L. Bull, personal observation). A large number of the frogs exhibiting these conditions were recaptured 2 mo later; none showed signs of any of the previous conditions. It was surmised that the low dissolved oxygen levels beneath the ice stressed the frogs, allowing fungal establishment. In Nevada, spotted frogs often exhibit an inflammation response to toe clipping (Reaser and Dexter, 1996b; J.K.R., D.L.Drake, and M.A. Hagerty, unpublished data). The presence of a chytrid fungus has been confirmed in sickly frogs at one pond and suspected at several others in the Owyhees of southwestern Idaho (J.C. Munger, personal communication; populations apparently stable at present) and in the Heber Valley population of Utah (Green and Converse, 2002; Green and Sohn, 2002).

R. Parasites. Parasites previously reported from Columbia spotted frogs include *Spironoura pretiosa* (Ingles, 1936, may have been *Rana cascadae*; J.K.R., unpublished data), *Gorgoderina multilobata* (Ingles, 1936), *Haplometrana utahensis*, *Gorgoderina tanneri* (Olsen, 1937), and *Aplectana gigantica* (Olsen, 1938). "Ascarid parasitism" was reported by Schonberger (1945). Turner (1958b) reports that of 62 hosts examined, 32 were parasitized by *Spironoura pretiosa*, 30 by lung flukes (*Haematolechus similiplexus* and an unknown species of this genus), and 10 by an undetermined species of *Glypthelminus*. *Gorgoderina tanneri* infected 8 of 27 frogs examined, and an undetermined species of *Halipegus* was found in 4 of 46 hosts. No cestodes were found. There were no differences in infections of males versus females. Three trematodes, including *Halipegus occidualis*, *Haematoloechus varioplexus*, and *Haplometrana intestinalis*, were found in 14–51% of 59 spotted frogs examined from five ponds in northern Idaho (Russell and Wallace, 1992). Large adult hosts had the largest infections. Upon necropsy of a sickly specimen from Nevada, D. E. Green (personal communication, 1996) identified a chronic *Haplipegus* infection of one ear and noted an accumulation of inflammatory cells, fluids, and fluke eggs within the middle ear. Various lesions (skin, ears, lungs, esophagus, stomach, intestines, and urinary bladder) in this animal appeared to have been caused by a variety of parasites (species undetermined). Columbia spotted frogs captured in southwestern Idaho frequently have leeches attached; one frog in 1998 had 14 small leeches (J.C. Engle, personal communication). A few specimens in Utah have been found with the leech *Helobdella stagnalis* attached to the ventral surface (Ross and Richardson, 1995). This is the third report of this leech species attached to vertebrates, but without further gastric analysis, it is unknown if this leech is actually feeding on the vertebrates (P. Hovingh, personal communication). Johnson et al. (2002) examined Columbia spotted frogs from the Owyhees in Idaho and found low rates of infection by trematodes (*Ribeiroia ondatrae*) and few individuals with limb deformities.

4. Conservation.

Columbia spotted frogs are one of the many species of amphibians in the western United States experiencing declines. Due to documented rarity, loss of populations, and increasing land-use pressures, Columbia spotted frogs of the Great Basin (Nevada and southern Idaho), Wasatch Front (Utah), and West Desert (Utah) were declared warranted (but precluded by work on other species having higher priority for listing) of protection under the Federal Endangered Species Act (Worthing, 1993). It is important to note, however, that this finding pre-dated the availability of molecular data on the distinct populations of *R. luteiventris*, and, thus, may have underestimated the distinctiveness and vulnerability of some populations. For example, isolated, genetically distinct populations in the Bighorn Mountains of Wyoming have yet to be considered in conservation efforts (Bos and Sites, 2001). It is also possible that efforts to conserve the species since 1993 have improved the status of some populations. For example, according to a report released by the U.S. Fish and Wildlife Service (USFWS, 2002d), habitat protection and conservation measures have minimized or removed potential threats such as urbanization, predation, and water depletion as stressors of Utah's Heber Valley population (the largest and best protected on the Wasatch Front).

The "heritage rank status" for the *R. luteiventris* was last assigned in July 1997 and took into consideration the molecular studies of Green et. al. (1996, 1997). The results are as follows: Global Heritage Status—G4; Rounded Global Heritage Status Rank—G4; U.S. National Heritage Rank—N4 (see www.natureserve.org).

Pooled water with strong sun exposure, springs, and floating vegetation (and/or other basking sites) are critical components for persistence of Columbia spotted frogs. Any stochastic event or land-use activity (e.g., grazing, removal of beaver and destruction of dams, fragmentation of migration corridors by road building, spring development and water diversion, climate change) that has a negative impact on these landscape elements, or access to them, poses a threat to Columbia spotted frog populations (e.g., Reaser, 1997; Patla and Peterson, 1999; Bull and Hayes, 2000; Maxell, 2000; Engle, 2001). Columbia spotted frogs are also susceptible to displacement or predation by introduced species (e.g., Werner et al., 1998; Pilliod and Peterson, 2001; Bull and Marx, 2002), disease from pathogens and parasites (e.g., Turner, 1958b; Johnson et al., 2002), and chemical contaminants (e.g., Kirk, 1988; Lefcort et al., 1998). Studies conducted by Blaustein et al. (1999) suggest that, at least at the embryonic stage, Columbia spotted frog populations are not presently being limited by UV-B radiation.

In the last decade, the plight of Columbia spotted frogs has received the attention of many biologists, and a considerable amount of work has gone into learning about their ecology and evolution, documenting their current distribution and status, and developing management and monitoring plans. In addition, two habitat restoration projects show promising results for spotted frog conservation. The first project, undertaken by the Bureau of Reclamation in Utah, used the best available information on spotted frog habitat requirements to design and construct wetlands for Columbia spotted frogs as part of a habitat restoration and mitigation project along the Provo River (Ammon et al., 2003). Within 5 yr of construction, approximately 90% of the 17 original constructed wetlands had been colonized and were being used for breeding by Columbia spotted frogs. A population repatriated via egg masses began to reproduce after 2 yr. Approximately 70 additional wetlands have since been constructed and are being monitored for frog use (E. Ammon, personal communication). The second notable conservation project is being conducted on Stoneman Creek in the Owyhee Mountains of southwestern Idaho, where the elimination of beaver had resulted in the loss of breeding habitat for Columbia spotted frogs (J.C. Munger, personal communication). Coincident with the erosion of an inactive beaver dam, a population of >100 frogs declined to near zero as water tables dropped and oxbows dried. As a restoration effort for spotted frogs, the eroded breech in the dam was repaired and beaver were introduced. One year later, 6 adult and >200 juvenile spotted frogs were captured in the slack water and oxbows above the dam, and beaver had created several more dams upstream (J.C. Munger, personal communication).

Acknowledgments. We thank Elizabeth Ammon, David Bos, Evelyn Bull, Tim Carrigan, Janice Engle, Kent Hatch, Marc Hayes, Peter Hovingh, Jim Munger, Deb Patla, Chuck Peterson, David Ross, and Jack Sites for their contributions to this account and their ongoing dedication to the conservation of the Columbia spotted frog. Support for DSP was provided by USGS Amphibian Research and Monitoring Initiative during this project.

Rana muscosa Camp, 1917b
MOUNTAIN YELLOW-LEGGED FROG

Vance Vredenburg, Gary M. Fellers, Carlos Davidson

1. Historical versus Current Distribution.
Mountain yellow-legged frogs (*Rana muscosa*) are endemic to two disjunct areas: (1) the Sierra Nevada in California and Nevada, and (2) the San Gabriel, San Bernardino, and San Jacinto Mountains in southern California. This species was originally described as two subspecies of *R. boylii* (*R. boylii muscosa* and *R. boylii sierrae*; Camp, 1917b). On the basis of morphological data, the two subspecies were separated from *R. boylii* and raised to the species level (*R. muscosa*; Zweifel, 1955). Recent molecular data (Macey et al., 2001) suggest that there may be large differences between the frogs in these two disjunct areas; therefore, we will refer to Sierran populations and southern California populations separately whenever possible.

In the Sierran range, mountain yellow-legged frogs occur from near Antelope Lake (northern Plumas County; (G.M.F., in preparation), south 490 km to Taylor Meadow (southern Tulare County; Zweifel, 1955) and range from 1,370–3,660 m (Camp, 1917b; Grinnell and Camp, 1917b; Storer, 1925; Zweifel, 1955; Mullally and Cunningham, 1956). They are not known from east of the Sierra Crest except for two regions: in the vicinity of Mt. Rose, near Lake Tahoe (Zweifel, 1955) where populations are now extirpated, and in the Glass Mountains south of Mono Lake, where there are two small populations. Currently, mountain yellow-legged frogs are found scattered throughout nearly all their historical range in Sierra Nevada, but the number of populations is greatly reduced. This is most notable in the northernmost 125 km of the range (north of Lake Tahoe) and the southernmost 50 km, where only a few populations have been found in the last few years (see "Historical versus Current Abundance" below; Jennings and Hayes, 1994a; (G.M.F., in preparation).

In southern California, the historical range included the San Gabriel, San Bernardino, and San Jacinto Mountains with an isolated population at Mt. Palomar (in northern San Diego County; Camp, 1917b) and ranged from 300–2,300 m (Camp, 1917b; Grinnell and Camp, 1917; Storer, 1925; Zweifel, 1955). The frogs were thought to have gone extinct in the San Bernardino Mountains (none were

Mountain Yellow-legged Frog (*Rana muscosa*)

found between 1970–93; Jennings and Hayes, 1994a), but a small population was recently discovered (M. Jennings, personal communication). In the southern California portion of their range, nearly all populations of mountain yellow-legged frogs have disappeared (Jennings and Hayes, 1994a; see below) and, given their status as a distinct vertebrate population segment, were given Endangered species as of 1 August 2002 (USFWS, 2002b).

2. Historical versus Current Abundance.
A large number of Sierran populations have disappeared, but the extent of decline is unclear due to the lack of systematic surveys (Jennings and Hayes, 1994a). Between 1989–93, Bradford et al. (1994a) resurveyed mountain yellow-legged frog "historical sites" (documented between 1959–79). In the western portion of Sequoia National Park (Kaweah River drainage), they resurveyed 27 historical sites and found no frogs at any of these locations (Bradford et al., 1994a). Elsewhere in Sequoia and Kings Canyon National Parks (Kern, Kings, and San Joaquin River drainages), they resurveyed 22 historical sites and only 11 contained frogs (Bradford et al., 1994a). Beginning just north of Kings Canyon National Park and running up into Yosemite National Park, they resurveyed 24 historical sites and found frogs present at only 3 sites (Bradford et al., 1994a). In another study, Drost and Fellers (1996) compared the presence of Sierran mountain yellow-legged frogs at historical sites (surveyed in 1915 by Grinnell and Storer) to distributions in 1995. Grinnell and Storer (1924) stated that "the yellow-legged frog is the commonest amphibian in most parts of the Yosemite section." Drost and Fellers (1996) report finding frogs in only 2 of 14 historical sites (a single tadpole at one site, and an adult female at another). If we combine the data from the two resurvey studies in the Sierra Nevada (Bradford et al., 1994a; Drost and Fellers, 1996), there are 86 historical sites (data from 1915–59), and only 16 contained frogs when they were revisited between 1989–95. At the northernmost and southernmost part of the Sierran range (Butte and Plumas counties in the north, and Tulare County in the south), few populations have been seen since 1970 (Jennings and Hayes, 1994a; (G.M.F., in preparation).

In southern California, mountain yellow-legged frog populations have declined nearly to extinction (Jennings and Hayes, 1994a; Stebbins and Cohen, 1995; Drost and Fellers, 1996; USFWS, 2002b). With only 6–8 extant populations, the largest having fewer than 100 adults (M. Jennings, personal communication), the situation is tenuous at best for mountain yellow-legged frogs in southern California.

3. Life History Features.
A. Breeding. Reproduction is aquatic.

i. Breeding migrations. Sierran mountain yellow-legged frogs do not have a distinct breeding migration because adults spend most of their time in the vicinity of suitable breeding habitat (Zweifel, 1955; Bradford, 1983). In some areas, there is a seasonal movement from lakes that are more favorable for overwintering (e.g., deeper water, see "Torpor" below) to nearby areas that are more favorable for breeding. Frogs typically move only a few hundred meters, but distances up to 1 km have been observed (V. V., in preparation). Breeding activity begins early in the spring, soon after ice-melt, and can range from April at lower elevations to June–July in higher elevations (Wright and Wright, 1933; Stebbins, 1951; Zweifel, 1955). The timing of the onset of breeding is dependent upon the amount of snowfall and subsequent ice-out dates of ponds, lakes, and streams (V. V., in preparation).

Almost no data exist on dispersal of juvenile mountain yellow-legged frogs from either Sierran populations (Bradford, 1991) or southern California. In the Sierra Nevada, juveniles have been observed in small intermittent streams and may have been dispersing to permanent water (Bradford, 1991). Although it has been reported that frogs avoid crossing even short distances of dry ground (Mullally and Cunningham, 1956), we have documented overland movement of Sierran frogs well away from water (≤400 m; Vredenburg and Fellers, in preparation).

ii. Breeding habitat. Southern California populations are almost exclusively stream-dwelling, perhaps reflecting the general lack of ponds and lakes in the area. It is not known whether southern California populations overwinter as tadpoles before metamorphosis, therefore, the restriction for breeding habitat to deeper, more permanent bodies of water (or habitats connected to them) may not apply to these populations.

B. Eggs.

i. Egg deposition sites. In Sierran populations, eggs are deposited underwater in clusters under banks or attached to rocks, gravel, or vegetation in streams or lakes (Wright and Wright, 1949; Stebbins, 1951; Zweifel, 1955). In the Sierra Nevada, tadpoles overwinter at least once before metamorphosis (Bradford, 1983; see below). Therefore, egg-laying sites must either be in or connected to lakes and ponds that do not dry in the summer and are sufficiently deep that they do not freeze to the bottom in the winter (>1.7 m deep, and preferably >2.5 m deep; Bradford, 1983; see below). Southern California populations are almost exclusively stream dwelling, perhaps reflecting the general lack of ponds in the area. It is not known whether southern California populations overwinter as tadpoles before metamorphosis, therefore the

restriction for breeding habitat to deeper, more permanent bodies of water (or habitats connected to them) may not apply to these populations.

ii. Clutch size. Livezey and Wright (1945) report an average of 233 eggs/egg mass (n = 6, range 100–350) for Sierran frogs, but we have observed egg masses with as few as 15 eggs, so the average is probably less (V. V., in preparation). In laboratory breeding experiments on Sierran frogs, egg hatching times ranged from 18–21 d at temperatures ranging from 5–13.5 °C (Zweifel, 1955).

C. Larvae/Metamorphosis.

i. Length of larval stage. The length of the larval stage depends upon the elevation. At lower elevations where the summers are longer, tadpoles are thought to be able to grow to metamorphosis in a single season (Storer, 1925). However, throughout most of their range in the Sierra Nevada, populations are clearly composed of tadpoles of three size classes that may correspond to year classes (G.M.F., in preparation). Hence, metamorphosis would occur at the end of the third summer when the tadpoles are 2.5 yr old (Wright and Wright, 1933; Zweifel, 1955). At higher elevations (>2,500 m) or after winters where ponds and lakes remain ice-covered for ≥9 mo, we believe that tadpoles may not metamorphose until the end of their fourth summer. There is no information on the length of the larval stage in southern California populations.

ii. Larval requirements.

a. Food. Not reported, but tadpoles are most likely herbivorous and detritivorous.

b. Cover. Tadpoles burrow in mud, under rocks, under banks, or into submerged vegetation (Stebbins, 1951). Before the spring overturn (thermal mixing in lakes), Sierran tadpoles remain in the warmer water below the thermocline. After mixing occurs, they move each day from the deeper water where they take refuge at night to warm shallow areas near shore where they aggregate in large numbers (Bradford, 1984). There are no reports from southern California populations.

iii. Larval polymorphisms. None.

iv. Features of metamorphosis. Tadpoles transform in July and August (22–27 mm SVL; Wright and Wright, 1933).

v. Post-metamorphic migrations. Not reported.

D. Juvenile Habitat. Believed to be similar to adults.

E. Adult Habitat. In the Sierra Nevada, adult mountain yellow-legged frogs occupy wet meadows, streams, and lakes; adults typically are found sitting on rocks along the shoreline, usually where there is little or no vegetation (Wright and Wright, 1933). In southern California, mountain yellow-legged frogs occupy streams in narrow rock-walled canyons (Grinnell and Camp, 1917) and streams in the chaparral belt (Zweifel, 1955).

In the Sierra Nevada, most frogs are seen on a wet substrate within 1 m of the water's edge. Both adults and larvae are found most frequently in areas with shallow water, in part because these were the warmest areas (Bradford, 1984). Bradford (1984) reported seeing aggregations of Sierran frogs of up to 58 individuals. Aggregations occurred in the early afternoon in situations where the head and back of individual frogs were in full sunlight.

F. Home Range Size. Using radio tracking, home range size was estimated for Sierran mountain yellow-legged frogs in Kings Canyon National Park (Matthews and Pope, 1999). They used adaptive kernal 90% contours to estimate home range separately by month for August, September, and October 1998. In August, ranges for ten females varied from 19.4–1,028 m^2 (x = 385.5 m2; s.e. = 113.4 m2). In September, ranges for seven females varied from 53–9,807 m^2 (x = 5,099 m^2; s.e. = 1,506 m^2), and one female was calculated at 6,990 m^2. In October, the calculated home range for one female was 3.2 m^2; for two males the values were 73 m^2 and 82 m^2. This study was conducted in an area dominated by introduced trout (Matthews and Pope, 1999); it is not known if these calculations reflect mountain yellow-legged frog movements in natural conditions (those that lack all fish predators).

G. Territories. Unknown, but other ranid frogs are well known to defend breeding areas (Wells, 1977). In the Sierra Nevada, where the largest populations remain, aggregations of 2–15 adult frogs (with a maximum of 58) can be seen sunning on warm days (Bradford, 1984; Fellers, in preparation).

H. Aestivation/Avoiding Desiccation. None.

I. Seasonal Migrations. See "Breeding migrations" above and Pope and Matthews (2001).

J. Torpor (Hibernation). As the temperatures drop to freezing or below (generally October–November), frogs become inactive for the winter (Zweifel, 1955; Bradford, 1983). Sierran frogs apparently spend the winter at the bottom of lakes or in rocky streams (V.V. and G.M.F., in preparation). One study (Matthews and Pope, 1999) reports Sierran frogs overwintering in rock crevices, but this behavior may be in response to the presence of introduced fish (see "Interspecific Associations/Exclusions" below).

Because most Sierran mountain yellow-legged frogs overwinter in lakes, they require lakes that do not freeze to the bottom (>1.7 m deep and preferably >2.5 m deep; Bradford, 1983). In the Sierra Nevada, tadpoles seem better able to survive long winters than juvenile and adult frogs (Bradford, 1983). In 1978, winterkill was responsible for the mortality of all but one Sierran adult in 21 of 26 lakes, while

tadpoles survived in all 26 lakes (Bradford, 1983). In laboratory studies, Bradford (1983) confirmed that Sierran mountain yellow-legged tadpoles have a greater tolerance of hypoxia and a reduced consumption of energy and oxygen during hibernation when compared to metamorphosed individuals (juveniles and adults).

In the Sierra Nevada, adults emerge as soon as the ponds and lakes begin to thaw and ice is clear from at least part of the water surface. As with Yosemite toads (*Bufo canorus*), adults sometimes travel over snow to reach preferred breeding sites early in the season. In years with particularly cold winters, high elevation populations (> 3,000 m) of Sierran mountain yellow-legged frogs may only be active for 90 d during the warmest part of the summer (V.V., personal observations).

K. Interspecific Associations/Exclusions. The native habitat for Sierran populations of mountain yellow-legged frogs is almost entirely outside the range of native fish (Knapp, 1996). This is largely due to the presence of impassable waterfalls in nearly all the Sierran drainages (from past glaciation and uplift processes). Beginning in the late 1800s, trout were introduced to most permanent bodies of water throughout the Sierra Nevada (Knapp, 1996). Accounts of introduced trout eating Sierran mountain yellow-legged frogs go back many years (Grinnell and Storer, 1924; Needham and Vestal, 1938), yet introductions of trout continue throughout much of the Sierran Range (Knapp, 1996). Sierran mountain yellow-legged frogs are found in substantially lower densities in ponds and lakes with trout compared with similar habitats that lack fish (Grinnell and Storer, 1924; Bradford, 1989), and those places of co-occurrence probably represent sink populations of frogs (Bradford, 1989; V.V., in preparation). A recent survey of Sierra Nevada sites comparing 669 bodies of water in U.S. Forest Service land (where fish are still routinely introduced) to 1,059 in National Park Service land (where fish introductions ceased in 1977) showed a dramatic difference in the occurrence of mountain yellow-legged frogs in the two areas (3% and 20%, respectively; Knapp and Matthews, 2000).

Little is known about the effect of introduced trout on southern California populations of mountain yellow-legged frogs. It is clear that trout are playing a substantial role in the declines of Sierran populations; however, there may be multiple explanations. For example, frogs from an entire Sierra Nevada watershed went extirpated in an area where trout were never introduced (27 locations; Bradford, 1991).

In a single drainage in the northernmost part of the Sierran range, mountain yellow-legged frogs historically occurred sympatrically with Cascade frogs (*R. cascadae*) and foothill yellow-legged frogs (*R.

boylii*; Zweifel, 1955). More recently, Stebbins and Cohen (1995) report that all three species of frogs have disappeared from that area; there are now no known sites where all three of these congeners occur together.

In southern California, there are historical accounts of overlap between foothill and mountain yellow-legged frogs along 1.6 km (1 mi) of the North Fork San Gabriel River in Los Angeles County (Zweifel, 1955), but they have both since disappeared from that area (Jennings and Hayes, 1994a). Sierran and southern California mountain yellow-legged frogs continue to co-occur frequently with Pacific treefrogs (*Pseudacris regilla*; Stebbins, 1985). Rarely, Sierran mountain yellow-legged frogs co-occur with Yosemite toads, western toads (*B. boreas*; Stebbins, 1985), and long-toed salamanders (*Ambystoma macrodactylum*).

L. Age/Size at Reproductive Maturity. Females reach sexual maturity at 45–50 mm SVL; males mature at a smaller size (Zweifel, 1955). There are few data on age at reproduction, however Zweifel (1955) reports age at first reproduction at 3 yr following metamorphosis. Three or 4 yr after metamorphosis for high elevation populations seems like a reasonable estimate, but studies of known-age individuals are needed to be certain.

M. Longevity. Unknown.

N. Feeding Behavior. In southern California, mountain yellow-legged frogs prey on a wide variety of invertebrates including beetles (Coleoptera), ants, bees, wasps (Hymenoptera), flies (Diptera), true-bugs (Hemiptera), and dragonflies (Odonata; Long, 1970). This is probably true of Sierran frogs, but there are no reports. Sierran frogs have been observed eating Yosemite toad tadpoles (Mullally, 1953), and Pacific treefrog tadpoles (Pope, 1999). There is one report of Sierran mountain yellow-leg cannibalism—tadpoles eating thousands of conspecific eggs (V.V., in review). In addition, these tadpoles have been seen feeding on carcasses of dead metamorphosed mountain yellow-legged frogs (V.V., in preparation). Pope and Matthews (2002) address the influence of prey on mountain yellow-legged frog condition and distribution.

O. Predators. Native predators of mountain yellow-legged frogs include western terrestrial garter snakes (*Thamnophis elegans*; Grinnell and Storer, 1924; Mullally and Cunningham, 1956; Jennings et al., 1992), Brewer's blackbirds (*Euphagus cyanocephalus*, in Sierran populations; Bradford, 1991), Clark's nutcrackers (*Nucifraga columbiana*, in Sierran populations; Camp, 1917b), and coyotes (*Canis latrans*; in the Sierra Nevada; Moore, 1929). There are two anecdotal reports of black bears (*Ursus americanus*) feeding on these frogs (Sierran populations; G.M.F., personal communication). Garter snakes in the Sierra

Nevada feed extensively on these frogs and commonly are found near large numbers of tadpoles (Jennings et al., 1992). Introduced trout (rainbow trout [*Oncorhynchus mykiss*], golden trout [*O. aguabonita*], brook trout [*Salvelinus fontinalis*], and brown trout [*Salmo trutta*]) have been observed to prey on Sierran mountain yellow-legged frogs (see above; Grinnell and Storer, 1924; Needham and Vestall, 1938; Knapp, 1996).

P. Anti-Predator Mechanisms. When alarmed, adults dive into streams, kick up silt with their hind legs, and bury themselves into the mud (southern California, Camp, 1917b). Similar behaviors have been seen in the Sierra Nevada (Grinnell and Storer, 1924; Wright and Wright, 1933).

Q. Diseases. In a population of Sierran frogs, Bradford (1991) observed a large-scale die-off of frogs that were infected with a bacterium (red-leg disease, *Aeromonas hydrophila*). Recently, a chytridiomycete fungus (likely *Batrachochytrium dendrobatidis*; Longcore et al., 1999) has been identified and cultured from both tadpoles and subadults from the central Sierra Nevada (Fellers et al., 2001). This same chytridiomycete fungus has been implicated in amphibian declines in the rain forests of Australia and Central America (Berger et al., 1998); however, at this time it is not known if this fungus has played a role in the decline of mountain yellow-legged frogs.

R. Parasites. Goodman (1989) described a trematode species from southern California mountain yellow-legged frogs.

4. Conservation.
Currently, mountain yellow-legged frogs are found scattered throughout nearly all their historical range in the Sierra Nevada, but the number of populations is greatly reduced. This is most notable in the northernmost 125 km of the range (north of Lake Tahoe) and the southernmost 50 km, where only a few populations have been found in the last few years (Jennings and Hayes, 1994a; G.M.F., in preparation; see "Historical versus Current Distribution" and "Historical versus Current Abundance" above). If we combine the data from the two resurvey studies in the Sierra Nevada (Bradford et al., 1994a; Drost and Fellers, 1996), of the 86 historical sites (data from 1915–59), only 16 contained frogs when they were revisited between 1989–95.

In southern California, mountain yellow-legged frog populations have declined nearly to extinction (Jennings and Hayes, 1994a; Stebbins and Cohen, 1995; Drost and Fellers, 1996; USFWS, 2002b; see also "Historical versus Current Distribution" and "Historical versus Current Abundance" above). With only six to eight extant populations, the largest having fewer than 100 adults (M. Jennings, personal communication), and given their status as a distinct vertebrate popula-

tion segment, southern California populations were given Endangered species status as of 1 August 2002 (USFWS, 2002b). In a study of displaced mountain yellow-legged frogs, Matthews (2003) concluded that stress due to a homing response may preclude translocation as an effective conservation tool.

Rana okaloosae Moler, 1985
FLORIDA BOG FROG
Paul E. Moler

1. Historical versus Current Distribution.
Florida bog frogs (*Rana okaloosae*) are locally distributed in Walton, Okaloosa, and Santa Rosa counties, Florida (Moler, 1985, 1993). They are known to occur only along small streams draining to Titi Creek, the East Bay River, or the lower Yellow River, all of which ultimately drain to Escambia Bay. The Titi Creek populations in Walton County, Florida, appear to be isolated by approximately 30 km from populations in the lower Yellow River basin. Most known populations occur on Eglin Air Force Base. Florida bog frogs were unknown prior to 1982, and nothing is known about their historical distribution. This species was almost certainly more widely distributed during glacial periods when sea levels were lower than present, but there is no reason to believe that its historical distribution was substantially greater than present.

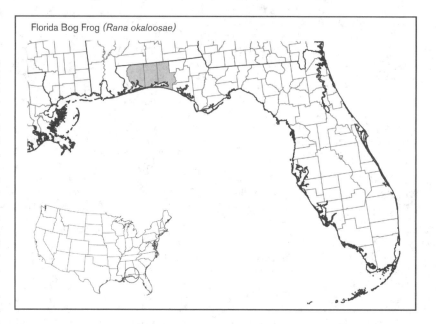

Florida Bog Frog (*Rana okaloosae*)

2. Historical versus Current Abundance.
Rare, but may be common locally. Historical abundance is unknown.

3. Life History Features.
A. Breeding. Reproduction is aquatic.
i. Breeding migrations. Florida bog frogs do not migrate. They breed and permanently reside in the same habitats.

ii. Breeding habitat. Live and breed in shallow, acidic, non-stagnant seeps or seepage-fed streams. Breeding occurs from April–August, with occasional calling heard into early September (Moler, 1985, 1992b).

B. Eggs.
i. Egg deposition sites. Eggs are laid as a film on the surface of quiet pools (Moler, 1992b).

ii. Clutch size. Has not been determined.

C. Larvae/Metamorphosis. At least some tadpoles overwinter (Moler, 1985, 1992b).

D. Juvenile Habitat. Same as adult habitat.

E. Adult Habitat. Adult Florida bog frogs are associated with shallow, non-stagnant, acid (pH 4.1–5.5) seeps and shallow, boggy overflows of seepage fed streams, usually in association with lush beds of sphagnum moss (Moler, 1985).

F. Home Range Size. Unknown.

G. Territories. Unknown.

H. Aestivation/Avoiding Desiccation. Not applicable.

I. Seasonal Migrations. None.

J. Torpor (Hibernation). Unknown.

K. Interspecific Associations/Exclusions. Bronze frogs (*R. clamitans clamitans*) and southern cricket frogs (*Acris gryllus*) are common in habitats occupied by Florida bog frogs (Moler, 1985).

L. Age/Size at Reproductive Maturity. Unknown.

M. Longevity. Unknown.

N. Feeding Behavior. Unknown. At night, frogs have taken moths attracted by photographers' lights.

O. Predators. Unknown. Cottonmouths (*Agkistrodon piscivorous*) and southern water snakes (*Nerodia fasciata*) are common associates, and both likely prey on bog frogs.

P. Anti-Predator Mechanisms. Fleeing; otherwise unknown.

Q. Diseases. Unknown.

R. Parasites. Unknown.

4. Conservation.
Most known populations occur on Eglin Air Force Base. Because Florida bog frogs were unknown prior to 1982, almost nothing is known about their historical distribution. Florida bog frogs are considered a Species of Special Concern in Florida.

Rana onca Cope, 1875(b)
RELICT LEOPARD FROG

David F. Bradford, Randy D. Jennings, Jef R. Jaeger

1. Historical versus Current Distribution.
Our taxonomy follows Jaeger et al. (2001), who ascribe recently extant leopard frog populations along the Virgin River from Littlefield, Arizona, downstream into the Black Canyon of the Colorado River below Hoover Dam in Nevada to relict leopard frogs *(Rana onca)*. Some of these populations were formerly considered to be lowland leopard frogs *(R. yavapaiensis;* Platz, 1988). Populations of extinct leopard frogs from the Las Vegas Valley, Nevada, often synonymized with *R. onca* (see review in Jennings, 1988), are treated in this volume as Vegas Valley leopard frogs *(R. fisheri).*

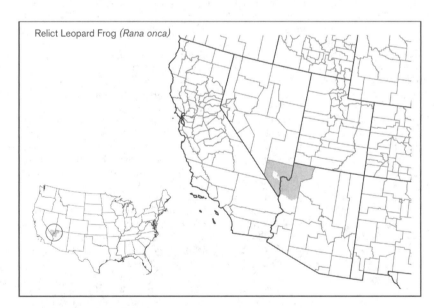

Relict Leopard Frog *(Rana onca)*

The historical distribution of relict leopard frogs is not well documented. This species was first identified from a single individual probably collected in the Virgin River Valley, Washington County, Utah (Cope, 1875, in Tanner, 1929). Their distribution historically has been characterized as springs, streams, and wetlands within the Virgin River drainage from the vicinity of Hurricane, Utah, to the Overton Arm of what is now Lake Mead,

Nevada, and along the Muddy River in Nevada. Jennings and Hayes (1994b) reported that the species was never recorded from "the Colorado River proper," but one observation at a site now inundated by Lake Mead indicates that it probably occurred in marsh habitat adjacent to the river (Cowles and Bogert, 1936). Leopard frogs also were collected from populations within the Black Canyon along the Colorado River, immediately downstream from Lake Mead. Populations in Utah appear to have been extirpated since the 1950s (Platz, 1984). Surveys in the 1990s revealed extant populations at six sites in three general areas: a spring-fed wetland adjacent to the Virgin River near Littlefield, Arizona; Rogers and Blue Point Springs near the Overton Arm of Lake Mead, Nevada; and three springs in Black Canyon below Lake Mead, Nevada (Jennings et al., 1995; Jaeger et al., 2001). Repeated surveys in 2001 and 2002, however, indicate that the Littlefield population has been extirpated, leaving only two general areas with extant populations. An additional population at the smallest site, Corral Spring, near Rogers Spring, was extirpated in 1995 (Bradford et al., 2004).

2. Historical versus Current Abundance.
The former abundance of relict leopard frogs is unknown. The species was once considered to be extinct (Platz, 1984; Jennings, 1988), but this designation resulted in part from taxonomic confusion with lowland leopard frogs (Jaeger et al., 2001). The current abundance of relict leopard frogs is quite low. Counts and limited efforts at mark–recapture estimation suggest that the total number of metamorphosed individuals at all sites in the late 1990s was little more than 1,100 (Bradford et al., 2004).

3. Life History Features.
The life history of relict leopard frogs has not been well studied and much remains unknown. The information presented here is based on our unpublished observations and those of other researchers.

A. Breeding. Reproduction is aquatic.

i. Breeding migrations. Adults reside in and near breeding habitat; breeding migrations are unknown. Time of oviposition varies among sites and years. Observations of eggs and tadpoles imply favored breeding times from approximately February–April and November.

ii. Breeding habitat. Pools or slow-moving side areas of streams, with or without emergent vegetation, appear to be favored breeding sites. Oviposition sites have been in shallow water within a few centimeters of the water surface.

B. Eggs.

i. Egg deposition sites. Clutches are typically attached to vegetation or sticks at the water surface or submerged within a few centimeters of surface.

ii. Clutch size. Clutches consist of a globular cluster of many hundred eggs.

C. Larvae/Metamorphosis. Limited observations suggest that several months are required to reach metamorphosis.

D. Juvenile Habitat. Juveniles have been observed in the same areas as adults, and their habitat characteristics are likely similar to that of adults.

E. Adult Habitat. Adult frogs formerly inhabited permanent streams, springs, and spring-fed wetlands below approximately 600 m (Linsdale, 1940), wetlands adjacent to the Virgin River, and at least one wetland adjacent to the Colorado River at a site now inundated by Lake Mead (Cowles and Bogert, 1936). Observations suggest that adults prefer relatively open shorelines where dense vegetation does not dominate. The three general areas where frogs remained in the 1990s are quite different in many features. The Littlefield site is a small, marshy wetland formed by a spring near the shore of the Virgin River. Frogs there were found mostly near the spring source. The sites around the Overton Arm of Lake Mead are fast-moving streams formed by geothermal springs. The stream channels are cut into the gypsiferous soil and are mostly overgrown with dense stands of emergent vegetation. Habitats in the Black Canyon are geothermal springs that flow over more rocky substrates where mesquite and *Tamarix* sp. dominate the over-story vegetation, where present. Water sources for all five of the sites where frogs remain are geothermally influenced, with relatively constant water temperatures between 30–55 °C (Pohlmann et al., 1998).

F. Home Range Size. Adult frogs are usually found at the water's edge and occasionally in adjacent low riparian vegetation or a few meters beyond. In a 3-yr mark-recapture study of an isolated

550-m stream reach at Blue Point Spring, the mean distance moved between captures averaged 18 m, and the longest distance recorded between recaptures was 120 m (Bradford et al., 2004).

G. Territories. Unknown.

H. Aestivation/Avoiding Desiccation. Species is found only in permanently wet sites, and individuals show no evidence of diminished activity during summer or dry weather periods.

I. Seasonal Migrations. None evident.

J. Torpor (Hibernation). There is no evidence that adults enter refugia or become inactive during part of the year. However, adults are more difficult to find during the coldest months (December–January), even in warm, geothermally influenced waters.

K. Interspecific Associations/Exclusions. Relict leopard frogs coexist with red-spotted toads (*Bufo punctatus*), Woodhouse's toads (*B. woodhousii*), Pacific treefrogs (*Pseudacris regilla*), and (at least historically) Arizona toads (*B. microscaphus*). Relict leopard frogs will also coexist with exotic fishes (mostly small and tropical) in Blue Point and Rogers Springs (Courtenay and Deacon, 1983). Exotic American bullfrogs (*R. catesbeiana*), crayfish, and game-fishes inhabit areas where relict leopard frogs have been extirpated, suggesting that these introduced species may have been a factor in population extinctions (Jennings and Hayes, 1994b).

L. Age/Size at Reproductive Maturity. Unknown for females. Males reach reproductive maturity at approximately 42 mm SVL, based on the appearance of pigmented thumb pads. At Blue Point Spring, this size appears to be reached during the first year because smaller individuals are rarely found a year after oviposition.

M. Longevity. In a 4-yr mark-recapture study at one site, population turnover was relatively high, with survivorship of adults averaging 0.27/yr (Bradford et al., 2004). Longevity may be greater for a few individuals; a marked individual at a different site was at least 4 yr old.

N. Feeding Behavior. Diet unknown; presumably similar to other *Rana* species.

O. Predators. Predators not documented. Probably eaten by American bullfrogs, where the two species co-occur (recently, Virgin River near Littlefield; historically, in Utah and along the Muddy River, Nevada), and western terrestrial garter snakes (*Thamnophis elegans*) in portions of Utah.

P. Anti-Predator Mechanisms. Frogs are drab in color and cryptic in appearance. When approached, individuals typically remain motionless until flight to water, where they often seek shelter among submerged vegetation or under rocks. During daylight, individuals are found in less conspicuous locations than at night.

Q. Diseases. Unknown.

R. Parasites. Unknown.

4. Conservation.
The extinction of relict leopard frog populations throughout the species' range occurred concomitantly with the elimination or dramatic alteration of aquatic habitat due to agriculture, marsh draining, and water development. The introduction and spread of American bullfrogs, crayfish, and predaceous fishes have also been implicated in population declines (Jennings and Hayes, 1994b).

At present, the remaining populations appear to be vulnerable to extinction; in May 2002, this species was petitioned by the Center for Biodiversity and the Southern Utah Wilderness Alliance to the U.S. Fish and Wildlife Service for listing under the U.S. Endangered Species Act. *Rana onca* currently is designated by the U.S. Fish and Wildlife Service as a federal candidate for listing, which means that sufficient information exists to list it but this action is precluded by higher priorities. The sites occupied by extant populations are small, population sizes are small, and population turnover appears to be relatively rapid. Moreover, remaining populations occur in only a few sites that are largely isolated from one another. Historically, relict leopard frogs were probably distributed in patches of suitable habitat throughout the regional river system. Extinction and subsequent recolonization of patches may have been an important and dynamic process in maintaining long-term population viability. Currently, the river system in the historical range of relict leopard frogs now forms Lake Mead and Lake Mojave or is otherwise degraded and occupied by American bullfrogs and other exotic species. Consequently, the opportunities for colonization of new sites, recolonization of former sites, and dispersal among the two areas containing extant populations appears to be largely precluded.

Specific potential threats to the few remaining populations include the spread of American bullfrogs and other exotic species into remaining sites, the spread of disease (Carey et al., 1999, Daszak et al., 1999), and the possibility of short-term demographic declines causing extinctions in small populations. Also, vegetation encroachment may be reducing the quality of remaining habitats because of changes in disturbance regimes and/or the establishment of exotic vegetation. Unfortunately, critical information on habitat characteristics is not known. These frogs appear to favor habitats that are somewhat disturbed with reduced vegetation cover. Thus, control of disturbance factors such as burro grazing may actually result in population declines. On the other hand, dense vegetation may be a critical factor in keeping bullfrogs from colonizing some remaining sites. All the remaining five sites occupied by relict leopard frogs are within Lake Mead National Recreation Area. Developing meaningful management options for these populations obligates the need for a clear understanding of habitat characteristics and factors that have maintained the viability of the remnant populations while all others have gone extinct.

Rana palustris LeConte, 1825
PICKEREL FROG

Michael Redmer

1. Historical versus Current Distribution.
Pickerel frogs (*Rana palustris*) are distributed from Nova Scotia and New Brunswick, west through southern Québec and Ontario, to Michigan, Wisconsin, and extreme southeastern Minnesota. From southeast Minnesota south along the Mississippi Drainage (west to extreme southeast Kansas and eastern Oklahoma, but absent from interior prairies of Illinois) to the Gulf Coast in eastern Texas. Absent from most of the eastern Gulf Coastal Plain (except Conecuh County, Alabama, and Escambia County, Florida). In the eastern United States, they are generally distributed from New England south to South Carolina and Georgia, where they are apparently absent from the Atlantic Coastal Plain (Smith, 1961; Schaaf and Smith, 1971; Pace, 1974; Mount, 1975; Ashton and Ashton, 1988; Conant and Collins, 1991).

One subspecies, *R. p. mansuetti*, was applied to populations along the Atlantic Coastal Plain by Hardy (1964), who noted selection pressures of floodplain swamps probably affect phenotype. That subspecies was synonymized by Schaaf and Smith (1970), who noted that the phenotype was not geographically restricted.

The similarity to northern leopard frogs (*R. pipiens*, Schaaf and Smith, 1970; Pace, 1974; Hunter et al., 1992; Redmer and Mierzwa, 1994) has lead to erroneous records. Around the peripheries of their range, pickerel frog populations may be localized in suitable habitat, and special efforts may be needed to detect new localities or to relocate them at or near historical localities (Redmer, 1998b). For example, during surveys in Illinois, pickerel frogs were found in 10 (about 59%) of 17 counties with previous historical records, but five new county records also were reported (Walley, 1991; Redmer and Mierzwa, 1994; Redmer and Ballard, 1995; Redmer et al., 1995; Redmer, 1998b; Petzing et al., 1998). They may be extirpated in the urban Chicago region and at a locality in extreme southern Illinois (Mierzwa, 1998a; Redmer, 1998b). Vogt (1981) reported only 15 active localities in Wisconsin, but Johnson (1984) identified 61 active localities (a 407% increase) in the state.

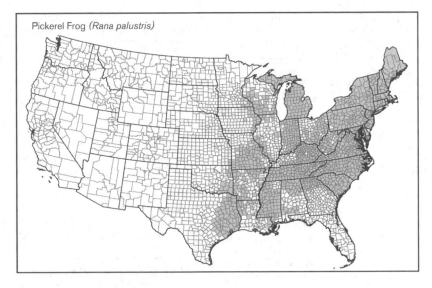

Pickerel Frog (*Rana palustris*)

2. Historical versus Current Abundance.
Unknown. It is unclear whether pickerel frogs have suffered range-wide decline, though their somewhat specialized habitat characteristics and intolerance of pollution may make them vulnerable to human activities (Redmer and Mierzwa, 1994; Harding, 1997). Because pickerel frogs are sometimes difficult to detect (Mossman et al., 1998; Redmer, 1998b), periodic species-specific surveys may be needed to monitor populations.

3. Life History Features.
   A. Breeding. Reproduction is aquatic.
   i. Breeding migrations. Timing and length of breeding season follow a south–north cline. The reported breeding season ranges from December–May in the south (Mount, 1975; Garrett and Barker, 1987; Dundee and Rossman, 1989; Trauth et al., 1990; Hardy and Raymond, 1991), March–May in the middle of their range (Walker, 1946; Minton, 1972; Vogt, 1981; Johnson, 1984; Green and Pauley, 1987; Johnson, 1987; Shaffer, 1991; Redmer and Mierzwa, 1994; Redmer, 1998b), and May–June in the north (Hunter et al., 1992; Oldfield and Moriarty, 1994).

   Immigration takes place after one or a combination of the following occurs: water temperatures reach 7–18 °C (Wright, 1914; Moore, 1939; Pope, 1944; Johnson, 1984; Hardy and Raymond, 1991), air reaches 10–26 °C (Wright, 1914; Pope, 1944; Johnson, 1984), or surface soil temperature reaches 7–12 °C (Hardy and Raymond, 1991). Weather conditions greatly affect emergence from caves where frogs winter (Resetarits, 1986). In southwestern Illinois, males and gravid females are found on roads as they move towards breeding ponds during late March and April (unpublished data).

   ii. Breeding habitat. Breed in a variety of aquatic habitats adjacent to adult habitat, including woodland pools and ponds,

stream overflow pools, farm ponds, sinkhole ponds, floodplain wetlands, marshes, and flooded quarries (Smith, 1961; DeGraaf and Rudis, 1983; Johnson, 1984; Resetarits, 1986; Adams and Lacki, 1993; Redmer and Mierzwa, 1994; Harding, 1997; Redmer, 1998b). Brown (D. R., 1984) reported oviposition in a cave pool in Indiana.

   B. Eggs.
   i. Egg deposition sites. Oviposition takes place more often in eutrophic zones than oligotrophic ones, but dissolved oxygen was high and water temperature was cool at breeding sites in Wisconsin (Johnson, 1984). Eggs are attached to dead or submerged vegetation at or near the water surface and are often concentrated in portions of the pool that receive greatest sunlight (Johnson, 1984; Harding, 1997; unpublished data).

   ii. Clutch size. Unknown. Egg masses are spherical; while the inner most embryos develop in a more hypoxic environment than those in the outer egg mass, ciliated epithelia of the embryos are used to create convective ventilation (Burggren, 1985). Under natural conditions, larvae hatch within 10–24 d (Johnson, 1984; Harding, 1997). In the lab, eggs hatch in 6–8 d (unpublished data).

   C. Larvae/Metamorphosis. Under natural conditions, larvae begin to metamorphose within 60–90 d after hatching (Johnson, 1984; Harding, 1997). In the lab, metamorphosis occurs from 75–90 d following hatching (unpublished data).

   D. Juvenile Habitat. Specific data are not available, presumably similar to that of adults.

   E. Adult Habitat. Often reported to prefer streams or ponds with cool, unpolluted water (Dickerson, 1906; Toner and St. Remy, 1941; DeGraaf and Rudis, 1983; Cook, 1984; Conant and Collins, 1991; Harding, 1997), but Johnson (1984) found no evidence to support this. Stream corri-

dors may be important conduits for movements in forests (Gibbs, 1998a). Some coastal and floodplain populations are reported to occupy swamps (Smith, 1961; Hardy, 1964; Schaaf and Smith, 1970). The most cave-adapted North American anuran, they are often abundant in areas of karst topography (Smith, 1961; Schaaf and Smith, 1970; McDaniel and Gardner, 1977; Resetarits, 1986). Individuals occurred in a Missouri cave from July–March (Resetarits, 1986). They may also enter abandoned mines (Heath et al., 1986). Other habitats include wooded wetlands, bogs, and shrubby, open meadows (Dickerson, 1906; DeGraaf and Rudis, 1983; Johnson, 1984; Redmer and Mierzwa, 1994; Harding, 1997). They prefer the margins of aquatic habitats with dense herbaceous vegetation (Pope, 1944; DeGraaf and Rudis, 1983; Johnson, 1984; Redmer and Mierzwa, 1994). Rural land use proximal to occupied habitats may or may not influence occurrence, though adults may be less abundant where stream bank vegetation is mowed or grazed (Johnson, 1984) or absent from areas that are logged (Cook, 1984).
   F. Home Range Size. Unknown.
   G. Territories. Unknown.
   H. Aestivation/Avoiding Desiccation. True aestivation is unknown. Warm summer temperatures may cause individuals to become nocturnal (Harding, 1997). In Missouri, some individuals retreat to caves during summer (Resetarits, 1986). Critical thermal maxima and capacity for acclimation to changing thermal conditions have been reported (Moore, 1939; Brattstrom and Lawrence, 1962; Brattstrom, 1968).
   I. Seasonal Migrations. Pickerel frogs migrate from breeding areas (pools or ponds) to summer habitat (stream sides) where individuals may remain sedentary during spring and summer until they return to hibernacula in the fall (Johnson, 1984).
   J. Torpor (Hibernation). Overwinter in the mud bottoms of ponds (DeGraaf and Rudis, 1983; Green and Pauley, 1987; Harding, 1997), ravines (Wright, 1914; DeGraaf and Rudis, 1983), under mud, rock, or debris in spring seeps and pools (Johnson, 1984), and in or around the edges of pools in caves (Resetarits, 1986). In spring-fed habitats, individuals may remain active (Pope, 1944; Harding, 1997).
   K. Interspecific Associations/Exclusions. Often reported to exclude northern leopard frogs (Manion and Cory, 1952; Johnson, 1984; Conant and Collins, 1991), but the reasons are unknown. It is believed that skin toxins (see "Anti-Predator Mechanisms" below) may kill other anurans kept in the same collecting container (Behler and King, 1998). Because of their anti-predator effect, it has been suggested that toxic skin secretions may be an adaptive advantage that makes pickerel frogs more abundant than northern leopard

frogs where they are sympatric (Dunn, 1935). Experimental evidence indicates that larvae compete with those of American toads (*Bufo americanus*; Wilbur and Fauth, 1990), but this is not affected by reproductive phenology (Alford, 1989b).

L. Age/Size at Reproductive Maturity. Age at reproductive maturity is unknown. Size at maturity is sexually dimorphic, with females averaging larger than males (Walker, 1946; Smith, 1961; Johnson, 1984; Resetarits and Aldridge, 1988; Hardy and Raymond, 1991). Minton (2001) reports a size range of 43.0 to 56.5 mm (mean 52.0) for adult males and from 54 to 78 mm (mean 64.0) for adult females.

M. Longevity. Unknown.

N. Feeding Behavior. Detailed studies of diet are unknown. Adults are reported to feed on insects, spiders, and other invertebrates (Pope, 1944; Harding, 1997). Tadpoles feed on algae and detritus (Pope, 1944; Harding, 1997).

O. Predators. Few observations of natural predation have been reported. Applegate (1990) reported predation by bald eagles *(Haliaeetus leucocephalus)*, and Beane (1990) reported two frogs from the stomach of a mink *(Mustela vison)*. Captive American bullfrogs *(R. catesbeiana)* and green frogs *(R. clamitans)* will eat small adults (Pope, 1944). A number of other vertebrate species (such as snakes, birds, and raccoons) may prey on adults. Some observations and experiments suggest that this species is distasteful to some predators (see "Anti-predator Mechanisms" below). Humans use adults as fishing bait (Cook, 1984). Several fish species were listed as potential tadpole predators in a stream (Holomuzki, 1995). Predators in experimental studies of larvae included eastern newts *(Notophthalmus viridescens)*, dragonfly naiads *(Anax* sp.), and diving beetles *(Dytiscus verticalis*; Brodie and Formanowicz, 1983; Wilbur and Fauth, 1990). Various other aquatic predators may prey on tadpoles.

P. Anti-Predator Mechanisms. Adults may have skin secretions that are toxic or distasteful to predators (Dickerson, 1906; Schaaf and Smith, 1970), and the bright yellow-orange colors on the concealed surface of the hind legs may serve to warn predators (Wright and Wright, 1949). Published observations and experimental evidence both support (Wright, 1932; Pope, 1944; Wright and Wright, 1949; Formanowicz and Brodie, 1979) and question (Mulcare, 1965) whether skin secretions are toxic or distasteful to predators. A combination of defensive posture and distasteful skin secretions may make this species unpalatable to shrews (Formanowicz and Brodie, 1979). Defensive behaviors and postures may be used in response to snakes (Marchisin and Anderson, 1978). In streams, choice of oviposition sites may make tadpoles unavailable to predatory fishes (Holomuzki, 1995). Unpalatability,

rapid growth/size, or behavior (reduced activity or use of fish inaccessible areas) have been suggested as anti-predator mechanisms of tadpoles (Formanowicz and Brodie, 1982; Brodie and Formanowicz, 1983; Holomuzki, 1995).

Q. Diseases. Little is known about natural disease and non– predation-caused mortality. A "red-leg" bacterial infection was reported in a cave (Lee and Franz, 1973). Johnson (1984) speculated that disease may have caused declines in Wisconsin during the 1970s. Devillars and Exbryant (1992) listed this species as one affected by a residue of the toxic pesticide dichlorodiphenyl trichloroethane (DDT).

R. Parasites. Reported parasites include trematodes (Bosma, 1934; Kuntz, 1941; Rankin, 1945; Goodchild, 1948; Walton, 1949; Bouchard, 1951; Coggins and Sajdak, 1982; McAllister et al., 1995b), cestodes (McAllister et al., 1995b), nematodes (Walton, 1929; Harwood, 1930, 1932; Coggins and Sajdak, 1982; Rankin, 1945; Baker, 1978a; McAllister et al., 1995b), and mites (Loomis, 1956; Murphy, 1965; McAllister et al., 1995b). Parasitic and commensal protozoans have also been reported (Metcalf, 1923; Walton, 1963; McAllister et al., 1995b; McAllister and Trauth, 1996).

4. Conservation.
Pickerel frogs are listed as Declining in Iowa (Christiansen, 1981), as Declining and a Species of Special Concern in Wisconsin (Casper, 1998; Mossman et al., 1998) and (due to restricted range) a Species of Special Concern in Minnesota (Oldfield and Moriarty, 1994; Lannoo, 1998d). Pickerel frogs may be extirpated in Kansas (Platt et al., 1974; Collins, 1993). Some authors report that they are common regionally (DeGraaf and Rudis, 1983; Green and Pauley, 1987; Johnson, 1987; Hunter et al., 1992), while others report that they are rare, uncommon, or localized (Minton, 1972; Pentecost and

Vogt, 1976; Vogt, 1981; Christiansen and Bailey, 1991; Redmer and Mierzwa, 1994; Harding, 1997; Redmer, 1998b).

Rana pipiens Schreber, 1782
NORTHERN LEOPARD FROG

James C. Rorabaugh

1. Historical versus Current Distribution.
Northern leopard frogs *(Rana pipiens)* occurred historically from Newfoundland and southern Québec, south through New England to West Virginia, and west across the Canadian provinces and northern and central portions of the United States to British Columbia, Oregon, Washington, and northern California (Stebbins, 1985; Conant and Collins, 1998). The species has been introduced at Lake Tahoe, California (Jennings and Hayes, 1994b), western Newfoundland (Maunder, 1997; Conant and Collins, 1998), Vancouver Island (Alberta Fish and Wildlife Division, 1991), and elsewhere. In the southwestern states, the species generally occurs at higher elevations, such as the mountains of northern and central Arizona and New Mexico. Geographic variation is discussed by Moore (1944, 1949a,b), Pace (1974), and Hoffman and Blouin (2004), however some morphs described by these authors have since been erected as separate species. Both burnsi (lacking dorsal spots) and kandiyohi (speckled appearance) color morphs are known from areas centered in western Minnesota (Merrell, 1965; see also Dapkus, 1976; McKinnell et al., this volume, Part One).

Although northern leopard frogs presently occur throughout most of their historical range, population declines and loss since the 1960s (Gibbs et al., 1971) or earlier have resulted in extirpation from some areas, particularly in the western 2/3 of the species' range. In Canada, populations of northern leopard frogs have declined dramatically in British Columbia (Orchard, 1992), and as of 2002 they were

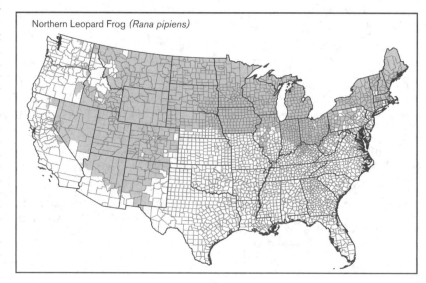

Northern Leopard Frog *(Rana pipiens)*

only known to be extant at a single site (Creston Valley; British Columbia Frog-watch Program, 2002). The species has disappeared from central Alberta and is greatly reduced in southern Alberta (Roberts, 1992; Seburn, 1992; Wagner, 1997; Takats and Willis, 2000), where declines were first noted in 1979 (Roberts, 1981). In Saskatchewan, populations reached a low in the early to mid-1970s (Seburn, 1992; Didiuk, 1997). The range of the species has contracted in Manitoba (Vial and Saylor, 1993), where a die-off in 1975–76 resulted in "heaps" of dead and dying frogs up to 1 m high on the shores of frog ponds (Koonz, 1992). Declines have occurred in northern Ontario (Old-ham and Weller, 1992) and the Richelieu River Valley of Québec (Gilbert et al., 1994). However, populations in New Brunswick, Nova Scotia, Prince Edward Island, and Northwest Territories show no evidence of decline (Green, 1997c; McAlpine, 1997a). Introduced populations in western Newfoundland apparently have disappeared (Maunder, 1997a). Reports of extirpations and range contractions are common in the western United States, where the species has disappeared from about 95% of their historical range in California (M.R. Jennings, 1995) and most historical localities in Washington, where they are probably now limited to two sites in the Crab Creek drainage of Grant County (Leonard et al., 1999; McAllister et al., 1999). In Arizona, northern leopard frogs are absent from most historical localities (Clarkson and Rorabaugh, 1989; Sredl et al., 1997; Meyer and Mikesic, 1998). Northern leopard frogs are extirpated from most historical localities in Oregon (St. John, 1985; Storm, 1986; Leonard et al., 1993), apparently are absent over much of western Montana (Reichel and Flath, 1995; Werner and Plummer, 1998), and may be extirpated from the middle Rio Grande Valley of New Mexico and Texas (Jackson, 1992; Degenhardt et al., 1996) and portions of eastern and north-central Colorado, including Rocky Mountain National Park (Corn et al., 1989, 1997; Stebbins and Cohen, 1995). In the western states, declines or extirpations have also been noted in Nevada (Panik and Barrett, 1994; R. Jennings, personal communication), south-eastern Wyoming (Corn et al., 1989; Stebbins and Cohen, 1995), Grand Teton National Park (Koch and Peterson, 1995), areas of Utah outside Cache and Kane counties (L. Colburn, personal communication), Targhee and Caribou National Forests, Idaho (Clark et al., 1993; Burton, 2001), and southwest of Twin Falls, Idaho (McDonald and Marsh, 1995). Northern leopard frogs are relatively rare and appear to have declined in the Greater Yellowstone Ecosystem (Patla et al., 2000). Call count survey results suggest that either the range of northern leopard frogs has

contracted in North Dakota or current sampling techniques are inadequate to detect frogs (Bowers et al., 1997). No reports of declines or extirpations are known from South Dakota or Nebraska (Vial and Saylor, 1993; but see McLeod, this volume, Part One). Declines and extirpations were first observed in portions of the Midwest (Wisconsin, Michigan, and northeastern Illinois) in the late 1960s or early 1970s (Rittschof, 1975; Hine et al., 1981; Mierzwa, 1998a), at the same time declines were noted in Manitoba and Saskatchewan. Northern leopard frogs apparently declined in Indiana from 1948–93 (Minton, 1998) and in northwestern Indiana, possibly due in part to habitat degradation (Brodman, 2002). The species has declined in Ohio in Hamilton County (Davis et al., 1998) and east of Cleveland away from the lake plain (Matson, 1999). In the northeastern states, northern leopard frogs are still locally common, but have been extirpated in developed areas such as New Haven, Connecticut; Cuttyhunk Island, Massachusetts; and Providence, Rhode Island (Klemens, 1993). Populations may be recovering in some areas (Seburn, 1992; Stebbins and Cohen, 1995; Didiuk, 1997; Casper, 1998). Northern leopard frog populations and habitats are dynamic (Lannoo, 1998a; Skelly et al., 1999), and local extirpation or apparent declines may not be permanent. Thus, the relationship of reported declines or local extirpation to the long-term viability of the species on a regional or rangewide basis is often unclear (Corn, 1994a; Green, 1997c). However, consistent, widespread reports of population loss are alarming, particularly in the western but also the central portions of the species' range.

2. Historical versus Current Abundance.

Much information exists concerning historical versus current distribution, however, relatively few data are available to compare historical versus current abundance at extant localities. At the two localities at which northern leopard frogs have been seen in recent years in California, only one individual was observed at one site and 8–10 at the other during 1990–94 (Jennings and Hayes, 1994b). In Arizona, northern leopard frogs are relatively common in livestock tanks at and near Stoneman Lake, Coconino County, but are scarce elsewhere (M. Sredl, personal communication). At Rocky Flats Environmental Technology Site, Colorado, northern leopard frogs were only recorded at one site, where they exhibited a relatively low vocalization index (Nelson, 1999). Lannoo et al. (1994) report that since 1900, northern leopard frogs have declined by 2–3 orders of magnitude in Dickinson County, Iowa. However, in northeastern Illinois, the species was considered "exceedingly abundant" (Kennicott, 1855), is still common today, and numbers of

northern leopard frogs at Glacial Park, Illinois, appear to be on the increase (Mierzwa, 1998a). The species is rare in Missouri (Vial and Saylor, 1993). Leopard frog harvests in Minnesota have declined from 100,000 lbs/yr in the early 1970s to approximately 1,000–2,000 lbs/yr in recent years (Moriarty, 1998), suggesting population declines. Breckenridge (1944) noted northern leopard frogs were scarce in the prairies of western Minnesota during a drought in 1931–34, but populations rebounded rapidly with the return of rain. At the Edwin S. George Reserve in Michigan, northern leopard frogs were abundant and green frogs (R. clamitans) were rare in the 1950s; from the 1970s through 1992, the reverse was true (Collins and Wilbur, 1979; Skelly et al., 1999). Northern leopard frogs are one of the least abundant amphibians in Jasper County, Indiana (Brodman and Kilmurry, 1998). Northern leopard frogs are common, but declining in Wisconsin (Mossman et al., 1998). In the Great Lakes region, northern leopard frogs currently are uncommon or rare in many places, but can be locally abundant (Harding, 1997). In New England, northern leopard frogs may be the predominant species in a narrow habitat strip, but are rare regionally (Klemens, 1993). Hinshaw (1992) notes that in Maine the species is typically less abundant than pickerel frogs (R. palustris). Northern leopard frogs were common in southwestern Ontario, but declined from 1992–93 (Hecnar, 1997). They are considered uncommon and of restricted distribution in the Northwest Territories (Fournier, 1997); and in Québec, northern leopard frogs are widespread but not abundant (Lepage et al., 1997). Predicting local abundance and distribution requires knowledge of the landscape in terms of all the habitat needs of northern leopard frogs, which include overwintering, breeding, and upland post-breeding habitats, as well as corridors among them (Pope et al., 2000).

3. Life History Features.

A. Breeding. Reproduction is aquatic.

i. Breeding migrations. In New England and Michigan, northern leopard frogs occupy grassy areas or damp wooded areas well away from water in summer, but breed and hibernate in aquatic sites (Wright, 1914; Noble and Aronson, 1942; Dole, 1965a; Merrell, 1970; Hinshaw, 1992; Klemens, 1993). Frogs migrate to breeding areas following emergence from hibernation (Dole, 1968). In Minnesota, sexually mature frogs move up to 1.6 km or more from hibernacula to breeding sites. Migration occurs at night or during the day when air temperatures approach 10 °C. Juvenile frogs tend to remain at hibernacula, which are relatively large, deep bodies of water (Merrell, 1970). Breeding and overwintering may occur in the same pond (Wagner, 1997). There is little evidence for breeding migrations in western North America (but see Simmons, 2000).

ii. Breeding habitat. Northern leopard frogs breed in a variety of aquatic habitats, such as quiet or slow-moving water along streams and rivers, wetlands associated with lakes or tidal areas, permanent or temporary pools, beaver ponds, and human-constructed habitats such as borrow pits, agriculture, and cattle ponds (Zenisek, 1963; Roberts, 1981; Klemens, 1993; Degenhardt et al., 1996; Wagner, 1997; Orr et al., 1998; Werner and Glennemeier, 1999). Permanent water is not necessary for breeding (Collins and Wilbur, 1979; Bonin et al., 1997b), and agricultural ditches and flooded fields may serve as breeding sites in the Midwest (Walker, 1946; Smith, 1961). However, at the Edwin S. George Reserve, Michigan, northern leopard frogs were never found to breed in temporary ponds and were almost never found at ponds in closed-canopy forests. In that area, location and number of breeding populations and habitats were dynamic. However, ponds occupied during 1967–74 and 1988–92 tended to be closer to potential source populations than ponds that were never occupied (Skelley et al., 1999). Near Ottawa, Ontario, mean pH, the amount of spawning habitat, amount of post-breeding summer habitat within 1 km, and the number of sites with leopard frogs calling within 1.5 km of a core breeding pond all contributed substantially to explaining observed variability in relative frog abundance at core ponds. Relative abundance was higher in core ponds with a neutral pH (Pope et al., 2000).

Northern leopard frogs may be absent or rare where large populations of American bullfrogs (R. catesbeiana) occur (Hammerson, 1982b; Jennings and Hayes, 1994b; Lannoo et al., 1994; Davis et al., 1998). Emergent or submergent vegetation may enhance habitat for oviposition and cover during the breeding season (Jennings and Hayes, 1994b). Waters no more acidic than pH 6.0 are optimal for fertilization and early development (Schlichter, 1981). In the Pacific Northwest, northern leopard frogs do not breed in bodies of water devoid of vegetation (Nussbaum et al., 1983). In Arizona, these frogs breed most often in constructed earthen cattle tanks (Sredl and Saylor, 1997). In New Mexico, northern leopard frogs breed in irrigation ditches in addition to natural aquatic sites (Degenhardt et al., 1996). In the Rocky Mountains of Colorado, breeding occurs in natural or manmade lakes and ponds (Corn and Fogelman, 1984). In Wisconsin, northern leopard frogs breed in many habitats, but are most common in open country (Mossman et al., 1998). Orr et al. (1998) found breeding northern leopard frogs in northeastern Ohio often associated with pools

or ponds in early stages of succession and typically where fish were absent. Increasing beaver populations and the ponds they create in New Brunswick have benefited northern leopard frog populations (McAlpine, 1997a). In southwestern Ontario, absence of fish was associated with presence of northern leopard frogs (Hecnar, 1997). In Québec, habitats with the greatest numbers of egg masses, in decreasing order of importance, were a wet meadow, a shallow marsh, and an abandoned (unplowed) meadow (Gilbert et al., 1994). In agricultural areas of southern Québec, northern leopard frogs were closely associated with wooded habitat strips (Maisonneuve and Rioux, 2001). At Lac Saint Pierre, Québec, high densities of frogs were found in habitats that were close to the marsh line, had a tall herbaceous stratum and high richness, and had low moss cover (Beauregard and Leclair, 1988). Calling northern leopard frogs are associated with both permanent and intermittent water bodies in Québec (Bonin et al., 1997b).

Breeding is generally restricted to a relatively short period in the spring, but Scott and Jennings (1985) reported eggs and small tadpoles during April–July and September–October in New Mexico. Breeding occurs later at higher elevations (Corn and Livo, 1989) and in northern areas as well (Breckenridge, 1944). In Colorado, initial breeding activities seemed to be related more to temperature than to precipitation. Oviposition followed the onset of male chorusing by 2–3 d and corresponded to periods of warm weather (Corn and Livo, 1989). In Wisconsin, eggs are found from early April to May (Watermolen, 1995). Breeding in the northeastern states occurs in March–May (Klemens, 1993). In Québec, northern leopard frogs were found calling from 2 May–5 July, with peak calling from 2–23 May (Lepage et al., 1997). In Glen Canyon, Arizona, northern leopard frogs probably deposit egg masses during late April to early May (Drost and Sogge, 1993).

B. Eggs.

i. Egg deposition sites. Egg masses are deposited in a tight, oval mass, and several egg masses may be communally attached to emergent vegetation or, less commonly, may lie on the pond bottom in shallow, slow-moving, or still water (Wright, 1914; Breckenridge, 1944; Wright and Wright, 1949; Corn and Livo, 1989; Gilbert et al., 1994; Degenhardt et al., 1996). Based on a review of pertinent literature, Pope et al. (2000) characterized northern leopard frog preferred spawning habitat as sites with non-acidic water, 10–65 cm water depth in full sun, usually on the north side of the pond, and with emergent, non-broad-leaved vegetation for attachment of egg masses.

ii. Clutch size. Egg masses are reported to contain between 645–7,648 eggs (Livezey

and Wright, 1947; Hupf, 1977; Corn and Livo, 1989; Gilbert et al., 1994; Watermolen, 1995). Post-spawning ovarian quiescence lasts 1–2 mo in the northeastern states. In Wyoming and Colorado, females probably do not produce multiple clutches (Corn and Livo, 1989). Time to hatching is correlated with temperature and ranges from 2 d at 27 °C to 17 d at 11.4 °C (Nussbaum et al., 1983; see also Volpe, 1957b).

C. Larvae/Metamorphosis.

i. Length of larval stage. Variable, depending on latitude and weather, with metamorphosis typically occurring approximately 3–6 mo following egg deposition (Merrell, 1977; Hinshaw, 1992). Time from hatching to metamorphosis is about 50 d (7 wk) in northwestern Iowa (Lannoo, 1996). Crowding and interaction may reduce growth rates of tadpoles where they are especially dense, in hatchling aggregations, and in pools in the final stages of drying (Richards, 1958; West, 1960; Gromko et al., 1973; John and Fenster, 1975; Alford, 1999). In the laboratory, high tadpole density or limited availability of food slowed both growth and developmental rates. Blocking corticoid synthesis reversed growth suppression caused by high density, but did not alter effects of density on developmental rates (Glennemeier and Denver, 2002). Overwintering of tadpoles has been documented in Arizona (Collins and Lewis, 1979) and Nova Scotia (DeGraaf and Rudis, 1983).

ii. Larval requirements.

a. Food. Tadpoles feed on attached and suspended algae, other attached plant and animal material, and occasionally on dead animals (e.g., Hendricks, 1973; Seale, 1982a).

b. Cover. In Michigan, northern leopard frog tadpoles survive and grow much better in open-canopy ponds as opposed to closed-canopy ponds. Addition of food to closed-canopy ponds dramatically increases tadpole survival and growth. Closed-canopy ponds tended to be slightly cooler, and pH and dissolved oxygen are lower (Werner and Glennemeier, 1999). Northern leopard frog tadpoles are poorly adapted for life in fast-moving water, and thus are most often found in quiet backwaters and pools (Roberts, 1981). Gulping of air at the pond surface begins to increase when dissolved oxygen drops below about 4 ppm (Wassersug and Seibert, 1975). Aerial respiration is more energetically expensive than aquatic respiration and can slow growth and development if food is not abundant (Feder and Moran, 1975). Tadpoles are active during the day and exhibit behavioral thermoregulation (Casterlin and Reynolds, 1978; see also Noland and Ultsch, 1981). Tadpoles do not school per se, and laboratory observations using spatial affinity as a recognition assay could not demonstrate sibling recognition (Fishwald et al., 1990). Determinants of larval behavior include responses to both intrinsic (body size, age) and extrinsic (group size) cues. Tadpoles in pairs or groups generally are more active and occur more often in the open water column than lone tadpoles. Younger tadpoles tend to be less active and use the open water column less than older tadpoles (Golden et al., 2001).

iii. Larval polymorphisms. None described.

iv. Features of metamorphosis. Metamorphosed frogs are 18–50 mm SUL (Merrell, 1977; Nussbaum et al., 1983; Leclair and Castanet, 1987; Hinshaw, 1992; Degenhardt et al., 1996).

v. Post-metamorphic migrations. Newly metamorphosed northern leopard frogs disperse away from their natal wetlands. Frogs can move up to 800 m in 2–3 d (Dole, 1971) and have a tendency to move to the edges of permanent bodies of water (Merrill, 1977; Cochran, 1982). During the metamorphic period, at least a few frogs will leave wetlands every night; mass emigrations can occur following heavy rains (Bovbjerg, 1965; Dole, 1971). Dole (1972a) cites evidence for celestial orientation in newly metamorphosed northern leopard frogs.

D. Juvenile Habitat. During post-metamorphic dispersal in summer, juvenile frogs can be found in upland forests and meadows, whereas adult frogs remain closer to water (Dole, 1971; see also Dole, 1972a,b). Juvenile frogs are also sometimes encountered at temporary bodies of water (Smith, 1961).

E. Adult Habitat. Adults require suitable habitats for breeding, upland foraging areas after leaving the breeding ponds, as well as sites for overwintering (see "Breeding habitat," and "Seasonal migrations" above, "Torpor [Hibernation]" below), and, where these habitats are different, require habitat corridors connecting them. Upland areas used as post-breeding habitat in summer are typically grassy areas, meadows, or fields but can include other sites such as peat bogs and perennial forage crops (e.g., grass, alfalfa, and clover). Post-breeding summer habitats usually do not include barren ground, open sandy areas, heavily wooded areas, cultivated fields, especially those that have been cut recently, heavily grazed pastures, or closely mowed lawns (Wright and Wright, 1949; Werner and Glennemeier, 1999; Pope et al., 2000; Mazzerole, 2001). Northern leopard frogs in the relatively arid landscapes of southwestern Alberta may be limited by lack of suitable upland post-breeding habitats and lack of dispersal opportunities (Roberts, 1992). Near Chicago, despite presence of abundant suitable breeding ponds, northern leopard frog abundance was low. However, after grasslands were restored around ponds, frog populations increased dramatically (K.S. Mierzwa, personal communication, in

Pope et al., 2000). Orr et al. (1998) suggest that in northeastern Ohio, the northern leopard frog is a colonizing species that takes advantage of habitats in early stages of succession but does poorly in later successional stages where predation and competition are more intense. Allozyme frequency and rapid amplified polymorphic DNA analysis of northern leopard frog populations in Utah and Arizona showed relatively large differences among populations, suggesting microgeographic differentiation was occurring and that each population may be locally adapted (Kimberling et al., 1996a,b). In a rural landscape of agriculture and fragmented wetlands and forests in Indiana, northern leopard frogs were likely to occur in more isolated forest patches that were in close proximity to wetlands. A sensitivity to isolation from wetlands suggested an important role of recolonization in the distribution of the species (Kolozsvary and Swihart, 1999). Larger leopard frogs may be less susceptible to desiccation and more capable of using arid environments such as mined peat bogs (Mazerolle, 2001). In agricultural and urbanized regions of southwestern Minnesota, reduced landscape connectivity caused by habitat fragmentation and loss negatively affected amphibian assemblages that include northern leopard frogs (Lehtinen et al., 1999).

F. Home Range Size. In Michigan, Dole (1965b) found that daily movements of adults are usually <5–10 m in wet pastures and marsh, but home range was not calculated. Both adults and juveniles wander widely during wet weather (Dole, 1971). Home ranges may include breeding sites, hibernacula, and upland foraging and dispersal areas (Merrell, 1970). Displaced northern leopard frogs will home and apparently use olfactory and auditory cues, and possibly celestial orientation, as guides (Dole, 1968, 1972a).

G. Territories. Because egg masses are typically confined to an area much smaller than that occupied by calling males, Pace (1974) suggested that relatively few males are involved in fertilizing egg masses. Short, terminal sounds in the call sequence have an aggressive or spacing function among males (Pace, 1974). These observations suggest territoriality among calling males.

H. Aestivation/Avoiding Desiccation. Northern leopard frogs apparently are active throughout warm periods and seasons. No periods of aestivation have been described. Dole (1967) described the role of dew and substrate moisture in the water balance of northern leopard frogs. Maximum thermal tolerances were addressed by Hutchison and Ferrance (1970).

I. Seasonal Migrations. Four types of seasonal migrations have been described, as follows: (1) spring migration of adult frogs from hibernacula to breeding sites;

(2) movements of newly metamorphic animals from breeding sites to new areas; (3) movements from summer locales to hibernacula; and (4) passive migration of tadpoles in waterways following heavy rains that carry them to new areas (Merrell, 1970). Both adult and juvenile frogs may also move in response to warm rains. Pope et al. (2000) characterize northern leopard frogs as a species dependent upon landscape complementation, that is, linking together different landscape elements through movement to complete their life cycles. This is well documented in eastern populations, however, descriptions of landscape complementation in western North America are lacking. Simmons (2000) notes that in Washington, northern leopard frogs are believed to move among overwintering, spring breeding, and summer feeding sites, but she finds little supporting documentation. In the Great Lakes region, adult frogs hibernate in deep water, which protects them from freezing, but move to shallow water for breeding (Harding, 1997). In Michigan, newly metamorphosed animals disperse overland and nocturnally to new locales, particularly during warm, wet weather. Young leopard frogs commonly move ≤800 m from their place of metamorphosis; three young males established residency up to 5.2 km from their place of metamorphosis (Dole, 1971). In the Cypress Hills, southern Alberta, young-of-the-year northern leopard frogs successfully dispersed to downstream ponds 2.1 km from the source pond, upstream 1 km, and overland 0.4 km. At Cypress Hills, a young-of-the-year northern leopard frog moved 8 km in 1 yr (Seburn et al., 1997). Streams are important dispersal corridors for young frogs (Seburn et al., 1997), and vegetated drainage ditches can enhance connectivity among seasonal habitats (Pope et al., 2000). Rainfall or humidity may be an important factor in dispersal, because odors carry well in moist air, making it easier for frogs to find other wetland sites (U. Sinsch, 1991). In Iowa, post-metamorphic frogs emigrated from breeding ponds in July, possibly in response to internal cues (Bovbjerg, 1965). During heavy, prolonged rains in Michigan, northern leopard frogs moved in a direct line until daybreak, in contrast to the short, winding movements at other times. Following such excursions, frogs usually returned to their typical areas of use (Dole, 1965b). In Minnesota, post-breeding frogs disperse to a variety of wetland types. In the fall, frogs move once again back to hibernacula (Breckenridge, 1944). In New Brunswick, adult and juvenile northern leopard frogs' use of peat bogs peaked in August, corresponding to juvenile dispersal from breeding ponds (Mazerolle, 2001). During migrations in Minnesota, large numbers of frogs are crushed by automobiles on roadways (Breckenridge,

1944; Merrell, 1970). Northern leopard frog populations within 1.5 km of roads are negatively affected by road mortality (Carr and Fahrig, 2001). At Ithaca, New York, of 44 recaptured northern leopard frogs, the maximum recorded distance from the first place of capture was 137 m (Ryan, 1953).

J. Torpor (Hibernation). Northern leopard frogs typically hibernate in ponds and lakes (Nussbaum et al., 1983), where they may sit on the bottom under rocks or logs or in shallow pits in silt substrates. Overwintering frogs may bury themselves in the mud (DeGraaf and Rudis, 1983; Cunjak, 1986; Harding, 1997) or they may aggregate into mounds over underwater spring heads (Lannoo, 1996). Northern leopard frogs overwintering in caves in Indiana remained active (Rand, 1950). In New England, northern leopard frogs hibernate from October or November to February or March, but may emerge on warm days in winter (DeGraaf and Rudis, 1983). In Minnesota, northern leopard frogs were found hibernating in January in a shallow (46–61 cm deep) outflow from a dam (Breckenridge, 1944); however, northern leopard frogs typically hibernate in deep, well-oxygenated water that does not freeze solid (Cory, 1952; Wagner, 1997). In the Lamoille River, Vermont, northern leopard frogs were found in well-oxygenated water in a common map turtle (*Graptemys geographica*) hibernaculum (Ultsch et al., 2000). In a laboratory setting, northern leopard frogs held in water at 1.5 °C remained submerged and motionless with limbs held loosely from the body. Frogs began to move and resurfaced at water temperatures of 7–10 °C (Licht, 1991). Northern leopard frogs are freeze intolerant. Frogs frozen at -2 °C survived no longer than 8 hr (Layne, 1992). Overwintering northern leopard frogs are intolerant of severely hypoxic or anoxic waters (Pinder, 1985) and will winterkill (Manion and Cory, 1952). Burnsi morphs appear to be more tolerant of temperature extremes than normal color patterned animals (Merrell and Rodell, 1968; Dapkus, 1976).

K. Interspecific Associations/Exclusions. In a competitive exclusion experiment, northern leopard frog tadpoles exhibited higher survival rates than Oregon spotted frogs (*R. pretiosa*), suggesting that the former might be capable of displacing the latter (Dumas, 1964). In the Pacific Northwest, spotted frogs historically co-occurred with northern leopard frogs at many localities, but the former has since disappeared from most sites (Nussbaum et al., 1983). Water conditioned by other taxa did not inhibit the growth of northern leopard frog tadpoles (Akin, 1966). Northern leopard frog populations may be eliminated by predators such as American bullfrogs (*R. catesbeiana*; Lannoo et al., 1994) and fishes (Bovbjerg, 1965;

Lannoo, 1996). However, at Point Pelee National Park, Ontario, after extinction of American bullfrogs, observations of northern leopard frogs declined, while those of green frogs increased (Hecnar and M'Closkey, 1997). When European carp (*Cyprinus carpio*) invaded a pond in Washington, they ate all aquatic vegetation and left the site denuded; northern leopard frogs subsequently disappeared (Corkran and Thomas, 1996). In the Caribou National Forest, southeastern Idaho, northern leopard frogs were not observed to commonly breed in association with tiger salamanders (*Ambystoma tigrinum*) or boreal chorus frogs (*Pseudacris maculata*). In comparison to these species, northern leopard frogs were more often associated with ponds that were less isolated and more likely to be seasonal or permanent (Burton, 2001). In Arizona, northern leopard frogs occur with Woodhouse's toads (*Bufo woodhousii*), red-spotted toads (*B. punctatus*), Mexican spadefoot toads (*Spea multiplicata*), Chiricahua leopard frogs (*R. chiricahuensis*), tiger salamanders, canyon treefrogs (*Hyla arenicolor*), and western chorus frogs (*P. triseriata*; Clarkson and Rorabaugh, 1989; Drost and Soggee, 1993), and, until recently, occurred with American bullfrogs at one site (personal observations). In areas of sympatry between northern leopard frogs and plains leopard frogs (*R. blairi*) in Nebraska, the former is found in clear, sandy bottom streams, the latter in turbid, silty streams (Lynch, 1978; see also Cousineau and Rogers, 1991; see also Hardy and Gillespie, 1976). Hybridization (usually <5%) occurs between northern leopard frogs and plains leopard frogs in areas of sympatry in Colorado, Nebraska, and South Dakota (Lynch, 1978). At Pawnee, Weld County, Colorado, northern leopard frogs occur with chorus frogs (*Pseudacris* sp.) and plains spadefoot toads (*S. bombifrons*; Corn et al., 2000). In Cuyahoga Valley National Recreation Area, Ohio, relatively large breeding choruses of northern leopard frogs occurred in areas with large choruses of American toads (*B. americanus*), spring peepers (*P. crucifer*), western chorus frogs, green frogs (*R. clamitans*), and wood frogs (*R. sylvatica*; Varhegyi et al., 1998). In the 1940s–50s at Bacon Swamp in Indiana, northern leopard frogs were common, as were eastern gray treefrogs (*H. versicolor*) and spring peepers (Minton, 1998). At the Edwin S. George Reserve, Michigan, northern leopard frogs co-occurred in wetlands with American toads, American bullfrogs, green frogs, wood frogs, spring peepers, eastern gray treefrogs, four species of mole salamanders (*Ambystoma*), and eastern newts (*Notophthalmus viridescens*), but bred after salamanders in sympatric habitats (Collins and Wilbur, 1979). Near Ithaca, New York, northern leopard frogs occurred with green frogs and bullfrogs

(Ryan, 1953). Wright (1914) found northern leopard frogs in amplexus with pickerel frogs (*R. palustris*) and American toads near Ithaca. Pace (1974) found leopard frogs often calling in association with chorus frogs and spring peepers. At Lac Saint Pierre, Québec, northern leopard frogs occurred with wood frogs, American toads, and American bullfrogs (Beauregard and Leclair, 1988). Wood frogs typically breed in woodland ponds and northern leopard frogs breed in more open areas; however, within agricultural areas of southern Québec, wood frogs typically occur in shrubby strips, whereas northern leopard frogs are more closely associated with wooded strips of habitat (Maisonneuve and Rioux, 2001). Where the two occur in sympatry, competition may occur (Smith-Gill and Gill, 1978; Werner, 1992), but the importance of competition varies among years (DeBenedictis, 1974). In the laboratory, when northern leopard frog and wood frog tadpoles were reared separately in the absence of predators, the former grew faster. However, when reared together, wood frog tadpoles grew faster. Addition of caged predators (larval dragonflies or mudminnows) to pens containing both species reversed competitive interactions—the leopard frog larvae grew faster. Competition also resulted in increased mouth width in northern leopard frog tadpoles, although this effect was eliminated when predators were introduced (Relyea, 2000). In open-canopy ponds in Michigan, growth rates of leopard frog tadpoles in the presence of wood frog tadpoles was only 60–63% of that in ponds without wood frogs. There was no indication that northern leopard frogs influenced growth rates of wood frog tadpoles. Wood frogs bred earlier, so their tadpoles were larger, and leopard frog tadpoles are probably less active than wood frog tadpoles. Werner and Glennemeier (1999) suggest these factors may explain the competitive dominance of the wood frog in their experiments. Overlap in prey species length suggests a potential for competition between northern leopard frogs and green frogs in New Brunswick. In that study, northern leopard frogs were more terrestrial and occupied denser vegetation than green frogs (McAlpine and Dilworth, 1989).

L. Age/Size at Reproductive Maturity. In southeastern Québec, female northern leopard frogs reached sexual maturity at 2 yr and ≥60 mm SUL; however, 55% of males were mature at age 1 (Gilbert et al., 1994). In Wisconsin, males may mature within 1 yr of metamorphosis, but probably more commonly after 2 yr (Hine et al., 1981). Females probably do not mature until 2–3 yr post-metamorphosis (Rittschof, 1975; Merrell, 1977; Hine et al., 1981; Hinshaw, 1992). Force (1933) suggested northern leopard frogs reach sexual maturity at age 3 in northern Michigan. In Wyoming, females matured at age 3 in high elevation populations and at age 2 in lower elevation sites (Baxter, 1952). In Minnesota, frogs matured more rapidly in populations of relatively low density (Merrell, 1977). Ryan (1953) suggested that near Ithaca, New York, a few frogs matured in the same year in which they metamorphosed.

M. Longevity. In southwestern Québec, 51 of 53 frogs collected were 2 yr old or less. The two other frogs were 3 and 4 yr old (Leclair and Castanet, 1987). Dole (1971) reported a 5-yr-old individual from Michigan. Captives have lived as long as 9 yr (Russell and Bauer, 1993).

N. Feeding Behavior. Prey of metamorphosed frogs include a variety of terrestrial invertebrates such as insect adults and larvae, spiders, slugs, snails, leeches, sowbugs, and earthworms (e.g., Knowlton, 1944; Linzey, 1967; Hedeen, 1970). As well, larger frogs take vertebrates such as spring peepers and western chorus frogs (Harding, 1997), small northern leopard frogs (Russell and Bauer, 1993), small birds and snakes (Breckenridge, 1944; DeGraaf and Rudis, 1983), fish (Leonard et al., 1993), and bats (Creel, 1963). Prey items correlate with peak abundances in insect species (Linzey, 1967). Drake (1914) reported that almost 90% of stomach contents of Ohio specimens consisted of insects and spiders. Foraging success of northern leopard frogs in Itasca State Park, Minnesota, was correlated with both the total distance traveled during the foraging bout and the distance to the nearest conspecific neighbor (Wiggins, 1992). Northern leopard frogs, as with most frog species, typically use visual cues to detect and locate prey, but Shinn and Dole (1978) provide evidence that olfactory cues are also used. Gut passage times vary from 12–24 h in active adults, 48–96 h in overwintering animals (Gossling et al., 1980).

O. Predators. In the Pacific Northwest and Québec, northern leopard frog tadpoles are preyed upon heavily by garter snakes (*Thamnophis elegans* and *T. sirtalis*), while adults are preyed upon by snakes, birds, and other small carnivores (Nussbaum et al., 1983; Russell and Bauer, 1993). Cannibalism has been documented (Borland and Rugh, 1943; Nussbaum et al., 1983). Predators in the Great Lakes region include green frogs, American bullfrogs, various species of snakes, hawks, waterfowl, herons, raccoons, foxes, mink, otters, and humans (Breckenridge, 1944; Harding, 1997). American bullfrogs prey on northern leopard frogs in New Brunswick (McAlpine and Dilworth, 1989). Invasion of American bullfrogs in Iowa is correlated with decline of northern leopard frogs (Lannoo et al., 1994; Lannoo, 1996). Northern leopard frogs apparently have disappeared from the Columbia National Wildlife Refuge, Washington, possibly as a result of American bullfrog predation (Leonard et al., 1993). In mesic regions, northern leopard frogs use a variety of seasonal habitats, including upland sites, and may wander far from water after breeding; American bullfrogs are more likely to remain close to the water's edge (Hecnar and M'Closkey, 1997). In arid areas, northern leopard frogs are more likely to stay close to water and therefore are probably more susceptible to predation by American bullfrogs. A large population of northern leopard frogs has been severely reduced by predation at Garlock Slough, Iowa, after a variety of fish species was stocked there by state fisheries biologists (Bovbjerg, 1965). Largemouth bass (*Micropterus salmoides*) feed on young-of-the-year northern leopard frogs inhabiting the littoral regions of permanent wetlands and lakes (Cochran, 1982; see also Cochran, 1983). Crayfish may feed on northern leopard frogs and may have adversely affected populations in California (Jennings and Hayes, 1994b) and other western states. Under laboratory conditions, northern leopard frog eggs and tadpoles were readily eaten by marbled salamanders (*A. opacum*) and eastern newts (Walters, 1975).

P. Anti-Predator Mechanisms. Northern leopard frogs seek shelter in water when threatened (Russell and Bauer, 1993); if in upland areas, they hop in an erratic pattern and then conceal themselves in vegetation (Harding, 1997). The resemblance of the species to pickerel frogs, which have skin secretions that repel predators, may deter some predators from pursuing northern leopard frogs (Harding, 1997). During summer in Minnesota, frogs that are disturbed are less likely to jump into water and remain submerged (Merrell, 1970). Recently metamorphosed northern leopard frogs crouch or cease to move in the presence of active eastern garter snakes (Heinen and Hammond, 1997). In the laboratory, as predator density increased, the proportion of time tadpoles are active and swimming speed declined, presumably as a means to avoid predation. Larger tadpoles, which are less vulnerable to predation, were more active in the presence of predators than small tadpoles (Anholt et al., 2000). In the presence of caged predatory dragonfly (*Anax* sp.) larvae, northern leopard frog tadpoles developed a deeper tail fin, deeper tail musculature, and shorter body in more natural pool and tank experiments, but not in a laboratory setting. Deeper tail fins are correlated with greater swimming speeds and improved ability to escape predators (Relyea and Werner, 2000; Van Buskirk, 2001).

Q. Disease. Massive die-offs attributable to disease and winter freezing and associated oxygen deprivation have been recorded (Cory, 1952; Koonz, 1992; Harding, 1997). Dead and moribund frogs are often symptomatic for "red-leg" (*Aeromonas hydrophila*), a bacterial infection and other gram-negative and occasionally gram-positive bacteria (Glorioso et al., 1973; Koonz, 1992; Harding, 1997; Faeh et al., 1998); however, the infectious agents are normal inhabitants of frog environments, and frogs may only become symptomatic when immune competence is compromised (Crawshaw, 1992). Intestines of mostly healthy northern leopard frogs and tadpoles from North Dakota and Minnesota contained *Aeromonas hydrophila* and 29 species of *Enterobacteriaceae* (Hird et al., 1983). Faeh et al. (1998) report isolation of polyhedral cytoplasmic amphibian viruses (in the iridovirus family) from northern leopard frogs, but they were not identified as a cause of lesions or illness. Renal carcinomas caused by a herpes virus are most evident during the winter and spring, and transmission is thought to occur via frog urine when adults congregate at breeding ponds (Hunter et al., 1989; Faeh et al., 1998). Hunter et al. (1989) believed the prevalence of tumors was increasing in the 1980s in Minnesota. Gibbs et al. (1966) reported intestinal tract cancer in northern leopard frogs. During a period from hibernation emergence until immune function is restored, frogs are particularly susceptible to diseases (Maniero and Carey, 1997).

Low pH affects fertilization and development of northern leopard frog eggs. Adult northern leopard frogs exhibited high mortality when maintained for 10 d in water of pH 5.5. Immune function declines during hibernation (Cooper et al., 1992), leaving northern leopard frogs more susceptible to low pH (Vatnick et al., 1999). Northern leopard frog tadpoles exposed to pH <4 died, and those exposed to pH <5.6 for >24 hr experienced high mortality. Prior to death, tadpoles exposed to low pH exhibited caudal curling, thoracic swelling, and failure to retract the yolk plug (Watkins-Colwell and Watkins-Colwell, 1998). Leopard frog abundance may be low in areas where waters are either acidic or basic (Pope et al., 2000). Endogenous gut bacteria persist through hibernation and may be a source of systematic infection when frogs emerge in the spring in an acidic environment (Leonard et al., 1999).

Large numbers of malformed northern leopard frogs (mostly limb abnormalities) and some other malformed anurans recently have been observed in Minnesota (Helgen et al., 1998; Souder, 2000; Lannoo et al., 2003). Malformed northern leopard frogs have also been reported from Arizona, Québec, Wisconsin, Ohio, South Dakota, Vermont, Maine, and other states. Meteyer et al. (2000) report that 86% of recently metamorphosed northern leopard frogs from Minnesota, Vermont, and Maine exhibited hindlimb deformities.

They describe these malformations in detail and suggest that "developmental events may produce a variety of phenotypes depending on the timing, sequence and severity of the environmental insult." In five Vermont counties, 7.5% of recently metamorphosed northern leopard frogs exhibited external malformations; observed malformation rates varied seasonally and annually (Levey, 2000). Some malformed frogs in Le Sueur County, Minnesota, had apparent parasitic cysts in the thigh muscles (Helgen et al., 1988). Infections of a parasitic trematode (*Ribeiroia* sp.) have been implicated in limb malformations of Pacific treefrogs *(P. regilla)* in California and may be a contributing factor in malformations observed in other amphibians (Johnson et al., 1999; Johnson and Lunck, this volume, Part One), including northern leopard frogs in Arizona (Sessions et al., 1999). However, Gilliland and Muzzall (1999, 2002) concluded that trematodes were not the cause of deformities in southern Michigan. Visiting the "hottest of the Minnesota malformed frog hotspots," and control sites, Lannoo et al. (2003) concluded that where *Ribeiroia* metacercariae were found, they likely cause malformations, but there were two important disconnects: some "control" sites contained *Ribeiroia*, but malformations were not present in high numbers, and at some "hotspots," malformations were present in the absence of *Ribeiroia* metacercariae.

Possible causes of malformations include pesticides, retinoids, heavy metals, increased UV-B light, mechanical perturbation including parasitic infestations and predation attempts, and estrogen mimics (Sessions and Ruth, 1990; Kao and Danilchik, 1991; Ankley, 1997; Ouellet et al., 1997b; Helgen et al., 1998; Johnson et al., 1999; Sessions et al., 1999). Agricultural contaminants are the suspected cause of deformities observed on the St. Lawrence River Valley, Québec (see discussion below), although variation in the proportion of deformities among sites was too large to conclude that there was a difference between control and pesticide-exposed habitats. Conspicuous deformities interfered with swimming and hopping and likely constituted a survival handicap (Ouellet et al., 1997). Deformed frogs from Minnesota exhibited normal chromosome number and morphology, and selectively stained metaphase plates had the normal number of nucleolar organizer regions, providing no evidence that affected frogs had damaged genetic material (Reister et al., 1998; Horner et al., 2000). Deformed frogs in Canada weighed less than normal frogs, while both deformed and normal frogs from sites of high deformity rates exhibited shorter body length, head width, and femur and forelimb length than frogs from sites with low incidence of deformities (Gallant and Teather, 2001).

A growing body of evidence suggests northern leopard frogs can be adversely affected both acutely and via sublethal symptoms by pesticides and other chemicals. However, with a few notable exceptions, as yet there is little evidence that concentrations of chemicals present in the environment are contributing to population decline or increased levels of malformations. Northern leopard frog embryos exposed to low levels of three insecticides and six herbicides commonly used in Canadian forests and croplands hatched at the same time and with the same hatching success as control animals. Experimental embryos did not exhibit higher levels of tadpole deformities than controls. Experimental tadpoles were paralyzed, but gradually recovered after exposure was terminated (Berril et al., 1997). Northern leopard frogs may be vulnerable to agricultural chemicals during spring runoff due to their habitat of overwintering in permanent water bodies in close contact with bottom sediments (Kaplan and Overpeck, 1964; Didiuk, 1997). During metamorphosis, resorption of the tail may cause mobilization of bioaccumulated pesticides and subsequent toxicity (Cooke, 1970). Streams, rivers, and reservoirs in mid-western corn-growing regions may be exposed to high levels of the herbicide atrazine. In the laboratory, high doses of atrazine caused deformities in northern leopard frog larvae and respiratory distress and cessation of feeding in adult frogs, but these doses were considerably higher than concentrations found in North American surface waters. Thus, direct toxicity of atrazine may not be a significant factor in recent declines of northern leopard frogs (Allran and Karasov, 2001). Furthermore, atrazine and nitrate fertilizers at concentrations found in the environment, and interactions between the two, do not appear to pose a substantial direct toxicity threat to northern leopard frog tadpoles (Allran and Karasov, 2000). Atrazine may not be present in the environment at levels that result in acute toxicity, but there is new evidence that sublethal effects to frogs are occurring. Atrazine disrupts endocrine function and, even at very low concentrations of 0.1 ppb, cause retarded gonadal development, hermaphroditism, and oocyte growth in male northern leopard frogs. Atrazine contamination is widespread in the United States and can be present in excess of 1 ppb in precipitation and even in areas where it is not used (Hayes et al., 2002b,c). Glennemeier (2001) and Glennemeier and Denver (2001) examined the effects of a polychlorinated biphenyl (PCB) congener, 77-TCB, on northern leopard frog tadpoles. Effects included decreased feeding rates, reduced whole-body corticosterone content, and altered

competitive interactions with wood frog tadpoles. Corticosterone is important in mediating the negative growth response of northern leopard frog tadpoles to increasing larval densities, and also affects development, morphology, and response to adrenocorticotropic hormone. The results suggest negative population-level consequences from sublethal effects in PCB or other pollutant-contaminated environments. However, no significant correlation between frog densities and severity of PCB contamination was found in contaminated wetlands. Tissue concentrations of PCBs in the field were much lower than in sediments (Glennemeier and Begnoche, 2002). Northern leopard frog tadpoles exposed to high levels of PCB 126 exhibited elevated incidence of edema, reduced swimming speed and growth, and 100% mortality before metamorphosis. Few deformities were observed. However, sublethal effects were not apparent at PCB concentrations that occur in the Green Bay ecosystem, Wisconsin (Rosenshield et al., 1999). The pre-emergent herbicide acetochlor interacts with thyroid hormone in northern leopard frog tadpoles, accelerating thyroid hormone induced metamorphosis and countering the effects of corticosterone (Cheek et al., 1999). Survival declined, the prevalence of deformities increased, and growth and development slowed in northern leopard frog tadpoles exposed to un-ionized ammonia concentrations in excess of 1.5 mg/L. This level is higher than that measured in waters of the Fox River-Green Bay ecosystem, but lower than for pore sediment water. Leopard frogs may be exposed to hazardous levels of un-ionized ammonia when they hibernate on the bottom or while buried in sediments (Jofre and Karasov, 1999). Among northern leopard frog tadpoles, growth rates slow and mortality rates increase when exposed to copper sulfate, used to control nuisance algal blooms (Lande and Guttman, 1973).

Intensive UV exposure in the laboratory can result in embryonic mortality and abnormal development (Higgins and Sheard, 1926). However, larvae are more sensitive than embryonic stages; ambient UV-B levels were found to be lethal to northern leopard frog tadpoles in the absence of shade or refuge (Ankley et al., 2000; Tietge et al., 2001). At 50–60% ambient sunlight and in laboratory exposure to UV radiation, incidence of hindlimb malformations increased. However, due to uncertainties in dose extrapolation, the significance of the results in explaining malformations observed in the field is unclear (Ankley et al., 2000). There is growing evidence that the deleterious effects of UV radiation and chemicals may interact or be additive. In the laboratory, northern leopard frog tadpoles exposed to the pesticide s-methoprene exhibited a deformity rate of 2.1%, whereas those exposed to both UV and s-methoprene had a deformity rate of 8.7%. No deformities were observed in the control group (Akins and Wofford, 1999). Exposure of northern leopard frog tadpoles to UV-A, simulating a fraction of summertime, midday sunlight in the northern latitudes, significantly increased the toxicity of fluoranthene (Monson et al., 1999).

A fungal disease, chytridiomycosis, implicated in declines of anurans in Australia, Central America (Berger et al., 1998), and elsewhere, has recently been documented in the United States (Milius, 1998) and as a contributing factor in mass mortality of northern leopard frogs in the Colorado Rockies in the 1970s (Carey et al., 1999). Additional work is needed to clarify the role of chytridomycosis in this and possibly other observed declines of northern leopard frogs.

Diana and Beasley (1998) discuss the toxicology of northern leopard frogs and other amphibians. Underhill (1966) noted an incidence of spontaneous caudal scoliosis in northern leopard frog tadpoles. One of us (M.J.L.) knows of a spring-fed wetland in northwestern Iowa where scoliotic animals are found regularly. The physiology and developmental biology of these animals have been studied extensively (Gibbs et al., 1971; Feder and Burggren, 1992); northern leopard frogs are commonly used as a model organism in laboratory studies.

R. Parasites. Parasitic cysts have been noted in the thigh muscles of northern leopard frogs in Minnesota (Helgen et al., 1998), and trematodes may be contributing to observed limb malformations in some regions (Sessions et al., 1999; see "Disease" above). Brooks (1976b) examined helminth faunas in northern leopard frogs (see also Pollack, 1971). Gilliland and Muzzall (1999) found 12 species of helminths in 43 northern leopard frogs from southern Michigan. Helminth infestations have also been investigated in North and South Dakota (Goldberg et al., 2001) and New Brunswick (McAlpine, 1997c; McAlpine and Burt, 1998). In New Brunswick, helminth species richness was greatest in adults. Helminths that infect the host via skin penetration are most abundant in larger frogs with greater epidermal area (McAlpine, 1997c). A tetrathyridia (*Mesocestoides* sp.) was found in a single specimen of northern leopard frog from Jefferson County, New York (McAllister and Conn, 1990). The respiratory tracts of ranid frogs are susceptible to infection by *Rhabdias* sp., a group of lung worms (Baker, 1978a). Gibbs et al. (1971) suggested that northern leopard frogs remain healthy despite serious parasite loads (Sutherland, this volume, Part One).

4. Conservation.
In Canada, the Southern Mountain population of northern leopard frogs (British Columbia) is listed as Endangered and the prairie population is designated a Species of Special Concern, but the eastern population is not considered to be at risk (Committee on the Status of Endangered Wildlife in Canada, 2002). Northern leopard frogs have no status under the U.S. Endangered Species Act. States and provinces often have species designations that include northern leopard frogs (e.g. the species is Threatened in Alberta and redlisted in British Columbia) or protect the species from hunting.

Causes of decline and extirpation are many, and several probably interact to exacerbate adverse effects. In the western states and provinces, where declines have been most evident, introduction and spread of non-native predators has played an important role. But frogs are often absent from seemingly pristine habitats in the west, particularly at high elevation, suggesting chytridiomycosis, UV radiation, or other less obvious causes. Relationships among stressors are complex; susceptibility to and virulence of diseases can be affected by many factors. Stressors such as contaminants, acidic rainfall, changes in climate or microclimate, or increased UV radiation can cause immunosuppression (Carey et al., 1999, 2001; Lips, 1999). Also, seemingly pristine forests of the western states are often quite altered from predevelopment conditions (Dahms and Geils, 1997) due to a long history of logging, fire suppression, and grazing. The intensity of wildfire in the forests of the western states in recent years highlights just how altered western forest ecosystems have become. How these changes have affected northern leopard frogs is unclear.

In Saskatchewan, Manitoba, Ontario, Québec, and the Midwestern states, widespread declines were evident in the 1960s or 70s, and observations of large die-offs during that time suggest disease or chemical insults as the proximate cause. However, in these areas, leopard frogs clearly depend on a varied landscape that includes overwintering aquatic habitats, breeding ponds, upland foraging areas, and corridors to move among these habitats. This dependence, combined with likely metapopulation structure, makes this species especially susceptible to fragmentation and loss or alteration of habitats on a landscape level. In the Midwest, massive land-clearing and draining of wetlands for cultivation has occurred over the last century and a half. For example, Indiana's forests and wetlands have been reduced by about 78% and 86%, respectively (Miller, 1993; Hartman, 1994; Kolozsvary and Swihart, 1999). On the other hand, closed-canopy forest has increased significantly at the Edwin S. George Reserve in Michigan since 1937, and similar ecological

succession is occurring over other portions of eastern North America due to abandonment of agricultural lands and possibly other causes (Skelley et al., 1999; Werner and Glennemeier, 1999). Northern leopard frogs do poorly in closed-canopy forest (Werner and Glennemeier, 1999). Alteration of habitats has, and no doubt will, continue to contribute to changes in northern leopard frog distribution and abundance. Recent widespread observations of malformed frogs in this region suggests frogs are also under assault from chemical or other stressors. The species is faring better in the eastern states and provinces, but habitat loss, degradation, and fragmentation have caused localized declines and extirpations in that region, and malformed frogs suggest additional stressors are present as well.

Crafting a comprehensive rangewide conservation plan for northern leopard frogs would be a difficult task due to incomplete knowledge of why the species is declining, regional differences in biology and threats, and the daunting challenge of initiating and coordinating conservation across many jurisdictions and the large range of the species. As a result, conservation would probably best be addressed at a regional or population level with coordination and communication among regions.

Rana pretiosa Baird and Girard, 1853c
OREGON SPOTTED FROG

Christopher A. Pearl, Marc P. Hayes

1. Historical versus Current Distribution.
Biochemical, morphological, and ecological differences between Oregon spotted frogs (*Rana pretiosa*; largely the former *R. pretiosa pretiosa*) and Columbia spotted frogs (*R. luteiventris*; mostly the former *R. pretiosa luteiventris*; see account, this volume) confirm these taxa deserve species designation (Green et al., 1996, 1997; M.P.H. and C.A.P., personal observations). Oregon spotted frogs historically ranged from northeastern California to southwestern British Columbia, occurring west of the Cascade crest in British Columbia, Washington, and the Willamette hydrographic basin of Oregon, and east of the crest in the Deschutes, Klamath, and Pit River basins of Oregon and California (Nussbaum et al., 1983; K.R. McAllister, 1995; Green et al., 1996; Hayes, 1997).

Oregon spotted frogs likely occurred patchily in larger emergent wetlands from sea level–1,635 m, with maximum elevations increasing in the southern portion of their range (Nussbaum et al., 1983; Hayes, 1994, 1997). As of 1999, Oregon spotted frogs were known to occupy 31 sites within this range—24 in Oregon, 4 in Washington, and 3 in British Columbia (Hayes, 1997; McAllister and Leonard, 1997; Pearl, 1999; R. Haycock, personal communication).

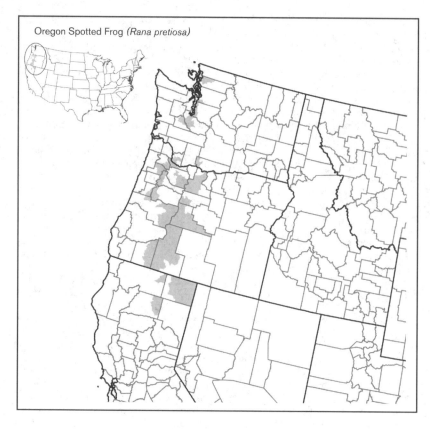

Oregon Spotted Frog *(Rana pretiosa)*

ii. Breeding habitat. Oregon spotted frogs breed in warm, vegetated shallows of open, freshwater marshes and lake margins with little flow (Licht, 1971). Breeding micro-environments are often located in seasonally inundated shallows and are usually hydrologically connected to permanent water (Licht, 1971). Breeding is usually concentrated within a 1–2 wk period (may extend to 4 wk, especially where population sizes are large or suboptimal temperatures alter breeding phenology) and is initiated when water temperatures reach 6–7 °C (Licht, 1969b, 1971; McAllister and Leonard, 1997). Timing of oviposition varies from late February at lowland sites (McAllister and Leonard, 1997) to late June at montane sites in Oregon (C. A. P., personal observations).

B. Eggs.
i. Egg deposition sites. Egg masses are often deposited communally, and aggregations can contain >120 individual masses (McAllister and Leonard, 1997; M. P. H., personal observations). At a lowland Washington site, mean water depth near time of oviposition ranged between 5.8–11.5 cm (McAllister and Leonard, 1997). Desiccation of shallowly deposited egg masses can be a major source of embryonic mortality (Licht, 1974; McAllister and Leonard, 1997).

ii. Clutch size. Egg masses (n = 18) averaged 643 eggs (Licht, 1971) in lowland British Columbia and 598 eggs (n = 6) in western Washington lowlands (McAllister and Leonard, 1997).

C. Larvae/Metamorphosis.
i. Length of larval stage. Larvae hatch in 18–30 d (McAllister and Leonard, 1997). Licht (1975a) reported tadpoles reaching metamorphosis between 110–130 d after hatching in British Columbia, although development can occur in as short as 95 d in Oregon (C. A. P., personal observation).

ii. Larval requirements.

a. Food. Larvae are thought to consume bacteria, algae, and organic detritus (Licht, 1974).

b. Cover. Larvae utilize warm shallows with relatively dense aquatic vegetation and can be cryptic where predatory fish are present (personal observations).

iii. Larval polymorphisms. None reported. Anerythristic larvae and post-metamorphic Oregon spotted frogs have been observed in one Oregon population (C. A. P., personal observation).

iv. Features of metamorphosis. Size at metamorphosis varies with elevation and growing season. In lowland British Columbia, metamorphosis began the first week of August, and transforming Oregon spotted frogs averaged 33 mm (Licht, 1974). At one site near 1,500 m elevation in the Oregon Cascade Range, 24 transforming frogs averaged 26.1 mm in late September 1999, a particularly large snow year (C. A. P., unpublished data). Survivorship to metamorphosis can be extremely

Surveys across the geographic range have not been absolutely comprehensive, but current understanding of the species' habitat characteristics suggests that limited potential habitat exists in unexamined areas and additional sizable populations are probably few. Surveys of historically occupied sites suggest that Oregon spotted frogs are extirpated from 70–90% of their native range, including their entire range in northeastern California (K. R. McAllister et al., 1993; Hayes et al., 1997). Populations west of the Cascade axis appear to have been lost disproportionately: no Oregon spotted frog populations are known to persist across their historical lowland range in western Oregon, and only five extant sites are known from western Washington and British Columbia (K. R. McAllister et al., 1993; Hayes, 1994; R. Haycock, personal communication).

2. Historical versus Current Abundance.
Concern has been voiced since the 1960s regarding Oregon spotted frog declines (since 1997 as Oregon spotted frogs) west of the Cascade crest (Dumas, 1966; Storm, 1966; Shay, 1973; Licht, 1974; Nussbaum et al., 1983; St. John, 1987). Existing data suggest that a sizable portion of extant populations are small, geographically isolated, or restricted to higher elevations in drainage basins that were historically occupied (Hayes, 1997; Pearl, 1999). At least eight (ca. 20%) of the extant populations

are currently thought to be <200 breeding females based on egg mass surveys (McAllister and Leonard, 1997; Pearl, 1999; R. Haycock, personal communication). Licht (1969b) reported similarly small numbers of egg masses (30 and 54 in consecutive years) in a population in British Columbia that is now thought to be extirpated. Surveys of adult abundance suggest seven additional Oregon sites are of similarly limited size (Hayes, 1997).

3. Life History Features.
A. Breeding. Reproduction is aquatic.

i. Breeding migrations. Initiation of breeding activity occurs soon after winter thaw, and adults are thought to move into breeding sites as early as 1–2 wk prior to oviposition (Licht, 1969b, 1971). Recent submerged trapping and telemetry work has revealed that underwater movements that would be undetected during standard visual surveys occur between an aquatic overwintering site (see "Torpor [Hibernation]" below) and the breeding site (Hayes et al., 2001; M. P. H., J. Bowerman, unpublished data). Adults have been observed to be active under ice (Leonard et al., 1997). Emergence of adults has been documented in early February in Oregon's Willamette Valley (Nussbaum et al., 1983) and in late February to early March in southwestern British Columbia and western Washington (Licht, 1969b; Watson et al., 1998, 2003).

low (as low as 1%) due to desiccation of egg masses and high larval predation rates (Licht, 1974).

v. Post-metamorphic migrations. Juveniles may remain around breeding ponds after transformation. Emigration patterns are unknown.

D. Juvenile Habitat. Thought to be similar to adult habitats, but poorly known.

E. Adult Habitat. Adult Oregon spotted frogs are highly aquatic and are rarely found >2 m from surface water (Licht, 1986b). In one shallow wetland in western Washington, adults were associated with microhabitats characterized by deeper water and more open canopy than randomly selected points (Watson et al., 1998). Adults remained associated with standing water as inundation limits receded through the summer season (Watson et al., 1998). Adults may also utilize seasonal pools within 300 m of the breeding site (McAllister and Leonard, 1997; Pearl, 1999).

F. Home Range Size. Watson et al. (2000) report home ranges of adult Oregon spotted frogs averaged between 5.4 and 6.4 ac in one western Washington population.

G. Territories. Not reported to be territorial.

H. Aestivation/Avoiding Desiccation. None documented.

I. Seasonal Migrations. See "Breeding migrations" and "Post-metamorphic migrations" above. Watson et al. (1998) reported that adults were found increasing distances from the breeding site with onset of substantial fall rains, but whether this represents movement toward overwintering sites is unclear. Recent telemetry work has shown the movement between aquatic active-season habitat and overwintering habitat at one western Washington site involves movements on a scale of 50–400 m (Hayes et al., 2001). Submerged trapping efforts suggest that movements between overwintering and breeding sites are relatively similar in scale (M.P.H., unpublished data).

J. Torpor (Hibernation). Dickerson (1906) reported one overwintering site in a mud bottom of a marshy lake fringe that was at least 30 cm (1 ft) deep. Recent telemetry work has revealed that adults use aquatic overwintering sites that appear to be spent in freeze-free low flow-velocity channels or seep and spring areas that are hydrologically linked to breeding sites (see also "Breeding migrations" above; Hayes et al., 2001; C.A.P, personal observation).

K. Interspecific Associations/Exclusions. Oregon spotted frogs can share breeding sites with northwestern salamanders (*Ambystoma gracile*), long-toed salamanders (*A. macrodactylum*), Pacific treefrogs (*Pseudacris regilla*), rough-skinned newts (*Taricha granulosa*), and western toads (*Bufo boreas*). At higher elevations in Oregon, Oregon spotted frogs co-occur with Cascade frogs (*R. cascadae*), and the two have been observed to hybridize in laboratory and field (Haertel and Storm, 1970; Green, 1985b). Oregon spotted frogs historically co-occurred with northern red-legged frogs (*R. aurora*) in lowland western Washington and southwestern British Columbia (Licht, 1971, 1974, 1986a,b; McAllister and Leonard, 1997), but areas of sympatry are now restricted due to loss of lowland spotted frog populations. The occurrence of non-native American bullfrogs (*R. catesbeiana*) at low and mid-elevation historical Oregon spotted frog sites now lacking spotted frogs may suggest a negative interaction (Dumas, 1966; Nussbaum et al., 1983; Hayes, 1997). Negative interactions have been suggested between Oregon spotted frogs and introduced salmonids and centrarchids that are now established at higher elevations and lower elevations, respectively (Hayes and Jennings, 1986; Hayes, 1997).

L. Age/Size at Reproductive Maturity. At their northern range limit in lowland British Columbia, male Oregon spotted frogs are thought to become sexually mature in their second year (mean size 45 mm) and females in either their second or third year (mean size 62 mm; Licht, 1975a). Licht (1974) suggested females at that site may breed each year.

M. Longevity. Not well understood, but adults are thought to be relatively short lived (i.e., generally 2–5 yr) compared to other native western ranids (see Licht, 1975a).

N. Feeding Behavior. Oregon spotted frogs consume a variety of invertebrate prey, many of which are captured in aquatic or near-shore environs (Licht, 1986). Adult Oregon spotted frogs will consume vertebrate prey, including Pacific treefrogs and juvenile conspecifics, northern red-legged frogs, and western toads (Licht, 1986; Pearl and Hayes 2002). Aquatic ambush behavior is described by Pearl and Hayes (2002).

O. Predators. Documented predators of larval Oregon spotted frogs include red-spotted garter snakes (*Thamnophis sirtalis concinnus*), larval diving beetles (Dytiscidae; C.A.P., personal observations), and water scorpions (Nepidae; M.P.H., personal observations). Suspected predators of larvae include sandhill cranes, a variety of introduced fish, rough-skinned newts, northwestern salamanders, backswimmers (Notonectidae), giant water bugs (Belostomatidae), and anisopteran odonates (Licht, 1974). Documented predators of post-metamorphic Oregon spotted frogs include American bullfrogs, adult Oregon spotted frogs, red-spotted garter snakes, and mink (Licht, 1974; Watson et al., 1998; M.P.H., personal observations). Suspected predators of post-metamorphic Oregon spotted frogs include raccoons, great blue herons, kingfishers, red fox, and striped skunks (Licht, 1974, 1986).

P. Anti-Predator Mechanisms. Larvae can reduce activity and distance themselves from known odonate predators in the laboratory (Barnett and Richardson, 2002). Larval Oregon spotted frogs occupy densely vegetated shallows, which may afford them some protection from visually oriented predators (C.A.P., personal observation). When not in water, adults usually position themselves within a few jumps of water. Once in water, adults bury themselves in organic sediments and remain still (Licht, 1986b).

Q. Diseases. Not well known, but in related and sympatric Cascade frogs, a similar communal breeding pattern may be associated with increased occurrence of the oomycete fungus *Saprolegnia* sp. in egg masses (Kiesecker and Blaustein, 1997b).

R. Parasites. Trematodes of the genus *Haplometrana* have been reported in Washington spotted frogs (Lucker, 1931; Pratt and McCauley, 1961), although these observations preceded the taxonomic revision of western spotted frogs. Clark et al. (1969) reported the blood parasites *Trypanosoma* and *Lankesterella* in Oregon spotted frogs, although this determination also preceded taxonomic revision and may include Columbia spotted frogs.

4. Conservation.

Direct loss and alteration of suitable marsh habitat to agriculture, flood control, and urbanization have been severe throughout large areas of the species' range, but are particularly acute in their range west of the Cascade Mountains. Loss of springs through development and spring capping may have resulted in loss of overwintering sites (Hayes, 1997). Introduction of a variety of non-native predators may have played important roles in reducing abundance of Oregon spotted frogs (Hayes, 1997; Hayes et al., 1997). Non-native bullfrogs and introduced fishes are widely distributed over much of the extant range of Oregon spotted frogs, and both have been posited as detrimental to spotted frog populations (Dumas, 1966; Nussbaum et al., 1983; St. John, 1987). Still, when determining the causes of Oregon spotted frog declines, it is difficult to separate the role of these introduced predators from that of widespread habitat modification, but additive or synergistic effects are also suspected (Hayes and Jennings, 1986; also see Adams, 1999). Water quality degradation associated with agricultural and urban land uses is also a stressor (Nussbaum et al., 1983), and laboratory trials suggest that larval Oregon spotted frogs may be more sensitive to nitrogen compounds associated with fertilizers than other northwestern wetland-breeding amphibians (Marco et al., 1999). Field trials and assessment of enzymatic repair activity suggest Oregon spotted frog embryos are relatively resistant to current incident ultraviolet radiation (Blaustein et al., 1999). Efforts to establish or enhance populations afield are underway in Oregon, United States (C. A. P. and R. B. Bury, unpublished data), and British Columbia, Canada (R. Haycock, personal communication).

Oregon spotted frogs are considered a candidate for Federal listing by the U.S. Fish and Wildlife Service (1997a). They have some designation indicating concern about their status in every political entity that encompasses their historical range. They are considered Endangered by the State of Washington, Sensitive-Critical in Oregon (Oregon Natural Heritage Program, 1995), and a Species of Special Concern in California, where they have been recommended for Endangered status (Jennings and Hayes, 1994a). Oregon spotted frogs are considered Red-listed (COSEWIC Endangered) in British Columbia.

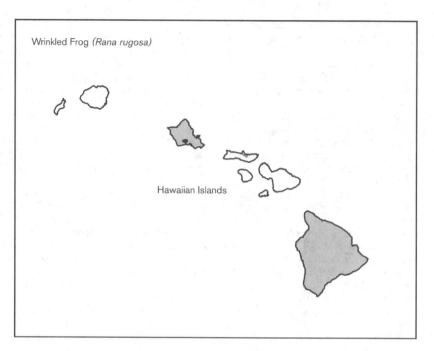

Wrinkled Frog (*Rana rugosa*)

Hawaiian Islands

Rana rugosa Temminck and Schlegel, 1838
WRINKLED FROG

Fred Kraus

1. Historical versus Current Distribution.
Wrinkled frogs (*Rana rugosa*) were probably introduced to Hawaii from Japan in 1895 or 1896 by Albert Koebele to assist in controlling introduced insects (Bryan, 1932; Oliver and Shaw, 1953). In their native range, wrinkled frogs are widely distributed in Japan, Korea, northeastern China, and parts of the Russian Far East (Maeda and Matsui, 1989; Zhao and Adler, 1993; Hasegawa et al., 1999), although the taxonomic status of populations on the mainland may need further assessment (Zhao and Adler, 1993). In Hawaii, they are presently found on Kauai, Oahu, Maui, and Hawaii Island (Cochran and Goin, 1970; McKeown, 1996), although vouchers are apparently absent from Hawaii Island. It is unclear whether wrinkled frogs are still expanding their range in Hawaii, although introduction to additional islands would provide them this opportunity. The species was apparently first released on Oahu (Oliver and Shaw, 1953), was reported from Maui by 1936 (Svihla, 1936; Fisher, 1948), and from Kauai and Hawaii Island by 1953 (Oliver and Shaw, 1953). Wrinkled frogs currently range from near sea level to at least 1,100 m elevation in Hawaii.

2. Historical versus Current Abundance.
In Hawaii, abundant populations have been established for decades (Svihla, 1936; Hunsaker and Breese, 1967), and wrinkled frogs remain common (McKeown, 1996; personal observations); they also are common in their native range (Shannon, 1959).

3. Life History Features.
A. Breeding. Reproduction is aquatic.

i. Breeding migrations. In Hawaii, wrinkled frogs breed from February–August (Tinker, 1938; Oliver and Shaw, 1953), whereas in their native range, breeding is from mid-May to 1 September (Okada, 1938, 1966; Maeda and Matsui, 1989; Hirai and Matsui, 2001). Breeding occurs only after daily average water temperatures >18 °C (Chang, 1994). Females may produce multiple clutches per year (Maeda and Matsui, 1989; Chang, 1994).

ii. Breeding habitat. In Hawaii, adults breed in the pools and slow-moving waters of mountain streams and in lowland ponds (Tinker, 1938; McKeown, 1996). In their native range, wrinkled frogs breed in rice paddy fields, ponds, ditches, or pools in dry riverbeds (Okada, 1966; Maeda and Matsui, 1989) and presumably in pools of mountain streams as well. Most breeding activity occurs at night (Okada, 1938, 1966).

B. Eggs.

i. Egg deposition sites. Females deposit eggs in loose, unattached masses in pools, either among protruding sticks and vegetation or in bare areas (Svihla, 1936; Okada, 1966; McKeown, 1996). Eggs are laid in clusters of 10–200 (Okada, 1938; Chang, 1994).

ii. Clutch size. Females spawn from 400–1,350 eggs at a time, depending on their body size (Okada, 1938, 1966; Chang, 1994). Females can spawn up to three times per year (Chang, 1994).

C. Larvae/Metamorphosis.

i. Length of larval stage. Eggs hatch approximately 5 d after laying (Chang, 1994). Newly hatched tadpoles are approximately 8 mm TL (Oliver and Shaw, 1953) and grow to between 38–80 mm TL prior to metamorphosis (Okada, 1938; Gilbertson and Watermolen, 1998). Juveniles transform at lengths of 19.5–26.8 mm SVL in Hawaii (Oliver and Shaw, 1953). In their native range in Japan, tadpoles do not metamorphose in the same year they hatch, but from April–October of the following year (Okada, 1966; Maeda and Matsui, 1989; Hirai and Matsui, 2001). However, Khonsue et al. (2001) state that intra-populational variation occurs in larval overwintering, with some animals metamorphosing prior to their first winter. It seems possible that tadpoles may metamorphose in their first year in Hawaii, because Tinker (1938) notes that "frogs may be found in Hawaii in all stages of development from early spring until late in the fall."

ii. Larval requirements.

a. Food. Tadpoles presumably feed on algae and microorganisms that use rocks and aquatic vegetation for a substrate, but to date this has not been investigated directly.

b. Cover. In Japan, tadpoles overwinter in the mud of rice paddy fields or streams (Okada, 1938, 1966).

iii. Larval polymorphisms. Unknown and unlikely.

iv. Features of metamorphosis. Metamorphosis occurs during the spring and summer following hatching in their native range (Okada, 1938, 1966; Hahn, 1960), but apparently occurs during much of the year in Hawaii (Tinker, 1938).

v. Post-metamorphic migrations. Unknown and unlikely.

D. Juvenile Habitat. Probably similar to adult habitats, but not investigated.

E. Adult Habitat. In Hawaii adults occur in both lentic and lotic habitats, ranging from low elevation taro ponds to cool, clear, mid-elevation streams (Bryan, 1932;

Svihla, 1936; personal observations). In the smaller streams they typically inhabit pools; in larger streams with more current, they occur along the sides of quiet backwaters. Adults frequently bask in the sun in open grassy areas (personal observations) or on rocks protruding from streams (McKeown, 1996). In their native range, wrinkled frogs are widely distributed in plains and low mountain valleys, being found in rice paddies, ponds, ditches, and reservoirs, and along small, fast-flowing streams (Hallowell, 1860; Stejneger, 1907; Okada, 1938, 1966; Stewart, 1953; Shannon, 1959; Hahn, 1960; Webb et al., 1962; Maeda and Matsui, 1989). They frequently bask in tufts of grass or reeds along stream or pond margins (Hahn, 1960; Webb et al., 1962).

F. Home Range Size. Unknown.

G. Territories. Unknown.

H. Aestivation/Avoiding Desiccation. Apparently absent inasmuch as frogs are active from March to early December and breed during the summer (Okada, 1938, 1966; Chang, 1994).

I. Seasonal Migrations. Although wrinkled frogs are typically found in or near water and hibernate under water, in wet forests in Hawaii, frogs can be found far from any permanent water body (personal observations).

J. Torpor (Hibernation). In their native range, adult wrinkled frogs will overwinter underwater (Maeda and Matsui, 1989), while tadpoles overwinter in the mud of paddy fields or streams (Okada, 1966). Hibernation season is variable and depends on latitude (Okada, 1938). In Japan and Korea, wrinkled frogs commence hibernation from late October to early December and emerge from hibernation beginning in mid-March to late April (Okada, 1938; Shannon, 1959; Chang, 1994). Hibernation in Hawaii seems unlikely because of the mild climate, but this has not been investigated directly.

K. Interspecific Associations/Exclusions. Not reported.

L. Age/Size at Reproductive Maturity. Adult males range in size from 30–47 mm and females from 45–60 mm (Okada, 1938, 1966; Chang, 1994). As with most ranids, males have swollen nuptial pads during the breeding season (Okada, 1938, 1966). Age at attainment of sexual maturity is usually 1–2 yr post-metamorphosis in males and 2–3 yr post-metamorphosis in females, although a few females mature at 1 yr post-metamorphosis (Khonsue et al., 2001).

M. Longevity. Males can live up to 4 yr post-metamorphosis; females up to 5 yr (Khonsue et al., 2001).

N. Feeding Behavior. Wrinkled frogs eat primarily a variety of insects (in 98% of examined stomachs), but also are known to consume arachnids, crustaceans, chilopods, diplopods, mollusks, oligochaetes, and small frogs (Okada, 1938, 1966; Maeda and Matsui, 1989;

Hirai and Matsui, 2000, 2001). They feed on ants to a larger extent than most other frogs (Hirai and Matsui, 2000), including syntopic congeners (Hirai and Matsui, 2001).

O. Predators. Chang (1994) reports goldfish (Carassius auratus) eating wrinkled frog eggs. McKeown (1996) implies American bullfrogs (R. catesbeiana) are predators, but presents no evidence. Humans eat wrinkled frogs in Japan (Oliver and Shaw, 1953).

P. Anti-Predator Mechanisms. When alarmed, wrinkled frogs leap into the water and hide in leaf litter, bottom muck, or among rocks for ≤10 min before resurfacing (Shannon, 1959; Hahn, 1960; McKeown, 1996). According to Choi et al. (1999), wrinkled frogs combine a crypsis-escape predator avoidance mechanism with a crouching posture and noxious skin secretions to deter predators. These authors note that by simply crouching, wrinkled frogs increased their survival rate by 37% and delayed the time to predation by 23 min when compared to a congener that did not crouch.

Wrinkled frogs may be unpalatable to some snakes (Mori, 1989, in Chang, 1994). Nine small peptides, named gaegurins and rugosins, have been isolated from R. rugosa skin. These compounds have antimicrobial properties against an array of bacteria, fungi, and protozoa (Park et al., 1994; Suzuki et al., 1995).

Q. Diseases. Unknown. Eggs can be infected by an unidentified fungus (Chang, 1994).

R. Parasites. The nematodes Angiostoma bufonis, Hedrurus ijimai, Oswaldocruzia insulae, Rhabdias nipponica, Spinitectus ranae, and Spiroxys japonica and the digenean helminth Opisthioglyphe japonicus have been reported from the digestive tracts or lungs of Japanese specimens (Morishita, 1926; Yamaguti, 1935; Hasegawa and Otsuru, 1977, 1978, 1979; Uchida et

al., 1980). Over their native range, wrinkled frogs appear resistant to parasitism by the nematode Gnathostoma nipponicum (Kim, 1983; Oyamada et al., 1998) and by parasites in the genus Sparganum (Kim, 1983). An investigation of 39 animals from 1,100 m elevation on Kauai revealed no helminth parasites (S. Goldberg, personal communication).

4. Conservation.
Wrinkled frogs have been introduced to Hawaii and have invaded native forest. They have no legal protection, and their demise in Hawaii would be a positive development.

Rana septentrionalis Baird, 1854(b)
MINK FROG

Gary S. Casper

1. Historical versus Current Distribution.
The southern limit of distribution of mink frogs (Rana septentrionalis), or "the frogs of the north," is at the highest latitude of any North American anuran (43 °N; Hedeen, 1986). Historically, mink frogs were distributed from southern Labrador and the Maritime Provinces to Minnesota and southeastern Manitoba, south to northern New York and northern Wisconsin, with isolated colonies in northern Québec and northern Labrador. Hedeen (1977) reviews erroneous locality records. Western range limits are thought to be influenced by sufficient moisture (Hedeen, 1986). Hypotheses to explain the southern limits of the distribution include predation pressures from American bullfrogs (R. catesbeiana; Moore, 1952; Bleakney, 1958; Schueler, 1975) and the limited tolerance of embryos to warmer water temperatures and consequent lower oxygen diffusion rates (Hedeen, 1986). No substantial changes in distribution have been reported, although global warming trends can be expected to have an effect in the future.

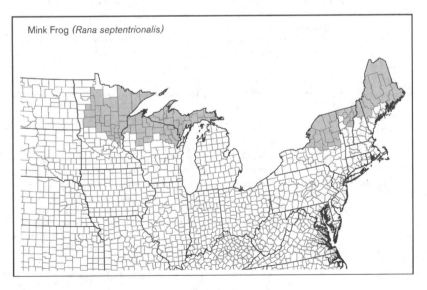

Mink Frog (*Rana septentrionalis*)

2. Historical versus Current Abundance.

Mink frogs generally are considered to be locally common in suitable habitats, and range-wide abundance appears to be unchanging. Judge et al. (1999) performed a mark–recapture study in an Ontario population and found wide annual fluctuations in population estimates of mink frogs, American bullfrogs, and northern green frogs (*R. clamitans melanota*). No directional trends could be discerned. Shirose and Brooks (1997) likewise detected no trend in a study spanning 9 yr in a central Ontario wilderness area, but noted marked short-term fluctuations in population sizes, with the percent of transforming individuals varying among years from 21.6–60.2% (mean = 38.4, s.d. = 17.2, among 6 yr). Mink frog density estimates in New Brunswick ranged from 0–88.8 frogs/100 m^2 (McAlpine, 1997b). Mink frogs may benefit from utilizing aquatic habitat created by beaver (*Castor canadensis*) ponds, and recent increases in beaver may have increased breeding habitat availability in some areas (New Brunswick, McAlpine, 1997a; Apostle Islands, Wisconsin, personal observations, 1999).

3. Life History Features.

A. Breeding. Reproduction is aquatic.

i. Breeding migrations. Mink frogs do not have breeding migrations. Their entire life cycle is completed in relatively close proximity in aquatic shoreline habitats, typically in areas with high macrophyte abundance.

ii. Breeding habitat. Breeding takes place in permanent waters with a high abundance of aquatic macrophytes. Shortly after males begin calling, females arrive to mate and deposit egg masses on the submerged stems of macrophytes. The breeding period extends from the first week in June to the first week in August (Garnier, 1883; Wright and Wright, 1949; Bleakney, 1958; Hedeen, 1971a, 1972c). In Minnesota, male choruses run from late May to early August, calling males are not restricted to fixed areas, and males spend more time at breeding sites than females (Hedeen, 1972c; Hunter et al., 1992).

B. Eggs.

i. Egg deposition sites. The globular egg mass is attached to stems of submerged vegetation, often ≥1 m below the water surface (Vogt, 1981), in permanent aquatic habitats such as rivers, lakes, and ponds (Hedeen, 1971a; Hedeen, 1986). The subsurface placement of the egg mass avoids subsequent freezing of surface waters in northern latitudes, which would kill eggs (Moore, 1940; Herreid and Kinney, 1967). Egg masses may detach from vegetation and drop to the bottom of the pond or stream some time after laying, usually remaining viable and completing development (DeGraaf and Rudis, 1983; Hunter et al., 1992).

ii. Clutch size. Clutch size ranges from 500–4,000 eggs (Vogt, 1981), although Shirose and Brooks (1995b) report a maximum of 2,000 eggs/clutch. The upper limiting embryonic temperature is 30 °C (Moore, 1952). At warmer temperatures, oxygen diffusion is inadequate to supply embryos in the center of the egg mass, and dying embryos may produce decomposition products lethal to neighboring embryos (Moore, 1949a,b). The length of time needed for eggs to develop and hatch is apparently variable and more data are needed (Bleakney, 1958).

C. Larvae/Metamorphosis.

i. Length of larval stage. Metamorphosis usually takes place in July–August following a 1-yr larval period at body lengths ranging from 25–42 mm, although sometimes 2 yr are required and tadpoles can reach total lengths exceeding 72 mm (Garnier, 1883; Wright, 1932; Wright and Wright, 1949; Hedeen, 1971a, 1972c; Vogt, 1981). Tadpole growth slows during the winter (Hedeen, 1971a), and northern populations exhibit a prolonged larval period (Leclair and Laurin, 1996).

ii. Larval requirements.

a. Food. Larvae feed primarily on algae. Garnier (1883) reported mink frog tadpoles feeding on dead fishes and dead green frog tadpoles, but this observation has not been corroborated. Hedeen (1972b) reports a primarily algal diet in Minnesota and no observations of carnivory despite carcasses being available. Feeding is suspended during metamorphosis when forelimbs first appear, and an adult diet begun when the tail is <2.1 cm long (Hedeen, 1972b).

b. Cover. Tadpoles use aquatic vegetation for cover.

iii. Larval polymorphisms. Unknown and unlikely.

iv. Features of metamorphosis. As mentioned above (see "Length of larval stage"), metamorphosis usually takes place in July–August at least 1 yr after eggs are laid.

v. Post-metamorphic migrations. Rare; juveniles and adults share the same habitats.

D. Juvenile Habitat. Similar to adults.

E. Adult Habitat. Mink frogs are highly aquatic. Tolerance to desiccation is substantially lower than the tolerances of other northern anurans, and tolerance to hydration is greater than in other anurans (Schmid, 1965b). Terrestrial activity is not commonly reported, and usually restricted to periods of nocturnal precipitation (Hedeen, 1986; Schueler, 1987; personal observations, 1990s). Adults typically occupy rivers, lakes, ponds, pools, puddles, ditches, and streams, avoiding rapid currents and large wave activity, and preferring quiet bays and protected areas with a high abundance of aquatic macrophytes, especially floating water-lily (Nymphaeaceae) and pickerel weed (*Pontederia cordata*) or the edges of sphagnum mats (Garnier, 1883; Jackson, 1914;

Wright, 1932; Moore and Moore, 1939; Wright and Wright, 1949; Gorham, 1970; Hedeen, 1971a, 1972a,b; Stewart and Sandison, 1972; Vogt, 1981; Hedeen, 1986). A mean body temperature of 28.8 °C is reported for four frogs basking in the sun on 25.0 °C rocks in 22.0 °C air (Brattstrom, 1963). A median lethal temperature of 36 °C is reported for Minnesota mink frogs submerged in warm water for 1 hr (Dean, 1966). Field body temperatures of 27 Minnesota mink frogs ranged from 16.2–27.1 °C, and thermoregulation (basking and posturing to control body temperatures) is utilized (Hedeen, 1971b). The maximum voluntary temperature recorded is 30.5 °C (Brattstrom, 1963).

F. Home Range Size. Unknown. Dispersal movements through forested habitats typically occur on rainy nights, but dispersal distances are unknown (Schueler, 1987; personal observations, 1990s).

G. Territories. Territoriality has not been reported, although investigations are probably incomplete.

H. Aestivation/Avoiding Desiccation. Aestivation has not been reported. Because mink frogs are highly aquatic and occupy the permanent waters of northern latitudes, aestivation is probably not a necessary behavior.

I. Seasonal Migrations. Schueler (1987) observed overland movements during rainy October nights in Ontario, which he interpreted as movements towards hibernation sites. Most mink frogs probably hibernate within the same waters they inhabit the rest of the year, however.

J. Torpor (Hibernation). Mink frogs are not freeze tolerant (Schmid, 1982); to escape freezing they hibernate in the bottom mud of permanent lakes and streams (Schmid, 1965b; Hunter et al., 1992; Harding, 1997). In the Great Lakes region, they may enter dormancy by late September and remain inactive until well into May (Harding, 1997).

K. Interspecific Associations/Exclusions. Sympatry is reported with green frogs (Kramek, 1972; Stewart and Sandison, 1972; Shirose and Brooks, 1995b; Leclair and Laurin, 1996), American bullfrogs (Bleakney, 1958; Stewart and Sandison, 1972; Werner and McCune, 1979; Shirose and Brooks, 1995b; Leclair and Laurin, 1996), northern leopard frogs (*R. pipiens*; Hedeen, 1972b; Leclair and Laurin, 1996), and wood frogs (*R. sylvatica*; Hedeen, 1972b). Hedeen (1972b) described seasonal niche overlap, where northern leopard frogs moved into the aquatic habitat of mink frogs in late summer. Diets differed, with mink frogs ingesting a greater proportion of aquatic insects, and northern leopard frogs ingesting more terrestrial items when terrestrial foraging habitat was available; however, northern leopard frog diets became similar to mink frog diets at sites where no terrestrial foraging habitat was available. Kramek (1972) suggested that interspecific competition with American bullfrogs and green frogs may influence diet through feeding niche separation. Stewart and Sandison (1972) found differences in habitat preferences between mink frogs, bullfrogs, and green frogs in New York; these differences were reflected in their diets. Mink frogs primarily inhabited the aquatic zone, sitting on lily pads or other floating vegetation when not submerged, while green frogs and American bullfrogs are found primarily on water margins, with green frogs moving rarely into the aquatic zone and often into the terrestrial zone. Mink frogs ingested fewer food types than bullfrogs (26 versus 46, with only 6 items shared), ingested fewer aquatic invertebrate food types than bullfrogs (26% versus 32%), and frogs comprised a major component of the bullfrog diet while no frogs were recorded from the mink frog diet. Dietary overlap of food items ingested by mink frogs with sympatric green frogs was 48%, with volume of aquatic food groups representing 36% in mink frogs and 24% in green frogs. Courtois et al. (1995) found that mink frogs and green frogs were more abundant and had a more even distribution across different habitats when American bullfrogs were absent. Viable triploid hybrids have resulted from lab crosses of mink frog eggs with American bullfrog sperm (Elinson, 1993).

L. Age/Size at Reproductive Maturity. Mink frogs transform at a relatively larger size than other ranids; immediately following metamorphosis, froglets already have a body size representing nearly 60% of the mean adult size, whereas this value is somewhere between 30–40% in other ranids (Leclair and Laurin, 1996). Maturity is reached after 1 yr of post-metamorphic life in southern populations, and after 2 yr in the North (Leclair and Laurin, 1996). Males become sexually mature when about 45–50 mm long, approximately 1 yr after transformation, while females become sexually mature when about 54–59 mm long, 1–2 yr after transformation (Hedeen, 1972c). In central Ontario, sexual maturity is reached at 3 yr post-transformation (mean standard length = 63.8 mm) in females, and 2 yr post-transformation (mean standard length = 56.6 mm) in males (Shirose and Brooks, 1995a). Females grow faster and attain larger sizes than males (Shirose and Brooks, 1995a; 11% larger than males, Leclair and Laurin, 1996). Mean age and maximum longevity of females is higher than males in southern populations (Leclair and Laurin, 1996). Hedeen (1972c) concluded that only some females bred 1 yr after transforming, at 54–59 mm. Wright (1932) measured size classes of mink frogs from New York and Ontario without discriminating between sexes

and localities and reported sizes of 44 mm (40–48 mm) in the first year after transformation, 53.5 mm (48–58 mm) in the second year, and 63 mm (59–72 mm) in the third year. Average SULs in Québec were 41.3–48.9 mm for first-year females, 41.2–46.5 mm for first-year males, 58.4–66.1 mm for second-year females, 52.8– 60.6 mm for second-year males, 65.4–70.4 mm for third-year females, 63.1 mm for third-year males, and 70.0 mm for fourth-year females (Leclair and Laurin, 1996). Schueler (1975) demonstrated geographic variation in size, with mink frogs much larger in the North, and postulated that individuals in southern populations attained sexual maturity earlier and at a smaller size due to predation pressures. Leclair and Laurin (1996) found that specimens from a northern population were 17% larger than specimens from a southern population and had similar annual growth rates despite the longer growing season in the South. Growth rate, delayed maturity, greater mean ages, and size at transformation all contribute to larger size in adult mink frogs at northern localities (Leclair and Laurin, 1996). Sex ratios favored females in southern populations where males incurred greater mortality (Leclair and Laurin, 1996).

M. Longevity. In populations studied by Leclair and Laurin (1996), most mink frogs survived only 1–2 yr past metamorphosis, with females sometimes reaching 3–4 yr of age and males only rarely entering a third year. In central Ontario, Shirose and Brooks (1995a,b) found mortality to be lowest at transformation and then increasing gradually, with an estimated longevity (maximum life span) of 5–6 yr past transformation and a mean life expectancy of 1.7–4.0 yr past transformation.

N. Feeding Behavior. As with most adult ranids, mink frogs are opportunistic feeders and prey upon anything of a proper size that moves, thus diet generally reflects the availability of prey within the foraging habitat, which is mostly the water surface. The diet is therefore dominated by organisms such as dragonflies and damselflies, diving and whirlygig beetles, waterbugs, and aphids (Hedeen, 1972b; Kramek, 1976). Kramek (1976) studied feeding behavior of the mink frog. Actively feeding frogs usually assume an erect, "head up" posture on floating vegetation and wait for prey items to come within striking distance. Hunting frogs will orient toward prey before striking, and stalking of prey occurs if the prey is outside of the striking distance, sometimes including an underwater approach to within striking distance. If no prey approaches, the frog may change position within a 1–3 m hunting area. Mink frogs exhibit strong tendencies to return to certain places within a hunting area, and capture suc-

cess is high (84%) for slow-moving prey (i.e., aphids) and low (16%) for fast-moving prey (i.e., aerial insects). In Minnesota, Hedeen (1972b) found plant material in 90.5% of the stomachs he examined (mostly duckweeds, *Lemna* sp.), but considered it to have been ingested incidental to predation on animals. Stomachs of Minnesota mink frogs contained Collembola, Odonata, Hemiptera, Homoptera, Neuroptera, Coleoptera, Trichoptera, Lepidoptera, Diptera, Hymenoptera, Araneida, Acarina, and Pulmonata (Hedeen, 1972b). Garnier (1883) reported food items of water insects, beetles (Carabidae), millepedes (Julidae), and small fish (chubs). Kramek (1972) reported leeches, snails, spiders, and 35 insect families from mink frog stomachs in New York. Odonata, Coleoptera, and Homoptera predominated, with aphids being consumed most often and in greatest volume (21.7%). There was no difference between the diets of similarly sized males and females. Stewart and Sandison (1972) recorded the following stomach contents from New York: Acarina, Araneae (7 families), Collembola, Plecoptera (larva), Odonata (2 families), Coleoptera (13 families), Hemiptera (5 families), Lepidoptera (larva), Mecoptera (larva), Diptera (5 families), and Hymenoptera (4 families). Dytiscids (diving beetles) and gyrinids (whirlygig beetles) were the most important food items. Frequency of both plant and animal matter was 100%; volume of plant and animal matter was 10.5% and 89.5%, respectively. The occurrence of plant matter in stomachs was interpreted as accidental ingestion of floating and moving material.

O. Predators. Schueler (1975) concluded that American bullfrogs are major predators on mink frogs. Stewart and Sandison (1972) also report bullfrog predation on mink frogs. Other predators include raccoons (*Procyon lotor*; Hedeen, 1972a), great blue herons (*Ardea herodius*) and other Ciconiiformes (Bent, 1926; Wright, 1932; adults and tadpoles, Hedeen, 1967), wood ducks (*Aix sponsa*; Mallory and Lariviere, 1998), the spruce grouse (*Dendragapus canadensis*; Applegate, 1978), eastern newts (*Notophthalmus viridescens*; eggs and embryos, Wright, 1932), green frogs (Moore, 1952), larvae of eastern tiger salamanders (*Ambystoma tigrinum*; tadpoles, Hedeen, 1972a), snakes (probably common garter snakes, *Thamnophis sirtalis*; Hoopes, 1938), five-spined sticklebacks (*Culaea inconstans*; tadpoles, Hedeen, 1972a), giant water bugs (*Lethocerus americanus*; tadpoles, Hedeen, 1972a), and leeches (*Macrobdella decora*; tadpoles, Hedeen, 1972a).

P. Anti-Predator Mechanisms. Hedeen (1972a) studied escape behavior in Minnesota and concluded that the usual response (+ 85%) of frogs to disturbance was to dive underwater and conceal themselves in mud or vegetation, remaining hidden for 30 s to 26 min. Upon returning to the surface, frogs remained wary, exposing only their eyes and snout above the water surface and diving again at the slightest disturbance. Vogt (1981) suggested that mink frogs give a warning cry when startled, but this vocalization, while characteristic of American bullfrogs, has not been observed by this author or by other mink frog researchers (D.F. McAlpine and S.E. Hedeen, personal communications, 2001). Presumably, the musky odor of the skin secretions is offensive to some predators (Vogt, 1981).

Q. Diseases. Red-leg (usually *Aeromonas* bacteria) has been reported (Hedeen, 1972a). The icosahedral frog erythrocytes virus (Iridoviridae) has been isolated from the cytoplasm of Ontario mink frogs (Gruia-Gray et al., 1989), as well as an icosahedral cytoplasmic virus in leukocytes (Desser, 1992). Malformations have been reported from several locations, with a marked increase in this phenomenon in the past decade (Wisconsin—supernumerary limbs, unresorbed or deformed tails, misplaced bony projections, underdeveloped digits, Robert DuBois, personal communication, 1996; Minnesota—over 40% malformation rates recorded, Hoppe, 1996). Investigations into causal agents of these malformations remain equivocal, but trematode parasites (Sessions and Ruth, 1990; Johnson et al., 1999; Sessions et al., 1999), chemical contaminants (Ouellet et al., 1997a; Gardiner and Hoppe, 1999), and UV-B light (Worrest and Kimeldorf, 1975; Long et al., 1995; Blaustein et al., 1997) are implicated.

R. Parasites. Parasites reported include leeches (*M. decora* on adults and tadpoles, Hedeen, 1972a; Barta and Desser, 1984), opalinid ciliate infusorians (*Opalina* sp.; Metcalf, 1923), protozoans (coccidians, Chen and Desser, 1989; the hematozoan *Lankesterella* sp., Desser et al., 1990; *Aegyptianella ranarum* [Rickettsiales, Anaplasmataceae], Desser, 1987), and nematodes (giant kidney worm *Dioctophyma renale*, Mace and Anderson, 1975; microfilaria, presumably *Foleyella* sp., Barta and Desser, 1984).

4. Conservation.
Range-wide, mink frog abundance appears to be unchanging. Threats include pesticides, acidification, and increased UV-B radiation. Repeated applications of the broad spectrum insecticide fenitrothion (0, 0-dimethyl 0-[4-nitor-m-tolyl] phosphorothioate), commonly used for spruce budworm (*Choristoneura fumiferana*) control, may adversely affect mink frog populations in forest ponds (McAlpine et al., 1998). The susceptibility of mink frog breeding ponds to pH depression resulting from atmospheric acid deposition is considered moderate (Schreiber and Newman, 1988). Because mink frogs thermoregulate

by basking in the sun, they may be susceptible to damage to the skin and eyes from UV-B radiation, especially as UV-B levels increase in northern latitudes with continuing ozone depletion in the upper atmosphere. However, no data are available on UV-B susceptibility.

Rana sevosa Goin and Netting, 1940
DUSKY GOPHER FROG

Stephen C. Richter, John B. Jensen

A recent genetic study of gopher frog (*Rana capito*) populations across the current geographic distribution (Mississippi–North Carolina) by Young and Crother (2001) indicated that the Mississippi population was genetically distinct. Young and Crother (2001) therefore elevated the Mississippi population to specific status by resurrecting *R. sevosa* Goin and Netting (1940; dusky gopher frog), because this population is the only one remaining in the historical geographic range of *R. sevosa* (Louisiana to Mobile County, Alabama) as described by Goin and Netting (1940). Assigning the population discovered in Baldwin County, Alabama, after the publication of Goin and Netting (1940) to *R. sevosa* or *R. capito* is difficult because this population was not included in their study and has since become extinct. Netting and Goin (1942a) assigned this population to *R. sevosa* based on the then-known distribution of *R. sevosa* (Louisiana to Mobile County, Alabama) and *R. capito* (Florida to North Carolina). However,

tions sampled east of the Mobile Bay (thus remaining *R. capito*). Based on this evidence, we assign the Baldwin County population to *R. capito* (as indicated by the geographic range maps) and note the inherent uncertainty. It currently is not possible to determine the location of the contact zone between *R. sevosa* and *R. capito*, but the extensive Mobile Basin creates a logical barrier to dispersal and thus probably separates the two species.

1. Historical versus Current Distribution.
Dusky gopher frogs (*R. sevosa*) are endemic to the coastal plain of Louisiana (Saint Tammany, Tangipahoa, and Washington parishes), Mississippi (Hancock, Harrison, Jackson, and Pearl River counties), and southwestern Alabama (Mobile County; Goin and Netting, 1940; Netting and Goin, 1942a; Altig and Lohoefener, 1983; Dundee and Rossman, 1989). They were last documented in Louisiana in 1967, although they are now thought to be extirpated (R. Thomas, personal communication). Populations from historical localities in Mobile County, Alabama, are also thought to be extinct (Mount, 1990; M. Bailey, personal communication). Two populations were found in Mississippi (Harrison and Jackson counties) following extensive surveys of historical localities and suitable habitat in the late 1980s, (S.C.R. unpublished; G. Johnson, personal communication).

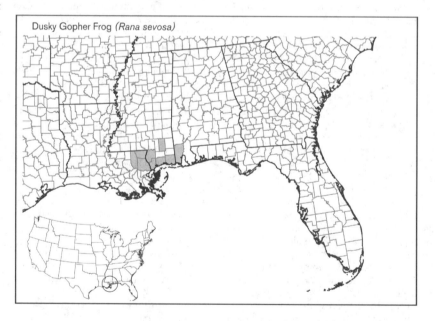

Dusky Gopher Frog (*Rana sevosa*)

populations later discovered in counties proximate to Baldwin County were included in the genetic study of Young and Crother (2001) and were not genetically distinct from all other popula-

2. Historical versus Current Abundance.
Although once abundant in coastal Mississippi (Allen, 1932), breeding populations of dusky gopher frogs currently are known from only a single pond (Glen's

Pond) located in DeSoto National Forest in Harrison County, Mississippi. Population size at Glen's Pond is relatively small; an average of 57 females/year bred from 1988–2001 (maximum = 130 females; Richter et al., 2003).

3. Life History Features.

A. Breeding.

i. Breeding migrations. Males typically arrive at the breeding ponds earlier than females and begin chorusing (Richter and Seigel, 2002). Breeding occurs from December–April in Louisiana and Mississippi (Dundee and Rossman, 1989; Richter et al., 2003), although earlier chorusing and breeding has occurred following tropical storms and hurricanes (e.g., Seigel and Kennedy, 2000).

ii. Breeding habitat. Dusky gopher frogs typically breed in relatively shallow, ephemeral ponds that contain both emergent and submergent vegetation and are located in upland longleaf pine (Pinus palustris) forests or pine flatlands (Wright and Wright, 1949; Dundee and Rossman, 1989; S. C. R., personal observations).

B. Eggs.

i. Egg deposition sites. Eggs typically are deposited on emergent herbaceous vegetation (Goin and Netting, 1940; Dundee and Rossman, 1989), but have been observed attached to small trees, submergent vegetation, and floating woody debris (S. C. R., personal observations). Egg masses are deposited singly and in more centrally located sites (as opposed to the clumped and peripherally located egg masses of syntopic southern leopard frogs [R. sphenocephala]; Richter, 2000).

ii. Clutch size. Females deposit a single clutch ranging in size from 500–2,800 eggs in Mississippi (Richter and Seigel, 1997, 1998; Richter, 1998). Clutch size estimates in Louisiana ranged from 3,000– 7,000 (Volpe, 1957; Dundee and Rossman, 1989). Volpe (1957) measured a mean egg diameter (vitellus) of 1.84 mm (range = 1.67–2.18 mm; n = 100; see Volpe [1957] for a detailed account of embryonic development).

C. Larvae/Metamorphosis. Developmental times range from 141–155 days at 20 ± 3 °C in the laboratory (Volpe, 1957) and from 81–179 days in the field (Richter et al., 2003). Volpe (1957) reported a maximum total length of 74 mm for larvae reared in the laboratory (see Volpe, 1957 for a detailed account of larval development to metamorphosis). Size at metamorphosis can vary widely among years, primarily as a result of hydroperiod length (Richter et al., 2003). Smith and List (1955) reported the body size (SVL) of one metamorphic frog in Mississippi as 31 mm. Volpe (1957) reported the mean size of laboratory-reared frogs in Louisiana to be 28.2 mm SVL (range = 26.5–30.1 mm). Richter and Seigel (2002) found variation across years in metamorphic body sizes primarily due to hydroperiod—1997:

mean SVL = 35.7 mm (n = 213; range = 31.8–42.0 mm) and mean mass = 5.0 g (n = 213; range = 3.2–6.8 g); 1998: mean SVL = 29.8 mm (n = 813; range = 24.8–34.6 mm) and mean mass = 2.6 g (n = 813; range = 1.5–3.9 g). Reproductive success (measured as proportion of metamorphic animals emerging compared to number of eggs deposited) is low (0.35–4.9%; Richter et al., 2003). Estimates for other ranid frogs with similar life histories range from 4.3–5% (Turner, 1960; Herreid and Kinney, 1966; Calef, 1973a).

D. Juvenile Habitat. Juvenile habitat use has not been studied but is assumed to be similar to that of adults.

E. Adult Habitat. Typical non breeding-season habitat consists of both upland and flatland longleaf pine forests with relatively open canopies (Dundee and Rossman, 1989). Adults reside in underground retreats associated with gopher tortoise (Gopherus polyphemus) and small mammal burrows, stump holes, and root mounds of fallen trees (Allen, 1932; Wright and Wright, 1949; Richter et al., 2001). They often can be seen outside of these retreats during the day (Wright and Wright, 1949; Richter et al., 2001). See "Breeding habitat" above, for further information on adult habitat characteristics.

F. Home Range Size. Unknown.

G. Territories. Doody et al. (1995) observed male–male combat at the breeding pond in Mississippi, which suggests that calling territories are established.

H. Aestivation/Avoiding Desiccation. Unknown.

I. Seasonal Migrations. No other migratory events other than breeding migrations are known (see "Breeding migrations" above).

J. Torpor (Hibernation). Dusky gopher frogs typically are active and breed during the winter (Allen, 1932; Dundee and Rossman, 1989; Richter et al., 2003), so torpor likely does not occur.

K. Interspecific Associations/ Exclusions. Dusky gopher frogs inhabit the burrows of gopher tortoises, (Allen, 1932; Wright and Wright, 1949) and small mammals (Richter et al., 2001). Amphibian species currently known to breed in the same pond as dusky gopher frogs, although not necessarily during the same season, include mole salamanders (Ambystoma talpoideum), southern cricket frogs (Acris gryllus), southern toads (Bufo terrestris), eastern narrow-mouthed toads (Gastrophryne carolinensis), green treefrogs (Hyla cinerea), pine woods treefrogs (H. femoralis), barking treefrogs (H. gratiosa), squirrel treefrogs (H. squirella), spring peepers (Pseudacris crucifer), southern chorus frogs (P. nigrita), ornate chorus frogs (P. ornata), green frogs (R. clamitans), southern leopard frogs, and eastern spadefoot toads (Scaphiopus holbrookii; Allen, 1932; G. Johnson and S.C.R., personal observations).

L. Age/Size at Reproductive Maturity. Richter and Seigel (2002) found a minimum age at maturity for males of 4–6 mo, with a mean length of 56.3 mm SVL and mass of 18.8 g. Average post-metamorphic growth was 20.6 mm SVL (37% increase) and 13.8 g (73% increase). No females originally marked when newly metamorphosed were captured, but age at maturity for females is estimated to be 2–3 yr (Richter and Seigel, 2002).

Adult body size ranges from 56–105 mm and varies between the sexes. Goin and Netting (1940) reported body size for males of 62–84 mm (mean = 73.6 mm; n = 21) and for females of 73–92.5 mm (mean = 82.3 mm; n = 29). Wright and Wright (1949) reported body size for males of 71–92 mm (mean = 82.4 mm; n = 19) and for females of 82–105 mm (mean = 87.5 mm; n = 11). Richter and Seigel (2002) reported variation in body sizes of adults—1996: males 54.0–81.6 mm (mean = 63.2 mm; n = 47), females 69.4–94.4 mm (mean = 82.7 mm; n = 50); 1997: males 59.0–84.0 mm (mean 70.2 mm; n = 50), females 64.6–93.6 mm (mean = 78.0 mm; n = 60); 1998: males 51.4–84.6 mm (mean = 67.7 mm; n = 29), females 69.0–88.3 mm (mean = 79.3 mm; n = 38).

M. Longevity. No data exist on longevity, although Richter et al. (2003) estimated maximum longevity based on recapture data to be 6–10 yr; most adults probably do not exceed 4–5 yr.

N. Feeding Behavior. Dusky gopher frog adults are carnivorous, especially insectivorous, with known gut contents including carabid (Pasimachus sp.) and scarabaeid (genera Canthon sp. and Ligyrus sp.) beetles (Netting and Goin, 1942a). They probably have a diet similar to that reported for gopher frogs—frogs, toads, beetles, hemipterans, grasshoppers, spiders, roaches, and earthworms (Dickerson, 1906; Deckert, 1920; Carr, 1940a).

O. Predators. No records of predation on adults or juveniles exist, but predators would be similar to those of other gopher frogs, and other ranid frogs (e.g., snakes, birds, and mammals; Jensen and Richter, this volume). Caddisfly (Trichoptera) larvae are known to prey on eggs and larvae (Richter, 2000). No other documentation of larval predation exists, but potential predators include those of other gopher frogs, and those observed in Glen's Pond feeding on southern leopard frog eggs and larvae—dragonfly naiads (Odonata), backswimmers (Hemiptera), giant water bugs (Hemiptera), predaceous diving beetles (Coleoptera), fish, salamanders, snakes, turtles, and birds (Jensen and Richter, this volume; S. C. R., personal observations).

P. Anti-Predator Mechanisms. Dusky gopher frogs inflate their bodies and cover their eyes when harassed or grasped by potential predators (S. C. R., personal observations), as is known for crawfish frogs (R. areolata; Altig, 1972a), a sister species. This

behavior in dusky gopher frogs is coupled with a milky secretion having a distinct musky odor and bitter taste, which exudes from the dorsal warts (Dickerson, 1906; Goin and Netting, 1940; S.C.R., personal observations). The secretion is composed of a variety of peptides that have a wide range of bioactive properties, some of which are thought to be associated with predator deterrence (C. Graham, S.C.R., P. Flatt, and C. Shaw, unpublished data).

Q. Diseases. *Perkinsus*-like infectious organism (M. Sisson, personal communication).

R. Parasites. Undescribed. Dusky gopher frogs are presumably susceptible to ticks *(Ornithodoros turicata)* commonly found on gopher tortoises and their burrow commensals, including gopher frogs (Blihovde, 2000a).

4. Conservation.
Dusky gopher frogs are thought to be extirpated in Louisiana and Alabama and currently are known only from a single population in southern Mississippi (see "Historical versus Current Distribution" above). They were listed under the U.S. Endangered Species Act as Endangered on 4 December 2001 (effective 3 January 2002; LaClaire, 2001). Threats to populations are similar to other gopher frogs, primarily involving human-induced stressors (see Jensen and Richter, this volume). Currently, conservation efforts are being considered, tested, and/or implemented, including the reintroduction of gopher tortoises to the habitat adjacent to Glen's Pond, artificial lengthening of hydroperiod at Glen's Pond, translocation of eggs to other suitable and/or historical ponds, creation of artificial breeding sites, and alteration of existing ponds (R. Seigel, J. Pechmann, personal communication). In addition, pertinent population genetic data are being collected in order to fully understand the status of this last known population (S.C.R., unpublished data).

Rana sphenocephala Cope, 1886
SOUTHERN LEOPARD FROG

Brian P. Butterfield, Michael J. Lannoo, Priya Nanjappa

1. Historical versus Current Distribution.
Southern leopard frogs *(Rana sphenocephala)* are distributed throughout the southeastern quarter of the continental United States, exclusive of the Appalachian Highlands. Their range extends from central Texas and Oklahoma and eastern Kansas eastward to southeastern Iowa, central Illinois, and eastern Kentucky, south to include most of Missouri, Arkansas, Tennessee, Louisiana, Mississippi, Alabama, Georgia, South Carolina, and Florida (Garrett and Barker, 1987; Hoffman, 1990). Their range also extends up the Atlantic Coast to include the eastern 3/4 of North Carolina (see also Myers, 1924), east-

ern Virginia, Maryland, eastern Pennsylvania, and southeastern New York, including Long Island. Southern leopard frogs are common throughout most of Tennessee, but are limited in the east by the higher elevations of the Blue Ridge Mountains (Tennessee Animal Biogeographic System, 2000; fwie.fw.vt.edu/TN). Two subspecies of southern leopard frogs *(R. sphenocephala)* are recognized: Florida leopard frogs *(R. s. sphenocephala* Cope, 1886) are found in Florida, especially on the peninsula; southern leopard frogs *(R. s. utricularia)* occupy the remainder of the distribution (Crother et al., 2000). Male Florida leopard frogs retain vestigial oviducts (Bartlett and Bartlett, 1999a). While a comparison of historical and current ranges has not been made, Hudson

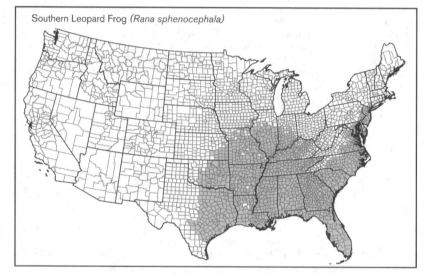

Southern Leopard Frog *(Rana sphenocephala)*

(1956) comments that southern leopard frogs occur "...only in isolated localities now undergoing modifications due to industrial expansion."

2. Historical versus Current Abundance.
Bartlett and Bartlett (1999a) state that these are the most abundant frogs in Florida. Mount (1975) notes that they are the most ubiquitous frogs in Alabama, but recent declines have been noted. The Tennessee Animal Biogeographic System (2000) states that southern leopard frogs are common throughout most of the state, but apparently are limited in the east by the higher elevations of the Blue Ridge Mountains.

3. Life History Features.
A. Breeding. Reproduction is aquatic.
i. Breeding migrations. Southern leopard frogs breed in early spring in the North, and at any month of the year in the South (Mount, 1975; Conant and Collins, 1991; Doody and Young, 1995) following rainfalls (Doody and Young, 1995), although Mount (1975) notes that most breeding in Alabama occurs from December–March. Johnson (1992) states that southern leopard frogs in Missouri are known to breed in

the autumn. Caldwell (1986) summarizes what is known by stating that although southern leopard frogs will breed throughout the year, in the South there are usually two major breeding periods—the first in early fall in September–October and the second from November–February or March. In the North, only one breeding season is typical, in the spring following ice melt. Caldwell (1986) notes exceptions: in Kansas southern leopard frogs breed in September and March, while in Missouri (D. Seale, personal communication) they breed in autumn and spring. Males occasionally will call when submerged or from crayfish burrows (Palis, 2000b). Following breeding, southern leopard frogs disperse throughout upland habitats (Brandt, 1936a).

ii. Breeding habitat. Males call while floating in open water from emergent vegetation or while perched on logs or sticks (Wright, 1932). Egg masses are laid in shallow, non-flowing waters (Hillis, 1982; Behler and King, 1998), which are usually fishless. Masses are typically partly floating and attached to vegetation (Wright, 1932).
B. Eggs.
i. Egg deposition sites. Eggs are found most often adhering to submergent or emergent vegetation (Caldwell, 1986; Ashton and Ashton, 1988; Bartlett and Bartlett, 1999a). Eggs hatch in 4–5 d in Florida (Ashton and Ashton, 1988) and ≤2 wk in Missouri (Johnson, 1992). Caldwell (1986) documented communal nesting during the cold breeding season, and suggested that this confers a thermal advantage to the eggs that is necessary to develop and survive in cold temperatures. Eggs hatch in 3–5 d (Wright, 1932). Eggs exhibit accelerated hatching times in the presence of crayfish (*Procambarus nigrocinctus*) predators (Saenz et al., 2003). Rapid pond drying is a threat to eggs (Wright, 1932).
ii. Clutch size. "The complement of a large female might reach 1,200–1,500 eggs

or even more" (Wright, 1932; 5,000 according to Johnson, 1992). Egg diameter ranges from 1.5–1.8 mm (vitellus).

C. Larvae/Metamorphosis.

i. Length of larval stage. From 50–75 d (Wright, 1932); "About three months" (Ashton and Ashton, 1988). Johnson (1992) notes that in Missouri, metamorphosis usually takes place from mid-June to late July.

ii. Larval requirements.

a. Food. Where southern leopard frogs and Rio Grande leopard frogs (*R. berlandieri*) co-occur, southern leopard frog tadpoles ingest mainly green algae and many fewer diatoms than Rio Grande leopard frog tadpoles (Hillis, 1982). Low food levels elicit interference mechanisms (production of growth inhibitors) among southern leopard frog tadpoles, while high food levels elicit exploitative competition (Steinwascher, 1978b; Alford and Crump, 1982).

b. Cover. Aquatic macrophytes and emergent vegetation provide cover for tadpoles.

iii. Larval polymorphisms. Do not occur.

iv. Features of metamorphosis. Across their range, metamorphosis occurs during every month from April–October (Wright, 1932). Metamorphosing animals generally are small, but sizes vary and range from 20–33 mm (Wright, 1932). Southern leopard frog tadpoles will shorten their time as larvae in response to wetland drying (Parris, 1997).

v. Post-metamorphic migrations. From breeding sites to upland feeding sites.

D. Juvenile Habitat. Similar to adults, although juveniles may seek wetter habitats.

E. Adult Habitat. All types of shallow freshwater habitats, including temporary pools, cypress ponds, ponds, lakes, ditches, irrigation canals, and stream and river edges; will inhabit slightly brackish coastal wetlands (Wright and Wright, 1949; Garrett and Barker, 1987; Hoffman, 1990; Conant and Collins, 1991; Bartlett and Bartlett, 1999a; Tennessee Animal Biogeographic System, 2000). In North Carolina, Pearse (1911) reported southern leopard frogs in 2.14‰ saltwater (see also Voice, 1923; Liner, 1954; Dundee and Rossman, 1989). Bartram (1791, in Wright, 1932) notes that in southern Pennsylvania, southern leopard frogs abound in rivers, swamps, and marshes. Hoffman (1990) concluded that southern leopard frogs occur chiefly, if not exclusively, along the floodplains of larger rivers in the western Piedmont. Southern leopard frogs will move into terrestrial habitats to feed during the summer, when vegetation in pastures, fields, and sod lands affords shade and shelter (Brandt, 1936a; Conant and Collins, 1991; Johnson, 1992; Bartlett and Bartlett, 1999a). Primarily nocturnal. Annual periods of activity range from nearly year-round in the south to winter dormancy in the north (Wright, 1932).

F. Home Range Size. Unknown.

G. Territories. Unknown and unlikely.

H. Aestivation/Avoiding Desiccation. Similar to northern leopard frogs (*R. pipiens*), southern leopard frogs seek moist areas, such as springs, river edges, and lake margins during dry periods.

I. Seasonal Migrations. To and from breeding sites in the spring, post-metamorphic migrations from breeding wetlands to upland feeding sites, and, where necessary, to winter refugia in permanent water bodies.

J. Torpor (Hibernation). Occurs in northern populations. Overwintering sites include permanent, well-oxygenated bodies of water.

K. Interspecific Associations/Exclusions. In Missouri, southern leopard frogs will occur and sometimes hybridize with plains leopard frogs (*R. blairi*; Johnson, 1992). In Louisiana, Dundee and Rossman (1989) note that bronze frogs (*R. clamitans clamitans*) will displace southern leopard frogs in wet woods and swamps. In Virginia, Pague and Buhlman (1992) note southern leopard frogs in association with carpenter frogs (*R. virgatipes*), northern leopard frogs, pickerel frogs (*R. palustris*), and green frogs (*R. clamitans*). Southern leopard frogs and Rio Grande leopard frogs breed in the same wetlands—Rio Grande leopard frogs usually breed several months before southern leopard frogs—and tadpoles co-occur (Wright and Wright, 1949). Pickens (1927b) suggests intergrades between northern and southern leopard frogs.

L. Age/Size at Reproductive Maturity. From 51–127 mm SVL (Conant and Collins, 1991; Johnson, 1992; see also Springer, 1938).

M. Longevity. Unknown; Snider and Bowler (1992) do not list southern leopard frogs in their compilation. A scan of Snider and Bowler's (1992) data for similarly sized ranids reveals great variation, from <2 to 6 yr, in longevity, and so is not helpful in formulating an estimation for southern leopard frogs.

N. Feeding Behavior. Southern leopard frogs will feed in upland habitats during the summer, where they take insects and a variety of other invertebrate prey (Johnson, 1992). In general, the food habits of southern leopard frogs resemble those of northern leopard frogs (Wright, 1932) and include a variety of insects and other aquatic invertebrates including crayfish (Force, 1925).

O. Predators. Southern leopard frogs are often hunted for their legs (Mount, 1975; Helfrich et al., 1997; Behler and King, 1998). Ashton and Ashton (1988) state that southern leopard frogs are too small to be taken for their legs, but are captured in large numbers for use in the bait industry, by scientific researchers, and for classroom teaching. They also note that southern leopard frogs, along with southern cricket frogs (*Acris gryllus*), are "undoubtedly an important food item in the diet of many aquatic predators." Wright (1932) records predators as including great blue herons (*Ardea herodias*), grackles (*Quiscalus* sp.), southern water snakes (*Nerodia fasciata*), brown water snakes (*N. taxispilota*), northern black snakes (*Coluber constrictor constrictor*), peninsular ribbon snakes (*Thamnophis sauritus sackenii*), and water moccasins (*Agkistrodon piscivorus*).

P. Anti-Predator Mechanisms. When frightened and when adjacent to water, southern leopard frogs will dive into the water, make a sharp angled turn while submerged, and surface among vegetation away from the predator's gaze (Behler and King, 1998). When frightened on land, they resemble northern leopard frogs in making a series of low leaps, each in a different direction (Bartlett and Bartlett, 1999a). Wright (1932) comments that southern leopard frogs are alert, active, agile, and hard to capture in water or on land. Further, when they are alarmed while resting on a bank of a stream or a pond, southern leopard frogs usually escape inland.

Q. Diseases. Southern leopard frogs are susceptible to agrichemical applications in combination with UV-B exposure in mesocosm experiments (Britson and Threlkeld, 1998b). They are also sensitive to the interactive toxicity of petroleum hydrocarbons and solar radiation (Little et al., 1998). Sparling (1998) reported an increase in hindlimb, orbital, and pigment malformations in southern leopard frogs in response to altosid applications. McCallum (1999) notes a site in southern Illinois characterized by malformed southern leopard frogs with multiple malformations, including missing limbs, partial limbs, complete but malformed limbs, duplicated limb segments, and missing eyes.

R. Parasites. Apparently not well studied. The respiratory tracts of some ranid frogs are susceptible to infection by a group of lung worms in the genus *Rhabdias* (Baker, 1978a).

4. Conservation.
Southern leopard frogs exhibit a broad distribution throughout the southeastern United States and are not considered a species of conservation concern by any state or the Federal Government. They are considered the most abundant frogs in Florida (Bartlett and Bartlett, 1999a) and the most ubiquitous frogs in Alabama (Mount, 1975). The Tennessee Animal Biogeographic System (2000) states that southern leopard frogs are common throughout much of this state, but are apparently limited in the east by the higher elevations of the Blue Ridge Mountains. Hudson (1956) notes that habitat destruction may be restricting southern leopard frogs to isolated localities.

Rana subaquavocalis Platz, 1993
RAMSEY CANYON LEOPARD FROG

Michael J. Sredl

1. Historical versus Current Distribution.

Little is known of the historical distribution of Ramsey Canyon leopard frogs (*Rana subaquavocalis*). As first described, the range of this species was limited to Ramsey and Brown canyons on the east side of the Huachuca Mountains, Cochise County, Arizona (Platz, 1993). There is speculation that its historical range included the San Pedro River valley and parts of Chihuahua, Mexico (Platz, 1997). Leopard frogs that may have been Ramsey Canyon leopard frogs were noted at 11 canyons in the Huachuca Mountains, including: Ash, Bear, Carr, Copper, Garden, Hunter, Miller, Montezuma, Parker, Scotia, and Sunnyside canyons (Slevin, 1928; Wright and Wright, 1949; Platz and Mecham, 1979; Holm and Lowe, 1995; Rorabaugh, personal communication; Beatty, unpublished data; Belfit, unpublished data). There are no recent records from these sites.

Ramsey Canyon Leopard Frog *(Rana subaquavocalis)*

Recent surveys of suitable sites in and near the Huachuca Mountains have found Ramsey Canyon leopard frogs at nine sites in two canyons on the east side of the range; no other leopard frog species were found (Platz, 1993, 1997; Sredl et al., 1997). One of these populations was not diagnosable as Ramsey Canyon leopard frogs (Platz, personal communication), and the identity of these animals remains unclear. Elevations of these localities range from 1,501–1,829 m (Sredl et al., 1997).

2. Historical versus Current Abundance.

Little is known of abundance prior to 1993. Platz et al. (1997) examined census data collected between 1990–95 from Ramsey Canyon and noted a decline in the population from over 90 in 1990 to 19 in 1995. Additionally, Platz (1997) summarized census data collected intermittently between 1991–96 from Ramsey Canyon and three other localities. Censuses varied from 1–38 individuals, and consistent reproduction was noted at only two localities.

3. Life History Features.

A. Breeding. Reproduction is aquatic.

i. Breeding migrations. Males call underwater while submerged as deep as 1.4 m (Platz, 1997). Although they are thought to produce calls that are inaudible in air (Platz, 1993, 1997), workers have heard calling males at the type locality (personal observations). Whether these calls are produced while the male is above or below the water's surface is unclear. Amplexus may last as long as 8–20 hr, but oviposition is brief (Platz, 1997).

Mating begins in late March to early April once water temperatures exceed 10 °C and continues through early October (Platz, 1997; unpublished data). In 1995, breeding activity at one site peaked during early May then again in mid-June, with a total of 19 egg masses produced (Platz, 1997). It is likely that females are capable of double clutching (Platz, 1997). Oviposition does not appear to be correlated with rain, but instead may be correlated with changes in water temperature. Platz (1997) noted that oviposition occurred on 10 of 11 nights shortly before or slightly after a decrease in water temperature.

ii. Breeding habitat. The aquatic systems such as springs, cienegas (= wetlands), earthen cattle tanks, small creeks, and slack water of main-stem rivers that support adult populations also provide reproduction sites.

B. Eggs.

i. Egg deposition sites. Females deposit eggs in spherical masses attached to submerged vegetation at a mean depth of 269 mm (n = 19, range = 110–710 mm; Platz, 1997).

ii. Clutch size. Egg masses contain an average of 1,518 ova (n = 7, range = 1,200–2,040; Platz, 1997). In nature, eggs hatch in approximately 14 d, depending on temperature (Platz, 1997); in captivity they take approximately 10 d at 23–25 °C (M. J. Demlong, unpublished data).

C. Larvae/Metamorphosis. Altig et al. (1998) describe the tadpoles of Ramsey Canyon leopard frogs. Length of larval period has not been well studied, but larvae may metamorphose in the year they were oviposited or overwinter as tadpoles (Platz and Grudzien, 1993; Platz, 1996; Platz et al., 1997). In captivity, larvae metamorphose as quickly as 100 d, but more commonly take 160–200 d (M. J. Demlong, unpublished data).

No comprehensive study of larval use of cover has been conducted, although Platz (1996) noted they may avoid predators by hanging motionless in algal mats. Given that many populations of Ramsey Canyon leopard frogs are small, recent observations of high densities of carnivorous giant water bugs (Belostomatidae) and low larval recruitment (personal observations) underscore the need for a greater understanding of predator avoidance and cover needs.

D. Juvenile Habitat. Juvenile habitats of Ramsey Canyon leopard frogs are unknown. Seim and Sredl (1994) studied the association of life stage and pool size in lowland leopard frogs (*R. yavapaiensis*) and found that juveniles were more frequently associated with small pools and marshy areas, while adults were more frequently associated with large pools.

E. Adult Habitat. Although no studies examining habitat use by Ramsey Canyon leopard frogs have been conducted, adults are habitat generalists found in aquatic systems in pine-oak and oak woodland and semi-desert grassland habitats in extreme southeastern Arizona. The aquatic systems from which Ramsey Canyon leopard frogs are known or likely to have occurred include springs, cienegas, earthen cattle tanks, small creeks, and slack water of main-stem rivers. These habitats are perennial to near perennial (unpublished data).

One of the nine known habitats inhabited by Ramsey Canyon leopard frogs is a natural aquatic system, the remainder are artificial or highly modified aquatic systems (Sredl et al., 1997). Sredl and Saylor (1998) found artificial aquatic systems, such as earthen cattle tanks, to be extremely important for the continued viability of Ramsey Canyon leopard frog populations and other members of the leopard frog (R. pipiens) complex in Arizona.

The role of habitat heterogeneity within aquatic and terrestrial environments is unknown, but likely important. Vegetation at extant populations varies greatly. Some sites are sparsely or seasonally vegetated with introduced or native grasses such as deer grass *(Muhlenbergia rigens)*, Bermuda grass *(Cynondon dactylon)*, and Johnson grass *(Sorghum halepense)*. More aquatic sites support wetland plants such as horsetail *(Equisitum* sp.), spikerush *(Eleocharis* sp.), monkey flowers *(Mimulus* sp.), watercress *(Rorippa* sp.), and cattail *(Typha* sp.). Shallow water with emergent vegetation and deeper water or root masses provide refuge from predators (Sredl, unpublished data).

Severe fragmentation and alteration of aquatic habitats in the southwestern United States has likely constricted wide ranging aquatic species into isolated pockets. Sredl and Howland (1995) speculated that the currently restricted distributions of leopard frog populations in Arizona may reflect habitat fragmentation producing extinction without recolonization, as well as impacts on habitat quality. Effects of grazing or logging on Ramsey Canyon leopard frog populations are unknown. Both may degrade ranid frog habitat by removing cover or increasing runoff and sedimentation rates (Jennings, 1988; Belsky and Blumenthal, 1997). Decreased frequency of widespread fire, effects of human uses, and variation in climatic patterns have resulted in altered forest communities and occurrence of infrequent but intense crown fires in the Huachuca Mountains (Danzer et al., 1996). Severe flooding and scouring of canyon bottoms following intense fire has probably contributed to the loss of wetland habitats and species in the Huachuca Mountains (Bowers and McLaughlin, 1994). Mining also historically resulted in habitat modification, including the upper portions of Ramsey Canyon. While the relationships between Ramsey Canyon leopard frogs and non-native predators (e.g., American bullfrogs *[R. catesbeiana]*, crayfish, and predatory fish) have not been studied in detail, there is a negative co-occurrence between other native leopard frog species and introduced predators (Rosen et al., 1998). Jennings and Scott (1991) questioned the importance of the role that American bullfrogs have played in declines of native amphibians in the southwestern United States, arguing that many populations of amphibians in New Mexico have declined without being impacted by bullfrogs.

F. Home Range Size. Unknown.

G. Territories. Platz (1996) suggests that Ramsey Canyon leopard frogs form a *lek* (mating priority determined by dominant status of males on communal mating grounds; Krebs and Davies, 1981), but a detailed analysis of the mating system has not been performed. Systematic study of within-site movement has not been done, although marked frogs have been observed moving hundreds of meters up and down Ramsey Canyon (unpublished data).

H. Aestivation/Avoiding Desiccation. Mechanisms that Ramsey Canyon leopard frogs use to survive the loss of surface water are unknown. However, other species of leopard frogs in the southwestern United States survive by burrowing into mud cracks (Howland et al., 1997).

I. Seasonal Migrations. Migrations of juvenile leopard frogs that may be Ramsey Canyon leopard frogs have been observed during summer rains (J. E. Wallace, personal observations).

J. Torpor (Hibernation). Although metamorphosed Ramsey Canyon leopard frogs generally are inactive between November–February, a detailed wintertime study has not been done.

K. Interspecific Associations/Exclusions. Other anurans that occur with Ramsey Canyon leopard frogs include Canyon treefrogs *(Hyla arenicolor)*, Woodhouse's toads *(Bufo woodhousii)*, red-spotted toads *(B. punctatus)*, Mexican spadefoot toads *(Spea multiplicata)*, and American bullfrogs (unpublished data). Couch's spadefoot toads *(Scaphiopus couchii)*, Sonoran Desert toads *(B. alvarius)*, Arizona treefrogs *(H. wrightorum)*, and lowland leopard frogs could either occur incidentally or may have occurred historically with Ramsey Canyon leopard frogs (unpublished data).

Sympatry with other species of native Arizona leopard frogs has been examined but not found (Platz, 1997; Sredl et al., 1997). Association with Chiricahua leopard frogs *(R. chiricahuensis)* is most likely (Platz, 1996).

L. Age/Size at Reproductive Maturity. Age structure between extant populations differs. Skeletochronology by Platz et al. (1997) indicated that 78% of the individuals from one population had lived at least 4 yr after metamorphosis, compared to 48% in a second population. Further, they found size to be a poor indicator of age. Growth rate for males was lower than that of females (Platz et al., 1997).

Although Platz et al. (1997) speculate that Ramsey Canyon leopard frogs do not generally reach sexual maturity earlier than 6 yr after metamorphosis, recently metamorphosed individuals reared in captivity and released August 1999 produced 31 egg masses between July and October 2000 (unpublished data). Captive-reared Ramsey Canyon leopard frogs also grew faster than those studied by Platz et al. (1997). Two females and one male released in October 1995 reached 102, 110, and 105 mm SUL, respectively, by the end of June 1998.

M. Longevity. In the wild, both sexes are known to live ≤10 yr following metamorphosis (Platz et al., 1997). In one population, a higher proportion of females (15 of 20) lived ≥ 5 yr compared with males (11 of 22; Platz et al., 1997). In this same population, males were better represented among the 3–5-yr age classes, while females were best represented among the 6–8-yr age classes. Of all individuals examined, 47% were 6 yr or older.

N. Feeding Behavior. No comprehensive studies of the feeding behavior or diet of Ramsey Canyon leopard frog larvae or adults have been completed. Platz (1996) states that like many ranid tadpoles, larval Ramsey Canyon leopard frogs are herbivorous. Available food resources at one site examined include bacteria, diatoms, phytoplankton, filamentous green algae, water milfoil *(Myriophyllum* sp.), duckweed *(Lemna minor)*, and detritus (Marti and Fisher, 1998). The diet of Ramsey Canyon leopard frog adults has not been determined. Stomach analyses of other members of the leopard frog complex from the western United States show a wide variety of prey items, including many types of aquatic and terrestrial invertebrates (e.g., snails, spiders, and insects) and vertebrates (e.g., fish, other anurans [including conspecifics], and small birds; Stebbins, 1951). Field et al. (2003) recently reported a Ramsey Canyon leopard frog taking a hummingbird. Captive larvae ate spinach, romaine lettuce, cucumber slices, frozen trout, duckweed, spirulina-type fish foods, and rabbit pellets. Captive juvenile frogs ate crickets (Demlong, 1997).

Oophagy (egg eating) may occur. Larger tadpoles have been observed ingesting the gelatinous envelopes of eggs, but have not been reported to consume the ovum (Platz, 1996). Cannibalism, primarily large adults eating juvenile frogs or large larvae, is likely but has not been studied.

O. Predators. Detailed studies of predators of Ramsey Canyon leopard frogs have not been done. Tadpoles are likely preyed upon by insects including belostomatids, notonectids, dytiscids, and anisopterans, and vertebrates including garter snakes *(Thamnophis* sp.), great blue herons *(Ardea herodias)*, and other birds. Predators of juvenile and adult frogs likely include American bullfrogs, garter snakes, great blue herons, and mammals, including rats, coyotes, gray foxes, raccoons, ringtail cats, coatis, black bears, badgers, skunks, bobcats, and mountain lions (Platz, 1996). Species of native and non-native fish likely would prey upon some or all life stages of Ramsey Canyon leopard frogs, but these taxa have not been documented to co-occur with extant leopard frog populations.

P. Anti-Predator Mechanisms. Antipredator mechanisms of tadpoles have not been studied. Metamorphosed frogs typically escape by jumping into water and seeking cover. Fright or distress calls have not been noted (Platz, 1996).

Q. Diseases. In 1998, chytrid fungus was implicated in declines of amphibians in Australia and Panama (Berger et al., 1998). That same year, it was first identified in Arizona (Milius, 1998). Presently, one salamander *(Ambystoma tigrinum stebbinsi)*, seven species of ranid frogs *(R. berlandieri, R. blairi, R. catesbeiana, R. chiricahuensis, R. subaquavocalis, R. tarahumarae, R. yavapaiensis)*, and one canyon treefrog *(H. arenicolor)* have been affected by this fungus in Arizona (Sredl et al., 2000; J. P. Collins, unpublished data;). All outbreaks have been cool season phenomena, and the pathogen appears widely distributed throughout central and southeastern Arizona (Sredl et al., 2000). Demlong (1997) described a "bloating malady" and malformed tails in captive tadpoles.

R. Parasites. An unknown fungus commonly infects egg masses, but whether this infection is the primary or secondary cause of mortality is unknown (Platz, 1995). Goldberg et al. (1998b) examined parasites of two closely related native ranids (Chiricahua and lowland leopard frogs) and one introduced ranid (American bullfrogs) collected in Arizona and found that none of the helminths identified from the two native species were found in American bullfrogs.

4. Conservation.
Due to the small number and size of populations, the U.S. Fish and Wildlife Service considered Ramsey Canyon leopard frogs be a candidate species for Federal listing from 1994–97. In 1997, they were removed from the list of Federal candidates (USFWS, 1997b) after a conservation assessment was developed (Platz, 1996) and a multiparty conservation agreement signed in 1996. Implementation of this agreement began in 1996. Platz and Grudzien (2003) have recently described limited genetic heterozygosity in Ramsey Canyon leopard frogs.

Rana sylvatica LeConte, 1825
WOOD FROG

Michael Redmer, Stanley E. Trauth

1. Historical Versus Current Distribution.
Although their range is primarily boreal, wood frogs *(Rana sylvatica)* are the most widespread North American amphibian species (Martof, 1970). They occur from the southern Appalachian Mountains of Georgia, north into Canada above the Arctic Circle, and west to Alaska (Martof, 1970; Conant and Collins, 1998). The western edge of their range runs roughly diagonally from Alaska and the Northwest Territories southeast through eastern

North Dakota, the Upper Midwestern states (Minnesota, Wisconsin, and Illinois), and to northeastern Alabama. Disjunct populations occur in Colorado, Wyoming, and the Ozark Plateau (Martof, 1970; Conant and Collins, 1998), but reports of this species in Kansas are probably erroneous (Collins, 1993).

Based on geographic isolation and reported genetic incompatibility with other populations, wood frogs from the Rocky Mountains of Colorado and Wyoming were named as a distinct species *R. maslini* by Porter (1969a,b). Bagdonas and Pettus (1976) showed that Rocky Mountain wood frogs were capable of breeding with

those from other populations and suggested that *R. maslini* remain a synonym of *R. sylvatica*.

Wood frogs have been the subject of at least two projects/studies of translocation as a conservation tool (Sexton and Phillips, 1986; Thurow, 1994). In both studies the preliminary results indicate at least short-

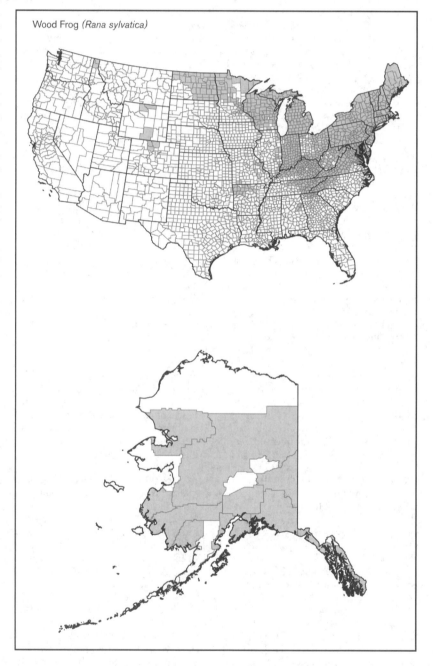

Wood Frog *(Rana sylvatica)*

term success, though the latter author has been criticized for translocating frogs from a donor area several hundred km away into an area for which historical presence of this species is not confirmed (Redmer, 1998b; Szafoni et al., 1999).

There have been no systematic studies done to compare current and historical distribution of wood frogs. No doubt populations have been lost as native landscapes have been anthropogenically altered, but it is likely that the current distribution resembles the historical distribution.

2. Historical versus Current Abundance.

There are few quantified data against which historical and current abundance can be compared. Based on 11 yr of volunteer surveys of chorusing frogs, Mossman et al. (1998) reported a strong increase in numbers of wood frogs detected in Wisconsin, but noted that this was likely due to improvements in observer ability to detect them over the course of their study. Casper (1998) also noted that coordinated statewide Wisconsin atlas surveys conducted between 1981 and 1996 documented an increase of county records from 46 to 65 (41%). Fluctuations in adult populations and reproductive traits may be strongly influenced by larval fitness, mortality, metamorphosis, and recruitment (Berven, 1982a,b, 1988, 1990). Other studies have used museum specimens and recent studies to compare historical and current presence and absence (Davis et al., 1998; Pfingston, 1998; Redmer, 1998b).

3. Life History Features.

A. Breeding. Reproduction is aquatic.

i. Breeding migrations. Breeding is well known to be "explosive" and occurs after the first warm rains of late winter or early spring (Banta, 1914; Wright and Wright, 1949; Howard, 1980; Seale, 1982b). Adults migrate from terrestrial overwintering sites to seasonal and semipermanent breeding wetlands. Timing of breeding season roughly follows latitudinal or elevational gradients. Examples (by state) of reported months in which breeding may occur include Alabama: January–February (Davis and Folkerts, 1979); Alaska: May–June (Herreid and Kinney, 1966, 1967); Arkansas: January–February (Trauth et al., 1989b, 1995a); Colorado and Wyoming: May (Corn and Livo, 1989); Georgia: February (Camp et al., 1990); Illinois (north): March–April (Pope, 1944); Illinois (south): February–March (Redmer, 2002); Maine: March–April (Hunter et al., 1999); Maryland: February (Berven, 1982a,b); Minnesota: March–April (Oldfield and Moriarty, 1994); Missouri: February (Guttman et al., 1991); New York: April (Wright, 1914; Wright and Wright, 1949); North Carolina: February–March (Hopey and Petranka, 1994); Pennsylvania: March–April (Seale, 1982b; Shaffer, 1991); Virginia: February (Berven, 1982a,b); Tennessee: February (Meeks and Nagel, 1973); Wisconsin: April–May (Mossman et al., 1998; Vogt, 1981).

ii. Breeding habitat. Usually described as fish-free, ephemeral woodland pools or ponds or cut-off sections of streams within or adjacent to the adult habitat, though human-made aquatic habitats

(including ditches, "wildlife" ponds, and tire ruts in dirt or clay roads) are commonly used as well (Trauth et al., 1989b, 1995a; Camp et al., 1990; Adams and Lacki, 1993; Hopey and Petranka, 1994; Cartwright et al., 1998; Trauth et al., 2000; Redmer, 2002). Assuming they are fish-free, colonization of small human-made pools such as tire ruts may benefit populations by increasing available breeding habitat, but such small habitats often have an insufficient hydroperiod to sustain eggs and tadpoles. Although the eggs can withstand some desiccation caused by temporary terrestrial standing (Forester and Lykens, 1988), tadpoles cannot. Eggs and tadpoles in small habitats may also be exposed to other risks including decreased food and increased exposure to predation and off-road vehicle traffic (Redmer, 2002). Presence of breeding ponds in open- versus closed-canopy ponds did not affect growth and survivorship (Werner and Glennemeier, 1999). Adults demonstrate considerable fidelity to certain breeding ponds, though a substantial number of juveniles disperse to breed in ponds other than their natal ponds (Berven and Grudzien, 1990). Recently, Paton et al. (2003) described breeding behavior during a drought.

B. Eggs.

i. Egg deposition sites. Oviposition sites have been characterized by many authors. Egg masses are usually deposited communally and are thus conspicuous. Communal oviposition is a possible function of explosive breeding, and additional studies have examined the relationships between mate choice, thermal advantage (freezing tolerance), and communal oviposition (Wells, 1977; Howard, 1980; Berven, 1981; Seale, 1982b; Waldman, 1982; Waldman and Ryan, 1983; Howard and Kluge, 1985; Corn and Livo, 1989; Trauth et al., 1995a; Cartwright et al., 1998). In particular, Waldman and Ryan (1983) point out a thermal advantage of up to 5 °C depending on position of egg, size of egg clump, and environmental conditions to communal egg laying in wood frogs. Egg masses are loose, and water convection between eggs is important for delivery of oxygen to developing embryos (Pinder and Friet, 1994).

ii. Clutch size. Some older texts (Pope, 1944; Wright and Wright, 1949; Smith, 1961; Minton, 1972; Mount, 1975) cite a general range of 1,000–3,000 eggs/clutch, a figure probably based on imprecise estimates. However, some more-recent studies have consistently reported smaller numbers, from approximately 300–1,500 eggs/clutch (Berven, 1982a; Meeks and Nagel, 1973; Davis and Folkerts, 1986; Corn and Livo, 1989; Trauth et al., 1989b; Camp et al., 1990; Redmer, 2002).

C. Larvae/Metamorphosis.

i. Length of larval stage. Tadpoles reach 50–60 mm TL at between 65–130 d

(Wright and Wright, 1949; Herreid and Kinney, 1967; Meeks and Nagel, 1973; Berven, 1982b; Berven and Chadra, 1988; Camp et al., 1990; Redmer, 2002).

ii. Larval requirements. Larvae feed opportunistically on detritus and plant material, but may also cannibalize conspecifics and prey upon other aquatic animals (Petranka et al., 1994; Petranka and Thomas, 1995; Petranka and Kennedy, 1999). Experimentally lengthened hydroperiods did not positively affect tadpole growth (Rowe and Dunson, 1995). Laboratory observations using spatial affinitiy as a recognition assay demonstrate sibling recognition (Fishwald et al., 1990).

iii. Larval polymorphisms. Distinctive cannibal morphs are unknown, though cannibalism by large tadpoles upon eggs and smaller conspecific tadpoles is well known (Bleakney, 1958; Petranka and Thomas, 1995). Such predation increases with size of the cannibalistic tadpole (Petranka and Thomas, 1995). Phenotypic plasticity has been demonstrated experimentally in tadpoles raised in the presence of dragonfly larvae (Van Buskirk and Relyea, 1998; Relyea and Werner, 2000).

iv. Features of metamorphosis. Metamorphosis occurs within 65–130 d (Wright and Wright, 1949; Herreid and Kinney, 1967; Meeks and Nagel, 1973; Berven, 1982b; Berven and Chadra, 1988; Camp et al., 1990; Redmer, 2002).

v. Post-metamorphic migrations. In a Virginia study of the effects of post-metamorphic dispersal on population genetic structure, up to 18% of juveniles were found to breed in ponds other than their own natal ponds (Berven and Grudzien, 1990).

D. Juvenile Habitat. Microhabitat preferences of juveniles are not specifically known, though observations of post-metamorphic juveniles indicate similarity to adults.

E. Adult Habitat. Wood frogs are found in a variety of habitats including tundra, subalpine woodlands, willow thickets, wet meadows, bogs, and temperate forests (both coniferous and deciduous) of various canopy species associations (Martof and Humphries, 1959; Martof, 1970; Pentecost and Vogt, 1976; Vogt, 1981; Davis and Folkerts, 1986; Hammerson, 1986; Corn and Livo, 1989; DeGraaf and Rudis, 1990; Russell and Bauer, 1993; Oldfield and Moriarty, 1994; Trauth et al., 1995a; Harding, 1997; Behler and King, 1998; Conant and Collins, 1998; Redmer, 2002). Wood frogs are sensitive to edge effects and reduced canopy cover created by forest cutting (deMaynadier and Hunter, 1998; Gibbs, 1998a,b). The relationship between landscape level distribution and population cycles was examined in southwest Ontario (Hecnar and McCloskey, 1996). Potential habitat has recently been described by Baldwin and Vasconcelos (2003).

F. Home Range Size. The only extensive study of this topic was a mark–recapture study conducted in Minnesota by Bellis (1961b), who found most recaptured adults near the initial capture site. The mean distance moved was 13.3 m² (n = 298). Home range sizes were from 3.8–477.2 m² (mean = 83.6 m²). Several studies examine movements (see "Seasonal Migrations" below). Though not specifically addressing home range size, one study of a Virginia population contributed extensive data on post-metamorphic dispersal and gene flow (Berven and Grudzien, 1990). In this study, up to 18% of juveniles dispersed to breed in ponds other than those of their origin, and maximum movement distances were reported to be 800–1,513 m. Cohorts formed mean genetic neighborhood areas of 5,035 ha, and gene flow between ponds was restricted beyond distances of 1,000 m (Berven and Grudzien, 1990).

G. Territories. Unknown.

H. Aestivation/Avoiding Desiccation. Unknown, but several authors provide evidence that in the warmer regions, cool or moist microhabitats are sought as temperature increases or humidity decreases (Heatwole, 1961a; Trauth et al., 1995a; Redmer, 2002).

I. Seasonal Migrations. Wood frogs exhibit explosive, synchronous migrations to breeding ponds in late winter to early spring (Wright and Wright, 1949; Meeks and Nagel, 1973; Howard, 1980; Seale, 1982b). Movements within a forest habitat are restricted by road corridors (Gibbs, 1998a).

J. Torpor (Hibernation). Wood frogs hibernate terrestrially near the soil surface (Heatwole, 1961a; Bellis, 1962; Howard, 1980; Schmid, 1982; Storey, 1984; Storey and Storey, 1987; Zweifel, 1989; Licht, 1991). They are resistant to dehydration (Schmid, 1965b), but tend to hibernate nearer to breeding ponds; this tendency is more pronounced in males than females (Regosin et al., 2003).

Wood frogs tolerate sub-freezing temperatures by producing cryoprotectants. The literature on this and related physiological responses is extensive (see Costanzo and Lee, 1994, for a review). Adults from southern populations have higher critical thermal maxima (CTM) then non-acclimated individuals from more northern populations, but there is no difference in tolerance to low temperatures (Lotshaw, 1977; Manis and Claussen, 1986).

K. Interspecific Associations/Exclusions. Although they are broadly sympatric, wood frogs and northern leopard frogs *(R. pipiens)* are reported to infrequently breed in the same ponds, and while no single cause for this segregation has been determined, food, predation, and habitat segregation may all contribute (DeBenedictis, 1974; Werner and Glennemeier, 1999). Wood frog tadpoles are known to prey

upon eggs and tadpoles of American toads *(Bufo americanus)*, and female American toads avoid oviposition in ponds where wood frogs are present (Petranka et al., 1994). In turn, the larvae of several species of ambystomatid salamanders are important predators of tadpoles and egg masses of wood frogs (see "Predators," below).

L. Age/Size at Reproductive Maturity. Males mature within 1–2 yr of metamorphosis, females usually at 2–3 yr (Berven, 1982a; Bastien and Leclair, 1992; Sagor et al., 1998; Redmer, 2002). Differential rates of maturity are a suggested cause of different sex ratios usually seen in breeding ponds (Berven, 1990). Sizes of mature individuals are extremely variable across their range (Martof and Humphries, 1959; Smith, 1961; Minton, 1972; Meeks and Nagel, 1973; Berven, 1982a; Davis and Folkerts, 1986; Trauth et al., 1989b, 1995a; Bastien and Leclair, 1992; Redmer, 2002). Geographic variation in growing season, elevation, and tadpole fitness (size at metamorphosis) are all known to affect life history and demographic traits such as age/size at which maturity is attained (Martof and Humphries, 1959; Berven, 1982a,b, 1988, 1990; Berven and Gill, 1983; Davis and Folkerts, 1986; Sagor et al., 1998).

M. Longevity. In Québec and southern Illinois, estimated ages of adults are 4 yr for males and 5 yr for females (Bastien and Leclair, 1992; Redmer, 2002). In Minnesota (Bellis, 1961a), Maryland, and Virginia (Berven, 1982a), and another Québec study (Sagor et al., 1998), maximum reported ages are 3 yr for males and 4 yr for females.

Bastien and Leclair (1992) suggested mean physiological longevity (MPL = mean age in years multiplied by the average number of frost-free days in the region) as an alternative to expressing longevity in years, and commented on possible variation in MPL of different wood frog populations. From their data, and other published data (Berven, 1982a), they calculated MPL for three populations (Québec, Maryland, and Virginia). They speculated that MPL was relatively constant in females but geographically more variable in males, and that earlier maturity curtailed the potential life span of males. However, MPL calculated for southern Illinois wood frogs was much longer than those that Bastien and Leclair (1992) calculated (Redmer, 2002).

N. Feeding Behavior. Tadpoles may prey heavily upon eggs and embryos of some salamanders (Petranka et al., 1998).

O. Predators. The larvae of several ambystomatid salamanders are important predators of tadpoles and egg masses of wood frogs (Wilbur, 1972; Walters, 1975; Rowe and Dunson, 1995). Leeches are also known predators of embryos and tadpoles (Cory and Manion, 1953). Adults are preyed upon by a number of terrestrial and aquatic predators.

P. Anti-Predator Mechanisms. Eggs are reportedly highly palatable to some aquatic predators (Walters, 1975). Although wood frogs exhibit a high degree of philopatry (Berven and Grudzien, 1990), newly created ponds are rapidly colonized, and females select fish-free ponds as oviposition sites (Hopey and Petranka, 1994). Experiments have shown that tadpoles raised in the presence of certain predators adopt alternate phenotypes or behaviors to avoid predation (Van-Buskirk and Relyea, 1998; Relyea and Werner, 1999, 2000; Relyea, 2000; but see Anderson and Petranka, 2003). In laboratory trials, post-metamorphic wood frogs (along with pickerel frogs *[R. palustris]*) were found to survive attacks by short-tailed shrews *(Blarina brevicauda)* more often than American bullfrogs *(R. catesbeiana)* or green frogs *(R. clamitans*; Formanowicz and Brodie, 1979).

Q. Diseases. Infection by *Aeromonas* sp. bacteria has been reported (Nyman, 1986). Mortality sometimes occurs in breeding choruses when multiple males attempt to amplex single females, resulting in "mating balls" that may drown the female (Phillips and Wade, 1990).

R. Parasites. The extensive literature on wood frog parasites was reviewed by McAllister et al. (1995e). In Canada alone, studies date back nearly 100 yr (Stafford, 1905; McAllister et al., 1995e). In the United States, data are available from Alaska (Metcalf, 1923), Arkansas (McAllister et al., 1995e), Maine (Bouchard, 1951), Maryland (Walton, 1931), Massachusetts (Rankin, 1945), Michigan (Najarian, 1955; Muzzall and Peebles, 1991), New York (Harwood, 1930, 1932), North Carolina (Metcalf, 1923), Ohio (Metcalf, 1923; Odlaug, 1954), and Wisconsin (Williams and Taft, 1980).

Protozoan parasites were reviewed in McAllister et al. (1995e). Metazoans infecting wood frogs include adult and/or larval stages of at least 10 trematodes, 3 cestodes, 8 nematodes, 1 acanthocephalian, and 1 hirudinean (McAllister et al., 1995e). A brief summary of the parasites from wood frogs in the United States is presented below; a listing of Canadian genera of helminth parasites (along with their references) can be found in McAllister et al. (1995e).

Protozoan parasites of U.S. populations, including *Opalina* sp., were found in the rectum of wood frogs from Arkansas (McAllister et al., 1995e) and Ohio (Metcalf, 1923), and *Cepedea* sp. were found from specimens in Alaska and Michigan. Wood frogs were found to host the myxosporean *Myxidium serotinum* in Arkansas (McAllister et al., 1995e). A previously unnamed eimerian *(Eimeria fitchi)* has been found in the feces of specimens from Arkansas (McAllister et al., 1995e).

Trematodes documented from wood frogs include the plagiorchid, *Brachycoelium salamandrae* (Rankin, 1938;

Odlaug, 1954; McAllister et al., 1995e; Najarian, 1955). Unidentified trematode metacercariae were mentioned in specimens from Arkansas (McAllister et al., 1995e), Massachusetts, and Michigan (Rankin, 1945; Najarian, 1955; Muzzall and Peebles, 1991), while echinostome cysts were found in Michigan populations (Najarian, 1955). The genus *Gorgoderina* and gorgoderid cysts have been found in several regions (Rankin, 1945; Bouchard, 1951; Najarian, 1955; Waitz, 1961). Four species of *Haematoloechus* have been reported (Waitz, 1961; Catalano and White, 1977; Williams and Taft, 1980; Muzzall and Peebles, 1991).

Tetrathyridia of the cyclophyllidean tapeworm genus *Mesocestoides* reported from Arkansas (McAllister et al., 1995e) also are found in wood frogs. Rankin (1945) observed plerocercoid larvae in Massachusetts wood frog populations.

Nematode genera known from wood frogs include *Abbreviata*, *Cosmocercoides*, *Megalobatrachonema*, *Oswaldocruzia*, *Rhabdias*, and *Spiroxys* from a number of U.S. states and Canada (Walton, 1931; Harwood, 1932; Rankin, 1945; Baker, 1978a, 1987; Muzzall and Peebles, 1991; McAllister et al., 1995e).

Odlaug (1954) found a cystacanth of an unidentified acanthocephalan from a wood frog in Ohio, and McAllister et al. (1995e) reported a widely distributed glossiphoniid leech, *Desserobdella* (= *Batracobdella*) *picta*, attached to a single wood frog from Arkansas.

4. Conservation.
Throughout much of their range, wood frogs are a common and familiar species. Many authors have made general comments concerning regional abundance or rarity (Dickerson, 1906; Wright and Wright, 1949; Mount, 1975; Pentecost and Vogt, 1976; Vogt, 1981; DeGraaf and Rudis, 1983; Green and Pauley, 1987; Johnson, 1987; Oldfield and Moriarty, 1994; Harding, 1997; Hunter et al., 1999). However, in several states along the periphery of their range they are either considered by local experts to be Restricted, Uncommon, or Rare, or they are afforded legal protection. These include Alabama (Mount, 1975), Arkansas (Trauth et al., 1989b, 1995a), Colorado (Corn and Livo, 1989; Levell, 1997), Illinois (Ackerman, 1975; Mierzwa, 1998a; Redmer, 1998b), Missouri (Johnson, 1987), and New Jersey (Levell, 1997).

Rana tarahumarae Boulenger, 1917
TARAHUMARA FROG

James C. Rorabaugh, Stephen F. Hale

1. Historical versus Current Distribution.
Tarahumara frogs (*Rana tarahumarae*) are known from 63 localities in montane canyons in extreme southern Arizona,

south to northern Sinaloa and southwestern Chihuahua, Mexico (Campbell, 1931a; Zweifel, 1968c; Hale et al., 1977, 1995; Hale, 2001). Reports of the species from New Mexico (Linsdale, 1933b; Little and Keller, 1937; Wright and Wright, 1949) and at Rose Creek near Roosevelt Reservoir, Arizona, were based on misidentified specimens of American bullfrogs (*R. catesbeiana*; Stebbins, 1951; Zweifel, 1968c). The range of Tarahumara frogs is centered in the northern Sierra Madre Occidental of Mexico (McCranie and Wilson, 1987); however, the eastern and southern distributional limits are not clear. Tarahumara frogs may not occur south of the Sierra Surutato in Sinaloa. Apparently suitable habitat occurs in Durango, well south of the current known range, but other ranid frogs occur there, and Tarahumara frogs have not been collected in Durango (Hale and Jarchow, 1988; Hale et al., 1995). Rangewide, most localities are in the mountains of eastern Sonora. In the United States, Tarahumara frogs were known historically from six locales, including three from the Santa Rita Mountains and three from the Atascosa-Pajarito–Tumacacori Mountains complex, which are located north and west, respectively, of Nogales in Santa Cruz County, Arizona (Campbell, 1931a; Zweifel, 1968c; Hale et al., 1977, 1995). By May 1983, Tarahumara frogs were extirpated from all localities in Arizona. However, in June 2004, the species was experimentally re-established to Big Casa Blanca Canyon, Santa Rita Mountains (see "Conservation" below). Surveys from May 1998–May 2000 in Sonora yielded Tarahumara frogs at 6 of 11 historical localities and 3 new localities (Hale et al., 1998; Hale, 2001).

2. Historical versus Current Abundance.
From 1975–77, the mean number of adult Tarahumara frogs was 509 in a 4.8-km reach of Big Casa Blanca Canyon, Santa

Rita Mountains, Arizona. Thirty Tarahumara frogs were reported in a 300-m reach of Arroyo el Cobre, southern Sonora, and 22 were found in a 50-m reach of Sycamore Canyon, Pajarito Mountains, Arizona (Hale and May, 1983). Based on qualitative observations, generally, where the species is currently extant, no long-term declines are apparent. However, Hale and May (1983), Hale and Jarchow (1988), and Hale et al. (1995, 1998) describe population declines in progress in Sycamore Canyon in the Pajarito Mountains in 1974, in Big Casa Blanca and Gardner canyons in the Santa Rita Mountains in 1977–83, and in northeastern Sonora in the lower reach of Arroyo La Carabina (La Bota) in the Sierra El Tigre in 1981. These declines resulted in extirpation from these locales. Occasional frogs were observed in lower Arroyo La Carabina from 1982–84 and larvae through 1986, apparently as a result of immigration from the upper canyon where healthy frogs were observed in 1982 and 1983. Healthy, affected, and dead frogs were found during every visit to a site between the lower and upper portions of the canyon from 1982–86 (Hale and Jarchow, 1988). Frogs were not found during a re-survey of lower Arroyo La Carabina in May 1998, suggesting that recolonization of the lower canyon by upstream immigrants had not occurred, despite the presence of apparently excellent habitat. In Arroyo El Tigre, Sierra El Tigre, flooding during the summer of 1998–99 may have eliminated most Tarahumara frog tadpoles (Hale, 2001).

3. Life History Features.
A. Breeding. Reproduction is aquatic.
i. Breeding migrations. Unknown.
ii. Breeding habitat. Breeding habitat is located within oak and pine-oak woodland and the Pacific coast tropical area (foothill thornscrub and tropical deciduous forest; Hale and May, 1983; McCranie

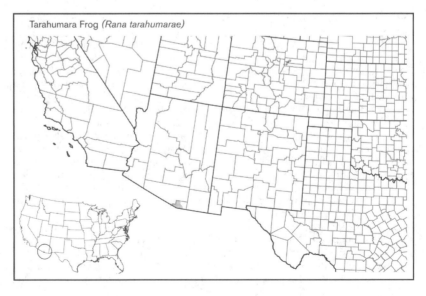

Tarahumara Frog (*Rana tarahumarae*)

and Wilson, 1987). Breeding occurs primarily toward the end of the dry season (April–May), when permanent water is often restricted to springs and "plunge pools" (deep [>1 m] pools in bedrock or among boulders) with deep underwater and streamside retreats. Plunge pools in canyons with low mean flows (<0.2 cubic ft/s) and relatively steep gradients (>60 m/km of stream) provide the best breeding sites (Hale and May, 1983; Hale, 2001). Permanent water is necessary for metamorphosis. At Pena Blanca Spring and Tinaja Canyon, Arizona, and Arroyo El Salto, northeastern Sonora, Tarahumara frogs inhabited artificial impoundments (Hale and May, 1983; Hale, 2001).

B. Eggs.

i. Egg deposition sites. Richard Zweifel, in a personal communication to Robert Stebbins (Stebbins, 1951), reported an egg mass from Alamo Canyon, Arizona, that was located in a pool in an intermittent stream. The eggs were on the bottom in about 18 cm of water, covered by a layer of sediment. Egg masses in Big Casa Blanca Canyon, Arizona, were found in calm shallows of bedrock pools, attached to the bedrock surface (Hale and May, 1983). A free-floating egg mass was found in the sandy shallows of a bedrock and boulder pool in Arroyo El Chorro, northern Sonora (Hale, 2001).

ii. Clutch size. Based on seven egg masses, Hale and May (1983) report number of eggs/egg mass ranged from 527–1,635 (mean = 1,083, s.e. = 161.4). Approximately 1/2 of an egg mass collected in the Sierra de la Madera, Sonora, contained about 900 eggs (Hale, 2001; J.C.R., personal observation).

In Arizona, amplectic frogs and egg masses were typically observed from April to mid-May (Hale and May, 1983); however, an egg mass was reported by Zweifel (1955) in late August in Alamo Canyon, Arizona. Hale (2001) reported small tadpoles at Arroyo El Tigre, northeastern Sonora, on 26 October 1999, suggesting egg deposition in late summer or fall.

C. Larvae/Metamorphosis.

i. Length of larval stage. Hale and May (1983) reported that tadpoles metamorphose in 2 yr (Hale and May, 1983). Tadpoles reared in semi-wild conditions in Arizona metamorphosed in as little as 86 d, but most took longer than 10 mo (J. C. R. personal observations). Compared to other ranids, Tarahumara frog larvae are large at all stages of development (Webb and Korky, 1977). Larvae grow as large as 106 mm before metamorphosing (Campbell, 1931a). At metamorphosis, frogs are as small as 21 mm SVL, but most are 35–40 mm. Size of larvae vary within and across populations, with larvae of the same stage varying in size by ≤20 mm (Webb and Korky, 1977).

ii. Larval requirements. Permanent water.

a. Food. As with most ranid tadpoles, larvae likely are omnivorous with a strong tendency towards algivory. Tadpoles reared in captivity ate spinach, sliced vegetables, fish food, algae, and boiled egg whites (USFWS, 2001a).

b. Cover. Probably necessary to avoid currents. Tadpoles are likely transported passively downstream in drainages (Hale and May, 1983).

iii. Larval polymorphisms. None.

iv. Features of metamorphosis. Moderately long; metamorphosed frogs were typically found from June–August in Big Casa Blanca Canyon, Arizona (Hale and May, 1983).

v. Post-metamorphic migrations. Hale and May (1983) report movements along stream courses of ≤1,885 m for juveniles (n = 23) in Big Casa Blanca Canyon.

D. Juvenile Habitat. No differences between adult and juvenile frog habitat use have been described.

E. Adult Habitat. Same as breeding habitats. The presence of hibernacula where frogs can remain moist and protected from predators and freezing temperatures is an important habitat feature (Hale and May, 1983), particularly in the northern portion of the species' range or at higher elevation sites. Hibernacula may include moist refugia among rocks and boulders along streams and at plunge pools.

F. Home Range Size. Hale and May (1983) report movements of ≤1,885 m for juveniles (n = 23) and males (n = 21), and ≤651 m for females (n = 9) in Big Casa Blanca Canyon. Movements apparently were along stream courses, suggesting linear home ranges, although the authors suggest limited overland movement also occurred. The longest movements occurred between the end of the dry season (June) and the middle of the summer rainy season (August).

G. Territories. Establishment and defense of territories has not been observed, but it is apparent that some individuals prefer to remain at certain pools or stream reaches (Hale and May, 1983).

H. Aestivation/Avoiding Desiccation. Hale and May (1983) speculated that some Tarahumara frogs may aestivate during the dry season before the onset of summer rains in July; however, most probably congregate at plunge pools.

I. Seasonal Migrations. Tadpoles are probably transported passively downstream in drainages. Hale and May (1983) observed upstream movements by a majority of frogs during the summer rainy season, followed by mostly downstream movements as fall approached. Juveniles may move upstream in late summer or fall, offsetting downstream dispersal by tadpoles. Frogs may also move upstream in the fall to access hibernacula.

J. Torpor (Hibernation). Tarahumara frogs have been observed from March–

October (Zweifel, 1955; Hale and May, 1983), but were not reported during February in Big Casa Blanca Canyon. Whether Tarahumara frogs remain active through the winter in the southern portion of their range is unknown.

K. Interspecific Associations/Exclusions. Tarahumara frogs have been observed in association with canyon treefrogs (Hyla arenicolor) at most localities. In southern Arizona, Tarahumara frogs usually were found in association with Chiricahua leopard frogs (R. chiricahuensis), while lowland leopard frogs (R. yavapaiensis) were also present in Sycamore Canyon and at Alamo Spring. Lowland leopard frogs usually are found in association with Tarahumara frogs in northern and central Sonora and are replaced by big-eyed leopard frogs (R. magnaocularis) in southern Sonora and adjacent areas in western Chihuahua. Arroyo El Cobre in southern Sonora is unique in that it supports both Tarahumara frogs and pustulose frogs (R. pustulosa). Possible hybrids or intergrades of these species were observed in Arroyo El Cobre by S. Hale in 1999. Erosion from hillsides where tropical deciduous forest has been cut to plant buffelgrass (Pennisetum ciliare) has resulted in siltation of plunge pools at Arroyo El Cobre. This conversion of habitat has favored pustulose frogs (Hale, 2001). In the Sierra Surutato in northern Sinaloa, pustulose frogs apparently replace Tarahumara frogs at low elevation sites and Tarahumara frogs occur at higher elevations (Hale and Jarchow, 1988). Rosy salamanders (Ambystoma rosaceum) are present at most Tarahumara frog sites in Mexico ≥ 1,000 m (Hale and Jarchow, 1988), but also have been found in association with Tarahumara frogs at 700 m in the Sierra Aconchi, northern Sonora (Hale et al., 1998). Tarahumara frogs may be excluded in habitats that support large populations of nonnative predators, such as American bullfrogs and fishes (i.e., green sunfish [Lepomis cyanellus], largemouth bass [Micropterus salmoides]; Hale and May, 1983). Hale (2001) suggested predation by a large chub species (Gila sp.) may have eliminated Tarahumara frogs from a site in Sonora.

L. Age/Size at Reproductive Maturity. Reproduction probably begins in the second spring following metamorphosis. In Big Casa Blanca Canyon, males matured at ± 64 mm SUL, females at ± 67 mm SUL (Hale and May, 1983).

M. Longevity. The oldest frogs in Big Casa Blanca Canyon were estimated to be at least 6 yr post-metamorphosis (Hale and May, 1983).

N. Feeding Behavior. Prey items are diverse and include juvenile Sonoran mud turtles (Kinosternon sonoriense); Sonora chubs (Gila ditaenia); snakes, including black-headed snakes (Tantilla atriceps); beetles (including Tenebrionidae, Scarabaeidae, and Buprestidae); moths (Lepidoptera); water bugs (Belostomatidae); scorpions (Scorpionida); centipedes (Chilopoda); grasshoppers (Agrididae); mantids (Mantidae); wasps (Hymenoptera); spiders (Lycosidae); crickets (Gryllidae); caddisflies (Tricoptera); and katydids (Tettigoniidae; Zweifel, 1955; Hale and May, 1983). Both diurnal and nocturnal feeding is indicated based on the activity patterns of prey species.

O. Predators. Hale and May (1983) report a possible predation attempt on a Tarahumara frog by a pair of ringtail cats (*Bassariscus astutus*) in Sonora. A variety of animals likely prey on frogs, tadpoles, and/or egg masses, including garter snakes (*Thamnophis* sp.) and other snakes, birds, other frogs, rosy salamanders, fishes, and invertebrates, particularly species of water bugs, *Belostoma* and *Lethocerus* (Hale and May, 1983; Hale, 2001). Predation by non-native fishes, particularly sunfishes (Centrarchidae) and American bullfrogs, may have contributed to the disappearance of the species from Pena Blanca Spring and portions of Pena Blanca Canyon, Arizona.

P. Anti-Predator Mechanisms. When disturbed, frogs typically hop into water where they take refuge under leaf litter, rocks, or other debris. In addition, Tarahumara frogs have skin secretions that can cause mild skin irritation and may be noxious tasting.

Q. Diseases. Hale and Jarchow (1988) list the following possible causal mechanisms in the extirpation of Tarahumara frog populations: (1) winter cold; (2) flooding or severe drought; (3) competition; (4) predation; (5) disease; and (6) heavy metal poisoning. Metals occur naturally in streamside deposits and may be mobilized by acid precipitation events. Acidic rainfall in southeastern Arizona and northern Sonora may have occurred as a result of atmospheric emissions from copper smelters at Cananea and Nacozari, northeastern Sonora, and Douglas, Arizona (Blanchard and Stromberg, 1987; Hale et al., 1995). Cadmium toxicity is a possible cause of observed Tarahumara frog die-offs in Arroyo La Carabina, Arroyo Pinos Altos, and Arroyo La Colonia in northeastern Sonora, and Big Casa Blanca and Sycamore canyons in Arizona (Hale and Jarchow, 1988; Hale et al., 1998). Cadmium is highly toxic due to its propensity to substitute for zinc and/or copper in enzymes (Coombs, 1979). Absorption through the skin or ingestion of zinc by frogs may act to reduce cadmium toxicity. Thus, in areas of relatively high zinc to cadmium ratios, frogs may be less affected (Hale and Jarchow, 1988; Hale et al., 1998). The solubility of zinc in weak acids may account for its depletion from decades of acidic precipitation on rhyolitic stream banks. Elevated levels of cadmium occur in and near tailings of copper, lead, and zinc mines (Peterson and Alloway, 1979). Cumulative sedimentation from physical erosion and deposition in drainages likely result in elevated concentrations of cadmium in downstream reaches. Thus, stream headwaters and springs may be important refuges for frogs when toxic conditions exist in downstream reaches (Hale et al., 1998). Die-offs of ranid frogs (Tarahumara frogs, lowland leopard frogs, and Chiricahua leopard frogs) in Sycamore Canyon, Arizona, are similar to die-offs of Chiricahua leopard frogs reported (Scott, 1993) in New Mexico, which were attributed to "postmetamorphic death syndrome." These die-offs are also consistent with chytridiomycosis, a fungal disease implicated in declines of anurans in Australia, Central America, (Berger et al., 1998) and elsewhere. Tarahumara frogs collected during a die-off in Sycamore Canyon in 1974 were infected with chytridiomycosis (T.R. Jones and P.J. Fernandez, personal communication). Chytridiomycosis was also confirmed via histology from frogs collected in northeastern and east-central Sonora at Arroyo La Carabina (1981, 1982), Arroyo El Tigre (1999), Arroyo La Colonia (1982), Arroyo El Trigo (1982), and Arroya El Aguaje (1999); and from Arroya El Cobre (1985) in southern Sonora (Hale, 2001). Tarahumara frogs are apparently extirpated from Sycamore Canyon, the lower reaches of Arroyo La Carabina, and perhaps Arroyo El Cobre; however, they have persisted at the other sites despite the presence of chytridiomycosis. (see also Rollins-Smith et al., 2002).

R. Parasites. Hale and Jarchow (1988) report mild infestations of nematodes and trematodes in Tarahumara frogs from Arroyo La Carabina and Arroyo El Tigre, northeastern Sonora. Fifteen species of helminths were identified from 42 Tarahumara frogs from Sonora, including a new species (*Falcaustra lowei*; Bursey and Goldberg, 2001).

4. Conservation.
Tarahumara frogs are considered an Endangered species by the Arizona Game and Fish Department (1988) and are included on a draft state list of Species of Concern (Arizona Game and Fish Department, 1996b). Tarahumara frogs have no status under the Federal Endangered Species Act, C.I.T.E.S., or Mexican law. From June–October 2004, a team of state, federal, and other partners experimentally released 56 adult, 229 juvenile, and 327 larval Tarahumara frogs to Big Casa Blanca Canyon, Santa Rita Mountains, Arizona. The source of the frogs was Arroyo el Chorro, Sierra de la Madera, northern Sonora (see Field et al., 2000; USFWS, 2001a; Rorabaugh and Humphrey, 2002).

Rana virgatipes Cope, 1891
CARPENTER FROG

Joseph C. Mitchell

1. Historical versus Current Distribution.
The historical range of carpenter frogs (*Rana virgatipes*) is unknown. Current distribution is in the Atlantic Coastal Plain from southern New Jersey to southeastern Georgia and two counties in northeastern Florida (Gosner and Black, 1968; Christman et al., 1979a; Ashton and Ashton, 1988; Conant and Collins, 1998).

2. Historical versus Current Abundance.
These frogs may be common locally in some areas, but most populations consists of small numbers of individuals (personal observations).

3. Life History Features.
The life history and ecology of this species were summarized by Wright (1932), Wright and Wright (1949) and Standaert (1967).

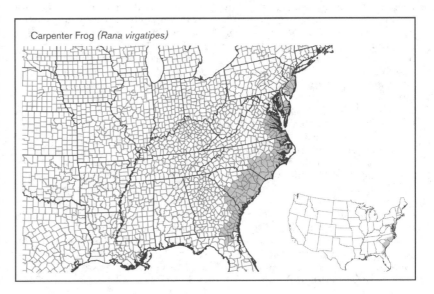

Carpenter Frog (*Rana virgatipes*)

A. Breeding. Breeding is aquatic.

i. Breeding migrations. Breeding migrations do not occur because adults live in the breeding ponds. Initiation of the calling period in Virginia occurs in April, but I have heard calls in March during warm spells. Calling season in New Jersey, Virginia, and North Carolina is late April to early August (Given, 1987; Mitchell, 1998; personal observations). Males call from open water usually while sitting on aquatic vegetation. They have a prolonged breeding season, generally May–July. Depending on the breeding site (Given, 1988a), males call on 4–94 nights of the generally 3-mo breeding period (Given, 1988b). Observations of amplectant pairs are rare, but have been observed from 26 April–27 July in New Jersey (Given, 1988b).

ii. Breeding habitat. Carpenter frogs breed in permanent, low to high acidity sphagnum ponds, beaver ponds, freshwater marshes, interdunal cypress swales, and pocosins (Wright, 1932; Wright and Wright, 1949; Mitchell, 1998; personal observations; Zampella and Bunnell, 2000). Number of submerged stems is positively correlated with the number of calling sites (Given, 1988b), indicating that vegetation structure of wetlands is an important component of breeding habitats.

B. Eggs.

i. Egg deposition sites. Eggs are laid in small masses attached to vegetation slightly submerged below the water's surface (Livezey and Wright, 1947). Carpenter frog embryos tolerate more strongly acidic water than most eastern temperate zone frogs. Successful hatching occurs as low as pH 3.8, but that concentration produces some abnormal development (Gosner and Black, 1957b).

ii. Clutch size. The number of eggs/mass is 200–600 (Wright, 1932; Livezey and Wright, 1947).

C. Larvae/Metamorphosis. The extended larval period lasts about 1 yr, and tadpoles overwinter in breeding ponds. Newly metamorphosed animals emerge in August–September from larvae hatched from eggs laid the previous summer. Size at metamorphosis was reported to be 23–31 mm SVL in Georgia (Wright, 1932) and 28–36 mm SVL in New Jersey (Standaert, 1967).

D. Juvenile Habitat. Apparently similar to that of adults. Juveniles are not known to disperse from breeding sites, and this size class is caught frequently in aquatic habitats (unpublished data).

E. Adult Habitat. Adults do not venture far from water and apparently remain in wetlands in all seasons. Buhlmann et al. (1994) found only three carpenter frogs in pitfall traps in upland habitats during a 6-mo trapping period in southeastern Virginia, suggesting that few move overland.

F. Home Range Size. Unknown.

G. Territories. Male carpenter frogs are territorial and use physical interactions and vocalizations to defend territories (Given, 1987). Territory size is 0.5–6.5 m in diameter (Given, 1988b). Larger males are aggressive to smaller males and exclude them from territories. Males exhibit high site fidelity to their territories.

H. Aestivation/Avoiding Desiccation. Unknown.

I. Seasonal Migrations. Carpenter frogs apparently do not migrate. Standaert (1967) noted that several adults moved from a large breeding pond to a series of smaller ponds nearby, where they overwintered.

J. Torpor (Hibernation). Overwintering sites are under water, usually in permanent breeding ponds.

K. Interspecific Associations/Exclusions. Carpenter frogs are sympatric with northern cricket frogs (*Acris crepitans*), southern cricket frogs (*A. gryllus*), American toads (*Bufo. americanus*), Fowler's toads (*B. fowleri*), southern toads (*B. terrestris*), eastern narrow-mouthed toads (*Gastrophryne carolinensis*), Pine Barrens treefrogs (*Hyla andersonii*), Cope's gray treefrogs (*H. chrysoscelis*), eastern gray treefrogs (*H. versicolor*), spring peepers (*Pseudacris crucifer*), American bullfrogs (*R. catesbeiana*), green frogs (*R. clamitans*), pickerel frogs (*R. palustris*), and southern leopard frogs (*R. sphenocephala*; Gosner and Black, 1957b; Given, 1987; Buhlmann et al., 1994; Bunnell and Zampella, 1999; Mitchell, 1998).

L. Age/Size at Reproductive Maturity. The smallest mature male carpenter frog reported was 39 mm SVL, 1 yr following metamorphosis (Standeart, 1967; Given, 1988b). Males ≥47 mm SVL are mature. Corresponding data for females are not available.

M. Longevity. This species has lived as long as 6 yr, 2 mo in captivity (Snider and Bowler, 1992). Standeart (1967) and Given (1988a) noted that few adults marked in 1 yr were recaptured in subsequent years and suggested that adults in nature live ≤3 yr following metamorphosis.

N. Feeding Behavior. The diet of carpenter frogs has not been published.

O. Predators. Known snake predators are southern water snakes (*Nerodia fasciata*) and northern water snakes (*N. sipedon*; Wright, 1932; Kauffeld, 1957; Gosner and Black, 1968; Palmer and Braswell, 1995). American bullfrogs are usually absent where carpenter frogs are found (Zampella and Bunnell, 2000) or they only occur in small numbers (personal observations), suggesting that this predator may influence carpenter frog distribution and abundance.

P. Anti-Predator Mechanisms. Unknown.

Q. Diseases. Unknown.

R. Parasites. Unknown.

4. Conservation.
Carpenter frogs are listed as a Species of Special Concern in Virginia due to the low number of locations within the state (Levell, 1997; Mitchell and Reay, 1999). Ecologically, this species is one of the poorest known ranid frogs.

Rana yavapaiensis Platz and Frost, 1984
LOWLAND LEOPARD FROG

Michael J. Sredl

1. Historical versus Current Distribution.
Historically, lowland leopard frogs (*Rana yavapaiensis*) were distributed from northwestern Arizona through central and southeastern Arizona, southwestern New Mexico, and northern Sonora, Mexico. Populations also were known from southwestern Arizona and southeastern California along the lower Colorado River and in the Coachella Valley (Platz and Frost, 1984; Platz, 1988; Jennings, 1995a). Identity of leopard frogs in southwestern Utah, southeastern Nevada, and extreme northwestern Arizona has been problematic (Platz, 1984; Jennings, 1988a; Jaeger et al., 2000). This account follows the taxonomy of Jaeger et al. (2000) and considers frogs of the Virgin River downstream into the Black Canyon of the Colorado River below Hoover Dam in Nevada to be relict leopard frogs (*R. onca*). Distribution of lowland leopard frogs, like other leopard frogs in the western United States, is fragmented and discontinuous due to the aridity of the region (Mecham, 1968c).

Interpreting historical distribution and abundance data for species of leopard frogs in the western United States is difficult for several reasons: (1) many of them are recently described and similar in appearance (Jennings, 1994); (2) multiple species can inhabit the same locality (Platz and Platz, 1973); and (3) their populations are prone to large fluctuations (Sredl et al., 1997). Historical records for lowland leopard frogs are known from Mohave, Yavapai, Coconino, La Paz, Maricopa, Gila, Pinal, Graham, Greenlee, Yuma, Pima, Santa Cruz, and Cochise counties, Arizona; Catron, Grant, and Hidalgo counties, New Mexico; and San Bernardino, Riverside, and Imperial counties, California (Vitt and Ohmart, 1978; Clarkson and Rorabaugh, 1989; Jennings and Scott, 1991; Jennings and Hayes, 1994b; Sredl et al., 1997). Historical records for lowland leopard frogs also exist from northern Sonora, Mexico, although the range of this species in Mexico is poorly known (Platz, 1988). Elevations of lowland leopard frog localities range from near sea level to 1,817 m (Jennings and Hayes, 1994a; Sredl et al., 1997).

Vitt and Ohmart (1978) surveyed numerous localities along the lower Colorado River and concluded that

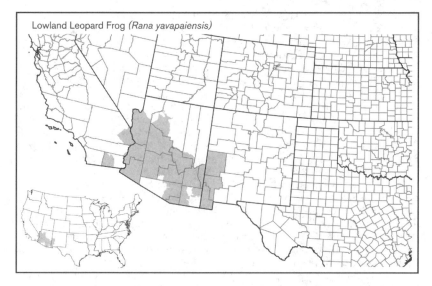

Lowland Leopard Frog *(Rana yavapaiensis)*

populations of leopard frogs, which would now be considered lowland leopard frogs, in that area may be extinct. All post-1980 records from the lower Colorado River and in the vicinity of the Salton Sea have turned out to be Rio Grande leopard frogs *(R. berlandieri)*, which have established themselves in the lower Colorado River and Gila River to Phoenix, Arizona (Platz et al., 1990; Jennings and Hayes, 1994a; Rorabaugh et al., 2004). During surveys conducted between 1983 and 1987 of the Arizona and California portions of its range, Clarkson and Rorabaugh (1989) found lowland leopard frogs at only 5 of 10 historical sites and 3 new sites; no extant population was found in California. The most recent California record for lowland leopard frogs was collected in 1965 from an irrigation ditch east of Calexico (Jennings and Hayes, 1994b). Lowland leopard frogs, if they are still present in California, are extremely rare (Jennings and Hayes, 1994a). Sredl et al. (1997) conducted 1,104 surveys within the range of lowland leopard frogs in Arizona and confirmed their presence at 43 of 115 sites that had prior records and 61 new sites; many of the extant populations found were in central Arizona. Jennings (1995a) found lowland leopard frogs at none of six historical sites visited in 1995. A single lowland leopard frog (specimen confirmed by R.D. Jennings) observed in August 2000 in Hidalgo County (Christman and Painter, unpublished data) was the first confirmed observation in New Mexico since 1985 (Jennings, 1995a; R.D. Jennings, personal communication). Additional populations may still occur at or near historical localities in Hidalgo County (Scott, 1992), but their status in New Mexico seems dire (Jennings, 1995a). Little is known of the status of populations in Mexico.

2. Historical versus Current Abundance.
Little is known of historical abundance of lowland leopard frogs. Zweifel (1968b) characterized leopard frogs as "abundant" at well-watered sites in valleys of southeastern Arizona. In other parts of southeastern and east-central Arizona, frogs that were likely lowland leopard frogs were abundant and widespread (Wright and Wright, 1949; J.T., Bagnara, personal communication). Like other leopard frogs in the Southwest, these frogs can be abundant where they occur (10–20 frogs jump with each step along the shoreline) or they can be quite scarce (personal observations).

3. Life History Features.
    A. Breeding. Reproduction is aquatic.
    i. Breeding migrations. Breeding migrations as have been described for some amphibians have not been noted in lowland leopard frogs. In most populations, breeding usually follows springtime emergence after a period of winter inactivity and continues through summer and into fall. Populations occupying geothermal springs or at low elevations are likely active year-round (R.D. Jennings, unpublished data).
    Egg masses have been observed from January through late April and October (Ruibal, 1959; Collins and Lewis, 1979; Frost and Platz, 1983). Reproductive activity may decrease between the time temperatures warm in mid-May to prior to the onset of the summer rains in early July (unpublished data).
    In populations examined, sex ratios generally do not differ from 1:1 (Sredl et al., 1997). Male lowland leopard frogs attract a potential mate by emitting an airborne call consisting of a series of low pulses lasting 3–8 s (Platz and Frost, 1984). Proximate cues that stimulate mating in

lowland leopard frogs are not well studied, although rainfall and water temperature have been mentioned as cues for other leopard frog species in the Southwest.
    Ruibal (1959, 1962) studied the physiological ecology of a brackish water population of lowland leopard frogs in southern California. Salinities at this site ranged between 6.0 and 9.0‰. Minimum lethal salinity of eggs was 5‰, while that of adult frogs was 13.0‰ (Ruibal, 1959). Reproduction in a tributary, which was less saline, likely explained persistence of the brackish water frog population (Ruibal, 1959). The upper and lower thermal limit of developing eggs in this population was 29 °C and 11 °C, respectively (Ruibal, 1962).
    ii. Breeding habitat. Lowland leopard frogs inhabit aquatic systems in desert scrub to pinyon-juniper (Platz and Frost, 1984). They are habitat generalists and breed in a variety of natural and man-made aquatic systems. Natural systems include rivers, permanent streams, permanent pools in intermittent streams, beaver ponds, cienegas (= wetlands), and springs; while manmade systems include earthen cattle tanks, livestock drinkers, canals, irrigation sloughs, wells, mine adits, abandoned swimming pools, and ornamental backyard ponds (Platz and Frost, 1984; Scott and Jennings, 1985; Sredl and Saylor, 1998). The preponderance of historical localities are small to medium-sized streams and rivers (R.D. Jennings, 1987; Sredl and Saylor, 1998). In lotic habitats, they are concentrated at springs, near debris piles, at heads of pools, and near deep pools associated with root masses (R.D. Jennings, 1987; unpublished data).
    The constant flow and warm water temperature of thermal springs, which permit year-round adult activity and winter breeding, and depauperate fish communities make these sites particularly important breeding habitat for leopard frogs in New Mexico (Scott and Jennings, 1985).

    B. Eggs.
    i. Egg deposition sites. Females deposit spherical masses attached to submerged vegetation, bedrock, or gravel. Eggs usually are deposited near the surface of the water (Sartorius and Rosen, 2000).
    ii. Clutch size. Has not been studied.
    C. Larvae/Metamorphosis. Altig et al. (1998) describe the tadpoles of lowland leopard frogs. In the wild, egg masses have been observed to hatch in 15–18 d (Sartorius and Rosen, 2000). Tadpoles may metamorphose in the same year they were oviposited or overwinter (Collins and Lewis, 1979), and length of larval period may be as short as 3–4 mo or as long as 9 mo (unpublished data). Jennings (1990) found that tadpoles of Chiricahua leopard frogs *(R. chiricahuensis)* in warm springs appeared to grow continuously, while

growth of those in cold water sites appeared to be arrested during the winter. Lowland leopard frog tadpoles would likely exhibit the same pattern (R. Jennings, personal communication).

D. Juvenile Habitat. Some spatial and temporal separation of adult and juvenile lowland leopard frogs may enhance survivorship. Seim and Sredl (1994) studied the association between juvenile and adult stages and pool size and found that juveniles were associated more frequently with small pools and marshy areas while adults were associated more frequently with large pools.

E. Adult Habitat. Although no study examining habitat use by adult lowland frogs has been conducted, lowland leopard frogs are found in a variety of natural and manmade aquatic systems within its range. Duration of water in these habitats ranges from semi-permanent to permanent. In semi-permanent aquatic systems, lowland leopard frogs may survive the loss of surface water by retreating into deep mud cracks, mammal burrows, or rock fissures (Howland et al., 1997).

The role of habitat heterogeneity within the aquatic and terrestrial environment is unknown but likely important. Shallow water with emergent and perimeter vegetation provide basking habitat, and deep water, root masses, undercut banks, and debris piles provide refuge from predators and potential hibernacula (R.D. Jennings, 1987; Platz, 1988; Jennings and Hayes, 1994b; unpublished data).

Sredl et al. (1997) estimated populations at six lowland leopard frog sites in Arizona. These sites ranged in elevation from 658–1,134 m and 30-yr average rainfall of 233–480 mm/yr. Common riparian overstory at these sites consisted of Fremont cottonwoods (*Populus fremonti*), willows (*Salix* sp.), seepwillows (*Baccharis glutinosa*), mesquite (*Prosopis* sp.), and introduced salt cedars (*Tamarix chinensis*). Common ground cover in moist areas included yerba-mansa (*Anemopsis californica*), canyon ragweeds (*Ambrosia ambrosioides*), and arrow-weeds (*Tessaria sericea*). Three-square rushes (*Scirpus americanus*), spike rushes (*Eleocharis* sp.), and introduced Bermuda grass (*Cynodon dactylon*) lined the banks or perimeter of ponds and slackwater pools. The largest, deepest pools had stands of narrow-leafed cattails (*Typha angustifolia*); large ponds in addition to having cattails, had pondweeds (*Potomageton* sp.).

Severe fragmentation and alteration of aquatic habitats in the southwestern United States have likely constricted wide ranging aquatic species such as the lowland leopard frog into isolated pockets, and maintenance of aquatic corridors may be critical to their future viability (Jennings and Scott, 1991). Sredl and Howland (1995) speculate that distribution of extant leopard frog populations in Arizona may be reflective of habitat fragmentation and extinction without recolonization as well as habitat quality.

Effects of livestock grazing on amphibian populations may be positive or negative (Jennings, 1988b; Rosen and Schwalbe, 1998; Sredl and Saylor, 1998). In the late 1800s and early 1900s, construction of earthen cattle tanks in upland drainages became a common range management practice (U.S. General Accounting Office, 1991), one that continues to this day. Because they were built primarily to water livestock, a positive secondary benefit of these systems is the water and aquatic habitat they provide to many species of wildlife, including amphibians.

Overgrazing negatively impacts amphibian habitat by removing bankside cover, increasing ambient ground and water temperatures, destroying bank structure (e.g., eliminating undercut banks), trampling egg masses, and adding high levels of organic wastes (Jennings, 1988b). Overgrazing in upland habitats also may degrade amphibian habitat by increasing runoff, which can lead to increased sedimentation of pool habitat in the drainages below (Jennings, 1988b; Belsky and Blumenthal, 1997).

Although the relationship between the lowland leopard frog and non-native predators (e.g., American bullfrogs [*R. catesbeiana*], crayfish, and predatory fish) has not been studied in detail, there is a negative co-occurrence between them (Rosen et al., 1995; Fernandez and Rosen, 1996). Among other factors, Sredl et al. (1997) attributed the low frequency of native leopard frog populations in mainstem rivers to the presence of large populations of non-native organisms.

F. Home Range Size. No study of home range has been completed. Sredl (1996) used repeated captures of lowland leopard frogs from a site that had been divided into 50 m sections. Of all captures, 46% were from the same section of initial capture, and 86% of all captures were within two sections of initial capture.

G. Territories. Unknown.

H. Aestivation/Avoiding Desiccation. Unknown.

I. Seasonal Migrations. Little is known of seasonal migrations. In one case, following drying of a pond, 154 frogs moved about 250 m upstream to a pond that did not dry, while 4 frogs moved 900 m downstream (Sredl, 1996).

J. Torpor (Hibernation). Although metamorphosed lowland leopard frogs generally are inactive between November and February, a detailed study of wintertime activity and habitat use has not been done.

K. Interspecific Associations/Exclusions. Throughout their range, lowland leopard frogs have been observed to occur with tiger salamanders (*Ambystoma tigrinum*), Colorado River toads (*Bufo alvarius*), Great Plains toads (*B. cognatus*), Arizona toads (*B. microscaphus*), red-spotted toads (*B. punctatus*), Woodhouse's toads (*B. woodhousii*), canyon treefrogs (*Hyla arenicolor*), Pacific treefrogs (*Pseudacris regilla*), Chiricahua leopard frogs, Tarahumara frogs (*R. tarahumarae*), and western narrowmouthed toads (*Gastrophryne olivacea*; Campbell, 1934; Ruibal, 1959; Platz and Frost, 1984; R.D. Jennings, 1987; Hale and Jarchow, 1988; Clarkson and Rorabaugh, 1989; unpublished data).

Lowland leopard frogs can be sympatric with another member of the *R. pipiens* complex, Chiricahua leopard frogs; but where sympatry occurs, F_1 hybrids are rare and presumed backcross individuals were not detected (Platz and Frost, 1984).

Rosen et al. (1995) noted a strong negative co-occurrence between lowland leopard frogs and introduced American bullfrogs. Jennings and Scott (1991) questioned the importance of the role that American bullfrogs have played in declines of native amphibians in the southwestern United States, arguing that many populations in New Mexico that declined were impacted by factors other than American bullfrogs.

L. Age/Size at Reproductive Maturity. Size at metamorphosis for lowland leopard frogs ranges from 25–29 mm SUL (Platz, 1988). The smallest males to exhibit secondary sexual characteristics from study sites in Graham and Yavapai counties, Arizona, were 53.5 mm and 56.2 mm SUL, respectively (Sredl, unpublished data). Size at which females reach sexual maturity is not known. Females have a mean asymptotic SUL of 76.4 mm, while that of males is 63.1 mm (Sredl et al., 1997).

M. Longevity. Preliminarily, skeletochronology of lowland leopard frogs indicates that they can live as long as 3 yr (M.J.S. and P. Fernandez, unpublished data). Estimates of survivorship of the adult and juvenile "age classes" appear to follow a seasonal pattern (Sredl et al., 1997)—high in the spring and summer and lower in the fall and winter. Within any given year, survivorships were always lowest in the winter. In 3 of 4 yr for which there were estimates for all four intervals, wintertime survivorship was by far the lowest; this pattern held for both adults and juveniles.

N. Feeding Behavior. No comprehensive studies of the feeding behavior or diet of lowland leopard frog larvae or adults have been conducted. Larval lowland leopard frogs are herbivorous. The diet of lowland leopard frog adults likely contains a wide variety of insects and other arthropods (Degenhardt et al., 1996). Stomach analyses of other members of the leopard frog complex from the western United States show a wide variety of prey items, including many types of aquatic and terrestrial invertebrates (e.g., snails, spiders, and insects) and vertebrates (e.g., fish, other anurans [including conspecifics], and small birds; Stebbins, 1951).

O. Predators. Detailed studies of predators of lowland leopard frogs have not been done. Tadpoles are likely preyed upon by insects, including belostomatids, notonectids, dytiscids, and anisopterans, and vertebrates, including native and non-native fish, tiger salamanders, garter snakes (*Thamnophis* sp.), Sonoran mud turtles (*Kinosternon sonoriense*), great blue herons (*Ardea herodias*), and other birds. Potential predators of juvenile and adult frogs likely include native and non-native fish, American bullfrogs, mud turtles, garter snakes, great blue herons, black hawks (*Buteogallus anthracinus*), and mammals including rats, coyotes, gray foxes, raccoons, ringtail cats, coatis, black bears, badgers, skunks, bobcats, and mountain lions. Cannibalism, primarily large adults eating juvenile frogs or large larvae, is likely but has not been studied. Jones (1990) studied the diet of black-necked garter snakes (*Thamnophis cyrtopsis*) in two desert streams in Arizona and found adult and larval lowland leopard frogs to be the most frequently consumed prey of adult and subadult snakes.

P. Anti-Predator Mechanisms. Adult lowland leopard frogs are cryptically colored and will sometimes remain motionless to escape detection. Other times, they rely on saltation, escaping to deep water or shoreline cover (personal observations). Jennings and Hayes (1994a) showed a picture taken by R. Ruibal of a lowland leopard frog in a defensive posture. Although others have noted this posture, its specific purpose and effectiveness have not been investigated.

Q. Diseases. In 1998, chytrid fungus was implicated in declines of amphibians in Australia and Panama (Berger et al., 1998). That same year, it was first identified in Arizona (Milius, 1998). Presently, one salamander (Sonoran tiger salamander *[A. tigrinum stebbinsi]*), seven species of ranid frogs (Rio Grande leopard frogs, plains leopard frogs *[R. blairi]*, American bullfrogs, Chiricahua leopard frogs, Ramsey Canyon leopard frogs *[R. subaquavocalis]*, Tarahumara frogs, and lowland leopard frogs), and one treefrog, canyon treefrogs, have been affected by this fungus in Arizona (Sredl et al., 2000; Collins, unpublished data). All outbreaks have been cool-season phenomena, and the pathogen appears widely distributed throughout central and southeastern Arizona (Sredl et al., 2000).

R. Parasites. Goldberg et al. (1998b) examined parasites of lowland and Chiricahua leopard frogs and American bullfrogs collected in Arizona and found that none of the helminths identified from the two native species were found in American bullfrogs.

4. Conservation.
Lowland leopard frog populations appear vulnerable to large-scale mortality, on a frequent basis, at the hands of a variety of causative factors. In 1991, Sredl et al. (1997) used mark–recapture to study six populations of lowland leopard frogs. In only 2 yr, each of six study sites showed substantial mortality events, sometimes followed by recovery (Sredl et al., 1997). At the site with the largest population, high rainfall and subsequent flooding during the winter of 1993 caused approximately 90% mortality. By the end of 1993, however, the population was nearly back to pre-flood size, although age structure had shifted toward younger age classes. In the second largest population, an outbreak of chytridiomycosis, which was retrospectively diagnosed, caused 50–80% mortality of metamorphosed frogs in late 1992 and early 1993. Adult frogs were affected to a lesser degree than juveniles. Reduction in total habitat area due to siltation of several important pools preceded and may have exacerbated the die-off. Population recovery from the epidemic has been modest and siltation continues to degrade the remaining habitat (unpublished data). Chytridiomycosis affected another population. In 1992–93 at two other small populations, declines also took place, and these populations were apparently extirpated. A fourth small population was apparently extirpated in 1992, but recolonized in 1994. It is likely that similar changes in abundance occurred in lowland leopard frog populations historically, although the role chytridiomycosis played is unknown at this time.

Because of suspected declines and extirpation of historical localities, lowland leopard frogs were added to the list of Category 2 candidate species (U.S.F.W.S., 1991). Later, the U.S. Fish and Wildlife Service discontinued the designation of multiple categories of candidates, and only those taxa meeting the definition of the former Category 1 candidate were considered candidates. At this time, lowland leopard frogs were dropped from consideration as a candidate species for Federal listing (U.S.F.W.S., 1996).

Family Rhinophrynidae

Rhinophrynus dorsalis Duméril and Bibron, 1841
BURROWING TOAD (SAPO BORRACHO)

M. J. Fouquette, Jr.

1. Historical versus Current Distribution.
Burrowing toads (*Rhinophrynus dorsalis*) have been a well-known, if infrequently encountered, burrowing anuran of lowland coastal areas in Mesoamerica for some 150 yr (Fouquette, 1969), but were not discovered as a component of the U.S. herpetofauna until 1964, when James (1966) found breeding populations in southern Texas. These toads surface only to breed during heavy rains, perhaps only once per year; thus, they are encountered infrequently by humans. Few additional reports of this species in Texas have been published since the first one.

Burrowing toads have been documented in a limited portion of Starr and Zapata counties in extreme southwestern Texas (James, 1966; Garrett and Barker, 1987; Dixon, 1987, 2000). Across their range, burrowing toads occur from sea level to 600 m, from southern Texas and northern Tamaulipas, Mexico, to Guanacaste Province in Costa Rica. In Texas, they occur in the Tamaulipan Biotic Province (Edwards et al., 1989), in the arid subtropical zone (Matamoran District;

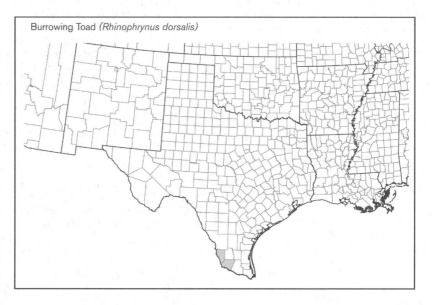

Burrowing Toad (*Rhinophrynus dorsalis*)

Blair, 1952). Duellman (1971) and Villa (1984) extended the southward range in Honduras and Nicaragua, beyond that indicated by the map of Fouquette (1969), and Foster and McDiarmid (1983) noted additional Costa Rican localities. Dixon (1987) commented that burrowing toads have not been found in Texas since about 1972; however, Dixon (2000) indicated specimens had been taken in 1984 and calls heard in 1998, presumably in the same limited area (although no specific literature records were reported).

2. Historical versus Current Abundance. No basis exists for evaluating any change in abundance over the 35 yr they have been known from the Texas localities. These animals are ephemerally active in limited arid habitats following heavy rains with flooding. James (1966) noted one moderate chorus that was active one night and had no activity on subsequent nights, but a few nights later (following another heavy rain), new choruses were found. There are no reports evaluating changes in abundance throughout the rest of their range.

3. Life History Features.
A. Breeding. Reproduction is aquatic.
i. Breeding migrations. Burrowing toads will breed any time of the year after rains create sufficiently deep breeding pools. Males will call from burrows, then emerge to breed in groups that can be large (Garrett and Barker, 1987). Calling and breeding may occur at the surface of the water or on soil along the flooded sites (Fouquette and Rossman, 1963; James, 1966). The distinctive advertisement call has been described by several authors (Fouquette and Rossman, 1963; James, 1966; Fouquette, 1969), and brief analysis with audiospectrogram was made by Fouquette (1969). Individuals occasionally are encountered on roads (Fouquette and Rossman, 1963; James, 1966) when there is breeding activity nearby, but there is no information available on dispersal or migratory movements.
ii. Breeding habitat. Seasonal pools and flooded areas created by heavy rainfall (James, 1966; Garrett and Barker, 1987).
B. Eggs.
i. Egg deposition sites. In ephemeral wetlands, females lay eggs in small clumps that soon separate and float to the water surface (Stuart, 1935).
ii. Clutch size. No information is available.
C. Larvae/Metamorphosis.
i. Length of larval stage. At least 2 mo, likely longer for most animals (Stuart, 1961). Growth rates, and thus times to

metamorphosis, are dependent on food availability (Stuart, 1961). Larvae were not observed by James (1966) in the Texas localities, although she searched for them, and those of other species were abundant.
ii. Larval requirements.
a. Food. Tadpoles are phytoplankton feeders and their guts are lined with large caecum-like areas (Altig and Kelly, 1974). Occasionally, tadpoles may feed on conspecifics (Starrett, 1960), although cannibalism was not observed by Stuart (1961).
b. Cover. Tadpoles school (Stuart, 1961; Foster and McDiarmid, 1982). When schooling, tadpoles form massed aggregates of varying sizes (Stuart, 1961). Schools may be mediated by either lateral line mechanoreceptive cues or olfaction, but apparently not by vision; they also may be influenced by socialization (Foster and McDiarmid, 1982). Generally, these masses are disc-like, with one estimated to be 25 cm in diameter and 10 cm deep that could have contained thousands of individuals (Stuart, 1961). Smaller schools may join to form larger schools, larger schools may break up (Stuart, 1961). Occasionally, larvae in schools will suddenly form a column and swim swiftly towards the surface (surfacing behavior; Stuart, 1961).
iii. Larval polymorphisms. Unknown and unlikely, although Starrett (1960) reported occasional cannibalism.
iv. Features of metamorphosis. No information available.
v. Post-metamorphic migrations. No information available.
D. Juvenile Habitat. No information available, but juveniles presumably burrow in the soil, similar to adults.
E. Adult Habitat. Friable soils that permit burrowing, including cultivated fields and gardens, are required (Garrett and Barker, 1987). Duellman (1971) pointed out that habitats for this species are typically arid or savanna, tropical or subtropical, non-forested areas. Villa (1984) also emphasized that habitats are in dry rather than moist environments, and activity is opportunistic with heavy rainfall. Fenolio and Ready (1995) indicated a strong correlation between breeding and rainfall. James (1966) described in detail the habitat in Texas.
F. Home Range Size. No information available.
G. Territories. No information available.
H. Aestivation/Avoiding Desiccation. Fouquette and Rossman (1963) reported a captive individual that did not feed for ≥ 18 mo, and during at least the last 5 mo, it apparently was aestivating in dried soil; upon rehydrating, the toad recovered and was quite active. No information is avail-

able for natural populations, but the soils where they burrow are often dry for weeks or months, so aestivation is presumed during such periods.
I. Seasonal Migrations. No information available.
J. Torpor (Hibernation). See "Aestivation/Avoiding Desiccation" above.
K. Interspecific Associations/Exclusions. James (1966) noted western narrow-mouthed toads (*Gastrophryne olivacea*), Texas toads (*Bufo speciosus*), and Coastal-Plain toads (*B. nebulifer*) calling abundantly in the choruses with burrowing toads in 1964, and Couch's spadefoot toads (*Scaphiopus couchii*) calling along with the Mexican burrowing toads in 1965.
L. Age/Size at Reproductive Maturity. Adults reach sexual maturity at 60–65 mm (Fouquette, 1969) with larger animals to 88 mm (Nelson and Nickerson, 1966). Females tend to be larger than males.
M. Longevity. No information available.
N. Feeding Behavior. Burrowing toads specialize in feeding on ants and termites in their underground nests. Hindlimbs have keratinized tubercles used in burrowing, and front limbs and snout are specially adapted for entering termite and ant tunnels to access prey (Trueb and Gans, 1983; Garrett and Barker, 1987). The tongue protrusion mechanism is uniquely designed for feeding on these insects underground (Trueb and Gans, 1983).
O. Predators. No information available.
P. Anti-Predator Mechanisms. Burrowing toads will grossly inflate their bodies, nearly concealing their head and limbs (Garrett and Barker, 1987). This is useful for subterranean burrowing (Trueb and Gans, 1983) and may also discourage predation on the surface. Other burrowing anurans (e.g., pelobatids and microhylids) may have noxious skin secretions that repel ants and termites from biting or stinging and are effective in discouraging many predators; no information is available on skin secretions of burrowing toads.
Q. Diseases. No information available.
R. Parasites. Santos-Barrera (1994) reported opalinid protozoans (*Zelleriela* sp.) from the gut of burrowing toads.

4. Conservation.
Burrowing toads are considered Threatened by Texas, the only U.S. state where they occur, but are not listed by the U.S. Federal Government. Because these are secretive animals, determining their distribution and abundance, and therefore their conservation status, is a problem. There are no data that indicate changes in abundance or distribution over the 35 yr they have been known from the Texas localities.

CAUDATA

Family Ambystomatidae

Ambystoma annulatum Cope, 1886

RINGED SALAMANDER

Stanley E. Trauth

1. Historical versus Current Distribution.

Ringed salamanders *(Ambystoma annulatum)* are endemic to the Interior Highlands (Ozark and Ouachita mountains) in Arkansas, Missouri, and Oklahoma (Anderson, 1950; Smith, 1950; Dowling, 1956, 1957; Anderson, 1965; McDaniel, 1975; Funk, 1979; Minter, 1979; Schuette, 1980; Trauth, 1980; Johnson, 1987; Turnipseed and Altig, 1991). No specimens have been reported from the Ozark Plateau of Kansas (Collins, 1993). There are no data to suggest that the current distribution differs from the pre-settlement distribution. Phillips et al. (2000) note that populations in the northeastern portion of the range (central Missouri) have less variable mitochondrial DNA components than populations to the southwest (southern Missouri, northwestern Arkansas, and eastern Oklahoma). This suggests that northern populations have been established more recently, perhaps with reforestation following the warm and dry Hypsithermal

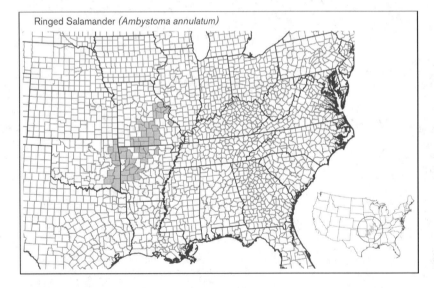

Ringed Salamander *(Ambystoma annulatum)*

Interval (8,000–4,000 yr before present; Phillips et al., 2000).

2. Historical versus Current Abundance.

Unknown. Published reports range from few to large numbers of animals. For example, while several early published records (prior to 1924) indicated few (<10) specimens (Black and Dellinger, 1938, in Arkansas), large populations (unspecified numbers) have been mentioned (Noble and Marshall, 1929, in Missouri; Trapp, 1956; Brussock and Brown, 1982,

in Arkansas). The first record reported in Oklahoma was a single specimen (Firschein and Miller, 1951). Spotila and Beumer (1970) observed 46 salamanders crossing a highway on 18 October 1966 near Fayetteville (Washington County), Arkansas, and observed 155 adults over a 3-yr period (1965–67). Several hundred ringed salamanders were counted by McDaniel and Saugey (1977) crossing a highway on 22 October 1976 near Blanchard Springs Caverns (Stone County, Arkansas). During a salamander sting operation conducted by the Arkansas Game and Fish Commission (Arkansas Democrat, 22 October 1987), a poacher was arrested while attempting to sell about 1,060 specimens for fish bait in Hot Springs, Arkansas. Peterson et al. (1991) estimated the number of females ovipositing in each of two ponds (Stone County, Missouri) to be between 150–230 individuals, whereas Peterson et al. (1992) captured 230 males and 237 females in one of these ponds using pit fall traps. In a recent study, Briggler et al. (1999) found 1,096 specimens migrating across a highway to a single pond in northwestern Arkansas from 22 September–14 November 1998. Trauth (2000) found 17 specimens in a woodland pond in the Ozark National Forest (Stone County, Arkansas) on 26 February 2000.

3. Life History Features.

A. Breeding. Reproduction is aquatic.

i. Breeding migrations. Ringed salamanders typically breed during autumn, primarily between September and early November (Noble and Marshall, 1929; Trapp, 1956; Anderson, 1965). Trauth et al. (1989c) observed the first incidence of winter breeding; apparently, winter breeding may be a common activity in ringed

salamanders in the Ozark National Forest of north-central Arkansas (Trauth, 2000). Adults are stimulated to migrate by medium to heavy rains, cool temperatures, and nighttime conditions. They travel to fishless, woodland ponds to breed; however, farm ponds (heavily used by livestock) in open pastures may also be utilized (Brussock and Brown, 1982). Hundreds of adults may be present during a single breeding episode, which normally lasts several days. Males apparently arrive before females (Spotila and Beumer, 1970). Each gravid female may be courted by 2–25 males. Males may also deposit spermatophores with or without attendant females (Spotila, 1976). Egg laying begins shortly after courtship and can last for 2 d (Johnson, 1987; Conant and Collins, 1998).

ii. Breeding habitat. Fishless woodland ponds as well as livestock ponds are suitable breeding sites for ringed salamanders (Brussock and Brown, 1982; Johnson, 1987).

B. Eggs.

i. Egg deposition sites. Eggs are laid on submerged branches, aquatic plant stems, or on the pond bottom in loose masses (Anderson, 1965; Spotila and Beumer, 1970; Johnson, 1987). An observation of terrestrial egg laying in March as reported by Strecker (1908a) has been discounted as being erroneous (see Trauth et al., 1989c; Peterson et al., 1992). Each female either lays one or two large clusters of eggs with 75–150 eggs/cluster (Johnson, 1987) or lays several smaller clusters of from 4–31 or 2–17 eggs (Trapp, 1956).

ii. Clutch size. Average clutch size is estimated to range from 205–390 eggs/female (Trauth, 2000; Peterson et al., 1992, respectively). Eggs hatch in 2–3 wk, depending on water temperature (Johnson, 1987).

C. Larvae/Metamorphosis. Larvae hatch at 12–15 mm SVL. Coloration in newly metamorphosed individuals is olive to black dorsally and grayish white ventrally (Hutcherson et al., 1989) with a fairly broad pigment-free band on the sides of the trunk (Bishop, 1943).

i. Length of larval stage. Larvae remain in the ponds throughout the winter and metamorphose the following summer. The length of the larval period varies from ~6–8.5 mo (Hutcherson et al., 1989).

ii. Larval requirements.

a. Food. Small larvae feed on cladocerans, copepods, and dipteran larvae, whereas older larvae primarily consume dipteran larvae; additional prey items included ostracods, hemipterans, snails, and dragonfly and damselfly naiads (Hutcherson et al., 1989). Trapp (1959) mentions cladocerans and *Chironomus* sp. (Diptera) as the major components of their diet. Cannibalistic larvae are facultative and will consume the same array of invertebrate food items as non-cannibals.

b. Cover. Little information is available on the terrestrial ecology of juveniles and adults (Petranka, 1998).

iii. Larval polymorphisms. Cannibalistic larvae were reported in a Missouri pond (Nyman et al., 1993). Other than having a relatively larger body size and somewhat broader heads, most cannibals were not different in appearance from non-cannibals. Cannibals exhibited mean body lengths twice that of their conspecific prey. Nyman et al. (1993) further pointed out that cannibalism in ringed salamander larvae appears to be an opportunistic behavior, an activity related to larger body size in older cohorts that tend to metamorphose at a larger size and at an earlier date.

iv. Features of metamorphosis. Larvae average 16 mm SVL 1 mo after hatching; average length at metamorphosis varies from 34–40 mm SVL and normally occurs from late April to early June in Missouri (Hutcherson et al., 1989; Nyman et al., 1993). Large ringed salamander larvae have been observed in September in the Ozark National Forest in Arkansas (unpublished data).

v. Post-metamorphic migrations. None observed.

vi. Neoteny. Has not been observed.

D. Juvenile Habitat. Juvenile habitat characteristics are unknown, although one juvenile reported by Trauth (1980; unpublished data) was collected on a forested hillside beneath a small flat rock.

E. Adult Habitat. Ringed salamanders are found in forested areas. Noble and Marshall (1929) found adults beneath piles of vines in a field as well as under leaves near a pond. They probably live hiding under logs and rocks or burrowing in the soil and are seldom found out or in the open (Johnson, 1987).

F. Home Range Size. Unknown.

G. Territories. Unknown.

H. Aestivation/Avoiding Desiccation. Aestivation is unknown; adults likely seek shelter under cover objects or burrow when faced with desiccating conditions.

I. Seasonal Migrations. Medium to heavy autumn rains with cooler temperatures initiate breeding movements (Spotila and Beumer, 1970). The triggering mechanism for winter breeding remains unknown (Trauth et al., 1989c), although Trauth (2000) mentioned the lack of adequate rainfall during the normal breeding cycle (due to a summer/fall drought) as a possible stimulus.

J. Torpor (Hibernation). No reports on hibernating animals are available.

K. Interspecific Associations/Exclusions. Ecological associates observed by Trauth et al. (1989c) and Trauth (2000) at a winter breeding pond include spotted salamanders (*A. maculatum*), central newts (*Notophthalmus viridescens louisianensis*), wood frogs (*Rana sylvatica*), and spring peepers (*Pseudacris crucifer*).

L. Age/Size at Reproductive Maturity. Males 74–104 mm SVL; females 82–112 mm SVL (Spotila and Beumer, 1970). Sixteen females that laid eggs during winter averaged 89.6 mm SVL (range, 80–98 mm; Trauth, 2000).

M. Longevity. Spotila and Beumer (1970) suggested that breeding cohorts included at least four size classes, making the oldest animals at least 4 yr old. Captive individuals live nearly 5 yr (Snider and Bowler, 1992).

N. Feeding Behavior. Ringed salamander adults probably feed on earthworms, insects, and snails (Johnson, 1987).

O. Predators. Petranka (1998) commented that although no known predators have been identified, potential predators include owls, woodland snakes, shrews, skunks, raccoons, opossums, and other mammals.

P. Anti-Predator Mechanisms. Brodie (1977) described the anti-predator display of the ringed salamander as being similar to that of flatwoods salamanders (*A. cingulatum*), Mabee's salamanders (*A. mabeei*), and small-mouthed salamanders (*A. texanum*), which are more passive than other species of *Ambystoma*. In these animals, anti-predator behaviors involve lashing the tail. Ringed salamanders will also elevate the proximal portion of their tail, coil their body, and position their head under the base of the tail.

Q. Diseases. No studies have focused on diseases in ringed salamanders.

R. Parasites. During a survey of new host records for the myxosporean *Myxidium serotinum* (from 28 amphibian species), McAllister and Trauth (1995) noted the absence of this gall bladder parasite in two ringed salamanders from Arkansas; however, in a follow-up study, McAllister et al. (1995c) for the first time found a single ringed salamander infected (6% of the salamanders examined). McAllister et al. (1995c) also identified what appeared to

be larval nematodes ("spirurid cysts") encysted in the stomach wall of seven salamanders; the rhabditid nematode, *Rhabdias ranae*, was found in 66% of the salamanders examined, whereas the ascarid nematode, *Cosmocercoides variabilis*, was identified in 71% of the salamanders.

4. Conservation.
Ringed salamanders are not protected in any of the states where they occur, nor are they Federally listed (Levell, 1997). There appears to be little difference between their current distribution and their historical distribution, and some populations have large (>1,000) breeding adults. Threats included habitat destruction and commercial collecting (for the bait industry).

Ambystoma barbouri Kraus and Petranka, 1989
STREAMSIDE SALAMANDER

Mark B. Watson, Thomas K. Pauley

1. Historical versus Current Distribution.
The current range of streamside salamanders (*Ambystoma barbouri*) includes central and western Kentucky, central Tennessee, southeastern Indiana, southwestern Ohio, and western West Virginia along the Ohio River (Kraus and Petranka, 1989; Longbine et al., 1991; Petranka, 1998). Kraus and Petranka (1989) first described streamside salamanders as a sibling species of small-mouthed salamanders (*A. texanum*). Streamside salamanders and small-mouthed salamanders are allopatric throughout most of their range, but parapatric or sympatric in portions of Kentucky and Tennessee (Petranka, 1982b). Streamside salamanders can be distinguished from small-mouthed salamanders by differences in range, ecology, and slight morphological characters such as different dentition. For example, small-mouthed salamanders have maxillary and

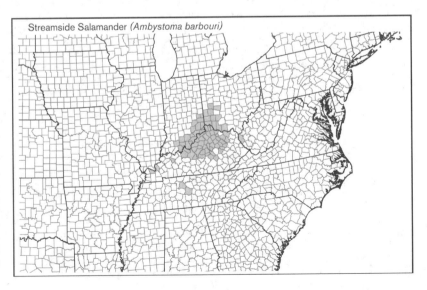

Streamside Salamander (*Ambystoma barbouri*)

premaxillary teeth with long pointed cusps, whereas streamside salamanders have more rounded cusps similar to other ambystomatids. All reports prior to 1989 considered streamside salamanders to be stream-breeding small-mouthed salamanders. Therefore, changes in historical distribution may not be apparent. New populations of streamside salamanders may be discovered in the states within their range after a thorough examination of small-mouthed salamander specimens.

2. Historical versus Current Abundance.
Reports of stream-breeding small-mouthed salamanders prior to the description of streamside salamanders as a separate species were probably of streamside salamanders. For example, Green (1955) stated that the only known population of small-mouthed salamanders in West Virginia breeds in streams. Subsequently, Longbine et al. (1991) showed that this population consists of streamside salamanders.

3. Life History Features.
A. Breeding. Reproduction is aquatic.

i. Breeding migrations. Streamside salamanders have an extended breeding season that lasts from late fall to early spring. Breeding migrations occur on rainy nights and individuals usually move from forests to first- and second-order streams, although streamside salamanders have been also found in ponds (Petranka, 1984a). Streamside salamanders do not migrate en masse as seen in congeneric species such as spotted salamanders (A. maculatum). Breeding in streamside salamanders begins 4–5 wk earlier than in small-mouthed salamanders where the species occur sympatrically. Additionally, streamside salamanders migrate to their breeding streams in the fall and undergo low levels of mating activity from early winter to early spring (Kraus and Petranka, 1989).

ii. Breeding habitat. Streamside salamanders were first described as a stream-breeding type of small-mouthed salamanders (Petranka, 1982a). They breed in first- and second-order streams that are usually devoid of fishes (Petranka, 1983, 1984d; Kats and Sih, 1992). Petranka (1982a) described the courtship behavior of stream-breeding small-mouthed salamanders in the laboratory. He described courtship in four phases. During Phase 1, males exhibited exaggerated undulations of their bodies and tails. They also circled in tight groups for 18–35 min. Phase 2 was described as spermatophore deposition, which lasted about 10 s. The males grasped the substrate with their legs and deposited 1–5 spermatophores on the top of gravel particles or other substrate. He also observed that males might try to dislodge spermatophores of conspecifics. In Phase 3, females entered a field of spermatophores. Females mounted 18–27 spermatophores, some more than once. During Phase 4,

females became dormant or motionless and laid eggs within 48 hr. McWilliams (1992) observed that breeding activities of streamside salamanders and small-mouthed salamanders are similar and that the location for breeding is the major difference. McWilliams (1992) also noted that unlike small-mouthed salamanders, breeding streamside salamander males are less likely to interfere with other males.

B. Eggs.

i. Egg deposition sites. Eggs are attached in a single row on the lower surface of flat rocks in flowing streams (Petranka, 1984d). Subsequent rows of eggs may be deposited, which may form clumps. Eggs are usually deposited in hidden or cryptic sites but may be in exposed locations (Sih and Maurer, 1992). Egg deposition occurs from January to early April. Oviposition generally occurs in pools, as opposed to riffles, in streams (Petranka, 1984a; Holomuzki, 1991).

ii. Clutch size. While clutch size varies from <10 to >1,000 eggs, the average number of ova in gravid females is around 260, suggesting communal egg deposition (Petranka, 1984a,d).

C. Larvae/Metamorphosis. Petranka (1984d) and Petranka and Sih (1986) observed that eggs hatch during late April, and larvae transform 6–9 wk after hatching. Brandon (1961) described the larval development of stream-breeding small-mouthed salamanders from central Kentucky. Newly hatched larvae are 12 mm in length. Individuals begin to metamorphose at 37–41 mm. He also observed two individuals in that size range that had completely lost their gills. Petranka (1984c,d) noted that temperature and food availability could influence larval growth rates. At warm temperatures (24 °C) larvae metamorphosed in 27 d on average, but at cooler temperatures (15 °C) metamorphosis was delayed to an average of 72 d. He also noted that larvae grew to a larger size when metamorphosis occurred at cooler temperatures. Maurer and Sih (1996) compared larval growth rates of streamside salamanders and small-mouthed salamanders. Streamside salamanders showed increased feeding and activity levels and developed faster than small-mouthed salamanders. Streamside salamanders also reduced their activity in response to food deprivation. These differences were attributed to differences in habitat duration between the stream-dwelling and pond-dwelling species.

In some streams, competition for food may increase the length of the metamorphic period (Petranka and Sih, 1986). Chemical cues from green sunfish (Lepomis cyanellus) could delay hatching of streamside salamanders (Moore et al., 1996). Similarly, Sih and Moore (1993) reported experimental evidence to show that flatworm (Pagocotus gracilis) predation

can induce streamside salamander eggs to delay hatching. Increased size of hatchlings decreased the likelihood of predation by flatworms.

D. Juvenile Habitat. Juvenile streamside salamanders utilize the same habitat as larvae. Streamside salamander juveniles feed on various macroinvertebrates, including insect larvae, zooplankton, isopods, and amphipods (Huang and Sih, 1990; Sparks, 1996).

E. Adult Habitat. Adult streamside salamanders are extremely fossorial and above ground activity is mostly observed during breeding migrations (Petranka, 1982a; Holomuzki, 1991). They are usually found in upland deciduous forests and are most common in regions with exposed limestone (Petranka, 1998).

F. Home Range Size. Unknown.

G. Territories. Unknown.

H. Aestivation/Avoiding Desiccation. Aestivation is unknown; animals may seek shelter under cover objects or burrow when facing desiccating conditions.

I. Seasonal Migrations. Petranka (1984a) found that migrations to breeding sites in central Kentucky occur from late October to March. Males reach breeding sites before females (Petranka, 1984a).

J. Torpor (Hibernation). Unknown.

K. Interspecific Associations/Exclusions. Streamside salamanders breed in first- and second-order streams and may come into contact with individuals in the genera *Desmognathus*, *Eurycea*, *Pseudotriton*, and *Gyrinophilus*.

L. Age/Size at Reproductive Maturity. Unknown.

M. Longevity. Unknown.

N. Feeding Behavior. Streamside salamander larvae can have substantial effects on the density of benthic isopods. Huang and Sih (1991a) observed a large negative relationship between density of salamander larvae and isopods in stream experiments. Sih and Petranka (1988) showed that small-mouthed salamander larvae were not selective when prey items were in low abundance, but would release previously acquired smaller prey in order to acquire larger prey.

O. Predators. Fishes are the major predators of streamside salamander larvae. Petranka (1983) showed that fish predation or the presence of fish restricted stream-breeding small-mouthed salamander larvae to upper regions of streams that are devoid of fishes. Kats et al. (1988) demonstrated that stream-breeding small-mouthed salamanders were palatable to fish. Sih et al. (1992) reported that fish preyed heavily upon larvae of streamside salamanders. They observed that 30–40% of larvae drifted into a pool with fish, and of those only 6–8% survived to drift out. Sih (1992) suggested that ineffective predator defense against sunfish (Centrarchidae) may be important in the evolution of small-mouthed salamanders. Sunfish in stream pools could be a barrier to gene flow by preventing movement of larvae between pools (Storfer, 1999a). Populations of streamside salamanders separated by fish within the same stream could be as genetically different as those from different streams (Storfer, 1999b). Flatworms prey on small streamside salamander larvae in streams (Petranka et al., 1987a; Sih and Moore, 1993). Three small-mouthed salamander larvae and one mud salamander *(Pseudotriton montanus)* larva were regurgitated by a northern water snake *(Nerodia s. sipedon)* that was found foraging in a small stream in central Kentucky (Kats, 1986).

P. Anti-Predator Mechanisms. Adult small-mouthed salamanders, a close relative of streamside salamanders, use biting and immobility as anti-predator defenses. They also elevate the tail and roll their bodies toward a grasped limb, right themselves, and coil upon release (Brodie et al., 1974a; Brodie, 1977). Members of the genus *Ambystoma* have many anti-predator adaptations including tail movements, coiling and elevating the body, aposematic coloration, and noxious skin secretions (Brodie, 1977). Huang and Sih (1990) found stream-dwelling small-mouthed salamander larvae increased the use of refugia in the presence of predatory sunfish.

In laboratory studies that simulated fish predation, streamside salamanders took refuge under artificial substrates (Huang and Sih, 1991b). Several studies have shown that streamside salamander larvae respond to chemical cues from predatory fish. In these studies, larvae exhibited anti-predator behavior when fish were kept isolated and not visible to the larvae (Petranka et al., 1987b; Sih and Kats, 1991, 1994). Kats et al. (1988) observed that stream-dwelling small-mouthed salamander larvae responded to chemical cues from green sunfish by spending less time in the open and more time under cover. They also observed that small-mouthed salamanders that breed in ponds normally devoid of fish did not exhibit this behavior. Other studies have shown larvae that increase time spent under substrates, such as rocks and algae, become immobile or shift to more nocturnal feeding behavior in the presence of fish (Kats et al., 1988; Holomuzki, 1989a; Sih et al., 1992; Sih and Kats, 1994). Storfer and Sih (1998) noted that larvae from populations most isolated from fishless populations showed stronger anti-predator behavior. Storfer et al. (1999) observed that streamside salamander larvae from streams with fish were a more cryptic, lighter color than those found in shallow ephemeral streams without fish.

Q. Diseases. Unknown.

R. Parasites. Nematodes *(Cosmoceroides dukae* or *C. variabilis)* are known to occur in the digestive tract (Baker, 1987).

4. Conservation.
Streamside salamanders usually are not found in streams where surrounding forest land has been timbered (Petranka, 1998). This suggests that deforestation and development around streams and ravines within their range will be detrimental to this species.

Only two populations are known in West Virginia, and one of these may have been destroyed recently by development. Mitchell et al. (1999) lists streamside salamanders as a species that may need to be monitored in West Virginia because of few verified populations and lack of data on the status of the known populations.

Ambystoma californiense Gray, 1853
CALIFORNIA TIGER SALAMANDER

H. Bradley Shaffer, Peter C. Trenham

1. Historical versus Current Distribution.
California tiger salamanders *(Ambystoma californiense)* are a California endemic with their range centered in the Central Valley from Tulare and San Luis Obispo counties in the south, to Sacramento and Solano counties in the north. They are known from sites in the bottom of the Central Valley up to a maximum elevation of roughly 1,200 m in the Coast Ranges, and 500 m in the Sierra Nevada foothills. Although the historical distribution of California tiger salamanders is not known in detail, their current distribution and genetic data suggest that they were distributed continuously in the vernal pool/grassland habitat that dominated much of

this region. In addition, there are two well-characterized disjunct populations of California tiger salamanders. To the northwest of their main range there are populations near Santa Rosa, Sonoma County; to the southwest there are populations in the Santa Maria region of northwestern Santa Barbara County (Storer, 1925; Jennings and Hayes, 1994a; Fisher and Shaffer, 1996). Genetic data indicate that the Sonoma and Santa Barbara populations have been isolated from the remainder of the range for sufficient time (on the order of 1,000,000 yr) that they may represent distinct species (H.B.S., unpublished data).

The northernmost California tiger salamander specimens on record are from the California Department of Fish and Game Refuge at Grey Lodge, Butte and Sutter counties (California Department of Fish and Game, 2002). The southernmost record is for specimens collected in 1892 in Riverside County, although this record is >300 km from the nearest known population and may represent a misidentification. In both cases, repeated surveys over the past two decades have failed to relocate these populations and they are presumed extirpated or in error (Jennings and Hayes, 1994a). Presently the northernmost population of California tiger salamander is in the vicinity of Dunnigan, Yolo County, and the southernmost populations are in Santa Barbara County. In addition to the loss of populations at the extremes of the species' range, extensive habitat conversion has extirpated many populations in the continuous portions of

California Tiger Salamander *(Ambystoma californiense)*

their range (Barry and Shaffer, 1994; Jennings and Hayes, 1994a; Fisher and Shaffer, 1996). Holland (1978) estimated that by 1973, 60–85% of natural vernal pool habitats had been lost, and what remains is increasingly fragmented by urban and agricultural development. Presently, few populations are known from sites in the Central Valley, and most recent records are for sites in the surrounding hills to the east and west (California Department of Fish and Game, 2002), presumably due to the nearly complete conversion of the Central Valley to intensive agriculture. Surveys conducted in the early 1990s failed to find any California tiger salamanders at 56% of 86 historical localities (H.B.S., unpublished data). Due to the rapid expansion of agricultural, urban development and the loss of breeding ponds in Sonoma and Santa Barbara counties, the U.S. Fish and Wildlife Service listed these distinct population segments as Endangered (USFWS, 2000b, 2002c).

2. Historical versus Current Abundance.
California tiger salamander populations can fluctuate widely both among years and among ponds within years. At one study site in Monterey County, the number of breeding adults visiting a pond varied from 57–244 individuals, with female breeders varying by an order of magnitude among years (16–140; Trenham et al., 2000). A breeding site approximately 200 km to the north (Contra Costa County) showed a similar pattern of variation, suggesting that such fluctuations are typical (Loredo and Van Vuren, 1996). At the local landscape level, nearby breeding ponds can vary by at least an order of magnitude in the number of individuals visiting a pond, and these differences appear to be stable across years (Trenham et al., 2001). Virtually nothing is known concerning the historical abundance of the species. Twitty (1941) reported that on two rainy nights in January 1940, 28 and 45 migrating adults were obtained on the road bordering Lake Lagunita, Stanford University, Palo Alto; similar breeding migrations have been observed in recent years (A. Launer, personal communication).

3. Life History Features.
    A. Breeding. Reproduction is aquatic.
    i. Breeding migrations. Migrations to and from breeding ponds occur during the rainy season (November–May), with the greatest activity from December–February (Storer, 1925; Loredo and Van Vuren, 1996; Trenham et al., 2000). Breeding may occur in one major bout or during a prolonged period of several months, depending on the rainfall pattern (Loredo and Van Vuren, 1996; Trenham et al., 2000). Breeding migrations are strongly associated with rainfall events (Loredo

and Van Vuren, 1996; Trenham et al., 2000). During drought years, adults (particularly females) migrate in low numbers. Males consistently arrive at the breeding pond before females and stay approximately four times longer—over a 7-yr period the average time spent in one Monterey County breeding pond was 44.7 d for males and 11.8 d for females (Trenham et al., 2000); Loredo and Van Vuren (1996) found similar results in a 2-yr study in Contra Costa County (males—37 d, females—10 d).
    ii. Breeding habitat. California tiger salamanders breed in fishless, seasonal and semi-permanent wetlands (Barry and Shaffer, 1994; Petranka, 1998). Historically, California tiger salamanders probably relied exclusively on shallow vernal pools for breeding habitat, but they now make extensive use of ponds constructed for cattle, particularly in foothill habitat. In Monterey County, ponds utilized as breeding habitat were natural vernal pools and artificial cattle ponds ranging in depth from 30 cm to >2 m and ranging in annual hydroperiod from 10–52 wk (Trenham et al., 2001).

    B. Eggs.
    i. Egg deposition sites. Females lay eggs singly or, more rarely, in clusters of 2–4 eggs, attached to live vegetation or detritus (Storer, 1925; Twitty, 1941). Eggs are about 3.5 mm in diameter and hatch 2–4 wk after being laid (Storer, 1925).
    ii. Clutch size. Eleven females from Monterey County contained an average of 814 eggs (range: 413–1,340; Trenham et al., 2000).

    C. Larvae/Metamorphosis.
    i. Length of larval stage. Average is 4–5 mo (range 3–6), with peak laying in January and peak emergence of newly metamorphosed animals from mid-June to mid-July (Loredo and Van Vuren, 1996; Trenham et al., 2000). Total annual production of metamorphic animals ranged from 122–775 (4.7–21.9/female) at a Monterey County pond (Trenham et al., 2000). Newly metamorphosed individuals from three sites across the species range are fairly consistent in size (mean SVLs are 58 mm, 64 mm, and 62.4 mm; Holland et al., 1990; Loredo and Van Vuren, 1996; Trenham et al., 2000).
    ii. Larval requirements.
    a. Food. Newly hatched larvae begin feeding after a few days. Larvae are gape-limited predators. Smaller larvae feed primarily on zooplankton (cladocerans and copepods); older larvae feed on tadpoles (primarily of Pacific treefrogs; *Pseudacris regilla*), ostracods, amphipods, midge larvae, water boatmen (Corixidae), and pond snails (Anderson, 1968).
    b. Cover. When not feeding, larvae are sedentary on the bottom of wetlands. Storer (1925) noted that larvae are wary, seeking cover in vegetation when disturbed.

    iii. Larval polymorphisms. Have not been reported, although large, late-metamorphosing larvae sometimes develop large heads reminiscent of cannibal morph eastern tiger salamanders (*A. t. tigrinum*; H.B.S., personal observation).
    iv. Features of metamorphosis. Metamorphosis typically occurs during the dry summer, and newly metamorphosed larvae emigrate from their breeding wetlands under both wet and dry conditions (Loredo and Van Vuren, 1996; Loredo et al., 1996). Median emergence date can vary by at least 2 mo across years and invariably precedes complete pond drying by several weeks (Trenham et al., 2000). At least a few metamorphic animals continue to emigrate after the pond has completely dried (Trenham et al., 2000), presumably by using dense vegetation or cracks in the substrate as temporary daytime retreats (Loredo et al., 1996).
    v. Post-metamorphic migrations. In the first night, newly metamorphosed animals move an average of about 26 m (range 6–57 m; Loredo et al., 1996). Of 69 individuals followed, roughly half found shelter in soil cracks and half in California ground squirrel (*Spermophilus beecheyi*) burrows (Loredo et al., 1996). Laboratory studies of burst and endurance locomotor capabilities demonstrate that juveniles can travel at least 25 m without resting (Shaffer et al., 1991).
    vi. Neoteny. Has not been reported. For much of their evolutionary history, California tiger salamanders were presumably restricted to vernal pool habitats and did not have the opportunity to express a fully paedomorphic life history (Storer, 1925). As breeding sites have been enlarged and made permanent, over-summering larvae have occasionally been found (H.B.S., unpublished observations). The complete absence of sexually mature paedomorphic animals suggests that this species lacks the ability to express this life history pattern. In central California, recently discovered populations containing paedomorphic animals invariably contained introduced genotypes from non-native eastern tiger salamanders, based on mitochondrial and nuclear genetic analyses (H.B.S. and colleagues, unpublished data).
    D. Juvenile Habitat. Newly metamorphosed animals emerging from one breeding pond settled in California ground squirrel burrows and soil cracks in roughly equal proportions (Loredo et al., 1996). Near Lake Lagunita at Stanford University, juveniles use the crevices in a sandbag retaining wall as terrestrial retreats (Barry and Shaffer, 1994).
    E. Adult Habitat. Adults clearly rely on rodent burrows for underground retreats—except for the brief breeding season, they spend the entire year in or near these retreats. Animals unable to gain access to underground burrows may be prone to

desiccation (Loredo et al., 1996). California ground squirrel and valley pocket gopher *(Thommomys bottae)* burrows are the primary source of these retreats (Barry and Shaffer, 1994; Trenham, 2001). Unlike eastern tiger salamanders, adult California tiger salamanders are able to burrow through moist soil in blocked mammal burrows (Jennings, 1996a). Adults rely heavily on ground squirrel burrows, with 83% of adults using them on their first night migrations from breeding ponds (Loredo et al., 1996). Radio-tracked adults most commonly settled in burrows in open grassland areas or beneath large oaks, with burrows in woodland areas less commonly occupied (Trenham, 2001).

F. Home Range Size. Following breeding, radio-tracked adults migrated away from breeding ponds and initially settled in ground squirrel burrows 3–158 m away (Trenham, 2001); Loredo et al. (1996) observed similar first night emigration distances. Most of the radio-tracked salamanders moved to another or several different burrow systems farther from the pond during the 1–4 mo tracking interval. For 11 salamanders, the average final distance traveled from the pond was 114 ± 83 m (Trenham, 2001). Within individual burrow systems, salamanders frequently made short moves of <10 m, apparently without surfacing (Trenham, 2001).

G. Territories. Generally unknown. However, using a fiber optic scope, Semonsen (1998) observed multiple juvenile salamanders in close proximity to one another within individual burrow systems near Santa Maria, suggesting that they are not strongly territorial.

H. Aestivation/Avoiding Desiccation. Summertime conditions are extremely hot (frequently >40 °C) and dry (essentially zero precipitation from June–October), and aestivation may be occurring in underground retreats. However, continued within-burrow movements through at least June by some radio-equipped adult salamanders suggest that aestivation, if it occurs, is not obligate (Trenham, 2001).

I. Seasonal Migrations. Seasonal migrations in California tiger salamanders appear to occur only for the purpose of breeding. In years when rainfall is sparse or late, larger proportions of surviving adults, especially females, fail to migrate to breeding ponds (Loredo and Van Vuren, 1996; Trenham et al., 2000). Subadults do not appear to make any regular seasonal migrations, as they are rarely captured at or near breeding ponds (P.C.T., personal observation). There is one report of juveniles migrating to a breeding site *en masse* following a rare August storm (Holland et al., 1990; see also Petranka, 1998). While most individuals breed for the first time in their natal pond, 31% of males and 27% of females marked at a pond were recaptured breeding for the first time at a second

pond 580 m away (Trenham et al., 2001). Within a system of 11 breeding ponds, 26% of surviving adults also dispersed to other breeding ponds in subsequent years (Trenham et al., 2001). Following breeding, adults move away from breeding ponds (see "Home Range Size" above).

J. Torpor (Hibernation). None recorded; none expected given the mild winters that California tiger salamanders encounter.

K. Interspecific Associations/Exclusions. According to Loredo et al. (1996), California tiger salamanders may have a commensal relationship with California ground squirrels. Radio-tracked adult salamanders were always located in close association with ground squirrel burrows (Trenham, 2001). Some burrows occupied by salamanders are simultaneously in use by ground squirrels (Semonsen, 1998; P.C.T., personal observations). Burrows of pocket gophers are also used (Barry and Shaffer, 1994; Jennings, 1996a; Trenham, 2001). Because of concerns to cattle and agriculture, California ground squirrels are currently being controlled on over 4 million ha, a management practice that may indirectly threaten California tiger salamanders (Loredo et al., 1996). Associations with burrowing mammals in flat, floodplain regions of the central valley are unknown and may differ from published studies in better drained, upland habitats.

Ponds that contain populations of exotic fishes and American bullfrogs *(Rana catesbeiana)* appear unsuitable as breeding habitat (Fisher and Shaffer, 1996; H.B.S. and colleagues, unpublished data). In a related experiment, California tiger salamander embryos suffered complete predation in a fish- and American bullfrog-free permanent pond that contained large numbers of resident adult California newts *(Taricha torosa)* and dragonfly larvae, but only light predation in a nearby temporary pond lacking these predators (P.C.T., unpublished data).

L. Age/Size at Reproductive Maturity. Minimum age at first reproduction based on recaptures of marked juveniles is 2 yr for males and 2–3 yr for females (Loredo and Van Vuren, 1996; Trenham et al., 2000). However, most individuals at one pond in Monterey County did not reach sexual maturity until 4–5 yr of age (Trenham et al., 2000); skeletochronology estimates of breeding adult age structure confirmed these estimates. Interestingly, <50% of individuals breed a second time during their lifetime at either their natal or any other pond (Trenham et al., 2000). Trenham et al. (2000) found sizes of breeding adults to be extremely variable, ranging from 75–130 mm SVL, and only a weak positive relationship between SVL and skeletochronological age (R < 0.4).

M. Longevity. Skeletochronology-based age estimates for the Monterey County breeding site ranged from 2–11 yr for sexually mature individuals and did not

differ between sexes (Trenham et al., 2000). Most breeding adults are 4–6-yr old; confirmation with known-age animals demonstrated that these estimates are accurate (Trenham et al., 2000).

N. Feeding Behavior. Nothing has been published on feeding ecology of post-metamorphic juveniles or adults. Based on captive individuals, they are presumed to take a wide variety of invertebrate and small vertebrate prey.

O. Predators. California red-legged frog *(R. draytonii)* adults are known to eat California tiger salamander larvae (Baldwin and Stanford, 1987; see also Petranka, 1998). California ground squirrels may eat adults, although salamanders do not appear to avoid occupied ground squirrel burrows (Loredo et al., 1996; Semonsen, 1998; see also Petranka, 1998). Garter snakes *(Thamnophis* sp.) will sometimes prey heavily on larvae, and at least one adult American bullfrog was found with a newly metamorphosed tiger salamander in its stomach (H.B.S., unpublished observations). Striped skunks *(Mephitis mephitis)* and garter snakes have been observed preying on adult salamanders in pitfall traps (P.C.T., personal observations). Introduced predatory fishes and California tiger salamander larvae do not co-occur in the same ponds (Fisher, 1995), suggesting that these fishes prey heavily on larvae. In a controlled field experiment, low densities of mosquito fish *(Gambusia* sp.; 0.5 fish/m^2 pond surface area) had no discernible effect on California tiger salamander hatchling growth or survival to metamorphosis. However, densities more typical of many permanent ponds (12.5 fish/m^2) significantly reduced growth and survival to metamorphosis of California tiger salamander larvae (K. Leyse, unpublished data). Presumably the same types of predators that occur in eastern tiger salamander populations also influence California tiger salamanders, including predatory birds and small mammals.

P. Anti-Predator Mechanisms. Unknown.

Q. Diseases. Unknown.

R. Parasites. Unknown.

4. Conservation.
California tiger salamanders are considered a Species of Special Concern across its range by the State of California. They are a candidate for listing under the U.S. Endangered Species Act, and in 1994, listing was determined to be "warranted" by the U.S. Fish and Wildlife service, but "precluded" due to higher priority species (USFWS, 1994a). In January 2000, the Santa Barbara populations received emergency listing under the Endangered Species Act as an Endangered species (USFWS, 2000b). In March 2003, the Sonoma County distinct population segment was listed as Endangered (Federal Register, 2003). Habitat destruction (both of breeding pools and upland terrestrial

habitat) and introduced exotic predators are widely considered to be the primary causes of decline (Stebbins and Cohen, 1995; Fisher and Shaffer, 1996; Davidson, 2000; Davidson et al., 2002). Introduced *A. tigrinum* also hybridize with native *A. californiense*, causing genetic "biopollution" problems in central California (Riley et al., 2003).

Ambystoma cingulatum Cope, 1867 (1868)

FLATWOODS SALAMANDER

John G. Palis, D. Bruce Means

1. Historical versus Current Distribution.

Flatwoods salamanders (*Ambystoma cingulatum*) are restricted to the Coastal Plain of the southeastern United States; ranging from southern South Carolina south to Marion County, Florida, and west through southern Georgia and the Florida Panhandle to southwestern Alabama (Conant and Collins, 1991). Formerly, Louisiana, Mississippi, and North Carolina were included in their range (Martof, 1968); however, these records have been discounted as misidentifications (Goin, 1950; Hardy and Olmon, 1974; P. Moler, personal communication).

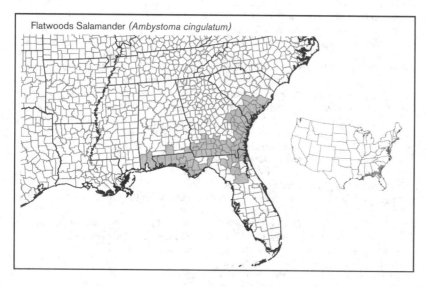

Flatwoods Salamander (*Ambystoma cingulatum*)

Flatwoods salamanders may no longer occur at many historical locations. The last observation of flatwoods salamanders in Alabama was 1981 (Jones et al., 1982). Palis (1997b) was unable to confirm the continued existence of flatwoods salamanders at/near 2/3 of historical breeding sites sampled in Florida. Recent (post-1989) surveys failed to detect flatwoods salamanders at 33 historical locations in Georgia and at 26 of 29 (90%) historical sites in South Carolina (USFWS, 1997c). However, recent surveys documented 102 previously unknown breeding sites in Florida (n =75) and Georgia (n = 27; USFWS, 1997c). Despite intense efforts,

no new breeding sites have been located in Alabama or South Carolina (USFWS, 1997c).

2. Historical versus Current Abundance.

Historical (Goin, 1950) and recent (Palis, 1997a) observations indicate that flatwoods salamanders can be common locally. However, due to habitat alteration, they may be less common now than they were historically. For example, Means et al. (1996) observed a 99% decline in the breeding migration of adults at a Florida site from the 1970s–90s.

3. Life History Features.

A. Breeding. Terrestrial, in wetland basins prior to pond filling.

i. Breeding migrations. Flatwoods salamanders are autumn breeders; most individuals migrate to breeding sites in October–November and emigrate in December–January (Means, 1972; Anderson and Williamson, 1976; Means et al., 1996; Palis, 1997a). Adults typically move during rains associated with passing cold fronts, but will also migrate on rainless nights when soils are near saturation (Palis, 1997a). Males and females migrate simultaneously (Palis, 1997a). Post-breeding

salamanders tend to emigrate in the direction of immigration, suggesting the ability to return to a particular terrestrial retreat (Palis, 1997a). In an all-night (29–30 October 1993) investigation of a flatwoods salamander breeding migration at a site in Okaloosa County, Florida (see Palis, 1997a), nearly equal numbers of salamanders were observed between 1800–2400 (n = 27) and 2401–0700 (n = 26; J.G.P., unpublished data). Rain fell continuously and the temperatures remained nearly constant across both observation periods. Adults may move as far as 1,700 m from breeding sites (Ashton, 1992). However, this distance may be exceptional inas-

much as other *Ambystoma* are reported to travel shorter distances from breeding sites (Semlitsch, 1998; P.K. Williams, 1973).

ii. Breeding habitat. Breeding sites include isolated swamps where pond cypress or blackgum predominate, marshy pasture ponds, roadside ditches, or small, shallow borrow pits (Anderson and Williamson, 1976; Palis, 1995b, 1997b; D. Stevenson, personal communication). Where woody regeneration is poor, breeding sites that were logged may now be dominated by shrubs or graminaceous vegetation.

B. Eggs.

i. Egg deposition sites. Eggs are deposited singly in small groups (1–34 eggs) in dry pond basins either in the open on bare soil or under cover beneath logs, leaf litter, dead grass, *Sphagnum* mats or within crayfish burrows (Anderson and Williamson, 1976). Anderson and Williamson's (1976) report that most eggs are deposited in the lowest elevations of depressions may be an artifact of sampling. Eggs are more easily located in the open, lower elevations of breeding basins than in the more densely vegetated grassy edges (G. Williamson, personal communication). Captures of recently hatched larvae suggest that many eggs are deposited in these upper grassy zones (J.G.P., personal observations). Cryptic oviposition sites offer protection from fire (J. Jensen, personal communication). In Florida, flatwoods salamanders may oviposit on vegetation in water (R. Ashton, personal communication).

ii. Clutch size. Based on ovarian counts, from 97–222 (mean 163) in South Carolina and Georgia (Anderson and Williamson, 1976) and about 225 in Florida (Ashton, 1992). Egg number correlates with female size (see also Petranka, 1998).

C. Larvae/Metamorphosis.

i. Length of larval stage. Palis (1995b) determined the larval period at two Florida breeding sites. At a site containing two distinct size classes of larvae, larvae grew 1.7–1.9 mm/wk, and the larval period encompassed 15–18 wk. At the second site, larvae grew 2.5 mm/wk and transformed in 11–13 wk. Sekerak et al. (1996) observed metamorphosis after an approximate 12-wk larval period.

ii. Larval requirements.

a. Food. Food habits have not been studied, but larvae likely prey on a variety of aquatic invertebrates and perhaps small vertebrates (e.g., other amphibian larvae, smaller conspecifics). Larvae have been raised in the laboratory on earthworms (*Sparganophilus* sp., *Diplocardia eiseni*), fairy shrimp (Eubranchiopoda), aquatic amphipods (*Hyallela* sp.), and mosquito (dipteran) larvae.

b. Cover. Larvae hide in submerged herbaceous vegetation during the day. In a quantitative study of ten larval sites in Florida, Sekerak et al. (1996) found larvae most often in sedge- and pipewort-dominated plots. In a daytime survey of 82 breeding sites in Florida, Palis (1997b) captured larvae principally in graminaceous vegetation. At night, however, larvae enter the water column (Palis, 1995b).

iii. Larval polymorphisms. None known.

iv. Features of metamorphosis. Larvae typically metamorphose within a 5–10 d span in March–April (Mecham and Hellman, 1952; Palis, 1995b). Metamorphic animals range in size from 35.5–46.0 mm SVL (Goin, 1950; Mecham and Hellman, 1952; Palis, 1995b; J.G.P., unpublished data).

v. Post-metamorphic migrations. Have not been studied. However, at a Florida site, 28% of the yearlings first observed at a drift fence in the fall were captured emigrating from the pond, suggesting that they summered at the breeding site (Palis, 1997a). Metamorphic animals have been found within 15 m of a breeding site in Liberty County, Florida, in April (J.G.P., personal observations) and collected while emigrating from a breeding site in Jasper County, South Carolina, in mid-April (data with specimens in collection of Savannah Science Museum).

vi. Neoteny. Not known to occur.

D. Juvenile Habitat. Unknown, but presumably similar to that of adults. Because small salamanders are susceptible to desiccation (Spotila, 1972), juveniles may be restricted to the vicinity of the breeding site in dry years (Palis, 1997a).

E. Adult Habitat. Often described as slash pine flatwoods (Goin, 1950; Martof, 1968; Conant and Collins, 1991), but Palis (1996a), Means (this volume, Part One), and Means et al. (1996) argued that the primary habitats of post-metamorphic animals are longleaf pine flatwoods and savannas. The distinction is important because longleaf pine-dominated uplands typically have a more open canopy and a greater herbaceous component to the ground cover than slash pine-dominated uplands. Neill (1951b) and Ashton (1992) have observed adults in crawfish burrows.

F. Home Range Size. Observations of home range size are limited to Ashton (1992) who determined that the activity range of one individual encompassed 1,500 m².

G. Territories. The potential for territorial behavior has not been studied.

H. Aestivation/Avoiding Desiccation. Unknown. However, we are not aware of any specimens collected in July–August, suggesting that either flatwoods salamanders are active only underground or aestivate during these months.

I. Seasonal Migrations. Documented migrations involve breeding (adults to/from breeding sites in the fall) and migrations of newly metamorphosed animals in the spring.

J. Torpor (Hibernation). Unknown and unlikely, given that the species is active during the winter (Goin, 1950; Palis, 1997a).

K. Interspecific Associations/Exclusions. Larvae co-occur with a wide variety of invertebrates and vertebrates in breeding sites, including other salamander species (Anderson and Williamson, 1976; Palis, 1997b). Competitive interactions with other salamander species, either in larval or post-metamorphic stages, is unknown.

L. Age/Size at Reproductive Maturity. In a western Florida population, males attained sexual maturity in 1 yr, but most females apparently did not mature for at least 2 yr (Palis, 1997a). In this population, the smallest mature male was 44 mm SVL, the smallest gravid female 53 mm. In the laboratory, Means (1972) raised larvae to adult sizes in 1 yr.

M. Longevity. Flatwoods salamanders were not included among the species for which captive longevity records were available to Bowler (1977). One of us (D.B.M.) and Clive Longden (personal communication) kept adults raised from larvae in captivity for ≤ 4 yr. Longevity of individuals in the wild is unknown.

N. Feeding Behavior. Although feeding behavior was not observed, Goin (1950) noted that stomachs of adults contained earthworms (*Diplocardia* sp.). Prey capture kinematics were investigated by Beneski et al. (1995). In the laboratory, newly metamorphosed animals were raised to adulthood on earthworms (*Diplocardia eiseni*, *Pheretima* sp.) and small slugs (D.B.M., personal observations).

O. Predators. Unknown, but some associated species are potential predators (Anderson and Williamson, 1976; Palis, 1997b).

P. Anti-Predator Mechanisms. As with other members of the genus *Ambystoma*, flatwoods salamanders have concentrations of granular glands along the length of their tail that produce distasteful secretions (Brodie, 1977). When threatened, adults coil with their head positioned under the base of their tail (Brodie, 1977). Occasionally individuals will weakly lash their tail (Brodie, 1977). Juveniles will raise their tail and slowly undulate it from side to side (J.G.P., personal observations).

Q. Diseases. Unknown; however, one of us observed a dying larva at a breeding site in Liberty County, Florida, that appeared to have red-leg disease (J.G.P., personal observations).

R. Parasites. Unknown.

4. Conservation.

Flatwoods salamanders may no longer occur at many historical locations and may be extirpated in Alabama (Jones et al., 1982). Flatwoods salamanders are listed as Threatened by the Federal Government (USFWS, 1997c), Protected in Georgia (http://georgiawildlife.dnr.state.ga.us), and State Endangered in South Carolina (www.dnr.state.sc.us). The main threat to this species appears to be habitat alteration (Means et al., 1996).

Acknowledgments. We thank Ray Ashton, John Jensen, Clive Longden, Dirk Stevenson, and Gerald Williamson for sharing their observations on flatwoods salamanders with us.

Ambystoma gracile (Baird, 1859)
NORTHWESTERN SALAMANDER

H. Bradley Shaffer

1. Historical versus Current Distribution.
Northwestern salamanders (*Ambystoma gracile*) are found from northern California (extreme northern Sonoma County) north to southeastern Alaska. Two subspecies are often recognized (Petranka, 1998), although recent allozyme data suggest that this may not be warranted (Titus, 1990). Within this range, the species is limited to mesic habitats of the Coast Range and Cascade Mountains (Nussbaum et al., 1983; Stebbins, 1951). Habitats range in altitude from sea level to 3,110 m (Nussbaum et al., 1983). The current distribution presumably resembles the historical distribution, but no serious analysis of the species has been undertaken. Almost certainly, populations have been lost as the Pacific Northwest has been developed for urban and agricultural uses; however, ponds constructed for stock and other uses have probably improved habitat in some instances.

2. Historical versus Current Abundance.
Few data exist and those that do are contradictory. Aubry and Hall (1991) found northwestern salamander numbers to be substantially lower in regrowing forest, while Corn and Bury (1991) found little correlation between salamander abundance and stand age once clearcuts began to grow over (see also Petranka, 1998).

3. Life History Features.
A. Breeding. Reproduction is aquatic.

i. Breeding migrations. The breeding season for northwestern salamanders varies based on latitude and altitude. In low-elevation Pacific Northwest populations, egg deposition occurs from January–April (Henry and Twitty, 1940; Watney, 1941; Snyder, 1956; Licht, 1969c, 1975b; H. A. Brown, 1976b; Eagleson, 1976; see also Petranka, 1998), with peak activity in late February in the Seattle area (Snyder, 1956). Breeding occurs from June to late August in higher elevation (1,300–1,676 m) lakes in Oregon (Snyder, 1956) and British Columbia (Eagleson, 1976). Breeding migrations can occur when ice remains on ponds (Snyder, 1956; Eagleson, 1976; Nussbaum et al., 1983). Breeding lasts 1–7 wk; at any given site the length of the breeding season is variable and apparently linked to the rate at which water temperature increases in the pond or lake (Snyder, 1956; Eagleson, 1976).

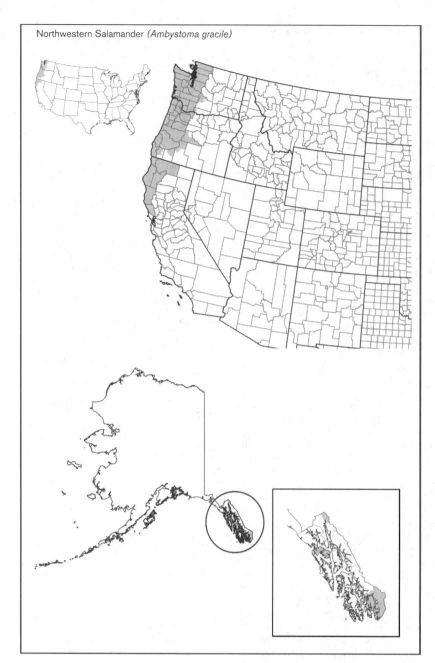

Northwestern Salamander (*Ambystoma gracile*)

phose at 12–14 mo, but some overwinter a second year; montane populations may overwinter a third year (Watney, 1941; Snyder, 1956; Licht, 1975b; Eagleson, 1976; see also Petranka, 1998). Some animals are neotenic and apparently never metamorphose (see "Neoteny" below). Under laboratory conditions, high food levels increase larval growth rates and decrease time to metamorphosis but do not affect size at metamorphosis or the tendency to be neotenic (Licht, 1992). Larvae of both sexes have identical growth rates.

ii. Larval requirements.

a. Food. Hatchlings initially feed on cladocerans, copepods, and ostracods, gradually adding prey items to their diet as they can swallow them. Larger prey include insect larvae, annelids, mollusks, amphipods, anostracans, and snails (Efford and Tsumura, 1973; Licht, 1975b). Northwestern salamander larvae will eat anuran tadpoles, but apparently avoid toxic bufonid larvae (Peterson and Blaustein, 1991).

b. Cover. Larvae seek cover in terrestrial vegetation or in shallows in the presence of predatory fishes (Neish, 1971). Sexually mature paedomorphic animals tend to burrow in the substrate and flee into deeper water in fishless lakes, but flee into shallow water and burrow less often in fish-containing lakes (Taylor, 1983b).

iii. Larval polymorphisms. Unknown and unlikely given the attention dedicated to northwestern salamanders.

iv. Features of metamorphosis. Larvae metamorphose at sizes ranging from 47–74 mm SVL under laboratory conditions (Eagleson, 1976), and size frequency analysis from field populations suggest that this accurately reflects sizes in the field.

v. Post-metamorphic migrations. For terrestrial adults, migration takes place from their breeding ponds to upland sites. Adults and terrestrial juveniles are assumed to spend most of the year in rodent burrows (Nussbaum et al., 1983).

vi. Neoteny. Nearly all populations contain neotenic (paedomorphic) individuals (Licht, 1992). The percentage of neotenic animals has been suggested to be linked to altitude, with higher elevation populations metamorphosing less frequently (Snyder, 1956; Sprules, 1974a,b; Eagleson, 1976; see also Petranka, 1998); recent experimental data has brought this correlation into question (Licht, 1992). There are no anatomical differences in cloacal anatomy between transforming and nontransforming adults, suggesting that the two forms should be capable of interbreeding (Licht and Sever, 1991).

D. Juvenile Habitat. Juvenile northwestern salamanders are thought to become sexually mature 1–2 yr after metamorphosis (Efford and Mathias, 1969), although this has not been validated with

ii. Breeding habitat. Adults breed in semipermanent and permanent wetlands, lakes, and slow-flowing streams and rivers.

B. Eggs.

i. Egg deposition sites. Vary depending on whether adults have metamorphosed or are neotenic. Metamorphosed females attach egg masses to submerged objects such as tree limbs or branches and cattails, from 0.5–1 m below the water surface. Over time, symbiotic algae invade egg masses, providing eggs with oxygen (Patch, 1922; Carl and Cowan, 1945; see also Petranka, 1998). Neotenic adults tend to lay eggs in smaller, looser masses (Henry and Twitty, 1940; Snyder, 1956; but see Knudsen, 1960) directly on the wetland bottom. Depending on water temperature, hatching occurs 2–9 wk after

laying (Slater, 1936; Watney, 1941; Licht, 1975b; H. A. Brown, 1976b). Egg masses are firm gelatinous structures 80–150 mm in diameter, and individual ova are 1.5–2.5 mm in diameter (Nussbaum et al., 1983).

ii. Clutch size. Egg masses are large (5–15 cm long, 5–8 cm wide), contain anywhere from 30–270 eggs (Slater, 1936), and generally have a firm jelly layer. There is some discrepancy concerning differences in egg masses produced by paedomorphic versus transformed individuals (Petranka, 1998).

C. Larvae/Metamorphosis.

i. Length of larval stage. Hatchlings average about 8 mm (Licht, 1975b), and larvae 1–7 d old were 15–20 mm (Bishop, 1943). Most larvae overwinter and metamor-

accurate field studies. Terrestrial juveniles likely have habitat characteristics similar to terrestrial adults.

E. Adult Habitat. Varies depending on whether adults are terrestrial or paedomorphic. Little is known about the habits of terrestrial adults. As is typical of many ambystomatids, males migrate to breeding sites before females (Nussbaum et al., 1983). The habitat characteristics for neotenic adults are similar to those for larvae, but the larger neotenic adults can presumably take a greater range of prey.

F. Home Range Size. Unknown.

G. Territories. Unknown.

H. Aestivation/Avoiding Desiccation. Unknown; may not be necessary in the moist coastal climate of the Pacific Northwest.

I. Seasonal Migrations. Terrestrial adults migrate when snow and ice are still present at high elevation sites (Nussbaum et al., 1983). The terrestrial ecology of northwestern salamanders has not been well studied.

J. Torpor (Hibernation). Important for terrestrial individuals but apparently unstudied.

K. Interspecific Associations/Exclusions. Unlike many ambystomatid salamanders, northwestern salamanders can coexist with predatory fishes; they apparently monitor fishes and alter their behavior to avoid predation (Efford and Mathias, 1969; Taylor, 1983a,b). Salamanders in fishless habitats are diurnal and exhibit weaker escape responses (Taylor, 1983b). In the presence of fish, larvae concentrate in shallow water where they seek refuge (Taylor, 1983a). Larger larvae and neotenic adults escape into shallow areas when fish are encountered. Both larvae and eggs are eaten by rainbow trout (Efford and Mathias, 1969). Rough-skinned newts *(Taricha granulosa)* co-occur in ponds with northwestern salamanders in some sites (Efford and Mathias, 1969) where they have similar diets and densities. In mountain ponds and lakes in Mount Ranier National Park, northwestern salamander larvae tend to occur in sites that are larger, deeper, lower in elevation, higher in organic content, with more flocculent sediment and coarse woody debris, when compared with long-toed salamander *(A. macrodactylum)* larvae.

L. Age/Size at Reproductive Maturity. Variable depending on whether animals are at low or high latitudes and altitudes and whether adults are terrestrial or neotenic. The youngest age at which animals could be reproductively mature appears to be 2 yr (after 1 yr as a larva and 1 yr as a juvenile); some animals take several years. Animals become sexually mature at about 70–75 mm SVL (Petranka, 1998). If the lifespan of 5 yr inferred by Efford and Mathias (1969) is accurate, it implies that most animals breed once or at most a few times.

M. Longevity. Unknown, for terrestrial animals. Neotenic animals live about 5 yr (Efford and Mathias, 1969).

N. Feeding Behavior. Neotenic adults feed on a wide range of aquatic invertebrates and tadpoles (Efford and Tsumura, 1973; Licht, 1975b; Peterson and Blaustein, 1991). Feeding behavior for terrestrial adults is apparently unstudied but probably includes insects and other invertebrates that inhabit grasslands and forest floors.

O. Predators. Predatory fishes (Efford and Mathias, 1969; Sprules, 1974a; Taylor, 1983b). Rough-skinned newts have been observed attempting to eat northwestern salamander eggs (Pearl, 2003).

P. Anti-Predator Mechanisms. Aquatic animals will seek shelter in vegetation and littoral zones (Neish, 1971; Taylor, 1983a,b; see also Petranka, 1998). Metamorphosed individuals have conspicuous parotid glands and poison glands along the tail. They assume a characteristic "head-down" defensive posture when challenged by predators (Brodie and Gibson, 1969).

Q. Diseases. Unknown. In experimental containers, embryos are affected by ambient UV-B radiation (Blaustein et al., 1995).

R. Parasites. Larvae, including paedomorphic adults carry heavy burdens of the nematode *Chabaudgolvania waldeni* (Quimperiidae; Seuratoidea), with 40–100% infection rates in four British Columbia lake populations (Adamson and Richardson, 1991).

4. Conservation.
Northwestern salamanders remain common over many parts of their range. Their current distribution appears to resemble their historical distribution, but undoubtedly populations have been lost as the Pacific Northwest has been developed. Ponds constructed for livestock and other uses have probably improved

habitat in some instances. While logging may reduce populations (Aubry and Hall, 1991), once clearcuts began to grow over, there is little correlation between salamander abundance and stand age (Corn and Bury, 1991).

Ambystoma jeffersonianum (Green, 1827)
JEFFERSON SALAMANDER

Robert Brodman

1. Historical versus Current Distribution.
Because of the difficulty of distinguishing Jefferson salamanders *(Ambystoma jeffersonianum)* from those of unisexual *Ambystoma* populations of hybrid origin, the status of Jefferson salamanders is uncertain and their ecology poorly understood throughout much of their range. Early studies (prior to the mid 1960s) need to be viewed with caution because many may include blue-spotted salamanders *(A. laterale)* or unisexual populations. Jefferson salamanders are distributed in the United States from eastern Illinois and south-central Kentucky northeast to northern Virginia and southwestern New England (Petranka, 1998). There is no evidence to indicate that current distributions differ from historical distributions, but populations have been lost due to habitat destruction, alteration, and acidification (Sadinski and Dunson, 1992; Rowe and Dunson, 1993).

2. Historical versus Current Abundance.
The relative abundance of Jefferson salamanders is uncertain because they often coexist with unisexual *Ambystoma* hybrids. Recent studies indicate that unisexual populations have a larger range than previously thought (Rye et al., 1997) and often outnumber Jefferson salamanders when they are syntopic (Uzzell, 1964; Nyman et al., 1988; Bogart and Klemens, 1997; unpublished data). Jefferson salamanders are considered Locally

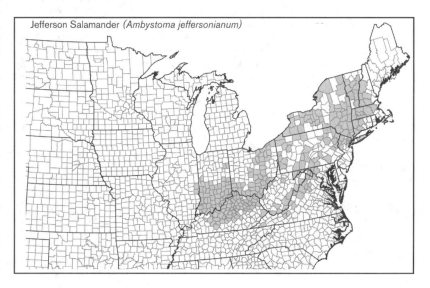

Jefferson Salamander *(Ambystoma jeffersonianum)*

Common–Rare in New England (DeGraaf and Rudis, 1983) and are on the Special Concern list in Vermont, Massachusetts, and Connecticut (Hunter et al., 1999). They are only known from two counties in eastern Illinois (Phillips, 1991) and have recently been listed as state Threatened. The relative abundance of Jefferson salamanders in Indiana is considered occasional (Simon et al., 1992) or uncommon (unpublished data). Long-term studies from Indiana (S. Cortwright, personal communication) and Ohio (Brodman, 1995, 2002) indicate that some populations are stable. From 1989–95 there were 36 new township records of Jefferson salamanders in Ohio. In Pennsylvania, the total number of eggs that females deposit is positively correlated with pH levels and negatively correlated with aluminum levels (Rowe and Dunson, 1993; see also Petranka, 1998). At a pH of < 4.5, eggs and larvae will perish. Given that over half of the ponds studied in a region of Pennsylvania had a pH of ≤ 4.5, the abundance of Jefferson salamanders is likely much lower than it was historically. Of the specimens examined from 106 sites in New York and New England, 70% are unisexual hybrids and most of the rest (23%) are blue-spotted salamanders (Bogart and Klemens, 1997). Fewer than 7% of the specimens collected were Jefferson salamanders and < 3% of the sites contain Jefferson salamander populations that were not syntopic with unisexual populations (Bogart and Klemens, 1997). Jefferson salamanders are common in the western panhandle of Maryland (Thompson et al., 1980; Thompson and Gates, 1982). Little is known about the relative abundance of southern populations in Kentucky, West Virginia, and Virginia.

3. Life History Features.

A. Breeding. Reproduction is aquatic.

i. Breeding migrations. Jefferson salamanders are among the first amphibians to breed and are the earliest *Ambystoma* species to breed, with the exception of the fall-breeding marbled salamanders (*A. opacum*; Brodman, 1995; Petranka, 1998, Minton, 2001). Jefferson salamanders breed as early as December–January in southern Indiana (P. K. Williams, 1973; Minton, 2001) and Kentucky (Smith, 1983) or as late as March in northern Ohio (Downs, 1989a; Brodman, 1995) and Pennsylvania (Mohr, 1931). They migrate from upland overwintering sites to wetland breeding sites. Males typically migrate first and will move while the ground remains frozen (Bishop, 1941a, 1943; Douglas, 1979; Douglas and Monroe, 1981; Downs, 1989a,b; Brodman, 1995; see Petranka, 1998). Warm nighttime rains or heavy snowmelts typically trigger spring breeding migrations; adults will migrate during the day under overcast

skies (Bishop, 1941a). Adults breed early enough in the season that they can be caught out and killed by cold snaps (Petranka, 1998). Adults in southern populations occasionally will migrate to breeding wetlands in autumn, where they overwinter (Douglas and Monroe, 1981; Petranka, 1998).

Breeding in northern populations tends to occur in single bouts that usually last only a few days while southern populations breed in several bouts interrupted by cold weather (Bishop, 1941a; Douglas, 1979; Brodman, 1995). Sex ratios of males:females at any one time are usually greater than 3:1, with males staying for 16–30 d and females for 19–21 d (Collins, 1965; Downs, 1989b; Petranka, 1998). In some populations males breed annually while females skip ≥1 yr before returning to breed (P. K. Williams, 1973; Petranka, 1998).

After breeding, adults migrate from wetlands back to underground retreats in the forest floor. Douglas and Monroe (1981) show that it takes about 45 d to move ≥250 m to these sites. Jefferson salamanders have been found between 250–1,600 m from breeding sites (Bishop, 1941a; P. K. Williams, 1973; Downs, 1989b).

Jefferson salamanders are one of four species that participate in the unisexual *Ambystoma* complex present in recently glaciated areas of the northeastern United States. In some ponds supporting this complex, unisexual female salamanders may outnumber males (Uzzell, 1964; Wilbur, 1971; Nyman et al., 1988; see also Petranka, 1998), and in these ponds Jefferson salamander females will interfere with amplexing pairs (Bishop, 1941a). Jefferson salamander males lay fewer spermatophores when courting females of the unisexual *Ambystoma* complex than when courting conspecific females (Petranka, 1998).

ii. Breeding habitat. Adults typically breed in vernal and semi-permanent woodland pools but also occasionally in permanent, fishless woodland ponds (Bishop, 1941a; Douglas and Monroe, 1981; Brodman, 1995). Breeding ponds tend to be isolated from larger water bodies such as oxbows and lakes and tend to be cooler, more turbid and contain more aquatic vegetation than ponds that are not used (Thompson et al., 1980).

B. Eggs.

i. Egg deposition sites. Females begin laying eggs on submerged twigs, tree branches, and emergent vegetation 1–2 d following mating. Egg masses on firm supports such as twigs are ovoid; eggs laid on grasses and other vegetation are scattered. Females preferentially lay their eggs on the thin twigs of fallen branches and submerged ends of live willow branches (personal observations). Eggs are 2–2.5 mm in diameter. In time, many egg masses are

colonized by symbiotic green algae *Oophila* sp. (Gatz, 1973). Average eggs/ mass range from 14–31 (Smith, 1911a; Bishop, 1941a; Seibert and Brandon, 1960; Smith, 1983; Brodman, 1995).

ii. Clutch size. Ovarian egg counts vary from 140–280 (Bishop, 1941a; Uzzell, 1964). Egg densities range from 57 eggs/m^2 in Ohio (Brodman, 1995) to 123 eggs/m^2 in Indiana (Cortwright, 1988) to 469 eggs/m^2 in Kentucky (Douglas, 1979).

Embryonic periods range from 3–14 wk, depending on temperature and latitude (Bishop, 1941a; Smith, 1983; Brodman, 1995). Embryonic survivorship tends to be high (71–96%) when the pH is ≥6.0 (Rowe and Dunson, 1993; Brodman, 1995).

C. Larvae/Metamorphosis.

i. Length of larval stage. From 2–4 mo and at sizes ranging from 52–78 mm (Bishop, 1941a; Minton, 1954; Downs, 1989b). Survival is reported to be <1% in some populations (Thompson et al., 1980; Downs, 1989b), but can be higher in others (Pough and Wilson, 1977; Brodman, 1996; S. Cortwright, personal communication).

ii. Larval requirements.

a. Food. Jefferson salamander larvae are opportunistic and size-selective feeders that are gape-limited because they swallow most prey whole. Larvae take a variety of prey including ostracods, cladoceran and copepod zooplankton, nematodes, snails, and a range of aquatic and adult insects including chironomid (Diptera) larvae (Smith and Petranka, 1987; Petranka, 1998). Larvae are aggressive feeders and will cannibalize smaller larvae and prey upon spotted salamander (*A. maculatum*) larvae (Brandon, 1961; Smith and Petranka, 1987; Brodman, 1996, 1999b). Smaller Jefferson salamander larvae are sometimes found with missing limbs, gills, or tails (Petranka, 1998; personal observations).

b. Cover. Larvae are nocturnal, emerging from vegetation, where they are found during the day, to feed either on the wetland bottom or to float and feed in the water column. Nocturnal vertical migration and stratification has been observed in some, but not all, Jefferson salamander populations (Anderson and Graham, 1967; Brodman, 1995). Jefferson salamander larvae will use leaf litter and algae patches as refuges in the presence of predatory eastern tiger salamander (*A. tigrinum*) or marbled salamander larvae (Brodman and Jaskula, 2002).

iii. Larval polymorphisms. While cannibalism is known, cannibal morphs (*sensu* Powers, 1907) have not been documented. Clanton (1934) and Bishop (1943) noted that there were dark and light forms in many populations of what was then thought to be Jefferson salamanders. Minton (1954) clearly distinguished these forms as two species (Jefferson and blue-spotted salamanders). Uzzell (1963, 1964) later found that the Jefferson salamanders and blue-spotted salamanders were distinct diploid species and that there were two sympatric all-female populations that were unisexual hybrids with triploid chromosome numbers. These unisexual *Ambystoma* populations typically use gynogenetic reproduction in which the unisexual females use male blue-spotted salamanders, Jefferson salamanders, small-mouth salamanders (*A. texanum*), or tiger salamanders to stimulate egg laying without the incorporation of male genes in the progeny (see the unisexual *Ambystoma* account by Phillips and Mui, this volume, for an explanation of this phenomenon). Many genomic combinations of unisexual hybrids are now known between blue-spotted salamanders and Jefferson salamanders (JLL, JJL, JL, JLLL, LLLL, and JLLLL [with initials denoting the species name of the parental genetic component as follows: J = jeffersonianum; L = laterale]), small-mouth salamanders (LT, LLT, LTT, JLT, TTT, JJLT, LLLT, LLTT, and LTTT [T = texanum]), and tiger salamanders (JLTi, LTTi, and LTTTi [Ti = tigrinum]). Because of this confusion, little research has been done on the autecology of Jefferson salamanders. The historical literature can also be confusing. Clanton (1934) and Wacasey (1961) performed detailed ecological studies on "Jefferson salamanders" in Michigan. However, the range of the Jefferson salamanders does not extend into Michigan, so these widely cited studies describing Jefferson salamander ecology must refer to either blue-spotted salamanders or unisexual hybrids.

iv. Features of metamorphosis. Jefferson salamander larvae grow fast and can complete development in 2–3 mo. Metamorphosis usually occurs in late July to early August in northern Ohio, but can occur as early as June if the pond dries (personal observations). In southern Indiana, larvae have transformed as early as late May (Minton, 2001), and larvae have been found in ponds as late as November, indicating that larvae may overwinter in some populations (Cortwright, 1988).

v. Post-metamorphic migrations. Juveniles move a mean of 92 m (range 3–247 m) away from breeding wetlands into floodplain and upland forest floor habitats during their first 10 d (P.K. Williams, 1973).

vi. Neoteny. Not known to occur.

D. Juvenile Habitat. Little is known about the terrestrial ecology of Jefferson salamanders (Petranka, 1998). In northern populations, juveniles are more active on the forest floor than are adults or juveniles in more southern populations (Green and Brant, 1966; Petranka, 1998). Juveniles appear to spend most of their time in burrows, where they feed on soil invertebrates (Petranka, 1998).

E. Adult Habitat. Jefferson salamanders are rarely caught above ground outside of breeding migrations. When encountered, they are typically scattered in deciduous forests near suitable breeding wetlands (Petranka, 1998). Adults live in burrows, including rodent burrows (P.K. Williams, 1973; Douglas and Monroe, 1981), and are found more often in well-drained upland forest sites than are other species of forest-dwelling *Ambystoma* (Downs, 1989b; Petranka, 1998).

F. Home Range Size. Observations of post-breeding adults returning to burrows (Bishop, 1941a; P.K. Williams, 1973; Douglas and Monroe, 1981) over long distances (250–1,600 m) suggest that Jefferson salamanders have home ranges that they retain from year to year.

G. Territories. Unknown.

H. Aestivation/Avoiding Desiccation. Unknown, but individuals likely respond to dry conditions by going deeper into burrows.

I. Seasonal Migrations. Jefferson salamanders migrate to their wetland breeding ponds from the underground burrows they use for overwintering (Douglas and Monroe, 1981). Following breeding, 3–4 wk later, they migrate back to feed and grow, often to their overwintering burrow (P.K. Williams, 1973; see also Faccio, 2003). Populations in south-central Kentucky have been observed in autumn migrations (Douglas and Monroe, 1981). Other migrations are unknown, although animals probably move nearer or farther from the soil surface with increasing or decreasing moisture conditions.

J. Torpor (Hibernation). In burrows below the frost line.

K. Interspecific Associations/Exclusions. Jefferson salamander populations frequently coexist and share breeding sites with unisexual *Ambystoma* hybrid salamanders, marbled salamanders, spotted salamanders, small-mouthed salamanders, eastern newts (*Notophthalmus viridescens*), wood frogs (*Rana sylvatica*), and spring peepers (*Pseudacris crucifer*; Thompson and Gates, 1982; Cortwright, 1988; Downs, 1989b; Sadinski and Dunson, 1992; Brodman, 1995; personal observations). They can share habitat with spotted salamanders as often as 40% of the time in some parts of their range (Thompson and Gates, 1982). In regions of sympatry, Jefferson salamanders and blue-spotted salamanders are not known to share breeding sites (Anderson and Giacosie, 1967; Nyman et al., 1988; Bogart and Klemens, 1997). Studies on the interactions among Jefferson salamanders, marbled salamanders, and spotted salamanders indicate a complex density-dependent set of interactions among larvae involving both competition and predation, including cannibalism (Cortwright, 1988; Brodman, 1996,

1999b). The effect on Jefferson salamander population dynamics of interactions with unisexual *Ambystoma* populations is unknown and may be important with regards to the conservation and status of this species.

L. Age/Size at Reproductive Maturity. Juveniles mature in 2–3 yr (Bishop, 1941a; P.K. Williams, 1973) and at SVLs of 62–68 mm for males and 76–78 mm for females (Uzzell, 1967a; P.K. Williams, 1973).

M. Longevity. Jefferson salamanders were not included among the species that have captive longevity records (Bowler, 1977); no studies using skeletochronology have been conducted. Mark–recapture studies indicate that 10–18% of marked adults survived for 3 yr (Collins, 1965; P.K. Williams, 1973).

N. Feeding Behavior. Adults and juveniles are thought to feed on earthworms and other soil invertebrates (Downs, 1989b; Petranka, 1998).

O. Predators. During breeding migrations, adults are vulnerable to predation from mammals such as striped skunks (*Mephitis mephitis*) and raccoons (*Procyon lotor*); shrews will feed on adults during other times of the year (Petranka, 1998). Petranka (1998) also suggests owls and woodland snakes will feed on adults and juveniles.

P. Anti-Predator Mechanisms. Adults elevate the pelvis and tail, undulate the tail and coil the body (Brodie, 1977), and will respond to tongue-flicks from snakes (Dodd, 1977b; Ducey and Brodie, 1983; Brodie, 1989). The dorsal surface of the tail contains glands that produce a noxious, adhesive secretion. In tiger salamanders, the noxious component of this secretion has neurotoxic effects (Hamning et al., 2000). Larvae seek refuge and reduce activity when in the presence of large tiger salamander larvae (Brodman and Jaskula, 2002) and adult eastern newts (personal observations).

Q. Diseases. Unknown.

R. Parasites. Unknown.

4. Conservation.

The main threats to Jefferson salamanders are habitat destruction and acidification of breeding ponds. They need undisturbed well-drained upland forest sites (Downs, 1989; Petranka, 1998) that are within 200–250 m of seasonal, semipermanent, and fish-free permanent wetlands (Bishop, 1941; Douglas and Monroe, 1981; Semlitsch, 1998) that are not acidic (Sadinski and Dunson, 1992; Rowe and Dunson, 1993). Because unisexual *Ambystoma* hybrids often outnumber Jefferson salamanders when they are syntopic (Uzzell, 1964; Nyman et al., 1988; Bogart and Klemens, 1997; R.B., unpublished data), their interactions, and the conservation implications of these interactions, need to be studied.

Ambystoma laterale Hallowell, 1856
BLUE-SPOTTED SALAMANDER
Robert Brodman

1. Historical versus Current Distribution.

The distribution of blue-spotted salamanders *(Ambystoma laterale)* has been uncertain because of the difficulty in distinguishing these animals from unisexual *Ambystoma* hybrids. Recent studies indicate that unisexual populations have a larger range than previously thought (Rye et al., 1997). Blue-spotted salamanders, the most northern species of *Ambystoma*, are found primarily north of the Wisconsinian glacial border (Downs, 1989c) and were primary invaders of post-glacial habitat (Holman, 1998). Today, blue-spotted salamanders are found across southern Canada and the northern United States from eastern Manitoba and Iowa to the Gulf of Saint Lawrence and northern New Jersey (Conant and Collins, 1991; Petranka, 1998).

eight new township records have been made of unisexual *Ambystoma* populations (Downs, 1989c; Pfingsten, 1998a). There were no new Indiana populations found from 1971–94 (Minton et al., 1982; Minton, 2001).

Remnant populations exist along the periphery of the distribution with isolated populations in Manitoba, Iowa, and New Jersey. Brownlie (1988) recently found a disjunct population in Nova Scotia. A population in Jay County, Indiana, was recently discovered that extends the range 110 km south (Brodman, 1999a, 2001). In New Jersey, blue-spotted salamanders are known only from four wetland regions of the Passaic River basin (Anderson and Giacosie, 1967; Nyman et al., 1988).

There is evidence of anthropogenic range contraction. While blue-spotted salamanders were once considered common in the Chicago area (Grant, 1936), Smith (1961) reported only seven populations in northern Illinois.

in the northern part of the state (Harding and Holman, 1992). Populations are stable in northeast Minnesota (Moriarty, 1998). Only two populations remain in Iowa (Camper, 1988).

Blue-spotted salamanders are the most abundant salamander species in Wisconsin deciduous and mixed woodlands (Casper, 1998). They are uncommon and local in areas of Wisconsin where woodland ponds are scarce and populations are becoming more isolated by development in the southeast (Casper, 1998). Blue-spotted salamanders are common in wooded moraines in northeast Illinois where populations fluctuate, tending to recover in wet years following dry years (Mierzwa, 1998a).

While found in 11 counties in northern Indiana, blue-spotted salamanders are most common in the Indiana Dunes along Lake Michigan and rare elsewhere (Minton, 2001). At Jasper-Pulaski Fish and Wildlife Area in northwest Indiana, blue-spotted salamanders were considered plentiful in oak woods from 1946–71, but became uncommon by 1995 (Minton, 1998). They are no longer found in adjacent habitats outside of Jasper-Pulaski where agriculture has isolated the few remaining oak stands (Brodman and Killmurry, 1998). Blue-spotted salamanders are found throughout New York (Bishop, 1941a; R. Ducey, personal communication) and New England, but are Rare and Threatened in southern New England (DeGraaf and Rudis, 1983; McCollough, 1999) and New Jersey (Nyman et al., 1988). Acid rain is considered a threat to blue-spotted salamanders in the Northeast (DeGraaf and Rudis, 1983; Knox, 1999).

Most (70%) of the specimens examined from New York and New England are associated with unisexual *Ambystoma* hybrids. Few sites contain only diploid bisexuals (Bogart and Klemens, 1997). There are only two known pure diploid populations in New England (Knox, 1999). In northern Indiana, most populations with a blue-spotted salamander morphology consist of unisexual hybrids (including diploid, triploid, and tetraploid animals) that use male small-mouth salamanders (*A. texanum*) for breeding (Brodman, 1999a, 2001). Unisexual *Ambystoma* salamanders are replacing blue-spotted salamanders and small-mouth salamanders in the western Lake Erie basin (Kraus, 1985b). Some unisexual populations have restricted ranges and are vulnerable to extinction (Petranka, 1998), but conservation of these unisexual populations is difficult because these forms do not fit the biological or evolutionary species concepts on which federal and state conservation laws are based (Kraus, 1995b; see also Goebel, this volume, Part One; Minton, this volume, Part One). However, New Jersey and Illinois give protective status to unisexual *Ambystoma* populations.

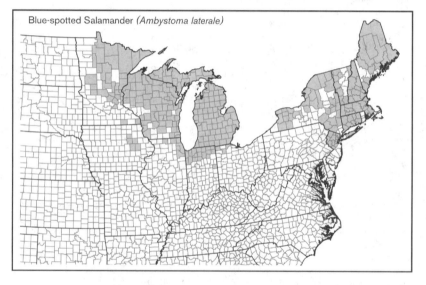

Blue-spotted Salamander *(Ambystoma laterale)*

Specifically, blue-spotted salamanders are known throughout New York. Blue-spotted salamanders are widespread in northeast Minnesota with isolated populations in remnant maple/basswood forests in southern counties (Moriarty, 1998). They previously were known from just 13 Minnesota counties (Breckenridge, 1944), but now are known from 30 counties (Oldfield and Moriarty, 1994). Blue-spotted salamanders are found throughout Wisconsin except for in the driftless area (Vogt, 1981). In Ohio, they are known from only four localities in the extreme northwest part of the state (Pfingsten, 1998a). In Ohio, a search for new populations from 1981–82 failed to find any new populations, and no new records of blue-spotted salamanders have been reported in Ohio since then, even though during this time

2. Historical versus Current Abundance.

Blue-spotted salamanders are relatively common in many areas of their range but have declined with the loss of native forests and wetland breeding sites (Petranka, 1998), especially in southern regions. They are considered Endangered in New Jersey, Iowa, and Ohio, Threatened in Connecticut, and Of Special Concern in Indiana, Vermont, and Massachusetts (Lannoo, 1998d; McCollough, 1999).

Although rare in southwestern Ontario (Hecnar, 1997), there is no evidence of a blue-spotted salamander decline in Canada (Weller and Green, 1997). In the United States, they have been considered common throughout the Lake Michigan basin (Pentecost and Vogt, 1976). Today, blue-spotted salamanders remain common in Michigan, but are abundant only

3. Life History Features.

A. Breeding.

i. Breeding migrations. Adults migrate to breeding ponds from late March to early April in the southern parts of their range in southern Michigan (Clanton, 1934), northern Indiana (Brodman and Killmurry, 1998), and northern Illinois (Stille, 1954; Uzzell, 1969). This migration occurs later in more northern habitats. Migration takes place in mid-April in Wisconsin (Vogt, 1981) and central Ontario (Lowcock et al., 1991) and from late April to early June in more northern parts of the range (Bleakney, 1957; Gilhen, 1974) and on Isle Royale (Van Buskirk and Smith, 1991). Blue-spotted salamanders typically migrate at the same time as spotted salamanders (*A. maculatum*) when sympatric (Wacasey, 1961; Nyman, 1991; Talentino and Landre, 1991). Blue-spotted salamanders tend to be explosive breeders with a breeding season that lasts from a few days (Talentino and Landre, 1991; Brodman and Killmurry, 1998) to 2–3 wk (Clanton, 1934; Uzzell, 1969; Lowcock et al., 1991). The courtship of blue-spotted salamanders is similar to Jefferson salamanders (*A. jeffersonianum*), and interspecific breeding isolation is based on chemical cues (Storez, 1969).

ii. Breeding habitat. Blue-spotted salamanders breed in temporary woodland ponds (Wilbur, 1977c; Van Buskirk and Smith, 1991) and are absent where pH is <4.5 (Karns, 1992).

B. Eggs.

i. Egg deposition sites. Females attach eggs singly or as masses of 2–15 eggs to aquatic vegetation (Stille, 1954; Bleakney, 1957; Uzzell, 1964; Gilhen, 1974; Wilbur, 1977c; Talentino and Landre, 1991).

ii. Clutch size. Females carry an average of 196–250 ova (Clanton, 1934; Wilbur, 1977c; Minton, 1972, 2001; Gilhen, 1974).

C. Larvae/Metamorphosis.

i. Length of larval stage. Eggs hatch in 3–4 wk (Smith, 1961; Minton, 1972, 2001; Talentino and Landre, 1991). Larvae metamorphose 2–3 mo following hatching (Smith, 1961; Talentino and Landre, 1991). Blue-spotted salamander larvae develop faster than sympatric spotted salamander larvae (Wacasey, 1961; Nyman, 1991; Talentino and Landre, 1991). On Isle Royale, average hatchling densities were 12–15/m^2 with a high of 158/m^2.

ii. Larval requirements.

a. Food. Larvae prey primarily upon microcrustaceans (cladocerans, copepods, ostracods) and aquatic dipteran larvae (chironomids, chaoborids, mosquito larvae; Nyman, 1991; personal observation). They also eat western chorus frog (*Pseudacris triseriata*; Smith, 1983) and northern leopard frog (*Rana pipiens*) tadpoles (personal observations).

b. Cover. Larvae spend more time in leaf litter and aquatic vegetation than out in the open or in the water column, especially in the presence of larger species of *Ambystoma* larvae (Jaskula and Brodman, 2000; Brodman and Jaskula, 2002).

iii. Larval polymorphisms. Polymorphisms are currently unknown. Clanton (1934) and Bishop (1943) noted that there were dark and light forms in many populations of what was then thought to be Jefferson salamanders. Uzzell (1963, 1964) found that blue-spotted salamanders and Jefferson salamanders were distinct diploid species and that there were two sympatric all-female populations that were unisexual hybrids with triploid chromosome number. These unisexual *Ambystoma* populations typically use gynogenetic reproduction, in which the unisexual females use male blue-spotted, Jefferson, or small-mouth salamanders to stimulate egg laying without the incorporation of male genes in the progeny (see the unisexual *Ambystoma* account by Phillips and Mui, this volume, for an explanation of this phenomenon). Many genomic combinations of unisexual hybrids are now known between blue-spotted salamanders and Jefferson salamanders (JLL, JJL, JL, JLLL, LLLL, and JLLLL [with initials denoting the species name of the parental genetic component as follows: J = jeffersonianum; L = laterale]), small-mouth salamanders (LT, LLT, LTT, JLT, TTT, JJLT, LLLT, LLTT and LTTT [T = texanum]), and tiger salamanders (JLTi, LTTi and LTTTi [Ti = tigrinum]). Because of this confusion, little research has been done on the autecology of blue-spotted salamanders. The historical literature can also be confusing. Clanton (1934) and Wacasey (1961) performed detailed ecological studies on "Jefferson salamanders" in Michigan. However, the range of the Jefferson salamanders does not extend into Michigan, so these widely cited studies describing Jefferson salamander ecology must refer to either blue-spotted salamanders or unisexual *Ambystoma* hybrids.

iv. Features of metamorphosis. Metamorphosis can occur as early as June in the southern parts of their range (Smith, 1961; personal observations) and as late as September in northern locations (Edgren, 1949; Gilhen, 1974).

v. Post-metamorphic migrations. Little is known about post-metamorphic migrations.

vi. Neoteny. Not known to occur.

D. Juvenile Habitat.
Little is known about juvenile habitat characteristics; they are presumably similar to those of adults.

E. Adult Habitat.
Adult blue-spotted salamanders are fossorial and most abundant in flatwoods with swamp white oak, wooded moraines (Mierzwa, 1998a,b), moist woodlands with sandy soil (Minton, 1972, 2001), and in remnant maple-basswood forest (Moriarty, 1998). Blue-spotted salamander numbers are reduced in clearcuts (deMaynadier and Hunter, 1995) and absent from open areas (Mierzwa, 1998b). Unlike most other members of the genus *Ambystoma*, blue-spotted salamanders are found regularly under logs (Downs, 1989c; Minton, 1972, 2001; Vogt, 1981). Blue-spotted salamanders are more tolerant of dry, sandy conditions than other salamanders in their range (Minton, 1972, 2001; Vogt, 1981). Blue-spotted salamanders are considered to be a forest management-sensitive species (deMaynadier and Hunter, 1998). In Maine, they occur in habitats perpendicular to silviculture edges that have good conifer and hardwood canopy, litter, bole-root, and nonvascular plants. They are less common in areas with good ambient light, mid story cover, and woody cover (deMaynadier and Hunter, 1998). The depth of edge effects was estimated to be 25–35 m from areas of silviculture.

F. Home Range Size.
Home range size has not been studied in blue-spotted salamanders. Because of their smaller size, their range may be smaller than of other *Ambystoma*.

G. Territories.
Unknown. Adult LLT unisexual salamanders will aggressively defend burrows from intruders (Ducey, 1989; Ducey and Heuer, 1991).

H. Aestivation/Avoiding Desiccation.
Aestivation is unknown; animals likely avoid desiccating conditions by seeking shelter under cover objects or burrowing.

I. Seasonal Migrations.
The only documented migrations are those associated with breeding (adults to/from breeding sites in the spring) and migrations of newly metamorphosed animals from breeding sites in the summer.

J. Torpor (Hibernation).
Blue-spotted salamanders are a terrestrially hibernating species. Individuals are freeze intolerant and are thought to overwinter by burrowing below the frost line (Storey and Storey, 1986).

K. Interspecific Associations/Exclusions.
Larvae co-occur with a wide variety of amphibians at breeding sites. They commonly breed at sites with spotted salamanders, wood frogs (*R. sylvatica*), spring peepers (*P. crucifer*), and western chorus frogs in Wisconsin (Vogt, 1981); they coexist at breeding sites with all of the amphibian species found in northern Indiana, including potential predators such as American bullfrogs (*R. catesbeiana*), eastern tiger salamanders, eastern newts (*Notophthalmus viridescens*), and lesser sirens (*Siren intermedia*; personal observations). Blue-spotted salamanders often share breeding ponds with unisexual *Ambystoma* populations in the southern half of their range. Mechanisms that allow unisexual salamanders to coexist with blue-spotted salamanders are poorly understood (Petranka, 1998). Blue-spotted salamanders and JLL unisexual salamanders experience

density-dependent effects due to interspecific competition among larvae in experimental enclosures (Wilbur, 1971, 1972; R.B. and H.D. Krouse, unpublished data). Larval aggression in natural populations is the primary mechanism regulating density-dependent growth and survival (Wilbur, 1971, 1972; Van Buskirk and Smith, 1991). Blue-spotted salamander and eastern tiger salamander larvae facilitate their coexistence by partitioning microhabitats; blue-spotted salamander larvae will use refugia (Jaskula and Brodman, 2000; Brodman and Jaskula, 2002). Little is known of interspecific interactions among juveniles and adults. When placed together, adult tiger salamanders will attempt to eat adult blue-spotted salamanders (personal observation).

L. Age/Size at Reproductive Maturity. In Michigan, juveniles mature in 2 yr (Wilbur, 1977c). Males mature at a minimum of 42–45 mm SVL, females at 51–52 mm SVL (Uzzell, 1967b; Gilhen, 1974; Licht, 1989).

M. Longevity. Blue-spotted salamanders were not included among the species that have captive longevity records (Bowler, 1977). No studies using skeletochronology have been conducted on these animals.

N. Feeding Behavior. Larvae are opportunistic, gape-limited predators; active foragers that stalk and pursue prey (Jaskula and Brodman, 2000; personal observations). Stomach contents of juveniles and adults indicate that blue-spotted salamanders eat a wide variety of invertebrates, but most commonly beetles, spiders, centipedes, earthworms, and slugs (Judd, 1957; Minton, 1972, 2001; Gilhen, 1974; Vogt, 1981; Bolek, 1997).

O. Predators. Larval activity is reduced at low pH levels (<5.0–4.5), which may make larvae more susceptible to predation (Brodman, 1993; Kutka, 1994). Predaceous diving beetles (Coleoptera), adult eastern newts, lesser sirens (personal observations) and tiger salamander larvae (Jaskula and Brodman, 2000) eat blue-spotted salamander larvae. Juveniles have been observed to be preyed upon by wolf-spiders (Arachnida; McLister and Lamond, 1991).

P. Anti-Predator Mechanisms. Adults curl their bodies and elevate and lash with their tails (Brodie, 1977). Larvae seek refuge and reduce activity when in the presence of larger *Ambystoma* larvae (Jaskula and Brodman, 2000; Brodman and Jaskula, 2002) and adult newts (personal observations).

Q. Diseases. Little is known of diseases of blue-spotted salamanders.

R. Parasites. Trypanosome protozoans have been detected in blue-spotted salamanders and JLL hybrids where as many as 26% of the salamanders in a sample from a given year may be infected (Woo and Bogart, 1986). An apicomplexian protozoan *(Hepatozoon clamatae)* that infects several sympatric amphibian species

does not infect blue-spotted salamanders (Kim et al., 1998).

Other parasites include the flukes *Rhabdias ranae, Spiroxys* sp., *Thelandros magnavulvaris, Brachycoelium salamandrae,* echinostomes (Muzzall and Schinderle, 1992), and the nematode *Cosmocercoides dukae* (Bolek, 1997). These nematodes can be present in as many as 50% of the salamanders in a population and are correlated with the number of slugs in the stomach contents (Bolek, 1997).

4. Conservation.
The main threats to blue-spotted salamanders are habitat destruction, land use, and acidification of breeding ponds. They need undisturbed upland forest sites (Downs, 1989; deMaynadier and Hunter, 1998; Petranka, 1998) with temporary woodland ponds (Wilbur, 1977; Van Buskirk and Smith, 1991) and pond pH >4.5 (DeGraaf and Rudis, 1983; Karns, 1992; Knox, 1999). A radius of 164 m around breeding ponds needs to be protected as core upland habitat (Semlitsch, 1998). A study on a population of blue-spotted salamanders in Wisconsin suggests a 147 m core of upland habitat (J.M. Jaskula, personal communication). Blue-spotted salamanders are sensitive to forestry management (deMaynadier and Hunter, 1998) and agriculture (Brodman and Kilmurry, 1995; Petranka, 1998) practices. Silviculture produces an edge effect ≤35 m that impacts blue-spotted salamander populations in adjacent undisturbed habitat (deMaynadier and Hunter, 1998). Therefore a buffer of at least 35 m should be added to protect core upland habitat, suggesting a radius of 182–199 m is necessary to conserve blue-spotted salamander populations.

Because unisexual larvae often outnumber blue-spotted salamanders in the

southern portions of their range where they are syntopic (Uzzell, 1964; Nyman et al., 1988; Bogart and Klemens, 1997; personal observations), competition with unisexual populations may limit the distribution and abundance of blue-spotted salamanders (R.B. and H.D. Krouse, unpublished data).

Ambystoma mabeei Bishop, 1928
MABEE'S SALAMANDER
Joseph C. Mitchell

1. Historical versus Current Distribution.
The historical distribution of Mabee's salamanders *(Ambystoma mabeei)* is unknown. Mabee's salamanders occur entirely in the Atlantic Coastal Plain and extend from Gloucester County, Virginia, to the southern tip of South Carolina (Mosimann and Rabb, 1948; Mitchell and Hedges, 1980; Ensley and Cross, 1984; Petranka, 1998; Mitchell and Reay, 1999). There are no records west of the Fall Line.

2. Historical versus Current Abundance.
There are no published estimates of population size for Mabee's salamanders.

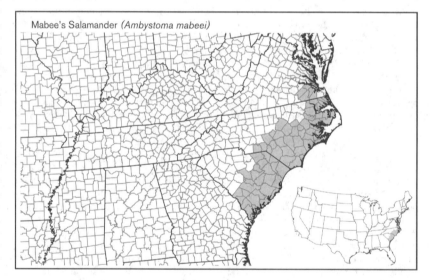

Mabee's Salamander *(Ambystoma mabeei)*

Maximum number of larval salamanders caught in 1963 with a minnow seine in four ponds in North Carolina was 51, 97, 115, and 212 (Hardy, 1969b). Roble (1998) reported that Mabee's salamanders were confirmed in 17 of 50 natural sinkhole ponds in the Grafton Plains area in the city of Newport News and York County, Virginia. Up to 100 larvae were caught on single visits to three of these ponds. A total of 183 individuals was caught in 1995–97, compared to 821 marbled salamanders *(A. opacum)* encountered during that period, suggesting that the best known population of Mabee's salamanders in Virginia exists in low densities (Roble, 1998).

3. Life History Features.

The life history and biology of Mabee's salamanders have been summarized by Pague and Mitchell (1991a) and Petranka (1998). Mabee's salamanders possess an aquatic larval stage and terrestrial juvenile and adult stages.

A. Breeding. Reproduction is aquatic.

i. Breeding migrations. During rains, adults migrate *en masse* from terrestrial retreats to breeding ponds during the winter (Hardy, 1969a). Movement distances are unknown. Hardy (1969b) found 91 recently transformed Mabee's salamanders under boards approximately 800 meters from the nearest water. Roble (1998), however, found no mass migrations in Virginia using drift fences and pitfall traps.

ii. Breeding habitat. Mabee's salamanders breed in small, shallow, typically ephemeral to semi-permanent wetlands that are usually free of fishes (Petranka, 1998). A wide variety of pools support breeding populations, including farm ponds, water-filled foxholes, vernal pools in pine and hardwoods, Carolina bays, sinkhole ponds, and cypress-tupelo ponds in pinewoods (Hardy, 1969a; Hardy and Anderson, 1970; Roble, 1998).

B. Eggs.

i. Egg deposition sites. Mating and egg laying occur in winter, extending from December–March depending on weather conditions. Courtship occurs during winter rains. Eggs vary from 5.1–5.9 mm in diameter and are deposited singly or in loosely connected clusters of 2–6 eggs (Hardy, 1969a). Eggs are attached to vegetation on the pond substrate.

ii. Clutch size. Undescribed. Hatching occurs in 9–14 d at about 9 mm TL.

C. Larvae/Metamorphosis. The larval stage lasts a few months (Petranka, 1998). Metamorphosis occurs in April–May depending on when eggs were deposited. Size at metamorphosis in North Carolina was estimated at 55–60 mm total length (Hardy, 1969b). Metamorphic animals with gill buds from Virginia were 60.1–78.6 mm total length (mean = 70.4; Mitchell and Hedges, 1980).

D. Juvenile Habitat. Juvenile habitat appears to be similar to adult habitat, but this aspect of the ecology of Mabee's salamanders has not been studied. Hardy (1969b) noted that a large number of recently transformed juveniles were found under coverboards in an overgrown tree-shaded lawn adjacent to an open field.

E. Adult Habitat. Adults use terrestrial habitats extensively outside of the breeding period. Terrestrial habitat includes open fields, pine forest, and hardwood forest (Hardy, 1969b).

F. Home Range Size. Unknown. Movements of adults have not been studied but observations by Hardy (1969b) suggest that some individuals remain close to breeding ponds during the remainder of the year.

G. Territories. Territoriality has not been described in Mabee's salamanders.

H. Aestivation/Avoiding Desiccation. Adults and juveniles are fossorial during the non-breeding months. We do not know if individuals are active year-round underground or if they aestivate during dry periods. Adults have been found under logs in the bottom of breeding ponds during a late August to November drought (Hardy, 1969b).

I. Seasonal Migrations. Recently metamorphosed juveniles migrate from natal ponds and are known to disperse ≤0.8 kilometers (Hardy, 1969b).

J. Torpor (Hibernation). Because Mabee's salamanders breed in the winter, they are unlikely to exhibit winter torpor.

K. Interspecific Associations/Exclusions. In aquatic habitats, Mabee's salamanders have been found in association with tiger salamanders (*A. tigrinum*), marbled salamanders, mole salamanders (*A. talpoideum*), lesser sirens (*Siren intermedia*), southern two-lined salamanders (*Eurycea cirrigera*), dwarf salamanders (*E. quadridigitata*), and eastern newts (*Notophthalmus viridescens*; Hardy, 1969a; Roble, 1998). As many as ten species of frogs and one fish, eastern mudminnows (*Umbra pygaea*), have been found in ponds containing Mabee's salamanders (Roble, 1998).

L. Age/Size at Reproductive Maturity. Age at maturity is unknown for either sex. The smallest male known from Virginia was 55 mm SVL, the smallest female was 53 mm SVL (unpublished data).

M. Longevity. Snider and Bowler (1992) reported a longevity record for a captive individual of unknown sex as 8 yr, 9 mo, 23 d. Longevity in nature is unknown.

N. Feeding Behavior. Larval Mabee's salamanders feed on zooplankton and other small invertebrates (Petranka, 1998). Hardy (1969b) noted that adults caught on land in March–April had eaten earthworms.

O. Predators. Known predators are tiger salamander larvae and lesser sirens (Hardy, 1969b). Petranka (1998) added odonate naiads and dytiscid beetle larvae.

P. Anti-Predator Mechanisms. Mabee's salamanders possess passive anti-predator displays that include lashing the tail weakly toward a touch, coiling the body with the head under the base of the tail, tail undulation, and, when touched, assuming an immobile position with the forelimbs clasped along the body (Brodie, 1977). Individual salamanders exhibited these display behaviors for 8.5–87 s (mean = 40.5 s) when approached by a small snake (Dodd, 1977b).

Q. Diseases. No diseases have been reported.

R. Parasites. Parasites have not been reported.

4. Conservation.

Mabee's salamanders are listed as a Threatened species in Virginia (Mitchell and Reay, 1999). Threats include habitat fragmentation, aquatic and terrestrial habitat loss, road mortality, and alteration of hydrology mostly due to urbanization. One Mabee's salamander site, a 151 ha portion of the Grafton area sinkhole pond complex in the city of Newport News, was dedicated in 1995 as a Virginia Department of Conservation and Recreation Natural Area Preserve (Clark, 1998).

Ambystoma macrodactylum Baird, 1850 "1849"
LONG-TOED SALAMANDER

David S. Pilliod, Julie A. Fronzuto

1. Historical versus Current Distribution.

Historical and current distributions of long-toed salamanders (*Ambystoma macrodactylum*) are similar; there is no evidence of a change in distribution. The type-locality for this common, broadly distributed northwestern species is Astoria, Oregon (Baird, 1849), with the syntype deposited at the U.S. National Museum 4042 (Ferguson, 1963). Populations occur from the Alaskan Peninsula across British Columbia; south through Washington into Oregon and the Sierra Nevada of California; and across the Rocky Mountains into eastern Alberta, western Montana, and central Idaho. Isolated populations occur in central California (Russell and Anderson, 1956) and in southeastern Oregon (Petranka, 1998). An erroneous record exists from Iowa (Ruthven, 1912). There are five subspecies recognized, based primarily on morphologically distinguishable dorsal banding patterns (Mittleman, 1948; Ferguson, 1961a; Crother et al., 2000). Generally, each subspecies is distributed allotopically along north–south ranges (see Petranka, 1998) as follows.

Western long-toed salamanders (*A. m. macrodactylum*; Baird, 1849) range from Vancouver Island and southwestern British Columbia through western Washington (west of the Cascade Mountains) and north of the Calapooya Divide in western Oregon (Nussbaum et al., 1983).

Eastern (also cited as central or Columbia) long-toed salamanders (*A. m. columbianum*; Ferguson, 1961a) range from southeastern Alaska through British Columbia, eastern Washington, Oregon (east of the Willamette Valley), and into central Idaho (Nussbaum et al., 1983; Petranka, 1998).

Northern (also erroneously cited as eastern or western) long-toed salamanders (*A. m. krausei*, Peters, 1882) range from eastern British Columbia and Idaho through western Alberta and Montana (Nussbaum et al., 1983; Petranka, 1998; Walsh, 1998; Graham and Powell, 1999).

Southern long-toed salamanders (*A. m. sigillatum*; Ferguson, 1961a) range from southwestern Oregon (south of Calapooya Divide) into the Sierra Nevada as far south as Carson Pass in California (R. Cutter, personal observations).

Santa Cruz long-toed salamanders (*A. m. croceum*; Russell and Anderson, 1956) represent an isolated subspecies found at 11 locations in Santa Cruz and Monterey counties, California (USFWS, 1999c). Listed in 1967, this is the only federally Endangered subspecies of *A. macrodactylum* (Bury et al., 1980). Recent efforts to halt habitat threats to this narrowly distributed population are being evaluated (Ruth, 1988; USFWS, 1999c).

Long-toed Salamander (*Ambystoma macrodactylum*)

2. Historical versus Current Abundance.

Little information exists regarding the historical abundance of long-toed salamanders; it is therefore difficult to determine if abundance has changed over time. In addition, determining the abundance of

adults in an area is usually difficult. Even when populations are large, long-toed salamanders are rarely found outside of the breeding season. When encountered in ponds during spring breeding aggregations, long-toed salamanders have been considered locally abundant (Anderson, 1967a; Beneski et al., 1986; Powell et al., 1997; Fukumoto and Herrero, 1998). Historical descriptions from Crater Lake, Oregon (Evermann, 1897; Bishop, 1943), Corvallis, Oregon (Storm and Pimentel, 1954), and Mt. Rainier, Washington (Slater, 1936b; Stebbins, 1966), indicate long-toed salamanders were abundant in those areas. In northwestern Idaho, over 2,030 adults were captured by a drift fence

encompassing a small (0.3 ha), fishless pond, providing habitat for an estimated 3,141 breeding adults (Beneski et al., 1986). Recent mark–recapture studies in Alberta found similarly large population sizes (Powell et al., 1997; Fukumoto and Herrero, 1998). However, the effective population sizes (estimated from allozyme data) of six high elevation populations in Idaho and Montana were considerably smaller (mean = 123; Funk et al., 1999). A five-year census of 11 high elevation basins in central Idaho indicated that long-toed salamanders may have been extirpated or reduced to low numbers in six of these basins, possibly due to the stocking of non-native trout into deep, fishless breeding habitats (Pilliod and Peterson, 2001).

3. Life History Features.

A. Breeding. Reproduction is aquatic.

i. Breeding migrations. In the Pacific Northwest, long-toed salamanders are the earliest breeding amphibians (Leonard et al., 1993; Corkran and Thoms, 1996), often migrating across snow and depositing eggs before complete ice-out. In the Willamette Valley, Oregon, adults migrate to breeding ponds in late October to early November (Stebbins, 1966; Nussbaum et al., 1983; Leonard et al., 1993), and as late as June–July at higher elevations in the Cascades, Rockies, Sierra Nevada, and Wallowas (Kezer and Farner, 1955; Stebbins, 1966; Howard and Wallace, 1985; Leonard et al., 1993; Walls et al., 1993a; Pilliod, 2001; Thompson, 2001). Males are the first to arrive at breeding ponds (Nussbaum et al., 1983; Beneski et al., 1986), probably to court arriving females (Slater, 1936a; Knudsen, 1960; Anderson, 1961) and compete with other males (Verrell and Pelton, 1996). While females as a group spend approximately 3 wk at a breeding site (Beneski et al., 1986), depositing eggs over a 6–7 d period (Anderson, 1967a), individual females spend only 1–2 d (Verrell and Pelton, 1996). Males generally leave the breeding site ~1 wk after the females (Beneski et al., 1986), but may remain in the ponds for the entire breeding season (up to 2 mo or more; Anderson, 1968; Beneski et al., 1986). In ephemeral habitats, breeding may only last 1 – 2 d (Walls et al., 1993a).

ii. Breeding habitat. Long-toed salamanders are opportunistic breeders, depositing eggs in a variety of habitats, including seeps (Hamilton et al., 1998); along the backwaters of slow-flowing streams (Beneski et al., 1986; Hamilton et al., 1998; Llewelyn and Peterson, 1998), temporary pools at lower elevations (Leonard and Klaus, 1994); and small to large permanent lakes and ponds at higher elevations (Anderson, 1967a; Howard and Wallace, 1985; Leonard et al., 1993; Hamilton et al., 1998; Pilliod, 2001) and higher latitudes (Green and Campbell, 1984). Eggs and larvae have also been found in disturbed areas, such as newly formed (Hamilton et al., 1998), recently disturbed (Corkran and Thoms, 1996), and human-influenced (Beneski et al., 1986; Llewelyn and Peterson, 1998; Monello and Wright, 1999) pools down to the size of tire ruts (K. R. McAllister, personal communication).

B. Eggs.

i. Egg deposition sites. Eggs are deposited in shallow water (<0.5 m) with silt-mud substrates, but also along rocky shorelines (Hamilton et al., 1998). Eggs are attached to vegetation, floating or submerged woody debris (logs, branches), and rocks, or placed unattached on the bottom in shallow (<20 cm) water (Slater, 1936a,b; Stebbins, 1954a; Nussbaum et al., 1983; Howard and Wallace, 1985; Corkran and Thoms, 1996).

ii. Clutch size. Clutch size is geographically variable, from 90–411 eggs (Slater, 1936b; Gordon, 1939; Ferguson, 1961a; Anderson, 1967a; Howard and Wallace, 1985; see also Petranka, 1998). Females deposit eggs over several hr, releasing 1–81 eggs in a cluster before moving to a new location in the pond (Petranka, 1998; Maxell, 2000).

C. Larvae/Metamorphosis.

i. Length of larval stage. Length of larval period varies with elevation, latitude, and pool permanence (Slater, 1936b; Kezer and Farner, 1955; Anderson, 1967a; Howard and Wallace, 1985; Watson, 1997). Eggs hatch in 5–35 d, depending on water temperature (Anderson, 1967a; Leonard et al., 1993). The larval period can be as short as 50 d in some temporary ponds (Nussbaum et al., 1983) or last ≤3 yr in permanent lakes at higher elevations (Bishop, 1943; Stebbins, 1966; Leonard et al., 1993; Pilliod, 2001).

ii. Larval requirements.

a. Food. Larvae are opportunistic carnivores and begin feeding shortly after hatching (Petranka, 1998). Prey size generally increases with salamander body size and includes crustaceans (amphipods, cladocerans, copepods), a variety of aquatic and terrestrial insects (coleopterans, dipterans, ephemeropterans, plecopterans, trichopterans), mollusks (gastropodans, pelecypodans), annelids (hirudineans, oligochaetes), and ranid frog tadpoles (Anderson, 1968; Tyler et al., 1998a). Larger larvae may cannibalize smaller larvae (Anderson, 1967a; Walls et al., 1993a), possibly resulting in increased growth and size at metamorphosis of the cannibals (Wildy et al., 1998, 1999).

b. Cover. Because of their diverse diets, feeding larvae are found in the open water column and within cover. During the day, larvae may use cover to avoid predation from vertebrate and invertebrate predators and are often found in or under bottom detritus (rotting leaves, woody debris), submerged logs, rocks, and aquatic vegetation (Anderson, 1967a; Green and Campbell, 1984; Liss et al.,

1995; Corkran and Thoms, 1996; Munger et al., 1997b; Hamilton et al., 1998; Petranka, 1998). In fishless lakes, larvae may move more freely across open substrates or in the water column (Liss et al., 1995; Tyler et al., 1998a). However, in fishless ponds in southeastern Washington that contain a variety of other vertebrate and invertebrate predators, larvae are seldom observed in open water and even less often captured in minnow traps (J. A. F. and P. Verrell, unpublished data). At high elevations, second-year larvae may use more open habitat compared to first-year and metamorphosing larvae that remain under cover (Anderson, 1967a).

iii. Larval polymorphism. Morphologically distinct cannibalistic larvae have been reported from at least one small, subalpine pond in Oregon (Walls et al., 1993a,b; Petranka, 1998). These cannibalistic larvae have longer, wider heads and larger vomerine teeth compared to conspecifics of the same size and from the same population that were reared in the laboratory on a diet of live *Tubifex* (Walls et al., 1993a). Diet is one of several intrinsic and extrinsic factors that influence the expression of the cannibalistic larval morphology (Walls et al., 1993b).

iv. Features of metamorphosis. The timing of metamorphosis varies with environmental conditions and is triggered by either intrinsic factors (possibly size of the animal) or extrinsic factors such as temperature and pond drying (Anderson, 1967a). Size at metamorphosis is highly variable, ranging from 23–48 mm SVL (40.5–90 mm TL; Carl, 1942; Howard and Wallace, 1985). Although size at metamorphosis does not appear to be associated with elevation (Anderson, 1967a), larvae that take 2–3 yr to transform are generally larger at metamorphosis (Kezer and Farner, 1955; Howard and Wallace, 1985).

v. Post-metamorphic migrations. Long-toed salamanders exhibit a strong breeding-site fidelity and generally will only migrate within 100 m of breeding ponds (Anderson, 1967a; Sheppard, 1977; Powell et al., 1997). However, outside of the breeding season, longer migrations may occur. For example, in Montana, adults have been captured in pitfall traps at least 600 m from the nearest breeding site (J. Pierson, unpublished data, as cited in Maxell, 2000). Terrestrial post-metamorphic migrations are generally associated with rains, high soil moisture, and air temperatures above 0 °C (Anderson, 1967a; Howard and Wallace, 1985; Beneski et al., 1986; Powell et al., 1997; Fukumoto and Herrero, 1998). When moisture levels are sufficient, metamorphic animals disperse away from breeding sites shortly after transforming (Anderson, 1967a). Dispersal of newly metamorphosed animals may be spread out over several months (May–August in southeastern Washington; J. A. F. and P. Verrell, unpublished data)

or occur as a mass migration when conditions permit (Anderson, 1967a; Marnell, 1997).

vi. Neoteny. Although the prevalence of delayed metamorphosis in long-toed salamanders indicates a potential for neoteny (Sprules, 1974a), paedomorphosis—retention of larval characteristics in reproductively active adults—has not been observed.

D. Juvenile Habitat. As far as we know, juvenile habitats are similar to adults. Recently transformed juveniles may remain close to the breeding pond (under cover objects) until conditions for migration are favorable (Anderson, 1967a).

E. Adult Habitat. Adult long-toed salamanders occur in a wide range of habitats ranging from sea level to 3,000 m in California (Stebbins, 1966) to ≤2,030 m in Washington (Leonard et al., 1993) to 2,470 m in Oregon (Howard and Wallace, 1981, 1985) and 2,725 m in Idaho (Munger et al., 1997b; Pilliod, 2001). Suitable habitats include semiarid grasslands and sagebrush steppes, alpine meadows, dry oak woodlands, humid coniferous forests, rocky shorelines of subalpine lakes, beaver ponds (Reichel, 1996), and even disturbed agricultural areas (Nussbaum et al., 1983; Monello and Wright, 1999), timber harvest areas (Hamilton et al., 1998; Naughton et al., 2000), pastures (Leonard et al., 1993), and residential green belts (Leonard et al., 1993). Adults are typically subterranean outside of the breeding season, hiding under logs, bark, rocks, and within rotten wood or rodent burrows, generally within 100 m of water (Gordon, 1939; Bishop, 1943; Stebbins, 1954a; Stebbins, 1966; Green and Campbell, 1984; Corkran and Thoms, 1996; Powell et al., 1997). In a comprehensive search conducted in August, of all cover objects within 10 m of the shoreline of three high elevation lakes in central Idaho, the majority (82%) of adults were found under logs (5–50 cm diameter) within 5 m of water (D. S. P. and M. Reed, unpublished data). Microhabitats are typically associated with higher substrate moisture (Anderson, 1967a). When soil moisture levels are low, juveniles and adults may aggregate (≤43 individuals have been observed in close proximity) and entwine (Anderson, 1967a), a behavior that reduces water loss (Alvarado, 1967). Adults can be found above ground at night or during rains in the summer months not associated with the breeding season. Migrations do not necessarily occur along stream corridors or within obvious habitat types (Beneski et al., 1986).

F. Home Range Size. Due to the limited vagility of long-toed salamanders, home range sizes are relatively small. Sheppard (1977) monitored 25 salamanders with implanted radioactive tags from July–October in Alberta and estimated (minimum-area convex polygon) home ranges to be 115.6 m² for females, 167.5 m² for males, and 281.6 m² for juveniles. The rugged topography in which long-toed salamanders are often found may reduce movements between distant populations, potentially reducing gene flow among populations (Howard and Wallace, 1981). However, in the Bitterroot Mountains in Montana, allozyme data indicate populations within basins are panmictic, with salamanders moving among breeding populations (Tallmon et al., 2000). These allozyme data also suggest that salamanders move across mountain ridges more frequently than they move across valley bottoms. In Idaho, a salamander was observed in June crossing a snow-covered ridge that separated two cirque basins at an elevation of 2,600 m (D.S.P., personal observations).

G. Territories. Although larvae are not known to be territorial, they are aggressive, possibly resulting in the spacing of individuals (Anderson, 1967a). Both juveniles and adults can be aggressive in competition over food (J.A.F., personal observations). Non-breeding adults may be social, rather than territorial—conspecifics aggregate rather than exclude each other (Verrell and Davis, 2003).

H. Aestivation/Avoiding Desiccation. Aestivation has not been documented, but low precipitation may inhibit migratory behavior and result in reduced surface activity (Howard and Wallace, 1985).

I. Seasonal Migrations. Long-toed salamanders migrate to breeding habitats in the spring and to overwintering habitats in the fall. Individuals home to breeding and wintering locations, but do not appear to follow population-level migratory routes (Beneski et al., 1986).

J. Torpor (Hibernation). Although little is known about long-toed salamander hibernation, adults probably hibernate terrestrially. Sheppard (1977) found three terrestrial hibernacula aggregations located in gravel substrate at 45–70 cm below the ground surface (frost line was estimated at 45 cm). When water temperatures drop and surface ice forms, overwintering larvae become less active and retreat under logs and bottom debris (Anderson, 1967a). In shallow (<1 m), high elevation ponds and ephemeral pools, larvae may move into subsurface springs when pools freeze solid (D.S.P., personal observations).

In lowland areas, adults can remain active year-round (Stebbins, 1966). In southeastern Washington, adults have been observed migrating to breeding ponds during the coldest winter months and have been retrieved from submerged minnow traps in early January, below ≤ 15 cm of ice, with air temperatures from 0 to −16°C (J.A.F. and P. Verrell, unpublished data).

K. Interspecific Associations/Exclusions. Long-toed salamanders occur in habitats used by other amphibians, including blotched tiger salamanders (*A. tigrinum melanostictum*), California slender salamanders *(Batrachoseps attenuatus)*, arboreal salamanders *(Aneides lugubris)*, rough-skinned newts *(Taricha granulosa)*, Pacific treefrogs *(Pseudacris regilla,)* mountain yellow-legged frogs *(Rana muscosa)*, Oregon spotted frogs *(R. pretiosa)*, and western toads *(Bufo boreas)*. Long-toed salamanders co-occur with Columbia spotted frogs *(R. luteiventris)* at >80% of survey sites in Idaho and Montana (Werner and Reichel, 1994; O'Siggins, 1995; Munger et al., 1997b; Llewelyn and Peterson, 1998; Pilliod, 2001). This is probably due to a similarity in habitat characteristics, but adult and larval long-toed salamanders may also utilize frequently abundant Columbia spotted frog tadpoles as a food source (Munger et al., 1997b). In Washington, long-toed salamanders and northwestern salamanders *(A. gracile)* have generally allotopic distributions, possibly resulting from competition and predation (Hoffman et al., 2003). Introduced trout (*Oncorhynchus* sp.; Liss et al., 1995) and goldfish (*Carassius auratus*; Monello and Wright, 2001) prey on long-toed salamander eggs and larvae, substantially reducing their numbers, sometimes to the point of excluding them from breeding sites (Liss et al., 1995; Munger et al., 1997b; Beck et al., 1998; Tyler et al., 1998a; Yeo and Peterson, 1998; Funk and Dunlap, 1999; Hoffman and Pilliod, 1999; Monello and Wright, 1999; Pilliod and Peterson, 2001). Artificial-pond experiments indicate that trout reduce growth and survivorship of larvae, presumably due to both indirect and direct effects of predation, such as limited foraging activity associated with refuge use and increased predation (Tyler et al., 1998b).

L. Age/Size at Reproductive Maturity. Sexual maturity is reached in 1–3 yr (at 50–55 mm SVL) after metamorphosis for both sexes (Anderson, 1967a; Nussbaum et al., 1983; Green and Campbell, 1984; Howard and Wallace, 1985; Russell et al., 1996).

M. Longevity. Skeletochronological techniques indicate that long-toed salamanders live ≤10 yr (Russell et al., 1996).

N. Feeding Behavior. Larvae are carnivorous, opportunistic predators, consuming small (zooplankton) to large (tadpoles) aquatic and terrestrial prey depending on size and availability. Young larvae use a sit-and-wait technique, lunging at approaching prey. Older larvae stalk or pursue prey (Anderson, 1968). Adults are also carnivorous, preying on a variety of terrestrial organisms, such as annelids, mollusks, and a variety of arthropods: arachnids, coleopterans, collembolans, dipterans, formicids, lepidopterans, and orthopterans (Schonberger, 1944; Farner, 1947; Anderson, 1968). Males feed at breeding sites, taking similar aquatic organisms as larvae (e.g., aquatic dipterans; Anderson, 1968); females do not feed at breeding sites. This difference may result from the longer time males spend at the breeding sites.

O. Predators. Invertebrate predators of larvae include predaceous diving beetles (*Dytiscus* sp.; Marnell, 1997), odonate naiads, and belostomatids (J.A.F., personal observations). Larvae and adults are also preyed upon by vertebrates, such as salmonid fish (Liss et al., 1995), goldfish (Monello and Wright, 2001), northwestern salamanders (Hoffman and Larson, 1999), blotched tiger salamanders (J.A.F., unpublished data), non-native American bullfrogs (*R. catesbeiana*; Nussbaum et al., 1983; M.P. Hayes, personal communication), western terrestrial and common garter snakes (*Thamnophis elegans* and *T. sirtalis*, respectively; Ferguson, 1961a; Marnell, 1997; Pilliod, 2001), and belted kingfishers (*Ceryle alcyon*; P. Murphy, personal communication).

P. Anti-Predator Mechanisms. The highly secretive, subterranean life-style of long-toed salamanders may be one of their most successful avoidance mechanisms (Ferguson, 1961a). However, in the presence of predators, such as tiger salamanders, adult long-toed salamanders may increase their activity and use cover objects less often (J.A.F. and P. Verrell, unpublished data). The yellow dorsal stripe may serve as a warning to predators. When attacks are simulated in a laboratory, long-toed salamanders demonstrate a combination of behavioral and chemical defenses, including coiling, tail undulations and lashing, and production of skin secretions (Anderson, 1963; Brodie, 1977; Williams and Larsen, 1986). Adult long-toed salamanders can vocalize with squeaks and clicks, possibly to startle predators once captured (Hossack, 2002). Adults will avoid areas occupied by a damaged conspecific (Chivers and Kiesecker, 1996), indicating the importance of chemical signals for avoiding predators. Larval long-toed salamanders use chemical cues and learned recognition (non-lethal encounters) to avoid inter- and intraspecific predators (Tyler et al., 1998b; Wildy et al., 2001). The presence of methoxychlor increases vulnerability to predators (Verrell, 2000).

Q. Diseases. Little is known about the susceptibility of long-toed salamanders to diseases. Two types of water molds, *Saprolegnia ferax* and *Achlya racemosa*, have been found growing on long-toed salamander eggs in Montana and Oregon (D.S.P. and others, unpublished data). Water molds, such as *Saprolegnia* sp., have been reported to increase mortality of injured ambystomatid salamanders (Walls and Jaeger, 1987) and may infect eggs that have been stressed by environmental conditions such as low water levels, cold temperatures, or increased UV-B radiation (Blaustein et al., 1994c; Kiesecker and

Blaustein, 1995; Kiesecker et al., 2001a). Investigations into the susceptibility of long-toed salamanders to the recently discovered chytrid fungus are warranted.

R. Parasites. Trematode parasites (*Ribeiroia* sp.) may be responsible for supernumerary limbs and related deformities observed in >1,600 long-toed salamanders collected from Oregon in the late 1980s (Sessions and Ruth, 1990; Johnson et al., 1999; Sessions et al., 1999).

4. Conservation.
Long-toed salamanders are widespread across their historical range. Santa Cruz long-toed salamanders are federally listed as Endangered and occur in three population clusters (metapopulations) in coastal areas of Santa Cruz and Monterey counties, California (USFWS, 1999). The species is considered "Secure" in Washington, Oregon, Idaho, Montana, and British Columbia according to the National and State Heritage Status Ranks. Long-toed salamanders are on the "Yellow B List" in Alberta, meaning that the species is not at risk but vulnerable to limiting factors such as habitat alteration, destruction of critical habitats, and non-native predatory fish (Graham and Powell, 1999). In Alaska, long-toed salamanders are ranked as "Imperiled" by the State Heritage system, but insufficient information exists to adequately assess the species' status there.

Long-toed salamanders may suffer local and possibly regional threats, but apparently these threats are not resulting in widespread declines. In the vicinity of urban and agricultural areas, lowland populations may be impacted by the loss of wetlands (Bury and Ruth, 1972; Ruth, 1974, 1988), road mortality during breeding migrations (Fukumoto and Herrero, 1998), chemical contaminants (Ingermann et al., 1997; Fukumoto and Herrero, 1998; Nebeker et al., 1998; Verrell, 2000), and predation from introduced goldfish (Monello and Wright, 2001). In relatively undisturbed mountain habitats, a number of possible threats have been identified, including increased UV-B radiation (Blaustein et al., 1997; Belden et al., 2000; Belden and Blaustein, 2002b), timber harvesting (McGraw, 1998; Naughton et al., 2000), and introduced salmonid fish (Liss et al., 1995; Tyler et al., 1998a; Funk and Dunlap, 1999; Hoffman and Pilliod, 1999; Pilliod and Peterson, 2001), but not increased acid deposition or mobilization of aluminum (Bradford and Gordon, 1992; Bradford et al., 1994c). Elasticity analyses of demographic models for long-toed salamanders suggest that populations are more likely to decline when environmental stressors result in higher mortality of post-metamorphic life stages as compared to stressors that result in mortality of embryos or larvae (Vonesh and De la Cruz, 2002).

Acknowledgments. We thank several people for commenting on this account and providing helpful natural history observations and access to unpublished documents, including J. Howard, B. Leonard, K. McAllister, G. McLaughlin, C. Peterson, L. Powell, and P. Verrell. Support for D.S.P. was provided by the USGS Amphibian Research and Monitoring Initiative during the final phase of this project.

Ambystoma maculatum (Shaw, 1802)
SPOTTED SALAMANDER

Wesley K. Savage, Kelly R. Zamudio

1. Historical versus Current Distribution.
Spotted salamanders (*Ambystoma maculatum*) are distributed throughout the interior and Atlantic Coastal Plains of the eastern United States, primarily in hardwood and mixed coniferous-deciduous forest habitats (Bishop, 1943; Pope, 1944; Anderson, 1967b; Upchurch, 1971; Lang, 1972; Schuette, 1980; Kats et al., 1988; Lovich and Fisher, 1988; Beane and Gaul, 1991; C. Phillips, 1991a,b, 1992, 1994; Schwaner and Anderson, 1991; E.E. Brown, 1992; Petranka, 1998). This species prefers lowland forests; they are occasionally found in more open habitats such as meadows, but usually near forest edges. Although not typically a montane species, spotted salamanders can occur at higher elevations when suitable breeding sites are available (Thompson and Gates, 1982).

however, range reductions undoubtedly have occurred, and continue to occur, due to habitat alteration. Despite these changes, new county records (e.g. Brinkman et al., 2001; Cochran and Cochran, 2001; Scott et al., 2001) documenting the geographic distribution of spotted salamanders have been published in recent years. Moreover, two new county records extend the range of this species into Minnesota (Hall, 2002), indicating that the extent of the range is still being identified.

Smith (1961) mentioned geographic variation between populations in northeast and southern Illinois in the average sizes of adults, number of teeth, number of dorsal spots, costal groove counts, and contrast between head spot and body spot color. Likewise, DuShane and Hutchinson (1944) raised larvae from independent populations in a common garden experiment and demonstrated significant differences between these populations in developmental rates and size at metamorphosis. This variation has led some to suggest the presence of two distinct subspecies of spotted salamanders (Smith, 1961).

There is indirect evidence that ranges may have been reduced substantially during the Pleistocene glacial cycles, causing geographic discontinuity and resulting in genetic differentiation between two highly differentiated lineages (C. Phillips, 1994; Zamudio and Savage, 2003). Post-glacial recolonization resulted in the expansion

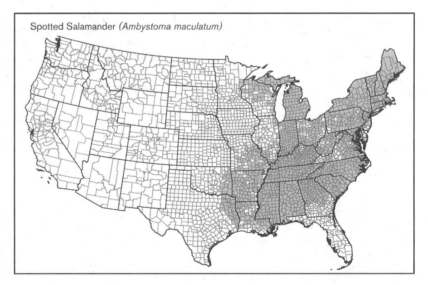

Spotted Salamander *(Ambystoma maculatum)*

With the exception of a few isolated populations, spotted salamanders are also absent from the Coastal Plain of North Carolina, as well as southeastern portions of Virginia, Maryland, Delaware, and New Jersey. No specimens have been recorded from the Florida Peninsula or the Coastal Plain of the Florida Panhandle to southern Georgia.

There is no evidence to suggest that current and historical distributions differ;

and recontact of these two lineages in at least three major regions: (1) the southern Interior Lowlands of Missouri and Illinois; (2) the Interior Lowlands of Ontario, Canada; and (3) from the southern Appalachian Highlands to the coast of South Carolina (Zamudio and Savage, 2003). Three fossil vertebral elements assigned to spotted salamanders date from the late Pleistocene and were collected in a cave near St. Louis, Missouri (Holman, 1965b), near the western extent of the species' current range. A second fossil form (*A. minshalli*) has been proposed as ancestral to *A. maculatum* (Holman, 1975) and is clearly a member of the *maculatum* group. These fossils, collected in Nebraska and Kansas (Estes and Tihen, 1963; Tihen and Chantell, 1963), date back to the Upper Miocene, suggesting that this group was well differentiated before Plio-Pleistocene climatic changes. The presence of fossils in these regions indicates that the historical range of spotted salamanders coincides in part with the contemporary range and that members of the spotted salamander lineage once extended farther west into the Great Plains.

2. Historical versus Current Abundance.

Although there are no quantitative assessments of declines in abundance for spotted salamanders, their habitat requirements suggest that, as with other pond-breeding species, they are adversely affected by deforestation and wetland destruction (Petranka, 1998). Habitat loss, acidification, metal concentrations, environmental contaminants, and fish introductions are certainly factors that could contribute to declines of spotted salamanders, and some have been shown to negatively affect this species (Pough, 1976; Clark, 1986b; Blem and Blem, 1989, 1991; Sadinski and Dunson, 1992; Brodman, 1993; Rowe and Dunson, 1993; but see Cook, 1983). Acid rain has received particular attention in the Northeast as a factor reducing survivorship (e.g., Portnoy, 1990). Future spotted salamander declines are possible because these animals have not evolved mechanisms to tolerate acidic conditions or introduced fishes, suggesting they will respond negatively to these threats (Tome and Pough, 1982; Clark, 1986b; Sexton et al., 1994; see also Petranka, 1998). Because overall range reductions caused by changing landscape features coincide with human land use, we expect that current abundances are lower than historical numbers.

3. Life History Features.

A. Breeding. Reproduction is aquatic.

i. Breeding migrations. Many life history aspects of spotted salamanders vary tremendously across their geographic range, most likely due to large differences in environmental conditions. The pattern of variation is complex and does not seem to be determined exclusively by latitude or geographic locality. Throughout their range, individuals migrate from terrestrial overwintering sites into seasonally available aquatic habitats to breed. Although the vast majority of individuals are reproductively active, a small percentage of non-reproductive individuals also migrate (Shoop, 1967). Why these non-reproductive animals migrate is not known. Breeding individuals are easily identified by

body size and characteristics of the vent: males have a conspicuously swollen vent covered with rough papillae; females have a rounded, smooth vent, and are generally larger in body size (Wilson, 1976; Sexton et al., 1986). Finkler et al. (2003) identified sex-related differences in terrestrial and aquatic locomotor performance in breeding spotted salamanders.

Males typically arrive earlier, anywhere from 1–6 d, than females at ponds (Bishop, 1941b; Hillis, 1977; Douglas, 1979; Sexton et al., 1986). This appears to be due to a differential response between the sexes to temperature cues for the onset of migration (Sexton et al., 1990) or to locomotor performance (Finkler et al., 2003). Migrations take place at night during or following early spring rains correlated with increasing air temperatures; in more northern regions, migrations follow periods of increased temperature and heavy snowmelt or rain (Wright, 1908; Wright and Allen, 1909; Blanchard, 1930; Bishop, 1941b, 1943; Baldauf, 1952; Whitford and Vinegar, 1966; Wilson, 1976; Hillis, 1977). The exact role of each of these ecological features in the onset of migrations is still unclear. A detailed study by Sexton et al. (1990) showed a threshold system of rainfall and temperature cues and a correlation of immigration with soil temperatures (4.5 °C, 30 cm below the surface) when the thermal profile was reversed.

The onset of breeding varies greatly across their range and can occur as early as December in southern populations or as late as April at higher latitudes (Brimley, 1921a; Blanchard, 1930; Dempster, 1930; King, 1939; Pawling, 1939; Welter and Carr, 1939; Gray, 1941; Baldauf, 1952; Hardy, 1952; Moulton, 1954b; Peckham and Dineen, 1954; Green, 1955; Seibert and Brandon, 1960; Whitford and Vinegar, 1966; Worthington, 1968, 1969; Minton, 1972, 2001; Keen, 1975; Mount, 1975; Harris, 1980; Walls and Altig, 1986; Talentino and Landre, 1991; Pinder and Friet, 1994; Brodman, 1995). There is a rough latitudinal pattern in the degree of breeding synchrony within populations: in more southern populations the breeding season may last over 2 mo, usually with 2–3 major bouts of activity following heavy rains (Peckham and Dineen, 1954; Mount, 1975; Harris, 1980; Sexton et al., 1986); breeding in northern regions is more synchronized and can be restricted to a few days (2–5 d; Wilson, 1976; Talentino and Landre, 1991; Brodman, 1995; Petranka, 1998). In a Missouri population, Sexton et al. (1986) estimated that the breeding season lasts 45 d for males, 37 d for females. In North Carolina, Stenhouse (1985a) reported 56 and 73 d in consecutive years. Husting (1965) recorded breeding seasons of 9–29 d over 5 yr in Michigan; in an Indiana population, breeding averaged 25 d (Peckham and

Dineen, 1954). Over 4 yr, Wilson (1976) recorded breeding seasons that averaged 18.5 d. By contrast, breeding seasons in Ohio lasted only 3–5 d (Brodman, 1995); 2–3 d in Massachusetts (Talentino and Landre, 1991). During the breeding season both sexes lose weight. In Missouri, Sexton et al. (1986) reported average weight losses of 13% and 38%, for males and females, respectively; whereas in Massachusetts, Windmiller (1996) found values of 4.5% and 24.1%.

Return rates of adults to breeding ponds suggest that while individuals in some populations breed annually, others may skip years; however, mark–recapture studies may underestimate the number of animals breeding in consecutive years because they do not account fully for dispersal of adults to other sites. Thus it is unknown whether adults are capable of sustained annual breeding or if the sexes differ in that regard. High annual return rates were recorded at a Rhode Island breeding pond (89.5%; Whitford and Vinegar, 1966). Douglas and Monroe (1981) reported six of eight (75%) tagged individuals returning to a Kentucky breeding pond in 2 consecutive years; however, these samples sizes were relatively small. In Missouri, 38% of marked females and 30% of marked males returned to breed the following year (Phillips and Sexton, 1989). Less consistent breeding patterns were found in a Michigan population, where annual return rates ranged from 18–55% over the course of 5 yr (Husting, 1965). In New York, Wilson (1976) recaptured a total of only 12 males and 1 female in the four years of his study, representing 8.7% of males and 0.7% of females. These longer-term studies suggest that there may be large temporal as well as inter-population differences in breeding patterns.

Numerous studies have investigated migratory orientation of spotted salamanders during the breeding season. Adults usually exit ponds near where they entered, suggesting that they use the same migratory route between terrestrial home ranges and breeding sites (Shoop, 1965a, 1968; Wilson, 1976; Douglas and Monroe, 1981; Kleeberger and Werner, 1983; Stenhouse, 1985a; Sexton et al., 1986; Phillips and Sexton, 1989; see also Petranka, 1998). In addition, the order of male arrival was correlated between years in one population in North Carolina (Stenhouse, 1985a), suggesting that males may be returning to the same home ranges after breeding. Downs (1989d) observed some adults returning to the same home burrow after emigrating from breeding ponds in Ohio.

Post-breeding adults move on average 115.7 m from the pond edge to their terrestrial home ranges (range, 0–249 m; average of adult movement distances reported from six studies; Wacasey, 1961;

P.K. Williams, 1973; Douglas and Monroe, 1981; Kleeberger and Werner, 1983; Madison, 1997). Semlitsch (1998) reviewed adult movement patterns in six species of ambystomatids (including spotted salamanders) and concluded that terrestrial buffer zones of breeding ponds would have to extend 164.3 m from the edge to encompass movements of 95% of the adults in a population.

Mark–recapture studies and displacement experiments suggest that two cues may be important in migratory orientation. Olfaction was suggested as an orientation mechanism in a laboratory choice experiment where adult salamanders preferred substrate saturated with their home pond odors (McGregor and Teska, 1989). Shoop (1968) and Whitford and Vinegar (1966) displaced adult individuals 500 m and 128 m from their home ponds. Most individuals successfully oriented back to their home ponds; a higher percentage of those released in and adjacent to a seepage returned to their home pond, suggesting they may be using rheotaxic orientation in addition to olfaction (Whitford and Vinegar, 1966). Rheotropism was also suggested in individuals migrating through runoff in two Michigan ponds following heavy rains (Finneran, 1951). These data are suggestive but not conclusive and underscore the need for a rigorously controlled experimental test of orientation mechanisms.

ii. Breeding habitat. In addition to ephemeral, fishless wetlands, spotted salamanders will breed in roadside ditches and tire ruts in dirt roads, artificial ponds, floodplain wetlands, and marshes. Although they typically use ephemeral habitats, breeding has been documented in permanent ponds containing fishes (Husting, 1965; Harris, 1984; Figiel and Semlitsch, 1990; see also Petranka, 1998), as well as in a fish-free lake in New York (Bahret, 1996). Populations in the Atlantic Coastal Plain sometimes breed in stream backwaters containing predatory fishes; larvae typically will not survive in these habitats unless there is sufficient submergent vegetation (Husting, 1965; Harris, 1984; Figiel and Semlitsch, 1990).

In breeding ponds, adults aggregate in large polyandrous groups for courtship and mating, referred to as a nuptial dance or *liebesspiel* (O'Donnell, 1937). In the presence of females, courting males deposit spermatophores in an area of approximately 1 m^2 on the pond bottom. Spermatophores are usually attached to leaves, twigs, and other submerged vegetation (Breder, 1927; Bishop, 1943; Petranka, 1998). Arnold (1976) showed in experimental field enclosures that males produce on average 40 spermatophores the first night when in the presence of a female (range, 10–81), and spermatophore production decreases rapidly on successive nights. A male vigorously

courts a female by repeatedly nudging her with his snout, swinging his head back and forth along her dorsum, and lifting his head under her chin (Arnold, 1976). A sexually receptive female will also nudge males and search for spermatophores by stepping from side to side with her hindlimbs while moving forward. Sound production has been reported during lulls in courtship (Breder, 1927), although it is unknown whether there is any courting significance because sounds were emitted more often in the presence of same-sex individuals (Wyman and Thrall, 1972). Once a female finds a spermatophore, she orients towards it tactually, squats, arches the base of her tail and removes the sperm mass from the gelatinous base with her vent. Females pick up multiple sperm masses, and paternity analyses of experimental matings (Tennessen and Zamudio, 2003) and wild-collected clutches (E.M. Myers and K.R.Z., unpublished data) confirm that multiple paternity is common in this species and that eggs are frequently fertilized by sperm storage.

Arrival time at breeding ponds has significant impact on male reproductive success. In experimental enclosures in the field, Tennessen and Zamudio (2003) enclosed mating trios (one female and two males) in control enclosures and both males were allowed to mate at the same time; in experimental enclosures one male was introduced in the mating chamber 2 hr before the second male. Paternity analyses of the resulting offspring indicate that early arrival at a mating aggregation is an important determinant of reproductive success. In control enclosures, the two males fathered similar proportions of offspring; in experimental enclosures, the early arriving male sired significantly more offspring than the late-arriving male (Tennessen and Zamudio, 2003).

Sex ratios of spotted salamanders are always male-biased in breeding ponds and vary from 1.5–3.5:1 (Peckham and Dineen, 1954; Husting, 1965; Whitford and Vinegar, 1966; Hillis, 1977; Stenhouse, 1985a; Sexton et al., 1986; Downs, 1989d; Flageole and Leclair, 1992). Differences in age at maturity, reproductive frequency, and survival may explain the strongly male-biased sex ratios (Wacasey, 1961; Flageole and Leclair, 1992). This ratio, combined with the short breeding season, produces extreme sexual competition among males. Sexual interference by spermatophore "capping" is a common behavior for males. When males encounter other spermatophores, they deposit a secondary spermatophore upon it, but it is unknown whether males can discriminate their own spermatophores from those of other males (Arnold, 1976). Spermatophores can be found in stacks 2–6 high, and the female presumably picks up the top-most sperm mass.

B. Eggs.

i. Egg deposition sites. Egg masses are either scattered individually in breeding ponds or deposited in aggregates of 50 clutches or more. Single eggs have been found on land near a pond (Brimley, 1921a; Smith, 1921). Reported mean numbers of eggs/mass are highly variable across regions, ranging from 58–155 (Bishop, 1941b; Seibert and Brandon, 1960; Woodward, 1982; Stangel, 1988; Downs, 1989d; Talentino and Landre, 1991). Egg masses are deposited in firm and compact masses on sticks, submergent vegetation, small branches, or directly on the pond bottom. The thick, firm jelly coat on eggs serves to protect them from desiccation (Nyman, 1987), physical disturbances, and predation (Cory and Manion, 1953; Ward and Sexton, 1981; Semlitsch, 1988).

Some females deposit eggs containing hydrophobic proteins in the outer jelly layer producing white or milky egg masses in which the embryos are not visible (Hardy and Lucas, 1991). These white and clear forms are present among individuals within populations, but there is no evidence of differences in egg viability or hatching size related to this polymorphism. However, Brodman (1995) demonstrated that white egg masses in an Ohio population contained significantly more eggs than clear egg clutches (103.5 ± 11.7 and 66.1 ± 4.3, respectively), suggesting the advantage may lie in higher fecundity. Metts (2001) also found that white egg masses contain more eggs than clear ones, but that clear egg masses have significantly better hatching success than white masses. The functional role or advantage of the white protein is still unclear (Ruth et al., 1993).

ii. Clutch size. Females deposit 2–4 egg masses, each containing from 1–250 eggs (Bishop, 1941b; Shoop, 1974; Wilson, 1976; Petranka, 1998).

After egg deposition, chlamydomonad algae *(Oophila amblystomatis)* enter the inner jelly capsules of individual eggs and presumably aids in increased oxygen supply to developing embryos (Breder, 1927; Gilbert, 1942, 1944; Hutchison and Hammen, 1958; Hammen and Hutchison, 1962; Gatz, 1973a; Bachmann et al., 1986). In laboratory experiments, embryos with algal symbionts hatched sooner than clutches with reduced algal growth (Gilbert, 1942). Embryonic mortality is variable and can be high, ranging from 0–100% (Pough, 1976; Harris, 1980; Cook, 1983; Clark, 1986b; Stenhouse, 1987; Stangel, 1988; Ireland, 1989; Brodman, 1995). Eggs laid later in the breeding season had higher rates of survivorship in a North Carolina pond, suggesting that within-year timing of breeding can have important implications for individual fitness (Harris, 1980). Egg mortality is increased by low pH, predation, and freezing in shallow areas of ponds (Pough, 1976;

Wilson, 1976; Ireland, 1989). Incubation times range from 8–60 d (Bishop, 1941b, 1943; Whitford and Vinegar, 1966; Worthington, 1968, 1969; Shoop, 1974; Keen, 1975; Freda, 1983; Sexton et al., 1986; Stangel, 1988; Nyman, 1991; Talentino and Landre, 1991), and hatchlings from a Mississippi pond measured 14.8 mm ± 1.46 (Walls and Altig, 1986). However, all developmental features are temperature-dependent (Voss, 1993a) and vary depending on the population and region (DuShane and Hutchinson, 1944).

C. Larvae/Metamorphosis.

i. *Length of larval stage.* Duration of the larval stage is highly variable and lasts anywhere from 6 wk to >12 mo (Bishop, 1941b, 1943; Dundee, 1947; Worthington, 1968; Freda, 1983; Nyman, 1991; Talentino and Landre, 1991). Metamorphic patterns can vary within a population across years (Phillips, 1992), as well as among individuals within populations (Ireland, 1973). Most larvae transform from June–August (Petranka, 1998), but overwintering larvae have been found in spring-fed ponds (Bleakney, 1952; Whitford and Vinegar, 1966; Ireland, 1973; Hillis and Miller, 1976; Stangel, 1988; Phillips, 1992). Overwintering larvae in Missouri reach, on average, 69 mm TL and transform as late as May (Phillips, 1992). Larvae range in size from 27–75 mm TL at metamorphosis, and these lengths vary greatly across populations (Bishop, 1941b, 1943; Dundee, 1947; Worthington, 1968; Freda, 1983; Harris, 1984; Nyman, 1991; Talentino and Landre, 1991; Phillips, 1992; see also Petranka, 1998). Survival to metamorphosis is low, usually ≤ 13% of hatched larvae. In some cases, entire cohorts die due to intense predation and wetland drying (Shoop, 1974; Stenhouse, 1985b, 1987; Stangel, 1988; Ireland, 1989; Figiel and Semlitsch, 1990; Rowe and Dunson, 1995).

ii. *Larval requirements.*

a. Food. Larvae are gape-limited and size-selective feeders, ingesting a range of aquatic invertebrates including zooplankton, coleopterans, isopods, ostracods, odonates, and trichopterans. Smaller larvae feed primarily on zooplankton; as they grow in size, consumption shifts towards larger prey such as chironomids, chaoborids, and isopods (Nyman, 1991). Branch and Altig (1981) and Freda (1983) reported that the predominant prey items of larvae were cladocerans and copepods. However, Harris (1995) observed a higher reduction of isopods and amphipods relative to zooplankton under experimental conditions. Freda (1983) also described stomach contents containing eastern newt (*Notophthalmus viridescens*) larvae with mean SVL 63% of the body size of the larval spotted salamander predators.

b. Cover. In the presence of predators, refugia such as submerged vegetation are necessary for larval survivorship (Walls,

1995). Spotted salamanders are vulnerable to predation by other ambystomatids, such as marbled salamanders (*A. opacum*). Interestingly, while spotted salamander larvae require cover, their use of cover does not increase in the presence of competing ambystomatids (Walls, 1995) or fishes (Ireland, 1989; but see Figiel and Semlitsch, 1990). In contrast, Brodman et al. (2002) found that the presence of other *Ambystoma* caused spotted salamander larvae to occupy refuges more and decrease their activity, and in the presence of *A. laterale* (blue-spotted salamanders; a potential competitor), to change their activity and use of microhabitat in opposite directions from that of their congeners. Stratification (floating in the water column, usually at night) is present in some populations, although its function is not well understood (but see Lannoo and Bachmann, 1984b). Branch and Altig (1981) reported larval stratification in Mississippi ponds. In New Jersey, small larvae tend to remain in leaf litter, while larger larvae stratify in the water column (Nyman, 1991; but see Anderson and Graham, 1967).

iii. *Larval polymorphisms.* Have not been described for spotted salamanders. Given the widespread distribution and numerous population studies on this species, it is unlikely that polymorphisms exist.

iv. *Features of metamorphosis.* Timing of metamorphosis varies greatly, even within populations, and larval overwintering occurs in some populations that breed in permanent ponds (Phillips and Sexton, 1989; Whitford and Vinegar, 1966). Metamorphosis typically occurs 8–16 wk after hatching, when larvae reach 27.0–32.0 mm SVL (Worthington, 1968; Talentino and Landre, 1991). In many populations that breed in seasonal and semipermanent wetlands, larval survivorship is correlated with wetland duration (Ball, 1937; Ling et al., 1986; Stangel, 1988; see also Petranka, 1998). Abundance and size at metamorphosis are also positively correlated with length of the hydroperiod (Rowe and Dunson, 1995; Phillips et al., 2002). However, in a short hydroperiod year, size at metamorphosis is negatively correlated with time to emergence from the pond, suggesting that faster growing larvae reach an optimal size when compared with other larvae that take longer to metamorphose at smaller sizes (Phillips et al., 2002). Likewise, mean size at metamorphosis in a year when hydroperiod is shorter than usual is significantly smaller than mean size in a longer hydroperiod year (Phillips et al., 2002). At high densities or in the presence of other ambystomatid larvae, spotted salamander larvae have higher mortality, grow more slowly, and metamorphose at smaller sizes (Wilbur, 1972; Wilbur and Collins, 1973). At a Massachusetts pond, larvae that remain in their natal pond longer metamorphose at greater weights

than earlier-metamorphosing individuals (Windmiller, 1996).

v. *Post-metamorphic migrations.* Newly metamorphosed animals dispersing from breeding ponds move beneath rocks and logs near the pond margin where they are sometimes exposed to high temperatures (Pough and Wilson, 1970) and desiccation (Shoop, 1974). Because metamorphosis varies among individuals, the seasonal migration of juveniles away from the ponds is not as synchronized as the adult breeding migrations. Wilson (1976) captured newly metamorphosed animals exiting a New York pond over a period of 5 wk from 27 July–29 August. He recorded sporadic movements ≤25 m a night over the course of several nights. In Massachusetts, Windmiller (1996) recorded episodic dispersal of post-metamorphic animals from 1 July to autumn. However, there is one report of a concentration of "hundreds" of juveniles migrating along a dirt road adjacent to a wooded area after a summer storm in Maryland (Hardy, 1952).

Orientation of post-metamorphic individuals in five of six cohorts was highly non-random and biased towards a northward direction in five Massachusetts ponds (Windmiller, 1996). These orientation patterns seemed to reflect the emigration routes of post-breeding parental cohorts and the location of upland forest habitat. In the one post-metamorphic cohort that did not disperse in a northward direction, no forested land occurred to the north of the pond. Most individuals (50%) remained within 40 m of their natal pond as late as 8 December, while 36% were between 40–100 m and 14% dispersed beyond 100 m. Later-metamorphosing individuals were heavier in weight and dispersed on average 18 m farther than those metamorphosing earlier (Windmiller, 1996).

vi. *Neoteny.* There are no reports of non-transforming individuals in spotted salamanders. Given the widespread distribution and numerous population studies on this species, it is unlikely that neotenic adults exist.

D. Juvenile Habitat.
Habitat characteristics of juveniles are similar to those of adults. Juvenile abundance declines sharply across a gradient running from relatively mature forest to interior habitat (70–90 yr old) to recently clearcut habitat (2–11 yr old) in Maine (deMaynadier and Hunter, 1999). Juveniles are found more frequently closer to the natal pond edge (<75 m) and seek refuge in burrows and under rocks and fallen logs (Wilson, 1976; Windmiller, 1996). Although usually found only in moist forest habitats, some juveniles were collected from under driftwood on the sandy shore of the Chesapeake Bay in Maryland (Hardy, 1952). Juveniles typically emerge at times beginning in midsummer (Parmalee, 1993). Detailed data for the period of juvenile

development are sparse, but presumably they forage on forest floor invertebrates in underground burrows while they mature.

E. Adult Habitat. Spotted salamanders occur within eastern mixed-deciduous forest habitats and are common in mesic to floodplain habitats. They primarily rely on cutaneous respiration (Whitford and Hutchison, 1966a) and are more commonly found under large, moist cover objects (Parmalee, 1993). As with juveniles, adult abundance declines sharply from relatively mature forest to interior habitat to recently clearcut habitat in Maine (deMaynadier and Hunter, 1999). Along the Atlantic Coastal Plain, spotted salamanders are not common in the drier, sandy, upland environments and are mostly found in hardwood, bottomland habitats with lower soil temperatures and higher soil moisture. These habitats frequently harbor fish populations that significantly decrease the reproductive success of spotted salamanders but do not exclude them from breeding there (Semlitsch, 1988). Potential habitat has been described by Baldwin and Vasconcelos (2003).

F. Home Range Size. Burrow systems of individual adults in Kentucky and Michigan encompass 12–14 m² and an average of 9.8 m² of forest floor, respectively (Douglas and Monroe, 1981; Kleeberger and Werner, 1983). In Michigan, average home range movements of 14 ± 3 m were recorded over a 6–7 wk period. During this time, adults were mostly found in burrows (72% of the time), but also in decaying logs (21%) and leaf litter (7%; Kleeberger and Werner, 1983).

G. Territories. Although territoriality has not been documented in larval salamanders, larvae exhibit visual and movement displays that function as aggressive behavior in crowded environments (Walls and Semlitsch, 1991). In the larval stage, refuge use and stratification may minimize such aggressive interactions through habitat partitioning (Branch and Altig, 1981; Nyman, 1991). Adults are almost always the sole residents under cover objects and are rarely found in close proximity in their forest habitats (Parmelee, 1993). In laboratory trials, adults will alter their normal movement patterns to avoid conspecifics (Ducey and Ritsema, 1988). In other laboratory trials, adults actively defended experimental burrows and feeding areas against conspecific and heterospecific intruders (Ducey and Ritsema, 1988; Walls, 1990; Ducey and Heuer, 1991). Conspecific interactions involve biting, head butting, and forebody raising (Ducey and Ritsema, 1988). In one case of interspecific interactions, spotted salamanders consumed 9% of an eastern red-backed salamander (*Plethodon cinereus*), but this aggression may have been due to predation rather than territorial defense (Ducey et al., 1994).

H. Aestivation/Avoiding Desiccation. In some areas of the range, adults are likely to experience surface temperatures above their critical thermal maximum (34.9–35.8 °C; Gatz, 1971, 1973b) and will usually seek refuge from heat and desiccation in small mammal burrows. Adults have been found in burrows as deep as 1.3 m (Semlitsch, 1983c).

I. Seasonal Migrations. Adults move to and from breeding ponds (see "Breeding migrations" above). Post-metamorphic animals migrate away from wetlands from mid to late summer (see Faccio, 2003).

J. Torpor (Hibernation). Adults and post-metamorphic animals hibernate in subterranean burrows and likely keep from freezing by moving below the frost line.

K. Interspecific Associations/Exclusions. The structure of larval communities has been the focus of numerous studies on the effects of phenology, density, and interspecific competition among sympatric ambystomatids. Spotted salamanders often breed syntopically with congeners, including Jefferson salamanders (*A. jeffersonianum*), marbled salamanders, mole salamanders (*A. talpoideum*), eastern tiger salamanders (*A. t. tigrinum*), blue-spotted salamanders, and small-mouthed salamanders (*A. texanum*). Many studies have investigated the mechanisms of coexistence among these species, and in general, spotted salamander larvae are inferior direct competitors because they generally are smaller than their interspecific competitors. However, female Cope's gray treefrogs (*Hyla chrysoscelis*) avoid depositing eggs in pools containing larval spotted salamanders, suggesting that larvae prey on hatchling tadpoles (Resetarits and Wilbur, 1989).

Brodman (1995, 1996) found that density of each species was an important component determining the outcome of interactions with conspecific and heterospecific competitors. Species show negative responses, such as reduced growth rate and smaller body size at metamorphosis to increased density (Semlitsch and Walls, 1993). Walls and Jaeger (1987) and Walls (1996) demonstrated that spotted salamander larvae were superior exploitative competitors when in competition with mole salamanders. However, Brodman (1999b) argued that coexistence between spotted salamanders and Jefferson salamanders could not be explained by the interference-exploitative competition trade-off, because spotted salamanders did not reduce the growth of congeners at high densities. Other factors influence interspecific competition, such as spatial or dietary partitioning through refugium selection (Walls, 1995; Brodman, 1996), temperature preference (Anderson and Graham, 1967; Stauffer et al., 1983), or dietary differences (Nyman, 1991). Larval spotted salamander behavior is highly variable and may explain their success in a variety of competitive larval environments throughout their range (Brodman, 1996). For example, spotted salamanders reduce the amount of time they spend in the water column (stratification) in the presence of interspecific predators (Walls and Williams, 2001). Predator-mediated reduction in activity patterns potentially limits foraging opportunities and hence potentially decreases survival in this species. Thus, in some cases the composition of larval amphibian assemblages could exclude spotted salamanders from communities of pond-breeding amphibians (Walls and Williams, 2001), yet they continue to persist in many highly competitive larval environments.

Juvenile and adult spotted salamanders live under cover objects and in burrows created by shrews, moles, and other small mammals (Gordon, 1968; Douglas and Moore, 1981; Madison, 1997). However, the associations among co-inhabitants of burrow systems are not well known. Laboratory experiments show that juvenile spotted salamanders exhibit preference for unoccupied burrows and will leave burrows already inhabited by conspecific or heterospecific (marbled salamander) juveniles (Smyers et al., 2002). Adults also share home ranges with other salamanders, including eastern red-backed salamanders (Ducey et al., 1994) and mole salamanders (Walls, 1990). Spotted salamanders share breeding habitats with numerous species, including fishes. Naive adults do not adjust their breeding behavior in the presence of fish, nor do females reduce the number of eggs they deposit (Sexton et al., 1994). However, larvae do seem to adopt behaviors that protect them from fish predation (Figiel and Semlitsch, 1990).

L. Age/Size at Reproductive Maturity. Adults generally range in size from 15–25 cm TL (Petranka, 1998). In southern populations, individuals may take <3 yr to reach sexual maturity (Bishop, 1941b, 1943; Minton, 1972, 2001). In northern populations, males reach maturity earlier than females. In a Québec population examined using skeletochronology, the majority of animals were 2–18 yr old with the mode at 6–8 yr. Most females matured by 7 yr at ≥78 mm SVL. Males reached maturity between the ages of 2 and 6 and at ≥63 mm SVL (Flageole and Leclair, 1992). In Michigan populations, males reach sexual maturity in 2–3 yr and females in 3–5 yr (Wilbur, 1977c; see also Petranka, 1998). In Massachusetts, some males attain sexual maturity in 2 yr, while females require an additional year (Windmiller, 1996). At this locality, the smallest sexually mature individuals captured were 56.5 and 60.2 mm SVL for males and females, respectively. Growth in spotted salamanders has been analyzed by Blackwell et al. (2003) and Homan et al. (2003).

M. Longevity. Individuals live ≤ 22–25 yr in captivity (P. H. Pope, 1928a, 1937; Snider and Bowler, 1992), and skeletochronology indicated the oldest individual was 32 yr in a Québec population (Flageole and Leclair, 1992). Several mark–recapture studies at breeding ponds suggest that adult survivorship is high from one breeding season to the next, ranging from 70–100% (Husting, 1965; Whitford and Vinegar, 1966; Douglas and Monroe, 1981; see also Petranka, 1998). Females appear to suffer higher mortality than males (Harris, 1980).

N. Feeding Behavior. During the breeding season adults apparently do not feed (Smallwood, 1928). In terrestrial sites, adults and juvenile spotted salamanders feed by either protruding their heads from burrows using a sit-and-wait strategy or by foraging more actively in wet weather on the forest floor. They are generalists on forest floor invertebrates, consuming mollusks, earthworms, centipedes, millipedes, spiders, and a wide variety of insects (Bishop, 1941b; Pope, 1944). Given the amount of time adults spend in burrows and the depths at which they are found (1.3 m; Gordon, 1968; Semlitsch, 1983c), they may also feed underground.

O. Predators. Larval wood frogs (*Rana sylvatica*), centrarchid and cyprinid fishes, and various larval invertebrate species prey on spotted salamander eggs (Ward and Sexton, 1981; Semlitsch, 1988; Petranka, 1998; Baldwin and Calhoun, 2002). In a Virginia population, high reproductive failure was attributed to predation by centrarchid fishes (Ireland, 1989). Eastern newt adults will eat spotted salamander eggs (Hamilton, 1932; Wood and Goodwin, 1954), as will caddisfly and midge (*Parachironomus* sp.) larvae (Murphy, 1961; Leclair and Bourassa, 1981; Stout et al., 1992).

Larvae are preyed upon by caddisflies (*Ptilostomis postica*; Murphy, 1961; Rowe and Dunson, 1993; Rowe et al., 1994; and *Banksiola dossuaria*; Stout et al., 1992) and other predatory aquatic insects such as midge larvae (Petranka, 1998). Wood frog tadpoles will prey on larvae (Petranka, 1998; Petranka et al., 1998; Baldwin and Calhoun, 2002), as will least sandpipers (*Calidris minutilla*; Stangel, 1983). Larval predators also include other *Ambystoma* species; in regions of sympatry, marbled salamander larvae are already present and will prey on spotted salamander larvae when they hatch (Stewart, 1956; Husting, 1965; Stenhouse, 1985b). Thus, it is not surprising that spotted salamander larval densities are inversely correlated with marbled salamander larval densities (but see Worthington, 1968; Cortwright, 1988; Petranka, 1998). Jefferson salamander and silvery salamander (*A. platineum*) larvae also prey on spotted salamander larvae (Noble, 1931; Nyman, 1991; Brodman, 1996). Newly metamorphosed animals and breeding adults are preyed on by rac-

coons (Huheey and Stupka, 1967) and probably other mammals such as opossums, weasels, and minks (Beachy, 1991a).

P. Anti-Predator Mechanisms. Egg masses have a thick jelly coat that may afford some protection from centrarchid fishes and other aquatic invertebrate predators (Ward and Sexton, 1981; Semlitsch, 1988). There are conflicting reports on whether or not leeches prey on eggs (Cory and Manion, 1953; Cargo, 1960). Cunnington and Brooks (2000) tested whether adults breeding in harsher environments containing predatory fishes produced larger eggs—which might offer some protection to larvae against predation—but found no significant differences. Larvae possess no direct anti-predator mechanisms (e.g., Anderson and Petranka, 2003) but are less prone to predation when refuges are available. One experiment of microhabitat partitioning demonstrated that spotted salamander larvae tend to occupy refuges at a higher rate and with lower activity levels in the presence of predators (Brodman et al., 2002).

Spotted salamanders behave aggressively toward potential predators, head butting and biting attackers (Dodd, 1977b; Brodie et al., 1979). Adults can assume defensive postures, vocalize, and produce noxious skin secretions from granular glands located primarily on the dorsolateral surface of the tail and the paratoid gland on the head (Barach, 1951; Howard, 1971; Wyman and Thrall, 1972; Brodie, 1978; Brodie et al., 1979). These secretions are effective deterrents to predation by short-tailed shrews (*Blarina brevicauda*; Brodie et al., 1979) and potentially against some reptilian predators (Barach, 1951; Howard, 1971), presumably because they are distasteful. Brodie (1971a) reported that the secretions "cause a drying and burning sensation to the tongue." Nonetheless, adult spotted salamanders have been found in the stomachs of common garter snakes (*Thamnophis sirtalis*, Klemens, 1993), eastern hog-nosed snakes (*Heterodon platirhinos*; Babbitt, 1932), and fishes (e.g., *Salmo trutta*; Bishop, 1941b).

Q. Diseases. In egg clutches collected from several Michigan ponds and raised in the laboratory, protozoans (*Tetrahymena* sp.) were found to infect and kill the developing embryos (Ling and Werner, 1988; see also Petranka, 1998). Evidence of infection was characterized by swelling of the epidermis, and histological preparations revealed *Tetrahymena* sp. concentrations in nervous tissue. Corliss (1954) originally found *Tetrahymena* sp. in egg masses and developing embryos, but no effects were noted. Fungal infections have been observed in egg masses maintained in the laboratory, as well as in nature, but have only been mentioned anecdotally in the literature (e.g., Ward and Sexton, 1981).

Kingsley (1880) first reported the occurrence of natural limb anomalies in spotted salamanders. Polydactyly (particularly of the forelimbs), syndactyly of the proximal first digits, and missing digits were common abnormalities observed in juveniles and adults in a New York population (Wilson, 1976). Cases of melanism, albinism, and partial albinism are documented in adults, but usually occur in populations at low frequencies (Brandt, 1952; Hensley, 1959; Smith and Michener, 1962; Husting, 1965; Easterla, 1968; Mount, 1975; Dyrkacz, 1981). In a Mississippi population, Worthington (1974) found a high incidence of abnormal coloration, deviant counts of vertebral trunk elements, and limb anomalies; he suggested high temperatures as a cause, although genetic and other environmental factors could not be ruled out.

R. Parasites. Rankin (1937) reported the following parasites from spotted salamanders in North Carolina: Protozoa—*Cytamoeba bactifera*, *Euglenamorpha hegneri*, *Hexamitus intestinalis*, *Prowazekella longifilis*, *Tritrichomonas augusta*; Trematoda—*Brachycoelium hospitale*, *Diplostomulum ambystomae*; Nematoda—spirurid cysts; and Acarina—*Hannemania dunni*.

Bolek and Coggins (1998) reported helminth species infecting the small and large intestine of spotted salamanders from northwest Wisconsin. Of 20 individuals sampled, nine were infected by the nematode *Batracholandros magnavulvaris*; the trematode *Brachycoelium salamandrae* was found in only three individuals but at a higher mean intensity. Kuzmin et al. (2001) also described a nematode in the lungs and body cavity of spotted salamanders from northwestern Wisconsin; this nematode, *Rhabdias ambystomae*, is the first species of the genus described from salamanders in North America. Leeches have also been commonly observed on adults within and exiting breeding ponds. The impact, if any, of helminth or leech parasitism on spotted salamander population dynamics has not been documented.

4. Conservation.
Spotted salamander populations are seemingly widespread throughout the eastern United States, frequently occurring in nature preserves and state and federal parks, as well as in relatively undisturbed forests and private lands. Although present in fragmented suburban areas (Klemens, 1993; Gibbs, 1998a; Wright and Zamudio, 2002) and sometimes in highly disturbed urban areas (Klemens, 1993; DiMauro and Hunter, 2002), spotted salamander populations decline with increasing urbanization (Windmiller, 1996) and mere presence indicates little about long-term population viability in these habitats. Because populations are vulnerable to human developments that alter or eliminate habitats and corridors, they

are at risk of local extinction or even local population genetic consequences. Population comparisons using fluctuating asymmetry in dorsal spot patterns as a correlate of habitat disturbance showed that a population of breeding adults was significantly more asymmetrical in a disturbed landscape (in this case, a golf course) than in a nature reserve (Wright and Zamudio, 2002). These findings suggest that populations in the disturbed habitats may experience reductions in fitness, potentially due to demographic/genetic bottlenecks or developmental perturbations (due to chemical applications) as measured by fluctuating asymmetry, despite their continued persistence in the disturbed habitat.

Connectivity among patches of breeding habitat will decrease with habitat alteration. Thus, the importance of individual populations increases because viable populations can rescue nearby sinks in a metapopulation framework. Semlitsch and Bodie (1998) examined frequency distribution of wetland sizes and concluded that even small, isolated wetlands serve as important sources of juvenile recruits. Their loss could result in a direct reduction of connectivity among remaining populations. Spotted salamanders are likely to benefit from efforts to maintain connectivity among patches of upland forest habitats and aquatic breeding sites with sufficient buffer zones (Semlitsch, 1998). Spotted salamanders are more likely to occupy ponds in more forested areas and those that are adjacent to forest (Guerry and Hunter, 2002). Thus, the composition and configuration of the landscape surrounding breeding ponds is associated with the likelihood of spotted salamander presence and population persistence.

Environmental contamination in breeding ponds may negatively affect juvenile recruitment in spotted salamanders. Habitat acidification is a concern for this species, as hatching success and larval development can be reduced in pH ranges of 4.5–5.5 (Pough, 1976; Pough and Wilson, 1977; Ling et al., 1986; Sadinski and Dunson, 1992; Brodman, 1993), levels that are not uncommon in the eastern United States (Clark, 1986b; Cook, 1983; Ling et al., 1986; Sadinski and Dunson, 1992; Petranka, 1998). The number of eggs present in a pond is positively correlated with alkalinity, as is hatching success (Clark, 1986b; Petranka, 1998). Ireland (1991) demonstrated that significantly reduced growth rates were not the result of pH alone but may be a function of anion concentration or the combined effect of anion and hydrogen ions. Cook (1983) found no correlation between low pH and embryonic mortality in 13 Massachusetts ponds; in contrast, Portnoy (1990) found reduced survivorship correlated with acidification in the Cape Cod region.

Low pH levels can also reduce growth and survival rates, as well as alter competitive interactions between larval spotted salamanders and Jefferson salamanders (Brodman, 1993). Acidification is not the only threat to adult reproductive fitness. Over an 8-yr period in eastern Virginia, reproductive activity of spotted salamanders declined severely in breeding ponds with low pH levels and high aluminum, copper, silicon, and zinc concentrations (Blem and Blem, 1989, 1991). Elevated hydrogen ion and aluminum concentrations in breeding ponds may also negatively affect hatching success (Clark and Hall, 1985). In Pennsylvania, the number of eggs present is positively correlated with pH and pond size, but negatively correlated with cations and silica levels (Rowe and Dunson, 1993).

During the breeding season, road mortality is a considerable threat to migrating amphibians. Although the direct effects of road mortality on population numbers have not been thoroughly evaluated, several studies show that amphibian densities decrease in response to roadways (Fahrig et al., 1995; Vos and Chardon, 1998; Trombulak and Frissell, 2000). The direct effects of roads on mortality may not be the only consideration. In New Hampshire, embryonic survivorship is reduced in breeding ponds contaminated with road salt runoff (Turtle,

2000). Roads also affect animal communities by creating habitat edges that promote invasion of exotic plants (Gaddy and Kohlstaat, 1987) and exposing salamanders to predation. Moreover, heavily trafficked roads may entirely block migrating salamanders from reaching a breeding pond (Trombulak and Frissell, 2000). In Massachusetts, Jackson and Tyning (1989) reported that drift fences

and two tunnels installed under a road aided migrating spotted salamanders and substantially reduced mortality. Migratory routes under roads are not difficult to implement, and amphibian populations would benefit from local efforts to reduce road mortality during their short breeding seasons.

Ambystoma opacum (Gravenhorst, 1807)
MARBLED SALAMANDER

David E. Scott

1. Historical versus Current Distribution.
Marbled salamanders (*Ambystoma opacum*) range throughout much of the eastern United States from eastern Texas and Oklahoma, northeast through Illinois and Indiana to southern New Hampshire and central Massachusetts, and south to north Florida. Disjunct populations occur along the southern edge of Lake Michigan; locality data are summarized by Anderson (1967b). Additional localities are reported for east Texas (Baldauf and Truett, 1964), Louisiana (Dundee and Rossman, 1989), southeastern Oklahoma (Trowbridge, 1937), Missouri (Johnson, 1987), Mississippi (Ferguson, 1961b), Indiana (LaPointe, 1953), Alabama (Mount, 1975), north Georgia (Martof, 1955), North Carolina (E.E. Brown, 1992), and Rhode Island (Doty, 1978).

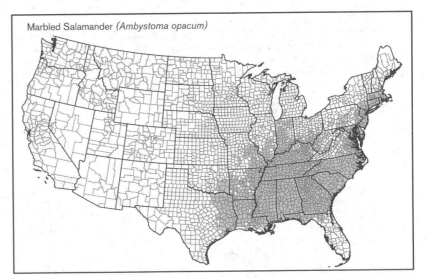

Marbled Salamander (*Ambystoma opacum*)

2. Historical versus Current Abundance.
Early accounts of marbled salamanders describe them as common but secretive (e.g., Noble and Brady, 1933), although prior to the early 1960s no data on population sizes were collected (see Murphy, 1962; Graham, 1971). Currently, marbled salamanders are common and may be locally abundant in some areas. Population sizes range from dozens of individuals

(Murphy, 1962) to hundreds (Graham, 1971; Shoop and Doty, 1972; Stenhouse, 1985a), ~1,000 (Pechmann et al., 1991; Semlitsch et al., 1996) to >10,000 (Taylor and Scott, 1997). However, given the reliance of marbled salamanders on small isolated seasonal wetlands and intact forested floodplain habitats, their abundance presumably has declined as wetland habitats have been destroyed (Petranka, 1998). For example, from the 1950s–70s the loss of wetlands in the Southeast was greater than in any other region of the country, with a net annual loss of 386,000 ac/yr (Hefner and Brown, 1985); in North Carolina approximately 51% of all wetland acreage on the Coastal Plain has been lost (Richardson, 1991), including 70% of the pocosins that have been "developed" or partially altered (Richardson, 1983); approximately 97% of the Carolina bays on the Coastal Plain of South Carolina have been altered or severely impacted; and <200 bays of the original thousands are "relatively unimpacted" (Bennett and Nelson, 1991).

3. Life History Features.

A. Breeding. Reproduction is terrestrial, in or near the wetland breeding sites prior to pond filling.

i. Breeding migrations. Onset of breeding migrations occurs from September–November. Timing varies geographically and may occur 1 mo or more earlier at southern latitudes compared with northern latitudes (Anderson and Williamson, 1973). On a broad scale, seasonal migrations are probably linked to regional climatic and hydrological cycles (Salthe and Mecham, 1974). Adult salamanders move to breeding sites on rainy nights and tend to enter and exit the site at approximately the same point (Shoop and Doty, 1972; P.K. Williams, 1973; Douglas and Monroe, 1981; Stenhouse, 1985a). Males generally arrive at the breeding site before females (Noble and Brady, 1933; Graham, 1971; Krenz and Scott, 1994). In a 25-yr study at Rainbow Bay in South Carolina (see Semlitsch et al., 1996), the mean date of arrival of males at the breeding site was 10 d earlier than females (unpublished data), perhaps due solely to the combination of a limited number of nights suitable for migration and slower nightly movements by fat, gravid females (Blanchard, 1930; personal observations). The sex ratio of the breeding population is often biased toward males (Graham, 1971; Stenhouse, 1987; Krenz and Scott, 1994), in part because males mature at an earlier age (Scott, 1994; Pechmann, 1994). The sex ratio in one study (Parmelee, 1993) during the non-breeding season did not differ from 1:1.

a. Courtship activity. At the time of autumn migration, males are at maximal testosterone levels (Houck et al., 1996; unpublished data). Courtship is terrestrial,

with males exhibiting nudging, head-swinging, lifting, and body-flexing behaviors (Arnold, 1972). Spermatophore deposition follows lateral undulations of the tail. Spermatophores are 4–5.5 mm tall (Lantz, 1930; illustrated in Noble and Brady, 1933). Typical and secondary spermatophore deposition may occur (Arnold, 1972, 1976; personal observation); a male may deposit over 10 spermatophores in 30–45 min (L. Houck, personal communication). Males will mate with females outside what is typically thought of as the wetland margin (Krenz and Scott, 1994). Males often will court other males (Noble and Brady, 1933), including spermatophore deposition in the absence of a female (L. Houck, personal communication). Females may follow a male to pick up a spermatophore (Noble and Brady, 1933) or simply move about an area until a spermatophore is located (Arnold, 1972). When a spermatophore contacts a female's vent she will lower herself onto it and insert it into her cloaca (Arnold, 1972). Sperm are stored in exocrine glands called spermathecae in the roof of the cloaca (Sever and Kloepfer, 1993). Eggs are fertilized internally by sperm released from spermathecae during oviposition (Sever et al., 1995). Females may pick up multiple spermatophores (Arnold, 1972), but sperm competition has not been definitively demonstrated. Sperm in the spermathecae do not persist for > 6 mo after oviposition (Sever et al., 1995).

ii. Breeding habitat. Marbled salamanders are one of two species of *Ambystoma* that breed on land (Petranka, 1998), and they are the only *Ambystoma* species that exhibit parental care (Nussbaum, 1985, 1987). Due to the terrestrial reproductive habits of marbled salamanders, breeding is restricted to fish-free wetlands with seasonally fluctuating water levels that include upland hardwood "swamp forests" (King, 1935), bottomland hardwood pools (Viosca, 1924a; Petranka and Petranka, 1981a,b), quarries (Graham, 1971), vernal ponds (Doty, 1978), Carolina bays (Jackson et al., 1989; Gibbons and Semlitsch, 1991), and floodplain pools (Petranka, 1990). Females remain with eggs (Noble and Brady, 1933) for varied lengths of time (Petranka, 1998); they may leave before eggs are inundated (McAtee, 1933; Jackson et al., 1989; Petranka, 1990). Nest brooding appears to enhance embryonic survival (Petranka and Petranka, 1981b; Jackson et al., 1989), although the mechanism is unknown. Opinions differ on whether there is an energetic cost to brooding by females (Kaplan and Crump, 1978; versus D.E.S., unpublished data). Occasionally nests are communal (Graham, 1971; Petranka, 1990), especially if cover items are scarce (Palis, 1996b).

B. Eggs.

i. Egg deposition sites. Breeding sites are generally the dried beds of temporary ponds, the margins of reduced ponds, or dry floodplain pools. Female marbled salamanders construct nests and lay eggs under virtually any cover in situations where the nest is likely to be flooded by subsequent winter rains (Noble and Brady, 1933). Eggs are laid on the edges of pools (Dunn, 1917b) and in dry basins under vegetation (Jackson et al., 1989), logs (Bishop, 1924; Doody, 1996), and leaf debris (Deckert, 1916; Petranka and Petranka, 1981b). Eggs are laid occasionally on non-soil substrate (Brimley, 1920a). Nest site selection by females is influenced by microsite elevation within the pond bed, site hydrologic regime, cover availability, and soil moisture (Petranka and Petranka, 1981a,b; Jackson et al., 1989; Figiel and Semlitsch, 1995; Wojnowski, 2000; but see also Marangio and Anderson, 1977). Females actively excavate oblong to ovoid-shaped depressions (King, 1935; Petranka and Petranka, 1981b).

ii. Clutch size. Of the three reproductive modes of salamanders outlined by Salthe (1969), marbled salamanders have an atypical type I mode (Salthe and Mecham, 1974; Kaplan and Salthe, 1979). Clutch size ranges from approximately 30 to <200 eggs (see Petranka, 1998) and usually is positively correlated with female body size (Kaplan and Salthe, 1979; Walls and Altig, 1986; Petranka, 1990; Scott and Fore, 1995), although not always (Kaplan and Salthe, 1979)—larger females may have larger eggs (Kaplan and Salthe, 1979).

Compared to other species of *Ambystoma*, females in some populations of marbled salamanders may have fewer, larger eggs than would be expected for an animal of their size (Kaplan and Salthe, 1979; D.E.S., unpublished data; for a different opinion see Nussbaum, 1985, 1987). For example, comparing female marbled salamanders and mole salamanders (*A. talpoideum*) of equal body size, marbled salamanders have 3–4 times fewer eggs, but each egg is 3–4 times larger with 3–4 times higher lipid amounts (Komoroski, 1996; D.E.S., unpublished data). Mean egg diameter is greater in marbled salamanders than in flatwoods salamanders (*A. cingulatum*; 2.8 vs. 2.3. mm; Anderson and Williamson, 1976). Mean egg dry mass is greater in marbled salamanders than in either mole salamanders or spotted salamanders (*A. maculatum*; Komoroski, 1996). The caloric content (cal/mg dry mass) of marbled salamander eggs is greater than the energy content of spotted salamanders and tiger salamander (*A. tigrinum*) eggs (Kaplan, 1980b). Relatively few, large eggs with lipid stores in excess of the amount needed for embryogenesis probably reflects a response to the terrestrial breeding habits of marbled salamanders and the extreme variability and unpredictability in the timing of hatching (i.e., the duration of the egg stage). Substantial (15–30%) variation in egg diameter occurs within and among populations (Kaplan, 1980a). Egg size is positively correlated with hatchling size and early larval size (Kaplan, 1980a).

In spite of terrestrial egg laying, egg structure in marbled salamanders is similar to aquatic *Ambystoma* species (Salthe, 1963). Egg development is temperature-dependent (Noble and Brady, 1933); development (at similar temperatures) is slower than for some other ambystomatids (Moore, 1939). The prospective neural tissue of marbled salamanders has a lower density and higher water-holding capacity than the embryonic tissue of aquatic breeders such as spotted salamanders and tiger salamanders (M.G. Brown, 1942). Embryos develop to a hatching stage, but do not hatch until stimulated by hypoxia when the nest is flooded (Petranka et al., 1982). Some eggs may remain viable several months after oviposition (Noble and Brady, 1933), but often eggs laid in October will die by December if the nest has not been flooded (McAtee, 1933). An embryo's lipid reserves in excess of reserves required for embryogenesis constrain the maximum time an embryo can survive in the egg (unpublished data). Hatching under natural conditions may occur at a wide range of developmental stages (Noble and Brady, 1933; Graham, 1971), probably due to the hydration state during development and the timing of nest inundation (Noble and Brady, 1933; S. Dooty, personal communication). Larvae will hatch from early stages (10.5 mm), with much yolk and little swimming ability, to late stages (20 mm; Brimley, 1920a; McAtee, 1933; Noble and Brady, 1933). By 18 mm the balancers are usually lost whether or not the egg has hatched (Brandon, 1961). Embryo mortality can be high due to freezing, dehydration, predation, or fungus (Stenhouse, 1987; Jackson et al., 1989) and is dependent on the timing and extent of pond-filling (Wojnowski, 2000).

Timing of hatching varies among sites and years (Petranka and Petranka, 1980). Within a site, pond-filling may be incremental or sudden; gradual pond-filling may result in staggered hatching of eggs and substantial size variation of larvae within a pond (Smith, 1988).

C. Larvae/Metamorphosis.

i. Length of larval stage. Hatchling densities average as high as 47 larvae/m^2 (Smith, 1988). Catastrophic larval mortality may result from winter kill due to extreme cold (Heyer, 1979; Cortwright and Nelson, 1990), incomplete pond filling and subsequent drying (King, 1935; Petranka and Petranka, 1981a), and early pond drying (Pechmann et al., 1991).

Larval growth rates of marbled salamanders vary considerably (see Doody,

1996; table 1), depending upon their density, food levels, and temperature. Growth rates are comparable to spotted salamanders (Walls and Altig, 1986), but comparisons to mole salamanders differ (Keen et al., 1984; Walls and Altig, 1986). Although larval growth is temperature dependent (Stewart, 1956), temperature effects may not be as pronounced as in some other *Ambystoma* species (Keen et al., 1984). Food level, temperature, hatching time, and larval density affect traits of newly metamorphosed animals (Stewart, 1956; Boone et al., 2002). Early hatching larvae are larger at metamorphosis, have higher survival, and metamorphose earlier than late-hatching larvae (Boone et al., 2002). Higher food levels and warmer temperature promote earlier metamorphosis; increased prey density promotes larger size at metamorphosis (Stewart, 1956). The larval period may range from 2–9 mo (unpublished data; Jenkins et al., 2002), and is strongly influenced by the timing of pond filling and drying. Intraspecific larval density affects larval growth and a suite of larval traits (Stenhouse et al., 1983; Stenhouse, 1985b; Smith, 1988, 1990; Petranka, 1989c; Scott, 1990). At high densities, larvae have slower growth, smaller size at metamorphosis, and reduced survival (Petranka, 1989c); they may also have longer larval periods (Scott, 1990). As ponds dry, larval densities may become extremely high (e.g., 237 m^2; Smith, 1988).

ii. Larval requirements.

a. Food. Larvae eat primarily macrozooplankton, beginning with copepod nauplii in hatchlings (Petranka and Petranka, 1980). Ostracod, cladoceran, and copepod zooplankton feed larger larvae. Other invertebrate prey include chironomids, amphipods, chaoborids, and isopods (Petranka and Petranka, 1980; Branch and Altig, 1981).

b. Cover. Larvae may remain mostly hidden on the pond bottom during the day and move into the water column at night (Anderson and Graham, 1967; Petranka and Petranka, 1980; Branch and Altig, 1981). Both the limited diurnal movements and increased nocturnal activity may serve to enhance feeding and avoid vertebrate predation (Hassinger et al., 1970; Branch and Altig, 1981), although larvae floating in the water column at night did not capture more prey than those feeding on benthos during daylight (Petranka and Petranka, 1980). Movements of larvae into the water column are associated with decreased light intensity (Hassinger and Anderson, 1970) as well as vertical stratification of some prey species (Anderson and Graham, 1967; Petranka and Petranka, 1980). Larval activity may also vary seasonally; larvae remain near the bottom of the water column early in the season and utilize the entire column for feeding on zooplankton as the season progresses (Hassinger et al.,

1970). Larvae nearing metamorphosis remain near the bottom at night (Petranka and Petranka, 1980).

iii. Larval polymorphisms. None reported, although behavior differences are known. Laboratory assays have demonstrated two divergent aspects of kin recognition. In some contexts, kin recognition may reduce aggression and cannibalism among siblings in larval marbled salamanders (Walls and Roudebush, 1991); whereas in other contexts, large larvae may eat siblings preferentially (Walls and Blaustein, 1995). Hokit et al. (1996) further demonstrated that kin discrimination is context dependent. Under more natural conditions, kinship effects on larval performance did not occur (Walls and Blaustein, 1994).

iv. Features of metamorphosis. As noted above, at high larval densities individual larvae have slower growth, a smaller size at metamorphosis, and reduced survival (Petranka, 1989c); they may also have longer larval periods (Scott, 1990). In field experiments, environmental conditions (i.e., larval density) accounted for more of the variation in body size at metamorphosis than did an individual's level of multilocus genetic heterozygosity (Chazal et al., 1996). In an artificial pond study, more heterozygous individuals had shorter larval periods than did less heterozygous larvae (Krenz, 1995). Juvenile recruitment into the terrestrial population may vary dramatically among sites and years (Stenhouse, 1984, 1987; Pechmann et al., 1991; Taylor and Scott, 1997), which probably reflects broad variation in abiotic (e.g., hydroperiod) and biotic (e.g., productivity, competition, and predation) conditions (Petranka, 1989c; Semlitsch et al., 1996). Effects initiated by aquatic conditions persist in terrestrial adults (Scott, 1994).

Timing of metamorphosis may vary geographically, although recently metamorphosed juveniles generally disperse from the breeding site in late spring. Metamorphosis occurred in June–July in Illinois (Smith, 1961); June in New York (Bishop, 1941b); late May to early June in Maryland (Worthington, 1968), New Jersey (Hassinger et al., 1970), and north Georgia (Martof, 1955); mid-May in West Virginia (Green, 1955); mid-April to May in North Carolina (Stewart, 1956; Smith, 1988); March–April in Alabama (Petranka and Petranka, 1980); and as early as mid-March in Louisiana (Dundee and Rossman, 1989). Although marbled salamanders will metamorphose in response to pond drying, timing also appears to be triggered by intrinsic factors (Hassinger et al., 1970; Scott, 1994). Larvae that hatch 2–3 mo later than others will nonetheless metamorphose within a few weeks of early hatching larvae, but at a smaller body size (Boone et al., 2002). Stages of metamorphosis are described by Grant (1931).

v. Post-metamorphic migrations. Juveniles may not disperse far from the edge of wetlands (P.K. Williams, 1973) and therefore require intact terrestrial habitats surrounding the breeding sites (Semlitsch, 1998).

vi. Neoteny. There are no reports of non-transforming marbled salamanders. Given the widespread distribution and numerous population studies on this species, it is unlikely that neotenic adults exist.

D. Juvenile Habitat. Same as adult habitat, although juveniles tend to occur under smaller cover objects (Parmelee, 1993). Juveniles retain the ability to discriminate their siblings, presumably by chemoreception, for ≥8 mo after metamorphosis (Walls, 1991). Juvenile marbled salamanders experienced low first-year survival (4.5%) in old field terrestrial enclosures when compared to survival in forest enclosures (45%; Rothermel, 2003). Juveniles >1 yr old experienced near zero annual survivorship in old field enclosures compared to >70% in forest enclosures (Rothermel, 2003).

E. Adult Habitat. Most reports of terrestrial habitats indicate that mature deciduous forests are preferred (Petranka, 1998). Mixed hardwood and pine stands (Smith, 1988; Pechmann et al., 1991), floodplains (Petranka, 1989c; Parmelee, 1993), and uplands (Smith, 1961) are also utilized. Of 15 radioactively tagged individuals released near a woodland pond in southern Indiana, 14 were tracked in hardwood forest, 1 in an old field (P.K. Williams, 1973). Microhabitats within the forest include under leaf litter and small mammal burrows (P.K. Williams, 1973; Douglas and Monroe, 1981). Salamanders do not actively dig their own burrows, but enlarge existing openings (Semlitsch, 1983a). Although generally described as woodland salamanders, marbled salamanders may also be tolerant of relatively dry conditions (Cagle, 1942; Smith, 1961; Mount, 1975; Dundee and Rossman, 1989) and can be found on rocky hillsides (Johnson, 1987). One laboratory experiment indicated a preference for relatively basic substrates (pH 7.7), although animals in the field were found on more acidic (pH 5.5) substrates (Mushinsky, 1975). Compared to other ambystomatids, marbled salamanders may use substantially drier habitat and tolerate higher substrate temperatures (Parmelee, 1993). Adults dispersed an average of 194 m from the wetland breeding site (P.K. Williams, 1973). Consequently, post-metamorphic individuals require intact terrestrial habitats surrounding the breeding sites (Semlitsch, 1998). Survivorship of marbled salamander adults and recently metamorphosed animals was low in 100 m² enclosures in clearcuts compared to enclosures in adjacent forests (P. Niewiarowski and A. Chazal, personal communication).

F. Home Range Size. Williams (P.K., 1973) examined home range size for mar-

bled salamanders (n = 8) by using radioactive wire tags. The spring and summer home range size varied from 1–225 m², with a median of 14.5 m². There was a tendency for home range size to increase as individuals were followed for longer periods. A laboratory study indicated that juvenile marbled salamanders tend to stay on their own marked substrate, which may be a mechanism to detect home areas by chemical cues (Smyers et al., 2001).

G. Territories. Although Martin et al. (1986) found no evidence for territoriality in small-mouthed salamanders (*A. texanum*), they suggested that territoriality might be expected in marbled salamanders. Individuals of some *Ambystoma* species may return to their summer home range from the previous year (Semlitsch, 1983b), and this may be true in marbled salamanders (P.K. Williams, 1973). The orientation behavior exhibited by marbled salamanders is a necessary precursor to territoriality, although territoriality itself has not been documented. In laboratory studies, "resident" individuals tend to bite conspecific "intruders;" however, animals housed together for long periods did not avoid each other (Ducey, 1989). Juvenile salamanders (<8 mo old) are less aggressive to familiar "neighbors" than to unfamiliar "strangers," especially among siblings (Walls, 1991). Adult marbled salamanders maintained on a low-food diet were more prone to bite an intruding salamander than those on a high-food diet (Ducey and Heuer, 1991), which may indicate that aggression functions to repel an intruder from an individual's feeding area/burrow refuge. Marbled salamanders almost always occur alone under cover objects (Parmelee, 1993).

H. Aestivation/Avoiding Desiccation. Marbled salamanders likely undergo prolonged periods of summer inactivity, corresponding to periods of little or no rainfall. Despite reports that marbled salamanders can occur in unusually dry habitats (e.g., Bishop, 1943), there is no evidence that they differ from more aquatic species in terms of their water exchange with soil (Spight, 1967b). However, a laboratory study of water loss rate showed marbled salamanders lose water more slowly than the other species examined (which were all plethodontids) and were able to withstand dehydration ≤ 30% of initial body weight (Spight, 1968). Dehydrated salamanders incur substantial metabolic costs, however (Sherman and Stadlen, 1986). To minimize water loss, marbled salamanders likely avoid desiccating conditions; as soils dry in late summer, animals may retreat to deeper burrows (P.K. Williams, 1973). Rehydration rates were faster in marbled salamanders than in more aquatic species, and faster in severely dehydrated animals than in less-dehydrated individuals (Spight, 1967a).

i. Heat stress. Larvae of marbled salamanders have less resistance to high temperatures (i.e., have a lower Critical Thermal Maximum, CTM) than either small-mouthed salamanders or spotted salamanders (Keen and Schroeder, 1975). Smaller adult salamanders reach their CTM faster than larger adults (Hutchison, 1961). The possible relationship between CTM in eggs, larvae, and adults, and geographic distribution or timing of breeding (*sensu* Gatz, 1971) is unknown.

ii. Water stress. In general, post-metamorphic marbled salamanders do not appear to respond well to prolonged immersion in water (personal observation). Interestingly, the total oxygen uptake through pulmonary surfaces is relatively low (34%; Whitford and Hutchison, 1966b), although lung sacs, ridges, and vascularization are well developed in marbled salamanders (Czopek, 1962). Under anoxic conditions, larvae may exhibit anaerobic glycolysis (Weigmann and Altig, 1975).

iii. Metabolic rate. Lunged salamanders, including marbled salamanders, increase levels of oxygen consumption with increasing body size (Whitford and Hutchison, 1967; Krenz, 1995). A daily cycle also occurs, with resting metabolic rate increasing by 50% during the early evening (Krenz, 1995). Metabolic rates increase by 119% following dehydration (Sherman and Stadlen, 1986). Resting metabolic rate is positively correlated with multi-locus heterozygosity (Krenz, 1995); more heterozygous females with higher metabolic demands allocated less energy to their clutches of eggs (Krenz, 1995).

I. Seasonal Migrations. Restricted to times of breeding (adults; see "Breeding migrations" above) and following metamorphosis (juvenile; see "Features of metamorphosis" above). Post-metamorphic dispersal is restricted to rainy nights. The period between metamorphosis and dispersal may be several weeks or more (depending on occurrence of nighttime rainfall) and is likely a period of high mortality for juveniles (personal observation).

J. Torpor (Hibernation). In the north, post-reproductive adult marbled salamanders move ≤30 m from the breeding site (Douglas and Monroe, 1981), where they remain for the winter. Hibernation in the southern portions of their range is unknown.

K. Interspecific Associations/Exclusions. Due to the terrestrial breeding habits and early egg hatching, larval marbled salamanders are often much larger than other amphibian larvae (Worthington, 1968, 1969; Keen, 1975; Stenhouse et al., 1983; Walls and Altig, 1986; Smith, 1988; Scott, 1993). Larval marbled salamanders will feed on other amphibian eggs and larvae (Walters, 1975), including *Ambystoma* larvae. Where they co-occur, marbled salamanders eat smaller spotted salamander larvae (Stewart, 1956; Stenhouse et al.,

1983; Stenhouse, 1985b), small-mouthed salamanders (Walters, 1975; Doody, 1996), Jefferson salamanders (*A. jeffersonianum*; Cortwright, 1988), tiger salamanders (Stine et al., 1954), and mole salamanders (Walls, 1995). Spotted salamander larvae may be more susceptible than mole salamander larvae to this predation due to increased use of refugia by mole salamanders (Walls, 1995). The size of marbled salamander larvae at the time when other *Ambystoma* eggs are hatching varies among ponds and years by 30–40% (Stenhouse et al., 1983). Consequently, although larval marbled salamanders are often predators, they may also be competitors (Wilbur, 1984; Stenhouse, 1985b; Cortwright, 1988; Semlitsch et al., 1996). Predation by marbled salamander larvae may substantially affect community dynamics (Cortwright and Nelson, 1990; Morin, 1995; Boone et al., 2002). Juvenile marbled salamanders that were tested under laboratory conditions with conspecifics and with juvenile mole salamanders did not show any overt aggression, perhaps indicating that such behavioral interactions are not important for juveniles (Smyers et al., 2001). Additional experiments with juvenile spotted salamanders indicated that juvenile marbled salamanders may defend burrow space by excluding heterospecific salamanders (Smyers et al., 2002).

L. Age/Size at Reproductive Maturity. Age and size at reproductive maturity are traits that vary and are highly dependent on size at metamorphosis (Scott, 1994), which in turn is influenced by intraspecific larval densities and the timing of pond drying (Petranka, 1989c; Scott, 1990). Males tend to mature at an earlier age than females (Scott, 1994; Pechmann, 1995); average age at first reproduction for males is 2.5–3.1 yr (Scott, 1994) to 3.3 yr (Pechmann, 1995), and for females, 2.8–3.4 yr (Scott, 1994) to 4.0 yr (Pechmann, 1995). The range of age at first reproduction for both sexes is 1–7 yr. Mean size at first reproduction is approximately 53–60 mm SVL for both sexes (Scott, 1994; Pechmann, 1995); the minimum size at first reproduction may be smaller for males (~42.0 mm) than for females (~45.0 mm).

M. Longevity. Survival to first reproduction can be low and is influenced by size at metamorphosis. Variation in body size at metamorphosis is coupled with variation in lipid stores (ranging from 2–16.5% of dry mass; unpublished data). Small, lean animals may suffer the highest mortality immediately following metamorphosis (Scott, 1994). Survival from metamorphosis to first reproduction ranges from 3–60% (Scott, 1994; Pechmann, 1994, 1995). Males may exhibit higher survivorship than females due to their earlier age at first reproduction (Scott, 1994). Maximum lifespan in the field appears to be 8–10 yr (Graham, 1971; Taylor and Scott, 1997).

N. Feeding Behavior. Stomach contents of juveniles and adults include millipedes, centipedes, spiders, insects, and snails (Dundee and Rossman, 1989); arthropods, annelids, and mollusks (Smith, 1961).

O. Predators.

i. Eggs. Eggs may be preyed upon by beetles, salamanders, frogs (Noble and Brady, 1933), and possibly a millipede species (*Uroblaniulus jerseyi*; Mitchell et al., 1996a).

ii. Larvae. Larvae are palatable to fishes (Kats et al., 1988), but usually do not inhabit ponds where fish occur. Larval marbled salamanders are prey for numerous species, especially invertebrates including dragonfly naiads (Odonata), spiders (Arachnida), dytiscid beetle larvae and adults (Coleoptera), and giant water bugs (Belostomatidae). Larval survivorship decreased from 60 to 70% to <20% when hatchlings inhabited experimental enclosures in a wetland replete with invertebrate predators (unpublished data); survivorship decreased to zero in a year when chain pickerel (*Esox niger*) colonized the wetland. Adult eastern newts (*Notophthalmus viridescens*) and paedomorphic mole salamanders also feed on larval marbled salamanders. Cannibalism may occur (Walls and Roudebush, 1991) when incremental pond-filling staggers dates of hatching and increases size variation (Smith, 1990). Wading birds and kingfishers (*Megaceryle alcyon*) are also likely predators (personal observations).

iii. Juveniles and adults. Raccoons (*Procyon lotor*), opossums (*Didelphis virginiana*), skunks (Mustelidae), and shrews (Soricidae) are known to kill adult marbled salamanders (DiGiovanni and Brodie, 1981; Petranka, 1998). Often the tails are not eaten (personal observation). Newly metamorphosed animals may be susceptible to mammalian predators as well as some snakes; one southern water snake (*Nerodia fasciata*) had eaten 34 recently metamorphosed marbled salamanders (unpublished data). Liner (1954) reported ingestion of two recently metamorphosed marbled salamanders by a western ribbon snake (*Thamnophis proximus*).

P. Anti-Predator Mechanisms.

i. Eggs. Protection of eggs from predators is possibly one function of nest-brooding by females (Petranka, 1990).

ii. Larvae. Limited diurnal movements and hiding in benthic debris may reduce predation (Hassinger et al., 1970; Petranka and Petranka, 1980; but see Marangio, 1975, for report of positive phototaxis in small larvae). Hatchlings and small larvae may use the sun as a cue to orient toward deep water (Tomson and Ferguson, 1972). Larvae do not change behavior (i.e., increase use of refugia) in the presence of fishes (Kats et al., 1988).

iii. Adults. Animals under attack by short-tailed shrews (*Blarina brevicauda*) exhibit tail lashing, body coiling, and head-butting behaviors, and/or may become immobile (Brodie, 1977). Such behaviors may draw the attacks toward the tail, which has concentrations of granular glands on dorsum that produce noxious secretions. Adults are unpalatable to common ribbon snakes (*T. s. sauritus*; T. Mills, personal communication). Secretions generally confer protection from a single attack by shrews (Brodie et al., 1979). Secretions in marbled salamanders are reduced after multiple attacks by shrews, resulting in increased vulnerability (DiGiovanni and Brodie, 1981).

Q. Diseases. An aquatic fungus (*Saprolegnia* sp.) may develop on the injured portions, especially limbs, of bitten larvae and may be lethal (Petranka, 1989c).

Marbled salamanders have been used in toxicological tests of hydrazine compounds (Slonim, 1986), beryllium sulfate (Slonim and Ray, 1975), pesticides (Hall and Swineford, 1981), and motor oil (Lefcourt et al., 1997).

R. Parasites. Rankin (1937) reported the following parasites from marbled salamander larvae in North Carolina: Protozoa—*Cryptobia borreli, Eutrichomastix batrachorum, Hexamitus intestinalis, Prowazekella longifilis, Tritrichomonas augusta*; Trematoda—*Diplostomulum ambystomae*; Acanthocephala—*Acanthocephalus acutulus*. Rankin (1933) reported the following parasites from marbled salamander adults in the same populations: Protozoa—*Cryptobia borreli, Cytamoeba bacterifera, Eimeria ranarum, Eutrichomastix batrachorum, Haptophyra michiganensis, Hexamastix batrachorum, Hexamitus intestinalis, Prowazekella longifilis, Tritrichomonas augusta*; Trematoda—*Brachycoelium hospitale, Diplostomulum ambystomae*; *Gorgoderina bilobata, Megalodiscus temperatus, Plagitura* sp.; Nematoda—*Capillaria inequalis, Cosmocercoides dukae, Filaria* sp., spirurid cysts; Acarina—*Hannemania dunni*.

The trematode *Brachycoelium ambystomae* was reported from marbled salamanders by Couch (1966), and an unidentified immature trematode by Malewitz (1956). The gall bladder myxosporean (*Myxidium serotinum*) has been reported in marbled salamanders in Arkansas and Texas (McAllister and Trauth, 1995).

4. Conservation.

Marbled salamanders are listed as Threatened in Massachusetts and Michigan, and Protected in New Jersey (Levell, 1997). In each of these states, permits are required for any activity involving marbled salamanders.

Given the reliance of marbled salamanders on small isolated seasonal wetlands and intact forested floodplain habitats, their abundance presumably has declined as wetland habitats have been destroyed (Petranka, 1998). Small isolated wetlands are the most valuable wetlands for maintaining amphibian biodiversity, but it is precisely these wetlands that are unprotected by current wetlands regulations and that are most "at risk" (Semlitsch and Bodie, 1998). Further loss of small wetlands such as Carolina bays will likely be accelerated by the U.S. Supreme Court's SWANCC decision in January 2001 (Sharitz, 2003), unless individual states pass legislation that protects small isolated wetlands. As isolated wetland habitats disappear and remaining wetlands become increasingly separated, the cumulative impact on amphibian populations such as marbled salamanders will likely be substantial and perhaps non-linear, as elimination of remaining wetlands results in proportionally larger and larger effects on pond-breeding amphibian populations. For conservation efforts to succeed it will also be critical that the wetland ecosystem be viewed not solely as the wetland itself, but also the adjacent terrestrial habitat that is essential to the persistence of pond-breeding amphibians (Scott, 1999; Gibbons, 2003).

Acknowledgments. Whit Gibbons reviewed early drafts of the manuscript. Manuscript preparation was supported by the Environmental Remediation Sciences Division of the Office of Biological and Environmental Research, U.S. Department of Energy, through Financial Assistance Award No. DE-FC09-96-SR18546 to the University of Georgia Research Foundation.

Ambystoma platineum Cope, 1867 (1868)
SILVERY SALAMANDER

Christopher A. Phillips, Jennifer Mui

See C. Phillips and J. Mui's account "Unisexual members of the *Ambystoma jeffersonianum* complex" for a full consideration of *Ambystoma platineum*, *Ambystoma tremblayi*, and unisexual hybrid ambystomas.

Ambystoma talpoideum (Holbrook, 1838[b])
MOLE SALAMANDER

Stanley E. Trauth

1. Historical versus Current Distribution.
Mole salamanders (*Ambystoma talpoideum*) are distributed along the southern Atlantic and Gulf Coastal Plains from central South Carolina to eastern Texas, and north along the Mississippi River Valley to southern Illinois (Owens, 1941; Baldauf and Truett, 1964; Robison and Winters, 1978; Erwin, 1979; Sutton and Paige, 1980; Bader and Mitchell, 1982; Redmond et al., 1982; Braswell, 1985; Murdock and Braswell, 1985; Seyle, 1985a; Meshaka and McLarty, 1988; Meshaka et al., 1989; Braswell et al., 1990; Somers, 1990; Teska, 1990; Trauth et al., 1993a; Conant and Collins, 1998; Jensen, 1998; Petranka, 1998). They do not occur in southern peninsular Florida or in southern Louisiana. Disjunct populations are scat-

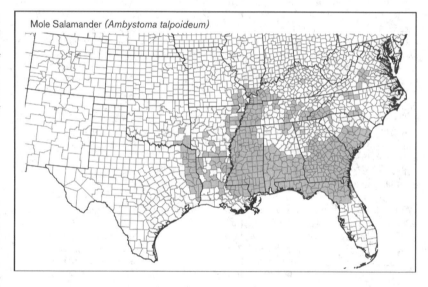

tered in Kentucky, Virginia, Tennessee, North Carolina, northern South Carolina, northern Georgia, and northern Alabama. There is no evidence that this current distribution differs substantially from the historical distribution.

2. Historical versus Current Abundance.
According to Petranka (1998) a large number of local ambystomatid populations have been lost in the conversion from mesic forest habitats to agriculture and urban areas. Clearcutting has been shown to affect mole salamander survival (Raymond and Hardy, 1991).

3. Life History Features.
  A. Breeding. Reproduction is aquatic.
  i. Breeding migrations. Breeding occurs primarily from December–March (Allen, 1932; Carr, 1940a; Gentry, 1955; Shoop,

1960; Mount, 1975; Hardy and Raymond, 1980; Walls and Altig, 1986; Trauth et al., 1993a, 1995b). Smith (1961) reported a November breeding aggregation. Variation in the timing of breeding migrations is related to annual variation in meteorological conditions (Gibbons and Semlitsch, 1991).

Neotenic adults do not migrate. Terrestrial adults migrate on rainy nights (Patterson, 1978; Semlitsch, 1981, 1985a,b, 1987). Breeding is most intense during periods of heavy, sustained rains and cold temperatures (Shoop, 1960; Petranka, 1998). Adults move along hardwood corridors, avoiding open grassy areas (Patterson, 1978).

Shoop (1960) reported that breeding of metamorphic adults lasted 7–15 d in any year. In contrast, Hardy and Raymond (1980) report that individuals remain in or near wetlands for anywhere between

8–108 d. They note that males arrive earlier and remain longer; however, both sexes leave the ponds at about the same time.

The timing of breeding appears to vary between neotenic and metamorphic animals. Peak breeding occurs earlier in neotenic animals—early November—compared with metamorphosed adults—middle November to early January . This temporal separation may produce partial reproductive isolation (see Petranka, 1998). Sperm are present within the spermathecae of metamorphosed adults from November–February (Trauth et al., 1994). These authors also mention the possibility of intermorph mating.

ii. *Breeding habitat.* In general, adults breed in forested, fishless wetlands (Semlitsch, 1988). These wetlands include a range of wetland classes—seasonal, semipermanent, and permanent (including Carolina bays)—gravel pits, and roadside ditches (see Petranka, 1998).

B. Eggs.

i. *Egg deposition sites.* Egg clusters are laid on small twigs or other submerged structures. When laying, females grasp a twig and lay eggs along its length (Shoop, 1960). Eggs are deposited at night, and females require several days to complete laying (Shoop, 1960; Krenz and Sever, 1995; see also Petranka, 1998). Most adults breed annually (Raymond and Hardy,1990; Semlitsch et al., 1993). Adults exhibit philopatry (Raymond and Hardy, 1990).

Mole salamanders from the Atlantic and Gulf Coastal Plains also differ in their mode of egg deposition. Atlantic Coastal Plain populations lay their eggs singly and scatter them throughout the wetland (Semlitsch and Walls, 1990), while Gulf Coastal Plain populations lay their eggs in small clusters.

Breeding adult numbers (100–6,000 individuals; see Gibbons and Semlitsch, 1991) are related to recent rainfall, not the number of recently produced offspring (Semlitsch, 1987; Semlitsch et al., 1996).

ii. *Clutch size.* Clutch size in mole salamanders averages higher in Atlantic Coastal Plain populations. For example, in South Carolina eight clutches from metamorphosed females ranged from 481–696 (mean=590) eggs, while clutches from seven metamorphosed females from Louisiana ranged from 208–504 (mean = 331) eggs (Raymond and Hardy, 1990). Similarly, Shoop (1960) found a range from 226–441 (mean about 300) eggs from 14 Louisiana female mole salamanders. Within regions, both clutch size and egg size increased with the size (age) of females (Semlitsch, 1985b; Semlitsch and Gibbons, 1990). Neotenic females are smaller than metamorphosed females and correspondingly lay fewer (mean = 173), smaller eggs (Petranka, 1998).

The number of eggs/cluster varies depending on the population: Bishop (1943) reports 4–20 eggs, Raymond and Hardy (1990) report both 5–50 eggs (38 clusters) and a mean of 18 eggs (107 clusters). In Arkansas, clutches from metamorphosed females contain from 12–99 (mean=41) eggs. Neotenic females lay eggs singly (Trauth et al., 1995b; see also Petranka, 1998).

C. Larvae/Metamorphosis.

i. *Length of larval stage.* Larval growth is positively related to egg size and hatchling size (Walls and Altig, 1986). Smith (1961) reports a larval period of 3–4 mo. Recently metamorphosed individuals range from 35–38 mm SVL in Louisiana populations (Raymond and Hardy, 1990). Parmalee (1993) first noted newly metamorphic animals in July; numbers peaked in September. At the Savannah River Ecology Lab (see Semlitsch, 1985b), larvae grow rapidly after hatching, begin to mature as neotenic adults by September, and can breed as 1 yr olds in December–January. Here, animals that metamorphose do so from May–September and become sexually mature a few months later. A subset of animals overwinter as larvae and metamorphose the following spring. Pond drying triggers metamorphosis in large larvae and neotenic adults (Semlitsch and Gibbons, 1985; Semlitsch and Wilbur, 1988; Semlitsch et al., 1990).

ii. *Larval requirements.*

a. Food. As with all ambystomatids, mole salamander larvae are gape-limited. They feed primarily on zooplankton when young (small) and add larger prey items to their diet as they are able to ingest them (Taylor et al., 1988). Larvae are size-selective, choosing larger prey. While feeding at night, larvae stratify (move into the water column; Anderson and Williamson, 1974). In Mississippi, Branch and Altig (1981) report the consumption of copepods, cladocerans, *Ambystoma* larvae, and ostracods. In Arkansas, neotenic individuals consume mostly midge larvae and scuds (amphipods), but will also cannibalize eggs and take the larvae of other *Ambystoma* species (McAllister and Trauth, 1996b).

b. Cover. During daylight hours, larvae remain hidden in leaf litter, vegetation, and debris on the bottom of ponds. At night, larvae leave these retreats to search for food (Anderson and Williamson, 1974).

iii. *Larval polymorphisms.* Not reported.

iv. *Features of metamorphosis.* Animals that metamorphose do so from May–September and become sexually mature a few months later. A subset of animals overwinter as larvae and metamorphose the following spring. Pond drying triggers metamorphosis in large larvae and neotenic adults (Semlitsch and Gibbons, 1985; Semlitsch and Wilbur, 1988; Semlitsch et al., 1990). In Louisiana, larval transformation occurs at 72–82 mm TL (Dundee and Rossman, 1989).

v. *Post-metamorphic migrations.* Undescribed.

vi. *Neoteny.* Populations that breed in permanent, fishless wetlands tend to be paedomorphic (see Petranka, 1998). In Arkansas, seasonal, temporary gravel pits of permanent, murky water also contain paedomorphic individuals (Trauth et al., 1993a).

D. Juvenile Habitat. Likely to be similar to adults. In contrast with adults, juveniles exhibit little agonistic behavior either towards conspecifics or heterospecifics (Walls, 1990). Most post-metamorphic growth occurs during the year following metamorphosis and prior to sexual maturity (Raymond and Hardy, 1990).

E. Adult Habitat. Adults can be metamorphosed or neotenic, and local populations can consist of either morphotype. Terrestrial adults are found in expansive floodplain forests—in areas near gum and cypress ponds (Shoop, 1960; Semlitsch, 1981). Outside the Atlantic and Gulf Coastal Plains, mole salamanders inhabit forested uplands, including mixed conifer-hardwood forests. The presence of fishless wetlands for breeding is important. Populations associated with seasonal and/or semipermanent wetlands produce terrestrial adults, populations associated with permanent wetlands produce neotenic adults (Semlitsch and Gibbons, 1985; Semlitsch et al., 1990; Scott, 1993; Heintzel and Rossell, 1995; see also Petranka, 1998).

Adult mortality varies from 16–37% (mean=26%) and is the same for males and females (Raymond and Hardy, 1990). According to Patterson (1978), 45% of adults die before leaving the breeding wetland.

F. Home Range Size. Home range size is based upon the number of activity centers within a variety of terrestrial habitats; of 22 monitored individuals, 66% of their time was spent in pine forests and 33% in hardwood forests (Semlitsch, 1981). Males and females average 5.0 (range 1–6) activity centers and 3.0 (range 3) activity centers, respectively. In juveniles, the number is 2 (range 1–3). Areas of minimal home range size for males, females, and juveniles average 3.61, 5.29, and 0.25 m^2, respectively (Semlitsch, 1981).

G. Territories. Unknown.

H. Aestivation/Avoiding Desiccation. Aestivation is unknown; animals likely avoid desiccating conditions by seeking shelter under cover objects or burrowing.

I. Seasonal Migrations. Environmental conditions have the greatest influence on seasonal migrations to breeding ponds; adults migrate only at night or shortly after rainfall (Semlitsch et al., 1993). Peak immigration is associated with the coldest period of the year in South Carolina (Semlitsch, 1985a). Emigration of breeding adults does not occur until March. Emigration of metamorphosing juveniles occurs from June–November (Semlitsch,

1985a). Average migration distance for adults is 178 m, juveniles, 47.0 m (Semlitsch, 1981, 1988).

J. Torpor (Hibernation). Unknown.

K. Interspecific Associations/Exclusions. Adult metamorphic animals breed in ponds utilized by marbled salamanders (*A. opacum*), spotted salamanders (*A. maculatum*), and central newts (*Notophthalmus viridescens louisianensis*) in Arkansas (Trauth et al., 1993a, 1995b). In Carolina bays (see Gibbons and Semlitsch, 1991) of South Carolina, several salamander and anuran species inhabit the same breeding sites as mole salamanders, including marbled salamanders, central newts, tiger salamanders (*A. tigrinum*), dwarf salamanders (*E. quadridigitata*), southern toads (*Bufo terrestris*), spring peepers (*Pseudacris crucifer*), southern chorus frogs (*P. nigrita*), little grass frogs (*P. ornata*), green frogs (*Rana clamitans*), southern leopard frogs (*R. sphenocephala*), and eastern spadefoot toads (*Scaphiopus holbrookii*; Semlitsch et al., 1996).

L. Age/Size at Reproductive Maturity. Individuals first breed at about 2 yr old, at ≥44 mm SVL (Shoop, 1960; Raymond and Hardy, 1990). In South Carolina, maturity in aquatic and terrestrial morphs is reached at 30 mm SVL (Semlitsch, 1985b).

M. Longevity. Individuals can live >9 yr (Gibbons and Semlitsch, 1991).

N. Feeding Behavior. Mole salamanders feed on a variety of invertebrates (Petranka, 1998). Gibbons and Semlitsch (1991) report zooplankton, aquatic insects, and tadpoles in their diet.

O. Predators. Semlitsch (1988) reports that bluegills (*Lepomis macrochirus*) will feed heavily on mole salamander eggs. Exposure to caged fish produces a higher rate of metamorphosis (Jackson and Semlitsch, 1993).

P. Anti-Predator Mechanisms. When attacked, metamorphosed juveniles and adults exhibit a head-down posture and expose their well-developed parotoid glands to predators. The parotoid glands secrete noxious chemicals. Mole salamanders will also lash their tails at predators (Brodie, 1977). Petranka (1998) describes head butting, biting, and body flipping followed by immobility or fleeing when artificially exposed to a smooth earth snake (*Virginia valeriae*).

Q. Diseases. Unknown.

R. Parasites. Baker (1987) reports the occurrence of nematodes. Small encapsulated nematodes were found in one of five adults in an Arkansas population (C.T. McAllister, personal communication). Upton et al. (1993) mention the absence of coccidian parasites in mole salamanders from 12 specimens in three Arkansas counties.

4. Conservation.
Mole salamanders have a core distribution centered on the Atlantic and Gulf Coastal Plains, with disjunct populations occurring peripherally. They are listed as a species of Special Concern in North Carolina, and a permit is required for all activities involving this species. Similarly, mole salamanders are classified as a species In Need of Management in Tennessee, where a permit is also required for all activities involving these animals. While the conversion of mesic forests to agricultural and urban/suburban areas has destroyed many populations of ambystomatid salamanders (Petranka, 1998), and while clearcutting reduces mole salamander numbers (Raymond and Hardy, 1991), there are no data showing that the current distribution of mole salamanders differs from their historical distribution.

Ambystoma texanum (Matthes, 1855)
SMALL-MOUTHED SALAMANDER
Stanley E. Trauth

1. Historical versus Current Distribution.
Small-mouthed salamanders (*Ambystoma texanum*) are distributed throughout the south-central United States, except the Ozark Plateau and the Louisiana Gulf Coastal Plain (Scott and Johnson, 1972; Lardie and Glass, 1975; Neck, 1980b; Saugey et al., 1986; Cochran and Cochran, 1989; Fletcher and Cochran, 1990; Conant and Collins, 1998; Petranka, 1998; Petzing and Phillips, 1998b). Populations range from southeastern Nebraska and Oklahoma, eastern Oklahoma and Texas to southeastern Michigan and northern Ohio, and south to eastern Kentucky, Tennessee, and western Alabama. Disjunct populations occur in southern Ohio and Indiana, primarily along the Ohio River. There is little evidence that the current distribution differs from the historical distribution, although interpreting this is complicated because some populations historically considered small-mouthed salamanders have been determined to be streamside salamanders (*A. barbouri*; Kraus and Petranka, 1989; see also Petranka, 1998).

2. Historical versus Current Abundance.
Typically reported as abundant, especially from observations made during the breeding season (e.g., Bragg, 1955b). According to Petranka (1998), populations of small-mouthed salamanders have been eliminated throughout their range as floodplain and mesic forests have been clearcut and converted to agricultural uses.

3. Life History Features.
A. Breeding.
i. Breeding migrations. Adult small-mouthed salamanders migrate from upland sites to breeding wetlands during warm rains from late winter to early spring (B.A. Brown et al., 1982; Petranka, 1984a; Kraus and Petranka, 1989), although adults will migrate in the absence of rains during drought conditions (Bailey, 1943). Breeding occurs earlier in the south (for breeding dates see Hay, 1892; Gloyd, 1928; Smith, 1934; Cagle, 1942; Bailey, 1943; Bragg, 1949; Ramsey and Forsyth, 1950; Smith, 1961; Brandon, 1966a; Minton, 1972, 2001; Plummer, 1977; B.A. Brown et al., 1982; Downs, 1989f; Doody, 1996).

ii. Breeding habitat. Adults breed in fishless seasonal and semipermanent wetlands including prairie potholes, forested wetlands, oxbows, ditches, borrow pits, flooded fields, and occasionally stream pools (Gloyd, 1928; Burt, 1938a; Cagle, 1942; Bailey, 1943; Ramsey and Forsyth, 1950; Petranka, 1982a, 1998; Kraus and Petranka, 1989; Collins, 1993). As with many species of ambystomatids, small-mouthed salamanders are explosive breeders.

B. Eggs.
i. Egg deposition sites. Typically in water, but Moore and Matson (1997) found eggs suspended from buttonbush (*Cephalanthus occidenatils*) branches. Females lay

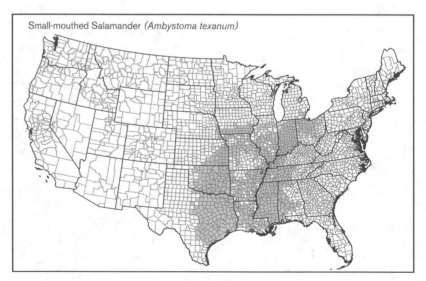

Small-mouthed Salamander (*Ambystoma texanum*)

eggs either singly or in clusters ranging from 2–15 eggs (Bailey, 1943; Minton, 1972, 2001; Petranka, 1982a; Licht, 1989; Trauth et al., 1990; see also Petranka, 1998). Surprisingly, the courtship pattern of small-mouthed salamanders has been controversial, which means it is probably variable. Working in McLean County, Illinois, Wyman (1971) observed that males combine rubbing and amplexus to stimulate females. Working in Jackson County, Illinois, Garton (1972) reports that similar to spotted salamanders (*A. maculatum*) and ringed salamanders (*A. annulatum*), small-mouthed salamander males deposit a large number of spermatophores in a small area, into which females enter and then mount the spermatophores. Garton (1972) does not observe amplexus nor males leading females over spermatophores and attributed these differences to introgression with blue-spotted salamanders (*A. laterale*). Licht and Bogart (1990) confirmed Garton's (1972) report of the absence of amplexus in small-mouthed salamanders, but rejected his hypothesis that this was due to introgression with blue-spotted salamanders.

ii. *Clutch size.* Females are reported to carry 550–700 eggs (Smith, 1934; Cagle, 1942; Burger, 1950; Camper, 1990). Eggs range from 1.6–2.5 mm. Incubation lasts from 2–8 wk, depending on temperature (Minton, 1972, 2001; see also Petranka, 1998).

C. **Larvae/Metamorphosis.**

i. *Length of larval stage.* Newly hatched larvae vary from 7–14 mm TL (Smith, 1934; Liner, 1954; Minton, 1972, 2001; see also Petranka, 1998) and metamorphose between 2–4 mo later at about 60 mm TL (Hay, 1892; Bragg, 1949; Rossman, 1960; Minton, 1972, 2001; Parmalee, 1993; see also Petranka, 1998).

ii. *Larval requirements.*

a. *Food.* Larvae are gape-limited and feed on cladocerans and ostracods when young (small) and larger prey such as insect larvae, isopods, and gastropods when older. Larvae feed dielly (day and night) and will occasionally cannibalize (Minton, 1972, 2001; Whitaker et al., 1982; McWilliams and Bachmann, 1989a). While Petranka (1998) emphasized that larvae feed on the surface of the leaf litter on the pond bottom, McWilliams and Bachmann (1989b) noted foraging in the water column, especially by smaller larvae.

b. *Cover.* During the day, larvae seek cover under leaf litter on the wetland bottom (Petranka, 1998). Kats (1988) noted that small-mouthed salamander larvae detect potential predators using olfactory cues.

iii. *Larval polymorphisms.* Unknown and unlikely per se, given the extent of the work done on small-mouthed salamanders. However, Underhill (1968b) suggests that there is a recessive gene for albinism maintained at a low frequency in central Illinois populations that produce albino eggs and larvae.

iv. *Features of metamorphosis.* Size at metamorphosis averages 48 mm TL (Bragg, 1949). Metamorphosis generally occurs from late May to July (see Petranka, 1998). Brandt (1947) reports that associative memory related to larvae feeding survives metamorphosis and is present in adults.

v. *Post-metamorphic migrations.* Within a few weeks of metamorphosis, small-mouthed salamanders migrate to cover objects on the floors of floodplain and mesic forests. They move to nearby underground sites in response to hot, dry conditions (Parmalee, 1993; see also Petranka, 1998). Juveniles do not move far from their breeding ponds, and adult small-mouthed salamanders remain closer to their breeding ponds than will other ambystomatid species (Parmalee, 1993).

vi. *Neoteny.* Unknown and unlikely.

D. **Juvenile Habitat.** According to Petranka (1998) juveniles use leafy cover objects more than will adults.

E. **Adult Habitat.** Adult small-mouthed salamanders spend most of their time in burrows in mesic forest floors, including crayfish burrows, near the breeding wetlands (Strecker and Williams, 1928; Cagle, 1942; Minton, 1972, 2001; Parmalee, 1993; see also Petranka, 1998). They surface during rainy nights. During breeding early in the spring, adults can be found beneath logs or other cover objects near the wetland margins. Adults tend to be found under larger cover objects and frequently share cover objects with conspecifics, infrequently with congenerics. Adults are sometimes unearthed by plows and excavation equipment.

F. **Home Range Size.** Data show that adults remain in the same area (maximum distance moved by a female = 20 m) for long periods (Petranka, 1998).

G. **Territories.** Adult small-mouthed salamanders do not appear to establish territories (Martin et al., 1986; Parmalee, 1993); they do not respond to substrates marked by conspecifics nor show aggression to conspecifics (see also Petranka, 1998).

H. **Aestivation/Avoiding Desiccation.** Adults respond to hot, dry conditions by burrowing underground (Parmalee, 1993; see also Petranka, 1998).

I. **Seasonal Migrations.** Adults migrate to and from breeding ponds; juveniles exhibit post-metamorphic migrations (see Petranka, 1998).

J. **Torpor (Hibernation).** Unknown.

K. **Interspecific Associations/Exclusions.** Adults share breeding wetlands with spotted salamanders, marbled salamanders (*A. opacum*), and mole salamanders (*A. talpoideum*). Adults appear to segregate by moisture levels, with small-mouthed salamanders and mole salamanders found in wetter sites. Larval numbers of small-mouthed salamanders are reduced in the presence of tiger salamander (*A. tigrinum*) larvae.

Populations of silvery salamanders (*A. platineum*) from east-central Illinois and central Indiana occur where Jefferson salamanders (*A. jeffersonianum*) are absent and small-mouthed salamanders are the sexual host (Uzzell and Goldblatt, 1967). In at least one of these populations, JJL eggs of silvery salamanders are regularly fertilized by small-mouthed salamander sperm resulting in tetraploid hybrids that are JJLT (Morris and Brandon, 1984; Spolsky et al., 1992). Phillips et al. (1997) also documented pentaploids in this population that are presumably the result of fertilization of JJLT eggs by *A. texanum* sperm (JJLTT; see Phillips and Mui's unisexual *Ambystoma* account in this volume).

L. **Age/Size at Reproductive Maturity.** Small-mouthed salamanders reach sexual maturity at 60–70 mm SVL, with a maximum size for females of 106 mm SVL (Redmer, 1995), but time to maturity is unknown (Petranka, 1998).

M. **Longevity.** Unknown.

N. **Feeding Behavior.** Small-mouthed salamander adults eat a wide variety of invertebrate prey including annelids, isopods, centipedes, arachnids, lepidopterans, coleopterans, and other insects. Breeding adults occasionally feed on aquatic invertebrates (Whitaker et al., 1982; see also Petranka, 1998).

O. **Predators.** Include predaceous aquatic insects, tiger salamander larvae (Wilbur, 1972), garter snakes (*Thamnophis* sp.), and water snakes (*Nerodia* sp).

P. **Anti-Predator Mechanisms.** In response to attacks by snakes and other predators, individual adults assume a defensive posture that includes lowering their head, curling their body, and raising and waving their tail, which secretes noxious substances (Hay, 1892; Minton, 1972, 2001; Dodd, 1977b).

Q. **Diseases.** No data are available on diseases in small-mouthed salamanders.

R. **Parasites.** Parasites of small-mouthed salamanders include a variety of protozoan and helminth species (Harwood, 1932; Walton, 1942; Landewe, 1963; Rosen and Manis, 1976; Price and St. John, 1980; Baker, 1987; McAllister and Upton, 1987b). No intraerythrocytic or trypanosomal hematozoans were observed in blood samples examined by McAllister and Upton (1987b). Protozoans, however, have been reported from the colon, rectum, gall bladder, intestinal mucosa, and feces (see studies cited above). A cyclophyllidean cestode (*Cylindrotaenia americana*) that is common in several plethodontid salamander species in the United States was also found in the small intestine of small-mouthed salamanders (McAllister and Upton, 1987b).

4. Conservation.

Small-mouthed salamanders are listed as Endangered by the State of Michigan, and legally protected under Michigan's Natural Resources and Environmental Protection Act (Levell, 1997). These populations represent the northern extreme of small-mouth salamander distribution. There are few data to suggest that their current distribution differs from their historical distribution, although undoubtedly many populations have been lost due to wetland drainage and deforestation due to agricultural practices and urban/suburban development.

Acknowledgments. Thanks to Robert Fiorentino for providing additional literature sources.

Ambystoma tigrinum (Green, 1825)
TIGER SALAMANDER

Michael J. Lannoo, Christopher A. Phillips

1. Historical versus Current Distribution.

Tiger salamanders *(Ambystoma tigrinum)* are the most widespread salamander species in North America. They occur along the Atlantic and Gulf Coastal Plains from Long Island, New York, to southeastern Louisiana and north into Mississippi, Alabama, Tennessee, and Kentucky. They are found throughout the Midwest, Great Plains, the eastern Front of the Rocky Mountains, and on the Columbia Plateau. Disjunct populations occur both in western and in Appalachian states. References to tiger salamander distributions include Engelhardt (1916a,b), Duellman (1955), Rossman (1965a), Cook (1957), Smith (1961), Cliburn (1965), Barbour (1971), Hudson (1972), Minton (1972, 2001), Harris (1975), Mount (1975), Hodge (1976), Williamson and Moulis (1979), Harker et al. (1980), Martof et al. (1980), DeGraaf and Rudis (1981, 1983), Vogt (1981), McCoy (1982), Hammerson (1986, 1999), Dixon (1987, 2000), Green and Pauley (1987), Pearson et al. (1987), Ashton and Ashton (1988), Bury and Corn (1988b), Black and Sievert (1989), Dundee and Rossman (1989), Hoberg and Gause (1989), Pfingsten and Downs (1989), Christiansen and Bailey (1991), Gibbons and Semlitsch (1991), Harding and Holman (1992), Hulse and Hulse (1992), Johnson (1992), Collins (1993), Klemens (1993), Leonard et al. (1993), Nussbaum et al. (1983), Oldfield and Moriarty (1994) McAllister (1995), Redmer and Ballard (1995), Casper (1996, 1997), Degenhardt et al. (1996), Esco and Jensen (1996), Killebrew et al. (1996), Redmond and Scott (1996), Blair et al. (1997), Dvornich et al. (1997), Luce et al. (1997), Oliver (1997), Brodman and Kilmurry (1998), Brodman (1999a), Kunzmann and Halvorson (1998), Petranka (1998), Petzing et al. (1998), Pfingsten (1998a), Arizona Game and Fish Department (1999a,b), Bartlett and Bartlett

(1999a), Fischer et al. (1999), Livo et al. (1999), McLeod (1999), Maxell (1999), Mitchell and Reay (1999), Nevada Natural Heritage Program (1999), Phillips et al. (1999), and Andrews (2000). Altitudinal variation extends from sea level to >3,300 m (Gehlbach, 1967a).

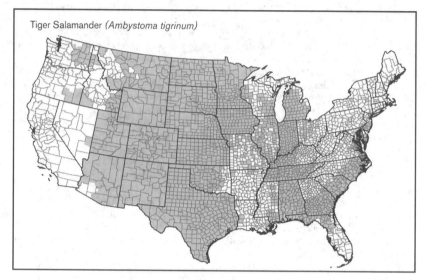

Tiger Salamander *(Ambystoma tigrinum)*

Six subspecies are currently recognized (Crother et al., 2000; see Petranka, 1998 for subspecies distribution maps): gray tiger salamanders *(A. t. diaboli* Dunn, 1940), barred tiger salamanders *(A. t. mavortium* Baird, 1850), blotched tiger salamanders *(A. t. melanosticum* Baird, 1860, see Beltz, 1995), Arizona tiger salamanders *(A. t. nebulosum* Hallowell, 1952), Sonoran tiger salamanders *(A. t. stebbensi* Lowe, 1954), and eastern tiger salamanders *(A. t. tigrinum* Green, 1825). There has been disagreement about whether or not tiger salamander subspecies deserve separate species status (Collins et al., 1980; Pierce and Mitton, 1980; Routman, 1993; Templeton et al., 1995; Shaffer and McKnight, 1996) and whether or not populations deserve subspecies status (Dunn, 1940; Gehlbach, 1967a; Jones and Collins, 1992; Jones et al., 1988, 1995). For descriptions of subspecies see Bishop (1941a), Collins (1981), Gehlbach (1967a), Lowe (1954), Stebbins (1985), Shaffer and McKnight (1996), Conant and Collins (1998), Petranka (1998), and Larson et al. (1999).

The current distribution of tiger salamanders differs from the historical distribution. Many populations within the historical distribution have been extirpated (e.g., Lannoo et al., 1994; Lannoo, 1996). Some extralimital populations, especially in the southwestern United States, have become established through activities associated with the fish bait industry (e.g., Carpenter, 1953; Espinoza et al., 1970; Collins et al., 1988; Lannoo, 1996; Petranka, 1998). The overall pattern is fragmentation and loss of populations

within their historical distribution, with sporadic, anthropogenically assisted invasions of new habitats.

2. Historical versus Current Abundance.

While remaining locally common in many regions, tiger salamander numbers have plummeted compared with historical levels (e.g., Lannoo, 1996). Lannoo et al. (1994) suggest that numbers of amphibians, including tiger salamanders, on the landscape have decreased more rapidly than has habitat loss, and in portions of the Midwest, EuroAmerican settlement has produced nearly 99% habitat (breeding wetland) loss (Leja, 1998).

Corn and Vertucci (1992) describe the role of acid rain in western tiger salamander declines. Larson (1998) examines the possible effect of agricultural pesticides in endocrine disruption in tiger salamanders.

Over a 7-yr period during the 1980s, Harte and Hoffman (1989) demonstrated a 65% decline in tiger salamander adults that led to a decline in larval recruitment. During the 1980s and early 1990s, Semlitsch et al. (1996) demonstrated a similar decline at the Savannah River Ecology Laboratory (see also Petranka, 1998). Pechmann et al. (1991) and Wissinger and Whiteman (1992) address the problem of discerning natural population fluctuations from actual declines.

3. Life History Features.

A. Breeding. Breeding is aquatic.

i. Breeding migrations. From upland overwintering sites to breeding wetlands. These migrations typically take place in March in Iowa, earlier in southern and coastal populations, and later in northern populations (Bishop, 1941a; Stine et al., 1954; Cooper, 1955; Gentry, 1955; Brandon and Bremer, 1967; Hassinger et al., 1970; Anderson et al., 1971; Peckham and Dineen, 1954; Seale, 1980; Morin, 1983a; Semlitsch,

1983a; Lannoo and Bachmann, 1984a; Downs, 1989g; Trauth et al., 1990; Lannoo, 1996). Migrations are triggered by warm spring rains and typically occur within a few weeks of ice-off in northern wetlands (Sever and Dineen, 1978; Semlitsch and Pechmann, 1985; Lannoo, 1996, 1998a).

Males migrate to breeding sites earlier than females. Semlitsch (1983a) reported males migrated 2–8 wk earlier than females at the Savannah River Ecology Laboratory site. This temporal separation is reduced at northern latitudes. Peckham and Dineen (1954) report that males and females arrive at about the same time at an Indiana site. Males tend to stay longer than females. Following breeding, Semlitsch (1983a) reported that one tagged male moved 162 m to an upland site and returned to the same site to breed the following autumn (see also Petranka, 1998).

While Sever and Dineen (1978) report that females breed annually, Pechmann et al. (1991) and Semlitsch et al. (1996) observed substantial variation in the number of breeding adults at Rainbow Bay on the Savannah River Ecology Laboratory site. Males tend to outnumber females at breeding sites, with reported male:female ratios ranging from 1:1–5.3:1 (Peckham and Dineen, 1954; Sever and Dineen, 1978; Semlitsch, 1983a).

Breeding times vary among morphotypes within populations. Rose and Armentrout (1976) report that large barred tiger salamander morphs breed from January–May, but small morphs will breed at any time of the year following sufficient rains.

Breeding populations are known to vary in size (Pechmann et al., 1991). An Indiana population has been estimated to consist of between 1,100–2,000 adults (Peckham and Dineen, 1954; Sever and Dineen, 1978; see also Petranka, 1998). In a New Jersey population, 540 breeding animals were counted (Hassinger et al., 1970; Anderson et al., 1971).

ii. Breeding habitat. Seasonal, semipermanent, and fishless permanent wetlands (Bishop, 1943; Blair, 1951c; Woodbury, 1952; Carpenter, 1953; Smith, 1961; Minton, 1972, 2001; Brandon and Bremer, 1967; Werner and McPeek, 1994; Lannoo, 1996; see also Petranka, 1998). Tiger salamanders will also breed in roadside ditches, quarry ponds, cattle tanks, subalpine lakes, and sluggish streams (e.g., Leonard and Darda, 1995; Petranka, 1998).

B. Eggs.

i. Egg deposition sites. Eastern tiger salamanders lay eggs in clusters attached to aquatic substrates such as the stems of emergent vegetation and larger detritus including submerged twigs and branches. The remaining subspecies lay eggs singly, in small clusters, or in strings. Diameters of ova are 2–3 mm (Hamilton, 1948; Collins et al., 1980; Kaplan, 1980a).

ii. Clutch size. Numbers of eggs/cluster from a number of geographically distinct populations ranged from an average of 38–59 eggs/mass (ranges from 5–122; Bishop, 1941a, 1943; Stine et al., 1954; Hassinger et al., 1970; Anderson et al., 1971; Morin, 1983a; Trauth et al., 1990).

Tiger salamanders show a wide range of clutch sizes, from an average of 421 ova reported in an eastern tiger salamander population from Michigan (Wilbur, 1977c) to 7,631 eggs from neotenic barred tiger salamander adults from Texas (Rose and Armentrout, 1976; see also Petranka, 1998).

Incubation times for eastern tiger salamanders range from 19–50 d, depending on water temperatures (Engelhardt, 1916b; Bishop, 1943; Stine et al., 1954; Sever and Dineen, 1978; Couture and Sever, 1979; Morin, 1983a; Trauth et al., 1990). Incubation times for barred tiger salamanders are 8.5 d at 25 °C (Webb and Rouche, 1971; see also Petranka, 1998). Incubation times for Arizona tiger salamanders range from 6.5 d at 19 °C and between 14–21 d at a natural wetland (Tanner et al., 1971; see also Petranka, 1998).

Hatchlings are 9–10 mm TL for barred tiger salamanders (Webb and Rouche, 1971), 9–14 mm for Arizona tiger salamanders (Tanner et al., 1971), and 13–17 mm for eastern tiger salamanders (Bishop, 1941a; see also Petranka, 1998).

C. Larvae/Metamorphosis.

i. Length of larval stage. Growth rate, length of larval period, size at metamorphosis, and duration of metamorphosis vary with environmental factors such as temperature, food level, salamander density, the presence of competitors, and wetland persistence (Wilbur and Collins, 1973; Bizer, 1978; Brunkow and Collins, 1996; Lannoo, 1996).

The larval stage lasts a minimum of 10 wk but can last longer, even within populations (Engelhardt, 1916a; Bishop, 1943; Stine et al., 1954; Hassinger et al., 1970; Sever and Dineen, 1978; Sexton and Bizer, 1978; Seale, 1980; Lannoo and Bachmann, 1984a; Trauth et al., 1990). Larvae inhabiting permanent wetlands can overwinter and metamorphose the following spring (Brandon and Bremer, 1966; unpublished observations). These animals tend to be sexually immature. Larvae may reach sexual maturity (technically becoming neotenic; see Larson et al., 1999 for a discussion) and metamorphose or not depending on a combination of genetic and environmental factors (Hensley, 1964; Larson, 1968; Dodson and Dodson, 1971; Tanner et al., 1971; Sexton and Bizer, 1978; Collins et al., 1980, 1988; Jones et al., 1993; Larson et al., 1999).

Size at metamorphosis varies from 49–75 mm SVL and may be considerably larger (≤150 mm SVL) in animals transforming after reaching sexual maturity (Dundee, 1947; Carpenter, 1953; Gehlbach, 1965, 1967b; Rose and Armentrout, 1976;

Sever and Dineen, 1978; Lannoo and Bachmann, 1984a; see also Petranka, 1998).

Tiger salamander larvae grow faster than larvae of all other *Ambystoma* species (Webb and Rouche, 1971; Rose and Armentrout, 1976; Keen et al., 1984; see also Petranka, 1998).

ii. Larval requirements.

a. Food. Tiger salamander larvae are gape-limited, size selective feeders. They remain size selective in the absence of visual and olfactory cues, suggesting that nocturnal feeding is mediated by lateral line mechanoreceptive and electroreceptive organs (Lannoo, 1986, 1987). Larvae are generalists, feeding on a wide range of invertebrate prey items including zooplankton, ostracods, aquatic insect larvae and adults, mollusks, oligochaetes, leeches, and crayfish, as well as anuran tadpoles, small fishes, and conspecifics (Little and Keller, 1937; Gehlbach, 1965; Black, 1969b; Buchli, 1969; Dodson, 1970; Dodson and Dodson, 1971; Wilbur, 1972; Rose and Armentrout, 1976; Sever and Dineen, 1978; Brophy, 1980; Zaret, 1980; Collins and Holomuzki, 1984; Lannoo and Bachmann, 1984a; Leff and Bachmann, 1986; Holomuzki and Collins, 1987; Zerba and Collins, 1992; Brunkow and Collins, 1996). Smaller larvae eat smaller prey. While larger larvae eat larger prey, they will also air gulp to fill their lungs and acquire buoyancy ("stratify" *sensu* Branch and Altig, 1981; see also Lannoo and Bachmann, 1984b) and feed on pelagic zooplankton (Rose and Armentrout, 1976; Lannoo and Bachmann, 1984b) and other smaller prey (Lee and Franz, 1974; Brophy, 1980; Tyler and Buscher, 1980). The kinematics of prey capture by *Ambystoma* larvae were described by Hoff et al. (1985), Shaffer and Lauder (1985), and Reilly et al. (1992).

Cannibal morph larvae occur in tiger salamanders (see "Larval polymorphisms" below). Cannibal morphs tend to eat larger prey than typical morphs. In populations of barred tiger salamanders, cannibal morphs may preferentially take conspecifics (Rose and Armentrout, 1976; Collins and Holomuzki, 1984; Holomuzki and Collins, 1987; Pfennig et al., 1994; Whiteman et al., 2003), while cannibal morph eastern tiger salamanders take conspecifics as a component of a generally broader diet (Lannoo and Bachmann, 1984a; Loeb et al., 1994). Large typical larvae may also be cannibalistic (Crump, 1983; Lannoo et al., 1989).

The diets of neotenic adults generally resemble that of large larvae (Sprules, 1972; Norris, 1989).

b. Cover. Larvae move vertically and horizontally within wetlands. Horizontal diel movements are thought to be related to temperature regulation (Prosser, 1911; Whitford and Massey, 1970; Heath, 1975; Holomuzki and Collins, 1983; Holomuzki, 1989b; see also Petranka, 1998). Larvae show site preferences, moving to the same general vicinity of the littoral zone every day following nocturnal migrations to deeper water (Gehlbach, 1967b; see also Petranka, 1998). Causes underlying vertical migrations are more complicated. Collins and Cheek (1983) and Holomuzki (1989b) observed larvae moving into the open water, pelagic zone, during the day and retreating to the pond bottom at night. These movements are thought to be related to thermoregulation and/or predator (dytiscid beetle) avoidance. Tiger salamander larvae, similar to other wetland-dwelling *Ambystoma*, migrate off the bottom and into the pelagic zone at night to feed on zooplankton (Anderson and Graham, 1967; Branch and Altig, 1981; Brophy, 1980; Lannoo and Bachmann, 1984b).

iii. Larval polymorphisms. Cannibal morphs, animals with proportionally large heads and large teeth, have been reported in nature from three of the six tiger salamander subspecies: barred tiger salamanders, Arizona tiger salamanders, and a single metapopulation of eastern tiger salamanders from northwestern Iowa (Powers, 1907; Rose and Armentrout, 1976; Lannoo and Bachmann, 1984a; Pierce et al., 1981, 1983; Reilly et al., 1992; Whiteman et al., 2003; see also Larson et al., 1999). Cannibal morphs in Sonoran tiger salamanders and in eastern tiger salamanders from Indiana have been induced in the laboratory (J.P. Collins and D. Pfennig, respectively, personal communication; see also Larson et al., 1999).

In nature, populations with the genetic capacity to express cannibal morphs may not do so every year (Collins and Cheek, 1983; Lannoo and Bachmann, 1984a; Pfennig et al., 1991). In these populations, conspecific density appears to trigger the expression of cannibal morphs (Collins and Cheek, 1983; Lannoo and Bachmann, 1984a; Pfennig et al., 1991, 1994; Pfennig and Collins, 1993; Maret and Collins, 1994).

Interestingly, the tooth morphology differs between cannibal morphs from eastern and barred tiger salamander populations, suggesting either separate evolutionary inventions or divergence from a common ancestor (Larson et al., 1999). Intermediate forms between cannibal and typical morphs appear under conditions that produce fully developed cannibal morphs (Pedersen, 1993; Larson et al., 1999).

iv. Features of metamorphosis. Metamorphosis can, but does not always, involve large numbers of individuals (M.J.L., personal observations). Semlitsch (1983a) noted that numbers of newly metamorphosed animals/breeding female ranged from 0–24.

v. Post-metamorphic migrations. From breeding wetlands to upland sites. In some populations newly metamorphosed animals remain near wetlands, in others

they migrate some distance. The longest recorded distance moved from a wetland is 229 m (Gehlbach, 1967b). These emigrations tend to occur during rainy nights (Sever and Dineen, 1978; personal observations); when wetlands occur near highways, mortality can be high (Duellman, 1954a; Lannoo, 1996).

vi. Neoteny. Known in eastern tiger salamanders from populations in Michigan (Hensley, 1964; Larson, 1968; Collins et al., 1980; Jones et al., 1993), Wisconsin (G.S. Casper and M. Mossmann, personal communication), Illinois (Whiteman et al., 1998), and in each of the western subspecies (Powers, 1903; Burger, 1950; Glass, 1951; Knopf, 1962; Buchli, 1969; Sprules, 1974a; Wiedenheft, 1983; Whiteman, 1994c; Whiteman and Howard, 1998; Petranka, 1998; Larson et al., 1999; see also Gould, 1977). A neotenic, cave dwelling population of tiger salamanders occurs in New Mexico (Thompson and Jones, 1992).

D. Juvenile Habitat. Juveniles may remain in or near wetlands to feed or may migrate to upland sites and burrow.

E. Adult Habitat. Tiger salamander adults occupy a wide variety of habitats, from the moist lowland coastal plains of the southeastern United States to the arid Great Plains, desert southwest, and altitudes in the Rocky Mountains (Burger, 1950; Gehlbach, 1969; Webb, 1969; Webb and Rouche, 1971). Tiger salamanders are tolerant of agriculture (but see Larson, 1998) and are the most common salamanders throughout the Midwest. They tend to be associated with grasslands, savannas, and woodland edges, and less so with forests.

Adult tiger salamanders can be terrestrial or aquatic (neotenic). Terrestrial adults burrow and require deep friable soils. Tiger salamanders actively burrow by using their forelimbs (Gruberg and Stirling, 1972; Semlitsch, 1983c). Animals tend to live near the surface (12 cm deep; Semlitsch, 1983b) but can be found deeper; Gehlbach (1965) found an animal 2 m below the soil surface. Other animals often live in or are associated with mammal burrows and are active on the surface during rainy nights (Hamilton, 1946; Calef, 1954; Duellman, 1954a; Carpenter, 1955; Gehlbach, 1967a; Collins et al., 1993).

Aquatic, neotenic adults generally require fishless permanent wetlands, where they are the top aquatic carnivores. These animals are threatened during droughts and may or may not be able to metamorphose; with increasing age, neotenic adults become less able to transform.

F. Home Range Size. Unknown for either terrestrial or neotenic adults.

G. Territories. Territories have not been documented in either terrestrial or neotenic adults.

H. Aestivation/Avoiding Desiccation. During dry conditions, terrestrial tiger salamanders can burrow deeper into moister

soil. One tiger salamander was found 2 m below the soil surface (Gehlbach, 1965).

Neotenic adults must metamorphose to avoid drying wetland conditions. There is no evidence that they have the capacity to burrow into the mud and aestivate.

I. Seasonal Migrations. Aside from migrations to and from breeding sites by terrestrial adults (see "Breeding migrations" and "Post-metamorphic migrations" above), no other migrations have been documented.

J. Torpor (Hibernation). Terrestrial adults overwinter by burrowing below the frost line. Aquatic adults overwinter in permanent, sometimes in semi-permanent, bodies of water.

K. Interspecific Associations/Exclusions. Predation by neotenic adults eliminates or controls cladocerans (*Daphnia pulex*), fairy shrimp (*Branchinecta shantzi*), and other invertebrates in subalpine lakes in Colorado (Sprules, 1972; see also Petranka, 1998) and arid wetlands (Holomuzki et al., 1994). Tiger salamander larvae lower survivability of larvae of syntopic *Ambystoma*, including blue-spotted salamanders (*A. laterale*), small-mouthed salamanders (*A. texanum*), and Tremblay's salamanders (*A. tremblayi*).

Egg mortality by eastern newts (*Notophthalmus viridescens*) can be sufficient to exclude tiger salamanders from wetlands (Morin, 1983a; see also Petranka, 1998). However, eastern newts and tiger salamanders co-exist in ponds, including in southern Illinois, where they have similar diets (Brophy, 1980).

L. Age/Size at Reproductive Maturity. Terrestrial adults can reach 35 mm TL (see also Petranka, 1998), with males either similarly sized or slightly smaller than females. Males tend to have longer tails. Semlitsch (1983a) noted that 6 of 1,041 marked post-metamorphic animals returned to breed 2 yr later; the following year, 52 marked animals were observed (Glass, 1951).

M. Longevity. Nigrelli (1954) reports that in captivity, neotenic adults live as long as 25 yr; terrestrial adults live for 16 yr (see also Petranka, 1998).

N. Feeding Behavior. Terrestrial adults feed upon a variety of invertebrates associated with the soil and the soil surface. These include annelids and insect larvae and adults. Adults are also known to feed on small vertebrates, including field mice (*Peromyscus* sp.; Bishop, 1941a) and a hatchling six-lined racerunner (*Cnemidophorus sexlineatus*; Camper, 1986). Captive adults will eat a wide variety of food items, including small frogs, snakes, and lizards (Duellman, 1948; Camper, 1986), that they may or may not eat in nature (Camper, 1986).

Non-breeding terrestrial adults and newly metamorphosed animals will forage in fishless wetlands (Whiteman et al., 1994; personal observations). Adult barred tiger salamanders from high elevation ponds can be terrestrial or may move to wetlands to feed. Terrestrial adults in permanent ponds often move to semipermanent ponds where competition is reduced and food density higher (Whiteman et al., 1994; see also Petranka, 1998).

O. Predators. A large number of predators are known to feed on tiger salamander eggs, larvae, and adults, including insects such as caddisflies (Dalrymple, 1970), dragonfly naiads, predaceous diving beetles (Holomuzki, 1985a,b, 1986a), and giant water bugs; amphibians such as eastern newts (Morin, 1983a), conspecifics (cannibal morphs and cannibalistic typical morphs), and marbled salamander *(A. opacum)* larvae; snakes including garter snakes (*Thamnophis* sp.) and eastern hog-nosed snakes (*Heterodon platirhinos)*; a variety of predatory birds, shorebirds, and wading birds including killdeer, bitterns, grackles, loggerhead shrikes (Jensen, 2003), gray jays, kingfishers, great blue herons, and egrets; and mammals including badgers (Long, 1964), bobcats, raccoons, coyotes, opossums, and humans (see also Petranka, 1998).

P. Anti-Predator Mechanisms. Tiger salamander larvae and neotenic adults exhibit a fast start movement that permits escape in aquatic environments. Terrestrial adults will assume a defensive posture by raising their hind legs and arching and waving their tail (Brodie, 1977; Smith, 1985). As with other *Ambystoma* species, terrestrial tiger salamander adults produce secretions from granular skin glands along the dorsal surface of their tail. These secretions have two components: a sticky component that adheres to a potential predator and can encumber their movement and a toxic/noxious component that can repel, even kill, predators (Brodie and Gibson, 1969; Arnold, 1982; Brodie, 1983; Evans, 1993; Hamning et al., 2000). Hamning et al. (2000) demonstrate that the adhesive component of these glands is supplied by disulfide linkages, and that the noxious component is provided by neurotoxins. Hamning et al. (2000) speculate that at least two neurotoxins are present: one that binds to a protein, inhibits neurotransmission, and is reversible, and a second that is either a protease or lipase, causes cellular damage, and is effective over a longer time period.

Larvae tend to be nocturnally active in warmer climates and temperatures (see Petranka, 1998), thus avoiding predation by visual predators. Barred tiger salamander larvae move into open water at night to avoid being eaten by predaceous diving beetles (*Dytiscus dauricus*; Holomuzki, 1986a; see also Petranka, 1998). Small larvae do not move to areas to avoid larger cannibal morph larvae (Holomuzki, 1986b; Skelly, 1992).

Q. Diseases. The parasite loads cannibal morph larvae acquire (see "Parasites" below) may make them more susceptible to disease than typical morphs (Pfennig et al., 1991). In wetlands, densities of cannibal morphs are inversely related to disease incidence.

In desert and grassland populations, epizooic bacteria (*Acinetobacter* sp.) blooms, perhaps caused by accumulations of livestock, are thought to be responsible for mass die-offs of tiger salamanders (Worthylake and Hovingh, 1989). Red-leg outbreaks can occur under stressful conditions. One outbreak was caused by a zooplankton (prey) crash following heavy siltation due to a nearby construction project, a failure to erect erosion fences, and a midwestern thunderstorm—the only year that red-leg disease has been observed to be pervasive and fatal in this population (M. J. L., personal observation).

Chytridiomycosis transmission and pathogenicity in desert populations of tiger salamanders is detailed by Davidson et al. (2003).

R. Parasites. Presumably due to their diet of conspecifics, cannibal morph larvae carry more parasitic nematodes than do typical larvae.

Larvae in some populations serve as hosts to a large number of leeches, but it is uncertain whether leeches cause direct mortality (Carpenter, 1953; Holomuzki, 1986c; see also Petranka, 1998).

4. Conservation.
As with many amphibian species that breed in semipermanent wetlands, larval tiger salamanders often experience mass mortality associated with pond drying (Sever and Dineen, 1978; Lannoo, 1998a).

This is a natural phenomenon. The biggest threat to existing tiger salamander populations comes from continued wetland destruction and wetland alteration through aquacultural activities such as those promoted officially by the states of Iowa and Wisconsin (Lannoo et al., 1994; Lannoo, 1996). Introduced fishes have long been known to reduce or eliminate tiger salamander populations (Carpenter, 1953; Espinoza et al., 1970; Collins et al., 1988). To quote Petranka (1998): "The ecological effects of fish introductions on native amphibians should be carefully considered by fish and wildlife managers when deciding whether or not to stock natural, fish-free habitats."

According to Petranka (1998), deforestation is also a problem in southeastern and midwestern populations. Northeastern populations are affected by acid deposition. Tiger salamander adults avoid waters with a low pH; at a pH of 4.2, 50% of embryos suffer mortality (Whiteman et al., 1995). At a low pH, tiger salamanders also experience reduced growth and longer larval periods (Kiesecker, 1996).

Tiger salamanders are listed as Endangered in Delaware, New York, New Jersey, and Maryland; Protected in Arizona; and Of Special Concern in North Carolina and South Carolina.

Ambystoma tremblayi Comeau, 1943
TREMBLAY'S SALAMANDER

Christopher A. Phillips, Jennifer Mui

See C. Phillips and J. Mui's account "Unisexual members of the *Ambystoma jeffersonianum* complex" for a full consideration of *Ambystoma tremblayi, Ambystoma platineum,* and unisexual hybrid ambystomas.

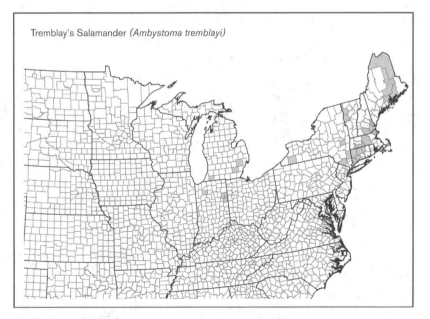

Tremblay's Salamander *(Ambystoma tremblayi)*

Unisexual members of the *Ambystoma jeffersonianum* complex

Ambystoma platineum Cope, 1868 "1867"
SILVERY SALAMANDER

Ambystoma tremblayi Comeau, 1943
TREMBLAY'S SALAMANDER

And other hybrids
Christopher A. Phillips, Jennifer Mui

The occurrence of all-female morphs of salamanders associated with Jefferson salamanders (*Ambystoma jeffersonianum*) and blue-spotted salamanders (*A. laterale*) has been recognized for more than 50 yr (see Uzzell, 1964 for a review). Uzzell (1964) was able to identify two distinct forms of these all-female morphs based on the same morphological characters that distinguish the two diploid species and suggested that the all-female morphs resulted from a post-Wisconsin hybridization between *A. jeffersonianum* and *A. laterale*. He resurrected the name *A. tremblayi* (Tremblay's salamanders) for the smaller species that is most often found associated with *A. laterale* and the name *A. platineum* (silvery salamanders) for the larger one found most often with *A. jeffersonianum*. Biochemical and karyological studies demonstrated that *A. platineum* has two chromosome sets from *A. jeffersonianum* and one from *A. laterale* (JJL); *A. tremblayi* has the reverse, one chromosome set from *A. jeffersonianum* and two from *A. laterale* (JLL; Uzzell and Goldblatt, 1967; Sessions, 1982). Uzzell (1963) demonstrated that the all-female morphs and their offspring are triploid and, based on observations of lampbrush chromosomes, Macgregor and Uzzell (1964) proposed pre-meiotic chromosomal duplication and gynogenesis as a mechanism to explain the perpetuation of the all-female triploid lineages. Under this hypothesis, sperm from spermatophores

deposited by males of *A. laterale* and *A. jeffersonianum* activate the unreduced eggs of *A. tremblayi* and *A. platineum*, respectively, but normally do not fertilize them.

Several exceptions to this scenario have been reported, including (1) Populations of silvery salamanders from east-central Illinois and central Indiana where Jefferson salamanders are absent and small-mouthed salamanders (*A. texanum*) are the sexual host (Uzzell and Goldblatt, 1967). In at least one of these populations, JJL eggs of silvery salamanders are regularly fertilized by small-mouthed salamander sperm resulting in tetraploid hybrids that are JJLT (Morris and Brandon, 1984; Spolsky et al., 1992). Phillips et al. (1997) also documented pentaploids in this population that are presumably the result of fertilization of JJLT eggs by *A. texanum* sperm (JJLTT). (2) Unisexual tetraploid and pentaploids that result from fertilization rather than simple activation of unreduced triploid ova (Servage, 1979; Bogart, 1989; Lowcock et al., 1991; Lowcock, 1994). (3) Diploid hybrids that have one genome from each of the parental species, blue-spotted salamanders and Jefferson salamanders (Bogart, 1989; Spolsky et al., 1992; Bogart and Klemens, 1997). (4) Diploid and triploid hybrids between small-mouthed salamanders and blue-spotted salamanders in northwestern Ohio and southeastern Michigan (Downs, 1978; Kraus, 1985b). This account deals with all but the last group of populations.

Unisexual members of the *A. jeffersonianum* complex can only be reliably identified using techniques such as karyology or protein electrophoresis. Depending on the geographic locality, some unisexual salamanders may be identified using morphological characters such as snout-vent length, inter-narial distance, head width, and dentition (Uzzell, 1964; Kraus, 1985b; Lowcock et al., 1992; Spolsky et al., 1992).

Partial identification (ploidy only) can be achieved by measuring the area of red blood cells or the total amount of DNA/cell (Uzzell, 1964; Lowcock et al., 1991). Because these methods are not widely available, much of the literature does not differentiate among members of the complex and the few studies that attempt to differentiate are usually not explicit concerning the methods they employed. However, studies referring to Jefferson salamanders and blue-spotted salamanders are likely based in part on the unisexuals and, because the unisexuals are derived from these two species, much of the life history is similar.

1. Historical versus Current Distribution.
Because of the potential confusion among members of the complex, no reliable historical distribution data exist for the unisexual *Ambystoma* salamanders, and even our knowledge of their current distribution is incomplete. In the United States, silvery salamanders are distributed sporadically throughout east-central Illinois, Indiana, the northwestern 3/4 of Ohio, extreme southern Michigan, Connecticut, Massachusetts, New York, and northern New Jersey (Uzzell, 1964; Uzzell, 1967a,b; Morris and Brandon, 1984; Nyman et al., 1988; Downs, 1989e; Bogart and Klemens, 1997; Petranka, 1998). In 1968, Creusere (1971) reported an individual identified by internarial distance and blood cell size as a silvery salamander from northern Kentucky. As noted above, tetraploid and pentaploid hybrids resulting from the fertilization of silvery salamander eggs by small-mouthed salamander sperm are known from east-central Illinois, and central and south-central Indiana (Uzzell, 1964; Morris and Brandon, 1984; Spolsky et al., 1992).

In the United States, Tremblay's salamanders are known from locations in Maine, Vermont, New Hampshire, Connecticut, Massachusetts, New York, northern New Jersey, southern Michigan, northern Indiana, and Wisconsin. (Uzzell, 1967a,b; Vogt, 1981; Nyman et al., 1988; Downs, 1989h; Bogart and Klemens, 1997; Petranka, 1998).

2. Historical versus Current Abundance.
Phillips et al. (1997) reported on abundance of breeding adult silvery salamanders and silvery salamander–small-mouthed salamander (*A. platineum* x *A. texanum*) hybrids over 4 yr at a pond in east-central Illinois. They used a drift fence and mark–recapture models to estimate breeding population size in a fifth year at the same site (Phillips et al., 2001).

3. Life History Features.
   A. Breeding. Reproduction is aquatic.
   i. Breeding migrations. Unisexuals require the spermatophores of male Jefferson salamanders, blue-spotted salamanders, small-mouthed salamanders, or tiger

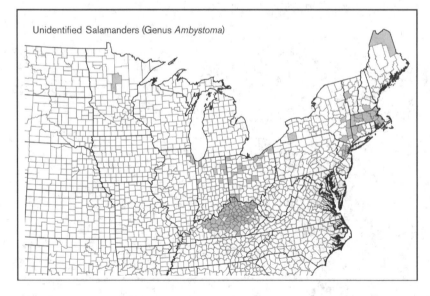

Unidentified Salamanders (Genus *Ambystoma*)

salamanders *(A. tigrinum)* to trigger gyno-genetic reproduction (i.e., sperm are used to activate cleavage of the egg but do not penetrate the egg or contribute genetic information) or fertilize their eggs. Therefore, the timing of unisexual breeding corresponds to breeding in these other ambystomatids and occurs in late winter or early spring—shortly after ice-off—during or following precipitation events. Uzzell (1969) observed breeding of silvery salamanders on 25 February in northern Ohio. Collins (1965) reported breeding of silvery salamanders in mid-February in southern Ohio. In east-central Illinois, Phillips et al. (2001) recorded breeding of adult silvery salamanders and silvery salamander–small-mouthed salamander hybrids as early as 11 February. Nyman (1991) reported 13 March as the date of the first large migration of silvery salamanders in New Jersey.

In southeastern Michigan, Clanton (1934) and Uzzell (1964) recorded breeding of Tremblay's salamanders from mid-March to early April. Lowcock et al. (1991) reported first breeding dates ranging from 3 April–28 April for Tremblay's salamanders at six breeding ponds in central Ontario.

ii. Breeding habitat. Fishless ponds in a variety of wooded and semi-wooded habitats including ponds, wetlands, ditches, and sloughs.

B. Eggs.

i. Egg deposition sites. Eggs of silvery salamanders are typically laid on the pond bottom or attached to submerged sticks (Uzzell, 1964; Morris and Brandon, 1984). Two silvery salamanders from Lorain County, Ohio, laid an average of 129.7 eggs; two from Franklin County, Massachusetts, averaged 156.6 eggs (Uzzell, 1964).

Eggs of Tremblay's salamanders are laid in small bunches (Clanton, 1934) attached to twigs, leaves, or stems (Wilbur, 1977c).

ii. Clutch size. Silvery salamanders from Illinois laid from 124–207 eggs in masses of 2–50 eggs/mass, each mass measuring 11–17 mm by 16–40 mm (Morris and Brandon, 1984). Downs (1989e) reported on a female silvery salamander that laid 137 eggs in a laboratory aquarium. The eggs were deposited in seven masses, each containing from 8–31 eggs; egg diameter ranged from 2.7–2.8 mm (n=10). Nyman (1991) reported an embryonic period of 28 d for silvery salamander eggs in a New Jersey pond. Under laboratory conditions, Panek (1978) demonstrated that eggs of silvery salamanders from a New Jersey pond developed from 6–37% faster than those of Jefferson salamanders, depending on incubation temperature.

Phillips et al. (1997) set up crosses between tetraploid silvery salamander–small-mouthed salamander hybrids and male small-mouthed salamanders from Illinois using the mating cage design of Uzzell

(1964). They found that seven females laid an average of 110.8 eggs each (range=70–137). Hatching success for the eggs of these crosses averaged 35.6% compared to 61.7% for the eggs of control crosses involving small-mouthed salamander males and females.

The number of dissected ovarian eggs/female has been reported several times for Tremblay's salamanders in Washtenaw County, Michigan; Clanton (1934) reported an average of 142 freshly dissected ovarian eggs/female (range=50– 216) for 66 females, Uzzell's (1964) average was 156.2 for 63 females, and Wilbur (1977c) reported an average of 136.2 ovarian eggs/female for 21 females. Wilbur (1971) also reported an average of 4.8 eggs/mass for 200 masses from the same area. The diameter of the eggs in Clanton's (1934) study averaged 2.1 mm with a range of 1.8–2.2 mm. Females usually deposit their eggs within 48 hr of insemination. Uzzell (1964) confined various combinations of Tremblay's salamanders from Washtenaw County, Michigan, and male Jefferson salamanders and blue-spotted salamanders in mating cages in a natural pond. He found that the embryonic period for these crosses lasted approximately 1 mo. Lowcock et al. (1991) set up "test crosses" between male blue-spotted salamanders and female blue-spotted salamanders, Tremblay's salamanders, and LLLJ. No details concerning the environmental conditions of the crosses are given. Two female blue-spotted salamanders laid 153 and 212 eggs, two female Tremblay's salamanders laid 167 and 143 eggs, and one LLLJ laid 84 eggs. Hatching success for these clutches was 75.8%, 72.6%, 61.1%, 68.5%, and 20.2%, respectively.

C. Larvae/Metamorphosis.

i. Length of larval stage. In New Jersey, Nyman (1991) found that silvery salamander larvae that hatched from 12 April–20 April metamorphosed from 21 June to late July. In the same study, the mean SVL of 14 larvae collected approximately 18 d post hatching was 9.9 mm. Phillips et al. (2002) documented transformation of silvery salamander and JJLT tetraploids in 2 yr in east-central Illinois. In 1997, transformation occurred from 25 June–20 July at a mean SVL of 31.4 mm; in 1998, it occurred from 28 June–4 August at a mean SVL of 39.3 mm.

Larvae of Tremblay's salamanders from Uzzell's (1964) mating cage experiments in southeastern Michigan metamorphosed under laboratory conditions approximately 100 d after the first hatchlings appeared. Wilbur (1971) followed the metamorphosis of larval Tremblay's salamanders from southeastern Michigan in window screen pens placed in a natural pond. The mean larval period for 134 individuals was 94.2 d compared to 91.5 d for larvae of blue-spotted salamanders raised under similar conditions. The mean

SVL of the newly metamorphosed Tremblay's salamander larvae was 34.1 mm (using the conversion formula provided by Wilbur). Lowcock (1994) reported that juvenile Tremblay's salamanders and tetraploids (presumably LLLJ) metamorphosed from late July to early October from a pond in central Ontario. Mean SVL at metamorphosis was 37.3 mm (range= 31.9–41.4; n=9) for Tremblay's salamanders and 38.5 mm (range=37.2–40.0; n=80) for tetraploids. They also reported a mean larval period of 75.7 d for 129 Tremblay's salamanders raised in the lab compared to 90.2 d for 22 tetraploids. Mean SVL was 34.6 mm (range=33.4–37.1; n=139) for Tremblay's salamanders and 33.7 mm (range=33.4–37.8; n=21) for the tetraploids.

ii. Larval requirements.

a. Food. Nyman (1991) examined the stomach contents of 124 larval silvery salamanders from 30 April–14 July at a New Jersey pond; the main prey items found (by frequency) were cladocerans, copepods, ostracods, isopods, and insect adults and larvae. Chaoborids were the most important food items by volume. Less important prey items included snails, water mites, leeches, volocine algae, and other *Ambystoma* larvae (Petranka, 1998). No data are available concerning the diets of the other unisexual *Ambystoma*, but they are thought to be similar to silvery salamanders (Petranka, 1998).

b. Cover. No data are available for the unisexual salamanders.

iii. Larval polymorphisms. Cannibalistic morphs have not been reported.

iv. Features of metamorphosis. As stated above, two studies in different geographic regions found that silvery salamander larvae metamorphosed from late June to late July.

v. Post-metamorphic migrations. No data available.

vi. Neoteny. Neoteny is unknown among unisexual *Ambystoma*.

D. Juvenile Habitat. No data are available for unisexual *Ambystoma*.

E. Adult Habitat. Silvery salamanders are found in upland forests as well as bottomland forests (Downs, 1989e). Tremblay's salamanders are primarily found in wooded areas with sandy soils, remaining near the surface most of the year until late autumn (Downs, 1989h).

F. Home Range Size. No data are available for unisexual *Ambystoma*.

G. Territories. Although territoriality has not been explicitly demonstrated in the wild for any of the unisexuals, Ducey (1989) showed that resident Tremblay's salamanders bit conspecific intruders in 80% of encounters staged in the laboratory. Ducey and Heuer (1991) extended these observations by showing that the level of aggression exhibited by resident Tremblay's salamanders was negatively correlated with food availability.

H. Aestivation/Avoiding Desiccation. No data on aestivation are available for unisexual *Ambystoma*; animals likely avoid desiccating conditions by seeking shelter under cover objects or burrowing.

I. Seasonal Migrations. No seasonal movements have been documented other than movement associated with breeding migrations (see above).

J. Torpor (Hibernation). No data are available for unisexual *Ambystoma*, but they probably overwinter underground in rodent tunnels and crayfish burrows.

K. Interspecific Associations/Exclusions. In the majority of populations, unisexual salamanders require the spermatophores of one of the parental, bisexual species either to activate cleavage (Uzzell, 1964) or fertilize eggs. Uzzell (1964) and Uzzell and Goldblatt (1967) set up crosses in aquaria in the lab or in mating cages in a pond and found that blue-spotted salamander and Jefferson salamander males were more likely to court conspecific females than the triploid species most closely related to them. Uzzell (1964) repeated these courtship experiments using the number of spermatophores deposited by males as a measure of their discriminatory ability. He found that Jefferson salamander males deposited 48.3% fewer spermatophores when confined with silvery salamanders than with female Jefferson salamanders; blue-spotted salamander males deposited 42.8% fewer spermatophores for Tremblay's salamanders than for female blue-spotted salamanders.

In the Illinois population of silvery salamanders where Jefferson salamanders are absent and small-mouthed salamanders are the sexual host, Phillips et al. (1997) used the same mating cage design to cross male small-mouthed salamanders with silvery salamanders and silvery salamander–small-mouthed salamander hybrids. They found that male small-mouthed salamanders deposited significantly more spermatophores in crosses with conspecific females than with either of the other two females. The number of spermatophores deposited for the hybrids was greater than that for silvery salamanders, but the difference was not significant at the 0.05 level.

Morris and Brandon (1984) set up artificial insemination experiments using salamanders from the same population and found that no embryos of crosses between silvery salamanders and spotted salamanders (*A. maculatum*) hatched; 4.1% of eggs from crosses between small-mouthed salamanders and silvery salamanders hatched.

Dawley and Dawley (1986) used a flow-through Y-maze to investigate whether male Jefferson salamanders could differentiate chemically between conspecific females and silvery salamanders. Eleven of 12 males chose the arm of the Y with the conspecific female.

L. Age/Size at Reproductive Maturity. Silvery salamanders and silvery salamander–small-mouthed salamander hybrids as small as 72 mm SVL were found migrating to breeding ponds in east-central Illinois (Morris, 1981). Uzzell's (1964) smallest sexually mature female was 72.7 mm SVL.

Uzzell (1964) found a sexually mature Tremblay's salamander that was 55.7 mm SVL. Downs (1989h) speculated that some individuals reach maturity at 2 yr, while others mature at 3 yr.

M. Longevity. Collins (1965) documented that 18% of the adult Jefferson salamanders and silvery salamanders that were marked in 1962 at a pond in southern Ohio returned to breed in 1965. Morris (1981) speculated that silvery salamanders and silvery salamander–small-mouthed salamander hybrids with an SVL of 99–104 mm were 7–10-yr old.

N. Feeding Behavior. The diet of adults has not been documented specifically, but it is probably similar to other *Ambystoma* of the eastern United States and therefore includes beetles, centipedes, worms, slugs, and other invertebrates (Phillips et al., 1999).

O. Predators. Known predators of larvae include other *Ambystoma* larvae. Adults returning to breeding pools were killed by raccoons (Morris, 1981). A newly transformed silvery salamander from the Illinois population was observed being consumed by a fishing spider (*Dolomedes* sp.; J.E. Petzing, personal communication). Newly transformed silvery salamanders were observed at a drift fence that were partially eaten, probably by raccoons, skunks, and short-tailed shrews, which were all abundant in the area.

P. Anti-Predator Mechanisms. Unisexual members of the complex show the same defense displays as the bisexual members of the complex (Uzzell, 1967a,b).

Q. Diseases. No data are available for unisexual *Ambystoma*.

R. Parasites. Morris (1981) found an unidentified fluke in the large intestine of a tetraploid silvery salamander–small-mouthed salamander hybrid.

4. Conservation.
Because of their genetic complexity, these animals are not accommodated under either biological or evolutionary species concepts (see Minton, this volume, Part One; Goebel, this volume, Part One). The conflicts surrounding the nomenclature of the unisexual hybrids and the papers arguing for the removal of formal binomial epithets (e.g., Lowcock et al., 1987) have resulted in much confusion and, unfortunately, the loss of protected status in several states. However, as of 2002, they continue to be recognized as animals Of Special Concern in Connecticut (http://dep.state.ct.us), and as Endangered (*A. platineum*) in Illinois (http://dnr.state.il.us) and (*A. tremblayi*) New Jersey (www.state.nj.us).

Family Amphiumidae

Amphiuma means Garden, 1821
TWO-TOED AMPHIUMA

Steve A. Johnson, Richard B. Owen

1. Historical versus Current Distribution.
The range of two-toed amphiumas (*Amphiuma means*) includes the Gulf and Atlantic Coastal Plains from about New Orleans to southeastern Virginia and all of Florida (e.g., Salthe, 1973a; E.E. Brown, 1992; Conant and Collins, 1998; Petranka, 1998). Two-toed amphiumas are absent from the Florida Keys (Duellman and Schwartz, 1958). There is no evidence to suggest that recent historical and current distributions differ. However, fossil evidence indicates that amphiumids were present by the late Cretaceous, and fossils identified as *A. means* have been found at

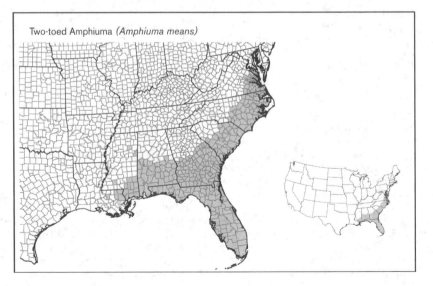

Two-toed Amphiuma (*Amphiuma means*)

three late Pleistocene sites in Florida and one site in Texas (Holman, 1995). Therefore, the range of two-toed amphiumas has contracted at least in the western portion of their range during the past 20,000 yr.

2. Historical versus Current Abundance.
In historical accounts, two-toed amphiumas were considered common throughout their range (Loennberg, 1894; Brimley, 1939; Carr, 1940a; Bishop, 1943; Hamilton, 1950). Although they are still considered common, destruction and degradation of wetlands throughout the Southeastern Coastal Plain has certainly caused the loss or decline of many local populations (Petranka, 1998). For example, Bancroft et al. (1983) reported that the density of two-toed amphiumas declined over a 3-yr period in a central Florida lake. They attributed the decline in part to destruction of littoral zone (shallow water) habitat. No data are available to accurately assess local long-term population or range-wide trends.

3. Life History Features.
In general, the life history features of two-toed amphiumas have not been studied in detail, and much research is needed on this species.

A. Breeding. Courtship and copulation presumably take place in the water, and fertilization is internal (Rose, 1967). However, females appear to deposit eggs in moist, terrestrial sites (e.g., Gunzburger, 2003). Two-toed amphiumas mate during the late winter, at least in the western portion of their range (Rose, 1967).

i. Breeding migrations. No pronounced breeding migrations are known.

ii. Breeding habitat. Females of this aquatic salamander may leave the water and slither to moist terrestrial sites to deposit their eggs (see below and Gunzburger, 2003).

B. Eggs.
i. Egg deposition sites. The few two-toed amphiuma nests observed have been in moist microhabitats close to standing water (Davison, 1895; Bancroft et al., 1983; Hayes and Lahanas, 1987; Gunzburger, 2003; K. Enge, personal communication; R.B.O., personal observations) or in the basins of dried or drying ponds (Brimley, 1910; Weber, 1944; Seyle, 1985b; Gunzburger, 2003). Those nests found near standing water have been 1.5 m (Hayes and Lahanas, 1987), 4 and 7 m (Bancroft et al., 1983), and 36 m (R.B.O., personal observations) from the water. Eggs are laid under objects such as logs and rocks. One clutch of eggs was found in the nest of an American alligator (K. Enge, personal communication). Although nests have all been found out of the water, Petranka (1998) suggests that the females probably laid the eggs in the water and the water subsequently receded, exposing the eggs. While this is possible for nests found near drying or dry ponds, the nests observed by Bancroft et al. (1983) were found several meters from the edge of a large lake. It seems unlikely that the females initially laid their eggs underwater and the lake receded enough to expose them. Inundation of the eggs stimulates hatching of fully developed embryos (Weber, 1944; Baker, 1945; Seyle, 1985b; Gunzburger, 2003; R.B.O., personal observations).

In most published accounts of nests, females were found with the eggs. Photographs of females attending their eggs can be found in Baker (1945, fig. 2), Neill (1971, fig. 29), and Behler and King (1998, color plate 18). The fact that females do not flee when nests are discovered suggests that they may defend their eggs as well as prevent desiccation of the clutch. Individual eggs desiccate quickly on a dry substrate (Hayes and Lahanas, 1987).

Eggs resemble a string of beads and are connected by thin constrictions of an outer sheath that covers the eggs (Petranka, 1998). Eggs of one clutch were laid in a string with 5–10 mm separating each egg in the string, and they averaged 10 mm × 10.5 mm (Weber, 1944; see fig. 78 in Petranka, 1998, for a photo of a clutch). Eggs examined by Gunzburger (2003) averaged 8.2, 9.8, and 10.1 mm in diameter for each of three clutches. She measured 20 eggs/clutch and the diameter included the outer sheath.

ii. Clutch size. Clutch size varies considerably in two-toed amphiumas. Average clutch size (based on dissections) of 22 females collected during February and April in Louisiana was 201 and ranged from 106–354 eggs (Rose, 1966a). Clutch size was positively correlated with female body size. Clutch sizes of six nests observed by various investigators were 210 and 97, 26, 42, 33, and 49 (Weber, 1944; Bancroft et al., 1983; Seyle, 1985b; Hayes and Lahanas, 1987; R.B.O., personal observations).

Duration of incubation is poorly studied, but the incubation period appears to be long. Weber (1944) estimated 5 mo as the incubation period for a clutch found in south Florida that was presumed to have been laid in January. Gunzburger (2003) reported three clutches collected in late July and early September from Lake Iamonia, which had dried several months earlier. Eggs in each of the clutches contained clearly visible embryos. The timing of inundation after eggs are deposited certainly impacts duration of the incubation period. Egg clutches have been found in February, May–July, and September. Based on the developmental condition of the embryos in these nests, two-toed amphiumas deposit eggs in the winter and spring.

C. Larvae/Metamorphosis.
i. Length of larval stage. Larvae hatch with external gills, but the gills are lost shortly after (Baker, 1945; Duellman and Trueb, 1986; Petranka, 1998). Gunzburger (2003) found that hatchlings from three clutches retained external gills an average of 2 wk before transformation. Juveniles (without external gills) as small as 60 mm TL (Neill, 1947a) and "but three inches in length" (Harlan, 1825a) have been reported.

ii. Larval requirements.
a. Food. Food requirements of larvae are unknown. Given the short time period between hatching and metamorphosis, larval two-toed amphiumas may not feed. Larvae can survive to metamorphosis exclusively on yolk reserves and juveniles are able to survive for several months without feeding (Gunzburger, 2003).

b. Cover. Presumably the same as transformed juveniles and adults.

iii. Larval polymorphisms. Larval polymorphisms have never been reported.

iv. Features of metamorphosis. Weber (1944) reports a length of 55 mm for hatchlings; it is not clear how many individuals were measured or if the length was SVL or TL. Average length for 37 recently hatched two-toed amphiumas from a clutch found in north Florida was 45.6 mm SVL with a mode of 45 mm (R.B.O., unpublished data). Gunzburger (2003) reported that hatchlings from three clutches ranged from 57–64 mm TL. In experiments she conducted with eggs from these clutches, Gunzburger (2003) found that, within a clutch, eggs that hatched early produced shorter hatchlings than eggs that hatched later.

Hatchlings have three finely branched gills, whitish in color, on either side of the head (Weber, 1944, S.A.J., personal observations). As noted (see "Length of larval stage" above), these gills are resorbed quickly. Hatchlings have three pairs of external gill slits, but only one gill slit on each side of the head is retained in adults (Hay, 1888; Duellman and Trueb, 1986; Behler and King, 1998).

v. Post-metamorphic migrations. Post-metamorphic migrations are unknown. Such movements would seem unlikely since larvae presumably require the same habitats as recently transformed juveniles and adults.

vi. Neoteny. Two-toed amphiumas, as well as the other two species in the family Amphiumidae, are considered to be obligate neotenes (Duellman and Trueb, 1986). Adults retain a single pair of gill slits (see "Features of metamorphosis" above).

D. Juvenile Habitat. Presumably the same as adults. However, in a central Florida lake, juvenile two-toed amphiumas were "unusually abundant" in deeper detrital substrates (>20 cm deep) while "adults were collected more frequently than expected at detrital depths of 16–20 cm" (Bancroft et al., 1983).

E. Adult Habitat. Two-toed amphiumas occupy a great variety of aquatic habitats. They occur in permanent ponds and lakes, preferring relatively shallow, heavily vegetated habitats (Loennberg, 1894; Goin, 1943; Telford, 1952; Bancroft et al., 1983; Franz, 1995). They may also be found in isolated, ephemeral wetlands (Smith and Franz, 1994; Snodgrass et al., 1999; S.A.J.; R.B.O., personal observations). They inhabit wet prairies and marshes (Harper, 1935; Dye, 1982; Smith and Franz, 1994), and they have been taken in the flooded houses of round-tailed muskrats (Harper, 1935; Smith and Franz, 1994). Two-toed amphiumas may be found in swamps (Wright, 1926; Harper, 1935; Bishop, 1943) and the Florida Everglades (Duellman and Schwartz, 1958; Machovina, 1994; Barr, 1997), as well as in small streams (Viosca, 1923; Harper, 1935; Martof et al., 1980; Gibbons and Semlitsch, 1991). They are common in canals and drainage ditches (they have been referred to as *ditch eels*), preferring to burrow in mucky substrates (Loennberg, 1894; Brimley, 1920b; Carr, 1940a; Bishop, 1943; Baker, 1947; Funderburg, 1955; Duellman and Schwartz, 1958; Lee, 1969c). They are often found inhabiting crayfish burrows (Carr, 1940a; Bishop, 1943; Dundee and Rossman, 1989). Snodgrass et al. (1999) found that the occurrence of two-toed amphiumas in depression wetlands at the Savannah River Site decreased as the distance from the nearest intermittent habitat increased.

F. Home Range Size. Results of recaptures of 25 marked individuals and radiotelemetry data for two adult individuals in a central Florida lake indicate that individual two-toed amphiumas have a small home range (Bancroft et al., 1983). One of the radio-telemetered individuals (a male) had a home range size of 12.4 m² over a 4-mo period. Juveniles are presumed to have smaller home range sizes than adults (Bancroft et al., 1983).

G. Territories. Unknown.

H. Aestivation/Avoiding Desiccation. Two-toed amphiumas are able to survive droughts by either burrowing into the substrate or by occupying crayfish burrows (Mount, 1975). They have been excavated from ≤1 m below the ground in dried wetlands (Harlan, 1825a; Brimley, 1920b; Baker, 1945; Knepton, 1954). Aresco (2001) observed two-toed amphiumas aestivating in organic/silt sediments of the shore of Lake Jackson, Leon County, Florida. Although the water level of the lake was low because of drought, a relatively large area of open water remained.

The maximum length of time two-toed amphiumas can survive buried in the substrate is unknown, but Brode and Gunter (1959) reported that these salamanders could survive starvation for a period of 3 yr. However, Rose (1966b) did not find support for survival over such a long period. Ten animals held without food were all dead after 13 mo. Live individuals were found buried at a bayhead

that had been dry for 2 yr (Knepton, 1954). At Lake Jackson, the location of aestivating two-toed amphiumas had not been covered with water for 1–2 yr (Aresco, 2001).

I. Seasonal Migrations. No distinct seasonal migrations are known. Two-toed amphiumas are primarily aquatic but will move overland, especially during rains (Carr, 1940a; Gibbons and Semlitsch, 1991; Conant and Collins, 1998; S.A.J., personal observations). Overland movements may be to disperse or to flee drying ponds (Snodgrass et al., 1999; Aresco, 2002). Females may migrate short distances from water to lay eggs (see "Egg deposition sites" above).

J. Torpor (Hibernation). Two-toed amphiumas are less active during the winter and in some habitats may burrow to overwinter. Carr (1940a) reported finding a two-toed amphiuma in Florida during January that was "under two feet of sphagnum and mud" and was "apparently hibernating." The capture rate of these salamanders in a central Florida lake was positively correlated with mean monthly water temperature; fewer animals were captured during the cooler winter (Bancroft, 1989). In the Florida Everglades there was also a significant positive correlation between average monthly temperature and the number of two-toed amphiumas captured (Machovina, 1994). The absence of seasonal differences in sizes of fat bodies led Machovina (1984) to conclude that two-toed amphiumas in the Florida Everglades do not hibernate.

K. Interspecific Associations/Exclusions. Two-toed amphiumas are sympatric with three-toed amphiumas (*A. tridactylum*), their closest relative (Karlin and Means, 1994), in extreme southeastern Louisiana, and can be syntopic (Dundee and Rossman, 1989). Juvenile two-toed amphiumas have been found occasionally with one-toed amphiumas (*A. pholeter*; see *A. pholeter* account by B. Means, this volume). Interactions among two-toed amphiumas and sirens (*Siren* sp.) may limit the distributions of these species in depression wetlands (Snodgrass et al., 1999).

L. Age/Size at Reproductive Maturity. There are few estimates of size and age at reproductive maturity for two-toed amphiumas. According to Machovina (1994), males and females from the Florida Everglades mature at about 260 mm SVL. Based on limited growth rate data from a central Florida lake, age at sexual maturity was estimated to be 3 yr for males and 4 yr for females (Bancroft et al., 1983). Mean size for confirmed females at this site was 398.1 mm SVL (n=147) and 404.5 mm (n=117) for males. Males averaged 168.7 g while the females averaged 148.6 g (Bancroft et al., 1983). Various authors provide lengths of what were assumed to be adult specimens (Bishop, 1943; Hamilton, 1950; Hill, 1954; Duellman and Schwartz, 1958). A female

collected in Nassau County, Florida, tending a clutch of 42 eggs was 245 mm SVL (325 mm TL; R.B.O., unpublished data). Seyle (1985) discovered a 289-mm female tending a clutch of 33 eggs, while Weber (1944) found a 390-mm female with eggs. It is unclear if Weber reported SVL or TL. A dissected female of 478 mm SVL contained >200 enlarged ovarian eggs (Bancroft et al., 1983). The smallest male with enlarged reproductive organs examined by Bancroft et al. (1983) was 365 mm SVL. Machovina (1994) found that the number of follicles in females increased with SVL. The maximum size attained by two-toed amphiumas appears to be 1,162 mm TL (Behler and King, 1998; Dundee and Rossman, 1989; Conant and Collins, 1998).

M. Longevity. Bowler (1977) reported an individual two-toed amphiuma maintained at the Philadelphia Zoo to be almost 15 yr old. This animal was collected as an adult. Another captive individual lived for 27 yr at the London Zoological Gardens (Flower, 1936). Longevity of wild animals is unknown.

N. Feeding Behavior. Two-toed amphiumas are chiefly nocturnal (Carr, 1940a; Bishop, 1943; Bancroft et al., 1983), and they emerge at night to actively forage in shallow water (Funderburg, 1955; Dundee and Rossman, 1989). However, fishermen are known to catch two-toed amphiumas during the day (Funderburg, 1955). Two-toed amphiumas are also said to employ a sit-and-wait feeding strategy where individuals wait for food to come near as they remain hidden in burrows or debris (Conant and Collins, 1998). Olfaction likely plays an important role in the ability of two-toed amphiumas to secure prey (Hargitt, 1892). Although the diet of two-toed amphiumas is well documented (see below), feeding behavior remains enigmatic, as does much of the life history of these salamanders. They are best described as carnivorous opportunists. According to Carr (1940a), "apparently any animal that can be captured and swallowed is eaten by the adults." Carr's supposition was shared by Hamilton (1950), who concluded, "the [amphiuma] eats that which it can master." Adults are known to feed on a variety of aquatic insect adults and larvae (e.g., odonate naiads, dytiscid and hydrophilid beetles, chironomid and syrphid larvae; Hamilton, 1950; Duellman and Schwartz, 1958; Lee, 1969c; Bancroft et al., 1983). They also have been documented to take a variety of amphibians and reptiles, including southern cricket frogs (*Acris gryllus*), southern leopard frogs (*Rana sphenocephala*), unidentified tadpoles, salamanders, smaller conspecifics (Carr, 1940a; Hamilton, 1950), greater sirens (*Siren lacertina*), peninsula newts (*Notophthalmus viridescens piaropicola*; Machovina, 1994), water snakes (*Nerodia* sp.; Hamilton, 1950; Lee, 1969c), anoles (*Anolis* sp.), and small mud turtles (*Kinosternon* sp.; Hamilton, 1950). Numerous species of fish, mollusks, and spiders have been observed in the digestive tracts of two-toed amphiumas (Hargitt, 1892; Hamilton, 1950; Duellman and Schwartz, 1958; Bancroft et al., 1983). Crayfish comprise a large part of their diet (Brimley, 1939; Carr, 1940a; Hamilton, 1950; Duellman and Schwartz, 1958; Mount, 1975; Bancroft et al., 1983; Dundee and Rossman, 1989). Individuals may ingest vegetation incidental to capturing prey (Duellman and Schwartz, 1958; Bancroft et al., 1983). Juvenile two-toed amphiumas feed mainly on amphipods, aquatic insects, and aquatic insect larvae (Carr, 1940a; Bancroft et al., 1983).

O. Predators. Known predators of two-toed amphiumas include snakes, birds, alligators, larger conspecifics, and possibly mammals. Mud snakes *(Farancia abacura)* and rainbow snakes *(F. erythrogramma)* are major predators (Brimley, 1939; Funderburg, 1955; Schwartz, 1957b; Duellman and Schwartz, 1958; Salthe, 1973a; Mount, 1975). Kingsnakes *(Lampropeltis getula*; Harper, 1935) and plain-bellied water snakes *(Nerodia erythrogaster*; Funderburg, 1955) also prey on amphiumas. Sandhill cranes *(Grus canadensis)* will eat two-toed amphiumas (Dye, 1982). In the Florida Everglades, amphiumas and sirens are important food items for alligators during the wet season (Barr, 1997). Hamilton (1950) reported on two two-toed amphiumas that had eaten smaller conspecifics. Bancroft et al. (1983) speculated that otters *(Lutra canadensis)* were predators of amphiumas; these mammals may have been partially responsible for a decline of two-toed amphiumas in a central Florida lake.

P. Anti-Predator Mechanisms. Their cryptic habits (e.g., nocturnal behavior and inhabiting dense vegetation) are probably their major anti-predator mechanism. Nevertheless, two-toed amphiumas will defend themselves and can deliver a painful bite (Carr, 1940a; Baker, 1945; Conant and Collins, 1998; Petranka, 1998). Bite scars found on mud snakes may be due to attempted predation or defense by amphiumas (Dundee and Rossman, 1989).

Q. Diseases. Unknown.

R. Parasites. The trematodes *Telorchis stunkardi*, *Cephalogonimus amphiumae*, and *Megalodiscus americanus*, as well as a nematode, presumably *Filaria amphiumae*, were found in a large male two-toed amphiuma from Louisiana (Chandler, 1923). Lee (1969c) found a dozen unidentified parasitic nematodes in an individual from Polk County, Florida.

4. Conservation.
Two-toed amphiumas have no federal protection and are not listed at any level in the eight states where they occur. They are considered common, but destruction

and degradation of wetlands throughout the Southeastern Coastal Plain has likely caused the loss or decline of many local populations. Nonetheless, the biological status of two-toed amphiumas is unknown because there are no data available to assess local population or range-wide biological status or trends.

Amphiuma pholeter Neill, 1964(b)
ONE-TOED AMPHIUMA

D. Bruce Means

1. Historical Versus Current Distribution.
One-toed amphiumas *(Amphiuma pholeter)* are found in a narrow distribution of only about 80–120 km (50–75 mi) inland from the seashore in the eastern Gulf Coast Plain of the southeastern United States, from Jackson County, Mississippi (Floyd et al., 1998), to Levy and Hernando counties, Florida (Stevenson, 1967; Means, 1996a). They are known from two localities in the Ochlockonee River drainage of Georgia (Means, 1996a) and only two localities in Alabama (Carey, 1984, 1985). In Florida, one-toed amphiumas are known from about 40 localities (Means, in preparation; Florida Natural Areas Inventory, personal communication). Knowledge of the species' current distribution has accumulated recently; because the species was first recognized by Neill (1964b), there are no historical data with which to make a comparison.

on a regular basis (personal observations). However, no declines in current abundance at these sites have been noted over the past three decades of monitoring (personal observations).

3. Life History Features.
Life history is poorly known, but see Means (1977, 1992b, 1996a, in preparation).

    A. Breeding.
    i. Breeding migrations. None reported.
    ii. Breeding habitat. Presumably the same as that of juveniles and adults; gravid females have been found co-occurring with juveniles and adults.

    B. Eggs.
    i. Egg deposition sites. Eggs and hatchlings have not been found (Means, 1996a). As with two-toed amphiumas *(A. means)*, brooding females may coil around their eggs during development.
    ii. Clutch size. Unknown.

    C. Larvae/Metamorphosis.
    i. Length of larval stage. Poorly known; larval life is thought to be short (personal observations).
    ii. Larval requirements.
    a. Food. Food requirements of larvae are unknown.
    b. Cover. Presumably the same as transformed juveniles and adults.
    iii. Larval polymorphisms. None reported.
    iv. Features of metamorphosis. Hatchlings may have thin, feathery gills and a short larval life before metamorphosing into air-breathing juveniles.

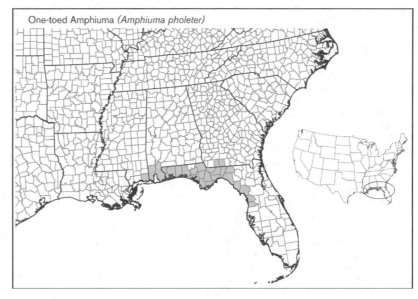

One-toed Amphiuma (*Amphiuma pholeter*)

2. Historical Versus Current Abundance.
Current abundance of one-toed amphiumas is difficult to assess because their microhabitats are difficult to sample. Usually, at most localities, only one or two specimens are found during several person-hours of vigorous searching. At only three or four localities in Florida is it possible to find one-toed amphiumas

    v. Post-metamorphic migrations. None reported.

    vi. Neoteny. One-toed amphiumas, as well as the other two species in the family Amphiumidae, are considered to be obligate neotenes (Duellman and Trueb, 1986).

    D. Juvenile Habitat. Not known to differ from that of adults because small, thin individuals (that upon dissection are obviously juveniles and probably recent hatchlings) have been collected in several localities with adults (personal communication).

E. Adult Habitat. Means (1977) analyzed the habitat qualities of 13 localities of one-toed amphiumas and found that individuals are primarily found in deep, liquid, amorphous mucks derived from hardwood and cypress litter. The most important habitat variables associated with one-toed amphiumas are (1) streams of low–moderate gradient; (2) swampy and periodically inundated floodplains; (3) mixed bottomland hardwoods and cypress; (4) seepage; and (5) vulnerability to drought. Muck, as compared with peat, is usually liquid, and decomposition of the organic material in it has progressed so far that it is relatively amorphous, not having large pieces of wood and leaf litter. Amphiumas cannot locomote through fibrous peat and are rarely found in shallow muck deposits of <15 cm (6 in) deep (personal observations), presumably because it increases their vulnerability to predators such as raccoons.

F. Home Range Size. Unknown.

G. Territories. Have not been reported.

H. Aestivation/Avoiding Desiccation. Means (1992b, in preparation) reported excavating individuals from the bottom of moist muck, where they had obviously sought refuge from drought. Animals were coiled in small chambers made by their bodies and were stimulated during excavation to move into small tunnels they appeared to have excavated.

I. Seasonal Migrations. Means (in preparation) noted that in winter, one-toed amphiumas were occasionally found under large logs buried along stream courses in first-order stream valleys where the species is not found in the spring, summer, or fall. Means (in preparation) speculated that some individuals migrated upstream into seepage heads of first-order valleys to find protection from cold weather in warm seeps.

J. Torpor (Hibernation). Individuals found in November–March are cold to the touch and sluggish (personal observations). In winter, one-toed amphiumas are less abundant in muck beds, indicating that they may migrate into either different microhabitats (i.e., the peaty soil held together by dense root mats of wetland shrubs) or upstream into the warmer headwater seepages.

K. Interspecific Associations/Exclusions. On several occasions, Means (in preparation) has taken a few juvenile two-toed amphiumas in the same muck beds with one-toed amphiumas, but adult two-toed amphiumas have not been found at these sites.

L. Age/Size at Sexual Maturity. Means (in preparation) determined that total

length of a large sample of mixed sexes and juveniles was 218.3 mm (s.d. ± 41.8 mm, range 89.3–314.2 mm, n=204), and there was no sexual dimorphism in body size. The smallest of eight gravid females was 247.3 mm in total length, but it is not known how long is required for individuals to achieve sexual maturity.

M. Longevity. Juveniles (probably recent hatchlings) were raised in the laboratory on a diet of earthworms. They reached average size in about 1 yr and lived in captivity for 5 yr before dying accidentally (Means, in preparation).

N. Feeding Behavior. Natural food contents of the guts of 100 one-toed amphiumas included sphaeriid clams, physellid snails, aquatic earthworms (*Sparganophilus* sp.), asellid isopods, the larvae of mayflies, tipulid flies, chironomid midges, culicid mosquitoes, stoneflies, megalopterans, tabanid flies, adults and larvae of small aquatic beetles, planarians, and occasional terrestrial beetles and lepidopteran larvae that drop onto the surface of the muck (Means, in preparation). Notably lacking from the diet are dragonfly naiads, salamander larvae, frog tadpoles, and small fish that cohabit the same mucky sites (Means, in preparation).

O. Predators. Potential predators found in the same muck beds with one-toed amphiumas are common snapping turtles (*Chelydra serpentina*), eastern mud turtles (*Kinosternon subrubrum*), mud snakes (*Farancia abacura*), red-bellied water snakes (*Nerodia c. erythrogaster*), brown water snakes (*N. taxispilota*), queen snakes (*Regina septemvittata*), ring-necked snakes (*Diadophis punctatus*), cottonmouths (*Agkistrodon piscivorus*), two-toed amphiumas, southern leopard frogs (*Rana sphenocephala*), and bronze frogs (*R. clamitans*; Means, in preparation).

P. Anti-Predator Mechanisms. One-toed amphiumas have a noxious skin secretion that produces foam when animals

are collected. This secretion is bitter tasting and slightly numbing (personal observations).

Q. Diseases. None observed.

R. Parasites. Parasites or gut symbionts taken from the alimentary canal include unidentified trematodes, cestodes, nematodes, and acanthocephalans (Means, in preparation).

4. Conservation.
One-toed amphiumas are considered Rare in Florida (Means, 1992b), Rare in Georgia (J. Jensen, Georgia Department of Natural Resources, personal communication), under consideration for Endangered status in Mississippi (T. Mann, personal communication), and poorly known in Alabama (Means, 1986c). This Rare status may be a matter of their secretive nature and proclivity for mucky and swampy environments, which makes effective sampling difficult.

Amphiuma tridactylum Cuvier, 1827
THREE-TOED AMPHIUMA
Jeff Boundy

1. Historical versus Current Distribution.
The distribution of three-toed amphiumas (*Amphiuma tridactylum*) is within the Gulf Coast Plain centered around the lower Mississippi River from the Brazos River Valley in Texas to central Alabama, and north to southern Illinois and extreme southwestern Kentucky.

2. Historical versus Current Abundance.
According to Petranka (1998), three-toed amphiumas are locally common and currently do not appear to need protection. The loss of swamps and other forms of wetland habitat have been offset by the creation of canals, ditches, and permanent ponds (Petranka, 1998).

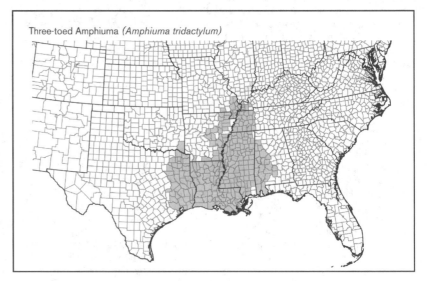
Three-toed Amphiuma (*Amphiuma tridactylum*)

3. Life History Features.

A. Breeding. Reproduction is aquatic.

i. Breeding migrations. Unknown and unlikely.

ii. Breeding habitat. All records of amphiuma nests have been out of water (Salthe, 1973b). However, Baker (1945, in Salthe, 1973b) felt that these observations were artifactual, that nests generally occur underwater and that observed nests were those stranded by drought conditions. Cagle (1948) suggested that females prefer to lay eggs underwater in submerged debris (in Salthe, 1973b).

B. Eggs.

i. Egg deposition sites. Three-toed amphiumas mate using internal fertilization. Baker (1937, in Salthe, 1973b) outlined courtship behaviors, and copulation has been observed by Baker and colleagues (1947). In Tennessee, Sturdivant (1949) reported mating from December–June (see also Salthe, 1973b; Behler and King, 1998), and nesting from August to mid-winter (Parker, 1937, in Salthe, 1973b). In Louisiana, mating was reported from April–September, depending on weather conditions (Wilson, 1940; Cagle, 1948; Rose, 1966a; Salthe, 1973b; and Dundee and Rossman, 1989). In Louisiana, males are in reproductive condition from January–April (Fontenot, 1999). Ovum diameter varies between 5–7 mm (Cagle, 1948; Salthe, 1973b; Petranka, 1998; Fontenot, 1999).

ii. Clutch size. About 200 eggs are laid in a strand 5–12 cm long that becomes tangled in the nesting cavity (Behler and King, 1998; Petranka, 1998). Cagle (1948) counted follicles and found an average of 98 eggs (range 42–131). However, Rose (1966a) examined larger females and found an average of 201 eggs with a maximum of 354 (see also Dundee and Rossman, 1989; Bartlett and Bartlett, 1999a). Similarly, working in Louisiana, Fontenot (1999) found a mean of 201.4 eggs (range 44–282, n=14). Cagle (1948) and Fontenot (1999) suggest that the incubation period may be about 5 mo. Eggs ready to hatch have been found in August in Arkansas (Hay, 1888; Salthe, 1973b) and November in Louisiana (Baker, 1945, Salthe, 1973b). In Louisiana, oviposition begins in early July (Fontenot, 1999). Many reports suggest that females may be biennial (Wilson, 1942; Cagle, 1948; Salthe, 1973b; Dundee and Rossman, 1989) or triennial (35% gravid females; Fontenot, 1999) reproductive cycles. Dundee and Rossman (1989) repeat Viosca's findings that three-toed amphiumas may be ovoviviparous.

C. Larvae/Metamorphosis.

i. Length of larval stage. Apparently short; gills are lost about 3 wk after hatching (Ultsch and Arceneaux, 1988; Petranka, 1998). Cagle (1948) and Liner (1954) found young with vestiges of external gills in November in Louisiana and Alabama. Hatchlings average 64 mm in length (Dundee and Rossman, 1989).

ii. Larval requirements. Found in the same habitats as adults.

a. Food. Unknown.

b. Cover. Young animals have been found schooling (Parker, 1939; Baker, 1945, in Salthe, 1973b).

iii. Larval polymorphisms. Have not been reported.

iv. Features of metamorphosis. Gills are lost about 3 wk after hatching (Ultsch and Arceneaux, 1988; Petranka, 1998). Metamorphosing animals are about the same size as hatchlings, approximately 60–75 mm TL (Barbour, 1971; Dundee and Rossman, 1989).

v. Post-metamorphic migrations. Unknown and unlikely.

vi. Neoteny. Three-toed amphiumas, as well as the other two species in the family Amphiumidae, are considered to be obligate neotenes (Duellman and Trueb, 1986).

D. Juvenile Habitat. Juveniles use smaller retreats (e.g., tiny holes in mud, discarded cans, etc.) and floating mats of vegetation (C. Fontenot, personal communication) more often than adults.

E. Adult Habitat. Adults are nocturnal; according to Petranka (1998) peak activity occurs 3–4 hr after sunset. Adult three-toed amphiumas occupy bottomland swamps, bayous, cypress swamps, and streams (Behler and King, 1998; Conant and Collins, 1998). They are especially abundant in drainage ditches in suburban and agricultural areas of the lower Mississippi River (Behler and King, 1998). Three-toed amphiumas will frequently inhabit crayfish burrows and can co-exist with fishes (Petranka, 1998). Three-toed amphiumas will move overland during and following heavy rains, and Petranka (1998) reports they have been observed ≤12 m away from the water's edge.

F. Home Range Size. Usually remain within a restricted area, but Cagle (1948) reports dispersal ≤396 m (see also Petranka, 1998).

G. Territories. Adult males will fight, suggesting that either territories are defended or there is competition for females (Dundee and Rossman, 1989).

H. Aestivation/Avoiding Desiccation. According to Petranka (1998), adult three-toed amphiumas are capable of burrowing through soft mud and utilizing crayfish burrows. They can remain in these refugia for many months without feeding.

I. Seasonal Migrations. Barbour (1971) reports that on rainy nights, amphiumas may be seen wandering on land in swampy areas. They are also known to cross roads, especially during rains (J.B., personal observations).

J. Torpor (Hibernation). Unknown.

K. Interspecific Associations/Exclusions. Three-toed amphiumas occur sympatrically with two-toed amphiumas (*A. means*) in parts of their range.

L. Age/Size at Reproductive Maturity. From 45.7–106 cm (18–41 3/4 inches;

Behler and King, 1998), although Fontenot (1999) found the minimum size of gravid females to be 33 cm body length.

M. Longevity. Not reported.

N. Feeding Behavior. Crawfish constitute a large proportion of the diet, followed by annelids, insects, mollusks, fishes, and even skinks. Fontenot and Fontenot (1989) found the remains of a common snapping turtle (*Chelydra serpentina*; see also Petranka, 1998). Vegetation is probably ingested incidentally. Seasonal differences reflect prey availability (Chaney, 1951; Dundee and Rossman, 1989). After capturing prey, amphiumas often twist violently, which may assist in incapacitating the prey or reduce the risk of the prey attacking them (see Petranka, 1998). The biomechanics of feeding have been detailed by Erdman and Cundall (1984).

O. Predators. Preyed upon by mudsnakes (*Faracina abacura*) and cottonmouths (*Agkistrodon piscivorous*; Behler and King, 1998).

P. Anti-Predator Mechanisms. Amphiumas are known to inflict a painful bite and are also known to flee when capture is attempted (Barbour, 1971; Dundee and Rossman, 1989; Petranka, 1998). Amphiumas may whistle when agitated (Baker, 1937; see also Petranka, 1998). Amphiumas are covered with a mucus slime and have thick skin, which makes them extremely difficult to grasp.

Q. Diseases. None reported.

R. Parasites. The trematodes *Telorchis stunkardi*, *Cephalogonimus amphiumae*, and *Megalodiscus americanus*; as well as the nematode, presumably *Filaria amphiumae*, were documented in amphiumids from Louisiana (Chandler, 1923). Turtle leeches (*Placobdella* sp.) were found on an amphiuma in Louisiana (Saumure and Doody, 1998).

4. Conservation.

Because these animals breathe through their skin, certain chemicals can adversely affect their respiration, forcing them from the water. For example, rotenone, a widely used poison that causes fish gills to constrict, is thought to similarly constrict the integumentary vessels of amphiumids (Dundee and Rossman, 1989). The use of such chemicals as a management technique in amphiumid habitat could contribute to the decline of the species. Large-scale amphiuma kills were noted in Louisiana following flushing of chlorine into a drainage ditch and after a gasoline spill in a bayou and swamp (J.B., personal observations). Three-toed amphiumas are considered widespread and abundant in Louisiana. Several dozen per year are sold from Louisiana as pets, while from several hundred to about 1,000 are sold annually to biological suppliers for laboratory experiments. Most are taken as off-catch during crayfish trapping. Commercial take dropped drastically during a

3-yr drought, but this was due to a reduction in crayfish trapping (J.B., personal observations).

Family Cryptobranchidae

Cryptobranchus alleganiensis (Daudin, 1803)

HELLBENDER

Christopher A. Phillips, W. Jeffrey Humphries

1. Historical versus Current Distribution.
Historically, eastern hellbenders (*Cryptobranchus a. alleganiensis*) were found in the Susquehanna system (Atlantic drainage) in New York, Pennsylvania, and Maryland; tributaries of the Savannah River (Atlantic drainage) in South Carolina and Georgia; the Tennessee system in Georgia, Virginia, Alabama, Mississippi, Tennessee, North Carolina, and Kentucky; and the Ohio system in New York, Maryland, Pennsylvania, Virginia, West Virginia, Ohio, Indiana, Kentucky, and Illinois. A second cluster of populations inhabits portions of the Missouri drainage in south-central Missouri and the Meramec (Mississippi drainage) in eastern Missouri. Ozark hellbenders (*C. a. bishopi*) are found in the White River system in southern Missouri and north-central Arkansas. Cope (1889) listed a specimen in the U.S. National Museum from Des Moines, Iowa, and Firschein (1951b) mentioned an unverified record from the Skunk River (Mississippi drainage) in southeastern Iowa. Others have referred to the hellbender's presence in Iowa (e.g., Hay, 1892; McMullen and Roudabush, 1936), suggesting that Iowa might be within the historical range. Firschein (1951b) convincingly discredited a specimen from Vernon County, Missouri (Arkansas drainage). Two specimens from the Neosho River (Arkansas drainage) in southeastern Kansas (Hall and Smith, 1947) have come under

scrutiny. Based on the extreme hiatus between the Kansas records and the nearest verified records to the east, several authors have speculated that these specimens were either introduced (Smith and Kohler, 1977) or are otherwise invalid (Dundee, 1971). Records from the Great Lakes, Louisiana, and New Jersey are certainly invalid and represent introductions or confusion with other species (see summary in Nickerson and Mays, 1973a).

Because of the secretive nature of hellbenders and their confusion with mudpuppies (*Necturus maculosus*), the present range is not known with certainty. They are no longer present in Iowa (if they ever occurred there), and they are almost certainly extirpated from the Ohio drainage in Illinois, although there is a verified 1991 record from the Wabash River in White County. Hellbenders have been eliminated from Indiana except for a small population in the Blue River and the lower portions of the South Fork of the Blue River (Kern, 1986a). In Ohio, Pfingsten (1989a) spent 2,000 person-hours searching for hellbenders from 1985–88 and failed to find any in the Miami River or its tributaries, but did locate them in the other main drainages of the Ohio River. Populations in the remainder of the Ohio drainage are extant, as are most of those in the Tennessee drainage. Green (1934) reported hellbenders to be common in the Ohio River, but not so common in the tributaries near Huntington, West Virginia. Hellbenders were also reported from the Ohio River near Marietta, Ohio (Krecker, 1916). Records for the hellbenders in the Ohio River have not been reported since these early sightings. Humphries (1999) reported hellbenders to still be common in many high elevation streams in West Virginia. Bothner and Gottlieb (1991) studied the distribution and abundance of hellbenders in

New York and found the species in both the Allegheny and Susquehanna drainages. The same is true for both systems in Pennsylvania and Maryland (Gates, 1983). No recent data are available for the Savannah drainage populations in Georgia and South Carolina. Populations are also still found throughout the species' historical range in Missouri and Arkansas (Trauth et al., 1992a,b; Trauth et al., 1993b; LaClaire, no date).

2. Historical versus Current Abundance.
Historical data on abundance are available for only a few populations and even in those cases, rigorous quantification of effort is lacking. For example, Green (1935) reported catching 34 hellbenders between the hours of 8 (PM) and midnight on 21 June 1934 in the headwaters of the Shavers Fork of the Cheat River (New River drainage of West Virginia). Hellbenders were detected using an acetylene light, but the number of observers or the length of the stream surveyed was not specified. Swanson (1948) reported collecting (and permanently removing) over 650 hellbenders from a 4.8 km (3 mi) stretch of Big Sandy Creek (Allegheny drainage) in Venango County, Pennsylvania, from 1932–48. Other vague accounts are available for Bear Creek in the Tennessee drainage of northeastern Mississippi (Ferguson, 1961b), French Creek in the Allegheny drainage of northwestern Pennsylvania (Hillis and Bellis, 1971), and the streams in the Ohio River drainage of Ohio (Pfingsten, 1989a).

More rigorously documented abundance data are available for the Allegheny River drainage of New York State where Bothner and Goettlieb (1991) performed mark–recapture studies to estimate both abundance and density at eight sites along Ischua Creek, Oswayo Creek, and the Allegheny River. Abundance estimates ranged from 3–58 individuals in study areas that ranged from 424–14,003 m^2 of stream bed. A series of mark–recapture studies has been conducted in Missouri starting with Nickerson and Mays (1973b), who estimated abundance at 1,142 hellbenders in a 2.67-km stretch of the North Fork River (White River drainage) in 1969 and 269 hellbenders in a single riffle (4,600 m^2) in the same river in 1970. Effort, in person-hours, was recorded as 750 in 1969 and 108 in 1970. Peterson et al. (1983) conducted a mark–recapture study in the same riffle during 1977–78, and their estimate of 231 fell within the 95% confidence interval of Nickerson and Mays, indicating no change in abundance in that riffle during a 7-yr period. Mark–recapture estimates of hellbender abundance ranging from 0.9–6.1 hellbenders/100 m^2 were reported for the Spring and Eleven Point rivers (White River drainage of Missouri and Arkansas) and Big Piney and Gasconade

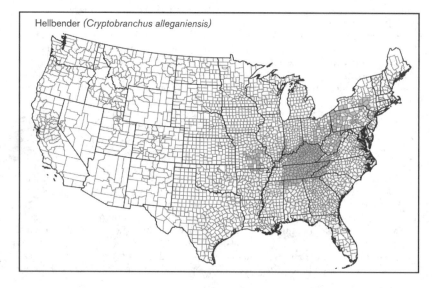

Hellbender (*Cryptobranchus alleganiensis*)

rivers (Missouri River drainage of Missouri) during 1980–82 (Peterson et al., 1988). Peterson (1987) captured 110 adults in a 2600-m^2 study site in the Niangua River in Missouri during 1985. Humphries (1999) conducted a mark–recapture study from 1998–99 in the West Fork of the Greenbrier River in West Virginia. An abundance estimate of 31 individuals was found within a 216 m stretch of stream. The density in this section was 0.80 individuals/100 m^2.

Four streams in the White River drainage of Missouri were surveyed in 1992 (Ziehmer, 1992). Numbers encountered and effort were recorded as follows: Jack's Fork, 4 in 66 person-hours; Current, 12 in 60 person-hours; North Fork, 122 in 49 person-hours; Bryant's Creek, 0 in 22 person-hours. Trauth et al. (1992a) surveyed seven sites on the Spring River (White River drainage of Arkansas) in 1991, including two of the same sites that Peterson (1985) had studied between 1980 and 1982. Trauth et al. (1992a) did not encounter any hellbenders at a site where Peterson had marked 60 and encountered only 5 where Peterson had marked 310. This is the only evidence of decline in a hellbender population that has been documented rigorously in the literature.

3. Life History Features.

A. Breeding. Reproduction is aquatic.

i. Breeding migrations. The suggestion by Alexander (1927) that male hellbenders move many km to reach their breeding grounds in the fall has not been supported by recent research. Most authorities agree that no actual breeding migration takes place, although males may move short distances within their home ranges to brooding sites.

ii. Breeding habitat. The breeding season is variable but occurs mainly in September and October, although evidence of breeding activity as late as December and January has been reported for the Spring River in Arkansas (Peterson et al., 1989). Only a few specific breeding dates are given in the literature. Smith (1907) reported egg-laying dates from 28 August–8 September in 6 consecutive years starting in 1906 in northwestern Pennsylvania. Swanson (1948) stated that egg-laying takes place about the 1 September in Venango County, Pennsylvania. The release of milt from captured males and the presence of gravid females was documented between 7 September–11 October during 2 years of study in the Blue River of Indiana (Kern, 1986b). Nests with eggs have been reported in the North Fork River, Missouri, on 13 September (Nickerson and Mays, 1973a), and 2 and 8 October (Nickerson and Tohulka, 1986). Dundee and Dundee (1965) noted a nest containing eggs in the Niangua River, Missouri, on 14 November, and Johnson (1981) noted a clump of eggs in the same river on

19 September. Bothner and Gottlieb (1991) reported nests in the Susquehanna River in New York on 10 and 11 September. Green (1934) reported the spawning season of the hellbender in the vicinity of Elkins, West Virginia, to be from the middle of August to early September. The release of milt from captured males was reported from 20 August–11 September in the West Fork of the Greenbrier River in West Virginia (Humphries, 1999). A nest with eggs has also been reported from the Williams River in Webster County, West Virginia, during September 1997 (S. Blackburn, personal communication).

B. Eggs.

i. Egg deposition sites. The beginning of the breeding season is marked by a change in behavior of hellbenders, especially males. They leave their routine hiding places and move around the stream bottom, even during daylight, exploring cavities under flat rocks and crevices or holes in the bedrock (Smith, 1907). Eventually a male occupies a suitable site and may actively prepare a nest by moving gravel to create a saucer-shaped depression (Bishop, 1941b). Peterson (1988) also reported males using a hole in a mud-gravel bank for nesting. The males lie at the opening of their nests, frequently with their heads exposed, waiting for gravid females. Females may enter nest sites voluntarily or they may be forced into the cavity by the male. As soon as the female starts to deposit eggs, the male moves alongside or slightly above the female and sprays the eggs with snowy-white seminal fluid that may take the form of a cloudy mass or ropy chunks (Smith, 1907).

ii. Clutch size. A single female may deposit from 200 to >400 eggs (Smith, 1907, 1912a; Bishop, 1941b, Nickerson and Mays, 1973a), but this may not represent all the eggs available for oviposition, as >20% of a female's compliment may be resorbed (Topping and Ingersol, 1981; Petranka, 1998). Males may accept several females into their nest cavity, so the total number of eggs in a single nest may be >2,000 (Bishop, 1941b). Deposited eggs are often eaten by both males and females (Smith, 1907; Bishop, 1941b). The eggs are yellow, round, approximately 6 mm in diameter, and surrounded by two transparent envelopes. The inner envelope is attached as a solid rope from egg to egg resulting in long egg strings (Nickerson and Mays, 1973a). The eggs swell with water and eventually increase to 18 mm in diameter (Smith, 1912a). It may take 2 d for a female to expel her eggs, at which time she either leaves the nest or is expelled by the male (Bishop, 1941b). Males usually remain in the nest cavity with the eggs, and both Smith (1907) and Bishop (1941b) witnessed episodes of active nest guarding by males. Bishop (1941b) also observed a brooding male swaying from side to side over the eggs, which may

increase the oxygen supply to the eggs. The duration of this brooding period varies, but Smith (1907, 1912a,b) found males attending nests that contained embryos about 3 wk old.

Bishop (1941b) estimated the incubation period at 68–84 d for western New York and Pennsylvania. Peterson (1988) encountered hatchlings in the Niangua River, Missouri, that he believed to be no more than 45 d old. Temperature undoubtedly plays a major role in determining length of embryonic period. Smith (1912) provides the most comprehensive data on embryonic development and should be consulted for details.

C. Larvae/Metamorphosis.

i. Larval stage. Newly hatched larvae are approximately 30 mm TL and are well pigmented dorsally and on the tail. The venter is unpigmented except for the yellow of the yolk sac. The mouth and eyes are conspicuous, the gills are short and flattened, the front limbs terminate in two lobes, and the hindlimbs are paddle-shaped and unlobed (Bishop, 1941b). Development is rapid, and hatchlings double their size during the first year (Bishop, 1941b). Larvae normally lose their external gills in the second summer after hatching, at 100–130 mm TL (Smith, 1907; Bishop, 1941b; Nickerson and Mays, 1973a).

ii. Larval requirements.

a. Food. The diet of larval hellbenders has not been studied but probably includes invertebrates.

b. Cover. Nickerson and Mays (1973a) reported that larval hellbenders utilize small stones and chert for cover. They also reported an anecdotal account of a larval hellbender taken from the interstices of a gravel bed in an area of subsurface percolation. The scarcity of records for larval hellbenders compared to adults supports this suggestion (Kern, 1986c; Petranka, 1998).

iii. Larval polymorphisms. None reported.

iv. Features of metamorphosis. The major morphological features of hellbender metamorphosis are the loss of the external gills and the attainment of adult color pattern.

v. Post-metamorphic migrations. None known.

vi. Neoteny. Not reported.

D. Juvenile Habitat. Same as for adults.

E. Adult Habitat. Adult hellbenders are found in fast-flowing streams containing abundant cover in the form of large flat rocks, bedrock shelves and crevices, and logs (Bishop, 1941b; Nickerson and Mays, 1973a).

F. Home Range Size. Home range has been reported in various forms for several populations of hellbenders. Using minimum area convex polygon in Missouri, average home range size was 28 m^2 for females and 81 m^2 for males (Peterson and Wilkinson, 1996). Coatney (1982) calculated an elliptical home range of 90 m^2 for seven Ozark hellbenders radio-tracked nocturnally for 2 wk. In Pennsylvania, average inter-capture distance was 18.8 m for males and 18.7 m for females (Hillis and Bellis, 1971). The mean activity radius for this population was 10.5 m. Calculated as a circular home range, the average home range was 346.4 m^2. Linear distance between captures in Tennessee ranged from 5–60 m (Casey et al., 1993). Topping and Peterson (1985) provided evidence for size-specific movement in hellbenders in Missouri. They demonstrated a tendency for upstream movements ranging from 2.3–25.7 m/d. In contrast, Peterson (1987) detected no net movement upstream or downstream in the Niangua River, Missouri. Mean linear movement of hellbenders in a West Virginia stream was 20.1 m, ranging from 0.8–70.2 m between captures at least 1 mo apart (Humphries, 1999).

G. Territories. Home ranges of hellbenders overlap (Peterson and Wilkinson, 1996), but they apparently avoid being in the area of overlap at the same time (Coatney, 1982). However, hellbenders have been observed in close proximity to each other at night without conflict between individuals (Humphries, 1999). Rarely is >1 hellbender found beneath the same rock except during the breeding season (Smith, 1907; Hillis and Bellis, 1971; Nickerson and Mays, 1973a; Peterson, 1988), and they are known to defend shelter rocks (Peterson and Wilkinson, 1996; Hillis and Bellis, 1971). Hellbenders will utilize rocks recently vacated by other individuals (Hillis and Bellis, 1971; Peterson and Wilkinson, 1996; Humphries, 1999). Male hellbenders become extremely territorial during the breeding season and will defend nest holes or rocks (Smith, 1907; Bishop, 1941b; Peterson, 1988). Blais (1996) reported that during the breeding season in New York, male hellbender's home ranges tended to overlap more than those of females.

H. Aestivation/Avoiding Desiccation. Not reported; however, Green (1934) stated that hellbenders in West Virginia moved to deeper holes in summer to find colder water.

I. Seasonal Migrations. See "Breeding migrations" above. Seasonal change in nocturnal activity has been reported in high elevation populations in West Virginia (Humphries and Pauley, 2000). Hellbenders were most active during early summer (May–June), with decreased activity in later months. Nocturnal searches with flashlights were most productive in early summer in these populations; however, nocturnal activity shifts have not been reported in other parts of the hellbender's range. Noeske and Nickerson (1979) reported on seasonal changes in activity rhythms in the laboratory.

J. Torpor (Hibernation). In most streams, hellbenders likely become inactive during winter. Overwintering sites in New York included deep pools >2 m deep, fast-flowing riffles that remained free of ice cover,

and deep water pockets within riffles 1.5–2 m deep (Blais, 1996). However, hellbenders sometimes breed in Missouri and Arkansas during winter (Dundee and Dundee, 1965; Peterson et al., 1989).

K. Interspecific Associations/Exclusions. None known.

L. Age/Size at Reproductive Maturity. Sexual maturity is reached from 300–400 mm TL, with males normally maturing at a smaller size than females (Taber et al., 1975), although there is much variation reported in the literature (see Petranka, 1998, for a review). Age at sexual maturity has been estimated at 3–4 yr (Smith, 1907) and 5–6 yr (Bishop, 1941b) for eastern populations and 5–6 yr (Dundee and Dundee, 1965; Nickerson and Mays, 1973a) for Ozark populations.

M. Longevity. Hellbenders have survived as long as 29 yr in captivity (Nickerson and Mays, 1973a). Extrapolations from growth rate data suggest that some large individuals may live as long as 30 yr in nature (Taber et al., 1975; Peterson et al., 1983; Petranka, 1998).

N. Feeding Behavior. Crayfish are the most important food item for hellbenders, as indicated by their position at the top of most food lists in the literature (Smith, 1907; Green, 1935; Bishop, 1941b; Swanson, 1948). Other items that have been recorded include fish, insects, earthworms, snails, tadpoles, fish eggs, other hellbenders, and hellbender eggs (Nickerson and Mays, 1973a).

O. Predators. Fishes, turtles, and water snakes are important predators (Nickerson and Mays, 1973a). As noted above, hellbenders and their eggs are eaten by conspecifics (Nickerson and Mays, 1973a). Man is an important predator, as a result of both commercial collecting and scientific research. Swanson (1948) reported taking over 650 individuals from a 4.8-km stretch of Big Sandy River, Pennsylvania. Peterson (1989) killed 108 hellbenders in the Niangua River, Missouri, in 1974 and 62 from the Spring River, Arkansas, in 1985–86 for a study of food habits. Ingersol et al. (1991) killed 127 from the Niangua River in 1979–80 to document their reproductive cycle.

P. Anti-Predator Mechanisms. Hellbenders produce skin secretions that are lethal when injected into white mice and are probably unpalatable to some predators (Brodie, 1971a).

Q. Diseases. Captives are often infected by water mold (*Saprolegnia* sp.).

R. Parasites. Protozoans, nematodes, trematodes, cestodes, acanthocephalans, and leeches have been found associated with hellbenders (see Rankin, 1937 and Nickerson and Mays, 1973a, for a thorough review).

4. Conservation.

Hellbenders are classified as Endangered in Illinois, Indiana, Maryland, and Ohio;

Rare in Georgia; Of Special Concern or Species of Concern in New York, North Carolina, and Virginia; Watch List in Missouri; and Deemed in Need of Management in Tennessee. The actual degree of protection each of these designations afford varies by state, but generally, Endangered status requires that a permit be secured before a hellbender can be captured and provides penalties for possessing hellbenders without such a permit. The other categories listed above do not afford this level of protection, but do allow for some acknowledgment that the future of the species within their boundaries is not totally secure. Other states such as Alabama, Arkansas, Kentucky, Mississippi, South Carolina, and West Virginia track hellbender distribution records in a database, but do not generally afford them protection from take. Pennsylvania apparently neither tracks hellbender records nor protects them from take.

The U.S. Fish and Wildlife Service performed a status review of Ozark hellbenders *(C. a. bishopi)* in the early 1990s (LaClaire, no date). This review concluded that "populations of the Ozark hellbender in the majority of its range (in Missouri) are apparently stable and new populations of the species have been found during the recent status surveys." The final recommendation was that the Ozark hellbender did not warrant protection at the time.

As early as 1957, it was noted that the hellbender's range was rapidly shrinking as a result of human modification of stream habitats (Smith and Minton, 1957). Dundee (1971) listed "siltation, general pollution, and thermal pollution" as being responsible for eliminating the hellbender from "much of the Ohio River drainage, and from other industrialized regions." Bury et al. (1980) mentioned channelization and impoundment of streams and rivers as an agent of decline specifically for Alabama, Maryland, Missouri, Tennessee, and West Virginia. They also cited Nickerson and Mays (1973) when implicating industrialization, agricultural runoff, and mine wastes as contributing factors in Ohio, Pennsylvania, and West Virginia. Other authors have alluded to the range-wide decline in hellbender numbers (Williams et al., 1981; Gates et al., 1985). However, rigorous quantification of effort is lacking in most hellbender surveys, so there are few data to back up these claims. The sole exception of which we are aware, that of Trauth et al. (1992a), documented a drastic decline in hellbenders along the Spring River of Arkansas. They attributed the decline to over-collection of specimens for scientific purposes (see "Predators" above), habitat alteration related to recreational activities, elimination of riparian habitats leading to an increase in the silt burden, and water pollution associated with human occupation and development along the river. Rigorous historical abundance data exist for other streams (see "Historical versus Current Abundance" above) and these areas should be targeted for resurvey.

Family Dicamptodontidae

Dicamptodon aterrimus (Cope, 1867)
IDAHO GIANT SALAMANDER

Kirk Lohman, R. Bruce Bury

1. Historical versus Current Distribution.
Idaho giant salamanders (*Dicamptodon aterrimus*) are found in forested watersheds in north-central Idaho from the Coeur d'Alene River south to the Salmon River (Maughan et al., 1976; Nussbaum et al., 1983) and from two locations in Mineral County in extreme western Montana (Reichel and Flath, 1995). Formerly considered California giant salamanders (*D. ensatus*), Idaho giant salamanders are now regarded as genetically distinct from other species in the genus (Daugherty et al., 1983; Good, 1989). Although the general outline of their range has not changed in recent years, the distribution of Idaho giant salamanders has likely been much reduced within heavily logged watersheds (Fisher, 1989; Hamilton et al., 1998; Hossack, 1998).

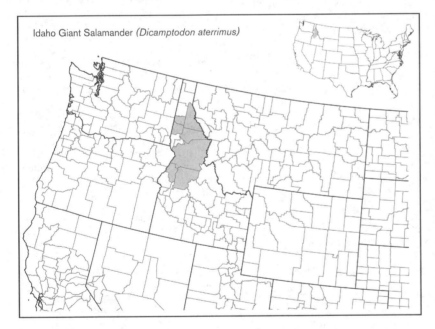

Idaho Giant Salamander (*Dicamptodon aterrimus*)

2. Historical versus Current Abundance.
Locally abundant in headwater streams in coniferous forest watersheds. Knowledge of historical abundance is scarce, but numbers have likely been reduced in areas of intense timber harvest where larvae may be adversely affected by sedimentation and increases in water temperature (Corn and Bury, 1989a; Bury et al., 1991).

3. Life History Features.
A. Breeding. Reproduction is aquatic.

i. Breeding migrations. The reproductive biology of Idaho giant salamanders is presumed to be similar in most respects to that described for other members of the genus, although breeding behavior of all dicamptodontids is poorly understood. There is circumstantial evidence from population structure that breeding occurs during both spring and fall (Nussbaum, 1969; Nussbaum and Clothier, 1973). Because neoteny is common, asynchronous breeding may also be characteristic of many populations with a high proportion of neotenes (Nussbaum and Clothier, 1973).

ii. Breeding habitat. Courtship likely takes places in hidden nest chambers beneath logs, stones, and crevices in small streams (Nussbaum et al., 1983).

B. Eggs.

i. Egg deposition sites. Nussbaum (1969a) discovered a gravid female buried 60 cm (2 ft) deep in a rock pile at the base of a small waterfall that he believed to be a nest site. Eggs are attached singly on the undersides of rocks within the nest chamber (Nussbaum, 1969a; Nussbaum et al., 1983).

ii. Clutch size. Females lay 135–200 eggs (Nussbaum, 1969a; Nussbaum et al., 1983). Clutch frequency is presumed to be biennial (Nussbaum et al., 1983; Blaustein et al., 1995b). Females remain with the eggs throughout development, apparently guarding them from predators (Nussbaum, 1969a). An incubation period of nearly 9 mo was reported by Henry and Twitty (1940).

C. Larvae/Metamorphosis.

i. Length of larval stage. Metamorphosis occurs during or after the third year (Nussbaum and Clothier, 1973).

ii. Larval requirements.

a. Food. Primarily benthic invertebrates and tailed frog (*Ascaphus* sp.) tadpoles (Metter, 1963). Some predation on juvenile fishes by other *Dicamptodon* sp. has been reported (Antonelli et al., 1972; Parker, 1993a).

b. Cover. Stones and submerged logs (Nussbaum et al., 1983).

iii. Larval polymorphisms. None.

iv. Features of metamorphosis. Transforming individuals are 92–166 mm TL (Nussbaum and Clothier, 1973). Larvae (145–150 mm TL) observed in process of metamorphosis during August in northern Idaho (K.L., personal observations). Complete metamorphosis of Idaho giant salamander larvae required from 11 d to 6 mo in aquaria (Kessel and Kessel, 1944). Mixed populations of neotenic and terrestrial adults are common (Nussbaum, 1976).

v. Post-metamorphic migrations. Move from streams to moist coniferous forest floors (Nussbaum et al., 1983).

vi. Neoteny. Common.

D. Juvenile Habitat. Same as adults.

E. Adult Habitat. Under logs and bark in coniferous forests (Nussbaum et al., 1983). Adults are rarely encountered and knowledge of their habits is scarce.

F. Home Range Size. Unknown.

G. Territories. Large larvae and adults are aggressive toward conspecifics, but whether larvae or adults defend territories is unknown.

H. Aestivation/Avoiding Desiccation. These behaviors have not been reported.

I. Seasonal Migrations. Unknown.

J. Torpor (Hibernation). Unknown.

K. Interspecific Associations/Exclusions. Often occur in the same streams as tailed frogs. Tailed frog tadpoles are an important prey item for larval Idaho giant salamanders.

L. Age/Size at Reproductive Maturity. Sexual maturity in both larval and terrestrial forms usually occurs at sizes greater than 115 mm SVL (Nussbaum et al., 1983).

M. Longevity. Unknown. Based on a 3-yr larval period and a maximum size of 250–300 mm, a lifespan of at least 6–10 yr would not be an unreasonable estimate.

N. Feeding Behavior. Larvae are sit-and-wait predators (Parker, 1994). Adult *Dicamptodon* feed on a wide variety of prey, including terrestrial invertebrates, small snakes, shrews, and mice (Nussbaum et al., 1983).

O. Predators. Predators include fishes, garter snakes, weasels, and water shrews (*Sorex palustris*) (Nussbaum et al., 1983; Blaustein et al., 1995b).

P. Anti-Predator Mechanisms. Toxic skin secretions, warning postures, and biting (Nussbaum et al., 1983). Adults are also known to growl or squawk (Nussbaum et al., 1983).

Q. Diseases. Unknown.

R. Parasites. Unknown.

4. Conservation.

The general outline of the range of Idaho giant salamanders has not changed in recent years, and they can be locally abundant in headwater streams in coniferous forest watersheds. However, numbers have likely been reduced in areas of heavily logged watersheds where larvae may be adversely affected by sedimentation and increases in water temperature (Corn and Bury, 1989a; Fisher, 1989; Bury et al., 1991; Hamilton et al., 1998; Hossack, 1998).

Dicamptodon copei Nussbaum, 1970
COPE'S GIANT SALAMANDER
Lawrence L.C. Jones, R. Bruce Bury

1. Historical versus Current Distribution.

Cope's giant salamanders (*Dicamptodon copei*) are known from western Washington and extreme northwestern Oregon (Nussbaum, 1970, 1976, 1983; Jones and Aubry, 1984; McAllister, 1995; Leonard et al., 1998). In Washington, they are known from the Cascades (south of Mt. Rainier), Willapa Hills, and Olympic Peninsula (except for most of the northeast portion; R.B.B., unpublished data). In Oregon, Cope's giant salamanders are confined to the vicinity of the Columbia River Gorge (Cascades and Coast Range). They can be locally abundant and are widely distributed through much of their range, but tend to have a more spotty distribution where they are sympatric with coastal giant salamanders (*D. tenebrosus*). Although it is likely that development and deforestation may have resulted in reduction in some populations, there has been no apparent reduction in range.

2. Historical versus Current Abundance.

Little is known about the effects of timber harvest on Cope's giant salamanders. Ruggiero et al. (1991) state that this species is associated with old-growth forests and/or stands with similar structural attributes. However, Bury et al. (1991) found densities somewhat lower in old forests than in young, naturally regenerated forest streams in the Washington Cascades. Unfortunately, results of investigations on this species may be obscured by combining Cope's giant salamander and coastal giant salamander data sets, as these species are phenotypically similar when young (e.g., Antonelli et al., 1972; Wilkins and Peterson, 2000). Ongoing investigations in the Olympic Peninsula and elsewhere will help clarify habitat relationships.

3. Life History Features.

A. Breeding. Reproduction is aquatic.

i. Breeding migrations. Unknown.

ii. Breeding habitat. Likely to be the same as egg deposition sites, which include hidden chambers under rocks and logs and in cutbanks of streams (Nussbaum et al., 1983; Steele et al., 2003).

B. Eggs.

i. Egg deposition sites. Nussbaum et al. (1983) reported on nine nests found in nature. Eggs are deposited in spring–fall. Females placed their eggs in hidden chambers under rocks and logs and in cutbanks. Steele et al. (2003) recently reported on two clutches.

ii. Clutch size. Clutches range from 25–115 eggs, averaging about 50. Females

guard the eggs, and conspecifics are often found with bite marks, which suggests nest defense. The two nests discovered by Steele et al. (2003) contained 23 and 28 eggs.

C. Larvae/Metamorphosis.

i. Length of larval stage. In general, the species is larviform throughout its life and growth rates are unknown.

ii. Larval requirements.

a. Food. Antonelli et al. (1972) studied food habits of Cope's giant salamanders and coastal giant salamanders, but did not distinguish between the two species or age classes, except transformed individuals were not included. They considered Cope's giant salamanders to be opportunistic, feeding primarily on invertebrate stream benthos.

b. Cover. Bury et al. (1991) found that Cope's giant salamanders (paedomorphic animals and larvae combined) primarily inhabited pools in mountain streams, using large stones for cover.

iii. Larval polymorphisms. None.

iv. Features of metamorphosis. Metamorphosis in Cope's giant salamanders is extremely rare. To date, six naturally metamorphosed individuals have been reported in the literature (Nussbaum, 1983; Jones and Corn, 1989; Loafman and Jones, 1996). All were from the Olympic Peninsula, except one from the Washington Cascades. All are extremely similar in coloration and external morphology (see color photograph in Jones and Corn, 1989), although we did not examine the Cascades specimen. This species has little proclivity for metamorphosis, as demonstrated by intensive thyroxin treatment to induce transformation (Nussbaum, 1976).

v. Post-metamorphic migrations. Unknown, but transformed individuals were usually found near streams.

vi. Neoteny. This species appears to be a near-obligate paedomorphic species.

D. Juvenile Habitat. Not applicable.

E. Adult Habitat. Unknown for metamorphosed individuals and little studied for larviform adults. Paedomorphic adults and larvae are usually associated with pools in small to moderately sized rocky mountain streams and, occasionally, montane lakes (Nussbaum et al., 1983; Bury et al., 1991).

F. Home Range Size. Unknown.

G. Territories. Unknown, but individuals often have scars inflicted by conspecifics (Nussbaum et al., 1983), which may be due to nest protection, territoriality, or both.

H. Aestivation/Avoiding Desiccation. Surface activity is normally high during the summer (Antonelli et al., 1972), but individuals probably seek refuge during desiccating conditions.

I. Seasonal Migrations. Unknown.

J. Torpor (Hibernation). Surface activity is absent or reduced in the winter, at least in areas where temperatures are near freezing (Antonelli et al., 1972).

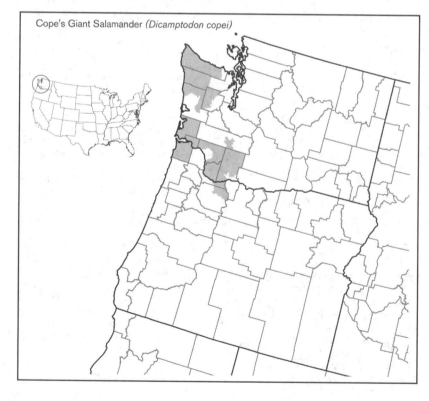

Cope's Giant Salamander (*Dicamptodon copei*)

K. Interspecific Associations/Exclusions.
They are sympatric with coastal giant salamanders, except on the Olympic Peninsula, where Cope's giant salamanders are the only members of the genus. Nussbaum (1976) listed eight sites where these two species coexist. They are not known to hybridize (Nussbaum, 1976; Good, 1989). Recent studies have yielded little information on the extent or mechanism of co-existence or interactions of sympatric giant salamanders. Torrent salamanders (*Rhyacotriton* sp., including Olympic torrent salamanders [*R. olympicus*], Columbia torrent salamanders [*R. kezeri*], and Cascade torrent salamanders [*R. cascadae*]) and tailed frogs (*Ascaphus* sp.) often co-occur with Cope's giant salamanders. In wet edge habitat, Van Dyke's salamanders (*Plethodon vandykei*) may occur throughout their Washington range and Dunn's salamanders (*P. dunni*) in its Oregon and southwestern Washington range.

L. Age/Size at Reproductive Maturity.
From 65–77 mm SVL (Nussbaum, 1976).

M. Longevity. Unknown.

N. Feeding Behavior. Branchiate adults may feed on larger prey, including tailed frog larvae and juveniles, small fish, and smaller congeners or their eggs (Antonelli et al., 1972; Kaplan and Sherman, 1980; Nussbaum et al., 1983; Jones and Raphael, 1998). They generally feed in streams, but sometimes venture out of the stream in wet weather (Nussbaum et al., 1983). Diet for metamorphosed forms is unknown.

O. Predators. Cope's giant salamanders, coastal giant salamanders, common garter snakes (*Thamnophis sirtalis*), and water shrews (*Sorex palustris*) have been reported (Nussbaum et al., 1983; Loafman and Jones, 1996).

P. Anti-Predator Mechanisms. Unreported, but flight is evident (personal observations).

Q. Diseases. Unknown.

R. Parasites. Unknown.

4. Conservation.
There has been no apparent reduction in the range of Cope's giant salamanders. They are widely distributed through much of their range and can be locally abundant. While little is known about the effects of timber harvesting on Cope's giant salamanders, it is likely that development and deforestation may have resulted in the reduction of some populations. According to Levell (1997), Cope's giant salamanders are listed as Protected in Oregon.

Dicamptodon ensatus (Eschscholtz, 1833)
CALIFORNIA GIANT SALAMANDER

R. Bruce Bury

1. Historical versus Current Distribution.
California giant salamanders (*Dicamptodon ensatus*) are the southernmost taxa of four species in the genus (Good, 1989). They occur in the Pacific Coastal region around San Francisco Bay north to Mendocino, Sonoma, and Napa counties, and south to Santa Cruz County. A disjunct population exists farther to the south in Monterey County (Anderson, 1969; Nussbaum, 1976). There is little evidence to suggest the current distribution differs from the historical distribution, but populations have undoubtedly been lost with the intense urbanization and habitat alteration characteristic of many parts of their range.

California Giant Salamander (*Dicamptodon ensatus*)

2. Historical versus Current Abundance.
Unknown, although it is likely that abundance is reduced in populations experiencing habitat modifications. In related Pacific giant salamanders (*D. tenebrosus*), larvae may be reduced in numbers where there has been clear-cut logging (Corn and Bury, 1989a) or siltation from roads (Welsh and Ollivier, 1998). However, opening of forest canopies over streams may lead temporarily to higher primary productivity that in turn increases the body sizes of larval Pacific giant salamanders (Murphy and Hall, 1981).

3. Life History Features.
A. Breeding. Reproduction is aquatic.
i. Breeding migrations. Unknown.
ii. Breeding habitat. During the breeding season, adults are found under stones in streams, but information on courtship behavior is not available.

B. Eggs.
i. Egg deposition sites. Subterranean habitats in running water (Henry and Twitty, 1940; Dethlefsen, 1948). On 17 June 1937, one nest of about 70 eggs was found attached to a submerged timber in a stream in the Santa Cruz Mountains, San Mateo County, California (Henry and Twitty, 1940); some eggs may have been lost in removing the timber from the stream bed. Dethlefsen (1948) reported on about 100 eggs and dismembered parts of two adult *Dicamptodon* that were ejected from a pipe drilled 6.1 m (20 ft) into a hillside spring in Santa Cruz County, California. Another live adult (243 mm TL) was in the pool outlet about 1 hr later. Three weeks later, another five adults were located under the same drilling station. No eggs appeared, but upon dissection, about 80 eggs were found in one female. All adults were weakly pigmented, suggesting they were underground for some time. Eggs are about 5.5 mm in diameter and attached singly to substrata such as rocks and logs by pedicels about 6 mm long.

ii. Clutch size. From 70 (Henry and Twitty, 1940) to about 100 eggs (Dethlefsen, 1948). Incubation time has been suggested as extremely long at almost 5 mo (Henry and Twitty, 1940), but this occurred in the laboratory.

C. Larvae/Metamorphosis.

i. Length of larval stage. Perhaps 18 mo (Kessel and Kessel, 1943a,b, 1944; see also Petranka, 1998). Larvae are mountain brook type with a reduced tail fin and no balancers.

ii. Larval requirements.

a. Food. No data exist for food of California giant salamanders. Diet is most likely similar to related coastal giant salamanders that feed on aquatic invertebrates, with a shift toward larger prey items with growth (Johnson and Schreck, 1969; Parker, 1994).

b. Cover. Younger larvae are found in slowly moving water near the shoreline, while older larvae tend to be found in the main stream channel (Kessel and Kessel, 1943a,b, 1944). Larvae often hide under gravel and cobble or other objects in streams.

iii. Larval polymorphisms. Unknown.

iv. Features of metamorphosis. Larvae metamorphose at about 130–140 mm TL in their second year of life (Kessel and Kessel (1943a,b).

v. Post-metamorphic migrations. Undocumented.

vi. Neoteny. Neotenic animals are present in many populations (De Marco, 1952; Nussbaum, 1976).

D. Juvenile Habitat. Probably similar to adult habitat.

E. Adult Habitat. California giant salamanders are associated with permanent and semipermanent streams. Adults occasionally are found under cover objects, such as rocks and logs, or under stones in streams during the breeding season (see Petranka, 1998).

F. Home Range Size. Unknown.

G. Territories. Unknown.

H. Aestivation/Avoiding Desiccation. Aestivation is unlikely; animals likely avoid desiccating conditions by seeking shelter under cover objects.

I. Seasonal Migrations. Adults return to streams with the advent of fall rains (Kessel and Kessel, 1943a). In related coastal giant salamanders, newly metamorphosed juveniles (Nussbaum et al., 1983) and sometimes neotenic adults (Welsh, 1986) move away from streams in rainy periods.

J. Torpor (Hibernation). Unknown, but unlikely.

K. Interspecific Associations/Exclusions. There is a hybrid zone between *D. ensatus* and *D. tenebrosus* about 10 km north of Gualala, Mendocino County, California (Good, 1989).

L. Age/Size at Reproductive Maturity. Adults measure 170–305 mm TL (Nussbaum, 1976; see also Petranka, 1998).

M. Longevity. Unknown.

N. Feeding Behavior. Includes a range of invertebrate prey and small vertebrate prey (Bury, 1972), including conspecifics (Anderson, 1960).

O. Predators. Likely include large birds, small mammals, and other vertebrates, including aquatic garter snakes (*Thamnophis atratus*; Lind and Welsh, 1990).

P. Anti-Predator Mechanisms. Anti-predator posturing has not been documented in California giant salamanders, but coastal giant salamanders have an arched posture and will release toxins when disturbed (Nussbaum et al., 1983). Unlike most salamanders, California giant salamanders can emit a vocalization that resembles a bark (Stebbins, 1951).

Q. Diseases. Unknown.

R. Parasites. Unknown.

4. Conservation.

The current distribution of California giant salamanders likely does not differ from the historical distribution, but populations have undoubtedly been lost due to the intense habitat alterations characteristic of many parts of their range. According to Petranka (1998), because California giant salamanders have a small range that exists within an area of intense human activity, they are at threat from siltation of their stream habitats and urbanization.

Dicamptodon tenebrosus (Baird and Girard, 1852[b])

COASTAL GIANT SALAMANDER

Lawrence L. C. Jones, Hartwell H. Welsh, Jr.

1. Historical versus Current Distribution.
Coastal giant salamanders (*Dicamptodon tenebrosus*) occur from Mendocino County, California, north to extreme southwestern British Columbia, Canada, along the Coast and Cascade Ranges, exclusive of the Olympic Peninsula (Nussbaum, 1976; Good, 1989; McAllister, 1995). Coastal giant salamanders are widely distributed throughout most of their range; however, in British Columbia, they have a limited range (Cook, 1970) and are listed as Endangered. There have been no indications of declines or increases in historical distribution, but there has probably been some fragmentation within their range resulting from habitat alterations, mostly due to forestry practices.

2. Historical versus Current Abundance.
Throughout most of their range, coastal giant salamanders are one of the most abundant stream vertebrates. Larvae and paedomorphic animals may reach high densities and are often the dominant vertebrate predators because of a relatively large biomass (Hall et al., 1978; Murphy and Hall, 1981; Hawkins et al., 1983; Corn and Bury, 1989b). Numerous studies have investigated the relationships of the aquatic life stages of coastal giant salamanders to timber harvesting (Hall et al., 1978; Murphy and Hall, 1981; Murphy et al., 1981; Hawkins et al., 1983; Bury and Corn, 1988a; Corn and Bury, 1989a; Bury et al., 1991; Welsh and Lind, 1991; Welsh, 1993;

Coastal Giant Salamander (*Dicamptodon tenebrosus*)

Kelsey, 1995). These studies have produced variable results, with aquatic coastal giant salamanders usually found in higher densities in streams in unlogged mature and old-growth forests; in several studies, higher densities were reported in streams crossing recently logged sites, presumably the result of an increase in primary productivity. Welsh and Ollivier (1998) studied the effects of increased sedimentation and found that coastal giant salamanders occurred in higher densities in streams unimpacted by fine sediments. Increased siltation negatively affects these animals by filling rocky interstices under cover objects, particularly in low-gradient streams that do not adequately flush sediments. Despite the negative effects of habitat alteration on coastal giant salamanders, they seem to be more resilient than other stream-associated amphibians in the coastal Northwest and are likely to persist in streams that are moderately disturbed.

Metamorphosed coastal giant salamanders are encountered less frequently in most field investigations (e.g., Bury, 1983; Bury and Corn, 1988b; Raphael, 1988; Welsh and Lind, 1988, 1991; McComb et al., 1993), even though they are probably locally abundant in inaccessible subterranean retreats. Consequently, stand-scale habitat relationships for the terrestrial life stage are poorly known. Habitat use patterns from a telemetry study of terrestrial animals (Johnston, 1998) were somewhat nebulous, but it appeared that good-quality, near-stream habitats (e.g., old forests or younger forest stands with wide stream buffers) were selected.

3. Life History Features.

A. Breeding. Reproduction is aquatic.

i. Breeding migrations. Adult females migrate from upland habitats to streams to oviposit.

ii. Breeding habitat. Pools or slow-moving portions of streams (Nussbaum, 1969a; Jones et al., 1990).

B. Eggs.

i. Egg deposition sites. Courtship sites are unknown. Eggs are attached singly to the undersurface of rock or wood, and females attend their eggs. Three nest sites have been reported from the Oregon Coast Range. Nussbaum (1969a) reported two subterranean nests in shallow, slow-moving portions of two streams; Jones et al. (1990) described the nest of a paedomorphic female in a small pool of a third stream. No nests are known from lentic habitats.

ii. Clutch size. Nussbaum (1969a) reported 83 and 146 eggs from the two subterranean nests he discovered, and Jones et al. (1990) reported 129 eggs from the paedomorphic female they found.

C. Larvae/Metamorphosis.

i. Length of larval stage. Larvae typically metamorphose 18–24 mo after hatching (Nussbaum et al., 1983). Some individuals may overwinter and transform in the third year (Nussbaum and Clothier, 1973). Metamorphosis occurs from 92–166 mm TL (Nussbaum and Clothier, 1973).

ii. Larval requirements.

a. Food. Larval coastal giant salamanders are known to feed on a variety of invertebrates, mostly benthic, including insect larvae and adults, amphipods, ostracods, trematodes, mollusks, and crayfish, as well as some vertebrates such as salmonids (Salmonidae), sculpins (Cottidae), northwestern salamanders (*Ambystoma gracile*), and congeners (Fitch, 1936; Schonberger, 1944; Johnson and Schreck, 1969; Antonelli et al., 1972; Parker, 1993a,b, 1994; Esselstyn and Wildman, 1997).

b. Cover. Larvae and paedomorphic animals are associated with coarse substrates of streams, rivers, and mountain lakes (Nussbaum et al., 1983; Leonard et al., 1993). Welsh (1993) found coastal giant salamanders most abundant in late-successional forests, but also noted that within-stream conditions were better predictors of presence and abundance than surrounding forest conditions. Parker (1991) reported that salamander density increased with density of large stones (>7.5 cm). Welsh (1993) found salamanders most abundant in reaches of intermediate gradient due to a combination of microhabitat diversity (i.e., pools, riffles, runs) and sediment flushing. Several studies have concluded that fine sediments fill and eliminate critical cover interstices within the streambed matrix (e.g., Hall et al., 1978; Hawkins et al., 1983; Corn and Bury, 1989a; Welsh and Ollivier, 1998). Welsh (1986) reported a paedomorphic individual in a pitfall trap 3 m from a small stream. Larvae increase cover use in the presence of chemical cues from cutthroat trout (*Oncorhynchus clarki*; Rundio and Olson, 2003).

iii. Larval polymorphisms. None.

iv. Features of metamorphosis. Paedomorphosis is typical in perennially aquatic sites (large streams, rivers, lakes, and ponds), while small streams that may dry up tend to harbor predominantly metamorphic populations (Nussbaum and Clothier, 1973).

v. Post-metamorphic migrations. Newly metamorphosed animals may venture into upland habitats during rainy periods, ≤400 m or more from water (McComb et al., 1993).

vi. Neoteny. Coastal giant salamanders exhibit facultative paedomorphosis (Nussbaum, 1976).

D. Juvenile Habitat. Similar to adults and larvae.

E. Adult Habitat. Terrestrial adults are more abundant in forested habitats than in pre-canopy sites, where they are found under rocks and in logs, root channels, and burrows (Johnston, 1998). Paedomorphic animals require perennial water, where they seek cover in coarse substrates. They are found primarily in lotic waters, from headwaters to rivers, but may also be found in some lentic habitats (Nussbaum et al., 1983; Leonard et al., 1993). Paedomorphic animals have habitat affinities similar to large larvae.

F. Home Range Size. Johnston (1998) found an average home range of 3,047–5,196 m^2 (modified convex polygon and 95% adaptive kernal methods, respectively) for 20 radio-telemetered, metamorphosed animals in British Columbia.

G. Territories. Not documented.

H. Aestivation/Avoiding Desiccation. Unknown; during the summer, larvae and paedomorphic animals are active in aquatic habitats (personal observations).

I. Seasonal Migrations. Newly metamorphosed (Nussbaum et al., 1983; Johnston, 1998), and sometimes paedomorphic (Welsh, 1986) animals move out of streams to the surrounding habitat during rainy and wet periods. Some metamorphosed individuals remain in the vicinity of the stream, others prefer upland areas (Johnston, 1998).

J. Torpor (Hibernation). Larvae (Antonelli et al., 1972) and adults probably seek refuge from temperature extremes in the winter, at least in areas with freezing temperatures.

K. Interspecific Associations/Exclusions. In a few streams of southern Mendocino County, coastal giant salamanders are known to hybridize with California giant salamanders (*D. ensatus*; Good, 1989). Coastal giant salamanders are sympatric with Cope's giant salamanders (*D. copei*) across the range of the latter, except for the Olympic Peninsula, and may occur in the same stream (Nussbaum and Clothier, 1973; Nussbaum, 1976; Daugherty et al., 1983). The extent and mechanisms of sympatry are not well known, as small larvae are phenotypically similar and are often grouped together as *Dicamptodon* sp. in scientific investigations (e.g., Antonelli et al., 1972; Wilkins and Peterson, 2000). The two species are not known to hybridize (Nussbaum, 1976; Good, 1989). They are often sympatric with torrent salamanders (*Rhyacotriton* sp.) and tailed frogs (*Ascaphus* sp.; Bury et al., 1991). Welsh (1993) found larval coastal giant salamander numbers higher in stream reaches with larval tailed frogs and southern torrent salamanders (*R. variegatus*) present, both of which are known to be prey of coastal giant salamanders. They are sometimes found in streams with sculpins or salmonids, with which they may compete for food or function as predator or prey (Antonelli et al., 1972; Parker, 1993a,b).

L. Age/Size at Reproductive Maturity. Except for paedomorphic animals, larvae mature after metamorphosis, generally when they are 115 mm SVL or longer (Nussbaum, 1976).

M. Longevity. Nothing is known about longevity in coastal giant salamanders, but other large aquatic salamanders can be long lived (Hairston, 1987).

N. Feeding Behavior. Coastal giant salamanders are sit-and-wait predators, lunging short distances with surprising speed to procure live prey (Bury, 1972). Adults may consume small mammals and other vertebrates and have even been seen with snakes in their jaws (Diller, 1907; Graf, 1949; Wilson, 1970; Bury, 1972).

O. Predators. Weasels and river otters (Mustelidae), water shrews (Soricidae), garter snakes (*Thamnophis* sp.), salmonids, and conspecifics (Fitch, 1941; Nussbaum and Maser, 1969; Nussbaum et al., 1983; Lind and Welsh, 1990, 1994; Parker, 1993a).

P. Anti-Predator Mechanisms. A variety of defenses may be employed, including biting, arching, tail-lashing, and exudation of noxious skin secretions (Brodie, 1977; Nussbaum et al., 1983).

Q. Diseases. Unknown.

R. Parasites. Known parasites include helminths (Lynch, 1936; Lehmann, 1954; Anderson et al., 1966).

4. Conservation.
There have been no indications of declines or increases in the historical distribution of coastal giant salamanders, but there has been some fragmentation within their range resulting from habitat alterations, mostly due to forestry practices. Coastal giant salamanders seem to be more resilient than other stream-associated amphibians in the coastal Northwest and are likely to persist in streams that are moderately disturbed.

Family Plethodontidae

Aneides aeneus (Cope and Packard, 1881)
GREEN SALAMANDER

Thomas K. Pauley, Mark B. Watson

1. Historical versus Current Distribution.
Dunn (1926) described the range of green salamanders (*Aneides aeneus*) as extending from Nickajack Cave, Tennessee, north to Baileysville, West Virginia, and through the Cumberland and Kanawha sections of the Allegheny Plateau. Bishop (1943) shows them from northern West Virginia south through southwestern Ohio, eastern Kentucky, western Virginia, North Carolina, South Carolina, and eastern Tennessee into northern Georgia and Alabama. Later range extensions included the extreme northeast of Mississippi (Ferguson, 1961b), southeastern Pennsylvania (Richmond, 1952; Bier, 1985), and southeastern Maryland (Harris and Lyons, 1968). Populations in the tri-state area of southwestern North Carolina, northwestern South Carolina, and northeastern Georgia are disjunct from the main stem

of the range (Petranka, 1998). Disjunct populations have also been found in extreme southern Indiana (Madej, 1998). In Virginia, they are known only from the extreme southwestern part of the state in the Appalachian Plateau and Ridge and Valley physiographic provinces (Mitchell and Reay, 1999). Green salamanders are found in the Cumberland Mountains, Cumberland Plateau, and Eastern Highland Rim in Tennessee (Redmond and Scott, 1996). There may be disjunct populations in the Bays Mountains, on Clinch Mountain, in Appalachian Ridge and Valley, and in the Inner Central Basin of Tennessee (Redmond and Scott, 1996). A disjunct population was reported on the eastern slope of Mount LeConte in the Great Smoky Mountains National Park (Weller, 1931), however, animals have not been located since, so it is probably a doubtful record (Redmond and Scott, 1966; Petranka, 1998).

High elevation populations range from 762 m on Clinch Mountain, Virginia (Fowler, 1947), to 915 m on Droop and Backbone mountains in West Virginia (Pauley, 1993b), 1,250 m at Highlands, North Carolina (Gordon and Smith, 1949), 1,265 m on Black Mountain, Kentucky (Barbour, 1953), and to 1,341 m on Cold Mountain, North Carolina (Gordon, 1967).

Systematic studies show there is more than one form or species of *Aneides* in the eastern United States. Morescalchi (1975) and Sessions and Kezer (1987) suggest that there are three chromosomally distinct forms. Morescalchi (1975) proposed one form in eastern Tennessee. Sessions and Kezer (1987) described one form in southern Pennsylvania, southern Kentucky, West Virginia, and southwestern North Carolina, and a second form in southern Tennessee and northern Alabama. Jeff Corser and his colleagues (personal communication) recently examined nucleotide

sequences from mitochondrial and nuclear genes, as well as ecological, life history, and morphological traits, and found that there are four species-level clades of green salamanders.

2. Historical versus Current Abundance.
Weller (1931) reported green salamanders on Mount Le Conte. King (1939) verified the identification of this specimen, but numerous attempts to confirm this locality and the occurrence of green salamanders in the Great Smoky Mountains have failed. Green salamanders declined for some unknown reason in the 1970s in North Carolina (Mitchell et al., 1999). Snyder (1991) discussed several potential reasons for this decline but concluded that the probable cause was that unusually prolonged cold periods may have frozen many hibernating green salamanders in a torpid condition, which prevented them from moving into deeper

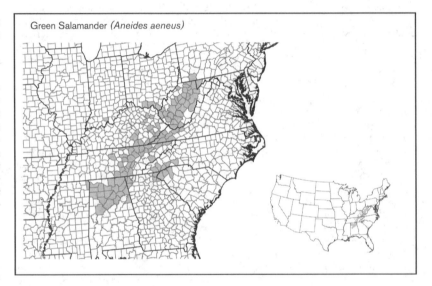

Green Salamander *(Aneides aeneus)*

and safer crevices. Corser (2001) monitored seven historical populations of green salamanders throughout the 1990s on the Blue Ridge Escarpment in northeastern Georgia, northwestern South Carolina, and southwestern North Carolina and found a 98% decline in relative abundance compared to numbers observed in the 1970s. He contributed this sharp decline to habitat loss, overcollecting, epidemic disease, and climate changes.

Barbour (1971) stated that green salamanders reached tremendous population numbers in the 1930s under bark of the millions of dead chestnut trees in eastern Kentucky. Most records of green salamanders beneath bark of downed trees are from the 1920s–50s, before the original forests were cut (C.H. Pope, 1928; Welter and Barbour, 1940; Fowler, 1947; Gordon, 1952; Barbour, 1953). Petranka (1998) concluded that large, thick slabs of bark on large old-growth logs probably provided more favorable microhabitats for foraging and nesting than the smaller logs found in secondary forests today.

3. Life History Features.

A. Breeding. Reproduction is terrestrial.

i. Breeding migrations. Migrations have been reported from hibernation crevices to outlying rock outcrops during the breeding period (Gordon, 1952). Dispersal of adults as well as juveniles occurs during prehibernation periods (Gordon, 1952; Canterbury, 1991). In the southern Appalachians, Gordon (1952) reported that mating occurs in late May to early June. Snyder (1971) and Cupp (1971, 1991) observed breeding activity in autumn and spring. Cupp (1971) observed courtship and spermatophore deposition on 17 October 1970 in Kentucky. In West Virginia, Brooks (1948a) reported a courting pair on a cliff face between midnight and 0100 h on 13 June 1941. Canterbury and Pauley (1994) determined that mating occurs in late May to early June and possibly in September–October in West Virginia. Dates of egg deposition vary from late May to early July in North Carolina (Gordon, 1952; Eaton and Eaton, 1956; Snyder, 1971); mid-July in Kentucky (Cupp, 1991); the third week of July in Mississippi (Woods, 1969); and the first 2 wk of June in West Virginia (Canterbury and Pauley, 1994).

ii. Breeding habitat. Breeding occurs in crevices or rock ledges of emergent rocks, rock outcrops, and rock cliffs in mesophytic forests. Cupp (1971) observed the courtship ritual and subsequent mating of a pair of green salamanders in a crevice of a sandstone rock outcrop in Kentucky. Brooks (1948a) reported watching a courting pair on a cliff face in West Virginia. Gordon (1952) reported that he found single pairs of salamanders in a crevice prior to egg deposition, but he did not observe courtship. Snyder (1971) observed courtship in late May, presumably in a rock crevice because green salamander nests have only been found in rock crevices in the southern Appalachians.

B. Eggs.

i. Egg deposition sites. Females attend clusters of unpigmented eggs that are held together and attached to the roof of a crevice by strands of mucus. Green and Pauley (1987) reported that several clusters of eggs were observed in central West Virginia on 4 July 1976.

ii. Clutch size. Number of eggs/clutch ranges from 10–30 in North Carolina (Gordon, 1952); 14–20 in Kentucky (Cupp, 1991); 12–27 (Canterbury and Pauley, 1994); and 20–32 (Lee and Norden, 1973) in West Virginia.

C. Direct Development. Females brood. Incubation period of eggs is 84–91 d in North Carolina (Gordon, 1952), 73 d in Kentucky (Cupp, 1991), 82 d in Mississippi (Woods, 1969), and 82–90 d in West Virginia (Canterbury and Pauley, 1994).

Norden and Groves (1974) found a female attending newly hatched young on 4 September 1972 in northern West Virginia. Females produce eggs on a biennial cycle (Snyder, 1991; Canterbury and Pauley, 1994).

D. Juvenile Habitat. Juveniles remain in the same habitat as adults. Walker and Goodpaster (1941) observed two groups of hatchlings, each with an attending female, on 30 September 1939 in Ohio. Post-hatching juveniles remain with the female for about 1 mo and then move upward on rock faces toward moss-covered ledges and crevices. Juveniles move about more on rock outcrops than do adults (Canterbury, 1991). Gordon (1952) found that after a brooding period of a few months, juveniles leave the nests and move to cracks and crevices. Cupp (1991) observed that dispersing hatchlings and juveniles were found occasionally in places (interfaces between rock cliffs and tree limbs and on more moist rock faces and crevices) where adults usually do not occur.

E. Adult Habitat. Green salamanders have been found in caves (Cope and Packard, 1881), under bark on logs (C.H. Pope, 1928), on cliffs and rock faces (Petranka, 1998) and under logs in forests (T.K.P., personal observation). Bishop (1928) considered green salamanders to be a weak arboreal species. Green salamanders are most commonly encountered in the crevices of emergent rocks, on rock outcrops, and on rock cliffs in forests. Primarily, such rocks do not have running water but are usually in cool, well-shaded, high humidity forests. Occasionally, green salamanders are found on dry rock outcrops (Pauley, 1993b; Waldron et al., 2000). Rock types include sandstone, limestone, dolomite, granite, and quartzite. Type of rock may be less important than crevice size and moisture (Gordon and Smith, 1949).

F. Home Range Size. Home range size has not been determined for green salamanders, but several studies have examined linear movement patterns. In an experiment on the movement of displaced green salamanders, Gordon (1961) found evidence for homing. However, most green salamander movements involve leaving crevices to forage for food and to disperse during the spring and fall. Gordon (1952) found that adults he followed in a study at Highlands, North Carolina, moved <3.7 m from the point where he originally captured them. He found two males (of 26 marked) that moved 76.2 m and 91.4 m, and one immature that moved 106.7 m. Most movements were the result of a prehibernation dispersal period that extended from late October to mid-December. Canterbury (1991) found that juveniles move more than adults. In central West Virginia, both sexes and all age classes move during the

fall dispersal, with some movement in August–September, but most dispersal occurs in October (Canterbury, 1991). Distance moved on single rock outcrops ranged from 3.4–15.8 m (n=15). Distances moved between rock outcrops ranged from 2.6 m for one female, 16.8–46.3 m for three males, and 6.4–49.4 m for five subadults. Waldron (2000) found in northern West Virginia that the mean linear distance that females travel is 1.86 m, males 3.18 m, and juveniles 3.34 m. Movement only occurred between crevices on a single rock outcrop and not between rock outcrops. Woods (1969) observed that subadults in Mississippi move ≤ 31 m. Green salamanders have been observed crossing roads (Williams and Gordon, 1961; Cupp, 1991).

G. Territories. Males, but not females and subadults, are territorial and aggressively defend their crevices against other males (Cupp, 1980; Canterbury and Pauley, 1991). Gravid females probably mate with males in optimal crevices, and females either use these crevices as nesting sites or move to other favorable crevices (Cupp, 1991). Cupp (1980) observed that resident males won 95.6% of all encounters, and he concluded that size was not a significant factor in territorial defense.

H. Aestivation/Avoiding Desiccation. Green salamanders become less active and tend to move deeper in rock crevices during high temperatures and dry conditions of July–August in the lower elevations in West Virginia (Pauley, 1993b). In central West Virginia, Canterbury (1991) found that green salamanders are less active in June–August than in September–October. In higher elevations in northern West Virginia, Waldron (2000) found green salamanders are the most active in June, with little activity from the first week in July to the following May. Snyder (1991) examined several potential causes for the decline of the Blue Ridge populations of green salamanders. While he did not believe drought was the major reason for the declines, he did suggest that the drying effect of droughts could eliminate safe refuges.

I. Seasonal Migrations. Seasonal migrations of green salamanders involve pre-hibernation and post-hibernation dispersals. Gordon (1952) reported that salamanders not observed during the summer on his study sites in North Carolina appeared in October just prior to hibernation. He considered this to be a pre-hibernation aggregation. He also noted a post-hibernation dispersal in late April to early May. By late May to early June, salamanders had moved into crevices. In southeastern Kentucky, Cupp (1991) recorded the major dispersal and aggregation period to be from early November to mid-December (pre-hibernation period). Cupp found the post-hibernation dispersal period extended from mid-March to late April. In central West Virginia males, females, and juveniles disperse from August–September and into October (Canterbury, 1991).

J. Torpor (Hibernation). Hibernation probably extends from mid-December to early March (Gordon, 1952; Canterbury, 1991; Cupp, 1991). Hibernation occurs in deep interconnecting crevices (Gordon, 1952). Gordon (1952) was only able to locate four individuals in deep anastomosing crevices from November–April in North Carolina. There are some reported occurrences during the winter. In central West Virginia, Canterbury (1991) found specimens at the mouths of crevices in November (n=2), December (n=9), and March (n=7). In the lower elevations of the western section of the state, he found individuals at the mouths of crevices in January (n=1), February (n=2), and March (n=2).

K. Interspecific Associations/Exclusions. Cliff-dwelling plethodontid salamanders display a stratification hierarchy (Baltar, 1983; Cliburn and Porter, 1987; Waldron, 2000). Baltar (1983) found that green salamanders occupied the highest crevices on rock faces; northern zigzag salamanders *(Plethodon dorsalis)* and northern slimy salamanders *(P. glutinosus)* occupied lower crevices. Waldron (2000) observed green salamanders more frequently on high areas of rock faces than either northern slimy salamanders or Allegheny Mountain dusky salamanders *(Desmognathus ochrophaeus).* Because green salamanders have a greater climbing ability than northern slimy salamanders (Cliburn and Porter, 1986; Canterbury, 1991) and Allegheny Mountain dusky salamanders (Canterbury, 1991), they may avoid competition with these two sympatric species by moving to upper crevices on rock faces. Canterbury and Pauley (1991) observed Cumberland Plateau salamanders *(P. kentucki)* in higher crevices on rock faces than northern slimy salamanders and concluded that Cumberland Plateau salamanders are probably keener competitors with green salamanders than are northern slimy salamanders.

Laboratory studies have shown that interspecific associations exist between green salamanders and Cumberland Plateau salamanders, Allegheny Mountain dusky salamanders, and Wehrle's salamanders *(P. wehrlei;* Cupp, 1980; Canterbury and Pauley, 1991). Canterbury and Pauley (1991) observed aggressive behaviors between male green salamanders, Cumberland Plateau salamanders, and northern slimy salamanders.

L. Age/Size at Reproductive Maturity. Both sexes of green salamanders become sexually mature at about 45 mm in Ohio (Juterbock, 1989). In central West Virginia, both sexes probably become sexually mature during the spring–summer of their third year (Canterbury, 1991).

M. Longevity. Snider and Bowler (1992) report a lifespan of 3 yr, 4 mo for a captive specimen. Since sexual maturity is not reached in both sexes until their third year (Canterbury, 1991), they probably live much longer than 3 yr. Corser (2001) reported that Castanet (unpublished data) found that green salamanders may live 10 yr or longer.

N. Feeding Behavior. As with most salamanders, green salamanders are opportunistic feeders. Gordon (1952) found that prey items included snails, slugs, spiders, and small insects. More specific food items include coleopterans, dipterans, and hymenopterans (Lee and Norden, 1973; Canterbury and Pauley, 1990). All age classes of green salamanders prey on the same items, but hatchlings consume more mites than other invertebrates (Canterbury and Pauley, 1990).

O. Predators. Predators include ring-necked snakes *(Diadophis punctatus)* and common garter snakes *(Thamnophis sirtalis;* T.K.P., personal observations) and probably other species of snakes. Juveniles could be prey items of adult Allegheny Mountain dusky salamanders in areas where Allegheny Mountain dusky salamanders inhabit crevices in rock faces.

P. Anti-Predator Mechanisms. Green salamanders avoid most predators through their nocturnal activity, but there is some diurnal activity (Gordon, 1952). Green salamanders are cryptically hidden from diurnal predators by their color pattern that resembles the lichen-covered rocks of their habitat (Brandon and Huheey, 1975). Hatchlings become immobile when touched (Netting and Richmond, 1932; Gordon, 1952; Brodie et al., 1974a). When adult green salamanders are touched with a twig, they turn their tail toward the opening of the crevice and undulate it from side to side (Brodie, 1977).

Q. Diseases. Diseases have not been reported.

R. Parasites. Baker (1987) has reported a nematode, *Batracholandros magnavulvaris.*

4. Conservation.
Several types of perturbations can negatively impact green salamanders. Obviously, removal of emergent rocks in forests eliminates this species in such areas. Road cuts and other larger corridors adjacent to emergent rocks and outcrops can result in an increase in airflow and greater insolation, thus increasing temperatures and decreasing moisture. Removal of trees that shade rock outcrops also results in greater insolation that ultimately dries nesting and foraging crevices. In clearcuts, Petranka (1998) suggests a buffer zone of 100 m or more around rock outcrops with known green salamander populations. Blue Ridge populations in North Carolina experienced declines in the late 1970s (Snyder, 1991),

and there appears to have been some local extinctions in that area (Mitchell et al., 1999). The total range of green salamanders is fragmented into several disjunct populations, and such isolated populations are always at risk of extinction. Over-collecting in some areas may be a threat (Mitchell et al., 1999; Corser, 2001).

Green salamanders are listed as Endangered in Indiana, Ohio, Maryland, and Mississippi, Threatened in Pennsylvania, Protected in Georgia (http://georgiawildlife.dnr.state.ga.us), Rare in North Carolina, and Of Special Concern in South Carolina and West Virginia.

Aneides ferreus Cope, 1869
CLOUDED SALAMANDER

Nancy L. Staub, David B. Wake

1. Historical versus Current Distribution.
Clouded salamanders *(Aneides ferreus)* are found in the coastal forests of Oregon and northern California. Their range extends from the Columbia River south through the Siskiyou and Coast mountains and western Cascades of Oregon (but they are not found in the extreme northwestern part of the Coast Mountains) and northwestern California. In California, their range extends to near the junctions of Hurdygurdy Creek and Goose Creek with the South Fork of the Smith River near the coast, and north of the junction of the Salmon and Klamath rivers further inland (Leonard et al., 1993; Wake and Jackman, 1999 [1998]). The current distribution of the clouded salamander is much more restricted than previously thought because all British Columbia specimens and most specimens from California have been assigned to the new species *Aneides vagrans* (wandering salamander; Wake and Jackman, 1999 [1998]).

Clouded salamanders occur in high densities in old growth forests (Corn and Bury, 1991) and in recently cut or burned areas in association with stumps, decaying logs, and coarse, woody debris (Van Denburgh, 1916; Fitch, 1936; McKenzie and Storm, 1970; Whitaker et al., 1986; Bury and Corn, 1988b; Welsh and Lind, 1988, 1991). Welsh and Lind (1988) found that mesic forest stands had significantly higher capture rates of clouded salamanders than did drier stands. This species tends not to be found in dense forests (Wood, 1939). Individuals have been founds ≤1,525 m elevation (Beatty, 1979). Populations have certainly been lost due to forestry management practices and urban sprawl.

2. Historical versus Current Abundance.
Clouded salamanders are typical inhabitants of old growth forests (Corn and Bury, 1991), especially in edge habitats (e.g., forest clearings) with downed or standing decaying trees and stumps.

Clouded Salamander *(Aneides ferreus)*

Populations associated with post-logged areas eventually decline, and it is doubtful that this species can survive in areas where forests are intensively managed on short rotation cycles because of severe reductions in moisture conditions and the amount of large woody debris (Corn and Bury, 1991). The amount of coarse, woody debris retained after timber harvesting under current forest management guidelines probably does not provide adequate habitat for clouded salamanders (Butts and McComb, 2000).

3. Life History Features.

    A. Breeding. Reproduction is terrestrial.

    i. Breeding migrations. Do not occur. Females lay eggs in late June and July, and hatchlings emerge in late August or September (Petranka, 1998).

    i. Breeding habitat. In rotting logs and possibly in the forest canopy.

    B. Eggs.

    i. Egg deposition sites. Egg clutches have been found in rotting logs (Storm, 1947).

    ii. Clutch size. Reported clutch sizes vary from 9–17 eggs (Storm, 1947). An ovarian count reported 12 eggs (Fitch, 1936).

    C. Direct Development.

    i. Brood sites. If the reproductive biology of the clouded salamander is similar to that of the wandering salamander, brood sites occur in the forest canopy as well as in decaying logs (Welsh and Wilson, 1995).

    ii. Parental care. Apparently variable. Clutches have been found without a parent in attendance (*A. ferreus*; Storm, 1947), with a female in attendance (*A. vagrans*; Dunn, 1942), and with both a male and a female in attendance (*A. ferreus*; Storm, 1947).

    D. Juvenile Habitat. Juvenile clouded salamanders prefer bark litter over rock or leaf litter (McKenzie and Storm, 1970). Subadults select bark litter at higher temperatures (20 °C, 25 °C) and show no preference between rock or bark litter at lower temperatures (10 °C; McKenzie and Storm, 1970). McKenzie and Storm (1971) demonstrate that as juvenile animals become older their color patterns become darker and duller.

    E. Adult Habitat. Adults are found associated with stumps and logs in Douglas fir forests, crevices in rock outcrops and road cuts, and talus (Van Denburgh, 1916; Wood, 1939; Lowe, 1950; Nussbaum, 1983; Maser and Trappe, 1984; Leonard et al., 1993; Jackman, 1999 [1998]). Animals are most commonly found in decaying logs and stumps with intact bark—either under the bark, within the log, or under the log (Bury and Corn, 1988b; Corn and Bury, 1991). Habitats with such logs are often associated with forest edges (Myers and Maslin, 1948; Stebbins, 1951). For example, McKenzie and Storm (1970) found clouded salamanders in stumps and fire-charred logs of Douglas fir associated with forest clearings and road banks. The preferred type of wood is Douglas fir;

individuals are found less commonly in Port Orford cedar, alder, and redwood (Myers and Maslin, 1948; Wake, 1965). Individuals have been found ≤6.1 m above the ground in dead stumps (Van Denburgh, 1916; Slevin, 1928). Based on recent work documenting the presence of wandering salamanders in the forest canopy, it is probable that clouded salamanders occupy similar habitat in their range (J.C. Spickler, personal communication). In a microhabitat selection experiment, adults consistently selected against leaf litter and showed equal preference for bark litter and rock (McKenzie and Storm, 1970). Jackman (1999 [1998]) suggests that there may be a difference in habitat preferences between clouded and wandering salamanders. In northwestern California, clouded salamanders are commonly found associated with both decaying logs and rocky slopes, while wandering salamanders are almost exclusively found associated with decaying logs.

    F. Home Range Size. Unknown.

    G. Territories. In contrast to wandering salamanders, clouded salamanders appear to be relatively aggressive. Thirty-two percent of specimens from an Oregon population had scars, presumably from conspecific attacks, and males had a higher percentage of scars than did females (Staub, 1993). Studies on *A. vagrans* indirectly suggest that this species may not be as aggressive as is *A. ferreus*—*A. vagrans* does not use the chemical signals of fecal pellets to delimit territory boundaries as other plethodontids tend to do (Ovaska and Davis, 1992).

    H. Aestivation/Avoiding Desiccation. Aestivation is unknown; animals likely avoid desiccating conditions by seeking shelter under cover objects or in burrows.

    I. Seasonal Migrations. Not known to occur.

    J. Torpor (Hibernation). Not known to occur.

    K. Interspecific Associations/Exclusions. Clouded and wandering salamanders overlap in a zone <15 km wide in northwestern California. Jackman (1999 [1998]) reports slight evidence of introgression, but clear hybrid individuals have never been identified. The range of clouded salamanders also overlaps with congeneric black salamanders (*A. flavipunctatus*; Fitch, 1936; Myers and Maslin, 1948; Stebbins, 1985), especially in the Klamath River valley, but black salamanders occur in hotter, drier regions (for example, further up the Klamath River Valley).

    Mitochondrial, allozyme, and karyotypic data distinguish *A. ferreus* from *A. vagrans* (Sessions and Kezer, 1987; Jackman, 1999 [1998]). Morphologically, these two species are similar (Beatty, 1979).

    L. Age/Size at Reproductive Maturity. Data from McKenzie (1970) suggest that males mature during their second year at lengths greater than 36 mm SVL, and that females first reproduce in their third year when they are approximately 55 mm SVL.

M. Longevity. Unknown.

N. Feeding Behavior. As with most salamanders, clouded salamanders are generalist feeders. Adults primarily eat isopods (sowbugs), hymenopterans (ants), and coleopterans, but their diet also includes an important assortment of other insects (e.g., dipterans, isopterans [termites]), and mites, spiders, pseudoscorpions, centipedes, and millipedes (Fitch, 1936; Storm and Aller, 1947; Bury and Martin, 1973; Whitaker et al., 1986). Whitaker et al. (1986) found no substantial differences between the diets of adult males and females. Hatchlings (<20 mm SVL) eat small prey, primarily mites, springtails, flies, and small beetles (Whitaker et al., 1986). As juveniles get larger, they switch to eating larger prey items such as sowbugs, larger beetles, and earwigs (Whitaker et al., 1986). Adults occasionally consume shed skin (Whitaker et al., 1986).

O. Predators. Poorly documented, but Petranka (1998) suggests predators of *Aneides* sp. include mammals, woodland birds, and snakes.

P. Anti-Predator Mechanisms. Several anti-predatory behaviors have been observed in the clouded or wandering salamander when individuals are startled or attacked: crawling away rapidly, immobility, a defensive posture (raising the body and undulating the tail), and flipping around followed by immobility (Fitch, 1936; Brodie, 1977). Skin secretions are thought to be noxious (Brodie, 1977).

Q. Diseases. Unknown.

R. Parasites. One species of nematode has been identified in clouded salamanders (Lehman, 1954; Goldberg et al., 1998c).

4. Conservation.

Clouded salamanders occur in high densities in old growth forests, and populations have certainly been lost due to forestry management practices and urban sprawl. It is doubtful that this species survives in areas where forests are intensively managed on short rotation cycles because of severe reductions in moisture conditions and the amount of large woody debris. According to Levell (1997), clouded salamanders are listed as Protected in Oregon.

Aneides flavipunctatus (Strauch, 1870)
BLACK SALAMANDER

Nancy L. Staub, David B. Wake

1. Historical versus Current Distribution.

Black salamanders (*Aneides flavipunctatus*) occur in western lowland forests and meadows in northern California and extreme southern Oregon where annual precipitation is >75 cm (Lynch, 1974). Populations generally are found at elevations below 600 m, but occur as high as 1,700 m (Lynch, 1974, 1981; Nussbaum et al., 1983). The distribution of black salamanders is disjunct; the southernmost populations (Santa Cruz Mountains) are separated from more northern populations by a gap that includes the northern part of the San Francisco Peninsula, the Marin Peninsula, and the nearly treeless area in southern Sonoma County, California. Populations south of Mt. Shasta and east of the Trinity Mountains appear to be separated from populations to the west (Larson, 1980; Lynch, 1981), although this may represent a collecting artifact. Analysis of protein variation among populations indicates a high level of genetic subdivision (Larson, 1980). Populations have been isolated from one another since the late Pliocene. Northern populations are paedomorphic in color pattern—adults retain the typical juvenile green-gray color pattern. Interestingly, in this part of the range, the salamander's coloration matches the greenish-gray talus substrate. In other parts of the range, animals are found on dark soil (Larson, 1980; Lynch, 1981).

2. Historical versus Current Abundance.

Black salamanders were once considered common in many areas of their range but have become rare in recent years (D.B. Wake, in Petranka, 1998). The proliferation of vineyards in northern California has destroyed much of the black salamander's prime habitat (N.L.S. personal observation).

3. Life History Features.

A. Breeding. Reproduction is terrestrial.

i. Breeding migrations. Breeding migrations do not occur.

ii. Breeding habitat. There are no published data available on the courtship and breeding behavior of black salamanders (Lynch, 1974).

B. Eggs.

i. Egg deposition sites. Females probably lay eggs in July or early August in cavities below ground. Clutches have been found underground at depths of 23 and 38 cm (Van Denburgh, 1895; Storer, 1925). Eggs are attached by peduncles to moist earth. In the lab, eggs have been attached to the underside of cover objects (broken clay flowerpot pieces; N.L.S., personal observation).

ii. Clutch size. Van Denburgh (1895) described a partial clutch of 15 eggs (about twice as many eggs composed the original clutch in the field) found next to a barn in soil with numerous spaces and pieces of rotten wood. Ovarian complements range

Black Salamander (*Aneides flavipunctatus*)

from 8–25 (mean about 12; Van Denburgh 1895; Stebbins, 1951).

C. Direct Development.

i. Brood sites. The same as egg deposition sites (underground and, in the lab, to the underside of cover objects).

ii. Parental care. Typically an egg clutch is found with a female in attendance. In the laboratory, females stayed with their clutches until eggs hatched (N.L.S., personal observation).

D. Juvenile Habitat. Juveniles are found in the same microhabitats as adults (e.g., under rocks and logs; Myers, 1930b).

E. Adult Habitat. Black salamanders occur in areas that receive >75 cm annual precipitation (Lynch, 1974). Specific habitats include lowland forests, under rocks and logs or in wet soil along streams, under logs and rocks in grassy meadows, pastures, and burned areas, and in talus slopes (Wood, 1936; Myers and Maslin, 1948; Stebbins, 1951; Lynch, 1974, 1981; Staub, 1993). The populations in the Santa Cruz Mountains appear to prefer moister microhabitats than more northern populations. Unlike their more arboreal congeners, black salamanders are primarily ground dwellers (Myers and Maslin, 1948). Despite their ground-dwelling habits, black salamanders have a prehensile tail (Van Denburgh, 1895).

F. Home Range Size. Unknown.

G. Territories. In captivity, adults often bite one another (e.g., Myers, 1930b), and adult males and females show agonistic behavior toward conspecific intruders (Staub, 1993). Animals captured in the field are frequently scarred; males show a higher frequency of scarring than do females (Staub, 1993). This species may be territorial in the field.

H. Aestivation/Avoiding Desiccation. In southern populations that are associated with streamside habitats, black salamanders are active year-round. In habitats that are not associated with permanent water, salamanders move underground during the dry season (mid-April to mid-October; Lynch, 1974).

I. Seasonal Migrations. Not known to occur.

J. Torpor (Hibernation). Not known to occur.

K. Interspecific Associations/Exclusions. Black salamanders occur syntopically with clouded salamanders (*A. ferreus*), wandering salamanders (*A. vagrans*), arboreal salamanders (*A. lugubris*), ensatinas (*Ensatina eschscholtzii*), and California slender salamanders (*Batrachoseps attenuatus*; Wood, 1936; Myers and Maslin, 1948; Lynch, 1974, 1985). Black salamanders also occur extensively with Del Norte salamanders (*Plethodon elongatus*) in the Klamath River Valley and Trinity River drainage.

L. Age/Size at Reproductive Maturity. Reproductively mature black salamanders range in size from 60–75 mm (Lynch, 1974).

M. Longevity. In the lab, black salamanders have lived 20 yr (N.L.S., personal observation).

N. Feeding Behavior. Juveniles and adults feed on a wide variety of prey. The diet of adult salamanders consists primarily of diplopods (millipedes), coleopterans, formicans (primarily ants), and isopterans (primarily termites; Lynch, 1985). The diet

of juveniles includes these prey as well as dipterans and collembolans (Lynch, 1985). Larger individuals consume larger prey items; mean and maximum prey size is correlated with body size. This correlation suggests that larger animals are selecting larger prey items and are ignoring smaller prey items. The number of prey items decreases as body size increases (Lynch, 1985).

O. Predators. Predators include western terrestrial garter snakes (*Thamnophis elegans*; Lynch, 1981).

P. Anti-Predator Mechanisms. When startled, juveniles generally remain immobile and adults flee (Van Denburgh, 1895; Jones, 1984). Other escape or defense behaviors include jumping (Van Denburgh, 1895), the production of sticky skin secretions (Lynch, 1981), an agonistic posture, and agonistic behaviors including biting (Lynch, 1981; Staub, 1993). The agonistic posture of black salamanders is distinctive. The animal raises its body off the substrate with the legs fully extended, the back is arched, the head elevated with the snout pointed slightly downward, and the tail undulates (Jones, 1984; Staub, 1993; Stebbins, 1954a). In the laboratory, black salamanders will bite

much of their prime habitat. According to Levell (1997), black salamanders are listed as Protected in Oregon.

Aneides hardii (Taylor, 1941)
SACRAMENTO MOUNTAIN SALAMANDER
Cindy Ramotnik

1. Historical versus Current Distribution. There is no evidence to support either an expansion or a contraction of the historical range of Sacramento Mountain salamanders (*Aneides hardii*) within the three disjunct areas where they occur: the Capitan, White, and Sacramento mountains of south-central New Mexico, in Lincoln and Otero counties (Ramotnik, 1997). Researchers have revisited some historical localities known from the 1950s and 1960s (these salamanders were discovered in 1940) in the Capitan and Sacramento mountains and have continued to find salamanders (Meents, 1987; Scott et al., 1987; C. Painter, personal communication). The increase in new locality records from the Sacramento Mountains in the last 10 yr most likely reflects increased survey effort and not range expansions (Ramotnik, 1997).

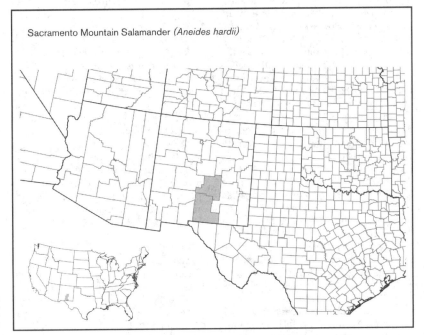

Sacramento Mountain Salamander *(Aneides hardii)*

western terrestrial garter snakes, which can result in serious injuries to the snakes (Lynch, 1981).

Q. Diseases.

R. Parasites. Nematodes have been found in the black salamander (Lehmann, 1954; Schad, 1960).

4. Conservation. Black salamanders have become rare in recent years due in large part to the proliferation of vineyards in northern California that has destroyed

2. Historical versus Current Abundance. Historical abundance unknown. Sacramento Mountain salamanders are abundant in appropriate habitat (above 2,800 m), but populations can be patchy within relatively uniform habitat (Ramotnik and Scott, 1988).

3. Life History Features.
   A. Breeding. Reproduction is terrestrial.
   i. Breeding migrations. None reported.

ii. Breeding habitat. Courtship and mating take place in underground sites (Johnston and Schad, 1959).

B. Eggs.

i. Egg deposition sites. Clutches have been found from mid-July to September, usually within large, decaying Douglas fir logs or stumps, but eggs also are believed to be oviposited below ground (Johnston and Schad, 1959).

ii. Clutch size. Clutch size ranges from 1–10 eggs (mean = 5.9; Staub, 1986), the lowest of any North American plethodontid salamander.

C. Direct Development. As previously noted (see "Egg deposition sites" above), clutches have been found from mid-July to September, usually within large, decaying Douglas fir logs or stumps, but eggs also are believed to be oviposited below ground (Johnston and Schad, 1959). Newly emerged salamanders from a clutch hatched in captivity measured 11–13 mm (SVL) and had color patterns similar to adults (Weigmann et al., 1980).

D. Juvenile Habitat. Presumably the same as adult habitat.

E. Adult Habitat. Within mesic mixed-habitat forest above 2,400 m, salamanders can be found within and under logs and moist litter and under rockslides and surface objects such as bark, rocks, and small woody debris. Above timberline they are associated with rocks and mats of mosses and lichens (Moir and Smith, 1970). Salamanders are often found within coniferous logs (primarily Douglas fir) that are in advanced stages of decay. They can be found under bark or in small cracks and chambers near the inner bark surface in less decayed logs (Johnston and Schad, 1959). Sacramento Mountain salamanders are more frequently associated with an understory of Rocky Mountain maple, less frequently with a sparse understory, and rarely with Gambel oak (Weigmann et al., 1980). However, Johnston and Schad (1959) noted that small oaks (*Quercus* sp.) were present among dominant conifers at all collecting sites.

F. Home Range Size. Unknown. Fifteen marked salamanders moved an average of 22.7 m (range = 0–50 m) between July and September (Staub, 1986).

G. Territories. Unknown.

H. Aestivation/Avoiding Desiccation. During periods of drought, salamanders retreat beneath surface objects (e.g., large decayed logs) or to subterranean retreats; they respond to decreased temperature and precipitation in September by reducing surface activity (Ramotnik, 1997).

I. Seasonal Migrations. None reported.

J. Torpor (Hibernation). Believed to be inactive below ground from October–May (Johnston and Schad, 1959; Scott and Ramotnik, 1992).

K. Interspecific Associations/Exclusions. Interspecific competition is unlikely within their range because the only salamanders

that occur sympatrically are tiger salamanders (*Ambystoma tigrinum*), which have distinctly different microhabitat preferences (Ramotnik, 1997).

L. Age/Size at Reproductive Maturity. Sexual maturity is reached at 43 mm SVL when females are 4 yr of age (based on growth-rate estimates by Staub, 1986). Williams (1976) reports that females reach sexual maturity from 2–3 yr of age but do not oviposit for another year in order to allow ova to reach sufficient size.

M. Longevity. At least 10 yr of age (based on a growth rate of 3.2 mm/yr; Staub, 1986).

N. Feeding Behavior. Salamanders feed on ground-dwelling invertebrates, primarily arthropods, especially spiders and insects, some mollusks (Johnston and Schad, 1959; Staub, 1986), and earthworms (Scott, 1990). In a sample of 83 stomachs of Sacramento Mountain salamanders, ants (Formicidae) occurred most frequently and were found in 55% of the stomachs, followed in frequency by rove beetles (Staphylinidae: 20%) and springtails (Sminthuridae: 18%; unpublished data). Males and females exhibit little difference in the sizes of prey taken (Staub, 1989). Captive salamanders initially foraged during daylight hours but after acclimation actively foraged at night or under low-light conditions (Johnston and Schad, 1959).

O. Predators. Western terrestrial garter snakes (*Thamnophis elegans*) are the only known predators (Painter et al., 1999).

P. Anti-Predator Mechanisms. Immobile when initially exposed; body coiled and body flipped (*sensu* Brodie, 1983); will spin and writhe in a continuous motion when held by hand (Ramotnik, 1997).

Q. Diseases. None reported.

R. Parasites. A sample of 30 adult salamanders collected in July showed a high infection rate, 83% and 90%, by two species of nematodes, *Oswaldocruzia* sp. and *Thelandros* sp., respectively (Johnston and Schad, 1959).

4. Conservation.
Sacramento Mountain salamanders are listed as Threatened in New Mexico (Levell, 1997; New Mexico Department of Game and Fish, 2000) and are potentially vulnerable to actions such as logging and fire that dry the habitat and reduce the amount of moisture available to them for respiration. However, salamanders have endured bouts of logging activities (sometimes intense) in the Sacramento Mountains over the past 60–90 yr, and there is no evidence that populations have been eliminated. Numbers of salamanders decreased the first year following logging in 1986 and 1987 in the Sacramento Mountains, but their numbers approached or exceeded pre-logging values after 5–7 yr (unpublished data). Salamanders apparently survived the frequent, low-intensity fires that occurred

historically in the Sacramento Mountains, but fire suppression has created opportunities for large, catastrophic, "stand-reducing" fires. The intensity of these fires compared to historical fires makes it difficult to predict how salamanders will respond. An opportunity to study a recent large-scale fire occurred in May 2000, when the Scott Able Fire burned over 16,000 ac within the habitat of Sacramento Mountain salamanders. Second year results from the five-year study indicate that salamanders minimize moisture loss on burned sites by aggregating in large numbers beneath logs, and that arthropod prey diversity is greatest on low-burned sites (unpublished data).

Aneides lugubris Hallowell, 1949
ARBOREAL SALAMANDER
Nancy L. Staub, David B. Wake

1. Historical versus Current Distribution.
Arboreal salamanders (*Aneides lugubris*) occur in coastal oak woodlands from northern California (Humboldt County) to approximately Valle Santo Tomás, Baja California del Norte, Mexico. Their range includes South Farallon, Santa Catalina, Los Coronados, and Año Neuvo Islands (Lynch and Wake, 1974; McPeak, 2000). In the foothills of the Sierra Nevada, a geographically isolated cluster of populations occurs in black oak and yellow pine forests (Lynch and Wake, 1974). This group of populations is genetically distinct from coastal populations (Jackman, 1993). The range of arboreal salamanders is similar to the range of the oaks (*Quercus agrifolia* and *Q. wislizenii*), presumably a consequence of shared moisture and soil characteristics (Rosenthal, 1957). However, in southern California, arboreal salamanders are frequently associated with sycamores (*Platanus racemosa*) bordering seasonal streams.

Populations have certainly been eliminated as coastal California habitats have been developed, but the species survives in many urbanized regions where adequate cover is present.

2. Historical versus Current Abundance.
Although arboreal salamanders remain common in many areas, in some areas populations have declined in the past 20 yr (D.B.W., personal observations). Petranka (1998) suggests that large oaks used for nesting and aestivation should be preserved. The current decline of live oaks in California will have negative effects on arboreal salamander populations.

3. Life History Features.
Reproduction is terrestrial.

A. Breeding.

i. Breeding migrations. Breeding migrations do not occur, but seasonal or daily vertical migrations into and out of trees are possible.

Arboreal Salamander (*Aneides lugubris*)

ii. **Breeding habitat.** Unknown. During courtship a courting male places his mental gland on the female's dorsum and, in a succession of quick strokes, draws the mental gland across the female's back (Arnold, 1977). During this behavior, the male's large and monocuspid premaxillary teeth may scratch the female's skin and enable efficient delivery of mental gland pheromones to the female's circulation.

B. Eggs.

i. Egg deposition sites. Reported oviposition sites include in decay holes of live oak trees (most common), under rocks set deeply in the ground, in logs, under surface cover objects (e.g., stone slabs, flower box), and beneath the ground surface (Ritter and Miller, 1899; Ritter, 1903; Storer, 1925). Egg clutches have been found ≤9 m above ground in live oak trees (Ritter, 1903). Most females oviposit in late spring or early summer (Stebbins, 1951; Anderson, 1960).

ii. Clutch size. The number of eggs in a clutch varies from 5–24 (Ritter, 1903; Storer, 1925; Stebbins, 1951), with larger females producing more eggs (Anderson, 1960). Ovarian counts range from 5–26 maturing oocytes (Anderson, 1960). Females on densely populated islands tend to produce fewer eggs than females in less dense mainland populations (Anderson, 1960). Eggs are large, 7–9.5 mm in diameter.

C. Direct Development.

i. Brood sites. The same as egg deposition sites.

ii. Parental care. Females are often found coiled around the eggs; males are often in attendance as well (Ritter and Miller, 1899; Ritter, 1903; Stebbins, 1951). After an approximately 3–4-mo developmental period, embryos hatch in August–September at between 26 and 32 mm TL (Storer, 1925; Stebbins, 1951). Presumed family groups may stay associated after hatching (Ritter, 1903).

D. Juvenile Habitat. Similar to that of adults.

E. Adult Habitat. Arboreal salamanders are found in a variety of terrestrial and arboreal habitats, including under rocks and woody surface cover, in decaying stumps and logs, in decay holes in trees, and in rock crevices (Ritter and Miller, 1899; Ritter, 1903; Storer, 1925; Miller, 1944; Stebbins, 1951; Rosenthal, 1957; Anderson, 1960).

Climbing is facilitated by expanded tips of terminal phalanges and large subdigital pads, as well as by the prehensile tail of the arboreal salamander (Ritter and Miller, 1889). Individuals have been found over 18 m above ground in trees (Ritter and Miller, 1899; Ritter, 1903; Stebbins, 1951). Arboreal salamanders can be found in microhabitats that are drier than those of sympatric salamanders (Storer, 1925; Cohen, 1952; Ray, 1958; Anderson, 1960). This species is generally absent from regions receiving <25 cm of precipitation per year (Rosenthal, 1957).

F. Home Range Size. Unknown.

G. Territories. Arboreal salamanders are well known for their aggressive tendencies and weaponry. This species has a suite of

morphological features that enable a strong, wound-inflicting bite. The jaw muscles are hypertrophied, the skull is heavily ossified with especially strong jaws, and both juveniles and adults possess enlarged and flattened, blade-like, monocuspid teeth (other plethodontids possess weaker bicuspid teeth as juveniles; Wake, 1966; Wake et al., 1983). In his description of arboreal salamanders, Cope (1889) writes: "On the whole, the physiognomy is not unlike that of a snapping tortoise." Scarred individuals are often found in the field (Miller, 1944; N.L.S., personal observation), and Myers (1930b) observed that salamanders housed in the same container bite each others' tails. In a study of museum specimens, Staub (1993) found that 15% of the examined individuals were scarred, presumably from conspecific attacks. The frequency of scarring did not differ significantly between males and females (Staub, 1993).

H. Aestivation/Avoiding Desiccation. Arboreal salamanders are more tolerant of dry conditions than are many species of salamanders and are often among the last salamanders to retreat underground or into tree holes to avoid desiccation (Miller, 1944; Cohen, 1952; Ray, 1958; Petranka, 1998). This species has relatively low rates of water loss compared with other salamanders, possibly due to postural adaptations (curled body and tightly coiled tail) and a rapid rate of water uptake (Cohen, 1952; Ray, 1958).

I. Seasonal Migrations. Not known to occur.

J. Torpor (Hibernation). Not known to occur.

K. Interspecific Associations/Exclusions. Arboreal salamanders are syntopic with California slender salamanders (*Batrachoseps attenuatus*), wandering salamanders (*A. vagrans*), and black salamanders (*A. flavipunctatus*) in regions north of San Francisco Bay. Throughout most of the rest of their range, arboreal salamanders occur in sympatry with *Ensatina* and a number of species of *Batrachoseps*. Ecological interactions between these species are not well understood. Maiorana (1978a) showed that there may be competition for food between California slender salamanders and arboreal salamanders when large prey items are limited. When large prey items are not limited however, Lynch (1985) found broad dietary differences between these two species—arboreal salamanders tend to eat a few, large-bodied prey in addition to a diverse assortment of other prey items. Arboreal salamanders occasionally prey on *Batrachoseps* (Storer, 1925; Miller, 1944).

L. Age/Size at Reproductive Maturity. Arboreal salamanders are the largest species of *Aneides*; mature individuals range in size from 65–100 cm SVL (Lynch and Wake, 1974). Age–size relationships suggest that 3 yr are required to reach maturity (Anderson, 1960). Anderson (1960) reported that the minimum size of sexual maturity was 34 mm SVL for females, though this size seems small for typical females reaching sexual maturity.

M. Longevity. Unknown.

N. Feeding Behavior. Arboreal salamanders are nocturnal and feed most actively under moist/wet conditions. Adults tend to feed on larger prey than juveniles, although Wake et al. (1983) point out that arboreal salamanders of all sizes take a range of prey items. The diet of arboreal salamanders includes millipedes, annelids, snails, and especially coleopterans, hymenopterans (ants), isopterans (termites), isopods (sowbugs), chilopods (centipedes), and lepidopterans (Miller, 1944; Zweifel, 1949; Bury and Martin, 1973; Lynch, 1985). Miller (1944) suggests that fungus is an important component of the diet, but other authors have not confirmed this observation. Relative to syntopic species, arboreal salamanders consume disproportionately larger prey items than expected for individuals of a given body size (Lynch, 1985), and comparative data suggest that arboreal salamanders consume numerous large-sized prey that other species are unable to capture (Bury and Martin, 1973). The structural components of the feeding apparatus are well developed (e.g., well-ossified skull; Wake et al., 1983). Despite the large jaws and teeth, prey are typically captured by the tongue and brought fully into the mouth, usually without contacting the marginal dentition (personal observation).

O. Predators. Pacific rattlesnakes *(Crotulus viridis helleri)* are known predators of arboreal salamanders (Mahrdt and Banta, 1997), and a California scrub jay *(Aphelocoma coerulescens)* has been observed trying to eat a juvenile arboreal salamander (Rubinoff, 1996).

P. Anti-Predator Mechanisms. Several anti-predatory behaviors have been observed when individuals are startled or attacked: a defensive posture (raising the body stiffly off the ground; Cohen, 1952; Stebbins, 1951), squeaking (Ritter and Miller, 1899; Storer, 1925), rapid movement and jumping (Ritter and Miller, 1899), and biting (Ritter, 1903; Storer, 1925; Stebbins, 1951; Lynch, 1981). Arboreal salamanders will bite western terrestrial garter snakes *(Thamnophis elegans)*, and in some cases the snakes can die from the inflicted wounds (Lynch, 1981). *Micturition*, the act of voiding the bladder when startled, is a novel putative antipredatory behavior that has been documented for arboreal salamanders (Staub and Anderson, 2001).

Q. Diseases. Not reported.

R. Parasites. Two species of nematode *(Batracholandros salamandrae, Oswaldocruzia pipiens)* have been found in arboreal salamanders (Schad, 1960; Goldberg et al., 1998c).

S. Comments. There are two geographically segregated groups of chromosomally differentiated populations of arboreal salamanders (Sessions and Kezer, 1987). These two karyotypes intergrade in south and east-central Mendocino County (Sessions and Kezer, 1987). Unpublished genetic analyses (allozymes and mitochondrial DNA sequences) show that the chromosomal units do not correlate with patterns of genetic variation (Jackman, 1993). The Farallon Island population is most similar genetically to the nearest mainland population, not populations in the Gabilan mountains to the south as suggested by Morafka (1976; see also Jackman, 1993).

4. Conservation.
The range of arboreal salamanders is similar to the range of their oak habitat; oaks are used for nesting and aestivation. The current decline of live oaks in California will have negative effects on arboreal salamander populations.

Populations of arboreal salamanders have certainly been eliminated as coastal California habitats have been developed, but the species survives in many urbanized regions where adequate cover is present.

Aneides vagrans Wake and Jackman, 1999 "1998"

WANDERING SALAMANDER

Nancy L. Staub, David B. Wake

1. Historical versus Current Distribution.
Wandering salamanders *(Aneides vagrans)* have an unusual distribution. They are found in the coastal forests of northern California and in British Columbia. They are not found in either Oregon or Washington (Jackman, 1999 [1998]). The distribution of wandering salamanders in California extends from northern Siskiyou and Del Norte counties, south through extreme western Trinity, Humboldt, and Mendocino counties in an increasingly narrow coastal strip to the vicinity of Stewart's Point, northwestern Sonoma County. In British Columbia, wandering salamanders are widespread on Vancouver Island and on neighboring islands and occur in a small area of mainland British Columbia (Green and Campbell, 1984). Jackman (1999 [1998]) hypothesizes that the British Columbia populations are derived from human-mediated introductions from California during the mid-to-late nineteenth century. In particular, genetic and historical evidence suggests that wandering salamanders were introduced to British Columbia from California in shipments of tan oak *(Lithocarpus densiflorus)* bark, used extensively in the tanning of leather at that time (Jackman, 1999 [1998]).

The California populations now assigned to *A. vagrans* previously were considered to be *A. ferreus* (clouded salamanders) but have been reassigned based on allozyme, mitochondrial DNA, and karyotypic evidence (Sessions and Kezer, 1987; Jackman, 1999 [1998]; Wake and Jackman, 1999 [1998]). Nearly all of the literature dealing with *A. ferreus* from

Wandering Salamander *(Aneides vagrans)*

California and British Columbia refers instead to *A. vagrans*.

Studies of salamander abundance in different forest types and stages in Oregon and northwestern California (includes both *A. ferreus* and *A. vagrans*) show that this species has higher densities in older forests (Welsh and Lind, 1991). In British Columbia, wandering salamanders are found in microhabitats typical of clouded salamanders—in cavities of decaying logs and stumps or beneath loose bark (Stelmock and Harestad, 1979). In contrast to studies on clouded salamanders that showed that logging is detrimental to salamander abundance (Corn and Bury, 1991; Butts and McComb, 2000), forest harvesting did not appear to have long-term effects on the abundance of wandering salamanders (Stelmock and Harestad, 1979). Stelmock and Harestad (1979) found that in British Columbia, old-growth forest and regenerating conifer seral stages were equally favorable habitats for wandering salamanders, although their sample sizes of collected salamanders were small. It has not yet been determined whether the persistence of wandering salamanders in logged areas in British Columbia reflects a difference in moisture regimes between the habitats, a difference in logging practices and the amount of coarse woody debris left behind, an artifact of small sample sizes, or a difference in salamander biology.

Populations have certainly been lost due to forestry management practices and to urban sprawl.

2. Historical versus Current Abundance. Populations of clouded salamanders associated with post-logged areas eventually decline, and Corn and Bury (1991) doubt that they can survive in areas where forests are intensively managed on short rotation cycles because of the severe reductions in moisture conditions and in the amount of large woody debris. Presumably this reasoning holds true for wandering salamanders as well (although see "Historical and Current Distribution" above). Ironically, wandering salamanders now have a much wider distribution than in the past as a result of their introduction into coastal British Columbia, where they apparently rapidly expanded their range so that they now occur widely on Vancouver Island and many neighboring islands and have been found on the mainland (Green and Campbell, 1984).

3. Life History Features.
    A. Breeding. Reproduction is terrestrial. Two albino specimens from two different egg clutches were found in Humboldt County, California (Houck, 1969).
    i. Breeding migrations. Do not occur. Oviposition is thought to occur in spring or early summer in British Columbia (Stelmock and Harestad, 1979).
    ii. Breeding habitat. See "Brood sites" below.

    B. Eggs.
    i. Egg deposition sites. Egg clutches have been found under the bark of a rotting Douglas fir log (Wood, 1939; Dunn, 1942) and at the base of a tree limb 30–40 m above the forest floor (Welsh and Wilson, 1995). Davis (2003) has recently described the characteristics of three egg masses.
    ii. Clutch size. Reported clutch sizes vary from 6–9 eggs (Dunn, 1942; Welsh and Wilson, 1995). The average for the ovarian complement of eggs is 18 (range 14–26, Stelmock and Harestad, 1979).
    C. Direct Development.
    i. Brood sites. The same as egg deposition sites.
    ii. Parental care. If similar to the clouded salamander, variable. Clutches have been found without a parent in attendance (*A. ferreus*; Storm, 1947), with a female in attendance (*A. vagrans*; Dunn, 1942), and with both a male and a female in attendance (*A. ferreus*; Storm, 1947).
    D. Juvenile Habitat. V. Brown (1948) found three juveniles (presumably *A. vagrans*) under bark on a log. If habitat selection preferences are similar to the clouded salamander's preferences, then juveniles prefer bark litter over rock or leaf litter; subadults prefer bark litter at higher temperatures (20 °C, 25 °C) and show no preference between rock or bark litter at lower temperatures (10 °C; McKenzie and Storm, 1970).
    E. Adult Habitat. Adults are found associated with stumps and logs in Douglas fir forests (Welsh and Lind, 1988, 1991), under rocks (V. Brown, 1948; Jackman, 1999 [1998]), under redwood slabs in redwood forests (Storer, 1925), and in the forest canopy (J.C. Spickler, personal communication). Wood (1939) found that adults have a tendency to be in, but not under, rotten logs. Bury and Martin (1973) found the wandering salamander associated with decaying stumps and logs in edge and open habitat (e.g., abandoned pastures, logging platforms in forest clearings), secondary forests, and in redwood forests. In coastal areas, Welsh and Lind (1988) captured more *A. ferreus* or *A. vagrans* in slightly younger forests than in the older forests, but inland, the opposite trend was evident. They suggest the variation is due to differences in moisture regimes between the different sites. In general, *A. ferreus* or *A. vagrans* were found in the more mesic habitats. Recent work suggests that the wandering salamander inhabits the canopy of old-growth redwood forests year-round (J.C. Spickler, personal communication). In the canopy, salamanders have been found in humus accumulations in trunk crotches, on limbs, under bark, and in cracked and rotting wood of broken limbs and trunks. They have been found ≤87 m above ground; at some sites, the canopy salamanders can be abundant (J.C. Spickler, personal communication).

Jackman (1999 [1998]) suggests that there may be a difference in habitat preference between clouded and wandering salamanders—in northwestern California clouded salamanders are commonly found associated with both decaying logs and rocky slopes, while wandering salamanders are almost exclusively found associated with decaying logs.
    F. Home Range Size. Unknown.
    G. Territories. Studies on wandering salamanders indirectly suggest that they may not be as aggressive as clouded salamanders; wandering salamanders do not use chemical signals in fecal pellets to delimit territory boundaries, as with other plethodontids (Ovaska and Davis, 1992; see also Davis, 2002a,b).
    H. Aestivation/Avoiding Desiccation. Aestivation is unknown; animals likely avoid desiccating conditions by seeking shelter under cover objects or in burrows.
    I. Seasonal Migrations. Not known to occur.
    J. Torpor (Hibernation). Not known to occur.
    K. Interspecific Associations/Exclusions. Clouded and wandering salamanders overlap in a zone <15 km wide in northwestern California. Jackman (1999 [1998]) reports evidence of introgression, but clear hybrid individuals have not been found. Mitochondrial, allozyme, and karyotypic data distinguish the clouded from the wandering salamander (Sessions and Kezer, 1987; Jackman and Wake, 1999 [1998]).

In California, the range of wandering salamanders also overlaps with its congeners, arboreal salamanders (*A. lugubris*) and black salamanders (*A. flavipunctatus*; Stebbins, 1985; Wake and Jackman, 1999 [1998]).
    L. Age/Size at Reproductive Maturity. Based on follicle size, Stelmock and Harestad (1979) suggest that females greater than 50 mm SVL are sexually mature. This size is slightly smaller than the data on clouded salamanders (McKenzie, 1970) that suggest that females first reproduce in their third year when they are approximately 55 mm SVL. McKenzie (1970) estimated that male clouded salamanders mature during their second year at lengths greater than 36 mm SVL.
    M. Longevity. Unknown.
    N. Feeding Behavior. As with most salamanders, the wandering salamander is a generalist feeder. In California, wandering salamanders have a diverse diet consisting primarily of hymenopterans (ants), coleopteran adults and larvae, isopods, and collembolans (Bury and Martin, 1973). In a population in British Columbia, hymenopterans (ants), coleopterans, and gastropods were significant prey items in terms of both frequency and prey volume (Stelmock and Harestad, 1979). Acarinians (mites) and collembolans had a high frequency of occurrence, but their

contribution to total prey volume was slight (Stelmock and Harestad, 1979). Juveniles eat smaller prey than do adults (Stelmock and Harestad, 1979).

O. Predators. Poorly documented, but Petranka (1998) suggests predators of *Aneides* sp. include mammals, woodland birds, and snakes.

P. Anti-Predator Mechanisms. Several anti-predatory behaviors have been observed in the clouded or wandering salamander when individuals are startled or attacked: immobility, a defensive posture (raising the body and undulating the tail), and flipping around followed by immobility (Brodie, 1977). Skin secretions are thought to be noxious (Brodie, 1977).

Q. Diseases. Unknown.

R. Parasites. Infestation rates of the nematode *Thelandros salamandrae* are greater than 50% in wandering salamanders (Stelmock and Harestad, 1979). These nematodes infect the large intestine. There was no difference in infection rate between males and females. Small animals (<30 mm SVL) had no parasites (Stelmock and Harestad, 1979).

4. Conservation.
Wandering salamanders have higher densities in old-growth forests, and while populations certainly have been lost due to forestry management practices and urban sprawl, animals can be found at high densities in regenerating forest. Wandering salamanders now have a much wider distribution than in the past as a result of their introduction into coastal British Columbia.

Batrachoseps aridus Brame, 1970
DESERT SLENDER SALAMANDER
Robert W. Hansen, David B. Wake

1. Historical versus Current Distribution.
Desert slender salamanders (*Batrachoseps aridus*) were discovered in 1969 from Hidden Palm Canyon (760 m elevation), a tributary of Deep Canyon on the eastern flank of the Santa Rosa Mountains, Riverside County, California (Brame, 1970). This remained the only known population until the discovery of salamanders along Guadalupe Creek (945–1,097 m elevation), about 4.5 air miles from the type locality (Giuliani, 1981; Wake and Jockusch, 2000). These two sites constitute the only known localities for the species (Wake, 1996). Available habitat for Hidden Palm Canyon (about 0.2 ha; Bleich, 1978) and Guadalupe Canyon (about 0.6 ha; Duncan and Esque, 1986) combines for one of the most restricted distributions of any North American salamander.

Portions of the Hidden Palm Canyon habitat were destroyed by a severe storm in 1976. Damage included loss of limestone sheeting and underlying soil on the canyon wall, as well as scouring of the

Desert Slender Salamander (*Batrachoseps aridus*)

canyon bottom to a depth of about 2 m (Bleich, 1978). Other parts of the canyon retain viable habitat (K. Nicol, personal communication).

2. Historical versus Current Abundance.
Brame (1970) and subsequent workers found desert slender salamanders to be extremely localized but common within Hidden Palm Canyon. For example, during single visits Brame and colleagues found 8 specimens in February 1970, 11 specimens in March 1970, 10 specimens in July 1970, 39 specimens in March 1971, and 25 specimens in March 1972 (Brame, 1970; Brame et al., 1973). Bleich (1978) found as many as 21 salamanders in one night in the 1977–78 field season. Giuliani (1981) observed 16 individuals in one night in January 1981, although most of these were juveniles. More recent searches (since 1995) confirm the presence of salamanders in Hidden Palm Canyon, although estimates of abundance are lacking (Brame and Hansen, 1994; K. Nicol, personal communication). Comparing the field results of Brame and colleagues (who relied on movement of surface cover to locate specimens) with more recent searches is difficult owing to the passive search techniques employed by later workers to minimize disturbance of the fragile habitat.

Within Guadalupe Canyon, salamanders have been found at discrete sites over approximately 1.6 linear km (1 mi) of riparian canyon-bottom habitat. However, numbers of salamanders at a given locality are much lower than in Hidden Palm Canyon. Duncan and Esque (1986), for example, never found more than four salamanders at any one locality/visit.

3. Life History Features.
A. Breeding. Reproduction is terrestrial.
i. Breeding migrations. Unknown. We expect that breeding movements are

similar to garden slender salamanders (*B. major*), their close phylogenetic relatives.
ii. Breeding habitat. Unknown.

B. Eggs.
i. Egg deposition sites. Oviposition sites are unknown. Presumably, desert slender salamander females deposit egg clutches deep within limestone crevices (Hidden Palm Canyon) where moisture may be present year-round. Egg attendance by females and communal nesting are unknown for this species.
ii. Clutch size. Unknown.

C. Direct Development. Although unreported, we presume that desert slender salamanders undergo direct development, as is the case with other species of bolitoglossine plethodontids. Dates of hatching are unknown.

D. Juvenile Habitat. It is unknown how this may differ from adult habitats.

E. Adult Habitat. Desert slender salamanders exist in seep formations and other localized moist areas within steep-sided, protected canyons on the eastern (desert-facing) slope of the Santa Rosa Mountains. Salamander sites are beyond the reach of direct sunlight year-round. Mesic-associated vegetation at the Hidden Palm Canyon site includes sugar bush, willow, Washington palms, creosote, mesquite, and various grasses and mosses. Bordering slopes display typical desert plant species (e.g., cholla, agave, barrel cactus, prickly pear cactus, creosote, juniper, and mesquite). Within Guadalupe Canyon, the following plants are characteristic of salamander sites: maidenhair fern, Fremont cottonwood, willows, waterweed, deer grass, California fuschia, wild grape, sugar bush, and scrub oaks.

Although individual desert slender salamanders have been found beneath rocks resting on damp substrates, perhaps the most important element of the Hidden Palm Canyon habitat is the numerous crevices and fissures present in the limestone rock formations (USFWS, 1982). These permit moisture to move through crevices and provide critical refugia for salamanders. This may be critical in sustaining the population following stochastic events, such as severe storms, that result in significant modification to surface habitat.

During the active season, desert slender salamanders move at night from locations deep in the recesses of crevices to the crevice openings to capture food. Thus, diurnal searches for this species that rely on examination of surface cover may seriously underestimate the number of individual salamanders present. Nighttime searches using a flashlight to examine moist crevices, together with use of artificial cover objects (plastic garbage bags), are more effective and minimize potential impact to the habitat (Duncan and Esque, 1986; Palazzo, 1994; Duncan, 1998).

F. Home Range Size. Unknown.

G. Territories. Unknown.

H. Aestivation/Avoiding Desiccation. Salamanders have been observed surface active or under cover mostly during the winter, although Brame (1970) found animals in July under porous limestone, suggesting that activity continues into the summer at Hidden Palm Canyon. Conditions were too dry in Guadalupe Canyon in July for surface activity (Duncan, 1998).

I. Seasonal Migrations. Unknown.

J. Torpor (Hibernation). Unknown. Potentially, low winter temperatures ($\leq 5\,^{\circ}\text{C}$) would preclude activity, but this remains unstudied.

K. Interspecific Associations/Exclusions. Unknown. This is the only salamander species within the range.

L. Age/Size at Reproductive Maturity. Unknown. Based upon measurements of a small series and comparisons with other species of *Batrachoseps*, we estimate that sexual maturity is attained at 31 mm SVL in males and at a slightly larger size in females. The largest individuals are females (maximum SVL 48.4 mm).

M. Longevity. Unknown.

N. Feeding Behavior. Bleich (1978) observed feeding at night on two occasions, with salamanders capturing small insect prey using their projectile tongue.

O. Predators. Bleich (1978) speculated that western skinks *(Eumeces skiltonianus)* and ring-necked snakes *(Diadophis punctatus)* were potential predators, although the latter species is unrecorded from desert slender salamander sites.

P. Anti-Predator Mechanisms. Brame et al. (1973) reported that when uncovered in the field, about 75% of desert slender salamanders would coil their body while elevating their tail. Although coiling is a common defensive response in several species of *Batrachoseps*, the combination of body coiling and tail elevation is apparently unique to desert slender salamanders.

Most of the specimens encountered by Bleich (1978) coiled tightly with the tail in a horizontal position or slightly elevated; the head was either on top of the body coil or hidden beneath.

Q. Diseases. Unknown.

R. Parasites. Unknown.

S. Comments. The taxonomic status of desert slender salamanders has been called into question by recently published analyses of allozymes and mtDNA (cytochrome b) sequences (Wake and Jockusch, 2000). With respect to allozymes, desert slender salamanders are no different than a local population of garden slender salamanders and are deeply nested within that taxon. Thus, continued recognition of desert slender salamanders *(B. aridus)* would render garden slender salamanders *(B. major)* paraphyletic. With respect to mtDNA sequences, desert slender salamanders have a unique haplotype (shared by the two populations) that is most similar to those found in a population of undetermined status (but which appears to be identical to *B. major* from the mainland) from Isla Todos Santos off the northwestern coast of Baja California. In turn, all of these haplotypes cluster with a phylogeographic segment of *B. major* from San Diego County. Wake and Jockusch (2000) recommend that *B. aridus* be treated as a subspecies of *B. major* because they differ from that taxon in morphology and possibly in ecology. We make no judgment about taxonomy, but choose to handle *B. aridus* as a separate entity for this work because of its status as a federally protected Endangered species, which would not change regardless of its formal taxonomy.

4. Conservation.
Desert slender salamanders are listed as Endangered at both state and federal levels. All parts of their known range occur on public lands administered by the U.S. Bureau of Land Management or the California Department of Fish and Game. The Hidden Palms Ecological Reserve, harboring the largest population, is closed to public access.

Acknowledgments. Todd Esque (U.S. Geological Survey), Derham Giuliani, and Kim Nicol (California Department of Fish and Game) provided unpublished reports or other information derived from their fieldwork with *Batrachoseps* in the Santa Rosa Mountains. Gavin Wright and Larry Foreman (Bureau of Land Management) also provided copies of reports from BLM files.

Batrachoseps attenuatus (Eschscholtz, 1833)
CALIFORNIA SLENDER SALAMANDER
Robert W. Hansen, David B. Wake

1. Historical versus Current Distribution.
California slender salamanders *(Batrachoseps attenuatus)* occur in two principal

California Slender Salamander *(Batrachoseps attenuatus)*

areas: along the California coast and adjacent Inner Coast Ranges from north and east of Monterey Bay (extreme western Merced, Monterey, and San Benito counties) northward to extreme southwestern Oregon (south side of the Rogue River, Curry County; Yanev, 1978, 1980; Leonard et al., 1993); and in the western foothills of the northern and central Sierra Nevada, from Calaveras County north to at least Butte County (Hayes and Cliff, 1982; Jockusch et al., 1998). Inner Coast Range populations extend as far north as Tehama County (Jennings, 1982; Boundy, 2000). Isolated records of *Batrachoseps* from the floor of California's Central Valley (Stanislaus, San Joaquin, and Yolo counties) probably represent California slender salamanders, although confirmation from genetic analysis is lacking. Additionally, there are isolated populations in Shasta County (north of California's Central Valley) in the Little Cow and Clipkapudi Creek drainages (Stebbins, 1985).

The Sierra Nevada portion of the range is poorly defined, and some of the museum specimens currently allocated to California slender salamanders on the basis of morphology may be misidentified. Two similar-looking species approach or overlap the range of California slender salamanders in the central Sierra Nevada: gregarious slender salamanders *(B. gregarius)* range as far north as Mariposa County, just south of the southernmost range limits for California slender salamanders; and Hell Hollow slender salamanders *(B. diabolicus)* may be broadly sympatric with California slender salamanders between El Dorado and Calaveras counties, although the two have been found together at only one site (north slope of the American River; Jockusch et al., 1998).

Populations of California slender salamanders and the morphologically similar Gabilan Mountains slender salamander *(B. gavilanensis)* narrowly (by a few hundred m to a few km) overlap in geographic distribution along the southern border of the range of California slender salamanders in Santa Cruz and San Benito counties (Yanev, 1978, 1980; Jockusch et al., 2001).

2. Historical versus Current Abundance.
Early reports (e.g., Stebbins, 1954a; Anderson, 1960) and more recent fieldwork by R.W.H., D.B.W., and colleagues suggest that California slender salamander population densities in the coastal portion of the range remain high. Estimates of densities range from 4,470/ha (Contra Costa County; Anderson, 1960, citing data from Hendrickson, 1954) to 17,290/ha on Red Rock Island in San Francisco Bay (Anderson, 1960). Stebbins (1954a) reported that a Contra Costa County population of California slender salamanders (presumably the same site studied by Hendrickson, 1954) occurred at densities relative to ensatinas *(Ensatina*

eschscholtzii) of 7:1; he estimated the latter to occur at 170–200 adults/ac. This suggests a population density of 1,190–1,400/ac (2,939–3,458/ha) for California slender salamanders.

3. Life History Features.
A. Breeding. Reproduction is terrestrial.

i. Breeding migrations. Movements associated with reproductive activity are virtually unknown for any species of *Batrachoseps*. However, Anderson (1960) observed 69 adult California slender salamanders surface active during a heavy rain at night on 26 October (San Francisco Bay area); nearly all were gravid females moving uphill and across a road, a minimum distance of about 9 m. Coastal populations lay eggs shortly after the onset of fall rains, generally in October–November (Stebbins, 1951). Periods of oviposition for Sierra Nevada populations are unknown, but likely later in the year (December–January).

ii. Breeding habitat. Courtship behavior probably occurs underground (Houck, 1977b), but the timing is unknown.

B. Eggs.

i. Egg deposition sites. Females deposit eggs beneath rocks, logs, or other objects, although the paucity of nests discovered in the field suggests that most females are ovipositing underground (Burke, 1911; Storer, 1925; Myers, 1930b). Females apparently abandon eggs to communal nests, although female association with eggs has been observed (Snyder, 1923; Maslin, 1939). Storer (1925) described one nesting site as consisting of 53 eggs found "under a plank in a moist, springy place near a brush pile." Maslin (1939) described finding a female in the act of egg laying; the female "was in a small depression in damp soil beneath a strip of tin" in the shade of a tree. Three other adult females occupied this depression, along with 74 eggs. Burke (1911) reported the discovery of 35 eggs (in groups of 21, 10, and 4) under a log in a moist ravine, unattended by any adults. Sixty-eight eggs were found in a small depression (ca. 25 mm diameter) beneath a board in late December in Contra Costa County; three adult salamanders were present under the board as well, but none was in contact with the egg cluster (R.F. Hoyer, personal communication).

ii. Clutch size. Clutch size is positively correlated with female body size. Jockusch and Mahoney (1997) reported mean clutch sizes of 8.3–13.4 eggs for four coastal populations and 6.8 eggs for a Sierra Nevada population. Maiorana (1976) found clutch sizes of 4–5 eggs for females occupying a forested area, while those from an oak woodland population produced larger clutches (7–8 eggs), with 82% and 95%, respectively, breeding annually. Mean egg diameter is 4.1 mm (Jockusch, 1997a).

C. Direct Development. Mean incubation periods for lab-incubated eggs (at 13°C) ranged from 72–86 d (Jockusch, 1997a; Jockusch and Mahoney, 1997). Hatchlings have been observed in late January to February (San Francisco Bay area), and in late December at Walnut Creek (Contra Costa County) and St. Helena (Napa County; R.F. Hoyer, personal communication).

D. Juvenile Habitat. Maiorana (1976) found that juvenile California slender salamanders were more likely than adults to be active under sub-optimal surface conditions. Differential use of surface cover elements is unstudied, although juveniles often occur under relatively smaller pieces of cover.

E. Adult Habitat. Along the coast of northwestern California, California slender salamanders are largely restricted to low elevation, coastal redwood forest (Bury and Martin, 1973; Bury, 1983; Welsh and Lind, 1988). In southwestern Oregon, they are closely associated with humid coastal mixed evergreen forests (Leonard et al., 1993). Farther south, populations occur in a broader range of habitats, from moist coastal forests to oak woodlands. Sierra Nevada records are principally associated with pine-oak woodland and chaparral of the foothills (Block and Morrison, 1998), although populations may extend onto the floor of the Central Valley along riparian corridors (Hayes and Cliff, 1982). Ecological and behavioral aspects of water economy are described by Wisely and Golightly (2003).

Individuals occur under logs, bark, rocks, boards, and other surface cover, and in damp leaf litter (Stebbins, 1985). Periods of surface activity correspond closely to the rainy season (November–April/May for most of the range, later in moist, coastal forests). Maiorana (1977a) observed salamanders dead and dying in the field in the San Francisco Bay region during March and April; she attributed mortality to thermal stress. Field body temperatures for salamanders found under surface cover ranged from 2.2–15.8°C (mean of 7.7°C, n = 82; Feder et al., 1982; R.W.H., unpublished data).

F. Home Range Size. Aside from longer movements associated with breeding activity, individuals tend to remain within a small area over most of their lives. Anderson (1960) found short-term movements averaging 1.5 m, and Maiorana (1978a) recorded movements to 2 m. There is a strong tendency for individuals to occupy the same cover objects over several seasons; Hendrickson (1954) recaptured 59% of his marked animals under their original cover.

G. Territories. Unknown.

H. Aestivation/Avoiding Desiccation. Surface activity corresponds to moist surface conditions, which for most of the range means a period of November–March/April, with local differences related to elevation, slope exposure, and recent precipitation. Surface activity is extended in some areas, such as those receiving daily on-shore fog or in closed canopy redwood forests, where moist conditions prevail outside the rainy season. In the San Francisco Bay area, surface activity begins in October and extends to May (or June in favorable years). In the moist redwood forests of the Santa Cruz Mountains, activity may occur in all months (Brame, 1959). During the dry season, individuals apparently retreat underground, using old root channels, earthworm burrows, or deep talus.

I. Seasonal Migrations. Unknown (but see "Breeding migrations" above).

J. Torpor (Hibernation). Rarely found on the surface during periods of low winter temperatures (e.g., soil temperatures below 5 °C).

K. Interspecific Associations/Exclusions. Over most of their range, California slender salamanders are the only species of *Batrachoseps* present. Along the southern slopes of the Santa Cruz Mountains, the ranges of California slender salamanders and morphologically similar Gabilan Mountains slender salamanders overlap, and they are locally sympatric (e.g., Hecker Pass on the Santa Cruz-Santa Clara County line; Jockusch et al., 2001). In general, California slender salamanders are found in more mesic and upland sites than the other species. Habitat associations for California slender salamanders in the Sierra Nevada are not well understood, and only a single area of sympatry with Hell Hollow slender salamanders is known (Jockusch et al., 1998).

Maiorana (1978a) suggested that differing patterns of fine-scale distribution between sympatric arboreal salamanders (*Aneides lugubris*) and California slender salamanders were due to differential cover use—arboreal salamanders prefer structurally diverse microhabitats (e.g., rock piles) with larger openings and retreats relative to those preferred by California slender salamanders. Further, she suggested that competition with arboreal salamanders for burrows, rather than prey availability, may limit the density of *Batrachoseps* in some areas.

L. Age/Size at Reproductive Maturity. Age estimates at sexual maturity range from 2–4 yr (Hendrickson, 1954) to 2.5–3.5 yr (Anderson, 1960; Maiorana, 1976). Adult females from the San Francisco Bay region were 32–52 mm SVL (mean = 41.6, n = 79); adult males are slightly smaller (Maslin, 1939). Maximum size is 59 mm SVL (Fisher, 1953). Once sexual maturity has been attained, there is little correlation between age and size (Wake and Castanet, 1995).

M. Longevity. Hendrickson (1954) estimated longevity at <10 yr. Based on skeletochronological data, Wake and Castanet (1995) showed that individuals live to be at least 8 yr old.

N. Feeding Behavior. Prey consist of small insects (especially springtails and small beetles), snails, isopods, mites, and spiders (Schonberger, 1944; Adams, 1968; Maiorana, 1978a,b; Lynch, 1985; Bury and Martin, 1973), captured by a projectile tongue (Lombard and Wake, 1977).

O. Predators. Several workers document predation under conditions of captivity (e.g., Hubbard, 1903; Storer, 1925; Stebbins, 1954a). Stebbins (1951) and Storer (1925) state that arboreal salamanders feed on California slender salamanders, but details are not provided and it is not possible to determine the prey species involved. The only unambiguous records of predation in the wild we can locate are by sharp-tailed snakes (*Contia tenuis*), Santa Cruz garter snakes (*Thamnophis atratus*; Stebbins, 1954a; Boundy, 1999), California giant salamanders (*Dicamptodon ensatus*; Bury, 1972; T. Burkhardt, personal communication), and scrub jays (*Aphelocoma coerulescens*; Reaser, 1997b). Several authors have suggested ring-necked snakes (*Diadophis punctatus*) as likely predators, and captive individuals readily consume California slender salamanders as well as other species of *Batrachoseps* (R.W.H., personal observations). We regard screech owls (*Otus kennicottii*) as likely predators as well, although this remains undocumented.

P. Anti-Predator Mechanisms. The following behaviors have been documented: coiling, violent thrashing (rapid coiling/uncoiling or "flipping" behavior), immobility/crypsis, release of adhesive skin secretions, and tail autotomy (Hubbard, 1903; Storer, 1925; Brodie et al., 1974a; Garcia-Paris and Deban, 1995). Flipping behavior may propel an individual salamander 10–20 cm where it may remain motionless (Brodie et al., 1974a). Arnold (1982) reported that in a lab-staged encounter between a California slender salamander and a garter snake, the salamander prevented ingestion by looping its tail around the snake's head to form a knot. Moreover, adhesive skin secretions released by the salamander caused the snake's jaw to remain glued shut at least 48 hr later. The drab brown to reddish-brown dorsal coloration often closely matches native substrates. Maiorana (1977b) reported that tail break frequencies of adult California slender salamanders from a population near Berkeley ranged from 28–48% annually, with an equal distribution between sexes.

Q. Diseases. Unknown.

R. Parasites. Two helminths are recorded from California slender salamanders: *Batracholandros salamandrae* (Lehmann, 1954; Goldberg et al., 1998c) and *Cylindrotaenia diana* (Helfer, 1949; Lehmann, 1960).

S. Comments. Until Yanev's work (1978, 1980), California slender salamanders included several other forms now recognized as distinct species: black-bellied slender salamanders (*B. nigriventris*), gregarious slender salamanders, Hell Hollow slender salamanders (*B. diabolicus*, in part), as well as the recently described *B. gavilanensis* and *B. luciae* from the Coast Ranges of central California. Collectively, these taxa exhibit a derived, attenuate morphology; however, recent studies of mtDNA gene sequences (Jockusch, 1997b) support Yanev's (1978, 1980) finding that California slender salamanders are phylogenetically isolated with respect to other members of the genus. For papers published prior to Yanev (1980), it may be impossible to determine whether references to California slender salamanders actually pertain to *attenuatus* or to sympatric, visually indistinguishable (and as yet undescribed) species of the Pacific slender salamander (*B. pacificus*) complex.

4. Conservation.

Large portions of the historical range have been modified by development for housing, agriculture, and other activities, although California slender salamanders are still present over most of their original range. In some places, such as the San Francisco Bay region, they continue to be locally abundant in urban or suburban edge settings (e.g., gardens, vacant lots). Habitat loss is perhaps most important in areas where the species has only a limited presence, such as the floor of California's Central Valley. California slender salamanders live in narrow strips of riparian oak woodlands along small creeks draining into the Central Valley. As modern agriculture replaces these residual habitats with drainage ditches, the salamanders survive only by becoming symbiotic with humans (they now occur in both Sacramento and Woodland).

Acknowledgments. We are indebted to Carla Cicero and Shawn Kuchta for assistance with literature searches. Stephen Goldberg provided key references for parasites. Bruce Bury and Tim Burkhardt extracted predation data from their field notes. Richard Hoyer shared information concerning his discovery of a communal nest.

Batrachoseps campi Marlow, Brode, and Wake, 1979
INYO MOUNTAINS SALAMANDER
Robert W. Hansen, David B. Wake

1. Historical versus Current Distribution.

Inyo Mountains salamanders (*Batrachoseps campi*) were discovered in 1973 from two desert springs (French Spring and Long John Canyon) on the western slopes of the Inyo Mountains in California's northern Mojave Desert (Marlow et al., 1979). Subsequent field surveys revealed 14 additional localities, all within the Inyo Mountains, including 10 on the eastern slopes (Giuliani, 1977, 1988, 1990; Yanev

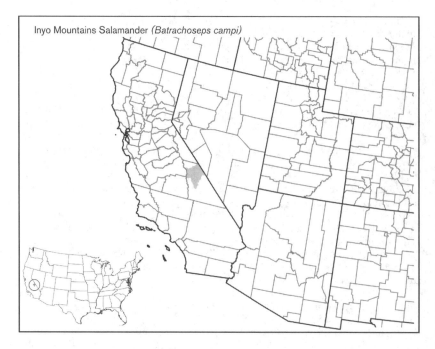

Inyo Mountains Salamander *(Batrachoseps campi)*

and Wake, 1981; Papenfuss and Macey, 1986; Jennings and Hayes, 1994a). Northern range limits are Waucoba Canyon (2,285 m, east slope) and Barrel Spring (1,950 m, west slope); southernmost sites are Long John Canyon (1,695–1,830 m, west slope) and Hunter Canyon (490–915 m, east slope). The elevational range is from 490 (Hunter Canyon, east slope) to 2,590 m (upper Lead Canyon, east slope; Giuliani, 1977; Papenfuss, 1986).

All inhabited sites are highly localized—small areas of suitable habitat bordered by large expanses of inhospitable desert or semi-desert terrain. Inyo Mountains salamanders and desert slender salamanders *(B. aridus)* are the only salamanders that occur exclusively in desert ecosystems.

Occupied habitat for Inyo Mountains salamanders totals <20 ha (Papenfuss and Macey, 1986; Giuliani, 1977). However, a few potential sites within the Inyo Mountains remain unsurveyed, and it is likely that additional populations will be found. Following surveys of eastern slope canyons of the Inyo Mountains, Giuliani (1977) estimated that potential habitat was present along about 19.2 linear km (11.9 mi).

2. Historical versus Current Abundance.
Water diversion for mining or other activities has occurred at some sites, resulting in habitat degradation and possible extirpation of some populations (Papenfuss and Macey, 1986). This is especially severe for Barrel Spring (1,950 m), where the original 0.8-km- (0.5-mi-)long corridor of riparian habitat has been reduced 90% by water diversion and road construction (Giuliani, 1990). At Long John Canyon, flash flooding in 1985 caused a scouring

of the canyon bottom, resulting in complete loss of riparian vegetation (Giuliani, 1990); the population appeared to be recovering slowly by 1995 (Giuliani, 1996). Considerable loss of riparian vegetation occurred at French Spring during flash flooding in 2001, although nearby salamander habitat was less affected (T. Russi, personal communication). Feral burros have degraded habitat at some sites (Giuliani, 1977; Papenfuss and Macey, 1986).

3. Life History Features.
A. Breeding. Reproduction is terrestrial.
i. Breeding migrations. Unknown.
ii. Breeding habitat. Unknown.
B. Eggs.
i. Egg deposition sites. Nests are unknown. Eggs are probably laid in moist crevices within the outcrops characteristic of Inyo Mountains salamander habitat.
ii. Clutch size. Unknown.
C. Direct Development. Eggs have not been recorded, but all species of *Batrachoseps* undergo direct development. A female found on 15 April at French Spring contained eggs visible through the abdomen.
D. Juvenile Habitat. Unknown how this may differ from adults. Both adults and juveniles have been found under rocks on wet substrates.
E. Adult Habitat. The Inyo Mountains are a north–south trending range, with east–west oriented canyons dissecting both slopes. Northern Mojave Desert vegetational associations (e.g., creosote bush scrub) occur at lower elevations, grading into Great Basin communities (sagebrush steppe and pinyon-juniper woodland) at higher elevations. Rainfall amounts range from 13 cm/yr at Independence in the adjacent Owens Valley

(S. Hsu, personal communicaton) to approximately 22 cm/yr on the western slopes of the Inyo Mountains (Marlow et al., 1979). Perennial springs and seepages and their associated riparian plant species occupy discrete sections of steep-sided canyons; these areas are grown to willow, wild rose, and coyote brush, often forming dense thickets such that direct sunlight rarely reaches ground level. Salamander populations are largely restricted to such places, especially where solid-rock cliffs, outcrops, or talus are in contact with surface flow (Giuliani, 1977; Marlow et al., 1979; Macey and Papenfuss, 1991a). Salamanders have been found under rocks resting on wet substrates, sometimes under woody debris (which is relatively scarce), and occasionally within clumps of moist ferns growing in waterfall spray zones (Giuliani, 1977; Papenfuss and Macey, 1986). Fissures and crevices within the adjacent granitic or limestone outcrops also harbor salamanders and presumably serve as refugia during periods of unfavorable surface conditions. All records are from riparian sites except one—several specimens were found in pitfall traps positioned near openings in rock formations on a ridge (2,285 m) between two canyons. This suggests that at least at higher elevations within pinyon-juniper woodland, salamanders may not be restricted to canyon bottom, riparian habitats.

Salamanders have been found in every month, although snowfall and extreme cold undoubtedly restrict activity at higher elevation sites. Field body temperatures (inferred from substrate temperatures) ranged from 10.5–18.0 °C (Giuliani, 1977; R.W.H., unpublished data). Surface movements presumably occur at night.

F. Home Range Size. Unknown, but presumably small. Surface habitat at some sites is extremely limited, although including the moist crevices in adjacent outcrops may considerably expand the total available habitat.

G. Territories. Unknown.

H. Aestivation/Avoiding Desiccation. Perennial springs and seepages provide a moist, thermally buffered environment, such that surface activity is possible at most sites year-round. Water temperatures recorded from several springs where salamanders were found (at elevations of 490–2,590 m) ranged from 10.5–17.5 °C, with only modest seasonal variation (Giuliani, 1977). There was no correlation between elevation and water temperature.

I. Seasonal Migrations. Unknown.

J. Torpor (Hibernation). Unknown. Winter snowfall and low temperatures at higher elevation sites probably limit winter activity, although salamanders have been observed in December and January at moderately high elevations (1,765–1,950 m). Higher elevation sites are not accessible in winter.

K. Interspecific Associations/Exclusions. Inyo Mountains salamanders are the only salamanders within their range.

L. Age/Size at Reproductive Maturity. Females attain larger average and maximum size. The smallest sexually mature male known is 41.3 mm; average size of 27 sexually mature males was 46.9 mm SVL (maximum = 53.7 mm, Waucoba Creek, east slope). Smallest sexually mature female known is 43.8 mm SVL; average size of 29 sexually mature females was 51.6 mm SVL (maximum = 60.8 mm, Cove Creek, east slope).

M. Longevity. Unknown.

N. Feeding Behavior. Has not been described, although all *Batrachoseps* species observed thus far use a projectile tongue to capture small invertebrates.

O. Predators. Unknown.

P. Anti-Predator Mechanisms. Defensive behaviors include immobility (Marlow et al., 1979) and coiling in both juveniles and adults (R.W.H., unpublished data).

Q. Diseases. Unknown.

R. Parasites. Unknown.

S. Comments. Considerable genetic differentiation exists among populations currently recognized as *B. campi*, likely due to genetic drift and restricted gene flow together with small population sizes (Yanev and Wake, 1981). The genetic distance data, in conjunction with the extremely localized nature of potential habitat (Giuliani, 1977; Papenfuss and Macey, 1986), indicate that populations are currently genetically isolated from one another. The extreme fragmentation of the species is indicated by the high levels of F_{ST} (0.59) and low levels of N_m (on the order of 0.025 for the species; Larson et al., 1984). Extrapolating from these analyses, the collective effective population size of the populations sampled by Yanev and Wake (1981) was estimated at 14,000. Studies of mtDNA gene sequences, allozymes, and morphological data (Yanev, 1980; Jockusch, 1996; Jackman et al., 1997) confirm that *B. campi*, Oregon slender salamanders (*B. wrighti*), and Kern Plateau salamanders (*B. robustus*) from the southern Sierra Nevada are phylogenetically distant from other species of *Batrachoseps*.

4. Conservation.
Papenfuss and Macey (1986) carefully detailed specific threats to each of the then-known populations of Inyo Mountains salamanders. Among these were mining activities (>360 mining claims near *B. campi* populations), damage from livestock and feral burros, and water diversions. All localities occur on federal lands managed by the U.S. Bureau of Land Management or USDA Forest Service. Concerted efforts have been made by federal agencies to reduce the number of feral burros in the Inyo Mountains region. This is a Species of Special Concern (California

Department of Fish and Game) and is listed as a Forest Service and BLM Sensitive Species.

Acknowledgments. Darrell Wong (California Department of Fish and Game) and Amy Kuritsubo (Bureau of Land Management) provided copies of unpublished reports. Simon Hsu of the Los Angeles Department of Water and Power provided rainfall data for the Owens Valley.

Batrachoseps diabolicus Jockusch, Wake, and Yanev, 1998
HELL HOLLOW SLENDER SALAMANDER
Robert W. Hansen, David B. Wake

1. Historical versus Current Distribution.
Hell Hollow slender salamanders (*Batrachoseps diabolicus*) are found on the western slope foothills of the Sierra Nevada of California, from the lower Merced River Canyon (Mariposa County) north to the American River (Placer County) (Jockusch et al., 1998), at elevations below 620 m. Recent work by Jockusch et al. (1998) partitioned the formerly wide-ranging *Batrachoseps relictus* of the Sierra Nevada into four species (from north to south in the central and southern Sierra Nevada): *B. diabolicus*, Kings River slender salamanders (*B. regius*), Sequoia slender salamanders (*B. kawia*), and relictual slender salamanders (*B. relictus*). Of these species, *B. diabolicus* has a relatively extensive distribution and is known from at least a dozen localities.

2. Historical versus Current Abundance.
Based upon limited sampling, populations from the Merced River Canyon appear stable. There is little information concerning the status of more northern populations, as these have been recognized only recently as *B. diabolicus*.

3. Life History Features.
A. Breeding. Reproduction is terrestrial.

i. Breeding migrations. Unknown.

ii. Breeding habitat. Unknown. Presumably, courtship occurs after the start of the rainy season in the fall; egg-laying probably takes place from November–January, depending on local rainfall.

B. Eggs.

i. Egg deposition sites. Nest sites have not been found for Hell Hollow slender salamanders. At several sites (for example at the type locality), salamanders are associated with extensive metamorphic rock talus; presumably eggs are laid well underground.

ii. Clutch size. Unknown.

C. Direct Development. Newly hatched salamanders were observed on 8 February at Hell Hollow (R.W.H., unpublished data). Onset of fall/winter rains in this part of the Sierra Nevada is unpredictable and varies from year to year; thus, timing of hatching is expected to vary both geographically and annually.

D. Juvenile Habitat. Hatchling-sized individuals have been found in the same general areas as adults, but often under smaller pieces of cover.

Hell Hollow Slender Salamander (*Batrachoseps diabolicus*)

E. Adult Habitat. Hell Hollow slender salamanders are found mostly in mixed pine-oak woodland and chaparral communities of the foothills of the Sierra Nevada, often at low elevations where the foothills first appear, in areas of extreme summer heat and drought.

At Hell Hollow in the lower part of the Merced River Canyon (ca. 300 m elevation), salamanders are found beneath metamorphic rocks (often occurring in patches of talus), bark rubble, and downed logs, usually in areas that receive little direct sunlight during the winter. Prominent components of the local vegetation include foothill pine, interior live oak, California buckeye, California bay, and toyon.

At one site at the northern end of their range (near the Middle Fork of the American River, El Dorado County; 245 m elevation), salamanders were found in a narrow, shaded ravine containing a small stream, with limestone outcrops and rubble. Vegetation included canyon live oak, bigleaf maple, California bay, and Douglas fir; the ground was covered with mosses and ferns. Salamanders were obtained from under bark layers on fallen logs, as well as beneath rocks, in areas of shade or filtered sunlight. At lower elevations in the same region (ca. 100 m elevation), the salamanders have been taken in open pastures under rocks.

F. Home Range Size. Unknown.

G. Territories. Unknown.

H. Aestivation/Avoiding Desiccation. Salamanders are present under surface cover only during periods of adequate soil moisture—generally from November–March/April. Timing of surface activity varies depending on arrival of fall/winter rains. Individual salamanders presumably move beneath the surface in burrows or rock rubble during dry periods and during episodes of extreme cold. Field body temperatures of salamanders found under cover averaged 8.5 °C (range = 6.0–12.9 °C; n = 11; R.W.H., unpublished data).

I. Seasonal Migrations. Unknown.

J. Torpor (Hibernation). We have not recorded surface activity at substrate temperatures below 6 °C.

K. Interspecific Associations/Exclusions. Hell Hollow slender salamanders occur in association with limestone salamanders *(Hydromantes brunus)*, arboreal salamanders *(Aneides lugubris)*, yellow-eyed ensatinas *(Ensatina eschscholtzii xanthoptica)*, and California newts *(Taricha torosa)* within the Merced River Canyon, and with these species (except for limestone salamanders) and California slender salamanders *(B. attenuatus)* at localities farther north. At the northern end of their range at the American River, Hell Hollow slender salamanders are sympatric with California slender salamanders (Jockusch et al., 1998)—the only confirmed instance of sympatry between these two species—although information concerning their local distribution and habitat characteristics is unknown. The ranges of these congeners overlap considerably and they undoubtedly occur in local sympatry at many places. Within the Hell Hollow area of the Merced River Canyon, Hell Hollow slender salamanders are often found in proximity to individuals of limestone salamanders; however, the latter are nearly always associated with moist talus, while Hell Hollow slender salamanders range into the surrounding chaparral.

L. Age/Size at Reproductive Maturity. This is a moderately small species of *Batrachoseps*, with adults ≤45 mm SVL (Jockusch et al., 1998). Age and size at sexual maturity are unknown.

M. Longevity. Unknown.

N. Feeding Behavior. Has not been described, although all *Batrachoseps* species observed thus far use a projectile tongue to capture small invertebrates.

O. Predators. Predation has not been observed. Ring-necked snakes *(Diadophis punctatus)* occur throughout the range of Hell Hollow slender salamanders, and the two species have been found meters apart at Hell Hollow.

P. Anti-Predator Mechanisms. Immobility, coiling, and rapid crawling have all been observed when salamanders are first uncovered (R.W.H., unpublished data). There does not seem to be a prevalent, stereotypical response. The dark dorsal and ventral coloration renders them effectively cryptic when uncovered.

Q. Diseases. Unknown.

R. Parasites. Unknown.

S. Comments. Until the work of Yanev (1978, 1980), Jockusch (1996), and Jockusch et al. (1998), taxonomic relationships among *Batrachoseps* in the central and northern Sierra Nevada were poorly resolved; thus, some earlier references (e.g., Basey, 1969, 1976; Basey and Sinclear, 1980) did not make distinctions among the three species of *Batrachoseps* now known to occur in this region: *B. attenuatus*, *B. diabolicus*, and *B. gregarius*. Moreover, specimens from this part of the Sierra Nevada in some institutional collections have typically been identified as *B. attenuatus*, and it is not clear which species are actually represented; reliance on locality for species identification is not possible in many instances.

Batrachoseps diabolicus shows substantial geographic variation in mtDNA sequences (cytochrome b). Jockusch et al. (1998) found five haplotypes falling into two clades which differ by 8.0–11.8%. This is substantial divergence, suggesting that more than one species might be represented. One of these clades has been found only in Calaveras County, whereas the other clade occurs throughout the entire range of the species.

4. Conservation.
Salamanders of the genus *Batrachoseps* can be difficult to locate in the central and northern Sierra Nevada, a region home to three species: *B. attenuatus*, *B. diabolicus*, and *B. gregarius* (which reaches its northern range limits near the southern boundary of *B. diabolicus*). Salamanders are absent from large areas of apparently suitable habitat and are seldom common where encountered. These factors, together with the difficulty of making species identifications in the field, have hampered field studies. Thus, detailed knowledge of distribution, habitat associations, and other aspects of the natural history of these species is fragmentary.

Batrachoseps gabrieli Wake, 1996
SAN GABRIEL MOUNTAINS SLENDER SALAMANDER

Robert W. Hansen, Robert H. Goodman, Jr., David B. Wake

1. Historical versus Current Distribution.
San Gabriel Mountains slender salamanders *(Batrachoseps gabrieli)* recently have been described (Wake, 1996) and are known from 13 discrete localities along the southern flanks of the San Gabriel and San Bernardino mountains of southern California (Wake, 1996; Goodman et al., 1998; R.H.G., unpublished data). Their distribution is discontinuous, from San Gabriel Canyon in the San Gabriel Mountains, Los Angeles County, east to Waterman Canyon in the San Bernardino Mountains, San Bernardino County (Stewart et al., in press). Elevation ranges from near 850–2,380 m. All sites occur on public lands administered by Angeles and San Bernardino National Forests.

2. Historical versus Current Abundance.
San Gabriel Mountains slender salamanders are difficult to find at most sites. The number of animals found/visit has ranged from 0–31, but a more typical yield is 1–4 specimens. The difficulty of discovering animals under surface cover may be attributable to their occurrence in often-extensive talus, where salamanders may be present well below the top layer of rocks and thus are not readily observed.

3. Life History Features.
A. Breeding. Reproduction is terrestrial.
i. Breeding migrations. Unknown.
ii. Breeding habitat. Courtship and oviposition may occur following the first heavy rains in the fall.
B. Eggs.
i. Egg deposition sites. Nest sites have not been located. Considering the close association of San Gabriel Mountains slender salamanders with talus, it seems likely that eggs are deposited deep in talus piles.
ii. Clutch size. Unknown.
C. Direct Development. Egg laying and hatching have not been observed. San Gabriel Mountains slender salamanders

San Gabriel Mountains Slender Salamander *(Batrachoseps gabrieli)*

occur at mid-to-high elevations with surface activity restricted to southern California's winter precipitation season from November–April (Wake, 1996; Stewart et al., in press). A single hatchling-sized salamander was discovered on 15 April in the South Fork Lytle Canyon (San Gabriel Mountains).

D. Juvenile Habitat. Differences between juvenile and adult habitat use have not been observed, although we have found subadults more often under branches and other small cover.

E. Adult Habitat. San Gabriel Mountains slender salamanders are southern California talus specialists. However, there are considerable differences in local habitats owing to variation in elevation, vegetation, and exposure. Most salamanders have been discovered in and around stable talus accumulations of various sizes, under rotting logs, bark, downed branches, fern fronds, and rocks (Wake, 1996; Goodman et al., 1998; Stephenson and Calcarone, 1999; Stewart et al., in press). The type locality (1,550 m, 1 km ESE of Crystal Lake, along Soldier Creek, San Gabriel Mountains) consists of a steep northwest-facing talus slope within mixed conifer forest (various species of pines, white fir, big-cone spruce, incense cedar, canyon live oak; Wake, 1996). Other populations occur at lower, drier, more exposed sites within chaparral communities. At Alpine Canyon (1,150 m, San Gabriel Mountains), San Gabriel Mountains slender salamanders are associated with isolated stands of big cone spruce in talus bordered by chaparral vegetation on a southwestern exposure.

F. Home Range Size. Unknown.

G. Territories. Unknown.

H. Aestivation/Avoiding Desiccation. Surface activity is limited to the rainy season during the winter and early spring (November–April). A large majority of specimens have been found following significant rain/snow events. Animals have been observed and collected while snow was present on the ground (Wake, 1996; R.H.G., unpublished data). As the talus habitat dries, there is a sharp decline in surface activity and salamanders become increasingly difficult to find. A substrate temperature of 4.2 °C was recorded for three salamanders found beneath a rock in late March (Wake, 1996) and is probably close to the low temperature threshold for surface presence.

I. Seasonal Migrations. Unknown.

J. Torpor (Hibernation). Unknown.

K. Interspecific Associations/Exclusions. San Gabriel Mountains slender salamanders occur in sympatry with black-bellied slender salamanders *(B. nigriventris)* at some sites in the San Gabriel Mountains (Wake, 1996; R.H.G., unpublished data). Within the San Gabriel Canyon area, black-bellied slender salamanders are common in oak woodlands and in riparian settings (with alders and sycamores). They have not been found in the forested areas east of San Gabriel Canyon, where San Gabriel Mountains slender salamanders occur.

Garden slender salamanders *(B. major)* occur within 5.5 km of the nearest population of San Gabriel Mountains slender salamanders in San Gabriel Canyon, but occupy distinctly different habitats at generally lower elevations in this region.

L. Age/Size at Reproductive Maturity. Adult males are 39.8–46.3 mm SVL (mean = 42.4 mm); females are 41.0–50.0 mm (mean = 46.1 mm; Wake, 1996).

M. Longevity. Unknown.

N. Feeding Behavior. Small invertebrates are captured with a projectile tongue. Field examination of San Gabriel Mountains slender salamander scat indicates at least some consumption of ants *(Crematogaster* sp.; R.H.G., unpublished data).

O. Predators. Unknown.

P. Anti-Predator Mechanisms. Various reactions to discovery include coiling (Wake, 1996), immobility, or slowly crawling away; less often, salamanders have escaped by moving into deeper talus layers (R.H.G., unpublished data).

Q. Diseases. Unknown.

R. Parasites. Unknown.

S. Comments. Following the description of San Gabriel Mountains slender salamanders in 1996, careful searching in rockslide habitats has revealed additional populations. Although we provisionally include them in *B. gabrieli,* some of these recently discovered populations appear to have diverged genetically following long periods of isolation and may warrant recognition as a distinct species (D.B.W. and colleagues, unpublished data).

4. Conservation.
The ecology of San Gabriel Mountains slender salamanders remains poorly known. All but two of the localities are represented by fewer than five specimens. Further searches of rockslide and talus habitats in the San Gabriel, San Bernardino, and San Jacinto mountains are needed. San Gabriel Mountains slender salamanders are a U.S. Forest Service Sensitive Species.

Batrachoseps gavilanensis Jockusch, Yanev, and Wake, 2001
GABILAN MOUNTAINS SLENDER SALAMANDER

Robert W. Hansen, David B. Wake

1. Historical versus Current Distribution.
Gabilan Mountains slender salamanders *(Batrachoseps gavilanensis)* were described in 2001 on the basis of differences in allozymes and in DNA haplotypes in the mitochondrial gene cytochrome b (Jockusch, 1996; Jockusch et al., 2001). Gabilan Mountains slender salamanders are broadly distributed and currently known from many locations in central coastal California (Jockusch et al., 2001). The northernmost known location is Rodeo Gulch, Santa Cruz County. From there, populations are found southward to the Pacific Coast, and along the coast to the eastern edge of the city of Monterey. They are found in Monterey on Jack's Peak, Monterey County. South of the Monterey Peninsula, populations do not occur along the coast, but instead extend southeast to the south-central portions of Monterey County.

The northern boundary of the range of Gabilan Mountains slender salamanders extends from Santa Cruz east along the base of the Santa Cruz Mountains. From

Gabilan Mountains Slender Salamander (*Batrachoseps gavilanensis*)

there populations extend southeast into the Gabilan Range and other mountains in San Benito County, onto Mustang Ridge in extreme eastern Monterey County. Inland locations include Coalinga Mineral Springs, western Fresno County, and the Cholame region of southeastern Monterey and northeastern San Luis Obispo counties, extending into the northwestern tip of Kern County. Preserved specimens from the Temblor Range in Kern County are tentatively assigned to this species (Jockusch et al., 2001). The elevational range is from near sea level to about 880 m (headwaters of San Benito River at Fresno/Monterey county line). Differences between historical and current distributions are not apparent.

2. Historical versus Current Abundance.
Aside from local extirpations associated with development and agriculture, populations appear stable.

3. Life History Features.
A. Breeding. Reproduction is terrestrial.
i. Breeding migrations. Unknown.
ii. Breeding habitat. Unknown.
B. Eggs.
i. Egg deposition sites. Oviposition sites are unknown. Egg attendance by females and communal nesting are unknown for this species.
ii. Clutch size. Unknown.
C. Direct Development. This species lays eggs that undergo direct development. Eggs have been laid in the laboratory but no field observations have been made as yet.
D. Juvenile Habitat. Differences between juvenile and adult habitat use are not apparent.

E. Adult Habitat. Gabilan Mountains slender salamanders are found in a variety of habitats from deeply shaded, moist redwood and mixed coniferous forests through oak woodland and chaparral to open grassland with widely scattered small oaks. The northwesternmost localities have cool, equable climates and rainfall that can exceed 100 cm annually; the southeasternmost localities are hot and dry during the summer, with annual rainfall <20 cm. Indeed, populations occurring in the southern Diablo Range (e.g., Coalinga Mineral Springs, Fresno County) and the northern Temblor Range (e.g., Cottonwood Pass, Kern County) occur in climatically harsh settings rarely occupied by members of this genus; these populations appear to be widely scattered and localized and are restricted to relatively mesic north-facing slopes in otherwise arid oak woodlands or chaparral.
F. Home Range Size. Unknown.
G. Territories. Unknown.
H. Aestivation/Avoiding Desiccation. Salamander activity closely tracks the rainy season (largely December–April), especially in the warmer and drier southern and eastern parts of the range. Northern and coastal populations may remain surface active into the summer. Body temperatures (inferred from substrate temperatures) for a small series of adult and juvenile Gabilan Mountains slender salamanders found on 24 March in Fresno County averaged 12.1 °C (range 11.5–12.5 °C; n = 16; R.W.H., unpublished data).
I. Seasonal Migrations. Unknown.
J. Torpor (Hibernation). Unknown.
K. Interspecific Associations/Exclusions. North of Monterey Bay, Gabilan Mountains

slender salamanders are found in sympatry with California slender salamanders (*B. attenuatus*) along an ecotone between upland redwood forests and lowland oak habitats. One such area is Hecker Pass at the southern end of the Santa Cruz Mountains (near Santa Cruz-Santa Clara county line), where both species are locally sympatric. Along the southwestern edge of their range, on Mustang Ridge and Peachtree Valley, Monterey County, and Coalinga Mineral Springs, Fresno County, Gabilan Mountains slender salamanders and black-bellied slender salamanders (*B. nigriventris*) are found in microsympatry except on steep slopes with scattered foothill pine and juniper (Jockusch et al., 2001). On Jacks Peak, along Carmel Valley and near Arroyo Seco, Monterey County, sympatry is expected between Gabilan Mountains slender salamanders and Santa Lucia Mountains slender salamanders (*B. luciae*) but has not been confirmed. These two species have been taken within 200 m of each other in southwestern Monterey County.

Gabilan Mountains slender salamanders are ecological equivalents of *B. attenuatus* and *B. nigriventris*; all three species are relatively widely distributed and occur over a broad range of ecological settings. These species replace one another geographically, with only limited range overlap, suggesting that range expansion might be limited by competition.

L. Age/Size at Reproductive Maturity. This is a moderately small species of *Batrachoseps*, although individuals attain larger sizes than other central coastal species of the *B. pacificus* complex. Based upon measurements of a small series, males ranged from 39.4–46.7 mm SVL (mean 43.1 mm; n = 10), females from 38.8–50.6 mm SVL (mean 43.9 mm; n = 10) mm.
M. Longevity. Unknown.
N. Feeding Behavior. Gabilan Mountains slender salamanders, similar to other species of *Batrachoseps*, capture small insect prey using their projectile tongue.
O. Predators. Unknown, although snakes, birds, and small mammals are likely predators.
P. Anti-Predator Mechanisms. Coiling is a common defensive response in several species of *Batrachoseps*.
Q. Diseases. Unknown.
R. Parasites. Unknown.
S. Comments. Although only recently described, specimens now referred to this species have been in scientific collections for many years though assigned to other taxa (e.g., *B. attenuatus* [Hendrickson, 1954] and later, *B. relictus* [Brame and Murray, 1968] and *B. pacificus* [Yanev, 1980]). Yanev (1978, 1980), on the basis of allozymes, first identified the major lineages within the central coastal *B. pacificus* complex. Subsequently, these have been recognized as four distinct species (*gavilanensis, incognitus, luciae,* and *minor*

[Jockusch et al., 2001]). Based on studies of mtDNA and allozymes, it seems likely that *B. gavilanensis* is the most basal member of the *pacificus* group (Jockusch and Wake, 2002). There is some evidence that *B. gavilanensis* and *B. luciae* might have been in genetic contact early in their history, although there is no indication of present gene exchange despite sharing a long border through Carmel Valley (Jockusch et al., 2001).

4. Conservation.
Gabilan Mountains slender salamanders are moderately widespread throughout their range and appear to be common at a number of localities. Portions of their range occur on publicly owned lands or other large land holdings that are likely to remain relatively undisturbed for the foreseeable future. Aside from local extirpations associated with human development, there are no known critical conservation concerns. Surveys for this species might be complicated by the closely similar appearance of *B. attenuatus* and *B. nigriventris*—species that have ranges that partly overlap that of *B. gavilanensis* and are extremely difficult to distinguish in the field.

Batrachoseps gregarius Jockusch, Wake, and Yanev, 1998
GREGARIOUS SLENDER SALAMANDER
Robert W. Hansen, David B. Wake

1. Historical versus Current Distribution.
Gregarious slender salamanders (*Batrachoseps gregarius*) are found on the western slopes of the central and southern Sierra Nevada of California, from the southern boundary of Yosemite National Park south to just north of the Kern River (Jockusch et al., 1998; Wake and Jockusch, 2000). Northernmost records are for Feliciana Mountain and Jerseydale Ranger Station (Mariposa County). Southern range limit is Rancheria Road, 3 mi north of the Kern River (Kern County). Especially abundant along riparian corridors of the White River–Arrastre Creek drainages in southern Tulare and northern Kern counties. Elevational range is from <100 m to about 1,800 m, with southern populations (e.g., Tulare and Kern counties) generally occurring below 900 m. Although mostly confined to foothills and middle-elevation mountainous regions, gregarious slender salamanders extend onto the floor of the Central Valley along riparian corridors (e.g., San Joaquin River north of Fresno, Kings River at Centerville). There is little evidence of modification of the historical range. Some range-margin populations along the eastern edge of the Central Valley may have been affected by housing, agriculture, or other development.

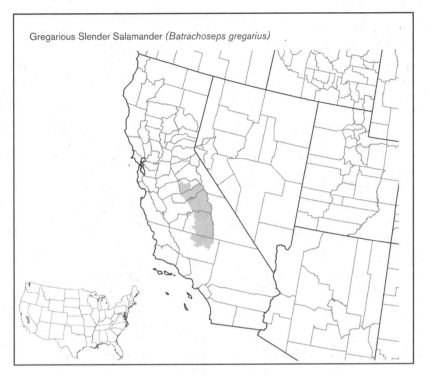

Gregarious Slender Salamander *(Batrachoseps gregarius)*

2. Historical versus Current Abundance.
In our experience, gregarious slender salamanders may be locally abundant at some sites and difficult to find at others, but long-term trends are not evident.

3. Life History Features.
 A. Breeding.
  i. Breeding migrations. Unknown. However, based upon their observations of the local distributions of adult females, Jockusch and Mahoney (1997) suggest that surface movements related to egg laying are more extensive than previously thought.
  ii. Breeding habitat. Unknown.
 B. Eggs.
  i. Egg deposition sites. The following information is taken from Jockusch and Mahoney (1997), who reported on numerous nest sites discovered at both the northern end of the range (near the Madera–Mariposa County line within mixed coniferous forest, 1,400 m elevation) and from the southern end of the range (in foothill oak woodland, Tulare County, 430 m elevation). Most females deposit eggs in communal nests that are then abandoned, although adult males and subadults may be found under the same cover objects (spent females also have been found with eggs but apparently do not remain under these cover objects for extended periods). Eggs are laid beneath downed logs, bark slabs, rocks, or within damp leaf litter usually associated with surface water—stream or seep margins. Nests at the foothill site were located under deeply imbedded rocks in a grassy pasture, while those from the higher elevation forested area were found under a variety of surface objects.

  ii. Clutch size. Egg counts ranged from <10 eggs (the product of a single female) to >300 eggs, which were attributed to >18 females. Clutch size is correlated with female body size and also varies geographically. Females from the northern end of the range produce larger clutches (mean = 15.3 eggs for 50 clutches; record = 29 eggs), while those from the southern part of the range produce fewer eggs (mean = 7.3 eggs for 20 clutches; maximum = 14 eggs; Jockusch, 1997b; Jockusch and Mahoney, 1997).
 C. Direct Development. Timing of egg laying varies with elevation and latitude. At the southern end of the range and at lower elevations, egg laying coincides with the start of the rainy season (Jockusch and Mahoney, 1997). Onset of fall rains in this part of the Sierra Nevada is unpredictable and varies from year to year; thus, egg-laying occurs from mid-November to early January, depending on local rainfall. At higher elevation sites, egg laying occurs later (late March to late April; Jockusch and Mahoney, 1997; R.W.H., unpublished data).

  Developmental times for lab-incubated (at 13 °C) eggs ranged from 65 d (for 50 eggs from a southern population) to 73 d (for 222 eggs from northern sites; Jockusch and Mahoney, 1997). Hatchlings were observed 6 March in the foothills of Fresno County in an area where females lay eggs in December–January.
 D. Juvenile Habitat. Unknown.
 E. Adult Habitat. Gregarious slender salamanders occur principally in the oak woodlands of the Sierra Nevada foothills, dominated by blue oak, interior live oak,

and foothill pine, but also range into mixed conifer forests grown to Ponderosa pine, incense cedar, white fir, and black oak. At their northern range limits, gregarious slender salamanders are abundant within a closed canopy conifer forest with sugar pine and giant sequoias. Annual precipitation at higher elevation sites may exceed 110 cm, with snow often present from November–April (Jockusch et al., 1998). Also recorded from low elevation riparian corridors associated with interior live oak (San Joaquin River north of Fresno) or valley oak (Kings River near Centerville). Gregarious slender salamanders from the east side of the Central Valley along Ash Slough (Madera County) occur in annual grassland. They generally are absent from steep slopes.

The southernmost population, just north of the Kern River, occupies arid rolling grassland with scattered rocks; rainfall averages only 27 cm/yr. Salamanders at this site have been found under rocks with desert night lizards (*Xantusia vigilis*), night snakes (*Hypsiglena torquata*), side-blotched lizards (*Uta stansburiana*), western fence lizards (*Sceloporus occidentalis*), and Gilbert skinks (*Eumeces gilberti*).

At some higher elevation sites in Fresno and Madera counties, gregarious slender salamanders may occur in settings that are similar ecologically to those of the allopatric relictual slender salamanders (*B. relictus*), namely under or within woody debris adjacent to seeps or other moist places within coniferous forests. This is in marked contrast to the large majority of gregarious slender salamander collection sites, most of which are in relatively open oak woodlands subject to high summer temperatures.

F. Home Range Size. Unknown.

G. Territories. Unknown.

H. Aestivation/Avoiding Desiccation. Surface activity is strongly correlated with soil moisture and, to some extent, temperature. For foothill areas, salamanders are surface active from the onset of fall rains (late October to November) to March–April. At higher elevations (>1,200 m), activity may occur in all but the coldest months. Body temperatures of salamanders found under surface cover averaged 12.3 °C (range = 5.0–19.0 °C, n = 280; R.W.H., unpublished data).

I. Seasonal Migrations. Unknown.

J. Torpor (Hibernation). We have not recorded surface activity at substrate temperatures below 5°C. Presumably, salamanders retreat underground below the frost line.

K. Interspecific Associations/Exclusions. Sympatric with Sequoia slender salamanders (*B. kawia*) in the South Fork Kaweah River drainage (Jockusch et al., 1998). Their distribution is allopatric to that of relictual slender salamanders, which

generally occur at higher elevations and wetter sites, although their ranges converge in several drainages in southern Tulare County and northern Kern County (Brame and Murray, 1968; Jockusch et al., 1998). Just beyond the northern range limits for relictual slender salamanders, gregarious slender salamanders are found at higher elevations and occur in moist coniferous forest settings that are typical of relictual slender salamander sites farther south. The ranges of Kings River slender salamanders (*B. regius*) and gregarious slender salamanders closely approach, but do not overlap, in the Kings River drainage (Fresno County), where Kings River slender salamanders are restricted to moist, north-facing slopes.

The southernmost locality for gregarious slender salamanders (just north of the Kern River) is <10 km from populations of Kern Canyon slender salamanders (*B. simatus*) in lower Kern River Canyon. Salamanders assigned to black-bellied slender salamanders (*B. nigriventris*) occur about 37 km south of the southernmost gregarious slender salamander locality, although we suspect that populations belonging to one or both species may occur in the intervening semi-arid foothills of the Tehachapi Mountains.

In the central Sierra Nevada (between Calaveras and Mariposa counties), the ranges of gregarious slender salamanders, California slender salamanders (*B. attenuatus*), and Hell Hollow slender salamanders (*B. diabolicus*) converge, but sympatry is recorded only for the latter two species. Fieldwork is needed in that area to better define the distributional relationships among those species.

L. Age/Size at Reproductive Maturity. Nine adult males from the northern end of the range measured 29.9–40.2 mm SVL (mean = 35.9 mm); 10 females ranged from 38.4–46.0 mm SVL (mean = 42.5 mm; Jockusch et al., 1998). An adult female from Fresno County measured 50 mm SVL.

M. Longevity. Unknown.

N. Feeding Behavior. Has not been described, although all *Batrachoseps* species observed thus far use a projectile tongue to capture small invertebrates.

O. Predators. Predation in the wild is undocumented, although captive night snakes and ring-necked snakes (*Diadophis punctatus*) have eaten gregarious slender salamanders (R.W.H., unpublished data).

P. Anti-Predator Mechanisms. Defensive behaviors include tail autotomy, coiling, immobility, and flipping followed by immobility. Most adults coil tightly or remain immobile when discovered under surface cover; their dark brown or reddish brown coloration is effectively cryptic.

Q. Diseases. Unknown.

R. Parasites. Unknown.

S. Comments. Until the work of Brame and Murray (1968) and Yanev (1978, 1980), Sierra Nevada populations of attenuate *Batrachoseps* were included within California slender salamanders (*B. attenuatus*), a wide-ranging composite species. Yanev restricted the name *B. attenuatus* in the Sierra Nevada to the northern and central portion of that range, allocating Sierran populations of *B. attenuatus* from Mariposa County southward to black-bellied slender salamanders (*B. nigriventris*). Jockusch et al. (1998) demonstrated that Sierra Nevada *B. nigriventris* were specifically distinct from southern California and coastal populations of that species and described gregarious slender salamanders to accommodate the Sierran portion of the range. Thus, some earlier references (e.g., Stebbins, 1951; Basey, 1969, 1976; Basey and Sinclear, 1980) provide information about *B. gregarius* under the name *B. attenuatus*. Additionally, many museum specimens of "*B. attenuatus*" from the Sierra Nevada are mis-identified and could represent any of a number of species, including *B. gregarius*, *B. diabolicus*, *B. relictus*, or *B. kawia*. Wake and Jockusch (2000) include *B. gregarius* in a *B. nigriventris* complex, consisting of *B. nigriventris*, *B. gregarius*, *B. simatus*, and *B. stebbinsi*, as well as undescribed taxa from the southern Sierra Nevada.

Substantial variation exists in morphology, allozymes, and mtDNA gene sequences between northern and southern populations of gregarious slender salamanders (Jockusch et al., 1998). Features of reproductive biology (spring versus fall egg-laying and pronounced clutch size differences) also vary, suggesting that *B. gregarius* as currently recognized may comprise more than one species. Jockusch et al. (1998) considered this possibility, but tentatively regarded intermediate populations as having undergone recent local gene flow and thus chose to recognize a single species.

4. Conservation.

Gregarious slender salamanders are especially abundant along riparian corridors of the White River–Arrastre Creek drainages in southern Tulare and northern Kern counties. There is little evidence of changes in distribution compared with the historical range, and long-term trends in abundance are not evident. Some range-margin populations along the eastern edge of the Central Valley may have been affected by housing, agriculture, or other development.

In general, the distributions of *Batrachoseps* species in the southern Sierra Nevada are complex and incompletely known. Additional field and laboratory studies are underway to resolve outstanding taxonomic and distributional questions.

Batrachoseps incognitus Jockusch, Yanev, and Wake, 2001
SAN SIMEON SLENDER SALAMANDER
Robert W. Hansen, David B. Wake

1. Historical versus Current Distribution.
San Simeon slender salamanders *(Batrachoseps incognitus)* were described in 2001 on the basis of differences in proteins and in DNA sequences of the mitochondrial gene cytochrome b (Jockusch, 1996; Jockusch et al., 2001). San Simeon slender salamanders are found in the Santa Lucia Range in extreme southwestern Monterey County and northern San Luis Obispo County of central coastal California. Although generally found in the mountains, they occur near sea level in the northwesternmost part of their range, where the mountains abruptly meet the ocean just north of the Monterey–San Luis Obispo county line. They range as high as 1,000 m on Pine Mountain and Rocky Butte in northern San Luis Obispo County.

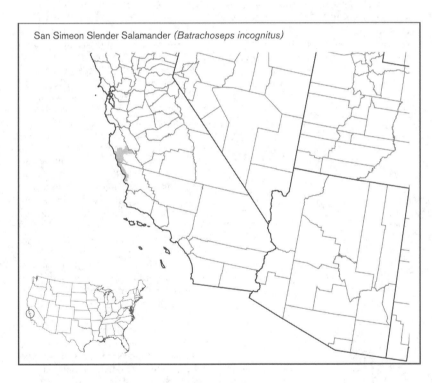

San Simeon Slender Salamander *(Batrachoseps incognitus)*

2. Historical versus Current Abundance. Unknown.

3. Life History Features.
　A. Breeding. Reproduction is terrestrial.
　i. Breeding migrations. Unknown.
　ii. Breeding habitat. Unknown.
　B. Eggs.
　i. Egg deposition sites. Oviposition sites are unknown. Egg attendance by females and communal nesting are unknown for San Simeon slender salamanders.
　ii. Clutch size. Unknown.
　C. Direct Development. Although unreported, we presume that San Simeon

slender salamanders undergo direct development, as is the case with other species of bolitoglossine plethodontids. Dates of hatching are unknown.
　D. Juvenile Habitat. Unknown how this may differ from adult habitat.
　E. Adult Habitat. San Simeon slender salamanders occur in a range of habitat types including closed and open forest (Jockusch et al., 2001). The type locality is characterized by open forest with yellow pine predominant. Near the Pacific Coast they occur in leaf litter under a closed canopy of laurel and sycamore. Near Rocky Butte in the San Simeon Creek drainage, individuals were found in a forest edge setting, being absent from the well developed oak forest (T. Burkhardt, personal communication).
　F. Home Range Size. Unknown.
　G. Territories. Unknown.
　H. Aestivation/Avoiding Desiccation. Unknown. Especially at the northern edge of the range, local weather conditions are strongly influenced by the convergence

of the Santa Lucia Range and Pacific Ocean, potentially extending periods of surface activity beyond the rainy season. San Simeon slender salamanders have been found under surface cover from mid-January to early June (upper elevations of San Simeon Creek drainage), but undoubtedly this species is active with the first cool, moist weather in the fall.
　I. Seasonal Migrations. Unknown.
　J. Torpor (Hibernation). Unknown.
　K. Interspecific Associations/Exclusions. San Simeon slender salamanders occur in sympatry with black-bellied slender salamanders *(B. nigriventris)* in the Santa Lucia

Range. The overall smaller range of *B. incognitus* within fragmented, relatively mesic uplands suggests a relict distribution together with possible ecological replacement by *B. nigriventris*, which enjoys a broader geographical and ecological distribution and occurs in higher densities (Yanev, 1978). However, only *B. incognitus* occupies the highest elevations within the San Simeon Creek drainage (Pine Mountain, Rocky Butte) in San Luis Obispo County. The ranges of San Simeon slender salamanders and Santa Lucia Mountains slender salamanders *(B. luciae)* approach one another but do not overlap. Near the Pacific Ocean, the two species are separated by ca. 25 km, while at higher elevation, the gap is ca. 50 km (Jockusch et al., 2001). However, the intervening areas in northern San Luis Obispo and southern Monterey counties have not been adequately explored and distributional details for both species are lacking. Additionally, the ranges of San Simeon slender salamanders and lesser slender salamanders *(B. minor)* approach to within ca. 15 km in northern San Luis Obispo County (Jockusch et al., 2001). Other plethodontid salamanders occurring within the range of San Simeon slender salamanders include ensatinas *(Ensatina eschscholtzii)* and arboreal salamanders *(Aneides lugubris)*.
　L. Age/Size at Reproductive Maturity. Unknown. Based upon measurements of a small series of San Simeon slender salamanders, males range from 38.3–46.3 (mean 41.7; n = 10) mm SVL, females from 39.2–48.0 (mean 42.8; n = 10) mm.
　M. Longevity. Unknown.
　N. Feeding Behavior. San Simeon slender salamanders are likely similar to other species of *Batrachoseps*, which capture small arthropod prey using their projectile tongue.
　O. Predators. Unknown, although snakes are likely predators.
　P. Anti-Predator Mechanisms. Coiling has been observed and is a common defensive response in *Batrachoseps*.
　Q. Diseases. Unknown.
　R. Parasites. Unknown.
　S. Comments. Although only recently described, specimens now referred to this species were first collected in 1972. Based on studies of mtDNA and allozymes, San Simeon slender salamanders are placed in a clade containing lesser slender salamanders and northern populations of the garden slender salamander *(B. major;* Jockusch et al., 2001; Jockusch and Wake, 2002).

4. Conservation.
San Simeon slender salamanders occupy a small range and are confirmed from only a few localities. Surveys for this species are potentially difficult owing to its general similarity to the microsympatric and more common black-bellied

slender salamander; surveys may be further complicated by limited habitat and highly restricted access to privately owned land (Jockusch et al., 2001). This species lacks protected status at state or federal levels.

Acknowledgments. We thank Tim Burkhardt for sharing field observations.

Batrachoseps kawia Jockusch, Wake, and Yanev, 1998

SEQUOIA SLENDER SALAMANDER

Robert W. Hansen, David B. Wake

1. Historical versus Current Distribution.
The known distribution of Sequoia slender salamanders (*Batrachoseps kawia*) lies entirely within the Kaweah River drainage (Tulare County) on the western slope of the Sierra Nevada of California, at elevations of 430–2,205 m (Jockusch et al., 1998). The six known sites are located within the South, East, and Middle Fork Kaweah River drainages; additional field-work will likely reveal populations of Sequoia slender salamanders along the North and Marble Forks as well. Most of these populations were discovered many years ago but were regarded at the time as belonging to more wide-ranging species (see "Comments" below).

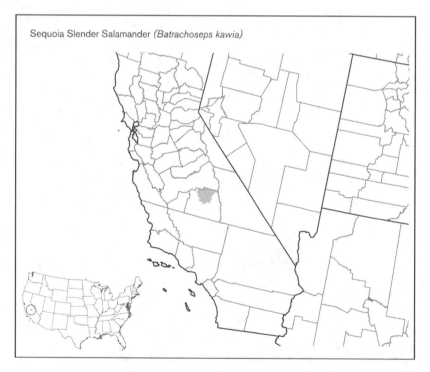

Sequoia Slender Salamander *(Batrachoseps kawia)*

2. Historical versus Current Abundance.
There is little information concerning the status of Sequoia slender salamander populations, although all known localities appear to have undergone little change in recent decades. The two high-elevation sites near Silver City (East Fork

Kaweah River drainage), first discovered in 1982, were revisited 18 yr later, but salamanders were not found (R.W.H., unpublished data).

3. Life History Features.
A. Breeding. Reproduction is terrestrial.
i. Breeding migrations. Unknown. For the lower elevation populations, courtship presumably occurs after the start of the rainy season in the fall (November–December), and egg-laying probably takes place from December–January, depending on local rainfall. The high elevation sites lie above 2,000 m and experience heavy snowfall during the winter and below freezing temperatures well into spring. Presumably, breeding activity commences in May–June, but this remains to be established.
ii. Breeding habitat. Nest sites have not been found.
B. Eggs.
i. Egg deposition sites. Unknown.
ii. Clutch size. Unknown.
C. Direct Development. Hatchlings have not been observed, but should be expected to appear in late winter to early spring at low elevation sites, and perhaps in mid-to-late summer at higher elevations.
D. Juvenile Habitat. Unknown.

E. Adult Habitat. Over their extensive elevational range, Sequoia slender salamanders occur in diverse habitats. Along the South Fork Kaweah River within foothill oak woodland at 430 m elevation, salamanders are present on or at the base of a mesic, north-facing slope

bordering a stream. Local vegetation includes California buckeye, California sycamore, white alder, Fremont cottonwood, western redbud, and interior live oak, with blue oak characteristic of the bordering exposed slopes (Jockusch et al., 1998). Salamanders occupy moist, moss-covered talus. Both Sequoia slender salamanders and gregarious slender salamanders (*B. gregarius*) occur here, but the latter's range extends into drier habitats.

At somewhat higher elevations (1,200 m) along the South Fork of the Kaweah River, Sequoia slender salamanders occur under fallen tree limbs and surface litter in mixed conifer forest.

At one high-elevation site (2,205 m) near Silver City, Sequoia slender salamanders were found under rocks on wet gravel at a perennial spring flowing over a west-facing slope. Vegetation in and near the spring includes willows, currant, white fir, and manzanita.

At a second high-elevation site (2,200 m), salamanders were found beneath wet logs resting on wet soil in a shaded draw about 5 m from a brook. The heavily forested slope was grown to sugar pine, incense cedar, giant sequoia, alders, and ferns.

F. Home Range Size. Unknown.
G. Territories. Unknown.
H. Aestivation/Avoiding Desiccation. Sequoia slender salamanders occur over a wide elevational range and periods of surface activity vary considerably with local rainfall and temperature. At low-elevation sites, salamanders are present under surface cover from November to March/April. This window of activity is extended for middle elevations, but with salamanders retreating underground during the coldest periods of December–January. At high-elevation sites (≥2,000 m), surface activity probably is confined to May–October. Field body temperatures of salamanders found under cover at two high-elevation sites averaged 13.2 °C (range = 12.0–15.0 °C, n = 9; R.W.H., unpublished data).
I. Seasonal Migrations. Unknown.
J. Torpor (Hibernation). Unknown.
K. Interspecific Associations/Exclusions. There is extensive range overlap between Sequoia slender salamanders and gregarious slender salamanders within the Kaweah River system, and cases of sympatry are known. Although confirmed from only six localities, Sequoia slender salamanders likely have a larger distribution within the Kaweah River drainage than present records indicate.

Sequoia slender salamanders are found occasionally with Sierra Nevada ensatinas (*Ensatina eschscholtzii platensis*); California newts (*Taricha torosa*) are widespread at lower elevations in the Kaweah River drainage and commonly are found at sites where Sequoia slender salamanders occur.

We expect to find Sequoia slender salamanders in sympatry with both Kings River slender salamanders *(B. regius)* and relictual slender salamanders *(B. relictus)*, but as yet these members of the same species group are not known to occur together.

L. Age/Size at Reproductive Maturity. This is a moderately small species of *Batrachoseps*, with maximum adult sizes ≤50 mm SVL (Jockusch et al., 1998). Age and size at sexual maturity are unknown.

M. Longevity. Unknown.

N. Feeding Behavior. Has not been described, although all *Batrachoseps* species observed thus far use a projectile tongue to capture small invertebrates.

O. Predators. Unknown. Ring-necked snakes *(Diadophis punctatus)* occur in the lower elevation portion of the range of Sequoia slender salamanders and should be considered likely predators.

P. Anti-Predator Mechanisms. Immobility, coiling, and flipping have been observed when salamanders are first uncovered (R.W.H., unpublished data). The dark dorsal coloration presumably is cryptic when viewed against dark substrates and in low-light environments.

Q. Diseases. Unknown.

R. Parasites. Unknown.

S. Comments. Populations now identified as *B. kawia* previously were placed within *B. relictus* (Brame and Murray, 1968) or *B. pacificus relictus* (Yanev, 1978, 1980). Although Yanev noted the distinctiveness of the Kaweah River populations, these were not elevated to full species status until recently (Jockusch et al., 1998), when the formerly wide-ranging *B. relictus* was partitioned into four species (from north to south in the central and southern Sierra Nevada): *B. diabolicus, B. regius, B. kawia,* and *B. relictus.* Recent mtDNA studies suggest a sister-taxon relationship between *B. kawia* and *B. relictus* (Jockusch, 1996; Jockusch et al., 1998).

Jockusch et al. (1998) tentatively assigned specimens from Summit Meadow, Fresno County, to *B. kawia,* but we now believe that this population represents *B. regius.*

4. Conservation.

The majority of Sequoia slender salamander populations occur on public lands administered by the USDA Forest Service or National Park Service. They thus enjoy some measure of habitat protection, although Forest Service lands are subject to various uses including timber harvest and grazing. Based upon available data, this species appears to have a very restricted range comprised of disjunct, quite localized populations. Additional field surveys are needed throughout the Kaweah River drainage to better define the distributional limits of this species and thereby offer informed opinions regarding possible protection measures.

Batrachoseps luciae Jockusch, Yanev, and Wake, 2001

SANTA LUCIA MOUNTAINS SLENDER SALAMANDER

Robert W. Hansen, David B. Wake

1. Historical versus Current Distribution.

Santa Lucia Mountains slender salamanders *(Batrachoseps luciae)* were described in 2001 on the basis of differences in allozymes and in DNA haplotypes in the mitochondrial gene cytochrome b (Jockusch, 1996; Jockusch et al., 2001). Santa Lucia Mountains slender salamanders are distributed throughout the northern Santa Lucia Mountains along the Pacific Coast in Monterey County (Jockush et al., 2001). They currently are known from a number of Monterey County locations in central coastal California. The northernmost known location is on the Monterey Peninsula and the southernmost is near the Monterey/San Luis Obispo County line. Inland, they are found along the Carmel Valley and the eastern slopes of the Santa Lucia Mountains from Arroyo Seco at least as far south as 36 °N. Differences between historical and current distributions are not apparent.

2. Historical versus Current Abundance.

Unknown. Santa Lucia slender salamanders have been found abundantly at only a few sites, including the type locality, a park in the city of Monterey (Jockusch et al., 2001).

3. Life History Features.

A. Breeding. Reproduction is terrestrial.

i. Breeding migrations. Unknown.

ii. Breeding habitat. Unknown.

B. Eggs.

i. Egg deposition sites. A cluster of 19–20 eggs was found beneath a log in Pacific Grove on 10 February, along with several adult Santa Lucia Mountains slender salamanders, one of which was positioned ca. 5 cm from the cluster (D. Roberson; http://montereybay.com/creagrus/CABatrachoseps.html). Almost certainly these eggs were deposited by more than one female; this appears to be the first example of communal nesting (Jockusch and Mahoney, 1997) for this species.

ii. Clutch size. Jockusch (1997b) induced oviposition in gravid females collected from two localities. Average clutch sizes were 5.1 eggs (Carmel Valley, n = 10 clutches) and 10.6 eggs (Monterey, n = 13 clutches).

C. Direct Development. Developmental times for eggs incubated in the lab at 13 °C averaged 78 d (Jockusch 1997b). Dates of hatching in the wild are unknown.

D. Juvenile Habitat. Unknown how this may differ from adult habitat.

E. Adult Habitat. San Lucia Mountains slender salamanders are found predominantly in moist redwood and mixed coniferous forests. Inland, they occur mainly on wooded (with predominant vegetation being tan bark oaks and maples), north-facing slopes. During favorable climatic conditions, San Lucia Mountains slender salamanders also can be found under suitable cover in open, disturbed habitats, including the type locality, a park in the city of Monterey (Jockusch et al., 2001). In the Big Sur area, they have been found in wet, creekside situations.

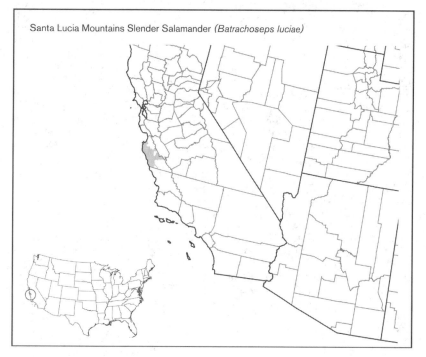

Santa Lucia Mountains Slender Salamander *(Batrachoseps luciae)*

F. Home Range Size. Unknown.

G. Territories. Unknown.

H. Aestivation/Avoiding Desiccation. Unknown, but likely in populations occurring on the eastern flanks of the Santa Lucia Range. On the western flank of these mountains, Santa Lucia Mountains slender salamanders have been found under surface cover year-round. The presence of sharply rising mountain slopes adjacent to the Pacific Ocean results in frequent summer fogs that in turn allow salamanders to remain surface active outside the rainy season. We have recorded Santa Lucia Mountains slender salamanders under surface cover at substrate temperatures of 7.4–8.4 °C (mean 7.9 °C; n = 14) in January, although we expect activity across a broader range of temperatures.

I. Seasonal Migrations. Unknown.

J. Torpor (Hibernation). Unknown.

K. Interspecific Associations/Exclusions. Santa Lucia Mountains slender salamanders and Gabilan Mountains slender salamanders *(B. gavilanensis)* are parapatric, their ranges largely interdigitating for ca. 80 km along a broad ecotone in Carmel Valley. Populations of these species occur to within a few km of each other in Carmel Valley, but sympatry has not been discovered (Yanev, 1978; Jockusch et al., 2001). Similarly, their ranges converge in the Monterey Peninsula in an area with continuous habitat; further collecting in this region seems likely to reveal sympatry. They also occur within a few hundred meters of each other in the Nacimiento River valley in extreme southern Monterey County.

At their southern range limits, Santa Lucia Mountains slender salamanders come close to populations of San Simeon slender salamanders *(B. incognitus)* and black-bellied slender salamanders *(B. nigriventris)*, but range overlap is unknown (Jockusch et al., 2001).

Other plethodontid salamanders occurring within the range of Santa Lucia Mountains slender salamanders include ensatinas *(Ensatina eschscholtzii)* and arboreal salamanders *(Aneides lugubris)*.

L. Age/Size at Reproductive Maturity. Unknown. Based upon measurements of a small series, males range from 32.0–40.1 (mean 36.3, n = 10) mm SVL, females from 36.8–44.8 (mean 41.2, n = 10) mm SVL.

M. Longevity. Unknown.

N. Feeding Behavior. Santa Lucia Mountains slender salamanders, similar to other species of *Batrachoseps*, likely capture small arthropod prey using their projectile tongue.

O. Predators. Unknown, although lizards and snakes are likely predators.

P. Anti-Predator Mechanisms. Coiling and tail autotomy are common defensive responses in several species of attenuate *Batrachoseps*.

Q. Diseases. Unknown.

R. Parasites. Unknown.

S. Comments. Although only recently described, specimens now referred to this species have been in scientific collections for many years though assigned to other taxa (e.g., *B. attenuatus* [Hendrickson, 1954]; and later, *B. relictus* [Brame and Murray, 1968] and *B. pacificus* [Yanev, 1980]). Yanev (1978, 1980), on the basis of allozymes, first identified the major lineages within the central coastal *B. pacificus* complex. Subsequently, these have been recognized as distinct species *(gavilanensis, incognitus, luciae, and minor)*; Jockusch et al., 2001).

Based on studies of mtDNA and allozymes, Santa Lucia Mountains slender salamanders are known to be close relatives of *B. incognitus*, *B. minor*, and *B. gavilanensis* (Jockusch et al., 2001; Jockusch and Wake, 2002). There is some evidence that *B. luciae* and *B. gavilanensis* may have been in genetic contact early in their history, although there is no indication of present gene exchange despite sharing a long border through Carmel Valley (Jockusch et al., 2001).

4. Conservation.
Santa Lucia Mountains slender salamanders are moderately widespread throughout the northern Santa Lucia Mountains and are abundant at a few sites. Portions of the range occur on publicly owned lands or other large land holdings that are likely to remain relatively undisturbed for the foreseeable future. Aside from local extirpations associated with human development, there are no known significant conservation concerns.

Acknowledgments. We thank Don Roberson for information concerning his discovery of a communal nest.

Batrachoseps major Camp, 1915
GARDEN SLENDER SALAMANDER
Robert W. Hansen, David B. Wake

1. Historical versus Current Distribution.
Garden slender salamanders *(Batrachoseps major)* are found in southern California from the southern foothills of the Santa Monica, San Gabriel, and San Bernardino mountains south along the Pacific Coast to the vicinity of El Rosario, Baja California (Wake and Jockusch, 2000). Populations extend eastward via San Gorgonio Pass to the margins of the Colorado Desert, occurring south of Cabazon (Wake and Jockusch, 2000), in Snow Creek Canyon (D.B.W., unpublished data), and within the city of Palm Springs (Cornett, 1981). They also are found on Santa Catalina, North, Middle, and South Coronados, and Todos Santos islands. Garden slender salamanders have been introduced at Hanford, Kings County, in the San Joaquin Valley of central California, 250 km north of their natural range (Jockusch, 1996). They occur from sea level to 1,500 m (Mt. Palomar, San Diego County), but most populations are found below 700 m elevation.

2. Historical versus Current Abundance.
While formerly widespread and common in southern California, garden slender salamanders have been extirpated from much of their historical range due to habitat destruction (Cunningham, 1960; Wake, 1996). DeLisle et al. (1986) recorded garden slender salamanders from only two sites along the southern base of the Santa Monica Mountains (lowland areas that have experienced substantial habitat loss) but found black-bellied slender salamanders *(B. nigriventris)* at 31 localities from higher elevations, where native habitats have been less affected by development.

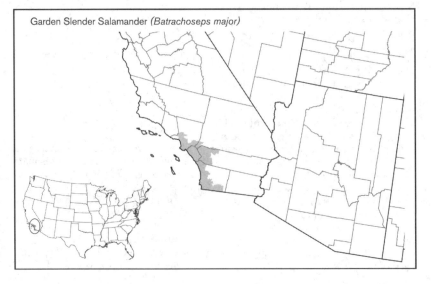

Garden Slender Salamander *(Batrachoseps major)*

3. Life History Features.

A. Breeding. Reproduction is terrestrial.

i. Breeding migrations. Seasonal movements associated with reproductive activity have not been described. Although unrecorded, courtship probably occurs in the fall (November–December) after the start of the rainy season. Groups of 15–20 adults have been found under single cover objects in spring (B. Campbell, in Stebbins, 1951), but such aggregations are unlikely to be related to breeding activity.

ii. Breeding habitat. Unknown. Eggs are laid underground in burrows (Cunningham, 1960), but the location of spermatophore deposition and capture is unrecorded.

B. Eggs.

i. Egg deposition sites. Eggs are laid underground. At night in late January, hatchlings were observed in groups of 3–12 near burrows following heavy rains (Cunningham, 1960), suggesting that eggs had been deposited underground in burrows. Communal nesting (Jockusch and Mahoney, 1997) is suggested by the mass emergence of 158 hatchlings following heavy rains in late January; salamanders had emerged from a 4–5 mm crack between the wall of a house and a concrete landing (Grant, 1958).

ii. Clutch size. Davis (1952) discovered a clutch of 13 freshly laid eggs; by 18 January, five additional eggs were observed, though it is unknown whether these were deposited by the same female. Two adult females collected in mid-December contained 15 and 20 fully developed ova (Stebbins, 1951).

C. Direct Development. In the northern part of the range (Los Angeles County), eggs are deposited in December–January (Stebbins, 1951; Davis, 1952), with hatchlings first appearing in late January. Hatchlings have been observed as late as mid-April in northern San Diego County (Holland and Goodman, 1998).

D. Juvenile Habitat. Differences in habitat use between juveniles and adults have not been noted.

E. Adult Habitat. Prior to extensive modification of southern California native landscapes, garden slender salamanders occupied areas of coastal sage scrub, chaparral, and coast live oak woodlands, mostly in open areas with little relief (Camp, 1915; Stebbins, 1951, 1954a; Wake, 1996). Such areas experience a mild, dry summer to wet winter climate, often with coastal fog. In the San Gabriel Mountains, garden slender salamanders occur mostly in the foothills, but occasionally are found in oak woodland on alluvial fans (Schoenherr, 1976). In northern San Diego County, garden slender salamanders may be locally abundant in oak woodland and chaparral (Holland and Goodman, 1998). In present-day southern California, garden slender salamanders frequently are associated with suburban landscapes that receive

regular irrigation (Cunningham, 1960; Cornett, 1981). At their southern range limits at the edge of the Vizcaino Desert (26 km east of El Rosario, Baja California), garden slender salamanders have been found beneath decaying desert agave, at a site characterized by cardon cactus, boojum tree, and several species of *Opuntia* (Grismer, 1982).

On Santa Catalina Island, garden slender salamanders are found in coastal sage scrub (Schoenherr et al., 1999). On South Todos Santos Island off the Pacific Coast of northern Baja California, garden slender salamanders were found under rocks on a northeast-facing slope in early March (Zweifel, 1958).

The eastward extension of garden slender salamanders through San Gorgonio Pass to the margins of the Colorado Desert at Cabazon and Palm Springs suggests that they are tolerant of both heat and drought. However, these populations remain poorly studied. Although the Palm Springs region receives only 13.7 cm of annual precipitation, salamanders are now associated with residential and commercial landscapes that receive regular irrigation (Cornett, 1981); presumably these populations formerly experienced much shorter periods of surface activity corresponding to the brief rainy season. Some salamanders near Cabazon were found on the desert floor, although a few cottonwood trees grew nearby at a spring; during a cold rain in mid-winter, one salamander was found under a dead stalk of a shrub on the open desert floor. The Cabazon site lies at the base of the San Jacinto Mountains in creosote bush desert (D.B.W., unpublished data).

The introduced San Joaquin Valley population is located in an area where summer daytime temperatures regularly exceed 37°C and winter temperatures occasionally drop below freezing. Salamanders are confined to residential gardens with ornamental trees and shrubs that receive daily watering during the dry season (R.W.H., unpublished data).

Salamanders use gopher burrows, soil crevices, and earthworm tunnels for retreats (Lowe and Zweifel, 1951). One specimen was found 0.8 m below ground in a gopher burrow in early June (von Bloeker, 1942). During periods of surface activity, salamanders may be found under rocks, logs, boards, and other surface cover, as well as within moist leaf litter, mostly on coarse, well-drained substrates. Individuals occasionally are found climbing in low vegetation (Cunningham, 1960).

F. Home Range Size. Of 539 salamanders marked during a study in Santa Monica (Los Angeles County), 141 animals were recaptured at least five times; the maximum distance from point of last capture averaged 6.0 m (Cunningham, 1960). Juveniles tend to wander and have larger home ranges than adults.

G. Territories. Unknown.

H. Aestivation/Avoiding Desiccation. Surface activity generally begins with the first significant rains of fall (October–November) and lasts into spring (April–May). However, in areas receiving coastal fog, activity may extend to June–July, and salamanders living in watered residential landscapes are active year-round (Cunningham, 1960). During a pitfall trapping survey on the western slope of Palomar Mountain (San Diego County), salamanders were captured from late December to late April and late December to late February over a 2-yr period (Powers and Banta, 1972).

Substrate temperatures for >400 salamanders found under surface cover ranged from 4–21°C, with most in the 9–11°C range (Cunningham, 1960). Body temperatures averaged 12.9°C for 53 salamanders (range 7.0–19.6°C; Brattstrom, 1963; Feder et al., 1982). A cluster of six aestivating salamanders was unearthed 1 m below ground in moist soil (Cunningham, 1960). Four salamanders collected in mid-February from the introduced San Joaquin Valley population had body temperatures of 12.0°C (R.W.H., unpublished data).

I. Seasonal Migrations. Unknown.

J. Torpor (Hibernation). Body temperatures below 4.0°C have not been recorded, and much of the range of garden slender salamanders rarely experiences nighttime temperatures below freezing.

K. Interspecific Associations/Exclusions. Garden slender salamanders are sympatric with black-bellied slender salamanders from southern Los Angeles County to the southern limits for black-bellied slender salamanders in Orange and Riverside counties (Brame, 1970; Glaser, 1970; Wake, 1996). In areas where both species occur, there is a general pattern of ecological segregation, with garden slender salamanders occurring in open, grassy areas, mostly absent from slopes, while black-bellied slender salamanders are largely restricted to areas with tree cover (Campbell, 1931b; Lowe and Zweifel, 1951; Cunningham, 1960). However, at some sites with northern exposures, both species have been taken under the same cover (Campbell, 1931b).

Garden slender salamanders are sympatric with Monterey ensatinas (*Ensatina eschscholtzii eschscholtzii*) and arboreal salamanders (*Aneides lugubris*) at many localities, although the presence of garden slender salamanders in suburban gardens is unique among southern California salamanders.

L. Age/Size at Reproductive Maturity. Minimum adult size unknown. Age at sexual maturity has not been established, but Campbell (in Stebbins, 1951) found three size classes, which probably correspond to young-of-the-year, 2-yr-olds, and animals in their third year or older. The 2-yr-old group (46–60 mm TL) is clearly too small to be mature; thus, we suggest that likely minimum age for adults is 3 yr.

Geographic variation in adult size remains unstudied, although the largest animals have been recorded from the northern part of the range. Brame and Murray (1968) reported on a small series from Los Angeles County; nine adult females ranged in size from 50.1–67.5 mm SVL, and a single adult male measured 50.4 mm SVL. For another series, also from Los Angeles County, average size of nine adult males was 46.5 mm SVL (maximum 54.5 mm SVL), while average size of 13 adult females was 45.6 mm SVL (maximum 55.3 mm SVL). At the southern limits of the range (near El Rosario, Baja California), the largest individual in a series of 14 measured 45.2 mm SVL (Grismer, 1982). Stebbins (1985) reports a maximum size of 69 mm SVL, sex not indicated. Female-biased sexual size dimorphism is characteristic of most species of *Batrachoseps* for which data are available; it is unclear whether adult size differences exist in garden slender salamanders.

M. Longevity. Unknown.

N. Feeding Behavior. Prey reported for a Los Angeles County population studied by Cunningham (1960) included annelid worms, sow bugs, earwigs, and other insects.

O. Predators. Despite their broad distribution and local abundance in southern California, reports of natural predators of garden slender salamanders are rare. Ring-necked snakes *(Diadophis punctatus)* have been observed feeding on garden slender salamanders (Holland and Goodman, 1998; D. Holland, personal communication). Ervin et al. (2003) reported predation by a two-striped garter snake *(Thamnophis hammondii).* Feral pigs are reported to prey on the salamander species occurring on Santa Catalina Island, although specific reference to *Batrachoseps* is lacking (Schoenherr et al., 1999).

P. Anti-Predator Mechanisms. Individuals of garden slender salamanders may exhibit coiling, vigorous lateral thrashing, or immobility upon discovery or when handled (Cunningham, 1960; Schoenherr, 1976). Cunningham (1960) observed individual salamanders partly emerged from home burrows, with tail still anchored within the burrow; the salamanders would quickly retreat upon his approach. Garden slender salamanders consistently are the lightest-colored species of *Batrachoseps*, perhaps an example of crypsis, as Stebbins (1954a) noted their frequent association with light-colored substrates. Captive snakes (two-striped garter snakes and common kingsnakes [*Lampropeltis getula*]) died after consuming garden slender salamanders, presumably because of the effects of skin toxins (Cunningham, 1960). The discovery of two bodies and six tail tips of garden slender salamanders in the stomach of a two-striped garter snake clearly attests to the efficacy of tail autotomy as a defensive response in this species (Ervin et al., 2003).

Q. Diseases. Unknown.

R. Parasites. Unknown.

S. Comments. Although Wake and Jockusch (2000) provisionally included salamanders from the Sierra San Pedro Martir (Baja California) within *B. major*, work in progress suggests that this population warrants recognition as a distinct species (D.B.W. and colleagues, unpublished data) and thus we exclude it from this account. Following Wake and Jockusch (2000), we also exclude from *B. major* those populations occurring in coastal central California previously included within a large *B. pacificus* complex (Yanev, 1978, 1980; now consisting of four recently described species [Jockusch et al., 2001]), as well as populations on the northern Channel Islands, now accorded full species status as *B. pacificus* (Channel Islands slender salamanders).

Recent studies of allozymes and mitochondrial DNA sequences have led Wake and Jockusch (2000) to reduce *B. aridus* to subspecific status within *B. major*. The population of *B. aridus* analyzed for allozymes is no different from *B. major* than any of a number of other local populations. There are no fixed differences and no unique alleles. However, there is a unique haplotype for the mitochondrial gene cytochrome b, and it forms a sister group with a population from Todos Santos Island, off the coast of Baja California. These two populations are in turn sister to a group of populations assigned to *B. major* from San Diego County. We make no judgment about taxonomy, but choose to deal with *B. aridus* as a separate entity for this work because of its status as a federally protected Endangered species, which would not change regardless of its formal taxonomy.

MtDNA sequences confirm a Los Angeles County origin for the extralimital population of *B. major* now established in the southern San Joaquin Valley (Jockusch, 1996). Several large commercial nurseries in southern California supply plants to retail nurseries in central California; this seems a likely method for the transport of salamanders and eggs, or both.

4. Conservation.
Much of the historical habitat of garden slender salamanders in southern California has been altered permanently by human development, thus fragmenting the range of this otherwise widespread species. Although portions of their range appear to be protected, examination of the relationships between genetically distinctive range segments of *B. major* and the long-term viability of native landscapes is needed.

Acknowledgments. We thank Dan Holland and Ted Papenfuss for sharing unpublished observations.

Batrachoseps minor Jockusch, Yanev, and Wake, 2001
LESSER SLENDER SALAMANDER
Robert W. Hansen, David B. Wake

1. Historical versus Current Distribution.
Lesser slender salamanders *(Batrachoseps minor)* were described in 2001. The species is distinguished from related and sympatric forms by differences in allozymes and mitochondrial DNA sequences, as well as small differences in morphology (Jockusch, 1996; Jockusch et al., 2001). They are distinguished from sympatric black-bellied slender salamanders *(B. nigriventris)* by being somewhat more robust and having longer limbs and larger hands and feet, but the differences are subtle. Lesser slender salamanders have a restricted distribution in the southern Santa Lucia Range of north-central San Louis Obispo County in central coastal California. In the north, they occur immediately north of Black Mountain and range south and east into the drainages of Paso Robles and Santa Rita Creeks. Populations found farther south and west of Atascadero and in the Cuesta Ridge Botanical Area have not been examined for DNA sequences or allozymes but are tentatively assigned to this species. The elevational range is generally 400–640 m.

2. Historical versus Current Abundance.
Lesser slender salamanders were once common, but today are difficult to find (Jockusch et al., 2001). Fewer than five of these salamanders have been seen in the past decade despite many attempts to find them. Although some areas within the historical range of this species have been modified for agriculture (i.e., conversion to vineyards), ample habitat remains and there is no obvious reason for this decline in abundnce.

3. Life History Features.
  A. Breeding. Reproduction is terrestrial.
  i. Breeding migrations. Unknown.
  ii. Breeding habitat. Unknown.
  B. Eggs.
  i. Egg deposition sites. Oviposition sites are unknown. Egg attendance by females and communal nesting are unknown for this species.
  ii. Clutch size. Unknown.
  C. Direct Development. Although unreported, we presume that lesser slender salamanders undergo direct development, as is the case with other species of bolitoglossine plethodontids. Dates of hatching are unknown.
  D. Juvenile Habitat. Unknown how this may differ from adult habitat.
  E. Adult Habitat. Lesser slender salamanders appear to be restricted to areas that are either higher in elevation or more

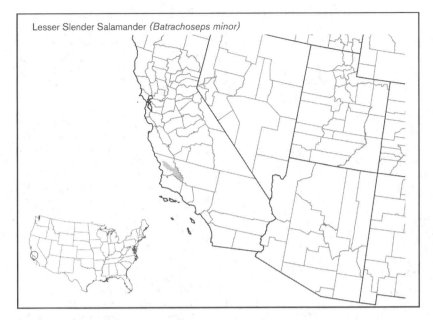

Lesser Slender Salamander *(Batrachoseps minor)*

Batrachoseps nigriventris Cope, 1869
BLACK-BELLIED SLENDER SALAMANDER
Robert W. Hansen, David B. Wake

1. Historical versus Current Distribution.
Black-bellied slender salamanders *(Batrachoseps nigriventris)* are found in the coastal mountains and valleys west of the Central Valley of California, from extreme southern Monterey County and western and southern Fresno County to the Santa Monica and San Gabriel mountains of Los Angeles County. They also are found in the Tehachapi Mountains at the southern and eastern margins of the Central Valley. They also occur in isolated upland areas in southern Los Angeles County and in Orange and extreme southwest Riverside counties. Black-bellied slender salamanders are widely distributed on Santa Cruz Island. They occur from near sea level to about 2,260 m on Mt. Pinos (Wake and Jockusch, 2000). Recent range modifications are not evident, except for areas of obvious habitat modification in many parts of their historical range in southern California.

2. Historical versus Current Abundance.
Black-bellied slender salamanders may be locally abundant at some times, but difficult to find at others. Long-term trends are not apparent.

3. Life History Features.
    A. Breeding. Reproduction is terrestrial.
    i. Breeding migrations. We have discovered seasonal aggregations of 12–20 black-bellied slender salamander adults (sex not determined) on Mt. Pinos, suggesting that adult females move to communal nesting sites. Adults have been observed moving across wet roads at night on Mt. Pinos (G. Keasler, personal communication).
    ii. Breeding habitat. Unknown.
    B. Eggs.
    i. Egg deposition sites. Nest sites have not been located. However, our observations of adult aggregations under and within downed logs in seepages on Mt. Pinos (2,260 m elevation) are strikingly similar to the descriptions of egg-laying sites reported for their sister species, gregarious slender salamanders *(B. gregarius* [Jockusch and Mahoney, 1997]; reported as *B. nigriventris)*.
    ii. Clutch size. Unknown.
    C. Direct Development. Timing of egg laying and hatching are poorly known. Black-bellied slender salamanders occur over a broad elevational range in diverse habitats—from lowlands that experience a mild, coastal climate to interior mountains that receive substantial winter snowfall, and it seems likely that periods of egg laying will vary accordingly. In southern California lowlands and foothills, eggs probably are laid in winter and hatch in winter and early spring.

mesic than surrounding areas (Jockusch et al., 2001). The type locality is a mesic canyon surrounded by relatively more xeric habitats. Here salamanders were collected from a deeply shaded slope with deep leaf litter. Canopy trees include tanbark oak, coast live oak, sycamore, and laurel. Dense shrubs, predominantly poison oak, were also present. A second site, at York Mountain, is characterized by blue oaks and coast live oaks and generally is less heavily shaded.

    F. Home Range Size. Unknown.

    G. Territories. Unknown.

    H. Aestivation/Avoiding Desiccation. Unknown but likely, especially in the southwestern portions of the range. Salamander activity closely tracks the rainy season and is strongly influenced by local conditions. Lesser slender salamanders have been found under surface cover from mid-November to mid-March.

    I. Seasonal Migrations. Unknown.

    J. Torpor (Hibernation). Unknown.

    K. Interspecific Associations/Exclusions. Everywhere lesser slender salamanders are found, they occur in microsympatry with black-bellied slender salamanders (Yanev, 1978; Jockusch et al., 2001). However, the overall smaller range of *B. minor* within fragmented, relatively mesic uplands suggests a relict distribution together with possible ecological replacement by *B. nigriventris*, which enjoys a broader geographical and ecological distribution and occurs in higher densities (Yanev, 1978). Other plethodontid salamanders occurring within the range of lesser slender salamanders include ensatinas *(Ensatina eschscholtzii)* and arboreal salamanders *(Aneides lugubris)*.

    L. Age/Size at Reproductive Maturity. Unknown. Based upon measurements of a small series of lesser slender salamanders, males range from 28.4–33.6 (mean 31.0,

n = 10) mm SVL, females from 26.5–32.8 (mean 30.1, n = 10) mm SVL. This is the smallest species of *Batrachoseps* (Jockusch et al., 2001).

    M. Longevity. Unknown.

    N. Feeding Behavior. Similar to other species of *Batrachoseps*, lesser slender salamanders likely capture small insect prey using their projectile tongue.

    O. Predators. Unknown, although snakes are likely predators.

    P. Anti-Predator Mechanisms. Coiling and tail autotomy are common defensive responses in several species of attenuate *Batrachoseps*.

    Q. Diseases. Unknown.

    R. Parasites. Unknown.

    S. Comments. Although only recently described, specimens now referred to this species were first collected in 1960. Based on studies of mtDNA and allozymes, lesser slender salamanders are placed in a clade containing San Simeon slender salamanders *(B. incognitus)* and northern populations of the garden slender salamander *(B. major*; Jockusch et al., 2001; Jockusch and Wake, 2002).

4. Conservation.
The naturally small range of this species, together with its apparent decline, merit conservation attention. The three known localities of lesser slender salamanders where specimen allocation to species has been confirmed by examination of DNA sequences occur on privately owned land. Surveys for this species are potentially difficult owing to its general similarity to the microsympatric and more common black-bellied slender salamander. Field surveys using artificial cover objects (plastic garbage bags) may be effective in locating this species and will minimize potential impact to the habitat (Palazzo, 1994; Duncan, 1998).

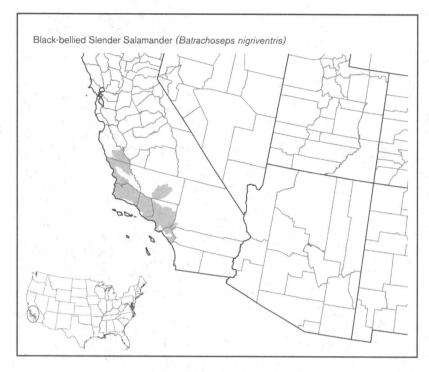

Black-bellied Slender Salamander *(Batrachoseps nigriventris)*

Adult aggregations have been found in July to early August at high elevations on Mt. Pinos, but eggs were not observed at this time.

D. Juvenile Habitat. Differences in habitat use between juveniles and adults have not been noted.

E. Adult Habitat. Black-bellied slender salamanders occupy a wide range of habitats, from semi-arid blue oak savannas at the northern limits of the range to moist, oak-filled canyons and pine-fir forest at higher elevations. Habitat extremes are illustrated in northern populations; near Comanche Point on the northern slopes of the Tehachapi Mountains, salamanders occur on semi-arid, grassy slopes, with only an occasional blue oak mixed with scattered granite outcrops; at Mt. Pinos (2,260 m elevation), salamanders are locally abundant in Jeffrey pine-white fir forests, especially in association with small streams and seepages. Elsewhere in the Tehachapi Mountains, the species is relatively common in canyons with extensive growths of canyon live and black oaks (Block and Morrison, 1998; R.W.H., unpublished data).

In southern California, south of the Tehachapi Mountains, black-bellied slender salamanders mostly are found in foothills and mountain canyons, usually within coast live oak woodlands and/or chaparral (Lowe and Zweifel, 1951; Cunningham, 1960; Schoenherr, 1976). These salamanders have been found on rocky, north-facing slopes covered with coastal sage scrub vegetation and some California buckeye at the ocean in southern Orange County.

Black-bellied slender salamanders are widely distributed on Santa Cruz Island, where they occur under rocks and scattered debris (e.g., old fence posts) in open grassland; under rocks, fallen branches and in leaf litter in oak woodland; in and under rotting branches and logs as well as under rocks; and in surface litter in pine forests, as well as under superficial surface cover in chaparral.

F. Home Range Size. Salamanders of the genus *Batrachoseps* generally are assumed to be sedentary (Hendrickson, 1954; Cunningham, 1960). The geographically restricted patterns of mtDNA haplotype distributions indicate that female movements are limited (Jockusch, 1996). Thus, although there may be some short-range surface movements associated with breeding and egg deposition (e.g., Jockusch and Mahoney, 1997), individual salamanders likely have small home ranges.

G. Territories. Unknown.

H. Aestivation/Avoiding Desiccation. Salamander activity closely tracks the rainy season and is strongly influenced by local conditions. For example, in the open oak woodlands on the northern flanks of the Tehachapi Mountains, the soil dries rapidly following spring rains and it is likely that surface activity declines abruptly after March. However, in adjacent canyons with extensive oak forests and north-facing slopes, seasonal activity may extend to late April to early May. On Mt. Pinos, salamanders have been found throughout the summer in association with perennially moist microhabitats. In the San Gabriel Mountains, we have found black-bellied slender salamanders

mostly during the winter, with surface activity usually associated with recent rainfall. In the Santa Monica Mountains, surface activity occurs mainly from January–April (De Lisle et al., 1986). Near the Pacific Ocean at the base of the Santa Ynez Mountains, marine air flows may extend surface activity to late May. Earthworm burrows and soil crevices are utilized as retreats when surface conditions deteriorate (Lowe and Zweifel, 1951). Body temperatures of salamanders found under surface cover averaged 9.8 °C (range = 6.0–14.0 °C, n = 40, mostly from Tehachapi Mountains populations; R.W.H., unpublished data) and 8.8 °C (range = 6.8–9.5 °C, n = 26, locality not stated; Feder et al., 1982). Stebbins (1951) reported finding 14 adults in early December in the Santa Monica Mountains and recorded body temperatures of 10.2–10.5 °C.

I. Seasonal Migrations. Unknown.

J. Torpor (Hibernation). Unknown.

K. Interspecific Associations/Exclusions. Sympatric with garden slender salamanders *(B. major)* from southern Los Angeles County to the southern limits for black-bellied slender salamanders in Orange County (Brame, 1970; Wake, 1996). In areas where both species occur, there is a general pattern of ecological segregation, with garden slender salamanders occurring in open, grassy areas, mostly absent from slopes, while black-bellied slender salamanders are largely restricted to areas with tree cover (Campbell, 1931b; Lowe and Zweifel, 1951; Cunningham, 1960). However, at some sites, both species have been taken under the same cover. Black-bellied slender salamanders and garden slender salamanders have been collected within a few meters of each other in southwestern Riverside County.

Black-bellied slender salamanders co-exist broadly with Channel Islands slender salamanders *(B. pacificus)* on Santa Cruz Island, where both species tend to occupy similar habitats (Schoenherr et al., 1999), and they frequently are found in microsympatry.

Along the central California coast and interior uplands west of the Central Valley, in southern Monterey and northern San Luis Obispo counties, some overlap occurs with three recently described species that formerly were included in *B. pacificus (B. gavilanensis, B. incognitus,* and *B. minor;* Jockusch, 1996; Yanev, 1978, 1980; Jockusch et al., 2001).

Black-bellied slender salamanders are sympatric with Tehachapi slender salamanders *(B. stebbinsi)* in the Pastoria and Tejon Creek drainages of the Tehachapi Mountains, as well as at Fort Tejon (Jockusch, 1996; R.W.H., unpublished data). Although both species have been found under the same cover, Tehachapi slender salamanders are restricted to moist, north-facing slopes, where they favor areas of talus or downed logs; black-bellied slender salamanders exhibit a broader ecological distribution in these areas and are mostly absent from steep slopes (R.W.H., unpublished data). The northernmost Tehachapi Mountains population of black-bellied slender salamanders lies about 37 km south of the nearest locality, near the Kern River, of its close relative, gregarious slender salamanders; it seems likely that populations belonging to one or both species will be found in the intervening foothills.

Black-bellied slender salamanders are sympatric with talus-dwelling San Gabriel Mountain slender salamanders (*B. gabrieli*) at a few sites in the San Gabriel Mountains at elevations of 1,158–1,200 m, and the two species have been taken under the same cover (Wake, 1996).

Black-bellied slender salamanders are broadly sympatric with yellow-blotched ensatinas (*Ensatina eschscholtzii croceater*) in the Tehachapi Mountains and Mt. Pinos regions, and with Monterey ensatinas (*E. e. eschscholtzii*), arboreal salamanders (*Aneides lugubris*), and California newts (*Taricha torosa*) at many sites in coastal and southern California.

L. Age/Size at Reproductive Maturity. Stebbins (1985) lists adult size as 31–47 mm SVL, but this information should be regarded as preliminary, considering that black-bellied slender salamanders then included populations that are now referable to *B. gregarius*. In a small series of adult males from San Luis Obispo County, the average was 38.0 mm SVL (range 35.3–42.0 mm, n = 7; reported as *B. attenuatus*; Brame and Murray, 1968). A female from Big Oak Flat, Los Angeles County, measured 49.1 mm SVL.

M. Longevity. Unknown.

N. Feeding Behavior. Has not been described, although all *Batrachoseps* species observed thus far use a projectile tongue to capture small invertebrates.

O. Predators. Predation in the wild is undocumented, although captive ring-necked snakes (*Diadophis punctatus*) and night snakes (*Hypsiglena torquata*) have consumed black-bellied slender salamanders (R.W.H., unpublished data).

P. Anti-Predator Mechanisms. Black-bellied slender salamanders exhibit coiling, rapid crawling, and immobility upon discovery (Stebbins, 1951; Schoenherr, 1976; R.W.H., unpublished data), and tail autotomy may occur when the tail is grabbed or pinched.

Q. Diseases. Unknown.

R. Parasites. A single species of helminth, *Batracholandros salamandrae*, is recorded from black-bellied slender salamanders (Goldberg et al., 1998c).

S. Comments. Jockusch et al. (1998) recently demonstrated that black-bellied slender salamanders (as defined by Yanev, 1978, 1980) comprised, at minimum, two species: *B. gregarius* of the Sierra Nevada and *B. nigriventris* of the central coast and

Tehachapi Mountains into southern California. However, based on studies of allozymes and mtDNA sequences, even this restricted *B. nigriventris* appears to consist of three distinctive lineages: a northern form including the Tehachapi Mountains and central coast north to Monterey County, an island form from Santa Cruz Island, and a southern form occurring from Ventura and Los Angeles counties to the southern and eastern margins of the species range (Wake and Jockusch, 2000). The southern form shows extensive phylogeographic structure.

4. Conservation.
Black-bellied slender salamanders have a relatively wide distribution (Wake and Jockusch, 2000). Recent range changes are not evident, except for areas of obvious habitat modification in many parts of their historical range in southern California. Long-term trends in abundance are not apparent.

Acknowledgments. Stephen Goldberg provided information concerning parasites.

Batrachoseps pacificus (Cope, 1865)
CHANNEL ISLANDS SLENDER SALAMANDER
Robert W. Hansen, David B. Wake, Gary M. Fellers

1. Historical versus Current Distribution.
Channel Islands slender salamanders (*Batrachoseps pacificus*) are restricted to the northern Channel Islands off the Pacific Coast of south-central California: East Anacapa, Middle Anacapa, West Anacapa, Santa Cruz, Santa Rosa, and San Miguel islands (Van Denburgh and Slevin, 1914; Brame and Murray, 1968; Wake and Jockusch, 2000). The elevational range is

from sea level to around 430 m (Mt. Pleasant, Santa Cruz Island). This is the only amphibian endemic to the California islands. There is no indication that the distribution of Channel Islands slender salamanders has changed in historical times.

2. Historical versus Current Abundance.
This species appears to occupy all parts of its potential range, and there is no indication of any changes in abundance. Channel Islands slender salamanders can be abundant under surface cover when the substrate is especially wet, but Schoenherr et al. (1999) indicate that *Batrachoseps* on San Miguel Island are never particularly abundant. This is probably because there is less surface cover and because the soils on San Miguel Island are less suitable for salamanders (e.g., too rocky or too sandy) on many parts of the island.

3. Life History Features.
A. Breeding. Reproduction is terrestrial.
i. Breeding migrations. The life history of Channel Islands slender salamanders is virtually unstudied. Ecological conditions on the northern Channel Islands are generally equivalent to those encountered by their sister species, garden slender salamanders (*B. major*), on the southern California mainland, though summer fogs allow for greatly extended periods of activity. Presumably, courtship occurs after the start of the rainy season, and egg laying is probably associated with the first fall or winter rains.
ii. Breeding habitat. Unknown.
B. Eggs.
i. Egg deposition sites. Eggs of Channel Islands slender salamanders have not been discovered in the field. As with garden slender salamanders, Channel Islands slender salamander eggs probably are deposited underground.

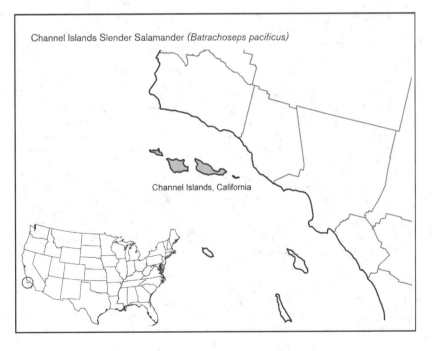

Channel Islands Slender Salamander *(Batrachoseps pacificus)*

Channel Islands, California

ii. Clutch size. Unknown; Stebbins (1954a) reported that one adult female from Santa Rosa Island contained 20 large ova.

C. Direct Development. Stebbins (1954a) reported finding a gravid female from Santa Rosa Island on 8 December. Hilton (1945) noted the presence of ovarian eggs (2 mm diameter) in a female from San Miguel Island on 20 May. Timing of hatchling emergence is unknown.

D. Juvenile Habitat. Differences in habitat use between juvenile and adult Channel Islands slender salamanders have not been noted.

E. Adult Habitat. Channel Islands slender salamanders occur in grassland, coastal sage scrub, chaparral, riparian, oak woodlands, and pine forest communities. They have been found under rocks and logs, especially near streams. Dense populations have been found in open areas near the ocean; in February at the west end of Santa Cruz Island, salamanders were abundant under driftwood on sand within 50–60 m of the ocean. Periods of surface activity correspond generally to the rainy season, especially in drier inland valleys (e.g., Santa Cruz Island). However, the moderating influence of cool, marine air, combined with daily fog, extends activity throughout the summer.

F. Home Range Size. Unknown.

G. Territories. Unknown.

H. Aestivation/Avoiding Desiccation. Channel Islands slender salamanders have been collected in every month, a reflection of the mild marine climate that prevails in the northern Channel Islands. Summer fog is not uncommon, especially on San Miguel Island. Unlike other species of *Batrachoseps* occurring at low elevations on the mainland in which surface activity declines following the rainy season, Channel Islands slender salamanders remain surface active at some sites throughout the year. Indeed, some of the largest collections have been made during the summer (e.g., 58 individuals found on 26 August on Anacapa Island). However, in the interior portions of the larger islands (e.g., the Central Valley of Santa Cruz Island), conditions become too warm and dry for activity during the summer, and salamanders retreat into ground cracks or other refuges (Schoenherr et al., 1999).

I. Seasonal Migrations. Unknown.

J. Torpor (Hibernation). Surface activity of Channel Islands slender salamanders in winter is limited by moisture rather than low temperatures, as freezing conditions are rare on the northern Channel Islands (Schoenherr et al., 1999).

K. Interspecific Associations/Exclusions. Channel Islands slender salamanders are broadly sympatric with black-bellied slender salamanders (*B. nigriventris*) on Santa Cruz Island, where both species tend to occupy similar habitats (Campbell, 1931b; Schoenherr et al., 1999). In one February survey, collectors obtained 155 Channel Islands slender salamanders and 152 black-bellied slender salamanders (Brame and Murray, 1968). During February 1974, both species were abundant, and no microhabitat differences were detected where the species occurred in sympatry. However, only Channel Islands slender salamanders were found under driftwood on a sandy substrate near the ocean (106 specimens observed, 16 February 1974; air temperature 13 °C, exposed sand temperature 31.8 °C, microhabitat under cover 13.2–17.6 °C, mean 16.4 °C; D.B.W., unpublished data). In general, when only one species is found at a more open site, it is usually Channel Islands slender salamanders. Only Channel Islands slender salamanders were found under dried cow pies in open grassland. In scrub oak habitat, black-bellied slender salamanders seem to be the predominant species.

L. Age/Size at Reproductive Maturity. Age and size at sexual maturity are unknown. Considering only animals ≥40 mm SVL, the average size for males is 50.1 mm SVL (range 43.3–58.9 mm, n = 21), and average size for females is 48.8 mm SVL (range 44.0–65.5 m, n = 20; Brame and Murray, 1968; R.W.H., unpublished data). Further restricting this comparison to those specimens ≥50 mm, average size of 11 males was 52.5 mm SVL; average size of 11 females was 55.5 mm SVL. In a large sample that also included subadults, Goldberg et al. (2000) reported mean sizes of 40.2 mm SVL for males and 40.5 mm SVL for females (n = 78, 96, respectively). Female-biased sexual size dimorphism is present in most species of *Batrachoseps* for which data are available. Although females may reach a larger maximum size in Channel Islands slender salamanders, it is unclear whether average adult sizes vary significantly between sexes.

M. Longevity. Unknown.

N. Feeding Behavior. Has not been described in Channel Islands slender salamanders, although all *Batrachoseps* species observed thus far capture prey using a projectile tongue and feed on small invertebrates.

O. Predators. Predation is unreported in this species. As with most islands, the northern Channel Islands contain a depauperate terrestrial vertebrate fauna, and it is possible that predation pressures are much reduced here. Southern alligator lizards (*Elgaria multicarinata*) are present on all the northern Channel Islands with *Batrachoseps* and may prey on these salamanders.

P. Anti-Predator Mechanisms. Occasionally, individual Channel Islands slender salamanders coil when cover objects are removed. This is a widely observed behavior in the genus.

Q. Diseases. Unknown.

R. Parasites. Goldberg et al. (2000) reported an infection rate of 57% (99 of 174 animals examined) in Channel Islands slender salamanders and recorded the following helminths: *Mesocestoides* sp. (a cestode) and *Batracholandros salamandrae* and *Oswaldocruzia pipiens* (both nematodes).

S. Comments. Wake and Jockusch (2000) restricted *B. pacificus* to the populations occurring on the northern Channel Islands. Thus, all mainland populations, as well as those on more southern islands, formerly included in a more broadly defined *B. pacificus* (Yanev, 1978, 1980) are now placed in other taxa. Among these are *B. major*, an undescribed species from the Sierra San Pedro Martir (Baja California), and four recently described species in coastal central California (Jockusch et al., 2001). Examination of mtDNA sequences and allozymes of *B. pacificus* from different islands reveals significant phylogeographic structure (Jockusch, 1996; D.B.W., unpublished data).

4. Conservation.

There are no known threats to Channel Islands slender salamanders. Santa Rosa and Santa Cruz Islands have a variety of introduced mammals (e.g., deer, elk, pigs), but there are active programs or plans to eliminate all three species from the islands. Habitat for *Batrachoseps* should improve with the absence of these large ungulates. The other four islands where these salamanders live (East Anacapa, Middle Anacapa, West Anacapa, and San Miguel islands) are managed entirely as natural areas and appear to provide good salamander habitat.

Acknowledgments. We thank Stephen Goldberg for providing information concerning parasites.

Batrachoseps regius Jockusch, Wake, and Yanev, 1998
KINGS RIVER SLENDER SALAMANDER
Robert W. Hansen, David B. Wake

1. Historical versus Current Distribution.

Kings River slender salamanders (*Batrachoseps regius*) are known from the drainage of the Kings River (Fresno County) on the western slope of the Sierra Nevada of California. The type locality and nearby sites are located on the south and east sides of the North Fork of the Kings River, at elevations of 335–440 m (Jockusch et al., 1998). A second population, provisionally assigned to this species, is known from Summit Meadow, within the South Fork Kings River drainage, at an elevation of 2,470 m and about 37 kilometers east-southeast (direct distance) of the lower elevation sites. The intervening areas within the Kings River drainage (a region of difficult terrain and few roads) have not been surveyed adequately for *Batrachoseps*, and it seems likely that additional populations of Kings River slender salamanders will be found.

More recently, salamanders discovered in the Middle Fork Kaweah River drainage (610 m elevation, Sequoia National Park, Tulare County) have been referred to this species (Jockusch and Wake, 2002); this new find extends the range ca. 29 km south from the nearest Kings River drainage population.

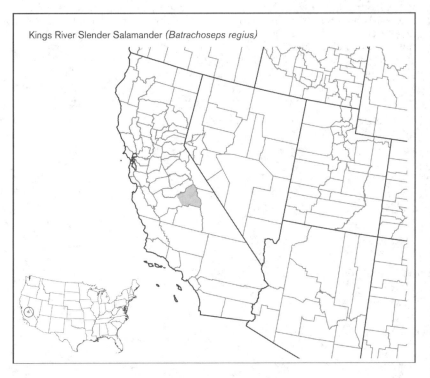

Kings River Slender Salamander (*Batrachoseps regius*)

2. Historical versus Current Abundance.
The small cluster of sites from the lower elevation Kings River appears to be stable; salamanders have been found here intermittently for the last 25 yr. However, these sites occupy localized habitat immediately adjacent to roads, and thus should be considered vulnerable to alteration. A total of seven specimens have been found at the single high-elevation site on two occasions over a 45-yr period.

3. Life History Features.
A. Breeding. Reproduction is terrestrial.

i. Breeding migrations. Unknown.

ii. Breeding habitat. Unknown. For the lower elevation populations, courtship presumably occurs after the start of the rainy season in the fall (November–December), and egg-laying probably takes place from December–January, depending on local rainfall.

B. Eggs.

i. Egg deposition sites. Nest sites have not been found. The association of the lower elevation populations with talus suggests that eggs may be laid well below ground within the rock/litter matrix. The high-elevation population is associated with the wet margins of a seasonally flooded meadow, and it is likely that eggs

are laid under or within moist, decomposing logs.

ii. Clutch size. Unknown.

C. Direct Development. Hatchlings have not been observed, but should be expected to appear in late winter to early spring (low elevation sites).

D. Juvenile Habitat. Unknown.

E. Adult Habitat. The lower Kings River sites lie within a mixed pine-oak/chaparral association, characterized by interior live oak, blue oak, foothill pine, and western redbud, with California bay and western sycamore in moist side canyons. Salamanders have been found under scattered granitic rocks (sometimes as talus), downed logs, or within moist sycamore/oak litter at the base of shaded, north-facing slopes and ravines; ferns and mosses are present at some sites (Jockusch et al., 1998; R.W.H., unpublished data).

The lone high-elevation record is for Summit Meadow (2,470 m elevation), located within a moist coniferous forest of lodgepole pine and red fir. This area receives considerable snowfall between November and April. In contrast to the lower elevation sites, surface activity is confined to summer to early fall. One salamander was found inside a rotted lodgepole pine log saturated from surface moisture and lying in deep shade at the margin of a boggy meadow (E.L. Karlstrom field notes, on file at the Museum of Vertebrate Zoology [MVZ]). Specimens have been found here on only two occasions, in May and June, separated by an interval of 45 yr.

F. Home Range Size. Unknown.

G. Territories. Unknown.

H. Aestivation/Avoiding Desiccation. At the low elevation sites, salamanders are present under surface cover only during periods of adequate soil moisture—generally from November to March–April. Timing of surface activity varies depending on arrival of fall/winter rains. Individual salamanders presumably move beneath the surface in burrows or rock rubble during dry periods and during the coldest periods in winter. Field body temperatures of 10 salamanders found under cover at the low elevation sites averaged 7.0 °C (range 5.2–10.1 °C; R.W.H., unpublished data). A single individual found on 9 June at Summit Meadow had a body temperature of 10.2 °C (E.L. Karlstrom field notes, on file at MVZ).

I. Seasonal Migrations. Unknown.

J. Torpor (Hibernation). Unknown.

K. Interspecific Associations/Exclusions. Kings River slender salamanders are found with Sierra Nevada ensatinas (*Ensatina eschscholtzii platensis*), which occupy a comparatively broad range of habitats. The ranges of gregarious slender salamanders (*B. gregarius*) and Kings River slender salamanders approach, but apparently do not meet, near Pine Flat Reservoir (Fresno County) along the Kings River (Jockusch et al., 1998). These two species undoubtedly come into contact within the Kaweah River drainage. Gregarious slender salamanders have also been found at higher elevations in the Kings River drainage within coniferous forest at sites that appear suitable for Kings River slender salamanders. California newts (*Taricha torosa*) are widespread at lower elevations in the Kings River drainage and commonly are found at sites where Kings River slender salamanders occur.

L. Age/Size at Reproductive Maturity. This is a moderately small species of *Batrachoseps*, with maximum adult size ≤45 mm SVL (Jockusch et al., 1998). Age and size at sexual maturity are unknown.

M. Longevity. Unknown.

N. Feeding Behavior. Has not been described, although all *Batrachoseps* species observed thus far use a projectile tongue to capture small invertebrates.

O. Predators. Unknown. Ring-necked snakes (*Diadophis punctatus*) occur in the lower elevation portion of the range of Kings River slender salamanders and should be considered likely predators.

P. Anti-Predator Mechanisms. Immobility and coiling have been observed when salamanders are first uncovered (R.W.H., unpublished data). The dark dorsal coloration presumably is cryptic against dark substrates and in low-light environments.

Q. Diseases. Unknown.

R. Parasites. Unknown.

S. Comments. Although Yanev (1978, 1980) clearly identified the lower Kings River populations of *Batrachoseps* as distinctive, formal species recognition did not come until recently (Jockusch et al., 1998), when the formerly wide-ranging *B. relictus* was partitioned into four species (from north to south in the central and southern Sierra Nevada): *B. diabolicus*, *B. regius*, *B. kawia*, and *B. relictus*. Recent mtDNA studies (Jockusch, 1996; Jockusch et al., 1998) suggest a sister-taxon relationship between *B. regius* and *B. diabolicus*, despite the geographic proximity of *B. regius* to *B. kawia*, both of which are now known from the Kaweah River drainage to the south.

Although Jockusch et al. (1998) provisionally assigned specimens collected in 1955 from Summit Meadow (Kings Canyon National Park, elevation 2,470 m) to *B. kawia*, more recent studies of freshly collected material suggest that this is a high elevation population of *B. regius*. This leaves a substantial gap in the known distribution of *B. regius*. Further, the recent discovery of *B. regius* within the Middle Fork Kaweah River drainage brings this species much closer to populations of its southern relative, *B. kawia*, which is known from the adjoining East Fork Kaweah River drainage. Additional fieldwork is needed in the upper reaches of the Kings and Kaweah drainages, an area where the ranges of three species *(B. regius, B. gregarius,* and *B. kawia)* converge and potentially overlap.

4. Conservation.

Kings River slender salamanders occupy a geographically small range and are confirmed from only three areas (one of which [the lower Kings River] consists of a cluster of several sites). Each of these populations appears to be quite localized, and the degree of genetic subdivision suggests that these have been isolated from one another for a long time (Jockusch and Wake, 2002). The lower Kings River sites are located immediately adjacent to a road and likely would be affected by road construction. All localities for Kings River slender salamanders occur on public lands administered by the USDA Forest Service or National Park Service.

Acknowledgments. We thank John Romansic for providing copies of his field notes.

Batrachoseps relictus Brame and Murray, 1968
RELICTUAL SLENDER SALAMANDER
Robert W. Hansen, David B. Wake

1. Historical versus Current Distribution.

Relictual slender salamanders (*Batrachoseps relictus*; *sensu* Jockusch et al., 1998) are restricted to the west slopes of the southern Sierra Nevada of California, ranging from the lower Kern River Canyon (Kern County) to highlands drained by the Tule and Kern rivers (Tulare County). Two principal distributional units are present: (1) lower Kern River Canyon, where relictual slender salamanders have been found at six sites at elevations of 485–730 m (Brame and Murray, 1968); and (2) higher elevations in the Greenhorn Mountains north to the Tule River drainage, at elevations of 1,125–2,440 m. There is a single record for the western margin of the Kern Plateau, east of the Kern River, at 2,440 m (Hansen, 1980; Jockusch et al., 1998). Despite repeated and careful searches, the species has not been found in the lower Kern River Canyon since 1971 and is presumed extirpated from those localities (Jennings and Hayes, 1994a; Hansen, 1997; Jockusch et al., 1998).

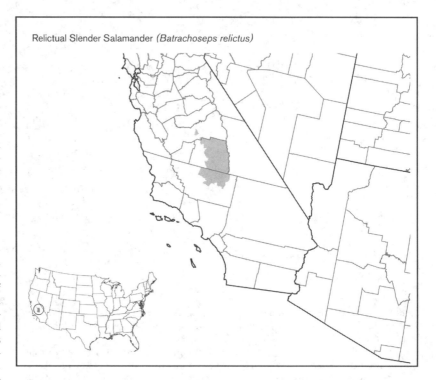

Relictual Slender Salamander *(Batrachoseps relictus)*

2. Historical versus Current Abundance.

Populations of relictual slender salamanders from throughout the range appear to be stable (with the aforementioned exception of the type locality and associated sites in the lower Kern River Canyon, where no individuals of this species have been found since 1971, despite concerted efforts to find them). The construction of State Route 178 along the lower, southern slopes of the Kern River Canyon severely impacted seepages and springs that once harbored relictual slender salamanders. Although specimens were collected at these sites for a number of years following highway construction, this loss of habitat may have initiated local population declines that finally reached non-sustainable levels (Hansen, 1988).

3. Life History Features.
    A. Breeding. Reproduction is terrestrial.
    i. Breeding migrations. Unknown.

    ii. Breeding habitat. For the mid-elevation populations (e.g., Tule River drainage, ≤1,250 m), courtship presumably occurs after the start of the rainy season in the fall (November, perhaps earlier), and egg-laying probably takes place in late November to December, depending on local rainfall and temperatures. Higher elevation sites (1,600–2,440 m) experience a wide range of winter conditions, including moderate snowfall and below freezing temperatures well into spring. Breeding phenology of these populations has not been studied.

    B. Eggs.
    i. Egg deposition sites. Nest sites of relictual slender salamanders have not been found. A large majority of populations are associated with seepages or springs, and we expect that nests will be found under rocks and logs at the margins of such wet microhabitats. Communal nests (Jockusch and Mahoney, 1997) are unknown for any member of the *B. relictus* species group.
    ii. Clutch size. Unknown.

    C. Direct Development. Timing of hatchling emergence is poorly known. Hatchlings were observed at one high-elevation site (1,820 m) in the Greenhorn Mountains on 28 May.

    D. Juvenile Habitat. Unknown how this may differ from adult habitats.

    E. Adult Habitat. At most localities where relictual slender salamanders have been found, they are associated with downed logs and bark rubble in moist conifer forest, frequently near seepages and springs where surface moisture persists through the summer. In the Greenhorn Mountains (Kern and Tulare counties), they are especially common in roadside seepages at elevations of 1,675–2,130 m. Typical forest components include Ponderosa pine, sugar pine, incense cedar, white fir, and black oak. We occasionally have found salamanders here beneath large sugar pine cones.

Farther north, near their lower elevation limits within the Tule River drainage (Tulare County, 1,215 m), relictual slender salamanders occur under rocks and logs beneath canyon live oaks in the transition to lower Ponderosa pine forest. Interestingly, several other species appear to reach their low elevation range limits within the Tule River Canyon at this site, including western skinks (*Eumeces skiltonianus*), northern alligator lizards (*Elgaria coerulea*), and rubber boas (*Charina bottae*); Gilbert's skinks (*Eumeces gilberti*) reach their uppermost point in the canyon here as well (R.W.H., unpublished data). At slightly higher elevation (1,265 m) in the same river canyon, relictual slender salamanders were found near a small creek in a forest of incense cedar, ponderosa and sugar pine, whitebark alder, bigleaf maple, and canyon live oak, under rocks, logs, and moist pine needles.

The single population east of the Kern River, located on the western flank of the Kern Plateau at an elevation of 2,440 m, is associated with a large seepage on a moderately steep slope, surrounded by a dense forest of Jeffrey pine and white fir. Relictual slender salamanders were found beneath rocks and downed logs that rested on saturated gravel substrates, some with surface flow (R.W.H., unpublished data).

The lower Kern River Canyon populations, now presumed extirpated, were associated with perennial springs, seepages, and the margins of small creeks (Hilton, 1948; Brame and Murray, 1968). In this area, the species was taken in sympatry with yellow-blotched ensatinas (*Ensatina eschscholtzii croceater*) and Kern Canyon slender salamanders (*B. simatus*), though these two species are usually found away from water, whereas relictual slender salamanders frequently were found in water.

F. Home Range Size. Unknown.

G. Territories. Unknown.

H. Aestivation/Avoiding Desiccation. Relictual slender salamanders are present under surface cover only during periods of adequate soil moisture. In mid-elevation conifer forest, this may extend from April–November, especially for seep-margin populations. At lower elevations, the period of seasonal activity is shorter, beginning with rains in the fall and extending until May–June. Individual salamanders presumably move beneath the surface in burrows or rock rubble during dry periods and the coldest winter months. In the lower Kern River Canyon, where salamanders were closely associated with perennial

seepages and springs, surface activity may have been possible over much of the year, as indicated by collection dates ranging from January–May. Field body temperatures for salamanders found under surface cover averaged 12.0 °C (range = 7.0–15.0 °C, n = 112; all data taken from populations ≥1,200 m elevation; R.W.H., unpublished data).

I. Seasonal Migrations. Unknown.

J. Torpor (Hibernation). We have not recorded surface activity of relictual slender salamanders at substrate temperatures below 7 °C.

K. Interspecific Associations/Exclusions. Over most of their range, relictual slender salamanders do not occur with any other species of *Batrachoseps*. Exceptions occur in the lower Kern River Canyon, at least historically, where relictual slender salamanders were sympatric with Kern Canyon slender salamanders at two known sites (Brame and Murray, 1968). The ranges of relictual slender salamanders and gregarious slender salamanders (*B. gregarius*) converge in a number of drainages in the southern Sierra Nevada, but cases of sympatry are unknown. The single record for east of the Kern River is close to populations of Kern Plateau salamanders (*B. robustus*), and sympatry is expected (Wake et al., 2002).

Sierra Nevada ensatinas (*E. e. platensis*) occur at or near most of the relictual slender salamander sites from the Greenhorn Mountains northward. Yellow-blotched ensatinas occur in the lower Kern River Canyon at historical localities for relictual slender salamanders. The ranges of California newts (*Taricha torosa*) and relictual slender salamanders overlap below 1,500 m in Tulare County.

L. Age/Size at Reproductive Maturity. Relictual slender salamanders are one of the smallest species in the genus. For a small series from the lower Kern River Canyon, sexually mature males averaged 31.2 mm SVL (range = 26.3–36.9 mm, n = 8), and females averaged 36.2 mm SVL (range = 33.3–41.7 mm, n = 9; Brame and Murray, 1968; Wake and colleagues, unpublished data). For the high-elevation populations (Greenhorn Mountains, Kern and Tulare counties), adult sizes appear to be somewhat larger (mean = 39.4 mm SVL; range = 34.4–48.1, n = 16, sexes lumped; Brame and Murray, 1968; Wake and colleagues, unpublished data), but more data are needed from throughout the range.

M. Longevity. Unknown.

N. Feeding Behavior. Has not been described for relictual slender salamanders, although all *Batrachoseps* species observed thus far use a projectile tongue to capture small invertebrates.

O. Predators. Predation on relictual slender salamanders is undocumented. Ring-necked snakes (*Diadophis punctatus*) are common in some parts of their range (e.g., Tule River) and are likely predators.

P. Anti-Predator Mechanisms. Defensive behaviors include coiling, immobility, thrashing, crypsis, and tail autotomy. Coiling behavior upon discovery is nearly universal among salamanders in a Tule River population. This behavior is uncommon among Greenhorn Mountains individuals, which are more likely to remain immobile or thrash vigorously when uncovered or touched (R.W.H., unpublished data). One individual escaped by crawling quickly down a tunnel beneath its rock.

There is a strong resemblance between Greenhorn Mountains relictual slender salamanders and a sympatric millipede. Although the salamanders display some variation, one common color pattern in the Greenhorns is that of a tan/brown dorsum with a thin, gray/black dorsal line—remarkably similar to that of the millipedes. Millipedes appear to prefer slightly drier substrates, but often both salamanders and millipedes are encountered under the same piece of cover (R.W.H., unpublished data). The noxious discharges of millipedes are well known; the effects of relictual slender salamander skin secretions are unstudied. Examples of vertebrate mimicry of millipedes have been offered by Vitt (1992) and Leonard and Stebbins (1999), and this may represent an additional occurrence. Alternatively, this may be a case of convergent crypsis, as the color patterns of both salamanders and millipedes well match the mixture of pine needles and incense cedar debris at seep margins. Mimicry of a noxious model and pattern selection for crypsis are not mutually exclusive hypotheses.

Q. Diseases. Unknown.

R. Parasites. Unknown.

S. Comments. The taxonomic composition of *B. relictus* has changed considerably following its description in 1968. Yanev (1978, 1980) removed coastal California populations from this species and restricted *B. relictus* to the Sierra Nevada, while noting extensive genetic subdivision across various Sierran drainages. Later, Jockusch et al. (1998) further partitioned this still wide-ranging group into four species (from north to south in the central and southern Sierra Nevada): *B. diabolicus*, *B. regius*, *B. kawia*, and *B. relictus*.

The lack of live material from the type locality and nearby areas in the lower Kern River Canyon, where *B. relictus* is presumed extinct, has precluded genetic comparisons with high-elevation Greenhorn Mountains populations included in the species. Morphological and ecological differences exist between these two segments of the range, and it is possible that more than one species is represented. Genetic divergence among high-elevation populations of *B. relictus* is low relative to other species within the *relictus* group (Jockusch and Wake, 2002), suggesting that these populations have been in recent genetic contact.

4. Conservation.

The lower Kern River Canyon populations (inclusive of the type locality), as described above, appear to have been extirpated. Despite our failure to locate specimens from this area over the last 30 + yr, additional search efforts are warranted. Although these historical localities occur on public lands administered by the USDA Forest Service, road margin habitat has been severely degraded by road maintenance and related construction activities. Relictual slender salamanders are a Species of Special Concern (California Department of Fish and Game), a Forest Service Sensitive Species, and a Federal Species of Concern.

Batrachoseps robustus Wake, Yanev, and Hansen, 2002
KERN PLATEAU SALAMANDER
Robert W. Hansen, David B. Wake

1. Historical versus Current Distribution.

Kern Plateau salamanders *(Batrachoseps robustus)* are endemic to the southeastern Sierra Nevada of California and have only recently been described (Wake et al., 2002). Their range consists of three principal units: the Kern Plateau (Tulare County), where they have been recorded from a number of sites; the eastern slopes of the Sierra Nevada draining into Owens and Indian Wells valleys (Inyo County), where they are restricted to steep, east-facing canyons; and the Scodie Mountains (Kern County), south of the main body of the species' range. The elevational range is from 1,700–2,800 m on the Kern Plateau (Tulare County); 1,430–2,440 m on the Sierran east slope (Inyo County); and 1,980–2,025 m in the Scodie Mountains (Kern County).

2. Historical versus Current Abundance.

Populations from the Kern Plateau and east slope Sierran canyons appear stable. An extensive wildfire burned through the Scodie Mountains within the last decade, and the status of populations there is uncertain as a recent survey (2002) failed to locate any salamanders (R. W. H., personal observation).

3. Life History Features.

A. Breeding. Reproduction is terrestrial. Timing of courtship and oviposition are unknown, but are likely to vary somewhat with elevation and perhaps with levels of seasonal precipitation. Adult females (gravid though physiologically not ready for egg laying) were found in early May at 2,070 m elevation on the Kern Plateau (Wake et al., 2002). This site is a seasonal seepage that may disappear by May–June and is located in an otherwise dry singleleaf pinyon pine forest.

Many of the females collected from various high-elevation sites (≥2,200 m)

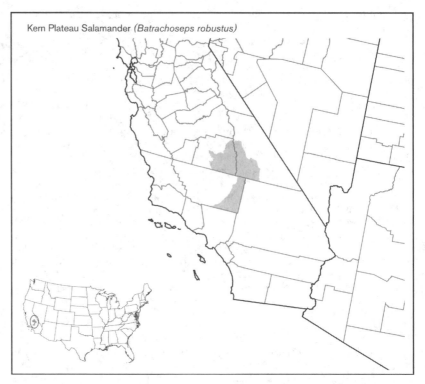

Kern Plateau Salamander *(Batrachoseps robustus)*

on the Kern Plateau during July have contained small to large ova (this variation in development of ova is usually present even within a local population).

  i. Breeding migrations. Unknown.

  ii. Breeding habitat. Unknown.

B. Eggs.

  i. Egg deposition sites. Eggs of this species have not been found in the wild. However, we have found aggregations of gravid females in association with wet substrates—under cover at margins of seeps, springs, or creeks. At one such site (near Osa Meadow, Tulare County, 2,680 m elevation, 22 July), we found 21 large adults, most of which were females. Nine of these were found within a rotted lodgepole pine log saturated from snowmelt. We suspect that such localized wet areas probably serve as oviposition sites. However, nests have not been discovered in the field and communal nesting is unknown for members of the robust clade of slender salamanders (subgenus *Plethopsis*; Jockusch and Mahoney, 1997).

  ii. Clutch size. Uncertain, although five gravid females induced to oviposit in the lab laid three eggs each while retaining 1–3 eggs (Wake et al., 2002).

C. Direct Development. Eggs incubated in the lab at 13 °C hatched after 96 and 103 days (Wake et al., 2002). Hatchlings are black, usually with gold or silver flecking, and measure ca. 12–13 mm SVL.

D. Juvenile Habitat. Hatchlings often are found under smaller cover objects (small rocks and pieces of bark) not utilized by adults.

E. Adult Habitat. Kern Plateau salamanders occur in a broad range of ecological settings, from high-elevation coniferous

forests to semiarid pinyon pine/sagebrush associations. In general, salamanders occurring in mesic pine-fir forests are more broadly distributed on a local scale and are less closely associated with surface moisture. By contrast, populations occupying the southeastern Kern Plateau, eastern slope Sierra Nevada canyons, and Scodie Mountains are restricted to areas of permanent or seasonal surface moisture. During favorable periods of surface moisture and temperature, Kern Plateau salamanders may be found under rocks, under or within downed logs, or among bark rubble.

Detailed descriptions and photographs of representative localities throughout the range of this species are provided by Wake et al. (2002).

F. Home Range Size. Unknown.

G. Territories. Unknown.

H. Aestivation/Avoiding Desiccation. Timing and duration of seasonal activity vary with elevation, availability of surface moisture, and annual variation in precipitation (Wake et al., 2002). Lower elevation populations (≤2,000 m) may be active in all but the coldest periods, retreating below ground by late summer or early spring; exceptions occur where a local population is associated with perennial surface moisture, in which case activity may continue during summer. At higher elevation sites, activity on the surface is confined to a few months, perhaps May–June to October (Wake et al., 2002). Substrate temperatures for salamanders found under surface cover ranged from 5.2–25.0 °C (mean = 13.5 °C, n = 217; Wake et al., 2002). As local surface conditions dry, adult salamanders are scarce, but juveniles may still be present; perhaps there is pressure to accumulate fat reserves prior to aestivation (Wake et al., 2002).

I. Seasonal Migrations. Unknown.

J. Torpor (Hibernation). The lower threshold for surface activity appears to be about 5 °C. Some high-elevation populations of this species occur in areas subject to long periods of cold winter weather and although activity may be reduced, we suspect that some activity might occur where snow-covered soil and logs remain above freezing.

K. Interspecific Associations/Exclusions. At the western margin of the Kern Plateau, relictual slender salamanders *(B. relictus)* have been found at a single high-elevation site only 1.6 km from the nearest locality for *B. robustus*, and sympatry is expected (Hansen, 1997; Wake et al., 2002). On the Sierra Nevada eastern slope (Inyo County), the range of *B. robustus* extends as far north as Olancha Creek, where it meets the southern range limits of a population of web-toed salamanders tentatively allocated to *Hydromantes platycephalus* (Mt. Lyell salamanders; Wake and Papenfuss, this volume). Within Olancha Creek Canyon, *Hydromantes* occupy higher elevations (≥1,800 m), and Kern Plateau salamanders have not been found here above 1,800 m elevation. Both species have been taken under adjacent rocks at 1,800 m elevation, but this is the only known instance of sympatry (Giuliani, 1989; Wake et al., 2002). The distributional pattern is one of geographical replacement with virtually no overlap, suggesting that range expansion might be limited by competition.

L. Age/Size at Reproductive Maturity. This is a large, robust species of *Batrachoseps*, with females attaining larger average and maximum size. Average size of 10 adult males was 51.9 mm SVL (maximum 58.9 mm); 10 adult females averaged 58.2 mm (maximum 61.3 mm; Wake et al., 2002). There is little variation in adult size among Kern Plateau populations. Average adult size may be smaller in east slope Sierra Nevada canyons, although samples are small. The smallest gravid female was 44 mm SVL (Wake et al., 2002).

M. Longevity. Unknown.

N. Feeding Behavior. In the laboratory, Kern Plateau salamanders have been observed to use a projectile tongue to capture small invertebrates.

O. Predators. Unknown, although garter snakes are likely predators.

P. Anti-Predator Mechanisms. We recorded responses of 79 individuals in the field as they were discovered under surface cover: attempt to crawl away (73%), immobility (16.5%), thrashing (5%), and coiling (3.8%; Wake et al., 2002). García-París and Deban (1995) observed partial coiling (body coiled, but tail extended) in all three individuals tapped on the back in a lab setting. Rarely, sticky skin secretions are produced. Tail autotomy is rare in robust members of this genus *(B. campi,*

robustus, and *wrighti*—the subgenus *Plethopsis*), in strong contrast to attenuate species (subgenus *Batrachoseps*) in which tail loss is a common defensive strategy.

Dorsal coloration varies in this species. In general, salamanders from low-rainfall areas are relatively lighter in coloration, while those from more mesic portions of the range, such as the red fir forests on the northern Kern Plateau, display dark dorsal patterns that appear cryptic against darker substrates (Wake et al., 2002).

Q. Diseases. Unknown.

R. Parasites. Unknown.

S. Comments. Although only recently described, specimens now allocated to *B. robustus* were first collected in 1972 on the Kern Plateau, but misidentified as Tehachapi slender salamanders *(B. stebbinsi*; Richman, 1973). Detailed field surveys initiated in 1979 revealed additional populations of this species on the Kern Plateau and in the Scodie Mountains; in the mid-1980s, populations were discovered on the eastern flank of the Sierra Nevada in Inyo County. References to this species (as *Batrachoseps* sp. or "Kern Plateau salamander") prior to its formal description appear in Hansen (1980, 1997; distribution; regional species diversity), Stebbins (1985; distribution, illustration), Macey (1986; regional biogeography), Giuliani (1988, 1990, 1996; field survey reports), Macey and Papenfuss (1991a,b; distribution and regional biogeography), Wake (1993; membership in *Plethopsis*), García-París and Deban (1995; defensive behavior), Jennings (1996b; distribution map), Jockusch (1996, 1997b, 2001; phylogeography; number of trunk vertebrae; similarity to *B. campi*), Jackman et al. (1997; confirmation of membership in *Plethopsis*), Jockusch and Mahoney (1997; reproduction), Jockusch et al. (1998; regional species diversity), and Jockusch and Wake (2002; phylogeography).

Studies of mtDNA gene sequences, allozymes, and morphological data (Yanev, 1980; Jockusch, 1996; Jackman et al., 1997) confirm that *B. robustus*, together with Oregon slender salamanders *(B. wrighti)* and Inyo Mountains salamanders *(B. campi)*, are phylogenetically distant from other species of *Batrachoseps*. Despite the relative isolation of the Scodie Mountains populations, they are relatively little differentiated from populations in the main body of the species' range on the Kern Plateau and western Owens Valley based on allozyme studies (Wake et al., 2002). Despite their proximity to populations of Inyo Mountains salamanders *(B. campi*; 43 km), Kern Plateau salamanders are as differentiated (or more so) from them as from Oregon slender salamanders *(B. wrighti)*.

4. Conservation.
Kern Plateau salamanders have been found to be surprisingly widespread on

the Kern Plateau, especially on the more mesic portions, as well as being present in virtually every stream-bearing canyon on the eastern flank of the Sierra Nevada in Inyo County (from Olancha Creek south to Ninemile Canyon). The two isolated populations in the Scodie Mountains are quite small and are comparatively more vulnerable to habitat disturbance. Nearly all populations occur on public lands administered by the USDA Forest Service or U.S. Bureau of Land Management. Some sites have been affected by road construction, timber harvesting activities, or forest fire suppression efforts. A critical habitat feature for this species at most localities is surface moisture in the form of springs, seepages, or creek margins—we suspect these are oviposition sites based upon the discovery of small aggregations of gravid females. This is a Forest Service Sensitive Species.

Batrachoseps simatus Brame and Murray, 1968
KERN CANYON SLENDER SALAMANDER
Robert W. Hansen, David B. Wake

1. Historical versus Current Distribution.
Kern Canyon slender salamanders *(Batrachoseps simatus)* are known only from the lower Kern River Canyon (Kern County) at the southern end of the Sierra Nevada of California. They have been recorded from nine sites within the lower Canyon, ranging from Stork Creek (455 m) to Clear Creek (685 m; Brame and Murray, 1968; Hansen, 1988). They also occur within Erskine Creek Canyon (a tributary to the Kern River) on the northern flank of the adjacent Piute Mountains; here they have been found in three small areas ranging in elevation from 890–1,220 m.

Populations of Kern Canyon slender salamanders appear to be confined to small, disjunct patches of habitat, suggesting that population sizes are quite small. Despite an earlier estimate that total available habitat for Kern Canyon slender salamanders comprised approximately 820 acres (332 ha), salamanders have been recorded only from roughly 17 acres (6.9 ha), about 2% of the potential habitat (Hansen, 1988). Substantial genetic differences between adjacent populations suggest that the apparent isolation is not an artifact of inadequate sampling (D.B.W., unpublished data).

2. Historical versus Current Abundance.
The southern end of the Sierra Nevada may experience multi-year droughts, making it nearly impossible to locate specimens of Kern Canyon slender salamanders under surface cover during such periods. Even under favorable conditions, this species can be difficult to find, rendering efforts to monitor the status of local populations problematic. Despite such obstacles, fieldwork over the last two

Kern Canyon Slender Salamander *(Batrachoseps simatus)*

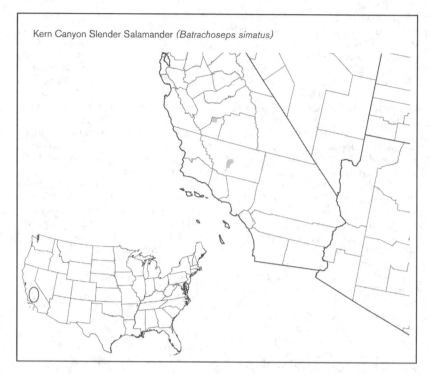

decades suggests that populations at all known localities are extant.

3. Life History Features.

A. Breeding. Reproduction is terrestrial.

i. Breeding migrations. Unknown. Onset of the fall/winter rainy season is especially unpredictable at the extreme southern end of the Sierra Nevada. An initial heavy rainstorm in October or November, which may stimulate salamander surface activity, may be followed by dry conditions lasting 1–2 mo before rains resume. We suspect that timing of courtship and egg laying is similarly variable.

ii. Breeding habitat. Unknown.

B. Eggs.

i. Egg deposition sites. Nest sites have not been found for this species. Communal nests have been reported for other members of the *B. nigriventris* species group (Jockusch and Mahoney, 1997) and might be expected in *B. simatus*. The discovery of a communal nest on nearby Breckenridge Mountain (Kern County, at an elevation of 1,920 m; R.W.H., unpublished data) was cited by Stebbins (1985) as representing *B. simatus*; however, the taxonomic status of the Breckenridge Mountain population is unresolved and it is doubtful that these animals are conspecific with *B. simatus*.

ii. Clutch size. Unknown.

C. Direct Development. Timing of hatchling emergence is unknown. A single hatchling was observed on 4 March along Clear Creek (685 m).

D. Juvenile Habitat. Unknown how this may differ from adult habitats.

E. Adult Habitat. Within the lower Kern River Canyon, populations of Kern Canyon slender salamanders occur mostly in smaller tributary canyons or ravines at the base of the north-facing slope within a pine-oak woodland. These areas receive little or no direct sunlight during the winter, and cool, moist conditions are likely to persist later in the spring. Predominant plant species include foothill pine, interior live oak, sycamore, California buckeye, Fremont cottonwood, and willow. Understory vegetation commonly includes poison oak, nettles, miner's lettuce, *Nemophila* sp., *Phacelia* sp., and grasses (Hansen, 1988). Salamanders have been found under rocks, under and within logs, or in moist oak and sycamore litter. Collection sites range from the wet margins of creeks and seepages to fairly exposed hillsides among chaparral vegetation (Brame and Murray, 1968; Hansen, 1988).

Along Clear Creek near Miracle Hot Springs (685 m elevation), Kern Canyon slender salamanders are associated with talus derived from metamorphic outcroppings. The surrounding slopes support interior live oak, California juniper, foothill pine, yucca, and beavertail cactus.

In Erskine Creek Canyon (890–1,220 m elevation), Kern Canyon slender salamanders are closely associated with localized groves of canyon live oak, in areas bordered by dry slopes of foothill pine, interior live oak, and chaparral shrubs such as manzanita, deer brush, and Great Basin sagebrush.

F. Home Range Size. Unknown.

G. Territories. Unknown.

H. Aestivation/Avoiding Desiccation. Salamanders are present under surface cover only during periods of adequate soil moisture—November to March–April. However, the timing of winter rains is erratic in the southern Sierra Nevada, such that conditions suitable for surface activity are shorter, often February–March in the lower Kern River Canyon and somewhat later for the higher elevation sites in Erskine Creek Canyon. Within the rainy season, periods of surface activity tend to be brief, and salamanders can be difficult to find. Field body temperatures for salamanders found under surface cover averaged 10.1 °C (range = 8.8–15.0 °C, n = 40; R.W.H., unpublished data).

I. Seasonal Migrations. Unknown.

J. Torpor (Hibernation). We have not recorded surface activity at substrate temperatures below 8.8 °C.

K. Interspecific Associations/Exclusions. Relictual slender salamanders *(B. relictus)* formerly were sympatric with Kern Canyon slender salamanders at two sites in the lower Kern River Canyon (Brame and Murray, 1968), prior to their apparent extirpation there. Although these two species were sometimes found in close proximity, relictual slender salamanders were invariably restricted to aquatic microhabitats. Populations assignable to gregarious slender salamanders *(B. gregarius)* occur just below the mouth of the Kern River Canyon in semiarid grassland (Jockusch et al., 1998), and suitable habitat extends into the Canyon as well, especially along the northern slopes. Yellow-blotched ensatinas *(Ensatina eschscholtzii croceater)* have been found at most Kern Canyon slender salamander localities; they occupy similar habitats but have a more extensive geographical, elevational, and ecological distribution. Sympatry with California newts *(Taricha torosa)* occurs along Mill Creek just above its confluence with the Kern River.

L. Age/Size at Reproductive Maturity. This is a moderately large species of *Batrachoseps*, with adults commonly exceeding 50 mm SVL. Including only animals ≥ 40 mm SVL, a series of adult males from the lower Kern River Canyon averaged 47.3 mm SVL (maximum = 53.8 mm SVL, n = 17), while the average SVL for females was 48.1 mm (maximum = 58.5 mm SVL, n = 9; Brame and Murray, 1968; Wake et al., unpublished data). Animals from Erskine Creek Canyon are markedly smaller: three adult males averaged 39.2 mm SVL (range 38.7–40.2), while four adult females averaged 39.6 mm SVL; one unsexed individual measured 43.0 mm SVL. The mean SVL for the 20 largest animals (sexes combined) from lower Kern River Canyon localities was 52.2 mm, while the mean SVL for the 10 largest specimens from Erskine Creek Canyon was 39.3 mm (D.B.W. and colleagues, unpublished data).

M. Longevity. Unknown.

N. Feeding Behavior. Has not been described, although all *Batrachoseps* species observed thus far use a projectile tongue to capture small invertebrates.

O. Predators. Predation is undocumented, although ring-necked snakes *(Diadophis punctatus)* occur throughout the southern Sierra Nevada and presumably prey on this species.

P. Anti-Predator Mechanisms. Defensive behaviors may include coiling, immobility (crypsis), rapid crawling, and tail autotomy.

Q. Diseases. Unknown.

R. Parasites. Unknown.

S. Comments. For this account, we have restricted *B. simatus* to those populations occurring within the lower Kern River Canyon and adjacent Erskine Creek Canyon (Piute Mountains). Even within this relatively small area, considerable genetic differences are evident, suggesting that these populations have long been isolated (Jockusch, 1996; D.B.W., unpublished data). Moreover, there are morphological differences between the Erskine Creek Canyon specimens and those from the lower Kern River Canyon, indicating that further work is needed in clarifying their relationship.

There are additional, undescribed populations of *Batrachoseps* known from within the Kern River drainage. One of these, near Fairview, adjacent to the Kern River (1,065 m, Tulare County), was tentatively allocated to *B. simatus* by Brame and Murray (1968); however, more recent studies of allozymes and mtDNA sequences indicate that this population and others nearby are not conspecific with *B. simatus* (Jockusch, 1996; Jockusch and Wake, unpublished data). Although these populations clearly belong within the *B. nigriventris* group (Jockusch and Wake, 2000; Wake and Jockusch, 2000), their relationships to *B. simatus* are unresolved but under study (D.B.W. and colleagues, unpublished data).

4. Conservation.
The construction of State Route 178 along the lower south slopes of the Kern River Canyon affected much of the habitat for this species (Hansen, 1988). Nearly all the known populations occur on public lands administered by the Sequoia National Forest. Cattle grazing has severely degraded salamander habitat, particularly in the narrow ravines where salamanders are often concentrated (Hansen, 1988). Another potential threat includes the proposed development of water storage facilities within the Kern River Canyon (Hansen and Stafford, 1994a; Jennings, 1996b). Kern Canyon slender salamanders are listed as Threatened by the State of California, a Federal Species of Concern, and a USDA Forest Service Sensitive Species.

Batrachoseps stebbinsi Brame and Murray, 1968

TEHACHAPI SLENDER SALAMANDER

Robert W. Hansen, David B. Wake

1. Historical versus Current Distribution.
Tehachapi slender salamanders *(Batrachoseps stebbinsi)* are known from two small areas in south-central California, both in Kern County. Within Caliente Canyon (site of the type locality), at the junction of

Tehachapi Slender Salamander *(Batrachoseps stebbinsi)*

the Sierra Nevada and Tehachapi Mountains, Tehachapi slender salamanders have been recorded from seven discrete localities at elevations of 550–790 m (Brame and Murray, 1968; R.W.H., unpublished data).

Populations tentatively allocated to *B. stebbinsi* also occur in several isolated canyons on the northern slopes of the Tehachapi Mountains, ranging from Tejon Canyon southwest to Fort Tejon, at elevations of 945–1,430 m (Yanev, 1980; Stebbins, 1985; Jockusch, 1996; Wake, 1996; Wake and Jockusch, 2000; R.W.H., unpublished data).

A single specimen was collected in 1957 from the north slope of Black Mountain (914 m), in the vicinity of Tehachapi Pass, an area that is geographically intermediate between the Tehachapi Mountains populations and Caliente Canyon sites (Brame and Murray, 1968). The Tehachapi Pass site, along the route of old U.S. Highway 466, was thought to have been buried by new highway construction (Brame and Murray, 1968). However, the original collector reports that the site remains relatively undisturbed (T.J. Papenfuss, personal communication), although no *Batrachoseps* have been found anywhere in Tehachapi Pass since 1957.

2. Historical versus Current Abundance.
Populations of Tehachapi slender salamanders are confined to seasonally shaded, north-facing slopes of canyons located in otherwise arid to semi-arid terrain. Individual populations are small and localized. Within Caliente Canyon, much of the known salamander habitat occurs on public lands administered by the U.S.

Bureau of Land Management. Some sites have been affected by road construction, mining, and cattle grazing, and potentially by flood control projects (Hansen and Stafford, 1994b; Jennings, 1996b). Portions of the Tehachapi Mountains (notably Bear, Cummings, and Tehachapi Valleys) are experiencing rapid human population growth, with much development occurring in the foothills. Plans exist for the development of several new communities on the vast Tejon Ranch property. Owing to the small size and localized nature of Tehachapi slender salamander populations, the Tejon Ranch sites appear especially vulnerable to habitat disturbance.

The construction of a major freeway through Grapevine Canyon (Tejon Pass) at the western edge of the Tehachapi Mountains undoubtedly has impacted populations of Tehachapi slender salamanders. Although the occurrence of Tehachapi slender salamanders within the Tejon Pass region has been confirmed only from the vicinity of Fort Tejon, potential habitat is present at additional sites to the north.

Much of the Tehachapi Mountains and adjoining areas are inaccessible owing to a combination of rugged terrain and private ownership, and knowledge of the distribution of populations allied to *B. stebbinsi* is sketchy. Preliminary ground and aerial surveys indicate the presence of potential habitat in a number of unexplored canyons on the south side of Cummings Valley, the northwest slope of Bear Mountain (e.g., Clear and Sycamore creeks), elsewhere on the north slopes of the Tehachapi Mountains (e.g., Cedar, Chanac, and Tunis creeks), and possibly in some of the canyons on the north slopes of the San Emigdio Range (e.g., Black Bob Canyon) to the west of Fort Tejon.

3. Life History Features.

A. Breeding. Reproduction is terrestrial.

i. Breeding migrations. Unknown. Extensive surface movements within the breeding season, as reported for California slender salamanders (*B. attenuatus*; Anderson, 1960), seem implausible for this species given that most populations are associated with small, discrete patches of suitable habitat.

Onset of the fall/winter rainy season is especially unpredictable at the extreme southern end of the Sierra Nevada and adjacent Tehachapi Mountains. Periods of surface activity similarly vary from year to year. A possibly gravid female was discovered on 13 February in Caliente Canyon, and another was found on 1 April at Fort Tejon (T. Manolis, personal communication).

ii. Breeding habitat. Unknown.

B. Eggs.

i. Egg deposition sites. Nest sites have not been found for this species, but it seems likely that eggs are deposited deep within the rock talus/litter matrix characteristic of Tehachapi slender salamander microhabitat. Communal nests have been reported for other members of the *B. nigriventris* species group (Jockusch and Mahoney, 1997) and might be expected in Tehachapi slender salamanders.

ii. Clutch size. Unknown.

C. Direct Development. Timing of hatchling emergence is unknown.

D. Juvenile Habitat. Unknown how this may differ from adult habitats. Curiously, juvenile Tehachapi slender salamanders rarely are found, suggesting that hatching occurs in spring as surface activity declines and that juveniles may remain well underground.

E. Adult Habitat. Caliente Canyon, lying at the southern end of the Sierra Nevada, is situated in a moderately arid region. Salamanders are restricted to lower margins of north-facing slopes bordering Caliente Creek, as well as a few small side canyons, and are associated with granitic or limestone talus and scattered rocks. Vegetation here consists of foothill pine, interior live oak, canyon oak, blue oak, Fremont cottonwood, sycamore, and California buckeye (Brame and Murray, 1968). At more exposed locations, California juniper, yucca, bush lupine, and buckwheats grow. Substrates range from sandy-gravelly loam to decomposed granite.

In the canyons of the Tehachapi Mountains, Tehachapi slender salamander populations are likewise restricted to north-facing slopes, although they occur at higher elevations. Unlike the Caliente Canyon populations, where salamanders are nearly always associated with rocks,

the Tehachapi Mountains salamanders occur in areas of downed wood or talus.

F. Home Range Size. Unknown.

G. Territories. Unknown.

H. Aestivation/Avoiding Desiccation. Tehachapi slender salamanders occur in a region where the timing and amount of winter precipitation is erratic. Surface activity closely tracks the onset of the rainy season (generally November–December), and at lower elevations may be of brief duration (2–3 mo). In years of below average rainfall or consecutive years of drought (not unusual in this region) salamanders may not appear under surface cover at all. Although most of the winter precipitation occurs as rainfall, higher elevation portions of the range regularly receive snow.

Peak surface activity occurs in February–March, extending into April in wet years or early May at higher elevations (e.g., upper reaches of Pastoria and Tejon Creek drainages, Tehachapi Mountains). Field body temperatures for salamanders found under surface cover averaged 10.4 °C (range = 5.0–12.0 °C, n = 47; R. W. H., unpublished data).

I. Seasonal Migrations. Unknown.

J. Torpor (Hibernation). Most of the range of Tehachapi slender salamanders experiences below-freezing temperatures during the winter, and salamanders rarely are found under surface cover (i.e., they are likely underground) during such episodes.

K. Interspecific Associations/Exclusions. Yellow-blotched ensatinas (*Ensatina eschscholtzii croceater*) have been found at all Tehachapi slender salamander localities; they occupy similar habitats, but have a more extensive geographical, elevational, and ecological distribution. Within Caliente Canyon, Tehachapi slender salamanders and yellow-blotched ensatinas are the only salamanders present. However, black-bellied slender salamanders (*B. nigriventris*) range as far north as Comanche Point on the northern slopes of the Tehachapi Mountains, and a sight record for a *Batrachoseps* in the semi-arid, open oak woodland along Caliente Creek below Caliente Canyon (J. M. Brode, personal communication) may pertain to this species or possibly gregarious slender salamanders (*B. gregarius*).

Within the Tehachapi Mountains segment of the range, Tehachapi slender salamanders and black-bellied slender salamanders are sympatric in the Pastoria and Tejon Creek drainages, at Fort Tejon in Grapevine Canyon, and probably elsewhere (Jockusch, 1996; Wake and Jockusch, 2000). Although the ranges of these two species are broadly overlapping, Tehachapi slender salamanders are habitat specialists, confined to moist, north-facing slopes within canyons or ravines, and are associated with scattered rock, talus, or woody debris (Brame and Murray, 1968; Wake, 1996; R. W. H., unpublished

data). By contrast, black-bellied slender salamanders enjoy a broader distribution, occurring both in moist, oak-filled canyons, as well as in drier oak woodlands on open hillsides. Both species occasionally have been collected from under the same log. The extensive range overlap between Tehachapi slender salamanders and black-bellied slender salamanders in the Tehachapi Mountains is notable in that it represents the only case of sympatry involving members of the same species group of *Batrachoseps* (Wake and Jockusch, 2000).

Within the Tehachapi Mountains, yellow-blotched ensatinas occur at all known Tehachapi slender salamanders localities. However, at sites where we have found yellow-blotched ensatinas to be abundant, Tehachapi slender salamanders do not occur. Conversely, at 1,400 m elevation along Pastoria Creek, Tehachapi slender salamanders are locally abundant, and yellow-blotched ensatinas are present but in much lower numbers.

L. Age/Size at Reproductive Maturity. This is a relatively large species of *Batrachoseps*, with modest female-biased sexual size dimorphism. For a series of 10 large females, mean was 57.0 mm SVL (maximum SVL = 60.4 mm); 10 large males averaged 54.0 mm SVL (maximum SVL = 59.3 mm; both series from Caliente Canyon; data in part from Brame and Murray, 1968). Minimum age and size at sexual maturity are unknown.

M. Longevity. Unknown.

N. Feeding Behavior. Has not been described, although all *Batrachoseps* species observed thus far use a projectile tongue to capture small invertebrates.

O. Predators. In Caliente Canyon, a small adult ring-necked snake (*Diadophis punctatus*) was observed attempting to subdue an adult Tehachapi slender salamander by constriction. The salamander forced the snake to release its hold by moving into rock rubble, where the salamander escaped (Burkhardt et al., 2001).

P. Anti-Predator Mechanisms. Defensive behaviors may include coiling, immobility (crypsis), rapid crawling, and tail autotomy. These behaviors have been observed in populations throughout the range.

Q. Diseases. Unknown.

R. Parasites. Unknown.

S. Comments. For this account, we have included populations currently allocated to *B. stebbinsi* ranging from Caliente Canyon to Fort Tejon (Wake, 1996). However, high levels of genetic differentiation, as well as differences in coloration and size, between the two principal distributional units of *B. stebbinsi* (Caliente Canyon versus Tehachapi Mountains) strongly suggest that two species are represented (Jockusch, 1996; Jockusch and Wake, 2002; D.B. Wake and colleagues, unpublished data). Ecological information presented in this account has been delineated for the most part with future taxonomic changes in mind.

An earlier report (Richman, 1973) of *B. stebbinsi* from high elevation on the Kern Plateau of the southern Sierra Nevada (Tulare County) actually represented the first collection of the then-undescribed Kern Plateau salamander (*B. robustus*; Wake et al., 2002).

Although the description of *B. stebbinsi* as a new species was relatively recent (Brame and Murray, 1968), recognition of the existence of two species of *Batrachoseps* in the Tehachapi Mountains appears to have occurred as early as 1858 (Wake and Jockusch, 2000). The naturalist John Xántus, then stationed at Fort Tejon, reported in correspondence to the Smithsonian Institution (reprinted in Zwinger, 1986) that three species of salamanders occurred there, one of which was yellow-blotched ensatinas (undescribed at the time) and the other two obviously referable to *Batrachoseps* (Wake and Jockusch, 2000).

4. Conservation.
Populations of this narrowly distributed species occur on both private and public lands (the latter including portions of the Caliente Canyon range segment and administered by U.S. Bureau of Land Management) and face a variety of threats as noted above (see "Historical vs. Current Abundance" above). Although listed as Threatened by the State of California, Tehachapi slender salamanders probably warrant some measure of federal protection, given the near-term development pressures in the Tehachapi Mountains. Tehachapi slender salamanders are listed as a Forest Service Sensitive Species and a Federal Species of Concern.

Acknowledgments. We thank John Brode, Amy Kuritsubo, Tim Manolis, and Ted Papenfuss for sharing field observations.

Batrachoseps wrighti (Bishop, 1937)
OREGON SLENDER SALAMANDER

R. Bruce Bury

1. Historical versus Current Distribution.
Oregon slender salamanders (*Batrachoseps wrighti*) only occur in western Oregon from the south side of the Columbia River Gorge southward in the Cascade Mountains to southern Lane County. Populations typically are restricted to (but not exclusively) the west slopes of the Cascades (Kirk, 1991; Kirk and Forbes, 1991). Animals are found from about 15 m elevation (in the Columbia Gorge) to 1,340 m (Nussbaum et al., 1983; Leonard et al., 1993; Petranka, 1998).

2. Historical versus Current Abundance.
Declines have been associated with the clearcutting of ancient and mature forests (Nussbaum et al., 1983; Bury and Corn, 1988b). Oregon slender salamanders are

Oregon Slender Salamander *(Batrachoseps wrighti)*

considered by the State of Oregon to be a Sensitive Species of vulnerable status. Results of time-constrained searches (Gilbert and Allwine, 1991) showed that Oregon slender salamanders were second only to ensatinas in abundance in naturally regenerated stands: 1.26 salamander/site in old growth; 2.10 in mature; and 2.1 in young. Earlier, Bury and Corn (1998) found that Oregon slender salamanders were associated most closely with well-decayed pieces of coarse woody debris on forest floors, which is characteristic of old-growth and mature forests.

3. Life History Features.
A. Breeding. Reproduction is terrestrial.
i. *Breeding migrations.* Unknown and unlikely.
ii. *Breeding habitat.* Unknown. As judged from females with spermatophores, courtship and breeding take place in April, May, and early June (Nussbaum et al., 1983).

B. Eggs.
i. *Egg deposition sites.* Likely to be underground or deep within large, decaying logs. Ovipositing occurs from April–June.
ii. *Clutch size.* Average ovarian egg count is 6.3 (range = 3–11, n = 38; Nussbaum et al., 1983). Tanner (1953) found a clutch with eight eggs. Jockusch (1997b) and Jockusch and Mahoney (1997) report a mean of 5.1 ± 2.0 (s.d.) eggs with a maximum of 10 eggs for 22 clutches. Eggs are about 4 mm in diameter.

C. Direct Development.
i. *Brood sites.* Same as egg deposition sites. Incubation time is about 4.5 mo at

12 °C (Stebbins, 1949a). Jockusch (1997b) and Jockusch and Mahoney (1997) report a mean of 126 d ± 8 (s.d.) to hatching at 13 °C for 28 clutches.
ii. *Parental care.* Tanner (1953) found a female attending eight eggs.

D. Juvenile Habitat. Hatchlings measure 13–14 mm SVL (19–20 mm TL; Stebbins, 1949a; see also Petranka, 1998). Juvenile habitat is unknown, but is likely similar to adult habitat.

E. Adult Habitat. Adults are most often found within large, well-decayed logs and stumps in old growth and mature forests. Bury and Corn (1988b) report that they occur more often than expected in older decay classes of coarse woody debris, and most (62%) were in logs, under bask (26%), or on/under logs (7%). They have also been found on lava flows (Nussbaum et al., 1983; Petranka, 1998). They occasionally are found under large rocks that are moss covered and in stabilized talus (Leonard et al., 1993). Animals are active after snowmelt in the spring, but at other times of the year inhabit underground burrows or sites deep within large logs.

F. Home Range Size. Unknown.

G. Territories. Unknown. Individuals often are found clumped together, with two or more under the same object (Leonard et al., 1993). This suggests little territoriality. However, there are no detailed ecological studies on this species.

H. Aestivation/Avoiding Desiccation. Aestivation is unknown; animals likely avoid desiccating conditions by seeking shelter under cover objects or in burrows.

I. Seasonal Migrations. Unknown.

J. Torpor (Hibernation). Unlikely.

K. Interspecific Associations/Exclusions. Not documented.

L. Age/Size at Reproductive Maturity. Males mature at 33 mm SVL, females at 35. Adults typically measure 85–120 mm TL, with females about 12% greater in length (SVL; Petranka, 1998).

M. Longevity. Unknown.

N. Feeding Behavior. Prey include small invertebrates such as pseudoscorpions, centipedes, mites, spiders, snails, annelids, and insects such as dipterans and collembolans.

O. Predators. Undocumented, but likely to include reptiles, birds, and small mammals.

P. Anti-Predator Mechanisms. Oregon slender salamanders are secretive, and in the presence of potential predators they remain immobile or coil (Brodie, 1977). They may rapidly uncoil and flip away (Stebbins and Lowe, 1949).

Q. Diseases. Unknown.

R. Parasites. Unknown.

4. Conservation.
Oregon slender salamanders are listed as Protected by the State of Oregon (Levell, 1997), which means that a permit is required to possess or collect these animals.

Desmognathus abditus Anderson and Tilley, 2003
CUMBERLAND DUSKY SALAMANDER

Editor's note: The publication (Anderson and Tilley, 2003) describing Cumberland dusky salamanders *(Desmognathus abditus)* appeared as the present manuscript was being prepared for publication. I have extracted the pertinent life history and natural history data from this publication.

1. Historical versus Current Distribution.
Unknown. The current distribution of this recently described salamander includes the Cumberland Plateau of Tennessee, in Cumberland, Morgan, and Grundy counties (Anderson and Tilley, 2003). The northernmost known sites are just south of the Cumberland Mountains section of the Plateau in the vicinity of Wartburg, Morgan County, Tennessee. The southern- and westernmost known sites are on the southern ridge of Walden Mountian near Tracey City, Grundy County, Tennessee.

2. Historical versus Current Abundance.
Historical data are unavailable. Anderson and Tilley (2003) note how difficult it can be to locate these salamanders (and by extension to do population surveys), a feature contributing to the specific name of this species.

3. Life History Features.
A. Breeding.
i. Breeding migrations. Unknown.
ii. Breeding habitat. Unknown, gravid females have not been collected.

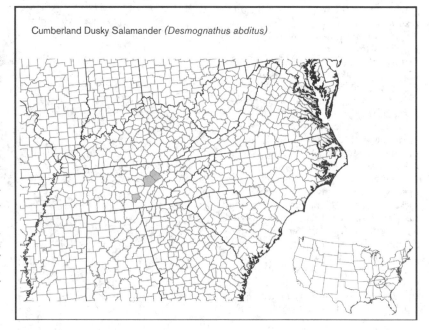

Cumberland Dusky Salamander *(Desmognathus abditus)*

B. Eggs.
i. Egg deposition sites. Unknown; eggs have not been observed or collected.
ii. Clutch size. Unknown, neither gravid females nor eggs have been discovered.

C. Larvae/Metamorphosis. Larvae have not been collected.

D. Juvenile Habitat. Likely to be the same as adults.

E. Adult Habitat. Specimens have been found "... under rocks along small streams and under moss and debris on vertical rock surfaces behind cascades. Most individuals have been located on land within one meter of surface water."

F. Home Range Size. Unknown.

G. Territories. Unknown.

H. Aestivation/Avoiding Desiccation. Unknown.

I. Seasonal Migrations. Unknown, but probably do not occur.

J. Torpor (Hibernation). Unknown.

K. Interspecific Associations/Exclusions. Cumberland dusky salamanders are sympatric with larger desmognathines, including northern dusky salamanders *(D. fuscus)*, seal salamanders *(D. monticola)* and Black Mountain salamanders *(D. welteri)*. They will hybridize with Allegheny Mountain dusky salamanders *(D. ochrophaeus)* "at the escarpment of the Cumberland Mountains in the vicinity of Frozen Head, just north of Wartburg, Morgan County," Tennessee (Anderson and Tilley, 2003). They will also hybridize with Ocoee salamanders *(D. ocoee)* along Walden Ridge, near Tracy City, Grundy County, Tennessee (Anderson and Tilley, 2003).

L. Age/Size at Reproductive Maturity. Unknown.

M. Longevity. Unknown.

N. Feeding Behavior. As with other species of *Desmognathus*, prey likely include leaf litter invertebrates.

O. Predators. Unknown, but individuals could be preyed upon occasionally by large sympatric *Desmognathus*, other salamanders, snakes, small mammals, and possibly species of birds that forage in leaf litter.

P. Anti-Predator Mechanisms. Anderson and Tilley (2003) write: "When disturbed, they often run rapidly toward water and are extremely adept at escape."

Q. Diseases. Unknown.

R. Parasites. Unknown.

4. Conservation.
Maintaining intact habitat for Cumberland dusky salamanders should be a top priority. This recently described species has had no studies done on population size or viability; they do not receive protection by the state of Tennessee or by the federal government.

Desmognathus aeneus Brown and Bishop, 1947
SEEPAGE SALAMANDER
Julian R. Harrison

1. Historical versus Current Distribution.
Seepage salamanders *(Desmognathus aeneus)* are distributed locally in southwestern North Carolina, southeastern Tennessee, northern Georgia, and north-central Alabama (Harrison, 1992; Conant and Collins, 1998; Petranka, 1998). Western populations in the Fall Line Hills region of Alabama are separated by an apparent hiatus from eastern populations of that state in the Blue Ridge and adjacent Piedmont regions. Another apparently disjunct population is present in the Piedmont of northeastern Georgia. Camp and Payne (1996) recorded a slight extension of the range in Georgia,

and Livingston et al. (1995) provided the first record for the species in northwestern South Carolina. Despite statements to the contrary (Harrison, 1967; Petranka, 1998), seepage salamanders do occur north of the Little Tennessee River, as a population is known in the Hazel Creek area of Great Smoky Mountains National Park (R. Highton, personal communication; see also Dodd, 2004).

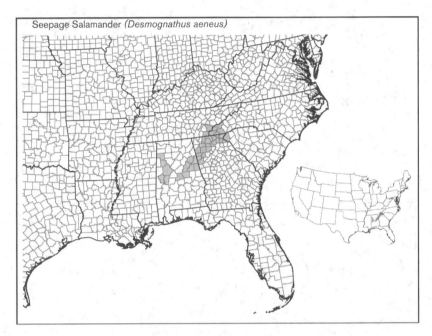

Seepage Salamander (*Desmognathus aeneus*)

2. Historical versus Current Abundance.
Few historical data are available. Folkerts (1968) found seepage salamanders to be abundant within their habitat in Alabama. For example, in an estimated 65 square feet of suitable habitat in Clay County, 131 specimens were collected. The writer's experience with this species in Georgia and North Carolina during 1958–62 was similar. The salamanders were common to abundant at virtually all localities visited (Dodd, 2004). On one occasion in Union County, Georgia, 64 specimens were collected within a 2-hr period from beneath leaf litter and mosses along a small stream.

3. Life History Features.
A. Breeding. Reproduction is terrestrial.
i. Breeding migrations. Unknown as such, but females move from beneath leaf litter retreats to the vicinity of streams and seepage areas to lay eggs. Courtship and mating behavior are unknown, except in laboratory studies (Promislow, 1987). There is a striking similarity between the courtship of this species and that of pigmy salamanders (*D. wrighti*; Verrell, 1997, 1999).

In Alabama, clutches of eggs are found from early February to late May, but most oviposition in west Alabama occurs in February and March (Valentine, 1963c; Folkerts, 1968). In east Alabama, most clutches

are found in April, and there is also a fall oviposition period with egg deposition occurring from mid-July to early October (Folkerts, 1968). In Georgia and North Carolina, oviposition occurs during late April to early May; hatching takes place from late May to early August (Harrison, 1967). In Tennessee, oviposition occurs from late April to early May and hatching from mid-June to mid-July (Jones, 1981).

ii. Breeding habitat. Forest floor in seepage areas.

B. Eggs.
i. Egg deposition sites. Eggs are deposited under mosses, logs, leaf litter, and root mats or other objects in seepage areas or near streams (Valentine, 1963c; Harrison, 1967; Folkerts, 1968; Jones, 1981).

ii. Clutch size. Females lay 3–19 eggs (Bishop and Valentine, 1950; Harrison, 1967; Folkerts, 1968; Jones, 1981; Beachy, 1993a; Collazo and Marks, 1994). For a description of eggs, see Brown and Bishop (1948); for a staging table, see Marks and Collazo (1998).

C. Direct Development. The developmental period of laboratory clutches lasts an estimated minimum of 60 d (Valentine, 1963c) or 34–45 d (Harrison, 1967). In field clutches in Tennessee, approximately 45 d elapsed between the earliest date observed for uncleaved eggs and the earliest date observed for hatching (Jones, 1981).

D. Juvenile Habitat. Same as adults. This species is terrestrial; there is no free-living aquatic larval stage. Seepage salamanders were included in a key to salamander larvae and larviform adults (Altig and Ireland, 1984), probably because of the presence of short, unpigmented gills and a partial dorsal fin in some hatchlings.

Though, as the authors note, they probably never enter the water.

E. Adult Habitat. Adults are terrestrial and live at the interface between the leaf or leaf mold layer and the underlying soil in the vicinity of seepages and small streams in heavily shaded hardwood or mixed forests (Harrison, 1967, 1992; Folkerts, 1968; Jones, 1981; Hairston, 1987; Bruce, 1991). When active, adults are nocturnal (Brandon and Huheey, 1975), typically remain under surface cover, and are not known to climb or actively burrow (Harrison, 1967; Folkerts, 1968; Jones, 1981; and Hairston, 1987). One apparent exception to the lack of scansorial activity by this species was given by Wilson (1984), who observed seepage salamanders climbing on grasses and bushes and jumping from branch to branch at a locality in central Alabama. Harper and Guynn (1999) studied the factors affecting salamander density and distribution within four forest types in the Southern Appalachian Mountains. They found that seepage salamanders and other salamanders preferred moist microsites within each forest type, with the highest densities occurring on sites with a northern and/or eastern exposure and within northern hardwood forests. They noted that densities were lowest on 0–12-yr plots but equal on 13–39 and ≥40-yr plots, suggesting a quicker recovery from clearcutting than reported by previous researchers. As plots with salamanders had significantly higher numbers of snails than plots without them, the authors also suggested that snails may be a necessary source of calcium for salamanders and may have a substantial impact on salamander distribution.

F. Home Range Size. Unknown.

G. Territories. Unknown. However, under laboratory conditions seepage salamanders displayed little or no aggression toward conspecifics or heterospecific Allegheny Mountain dusky salamanders (*D. ochrophaeus*) and pigmy salamanders (Peele, 1992).

H. Aestivation/Avoiding Desiccation. Seepage salamanders do not aestivate; major activity occurs in the spring and summer, but individuals can be observed during any time of the year.

I. Seasonal Migrations. Unknown, but probably do not occur. The possible exception is the movements of females to oviposition sites within seepage areas in late spring and early summer.

J. Torpor (Hibernation). Unknown.

K. Interspecific Associations/Exclusions. Seepage salamanders are sympatric with pigmy salamanders in the vicinity of Deep Gap and Wayah Bald, Macon County, North Carolina, and the two species also occur in close proximity in forest floor habitats along the Nantahala River, North Carolina (Tilley and Harrison, 1969; Rubin, 1971; Bruce, 1991). At most known localities within their range, seepage salamanders coexist with one to several other congeneric species, though not completely syntopically. Where studied in the Nantahala and Unicoi mountains, they occupy the forest floor habitat, occurring farther from streams or seepages than other species (Hairston, 1973; Jones, 1981). Seepage salamanders are extremely secretive, apparently unaffected by competitive or predatory interactions with sympatric, congeneric seal salamanders (*D. monticola*) and Allegheny Mountain dusky salamanders (Hairston, 1987). Hairston speculated that this is perhaps attributable to an original predatory relationship that is no longer detectable. Sever et al. (1976) list seepage salamanders as an associate of Junaluska salamanders (*Eurycea junaluska*) in Graham County, North Carolina.

L. Age/Size at Reproductive Maturity. Both sexes reach sexual maturity at 18–19 mm SVL (measured to anterior margin of vent) at 2 yr (Harrison, 1967).

M. Longevity. A wild-caught specimen in the Cincinnati Zoo survived 4 yr, 14 d (Bowler, 1977). Survivorship in nature is unknown.

N. Feeding Behavior. Arthropods are the principal foods, primarily insects, but arachnids, isopods, amphipods, centipedes, millipedes, nematodes, earthworms, and land snails are also eaten (Folkerts, 1968; Donovan and Folkerts, 1972; Jones, 1981). Donovan and Folkerts (1972) also reported the occurrence of a recently hatched seepage salamander in the stomach of a small adult male. Most of the items consumed are leaf litter species, indicating the confinement of foraging activity to that microhabitat rather than the forest floor surface (Jones, 1981).

O. Predators. Unknown, but individuals could be preyed upon occasionally by large sympatric *Desmognathus* of other species, spring salamanders (*Gyrinophilus* sp.), ring-necked snakes (*Diadophis punctatus*), and possibly species of birds that forage in leaf litter. Folkerts (1968) listed ring-necked snakes as an associate of seepage salamanders in his study of Alabama populations.

P. Anti-Predator Mechanisms. Most individuals attempt predator avoidance by remaining immobile when exposed, primarily in a linear posture (Dodd, 1990a; personal observation).

Q. Diseases. Unknown.
R. Parasites. Unknown.

4. Conservation.
There is evidence that logging activities are responsible for the extirpation of some Alabama populations (Folkerts, 1968). In that state, seepage salamanders are currently ranked as S2, Imperiled because of rarity (6–20 populations) or vulnerability to extirpation (Alabama Natural Heritage Program, 1996). Approximately half of the populations known in 1976 no longer exist, primarily because of various forestry practices (G.W. Folkerts, personal com-

munication). The species' present status in North Carolina is SR (Significantly Rare) with a rank of S3, Rare or Uncommon (21–100 extant populations; LeGrand and Hall, 1995). At present, however, populations appear to be stable and the species is common locally (R.C. Bruce, personal communication). Hairston and Wiley (1993) conducted identical observations 1–4 times per year for 15–20 yr at two locations in the southern Appalachians that showed fluctuations in numbers but no trend among several salamanders species, including seepage salamanders. Georgia accords seepage salamanders no legal status but ranks them S3 (Georgia Natural Heritage Program, 1996). Three populations in Stephens and Union counties, Georgia, monitored by Carlos Camp (personal communication) appear to be stable, but these have incidental protection either because they exist in areas set aside as special botanical areas or occur adjacent to the Appalachian Trail. In Ten-

nessee, seepage salamanders are regarded as a species in need of management and are ranked S1, "Extremely Rare and Critically Imperiled with five or fewer occurrences, or very few remaining individuals, or because of some special condition where the species is particularly vulnerable to extinction" (Withers, 1996). The U.S. Fish and Wildlife Service (1994c) listed seepage salamanders as a Category 2 candidate for listing. However, a more recent list (USFWS, 1996a) removed that classification.

Maintenance of wide buffer zones adjacent to seepages and streams would protect this species' habitat (Wilson, 1995).

Desmognathus apalachicolae Means and Karlin, 1989
APALACHICOLA DUSKY SALAMANDER

D. Bruce Means

1. Historical versus Current Distribution.
Apalachicola dusky salamanders (*Desmognathus apalachicolae*) have a relatively small geographic distribution in the Coastal Plain of the southeastern United States. The entire range of the species is confined principally to tributaries of the Apalachicola and Chattahoochee rivers below the Fall Line at Columbus, Georgia, but populations are alsoknown from the Chipola River basin in Florida, the Ochlockonee River basin in Georgia and Florida, and from the upper Choctawhatchee River drainage in Alabama (Means and Karlin, 1989; Means, 1993). So far as known, the species' current distribution is identical with its historical range.

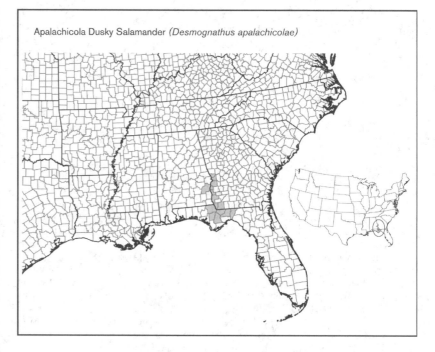

Apalachicola Dusky Salamander *(Desmognathus apalachicolae)*

2. Historical Versus Current Abundance.
Currently abundant in two types of deep, moist ravines. One type is the classic gully-eroded ravine in sandy clay "red hills" soils, which are common all along the Chattachoochee River from Columbus to Fort Gaines in Georgia and on the eastern side of the Apalachicola River from the town of Chattahoochee to Big Sweetwater Creek in Florida (Means and Karlin, 1989). The second type of ravine, called a *steephead*, is found in a wide band of deep Pleistocene sands in parts of the Chipola, Apalachicola, and Ochlockonee rivers (Means, 1991). Steepheads are unique in having permanent, copious, cold-water

(20–21 °C) flow from numerous hillside seepages (Means, 1975a, 2000). Historical abundance is probably not much different from that of today because principal habitats are in high relief terrain and many are protected by state parks and The Nature Conservancy.

3. Life History Features.

A. Breeding. Reproduction is semi-aquatic. Adults lay eggs on land near water and brood them. Eggs hatch and larvae are aquatic.

i. Breeding migrations. Probably not present in this territorial species.

ii. Breeding habitat. Same as adult habitat.

B. Eggs.

i. Egg deposition sites. Beneath stones, small logs, and other cover, or in or along seeps. Eggs have been found in the field from the third week in May to the second week in November. Females presumably brood (Means and Karlin, 1989).

ii. Clutch size. Unknown.

C. Larvae/Metamorphosis. No specific studies of the larval development of Apalachicola dusky salamanders have been conducted.

i. Length of larval stage. From 9–10 mo. Larvae have been found from the second week in July to the second week in March (Means and Karlin, 1989).

ii. Larval requirements.

a. Food. Unknown, but larvae are undoubtedly carnivorous.

b. Cover. Larvae are found in organic litter such as twigs, leaf fragments, and pieces of bark lying on seepage slopes (personal observations).

iii. Larval polymorphisms. Unknown.

iv. Features of metamorphosis. Unknown.

v. Post-metamorphic migrations. Unknown.

vi. Neoteny. Not present. All larvae metamorphose into the fully terrestrial morphology (personal observations).

D. Juvenile Habitat. Similar to adult habitats in mesic ravines except that juveniles are excluded from the choice microhabitats by the aggressive, territorial adult males (personal observations). Juveniles are found under litter and other debris, rocks when they occur, and wet organic matter in mesic ravines in or adjacent to seepages and water (personal observations).

E. Adult Habitat. The principal habitats are first-order streams in highly relieved areas—gully-eroded ravine valley heads or steepheads (Means, 2000). Adults are found in daytime under suitable debris including logs, rocks when they occur, and leaf packs in mesic ravines. Individuals usually are situated at the ground/water/air interface lying under debris with their bodies immersed in water and their heads protruding into the air column. Debris at the edge of water is best. When suddenly exposed, Apalachicola dusky salamanders usually dive into water to escape. At night, adults are often found out foraging in leaf litter or over the moist soil at the edge of seepages and on the banks of small streams. Individuals are rarely found >2 m from water.

F. Home Range Size. Not known.

G. Territories. Adult males are almost never found under debris with conspecifics, except for an occasional adult female (personal observations). Behavioral experiments indicate that adult males actively defend their daytime refugia from conspecific males by chasing and biting (unpublished data). This behavior suggests that the species is highly territorial.

H. Aestivation/Avoiding Desiccation. Apalachicola dusky salamanders are found least commonly during daytime in midsummer, when air temperatures are hot, and especially during regional droughts. It is presumed that they retire to deeper recesses in the ground during these times, although individuals are active on warm summer nights, especially during and after rains.

I. Seasonal Migrations. Not known, but probably do not occur in this relatively sedentary, territorial species.

J. Torpor (Hibernation). Individuals are sluggish during winter when found in cold microhabitats, but become active immediately upon warming in one's hand. In microhabitats bathed by seepage water, which is relatively warm in midwinter, individuals are as active as at other times of the year (personal observation).

K. Interspecific Associations/Exclusions. Means (1974), Karlin and Guttman (1986), and Verrell (1990c) demonstrated that Apalachicola dusky salamanders are most closely related to Allegheny mountain dusky salamanders *(D. ochrophaeus)* of the southern Appalachian Mountains, but the two species are allopatric in the Chattahoochee River drainage in Georgia. Means (1975a) used parapatric distributional relationships to demonstrate that Apalachicola dusky salamanders probably competitively exclude southern dusky salamanders *(D. auriculatus)* from their habitats. Wherever the two species are sympatric geographically, Apalachicola dusky salamanders occupy first- and second-order ravines, while southern dusky salamanders live downstream in larger order, swampier reaches of drainages (Means, 1974, 1975a). Two other plethodontid salamanders, red salamanders *(Pseudotriton ruber)* and southern two-lined salamanders *(Eurycea cirrigera)*, are always allotopic with Apalachicola dusky salamanders (Means, 2000).

L. Age/Size at Reproductive Maturity. The smallest males with two lobes/testis measured 40 mm SVL, but smaller males with one lobe/testis can be sexually mature; the smallest gravid female measured 33.0 mm SVL (Means, 1974, 1993). Age at sexual maturity may be 2 yr for males and 3 yr for females, but a definitive study has not been published.

M. Longevity. Males with three lobes/testis are common in populations, occupying the larger size classes of males. Allowing for 1 yr to metamorphosis and 1 yr for each lobe/testis, some males readily may reach 4 yr of age. There is no physiological/anatomical size-correlated marker for females, but there is no obvious reason why females do not live as long as males. These estimates are probably conservative.

N. Feeding Behavior. No food studies have been published.

O. Predators. Potential predators often found in the habitat of Apalachicola dusky salamanders are copperheads *(Agkisrodon contortrix)*, cottonmouths *(A. piscivorus)*, plainbelly water snakes *(Nerodia erythrogaster)*, southern water snakes *(N. fasciata)*, eastern ribbon snakes *(Thamnophis sauritus)*, snapping turtles *(Chelydra serpentina)*, red salamanders, two-toed amphiumas eels *(Amphiuma means)*, green frogs *(Rana clamitans)*, and mammals including shrews (Soricidae), opossums *(Didelphis virginiana)*, raccoons *(Procyon lotor)*, gray fox *(Urocyon cinereoargenteus)*, river otters *(Lutra canadensis)*, and armadillos *(Dasypus novemcinctus)*.

P. Anti-Predator Mechanisms. Apalachicola dusky salamanders use cryptic coloration to go unnoticed by their predators. There is a high degree of intraspecific variation and sexual dimorphism in color pattern (Means, 1974). Adult males become uniformly dark brown in color, but females retain the juvenile pattern of bright dorsal blotches. Color pattern varies with ontogeny, and individuals also exhibit metachrosis to effect background color matching (Means, 1974). As with other members of the genus, Apalachicola dusky salamanders can flip themselves out of a grasp by the action of strong trunk, leg, and tail muscles. The hind legs are twice the thickness of the front legs, allowing individuals to make impressive leaps.

Q. Disease. None recorded.

R. Parasites. None recorded.

4. Conservation.

Apalachicola dusky salamander populations seem to be stable throughout their range. This is probably due, in part, to their proclivity for deep, shaded, wet ravines which are unsuitable for human development. Pollution from stormwater runoff, however, could be a threat to water quality and the integrity of streamside microhabitats, but few towns exist in the rugged topography in which their ravines are located. Apalachicola dusky salamanders are potentially protected in Torreya State Park and the Apalachicola Bluffs and Ravines Preserve of The Nature Conservancy in Florida, and in Kolomoki Mounds State Park in Georgia.

Desmognathus auriculatus (Holbrook, 1838[b])
SOUTHERN DUSKY SALAMANDER
D. Bruce Means

1. Historical versus Current Distribution.
The geographic distribution of southern dusky salamanders *(Desmognathus auriculatus)* has been mapped according to museum records and the literature (Means, 1999a). This distribution includes the Coastal Plain of the southeastern United States from extreme southeastern Virginia to mid-peninsular Florida, then west through the Florida Panhandle to the Escambia River and barely entering southern Alabama, except at Florala and Mobile Bay. There appears to be a hiatus in the range between the Escambia and Perdido rivers, but the species ranges westward from Mobile Bay into southern Mississippi, Louisiana, and eastern Texas to the Trinity River basin. Many areas should be further investigated, such as northeastern Florida east of the St. Johns River; Escambia and Santa Rosa counties, Florida; southern Alabama; southern Louisiana; and Texas. Because there has been an alarming decline in the number of historical localities throughout the range of the species (Dodd, 1998; Means, submitted), it may not be possible now to find southern dusky salamanders where they once might have ranged.

salamander in their habitat, but Dodd (1998) and Means (submitted) have reported wholesale extinctions in > 50 sites that appear to be relatively undisturbed. Likewise, the species has declined dramatically in Georgia, South Carolina, and North Carolina and may have been extirpated entirely from Virginia (Means, submitted).

3. Life History Features.
    A. Breeding. Reproduction is semi-aquatic. Eggs are laid and incubated on land. Hatchlings move into water and larvae are aquatic.
    i. Breeding migrations. Probably do not occur. In >30 yr of road cruising on rainy nights at all seasons, only one specimen was ever found crossing a highway at a creek swamp inhabited by the species (personal observation).
    ii. Breeding habitat. Same as adult habitat.
    B. Eggs.
    i. Egg deposition sites. Females lay eggs in grape-like clusters in cavities or small depressions. Nests also have been found in *Sphagnum* moss, in cypress logs, beneath logs and bark, and in stumps within 1–2 m from water (see Petranka, 1998). Eggs with developing embryos were found in the field from 4 September–12 October in Alabama (Folkerts, 1968); from early September to late October in North Carolina (Robertson and Tyson, 1950; Eaton, 1953), and on 28 October in Florida

eggs in late October in north-central Florida, and eggs were estimated to have been deposited in Devil's Millhopper in north-central Florida from August–December (Dodd, 1998). Tiny hatchlings (9.2–10.1 mm SVL) were found on 7 October in Devil's Millhopper and late October in North Carolina (Eaton, 1953).
    ii. Clutch size. From 9–26 eggs (Robertson and Tyson, 1950; Goin, 1951; Wood and Clarke, 1955).
    C. Larvae/Metamorphosis. Older larvae have been found in all months except March, July, and September at Devil's Millhopper (Dodd, 1998), and close by at other localities in October, December, and January (Neill, 1951c). Metamorphosis was reported in January in Louisiana (Chaney, 1949) and January and February in North Carolina (Eaton, 1953).
    D. Juvenile Habitat. Same as for adults.
    E. Adult Habitat. Southern dusky salamanders have been collected in shallow water under logs along the edge of a creek in Texas (Dial, 1965); in swampy muck-lands or ravines, especially around seepages, beneath logs and debris, and in shallow water or adjacent to the water in Louisiana (Dundee and Rossman, 1989); in muddy, bottomland swamps and sloughs in Mississippi (Valentine, 1963a); from under debris at the edge of mucky floodplain sloughs, at mucky edges of swampy lakes, and in other sites of muck or wet peat associated with black waters in Florida; when in steephead ravines, they were found only where organic debris had accumulated along seepages (Means, 1974, 2000). In Alabama, they have been found at the edges of cypress ponds and sloughs, swampy places, and along the edges of springs and spring-runs under leaf mold, logs, or mats of marginal aquatic vegetation (Folkerts, 1968; Means, 1986d). In Georgia, the species was common in mucky, swampy black-water streams of the Lower Coastal Plain (Means, 1974). Martof et al. (1980) reported southern dusky salamanders to be abundant under leaf litter and rotten logs in swamps and bottomland forests in the Carolinas and Virginia; Eaton (1953) found a female with unhatched eggs and hatchlings under a log in wet mud in a cypress swamp in North Carolina; and Wood and Clarke (1955) found a nest and attending female in a small crevice beneath a cypress limb embedded in the mud in a dense cypress forest in Virginia.
    F. Home Range Size. Unknown.
    G. Territories. Because some congeners are highly territorial and southern dusky salamanders do not migrate, it is possible that adult males, at least, defend territories or choice hiding places, but no studies have been conducted.
    H. Aestivation/Avoiding Desiccation. During droughts, which may occur at any time of the year, individuals may seek refuge by burrowing deeply into peat

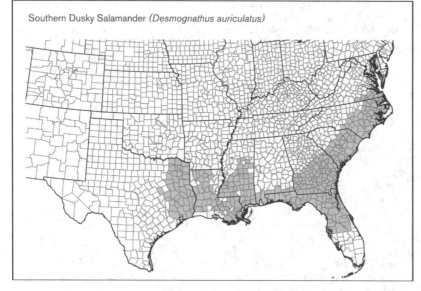

Southern Dusky Salamander *(Desmognathus auriculatus)*

2. Historical Versus Current Abundance.
Chaney (1949) found southern dusky salamanders abundant in the Florida Parishes of Louisiana, but Dundee and Rossman (1989) and Jeff Boundy (in Means, submitted) report that they have disappeared or severely declined in known sites there. In Florida, southern dusky salamanders were the most abundant

(Goin, 1951). Chaney (1949) reported egg clutches in many stages of development on 26 September in Louisiana, indicating to him that the eggs were laid in late August or early September. The mid-June record from near Augusta, Georgia, probably is for northern dusky salamanders *(D. fuscus;* Neill and Rose, 1949). Neill (1951c) reported a female guarding her

(personal observation). Whether any specific physiological changes take place during these times is unknown.

I. Seasonal Migrations. Unknown, but in swampy flatwoods and river floodplains during drought, individuals may burrow deep into peat or follow the edge of receding or rising water (personal observation).

J. Torpor (Hibernation). Neill (1948b) found this species to be active in air temperatures of −2–0 °C (28–32 °F) in Richmond County, Georgia. In northern Florida steepheads, in which the preferred microhabitats are kept constantly at the temperature of the seeping groundwater (19–21 °C; 67–70 °F), specimens normally were active when found throughout the winter; in swampy habitats subject to extremes of temperature, southern dusky salamanders were not so commonly found as in spring and fall and were cold and sluggish when handled (personal observations).

K. Interspecific Associations/Exclusions. Means (1975a) demonstrated through geographic and habitat distributional analysis that the southern dusky salamander is probably competitively excluded from first-and second-order ravine habitats by Apalachicola dusky salamanders (*D. apalachicolae*) and spotted dusky salamanders (*D. conanti*) in western Florida. Southern dusky salamanders are almost always associated with mud salamanders (*Pseudotriton montanus*), dwarf salamanders (*Eurycea quadridigitata*), three-lined salamanders (*E. guttolineata*), many-lined salamanders (*Stereochilus marginatus*), and one-toed amphiumas (*Amphiuma pholeter*) where they are sympatric with these species (Means, 1974, 2000).

L. Age/Size at Reproductive Maturity. In Louisiana, Chaney (1949) reported that males attain sexual maturity in their second or third year after hatching; the majority of females reach sexual maturity in their third year of growth, but some may do so in their second year. In Florida, one population of males reached sexual maturity by 44 mm SVL, and females were gravid by 39 mm SVL (Means, 1974).

M. Longevity. Wild-caught southern dusky salamanders have been maintained in captivity nearly 7 yr (Snider and Bowler, 1992).

N. Feeding Behavior. Carr (1940a) reported aquatic beetle larvae, lumbricid worms, beetles, tabanid larvae, lycosid spiders, and tipulid larvae. Folkerts (1968) found larval and adult insects, arachnids, and annelids in the stomachs of eight specimens.

O. Predators. Only banded water snakes (*Nerodia fasciata*) and red-fin pickerels (*Esox americanus*) have been reported as predators, but Means (submitted) implicated feral pigs (*Sus scrofa*). Oophagy and cannibalism were reported by Chaney

(1949), Rose (1966c), and Wood and Clarke (1955).

P. Anti-Predator Mechanisms. Individuals thrash about and dive into black, saturated organic soils when exposed, disappearing quickly down crayfish burrows or among masses of rootlets and wet peat (personal observations).

Q. Diseases. None known, except that the widespread and pervasive disappearance of southern dusky salamanders throughout their range since the mid 1970s may have been caused by some unknown pathogen (Means, submitted).

R. Parasites. Chigger mites were found on southern dusky salamanders in Texas (Loomis, 1956).

4. Conservation.
No detailed population surveys or studies of the ecology of southern dusky salamanders have been undertaken since the work of Cook and Brown (1974) and Means (1974, 1975a), until Dodd (1998) and Means (submitted) discovered that the species had experienced a widespread decline throughout its range beginning about the mid-1970s. Considering how widespread and severe this decline has been (Means, submitted), it is impera-

Desmognathus brimleyorum Stejneger, 1894
OUACHITA DUSKY SALAMANDER
D. Bruce Means

1. Historical versus Current Distribution.
The current distribution of Ouachita dusky salamanders (*Desmognathus brimleyorum*) is highland areas south of the Arkansas River, including all the Ouachita Mountains and Petit Jean Mountain in western Arkansas; Rich, Winding Stair, Blackfork, and Kiamichi mountains, and the Potato Hills in eastern Oklahoma (Means, 1999b). This species is also found in small streams in the rugged topography south of Kiamichi Mountain along the Arkansas–Oklahoma border to Beaver Bend State Park in McCurtain County, Oklahoma, and also southeast of Little Rock, Arkansas (Means, 1999b). They have not been found on Iron Fork Mountain, the westernmost of the Ouachita Mountains in Oklahoma, but Chaney (1958) reported the only population known north of the Arkansas River at Russellville in Arkansas. It is not known whether the historical distribution was different from the current distribution.

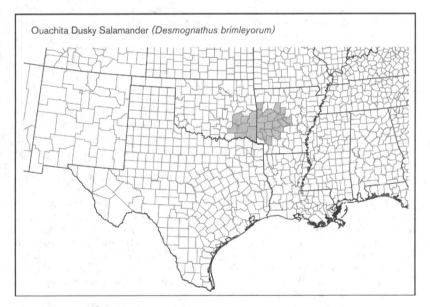

Ouachita Dusky Salamander (*Desmognathus brimleyorum*)

tive that studies be undertaken to determine the cause of the decline. Also, surveys are urgently needed to ascertain the population status in every state in the geographic distribution of southern dusky salamanders. Because the disappearance of many local populations has taken place in relatively undisturbed, and in some cases pristine, native habitat (Dodd, 1998; Means, submitted), management recommendations must await determination of the cause(s) of the decline.

2. Historical versus Current Abundance.
No studies have been undertaken specifically to assess changes over time of known populations, but three robust populations on Rich Mountain in the 1970s were nearly devoid of salamanders in 1998 (personal observation). The reason, however, may have been a severe drought that had dried up all surface water and may have caused Ouachita dusky salamanders to retreat underground into the interstices of the rocky soils there. The practice of harvesting

Ouachita dusky salamanders for fish bait has increased in the past two decades (personal observation) and could have a severe effect on local populations.

3. Life History Features.

A. Breeding. Reproduction is semi-aquatic. Eggs are laid on land and incubated by the female. Hatchlings move into water, and larvae are strictly aquatic.

    i. Breeding migrations. Unknown, but see below. Oviposition occurs in late June and early July on Rich, Kiamichi, and the Cossatot mountains (Means, 1975b). Strecker (1908a) reported eggs in late August and early September near Hot Springs, Arkansas. Chaney (1958) found eggs in late stages of development at Russellville, Arkansas, on 12 October. Trauth (1988) and Taylor et al. (1990) reported finding eggs in July and August on Fourche, Winding Stair, and the Cossatot mountains, and Trauth (1988) described three stages of development in three egg clutches he excavated on Rich Mountain. Means (1975b) reported hatching on Rich, Kiamichi, and the Cossatot mountains in the second week of September, when 42 of 92 clutches collected at that time hatched immediately or after only a few minutes post-collection; the remaining 50 clutches were about 3/4 developed. No egg clutches were found after late November (Means, 1975b). Hatching was reported on 12 October in the Russellville, Arkansas, population, but large larvae were also present and common, indicating that many had hatched sometime earlier (Chaney, 1958).

    ii. Breeding habitats. Same as adult habitat.

B. Eggs.

    i. Egg deposition sites. Eggs are laid most commonly under the surface of rocks, sometimes totally immersed in water and sometimes in air, and always with a brooding female (Chaney, 1958; Means, 1975b). Trauth (1988) reported on three egg masses he excavated 0.5 m deep near the water table of a dry seepage site in mud during a summer drought. The clutches were deposited in grape-like clusters on the ceiling of mud chambers made by females.

    ii. Clutch size. Reported numbers range from 20–29 eggs (Trauth, 1988; Taylor et al., 1990; Trauth et al., 1990).

C. Larvae/Metamorphosis.

    i. Length of larval stage. Perhaps 1 yr (Means, 1974; Trauth et al., 1990).

    ii. Larval requirements.

    a. Food. Unknown, but larvae are undoubtedly carnivorous.

    b. Cover. Larvae use interstitial spaces among gravel, rocks, and organic litter in the streambed.

    iii. Larval polymorphisms. None known.

    iv. Features of metamorphosis. Unknown.

    v. Post-metamorphic migrations. Unknown.

    vi. Neoteny. All larvae completely metamorphose into morphologically terrestrial adults.

On Rich, Kiamichi, and the Cossatot mountains, collections in March and April contained a larger percentage of larvae than newly metamorphosed animals; by the end of June, metamorphosed animals made up the youngest age/size class, and only a few larvae remained (Means, 1975b). Trauth et al. (1990) stated that by mid-August, transformation had occurred. Metamorphosis occurred from 15 July–1 September in Russellville (Chaney, 1958).

D. Juvenile Habitat. Same as adult habitat, but animals are found wedged in between smaller crevices in gravel and rocks (Means, 1974, 1975b). Post-metamorphic growth was described by Chaney (1958) in the Russellville population and Means (1975b) from Rich, Kiamichi, and the Cossatot mountains.

E. Adult Habitat. Ouachita dusky salamanders are found most often and abundantly in first- and second-order mountain streams and adjacent ravine woodlands, especially where there is a rocky, gravelly, porous substrate. Rocky, coarse-gravelled stream benches, islands, and cascades are most densely populated. Where hillsides near streams are composed of coarse, rocky talus, adults may be found at night up to several tens of meters away from the streamside with their heads and anterior bodies protruding from crevices between rocks (Means, 1974, 1975b). Animals have also been collected from beneath rocks and from the banks of a spring (Chaney, 1958).

F. Home Range Size. Unknown, but densities during oviposition season (late June to July) ranged from 1.8 to an astounding 61.7 individuals/m^2 of habitat (Means, 1975b).

G. Territories. Probably exist in this sedentary species, but have not been studied.

H. Aestivation/Avoiding Desiccation. During droughts when surface water dries up in mountain streams, Ouachita dusky salamanders become difficult to find, presumably because they move downward into rocky substrata and talus slopes to avoid desiccation (personal observations). It is not known whether physiological changes take place during these periods.

I. Seasonal Migrations. At night, gravid females were observed moving over a wooded talus hillside and a gravel road on Rich Mountain and across a paved, macadam road in the Cossatot Mountains, moving both away and toward streamsides and up and down slopes (Means, 1975b).

J. Torpor (Hibernation). Specimens collected in midwinter were sluggish because their bodies were immersed in near-freezing water, but they became active when warmed (personal observation).

K. Interspecific Associations/Exclusions. Ouachita dusky salamanders are often found in mountain brooks with many-ribbed salamanders (*Eurycea multiplicata*), and the adjacent terrestrial habitat usually contains northern slimy salamanders (*Plethodon glutinosus*) and either Caddo Mountain salamanders (*P. caddoensis*) or Rich Mountain salamanders (*P. ouachitae*) (personal observation).

L. Age/Size at Reproductive Maturity. Males and females at Russellville, Arkansas, attain sexual maturity in their fourth year of growth and had attained a minimum SVL of 50.7 mm and 52.0 mm, respectively (Chaney, 1958). Means (1975b) reported age at sexual maturity as 3 yr for males and 4 yr for females, on each of Rich, Kiamichi, and the Cossatot mountains.

M. Longevity. Unknown.

N. Feeding Behavior. Food items found in the stomachs of 48 out of 64 Ouachita dusky salamanders from Russellville, Arkansas, were *Lirceus* sp. isopods, spiders, lepidopteran larvae, buprestid beetles, collembolans, larval chironomids, plecopterans, hydrophilid beetles, centipedes, and adult dipterans (Chaney, 1958).

O. Predators. Documented predators are western cottonmouths (*Agkistrodon piscivorus conanti*; reported by Means, in Gloyd and Conant, 1990) and speckled kingsnakes (*Lampropeltis getula holbrooki*; reported by Trauth and McAllister, 1995). Potential other predators observed in the habitat were spotted skunks, raccoons, garter snakes, and large crayfish (Means, 1975b). Cannibalism on larvae was documented by Chaney (1958).

P. Anti-Predator Mechanisms. None known except that, as in desmognathine salamanders, in general, Ouachita dusky salamanders twist and flip their bodies with great power and agility when grasped, and jump and burrow rapidly into the substrate to escape and hide (personal observation).

Q. Diseases. An intraerythrotic inclusion thought to be a rickettsia or virus of undetermined taxonomic status was reported in an Ouachita dusky salamander by McAllister et al. (1995f).

R. Parasites. Few salamanders have been investigated for parasites, but only 41 Ouachita dusky salamanders produced *Chloromyxum salamandrae*, *Brachycoelium salamandrae*, *Cylindrotaenia americana*, *Mesocestiodes* sp., *Batracholandros magnavulvaris*, *Desmognathinema nantahalensis*, *Hedrurus pendula*, *Omeia papillocauda*, Ascaridoidea larvae, an acanthocephalan, and a *Hannemania* sp. mite (McAllister et al., 1995f). Loomis (1956), Means (1974), Winter et al. (1986), and Anthony et al. (1994) also reported the presence of trombiculid mites of the genus *Hannemania*. Winter et al. (1986) reported cestodes, nematodes, and acanthocephalans.

4. Conservation.
Ouachita Mountain dusky salamanders were sampled intensively over their entire geographic range involving about fifty localities in the early 1970s (Means, 1974, 1975b), but there has been no follow-up. Most of the geographic range of the species, however, is on publicly owned lands such as the Ouachita National Forest, part of the Ozark National Forest, Hot Springs National Park, and several state parks in both Arkansas and Oklahoma. The most important conservation measures are to ensure that the ecological integrity of the mountain brook habitats of the species are maintained. Careful attention should be paid to avoiding silvicultural practices, such as clearcutting, that would increase sedimentation or alter the hydrology of mountain brook watersheds or deliver pesticides, herbicides, or fertilizers to the aquatic habitat. Impacts of the fish bait industry should be assessed, including whether populations can recover from different levels of harvesting and, particularly, whether local gene pools are being altered by the artificial movement of animals across drainages and among mountain masses. This latter is important because Karlin et al. (1993) showed that populations on different mountain masses were genetically distinct, indicating a high degree of local adaptation. Means (1975b) verified this through morphological analyses.

Desmognathus carolinensis Dunn, 1916
CAROLINA MOUNTAIN DUSKY SALAMANDER
Carlos D. Camp, Stephen G. Tilley

1. Historical versus Current Distribution.
Elevated to species status only recently (Tilley and Mahoney, 1996), Carolina mountain dusky salamanders *(Desmognathus carolinensis)* have a restricted distribution along the Tennessee–North Carolina border in the southwestern Blue Ridge Physiographic Province. They are found on the Blue Ridge, Black, Bald, and Unaka mountains. Their range extends from between Linville Falls and McKinney Gap on the Blue Ridge Divide and the valley of the Doe River on the Tennessee–North Carolina border southwest to the valley of the Pigeon River in North Carolina (Tilley and Mahoney, 1996; Petranka, 1998; Mead and Tilley, 2000). Populations range to the peaks of the highest mountains (approximately 2,000 m) in the Appalachian Mountains (Tilley, 1980). Tilley (1997) speculated that interactions with seal salamanders *(D. monticolu)* may have contributed to the isolation and genetic differentiation of units of the *D. ochrophaeus* complex.

2. Historical versus Current Abundance.
Within their restricted range, Carolina mountain dusky salamanders are one of

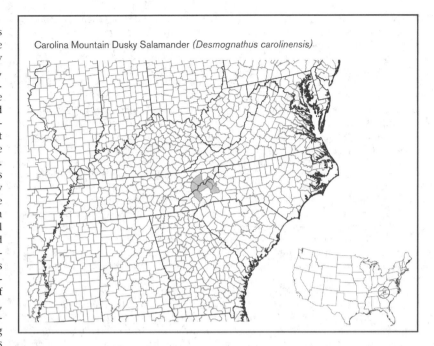

Carolina Mountain Dusky Salamander *(Desmognathus carolinensis)*

the most common Appalachian salamanders. Petranka and Murray (2001) sampled a cove forest at approximately 1,200 m and estimated the density of this species to exceed 10,000 individuals/ha and 7.5 kg/ha. Petranka et al. (1993) demonstrated that clearcut timber harvesting negatively affects the number of total *Desmognathus* individuals, including members of the *D. ochrophaeus* complex. They estimated that, at the rate of clearcutting carried out during the 1980s and early 1990s, the Appalachian forests of North Carolina lost as many as 14 million salamanders of all species each year during that time. However, Ash and Bruce (1994) strongly disagreed with these estimates and did not consider clearcutting to have as strong an impact on native salamanders.

3. Life History Features.
A. Breeding. Courtship occurs on land. Eggs are deposited in or near flowing water.

i. Breeding migrations. Females may move short distances from terrestrial foraging areas into semi-aquatic nesting sites. Closely related Ocoee salamanders *(D. ocoee)* move to nesting sites several weeks prior to oviposition (Forester, 1981), and Carolina mountain dusky salamanders presumably exhibit similar behavior.

ii. Breeding habitat. Mating occurred among captive individuals during July and September under experimental conditions (Mead and Tilley, 2000). Females apparently breed annually (Tilley, 1973b), similar to other members of the *D. ochrophaeus* complex (Forester, 1977). Inseminations of females by more than one male are relatively common, with numbers of multiple inseminations varying among populations (estimated

to average 7% of matings; Tilley and Hausman, 1976).

B. Eggs.

i. Egg deposition sites. Eggs were deposited at several localities on Mt. Mitchell, North Carolina, from late May to early July (Tilley, 1973b). Oviposition during the winter has been inferred from the abundance of small, yolk-laden larvae during the early spring; wintertime clutches were not found during diligent searches, however (Tilley, 1973b). Females that oviposit in summer brood their eggs under moss on rocks and logs associated with seepage areas and small streams. Tilley (1973b) speculated that winter-brooding females may lay their eggs deep below the surfaces of respective seeps. Eggs generally are laid in lacunae hollowed out by the female within or under moss (Martof and Rose, 1963; Tilley, 1973b). Martof and Rose (1963) found clutches on Mt. Mitchell under thick moss beside seepages or in rotting logs in the forest. Eggs are deposited in small, grape-like clusters, each egg attached singly or in twos and threes via respective stalks (Martof and Rose, 1963). As with all amphibians, time to hatching varies with temperature. In laboratory observations, eggs kept at 16 °C hatched in 71 d; clutches observed in nature on Mt. Mitchell hatched in 58–69 days (Tilley, 1972). Eggs and/or hatchlings are sometimes consumed by the attending female (Tilley, 1972).

ii. Clutch size. Clutch sizes determined from developing oocytes counted in dissected females ranged from 12–40 eggs (Martof and Rose, 1963). Some females fail to lay their entire clutch complements, and a small number of oocytes may be resorbed (Tilley, 1973a,b); clutch sizes taken from natural nests on Mt. Mitchell ranged from 16–26 (Tilley, 1973b). Clutch size varies with elevation on Mt. Mitchell as a function of variable female body size (Tilley, 1973a). Ova are approximately 3 mm in diameter (Martof and Rose, 1963; Tilley, 1973b). Females brood their eggs (Bishop, 1941; Tilley, 1972). Hatchlings measure 13–18 mm TL and have prominent yolk sacs (Bishop, 1941). Timing of egg laying appears highly variable, with evidence of egg deposition ranging from mid-to-late fall or winter to late July (Tilley and Tinkle, 1968; Tilley, 1973b).

C. Larvae/Metamorphosis.

i. Length of larval stage. Anywhere from 2–8 mo, depending on time of oviposition. Larvae from late winter and early spring eggs apparently metamorphose in 2–3 mo, while those from the summer may take 4–8 mo (Tilley, 1973b).

ii. Larval requirements.

a. Food. Probably small, aquatic invertebrates.

b. Cover. Larvae inhabit shallow water in seepage areas and water films on vertical rock faces. They may take cover during the day under leaf litter and other detritus in shallow water and seepage areas.

iii. Larval polymorphisms. Individuals that overwinter as larvae reach significantly larger body sizes than those that metamorphose prior to winter (Tilley, 1973b). Variability in dorsolateral stripe configuration is evident in larval specimens (Tilley, 1969).

iv. Features of metamorphosis. SVL of newly metamorphosed animals are typically 10–12 mm (Tilley, 1973a).

v. Post-metamorphic migrations. Juveniles and adults may move into forest-floor habitats.

vi. Neoteny. Neoteny is not known in this species.

D. Juvenile Habitat. Generally similar to adult habitats. Growth rates vary depending on elevation, age, and microhabitat features. Juveniles can grow 8–10 mm/yr in SVL (Tilley, 1973a, 1974).

E. Adult Habitat. Adults are found in seepage areas, on wet rock faces, and in the forest-floor litter in association with streams, particularly headwater seepages (Hairston, 1949; Tilley, 1973a, 1974, 1997). The distribution of this species is directly related to moisture conditions; individuals living at moist, high elevation sites may occur on the forest floor far from streams (Hairston, 1949; Petranka and Murray, 2001). Individuals at low elevations are found in close association with seeps and streams (Hairston, 1949).

Carolina mountain dusky salamanders are largely nocturnal, remaining beneath cover during the day and emerging to feed at night. They may be active during the day when it is heavily overcast. On rainy or foggy nights, Carolina mountain dusky salamanders frequently climb understory plants and tree trunks and perch >1 m above the forest floor (Hairston, 1949; Petranka, 1998). Activity in members of the *D. ochrophaeus* complex is directly related to humidity (Feder and Londos, 1984).

F. Home Range Size. Unknown, but probably small. Movements of similar Ocoee salamanders center around a small home range (Huheey and Brandon, 1973).

G. Territories. Adults are aggressive and actively defend space from other Carolina mountain dusky salamanders (Bennett and Houck, 1983; Petranka, 1998).

H. Aestivation/Avoiding Desiccation. Unlikely. Individuals are active throughout the summer. Activity levels, however, are directly related to humidity (Feder and Londos, 1984; Petranka and Murray, 2001).

I. Seasonal Migrations. Metamorphosed individuals congregate in seepage areas to escape freezing during winter and disperse into the surrounding forest in the spring (Tilley, 1973b; Petranka, 1998).

J. Torpor (Hibernation). Unknown.

K. Interspecific Associations/Exclusions. Carolina mountain dusky salamanders often occur sympatrically with black-bellied salamanders (*D. quadramaculatus*), seal salamanders (*D. monticola*), and pigmy salamanders (*D. wrighti*; Hairston, 1949). These species are organized similar to other desmognathine communities throughout the southern Appalachian Mountains, where different species predictably sort by body size along the stream–forest interface. Larger species are more aquatic, and smaller ones occur more terrestrially. This pattern is evident both along a horizontal gradient from stream–stream bank–forest and along a vertical gradient from stream–seepage–forest (Hairston, 1949). As with other members of the *D. ochrophaeus* complex, Carolina mountain dusky salamanders are relatively small and occur more terrestrially than most of their sympatric congeners. The observed pattern of desmognathine assortment initially was explained as niche partitioning among competitors (Hairston, 1949; Organ, 1961a). Tilley (1968) and Hairston (1980c), however, suggested that interspecific predation was a more likely cause. A number of studies attempted to determine which was the more probable factor (e.g., Kleeberger, 1984; Carr and Taylor, 1985; Hairston, 1986; Southerland, 1986a,b,d). They generally concluded that both predation and aggressive interference were important factors in interspecific desmognathine interactions. Hairston (1986) made the strongest case for predation with competition being a secondary factor. His statistical methods have been criticized, however (Jaeger and Walls, 1989). Although large desmognathines readily eat small ones in artificial environments, dietary studies have demonstrated that neither black-bellied salamanders nor seal salamanders are important predators of heterospecific congeners (Camp, 1997b). The lack of predation under natural conditions is probably a result of differential habitat selection and behavioral avoidance (perhaps involving chemical cues) of larger congeners by small individuals. Predation by large species may have been important historically in the organization of desmognathine communities. Alternative hypotheses based on abiotic factors, rather than biotic ones such as competition and predation, recently have been proposed to explain patterns of habitat preference among desmognathines (Bruce, 1996; Camp et al., 2000). In forest-floor habitats, Carolina mountain dusky salamanders are often syntopic with Blue Ridge two-lined salamanders (*Eurycea wilderae*) and several species of woodland salamander (genus *Plethodon*; Petranka and Murray, 2001).

The Pigeon River forms a southern boundary between Carolina mountain dusky salamanders and the closely related Ocoee salamanders. No such physical barrier exists, however, between Carolina mountain dusky salamanders and Blue Ridge dusky salamanders (*D. orestes*) to the north. Limited hybridization occurs at points of contact between these two species (Mead and Tilley, 2000). Carolina mountain dusky salamanders exhibit considerable sexual isolation from their sibling, parapatric neighbors (Verrell and Arnold, 1989; Tilley et al., 1990).

L. Age/Size at Reproductive Maturity. Populations on Mt. Mitchell vary in size and age at maturation with elevation. Those at high elevations mature at a relatively late age and large size and reach larger maximum sizes (>65 mm SVL) than elsewhere in the species' range (Tilley, 1973a). Adult females at 1,800–1,900 m average 44–48 mm SVL; those at <1,000 m average 37–38 mm. Populations inhabiting wet rock face versus woodland habitats at the same elevation also differ in size at maturity, with woodland individuals being larger. This difference in maturation size is the result of differential juvenile growth rates (Tilley, 1974).

M. Longevity. Unknown. Similar Ocoee salamanders are known to live at least 7–10 yr (Castanet et al., 1996). Large, dark individuals may be considerably older.

N. Feeding Behavior. Both juveniles and adults are generalists, feeding on stream- and forest-floor–associated invertebrates including pseudoscorpions, snails, annelids, and insects such as beetles, dipteran and lepidopteran larvae, homopterans, and parasitic wasps (Hairston, 1949). Females sometimes cannibalize their own eggs and hatchlings (Tilley, 1972). Bernardo (2002) reported that an adult Carolina mountain dusky salamander ate a Weller's salamander (*Plethodon welleri*).

O. Predators. Woodland birds, snakes, and small mammals, particularly shrews (Brannon, 2000), undoubtedly prey on Carolina Mountain dusky salamanders. They may also be preyed upon by spring salamanders (*Gyrinophilus porphyriticus*). The latter species is a major predator of other members of the *D. ochrophaeus* complex (Bruce, 1979).

P. Anti-Predator Mechanisms. Unknown. They are probably similar to other desmognathines, i.e., flight, writhing, and biting (Brodie et al., 1989).

Q. Diseases. Unknown.

R. Parasites. Rankin (1937) reported the following parasites from Carolina mountain dusky salamanders in North Carolina: Protozoa—*Cryptobia borreli, Eutrichomastix batrachorum, Hexamastix batrachorum, Hexamita intestinalis, Karotomorpha swezi, Prowazekella longifilis, Tritrichomonas augusta*; Trematoda—*Brachycoelium hospitale*, Cestoda—*Crepidobothrium cryptobranchi*; Nematoda—*Capillaria inequalis, Oxyuris magnavulvaris.*

Closely related Ocoee salamanders occupy similar habitats and harbor a helminth fauna that likely is similar to Carolina mountain dusky salamanders and includes nematodes, flukes, and tapeworms (Goater et al., 1987). Leeches occasionally occur on Ocoee salamanders (Goater, 2000).

4. Conservation.
Carolina mountain dusky salamanders are among the most common salamanders within their range. Because of their reliance on moist habitats, the greatest potential threat is likely the removal of the protective forest canopy through the harvesting of timber and resulting habitat desiccation. Petranka et al. (1993) estimated that the clearcut logging of Appalachian forests of North Carolina during the 1980s and early 1990s killed millions of salamanders each year, including members of the *D. ochrophaeus* complex. Ash and Bruce (1993), however, disputed these estimates and considered them to be exaggerations of the actual number killed. Desmognathine salamanders currently are abundant in areas of the southern Appalachians (the Great Smoky Mountains) that have been logged extensively in the past (S.G.T., personal observations). Appalachian seepages occasionally dry up, negatively affecting both reproduction (Camp, 2000) and survival in closely related Ocoee salamanders (C.D.C., unpublished data). It is not known how periodic drought may interact with techniques of timber harvesting to affect populations of Carolina mountain dusky salamanders and other seepage-dwelling salamanders.

Desmognathus conanti Rossman, 1958
SPOTTED DUSKY SALAMANDER

D. Bruce Means, Ronald M. Bonett

1. Historical versus Current Distribution.
The geographic distribution of spotted dusky salamanders *(Desmognathus conanti)* extends south of a line beginning at the confluence of the Cumberland and Ohio rivers on the border of Illinois and Kentucky and continuing southeast through western Kentucky, central Tennessee, and into northwestern South Carolina (Karlin and Guttman, 1986; Petranka, 1998). The species also occurs in a few colonies at the southern tip of Illinois (Smith, 1961). Spotted dusky salamanders occur as far south as southern Mississippi and eastern Louisiana (K. Kozak and colleagues, unpublished data) and may occur in eastern Alabama (Rossman, 1958), but they do not reach western Florida (Means, 1998). Still, many aspects of the southern distributional limits of spotted dusky salamanders presently are unclear. Chaney (1949) did a life history study of six localities in southern Louisiana, believing he was working with a single species he called *D. fuscus auriculatus*. Later, he conducted additional intensive studies of Louisiana *Desmognathus* but only considered that he was working with one species, when in fact he may have been working with at least two (Chaney, 1958), including spotted dusky salamanders. Dundee and Rossman (1989) did not question that there were two morphological types in Louisiana, one a ravine dweller with light-colored, heavily patterned adult and larval dorsum and fewer gill filaments in larvae (which could be spotted dusky salamanders), and the other a muckland form with darker coloration and bushy gills (which are probably southern dusky salamanders [*D. auriculatus*]). They did not, however, reach a conclusion about the taxonomic status of Louisiana *Desmognathus*. They simply discussed all Louisiana *Desmognathus* as either "*Desmognathus fuscus-auriculatus* complex, *D. fuscus*, or *D. auriculatus*," with no reference to *D. conanti*. The ravine-inhabiting form represents either spotted dusky salamanders or some unrecognized species. The only known occurrence of spotted dusky salamanders west of the Mississippi River is a small group of populations on Crowley's Ridge, an isolated gravel ridge in an ancient part of the Mississippi River valley in northeastern Arkansas. These may have been extirpated (Means, 1974; Nickerson et al., 1979; Bonett, 2002; Trauth et al., in preparation).

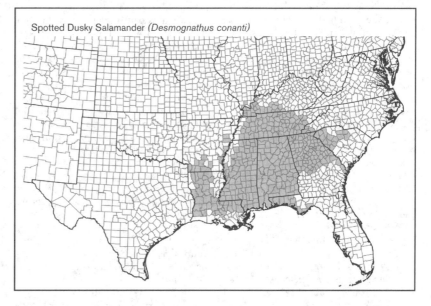

Spotted Dusky Salamander *(Desmognathus conanti)*

2. Historical versus Current Abundance.
In addition to Crowley's Ridge, where populations seem to have disappeared, population densities in streams near Atlanta, Georgia, were found to be inversely proportional to the degree of stream disturbance associated with urbanization (Orser and Shure, 1972). Siltation and sedimentation of their small stream habitats from runoff following construction and farming probably have extirpated or severely reduced populations of this species throughout its range, but no specific studies have been undertaken to assess historical versus current abundance relationships.

3. Life History Features.
A. Breeding. Reproduction is semiaquatic.

i. Breeding migrations. Probably do not occur. No breeding migrations have been reported.

ii. Breeding habitat. Desmognathine salamanders are highly territorial and breed in the same habitats in which their eggs are laid.

B. Eggs.

i. Egg deposition sites. In secretive habits such as the undersides of rocks, leaf mats, moss, and cavities in rotting logs near streams and seeps.

ii. Clutch size. Eggs (13–24) with developing embryos were found in the field from mid-July to mid-October in Alabama (Mount, 1975), and in the fall in northern Mississippi (Brode, 1961). Eggs hatch in 5–7 wk in Alabama (Mount, 1975).

C. Larvae/Metamorphosis.

i. Length of larval period. In Alabama, larval development may take ≤13 mo in water or is greatly accelerated if on a damp place on land (Mount, 1975). Life history information in Dundee and Rossman (1989) for Louisiana was based upon that of Chaney (1949, 1958), who may have reported mixed data from southern and spotted dusky salamanders. Because of the uncertainty about the taxonomic status and geographic limits of this species (e.g., Dundee and Rossman, 1989; Petranka, 1998), few studies have been undertaken on their life history, reproduction, and development.

D. Juvenile Habitat. Same as for adults, except that choice microhabitats, occupied by the aggressive adult males, are not occupied by juveniles.

E. Adult Habitat. In the deeply shaded, heavily wooded ravines of the Alabama Red Hills where they are syntopic with the Red Hills Salamanders (Phaeognathus hubrichti), spotted dusky salamanders were found on the wetter portions of the stream banks and under the rocks of the stream itself (Brandon, 1966b). Elsewhere in Alabama, Mount (1975) reported them from a variety of damp situations including seepage areas and the edges of small rocky streams, where they were found during the daytime under rocks, damp leaf litter, and in burrows or rotting logs. In Louisiana, populations with boldly patterned individuals that may be spotted dusky salamanders were reported from spring seepages in ravines under logs and debris in shallow water or adjacent to water (Dundee and Rossman, 1989). Smith (1961) reported that spotted dusky salamanders are found in cold springs under wet leaves or rocks at the margin of springs or streams in southern Illinois. In north Georgia, the species was reported from spring-fed streams impacted by different degrees of disturbance from urbanization in and around relieved terrain in Atlanta (Orser and Shure, 1972).

F. Home Range Size. Unknown.

G. Territories. Because some congeners are highly territorial and the species does not migrate, it is probable that adult males, at least, defend territories or choice hiding places, but no studies have been conducted.

H. Aestivation/Avoiding Desiccation. Unknown and unstudied.

I. Seasonal Migrations. Unknown. Short movements associated with drought or floods might take place toward or away from the streams along which the species is associated, but as a normal, regular activity, seasonal migrations probably do not exist in this territorial species.

J. Torpor (Hibernation). Unknown.

K. Interspecific Associations/Exclusions. Lake Barkley (Cumberland River) in western Kentucky forms a relatively discrete boundary between northern dusky salamanders (Desmognathus fuscus) and spotted dusky salamanders, with their populations inhabiting tributaries on the eastern and western shores of the lake, respectively (Bonett, 2002). Fine scale examination of populations at the interface of these species revealed only one population that contained few putative hybrid individuals (Bonett, 2002). In Tennessee, populations of dusky salamanders (D. fuscus and D. conanti) previously have been illustrated as continuous across the state (Petranka, 1998), however the well-drained Central Basin Region actually appears to geographically separate the species (see D. fuscus spot map in Redmond and Scott, 1996). Bonett (2002) identified populations of spotted dusky salamanders in northern Alabama and populations of northern dusky salamanders approximately 150 km north in east-central Tennessee; it is unclear if contact between the species exists in the intermediate region. This leaves an ambiguous identification on specimens observed in studies of courtship behavior (Verrell, 1995), reproductive ecology (Hom, 1987), and cover object choice (Hom, 1988) from the Reed Creek drainage of eastern Tennessee, which occurs between the known species boundaries of D. fuscus and D. conanti.

In the Cheaha Mountains of east-central Alabama, spotted dusky salamanders are syntopic with seal salamanders (D. monticola), red salamanders (Pseudotriton ruber), and southern two-lined salamanders (Eurycea cirrigera; Rubenstein, 1969, personal observation). In the Alabama Red Hills, they are found in deep ravines with Red Hills salamanders, seal salamanders, and southern two-lined salamanders (Brandon, 1966b). In southern Mississippi they are found with dwarf salamanders (E. quadridigitata) in leaf packs along swamp edges (R.M.B., personal observations). In southern Louisiana, spotted dusky salamanders may be syntopic with southern dusky salamanders (Chaney, 1949; Dundee and Rossman, 1989). In western Tennessee, spotted dusky salamanders were found with long-tailed salamanders (Eurycea longicauda) and southern two-lined salamanders (Sites, 1978). Verrell (1995) may not have described the courtship behavior of spotted dusky salamanders because his experimental specimens came from north of the hypothesized "contact zone" in the range of northern dusky salamanders.

L. Age/Size at Reproductive Maturity. Unstudied.

M. Longevity. Unknown.

N. Feeding Behavior. Sites (1978) reported food habits and feeding behavior of animals from three sites in western Tennessee. In the diet he reported amphipods, isopods, pulmonate snails, oligochaetes, chilopods, diplopods, insect larvae, coleopterans, dipterans, collembolans, homopterans, and hymenopterans.

O. Predators. None reported.

P. Anti-Predator Mechanisms. Unstudied.

Q. Diseases. None reported and unstudied.

R. Parasites. Dyer et al. (1980) reported two species of nematodes (Thelandros magnavulvaris, Cosmocercoides dukae), one trematode (Brachycoelium obesum), and unidentified acanthocephalans in the gastrointestinal tracts of 149 (34%) of 442 spotted dusky salamanders from Pulaski County, Illinois.

4. Conservation.

With the exception of the studies of Orser and Shure (1972, 1975) on populations in the vicinity of Atlanta, Georgia, no detailed population surveys or studies of the ecology of the spotted dusky salamander have been undertaken. Phylogenetic analyses of Desmognathus based on mitochondrial DNA and morphology suggest many of the characters that have previously been used are labile and can be rapidly acquired when populations enter new adaptive zones (K. Kozak et al., unpublished data). The most urgent study need is the identification of the geographical limits of the species. Unfortunately, spotted dusky salamanders represent one of the most poorly understood amphibians in eastern North America, despite its potentially large geographic distribution.

Desmognathus folkertsi Camp, Tilley, Austin, and Marshall, 2002
DWARF BLACK-BELLIED SALAMANDER
Carlos D. Camp, Stephen G. Tilley

1. Historical versus Current Distribution.

Dwarf black-bellied salamanders (Desmognathus folkertsi) are cryptic and syntopic with black-bellied salamanders (D. quadramaculatus). Animals are currently known from the Nottely, Ocoee (Union Co.), Hiwassee (Towns Co.), and Chattahoochee (Lumpkin Co.) river systems in the Blue Ridge Physiographic Province of Georgia (Camp et al., 2004). Black-bellied salamanders are commonly collected and sold as fish bait in Georgia (Jensen and Waters, 1999) and have been introduced into new areas as a result (Martof, 1953b). It is possible that collectors in the range of dwarf black-bellied salamanders have also collected and sold them, possibly having introduced them into other areas as well.

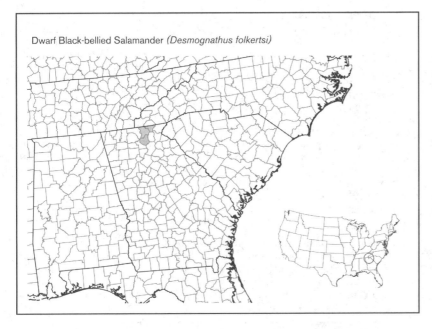

Dwarf Black-bellied Salamander (Desmognathus folkertsi)

2. Historical versus Current Abundance.
Dwarf black-bellied salamanders are abundant in the two streams where they occur. There seemed to be no discernible differences in abundance from collections made during 1998 versus those made a decade earlier (C.D.C., unpublished data).

3. Life History Features.
A. Breeding.
i. Breeding migrations. Unknown.

ii. Breeding habitat. Adult males and females are found throughout the year in and along high-gradient streams (C.D.C., unpublished data). They presumably use these habitats for mating as well as other activities (e.g., foraging).

B. Eggs.
i. Egg deposition sites. Unknown. Presumably similar to closely related black-bellied salamanders syntopic congeners that share similar habitats. Females of the latter species have been found attending clutches beneath stones in the middle of creek channels at one of the streams where dwarf black-bellied salamanders occur (C.D.C., personal observations).

ii. Clutch size. Oviposited clutches have not been found. Clutches estimated from developing follicles of mature females range from 2–62 eggs/clutch, averaging 39 eggs/clutch. Clutch size is positively related to female body size (C.D.C., unpublished data), similar to other desmognathines (Tilley, 1968).

C. Larvae/Metamorphosis.
i. Length of larval stage. Larvae of dwarf black-bellied salamanders and black-bellied salamanders currently cannot be distinguished. Collections of larvae of presumably both species indicated the presence of four distinct size classes (Austin and Camp, 1992). The largest size class corresponds with sizes of newly metamorphosed black-bellied salamanders (Camp et al., 2000). Newly metamorphosed dwarf black-bellied salamanders appear to correspond to the second size class. The probable larval period, therefore, is 2 yr (Camp et al., 2002).

ii. Larval requirements.

a. Food. Unknown. Presumably aquatic invertebrates.

b. Cover. Larvae of (presumably) both dwarf black-bellied and black-bellied salamanders occur in the interstices of riffle areas in streams (Austin and Camp, 1992).

iii. Larval polymorphisms. No distinct larval polymorphisms are known.

iv. Features of metamorphosis. Newly metamorphosed individuals average 36 mm SVL (range = 34–37; Camp et al., 2000, 2002).

v. Post-metamorphic migrations. Unknown.

vi. Neoteny. Complete neoteny is unknown. As with black-bellied salamanders, adults exhibit the neotenic retention of lateral line pores (sensu Hilton, 1947; Camp et al., 2002).

D. Juvenile Habitat.
Juveniles have been collected under cover objects within streams during the day and actively wandering within streams at night. Several juvenile specimens were collected while raking through wet leaves at the margins of streams (S.G.T., personal observation).

E. Adult Habitat.
Adults have been collected in high-gradient, turbulent first- and second-order streams. They are most abundant in shallow areas of streams where they take cover under rocks. At night they occasionally are seen wandering away from cover; however, they usually remain with only their heads and/or anterior parts of their bodies protruding from cover sites. No differences in habitat have been noted between the sexes (C.D.C., personal observations).

F. Home Range Size.
Unknown.

G. Territories.
Unknown in nature, although it is probable that individuals defend cover sites in the way that other semi-aquatic desmognathines do (e.g., seal salamanders [D. monticola], Keen and Sharp, 1984; black-bellied salamanders, Camp and Lee, 1996). A study of bite-scar patterns (Camp, 1996) in presumed black-bellied salamanders included dwarf black-bellied salamanders as well. Several dwarf black-bellied salamanders had distinct wounds inflicted by bites of other salamanders. It is unknown whether these bites were inflicted by conspecific and/or heterospecific salamanders.

H. Aestivation/Avoiding Desiccation.
Unlikely. Individuals are active throughout the summer (C.D.C., personal observations).

I. Seasonal Migrations.
Unknown, but unlikely.

J. Torpor (Hibernation).
Unknown.

K. Interspecific Associations/Exclusions.
Dwarf black-bellied salamanders are syntopic with four congeneric species: seepage salamanders (D. aeneus), Ocoee salamanders (D. ocoee), seal salamanders, and black-bellied salamanders. The extent of potential competitive interactions is unknown. Dwarf black-bellied salamanders are similar in size to seal salamanders but are more aquatic. Dwarf black-bellied salamanders occupy habitats similar to larger black-bellied salamanders. Gut analyses of syntopic adult black-bellied salamanders provided no evidence of predation on dwarf black-bellied salamanders (Camp, 1997b).

Dwarf black-bellied salamanders exhibited avoidance of substrates with chemical cues of individual black-bellied salamanders from both sympatric and allopatric populations in laboratory-based experiments. Black-bellied salamanders were neutral to the cues of dwarf black-bellied salamanders (S. Myers, M. Rubo, and J.L. Marshall, unpublished data).

Martof (1962b) noted the enigmatic absence of shovel-nosed salamanders (D. marmoratus) in areas of apparently favorable habitat in the Nottely and Hiwassee rivers. The presence of a similar-sized desmognathine of the quadramaculatus-marmoratus clade (sensu Titus and Larson, 1996) in streams of the Nottely River system may help explain this absence. Whether this potential parapatry is the result of ecological exclusion or evolutionary divergence, however, has yet to be investigated. Shovel-nosed salamanders are entirely aquatic, and dwarf black-bellied salamanders are semi-aquatic. Therefore, competitive exclusion, while possible, seems unlikely.

L. Age/Size at Reproductive Maturity.
Adult males average 72 mm SVL (range = 58–81 mm), and adult females average 66 mm SVL (range = 57–75 mm; Camp et al., 2002).

M. Longevity. Unknown. Similar-sized desmognathines are known to live at least 11 yr (Castanet et al., 1996).

N. Feeding Behavior. The guts of 12 adult individuals contained adult water bugs (Hemiptera) and crane-fly larvae (Tipulidae; C.D.C., unpublished data). Individuals with heads protruding from refugia readily seize earthworms when offered as bait on a hook (Camp and Lovell, 1989). This behavior and the observed tendency of most adult individuals to remain in refugia at night with only their heads (or anterior parts of bodies) exposed suggest an ambush foraging strategy similar to that of adult seal and black-bellied salamanders (Brandon and Huheey, 1971; Kleeberger, 1985; Camp and Lee, 1996).

O. Predators. Unknown. The streams where dwarf black-bellied salamanders occur are undoubtedly inhabited by predatory fish (e.g., sculpins, trout). Although no aquatic snakes have been observed at these sites, we presume that northern water snakes *(Nerodia sipedon)* occur, at least on lower stretches of the streams. Spring salamanders *(Gyrinophilus porphyriticus)* are natural predators of small salamanders (Bruce, 1979) and are relatively common in seepages that feed into streams containing dwarf black-bellied salamanders (Wharton, 1978; C.D.C., personal observations). Spring salamanders almost certainly prey on juvenile/larval dwarf black-bellied salamanders. Larvae are probably eaten by aquatic invertebrates, including insects.

P. Anti-Predator Mechanisms. Presumably the same as other desmognathines, that is, flight, writhing, and biting. A number of specimens have missing or regrown tails (C.D.C., personal observations), a possible indication of failed attempts at predation.

Q. Diseases. Unknown.

R. Parasites. Leeches have been observed on specimens (C.D.C., personal observations). Goater (2000) noted the prevalence of leeches *(Oligobdella biannulata)* on black-bellied salamanders and their probable use as vectors for the transmission of blood trypanosomes.

4. Conservation.
Dwarf black-bellied salamanders are not legally protected at present. Their known range is small, consisting of streams of the Nottely River system of Union County, Georgia. Whether dwarf black-bellied salamander populations extend beyond this range is unknown; they may occur in other stream systems such as the Ocoee and upper Hiwassee rivers of Georgia, North Carolina, and Tennessee. However, if they are restricted to the vicinity of the Nottely River system, dwarf black-bellied salamanders may be in need of legal protection.

Georgia does not require permits for the collection of desmognathine salamanders for fish bait, and local bait collectors are known to collect *Desmognathus* from streams inhabited by dwarf black-bellied salamanders (C.D.C., personal observations). Dwarf black-bellied salamanders, therefore, are undoubtedly exploited for this purpose. They remain abundant, however, at sites where they have been previously collected (C.D.C., unpublished data).

Desmognathus fuscus (Green, 1818)
NORTHERN DUSKY SALAMANDER

D. Bruce Means

1. Historical versus Current Distribution.
The geographic distribution of northern dusky salamanders *(Desmognathus fuscus)* extends in the United States southwest from Maine, through New England, New

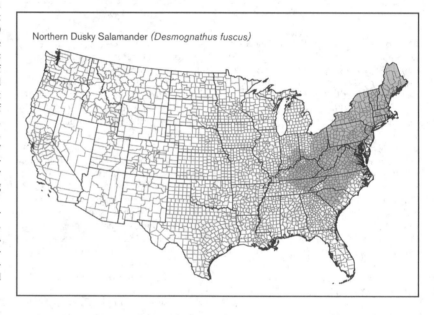

Northern Dusky Salamander *(Desmognathus fuscus)*

York, and Pennsylvania to Virginia, then west to southern and eastern Ohio, southeastern Indiana, eastern Kentucky, eastern Tennessee, and western North Carolina (Petranka, 1998). Northern dusky salamanders also occur in Canada, and are found in Niagara Gorge in Ontario, in southeastern Québec, and in southern New Brunswick. The southern geographic limits of northern dusky salamanders are a poorly delineated contact zone between northern and spotted dusky salamanders *(D. conanti)* that extends southeast from the Cumberland River in western Kentucky to western South Carolina (see Karlin and Guttman, 1986; Petranka, 1998). Spotted dusky salamanders range from the contact zone to the southwest of the range of northern dusky salamanders, extending from extreme southern Illinois, western Kentucky, and western Tennessee into northern Georgia, Alabama,

and Mississippi (Rossman, 1958; E.E. Brown, 1992; Petranka, 1998; Means, this volume). No reductions have been noted in the geographical distribution of northern dusky salamanders.

2. Historical versus Current Abundance.
In eastern Kentucky and Tennessee, northern dusky salamander larvae are absent from many streams that drain coal strip mines (Gore, 1983). Throughout their New England and midwestern ranges, urbanization has extirpated many populations, but the species is wide ranging and ubiquitous in small stream valleys. There have been no studies documenting declines.

3. Life History Features.
A. Breeding. Reproduction is semi-aquatic.

i. Breeding migrations. Northern dusky salamanders commonly live within a few feet of streams or in springs and are not known to undertake extensive migrations (Bishop, 1941b). Adults mate during the fall and spring in Virginia (Organ, 1961a) and New York (Bishop, 1941b).

ii. Breeding habitat. Same as adult habitat.

B. Eggs.

i. Egg deposition sites. Eggs are attached to the undersides of rocks, logs, or other substrates in or near water (see Petranka, 1998).

Maryland salamanders lay eggs from June to early August (Danstedt, 1975) and New York salamanders from early June to late August (Wilder, 1913). In Ohio, females oviposit in the last two weeks in July and the embryos hatch from early September to mid-October (Wood and Fitzmaurice, 1948; Dennis, 1962; Orr and Maple, 1978; Juterbock, 1986). Populations in eastern Kentucky and southeastern Ohio oviposit during July, and the eggs incubate for 46–61 d with a median incubation period of 47 d (Juterbock, 1986). In eastern Tennessee, females oviposit from the third week in June to the end of August, and embryos hatch from late August to mid-October (Hom, 1987). In other parts of their range, eggs have been found from 28 June–5 October in Indiana (Minton, 1972, 2001); early June to early August in Maryland (Danstedt, 1975, 1979); 11 June–24 September in Massachusetts (Wilder, 1913); July–September in southwestern Pennsylvania (Krzysik, 1980a,b); and late June to mid-August in Viriginia (Organ, 1961a). Females lay eggs every year in Maryland (Danstedt, 1975) and North Carolina (Spight, 1967c), but may be biennial in higher altitudes in southwestern Virginia (Organ, 1961a).

ii. Clutch size. There is variation in egg numbers, ranging from 13–34 with means ranging from 21–33 (see Petranka, 1998).

C. Larvae/Metamorphosis. Hatchlings were discovered between 17 June–1 September in Massachusetts (Wilder, 1913); in July and August in Maryland (Danstedt, 1975) and Pennsylvania (Krzysik, 1980a); in August and September in New York (Bishop, 1941b); and in mid-September in Ohio (Dennis, 1962). The number of eggs/nest was 5–33 in nine nests in Massachusetts (Wilder, 1913); 22–34 in 44 nests in Pennsylvania (Pawling, 1939); and an average of 26 eggs in seven nests in Ohio (Wood and Fitzmaurice, 1948). Enlarged ovarian ova counts averaged 27 in 18 females from western North Carolina (Spight, 1967c); 21 in 17 Pennsylvania females (Hall, 1977); and 24–33 ova in females from six populations from Maryland (Danstedt, 1975). Larvae transform in late spring and early summer when 9–12 mo old in Maryland (Danstedt, 1975); in July–October when 11–14 mo old in Virginia (Organ, 1961a); and between 11 May–17 June when 8–10 mo old in Massachusetts (Wilder, 1913).

D. Juvenile Habitat. Same as for adults.

E. Adult Habitat. In New York, northern dusky salamanders are found along the margins of small wooded streams, on seepage hillsides, in shallow weed-choked streams with sandy bottoms, and in low boggy places under stones, logs, bark, and other debris on the ground (Bishop, 1941b). In Pennsylvania, they are found along wet streambanks with substrates of medium or coarse gravel and small rocks, and under logs about 15–90 cm from open water (Krzysik, 1979). In Ohio, they frequent the margins of streams, springs, leaf-littered trickles, spring banks with constantly moist soil, and the beds of ephemeral streams; preferred hiding places are under flat rocks and debris that are partially submerged in streams (Karlin and Pfingsten, 1989). Indiana northern dusky salamanders are largely restricted to hillside springs and small, rocky brooks in limestone country, but are sometimes found in river bottomlands in spring seepage areas and are often common in the mouths of caves, but not within caves (Minton, 1972, 2001). Brimley (1944) stated that in North Carolina they are salamanders of small woodland brooks, found in both streams and under rocks, logs, rubbish or other shelter close by, but never in strictly terrestrial surroundings.

F. Home Range Size. Barbour et al. (1969) reported an average home range of 48.4 m^2 (25.2–114.5 m^2) in 14 northern dusky salamanders found along a creek in east-central Kentucky with a sandstone bed. In strong contrast, along a small limestone stream in Ohio, Ashton (1975) calculated an average home range for 14 northern dusky salamanders of 1.4 m^2 (0.1–3.6 m^2). Mean activity radii were calculated for 21 Pennsylvania northern dusky salamanders from which home range can be determined (Barthalmus and Bellis, 1972). Population densities ranging between 0.43 and 1.42/m^2 and biomass estimates were reported for salamanders in New Hampshire by Burton and Likens (1975b). Densities were calculated for populations at 0.8 m^2 in Pennsylvania (Hall, 1977), and between 0.4 and 1.4 animals/m^2 in a North Carolina stream bed (Spight, 1967c).

G. Territories. Aggressive behavior has been documented in northern dusky salamanders (Organ, 1961a), but this author speculated that aggressive behavior served as a method of sex recognition. Because of the aggressive behavior among males and high site tenacity of the species (Barthalmus and Bellis, 1969, 1972; Barthalmus and Savidge, 1974; Hom, 1987; Juterbock, 1987), territorial behavior is probably well developed, but no studies have specifically targeted this facet of its biology.

H. Aestivation/Avoiding Desiccation. Aestivation is unknown, and probably does not occur. Animals likely avoid desiccating conditions by seeking shelter under cover objects or in burrows.

I. Seasonal Migrations. In Ohio between 23 November–21 December, Ashton (1975) reported that salamanders moved into subterranean microhabitats that were warmer by several degrees centigrade upstream from their normal summer home.

J. Torpor (Hibernation). Ashton (1975), following the activity of five salamanders during winter in Ohio, discovered that all were active and burrowed quickly to escape through gravel and broken limestone of their winter streambed retreats when disturbed. The winter retreats were 30–50 cm below the soil surface, which was frozen to a depth of 14 cm. He concluded that at no time were northern dusky salamanders in an inactive state.

K. Interspecific Associations/Exclusions. Northern dusky salamanders have been reported in association (=locally sympatric or syntopic) with seal salamanders (*D. monticola*; Organ, 1961a; Krzysik, 1979; Southerland, 1986e), Allegheny Mountain dusky salamanders (*D. ochrophaeus*; Noble and Evans, 1932; Organ, 1961a; Stewart and Bellis, 1970; Hall, 1977; Orr and Maple, 1978; Krzysik, 1979; Karlin and Pfingsten, 1989; Holumuzki, 1980; Southerland, 1986e; Sharbel and Bonin, 1992; Sharbel et al., 1995), black-bellied salamanders (*D. quadramaculatus*; Organ, 1961a; Southerland, 1986e), pigmy salamanders (*D. wrighti*; Organ, 1961a), northern two-lined salamanders (*Eurycea bislineata*; Spight, 1967c; Stewart and Bellis, 1970; Holumuzki, 1980; Karlin and Pfingsten, 1989), long-tailed salamanders (*E. longicauda*; Karlin and Pfingsten, 1989), red salamanders (*Pseudotriton ruber*; Stewart and Bellis, 1970; Karlin and Pfingsten, 1989), mud salamanders (*P. montanus*; Spight, 1967c; Karlin and Pfingsten, 1989), spring salamanders (*Gyrinophilus porphyriticus*; Stewart and Bellis, 1970; Krzysik, 1979; Karlin and Pfingsten, 1989), southern ravine salamanders (*Plethodon richmondi*; Karlin and Pfingsten, 1989; Stewart and Bellis, 1970), northern slimy salamanders (*P. glutinosus*; Spight, 1967c; Karlin and Pfingsten, 1989), eastern red-backed salamanders (*P. cinereus*; Karlin and Pfingsten, 1989; Stewart and Bellis, 1970), and marbled salamanders (*Ambystoma opacum*; Spight, 1967c). Over their extensive geographical distribution, northern dusky salamanders are syntopic with more salamanders than listed above, but such associations rarely have been reported in the literature. Davic (1983) concluded that adult northern dusky salamander females were smaller because of syntopy with seal salamanders in Pennsylvania.

L. Age/Size at Reproductive Maturity. In New York, sexual maturity is reached when salamanders are 24–26 mo old; mating and egg laying follows 1–2 mo later (Bishop, 1941b). In Pennsylvania, age at maturity for males is 3 yr (Hall, 1977). In the Unicoi Mountains of Tennessee, males mature at approximately 2 yr and about 35 mm SVL; but females mature at 3 yr and about 40 mm SVL (Jones, 1986). In Maryland, males mature at 2 yr, females at 3 (Danstedt, 1975). At high elevation sites in Virginia, males become sexually mature at 3.5 yr and females 1 yr later, first ovipositing when 5 yr old (Organ, 1961a). Females in a Tennessee population reproduced on an annual cycle (Hom, 1987). Juterbock (1978) discussed maturity characteristics in northern dusky salamanders from Ohio and Kentucky.

M. Longevity. Snider and Bowler (1992) reported that a wild-caught adult of unknown sex lived 4 yr, 4 mo, and 11 d.

N. Feeding Behavior. Larvae in two studies were found to feed on copepods, chironomid midge larvae, plecopteran nymphs, collembolans, mites, and fingernail clams (Wilder, 1913; Burton, 1976).

In the stomachs of 18 adults, Wilder (1913) found the remains of caddisflies, larval and adult dipterans, ants, spiders, beetles, sowbugs, caterpillars, earthworms, amphipods, mites, and molted skins and larvae of northern dusky salamanders. Food items identified from 53 stomachs of northern dusky salamanders from eastern Kentucky were amphipods, chilopods, orthopterans, ephemeropterans, odonates, hemipterans, lepidopterans, coleopterans, hymenopterans, dipterans, spiders, and one gastropod (Barbour and Lancaster, 1946). From 83 stomachs from the White Mountains in New Hampshire, Burton (1976) reported many of the same groups plus mites, collembolans, plecopterans, thysanopterans, homopterans, trichopterans, and sphaeriid clams. In stomachs of Indiana northern dusky salamanders, Minton (1972, 2001) found beetles, lepidopteran larvae, ants, dipteran larvae, an ichneumonid wasp, and a stonefly nymph. Montague and Poinski (1976) and Wood et al. (1955) reported on brooding females that had ingested oligochaete annelids. Holumuzki (1980) reported the dietary overlap and synchronous foraging of northern dusky salamanders with two other plethodontids and showed that salamander surface activity was highly correlated with that of potential invertebrate prey. Krzysik (1979) reported niche overlap, niche width, and utilization spectra of prey volumes among Pennsylvania northern dusky salamanders and sympatric Allegheny Mountain dusky salamanders, seal salamanders, and spring salamanders, but did not list the prey species eaten. Krzysik (1980b) compared gut contents among brooding and non-brooding females, finding significantly more food in the latter, suggesting that brooding females do not actively forage.

O. Predators. Uhler et al. (1939) found that northern dusky salamanders were eaten by northern water snakes *(Nerodia sipedon)* and common garter snakes *(Thamnophis sirtalis)*. Black-bellied salamanders were reported as predators by Noble and Evans (1932). Raccoons, skunks, opossums, and other small mammals, snakes, and birds probably eat northern dusky salamanders (Karlin and Pfingsten, 1989). Brooding females will cannibalize their own eggs and those of others (Bishop, 1941b; Baldauf, 1947; Wood and Clarke, 1955; Jones, 1986). Cannibalism by adults on larvae was reported by Wilder (1913) and Hamilton (1943).

P. Anti-Predator Mechanisms. Brooding females are hypothesized to keep egg predators away from incubating clutches (Dennis, 1962). Dusky salamanders often bite their predators (garter snakes) and autotomize their tails and flee (Whiteman and Wissinger, 1991).

Q. Diseases. No diseases have been reported.

R. Parasites. Rankin (1937) conducted an exhaustive study of the parasites of northern dusky salamanders in North Carolina, finding the following major groups in both larvae and adults: protozoans, trematodes, cestodes, nematodes, and acanthocephalans. Possible parasitism by roundworms was reported in New Hampshire northern dusky salamanders by Burton (1976).

4. Conservation. Monetary values of $0.25/specimen were assessed for northern dusky salamanders for the purpose of establishing a system to quantify the loss of native amphibians by the activities of man and to provide guidelines for financial penalties in attempts to mitigate or repair damage should population losses occur (Society for the Study of Amphibians and Reptiles Monetary Value of Amphibians Subcommittee, 1989). The absence of larvae of northern dusky salamanders from streams draining coal strip mines in eastern Kentucky and Tennessee appear to be caused by siltation and high metal concentrations (Gore, 1983). Although studies are lacking, there is little doubt that many northern dusky salamander populations have been impacted severely or extirpated in the heavily urbanized and developed areas of their range, which coincide with the most densely human populated region of the United States. Another problem has been the difficulty recognizing the southern limits of the northern dusky salamanders' range, which has been shown to overlap or abut the ranges of several cryptic species, with which it has been confused (e.g., spotted dusky salamanders, Santeetlah dusky salamanders [D. santeetlah], Allegheny Mountain dusky salamanders, Black Mountain salamanders [D. welteri], and even south-

ern dusky salamanders [D. auriculatus]). More studies of the evolutionary relationships of northern dusky salamanders and their relatives are sorely needed to clearly map out their geographic ranges so that studies of their population health and assessment of their conservation status can be made.

Desmognathus imitator Dunn, 1927
IMITATOR SALAMANDER

Carlos D. Camp, Stephen G. Tilley

1. Historical versus Current Distribution. Imitator salamanders *(Desmognathus imitator)* are found in a restricted area of the Great Smoky, Plott Balsam, and Balsam mountains of eastern Tennessee and western North Carolina at elevations above 900 m (Tilley, 1985; Petranka, 1998). Their range extends from the Great Smoky Mountains to slightly east of Soco Gap on Balsam Mountain (Petranka, 1998; Tilley, 2000a). Populations inhabiting rock faces at Waterrock Knob are phenotypically distinct from populations elsewhere (Petranka, 1998; Tilley, 2000a). Tilley (2000a) noted, however, that recognition of this form as a distinct species would "obscure patterns of evolutionary diversification" indicated by allozyme data.

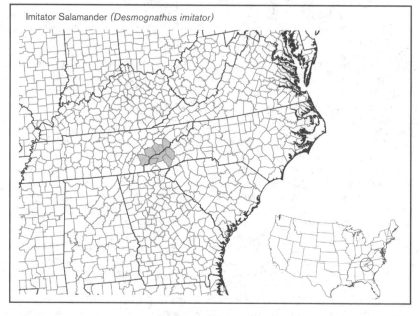

Imitator Salamander (*Desmognathus imitator*)

2. Historical versus Current Abundance. Unknown. Most populations are now protected, as the habitat of imitator salamanders lies almost entirely within the boundaries of the Great Smoky Mountains National Park and Blue Ridge Parkway.

3. Life History Features.
    A. Breeding. Reproduction is aquatic.
     i. Breeding migrations. Unlikely. Breeding probably occurs both in the autumn

and in the spring, with ovipositing in late spring to early summer, similar to other small species of *Desmognathus* (Petranka, 1998). Verrell (1994a) observed the mating of individuals from Waterrock Knob from July–October in a laboratory environment.

ii. Breeding habitat. Mating presumably occurs in the same habitats where adults consistently are found, including seepage areas, wet rock faces, and forest-floor habitats (Tilley et al., 1978; Petranka, 1998).

B. Eggs.

i. Egg deposition sites. Koenings et al. (2000) found six egg clutches in saturated soil within a seepage in late June in western North Carolina. All clutches were attached to the undersides of respective rocks. One clutch was still attended by a brooding female. Five of the clutches were attached to the rock as a monolayer. The sixth clutch had eggs arranged in two layers. Two clutches were found under rocks in saturated gravel in a spring at Double Springs Gap in the Great Smoky Mountains (Swain County, North Carolina; S.G.T., personal observation).

ii. Clutch sizes. The clutches observed by Koenings et al. (2000) averaged 19 eggs/ clutch (range = 13–24). Two presumed clutches of imitator salamander eggs reported by Petranka (1998) contained 23 and 30 eggs (Koenings et al., 2000).

C. Larvae/Metamorphosis.

i. Length of larval stage. Unknown. Presumed to be similar to that of Ocoee salamanders (*D. ocoee*; Bernardo, 2000), which is <1 yr (Huheey and Brandon, 1973; Bruce, 1989).

ii. Larval requirements.

a. Food. Probably small, aquatic invertebrates.

b. Cover. Under rocks, leaf litter, and moss in headwater seepage areas, which may serve as nurseries for this species (Bernardo, 2000).

iii. Larval polymorphisms. Unknown.

iv. Features of metamorphosis. A series of newly metamorphosed juveniles reported by Bernardo (2000) averaged 12 mm SVL (range = 10–14).

v. Post-metamorphic migrations. Unknown. However, adults may be found under cover in the forest floor at some distance from seepages (Tilley et al., 1978).

vi. Neoteny. Unknown.

D. Juvenile Habitat. Juveniles may be abundant in headwater seeps (Bernardo, 2000).

E. Adult Habitat. Imitator salamanders are found along streamsides, on wet rock faces, and associated with the forest floor at high elevations in spruce-fir and northern hardwood forests (Tilley et al., 1978). They usually are found closer to streams than Ocoee salamanders, which they resemble and overlap in distribution. A unique population that inhabits Waterrock Knob occurs on wet rock faces above 1,650 m (Tilley, 2000a). Other rock face populations, phenotypically distinct from those on Waterrock Knob, occur along the Blue Ridge Parkway on Balsam Mountain. Koenings et al. (2000) suggested that the nesting habits of imitator salamanders may be more similar to those of semi-aquatic seal salamanders *(D. monticola)* than to the more closely related Ocoee salamanders.

F. Home Range Size. Unknown.

G. Territories. Unknown.

H. Aestivation/Avoiding Desiccation. Unlikely. Active individuals have been found during summer (Tilley et al., 1978; Bernardo, 2000; Koenings et al., 2000).

I. Seasonal Migrations. Unknown.

J. Torpor (Hibernation). Unknown.

K. Interspecific Associations/Exclusions. Red-cheeked imitator salamanders are hypothesized Batesian mimics of Jordan's salamanders *(Plethodon jordani)* in the Great Smoky Mountains (Petranka, 1998; Tilley, 2000a). They occur syntopically with Ocoee salamanders at higher elevations, but may exclude them from some lower elevation areas (Tilley et al., 1978; Bernardo, 2000). Imitator salamanders and Ocoee salamanders exhibit sexual isolation via a reduced willingness of individuals to mate with members of the other species (Verrell, 1989, 1990a; Verrell and Tilley, 1992). Imitator salamanders are replaced by Santeetlah dusky salamanders *(D. santeetlah)* on lower elevation rock faces in the Plott Balsam Mountains (Tilley, 2000a).

L. Age/Size at Reproductive Maturity. Adult males and females reach 50 and 57 mm SVL, respectively (Tilley, 1985). Bernardo (2000) reported two size classes of juveniles, one averaging 12 mm SVL and a second averaging 18 mm SVL. He noted that a juvenile growth rate of 6 mm/yr is similar to that seen in juvenile Ocoee salamanders (Tilley, 1980; Bernardo, 1994).

M. Longevity. Unknown. An individual of the similar Ocoee salamander has been aged at 10 yr (Castanet et al., 1996).

N. Feeding Behavior. Imitator salamanders take cover during the day and come out at night (Tilley et al., 1978), presumably to forage. Foods are probably similar to those of Ocoee salamanders, that is small invertebrates (Huheey and Brandon, 1973).

O. Predators. Unknown, but probably include ring-necked snakes (*Diadophis punctatus*), northern water snakes (*Nerodia sipedon*), common garter snakes (*Thamnophis sirtalis*), larger salamanders including spring salamanders (e.g., *Gyrinophilus porphyriticus*), birds, and mammals. For the Batesian system to have evolved, predators would include those that find red-cheeked Jordan's salamanders to be noxious as prey and also have the ability to detect chromatic cheek patches.

P. Anti-Predator Mechanisms. Immobility followed by flight (Dodd, 1990a). The chromatic cheek patches have been hypothesized to represent expressions of Batesian mimicry (Petranka, 1998; Tilley, 2000a).

Q. Diseases. Unknown.

R. Parasites. Unknown. Similar Ocoee salamanders harbor a fauna of helminths including nematodes, flukes, and tapeworms (Goater et al., 1987).

4. Conservation.
Most of the geographic range of the imitator salamander falls within the boundaries of the Great Smoky Mountains National Park. Therefore, populations of this species are largely protected from such threats as mining and the harvesting of timber. Petranka (1998) suggested that the greatest long-term threat to this species may result from acid precipitation that may fall on high elevation sites in the Great Smoky Mountains and other areas of the southern Appalachians.

Desmognathus marmoratus (Moore, 1899)
SHOVEL-NOSED SALAMANDER
Carlos D. Camp, Stephen G. Tilley

1. Historical versus Current Distribution.
Shovel-nosed salamanders (*Desmognathus marmoratus*) are found from southwestern Virginia southwest through eastern Tennessee, western North Carolina, and extreme northwestern South Carolina and into northern Georgia at elevations from 300–1,680 m (Petranka, 1998). Drainage patterns, and consequentially rate of stream flow, rather than elevation, apparently limit distribution (Martof, 1962b). Populations are scattered, especially at the northeastern and southwestern extremes of the range. Shovel-nosed salamanders commonly were collected and sold as fish bait in Georgia during the mid-twentieth century (Martof, 1962b), and bait dealers have introduced shovel-nosed salamanders into areas outside their historical range (Martof, 1962b). A survey of bait shops in the Appalachian region of Georgia during 1997–98 showed that this species currently is not important to the bait industry in that state (Jensen and Waters, 1999).

Martof (1962b) noted the absence of shovel-nosed salamanders in streams that have been heavily silted. He also noted the enigmatic absence of this species in apparently favorable habitats in Georgia stream systems (e.g., Hiwassee, Ocoee, and Nottely rivers) to the west of its distribution. A newly discovered, dark-ventered desmognathine *(D. folkertsi)* that is similar in size to shovel-nosed salamanders occurs in these river systems (Camp et al., 2000, 2004). The recently found form is presumably a member of the *marmoratus–quadramaculatus* clade (*sensu* Titus and Larson, 1996) within the genus *Desmognathus* (Camp et al., 2002). Competitive exclusion between the two, while possible,

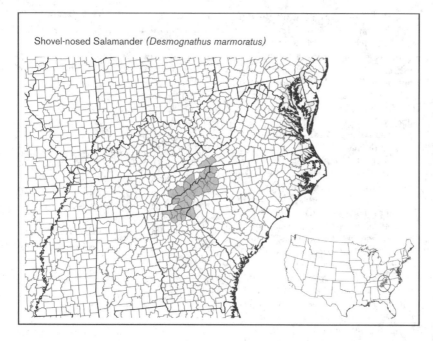

Shovel-nosed Salamander *(Desmognathus marmoratus)*

seems unlikely because the newly discovered form is semi-aquatic, occupying habitats similar to black-bellied salamanders *(D. quadramaculatus)*; shovel-nosed salamanders are nearly (see below) completely aquatic (Camp et al., 2002). Stream capture apparently has influenced the distributional history of shovel-nosed salamanders (Martof, 1962b; Voss et al., 1995).

2. Historical versus Current Abundance. Petranka (1998) noted that shovel-nosed salamanders generally are common in second- and third-order streams at elevations below 1,220 m. Shovel-nosed salamanders are nearly (see below) completely aquatic (Hairston, 1949; Martof, 1962b) and are most abundant in rapids and riffles where densities can be >6 animals/m² (Martof, 1962b). Populations have been shown to be vulnerable to low pHs and heavy metal contamination (Mathews and Morgan, 1982).

3. Life History Features.

   A. Breeding. Reproduction is aquatic.

   i. Breeding migrations. This species is not known to migrate. Martof (1962b) suggested that females may breed every other year. The bienniality of female oogenic cycles in other desmognathines has been questioned, however (Tilley, 1968, 1977; Tilley and Tinkle, 1968).

   ii. Breeding habitat. Mating presumably occurs in habitat that is favored for other activities, that is riffle areas of streams (Martof, 1962b). Mating behavior has not been described.

   B. Eggs.

   i. Egg deposition sites. Females oviposit and attend their clutches in late spring and summer. Eggs are laid on the undersides of large rocks in fast-flowing streams (Pope, 1924). Eggs are attached singly or in tight clusters of 2–4 eggs. Martof (1962b) found that clutches are deposited in the main currents of streams in an average depth of 20 cm (range = 8–36 cm).

   ii. Clutch size. Mean clutch size varies among populations. Clutch size, which varies positively with female body size, ranges from 20–65, and eggs average 4.1 mm in diameter (Martof, 1962b). Incubation times have been estimated to be 10–12 wk, so that eggs hatch from mid-August to mid-September; hatchlings measure about 11 mm SVL (Martof, 1962b).

   C. Larvae/Metamorphosis.

   i. Length of larval stage. In western North Carolina, Bruce (1985a) determined larval period to be 3 yr, whereas Martof (1962b) estimated a larval period of 10–20 mo in Georgia. One of Martof's (1962b) larval samples, however, exhibited three distinct size classes, similar to Bruce's (1985a) data. Both workers found maximum larval size to be similar (38 mm SVL), although Martof (1962b) found newly metamorphosed animals as small as 25 mm SVL.

   ii. Larval requirements.

   a. Food. Larvae feed on aquatic invertebrates. From an analysis of stomach contents, Martof and Scott (1957) reported that larvae fed on aquatic insects represented by five different orders (Plecoptera, Ephemeroptera, Trichoptera, Diptera, Coleoptera).

   b. Cover. Larvae inhabit interstices of rock and gravel on the bottoms of streams (Bruce, 1985a). Small larvae can be under collected in samples and may be more secretive in inter-gravel spaces. Bruce (1985a) noted that larvae readily burrow into aquarium gravel when placed into artificial tanks.

   iii. Larval polymorphisms. Larvae of this species are not known to exhibit distinct polymorphisms.

   iv. Features of metamorphosis. In western North Carolina, larvae reach a maximum size of 37–38 mm SVL (Bruce, 1985a); at a Georgia site, larvae reach a maximum size of 30–36 mm (Martof, 1962b). In other populations, larvae are smaller, with average sizes at metamorphosis between 26 and 33 mm SVL (Petranka, 1998).

   v. Post-metamorphic migrations. This species is not known to migrate.

   vi. Neoteny. Occasionally, larvae with enlarged gonads have been found (Martof, 1962b). All individuals exhibit the neotenic retention of lateral line pores (*sensu* Hilton, 1947).

   D. Juvenile Habitat. Following metamorphosis, juveniles remain in streams, where their habitat resembles that of adults (Martof, 1962b).

   E. Adult Habitat. Shovel-nosed salamanders inhabit cool, well-oxygenated, second- and third-order streams or low gradient first-order streams (Pope and Hairston, 1947; Martof, 1962b). Adults inhabit shallow waters in areas with rocks, loose gravel, and moderate- to fast-flowing water (Petranka, 1998). There is a greater density of salamanders in rapids and riffles than in pools. This may have to do with a physiological need to constantly flush the water next to the skin for aeration purposes. Booth and Feder (1991) showed that, because of the differential diffusion of oxygen, a thin, hypoxic zone develops around the skins of plethodontids in quiet water. Captive shovel-nosed salamanders have been observed to climb into air and rest on the sides of plastic bags when left in cool, quiet water for 12–24 hr (C.D.C., personal observations). Both males and females are found under cover objects during the day and emerge to feed at night (Martof, 1962b). Southerland (1986f) presented evidence that they may be able to climb onto land during rainy weather. One of us (S.G.T.) observed an individual during the daytime on a branch about 10 cm above flowing water of Mill Creek, Blount County, Tennessee. Bishop (1943) stated that *Leurognathus marmorata intermedia [sic]* sometimes occurs on land, but he did not provide corroborative evidence.

   F. Home Range Size. Unknown.

   G. Territories. Preliminary observations of laboratory-based, staged encounters suggest that individuals may not be aggressive toward conspecifics (Jaeger and Forester, 1993).

   H. Aestivation/Avoiding Desiccation. Unlikely. Martof (1962b) found individuals that were active throughout most of the year (April–November).

   I. Seasonal Migrations. Shovel-nosed salamanders are not known to migrate.

J. Torpor (Hibernation). Unknown. However, Martof and Scott (1957) reported collecting individuals with stomachs full of food in the early spring.

K. Interspecific Associations/Exclusions. Shovel-nosed salamanders often coexist with black-bellied salamanders. Individuals of the latter species are not fully aquatic and are often associated with the banks of streams where shovel-nosed salamanders are found (Martof, 1962b). In addition, black-bellied salamanders usually rest or forage with a large part of their body out of the water (Camp and Lee, 1996; Mills, 1996). Therefore, contact between the two species may not be frequent. Preliminary observations of laboratory-based, staged encounters suggest that shovel-nosed salamanders may not be aggressive toward black-bellied salamanders (Jaeger and Forester, 1993). Because of their unique habitat, shovel-nosed salamanders are more likely to compete with small fishes that have similar niche characteristics (e.g., sculpins [*Cottus* sp.]; Greenberg and Holtzman, 1987) than with other species of salamanders.

Shovel-nosed salamanders share habitats with larval members of the two-lined salamander *(Eurycea bislineata)* complex, and Martof and Scott (1957) reported a shovel-nosed salamander that had eaten a larval two-lined salamander. Female two-lined salamanders often nest in riffle areas of streams (Petranka, 1998; C.D.C., personal observations). Any interactions between adult two-lined and shovel-nosed salamanders may be seasonal and related to the timing of nesting by two-lined salamanders.

L. Age/Size at Reproductive Maturity. Adult males range 50–73 mm SVL, with males reaching sizes that average 6–13% larger than that attained by adult females (Martof, 1962b). The largest specimen (78 mm SVL) reported by Martof (1962b), however, was a female. Females mature at 55–59 mm SVL (Martof, 1962b). Maturation may occur at 4–5 yr (Tilley and Bernardo, 1993).

M. Longevity. Unknown. Other desmognathines are known to live for at least 13 yr (Castanet et al., 1996).

N. Feeding Behavior. Generally, adults feed on aquatic and stream-associated terrestrial invertebrates, especially insects (Martof, 1962b). Stomach contents showed prey items to include snails, crayfish, and insects such as ephemeropterans, trichopterans, dipterans, hymenopterans, and coleopterans (Martof and Scott, 1957). Martof and Scott (1957) found the remains of salamanders in 3 of nearly 200 stomachs examined. The two sets of remains that could be identified were a two-lined salamander larva and a conspecific larva.

Underwater observations indicate that shovel-nosed salamanders often lie under stones with their heads sticking out in an alert manner (Martof, 1962b). This sug-gests that at least some of these salamanders may use an ambush foraging strategy. Other individuals have been observed to leave cover objects and move along the stream bottom to forage, especially at night (Martof, 1962b). Observations of similar behaviors have been made of sculpins, small predaceous fishes that occupy the same riffle-type habitats as shovel-nosed salamanders (Greenberg and Holtzman, 1987).

O. Predators. Shovel-nosed salamanders live in streams with predaceous fish (e.g., Cottidae, Cyprinidae, and Salmonidae). Much of the habitat occupied by these salamanders is also prime habitat for trout (Salmonidae), and stock trout *(Oncorhynchus mykiss* and *Salmo trutta)* are often released in these streams by fish and wildlife agencies. Martof (1962b) reported a case of predation of a shovel-nosed salamander by a native brook trout *(Salvelinus fontinalis)*. Water snakes *(Nerodia sipedon),* which readily feed on desmognathines (C.D.C., personal observations), may be common along streams containing shovel-nosed salamanders. In a confined laboratory situation, large stone fly *(Acroneuria* sp.) nymphs attacked small shovel-nosed salamander larvae (Mathews, 1982). Shovel-nosed salamanders occasionally are cannibalistic (Martof, 1962b).

P. Anti-Predator Mechanisms. Individuals tend to take shelter under rocks on stream bottoms and leave them to forage at night (Martof, 1962b). When disturbed by humans, adults move slowly away or swim away for a short distance (Martof, 1962b).

Q. Diseases. Unknown.

R. Parasites. Goater et al. (1987) reported four species of mature nematodes in a survey of 50 shovel-nosed salamanders from southwestern North Carolina. These four species were *Capillaria inequalis, Thelandros magnavulvaris, Omeia papillocauda,* and *Falcaustra plethodontis.* They demonstrated that this nearly completely aquatic species of salamander harbors a less diverse helminth fauna than semi-aquatic congeners. Goater (2000) found no leeches on 50 shovel-nosed salamanders surveyed. He did, however, find trypanosomes, flagellates that live in the bloodstream and presumably introduced by the leech *Oligobdella biannulata,* in 4 of 17 surveyed salamanders.

4. Conservation.
Shovel-nosed salamanders depend heavily on flowing streams having rocky substrates with abundant interstices. They are vulnerable, therefore, to the degradation of these habitats. Shovel-nosed salamanders are largely absent from streams that have been heavily silted (Martof, 1962) and have been eliminated from many areas due to the impoundment of streams (Petranka, 1998). Because of the dependence of this species on aquatic insects, pol-lution or other degrading factors that affect insect populations may also affect populations of shovel-nosed salamanders through the loss of potential food resources. Shovel-nosed salamanders are vulnerable to pollution that results in high acidity and contamination of streams by heavy metals (Mathews and Morgan, 1982). Shovel-nosed salamanders were exploited as fish bait in some areas during the mid-twentieth century (Martof, 1962). However, a recent survey of bait shops in northern Georgia showed that this species is no longer important to the bait industry of that state (Jensen and Waters, 1999).

Desmognathus monticola Dunn, 1916
SEAL SALAMANDER

Carlos D. Camp, Stephen G. Tilley

1. Historical versus Current Distribution.
Seal salamanders *(Desmognathus monticola)* range throughout the central and southern Appalachians from western Pennsylvania to central Alabama. Disjunct populations occur in the coastal plain of southern Alabama and the Florida Panhandle (Rose and Dobie, 1963; Means and Longden, 1970; Mount, 1975; Caldwell and Folkerts, 1976). Seal salamanders are most commonly found below elevations of 1,219–1,372 m, although they occur as high as 1,555 m (Hairston, 1949; Organ, 1961a). Seal salamanders are commonly used as fish-bait in the southern Appalachians. In a recent survey of bait shops in northern Georgia, seal salamanders made up 67% of salamanders sold as "spring lizards" (Jensen and Waters, 1999). The introduction of other desmognathines into uninhabited areas has been attributed to the bait trade (Martof, 1953b). Because of the widespread use of seal salamanders as bait, this species may have been introduced into new areas by anglers.

2. Historical versus Current Abundance.
It is likely that certain populations of seal salamanders have suffered due to collection for fish bait and other human activities. One technique of bait collection, which may have a negative effect, involves pouring liquid bleach into small, high-gradient streams to drive out salamanders. Because salamanders thus collected have low survivabilities, wholesale purchasers (i.e., owners of bait shops) generally do not buy salamanders from such collectors more than once, and this attitude may restrict usage of the bleach (Jensen and Waters, 1999). Acidification of streams due to mine drainage may affect seal salamanders. Roudebush (1988) showed that feeding rates were depressed in salamanders exposed to low pH levels. Petranka et al. (1993) demonstrated that clearcut timber harvesting negatively

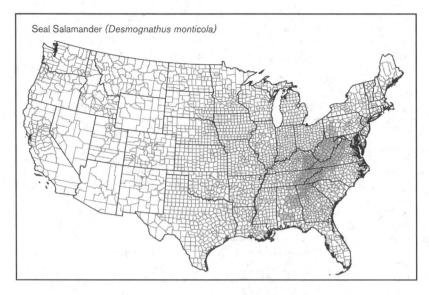

Seal Salamander (*Desmognathus monticola*)

affects the number of total *Desmognathus* individuals, including seal salamanders. They estimated that, at the rate of clearcutting carried out during the 1980s and early 1990s, the Appalachian forests of North Carolina lost as many as 14 million salamanders of all species each year during that time. However, Ash and Bruce (1994) strongly disagreed with these estimates and did not consider clearcutting to have as strong an impact on native salamanders. No significant change in abundance was noted in 20 yr of seal salamander surveys carried out in undisturbed habitats in the Appalachian Mountains of southwestern North Carolina (Hairston, 1996).

3. Life History Features.

A. Breeding. Reproduction is aquatic.

i. Breeding migrations. Seal salamanders are not known to migrate. Mating occurs in the fall and spring in Virginia (Organ, 1961a). Brock and Verrell (1994) observed mating in July–November in specimens collected in southwestern North Carolina. Organ (1961a) suggested a biennial breeding cycle in females; his conclusions of biennial oogenic cycles in desmognathines have been questioned, however (Tilley, 1968, 1977; Tilley and Tinkle, 1968). Eggs have been found from mid-June to August (Pope, 1924; Organ, 1961a; Folkerts, 1968; Bruce, 1990; Camp, 1997a).

ii. Breeding habitat. Mating presumably occurs in the same areas that are optimal for non-reproductive activities. Adult male and female seal salamanders have been observed occupying common refugia during late summer in Georgia (C.D.C., personal observations). Mating involves orientation of the male to the female, chemical and tactile stimulation of the female by the male, and the "tail-straddle walk" characteristic of other plethodontids (Organ, 1961a; Brock and Verrell, 1994).

B. Eggs.

i. Egg deposition sites. Eggs are attended by females and are laid in or near running water. Eggs may be buried in nests in streambeds (Bruce, 1990) >30 cm below ground (Organ, 1961a). Petranka (1998) reports finding eggs in leaf clumps. Egg clutches have been found in seepages flowing under rocks (Folkerts, 1968), under moss next to flowing water (Pope, 1924), and in crevices in wet cliffs (Camp, 1997a). Eggs are attached, typically to the underside of a rock, in either a monolayer or in a loose group, 2–3 eggs thick (Pope, 1924; Folkerts, 1968).

ii. Clutch size. Egg numbers range from 13–39 (Pope, 1924; Organ, 1961a; Folkerts, 1968; Camp, 1997a). Clutch sizes determined from eggs collected in the field averaged 27 and 22 in Virginia (Organ, 1961a) and Alabama (Folkerts, 1968), respectively. Clutch sizes determined from the numbers of developing oocytes are independent of female size (Tilley, 1968). Eggs range in diameter from 4.0–4.8 mm (Bruce, 1989; Camp, 1997a). Incubation time is 1–2 mo, and hatching occurs at 11–12 mm SVL (Folkerts, 1968; Bruce, 1990).

C. Larvae/Metamorphosis.

i. Length of larval stage. Larvae overwinter and metamorphose in either the spring or early to mid-summer, suggesting a larval period of 8–13 mo (Organ, 1961a; Juterbock, 1984; Bruce, 1989).

ii. Larval requirements.

a. Food. Unreported. Larvae probably feed on small aquatic arthropods.

b. Cover. Larvae typically occur along small gravelly bars and rocky seepage areas in streams (Folkerts, 1968).

iii. Larval polymorphisms. Larvae of this species do not exhibit distinct polymorphisms.

vi. Features of metamorphosis. Metamorphosis occurs at approximately 14–16 mm SVL (Organ, 1961a; Folkerts, 1968; Juter-

bock, 1984; Bruce, 1989, 1990, 1995; Bruce and Hairston, 1990).

v. Post-metamorphic migrations. This species is not known to migrate.

vi. Neoteny. This species is not known to exhibit neoteny.

D. Juvenile Habitat. Metamorphosed seal salamanders are nocturnal, emerging from under rocks, logs, or stream-bank burrows (Brandon and Huheey, 1971; Shealy, 1975; Hairston, 1986). Juveniles and adults use different microhabitats based on cover size, moisture levels, and coarseness. Adults are found under larger, moister, coarser objects; juveniles appear to avoid adults, and thus these sites (Krzysik and Miller, 1979; Colley et al., 1989). Juveniles shift their choices of substrates in the presence of adults (Colley et al., 1989). Intraspecific competition for cover objects may occur (Brandon and Huheey, 1971; Kleeberger, 1984). Juveniles are located closer to water in streambeds, on average, than adults (Hairston, 1986). Immature seal salamanders occasionally are found wandering on the faces of wet cliffs (C.D.C. and S.G.T., personal observations).

E. Adult Habitat. Seal salamanders most commonly are found in hardwood forests in association with small- to medium-sized streams containing cool, well-aerated water (Petranka, 1998). Seal salamanders are found under cover objects in association with streambanks and uninundated portions of streambeds, rather than in the stream channel proper (Hairston, 1949; Organ, 1961a; Krzysik, 1979). They are also common in seepages (Organ, 1961a), although they typically do not occur as close to the headwaters of seeps as members of the *D. ochrophaeus* complex (C.D.C. and S.G.T., personal observations). Seal salamanders occasionally use crevices associated with wet cliffs as refugia (Tilley, 1980; Camp, 1997a). Adult males have been observed to share crevices in wet cliffs with adult females (Camp, 1997a; C.D.C., personal observations).

Following heavy rains, seal salamanders may temporarily leave streams and stream banks and forage in surrounding forest (Kleeberger, 1985), occasionally climbing on tree trunks 1–2 m above the ground (Petranka, 1998).

F. Home Range Size. Generally small, estimated to be 0.07–0.45 m^2 in one study (Kleeberger, 1985) and 8.4 m^2 in another (Hardin et al., 1969).

G. Territories. Seal salamanders defend cover sites from conspecifics of either sex (Keen and Sharp, 1984; Keen and Reed, 1985). They defend feeding cover sites significantly more than they do non-feeding cover sites. Individuals of this species move among refugia, defending whichever refugia they happen to inhabit. Therefore, Keen and Sharpe (1984) suggested that seal salamanders possess mobile territories. A similar pattern is seen in black-bellied salamanders *(D. quadramaculatus)* that move among a small set of favored refugia, functionally maintaining a constant territory through a combination of aggressive defense and pheromonal advertisement (Camp and Lee, 1996). Camp (2003a) reports an instance of intraspecific aggression in seal salamanders.

H. Aestivation/Avoiding Desiccation. Unlikely. Surface activity peaks in April and declines with summer. Winter rains may trigger additional surface activity (Shealy, 1975; Petranka, 1998). Individuals are active at the surface in the southern part of their range during periods of warm temperatures throughout the winter (C.D.C., personal observations).

I. Seasonal Migrations. This species is not known to migrate.

J. Torpor (Hibernation). Unknown.

K. Interspecific Associations/Exclusions. Seal salamanders are sympatric with a variety of combinations of congeners throughout their range. Over much of their range, seal salamanders are syntopic with members of the dusky salamander *(D. fuscus)* and/or mountain dusky salamander *(D. ochrophaeus)* complexes. Members of *monticola-fuscus-ochrophaeus* assemblages in Pennsylvania assort positively by body size and substrate-particle size. Seal salamanders, being the largest members, choose the coarsest substrates. They are also the most aquatic species among this group, preferring moister substrates than either of their two sympatric congeners (Krzysik, 1979). Likewise, seal salamanders are more aquatic than their smaller, sympatric counterparts (Ocoee salamanders *[D. ocoee]* and spotted dusky salamanders *[D. conanti]*) in Alabama (Folkerts, 1968). Seal salamanders and Ocoee salamanders segregate spatially and temporally in the Piedmont of South Carolina, with seal salamanders being more aquatic and diurnal than Ocoee salamanders (Shealy, 1975). In experimental, laboratory-based trials, seal salamanders are more aggressive toward dusky salamanders than to conspecifics (Keen and Sharp, 1984). Choices of substrate moisture and sizes of cover objects by either of these two species were not affected by the presence of the other, although activity of dusky salamanders was depressed by presence of seal salamanders (Keen, 1982). Tilley (1997) speculated that interactions with seal salamanders may have contributed to the isolation and genetic differentiation of units of the *D. ochrophaeus* complex.

Throughout much of the southern Appalachian region, seal salamanders are sympatric with at least two congeneric species. They often occur syntopically with black-bellied salamanders and members of the *D. ochrophaeus* complex. They additionally may occur with shovel-nosed salamanders *(D. marmoratus)* and seepage salamanders *(D. aeneus)*, as well as dusky salamanders. Except for the

completely aquatic shovel-nosed salamander, these species assort by body size along the stream–forest interface with larger species being more aquatic and smaller ones occurring more terrestrially. This pattern is evident both along a horizontal gradient from stream–stream bank–forest and along a vertical gradient from stream–seepage–forest (Organ, 1961a). Seal salamanders are intermediate in both body size and habitat preference, occurring along stream banks, in uninundated parts of streambeds, and seepages. The observed pattern of desmognathine assortment was explained initially as niche partitioning among competitors (Hairston, 1949; Organ, 1961a). Tilley (1968) and Hairston (1980c), however, suggested that interspecific predation was a more likely cause. A number of studies attempted to determine which was the more probable factor (e.g., Kleeberger, 1984; Carr and Taylor, 1985; Hairston, 1986; Southerland, 1986a,b,d). They generally concluded that some combination of predation and aggressive interference were important factors in interspecific desmognathine interactions. Hairston (1986) made the strongest case for predation, with competition being a secondary factor. His statistical methods have been criticized, however (Jaeger and Walls, 1989). Although large desmognathines readily eat small ones in artificial environments, there is no evidence from extensive dietary studies that either black-bellied salamanders or seal salamanders are significant predators of heterospecific congeners (Camp, 1997b). Nevertheless, some researchers hold that predation by large species may have been important historically in the organization of desmognathine communities. Alternative hypotheses based on abiotic factors rather than biotic ones, such as competition and predation, have recently been proposed (Bruce, 1996; Camp et al., 2000). The lack of predation under natural conditions is probably a result of differential habitat selection and behavioral avoidance, perhaps involving chemical cues, of larger congeners by small individuals. The activity levels of dusky salamanders are depressed by the presence of larger seal salamanders (Keen, 1982), and both activity levels and substrate choices of seal salamanders are altered by the presence of larger black-bellied salamanders in experimental tanks (Carr and Taylor, 1985; Roudebush and Taylor, 1987a). Seal salamanders avoid chemical extracts of black-bellied salamanders under experimental conditions (Southerland, 1986a; Roudebush and Taylor, 1987b; Jacobs and Taylor, 1992). Keen (1985), however, suggested that interspecific aggression between these two species may not be more important than intraspecific interactions. The presence or absence of black-bellied salamanders did not affect the dispersion

of seal salamanders in a study by Grover (2000).

L. Age/Size at Reproductive Maturity. Adult size varies among populations. Bruce and Hairston (1990) reported size data on two populations in southwestern North Carolina. Adult males at Wolf Creek averaged 57 mm (range = 46–72) SVL, and those from Coweeta averaged 67 mm (range = 48–80) SVL. Adult females averaged 59 mm (range = 53–65) and 64 mm (range = 52–76) SVL, respectively, at the same two sites. Castanet et al. (1996) and Bruce et al. (2002) aged individuals of these same populations using skeletochronology. They found that males and females at each site mature at 4–6 yr and 5–7 yr, respectively. Bruce et al. (2002) showed, however, that more individuals at Coweeta matured at later ages than those at Wolf Creek. Adults measured from northern Alabama had respective average SVLs for males and females of 64 mm and 57 mm; those from the Alabama Coastal Plain averaged 68 mm and 59 mm for males and females, respectively (Folkerts, 1968). Duncan (1967) reported maximum sizes for seal salamanders in Virginia to be 78 mm for males, 67 mm for females. Series of salamanders examined from Kentucky and North Carolina indicated that maturity was reached at 42 mm in males, 48 mm in females (Juterbock, 1978).

Adult seal salamanders exhibit male-biased sexual size dimorphism. Although males mature at smaller sizes and earlier ages than females, males reach larger maximum sizes than females. This results from females having more depressed post-maturation growth rates than males (Bruce, 1993; Castanet et al., 1996).

M. Longevity. Castanet et al. (1996) and Bruce et al. (2002) aged individuals using skeletochronological techniques and reported that females live at least 9 yr and males live at least 11 yr. Bruce and Hairston (1990) suggested greater potential longevity for animals in a population characterized by delayed maturity and larger body size.

N. Feeding Behavior. Nocturnal. Animals emerge from daytime retreats under logs and rocks or from stream-bank burrows to forage. Juveniles actively move about to forage, whereas adults are more likely to sit in the entrances to their burrows and wait for prey (Brandon and Huheey, 1971; Kleeberger, 1985; C.D.C., personal observations). Surface activity peaks around midnight, with a second bout of activity near sunrise (Hairston, 1949, 1986; Shealy, 1975). Both juveniles and adults feed on aquatic and terrestrial invertebrates including true bugs, stoneflies, caddisflies, lepidopterans, beetles, mayflies, dipterans, wasps, ants, odonates, millipedes, and earthworms (Hairston, 1949; Duncan, 1967; Krzysik, 1979; Kleeberger, 1982). Larger animals eat larger prey items (Krzysik, 1979). Seal salamanders have been reported to occasionally eat other salamanders (Shealy, 1975; Bernardo, 2002); however, evidence from dietary studies suggests that predation on other salamanders occurs rarely under natural conditions (Camp, 1997b). Brown et al. (2003) report a seal salamander taking an especially large (35 mm, 74.5% of the salamander's SVL) hesperiid lepidopteran larva.

O. Predators. Natural predators are not reported (Petranka, 1998). Spring salamanders (*Gyrinophilus porphyriticus*) feed regularly on other salamanders (Bruce, 1979) and may prey on larval and juvenile seal salamanders. In Georgia, seal salamanders occasionally occur along the edges of streams with potentially predaceous fish (e.g., *Cottus carolinae*, *Nocomis leptacephalus*, *Semotilus atromaculatus*). Snake species known to feed on amphibians (e.g., *Nerodia sipedon*, *Diadophis punctatus*) have been seen foraging at night in streams inhabited by seal salamanders (C.D.C., personal observations). Predation pressure and/or intraspecific aggression may be high. Wake and Dresner (1967) report that 11% of a museum sample had broken tails.

P. Anti-Predator Mechanisms. Seal salamanders apparently use chemical cues to avoid contact with, and thus possible predation by, black-bellied salamanders (Roudebush and Taylor, 1987b; Jacobs and Taylor, 1992). Choices of egg-deposition sites may, in part, mitigate the effects of predation on eggs. Egg-attendance by maternal females apparently prevents at least some predation on egg clutches. Removal of a 65 mm female from her clutch resulted in the predation on the unattended eggs by a 67 mm female conspecific within 10 min (Camp, 1997a).

Q. Diseases. Unknown.

R. Parasites. Goater et al. (1987) reported the occurrence of six adult nematode species (*Capillaria inequalis*, *Thelandros magnavulvaris*, *Omeia papillocauda*, *Desmognathinema nantahalaensis*, *Falcaustra plethodontis*, and *Cosmocercoides dukae*) and one larval nematode species (an ascaridoidid) in seal salamanders. They found mature forms of three species of trematode (*Brachycoelium elongatum*, *Gorgoderina bilobata*, and *Phyllodistomum solidum*) and one species of tapeworm (*Cylindrotaenia americana*). They also found larval tapeworms (proteocephalan plerocercoids) and one species of larval acanthocephalan (*Centrorynchus conspectus*). Goater (2000) found a much lower incidence of leeches (*Oligobdella biannulata*) in seal salamanders than in more aquatic black-bellied salamanders. He similarly found low infection rates of trypanosomes, blood parasites that presumably use leeches as vectors, in seal salamanders.

4. Conservation.

Seal salamanders remain abundant in areas of preferred habitat over much of their geographic range. In 20 yr of surveys of seal salamanders carried out in undisturbed habitats in the Appalachian Mountains of North Carolina, Hairston (1996) found no significant changes in abundance. There may be local threats to viable populations, however. Stream acidification as a consequence of mine drainage may negatively affect some populations. Roudebush (1988) showed that low pH depressed feeding rates, but salamanders did not die when exposed to a pH of 3.5 for 3 wk.

Certain populations of seal salamanders may have experienced declines due to exploitation as fish bait. In a recent survey of bait shops in northern Georgia, this species made up 67% of salamanders sold as "spring lizards" (Jensen and Waters, 1999). The collecting technique of pouring liquid bleach into small, high-gradient streams to drive out salamanders may be particularly harmful to this and other species of streamside salamanders. Because salamanders collected using this method have low survivabilities, bait shops owners often refuse to purchase salamanders from such collectors more than once (Jensen and Waters, 1999).

Optimal habitats for seal salamanders are seepages and small- to medium-sized streams containing cool, well-aerated water, located within mesic, hardwood forests (Organ, 1961; Petranka, 1998). The greatest threat to populations of this species, therefore, may be timber harvesting techniques (e.g., clearcutting) that increase rates of evaporative water loss through the removal of the protective canopy. Petranka et al. (1993) argued that clearcut timber harvesting negatively affects the number of total *Desmognathus* individuals, including seal salamanders, and estimated that clearcutting killed millions of salamanders from the Appalachian forests of North Carolina during the 1980s and early 1990s. Ash and Bruce (1993), however, disputed these estimates and considered them to be exaggerations of the actual numbers lost. Desmognathine salamanders currently are abundant in areas of the southern Appalachians (the Great Smoky Mountains) that have been extensively logged in the past (S.G.T., personal observations). Appalachian seepages occasionally dry up (Camp, 2000), and it is not known how periodic drought may interact with techniques of timber harvesting to affect populations of seal and other seepage-dwelling salamanders. Because of restricted ranges/population sizes, inherently warmer climates, and intensive silviculture in the Coastal Plain, timber harvesting may be a greater threat to Coastal Plain populations of seal salamanders than to Appalachian populations.

Desmognathus ochrophaeus Cope, 1859
ALLEGHENY MOUNTAIN DUSKY SALAMANDER

Thomas K. Pauley, Mark B. Watson

1. Historical versus Current Distribution.
Dunn (1926) described the range of Allegheny Mountain dusky salamanders (*Desmognathus ochrophaeus*) from Clinton County, New York, south to Garrett County, Maryland, and west to Columbus, Ohio, through the Appalachian Plateau of Pennsylvania. Bishop (1941b) extended the range south to Highland County, Virginia. Bishop (1943) further extended the range west into eastern Kentucky, and south through West Virginia, southwestern Virginia, western North Carolina, eastern Tennessee, and into northern Georgia. Sharbel and Bonin (1992) described the northernmost point of the range of Allegheny Mountain dusky salamanders at the Chateauguay River drainage in Québec, Canada. Recent descriptions of several sibling species based upon geographic and molecular data have reduced the southern range of Allegheny Mountain dusky salamanders to the southern mountains of Tennessee (Tilley and Mahoney, 1996). Tilley and Mahoney (1996) determined that the range of Allegheny Mountain dusky salamanders extends from southwestern Virginia (Burmley, Clinch, Walker, and Potts mountains) west into the Cumberland Mountains and Cumberland Plateau of southeastern Kentucky, north through the Allegheny Plateau and Allegheny Mountains of West Virginia, Maryland, Pennsylvania, Ohio, and through the Adirondack Mountains in New York into southern Québec. While populations have undoubtedly been lost, there are no data to support recent range reductions.

2. Historical versus Current Abundance.
In the higher elevations of West Virginia (975–1,463 m), Pauley (1980a) found Allegheny Mountain dusky salamanders in forested habitats on 17 of 19 mountains surveyed from 1976–79. Of the five most common species observed, Allegheny Mountain dusky salamanders comprised 26.5% of the salamander assemblage. In this study, while Allegheny Mountain dusky salamanders ranged from 975–1,400 m, they were most abundant between 975 and 1,036 m.

Petranka et al. (1994) demonstrated that local populations of the sibling species Blue Ridge dusky salamanders (*D. orestes*) can be severely impacted by clearcutting. Pauley (unpublished data) found Allegheny Mountain dusky salamanders are slower than eastern red-backed salamanders (*Plethodon cinereus*) to recolonize early successional forests in clearcut areas in northern West Virginia. Pauley (1980b) found that Allegheny

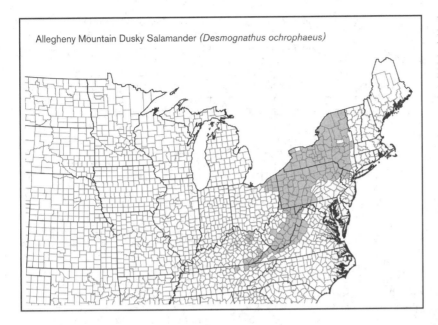

Allegheny Mountain Dusky Salamander (*Desmognathus ochrophaeus*)

Mountain dusky salamanders require cooler and more moist areas for foraging and nesting sites than do eastern red-backed salamanders. The inability of Allegheny Mountain dusky salamanders to recolonize clearcuts may be due to the lack of optimal foraging and nesting sites in young regrowth forests.

3. Life History Features.

    A. Breeding. Reproduction is aquatic.

    i. Breeding migrations. Allegheny Mountain dusky salamanders do not migrate.

    ii. Breeding habitat. Allegheny Mountain dusky salamanders occupy moist areas such as first-order streams, spring seeps, and wet rock faces. As with most plethodontids, courtship and mating occur at night. Barbour (1971) observed a pair in a courtship dance in Harlan County, Kentucky, on 14 July. Bishop and Crisp (1933) reported finding a spermatophore in the cloacal vent of a female from New York in May. Tactile and chemical stimulation are used by males during courtship (Verrell and Mabry, 2000).

    B. Eggs.

    i. Egg deposition sites. Nesting areas are usually under moist fallen logs, under rocks, and in mud banks. Nests have been found under rocks ≤ 0.5 m from a water source (Pauley, 1993b) and in interstices within seepage banks of first-order streams (Keen and Orr, 1980; Marcum, 1994).

    Eggs are deposited in grape-like clusters, and each egg is attached to a central stalk (Bishop and Crisp, 1933; Bishop, 1941b). Keen and Orr (1980) found spent females in northern Ohio populations in late April. The clutch size in Allegheny Mountain dusky salamanders corresponds to female body size, with larger females producing larger clutches (Keen and Orr, 1980).

    ii. Clutch size. Pfingsten (1966) found clutch sizes numbering around 19 eggs in Ohio. Wood and Wood (1955) reported 10–37 eggs/clutch in Virginia. Green and Pauley (1987) reported clutch sizes around 19 in West Virginia, although Marcum (1994) reported smaller clutch sizes in West Virginia, ranging from 3–19. In Pennsylvania, clutch size ranges from 8–24 (Hall, 1977). Allegheny Mountain dusky salamander females are able to recognize and brood eggs from related individuals (Masters and Forester, 1995).

    Bishop and Crisp (1933) found freshly laid eggs in August and newly hatched larvae in September and March in New York and northern Pennsylvania. Marcum (1994) determined that eggs hatch in northern West Virginia in early October. The average SVL of young at hatching is 8.7 mm. Keen and Orr (1980) noted that larvae emerged from a clutch of eggs on 19 April in Ashtabula County, Ohio. In Kentucky, eggs are deposited in August, September, and October, and larvae are about 2 cm (3/4 in) long (Barbour, 1971). Bishop (1941b) observed hatching in a New York population in mid-March. The timing of hatching varies with elevation and may be biphasic. Eggs and recently hatched young have been observed during autumn and spring (Bishop, 1943; Keen and Orr, 1980; Marcum, 1994).

    C. Larvae/Metamorphosis. The larval period of Allegheny Mountain dusky salamanders varies depending on moisture and temperature (Bishop, 1941b). Gills may be lost in several days to several weeks or up to 8–10 mo, depending on temperature and food availability (Bishop, 1941b). Allegheny Mountain dusky salamanders from northern West Virginia may have larval periods of only 1–3 wk. Marcum (1994) found transformed juveniles that still showed signs of yolk plug remnants and also observed transformed individuals with considerable amounts of yolk present. Allegheny Mountain dusky salamander larvae were not found in the same streams during a 5-yr study prior to Marcum's work (Pauley, 1995a). Bishop and Crisp (1933) suggested that larvae transformed at a length of 18 mm without entering water.

    i. Food. Orr and Maple (1978) found that Allegheny Mountain dusky salamander larvae used their yolk sacs by 140 d after hatching. Several reports indicate that the larval period is short and in some cases may only occur in the egg (Bishop, 1943; Marcum, 1994). Marcum (1994) found newly transformed individuals with yolk plugs or traces of yolk plugs that could supply them with sufficient nutrients until they completely transformed.

    D. Juvenile Habitat. The juvenile habitats of Allegheny Mountain dusky salamanders are similar to those of adults. Although juveniles can be found under the same types of cover objects as the adults and in the same habitat, they are observed more frequently under and between damp leaves on the forest floor and at the edges of first-order streams. Smaller individuals may utilize smaller cover objects and may be found at greater distances from low elevation streams, where larger congenerics are absent.

    E. Adult Habitat. Allegheny Mountain dusky salamanders occupy moist woodlands, seepage areas, wet rock faces, and small streams throughout most of their range (Keen, 1979; Green and Pauley, 1987). These salamanders are found under rocks, leaves, bark, and logs and can be found in forests some distance upslope from streams (Weber, 1928; Green and Pauley, 1987; Pauley, 1995a). In a 5-yr study of salamander composition along two, 100-m, horizontal transects located 20 m and 40 m upslope of first-order streams in four watersheds in northern West Virginia, Allegheny Mountain dusky salamanders (n = 1,787) outnumbered the typically more terrestrial eastern red-backed salamanders (n = 855) 2:1 (Pauley, 1995a). This result demonstrates that Allegheny Mountain dusky salamanders can be abundant in moist ravines, moist old logging roads, and close to seepages.

    F. Home Range Size. The average home range of Allegheny Mountain dusky salamanders from Ohio populations is <1 m^2 (Holomuzki, 1982). Holomuzki (1982) found that some Allegheny Mountain dusky salamanders return to their home territory when displaced ≤30 m. In Pennsylvania, Hall (1977) found the average movement of Allegheny Mountain dusky salamanders to be 1.8 m.

    G. Territories. Smith and Pough (1994) observed that when compared to eastern red-backed salamanders, Allegheny Mountain dusky salamanders were more successful defending their territories and intruding into other territories. Stewart and Bellis (1970) found that Allegheny Mountain dusky salamanders were arranged randomly along a stream bank and that disturbing the cover objects caused salamanders to disperse to other cover within 1–2 d. Individuals from West Virginia have been found sharing cover objects such as rocks and logs with up to three or more conspecifics (T.K.P., personal observations).

Evans et al. (1997) showed that female Allegheny Mountain dusky salamanders exhibited a significant preference for substrates that were marked with scents from males outside their population. Evans and Forester (1996) suggest that male Allegheny Mountain dusky salamanders use chemical clues to identify potential rivals in their territory, but chemical clues may be less important for female recognition of other female conspecifics. Geographic proximity may be important in recognition of conspecifics for courtship. Electrophoretic data demonstrated low levels of hybridization between Allegheny Mountain dusky salamanders with northern dusky salamanders (D. fuscus) in Québec (Sharbel et al., 1995).

H. Aestivation/Avoiding Desiccation. Allegheny Mountain dusky salamanders are active on the surface from March–October in Ohio and southern and northern West Virginia (Keen, 1979; Pauley, 1993b, 1995a). Locomotion and foraging are affected if a dehydration deficit exceeds 12% (Houck and Bellis, 1972).

I. Seasonal Migrations. Seasonal migrations of Allegheny Mountain dusky salamanders have been reported. Allegheny Mountain dusky salamanders will move from the surface to underground refugia with the onset of cold weather.

J. Torpor (Hibernation). Allegheny Mountain dusky salamanders apparently emerge from winter refugia in late March and April and remain active on the forest floor through October (Keen, 1979; Pauley, 1993b, 1995a). They are less active on the surface and ingest less prey when air temperatures drop from 5–0 °C, and they remain in underground refugia when temperatures drop below 0 °C (Keen, 1979). They will remain active during the winter around springs, seepages, and bogs or fens (Green and Pauley, 1987).

K. Interspecific Associations/Exclusions. Over a large portion of their range, Allegheny Mountain dusky salamanders are associated with three other desmognathine salamanders: black-bellied salamanders (D. quadramaculatus), seal salamanders (D. monticola), and northern dusky salamanders. These four species make up an assemblage within the mountain streams where they are found. In northern Ohio, Allegheny Mountain dusky salamanders are sympatric with northern dusky salamanders (Orr and Maple, 1978). At higher eleva-

tions and more northern latitudes, Allegheny Mountain dusky salamanders occupy more terrestrial habitats and can be found at considerable distances from the nearest water source. At high elevations in West Virginia, Allegheny Mountain dusky salamanders are sympatric with eastern red-backed salamanders, northern slimy salamanders (P. glutinosus), Wehrle's salamanders (P. wehrlei), and Cheat Mountain salamanders (P. nettingi; Santiago, 1999).

L. Age/Size at Reproductive Maturity. Bishop (1943) lists the length of males from Rochester, New York, at over 70 mm and females at 73 mm. Orr (1989) reports adult males from Ohio to be 37 mm SVL. Hall (1977) found males mature at 3 yr. Females from northern Ohio yolked their first clutches of eggs at around 27–30 mm SVL and at approximately 36 mo of age (Keen and Orr, 1980). Keen and Orr (1980) note that some females from Ohio do not complete their first clutch of eggs until 36–42 mo.

M. Longevity. Snider and Bowler (1992) report that a captive Allegheny Mountain dusky salamander lived for nearly 20 yr. It is not known if this specimen was truly D. ochrophaeus or a sibling species.

N. Feeding Behavior. Juvenile and adult Allegheny Mountain dusky salamanders, similar to other Desmognathus species, are ambush predators. Adult and larval dipterans, hymenopterans, snails, mites, earthworms, and collembolans are important food items for Allegheny Mountain dusky salamanders (Krzysik, 1979; Keen, 1979; Pauley, 1995a). Keen (1979) observed that temperature alters feeding activity of Allegheny Mountain dusky salamanders. Salamanders in seepage habitats were more active at lower temperatures than in streamside habitats, and foraging success was closely correlated with precipitation. They are most active around sunset and during the first hour thereafter when prey items (especially dipterans) become active (Holomuzki, 1980). Fitzpatrick (1973) found that female Allegheny Mountain dusky salamanders need only 2,940 cal/yr to survive and reproduce.

O. Predators. Female Allegheny Mountain dusky salamanders have been reported to prey upon conspecific eggs, larvae, and hatchlings (Fitzpatrick, 1973; Wood and Wood, 1955). Undoubtedly, Allegheny Mountain dusky salamander eggs and larvae serve as prey for larger streamside salamanders, such as other species of Desmognathus, and adults, juveniles, and larger larvae of Eurycea and Gyrinophilus.

Juveniles and adult predators include eastern garter snakes (Thamnophis s. sirtalis; Whiteman and Wissinger, 1991) and birds (Coker, 1931). Other predators probably include ring-necked snakes (Diadophis punctatus), small rodents, and other mammals, and turkeys and other birds. Ring-necked snakes may be less of a threat

because Allegheny Mountain dusky salamanders are known to avoid substrates that ring-necked snakes have marked with their scent (Cupp, 1994).

P. Anti-Predator Mechanisms.
i. Eggs. The most effective anti-predator mechanisms are probably cryptic sites for nests and maternal care of the eggs.

ii. Larvae. Little is known of the anti-predator mechanism of Allegheny Mountain dusky salamander larvae. As some populations have short larval stages, anti-predator mechanisms are probably similar to those of the eggs—cryptic sites with maternal brooding.

iii. Adults. Lutterschmidt et al. (1994) demonstrated that Allegheny Mountain dusky salamanders avoid skin extracts from conspecific and heterospecific salamanders. This suggests that Allegheny Mountain dusky salamanders respond to chemical signals from damaged salamander skin, and these chemicals may be important in predator avoidance. Brodie (1977) lists members of the genus Desmognathus as using biting and pseudoaposematic coloration as anti-predator defenses. Whiteman and Wissinger (1991) similarly report biting and tail autotomy as important anti-predator behaviors in response to garter snake attacks, although they suggest tail autotomy may be most important, especially in regard to larger snakes. Brodie and Howard (1973) suggested that Allegheny Mountain dusky salamanders resemble several species of Plethodon, including eastern red-backed salamanders and Cheat Mountain salamanders, and that Allegheny Mountain dusky salamanders may be a polymorphic Batesian mimic of these Plethodon species. Dodd et al. (1974) found that when predators were fed one Plethodon species (Peaks of Otter salamanders, P. hubrichti), they experienced what was interpreted as discomfort. The dorsal coloration of Allegheny Mountain salamanders within the range of eastern red-backed salamanders and Cheat Mountain salamanders in West Virginia varies from a red dorsal stripe to dark backs (T.K.P., personal observations). These dorsal patterns could be an effective mimic of these two potentially distasteful species of Plethodon.

Q. Diseases. There are no reports of diseases in Allegheny Mountain dusky salamanders.

R. Parasites. Rankin (1937) found the protozoans Prowazekella longifilis and Tritrichomonas augusta, and the nematode Capillaria augusta. Baker (1987) lists three nematode parasites of Allegheny Mountain dusky salamanders: Batracholandros salamandrae, B. magnavulvaris, and Falcaustra plethodontis. Goater et al. (1987) lists adult nematodes (Capillaria inequalis, Thelandros magnavulvaris, Omeia papillocauda, Falcaustra plethodontis, Cosmoceroides dukae), trematodes (Brachycoelium elongatum, Gorgoderina bilobata, Phyllodistomum *solidum)*, and cestodes *(Cylindrotaenia americana)* in Allegheny Mountain dusky salamanders. Goater et al. (1987) also found larval nematodes *(Ascaridoidea* sp.), cestodes *(proteocephalan plerocercoid)*, and acanthocephala *(Centrorynchus conspectus)* in Allegheny Mountain dusky salamanders.

4. Conservation.
Allegheny Mountain dusky salamanders are common terrestrial and semi-aquatic salamanders in the central and northern Appalachian Mountains. They occupy a variety of forested habitats and appear to survive silvicultural impacts and forest fragmentation. Isolated mountain populations in eastern Kentucky and Tennessee should be monitored more closely (Petranka, 1998). This may be particularly true for those populations that could be impacted by mountain top removal coal mining.

Desmognathus ocoee Nicholls, 1949
OCOEE SALAMANDER

Carlos D. Camp, Stephen G. Tilley

1. Historical versus Current Distribution.
Ocoee salamanders *(Desmognathus ocoee)* occur in two distinct regions. The larger region lies in the southwestern Blue Ridge and the adjacent Piedmont Physiographic Provinces. Blue Ridge populations include those in the Balsam, Blue Ridge, Cowee, Great Smoky, Nantahala, Snowbird, Tusquitee, and Unicoi mountains, as well as lower elevation populations in gorges of major rivers, including the Hiwassee, Ocoee, Tugaloo, and Tallulah (Valentine, 1961, 1964; Martof and Rose, 1963; Tilley and Mahoney, 1996). Populations also occur in adjacent areas of the Piedmont Province of Georgia (Martof and Rose, 1963) and South Carolina (Valentine, 1964). They occur as far south as the Gainesville Ridges (*sensu* Wharton, 1978) in the upper Piedmont of northeastern Georgia (Camp, 2000). A second cluster of populations occurs in the Appalachian Plateau of northeastern Alabama (Folkerts, 1968; Mount, 1975; Tilley and Mahoney, 1996). Ocoee salamanders occur over a greater elevational range (from low-lying gorges to mountain tops) than any other species of *Desmognathus* (Petranka, 1998). Populations of Ocoee salamanders in different mountain ranges have undergone considerable genetic differentiation and may ultimately be subdivided into several additional taxa (Tilley and Mahoney, 1996). Tilley (1997) speculated that interactions with seal salamanders may have contributed to the isolation and genetic differentiation of units of the *D. ochrophaeus* complex.

2. Historical versus Current Abundance.
Ocoee salamanders are widespread and abundant in seepage areas, on wet rock

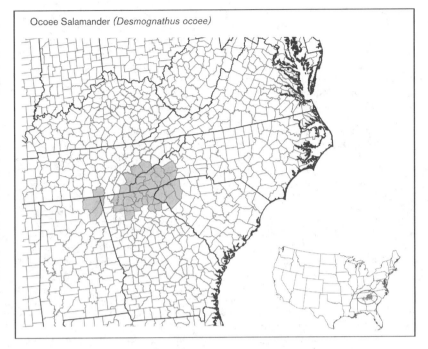

Ocoee Salamander *(Desmognathus ocoee)*

faces, along banks of small streams, and in mesic forest floor habitats. Individuals often are found some distance from water in mesic forests at higher elevations. Low elevation populations are less terrestrial than those at high elevations; those on the Georgia Piedmont are restricted to areas of streambed except during wet weather (C.D.C., personal observations). Population densities on rock faces generally exceed those along streams. Rock face densities have been reported to be six adults/m^2 and 18–25 total individuals/m^2 (Huheey and Brandon, 1973; Tilley, 1980; Bernardo, 1994).

Petranka et al. (1993) demonstrated that clearcut timber harvesting negatively affects the number of total *Desmognathus* individuals, including members of the *D. ochrophaeus* complex. They estimated that, at the rate of clearcutting carried out during the 1980s and early 1990s, the Appalachian forests of North Carolina lost as many as 14 million salamanders of all species each year during that time. However, Ash and Bruce (1994) strongly disagreed with these estimates and did not consider clearcutting to have as strong an impact on native salamanders. Ocoee salamanders are significantly more abundant in cove hardwood stands >85 yr of age than in younger stands (Ford et al., 2002). No significant change in abundance was noted in 20 yr of surveys of Ocoee salamanders carried out in undisturbed habitats in the Appalachian Mountains of southwestern North Carolina (Hairston and Wiley, 1993; Hairston, 1996).

3. Life History Features.
   A. Breeding. Courtship occurs on land. Eggs are laid in or near flowing water.

   i. Breeding migrations. Gravid females choose nesting sites as many as 2–3 wk before ovipositing (Forester, 1981).

   ii. Breeding habitat. Courtship occurs on land from September–June (Martof and Rose, 1963; Huheey and Brandon, 1973; Forester, 1977; Houck et al., 1985).

   B. Eggs.
   i. Egg deposition sites. Females oviposit in cavities beneath rocks, in or under decaying logs, in leaf litter or under moss, under cover objects (e.g., rocks) near seepages, springs, and small streams, and behind vegetation or in crevices associated with wet cliffs (Pope, 1924; Martof and Rose, 1963; Folkerts, 1968; Forester, 1977; Bruce, 1990; Bernardo and Arnold, 1999; Camp, 2000). As is the case in other plethodontid species, females brood their eggs. Brooding appears to serve in defense against predators, remove dead eggs from the nest, reduce egg desiccation, reduce fungal attacks, and perhaps help the hatchlings escape (Tilley, 1972; Huheey and Brandon, 1973; Forester, 1978, 1979, 1984). Oviposition occurs from June–September (Martof and Rose, 1963; Folkerts, 1968), and some females may brood eggs during the winter (Tilley, 1977). Salamanders that nest near the headwaters of seepages risk the loss of reproductive investment due to failure to lay or the loss of eggs to desiccation during periods of drought (Camp, 2000).

   ii. Clutch size. Females may congregate and oviposit their eggs in close proximity in areas of favorable habitat (Martof and Rose, 1963). Clutch sizes determined from ovarian follicles vary from 5–29 and are positively related to female body size in most populations (Martof and Rose, 1963). Clutch sizes determined from counting eggs laid in natural clutches averaged 12 (range = 5–18) in Alabama (Folkerts, 1968); 13 (range = 5–23) in North Carolina (Bruce, 1996); and 15 (range = 5–21) in Georgia (C.D.C., unpublished data). There is a strong, positive relationship between clutch size and the body size of the respective attending female in Georgia (C.D.C., unpublished data). Although clutch size and female body size are positively related, the same relationship may not be true between clutch size and female age (Tilley, 1980). Females are capable of oviposition in successive years (Forester, 1977; Tilley, 1977, 1980), but probably seldom reproduce annually throughout their lives (Tilley, 1977). Eggs measure about 2–3 mm in diameter (Martof and Rose, 1963; Folkerts, 1968; Huheey and Brandon, 1973; Bruce, 1990).

Clutches observed near Highlands, North Carolina, at 1,265 m hatched in 53–66 d (Tilley, 1972). Seven clutches observed at approximately 250 m on the Georgia Piedmont hatched in 32–39 days (C.D.C., unpublished data). Clutches near hatching have been found in August–September (Pope, 1924; Noble, 1927a; Martof and Rose, 1963; Tilley, 1972). Eggs on the Georgia Piedmont hatch from September–October (C.D.C., personal observations). Hatching occurs at approximately 11–12 mm SVL (Tilley, 1980; Bruce, 1989).

C. Larvae/Metamorphosis.

i. Length of larval stage. The larval period in montane populations lasts between 9–10 mo (Huheey and Brandon, 1973; Bruce, 1989). Larvae in a Georgia Piedmont population metamorphose in 7–8 mo (C.D.C., unpublished data). Beachy (1995) demonstrated that food and temperature can have environmental effects on the larval period under experimental conditions.

ii. Larval requirements.

a. Food. Larvae feed on small aquatic invertebrates. Individuals in captive environments readily fed on live *Tubifex* worms (Beachy, 1995).

b. Cover. Larvae are found in shallow water associated with seepages or in thin films of water flowing over rock faces (Huheey and Brandon, 1973).

iii. Larval polymorphisms. Ocoee salamanders are not known to exhibit distinct larval polymorphisms.

iv. Features of metamorphosis. Metamorphosis usually occurs in May and June, with animals averaging 13–14 mm SVL (Bruce, 1989; Bernardo, 1994). Newly metamorphosed individuals averaged 61–94 mg in weight under experimental conditions (Beachy, 1995).

v. Post-metamorphic migrations. Ocoee salamanders are not known to migrate.

vi. Neoteny. Neoteny is not known in this species.

D. Juvenile Habitat. Juveniles occur in seepages (Bernardo, 1994), on wet rock faces (Huheey and Brandon, 1973), and under cover along edges and in the streambeds of small streams (C.D.C., personal observations). Growth rates vary depending on elevation, microhabitat (rock faces versus forest floor), and size. Generally, growth rates are about 5–7 mm SVL/yr until animals reach sexual maturity, when growth slows (Tilley, 1977, 1980).

E. Adult Habitat. Ocoee salamanders have a strong affinity for the headwaters of first-order streams in montane regions (Tilley, 1997). Individuals are common in seepages, on wet rock faces, and in streambeds of larger streams (Petranka, 1998). Ocoee salamanders will move away from the streambeds under moist conditions, and in the mesic forests of higher elevations, individuals become components of the terrestrial salamander community (Hairston, 1987). Ocoee salamanders are substantially more abundant in cove hardwood forests of >85 yr of age than in younger stands. They are also more abundant in cove forests with a significant amount of emergent rock. This possibly reflects the positive effect that shallow and emergent rock has on soil moisture retention as well as the abundance of refugia (Ford et al., 2002). Camp (2000) reported a seepage that dried due to drought, resulting in the failure to oviposit and the loss of clutches by female Ocoee salamanders that inhabited the seepage. Following an extended drought during which this same seepage dried every summer for 4 yr, the subpopulation collapsed, apparently due to desiccation-induced mortality (Camp, 2003b; C.D.C., unpublished data).

F. Home Range Size. Home ranges are small. In a study of salamanders inhabiting a wet rock face, Huheey and Brandon (1973) marked individuals that moved an average of 47 cm between observations. Movements centered around a small home range, to which displaced individuals successfully homed. Individual salamanders were randomly distributed on the rock face.

G. Territories. Adults are aggressive and defend space against conspecific intruders (Jaeger, 1988). Males are aggressive towards other males (Verrell and Donovan, 1991).

H. Aestivation/Avoiding Desiccation. Unlikely. Individuals are active throughout the summer (Huheey and Brandon, 1973; Tilley, 1980).

I. Seasonal Migrations. Unknown. At lower elevations, individuals move into forest-floor habitats from streams during wet, mild weather (C.D.C., personal observations). It is possible that individuals in high elevation sites move between streams and surrounding forests in association with seasonal weather patterns.

J. Torpor (Hibernation). Adults and juveniles may aggregate in seepages or underground retreats during the winter (Shealy, 1975). Individuals become active in mild weather conditions throughout the winter; winter-collected specimens on a wet rock face will feed (Huheey and Brandon, 1973).

K. Interspecific Associations/Exclusions. Throughout their range, Ocoee salamanders are sympatric with several congeneric species. They often occur syntopically with black-bellied salamanders (*D. quadramaculatus*) and seal salamanders (*D. monticola*). They additionally may occur with seepage salamanders (*D. aeneus*). Except for the completely aquatic shovel-nosed salamander (*D. marmoratus*), southern Appalachian desmognathines sort by body size along the stream–forest interface with larger species being more aquatic and smaller ones occurring more terrestrially. This pattern is evident both along a horizontal gradient from stream–stream bank–forest and along a vertical gradient from stream–seepage–forest (Organ, 1961a). Ocoee salamanders are relatively small and occur more terrestrially than most of their sympatric congeners. The observed pattern of desmognathine assortment was explained initially as niche partitioning among competitors (Hairston, 1949; Organ, 1961a). Tilley (1968) and Hairston (1980c), however, suggested that interspecific predation was a more likely cause. A number of studies have attempted to determine which was the more probable factor (e.g., Kleeberger, 1984; Carr and Taylor, 1985; Hairston, 1986; Southerland, 1986a,b,d). They generally concluded that some combination of predation and aggressive interference was an important factor in interspecific desmognathine interactions. Hairston (1986) made the strongest case for predation, with competition being a secondary factor. His statistical methods have been criticized, however (Jaeger and Walls, 1989). Although large desmognathines readily eat small ones in artificial environments, dietary studies show that neither black-bellied salamanders nor seal salamanders are important predators of heterospecific congeners (Camp, 1997b). The lack of predation under natural conditions is probably a result of differential habitat selection and behavioral avoidance (perhaps involving chemical cues) of larger congeners by small individuals. Predation by large species may have been important historically in the organization of desmognathine communities. Alternative hypotheses based on abiotic factors, rather than biotic ones such as competition and predation, recently have been proposed to explain patterns of habitat preference among desmognathines (Bruce, 1996; Camp et al., 2000).

At elevations above 900 m in the Great Smoky Mountains, Ocoee salamanders are syntopic with imitator salamanders (*D. imitator*; Tilley, 1985), similar members of the *D. ochrophaeus* complex (Tilley et al.,

1978). In these areas, Ocoee salamanders are more terrestrial than imitator salamanders (Tilley et al., 1978). Imitator salamanders may exclude Ocoee salamanders from lower elevation sites (Tilley et al., 1978; Bernardo, 2000). At Waterrock Knob in North Carolina, Ocoee salamanders may be largely excluded from both high-and low-elevation rock faces by Santeetlah dusky salamanders (*D. santeetlah*) and imitator salamanders, respectively (Tilley, 2000a). Ocoee salamanders and spotted dusky salamanders (*D. conanti*) are broadly sympatric but apparently largely exclude each other in areas of the Georgia Piedmont (C.D.C., personal observations) and the Appalachian Plateau region of Alabama (Folkerts, 1968; Mount, 1975).

L. Age/Size at Reproductive Maturity. Variable, and may increase with elevation (Tilley, 1977, 1980; Bernardo, 1994). In Bruce's (1990) study, both sexes apparently reached sexual maturity at 3 yr old; females first oviposit at 4 yr old. Females mature at about 29–30 mm SVL, males at or slightly smaller than 28 mm (Martof and Rose, 1963; Valentine, 1964; Huheey and Brandon, 1973). In skeletochronological studies of North Carolina populations, Castanet et al. (1996) and Bruce et al. (2002) found that males mature at 3–4 (generally 3) yr old, and females one year later (4–5 yr, but generally at 4 yr old).

M. Longevity. Using skeletochronological techniques, Castanet et al. (1996) and Bruce et al. (2002) estimated the ages of several males and females to be as long as 7 yr. They aged one male at 10 yr. Tilley (1977) recaptured several males 5 yr after their initial captures as large adults. These individuals must have been in at least their fifth adult year (8 yr of age) and were probably considerably older than that.

N. Feeding Behavior. As with most salamanders, Ocoee salamanders are generalist feeders, taking an array of small invertebrates, especially insects. Huheey and Brandon (1973) surveyed the stomachs of 54 individuals and reported flies (both larval and adult), ants, wasps, beetles, spiders, mites, an isopteran, a trichopteran, and a larval salamander (*Desmognathus* sp.). Folkerts (1968) additionally reported collembolans, lepidopterans, and a grasshopper.

Female Ocoee salamanders occasionally cannibalize their own eggs, the feeding response often being elicited by the presence of dead eggs (Tilley, 1972; Forester, 1979). They also occasionally eat conspecific hatchlings (Forester, 1981). A brooding female consumed its entire clutch of healthy eggs after she had been disturbed and moved to laboratory conditions (Bruce, 1990).

O. Predators. Woodland birds are probable predators (Petranka, 1998), as are various snakes, including ring-necked snakes (*Diadophis punctatus*), common garter snakes (*Thamnophis sirtalis*), and northern water snakes (*Nerodia sipedon*; Huheey and Brandon, 1973). On one occasion, a young northern water snake was found in a pile of loose shingles that had been dumped into a seepage on Rabun Bald Mountain in Georgia (C.D.C., personal observation). Ocoee salamanders were extremely abundant among the shingles, and the snake appeared to be gorged with salamanders. Although large desmognathine salamanders have been implicated as predators in experimental work (Hairston, 1986; Formanowicz and Brodie, 1993), dietary studies of seal salamanders and black-bellied salamanders indicate that neither species is a significant predator of Ocoee salamanders (Camp, 1997b), perhaps as a consequence of predator avoidance mechanisms. Spring salamanders (*Gyrinophilus porphyriticus*), on the other hand, are important predators of Ocoee salamanders (Bruce, 1979; Bernardo, 2002), explaining why Ocoee salamanders are more likely to flee from spring salamanders than from black-bellied salamanders (Hileman and Brodie, 1994).

P. Anti-Predator Mechanisms. Ocoee salamanders autotomized their tails when attacked by chickens in experimental trials (Labanick, 1984). These salamanders flee from predators such as spring salamanders (Hileman and Brodie, 1994). Ocoee salamanders may remain immobile to avoid detection (Dodd, 1990a). When attacked by snakes, Ocoee salamanders flip their bodies and bite, directing their bites to the face of the snake (Brodie et al., 1989).

Q. Diseases. Unknown.

R. Parasites. Goater et al. (1987) reported eight species of adult helminths, including nematodes (*Capillaria inequalis, Thelandros magnavulvaris, Omeia papillocauda, Falcaustra plethodontis, Cosmocercoides dukae*), flukes (*Brachycoelium elongatum, Phyllodistomum solidum*), and a tapeworm (*Cylindrotaenia americana*). They also reported two species of larval helminth, an ascaridoidid nematode and the plerocercoid of a proteocephalan tapeworm. Leeches (*Oligobdella biannulata*) are apparently rare parasites of Ocoee salamanders (Sawyer and Shelley, 1976; Goater, 2000).

4. Conservation.
Ocoee salamanders are among the most common salamanders of the southern Appalachians (Petranka, 1998). In 20 yr of surveys of desmognathine salamanders carried out in undisturbed habitats in the Appalachian Mountains of North Carolina, Hairston (1996) found no significant change in the abundance of Ocoee salamanders. Petranka (1998) suggested that the isolated populations in northeastern Alabama may be more vulnerable to environmental degradation than other populations.

Because of the reliance of populations of this species on moist habitats, the greatest potential threat is probably the removal of the protective forest canopy through the harvesting of timber and the resulting desiccation of habitats. Petranka et al. (1993) estimated that the clearcut logging of Appalachian forests of North Carolina during the 1980s and early 1990s killed millions of salamanders, including members of the *D. ochrophaeus* complex, each year. Ash and Bruce (1993), however, disputed these estimates and considered them to be exaggerations of the actual number killed. Desmognathine salamanders currently are abundant in areas of the southern Appalachians (the Great Smoky Mountains) that have been logged extensively in the past (S.G.T., personal observations). Appalachian seepages occasionally dry up, negatively affecting both reproduction (Camp, 2000) and survival in Ocoee salamanders (C.D.C., unpublished data). It is not known how periodic drought may interact with timber harvesting to affect populations of Ocoee and other seepage-dwelling salamanders. Populations of Ocoee salamanders occupying high elevation sites in the southern Appalachians may also be vulnerable to the effects of acid precipitation.

Although Ocoee salamanders are abundant, their considerable genetic differentiation across mountain ranges indicates that relatively isolated genetic subunits may be vulnerable to declines and the potential loss of genetic diversity. Some of these subunits may ultimately be described as separate taxa (Tilley and Mahoney, 1996).

Desmognathus orestes Tilley and Mahoney, 1996
BLUE RIDGE DUSKY SALAMANDER
Carlos D. Camp, Stephen G. Tilley

1. Historical versus Current Distribution.
Recently described Blue Ridge dusky salamanders (*Desmognathus orestes*; Tilley and Mahoney, 1996) occur in the Blue Ridge Physiographic Province from southwestern Virginia (Floyd County) into northwestern North Carolina (southward to Burke and Mitchell counties) and northeastern Tennessee (Unicoi County; Tilley and Mahoney, 1996; Mead and Tilley, 2000). This species is currently considered to comprise two genetically distinct groups of populations (Tilley and Mahoney, 1996; Tilley, 1997; Mead and Tilley, 2000; Mead et al., 2002).

Like other members of the *D. ochrophaeus* complex, Blue Ridge dusky salamanders are strongly associated with the headwaters of first-order streams (Tilley, 1997). This species occurs to the tops of the highest mountains within its range (Organ, 1961a; Tilley, 1968; Petranka, 1998). Tilley (1997) speculated that interactions with low elevation seal

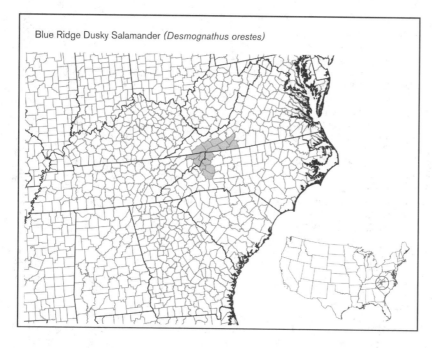

Blue Ridge Dusky Salamander *(Desmognathus orestes)*

salamanders may have contributed to the isolation and genetic differentiation of units of the *D. ochrophaeus* complex.

2. Historical versus Current Abundance.
Blue Ridge dusky salamanders are abundant in seepages, beds and banks of headwater streams, and wet cliffs. In mesic, high elevation forests, individuals may disperse widely into forest-floor habitats (Tilley, 1997). Petranka et al. (1993) demonstrated that clearcut timber harvesting negatively affects the number of total *Desmognathus* individuals, including members of the *D. ochrophaeus* complex. They estimated that, at the rate of clearcutting carried out during the 1980s and early 1990s, the Appalachian forests of North Carolina lost as many as 14 million salamanders of all species each year during that time. However, Ash and Bruce (1994) strongly disagreed with these estimates and did not consider clearcutting to have as strong an impact on native salamanders.

3. Life History Features.
A. Breeding. Courtship occurs on land. Eggs are deposited in or near flowing water.

i. Breeding migrations. Adult females of the closely related Ocoee salamander *(D. ocoee)* have been shown to be philopatric with regard to oviposition sites (Forester, 1977) and move to nesting sites several weeks prior to oviposition (Forester, 1981). Blue Ridge dusky salamanders presumably exhibit similar behaviors and may repeatedly return to headwater streams to lay eggs (Tilley, 1997). Mating occurs throughout the warmer months by individuals held in captivity for 1–2 d (Organ, 1961a).

ii. Breeding habitat.

B. Eggs.
i. Egg deposition sites. Females lay eggs and brood them in small chambers under moss, rocks, or logs in sites from which larvae can easily enter shallow water (Organ, 1961a; Tilley, 1997; Petranka, 1998). Eggs may be attached to the undersides of rocks ≤25 cm beneath stream banks in underground seepages (Organ, 1961a). Eggs have been observed in the field from June–October (Organ, 1961a).

ii. Clutch size. Three recently laid clutches averaged 11 eggs/clutch (Organ, 1961a). Clutch size determined from the dissection of specimens with developing oocytes averaged 14–16 (Martof and Rose, 1963). Clutch size is strongly related to female body size (Martof and Rose, 1963; Tilley, 1968). Eggs hatch after an incubation period of about 2 mo (Organ, 1961a).

Organ (1961a) hypothesized a biennial female reproductive cycle. This conclusion has since been questioned (Tilley, 1968, 1977; Tilley and Tinkle, 1968). Females probably are capable of nesting annually (Tilley and Tinkle, 1968), similar to other members of the *D. ochrophaeus* complex (Forester, 1977; Tilley, 1980).

C. Larvae/Metamorphosis.
i. Length of larval stage. Organ (1961a) determined an 8–9 mo larval period.

ii. Larval requirements.

a. Food. They presumably feed on small, aquatic invertebrates.

b. Cover. Similar to other members of the *Desmognathus ochrophaeus* complex, larvae are found in shallow water associated with seepages, small streams, and wet rock faces (Huheey and Brandon, 1973; Tilley, 1997).

iii. Larval polymorphisms. Variability in dorsolateral stripe configuration is evi-

dent in larval specimens of the related Carolina mountain dusky salamanders (*D. carolinensis*; Tilley, 1969).

iv. Features of metamorphosis. Metamorphosis occurs at 10–12 mm SVL (Organ, 1961a).

v. Post-metamorphic migrations. Similar to other members of the *D. ochrophaeus* complex, individuals disperse into forest habitats when temperature and moisture conditions are conducive to overland movement. In mesic, high elevation forests, they become functional components of the forest-floor salamander community during warm months (Hairston, 1949; Tilley, 1997; Petranka, 1998; Petranka and Murray, 2001).

vi. Neoteny. Neoteny is not known in this species.

D. Juvenile Habitat. Similar to adults, they are commonly associated with seepages of headwater streams where they may be found on stream banks, in streambeds, and on wet rock faces (Organ, 1961a; Martof and Rose, 1963).

E. Adult Habitat. Blue Ridge dusky salamanders are found associated with wet rock faces, seepages, and forest-floor habitats in the vicinity of streams and seeps (Organ, 1961a; Martof and Rose, 1963; Tilley, 1997). Adults are nocturnal, spending the day under cover objects. Animals sometimes are active during the day under overcast conditions (Petranka, 1998).

F. Home Range Size. Home ranges are presumably small, similar to those of Ocoee salamanders (Huheey and Brandon, 1973).

G. Territories. Territorial behavior has not been studied in this species. Individuals of other members of the *D. ochrophaeus* complex exhibit intraspecific aggression (Jaeger, 1988; Verrell and Donovan, 1991).

H. Aestivation/Avoiding Desiccation. Unlikely. Blue Ridge dusky salamanders are active throughout the summer (Organ, 1961a).

I. Seasonal Migrations. Individuals may move from streams into forest-floor habitats during warmer months when and where conditions are sufficiently moist. They aggregate in seepages, presumably to overwinter (Organ, 1961a).

J. Torpor (Hibernation). In winter, adults migrate to seepages and springs, or underground retreats (Organ, 1961a).

K. Interspecific Associations/Exclusions. Blue Ridge dusky salamanders often occur sympatrically with black-bellied salamanders (*D. quadramaculatus*), seal salamanders (*D. monticola*), northern dusky salamanders (*D. fuscus*), and pigmy salamanders (*D. wrighti*; Organ, 1961a). These species are organized like desmognathine communities throughout the southern Appalachians, where different species predictably assort by body size along the stream–forest interface. Larger species are more aquatic, smaller ones more terrestrial. This pattern is evident both along a horizontal gradient from stream–stream bank–forest and along a vertical gradient from stream–seepage–forest (Hairston, 1949; Organ, 1961a). As with other members of the *D. ochrophaeus* complex, Blue Ridge dusky salamanders are relatively small and occur more terrestrially than most of its sympatric congeners. The observed pattern of desmognathine assortment was explained initially as niche partitioning among competitors (Hairston, 1949; Organ, 1961a). Tilley (1968) and Hairston (1980c), however, suggested that interspecific predation was a more likely cause. A number of studies attempted to determine which was the more probable factor (e.g., Kleeberger, 1984; Carr and Taylor, 1985; Hairston, 1986; Southerland, 1986a,b,d). They generally concluded that some combination of predation and aggressive interference were important factors in interspecific desmognathine interactions. Hairston (1986) made the strongest case for predation with competition being a secondary factor. His statistical methods have been criticized, however (Jaeger and Walls, 1989). Although large desmognathines readily eat small ones in artificial environments, there is no evidence from extensive dietary studies that large desmognathine salamanders are important predators of heterospecific congeners (Camp, 1997b). The lack of predation under natural conditions is probably a result of differential habitat selection and behavioral avoidance (perhaps involving chemical cues) of larger congeners by small individuals. Predation by large species may have been important historically in the organization of desmognathine communities. Alternative hypotheses based on abiotic factors, rather than biotic ones such as competition and predation, recently have been proposed to explain patterns of habitat preference among desmognathines (Bruce, 1996; Camp et al., 2000).

In forest-floor habitats, Blue Ridge dusky salamanders may be syntopic with Blue Ridge two-lined salamanders (*Eurycea wilderae*) and several species of woodland salamanders (genus *Plethodon*). In and around seepages, they may be syntopic with Blue Ridge two-lined salamanders, red salamanders (*Pseudotriton ruber*), and spring salamanders (*Gyrinophilus porphyriticus*; Organ, 1961a). Spring salamanders are known to prey on other salamanders in the *D. ochrophaeus* complex (Bruce, 1979).

A 25-km wide valley forms a northern boundary between Blue Ridge dusky salamanders and the closely related Alleghany mountain dusky salamanders (*D. ochrophaeus*). No such physical barrier exists, however, between Blue Ridge dusky salamanders and Carolina mountain dusky salamanders (*D. carolinensis*) to the south. Limited hybridization occurs at points of contact between these two species (Mead

and Tilley, 2000; Mead et al., 2002). Blue Ridge mountain dusky salamanders exhibit considerable sexual isolation from their sibling, parapatric neighbors (Verrell and Arnold, 1989; Tilley et al., 1990) and from northern dusky salamanders (Uzendoski and Verrell [1993]).

L. Age/Size at Reproductive Maturity. Organ (1961a) hypothesized that maturity is achieved in 3.5 yr and 4.5 yr in males and females, respectively. His conclusions were based on size-class estimates derived from samples pooled over a number of different populations collected over a wide range of elevations. Members of this species complex are known to vary in maturation age with elevation (Tilley, 1973a, 1980), but Organ's (1961a) estimates fall within the ranges for Ocoee salamanders as determined by skeletochronology (Castanet et al., 1996) and capture–recapture studies (Tilley, 1977, 1980).

M. Longevity. Closely related Ocoee salamanders can live up to at least 10 yr (Castanet et al., 1996), and maximum life spans are probably considerably more than that (Tilley, 1977).

N. Feeding Behavior. According to Petranka (1998), no information is available on diet. Blue Ridge dusky salamanders probably feed heavily on small arthropods, as do the other members of the species complex (Hairston, 1949; Fitzpatrick, 1973; Keen, 1979; Krzysik, 1980b).

O. Predators. Blue Ridge dusky salamanders probably are eaten by various species of vertebrate predators, including birds, small mammals, snakes, and predatory salamanders such as spring salamanders (Petranka, 1998).

P. Anti-Predator Mechanisms. Undescribed. They are presumably the same as in other desmognathines, i.e., flight, writhing, biting, and tail autotomy (Labanick, 1984; Brodie et al., 1989; Hileman and Brodie, 1994).

Q. Diseases. Unknown.

R. Parasites. Unreported. Closely related Ocoee salamanders occupy similar habitats and harbor a helminth fauna that includes nematodes, flukes, and tapeworms (Goater et al., 1987). Leeches occasionally occur on Ocoee salamanders (Goater, 2000).

4. Conservation. Blue Ridge dusky salamanders are abundant at high elevation sites. Because of their reliance on moist habitats, the greatest potential threat is probably the removal of the protective forest canopy through the harvesting of timber and the resulting desiccation of habitats. Petranka et al. (1993) estimated that the clearcut logging of Appalachian forests of North Carolina during the 1980s and early 1990s killed millions of salamanders, including members of the *D. ochrophaeus* complex, each year. Ash and Bruce (1993), however, disputed these estimates and considered

them to be exaggerations of the actual number killed. Petranka et al. (1994) indicated that populations of this species at low elevations may take many years to recover from intensive timber harvesting. However, desmognathine salamanders currently are abundant in areas of the southern Appalachians (the Great Smoky Mountains) that have been logged extensively in the past (S.G.T., personal observations). Appalachian seepages occasionally dry up, negatively affecting both reproduction (Camp, 2000) and survival in the closely related Ocoee salamanders (C.D.C., unpublished data). It is not known how periodic drought may interact with timber harvesting to affect populations of Blue Ridge dusky and other seepage-dwelling salamanders. Populations of Blue Ridge dusky salamanders that occupy high-elevation sites in the southern Appalachians may also be vulnerable to the effects of acid precipitation.

Desmognathus quadramaculatus (Holbrook, 1840)
BLACK-BELLIED SALAMANDER

Mark B. Watson, Thomas K. Pauley, Carlos D. Camp

1. Historical versus Current Distribution. Bishop (1943) showed that black-bellied salamanders (*Desmognathus quadramaculatus*) range from northern Georgia in the southeastern United States northward through the mountains of western North Carolina, eastern Tennessee, and southeastern Virginia to southern West Virginia. Research since Bishop's work has extended the range north into Allegheny and Franklin counties, Virginia, and upstream in the New River in West Virginia to its confluence with the Gauley River in Fayette County (Valentine, 1974). Recent surveys in West Virginia have shown that the range of black-bellied salamanders extends farther north than previously known. Turner (1997) located black-bellied salamanders upstream in tributaries of the Gauley River (Nicholas County) and T.K.P. and W.J. Humphries (unpublished data) found them in tributaries 4.8 km upstream from the mouth of the Greenbrier River. They also have been located approximately 4.8 km upstream from the mouth of the Bluestone River (Waldron et al., 2000). Black-bellied salamanders have probably been introduced into drainages where they previously were not present due to fishermen releasing individuals used as bait. Isolated records from the Piedmont of Georgia and South Carolina have been attributed to such bait–release introductions (Martof, 1953, 1955). Recent collections on the Georgia Piedmont suggest that at least one disjunct record represents a breeding, persistent population (J.B. Jensen, unpublished data).

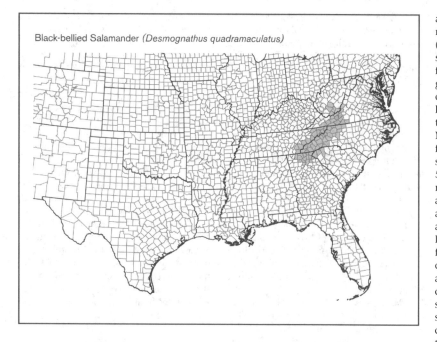

Black-bellied Salamander *(Desmognathus quadramaculatus)*

2. Historical versus Current Abundance.
Black-bellied salamanders are used widely as fish bait over the extent of their range (Camp and Lovell, 1989; Jensen and Waters, 1999). Petranka (1998) suggests that over-harvesting of black-bellied salamanders for bait may be detrimental to some local populations. Statements recorded by a local fisherman in West Virginia suggest that salamander populations have declined in that area in past years (Green and Pauley, 1987). Turner and Pauley (1992) found that black-bellied salamanders were absent at the mouth of a tributary where fishermen could easily search for salamanders as bait, but present farther upstream where fishermen had not searched.

Mitchell et al. (1999) concluded that stream pollution caused by acid mine drainage and sewage and over-collection for fish bait are threats to black-bellied salamanders. Stream acidification and contamination by metal pollutants have resulted in the elimination of black-bellied salamanders from one stream in the Great Smoky Mountains (Kucken et al., 1994). Petranka et al. (1993) demonstrated that clearcut harvesting of timber negatively affects populations of black-bellied salamanders. They estimated that timber-harvesting rates during the 1980s and early 1990s resulted in the loss of at least 14 million salamanders of all species annually in western North Carolina. Long-term (20 yr) monitoring has shown no significant changes in the abundance of this species in undisturbed locations in the Appalachian Mountains of western North Carolina (Hairston, 1996).

3. Life History Features.
A. Breeding. Reproduction is aquatic.
i. Breeding migrations. Black-bellied salamanders do not migrate.

ii. Breeding habitat. First-, second-, and third-order streams (Pauley, 1993b).
B. Eggs.
i. Egg deposition sites. Eggs of black-bellied salamanders are deposited under rocks in moving water. Austin and Camp (1992) found egg clutches in a low-elevation Georgia site (Habersham County) in May and hatchlings in July. Eggs with attending females have been found in July at a high-elevation site in Union County (C.D.C., personal observation). In Virginia, egg deposition occurs from June–September (Organ, 1961a). Pauley (1993b) reported a nest in West Virginia in July.

ii. Clutch size. Averages 45 eggs (range = 38–55) in black-bellied salamanders from North Carolina (Bruce, 1996). Bruce (1996) found eggs attached to large rocks in the middle of Wolf Creek, North Carolina, in June. Mills (1996) similarly found eggs in this time period from the northern portion of the range. Eggs range from an average of 3.9 mm in diameter and 188–203 mg dry weight (Tilley and Bernardo, 1993) to >5 mm in diameter (C.D.C., unpublished data). Oocytes require 2 yr to develop, and females probably oviposit biennially (Austin, 1993).

Marks (2000) describes the embryonic development of black-bellied salamanders. Turner (1997) observed hatchlings from April–September in tributaries of the Gauley River in West Virginia.
C. Larvae/Metamorphosis.
i. Length of larval stage. Bruce (1988a) found 3–4 age classes of larvae in southwestern North Carolina, depending on the population. Castanet et al. (1996) used skeletochronological aging techniques to show that larval development in one North Carolina population took 3 yr. Bruce et al. (2002) found that larvae in

another North Carolina population metamorphosed in 4 yr. Austin and Camp (1992) found differences in the age and size of newly metamorphosed animals from two populations in northern Georgia. Low-elevation populations had an average of 40–43 mm SVL at 3 yr, whereas hatchlings from a high-elevation population averaged 54 mm SVL when 4 yr old. Mills (1996) interpreted five size classes from collections of larvae in West Virginia streams and suggested that there may be a 54–60 mo larval period. Populations in northeastern Tennessee (Csanady, 1978) and northwestern North Carolina (Beachy and Bruce, 1998) may have larval periods as short as 1 yr. Larval period in black-bellied salamanders is strongly related to rainfall patterns and is the most important determinant of adult body size (Camp et al., 2000). Bruce (1985a) found a higher frequency of young larvae in small tributary streams and fewer young larvae in larger streams, suggesting that reproduction was concentrated in headwater streams and that older larvae move downstream. Orr and Maple (1978) found that black-bellied salamander larvae hatched earlier and had more rapid yolk absorption than seal salamanders *(D. monticola)*. They suggested this may have an adaptive advantage.

a. Food. Beachy (1997a) showed that black-bellied salamander larvae fed on Blue Ridge two-lined salamander *(Eurycea wilderae)* larvae in controlled experiments. Davic (1991) found that aquatic insects made up the bulk of the natural diet of larval black-bellied salamanders. In West Virginia, in the northern extreme of the range of black-bellied salamanders, larvae prey on larval dipterans and trichopterans, and plecopteran and ephemeropteran nymphs, as well as several other insect taxa (Mills, 1996). Growth of black-bellied salamander larvae is not regulated by their densities, but survival of first-year larvae is reduced by the presence of larger black-bellied salamanders under artificial, experimental conditions (Beachy, 1993b).

b. Cover. Larvae often take cover in the interstices of shallow riffle areas in streams (Austin and Camp, 1992; C.D.C., personal observation).

D. Juvenile Habitat. Juvenile habitats are similar to those of adults, although juveniles tend to be farther into the middle of streams than adults (Camp and Lee, 1996). Juveniles occupy smaller substrates than adults, avoid refugia containing large salamanders (Southerland, 1986c), and flee from large conspecifics (Camp and Lee, 1996). Juveniles spend more time actively foraging outside of refugia than do adults (Camp and Lee, 1996).

E. Adult Habitat. With the exception of shovel-nosed salamanders *(D. marmoratus)*, black-bellied salamanders are the most aquatic of the *Desmognathus* species. They are most often found in

cool, small-volume, high-gradient perennial streams. Whitford and Hutchison (1965) suggested that black-bellied salamanders may be limited to cool streams by their limited surface area for respiration. In the Great Smoky Mountains of Tennessee, they have been found from 375–1,725 m (1150–5200 ft; Huheey, 1966). They are associated with clean water and large cobbles and flat rocks (Pope, 1924; Davic and Orr, 1987; Mills, 1996; Turner, 1997). Davic and Orr (1987) suggested there may be a positive correlation between the density of rocks and pebbles and the density of black-bellied salamanders. Black-bellied salamanders may also use burrows in stream banks as refugia or for ambushing prey (Huheey, 1964; Brandon and Huheey, 1971; Camp and Lee, 1996). Southerland (1986c) found that increasing streamside cover might influence species composition due to competition for refugia in densely populated streams. Pollio (1993) observed that adult black-bellied salamander surface density decreased as water temperature increased in a southern West Virginia stream. Mills (1996) observed a seasonal shift in the foraging location of black-bellied salamanders in West Virginia. Most (97%) black-bellied salamanders forage above the water's edge; as summer progresses, more salamanders forage in the stream away from the bank. Post-metamorphic black-bellied salamanders forage in the water from October–April. Adults spend more time ambushing prey from refugia than do juveniles, which tend to forage more actively (Camp and Lee, 1996).

Black-bellied salamanders inhabit cool mountain streams. Bogert (1952) observed black-bellied salamanders in streams and springs with water temperatures that varied from 10–17.7 °C. Spight (1968) found that black-bellied salamanders lost water as a linear function of body weight (y = 2.0031+0.41x at 60–70% humidity, where y = log rate loss, x = log of body weight). Black-bellied salamanders lost water at an intermediate rate compared to other species tested. Whitford and Hutchison (1965) described various measurements of metabolism of black-bellied salamanders. Black-bellied salamanders had a lower metabolic rate than spotted salamanders (*Ambystoma maculatum*) of comparable size. Below 10 °C, 80–90% of the oxygen was absorbed through the skin. The optimum temperature is from 15–20 °C. Booth and Feder (1991) showed that low flow rates of water result in the buildup of hypoxic layers around black-bellied salamanders that depend entirely on cutaneous breathing for gas exchange. They hypothesized that this species can survive reduced aerobic conditions using anaerobic metabolic pathways. Field studies, however, have shown that most individuals occur with at least part of their bodies

above the water level (Camp and Lee, 1996; Mills, 1996), perhaps to avoid hypoxic conditions in quiet water.

F. Home Range Size. Individual black-bellied salamanders repeatedly used the same refugia or moved among a series of primary and secondary refugia (Camp and Lee, 1996). Camp and Lee (1996) found home ranges to be linear with a minimum area of 1,207 cm^2. They found no correlation between home range size and body size (SVL). Home ranges often overlapped with home ranges of neighboring salamanders.

G. Territories. Black-bellied salamanders aggressively defended refugia from conspecifics (Camp and Lee, 1996). Camp and Lee (1996) rarely found two conspecifics in the same refugium, and they also found intruders absent from refugia even though the resident was away. This suggested that chemical cues might be used to mark territories. These chemical cues may be detected by smaller conspecifics to avoid predation. Roudebush and Taylor (1987b) found black-bellied salamanders were attracted to extracts of like-sized salamanders. Camp (1996) examined bite scars on black-bellied salamanders and suggested that bites frequently are inflicted by conspecifics. A high portion of these bites are on the head in larger salamanders and towards the tail on smaller ones. Males and females had similar bite frequencies, and mature animals had more scars than immature ones. These data suggest that larger animals may tend to stand their ground more against aggressors while smaller individuals may flee.

H. Aestivation/Avoiding Desiccation. Aestivation is unknown. Black-bellied salamanders are associated with streams, which may buffer them from desiccating conditions.

I. Seasonal Migrations. Black-bellied salamanders do not migrate.

J. Torpor (Hibernation). Unknown.

K. Interspecific Associations/Exclusions. Black-bellied salamanders can be found associated with other members of the genus *Desmognathus* in streams. Seal salamanders, dusky salamanders (*D. fuscus* complex), and mountain dusky salamanders (*D. ochrophaeus* complex) are all associated with streams and seeps and typically sort by size and association with water. *Desmognathus* species that inhabit these streams are structured into size gradients with larger-bodied, more aquatic species found closer to water, and smaller-bodied salamanders found in more terrestrial locations (Hairston, 1949, 1987; Organ, 1961a). More terrestrial species used woody debris rather than rocks as cover objects (Southerland, 1986a). Larger black-bellied salamanders have been shown to displace smaller salamanders from refugia and feeding areas. Predation by black-bellied salamanders has been suggested to alter habitat selection of seal salamanders (Kleeberger,

1984; Carr and Taylor, 1985; Southerland, 1986b) and to be an important factor in the development of the community structure of streamside desmognathines (Hairston, 1980c). Grover (2000) reported that northern dusky salamanders (*D. fuscus*) and other salamanders (e.g., *Plethodon* species) were farther from water in stream transects where black-bellied salamanders were present than in streams where black-bellied salamanders were absent. There was not a significant difference in distance from water between seal salamanders inhabiting the same respective transects, and Grover (2000) could not rule out the possibility that abiotic differences among the streams caused the observed differences in salamander dispersion. Roudebush and Taylor (1987a,b) suggested that seal salamanders use chemical clues to avoid black-bellied salamanders. Jacobs and Taylor (1992) found that seal salamanders avoided substrates that were covered with secretions from mucous glands of black-bellied salamanders. Black-bellied salamanders did not avoid secretions from seal salamanders. Hairston (1986) suggested that predation and competition were important factors in determining species composition of streams containing desmognathine assemblages. Evidence from dietary studies, however, suggests that interspecific predation among black-bellied salamanders and other salamanders is extremely rare (Camp, 1997b).

Exceptions to the body-size/moisture gradient have been recently reported. Beachy and Bruce (1998) discovered a population of small-sized black-bellied salamanders in northwestern North Carolina that is syntopic with seal salamanders that are equivalent in size but occur more terrestrially. Black-bellied salamanders in Union County, Georgia, are syntopic with a cryptic, newly discovered species of dark-ventered salamander that is approximately half the SVL of black-bellied salamanders (Camp et al., 2000). The newly discovered form, though smaller, appears to be as aquatic as black-bellied salamanders (Camp et al., 2002).

L. Age/Size at Reproductive Maturity. Organ (1961a) estimated that male black-bellied salamanders from Virginia are sexually mature at 3.5 yr of age and females at 4.5 yr of age. However, because his estimates were based on specimens that were pooled from a number of different populations, their accuracy has been questioned (Austin and Camp, 1992). Bruce (1993) found that black-bellied salamander males were 64–94 mm SVL as adults compared to 73–85 mm for females in North Carolina. Males from Georgia populations may reach 120 mm SVL (Camp et al., 2000). Black-bellied salamanders in the northern portion of the range appear to reach maturity at a smaller size range (57.6–77.5 mm; Mills, 1996). Bruce (1988a) estimated that first mating occurred at

5–6 yr of age for males and 6–7 yr for females. Using data from skeletochronological analysis, Castanet et al. (1996) found that male black-bellied salamanders in North Carolina reached sexual maturity in 6–7 yr, while females matured at 7–8 yr of age. Using the same technique, Bruce et al. (2002) found that males and females at Coweeta, North Carolina, reached maturity at 7–8 yr and 9–10 yr, respectively.

M. Longevity. Skeletochronological studies have shown that natural longevity is at least 15 yr (Castanet et al., 1996; Bruce et al., 2002).

N. Feeding Behavior. Camp and Lee (1996) showed that black-bellied salamander adults are largely ambush predators using burrows and rock crevices to wait for prey. Smaller salamanders are more likely to actively forage among refugia than adults. Adult salamanders will move away from the stream at night to feed (Hairston, 1987). During warm seasons with high relative humidity, post-metamorphic black-bellied salamanders forage at the edge of the stream and ≤ 30 cm from the edge (Mills, 1996).

Davic (1991) compared the diets of larval, juvenile, and adult black-bellied salamanders and found that there was a shift in the diet from aquatic prey to aerial prey after metamorphosis. Over 82% of prey in larvae were aquatic compared to 35% of aquatic prey in juveniles. Adult black-bellied salamanders preyed upon crayfish as well as aquatic and terrestrial insects (Hairston, 1949; Martof and Scott, 1957; Kleeberger, 1982). Mills (1996) found that the major prey items of post-metamorphic black-bellied salamanders were dipteran larvae followed by adult coleopterans and a variety of other insect taxa. A number of studies have characterized black-bellied salamanders as important predators of other salamanders (e.g., Hairston, 1980c, 1986, 1987; Southerland, 1986b; Formanowicz and Brodie, 1993; Grover, 2000). However, these conclusions have been based on anecdotal reports (e.g., Hairston, 1980c, 1986) or predation that was induced under artificial, experimental conditions (e.g., Formanowicz and Brodie, 1993). In an analysis of 659 stomachs of black-bellied salamanders from Georgia, Camp (1997b) found no evidence of predation on heterospecific desmognathines, although larval or juvenile conspecifics occasionally were taken as prey. This analysis corroborated other dietary studies (e.g., Hairston, 1949; Martof and Scott, 1957; Kleeberger, 1982; Davic, 1991; Mills, 1996) that demonstrated that predation by this species on other species of salamanders is rare. The ambush–predator strategy (Brandon and Huheey, 1971; Camp and Lee, 1996) combined with the avoidance of pheromones by smaller species (Roudebush and Taylor, 1987a; Jacobs and Taylor,

1992) probably preclude predation of salamanders under natural conditions (Camp and Lee, 1996; Camp, 1997b).

O. Predators. Egg predators probably include aquatic insects, crayfish, fishes, and other salamanders.

Larger larvae of black-bellied salamanders occasionally are cannibalistic and will prey upon smaller black-bellied salamander larvae (Beachy, 1993b; Camp, 1997b). Beachy (1991b) found that microhabitat differences resulted from experimentally induced predation of species with larger larvae (e.g., black-bellied salamanders) on smaller larvae (e.g., Blue Ridge two-lined salamanders). Beachy (1994) experimentally demonstrated that spring salamander (*G. porphyriticus*) larvae are more efficient predators on Blue Ridge two-lined salamanders than are black-bellied salamander larvae.

Brodie et al. (1989) showed that garter snakes (*Thamnophis* sp.) prey on black-bellied salamanders. Adult black-bellied salamanders occasionally are cannibalistic on smaller conspecific juveniles and larvae (Camp, 1997b). In Georgia, black-bellied salamanders may occur in streams that are also occupied by potentially predaceous fishes (e.g., banded sculpins [*Cottus carolinae*], bluehead chubs [*Nocomis leptocephalus*], creek chubs [*Semotilus atromaculatus*]), and northern watersnakes (*Nerodia sipedon*) have been observed foraging in optimum habitat for black-bellied salamanders (C.D.C., personal observation).

P. Anti-Predator Mechanisms. Eggs of black-bellied salamanders frequently are deposited beneath large rocks in streambeds, obscuring them from potential predators. Females of many plethodontid salamanders exhibit brooding behavior (Jaeger and Forester, 1993). Female black-bellied salamanders have been observed tending clutches of eggs (T.K.P., personal observation).

Adults do not exhibit noxious skin secretions. They remain motionless when touched or show other passive defensive postures. They will bite and use pseudoaposematic coloration for defense. Members of the genus *Desmognathus* seem to be palatable to many species. The aggressive nature and biting ability of black-bellied salamanders are used as anti-predator defenses. Brodie (1978) showed that biting was used as a defense against shrew attacks. Brodie et al. (1989) observed three anti-predator responses during encounters with garter snakes (*Thamnophis* sp.): flipping, biting, and tail autotomy. Biting appeared to be a successful defense against snake attacks, and all salamanders that were bitten by a snake on the tail were not eaten by the snake (Brodie et al., 1989). Pope (1928) describes a black-bellied salamander from Highlands, North Carolina, as assuming a defensive posture and opening its mouth threateningly when he attempted to capture it. Brodie (1977) lists

members of genus *Desmognathus* as using biting and pseudoaposematic coloration as anti-predator defenses.

Q. Diseases. None reported.

R. Parasites. Rankin (1937) lists the following species collected from black-bellied salamanders: Protozoa—*Cryptobia borreli, Cytamoeba bacterifera, Eutrichomastix batrachorum, Hexamastix batrachorum, Hexamitus batrachorum, Hexamitus intestinalis, Karotomorpha swezi, Prowazekella longifilis,* and *Tritrichomonas augusta;* Trematoda—*Brachycoelium hospitale, Diplostomulum desmognathi,* and numerous unidentified metacercariae; Cestoda—*Crepidobothrium cryptobranchi* and proteocephalid cysts; Nematoda—*Capillaria inequalis, Omeia papillocauda, Oxyuris magnavulvaris,* and spirurid cysts; Acanthocephala—cysts.

Huheey (1966) observed leeches attached to black-bellied salamanders in the Great Smoky Mountains National Park, Tennessee, and stated that leeches were a common parasite. Leeches also are common on specimens from Georgia (C.D.C., personal observation). Baker (1987) lists *Batracholandros magnavulvaris, Falcaustra plethodontis, Desmognathinema nantahalaensis,* and *Omeia papillocauda* as parasitic nematodes in black-bellied salamanders. Goater et al. (1987) lists adult nematodes (*Capillaria inequalis, Thelandros magnavulvaris, Omeia papillocauda, Desmognathinema nantahalaensis, Falcaustra plethodontis,* and *Cosmoceroides dukae*), trematodes (*Brachycoelium elongatum, Gorgoderina bilobata,* and *Phyllodistomum solidum*), and cestodes (*Cylindrotaenia americana*) in black-bellied salamanders. They also found larval nematodes (*Ascaridoidea* sp.), cestodes (proteocephalan plerocercoids), and acanthocephalans (*Centrorynchus conspectus*) in black-bellied salamanders.

4. Conservation.
Stream pollution caused by acid mine drainage and sewage, clearcut forestry practices, and overcollection for fish bait are threats to black-bellied salamanders. In contrast, black-bellied salamanders are used widely as fish bait over the extent of their range, and because of this, they have been introduced into drainages where they previously were not present. While long-term (20 yr) monitoring has shown no substantive changes in the abundance of this species in undisturbed locations in the Appalachian Mountains of western North Carolina (Hairston, 1996), black-bellied salamanders are listed as a Species of Special Concern in West Virginia.

Desmognathus santeetlah Tilley, 1981
SANTEETLAH DUSKY SALAMANDER
Carlos D. Camp, Stephen G. Tilley

1. Historical versus Current Distribution.
Santeetlah dusky salamanders (*Desmognathus santeetlah*) occur at higher elevations

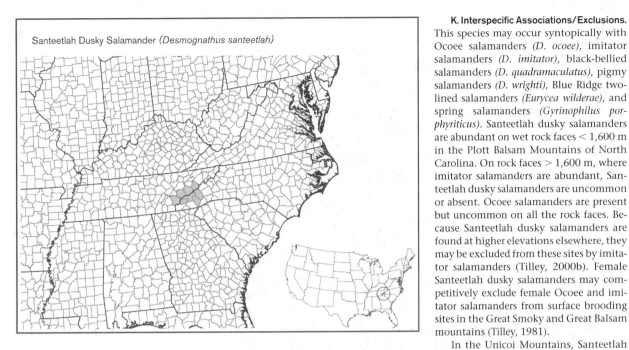

Santeetlah Dusky Salamander *(Desmognathus santeetlah)*

in the Unicoi, Great Smoky, Great Balsam, and Cheoah mountains in the southern Appalachians (Tilley, 2000a).

2. Historical versus Current Abundance.
Unknown. Currently, Santeetlah dusky salamanders may be abundant along small streams flowing through mesic, high-elevation forests (Jones, 1986) and on wet rock faces (Tilley, 2000b).

3. Life History Features.
    A. Breeding.
    i. Breeding migrations. Unknown.
    ii. Breeding habitat. Mating presumably takes place in habitats where other activities (e.g., foraging) take place. Courtship is virtually identical to that described for the closely related northern dusky salamander *(D. fuscus)* and involves the "tail-straddle walk" typical of plethodontids (Maksymovitch and Verrell, 1992).
    B. Eggs.
    i. Egg deposition sites. Females oviposit and brood their egg clutches in cavities excavated in moss, rotting logs, or soil within a few centimeters of shallow, flowing surface water. The most abundant nesting site observed by Jones (1986) in the Unicoi Mountains was under moss atop logs in seepage areas. He found nests from 16–83 cm from open water. These were often vertical distances, because clutches commonly were on top of logs lying in shallow water. Incubation time for eggs kept in the laboratory at 20 °C is approximately 7 wk (Jones, 1986).
    ii. Clutch size. Clutch size averages 17–20 eggs in the Great Smoky Mountains (Tilley, 1988) and 21 eggs in the Unicoi Mountains (Jones, 1986). In both areas, clutch sizes are slightly smaller than in nearby populations of spotted dusky

salamanders *(D. conanti).* Egg number is positively related to female body size. Santeetlah dusky salamanders, however, produce significantly more eggs/mm SVL than spotted dusky salamanders (Jones, 1986).
    C. Larvae/Metamorphosis.
    i. Length of larval stage. Jones (1986) estimated that larvae metamorphose in slightly <1 yr in the Unicoi Mountains.
    ii. Larval requirements.
    a. Food. Larvae probably feed on small, aquatic invertebrates.
    b. Cover. Larvae occur in shallow water in small streams and seepages, which are common nesting sites (Jones, 1986).
    iii. Larval polymorphisms. Unknown.
    iv. Features of metamorphosis. Newly metamorphosed Santeetlah dusky salamanders measure 9–10 mm in SVL (Jones, 1986).
    v. Post-metamorphic migrations. Unknown.
    vi. Neoteny. Unknown.
    D. Juvenile Habitat. Similar to that of adults.
    E. Adult Habitat. Santeetlah dusky salamanders occur under cover along small streams and seepages in mesic, high-elevation forests (Jones, 1986). They may be found several meters into the surrounding forest (Petranka, 1998), but usually occur in or within a few centimeters of surface water. They also may be abundant on wet cliff faces (Tilley, 2000b).
    F. Home Range Size. Unknown.
    G. Territories. Unknown.
    H. Aestivation/Avoiding Desiccation. Unlikely. Jones (1986) and Tilley (2000b) found them to be active during the summer.
    I. Seasonal Migrations. Unknown.
    J. Torpor (Hibernation). Unknown.

K. Interspecific Associations/Exclusions. This species may occur syntopically with Ocoee salamanders *(D. ocoee),* imitator salamanders *(D. imitator),* black-bellied salamanders *(D. quadramaculatus),* pigmy salamanders *(D. wrighti),* Blue Ridge two-lined salamanders *(Eurycea wilderae),* and spring salamanders *(Gyrinophilus porphyriticus).* Santeetlah dusky salamanders are abundant on wet rock faces < 1,600 m in the Plott Balsam Mountains of North Carolina. On rock faces > 1,600 m, where imitator salamanders are abundant, Santeetlah dusky salamanders are uncommon or absent. Ocoee salamanders are present but uncommon on all the rock faces. Because Santeetlah dusky salamanders are found at higher elevations elsewhere, they may be excluded from these sites by imitator salamanders (Tilley, 2000b). Female Santeetlah dusky salamanders may competitively exclude female Ocoee and imitator salamanders from surface brooding sites in the Great Smoky and Great Balsam mountains (Tilley, 1981).
    In the Unicoi Mountains, Santeetlah dusky salamanders and spotted dusky salamanders occur parapatrically (Tilley, 1981). The two species hybridize along the northwestern edge of the Great Smoky Mountains (Tilley, 1981, 1988). Santeetlah dusky salamanders exhibit complete sexual isolation with sympatric populations of Ocoee salamanders and imitator salamanders (Verrell, 1990a) and partial isolation from allopatric populations of spotted dusky salamanders (Verrell, 1990b).
    L. Age/Size at Reproductive Maturity. Santeetlah dusky salamanders reach sexual maturity at 30 mm and 33–35 mm SVL for males and females, respectively. Males mature at approximately 2 yr, and females mature a year later at 3 yr (Jones, 1986).
    M. Longevity. Unknown. However, Ocoee salamanders are similar in size and reach a maximum age of at least 10 yr (Castanet et al., 1996).
    N. Feeding Behavior. Santeetlah dusky salamanders probably feed heavily on small arthropods as do other medium-sized species of *Desmognathus* (Fitzpatrick, 1973; Keen, 1979), including members of the *D. fuscus* complex (Barbour and Lancaster, 1946; Sites, 1978).
    O. Predators. Unreported. Predators probably include woodland birds, small mammals, and snakes. Spring salamanders, which may occur syntopically with Santeetlah dusky salamanders at some sites, are known to feed heavily on small salamanders (Bruce, 1979).
    P. Anti-Predator Mechanisms. Unreported. They are presumably the same as in other desmognathines, that is flight, writhing, biting, and tail autotomy (Labanick, 1984; Brodie et al., 1989; Hileman and Brodie, 1994).
    Q. Diseases. Unknown.

R. Parasites. Unknown. Other species of semi-aquatic and semi-terrestrial desmognathines harbor a parasite fauna that includes endoparasitic nematodes, trematodes, and tapeworms (Goater et al., 1987) and ectoparasitic leeches (Goater, 2000).

4. Conservation.
A substantial portion of the geographic range of Santeetlah dusky salamanders falls within the boundaries of the Great Smoky Mountains National Park. These populations, therefore, are protected from such environmental disturbances as mining and timber harvesting. Populations outside park boundaries, however, are potentially vulnerable. Santeetlah dusky salamanders are largely restricted to stream-headwater habitats that can be degraded by logging, road building, construction, and the activities of feral hogs. The last of these disruptions is particularly important in the Unicoi Mountains (S.G.T., personal observations). Petranka et al. (1993) estimated that the clearcut logging of Appalachian forests of North Carolina during the 1980s and early 1990s killed millions of salamanders, including desmognathine salamanders, each year. Ash and Bruce (1993), however, disputed these estimates and considered them to be exaggerations of the actual number killed. Desmognathine salamanders currently are abundant in areas of the southern Appalachians (the Great Smoky Mountains) that have been logged extensively in the past (S.G.T., personal observations). Populations of Santeetlah dusky salamanders occupying high elevation sites in the southern Appalachians may also be vulnerable to the effects of acid precipitation.

Desmognathus welteri Barbour, 1950
BLACK MOUNTAIN SALAMANDER

J. Eric Juterbock, Zachary I. Felix

1. Historical versus Current Distribution.
Black Mountain salamanders (*Desmognathus welteri*) currently occur in the Cumberland Mountains and Cumberland Plateau in southwestern Kentucky, southern West Virginia, eastern Virginia, and north-central Tennessee (Redmond, 1980; Juterbock, 1984; Redmond and Scott, 1996; Petranka, 1998; Mitchell and Reay, 1999; Felix and Pauley, in preparation). Few data are available to allow an evaluation of present versus historical distributions. A specimen in the West Virginia Biological Survey at Marshall University was collected in 1938 by Neil Richmond at Blair Mountain in Logan County, West Virginia. Recent surveys in the area between this, the northernmost locality for the species, and known populations in McDowell and Wyoming counties, West Virginia, detected no new populations (Felix and Pauley, in preparation). This area has been mined extensively for coal, and it is

possible that habitat degradation has resulted in the isolation of populations in the area, effectively limiting their distribution. Blair Mountain has since been mined using mountain top removal, and it is doubtful that habitat remains for the species.

2. Historical versus Current Abundance.
There are no data available to properly address changes in abundance. That said, it is likely that there have been at least local decreases in the abundance of Black Mountain salamanders due to habitat alteration from the strip mining of coal and the widespread use of these salamanders as bait (see "Conservation" below).

3. Life History Features.
 A. Breeding. Courtship is unknown for this species, but well known and terrestrial for other congeneric species (see accounts for *D. orestes*, Blue Ridge dusky salamanders, in Petranka, 1998; Tilley and Camp, this volume). Nests and larvae are associated with streams (Juterbock, 1984; Smith et al., 1996; Petranka, 1998). A gravid female 73 mm SVL was found on 30 May, with a spermatophore in her vent, indicating breeding takes place in spring. Hedonic mental glands of males are most prominent in the same populations from April to late June. Sperm wave analysis showed production of sperm occurs in summer and fall, but males appear capable of producing spermatophores throughout the active season (Felix, 2001). Data from dissections and mark–recapture studies suggest a biennial breeding cycle for females (Felix, 2001).
 i. Breeding migrations. Unknown and unlikely. Black Mountain salamanders, as with other dusky salamanders, reproduce in their adult, stream or streamside, habitats.

 ii. Breeding habitat. Usually associated with permanent, small- to medium-sized streams with moderate to steep gradients located in mesic forests (Redmond, 1980; Petranka, 1998; Felix, 2001; J.E.J., personal observations). This apparent requirement for stream permanence has been cited (Redmond, 1980) as an explanation for the species' apparent absence from the southern Cumberland Mountains. However, a population of Black Mountain salamanders in Clay County, Kentucky, observed from 1970–77, remained reasonably abundant and continued to reproduce in spite of the periodic absence of running water at the surface during the summer. One of us (J.E.J., personal observations) was unable to determine whether this intermittent condition was normal for that habitat or a result of road-building activity on the opposite side of the mountain. It does, however, indicate that stream permanence, short term, may not be an absolute requirement for the species.
 B. Eggs. Egg clutches occur as a grape-like cluster, reminiscent of those of northern dusky salamanders (*D. fuscus*; Smith et al., 1996), and are guarded in nests by a brooding female. Individual eggs, averaging 4.5 mm in diameter (including capsules), are suspended by 4–12 mm long pedicles, all of which twist and attach to form the cluster. Eggs hatch in September (Juterbock, 1984; see Petranka, 1998).
 i. Egg deposition sites. Smith et al. (1996) reported on four egg masses deposited in packed leaves, 5–20 cm above the water surface in the stream's main channel. One of us (J.E.J., personal observations) has found Black Mountain salamander nests with eggs and attendant female in such situations as these, as well as under rocks on the adjacent stream

Black Mountain Salamander (*Desmognathus welteri*)

bank and in the wet interstices of rocks below where the stream tumbles over surface rocks. It appears that Black Mountain salamanders, which coexist with and are often confused with northern dusky and seal salamanders (*D. monticola*), broadly overlap these two species in nesting requirements.

ii. Clutch sizes. The number of eggs in four clutches averaged 26 (range = 18–33; Smith et al., 1996).

C. Larvae/Metamorphosis.

i. Length of larval stage. The larval period is apparently 20–24 mo (Juterbock, 1984). Hatchling larvae have been found in September, ranging in size from 11–12.5 mm SVL. At this time of year, there are two size classes of Black Mountain salamander larvae, with the larger cohort representing animals from the preceding summer. By this time, sympatric larval northern dusky and seal salamanders from the previous summer had undergone metamorphosis. Similar evidence suggests a 2-yr larval period in the northern portion of the species' range (McCleary, 1989; Felix, 2001).

ii. Larval requirements.

a. Food. Unknown.

b. Cover. One of us (J.E.J., personal observations) has typically observed small larvae among sand and gravel and leaf litter in shallow (usually ≤3 cm) water, usually out of the current. Larger larvae are found here or in deeper water (usually ≤10 cm) without much current and often associated with slightly larger stones. Larvae often are not seen in the middle of pools.

iii. Larval polymorphisms. Unknown and unlikely.

iv. Features of metamorphosis. In a previous study (Juterbock, 1984), the largest larval Black Mountain salamander observed was 28.5 mm SVL; several other animals approached this size. The smallest newly metamorphosed animal observed was 19 mm SVL, but 21 mm SVL was more typical. In this same study, in September when the eggs of the newest cohort were hatching, no 2-yr-old cohort of larvae was observed. Thus, it appears that metamorphosis occurs somewhat variably during the summer at the end of the second year of larval life. Some individuals in West Virginia populations appear to transform in July and August at 26–28 mm SVL (Felix, 2001).

v. Post-metamorphic migrations. Unknown and unlikely.

vi. Neoteny. Unknown and unlikely.

D. Juvenile Habitat. Juveniles are found associated with the same habitats as the adults: permanent, small- to medium-sized streams with moderate to steep gradients located in mesic forests (Redmond, 1980; Petranka, 1998; J.E.J., personal observations). Differences in micro-scale distribution between adults and juveniles remain unstudied, but juveniles and adults may partition microhabitats along

streams in terms of distance to the land/water interface. While it appears to one of us (J.E.J.) that juveniles are more terrestrial than adults, in other populations, juveniles were found substantially farther into streams than adults (Felix and Pauley, 2001). It is possible adults exclude juveniles from the land/water interface, and that in the population studied by Felix and Pauley (2001), juveniles move into the water to avoid competition or predation from more terrestrial seal and northern dusky salamanders. Because the cannibalistic (both intra- and interspecific) tendencies of congeneric black-bellied salamanders (*D. quadramaculatus*) are well known (see Petranka, 1998) and size dependent, a difference in microhabitat preferences between adult and juvenile Black Mountain salamanders would not be unexpected.

E. Adult Habitat. Adults are found along permanent, small- to medium-sized streams with moderate to steep gradients located in mesic forests at elevations from 300–800 m (Redmond, 1980; Petranka, 1998; Felix, 2001; J.E.J., personal observations). Adults frequently are associated with large rocks (Petranka, 1998). Although generally aquatic, individuals are also commonly found on the stream banks (J.E.J., personal observations). Sexual dimorphism in habitat characteristics is unknown.

F. Home Range Size. Limited data suggest Black Mountain salamander movement patterns and home range size are similar to other *Desmognathus* species (see Petranka, 1998). Measurements from 18 recaptures of 12 salamanders show individuals moved an average of 1.8 m, with the longest distance moved 7.5 m. Individuals recaptured multiple times showed an association with structures such as cascades or mid-stream gravel bars (Felix, 2001).

G. Territories. Unknown.

H. Aestivation/Avoiding Desiccation. Aestivation is unknown. Black Mountain salamanders are found adjacent to stream banks, which may buffer them from the effects of desiccating conditions.

I. Seasonal Migrations. Unknown and unlikely.

J. Torpor (Hibernation). Unknown.

K. Interspecific Associations/Exclusions. Syntopic with northern dusky salamanders and seal salamanders throughout their range, and with Allegheny Mountain dusky salamanders (*D. ochrophaeus*) in portions of their range. In southern West Virginia, Black Mountain salamanders occur sympatrically with three congeners in communities similar to those described by Bruce (1991) and others. Allegheny Mountain salamanders reside in forests adjacent to stream banks. Within the streambed, Black Mountain salamanders represent the large-bodied, aquatic species, followed by seal and northern dusky salamanders in order of increasing

terrestrial tendency (Felix and Pauley, 2001). It has been suggested that Black Mountain salamanders are excluded from some streams through competition with morphologically and ecologically similar black-bellied salamanders (Seeman, 1996).

L. Age/Size at Reproductive Maturity. Maturity is estimated to be at 4–5-yr old and at 50–55 mm SVL (Petranka, 1998). As one of us has discussed (Juterbock, 1978, p. 225), it is not clear how "effective sexual maturity" relates to the non-synchronous appearance of secondary sexual characteristics in male dusky salamanders. By the time they reach 50 mm SVL, virtually all male Black Mountain salamanders exhibit the adult condition for most secondary sexual traits and are presumably mature. The smallest male examined by Felix (2001) with mature sperm was 50 mm SVL. Females typically are found with yolked ovarian ova at the same size (Juterbock, 1978). With a 20–24-mo larval period, this is a minimum of 4 yr to maturity.

M. Longevity. Individuals can live at least 20 yr in captivity (Snider and Bowler, 1992).

N. Feeding Behavior. The five most numerically important prey groups in West Virginia specimens were adult dipterans, coleopterans, winged hymenopterans, larval dipterans, and larval lepidopterans. Four of these groups are terrestrial in origin, suggesting Black Mountain salamanders forage mainly out of water and capture both aerial and ground-dwelling insects. No remains of salamanders were observed in stomachs. The diet of Black Mountain salamanders is similar to sympatric seal salamanders, differing mostly in that the former consumed more aquatic prey (Felix and Pauley, in preparation). Captive individuals readily eat earthworms (J.E.J., personal observations).

O. Predators. Unknown.

P. Anti-Predator Mechanisms. Unknown, but presumably similar to the aggressive defense described for other dusky salamander species (Brodie, 1978). In the laboratory, black-bellied salamanders oriented their heads toward approaching short-tailed shrews (*Blarina brevicauda*), even if that required flipping their bodies around. This was accompanied by an open-mouth display. If the predator continued approaching, the salamander lunged toward it, loudly snapping the mouth closed. If bitten by the predator, they bit back, normally holding on (up to 47 s in one case) and sometimes twisting, especially if they bit the predator's snout. This defense kept four of five alive for 3–7.5 min, in a situation where escape was impossible. Smaller seal and Allegheny Mountain dusky salamanders (*D. ochrophaeus*) were less successful, but similarly defended themselves. Northern dusky salamanders also bite, apparently in defense (Noble, 1954; J.E.J., personal observations); Noble also notes that when grasped, they are very likely to "twist strenuously." Both behaviors have been observed in Black Mountain salamanders (J.E.J., personal observations).

Q. Diseases. Unknown.

R. Parasites. Intestinal parasites from West Virginia specimens were identified by J. Joy of Marshall University as the nematode *Batracholandros magnavulvaris*. This parasite has been reported from other species of *Desmognathus* salamanders (Joy et al., 1993).

4. Conservation.

Redmond (1980) indicates that both habitat alteration from the strip mining of coal and the widespread use of these salamanders as bait have led to local population declines in Black Mountain salamanders. During the 1970s, after the 1973 oil embargo revitalized the coal industry in this region, one of us (J.E.J., unpublished data) observed two kinds of negative mining effects on Black Mountain salamander populations. Siltation increased dramatically downstream of mining activity,which appeared to at least alter the micro-distribution of larvae, if not their abundance; mountain top removal, with the resultant deposition of the overburden in the valley below, completely eliminated salamander habitat. Black Mountain salamanders are now listed as Rare in West Virginia.

Desmognathus wrighti King, 1937
PIGMY SALAMANDER

Julian R. Harrison

1. Historical versus Current Distribution.

Pigmy salamanders (*Desmognathus wrighti*) occur in isolated populations in the high elevation forests of southwestern Virginia, western North Carolina, and eastern Tennessee, at elevations ranging from approximately 762–2,012 m. Pague (1984) reported a slight extension of the range in Virginia, at Pine Mountain, 2.4 km east of Mount Rogers. Pigmy salamanders are especially characteristic of spruce-fir forests, but populations also occur at lower elevations in mesophytic hardwoods; populations found east and south of the present range of spruce and fir may represent post-glacial relicts of a formerly more widespread distribution (Huheey, 1966; Tilley and Harrison, 1969; Rubin, 1971; Bruce, 1977). The historical distribution is unknown.

Crespi (1996) presented evidence that *D. wrighti*, as currently recognized, is not monotypic but instead a complex of two different allopatric species. She found that populations north of the Black Mountains in North Carolina (Roan, Grandfather, and Whitetop mountains, and Deep Gap, Virginia) differ from the

Pigmy Salamander (*Desmognathus wrighti*)

more southwestern populations by fixed alternative alleles at more than one locus, and a unique morphological characteristic, partial to complete loss of ventral iridophores. A description of the new species is in preparation.

2. Historical versus Current Abundance.

The extent to which pigmy salamanders have been affected by the decline of spruce-fir forests at high-elevation sites is unknown. Wilson (1995) offered the protection of spruce-fir forests and retention of ground litter in areas occupied by this species as management suggestions. Pigmy salamanders apparently were relatively common in 1936, as the original description was based on 100 carefully examined specimens of which 3/4 were seen in the field; an additional 50 specimens were observed but not collected (King, 1936). Also, Weller (1931) stated that this species (misidentified as a juvenile form of *D. fuscus carolinensis*) ". . . was found abundantly only under the tightly fitting bark of stumps or standing trees at high elevations." In his study of the local distribution and ecology of southern Appalachian plethodontids, Hairston (1949) found most species, including pigmy salamanders, to be abundant. Bogert (1952) noted that pigmy salamanders were relatively abundant in samples of salamanders from high elevations on White Top Mountain (n = 95) and Mount Rogers (n = 94), Virginia, constituting 20% of the sample from each locality. Organ (1961a) collected approximately 7,000 specimens of *Desmognathus* from Mount Rogers and Whitetop Mountain, Virginia, in the summer of 1957 and the summer, fall, and spring of 1958–59; none of the five species, including pigmy salamanders, was represented by < 1,000 individuals.

Pague and Mitchell (1987) regarded the species as abundant on Whitetop Mountain in Virginia, despite heavy collecting for at least 30 yr. In their study of the effects of timber harvesting on salamanders, Petranka et al. (1993) reported that captures of salamanders in plots from mature forests were five times greater than those from recent clearcuts and that 50–70 yr would be required for populations to return to pre-disturbance levels. Pigmy salamanders were represented by 26 specimens (3% of the total number of salamanders collected) and had a 30% frequency of occurrence in plots (see also Dodd, 2004) . However, Ash and Bruce (1994) questioned the methodology, analysis, and interpretation of the study, suggesting that <1% of the *Desmognathus* community may have been sampled (but see Petranka, 1994). Singer et al. (1982) reported no significant difference in numbers of pigmy salamanders and four other salamander species between northern hardwood stands rooted by hogs and stands that were not rooted. However, they did not report that only pigmy salamanders declined significantly as cited by Mathews and Echternacht (1984). Petranka (1998) has observed pigmy salamanders to be common in old-growth deciduous forests at relatively low elevations but uncommon in younger stands.

3. Life History Features.

A. Breeding. Reproduction is terrestrial.

i. Breeding migrations. Unknown.

ii. Breeding habitat. The six nests that are known were found beneath the bank of a creek (Organ, 1961b).

B. Eggs.

i. Egg deposition sites. Only six nests are known; these were found beneath the bank of a creek at a depth of about 30 cm in a pocket of gravel and mud through which water percolated (Organ, 1961b). Each nest was suspended from a small rock within a small cavity in the mud surrounding the rock and attended by an adult—spent females in the two instances where this could be determined. Courtship of captive adults occurs in both fall and spring (Organ, 1961b). Brooks (1948a), Organ (1961b), Houck (1980), and Verrell (1999) described courtship behavior. Brock and Verrell (1994), Verrell (1997), and Verrell and Mabry (2000) presented information concerning courtship pattern.

ii. Clutch sizes. The six known clutches were found in mid-October and contained 3–8 embryos in late stages of development (Organ, 1961b); hatching of the eggs occurred 19–25 October.

C. Direct Development. Length of the developmental period is unknown. The terrestrial hatchlings are essentially miniature versions of the adult; there is no aquatic larval stage.

D. Juvenile Habitat. Juveniles are terrestrial and have the same habitat characteristics as adults.

E. Adult Habitat. Typical habitat includes soil surfaces beneath small logs and stones in spruce-fir forests (King, 1936, 1939) and under bark of stumps or standing trees at high elevations (Weller, 1931). Pigmy salamanders are highly terrestrial, characteristic of spruce-fir forests, but present also at lower elevations in mesophytic hardwoods. All the pigmy salamanders observed by Hairston (1949) were in forest habitats with 76% of animals >61 m from the closest stream. Organ (1961a) confirmed Hairston's observations but also provided evidence that, in late fall, pigmy salamanders move into underground seepage areas. When active under conditions of complete darkness and a saturated atmosphere, individuals climb to heights of ≤ 1.2 m (7 ft) on living or dead spruces (Hairston, 1949; Organ, 1961a). On especially damp nights, pigmy salamanders may be found on leaves some distance above the ground (Mathews and Echternacht, 1984).

F. Home Range Size. Unknown.

G. Territories. Unknown. However, males are aggressive toward other males during the courtship season (Organ, 1961a). Larger males always dominate smaller ones. If both are about the same size, usually one male bites and the other flees.

H. Aestivation/Avoiding Desiccation. Unknown. Pigmy salamanders are active throughout the warmer parts of the year.

I. Seasonal Migrations. Pigmy salamanders abandon forest habitat in late fall and move into underground seepage areas for hibernation in winter (Organ, 1961a; Bruce, 1977).

J. Torpor (Hibernation). In the fall and spring of 1958–59, Organ (1961b) removed 649 specimens from a site measuring about 1 m (3 ft) on a side by 30 cm (1 ft) deep in saturated gravel and mud beneath the bank of a creek. He interpreted this as a hibernating aggregation.

K. Interspecific Associations/Exclusions. Pigmy salamanders occur syntopically with up to four or five different congeners in various parts of their range. These species are segregated ecologically by microhabitat based primarily on distance from open water, though considerable overlap in horizontal distribution occurs. Pigmy salamanders are the most terrestrial desmognathines in this streamside community, occurring exclusively in the forest, especially during the summer. They also exhibit greater scansorial activity than other species when active at night. The ecological differences between pigmy salamanders and their congeners, especially members of the *D. ochrophaeus* complex (Tilley and Mahoney, 1996), in part also forest species, have been interpreted as a consequence of interspecific competition (Hairston, 1949). Tilley (1968), however, suggested that selection has favored increased terrestrialism in the smaller species because of their susceptibility to predation by the larger ones. Subsequent studies by Hairston (1986, 1987) also pointed to predation rather than competition as the mechanism involved in streamside desmognathine communities. Pigmy salamanders are known to occur sympatrically with seepage salamanders (*D. aeneus*), an ecologically similar species, at several localities in the southern Nantahala Mountains of southwestern North Carolina (Tilley and Harrison, 1969; Rubin, 1971; Bruce, 1991), but the nature of the association between these two species in that area is unknown.

L. Age/Size at Reproductive Maturity. Males reach sexual maturity at 17–19 mm SVL (to anterior corner of vent), females at 20 mm SVL (Harrison, 1963). According to Organ (1961a), males mature sexually when they are 3.5 yr old and have attained a SVL >24 mm (to posterior corner of vent); females mature when 4.5 yr old but do not lay eggs until they are 5 yr old.

M. Longevity. Unknown. Males and non-brooding females have a mean annual survival rate of 91% in the early years of life and 29% in the later years (Organ, 1961a).

N. Feeding Behavior. Food items in 10 stomachs examined by Hairston (1949) included springtails, parasitic wasps, moths, beetles, thrips, flies, pseudoscorpions, spiders, and acarine mites, particularly orabatids. All or most of these prey items were species characteristic of the forest-floor habitat. Orabatid mites are abundant and ecologically important components of coniferous forests (Jacot, 1936).

O. Predators. Known predators include spring salamanders (*Gyrinophilus porphyriticus*) and a carabid beetle (Huheey and Stupka, 1967). Other presumed predators include larger, syntopic species of *Desmognathus*, small snakes such as ringnecked snakes (*Diadophis punctatus*), and birds. Of a pigmy salamander museum sample, 7% had broken tails (Wake and Dresner, 1967).

P. Anti-Predator Mechanisms. When exposed or touched, pigmy salamanders flip and become immobile (Brodie, 1977).

Q. Diseases. Unknown.

R. Parasites. Unknown.

4. Conservation.

In North Carolina, pigmy salamanders have been placed in category W5 of that state's "Watch List," a species facing "increasing amounts of threats to its habitat, whether or not populations are known to be declining" (LeGrand and Hall, 1995). Their rank in that state is S3, "Rare or Uncommon" (21–100 extant populations). Based on his experience with the species in North Carolina, R.C. Bruce (personal communication) has no reason to believe pigmy salamanders are in any jeopardy in that state, though he has not monitored populations in spruce-fir sites. Kucken et al. (1994) found that in a Great Smoky Mountains stream contaminated by sulfuric acid and heavy metals from the Anakeesta Formation, stream breeders were almost entirely eliminated, while terrestrial breeders, including pigmy salamanders, increased in numbers. Along contaminated sections of the stream where black-bellied salamanders (*D. quadramaculatus*) were almost entirely eliminated, pigmy salamanders were about four times more abundant than they were along uncontaminated sections where black-bellied salamanders were present. This was interpreted by the authors as reflecting the absence of predation on the smaller species by the larger one along contaminated sections of the stream.

In Tennessee, pigmy salamanders are considered to be in need of management and are given a rank of S2, "very Rare and Imperiled within the state" (Withers, 1996). Populations of pigmy salamanders in Virginia appear to be viable within their restricted range in the Mount Rogers National Recreation area, but they are considered to be a Species of Special Concern with a rank of S2 because of their limited distribution within the state (Pague, 1991). Logging activities and increased recreational development could threaten portions of this range, particularly at lower elevations (Pague, 1991). J.A. Organ (personal communication) has found that pigmy salamander populations on Whitetop Mountain, Virginia, below 1,525 m (5,000 ft) are stable, but above that elevation they may be declining. Further, there is no evidence of any local extinctions on Whitetop Mountain, Mt. Rogers, Beech Mountain, Bluff Mountain, or Pine Mountain in Virginia during the last half of the twentieth century (Organ, 1993, and personal communication). Organ's findings were based on intensive quantitative sampling in the Mt. Rogers National Recreational Area in the late 1950s, the late 1960s, 1970, and again in 1989–92. All the known Virginia localities for the species are situated within the Mt. Rogers Recreational Area, Jefferson National Forest (Mitchell and Reay, 1999).

Based on her recent and extensive experience with pigmy salamanders in the field, Erica Crespi (personal communication) observed that in the Great Smoky Mountains, Plott Balsams, Great Balsams, and southward, populations of pigmy salamanders were patchy in distribution, but that enough of them exist to provide long-term stability. In that part of the range, individuals were abundant in small patches (approximately 20 m × 50 m²), but such patches were difficult to find. In contrast, Crespi noted that the more northern populations of the species (Mt. Rogers, Whitetop Mountain, Roan Mountain, and Grandfather Mountain) are more isolated from one another and exist in fewer patches that are farther apart. In that area, pigmy salamanders were locally common in small patches, but patches were uncommon or rare. Thus, northern populations may be more vulnerable to extinction in the face of environmental change.

Ensatina eschscholtzii Gray, 1850
ENSATINA

Shawn R. Kuchta, Duncan Parks

1. Historical versus Current Distribution.
Ensatinas (*Ensatina eschscholtzii*) have the widest distribution of any West Coast plethodontid, ranging from the central coast of British Columbia and Vancouver Island, Canada (Kelson et al., 1999), to northern Baja California, Mexico (Mahrdt, 1975; Stebbins, 1985). Seven morphologically distinct subspecies—painted salamanders (*E. e. picta*), Oregon salamanders (*E. e. oregonensis*), yellow-eyed salamanders (*E. e. xanthoptica*), Monterey salamanders (*E. e. eschscholtzii*), Sierra Nevada salamanders (*E. e. platensis*), yellow-blotched salamanders (*E. e. croceater*), and large-blotched salamanders (*E. e. klauberi*)—are arranged parapatrically in a ring around the Central Valley of California (Stebbins, 1949b, 1985). Monterey salamanders (coastal) and large-blotched salamanders (inland) are locally sympatric at four sites in southern California, with little or no hybridization (Brown and Stebbins, 1964; Stebbins, 1957; Wake et al., 1986), yet intergradation occurs between geographically adjacent subspecies in a loop around the Central Valley of California. Thus, *Ensatina eschscholtzii* is

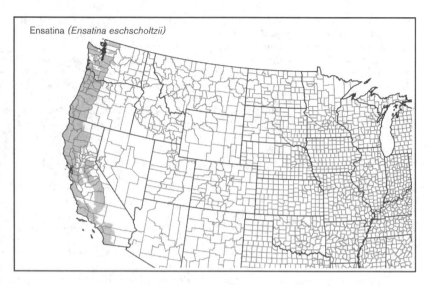

Ensatina (*Ensatina eschscholtzii*)

usually considered a ring species (Stebbins, 1949b; for recent studies of the ring species issue, see Wake et al., 1986; Wake and Yanev, 1986; Wake et al., 1989; Moritz et al., 1992; Jackman and Wake, 1994; Wake, 1997; Highton, 1998; Wake and Schneider, 1998). A second ring is completed in the Sierra Nevada where yellow-eyed salamanders and Sierra Nevada salamanders hybridize (Wake et al., 1989). Some researchers consider the complex to be comprised of many distinct species (Highton, 1998), and large-blotched salamanders in particular are sometimes treated as a separate species (Frost and Hillis, 1990; Collins, 1991; Grismer, 1994). Aside from localized extinctions, historical and current distributions appear similar.

2. Historical versus Current Abundance.
Ensatinas usually are common where present. Stebbins (1954b) estimated 1,730 yellow-eyed salamanders/ha in mature redwood forest in Contra Costa County, California. Gnaedinger and Reed (1948) estimated 2,833 Oregon salamanders/ha in a wooded canyon in Multnomah County, Oregon.

In the Tehachapi Mountains, yellow-blotched salamanders are positively associated with canyon live oak (*Quercus chrysolepis*) and negatively associated with blue oak (*Q. douglassi*; Block and Morrison, 1998). Development and the cutting of oak woodland in the Tehachapi Mountains may threaten yellow-blotched salamanders (Jennings and Hayes, 1994a). In Douglas fir forests in northern California, Oregon salamanders are more abundant in old-growth stands than in young or mature stands (Bury, 1983; Raphael, 1988; Welsh and Lind, 1988, 1991). In the industrial, managed forests in the Cascades of Washington, Oregon salamanders are more abundant in rotation age stands (from about 45–70 yr) than in younger stands (Aubry, 2000; Grialou et al., 2000). However, in Washington and

Oregon there is no correlation between stand age and Oregon salamander abundance in unmanaged Douglas fir forests (Aubry et al., 1988; Corn and Bury, 1991; Gilbert and Allwine, 1991). In old-growth Douglas fir forests in northern California, Oregon, and Washington, Oregon salamanders consistently are more abundant in drier sites (Aubry et al., 1988; Aubry and Hall, 1991; Gilbert and Allwine, 1991; Welsh and Lind, 1991). Finally, in managed Douglas fir forests in the Oregon Cascades, Oregon salamander abundance is positively correlated with the volume of coarse woody debris (Butts and McComb, 2000), but this correlation is not found in the Washington Cascades (Aubry, 2000).

Stebbins (1951) has observed that edge habitats support the highest abundances of ensatinas and notes that they appear more abundant on flat or gently sloping shelves above flood level than on steep terrain. Storer (1925) stated that yellow-eyed salamanders in central California are less abundant than arboreal salamanders (*Aneides lugubris*), a situation that is now reversed, probably in part because arboreal salamanders are now less common (D. B. Wake, personal communication).

3. Life History Features.
   A. Breeding. Reproduction is terrestrial.
   i. Breeding migrations. Ensatinas do not migrate and movements generally are limited (Staub et al., 1995). Males have abundant spermatozoa throughout the rainy season (September–April throughout much of the range, but extends into June in the Sierra Nevada and northern parts of the range; in the south, rains may not begin until December). Females with sperm capsules in their vents have been found from November–March (Stebbins, 1951).
   ii. Breeding habitat. Females lay eggs after retreating to aestivation sites at the end of the rainy season (Stebbins, 1951, 1954b; Jones and Aubry, 1985).

B. Eggs.

i. Egg deposition sites. Eggs masses are found in dark, moist, insulated habitats, such as under and inside logs, under bark, and inside animal burrows (summarized by Petranka, 1998; Stebbins, 1951, 1954b).

ii. Clutch size. Females lay from 3–25 eggs, but 9–16 are most common (data summarized by Petranka, 1998; Stebbins, 1951, 1954b).

C. Direct Development. Clutches are always attended by the female (Stebbins, 1954b). Under laboratory conditions, Collazo (1990) observed that most animals laid eggs in April and May. He monitored the development of five *E. eschscholtzii* clutches to hatching: painted salamander eggs took 113–119 d to hatch (5 eggs), Monterey salamander eggs took 177 d to hatch (4 eggs), and eggs from two Oregon salamander clutches took about 137–142 d (5 eggs) and 129 d (2 eggs), respectively. Time to hatching was negatively correlated with egg size.

D. Juvenile Habitat. Same as adult habitat.

E. Adult Habitat. Ensatinas typically inhabit coniferous forest, deciduous forest, oak woodland, coastal sage scrub, and chaparral (Stebbins, 1951). Individuals are found in thermally buffered, mesic microclimates, such as under logs, bark, and moss (Aubry et al., 1988; Bury and Corn, 1988b; Aubry and Hall, 1991; Corn and Bury, 1991; Gilbert and Allwine, 1991; Welsh and Lind, 1991), under leaf litter (Corn and Bury, 1991), in talus (Herrington, 1988), and in animal burrows (Storer, 1925). They quickly dehydrate on dry substrates (Cohen, 1952) and prefer moist, but unsaturated, soils. Recently, ecological and behavioral aspects of water economy have been described (Wisely and Golightly, 2003). Unlike many Pacific Northwest plethodontid salamanders, Oregon salamanders tend to be more abundant away from streams (McComb et al., 1993a,b). The range of habitats used by southern California ensatinas is greater than previously described by Stebbins (1951), Fisher and Case (1997), and Jennings and Hayes (1994a), but their habitat use and associations are in need of study (D. B. Wake, personal communication).

F. Home Range Size. Stebbins (1954b) estimated from a 122 × 49 m (400 × 160 ft) plot that home ranges had a maximum width of 10–41 m (mean=19.5 m) for males, and 6–23 m (mean=10 m) for females. Males moved about twice as far between recaptures as females, and the movement of young animals was similar to females. Jacqmotte (1992) estimated home ranges from 0.038–2.6 m² on 10 × 10 m plots, but the small plot size may have biased the home range estimate (Staub et al., 1995). Staub et al. (1995) found average distances between captures to be 26.5 m for males and 19.7 m for females in a grid of pit traps (100 × 300 m)

designed to detect long-distance dispersal (minimum detectable trapping distance= 7.1 m). The maximum movement was 120 m for males and 61 m for females.

G. Territories. Evidence for marking, recognizing, and defending home areas has been observed in laboratory settings outside of the breeding season, suggesting territoriality (Wiltenmuth, 1996; Wiltenmuth and Nishikawa, 1998).

H. Aestivation/Avoiding Desiccation. Surface activity by ensatinas is highly correlated with surface moisture (Stebbins, 1951, 1954b), and animals are not commonly encountered above ground during the summer dry season. Individuals aestivate in cool, moist areas, such as animal burrows and inside logs. Emerging individuals often appear dehydrated and usually have empty digestive tracts (Stebbins, 1954b).

I. Seasonal Migrations. None.

J. Torpor (Hibernation). Ensatinas retreat from the surface during periods of freezing temperatures but are not known to hibernate (Stebbins, 1954b).

K. Interspecific Associations/Exclusions. Ensatinas co-occur with salamanders in the genera *Ambystoma*, *Aneides*, *Batrachoseps*, *Dicamptodon*, *Plethodon*, and *Taricha* (Stebbins, 1985). In northern California, Oregon salamanders, black salamanders (*Aneides flavipunctatus*), arboreal salamanders, and California slender salamanders (*B. attenutatus*) differ in their mean and maximum prey size, despite considerable prey overlap (Lynch, 1985). Juvenile Oregon salamanders have a relatively broad head and eat larger prey than sympatric plethodontid juveniles, yet adult ensatinas eat smaller prey than arboreal salamanders (Lynch, 1985). Oregon salamanders eat larger prey than sympatric Larch Mountain salamanders (*P. larselli*) and Dunn's salamanders (*P. dunni*; Altig and Brodie, 1971). Both Lynch (1985) and Bury and Martin (1973) conclude that the diet of Oregon salamanders is generalized relative to sympatric California slender salamanders and climbing salamanders (genus *Aneides*).

L. Age/Size at Reproductive Maturity. Three to four years are required to achieve maturity. Males are sexually mature at about 48–55 mm SVL; females are sexually mature at >60 mm SVL (Stebbins, 1954b).

M. Longevity. Stebbins (1954b) estimated the oldest animals in his study plot to be >8.5 yr, and Staub et al. (1985) estimated ages up to 15 yr.

N. Feeding Behavior. Ensatinas are euryphagic predators of small animals, especially arthropods. Diet is known to include, among others, spiders, mites, beetles, sowbugs, crickets, springtails, centipedes, millipedes, termites, earthworms, and snails (Gnaedinger and Reed, 1948; Zweifel, 1949; Stebbins, 1951, 1954b; Altig and Brodie, 1971; Bury and Martin, 1973;

Lynch, 1985). Most feeding is sit-and-wait ambush, but ensatinas will stalk to get within range of prey items once they are spotted. The functional morphology of prey capture behavior is a stereotyped protrusion of the hyomandibular apparatus and partially attached tongue; timing and magnitude of tongue and jaw movements are modulated prior to protrusion to accommodate distance, prey type, and other factors (Deban, 1997).

O. Predators. Known predators include Steller's Jays (*Cyanocitta stellari*; Stebbins, 1954b), garter snakes (*Thamnophis* sp.; Fitch, 1940; Beneski, 1989), and raccoons (*Procyon lotor*; Wake et al., 1989). Stebbins (1954b) discusses other possible predators.

P. Anti-Predator Mechanisms. Harassed ensatinas stand on their toes stiff-legged, arch their back down, hold the neck erect with the head horizontal or downward, and arch and flip their tails in the direction of the attacker (Stebbins, 1951; Brodie, 1977). The dorsal surface of ensatina tails contains large, densely packed poison glands, and a sticky, milky secretion is exuded when the animal is threatened (Hubbard, 1903). Ensatinas can autotomize their tails at the constricted base, and autotomized tails bend wildly for several minutes (Stebbins, 1954b; Wake and Dresner, 1967). Beneski (1989) found that ensatina tails stick to the mouths of garter snakes, inhibiting further consumption. It takes about 2 yr to regenerate an autotomized tail (Stebbins, 1954b; Staub et al., 1985), and tails are rarely dropped unless conditions are life threatening (Beneski, 1989). Rarely, threatened ensatinas will vocalize with a hissing sound, similar to a snake (Stebbins, 1951; Brodie, 1978). Yellow-eyed salamanders are probably Müllerian mimics of the extremely toxic newts of the genus *Taricha* (Stebbins, 1949b; C. W. Brown, 1974; Wake et al., 1989; Kuchta, 2002).

Q. Diseases. Unknown.

R. Parasites. Helfer (1949), Lehmann (1954), and Goldberg et al. (1998c) examined the helminth communities in ensatinas. Two cestode and two nematode species were documented. Of 30 ensatinas from Del Norte and Humboldt counties, California (within the range of painted salamanders and Oregon salamanders), Goldberg et al. (1998c) found 17 contained at least one species of parasitic helminth (average number of helminth species/infected salamander = 1.1 ± 0.3). They concluded that the helminth infracommunities of ensatinas are depauperate relative to other vertebrates. The effect of parasites on ensatina fitness or ecology has not been examined.

4. Conservation.

Ensatinas usually are common where present. In Douglas fir forests in northern California, Oregon salamanders are more abundant in old-growth stands than in young or mature stands and are consistently more abundant in drier sites (see "Historical versus Current Abundance" above). In the industrial, managed forests in the Cascades of Washington, Oregon salamanders are more abundant in rotation-age stands from about 45–70 yr than in younger stands. However, in Washington and Oregon there is no correlation between stand age and Oregon salamander abundance in unmanaged Douglas fir forests (see "Historical versus Current Abundance" above).

Acknowledgments. We extend special thanks to Dr. D.B. Wake for valuable comments on the manuscript, and, more generally, for expertly and generously supporting our work on *Ensatina eschscholtzii*.

Eurycea aquatica Rose and Bush, 1963
DARK-SIDED SALAMANDER

Kenneth H. Kozak, Michael J. Lannoo

1. Historical versus Current Distribution.
Rose and Bush (1963) and Jones (1980) reported dark-sided salamanders (*Eurycea aquatica*) from small springs in the Ridge and Valley Physiographic Province of Alabama (Jefferson, Shelby, and Saint Clair counties) and northwestern Georgia (Catoosa, Chattooga, Floyd, Gordon, Murray, Walker, and Whitfield counties). They appear to be absent south of the Coosa River despite the availability of suitable spring habitats (Jones, 1980). Ashton (1966) reported specimens matching the description of dark-sided salamanders from Davidson County, Tennessee, in the Interior Low Plateaus Physiographic Province.

The species status of *E. aquatica* is controversial. Mount (1975) examined dark-sided salamanders from Alabama and found many individuals intermediate in morphology between southern two-lined salamanders (*E. cirrigera*) and dark-sided salamanders; he therefore considered these two taxa conspecific. Similarly, Jones (1980) conducted a comprehensive morphological study of the *E. bislineata* complex in the Ridge and Valley Province of Alabama and Georgia. At many of the examined localities, "intermediates" between southern two-lined salamanders and dark-sided salamanders were found, which Jones (1980) also interpreted as evidence of hybridization. Both Mount (1975) and Jones (1980) considered dark-sided salamanders to represent a "spring ecotype" of southern two-lined salamanders rather than a distinct species. Jacobs (1987) studied allozyme variation in the *E. bislineata* complex and found *E. aquatica* from the type locality in Jefferson County, Alabama, to be similar to Junaluska salamanders (*E. junaluska*) and some populations of *E. cirrigera*. Similarly, Wallace (1975) found that individuals possessing morphological characteristics of *E. aquatica* and *E. cirrigera* were genetically indistinguishable.

Redmond and Scott (1996) follow Wallace (1975) and place *E. aquatica* in Tennessee in synonomy with *E. cirrigera*. However, the status of *E. aquatica* in Alabama and Georgia remains controversial. Petranka (1998) considers *E. aquatica* to be synonymous with *E. cirrigera*, but Sever (1989) states that "the formal decision on the status of *E. aquatica* must await the comprehensive study of the species in Alabama and neighboring areas called for by Mount (1975) and Jones (1980)." Phylogeographic analyses of mtDNA variation indicate that populations of *E. aquatica* from the Ridge and Valley Province in Alabama form a monophyletic group that is divergent from any recognized species in the *E. bislineata* complex (Kozak and Larson, in preparation). This result corroborates Jacob's (1987) allozyme data indicating that *E. aquatica* from the type locality are genetically distinct from populations of *E. cirrigera* in Alabama, Georgia, and Tennessee. Thus, *E. aquatica* from the Ridge and Valley Province of Alabama should be considered an evolutionary unique lineage that warrants conservation (Moritz, 1999).

2. Historical versus Current Abundance.
Unknown. Rose (1971) states: "The springs in the area of the type locality of *Eurycea aquatica* are very small, and transformed animals in this area are nonexistent. Wholesale destruction of the springs was hastened by the discovery of endemic fish in the area and undisciplined herpetological collecting."

3. Life History Features.
Many aspects of the life history of dark-sided salamanders have not been described. Refer to the *Eurycea cirrigera* (Pauley, this volume) and *E. bislineata* (northern two-lined salamanders; Sever, this volume) accounts for information that can be used provisionally until data on dark-sided salamanders can be collected.

  A. Breeding.
    i. Breeding migrations. Unlikely. During wet periods and at higher altitudes, adults may wander far from breeding areas in springs, seeps, and streams. Dry conditions, however, will cause these animals to return to aquatic habitats. Eggs are laid in water, and larvae are aquatic, so females will congregate at appropriate nesting sites (see species accounts for *Eurycea cirrigera*; Pauley, this volume; and *E. bislineata*; Sever, this volume).
    ii. Breeding habitat. Springs and small woodland streams are the most likely nesting sites (Rose and Bush, 1963).

  B. Eggs.
    i. Egg deposition sites. Eggs are likely deposited on the undersides of rocks and logs in flowing water, and females brood (see Pauley, this volume; Sever, this volume).
    ii. Clutch sizes. The average clutch size in the vicinity of the type locality is 80 (range 60–96; Rose and Bush, 1963). Clutch sizes for northern two-lined salamanders and southern two-lined salamanders average from 20–50 (see Pauley, this volume; Sever, this volume), suggesting the reproductive potential of dark-sided salamanders is high relative to other species in the *E. bislineata* complex.

  C. Larvae/Metamorphosis. Larvae are aquatic.
    i. Length of larval stage. There is typically a bimodal distribution of larval SVLs, suggesting that the larval stage lasts 2 yr. The

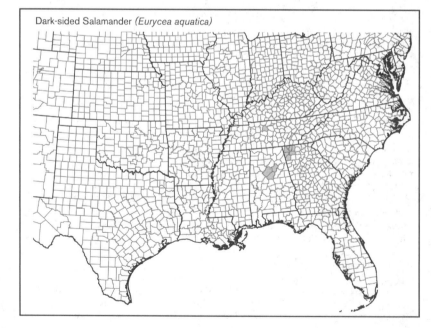

Dark-sided Salamander *(Eurycea aquatica)*

largest larvae are 30 mm SVL (Rose and Bush, 1963).

ii. Larval requirements.

a. Food. Larvae are carnivorous and likely eat a variety of insect and other invertebrate prey. Major prey items include isopods, amphipods, midges, dipteran pupae, stonefly nymphs, and chironomid larvae, as well as copepods, cladocerans, and other zooplankton (Caldwell and Houtcooper, 1973; Burton, 1976; Petranka, 1984b).

b. Cover. As with larvae of southern two-lined salamanders and northern two-lined salamanders, dark-sided salamander larvae generally are nocturnal, and likely prefer the bottoms of shallow, quiet pools that form below riffle areas in rocky streams. They may also be found by searching among the rocks and logs in riffles (see Pauley, this volume; Sever, this volume). At the type locality, larval dark-sided salamanders were found hidden on the bottom of the spring among aquatic vegetation (Rose and Bush, 1963).

iii. Larval polymorphisms. Not known to occur.

iv. Features of metamorphosis. There may be a large amount of variation in the time to metamorphosis, although this must be examined.

v. Post-metamorphic migrations. Unlikely.

vi. Neoteny. In the vicinity of the type locality seven larvae >28 mm with large black testes and coiled vasa deferentia were collected. Likewise, 20 larvae >28 mm with enlarged yolked ova have been collected from this same region. All these larvae also had adult pigmentation patterns (Rose and Bush, 1963).

D. Juvenile Habitat. Unknown, but likely the same as adults.

E. Adult Habitat. The primary habitat of dark-sided salamanders appears to be springs and streams in wooded areas (Rose and Bush, 1963). In the Ridge and Valley Province of Alabama and Georgia, populations are associated with limestone and dolomitic springs (Jones, 1980).

F. Home Range Size. Unknown.

G. Territories. Unknown.

H. Aestivation/Avoiding Desiccation. Aestivation is unknown.

I. Seasonal Migrations. Unknown.

J. Torpor (Hibernation). Unknown.

K. Interspecific Associations/Exclusions. Dark-sided salamanders occur syntopically with southern dusky salamanders (*Desmognathus auriculatus*), long-tailed salamanders (*E. longicauda*), spring salamanders (*Gyrinophilus porphyriticus*), red salamanders (*Pseudotriton ruber*), and northern dusky salamanders (*Desmognathus fuscus*). Detailed studies on the ecological interactions between dark-sided salamanders and sympatric species have not been determined.

L. Age/Size at Reproductive Maturity. Attained during or shortly after metamorphosis (Rose and Bush, 1963).

M. Longevity. Unknown.

N. Feeding Behavior. Dark-sided salamanders likely feed on a wide range of invertebrates. Potential prey include annelids, snails, arachnids, isopods, and insects such as beetles, bugs, roaches, springtails, dipterans, and hymenopterans (Hamilton, 1932; Burton, 1976).

O. Predators. Unknown.

P. Anti-Predator Mechanisms. Closely related species in the *E. bislineata* complex assume a defensive posture when attacked by small mammals and birds, and they may be unpalatable (Dowdy and Brodie, 1977). Tail autotomy is also known in this group (Whiteman and Wissinger, 1991).

Q. Diseases. Unknown.

R. Parasites. Unknown.

4. Conservation.

There appear to be at least three threats to dark-sided salamanders: (1) host springs are small, (2) endemic fish have drawn much attention, and (3) herpetological collections have been undisciplined.

Eurycea bislineata (Green, 1818)
NORTHERN TWO-LINED SALAMANDER

David M. Sever

1. Historical versus Current Distribution.

Northern two-lined salamanders (*Eurycea bislineata*) are the well-known "yellow salamanders" of the northeastern United States and eastern Canada. This species was one of the first North American salamanders formally given a Linnaean name (Green, 1818), and the numerous nineteenth century synonyms (see Harlan, 1826a; Holbrook, 1840; Dumeril et al., 1854; Cope, 1889) indicate that the species was widespread and well known to naturalists of that era. No subspecies are currently recognized. However, until Jacobs (1987) raised southern two-lined salamanders (*E. cirrigera*) and Blue Ridge two-lined salamanders (*E. wilderae*) to full specific status, these forms were considered subspecies of northern two-lined salamanders, so the historical distribution included the range of those species as well. Some workers, such as Petranka (1998), still recognize southern two-lined salamanders and Blue Ridge two-lined salamanders as subspecies of northern two-lined salamanders. I support the elevation of southern two-lined salamanders and Blue Ridge two-lined salamanders and, indeed, I believe both of those forms are polytypic (Sever, 1999b,c). In this report, *E. bislineata* is considered to include group C of Jacobs (1987), all Canadian populations (none were analyzed by Jacobs), and additional populations assigned to the species by morphological or genetic analysis in Ohio, West Virginia, and Virginia (Sever, 1989, 1999a, Mitchell and Reay, 1999). The distribution of northern two-lined salamanders, therefore, is

from Labrador and northern Québec and eastern Ontario, through New Brunswick and New England to northern Virginia, west through eastern Ohio and the Kanawha River valley of West Virginia (Sever, 1999a). In Canada, northern two-lined salamanders still have not been recorded from Nova Scotia or Prince Edward Island. Sever (1999a) stated that the distributional limits of the species in Canada are unclear, and that the range of the species may actually be expanding. The boundaries between northern two-lined salamanders and southern two-lined salamanders in Ohio (Guttman and Karlin, 1986) and Virginia (Mitchell and Reay, 1999) have been established by allozyme analysis, and some geographic overlap occurs in both states. The specific identity of Maryland populations has not been determined, but allozyme analysis indicates that two genetically distinct, geographically overlapping forms occur (Miller and Hallerman, 1994). The contact zone in West Virginia has not yet been investigated by genetic analysis, but overlap occurs in the Kanawha River Valley (Sever, 1999a,b). Ecological studies are necessary in areas of sympatry between northern two-lined salamanders and southern two-lined salamanders to determine how reproductive isolation is maintained. Southern two-lined salamanders replace northern two-lined salamanders in the southern Blue Ridge Mountains in southwestern Virginia (Dunn, 1926; Jacobs, 1987), and no areas of sympatry are known (Sever, 1999a,c). Some studies on two-lined salamanders published prior to Jacobs (1987) do not specify localities from which specimens were examined; thus, the specific identity of animals used cannot be determined. In the present account, an effort was made to restrict comments to studies on populations within the range of northern two-lined salamanders as defined above.

2. Historical versus Current Abundance.

Green (1818), in the original description of the species, simply reported: "Inhabits shallow waters; is found in numbers early in the spring, and is very active." DeKay (1842), however, stated: "Although this species is said to be very common, both by Green and Harlan, I have never had the good fortune to meet with it." In many late nineteenth-century and early twentieth-century reports, northern two-lined salamanders were considered common to very common (e.g., Allen, 1868; Morse, 1904; Reed and Wright, 1909). Cope (1889) stated that the species was "very abundant in Pennsylvania, and extends its range, with decreasing numbers, to Maine." I collected my first specimen in 1961; the animal was lying upside down under a log, along a rivulet at the edge of a dirt road that was bordered by a fence-row of Osage orange. In 40 yr of experience

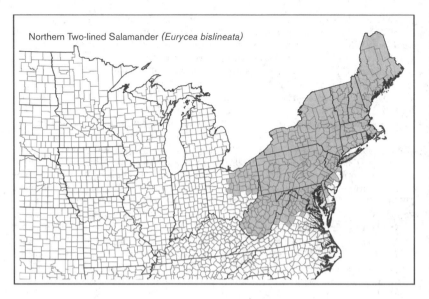

Northern Two-lined Salamander (*Eurycea bislineata*)

with the species, I find that northern two-lined salamanders remain common to seasonally abundant in appropriate habitats. Obviously, habitat destruction (deforestation, pollution, and siltation of streams) has eliminated many populations, but northern two-lined salamanders likely are present in most clean-water, small, rocky, woodland streams within their range. Weller and Green (1997) report no indication of decline in Canada where for most of their range, northern two-lined salamanders are the only stream-side salamanders. The post-glacial dispersal of the species into Labrador and nearly to Hudson Bay in Québec and Ontario indicates tremendous dispersal ability in such a small amphibian.

3. Life History Features.
Much literature occurs on these topics, from the first description of nesting in shallow streams (the usual situation) by H. H. Wilder (1899) to the most recent report involving nests at depths of 9–13.5 m in a mountain lake (Lake Minnewaska, Shawangunk Mountains, New York) by Bahret (1996). However, Bishop's (1941b) account on the natural history of northern two-lined salamanders in New York is still the most comprehensive available for the species.

A. Breeding. Reproduction is aquatic.

i. Breeding migrations. As noted by Bishop (1941b), no marked seasonal movements, such as those associated with mole salamanders (*Ambystoma* sp.), occur. In summer, especially during wet periods and at higher altitudes, adults may wander far from breeding areas in springs, seeps, and streams. Dry conditions, however, will cause these animals to return to aquatic habitats, independently of any mating or breeding behavior (see MacCulloch and Bider [1995] for an extensive study of summer movements of northern two-

lined salamanders in Québec). In the population that Bahret (1996) discovered nesting at considerable depths in a lake, the adults stayed in the lake the entire year. In any event, because the eggs are laid in water and the larvae are aquatic, females will congregate at appropriate nesting sites. Oviposition follows within 1 mo or so of mating, so males congregate at such sites as well. In Ohio, congregations of males and females in appropriate mating/nesting areas sometimes commence in fall prior to hibernation. The mating period is at its peak in March and April after hibernation. Northern two-lined salamanders are relatively easy to find in these generally wet, cool periods, and adults seem abundant at locales where they are hard to find in midsummer. Mating occurs on land (Noble, 1929a), but it is not unusual to find pairs under rocks in the middle of streams during daylight hours. All adult females at mating/nesting sites are gravid with large, vitellogenic eggs; this does not necessarily mean an annual breeding cycle. Non-breeding females have no compelling reasons to reside at mating/nesting sites and compete with sexually active individuals for food and shelter. The classic description of courtship and mating in northern two-lined salamanders is by Noble (1929a), which includes the first observations on any plethodontid of the tail straddle walk and use of courtship glands.

ii. Breeding habitat. Although small, swift-flowing, rocky, woodland streams are the most frequent nesting areas, nests can be found in a wide variety of aquatic habitats. The occurrence of nests in deep lakes already has been noted (Bahret, 1996), and I have found nests in gravelly spring-heads, wells, and boggy areas.

B. Eggs.

i. Egg deposition sites. In eastern Ohio, I commonly find the first nests during the

last week of April to the first week of May, and nesting is nearly synchronous among females in a population. The eggs most frequently are found attached to the undersides of rocks and logs in flowing water. In Lake Minnewaska, however, Bahret (1996) found clutches on the topmost leaves of water moss and not deposited cryptically.

ii. Clutch sizes. In Massachusetts, I. W. Wilder (1924b) found that the usual complement was 12–36 with an average of 18. In New York, Noble and Richards (1932) reported a maximum of 41 eggs, and Bishop (1941b) found a maximum of 43 eggs with a mean of about 30. Stewart (1968) examined the number of enlarged ovarian follicles in New York specimens and found 19–86, with a mean of 46 follicles. However, some ovarian follicles may not mature, so follicle counts probably overestimate the number of eggs ovulated and fertilized. Oftentimes, more than one nest can be found under a single large rock. One should consider strongly the possibility that any nest that contains >50 eggs likely represents the complements of more than one female. The female stays with the eggs; I have found females guarding eggs that were nearly ready to hatch. Incubation period is 4–10 wk (Petranka, 1998).

C. Larvae/Metamorphosis. Larvae are aquatic; excellent descriptions and illustrations are provided by Trapido and Clausen (1940) for specimens from Québec and by Bishop (1941b) for specimens from New York.

i. Length of larval stage. For New York stream populations, the larval period is generally 2 yr with a mean size of 30 mm SVL (45.7–80.0 mm TL) at metamorphosis (Stewart, 1968). In Pennsylvania, some individuals take 3 yr to metamorphose, but despite whether 2 or 3 yr are necessary, metamorphosis occurs at 27.1–34.1 mm SVL and 54.6–60.9 TL (Hudson, 1955). Trapido and Clausen (1940) reported that in Québec the larval period in streams may be 3 yr and the smallest metamorphosed individual was 77 mm TL. Larvae from still waters seem to attain a larger size than those from streams. For example, the largest larvae Bahret (1996) collected from Lake Minnewaska were 43–46 mm SVL and 84–92 mm TL. Whether the larger size of these larvae is due to a longer larval period is unknown.

ii. Larval requirements.

a. Food. Smallwood (1928) stated that the larvae of northern two-lined salamanders from the Onondaga Mountains of New York eat caddisfly larvae and beetle larvae. Burton (1976) reported the most important food items found in larvae from a New Hampshire stream are midges, stoneflies, cladocerans, and copepods.

b. Cover. At night I often have observed northern two-lined salamanders larvae at the bottoms of shallow, quiet pools that form below riffle areas in rocky streams. At these times, the larvae are either stationary on the substrate or slowly crawling along the bottom. Larvae can be found in such pools in the daytime as well, but usually then they are found under sticks, leaves, and other debris that accumulate in areas of such pools. Larvae also are collected frequently after scrapping through rocks and logs in riffles and holding a dipnet downstream from the disturbance. Hudson (1955) in southeastern Pennsylvania noted that larger larvae are less secretive than smaller larvae. Larvae are prone to drift downstream (Johnson and Goldberg, 1975), which might be an important factor in the extraordinary dispersal ability of this species.

iii. Larval polymorphisms. Unknown.

iv. Features of metamorphosis. Some observations suggest that a great deal of variation occurs in the time to metamorphosis within a given cohort, independent of environmental factors. In Massachusetts, I. W. Wilder (1924a) noted that the majority of larvae pass through their second winter in a pre-metamorphic state, whereas others lag behind a year. Transformation occurs as early as May in the 2-yr-old premetamorphic group, reaches its height in July, and continues into early fall. Laggards from this group may pass through a third winter. Elsewhere, I. W. Wilder (1924b) reported that two individuals, hatching at the same time in the laboratory and kept under the same conditions, metamorphosed at dates approximately 1 yr apart. My own observations lead me to believe that metamorphosis may occur more synchronously in some populations. In late summer and early fall, I often find a marked increase in the number of recently transformed juveniles under rocks along woodland streams.

v. Post-metamorphic migrations. Juveniles apparently stay close to streams through their first breeding season, which may occur in the spring following metamorphosis in the previous autumn (Bishop, 1941b). MacCulloch and Bider (1975) in Québec found that metamorphosing individuals in August remained closer to their natal streams than post-breeding adults.

vi. Neoteny. Neoteny is not known to occur. Cope (1889) noted, "It is one of those species whose metamorphoses are prolonged and which remains in the larval state until nearly fully grown." Thus, large larvae often possess well-developed gonads, and, as proposed by Bishop (1941b), maturity may be attained within a comparatively short time after transformation.

D. Juvenile Habitat. As indicated above, juveniles rapidly mature into adults, and if metamorphosis occurred in a given stretch of a stream, juveniles occupy that area through their first breeding season. A brief post-larval migration into terrestrial habitats <100 m from the stream may occur (MacCulloch and Bider, 1975).

Thus, the requirements during this short juvenile stage are not different from adults.

E. Adult Habitat. Adults from various populations of northern two-lined salamanders are 28.9–40.9 mm mean SVL; the record individual is a female 53 mm SVL and 123 mm TL from Ohio (Sever, 1999a). Stewart (1968) reported that females from New York are significantly longer than males. Two-lined salamanders usually are the smallest species when they occur in communities containing other salamanders in the northeastern United States. The primary habitats are unpolluted bogs, springs, streams, or lakes in wooded areas. Most small streams in the wooded, mountainous areas of the range are rocky, which provides a good substrate for nesting, and lack large predatory fish. Occasionally, specimens are found along larger streams and rivers, but nesting has not been reported in such habitats. Although members of the northern two-lined salamander complex can co-exist in streams with predatory fishes (Petranka et al., 1987), predation by fish still is a factor that likely limits extensive exploitation of rocky areas along large streams. Lake Minnewaska, where Bahret (1996) found nesting at depths of 9–13.5 m and adults living year-round at 18 m, lacks fish. The disappearance of fish is attributed to atmospheric acid deposition in the lake during the early 1900s, resulting in a pH of 4.5 (Bahret, 1996). This high acidity, however, seemingly does not affect success of northern two-lined salamanders.

F. Home Range Size. Stewart (1968) found as many as 11 adults/m^2 in New York populations, whereas Burton and Likens (1975b) reported only 0.02–0.04 individuals/m^2 in the Hubbard Brook Experimental Forest of New Hampshire. These densities seem quite disparate in populations that presumably occupy favorable habitats. Density studies on terrestrial northern two-lined salamanders, however, can be confounded by influx of numerous newly metamorphosed juveniles in late summer and fall, aggregations in spring of adults at favorable mating/nesting sites, interspecific interactions, and other factors.

G. Territories. Grant (1955) reported territorial defense (by biting intruders) in captive northern two-lined salamanders kept in relatively close quarters within a terrarium; whether this occurs in the wild is uncertain.

H. Aestivation/Avoiding Desiccation. I am unaware of any data.

I. Seasonal Migrations. Include breeding migrations and post-metamorphic migrations (see above).

J. Torpor (Hibernation). I am unaware of any data.

K. Interspecific Associations/Exclusions. As noted previously, in much of the Canadian range, especially north of the St. Lawrence Seaway and the Great Lakes,

northern two-lined salamanders are the only stream-side salamanders. In more southern portions of its range, northern two-lined salamanders share stream habitats with one or more of the following species: long-tailed salamanders (*E. longicauda*), northern dusky salamanders (*Desmognathus fuscus*), Allegheny Mountain dusky salamanders (*D. ochrophaeus*), red salamanders (*Pseudotriton ruber*), and/or spring salamanders (*Gyrinophilus porphyriticus*). The latter species is a known predator of northern two-lined salamanders (Wright and Haber, 1922; Bishop, 1941b). No detailed studies, however, have been done on ecological interactions between northern two-lined salamanders and these species. Studies are also necessary on the ecological and genetic relationships of northern two-lined salamanders and southern two-lined salamanders in areas of sympatry in Ohio, West Virginia, Virginia, and Maryland.

L. Age/Size at Reproductive Maturity. As mentioned earlier, metamorphosis generally occurs in the summer or fall of the second or third year, and these individuals may reach sexual maturity in time to participate in breeding activities the following spring. As a result, individuals are breeding at the beginning of their third or fourth year, depending upon length of the larval period. The smallest mature specimens I have examined were 30 mm SVL; however, I examined two males 25 mm SVL from Chittenden County, Vermont, that had evenly pigmented testes and may have been mature. Bishop (1941b) reported that sexual maturity is reached in males at 67 mm TL and some females when only 61 mm TL.

M. Longevity. In Québec, MacCulloch and Bider (1975) reported that only 25% of post-breeding adults who migrated >100 m from the mating/nesting area returned for the subsequent breeding season. They attributed this loss of 75% of these migrants to mortality, which indicates a rather high turnover in the adult population each year. I am unaware of any literature on maximum life span either in nature or in captivity.

N. Feeding Behavior. Feeding habits of metamorphosed northern two-lined salamanders have been studied by Surface (1913) in Pennsylvania, Smallwood (1928) and Hamilton (1932) in New York, and Burton (1976) in New Hampshire. As expected, a wide range of invertebrates are eaten, although Burton (1976) noted that prey items are primarily terrestrial. Terrestrial items include various insects (beetles, roaches, springtails, dipterans, and hymenopterans), earthworms, snails, spiders, and isopods, whereas aquatic prey includes stonefly nymphs, caddisfly larvae, midges, and mayflies. Thus, northern two-lined salamanders are generalist feeders that are likely to eat any small invertebrate within the appropriate size range.

O. Predators. Wright and Haber (1922) included northern two-lined salamanders among other salamander species that are preyed upon by spring salamanders. Bishop (1941b) noted that a number of recently captured spring salamanders from New York disgorged partly digested specimens of northern two-lined salamanders. Other known predators include shrews, owls, blue jays, garter snakes (*Thamnophis* sp.) and trout (Brodie, 1977; Petranka, 1998). I have observed screech owls *(Otus asio)* capturing northern two-lined salamanders on blacktop roads on rainy nights.

P. Anti-Predator Mechanisms. Brodie (1977) reported that northern two-lined salamanders assume a defensive posture when attacked by short-tailed shrews and blue jays. The body is coiled with the head near the vent, and the tail is elevated and undulated. In a test chamber, 10 of 11 northern two-lined salamanders attacked by shrews were first bitten on the tail. The shrew backed away, wiping its mouth, which in nature could give the salamander time to escape (Brodie, 1977). In a later study in which northern two-lined salamanders from New York were exposed to shrews, only 26% of the salamanders survived attacks (Brodie et al., 1979). In some cases, the tail was contacted by the shrew and avoidance followed, but in other cases, the head was bitten before the tail was contacted. The shrews, however, did not eat 47.8% of the northern two-lined salamanders killed, indicating that shrews find the salamander relatively unpalatable (Brodie et al., 1979). Dowdey and Brodie (1989) studied anti-predator response of northern two-lined salamanders from New York to garter snakes. Garter snakes are predators that are not repulsed by skin secretions, and salamanders do not assume the defensive postures used against blue jays and shrews. Instead, northern two-lined salamanders run or remain immobile during encounters with garter snakes. In a testing chamber, most salamanders ran when flicked by a snake's tongue, and 90% of these with tails survived (tailed individuals ran faster than tail-less ones). All salamanders that remained immobile after tongue flicks were eaten. The response to running was well developed only in a population subjected to heavy snake predation. However, when a salamander was merely contacted by the head or body of a garter snake, most remained immobile and 100% of these survived. Only 34% survived of those who ran. In an experiment using Pennsylvania specimens, Whiteman and Wissinger (1991) found that northern two-lined salamanders (59%) were almost twice as likely to autotomize their tails when attacked by garter snakes than northern dusky salamanders (24%) or Allegheny Mountain dusky salamanders (28%). Tail autotomy can increase the chance of the salamander escaping, although it makes the animal more susceptible to future predator attack (Dowdey and Brodie, 1989). Two-lined salamanders, therefore, have alternative anti-predator strategies, and the type and efficacy of each strategy varies with the type of predator and, in the case of garter snakes, the nature of contact.

Q. Diseases. I am not aware of any reports on diseases of northern two-lined salamanders.

R. Parasites. I am not aware of any reports on parasites of northern two-lined salamanders. Some literature on this subject (Burchett and Shoemaker, 1990) does exist for the sister taxon, southern two-lined salamanders.

S. Comments. A considerable literature exists on northern two-lined salamanders, and Sever (1999a) should be consulted for additional references, especially on topics not covered here such as morphology, physiology, and diagnostic characters. Note especially that several recent accounts (e.g., Petranka, 1998, Powell et al., 1998) continue to use characters like tail stripe length and number of costal grooves between toes of the adpressed limbs to separate species in the complex, even though such characters have long been discredited (Sever, 1972, 1989, 1999a,b,c). Because Petranka (1998) did not recognize southern two-lined salamanders and Blue Ridge two-lined salamanders as separate species, he combined data on these species with those on northern two-lined salamanders into one account; a careful reading, however, will reveal much material relevant to each taxon.

4. Conservation.

Northern two-lined salamanders are listed as Protected by the State of New Jersey (Levell, 1997). This categorization applies to all indigenous non-game species in this state and does not necessarily indicate a conservation concern. In general, northern two-lined salamanders are widespread and suitable habitat is available across their range. Indeed, northern two-lined salamanders may be spreading their range north.

Eurycea chamberlaini Harrison and Guttman, 2003
CHAMBERLAIN'S DWARF SALAMANDER
Julian R. Harrison

1. Historical versus Current Distribution.
Unknown. According to Harrison and Guttman (2003), Chamberlain's dwarf salamanders *(Eurycea chamberlaini)* presently are known from an isolated area in the western Piedmont of South Carolina, the lower Piedmont of North Carolina, the upper Coastal Plain of South Carolina, and the central portion of the Coastal Plain in North Carolina. They may also occur in Georgia and Florida (J. Jensen and D. Bruce Means, respectively, personal communications). The isolated area in the upper Piedmont of South Carolina depicted on the range maps of dwarf salamanders *(E. quadridigitata)* by Martof et al. (1980), Conant and Collins (1998), and Petranka (1998) is occupied only by populations of this new species. The record of *Manculus* (=*Eurycea*) *quadridigitatus* listed by Cope (1889) from "Abbeville, S.C." and depicted on the range map in Martof et al. (1980) may refer to Chamberlain's dwarf salamanders, but the specimen has not been re-examined and the record has not been verified. The paper by Brimley (1923) concerning the habits of *M. quadridigitatus* in North Carolina is based on populations of this new species.

Folkerts (1971) speculated that the isolated populations in the South Carolina Piedmont are remnants of a once more-widespread upland form of *Manculus* (= *Eurycea*) that disappeared or withdrew to the Coastal Plain when conditions became unfavorable.

2. Historical versus Current Abundance.
Unknown, but at present Chamberlain's dwarf salamanders appear to be relatively common throughout most of their known range.

3. Life History Features.
A. Breeding. Reproduction is aquatic. Chamberlain's dwarf salamanders have a biphasic life cycle: adults are semi-terrestrial, eggs and larvae are aquatic. Breeding occurs in winter in North Carolina (Brimley, 1923) but in the autumn at the Savannah River Site in Barnwell County, South Carolina (J.H. Pechmann and R.D. Semlitsch, personal communication).

i. Breeding migrations. Unknown, but these probably involve only limited movements to the breeding habitats from immediately adjacent uplands. At the Savannah River Site in Barnwell County, South Carolina, where the two species co-occur, Chamberlain's dwarf salamanders migrate to Carolina bays later in the autumn than do dwarf salamanders (J.H. Pechmann, personal communiction).

ii. Breeding habitat. Reported by Brimley (1923) to be seepages and runs from springs.

B. Eggs.

i. Egg deposition sites. Eggs are reported by Brimley (1923) to be scattered about singly or in groups of 3–6 among dead and decaying leaves in seepages and runs from springs.

ii. Clutch size. Unknown. However, Harrison and Guttman (2003) report that the mean number of large, yolk-laden ovarian eggs in Chamberlain's dwarf salamanders is 45.2 (35–64). Most or all of these probably are oviposited.

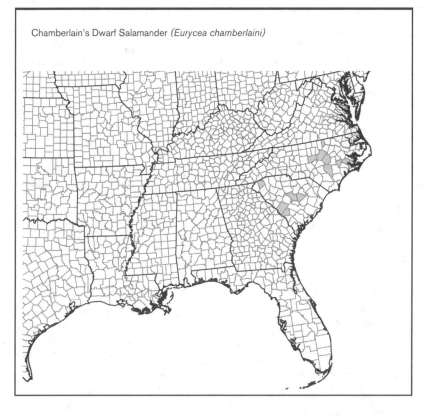

Chamberlain's Dwarf Salamander *(Eurycea chamberlaini)*

M. Longevity. Unknown.

N. Feeding Behavior. Unknown, but this is a species with a highly projectile (boletoid) tongue, and is therefore likely to be a sit-and-wait predator. Prey items eaten by adults are probably similar to those consumed by dwarf salamanders; these include earthworms, several kinds of insects, spiders, pseudoscorpions, mites, ticks, and millipedes (Carr 1940; McMillan and Semlitsch, 1980; Powders and Cate, 1980).

O. Predators. Unknown, but likely predators probably include crayfish, predaceous insects, large spiders, small snakes, and birds.

P. Anti-Predator Mechanisms. Unknown, but probably limited to immobility coupled with cryptic coloration.

Q. Diseases. Unknown.

R. Parasites. Unknown.

4. Conservation.

Chamberlain's dwarf salamanders have no legal protection but, with the possible and likely exception of some local populations, are probably not in any immediate jeopardy. Protection of larval and adult habitats (springs, seepage areas, and small streams in forested areas) is important for the species' survival.

C. Larvae/Metamorphosis.

i. Length of larval stage. According to Brimley (1923), most hatching occurs in March with metamorphosis occurring 2–3 mo later.

ii. Larval requirements.

a. Food. Unknown but presumably similar to that of congeneric dwarf salamanders (Carr, 1940; Taylor et al., 1988), their putative relative; items consumed include amphipods, ostracods, cladocerans, and chironomid larvae.

b. Cover. As reported by Brimley (1923) for populations in North Carolina, leaf-choked seepage areas and spring runs presumably provide adequate cover for larvae.

iii. Larval polymorphisms. Hatchling larvae of Chamberlain's dwarf salamanders have an average of 4.5 (1–7) dorsal spots; the number of spots declines over the larval period (Harrison and Guttman, 2003). Spots are absent in hatchling dwarf salamander larvae.

iv. Features of metamorphosis. Unknown.

v. Post-metamorphic migrations. Unknown, but probably limited to immediately adjacent upland habitats.

D. Juvenile Habitat Requirements. Similar to those of adults.

E. Adult Habitat Requirements. Chamberlain's dwarf salamanders are a semi-terrestrial species. Information from 33 records in the files of the North Carolina State Museum of Natural History (A. Braswell, personal communication) indicates that they normally occupy the margins of streams or seepages (70%) or floodplain or pond sites (30%).

F. Home Range Size. Unknown.

G. Territories. Unknown.

H. Aestivation/Avoiding Desiccation. Unknown. However, in North Carolina, Brimley (1923) apparently did not find Chamberlain's dwarf salamanders during the period from late spring to early autumn. Presumably, the salamanders were in underground retreats, but whether or not they were aestivating remains to be determined.

I. Seasonal Migrations. Unknown, but probably do not occur.

J. Torpor (Hibernation). Unknown. However, Brimley (1923) reported that he often found Chamberlain's dwarf salamanders from mid-October to late April, most frequently during the winter breeding season.

K. Interspecific Associations/Exclusions. Chamberlain's dwarf salamanders co-occur with congeneric dwarf salamanders at two locations: the Savannah River site in Barnwell County and a site in Allendale County, both in South Carolina (Harrison and Guttman, 2003).

L. Age/Size at Reproductive Maturity. The smallest mature adult measured by Harrison and Guttman (2003) is 22 mm SVL. Folkerts (1971) studied specimens of Chamberlain's dwarf salamanders (identified as *E. quadridigitata*) from the upper Piedmont of South Carolina and measured adults or subadults from 20.5–28.7 SVL.

Eurycea chisholmensis Chippindale, Price, Wiens, and Hillis, 2000
SALADO SALAMANDER

Paul T. Chippindale

1. Historical versus Current Distribution.

Salado salamanders *(Eurycea chisholmensis)* were described by Chippindale, Price, Wiens, and Hillis (2000) from springs at Salado, Bell County, Texas. Prior to the work of Chippindale (1995) and Chippindale et al. (2000), this population was known from a single juvenile specimen and was considered a peripheral isolate of Texas salamanders *(E. neotenes;* Sweet, 1978a, 1982). Salado salamanders are members of the "northern group" of Chippindale (1995, 2000) and Chippindale et al. (2000); this monophyletic group occurs northeast of the Colorado River in the Edwards Plateau region of central Texas. Based on molecular markers, this and other northern species are extremely divergent from *E. neotenes* and other *Eurycea* from the southern Edwards Plateau region (Chippindale et al., 2000).

2. Historical versus Current Abundance.

Almost nothing is known of the historical abundance of this species. Chippindale et al. (2000) collected most specimens in 1989–91, when several could sometimes be found on a single visit. Between 1991 and 1998, no additional animals could be located despite more than 20 visits to the type locality; one specimen was found in August 1998.

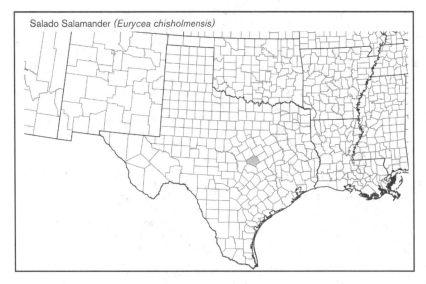

Salado Salamander *(Eurycea chisholmensis)*

several groundwater contamination incidents have occurred in the recent past. The potential remains for more incidents. The type locality for this species is located in a municipal park. Salado salamanders are listed as a Candidate species for federal listing (http://ecos.fws.gov).

Eurycea cirrigera (Green, 1830)
SOUTHERN TWO-LINED SALAMANDER
Thomas K. Pauley, Mark B. Watson

1. Historical versus Current Distribution.
The range of southern two-lined salamanders *(Eurycea cirrigera)* extends from Illinois and Indiana southeast through southern Ohio, southern West Virginia and Virginia, and south through Kentucky, Tennessee, North Carolina, and South Carolina to Georgia, Alabama, Mississippi, and Louisiana (E. E. Brown, 1992; Sever, 1999b). Southern two-lined salamanders have contact zones with northern two-lined salamanders *(E. bislineata)* in Ohio (Guttman and Karlin, 1986), Virginia (Mitchell and Reay, 1999), and West Virginia (Montani and Pauley, 1992; Brophy, 1995). Southern two-lined salamanders are absent in eastern Ohio and northern West Virginia where northern two-lined salamanders are found, and from the southern Blue Ridge Mountains where Blue Ridge two-lined salamanders *(E. wilderae)* occur (Sever, 1999b). There has been some debate among systematists on the validity of assigning full species status to members of the *E. bislineata* complex. For a summary of the taxonomic status of this group see Guttman and Karlin (1986), Jacobs (1987), Guttman (1989), Petranka (1998), Sever (1999b), Camp et al. (2000), and Highton (2000).

Thurow (1997) reports a successful translocation and subsequent establishment of southern two-lined salamanders from western Indiana to west-central Illinois.

3. Life History Features.
A. Breeding. Reproduction is aquatic.
i. Breeding migrations. Unlikely to occur.
ii. Breeding habitat. The aquatic habitats of adults.
B. Eggs.
i. Egg deposition sites. Unknown; some other spring-dwelling species of central Texas *Eurycea* are thought to deposit eggs in gravel substrates.
ii. Clutch size. Unknown.
C. Larvae/Metamorphosis. Salado salamanders are paedomorphic and natural metamorphosis is unknown.
D. Juvenile Habitat. Probably similar to adult habitat.
E. Adult Habitat. Completely aquatic. Salado salamanders are known only from the immediate vicinity of spring outflows, under rocks, and in gravel substrate. Water temperature in springs of the Edwards Plateau is relatively constant throughout the year and typically ranges from 18–20 °C, or slightly warmer near the fault zone at the Plateau's edge (Sweet, 1982). Sweet (1982) provided a comprehensive distributional analysis of the central Texas *Eurycea* and discussed hydrogeology of the region in relation to salamander distribution.
F. Home Range Size. Unknown.
G. Territories. Unknown.
H. Aestivation/Avoiding Desiccation. Unknown.
I. Seasonal Migrations. Unlikely to occur, although there may be seasonal variation in surface versus subsurface habitat use.
J. Torpor (Hibernation). Probably active throughout the year.
K. Interspecific Associations/Exclusions. Unknown.
L. Age/Size at Reproductive Maturity. Unknown. The average SVL of Salado salamanders measured by Chippindale et al. (2000) was 32.9 mm; all specimens measured were thought to be sexually mature, but this was verified for only some of the specimens.

M. Longevity. Unknown.
N. Feeding Behavior. Prey probably consists mainly of small aquatic invertebrates, but no detailed feeding studies of this species have been conducted. Captive specimens accepted amphipods *(Hyalella azteca;* personal observations).
O. Predators. Unknown.
P. Anti-Predator Mechanisms. Secretive.
Q. Diseases. Unknown.
R. Parasites. Unknown.

4. Conservation.
Chippindale et al. (2000) and Price et al. (1995) address some aspects of the conservation biology of Salado salamanders. They note that determining the conservation status of this species presents a problem because of the difficulty in acquiring specimens and determining the extent of its range. These authors point out that most of the spring outlets at Salado have been modified in the past 150 yr and that

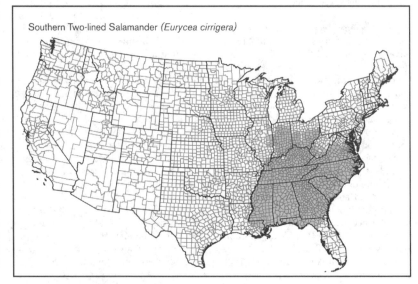

Southern Two-lined Salamander *(Eurycea cirrigera)*

2. Historical versus Current Abundance.

Southern two-lined salamanders appear to remain common throughout most of their range. There are no known reports of declines in abundance. Disjunct populations on the periphery of their range in northeastern Illinois apparently are the same as in the 1930s (Mierzwa, 1998). Numbers of specimens in museums from Hamilton County, Ohio, from pre-1940–95 show the number of southern two-lined salamanders to be about the same (Davis et al., 1998). While museum records cannot identify populations trend (but see Boundy, this volume), they can illustrate the occurrence of species. In Indiana, Minton (1998) compared historical and present abundance of amphibians and reptiles at eight sites. Southern two-lined salamanders initially were found at two sites in 1948, and their abundance was basically the same when Minton checked again in 1991 and 1993.

3. Life History Features.

A. Breeding. Reproduction is aquatic.

i. Breeding migrations. Weichert (1945) reported that during the spring breeding season, adult salamanders were found in water, but at all other times of the year, two-lined salamanders in the Cincinnati, Ohio, area were found away from water. In West Virginia during the summer, they can either be associated with aquatic habitats or found great distances from visible water sources. Individuals in upland, non-breeding habitats migrate to first-, second-, and third-order streams in the autumn where they remain through the winter (T. K. P., personal observations).

ii. Breeding habitat. Breeding and egg deposition occur in aquatic habitats, especially in streams. Courtship activity of southern two-lined salamanders in a laboratory setting was described by Noble and Brady (1930). They also reported that southern two-lined salamanders from North Carolina would readily mate with northern two-lined salamanders from New York. Weichert (1945) found the breeding season for southern Ohio was limited to the end of March to the first two weeks of April. Courtship occurs in the fall in North and South Carolina and Virginia (Martof et al., 1980). Brophy (1995) located a mature male and gravid female together at the end of March in southwestern West Virginia, and the female had a spermatophore visible in her cloaca, suggesting the breeding season is early spring for this region.

B. Eggs.

i. Egg deposition sites. Eggs may be deposited under rocks or leaves, attached to logs and sticks adjacent to streams (Richmond, 1945; Wood, 1953a; Baumann and Huels, 1982; Green and Pauley, 1987), or broadcast underwater among rocks and gravel in stream beds (T. K. P., personal observations). Females usually brood egg clusters (Green and Pauley, 1987; Marshall,

1996). Marshall (1996) found seven clusters of eggs in a single stream in Mississippi; three clutches were attached to the underside of logs and four were unattached and buried in the substrate. In Georgia, Martof (1955) found females in January–February to be distended with eggs and suggested that eggs were laid in February. Wood (1953a) reported that eggs were deposited over a 10-wk period from late January to mid-April on the Virginia Coastal Plain. Baumann and Huels (1982) found 1–3 egg masses of two-lined salamanders in Pine Creek in southeastern Ohio attached to the undersides of rocks. Brophy (1995) found egg masses in early development between 21 March–8 April in southwestern West Virginia. Masses were ovular in shape and attached to the underside of rocks in streams. Eggs are deposited in streams under rocks and other objects in winter and spring in North and South Carolina and Virginia (Martof et al., 1980) and in May in Illinois (Smith, 1961).

ii. Clutch sizes. Baumann and Huels (1982) reported that egg masses of southern two-lined salamanders averaged 16 cm^2 in total area. The number of eggs/nest reported in the literature varies greatly. The number of eggs in a clutch ranges from 12–110 and averages from 18–50 (Noble and Richards, 1932; Richmond, 1945; Wood and Duellman, 1951; Wood and McCutcheon, 1954; Smith, 1961; Mount, 1975). Differences in the number of ova reported are likely due to differences in the body size of females (Wood and McCutcheon, 1954). Barbour (1971) reported 30 eggs/clutch in Kentucky, and Brophy (1995) found clutch sizes from 36–59 eggs in southwestern West Virginia. Clutches were commonly made up of around 40 eggs. Large counts of 200 or more eggs in a nest may be the result of communal nests (Baumann and Huels, 1982).

In Illinois, eggs hatch about 1 mo after deposition (Smith, 1961). In northern Georgia, hatching occurs in early March (Martof, 1955). Duellman and Wood (1954) found that eggs hatch in late June in southwestern Ohio, where hatchlings are <9 mm SVL. Reports of newly hatched larvae vary from 10.5 mm (TL) in Louisiana (Dundee and Rossman, 1989) to 13 mm in Kentucky (Barbour, 1971).

C. Larvae/Metamorphosis.

i. Length of larval stage. The larval period of southern two-lined salamanders lasts from 1–3 yr. Lengths of the larval stage range from 2 yr (Petranka, 1984b) to 2–3 yr in Kentucky (Barbour, 1971); 1–3 yr in Alabama (Mount, 1975); 1–2 yr in Louisiana (Dundee and Rossman, 1989); 2 yr in Illinois (Phillips et al., 1999) and Ohio (Duellman and Wood, 1954); and possibly 3 yr in West Virginia (Pollio, 2000). A small number of larvae transform during the third year in Ohio (Duellman

and Wood, 1954). Brophy (1995) found two size classes of larvae from southern and southwestern West Virginia. First-year larvae from a pond-dwelling population grew significantly faster than a stream-dwelling population, but the stream-dwelling population grew significantly faster in the second year. Larvae from both populations grew little during cooler months (Brophy, 1995). Some larvae can reach the same size as sexually mature adults.

ii. Larval requirements.

a. Food. Larval southern two-lined salamanders are euryphagous feeders. Prey items usually are invertebrates and include ostracods, copepods, and insects such as dipterans (chironomids), ephemeropterans, and coleopterans (dytiscid beetle larvae; Caldwell and Houtcooper, 1973; Petranka, 1998). In a comparison of prey items between pond- and stream-dwelling larval populations in southern West Virginia, Brophy and Pauley (in press) found that pond larvae consumed 9 prey taxa, while stream larvae consumed 15. Primary prey for pond larvae included ostracods and chironomid larvae; stream larvae fed on copepods, isopods, and chironomids.

Southern two-lined salamander larvae also prey on vertebrates. Petranka (1984b) observed that large larvae will prey heavily on streamside salamander (*Ambystoma barbouri*) larvae.

b. Cover. Petranka (1984b) found that larvae showed diurnal movements to and from cover objects, and that larvae feed along streambeds during darkness. Smith and Grossman (2003) show that microhabitat use is correlated with habitat heterogeneity and the availability of cover.

iii. Larval polymorphisms. Unknown and unlikely.

iv. Features of metamorphosis. Size of larvae at transformation has been reported at 52 mm (TL) in Kentucky (Petranka, 1984b); 52 mm TL for males and 50.9 mm TL for females in southwestern Ohio (Duellman and Wood, 1954); 72 mm TL in Georgia (Martof, 1955); and 34–40.2 mm SVL in Ohio and Indiana populations (Sever, 1972).

v. Post-metamorphic migrations. Migrations occur from pond and stream habitats to adjacent uplands. Martof (1955) found transforming larvae in September in Georgia. Duellman and Wood (1954) discovered transformed individuals in southwestern Ohio during the summer. Transformation occurs in May to mid-June in Kentucky (Petranka, 1984b). Early metamorphic animals from a southern West Virginia pond in August were composed of two size classes, suggesting that these individuals had metamorphosed from 1–2-yr larvae (Brophy, 1995).

vi. Neoteny. Mount (1975) reported that neotenic individuals are not uncommon in Alabama. This condition has not been reported in other areas of the range of southern two-lined salamanders, but needs further study.

D. Juvenile Habitat. Juveniles are found under stones and other cover objects at the edges of aquatic habitats (Hudson, 1955; Petranka, 1998) and in surrounding forests (Petranka, 1998).

E. Adult Habitat. Southern two-lined salamanders are a semi-aquatic species and can be found in a variety of habitats throughout its range such as streams, pools, seeps, ditches, and damp woods (Smith, 1961; Barbour, 1971; Minton, 1972, 2001; Mount, 1975; Martof et al., 1980; Green and Pauley, 1987; Dundee and Rossman, 1989; Guttman, 1989; Bartlett and Bartlett, 1999a; Phillips et al., 1999). Ashton and Ashton (1978) reported that two-lined salamanders frequently used stream habitats with coarse sand and gravel, as well as broken limestone rock, leaf litter, and crayfish burrows. Brophy (1995) found southern two-lined salamanders mostly within the stream banks of two southwestern West Virginia streams. Martof (1955) suggested southern two-lined salamanders are abundant around springs and small streams in north-central Georgia. Grover (2000) reported that southern two-lined salamanders found along streams with black-bellied salamanders (*Desmognathus quadramaculatus*) were captured farther from streams than in an area without black-bellied salamanders. Means (2000) found southern two-lined salamanders most often around the heads of ravines in the coastal plain.

F. Home Range Size. In laboratory experiments, two-lined salamanders demonstrated a home range size that extended for a 5–6 inch radius from a central shelter site (Grant, 1955). In natural habitats, home range size is probably much larger. A mark–recapture study by Brophy (1995) suggested that the home range size of southern two-lined salamanders from southwestern West Virginia was around 14 m^2.

G. Territories. Grant (1955) showed in laboratory experiments that two-lined salamanders defend territories from conspecifics by advancing toward an intruder and placing their snout in contact with it. In other cases, the salamander defending a territory would bite the snout or tail, often resulting in autotomy of the antagonist's tail.

H. Aestivation/Avoiding Desiccation. Aestivation studies have not been reported in the literature, but southern two-lined salamanders, as with other plethodontids, probably remain in refugia underground during drought conditions.

Brooks and Sassaman (1965) measured the critical thermal maximum (CTM) of larval and adult *Eurycea* from the coastal plain of Virginia. The average CTM of larvae was 33.3 °C and 34.6 °C for adults.

Grover (2000) measured the dehydration and rehydration rates of *E. cirrigera*. The rehydration rate of *E. cirrigera* was intermediate to that of *Plethodon glutinosus* (northern slimy salamanders) and *D. monticola* (seal salamanders).

I. Seasonal Migrations. Ashton and Ashton (1978) reported that when stream temperatures fell below 7 °C, southern two-lined salamanders moved upstream to winter refugia but remained within close proximity to the stream. Brophy (1995) found no cyclic movements of southern two-lined salamanders in two populations from southwestern West Virginia.

J. Torpor (Hibernation). In an experimental hibernation study, Vernberg (1953) observed that southern two-lined salamanders, in response to an artificial temperature gradient, burrowed into smooth-walled hibernacula 25–30 cm into soil. Weichert (1945) found active southern two-lined salamanders on warm days during the winter in southern Ohio. Similarly, Ashton and Ashton (1978) observed movement during winter in refugia with moderate conditions.

K. Interspecific Associations/Exclusions. Southern two-lined salamanders are associated with other stream-dwelling salamanders including black-bellied salamanders, northern dusky salamanders (*D. fuscus*), seal salamanders, and spring salamanders (*Gyrinophilus porphyriticus*). Means (2000) described an assemblage of plethodontid salamanders that inhabit the steephead ravines of the coastal plain that included southern two-lined salamanders, red salamanders (*Pseudotriton ruber*), and either Apalachicola dusky salamanders (*D. apalachicolae*) or spotted dusky salamanders (*D. conanti*).

Northern two-lined salamanders and southern two-lined salamanders will hybridize (Noble and Brady, 1930), and broad regions of intergradation exist between these forms of two-lined salamanders (Howell and Switzer, 1953; Mittleman, 1966).

L. Age/Size at Reproductive Maturity. In Ohio, sexual maturity is reached in 2–4 yr (Guttman, 1989). Age and size at sexual maturity vary according to the length of the larval period and season when transformation of larvae occurs.

M. Longevity. Unknown.

N. Feeding Behavior. Weichert (1945) reported the stomach contents of southern two-lined salamanders from southern Ohio included small wood roaches, spiders, ticks, earthworms, beetles, isopods, millipedes, small snails, grubs, springtails, and dipteran and hymenopteran insects. Food was found in stomachs from every month of the year.

O. Predators. Wood (1953a) suggested that mosquitofish (*Gambusia affinis holbrookii*) ate two-lined salamander eggs. Resetarits (1991) showed that brook trout (*Salvelinus fontinalis*) and crayfish (*Cambarus bartonii*) were predators of

Eurycea larvae in a Virginia stream, and the presence or absence of trout could alter the salamander assemblages in the stream. Gustafson (1994) showed that spring salamander larvae preyed upon southern two-lined salamander larvae during laboratory experiments, and the efficiency of spring salamanders as predators of southern two-lined salamanders increased with the size of spring salamander larvae. Petranka (1984b) reported that larvae were palatable to sunfish and darters in streams.

Grover (2000) found that the presence of black-bellied salamanders in streams caused a shift of southern two-lined salamanders to drier sites farther from streams, suggesting that black-bellied salamanders were predators of southern two-lined salamanders. Other predators of two-lined salamanders (including other members of the *E. bislineata* complex) include screech owls, common garter snakes (*Thamnophis sirtalis*), ring-necked snakes (*Diadophis punctatus*), and rainbow trout (*Oncorhynchus mykiss*; Huheey and Stupka, 1967; Rising and Schueler, 1980; Beachy, 1993b).

P. Anti-Predator Mechanisms. Eggs of southern two-lined salamanders are deposited in cryptic sites under logs, leaves, and rocks. Females often are seen tending the eggs as a defense against predators (Baumann and Huels, 1982).

Petranka (1984b) suggested that diurnal behavior might be an anti-predator mechanism in larvae. Larvae may also use chemical cues from predatory fish to increase use of refugia (Petranka et al., 1987; Kats et al., 1988).

Several anti-predator mechanisms of northern two-lined salamander adults have been reported. The behavior of southern two-lined salamanders is probably similar to that of northern two-lined salamanders. In a laboratory setting, northern two-lined salamanders responded to common garter snakes with a protean, flipping escape rather than posturing or undulating the tail (Ducey and Brodie, 1983). Salamanders with tails could autotomize the tail during an encounter with a snake and were more successful in escaping capture than salamanders without tails. Dowdey and Brodie (1989) showed that different densities of predators can affect the response of two-lined salamanders to those predators. Salamanders in a high density of northwestern garter snakes (*T. ordinoides*) ran away more than salamanders from other areas. Salamanders that ran had a survival advantage. Whiteman and Wissinger (1991) reported that tail autotomy during predation experiments with common garter snakes as predators was nearly twice as frequent in northern two-lined salamanders as in northern dusky salamanders or Allegheny Mountain dusky salamanders (*D. ochrophaeus*). They found that two-lined salamanders with tails were more likely to escape a predator than those without tails. Two-lined salamanders were less aggressive during encounters and bit garter snakes less frequently than dusky salamanders.

Q. Diseases. Not known.

R. Parasites. Rankin (1937) lists the protozoan *Prowazekella longifilis*, the trematode *Brachycoelium hospitale*, and proteocephalid cestode cysts from southern two-lined salamanders.

4. Conservation.
Southern two-lined salamanders are abundant throughout most of the range. They are found in water polluted with sewage and other organic matter. As with most forest salamanders, major concerns are habitat destruction through activities such as clearcutting and habitat degradation, including acid mine drainage and acid deposition.

Eurycea guttolineata (Holbrook, 1838)
THREE-LINED SALAMANDER
Travis J. Ryan, Brooke A. Douthitt

1. Historical versus Current Distribution.
Three-lined salamanders (*Eurycea guttolineata*) are distributed throughout much of the southeastern United States. From west to east, the range of three-lined salamanders begins along the eastern bank of the Mississippi from Louisiana, north through all of Mississippi and Tennessee and into Kentucky. Eastward, they are distributed throughout Mississippi (except for the northeastern corner) and Alabama (except for the northern portion), all but the northwestern and southeastern extremes of Georgia, all of South Carolina, most of western and central North Carolina, and eastern Virginia. The absence of the three-lined salamanders from the bulk of North Carolina's Coastal Plain is particularly curious. Populations rarely are found above 800 m and almost always below 1,000 m (Fisher, 1887; Ireland, 1979; Freeman and Bruce, 2000).

There is no evidence to support a difference between the current and historical distributions, but Petranka (1998) points out that the loss of bottomland hardwood forests throughout the Southeast has undoubtedly resulted in the extirpation of many populations. However, links between habitat loss and any putative population declines have not been empirically demonstrated.

2. Historical versus Current Abundance.
Three-lined salamanders are fairly abundant throughout their range. While there have been a fair number of population studies of their sister species, *E. longicauda* (long-tailed salamanders), the population ecology of three-lined salamanders has not been studied nearly as well. Anecdotally, some of the populations first studied two and three decades ago (Bruce, 1970, 1982) apparently are still stable.

3. Life History Features.
A. Breeding. Courtship has not been described for three-lined salamanders, and relatively little is known of their reproductive activities.

i. Breeding migrations. Three-lined salamanders rarely stray far from aquatic habitats, but are found more frequently at the terrestrial/aquatic interface and in the water during the late fall, winter, and early spring. If there is a migration per se from terrestrial sites to aquatic ones, it occurs in the fall, probably coincident with the onset of the breeding season. Marshall (1999) suggested an extended breeding season (July–December).

ii. Breeding habitat. Breeding occurs in lentic and slow-moving lotic systems, such as sluggish streams and seeps, bogs, and cypress bays (Petranka, 1998).

B. Eggs. Oviposition occurs in the winter, but varies considerably in published reports and is likely a consequence of both spatial (geographic) and temporal

(year-to-year) variation. Gordon (1953) reports December oviposition. In North Carolina–South Carolina populations, a February oviposition date seems more likely, based on the appearance of hatchlings (Bruce, 1970, 1982; Freeman and Bruce, 2000). Marshall (1999) suggested that oviposition could occur as early as November in some populations.

i. Egg deposition sites. Few egg clutches have been observed. Mature ova are 2.5–3.0 mm in diameter. Bruce (1970) found hatchlings and advanced embryos scattered on the bottom of a cistern in mid-March. The eggs were not attached to any cover object, which is unusual for eastern *Eurycea*. Given the unconventional location of the embryos and the fact that no female was in attendance, Bruce (1970) speculated that eggs had washed into the cistern rather than having been oviposited there.

ii. Clutch size. Mount (1975) found groups of 8–14 eggs associated with several adults in a covered concrete reservoir associated with a shallow spring. Beyond this, we are aware of no reliable data regarding fecundity (Ryan and Bruce, 2000).

C. Larvae/Metamorphosis. The larval life history of three-lined salamanders is one of the best studied aspects of the species with at least five comprehensive studies.

i. Length of larval stage. In North Carolina, hatchlings emerge at 10–13 mm and undergo metamorphosis when they are 22–27 mm SVL after a 4–6 mo larval period (Bruce, 1970, 1982). Larvae in montane populations occasionally will overwinter and transform during the early summer, 16 mo after hatching and at between 30 and 32 mm SVL. The effect of elevation on larval periods was studied in a more comprehensive manner by Freeman and Bruce (2000). They investigated changes in timing of and size at metamorphosis over an elevational gradient within a single watershed. At the low-elevation populations (in Georgia and South Carolina), metamorphosis came 5–6 mo posthatching. Larvae overwintered and delayed metamorphosis until 14–15 mo post-hatching in the high-elevation populations in North Carolina. There were no differences between size at hatching or growth rate between the high- and low-elevation populations. Freeman and Bruce (2000) posited that differences in hydrological stability (which is greater in the high-elevation populations) likely accounts for these differences. Marshall (1999) also found metamorphosis occurring at 5–6 mo post-hatching. However, his analysis of variation in size and age at metamorphosis led him to conclude that three-lined salamander larvae are adapted to warm, stable aquatic environments and found no support for their adaptation to variable habitats.

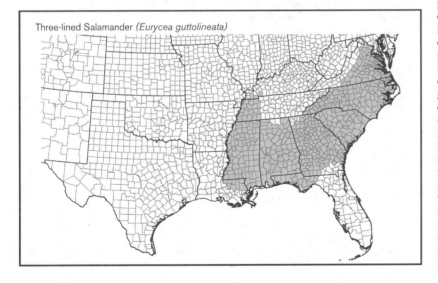

Three-lined Salamander *(Eurycea guttolineata)*

ii. Larval requirements. Larvae are found in the same slow-moving streams, bogs, and marshes as the adults.

a. Food. Larval three-lined salamanders most likely feed on small invertebrates (Petranka, 1998), but there are no detailed studies of foraging behavior or gut content analyses.

b. Cover. It is difficult to quantify or even accurately describe the cover objects of three-lined salamander larvae, as they most frequently inhabit waters that make direct observation extremely difficult. Larvae are captured most easily by thrusting a dipnet (e.g., Freeman and Bruce, 2000) rather blindly through the shallow water near the land–water interface or in deeper waters along the substrate. Larvae most likely seek refuge in decaying vegetation along the stream/pond/bog/marsh bottom (Bruce, 1982).

iii. Larval polymorphisms. Unknown.

iv. Features of metamorphosis. Metamorphosis occurs fairly early in three-lined salamanders (see "Length of larval stage" above). The first sign of metamorphosis is the adoption of an adult pigment pattern that frequently far precedes other signs of metamorphosis (such as the resorption of the tail fin and external gills; Bruce, 1970).

v. Neoteny. Paedomorphosis is not known in three-lined salamanders.

vi. Post-metamorphic migrations. Recently metamorphosed three-lined salamander juveniles are encountered most frequently at the land–water interface, but it is not altogether unlikely to find juveniles in the surrounding forest among adults. Coordinated migrations (in the manner of various *Ambystoma*, which may frequently be members of the same guild) per se are unknown and unlikely.

D. Juvenile Habitat. Same as adult habitat, see below.

E. Adult Habitat. Mainly terrestrial as adults, however, they rarely are found considerable distances from wetlands. Most abundant in river-bottom wetlands and in the vicinity of springs and streams (sometimes ditches, vernal ponds, and bogs) where seepage keeps the ground moist. Animals occasionally are found some distance from water, but are good swimmers and at home in the water. Like most other plethodontids, three-lined salamanders are primarily nocturnal, but may be found during the day under cover objects. Surface activity is closely tied with surface moisture; adults are likely to be encountered foraging on humid or rainy nights shortly after sunset (Petranka, 1998).

F. Home Range Size. Unknown. Because three-lined salamander adults do not defend territories (see below), the definition of individual home ranges is problematic. Furthermore, detailed autecology studies have not been published to date.

G. Territories. According to Jaeger (1988), three-lined salamanders are one of the few

plethodontid salamanders that are not territorial.

H. Aestivation/Avoiding Desiccation. Aestivation is unknown and unlikely.

I. Seasonal Migrations. Seasonal habitat shifts in response to temperature occur(toward the water in fall and winter as temperatures descend and away from aquatic habitats in the spring when air, ground, and water temperatures rise) but highly synchronized movements have not been described.

J. Torpor (Hibernation). Unknown and unlikely.

K. Interspecific Associations/Exclusions. Three-lined salamanders often are associated with southern two-lined salamanders (*E. cirrigera*) or Blue Ridge two-lined salamanders (*E. wilderae*). Little is known about how (Petranka, 1998) or whether these species compete. Bruce (1982) found three-lined salamanders and Blue Ridge two-lined salamanders inhabiting the same creeks in western North Carolina with no noticeable adverse effects on either; for example both species demonstrated life history patterns consistent with isolated populations (Bruce, 1970, 1988). The differences in larval life history (see Ryan and Bruce, 2000) may be sufficient to reduce competition prior to metamorphosis. As adults, three-lined salamanders tend to stay closer to aquatic habitats than do Blue Ridge two-lined salamanders or southern two-lined salamanders, particularly in the summer (T. J. R., personal observations); this may reduce competition in the post-metamorphic phase. At the higher elevation extreme, other stream-dwelling plethodontids may be encountered, such as Ocoee salamanders (*Desmognathus ocoee*) and more rarely seal salamanders (*D. monticola*), red salamanders (*Pseudotriton ruber*) and mud salamanders (*P. montanus*; Freeman and Bruce, 2000). However, because three-lined salamanders also inhabit more lentic waters, they also may be syntopic with members of the local pond-dwelling guilds (e.g., eastern newts [*Notophthalmus viridescens*] and spotted salamanders [*Ambystoma maculatum*]; Freeman and Bruce, 2000). Competitive interactions among these species, in either lentic or lotic environments, have not been evaluated.

Until relatively recently, three-lined salamanders were considered a subspecies of long-tailed salamanders, with which they were believed to hybridize (Bailey, 1937; Martof and Humphries, 1955; Valentine, 1962; Ireland, 1979). Carlin (1998) used morphological and genetic data to elevate *E. guttolineata* to full species. Furthermore, Carlin's analysis was capable of unambiguously identifying individuals from the putative zone of intergradation (located in northern Alabama and Georgia), making uncertain the status of hybrids that previously had been described based solely on morphological

characters. There are also reports of zones of sympatry along Blue Ridge escarpment that lack apparent intergradation (Ireland, 1979; T. J. R. and R. C. Bruce, personal observations).

L. Age/Size at Reproductive Maturity. Maturation is synchronous between the sexes, coming at about 2 yr post-hatching (Marshall, 1999; Ryan and Bruce, 2000). Females are slightly larger than males on average (Gordon, 1953).

M. Longevity. Unknown. Because of substantial differences in the size at maturation and average (not to mention maximum) adult sizes (Ryan and Bruce, 2000), it is reasonable to believe that there is the potential for substantial post-maturation growth. Marshall (1999) was able to measure post-metamorphic growth in a population of three-lined salamanders and found that following maturation, growth slowed considerably (from 1.7 mm SVL/mo to 0.11 mm SVL/mo). If this is consistent across populations, then it is likely that some adult three-lined salamanders must survive for upwards of a decade or more.

N. Feeding Behavior. Three-lined salamanders feed on a variety of invertebrate prey including snails, snail eggs, arachnids, millipedes, annelids, nematodes, and insects including hymenopterans (especially ants), dipterans, coleopterans, orthopterans, hemipterans, homopterans, lepidopterans, neuropterans, odonates, collembolans, and trichopterans (Tinkle, 1952; Petranka, 1998).

O. Predators. Unknown. It is easy to speculate that inhabitants of water edges and forest floors (e.g., semi-aquatic and/or semi-fossorial snakes and small mammals such as voles and shrews) and larger salamanders are likely important predators.

P. Anti-Predator Mechanisms. Three-lined salamanders assume a classic urodelean defensive posture that includes coiling their body, tucking their head beneath their tail, and raising and undulating their tail (Brodie, 1977).

Q. Diseases. Unknown.

R. Parasites. Rankin (1937) lists the following parasites from three-lined salamanders: Protozoa—*Cryptobia borreli*, *Cytamoeba bacterifera*, *Eutrichomastix batrachorum*, *Haptophyra michiganensis*, *Hexamastix batrachorum*, *Hexamitus batrachorum*, *Prowazekella longifilis*, and *Tritrichomonas augusta*; Trematoda—*Brachycoelium hospitale* and *Gorgoderina tenua*; Nematoda—*Oxyuris magnavulvaris* and spirurid cysts; Cestoda—proteocephalid cysts; Acarina—*Hannemania dunni*.

4. Conservation. Three-lined salamanders remain abundant throughout much of their range, and there is no evidence to support a difference between the current and historical distributions. While the loss of bottomland hardwood forests throughout the Southeast has undoubtedly resulted in the extirpation of many populations (Petranka, 1998), direct links between habitat loss and population declines have not been demonstrated.

Eurycea junaluska Sever, Dundee, and Sullivan, 1976
JUNALUSKA SALAMANDER

Travis J. Ryan, David M. Sever

1. Historical versus Current Distribution.
When described in 1976, Junaluska salamanders *(Eurycea junaluska)* were known "officially" from only three creeks, all located within Graham County, North Carolina. However, Junaluska salamanders had been collected in the Great Smoky Mountains National Park (GSMNP) as early as 1937. At that time, King (1939) recognized a unique form as *E. bislineata* x *cirrigera*, an intermediate between the northern *(bislineata)* and southern *(cirrigera)* subspecies of the widely distributed two-lined salamander. King was certain that the form he found in the GSMNP was distinct from the common Blue Ridge two-lined salamander (*E. bislineata wilderae*; the three subspecies of *E. bislineata* later were all elevated to species level by Jacobs, 1987). After describing *E. junaluska*, Sever (1976) examined a portion of King's collection and determined that they were in fact the earliest collected forms of the new species. This post hoc discovery expanded the range of Junaluska salamanders to two additional counties in Tennessee—Blount and Sever. Subsequent surveys resulted in new records from Monroe County, Tennessee (Sever, 1983a). An additional site in Graham County was later reported, but extensive surveys in the three North Carolina counties surrounding Graham County proved fruitless (Ryan, 1997). More recent fieldwork in Tennessee has resulted in new records from within the GSMNP and in Polk County (W. H. M. Gutzke, personal communication).

2. Historical versus Current Abundance.
Most of the other members of the *E. bislineata* complex (*E. bislineata*, *E. cirrigera*, and *E. wilderae)* are locally abundant throughout their ranges (Sever, 1989); but in our experience, Junaluska salamanders are an exception to this rule. Sever (1984) gives a particularly lucid and entertaining account of the difficulties that can be associated with studying this species. In the earliest account, King (1939) speculated that Junaluska salamanders were "a relic of a much older population" but did not mention their abundance. The type series of 23 individuals had to be assembled from collections made over a 3-yr period (Sever, 1984). A decade after the first members of the type series had been collected, Sever (1983a) commented on having "collected fewer than 50 metamorphosed individuals." Ryan (1998a) suggested that larvae are preferred indicators of population size because larvae are locally concentrated (i.e., confined to streams) whereas the more uncommon metamorphosed individuals may be spread out over a substantially larger area. Bruce (1982b) and Ryan (1997, 1998a) studied larval populations of Junaluska salamanders and Blue Ridge two-lined salamanders; in both cases, larvae of the latter were far more abundant than that of the former. Ryan (1998a) determined that in North Carolina, Junaluska salamanders accounted for fewer than 20% of the larval *Eurycea* assemblage at Santeetlah Creek (perhaps the most stable and best studied population). Junaluska salamanders represented ≤ 30% of the assemblage at Snowbird Creek in August 1994, but have not been collected at that site since later that year (Ryan, 1998a; T.J.R., unpublished data). Essentially, Junaluska salamanders were rare to begin with (perhaps contributing to their relatively late discovery), and remain so in North Carolina.

Recent collecting efforts at the type locality on the Cheoah River have been in vain (T.J.R., unpublished data; W.H.N. Gutzke, personal communication), and this population is feared to be extirpated, due to anthropogenic activities upstream. Tennessee populations appear stable and quite possibly larger than those in North Carolina.

3. Life History Features.
The life history of Junaluska salamanders is perhaps the best-known aspect of this otherwise enigmatic species. While many of Sever's Junaluska salamander papers (e.g., Sever et al., 1976; Sever, 1979, 1983a) contain valuable life history data, studies by Bruce (1982b) and Ryan (1998a) of the Santeetlah Creek population have resulted in the clearest picture of this species' life history. The life histories of Junaluska salamanders, as well as all other *Eurycea*, are summarized in Ryan and Bruce (2000).

A. Breeding. Reproduction is aquatic.

i. Breeding migrations. We are unaware of anything approaching a true breeding migration in Junaluska salamanders. However, the majority of adults that we have collected have come either in the fall, prior to when courtship and breeding occur in members of the *E. bislineata* complex, or in the spring, near the time of oviposition (e.g., Sever, 1983a; Ryan, 1998a; D.M.S. and T.J.R., unpublished data).

ii. Breeding habitat. Courtship of Junaluska salamanders has not been observed, but we believe that it likely occurs along streams where adults are found during the putative breeding season and also where eggs are deposited.

B. Eggs.

i. Egg deposition sites. Salamanders attend clutches of eggs around mid-May. Eggs typically are found attached to the underside of a large rock in water <0.5 m deep, with moderate stream flow, and between 1–12 m from the streambank. The location of the nests does not differ appreciably from where larvae have most frequently been collected.

ii. Clutch size. Clutch sizes observed in the field range from 30–49 (mean = 38, n = 5). Sever (1983a) reported gravid females contained between 41–68 (mean = 51, n = 10) mature ovarian follicles. About 1 mo is required for embryonic development, and hatchlings are about 7–8 mm TL (Bruce, 1982b). As a point of interest, each time Junaluska salamander nests have been discovered in the field, Blue Ridge two-lined salamander nests have been located syntopically, with as little as 1–2 m separating the two species.

C. Larvae/Metamorphosis. The larvae of Junaluska salamanders are superficially similar to those of Blue Ridge two-lined salamanders with regard to morphology, ecology, and life history. A photograph of

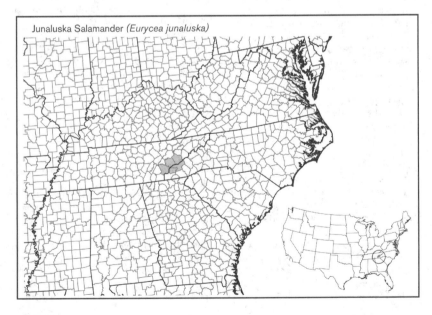

Junaluska Salamander *(Eurycea junaluska)*

a larva appears in Sever (1983a), and Ryan (1997) offers a description of the larvae and illustrations of larval Junaluska salamanders and Blue Ridge two-lined salamanders.

i. Length of larval stage. In North Carolina, the larval period appears to be 2 yr, possibly 3 (Bruce, 1982b; Ryan, 1998a); Ryan (1998a) estimated the age at metamorphosis to be 25.5 mo at Santeetlah Creek. Larvae grow at a faster rate in the first year of larval development than they do in the subsequent year(s) (Ryan, 1998a). Metamorphic individuals may be as small as 34 mm SVL, but most are closer to 40 mm SVL (Bruce, 1982b; Ryan, 1998a) making them the largest naturally metamorphosing larvae in the *E. bislineata* complex (Ryan and Bruce, 2000).

ii. Larval requirements.

a. Food. No data on the diet of larval Junaluska salamanders have been reported. The most commonly encountered macroinvertebrates in the streams where larvae are abundant are stonefly and caddisfly larvae. In the laboratory, we have observed that larvae feed readily on white worms (*Enchytraeus* sp.).

b. Cover. The ideal cover objects for Junaluska salamander larvae are large flat-bottomed rocks that come in close contact with the stream substrate. Most frequently these rocks are located in regions of relatively large (i.e., higher-order) streams where there is moderate water flow and low sedimentation. We also often find larvae in quiet pools on the margins of streams. Larvae are located by carefully displacing rocks. Less frequently, we have collected larvae in riffle areas of streams. In November, when the bulk of a stream's allochthonous input has been received, we have collected larvae using dipnets and searching through the leaf mats.

iii. Larval polymorphisms. Unknown.

iv. Features of metamorphosis. Based on the few field studies, it appears that metamorphosis is more or less synchronous within a population, occurring mostly in May–August, corresponding roughly to the time of hatching of a new cohort (Bruce, 1982b; Ryan, 1998a). The length of time to complete metamorphosis once it has been initiated (e.g., once gill resorption has begun) has not been studied rigorously, but Bruce (1982b) showed that while metamorphic individuals were common (approximately 1 in 4) during a July collection, none were found in the preceding (May) or following (September) collections.

v. Post-metamorphic migrations. Just as there appears to be no marked breeding migration, neither is there a record of recently metamorphosed juvenile migrations.

vi. Neoteny. Paedomorphosis is not known in Junaluska salamanders.

D. Juvenile Habitat. Juveniles are poorly recorded in the literature. The minimum size of adult Junaluska salamanders is not appreciably different from the size of metamorphic larvae; thus, the transition from juvenile to adult is likely minor, with the two classes overlapping to a wide degree. Unfortunately, little is known regarding the habitat characteristics of adults, as well (see below).

E. Adult Habitat. Little is known regarding the habitats of adults. Most adults have been collected on roads near creeks during warm, rainy nights (Sever, 1984) or on streambanks and in streams during early spring (Bruce, 1982b; Sever, 1983a, 1984; Ryan, 1998a). Most likely, adult Junaluska salamanders have seasonal activity patterns similar to Blue Ridge two-lined salamanders. In colder months, adults are most commonly found within streams, as this habitat is more thermally stable than the surrounding terrestrial habitat; in warmer months, adults are found predominantly in the forests surrounding the streams. Within the streams, adults have been collected within the same habitat as larvae, that is., beneath large flat rocks in regions of shallow water and moderate stream flow.

F. Home Range Size. Unknown.

G. Territories. Unknown.

H. Aestivation/Avoiding Desiccation. Aestivating behavior is unknown.

I. Seasonal Migrations. See "Breeding migrations" above.

J. Torpor (Hibernation). Unknown.

K. Interspecific Associations/Exclusions. The range of Junaluska salamanders is sympatric with that of Blue Ridge two-lined salamanders, with the latter being far more common. Nonetheless, the two species are frequently syntopic, with larvae and adults of both species utilizing identical habitats (Ryan, 1997, 1998a). We frequently have found larvae of both species beneath the same rock, and, as noted above, the nesting sites of both species are, for all intents and purposes, identical. It is unclear whether the presence of one species impacts the other. At one larval Junaluska salamander site along the Cheoah River (the type locality), eastern newts (*Notophthalmus viridescens*) were abundant (Ryan, 1998a). Other salamanders frequently encountered while collecting Junaluska salamanders include hellbenders (*Cryptobranchus alleganiensis*), red salamanders (*Pseudotriton ruber*), spring salamanders (*Gyrinophilus porphyriticus*), various dusky salamanders including shovel-nosed salamanders (*Desmognathus marmoratus*), seal salamanders (*D. monticola*), Ocoee salamanders (*D. ocoee*), and black-bellied salamanders (*D. quadramaculatus*; Bruce, 1982b; Sever, 1983a; Ryan, 1998a).

L. Age/Size at Reproductive Maturity. The age at first reproduction is unknown (Ryan and Bruce, 2000). The smallest mature female recorded in the literature (37 mm SVL; Sever, 1983a) falls well within the size range of metamorphic larvae (34–44 mm SVL; Ryan and Bruce, 2000), leading us to speculate that reproduction likely shortly follows metamorphosis. Thus, if most individuals metamorphose in their third summer, they likely breed initially at the end of the third year. The average size of females is approximately 43 mm SVL (Sever, 1983a) with maximum sizes of 47 and 49 mm SVL for females and males, respectively (Sever, 1983b).

M. Longevity. We are unaware of any reliable information regarding the age of Junaluska salamanders.

N. Feeding Behavior. Both the diet and feeding behavior of metamorphosed larval Junaluska salamanders are unknown. We speculate that they feed on a wide variety of invertebrates, as with other members of the *E. bislineata* complex (e.g., Burton, 1976).

O. Predators. There are no reports of predation specifically on larval Junaluska salamanders. They likely face threats similar to those experienced by other members of the *E. bislineata* complex: fish, birds, small mammals, snakes, and other salamanders (Petranka, 1998).

P. Anti-Predator Mechanisms. Unknown.

Q. Diseases. We are not aware of any reports of diseases of larval Junaluska salamanders. However, Ryan (1998b) described scoliosis (lateral spinal malformations) in two larvae from the Cheoah River. The cause of the condition is unknown and could be due to either exogenous (e.g., disease or parasites) or endogenous (e.g., genetic) causes.

R. Parasites. See "Diseases" above.

4. Conservation.
Tennessee populations of Junaluska salamanders appear stable. In North Carolina, however, they were rare to begin with and remain so. For example, they have not been collected at Snowbird Creek since late in 1994, and recent collecting efforts at the type locality on the Cheoah River have been in vain. This population is feared to be extirpated, due to anthropogenic activities upstream. In North Carolina, Junaluska salamanders are listed as a Species of Special Concern and has been proposed for Threatened status.

Eurycea latitans Smith and Potter, 1946
CASCADE CAVERNS SALAMANDER
Paul T. Chippindale

1. Historical versus Current Distribution.
Cascade Caverns salamanders (*Eurycea latitans*) were first described by Smith and Potter (1946) from Cascade Caverns, Kendall County, Texas. Brown (B.C., 1967a) provided a review of what was known about this species at that time. Sweet (1978a, 1984) demonstrated that this population includes individuals with a

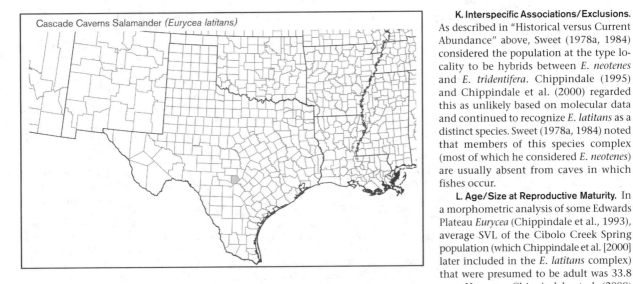

Cascade Caverns Salamander (*Eurycea latitans*)

spectrum of morphological features, ranging from highly cave-associated morphologies most similar to those of Comal blind salamanders (*E. tridentifera*), to surface-like morphologies most similar to those of what he considered Texas salamanders (*E. neotenes*). He hypothesized that this was the result of hybridization between a surface population of Texas salamanders and a population of the cave-dwelling Comal blind salamanders. Most recent authors have not recognized Cascade Caverns salamanders as a distinct species. Chippindale et al. (2000) found no evidence that salamanders from the Cascade Caverns system are hybrids. Based on molecular data, the Cascade Caverns salamander population appears most closely related to many other cave and spring populations in the southeastern Edwards Plateau region that do not include the type localities of either Texas salamanders or Comal blind salamanders. Chippindale (2000) and Chippindale et al. (2000) provisionally recognized the *E. latitans* complex, in which they included the population at the type locality, plus many others from Comal, Kendall, and eastern Kerr counties. Relationships among members of this group, and their relationships to other southeastern Edwards Plateau *Eurycea*, remain to be studied in more detail.

2. Historical versus Current Abundance.
As with many cave-dwelling populations of Texas *Eurycea*, it is difficult to assess population sizes. The most recent collection of topotypes of this species of which I am aware was in 1992, when P. Chippindale, A. Grubbs, and J. Hunter obtained five specimens from Pfeiffer's Water Cave, a subterranean extension of Cascade Caverns. Sweet (1978a, 1984) documented an apparent shift in phenotypes at the type locality, from predominance of individuals with cave-associated morphologies to predominance of individuals with surface-

associated morphologies, over a period of several decades. Members of the *E. latitans* complex (as broadly recognized by Chippindale et al., 2000) often are common at spring outflows, but their distribution appears to be limited and patchy.

3. Life History Features.
 A. Breeding. Reproduction is aquatic.
  i. Breeding migrations. Unlikely to occur.
  ii. Breeding habitat. Unknown within their cave ecosystem.
 B. Eggs.
  i. Egg deposition sites. Unknown; closely related species are thought to deposit eggs in gravel substrate.
  ii. Clutch size. Unknown.
 C. Larvae/Metamorphosis. Cascade Caverns salamanders are paedomorphic, and natural metamorphosis is unknown. Features of larval life are also unknown.
 D. Juvenile Habitat. Probably similar to adult habitat.
 E. Adult Habitat. Completely aquatic. Known only from caves that contain water and the immediate vicinity of spring outflows; individuals in caves are often seen in the open on submerged rock or mud substrate, whereas individuals from spring populations are found under rocks and leaves and in gravel substrate. Water temperature in springs and caves of the Edwards Plateau is relatively constant throughout the year and typically ranges from 18–20 °C or slightly warmer near the fault zone at the Plateau's edge (Sweet, 1982). Sweet (1982) provided a comprehensive distributional analysis of the central Texas *Eurycea* and discussed hydrogeology of the region in relation to salamander distribution.
 F. Home Range Size. Unknown.
 G. Territories. Unknown.
 H. Aestivation/Avoiding Desiccation. Aestivation is unknown.
 I. Seasonal Migrations. Unlikely.
 J. Torpor (Hibernation). Probably active throughout the year.

K. Interspecific Associations/Exclusions. As described in "Historical versus Current Abundance" above, Sweet (1978a, 1984) considered the population at the type locality to be hybrids between *E. neotenes* and *E. tridentifera*. Chippindale (1995) and Chippindale et al. (2000) regarded this as unlikely based on molecular data and continued to recognize *E. latitans* as a distinct species. Sweet (1978a, 1984) noted that members of this species complex (most of which he considered *E. neotenes*) are usually absent from caves in which fishes occur.

 L. Age/Size at Reproductive Maturity. In a morphometric analysis of some Edwards Plateau *Eurycea* (Chippindale et al., 1993), average SVL of the Cibolo Creek Spring population (which Chippindale et al. [2000] later included in the *E. latitans* complex) that were presumed to be adult was 33.8 mm. However, Chippindale et al. (2000) probably did not include the smallest reproductively mature specimens in their analysis, and no rigorous studies of reproductive biology have been conducted for this species. Barden and Kezer (1944) described eggs and egg deposition by a captive individual from a population that is likely (based on geographic location) to be part of this species complex.

 M. Longevity. Unknown.

 N. Feeding Behavior. Prey probably consists mainly of small aquatic invertebrates, but no detailed feeding studies of this species have been conducted.

 O. Predators. Unknown.

 P. Anti-Predator Mechanisms. Individuals from spring populations are secretive. Sweet (1978a, 1984) noted that individuals from the type locality (Cascade Caverns) show escape behaviors similar to those of animals from surface springs (i.e., movement toward substrate and cover items).

 Q. Diseases. Unknown.

 R. Parasites. Unknown.

4. Conservation.
As with many cave-dwelling populations of Texas *Eurycea*, it is difficult to assess population sizes of Cascade Caverns salamanders. Cascade Caverns salamanders can be common at spring outflows, but their distribution appears to be limited and patchy. They are listed as Threatened in Texas, but have no special recognition by the Federal Government.

Eurycea longicauda (Green, 1818)
LONG-TAILED SALAMANDER

Travis J. Ryan, Christopher Conner

1. Historical versus Current Distribution.
Long-tailed salamanders (*Eurycea longicauda*) are distributed throughout the Ozark Highlands, the Appalachian Highlands, and the Ohio River Valley. There is a narrow connection between the Ozarks

and the rest of their range through southern Illinois and western Kentucky. Two subspecies, dark-sided salamanders *(E. l. melanopleura)* and long-tailed salamanders *(E. l. longicauda)*, are recognized. Dark-sided salamanders are associated with the Ozark Highlands and are distributed from eastern Oklahoma and extreme southeastern Nebraska into central and eastern Missouri. Long-tailed salamanders range in a narrow band from southeastern Missouri through extreme southern Illinois, throughout most of Kentucky, central and western Tennessee, extreme northeastern Mississippi, northern Alabama, northern Georgia, extreme southwestern and northwestern North Carolina, western Virginia, West Virginia, Maryland, Pennsylvania, southern New York, and in the north from extreme eastern Illinois, west through southern Indiana, and into southern and eastern Ohio. Locally, distribution is somewhat dependant on the availability of suitable habitats.

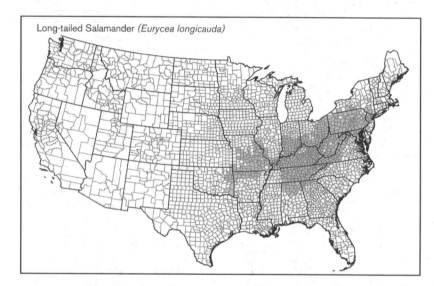

Long-tailed Salamander *(Eurycea longicauda)*

Long-tailed salamanders frequently are associated with caves, mines, and shale and limestone creek beds. They have a bi-phasic life cycle; aquatic habitats are necessary for breeding and embryonic/larval development, while terrestrial habitats, especially forests surrounding these aquatic habitats, support post-metamorphic individuals. Populations undoubtedly have been lost due to factors such as habitat loss, acid drainage from coal mining, and clearcutting. However, as with most other wide-ranging, stream-dwelling plethodontids, there are no robust distributional studies that document changes in gross distribution.

2. Historical versus Current Abundance.
Long-tailed salamanders can be locally abundant, with densities exceeding 10 adults/m^2 (Mohr, 1944; Guttman, 1989; T. J. R., unpublished data). No clear long-term population studies (c.f., Semlitsch et al., 1988) have been published, obfuscating any differences between historical and current local population sizes.

3. Life History Features.
The main aspects of the life history of long-tailed salamanders are typical of the lineage (i.e., subfamily Plethodontinae, tribe Hemidactyinii; Ryan and Bruce, 2000), and given the fairly broad distribution, there is relatively little variation in the pattern between subspecies or otherwise across the range.

A. Breeding. Oviposition is aquatic, as in all other *Eurycea*, and presumably courtship is aquatic as well. There are no published accounts of complete courtship encounters. One field observation (Cooper, 1960) mentions head-rubbing behavior, typical of plethodontids. There is only anecdotal evidence that females brood their clutches (Franz, 1964). Petranka (1998) speculated that the lone observation of brooding instead may well have been the lone observation of ovipositioning.

i. Breeding migrations. Adults and juveniles first become active above the surface in mid-spring (Anderson and Martino, 1966; Minton, 2001). At this time they become more prominent at the water–land interface. Courtship, however, may not occur until much later in the year. Cooper's (1960) field observation was made in October, and Ireland's (1974) analyses of reproductive tract gross morphology indicated that the bulk of breeding activity occurs in late fall to early spring. Thus, while there may be a mid-spring migration, it apparently is more connected with foraging and general surface activity than breeding per se.

ii. Breeding habitat. Egg laying occurs from late autumn to early spring, depending on latitude and altitude (Hutchison, 1956; Rossman, 1960; Anderson and Mar-

tino, 1966; Minton, 1972, 2001; Guttman, 1989) and also on temporal availability of suitable aquatic habitats. Eggs have been found in mid-autumn (November; Franz, 1964), late winter (March; Ireland, 1974), and in-between (January; Mohr, 1943).

B. Eggs. Egg size is typical for the genus, about 3 mm in diameter (Ryan and Bruce, 2000).

i. Egg deposition sites. Eggs generally are deposited in aquatic environments. Petranka (1998) notes that the discovery of eggs in the field is rare, but a trend is that oviposition is not only aquatic, but frequently subterranean as well (e.g., in caves, mine shafts, and cisterns). Non-aquatic eggs most likely are encountered in areas of high and constant humidity. For example, Franz (1964) found eggs suspended from the roof of a cave near a subterranean stream. Eggs have been found attached to undersides of stones in running water (Mohr, 1943), a pattern more typical for the lineage.

ii. Clutch sizes. Females produce between 61–106 eggs (Hutchison, 1956; Minton, 2001), apparently on an annual basis (Ireland, 1976). The incubation period ranges from 4–12 wk (Mohr, 1943; Ireland, 1974). Hatchlings are about 10 mm SVL (Hutchison, 1956; Anderson and Martino, 1966; Ireland, 1974). There is some discrepancy between the number of mature follicles in oviducal egg counts (Hutchison, 1956) and the number of eggs encountered in the field, indicating that a female may split her ovarian compliment among several clutches. This would be consistent with the apparent absence of brooding (see above).

C. Larvae/Metamorphosis.
i. Length of larval stage. The larval period of long-tailed salamanders is about 6 mo, but overwintering (with metamorphosis occurring at 12 mo post-hatching) occurs in some populations. In New Jersey populations, metamorphosis occurs after a 2–2.5 mo larval period and at a size of about 20 mm SVL (Anderson and Martino, 1966). Larvae taken in February in Arkansas measure 10 mm SVL and grow rapidly, up to 6 mm/mo during spring and summer, arriving at metamorphosis in 5–7 mo and at 23–28 mm SVL (Ireland, 1976). A similar pattern was recorded by Rudolph (1978) in Oklahoma populations—hatching at about 10 mm SVL, metamorphosis at 25–32 mm SVL after 4–7 mo. In Rudolph's populations, however, some portion of each cohort was observed to overwinter and metamorphose the following spring. He argues that overwintering is a response to lower invertebrate densities, and thus lower growth potentials, at the mouths of caves. There are reports of some populations requiring 2 yr for larval development (Smith, 1961, cited in Johnson, 1992).

ii. Larval requirements.

a. Food. Larvae ingest a variety of aquatic invertebrates, including ostracods, copepods, snails, and isopods, as well as insects such as dipteran and ephemeropteran larvae, and coleopterans (Rudolph, 1978).

b. Cover. Larvae most frequently are found beneath stones, limbs, and vegetation (rotting and emergent) in streams and ponds (Anderson and Martino, 1966; Petranka, 1998). Larvae may be active above cover objects and in the open at night (Petranka, 1998) and occasionally even in the middle of the day (T. J. R. and C. C., personal observations).

iii. Larval polymorphisms. None.

iv. Features of metamorphosis. Populations frequently breed in temporally variable aquatic habitats, such as classic *Ambystoma*-type temporary ponds (Anderson and Martino, 1966) and spring-fed intermittent streams in the Ozarks (Rudolph, 1978) and central Missouri (T. J. R., personal observations), and thus are subject to the pressure of completing metamorphosis prior to the completion of pond (or stream) drying. Obviously, overwintering is dependent on the persistence of suitable aquatic habitat throughout the year and is not possible in these ephemeral habitats.

v. Post-metamorphic migrations. Migrations of post-metamorphic juveniles are typically diffuse, with individuals gradually moving farther from the water's edge as time passes, but Franz and Harris (1965) report a mass migration of post-metamorphic animals from a Maryland population.

vi. Neoteny. Perennibranchism is not known in long-tailed salamanders; the species is sympatric with a pair of perennibranchiate congeners, Oklahoma salamanders (*E. tynerensis*) and many-ribbed salamanders (*E. multiplicata*; being perennibranchiate in some populations). The coincidence of their ranges indicates that long-tailed salamanders (at least members of the subspecies *E. l. melanopleura*) live in habitats that may favor perennibranchism, but long-tailed salamanders apparently lack the phenotypic plasticity in the timing of metamorphosis and maturation to adopt a neotenic life history pattern.

D. Juvenile Habitat. In New Jersey, post-metamorphic juveniles can be abundant near pond edges immediately following metamorphosis, taking refuge under rocks, fallen tree trunks, and even beneath tree bark (Anderson and Martino, 1966); this seems to be a standard pattern (Petranka, 1998). In general, juveniles are found closer to the water than are adults.

E. Adult Habitat. Adults are mainly terrestrial, found in and beneath old rotting logs and under stones. They are commonly found in crevices of shale and beneath stones and rock fragments near the margins of streams. Adults freely enter water and swim with ease. As with some other members of the genus *Eurycea*, they will enter caves. Adults emerge to feed on humid and rainy nights, where they are most active during the first few hours after dark (Hutchison, 1958; Smith, 1961; see also Petranka, 1998). Anderson and Martino (1966) found reduced densities surrounding permanent streams compared with populations surrounding temporary wetlands.

F. Home Range Size. Unknown. Adults can cover a considerable distance over the course of the year ($\geq$100 m to and from the breeding habitat), but how much of this is considered "home range" is not clear, and it is made even less clear by the suggestion that many juveniles and adults spend a great deal of time underground.

G. Territories. Adults frequently are found in large aggregations. For example, Mohr (1944) found over 300 adults near the rear of a mine shaft, and Guttman (1989) found 80 animals underneath a limestone slab and 23 adults under a 4-m-long log (see also Petranka, 1998). No territorial behavior was evident (S. A. Perrill, personal communication) when Indiana long-tailed salamanders were tested under protocols that have demonstrated territoriality in numerous other plethodontids (Jaeger and Marks, 1993). Also, territoriality is absent in three-lined salamanders (*E. guttolineata*; Jaeger, 1988), the sister species to long-tailed salamanders.

H. Aestivation/Avoiding Desiccation. Aestivation is unknown.

I. Seasonal Migrations. Adults exhibit marked seasonal patterns in habitat use. During periods of heavy rains, adults will migrate uphill to slopes. Adults are known to migrate into and out of caves and mineshafts. Mohr (1944) found that large numbers of (300+) adults aggregate in a mineshaft for about 8 mo of the year, beginning in August–September and emerging again in April–May. Because hatchling larvae were detected in the ponds prior to notable surface activity, Anderson and Martino (1966) speculated that the breeding migrations may be subterranean, or that eggs are deposited in subsurface waters. This notwithstanding, they found surface activity begins in April, and by May most adults were within 6 m (20 ft) of the breeding ponds.

J. Torpor (Hibernation). Juveniles and adults migrate to underground retreats in forests in October and emerge to breed in April to early May. Whether or not this subterranean period is marked by inactivity is unclear.

K. Interspecific Associations/Exclusions. Long-tailed salamanders are rarely the only plethodontid salamanders at a particular site. For example, they are known to exist in close association with cave salamanders (*E. lucifuga*) in the Ridge and Valley province in western Virginia and eastern Tennessee and Kentucky (Hutchison, 1956, 1958); in eastern Oklahoma they are also found with congeneric many-ribbed salamanders and Oklahoma salamanders, but also grotto salamanders (*Typhlotriton [Eurycea] spelaeus*, Rudolph, 1978); and in Indiana they are found syntopically with southern two-lined salamanders (*E. cirrigera*; T. J. R. and C. C., unpublished data). Furthermore, when breeding in temporary ponds, larvae may interact with members of this pond-dwelling salamander guild, such as marbled salamanders (*Ambystoma opacum*), Jefferson salamanders (*A. jeffersonianum*), spotted salamanders (*A. maculatum*), and eastern newts (*Notophthalmus viridescens*; Anderson and Martino, 1966). Some of these potential interactions appear to be ecologically important, others appear benign.

As larvae in ephemeral ponds, long-tailed salamanders and marbled salamanders are the first to appear (Anderson and Martino, 1966). Marbled salamanders are fall breeders and are likely well established by the time long-tailed salamander hatchlings become active in the spring. Furthermore, marbled salamander larvae can be important predators on and/or competitors with other larval salamanders (e.g., Boone et al., 2002). Perhaps the lower density of long-tailed salamanders near streams as opposed to ponds (see above) is a response to avoiding competition from stream-dwelling plethodontids in the region (e.g., northern two-lined salamanders [*E. bislineata*], northern dusky salamanders [*Desmognathus fuscus*], and red salamanders [*Pseudotriton ruber*]). However, the larvae of long-tailed salamanders and southern two-lined salamanders are found syntopically in some limestone creeks in southern Indiana (T. J. R. and C. C., unpublished data) in a manner analogous to their respective southern Appalachian sister species, three-lined salamanders and Blue Ridge two-lined salamanders (*E. wilderae*). The nature of potential competitive interactions has not been resolved.

Long-tailed salamander larvae appear to be competitive equals with larval cave salamanders (Wooley, 1971; Rudolph, 1978; see also Hutchison, 1956, 1958). Rudolph's (1978) study of larval plethodontid community ecology indicates that long-tailed salamanders are not equal to other species, however. Both long-tailed salamanders and cave salamanders inhabit waters at the mouths of streams with subterranean origins and as far downstream as the stream's hydrological stability permits (the likelihood of stream failure increases with distance from the stream origin). However, both species are displaced downstream when the cave-adapted grotto salamanders are present. Also, the diets of long-tailed salamanders and cave salamanders were more similar to each other than to other species (e.g., Oklahoma salamanders and three-lined salamanders) in field enclosures.

L. Age/Size at Reproductive Maturity. Long-tailed salamanders mature about 1–2 yr after metamorphosis (Ladd, 1947;

Anderson and Martino, 1966; Ireland, 1974). In New Jersey, males mature when they reach about 43 mm SVL and females at 46 mm SVL, almost uniformly at 2 yr post-hatching. Dark-sided salamanders in Arkansas mature at smaller sizes (31–43 mm SVL for males, 33–43 mm SVL for females) and as much as a year earlier than New Jersey populations (Ireland, 1974).

M. Longevity. Unknown.

N. Feeding Behavior. Adults feed on a wide variety of invertebrate prey. Specifically, Anderson and Martino (1966) documented annelids, isopods, diplopodans, chilipodans, arachnids (pseudoscorpions, spiders, phalangids, mites, and ticks), and various insects such as homopterans, coleopterans, dipterans, hymenopterans, lepidopterans, thysanurans, and orthopterans in the diet in their New Jersey populations. Hutchison (1958) found long-tailed salamanders in Virginia caves eat primarily dipterans, orthopterans, and coleopterans. A diet analysis of an Indiana population included > 20 types of invertebrates, with isopods, areneans, dipterans, coleopoterans, and collembolans being most numerous. Collectively, these reports indicate that long-tailed salamanders are invertebrate generalists; variations across adult habitats (e.g., caves versus forests) and within habitats across seasons produce different opportunities for feeding.

O. Predators. Larvae are preyed upon by sculpins (*Cottus* sp.) and sunfishes (*Lepomis* sp.; Rudolph, 1978).

P. Anti-Predator Mechanisms. This aspect of the long-tailed salamanders' biology has not been studied rigorously, but individuals discovered in the field have displayed the classic defensive posture with an elevated tail (T. J. R., personal observations). The tail autotomizes readily when handled; additionally, long-tailed salamanders are quick, bolting for cover when disturbed (Johnson, 1992).

Q. Diseases. Unknown.

R. Parasites. Unknown.

4. Conservation.

Long-tailed salamanders can be locally abundant, but populations have undoubtedly been lost due to habitat loss, effects of coal mining, and clearcutting. However, as with most other wide-ranging, stream-dwelling plethodontids, there are no robust distributional studies that document changes in gross distribution. Long-tailed salamanders are listed as Threatened in Kansas and New Jersey (Levell, 1997), and a Species of Special Concern in North Carolina.

Eurycea lucifuga Rafinesque, 1822å
CAVE SALAMANDER

J. Eric Juterbock

1. Historical versus Current Distribution.

Cave salamanders (*Eurycea lucifuga*) range from the southern half of Indiana and extreme southwestern Ohio in the North,

to the northern third of Alabama plus extreme northeastern Mississippi and northwestern Georgia in the South. They extend from northern Virginia in the East, to northeastern Oklahoma and extreme southeastern Kansas in the West. However, they are not always uniformly distributed within this range, due to their general (but not absolute) reliance on limestone caves and springs (see "Adult Habitat" below). Hutchison (1966) and Petranka (1998) contain additional information.

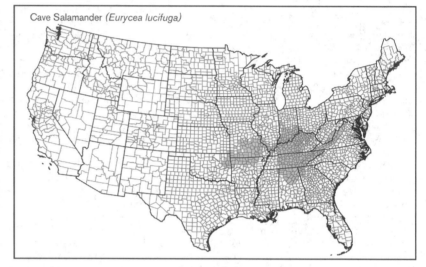

Cave Salamander (*Eurycea lucifuga*)

2. Historical versus Current Abundance.

It is impossible to determine accurately the general abundance of cave salamanders from available data. Although numerous accounts give some indication of the number of individuals collected, and some even indicate the number of individuals seen, only three accounts give any indication of population size/density. Hutchison (1958), after a mark–release–recapture study, presented what he suggested were "rough estimate[s]" that he thought "to be rather close" to actual population sizes in four Virginia caves: 36, 60, 62, and 63. He gave no estimate of density nor indication of cave size; there has been no subsequent study of these populations. Williams (1980), who collected and preserved the specimens he observed in a 1968 Illinois study, concluded that the investigator has an effect on the size of the visible population and compared his results with those of Hutchison. Whether or not such an effect exists, it seems likely that the removal of at least 68 adults from one cave, as he apparently did during his study, would have at least as great an effect upon the population as would disturbance through visitation.

Juterbock (1998) studied three ravine sites in Ohio and, after 3 yr of recapturing marked individuals, concluded that estimates of 30–35, 20–25, and 5–10 adults,

respectively, were reasonable at the sites, as well as 30–35, 5–10, and 25–30 post-larval immatures, respectively. Densities in the inhabited portion of each of these three habitats varied by a factor of 5. It is interesting that given the localized nature of their occurrence (and not knowing their population or metapopulation genetics), the sizes of all these localized groups are well within an order of magnitude. That said, none of these studies was long term, and none has been repeated. A further caution is evidenced by a removal study of salamanders in an Appalachian streamside community by Petranka and Murray (2001). They estimated that between 8 and 32 consecutive nights of sampling were required to remove 70% of the population for the six studied species. Thus, studies utilizing occasional sighting or capture–release–recapture techniques would probably underestimate population size, possibly substantially so.

To the degree that cave salamanders are dependent upon, or associated with, caves and similar limestone features, there is some reason for concern about their populations. Although the biggest threat to cave faunas may be their extremely localized occurrence, there are three recognized sources of vulnerability for cave organisms (Culver et al., 2000): (1) actions that directly degrade subsurface habitat; (2) actions that degrade surface terrestrial habitat and lead to degradation of subsurface habitat; and (3) actions that degrade surface aquatic habitat and lead to degradation of subsurface habitat. Minton (1998) contains one descriptive account of decline in cave salamander numbers, based upon his observations of Indiana sites from 1948–93. At a spring on a steep slope of exposed siltstone in Floyd County, cave salamanders, once "regularly found," ceased to be observed as the habitat was degraded over the years.

3. Life History Features.

For information beyond that in the following sections, Petranka (1998) is a good recent starting point. Hutchison's (1956, 1958, 1966) reports contain much original material and review previous work.

A. Breeding. Reproduction is aquatic.

i. Breeding migrations. Breeding migrations of the type typical of pond-breeding frogs and salamanders are unknown. Neither are such migrations suspected, given the life history of cave salamanders. However, that is not to say that such seasonal movements as occur (see "Seasonal Migrations" below) may not also be related to reproduction.

ii. Breeding habitat. There are few records for this species of naturally occurring egg clutches. The sparse evidence is summarized by Petranka (1998) and would indicate sites deep within surface or cave springs or cave streams. Minton (1972, 2001) reported finding large larvae in an Indiana cave approximately 1 km from the entrance.

B. Eggs.

i. Egg deposition sites. Myers (1958, p. 126) reported that eggs found in early January in a Missouri cave were "attached singly to the bottoms and sides of submerged rocks." The site in the stream was approximately 245 m from the entrance to the cave and about 1 m from where the stream flowed from the cave wall. Green et al. (1967) found eggs in three West Virginia caves attached to the sides of rimstone pools on the floor or sides of the caves, as well as unattached in the silt of the pools.

Banta and McAtee (1906) proposed, on the basis of finding 18 mm TL larvae in early February in Indiana, that oviposition had occurred around the end of December to early January. They also found "small larvae" as late as 20 March. Myers (1958, p. 126) found eggs "in various stages of development, from early cleavage to advanced embryos" in Missouri on 2 January. At the same time, he also found newly hatched larvae, measuring 11 mm TL (approximately 80% of which was SVL). Perhaps Hutchison's (1956) report of a 17 mm (TL or SVL not indicated) larva in Virginia during July could indicate a slow growth rate rather than a later time of oviposition. Myers (1958) "tentatively" interpreted the available data as indicating that the period from oviposition to hatching lasted from as early as October to as late as May. Green et al. (1967) slightly increased that range, reporting eggs in West Virginia caves from 24 September–5 November. The data do not exist to evaluate inter- and intra-population variation for this trait.

ii. Clutch size. Hutchison (1956) counted ovarian eggs in 17 adults collected during July and August in Virginia. They averaged 68.3 (median=67, range= 49–87). Trauth et al. (1990) counted ovarian eggs in 11 adults collected in Arkansas; these

averaged 77.7 (range=60–120). Clutch size in all likelihood varies with female body size, but insufficient data exist to evaluate such a hypothesis.

C. Larvae/Metamorphosis.

i. Length of larval stage. Banta and McAtee (1906) estimated a 12–15 mo larval period as typical for cave salamanders in Indiana. They suggested that some larvae undergo metamorphosis in autumn, when they collected larvae ranging between 31–56.5 mm TL. They concluded that most larvae transformed in March, when they collected an individual metamorphosing. Green et al. (1967) believed that a 30 mm TL larva in one of their West Virginia caves and a 41 mm TL larva reported by Myers (1958) from Tennessee, collected during late February to March, were 1 yr old. Sinclair's (1950) observation of three larvae measuring 22.5 mm, 31.5 mm, and 51.5 mm TL in mid March more clearly demonstrates overlapping generations.

ii. Larval requirements.

a. Food. Rudolph (1978) studied larval food habits of cave salamanders in northeastern Oklahoma and compared them to the larvae of four related salamander species. Of 370 cave salamander food items, 71.6% were ostracods and 12.2% were dipteran larvae; other food items, in order of abundance, included pulmonate snails, ephemeropteran nymphs, isopods, dipteran adults, trichopteran nymphs, adult coleopterans, larval coleopterans, plecopteran nymphs, copepods, and Araneae. As expected, most of these food items are aquatic. The food items of larvae of four syntopic species showed some overlap with larval cave salamanders. Most similar in their feeding habits were long-tailed salamander (*E. longicauda*) larvae, although Rudolph concluded that food competition was probably secondary to competition over space. The other species' food habits probably allow rejection of a hypothesis that cave salamander larvae were only eating things in the proportion in which they occurred in the habitat. Although ostracods were abundant in the diets of all 5 species, they were not the most numerous item for two species. Additionally, certain food items were considerably more abundant in the diets of one or more of the other species than in the diet of cave salamanders (e.g., copepods, amphipods, and isopods for grotto salamanders [*Typhlotriton (Eurycea) speleaus*]; copepods for long-tailed salamanders; and, isopods for Oklahoma salamanders [*E. tynerensis*]). Still, it is possible that microhabitat segregation may be at least partially responsible for the observed dietary differences.

b. Cover. Banta and McAtee (1906), working in Indiana, indicated that oviposition was in the deepest parts of caves and felt that larvae found at cave mouths and in outside streams were carried there by currents. Sinclair (1950)

reported counting hundreds of larvae, at all hours of the day, crawling about the bottom of a surface spring with no cover. He further indicates that he could not find larvae in caves, and that they seemed to prefer cover when they approached metamorphosis. Although the numbers of larvae sound impressive, it is worth noting that, with clutch sizes over 50 (probably averaging 70–75), "hundreds" of larvae, especially if young, might only represent a few clutches. Green et al. (1967) stated that larvae move out of the rimstone pools in which they hatch as those pools overflow in the winter and early spring. Larvae were found in the small, temporary overflow streams and the permanent main stream in the cave; no cover was mentioned. These authors note that larvae were always observed to move downstream, both under their own power and by the action of currents. This alone could explain the appearance of larvae in surface springs and streams, although more needs to be learned about oviposition sites and larval life history. In northeastern Oklahoma (Rudolph, 1978), larvae were found in some surface streams ≤ 45 m from the source spring under unspecified cover. In southwestern Ohio ravines, larvae were rarely seen and presumed to remain underground (unpublished data).

iii. Larval polymorphisms. None known.

iv. Features of metamorphosis. In southwestern Ohio, the smallest metamorphic animals appear in the surface-active population in late summer, at approximately 35–40 mm SVL and probably 18–21 mo old. Metamorphosis in Ohio does not appear to be at all synchronous. Of individuals captured while undergoing metamorphosis, 10 individuals were captured at two sites between May and September, with six of these in August (unpublished data). The smallest two metamorphosing salamanders were 27 mm SVL (captured in July and September); the largest one was 41 mm SVL in May (the second smallest of its age/size class [n=8, SVL range=38–50] in a collection containing a 23 mm SVL larva); the median size was 32 mm SVL. Only three larvae were captured at these sites, with SVLs of 15 mm (2 May), 23 mm (24 May), and 24 mm (13 June). I am aware of no other data on the timing of metamorphosis that is this complete (although individuals at these sites could only be found on the surface from April–September).

The smallest metamorphosed specimen reported by Williams (1980) was a 31 mm SVL female, and the largest larva was 33 mm SVL; TLs were 68 mm and 70 mm, respectively. He concluded that metamorphosis usually occurred between 25–35 mm SVL. He reported 26 larval specimens, but indicated that this was too small a sample to determine age classes; no data beyond those of the previous sentence were presented. Green et al. (1967) stated that metamorphosis occurred between 50–56 mm TL. Sinclair (1950) found two recently metamorphosed animals (59 and 60.5 mm TL) in a Tennessee cave on 2 June.

Rudolph (1978) studied northeastern Oklahoma populations found in surface springs, as well as caves, and stated that metamorphic size for cave salamanders in these populations was approximately 25 mm SVL. He indicated that recently hatched larvae appeared in springs during winter and early spring, and their spring and summer growth usually allowed them to metamorphose between July–October of the same year. Metamorphosis at this time was also supported by the observation of recently metamorphosed animals in the nearby terrestrial habitat. However, he adds that during winter one can usually find a few large larvae, because some individuals overwinter and metamorphose during their second spring. This implies that metamorphosis could occur as early as about 6 mo or as late as perhaps 18 mo.

Apparently, the only indication of growth in larval cave salamanders is contained in the two samples Rudolph (1978) collected from the same surface stream in northeastern Oklahoma. In those, there was a median size (SVL) difference of 10.5 mm (8 June 1976 collection: n=17, median=14 mm, range 10–22 mm; 1 August 1976 collection: n=26, median=24.5, range 19–29 mm).

v. Post-metamorphic migrations. None known.

vi. Neoteny. Unknown, although some larvae will overwinter and metamorphose the following year (Minton, 1972; Rudolph, 1978; see also Petranka, 1998).

D. Juvenile Habitat. There is no evidence that juvenile habitats differ from those of adults. One possible exception involves the timing of habitat use. In southwestern Ohio, I found that adults were much more likely to be active on the surface in late spring as compared to juveniles, whereas juveniles were much more likely to be active on the surface in late summer. For example, at one site over three years, 27 individual adults were captured in May, but only two in August or September; for juveniles, the numbers were 19 and 22, respectively (chi-squared=16.7, p < 0.001). The surface habitat at these sites generally is drier in late summer.

E. Adult Habitat. Adults are essentially terrestrial and/or associated with caves in limestone regions (e.g., Peters, 1946). Although most records are from, and the species appears to be most abundant in, the twilight regions of caves, where they climb over walls and ledges, they are also found outside of caves, under stones, logs, and other surface matter, as well as deeper in caves. Hutchison (1958) reported one locality where the species is associated with a non-calcareous cave. Banta and McAtee (1906) indicated that cave salamanders were found on the walls of the caves and rarely on the cave floor or in the water. Green et al. (1967) agree with this assessment, but Williams (1980) collected 16% of his adult specimens from the stream; most were under rocks. Petranka (1998) summarizes the variety of relevant reports, as well as cautioning that the species' restriction to cave habitats is over-emphasized. Certainly there are no caves in Hamilton County, Ohio, where terrestrial stage individuals are infrequent to common in at least six county and one city park units. Here the habitat consists of forested limestone ravines, at least some of which appear to have subsurface water flow (Davis et al., 1998; Juterbock, 1998). In these situations, adults seldom are seen free of cover, regardless of whether or not it was during the day or night (personal observations). Metamorphosed individuals of all ages are found in the stream bed, but rarely in the water; they are under rocks (mostly), logs, and debris. They only are present on the surface when there is water present or when the soil is muddy (unpublished data). Smith (1961) reported that spring-fed cypress swamps, located "well away from" rock bluffs, were the Illinois sites where cave salamanders were most abundant. Adults and larvae were commonly found there under leaves and logs. I have also seen numerous individuals of various sizes active on the surface of a roadcut on a hillside above a stream in Kentucky (personal observations).

F. Home Range Size. In one southwestern Ohio ravine, over the course of three surface-active seasons, 31 adults were recaptured at least once. Of these, 22 (71%) had maximum ranges along the ravine (at least 50 m of habitat) of ≤10 m, and mean distances between captures of ≤10 m (unpublished data). Thirteen of the 22 (42% of the total 31) were recaptured after overwintering below the surface at least once (and thus at least seasonally shifting the area in which they were active).

G. Territories. Territoriality has not been reported and seems unlikely. Smith (1961), for example, reported that a single large rock might contain "a number" of individuals.

H. Aestivation/Avoiding Desiccation. Aestivation has not been reported and seems unlikely, especially given the cave/spring habitat of this species. However, it is not clear what effects drought may have on the species. Hutchison (1958) found at least one individual in at least one of his four Virginia study caves each month except January.

I. Seasonal Migrations. Hutchison (1958) measured the distance from the mouth of one of his Virginia study caves to the site of each salamander's capture during the year. The salamanders were closest to the mouth in June–July and farthest from the mouth in February–March. He concluded that these data were evidence that migration from the cave did not occur. They do, however, clearly indicate seasonal movements within the cave ecosystem, from a June mean of 4.7 m from the mouth to a March mean of 26.6 m from the mouth. Hutchison (1958) also found that the visible population in all caves that he studied increased from late February to March (in different caves) to a peak in June and then declined dramatically by September. These data mirror the distance data and support a hypothesis of seasonal movements within the habitat. Williams (1980) collected almost 4–5 times as many individuals in May–June as he had in March–April, after which the July and August numbers returned to the level he had seen before the peak. As noted above (see "Historical versus Current Abundance" above), removing 60 or more individuals from the population during the spring should have affected summer counts. Juterbock (unpublished data) found a May–June peak in the surface-active population in southwestern Ohio limestone ravines, with a lesser, secondary peak (comprised primarily of juveniles) in late August to September. Although surface activity is unlikely in the winter in these ravine habitats, the overall pattern is similar to that described for caves and indicates at least seasonal movements from surface (active) to subsurface (inactive?) habitat.

J. Torpor (Hibernation). Considered unlikely by Hutchison (1958), who saw salamanders that were active during the winter while deep in the caves.

K. Interspecific Associations/Exclusions. Hutchison studied cave salamanders and long-tailed salamanders from four caves in Virginia. He found that the two species often shared the twilight zones of caves, with cave salamanders usually more abundant. However, because he found long-tailed salamanders in "comparatively larger numbers" in areas where cave salamanders did not occur (Hutchison, 1958, p. 11), he concluded that interspecific competition between the two species may occur. The similarities of diet (see "Larval requirements" above and "Feeding Behavior" below) may offer a hint as to the mechanism involved.

L. Age/Size at Reproductive Maturity. Hutchison (1958) reported that Virginia males were mature at >46 mm SVL, and females at >48 mm SVL. In southern Illinois, Williams (1980) reported that females >49 mm SVL were mature. Juterbock (1998), sexing recaptured individuals by means of externally visible characteristics, found the smallest mature males in Ohio to be 54 mm SVL and the smallest mature females to be 56 mm SVL. One of the smallest individuals was 53 mm and could not be sexed on 3 May, but was clearly a male when recaptured on 31 May of the same year. Although not nearly as dramatic, many of the smallest mature individuals he recorded were previously observed as unsexable immatures. The differences in size indicated by these three accounts may be real variation or an artifact of using different techniques.

Mostly because of the uncertainty surrounding larval age classes, age at maturity is problematic. The recapture data from southwestern Ohio (unpublished data) leave little doubt that at least 2 yr are required before maturity is attained after the summer of metamorphosis. This would be either the third or fourth summer after the winter of hatching (or approximately 2.5 or 3.5 yr of age), depending upon the length of the larval period. As noted above, it is likely that the duration of the larval period varies, probably both within and between populations. From a conservation perspective, this is one of the more glaring gaps in our knowledge of the species.

Hutchison (1958) observed an overall male:female sex ratio of 1.51:1 in his four Virginia caves. Three of those caves, with total sample sizes of 38, 54, and 59, had individual ratios of 0.65, 1.6, and 1.1, respectively (so the largest sample was the closest to 1:1). Williams (1980) reported a ratio of 1.125:1 (n=70) with males dominant. In contrast, Juterbock (unpublished) found that (based upon externally detectable characteristics of marked salamanders) mature females insignificantly outnumbered males 39:26 (chi-squared=2.6, p=0.11) in southwestern Ohio. Only at the site with the most adults captured (52%) was the difference significant (11 males, 23 females; chi squared=4.24, p=0.04). It is not known whether these differences represent a real result of interspecific variation in life history or an artifact of what are actually rather small samples.

M. Longevity. Not known.

N. Feeding Behavior. Peck (1974) surveyed for food in the guts of 112 cave salamanders (11 were empty) from nine (mostly southeastern) states and found a minimum of 73 prey species. These included annelid worms, snails, crustaceans, millipedes, various arachnids, and 14 orders of insects (14 families of beetles, 12 families of flies, and 4 families of hymenopterans). Spiders, crickets, and at least four families of flies were found in at least 10% of those guts containing food items. Peck and Richardson (1976) studied the diets of an additional 213 cave salamanders from four southeastern states, primarily to elucidate any differences in feeding ecology with respect to location within the cave. They found that the salamanders were best fed within the twilight zone of the cave and least well-fed in the zone of permanent darkness. The major dietary difference discovered was the importance of trichopteran insects, but these were an ephemeral resource at only a few of the studied caves. They identified at

least 101 taxa of prey in the study. Hutchison (1958) compared the food items in guts of 13 cave salamanders and 10 long-tailed salamanders. He found seven orders of insects (more fly taxa and individuals), three orders of arachnids (but no spiders), and isopods. The frequency of occurrence of most taxa was slightly greater for cave salamanders than for long-tailed salamanders, but their diets overlapped greatly.

O. Predators. There appear to be no records of specific predators attacking or consuming cave salamanders, but their responses to disturbance (see "Anti-Predator Mechanisms" below) presumably evolved for a reason. Numerous authors have suggested potential predators (e.g., Hutchison, 1958).

P. Anti-Predator Mechanisms. As do most plethodontid salamanders, cave salamanders possess skin glands that secrete noxious substances. Cave salamanders and their relatives raise and undulate the tail over the head, which, because the body is coiled, rests near the vent (Brodie, 1977). Brodie (1977) has witnessed congeneric species use this posture when attacked by short-tailed shrews (Blarina brevicauda) and blue jays (Cyanocitta cristata). In tests with shrews, he noted that 12 of 13 attacks resulted in bites to the tail, with the shrew briefly retreating and wiping its mouth; this should allow the salamander a brief opportunity to escape.

Under this scenario, one would expect that one explanation of broken tails would be failed predation attempts. The percentage of individuals with broken or obviously regenerating tails varies widely in different populations: approximately 4% in Virginia (Hutchison, 1956); 28.3% in Illinois (Williams, 1980); and 59.5% of 74 adults, 16.4% of 61 immatures in Ohio (unpublished data). This difference between tail damage rates of adults and immatures is significant (chi squared = 24.5, p < 0.001) and presumably relates to the accumulation of time spent in the terrestrial environment. It is worth noting that the Ohio populations occur in ravines, not caves, in a generally urban area. Whatever else may be harassing cave salamanders at these sites (and I have observed children turning rocks at these places), there also are urban population levels of raccoons, and all the Ohio sites studied did occasionally exhibit signs of raccoon foraging.

Q. Diseases. No records.

R. Parasites. McAllister et al. (1995d), in a study of another species of Eurycea in Arkansas, noted the first record of the nematode Desmognathinema nantahalaensis in cave salamanders.

4. Conservation.
It is impossible to accurately determine from available data the general abundance, and therefore the conservation

status, of cave salamanders. Because cave salamanders are dependent upon, or associated with, caves and similar limestone features, there is some reason for concern. Although the biggest threat to cave faunas may be their extremely localized occurrence, actions that directly degrade subsurface habitat or surface terrestrial and/or aquatic habitats negatively affect populations. Cave salamanders are listed as Endangered in Ohio, Mississippi, and Kansas (Levell, 1997), and considered Rare in West Virginia.

Eurycea multiplicata (Cope, 1869)
MANY-RIBBED SALAMANDER

Stanley E. Trauth, Harold A. Dundee

1. Historical versus Current Distribution.
Many-ribbed salamanders (Eurycea multiplicata) occur in the Ozark Plateaus and the Boston and Ouachita mountains and associated lowland rocky formations in southwestern Missouri, eastern Oklahoma, and northwestern Arkansas, at elevations of 107–763 m (Dundee, 1965a; but see Bonett and Chippendale, 2004). Two subspecies are recognized: many-ribbed salamanders (E. m. multiplicata) occur in southeastern Oklahoma and west central Arkansas, while gray-bellied (or graybelly) salamanders (E. m. griseogaster) occur in northeastern Oklahoma, southwestern Missouri, and northwestern Arkansas. The type locality of E. m. multiplicata is in question (Dundee, 1950), but the type locality of E. m. griseogaster is clearly stated (Moore and Hughes, 1941). Cope (1889) reported many-ribbed salamanders from Kansas, but this record apparently is not backed by a voucher specimen (Dundee, 1965). If Cope was correct, a range contraction for this species has occurred. Besides this mention, no evidence supports declines or shifts in distributions.

2. Historical versus Current Abundance.
Generally unknown, but Dundee (1947) reports finding gray-bellied salamanders "in abundance" and "in large numbers" at certain sites and certain times of the year in Oklahoma, particularly in winter and early spring.

3. Life History Features.
A. Breeding. Reproduction is aquatic.
i. Breeding migrations. Long-distance migrations are unlikely.
ii. Breeding habitat. The mating season varies among populations of gray-bellied salamanders. Populations that inhabit thermally stable springs have a prolonged mating season compared with populations inhabiting surface streams with more variable temperatures. Based on the times when females contained spermatozoa in their reproductive tracts, mating activity could be from July–May (Ireland, 1976). Ireland (personal communication) agrees that he meant spermatheca for the term reproductive tracts.

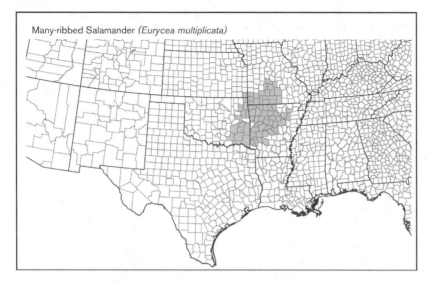

Many-ribbed Salamander *(Eurycea multiplicata)*

B. Eggs.

i. Egg deposition sites. Females lay their eggs in springs, spring-fed pools, and spring-fed ponds from autumn to early spring (Spotila and Ireland, 1970; Ireland, 1976). Eggs usually are laid on the undersides of submerged stones or beneath several layers of smaller stones (Trauth et al., 1990).

ii. Clutch size. Egg counts range from 2–21, but average 10–13 (Ireland, 1976; Trauth et al., 1990). Eggs range from 2.0–2.6 mm in diameter. Females do not brood. Hatchlings average about 10 mm SVL (Petranka, 1998), but Petranka does not document this size.

C. Larvae/Metamorphosis.

i. Length of larval stage. Larvae grow quickly, range in size from 23–85 mm TL, and metamorphose between 33–48 mm SVL, 5–8 mo after hatching.

ii. Larval requirements.

a. Food. Larvae feed most actively at night. Although larvae are described as benthic feeders and will eat isopods, ostracods, and zooplankton, aquatic insects usually are consumed (Rudolph, 1978; see also Petranka, 1998).

b. Cover. During daylight, many-ribbed salamanders are usually found under stones in slow-moving streams. Larvae will inhabit more ephemeral portions of streams than will adults (Loomis and Webb, 1951). Bragg (1955b) observed that small (<18 mm) gray-bellied salamanders can be found in shallow water, in the open, both day and night. Larger larvae spend more time burrowed in gravel.

c. Temperature. Gray-bellied salamander larvae have been found in water as warm as 22 °C in summer (August) and as cold as 9.5 °C in winter (Dundee, 1958).

iii. Larval polymorphisms. Undescribed and unlikely.

iv. Features of metamorphosis. At metamorphosis, animals are approaching sexual maturity. Ireland (1976) found that

16% of males in the process of metamorphosing contained spermatozoa.

v. Post-metamorphic migrations. Bragg (1955b) observed that when gray-bellied salamanders begin to metamorphose, they leave the stream and complete their development along the streambank in moist leaves or under stones.

vi. Neoteny. At metamorphosis, animals are approaching sexual maturity. Gilled adults are common in gray-bellied salamander populations that inhabit caves or streams draining caves, especially on the Salem Plateau (Dundee, 1965; Trauth et al., 2004; see also Dundee, 1947, for many-ribbed salamanders, and Petranka, 1998). Neotenic animals that inhabit caves often are pale and lack components of the normal pigment pattern, whereas neotenic animals in the vicinity of caves retain normal larval pigment patterns.

D. Juvenile Habitat. Similar to that of adults (cool, moist habitats near water under stones, logs, and other large materials in streams and springs, both in the open and in the twilight zone of caves; see "Adult Habitat" below), or if neotenic, similar to that of larvae (see "Neoteny" above).

E. Adult Habitat. Adults are essentially aquatic but occasionally are found on land. Many-ribbed salamanders are found in cool, moist habitats near water under stones, logs, and other large materials in streams and springs, both in the open and in the twilight zone of caves (Moore and Hughes, 1941; Dundee, 1947, 1958, 1965a; Loomis and Webb, 1951; Ireland, 1976; see also Petranka, 1998). Large numbers of gray-bellied salamanders were found under cover objects when the adjacent water had a thin film of ice after overnight subfreezing temperatures (H. A. D., field notes).

In the Ozark highlands, adults are found in caves (Dundee, 1965a). At a study site in Cherokee County, Oklahoma, gray-bellied salamanders were common where water flowed over shallow soils and exposed solid limestone in cool months,

but which were dry during summer and early autumn (Dundee, 1958). Bragg (1955b) noted habitat differences between many-ribbed and gray-bellied salamanders:

> Both [subspecies] occur in Cherokee County [Oklahoma] but apparently they tend to occupy different habitats. Wherever very small streams traverse solid limestone rock or soil, [many-ribbed salamanders are] very likely to be present. If a similar stream cuts through chert, flint, or granite, [gray-bellied salamanders are] the form to be expected, if either form occurs. My experience is not wide enough to say that there never are exceptions, but I can say that so far I have found none in Oklahoma.

One exception is known,—one of Dundee's (1958) study sites for gray-bellied salamanders was over solid limestone rock with grassy hummocks on it.

F. Home Range Size. Unknown but likely to be small. In the first collections of gray-bellied salamanders (Moore and Hughes, 1941), adults and larvae were collected together.

G. Territories. Unknown but unlikely. Loomis and Webb (1951) reported that frequently two or three, and up to four, many-ribbed salamander adults were found under a single rock. Much higher numbers of gray-bellied salamanders have been found under single cover objects (H. A. D., field notes).

H. Aestivation/Avoiding Desiccation. Adult gray-bellied salamanders tend to be found in association with springs and permanent streams and within the wetter portions of such streams during drier periods (Loomis and Webb, 1951). Animals may aestivate, as suggested from at least one site in Oklahoma that would dry in the summer when no salamanders could be found (Dundee, 1958). Bragg (1955b) noted terrarium observations on many-ribbed salamanders that support the possibility of aestivation.

I. Seasonal Migrations. Individuals do not disperse far from their natal streams.

J. Torpor (Hibernation). Animals apparently stay active near the surface, being found under rocks, logs, and moss in or near the edges of streams, except during periods of extreme winter weather (Dundee, 1947).

K. Interspecific Associations/Exclusions. Many-ribbed salamanders are found in association with Ouachita dusky salamanders *(Desmognathus brimleyorum)* in Arkansas (Strecker, 1908a). Gray-bellied salamanders are found with grotto salamanders *(Typhlotriton [Eurycea] spelaeus)* in Missouri (Noble, 1927b). Loomis and Webb (1951) noted that oftentimes stones that held many-ribbed salamanders provided a substrate for small- to medium-sized tarantulas *(Aphonopelma hentzii),* but tarantulas preferred the drier area under the stone.

Spring-fed headwaters in eastern Oklahoma often contain assemblages of 3–5 species of salamander larvae, including long-tailed salamanders (*E. longicauda*), cave salamanders (*E. lucifuga*), many-ribbed salamanders, Oklahoma salamanders (*E. tynerensis*), and grotto salamanders that segregate by distance from the spring head. This habitat segregation may be due to competitive factors (Rudolph, 1978; see also Petranka, 1998).

L. Age/Size at Reproductive Maturity. Reproductive maturity is reached at metamorphosis, or shortly thereafter (Ireland, 1976), although gilled adults are common in some populations (Dundee, 1947; see also Petranka, 1998). Ireland (1976) found that all metamorphosed males have spermatozoa in their seminiferous tubules; 16% of males in the process of metamorphosing also contained spermatozoa. Size varies across populations, with transformed animals averaging 28–45 mm SVL (Moore and Hughes, 1941; Loomis and Webb, 1951; Dundee, 1965). Males and females vary little in size, form, or color (Moore and Hughes, 1941). Little growth occurs after metamorphosis (Dundee, 1965a). Neotenic animals can exceed metamorphosed animals in size, reaching 54 mm SVL and 160 mm TL (Dundee, 1965a).

M. Longevity. Unknown.

N. Feeding Behavior. Undescribed, but adults likely feed on a variety of aquatic and semi-aquatic vertebrates associated with springs and permanent streams. The diet of neotenic adults likely resembles that of larvae (see "Larval requirements" above), although adults, being larger, may take larger prey. Terrestrial adult gray-bellied salamanders consume some aquatic arthropods but primarily eat terrestrial arthropods, snails, and oligochaetes (Dundee, 1958).

O. Predators. According to Petranka (1998), few data are available on natural predators, but they undoubtedly include crayfish and raccoons. Terrestrial stages probably are preyed upon by large beetles, other salamanders, and frogs. Fishes will feed on larvae and appear to exclude them in downstream sections of streams (Petranka, 1998).

P. Anti-Predator Mechanisms. In daylight, both larvae and adults seek cover under stones in slowly moving water. If their stone is removed and animals feel threatened, they will seek cover under a nearby stone (Moore and Hughes, 1941).

Q. Diseases. Undescribed.

R. Parasites. McAllister et al. (1995d) described the metazoan parasites of 50 larval and adult gray-bellied salamanders from seven Arkansas counties. In general, these parasites are typical of parasites reported from other plethodontid salamanders and exhibit little or no host specificity. Seven (14%) animals were infected with ≥1 parasite as follows:

a seuratoid nematode (*Desmognathinema nantahalaensis*; three animals), an ancanthocephalan (*Fessisentis vancleavei*; two animals [this species also reported by Malewitz (1956) from Oklahoma specimens and from Madison and Benton counties, Arkansas, by Saltarelli (1977) and Buckner and Nickol (1978)]), larval intradermal mites (*Hannemania* sp.; two animals), a plagiorchid nematode, a trematode (*Brachycoelium salamandrae*; one animal). Ectoparasitic flukes (*Sphyranra euryceae*) occur in larval and neotenic gray-bellied salamanders (Dundee, 1958; McAllister et al., 1991) and nematodes in the larvae (Dundee, 1958). In terrestrial gray-bellied salamanders, cysts, probably of nematodes and flukes, are found in the skins of terrestrial individuals, and cestodes occur in the gut (Dundee, 1958).

4. Conservation.
Many-ribbed salamanders inhabit a fairly large region of rugged, hilly, and mountainous terrain, most of which is not amenable to agriculture or developing urban settings. Despite considerable logging over much of the terrain and use of flatter areas for cattle grazing, substantial populations of this salamander are found in many settings, even within properties that people have developed. The only threats to the species are an increasing number of recreational homes scattered throughout the region and some improvements to caves that they inhabit being turned into commercial caves. Some of the stream areas that the species inhabits might become polluted if improper sanitary facilities are constructed. Overall, we do not visualize much impact on the species.

Eurycea nana Bishop, 1941(a)
SAN MARCOS SALAMANDER

Paul T. Chippindale, Joe N. Fries

1. Historical versus Current Distribution.
San Marcos salamanders (*Eurycea nana*) were described from outflows of San Mar-

cos Springs in the city of San Marcos, Hays County, Texas, by Bishop (1941a). Some authors (Sweet, 1978a; Dixon, 1987) also have considered the population of *Eurycea* at Comal Springs, Comal County, to be this species, but morphological and molecular evidence strongly reject this hypothesis (Chippindale, 2000; Chippindale et al., 1998, 2000). Their historical distribution probably is similar to their current distribution, although San Marcos Springs has been heavily modified by humans in the past century to form a small lake. Salamanders occur throughout much of this lake and extend about 150 m into the most upstream portion of the San Marcos River (Nelson, 1993). Based on phylogenetic analyses, *E. nana* appears to be the sister taxon to the southeastern Edwards Plateau subgroup of Texas *Eurycea* (Chippindale, 1995, 2000; Chippindale et al., 2000). Although Schmidt (1953) regarded this taxon as a subspecies of Texas salamanders (*E. neotenes*), few others (and no recent authors) have followed this approach. Molecular and morphological data strongly support their recognition as a distinct species (Chippindale, 1995, 2000; Chippindale et al., 1998, 2000).

2. Historical versus Current Abundance.
Extremely abundant within their severely limited range. Population densities are estimated to be about 116–129 individuals/m^2 in vegetation mats (Tupa and Davis, 1976; Nelson, 1993). The entire population has been estimated to be about 53,200 individuals in vegetation mats and suitable rocky substrates (USFWS, 1996b).

3. Life History Features.
  A. Breeding. Reproduction is aquatic.
    i. Breeding migrations. Unlikely to occur.
    ii. Breeding habitat. A subset of adult habitat.

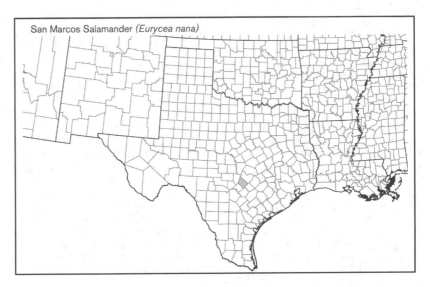

San Marcos Salamander (*Eurycea nana*)

B. Eggs.

i. Egg deposition sites. Eggs have never been observed in the wild. In captivity, ovipositioning has occurred on aquatic moss, filamentous algae, rocks, and glass marbles.

ii. Clutch size. In captivity, an average of 33 eggs/female has been oviposited during a single egg-laying event. Egg size is about 1.5–2.0 mm (Tupa and Davis, 1976). Eggs hatched at 16–35 d post-oviposition; total lengths of larvae were 9–12 mm.

C. Larvae/Metamorphosis. This species is paedomorphic, and natural metamorphosis is unknown. Transformation has been induced artificially through use of thyroid hormone (Potter and Rabb, 1960).

D. Juvenile Habitat. Probably similar to adult habitat.

E. Adult Habitat. Completely aquatic. Found in mats of blue-green algae (*Lyngbya* sp.), under rocks, and in gravel substrate at water depths of <1 m to several meters. Water temperature is relatively constant at approximately 22°C throughout the year; experimental studies show a critical thermal maximum of 36–37°C (Berkhouse and Fries, 1995).

F. Home Range Size. Unknown.

G. Territories. Unknown.

H. Aestivation/Avoiding Desiccation. Aestivation is unknown.

I. Seasonal Migrations. Unlikely to occur.

J. Torpor (Hibernation). Active throughout the year.

K. Interspecific Associations/Exclusions. Little known; fountain darters (*Etheostoma fonticola*) are common in the same habitats in which this species is found.

L. Age/Size at Reproductive Maturity. Tupa and Davis (1976) noted size at sexual maturity as 19–23.5 mm SVL for males and 21 mm SVL for females. In captivity, eggs were first observed in females at 250 d of age.

M. Longevity. At least 3.7 yr in captivity.

N. Feeding Behavior. Prey consists primarily of invertebrates, particularly chironomids and amphipods (Tupa and Davis, 1976). Oligochaete worms, snails, and zooplankton also are fed in captivity.

O. Predators. Suspected predators include catfishes, centrarchid fishes, and crayfishes (Tupa and Davis, 1976).

P. Anti-Predator Mechanisms. Secretive. Although tails do not autotomize, individuals sometimes exhibit partially missing or partially regrown tails and limbs.

Q. Diseases. Unknown.

R. Parasites. Unknown.

4. Conservation.

The entire population of San Marcos salamanders has been estimated to be about 53,200 (USFWS, 1996b). They are listed as Threatened both by the State of Texas (www.tpwd.state.tx.us) and the federal government.

Eurycea naufragia Chippindale, Price, Wiens, and Hillis, 2000
GEORGETOWN SALAMANDER

Paul T. Chippindale

1. Historical versus Current Distribution.

Georgetown salamanders (*Eurycea naufragia*) were described by Chippindale, Price, Wiens, and Hillis (2000) from springs of the San Gabriel River drainage in the vicinity of Georgetown, Williamson County, Texas; they also provisionally included one spring and one cave population from further north in Williamson County in this species. A cave population that probably represents this species was recently discovered west of Georgetown (J. Reddell, personal communication). Most of the known populations were discovered recently; the few populations known prior to the work of Chippindale (1995) and Chippindale et al. (2000) had been considered peripheral isolates of Texas salamanders (*E. neotenes*; Sweet, 1978a, 1982). Georgetown salamanders are members of the "northern group" of Chippindale (1995, 2000) and Chippindale et al. (2000); this monophyletic group occurs northeast of the Colorado River in the Edwards Plateau region of central Texas. Based on molecular markers, this and other northern species are extremely divergent from *E. neotenes* and other *Eurycea* from the southern Edwards Plateau region (Chippindale et al., 2000).

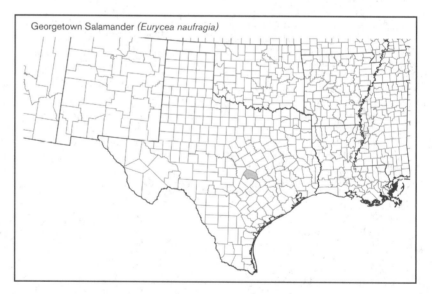

Georgetown Salamander (*Eurycea naufragia*)

2. Historical versus Current Abundance.

Little is known of the historical abundance of Georgetown salamanders. Several of the spring populations occur adjacent to Lake Georgetown, and it is likely that others were submerged when this manmade lake was created. Springs in Georgetown's San Gabriel Park, a historical locality for this species (Sweet, 1978a, 1982) have been heavily modified. One juvenile specimen

was discovered at this site in 1991 (Chippindale et al., 2000), but existence of this species at these springs appears precarious. Chippindale et al. (2000) briefly addressed some aspects of the conservation biology of this species; see Price et al. (1995) for a more detailed discussion.

3. Life History Features.

A. Breeding. Reproduction is aquatic.

i. Breeding migrations. Unlikely to occur.

ii. Breeding habitat. Same as adult habitat.

B. Eggs.

i. Egg deposition sites. Unknown; some other spring-dwelling species of central Texas *Eurycea* are thought to deposit eggs in gravel substrates.

ii. Clutch size. Unknown.

C. Larvae/Metamorphosis. Georgetown salamanders are paedomorphic, and natural metamorphosis is unknown.

D. Juvenile Habitat. Probably similar to adult habitat.

E. Adult Habitat. Completely aquatic. Georgetown salamanders are known only from the immediate vicinity of spring outflows, under rocks and leaves and in gravel substrate, and from two water-containing caves. Water temperatures in springs of the Edwards Plateau are relatively constant throughout the year and typically range from 18–20°C or slightly warmer near the fault zone at the Plateau's edge (Sweet, 1982). Sweet (1982) provided a comprehensive distributional analysis of the central Texas *Eurycea* and discussed hydrogeology of the region in relation to salamander distribution.

F. Home Range Size. Unknown.

G. Territories. Unknown.

H. Aestivation/Avoiding Desiccation. Unknown.

I. Seasonal Migrations. Unlikely to occur, although there may be seasonal variation in surface versus subsurface habitat use.

J. Torpor (Hibernation). Probably active throughout the year.

K. Interspecific Associations/Exclusions. Unknown.

L. Age/Size at Reproductive Maturity. Unknown. Average SVL of specimens measured by Chippindale et al. (2000) was 29.0 mm; all measured were thought to be sexually mature, but this was only verified for some of the specimens.

M. Longevity. Unknown.

N. Feeding Behavior. Prey probably consists mainly of small aquatic invertebrates, but no detailed feeding studies of this species have been conducted.

O. Predators. Unknown.

P. Anti-Predator Mechanisms. Secretive.

Q. Diseases. Unknown.

R. Parasites. Unknown.

4. Conservation.
Georgetown salamanders were described only recently (Chippindale et al., 2000) and most known populations were discovered within the past decade. Little is known of their historical abundance, although several populations occurred adjacent to Lake Georgetown, and it is likely that they were submerged when this man-made lake was created. Springs in Georgetown's San Gabriel Park have been heavily modified, and the existence of Georgetown salamanders at these springs appears precarious. They currently are considered as a Candidate species for federal listing (http://ecos.fws.gov), but they have not been protected by the state of Texas (www.tpwd.state.tx.us).

Eurycea neotenes Bishop and Wright, 1937
TEXAS SALAMANDER

Paul T. Chippindale

1. Historical versus Current Distribution.
Texas salamanders *(Eurycea neotenes)* were described by Bishop and Wright (1937) from a spring at Helotes, Bexar County, north of the city of San Antonio. In subsequent years, many spring and cave populations from throughout the Edwards Plateau region of central Texas were assigned to this species (e.g., B.C. Brown, 1942, 1950, 1967a; Schmidt, 1953; Conant, 1958a, 1975; Baker, 1961; Mitchell and Smith, 1972; Sweet, 1977a, 1978a,b, 1982, 1984; Dixon, 1987; Conant and Collins, 1991; Behler and King, 1998; Petranka, 1998). These identifications were based primarily on the high degree of morphological similarity among individuals from many populations, especially those inhabiting springs.

Chippindale (1995, 2000) and Chippindale et al. (1993, 1998, 2000) used molecular and morphological data to assess species boundaries in the central Texas *Eurycea*. Allozymes and mitochondrial sequences revealed extensive genetic subdivision within what had been considered *E. neotenes*, and Chippindale et al. (2000) restricted the distribution of this species to several springs at and near the type locality. Most references to *E. neotenes* in the literature involve populations that Chippindale et al. (2000) considered Fern Bank salamanders *(E. pterophila)* or members of the *E. latitans* or *E. troglodytes* species complexes. Other species that formerly were considered *E. neotenes* are Barton Springs salamanders *(E. sosorum)*, *E.* sp. 1 (Comal Springs), Jollyville Plateau salamanders *(E. tonkawae)*, Georgetown salamanders *(E. naufragia)*, and Salado salamanders *(E. chisholmensis)*. Based on phylogenetic analyses, *E. neotenes* is a member of the southeastern Edwards Plateau subgroup of Texas *Eurycea* (Chippindale, 1995, 2000; Chippindale et al., 2000), and appears to be restricted to several springs in Bexar and Kendall counties.

2. Historical versus Current Abundance.
Texas salamanders may be common at spring outflows, but their distribution appears to be limited and patchy.

3. Life History Features.
A. Breeding. Reproduction is aquatic.
i. Breeding migrations. Unlikely to occur.
ii. Breeding habitat. Unknown; closely related species are thought to deposit eggs in gravel substrates. Bogart (1967) described courtship and oviposition in Texas salamanders; in the laboratory, eggs were deposited on a variety of substrates.

B. Eggs.
i. Egg deposition sites. In the laboratory, eggs from an individual that might represent this species (see "Clutch size" below) were either free or attached to twigs and glass surfaces (Barden and Kezer, 1952).
ii. Clutch size. Barden and Kezer (1952) artificially induced egg laying in an individual that may represent this species; 12 eggs were produced in a 10-d period.

C. Larvae/Metamorphosis. Texas salamanders are paedomorphic, and natural metamorphosis is unknown. Kezer (1952a) described thyroxin-induced metamorphosis in individuals from a locality in Bexar County, which he considered to be Texas salamanders. It is likely that the populations with which he worked actually belong to the *E. latitans* complex, based on their geographic location. Barden and Kezer (1944) described eggs and egg-laying by a captive individual from one of these populations. Bogart (1967) described oviposition and egg development.

D. Juvenile Habitat. Probably similar to adult habitat.

E. Adult Habitat. Completely aquatic. Known only from the immediate vicinity of spring outflows, under rocks and leaves and in gravel substrate. Water temperatures in springs of the Edwards Plateau are relatively constant throughout the year and typically range from 18–20 °C or slightly warmer near the fault zone at the Plateau's edge (Sweet, 1982). Sweet (1982) provided a comprehensive distributional analysis of the central Texas *Eurycea* and discussed hydrogeology of the region in relation to salamander distribution.

F. Home Range Size. Unknown.

G. Territories. Unknown.

H. Aestivation/Avoiding Desiccation. In 1990, P. Chippindale, D. Hillis, A. Price, and D. Bell visited the type locality of this species, a spring at the headwaters of Helotes Creek, Bexar County. The landowner informed us that the spring had been dry for approximately 2 yr and had only started to flow again days earlier. We found dozens of extremely thin salamanders in the spring pool (some dead and dying); presumably they had retreated into subterranean habitat while the spring was dry. Although this may not constitute true aestivation, it indicates that this species can survive temporary drying of surface springs.

I. Seasonal Migrations. Unlikely to occur.

J. Torpor (Hibernation). Probably active throughout the year.

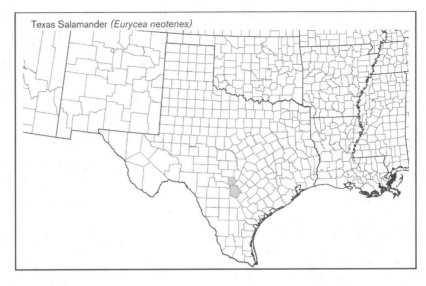
Texas Salamander (*Eurycea neotenes*)

K. Interspecific Associations/Exclusions. Unknown.

L. Age/Size at Reproductive Maturity. In a morphometric analysis of some Edwards Plateau *Eurycea* (Chippindale et al., 1993), average SVL of topotypical Texas salamanders that were presumed to be adult was 32.2 mm. However, Chippindale et al. (1993) probably did not include the smallest reproductively mature specimens in their analysis, and no rigorous studies of reproductive biology have been conducted for this species.

M. Longevity. Unknown.

N. Feeding Behavior. Prey probably consists mainly of small aquatic invertebrates, but no detailed feeding studies of this species have been conducted. Whiteworms were accepted in the laboratory (Bogart, 1967).

O. Predators. Unknown.

P. Anti-Predator Mechanisms. Secretive.

Q. Diseases. Hunsaker and Potter (1960) documented mortality due to "redleg" disease caused by infection with bacteria *(Pseudomonas hydrophila)* for a population in the vicinity of the type locality.

R. Parasites. Unknown.

4. Conservation.
Most references to Texas salamanders in the literature involve populations that are now considered members of the *E. latitans* or *E. troglodytes* species complexes. In fact, Texas salamanders appear to be restricted to several springs in Bexar and Kendall counties. These salamanders may be common at spring outflows, but their distribution appears to be limited and patchy. Despite this, they receive no protection by either the State of Texas (www.tpwd.state.tx.us) or by the federal government.

Eurycea pterophila Burger, Smith, and Potter, 1950
FERN BANK SALAMANDER

Paul T. Chippindale

1. Historical versus Current Distribution.
Fern Bank salamanders *(Eurycea pterophila)* were described originally from Fern Bank (Little Arkansas) Spring, Hays County, Texas, by Burger, Smith, and Potter (1950). Schmidt (1953) considered Fern Bank salamanders to be a subspecies of the supposedly widespread Texas salamanders *(E. neotenes)*, and Sweet (1978a,b) synonymized this taxon under *E. neotenes* without recognizing subspecies. Chippindale (1995), Chippindale (2000), and Chippindale et al. (2000) resurrected the name *E. pterophila* for populations of Edwards Plateau *Eurycea* in springs and caves of the Blanco River drainage of Blanco, Hays, and Kendall counties, Texas. Their status remains open to question, pending further studies of relationships among populations of southeastern Edwards Plateau *Eurycea*.

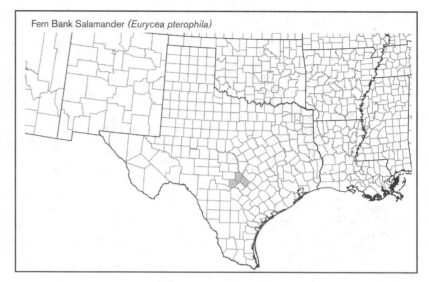

Fern Bank Salamander *(Eurycea pterophila)*

2. Historical versus Current Abundance.
Fern Bank salamanders may be common at spring outflows, but their distribution appears to be extremely limited and patchy.

3. Life History Features.

A. Breeding.
i. Breeding migrations. Unlikely to occur.
ii. Breeding habitat. Probably similar to adult habitat.

B. Eggs.
i. Egg deposition sites. Unknown; closely related species are thought to deposit eggs in gravel substrates.
ii. Clutch size. Unknown.

C. Larvae/Metamorphosis. Fern Bank salamanders are paedomorphic, and natural metamorphosis is unknown.

D. Juvenile Habitat. Probably similar to adult habitat; however, Conrads (1969) found that small juveniles often occurred in shallow (<1 cm) water.

E. Adult Habitat. Completely aquatic. Spring populations are known only from the immediate vicinity of spring outflows, under rocks and leaves, and in gravel substrate. This species also has been found in Grapevine Cave and T Cave, Blanco County (Chippindale et al., 2000). Water temperatures in springs of the Edwards Plateau are relatively constant throughout the year and typically range from 18–20 °C or slightly warmer near the fault zone at the Plateau's edge (Sweet, 1982). Sweet (1978a, 1982) provided a comprehensive distributional analysis of the central Texas *Eurycea* and discussed the hydrogeology of the region in relation to salamander distribution.

F. Home Range Size. Unknown.

G. Territories. Unknown.

H. Aestivation/Avoiding Desiccation. Hamilton (1973) reported that the population at the type locality survived an episode in which the springs ceased to flow. Although this may not constitute true aestivation, it indicates that this species

can survive temporary drying of surface outflows.

I. Seasonal Migrations. Unlikely to occur.

J. Torpor (Hibernation). Probably active throughout the year.

K. Interspecific Associations/Exclusions. Unknown.

L. Age/Size at Reproductive Maturity. In a morphometric analysis of some Edwards Plateau *Eurycea* (Chippindale et al., 1993), the average SVL of Fern Bank salamanders that were presumed to be adults was 30.6 mm for specimens from the type locality and 36.6 mm from another site, Boardhouse Spring. However, Chippindale et al. probably did not include the smallest reproductively mature specimens in their analysis, and no rigorous studies of reproductive biology have been conducted for this species. Bogart (1967) described oviposition in the laboratory.

M. Longevity. Unknown.

N. Feeding Behavior. Prey probably consist mainly of small aquatic invertebrates, but no detailed feeding studies of this species have been conducted.

O. Predators. Unknown.

P. Anti-Predator Mechanisms. Secretive.

Q. Diseases. Sweet (1978b) demonstrated that occurrence of short digits, one of the features that Burger et al. (1950) considered diagnostic in their original description of this species, was probably the result of tissue loss due to infection by bacteria *(Aeromonas* sp.).

R. Parasites. Unknown.

4. Conservation. The status of Fern Bank salamanders remains open to question, pending further studies of relationships among populations of southeastern Edwards Plateau *Eurycea*. Fern Bank salamanders may be common at spring outflows, but their distribution appears to be extremely limited and patchy. They are not listed by either the State of Texas or the federal government.

Eurycea quadridigitata (Holbrook, 1842)
DWARF SALAMANDER

Ronald M. Bonett, Paul T. Chippindale

1. Historical versus Current Distribution.
Dwarf salamanders *(Eurycea quadridigitata)* were described (as *Salamandra quadridigitata*) by Holbrook (1842); the type locality was restricted by Schmidt (1953) to the vicinity of Charleston, South Carolina. Cope (1869) transferred this taxon to the genus *Manculus* (as *M. quadridigitatus*). Dunn (1923) considered *Manculus* a junior synonym of *Eurycea*, and Wake (1966) concurred. However, Mittleman (1947, 1967) supported recognition of the genus *Manculus*, and recognized three subspecies: *M. q. quadridigitatus* throughout the southeast and much of the Gulf Coastal Plain; *M. q. paludicolus* from Louisiana and eastern Texas; and *M. q. uvidus* from northeastern Texas, western Arkansas, and southwestern Missouri. Nearly all recent authors consider this taxon a single species—*Eurycea quadridigitata*. Molecular work supports inclusion of this taxon in *Eurycea* under the Linnean system of nomenclature (Chippindale et al., 2000; C. Hass and R. Highton, unpublished data; P.T.C., unpublished data). In the broad sense, dwarf salamanders have a wide range from North Carolina south into much of peninsular Florida, and west into eastern Texas and southern Arkansas; they occur primarily in the Atlantic and Gulf Coastal Plain regions (see map in Petranka, 1998). However, recent molecular work has revealed deep divergences within this "species," and also suggests that the taxon may not be monophyletic (Chippindale et al., 2000; C. Hass and R. Highton, unpublished data; P.T.C., unpublished data). *Eurycea quadridigitata* probably consists of at least four distinct species, some of which occur sympatrically but apparently do not interbreed (R. Highton, personal communication). In fact, Harrison and Guttman (2003) have recently described Chamberlain's dwarf salamanders *(E. chamberlaini)* from an isolated area in the western Piedmont of South Carolina, the lower Piedmont of North Carolina, the upper Coastal Plain of South Carolina, and the central portion of the Coastal Plain in North Carolina (see account, this volume). Even in the area of the type locality, at least two species are present and sometimes sympatric (R. Highton, personal communication). A formal taxonomic treatment of this complex is expected in the near future. Given this situation, the information for "*E. quadridigitata*" summarized here almost certainly represents a composite of data for several, perhaps distantly related, species.

2. Historical versus Current Abundance.
Generally unknown, but numbers are undoubtedly lower in areas impacted by human activities (Petranka, 1998). Local

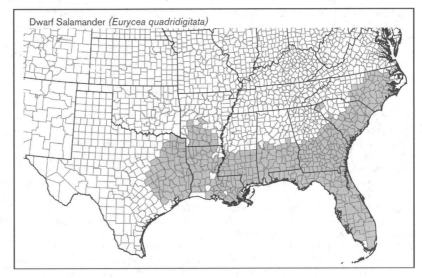

Dwarf Salamander *(Eurycea quadridigitata)*

abundance of adult members of the *E. quadridigitata* complex was monitored during a 16-yr study (1979–94) of a Carolina bay in South Carolina (Semlitsch et al., 1996). Breeding adult numbers were not influenced by seasonal rainfall amounts, and recruitment of juveniles was not influenced by larval densities. Juvenile recruitment was, however, influenced by the number of breeding adults. Members of this complex appear to be common in suitable habitat (Petranka, 1998), but given the current confusion regarding species boundaries and the probable occurrence of multiple, sometimes sympatric species, estimates of distribution and abundance should be considered questionable.

3. Life History Features.
   A. Breeding. Reproduction is aquatic.
   i. Breeding migrations. In South Carolina, dwarf salamanders migrate to breeding ponds from August–October (Gibbons and Semlitsch, 1991) and have been collected during migration until late November (McMillan and Semlitsch, 1980). South Carolina breeding populations can include ≤10,000 adults, with approximately equal representation of males and females (Gibbons and Semlitsch, 1991). In South Carolina, breeding migrations occur primarily at night, but individuals will also migrate crepuscularly and diurnally (Semlitsch and Pechmann, 1985; Gibbons and Semlitsch, 1991). It has been suggested that dwarf salamanders avoid predation and desiccation during diurnal migrations by traveling beneath the leaf litter (Semlitsch and Pechmann, 1985). Individuals were collected during apparent breeding migrations from late autumn to early winter in east-central Alabama (Trauth, 1983).
   ii. Breeding habitat. In the Atlantic Coastal Plain of South Carolina, dwarf salamanders use lentic habitats such as

Carolina bays and ephemeral ponds (Gibbons and Semlitsch, 1991). Similarly, in Florida, dwarf salamanders were found only to be associated with ponds (Goin, 1951). In east-central North Carolina, however, adult dwarf salamanders and their eggs have been found along small streams (Brimley, 1923), suggesting that members of this complex will also breed in lotic habitats (Goin, 1951).
   B. Eggs.
   i. Egg deposition sites. Females attach their eggs singly to vegetation and other substrates such as twigs, rootlets, and debris. Eggs usually are laid so that they are associated with flowing water (Brimley, 1923; Carr, 1940a; Harrison, 1973; Trauth, 1983). Eggs may also be laid in shallow depressions beneath cover objects along wetland margins (Goin, 1951) or in dry depressions that will fill with spring rains (Taylor et al., 1988).
   ii. Clutch size. Developmental data have been collected on members of the *E. quadridigitata* complex from several localities throughout the range. Clutches containing from 7–48 eggs are oviposited by South Carolina females from November–December when breeding ponds fill (Gibbons and Semlitsch, 1991). Females from east-central Alabama populations had previtellogenic follicles within their ovaries from March–September and enlarged ovarian follicles (14–59) from October–February; oviposition probably occurs over a broad time period in this region (Trauth, 1983). Numerous eggs have been discovered in the field in east-central North Carolina during early February (Brimley, 1923). In Florida, single clutches containing 20 eggs and 62 eggs were recovered from the field in November and February, respectively (Goin, 1951). This lack of synchrony in oviposition by populations in different locations may actually reflect interspecific differences among members of this complex. In the lab, eggs

from Florida populations took approximately 3–4 wk to hatch, with hatchlings measuring from 7.5–8.3 mm in total length (Goin, 1951).

C. Larvae/Metamorphosis. As with all hemidactyliines, members of the *E. quadridigitata* complex have an aquatic larval stage (Petranka, 1998). Goin (1951) described larvae from Florida populations to be morphologically intermediate between typical pond-type and stream-type salamander larvae, with a dorsal caudal fin that extends anteriorly to the mid dorsum (pond characteristic) and relatively few gill filaments (stream characteristic). A photograph of a larval dwarf salamander from southeastern North Carolina (R. W. VanDevender, in Petranka, 1998) matches the description of dwarf salamander larvae from Florida by Goin (1951). Larvae of members of the *E. quadridigitata* complex have been found to inhabit temporary ponds and Carolina bays in South Carolina (Semlitsch, 1980a; Taylor et al., 1988; Gibbons and Semlitsch, 1991). Additionally, it is likely that larvae develop in the aquatic habitats immediately adjacent to oviposition sites. This suggests that hammock ponds (Goin, 1951) and streams (Brimley, 1923) should also be included as larval habitats for members of this complex. It is unclear whether these reported differences in larval habitat reflect interspecific differences among members of this complex or the ability of larvae of these species to use a variety of aquatic habitats.

i. *Length of larval stage.* Most larvae metamorphose 2–6 mo after hatching (Brimley, 1923; Harrison, 1973; Semlitsch, 1980a; Petranka, 1998). Harrison (1973) and Semlitsch (1980a) found small larvae in January and February in South Carolina. Mount (1975; see also Petranka, 1998) noted small larvae in pools and ditches in March in Alabama.

ii. *Larval requirements.*

a. *Food.* According to Petranka (1998), larvae are benthic feeders. Prey include small invertebrates such as zooplankton, ostracods, and insect larvae (Taylor et al., 1988; see also Petranka, 1998).

b. *Cover.* Larvae are likely benthic.

iii. *Larval polymorphisms.* Unknown and unlikely.

iv. *Features of metamorphosis.* Newly metamorphosed dwarf salamanders have been found from April (Harrison, 1973) to early July (Semlitsch, 1980a; Taylor et al., 1988; see also Petranka, 1998).

v. *Post-metamorphic migrations.* Unknown and unlikely.

vi. *Neoteny.* Unknown.

D. Juvenile Habitat. Similar to adult habitat.

E. Adult Habitat. Most studies of adult members of the *E. quadridigitata* complex have involved individuals intercepted during migrations to and from breeding ponds. Therefore, aside from general notes on habitats where adult dwarf salamanders have been uncovered in the field, little is known about their terrestrial habitat characteristics outside of the breeding season. In general, members of the *E. quadridigitata* complex have been found beneath cover objects at the edges of ponds and swamps as well as in seeps and amongst leaf litter in springs (Mount, 1975; Petranka, 1998).

F. Home Range Size. Unknown. However, Carr (1940a) notes that dwarf salamanders from Florida can be found at considerable distances from aquatic habitats outside of the breeding season.

G. Territories. Unknown.

H. Aestivation/Avoiding Desiccation. Aestivation is unknown.

I. Seasonal Migrations. The only known migrations for the species of the *E. quadridigitata* complex are apparently for breeding purposes.

J. Torpor (Hibernation). Unknown.

K. Interspecific Associations/Exclusions. Studies in South Carolina have examined diel patterns of adult migratory activity (Semlitsch and Pechmann, 1985), and larval trophic relations of dwarf salamanders and other sympatric salamanders such as eastern newts *(Notophthalmus viridescens)* and various ambystomatids (Taylor et al., 1988). These studies have revealed a host of interspecific differences, but none that can be directly attributed to interactions between the species examined. Some members of the *E. quadridigitata* complex are known to occur sympatrically and are distinguishable based on both external morphology and molecular markers (R. Highton, personal communication). Other plethodontids that have been found syntopic with members of the *E. quadridigitata* complex in southern Mississippi and southeastern Louisiana include three-lined salamanders *(E. guttolineata)*, southern red salamanders *(Pseudotriton ruber vioscai)*, and spotted dusky salamanders *(Desmognathus conanti;* R. M. B., personal observation).

L. Age/Size at Reproductive Maturity. In east-central North Carolina, dwarf salamanders are reported to hatch in March and transform 2–3 mo later (Brimley, 1923). Larval periods of approximately 5–6.5 mo are reported for South Carolina populations (Harrison, 1973; Semlitsch, 1980a), and both males and females reach sexual maturity about 8–9 mo after hatching (Semlitsch, 1980a). Semlitsch (1980a) proposes that males are likely to reproduce during their first year, while females likely take an additional year for ova to develop. There has been considerable variation in size at maturity reported among populations of this complex, and the different species that currently are considered *E. quadridigitata* may mature at different sizes. Monitoring natural populations in South Carolina, Semlitsch (1980a) found larval periods to last from 5–6 mo when individuals attain lengths of 20.7–25.6 mm SVL. Other measurements of newly metamorphosed individuals from South Carolina showed a smaller size at metamorphosis (17–20 mm SVL; Harrison, 1973).

M. Longevity. Unknown.

N. Feeding Behavior. In South Carolina, dwarf salamander larvae feed diurnally, with small individuals taking primarily zooplankton and large larvae consuming small invertebrates (Taylor et al., 1988). Metamorphosed individuals in South Carolina were found to feed upon acarinans, arachnids, coleopterans, collembolans, diplopods, dipterans, hemipterans, homopterans, hymenopterans, and pseudoscorpionids (McMillan and Semlitsch, 1980). Acarinans, arachnids, coleopterans, collembolans, hymenopterans, and larval insects were recovered from the stomachs of terrestrial individuals from southern Georgia (Powders and Cate, 1980). The individuals examined by Powders and Cate (1980) were from two distinct size classes, but showed no differences in preferred prey items. Individuals from Florida populations were found to feed upon coleopterans (larval and adult), annelids, and amphipods (Carr, 1940a).

O. Predators. Examination of stomach contents of 122 pig frogs *(Rana grylio)* from southwest Georgia revealed four dwarf salamanders (Lamb, 1984). This is the only evidence of predation on dwarf salamanders, although it has been speculated that birds, snakes, and large invertebrates are likely predators (Petranka, 1998).

P. Anti-Predator Mechanisms. Unknown.

Q. Diseases. Unknown.

R. Parasites. Unknown.

4. Conservation.

Members of the *E. quadridigitata* complex currently are given no special protective status. Given the existence of several putative, currently undescribed species in the group, their conservation status should be re-examined after species boundaries and geographic ranges have been accurately delineated.

Eurycea rathbuni Stejneger, 1896
TEXAS BLIND SALAMANDER
Paul T. Chippindale

1. Historical versus Current Distribution.
Texas blind salamanders *(Eurycea rathbuni)* were first described by Stejneger (1896; as *Typhlomolge rathbuni*) from a 58 m-deep artesian well drilled in 1895 in the city of San Marcos, Hays County, Texas, on what is now the campus of Southwest Texas State University. These salamanders are known from several caves, wells, and pipes that intersect the San Marcos Pool of the Edwards Aquifer in San Marcos and are unlikely to range beyond this region

(for more detailed discussions of this species with respect to hydrogeology of the region, see also Uhlenhuth, 1919; Russell, 1976; Longley, 1978; Potter and Sweet, 1981; Chippindale et al., 2000). Status of the genus *Typhlomolge* as distinct from *Eurycea* has been controversial (e.g., Mitchell and Reddell, 1965; Wake, 1966; Mitchell and Smith, 1972; Potter and Sweet, 1981), but it is now clear that "*T.*" *rathbuni* and its presumed sister species "*T.*" *robusta* are phylogenetically nested within the central Texas *Eurycea* and should be considered species of *Eurycea* (Chippindale, 1995, 2000; Chippindale et al., 2000). Petranka (1998) recently followed this taxonomic approach. *Eurycea rathbuni* plus *E. robusta*, and a newly described species from Austin (Hillis et al., 2001) appear to represent the sister group to other central Texas *Eurycea* from south of the Colorado River.

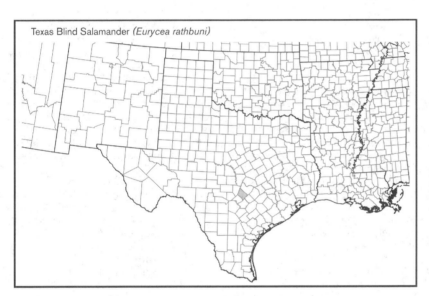

Texas Blind Salamander *(Eurycea rathbuni)*

2. Historical versus Current Abundance.
Densities of cave-dwelling populations of the central Texas *Eurycea* are difficult to assess. For several years after drilling of the artesian well at San Marcos (in the late 1800s), over 100 individuals emerged annually; this number soon dropped to a few per year (Uhlenhuth, 1921). Individuals of this species still appear common in outflows of Diversion Spring, a pipe that carries outflows from the Edwards Aquifer at San Marcos (Aquarena) Springs; most individuals that emerge probably are eaten by fishes, but salamanders can be captured if a net is placed over the pipe's outflow. However, numbers collected vary widely from year to year; currently, most individuals recovered are juveniles (J. Fries, personal communication). The National Fish Hatchery at San Marcos, Texas, currently maintains approximately 180 individuals, almost all obtained as juveniles, captured over a 4-yr period by

netting two wells that provide outflow from the Edwards Aquifer (J. Fries, personal communication). Individuals of this species can reliably be observed at a tiny cave opening into the Edwards Aquifer in San Marcos (Russell, 1976; personal observations, early 1990s). When the cave floods, Texas blind salamanders are active on the surface in broad daylight (personal observations).

3. Life History Features.
    A. Breeding. Reproduction is aquatic.
    i. Breeding migrations. Unknown.
    ii. Breeding habitat. Unknown in nature. However, this species has laid eggs on numerous occasions in captivity at the Dallas Aquarium at Fair Park, the Cincinnati Zoo, the Aquarena Center (San Marcos), and the San Marcos National Fish Hatchery and Technology Center (L. Ables, Dallas Aquarium at Fair Park, personal communication).
    B. Eggs.
    i. Egg deposition sites. In nature, unknown.
    ii. Clutch size. Unknown.
    C. Larvae/Metamorphosis. Texas blind salamanders are paedomorphic, and natural metamorphosis is unknown. Attempts to artificially induce transformation through use of thyroid hormone resulted in only partial metamorphosis (Dundee, 1957). Captive-hatched individuals at the Dallas Aquarium at Fair Park grew from approximately 10 mm to 80–90 mm TL in about 14–16 mo (L. Ables, personal communication). Grobman (1957) investigated the thyroid gland of this species, and Sever (1985) provided information of sexual dimorphism of the cloacal glands of this species.
    D. Juvenile Habitat. Probably similar to adult habitat.

    E. Adult Habitat. Completely aquatic. Found in caverns of the San Marcos Pool of the Edwards Aquifer, where they have been observed climbing rock surfaces or swimming in open water. Water temperature of the spring outflows of this region of the Edwards Aquifer is relatively constant at approximately 21–21.5 °C throughout the year (Berkhouse and Fries, 1995).
    F. Home Range Size. Unknown.
    G. Territories. Unknown.
    H. Aestivation/Avoiding Desiccation. Unknown.
    I. Seasonal Migrations. Unknown.
    J. Torpor (Hibernation). Probably active throughout the year.
    K. Interspecific Associations/Exclusions. No other salamanders are known from the subterranean habitat of this species, although San Marcos salamanders *(E. nana)* are abundant in springs directly above the caves occupied by this species.
    L. Age/Size at Reproductive Maturity. Petranka (1998) listed the size of adults as 90–135 mm in total length. Brandon (1971b) found that males mature at about 40 mm SVL, and females at about 40–50 mm SVL. Captive-raised females at the Dallas Aquarium at Fair Park displayed visible eggs (and deposited infertile eggs) at approximately 35 mm SVL (L. Ables, personal communication). Courtship and reproduction was documented by Belcher (1988), and captive reproduction has occurred at the Dallas Aquarium at Fair park, the Cincinnati Zoo, the Aquarena Center (San Marcos), and the San Marcos National Fish Hatchery and Technology Center (L. Ables, personal communication). Reproduction in the wild probably occurs throughout the year (Longley, 1978).
    M. Longevity. The longest officially recorded period in captivity was 10 yr, 4 mo, with this individual still living at the time of the report (Snider and Bowler, 1992).
    N. Feeding Behavior. Prey probably consists primarily of subterranean invertebrates; Longley (1978) reported amphipods, snails, and cave shrimp *(Palaemonetes antrorum)* as food items. One individual was seen skimming the water surface in a cave, perhaps seeking insects on the water's surface (personal observations). An individual outside a flooded cave was seen feeding on an earthworm (personal observation), and captive specimens will eat meat (Norman, 1900). This species will readily enter traps baited with potato peels; this may be due to the bait's attraction for aquatic invertebrates (Russell, 1976; personal observations).
    O. Predators. Unknown.
    P. Anti-Predator Mechanisms. Unknown.
    Q. Diseases. Unknown.
    R. Parasites. Unknown.

4. Conservation.
Texas blind salamanders are known from several caves, wells, and pipes that intersect the San Marcos Pool of the Edwards Aquifer in San Marcos and are unlikely to range beyond this region. Populations have been lost, and they are listed as Endangered by both the State of Texas (www.tpwd.state.tx.us) and the federal government (http://ecos.fws.gov).

Eurycea robusta Longley, 1978
BLANCO BLIND SALAMANDER
Paul T. Chippindale

1. Historical versus Current Distribution.
Blanco blind salamanders *(Eurycea robusta)* were described based on a single specimen collected in 1951, when workers drilled a hole for water in the bed of the then-dry Blanco River just east of San Marcos, Hays County, Texas (additional specimens were lost; apparently some were eaten by a heron shortly after capture). Authorship of the name is problematic. Potter (1963) described the species as *Typhlomolge robusta* in a Master's thesis, but this description is invalid under current rules of zoological nomenclature. Longley (1978) must be credited with description of the species, although he did so unintentionally in a government report. Potter and Sweet (1981) redescribed this taxon and discussed biogeographic history of salamanders in the Edwards Plateau region and status of the genus *Typhlomolge* (note that Dixon [1987] argued that Potter and Sweet should be credited with description of this species). Potter and Sweet (1981) and Russell (1976) provided evidence that the geological formation in which this species occurs is hydrologically isolated from that in which the geographically proximal species *E.* (formerly *T.*) *rathbuni* is found, supporting recognition of this population as a distinct species. Based on molecular data, Chippindale (1995, 2000) and Chippindale et al. (2000) agreed with Mitchell and Reddell (1965) and Mitchell and Smith (1972), who recommended synonymy of the genus *Typhlomolge* under *Eurycea*. Because *"T."* *robusta* appears to be closely related to *"T."* *rathbuni* based on morphology (no molecular data are available for *"T." robusta*), this renders the species *E. robusta*. Petranka (1998) followed this designation. No further specimens have been collected.

2. Historical versus Current Abundance.
Nothing is known; a 1995 petition to list this species as Federally Endangered was rejected due to lack of information on its status and distribution (O'Donnell, 1998).

3. Life History Features.
  A. Breeding. Reproduction is aquatic.
    i. Breeding migrations. Unknown.
    ii. Breeding habitat. Unknown.

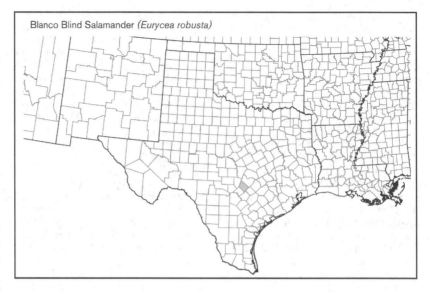

Blanco Blind Salamander *(Eurycea robusta)*

  B. Eggs.
    i. Egg deposition sites. Unknown.
    ii. Clutch size. Unknown.
  C. Larvae/Metamorphosis. Unknown.
  D. Juvenile Habitat. Unknown.
  E. Adult Habitat. Completely aquatic and subterranean; nothing else is known.
  F. Home Range Size. Unknown.
  G. Territories. Unknown.
  H. Aestivation/Avoiding Desiccation. Unknown.
  I. Seasonal Migrations. Unknown.
  J. Torpor (Hibernation). Unknown.
  K. Interspecific Associations/Exclusions. Unknown.
  L. Age/Size at Reproductive Maturity. Unknown; the type specimen is a reproductively mature female, 57.1 mm SVL, 100.8 mm total length.
  M. Longevity. Unknown.
  N. Feeding Behavior. Unknown.
  O. Predators. Unknown (excluding the heron mentioned above, certainly not a natural predator).
  P. Anti-Predator Mechanisms. Unknown.
  Q. Diseases. Unknown.
  R. Parasites. Unknown.

4. Conservation.
Little is known about the conservation status of Blanco blind salamanders. A 1995 petition to list this species as Federally Endangered was rejected due to lack of information on its status and distribution. However, the State of Texas lists them as Threatened (www.tpwd.state.tx.us).

Eurycea sosorum Chippindale, Price, and Hillis, 1993
BARTON SPRINGS SALAMANDER
Paul T. Chippindale, Robert Hansen

1. Historical versus Current Distribution.
Barton Springs salamanders *(Eurycea sosorum)* were described by Chippindale, Price, and Hillis (1993); the type locality is Barton Springs pool, a spring-fed swimming hole in the city of Austin, Travis County, Texas. This species is known only from the pool and three other springs that are immediately adjacent (Chippindale et al., 1993; City of Austin, 1997; Hansen et al., 1998). The spring outflows inhabited by this species are fed by the Barton Springs segment of the Edwards Aquifer. *Eurycea sosorum* was first recognized as a distinct, undescribed species by Sweet (1978a, 1984), based on its morphological differentiation from other surface and subterranean species of central Texas *Eurycea*. Chippindale et al. (1993) also demonstrated that this species is distinct based on morphology and allozymes. Chippindale et al. (2000) support recognition of this species and determined its relationships to other central Texas *Eurycea* using allozyme and mitochondrial DNA sequence data. Hillis et al. (2001) confirmed the phylogenetic position of this species using additional sequence data. *Eurycea sosorum* is a member of the "southeastern" subset of the "southern group" of Chippindale (1995, 2000) and Chippindale et al. (2000); this monophyletic group occurs south of the Colorado River in the Edwards Plateau region of central Texas.

2. Historical versus Current Abundance.
Little is known of the historical abundance of this species. The first known specimens were collected in 1946 by B.C. Brown from among plants in Barton Springs Pool (Chippindale et al., 1993). Observations by others (summarized by Chippindale et al., 1993) indicate that this species was abundant in the 1960s and 1970s. Considerable evidence (summarized by Chippindale et al., 1993; O'Donnell, 1994, 1997; Hansen et al., 1998; Hillis et al., 2001) indicates that the population underwent a major decline in the 1980s–90s, probably due in part to cleaning procedures used by the

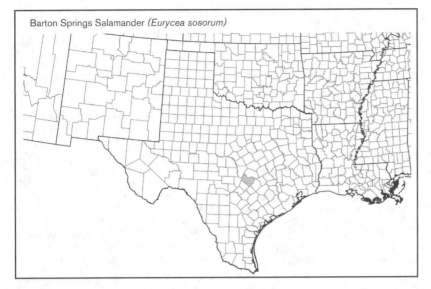

Barton Springs Salamander (*Eurycea sosorum*)

City of Austin at Barton Springs Pool. The City was made aware of the problems in the early 1990s and has since cooperated closely with biologists and conservation agencies to balance the need for pool maintenance with protection of Barton Springs salamanders. Details of the City's actions and the conservation history of this species are provided by Chippindale and Price (this volume, Part One). Abundance of salamanders appeared to increase substantially following modification of pool maintenance procedures, but the number of individuals located have been highly variable from year to year and the most recent data suggest a decline in numbers in 2000. Considerable concern remains regarding the impacts of human activities on water quality in the Barton Springs Aquifer, summarized by Chippindale and Price (this volume, Part One).

3. Life History Features.

   A. Breeding. Reproduction is aquatic.

   i. Breeding migrations. Unlikely to occur.

   ii. Breeding habitat. Unknown; a subset of adult habitat.

   B. Eggs.

   i. Egg deposition sites. Unknown in the wild; some other spring-dwelling species of central Texas *Eurycea* are thought to deposit eggs in gravel substrate. In captivity, oviposition has been observed at the Dallas Aquarium and the city of Austin facility. Females appear to deposit the eggs randomly on cobble, gravel, aquatic macrophytes, and even the glass sides and bottom of the aquaria. Gravid females may retain the eggs for periods longer than 12 mo (L. Ables, D. Chamberlain, personal communication). With some gravid females, oviposition does not occur and the eggs are resorbed by the female.

   ii. Clutch size. Twenty-nine ovipositions have been reported in captivity as of August 2000. The Dallas Aquarium has reported 24 egg-laying events. Of these, 20 have occurred in one particular tank containing one male and four females. In captivity, eggs hatch in approximately 25–35 d (L. Ables, personal communication). The high mortality rate of small juveniles at the Dallas Aquarium has limited the number of surviving young Barton Springs salamanders to about 50 juveniles (L. Ables, personal communication). The City of Austin has recorded three ovipositions of 29, 26, and 28 eggs with hatch rates of 0%, 27%, and 7%, respectively. Survival rates for the three ovipositions are 0%, 11%, and 4% (City of Austin, unpublished data). The San Antonio Zoo reported one oviposition of 18 eggs and a 0% hatch rate (G. Stettner, personal communication). The USGS's Midwest Science Center reported one oviposition of 29 eggs with a 10% hatch rate (J. Dwyer, personal communication). Clearly, further study is necessary to determine what cues (if any) trigger breeding and what conditions are optimal for development of eggs and young. The City of Austin is currently expanding their captive breeding program, and the facility has been relocated to the University of Texas at Austin campus.

   C. Larvae/Metamorphosis. Barton Springs salamanders are paedomorphic, and natural metamorphosis is unknown.

   D. Juvenile Habitat. Probably similar to those of adults. City of Austin field studies (unpublished data) indicate that larvae, juveniles, and adults utilize similar substrate types (cobble, gravel, aquatic macrophytes).

   E. Adult Habitat. Completely aquatic. Barton Springs salamanders are known only from the vicinity of spring outflows, under rocks and leaves, and in gravel substrate at depths ranging from a few cm to about 5 m. The first known specimens were collected in 1946 among aquatic plants. Water temperature of the spring outflows remains relatively constant at about 20–22 °C throughout the year.

   A recent study conducted by Alan Plummer Associates, Incorporated (2000), for the City of Austin summarizes water quality conditions at Barton Springs Pool. The analysis was based on City of Austin and U.S. Geological Survey data and concluded the following: (1) no trends of increasing concentration were found for several of the parameters commonly associated with nonpoint source pollution—nutrients, total suspended solids, and pesticides/herbicides; (2) trends of increasing concentrations were found for conductivity, sulfate, turbidity, and total organic carbon; (3) a trend of decreasing dissolved oxygen concentration was identified; and (4) the ratio of total nitrogen-to-total phosphorus was found to vary widely over time (from ≤20 to ≥100) and no observable trend was identified.

   The City of Austin has also conducted contaminated sediment studies and analysis at Barton Springs Pool and sites in Barton Creek above and below the pool. Polycyclic aromatic hydrocarbons (PAHs) were detected in sediment in or near Barton Springs at levels that may have biological effects. High levels of several pesticides were also detected in sediment directly upstream of Barton Springs Pool. These pesticides include aldrin, DDD, DDE, DDT, BHC, endosulfan, endrin, heptachlor epoxide, heptachlor, and lindane. Heavy metals also were detected in sediment at sites in and near Barton Springs. These metals include arsenic, cadmium, chromium, copper, lead, and zinc (City of Austin, 1997).

   Barton Springs salamanders appear to be primarily surface dwelling, but may also use subterranean habitat; the extent of their occurrence underground is uncertain (Chippindale et al., 1993). Sweet (1982) provided a comprehensive distributional analysis of the central Texas *Eurycea* and discussed hydrogeology of the region in relation to salamander distribution.

   F. Home Range Size. Unknown.

   G. Territories. Unknown.

   H. Aestivation/Avoiding Desiccation. Unknown.

   I. Seasonal Migrations. Very unlikely to occur; recent field studies by City of Austin staff indicate that no seasonal variation occurs for habitat use.

   J. Torpor (Hibernation). Individuals have been found active throughout the year.

   K. Interspecific Associations/Exclusions. Partially sympatric with the recently discovered Austin blind salamander (*E. waterlooensis*), which inhabits the subterranean portion of the Barton Springs Aquifer (Hillis et al., 2001). Individuals of this new species occasionally wash out of the spring outflows where *E. sosorum* occur. Specimens of *Eurycea* have been collected at Barton Springs Pool, Eliza Spring, and Old Mill Spring. No specimens have been observed at the Upper Barton Springs site (City of Austin, unpublished data). *Eurycea waterlooensis* is much more closely related to Texas and Blanco blind salamanders *(E. rathbuni* and *E. robusta)* than to *E. sosorum* (Hillis et al., 2001).

L. Age/Size at Reproductive Maturity. Unknown. Average SVL of specimens measured by Chippindale et al. (1993) was 29.2 mm; all animals measured were thought to be sexually mature, but this was only verified for some of the specimens. Maximum size of specimens examined by Chippindale et al. (1993) was 36.5 mm SVL (62.6 mm TL).

M. Longevity. Individuals (still living) have been maintained for over 6 yr in captivity.

N. Feeding Behavior. Prey probably consists mainly of small aquatic invertebrates, especially amphipods *(Hyallela azteca),* which are abundant in the habitat of this species. Gut analyses have revealed the following prey items: mayfly larvae, midge larvae, ostracods, copepods, physid snails, planorbid snails, and leeches (City of Austin, unpublished data). Captive specimens have accepted amphipods, earthworms, brine shrimp, bloodworms, blackworms, mosquito larvae, and commercial fishfood pellets (Chippindale et al., 1993; L. Ables, D. Chamberlain, personal communication; personal observations). Small larvae have disappeared when kept in aquaria with large juveniles and adults, so cannibalism may occur (L. Ables, personal communication).

O. Predators. Bass and sunfish are known predators (City of Austin, unpublished data; D. Hillis, personal communication). Crayfish have been suggested as potential predators, but most local experts think that they are not major predators.

P. Anti-Predator Mechanisms. Secretive.

Q. Diseases. Unknown.

R. Parasites. Unknown.

4. Conservation.
While Barton Springs salamanders were abundant in the 1960s–70s, they underwent a major decline in the 1980s–90s, probably due in part to cleaning procedures used by the City of Austin at Barton Springs Pool. Abundance of salamanders appeared to increase substantially following modification of pool maintenance procedures, but the number of individuals located have been highly variable from year to year. The City of Austin has established a captive breeding program for this species, and considerable concern remains regarding the impacts of human activities on water quality in the Barton Springs Aquifer. Barton Springs salamanders have been listed as a Federally Endangered species since 1997 (O'Donnell, 1997).

Eurycea tonkawae Chippindale, Price, Wiens, and Hillis, 2000
JOLLYVILLE PLATEAU SALAMANDER

Paul T. Chippindale

1. Historical versus Current Distribution.
Jollyville Plateau salamanders *(Eurycea tonkawae)* were described by Chippindale, Price, Wiens, and Hillis (2000); the type locality is a spring at the margin of the Jollyville Plateau in the city of Austin, Travis County, Texas. Other spring populations of this species are known from the Jollyville Plateau and Brushy Creek areas of Travis and Williamson counties (Chippindale et al., 2000; Davis et al., 2001). Chippindale et al. (2000) provisionally considered populations from several caves in the area, including the recently discovered Buttercup Creek Cave system in the Cedar Park area of Williamson County, to represent this species. However, they emphasized that some of these cave forms may prove to be distinct species. Most of the known populations were discovered recently; the few populations known prior to the work of Chippindale (1995) and Chippindale et al. (2000) had been considered peripheral isolates of *E. neotenes* (Baker, 1961; B.C. Brown, 1967a,c; Sweet, 1978a, 1982). *Eurycea tonkawae* is a member of the "northern group" of Chippindale (1995, 2000) and Chippindale et al. (2000); this monophyletic group occurs northeast of the Colorado River in the Edwards Plateau region of central Texas. Based on molecular markers, this and other northern species are extremely divergent from *E. neotenes* and other *Eurycea* from the southern Edwards Plateau region (Chippindale et al., 2000).

2. Historical versus Current Abundance.
Little is known of the historical abundance of this species. Individuals may be common at some spring outflows. Krienke Spring, a site in the Brushy Creek drainage

of Williamson County from which a large series was collected (preserved at the Texas Memorial Museum, Austin, Texas), apparently was destroyed by quarrying operations in the 1960s (Sweet, 1978a). An office building recently was built directly above the one other known Brushy Creek locality (Chippindale et al., 2000). A morphologically unusual cave population on the Jollyville Plateau, which may represent this species (Salamander Cave; Sweet, 1978a), now lies beneath an apartment complex (J. Reddell, personal communication). Chippindale et al. (2000) briefly addressed some aspects of the conservation biology of this species; Price et al. (1995) provided a more detailed discussion. Currently, the City of Austin is conducting comprehensive studies of spring populations in the Jollyville Plateau region and formulating policies for protection of this species (Davis et al., 2001; unpublished data). This work involves detailed analyses of habitat and water quality parameters and their relationship to salamander distribution and abundance. Preliminary results indicate an inverse correlation between the degree of urbanization and salamander abundance at spring outflows.

3. Life History Features.
A. Breeding. Reproduction is aquatic.
i. Breeding migrations. Unlikely to occur.
ii. Breeding habitat. Unknown; a subset of the adult habitat.
B. Eggs.
i. Egg deposition sites. Unknown; some other spring-dwelling species of central Texas *Eurycea* are thought to deposit eggs in gravel substrate. The highest ratios of small juveniles to large juveniles and adults have been observed in March–August (City of Austin, 2000).
ii. Clutch size. Unknown.
C. Larvae/Metamorphosis. This species is paedomorphic, and natural metamorphosis is unknown.

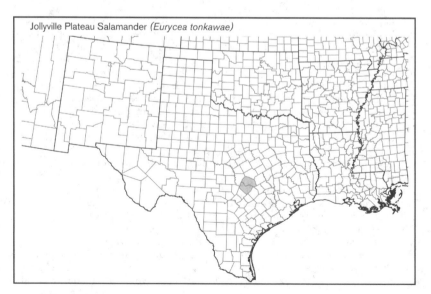

Jollyville Plateau Salamander *(Eurycea tonkawae)*

D. Juvenile Habitat. City of Austin personnel found that juveniles are more likely than adults to occur in shallow water near stream edges; they appear to prefer substrates with smaller particles than do large juveniles or adults (Davis et al., 2001).

E. Adult Habitat. Completely aquatic. Jollyville Plateau salamanders are known only from the vicinity of spring outflows, under rocks and leaves, and in gravel substrate. City of Austin personnel have found a positive correlation between abundance of large juvenile and adult salamanders and area of cobble available (Davis et al., 2001). Individuals in spring populations of the Jollyville Plateau region were found behaving (apparently) normally at water temperatures that ranged from 10.6–30.0 °C, and no correlation was found between water temperature and salamander abundance (Davis et al., 2001). Preliminary results indicate that salamander abundance decreases as the degree of urbanization increases (Davis et al., 2001). Nitrate levels are a particular concern; during a 2-yr study, the highest observed incidences of dead individuals and animals with spinal deformities occurred at one site (the type locality) for which nitrate levels were particularly high (Davis et al., 2001). Sweet (1982) provided a comprehensive distributional analysis of the central Texas *Eurycea* and discussed hydrogeology of the region in relation to salamander distribution.

F. Home Range Size. Unknown.

G. Territories. Unknown.

H. Aestivation/Avoiding Desiccation. Unknown. However, available evidence suggests that this species makes extensive use of subterranean aquatic habitat, especially when surface spring flows decrease (Davis et al., 2001). On several occasions, City of Austin personnel found apparently healthy adult salamanders with the return of spring flow at sites that had been dry for months.

I. Seasonal Migrations. Unlikely to occur, although there may be seasonal variation in surface versus subsurface habitat use (Davis et al., 2001).

J. Torpor (Hibernation). Animals are probably active throughout the year.

K. Interspecific Associations/Exclusions. Other species of salamanders are not known from the habitat of this species.

L. Age/Size at Reproductive Maturity. Unknown. Average length of specimens measured by Chippindale et al. (2000) was 30.5 mm SVL; all measured were thought to be sexually mature, but this was only verified for some of the specimens.

M. Longevity. Unknown.

N. Feeding Behavior. Prey probably consist mainly of small aquatic invertebrates, but no detailed studies of feeding in this species have been conducted. City of Austin personnel analyzed gut contents of individuals that were found dead and found a wide range of invertebrate prey items representative of the diversity found in salamander habitats (Davis et al., 2001).

Particularly common prey items included chironomid larvae, physid snails, copepods, and ostracods.

O. Predators. Circumstantial evidence suggests that centrarchid fishes may substantially reduce numbers of salamanders at some sites (Davis et al., 2001).

P. Anti-Predator Mechanisms. Secretive.

Q. Diseases. Unknown.

R. Parasites. Analyses of specimens found dead revealed encysted metazoan parasites (species unknown) in ova, pancreas, gut, and skeletal muscle. One individual had nematode parasites in the intestine (Davis et al., 2001).

4. Conservation.

Because Jollyville Plateau salamanders were described only recently and most of the known populations were discovered recently, little is known of the historical abundance of this species. Individuals may be common at some spring outflows, although development is known to have affected some populations and may be affecting others. Currently, the City of Austin is conducting comprehensive studies of spring populations in the Jollyville Plateau region and formulating policies for protection of this species. This work involves detailed analyses of habitat and water-quality parameters and their relationship to salamander distribution and abundance. Not surprisingly, preliminary results indicate an inverse correlation between the degree of urbanization and salamander abundance at spring outflows.

Eurycea tridentifera Mitchell and Reddell, 1965

COMAL BLIND SALAMANDER

Paul T. Chippindale

1. Historical versus Current Distribution.

Comal blind salamanders (*Eurycea tridentifera*) were described by Mitchell and Reddell

(1965) from Honey Creek Cave, Comal County, Texas. Sweet (1977b, 1978a, 1984) extended the distribution of this species to include several other caves in the Cibolo Sinkhole Plain of the southeastern Edwards Plateau region (Comal and Bexar counties). Chippindale et al. (2000) listed additional cave localities for this species in the same area and suggested that this species probably extends into Kendall County. Sweet (1978a, 1984) demonstrated that populations assigned to this species cluster together based on morphometric analyses. Some authors (Mitchell and Reddell, 1965; Bogart, 1967; Mitchell and Smith, 1972) have suggested that additional species may be present within what is considered *E. tridentifera*. There has been some difference of opinion regarding the generic allocation of this species. Wake (1966) considered it a member of the genus *Typhlomolge*, but despite some external and osteological similarity to the two species that were once included in this genus, it is not as closely related to either "*T.*" (now *Eurycea*) *rathbuni* or *robusta* as it is to other southeastern Edwards Plateau *Eurycea*. Bogart (1967), Mitchell and Smith (1972), and Sweet (1978a, 1984) discuss the placement of this species; Chippindale (1995, 2000), Chippindale et al. (2000), and Wiens et al. (2003) address relationships of *E. tridentifera* in an explicitly phylogenetic context. Generic boundaries in the central Texas hemidactyliine plethodontid salamanders have been controversial, but all species now are considered members of the genus *Eurycea* (see Chippindale, 1995, 2000; Chippindale et al., 2000; Chippindale and Price, this volume, Part One). All recent authors have considered this species a member of the genus *Eurycea*.

2. Historical versus Current Abundance.

Assessment of abundance of Texas cave *Eurycea* is difficult, so no reliable assessments of past versus current abundance

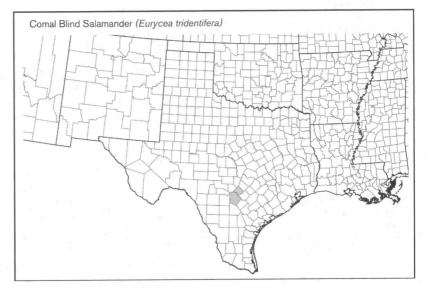

Comal Blind Salamander (*Eurycea tridentifera*)

can be made. Individuals of this species appeared scarce during visits to the type locality in the early 1990s (personal observations), although Mitchell and Reddell (1965), Bogart (1967), and Sweet (1978a, 1984) were able to collect fairly large series at this site.

3. Life History Features.
A. Breeding. Reproduction is aquatic.

i. Breeding migrations. Unlikely to occur.

ii. Breeding habitat. Likely the same as adult habitat.

B. Eggs.

i. Egg deposition sites. Unknown; closely related species are thought to deposit eggs in gravel substrates.

ii. Clutch size. Bogart (1967) artificially induced hybridization between this species and Valdina Farms salamanders (*E. troglodytes*), and the resulting eggs developed and hatched. Approximately 7–18 mature ova are produced/clutch, and eggs hatch when embryos are about 7 mm SVL (Sweet, 1977b).

C. Larvae/Metamorphosis. Comal blind salamanders are paedomorphic, and natural metamorphosis is unknown. Bogart (1967) reported that Comal blind salamanders fail to metamorphose when treated with either thyroid hormone or pituitary implantation.

D. Juvenile Habitat. Probably same as adult habitat.

E. Adult Habitat. Comal blind salamanders are found on rock and mud substrates in caves. Water temperature in waters of the Edwards Plateau is relatively constant throughout the year and typically ranges from 18–20 °C (Sweet, 1982).

F. Home Range Size. Unknown.

G. Territories. Unknown.

H. Aestivation/Avoiding Desiccation. Unknown.

I. Seasonal Migrations. Unlikely to occur.

J. Torpor (Hibernation). Probably active throughout the year.

K. Interspecific Associations/Exclusions. At the outflow of the type locality, where Honey Creek Cave becomes a spring, surface *Eurycea* occur (Mitchell and Reddell, 1965; Bogart, 1967; Sweet, 1978a, 1984). Sweet (1978a, 1982, 1984) considered these to be Texas salamanders (*E. neotenes*). Chippindale (1995) and Chippindale et al. (2000) considered these a member of the *E. latitans* complex. At the cave entrance, individuals intermediate in morphology between *E. tridentifera* and the surface species have been found and hybridization between the two species seems likely (Sweet, 1978a, 1984).

L. Age/Size at Reproductive Maturity. Brandon (1971b) found that males are mature at 22 mm SVL and females at 25 mm; Sweet (1977b) reported maturity at 25–27 mm for males and 28–32 mm for females.

M. Longevity. Unknown.

N. Feeding Behavior. Prey probably consists mainly of small aquatic invertebrates, but no detailed feeding studies of this species have been conducted. Individuals often have large amounts of detritus in their stomachs, suggesting they may graze the substrate for tiny invertebrates (personal observations). Bogart (1967) found remains of insects in fecal matter and suggested that this species may eat bat guano. Specimens maintained by Bogart (1967) accepted liver as food.

O. Predators. Sweet (1978a, 1984) noted that *Eurycea* usually are absent from caves where fishes are present in the general area of the Edwards Plateau inhabited by Comal blind salamanders. Bogart (1967) observed crayfish at the type locality, Honey Creek Cave, and suggested these as possible predators.

P. Anti-Predator Mechanisms. Sweet (1978a, 1984) reported that Comal blind salamanders usually swim upward when disturbed. This contrasts with behavior of most other members of the group that swim toward the substrate or to cover when frightened. Sweet suggested that this may reflect the absence of fish predators in caves inhabited by Comal blind salamanders.

Q. Diseases. Unknown.

R. Parasites. Unknown.

4. Conservation.
Assessment of abundance of Texas cave *Eurycea*, including Comal blind salamanders, is difficult, so no reliable assessments of past versus current abundance can be made. While historical collectors were able to collect fairly large series of individuals (Mitchell and Reddell, 1965;, Bogart, 1967; and Sweet, 1978a, 1984), Comal blind salamanders appeared scarce during visits to the type locality in the early 1990s. This species is listed as Threatened by the State of Texas, although it has not attracted federal attention.

Eurycea troglodytes Baker, 1957
VALDINA FARMS SALAMANDER
Paul T. Chippindale

1. Historical versus Current Distribution.
Valdina Farms salamanders (*Eurycea troglodytes*) were described by Baker (1957) from Valdina Farms Sinkhole, Medina County, Texas. Sweet (1978a, 1984) demonstrated that this population includes individuals with a spectrum of morphological features, ranging from highly cave-associated morphologies most similar to those of Comal blind salamanders (*E. tridentifera*) to surface-like morphologies most similar to those of what he considered Texas salamanders (*E. neotenes*). He hypothesized that this range of morphologies was the result of hybridization between surface Texas salamanders and a cave-dwelling species, perhaps Comal blind

salamanders (note that the known range of Comal blind salamanders is far to the east of the type locality for Valdina Farms salamanders; Sweet suggested that Comal blind salamanders might have a more extensive subterranean range than was recognized). Most recent authors have not recognized *E. troglodytes* as a distinct species. However, Chippindale et al. (2000) found that salamanders from the Valdina Farms Sinkhole system were phylogenetically nested (based on mitochondrial DNA analysis) within a group of spring and cave populations of *Eurycea* in the southwestern Edwards Plateau region, where Valdina Farms Sinkhole is located. Combined analyses of allozyme and mitochondrial sequence data support monophyly of this group and reveal deep divergences among many populations. Chippindale (2000) and Chippindale et al. (2000) included all these southwestern populations in the *E. troglodytes* complex, but noted that additional undescribed species may exist. At present, the *E. troglodytes* complex encompasses a large and wide-ranging array of spring and cave populations in Bandera, Edwards, Gillespie, western Kerr, Medina, Real, and Uvalde counties. Populations from Val Verde County probably represent the *E. troglodytes* complex as well, but have not yet been examined for molecular markers. It is likely that salamanders from the *E. troglodytes* complex occur in Kinney County, but no populations have yet been discovered.

2. Historical versus Current Abundance.
Salamanders of the *E. troglodytes* complex are often abundant at spring outflows, especially in canyons of Bandera, Kerr, and Real counties. It is difficult to assess densities for cave populations. Construction of a diversion dam in the 1980s temporarily submerged Valdina Farms Sinkhole, the type locality of Valdina Farms salamanders, and introduced catfish and other predators. Subsequent surveys (Veni and Associates, 1987; G. Veni, personal communication) have failed to reveal any Valdina Farms salamanders, even in areas of the cave where they once were common.

3. Life History Features.
A. Breeding. Reproduction is aquatic.

i. Breeding migrations. Unlikely to occur.

ii. Breeding habitat. Likely the same as adult habitat.

B. Eggs.

i. Egg deposition sites. Unknown; closely related species are thought to deposit eggs in gravel substrate.

ii. Clutch size. Unknown.

C. Larvae/Metamorphosis. Most populations in this species complex are paedomorphic, and metamorphosis is unknown for these. However, natural metamorphosis has been observed in populations from several springs in the Sabinal River drainage of Bandera County (Bogart, 1967; Sweet,

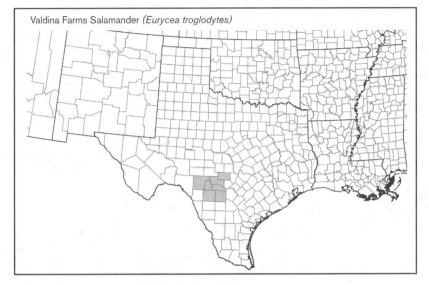

Valdina Farms Salamander (*Eurycea troglodytes*)

1977a). Sweet (1978a) observed a trans-formed individual in a Uvalde County cave and reported on a transformed individual observed by B.C. Brown in another cave in Uvalde County. Bogart (1967) induced transformation in animals from the type locality through implantation of frog pituitary glands.

D. Juvenile Habitat. Probably the same as adult habitat.

E. Adult Habitat. Individuals in most populations are completely aquatic, but Sweet (1977a, 1978b) found remains of terrestrial invertebrates in stomachs of some transformed individuals that were captured in water; this suggests that transformed animals may venture short distances onto land. Members of the *E. troglodytes* complex are known only from caves that contain water and the immediate vicinity of spring outflows; individuals in caves are often seen in the open on submerged rock or mud substrate, whereas individuals from spring populations are found under rocks and leaves and in gravel substrate. Water temperature in springs and caves of the Edwards Plateau is relatively constant throughout the year and typically ranges from 18–20 °C in the areas inhabited by this species (Sweet, 1982). Sweet (1982) provided a comprehensive distributional analysis of the central Texas *Eurycea* and discussed hydrogeology of the region in relation to salamander distribution.

F. Home Range Size. Unknown.

G. Territories. Unknown.

H. Aestivation/Avoiding Desiccation. Unknown.

I. Seasonal Migrations. Unlikely to occur.

J. Torpor (Hibernation). Probably active throughout the year.

K. Interspecific Associations/Exclusions. As described above, Sweet (1978a, 1984) considered the population at the type locality to be hybrids between *E. neotenes*

and *E. tridentifera*. Chippindale (1995) and Chippindale et al. (2000) regarded this as unlikely, based on molecular data and geographical considerations, and continued to recognize *E. troglodytes* as a distinct species.

L. Age/Size at Reproductive Maturity. Sweet (1977a, 1978a) found that in transforming populations (which he considered to be *E. neotenes*), sexual maturity is concurrent with transformation at 30–32 mm SVL. Bruce (1976) studied Kerr County populations, which almost certainly are part of this species complex (he also considered them *E. neotenes*). He found that individuals under 25 mm were invariably immature and concluded that males become reproductively active early in their second year, while females mature at the same time but first oviposit at 2 yr of age. Bogart (1967) artificially induced hybridization between *E. troglodytes* from the type locality and *E. tridentifera*.

M. Longevity. Unknown.

N. Feeding Behavior. Prey probably consists mainly of small aquatic invertebrates, but no detailed feeding studies of this species have been conducted. Sweet (1977a, 1978a) found remains of terrestrial collembolans and isopods in stomachs of transformed specimens. Transformed individuals maintained on wet moss by Bogart (1967) accepted *Drosophila* as food.

O. Predators. Unknown.

P. Anti-Predator Mechanisms. Spring-dwellers are secretive.

Q. Diseases. Sweet (1978a,b) found evidence of "red-leg" disease, thought to be caused by bacteria of the genus *Aeromonas*, in some populations (note that Sweet considered these populations to be *E. neotenes*).

R. Parasites. Unknown.

4. Conservation.
Valdina Farms salamanders were described from Valdina Farms Sinkhole, Medina

County, Texas. While it is difficult to assess densities for cave populations, salamanders of the *E. troglodytes* complex often are abundant at spring outflows, especially in canyons of Bandera, Kerr, and Real counties. Construction of a diversion dam in the 1980s temporarily submerged the type locality of Valdina Farms salamanders and introduced catfish and other predators. Subsequent surveys have failed to reveal any Valdina Farms salamanders, even in areas of the cave where they once were common. This species has not been given special conservation status by either the State of Texas or the federal government.

Eurycea tynerensis Moore and Hughes, 1939
OKLAHOMA SALAMANDER
Ronald M. Bonett

1. Historical versus Current Distribution.
Oklahoma salamanders (*Eurycea tynerensis*) were discovered in 1939 by ichthyologists seining in Tyner Creek, Adair County, Oklahoma (Moore and Hughes, 1939). The geographic distribution of Oklahoma salamanders currently is uncertain. They have been considered to range throughout the Springfield Plateau of northwestern Arkansas, northeastern Oklahoma, and southwestern Missouri (Dundee, 1965b; Conant and Collins, 1998; Petranka, 1998). Due to the subterranean tendencies of this neotenic species, the putative distribution described above includes drainage systems and physiographic regions that may potentially harbor populations of Oklahoma salamanders (Dundee, 1965b). The distribution of known localities, however, appears to be considerably smaller than their proposed range. The entire distribution of Oklahoma salamanders is nested within that of the extremely similar gray-bellied salamanders (*E. multiplicata griseogaster*), which have transforming adults (Petranka, 1998). Multivariate morphological analyses by Tumlison et al. (1990a) attempted to provide a mechanism for discerning Oklahoma salamanders from larval gray-bellied salamanders. However, it is possible that the morphometric differences that they found are associated with life history mode (paedomorphosis versus transformation), and do not represent species-specific differences. Furthermore, unpublished data show some populations of Oklahoma salamanders and gray-bellied salamanders to be indistinguishable based on allozymes (R. Wilkinson, personal communication). For this reason, all populations in Missouri formerly considered *E. tynerensis* are now considered to be *E. m. griseogaster* (Johnson, 2000). I am currently undertaking a study that includes delineating the species boundaries and revising the taxonomic status of the

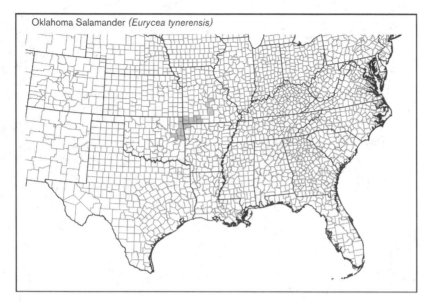

Oklahoma Salamander *(Eurycea tynerensis)*

members of this complex using molecular and morphological data.

2. Historical versus Current Abundance.
No extirpations of populations or range extensions in previously uninhabited regions have been reported for Oklahoma salamanders.

3. Life History Features.
   A. **Breeding.** Reproduction is aquatic.
   i. *Breeding migrations.* Natural populations of Oklahoma salamanders currently are thought to include only permanently aquatic, non-transforming individuals (Petranka, 1998). As explained in other sections, Oklahoma salamanders exhibit considerable movement within their aquatic habitat, but the degree of migration for breeding purposes is unknown.
   ii. *Breeding habitat.* Presumably the same as adult habitat.
   B. **Eggs.**
   i. *Egg deposition sites.* Egg clutches of Oklahoma salamanders have yet to be recovered in the field.
   ii. *Clutch size.* In mid-May and mid-November, 1–11 eggs were harvested from three gravid females collected in Arkansas (Trauth et al., 1990), and gravid females collected in Oklahoma in late May contained as many as seven eggs measuring 1.8 × 1.5 mm (Moore and Hughes, 1939).
   C. **Larvae/Metamorphosis.** Hatchlings measure from 9–13 mm TL (Dundee, 1965b). Morphological differences between larvae and adults are slight, with larvae having fewer ampullary (electroreceptor) organs and larger tail fins and gills (Moore and Hughes, 1939). Four Oklahoma salamanders were induced to metamorphose (undergoing a loss of gills and tail fin, alterations in the morphology of the skull and eyes, and exhibiting an affinity for terrestriality) when treated with thyroxin (Kezer, 1952a).

   D. **Juvenile Habitat.** Probably are the same as for adults, although studies that examine micohabitat usage by Oklahoma salamanders at various life stages have not been conducted.
   E. **Adult Habitat.** Traditionally, Oklahoma salamanders were thought to inhabit only cool, clear, swift streams that contain coarse gravel, where this species hides (Moore and Hughes, 1939; Dundee, 1958). Tumlison et al. (1990c) conducted thorough examinations of surface habitat parameters that are most preferable to Oklahoma salamanders. Abiotically, they found shallow, slowly moving streams containing medium-sized rocks that are only partially embedded to reliably contain Oklahoma salamanders. Oklahoma salamanders also were found to be most abundant in areas where aquatic invertebrate densities are high. However, it is unclear if this biotic factor results from Oklahoma salamanders preferring areas of high prey density or a mutual response of both the aquatic invertebrates and Oklahoma salamanders to water conditions or predator avoidance. Additional observations on the habitat of Oklahoma salamanders reveal that they also occur in small springs and seeps amongst moist leaf litter over a mud-and-detritus substrate (Tumlison and Cline, 1997).
   F. **Home Range Size.** Unknown.
   G. **Territories.** Unknown.
   H. **Aestivation/Avoiding Desiccation.** Dundee (1958) reported drought conditions to cause mass migrations of Oklahoma salamanders to more hospitable subsurface environments. A persistent drought dried Tyner Creek, Adair County, Oklahoma (the type locality for *E. tynerensis*), from 1951–55, but excavation of the stream bed revealed water 2.4 m (8 ft) below the surface, where Oklahoma salamanders apparently sought refuge (Dowling, 1956).

   I. **Seasonal Migrations.** The discovery of subterranean isopods (*Caecidotea* sp.) in the stomachs of two specimens and the location of many individuals in small, isolated springs distant from a main stream course led Tumlison and Cline (1997) to propose that Oklahoma salamanders may be migrating along subterranean corridors to reach resource-rich habitats on the surface. However, high densities of Oklahoma salamanders in rather atypical habitats might also be interpreted as a sequestering of all individuals into the last remaining moist habitats to survive drought conditions.
   J. **Torpor (Hibernation).** Unknown and unstudied.
   K. **Interspecific Associations/Exclusions.** Rudolph (1978) reported that ≤4 additional species of plethodontids with stream-dwelling larval stages are sympatric with Oklahoma salamanders. Among those are cave salamanders (*E. lucifuga*), dark-sided salamanders (*E. longicauda melanopleura*), gray-bellied salamanders; and grotto salamanders *(Typhlotriton [Eurycea] spelaeus)*. Rudolph (1978) found substantial differences between Oklahoma salamanders and other species in their ability to survive flood conditions, probably by seeking refuge within the gravel. He also noted that Oklahoma salamanders were the only species that are able to coexist with grotto salamanders at the heads of springs.
   L. **Age/Size at Reproductive Maturity.** Oklahoma salamanders reach sexual maturity in 2–3 yr (Dundee, 1958) at approximately 26 mm SVL (Dundee, 1965b). Sexual size dimorphism has not been noted.
   M. **Longevity.** Unknown.
   N. **Feeding Behavior.** A wide array of prey items has been identified from the digestive tracts of Oklahoma salamanders, including dipterans, ephemeropterans, plecopterans, coleopterans, trichopterans, hymenopterans, thysanopterans, odonates, ostracods, isopods, amphipods, decapods, hydracarians, and pulmonates (Tumlison et al., 1990b). In addition, pulmonates, copepods, and homopterans were identified from the feces of Oklahoma salamanders (Rudolph, 1978).

   Specimens used in an attempt to observe feeding postures of Oklahoma salamanders (Dodd, 1980) were later re-identified as grotto salamanders (Dodd, 1982).
   O. **Predators.** Oklahoma salamander larvae were consumed by fishes, including banded sculpins (*Cottus carolinae*), black bullheads (*Ameiurus melas*), and green sunfish (*Lepomis cyanellus*) under laboratory conditions (Rudolph, 1978), but predation by fish in the wild has not been reported. Of the Oklahoma salamanders collected by Tumlison et al. (1990c), 22% had autotomized tails, which was suggested might be a result of predation by crayfish that occupy the same rocky substrate.

P. Anti-Predator Mechanisms. In laboratory trials, fish predation was substantially less on larval Oklahoma salamanders than on grotto salamander, cave salamander, and long-tailed salamander larvae. This was attributed to the tendency of Oklahoma salamanders to seek refuge beneath the gravel substrate (Rudolph, 1978). In additional trials using leafy substrates, Oklahoma salamanders still evaded fish predation best, but there were more individuals consumed than in the trials on gravel substrate (Rudolph, 1978). Tumlison et al. (1990c) hypothesized that the shallow water preference of Oklahoma salamanders may reflect the exclusion of some fish (i.e., *Cottus* sp.) from such shallow depths.

Q. Diseases. Have not been reported or systematically studied.

R. Parasites. Two parasites have been described from Oklahoma salamanders: an echinorynchid worm, *Acanthocephalus van cleavi* (Hughes and Moore, 1943a), and a polystomatid fluke, *Sphyranura euryceae* (Hughes and Moore, 1943b).

4. Conservation.
Eurycea tynerensis, once listed as Rare on Missouri's Rare and Endangered Species List (Johnson, 1987), is no longer recognized as a valid taxon in Missouri (Johnson, 2000). The name *E. tynerensis* is still used in Oklahoma, but it is provided with no special protected status there. In Arkansas, *E. tynerensis* is considered to be a Species of Special Concern, and collecting permit requests are closely monitored (K. Irwin, personal communication).

Being permanently aquatic leaves this species particularly vulnerable to alterations in water quality and pollutants. Much needed studies using molecular techniques to determine the differences between *E. tynerensis* and *E. multiplicata*, and to define their distributions, are currently in progress. The seasonal movements of Oklahoma salamanders, in particular their tendency to follow stream levels to subsurface habitats, may complicate monitoring studies of this species (Dowling, 1956). An understanding of the subterranean abundance and activity of Oklahoma salamanders will provide valuable insights into their status and conservation requirements.

Eurycea waterlooensis Hillis, Chamberlain, Wilcox, and Chippindale, 2001
AUSTIN BLIND SALAMANDER
Paul T. Chippindale

1. Historical versus Current Distribution.
Austin blind salamanders (*Eurycea waterlooensis*) are a newly discovered species known only from the outflows of Barton Springs in the city of Austin, Travis County, Texas. A formal description is given in

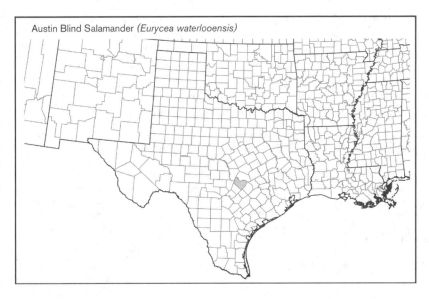

Austin Blind Salamander *(Eurycea waterlooensis)*

Hillis et al. (2001). This subterranean species is most closely related to Texas blind salamanders (*E.* [formerly *Typhlomolge*] *rathbuni*) and Blanco blind salamanders (*E.* [formerly *Typhlomolge*] *robusta)*; both species occur in or near San Marcos, Hays County, Texas (Hillis et al., 2001). Only a few specimens are available, and Austin blind salamanders are much more rarely encountered than Barton Springs salamanders (*E. sosorum*), which occur on the surface at Barton Springs.

2. Historical versus Current Abundance. Unknown.

3. Life History Features.
    A. Breeding.
    i. Breeding migrations. Unknown.
    ii. Breeding habitat. Unknown.
    B. Eggs.
    i. Egg deposition sites. Unknown.
    ii. Clutch size. Unknown.
    C. Larvae/Metamorphosis. This species is paedomorphic, and natural metamorphosis is unknown. Several specimens obtained as juveniles grew in captivity from about 15–60 mm TL in 8 mo, at which time they were presumed to be sexually mature (Hillis et al., 2001).
    D. Juvenile Habitat. Probably similar to those of adults.
    E. Adult Habitat. Completely aquatic. Austin blind salamanders are known only from spring outflows (juveniles that probably washed out accidentally); this species almost certainly is a cave dweller. Water temperature in the Barton Springs Aquifer is relatively constant at about 20 °C. Sweet (1982) provided a comprehensive distributional analysis of the central Texas *Eurycea* and discussed hydrogeology of the region in relation to salamander distribution.

    F. Home Range Size. Unknown.
    G. Territories. Unknown.
    H. Aestivation/Avoiding Desiccation. Unknown.
    I. Seasonal Migrations. Unlikely to occur.
    J. Torpor (Hibernation). Probably active throughout the year.
    K. Interspecific Associations/Exclusions. Partially sympatric with Barton Springs salamanders.
    L. Age/Size at Reproductive Maturity. Unknown, but based on growth patterns for captive specimens, this species probably matures at about 60 mm total length. The largest known specimen (maintained alive by the City of Austin) had a total length of about 79 mm in July 2000. The largest known preserved specimen is 35.6 mm SVL with a total length of 66.6 mm (2 mm of tail tip missing).
    M. Longevity. Unknown.
    N. Feeding Behavior. Prey probably consists mainly of aquatic invertebrates, but no feeding studies of this species have been conducted.
    O. Predators. Unknown.
    P. Anti-Predator Mechanisms. Unknown.
    Q. Diseases. Unknown.
    R. Parasites. Unknown.

4. Conservation.
Austin blind salamanders are a newly discovered species known only from the outflows of Barton Springs in the city of Austin, Travis County, Texas. Only a few specimens are available, and Austin blind salamanders are much more rarely encountered than syntopic Barton Springs salamanders, which occur on the surface at Barton Springs. Austin blind salamanders are currently listed by the federal government as a Candidate species; they receive no special protection in Texas.

Eurycea wilderae Dunn, 1920
BLUE RIDGE TWO-LINED SALAMANDER
David M. Sever

1. Historical versus Current Distribution.

Until raised to full species status by Jacobs (1987), Blue Ridge two-lined salamanders (*Eurycea wilderae*) were considered a subspecies of northern two-lined salamanders (*E. bislineata*) as described by Dunn (1920). Dunn (1920) stated that the range of his new taxon is the "southern division of the Blue Ridge," and Dunn (1926) reported that the distribution is "from White Top Mountain, Virginia; south in mountains to Clayton, Rabun County, Georgia, and Cherry Log, Gilmer County, Georgia. They inhabit the whole Southern Blue Ridge region." The range is sometimes loosely given as "the southern Appalachian Mountains" (Jacobs, 1987; Conant and Collins, 1998), but essentially the range as defined by Dunn (1926) is restricted to the Southern Blue Ridge Mountain physiographic province. Blue Ridge two-lined salamanders occur from base-level streams to the tops of the highest peaks (about 1,900 m). The most remarkable characteristic of this species is the presence of two male morphs that differ dramatically in morphology, especially in regards to the male secondary sexual characters (Sever, 1979, 1999c). One of these is considered the typical "*wilderae*" morph because it possesses labial cirri, which also characterize males in the type series (Dunn, 1920). This gracile form also possesses a mental gland, seasonally enlarged premaxillary teeth, and 0–2 costal grooves between toes of the adpressed limbs. Males of the other morph, called "morph A" by Sever (1979), lack cirri, mental glands, and seasonally enlarged premaxillary teeth, and they possess 2–3 costal grooves between toes of the adpressed limbs. During the breeding season, the muscles comprising the jaw adductors hypertrophy, making morph A the "big-headed" form in the Southern Blue Ridge. Hypertrophy of the jaw adductors is also known in populations of northern two-lined salamanders and Junaluska salamanders (*E. junaluska*), but these species all possess mental glands and enlarged premaxillary teeth, and cirri also occur in Junaluska salamanders (Sever, 1979). Females associated with the two male morphs of Blue Ridge two-lined salamanders are indistinguishable but generally have 2–4 costal grooves between adpressed limbs. Morph A has been reported from Blount, Monroe, and Sevier counties, Tennessee, and Graham, Haywood, Macon, and Watuga counties, North Carolina (Sever, 1989), and probably occurs throughout the Southern Blue Ridge. Whether the morphs represent separate species or a polymorphism in Blue Ridge two-lined salamanders still is unresolved,

but morph A also occurs in populations in the Piedmont and Coastal Plain of North Carolina within the defined range of southern two-lined salamanders (*E. cirrigera*; Sever, 1999b). Note that in some other populations of southern two-lined salamanders, such as in northern Alabama and the Cumberland Plateau of southeastern Tennessee, "big-headed" males occur (Mount, 1975; Sever, 1999b), but these individuals possess mental glands and cirri and therefore are not identifiable as morph A. Specimens resembling the male *wilderae* morph also have been found outside of the Southern Blue Ridge. Individuals morphologically similar to the *wilderae* morph have been reported by Chermock (1952) from Mount Cheaha in Alabama (which Chermock, as well as Mount, 1975, considers an extension of the Blue Ridge) and by Rossman (1965b) from Wilcox County in the Coastal Plain of Alabama. Allozyme analysis of specimens from these localities, however, align them with *E. cirrigera* (Jacobs, 1987). I have seen specimens that resemble the male *wilderae* morph from Cloudland Canyon State Park, Dade County, Georgia, and Kings Mountain State Park, York County, South Carolina. Howell and Switzer (1953) reported "integrades" between *E. cirrigera* and *E. wilderae* in the Piedmont of Georgia. An allozyme analysis of the contact zone between *E. wilderae* and *E. cirrigera* in northwestern South Carolina revealed the presence of populations of *E. wilderae* in the Piedmont of Anderson, Oconee, and Pickens counties (Kozak, 1999). The limits of the range of *E. wilderae* and interactions with *E. bislineata* and *E. cirrigera* in areas of contact or sympatry need much more study (Sever, 1999a,b,c). Finally, Jacob's (1987) group E of *E. wilderae* included samples from the type locality (Grayson County, Virginia) and two samples from Watuga County, North Carolina, whereas his group F contains all other samples of

E. wilderae from more southern areas of the Blue Ridge. Mean $D = 0.30$ between the E and F groups, far exceeding levels (as low as $D > 0.15$) commonly used to denote species level differences (Highton, 1998; Thorpe, 1982). Thus, *E. wilderae* is likely polyspecific; the northern taxon (group E) is referable to *E. wilderae*, but the southern taxon (group F) requires a new name.

2. Historical versus Current Abundance.

Early reports indicate that Blue Ridge two-lined salamanders were frequently encountered during fieldwork. Dunn (1917a) reported on collections of the species near Brevard, Transylvania County, and Linville, Avery County, North Carolina. He stated,

> We found the larvae of this species common in every small stream examined, and even in those as large as the Linville River. Adults were found rarely at Brevard. Only 1 adult and 1 transforming specimen were taken near there. At Linville both adults and larvae were common, but adults were found only on land under logs in situations similar to those chosen by various *Plethodons* and by *Desmognathus o. carolinensis* (*D. carolinensis*), in fact often in company with these species.

In Rutherford County, North Carolina, Weller (1930) found Blue Ridge two-lined salamanders to be "very common, both at camp and in the surrounding mountains." King (1939) stated, "This is one of the common salamanders in the Great Smokies. It may be found in nearly every spring, seep and permanently damp place at suitable elevations." In another account on Blue Ridge two-lined salamanders in the Smokies, Huheey and Stupka (1967) reported, "This common salamander ranges throughout the Park, occurring at all altitudes." I have been collecting the species for over 35 yr. I believe that historically the species probably occurred in

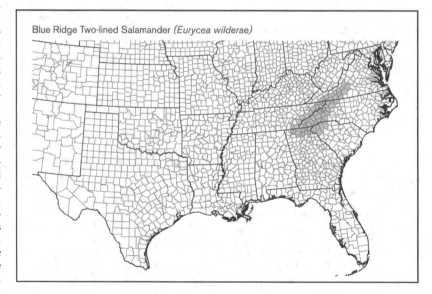

Blue Ridge Two-lined Salamander (*Eurycea wilderae*)

every rocky mountain stream in the Southern Blue Ridge. Today, Blue Ridge two-lined salamanders are still likely to be found in every stream that has not been damaged by pollution, siltation, deforestation, channeling, and other factors. The species, however, appears rather resilient, and one should not be surprised to find it almost anywhere in the Southern Blue Ridge, even in seemingly inhospitable habitats. For example, Tullulah Creek in Graham County, North Carolina, is historically a clear-flowing rocky, base-level stream. Large samples of Blue Ridge two-lined salamanders could be collected in the 1970s where Tullulah Creek runs through Robbinsville. In the past 15 yr, the stretch through Robbinsville has become increasingly murky and exposed as businesses and homes along the creek have flourished. Other formerly common salamanders, such as three-lined salamanders (*E. guttolineata*), Junaluska salamanders, and black-bellied salamanders (*D. quadramaculatus*), are now rare along Tullulah Creek in Robbinsville, but Blue Ridge two-lined salamanders are still common. Adults aggregate at mating/nesting sites from October–April and may be abundant in streams where few adults can be found in mid summer. However, mass metamorphosis of larvae can again make the species seem incredibly abundant along rocky mountain streams in summer, but these individuals are mostly juveniles.

3. Life History Features.
Consult Sever (1999a) for additional references on Blue Ridge two-lined salamanders.

A. Breeding. Reproduction is aquatic.

i. Breeding migrations. Many authors have noted that adults can be found considerable distances from water (e.g., King, 1939; Huheey and Stupka, 1967). In midsummer on Wayah Bald, Macon County, North Carolina, I have found dozens under single strips of bark on fallen trees hundreds of meters from the nearest streams. Because the eggs are laid in water and the larvae are aquatic, however, migration must occur to suitable nesting sites. Also, because both males and females migrate, nesting sites constitute mating areas as well. Aggregations start in the fall, and at higher elevations, the mating/nesting areas also serve as hibernation sites. The peak concentration of adults in mating/nesting areas is in spring. Courtship and mating occur on the banks of streams adjacent to nesting areas. The projecting premaxillary teeth of males of the *wilderae* morph scrape the female's skin during courtship, allowing secretions of the male's mental gland to enter the superficial circulation of the female (Arnold, 1977).

ii. Breeding habitat. Usually the eggs are attached to the underside of rocks in flowing water. I have not noticed any consistency in the size of the rock (large or small) or shape (flat or round). Females remain with the eggs. The gelatinous matrix of the eggs is naturally adhesive; eggs may be in one rather discreet cluster or more scattered. A large, favorable rock may have nests of several females. I occasionally have found eggs (with attendant females) by digging through gravelly spring heads; an individual egg in these situations may adhere to several small pieces of gravel.

B. Eggs.

i. Egg deposition sites. As reported above, nests most frequently have been found under rocks in streams. A considerable amount of variation, however, occurs in the timing of oviposition in this species. Dunn (1920) reported: "At Linville a batch of eggs was found hatching on July 19. They were attached to the under side of a stone in a brook just as are the eggs of *bislineata*." Wood (1949) stated that late summer deposition occurs at elevations of 1,525 m, and egg laying occurs earlier at lower elevations. At 1,220 m on Mount Mitchell on 4 May, Wood (1949) found a batch of 87 eggs suspended from the lower surface of a flat stone in a small seepage spring. Bruce (1982a) reported numerous egg clusters in late winter and early spring in streams in the Tuckasegee River basin (695–1,050 m) in Jackson County, North Carolina. At Santeetlah Creek (650 m), Graham County, North Carolina. Bruce (1982b) reported finding nests on 13 and 15 May; at the same site, however, I have found nests in mid-March.

ii. Clutch size. The batch of 87 eggs reported by Mitchell from the lower surface of a rock in a small spring on Mount Mitchell probably represents the clutches of more than one female. Clutch sizes in various populations have been reported to range from 8–34 and 28–56 (Ryan and Bruce, 2000).

C. Larvae/Metamorphosis. Ryan (1997) provides an excellent drawing of the larva. The larvae of Blue Ridge two-lined salamanders are a pale yellow to yellow-green dorsally with a thin, broken dorsolateral stripe, ventral to which is fine mottling and three rows of unpigmented lateral line spots (Eaton, 1956; Ryan, 1997). The tail is flattened and mottled while the venter is clear and a light cream color. Bruce (1986) reported on drift movements of Blue Ridge two-lined salamanders in a stream at 1,170 m in Macon County, North Carolina. He found that downstream movements are dominated by first-year larvae. Upstream movements are not sufficient to compensate for downstream drift, so drift may represent a density-dependent means of population regulation.

i. Length of larval stage. Bruce (1982a, 1985b) reported metamorphosis usually occurs in late spring and early summer after 1–2 yr at a mean 18.5–23.9 mm SVL in stream populations, and at 26.4 mm SVL in a pond. Although a tendency exists for growth rates to be lower at higher elevations, no corresponding tendency occurs for the larval period to be prolonged (Bruce, 1985b). At Santeetlah Creek, Graham County, North Carolina, the larval period typically is 2 yr with mean 31.8 mm SVL in the oldest cohort to metamorphose (Bruce, 1982b). Voss (1993b) found that larvae metamorphose after 1 yr in first-order streams, whereas in higher-order streams metamorphosis may be delayed for an additional year. This variation is due to warmer temperatures in first-order streams (Voss, 1993b). Beachy (1994) found that survival and growth of Blue Ridge two-lined salamander larvae raised in the laboratory together were independent of density, suggesting a lack of competition.

ii. Larval requirements.

a. Food. I am unaware of any study specifically dealing with food habits of larval Blue Ridge two-lined salamanders, although such studies do exist for the sibling species *E. bislineata* (Smallwood, 1928; Burton, 1976) and *E. cirrigera* (Petranka, 1984b).

b. Cover. As Petranka (1984b) reported for the sibling species *E. cirrigera*, larval Blue Ridge two-lined salamanders are found under rock cover during the day and move about stream beds feeding continuously at night (Wiltenmuth, 1997a).

iii. Larval polymorphisms. None are known.

iv. Features of metamorphosis. As reported in "Length of larval stage" above, metamorphosis occurs after 1–2 yr of larval development. Newly metamorphosed juveniles are often abundant around breeding areas in late spring and summer, indicating some synchrony in metamorphosis within a population.

v. Post-metamorphic migrations. Individuals that metamorphose in the late spring or summer may participate in breeding activities the following spring, at the beginning of their third or fourth year (Bruce, 1988b). Thus, the juvenile stage is short, and juveniles may not move far from streamside habitats along the natal area. I have found mature gonads in dissected individuals that are only 25–28 mm SVL, within the range of body sizes characterizing newly metamorphosed animals in some populations (Bruce, 1982a,b, 1985b).

vi. Neoteny. Not known to exist.

D. Juvenile Habitat. As mentioned above, juveniles often are found in streamside habitats. They may be found under rocks and logs, by scraping through leaves and other detritus. Bruce (1986) did not find significant differences between upstream and downstream movements in second-year larvae and metamorphosed individuals.

E. Adult Habitat. Males frequently are found with females under rocks in streams during the spring mating period. Females subsequently stay in the water with their nests, whereas the males move into terrestrial habitats. After eggs hatch, females must follow males into more terrestrial habitats because adults of either sex are usually uncommon along streams in midsummer.

F. Home Range Size. I am unaware of any literature on whether individuals of Blue Ridge two-lined salamanders establish home ranges during any period of the year or stage of life. The study done by Bruce (1986) on upstream and downstream movements suggests that adults and second-year larvae move upstream and downstream in equal frequencies, which no doubt contributes to maintaining a certain density at a locale. First-year larvae, however, move downstream more frequently, resulting in a density dependent mechanism of regulation of excess production (Bruce, 1986).

G. Territories. Wiltenmuth (1997a) reported aggression in larval Blue Ridge two-lined salamanders, but whether this behavior is due to territorial or nonterritorial interference competition requires further investigation.

H. Aestivation/Avoiding Desiccation. Not known to occur. Hutchison (1961) reported on critical thermal maxima in a number of salamanders, including three juvenile Blue Ridge two-lined salamanders. He found a CTM of 32.1 °C, the lowest of any salamander tested (Hutchison, 1961).

I. Seasonal Migrations. Movements between terrestrial and stream habitats associated with mating/nesting activities are discussed in "Breeding migrations" above.

J. Torpor (Hibernation). Numerous individuals can be found by digging through gravelly spring heads in mid-winter, even at high elevations where harsh winter weather surely precludes much surface activity. At lower elevations, however, activity may occur through mid-winter.

K. Interspecific Associations/Exclusions. The Southern Blue Ridge physiographic province is a center of salamander diversity in North America (Bruce et al., 2000). Blue Ridge two-lined salamanders commonly are found in association with a dozen or more other species. Brodie (1981) reported that yellow-striped Ocoee salamanders (*Desmognathus ocoee*; called *D. ochrophaeus* by Brodie) from several North Carolina localities are mimics of Blue Ridge two-lined salamanders. However, this model–mimic relationship is not as prevalent as in New York between Allegheny Mountain dusky salamanders (*D. ochrophaeus*) and northern two-lined salamanders, which are relatively more common in association with Allegheny Mountain dusky salamanders than Blue Ridge two-lined salamanders are with Ocoee salamanders in North Carolina.

L. Age/Size at Reproductive Maturity. The smallest individuals I have dissected that definitely possess mature gonads are 25 mm SVL for a female and 28 mm SVL for a male. However, some individuals as small as 23 mm SVL may be mature, and many are still immature at 30 mm SVL. Bruce (1988b) reported that individuals spend ≥1 yr as juveniles after 1–2 yr as larvae. Age at first reproduction in both sexes is estimated to be 3–4 yr, but usually the latter, since most individuals metamorphose at 2 yr (Bruce, 1988b). Mean SVL of adults from various populations ranges from 30.3–49.0 mm (Sever, 1999a), with total lengths of 60–90 mm (Bishop, 1943). The record specimen came from Indian Gap in the Great Smoky Mountains and is 120 mm TL (King, 1939).

M. Longevity. Bruce (1988b) constructed a life table for a population of Blue Ridge two-lined salamanders from 1,100 m in Macon County, North Carolina. He found that R0=0.821, indicating a declining population and an unstable age distribution. Bruce believed that this R value was a result of procedural errors and used alternative methods (that do not rely upon any assumptions concerning R) to calculate an estimate mean generation time of 4.4 yr; an annual survivorship of 0.408 was calculated for females. Few animals survive beyond 5 yr and none beyond 10 yr.

N. Feeding Behavior. I am not aware of any studies that specifically address feeding behavior of Blue Ridge two-lined salamanders.

O. Predators. Huheey and Stupka (1967) mention that spring salamanders (*Gyrinophilus porphyriticus*; see also Bruce, 1979; Beachy, 1994) and common garter snakes (*Thamnophis sirtalis*) are predators. Larval (Beachy, 1994, 1997) and metamorphosed (Davic, 1991) black-bellied salamanders are known predators, and it is likely that several other larger plethodontids (including red salamanders, [*Pseudotriton ruber*] and shovel-nosed salamanders, [*D. marmoratus*]) eat Blue Ridge two-lined salamander larvae or adults as well (Bruce, 1982a). Beachy (1994) raised larval Blue Ridge two-lined salamanders in the laboratory for 30 d with larvae of spring salamanders and/or black-bellied salamanders. He found that both predators significantly reduced survivorship of Blue Ridge two-lined salamanders, but that spring salamanders were more effective. Larvae of spring salamanders exposed to high prey densities grew more than those exposed to low prey densities, but prey density had no effect on prey survival. When grouped together with Blue Ridge two-lined salamanders, predator effects on prey survival were additive, indicating neither a mutualistic nor a competitive interaction between spring salamanders and black-bellied salamanders. Beachy (1997a) conducted additional experiments in which he exposed larval Blue Ridge two-lined salamanders to larval black-bellied salamanders. Risk of predation caused significant variation in growth rate, with larvae under highest predation risk growing faster during certain periods. However, larvae metamorphosed at the same time regardless of different growth rates and predation risks. Beachy (1997a) hypothesized that no advantage exists in varying time or size at metamorphosis due to the permanency and low productivity of mountain streams.

P. Anti-Predator Mechanisms. Wiltenmuth (1997a) conducted experiments to determine whether differences in body size among age classes of Blue Ridge two-lined salamander larvae affect intraspecific agonistic behavior, and whether cover availability in the presence of a predator species (larval black-bellied salamanders) affects spacing behavior of larvae. She found that larvae of Blue Ridge two-lined salamanders prefer the vicinity of rock cover and that larger larvae are dominant to smaller larvae. However, Blue Ridge two-lined salamanders did not increase use of rock cover in the presence of visual or chemical cues from black-bellied salamanders. She concluded that either black-bellied salamanders do not pose a substantial threat to Blue Ridge two-lined salamanders or that Blue Ridge two-lined salamanders do not hide under cover to avoid black-bellied salamanders (Wiltenmuth, 1997a).

Q. Diseases. None are known.

R. Parasites. I am not aware of any reports on parasites of Blue Ridge two-lined salamanders.

4. Conservation.

There are no current conservation concerns associated with Blue Ridge two-lined salamanders. They are not state or federally listed (Levell, 1997). Habitat remains plentiful throughout their range, and they remain abundant in suitable streams.

Eurycea sp. 1
COMAL SPRINGS SALAMANDER
Paul T. Chippindale, Joe N. Fries

1. Historical versus Current Distribution.
This putative species is known only from Comal Springs in the city of New Braunfels, Comal County, Texas. Some authors (Sweet, 1978a; Dixon, 1987) have considered this population conspecific with *E. nana* from San Marcos Springs, Hays County, but morphological and molecular evidence strongly supports a rejection of this hypothesis (Chippindale, 1995, 2000; Chippindale et al., 1998, 2000). This taxon appears to be most closely related to the southeastern Edwards Plateau clade of *Eurycea* recognized by Chippindale (1995, 2000) and Chippindale et al. (2000). Systematic studies are in progress.

2. Historical versus Current Abundance.
Unknown. Currently common at and near spring outflows.

3. Life History Features.
A. Breeding. Reproduction is aquatic.

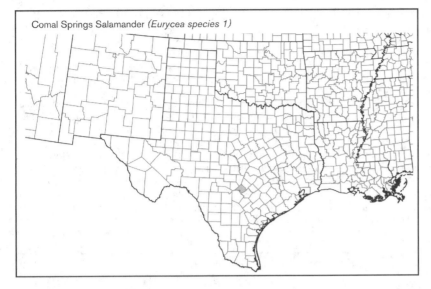

Comal Springs Salamander *(Eurycea species 1)*

i. Breeding migrations. Unlikely to occur.

ii. Breeding habitat. Roberts et al. (1995) successfully bred this species in acrylic columns filled with large gravel through which artesian water was pumped.

B. Eggs.

i. Egg deposition sites. Roberts et al. (1995) provided detailed information on egg laying. Based on their observations, they suggested that individuals travel downward into spring upwellings to deposit eggs.

ii. Clutch size. From 19 to >50 eggs (Roberts et al., 1995).

C. Larvae/Metamorphosis. This species is paedomorphic, and natural metamorphosis is unknown. Captive-hatched individuals at the Dallas Aquarium at Fair Park reached total lengths of 26 mm by 2 mo of age; by 6 mo of age they were 60 mm, and females showed signs of ova development; at 1 yr, two individuals were gravid (Roberts et al., 1995). Roberts et al. (1995) described development of eggs and larvae.

D. Juvenile Habitat. Probably the same as adult habitat.

E. Adult Habitat. Completely aquatic. This species is known only from the vicinity of spring outflows, under rocks, and in gravel substrate. Water temperature in Comal Springs typically is 23 °C (George et al., 1952; USFWS, 1996b). Roberts et al. (1995) provided detailed information on water chemistry for captive specimens. Sweet (1982) provided a comprehensive distributional analysis of the central Texas *Eurycea* and discussed hydrogeology of the region in relation to salamander distribution.

F. Home Range Size. Unknown.

G. Territories. Unknown.

H. Aestivation/Avoiding Desiccation. Unknown. However, Comal Springs is known to have ceased flowing during a drought in the 1950s; therefore, some members of this species are able to retreat underground when necessary.

I. Seasonal Migrations. Unlikely to occur, although there may be seasonal variation in surface versus subsurface habitat use.

J. Torpor (Hibernation). Probably active throughout the year.

K. Interspecific Associations/Exclusions. Fountain darters *(Etheostoma fonticola)* are common in the same areas where salamanders are found.

L. Age/Size at Reproductive Maturity. In morphometric analyses of some Edwards Plateau *Eurycea* (Chippindale et al., 1993, 1998), the average SVL of individuals that were presumed to be adults was 27.8 mm (in their 1993 paper, Chippindale et al. tentatively assigned this population to *E. neotenes*). Roberts et al. (1995) reported that 6-mo-old, captive-hatched individuals were approaching maturity at about 60 mm TL; at 1 yr, some were gravid. Breeding animals at the Dallas Zoo Aquarium were 91 mm TL (male) and 74 mm

(female) for a wild-caught pair, and 76 mm (male) and 65 mm (female) for an F1 pair (Roberts et al., 1995).

M. Longevity. At least 3 yr in captivity.

N. Feeding Behavior. Prey probably consists mainly of small aquatic invertebrates. Captive individuals fed on brine shrimp, amphipods, oligochaete worms, snails, and zooplankton.

O. Predators. Unknown.

P. Anti-Predator Mechanisms. Secretive.

Q. Diseases. Unknown.

R. Parasites. Unknown.

4. Conservation.

This putative species is known only from Comal Springs in the city of New Braunfels, Comal County, Texas. Their conservation status is unknown; they currently are considered common at and near spring outflows.

Eurycea sp. 2
PEDERNALES SPRINGS SALAMANDER
Paul T. Chippindale

1. Historical versus Current Distribution.
Pedernales Springs salamanders (unnamed *Eurycea* [*Eurycea* sp.]) are a putative species first found in 1989 by D. Hillis and P. Chippindale in a spring along the Pedernales River in extreme western Travis County, Texas; the only other known locality is a second nearby spring. These animals have never been formally assigned to any species, but appear to be distinct based on molecular markers, and are well separated geographically, geologically, and hydrologically from other members of the group (Chippindale, 1995, 2000; Chippindale et al., 2000). This taxon appears to be most closely related to the southeastern Edwards Plateau clade of *Eurycea* recognized by Chippindale (1995, 2000) and Chippindale et al. (2000). Systematic studies are in progress.

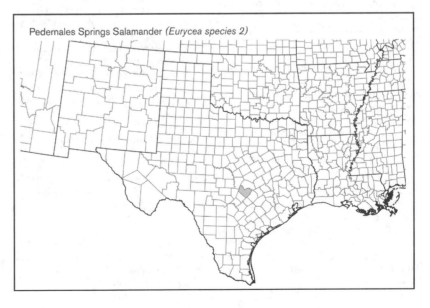

Pedernales Springs Salamander *(Eurycea species 2)*

2. Historical versus Current Abundance.
May be common at spring outflows, but their distribution appears to be extremely limited and patchy.

3. Life History Features.
A. Breeding. Reproduction is aquatic.

i. Breeding migrations. Unlikely to occur.

ii. Breeding habitat. Probably the same as adult habitat.

B. Eggs.

i. Egg deposition sites. Unknown; some other central Texas *Eurycea* are thought to deposit eggs in gravel substrate.

ii. Clutch size. Unknown.

C. Larvae/Metamorphosis. This species is paedomorphic, and natural metamorphosis is unknown.

D. Juvenile Habitat. Probably the same as adult habitat.

E. Adult Habitat. Completely aquatic. These animals are known only from the immediate vicinity of spring outflows, under rocks and leaves and in gravel substrate. Water temperature in springs of the Edwards Plateau is relatively constant throughout the year and typically ranges from 18–20 °C or slightly warmer near the fault zone at the Plateau's edge (Sweet, 1982). Sweet (1982) provided a comprehensive distributional analysis of the central Texas *Eurycea* and discussed hydrogeology of the region in relation to salamander distribution.

F. Home Range Size. Unknown.

G. Territories. Unknown.

H. Aestivation/Avoiding Desiccation. Unknown.

I. Seasonal Migrations. Unlikely to occur.

J. Torpor (Hibernation). Probably active throughout the year.

K. Interspecific Associations/Exclusions. Unknown.

L. Age/Size at Reproductive Maturity. Unknown, but Chippindale et al. (2000) noted that this species may mature at a slightly smaller size compared with other spring-dwelling central Texas Eurycea.

M. Longevity. Unknown.

N. Feeding Behavior. Prey probably consists mainly of small aquatic invertebrates, but no feeding studies of this species have been conducted.

O. Predators. Unknown.

P. Anti-Predator Mechanisms. Secretive.

Q. Diseases. Unknown.

R. Parasites. Unknown.

4. Conservation.
Pedernales Springs salamanders are a putative species first found in two springs near the Pedernales River in extreme western Travis County, Texas. They may be common at spring outflows, but their distribution appears to be extremely limited and patchy.

Gyrinophilus gulolineatus Brandon, 1965(a)
BERRY CAVE SALAMANDER
Christopher K. Beachy

1. Historical versus Current Distribution.
Berry Cave salamanders (*Gyrinophilus gulolineatus*) were originally a subspecies of Tennessee cave salamanders (*G. palleucus*). Collins (1991) suggested their elevation to species status based on allopatry and substantial morphometric differentiation (e.g., unique throat stripe, large size, and fewer trunk vertebrae in *G. gulolineatus* [Brandon, 1965a]) compared to other members of the *G. palleucus* complex.

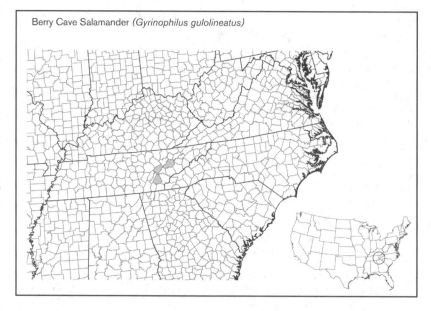

Berry Cave Salamander *(Gyrinophilus gulolineatus)*

Berry Cave salamanders are known only from sites in the Ridge and Valley Province in Knox, McMinn, and Roane counties, Tennessee (Brandon, 1965a, 1966c, 1967a; Petranka, 1998). The data necessary to compare current versus historical distributions have not been collected.

2. Historical versus Current Abundance.
Berry Cave salamander populations are declining (Caldwell and Copeland, 1992), likely due to above-ground habitat destruction and subsequent effects on water quality; and Caldwell and Copeland (1992) have suggested that Berry Cave salamanders should be given Endangered status.

3. Life History Features.
A. Breeding. No aspect of breeding has been observed. However, reproduction is undoubtedly aquatic, because Berry Cave salamanders are neotenic.

i. Breeding migrations. Unlikely. Given that Berry Cave salamanders are neotenic, breeding habitat is likely to be the same as, or a subset of, adult habitat.

ii. Breeding habitat. Berry Cave salamanders most likely breed in the caverns and passages they occupy.

B. Eggs.

i. Egg deposition sites. Microhabitat characteristics of egg deposition sites are unknown. Extending what is known about egg deposition sites in spring salamanders (*G. porphyriticus*), clutches will be attached as a single mass to the undersides of large stones.

ii. Clutch size. Unknown. However, a large clutch size (compared to other species of *Gyrinophilus*) is predicted, based on well-established relationships between salamander body size and clutch size (Kaplan and Salthe, 1979).

C. Larvae/Metamorphosis.

i. Length of larval stage. Unknown. Berry Cave salamanders are neotenic and the transition from larvae to reproductive adults has not been documented.

ii. Larval requirements.

a. Food. Unknown, although presumably larvae feed on aquatic, primarily benthic, invertebrates that are small enough to ingest whole (see Brandon, 1967b).

b. Cover. Unknown.

iii. Larval polymorphisms. Unknown.

iv. Features of metamorphosis. Unknown.

v. Post-metamorphic migrations. Unlikely in these neotenic animals.

vi. Neoteny. Berry Cave salamanders are obligate neotenes (Brandon, 1965a, 1966c; Simmons, 1975).

D. Juvenile Habitat. Juveniles live in the same cave systems occupied by adults and are therefore likely to have similar habitat characteristics.

E. Adult Habitat. Berry Cave salamanders either inhabit, or are associated with, caves. Caldwell and Copeland (1992) suggest that inflow (sinkhole) caves versus outflow caves may provide the best habitat. Inflow caves provide a detritus base that appears to be necessary for Berry Cave salamanders.

F. Home Range Size. Unknown, but possibly extremely small. In mark–recapture studies, animals are found in exactly the same location (e.g., Simmons, 1975).

G. Territories. Unknown.

H. Aestivation/Avoiding Desiccation. Aestivation is unknown and unlikely.

I. Seasonal Migrations. Unknown but unlikely. Unstudied, but if they occur, migrations occur either within their cave system or from caves to the immediate vicinity of cave openings (where animals were first collected; Brandon, 1965a).

J. Torpor (Hibernation). Unknown and unlikely.

K. Interspecific Associations/Exclusions. Berry Cave salamanders are not syntopic with any other amphibian species.

L. Age/Size at Reproductive Maturity. Berry Cave Salamanders are extremely large plethodontids. The holotype is an apparently reproductively mature female measuring 122 mm SVL (preserved; Brandon, 1965a).

M. Longevity. Unknown.

N. Feeding Behavior. Berry Cave salamanders likely feed on isopods, annelids, and aquatic invertebrates, similar to other troglobitic *Gyrinophilus* (see Brandon, 1967b; Simmons, 1975, 1976). Individuals have larger heads than Tennessee cave salamanders. Brandon (1965a) suggests that this is a feeding specialization, noting that among salamanders the most highly modified snouts are found on the most highly specialized cave salamanders, and speculates that because cave-dwelling salamanders tend to feed on bottom-dwelling invertebrates, a broad, spatulate snout may be effective in detecting and capturing food under dark conditions.

O. Predators. Unknown.

P. Anti-Predator Mechanisms. Unknown.

Q. Diseases. Unknown.

R. Parasites. Brandon (1967b) noted intestinal parasites (e.g., nematodes, cestodes, and acanthocephalans) in closely related Tennessee cave salamanders.

4. Conservation.

Berry Cave salamanders are known only from sites in the Ridge and Valley Province in Knox, McMinn, and Roane counties, Tennessee. These populations are declining due to above-ground habitat destruction and subsequent effects on water quality. The Tennessee Wildlife Resources Agency (1994) has listed *G. palleucus* as Threatened; because *G. gulolineatus* was recognized as a subspecies of Tennessee cave salamanders at the time of listing and only occurs in Tennessee, the arguments for listing *G. gulolineatus* are equally valid. Caldwell and Copeland (1992) have suggested that Berry Cave salamanders should be given Endangered status.

Gyrinophilus palleucus McCrady, 1954
TENNESSEE CAVE SALAMANDER

Christopher K. Beachy

1. Historical versus Current Distribution.
Tennessee cave salamanders (*Gyrinophilus palleucus*) exhibit a spotty distribution associated with cave systems throughout central Tennessee, northern Alabama, and northwestern Kentucky (Brandon, 1967a,b; Cooper, 1968; Cooper and Cooper, 1968; Redmond and Scott, 1996). Two subspecies of Tennessee Cave salamanders are recognized: Sinking Cove Cave salamanders (*G. p. palleucus*) and Big Mouth Cave salamanders (*G. p. necturoides*). The current distribution of Tennessee cave salamanders is probably similar to the historical distribution—there is no evidence that populations have been lost.

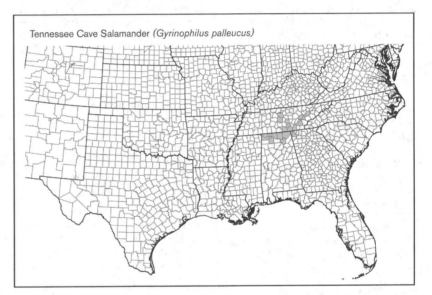

Tennessee Cave Salamander *(Gyrinophilus palleucus)*

2. Historical versus Current Abundance.
Petranka (1998) notes that population surveys rarely reveal >10–20 animals/cave visit, suggesting that populations are small. Population estimates from various caves reveal sizes of 25, 32, 48, and 88 animals, with densities ranging from 0.06–0.15 animals/m^2. The abundance of animals in some populations has been affected by siltation and increased water flows associated with deforestation (see Petranka, 1998). Most populations appear to be declining (Simmons, 1975; Caldwell and Copeland, 1992; Redmond and Scott, 1996).

3. Life History Features.

A. Breeding.

i. Breeding migrations. Do not occur. The presence of males with spermatophores in August (Lazell and Brandon, 1962) and the occurrence of small hatchlings in caves in December–February suggest that females lay eggs in autumn or early winter (Simmons, 1975; see also Petranka, 1998).

ii. Breeding habitat. The sinkhole-type caves characteristic of adult habitats.

B. Eggs.

i. Egg deposition sites. Unknown. It is expected that eggs will be deposited in a manner similar to that of spring salamanders (*G. porphyriticus*), that is attached to the undersides of large rocks. Ova (from one female) averaged 3.2 mm in diameter.

ii. Clutch size. Unknown. Clutch size/SVL relationships for spring salamanders provide a basis for estimating clutch size in Tennessee cave salamanders (Bruce, 1972).

C. Larvae/Metamorphosis.

i. Length of larval stage. Unusually long, which led to their genus name, which in Greek means "tadpole loving" (Brandon, 1967a). Naturally metamorphosed animals have only rarely been found (e.g., Simmons, 1975, 1976; Yeatman and Miller, 1985); populations typically consist of only neotenic forms (Lazell and Brandon, 1962; Brandon, 1966c, 1967a; Simmons, 1975, 1976; Caldwell and Copeland, 1992).

ii. Larval requirements.

a. Food. Tennessee cave salamanders consume benthic invertebrates and are constrained primarily by gape limitations (Brandon, 1966c; Simmons, 1975).

b. Cover. Animals can be found under rocks. However, most animals are found by direct observation without removal of cover objects (see Simmons, 1975). It is likely that cover is not used. Simmons (1975) describes the "disconcerting" habit of Tennessee cave salamanders to be found in exactly the same spot as months earlier.

iii. Larval polymorphisms. Unknown.

iv. Features of metamorphosis. Metamorphosis does not occur in Tennessee cave salamanders. Animals can be stimulated to metamorphose with thyroxin treatment (Dent and Kirby-Smith, 1963), and animals occasionally will metamorphose after collection and transport to the laboratory.

v. Post-metamorphic migrations. Unlikely.

vi. Neoteny. Most populations of Tennessee cave salamanders consist of only neotenic animals. Naturally metamorphosed specimens are found occasionally (e.g., Simmons, 1975, 1976; Yeatman and Miller, 1985).

D. Juvenile Habitat. Juvenile habitats are the same as adults.

E. Adult Habitat. Tennessee cave salamanders are found in sinkhole-type caves or phreatic cave systems in the vicinity of sinkholes. This association is due to the nutrients that flow into these systems and the prey base they support. Caldwell and Copeland (1992) suggest that inflow (sinkhole) caves versus outflow caves may provide the best habitat. Animals are found under rocks in rocky and sandy substrates in quiet, shallow pools (McCrady, 1954; Simmons, 1975; see also Petranka, 1998).

F. Home Range Size. Petranka (1998) notes that individuals are highly sedentary, rarely moving >3–4 m between surveys, with many individuals repeatedly found in the same locations (Simmons, 1975).

G. Territories. Unknown.

H. Aestivation/Avoiding Desiccation. Unknown and unlikely.

I. Seasonal Migrations. Unknown and unlikely.

J. Torpor (Hibernation). Unknown.

K. Interspecific Associations/Exclusions. There are no other amphibian species in habitats where Tennessee cave salamanders are found.

L. Age/Size at Reproductive Maturity. Growth rates are slow, and animals may be larvae for many years (Brandon, 1967a,b; Petranka, 1998). Males reach sexual maturity at 66 mm SVL (Petranka, 1998); in Sinking Cove cave salamanders, size at sexual maturity is 70–100 mm SVL (Brandon, 1967b). The inner contour of the vent is sexually dimorphic (Brandon, 1967a).

M. Longevity. Unknown.

N. Feeding Behavior. Tennessee cave salamanders feed on invertebrates and conspecifics. Invertebrates include amphipods, annelids (oligochaetes and earthworms), cladoceran zooplankton, crayfish, and insects such as coleopterans, plecopterans, ephemeropterans, trichopterans, dipterans (chironomid larvae), and thrips. The invertebrate (potential prey) fauna associated with caves has been described by Cooper and Cooper (1968).

O. Predators. Known predators include conspecifics (Lazell and Brandon, 1962; Simmons, 1975) and American bullfrogs (*Rana catesbeiana*), which can inhabit the mouths of cave entrances (Lee, 1969b). Petranka (1998) suspects that crayfish feed on small larvae.

P. Anti-Predator Mechanisms. Being troglobytic assists in the avoidance of most amphibian predators.

Q. Diseases. Unknown.

R. Parasites. Unknown.

4. Conservation.
The current distribution of Tennessee cave salamanders probably is similar to the historical distribution, although populations have been affected by the indirect effects of deforestation and most appear to be declining. Petranka (1998) makes a plea for conservation through water quality and protective land management initiatives. The Tennessee Wildlife Resources Agency (1994; see also www.state.tn.us) has listed Tennessee cave salamanders as Threatened. Although the U.S. Fish and Wildlife Service (1994c) listed Tennessee cave salamanders as a Category 2 candidate for federal listing, they were not included in a more recent federal list (USFWS, 1996a).

Gyrinophilus porphyriticus (Green, 1827)
SPRING SALAMANDER

Christopher K. Beachy

1. Historical versus Current Distribution.
Spring salamanders (*Gyrinophilus porphyriticus*) range from the middle of Maine southwest along the Appalachian spine (Dunn, 1926; Brandon, 1967c; French, 1976; Petranka, 1998). Brandon (1966c) recognized four subspecies. Northern spring salamanders (*G. p. porphyriticus*) are found throughout most of New England, New York, and Pennsylvania, and in portions of Ohio, West Virginia, Virginia, Tennessee, North Carolina, Georgia, Alabama, and extreme northwestern Mississippi (Engelhardt, 1919; Warfel, 1937; Fowler and Sutcliffe, 1952; Thurow, 1954; Brandon, 1966c; Graham, 1981; Graham and Stevens, 1982; Lazell and Raithel, 1986; Petranka, 1998). A disjunct population occurs near Cincinnati in southwestern Ohio. Kentucky spring salamanders (*G. p. duryi*) are found in western West Virginia, northeastern Kentucky, and south-central Ohio (Brandon, 1967c; Petranka, 1998), with a single record documented in Tazewell County, Virginia (Newman, 1954a). Carolina spring salamanders (*G. p. dunni*) are found in southwestern North Carolina, northwestern South Carolina, northern Georgia, and northeastern Alabama (Brandon, 1966c, 1967c). Blue Ridge spring salamanders (*G. p. danielsi*) occur in extreme western North Carolina (Brandon, 1966c, 1967c).

The range of the species apparently is the same currently (Petranka, 1998) as when Dunn (1926) first summarized the range of spring salamanders. It is probable that *G. porphyriticus* consists of several cryptic species. Southern Appalachian populations exhibit significant life history variation, morphometric differentiation, and ethological isolation among parapatric populations (Bruce, 1972, 1978; Beachy, 1996; Adams and Beachy, 2001).

Petranka (1998) notes that deforestation is a threat to many populations of spring salamanders.

2. Historical versus Current Abundance.
Spring salamanders are well known for being difficult to find. Repeated trips to classic salamander localities usually results in finding one or two spring salamanders, but often none at all (Bruce, 1972a, 1978a; Beachy, 1996). The habitat (see "Adult Habitat" below) simply proves difficult to penetrate, and the salamanders that are obtained seem to be the occasional animals that are active on the surface. Current densities seem in line with historical densities. This means that in most of their range, spring salamanders have always been difficult to obtain.

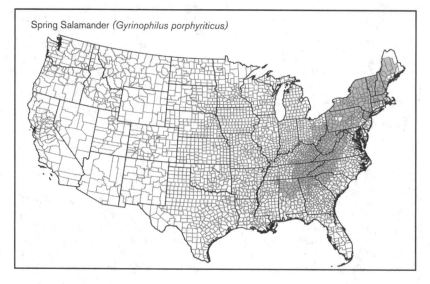

Spring Salamander (*Gyrinophilus porphyriticus*)

3. Life History Features.

A. Breeding. Courtship is in streamside conditions with oviposition being in headwater seeps (Beachy, 1996, 1997b).

i. Breeding migrations. Do not occur. Courtship occurs during the winter from December–February (Bruce, 1972a; Beachy, 1996, 1997b). Oviposition occurs during the late spring and summer (Green, 1925; Bishop, 1941b; Organ, 1961c; Bruce, 1972a, 1980).

ii. Breeding habitat. Same as adult habitats. Most females lay their eggs during the summer; embryos hatch in late summer or autumn (Green, 1925; Organ, 1961c; Bruce, 1972a, 1978a, 1980).

B. Eggs.

i. Egg deposition sites. Few egg masses have been found. Females likely lay their eggs deep in underground recesses in streams and seeps. Females attach their eggs to the undersides of rocks or other cover objects (Noble and Richards, 1932). Eggs are 3.5–4.0 mm in diameter (Bishop, 1941b; Bruce, 1969, 1972a).

ii. Clutch sizes. Ova numbers range from 39–63; clutch sizes vary from 16–106 and are related to female body size. Females brood (Bishop, 1924; Organ, 1961c; Bruce, 1978a). Hatchlings range in size from 18–22 mm TL in the southern Appalachians (Bruce, 1978a), and to 26 mm TL in New York (Bishop, 1924; see also Petranka, 1998).

C. Larvae/Metamorphosis.

i. Length of larval stage. Bruce (1980) estimates a modal larval length of 4 yr, with some individuals metamorphosing after 3–5 yr. Weber (1928) and Bishop (1941b) suggest a 3-yr larval length. Estimates of larval period are difficult because large samples of larvae are difficult to obtain.

ii. Larval requirements.

a. Food. Larvae feed at night. Spring salamander larvae feed on a variety of prey, including the following invertebrates: oligochaetes, arachnids, isopods, centipedes, crayfish, and insects including mayflies, odonates, stoneflies, and dipterans. Spring salamanders also will feed on vertebrates including salamander eggs, two-lined salamander (*Eurycea bislineata* complex) adults and larvae, and adult Ocoee salamanders (*Desmognathus ocoee*; Bruce, 1979; Resetarits, 1991; Beachy, 1994; Gustafson, 1994). Spring salamanders are cannibalistic and will feed on smaller conspecific larvae (Burton, 1976; Bruce, 1979).

b. Cover. Larvae are found most frequently beneath stones and logs or in gravel beds in springs, seeps, or spring-fed streams. Spring salamanders occasionally are found in lakes (Bishop, 1941b). Larvae are nocturnal; they are secretive during the day, where they can be found in subterranean cracks and crevices, sometimes far below the surface (Bruce, 1980; 2003). At night, individuals emerge to forage (Resetarits, 1991). Spring salamander larvae

generally do not occur in large numbers (Bruce, 1972a, 1978a), but densities can reach as high as 5–10/m^2 in streambeds in Virginia (Resetarits, 1991, 1995; see also Petranka, 1998).

Larvae have been found in caves in Kentucky, Virginia, and West Virginia (Green and Brant, 1966; Cooper and Cooper, 1968; see also Petranka, 1998).

iii. Larval polymorphisms. Unknown and unlikely.

iv. Features of metamorphosis. Metamorphosis occurs at about 55–65 mm SVL in populations below 1,200 m, and 61–70 mm in montane populations. Metamorphosis generally occurs from July–August, but has been reported from March–October (Bishop, 1941b). Bruce (1979) hypothesized that time to metamorphosis reflects an evolutionary response to food resources available to larvae and adults.

v. Post-metamorphic migrations. Unknown.

vi. Neoteny. Unknown.

D. Juvenile Habitat. The juvenile habitat is the same as for adults.

E. Adult Habitat. Adults are most abundant in the headwater sections of small tributaries and small streams that lack fishes, in seepages and caves, and can sometimes be found in roadside ditches (Petranka, 1998). Bruce (1972a) notes that in the Piedmont of South Carolina, populations are associated with springs and small streams in deep ravines covered with mature hardwood forest (see also Petranka, 1998). Citing Cooper and Cooper (1968), Besharse and Holsinger (1977) note that while spring salamanders are found in springs and cave streams in the south-central Appalachians, they are more common in caves than in springs in limestone areas.

F. Home Range Size. Unknown.

G. Territories. Unknown.

H. Aestivation/Avoiding Desiccation. Unknown and unlikely.

I. Seasonal Migrations. Unknown and unlikely.

J. Torpor (Hibernation). Spring salamanders in the southern Appalachians remain active throughout the year. Less is known about these animals in the northern part of their range. Despite ice cover, it is likely that spring salamanders remain active below ground.

K. Interspecific Associations/Exclusions. Spring salamanders are voracious predators of other salamanders (see "Feeding Behavior" below). Although there is no evidence that the presence of spring salamanders excludes other species, spring salamanders restrict two-lined salamander nocturnal feeding activity, causing slower growth rates and increased mortality in two-lined salamander larvae in regions where they co-occur (Resetarits, 1991; Beachy, 1994; Gustafson, 1994). Larger spring salamanders can also reduce the growth rates of smaller conspecifics

(Gustafson, 1994; see also Petranka, 1998). Where spring salamanders co-exist with black-bellied salamanders (*D. quadramaculatus*), there is no evidence that they compete for food (Beachy, 1994).

Spring salamanders co-occur in streams with fishes, but reach their highest densities in the absence of fishes (Petranka, 1998). Resetarits (1995) demonstrates that in the presence of trout fingerlings, spring salamanders use shallower habitats.

L. Age/Size at Reproductive Maturity. Males become reproductively mature at about 55 mm SVL, with no obvious sexual dimorphism (Bruce, 1972a; see also Petranka, 1998). Males at low to intermediate elevations become sexually mature shortly after metamorphosing (3–4 yr, see above); males in high-elevation populations require up to 1 yr longer (Bruce, 1972a), maturing at as large as 81 mm SVL.

Females in low-elevation populations can mature shortly after metamorphosing (3–4 yr), as small as 61 mm SVL. Females at higher elevations mature when they are older and larger (Bruce, 1972a; see also Petranka, 1998).

M. Longevity. Unknown. Assuming a median larval period of 4 yr, most animals attain sexual maturity at anywhere from 4–6 yr (Bruce, 1972a, 1978a, 1980).

N. Feeding Behavior. While spring salamander adults are feeding generalists, according to Petranka (1998), food habit tendencies vary regionally. Adults in northern populations tend to feed on invertebrates, including annelids, snails, centipedes, millipedes, arachnids (spiders and mites), and insects. Insect prey includes mayflies, caddisflies, stoneflies, dipterans, hymenopterans, and hemipterans (Bishop, 1941b; Culver, 1973). In northern populations, cannibalism and preying on other salamander species such as northern dusky salamanders (*D. fuscus*) is known, but thought to be uncommon (Hamilton, 1932).

In southern populations, spring salamander adults are voracious consumers of salamanders (Wright and Haber, 1922; King, 1939; Bishop, 1941b; Martof, 1955; Huheey and Stupka, 1967; Bruce, 1972a, 1979). They are known to feed on the following salamander species: pigmy salamanders (*D. wrighti*), adult and larval northern two-lined salamanders, Ocoee salamanders, Jordan's salamanders (*Plethodon jordani*), southern red-backed salamanders (*P. serratus*), southern Appalachian salamanders (*P. Teyahalee*), and red salamanders (*Pseudotriton ruber*). They also are known to be cannibalistic. Invertebrate prey include annelids, centipedes, and insects such as coleopteran larvae. Adult spring salamanders have a higher tendency than larvae to feed on other salamanders.

O. Predators. Northern water snakes (*Nerodia sipedon*) and common garter snakes

(Thamnophis sirtalis) prey on spring salamanders (Uhler et al., 1939). Smaller individuals are cannibalized by larger
individuals (Burton, 1976).

P. Anti-Predator Mechanisms. Spring salamanders use defensive postures. Adults
produce noxious skin secretions that are
known to repel shrews (Brodie et al., 1979).
Spring salamanders are thought to be part
of the Batesian mimicry complex that also
involves red salamanders and the red eft
stage of eastern newts (*Notophthalmus viridescens*; Howard and Brodie, 1973; Brandon and Huheey, 1975, 1981).

Q. Diseases. Unknown.

R. Parasites. Ranik (1937) studied the
parasites of Blue Ridge spring salamanders
and found two protozoans, *Hexamastix
batrachorum* and *Prowazekella longifilis*,
and one nematode, *Omeia papillocauda*.

4. Conservation.
With the caveat that spring salamanders
are difficult to find, current densities
appear to be in line with historical densities. Petranka (1998) notes that deforestation is a threat to many populations of
spring salamanders. Spring salamanders
are considered Endangered in Mississippi, Threatened in Connecticut (http://
dep.state.ct.us), of Special Concern in
Massachusetts (www.state.ma.us), and of
Concern in Rhode Island.

Gyrinophilus subterraneus Besharse and
Holsinger, 1977
WEST VIRGINIA SPRING SALAMANDER

Christopher K. Beachy

1. Historical versus Current Distribution.
The distribution of West Virginia spring
salamanders *(Gyrinophilus subterraneus)*
is limited to General Davis Cave, Greenbrier County, West Virginia (Besharse
and Holsinger, 1977; but see Blaney and
Blaney, 1978, and discussions in Green
and Pauley, 1987; Petranka, 1998). It is
likely that this species arose in General
Davis Cave, and their current distribution
reflects the historical distribution, although Besharse and Holsinger (1977)
consider the possibility that the current
distribution of West Virginia spring salamanders is relictual and reflects the remnants of a much broader distribution.

2. Historical versus Current Abundance.
Unknown, but Besharse and Holsinger
(1977) note they found 15 *Gyrinophilus*
specimens comprising West Virginia
spring salamanders and spring salamanders (*G. porphyriticus*) along a 180-m length
of cave stream. In these caves, West Virginia spring salamanders outnumbered
spring salamanders by about 2:1. Besharse
and Holsinger (1977) note that spelunkers have observed West Virginia spring
salamanders 1,800 m beyond the cave
entrance.

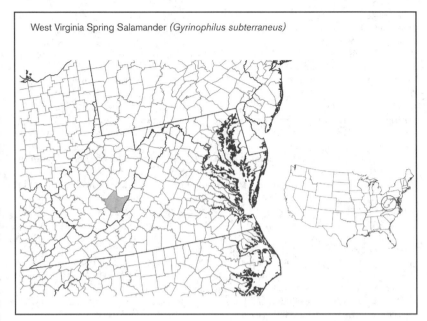

West Virginia Spring Salamander *(Gyrinophilus subterraneus)*

Because of the restricted distribution of
West Virginia spring salamanders, Besharse
and Holsinger (1977) strongly recommend
that future collecting of specimens be done
sparingly.

3. Life History Features.
Because West Virginia spring salamanders
are probably a sister taxon of spring salamanders, they may share certain life history features. In fact, Green and Pauley
(1987) state: "Nothing has been discovered
about life history or reproductive behavior
to indicate that [their] habits are different
from those of *Gyrinophilus p. porphyriticus.*"

A. Breeding. Reproduction is aquatic.

i. Breeding migrations. Probably do not
occur. West Virginia spring salamanders
are restricted to General Davis Cave and its
vicinity.

ii. Breeding habitat. The aquatic habitats
in and near General Davis Cave.

B. Eggs.

i. Egg deposition sites. Unknown.

ii. Clutch size. Unknown.

C. Larvae/Metamorphosis.

i. Length of larval stage. Unknown. Larvae metamorphose at about 95 mm SVL.
The largest larvae are either near or have
attained sexual maturity.

ii. Larval requirements.

a. Food. Likely to be the small aquatic
invertebrates found in the cave system
(see "Feeding Behavior" below).

b. Cover. Unknown, other than being
troglobitic.

iii. Larval polymorphisms. Unknown, but
unlikely.

iv. Features of metamorphosis. Larvae
metamorphose at about 95 mm SVL.

v. Post-metamorphic migrations. Unlikely.

vi. Neoteny. The largest larvae are at or
near sexual maturity. It is not known
whether individuals will reproduce as gilled

adults (Petranka, 1998). After a thorough
consideration of the morphology of West
Virginia spring salamanders, Besharse and
Holsinger (1977) note that their "large
body size, retention of the larval color pattern and fused premaxillae in transformed
adults, and development of mature gonads
in large larvae and transforming animals
all suggest an evolutionary trend towards
paedomorphosis." The massive size of larvae compared with the emaciated appearance of adults suggests a major difference
in the ability to obtain food between the
two life history stages of these animals,
which could lead to a selection pressure
for paedomorphosis (Bruce, 1979).

D. Juvenile Habitat. Unknown.

E. Adult Habitat. West Virginia spring
salamanders are restricted to General Davis
Cave or on the muddy banks along the
stream in the vicinity of the cave. Besharse
and Holsinger (1977) describe General
Davis Cave as a large, stream-passage cave
developed along the strike in the Union
Limestone of the Greenbrier Group (Mis-
sissippian). The entrance is large, at the
head of a short ravine. The cave passage
floor is damp clay that joins a stream passage 150 m beyond the entrance. This
cave is essentially a single, long, stream
passage that contains about 4,000 m of
surveyed length. The stream usually varies
in depth from 15–30 cm, but will flood
following rains. The primary source of
water is surface flow from a stream that
courses down Muddy Creek Mountain
(Besharse and Holsinger, 1977).

The banks of the stream are muddy,
steep, and contain a large amount of decaying leaf litter occasionally washed into
the cave by floods. This leaf litter offers a
source of nutrients for the cave invertebrates that West Virginia spring salamanders use as a prey base.

Spelunkers have observed West Virginia spring salamanders almost 2 km into General Davis Cave (Besharse and Holsinger, 1977).

F. Home Range Size. Unknown.

G. Territories. Unknown.

H. Aestivation/Avoiding Desiccation. Unknown and unlikely.

I. Seasonal Migrations. Unlikely.

J. Torpor (Hibernation). Unknown and unlikely.

K. Interspecific Associations/Exclusions. Co-occur with spring salamanders, but interactions between these two species have not been examined.

L. Age/Size at Reproductive Maturity. At about 95 mm SVL, near the time of metamorphosis.

M. Longevity. Unknown.

N. Feeding Behavior. Potential prey have been well documented and include the invertebrate prey base found in the cave. These prey include the following troglobitic species: crayfish (*Cambarus nerterius*), amphipods (*Gammarus minus; Stygobromus spinatus*), isopods (*Asellus holsingeri*), and carabid beetles (*Pseudanophthalmus grandis; P. laldermani*). Cavernous collembolans (*Pseudosinella gisini*), millipedes (*Trichopetalum weyeriensis* and *Pseudotremia* sp. [probably *fulgida*]), a pseudoscorpion (*Kleptochthonius henroti*), and an oligochaete (*Allobophora chlorotica*) also serve as potential prey.

Besharse and Holsinger (1977) note that the head of larval West Virginia spring salamanders is broader than the heads of either spring salamanders or Tennessee cave salamanders (*G. palleucus*), but does not have the spatulate snout characteristic of Berry Cave salamanders (*G. gulolineatus*).

O. Predators. Unknown.

P. Anti-Predator Mechanisms. Unknown other than cave dwelling.

Q. Diseases. Unknown.

R. Parasites. Unknown.

4. Conservation.
West Virginia spring salamanders are found only in General Davis Cave, Greenbrier County, West Virginia. Because of their restricted distribution they occur on West Virginia's Rare species list, and every attempt should be made to preserve this habitat and its water sources.

Haideotriton wallacei Carr, 1939
GEORGIA BLIND SALAMANDER
D. Bruce Means

1. Historical versus Current Distribution.
The name Georgia blind salamander (*Haideotriton wallacei*) originates from the first specimen available to science, which was pumped from a 60-m (200 ft) deep well at Albany, Georgia (Carr, 1939; Brandon, 1967d). Since that time only one other Georgia locality has been established, Climax Cave in Decatur County (Dundee,

1962), yet at least eight localities are known in the Florida Panhandle in the vicinity of Marianna in Jackson County (Pylka and Warren, 1958; Means, 1977, 1992c). All localities allow access to the aquatic habitat of Georgia blind salamanders, the Floridan Aquifer, which circulates in underground passageways in limestones of the Ocala and Suwannee formations. These passageways, and the vadose caves exposed in them, are part of a vast karst region called the Marianna Lowlands in Florida and the Dougherty Plain in Georgia.

2. Historical Versus Current Abundance.
The presence of Georgia blind salamanders in underground waters of the Dougherty Plain in Georgia has been re-confirmed recently (March, 1999) in Climax Cave (J. Jensen, personal communication). At least two caves in which Georgia blind salamanders were known to occur in the Marianna Lowlands have been destroyed by human activities; in five other undisturbed caves, the populations of Georgia blind salamanders seemed stable over the period 1970–92 (personal observations).

3. Life History Features.

A. Breeding. Reproduction is aquatic.

i. Breeding migrations. None known, although it is conceivable that adults move to sites of energy recharge (bat caves, sinkholes) to court and lay eggs. Gravid females with enlarged ova have been found in May and November (Carr, 1939; Means, 1977). Breeding, therefore, may be aseasonal.

ii. Breeding habitat. Not known to differ from habitat of the adults.

B. Eggs.

i. Egg deposition sites. Unknown.

ii. Clutch size. Unknown.

C. Larvae/Metamorphosis. These salamanders are thought to remain neotenic; no transformed animals have been collected, and larvae show no response to metamorphic agents (Dundee, 1962; Petranka, 1998).

D. Juvenile Habitat. Not known to be different from that of adults.

E. Adult Habitat. Knowledge of the dispersion of populations of Georgia blind salamanders in subterranean waters is biased toward places where air-breathing humans have easy access to those waters

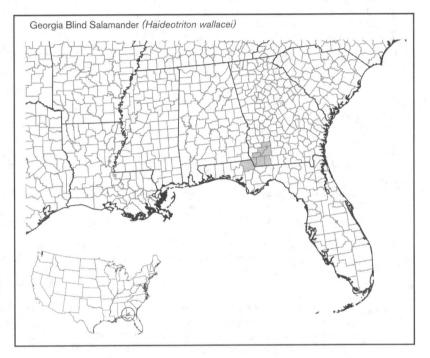

Georgia Blind Salamander *(Haideotriton wallacei)*

in vadose caves and sinkholes. In such sites, Georgia blind salamanders commonly are observed in pools and underground streams of the Floridan Aquifer, especially in caves where bats defecate over or near the water. Individuals move about slowly, resting on bottom sediments or climbing on limestone sidewalls and ledges underwater (Pylka and Warren, 1958; Means, 1992c). Farther back in subterranean tunnels away from air, salamanders become much less common (Means, 1992c), presumably because their food is scarce. The water is usually crystal clear and about 18–21 °C, but becomes turbid during heavy rains. Limestone forms the walls and ceiling of the phreatic habitat, and fine red clays and silts lie deep on the floors.

F. Home Range Size. Unknown. Needs study.

G. Territories. Unknown.

H. Aestivation/Avoiding Desiccation. These behaviors probably do not occur.

I. Seasonal Migrations. Not known, but possibly happens from distant reaches of passageways to food sources at breeding time.

J. Torpor (Hibernation). Probably does not exist because of the constant temperature of groundwater.

K. Interspecific Associations/Exclusions. Georgia blind salamanders are almost always found with Dougherty Plain blind crayfish (*Cambarus cryptodytes*; Pylka and Warren, 1958; Means, 1977, 1992c).

L. Age/Size at Reproductive Maturity. Adults measure 51–76 mm TL (Petranka, 1998), but no studies have been conducted on growth rates or age or size at maturity.

M. Longevity. Unknown.

N. Feeding Behavior. Ostracods and amphipods that occur in the water with Georgia blind salamanders were the most numerically common food items in the 40 specimens examined by Lee (1969d) and Peck (1973). Other prey included isopods, copepods, mites, and beetles.

O. Predators. Dougherty Plain blind crayfish probably prey on Georgia blind salamanders, as may freshwater eels (*Anguilla rostrata*), brown bullheads (*Ameiurus nebulosus*), and Florida chubs (*Notropis harperi*), all of which are commonly found in subterranean waters with Georgia blind salamanders (Means, 1992c).

P. Anti-Predator Mechanisms. Individuals bolt suddenly when any physical disturbance in the water is detected and commonly swim rapidly in an upward spiral, then become motionless in the water and float back down to the bed of the pool or underground passageway (personal observations).

Q. Diseases. None observed.

R. Parasites. Lee (1969d) observed "transparent parasitic nematodes" in the stomachs of 3 of 32 individuals he collected from one cave in the Marianna Lowlands between April and July. Another small specimen that died in captivity and was partially decomposed had "large numbers of live nematodes protruding from the body wall."

4. Conservation.

A review of the population status of Georgia blind salamanders has not been conducted. In the period 1969–92, populations seemed fairly abundant, in spite of heavy collecting in one or two localities (personal observations). Because of agricultural draws-down of the aquifer and possible pollution from agricultural runoff, a resurvey of the known localities and a search for new ones should be conducted. Only caves in the Marianna Caverns State Park are protected.

Hemidactylium scutatum Temminck and Schlegel, 1838

FOUR-TOED SALAMANDER

Reid N. Harris

1. Historical versus Current Distribution.

The current knowledge of distribution of four-toed salamanders (*Hemidactylium scutatum*) is summarized in Petranka (1998); four-toed salamanders range from south-eastern Canada south to the Gulf of Mexico and west to Oklahoma, Missouri, and Wisconsin. Populations appear to be disjunct in the southern and western parts of this range.

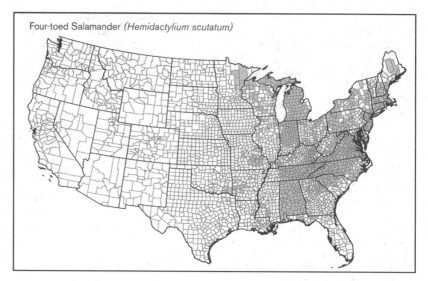

Four-toed Salamander (*Hemidactylium scutatum*)

2. Historical versus Current Abundance.

No published population-size data exist. In a study spanning 23 yr (1979–83 and 1994–2002; unpublished data) in the George Washington National Forest (GWNF), Virginia, the number of ovipositing females at one montane pond in Virginia fluctuated between 15 and 177; at another nearby pond, numbers fluctuated between 15 and 91. The highest number of females oviposited at both ponds in 1980, and numbers of females between 1994–2002 have averaged 48.31 (SD = 37.42, CV = 0.77) at one pond and 48.5 (SD = 23.94, CV = 0.49) at another. The population density of adult eastern newts (*Notophthalmus viridescens*) increased dramatically during the study period, and because adult newts prey on larval four-toed salamanders, the failure of the population to approach numbers seen in 1980 may be due to prey–predator dynamics. Surveys of apparently similar nearby ponds have revealed few nesting females.

Early concern was expressed by Wright (1918), who stated that four-toed salamanders were vanishing in New York due to draining of wetlands. Four-toed salamanders are considered Rare in Florida (Means, 1992d) and southern New England (Klemens, 1993), and worthy of Special Concern in Maine (Burgason, 1999) and Minnesota (Moriarty and Jones, 1997). Populations exist in fairly substantial numbers in the Great Smoky Mountains National Park (GSMNP; C. K. Dodd, personal communication), but their overall conservation status is unknown due to their secretive nature, scattered distribution, short larval period, and propensity to brood early in the season (Dodd, 1997;

C. K. Dodd, personal communication). Based on reports from two protected areas (GWNF and GSMNP), it appears that populations are doing well in these habitats in the upland south (see above). However, J. W. Petranka (personal communication) feels that current abundance is likely to be much lower than historical abundance due to loss of vernal ponds and other small wetlands across the eastern United States (see also Mitchell et al., 1999).

Inventories of four-toed salamanders will need to consider the following aspects of their biology, which will tend to lead to an underestimate of population size: a short larval period, secretive nature of adults, and a propensity of adult females to skip years of reproduction.

3. Life History Features.

A. Breeding. Insemination and brooding are terrestrial; larvae are aquatic.

i. Breeding migrations. Only females migrate from woodlands to the banks of ponds, bogs, or streams to oviposit. Migration date is related to altitude and latitude, occurring from mid-February to mid-April in lowland Virginia (Wood, 1955), and from April to mid-May in montane Virginia (Harris and Gill, 1980), lowland Michigan (Blanchard, 1934b), and New York (Bishop, 1941b).

ii. Breeding habitat. In Michigan, mating occurs on land from September–November (Blanchard, 1933a; Branin, 1935).

B. Eggs.

i. Egg deposition sites. Females prefer relatively steep banks, presumably to facilitate larvae making it from land to water. Nests are often found in moss, although nests are also found in grass and sedges and even inside rotted logs (Wood, 1955; Harris and Gill, 1980). It is frequently assumed that *Sphagnum* moss is preferred; however, females readily nest in other species of moss, such as *Thuidium* (Gilbert,

1941; Wood, 1955; Easterla, 1971; Petranka, 1998). Large differences in population sizes at apparently similar ponds in the GWNF suggest habitat preferences may be complex (unpublished data).

ii. Clutch size. Average clutch sizes based on dissection and counts of large ova range from 4–80 (Blanchard, 1936; Wood, 1953b). Joint nests are usually formed by two females; as many as 1,110 eggs have been found in one joint nest in Michigan, which suggests that about 20 females contributed eggs (Blanchard, 1934a; Harris et al., 1995). Females often skip years of reproduction (R. N. H. and P. M. Ludwig, unpublished data).

Females may lay eggs in a solitary nest and brood them. Alternatively, females may accept eggs from another female into her nest (Blanchard, 1934a; Harris and Gill, 1980; Breitenbach, 1982). Typically, joint nests have only one brooder (Blanchard, 1934a; Harris et al., 1995), meaning that some females desert their eggs. Females that desert their eggs in joint nests tend to be smaller (unpublished data). The percentage of females involved in joint nests can be over 70% (Blanchard, 1934a; Harris et al., 1995).

C. Larvae/Metamorphosis. For a plethodontid, four-toed salamanders have a short length of larval period of between 20–40 d (Blanchard, 1923; Berger-Bishop and Harris, 1996; O'Laughlin and Harris, 2000). Larvae require still waters (ponds, bogs, swamps, or sluggish streams) that are free of fish. Larvae have the ability to recognize kin (Harris et al., 2003).

D. Juvenile Habitat. Same as adult habitat.

E. Adult Habitat. Adults are found in forests that surround larval habitats. These can include mature hardwood or coniferous forests (Blanchard, 1923; Bishop, 1941b; Wood, 1955). Females have been found to nest in ponds that are near, but not directly adjacent to, a forest (Bishop, 1941b).

F. Home Range Size. Unknown.

G. Territories. Although reported as weakly territorial by Grant (1955), four-toed salamanders have not been shown to defend an area. Carreno and Harris (1998) reported no aggressive defense by brooding females against conspecifics.

H. Aestivation/Avoiding Desiccation. Aestivation is unknown; animals likely avoid desiccating conditions by moving under cover objects or into burrows.

I. Seasonal Migrations. Unknown.

J. Torpor (Hibernation). During the winter, males and females have been found in high density with conspecifics and other amphibians in holes, channels, and crevices in the ground (Blanchard, 1933b).

K. Interspecific Associations/Exclusions. The species is often assumed to be associated with Sphagnum moss for nesting and hardwood forests for juvenile and adult habitat, but see "Egg deposition sites" and "Adult Habitat" above. Four-toed salamanders co-occur with most of the pond-breeding amphibians of the east central and northeastern United States. For example Minton (1998) notes that in east-central Indiana, four-toed salamanders co-occur with eastern newts (*Notophthalmus viridescens*), spotted salamanders (*Ambystoma maculatum*), small-mouthed salamanders (*A. texanum*), tiger salamanders (*A. tigrinum*), unisexual *Ambystoma* hybrids, striped chorus frogs (*Pseudacris triseriata* complex), spring peepers (*P. crucifer*), northern cricket frogs (*Acris crepitans*), green frogs (*Rana clamitans*), and northern leopard frogs (*R. pipiens*).

L. Age/Size at Reproductive Maturity. Considerable geographic variation appears to exist in size at reproductive maturity for females. In montane populations in GWNF (1,000 m elevation), average SVL at reproductive maturity is 42.0 mm (SD = 11.3, n = 49); average total length is 92.8 mm (SD = 6.9, n = 45; R. N. H., unpublished data). In lowland Virginia populations, females can reproduce at SVLs of 29–33 mm (Wood, 1953b). In lowland Michigan populations, females can reproduce at total lengths of 62–72 mm (Blanchard and Blanchard, 1931).

For males in Michigan, individuals can mature as small as 50–52 mm total length, with maturity typically occurring at sizes of at least 54 mm total length (Blanchard and Blanchard, 1931).

Age at reproductive maturity is estimated as 2.3 yr in Michigan (Blanchard and Blanchard, 1931) and 22 mo in lowland populations of Virginia (Wood, 1953b). These estimates are based on analyses of size-frequency histograms and not on longitudinal studies or skeletochronology. As such, these estimates may be underestimated.

M. Longevity. Females appear to be long-lived. Based on longitudinal study in GWNF, a female first captured in 1992 as a breeding adult was recaptured in 1999, although it may have reproduced prior to 1992 (unpublished data). Using a minimum estimate of age at first reproduction as 2 yr, females can live to at least 9 yr of age.

N. Feeding Behavior. Larvae eat small planktonic organisms and small worms (personal observation). Adults also eat arthropods and worms (Bishop, 1919; personal observation). Brooding females do not eat (Wood, 1953b; unpublished data).

O. Predators. Eastern newts prey on larvae (Wells and Harris, 2001; Harris et al., 2003). It is likely that an array of other species prey on larvae, such as larval ambystomatids and odonate naiads. Adult four-toed salamanders will assume a defensive posture in the presence of ring-necked snakes, and skunks may prey on adults (personal observation).

P. Anti-Predator Mechanisms. Brodie (1977) has observed the following anti-predator mechanisms in adults: a glandular concentration on the tail dorsum; tail raised and undulated in the presence of a predator; body can flip or coil; the venter can be exposed; immobility; aposematic coloration; noxious skin secretions; and tail autotomy. Larvae are palatable to newts (Wells and Harris, 2001) and fish (Kats et al., 1988). Predator presence causes larvae to lower their activity level, thereby lowering their growth rate (Wells and Harris, 2001). Eggs are unpalatable to insect predators, such as carabid beetles (Hess and Harris, 2000).

Q. Diseases. Unknown.

R. Parasites. Fungus is known to infest the eggs (Harris and Gill, 1980), The ascomycete fungus *Gliocladium* is pathogenic to eggs. Small nematodes have been observed in the egg jelly (Bishop, 1919). A protozoan parasite has been reported in adults (Blanchard and Blanchard, 1931; Blanchard, 1934c).

4. Conservation.
Early concern was expressed by Wright (1918), who stated that four-toed salamanders were vanishing in New York due to draining of wetlands. Four-toed salamanders are considered Rare in Florida (Means, 1992) and southern New England (Klemens, 1993), and worthy of Special Concern in Maine (Burgason, 1999) and Minnesota (Moriarty and Jones, 1997). C. K. Dodd (1997, personal communication) reports populations in fairly substantial numbers in the Great Smoky Mountains National Park (GSMNP), but considers their overall conservation status as unknown due to their secretive nature, scattered distribution, short larval period, and propensity to brood early in the season. Based on reports from two protected areas (GWNF and GSMNP), it appears that populations are doing well in protected habitats in the upland south (see above). However, J. W. Petranka (personal communication) feels that current abundance is likely to be much lower than historical abundance due to loss of vernal ponds and other small wetlands across the eastern United States (see also Mitchell et al., 1999).

Inventories of four-toed salamanders will need to consider the following aspects of their biology, which will tend to lead to an underestimate of population size: a short larval period, secretive nature of adults, and a propensity of adult females to skip years of reproduction.

Hydromantes brunus Gorman, 1954
LIMESTONE SALAMANDER

David B. Wake, Theodore J. Papenfuss

1. Historical versus Current Distribution.
Limestone salamanders (*Hydromantes brunus*) were discovered in 1952 when an area of limestone outcrops in an unlikely region (relatively low elevation in a region

that becomes extraordinarily hot and dry in summer and experiences prolonged summer drought) was targeted for search. A previously unknown population was found in an area of limestone talus. The species occurs only along a short section (about 16–17 km) of the Merced River, from the vicinity of the type locality on the main highway to Yosemite National Park west to a region known as Hell Hollow, and a short distance up the North Fork of the Merced River, between elevations of 365–760 m. A fossil (Late Pleistocene) vertebra belonging to a species of *Hydromantes* was recovered from a woodrat midden found outside the southwestern margin of the range of this species in the lower Kings Canyon at 1,280 m. It likely represents either this species or Mt. Lyell salamanders (*H. platycephalus*). If it is this species, its range was more extensive in the past (Mead et al., 1985).

B. Eggs.

i. *Egg deposition sites.* Oviposition sites are unknown.

ii. *Clutch size.* Unknown.

C. Direct Development. Egg deposition has not been observed, nor has development, but its close relatives undergo direct development. Ovarian eggs were enlarged (4.6 mm in diameter) in the holotype collected in late February (Gorman, 1954). Eggs probably are laid in late spring and develop over the summer.

D. Juvenile Habitat. Unlikely to differ from adult habitat use. Juveniles and adults typically are active at the same time.

E. Adult Habitat. There is a general association with limestone, but salamanders have been found on the surface under both slate slabs and irregularly shaped pieces of limestone. They have been found in small areas of moss-covered or

steep slopes, where individuals use their tail to assist in locomotion (Gorman, 1954). Temperatures at which this species has been taken range from 10.0 °C–14.0 °C (mean 11.4 °C; Feder et al., 1982).

F. Home Range Size. Unknown.

G. Territories. Unknown.

H. Aestivation/Avoiding Desiccation. Salamanders are present on the surface when the soil is moist and temperatures moderate to cool. The species is thus active throughout the winter, but as spring progresses they disappear from the surface, typically by mid-April or earlier. However, they are probably active throughout the summer below the surface; an adult was found in July in an abandoned mine shaft near the type locality (D. B. W., unpublished observation).

I. Seasonal Migrations. Unknown and unlikely.

J. Torpor (Hibernation). Unknown and unlikely.

K. Interspecific Associations/Exclusions. Other amphibians known to be associated with this species include arboreal salamanders (*Aneides lugubris*), Hell Hollow slender salamanders (*Batrachoseps diabolicus*), yellow-eyed salamanders (*Ensatina eschscholtzii xanthoptica*), and Sierra newts (*Taricha torosa sierrae*; Gorman, 1954).

L. Age/Size at Reproductive Maturity. Unknown.

M. Longevity. Unknown.

N. Feeding Behavior. Members of this genus have highly specialized tongue projection capability and can project the tongue both far and fast (e.g., Deban et al., 1997).

O. Predators. Unknown.

P. Anti-Predator Mechanisms. Unknown.

Q. Diseases. Unknown.

R. Parasites. Unknown.

S. Comments. This is the largest of the California species of this genus and differs from the other members of the genus in its relatively uniform coloration, consisting of a rich brown. It is the most genetically distinct of the three species (Wake et al., 1978; Jackman et al., 1997). General species accounts have been provided by Stebbins (2003), Gorman (1954, 1964, 1988), and Petranka (1998).

Limestone Salamander *(Hydromantes brunus)*

2. Historical versus Current Abundance. There is no indication that the size of the range or the density of this species has changed, but neither have there been any significant new discoveries of populations for many years.

3. Life History Features.

A. Breeding. Reproduction is terrestrial.

i. *Breeding migrations.* Unknown and unlikely.

ii. *Breeding habitat.* Likely to be in cracks and crevices below the surface in limestone talus.

barren talus (Tordoff, 1980), as well as in rock crevices and even in abandoned mine tunnels. The vegetation in the region where salamanders are found is mainly chaparral, with a scattering of xeric-adapted trees such as gray pine (*Pinus sabiniana*), and with California laurel (*Umbellularia californica*) in more mesic sites. In general, the habitat appears marginal for salamander occupancy, and yet five species are sympatric in this seemingly hostile environment. The species has been found on relatively level ground, but it is more typically encountered on

4. Conservation. The type locality of this species is along the main access route to Yosemite National Park, and any widening of the road would destroy habitat, which is already highly restricted. Limestone salamanders are listed as a state Threatened species by the California Department of Fish and Game (www.dfg.ca.gov) and are protected. Apparently, they are in no immediate danger. In 1975, a Limestone Salamander Ecological Reserve was established by the Department on 48 ha of land, including the type locality (Anonymous, 1980).

Hydromantes platycephalus (Camp, 1916 [b])
MT. LYELL SALAMANDER

David B. Wake, Theodore J. Papenfuss

1. Historical versus Current Distribution.
Mt. Lyell salamanders (*Hydromantes platycephalus*) were discovered accidentally in the summer of 1915 when a male and a female were caught in a single snap-trap set for small mammals high on Mt. Lyell 3,292 m (10,800 ft) in Yosemite National Park, California (Camp, 1916b). For many years they were known only from the immediate vicinity of the park. Later they were discovered to the north at Sonora Pass and to the south at Silliman Gap in Sequoia National Park, but were considered to be restricted to the high southern Sierra Nevada (Adams, 1942). In recent years, isolated populations have been found at Smith Lake, in the Desolation Wilderness west of Lake Tahoe, and on the Sierra Buttes (Feder et al., 1982), near the northern end of the Sierra Nevada in Sierra County, the northernmost locality (ca. 39° 35'N). They have been found as far south as ca. 36° 25'N. The elevational range is from 1,220 m (at the base of Bridal Veil Falls, Yosemite National Park) to about 3,600 m. Salamanders tentatively assigned to this species occur in canyons along the east side of the Sierra Nevada as low as the floor of the Owens Valley (about 1,500 m).

2. Historical versus Current Abundance.
There is no indication that either the size of the range or the density of this species has changed recently. In fact, new discoveries continue to expand the known range at regular intervals. A fossil (Late Pleistocene) vertebra belonging to a species of *Hydromantes* was recovered from a woodrat midden found outside the southwestern margin of the range of this species in the lower Kings Canyon at 1,280 m. It may represent this species, and, if so, the range of the species was more extensive in the past (Mead et al., 1985).

3. Life History Features.
A. Breeding. Reproduction is terrestrial.
i. Breeding migrations. Unknown and unlikely.
ii. Breeding habitat. Likely to be in cracks and crevices below the surface in moist or wet granite talus or in other subterranean cavities.
B. Eggs.
i. Egg deposition sites. Oviposition sites are unknown.
ii. Clutch size. Unknown.
C. Direct Development. Egg deposition has not been observed, nor has development, but its close relatives undergo direct development. Ovarian eggs are enlarged in females collected from late June to early August (Stebbins, 1951). Dates of hatching are unknown, but apparent hatchlings are found in early summer (Adams, 1942).
D. Juvenile Habitat. Unlikely to differ from adult habitat use. The most detailed account of the natural history of this species discusses juveniles, but notes no differences in ecology from adults (Adams, 1942).
E. Adult Habitat. Mt. Lyell salamanders commonly are found in talus slopes of granite through which water is flowing. Granite boulders that are roughly quadrangular or flat, and resting on granite bedrock that have a thin film of flowing water are favored microhabitats. Favored habitats are downslope from melting snowfields, which persist long into or even through the summer in the high Sierra Nevada. Individuals also are encountered under rocks at the edge of streams; at low elevations, they may be found in direct contact with moist soil under rocks or occasionally under pieces of wood. Salamanders also are found under moss on wet rock faces (e.g., in Yosemite Valley). Usually the collecting sites are in the open, although stream-side localities may be shaded. Most of the range is in the elevational zone historically called "Hudsonian." Gorman (1988) reported that Mt. Lyell salamanders are active at lower temperatures than the other species of *Hydromantes* and that they become inactive as the temperature increases. Body temperatures have varied greatly, from $-2.0°C$ to $11.5°C$ (mean $5.7°C$) in the central Sierra Nevada (Brattstrom, 1963), and from $13.8-17.8°C$ (mean $15.0°C$) on Sierra Buttes at 2,125 m (Feder et al., 1982). In a temperature gradient in the laboratory, specimens preferred $13-14°C$ (Brattstrom, 1963).
F. Home Range Size. Unknown. Gorman (1988) believed local populations to be small (on the order of 60 individuals) and reported the population on top of Half Dome as consisting of 120–130 individuals, but no supporting data were published.
G. Territories. Unknown.
H. Aestivation/Avoiding Desiccation. As summer progresses into higher elevations, surface activity is curtailed as the animals move into below-ground microhabitats. Whether they aestivate or simply move away from the surface is unclear.
I. Seasonal Migrations. Unknown and unlikely.
J. Torpor (Hibernation). Unknown, but almost certainly occurs at high elevation as snow accumulates and temperatures drop.
K. Interspecific Associations/Exclusions. Unknown. In general, this is the only species of salamander in its range. However, Sierra Nevada salamanders (*Ensatina eschscholtzii platensis*) are known from Yosemite Valley and might be locally sympatric with low elevation populations of *Hydromantes*. At the southernmost localities in the Owens Valley, populations tentatively assigned to this species are locally sympatric with Kern Plateau salamanders (*Batrachoseps robustus*; Wake et al., 2002).
L. Age/Size at Reproductive Maturity. Unknown, but Adams (1942) presents size-frequency histograms for a population from the top Half Dome, Yosemite National Park, and argues that adults are

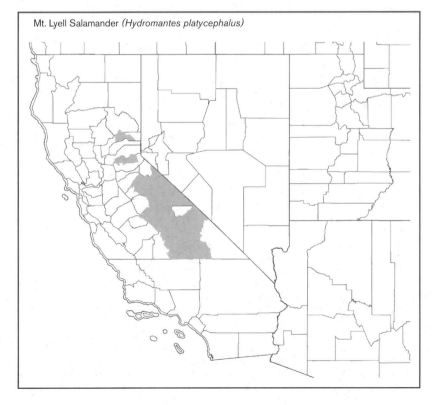

Mt. Lyell Salamander (*Hydromantes platycephalus*)

survivors of the first 2 yr of growth. He also argues that enlarged maxillary and premaxillary teeth, secondary sexual characters of males of this species, may serve as a guide to size and age at first reproduction, which he believes to be at approximately 2.5 yr of age.

M. Longevity. Unknown, but longer than 3 yr and possibly much longer.

N. Feeding Behavior. Members of this genus have highly specialized tongue projection capability and can project the tongue both far and fast (e.g., Deban et al., 1997). Adams (1942) reported that food includes mostly small, ground-dwelling insects, spiders, and centipedes.

O. Predators. Unknown.

P. Anti-Predator Mechanisms. Adults respond to being handled or being touched with a sharp object by raising the head and tail and depressing the body (Hansen, 1990; Stebbins, 1951). The skin contains toxins that have been reported to produce temporary (about 30 hr) blinding when introduced accidentally into human eyes (Hansen, 1990). García-París and Deban (1995) also reported that the salamanders produced a sticky, noxious secretion that burned the eyes. Adults also respond to simulated predator attacks by coiling the body and tail tightly, forming a spherical body that rolls downhill under favorable conditions (García-París and Deban, 1995). Mt. Lyell salamanders are adept at climbing and maneuvering on wet, sloping surfaces using their webbed feet and employing their short, blunt tail as an aid in locomotion (Stebbins, 1947).

Q. Diseases. Unknown.

R. Parasites. Unknown.

S. Comments. General species accounts have been provided by Stebbins (1985), Gorman (1964, 1988), Jennings and Hayes (1994a), and Petranka (1998). The three species of *Hydromantes* in California are close relatives, based on general similarities in allozymes, albumin (Wake et al., 1978), mitochondrial DNA sequences (Jackman et al., 1997), and morphology (Gorman, 1988, reported results from his unpublished doctoral dissertation), but Mt. Lyell salamanders differ markedly from the other species in occurring at much higher elevations and having more specialized habitat characteristics (being associated with granite rather than limestone).

4. Conservation.
Mt. Lyell salamanders enjoy protected status and appear to be in good condition. Much of their range is in National Parks and Wilderness Areas, so there are few threats from human activities, although road construction in the Sonora Pass region could harm excellent habitat, that occurs near roadside. They are protected as a species of Special Concern by the California Department of Fish and Game and may not be taken or possessed at any time.

Hydromantes shastae Gorman and Camp, 1953

SHASTA SALAMANDER

David B. Wake, Theodore J. Papenfuss

1. Historical versus Current Distribution.
Shasta salamanders *(Hydromantes shastae)* were discovered prior to 1915 but not described formally until their rediscovery in 1950 (Gorman and Camp, 1953), north of the current Shasta Reservoir in northern California. Subsequently, they have been found in a number of sites in the vicinity of the reservoir, east and west of the Sacramento River, and both north and south of the Pit River arm of the reservoir. However, the total range of the species is <35 km in greatest dimension. The species ranges between about 300 and 900 m in elevation.

2. Historical versus Current Abundance.
There is no indication that either the size of the range or the density of this species have undergone any substantive changes, although construction of Shasta Dam, road building, and mining all have impacted the species in the past. Development, mining, and other human activities continue to threaten the species, but most of the range of the species is on national forest lands that have little timber value.

3. Life History Features.

A. Breeding. Reproduction is terrestrial.

i. Breeding migrations. Unknown and unlikely.

ii. Breeding habitat. Caves have been the only sites where eggs have been found (Gorman, 1956); it is not known whether salamanders breed in other areas.

B. Eggs.

i. Egg deposition sites. Egg deposition has not been observed. The only eggs discovered were deposited in a cave (Gorman, 1956).

ii. Clutch size. Gorman (1956) studied two clutches of nine eggs each.

C. Direct Development. Gorman (1956) described late stages of development. Hatching of these clutches would have occurred in late October to early November had the eggs been left where found.

Shasta Salamander *(Hydromantes shastae)*

D. Juvenile Habitat. Unlikely to differ from adult habitat use. Juveniles and adults are found together on the surface during the winter.

E. Adult Habitat. Although the species was originally described from a cave and considered to be a cave species, individuals are commonly encountered on the surface from late autumn to early spring. Most populations occur in areas where limestone outcrops occur, and individuals are active on moist rock faces by night. Salamanders are found under rocks and logs and in leaf litter in lightly to densely forested areas, as well as in small to large caves. The forest is dominated by California black oaks *(Quercus kelloggii)* and pines *(Pinus sp.)*.

F. Home Range Size. Unknown.

G. Territories. Unknown.

H. Aestivation/Avoiding Desiccation. As summer progresses into higher elevations, surface activity is curtailed as the animals move into below-ground microhabitats. Whether they aestivate or simply move away from the surface is unclear. They have been found on the walls of caves during the summer. In August 1978, over 20 individuals were observed in a single small cave.

I. Seasonal Migrations. Unknown and unlikely.

J. Torpor (Hibernation). Unknown and unlikely, because the winter is mild and they are active throughout.

K. Interspecific Associations/Exclusions. A common associate are Oregon salamanders *(Ensatina eschscholtzii oregonensis)*. Less commonly encountered but with broadly overlapping distributions are black salamanders *(Aneides flavipunctatus)*, Sierra newts *(Taricha torosa sierrae)*, and coastal giant salamanders *(Dicamptodon tenebrosus)*.

L. Age/Size at Reproductive Maturity. Unknown.

M. Longevity. Unknown, but longer than 3 yr and possibly much longer.

N. Feeding Behavior. Members of this genus have highly specialized tongue projection capability, and can project the tongue both far and fast (e.g., Deban et al., 1997).

O. Predators. Unknown.

P. Anti-Predator Mechanisms. Adults respond to being handled or being touched with a sharp object by raising and undulating the tail (Brodie, 1977).

Q. Diseases. Unknown.

R. Parasites. Unknown.

S. Comments. This species shows great genetic substructuring, especially unusual given its small geographic range (Wake et al., 1978). The genetic data were analyzed by Larson et al. (1984), who found that the species conforms to a genetic structure and pattern of gene flow in accordance with an island model and a value of *Nm* (Slatkin, 1981) of 0.125, which means that effectively no species-wide gene flow is taking place now or in the recent past. Their distribution has been carefully mapped (Papenfuss and Brouha, 1979). General species accounts have been provided by Stebbins (1951, 1985), Gorman (1964, 1988), and Petranka (1998).

The three species of *Hydromantes* in California are closely related (Wake et al., 1978; Jackman et al., 1997), and this species appears to be a close relative of Mt. Lyell salamanders *(H. platycephalus)*, from

which they differ in elevational zonation and habitat characteristics. This species shows strong geographic differentiation in both allozymes and albumin (Wake et al., 1978).

4. Conservation.
Shasta salamanders are considered State Threatened by the California Department of Fish and Game (www.dfg.ca.gov); with this protection they may not be taken or possessed at any time. Much of the distribution of Shasta salamanders is on national forest land, and there is little reason to be concerned for their survival. However, a number of populations are small and isolated, and human activities could easily lead to their destruction by relatively small amounts of habitat disturbance.

Phaeognathus hubrichti Highton, 1961
RED HILLS SALAMANDER

C. Kenneth Dodd Jr.

1. Historical versus Current Distribution.
Red Hills salamanders *(Phaeognathus hubrichti)* are known from 13 discrete populations located in the Red Hills physiographic region between the Alabama and Conecuh rivers in south-central Alabama (Dodd, 1991). Electrophoretic evidence suggests weak separation of populations east and west of the Sepulga River (McKnight et al., 1991). Whereas salamanders have disappeared or declined at several sites within these populations because of adverse land-use practices, there is no evidence that the current range of Red Hills salamanders is any different from its historical range.

2. Historical versus Current Abundance.
Because of the secretive nature of this species, there are no range-wide estimates of abundance, either current or historical.

Areas occupied by Red Hills salamanders may be widely separated, and even within a site, burrows are not spaced uniformly. Presumably, burrow density and abundance are correlated with Red Hills salamander density and abundance, but the relationship is unclear. Dodd (1991) used time constraint sampling to determine the location of sites with relatively large numbers of burrows, then used line transect techniques to obtain statistically rigorous estimates of burrow abundance (Dodd, 1990b). In optimum habitat (n = 10 sites), he estimated 2.6–9.4 burrows/$100m^2$ (mean = 5.05 burrows/$100m^2$). Carroll et al. (2000) used mark–recapture data collected from a single 18-m transect to estimate a salamander abundance of 0.5 animals/m^2, again in optimal habitat. They estimated that 26 animals inhabited the 33 burrows observed, or 0.8 salamanders/burrow. Burrow abundance varies considerably among sites, however, particularly in areas affected by clearcutting and other forestry practices. In addition, there are no long-term data on the number of burrows occupied by an individual salamander; salamanders have been observed in >1 burrow, and >1 salamander has been observed in a single burrow. Finally, burrows do not seem to persist. Gunzburger and Guyer (1998) found that 50% of marked burrows remained open after 6 mo and inferred that Red Hills salamanders must continually repair old burrows and open new burrows because of the dynamic nature of the steep ravines they inhabit. These factors taken together suggest that local abundance varies considerably and that it will be difficult to obtain good estimates of relative abundance.

Although the factors discussed above make it difficult to obtain accurate estimates of Red Hills salamander abundance,

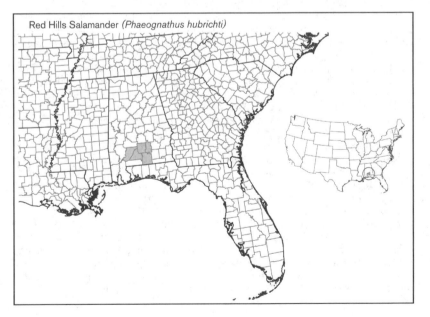

Red Hills Salamander *(Phaeognathus hubrichti)*

it is apparent that clearcutting and other forestry practices reduce or eliminate Red Hills salamanders in what was once optimal habitat. Red Hills salamander burrows may persist at reduced densities after cutting as long as burrow systems are not destroyed. It is unknown how long it might take for densities to reach pre-cut levels. If burrow systems are destroyed mechanically, as by plowing, tilling, or other forms of intensive site preparation, Red Hills salamanders disappear or are confined to small refugia unimpacted by site disturbance. Thus, current Red Hills salamander abundance is probably similar to historical abundance at some sites, but drastically reduced at other sites depending on the historical and current land-use practices.

3. Life History Features.

The life history and biology of Red Hills salamanders are summarized by Petranka (1998).

A. Breeding. Reproduction is terrestrial.

i. Breeding migrations. It is unknown how Red Hills salamanders find each other, but true breeding migrations probably do not occur.

ii. Breeding habitat. Courtship by Red Hills salamanders presumably takes place within the burrow system or in deep interconnecting rock fissures underground.

B. Eggs.

i. Egg deposition sites. Egg clusters of this species have not been found in nature, but egg deposition presumably takes place within the burrow system or in deep interconnecting rock fissures underground.

ii. Clutch size. In captivity, a single 115-mm Red Hills salamander produced 16 eggs (Brandon and Maruska, 1982).

C. Direct Development. Red Hills salamanders have direct development within the egg, one of only three desmognathines with this form of development. When hatched, the young resemble the adults.

i. Brood Sites. Eggs are presumably deposited within the burrow system or in deep interconnecting rock fissures underground. In captivity, a single 115-mm Red Hills salamander attached her egg cluster to an overhanging support via a stalk (Brandon and Maruska, 1982).

ii. Parental Care. Parental care is unknown, although many desmognathine salamanders remain with their eggs during development. In the single case of egg deposition in captivity, the female did not brood her eggs and none developed (Brandon and Maruska, 1982).

D. Juvenile Habitat. Very few juveniles have been observed, and all within the burrows of adult salamanders. Nothing is known concerning specialized habitat characteristics.

E. Adult Habitat. Males and females occupy similar habitats and are active year-round (Bakkegard, 2001, 2002). Burrows are key to the survival of Red Hills sala-

manders, and few individuals have been observed outside burrows (Gunzburger and Guyer, 1998). Red Hills salamander burrows are confined to steep (slope: mean = 50°; height: mean = 17 m) mesic ravines within the Tallahatta and Hatchetigbee geological formations. These geological formations consist of extremely porous rocks with a large water-holding capacity. Such ravines are humid and have friable soils ideal for burrow construction and large invertebrate populations. North-facing slopes are preferred, but salamanders may be found on slopes facing all directions. Most burrows are found on the lower 2/3 of a slope. The forest canopy consists of broad-leaved deciduous trees that provide shade and retain high humidity. These conditions combine to provide optimal habitat (see Valentine, 1963b; Schwaner and Mount, 1970; Jordan, 1975; French and Mount, 1978; and Dodd, 1991).

F. Home Range Size. Unknown.

G. Territories. Red Hills salamanders rarely leave their burrows, although movements away from the burrow have been observed (Bakkegard, 2001, 2002). It is unknown if Red Hills salamanders defend burrows, but sightings of different Red Hills salamanders within a single burrow suggest this is not the case. Males could defend territories (burrows?) during the breeding season, but it is not known if they do so.

H. Aestivation/Avoiding Desiccation. Red Hills salamanders probably retreat deep into the burrow system during periods of extreme cold or drought. However, they likely are active year-round when temperature and moisture conditions are favorable. The periods of greatest activity are the spring and summer seasons.

I. Seasonal Migrations. Not known to occur.

J. Torpor (Hibernation). See "Aestivation/Avoiding Desiccation" above.

K. Interspecific Associations/Exclusions. The burrows of the Red Hills salamanders are used by other salamanders, including southern two-lined salamanders (*Eurycea cirrigera*), three-lined salamanders (*E. guttolineata*), southeastern slimy salamanders (*Plethodon grobmani*), and eastern newts (*Notophthalmus viridescens*; see also Brandon, 1966b). Invertebrates also undoubtedly use the burrows.

L. Age/Size at Reproductive Maturity. Brandon (1965b) suggested that Red Hills salamander males attain sexual maturity at approximately 80 mm "body length" (presumably snout-vent length) and females at 100 mm body length. For females, this occurs in 5–6 yr of age (Parham et al., 1996). The age of maturity for males is uncertain, but males probably mature earlier than females. Egg deposition extends from early spring to September (Brandon, 1965; Schwaner and Mount, 1970).

M. Longevity. Based on skeletochronology, Parham et al. (1996) recorded a 121-mm SVL male at 11 yr. They suggested that it is unlikely that Red Hills salamanders live much longer.

N. Feeding Behavior. Red Hills salamanders sit at or just inside the burrow mouth and ambush invertebrates passing the entrance or entering the burrow. At night, they partially emerge from the burrow to attack prey. Prey include snails, millipedes, insects, insect larvae, spiders, mites, and probably any small animal that can be caught and swallowed (Brandon, 1965b; Gunzburger, 1999). They appear to avoid daddy longlegs (Phalangiidae; Bakkegard, 2002).

O. Predators. Direct predation has not been recorded, but small mammals (shrews; Soricidae) and reptiles likely eat this species. Feral pigs (*Sus* sp.) and armadillos (*Dasypus novemcinctus*) are common in the Red Hills and do considerable damage to the steep ravines inhabited by Red Hills salamanders. They undoubtedly dig out and eat salamanders whenever possible (Dodd, 1991).

P. Anti-Predator Mechanisms. The burrow would seem to be their primary defense. Red Hills salamanders are not known to possess granular skin glands, and hence apparently do not have skin secretions that repel predators. No defensive postures are known. However, Red Hills salamanders are capable of biting and will spin when grabbed (C. K. D. and C. Guyer, personal communication).

Q. Diseases. Nothing known.

R. Parasites. Intestinal trematodes (*Brachycoelium salamandrae*) and colic roundworms (*Oxyuris* sp.; Brandon, 1965b).

4. Conservation.

Red Hills salamanders are protected as Threatened under provisions of the Federal Endangered Species Act of 1973, as amended (USFWS, 1976). Six Habitat Conservation Plans (HCPs) for populations of Red Hills salamanders, covering approximately 25,169 ha, have been approved by the U.S. Fish and Wildlife Service in cooperation with timber companies in south Alabama. The goals of the Red Hills salamander HCPs are to allow for timber harvesting while promoting species conservation. Red Hills salamanders are listed as a protected non-game species by the State of Alabama.

Virtually all Red Hills salamander habitat is on private land, with only a small amount owned by the U.S. Army Corps of Engineers and the State of Alabama. Several (n = 25) excellent tracts of habitat remained in private ownership in 1988, and these were recommended for acquisition for conservation purposes (Dodd, 1988). To date, none have been acquired. Long-term protection is best assured through private landowner cooperation. With that in mind, Dodd (1988,

1991) recommended a series of management actions that would help to maintain the integrity of salamander habitat, especially in areas where forestry occurred. These included avoiding clearcutting on slopes containing salamander burrows; eliminating mechanical site preparation; maintaining woody leaf litter and an overstory hardwood tree canopy; leaving buffer zones above and below slopes containing salamander burrows; avoiding herbicides and other chemical applications; and allowing hardwoods to regenerate on previously cut and selectively cut slopes containing salamander burrows. Protection from collecting is also important for this species because of its restricted range and vulnerable populations; because it is the sole member of a monotypic genus, the collector value is high.

In areas unaffected by forestry (which are few in south Alabama), populations of Red Hills salamanders are likely stable. If habitat is degraded, Red Hills salamander populations will decline and may disappear. For this reason, periodic assessments of Red Hills salamander habitat must be carried out, especially to ensure that HCP provisions are being honored and that salamander habitat is protected during forestry operations.

Plethodon ainsworthi Lazell, 1998
BAY SPRINGS SALAMANDER

James Lazell

1. Historical versus Current Distribution.
Bay Springs salamanders (*Plethodon ainsworthi*) are known from only two specimens collected in 1964 from a single site in Jasper County, Mississippi (Lazell, 1998). Subsequent collections (1 March 1991; 23 April 1991; 17 July 1991; 4 June 1992; 20–21 May 1994; 14–16 December 1994; 8 and 20 February 1995; 11 and 23 March 1995; 3 and 13 April 1995; and 27 February 1997) attempted at and near the presumed collection site using visual searching and coverboard techniques revealed other salamander species (see "Interspecific Associations/Exclusions" below) but failed to discover additional Bay Springs salamander specimens. Bay Springs salamanders, combining features of the more derived groups within the genus *Plethodon*, may be close to the ancestral stock of the genus. If so, biogeographical comparisons to other groups such as cambarid crawfishes (Fitzpatrick, 1986) may be revealing. The Miocene path of the Tennessee River cut nearly straight southwest from the central Appalachian Mountains to cross the Bay Springs area. As with some crawfishes, the Bay Springs salamander may be a relictual isolate far downstream of the region of active evolutionary radiation in the highland headwater drainages.

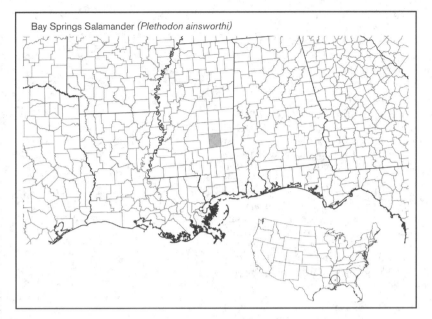

Bay Springs Salamander (*Plethodon ainsworthi*)

2. Historical versus Current Abundance.
Unknown, but not demonstrably less common than the *Eurycea* and *Pseudotriton* species that occur at the potential sites (see "Interspecific Associations/Exclusions" below)—each of these species is represented by one or two specimens to date. Ainsworth's original collections did not include any of these species, but the data are too sparse to indicate any change in relative abundance. Both the Ainsworth site and nearby Six Springs (Lazell, 1998, p. 970) are intact and in good ecological condition. The single night hunt conducted to date, on the dry night of 23 March 1995, produced a glimpse of a possible Bay Springs salamander (R. Highton, personal communication). Rainy night hunts and drift fences with pitfall traps, presumably the best collecting methods, have yet to be tried.

3. Life History Features.
Bay Springs salamanders do not fit into the usual morphological groups described by Highton (1962a) and Highton and Larson (1979), and because modern molecular techniques cannot be done on formalin-fixed animals, the phylogenetic relationships of this species will not be known until additional animals are discovered.

A. Breeding. Reproduction is terrestrial.
i. Breeding migrations. Unknown.
ii. Breeding habitat. Unknown.
B. Eggs.
i. Egg deposition sites. Unknown.
ii. Clutch size. Unknown.
C. Direct Development. Presumed.
i. Brood sites. Unknown.
ii. Parental care. Unknown.
D. Juvenile Habitat. Unknown, but likely is similar to adult habitat.

E. Adult Habitat. The presumed type locality is now a 2 ha woods. The predominant tree species are sweet gums (*Liquidambar styraciflua*), tulip trees (*Liriodendron tulipifera*), water oaks (*Quercus nigra*), white oaks (*Q. alba*), and loblolly pines (*Pinus taeda*). Understory species include sweet bay (*Magnolia virginiana*), magnolia (*M. grandiflora*), sourwood (*Oxydendron arboreum*), dogwood (*Cornus florida*), and holly (*Ilex opaca*). Ground cover includes ferns, especially royal ferns (*Osmunda regalis*) and sensitive ferns (*Onoclea sensibilis*), orchids (*Plantanthera clavilata* and *P. ciliaris*), greenbriars (*Smilax* sp. esp. *pumila*), and jack-in-the-pulpits (*Arisaema triphyllum*).

F. Home Range Size. Unknown, but likely to be small.

G. Territories. Unknown.

H. Aestivation/Avoiding Desiccation. Unknown.

I. Seasonal Migrations. Unknown, but unlikely.

J. Torpor (Hibernation). Unknown, but at this southern latitude, unlikely.

K. Interspecific Associations/Exclusions. Searches for Bay Springs salamanders have revealed the presence of Mississippi slimy salamanders (*P. mississippi*), southern dusky salamanders (*Desmognathus auriculatus*), southern two-lined salamanders (*Eurycea cirrigera*), and southern red salamanders (*Pseudotrition ruber vioscai*) at Ainsworth's springs, and three-lined salamanders (*E. guttolineata*; but not southern two-lined salamanders or red salamanders) at Six Springs. Southern dusky salamanders and Mississippi slimy salamanders are abundant at both sites.

L. Age/Size at Reproductive Maturity. Unknown.

M. Longevity. Unknown.

N. **Feeding Behavior.** Unknown.
O. **Predators.** Unknown.
P. **Anti-Predator Mechanisms.** Unknown.
Q. **Diseases.** Unknown.
R. **Parasites.** Unknown.

4. Conservation.
The conservation status of this species is unknown. David A. Beamer (personal communication) recently visited the type locality of *P. ainsworthi* and noted that the region has been planted to pine along the steep slopes, even across the tiny trickling streams in the ravines. Further, he notes that the type locality itself has a small amount of reasonable forest but it is surrounded by poor quality habitat. On one side is a cow pasture, another side has been recently clearcut, and yet another side is an old farmyard now covered with bamboo.

Acknowledgments. Richard Highton reviewed this account and provided valuable feedback.

Plethodon albagula Grobman, 1944
WESTERN SLIMY SALAMANDER

Carl D. Anthony

1. Historical versus Current Distribution.
Western slimy salamanders (*Plethodon albagula*) are widely distributed from Missouri south through the mountainous areas of Arkansas and extreme eastern Oklahoma. Disjunct populations occur in central Texas, with two isolates occurring in southeastern and northeastern Texas. There is some doubt concerning the presence of these populations, but museum records exist, as indicated below (R. Highton, personal communication). Western slimy salamanders were previously considered to be a subspecies of northern slimy salamanders (*P. glutinosus*), recognized as a species by Highton et al. (1989). Highton et al. (1989–addendum) noted that a population of Texas western slimy salamanders (the only light-chinned population examined) was the most divergent of *albagula* populations studied. Missouri, Arkansas, and Oklahoma populations are dark-chinned, as are the eastern Texas isolates and some of the central Texas populations (Highton, 1962a). If a new taxonomy is proposed, light-chinned populations in Texas will retain the name *albagula*. Though no recent specimens have been collected in eastern Texas, the southeastern Texas localities are indicated on some recent range maps (Conant and Collins, 1991; Petranka, 1998), but not others (Dixon, 1987, 2000; Garrett and Barker, 1987). Interestingly, the northeast Texas locality is reported as near the town of Big Sandy, while one of the southeastern Texas localities is near a town (Dallardsville) that was historically known as "Big Sandy."

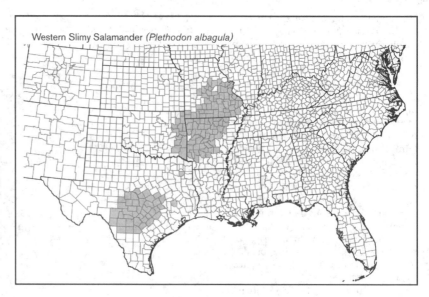

Western Slimy Salamander (*Plethodon albagula*)

2. Historical versus Current Abundance. Unknown.

3. Life History Features.
   A. Breeding. Reproduction is terrestrial.
   i. Breeding migrations. In Oklahoma and Arkansas, males have well-developed mental glands in late spring and late fall. Gravid females began entering an abandoned mine shaft near Hot Springs, Arkansas, in late August and began exiting the mine in late December, after brooding their eggs (S. Trauth, personal communication).
   ii. Breeding habitat. Unknown, but likely the damp ravines, wooded hillsides, cave entrances, wooded canyons, and ravines of adult habitat (Grobman, 1944; Crowell, 1981; Johnson, 1987) or the caves and abandoned mines known to be used as egg deposition sites (Noble and Marshall, 1929; Wells and Gordon, 1958; Heath et al., 1986; Trauth, et al., 2003).
   B. Eggs.
   i. Egg deposition sites. Egg clutches have been reported in caves (Noble and Marshall, 1929; Wells and Gordon, 1958) and abandoned mines (Heath et al., 1986).
   ii. Clutch size. Clutches with 18 and 10 eggs were found by Noble and Marshall (1929). Clutches with 8 and 11 eggs were reported by Wells and Gordon (1958). A clutch size range of 6–15 was reported from an abandoned mine shaft near Hot Springs, Arkansas (Trauth et al., 2003).
   C. Direct Development. Clutches were found with 18 eggs on 17 August and with 10 well-developed embryos on 3 September in an Arkansas cave (Noble and Marshall, 1929). On 27 October, clutches with 8 and 11 full-term embryos were reported from a Missouri cave (Wells and Gordon, 1958). In Arkansas, brooding females remain with their egg clutches for nearly 4 mo; hatching begins in late November to early December and continues until

late January to mid-February (Trauth, et al., 2003).
   D. Juvenile Habitat. Same as adult habitat.
   E. Adult Habitat. Damp ravines, wooded hillsides (Johnson, 1987), cave entrances (Crowell, 1981), wooded canyons, and ravines of the Edwards Plateau, Texas (Grobman, 1944).
   F. Home Range Size. Home range size is unknown.
   G. Territories. In Oklahoma, adults utilize the same cover objects from year to year (Anthony, 1995); in Arkansas, females use the same brooding sites from year to year (S. Trauth, personal communication). Adult males from Oklahoma (Anthony et al., 1997) and Arkansas (Forster, 1998) defend areas from conspecific intruders in laboratory chambers. Territory size is unknown.
   H. Aestivation/Avoiding Desiccation. Some Arkansas populations utilize caves during the summer (Crowell, 1981). Other populations remain underground from late May to mid-September.
   I. Seasonal Migrations. None reported.
   J. Torpor (Hibernation). Northern populations become inactive in late November and return to the surface in late March. Southern populations are probably surface active during cool, wet periods throughout the year.
   K. Interspecific Associations/Exclusions. Western slimy salamanders are found much less commonly than Rich Mountain salamanders (*P. ouachitae*) where these species co-occur in the Ouachita Mountains (Pope and Pope, 1951; Spotila, 1972; Duncan and Highton, 1979). Western slimy salamanders may be excluded from areas by Rich Mountain salamanders (Anthony et al., 1997), perhaps through interspecific territoriality. In laboratory encounters, adult males were largely unsuccessful in usurping cover objects from

smaller Rich Mountain salamanders (Anthony et al., 1997). Kuss (1986) found that habitat differed between the two species in Arkansas, with western slimy salamanders occurring more often at lower elevations, nearer to ravines, and under thinner vegetation than Rich Mountain salamanders. Spotila (1972) noted that western slimy salamanders are found in wetter microhabitats than are Rich Mountain salamanders. Dowling (1956) observed that western slimy salamanders become active on the forest floor later in the evening than do Rich Mountain salamanders, which may be able to withstand warmer conditions. Britton (1986) found that in northwestern Arkansas, where western slimy salamanders occur with northern zigzag salamanders *(P. dorsalis)*, the latter is restricted to seepage-type habitats while western slimy salamanders are found more often in less mesic hillside habitats.

L. Age/Size at Reproductive Maturity. Adults probably mature at about 50 mm SVL.

M. Longevity. Individuals that were sexually mature when collected have been kept in the laboratory for 3 yr. Longevity is probably >6 yr.

N. Feeding Behavior. Prey of specimens from central Texas consisted of hymenopterans (Formicidae), coleopterans, and isopods (Oliver, 1967). In northwestern Arkansas, Crowell (1981) found dipterans and formicids to be important prey of individuals utilizing cave entrances.

O. Predators. Unknown.

P. Anti-Predator Mechanisms. Nocturnal. All members of the genus *Plethodon* produce noxious skin secretions (Brodie, 1977). When handled, western slimy salamanders release an adhesive secretion that is difficult to remove from the fingers and hands (Johnson, 1987). Adults are agile and in some habitats flee to below ground retreats when disturbed (Pope and Pope, 1951).

Q. Diseases. Foot abnormalities, apparently developmental in origin, have been reported from a Hayes County, Texas, population (Lazell, 1995). The cause of these anomalies is unknown.

R. Parasites. Winter et al. (1986) found nematodes and protozoans in six adults from western Arkansas. McAllister et al. (C. T., 1993b) reported cestodes, nematodes, protozoans, and an acanthocephalon from 37 adults from throughout Arkansas. Intradermal mites (genus *Hannemania*) occasionally infest western slimy salamanders (Duncan and Highton, 1979), but they appear to be more resistant to mite infection than are other members of *ouachitae* complex (Anthony, 1995).

4. Conservation.
Western slimy salamanders are widely distributed (see "Historical versus Current

Distribution" above). Their historical abundance is unknown, but they are considered Protected by the State of Oklahoma, along the periphery of their range.

Acknowledgments. Richard Highton reviewed this account and provided valuable feedback.

Plethodon amplus Highton and Peabody, 2000
BLUE RIDGE GRAY-CHEEKED SALAMANDER
David A. Beamer, Michael J. Lannoo

1. Historical versus Current Distribution.
Blue Ridge gray-cheeked salamanders *(Plethodon amplus)* are known from Blue Ridge Mountain sites in Buncombe, Rutherford, and Henderson counties in North Carolina. With the exception of local extirpations due to habitat destruction and modification, their current distribution is likely similar to their historical distribution, but there has been no documentation of their historical distribution.

A. Breeding. Reproduction is terrestrial.
i. Breeding migrations. Unlikely; breeding migrations are unknown in any *Plethodon* species.
ii. Breeding habitat. Unknown.

B. Eggs.
i. Egg deposition sites. Unknown, but likely to be in underground cavities.
ii. Clutch size. Unknown.

C. Direct Development.
i. Brood sites. Unknown, but probably include underground cavities or chambers.
ii. Parental care. Unknown, but it is likely that females brood, as with other species of *Plethodon*.

D. Juvenile Habitat. Unknown, but likely to be similar to adults.

E. Adult Habitat. Blue Ridge gray-cheeked salamanders are reported from crevices in metamorphic rock and from the forest floor (Adler and Dennis, 1962; Rubin, 1969).

F. Home Range Size. Unknown, but small home ranges are typical for *Plethodon* species.

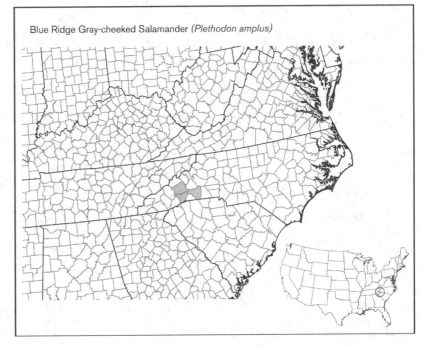

Blue Ridge Gray-cheeked Salamander *(Plethodon amplus)*

2. Historical versus Current Abundance. Unknown.

3. Life History Features.
Because Blue Ridge gray-cheeked salamanders have been described only recently (Highton and Peabody, 2000), little specific information is known about their life history and natural history features. As a portion of his larger research program, R. Highton has collected basic life history and natural history information on Blue Ridge gray-cheeked salamanders and has plans to publish these data in a monographic treatment.

G. Territories. At least some members of the *P. jordani* complex aggressively defend territories (Thurow, 1976), but it is unknown whether Blue Ridge gray-cheeked salamanders establish and defend territories.

H. Aestivation/Avoiding Desiccation. Conditions are almost always wet all summer, so aestivation is unlikely (R. Highton, personal communication). As with all *Plethodon*, their response to dry conditions is to move to moister (deeper) subterranean sites.

I. Seasonal Migrations. Unknown other than vertical movements from the forest floor to underground sites to avoid seasonally dry summer conditions and cold winter conditions.

J. Torpor (Hibernation). Unstudied, but in response to cold conditions, animals likely move to warmer, deep, subterranean sites.

K. Interspecific Associations/Exclusions. The following species were found on a cliff face with Blue Ridge gray-cheeked salamanders in Rutherford County, North Carolina: green salamanders (*Aneides aeneus*), Yonahlossee salamanders (*P. yonahlossee*), white-spotted slimy salamanders (*P. cylindraceus*), and Ocoee salamanders (*Desmognathus ocoee*; Adler and Dennis, 1962; Rubin, 1969).

White-spotted slimy salamanders are widely sympatric throughout the range of Blue Ridge gray-cheeked salamanders. There is no evidence of hybridization between these species (Highton and Peabody, 2000).

Blue Ridge gray-cheeked salamanders may contact southern gray-cheeked salamanders (*P. metcalfi*), although their geographic and genetic interactions have not been analyzed (Highton and Peabody, 2000).

Yonahlossee salamanders are widely sympatric throughout the range of Blue Ridge gray-cheeked salamanders. There is no evidence of hybridization between these species (Highton and Peabody, 2000).

L. Age/Size at Reproductive Maturity. Unknown. Females may be larger. The two largest animals (73 mm SVL) from the type locality were females (Highton and Peabody, 2000). The holotype is a male 63 mm SVL, the allotype a female 70 mm SVL. Other animals from the type locality ranged from 50–72 mm SVL.

M. Longevity. Unknown.

N. Feeding Behavior. The stomachs of four Blue Ridge gray-cheeked salamanders (average size 57.4 mm SVL) from Rutherford County, North Carolina, contained millipedes (32.5%), ants (18.8%), lepidopteran larvae (16.3%), spiders (13.8%), and centipedes (5.0%; Rubin, 1969).

O. Predators. Unknown, but likely to include small mammals, birds, snakes, and perhaps large predaceous invertebrates.

P. Anti-Predator Mechanisms. Nocturnal. Secretive. All members of the genus *Plethodon* produce noxious skin secretions (Brodie, 1977).

Q. Diseases. Unknown.

R. Parasites. Blue Ridge gray-cheeked salamanders from a cliff face population in Rutherford County, North Carolina, were infected with dermal mites (probably *Hannemania hegeneri*; Adler and Dennis, 1962).

Rankin (1937) lists parasites from at least two species of the *P. jordani* complex. It is not possible to determine which parasites were found in which salamander species, but it is possible that some of these salamanders were Blue Ridge gray-cheeked salamanders. The following parasites are noted: *Crytobia borreli, Eutrichomastix batrachorum, Hexamitus intestinalis, Karotomorpha swezi, Prowasekella longifilis, Tritrichomonas augusta, Brachycoelium hospitale*, and *Crepidobothrium cryptobranchi*.

4. Conservation.
Blue Ridge gray-cheeked salamanders are not protected in North Carolina, the only state within their range. Among members of the *P. jordani* complex, Blue Ridge gray-cheeked salamanders have one of the smallest distributions. Within this range there are a few federal and state properties that contain suitable habitat for these salamanders.

As with all species of *Plethodon*, Blue Ridge gray-cheeked salamanders do not migrate to breeding grounds and they do not have large home ranges. Thus, they can exist in habitats of smaller size than many other amphibian species. Conservation activities that promote mature closed-canopy forests should benefit this species.

Acknowledgments. Thanks to Richard Highton, who reviewed this account and gave us the benefit of his insight and experience.

Plethodon angusticlavius Grobman, 1944
OZARK ZIGZAG SALAMANDER

Walter E. Meshaka Jr.

1. Historical versus Current Distribution.
Ozark zigzag salamanders (*Plethodon angusticlavius*) are distributed throughout the Ozark Plateaus Province, from northwestern Arkansas to the eastern edge of Oklahoma and the southern tip of Missouri (Thurow, 1966). Their current distribution is not thought to be different from their historical distribution.

2. Historical versus Current Abundance. Unknown, although population declines have not been noted.

3. Life History Features.
A. Breeding. Reproduction is terrestrial.
i. Breeding migrations. In Arkansas, adults migrate to sandstone cedar glades from the forest during the courting season. Courtship activities among Ozark zigzag salamanders occur in the cool wet winter to spring. Vasa deferentia are largest from January–April, and yolked follicles are present January–May (Meshaka and Trauth, 1995).
ii. Breeding habitat. Clutches of closely related northern zigzag salamanders (*P. dorsalis*) are deposited within rock crevices, subterranean cavities, and in caves (Mohr, 1952; Smith, 1961; Mount, 1975; Miller et al., 1998). Logs appear to be avoided as nesting sites. Eggs rest on the substrate and are attended by the female (Mohr, 1952; Miller et al., 1998).

B. Eggs.
i. Egg deposition sites. Clutches of closely related northern zigzag salamanders are deposited within rock crevices, subterranean cavities, and in caves (Mohr, 1952; Smith, 1961; Mount, 1975; Miller et al., 1998). Logs appear to be avoided as nesting sites. Eggs rest on the substrate and are attended by the female (Mohr, 1952; Miller et al., 1998).
ii. Clutch size. Ozark zigzag salamander clutches in Arkansas average 5.3 eggs (range 3–9) and are produced annually (Meshaka and Trauth, 1995). Hatchlings appear in the fall (Wilkinson et al., 1993; Meshaka and Trauth, 1995).

C. Direct Development.
i. Brood sites. Within rock crevices and underground.
ii Parental care. Females attend the nest during the summer when they are absent from foraging sites (Wilkinson et al., 1993; Meshaka and Trauth, 1995).

D. Juvenile Habitat. Same as adults.

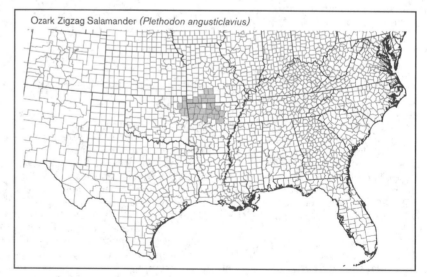

Ozark Zigzag Salamander (*Plethodon angusticlavius*)

E. Adult Habitat. Ozark zigzag salamanders are found in moist leaf litter, talus areas, and caves (Thurow, 1955; Meshaka and Trauth, 1995). Rocky habitat is preferred (Smith, 1961, Reinbold, 1979). This habitat serves to buffer them from weather extremes and enhance population sizes. The availability of these rocky habitats therefore limits the abundance of Ozark zigzag salamanders (Petranka, 1979). Rocky cedar glades are used extensively by adults during the courtship season (Meshaka and Trauth, 1995).

F. Home Range Size. Unknown.

G. Territories. Unknown.

H. Aestivation/Avoiding Desiccation. Aestivation is unknown; animals likely avoid desiccating conditions by moving under cover objects or into burrows.

I. Seasonal Migrations. In Arkansas, adults migrate to sandstone cedar glades from the forest during the courting season (see "Breeding migrations" above). Immature animals are not found on the glade (Meshaka and Trauth, 1995).

J. Torpor (Hibernation). Unknown.

K. Interspecific Associations/Exclusions. Centipedes in the genus *Scolopendra* sp., which are potential predators, were found in the same leaf litter as Ozark zigzag salamanders (Meshaka and Trauth, 1995).

L. Age/Size at Reproductive Maturity. Both sexes mature at 35 mm SVL and reproduce for the first time in the winter of their third year of life (Meshaka and Trauth, 1995).

M. Longevity. Unknown.

N. Feeding Behavior. Spiders and beetles comprised most of the diet of a fall and winter sample of closely related northern zigzag salamanders (Holman, 1955).

O. Predators. *Scolopendra* sp. are likely predators (D.A.B. personal observations).

P. Anti-Predator Mechanisms. Unknown.

Q. Diseases. Unknown.

R. Parasites. Chiggers of the genus *Hannemania* have been reported from Ozark zigzag salamanders (Duncan and Highton, 1979).

4. Conservation.
Ozark zigzag salamanders have a relatively small geographic range, which by itself is cause for concern. Additionally, little is known about their ecology, and therefore risks to populations through perturbations of their mesic woodland habitat can only be approximated.

Acknowledgments. Richard Highton reviewed this account and provided valuable feedback.

Plethodon aureolus Highton, 1984
TELLICO SALAMANDER
David A. Beamer, Michael J. Lannoo

1. Historical versus Current Distribution.
Tellico salamanders (*Plethodon aureolus*) occur on the western slopes of the Unicoi

Mountains and nearby lowlands, between the Little Tennessee and Hiwassee rivers, in northeastern Polk and eastern Monroe counties in Tennessee, and in northwestern Graham and northwestern Cherokee counties in North Carolina. They were first described by Highton (1984; see also Highton, 1986a). There is no evidence that the current distribution differs from the historical distribution.

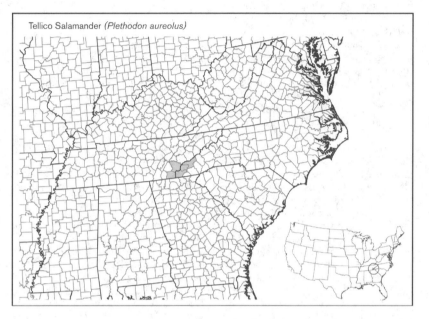

Tellico Salamander (*Plethodon aureolus*)

2. Historical versus Current Abundance.
A resurvey of a Tellico salamander population in Monroe County, Tennessee, by Highton (this volume, Part One) revealed a decrease from an observed mean of 7.3 salamanders from 1978–84 to a mean of 5.5 salamanders in 1997. Additional studies are necessary to determine whether this result reflects a real decline or simply natural variation.

3. Life History Features.
There are almost no published data on life history and natural history features of Tellico salamanders (Petranka, 1998). As a portion of his larger research program, R. Highton has collected basic life history and natural history information on Tellico salamanders and has plans to publish these data in a monographic treatment.

A. Breeding. Reproduction is terrestrial.

i. Breeding migrations. Undocumented, but breeding migrations are unknown for any *Plethodon* species.

ii. Breeding habitat. Unknown.

B. Eggs.

i. Egg deposition sites. Unknown, but likely to be in underground cavities.

ii. Clutch size. Unknown.

C. Direct Development.

i. Brood sites. Unknown.

ii. Parental care. Unknown, but it is likely that females brood, as with other mem-

bers of the slimy salamander (*P. glutinosus*) complex.

D. Juvenile Habitat. Juveniles have been found in abundance at the type locality in August under superficial cover such as twigs (D.A.B., personal observations).

E. Adult Habitat. Tellico salamanders occur mostly at lower altitudes. However, they have been found at elevations as high as 1,622 m in Graham County, North Carolina (Highton, 1984).

F. Home Range Size. Unknown, but small home ranges are typical for *Plethodon* species.

G. Territories. At least some members of the *P. glutinosus* complex aggressively defend territories (Thurow, 1976), but it is unknown whether Tellico salamanders establish and defend territories.

H. Aestivation/Avoiding Desiccation. Unknown, but it is likely that animals move to moist underground sites in response to dry conditions, similar to other members of the *P. glutinosus* complex (see above).

I. Seasonal Migrations. Unknown and unlikely, with the exception of vertical movements from underground sites to the forest floor and back again, in response to seasonal dry and cold conditions.

J. Torpor (Hibernation). Unknown, but in response to cold conditions, animals probably move to warmer underground sites, similar to other members of the *P. glutinosus* complex (see above).

K. Interspecific Associations/Exclusions. Tellico salamanders are sympatric with southern Appalachian salamanders (*P. teyahalee*) throughout the range of Tellico salamanders. There is no morphological or genetic evidence of hybridization (Highton and Peabody, 2000).

Tellico salamanders have a largely parapatric range with northern slimy salamanders *(P. glutinosus)*. They have been taken together at one site in Polk County, Tennessee, without any evidence of hybridization (Highton and Peabody, 2000).

An unusual association consisting of three members of the *P. glutinosus* complex occurs at a site in Polk County, Tennessee. Here southern Appalachian salamanders, Tellico salamanders, and northern slimy salamanders occur sympatrically. There is no evidence of hybridization at this location (Highton, 1984).

Tellico salamanders contact red-legged salamanders *(P. shermani)* on the west slope of the Unicoi Mountains, in Monroe County, Tennessee. A transect done at this site indicates a wide hybrid zone on Sassafras Ridge (Highton and Peabody, 2000).

Other salamander species associated with Tellico salamanders include southern red-backed salamanders *(P. serratus)*, Ocoee salamanders *(Desmognathus ocoee)*, black-bellied salamanders *(D. quadramaculatus)*, Santeetlah dusky salamanders *(D. santeetlah)*, northern dusky salamanders *(D. fuscus)*, seal salamanders *(D. monticola)*, seepage salamanders *(D. aeneus)*, and Blue Ridge two-lined salamanders *(Eurycea wilderae;* D.A.B., personal observations).

L. Age/Size at Reproductive Maturity. According to Highton (1986), adults range in size from 50–72 mm SVL.

M. Longevity. Unknown.

N. Feeding Behavior. Unknown, but as with other *Plethodon* species, feeding likely takes place at night under moist conditions. Prey items likely include a range of invertebrates, especially insects. Captives feed readily on earthworms and wax "worms" (i.e., lepidopteran larvae; D.A.B., personal observations).

O. Predators. Unknown, but likely to include small mammals, birds, snakes, and perhaps large predaceous invertebrates.

P. Anti-Predator Mechanisms. Members of the *P. glutinosus* complex produce large amounts of skin secretions that have an adhesive component (Brodie et al., 1979).

Q. Diseases. Unknown.

R. Parasites. Unknown.

4. Conservation.
Tellico salamanders are not protected in any of the states within their range. Among members of the *P. glutinosus* complex, Tellico salamanders have one of the smaller distributions. Within this range there are many federal and state properties that contain suitable habitat for these salamanders.

Tellico salamanders are relatively resilient to disturbances, such as those associated with timbering operations, and are frequently found in second-growth forests (D.A.B., personal observation).

As with all species of *Plethodon*, Tellico salamanders do not migrate to breeding grounds and they do not have large home ranges. Thus, they can exist in habitats of smaller size than many other amphibian species. Conservation activities that promote mature closed-canopy forests should benefit this species.

Acknowledgments. Thanks to Richard Highton, who reviewed this account and gave us the benefit of his insight and experience.

Plethodon caddoensis Pope and Pope, 1951
CADDO MOUNTAIN SALAMANDER
Carl D. Anthony

1. Historical versus Current Distribution.
Caddo Mountain salamanders *(Plethodon caddoensis)* are locally distributed in the Caddo and Cossatot mountains of western Arkansas. Knowledge of their geographic range has been expanded since the species was first collected on Polk Creek Mountain by Pope and Pope (1951; see Pope, 1964b; Blair and Lindsay, 1965; Duncan and Highton, 1979; Plummer, 1982). Blair (A.P., 1957b) reported specimens from the south side of Poteau Mountain near Oliver, Arkansas, but repeated field trips to this area have not yielded any specimens. Caddo Mountain salamanders are apparently not found outside the Novaculite Uplift (Plummer, 1982). Caddo Mountain salamanders are considered a Species of Special Concern in Arkansas. Their historical distribution is unknown.

2. Historical versus Current Abundance.
Abundant and widespread within their limited range. No changes in abundance have been noted.

3. Life History Features.
A. Breeding. Reproduction is terrestrial.
i. Breeding migrations. Breeding migrations are unknown.
ii. Breeding habitat. Breeding may occur in late fall, winter, or early spring (Taylor et al., 1990). Breeding habitat is unknown, but breeding may take place in deep crevices (see below) or perhaps on the forest floor under moist conditions.

B. Eggs.
i. Egg deposition sites. Nest sites have been found in abandoned mines (Heath et al., 1986; D. A. Saugey, unpublished data). Caves and deep crevices may be important breeding habitats as well. Ratios of reproductive to nonreproductive females suggest that females breed biennially (Taylor et al., 1990).
ii. Clutch size. A mean of 11.3 enlarged ovarian follicles from 22 mature females was reported by Taylor et al. (1990). A mean of 7.8 eggs (range = 4–11) was noted by Saugey (unpublished data).

C. Direct Development. Saugey (unpublished data) has observed females attending egg clutches. Eggs have been observed as early as 9 June and hatching as late as 5 November.

D. Juvenile Habitat. Same as adult habitat.

E. Adult Habitat. Caddo Mountain salamanders are most commonly found at higher elevations of mixed deciduous, north-facing wooded slopes (Pope and Pope, 1951; Plummer, 1982; Petranka, 1998). Rocks, logs, and other forest debris are common cover objects. Moisture conditions at the surface appear to influence activity greatly (Plummer, 1982), with salamanders retreating to lower levels of talus to escape hot and dry conditions (Spotila, 1972). Caves and abandoned

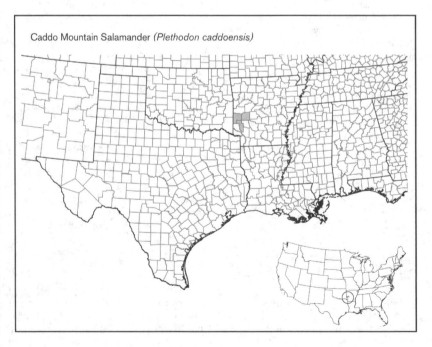

Caddo Mountain Salamander *(Plethodon caddoensis)*

mines are also utilized (Saugey et al., 1985; Heath et al., 1986).

F. Home Range Size. Unknown.

G. Territories. Adult Caddo Mountain salamanders recognize and respond to odors of conspecifics (Anthony, 1993), and they defend areas in laboratory chambers (Thurow, 1976; Anthony, 1995). Territory size is unknown.

H. Aestivation/Avoiding Desiccation. Populations move underground in late May, but may return to the surface during periods of rainfall and/or cool weather. By mid September, adults can again be found under cover objects at the surface.

I. Seasonal Migrations. Unknown.

J. Torpor (Hibernation). Probably hibernates from mid-November to late March.

K. Interspecific Associations/Exclusions. Occur syntopically with western slimy salamanders (*P. albagula*) and southern red-backed salamanders (*P. serratus*). Kuss (1986) found their habitat to be similar to that of western slimy salamanders. Dowling (1956) observed that Caddo Mountain salamanders become active on the forest floor earlier in the evening than do western slimy salamanders, which may prefer cooler and more humid conditions.

L. Age/Size at Reproductive Maturity. Maturity is reached at approximately 40 mm SVL (Highton, 1962a). Age at maturity is unknown.

M. Longevity. Individuals that were sexually mature when collected have been kept in the laboratory for at least 3 yr. Longevity is probably >6 yr.

N. Feeding Behavior. Unknown, but prey likely consists of small invertebrates such as worms, insects, and spiders.

O. Predators. Unknown.

P. Anti-Predator Mechanisms. Nocturnal. As well, all *Plethodon* produce noxious skin secretions (Brodie, 1977). Similar to other members of the *P. glutinosus* group, Caddo Mountain salamanders release an adhesive secretion when handled.

Q. Diseases. Unknown.

R. Parasites. Winter et al. (1986) found trematodes, cestodes, nematodes, and mites on 25 adults. Intradermal mites of the genus *Hannemania* are common (Saugey et al., 1985; Winter et al., 1986), with 7–75% of individuals infected (Duncan and Highton, 1979; Anthony et al., 1994). They appear as raised red pustules, typically found on the trunk (Winter et al., 1986; Anthony et al., 1994). Males incur greater infection intensities than do females (Anthony et al., 1994).

4. Conservation.
Caddo Mountain salamanders have a local distribution in the Caddo and Cossatot mountains of western Arkansas, and within this limited range they are abundant. No changes in abundance have been noted. Caddo Mountain salamanders are considered a Species of Special Concern in Arkansas.

Plethodon chattahoochee Highton, 1989
CHATTAHOOCHEE SLIMY SALAMANDER

David A. Beamer, Michael J. Lannoo

1. Historical versus Current Distribution.
Members of the *P. glutinosus* complex, Chattahoochee slimy salamanders (*Plethodon chattahoochee*) occur throughout much of the Chattahoochee National Forest in the Blue Ridge Physiographic Province of northern Georgia and southeastern Cherokee County, North Carolina. The type specimens were collected at an elevation of 1,353 m (Highton et al., 1989). There is no evidence to suggest that the current distribution differs from the historical distribution.

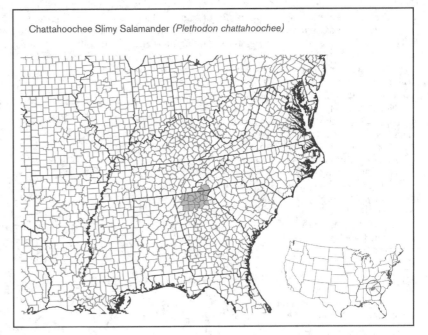

Chattahoochee Slimy Salamander (*Plethodon chattahoochee*)

2. Historical versus Current Abundance.
Highton (this volume, Part One) reports a decline in the one Chattahoochee slimy salamander population he sampled, when comparing data from 1961–84 with data from 1997, under similar sampling conditions and with a similar search effort. However, additional sampling must be done to determine whether this is the result of a true decline or a natural fluctuation.

3. Life History Features.
Chattahoochee slimy salamanders were recently described (Highton, 1989). In this time, there has been no published work done on this species. As a portion of his larger research program, R. Highton has collected basic life history and natural history information on Chattahoochee slimy salamanders and has plans to publish these data in a monographic treatment.

A. Breeding. Reproduction is terrestrial.

i. Breeding migrations. Undocumented, but breeding migrations are not known for any *Plethodon* species.

ii. Breeding habitat. Unknown.

B. Eggs.

i. Egg deposition sites. Unknown, but likely to be in underground cavities.

ii. Clutch size. Unknown.

C. Direct Development.

i. Brood sites. Unknown.

ii. Parental care. Unknown, but it is likely that females brood, as with other members of the slimy salamander complex.

D. Juvenile Habitat. Unknown, but likely to be similar to adults.

E. Adult Habitat. Has not been described. But, as with other *Plethodon* sp., forest floor habitats and cover objects are important. Animals are likely to be most active under moist conditions. Burrowing is likely to be important under inhospitable surface conditions.

F. Home Range Size. Unknown, but small home ranges are typical for *Plethodon* species.

G. Territories. At least some members of the *P. glutinosus* complex aggressively defend territories (Thurow, 1976), but it is unknown whether Chattahoochee slimy salamanders establish and defend territories.

H. Aestivation/Avoiding Desiccation. Generally unknown, but Chattahoochee slimy salamanders likely avoid desiccating conditions by seeking shelter in underground sites.

I. Seasonal Migrations. Animals likely make vertical migrations, moving from the forest floor to underground sites with the onset of seasonally related cold or dry conditions, then back up to the forest floor with the return of favorable surface conditions.

J. Torpor (Hibernation). Generally unknown, but Chattahoochee slimy salamanders likely avoid cold conditions by seeking shelter in underground sites.

K. Interspecific Associations/Exclusions. Chattahoochee slimy salamanders are parapatric with red-legged salamanders (*P. shermani*) near the Georgia–North Carolina border. A transect along the Tallulah River indicates that there is a wide hybrid zone (Highton and Peabody, 2000).

Chattahoochee slimy salamanders contact southern Appalachian salamanders (*P. teyahalee*) in Clay County, North Carolina. A transect in the vicinity of Hayesville indicates that there is a narrow hybrid zone (Highton and Peabody, 2000).

Atlantic Coast slimy salamanders (*P. chlorobryonis*) and Chattahoochee slimy salamanders contact in northeastern Georgia. A partially completed transect in White and Habersham counties has indicated that there is a parapatric hybrid zone (Highton and Peabody, 2000).

Northern slimy salamanders (*P. glutinosus*) and Chattahoochee slimy salamanders contact at the western edge of the Blue Ridge Province in Georgia; there is no published information on their interactions (Highton and Peabody, 2000).

L. Age/Size at Reproductive Maturity. Unknown. The holotype is a male, 62 mm SVL; the allotype a female 64 mm.

M. Longevity. Unknown.

N. Feeding Behavior. Unknown, but as with other *Plethodon* species, feeding likely takes place at night under moist conditions. Prey items likely include a range of invertebrates, especially insects.

O. Predators. Undocumented, but likely to include forest snakes, birds, and small mammals.

P. Anti-Predator Mechanisms. All *Plethodon* produce noxious skin secretions (Brodie, 1977). Members of the *P. glutinosus* complex frequently become immobile when initially contacted. Chattahoochee slimy salamanders were included in a field study on immobility; however, it is not possible to separate their behavior from the other members of this complex in this published data set. Immobility may increase survival by making salamanders less likely to be detected, especially by visually oriented predators (Dodd, 1989).

Q. Diseases. Unknown.

R. Parasites. Unknown.

4. Conservation.
Chattahoochee slimy salamanders are not listed in any of the states within their range. Among members of the *P. glutinosus* complex, Chattahoochee slimy salamanders have one of the smaller distributions. Within this range, there are many federal and state properties that contain suitable habitat for these salamanders.

As with all species of *Plethodon*, Chattahoochee slimy salamanders do not migrate to breeding grounds and they do not have large home ranges. Thus, they can exist in habitats of smaller size than many other amphibian species. Conservation activities that promote mature closed-canopy forests should benefit this species.

Acknowledgments. Thanks to Richard Highton, who reviewed this account and gave us the benefit of his insight and experience.

Plethodon cheoah Highton and Peabody, 2000
CHEOAH BALD SALAMANDER

David A. Beamer, Michael J. Lannoo

1. Historical versus Current Distribution. Cheoah Bald salamanders (*Plethodon cheoah*) are known from the Cheoah Bald in Graham and Swain counties in North Carolina. The type specimens were collected at an elevation of 1,445 m.

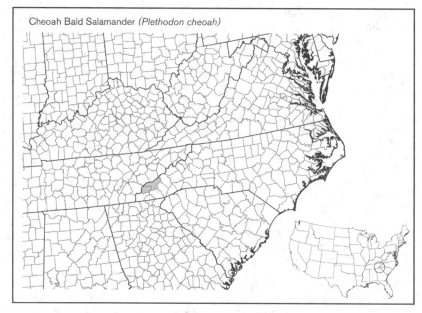

Cheoah Bald Salamander (*Plethodon cheoah*)

2. Historical versus Current Abundance. Generally unknown, but Highton (this volume, Part One) sampled a population in Graham County, North Carolina, in 1987–88 and again in 1992–97 and found a slight decrease from 8.5 to 6.7 animals/sampling effort. Additional sampling will be necessary to determine whether these data reflect a true decline or a natural population fluctuation.

3. Life History Features.
Until recently (Highton and Peabody, 2000), Cheoah Bald salamanders were considered to be a variant of Jordan's salamanders (*P. jordani*). Therefore, little specific information is known about the life history and natural history features of this

species. As a portion of his larger research program, R. Highton has collected basic life history and natural history information on Cheoah Bald salamanders and has plans to publish these data in a monographic treatment.

A. Breeding. Reproduction is terrestrial.
i. Breeding migrations. Unknown, but breeding migrations are unlikely in members of the genus *Plethodon*.
ii. Breeding habitat. Unknown.

B. Eggs.
i. Egg deposition sites. Unknown.
ii. Clutch size. Unknown.

C. Direct Development.
i. Brood sites. Unknown.
ii. Parental care. Unknown, but it is likely that females brood, as is true of other species of *Plethodon*.

D. Juvenile Habitat. Unknown, but likely to be similar to adults.

E. Adult Habitat. Unknown.

F. Home Range Size. Unknown.

G. Territories. At least some members of the *P. jordani* complex aggressively defend territories (Thurow, 1976); it is unknown whether Cheoah Bald salamanders establish and defend territories.

H. Aestivation/Avoiding Desiccation. Undocumented, but animals likely move from forest floor habitats to underground sites in response to desiccating surface conditions.

I. Seasonal Migrations. Animals likely make vertical migrations, moving from the forest floor to underground sites with the onset of seasonally related cold or dry conditions, then back up to the forest floor with the return of favorable surface conditions.

J. Torpor (Hibernation). Generally unknown, but Cheoah Bald salamanders likely avoid cold conditions by seeking shelter in underground sites.

K. Interspecific Associations/Exclusions. Cheoah Bald salamanders are sympatric with southern Appalachian salamanders (*P. teyahalee*) throughout their range, but these two species rarely hybridize (Highton and Peabody, 2000). Members of the *P. jordani* complex usually are not found with members of the *P. glutinosus* complex over wide elevational ranges and especially at high elevations. However, Cheoah Bald salamanders and southern Appalachian salamanders are an exception, as both species occur together in this area even at the highest elevations (1,543 m; Highton, 1970).

L. Age/Size at Reproductive Maturity. Unknown. The holotype is an adult male 50 mm SVL; the allotype is an adult female 52 mm. The largest individual from the type locality is a female 63 mm SVL; the range of adult sizes in 45–60 mm.

M. Longevity. Unknown.

N. Feeding Behavior. Unknown.

O. Predators. Unknown.

P. Anti-Predator Mechanisms. All *Plethodon* produce noxious skin secretions (Brodie, 1977).

Q. Diseases. Unknown.

R. Parasites. Unknown.

4. Conservation.

Cheoah Bald salamanders are not protected in North Carolina, the only state within their range. Among members of the *P. jordani* complex, Cheoah Bald salamanders have one of the smallest distributions. Within this range there are federal properties that contain suitable habitat for these salamanders.

As with all species of *Plethodon*, Cheoah Bald salamanders do not migrate to breeding grounds, and they do not have large home ranges. Thus, they can exist in habitats of smaller size than many other amphibian species. Conservation activities that promote mature closed-canopy forests should benefit this species.

Acknowledgments. Thanks to Richard Highton, who reviewed this account and gave us the benefit of his insight and experience.

Plethodon chlorobryonis Mittleman, 1951
ATLANTIC COAST SLIMY SALAMANDER

David A. Beamer, Michael J. Lannoo

1. Historical versus Current Distribution. Atlantic Coast slimy salamanders (*Plethodon chlorobryonis*) occur in the Coastal Plain Physiographic Province of southeastern Virginia, North Carolina, and northeastern South Carolina. They also enter the Piedmont Physiographic Province in southeastern Virginia and central and western South Carolina, and the Blue Ridge Physiographic Province in northeastern Georgia. Following surveys of historical sites, Highton (this volume, Part One) reports he could find no animals in three of ten populations—two populations in South Carolina (Aiken and

Atlantic Coast Slimy Salamander (*Plethodon chlorobryonis*)

Florence counties) and one in Virginia (Dinwiddie-Sussex County).

2. Historical versus Current Abundance. Highton (this volume, Part One) reports a decline in all 10 Atlantic Coast slimy salamander populations he sampled prior to 1985 and re-sampled after 1995 under similar sampling conditions and with a similar search effort. These populations should continue to be monitored to determine if Highton's data are indicative of true declines or natural population fluctuations.

3. Life History Features. Atlantic Coast slimy salamanders were recently elevated to species status (Highton, 1989). In this time, there has been no published work done on this species. As a portion of his larger research program, R. Highton has collected basic life history and natural history information on Atlantic Coast slimy salamanders and has plans to publish these data in a monographic treatment.

A. Breeding. Reproduction is terrestrial.

i. Breeding migrations. Undocumented, but breeding migrations are unknown for any *Plethodon* species.

ii. Breeding habitat. Unknown.

B. Eggs. Eggs in early gastrula stage were discovered on 30 May near Carrsville, Virginia. Two females collected at the same time had fully gravid ovaries indicating egg deposition was not complete by this date (Wood and Rageot, 1955).

i. Egg deposition sites. The eggs of this species were reported from a damp pile of paper cement bags in southeastern Virginia (Wood and Rageot, 1955).

ii. Clutch sizes. Seven eggs (average size 5.0–5.5 mm) were reported from a nest located within a damp pile of paper cement bags in southeastern Virginia (Wood and Rageot, 1955). A gravid female from

the Carrsville area of Virginia contained 19 ova; a second, slightly smaller individual contained 16 ova (Wood and Rageot, 1955).

C. Direct Development.

i. Brood sites. The same as egg deposition sites; the only known site was within a damp pile of paper cement bags in southeastern Virginia (Wood and Rageot, 1955).

ii. Parental care. Five Atlantic Coast slimy salamanders were found within a damp pile of paper cement bags, which contained a nest. One individual was a juvenile and the other four were adults. Three of the adults were females, of which two were gravid and one had "spent" ovaries. It is likely that the spent female was guarding the eggs, but the shifting of the bags during collection made it impossible to determine if the female was in attendance of the eggs (Wood and Rageot, 1955).

D. Juvenile Habitat. A juvenile Atlantic Coast slimy salamander was found in a damp pile of paper cement bags in southeastern Virginia (Wood and Rageot, 1955).

E. Adult Habitat. The type specimens were collected in dry bottomlands along a small creek. Populations in Pitt County, North Carolina, were found under logs on slopes above a cypress swamp (Robertson and Tyson, 1950; Eaton, 1953). According to Mittleman (1951), Atlantic Coast slimy salamanders have prehensile tails and show "a marked predilection for climbing." Highton (personal communication), however, disputes this claim. As with other members of the *P. glutinosus* complex, Atlantic Coast slimy salamanders likely seek shelter under cover objects during the day, and are active nocturnally, with activity levels proportional to moisture levels. Atlantic Coast slimy salamanders likely avoid dry and cold extremes by moving to underground sites.

F. Home Range Size. Unknown, but small home ranges are typical for *Plethodon* species.

G. Territories. At least some members of the *P. glutinosus* complex aggressively defend territories (Thurow, 1976); it is unknown whether Atlantic Coast slimy salamanders establish and defend territories.

H. Aestivation/Avoiding Desiccation. Generally unknown, but Atlantic Coast slimy salamanders likely avoid desiccating conditions by seeking shelter in underground sites.

I. Seasonal Migrations. It is likely that Atlantic Coast slimy salamanders move vertically from forest floor sites to underground sites in response to seasonal dry and cold conditions. For instance, on a transect in Pitt County, North Carolina, no Atlantic Coast slimy salamanders were found beneath cover objects during May or August. However, several individuals were discovered beneath these same cover objects in October (D.A.B., personal observations).

J. Torpor (Hibernation). Generally unknown, but Atlantic Coast slimy salamanders likely avoid cold conditions by seeking shelter in underground sites.

K. Interspecific Associations/Exclusions. Eaton (1953) reports occasionally finding Atlantic Coast slimy salamanders in wet situations shared with southern dusky salamanders *(Desmognathus auriculatus)*. In Pitt County, North Carolina, Atlantic Coast slimy salamanders have been found beneath the same cover objects with eastern red-backed salamanders *(P. cinereus)* and dwarf salamanders *(Eurycea quadridigitata;* D.A.B., personal observations).

The type specimen of Mabee's salamander *(Ambystoma mabeei)* was collected beneath a dead log near Dunn, Harnett County, North Carolina, in company with a marbled salamander *(A. opacum)* and a slimy salamander (Brimley, 1939). There has not been any work addressing the taxonomy of slimy salamanders from Harnett County; however, they appear to be within the range of Atlantic Coast slimy salamanders.

During a drift-fence study at the Savannah River Plant, Atlantic Coast slimy salamanders were captured with mole salamanders *(A. talpoideum)*, marbled salamanders, tiger salamanders *(A. tigrinum)*, eastern newts *(Notophthalmus viridescens)*, red salamanders *(Pseudotriton ruber)*, mud salamanders *(P. montanus)*, dwarf salamanders, and southern two-lined salamanders *(E. cirrigera;* Gibbons and Semlitsch, 1981).

In McCormick County, South Carolina, Atlantic Coast slimy salamanders were found with Webster's salamanders *(P. websteri)*, northern dusky salamanders *(Desmognathus fuscus)*, southern two-lined salamanders and long-tailed salamanders *(E. longicauda;* Semlitsch and West, 1983).

An unusual association of large eastern *Plethodon* occurs in the vicinity of Rabun Bald, Rabun County, Georgia. Here, a wide hybrid zone exists between Atlantic Coast slimy salamanders and southern gray-cheeked salamanders *(P. metcalfi)*, a member of the *P. jordani* complex. In this same area, Atlantic Coast slimy salamanders are also sympatric with another member of the *P. glutinosus* complex, southern Appalachian salamanders *(P. teyahalee)*, but there is no evidence of hybridization between these species. Thus, two members of the *P. glutinosus* complex contact without hybridization in the same area where one of them *(P. chlorobryonis)* hybridizes with a member of the *P. jordani* complex. Atlantic Coast slimy salamanders also contact southern Appalachian salamanders in northwestern South Carolina, where there is a parapatric hybrid zone in Anderson and Abbeville (Highton and Peabody, 2000).

Atlantic Coast slimy salamanders have a long contact with white-spotted slimy salamanders *(P. cylindraceus)* from southeastern Virginia to western South Carolina. Two transects, one in southeastern Virginia and one in northeastern North Carolina, indicate that there is a zone of parapatric hybridization (Highton and Peabody, 2000).

Atlantic Coast slimy salamanders contact Chattahoochee slimy salamanders *(P. chattahoochee)* in northeastern Georgia. A partially completed transect in White and Habersham counties indicates that there is a parapatric hybrid zone (Highton and Peabody, 2000).

L. Age/Size at Reproductive Maturity. Seven gravid or postpartum females examined by Mittleman (1951) ranged from 47.5–61 mm SVL; three sexually mature males ranged from 52–69 mm. The size of gravid or post partum females from the Virginia coastal plain ranged from 56.8–62.5 mm SVL (Wood and Rageot, 1955).

M. Longevity. Unknown.

N. Feeding Behavior. Unknown, but as with other *Plethodon* species, feeding likely takes place at night under moist conditions. Prey items likely include a range of invertebrates, especially insects.

O. Predators. Undocumented, but likely to include forest snakes, birds, and small mammals.

P. Anti-Predator Mechanisms. All *Plethodon* produce noxious skin secretions (Brodie, 1977).

Q. Diseases. Unknown.

R. Parasites. Rankin (1937) reported the following parasites from salamanders that were likely Atlantic Coast slimy salamanders: *Cryptobia borreli, Prowazekella longifilis, Tritrichomonas augustas,* and *Brachycoelium hospitale*.

4. Conservation.

Among members of the *P. glutinosus* complex, Atlantic Coast slimy salamanders have one of the wider distributions; they are not protected by any state. Within their range, there are many federal and state properties that contain suitable habitat for these salamanders.

Atlantic Coast slimy salamanders are relatively resilient to disturbances such as those associated with timbering operations, and they are frequently found in second-growth forests (D.A.B., personal observation). In the vicinity of the type locality, most of the area has been converted to agricultural uses but populations still exist along larger, forested streams corridors.

As with all species of *Plethodon*, Atlantic Coast slimy salamanders do not migrate to breeding grounds and they do not have large home ranges. Thus, they can exist in habitats of smaller size than many other amphibian species. Conservation activities that promote mature closed-canopy forests should benefit this species.

Acknowledgments. Thanks to Richard Highton, who reviewed this account and gave us the benefit of his insight and experience.

Plethodon cinereus (Green, 1818)
EASTERN RED-BACKED SALAMANDER
Gary S. Casper

1. Historical versus Current Distribution. Eastern red-backed salamanders *(Plethodon cinereus)* have undergone extensive taxonomic revision. For example, two southern subspecies *(P. c. serratus* and *P. c. polycentratus)* have been combined into southern red-backed salamanders *(P. serratus*; Highton and Larson, 1979). Range maps reflecting this taxonomic change do not imply a range retraction for eastern red-backed salamanders. A *P. cinereus* group has been recognized by Grobman (1944) and Highton and Larson (1979).

In the United States, eastern red-backed salamanders range throughout New England, southward to western and northeastern North Carolina, and northwestward to western Minnesota, where a disjunct population is found (Petranka, 1998). No range retractions have been reported, but local extirpations have been due to habitat changes (chiefly deforestation) and other, unknown causes (see Highton, this volume, Part One).

Eastern red-backed salamanders exhibit color variants that include red-backed and lead-backed morphs co-occurring in most populations (summarized in Petranka, 1998). Proportions of red-backed and lead-backed morphs vary predicably in many areas of the United States, but this variation is not correlated with any obvious environmental factor (Petranka, 1998).

2. Historical versus Current Abundance. Highton (this volume, Part One) documents recent widespread declines in most species of this complex. Jaeger's (1980a) data actually support Highton (this volume, Part One) in suggesting stable populations over a 14-yr period prior to 1980. There has been no evidence of declines in Canada (Weller and Green, 1997). However, positive correlations have been made between forest age, the quantity and quality of downed woody debris, and salamander abundance (Herbeck and Larsen, 1999); it is widely believed that abundance declined following European settlement, as virgin old-growth forests were first clearcut, then burned, then subjected to modern forestry practices with stand rotations of only a few decades. The occurrence of eastern red-backed salamanders is also positively associated with forest patch area, suggesting that forest fragmentation may result in declines (Kolozsvary and Swihart, 1999). Petranka et al. (1993) estimates that 14 million salamanders are lost annually in western North Carolina as a direct result of forestry practices (but see Ash, 1988, 1997; Ash and Bruce, 1994). Studies suggest that populations may recover from clearcutting within 30–60 yr (Pough et al., 1987; deMaynadier and Hunter, 1995). Forestry impacts demonstrated in other plethodontids probably apply to eastern red-backed salamanders as well, where five times more salamanders were found in Missouri old-growth stands (>120 yr old) than in second-growth stands (70–80 yr old); 20 times more salamanders were found in second-growth stands than in regenerating forests (<5 yr old; Herbeck and Larsen, 1999). Indeed, Waldick (1997) concluded that habitat modifications associated with standard forestry practices resulted in a decline of all forest amphibians in eastern North America, with terrestrial salamanders, such as eastern red-backed salamanders, being most susceptible. Selective logging appears to have less impact on this species than does clearcutting, with little difference in salamander densities detected between closed-canopy plots and 1-yr-old canopy gap plots (Messere and Ducey, 1998). In Virginia, abundance was far less in 2 to 7-yr-old clearcuts than in >60-yr-old forest (Blymyer

TABLE SP-5

Eastern Red-backed Salamander *(Plethodon cinereus)* Density Estimates.

Density (Salamanders/100 m²)	State	Source
280.0	Virginia	Mathis, 1991
220.0	Virginia	Jaeger, 1980b
89.0	Michigan	Heatwole, 1962
50.0	New York	Wyman, 1988b
27.6	New Hampshire	Burton and Likens, 1975a
21.0	Pennsylvania	Klein, 1960
8.38	Michigan	Kleeberger and Werner, 1982
3.0	New Brunswick	Waldick, 1997
0.15–3.85	New York	Pough et al., 1987
25.5[a]	Virginia	Blymyer and McGinnes, 1977

[a]relative abundance

and McGinnes, 1977). McAlpine (1997a) found no evidence of decline in New Brunswick, but suggests that clearcutting, conversions to conifer plantations, and shorter cutting cycles may have depleted populations. Conifer plantations are especially harmful; their drying, acidifying, and warming effects can permanently degrade salamander habitat (Waldick, 1997).

Eastern red-backed salamanders can be extremely numerous and play an important role in forest ecology, especially in energy flow and nutrient cycling, where they are 60% efficient at converting ingested energy into new tissue (Burton and Likens, 1975b). Digestive efficiencies of 83.57–90.49% and salamander tissue energy content values of 26.51 and 25.07 J/mg ash free dry weight were calculated by Crump (1979). In a New Hampshire study, the biomass of six species of salamanders (of which eastern red-backed salamanders comprised 93.5%) exceeded that for all

birds during the nesting season and was similar to the biomass estimate for all small mammals (Burton and Likens, 1975a). Available density estimates are given in Table SP-5. These estimates usually do not take into account the portion of the population that remains under the surface, which is probably greater than the number of individuals on the surface (e.g., Highton, this volume, Part One). Surface censuses are likely to encounter only 2–32% of the total population (Taub, 1961; Burton and Likens, 1975a; Highton, this volume, Part One). Monitoring protocols are being refined, and monitoring of some populations has begun (Carfioli et al., 2000).

3. Life History Features.

A. Breeding. Reproduction is terrestrial.

i. Breeding migrations. None reported. A prolonged mating season lasts from autumn to early spring.

ii. Breeding habitat. Spermatogenesis occurs from October–December in New York (Hood, 1934; Bishop, 1941b; Feder and Pough, 1975) and in late March in Michigan (Werner, 1969). Both temperature and photoperiod regulate the spermatogenic cycle (Werner, 1969). Ovipositing typically occurs in late spring and early summer. The earliest I have observed eggs is 28 April on Stockton Island in western Lake Superior (in an atypically early and warm spring). Unusually late dates are late October in New York (Sherwood, 1895) and 2 August in northern Michigan (Davidson and Heatwole, 1960).

B. Eggs.

i. Egg deposition sites. Females deposit eggs in moist natural cavities within leaf litter, soil burrows, or rotting logs (Test and Heatwole, 1962). Eggs are susceptible to dehydration, and the rehydration rate is slower than the rate of dehydration (Heatwole, 1961b). Egg-laying behavior is described by Madison et al. (1999). Where

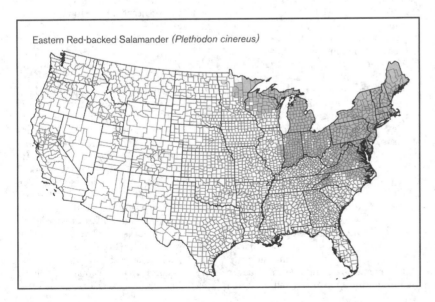

Eastern Red-backed Salamander *(Plethodon cinereus)*

logging activities have reduced the number of natural cavities available in downed woody debris, females may instead utilize cavities within matted leaf litter (Petranka, 1998). The grape-like clusters are usually suspended from the cavity roof by a short stalk. Eggs of this species can be confused with those of large gastropods that also nest within rotting logs. Females usually remain coiled with the eggs for about 60 d until hatching, and this behavior is thought to provide some protection for the eggs from predators and dehydration (Petranka, 1998), as well as accrue energetic and growth costs to the brooding female (Ng and Wilbur, 1995). There is no evidence that skin secretions from brooding females have antibiotic properties (however, see Vial and Preib, 1966; and Austin, 2000 [southern zigzag salamanders, *P. ventralis*]). Watermolen (1996) observed a female pick up an egg mass in her mouth, breaking it free from the pedicel, and carry it deeper into a log crevice when disturbed. Brooding females do not actively forage but will eat opportunistically (Ng and Wilbur, 1995).

　ii. Clutch size. Average clutch sizes are 6–9 relatively large eggs (range=1–14). Freshly laid ova are pale yellow to yellowish white, 3.0–4.0 mm in diameter, and surrounded by two jelly envelopes (Piersol, 1910; Cockran, 1911; Blanchard, 1928a; Bishop, 1941b; Lynn and Dent, 1941; Sayler, 1966; Nagel, 1977; Lotter, 1978; Petranka, 1998). The number of mature ova has been positively correlated with female length (Nagel, 1977; Lotter, 1978), as well as with female mass but not length (Fraser, 1980), suggesting that low food levels or quality may reduce clutch size. Brooding females will aggressively defend their eggs from conspecifics (Bachmann, 1984), and sometimes males are found with brooding females (Friet, 1995). In Virginia, 95% of nests were attended by brooding females (Highton and Savage, 1961).

　C. Direct Development. Embryonic gills are lost just before or shortly after hatching. The incubation period is about 6–8 wk (Burger, 1935; Davidson and Heatwole, 1960; Pfingsten, 1989b), with hatching usually taking place in August–September. Hatchlings are reported as 22 mm TL (Wisconsin; Vogt, 1981) and averaging 13.5 mm SVL (Ohio; Pfingsten, 1989b).

　D. Juvenile Habitat. Similar to adults. Juveniles often remain in the nest cavity with the mother for 1–3 wk after hatching before dispersing (Piersol, 1910; Burger, 1935; Test, 1955; Highton, 1959). Kin discrimination between mother and offspring may be context dependent (Gibbons et al., 2003).

　E. Adult Habitat. Eastern red-backed salamanders occupy deciduous, mixed conifer-deciduous, and sometimes northern conifer forests, where they inhabit leaf litter and utilize retreats under stones, within soil cavities, and in rotting logs.

Eastern red-backed salamanders have a limited ability to burrow, being effective only in soft substrates such as leaf litter or loose humus, and they prefer to use or enlarge existing retreats (Heatwole, 1960). They may also forage in bogs (Hughes et al., 1999). Soil moisture, soil pH, cover object availability, and light intensity all affect salamander distribution, with soil pH being the most influential factor (Wyman, 1988a,b; Frisbie and Wyman, 1992; Sugalski and Claussen, 1997; Grover, 1998). Eastern red-backed salamanders prefer cool, moist microhabitats and avoid temperature extremes and desiccating environments (Heatwole, 1960). A large percentage of the total population resides below the soil surface and is typically under-sampled in surface counts and mark–recapture methods (Test and Bingham, 1948; Taub, 1961). Their highest abundance occurs in mature hardwood forests, with deep soils and abundant downed woody debris in various stages of decomposition (Grover, 1998). Pfingsten (1989b) considers eastern red-backed salamanders to be an indicator organism of the beech-maple forest in Ohio, where cool, moist conditions prevail. Removal of dead and dying timber is likely to severely impact populations of terrestrial salamanders by reducing the availability of cover objects (Grover, 1998).

　Eastern red-backed salamanders avoid shallow soils, rocky substrates, hydric soils, and soils with pH < 3.7 (Wyman and Hawksley-Lescault, 1987; Petranka, 1998). Vernberg (1955) reported soil pH preferences of 6.0–6.8. Temperatures below 10 °C inhibit locomotion (Feder and Pough, 1975). Conifer-dominated forests often have litter temperatures of 39 °C (Heatwole, 1962), exceeding the maximum temperature tolerances for plethodontids (32.3–34.6 °C; Spotila, 1972), whereas litter in deciduous forests typically stays cooler (28 °C maximum; Heatwole, 1962). This phenomenon may have implications for eastern red-backed salamander distribution as global warming progresses, because increased physiological stress is likely in warm summer periods (Ovaska, 1997). There is some evidence that eastern red-backed salamander black morphs can tolerate warmer temperatures than can striped morphs, and maintenance of this polymorphism may therefore increase the species' tolerance to thermal variation (Moreno, 1989). Eastern red-backed salamanders have among the highest mean rates of dehydration and rehydration (4 mg/cm^2/hr) of all plethodontids (Grover, 2000). Preferred temperatures are higher in late summer and fall (maximum August mean selected temperature 21.0 °C) than in early summer (mean selected temperature 16.2 °C in early June), possibly facilitating surface activity in the summer, spermatogenesis during the fall mating season, and the selection of well-protected

hibernation sites (Feder and Pough, 1975). Field body temperatures range from 6.5–22.0 °C during the active season (Feder et al., 1982).

　Moisture requirements also influence microhabitat choice. Soils with an interstitial relative humidity <85% are probably unsuitable for this species (Heatwole and Lim, 1961; Heatwole, 1962).

　Eastern red-backed salamanders are largely nocturnal (Piersol, 1910; Cockran, 1911; Park et al., 1931; Heatwole, 1962). Stern and Mueller (1972) reported a diurnal rhythm, as measured by oxygen consumption, with activity peaking in the early morning hours, coinciding with that time of day when the lowest temperatures and highest humidity are likely to occur.

　F. Home Range Size. Kleeberger and Werner (1982) estimated home ranges in Michigan average 13 m^2 for males and juveniles, 24 m^2 for females.

　G. Territories. Eastern red-backed salamanders scent-mark territories on the forest floor with pheromones and fecal matter, which convey information concerning body size and gender. Kin recognition is suspected (Forester and Anders, 2000), and Gillette et al. (2000) provide evidence through behavioral experiments for social monogamy. Petranka (1998) gives home area (the defended territory) averages for males, females, and juveniles as 0.16–0.33 m^2. Peterson et al. (2000) emphasize that males and females can cohabit territories as pairs and allow juveniles to forage within their territories. When territorial sites are limited, pairs of females may defend sites (Peterson et al., 2000). These territories are defended aggressively against conspecific adults by threat displays and biting, and both males and females defend territories (Jaeger et al., 1982; Jaeger, 1984; Horne, 1988; Horne and Jaeger, 1988; Mathis, 1989, 1991; Simons et al., 1997; Lang and Jaeger, 2000; Maerz and Madison, 2000). The intensity of the defense varies depending on the quality of food resources contained within the territory (Gabor and Jaeger, 1999). Eastern red-backed salamanders can recognize individual neighbors by odors (McGavin, 1978), and exhibit considerable site tenacity, with 91% of recaptured, displaced individuals returning to within 1 m of their capture sites in Virginia (Gergits and Jaeger, 1990). Homing behavior has also been demonstrated in Michigan, where displaced salamanders returned to their territories after displacements of 30 m (90% return) and 90 m (25% return; Kleeberger and Werner, 1982). Jaeger et al. (1993) hypothesized that homing was accomplished by forming a cognitive map of the surrounding pheromone-marked territories of other salamanders in the home area of the forest. Territories appear to function primarily as feeding areas, but may also play a role in mating success (Jaeger et al., 1982;

Mathis, 1991). A large percentage of the population may be floaters (typically smaller animals that do not hold territories); up to 49% of the animals in a Virginia study may have been floaters (Mathis, 1991). Placyk et al. (2000) reported a probable breeding aggregation of five individuals in Michigan.

H. Aestivation/Avoiding Desiccation. Surface activity is reduced in midsummer (late June to August; Blanchard, 1928a; Test, 1955; Taub, 1961; Highton, 1972; Nagel, 1977; Maglia, 1996). Surface moisture and temperature affect the vertical distribution of individuals in the soil (Taub, 1961). Hot, dry conditions are avoided, but true aestivation has not been recorded.

I. Seasonal Migrations. Midsummer movements occur in response to rising temperature and falling humidity, resulting in salamanders moving to cooler, moister environments (i.e., deeper into soils and from hilltops to depressions in the forest floor; Heatwole, 1962). Waldick (1997) presents evidence of mass emigration away from a clearcut in New Brunswick.

J. Torpor (Hibernation). With the advent of freezing weather, individuals move into underground retreats, beneath stones, into ant mounds, or under and within rotting logs and stumps, where they usually remain until snowmelt (Cockran, 1911; Grizzell, 1949; Vernberg, 1953; Cooper, 1956; Sayler, 1966; Highton, 1972; Caldwell and Jones, 1973; Hoff, 1977; Lotter, 1978; Buhlmann et al., 1988). In Hoff's (1977) Massachusetts study of tree stump hibernacula, a decided preference for selecting the decayed root systems of white oak (*Quercus alba*) over other tree species was indicated. In Maryland, Cooper (1956) reported aquatic hibernation in 7.5–25 cm (3–10 in) of water. Regular surface activity during prolonged warm spells in winter has been observed (Highton, 1972), and based on full stomachs in January in Indiana, there is some evidence that feeding continues through the winter (Caldwell, 1975), but declines from December–February (Petranka, 1998). Eastern red-backed salamanders are not freeze tolerant and must avoid freezing temperatures using behavioral mechanisms (Storey and Storey, 1986).

K. Interspecific Associations/Exclusions. Competition between eastern red-backed salamanders and other plethodontids is recognized (e.g., Adams, 2000). Microspatial segregation occurs between eastern red-backed and Shenandoah salamanders (*P. shenandoah*), with possible competitive exclusion, but experimental results are inconclusive (Jaeger, 1970, 1971a,b, 1972, 1974a; Kaplan, 1977; Wrobel et al., 1980; Lancaster and Jaeger, 1995). Hybridization between eastern red-backed and Shenandoah salamanders is also reported, with concern that genetic swamping may be contributing to the decline of the Endangered Shenandoah salaman-

ders (Thurow, 1999). Eastern red-backed salamanders are aggressive toward northern slimy salamanders (*P. glutinosus*), defending territories against them (Lancaster and Jaeger, 1995). Allegheny Mountain dusky salamanders (*Desmognathus ochrophaeus*) behave aggressively towards eastern red-backed salamanders and can drive them from occupied sites (Smith and Pough, 1994). Eastern red-backed salamanders are replaced by the more drought resistant southern ravine salamanders (*P. richmondi*) on steep slopes in Ohio (Pfingsten, 1989b). However, in Virginia, Grover (2000) showed that eastern red-backed salamanders were displaced from moist habitats near streams and seeps by northern dusky salamanders (*D. fuscus*) and seal salamanders (*D. monticola*).

Spotted salamanders (*Ambystoma maculatum*) preyed upon eastern red-backed salamanders in 9% of lab trials (Ducey et al., 1994), which might best be interpreted as interspecific interference. The presence of spotted salamanders therefore can affect eastern red-backed salamander distribution on the forest floor in areas of sympatry. Possible competitive interactions between eastern red-backed salamanders and Valley and Ridge salamanders (*P. hoffmani*; Fraser, 1976b), and between eastern red-backed salamanders and Wehrle's salamanders (*P. wehrlei*; Pauley, 1978a,b,c), have been suggested but not conclusively demonstrated. In New Hampshire, eastern red-backed salamanders predominate in salamander assemblages, comprising 93.5% of the biomass of six salamander species (Burton and Likens, 1975a).

L. Age/Size at Reproductive Maturity. Sexual maturity is reached about 2 yr after hatching (Bausmann and Whitaker, 1987). Ohio males (n=904) averaged 40.5 mm SVL, and females (n=632) 41.2 mm (Pfingsten, 1989b). The smallest mature males from Ohio are reported to be 32–37 mm SVL, and females 34–39 mm (Pfingsten, 1989b). Nagel (1977) reported growth rates in an eastern Tennessee population averaging 15 mm SVL during the first year and 8 mm in the second year, with growth surprisingly not slowing during the winter (females in this study reproduced annually, so growth rates may be lower in biennially reproducing populations). Females often exhibit biennial breeding cycles in the North and annual cycles in the South (Sayler, 1966; Petranka, 1998). However, both biennial (Vogt, 1981) and annual (M. Bergeson, personal communication) breeding has been reported from Wisconsin. Females first oviposit about 3.5 yr after hatching, when they measure >34–38 mm SVL. Males breed annually throughout the range and are sexually mature upon reaching 32–37 mm SVL (Blanchard, 1928a; Sayler, 1966; Werner, 1971; Nagel, 1977; Lotter, 1978; Petranka, 1998).

M. Longevity. Unknown.

N. Feeding Behavior. Eastern red-backed salamanders are a top predator of the detritus food chain, feeding on any prey they can capture. Salamanders will climb on vegetation to forage at night (Cockran, 1911; Burton and Likens, 1975a; Jaeger, 1978). Small invertebrates are the staple of the diet. Wyman (1988b) estimated that eastern red-backed salamanders consume 1.5 million prey items/ha/yr in New York. Ants, termites, beetles, flies, earthworms, spiders, snails, slugs, mites, centipedes, millipedes, springtails, midges, pseudoscorpions, and other lepidopterans, thysanopterans, and hymenopterans are all reported as prey (Cockran, 1911; Murphy, 1918; Blanchard, 1928a; Hamilton, 1932; Jameson, 1944; Jaeger, 1972; Caldwell and Jones, 1973; Caldwell, 1975; Burton, 1976; Hoff, 1977; Pauley, 1978b; Mitchell and Woolcott, 1985; Bausmann and Whitaker, 1987; Maglia, 1996; Hughes et al., 1999). Ants and mites formed the bulk of the diet in a Canadian jack pine forest (Bellocq et al., 2000). Eastern red-backed salamanders are also reported to eat their own cast skins and occasionally will cannibalize conspecific eggs and juveniles (Surface, 1913; Piersol, 1914; Burger, 1935; Heatwole and Test, 1961; Highton and Savage, 1961; Burton, 1976). Maerz and Karuzas (2003) report an instance of an adult cannibalizing a juvenile.

O. Predators. A wide variety of animals and one plant will prey upon eastern red-backed salamanders, with ring-necked snakes (*Diadophis punctatus*) and short-tailed shrews (*Blarina brevicauda*) likely being the most common predators. Other reported predators include woodland snakes (i.e., garter snakes [*Thamnophis* sp.], copperheads [*Agkistrodon contortix*], and ring-necked snakes; Cockran, 1911; Uhler et al., 1939; Arnold, 1982; Mitchell, 1994a; Lancaster and Wise, 1996), spiders (Lotter, 1978), rove beetles (*Platydracus viduatus* [Staphylinidae]; Jung et al., 2000), spotted salamanders (Ducey et al., 1994), praying mantis (*Mantis religiosa*: Stein, 1989), mammals (shrews [Insectivora], voles and chipmunks [Rodentia], raccoons and foxes [Carnivora]; Brodie et al., 1979; Wyman, 1988b), and birds that forage in the leaf litter (Coker, 1931; Bent, 1949; Lotter and Scott, 1977; Brodie et al., 1979; Jaeger, 1981a; Fenster and Fenster, 1996). Most brooding female eastern red-backed salamanders will desert nests and flee when approached by ring-necked snakes (Petranka, 1998). Eastern red-backed salamanders have also been found dead within the insectivorous leaves of the bog-dwelling purple pitcher plant (*Sarracenia purpurea*; Hughes et al., 1999). Other likely predators include woodland mice (Cricetidae, Zapodidae), centipedes (Chilopoda), and ground beetles (Carabidae).

P. Anti-Predator Mechanisms. Eastern red-backed salamanders possess noxious skin secretions concentrated along the dorsum of the tail (Brodie et al., 1979; Petranka, 1998), which convey protection. When exposed, individuals may remain motionless to avoid detection, flee for protective cover, or assume a coiled position with the tail on top, presenting a dispensable body part to the predator (see below). The mean duration of immobility of disturbed salamanders is 39.4 s (range 1.0–169.5 s, n = 287; Dodd, 1989). Eastern red-backed salamanders release alarm pheromones from skin glands when attacked, which, unlike territorial pheromones, are short lived (about 2 min; Graves and Quinn, 2000; see also Hecker et al., 2003). Tail autotomy has also been reported as an anti-predator defense mechanism (Lancaster and Wise, 1996). When encountering shrews, eastern red-backed salamanders orient their tail toward the predator and arch and undulate this appendage, which contains glands thought to be distasteful to predators (Brodie et al., 1979). Shrews ate only 40% of red-backed salamanders offered in lab trials, which was attributed to distasteful glandular secretions (Brodie et al., 1979; see also Hecker et al., 2003). Mimicry has also been postulated as an anti-predator mechanism in the erythristic (all red) color morph of eastern red-backed salamanders, which are suspected of mimicking the red eft stage of eastern newts (*Notophthalmus viridescens*), a highly noxious species distasteful or poisonous to predators (Tilley et al., 1982).

Q. Diseases. No information is available, but eggs are susceptible to fungal infections (Pfingsten, 1989b).

R. Parasites. The following parasites and protozoans have been reported from eastern red-backed salamanders (Rankin, 1937; Ernst, 1974; Muzzall, 1990; Bursey and Schibli, 1995; Muzzall et al., 1997; Bolek and Coggins, 1998): protozoans—*Cryptobia borreli*, *Cytamoeba bacterifera*, *Eutrichomastix batrachorum*, *Haptophyra* (=*Cepedietta*) *michiganensis*, *Hexamastix batrachorum*, *Hexamitus* sp:, *Hexamitus batrachorum*, *Hexamitus intestinalis*, *Karatomorpha swezi*, *Monocercomonoides* sp., *Monocercomonas batrachorum*, *Octomitus* sp., *Proteromonas longifila*, *Prowazekella longifilis*, *Trimitus parvus*, *Tritrichomonas augusta*, and *Tritrichomonas batrachorum*; nematodes—*Angiostoma plethodontis*, *Batracholandros magnavulvaris*, *Cosmocercoides dukae*, *Cosmocercoides variabilis*, *Falcaustra* sp., *Oswaldocruzia pipiens*, *Oxyuris magnavulvaris*, and *Rhabdias ranae*; helminths—*Brachycoelium hospitale*, *Brachycoelium louisianai*, *Brachycoelium obesum*, *Brachycoelium salamandrae*, *Brachycoelium storeriae*, *Cylindrotaenia americana*, and *Cylindrotaenia idahoensis*.

4. Conservation.
No range retractions of eastern red-backed salamanders have been reported, but local extirpations have been due to habitat changes, chiefly deforestation, and other,

unknown causes (see Highton, this volume, Part One). Positive correlations have been made between forest age, the quantity and quality of downed woody debris, and salamander abundance (Herbeck and Larsen, 1999), and it is widely believed that abundance declined following European settlement. Despite this, eastern red-backed salamanders can be extremely numerous (see Table SP-5).

Plethodon cylindraceus (Harlan, 1825[b])
WHITE-SPOTTED SLIMY SALAMANDER
David A. Beamer, Michael J. Lannoo

1. Historical versus Current Distribution.
White-spotted slimy salamanders (*Plethodon cylindraceus*) occur in the Piedmont and Blue Ridge Physiographic Provinces of Virginia and North Carolina, west to the French Broad River and south to the Northern Piedmont of South Carolina. They also occur in parts of the Valley and Ridge Physiographic Province in western Virginia and extreme eastern West Virginia, and in a small area of the Coastal Plain Physiographic Province of eastern Virginia. Of 14 populations sampled prior to 1988 and resampled in 1993 or later, Highton (this volume, Part One) could find no animals in three, under similar sampling conditions and with a similar effort. These sites will need to be resurveyed to determine whether Highton's data reflect true declines or natural population fluctuations.

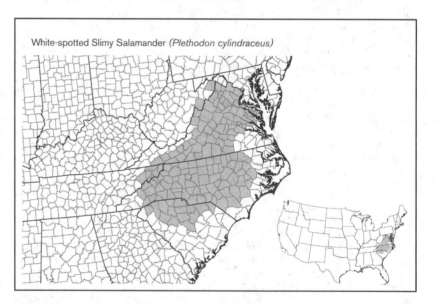
White-spotted Slimy Salamander (*Plethodon cylindraceus*)

2. Historical versus Current Abundance.
Highton (this volume, Part One) found evidence of declines in all 14 white-spotted slimy salamander populations he sampled during the past four decades. In 13 of these populations, declines appeared precipitous, with recent (1990s) numbers fewer than half of pre-1990s numbers. These sites must be resurveyed to

determine whether Highton's data reflect real declines or natural population fluctuations.

Gordon et al. (1962) compared their collection to a collection made by Dunn 44 yr earlier. The collections were made in the same area, though there had been at least one lumbering of the area between the collections. The number of white-spotted slimy salamanders was nearly identical in the two collections.

Clearcuts negatively affect populations of white-spotted slimy salamanders. Petranka et al. (1993, 1994) found that these salamanders occur in much higher numbers in mature forests than in clearcuts. Petranka (1998) states that deforestation and the conversion of mixed hardwood forests to pine monocultures has eliminated or reduced a large number of eastern *Plethodon* populations. However, Ash (1988, 1997) and Ash and Bruce (1994) have strongly disagreed with Petranka's estimates and do not consider forestry practices to be having as strong an impact on native salamanders.

3. Life History Features.
  A. Breeding. Reproduction is terrestrial. The frequency of males closely associated with females beneath cover objects is much higher in the spring than at other times of the year. Wells (1980) believes that this strongly suggests that courtship occurs in April and May in a

population from Durham County, North Carolina.

Organ (1960a) observed courtship of captive individuals from Whitetop Mountain, Virginia, throughout September. Males place their nasolabial grooves and mental glands in contact with the female's body. The male then begins a foot dance that involves raising and lowering the rear limbs simultaneously or alternately. The male then moves towards the female's head while repeatedly rubbing his nasolabial grooves on the female. Occasionally, the male will grasp the female's body or tail with his mouth. When the male reaches the female's head he rubs his mental gland over her head and nasolabial grooves. The male then pushes his head under her chin and passes beneath her, undulating his tail as it passes under the female's chin. When the male stops moving forward, the female straddles his tail and then the pair moves forward while the female maintains a straddled position. While moving forward, the male often flexes his body laterally and slaps the female with his mental gland. The pair then moves forward until the spermatophore is deposited. Just prior to depositing a spermatophore, the male begins a series of lateral sacral movements while the female moves her head laterally in synchronization with but counter to the male's movements. The male then lowers his vent to the substrate and deposits a spermatophore. After the spermatophore is deposited, the male withdraws his tail and bends it sharply to one side. The female keeps her chin against the sharply bent tail while the pair moves forward until the vent of the female reaches the spermatophore. The female then picks up the cap from the spermatophore with her cloacal lips.

i. Breeding migrations. Undocumented, but breeding migrations are not known for any *Plethodon* species.

ii. Breeding habitat. Unknown.

B. Eggs. Examination of preserved material from the Durham area suggests that oviposition probably occurs in late summer or early fall (Wells, 1980).

i. Egg deposition sites. Unknown, but likely to be in underground cavities.

ii. Clutch size. Unknown.

C. Direct Development.

i. Brood sites. Unknown.

ii. Parental care. Unknown, but it is likely that females brood, as with other members of the slimy salamander complex.

D. Juvenile Habitat. First-year young (20 mm SVL) were found beneath logs in Durham County, North Carolina, in early March (Wells and Wells, 1976).

E. Adult Habitat. White-spotted slimy salamanders have been recorded from both virgin and second-growth forest (Dunn, 1917a; Gordon et al., 1962).

White-spotted slimy salamanders in Durham County, North Carolina, were found in second-growth oak-hickory forest on a gently sloping hill. The ground had a thick layer of leaf litter and a large number of fallen logs were present while rocks and other cover was lacking (Wells and Wells, 1976). Salamanders were found frequently beneath logs but did not seem to occupy the leaf litter even in wet periods (Wells, 1980).

F. Home Range Size. The maximum distance moved by a white-spotted slimy salamander in Durham County, North Carolina, was 91.5 m, though most adults moved <9 m. Juveniles are more sedentary, and only one juvenile moved >9 m. Most juvenile movements were <6 m. White-spotted slimy salamanders between 55 and 65 mm SVL made the largest movements (30–90 m). Individuals >65 mm rarely moved >18 m (Wells and Wells, 1976). The most distant movements are made shortly after reaching sexual maturity.

G. Territories. Data from Wells (1980) show that juveniles and females do not avoid occupied cover objects. Males, however, may defend sites from other males.

H. Aestivation/Avoiding Desiccation. During droughts, groups of salamanders ranging from 10–25 animals are found under large logs. This behavior was observed in a period of drought and may reduce the rate of desiccation (Wells and Wells, 1976).

I. Seasonal Migrations. Animals likely move from surface sites to underground sites with seasonal changes in temperature and moisture conditions.

J. Torpor (Hibernation). Wells (1980) states that the entire white-spotted slimy salamander population disappeared into underground retreats for the winter shortly after 13 November.

K. Interspecific Associations/Exclusions. In Avery County, North Carolina, white-spotted slimy salamanders were found with eastern newts (*Notophthalmus viridescens*), Blue Ridge dusky salamanders (*Desmognathus orestes*), seal salamanders (*D. monticola*), red-backed salamanders (*P. cinereus*), Yonahlossee salamanders (*P. yonahlossee*), northern gray-cheeked salamanders (*P. montanus*), and Blue Ridge two-lined salamanders (*Eurycea wilderae*; Gordon et al., 1962).

Green salamanders (*Aneides aeneus*), Yonahlossee salamanders, and Blue Ridge gray-cheeked salamanders (*P. amplus*) were found on a cliff face with white-spotted slimy salamanders in Rutherford County, North Carolina (Rubin, 1969). Ocoee salamanders (*D. ocoee*) also have been reported from these same crevices (Adler and Dennis, 1962).

Hairston (1949) reports the following salamanders along with white-spotted slimy salamanders from the Black Mountains area, Yancey County, North Carolina: spring salamanders (*Gyrinophilus porphyriticus*), red salamanders (*Pseudotriton ruber*), Blue Ridge two-lined salamanders, black-bellied salamanders (*D. quadramaculatus*), seal salamanders, Carolina mountain dusky salamanders (*D. carolinensis*), pigmy salamanders (*D. wrighti*), shovel-nosed salamanders (*D. marmoratus*), Yonahlossee salamanders, and northern gray-cheeked salamanders.

At Limestone Cove, Unicoi County, Tennessee, white-spotted slimy salamanders were found with Yonahlossee salamanders, Weller's salamanders (*P. welleri*), eastern newts, seal salamanders, Carolina mountain dusky salamanders, and black-bellied salamanders (Thurow, 1963).

White-spotted slimy salamanders, Cow Knob salamanders (*P. punctatus*), and Shenandoah Mountain salamanders (*P. virginia*) are known from Shenandoah Mountain. There may be competitive exclusion between white-spotted slimy salamanders and Cow Knob salamanders, as usually only one or the other are found locally. There are no published data on the association of white-spotted slimy salamanders with Shenandoah Mountain salamanders there (Highton, 1972, 1989, 1999).

White-spotted slimy salamanders are found with red-backed salamanders, southern ravine salamanders (*P. richmondi*), and Yonahlossee salamanders at Comer's Rock, Grayson-Wythe counties, Virginia (Dodd, 1989).

In the Peaks of Otter region of Virginia, white-spotted slimy salamanders are sympatric with Peaks of Otter salamanders (*P. hubrichti*), red-backed salamanders, southern two-lined salamanders (*Eurycea cirrigera*), and seal salamanders (Thurow, 1957; Mitchell and Wicknick, this volume).

White-spotted slimy salamanders have a long contact from southeastern Virginia to western South Carolina with Atlantic Coast slimy salamanders (*P. chlorobryonis*). Two transects, one in southeastern Virginia and one in northeastern North Carolina, indicate that there is a zone of parapatric hybridization (Highton and Peabody, 2000).

White-spotted slimy salamanders have a long contact with northern slimy salamanders (*P. glutinosus*) from Maryland south through West Virginia, Virginia, and Tennessee, from the Potomac River to the French Broad River. Hybrid populations are known from Washington County, Maryland; and Highland, Washington, and Smyth counties, Virginia (Highton and Peabody, 2000).

White-spotted slimy salamanders probably contact southern Appalachian salamanders (*P. teyahalee*) along the North Carolina–South Carolina state line at the headwaters of the French Broad River. There are no data on their genetic interaction (Highton and Peabody, 2000).

White-spotted slimy salamanders have a long contact from southeastern Virginia to western South Carolina with Atlantic Coast slimy salamanders. Two transects, one in southeastern Virginia and one in northeastern North Carolina, indicate that there is a zone of parapatric hybridization (Highton and Peabody, 2000).

White-spotted slimy salamanders are sympatric with Yonahlossee salamanders at intermediate elevations throughout much of the Blue Ridge Province east of the French Broad River. There is no evidence of hybridization between these species (Highton and Peabody, 2000).

White-spotted slimy salamanders are sympatric throughout the range of South Mountain gray-cheeked salamanders (*P. meridianus*). There is no evidence of hybridization between these species (Highton and Peabody, 2000).

White-spotted slimy salamanders are widely sympatric throughout the range of Blue Ridge gray-cheeked salamanders. There is no evidence of hybridization between these species (Highton and Peabody, 2000).

White-spotted slimy salamanders are sympatric at the eastern edge of the Blue Ridge isolate of southern gray-cheeked salamanders (*P. metcalfi*). There is no evidence of hybridization between these species (Highton and Peabody, 2000).

White-spotted slimy salamanders are widely sympatric with northern gray-cheeked salamanders in the Blue Ridge province in the Black and Bald isolates, and most of the Roan isolate. There is no evidence of hybridization between these species (Highton and Peabody, 2000).

In the Black Mountains, Yancey County, North Carolina, white-spotted slimy salamanders and northern gray-cheeked salamanders generally replace each other altitudinally. On four of five transects there was no overlap of these species occurrences; at a fifth transect, the two occur together over an elevational range of about 60 m (Hairston, 1951).

L. Age/Size at Reproductive Maturity. The neotype is an adult female, 76 mm SVL. The smallest sexually mature males in Durham County, North Carolina, were 50 mm SVL (Wells and Wells, 1976).

M. Longevity. Unknown.

N. Feeding Behavior. The stomachs of five white-spotted slimy salamanders (average size 54.6 mm SVL) from Rutherford County, North Carolina, contained ants (22.0%), centipedes (11.0%), springtails (9.6%), camel crickets (7.0%), slugs (7.0%), millipedes (6.8%), snout beetles (6.0%), and earthworms (4.0%; Rubin, 1969).

The stomach contents of 58 white-spotted slimy salamanders from Johnston County, Tennessee, contained Collembola, Corrodentia, Membracidae, Cicadellidae, Miridae, lepidopteran larvae, Heterocera, Asilidae, culicid larvae, Diptera and dipteran larvae, Cantharidae, Carabidae, Curculionidae, Elateridae, Lathridiidae, Scaphidiidae, Scarabeidae, Staphylinidae, Tenebrionidae, coleopteran larvae, Formicidae, Vespidae, Phalangidea, Chelonethida, Araneida, Acarina, Diplopoda, Chilopoda, Lumbricidae, and Pulmonata (Pope, 1950).

The stomach contents of 94 white-spotted slimy salamanders from Augusta County, Virginia, included snails, sowbugs, earthworms, centipedes, millipedes, arachnids, phalangids, pseudoscorpions, a roach, Homoptera, Hemiptera, Diptera, lepidopteran larvae, Coleoptera, coleopteran larvae, and Hymenoptera (note the stomach contents of six northern slimy salamanders are also included in this list). Sowbugs and earthworms were commonly eaten in the spring while phalangids, pseudoscorpians, roaches, homopterans, hemipterans and lepidopteran larvae were eaten in the fall. The largest food item was a centipede (45 mm), and another stomach contained 10 buprestid larvae (Davidson, 1956).

O. Predators. Garter snakes (*Thamnophis* sp.) and copperheads (*Agkistrodon contortrix*) feed on white-spotted slimy salamanders in the George Washington National Forest in Virginia (Uhler et al., 1939).

P. Anti-Predator Mechanisms. All *Plethodon* produce noxious skin secretions (Brodie, 1977). Members of the *P. glutinosus* complex frequently become immobile when initially contacted. White-spotted slimy salamanders were included in a field study on immobility, however it is not possible to separate their behavior from the other members of this complex in this published data set. Immobility may increase survival by making salamanders less likely to be detected, especially by visually oriented predators (Dodd, 1989).

Q. Diseases. Unknown.

R. Parasites. Rankin (1937) reported the following parasites from salamanders that were likely white-spotted slimy salamanders: *Cryptobia borreli, Cytamoba bacterifera, Eutrichomastix batrachorum, Haptophrya gigantean, Haptophrya michiganensis, Hexamastix batrachorum, Hexamitus intestinalis, Karotomorpha swezi, Prowazekella longifilis, Tritrichomonas augusta, Brachycoelium hospitae, Capillaria inequalis, Cosmocercoides dukae, Oswaldocruzia pipiens, Oxyuris magnavulvaris, Acanthocephalus acutulus,* and *Hannemania dunni.*

4. Conservation.
White-spotted slimy salamanders are not protected by any state. Among members of the *P. glutinosus* complex, white-spotted slimy salamanders have one of the widest distributions. Within this range, there are many federal and state properties that contain suitable habitat for these salamanders.

White-spotted slimy salamanders are relatively resilient to disturbances, such as those associated with timbering operations, and frequently found in second-growth forests and relatively small, fragmented woodlots (Gordon et al., 1962; D.A.B., personal observations). However, clearcuts negatively affect populations (Petranka et al., 1993, 1994; but see Ash, 1988, 1997, and Ash and Bruce, 1994).

As with all species of *Plethodon*, white-spotted slimy salamanders do not migrate to breeding grounds and they do not have large home ranges. Thus, they can exist in habitats of smaller size than many other amphibian species. Conservation activities that promote mature closed-canopy forests should benefit this species

Acknowledgments. Thanks to Richard Highton, who reviewed this account and gave us the benefit of his insight and experience.

Plethodon dorsalis Cope, 1889
NORTHERN ZIGZAG SALAMANDER

Walter E. Meshaka Jr.

1. Historical versus Current Distribution.
Northern zigzag salamanders (*Plethodon dorsalis*) are distributed from southern Indiana southward through western and central Kentucky and central Tennessee (Thurow, 1966). The geographic range of northern zigzag salamanders was considered to include central Alabama and western Georgia until the discovery of Webster's salamanders (*P. websteri*), a sibling species (Highton, 1979) with a potentially narrow contact zone with southern zigzag salamanders (*P. ventralis*; Highton, 1985). Minton (2001) reports a reversal in relatively greater abundance from northern zigzag salamanders to eastern red-backed salamanders (*P. cinereus*) at an Indiana site.

2. Historical versus Current Abundance.
Northern zigzag salamanders are seasonally abundant under moist leaf litter and flat rocks, often on wooded hillsides and ravines, along shale banks, and associated with sinkholes (Minton, 2001). They have also been found in caves (Miller et al., 1998).

3. Life History Features.
  A. Breeding. Reproduction is terrestrial.
    i. Breeding migrations. Do not occur.
    ii. Breeding habitat. Same as adult habitat. In Indiana, spermatozoa are present in vasa deferentia in October and April (Sever, 1978a), and sperm are present in females in November and March (Minton, 2001) and in April (Sever, 1978b). Egg laying in Indiana occurs in the spring, collections indicate from April–June (Sever, 1978b).
  B. Eggs.
    i. Egg deposition sites. Clutches are deposited within rock crevices, subterranean cavities, and in caves (Mohr, 1952; Smith, 1961; Mount, 1975; Miller et al., 1998). Logs appear to be avoided as nesting sites. Eggs rest on the substrate and are attended by the female (Mohr, 1952; Miller et al., 1998).
    ii. Clutch size. Clutches of closely related Ozark salamanders in Arkansas average 5.3 eggs (range = 3–9) and are produced annually (Meshaka and Trauth, 1995).

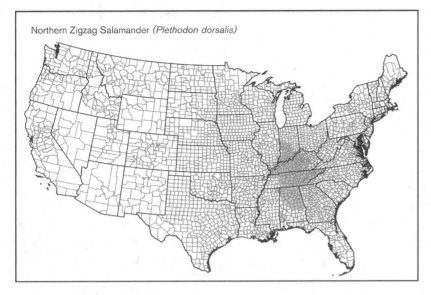

Northern Zigzag Salamander *(Plethodon dorsalis)*

C. Direct Development.

i. Brood sites. Within rock crevices and underground.

ii. Parental care. Females attend the nest during the summer, at which time they are absent from foraging sites (Wilkinson et al., 1993; Meshaka and Trauth, 1995). Hatchlings appear in the fall (Wilkinson et al., 1993; Meshaka and Trauth, 1995).

D. Juvenile Habitat. Same as adults.

E. Adult Habitat. Moist leaf litter, talus areas, and caves (Thurow, 1955; Minton, 2001). Northern zigzag salamanders prefer rocky habitat (Smith, 1961; Reinbold, 1979). Minton (2001) notes a preference for moist rocky slopes. This habitat serves to buffer salamanders from weather extremes, and the restricted availability of this habitat may limit the distribution and abundance of this species (Petranka, 1979).

F. Home Range Size. Unknown.

G. Territories. Unknown.

H. Aestivation/Avoiding Desiccation. Aestivation is unknown, however under desiccating conditions animals likely move under cover objects or into burrows.

I. Seasonal Migrations. Unknown. However, in Arkansas, closely related Ozark zigzag salamander *(P. augusticlavius)* adults migrate to sandstone cedar glades from the forest during the courting season. Immature individuals are not found on the glade (Meshaka and Trauth, 1995).

J. Torpor (Hibernation). Unknown.

K. Interspecific Associations/Exclusions. In areas of syntopy, there is habitat separation between northern zigzag salamanders and eastern red-backed salamanders (Reinbold, 1979). Minton (2001) notes segregation between these species in Indiana, with northern zigzag salamanders exhibiting a preference for moist rocky habitats. Bausmann and Whitaker (1987) note more earthworms (Annelida) and fewer beetles (Insecta: Coleoptera) in the stomachs of northern zigzag salamanders when compared to eastern red-backed salamanders.

L. Age/Size at Reproductive Maturity. Mature female northern zigzag salamanders in Indiana measure at least 35.6 mm SVL (Sever, 1978b; Minton, 2001). Both sexes of closely related Ozark zigzag salamanders are mature at 35 mm SVL and reproduce for the first time in the winter of their third year of life (Meshaka and Trauth, 1995).

M. Longevity. Unknown.

N. Feeding Behavior. Spiders and beetles comprised most of the diet of a fall and winter sample of northern zigzag salamander adults (Holman, 1955).

O. Predators. Centipedes in the genus *Scolopendra* sp. are likely predators of the closely related Ozark salamanders (personal observations).

P. Anti-Predator Mechanisms. Unknown.

Q. Diseases. Unknown.

R. Parasites. Unknown.

4. Conservation.
Northern zigzag salamanders are listed as a Species of Special Concern in North Carolina, at the eastern edge of their distribution (Levell, 1997). As with other members of the *P. dorsalis* complex, comprehensive life history and natural history studies of northern zigzag salamanders are necessary to obtain the knowledge required to make informed and effective management decisions.

Acknowledgments. Richard Highton reviewed this account and provided valuable feedback.

Plethodon dunni Bishop, 1934
DUNN'S SALAMANDER
R. Bruce Bury

1. Historical versus Current Distribution.
Dunn's salamanders *(Plethodon dunni)* are distributed from the southwestern corner of Washington State (but not in the Washington Cascade Mountains) southward through western Oregon, including the Cascade Range (Storm and Brodie, 1970a; Leonard et al., 1993), and barely into Del Norte County, California (Bury et al., 1969). In western Oregon, where their range is most extensive, Dunn's salamanders occur from the high tide line of the Pacific Ocean (Ferguson, 1956; Storm and Brodie, 1970a) throughout the Coast Range, and up the west flank of the Cascades to about 1,000 m (Nussbaum et al., 1983; Petranka, 1998). Dunn's salamanders are absent from the low parts of the Willamette Valley (Storm and Brodie, 1970a).

2. Historical versus Current Abundance.
Unknown. Unlike many Pacific northwestern species, Dunn's salamanders can be found in forest stands of all ages (Corn and Bury, 1991). However, they are absent or reduced from banks of streams that are in or downstream from clearcut forests in western Oregon (Corn and Bury, 1989a).

3. Life History Features.

A. Breeding. Reproduction is terrestrial.

i. Breeding migrations. Unknown for any species of *Plethodon*, although movements from forest floor habitats to underground brood sites occur. The breeding season for Dunn's salamanders is not well defined. Gravid females have been found throughout the year (Dumas, 1956). Spermatophores or remnants of spermatophores have been found in the cloacas of mature females in April and October, indicating that courtship may occur over a long time (Nussbaum et al., 1983).

ii. Breeding habitat. Unknown.

B. Eggs.

i. Egg deposition sites. See "Brood sites" below.

ii. Clutch size. Ranges from 4–15 (average = 9.4), and larger females contain more eggs than smaller females (Nussbaum et al., 1983). Mature ova are 4.5–5.5 mm in diameter (Nussbaum et al., 1983).

C. Direct Development.

i. Brood sites. On 6 July 1952, Dumas (1955) found a Dunn's salamander egg mass located about 2 m from Sugarbowl Creek, in Lincoln County, Oregon. The egg mass was attached by a 3.9 mm stalk to a slab, and located about 37 cm back in a crevice of a shale outcrop on a heavily shaded southwestern-facing slope. Nine eggs were arranged in a grape-like cluster. Eggs ranged from 4.8–5.3 mm in diameter. Dumas brought these into the lab and reared them until the last egg succumbed to a fungus. At this time (after an incubation period of about 70 d), the nearly full-term embryo was 16 mm SVL (in nature, hatchlings as small as 13 mm SVL have been found; Dumas, 1956).

ii. Parental care. As in other members of the *Plethodon* genus, the egg mass is probably protected by the female.

Dunn's Salamander (*Plethodon dunni*)

I. Seasonal Migrations. Coastal populations may be active year-round in wet weather, but in many areas surface activity is restricted by cold temperatures and dry conditions (Nussbaum et al., 1983; Petranka, 1998). During the winter, some populations remain associated with talus slopes but move to rocky streamsides with warmer spring temperatures. Under dry summer conditions, individuals move deep within talus, into deep cracks in rocks, or beneath stones and rocks in cool habitats along streams. With autumn rains, individuals return to the surface. Surface activity is greatest in the spring and autumn (Dumas, 1956; Petranka, 1998).

J. Torpor (Hibernation). Dunn's salamanders tolerate cold temperatures well and are found at slightly higher altitudes than congeneric western red-backed salamanders (*P. vehiculum*; Dumas, 1956).

K. Interspecific Associations/Exclusions. Co-occur with Del Norte salamanders (*P. elongatus*), Larch Mountain salamanders (*P. larselli*), Van Dyke's salamanders (*P. vandykei*), and western red-backed salamanders (Dumas, 1956; Storm and Brodie, 1970a). Dumas (1956) examined sympatric Dunn's salamanders and western red-backed salamanders and found that Dunn's salamanders prefer wetter sites and feed on a more diverse assemblage of prey items than western red-backed salamanders. Ovaska and Davis (1992) found that Dunn's salamanders and western red-backed salamanders can distinguish between each other's fecal pellets, but neither species avoids the other's burrow. Dumas (1956) suggested that Dunn's salamanders exclude western red-backed salamanders from preferred cover objects.

Other amphibians associated with Dunn's salamanders include the more aquatic coastal giant salamanders (*Dicamptodon tenebrosus*), torrent salamanders (*Rhyacotriton* spp.), coastal tailed frogs (*Ascaphus truei*), and the more terrestrial clouded salamanders (*Aneides ferreus*), rough-skinned newts (*Taricha granulosa*), and ensatinas (*Ensatina eschscholtzii*).

L. Age/Size at Reproductive Maturity. Animals become sexually mature at about 50–55 mm SVL, at an age of between 2–4 yr (Brodie, 1970; Nussbaum et al., 1983). Dumas (1956) concluded that for the population he studied there were more males than females.

M. Longevity. Unknown.

N. Feeding Behavior. Individuals forage for invertebrates on the forest floor on wet nights (Dumas, 1956). Prey include annelids, snails, centipedes, millipedes, isopods, scorpions, pseudoscorpions, mites, phalangids, and insects, including collembolans, coleopterans, dipterans, and hymenopterans (Dumas, 1956; Altig and Brodie, 1971). In the laboratory, adults can survive for >5 mo without food (Dumas, 1956).

D. Juvenile Habitat. Unknown.

E. Adult Habitat. Dunn's salamanders are almost always associated with rocks. Along the Coast Range, they occur in sandstone or shale outcrops near seepages, springs, and streams (Dumas, 1956). In the Cascades, they occur in basaltic talus near seepages, springs, and streams. They often are found in wetted rocky areas along with torrent salamanders (*Rhyacotriton* sp.; Nussbaum et al., 1983; Corn and Bury, 1989a). However, Bury et al. (1991) found that Dunn's salamanders were mostly along stream banks (about 90% of captures) with fewer (<10%) in riffles compared to torrent salamanders, with some (about 25%) along stream banks but most (about 50%) in riffles and the rest in pools. In general, torrent salamanders tend to occur in riffles or rocky substrata with fast waters nearby (often with their feet or venters wet), whereas Dunn's salamanders live under rocks just a little less wet and upward on talus slopes (personal observations). In rainy weather, they may be found in or under logs near streams or under surface debris (Corkran and Thoms, 1996).

Dumas (1956) collected animals when ambient temperatures ranged from 4–17°C (see also Storm and Brodie, 1970a). Storm (1955) describes locations where Dunn's salamanders were found as ". . . under well imbedded rocks, about 4 ft (1.3 m) from the water's edge, along the base of the steep bank of a small stream.

The area was well shaded by alders and second-growth hemlock, and many of the rocks were overgrown with moss." Further, at another site, "Two of the salamanders were taken from beneath shaded moss-covered rocks at the edge of a tumultuous mountain brook and 4 were collected among small stones and gravel in a spring seepage of the well shaded east slope . . . ." Dumas (1956, p. 485) provides additional details of this habitat and notes that under high humidity conditions Dunn's salamanders can be found on drier substrates.

While Dumas (1956) notes that Dunn's salamanders "are sluggish salamanders with rather low mobility," Petranka (1998) characterizes them as active and agile and rapidly fleeing when disturbed.

F. Home Range Size. Varies with the season (see "Seasonal Migrations" below).

G. Territories. Dumas (1956) found Dunn's salamanders under 113 cover objects. Of these, only four objects had two animals together, and one concealed three salamanders. This suggests that either individuals avoid each other or territories are established and defended. However, this needs further study.

H. Aestivation/Avoiding Desiccation. Ray (1958) points out that Dunn's salamanders generally live in wetter situations than most *Plethodon*, and that the single individual he tested for desiccation tolerance showed a low resistance to drying and demonstrated no protective behaviors.

O. Predators. Birds, such as dippers (*Cinclus mexicanus*) and Stellar's jays (*Cyanocitta stelleri*), and northwestern garter snakes (*Thamnophis ordinoides*) are reported to take Dunn's salamanders (Nussbaum et al., 1983). Potential predators include beetles (*Scaphinotus* sp.), shrewmoles (*Neurotrichus gibbsi*), and other shrews (*Sorex trowbridgei* and *S. pacificus*). While cannibalism is rare, Stebbins (1951) observed a female ingesting some newly laid eggs. Dissection of 21 adult Dunn's salamanders from the Oregon Coast Range revealed that two adults had cannibalized juveniles (Riesecrer et al., 1996).

P. Anti-Predator Mechanisms. Dunn's salamanders may remain motionless or rapidly flee when disturbed (Brodie, 1977).

Q. Diseases. Dumas (1956) has described a fungus attacking the eggs, and sometimes adults, in captivity.

R. Parasites. Unknown.

4. Conservation.
This is a fairly widespread species, but it appears to require wet rocky substrate and often is most abundant along streamside habitat. There is no clear evidence to indicate declines in populations, but they are absent or reduced in clearcut forests in western Oregon.

Plethodon electromorphus Highton, 1999
NORTHERN RAVINE SALAMANDER

David A. Beamer, Michael J. Lannoo

1. Historical versus Current Distribution.
Northern ravine salamanders (*Plethodon electromorphus*) are a cryptic species previously considered to be a part of *P. richmondi* (southern ravine salamanders; Highton, 1999a). Northern ravine salamanders differ from southern ravine salamanders in their protein composition and their distribution. Northern ravine salamanders range "from southwestern Pennsylvania, Ohio, and southeastern Indiana south to northwestern West Virginia, east of the New and Kanawha rivers, and northern Kentucky" (Highton, 1999a). While populations have undoubtedly been lost, the current distribution of northern ravine salamanders is likely similar to their historical distribution.

2. Historical versus Current Abundance.
Unknown.

3. Life History Features.
Information gathered under this scientific name likely holds for both northern ravine salamanders and southern ravine salamanders—until 1999 they shared the name "*P. richmondi*" and there are morphological and ecological similarities between the two species. As the older literature is sorted and a new literature is developed, it will be interesting to note whether behavioral and ecological differences between these sister species are discovered.

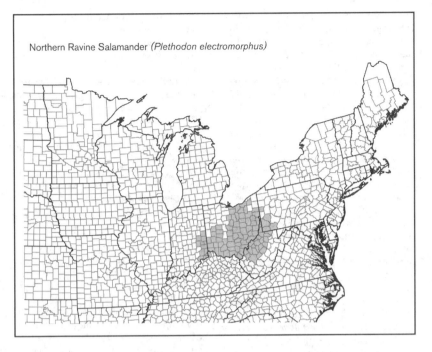

Northern Ravine Salamander (*Plethodon electromorphus*)

A. Breeding. Reproduction is terrestrial. A late winter breeding period is suggested, as a female collected in Indiana on 21 February dispelled a spermatophore while being anaesthetized (Minton, 1972, 2001). Wood (1945b) suggests that eggs are deposited from 21 April–14 May in Ohio.

i. Breeding migrations. Undocumented, but breeding migrations are not known for any *Plethodon* species.

ii. Breeding habitat. Unknown.

B. Eggs.

i. Egg deposition sites. A clutch of eggs presumably from a northern ravine salamander was found beneath a talus slope in Greene County, Ohio (Duellman, 1954b).

ii. Clutch size. A clutch of 12 eggs presumably from a northern ravine salamander was reported from Greene County, Ohio, on 14 July 1949 (Duellman, 1954b). Females from Ohio had from 9–15 large ovarian eggs (Wood, 1945b).

C. Direct Development. The eggs of the closely related southern ravine salamander hatched on 23 August in Kentucky. Hatchlings were 15 mm SVL (Wallace and Barbour, 1957).

i. Brood sites. Likely to be the same as egg deposition sites, but the single reported clutch of eggs did not have a female in attendance (Duellman, 1954b).

ii. Parental care. While the single clutch of eggs observed did not have a female in attendance (Duellman, 1954b), it is likely that females brood, as with other species of *Plethodon*.

D. Juvenile Habitat. Hatchling northern ravine salamanders remain underground, possibly at the nest site, for several months. They are first seen in the spring when they average 17 mm SVL (Pfingsten, 1989c).

E. Adult Habitat. Northern ravine salamanders have been found on forested slopes with friable soil and flat rocks. They avoid the dry crests of ridges and also stream margins and other excessively moist situations (Minton, 1972, 2001). Pfingsten (1989c) reports occasionally finding animals along the margins of streams. In Ohio, northern ravine salamanders show a strong preference for rocks and are rarely found under any other type of cover (see also Wood, 1945a; Minton, 1972, 2001). Talus slopes sometimes harbor large numbers of individuals.

F. Home Range Size. Unknown.

G. Territories. Thurow (1976) reported aggressive interactions between ravine salamanders (*P. richmondi*) in captivity. At that time, populations presently referred to as northern ravine salamanders were recognized as ravine salamanders (*P. richmondi*). It is possible that northern ravine salamanders were included in this experiment. The extent of aggressive behaviors is unknown in these salamanders. At times it may be minimal, as up to eight adults and subadults have been found beneath a single stone (Minton, 1972, 2001).

H. Aestivation/Avoiding Desiccation. Northern ravine salamanders move to underground sites in response to dry surface conditions during the summer (Minton, 1972, 2001; Pfingsten, 1989c). In Ohio, they have noticeably smaller tail diameters in the fall after they emerge from aestivation (Duellman, 1954b).

I. Seasonal Migrations. Animals likely move from the forest floor to underground sites in response to seasonal cold and dry extremes, and back again as surface conditions become tolerable (Minton, 1972, 2001; Pfingsten, 1989c).

J. Torpor (Hibernation). Unknown. Lethargic individuals have been found in Ohio when temperatures were below freezing and snow was on the ground (Duellman, 1954b).

K. Interspecific Associations/Exclusions. In Indiana, northern ravine salamanders have been found with eastern red-backed salamanders (*P. cinereus*), northern zigzag salamanders (*P. dorsalis*), and northern slimy salamanders (*P. glutinosus*). However, northern ravine salamanders were not found at a locality with both eastern red-backed salamanders and northern zigzag salamanders (Minton, 1972, 2001).

In Ohio, northern ravine salamanders are associated with eastern newts (*Notophthalmus viridescens*), eastern red-backed salamanders, northern slimy salamanders, southern two-lined salamanders (*Eurycea cirrigera*), northern two-lined salamanders (*E. bislineata*), long-tailed salamanders (*E. longicauda*), northern dusky salamanders (*Desmognathus fuscus*), spotted salamanders (*Ambystoma maculatum*), Jefferson salamanders (*A. jeffersonianum*), and green salamanders (*Aneides aeneus*; Pfingsten, 1989c).

In Wooster County, Ohio, northern ravine salamanders occasionally hybridize with eastern red-backed salamanders. Ten individuals among a collection of approximately 150 specimens were morphologically intermediate between these two species. When analyzed genetically, most were more similar to eastern red-backed salamanders, which may indicate that back-crosses to red-backed salamanders are more frequent than those to northern ravine salamanders (Highton, 1999a).

L. Age/Size at Reproductive Maturity. In Ohio, males with pigmented vasa and testes ranged in size from 38–56 mm, with an average SVL of 47.3 mm (measured to the anterior angle of the vent). Gravid females with pigmented eggs ranged from 40–57 mm SVL, with an average of 49.5 (measured to the anterior angle of the vent; Pfingsten, 1989c). The largest specimen from Ohio is a 144-mm TL female from Lawrence County (Hirschfield, 1962).

Maturity probably occurs during an animal's second summer. It is likely that females reproduce biennially (Pfingsten, 1989c).

M. Longevity. Unknown.

N. Feeding Behavior. Salamanders collected in Indiana on 23 September and 21 February contained (in approximate order of frequency) small dipteran and coleopteran larvae, ants, beetles, small spiders, earthworms, mites, and larval ticks (Minton, 1972, 2001).

The stomach contents of Ohio specimens collected at different times of the year included ants, sow bugs, earthworms, ground beetles, spiders, slugs, and land snails (Duellman, 1954b). Three adult females collected from Athens County, Ohio, in late March had eaten sow bugs,

spiders, and rove beetles (Seibert and Brandon, 1960).

O. Predators. Unknown, but likely include forest snakes, birds, and small mammals.

P. Anti-Predator Mechanisms. Nocturnal. All members of the genus *Plethodon* produce noxious skin secretions (Brodie, 1977). The closely related southern ravine salamanders frequently become immobile when initially contacted. Immobility may increase survival by making the salamander less likely to be detected, especially by visually oriented predators (Dodd, 1989).

Q. Diseases. Unknown.

R. Parasites. Unknown.

4. Conservation.

Northern ravine salamanders are not protected by any state. Within their range, there are many federal and state properties that contain suitable habitat for these salamanders.

As with all species of *Plethodon*, northern ravine salamanders can exist in habitats of smaller size than many other amphibian species. Conservation activities that promote mature closed-canopy forests should benefit this species.

Acknowledgments. Thanks to Richard Highton, who reviewed this account and gave us the benefit of his insight and experience.

Plethodon elongatus Van Denburgh, 1916
DEL NORTE SALAMANDER

Hartwell H. Welsh, Jr., R. Bruce Bury

1. Historical versus Current Distribution.
Del Norte salamanders (*Plethodon elongatus*) are restricted to the Klamath Province in northwestern California and southwestern Oregon (Brodie, 1970; Nussbaum et al., 1983; Stebbins, 1985; Leonard et al., 1993, Jennings and Hayes, 1994a; Bury and Pearl, 1999). In California, they range from central Humboldt County eastward, barely into Trinity County, and then northward through Del Norte and western Siskiyou counties. In Oregon, Del Norte salamanders occur along the coast (Curry and southernmost Coos counties) and eastward in the Siskiyou Mountains (Josephine County). They have recently been found inland along West Cow Creek in the Umpqua River watershed (southwestern corner of Douglas County; R. B. B., unpublished data). Their range is about 250 km long from north to south and extends about 95 km inland.

Van Denburgh (1916) first described Del Norte salamanders from a few specimens in Del Norte County, California. Individuals in a few more sites were found in coastal California and Oregon (Wood, 1934; Fitch, 1936), and further inland in

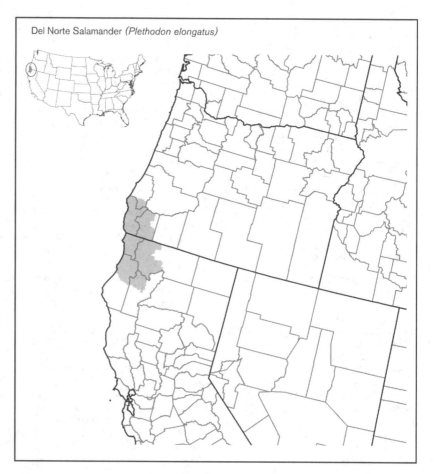

Del Norte Salamander (*Plethodon elongatus*)

California (Stebbins and Reynolds, 1947; Stebbins, 1951). Highton (1962a) shows 11 locality records for Del Norte salamanders, and Brodie (1970) reports 42 sites range wide. Currently, there are museum records and collected specimens from 100 sites (Bury, 1998) as follows: Oregon, 21 coastal localities and 16 inland sites; California, 39 coastal or somewhat inland sites, 12 inland sites along the Trinity River, and 12 inland sites along the Klamath River. Other records are reported in Welsh (1990), Diller and Wallace (1994), and Welsh and Lind (1995), mostly within the known range. However, these were ecological studies, and locality data are not shown by site. The number of new localities continues to increase, with mandated surveys under the Northwest Forest Plan (see below), but the limits of the range have expanded only slightly in recent years.

At the southwestern terminus of the range, two divergent lineages of Del Norte salamanders, both distinct from populations to the north, occur in the lower Klamath Basin and Trinity River Basin (a major tributary of the lower Klamath River); each may represent a new species (Mahoney, 2004). There is marked clinal variation in populations of Del Norte salamanders from coastal sites inland up the Klamath River corridor (Bury, 1973b, 1998). Recent phylogenetic analyses (Mahoney, 2001, 2004; Mead et al., 2004) indicate that *P. elongatus*, as currently recognized, is comprised of more than one genetically distinct lineage that could be recognized as separate taxa. The relationships among these entities and *P. stormi* (Siskiyou Mountains salamander, which also appears to be comprised of more than a single genetically distinct lineage) is under active investigation with new molecular data indicating that the population currently recognized as *P. stormi* in the Scott Bar Mountains region of Siskiyou County, California, is highly distinct and ancestral to both *P. elongatus* and *P. stormi* (Mead et al., 2004). They have now been found from sea level to 1,570 m in elevation (Ollivier and Welsh, 1999).

2. Historical versus Current Abundance.

Del Norte salamanders may be declining due to timber harvesting, especially in interior locales (Raphael, 1988; Welsh, 1990; Jennings and Hayes, 1994a; Bury, 1998). Prior to June 2002, this was a Survey and Manage Species that was afforded some protection on federal lands under the Northwest Forest Plan, where ground-disturbing activities were restricted on occupied sites and a surrounding 33 m buffer (H.H.W. and L.M. Ollivier, unpublished data). Recently (2002), all protections were been removed, and we can expect downward trends in numbers of occupied sites and population numbers to resume across federal lands as resource extraction

once again impacts and reduces suitable habitats in mature and late-seral forests (Welsh and Lind, 1995).

3. Life History Features.

A. Breeding. Reproduction is terrestrial.
i. Breeding migrations. Unknown.
ii. Breeding habitat. Breeding has not been observed but probably occurs in both spring and fall (L.M. Ollivier and H.H.W., unpublished data) on the forest floor of mature to late-seral forests. Lowe and Nieto (2003) report an instance of female-female aggression associated with the courtship season.

B. Eggs.
i. Egg deposition sites. Livezey (1959) reported a subterranean nest at the base of a redwood fence post. This is probably an atypical site, given the association of this species with rocky substrates. Typical sites are probably in subterranean cavities associated with unconsolidated talus and other hard substrates.
ii. Clutch size. One subterranean nest in the wild had 10 eggs in a grape-like cluster (Livezey, 1959). Two adult females were found with 10–11 eggs (Stebbins, 1951). Nussbaum et al. (1983) report that 18 mature females had an average of 7.0 large eggs with a range of 3–11.

C. Direct Development. Parental care is undocumented, but females likely remain with eggs.

D. Juvenile Habitat. Same as adults.

E. Adult Habitat. Adults are mostly associated with talus or rocky substrates (Nussbaum et al., 1983; Diller and Wallace, 1994; Jennings and Hayes, 1994a; Bury, 1998; Ollivier and Welsh, 1999), but they also are associated with downed woody debris in areas with nearby rock substrates (Welsh and Lind, 1991). Canopy closure of 60% or greater and older stand age are required to support populations (Welsh and Lind, 1995), except in coastal areas where talus is the primary predictor of occurrence (Diller and Wallace, 1994).

F. Home Range Size. Unknown. Sedentary, rarely moving over 7.5 m² in 3 yr (Welsh and Lind, 1992).

G. Territories. Unknown.

H. Aestivation/Avoiding Desiccation. Generally, this species is far underground during the hot summer.

I. Seasonal Migrations. Unknown.

J. Torpor (Hibernation). Del Norte salamanders become inactive in cold weather.

K. Interspecific Associations/Exclusions. Co-occur with ensatina salamanders *(Ensatina eschscholtzii)*, clouded salamanders *(Aneides ferreus)*, black salamanders *(A. flavipunctatus)*, wandering salamanders *(A. vagrans)*, and, in the northern portions of their range, Dunn's salamanders *(P. dunni;* Brodie, 1970; Bury and Pearl, 1999). Earlier thought to not occur with western red-backed salamanders *(P. vehiculum;* Nussbaum et al., 1983), but recently both species were found

together in southern Douglas County, Oregon (R.B.B., unpublished data).

L. Age/Size at Reproductive Maturity. Mature males range from 46–68 mm SVL and mature females, 57–71 mm SVL (Nussbaum et al., 1983).

M. Longevity. Unknown in the wild.

N. Feeding Behavior. Sit-and-wait predators that dart from cover to seize small prey such as collembolans, termites, mites, and beetles (Bury and Johnson, 1965; L.M. Ollivier and H.H.W., unpublished data).

O. Predators. Likely are prey to shrews, other carnivorous mammals, and garter snakes.

P. Anti-Predator Mechanisms. Unknown.

Q. Diseases. Unknown.

R. Parasites. Unknown.

4. Conservation.

Prior to June 2002, this was a Survey and Manage Species that was afforded protection on federal lands under the Northwest Forest Plan, where ground-disturbing activities were restricted on occupied sites and a surrounding buffer (H.H.W. and L.M. Ollivier, unpublished data). Removing this salamander from Survey and Manage protection was based on the large number of new sites with detections within the known range, as a result of pre-project surveys on federal lands. However, no considerations were given to population numbers or age structure at these sites, or whether individual sites actually represented new or distinct populations. More importantly, no consideration was given to the available molecular (Mahoney, 2004; Mead et al., 2004) and morphological (Bury, 1973; unpublished data) data that indicate high genetic diversity, and the probability that this taxon actually is more than a single species. By contrast, molecular data were instrumental during the same review process when considering the status of the related *P. stormi* to relax protections in the northern portion of their range. New protective measures for *P. elongatus* populations may prove necessary if there are two or more species in the complex. There is a critical need to map this newly discovered genetic diversity and to determine the habitat use of each taxon and how these entities are distributed relative to protected lands, and how each might respond to habitat loss and fragmentation. Protective measures should probably be directed first at the inland and more southern populations where the genetic diversity is highest. Salamanders here are also more likely to be extirpated because they occur in more isolated patches than populations on the coast and northward. This fragmentation is probably related to higher summer temperatures and reduced seasonal precipitation inland and southward.

Plethodon fourchensis Duncan and
Highton, 1979
FOURCHE MOUNTAIN SALAMANDER
Carl D. Anthony

1. Historical versus Current Distribution.
Fourche Mountain salamanders *(Plethodon
fourchensis)* were previously considered a
variant of Rich Mountain salamanders
(*P. ouachitae*; the Buck Knob variant; Blair
and Lindsay, 1965), and originally described by Duncan and Highton (1979).
They are found only on Fourche, eastern
Iron Fork, and Shut-In mountains (Blair
and Lindsay, 1965; Duncan and Highton,
1979, 1986b; Plummer, 1982; Lohoefener
and Jones, 1991). Fourche Mountain salamanders hybridize with Rich Mountain
salamanders in a narrow zone of sympatry
on western Fourche Mountain. Their historical distribution is unknown.

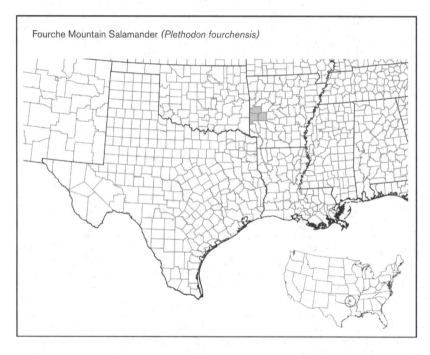

Fourche Mountain Salamander *(Plethodon fourchensis)*

2. Historical versus Current Abundance.
Fourche Mountain salamanders are abundant
and widespread within their limited range.
Lohoefener and Jones (1991) noted a reduction in abundance at a number of localities.

3. Life History Features.
   A. Breeding. Reproduction is terrestrial.
   i. Breeding migrations. Breeding may
occur in late fall, winter, or early spring
(Taylor et al., 1990). Breeding migrations
are unknown.
   ii. Breeding habitat. Unknown. Ratios of
reproductive to nonreproductive females
suggest that females breed biennially (Taylor et al., 1990).
   B. Eggs.
   i. Egg deposition sites. Unknown, but
most likely similar to that of other species
in the *P. ouchitae* complex.

   ii. Clutch size. Taylor et al. (1990) reported a mean of 14.1 enlarged ovarian
follicles from 14 mature females.
   C. Direct Development. Unknown, but
most likely similar to that of other species
in the *P. ouchitae* complex.
   D. Juvenile Habitat. Same as adult
habitat.
   E. Adult Habitat. Most commonly found
at higher elevations of mixed deciduous,
north-facing wooded slopes, especially in
deep ravines (Plummer, 1982). Lohoefener
and Jones (1991) noted a positive relationship between salamander abundance and
plant diversity, canopy cover, and overall
wetness of site. Rocks, logs, and other forest
debris are common cover objects.
   F. Home Range Size. Unknown.
   G. Territories. Adults actively defend
areas against conspecifics in laboratory
chambers (Anthony, 1995). Territory size
is unknown.

   H. Aestivation/Avoiding Desiccation. Individuals move underground in late May
but may return to the surface during periods of rainfall and/or cool weather. By
mid-September, adults can again be found
under cover objects at the surface.
   I. Seasonal Migrations. Unknown.
   J. Torpor (Hibernation). Unknown, likely
hibernate from mid-November to late March.
   K. Interspecific Associations/Exclusions.
Fourche Mountain salamanders occur syntopically with western slimy salamanders
(*P. albagula*) and southern red-backed salamanders (*P. serratus*). Kuss (1986) found
that habitat differed between western
slimy salamanders and Fourche Mountain
salamanders, with Fourche Mountain salamanders occurring more often at higher elevations, farther from ravines, and in
association with denser vegetation.

   L. Age/Size at Reproductive Maturity.
Adults typically range in size from 60–
78 mm SVL (Highton, 1986b). Age at maturity is unknown.
   M. Longevity. Individuals that were
sexually mature when collected have been
kept in the laboratory for at least 3 yr.
Longevity is probably greater than 6 yr.
   N. Feeding Behavior. Unknown, but
prey likely consists of small invertebrates
such as worms, insects, and spiders.
   O. Predators. Unknown.
   P. Anti-Predator Mechanisms. Nocturnal. All *Plethodon* produce noxious skin secretions (Brodie, 1977). When handled,
Fourche Mountain salamanders release an
adhesive secretion.
   Q. Diseases. Unknown.
   R. Parasites. Winter et al. (1986) found
trematodes, cestodes, nematodes, protozoans, and mites on six adults. Intradermal mites of the genus *Hannemania* are
common on the feet and toes and appear
as raised red pustules. Toe loss and damage
have been attributed to these mites (Conant and Collins, 1991). Duncan and
Highton (1979) found that 89–92% of individuals from three populations had
mite infestations.

4. Conservation.
Fourche Mountain salamanders have a limited distribution. Within this range they
can be abundant and widespread, although
Lohoefener and Jones (1991) noted a reduction in numbers at several localities.

Plethodon glutinosus Green, 1818
NORTHERN SLIMY SALAMANDER
David A. Beamer, Michael J. Lannoo

1. Historical versus Current Distribution.
Northern slimy salamanders *(Plethodon
glutinosus)* are present throughout Kentucky, West Virginia, and Pennsylvania,
extending northwest into southern Illinois, southern and western Indiana, and
eastern Ohio, south through Tennessee
into northeastern Alabama, northern
Georgia, and extreme southwestern North
Carolina, east into western Virginia,
Maryland, and New Jersey, and northeast
into southwestern Connecticut and
southern New York, with a disjunct population in southern New Hampshire. Populations range in elevation from sea level to
about 1,500 m (Petranka, 1998).
   Highton and his colleagues (Highton,
1989; Highton and Peabody, 2000) recognize 16 species in the *P. glutinosus* complex
as follows: western slimy salamanders
(*P. albagula*), Tellico salamanders (*P. aureolus*), Chattahoochee slimy salamanders
(*P. chattahoochee*), Atlantic Coast slimy
salamanders (*P. chlorobryonis*), white-
spotted slimy salamanders (*P. cylindraceus*),
northern slimy salamanders (*P. glutinosus*),
southeastern slimy salamanders (*P. grobmani*), Cumberland Plateau salamanders

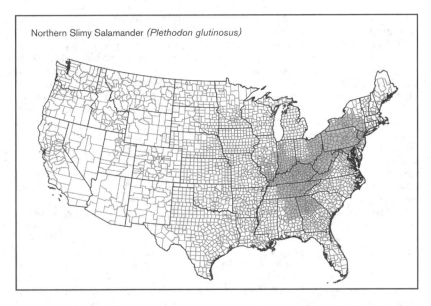

Northern Slimy Salamander *(Plethodon glutinosus)*

(*P. kentucki*), Kiamichi slimy salamanders (*P. kiamichi*), Louisiana slimy salamanders (*P. kisatchie*), Mississippi slimy salamanders (*P. mississippi*), Ocmulgee slimy salamanders (*P. ocmulgee*), Savannah slimy salamanders (*P. savannah*), Sequoyah slimy salamanders (*P. sequoyah*), southern Appalachian salamanders (*P. teyahalee*), and South Carolina slimy salamanders (*P. variolatus*). Information gathered on these species has been published under the name "*P. glutinosus*" (indeed, Petranka [1998] recognizes only Tellico salamanders, northern slimy salamanders, Cumberland Plateau salamanders, and southern Appalachian salamanders). Because of this, the literature on "*P. glutinosus*" must be interpreted carefully and with a consideration of locality data.

2. Historical versus Current Abundance. Density estimates in optimal habitats range from 0.52–0.81 individuals/m^2 (Semlitsch, 1980b). Densities can vary among nearby populations. Highton (this volume, Part One) found that of 30 populations sampled prior to 1990 and re-sampled in the 1990s, 26 (87%) had fewer animals. Several of these differences were alarming. These populations should continue to be monitored to determine whether these counts reflect true population declines or natural fluctuations.

3. Life History Features.
A. Breeding. Reproduction is terrestrial. Females from populations in Alabama had sperm present within the spermatheca in the spring (Trauth, 1984; note the species identities of the specimens in this study cannot be precisely determined, as exact locality data is unknown, although it is likely that at least some were northern slimy salamanders).

Males collected in October from Rochester and Ithaca, New York, had vas deferens packed with sperm (Bishop, 1941b). The vasa deferentia from northern slimy salamanders collected in Bedford County, Pennsylvania, and Frederick County, Maryland, were enlarged during September–October. One male collected in October and all those collected in March in southern Illinois had large vasa deferentia. (Highton, 1962b).

Northern slimy salamanders from Somerset County, New Jersey, courted in September and October. The early stages of courtship were not observed, but the tail-straddling stage was similar to that described for white-spotted slimy salamanders from Whitetop Mountain, Virginia (Organ, 1960a, 1968). A female collected in southern Illinois on 27 April had a spermatophore in her cloaca when she was preserved on 9 May (Highton, 1962b).

i. Breeding migrations. Undocumented, but breeding migrations are not known for any *Plethodon* species.

ii. Breeding habitat. The courtship of a pair of northern slimy salamanders in Giles County, Virginia, was observed on the bank of a road within a bare area of about 0.3 m^2 (Pope, 1950).

B. Eggs. Females collected from September–December in New York had large ovarian eggs, while large eggs were absent from specimens collected in May and through the summer (Bishop, 1941b). Females from Frederick County, Maryland, and Bedford County, Pennsylvania, were induced to oviposit in the laboratory during May and June (Highton, 1962b). One female collected on 23 August from DeKalb County, Georgia, was recently spent while another contained large ovarian eggs (5 mm) ready to be deposited (Highton, 1956). Females in southern Illi-

nois apparently oviposit in late spring to early summer (Highton, 1962b). Females from Virginia and West Virginia oviposit in late July to early August (Pope and Pope, 1949).

i. Egg deposition sites. A nest was discovered within a rotting stump in West Virginia (Fowler, 1940). A nest in Vigo County, Indiana, was found in a rotting log, while two recently hatched clutches in Parke County, Indiana, were beneath rocks (Rubin, 1965).

ii. Clutch size. A nest in Hampshire County, West Virginia, contained 15 eggs (Fowler, 1940). A female from Vigo County, Indiana, was found brooding 34 eggs (Rubin, 1965). Twenty-three gravid females from Bedford County, Pennsylvania, contained from 13–25 eggs while 10 females from Frederick County, Maryland, contained from 16–34 eggs (Highton, 1962b). Females from New York contained from 17–38 eggs (Bishop, 1941b). Females from Virginia and West Virginia contained from 17–33 eggs (Pope and Pope, 1949). Clutch sizes in Frederick County, Maryland, and Bedford County, Pennsylvania, are correlated with SVL (Semlitsch, 1980b).

C. Direct Development. A hatchling from Fulton County, Georgia, collected on 26 December was 18 mm SVL (Highton, 1956). Thirty-two hatchlings from Parke County, Indiana, ranged in size from 18.5–26 mm (Rubin, 1966).

i. Brood sites. A West Virginia female was found brooding a clutch inside a rotting stump (Fowler, 1940). A female from Vigo County, Indiana, was found brooding a clutch inside a rotting log, while two females with recently hatched young were found beneath rocks in Parke County, Indiana (Rubin, 1965).

ii. Parental care. Clutches in West Virginia and Indiana each were accompanied by a female (Fowler, 1940; Rubin, 1965).

D. Juvenile Habitat. Unknown, but likely to be similar to adults.

E. Adult Habitat. Northern slimy salamanders are 8–12 times denser in mature hardwood forests than in young pine monocultures (Bennett et al., 1980), and denser in old pine stands than younger pine stands (Grant et al., 1994).

In New York, Bishop (1941b) reports northern slimy salamanders as commonly being found beneath logs and stones in woods, in the crevices of shale banks, and along the sides of gullies and ravines. He also reports them from under moist humus and leaf mold or in manure piles.

Northern slimy salamanders in Cleburne County, Alabama, are found beneath logs, rocks and occasionally beneath leaf litter (Rubenstein, 1969).

Northern slimy salamanders are reported from caves in DeKalb, Jackson, and Marshall counties, Alabama, and Monroe County, Illinois (Peck, 1974).

F. Home Range Size. Unknown.

G. Territories. Northern slimy salamanders and members of their complex aggressively defend territories (Thurow, 1976).

H. Aestivation/Avoiding Desiccation. In New York, northern slimy salamanders disappear from their usual haunts and may be found only by digging deeply into the soil or following the crevices that extend far into banks (Bishop, 1941b).

I. Seasonal Migrations. Undocumented, but animals likely move from forest floor sites to underground sites in response to seasonally related dry and cold conditions.

J. Torpor (Hibernation). Bishop (1941b) reports finding northern slimy salamanders far below the surface during the winter.

K. Interspecific Associations/Exclusions. In a series of laboratory experiments with northern slimy salamanders and Cumberland Plateau salamanders, there was not a significant difference in frequency of occurrences for initiator, aggressor, escaper, and biter behaviors, but northern slimy salamanders were defenders significantly more often than were Cumberland Plateau salamanders. Cumberland Plateau salamanders were appeasers and intruders significantly more often than northern slimy salamanders (Bailey, 1992). Cumberland Plateau salamanders show an increase in territorial behavior relative to shelter availability and population density when involved in interactions with northern slimy salamanders (Marvin, 1998).

Male green salamanders (*Aneides aeneus*) exhibit aggressive behavior towards northern slimy salamanders (Canterbury and Pauley, 1991). Other interactions between these two salamanders likely include competition for space, nesting sites, and food (Bailey, 1992; Marvin, 1998). Green salamanders are found frequently to occupy higher crevices in rock faces than are northern slimy salamanders (Baltar, 1983; Cliburn and Porter, 1987; Waldron, 2000). This stratification may be due to superior climbing abilities of green salamanders (Cliburn and Porter, 1986) or to competition (Canterbury and Pauley, 1991).

Juvenile northern slimy salamanders and juvenile eastern red-backed salamanders (*P. cinereus*) compete reciprocally when resources are limiting. Interactions between adults are more obscure, but northern slimy salamanders may have reduced growth when eastern red-backed salamanders are present. In competition between these two species for territories, it appears that body size is the primary factor dictating territory ownership (Price and Shield, 2002).

The following salamanders were collected along with northern slimy salamanders on Mount Cheaha, Cleburne County, Alabama: spotted dusky salamanders (*Desmognathus conanti*), seepage salamanders (*D. aeneus*), seal salamanders (*D. monticola*), southern two-lined salamanders (*Eurycea cirrigera*), three-lined salamanders (*E. guttolineata*), spring sala-

manders (*Gyrinophilus porphyriticus*), red salamanders (*Pseudotriton ruber*), and Webster's salamanders (*P. websteri*; Rubenstein, 1969).

In Ohio, the following species are associated with northern slimy salamanders: eastern newts (*Notophthalmus viridescens*), Jefferson's salamanders (*Ambystoma jeffersonianum*), spotted salamanders (*A. maculatum*), small-mouthed salamanders (*A. texanum*), marbled salamanders (*A. opacum*), Allegheny Mountain dusky salamanders (*D. ochrophaeus*), northern dusky salamanders (*D. fuscus*), northern two-lined salamanders (*E.bislineata*), southern two-lined salamanders (*E. longicauda*), northern ravine salamanders (*P. electromorphus*), and eastern red-backed salamanders (Pfingsten, 1989d).

At a site in Randolph County, West Virginia, the following salamanders have been found in close association with northern slimy salamanders: Wehrle's salamanders (*P. wehrlei*), eastern red-backed salamanders, northern dusky salamanders, mountain dusky salamanders, spring salamanders, four-toed salamanders (*Hemidactylium scutatum*), and spotted salamanders (D.A.B., personal observations).

Northern slimy salamanders are found occasionally with northern zigzag salamanders (*P. dorsalis*), cave salamanders (*E. lucifuga*), and long-tailed salamanders on rock faces in the unglaciated sections of Indiana (D. A. B., personal observations).

Northern slimy salamanders were found with southern zigzag salamanders (*P. ventralis*), eastern newt efts, and northern dusky salamanders in Knox County, Tennessee (Powders, 1973).

Northern slimy salamanders are found with Pigeon Mountain salamanders (*P. petraeus*), southern zigzag salamanders, green salamanders, long-tailed salamanders, cave salamanders, and spring salamanders on Pigeon Mountain, Walker County, Georgia (Wynn et al., 1988).

Northern slimy salamanders are sympatric with Yonahlossee salamanders (*P. yonahlossee*) only in the vicinity of Skulls Gap, Smyth County, Virginia. A probable F1 hybrid was found at this location (Highton and Peabody, 2000).

Northern slimy salamanders contact southern Appalachian salamanders on the western side of the Blue Ridge Mountains in Tennessee, southwest of the French Broad River. At one site in Polk County, Tennessee, they occur sympatrically and probably do not hybridize. However, at two other transects in Monroe and Sevier counties, Tennessee, there are narrow hybrid zones (Highton and Peabody, 2000).

Northern slimy salamanders have a largely parapatric range with Tellico salamanders. They have been taken together at one site in Polk County, Tennessee, without any evidence of hybridization (Highton and Peabody, 2000).

An unusual association consisting of three members of the *P. glutinosus* complex occurs at a site in Polk County, Tennessee. Here, southern Appalachian salamanders, Tellico salamanders, and northern slimy salamanders occur sympatrically. There is no evidence of hybridization at this location (Highton, 1984).

Northern slimy salamanders and Chattahoochee slimy salamanders contact at the western edge of the Blue Ridge Province in Georgia; there is no published information on their interactions (Highton and Peabody, 2000).

Northern slimy salamanders have a long contact with white-spotted slimy salamanders from Maryland south through West Virginia, Virginia, and Tennessee, from the Potomac River to the French Broad River. Hybrid populations are known from Washington County, Maryland, and Highland, Washington, and Smyth counties, Virginia (Highton and Peabody, 2000).

The six isolates of northern gray-cheeked salamanders (*P. montanus*) in the Valley and Ridge Province are widely sympatric with northern slimy salamanders. The two species also are sympatric in a small section of the Roan isolate of northern gray-cheeked salamanders. There is evidence of occasional hybridization (Highton and Peabody, 2000).

Northern slimy salamanders occur within 0.4 km of Jordan's salamanders (*P. jordani*) on Parsons Bald, Swain County, North Carolina. There is no evidence of hybridization (Highton and Peabody, 2000).

Northern slimy salamanders are sympatric throughout most of the range of Cumberland Plateau salamanders in the Cumberland Plateau of eastern Kentucky and western West Virginia and in areas in adjacent Tennessee and Virginia. There is evidence that the two species occasionally hybridize (Highton and MacGregor, 1983; Highton and Peabody, 2000).

Northern slimy salamanders are found in association with Wehrle's salamanders throughout the entire range of Wehrle's salamanders (Highton, 1972).

Northern slimy salamanders are sympatric throughout the known range of Pigeon Mountain salamanders. There is some evidence that indicates hybridization may rarely take place (Wynn et al., 1988; Highton and Peabody, 2000).

L. Age/Size at Reproductive Maturity. The smallest mature female from Frederick County, Maryland, was 62 mm SVL; the smallest mature male was 60 mm. The smallest mature female from Bedford County, Pennsylvania, was 61 mm SVL and the smallest mature male was 54 mm SVL. Maturity in these populations is probably not reached until 4 yr old (but see also Semlitsch, 1980b). Females from these locations have a biennial breeding cycle (Highton, 1962b).

The smallest mature female from southern Illinois was 50 mm SVL and the smallest mature male was 47 mm SVL. Maturity in these populations may be reached at 3 yr old (Highton, 1962b). The smallest mature female from Giles County, Virginia, was 57 mm SVL and the smallest mature male was 45 mm SVL. Maturity in these populations is probably reached at the age of 3 or 4 (Pope and Pope, 1949; Highton, 1962b).

The smallest male with a well-developed mental gland in Ohio was 48 mm SVL (Pfingsten, 1989d).

M. Longevity. Northern slimy salamanders in Frederick County, Maryland, and Bedford County, Pennsylvania, live in excess of 5 yr (Semlitsch, 1980b).

N. Feeding Behavior. Adults are dietary generalists, preying on forest floor invertebrates including insects such as coleopterans, collembolans, dipterans, hemipterans, homopterans, hymenopterans, and lepidopterans. They also feed on arachnids, centipedes, isopods, millipedes, phalangids, pseudoscorpions, and snails (Surface, 1913; Hamilton, 1932; Pope, 1950; Davidson, 1956; Oliver, 1967).

The following food items were reported for northern slimy salamanders from Knox County, Tennessee: Annelida, Gastropoda, Diplopoda, Chilopoda, Isopoda, Phalangidea, Pseudoscorpionida, Araneae, Acarina, Collembola, Homoptera, Hemiptera, Coleoptera, Diptera, Formicidae, and non-formicid hymenopteran and other insect larvae (Powders and Tietjen, 1974).

Surface (1913) reports the following food items from 170 specimens from various locations in Pennsylvania: earthworms, snails, spiders, other invertebrates, and insect adults and larvae.

A large female northern slimy salamander (65 mm SVL) was captured with the tail of a juvenile northern slimy salamander (27.5 mm SVL) protruding from its mouth (Powders, 1973).

O. Predators. Undocumented, but likely to include forest snakes, birds, and small mammals.

P. Anti-Predator Mechanisms. Nocturnal. Secretive. Northern slimy salamanders and members of their complex produce large amounts of skin secretions that have an adhesive component (Brodie et al., 1979). These adhesives bind to potential predators and can compromise both mastication and locomotion. Individuals will body flip and lash their tails when attacked by shrews (Brodie et al., 1979). They will vocalize when physically disturbed (Mansueti, 1941).

Members of the *P. glutinosus* complex frequently become immobile when initially contacted. Northern slimy salamanders were included in a field study on immobility; however, it is not possible to separate their behavior from the other members of this complex in this published

data set. Immobility may increase survival by making the salamander less likely to be detected, especially by visually oriented predators (Dodd, 1989).

Q. Diseases. Unknown.

R. Parasites. Northern slimy salamanders are sometimes infected by the astomatous ciliate, *Cepedietta michiganensis* (Powders, 1970). For a considered list of the parasites of North Carolina animals now considered to be in the *P. glutinosus* complex, see Rankin (1937).

4. Conservation.
Northern slimy salamanders are listed as Threatened in Connecticut (http://dep.state.ct.us) and Protected in New Jersey, but are not listed in any of the other states within their range. Among members of the *P. glutinosus* complex, northern slimy salamanders have the widest distribution.

Within this range, there are many federal and state properties that contain suitable habitat for these salamanders.

Northern slimy salamanders are relatively resilient to disturbances, such as those associated with timbering operations, and are found frequently in second-growth forests and relatively small, fragmented woodlots. In Indiana, northern slimy salamanders are widespread and abundant in areas that had up to 99% surface erosion 100 yr ago (D.A.B., personal observations).

As with all species of *Plethodon*, northern slimy salamanders do not migrate to breeding grounds, and they do not have large home ranges. Thus, they can exist in habitats of smaller size than many other amphibian species. Conservation activities that promote mature closed-canopy forests should benefit this species.

Acknowledgments. Thanks to Richard Highton, who reviewed this account and gave us the benefit of his insight and experience.

Plethodon grobmani Allen and Neill, 1949
SOUTHEASTERN SLIMY SALAMANDER

David A. Beamer, Michael J. Lannoo

1. Historical versus Current Distribution.
Southeastern slimy salamanders (*Plethodon grobmani*) are distributed from southern Alabama and southern Georgia south to central Florida (Allen and Neill, 1949; Highton, 1989). There is no evidence to suggest that their current distribution differs from their historical distribution, but it seems likely that populations have been lost due to habitat loss and alteration (Lazell, 1994).

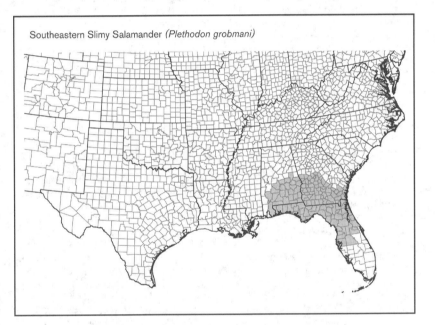

Southeastern Slimy Salamander *(Plethodon grobmani)*

2. Historical versus Current Abundance.
Four of five populations sampled by Highton (this volume, Part One) prior to 1987 and re-sampled in 1990 or later appear to have declined, although numbers of animals observed were high in Jackson County, Florida, and Schley County, Georgia. Additional sampling will be required to determine whether trends in these data reflect true declines or natural population fluctuations.

3. Life History Features.

A. Breeding. Reproduction is terrestrial. Spermatogenesis occurs from October–December in Florida. The vasa deferentia are filled with sperm until the midsummer breeding season. Courtship activity and insemination probably occur from mid-July to mid-August, several weeks before the females lay their eggs.

i. **Breeding migrations.** Undocumented, but breeding migrations are not known for any *Plethodon* species.

ii. **Breeding habitat.** Unknown.

B. Eggs. In Florida, females begin to lay their eggs at the end of August (Highton, 1956; but see also Trauth, 1984). A female collected between 11–13 August from Lowndes County, Alabama, contained large ovarian eggs ready to be deposited (Highton, 1962b).

i. **Egg deposition sites.** Southeastern slimy salamanders in Florida deposit their eggs in or under rotting logs (Highton, 1956, 1962b).

ii. **Clutch size.** Clutches discovered in the field in Florida contained from 5–11 eggs. Gravid females from Florida contained from 10–22 heavily yolked ovarian eggs (Highton, 1956, 1962b).

C. Direct Development. Embryonic gills are lost 2–3 d after hatching. The incubation period is about 2 mo, with hatching taking place in October–November. Hatchlings range in size from 12–14 mm SVL (20–22 mm TL; Highton, 1956).

i. **Brood sites.** Females brood their clutches in or under rotting logs (Highton, 1956).

ii. **Parental care.** The six known clutches of eggs were each accompanied by a female. However, recently spent females have been found without eggs; it is unknown if these females abandoned their eggs or if they were lost to predators (Highton, 1956, 1962b).

D. Juvenile Habitat. Similar to adults. Juveniles likely stay at the nest site for the first 2 mo after hatching. They are first found on the surface in late December (Highton, 1956).

E. Adult Habitat. Southeastern slimy salamanders are common in steephead ravines, maritime forests, and river bottom hardwood forests (Lazell, 1994; Enge, 1998). Much of the area within the range of southeastern slimy salamanders historically was covered with savanna and prairie, and in these areas they are rare or absent. Southeastern slimy salamanders do not require pristine habitats or old-growth forests and are often found under discarded rubbish. There may be a minimum size wood lot necessary as they are absent from small wood lots (Lazell, 1994). A southeastern slimy salamander was found in a cave in Alachua County, Florida (Peck, 1974).

F. Home Range Size. Individuals are sedentary, rarely moving >60 cm from their original capture-and-release sites (Highton, 1956).

G. Territories. At least some members of the *P. glutinosus* complex aggressively defend territories (Thurow, 1976); it is unknown whether southeastern slimy salamanders do the same.

H. Aestivation/Avoiding Desiccation. Southeastern slimy salamanders in all size groups were rarely encountered during the summer in peninsular Florida, despite the fact that there is increased rainfall at this time. Higher temperatures and higher evaporation rates may be factors inhibiting summer activity. During the summer, the growth rate of the young decreases (Highton, 1956).

I. Seasonal Migrations. Southeastern slimy salamanders probably move vertically from forest-floor sites to underground sites in response to seasonal dry and cold conditions. Southeastern slimy salamanders were much more abundant beneath logs in steephead ravines in the Florida Panhandle during the winter and early spring than during the summer (D. A. B., personal observations).

J. Torpor (Hibernation). Generally unknown, but southeastern slimy salamanders likely avoid cold conditions by seeking shelter in underground sites.

K. Interspecific Associations/Exclusions. Allen and Neill (1949) discuss hybrids with other members of the *P. glutinosus* complex.

During a drift-fence survey of steephead ravines in the Apalachicola and Ochlockonee River drainages, southeastern slimy salamanders were captured with one-toed amphiumas (*Amphiuma pholeter*), Apalachicola dusky salamanders (*Desmognathus apalachicolae*), southern two-lined salamanders (*Eurycea cirrigera*), three-lined salamanders (*E. guttolineata*), four-toed salamanders (*Hemidactylium scutatum*), eastern newts (*Notophthalmus viridescens*), mud salamanders (*Pseudotriton montanus*), and red salamanders (*P. ruber*; Enge, 1998).

Historically, southeastern slimy salamanders were found with southern dusky salamanders (*D. auriculatus*) at Devil's Millhopper State Geological Site, Alachua County, Florida (Dodd, 1998). However, most known populations of southern dusky salamanders within the range of southeastern slimy salamanders are now extirpated (Means, submitted).

In Alabama, southeastern slimy salamanders have been observed in the burrows of Red Hills salamanders (*Phaeognathus hubrichti*) along with the following salamanders: southern two-lined salamanders, three-lined salamanders, and eastern newts (Brandon, 1966b; Dodd, this volume). Seal salamanders (*D. monticola*) are common at some Alabama sites with southeastern slimy salamanders (Brandon, 1965b).

L. Age/Size at Reproductive Maturity. The holotype is an adult female, 49.9 mm SVL. Allen and Neill (1949) mention a series of their 20 largest specimens, which average 53.7 mm. The smallest mature female from Florida was 45 mm, the smallest mature male was 37 mm (Highton, 1956).

Maturity in female southeastern slimy salamanders is more closely related to age than size. Females do not mature until they are 2 yr old and cannot lay eggs until they approach the age of 3. Female southeastern slimy salamanders have an annual breeding cycle (Highton, 1956, 1962b).

Male southeastern slimy salamanders generally mature when they are 2 yr old, though some large individuals may mature when they are 1 yr old and breed at age 2. Many maturing 1-yr-old males do not produce functional spermatozoa, but instead undergo an abortive sexual cycle in which the maturing sex cells degenerate. These males do not breed until the age 3. Male southeastern slimy salamanders breed annually (Highton, 1956, 1962b).

M. Longevity. Unknown.

N. Feeding Behavior. The stomach contents of thirteen southeastern slimy salamanders from Alabama contained snails, millipedes, spiders, phalangids, beetles, Hymenoptera (mainly ants), and miscellaneous insect larvae (Brandon, 1965b).

O. Predators. Undocumented, but likely to include forest snakes, birds, and small mammals.

P. Anti-Predator Mechanisms. Members of the *Plethodon glutinosus* complex produce large amounts of skin secretions that have an adhesive component (Brodie et al., 1979). Individuals will vocalize when physically disturbed (Highton, 1956).

Q. Diseases. Unknown.

R. Parasites. Unknown.

4. Conservation.
Southeastern slimy salamanders are not protected in any of the states within their range. Among members of the *P. glutinosus* complex, southeastern slimy salamanders have one of the widest distributions. Within this range there are several federal and state properties that contain suitable habitat for these salamanders.

Southeastern slimy salamanders are relatively resilient to disturbances, such as those associated with timbering operations, and are frequently found in second-growth forests and relatively small, fragmented woodlots (Lazell, 1994; D. A. B., personal observation).

As with all species of *Plethodon*, southeastern slimy salamanders do not migrate or have large home ranges. Thus, they can exist in habitats of smaller size than many other amphibian species. However, there may be a minimum size habitat necessary, as they are absent from small wood lots (Lazell, 1994). Conservation activities that promote mature closed-canopy forests should benefit this species.

Acknowledgments. Thanks to Richard Highton, who reviewed this account and gave us the benefit of his insight and experience.

Plethodon hoffmani Highton, 1972 "1971"
VALLEY AND RIDGE SALAMANDER

David A. Beamer, Michael J. Lannoo

1. Historical versus Current Distribution.
Valley and Ridge salamanders (*Plethodon hoffmani*) inhabit the Valley and Ridge Physiographic Province from central

Pennsylvania south and southwest to the New River in Virginia and West Virginia. They also occur to the west in the adjacent Appalachian Plateau Physiographic Province and to the east in the adjacent Blue Ridge Physiographic Province (Highton, 1986c; Petranka, 1998). Some populations in northeastern West Virginia and western Virginia that were originally attributed to Valley and Ridge salamanders are now recognized as Shenandoah Mountain salamanders (*P. virginia*; Highton, 1999a, fig. 4). Their current distribution likely is similar to their historical distribution, but populations have likely been lost due to habitat destruction and alteration.

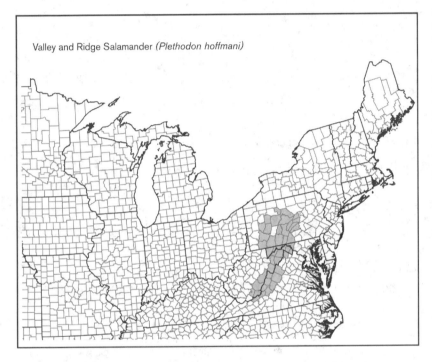

Valley and Ridge Salamander *(Plethodon hoffmani)*

2. Historical versus Current Abundance.
Largely unknown, but data collected by Highton (this volume, Part One) indicate that populations in Snyder County, Pennsylvania, and Augusta County, Virginia, may have crashed. Additional data from these sites will be needed to determine whether these are true declines or natural population fluctuations. Populations at other sites sampled by Highton (this volume, Part One) in Virginia and West Virginia appear to be more stable.

3. Life History Features.
   A. Breeding. Reproduction is terrestrial. The vasa deferentia are filled with sperm from late September to May. Valley and Ridge salamanders in Maryland and Pennsylvania primarily mate in the spring, although occasionally females are found with spermatozoa in their spermathecae during the fall (Angle, 1969).

   i. Breeding migrations. Undocumented, but breeding migrations are not known for any *Plethodon* species.
   ii. Breeding habitat. Unknown.
   B. Eggs. In Maryland and Pennsylvania, eggs probably are deposited in April–May (Angle, 1969).
   i. Egg deposition sites. Unknown, but as with other members of the *P. cinereus* group, likely include underground sites and perhaps sites under surface cover objects.
   ii. Clutch size. The number of mature ovarian eggs (maximum 4.0 mm diameter) in females from Maryland and Pennsylvania range from 3–8 with an average of 4.6 (Angle, 1969).

   C. Direct Development. Eggs probably are incubated for about 2 mo and hatch in late August to September (Angle, 1969).
   i. Brood sites. Unknown, but probably include underground cavities or chambers.
   ii. Parental care. Unknown, but it is likely that females brood, as with other species of *Plethodon*.
   D. Juvenile Habitat. Juveniles likely stay at the nest site for several months after hatching. They are first found on the surface in March (Angle, 1969).
   E. Adult Habitat. In West Virginia, Valley and Ridge salamanders inhabit hillside slopes of mixed deciduous forest with flat stones (Green and Pauley, 1987). Most of the range is dry and well drained (Highton, 1999a).
   Valley and Ridge salamanders have been found in two different caves in

Greenbrier County, West Virginia. In both caves they were within the twilight zone beneath cover (Cooper, 1961).
   F. Home Range Size. Unknown, but small home ranges are typical for *Plethodon* species.
   G. Territories. Unknown.
   H. Aestivation/Avoiding Desiccation. Few Valley and Ridge salamanders have been collected during the summer (Highton, 1962a; Angle, 1969).
   I. Seasonal Migrations. Animals likely make vertical migrations, moving from the forest floor to underground sites with the onset of seasonally related cold or dry conditions, then back up to the forest floor with the return of favorable surface conditions.
   J. Torpor (Hibernation). Valley and Ridge salamanders avoid cold conditions by moving to warmer underground sites (Highton, 1962a; Angle, 1969).
   K. Interspecific Associations/Exclusions. Valley and Ridge salamanders are usually not sympatric with other members of the *P. cinereus* group. However, they are sympatric with red-backed salamanders in several different areas including central Pennsylvania and near Monterey, Highlands County, Virginia (Highton, 1972). Most of the range occupied by Valley and Ridge salamanders may be too dry for red-backed salamanders to occupy (Highton, 1972).
   Populations of Valley and Ridge salamanders have been found proximate to other members of the *P. cinereus* group. They are found only 1.9 km away from a site that contains Peaks of Otter salamanders *(P. hubrichti)*. In the vicinity of Bell Knob in Tucker and Grant counties, West Virginia, Valley and Ridge salamanders are found within 3.7 km of Cheat Mountain salamanders *(P. nettingi)*. Southern ravine salamanders *(P. richmondi)* are known from a locality 0.6 km away from Valley and Ridge salamanders on the other side of the New River in Summers and Raleigh counties, West Virginia (Highton, 1999a).
   Valley and Ridge salamanders contact Shenandoah Mountain salamanders at the northern and southern extremities of Shenandoah Mountain salamanders range. In these areas there are some hybrid populations (Highton, 1999a).
   Valley and Ridge salamanders also occur in areas with northern slimy salamanders *(P. glutinosus)*, white-spotted slimy salamanders *(P. cylindraceus)*, and Wehrle's salamanders *(P. wehrlei*; Highton, 1962a, 1972, 1989).
   The following species were associated with Valley and Ridge salamanders in caves in Greenbrier County, West Virginia: cave salamanders *(Eurycea lucifuga)*, long-tailed salamanders, *(E. longicauda)* eastern red-backed salamanders *(P. cinereus)*, northern slimy salamanders, eastern newt efts *(Notophthalmus viridescens)*, and Jefferson's salamanders *(Ambystoma jeffersonianum*; Cooper, 1961).

L. Age/Size at Reproductive Maturity. Juveniles grow 10–12 mm SVL during their first year and become sexually mature in the autumn, at 2 yr old. At this time, males range in size from 38–44 mm; females are 39–47 mm. Most females will oviposit the year after becoming sexually mature, but a small percentage may require an additional year and first breed when they are nearly 4 yr old. Female Valley and Ridge salamanders have a biennial reproductive cycle (Angle, 1969).

M. Longevity. Unknown.

N. Feeding Behavior. Unknown, but as with other *Plethodon* species, feeding likely takes place at night under moist conditions. Prey items likely include a range of invertebrates, especially insects.

O. Predators. Unknown.

P. Anti-Predator Mechanisms. All *Plethodon* produce noxious skin secretions (Brodie, 1977). Valley and Ridge salamanders frequently become immobile when initially contacted. Valley and Ridge salamanders were included in a field study on immobility, however it is not possible to separate their behavior from Shenandoah Mountain salamanders in this published data set. Immobility may increase survival by making the salamander less likely to be detected, especially by visually oriented predators (Dodd, 1989).

Q. Diseases. Unknown.

R. Parasites. Unknown.

4. Conservation.

Valley and Ridge salamanders are not protected by any state. Within their range there are many federal and state properties that contain suitable habitat for these salamanders.

As with all species of *Plethodon*, Valley and Ridge salamanders do not migrate to breeding grounds, and they do not have large home ranges. Thus, they can exist in habitats of smaller size than many other amphibian species. Conservation activities that promote mature closed-canopy forests should benefit this species.

Acknowledgments. Thanks to Richard Highton, who reviewed this account and gave us the benefit of his insight and experience.

Plethodon hubrichti Thurow, 1957
PEAKS OF OTTER SALAMANDER
Joseph C. Mitchell, Jill A. Wicknick

1. Historical versus Current Distribution.

Peaks of Otter salamanders (*Plethodon hubrichti*) are endemic to the Blue Ridge Mountains, restricted to an approximately 16-km length of the mountain chain in Bedford and Botetourt counties, Virginia (Highton, 1972, 1986d; Mitchell and Reay, 1999). Information on their historical range is not available.

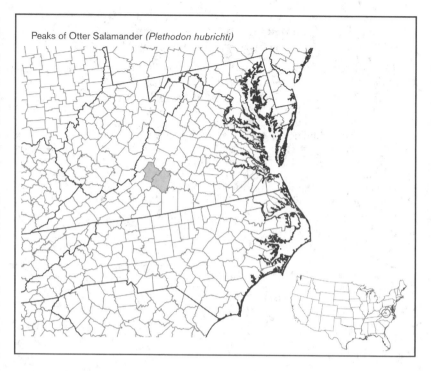

Peaks of Otter Salamander *(Plethodon hubrichti)*

2. Historical versus Current Abundance.

Historical abundance is unknown, but was probably less than today because of agricultural practices in the area in the 1800s and early 1900s. In optimal areas, current abundance can be high. Current abundance is affected by timber operations in parts of its range in the Jefferson National Forest. Kramer et al. (1993) marked 250 Peaks of Otter salamanders in a 10 × 10 m plot and estimated a population size of 450 individuals. Average density/field trip was 0.24 individuals/m². Mitchell and Wicknick (1996) found 0.13–0.18 salamanders/m² in transects in mature forests, 0.5–0.16/m² in forest stands that had been thinned, and 0.04–0.09/m² in recent clearcuts.

3. Life History Features.

The ecology and life history of Peaks of Otter salamanders has been summarized by Pague and Mitchell (1991b) and Petranka (1998); details are given below.

A. Breeding. Reproduction is terrestrial.

i. Breeding migrations. Peaks of Otter salamanders do not migrate to breeding locations.

ii. Breeding habitat. Breeding apparently occurs in and on leaf litter in hardwood forests and under logs and rocks in moist soil.

B. Eggs.

i. Egg deposition sites. Females lay eggs under logs and rocks in deep soil (Pague, 1989).

ii. Clutch sizes. Females lay 5–15 eggs (Pague, 1989).

C. Direct Development. Mating occurs in September and October when pairs of males and females may be found under the same log or rock (Wicknick, 1995). Females lay eggs in June (Pague, 1989). Hatching occurs in August and September. The few nests that have been observed were under rocks in moist soil. Females were in attendance on these nests (J.C.M., unpublished data).

D. Juvenile Habitat. Juvenile habitat is similar to adult habitat.

E. Adult Habitat. Within the range of this species, adults occupy the forest floor of mature Appalachian hardwood forest at elevations ≥550 m above sea level (Pague and Mitchell, 1991b). Salamanders have been found in a variety of vegetation types, including rhododendron thickets. Rocks and logs are usually in abundance. Salamanders are not distributed evenly throughout their range, instead highest densities occur in areas of deep, moist soil in mature hardwood forest stands.

F. Home Range Size. Median area for home range of Peaks of Otter salamanders has been measured at 0.6 m² in hardwood forest, and median distance moved was 1.0 m (Kramer et al., 1993).

G. Territories. Peaks of Otter salamanders are intraspecifically territorial. They exhibit most of the characteristics of a territorial salamander species (as defined by Gergits, 1982). They are site tenacious and can remain associated with a particular site (cover object: rock or log) within and between years (Wicknick, 1995). They may have the ability to home after displacement, although this is not yet clear (Wicknick, 1995). They are aggressive toward intruding conspecifics, retaining residency of an area and successfully expelling conspecific intruders (Wicknick, 1995). Negative intra-and intersex spatial associations indicate that intraspecific territoriality occurs within and between sexes (Wicknick, 1995). Competition for cover objects via intra- and interspecific territoriality has been only weakly supported by field experiments conducted in the fall (Wicknick, 1995).

H. Aestivation/Avoiding Desiccation. Surface activity ceases when the forest floor is dry. This can occur for several days to weeks during the normal activity period (usually April to early November).

I. Seasonal Migrations. This species does not migrate.

J. Torpor (Hibernation). Peaks of Otter salamanders usually are inactive during November–March, although periods of warm temperatures and moist surface conditions may cause these salamanders to become active on the forest floor for short periods of time during these months. The coldest soil temperature for surface-active individuals we measured was 10.0 °C on 14 October 1994 (unpublished data). None were observed in the same sites at 9.0 °C.

K. Interspecific Associations/Exclusions. Peaks of Otter salamanders occur sympatrically with southern two-lined salamanders (*Eurycea cirrigera*), seal salamanders (*Desmognathus monticola*), spring salamanders (*Gyrinophilus porphyriticus*), eastern red-backed salamanders (*P. cinereus*), white-spotted slimy salamanders (*P. cylindraceus*), red salamanders (*Pseudotriton ruber*), and eastern newt (*Notophthalmus viridescens*) efts. Eastern red-backed salamanders, the sibling species of Peaks of Otter salamanders, occur in a highly variable zone of overlap but generally narrow band of sympatry around the margins of the range of the Peaks of Otter salamanders (Wicknick, 1995). Wicknick (1995) found some support for competition for cover objects and their associated resources between Peaks of Otter salamanders and eastern red-backed salamanders. The two species occur in the same microhabitat but are negatively associated when in sympatry (Wicknick, 1995). It appears that Peaks of Otter salamanders hold interspecific territories, but despite a relatively high level of aggression, are only moderately successful at defending a site against invasion by red-backed salamanders (Wicknick, 1995). The two species may be in competitive equilibrium.

L. Age/Size at Reproductive Maturity. Age at maturity is unknown, but minimum size known for reproductive adults of both sexes is 39 mm SVL (Wicknick, 1995; Mitchell and Wicknick, 1996).

M. Longevity. Longevity is unknown.

N. Feeding Behavior. Peaks of Otter salamanders are opportunistic feeders along and above the forest floor on low herbaceous vegetation during warm, wet nights (Kramer et al., 1993). Prey include a wide variety of invertebrates, including ants, collembolans, isopods, including

beetles, spiders, centipedes, gastropods, worms, and lepidopteran larvae (Mitchell and Wicknick, 1996). Mitchell et al. (1996b) found 949 prey in 80 salamanders from 20 sites, of which 32.2% were ants and 67.8% were collembolans. They determined that there were no significant differences in numbers of ants consumed in four forest stand types (recent clearcut, old clearcut, mature, and shelterwood). However, salamanders in mature stands ate significantly more collembolans than those in other forest stand types.

O. Predators. Actual predation has not been observed. Sympatric snakes include ring-necked snakes (*Diadophis punctatus*), common garter snakes (*Thamnophis sirtalis*), juvenile black racers (*Coluber constrictor*), and milk snakes (*Lampropeltis triangulum*). Potential bird predators include wild turkeys (*Meleagris gallopavo*), towhees (*Pipilo* sp.), and brown thrashers (*Toxostoma rufum*) that forage in the leaf litter. Short-tailed shrews (*Blarina brevicauda*) and southeastern shrews (*Sorex longirostris*) occur sympatrically with Peaks of Otter salamanders.

P. Anti-Predator Mechanisms. Anti-predator behaviors include noxious skin secretions, immobility, and defensive postures (Brodie, 1977; Dodd et al., 1974). Dodd (1989) determined the length of time salamanders remained immobile when disturbed in the field by close approach, when their nearby microhabitat was disturbed, or when they were touched. Immobility times for Peaks of Otter salamanders averaged 16.3 s (range=1.5–99.5 s) and were not significantly different from times for other eastern *Plethodon*. Three of 28 salamanders had immobility times >180 s in duration.

Q. Diseases. Unknown.

R. Parasites. There are no published observations on parasites for this species.

4. Conservation.
Peaks of Otter salamanders are recognized as a federal species At Risk by the U.S. Fish and Wildlife Service and listed as a Species of Concern by the Virginia Department of Game and Inland Fisheries (Mitchell et al., 1999). The U.S. National Forest Service lists them as a Sensitive Species. Potential threats include logging activities, habitat fragmentation, and loss of forest canopy from the introduced gypsy moth (*Lymantria dispar*; Mitchell et al., 1996b; Sattler and Reichenbach, 1998). Mitchell et al. (1996b) determined that numbers of Peaks of Otter salamanders were highly variable and that clearcut sites consistently supported fewer individuals than mature forest or stands that had been thinned (shelterwood sites). This study also showed that diet quality may decrease following clearcutting. In another study, numbers of salamanders were stable over a 3-yr period in mature hardwood and shelterwood stands, but declined substantially

after clearcutting (Sattler and Reichenbach, 1998). In the clearcut site, 30% of the initial population remained after treatment; many adults and juveniles either likely emigrated or died.

The current conservation plan followed by the U.S. Forest Service and the U.S. National Park Service (George Washington and Jefferson National Forests, U.S. Fish and Wildlife Service, and Blue Ridge Parkway, 1997) specifies the entire range of the Peaks of Otter salamander as a Special Biological Area with a primary conservation area that allows no logging activities and a secondary conservation area that allows logging to take place under certain restrictions. These restrictions include the use of the shelterwood technique only, a 50 ft^2 basal area minimum, no cutting of remaining hardwood trees for 15 yr, location of the timber sale area away from previous sale areas, and seeding of the logged area with downed woody debris. In addition, cutting cannot occur during peak salamander surface activity time and no more than 100 ac may be harvested per year.

Plethodon idahoensis Slater and Slipp, 1940

COEUR D'ALENE SALAMANDER

Kirk Lohman

1. Historical versus Current Distribution.
Coeur d'Alene salamanders (*Plethodon idahoensis*) are locally distributed in northern Idaho, northwestern Montana, and extreme southeastern British Columbia (Corkran and Thoms, 1996; Wilson and Ohanjanian, 2002). Knowledge of their range has been greatly extended from the original description of one location in northern Idaho (Slater and Slipp, 1940; for example see Dumas, 1957; Teberg, 1963, 1965; Nussbaum et al., 1983; Leonard et al., 1993; Wilson and Ohanjanian, 2002). Coeur d'Alene salamanders previously were considered a subspecies of Van Dyke's salamanders (*P. vandykei*), which occur in western Washington (Howard et al., 1993; Leonard et al., 1993). Their historical distribution is unknown.

2. Historical versus Current Abundance.
Rare, but Coeur d'Alene salamanders may be abundant in small pockets (Wilson and Ohanjanian, 2002). Their historical abundance is unknown.

3. Life History Features.
A. Breeding. Reproduction is terrestrial.
i. Breeding migrations. Mating can occur from August–April, with peaks in the fall and early spring (Lynch, 1984). In areas with harsh springs, most animals breed in the fall, whereas breeding in the spring is more common in more temperate locales (Lynch, 1984). Migratory behavior is unknown.

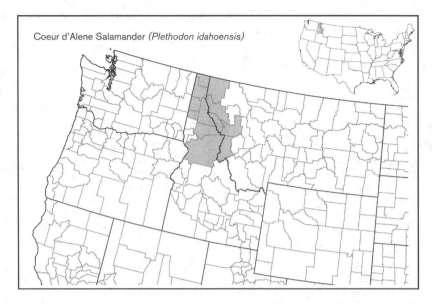

Coeur d'Alene Salamander *(Plethodon idahoensis)*

ii. Breeding habitat. Breeding apparently occurs just before and just after hibernation (Nussbaum et al., 1983; Lynch, 1984). Breeding and nest sites have been rarely observed.

B. Eggs.

i. Egg deposition sites. Have rarely been observed.

ii. Clutch size. Females typically lay 1–13 eggs (Lynch, 1984).

C. Direct Development. Hatching usually occurs in late summer and early fall (Nussbaum et al., 1983).

D. Juvenile Habitat. Same as adult habitat.

E. Adult Habitat. Most commonly found in semi-aquatic habitats: seepages, streamside talus, and splash zones (Nussbaum et al., 1983; Lynch, 1984); but may also occur in forest debris and damp talus (Slater and Slipp, 1940).

F. Home Range Size. Wilson and Larsen (1988) reported the movement of adults 5–10 m away from splash zones during nocturnal activities, but little is known concerning home range size.

G. Territories. Unknown.

H. Aestivation/Avoiding Desiccation. Nussbaum et al. (1983) reported that forest and talus-dwelling populations remain underground from June–September; populations along the Kootenai River, Montana, however, were inactive in midsummer only during periods of high daytime temperature and little rainfall (Wilson and Larsen, 1988).

I. Seasonal Migrations. None reported.

J. Torpor (Hibernation). Coeur d'Alene salamanders hibernate from late November to early March (Nussbaum et al., 1983).

K. Interspecific Associations/Exclusions. Unknown.

L. Age/Size at Reproductive Maturity. Age at first reproduction is 3.2–3.5 yr, with an SVL of 42–48 mm (Lynch, 1984).

M. Longevity. At least 5 yr (Lynch, 1984).

N. Feeding Behavior. Prey consists primarily of aquatic and semi-aquatic insects, including collembolans, dipterans, coleopterans, hemipterans, and acarinids, as well as homopterans and hymenopterans (Lindeman, 1993; Wilson and Larsen, 1988).

O. Predators. Generally unknown, but predation by American robins *(Turdus migratorius)* has been documented (Wilson and Simon, 1985), and predation by shrews (Soricidae) or invertebrates (possibly carabid beetles [*Scaphinotus* sp.]) has been inferred (Staub, 1995).

P. Anti-Predator Mechanisms. Nocturnal. Occasionally will coil as if to imitate a millipede; may rarely coil and uncoil to flip away if disturbed (Nussbaum et al., 1983).

Q. Diseases. Unknown.
R. Parasites. Unknown.

4. Conservation.
The historical distribution and abundance of Coeur d'Alene salamanders is unknown. They can be abundant in small pockets (Wilson and Ohanjanian, 2002). Coeur d'Alene salamanders are considered a Species of Special Concern in both Idaho and Montana.

Plethodon jordani Blatchley, 1901
JORDAN'S SALAMANDER, RED-CHEEKED SLAMANDER

David A. Beamer, Michael J. Lannoo

1. Historical versus Current Distribution.
Jordan's salamanders *(Plethodon jordani)* have a limited and disjunct distribution. The major cluster of populations occurs along the Tennessee–North Carolina border. A more southern set of populations occurs in Rabun County in extreme northeastern Georgia. They occur at elevations from 213–1,951 m (Grobman, 1944; see Petranka, 1998), but usually to elevations above 600 m (Hairston and Pope, 1948; Hairston, 1949; see Petranka, 1998; see also Highton and Peabody, 2000). There is no evidence to suggest that the current range of Jordan's salamanders differs from their historical range, although populations have undoubtedly been lost in association with habitat loss and alteration.

2. Historical versus Current Abundance.
Highton (this volume, Part One) sampled one population of Jordan's salamanders from 1961–75, then re-counted this population again in 1994 and found a precipitous decline in the number of animals found under similar sampling conditions

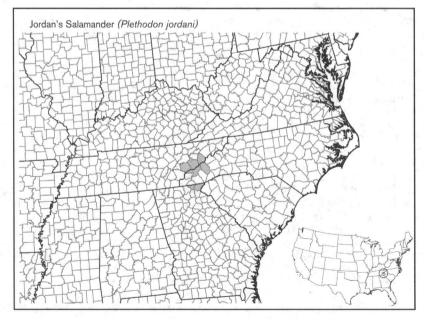

Jordan's Salamander *(Plethodon jordani)*

using a similar level of effort. While this is suggestive of declines, this population should continue to be monitored to determine whether these data are the result of true declines or natural population fluctuations.

Petranka (1998) states that deforestation and the conversion of mixed hardwood forests to pine monocultures has eliminated or reduced a large number of eastern *Plethodon* populations. However, Ash (1988, 1997) and Ash and Bruce (1994) have strongly disagreed with Petranka's estimates and do not consider forestry practices to be having as strong an impact on native salamanders.

3. Life History Features.

A. Breeding. Reproduction is terrestrial.

i. Breeding migrations. Unlikely; breeding migrations are unknown in any *Plethodon* species.

ii. Breeding habitat. Unknown.

B. Eggs.

i. Egg deposition sites. Unknown, but likely to be in underground cavities.

ii. Clutch size. Unknown.

C. Direct Development.

i. Brood sites. Unknown, but probably include underground cavities or chambers.

ii. Parental care. Unknown, but it is likely that females brood, as with other species of *Plethodon*.

D. Juvenile Habitat. Juveniles remain underground until the next summer, 10–12 mo following hatching. Once they surface, juveniles increase by an average of 12 mm SVL from May–October. They add about 9 mm SVL during the following summer.

Juveniles (18–34 mm SVL) were abundant in spruce-fir forest on 27 August at Indian Gap, Sevier County, Tennessee. They were found beneath rocks that were strewn about the forest floor in the midst of a thick carpet of fallen needles and occasionally beneath pieces of wood (Reynolds, 1959).

Juveniles (22.5–28.9 mm TL) were abundant in spruce-fir forest in May at Indian Gap, Sevier County, Tennessee. They were found beneath flat rocks protruding through a thick bed of fallen needles and beneath rotten logs and bark (Wood, 1947a).

E. Adult Habitat. Jordan's salamanders are most abundant in red spruce-Fraser's fir forest but are also found on hardwood covered ridges. The forest floor where this species is most abundant is covered with a heavy layer of moss with only a little soil over a mass of large boulders (King, 1939). They occur from the highest peak in the Great Smoky Mountains (Clingmans Dome, 2,024 m [6,643 ft]) down to an elevation of 853 m (2,800 ft; Huheey and Stupka, 1967).

Jordan's salamanders inhabit burrows or other subterranean passageways under rocks, logs, and other cover objects during warmer months. These burrow systems can be extensive (Chadwick, 1940). Animals are most active at night and during and following rains.

F. Home Range Size. Small; Madison (1969) found that the maximum distances moved by adults were ≤11 m for males and ≤4 m for juveniles. Individuals home after displacements of over 300 m (Madison, 1969, 1972).

Merchant (1972) estimates mean home ranges for males to be 11.4 m^2, for females to be 2.8 m^2, and for juveniles to be 1.7 m^2. Nishikawa (1990) found that adults occupy fixed home ranges with little overlap between individuals of the same sex or age group (see also Petranka, 1998).

Density estimates range from 0.18–0.86 animals/m^2 (Ash, 1988; Petranka, 1998).

G. Territories. At least some members of the *P. jordani* complex aggressively defend territories (Thurow, 1976), but it is unknown whether Jordan's salamanders establish and defend territories.

H. Aestivation/Avoiding Desiccation. Undocumented, but *Plethodon* species will move from forest-floor habitats to underground sites in response to desiccating surface conditions.

I. Seasonal Migrations. Jordan's salamanders have been found from April–November (Huheey and Stupka, 1967). Animals likely make vertical migrations, moving from the forest floor to underground sites with the onset of seasonally related cold or dry conditions, then back up to the forest floor with the return of favorable surface conditions.

J. Torpor (Hibernation). Subzero temperatures drive Jordan's salamanders deep into burrows, where they remain underground, even during thaws.

K. Interspecific Associations/Exclusions. In the Great Smoky Mountains, the following species were encountered on experimental plots with Jordan's salamanders: black-bellied salamanders (*Desmognathus quadramaculatus*), seal salamanders (*D. monticola*), Ocoee salamanders (*D. ocoee*), imitator salamanders (*D. imitator*), pigmy salamanders (*D. wrighti*), southern redbacked salamanders (*P. serratus*), southern Appalachian salamanders (*P. teyahalee*), spring salamanders (*Gyrinophilus porphyriticus*), red salamanders (*Pseudotriton ruber*), and Blue Ridge two-lined salamanders (*Eurycea wilderae*; Hairston, 1980b, 1981).

Southern Appalachian salamanders are widely sympatric at intermediate elevations with the central and western portions of the Great Smoky isolate and Gregory Bald isolate of Jordan's salamanders. In these areas, they rarely hybridize. In the eastern portion of the Great Smoky isolate of Jordan's salamanders, they replace each other altitudinally and hybridize extensively in a narrow contact zone (Highton and Peabody, 2000).

Jordan's salamanders occur within 0.4 km of northern slimy salamanders (*P. glutinosus*) on Parsons Bald, Swain County, North Carolina. There is no evidence of hybridization (Highton and Peabody, 2000).

Jordan's salamanders contact southern gray-cheeked salamanders (*P. metcalfi*) on Balsam Mountain and on Hyatt Ridge. Narrow hybrid zones are present at both contact zones (Hairston, 1950; Highton, 1970; Hairston et al., 1992; Highton and Peabody, 2000).

Hairston (1980b, 1981) demonstrated competition between Jordan's salamanders and southern Appalachian salamanders in the Great Smoky Mountains. The number of southern Appalachian salamanders increased significantly at plots in the Great Smoky Mountains where Jordan's salamanders were removed. The removal of southern Appalachian salamanders from plots resulted in a significant increase in the proportion of young Jordan's salamanders. Hairston did not identify what resources are limiting, but hypothesizes that southern Appalachian salamanders may compete with members of the *P. jordani* complex for nesting sites.

Powders and Tietjen (1974) hypothesize that competition for food during the fall occurs between southern Appalachian salamanders and Jordan's salamanders. Because competition occurs during the fall, they reason that it would act earlier at higher elevations. This fall competition may exclude southern Appalachian salamanders from higher altitudes that are occupied exclusively by Jordan's salamanders.

In the Great Smoky Mountains, Jordan's salamanders and southern Appalachian salamanders replace one another altitudinally. At two of four transects, there is no elevational overlap in the occurrence of these two species, while at the other two there is elevational overlap of 3–8 m (Hairston, 1951).

L. Age/Size at Reproductive Maturity. Unknown.

M. Longevity. Unknown.

N. Feeding Behavior. The following food items were reported for Jordan's salamanders from Great Smoky Mountains National Park: Annelida, Gastropoda, Diplopoda, Chilopoda, Isopoda, Phalangidea, Pseudoscorpionida, Aranae, Acarina, Collembola, Homoptera, Hemiptera, Coleoptera, Diptera, Formicidae, non-formicid-Hymenoptera, and insect larvae. Some seasonal variation was observed in the diet, which may indicate seasonal availability of food items. Millipedes appear to be more important in spring, and insect larvae increase during autumn; collembolans and annelids increase in importance with altitude (Powders and Tietjen, 1974).

Weller (1931) reported beetles, lepidopteran larvae, and mollusks from the stomachs of Jordan's salamanders.

Huheey (1959) described the feeding behavior of captive Jordan's salamanders on large earthworms (*Lumbricus* sp.). Salamanders would seize the worms firmly in the jaws and then rapidly rotate the body until the earthworm broke. It would then swallow the fragment and attack the remainder.

O. Predators. Jordan's salamanders are preyed upon by other salamanders, including black-bellied salamanders (Huheey and Brandon, 1962) and spring salamanders (Orr, 1962). Huheey (1960) reported predation by common garter snakes (*Thamnophis sirtalis*). Weller (1931) reported a beetle (*Cychrus* sp.) feeding on a freshly killed Jordan's salamander. In captivity, short-tailed shrews (*Blarina brevicauda*) will feed on Jordan's salamanders (Orr, 1962).

P. Anti-Predator Mechanisms. Jordan's salamanders have slimy tail secretions that are noxious to potential avian predators (Huheey, 1960; Brodie and Howard, 1973; Hensel and Brodie, 1976; see also Petranka, 1998). King (1939) reports that the mucous from Jordan's salamanders is extremely viscous and is removed from human skin with difficulty. Feder and Arnold (1982) report that when attacked by garter snakes (*Thamnophis* sp.), Jordan's salamanders may wrap their tail around the snake's head while releasing tail secretions, thrash wildly, bite, or autotomize their tail (see also Petranka, 1998). When attacked by shrews (*Blarina* sp.), individuals will flip and position their tail towards the predator (Brodie et al., 1979). Wake and Dresner (1967) report 28% of animals had damaged tails (although the species referred to here is in question [R. Highton, personal communication]).

Members of the *P. jordani* complex frequently become immobile when initially contacted. Jordan's salamanders were included in a field study on immobility, however, it is not possible to separate their behavior from the other members of this complex in this published data set. Immobility may increase survival by making the salamander less likely to be detected, especially by visually oriented predators (Dodd, 1989).

Q. Diseases. Unknown.

R. Parasites. Jordan's salamanders are sometimes infected by the astomatous ciliate, *Cepedietta michiganensis*. Infection rates seem to decrease at higher altitudes and during the spring (Powders, 1967, 1970).

4. Conservation.
Jordan's salamanders are not protected by any of the states within their range. Among members of the *P. jordani* complex, Jordan's salamanders have one of the wider distributions. The entire range of this species is contained within the Great Smoky Mountains National Park.

Jordan's salamanders are relatively resilient to disturbances such as those associated with timbering operations and often are common in areas that were logged heavily prior to the establishment of the national park (D.A.B., personal observation).

As with all species of *Plethodon*, Jordan's salamanders do not migrate to breeding grounds and they do not have large home ranges. Thus they can exist in habitats of smaller size than many other amphibian species. Conservation activities that promote mature closed-canopy forests should benefit this species.

Acknowledgments. Thanks to Richard Highton, who reviewed this account and gave us the benefit of his insight and experience.

Plethodon kentucki Mittleman, 1951
CUMBERLAND PLATEAU SALAMANDER
Thomas K. Pauley, Mark B. Watson

1. Historical versus Current Distribution.
Mittleman (1951) first described Cumberland Plateau salamanders (*Plethodon kentucki*) from specimens collected in Harlan and Pike counties, Kentucky. Highton and MacGregor (1983) showed that their range extends from eastern Kentucky east to Summers County, West Virginia, and Dickenson County, Virginia. Their range is now known to be in the Cumberland Plateau of eastern Kentucky, northeastern Tennessee, southwestern Virginia, and West Virginia west of the New and Kanawha rivers (Highton, 1986e; Highton and Peabody, 2000). The most southern point known is in Scott County, Tennessee (Redmond and Scott, 1996), and the easternmost locality is in Washington County, Virginia (Mitchell and Reay, 1999).

Prior to Mittleman's description, Cumberland Plateau salamanders were considered by biologists to be slimy salamanders (then *P. glutinosus*). Clay et al. (1955) could not distinguish *P. kentucki* from *P. glutinosus* using morphological features, and thus considered *P. kentucki* to be a local population variant of *P. glutinosus*. Highton and MacGregor (1983) resurrected *P. kentucki* as a separate species based on electrophoretic analyses. Other evidence that has verified *P. kentucki* as a valid species includes immunological studies by Maha et al. (1983) and behavioral evidence by Dawley (1986). Individuals observed in the known range of Cumberland Plateau salamanders prior to 1951, and between the time of the work by Clay et al. (1955) and Highton and MacGregor (1983), were recorded as a junior synonym of *P. glutinosus* (Highton, 1962a; Gorham, 1974).

2. Historical versus Current Abundance.
In western West Virginia, Bailey and Pauley (1993) reported that Cumberland Plateau salamanders comprised 70% of all salamander species active on the forest floor in March and April. They found few adults on the surface in February prior to the spring emergence in March–April and few in May–June after soil moisture decreased. Surface density was significantly correlated with soil moisture but not with soil temperature, air temperature, or soil pH. They also noted that Cumberland Plateau salamanders reached their greatest surface abundance between 2100 and 2200 h DST and that salamanders were most numerous on west-facing slopes, compared with southwest- and northwest-facing slopes. Cumberland Plateau salamanders are found in diverse habitats from floodplains to ridgetops in the

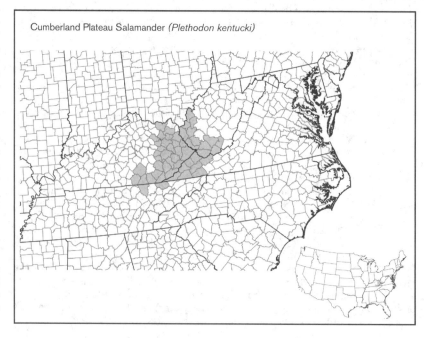

Cumberland Plateau Salamander (*Plethodon kentucki*)

Bluestone River Gorge (Bluestone National Scenic River) in southern West Virginia (Waldron et al., 2000). In the Bluestone River Gorge, the highest surface activity of Cumberland Plateau salamanders is from March–October. On Black Mountain, in eastern Kentucky, the greatest surface activity is from April–August. Differences in surface activity in western West Virginia reported by Bailey and Pauley (1993), Marvin (1990), and Waldron et al. (2000) may be attributed to elevational differences. The elevation in western West Virginia ranges from 197–247 m compared to 427–671 m in the Bluestone River Gorge and 1,265 m on Black Mountain.

In eastern Kentucky, Marvin (1996) estimated densities of 2–3-yr-olds, and found adults (both sexes) to be <0.20 animals/m². In Virginia, Roble and Hobson (1995) reported observing 30–35 Cumberland Plateau salamanders and a few northern slimy salamanders (the current *P. glutinosus*) on 13 September in Scott County. Two visits to the same area in 1993 (date not provided) revealed no Cumberland Plateau salamanders.

3. Life History Features.

A. Breeding. Reproduction is terrestrial.

i. Breeding migrations. This species does not migrate.

ii. Breeding habitat. Reproductive behavioral studies in the laboratory revealed courtship events similar to what has been shown in other large *Plethodon* (i.e., rubbing and twisting of both partners followed by a tail-straddle walk, spermatophore deposition, and retrieval of the spermatophore by females; Marvin and Hutchison, 1996). Mating most likely occurs from late July or early August to October (Marvin, 1996), although a pair was observed in courtship on the forest floor at the base of a rock outcrop in the Bluestone River Gorge, West Virginia, on 14 May 1998 (W. J. Humphries and T. K. Pauley, personal observation).

B. Eggs.

i. Egg deposition sites. Eggs are deposited in grape-like clusters in underground cavities (Marvin, 1996).

ii. Clutch size. In females studied in the laboratory, clutch sizes ranged from 9–12 eggs, and eggs hatched from 5–23 October (Marvin, 1996).

C. Direct Development. Mittleman (1951) concluded that oviposition probably occurs in early September, based on the large size of follicles found in females collected in August. Marvin (1996) found that females oviposit biennially and males probably mate annually. In Kentucky, eggs are deposited in July and hatch in October (Marvin, 1996). Marvin also found that upon oviposition, females lose an average of 37% (range from 24–50%) of their wet mass.

D. Juvenile Habitat. Juveniles have been found in the same habitat as adults (J.E. Bailey, 1992; Marvin, 1996).

E. Adult Habitat. Cumberland Plateau salamanders occupy a variety of woodland habitats such as moist ravines and hillsides, shale banks, cave entrances, and rock crevices (Green and Pauley, 1987; J. E. Bailey, 1992, Pauley, 1993b; Petranka, 1998). In West Virginia, they use rocks as primary cover objects and are found more frequently on west-facing slopes (Bailey and Pauley, 1993). Pauley (1993b) examined habitats of 51 Cumberland Plateau salamanders in the New River Gorge in West Virginia in 1991 and 1992 and found 30 under cover objects on the forest floor and 21 in rock crevices.

F. Home Range Size. The size of the home range is small, probably similar to that of northern slimy salamanders. One female was recorded to move 1.8 m, and one juvenile moved 1.3 m (Bailey and Pauley, 1993).

G. Territories. Bailey (J.E., 1992) examined interactions between Cumberland Plateau salamanders and northern slimy salamanders in a laboratory experiment. He found no significant differences in frequency of occurrences for the initiator (first salamander to move from its original position), aggressor (salamander that threatens or attacks first), escaper (salamander that attempts to flee), and biter (salamander that grasped or snapped at another). Cumberland Plateau salamanders were appeasers (positioned themselves to avoid or hinder an attack) and intruders (one that entered the territory of another) significantly more often than northern slimy salamanders. Northern slimy salamanders were defenders (those whose territory had been invaded) significantly more often than were Cumberland Plateau salamanders. In interactions with northern slimy salamanders, Cumberland Plateau salamanders show an increase in territorial behavior in accordance with shelter availability and population density (Marvin, 1998).

H. Aestivation/Avoiding Desiccation. Apparently, there is little surface activity in lower elevations after April when temperatures rise and soil moisture decreases (Bailey and Pauley, 1993). Populations on Black Mountain, Kentucky, move underground during summer droughts, and juveniles in this area probably grow slower and suffer higher mortality during summers with reduced rainfall (Marvin, 1996). During summers with sufficient rainfall, they remain active on the surface throughout the summer (Marvin, 1990).

I. Seasonal Migrations. This species does not migrate.

J. Torpor (Hibernation). A few adults have been observed on the surface in February (Bailey and Pauley, 1993). As with other *Plethodon*, some adults will emerge to the surface during warm periods in winter.

K. Interspecific Associations/Exclusions. Cumberland Plateau salamanders and northern slimy salamanders are sympatric throughout most of the range of Cumberland Plateau salamanders (Highton and Peabody, 2000). For a detailed description of the geographic protein variation of Cumberland Plateau salamanders and northern slimy salamanders, see Highton and Peabody (2000).

Competitive interactions for space, nesting sites, and food probably occur between green salamanders (*Aneides aeneus*) and northern slimy salamanders (J.E. Bailey, 1992; Marvin, 1998).

L. Age/Size at Reproductive Maturity. Mittleman (1951) suggested females reach sexual maturity at about 46 mm SVL; males at 40 mm. Marvin (1996) estimated that females become sexually mature in 4–5 yr when they exceed 52 mm SVL and males in 3–4 yr at about 47–48 mm.

M. Longevity. Probably similar to northern slimy salamanders, which have been reported to live over 20 yr (Snider and Bowler, 1992). Based on recapture data, Marvin (1996) estimates that annual survivorship of 2 yr olds is at least 48% and 68% in 3 yr olds. Annual survivorship of adults ranges from 72–92%.

N. Feeding Behavior. Major prey items include hymenopterans, collembolans, coleopterans, spiders, snails, pseudoscorpions, dipterans, and mites (J.E. Bailey, 1992).

P. Anti-Predator Mechanisms. Not reported, but as with slimy salamanders, Cumberland Plateau salamanders undoubtedly can secrete copious amounts of mucous when attacked.

Q. Diseases. Not reported.

R. Parasites. Not reported.

4. Conservation.

Optimal habitats for Cumberland Plateau salamanders are found in mature hardwood forests (Petranka, 1998). Harvesting of hardwoods and aggressive rotation schedules will have a negative impact on woodland salamanders, including Cumberland Plateau salamanders (but see Ash, 1988, 1997, and Ash and Bruce, 1994).

Plethodon kiamichi Highton, 1989
KIAMICHI SLIMY SALAMANDER
Carl D. Anthony

1. Historical versus Current Distribution.

Kiamichi slimy salamanders (*Plethodon kiamichi*) were described by Highton et al. (1989) as a member of the slimy salamander (*P. glutinosus*) species complex. Blair and Lindsay (1965) reported hybridization between Rich Mountain (*P. ouachitae*) and western slimy salamanders (*P. albagula [glutinosus]*), but this was not confirmed by protein comparisons in animals from the same localities (Duncan and Highton, 1979). Kiamichi slimy salamanders are known only from Kiamichi and

Round mountains in southeastern Oklahoma (Black and Seivert, 1989; Highton et al., 1989). Kiamichi slimy salamanders are listed as a Protected species by Oklahoma. Their historical distribution is unknown.

J. Torpor (Hibernation). Individuals likely hibernate from mid-November to late March.

K. Interspecific Associations/Exclusions. Occur syntopically with Rich Mountain salamanders (P. ouachitae) and southern red-backed salamanders (P. serratus).

produce noxious skin secretions (Brodie, 1977). When handled, Kiamichi slimy salamanders, as with other members of the P. glutinosus group, release an adhesive secretion.

Q. Diseases. Unknown.

R. Parasites. Unknown.

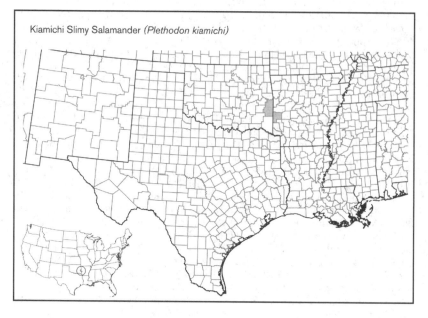

Kiamichi Slimy Salamander (Plethodon kiamichi)

4. Conservation.
Kiamichi slimy salamanders have a limited distribution, being known only from Kiamichi and Round mountains in southeastern Oklahoma. They can be locally common within their limited range. Kiamichi slimy salamanders are listed as Protected by Oklahoma.

Plethodon kisatchie Highton, 1989
LOUISIANA SLIMY SALAMANDER

Carl D. Anthony

1. Historical versus Current Distribution.
Louisiana slimy salamanders (Plethodon kisatchie) were described by Highton et al. (1989) as a member of the slimy salamander (P. glutinosus) species complex. Most similar to Sequoyah slimy salamanders (P. sequoyah), Louisiana slimy salamanders are known from the hill parishes of north-central Louisiana (Dundee and Rossman, 1989; Highton et al., 1989). They were first recorded in central Louisiana by Wilson (1966), and their known range was expanded by Warner (1971) and Boundy (1994a, 1998). Their historical distribution is unknown.

2. Historical versus Current Abundance.
2. Historical versus Current Abundance.
Locally common within their limited range.

L. Age/Size at Reproductive Maturity. Adults typically range in size from 48–75 mm SVL.

M. Longevity. Unknown.

N. Feeding Behavior. Unknown, but prey likely consists of small invertebrates such as worms, insects, and spiders.

Not common, but well established in central Louisiana (Warner, 1971). Historical abundance is unknown.

3. Life History Features.
Little is known of the life history of this species. Life history features of Kiamichi slimy salamanders presumably are similar to that of related forms.

A. Breeding. Reproduction is terrestrial.

i. Breeding migrations. Unknown.

ii. Breeding habitat. Breeding habitat is unknown.

B. Eggs.

i. Egg deposition sites. Unknown.

ii. Clutch size. No reports of egg sizes or numbers.

C. Direct Development. Details have not been reported.

D. Juvenile Habitat. Same as adult habitat.

E. Adult Habitat. Kiamichi slimy salamanders are most commonly found at higher elevations in moist woods (Black and Seivert, 1989) and ravines. Rocks and logs are common cover objects.

F. Home Range Size. Unknown.

G. Territories. Adults defend areas in laboratory chambers (Anthony, 1995). Territorial behavior is similar to that of western slimy salamanders.

H. Aestivation/Avoiding Desiccation. Aestivation is unknown, however animals likely avoid desiccating conditions by moving under cover objects or into burrows.

I. Seasonal Migrations. Unknown.

O. Predators. Unknown.

P. Anti-Predator Mechanisms. Nocturnal. All members of the genus Plethodon

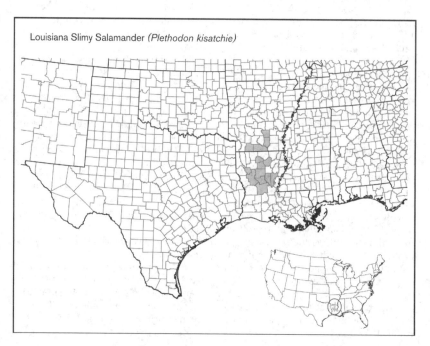

Louisiana Slimy Salamander (Plethodon kisatchie)

3. Life History Features.

Little is known of the life history of Louisiana slimy salamanders. Life history features presumably are similar to that of related forms.

A. Breeding. Reproduction is terrestrial.

i. Breeding migrations. Unknown.

ii. Breeding habitat. Breeding habitat is unknown.

B. Eggs.

i. Egg deposition sites. Unknown.

ii. Clutch size. Unknown; there are no reports of egg number or size.

C. Direct Development. The hatching period probably is around mid autumn (Dundee and Rossman, 1989).

D. Juvenile Habitat. Same as adult habitat.

E. Adult Habitat. Hardwood forests (Dundee and Rossman, 1989), but they also occur in forests that are composed primarily of pines (Warner, 1971).

F. Home Range Size. Unknown.

G. Territories. Territorial behavior has been undescribed.

H. Aestivation/Avoiding Desiccation. Aestivation is unreported. Animals likely avoid desiccating conditions by moving under cover objects or into burrows.

I. Seasonal Migrations. Unknown.

J. Torpor (Hibernation). Unreported.

K. Interspecific Associations/Exclusions. May be sympatric with southern red-backed salamanders *(P. serratus)*.

L. Age/Size at Reproductive Maturity. Adults typically range in size from 46–70 mm SVL (Warner, 1971).

M. Longevity. Unknown.

N. Feeding Behavior. Prey likely consists of small invertebrates such as worms, insects, and spiders.

O. Predators. Unknown.

P. Anti-Predator Mechanisms. Nocturnal. All members of the genus *Plethodon* produce noxious skin secretions (Brodie, 1977). When handled, Louisiana slimy salamanders, like other members of the *glutinosus* group, release an adhesive secretion (Dundee and Rossman, 1989).

Q. Diseases. Unknown.

R. Parasites. Rabalais (1970) reported trematodes from a sample that contained both Louisiana slimy salamanders and Mississippi slimy salamanders *(P. mississippi)*.

4. Conservation.

Louisiana slimy salamanders are found in the hill parishes of north-central Louisiana, where they are well established but not common.

Plethodon larselli Burns, 1954
LARCH MOUNTAIN SALAMANDER

Robert E. Herrington

1. Historical versus Current Distribution.

Larch Mountain salamanders *(Plethodon larselli)* were originally described as a subspecies of Van Dyke's salamanders *(P. vandykei;* Burns, 1954) and later elevated to full species status (Burns, 1962). They are endemic to a narrow region where the Cascade Mountains have been eroded by the Columbia River in the Columbia River Gorge. Nussbaum et al. (1983) indicated they had been reported from three counties (one in Washington and two in Oregon). Fieldwork conducted from 1981–84 (Herrington, 1985) revealed several new populations and expanded the Washington range eastward into Klickitat County. Aubry et al. (1987) extended the range northward into Lewis County, Washington. While the historical distribution is unknown, it is likely that road building, forestry practices, and human population growth in the limited range of this species has resulted in the loss of some populations. Because of the continued interest in this species by the Washington Department of Wildlife (1993), their distribution is better known in Washington. Intensive fieldwork in Oregon could yield additional populations.

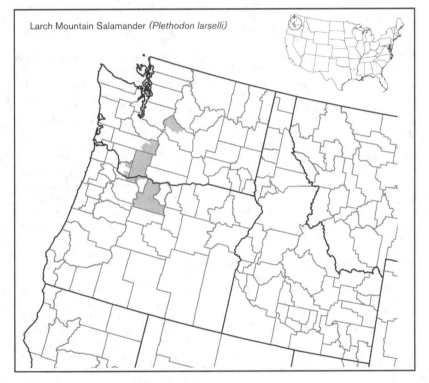

Larch Mountain Salamander *(Plethodon larselli)*

2. Historical versus Current Abundance.

Little is known about the historical or current abundance of Larch Mountain salamanders, and there are no detailed population studies available. It has been reported that Larch Mountain salamanders exhibit a remarkably low level of genetic diversity (Howard et al., 1983). Currently, Larch Mountain salamanders are known from isolated patches of talus habitat (with some exceptions) that appear to be separated by large expanses of unsuitable habitat. Many localities with

what appears to be suitable habitat do not contain this species. In optimal habitat, during ideal environmental conditions, Larch Mountain salamanders are relatively common. They are usually less abundant than sympatric western red-backed salamanders *(P. vehiculum;* in Washington) or Dunn's salamanders *(P. dunni;* in Oregon). However, searches conducted a few meters to either side of optimal habitat or during adverse weather conditions will often fail to detect them. Environmental alteration, such as clearcutting of trees on talus slopes, render the habitat unsuitable for Larch Mountain salamanders, while sustaining populations of more tolerant western red-backed salamanders (Herrington, 1985). During June 1998, I revisited most of the known collection sites in both Washington and Oregon (Herrington, 1985); generally, where habitat is intact, I found Larch Mountain salamanders present. However, several sites had been severely impacted over the ensuing 13 yr by anthropogenic factors.

In particular, clearcutting above one site (Mabee Mines) resulted in the erosional deposition of large amounts of soil over the exposed talus. This allowed the encroachment of hardwoods that have completely closed the canopy and greatly reduced the abundance of Larch Mountain salamanders. Also, along the Washougal River, private home developments have compromised several sites. Major road construction is planned for Oregon, which could jeopardize populations in that state.

3. Life History Features.

A. Breeding. Reproduction is terrestrial.

i. Breeding migrations. No breeding migrations are known.

ii. Breeding habitat. Courtship occurs on moist talus during suitable periods of temperature and moisture from September–April (Herrington and Larsen, 1987).

B. Eggs.

i. Egg deposition sites. Oviposition sites have not been located, but it is likely that gravid females descend to suitable moist chambers within the talus to oviposit and brood eggs.

ii. Clutch size. Egg numbers average 7.3 (range = 2–12, n = 43; Herrington and Larsen, 1987).

C. Direct Development. Although development has not been observed in either the field or laboratory, it is probable that young are produced directly from eggs deposited in talus slopes (Herrington, 1985). Hatchlings (18–21 mm SVL) appear in talus during the beginning of the fall rainy season (October–November), and the absence of yolk in these individuals suggests that hatching occurred weeks earlier.

D. Juvenile Habitat. Juvenile habitat is probably similar to that of adults. Juveniles are collected along with adults from the same habitats.

E. Adult Habitat. Larch Mountain salamanders typically are associated with steep, at least partially forested, talus slopes. Within the slopes, rock size, nutrient import, and moisture levels interact to produce optimal habitat (Herrington and Larsen, 1985). Larch Mountain salamanders are not tolerant of completely open canopies, such as that which occurs following clearcutting, although such areas frequently contain the more tolerant western red-backed salamanders (Herrington, 1985; Petranka, 1998). Larch Mountain salamanders occasionally have been found in other habitats such as at the entrance to an ice cave (Aubry et al., 1987) and in forested situations apparently no longer associated with exposed talus (Jones and Bury, 1983). However, it appears that in the latter case, these probably represent old talus slopes that have undergone successional processes associated with increased soil accumulation and subsequent tree growth.

F. Home Range Size. The home range is unknown.

G. Territories. Territoriality has not been studied in this species, but see "Interspecific Interactions" below.

H. Aestivation/Avoiding Desiccation. Highly seasonal precipitation patterns occur in the Pacific Northwest, with most rainfall occurring from October–March. Greatly reduced rainfall from May–September effectively reduces surface activity by Larch Mountain salamanders to periods immediately following substantial rainfalls. During dry periods, salamanders descend deeper in the talus to locate microhabitats with suitable moisture levels.

I. Seasonal Migrations. Migrations are unknown for this species.

J. Torpor (Hibernation). Salamanders are absent from the surface during snow cover, when temperatures are <4°C, or when conditions are dry. They presumably descend into the talus to where more favorable microclimates occur.

K. Interspecific Associations/Exclusions. The distribution pattern of members of the genus *Plethodon* within the range of Larch Mountain salamanders suggests that interspecific competition may play a role in the micro-geographic distribution of these species. Larch Mountain salamanders are sympatric with western red-backed salamanders throughout most of their range in Washington, and with Dunn's salamanders (*P. dunni*) throughout their range in Oregon (Nussbaum et al., 1983). However, even though Dunn's salamanders occur only a few miles west of the westernmost Larch Mountain salamander population in Washington, and western red-backed salamanders occur west of the westernmost population of Larch Mountain salamanders in Oregon, I have been unable to find any locality, in either state, where all three species are sympatric.

L. Age/Size at Reproductive Maturity. Based on a relatively large sample of size distributions, it appears that male Larch Mountain salamanders attain sexual maturity at 39–42 mm SVL and at 3.0–3.5 yr old. Female Larch Mountain salamanders are ≥44 mm SVL and are at least 4 yr old. Clutch size determined from ovarian/oviductal eggs ranged from 2–12 (mean = 7.3), and oviposition probably occurs only every other year after sexual maturity is reached. This is the only known species of salamander that lacks abdominal fat bodies, and excess energy is stored in the tail.

M. Longevity. Nothing is known about the longevity of this species in the wild.

N. Feeding Behavior. Larch Mountain salamanders feed on a variety of small invertebrates, with mites and collembolans making up a large part of their diet (Altig and Brodie, 1971). They are capable of feeding below the surface of the talus in laboratory experiments (Smith and Herrington, 1989).

O. Predators. I am unaware of studies on predation of this species. However, large salamanders, such as Dunn's salamanders and giant salamanders (*Dicamptodon* sp.), are known to feed on other salamanders. Garter snakes (*Thamnophis* sp.) are often abundant in the same habitat as Larch Mountain salamanders and are known to feed on a variety of amphibians.

P. Anti-Predator Mechanisms. Larch Mountain salamanders occasionally will coil, spring, and then lie motionless with their reddish venter exposed.

Q. Diseases. Unknown.

R. Parasites. Unknown.

4. Conservation.

Larch Mountain salamanders are known from isolated patches of talus habitat that appear to be separated by expanses of unsuitable habitat. Because of the continued interest in this species by the Washington Department of Wildlife (1993), their distribution is better known in Washington. Localities with what appears to be suitable habitat do not contain these salamanders. In optimal habitat, this species is relatively common. Little is known about the historical or current abundance of Larch Mountain salamanders, and there are no detailed population studies available. Clearcutting above one site (Mabee Mines) resulted in the erosional deposition of large amounts of soil over the exposed talus. This allowed the encroachment of hardwoods that have completely closed the canopy and greatly reduced the abundance of Larch Mountain salamanders. Along the Washougal River, private home developments have compromised several sites. Major road construction is planned for Oregon, which could jeopardize populations in that state.

Plethodon meridianus Highton and Peabody, 2000
SOUTH MOUNTAIN GRAY-CHEEKED SALAMANDER

David A. Beamer, Michael J. Lannoo

1. Historical versus Current Distribution.

South Mountain gray-cheeked salamanders (*Plethodon meridianus*) are known from Piedmont Province sites in Burke, Cleveland, and Rutherford counties in North Carolina. South Mountain gray-cheeked salamanders are members of the *P. jordani* complex. Because this species was described only recently, and because there have been no studies of "*Plethodon jordani*" from this region, historical distributional data that could be used to compare to current distributional data are lacking. Highton (1972) describes South Mountain gray-cheeked salamanders as "quite common throughout" the South Mountains.

2. Historical versus Current Abundance.

Generally unknown, but Highton (this volume, Part One) sampled two counties (Burke and Cleveland counties, North Carolina) in the decade between 1967 and 1977 and again in 1997, and found recent numbers to be reduced compared to historical numbers. Additional surveys will make it possible to determine whether these data reflect true declines or natural habitat fluctuations.

3. Life History Features.

Because this species has been described only recently, little specific information is known about the life history and natural history features of South Mountain

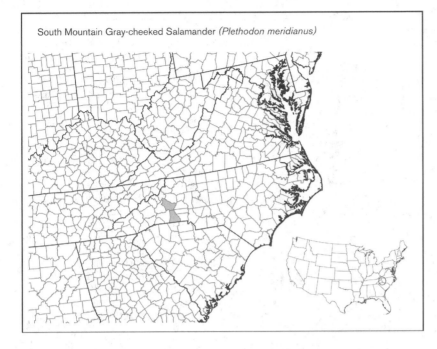

South Mountain Gray-cheeked Salamander *(Plethodon meridianus)*

gray-cheeked salamanders. Until recently, South Mountain gray-cheeked salamanders were considered to be Jordan's salamanders *(P. jordani).*

A. Breeding. Reproduction is terrestrial.

i. Breeding migrations. Unlikely; breeding migrations are unknown in any *Plethodon* species.

ii. Breeding habitat. Unknown.

B. Eggs.

i. Egg deposition sites. Unknown.

ii. Clutch size. Unknown.

C. Direct Development.

i. Brood sites. Unknown.

ii. Parental care. Unknown, but it is likely that females brood, as with other species of *Plethodon.*

D. Juvenile Habitat. Unknown, but likely to be similar to adults.

E. Adult Habitat. Unreported, but as with other species in the *P. jordani* complex, South Mountain gray-cheeked salamanders undoubtedly are active on the forest floor. They seek shelter under cover objects during the day and are nocturnally active, with activity levels proportional to moisture levels. Animals likely avoid dry and cold extremes by moving to underground sites.

F. Home Range Size. Unknown, but likely to be small.

G. Territories. At least some members of the *P. jordani* complex aggressively defend territories (Thurow, 1976); it is unknown whether South Mountain gray-cheeked salamanders establish and defend territories.

H. Aestivation/Avoiding Desiccation. Unstudied, but in response to dry conditions, animals likely move to moist (deep) subterranean sites.

I. Seasonal Migrations. Unstudied, but animals likely respond to seasonal shifts to dry and cold conditions by moving from forest-floor habitats to underground sites, and then back up to the surface when conditions become favorable.

J. Torpor (Hibernation). Unstudied, but in response to cold conditions, animals likely move to warm (deep) subterranean sites.

K. Interspecific Associations/Exclusions. South Mountain gray-cheeked salamanders are found in association with white-spotted slimy salamanders *(P. cylindraceus)* throughout their range. There is no evidence of hybridization between South Mountain gray-cheeked salamanders and white-spotted slimy salamanders (Highton and Peabody, 2000).

L. Age/Size at Reproductive Maturity. Unknown. Females may be larger. The two largest animals (73 mm SVL) from the type locality were females (Highton and Peabody, 2000). The holotype is a 63 mm male SVL; the allotype is a 70 mm female. Other animals from the type locality ranged from 50–72 mm.

M. Longevity. Unknown.

N. Feeding Behavior. Unreported, but as with other species of *Plethodon*, animals likely feed at night, with activity proportional to moisture levels. Prey items include small invertebrates, especially insects, that inhabit or are associated with the forest floor.

O. Predators. Likely to include forest-dwelling snakes, birds, and small mammals.

P. Anti-Predator Mechanisms. All *Plethodon* produce noxious skin secretions (Brodie, 1977).

Q. Diseases. Unknown.

R. Parasites. Unknown.

4. Conservation.

South Mountain gray-cheeked salamanders are not protected by North Carolina, the only state within their range. Among members of the *P. jordani* complex, South Mountain gray-cheeked salamanders have one of the smallest distributions, and within their range, there are only a few state properties that contain suitable habitat for these salamanders.

Presently, some habitat within the range of South Mountain gray-cheeked salamanders is preserved as game lands. However, recently much of this land has been converted to housing. Suitable habitat will likely remain in South Mountain State Park, but it is likely that habitat in this already restricted area will be lost in the near future. This area is isolated both geographically and physiographically from areas that contain other members of this complex. As such, the South Mountains are a likely candidate to contain other endemic species and should be a conservation priority.

As with all species of *Plethodon*, South Mountain gray-cheeked salamanders do not migrate to breeding grounds and they have small home ranges. Thus, they can exist in habitats of smaller size than many other amphibian species. Conservation activities that promote mature closed-canopy forests should benefit this species.

Acknowledgments. Thanks to Richard Highton, who reviewed this account and gave us the benefit of his insight and experience.

Plethodon metcalfi Brimley, 1912
SOUTHERN GRAY-CHEEKED SALAMANDER

David A. Beamer, Michael J. Lannoo

1. Historical versus Current Distribution.
Southern gray-cheeked salamanders *(Plethodon metcalfi)* are found in the southern Blue-Ridge Mountains, including Haywood and Macon counties, North Carolina, and Oconee County, South Carolina, at elevations above 750 m to the limits of large tree growth. One population occurs at an elevation of 256 m, the lowest elevation known for the entire *P. jordani* complex (Highton and Peabody, 2000). A portion of their original distribution is now underwater, covered by Lake Jocassee (Highton and Peabody, 2000).

2. Historical versus Current Abundance.
Generally unknown, but of four sites sampled by Highton (this volume, Part One) prior to 1990 and resurveyed in the 1990s, numbers of animals observed at three sites were lower, and at two of these sites numbers were substantially lower. Further sampling will be necessary to determine whether these data reflect true declines or natural population fluctuations. Southern gray-cheeked salamanders disappeared from clearcut areas in Macon County, North Carolina, within 4 yr of lumbering (Ash, 1988, 1997).

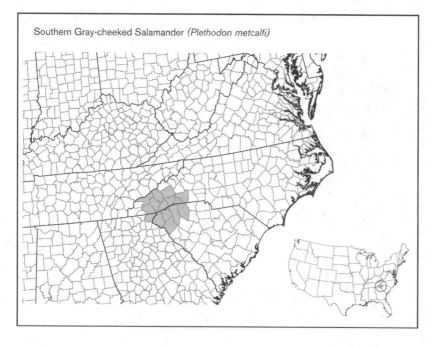

Southern Gray-cheeked Salamander *(Plethodon metcalfi)*

3. Life History Features.

A. Breeding. Reproduction is terrestrial.

i. Breeding migrations. Unlikely; breeding migrations are unknown in any *Plethodon* species.

ii. Breeding habitat. Unknown.

B. Eggs.

i. Egg deposition sites. Unknown, but likely to be in underground cavities.

ii. Clutch size. Unknown.

C. Direct Development.

i. Brood sites. Unknown.

ii. Parental care. Unknown, but it is likely that females brood, as with other species of *Plethodon*.

D. Juvenile Habitat. Juvenile southern gray-cheeked salamanders (17–25 mm SVL) from Macon and Jackson counties, North Carolina, were collected from 12–14 July. They were flushed from wet leaf litter within a cave-like opening and also were found in wet leaves at or above their intersection with loam. At night, juveniles were found on rock faces near the intersection of stone and leaf litter, and on rainy nights they were found on the leaves of herbaceous vegetation ≤30 cm above ground (Gordon, 1960).

E. Adult Habitat. King (1939) found southern gray-cheeked salamanders beneath logs in oak-chestnut forest in areas having good deep soil.

F. Home Range Size. Adult male southern gray-cheeked salamanders in the Balsam Mountains have an average home range of 5.04 m^2. Adult females have an average home range of 1.87 m^2. Two-and three-year-olds have average home ranges of 1.52 m^2 and 2.98 m^2, respectively (Nishikawa, 1990).

Experimentally displaced animals are known to climb trees immediately after being released, presumably to enhance olfactory cues (Madison and Shoop, 1970; see also Petranka, 1998; original observation recorded under *P. jordani*, but probably based on *P. metcalfi* [R. Highton, personal communication]).

G. Territories. At least some members of the *P. jordani* complex aggressively defend territories (Thurow, 1976); it is unknown whether southern gray-cheeked salamanders establish and defend territories.

H. Aestivation/Avoiding Desiccation. Large series of southern gray-cheeked salamanders have been collected in Macon County, North Carolina, from mid-August to early September (Houck et al., 1998).

I. Seasonal Migrations. Unknown.

J. Torpor (Hibernation). At lower elevations, animals remain active during warmer winter weather, retreating during freezes (see Petranka, 1998; original observation recorded under *P. jordani*, but probably based on *P. metcalfi* [R. Highton, personal communication]).

K. Interspecific Associations/Exclusions. In the Balsam Mountains, the following species were encountered on experimental plots with southern gray-cheeked salamanders: black-bellied salamanders *(Desmognathus quadramaculatus)*, seal salamanders *(D. monticola)*, Ocoee salamanders *(D. ocoee)*, pigmy salamanders *(D. wrighti)*, southern red-backed salamanders *(P. serratus)*, southern Appalachian salamanders *(P. teyahalee)*, spring salamanders *(Gyrinophilus porphyriticus)*, red salamanders *(Pseudotriton ruber)*, Blue Ridge two-lined salamanders *(Eurycea wilderae)*, and eastern newts *(Notophthalmus viridescens;* Hairston, 1980b, 1981).

Hairston (1949) reports the following salamanders along with southern gray-cheeked salamanders from near Highlands, Macon County, North Carolina:

black-bellied salamanders, seal salamanders, Ocoee salamanders, and red salamanders. Southern red-backed salamanders, southern Appalachian salamanders, and Blue Ridge two-lined salamanders also are found in this area and are often closely associated with southern gray-cheeked salamanders (Ash, 1997).

Southern Appalachian salamanders are widely sympatric with the Blue Ridge, Balsam, and Cowee isolates of southern gray-cheeked salamanders. There is no evidence of hybridization except at one site along Alarka Creek in the northern Cowee isolate where there is considerable hybridization (Highton and Peabody, 2000).

White-spotted slimy salamanders *(P. cylindraceus)* are sympatric at the eastern edge of the Blue Ridge isolate of southern gray-cheeked salamanders. There is no evidence of hybridization between these species (Highton and Peabody, 2000).

An unusual association of large eastern *Plethodon* occurs in the vicinity of Rabun Bald, Rabun County, Georgia. Here there is a wide hybrid zone between southern gray-cheeked salamanders and a member of the *P. glutinosus* complex, Atlantic Coast slimy salamanders *(P. chlorobryonis)*. In this same area, Atlantic Coast slimy salamanders are also sympatric with southern Appalachian salamanders, another member of the *P. glutinosus* complex, and there is no evidence of hybridization between these species. Thus two members of the *P. glutinosus* complex contact without hybridization in the same area in which one of them (Atlantic Coast slimy salamanders) hybridizes with a member of the *P. jordani* complex (Highton and Peabody, 2000).

Southern gray-cheeked salamanders contact Jordan's salamanders *(P. jordani)* on Balsam Mountain and on Hyatt Ridge. Narrow hybrid zones are present at both contact zones (Hairston, 1950; Highton, 1970; Hairston et al., 1992; Highton and Peabody, 2000).

Southern gray-cheeked salamanders may contact Blue Ridge gray-cheeked salamanders *(P. amplus)*. Their geographic and genetic interactions have not been analyzed (Highton and Peabody, 2000).

Hairston (1980b, 1981) demonstrated competition between southern gray-cheeked salamanders and southern Appalachian salamanders in the Balsam Mountains. The number of southern Appalachian salamanders increased significantly at plots in the Balsam Mountains where southern gray-cheeked salamanders were removed. The removal of southern Appalachian salamanders from plots resulted in a substantial increase in the proportion of young southern gray-cheeked salamanders. Hairston (1980b, 1981) did not identify which resources are limiting but hypothesizes that southern Appalachian salamanders may compete with members of the *P. jordani* complex for nesting sites.

In a portion of the Balsam Mountains, North Carolina, southern gray-cheeked salamanders and southern Appalachian salamanders replace each other altitudinally. In one transect there is no elevational overlap, while in two others there is overlap of 15 m and 46 m. However, in another section of the Balsam Mountains, the two species occur together over a wide elevational range with an overlap of 457–594 m (Hairston, 1951).

L. Age/Size at Reproductive Maturity. Males reach sexual maturity about 3 yr after hatching, at a size of about 50 mm SVL (Howell and Hawkins, 1954; original observation recorded under *P. jordani* but probably based on *P. metcalfi* [R. Highton, personal communication]).

M. Longevity. Unknown.

N. Feeding Behavior. In captivity, southern gray-cheeked salamanders fed on waxworms (lepidopteran larvae; Houck et al., 1998).

O. Predators. Bruce (1972a) reported that a spring salamander ate a member of the *P. jordani* complex (either a red-legged salamander [*P. shermani*] or a southern gray-cheeked salamander). Other predators likely include woodland mammals, birds, and snakes.

P. Anti-Predator Mechanisms. Nocturnal. Secretive. All members of the genus *Plethodon* produce noxious skin secretions (Brodie, 1977). Members of the *P. jordani* complex frequently become immobile when contacted initially. Southern gray-cheeked salamanders were included in a field study on immobility, however it is not possible to separate their behavior from the other members of this complex in this published data set. Immobility may increase survival by making the salamander less likely to be detected, especially by visually oriented predators (Dodd, 1989).

Q. Diseases. Unknown.

R. Parasites. Rankin (1937) lists parasites from at least two species of the *P. jordani* complex. It is not possible to determine which parasites were found in which salamander species, but it is likely that some of these salamanders were southern gray-cheeked salamanders. The following parasites are noted: *Crytobia borreli*, *Eutrichomastix batrachorum*, *Hexamitus intestinalis*, *Karotomorpha swezi*, *Prowasekella longifilis*, *Tritrichomonas augusta*, *Brachycoelium hospitale* and *Crepidobothrium cryptobranchi*.

4. Conservation. Among members of the *P. jordani* complex, southern gray-cheeked salamanders have one of the widest distributions, and they are not protected by any state within their range. While relatively wide ranging, southern gray-cheeked salamanders usually are restricted to higher elevations. Suitable habitat at these elevations may be separated by stretches of lower uninhabited

areas, and populations are often disjunct. Within their range there are many federal and state properties that contain suitable habitat for these salamanders.

Southern gray-cheeked salamanders are relatively resilient to disturbances, such as those associated with timbering operations, and frequently are found in second-growth forests and relatively small, fragmented woodlots (D.A.B., personal observation). However, clearcuts negatively affect populations (Petranka et al., 1993, 1994; Ash et al., 2003) and southern gray-cheeked salamanders disappeared from clearcut areas in Macon County, North Carolina, within 4 yr of lumbering (Ash, 1988, 1997).

As with all species of *Plethodon*, southern gray-cheeked slimy salamanders do not migrate to breeding grounds and do not have large home ranges. Thus, they can exist in habitats of smaller size than many other amphibian species. Conservation activities that promote mature closed-canopy forests should benefit this species.

Acknowledgments. Thanks to Richard Highton, who reviewed this account and gave us the benefit of his insight and experience.

Plethodon mississippi Highton, 1989
MISSISSIPPI SLIMY SALAMANDER
David A. Beamer, Michael J. Lannoo

1. Historical versus Current Distribution. Mississippi slimy salamanders (*Plethodon mississippi*) are distributed from southwestern Kentucky south through western Tennessee, eastern Mississippi, western Georgia, and southeastern Louisiana. The type specimens were described by Highton (1989) and collected at 177 m elevation. Their historical distribution is unknown, but likely to be similar to the current distribution.

2. Historical versus Current Abundance. Highton (this volume, Part One) noted declines in Mississippi slimy salamanders in each of the eight populations he sampled prior to 1986, then resampled under similar conditions using similar effort in or after 1992. In six of these populations, declines were precipitous, with fewer than half the original number of animals detected in the more recent census. Additional monitoring will be required to determine whether these data reflect true declines or natural fluctuations.

3. Life History Features.

A. Breeding. Reproduction is terrestrial. In males from Louisiana, the testes are large from September–February. The vasa deferentia are enlarged from March–July. Spermatophores probably are formed in July–August (Highton, 1962b).

i. Breeding migrations. Undocumented, but breeding migrations are unknown for any *Plethodon* species.

ii. Breeding habitat. Unknown.

B. Eggs. Females collected in late June had heavily yolked eggs. Eggs probably are deposited in late summer (Highton, 1962b).

i. Egg deposition sites. Mississippi slimy salamanders from Wayne County, Mississippi, deposited their eggs in a cave (Brode and Gunter, 1958). Caves are not available for nesting in large portions of the range of Mississippi slimy salamanders; in these areas, other subterranean areas or rotting logs may be used as nesting sites.

ii. Clutch size. Seventeen eggs (5–8 mm diameter) were reported from a cave in Mississippi (Brode and Gunter, 1958).

C. Direct Development. Hatchlings have been found in a Mississippi cave from early August to September (Brode and Gunter, 1958). A juvenile that was likely a few months old was collected from Harrison

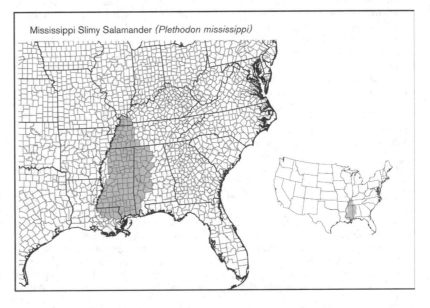

Mississippi Slimy Salamander *(Plethodon mississippi)*

County, Mississippi, and measured 17 mm (Highton, 1956).

i. Brood sites. Females have been found brooding their clutches in a cave in Mississippi (Brode and Gunter, 1958). Caves are not available for brooding in large portions of the range of Mississippi slimy salamanders; in these areas, other subterranean areas or rotting logs may be used as brood sites.

ii. Parental care. Females were found in attendance of their eggs in a Mississippi cave. Females remain with the hatchlings for 1–2 wk before the young disperse (Brode and Gunter, 1958).

D. Juvenile Habitat. Similar to adults. In a study of a Louisiana population, only a single juvenile was collected from December–February. It is likely they stay at the nest site for the first several months after hatching (Highton, 1962b).

E. Adult Habitat. Mississippi slimy salamanders are common in maritime forest and river bottom hardwood forests. Some of the area within the range of Mississippi slimy salamanders historically was covered with savanna and prairie, and in these areas they are rare or absent. Mississippi slimy salamanders do not require pristine habitats or old-growth forests and are often found under discarded rubbish. There may be a minimum size wood lot necessary, as they are absent from small wood lots (Lazell, 1994). In Mississippi, at Tishmingo State Park, Mississippi slimy salamanders are found in oak-hickory forest (Ferguson, 1961b). Mississippi slimy salamanders are reported from a cave in Lauderale County, Alabama (Peck, 1974).

F. Home Range Size. Unknown, but small home ranges are typical for *Plethodon* species.

G. Territories. At least some members of the *P. glutinosus* complex aggressively defend territories (Thurow, 1976); it is unknown whether Mississippi slimy salamanders establish and defend territories.

H. Aestivation/Avoiding Desiccation. Generally unknown, but Mississippi slimy salamanders likely avoid desiccating conditions by seeking shelter in underground sites.

I. Seasonal Migrations. Mississippi slimy salamanders are the most common salamander in Harrison County, Mississippi, during the winter but rarely are found during the summer (Allen, 1932). It is likely that Mississippi slimy salamanders move vertically from forest-floor sites to underground sites in response to seasonal dry and cold conditions.

J. Torpor (Hibernation). Generally unknown, but Mississippi slimy salamanders likely avoid cold conditions by seeking shelter in underground sites.

K. Interspecific Associations/Exclusions. The following salamanders, along with Mississippi slimy salamanders, are all reported from Tishmingo County, Mississippi:

marbled salamanders (*Ambystoma opacum*), spotted dusky salamanders (*Desmognathus conanti*), southern zigzag salamanders (*P. ventralis*), spring salamanders (*Gyrinophilus porphyriticus*), red salamanders (*Pseudotriton ruber*), green salamanders (*Aneides aeneus*), southern two-lined salamanders (*Eurycea cirrigera*), long-tailed salamanders (*E. longicauda*), and cave salamanders (*E. lucifuga*; Ferguson, 1961b; Wake and Woods, 1968).

In the vicinity of the presumed type locality of Bay Springs salamanders (*P. ainsworthi*), Mississippi slimy salamanders are found with red salamanders, southern dusky salamanders (*D. auriculatus*), southern two-lined salamanders, and three-lined salamanders (*E. guttolineata*; D.A.B., personal observations).

L. Age/Size at Reproductive Maturity. The holotype is an adult male, 65 mm SVL; the allotype is an adult female 71 mm. The smallest mature female from Louisiana was 42 mm, the smallest mature male 37 mm (Highton, 1962b).

Louisiana populations of Mississippi slimy salamanders probably reach maturity in 2 yr and breed for the first time at age 3. In both Louisiana and Mississippi at least some females may breed biannually (Highton, 1956, 1962b).

M. Longevity. Unknown.

N. Feeding Behavior. Unknown, but as with other *Plethodon* species, feeding likely takes place at night under moist conditions. Prey items likely include a range of invertebrates, especially insects.

O. Predators. Undocumented, but likely to include forest snakes, birds, and small mammals.

P. Anti-Predator Mechanisms. Nocturnal. All members of the genus *Plethodon* produce noxious skin secretions (Brodie, 1977). Individuals will vocalize when physically disturbed (Marshall, 1997). Members of the *P. glutinosus* complex frequently become immobile when initially contacted. Mississippi slimy salamanders were included in a field study on immobility; however, it is not possible to separate their behavior from the other members of this complex in this published data set. Immobility may increase survival by making the salamander less likely to be detected, especially by visually oriented predators (Dodd, 1989).

Q. Diseases. Unknown.

R. Parasites. Unknown.

4. Conservation.
Among members of the *P. glutinosus* complex, Mississippi slimy salamanders have one of the widest distributions and they are not protected by any state. Within their range there are several federal and state properties that contain suitable habitat for these salamanders.

Mississippi slimy salamanders are relatively resilient to disturbances, such as those associated with timbering operations, and are found frequently in second-

growth forests and relatively small, fragmented woodlots (Lazell, 1994; D.A.B., personal observations).

As with all species of *Plethodon*, Mississippi slimy salamanders do not migrate and have small home ranges. Thus, they can exist in habitats of smaller size than many other amphibian species. However there may be a minimum habitat size necessary, as they are absent from small wood lots (Lazell, 1994). Conservation activities that promote mature closed-canopy forests should benefit this species.

Acknowledgments. Thanks to Richard Highton, who reviewed this account and gave us the benefit of his insight and experience.

Plethodon montanus Highton and Peabody, 2000
NORTHERN GRAY-CHEEKED SALAMANDER
David A. Beamer, Michael J. Lannoo

1. Historical versus Current Distribution.
Northern gray-cheeked salamanders (*Plethodon montanus*) are recently described members of the *P. jordani* complex (Highton and Peabody, 2000). They tend to occur at higher elevations on Flat Top, Buckhorn, Burkes Garden, Knob, Clinch, and Brumley mountains in the Valley and Ridge Province of Virginia, and on Roan, Bald, Black, Max Patch, and Sandymush mountains in the Blue Ridge Province of Virginia, North Carolina, and Tennessee. There is no evidence to suggest that the current distribution differs from the historical distribution. A history of the nomenclature of this species is given in Highton and Peabody (2000). Much of the literature that deals with *P. metcalfi* (southern gray-cheeked salamanders) refers to this species.

2. Historical versus Current Abundance.
Highton (this volume, Part One) surveyed 12 northern gray-cheeked salamander populations prior to 1988 and again after 1994, and found decreased numbers of animals (under similar sampling conditions and with similar effort) in 10 populations. Highton (this volume, Part One) stresses that animals remain abundant, but these populations should continue to be monitored.

A possible decline in the numbers of northern gray-cheeked salamanders was noted after lumbering (Gordon et al., 1962). By comparing a sample made by Dunn 44 yr earlier, the authors determined that the post-lumbering population of salamanders was only 21% of the pre-lumbering population.

Clearcuts negatively affect populations of northern gray-cheeked salamanders. Petranka et al. (1993, 1994) found that these salamanders occur in much

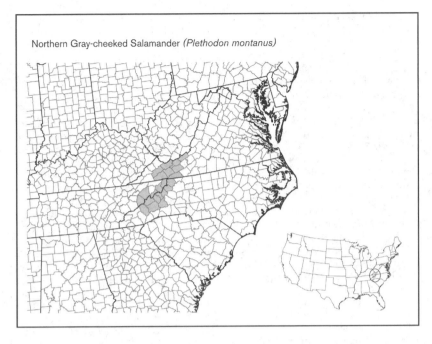

Northern Gray-cheeked Salamander *(Plethodon montanus)*

higher numbers in mature forests than in clearcuts. Petranka (1998) states that deforestation and the conversion of mixed hardwood forests to pine monocultures has eliminated or reduced a large number of eastern *Plethodon* populations. However, Ash (1988, 1997) and Ash and Bruce (1994) have strongly disagreed with Petranka's estimates and do not consider forestry practices to be having as strong an impact on native salamanders.

3. Life History Features.

A. Breeding. Reproduction is terrestrial. Courtship was observed on 3 August at Whitetop Mountain, Smyth County, Virginia (Organ, 1958), and on 31 July near Linville, Avery County, North Carolina (MacMahon, 1964).

Courtship is initiated when a male approaches a female and nudges her with his snout. The male then places his mental gland and nasolabial grooves in contact with the body of the female and begins a foot dance in which the limbs are raised and lowered off the substrate one at a time. The male moves forward to the female's head, at which point he presses his mental gland on her head and nasolabial grooves. The male then circles beneath the female's chin and laterally undulates his tail. The female then places her chin on his tail and moves forward to straddle the tail. The pair moves forward in a tail-straddle walk until the male stops and begins to make lateral sacral movements. The female makes synchronous lateral head movements counter to the sacral movements of the male. The male then lowers his vent to the substrate and deposits a spermatophore. He then flexes his tail to one side and leads the female forward until her vent is over the spermatophore. The female then picks up

the cap from the spermatophore with her cloacal lips (Organ, 1958; MacMahon, 1964).

i. Breeding migrations. Unlikely; breeding migrations are unknown in any *Plethodon* species.

ii. Breeding habitat. At a North Carolina site, courtship occurred within 1 m of a road in area almost covered with jewelweed (*Impatiens* sp.; MacMahon, 1964). The surrounding area is described as a second-growth mixed mesophytic woods at approximately 1,250 m.

B. Eggs.

i. Egg deposition sites. Unknown, but likely to be in underground cavities.

ii. Clutch size. Unknown.

C. Direct Development.

i. Brood sites. Unknown.

ii. Parental care. Unknown, but it is likely that females brood, as with other species of *Plethodon*.

D. Juvenile Habitat. Unknown, but presumably similar to adult habitat.

E. Adult Habitat. Northern gray-cheeked salamanders have been recorded from both virgin and second-growth forest (Dunn, 1917a; Gordon et al., 1962). Northern gray-cheeked salamanders in Avery County, North Carolina, are found in mixed mesophytic forest with a canopy consisting of read and white oak, maples, buckeyes, ironwood, gum, tulip trees, and a few scattered hemlocks. The understory at this site is sparse and composed of jewelweed, nettle shield fern, pipsissewa, goldenrod, twayblade, bellflowers, and pinesap (Gordon et al., 1962).

Northern gray-cheeked salamanders are reported from high elevations in spruce-fir forest (Walker, 1931). Here they are found under logs but not in stumps (Hoffman and Kleinpeter, 1948).

F. Home Range Size. Unknown, but small home ranges are typical for *Plethodon* species.

G. Territories. Selby et al. (1996) observed agonistic behavior in northern gray-cheeked salamanders.

H. Aestivation/Avoiding Desiccation. Undocumented, but *Plethodon* species will move from forest floor habitats to underground sites in response to desiccating surface conditions.

I. Seasonal Migrations. Unknown, other than vertical movements from the forest floor to underground sites to avoid dry summer and cold winter conditions.

J. Torpor (Hibernation). Unknown, but animals presumably move to warmer underground sites to escape cold weather.

K. Interspecific Associations/Exclusions. In Avery County, North Carolina, northern gray-cheeked salamanders were found with eastern newts (*Notophthalmus viridescens*), Blue Ridge dusky salamanders (*Desmognathus orestes*), seal salamanders (*D. monticola*), red-backed salamanders (*P. cinereus*), Yonahlossee salamanders (*P. yonahlossee*), white-spotted slimy salamanders (*P. cylindraceus*), and Blue Ridge two-lined salamanders (*Eurycea wilderae*; Gordon et al., 1962).

On Grandfather Mountain and in the Whitetop Mountain-Mount Rogers area, the following salamander species are reported in association with northern gray-cheeked salamanders: Weller's salamanders (*P. welleri*), red-backed salamanders, Blue Ridge two-lined salamanders, Blue Ridge dusky salamanders, and pigmy salamanders (*D. wrighti*; Walker, 1931; Hoffman and Kleinpeter, 1948; Bogert, 1952; Thurow, 1963).

Hairston (1949) reports the following salamanders along with northern gray-cheeked salamanders from the Black Mountains area, Yancey County, North Carolina: spring salamanders (*Gyrinophilus porphyriticus*), red salamanders (*Pseudotriton ruber*), Blue Ridge two-lined salamanders, black-bellied salamanders (*D. quadramaculatus*), seal salamanders, Carolina mountain dusky salamanders (*D. carolinensis*), pigmy salamanders, shovel-nosed salamanders (*D. marmoratus*), Yonahlossee salamanders, and white-spotted slimy salamanders.

Northern gray-cheeked salamanders are widely sympatric with white-spotted slimy salamanders in the Blue Ridge province in the Black and Bald isolates, and most of the Roan isolate. There is no evidence of hybridization between these species (Highton and Peabody, 2000).

Southern Appalachian salamanders (*P. teyahalee*) are widely sympatric with the Max Patch and Sandymush isolates of northern gray-cheeked salamanders. There is no known morphological or genetic evidence of hybridization (Highton and Peabody, 2000).

The six isolates of northern gray-cheeked salamanders in the Valley and Ridge Province are widely sympatric with northern slimy salamanders (*P. glutinosus*). The two species also are sympatric in a small section of the Roan isolate of northern gray-cheeked salamanders. There is evidence of occasional hybridization (Highton and Peabody, 2000).

Northern gray-cheeked salamanders are widely sympatric with Yonahlossee salamanders at the eastern edge of the Blue Ridge isolate. There is no evidence of hybridization (Highton and Peabody, 2000).

The ranges of northern gray-cheeked salamanders and Cumberland Plateau salamanders *(P. kentucki)* overlap in Russell, Smyth, and/or Washington counties in Virginia, but they have not yet been collected at the same sites (Highton and Peabody, 2000).

In the Black Mountains, Yancey County, North Carolina, northern gray-cheeked salamanders and white-spotted slimy salamanders generally replace each other altitudinally. On four of five transects, there was no overlap of these species occurrences; at a fifth transect, the two occur together over an elevational range of 61 m (Hairston, 1951).

L. Age/Size at Reproductive Maturity. Unknown.

M. Longevity. Unknown.

N. Feeding Behavior. Northern gray-cheeked salamanders sometimes forage above ground. Many have been observed on trunks and low branches of shrubs ≤1 m from the ground. Groups were observed feeding on fungal gnats around decaying fungi, or on fruit flies and other insects at the base of trees from which sap flowed (Gordon et al., 1962).

The following food items are reported from specimens from Grandfather Mountain, Avery County, North Carolina: Formicidae, Araneida, lepidopteran larvae, coleopteran larvae, Collembola, Diplopoda, Chilopoda, Acarina, Stylommatophora, dipteran larvae, Annelida, Tipulidae, Lepidoptera, shed skin, Ichneumonidae, Diptera, Gryllacrididae, Cicadellidae, Coleoptera, Miridae, Cantharidae, Chrysomelidae, Isopoda, Cynipodea, Hymenoptera, Phoridae, Psychodidae, Chironomidae, Cydnidae, Fulgoridae, Scarabaeidae, Reduviidae, Carabidae, Curculionidae, Chelonethida, Vespidae, Gryllidae, Simuli-idae, Histeridae, Pentatomidae, Mecoptera, Tingidae, Elateridae, Otitidae, Mycetophilidae, Cryptophagidae, Culicidae, Aphididae, Phalacridae, Dolichopodidae, Staphylinidae, Fungivoridae, and Hemiptera. Ants, mites, and springtails were eaten less frequently by larger individuals while millipedes, earthworms, craneflies, spiders, and centipedes were eaten more frequently. In general, the number of different food items increases in larger individuals (Whitaker and Rubin, 1971).

O. Predators. Undocumented, but probably include forest-dwelling snakes, birds, and small mammals.

P. Anti-Predator Mechanisms. Unknown specifically, but members of the *P. jordani* complex are known to assume defensive postures and to produce slimy tail secretions that predators find noxious (Brodie et al., 1979; Petranka, 1998). Members of the *P. jordani* complex frequently become immobile when initially contacted. Northern gray-cheeked salamanders were included in a field study on immobility; however, it is not possible to separate their behavior from the other members of this complex in this published data set. Immobility may increase survival by making the salamander less likely to be detected, especially by visually oriented predators (Dodd, 1989).

Q. Diseases. Unknown.

R. Parasites. Rankin (1937) lists parasites from at least two species of the *P. jordani* complex. It is not possible to determine which parasites were found in which salamander species, but some of these salamanders were northern gray-cheeked salamanders. The following parasites are noted: *Crytobia borreli, Eutrichomastix batrachorum, Hexamitus intestinalis, Karotomorpha swezi, Prowazekella longifilis, Tritrichomonas augusta, Brachycoelium hospitale* and *Crepidobothrium cryptobranchi*.

4. Conservation.
Northern gray-cheeked salamanders are not protected by any state. Among members of the *P. jordani* complex, northern gray-cheeked salamanders have one of the widest distributions. While relatively wide ranging, northern gray-cheeked salamanders are restricted to higher elevations. Suitable habitat at these elevations may be separated by stretches of lower uninhabited areas, and populations often are not continuous. Within this range there are many federal and state properties that contain suitable habitat for these salamanders.

Northern gray-cheeked salamanders are relatively resilient to disturbances, such as those associated with timbering operations, and frequently are found in second-growth forests and relatively small, fragmented woodlots (Gordon et al., 1962; D.A.B., personal observation).

As with all species of *Plethodon*, northern gray-cheeked salamanders do not migrate and have small home ranges. Thus, they can exist in habitats of smaller size than many other amphibian species. Conservation activities that promote mature closed-canopy forests should benefit this species.

Acknowledgments. Thanks to Richard Highton, who reviewed this account and gave us the benefit of his insight and experience.

Plethodon neomexicanus Stebbins and Riemer, 1950
JEMEZ MOUNTAINS SALAMANDER
Charles W. Painter

1. Historical versus Current Distribution.
Jemez Mountains salamanders were originally collected and described as *Spelerpes multiplicatus* (=*Eurycea multiplicata*; Bailey, 1913; VanDenburgh, 1924; Dunn, 1926; Bishop, 1943) but rediscovered and described as *P. neomexicanus* by Stebbins and Riemer (1950). Their range is limited to an area of approximately 971 km^2 (375 mi^2) within Sandoval, Los Alamos, and Rio Arriba counties in north-central New Mexico. Jemez Mountains salamanders are known only from high elevations of the mountains surrounding the caldera of the Valle Grande of the Jemez Mountains (Degenhardt et al., 1996; Petranka, 1998). Knowledge of their current distribution has been greatly expanded from early descriptions (Dunn, 1926; Stebbins, 1951; S.R. Williams, 1973). The current distribution is expected to be similar to their historical distribution.

2. Historical versus Current Abundance.
Jemez Mountains salamanders are rare and localized, but common in small areas of suitable habitat. Populations are highly fragmented due to subsurface geology. Historical abundance is unknown but expected to be higher in select areas of early heavy collecting and clearcut logging (Williams, 1972a).

3. Life History Features.
A. Breeding. Reproduction is assumed to be terrestrial, although it has not been observed in the wild.

i. Breeding migrations. Migratory behavior not reported or observed; not believed to occur.

ii. Breeding habitat. Males are sexually active throughout the rainy season (June–August). Females likely brood underground in fall or early spring every other year. Gravid females generally are not collected after 20 August. A gravid female collected in August and kept in the lab deposited eggs in early June the following spring. Fewer than 50% of mature females collected are gravid (Reagan, 1972; S.R. Williams, 1972b, 1973, 1976).

B. Eggs.

i. Egg deposition sites. Although no egg clutches have been discovered in the wild, it is believed they are laid in the deeply fractured interstices of subterranean, metamorphic rock. The composition of this subterranean habitat has not been fully investigated, although soils comprised of pumice or tuft generally are not suitable. Heavy equipment used during logging may cause compaction of soils and thus destroy suitable breeding habitat.

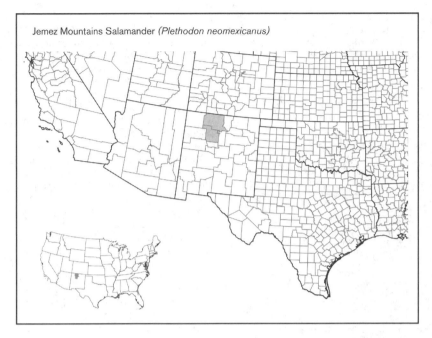

Jemez Mountains Salamander *(Plethodon neomexicanus)*

ii. Clutch size. Average clutch size is 7.7 (5–12; Reagan, 1972), with eggs 6.8–7.3 mm in diameter (S.R. Williams, 1978).

C. Direct Development. Females likely brood underground in fall or early spring every other year. Hatchlings appear on the surface during wet weather in mid-July to late August at 15–18 mm SVL. Size at hatching in wild individuals is unknown.

D. Juvenile Habitat. Similar to adult habitat.

E. Adult Habitat. Generally found on loose rocky soils between 2,200–2,900 m (Degenhardt et al., 1996), although individuals have been reported as high as 3,429 m (Whitford and Ludwig, 1976). Mostly found in and under rotting Douglas fir logs or under rocks on flat areas and steep slopes. Slope aspect may face any direction. Macrohabitat is coniferous forest dominated by Douglas fir, blue spruce, Engelmann spruce, ponderosa pine, and white fir with occasional aspen. Deciduous understory shrubs are Rocky Mountain maple, New Mexico locust, oceanspray, and various shrubby oaks (Degenhardt et al., 1996). Rarely observed on the surface at night or during the day and rarely encountered under bark, surface litter, or aspen logs. Subsurface geology, elevation, and moisture are key habitat features (Ramotnik, 1988; Ramotnik and Scott, 1988). Desiccation of habitat (Wiltenmuth, 1997b) and the species' low critical thermal maximum (33.5 °C; Whitford, 1968) are important considerations in any habitat manipulation.

F. Home Range Size. Unknown, although studies suggest home range may be small. Ramotnik (1986) tracked an individual that moved 13 m in 1.5 hr during a light drizzle. Two females implanted with Tantalum-182 wire showed strong site fidelity, returning to the same site after repeated movements of up to 5 m. Painter (2000) reported an individual marked on 3 August 1994 that was recaptured near the original capture site nearly 5 yr (1,831 d) later on 10 August 1999; maximum distance recorded from the original capture site was 1.5 m. Of all movements recorded from recaptured individuals, 37% were fewer than 2 m from the original capture site (range = 0–19.7 m). Data indicate a lack of homing ability in Jemez Mountains salamanders displaced from the original capture site (Ramotnik, 1986).

G. Territories. Unknown. Movements of up to 13 m (Ramotnik, 1986) or 19.7 m (Painter, 2000) are known from marked individuals. Of all movements recorded from marked individuals, 37% moved fewer than 2 m from the previous capture site, and 36% moved fewer than 2 m from the original capture site (Painter, 2000).

H. Aestivation/Avoiding Desiccation. Unknown. Individuals are not surface active unless sufficient moisture is available following summer monsoon rainfalls.

I. Seasonal Migrations. Individuals retreat underground in response to seasonal extremes of temperature and dryness.

J. Torpor (Hibernation). Retreat underground after temperatures fall below about 6 °C.

K. Interspecific Associations/Exclusions. Unknown.

L. Age/Size at Reproductive Maturity. Sexual maturity is reached at 2–3 yr of age in males, 3 yr in females. Mean SVL of mature males is 54.4 mm; females 55.5 mm (Williams, 1972b). Mean annual growth is approximately 3.7 mm SVL (Painter, 2000).

M. Longevity. Unknown, although based on reproductive data, longevity is expected to be about 10 yr.

N. Feeding Behavior. Feed on a wide variety of invertebrates, including annelid worms, spiders and mites (Arachnida), millipedes, centipedes, snails (Mollusca), and insects, including springtails (Collembola; Reagan, 1972; Painter and Altenbach, unpublished data). Ants (Hymenoptera) and mites comprise about 71% of the diet by percent prey items, while ants, beetles (Coleoptera), and non-ant hymenopterans comprise about 67% of the total volume (Painter, 2000). Most prey are taken while Jemez Mountains salamanders actively forage inside rotted logs.

O. Predators. Western terrestrial garter snakes (*Thamnophis elegans*; Painter et al., 1999). No other predators have been reported, but individuals are possibly preyed upon by black bears, shrews, and larger ground feeding birds.

P. Anti-Predator Mechanisms. None reported in the literature. Newly captured individuals often respond by tightly coiling the body into a loop with the tail covering the head.

Q. Diseases. Unknown.

R. Parasites. The nematode *Thelandros salamandrae* (Oxyuroidea) has been reported in 55% (Panitz, 1967) and 41% (Reagan, 1972) of specimens examined.

4. Conservation.
The range of Jemez Mountains salamanders is limited to the higher elevations of the mountains surrounding the caldera of the Valle Grande of the Jemez Mountains (Degenhardt et al., 1996; Petranka, 1998). Populations are highly fragmented due to subsurface geology; they are rare and localized, but common in small areas of suitable habitat. Jemez Mountains salamanders are listed as Threatened by the State of New Mexico and as a Species of Concern by the U.S. Forest Service and the U.S. Fish and Wildlife Service.

Plethodon nettingi Green, 1938(b)
CHEAT MOUNTAIN SALAMANDER

Thomas K. Pauley, Beth Anne Pauley, Mark B. Watson

1. Historical versus Current Distribution.
Cheat Mountain salamanders (*Plethodon nettingi*) are endemic to the eastern highlands of West Virginia. They inhabit what is presently or was historically red spruce forests. Initially, their range was thought to be limited to Cheat Mountain at elevations above 1,067 m in Randolph and Pocahontas counties (Green, 1938b; Brooks, 1945, 1948b). Later inventories expanded their range to include Pendleton and Tucker counties (Highton, 1971, 1986f;

Pauley, 1980a,b, 1981, 1986, 1987). More recent inventories have expanded their total range to include the western edge of Grant County and their elevational range down to 750 m (Pauley, 1991a).

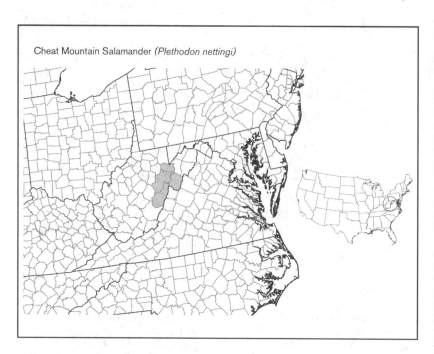

Cheat Mountain Salamander *(Plethodon nettingi)*

2. Historical versus Current Abundance.
Unknown historically. From 1976–00, Pauley (T., unpublished data) surveyed 1,300 sites within their known total range and has located Cheat Mountain salamanders in 60 disjunct populations. They are fairly abundant in some populations but appear to be scarce in others.

3. Life History Features.
Their life history and biology have been summarized by Green and Pauley (1987) and Petranka (1998). Mental hedonic gland-cluster is discussed by Dodd and Brodie (1976).

  A. Breeding. Reproduction is terrestrial.
  i. Breeding migrations. This species does not migrate.
  ii. Breeding habitat. Breeding activities of Cheat Mountain salamanders have not been observed, but most likely occur on the forest floor. Pairs of males and females have been found under rocks in the spring and autumn. Both sexes during these months are in breeding condition: males with swollen cloacas and squared-off snouts, females with mature follicles (gravid).

  B. Eggs.
  i. Egg deposition sites. Nests have been found in decayed logs and beneath rocks on the surface of the ground in red spruce forests or in deciduous forests (Green and Pauley, 1987).
  ii. Clutch size. Numbers of eggs/clutch vary from 4–17 (Brooks, 1948b).

  C. Direct Development. Spermatophores probably are deposited in late spring;

mating may also occur in the fall. Nests with attending females have been found from 28 April–15 August (Brooks, 1948b; Green and Pauley, 1987). Eggs hatch in late August to early September. Young of the year averaging 1.8 cm TL were observed in a nest with an attending female in September (Green and Pauley, 1987).

  D. Juvenile Habitat. Juveniles have been found in the same habitat as adults.

  E. Adult Habitat. Cheat Mountain salamanders are found in spruce, hemlock, or deciduous stands with scattered spruce and hemlock above 750 m in elevation. Their occurrence is not associated solely with any particular type of vegetation (Clovis, 1979), but is associated with boulder fields, rock outcrops, or steep ravines lined with a dense growth of *Rhododendron* sp. (Pauley, 1998).

  F. Home Range Size. Unknown. Movements probably are similar to eastern red-backed salamanders (*P. cinereus*; Kleeberger and Werner, 1982).

  G. Territories. Thurow (1976) noted that in all laboratory trials but one, Cheat Mountain salamanders did not show aggressive behavior toward other small *Plethodon* when they were introduced into the species' territory. Conversely, Pauley and Pauley (1990) found in laboratory trials that Cheat Mountain salamanders did display an aggressive behavior towards eastern red-backed salamanders introduced into their territory.

  H. Aestivation/Avoiding Desiccation. This species remains active on the surface from late March to mid-October (Santiago, 1999). Aestivation only occurs during unusual drought conditions.

  I. Seasonal Migrations. This species does not migrate.

  J. Torpor (Hibernation). Surface activity and abundance of Cheat Mountain salamanders are influenced by environmental conditions (Santiago, 1999). Depending on soil temperature, Cheat Mountain salamanders emerge from winter refugia at the end of March and retreat to underground refugia in mid October.

  K. Interspecific Associations/Exclusions. Pauley (1980a) and Pauley (1998) found that Cheat Mountain salamanders compete with eastern red-backed salamanders for nesting sites and primary and secondary food items. They also compete with Allegheny Mountain dusky salamanders *(Desmognathus ochrophaeus)* for cool, moist sites. Cheat Mountain salamanders have a narrow elevational overlap with red-backed salamanders (12.2 m–61.0 m) and Allegheny Mountain dusky salamanders (0–97.5 m; Pauley, 1980b). Elevational overlaps from 61 m–121.9 m result in keen competition (Hairston, 1951, 1980b). Pauley (1998) found that Cheat Mountain salamanders are more abundant adjacent to large emergent rocks where soil and litter are more moist and cooler than the surrounding hillsides. Pauley (T., this volume, Part One) has hypothesized that Cheat Mountain salamanders were protected in these refugia (emergent rocks) when the original forests were cut and in some areas burned. Because of competitive interactions with red-backed salamanders and Allegheny Mountain dusky salamanders, Cheat Mountain salamanders remain associated with these refugia.

  L. Age/Size at Reproductive Maturity. Mature females range from 38.2–53.4 mm SVL (average = 45.8 mm); males range from 35.5–49.4 mm (average = 42.5 mm; T. K. P., unpublished data).

  M. Longevity. Unknown. Probably similar to other small *Plethodon* (Snider and Bowler, 1992).

  N. Feeding Behavior. Cheat Mountain salamanders are opportunistic feeders. Major prey items are mites, springtails, beetles, flies, and ants (Pauley, 1980a).

  O. Predators. Unknown. Probably include shrews, common garter snakes *(Thamnophis sirtalis)*, and ring-necked snakes *(Diadophis punctatus)*.

  P. Anti-Predator Mechanisms. Unknown. As with other members of the genus *Plethodon*, mucus secretions in the tail may inhibit ingestion by small predators.

  Q. Diseases. Diseases of this species have not been reported.

  R. Parasites. Parasites have not been determined.

4. Conservation.
Cheat Mountain salamanders were listed as a Threatened species by the U.S. Fish and Wildlife Service in 1989b (Federal Register). The recovery plan for this species was developed by Pauley (1991b) for the Northeast Region of the U.S. Fish and Wildlife Service. Of the 60 known

populations, all or portions of 46 are located within the boundaries of the Monongahela National Forest, which provides protection from habitat disturbances. Cheat Mountain salamanders may be sensitive to logging practices such as clearcutting and development of roads, trails, and ski slopes (Pauley, 1994, this volume, Part One). Mitchell et al. (1999) listed existing and potential threats to Cheat Mountain salamanders. In 1981, Pauley (1997) attempted to relocate 47 salamanders from a population of Cheat Mountain salamanders that was going to be destroyed by a deep mine portal. All attempts to recover relocated specimens for the ensuing 4 yr were unsuccessful.

Plethodon ocmulgee Highton, 1989
OCMULGEE SLIMY SALAMANDER
David A. Beamer, Michael J. Lannoo

1. Historical versus Current Distribution.
Ocmulgee slimy salamanders *(Plethodon ocmulgee)* are found in the upper Coastal Plain and adjacent Piedmont Physiographic Provinces of central Georgia associated with the Ocmulgee River drainage (Highton, 1989). The type specimens were collected at an elevation of 49 m. There is little evidence that their current distribution differs from the historical distribution, although Highton (this volume, Part One) could not find animals at his Long County, Georgia, site. Further sampling will be required to determine whether or not this population is truly extirpated.

2. Historical versus Current Abundance.
Highton (this volume, Part One) sampled three populations of Ocmulgee slimy

salamanders from Bulloch, Long, and Wheeler counties in Georgia in 1953–88, then resampled them in or after 1995. He found evidence of precipitous declines (fewer than half the original number of animals observed in the resurvey) in two populations (Long and Wheeler counties, Georgia). Whether these data reflect true declines or natural population fluctuations can only be determined with additional sampling.

3. Life History Features.
Ocmulgee slimy salamanders have been only recently described (Highton, 1989). Since this time, there has been no published work done on this species. As a portion of his larger research program, Highton has collected basic life history and natural history information on Ocmulgee slimy salamanders and has plans to publish these data in a monographic treatment.

A. Breeding. Reproduction is terrestrial.

i. Breeding migrations. Undocumented, but breeding migrations are unknown for any *Plethodon* species.

ii. Breeding habitat. Unknown.

B. Eggs.

i. Egg deposition sites. Undocumented.

ii. Clutch size. Unknown.

C. Direct Development.

i. Brood sites. Unknown, but are likely to be the same as egg deposition sites.

ii. Parental care. Unknown, but it is likely that females brood, as with other members of the slimy salamander complex.

D. Juvenile Habitat. Unknown, but as with other *Plethodon*, likely to be similar to adult habitat.

E. Adult Habitat. Include forest-floor habitats. Logs and rocks are used as cover objects during the daytime. Animals tend

to be nocturnal, and activity levels likely are related to moisture levels.

F. Home Range Size. Unknown, but small home ranges are typical for *Plethodon* species.

G. Territories. At least some members of the *P. glutinosus* complex aggressively defend territories (Thurow, 1976); it is unknown whether Ocmulgee slimy salamanders establish and defend territories.

H. Aestivation/Avoiding Desiccation. Generally unknown, but Ocmulgee slimy salamanders likely avoid desiccating conditions by seeking shelter in underground sites.

I. Seasonal Migrations. Unknown, but vertical migrations from surface sites to underground sites and back again are likely to be important in surviving seasonally variable conditions.

J. Torpor (Hibernation). Generally unknown, but Ocmulgee slimy salamanders likely avoid cold conditions by seeking shelter in underground sites.

K. Interspecific Associations/Exclusions. Undocumented.

L. Age/Size at Reproductive Maturity. Unknown. The holotype is an adult male 56 mm SVL; the allotype is a 62 mm adult female.

M. Longevity. Unknown.

N. Feeding Behavior. Unknown, but as with other *Plethodon* species, feeding likely takes place at night under moist conditions. Prey items likely include a range of invertebrates, especially insects.

O. Predators. Undocumented, but likely to include forest snakes, birds, and small mammals.

P. Anti-Predator Mechanisms. Nocturnal. Secretive. All members of the genus *Plethodon* produce noxious skin secretions (Brodie, 1977).

Q. Diseases. Unknown.

R. Parasites. Unknown.

4. Conservation.
Ocmulgee slimy salamanders are not protected in Georgia, the only state within their range. Among members of the *P. glutinosus* complex, Ocmulgee slimy salamanders have one of the smaller distributions. There are few federal and state properties that preserve suitable habitat for these salamanders.

Acknowledgments. Thanks to Richard Highton, who reviewed this account and gave us the benefit of his insight and experience.

Plethodon ouachitae Dunn and Heinze, 1933
RICH MOUNTAIN SALAMANDER
Carl D. Anthony

1. Historical versus Current Distribution.
Rich Mountain salamanders *(Plethodon ouachitae)* are locally distributed in the

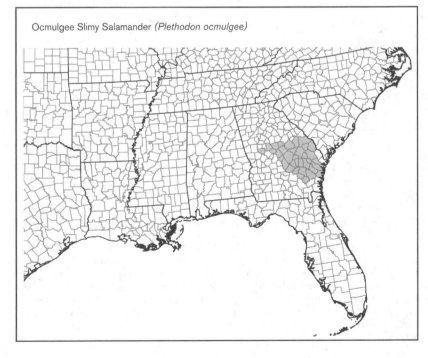

Ocmulgee Slimy Salamander *(Plethodon ocmulgee)*

Ouachita Mountains of eastern Oklahoma and western Arkansas (Pope and Pope, 1951; Blair and Lindsay, 1965; Duncan and Highton, 1979; Black and Sievert, 1989). Knowledge of their geographic range has been greatly expanded since the species was first collected on Rich Mountain east of Page, Oklahoma (Dunn and Heinze, 1933). Their current distribution in Oklahoma includes parts of Buffalo, Winding Stair, Coon, Spring, Honess, Rich, Blackfork, Kiamichi, Phillips, and Round mountains (A.P. Blair, 1957b; Blair and Lindsay, 1965; Duncan and Highton, 1979; Sievert, 1986; Anthony and Wicknick, 1993a; Anthony, 1995). Their current distribution in Arkansas includes parts of Rich, Fourche, and Blackfork mountains, with hybridization occurring between them and Fourche Mountain salamanders (*P. fourchensis*) in a narrow zone on Fourche Mountain (Blair and Lindsay, 1965; Duncan and Highton, 1979). Three variants are recognized (Winding Stair, Rich, and Kiamichi), each representing a distinct genetic group (Duncan and Highton, 1979). Historical distribution is unknown.

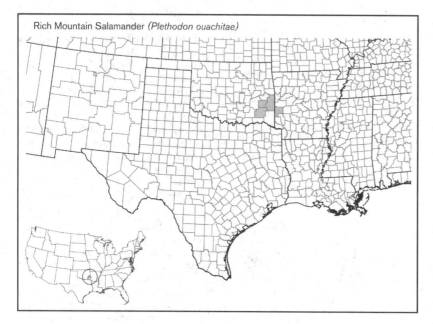

Rich Mountain Salamander (*Plethodon ouachitae*)

2. Historical versus Current Abundance. Locally abundant. Pope and Pope (1951) reported collecting 131 individuals in 4 hr at a disturbed site. Daytime collecting rates typically range from 3–17 salamanders/hr (Pope and Pope, 1951; Sievert, 1986), but may reach 30–40/hr under ideal conditions. No changes in abundance have been noted.

3. Life History Features.
  A. Breeding.
    i. Breeding migrations. Migratory behavior is unknown, but adults are active on

roadways on foggy nights in late spring and early fall.
    ii. Breeding habitat. Breeding may occur in late fall, winter, or early spring (Taylor et al., 1990).
  B. Eggs.
    i. Egg deposition sites. Breeding and nest sites are undescribed but likely occur deep in talus and in caves. Black (1974) reported a 7 mm SVL juvenile from a cave on Winding Stair Mountain. Surface active juveniles are considerably larger. Ratios of reproductive to nonreproductive females suggest that they breed biennially (Taylor et al., 1990).
    ii. Clutch sizes. Pope and Pope (1951) reported a mean of 16.7 enlarged ovarian follicles from 22 mature females; Taylor et al. (1990) reported a mean of 15.4 from 17 mature females.
  C. Direct Development. Juveniles averaging 21 mm SVL are active at the surface in May (Anthony and Wicknick, 1993b) and tend to remain at the surface through late May, after adults have moved underground.
  D. Juvenile Habitat. Same as adult habitat.

  E. Adult Habitat. Most commonly found at higher elevations of mixed deciduous wooded north-, northeast- (Blair, 1967) and northwest- (Black, 1974) facing slopes, often adjacent to seeps. Sandstone rocks, logs, and other forest debris are common cover objects (Black, 1974). Moisture conditions at the surface appear to influence activity greatly (Pope and Pope, 1951). Lower levels of talus are used to escape hot and dry conditions (Spotila, 1972). Cave entrances also are utilized (Black, 1974; Black and Puckette, 1974).

  F. Home Range Size. Little is known concerning home range size. Anthony (1995) reported that 23% of adults displaced 5 m returned to their original cover object. Adults are commonly found on roadways up to 20 m from suitable habitat.
  G. Territories. Individuals can be found under the same cover objects from year to year, and they home to those cover objects if displaced (Anthony, 1995). Adults can recognize and respond to odors of conspecifics (Anthony, 1993), and they actively defend areas in laboratory chambers (Thurow, 1976; Anthony and Wicknick, 1993b; Anthony et al., 1997). Territory size is unknown.
  H. Aestivation/Avoiding Desiccation. Populations move underground in late May but may return to the surface during periods of rainfall and/or cool weather. By mid-September, adults can again be found under cover objects at the surface.
  I. Seasonal Migrations. Adults are common on wet roads in spring and fall at temperatures to 23.8 °C. It is not known whether these salamanders are migrating, dispersing, or simply foraging.
  J. Torpor (Hibernation). Probably hibernate from mid-November to late March.
  K. Interspecific Associations/Exclusions. Occur syntopically with western slimy salamanders (*P. albagula*) and southern red-backed salamanders (*P. serratus*). Within their range, Rich Mountain salamanders are almost always more common than are western slimy salamanders (Pope and Pope, 1951; Spotila, 1972; Duncan and Highton, 1979). Western slimy salamanders are rare or absent at many localities and may be excluded by Rich Mountain salamanders (Anthony, 1995). Interspecific territoriality is apparently important. In laboratory encounters, adult males aggressively defended areas against conspecifics and against intruding western slimy salamanders. Despite the smaller size of Rich Mountain salamanders, defense was successful largely due to its extremely aggressive nature (Anthony et al., 1997). Kuss (1986) found that habitat differed between the two species, with Rich Mountain salamanders occurring more often at higher elevations, farther from ravines, and under denser vegetation than western slimy salamanders. Dowling (1956) observed that Rich Mountain salamanders become active on the forest floor earlier in the evening than do western slimy salamanders, which may have to wait for cooler and more humid conditions. Spotila (1972) noted that western slimy salamanders occur in wetter habitats than do Rich Mountain salamanders.
  L. Age/Size at Reproductive Maturity. Maturity is reached at an age of almost 3 yr (Pope and Pope, 1951). Females mature at 50–51 mm SVL; males mature at 47–49 mm SVL (Pope and Pope, 1951).
  M. Longevity. Individuals that were sexually mature when collected have been

kept in the laboratory for at least 3 yr. Longevity is probably greater than 6 yr.

N. Feeding Behavior. Prey consists of terrestrial invertebrates including annelids, chilopods, acarinids, coleopterans (adults and larvae), hemipterans, orthopterans, and hymenopterans (Black and Puckette, 1974). Feeding probably occurs in the leaf litter, but one adult was observed consuming an opilionid while beneath a cover object in the field.

O. Predators. Unknown.

P. Anti-Predator Mechanisms. Largely nocturnal, though individuals have been observed moving about on rainy days. All *Plethodon* produce noxious skin secretions (Brodie, 1977). When handled, Rich Mountain salamanders release a secretion that is initially lubricative then adhesive. The secretion is difficult to remove from the fingers and hands. Adults are agile and rapidly flee to below ground retreats when disturbed (Pope and Pope, 1951; Black, 1974).

Q. Diseases. Unknown.

R. Parasites. Winter et al. (1986) found cestodes, nematodes, protozoans, and mites in 29 adults. Intradermal mites of the genus *Hannemania* were first noted in the original species description (Dunn and Heinze, 1933) and appear as raised red pustules often on the feet and toes (Winter et al., 1986; Anthony et al., 1994). Up to 100% of a population can be infected (range 62–100%, Duncan and Highton, 1979). Males incur higher infection intensities than do females (Anthony et al., 1994). Toe loss is common and has been attributed to these mites (Winter et al., 1986; Conant and Collins, 1991). Anthony et al. (1994) reported structural damage to the nasolabial groove from mite attachment. Sympatric western slimy salamanders are seldom infected by mites (Pope and Pope, 1951; Duncan and Highton, 1979; Anthony et al., 1994) and have been shown to be more resistant to infection (Anthony, 1995).

4. Conservation.
Rich Mountain salamanders are locally distributed in the Ouachita Mountains. No changes in abundance have been noted. Three variants are recognized, each representing a distinct genetic group (Duncan and Highton, 1979). They are listed as a Protected species by Oklahoma.

Plethodon petraeus Wynn, Highton and Jacobs, 1988
PIGEON MOUNTAIN SALAMANDER
John B. Jensen, Carlos D. Camp

1. Historical versus Current Distribution.
Pigeon Mountain salamanders (*Plethodon petraeus*) are known only from the eastern slope of Pigeon Mountain in northwestern Georgia (Wynn et al., 1988). The original description reported them from only two sites, both in Walker County

(Wynn et al., 1988). Since that time, several additional sites have been discovered on Pigeon Mountain, including one in Chattooga County (Highton, 1995a; Buhlmann, 1996; J.B.J. and C.D.C., unpublished data).

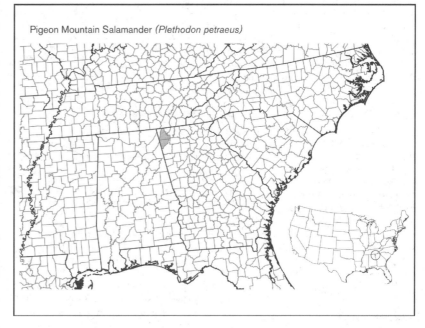

Pigeon Mountain Salamander *(Plethodon petraeus)*

2. Historical versus Current Abundance.
Wynn et al. (1988) reported the species as abundant, far outnumbering other syntopic salamander species. Recent surveys at two of the known sites indicated no detectable change in their abundance (Buhlmann, 1996; J.B.J., personal observations), however Pigeon Mountain salamanders have become uncommon at one locality, possibly due to disturbance created by increased cave visitation and/or perhaps scientific over-collection (Jensen, 1999a).

3. Life History Features.
A. Breeding. Reproduction is terrestrial.
i. Breeding migrations. Mating and migratory behavior unknown.
ii. Breeding habitat. Breeding and nest sites unknown.
B. Eggs.
i. Egg deposition sites. Unknown, but likely in caves (Jensen et al., 2002).
ii. Clutch size. Based on the number of developing oocytes found in dissected, mature females, Jensen et al. (2002) determined an average clutch size of 19.3.
C. Direct Development. As a member of the genus *Plethodon*, direct development of Pigeon Mountain salamanders is assumed, however eggs and hatchlings are unknown.
D. Juvenile Habitat. Thought to be the same as adult habitat.
E. Adult Habitat. Pigeon Mountain salamanders are associated with limestone

outcroppings, boulder fields, and caves (Wynn et al., 1988). Those found in caves are rarely deeper than the twilight zone. Individuals are most often found in and around cracks and crevices within rocks. These microhabitats are embedded within

mesic deciduous forests consisting of an overstory comprised primarily of oak and hickory (Jensen, 1999a). However, Pigeon Mountain salamanders are rarely encountered away from rock outcrops or caves (Jensen et al., 2002).
F. Home Range Size. Unknown.
G. Territories. Unknown.
H. Aestivation/Avoiding Desiccation. Little is known, although individuals are difficult to find during extremely dry conditions (J.B.J., personal observations).
I. Seasonal Migrations. Unknown.
J. Torpor (Hibernation). Little is known, although individuals are difficult to find during winter (Jensen, 1999a).
K. Interspecific Associations/Exclusions. Unknown, however aggressive interference is common among other species of *Plethodon*, and laboratory trials involving Pigeon Mountain salamanders and sympatric northern slimy salamanders (*P. glutinosus*) revealed territorial defense and aggression (J.L. Marshall, C.D.C., and R.G. Jaeger, unpublished data). In addition to northern slimy salamanders, other salamanders often found in the same microhabitats include cave salamanders (*Eurycea lucifuga*), long-tailed salamanders (*E. longicauda*), southern zigzag salamanders (*P. ventralis*), and green salamanders (*Aneides aeneus*; Jensen et al., 2002).
L. Age/Size at Reproductive Maturity. Females mature at a larger size (≥65 mm SVL) than males (≥56 mm) and grow to a greater maximum size (85 mm for

females, 80 mm for males). Males mature at a minimum of 3 yr, females at 4 yr (Jensen et al., 2002).

   M. Longevity. Unknown.

   N. Feeding Behavior. Jensen and Whiles (2000) found ants and beetles to be especially important prey items, though many other invertebrates also are consumed. While Pigeon Mountain salamanders are often opportunistic feeders "sitting and waiting" for prey (J.B.J., personal observations), the presence in dissected stomachs of larval beetles and flies associated with rotting logs, dung, and other moist substrates indicates that prey are also actively searched for and removed from such substrates. Larval insects associated with small, stagnant pools of water are also found in dissected stomachs, suggesting that Pigeon Mountain salamanders also forage in aquatic microhabitats.

   O. Predators. Unknown.

   P. Anti-Predator Mechanisms. Pigeon Mountain salamanders produce slimy tail secretions, similar to those found in the *P. glutinosus* complex, that may be noxious or irritating to predators (J.B.J., personal observations).

   Q. Diseases. Unknown.

   R. Parasites. Unknown.

4. Conservation.

Listed as Rare and thus protected by the State of Georgia. The restricted distribution of Pigeon Mountain salamanders makes them especially vulnerable to threats. The vast majority of their range is under public ownership as the Crockford-Pigeon Mountain Wildlife Management Area. However, mineral rights to a portion of this property are leased to a mining company that has proposed quarrying operations, which may threaten both this species and green salamanders, another rare amphibian. Over-collection for scientific study and possibly the illegal pet trade, as well as disturbance from recreational spelunkers, may threaten populations. Loss or reduction of moisture-trapping canopy cover as a result of timber removal on private lands could pose a future threat.

Plethodon punctatus Highton, 1972 "1971"
COW KNOB SALAMANDER

Joseph C. Mitchell, Thomas K. Pauley

1. Historical versus Current Distribution.

Cow Knob salamanders (*Plethodon punctatus*) are restricted to elevations >850 m on Shenandoah Mountain, Augusta and Rockingham counties, Virginia; Pendleton County and North Mountain in Hardy County, West Virginia; and Shenandoah County, Virginia (Highton, 1972, 1988a). Recent inventories have expanded the known range to include Hampshire County, West Virginia, down to 732 m elevation (Pauley, 1998; R. Highton, personal communication).

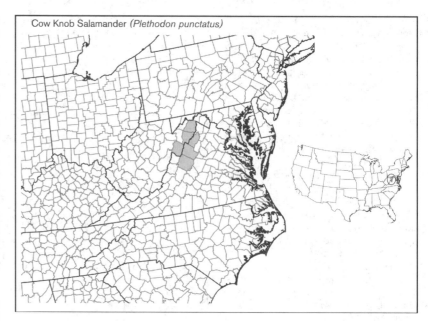

Cow Knob Salamander (*Plethodon punctatus*)

2. Historical versus Current Abundance.

Unknown historically. Cow Knob salamanders apparently are uncommon in many areas of their range but may be abundant at the surface in small pockets associated with ample rock cover and deep soils. Known density estimates range from 0.03–0.54/m^2 in Rockingham County, Virginia (Fraser, 1976b), to 1.62/m^2 in Pendleton County, West Virginia (Tucker, 1998).

3. Life History Features.

Petranka (1998) has summarized the life history and biology of Cow Knob salamanders.

   A. Breeding. Reproduction is terrestrial.

   i. Breeding migrations. This species does not migrate. Spermatophores are deposited in early spring. Mating probably occurs in spring and fall. Based on the presence of enlarged follicles, Tucker (1998) thought that egg laying occurs February–April (although R. Highton [personal communication] feels this is unlikely).

   ii. Breeding habitat. The breeding habitat of Cow Knob salamanders is unknown. This species may be more subterranean than other large *Plethodon* and may mate underground. Pairs of males and females have been found under rocks in spring and fall, suggesting that some mating occurs on the forest floor.

   B. Eggs.

   i. Egg deposition sites. Eggs are deposited beneath rocks or underground in deciduous forest habitats, but no nests have been reported for this species.

   ii. Clutch size. From 7–16 eggs.

   C. Direct Development.

   i. Brood sites.

   ii. Parental care. Unknown, but present in all known *Plethodon* in which nests have been found (R. Highton, personal communication).

   D. Juvenile Habitat. Young of the year emerge on the surface in September. No special habitat characteristics are known for juveniles, and they are apparently the same as adult habitat characteristics.

   E. Adult Habitat. Cow Knob salamanders have been observed in hemlock (*Tsuga canadensis*) stands, old-growth hardwoods, and mature hardwoods. They do not appear to be restricted to any one forest type as long as forest floor and subterranean features and canopy cover are present (Green and Pauley, 1987; Buhlmann et al., 1988; Tucker, 1998). Adults and juveniles are most commonly found under rocks in moist areas in deep soil on north-facing slopes above 914 m. This species had been found most often in mature and virgin hardwood forest patches. Individuals occasionally are found in younger hardwoods but not in pine forests, young hardwoods, or clearcuts.

   F. Home Range Size. Unknown. Movements are probably only a few meters in the lifetime of an individual. Buhlmann et al. (1988) reported movements of <2m to 17.4m. Tucker (1998) recaptured one individual that had moved 0.9m.

   G. Territories. The territorial behavior of this species has not been studied.

   H. Aestivation/Avoiding Desiccation. Surface abundance of Cow Knob salamanders is influenced by environmental conditions (Buhlmann et al., 1988; Tucker, 1998). Adults and juveniles are active during cool, wet periods in the spring (April–June) and autumn (September–October). Individuals remain underground during warm, dry months.

   I. Seasonal Migrations. This species does not exhibit seasonal migrations.

   J. Torpor (Hibernation). Cow Knob salamanders apparently are inactive November–March when surface temperatures are at or near freezing.

K. Interspecific Associations/Exclusions.
Other salamanders that occur sympatrically with Cow Knob salamanders are Jefferson salamanders (*Ambystoma jeffersonianum*), spotted salamanders (*A. maculatum*), Allegheny Mountain dusky salamanders (*Desmognathus ochrophaeus*), spring salamanders (*Gyrinophilus porphyriticus*), four-toed salamanders (*Hemidactylium scutatum*), eastern newt (*Notophthalmus viridescens*) efts, eastern red-backed salamanders (*P. cinereus*), white-spotted slimy salamanders (*P. cylindraceus*), Shenandoah Mountain salamanders (*P. virginia*), and red salamanders (*Pseudotriton ruber*; Mitchell et al., 1999). Fraser (1976a,b) determined that there was little competition between juvenile and adult Cow Knob salamanders and adult Shenandoah Mountain salamanders (then recognized as Valley and Ridge salamanders, *P. hoffmani*) for food and habitat characteristics. White-spotted slimy salamanders have been found syntopically with Cow Knob salamanders (Pauley, 1995b, 1998; Mitchell, 1996; Tucker, 1998), but interactions with this larger *Plethodon* have not been studied.

L. Age/Size at Reproductive Maturity.
Sexual maturity is reached in about 3 yr after hatching (Fraser, 1976a), when males are approximately 49 mm SVL and females are 59 mm SVL (Tucker, 1998).

M. Longevity. Length of life is unknown but is probably similar to that of other large Plethodon—about 15 yr (Snider and Bowler, 1992).

N. Feeding Behavior. Cow Knob salamanders are opportunistic carnivores on and above the forest floor during wet conditions. Most active foraging apparently occurs at night. Individuals have been observed to climb on tree trunks and rocks at night (Buhlmann et al., 1988). It is unknown whether foraging takes place under leaf litter or underground during dry periods, but it is likely that most energy consumption occurs during wet weather, as in other terrestrial salamanders.

Adults and juveniles prey on a wide variety of invertebrates, including ants, collembolans, beetles, dipterans, coleopterans, orthopterans, insect larvae, millipedes, centipedes, spiders, and mites (Fraser, 1976b; Tucker, 1998). The size of the prey item is positively correlated with the size of the salamander.

O. Predators. Not reported, but Petranka (1998) suggested shrews, small birds, woodland snakes, opossums, and skunks.

P. Anti-Predator Mechanisms. Mechanisms to fend off predators are unknown, but mucous secretions in the tail may inhibit swallowing by small predators such as snakes (e.g., *Diadophis* sp. and *Lampropeltis* sp.) and shrews.

Q. Diseases. No diseases have been reported in this species.

R. Parasites. *Cepedietta michiganensis* (a protozoan) and *Batracholandros magnavul-* *varis* (a nematode) are known to occur in the digestive tract (Tucker, 1998), but the etiology of these parasites is unknown.

4. Conservation.
The George Washington National Forest, in which most of the range of Cow Knob salamanders occurs, recognizes this as a Sensitive species. Most of the land >914 m has been allocated to Management Area 4 (areas off limits to logging) and has been designated as the Shenandoah Mountain Crest Special Biological Area. A formal Conservation Agreement exists, via a Memorandum of Understanding between the George Washington National Forest and the U.S. Fish and Wildlife Service, that affords the habitat of this species on public lands some protection from logging and other potentially damaging operations (Mitchell, 1994b). Cow Knob salamanders are listed as a Species At Risk by the U.S. Fish and Wildlife Service and a Species of Special Concern in Virginia and West Virginia (Mitchell et al., 1999). Cow Knob salamanders may be threatened by logging operations and the loss of hemlock trees by the introduced hemlock woolly adelgid (*Adelges tsugae*), and potentially threatened by defoliation of canopy hardwood trees by the introduced gypsy moth (*Lymantria dispar*).

Plethodon richmondi Netting
and Mittleman, 1938
SOUTHERN RAVINE SALAMANDER

Thomas K. Pauley, Mark B. Watson

1. Historical versus Current Distribution.
Netting and Mittleman (1938) described southern ravine salamanders (*Plethodon richmondi*) from specimens collected at Ritter Park in Cabell County, West Virginia. Early distribution records indicated that ravine salamanders were limited to Pennsylvania, Ohio, Kentucky, Indiana, and West Virginia (Netting and Mittleman, 1938; Netting, 1939; Bishop, 1943). Netting and Mittleman (1938) reported the range from Bedford and Allegheny counties in Pennsylvania, south to Grant, Cabell, and Wayne counties in West Virginia, and west into southeastern Ohio to Hamilton County. Bishop (1943) included eastern Kentucky within their range, and Grobman (1944) listed them in western Virginia. Netting (1939) described a limited distribution (in just three counties) in Pennsylvania. Highton (1962a) described the total range from Centre County, Pennsylvania, south through Maryland, West Virginia, western Virginia, northwestern North Carolina, and northeastern Virginia, west to Ohio, southeastern Indiana, and eastern Kentucky. Wallace (1969) reported specimens in 31 counties in eastern Kentucky. Barbour (1971) showed a slightly western expansion of the range, from Whitley County northeast to Jefferson County. Prior to 1972, the total range of southern ravine salamanders was considered to extend in the West from Lake Erie in north-central Ohio southwest into extreme eastern Indiana, eastern Kentucky, western Virginia, extreme northeastern Tennessee, and extreme northwestern North Carolina. In the East, they were thought to extend from the New River in West Virginia and Virginia north to the Susquehanna River Valley in Pennsylvania. Highton (1971) described individuals in the East as Valley and Ridge salamanders (*P. hoffmani*). Highton

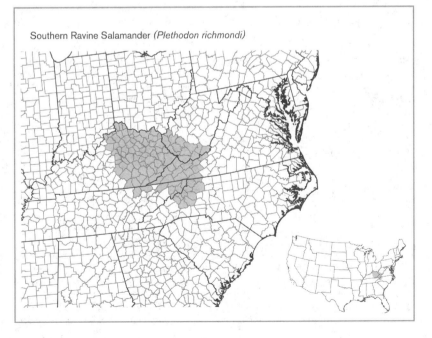

Southern Ravine Salamander (*Plethodon richmondi*)

(1999a) split the western populations into two species. Those north of the Kanawha and Ohio rivers were described as northern ravine salamanders *(P. electromorphus)* and those south of these rivers remained *P. richmondi* (Highton, 1999a; Regester, 2000a,b). Southern ravine salamanders reach 1,300 m in elevation on Big Black Mountain in Harlan County, Kentucky (Barbour, 1953).

The distribution of southern ravine salamanders has been updated for Tennessee (Redmond and Scott, 1996) and for Virginia (Mitchell and Reay, 1999).

2. Historical versus Current Abundance.

Little quantitative historical information is available for southern ravine salamanders. Welter and Barbour (1940) reported southern ravine salamanders to be as common as eastern red-backed salamanders *(P. cinereus)* in Rowan and Carter counties in eastern Kentucky. In Tazewell County, Virginia, Hoffman and Hubricht (1954) reported that they observed over 30 specimens in March in <1 hr. Ten days later they found another large number at a nearby locality. They reported that salamanders were particularly abundant in a nearly cleared pasture, under logs and split fence rails, but not flat stones. Studies of abundance of northern ravine salamanders in West Virginia were conducted by Jewell (1991) and Kramer (1996) prior to the recognition of northern ravine and southern ravine salamanders into separate species (Highton, 1999a).

3. Life History Features.
   A. Breeding. Reproduction is terrestrial.
   i. Breeding migrations. This species does not migrate. Females exhibit a biennial reproductive cycle (Nagel, 1979; Jewell and Pauley, 1995).
   ii. Breeding habitat. As with most *Plethodon*, breeding probably takes place at night on the forest floor. Southern ravine salamanders mate from fall to early spring. In the northeastern Tennessee populations, southern ravine salamanders mate from November–March, and egg deposition occurs in May (Nagel, 1979). Spermatophore deposition in West Virginia populations occurs in April–May, and egg deposition occurs in April (Jewell and Pauley, 1995).
   B. Eggs.
   i. Egg deposition sites. Nests are located in deep underground cavities (Petranka, 1979, 1998) or beneath rocks buried deeply in the soil (Wallace and Barbour, 1957). Egg deposition in Kentucky populations occurs from May to early June (Wallace, 1969). Eggs hatch in early fall, but most neonates do not migrate to the surface until the following spring (Nagel, 1979). A juvenile (23 mm TL) was reported in West Virginia in mid October (Netting

and Mittleman, 1938). Two adult southern ravine salamanders with two eggs and two newly hatched juveniles were found in late August in Kentucky (Wallace and Barbour, 1957).
   ii. Clutch size. Clutch sizes have been determined primarily on counts of mature follicles. Reports of mature follicles vary from 8.5 in Kentucky (Wallace, 1969) to 8.3 in Tennessee (Nagel, 1979).
   C. Direct Development.
   i. Brood sites. In deep underground cavities (Petranka, 1979, 1998) or beneath rocks buried deep in the soil (Wallace and Barbour, 1957).
   ii. Parental care. Likely; present in all *Plethodon* species where clutches have been observed.
   D. Juvenile Habitat. Juveniles have been found in the same habitat as adults (Netting and Mittleman, 1938).
   E. Adult Habitat. Habitat includes steep to sloping hillsides and ravines, and mesic forests with flat rocks, rock outcrops, logs, and abundant leaf litter (Wallace, 1969; Barbour, 1971; Martof et al., 1980; Green and Pauley, 1987; Redmond and Scott, 1996). Southern ravine salamanders are less common in floodplains or on dry ridge tops (Netting, 1939). They are found occasionally in pastures adjacent to wooded areas (Hoffman and Hubricht, 1954).
   F. Home Range Size. Unknown. Movements probably are similar to eastern red-backed salamanders (Kleeberger and Werner, 1982) and northern ravine salamanders (Jewell, 1991).
   G. Territories. In laboratory studies, southern ravine salamanders were found not to be as aggressive in defending territories (Thurow, 1976). Jewell (1991) found that in a laboratory setting, northern ravine salamanders are more territorial than sympatric eastern red-backed salamanders.
   H. Aestivation/Avoiding Desiccation. Southern ravine salamanders are most active above ground during the fall, winter, and spring. During the hot summer months of June–September, they move to underground refugia (Nagel, 1979; Green and Pauley, 1987).
   I. Seasonal Migrations. Southern ravine salamanders do not migrate.
   J. Torpor (Hibernation). Southern ravine salamanders remain active on the surface during mild winters. Petranka (1979) observed that they can be abundant during winter thaws at shallow depths beneath rocks near talus slopes. After a severe winter in central Kentucky, Petranka (1979) found few southern ravine salamanders at sites that lacked talus. During the previous autumn, he found them to be common at these sites. He postulated that they may have frozen during the severe weather. Fat for winter inactivity is known to be stored in the first 18 mm of the tail of most *Plethodon* (Fraser, 1980). Southern ravine salamanders found in the fall after emerg-

ing from summer refugia (aestivation) have been reported to have tails smaller in diameter than those salamanders collected in the spring (Netting, 1939; Green and Pauley, 1987). This would suggest that southern ravine salamanders feed more during the winter than summer, thus supporting the idea that they are active during the winter.
   K. Interspecific Associations/Exclusions. Southern ravine salamanders are sympatric with eastern red-backed salamanders, Wehrle's salamanders *(P. wehrlei)*, and Cumberland Plateau salamanders *(P. kentucki)* in West Virginia (Waldron et al., 2000). In Kentucky and eastern Tennessee, their distribution overlaps with northern zigzag salamanders *(P. dorsalis;* Conant and Collins, 1998) and with northern ravine salamanders in northern Kentucky (Highton, 1999b). In the southern part of their range, southern ravine salamanders have been taken with Weller's salamanders *(P. welleri)*, Yonahlossee salamanders *(P. yonahlossee)*, and northern gray-cheeked salamanders *(P. montanus;* R. Highton, personal communication).
   L. Age/Size at Reproductive Maturity. Male southern ravine salamanders in Tennessee populations become sexually mature in the fall of their third year of life and most females in their fourth year, but some may reach reproductive maturity in their third year (Nagle, 1979). Mature males in Tennessee had a mean SVL of 48.1 mm and gravid females 50.6 mm (Nagle, 1979). In Kentucky, Barbour (1953) reported body lengths of males from 34–50 mm and females from 34–50 mm. In another study in Kentucky, Wallace (1969) reported that the mean SVL of gravid females ranged from 47.6–53.1 mm. The smallest mature gravid female was 35 mm (SVL). Wallace and Barbour (1957) found a female (46 mm SVL) in Kentucky under a rock with two eggs and two hatchlings.
   M. Longevity. Unknown. Probably similar to other small *Plethodon* (Snider and Bowler, 1992).
   N. Feeding Behavior. Little is known about prey items of southern ravine salamanders. Several food studies have been conducted on northern ravine salamanders, a sibling species. Major prey items of northern ravine salamanders include ants, sow bugs, dipteran larvae, springtails, mites, snails, beetle larvae and adults, earthworms, termites, spiders, and larval ticks (Seibert and Brandon, 1960; Minton, 1972, 2001; Kramer, 1996).
   O. Predators. Undocumented, but undoubtedly include woodland snakes, birds, and small mammals.
   P. Anti-Predator Mechanisms. Southern ravine salamanders, like most small *Plethodon*, display an immobile, coiled, or contorted posture that makes them less obvious to predators when cover objects such as litter, logs, rocks, etc. are removed (Dodd, 1990c). Southern ravine salamanders also

avoid substrates with odors from ring-necked snakes (*Diadophis punctatus*; Cupp, 1994).

Q. Diseases. Diseases of this species have not been reported.

R. Parasites. Parasites that have been identified include *Batracholandros salamandrae* (Baker, 1987), *Thelandros salamadrae*, and *Cylindrotaenia americana* (Dunbar and Moore, 1979).

4. Conservation.
Mitchell et al. (1999) listed the conservation status of southern ravine salamanders as Scientific Interest in Tennessee because of their limited distribution in the Blue Ridge Mountain, Ridge and Valley and Cumberland Plateau. Petranka (1998) listed deforestation and urbanization as primary factors that have eliminated local populations of southern ravine salamanders throughout their range.

Plethodon savannah Highton, 1989
SAVANNAH SLIMY SALAMANDER

David A. Beamer, Michael J. Lannoo

1. Historical versus Current Distribution.
Savannah slimy salamanders (*Plethodon savannah*) are known only from Burke, Jefferson, and Richmond counties in Georgia (Highton, 1989). There is no evidence to suggest that their current distribution differs from their historical distribution.

2. Historical versus Current Abundance.
Highton (this volume, Part One) sampled a population in Richmond County, Georgia, between 1979 and 1988, and again

in 1995 under similar conditions using similar effort and found evidence of a precipitous decline. Whether this decline is a true decline or within the range of natural variation will only be known with additional monitoring. A housing development destroyed most of the habitat at the type locality (R. Highton, personal communication).

3. Life History Features.
Savannah slimy salamanders were recently described (Highton, 1989). In this time, there has been no published work done on this species. As a portion of his larger research program, R. Highton has collected basic life history and natural history information on Savannah slimy salamanders and has plans to publish these data in a monographic treatment.

A. Breeding. Reproduction is terrestrial.

i. Breeding migrations. Undocumented, but breeding migrations are unknown for any *Plethodon* species.

ii. Breeding habitat. Unknown.

B. Eggs.

i. Egg deposition sites. Undocumented.

ii. Clutch size. Unknown.

C. Direct Development.

i. Brood sites. Unknown, but are likely to be the same as egg deposition sites.

ii. Parental care. Unknown, but it is likely that females brood, as with other members of the slimy salamander complex.

D. Juvenile Habitat. Unknown, but as with other *Plethodon*, is likely to be similar to adult habitat.

E. Adult Habitat. Forest floor habitats are important. Logs and rocks are used as cover objects during daylight hours. Animals are active at night, and more active under moist conditions.

F. Home Range Size. Unknown, but small home ranges are typical for *Plethodon* species.

G. Territories. At least some members of the *P. glutinosus* complex aggressively defend territories (Thurow, 1976); it is unknown whether Savannah slimy salamanders establish and defend territories.

H. Aestivation/Avoiding Desiccation. Generally unknown, but Savannah slimy salamanders likely avoid desiccating conditions by seeking shelter in underground sites.

I. Seasonal Migrations. Unknown, but vertical migrations from surface sites to underground sites and back again are likely to be important in surviving seasonally variable conditions.

J. Torpor (Hibernation). Generally unknown, but Savannah slimy salamanders likely avoid cold conditions by seeking shelter in underground sites.

K. Interspecific Associations/Exclusions. Undocumented.

L. Age/Size at Reproductive Maturity. Unknown.

M. Longevity. Unknown.

N. Feeding Behavior. Unknown, but as with other *Plethodon* species, feeding likely takes place at night under moist conditions. Prey items likely include a range of invertebrates, especially insects.

O. Predators. Undocumented, but likely to include forest snakes, birds, and small mammals.

P. Anti-Predator Mechanisms. Nocturnal. Secretive. All members of the genus *Plethodon* produce noxious skin secretions (Brodie, 1977).

Q. Diseases. Unknown.

R. Parasites. Unknown.

4. Conservation.
Savannah slimy salamanders are not protected in Georgia, the only state within their range. Among members of the *P. glutinosus* complex, Savannah slimy salamanders have one of the smallest distributions. The entire range of this salamander is contained within Burke, Jefferson, and Richmond counties. Within this range, there do not appear to be any federal or state properties that would preserve suitable habitat for these salamanders.

The extremely small distribution and lack of preserved forested habitats are a concern for the conservation of Savannah slimy salamanders. Further field and laboratory work are required to determine the maximum extent of this salamander's range. Only then can an accurate assessment of their conservation status be made.

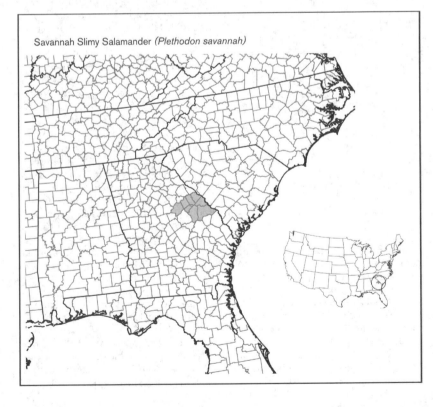

Savannah Slimy Salamander (*Plethodon savannah*)

Acknowledgments. Thanks to Richard Highton, who reviewed this account and gave us the benefit of his insight and experience.

Plethodon sequoyah Highton, 1989
SEQUOYAH SLIMY SALAMANDER

Carl D. Anthony

1. Historical versus Current Distribution.
Sequoyah slimy salamanders *(Plethodon sequoyah)* were previously considered a form of Rich Mountain salamanders (*P. ouachitae*; Dundee, 1947; Pope and Pope, 1951) or northern slimy salamanders (*P. glutinosus*; Blair and Lindsay, 1965). They were originally described by Highton et al. (1989) as a member of the *glutinosus* species complex most closely allied to Louisiana slimy salamanders (*P. kisatchie*). The only reported locality for Sequoyah slimy salamanders is Beavers Bend State Park in southeastern Oklahoma (Black and Sievert, 1989; Highton et al., 1989; Huntington et al., 1993), but specimens that presumably belong to this species have been taken from outside the park (Dundee, 1947; Blair and Lindsay, 1961), in the southern part of McCurtain County.

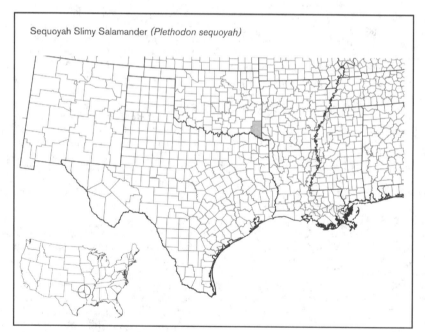

Sequoyah Slimy Salamander *(Plethodon sequoyah)*

2. Historical versus Current Abundance.
Unknown. Currently, they are locally common within their limited range.

3. Life History Features.
Little is known of the life history of Sequoyah slimy salamanders. Life history features are presumably similar to that of related forms.

A. Breeding. Reproduction is terrestrial.
i. Breeding migrations. Unknown.

ii. Breeding habitat. Breeding habitat is unknown.

B. Eggs.
i. Egg deposition sites. Unknown, but probably similar to that of other members of the species group.

ii. Clutch size. Unknown, but probably similar to that of other members of the species group.

C. Direct Development. No reports of egg size or number.

D. Juvenile Habitat. Probably the same as adult habitat.

E. Adult Habitat. Sequoyah slimy salamanders are most commonly found in moist woods (Black and Sievert, 1989) and ravines. Rocks and logs are common cover objects.

F. Home Range Size. Unknown.

G. Territories. Unknown.

H. Aestivation/Avoiding Desiccation. Aestivation is unknown. Animals likely avoid desiccating conditions by moving under cover objects or into burrows.

I. Seasonal Migrations. Unknown.

J. Torpor (Hibernation). Probably hibernate from November to late March.

K. Interspecific Associations/Exclusions. Occur sympatrically with southern red-backed salamanders (*P. serratus*).

L. Age/Size at Reproductive Maturity. Adults range in size from 46–70 mm SVL (Huntington et al., 1993).

M. Longevity. Unknown.

N. Feeding Behavior. Unknown, but prey likely consists of small invertebrates such as worms, insects, and spiders.

O. Predators. Unknown.

P. Anti-Predator Mechanisms. Nocturnal. All *Plethodon* produce noxious skin secretions (Brodie, 1977). When handled, Sequoyah slimy salamanders like other members of the *glutinosus* complex, release an adhesive secretion.

Q. Diseases. Unknown.

R. Parasites. Unknown.

4. Conservation.
Sequoyah slimy salamanders have an extremely restricted distribution; their only reported locality is Beavers Bend State Park in southeastern Oklahoma, but specimens that presumably belong to this species have been taken from outside the park (see "Historical versus Current Abundance" above). They are currently considered locally common within their limited range.

Plethodon serratus Grobman, 1944
SOUTHERN RED-BACKED SALAMANDER

David A. Beamer, Michael J. Lannoo

1. Historical versus Current Distribution.
Southern red-backed salamanders (*Plethodon serratus*) occur in four disjunct isolates: (1) southeastern Missouri and western Illinois; (2) northwestern Georgia, eastern Alabama, eastern Tennessee, and western North Carolina; (3) central Louisiana; and (4) southeastern Oklahoma and western Arkansas (Petranka, 1998; Smith, 1998). Populations occur as high as 1,690 m (Huheey and Stupka, 1967).

2. Historical versus Current Abundance.

Populations in areas where hardwood forests have been converted to intensively managed pine forests are compromised (Petranka, 1998). Highton (this volume, Part One) surveyed six populations in Arkansas, Missouri, and Oklahoma prior to 1987, and again during or after 1990 and found evidence of precipitous (<50% of original numbers) declines in five populations. Further monitoring of these populations will be necessary to determine whether these data reflect true declines or natural fluctuations.

3. Life History Features.
A. Breeding. Reproduction is terrestrial. Females in the Georgia Piedmont have spermatophores in their cloacae from February–March, while females in the southern Blue Ridge have spermatophores in their cloacae during December (Camp, 1988).

i. Breeding migrations. Undocumented, but breeding migrations are not known for any *Plethodon* species.

ii. Breeding habitat. Unknown.

B. Eggs. Eggs are laid in June and July (Camp, 1988; Taylor et al., 1990; Trauth et al., 1990).

i. Egg deposition sites. It is likely that eggs are deposited underground at depths of ≥ 1 m. In an experiment, two females deposited eggs when buried at a depth of

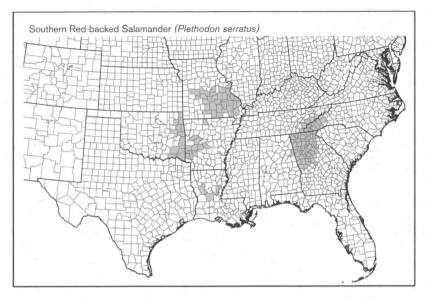

Southern Red-backed Salamander *(Plethodon serratus)*

1 m, but none of the females deposited eggs when buried at 0.5 m. These females resorbed the eggs and started developing new ova for the next season (Camp, 1988).

    ii. Clutch sizes. Females in Arkansas contained an average of 5.9 mature ova in one population and 7.0 in another (Taylor et al., 1990; Trauth et al., 1990). Females buried in boxes in the Georgia Piedmont each laid five eggs (4.5 mm diameter; Camp, 1988).

    C. Direct Development. A clutch from the Georgia Piedmont hatched on 11 September (Camp, 1988).

    i. Brood sites. It is likely that eggs are brooded underground at depths of ≥1 m (see "Egg deposition sites" above). In an experiment, two females deposited eggs when buried at a depth of 1 m, but none of the females deposited eggs when buried at 0.5 m . These females resorbed the eggs and started developing new ova for the next season (Camp, 1988).

    ii. Parental care. Unknown, but it is likely that females brood, as with other species of *Plethodon*.

    D. Juvenile Habitat. Juveniles likely stay at the nest site for several months after hatching. They are first found on the surface in autumn (Camp, 1988).

    E. Adult Habitat. In Natchitoches Parish, Louisiana, southern red-backed salamanders are found on slopes littered with sandstone rocks and with sandy/clay soils. Dominant vegetation includes longleaf pine *(Pinus palustris)*, yaupon *(Ilex vomitoria)*, oaks *(Quercus* sp.), and a variety of forbs and grasses (Keiser and Conzelmann, 1969).

    In the Great Smoky Mountains, southern red-backed salamanders are found in pine-oak or mixed hardwood forests between 457–1,686 m. Slopes having fairly dry soil seem to be preferred habitats (King, 1939; Huheey and Stupka, 1967).

In Arkansas, southern red-backed salamanders are found in leaf duff and under logs. They also are found in leaf packs along the margins of small streams (Thurow, 1957; D.A.B., personal observations).

    In the Georgia Piedmont, southern red-backed salamanders are found in hilly terrain with frequent steep bluffs covered by mesic deciduous forest. In this area, they are found beneath logs and rocks and in leaf litter, stump holes, and leaf packs in hillside seeps (Camp, 1986, 1988).

    F. Home Range Size. Unknown, but small home ranges are typical for *Plethodon* species.

    G. Territories. Unknown.

    H. Aestivation/Avoiding Desiccation. In eastern populations, southern red-backed salamanders are rarely found during the summer (Camp, 1988).

    I. Seasonal Migrations. Animals migrate from forest-floor sites to underground sites in response to seasonal drying and move back into forest-floor habitats in response to favorable surface conditions.

    J. Torpor (Hibernation). Southern red-backed salamanders frequently are active during the winter (Highton and Grobman, 1956; Camp, 1988).

    K. Interspecific Associations/Exclusions. Southern red-backed salamanders occur together with southern zigzag salamanders *(P. ventralis)* at Whiteoak Sink and the Sinks on Little River in Great Smoky Mountains National Park (King, 1939; Huheey and Stupka, 1967). However, in both of the areas the 2 seem to replace each other altitudinally between 518–619 m. Southern zigzag salamanders are largely restricted to the lower elevations and southern red-backed salamanders occur at higher elevations. There is evidence of character displacement in this area, as southern zigzag salamanders are all of the unstriped phase and southern red-backed salamanders are all striped (Highton, 1972).

Following experimental exclusion studies in the Great Smoky Mountains and Balsam Mountains, Hairston (1981) concluded that there is little competition between southern red-backed salamanders and other large eastern *Plethodon* species.

    In the Great Smoky Mountains, the following species were encountered on experimental plots with southern red-backed salamanders: black-bellied salamanders *(Desmognathus quadramaculatus)*, seal salamanders *(D. monticola)*, Ocoee salamanders *(D. ocoee)*, imitator salamanders *(D. imitator)*, pigmy salamanders *(D. wrighti)*, Jordan's salamanders *(P. jordani)*, southern Appalachian salamanders *(P. teyahalee)*, spring salamanders *(Gyrinophilus porphyriticus)*, red salamanders *(Pseudotriton ruber)*, and Blue Ridge two-lined salamanders *(Eurycea wilderae*; Hairston, 1980b, 1981).

    In the Balsam Mountains, the following species were encountered on experimental plots with southern red-backed salamanders: black-bellied salamanders, seal salamanders, Ocoee salamanders, pigmy salamanders, southern gray-cheeked salamanders *(P. metcalfi)*, southern Appalachian salamanders, spring salamanders, red salamanders, Blue Ridge two-lined salamanders, and eastern newts *(Notophthalmus viridescens*; Hairston, 1980b, 1981).

    Ash (1997) reports the following salamanders along with southern Appalachian salamanders from near Highlands, Macon County, North Carolina: Ocoee salamanders, southern red-backed salamanders, southern gray-cheeked salamanders, and Blue Ridge two-lined salamanders (Ash, 1997).

    Southern red-backed salamanders from Natchitoches and Rapides parishes, Louisiana, were found with dwarf salamanders *(E. quadridigitata*; Keiser and Conzelmann, 1969).

    In Polk County, Arkansas, southern red-backed salamanders were frequently found in wet leaf packs along the margins of small streams in May. Many-ribbed salamanders *(E. multiplicata)* and Ouachita dusky salamanders *(D. brimleyorum)* were also abundant in these leaf packs. Western slimy salamanders *(P. albagula)* and Rich Mountain salamanders *(P. ouachitae)* were found in nearby terrestrial habitat (D.A.B., personal observations).

    L. Age/Size at Reproductive Maturity. In populations from the Georgia Piedmont, males mature when 33–45 mm SVL, females between 33–47 mm. In populations from the Blue Ridge Mountains, adults are 33–39 m and 35–46 mm, respectively (Camp, 1988).

    M. Longevity. Unknown.

    N. Feeding Behavior. The following food items were found in 55 specimens (>30 mm SVL) collected during March–April from Fulton and Harris counties, Georgia: Gastropoda, Annelida, Acarina, Araneae, Diplopoda, Chilopoda, Isopoda, Collembola, Thysanura, Isoptera, Diptera,

Lepidoptera, Orthoptera, Coleoptera, Hymenoptera (ants), and unidentified larvae and adults (Camp and Bozeman, 1981).

O. Predators. Undocumented, presumably include forest mammals, birds, and snakes (Petranka, 1998).

P. Anti-Predator Mechanisms. All *Plethodon* produce noxious skin secretions (Brodie, 1977). Southern red-backed salamanders frequently become immobile when initially contacted. Immobility may increase survival by making the salamander less likely to be detected, especially by visually oriented predators (Dodd, 1989).

Q. Diseases. Unknown.

R. Parasites. Unknown.

4. Conservation.
Southern red-backed salamander populations occur in several isolates and are listed as a Species of Special Concern in Louisiana. Within the range of most isolates, there are federal and state properties that contain suitable habitat for these salamanders.

As with all species of *Plethodon*, southern red-backed salamanders do not migrate to breeding grounds and they do not have large home ranges. Thus, they can exist in habitats of smaller size than many other amphibian species. Conservation activities that promote mature closed-canopy forests should benefit this species.

Acknowledgments. Thanks to Richard Highton, who reviewed this account and gave us the benefit of his insight and experience.

Plethodon shenandoah Highton and Worthington, 1967
SHENANDOAH SALAMANDER

Joseph C. Mitchell

1. Historical versus Current Distribution.
Shenandoah salamanders (*Plethodon shenandoah*) are small members of the genus *Plethodon* known from only three small, isolated areas in Shenandoah National Park, Virginia. They occur on three separate mountains: Hawksbill, Stony Man, and The Pinnacles (Highton and Worthington, 1967; Jaeger, 1970; Highton, 1972, 1988b) that lie along the border of Madison and Page counties, Virginia. Their historical distribution is unknown.

2. Historical versus Current Abundance.
Historical abundance is unknown. Current abundance apparently varies depending on microhabitat location. In the area of sympatry with eastern red-backed salamanders (*P. cinereus*) outside of the talus, Shenandoah salamander density is low (about 0.1 individuals/m^2; Griffis and Jaeger, 1998). Within-talus habitat densities are unknown but are probably higher in allopatry.

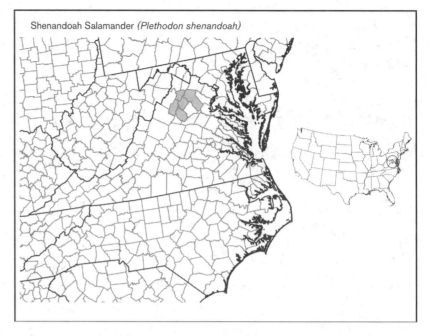

Shenandoah Salamander *(Plethodon shenandoah)*

3. Life History Features.
As with all members of the genus *Plethodon*, Shenandoah salamanders are terrestrial breeders and exhibit direct development (Wynn, 1991; Petranka, 1998).

A. Breeding. Reproduction is terrestrial.

i. *Breeding migrations.* Shenandoah salamanders do not migrate.

ii. *Breeding habitat.* The breeding microhabitat of this species is unknown. Mating probably occurs in spring and autumn.

B. Eggs.

i. *Egg deposition sites.* Eggs probably are laid in moist crevices underground (Jaeger, 1971a).

ii. *Clutch size.* Clutch size is 4–19 (average = 13), and females may produce eggs on a 2-yr cycle (Jaeger, 1980a, 1981b).

C. Direct Development. All larval development occurs in the egg. Young of the year emerge in September. Parental care of eggs is unknown.

D. Juvenile Habitat. Similar to adult habitat.

E. Adult Habitat. Shenandoah salamanders occupy relatively dry, north-facing talus slopes in each of the three known areas above 900 m elevation. Moist soil pockets and rocks characterize these habitats, and talus slopes support varying densities of hardwood trees. Jaeger (1970) described three microhabitats in the talus of Hawksbill Mountain that correlated with the distribution of Shenandoah salamanders. Type I talus is comprised of areas with rock and no soil; Type II talus has soil and leaf litter among rocks but not beneath them; and Type III talus has islands of soil among and under rocks. Shenandoah salamanders are most abundant in Type III talus, less so in Type II, and do not occur in Type I talus. Shenandoah

salamanders could occupy deeper, moister soils outside the talus areas were it not for direct behavioral competition with eastern red-backed salamanders. Shenandoah salamanders have a higher tolerance to desiccation and are able to survive within drier talus slopes that are inhospitable to eastern red-backed salamanders (Jaeger, 1971a). However, Shenandoah salamanders face a higher risk of mortality due to environmental perturbations that may cause drying in the talus (Jaeger, 1980a).

F. Home Range Size. Unknown.

G. Territories. Shenandoah salamanders are poor competitors compared to eastern red-backed salamanders (Griffis and Jaeger, 1998). Eastern red-backed salamanders competitively exclude Shenandoah salamanders from moist soil pockets and effectively restrict them to drier talus (Jaeger, 1970, 1971a,b, 1972). Competition is mediated through interspecific aggression and territoriality (Thurow, 1976; Jaeger and Gergits, 1979; Griffis and Jaeger, 1998). Shenandoah salamanders are less aggressive than eastern red-backed salamanders in territorial defense (Wrobel et al., 1980). Individual Shenandoah salamanders seldom invade territories defended by eastern red-backed salamanders, and they tend to snap at, rather than bite, intruders in their territories (Gergits, 1982).

H. Aestivation/Avoiding Desiccation. Shenandoah salamanders are inactive during the dry summer and during periods of drought at other times of the year. Patterns of rainfall strongly influence surface activity.

I. Seasonal Migrations. Shenandoah salamanders do not migrate.

J. Torpor (Hibernation). Overwintering occurs below the frost line presumably in moist areas in deep rock crevices. Surface activity terminates in October and usually resumes in April (W.L. Witt, personal communication).

K. Interspecific Associations/Exclusions. Shenandoah salamanders occur largely allopatrically from eastern red-backed salamanders on two of the three isolates. Both species occur throughout the isolate on The Pinnacle (Highton and Worthington, 1967). Eastern red-backed salamanders are the major competitor of Shenandoah salamanders on the periphery of the Hawksbill and Stony Man isolates and exclude this species to areas of talus that eastern red-backed salamanders cannot tolerate due to the drier microhabitat (Jaeger, 1972). White-spotted slimy salamanders *(P. cylindraceus)* occur in the area but they are encountered rarely (Jaeger, 1971b).

L. Age/Size at Reproductive Maturity. Age and size at maturity are unknown.

M. Longevity. Unknown, but probably similar to that for other small *Plethodon* (3–5 yr; Snider and Bowler, 1992).

N. Feeding Behavior. Shenandoah salamanders probably forage on the surface of the leaf litter and on low vegetation during moist nights, similar to sympatric eastern red-backed salamanders (Jaeger, 1978). They resemble other plethodontid salamanders in that they do not specialize and instead prey on a wide variety of invertebrates small enough to be ingested. Size of prey is limited to mouth size in these gape-limited predators. A list of prey species is in Jaeger (1972) and includes a wide variety of insects, arachnids, centipedes, millipedes, and worms. Kaplan (1977) determined in laboratory experiments that Shenandoah salamanders ingested *Drosophila* prey at rates of 0.006–0.012 prey/s. Prey assimilation efficiency is inversely related to temperature, as efficiency declined from 91% at 10 °C to 79% at 20 °C (Bobka et al., 1981).

O. Predators. Direct observation of predation has not been reported. Potential predators known to occur within the range include ring-necked snakes *(Diadophis punctatus)*, short-tailed shrews *(Blarina brevicauda)*, brown thrashers *(Toxostoma rufum)*, and towhees *(Pipilo* sp.; Jaeger, 1971b).

P. Anti-Predator Mechanisms. Anti-predator mechanisms for Shenandoah salamanders include noxious skin secretions, immobility, and defensive postures (Brodie, 1977). Dodd (1989) determined the length of time salamanders remained immobile when disturbed in the field by close approach, through disturbance of nearby microhabitat, or when touched. Immobility times for Shenandoah salamanders averaged 5.9 s (range 1–36 s) and were significantly lower from times for seven other species of eastern *Plethodon*. One of 21 salamanders had an immobility time >180 s in duration.

Q. Diseases. None known.

R. Parasites. None reported.

4. Conservation.
Shenandoah salamanders were listed as Endangered by the Virginia Department of Game and Inland Fisheries on 1 October 1987 (Wynn, 1991) and as Endangered by the U.S. Fish and Wildlife Service on 18 August 1989 (USFWS., 1989a). The primary threat is interspecific competition by eastern red-backed salamanders that limits their use of optimal habitat. Additional threats include alteration of forest canopy cover from defoliation by the introduced gypsy moth *(Lymantria dispar)*, acid precipitation, and succession of the talus habitat (USFWS., 1994d). Jaeger (1980a) documented the extinction of one subpopulation due to summer drought. Griffis and Jaeger (1998) described how interspecific competition, through interspecific aggression and territoriality, can lead to extinctions of sink populations by preventing dispersal of Shenandoah salamanders from the primary source population. The federal recovery plan outlines objectives to minimize human impacts in the national park while still allowing natural processes to occur (USFWS, 1994d).

Plethodon shermani Stejneger, 1906
RED-LEGGED SALAMANDER

David A. Beamer, Michael J. Lannoo

1. Historical versus Current Distribution.
Red-legged salamanders *(Plethodon shermani)* are found in the Unicoi and Nantahala mountains in North Carolina. They are considered by Highton and Peabody (2000) to be the Standing Indian, Wayah, Tusquitee, and Unicoi Mountain isolates of the *P. jordani* complex. There is no evidence that their current distribution differs from their historical distribution.

2. Historical versus Current Abundance.
Highton (this volume, Part One) sampled two populations of red-legged salamanders in or prior to 1973, then re-sampled them in or after 1994 and found evidence of declines in both populations. In the populations from Macon County, North Carolina, red-legged salamander numbers dropped from 31.8/person/visit to 6.0/person/visit (similar sampling effort under similar conditions). Additional sampling will be necessary to determine if these are true declines or natural population fluctuations.

3. Life History Features.
  A. Breeding. Reproduction is terrestrial.
    i. Breeding migrations. Unlikely; breeding migrations are unknown in any *Plethodon* species.
    ii. Breeding habitat. Unknown.
  B. Eggs.
    i. Egg deposition sites. Unknown.
    ii. Clutch size. Unknown.
  C. Direct Development.
    i. Brood sites. Unknown.
    ii. Parental care. Unknown, but it is likely that females brood, as with other species of *Plethodon*.
  D. Juvenile Habitat. A single juvenile was found with 40 adults in a maple and birch forest on 30 May at Wayah Bald, Macon County, North Carolina (Wood, 1947b).
  E. Adult Habitat. Six specimens were collected from beneath stones and logs in a *Rhododendreon* thicket bordering a little stream at the base of a mountain (Bishop, 1928).

On Wayah Bald, Macon County, North Carolina, red-legged salamanders were found inside and under rotten logs, under sticks and bark, and beneath solid logs. They were often found under the loose bark of prostrate sticks and logs (C.H. Pope, 1928).

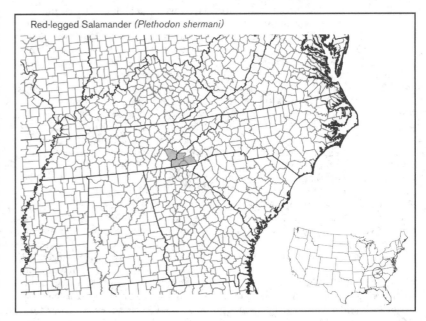

Red-legged Salamander *(Plethodon shermani)*

F. Home Range Size. Unknown, but small home ranges are typical for *Plethodon* species.

G. Territories. Selby et al. (1996) observed agonistic behavior in red-legged salamanders.

H. Aestivation/Avoiding Desiccation. Red-legged salamanders are active during the summer (C.H. Pope, 1928; Bailey, 1937; D.A.B., personal observations).

I. Seasonal Migrations. Animals likely make vertical migrations, moving from the forest floor to underground sites with the onset of seasonally related cold or dry conditions, then back up to the forest floor with the return of favorable surface conditions.

J. Torpor (Hibernation). During the winter when temperatures were below freezing, red-legged salamanders were not beneath surface cover such as flat stones and logs; in some instances, these exact stones and logs had sheltered salamanders during warmer periods. However, red-legged salamanders were found beneath deeply embedded rocks at this time (D.A.B., personal observations).

K. Interspecific Associations/Exclusions. The following salamanders are reported from Wayah Bald: red-legged salamanders, black-bellied salamanders *(Desmognathus quadramaculatus)*, and Ocoee salamanders *(D. ocoee;* Hairston, 1949). Blue Ridge two-lined salamanders *(Eurycea wilderae)* are also associated with red-legged salamanders on Wayah Bald (C.H. Pope, 1928; D.A.B., personal observations).

Red-legged salamanders contact Tellico salamanders *(P. aureolus)* on the west slope of the Unicoi Mountains, Monroe County, Tennessee. A transect on Sassafras Ridge indicates there is a wide hybrid zone (Highton and Peabody, 2000).

Red-legged salamanders are parapatric with Chattahoochee slimy salamanders *(P. chattahoochee)* near the Georgia–North Carolina border. A transect along the Tallulah River indicates there is a wide hybrid zone (Highton and Peabody, 2000).

Red-legged salamanders hybridize at all known contacts with southern Appalachian salamanders *(P. teyahalee).* In the area between the Standing Indian, Wayah, and Tusquitee isolates of red-legged salamanders, all populations appear to be hybrid swarms (Highton and Peabody, 2000).

In the Nantahala Mountains, North Carolina red-legged salamanders and southern Appalachian salamanders replace one another altitudinally. On two transects there is no elevational overlap, while at a third transect on this mountain there is vertical overlap of 61 m (200 ft; Hairston, 1951).

L. Age/Size at Reproductive Maturity. Unknown.

M. Longevity. Unknown.

N. Feeding Behavior. The following food items are reported from specimens from Grandfather Mountain, Avery County, North Carolina: Formicidae, Araneida, lepidopteran larvae, coleopteran larvae, Collembola, Diplopoda, Chilopoda, Acarina, Stylommatophora, dipteran larvae, Annelida, Tipulidae, Lepidoptera, shed skin, Ichneumonidae, Diptera, Gryllacrididae, Cicadellidae, Coleoptera, Cantharidae, Chrysomelidae, Isopoda, Cynipodea, Hymenoptera, Phoridae, Cydnidae, Scarabaeidae, Carabidae, Curculionidae, Chelonethida, Vespidae, Gryllidae, Simuliidae, Histeridae, Pentatomidae, Elateridae, Mycetophilidae, Culicidae, Aphididae, Phalacridae, Staphylinidae, Fungivoridae, Hemiptera, and Scaphididae. Ants, mites, and springtails were eaten less frequently by large individuals, while millipedes, earthworms, and craneflies were eaten more frequently. In general, the number of different food items increases in larger individuals (Whitaker and Rubin, 1971).

O. Predators. A spring salamander *(Gyrinophilus porphyriticus)* was found to have eaten a member of the *P. jordani* complex (either a southern gray-cheeked salamander *[P. metcalfi]* or a red-legged salamander; Bruce, 1972a).

P. Anti-Predator Mechanisms. Nocturnal. Secretive. All members of the genus *Plethodon* produce noxious skin secretions (Brodie, 1977). Members of the *P. jordani* complex frequently become immobile when initially contacted. Red-legged salamanders were included in a field study on immobility, however it is not possible to separate their behavior from the other members of this complex in this published data set. Immobility may increase survival by making the salamander less likely to be detected, especially by visually oriented predators (Dodd, 1989).

Q. Diseases. Unknown.

R. Parasites. Unknown.

4. Conservation.
Red-legged salamanders are not protected in North Carolina, the only state within their range. Among members of the *P. jordani* complex, red-legged salamanders have a relatively small and disjunct distribution. These salamanders are restricted to higher elevations and to suitable habitat at these elevations, and populations may be separated by stretches of lower uninhabited areas. Within their range, there are several federal and state properties that contain suitable habitat for these salamanders.

Red-legged salamanders are relatively resilient to disturbances such as those associated with timbering operations and frequently are found in second-growth forests and relatively small, fragmented woodlots (D.A.B., personal observations). Bishop (1928) reported red-legged salamanders to be scarce on the higher slopes of Wayah Bald, due to dry conditions that were exacerbated by razor-back hogs. Red-legged salamanders are now abundant there (D.A.B., personal observations), so it appears that populations on the upper slopes of Wayah Bald may be more robust than they were in 1926.

As with all species of *Plethodon*, red-legged salamanders do not migrate to breeding grounds and they do not have large home ranges. Thus, they can exist in habitats of smaller size than many other amphibian species. Conservation activities that promote mature closed-canopy forests should benefit this species.

Acknowledgments. Thanks to Richard Highton, who reviewed this account and gave us the benefit of his insight and experience.

Plethodon stormi Highton and Brame, 1965
SISKIYOU MOUNTAINS SALAMANDER
R. Bruce Bury, Hartwell H. Welsh, Jr.

1. Historical versus Current Distribution.
Siskiyou Mountains salamanders *(Plethodon stormi)* are restricted to the Siskiyou Mountains in the upper Klamath River, Siskiyou County, California, and the adjacent Applegate River watershed, Jackson and Josephine counties, Oregon (Brodie, 1970, 1971b; Highton and Larson, 1979; Bury, 1998; Bury and Pearl, 1999). Highton and Brame (1965) first described the species from five inland localities along the California–Oregon border. More recently, Siskiyou Mountains salamanders are known from 17 sites in Oregon and 10 in California (Bury, 1998). There are many new sites based on recent surveys (D. Clayton, R.B.B. and others; unpublished reports). The range of Siskiyou Mountains salamanders occupies an area of only about 35 × 75 km.

There is clinal variation in morphometric traits and color patterns of Del Norte salamanders *(P. elongatus)* from coastal to inland areas, and in the Upper Klamath River they closely resemble populations of Siskiyou Mountains salamanders (Bury, 1973b). Siskiyou Mountains salamanders appear to be the inland terminus of Del Norte salamanders and likely became separated from more coastal groups during glacial times, evolving in isolation into a new species (Bury, 1998). Earlier, Stebbins (1985) recognized these as two subspecies *(P. e. elongatus* and *P. e. stormi).* However, Highton and Larson (1979), Leonard et al. (1993), Highton (1995b), and Petranka (1998) continue to recognize Siskiyou Mountains salamanders as a distinct species; a view supported by recent molecular genetics analyses (Mahoney, 2001, 2004).

Thompson Creek is a drainage of the Applegate River, Oregon, where Siskiyou Mountains salamanders occur on the east side of the valley and Del Norte salamanders are 2 km away on the west side. There is no indication of intergradation in this area. However, on the southern flanks of

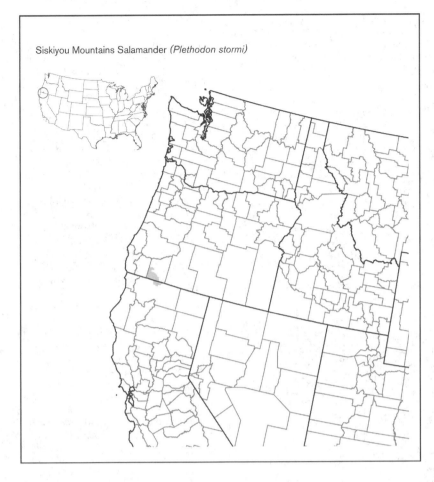

Siskiyou Mountains Salamander *(Plethodon stormi)*

the Siskiyou Mountains in California, Siskiyou Mountains salamanders may intergrade with Del Norte salamanders (Bury, 1998). Several studies on the genetic variation of *P. elongatus* and *P. stormi* are underway (Mead et al., in review; M. J. Mahoney, 2004).

2. Historical versus Current Abundance.
Siskiyou Mountains salamanders are listed as a Survey and Manage Species under the Northwest Forest Plan on federal lands, whereby ground-disturbing activities are not permitted where the species occurs or within a 33 m buffer around suitable habitat (Clayton et al., 1999). There has been considerable habitat loss and fragmentation due to past logging practices. This species, as with its sister species, Del Norte salamanders (Welsh and Lind, 1995), is highly associated with rocky talus slopes in areas of dense mature and late-seral forest (Welsh and Lind, 1995; Bury, 1998; Ollivier et al., 2001). Once considered a Federal Candidate Species for listing, Siskiyou Mountains salamanders are recognized as a species of Special Concern by Oregon and California.

3. Life History Features.
A. Breeding. Reproduction is terrestrial.
i. Breeding migrations. Unknown.
ii. Breeding habitat. Unknown.

B. Eggs.
i. Egg deposition sites. Unknown; no nests are known from the wild. Eggs are likely deposited deep in rocky substrates.
ii. Clutch size. Average clutch size of dissected females was 9 eggs with a range of 2–18 (Nussbaum et al., 1983).
C. Direct Development. Females may stay with nests until eggs hatch.
D. Juvenile Habitat. Same as adults.
E. Adult Habitat. Salamanders of all sizes and sexes occur in talus and rocky soils or slopes and occasionally are found under logs, leaf litter, and other cover if talus is nearby (Nussbaum et al., 1983; Bury, 1998). Heavily wooded, north-facing slopes with rocky talus contain the largest populations (Brodie, 1970; Nussbaum et al., 1983).
F. Home Range Size. Unknown, but related western red-backed salamanders *(P. vehiculum)* have a high site specificity, and mean distances between captures were 2.5 m for males and 1.7 m for females (Ovaska, 1988b). Similarly, the sibling species *P. elongatus* appears sedentary, rarely moving over 7.5 m^2 in 3 yr (Welsh and Lind, 1992).
G. Territories. Unknown.
H. Aestivation/Avoiding Desiccation. Salamanders are generally deep underground in dry summer conditions. During wet nights in summer, individuals may emerge from deeper locations to feed at the surface (Nussbaum et al., 1983).

I. Seasonal Migrations. Unknown.
J. Torpor (Hibernation). Inactive in cold winter months, but may appear on surface once soils warm above freezing.
K. Interspecific Associations/Exclusions. Co-occur with ensatina salamanders *(Ensatina eschscholtzii)*, clouded salamanders *(Aneides ferreus)*, and black salamanders *(A. flavipunctatus)*. No other species of *Plethodon* are known within the range of Siskiyou Mountains salamanders.
L. Age/Size at Reproductive Maturity. Mature males are 47–70 mm SVL, females are 56–70 mm SVL (Nussbaum et al., 1983).
M. Longevity. Unknown in the wild.
N. Feeding Behavior. Usually are sit-and-wait predators that dart out from cover to seize prey such as collembolans, termites, beetles, moths, spiders, and mites (Nussbaum et al., 1983).
O. Predators. None reported in literature, but likely are eaten by shrews, other carnivores, and garter snakes *(Thamnophis* sp.).
P. Anti-Predator Mechanisms. None reported.
Q. Diseases. Unknown.
R. Parasites. Unknown.

4. Conservation.
In 1993, Siskiyou Mountains salamanders were placed on the federal Survey and Manage species list, which was used for protecting Rare species under the Northwest Forest Plan. This status required surveying forest stands on federal lands before any proposed management activities and the protection of occupied sites with a 33-m or 1-tree-height (whichever is greater) protective buffer around suitable habitat where the species was detected. In June 2002, the status of the populations north of the crest of the Siskiyou Mountains (Applegate River drainage) was changed such that pre-project surveys are no longer required, but known sites are still supposed to be protected, as are any new sites that might be discovered serendipitously. South of the Siskiyou Mountain crest, pre-project surveys are still required. These changes were the result of new molecular genetics data indicating highly homogeneous populations north, and highly differentiated, as well as more geographically scattered, populations south, of the Siskiyou Mountain crest (Mead et al., in review). Mead et al. (in review) distinguish two lineages south of the Siskiyou Mountain crest that are distinct from the Applegate River drainage populations, with the southeasternmost populations (Scott Bar Mountain area of the Upper Klamath River) representing a lineage that is ancestral to all other *P. stormi* plus *P. elongatus*. Discovery and recognition of these new species will be of high conservation concern because they appear to occupy relatively small geographic pockets and occur in areas with resource extraction, primarily timber harvesting.

Plethodon teyahalee Hairston, 1950
SOUTHERN APPALACHIAN SALAMANDER

David A. Beamer, Michael J. Lannoo

1. Historical versus Current Distribution.
Southern Appalachian salamanders *(Plethodon teyahalee)* are distributed in the Blue Ridge Physiographic Province of southeastern Tennessee, southwestern North Carolina, northwestern South Carolina (Oconee, Pickens, and Anderson counties), and northeastern Georgia (Rabun County; Highton, 1987a; Petranka [as *P. oconaluftee*], 1998). Adults are found up to 1,550 m in elevation. There is no evidence that the current distribution differs markedly from the historical distribution.

Southern Appalachian Salamander *(Plethodon teyahalee)*

2. Historical versus Current Abundance.
Many southern Appalachian salamander populations occur on federal lands and thus receive some protection. Indeed, Hairston and Wiley (1993) monitored southern Appalachian salamander populations for almost two decades and found no evidence of declines. However, Highton (this volume, Part One), working more recently, has found evidence of declines in southern Appalachian salamander populations from Graham, Macon, and Madison counties in North Carolina, and from Monroe County in Tennessee, but not from Pickens County, South Carolina. Additional monitoring will be required to determine whether these data

reflect true declines or natural population fluctuations. Southern Appalachian salamanders disappeared from clearcut areas in Macon County, North Carolina (Ash, 1988).

3. Life History Features.
A. Breeding. Reproduction is terrestrial. Adults probably breed during the autumn (September–October, perhaps as early as August, but not earlier; R. Highton, personal communication) when Jordan's salamanders *(P. jordani)* breed, because these two species will hybridize.

i. Breeding migrations. Undocumented, but breeding migrations are not known for any *Plethodon* species.

ii. Breeding habitat. Unknown.

B. Eggs.
i. Egg deposition sites. Unknown, but likely to be in underground cavities.

ii. Clutch size. Unknown.

C. Direct Development.
i. Brood sites. Likely to be the same as egg deposition sites; presumably in chambers underground.

ii. Parental care. Unknown, but it is likely that females brood, as with other members of the slimy salamander complex.

D. Juvenile Habitat. During August at Farr Gap in Monroe County, Tennessee, juvenile southern Appalachian salamanders were commonly found in the same area as adults, but were frequently found under small superficial cover such as twigs

and bits of bark (D A. B., personal observations).

E. Adult Habitat. During the day, adults typically are found under cover objects (rocks and logs) in deciduous forests. At night, adults emerge to feed on the forest floor. As with other *Plethodon* species, activity corresponds with moisture levels.

F. Home Range Size. Adult male southern Appalachian salamanders in the Great Smoky Mountains have an average home range of 0.49 m^2. Adult females have an average home range of 1.03 m^2. Two- and three-year-olds have average home ranges of 0.37 m^2 and 0.06 m^2, respectively (Nishikawa, 1990).

G. Territories. At least some members of the *P. glutinosus* complex aggressively defend territories (Thurow, 1976); it is unknown whether southern Appalachian salamanders establish and defend territories.

H. Aestivation/Avoiding Desiccation. Undocumented, but *Plethodon* species will move from forest floor habitats to underground sites in response to desiccating surface conditions.

I. Seasonal Migrations. Huheey and Stupka (1967) reported finding slimy salamanders during every month of the year in the Great Smoky Mountains. These observations probably pertain to, at least in part, southern Appalachian salamanders. Animals likely make vertical migrations, moving from the forest floor to underground sites with the onset of seasonally related cold or dry conditions, then back up to the forest floor with the return of favorable surface conditions.

J. Torpor (Hibernation). Undocumented, but *Plethodon* salamanders move from forest-floor habitats to underground sites in response to cold surface conditions.

K. Interspecific Associations/Exclusions. In the Great Smoky Mountains, the following species were encountered on experimental plots with southern Appalachian salamanders: black-bellied salamanders *(Desmognathus quadramaculatus)*, seal salamanders *(D. monticola)*, Ocoee salamanders *(D. ocoee)*, imitator salamanders *(D. imitator)*, pigmy salamanders *(D. wrighti)*, southern red-backed salamanders *(P. serratus)*, Jordan's salamanders, spring salamanders *(Gyrinophilus porphyriticus)*, red salamanders *(Pseudotriton ruber)*, and Blue Ridge two-lined salamanders *(Eurycea wilderae)*. At plots in the Balsam Mountains, all the above species were also found, with the exception of imitator salamanders and Jordan's salamanders. However, southern gray-cheeked salamanders *(P. metcalfi)* and eastern newts *(Notophthalmus viridescens)* were also found in the Balsam Mountains (Hairston, 1980b, 1981).

In the vicinity of Santeetlah Creek, Graham County, North Carolina, southern Appalachian salamanders have been found with black-bellied salamanders, Ocoee salamanders, Santeetlah dusky salamanders

(D. santeetlah), seepage salamanders *(D. aeneus)*, southern red-backed salamanders, spring salamanders, and Blue Ridge two-lined salamanders,(D.A.B., personal observations).

Ash (1997) reports the following salamanders along with southern Appalachian salamanders from near Highlands, Macon County, North Carolina: Ocoee salamanders, southern red-backed salamanders, southern gray-cheeked salamanders, and Blue Ridge two-lined salamanders (Ash, 1997).

Southern Appalachian salamanders contact Chattahoochee slimy salamanders *(P. chattahoochee)* in Clay County, North Carolina. A transect in the vicinity of Hayesville indicates there is a narrow hybrid zone (Highton and Peabody, 2000).

Southern Appalachian salamanders are sympatric with Tellico salamanders *(P. aureolus)* throughout the range of Tellico salamanders. There is no morphological or genetic evidence of hybridization (Highton and Peabody, 2000).

Southern Appalachian salamanders contact northern slimy salamanders *(P. glutinosus)* on the western side of the Blue Ridge Mountains in Tennessee, southwest of the French Broad River. At one site in Polk County, Tennessee, they occur sympatrically and probably do not hybridize. However, at two other transects in Monroe and Sevier counties, Tennessee, there are narrow hybrid zones (Highton and Peabody, 2000).

An unusual association consisting of three members of the *P. glutinosus* complex occurs at a site in Polk County, Tennessee. Here, southern Appalachian salamanders, Tellico salamanders, and northern slimy salamanders occur sympatrically. There is no evidence of hybridization at this location (Highton, 1984).

Southern Appalachian salamanders probably contact white-spotted slimy salamanders *(P. cylindraceus)* along the North Carolina–South Carolina state line at the headwaters of the French Broad River. There are no data on their genetic interaction (Highton and Peabody, 2000).

An unusual association of large eastern *Plethodon* occurs in the vicinity of Rabun Bald, Rabun County, Georgia. Here, southern Appalachian salamanders are sympatric with Atlantic Coast slimy salamanders *(P. chlorobryonis)*; and there is no evidence of hybridization between these species. However, there is a wide hybrid zone between Atlantic Coast slimy salamanders and a member of the *P. jordani* complex, the southern gray-cheeked salamanders. Thus, two members of the *P. glutinosus* complex contact without hybridization in the same area in which one of them hybridizes with a member of the *P. jordani* complex. Southern Appalachian salamanders also contact Atlantic Coast slimy salamanders in northwestern South Carolina, where there is a parapatric hybrid

zone in Anderson and Abbeville (Highton and Peabody, 2000).

Southern Appalachian salamanders are widely sympatric with the Max Patch and Sandymush isolates of northern gray-cheeked salamanders *(P. montanus)*. There is no known morphological or genetic evidence of hybridization (Highton and Peabody, 2000).

Southern Appalachian salamanders are widely sympatric with the Blue Ridge, Balsam, and Cowee isolates of southern gray-cheeked salamanders. There is no evidence of hybridization except at one site along Alarka Creek in the northern Cowee isolate where there is considerable hybridization (Highton and Peabody, 2000).

Southern Appalachian salamanders are widely sympatric at intermediate elevations with the central and western portions of the Great Smoky isolate and Gregory Bald isolate of Jordan's salamanders. In these areas they rarely hybridize. In the eastern portion of the Great Smoky isolate of Jordan's salamanders, they replace each other altitudinally and hybridize extensively in a narrow contact zone (Highton and Peabody, 2000).

Southern Appalachian salamanders are sympatric throughout the range of Cheoah Bald salamanders *(P. cheoah)* and rarely hybridize. Instead of being replaced at higher elevations, southern Appalachian salamanders occur in sympatry to the top of the highest mountain in the Cheoah isolate (Highton and Peabody, 2000).

Southern Appalachian salamanders hybridize at all known contacts with red-legged salamanders *(P. shermani)*. In the area between the Standing Indian, Wayah, and Tusquitee isolates of red-legged salamanders, all populations appear to be hybrid swarms (Highton and Peabody, 2000).

Hairston (1980b, 1981) demonstrated competition between southern Appalachian salamanders and Jordan's salamanders in the Great Smoky Mountains, and between southern Appalachian salamanders and southern gray-cheeked salamanders in the Balsam Mountains. The number of southern Appalachian salamanders increased significantly at plots in the Great Smoky Mountains where Jordan's salamanders were removed. The removal of southern Appalachian salamanders from plots resulted in a significant increase in the proportion of young Jordan's salamanders. Similar results were obtained from the experiments in the Balsam Mountains involving southern Appalachian salamanders and southern gray-cheeked salamanders, except that it took a longer amount of time for the populations to respond to the absence of the other species. Hairston did not identify what resources are limiting, but hypothesizes that southern Appalachian salamanders may compete with members of the *P. jordani* complex for nesting sites.

Powders and Tietjen (1974) hypothesize that competition for food during the fall occurs between southern Appalachian salamanders and Jordan's salamanders. Because competition occurs during the fall, they reason that it would act earlier at higher elevations. This fall competition may exclude southern Appalachian salamanders from higher altitudes that are occupied exclusively by Jordan's salamanders.

In the Great Smoky Mountains, southern Appalachian salamanders and Jordan's salamanders replace one another altitudinally. At two of four transects there is no elevational overlap in the occurrence of these two species; while at two other transects there is elevational overlap of 3–8 m (Hairston, 1951).

In the Nantahala Mountains, North Carolina, southern Appalachian salamanders and red-legged salamanders replace one another altitudinally. On two transects there is no elevational overlap of the occurrence of these two species, while at a third transect on this mountain there is vertical overlap of 61 m (200 ft; Hairston, 1951).

L. Age/Size at Reproductive Maturity. Females first lay eggs when ≥5 yr old (Hairston, 1983).

M. Longevity. No published data exist on longevity (Petranka, 1998).

N. Feeding Behavior. Weller (1931) listed the following food items for an animal that may have been a southern Appalachian salamander: a millipede, an isopod, a mollusk, a beetle, a lepidopteran larva, ants, and flies (see Highton, 1987a, for a discussion).

The following food items were reported for southern Appalachian salamanders from Porters' Creek Trail in Great Smoky Mountains National Park: Gastropoda, Diplopoda, Chilopoda, Isopoda, Phalangidea, Pseudoscorpionida, Araneae, Acarina, Collembola, Homoptera, Hemiptera, Coleoptera, Diptera, Formicidae, non-formicid-Hymenoptera, and insect larvae. Variation observed in the diet may indicate seasonal availability of food items (Powders and Tietjen, 1974).

There is evidence that southern Appalachian salamanders occasionally eat other salamanders. The bones of a salamander were found in the gut of a specimen collected in Sevier County, Tennessee. The specific identity of the salamander could not be determined (Powders and Tietjen, 1974).

O. Predators. Undocumented, but likely include forest-dwelling snakes, birds, and small mammals.

P. Anti-Predator Mechanisms. Nocturnal. Secretive. All members of the genus *Plethodon* produce noxious skin secretions (Brodie, 1977). Members of the *P. glutinosus* complex frequently become immobile when initially contacted. Southern Appalachian salamanders were included in a field study on immobility; however, it is not possible to separate their behavior from the other members of this complex in this published data set. Immobility may increase survival by making the salamander less likely to be detected, especially by visually oriented predators (Dodd, 1989).

Q. Diseases. Unknown.

R. Parasites. Southern Appalachian salamanders are sometimes infected by the astomatous ciliate, *Cepedietta michiganensis*. Infection rates seem to decrease at higher altitudes and during the spring (Powders, 1970).

4. Conservation.
Southern Appalachian salamanders are not protected by any of the states within their range. Among members of the *P. glutinosus* complex, southern Appalachian salamanders have one of the smaller distributions. Within their range there are many federal and state properties that contain suitable habitat for these salamanders.

Southern Appalachian salamanders are relatively resilient to disturbances such as those associated with timbering operations and frequently are found in second-growth forests (D.A.B., personal observations).

As with all species of *Plethodon*, southern Appalachian salamanders do not migrate to breeding grounds and they do not have large home ranges. Thus, they can exist in habitats of smaller size than many other amphibian species. Conservation activities that promote mature closed-canopy forests should benefit this species.

Acknowledgments. Thanks to Richard Highton, who reviewed this account and gave us the benefit of his insight and experience.

Plethodon vandykei Van Denburgh, 1906
VAN DYKE'S SALAMANDER

David A. Beamer, Michael J. Lannoo

1. Historical versus Current Distribution.
The distribution of Van Dyke's salamanders (*Plethodon vandykei*) is discontinuous, with isolated localities in western and west-central Washington (Brodie, 1970; Wilson et al., 1995; Petranka, 1998). Four disjunct populations are known, including in (1) the Olympic Mountains of western Washington (Clallam, Jefferson, and north-eastern Gray's Harbor counties in Washington); (2) the Willapa Hills of southwest Washington (Pacific and Wahkiakum counties, Washington); (3) the Cascade Mountains of west-central Washington (Pierce, Mason, and Lewis counties, Washington); and (4) southern and west-central Washington (Skamania and Thurston counties). Populations are distributed from sea level to 1,560 m. There is no evidence to suggest that the current distribution differs from the historical distribution.

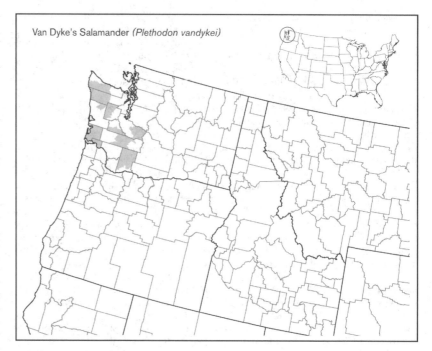

Van Dyke's Salamander (*Plethodon vandykei*)

2. Historical versus Current Abundance.
Van Dyke's salamanders are patchily distributed and generally uncommon (Brodie, 1970; Petranka, 1998). They do not fare well in intensively managed forests (Welsh, 1990).

3. Life History Features.
Most aspects of the natural history of adult and juvenile Van Dyke's salamanders have not been well documented (Petranka, 1998).

A. Breeding. Reproduction is terrestrial.

i. Breeding migrations. Undocumented, but breeding migrations are not known for any *Plethodon* species. Courtship and a description of the spermatophore are described by Lynch (1987).

ii. Breeding habitat. Unknown.

B. Eggs. A female collected on 17 April from Lewis County, Washington, contained eggs (2.6 mm in diameter), and a female collected on 1 July from Pierce County, Washington, contained eggs (2.4 mm in diameter; Stebbins, 1951). A recently deposited clutch (5.3 mm in diameter) was discovered on 21 May in Mason County, Washington (Jones, 1989).

i. Egg deposition sites. Only two nests of Van Dyke's salamanders have been discovered. The first was attached to a moss-covered stone in a damp location (Noble, 1925) and the second was located inside a partially rotted log (85 cm in diameter and 5 m in length) in an old-growth forest on a north-northeast facing slope (Jones, 1989).

ii. Clutch size. Ova counts were 11 and 14 from two females examined by Stebbins (1951). A nest in Mason County, Washington, contained seven eggs (Jones, 1989).

C. Direct Development. Hatchlings range in size from 15–18 mm SVL (Nussbaum et al., 1983).

i. Brood sites. A female in Mason County, Washington, was found brooding a clutch in a rotted log (Jones, 1989).

ii. Parental care. An adult female was found within 4 cm of a clutch discovered in Mason County, Washington (Jones, 1989).

D. Juvenile Habitat. A juvenile from Lewis County, Washington, was collected in rocks with little soil approximately 30 cm from the edge of a river (Stebbins, 1951).

E. Adult Habitat. Van Dyke's salamanders are found beneath moist stones and moss near running water and seeps (occasionally under bark) associated with coniferous forests (Slater, 1933; Bishop, 1943; Stebbins, 1951; Brodie, 1970). Adults are most active on the surface of the forest floor in spring and fall (Petranka, 1998). Populations generally are found in regions that receive >1.5 m of annual rainfall and have an upper elevational limit at the lower edge of subalpine forests (Wilson et al., 1995).

In coastal habitats, Van Dyke's salamanders are found associated with rocks or woody debris. In interior forests they generally are found associated with moist talus on north-facing slopes (Wilson et al., 1995).

F. Home Range Size. Unknown.

G. Territories. Unknown.

H. Aestivation/Avoiding Desiccation. Van Dyke's salamanders likely avoid desiccating conditions by seeking shelter in underground sites.

I. Seasonal Migrations. Van Dyke's salamanders probably move vertically from forest-floor sites to underground sites in response to seasonal dry and cold conditions.

J. Torpor (Hibernation). Generally unknown, but Van Dyke's salamanders likely avoid cold conditions by seeking shelter in underground sites.

K. Interspecific Associations/Exclusions. Olympic torrent salamanders (*Rhyacotriton olympicus*), Columbia torrent salamanders (*R. kezeri*), Cascade torrent salamanders (*R. cascadae*), Dunn's salamanders (*P. dunni*), Larch Mountain salamanders (*P. larselli*), and western red-backed salamanders (*P. vehiculum*) may be found with Van Dyke's salamanders (Leonard et al., 1993).

At a site in Mason County, Washington, Van Dyke's salamanders are sympatric with western red-backed salamanders and Olympic torrent salamanders (Jones, 1989).

At a site in Pacific County, Washington, Van Dyke's salamanders are sympatric with Dunn's salamanders and western red-backed salamanders (Ovaska and Davis, 1992).

Van Dyke's salamanders can distinguish between burrows marked with feces by western red-backed salamanders. However, Van Dyke's salamanders do not avoid marked burrows nor exhibit aggression towards western red-backed salamanders (Ovaska and Davis, 1992).

Van Dyke's salamanders are generally scarce. In suggesting causes for this rarity, Petranka (1998) offers either narrow habitat preferences or competitive/predatory interactions between Van Dyke's salamanders and other species.

L. Age/Size at Reproductive Maturity. Females generally are larger than males. Males mature at 44 mm SVL, females at 47 mm (Petranka, 1998).

M. Longevity. Unknown.

N. Feeding Behavior. Unknown.

O. Predators. Undocumented, but likely include forest snakes, birds, and small mammals.

P. Anti-Predator Mechanisms. All members of the genus *Plethodon* produce noxious skin secretions (Brodie, 1977).

Q. Diseases. Unknown.

R. Parasites. Unknown.

4. Conservation.
Van Dyke's salamanders are rare, and protected as a State Candidate species for listing by the State of Washington. They are federally listed as a Species of Concern (Petranka, 1998; Washington Department of Fish and Wildlife's Web site at www.wa.gov). Many populations of these salamanders are on National Park Service lands and are therefore afforded some degree of protection.

Acknowledgments. Thanks to Richard Highton, who reviewed this account and gave us the benefit of his insight and experience.

Plethodon variolatus (Gilliams, 1818)
SOUTH CAROLINA SLIMY SALAMANDER

David A. Beamer, Michael J. Lannoo

1. Historical versus Current Distribution.
South Carolina slimy salamanders (*Plethodon variolatus*) occur in the Southern Coastal Plain Physiographic Province of South Carolina and extreme southeastern Georgia. The neotype was collected at an elevation of 6 m in Berkeley County, South Carolina. There is little evidence to suggest that the current distribution differs from the historical distribution.

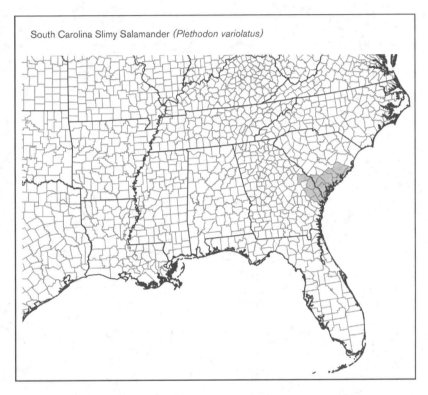

South Carolina Slimy Salamander *(Plethodon variolatus)*

2. Historical versus Current Abundance.
Generally unknown. Highton (this volume, Part One) found that in all six populations he studied (located in Chatham County, Georgia, and in Allendale, Berkeley, Charleston-Dorchester, and Jasper counties, South Carolina) numbers of animals collected in the 1990s were reduced compared with collections he had made during and prior to 1980. Additional surveys are necessary to determine whether these results indicate true declines or natural population fluctuations.

3. Life History Features.
South Carolina slimy salamanders were fairly recently described (Highton, 1989). Since this time, there has been no published work done on this species. As a portion of his larger research program, R. Highton has collected basic life history and natural history information on South Carolina slimy

---

salamanders and has plans to publish these data in a monographic treatment.

A. Breeding. Reproduction is terrestrial.

i. Breeding migrations. Undocumented, but breeding migrations are unknown for any *Plethodon* species.

ii. Breeding habitat. Unknown.

B. Eggs.

i. Egg deposition sites. Undocumented, but likely to be similar to other Coastal Plain members of the slimy salamander (*P. glutinosus*) complex, such as southeastern slimy salamanders (*P. grobmani*), which deposit their eggs in and under rotting logs (Highton, 1956).

ii. Clutch size. Unknown.

C. Direct Development.

i. Brood sites. Unknown, but are likely to be the same as egg deposition sites.

ii. Parental care. Unknown, but it is likely that females brood, as with other members of the slimy salamander complex.

D. Juvenile Habitat. Juveniles were commonly encountered in piles of pine bark surrounding naked pine boles during early spring in Berkeley County, South Carolina (D.A.B., personal observations).

E. Adult Habitat. Neill (1948c) reports South Carolina slimy salamanders from thick damp woods, where the ground was littered with bark scraps, fallen timber, and leafy debris. In Jasper County, Neill (1948c) stated that these salamanders were found only in the immediate vicinity of shrew burrows, which they used as retreats. South Carolina slimy salamanders appear to avoid low sandy areas with palmetto and pine flatwoods.

In the Francis Marion National Forest, Berkeley County, South Carolina slimy salamanders are found in mixed hardwood forests, bottomland forests, and longleaf pine savannas. They were frequently found in areas that were prescription burned as a land management practice. Unburned areas were always present nearby (usually across the road), so it is unknown if South Carolina slimy salamanders can persist in areas that burn regularly without refugia of nearby unburned areas (D.A.B., personal observations).

F. Home Range Size. Unknown, but small home ranges are typical for *Plethodon* species.

G. Territories. At least some members of the *P. glutinosus* complex aggressively defend territories (Thurow, 1976), but it is unknown whether South Carolina slimy salamanders establish and defend territories.

H. Aestivation/Avoiding Desiccation. Generally unknown, but South Carolina slimy salamanders likely avoid desiccating conditions by seeking shelter in underground sites.

I. Seasonal Migrations. Unknown, but vertical migrations from surface sites to underground sites and back again are likely to be important in surviving seasonally variable conditions.

J. Torpor (Hibernation). Generally unknown, but South Carolina slimy salamanders likely avoid cold conditions by seeking shelter in underground sites.

K. Interspecific Associations/Exclusions. In Berkeley County, South Carolina slimy salamanders have been found in association with marbled salamanders (*Ambystoma opacum*), mole salamanders (*A. talpoideum*), and dwarf salamanders (*Eurycea quadridigitata*; D.A.B., personal observations).

L. Age/Size at Reproductive Maturity. The neotype is an adult male, 61 mm SVL. Highton (1989) views South Carolina slimy salamanders as a small species.

M. Longevity. Unknown.

N. Feeding Behavior. Unknown, but as with other *Plethodon* species, feeding likely takes place at night under moist conditions. Prey items likely include a range of invertebrates, especially insects.

O. Predators. Undocumented, but likely include forest snakes, birds, and small mammals.

P. Anti-Predator Mechanisms. All *Plethodon* produce noxious skin secretions (Brodie, 1977).

Q. Diseases. Unknown.

R. Parasites. Unknown.

4. Conservation.
South Carolina slimy salamanders are not protected by either of the two states within their range. Among members of the *P. glutinosus* complex, South Carolina slimy salamanders have one of the smaller distributions. Within their range there are several federal and state properties that contain suitable habitat for these salamanders.

As with all species of *Plethodon*, South Carolina slimy salamanders do not migrate to breeding grounds and they do not have large home ranges. Thus, they can exist in habitats of smaller size than many other amphibian species.

South Carolina slimy salamanders are frequently found in woodlands with relatively sparse canopy due to frequent fire return intervals. Until data are accumulated about how South Carolina slimy salamander populations respond to fire frequencies, it is difficult to suggest land management plans that address burn regimes (see "Adult Habitat" above).

Acknowledgments. Thanks to Richard Highton, who reviewed this account and gave us the benefit of his insight and experience.

Plethodon vehiculum (Cooper, 1860)
WESTERN RED-BACKED SALAMANDER
R. Bruce Bury

1. Historical versus Current Distribution.
Western red-backed salamanders (*Plethodon vehiculum*) range throughout much of western Washington and Oregon, and northward to southern British Columbia and Vancouver Island. They occur on some coastal islands eastward to an altitude of about 800 m in the Washington Cascades (Brown and Slater, 1939; Slater and Brown, 1941; Slater, 1955). According to Storm and Brodie (1970b), the easternmost record in the Columbia Gorge is about 5 km east of Stevenson, Skamania County, Washington. In Oregon, except for a record near Boring, southeast of Portland, western red-backed salamanders appear to be absent from the northern portion of the Oregon Cascades. South of Oakridge, Lane County, salamanders again occur in the Cascades, but only to the North Umpqua River in Douglas County (Storm and Brodie, 1970b). At this location, western red-backed salamanders occur at about 1,250 m, one of the highest known elevations for any western *Plethodon* (Storm and Brodie, 1970b).

2. Historical versus Current Abundance.
Western red-backed salamanders are common in Douglas fir and other mixed forests throughout their range (Bishop, 1943; Nussbaum et al., 1983). Western red-backed salamanders are one of the few species of Pacific Northwest salamanders that are common in young forests (Corn and Bury, 1991; Petranka, 1998). They occur across all forest stand ages (Aubry et al., 1988; Aubry and Hall, 1991).

3. Life History Features.

A. Breeding. Reproduction is terrestrial. The mating season varies from November to early March, depending on geography. In Oregon, breeding activity

Western Red-backed Salamander (*Plethodon vehiculum*)

may begin as early as September but peaks from November–January (Dumas, 1956; Peacock and Nussbaum, 1973). Females in many populations have biennial reproductive cycles (Peacock and Nussbaum, 1973; Petranka, 1998). They may only oviposit every third year on Vancouver Island (Ovaska and Gregory, 1989).

  i. Breeding migrations. Do not occur.

  ii. Breeding habitat. Same as adult habitat.

  B. Eggs.

  i. Egg deposition sites. Nests have rarely been found, suggesting that females lay eggs in subsurface retreats.

  ii. Clutch size. Eggs number from 4–19 (mean usually near 10) and are large at 4.4–6.0 mm in diameter (Carl, 1943; Stebbins, 1951; Peacock and Nussbaum, 1973; Hanlin et al., 1979; Nussbaum et al., 1983; Norman and Swartwood, 1991).

  C. Direct Development.

  i. Brood sites. Females oviposit during the spring or early summer but nests have rarely been found, suggesting that females lay eggs in subsurface retreats. Eggs are laid in grape-like clusters attached to the roofs and sides of cavities under or in rocks and logs (Peacock and Nussbaum, 1973; Hanlin et al., 1979; Norman and Swartwood, 1991).

  ii. Parental care. Females brood; on rare occasions males are found with brooding females (Norman and Smartwood, 1991).

  D. Juvenile Habitat. Hatching occurs at about 13–15 mm SVL (Peacock and Nussbaum, 1973; Petranka, 1998). Hatchlings can be abundant on the forest floor in late autumn and early winter. Juveniles grow about 10 mm SVL annually for their first 3 yr. Microhabitat use varies with age (increased size). Juveniles (<30 mm SVL) tend to use smaller cover objects such as leaf litter and stones, while larger animals use larger objects. During drier conditions, juveniles tend to be more active on the forest floor than are adults (Ovaska and Gregory, 1989).

  E. Adult Habitat. Western red-backed salamanders are found beneath logs, bark, stones, and moss in cool, shaded areas (Dumas, 1956; Brodie, 1970; Nussbaum et al., 1983; Corn and Bury, 1991). They are sometimes extremely abundant in rocky, talus slopes. Bishop (1943) reports collecting many specimens from such areas and among loose stones at the foot of a cliff. Populations are found associated with rocky seeps, springs, and small streams in the northern Puget Trough and Cascade Range foothills in Washington (Leonard, 1996; Petranka, 1998).

  F. Home Range Size. According to Ovaska (1988b), individuals show a high degree of site specificity, with adults maintaining small home ranges on the order of a few m².

  G. Territories. According to Ovaska (1987), adults and particularly males competing for females are aggressive during their mating season. Individuals can detect chemical odors from the fecal pellets and bodies of conspecifics. Both males and females avoid odors of conspecific males, but not females (Ovaska, 1988a). Nevertheless, adults frequently share cover objects, are rarely aggressive toward each other, and do not avoid burrows with fecal pellets, suggesting that while males may compete for females, they are not territorial (Ovaska, 1987, 1988a,b, Ovaska and Davis, 1992).

  H. Aestivation/Avoiding Desiccation. During the drier summer, adults retreat to moist underground sites, although juveniles remain active on the forest floor (Ovaska and Gregory, 1989).

  I. Seasonal Migrations. Unlikely.

  J. Torpor (Hibernation). During the winter, western red-backed salamanders move deep into the talus (Dumas, 1956; see also Petranka, 1998).

  K. Interspecific Associations/Exclusions. Western red-backed salamanders are rarely found with Dunn's salamanders (*P. dunni*) and species of torrent salamanders (*Rhyacotriton* sp.), which are species that inhabit wet areas such as seeps or sides of streams. Recently, I found western red-backed salamanders for the first time in sympatry with Del Norte salamanders (*P. elongatus*). The locality was in the West Fork of Cow Creek, a tributary of the South Fork of the Umpqua River, southwest of Roseburg, Douglas County, Oregon (personal observations). This area is in the Oregon Coast Range. I captured both species, and a few Dunn's salamanders, in a rocky talus slope, 5–20 m from a small, north-flowing creek.

  L. Age/Size at Reproductive Maturity. Individuals become sexually mature between their second and third year, when males reach about 42 mm SVL and females reach about 44 mm SVL, although this varies geographically (Brodie, 1970). The largest specimens in most populations are females.

  M. Longevity. Unknown.

  N. Feeding Behavior. The diet of western red-backed salamanders consists of small invertebrates including annelids, isopods, snails, spiders, mites, millipedes, centipedes, pseudoscorpions, and insects such as collembolans, dipterans, lepidopterans, and hymenopterans—including ants (Dumas, 1956; Nussbaum et al., 1983; Petranka, 1998).

  O. Predators. Poorly documented but are known to (or probably) include common garter snakes (*Thamnophis sirtalis*; Norman, 1988b), shrewmoles (*Neurotrichus gibbsi*; Dumas, 1956), American dippers (*Cinclus mexicanus*; Dumas, 1956), and carabid beetles (Ovaska and Smith, 1988; Petranka, 1998).

  P. Anti-Predator Mechanisms. Western red-backed salamanders rarely assume anti-predator postures when approached by carabid beetles in laboratory settings (Ovaska and Smith, 1988).

  Q. Diseases. Unknown.

  R. Parasites. Unknown

4. Conservation.
This is a widespread species that occurs in a variety of habitats. Rocky substrate is a key habitat need, but forest age is not. The species appears not to be declining at the present time.

Plethodon ventralis Highton, 1997
SOUTHERN ZIGZAG SALAMANDER

David A. Beamer, Michael J. Lannoo

1. Historical versus Current Distribution.
Southern zigzag salamanders (*Plethodon ventralis*) occur in scattered sites from northern Mississippi east-northeast to southeastern Virginia, including sites in northern Alabama and Georgia, eastern Tennessee, and western North Carolina. With the exception of local extirpations due to habitat destruction and modification, their current distribution is likely similar to their historical distribution, but there has been no documentation of their historical distribution.

2. Historical versus Current Abundance.
Generally unknown. Highton (this volume, Part One) sampled two sites (Lawrence County, Alabama, and Blount County, Tennessee) prior to 1987 and again in 1996 and found that numbers of animals in both populations were reduced. Additional data are necessary to determine whether these data reflect true declines or natural population fluctuations.

3. Life History Features.
Southern zigzag salamanders were recently described (Highton, 1997). In this time, there has been no published work done on this species. As a portion of his larger research program, R. Highton has collected basic life history and natural history information on southern zigzag salamanders and has plans to publish these data in a monographic treatment.

  A. Breeding. Reproduction is terrestrial.

  i. Breeding migrations. Undocumented, but breeding migrations are not known for any *Plethodon* species.

  ii. Breeding habitat. Unknown.

  B. Eggs.

  i. Egg deposition sites. Unknown.

  ii. Clutch size. Unknown.

  C. Direct Development.

  i. Brood sites. Unknown.

  ii. Parental care. Unknown, but it is likely that females brood, as with other species of *Plethodon*.

  D. Juvenile Habitat. Unknown, but likely to be similar to adult habitat.

  E. Adult Habitat. In the Great Smoky Mountains, southern zigzag salamanders are found in flat, moist areas at elevations lower than 579 m (1,900 ft; King, 1939; Highton, 1972). Thurow (1963) reports

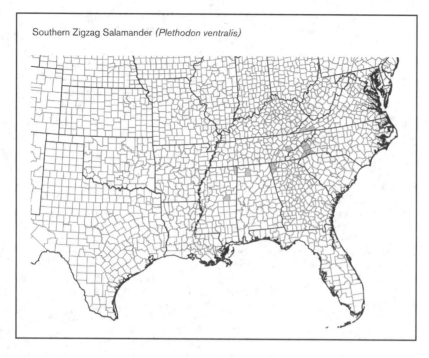

Southern Zigzag Salamander *(Plethodon ventralis)*

southern zigzag salamanders from a talus slope on the lower mountain slopes near Gatlinburg, Tennessee.

F. Home Range Size. Unknown, but small home ranges are typical for *Plethodon* species.

G. Territories. At least some members of the *P. dorsalis* complex aggressively defend territories (Thurow, 1976); it is unknown whether southern zigzag salamanders establish and defend territories.

H. Aestivation/Avoiding Desiccation. Southern zigzag salamanders likely avoid desiccating conditions by seeking shelter in underground sites.

I. Seasonal Migrations. Southern zigzag salamanders are rarely found on the surface during the summer (Highton, 1972).

J. Torpor (Hibernation). At low elevations in the Great Smoky Mountains, it is not unusual to find lethargic individuals under rocks and logs on mild days in the winter (Huheey and Stupka, 1967).

K. Interspecific Associations/Exclusions. Dodd (1989) reports southern ravine salamanders *(P. richmondi)* occurring with salamanders that are likely southern zigzag salamanders in Madison County, Kentucky.

Southern zigzag salamanders occur together with southern red-backed salamanders *(P. serratus)* at Whiteoak Sink and the Sinks on Little River in Great Smoky Mountains National Park (King, 1939; Huheey and Stupka, 1967). However, in both areas the two seem to replace each other altitudinally between 518–610 m (1,700–2,000 ft). Southern zigzag salamanders are largely restricted to the lower elevations, and southern red-backed salamanders occur at higher elevations. There is evidence of character displacement in

this area, as southern zigzag salamanders are all of the unstriped phase, and southern red-backed salamanders are all striped (Highton, 1972).

Southern zigzag salamanders are found with Pigeon Mountain salamanders *(P. petraeus)*, northern slimy salamanders *(P. glutinosus)*, green salamanders *(Aneides aeneus)*, long-tailed salamanders *(Eurycea longicauda)*, cave salamanders *(E. lucifuga)*, and spring salamanders *(Gyrinophilus porphyriticus)* on Pigeon Mountain, Walker County, Georgia (Wynn et al., 1988).

Southern zigzag salamanders were occasionally found with northern slimy salamanders, eastern newt *(Notophthalmus viridescens)* efts, and northern dusky salamanders *(Desmognathus fuscus)* in Knox County, Tennessee (Powders, 1973).

The following salamanders, along with southern zigzag salamanders, are all reported from Tishmingo County, Mississippi: hellbenders *(Cryptobranchus alleganiensis)*, marbled salamanders *(Ambystoma opacum)*, spotted dusky salamanders *(D. conanti)*, Mississippi slimy salamanders *(P. mississippi)*, spring salamanders, red salamanders *(Pseudotriton ruber)*, green salamanders, southern two-lined salamanders *(E. cirrigera)*, long-tailed salamanders, and cave salamanders (Ferguson, 1961b; Wake and Woods, 1968).

Southern zigzag salamanders contact northern zigzag salamanders *(P. dorsalis)* in the vicinity of Muldraugh's Ridge, Lincoln County, Kentucky. A transect in this area indicates that there is a narrow hybrid zone (Highton, 1997).

Southern zigzag salamanders contact Webster's salamanders *(P. websteri)* in Jefferson County, Alabama. The contact zone may be little more than 1 km in width

in this area. There is no evidence of hybridization between these two species. There is evidence of character displacement, as both species generally exhibit polymorphism for striped and unstriped morphs, but in this area southern zigzag salamanders are all of the unstriped morph and Webster's salamanders are all of the striped morph (Highton, 1976, 1985).

L. Age/Size at Reproductive Maturity. Unknown.

M. Longevity. Unknown.

N. Feeding Behavior. Unreported, but as with other species of *Plethodon*, animals likely feed at night, with activity proportional to moisture levels. Prey items include small invertebrates, especially insects, that inhabit or are associated with the forest floor.

O. Predators. Two southern zigzag salamanders were found in the stomach of a screech owl in the Great Smoky Mountains National Park (Stupka, 1953). Other predators are likely to include forest-dwelling snakes, birds, and small mammals.

P. Anti-Predator Mechanisms. All *Plethodon* produce noxious skin secretions (Brodie, 1977). Members of the *P. dorsalis* complex frequently become immobile when initially contacted. Southern zigzag salamanders were included in a field study on immobility; however, it is not possible to separate their behavior from the other members of this complex in this published data set. Immobility may increase survival by making the salamander less likely to be detected, especially by visually oriented predators (Dodd, 1989).

Q. Diseases. Unknown.

R. Parasites. Unknown.

4. Conservation.

Southern zigzag salamanders are listed as a Species of Special Concern in North Carolina. Within their range there are many federal and state properties that contain suitable habitat for these salamanders.

As with all species of *Plethodon*, southern zigzag salamanders do not migrate to breeding grounds, and they do not have large home ranges. Thus, they can exist in habitats of smaller size than many other amphibian species. Conservation activities that promote mature closed-canopy forests should benefit this species.

Acknowledgments. Thanks to Richard Highton, who reviewed this account and gave us the benefit of his insight and experience.

Plethodon virginia Highton, 1999
SHENANDOAH MOUNTAIN SALAMANDER
David A. Beamer, Michael J. Lannoo

1. Historical versus Current Distribution.
Shenandoah Mountain salamanders *(Plethodon virginia)* were recognized by Highton (1999a) on the basis of molecular and distributional data from animals previously

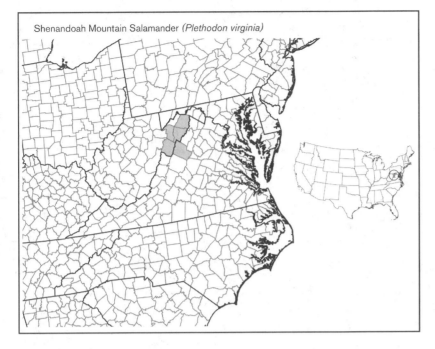

Shenandoah Mountain Salamander *(Plethodon virginia)*

recognized as Valley and Ridge salamanders *(P. hoffmani)*. The type specimen is an adult male collected within 1 km SE–SSE from the top of Cow Knob Mountain, at an elevation of 1,100–1,200 m, along the Pendleton County, West Virginia–Rockingham County, Virginia, state line. The known distribution of Shenandoah Mountain salamanders and its relation to the distribution of Valley and Ridge salamanders is presented in Highton (1999a, fig. 4). This distribution includes Shenandoah Mountain, plus South Branch and Nathaniel mountains, from central Rockingham County, Virginia, and Pendleton County, West Virginia, north to Hardy and Hampshire counties, West Virginia, and west to the South Fork of the South Branch of the Potomac River in western Hardy and southern Hampshire counties in West Virginia (Highton, 1999a).

2. Historical versus Current Abundance.
When compared with data collected prior to 1987, data collected since 1992 show that population declines of Shenandoah Mountain salamanders may have occurred at sites in Rockingham County, Virginia, and Pendleton County, West Virginia, while a population in Hampshire, West Virginia, contains a low but stable number of animals (Highton, this volume, Part One). Additional monitoring will be necessary to determine whether these data reflect true population trends or natural fluctuations.

3. Life History Features.
A. Breeding. Reproduction is terrestrial.
   i. Breeding migrations. Undocumented, but breeding migrations are not known for any *Plethodon* species.

ii. Breeding habitat. Unknown.
   B. Eggs.
   i. Egg deposition sites. Unknown, but as with other members of the *P. cinereus* group, likely to include underground sites and perhaps sites under surface cover objects.
   ii. Clutch size. The average number of eggs was 4.6, deposited by five female Shenandoah Mountain salamanders in Rockingham County, Virginia (Fraser, 1974).

C. Direct Development.
   i. Brood sites. Unknown, but probably include underground cavities or chambers.
   ii. Parental care. Unknown, but it is likely that females brood, as with other species of *Plethodon*.

D. Juvenile Habitat. Unknown.

E. Adult Habitat. The habitat of Shenandoah Mountain salamanders at a site in Rockingham County, Virginia, includes ridges and slopes. The ridges are characterized by deep soil with a predominant vegetation of white oak *(Quercus alba)*, pink honeysuckle *(Rhododendren nudifolium)*, and late low blueberry *(Vaccinium vacillans)*. The slopes are characterized by shallow, rocky soil with a predominant vegetation of chestnut oak *(Q. montanta)*, red maple *(Acer rubrum)*, late low blueberry, witch hazel *(Hamamelis virginiana)*, and mountain laurel *(Kalmia latifolia)*. Farther down the slopes, the soil is rocky with exposed bedrock and a predominate vegetation of chestnut oak, red oak *(Q. rubra)*, bitternut hickory *(Carya cordiformis)*, black gum *(Nyssa sylvatica)*, and witch hazel (Fraser, 1976a).

F. Home Range Size. Unknown, but small home ranges are typical for *Plethodon* species.

G. Territories. Unknown.

H. Aestivation/Avoiding Desiccation. Shenandoah Mountain salamanders were not found between August and September in Rockingham County, Virginia (Fraser, 1976a).

I. Seasonal Migrations. Animals make vertical migrations, moving from the forest floor to underground sites. Only a small proportion of the population is at the surface at any given time (Fraser, 1976a).

J. Torpor (Hibernation). Shenandoah Mountain salamanders in Rockingham County, Virginia, were not found at the surface during periods of below-freezing weather from December–February (Fraser, 1976a).

K. Interspecific Associations/Exclusions. The type locality of Shenandoah Mountain salamanders is only 1.5 km from the type locality of Cow Knob salamanders *(P. punctatus)*. Fraser (1976a) studied the coexistence of Shenandoah Mountain salamanders and Cow Knob salamanders at a site in Rockingham County, Virginia. There is high overlap of food resources and surface habitat utilization between adult Shenandoah Mountain salamanders and juvenile Cow Knob salamanders. Staggered feeding schedules and partitioning of structural habitat by adults appears to be important in reducing interspecific competition between these two species at this site.

Fraser (1976a) performed a study of behavioral interactions between captive Shenandoah Mountain salamanders and Cow Knob salamanders. Salamanders were introduced into large containers with surface cover objects. There was no tendency for either a positive or negative association when adult Valley and Ridge salamanders were added to an enclosure containing juvenile Cow Knob salamanders and vice versa. However, there was a significant negative association between Valley and Ridge salamanders when they were added to an enclosure containing adult Cow Knob salamanders. There was not a significant association when adult Cow Knob salamanders were added to an enclosure containing adult Valley and Ridge salamanders.

Shenandoah Mountain salamanders are sympatric with red-backed salamanders *(P. cinereus)* at only two sites. Highton (1999a) suggests that there might be strong competitive exclusion between these two species.

White-spotted slimy salamanders *(P. cylindraceus)* are known from Shenandoah Mountain (Highton, 1972, 1989), although there is not any published information on its association with Shenandoah Mountain salamanders there.

Shenandoah Mountain salamanders contact Valley and Ridge salamanders at the northern and southern extremities of Shenandoah Mountain salamanders' range.

In these areas there are some hybrid populations (Highton, 1999a).

L. Age/Size at Reproductive Maturity. The male holotype and the female allotype were the same size, with a 57 mm SVL.

M. Longevity. Unknown.

N. Feeding Behavior. The main food items in the stomachs of 94 Shenandoah Mountain salamanders included Hymenoptera, Collembola, Chilopoda, Diplopoda, Araneida, and larval Diptera, Oligochaeta, Coleoptera, and Pulmonata (Fraser, 1976a).

O. Predators. Unknown, but likely include forest birds, small mammals, and snakes.

P. Anti-Predator Mechanisms. All *Plethodon* produce noxious skin secretions (Brodie, 1977). Shenandoah Mountain salamanders frequently become immobile when initially contacted. Shenandoah Mountain salamanders were included in a field study on immobility, however it is not possible to separate their behavior from Valley and Ridge salamanders in this published data set. Immobility may increase survival by making the salamander less likely to be detected, especially by visually oriented predators (Dodd, 1989).

Q. Diseases. Unknown.

R. Parasites. Unknown.

4. Conservation.
Shenandoah Mountain salamanders are not conserved in either of the two states within their range. Within this range, however, there is a large amount of federal land that contains suitable habitat for these salamanders.

The range of Shenandoah Mountain salamanders is approximately the same as Cow Knob salamanders. Several conservation measures have been implemented to ensure the viability of Cow Knob salamander populations (Mitchell and Pauley, 2002) that also should benefit Shenandoah Mountain salamanders. Threats to Cow Knob salamanders that may also pertain to Shenandoah Mountain salamanders include logging operations and the loss of hemlock trees by the introduced hemlock woolly adelgid (*Adelges tsugae*), as well as defoliation of canopy hardwood trees by the introduced gypsy moth (*Lymantria dispar*; Mitchell and Pauley, this volume).

Acknowledgments. Thanks to Richard Highton, who reviewed this account and gave us the benefit of his insight and experience.

Plethodon websteri Highton, 1979
WEBSTER'S SALAMANDER

David A. Beamer, Michael J. Lannoo

1. Historical versus Current Distribution.
The distribution of Webster's salamanders (*Plethodon websteri*) is unusual. These animals occur in a relatively small number of scattered sites from eastern Louisiana, through Mississippi, Alabama, and Georgia,

Webster's Salamander *(Plethodon websteri)*

to western South Carolina. Historical data are lacking.

2. Historical versus Current Abundance.
Generally unknown. Highton (this volume, Part One) sampled two sites in Etowah County, Alabama, from 1971–88, and one site in Winston County, Mississippi, from 1976–84, then re-sampled these sites during the 1990s and found fewer animals at each site. Whether these data represent true declines or natural population fluctuations will only be determined by subsequent monitoring.

3. Life History Features.

A. Breeding. Reproduction is terrestrial. Males show maximum spermatogenic activity during the late summer. The vasa deferentia were packed with sperm from January–March, and the mental gland was most swollen at this time, indicating that courtship and insemination probably occur during this period (Semlitsch and West, 1983).

i. Breeding migrations. Undocumented, but breeding migrations are not known for any *Plethodon* species.

ii. Breeding habitat. Unknown.

B. Eggs. In South Carolina, females likely lay their eggs from June–September. Despite intensive searching, gravid females could not be found later than June (Semlitsch and West, 1983).

i. Egg deposition sites. Unknown, but likely to be in underground cavities.

ii. Clutch size. Mean number of ovarian follicles is 5.8 (69 females), with a range of 3–8 (Semlitsch and West, 1983). The number of follicles increases with female length. Follicle diameter is 3 mm.

C. Direct Development. Hatching probably takes place in October–November, as emerging juveniles with remnants of yolk have been found at this time. The smallest Webster's salamanders were found in

November and were 12.5 mm SVL (Semlitsch and West, 1983).

i. Brood sites. Females likely oviposit and brood underground from June–September (Semlitsch and West, 1983).

ii. Parental care. Unknown, but it is likely that females brood, as with other species of *Plethodon*.

D. Juvenile Habitat. Juveniles with yolk remnants begin to emerge onto the forest floor in October. Growth is about 1.3 mm/mo for the smallest animals, 1.1 mm/mo for second-year animals, and 0.33 mm/mo for adults.

E. Adult Habitat. Adult Webster's salamanders are found in mixed mesophytic forest bordering rocky, lower-order streams (Semlitsch and West, 1983). Semlitsch and West (1983) point out that at their study sites, predominant canopy species include sugar maples (*Acer saccharinum*), pignut hickory (*Carya glabra*), red oak (*Quercus rubra*), yellow poplar (*Liriodendron tulipifera*), sweetgum (*Liquidambar styraciflua*), and slippery elm (*Ulmus rubra*). The vegetation is considered to be a relict mesophytic community and contains disjunct populations of plant species.

On 4 February 1967, Webster's salamanders were collected in Clarke County, Alabama, under rocks and logs along the base of the slopes of an extensive ravine characterized by limestone outcroppings. Conspicuous floral elements at this site include blue beech, long-leaf magnolia, and sweet gum (Blaney and Relyea, 1967).

Webster's salamanders were collected from Winston County, Mississippi, in March–April 1958 under decaying logs near a small stream and in a pasture in a second-growth forest. The major tree species present were sweetgum, black willow, red and white oak, yellow popular, American elm, shagbark and mockernut hickory, beech, and red maple (Ferguson and Rhodes, 1958).

F. Home Range Size. Unknown, but as with other *Plethodon* species is likely to be small.

G. Territories. At least some members of the *P. dorsalis* complex aggressively defend territories (Thurow, 1976), it is unknown whether Webster's salamanders establish and defend territories.

H. Aestivation/Avoiding Desiccation. Webster's salamanders appear to remain underground during the summer. Forest floor activity was restricted to times when air temperatures were between 15–30 °C. Salamanders were not found in the summer even after heavy rainfall that left the leaf litter wet (Semlitsch and West, 1983).

I. Seasonal Migrations. Vertical migrations occur seasonally; Semlitsch and West (1983) report Webster's salamanders as being active from October–May and apparently retreating underground during the summer.

J. Torpor (Hibernation). Animals are active throughout the winter and are commonly found near the surface at this time (Semlitsch and West, 1983).

K. Interspecific Associations/Exclusions. Salamanders associated with Webster's salamanders at Semlitsch and West's (1983) study site include Atlantic Coast slimy salamanders (*P. chlorobryonis*), eastern newts (*Notophthalmus viridescens*), southern dusky salamanders (*Desmognathus auriculatus*), southern two-lined salamanders (*Eurycea cirrigera*), and long-tailed salamanders (*E. longicauda*).

In Clarke County, Alabama, the following species were collected with Webster's salamanders: Mississippi slimy salamanders (*P. mississippi*), spotted dusky salamanders (*D. conanti*; may actually have been *D. auriculatus*), and southern two-lined salamanders (Blaney and Relyea, 1967).

Webster's salamanders contact southern zigzag salamanders (*P. ventralis*) in Jefferson County, Alabama. The contact zone may be little more than 1 km wide in this area. There is no evidence of hybridization between these two species. There is evidence of character displacement, as both species generally exhibit polymorphism for striped and unstriped morphs, but in this area Webster's salamanders are all of the striped morph and southern zigzag salamanders are all of the unstriped morph (Highton, 1976, 1985).

The following salamanders were collected along with Webster's salamanders on Mount Cheaha, Cleburne County, Alabama: spotted dusky salamanders, seepage salamanders (*D. aeneus*), seal salamanders (*D. monticola*), southern two-lined salamanders, three-lined salamanders (*E. guttolineata*), spring salamanders (*Gyrinophilus porphyriticus*), red salamanders (*Pseudotriton ruber*), and northern slimy salamanders (*P. glutinosus*; Rubenstein, 1969).

L. Age/Size at Reproductive Maturity. The estimated age at sexual maturity for both males and females is between 21–26 mo. The estimated age when Webster's salamanders first court is between 29 and 31 mo, and the age of first oviposition is from 34–37 mo. Both sexes breed annually.

The smallest mature male in a McCormick County, South Carolina, study was 29 mm SVL and the smallest mature female was 27.5 mm (Semlitsch and West, 1983).

M. Longevity. Unknown.

N. Feeding Behavior. The following food items were found in specimens collected during March–April from Upson County, Georgia, and Lee County, Alabama: Gastropoda, Annelida, Acarina, Araneae, Pseudoscorpionida, Chilopoda, Isopoda, Collembola, Isoptera, Diptera, Lepidoptera, Thysanoptera, Coleoptera, ants, wasps, and unidentified larvae. Larger individuals generally feed on larger prey (ants and termites); smaller individuals feed on smaller prey (springtails and mites; Camp and Bozeman, 1981).

Webster's salamanders in South Carolina have a long (8 mo) continuous activity period, during which they forage and obtain resources (Semlitsch and West, 1983).

O. Predators. Undocumented, but likely include forest snakes, birds, and small mammals.

P. Anti-Predator Mechanisms. Nocturnal. Secretive. All members of the genus *Plethodon* produce noxious skin secretions (Brodie, 1977). Webster's salamanders frequently become immobile when initially contacted. Immobility may increase survival by making the salamander less likely to be detected, especially by visually oriented predators (Dodd, 1989).

Q. Diseases. Unknown.

R. Parasites. Unknown.

4. Conservation.

Webster's salamanders are listed as an Endangered species in South Carolina and as a Species of Special Concern in Louisiana. This species occurs in several isolates. Within some of these isolates there are not any federal and state properties that contain suitable habitat for these salamanders.

As with all species of *Plethodon*, Webster's salamanders do not migrate to breeding grounds and they do not have large home ranges. Thus, they can exist in habitats of smaller size than many other amphibian species. Conservation activities that promote mature closed-canopy forests should benefit this species.

Acknowledgments. Thanks to Richard Highton, who reviewed this account and gave us the benefit of his insight and experience.

Plethodon wehrlei Fowler and Dunn, 1917(a)
WEHRLE'S SALAMANDER
Thomas K. Pauley, Mark B. Watson

1. Historical versus Current Distribution.

Bishop (1943) and Grobman (1944) described the range of Wehrle's salamanders *(Plethodon wehrlei)* from southwestern New York through western Pennsylvania, eastern Ohio, and West Virginia, with disjunct populations in western North Carolina and extreme eastern Tennessee. Early distributional records include Pennsylvania (Netting, 1936a; Burt, 1942), Virginia (Netting et al., 1946; Grobman, 1949; Hoffman, 1967), and West Virginia (Netting, 1936b). Current work shows the range from Cattaraugus County, New York, south through western Pennsylvania, western Maryland, eastern Ohio, West Virginia, and southwestern Virginia to the northern edge of North Carolina in the East, and through eastern Kentucky to the northern edge of Tennessee in the West (Highton, 1971, 1987b). Wehrle's salamanders have not been found in Ohio since the 1930s (Pfingsten, 1989e). Yellow-spotted morphs are known to occur in Summers County, West Virginia (Cupp and Towles, 1983; Highton, 1987b; Waldron et al., 2000), Letcher County, Kentucky (Cupp and Towles, 1983), and Campbell County, Tennessee (Redman and Jones, 1985).

2. Historical versus Current Abundance.

No historical information is available. Hall and Stafford (1972) estimated that these are the most abundant salamanders in their habitat and have a high density and biomass. They estimated 1,000 salamanders/ha or 0.1 salamander/m^2 in one Pennsylvania site. Pauley (1980, 1993a) found them to be common in West Virginia at elevations from 305 m in mixed deciduous forests to 1,446 m in red spruce forests. Mitchell et al. (1999) determined existing and potential threats to Wehrle's salamander to be deforestation.

3. Life History Features.

A. Breeding. Reproduction is terrestrial.

i. Breeding migrations. Wehrle's salamanders do not migrate.

ii. Breeding habitat. Times of mating and egg deposition vary throughout the range. Mating occurs in September–October, and egg deposition occurs in January–March in Pennsylvania (Hall and Stafford, 1972); in New York, mating in captive salamanders occurs in September–October, and eggs are deposited in March–April (Johnson, 1961); in West Virginia, mating is in March–April and egg deposition prior to May (Pauley and England, 1969). Dodd and Brodie (1976) described the hedonic gland-cluster.

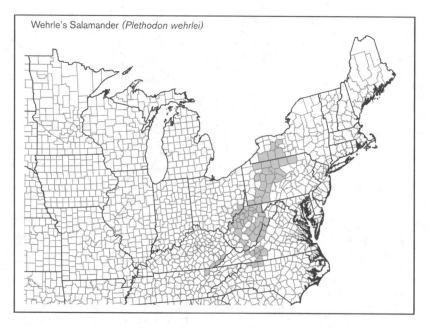

Wehrle's Salamander (*Plethodon wehrlei*)

K. Interspecific Associations/Exclusions. Highton (1971) discusses the geographical interactions between Wehrle's salamanders and Cow Knob salamanders. In West Virginia, habitat partitioning between Wehrle's salamanders and eastern red-backed salamanders is regulated by soil temperature, soil moisture, and burrow sizes (Pauley, 1978b,c,d; Pauley and Keller, 1993).

L. Age/Size at Reproductive Maturity. Males probably mature in 4 yr, females in 5 yr (Hall and Stafford, 1972). Total length of adults is 100–170 mm with males being slightly smaller than females (Petranka, 1998).

M. Longevity. Unknown. Probably similar to other large *Plethodon* (Snider and Bowler, 1992).

N. Feeding Behavior. In West Virginia specimens, Pauley (1978a) found the major prey items (in order of percent consumption) for adults to be ants, beetles, mites, spiders, and collembolans; for juveniles, prey items include ants, collembolans, mites, beetles, and spiders. In Pennsylvania, prey items included annelids, gastropods, millipedes, centipedes, isopods, phalangids, spiders, mites, collembolans, orthopterans, homopterans, hemipterans, beetles, flies, and ants (Hall, 1976). Bishop (1941b) found New York specimens to prey on ants, lepidopteran larvae, crickets, beetles, craneflies, hymenopterans, aphids, and mites.

O. Predators. Predators include ring-necked snakes (*Diadophis punctatus*; Hall and Stafford, 1972) and probably garter snakes, rodents, and some birds.

P. Anti-Predator Mechanisms. Unknown. As with other *Plethodon*, mucous secretions in the tail may inhibit ingestion by small predators.

Q. Diseases. Diseases of this species have not been reported.

R. Parasites. Parasites have not been determined.

4. Conservation.
Hall and Stafford (1972) estimated that Wehrle's salamanders may be the most abundant salamanders in their habitat. Deforestation may be the most important threat to Wehrle's salamanders (Mitchell et al., 1999). In fact, they have not been found in Ohio since the 1930s (Pfingsten, 1989e). Wehrle's salamanders are listed as In Need of Conservation in Maryland, In Need of Management in West Virginia, and Threatened in North Carolina (Levell, 1997).

Plethodon welleri Walker, 1931
WELLER'S SALAMANDER
David A. Beamer, Michael J. Lannoo

1. Historical versus Current Distribution. Weller's salamanders (*Plethodon welleri*) inhabit high-elevation regions (usually

B. Eggs.
i. Egg deposition sites. Because eggs have never been found on the surface of the ground, they are most likely deposited in underground nests in mixed deciduous and coniferous forests (Green and Pauley, 1987). Fowler (1952) found one nest of six eggs in a cavity in a cave in Virginia. Gross (1982) studied the morphology and physiology of Wehrle's salamanders in caves in West Virginia but was unable to locate nests.

ii. Clutch size. Hall and Stafford (1972) determined that the number of mature follicles varied from 7–24. This range of sizes is correlated to the sizes of the females.

C. Direct Development. Females likely tend nests.

D. Juvenile Habitat. Juveniles have been found in the same habitat as adults (T. K. P., unpublished data).

E. Adult Habitat. Habitat varies from mixed deciduous forests in the Allegheny Plateau and Cumberland Plateau to coniferous forests of the high elevations in the Allegheny Mountains. Ground cover in these habitats includes logs, rocks, and leaves (Hassler, 1932; Bishop, 1941b; Brooks, 1945; Pauley and England, 1969; Hall and Stafford, 1972; Green and Pauley, 1987). They also occur in caves (Netting, 1933; Reese, 1933; Netting et al., 1946; Pope and Fowler, 1949; Fowler, 1951; Newman, 1954b; Cooper, 1961; Gross, 1982; Holsinger, 1982; Pauley and Keller, 1993). In West Virginia, Wehrle's salamanders inhabit dry hillsides and are associated with plants indicative of xeric habitats when in sympatry with eastern red-backed salamanders (*P. cinereus*; Pauley, 1978b,c). Pauley and Keller (1993) found that when in sympatry with eastern red-backed salamanders, Wehrle's salamanders use larger

burrows than eastern red-backed salamanders, and there was a significant negative correlation between cloacal temperatures and burrow size for eastern red-backed salamanders but not Wehrle's salamanders.

F. Home Range Size. Unknown, but probably similar to other large *Plethodon* species such as *P. punctatus* (Cow Knob salamanders; Buhlmann et al., 1988) and northern slimy salamanders (*P. glutinosus*; Merchant, 1972).

G. Territories. Wehrle's salamanders defend their territories against conspecifics and other salamander species (Thurow, 1976).

H. Aestivation/Avoiding Desiccation. Members of this species remain active on the surface from March to early June in low elevations and April to early October in higher elevations (Pauley, 1978d, 1993a). In West Virginia, aestivation probably occurs from mid-June to late September in the low elevations of the Allegheny Plateau, and during unusual drought conditions from late April to mid-October in the Allegheny Mountains (Green and Pauley, 1987).

I. Seasonal Migrations. Wehrle's salamanders do not migrate across the forest floor, but will move from surface sites to underground sites in response to seasonal dry and cold conditions and back to the forest floor when surface conditions become favorable.

J. Torpor (Hibernation). In the lower elevations in West Virginia, Wehrle's salamanders remain active on the surface in late autumn until air and soil temperatures become too cold, and return to the surface in March as the temperatures increase. In the higher elevations, they are not active on the surface from late October to May (Green and Pauley, 1987; Santiago, 1999).

above 1,500 m, but occasionally down to about 700 m) from the Mt. Rogers and Whitetop Mountain areas of southwestern Virginia along the Unaka Mountain ridges to areas in extreme northeastern Tennessee (Johnson and Unicoi counties) and northeastern North Carolina (Yancey County), including Grandfather Mountain, North Carolina (Walker, 1934; Bishop, 1943; Hoffman, 1953; Thurow, 1956, 1964; Petranka, 1998). These populations are all associated with the Blue Ridge Physiographic Province. Populations are scattered and appear to be relictual from a previously more widespread distribution (Thurow, 1956).

September and secondary sex characters show greatest development at this time (Thurow, 1963).

Organ (1960b) observed courtship of captive individuals in October. Males first become restless and begin moving. When he approaches another salamander, the tail begins to undulate laterally. The male then brings his nasolabial projections and mental gland in contact with the body of the female and advances toward her head while holding his chin against her skin. When the male reaches the female's chin, he breaks contact and circles beneath her chin. The male then passes beneath her until his undulating tail reaches her chin.

ii. Breeding habitat. Unknown.

B. Eggs. A female collected on 26 May 1997 at Grandfather Mountain deposited eggs shortly after it was placed in a container to temporarily house it for photographic purposes. It is likely that the stress of capture may have caused early deposition of the eggs, and it is unknown if females begin depositing eggs at this time naturally (D. A. B., personal observations).

i. Egg deposition sites. Nests have been found in well-rotted conifer logs under moss mats from mid-August to September. Two nests are occasionally located within the same log within a few inches of one another (Hoffman and Kleinpeter, 1948; Organ, 1960b).

ii. Clutch size. In the Whitetop Mountain–Mount Rogers area, Virginia, the number of eggs range from 4–11. Eggs range in diameter from 3.6–6.5 mm and are deposited in tight clusters suspended by a stalk (Hoffman and Kleinpeter, 1948; Organ, 1960b).

C. Direct Development. As the eggs approach hatching, they enlarge slightly and become slightly ovoid. The eggs at the periphery of the mass are the first to hatch. Those that hatch first have small gill remnants and relatively little yolk. Meanwhile, those that hatch from more centrally located eggs produce smaller, less well-developed young with larger yolk supplies and larger gills. Hatchlings emerge from 12.5–15.0 mm SVL (Organ, 1960b).

i. Brood sites. Females have been found brooding eggs within well-rotted conifer logs under moss mats (Hoffman and Kleinpeter, 1948; Organ, 1960b).

ii. Parental care. Ten nests discovered in the Whitetop Mountain–Mount Rogers area each had attending females. Females guarding eggs were emaciated and had empty stomachs; it is likely that females do not feed while they are with the eggs (Hoffman and Kleinpeter, 1948; Organ, 1960b). A nest discovered within a rotting log at Grandfather Mountain, Avery County, North Carolina, had a female in attendance (D. A. B., personal observations).

D. Juvenile Habitat. Juveniles have been found in the same habitat with adults (Thurow, 1963; D. A. B., personal observations).

E. Adult Habitat. Generally found beneath logs, stones, and flakes of rock (course talus) in spruce-fir forests covering high slopes above 1,524 m (5,000 ft; Bishop, 1943; Hoffman and Kleinpeter, 1948; Organ, 1960b; Thurow, 1963, 1964). They are occasionally found at lower elevations in coves associated with limestone or in mixed deciduous forests below spruce-fir forests (Walker, 1934; Thurow, 1956; Organ, 1960b).

F. Home Range Size. Unknown, but small home ranges are typical for *Plethodon* species.

G. Territories. Weller's salamanders aggressively defend territories under captive

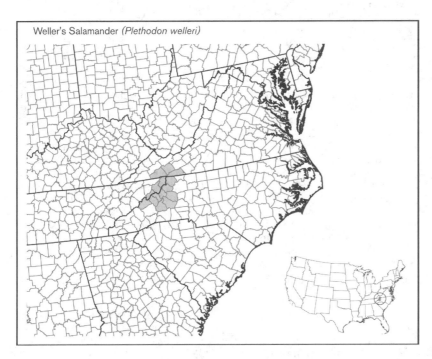

Weller's Salamander (*Plethodon welleri*)

2. Historical versus Current Abundance.
Bishop (1943) reports that in 1938, Weller's salamanders were abundant along the trail leading to Grandfather Mountain. In Petranka's (1998) view, the die-off of spruce-fir forests in the southern Appalachians constitutes the most obvious threat to populations of Weller's salamanders. Highton (this volume, Part One) sampled Weller's salamanders from a population on the North Carolina–Tennessee border (Mitchell and Unicoi counties, respectively) and from a population in Virginia (Grayson-Smyth County) prior to 1988 and again after 1993 and found precipitous declines in both populations. Whether these data reflect true declines or natural population fluctuations can only be determined with additional monitoring.

3. Life History Features.
A. Breeding. Apparently breeding takes place in spring and fall. Males had vasa deferentia swollen with sperm in April and

When the male's tail reaches the female's chin, he begins alternating vertical undulations with the lateral undulations of the tail. The female then raises her forelimb over the male's tail and maintains a straddled position. The pair then moves forward until the spermatophore is deposited. Just prior to depositing a spermatophore, the male begins a series of lateral sacral movements while the female moves her head laterally in synchronization with but counter to the male's movements. The male then lowers his vent to the substrate and deposits a spermatophore. The male then withdraws his tail and bends it sharply to one side. The female keeps her chin against the sharply bent tail while the pair moves forward until the vent of the female reaches the spermatophore. The female then picks up the cap from the spermatophore with her cloacal lips.

i. Breeding migrations. Undocumented, but breeding migrations are not known for any *Plethodon* species.

conditions. However, groups of male and female salamanders also remain closely associated with one another for long periods of time without showing aggression (Thurow, 1976).

Organ (1960b) reports that captive male Weller's salamanders did not establish definite territories, but during September, direct contact with other males usually resulted in fighting. In October, the mere sight of another male caused an aggressive attack.

H. Aestivation/Avoiding Desiccation. Weller's salamanders are well represented in collections made during the summer (Walker, 1934; Hoffman and Kleinpeter, 1948; Thurow, 1956, 1963; Organ, 1960b).

I. Seasonal Migrations. Unknown, surface migrations are unlikely, but animals likely move from forest-floor habitats to underground sites with the seasonal onsets of cold conditions, and back to forest-floor sites with the return of favorable conditions.

J. Torpor (Hibernation). Weller's salamanders were not found at a site on Grandfather Mountain, North Carolina, in December when temperatures were near freezing during the day and below freezing at night. This same locality harbored specimens earlier in the year (Thurow, 1963). It is likely that Weller's salamanders move into underground retreats during colder weather.

K. Interspecific Associations/Exclusions. In the Whitetop Mountain–Mount Rogers area, the following salamander species are reported in association with Weller's salamanders: northern gray-cheeked salamanders (*P. montanus*), red-backed salamanders (*P. cinereus*), Blue Ridge two-lined salamanders (*Eurycea wilderae*), Blue Ridge dusky salamanders (*Desmognathus orestes*), and pigmy salamanders (*D. wrighti*; Walker, 1934; Hoffman and Kleinpeter, 1948; Bogert, 1952). Hoffman and Kleinpeter (1948) report finding two pigmy salamanders with Weller's salamanders on Mount Rogers at an elevation of 1,707 m (5,600 ft) in rotting logs.

Weller's salamanders and southern ravine salamanders (*P. richmondi*) have both been found on Whitetop Mountain and Mount Rogers, Virginia. In this area, Weller's salamanders appear to be restricted to high elevations and southern ravine salamanders to low elevations; the two have not been collected together (Highton, 1972).

At Limestone Cove, Unicoi County, Tennessee, Weller's salamanders have been found with Yonahlossee salamanders (*P. yonahlossee*), white-spotted slimy salamanders (*P. cylindraceus*), eastern newts (*Notophthalmus viridescens*), seal salamanders (*D. monticola*), Carolina mountain dusky salamanders (*D. carolinesis*), and black-bellied dusky salamanders (*D. quadramaculatus*; Thurow, 1963).

At a low elevation near Carderview, Johnson County, Tennessee, Weller's salamanders occur together with southern ravine salamanders (Highton, 1972).

On Grandfather Mountain, North Carolina, Weller's salamanders have been found associated with northern gray-cheeked salamanders, Blue Ridge dusky salamanders, and pigmy salamanders (Walker, 1934; Thurow, 1963; D.A.B., personal observation). Near the summit of Grandfather Mountain, Weller's salamanders, Blue Ridge dusky salamanders, and pigmy salamanders have occasionally been found sheltering beneath the same log within a few inches of one another (D.A.B., personal observation). Pope (1950) reports a Yonahlossee salamander taken on Grandfather Mountain at an elevation of 1,737 m (5,700 ft). This altitude is within the elevational range that Weller's salamanders are known from in this area.

L. Age/Size at Reproductive Maturity. Males reach sexual maturity at about 30 mm SVL. Males probably reach breeding condition by 3 yr, though large males may breed at 2 yr of age. Females mature at about 35 mm, when they are over 3 yr old (Highton, 1962a; Thurow, 1963). The smallest female found attending a nest was 42.5 mm (Organ, 1960b).

M. Longevity. Unknown.

N. Feeding Behavior. The following food items are reported from twelve specimens collected at Grandfather Mountain, North Carolina: Chelonethida, Araneida, Acarina, Collembola, Hemiptera, Lepidoptera, Diptera, and Coleoptera. The animals had been captured 3 d prior to sacrificing them and examining their stomach contents, thus this list represents only part of the food consumed by these animals. A captive Weller's salamander from Limestone Cove, Unicoi County, Tennessee, consumed its own shed skin. The items in the diet are found largely in the forest floor suggesting that Weller's salamanders forage beneath the forest-floor surface or beneath surface cover (Thurow, 1963).

O. Predators. Undocumented, but likely to include forest snakes, birds, and small mammals.

P. Anti-Predator Mechanisms. Weller's salamanders, as do all *Plethodon* (Brodie, 1977), produce noxious skin secretions. Weller's salamanders frequently become immobile when initially contacted. Immobility may increase survival by making the salamander less likely to be detected, especially by visually oriented predators (Dodd, 1989).

Q. Diseases. Unknown.
R. Parasites. Unknown.

4. Conservation.
Weller's salamanders are listed as a Species of Special Concern in North Carolina and as Wildlife in Need of Management in Tennessee. Weller's salamanders are usually

restricted to elevations above 1,524 m (5,000 ft) in areas with spruce-fir forest. Thus, populations are generally separated by uninhabited lower elevations. Several of these mountain peaks are on state or federal property and enjoy some degree of protection. Petranka (1998) states that die-off of spruce-fir forests constitutes a long-term environmental threat to this species.

Acknowledgments. Thanks to Richard Highton, who reviewed this account and gave us the benefit of his insight and experience.

Plethodon yonahlossee Dunn, 1917(a)
YONAHLOSSEE SALAMANDER

David A. Beamer, Michael J. Lannoo

1. Historical versus Current Distribution.
Yonahlossee salamanders (*Plethodon yonahlossee*) are found in the Blue Ridge Mountains of western North Carolina, northeastern Tennessee, and southwestern Virginia (Dunn, 1917a; Bishop, 1943; Hairston, 1949; Newman, 1954b; Petranka, 1998). They occur between about 437–1,737 m in elevation (Petranka, 1998).

Highton (this volume, Part One) sampled four populations during or prior to 1987, and again in 1993 or later, and could not find animals at three sites (two along the North Carolina–Tennessee border and one in Grayson-Wythe County, Virginia). Additional sampling is required to determine whether these data reflect true declines (perhaps extirpations) or natural population fluctuations.

2. Historical versus Current Abundance.
Unknown, but threats include aggressive forestry practices (Petranka et al., 1993, 1994; Petranka, 1998; but see Ash, 1988, 1997, and Ash and Bruce, 1994).

Gordon et al. (1962) compared their collection to a collection made by Dunn 44 yr earlier. The collections were made in the same area, though there had been at least one lumbering of the area between the collections. The number of Yonahlossee salamanders was nearly identical in the two collections.

3. Life History Features.
A. Breeding. Reproduction is terrestrial. Spermatogenesis likely occurs after emergence from hibernation. A spermatogenic wave in the anterior third of the testes was found in a specimen collected on 25 May. Courtship probably occurs in early August, as at this time pairs of salamanders have been found under a single small cover object and males have conspicuous mental glands (Pope, 1950).

i. Breeding migrations. Undocumented, but breeding migrations are not known for any *Plethodon* species.

ii. Breeding habitat. Unknown.

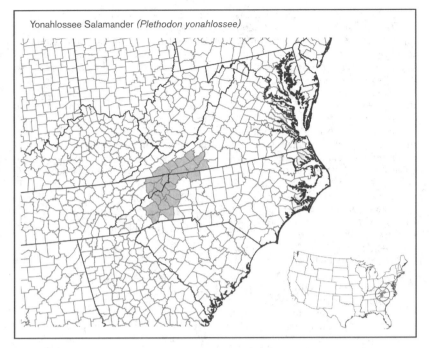

Yonahlossee Salamander *(Plethodon yonahlossee)*

B. Eggs. Females probably begin to lay their eggs in late August. Advanced ova were found in three females collected between 11–16 August at Whitetop Mountain and Comer's Rock, Virginia (Pope, 1950).

i. Egg deposition sites. Unknown, but likely to be in underground cavities.

ii. Clutch size. Three females (77–78 mm SVL) collected between 11–16 August at Whitetop Mountain and Comer's Rock, Virginia, had 19, 24, and 27 mature ova (Pope, 1950).

C. Direct Development.

i. Brood sites. Unknown.

ii. Parental care. Unknown, but it is likely that females brood, as with other species of *Plethodon*.

D. Juvenile Habitat. Juveniles have been collected in the same areas as adults (Pope, 1950; D.A.B., personal observations). Both juveniles and adults emerge from burrows at night to forage. Juveniles are most active for about 1 hr after sunset, while adult activity peaks 1 or 2 hr later (Gordon et al., 1962).

E. Adult Habitat. Yonahlossee salamanders have been recorded from both virgin and second-growth forest (Dunn, 1917a; Gordon et al., 1962). Bailey (1937) reports finding Yonahlossee salamanders common in a rock-filled ravine near Swannanoa, Buncombe County, North Carolina.

In the Iron Mountains, Johnson County, Tennessee, Yonahlossee salamanders have been found in a variety of habitats. Here they have been collected on bare, rock-covered road embankments exposed to the sun, in a talus slope, in deciduous forest, near a wet weather spring, and in an open pasture (Pope, 1950).

In Rutherford County, North Carolina, Yonahlossee salamanders are commonly found in deep crevices of metamorphic rock (Adler and Dennis, 1962; Rubin, 1969). Rock crevices are also occasionally occupied in Avery County and Watuga County, North Carolina (Adler and Dennis, 1962; Guttman et al., 1978).

Dunn (1926) reports that they create long burrows in the forest floor, with openings usually under a fallen log or piece of bark (Bishop, 1943; Martof et al., 1980). However, Pope (1950) suggests that Yonahlossee salamanders do not excavate burrows but may reopen partly obliterated passages.

Petranka (1998) challenged Hairston's (1949) claims that Yonahlossee salamanders in the Black Mountains of North Carolina are generally found within 30 m of streams and are most common in old growth forests.

Yonahlossee salamanders are often found in and under rotting logs, under bark on the ground or still on its log, and under stones. Preferred niches are old windfalls that have shed most of their bark and logs >25 cm (10 in) in diameter, with not more than 5–15 cm (1–3 in) of the log below the surface and a thick layer of leaf accumulation at the log–ground interface (Dunn, 1926; Pope, 1950; Gordon et al., 1962).

F. Home Range Size. Unknown, but small home ranges are typical for *Plethodon* species.

G. Territories. Both juveniles and adults aggressively defend territories in laboratory terraria (Thurow, 1976).

H. Aestivation/Avoiding Desiccation. Yonahlossee salamanders are well represented in collections made during the summer (Dunn, 1917a, 1920; Bailey, 1937; Pope, 1950; Adler and Dennis, 1962; Rubin, 1969; Guttman et al., 1978).

I. Seasonal Migrations. Animals likely move vertically from forest-floor sites to underground sites in response to seasonal dry and cold conditions.

J. Torpor (Hibernation). Unknown, but it is likely that Yonahlossee salamanders seek underground retreats during winter weather.

K. Interspecific Associations/Exclusions. In Avery County, North Carolina, Yonahlossee salamanders were found with eastern newts *(Notophthalmus viridescens)*, Blue Ridge dusky salamanders *(Desmognathus orestes)*, seal salamanders *(D. monticola)*, red-backed salamanders *(P. cinereus)*, northern gray-cheeked salamanders *(P. montanus)*, white-spotted slimy salamanders *(P. cylindraceus)*, and Blue Ridge two-lined salamanders *(Eurycea wilderae*; Gordon et al., 1962).

Pope (1950) reports a Yonahlossee salamander taken on Grandfather Mountain, North Carolina, at an elevation of 1,737 m (5,700 ft). Weller's salamanders *(P. welleri)*, northern gray-cheeked salamanders, pigmy salamanders *(D. wrighti)*, and Blue Ridge dusky salamanders are well known from this elevation at Grandfather Mountain.

Hairston (1949) reports the following salamanders along with Yonahlossee salamanders from the Black Mountains area, Yancey County, North Carolina: spring salamanders *(Gyrinophilus porphyriticus)*, red salamanders *(Pseudotriton ruber)*, Blue Ridge two-lined salamanders, black-bellied salamanders *(D. quadramaculatus)*, seal salamanders, Carolina mountain dusky salamanders *(D. carolinensis)*, pigmy salamanders, shovel-nosed salamanders *(D. marmoratus)*, white-spotted slimy salamanders, and northern gray-cheeked salamanders.

At Limestone Cove, Unicoi County, Tennessee, Yonahlossee salamanders have been found with Weller's salamanders, white-spotted slimy salamanders, eastern newts, seal salamanders, Carolina mountain dusky salamanders, and black-bellied salamanders (Thurow, 1963).

The following species were found on a cliff face with Yonahlossee salamanders in Rutherford County, North Carolina: green salamanders *(Aneides aeneus)*, Blue Ridge gray-cheeked salamanders *(P. amplus)*, white-spotted slimy salamanders, and Ocoee salamanders *(D. ocoee*; Adler and Dennis, 1962; Rubin, 1969).

At Whitetop Mountain, Virginia, Yonahlossee salamanders have been found with Wehrle's salamanders *(P. wehrlei*; Highton, 1962a). Pope (1950) reports Yonahlossee salamanders from elevations between 1,151–1,669 m (3,775–5,475 ft) on Whitetop Mountain. Organ (1961a) reports the following species from this same area: red-backed salamanders, white-spotted slimy salamanders, northern gray-cheeked salamanders, Weller's salamanders, southern ravine salamanders (*P. richmondi*), red salamanders, spring salamanders, Blue Ridge two-lined salamanders, black-bellied salamanders, seal salamanders, northern dusky salamanders (*D. fuscus*), Blue Ridge dusky salamanders, and shovel-nosed salamanders.

Yonahlossee salamanders are sympatric with white-spotted slimy salamanders at intermediate elevations throughout much of the Blue Ridge Province east of the French Broad River. There is no evidence of hybridization between these species (Highton and Peabody, 2000).

Yonahlossee salamanders are sympatric with northern slimy salamanders (*P. glutinosus*) only in the vicinity of Skulls Gap, Smyth County, Virginia. A probable F1 hybrid was found at this location (Highton and Peabody, 2000).

Yonahlossee salamanders are widely sympatric throughout the range of Blue Ridge gray-cheeked salamanders (*P. amplus*). There is no evidence of hybridization between these species (Highton and Peabody, 2000).

Yonahlossee salamanders are widely sympatric with northern gray-cheeked salamanders at the eastern edge of the Blue Ridge isolate. There is no evidence of hybridization (Highton and Peabody, 2000).

Yonahlossee salamanders tend to be much more altitudinally restricted than any of these associated species. A comparison between Yonahlossee salamanders and white-spotted slimy salamanders suggests a large degree of dietary overlap, although there is no evidence to suggest that competitive exclusion occurs (Pope, 1950).

L. Age/Size at Reproductive Maturity. Cloacal gland papillae and mental glands are evident in males at approximately 56 mm SVL. It is likely that these salamanders are mature. A 60-mm female appeared to be mature, though a 66-mm female was immature. Maturity in females is likely reached somewhere around 60–66 mm SVL. Sexual maturity in Yonahlossee salamanders probably occurs when they are almost 3 yr old (Pope, 1950).

A cliff-dwelling population of Yonahlossee salamanders from Rutherford County, North Carolina, apparently does not mature until a large size. Males do not mature until they are >65 mm and females do not mature until they are >61 mm (Adler and Dennis, 1962).

M. Longevity. Unknown.

N. Feeding Behavior. Juveniles forage most actively about 1 hr after sunset, while adult foraging activity peaks 1 or 2 hr later (Gordon et al., 1962).

An examination of the stomachs of 28 Yonahlossee salamanders during summer from Rutherford County, North Carolina, included representatives from the following invertebrate taxa: Gastropoda, Araneida, Diplopoda, Curculion-

idae, coleopteran adults and larvae, lepidopteran larvae, Collembola, Orthoptera, dipteran adults, larvae, and pupae, shed skins, Acarina, Hymenoptera, Coleoptera, Lepidoptera, Hemiptera, Chilopoda, Pseudoscorpianida, Staphylinidae, Annelida, Mecoptera, Neuroptera, and Nematoda (Rubin, 1969).

Food items from 50 Yonahlossee salamanders collected in the Iron Mountains included Collembola, Isoptera, lepidopteran larvae, dipteran larvae, coleopteran adults and larvae, Chelonethida, Araneida, Acarina, Diplopoda, Chilopoda, Isopoda, and Pulmonata. These specimens were kept alive for 1–3 d after capture, so it is possible soft-bodied prey that are digested more rapidly were absent. A female (77 mm SVL) from this series had eaten 15 invertebrates (Pope, 1950).

Rankin (1937) reported centipedes, ants, and beetles from the stomachs of three specimens. In captivity, a Yonahlossee salamander ate a small member of *P. jordani* complex (Thurow, 1976).

O. Predators. Undocumented, but likely to include forest snakes, birds, and small mammals.

P. Anti-Predator Mechanisms. Yonahlossee salamanders produce tail secretions that are noxious to birds and other potential predators (Petranka, 1998). Yonahlossee salamanders frequently become immobile when initially contacted. Immobility may increase survival by making the salamander less likely to be detected, especially by visually oriented predators (Dodd, 1989).

Q. Diseases. Unknown.

R. Parasites. Rankin (1937) reported the following parasites from Yonahlossee salamanders: Protozoa—*Cytamoeba bacterifera, Eutrichomastix batrachorum, Karotomorpha swezi, Prowazekella longifilis,* and *Tritrichomonas augusta;* Trematoda—*Brachycoelium hospitale;* Nematoda—*Oxyuris magnavulvaris.*

Yonahlossee salamanders from a cliff-face population in Rutherford County, North Carolina, were infected with dermal mites (*Hannemania hegeneri;* Adler and Dennis, 1962).

4. Conservation.
Yonahlossee salamanders are not protected in any of the states within their range. While relatively wide ranging, Yonahlossee salamanders are usually found from intermediate to high elevations. Suitable habitat at these elevations may be separated by stretches of lower uninhabited areas, and populations are often separated. Within this range, there are many federal and state properties that contain suitable habitat for these salamanders.

Yonahlossee salamanders are relatively resilient to disturbances, such as those associated with timbering operations, and are frequently found in second-growth

forests and relatively small, fragmented woodlots (Gordon et al., 1962; D. A. B., personal observations). However, aggressive forestry practices may negatively affect populations (Petranka et al., 1993, 1994; Petranka, 1998).

As with all species of *Plethodon,* Yonahlossee salamanders do not migrate to breeding grounds and they do not have large home ranges. Thus, they can exist in habitats of smaller size than many other amphibian species. Conservation activities that promote mature closed-canopy forests should benefit this species.

Acknowledgments. Thanks to Richard Highton, who reviewed this account and gave us the benefit of his insight and experience.

Pseudotriton montanus Baird, 1849
MUD SALAMANDER
Todd W. Hunsinger

1. Historical versus Current Distribution.
Mud salamanders (*Pseudotriton montanus*) are found from extreme southeastern Louisiana east to the Atlantic Coast, north to southern New Jersey, and westward to the Illinois boundary (Martoff, 1975b). Four subspecies are currently identified: eastern mud salamanders (*P. m. montanus*), midland mud salamanders (*P. m. diasticus*), Gulf Coast mud salamanders (*P. m. flavissimus*), and rusty mud salamanders (*P. m. floridanus*). Mud salamanders are naturally absent from the higher elevations of the Appalachian Mountains and the northern coast of North Carolina. Mud salamanders are reported from only two counties in New Jersey where they are listed as Threatened. Heckscher (1995) confirmed records for Sussex County, Delaware, where they were previously unreported. Miller (1990) verified their presence in Wilson County, Tennessee (see also Redmond and Scott, 1996). Mud salamanders are listed as Endangered in Pennsylvania after being verified in Franklin County in 1991 (McCoy, 1992). Bruce (1974) reported the loss of habitat in Oconee, South Carolina, due to the impoundment of the Little River to form Lake Keowee.

2. Historical versus Current Abundance.
Little is known about the abundance of this secretive, often subterranean species. It is not uncommon for decades to pass between sightings despite intense collection efforts at a given location. Many of the written accounts are based on observations of a single specimen or few individuals (Wright and Trapido, 1940; Fowler, 1946; Barbour, 1953; Hirschfield and Collins, 1963; Miller, 1990; Heckscher, 1995). The scarcity of records for the Delmarva Peninsula has been attributed in part to the secretive nature of mud salamanders, although wetland loss and

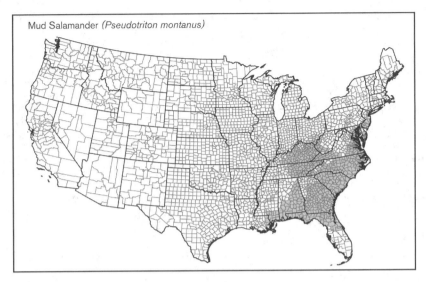

Mud Salamander (*Pseudotriton montanus*)

degradation are believed to have impacted this species (Heckscher, 1995). Mount (1975) reported a lack of abundance of specimens from known localities in Alabama. Mud salamanders were found in Louisiana in 1998 after a 31-yr absence of records (J. Boundy, personal communication). Mud salamanders were first described by Baird based on a specimen from South Mountain in Cumberland County, Pennsylvania. This was long believed to be a questionable account (McCoy, 1982), as repeated searching at this locality failed to locate mud salamanders. However, McCoy (1992) confirmed the species in neighboring Franklin County after an absence of records spanning approximately 140 yr.

3. Life History Features.

A. Breeding. Reproduction is aquatic.

i. Breeding migrations. There is no known breeding migration for this predominately aquatic and fossorial species. Mating has been reported in September in Kentucky (Robinson and Riechard, 1965). Mature sperm, indicative of the reproductive peak, were found in the vas deferens of adult males in South Carolina from mid-August to mid-September (Bruce, 1975). Adult males and females are often found together in the same burrow during this time.

ii. Breeding habitat. Springs, seeps, and bogs (Brimley, 1939; Fowler, 1946; Goin, 1947c).

B. Eggs.

i. Egg deposition sites. Bruce (1975) estimated oviposition to occur in November based on reports in the literature. Sexually mature females were gravid from mid-August to mid-September (Bruce, 1975). Fowler (1946) found eggs in various stages of development in late December.

Little data on breeding habitats exist for mud salamanders. Brimley (1939) reported egg masses attached to dead leaves in a small spring in North Carolina. Fowler (1946) found egg clusters in the rootlets of a hillside seep near a cypress swamp in Maryland. Goin (1947c) reported eggs in a boggy, sphagnum seepage. Eggs were gathered in clusters ranging from 2–8 ova and attached to rootlets that hung down into the water on the edge of an undercut bank.

ii. Clutch size. Dissection of the ovaries of mature females revealed 77–192 yolked oocytes, increasing in number as a function of body volume, for populations in South Carolina (Bruce, 1975). However, clutches exceeding 30 eggs have not been found.

Goin (1947c) reported recently hatched larvae on 14 January and 22 February in Florida. Hatching has been estimated to occur in February to early March in the Carolinas (Bruce, 1978b), with hatchlings ranging from 7.5–9.0 mm SVL (Goin, 1947c) to 10–13 mm (Bruce, 1974, 1978b).

C. Larvae/Metamorphosis.

i. Length of larval stage. Development from ovipositing to metamorphosis can be as quick as 1.5 yr or last as long as 2.5 yr, and may be a consequence of differences in temperature (Bruce, 1978b). Size at metamorphosis is consistent, with a SVL ranging from 36–48 mm, with a mean near 40 mm (Bruce, 1978b [It should be noted that Bruce (1978b) measured SVL to the anterior edge of the cloaca, not to the posterior edge, which is the currently accepted method.]). Goin (1947c) reported larvae ranging from 34–39 mm SVL in January. Netting and Goin (1942b) found transformed larvae ranging from 38–43 mm SVL with a total length of about 74 mm. Metamorphosis may occur from May–September.

ii. Larval Requirements.

a. Food. Larvae presumably feed on a variety of aquatic invertebrates.

b. Cover. Bruce (1974) found larvae in the mud and decaying vegetation of springs, small streams, and swampy pools and ponds; Bruce (2003) found mud salamanders in the bottomland swamp of a third order stream.

iii. Larval polymorphisms. Unknown.

iv. Time to metamorphosis. From 1.5–2.5 yr after ovipositing. Bruce's (1974) data for South Carolina populations suggest that metamorphosis occurs over a lengthy period, from mid-May to early September, with a concentration in July–August.

v. Post-metamorphic migrations. Bruce (1975) found recently transformed larvae along the margin of watercourses in the leaves and debris and observed movements through a series of burrows that connect their watercourse to upland habitats.

vi. Neoteny. Unknown.

D. Juvenile Habitat. Recently transformed individuals do not burrow as readily as older adults, but are found in leaves and debris at the edge of the water (Bruce, 1975).

E. Adult Habitat. Contrary to what the species name *montanus* might imply, mud salamanders are predominantly a lowland species (Wright and Trapido, 1940; Bruce, 1975; Heckscher, 1995). The preferred habitat has been described as consisting of lowland seeps, palustrine wetlands, muddy springs and streams, and swampy pools and ponds. Bruce (1975) reported the use of complexes of burrows concentrated near the water, but occasionally being located as far as 15–20 m away from surface water. These contained vertical channels to the surface from the water table below. Adults were located in the burrows with their heads near the surface. Laboratory observations revealed that individuals will construct their own burrows, rather than use those abandoned by other species (Bruce, 1975). Mount (1975) reported finding mud salamanders beneath logs in low, wooded flood plains. Funderburg (1955) observed that "it is a subterranean form and is seldom found under old logs as it is in other localities" in North Carolina. Barbour (1953) found a female under a stone in a dry, rocky pasture during August in Kentucky.

F. Home Range Size. Unknown.

G. Territories. Unknown.

H. Aestivation/Avoiding Desiccation. Undocumented.

I. Seasonal Migrations. Limited migrations away from the breeding habitat occur during the summer.

J. Torpor (Hibernation). Does not occur; mud salamanders are active throughout the year.

K. Interspecific Associations/Exclusions. Found in association with spring salamanders (*Gyrinophilus porphyriticus*), northern two-lined salamanders (*Eurycea bislineata*), long-tailed salamanders (*E. longicauda*), red salamanders (*P. ruber*), northern dusky salamanders (*Desmognathus fuscus*), four-toed salamanders (*Hemidactylium scutatum*), and eastern newts (*Notophthalmus viridescens*). Along the Coastal Plain of South Carolina they are found in association with southern dusky salamanders (*D. auriculatus*), long-tailed salamanders, two-lined salamanders, lesser sirens (*Siren intermedia*), and two-toed amphiumas (*Amphiuma means*; Semlitsch, 1983d).

L. Age/Size at Reproductive Maturity. Females reach sexual maturity at 4–5 yr of age (Bruce, 1975). Males reach sexual maturity in the first year following metamorphosis at a size exceeding 43 mm SVL, reproducing for the first time in the second, third, or fourth year of life, depending on the duration of the larval stage (Bruce, 1975).

M. Longevity. In captivity, individuals have lived past 15 yr (Snider and Bowler, 1992).

N. Feeding Behavior. Quantitative data are lacking on this aspect of the life history of the mud salamander. Dunn (1926) believed that adults may prey on smaller salamanders.

O. Predators. Snakes are the only documented predators of mud salamanders. Northern water snakes (*Nerodia sipedon*) feed on larval mud salamanders (E.E. Brown, 1979; Kats, 1986). Common garter snakes (*Thamnophis sirtalis*) were reported as predators of adult mud salamanders by Carr (1940a) and E.E. Brown (1979).

P. Anti-Predator Mechanisms. Defensive posturing includes extending the rear limbs, curling the body, and raising the tail or wrapping it around the head. A toxic substance is secreted along the back. Mud salamanders are currently believed to be part of a Mullerian mimicry complex with eastern newts and spring salamanders.

Q. Diseases. Unknown.

R. Parasites. Rankin (1937) described the parasites of mud salamanders from North Carolina as follows: Protozoa—*Haptophrya michiganensis*, *Prowazekella longifilis*, and *Tritrichomonas augusta*; Trematoda—*Allocreadium pseudotritoni*, and *Gorgoderina bilobata*; Nematoda—*Physaloptera* sp.; Acarina—*Hannemania dunni*.

4. Conservation.
Mud salamander populations have been lost (Bruce, 1974). They are listed as Rare and on the Watchlist in Maryland, as a Species of Special Concern in Louisiana and South Carolina, and are limited in range and considerd Rare in Ohio (Pfingsten, 1989f) and West Virginia. They are listed as Threatened in New Jersey (Levell, 1997), and as Endangered in Pennsylvania, after being verified in Franklin County in 1991 (McCoy, 1992).

Acknowledgments. The following people assisted with the preparation of this account: Jeff Boundy, Chris Heckscher, Scott Smith, Eric Stiles, Jim White, and Robert Zappalorti. This is contribution number

825 of the New York State Museum and Science Service.

Pseudotriton ruber (Latreille, 1801)
RED SALAMANDER
Todd W. Hunsinger

1. Historical versus Current Distribution.
Red salamanders (*Pseudotriton ruber*) are found from the Hudson River in New York southwestward to Indiana and southward to Louisiana and the Gulf Coast. Four subspecies are currently recognized: northern red salamanders (*P. r. ruber*), Blue Ridge red salamanders (*P. r. nitidus*),black-chinned

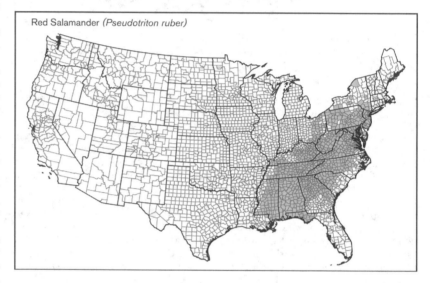

Red Salamander (*Pseudotriton ruber*)

red salamanders (*P. r. schencki*), and southern red salamanders (*P. r. vioscai*). Red salamanders are naturally absent from much of the Atlantic Coastal Plain and are uncommon on the Coastal Plain of Delaware (Martof, 1975c; J. White, personal communication). Red salamanders are known from only one county in Louisiana, where they are listed as Species of Special Concern. Red salamanders have vanished from places where they historically occurred. For example, Pfingsten (1989g) reported that this species is no longer found near Cincinnati, Ohio. In 1926, Bishop (1927) did not find red salamanders in streams in southwestern New York where they were taken in 1923. In a letter written during fieldwork in the area, Bishop lamented: "Some fine collecting places have been destroyed in the process of 'improving' the park and among them 1 of the 2 known streams for the Red Salamander." Recent attempts to document red salamanders in this location have failed (unpublished data). Overlooked in previous accounts of red salamanders is a specimen taken on Long Island in 1938 (NYSM3206; Hinderstein, 1968). No further records of red salamanders exist from this locality. Smith (1887) and Deckert (1914c) reported populations east of the

Hudson River in New York. The lack of vouchered specimens from these reports and inconsistencies with the known range make these identifications questionable. This sentiment was also reflected in a letter from G. Kingsley Noble to Sherman Bishop dated 23 April 1928. Noble wrote, "Dunn [Emmit Dunn of Smith College] has seen it [Noble, 1927c] and loudly protests against my having left out Deckert's references. I am convinced that his material was incorrectly identified, but that does not exclude the possibility of the red salamander . . . being yet found on this side of the Hudson in our region."

2. Historical versus Current Abundance.
Wilmott (1933) stated that red salamanders are not nearly as common as they formerly were on Staten Island, New York. They are also rare in the inner and outer central basins of Tennessee (Redmond and Scott, 1996).

3. Life History Features.
A. Breeding. Reproduction is aquatic.

i. Breeding migrations. There is no true breeding migration for this species. Breeding occurs in the same streams where adults spend much of their lives. Adults will return to the streams in the fall for breeding and egg laying.

ii. Breeding habitat. Ovipositing has rarely been observed. Eggs have been found deposited in springs, headwater streams, and seepage-fed mountain bogs (Bishop, 1941b; Bruce, 1972b, 1978c; Semlitsch, 1983d). Bishop (1925) reported a clutch of eggs found in a tamarack swamp in New York. A simple courtship pattern, including terrestrial components, is proposed for red salamanders and inferred to be ancestral for plethodontids (Organ and Organ, 1968).

B. Eggs.
i. Egg deposition sites. These eggs were on the lower surface of a large flat rock

15 cm (6 in) below the water surface and buried beneath the sod at the margin of a cold spring (Bishop, 1925).

ii. Clutch size. Clutch sizes range from 29–130 eggs (Bruce, 1978c), with a mean of 80. Bruce concluded that egg laying occurs in autumn. Bishop (1925) reported a single clutch of 72 eggs. Incubation lasts for 2–3 mo (Bishop, 1941b). Hatchlings and late embryos have been reported from November–March. Hatching was inferred to occur in January at 11.2–13.5 mm SVL along the Coastal Plain of South Carolina (Semlitsch, 1983d).

C. Larvae/Metamorphosis. Bruce (2003) confirmed the affinity of larvae for springs. Semlitsch (1983d) revealed that larval growth rates range from 1.2–2.0 mm/mo during the first six months.

i. Length of larval period. The larval period may last from 1.5–3.5 yr, depending on location. Semlitsch (1983d) reported a larval period of only 18–23 mo along the Coastal Plain of South Carolina and suggests that development may be correlated to water temperatures. Length ranged from 45.5–50.0 mm SVL. Bruce (1978c) reported metamorphosis at 31–33 mo in the Blue Ridge Mountains. Gordon (1966) estimated a larval period ranging from 29–33 mo in the Highland Plateau of North Carolina. Based on data from an Ohio population, Pfingsten (1989e) suggests a larval period of 27–31 mo, with metamorphosis at 44 mm SVL. Larvae at 54 mm SVL will probably metamorphose during their fourth year. Bishop (1941b) reported a larval period of 3.5 yr at the northern edge of their range in New York. Bell (1956) reported similar data in Pennsylvania, indicating the same development pattern. Larvae mature at 34–46 mm SVL in the Blue Ridge Mountains of the Carolinas (Bruce, 1972b [It should be noted that Bruce measured SVL to the anterior edge of the cloaca, not to the posterior edge, which is the usual method.]). Pfingsten (1989e) reported larvae ranging from 44–52 mm SVL in Ohio. Larvae in a South Carolina population averaged 47.1 ± 0.6 mm SVL (Semlitsch, 1983d). At the northern edge of the range, larvae have been known to exceed 85 mm TL (Bishop, 1941b; Bell, 1956; unpublished data).

ii. Larval requirements.

a. Food. Larvae likely feed on a range of aquatic invertebrates.

b. Cover. Larvae are found underneath logs and rocks or buried in the soft substrate in slower sections of streams. Bruce (1972b) found larvae among accumulations of decaying vegetation in springs, seepage areas, and pools along the course of small streams in the Blue Ridge Mountains. Bahret (1996) reported larvae in a deep acidic lake in New York.

iii. Larvae polymorphisms. Unknown.

iv. Time to metamorphosis. Variable; ranges from 1.5–3.5 yr, depending on location. (see "Length of larval period" above).

v. Post-metamorphic migrations. Unknown.

vi. Neoteny. Unknown.

D. Juvenile Habitat. Likely similar to adult characteristics.

E. Adult Habitat. Adults are aquatic and terrestrial. Aquatic habitat is most often slow-moving headwater springs and seeps in wooded lowland and upland areas. Tamarack wetlands are frequently mentioned in the habitat accounts for red salamanders in New York (Bishop, 1941b; Axtell and Axtell, 1948). Occasionally, individuals are found in swifter streams (Hunsinger and Morse, unpublished data). Adults have been reported from a deep acidic lake in New York (Bahret, 1996). Terrestrial adults are found under rocks, logs, or mats of sphagnum moss (personal observations) in wooded ravines, swamps, open fields, and meadows. Adults use burrows that connect to watercourses (Bishop, 1941b; Axtell and Axtell, 1948). Marvin (2003) examined aquatic and terrestrial burst speeds related to temperature acclimation.

F. Home Range Size. Unknown.

G. Territories. Unknown.

H. Aestivation/Avoiding Desiccation. Undocumented.

I. Seasonal Migrations. Adults migrate from the stream to terrestrial habitats on warm rainy nights in early April in New York (unpublished data) and return to the streams for breeding in the late summer or fall. Adults return to the stream in October in New York (Axtell and Axtell, 1948; unpublished data). Most activity occurs in the evening.

J. Torpor (Hibernation). Adults and juveniles are difficult to locate during the winter. For example, I could not locate larvae in a New York stream on 12 December 1998 in sections where they had been easily found on 16 November 1998 (unpublished data), indicating that larvae may cease activity and retreat to deeper underground springs during the winter.

K. Interspecific Associations/Exclusions. Red salamanders share habitat with Allegheny mountain dusky salamanders (*Desmognathus ochrophaeus*), northern dusky salamanders (*D. fuscus*), northern two-lined salamanders (*Eurycea bislineata*), and occasionally spring salamanders (*Gyrinophilus porphyriticus*) in the northern part of their range, and seal salamanders (*D. monticola*) and black-bellied salamanders (*D. quadramaculatus*) in southern localities (Bruce, 1974). Semlitsch (1983d) reported mud salamanders (*Pseudotriton montanus*), southern dusky salamanders (*D. auriculatus*), northern two-lined salamanders, and long-tailed salamanders (*E. longicauda*), as well as lesser sirens (*Siren intermedia*) and two-toed amphiumas (*Amphiuma means*), in association with northern red salamanders in the coastal plain of South Carolina. Goin (1939) reports association with northern slimy salamanders (*Plethodon glutinosus*)

and three-lined salamanders (*E. guttolineata*) in western Florida.

L. Age/Size at Reproductive Maturity. Bruce (1978c) found sexual maturity in males is reached during their fourth summer, just short of 4 yr old. Females attain sexual maturity at 5 yr and reproduce annually after that. The smallest breeding female was 55 mm SVL.

M. Longevity. Individuals are known to live >20 yr in captivity (Snider and Bowler, 1992).

N. Feeding Behavior. Diet includes smaller salamanders such as eastern red-backed salamanders (*P. cinereus*; Bishop, 1941b; Bock and Fauth, 1992), aquatic and terrestrial insects, earthworms, slugs, spiders, and millipedes. Axtell and Axtell (1948) reported water beetles (*Hydrophilus* sp.), sow bugs (*Oniscus* sp.) and crickets (*Gryllus* sp.), in the diet of New York specimens.

O. Predators. Little data are available on known predators. Uhler et al. (1939) found a red salamander in the stomach of a copperhead (*Agkistrodon contortrix*). I have documented predation by a common garter snake (*Thamnophis sirtalis*; NYSM color slide 20) in New York. Hunsinger and Morse (unpublished data) found red salamanders in association with fishes, including brown trout (*Salmo trutta*), which are known predators of salamanders (Bishop, 1941b).

P. Anti-Predator Mechanisms. When threatened, individuals assume a defensive posture of raising their hindlimbs and tail, slowly undulating their tail, and tucking their head under their tail. Red salamanders also produce a highly toxic secretion from their dorsal surface (Brandon et al., 1979b; Brandon and Huheey, 1981). Numerous studies have examined possible mimicry with the red eft stage of eastern newts (*Notophthalmus viridescens*) and spring salamanders (e.g., Brodie, 1976; Brandon et al., 1979b). It is believed that these species form a Mullerian mimicry complex (Brandon et al., 1979b).

Q. Diseases. Unknown.

R. Parasites. Rankin (1937) described the parasites of red salamander larvae from North Carolina as follows: Protozoa—*Cryptobia borreli*, *Cytamoeba bacterifera*, *Eutrichomastix batrachorum*, *Hexamastix batrachorum*, *Hexamitus intestinalis*, *Karotomorpha swezi*, *Prowazekella longifilis*, and *Tritrichomonas augusta*; Trematoda—*Allocreadium pseudotritoni*, and intestinal wall metacercariae; Cestoda—proteocephalid cysts. Adults contained the following parasites: Protozoa—*Cryptobia borreli*, *Cytamoeba bacterifera*, *Prowazekella longifilis*, and *Tritrichomonas augusta*; Trematoda—*Allocreadium pseudotritoni*, *Brachycoelium hospitale*, and *Gorgoderina bilobata*; Cestoda—*Crepidobothrium cryptobranchi*.

4. Conservation.

Red salamanders are listed as Endangered in Indiana (www.in.gov/dnr), as a Species of Special Concern in Louisiana, and as Protected in New Jersey (Levell, 1997). They are known to have vanished from places where they historically occurred (See "Historical versus Current Distribution" above). Other populations appear to be less robust than they were historically (Wilmott, 1933; Redmond and Scott, 1996).

Acknowledgments. The following people assisted with the preparation of this report: Kraig Adler, Jeff Boundy, Scott Smith, and Jim White. This is contribution number 824 of the New York State Museum and Science Service.

Stereochilus marginatus (Hallowell, 1856)
MANY-LINED SALAMANDER

Travis J. Ryan

1. Historical versus Current Distribution.

Many-lined salamanders (*Stereochilus marginatus*) inhabit wetlands, permanent lentic habitats, and slow-moving streams on the Atlantic Coastal Plain. The northern extreme of the distribution is southeastern Virginia, and they may be found as far south as northeastern Florida (Bishop, 1943; Schwartz and Etheridge, 1954; Neill, 1957e; Cooper, 1962; Rabb, 1966; Gerhardt, 1967; Christman et al., 1979b; Campbell et al., 1980; Petranka, 1998; Mitchell and Reay, 1999). Mitchell and Reay (1999) note that the northern portion of their range needs to be better determined; many-lined salamanders are one of the more poorly known plethodontid species in this fairly well-studied region. Loss of populations has undoubtedly occurred with wetland drainage (Petranka, 1998; Means, this volume, Part One), but this has not been documented or quantified to any appreciable degree. Mitchell (1991) describes their conservation status as Undetermined.

2. Historical versus Current Abundance.

Wetlands along the Atlantic Coastal Plain, where many-lined salamanders are found, are rapidly disappearing or being substantially altered (Means, 2000, this volume, Part One). How this is affecting the many-lined salamander, however, is not yet known. It is worth noting, however, that the range of many-lined salamanders overlaps substantially with flatwoods salamanders (*Ambystoma cingulatum*), a species federally listed as Threatened in 1999. While these species differ ecologically (e.g., flatwoods salamander adults are primarily terrestrial and migrate to breed in ephemeral wetlands), wetland conversion and destruction is likely a common primary threat to their long-term persistence.

3. Life History Features.

A. Breeding. Reproduction is aquatic and appears to follow the stereotypical behaviors associated with plethodontid salamanders with both tactile and chemical stimulation appearing mandatory. Noble and Brady (1930) make mention of a tail straddle walk and that males rub their mental glands over the nares of females. Beyond this, however, there is no detailed account of many-lined salamander courtship.

i. Breeding migrations. Negligible, if present at all. Because adults are primarily aquatic, they reside in the nuptial/natal site year-round. Many-lined salamanders are autumn breeders, and both sexes reproduce annually (Bruce, 1971).

ii. Breeding habitat. The same as general adult habitats: ponds, bogs, marshes, ditches, and slow-moving streams.

B. Eggs.

i. Egg deposition sites. Females oviposit during the winter in sites usually associated with aquatic habitats. Eggs are laid in either water or on land (Wood and Rageot, 1963; Bruce, 1971; see also Petranka,

1998). In water, eggs are attached singly to a support; in nature, to the lower surface of a stone, log, or piece of bark, 8–15 cm below the water surface in quiet pools (Bishop, 1943; Wood and Rageot, 1963); females do not brood. In the Dismal Swamp, where *Sphagnum* sp. moss is common, females show a preference for laying eggs on the water moss *Fontinalis* sp. (Wood and Rageot, 1963; see also Rabb, 1966). On land, eggs tend to be more clustered, and in about 30% of clutches females brood. Terrestrial nests have been found inside a decaying gum log near a pond margin (Rabb, 1956) and under a small log near a borrow pit (Schwartz and Etheridge, 1954; see also Petranka, 1998).

ii. Clutch size. Between 16–121 eggs are laid (Wood and Rageot, 1963; Bruce, 1971; see also Petranka, 1998; Ryan and Bruce, 2000) in most populations, but may average as many as 57 in a South Carolina population (Rabb, 1966). Eggs are 2–3.4 mm in diameter (Noble and Richards, 1932; Rabb, 1956). Hatchlings measure 8 mm SVL (Bruce, 1971). When compared with other generalized plethodontids, many-lined salamanders lay more eggs of smaller size in more ephemeral wet areas, likely a response to high larval mortality (Ryan and Bruce, 2000).

C. Larvae/Metamorphosis. Many-lined salamanders demonstrate a pond-like larval morphology, rare among plethodontids, including a high caudal fin extending onto the body, large filamentous gills, and a strongly invaginated gular fold (see Ryan and Bruce, 2000).

i. Length of larval stage. Larval size at hatching is about 8 mm, and larvae have functional limbs (Bruce, 1971; Ryan and Bruce, 2000). Growth is greater in the first year (about 15–16 mm SVL) than in the second (9–10 mm SVL). Larvae metamorphose either at the beginning of their second year (13–16 mo post hatching) or in the early part of their third year (25–28 mo post-hatching; Bruce, 1971). Size at metamorphosis ranges between 27–40 mm SVL (Foard and Auth, 1990; Ryan and Bruce, 2000).

ii. Larval requirements.

a. Food. Larvae are carnivorous, feeding on small invertebrates associated with vegetation beds (see Petranka, 1998). In the only published study of the diet of many-lined salamanders, Foard and Auth (1990) found isopods to be the most common food item in the gut of the larvae. Other aquatic invertebrate taxa documented were (in order of decreasing occurrence) chironomid larvae, ostracods, amphipods, copepods, dytiscid beetles, and cladocerans. The diet of larvae differs significantly from that of adults (see "Feeding Behavior" below).

b. Cover. Larvae are most commonly found in the aquatic vegetation at the periphery of wetlands, such as *Sphagnum*

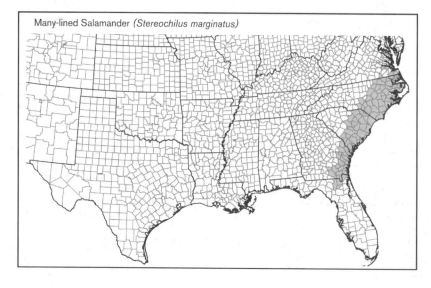

Many-lined Salamander (*Stereochilus marginatus*)

mats or *Fimbristylis* (Rabb, 1966; Bruce, 1971; Foard and Auth, 1990).

iii. Larval polymorphisms. Unknown and unlikely.

iv. Features of metamorphosis. In North Carolina populations, metamorphosis occurred after 1 or 2 yr of development when larvae were 27–42 mm SVL (Bruce, 1971). In Georgia, metamorphic individuals measured between 30 and 35 mm SVL (Foard and Auth, 1990), presumably also 1 or 2 yr post-hatching. While Bruce indicated that metamorphosis is most common in spring and summer, there are no corresponding data on timing of metamorphosis for the Georgia population.

v. Post-metamorphic migrations. Most pond-breeding salamanders (e.g., those of the genera *Ambystoma* and *Notophthalmus*) migrate from the nuptial/natal site following metamorphosis; many-lined salamanders are a clear exception to this trend. Juveniles and adults are primarily aquatic and appear to stay within or at least near the pond of origin.

vi. Neoteny. Perennibranchism, the retention of external gills throughout life, is not known, but these aquatic salamanders demonstrate some degree of reproductive acceleration when compared to their closest relatives in the genera *Gyrinophilus* and *Pseudotriton* (Ryan and Bruce, 2000).

D. Juvenile Habitat. Because post-metamorphic individuals are still predominantly aquatic, the larval, juvenile, and adult habitat characteristics are essentially the same (see Petranka, 1998).

E. Adult Habitat. Many-lined salamanders are usually aquatic, especially in permanent water, but occasionally found on land under logs in damp situations (Brimley, 1909, 1939; Bishop, 1943; Rabb, 1966; Bruce, 1971; Foard and Auth, 1990). Means (2000) refers to these animals as "technically a wetland species" living their entire larval and metamorphosed life in shallow, acid waters of Lower Coastal Plain swampy streams, coming onto land only occasionally. Most abundant in pools and slow streams such as gum and cypress swamps, woodland ponds, borrow pits, canals, and drainage ditches. Ryan and Bruce (2000), following Rabb (1966), note they are usually restricted to swamps, ditches, and sluggish streams of the Atlantic Coastal Plain. Animals can be collected by raking out dead leaves and detritus or by searching in and under *Sphagnum* sp. mats (Bishop, 1943; Rabb, 1966). However, adults occasionally are captured terrestrially.

F. Home Range Size. No studies to date have investigated home range size (or even the presence thereof) in many-lined salamanders.

G. Territories. Unknown. However, it may be safe to speculate that many-lined salamanders do not maintain distinct territories as opposed to many other plethodontid salamanders that are highly territorial. Because these are aquatic salamanders, marking territories via conventional methods (see Jaeger, 1988; Jaeger and Forrester, 1993) is problematic.

H. Aestivation/Avoiding Desiccation. Foard and Auth (1990) reported digging up many-lined salamander adults from exposed riverbeds during drought conditions. Individuals were found singly in small cavities measuring "about two-thirds their body length and twice their body width." It is unknown how long the individuals were in these cavities or how long they would be capable of remaining there.

I. Seasonal Migrations. Because of the almost exclusively aquatic life style, seasonal migrations are unlikely.

J. Torpor (Hibernation). Unlikely, given that many-lined salamanders live in a region that does not regularly experience extended sub-freezing temperatures.

K. Interspecific Associations/Exclusions. According to Petranka (1998), many-lined salamanders are found in association with other amphibians, but competitive and/or predatory interactions are poorly understood. Means (2000) notes that many-lined salamanders are found in association with mud salamanders (*Pseudotriton montanus*), dwarf salamanders (*Eurycea quadridigitata*), and southern dusky salamanders (*Desmognathus auriculatus*). Other species found by Bruce (1971) to be in association with many-lined salamanders include lesser sirens (*Siren intermedia*), two-toed amphiumas (*Amphiuma means*), and several anurans such as southern cricket frogs (*Acris gryllus*), southern leopard frogs (*Rana sphenocephala*), and carpenter frogs (*R. virgatipes*).

L. Age/Size at Reproductive Maturity. In males, maturation follows metamorphosis. In females, maturation is delayed until 1 yr following metamorphosis. Therefore, males reproduce in the autumn following metamorphosis, when about 21–33 mo old and 33–40 mm SVL (Bruce, 1971; Ryan and Bruce, 2000). Unlike males, females remain juveniles for >1 yr after metamorphosing and do not reproduce for the first time until 3 or 4 yr old and between 37–45 mm SVL (Bruce, 1971; Ryan and Bruce, 2000). Adults range from 63–112 mm TL, with slightly <1/2 length being tail (Rabb, 1966). Despite the asynchrony in the attainment of maturation, there appears to be no sexual size dimorphism (Bruce, 1971; see also Petranka, 1998).

M. Longevity. Unknown.

N. Feeding Behavior. While they share the same habitat, the prey of post-metamorphic many-lined salamanders differs from larvae in terms of the diversity of prey taken and the relative abundances. Only seven taxa were represented in the guts of larvae, whereas twice as many taxa were present in the gut of adults (Foard and Auth, 1990). The most common prey items were still isopods (52% of all stomachs investigated; only 37% in larvae), with amphipods a distant second (21%).

No other taxa were represented in >5% of the 161 transformed individuals examined. Most of the time, transformed individuals feed off the bottom, consuming larger prey items than larvae. However, they are capable of feeding off the surface, as long as their limbs are in contact with the substrate, and may in fact feed on terrestrial invertebrates (e.g., lepidopterans, coleopterans, chilipodans, and formicidians) during periods of heavy rainfall. Foard and Auth (1990) speculated that the terrestrial invertebrates were swept into the aquatic habitats during the rainfall, but because adults are known to occasionally leave wetlands (which would be most likely during periods of significant rainfall) it should not be ruled out that some terrestrial feeding occurs.

O. Predators. Unknown, but Petranka (1998) states that natural predators probably include aquatic snakes, fishes, wading birds, and invertebrates such as dragonfly naiads and dytiscid beetle larvae. Bruce (1971) documented several species of snakes in many-lined salamander habitats, including at least two species, southern water snakes (*Nerodia fasciata*) and black swamp snakes (*Seminatrix pygaea*), likely to feed on aquatic salamanders.

P. Anti-Predator Mechanisms. Many salamanders are well known for their defensive displays and postures (e.g., Brodie, 1977), but virtually all species for which such behaviors are known are at least partially, if not altogether, terrestrial. Defensive displays are likely ineffective in murky waters such as those inhabited by many-lined salamanders. Accordingly, there is not record of defensive posturing in this species, and it is unlikely that such behaviors exist.

Q. Diseases. Unknown.

R. Parasites. Gut parasites were common in Foard and Auth's (1990) samples; nearly half of all the guts examined contained at least one endoparasite, most commonly the acanthocephalan, *Pilum pilum*. They speculate that parasitism was a result of gastropod ingestion; however, they did not record gastropods in the gut contents of any animals investigated.

4. Conservation.
Many-lined salamanders are one of the more poorly known plethodontids. Loss of populations has undoubtedly occurred with wetland drainage, but this has not been documented. Mitchell (1991) describes their conservation status as Undetermined. Wetland destruction is likely a primary threat to their long-term persistence.

Typhlotriton (Eurycea) spelaeus Stejneger, 1893
GROTTO SALAMANDER
Dante B. Fenolio, Stanley E. Trauth

1. Historical versus Current Distribution.
The modern distribution of grotto salamanders (*Typhlotriton [Eurycea] spelaeus*;

sometimes called "ghost lizards" by local citizens; DiSilvestro, 1983) spans the Salem and Springfield regions of the Ozark uplift in Missouri, Oklahoma, extreme southeastern Kansas, and Arkansas. Grotto salamanders currently have no fossil record (Brandon, 1970).

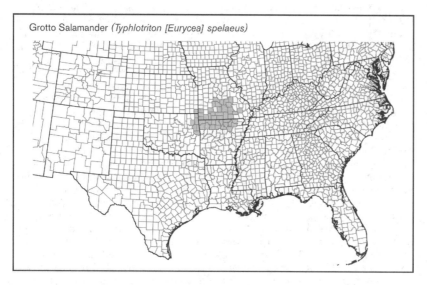

Grotto Salamander (*Typhlotriton [Eurycea] spelaeus*)

Grotto salamanders and other probable close relatives, including Texas blind salamanders (*Eurycea rathbuni*) and Georgia blind salamanders (*Haideotriton wallacei*), may have evolved from proto-*Eurycea* stock (Wake, 1966; Brandon, 1971b; see also Bowett and Chippindale, 2004). Origins of this species remain speculative, but at least one hypothesis suggests grotto salamanders evolved in the Ozark Plateau—the "Interior Highlands" (Dowling, 1956). The Interior Highlands were not covered by shallow seas of the Cretaceous and provided a refuge from the glaciers of the Pleistocene epoch because the southward push of the ice stopped at the base of the plateau. Toward the end of the Pliocene, an arid stage may have drastically affected Kansas and the Ozarks. According to the hypothesis, these climactic changes drove the ancestors of *Typhlotriton* below ground, along with the ancestors of this region's cave fish and cave crayfish fauna.

2. Historical versus Current Abundance.
Historical versus current abundance is generally unknown. Rudolph (1980) found larval grotto salamanders in densities of over 10/m2 in an epigeac spring, compared to fewer than 2/m^2 in a cave habitat. Smith (1959) examined several hundred larvae predominantly from a single spring in Independence County, Arkansas. Over a 7.5-yr period (1960–68), 220 adults were collected from a single cave in Shannon County, Missouri (Brandon, 1971a; Besharse and Brandon, 1974). Rudolph (1978) examined 111 larvae from a single spring run in northeast

Oklahoma primarily during a 2-yr period (1975–76).

Detailed records of sightings, and in some cases even numerical records of encounters, can be found in speleological society newsletters and peer-reviewed journals dating as far back as the 1930s (Blair, 1939). Some of these notes, combined with recent surveys of the same systems, demonstrate stable populations of grotto salamanders for >25 yr (Rimbach, 1968; Looney and Puckett, 1970; Looney, 1973; Graening and Brown, 2000; Graening et al., 2001).

3. Life History Features.
A. Breeding. Reproduction is likely aquatic.

i. Breeding migrations. Observations in some caves suggest that individuals move seasonally to and from small fissures, cracks, and crevices. Animals may aggregate in larger cave systems during and shortly after the wettest times of the year. These movements serve a dual purpose: reproduction and foraging. Prey densities in larger caverns are highest at this time (Hendricks and Kezer, 1958; Brandon, 1971a). Pyle (1964) indicated that breeding occurred year-round, although sexing of individuals in this study was questionable. Other authors have mentioned the ease of clearly observing light-colored eggs through the flanks and ventral surface of adult females (Barden and Kezer, 1944; Kezer, 1952b; Brandon, 1962; personal observations). Brandon (1971a) indicated that food may be limiting in these troglobitic systems and may provide sufficient nutrients for females to breed only every second or third year.

Breeding likely involves the transfer of a spermatophore from a male to a female, but neither this nor the courting behaviors have been observed.

ii. Breeding habitat. Reproduction and egg deposition occur within caves. Adults mate during the summer (May–August; Brandon, 1971a), and females oviposit from 1–4 mo after mating (Smith, 1960).

B. Eggs.
i. Egg deposition sites. Females oviposit in cryptic sites within caves, possibly in the water and within the rock rubble or immediately above a water source on a damp surface, where high humidity prevents the desiccation of the egg mass. In the laboratory, a female given pituitary gland implants deposited four eggs in the water and nine more attached to the moist surface of a rock above the water line (Barden and Kezer, 1944). Eggs, larvae in gelatinous egg membranes, and newly hatched larvae were attached singly to the edges of rocks in a cave waterway (Smith, 1960; but see Brandon, 1962).

ii. Clutch size. Thirteen eggs were produced by a female given pituitary implants; the eggs were 2.7–3.0 mm in diameter (Barden and Kezer, 1944; Brandon, 1966d; Trauth et al., 1990). Hatchlings average 13 mm SVL (range 10–16 mm; Brandon, 1965c, 1970) and were found from mid-December to January (Smith, 1959, 1960).

C. Larvae/Metamorphosis.
i. Length of larval stage. Larvae have been collected throughout the year (Pyle, 1964; Brandon, 1971a,b; Collins, 1993). Cold water temperatures and, in some cases, low food availability may hinder accurate assessment of larval age due to slower growth rates relative to above-ground *Eurycea*. The larval period apparently lasts 2–3 yr (Hendricks and Kezer, 1958; Brandon, 1966d, 1971a,b; Rudolph, 1980; Trauth et al., 1990), although data are limited.

Recently hatched larvae range from 10–16 mm SVL, and a second size class, representing approximately 1 yr of age, ranged from 22–30 mm SVL (Rudolph, 1978); maximum larval size is 60 mm SVL. Larvae are known to drift from cave waterways, out through the cave entrance, and into above-ground portions of the same stream. Larvae generally remain within 12 m of the mouth of natal caves (Rudolph, 1978). Whether they migrate back into natal caves or are swept downstream, perhaps to colonize other cave systems, is unknown (Brandon, 1971b; see also Petranka, 1998). Larvae collected from caves tend to be lighter than larvae from surface streams and can be solid pink or a silvery-blue color. Post-metamorphic individuals range from 38–57 mm SVL (Besharse and Brandon, 1974).

ii. Larval requirements. Epigean larvae are more pigmented than larvae from deep inside caves. Above-ground larvae occur in springs with clear water and sand or gravel substrates with little silt (Smith, 1960), as well as in streams associated with a nearby subterranean system; larvae can also be observed in waterways of the deepest accessible reaches of caves

(Hendricks and Kezer, 1958; Brandon, 1971b). Below-ground aquatic habitats where larvae have been observed include flowing waterways as well as rimstone pools (Bretz et al., 1983; personal observations). Preliminary evidence indicates larvae may move through phreatic passages (flooded caves, solution channels, and fault lines), possibly between cave systems and throughout the expanse of aquifers (V. Brahanna, personal communication). In a comparison of five spring-inhabiting species, grotto salamander larvae preferred cooler waters (Rudolph, 1980). Water characteristics of larval habitats have been collected at a number of locations: water temperature typically ranges from 5.5–16.5 °C in springs (Smith, 1960; Rudolph, 1980) and from 11–16 °C in caves (Pyle, 1964; Rudolph, 1980). The pH values collected for springs and caves range from 6.0–7.5 (Smith, 1960; Pyle, 1964; G.O. Graening, personal communication).

Small and intermediate-sized larvae have functional eyes that degenerate in older larvae and adults; eyelids of adults may grow over vestigial eyes in some adults (Smith, 1960; Stone, 1964a; Brandon, 1970; Besharse and Brandon, 1974, 1976). Data suggest that larvae regularly exposed to light may retain vision for longer periods than larvae maintained in darkness (Noble and Pope, 1928; Besharse and Brandon, 1976). Laboratory manipulation has shown that functional eyes of larvae that are surgically removed and then grafted back in place can regenerate the optic nerve and at least some degree of vision (Stone, 1964b).

a. Food. Larvae feed on small invertebrates and may employ a mix of sit-and-wait (Dodd, 1980; see also Petranka, 1998) and active foraging (personal observations) behaviors. Isopods *(Lirceus happinae)* form the bulk of the diet in some populations (Smith, 1948a,b). In other populations, diets are more diverse and include snails, dipteran larvae, annelids, arachnids, ostracods, copepods, amphipods, decapods (specifically small crayfish), diplopods, ephemeropteran nymphs, plecopteran nymphs, homopterans, hymenopterans, coleopterans, trichopterans, lepidopterans, centipedes, and other salamander *(Eurycea* sp.) larvae (Brandon, 1971a; Rudolph, 1980).

b. Cover. Surface larvae are secretive, being found during the day in gravel or under flat rocks (Smith, 1960). In some instances, larvae have been found at depths ≤0.75 m in rock rubble (Smith, 1960).

iii. Larval polymorphisms. Unknown.

iv. Features of metamorphosis. Larvae metamorphose from 85–96 mm TL.

v. Post-metamorphic migrations. Larvae approaching metamorphosis that live outside a cave are believed to return to subterranean retreats to complete metamorphosis; no adults have been found outside caves (Brandon, 1971b). The fact that larvae are readily found in springs and waterways, even a good distance from the mouth of a cave, may indicate that larval dispersion enables genetic exchange between populations (Brandon, 1971b).

vi. Neoteny. Some authors have indicated that gilled adults may occur in nature (Bishop, 1944; Mohr, 1950; Smith, 1960); although Brandon (1966d) found no evidence to support this idea. Reproductive maturity occurs near the time of metamorphosis.

D. Juvenile Habitat. Likely to be no different than adult habitats. The juvenile stage is short; sexual maturity appears to occur at, or shortly after, metamorphosis (Brandon, 1966d).

E. Adult Habitat. Adults are limited to limestone caves and underground passages in the karst formations of the Ozark Plateau and are most frequently found beyond the twilight zone on moist rock walls (Hendricks and Kezer, 1958; Brandon, 1971a).

F. Home Range Size. Unknown.

G. Territories. Unknown.

H. Aestivation/Avoiding Desiccation. Unlikely.

I. Seasonal Migrations. See "Breeding migrations" above.

J. Torpor (Hibernation). Unlikely.

K. Interspecific Associations/Exclusions. In a study investigating competition among five species of larval salamanders in the Ozarks, Rudolph (1980) found that grotto salamander larvae commonly inhabit spring headwaters and areas of subsequent drainages in conjunction with other salamander larvae. These spring habitats have the highest prey abundances, reduced temperature fluctuations (and cooler temperatures), a decreased likelihood of desiccation from receding waters, the least exposure to floods, an ease of reentering subterranean habitats, and the least exposure to predatory fish species. The study suggested that grotto salamander larvae occurring in low to moderate densities will displace larvae of at least two other salamander species, cave salamanders *(Eurycea lucifuga)* and long-tailed salamanders *(E. longicauda),* through aggressive behavior and predation. In some springs examined, larval grotto salamanders in moderate to high densities entirely eliminated larvae of other species. According to Rudolph (1980), only Oklahoma salamanders *(E. tynerensis)* can consistently coexist with larval grotto salamanders, but only at reduced densities and when well-developed gravel substrates provide cover. If grotto salamander larvae are absent or are removed from spring habitat, larvae from the other species will successfully colonize until grotto salamander populations become reestablished. The aggressive behavior of the larvae is not directed exclusively to other species; physical damage to conspecifics has been noted as well.

L. Age/Size at Reproductive Maturity. Adults reach sexual maturity between 36–60 mm SVL (Smith, 1960). Both males and females have cirri, but males in breeding condition have longer, enlarged cirri. The mental gland in reproductively active males is located beneath the lower jaw (Brandon, 1971a). Reproductive females of lighter color have cream to white eggs that can be observed through the walls of the flanks and the ventral surface (Barden and Kezer, 1944; Kezer, 1952; Brandon, 1962; D. B. F., personal observations).

M. Longevity. A captive specimen lived nearly 12 yr (Snider and Bowler, 1992).

N. Feeding Behavior. While adult grotto salamanders may be top predators in some caves (Schwartz, 1976), potential predators of the salamanders have been identified in others (see "Predators" below). Adults forage on land as well as in the water (Mohr, 1950; Brandon, 1971a; D.B.F., personal observations). In the terrestrial environment, they commonly forage directly on guano piles (Black, 1971c; D.B.F., personal observations). Adult prey includes gnats, mosquito larvae, beetles (Brandon, 1971a), and isopods (Smith, 1948a). Adults climb into the webs of mycetophilid gnat *(Macrocera nobilis)* larvae to eat them (D. B. F., personal observations). Grotto salamanders are opportunistic predators, taking advantage of seasonal and regional prey availability. Ninety-six percent of the stomach contents of larvae from one cave were composed of a single species of isopod *(Lirceus happinae)*; 80% of the prey of sympatric adults were the same isopod species (Smith, 1948b).

O. Predators. Larvae may be eaten by crayfish; adults appear to have few natural predators (Petranka, 1998); although green frogs *(Rana clamitans)* and pickerel frogs *(R. palustris)* are known to inhabit cave habitat at least seasonally, with many large enough to consume larvae and small, newly metamorphosed grotto salamanders. Raccoon *(Procyon lotor)* tracks in Ozark caves are common, even deep within caves, and it is possible that raccoons eat adult and larval grotto salamanders (D.B.F. and G.O. Graening, personal observations). Following flooding of their epigean streams, grotto salamander larvae can be consumed by predatory fishes (Rudolph, 1980).

P. Anti-Predator Mechanisms. Individuals on walls or ledges above pools or streams drop into the water below when approached (D. B. F., personal observations).

Q. Diseases. Unknown.

R. Parasites. Isopods *(Lirceus happinae)* serve as prey for both larval and adult grotto salamanders and harbor a cestode parasite, *Ophiotaenia cestodes,* that infects larvae (Smith, 1948a). The nematode *Falcaustra catesbeianae* is found in Missouri populations (Dyer, 1975).

4. Conservation.

Among natural threats, larval grotto salamanders are sensitive to flooding of epigean streams. Only 20% of a population of larval grotto salamanders remained in a surface stream after heavy flooding (Rudolph, 1980). Moreover, grotto salamander larvae are consumed by predatory fishes (Rudolph, 1980; see also "Predators," above).

There are several human threats to grotto salamanders. Both aquatic larvae and adults share subterranean living space with several critically endangered aquatic organisms including troglobitic isopods and amphipods, troglobitic crayfish of the genus *Cambarus*, and amblyopsid cavefish. Studies of these syntopic species have produced a considerable amount of information about human threats to the subterranean aquatic environments of the Ozark Plateau. The threats that face these endangered organisms equally endanger syntopic grotto salamanders. Among the threats are deforestation and development of critical above-ground recharge zones (Means and Johnson, 1995; Culver et al., 1999; Graening, 2000), human dumping of materials above ground that drastically affect groundwater quality (Means, 1990; Culver et al., 1999; Graening, 2000), manipulation of nutrients in groundwater allowing above-ground species to move into below-ground habitat and compete with or consume troglobites (A. V. Brown et al., 1994), and inexperienced cave explorers damaging critical aquatic habitat, trampling animals, and over-collecting regardless of protective laws (Willis and Brown, 1985). Grotto salamanders in particular seem sensitive to disturbance or impurities introduced into subterranean waterways and caves. During commercialization of caves in Camden County, Missouri, grotto salamanders disappeared from areas of the cave that were electrically lighted. When the lighting was removed, the salamanders returned (Weaver, 1987). Surveys after a 1981 spill of approximately 80,000 l of liquid ammonia nitrate in Missouri found that grotto salamanders 21 km from the spill were killed as the contaminant moved through a connected, subterranean aquifer (Crunkilton, 1984). Grotto salamanders also suffer from the introduction of sport-fishes into their habitat (Rudolph, 1980). Because bat guano supplies many terrestrial cave invertebrates with food, decline of colonial bats in the Ozarks may affect prey abundance.

Three of four states where grotto salamanders are found protect the species. The fourth state, Arkansas, is taking steps to protect the species as well (K. Irwin, Arkansas Game and Fish Commission, personal communication). Even with the state protection, habitat alteration poses the greatest threat to the species' survival. For example, a 1976 study of Cathedral Cave, Missouri, warned that the proposal

of an artificial lake in the region would eliminate all available habitat for the species in that cave system (Schwartz, 1976). The recovery of reduced colonies of colonial bats in the Ozarks is essential to sustain large deposits of guano in subterranean habitats, which in turn serve as food for many of the invertebrates that grotto salamanders rely on as prey. Currently, gray myotis (*Myotis grisescens*), only one of the three colonial bat species in the Ozarks, is showing signs of rebounding population numbers (S. Hensley, U.S. Fish and Wildlife Service, personal communication).

Family Proteidae

Necturus alabamensis Viosca, 1937
BLACK WARRIOR WATERDOG

Mark A. Bailey

1. Historical versus Current Distribution.

The type locality of black warrior waterdogs (*Necturus alabamensis*) is the Black Warrior River, tributary of the Alabama River, near Tuscaloosa, Tuscaloosa County, Alabama (Viosca, 1937; Bailey and Moler, 2003). Black warrior waterdogs range through a restricted segment of north-central Alabama. They apparently are confined to medium–large streams of the upper Black Warrior River system above the Fall Line (Bart et al., 1997). Black warrior waterdogs are known from nine stream segments in four counties: Sipsey Fork and Brushy Creek in Winston County, Locust Fork and Blackburn Fork in Blount County, Mulberry Fork, Blackwater Creek, and Lost Creek in Walker County, and Yellow Creek and North River in Tuscaloosa County (Ashton and Peavy, 1986; Bart et al., 1997). Although their geographic distribution has not been clearly delineated, their range is thought to essentially mimic

that of flattened musk turtles (*Sternotherus depressus*; Ashton and Peavy, 1986; Guyer, 1997, 1998). Additional comments on the identity and distribution of black warrior waterdogs can be found in Gunter and Brode (1964), Brode (1969), Mount (1975), and Guttman et al. (1990). Neill (1963) commented on the distribution of "*N. alabamensis*," but considered distant Coastal Plain *Necturus* populations to be conspecific.

2. Historical versus Current Abundance.

The historical abundance of black warrior waterdogs is poorly known, but a remarkably large series of 135 specimens was collected in late winter and spring of 1938 in pre-impoundment Mulberry Fork at Cordova, Walker County (Bart et al., 1997). Collection methods, effort, and collector are unknown. Black warrior waterdogs were recently documented (by single specimens) at two localities upstream from this site (Bailey, 1995; Guyer, 1997) and near the upper reaches of Bankhead Lake. However, there is no indication that waterdogs remain present anywhere in the densities that must have existed in 1938 to enable the collection of such a large number of animals. Mount (1981) estimated that sympatric flattened musk turtles no longer inhabited 27% of the stream miles of their historical occupation. The flattened musk turtle recovery plan (USFWS., 1990) suggests that only 142 out of 947 stream miles (15%) in the upper Black Warrior drainage may support flattened musk turtle populations, and there is no reason to assume conditions are different for black warrior waterdogs. The status of black warrior waterdogs in impoundments remains poorly known (Guyer, 1997). Bailey (1992, 1995) sampled for black warrior waterdogs at 77 sites scattered across the presumed range. Guyer (1997) re-sampled most of these sites

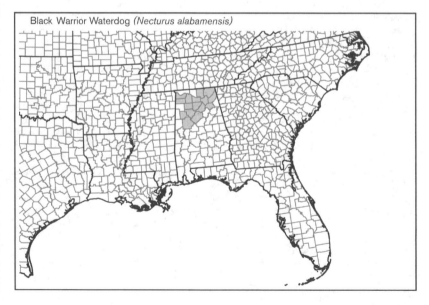

Black Warrior Waterdog *(Necturus alabamensis)*

and an additional 55 sites. Virtually all localities where roads cross or approach streams within the species' presumed range were sampled from 1992–97; of the 122 sites sampled by both researchers, 10 were found to support black warrior waterdogs. At one site, 17 individuals were collected from a 40 m² area (Guyer and Durflinger, 1999). Distribution, even within the best habitat, appears to be patchy, and abundance may fluctuate from year to year depending on the development of submerged leaf beds (Guyer and Durflinger, 1999).

3. Life History Features.

A. Breeding. Reproduction is aquatic.

i. Breeding migrations. Unknown. Post-hatchling larvae have been collected in December, suggesting late spring or summer nesting (Ashton and Peavy, 1986).

ii. Breeding habitat. Adults with swollen cloacal lips, indicating a sexually active condition, have been collected in the winter (Ashton and Peavy, 1986). Adults collected in winter are usually in or near leaf beds or rock crevices (personal observations).

B. Eggs.

i. Egg deposition sites. Unknown, but perhaps associated with leaf beds and rock crevices where sexually active adults are found.

ii. Clutch size. Unknown.

C. Larvae/Metamorphosis. Three size classes are evident from the samples taken to date (Guyer, 1997). These presumably correspond to larvae, subadults, and adults. Post-hatching animals (28–50 mm TL) are distinctly striped, closely resembling the larvae of mudpuppies *(N. maculosus)* of the Tennessee River drainage (Ashton and Peavy, 1986). Black warrior waterdog larvae are black to dark brown on the dorsum with two light dorsolateral stripes beginning at the nostril and extending posteriorly through the eye and terminating on the dorsal fin of the tail. Subadults (50–150 mm TL) lack the bold stripes, but retain traces. Adults (150–248 mm) generally lack stripes and have poorly developed dorsal spotting and a dark line beginning at the nostril, extending through the eye, and dissipating in the gill area. Large adults may be melanistic.

D. Juvenile Habitat. Poorly understood, but larvae typically are collected from submerged leaf beds with rich invertebrate faunas.

E. Adult Habitat. Guyer (1997) reported black warrior waterdogs associated with clay substrates lacking silt, wide and/or narrow stream morphology, increased snail and larval northern dusky salamander *(Desmognathus fuscus)* abundance, and decreased Asiatic mussel *(Corbicula* sp.) occurrence. The presence of leaf beds is important.

F. Home Range Size. Unknown.

G. Territories. Unknown.

H. Aestivation/Avoiding Desiccation. Aestivation is suspected, but not known. No specimens have been collected during the summer.

I. Seasonal Migrations. Unknown.

J. Torpor (Hibernation). Nonexistent. Adults are active throughout the winter.

K. Interspecific Associations/Exclusions. Unknown.

L. Age/Size at Reproductive Maturity. Not known with certainty, but the adult size class is probably attained in the third winter, or at 2.5 yr. No discernible size classes exist in adults.

M. Longevity. Unknown. The largest specimen on record is a 248 mm (TL) melanistic female, which exceeds the previously known maximum size (Ashton and Peavy, 1986) by almost 90 mm. Although the average growth rate is unknown, this was likely an unusually old individual.

N. Feeding Behavior. Adults are attracted to traps baited with fish flavored cat food, and a captive took a small fish *(Elassoma* sp.) that settled to the bottom. Captives take earthworms and are probably opportunistic feeders (Ashton and Peavey, 1986).

O. Predators. Unknown.

P. Anti-Predator Mechanisms. Unknown.

Q. Diseases. Unknown.

R. Parasites. Unknown.

4. Conservation.

Black warrior waterdogs are confined to medium–large streams of the upper Black Warrior River system above the Fall Line (Bart et al., 1997). There is no indication that waterdogs remain present anywhere in the densities that must have existed historically. Distribution, even within the best habitat, appears to be patchy, and abundance may fluctuate from year to year (Guyer and Durflinger, 1999).

Necturus beyeri Viosca, 1937
GULF COAST WATERDOG
Craig Guyer

1. Historical versus Current Distribution.

Gulf Coast waterdogs *(Necturus beyeri)* are thought to occur in two distinct regions: western populations are found in eastern Texas and west-central Louisiana, eastern populations are found in northern and western Mississippi (Gunter and Brode, 1964; Guttman et al., 1990; Conant and Collins, 1998; Petranka, 1998). Supporting this interpretation, waterdogs from the major rivers draining into Mobile, Alabama (except for the upper Black warrior Drainage, which contains Black warrior waterdogs, *N. alabamensis*), are genetically indistinguishable from the eastern form of *N. beyeri* (C. G. unpublished data).

2. Historical versus Current Abundance.

Petranka (1998) speculates that, similar to other river-dwelling fishes and amphibians, numbers of Gulf Coast waterdogs likely have been reduced due to siltation and pollution.

3. Life History Features.

A. Breeding. Reproduction is aquatic.

i. Breeding migrations. Unknown and unlikely; there is strong evidence for restricted home ranges that persist throughout the year (Shoop and Gunning, 1967; see "Home Range Size" below). Mating takes place in late autumn to early winter (Shoop, 1965b; see also Petranka, 1998). All females from a population in southern Louisiana collected from December–May had sperm stored in their cloacas, and females can apparently store sperm for ≤6 mo (Shoop, 1965b; see also Petranka, 1998).

Based on gravid females found throughout the winter and early spring, Bishop (1943) suggests an early spring breeding/

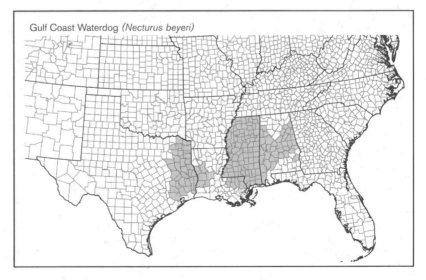

Gulf Coast Waterdog *(Necturus beyeri)*

nesting season. Shoop (1965b) found five nests without females in early May. Sever and Bart (1996) found four nests with brooding females in late May.

ii. Breeding habitat. Underwater.

B. Eggs.

i. Egg deposition sites. Females attach their eggs singly to the undersides of underwater substrates, such as pine logs, large boards, and railroad ties, at depths of 16–65 cm (Shoop, 1965b; see also Petranka, 1998). Neural fold-stage embryos measure 5.6–5.8 mm in diameter. Embryos maintained in the laboratory hatch after a 2-mo incubation time at a size from 13–16 mm SVL.

ii. Clutch size. Egg numbers in untended nests ranged from 4–40; in tended nests, eggs numbered from 26–37 (Shoop, 1965b; Sever and Bart, 1996; see also Petranka, 1998). Mature females have an average of 47.5 ova (range = 28–76; Sever and Bart, 1996).

C. Larvae/Metamorphosis.

i. Length of larval stage. Older larvae generally resemble adults in shape and color pattern (Bishop, 1943). An abrupt metamorphosis does not occur.

ii. Larval requirements.

a. Food. Unstudied, but larvae undoubtedly are carnivorous and feed on an array of aquatic invertebrates.

b. Cover. Unknown, but likely use debris and detritus as cover objects.

iii. Larval polymorphisms. Unknown and unlikely.

iv. Features of metamorphosis. Unknown.

v. Post-metamorphic migrations. Unknown.

vi. Neoteny. Gulf Coast waterdogs, as is true with all species of *Necturus*, are neotenic, retaining gills and an aquatic lifestyle through adulthood.

D. Juvenile Habitat. Young juveniles inhabit bottom debris, especially leaf litter, where currents are slow and prey aggregate (Bart and Holzenthal, 1985; see also Petranka, 1998).

E. Adult Habitat. Permanently aquatic. Gulf Coast waterdogs are found in sandy, spring-fed streams; they are not found in the turbid, sluggish waters characteristic of bayous, rivers, and lakes of the Lower Mississippi River System.

Most adults are found in slow-moving sections of streams under large objects such as logs, flood debris, and other obstructions. Some Gulf Coast waterdogs live in stream bank burrows (Shoop and Gunning, 1967; see also Petranka, 1998).

F. Home Range Size. Apparently relatively small. Adults are known to remain in the same stretch of stream for over 2.5 yr (724 d). In a mark–recapture study, mean movement between capture for males was 16 m; for females, 10 m; and for juveniles, 25 m (Shoop and Gunning, 1967; see also Petranka, 1998). All recaptures were within 64 m of the original capture and release site.

G. Territories. Unknown.

H. Aestivation/Avoiding Desiccation. Bart and Holzenthal (1985) suspect that Gulf Coast waterdogs aestivate in burrows during the summer and autumn, when invertebrate prey are scarce (see also Petranka, 1998).

I. Seasonal Migrations. Unlikely (Shoop and Gunning, 1967).

J. Torpor (Hibernation). Animals appear most active in the winter and spring, when breeding, nesting, and the bulk of feeding occur.

K. Interspecific Associations/Exclusions. The competitive relationships between Gulf Coast waterdogs and fishes have not been detailed; fish will prey on these mudpuppies (see Petranka, 1998).

L. Age/Size at Reproductive Maturity. From 112–123 mm SVL (males) and 115–135 mm SVL (females). According to Viosca (1937) and Bishop (1943), sizes average 203 mm TL (range = 160–223). Tails comprise 30.8% of TL (Bishop, 1943).

Time to reach sexual maturity is probably 4–6 yr, based on studies of other *Necturus* species (Cooper and Ashton, 1975; see also Petranka, 1998).

M. Longevity. Bart and Holzenthal (1967) suggest adults live at least 6–7 yr in nature (see also Petranka, 1998).

N. Feeding Behavior. Gulf Coast waterdogs feed on a variety of prey including crayfish, isopods, amphipods, freshwater clams, and insects including mayflies, caddisflies, dragonfly naiads, dytiscid beetles, and midges. Across species of *Necturus* inhabiting the Coastal Plain, diversity of prey ingested increases with prey availability during colder months (Bart and Holzenthal, 1985; Braswell and Ashton, 1985).

Adults appear to forage along logs, away from leaf-litter beds (Neill, 1963; Shoop and Gunning, 1967; Bart and Holzenthal, 1985; Petranka, 1998).

O. Predators. Predatory fishes (see Petranka, 1998).

P. Anti-Predator Mechanisms. Shoop and Gunning (1967) suggest that reduced feeding during summer and autumn is a predator avoidance mechanism, minimizing overlap between Gulf Coast waterdogs and predatory fishes.

Q. Diseases. Unknown.

R. Parasites. Unknown.

4. Conservation.
Gulf Coast waterdogs occur in western populations, found in eastern Texas and west central Louisiana, and eastern populations, found in northern and western Mississippi (Gunter and Brode, 1964; Guttman et al., 1990; Conant and Collins, 1998; Petranka, 1998). Petranka (1998) speculates that numbers of Gulf Coast waterdogs likely have been reduced due to siltation and pollution.

Necturus lewisi Brimley, 1924
NEUSE RIVER WATERDOG

Alvin L. Braswell

1. Historical versus Current Distribution.
Neuse River waterdogs (*Necturus lewisi*) are known only from the Neuse and Tar river systems in the Piedmont and Coastal Plain regions of North Carolina (Bishop, 1943; Ashton, 1990; Petranka, 1998). A survey conducted between 1978 and 1980 demonstrated that apparently healthy Neuse River waterdog populations remained and that the historical range had not been reduced, but severely polluted streams are known to have lost their populations (Braswell and Ashton, 1985; see also Petranka, 1998).

2. Historical versus Current Abundance.
Unknown, but if high levels of pollution eliminate populations (Braswell and Ashton, 1985; see also Petranka, 1998), low levels of pollution may reduce abundances within remaining populations.

3. Life History Features.
A. Breeding. Reproduction is aquatic. Details of a courtship are described by Ashton (1985). Sperm transfer has not been described. Testis morphology and spermatogenesis are reported on by Pudney et al. (1985).

i. Breeding migrations. Unknown. Bishop (1943) suggests a spring breeding season based on both male and female morphology (in March, one male had swollen cloacal glands; in April, one gravid female was found). Males have sperm in their vas deferens from November–May, and females, which store sperm following mating, have sperm in their spermathecae from December–May (Cooper and Ashton, 1985).

ii. Breeding habitat. Unknown, but probably not different from normal habitats occupied.

B. Eggs.

i. Egg deposition sites. One nest has been reported (Ashton and Braswell, 1979). In the spring, females lay eggs under large rocks in moderate currents (see Ashton, 1985; Cooper and Ashton, 1985). Seven additional nests (unpublished data) observed during May and June between 1980 and 1984 were all under rocks in water 25–41 cm deep. The rocks ranged in size from 27 cm length × 27 cm width × 5 cm height to ca. 80 cm × 50 cm × 25 cm. Nests were found at low-water conditions. At this time the minimum flow rate was ca. 2 cm/s. Nest sites were in areas that received only about 2 hr/d of direct sunlight.

ii. Clutch size. One nest observed on 2 July contained 35 egg capsules and had a male present (Ashton and Braswell, 1979). Male guarding of nests is unknown in *Necturus*, so there is some question as to

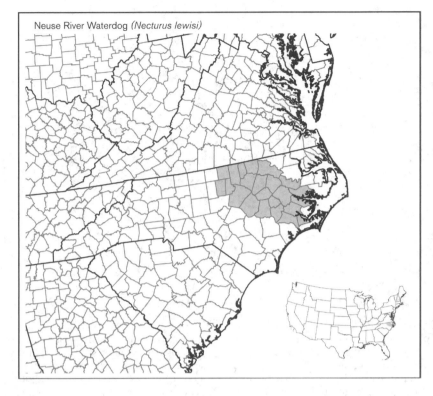

Neuse River Waterdog (*Necturus lewisi*)

whether or not this male was attending the eggs (Ashton and Braswell, 1979; see also Petranka, 1998). An attending adult was captured at six additional nests observed between 1980 and 1984 (unpublished data), and all were females that ranged in size from 93–120 mm SVL. The attending adult escaped from one other nest. Clutch sizes for these seven nests were 19, 21, 22, 20, 22, 36, and 32 (i.e., 31 eggs and 1 empty egg capsule).

C. Larvae/Metamorphosis.

i. Length of larval stage. Unknown.

ii. Larval requirements.

a. Food. Smaller larvae feed on invertebrates associated with leaf beds (Braswell and Ashton, 1985; see also Petranka, 1998).

b. Cover. Smaller larvae are found in quiet waters and in leaf beds, where they find cover and food (Braswell and Ashton, 1985; see also Petranka, 1998).

iii. Larval polymorphisms. Unknown and unlikely. Color patterns of larvae are described in Ashton and Braswell (1979).

iv. Features of metamorphosis. Unknown.

v. Post-metamorphic migrations. Unknown; long distance migrations unlikely.

vi. Neoteny. All mudpuppies (*Necturus* sp.), indeed all proteids, are neotenic, retaining larval features such as gills and tail fins through adulthood.

D. Juvenile Habitat. Ashton (1985) reported juveniles finding shelter under granite boulders on a sand/gravel substrate, and early spring use of leaf beds at an eastern Piedmont site (see Petranka, 1998).

E. Adult Habitat. Permanently aquatic. While Bishop (1943) notes animals generally are found in backwaters off the main current, where substrates are sandy or muddy, Braswell and Ashton (1985) found animals to be most abundant in streams greater than 15 m wide and 1 m deep, with flow rates of greater than 10 cm/s. Further, Braswell and Ashton (1985) found more animals associated with clay or hard soil substrates, while Brimley (1924) and Martof et al. (1980) found animals associated with leaf beds (see also Petranka, 1998). Neuse River waterdogs are distributed from larger headwater streams in the Piedmont to coastal streams up to the point of saltwater intrusion (Braswell and Ashton, 1985).

In the laboratory and in the field, juveniles and adults larger than 47 mm TL construct retreats under cover objects. In nature, adults construct retreat entrances on the downstream side of rocks (Ashton, 1985).

At night, adults become active and will leave their cover (Ashton, 1985; Braswell and Ashton, 1985). Daytime activity by adults is limited, as evidenced by no daytime trapping success (Braswell and Ashton, 1985). Neuse River waterdogs tend to become inactive away from cover when water temperatures are greater than 18 °C and cannot be trapped, but remain active at temperatures as low as 0 °C (Braswell and Ashton, 1985; Petranka, 1998).

F. Home Range Size. 16–19 m^2 for two females, 49–90 m^2 for three males (Ashton, 1985). Further, males tended to move greater distances between captures.

G. Territories. Retreats may constitute territories. Females actively defend their retreats; males also will defend retreats, but to a lesser extent (Ashton, 1985; Petranka, 1998).

H. Aestivation/Avoiding Desiccation. Unknown and unlikely.

I. Seasonal Migrations. Movements are highest in the spring and fall; increase following moderate rains but decrease following heavy rains; increase when barometric pressure is low or falls; and increase during new moons (Ashton, 1985; see also Petranka, 1998). In the spring, adults move from winter shelters (see "Torpor" below) to outcrops or boulders associated with fast currents and well-oxygenated water. Here they nest and remain for the summer.

J. Torpor (Hibernation). Unlikely. During the winter, adults find shelter in leaf beds, under rocks, or in burrows in river banks. Active feeding occurs throughout the winter and at near-freezing water temperatures; winter and early spring are the most productive trapping seasons (Braswell and Ashton, 1985).

K. Interspecific Associations/Exclusions. Resident Neuse River waterdogs inhabiting retreats exhibit threat displays that include flaring and pulsating their gills and curling their upper lip. When this fails they will attack intruders, including dwarf waterdogs (*N. punctatus*), by viciously biting them; intraspecific attacks are less vicious, whereas interspecific attacks occasionally resulted in death. Larvae and small juvenile Neuse River waterdogs may enter adult retreats without being attacked (Ashton, 1985).

Neuse River waterdogs are thought to compete with fishes for food (Petranka, 1998). The diets of Neuse River waterdogs and dwarf waterdogs also overlap, and these two mudpuppies likely compete for food where they are syntopic (Braswell and Ashton, 1985).

L. Age/Size at Reproductive Maturity. Males become sexually mature at 102 mm SVL, females at 100 mm SVL (Cooper and Ashton, 1985; see also Petranka, 1998). Bishop (1943) notes that the tail in males is about 28% of TL. Estimated age at maturity is 5.5 yr for males, 6.5 yr for females (Cooper and Ashton, 1985; see Petranka, 1998).

M. Longevity. Unknown.

N. Feeding Behavior. Adults will eat lampreys (Petromyzontidae; Nickerson et al., 1983), and can be caught using a baited hook and line and by dipnetting (C. S. Brimley and W. B. Mabee, in Bishop, 1943). Dietary analysis shows Neuse River waterdogs prey heavily on invertebrates such as ostracods, zooplankton such as copepods and cladocerans, snails, annelids, fishes, other species of salamanders, adult eastern worm snakes (*Carphophis amoenus amoenus*), isopods, slugs, spiders, crayfish, centipedes, millipedes, and insects such as mayflies, true flies, beetles, dragonfly and damselfly naiads, hellgrammites, caterpillars, and caddisflies (Braswell and Ashton, 1985; Petranka, 1998).

O. Predators. It is not known whether or not fish prey on Neuse River waterdogs, but workers have speculated that their inactivity during warmer months is in part due to the avoidance of fishes (Neill, 1963; Shoop and Gunning, 1967; Braswell and Ashton, 1985; see also Petranka, 1998).

P. Anti-Predator Mechanisms. Behavioral observations suggest that skin extracts from Neuse River waterdogs have a relatively low level of toxicity that causes moderate distress in mice (Brandon and Huheey, 1985).

Q. Diseases. Unknown.

R. Parasites. Unknown.

4. Conservation.

Pesticide and PCB residues were reported in Neuse River waterdogs from Piedmont and Coastal Plain localities (Hall et. al., 1985). Specifically, DDE, DDD, dieldrin, cis-chlordane, trans-nonachlor, and PCP 1254 were detected. Several other potential contaminants were not detected at the 0.01 ppm level of sensitivity. Neuse River waterdogs are considered a Species of Special Concern in North Carolina.

Necturus maculosus (Rafinesque, 1818)
MUDPUPPY

Timothy O. Matson

1. Historical versus Current Distribution.

Mudpuppies *(Necturus maculosus)* were historically distributed throughout eastern and middle North America (Eycleshymer, 1906). The distribution was further defined by Stejneger and Barber (1923) to include tributaries of the Great Lakes, the Mississippi and Hudson River systems, Lake Champlain, and rivers of Alabama, Georgia, and North and South Carolina. Two subspecies are recognized. The nominate form, *N. m. maculosus* (Rafinesque, 1818; Collins, 1990), occupies a broad distributional range extending from southern Québec, Lake Champlain, and eastern New York state, westward across southern Ontario to southeastern Manitoba and eastern Kansas, and south to northern Mississippi, Alabama, and Georgia. They are absent from the Adirondack Mountains, northern and far southern Minnesota, all but eastern Iowa, and parts of Missouri and Tennessee. Red River mudpuppies *N. m. louisianensis*; Viosca, 1937; Collins, 1990) occur from southern Missouri into south-central Kansas through eastern Oklahoma and Arkansas into northern Louisiana (Conant and Collins, 1991).

2. Historical versus Current Abundance.

Few quantitative data are available. Mudpuppies were reported as abundant in the Great Lakes region (Eycleshymer, 1906; Pearse, 1921). For example, Milner (1874) cited a fisherman at Evanston, Illinois, who set 900 hooks and caught 500 mud-

puppies in one night. Pearse (1921) credited Alexander Nielsen of Venice, Ohio, as taking large numbers of mudpuppies for years from Sandusky Bay of southern Lake Erie. Bishop (1941b) recorded mudpuppies as common in the St. Lawrence, Hudson, and Mohawk rivers and as abundant in several of the Finger Lakes of New York. Vogt (1981) relates that mudpuppies formerly were caught by the hundreds in Lake Mendota, Wisconsin. The current status of mudpuppies is unknown in many areas. They are reported to have declined in Lake Erie (Pfingsten and White, 1989), in Iowa (Lannoo, 1994), and in parts of the Great Lakes region (Harding, 1997).

3. Life History Features.

A. Breeding. Reproduction is aquatic.

i. Breeding migrations. Definite breeding migrations are unknown. In the autumn, males occupying both stream and lake habitats search for females within their shallow water retreats. Several males and females may share a communal retreat throughout the breeding season (Bishop, 1941b). Although most mating is believed to occur during the autumn, Bishop (1941b) located a male in reproductive condition beneath a rock slab with three females on 18 April, and occurrences of multiple males and females beneath the same slab become commonplace in August in northeastern Ohio (personal observations).

ii. Breeding habitat. Fall breeding activities take place in the shallow waters of lakes and streams at depths ranging from only decimeters to several meters, where retreats beneath rock slabs, logs, or planks occur. Sperm are stored in spermatheca over winter (Harris, 1959c). Ovulation followed by delayed fertilization and spawning occur in the spring, from April–June; timing is dependent upon water temperature. Locally, most females spawn within a 2–3 wk period (Bishop, 1926; Smith, 1911b).

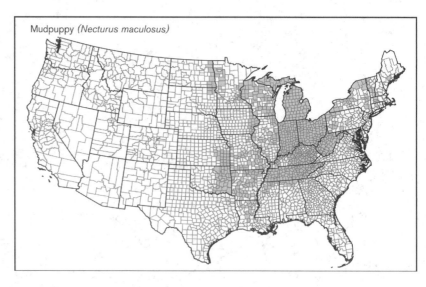

Mudpuppy *(Necturus maculosus)*

B. Eggs.

i. Egg deposition sites. Oviposition occurs on the roof of an excavated nesting cavern (Smith, 1911b; Bishop, 1926).

ii. Clutch size. Clutch size in mudpuppies was found to average 66 at Lake Monona, Wisconsin, in lake habitat (Smith, 1911b); 85 in New York (Bishop, 1941b); and 83 at northeastern Ohio in stream habitats (Matson, 1998). Clutch size for Red River mudpuppies was 36 in Big Creek, Louisiana (Shoop, 1965b). The ova measure 5–6 mm in diameter and are yellow spheres containing much yolk. While pendant from a pedicel in the nest chamber, three membranes encompass each egg increasing its width to 11 mm and its length to 14–16 mm. The period of incubation for the developing clutch extends from 38–63 d and is water temperature dependent (Bishop, 1941b).

C. Larvae/Metamorphosis. Larval mudpuppies average 22.5 mm TL at hatching (Bishop, 1941b). Most hatchlings remain in the nest cavity at least 6–8 wk until the large yolk sac has been absorbed. By the end of August, most hatchlings have left the brood site and have found retreats beneath objects in the stream channel (personal observations).

D. Juvenile Habitat. Larvae in their first winter or spring and juveniles through ages 3–4 yr are found in greater numbers in the substrate of pools where silt and organic debris have accumulated to a minimum thickness of several cm (Matson, 1990). Later in spring, many juveniles are found in portions of the stream lacking organic detritus, beneath retreats not occupied by adults or predatory fishes such as stonecats (*Noturus flavus*; Matson, 1990; personal observations). Cagle (1954) collected larvae during February in shallow holes where organic detritus had accumulated to many cm, and Bishop (1941b) reported that juveniles take up residence in deeper holes. Juvenile mudpuppies have been observed in riffle areas (Pfingsten and White, 1989).

E. Adult Habitat. Mudpuppies are found in both clear and silted waters of lakes, reservoirs, canals, ditches, and streams in the presence or absence of aquatic plants. Large populations may be present where retreats, such as flat rock slabs, logs, and planks are numerous. Lentic streams with mud substrates, crayfish burrows, undercut banks, and tree roots provide suitable habitat. Adults prefer well-aerated water downstream or to the sides of riffles (Bishop, 1941b). Streams or sections of streams coursing over exposed bedrock and providing apparently suitable rock slabs or other retreats will support few mudpuppies if a mud or mud-detritus substrate is absent from beneath the objects (personal observations). Mudpuppies have been reported from depths to 17 m in Lake Erie by White (Pfingsten and White, 1989) and to 27 m in Green Bay, Lake Michigan, by Reigle (1967).

F. Home Range Size. The summer activity ranges of adult mudpuppies in the Grand River of northeastern Ohio averaged 136.1 m^2 (Matson, 1998). No other reports of home range size are available. Both Shoop and Gunning (1967) and Matson (1998) cite evidence for long-term site fidelity.

G. Territories. Unknown.

H. Aestivation/Avoiding Desiccation. Active all year (Eycleshymer, 1906; Bishop, 1941b; Shoop and Gunning, 1967).

I. Seasonal Migrations. Vernal movements in lakes from deep water toward the shore were reported by Milner (1874), and Pearse (1921), Gibbons (1968); Pope (1947) reported annual migrations up tributaries to Lake Michigan in the Chicago area. Pope (1947) did not report any associated mortality. Gibbons and Nelson (1968) stated that these migrations did not appear to be related to reproduction because a 1:1 sex ratio was observed, mating occurs in fall or winter, and the movement included juveniles. Late mid-March movements in the Mud River and Twelvepole River were reported in West Virginia (Green and Pauley, 1987).

J. Torpor (Hibernation). Active all year (Eycleshymer, 1906; Evermann, 1920; Bishop, 1941b).

K. Interspecific Associations/Exclusions. Only known host for salamander mussels (*Simpsonaias ambigua*).

L. Age/Size at Reproductive Maturity. Age at first reproduction was reported as 5–6 yr at about 200 mm TL (Bishop, 1941b) and 7–8 yr at 175–200 mm TL (Pope, 1947).

M. Longevity. At least 29 yr (Bonin et al., 1995). Recently, a male has been aged at 34 yr (Gendron, in press).

N. Feeding Behavior. Prey include amphipods (*Gammarus* sp.), sculpins (*Cottus* sp.; Bishop, 1941b), crayfish, small fish, fish eggs, numerous aquatic insects including adults and their larvae or nymphs, ostracods, amphipods, plant material in

excess of what would be ingested by accident (Pearse, 1921), annelids, small salamanders (*Eurycea* sp., *Desmognathus* sp., *Notophthalmus* sp.; Bishop, 1941b), tadpoles (Harris, 1959b), and eggs of hellbenders (*Cryptobranchus alleganiensis*; Abbott, 1934). Evermann (1920) reported mudpuppies feeding upon small fish including brook silversides (*Labidesthes sicculus*), river shiners (*Notropis blennius*), and banded killifish (*Fundulus diaphanus*). Mudpuppies will also ingest lamprey ammocetes present in the water column during sea lampricide treatments (personal observations; J. Reblin, unpublished data). Cannibalism is not uncommon (Bishop, 1941b). Mudpuppies are primarily nocturnal but may be active diurnally in clouded, silted, or heavily vegetated waters (Bishop, 1926). Mudpuppies forage nocturnally in shallow waters affording little cover and do so more extensively during the colder months from December–April (Shoop and Gunning, 1967). Increased winter foraging activity may be a behavioral change due to reduced activity and feeding by predatory fishes.

O. Predators. Known predators of mudpuppies include predatory fishes, hellbenders, water snakes (Colubridae), herons, otters (Harris, 1959b), larger mudpuppies (Bishop, 1941b), and crayfish; fish prey upon eggs and hatchlings flushed from the nest cavity (personal observations).

P. Anti-Predator Mechanisms. Primarily nocturnal. Cryptic coloration (Bishop, 1941b). Mudpuppies have a lateral line system consisting of pressure, motion, and electroreceptor sensory cells triggering fast escape (Duellman and Trueb, 1986). Escapes are accomplished through C-starts—a rapid fishlike swimming accomplished by folding legs against the flanks and vigorously lashing the laterally compressed tail (Harris, 1959a).

Q. Diseases. *Saprolegnia*, water mold (Bishop, 1926).

R. Parasites. *Sphyranura osleri*, trematode attached to the gills and skin (Bishop, 1941b; Harris, 1959a); *Ophiotaenia lonnbergii*, a proteocephalid of the intestine; *Simpsonaias ambigua*, a salamander mussel that attaches to the gills (Pearse, 1921).

4. Conservation.
Mudpuppies have a broad geographical range, and their survival over the short term is probably secure, but their status over much of their range is poorly understood. Chemical water pollutants and heavy siltation have reduced habitat quality in many regions and have contributed to declines in population size. Agricultural, industrial, and residential practices targeted toward the reduction of pollution and siltation will aid in the conservation of this species. Less reliance upon the lampricide TFM in the Great Lakes and some eastern states for control of sea lampreys

(*Petromyzon marinus*) would allow chemically depressed populations to recover and approach carrying capacity (see Matson, 1998). Education must overcome the ignorance and misunderstanding regarding the existence and niche of mudpuppies. Erroneous information remains pervasive among biologists, fishers, and others who continue to persecute mudpuppies; consequently, mudpuppies are often used for fish bait, or commonly, hooked animals are discarded to die on the ground or on the ice.

Mudpuppies are considered Endangered/Extirpated by Maryland, Threatened in Iowa, and a Species of Special Concern in Indiana and North Carolina.

Necturus punctatus (Gibbes, 1850)
DWARF WATERDOG

Harold A. Dundee

1. Historical versus Current Distribution.
Dwarf waterdogs (*Necturus punctatus*) are found in Atlantic Coastal Plain streams from the Chowan River in southeastern Virginia through North Carolina, South Carolina, and Georgia to the Ocmulgee–Altamaha River systems in southeastern Georgia (Bishop, 1943; Seyle, 1985c; Dundee, 1998; Petranka, 1998). Dundee (1998) mentions two small gaps within the distribution: in the lower reaches of the Pee Dee River in South Carolina and in several streams in the southeastern corner of South Carolina. These may be true absences or reflect a lack of collecting effort. Dwarf waterdogs typically occur in small streams and relatively shallow water, whereas the lower reaches of streams normally are deep.

2. Historical versus Current Abundance.
Unknown. Populations probably persist in most areas of the historical range. Recent records from several Virginia counties (Tobey, 1985; Roble et al., 1999) suggest that the species occupies areas that probably have long harbored dwarf waterdogs. Despite stream pollution, they occur widely throughout their range, which also suggests that they occur where earliest records were noted. Traditional collecting methods of seining, dip netting, hook and line fishing, and minnow trapping have been augmented by the use of rotenone and electroshocking, thus allowing a more comprehensive discovery of distribution and habitat.

3. Life History Features.
A. Breeding. Reproduction presumably is aquatic because that is the typical pattern for *Necturus* sp., and all dwarf waterdogs have been found in aquatic habitats. Oocyte number increases with size (Meffe and Sheldon, 1987). Mid-autumn ovaries contained yolked and unyolked oocytes; males showed signs of sexual activity—large, firm, yellow testes with heavy dark

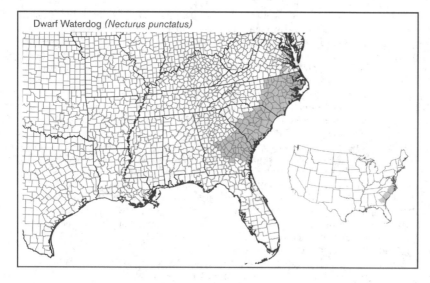

Dwarf Waterdog (Necturus punctatus)

pigmentation, involuted vasa deferentia, and swollen cloacal glands (Meffe and Sheldon, 1987). Bishop (1943) suggests a long breeding season, but likely most hatching takes place in winter. Neill (1963) found that dwarf waterdogs are most active during the winter when water is cold and streams are swollen.

i. Breeding migrations. Little is known of breeding in dwarf waterdogs; nests have not been discovered and no study has been made of tagged individuals.

ii. Breeding habitat. Unknown, but surely in water.

B. Eggs.

i. Egg deposition sites. Nests have not been discovered. Females likely attach their eggs to the undersides of logs and other objects lying in the water.

ii. Clutch size. Gravid females contain 15–55 ova (most between 20 and 40), and ova number is positively correlated with female size. Folkerts (1971) reported a gravid female collected on 12 April with 33 ovarian eggs; average egg diameter was 4.2 mm (other females in this sample had previously oviposited). Meffe and Sheldon (1987) found a single female on 20 February that had ovarian eggs averaging about 4 mm diameter. From these data, mating presumably occurs during the winter (Meffe and Sheldon, 1987; Petranka, 1998).

C. Larvae/Metamorphosis.

i. Length of larval stage. At least 2 yr (Bishop, 1943). The smallest known larva was 28 mm TL and was collected on 30 November; 40 mm TL larvae, presumably young of the year, were also collected at that time (Bishop, 1943).

ii. Larval requirements. Shallow water and leaf beds and a suitable source of food.

a. Food. Oligochaetes, aquatic Crustacea, especially ostracods (24.6%), various Insecta, particularly the dipterans Simuliidae (12.3%), Ceratopogonidae (24.6%), and beetles (Dytiscidae, 12.3%;

Braswell and Ashton, 1985). Similarly, Folkerts (1971) examined six immature specimens and found they had eaten annelids, millipedes, amphipods, and insects, especially caddisfly larvae.

b. Cover. Larvae live in both shallow and deeper waters (Brimley, 1924) and will burrow in silt (Martof et al., 1980).

iii. Larval polymorphisms. Most larvae are brown but gradually change to gray when they are 40–50 mm SVL (Folkerts, 1971).

iv. Features of metamorphosis. Metamorphosis does not occur; only maturation of the reproductive system signifies change from larva to adult.

v. Post-metamorphic migrations. Unknown.

vi. Neoteny. Dwarf waterdogs have a larval structure, but no experiments have checked their response to metamorphic stimulating hormones. Thyroxin will cause reduction of gill and fin size in mudpuppies *(N. maculosus)*; although these salamanders produce substantial amounts of thyroxin, they are essentially immune to this metamorphic agent and are thus permanently neotenic. The same likely applies to dwarf waterdogs.

D. Juvenile Habitat. Juveniles apparently prefer shallower water than adults (Folkerts, 1971) and are most common in mats of leaves.

E. Adult Habitat. Permanently aquatic and apparently only in streams. Bishop (1943) reported that they are usually found in the slower regions of streams, including side ditches, having mud or sandy banks/bottoms. These streams are usually small to medium sized. Dwarf waterdogs are most common in deeper sections with reduced flows and an accumulation of mud, silt, and leaves (Folkerts, 1971; Meffe and Sheldon, 1987). Individuals are rarely found in the main channels of the Neuse and Tar rivers (Braswell and Ashton, 1985; Petranka, 1998). A univariate analysis of habitat use based on water depth,

water velocity, stream width, and substrates is presented in Meffe and Sheldon (1987).

F. Home Range Size. Unknown.

G. Territories. Unknown, but the numbers collected at given sites and habitats suggest that territorial behavior is not exhibited.

H. Aestivation/Avoiding Desiccation. Unknown, but most animals reported were taken in the cooler months of October–April.

I. Seasonal Migrations. Unknown.

J. Torpor (Hibernation). Adults are active during wintertime and will aggregate in leaf beds (Brimley, 1924; Martof et al., 1980).

K. Interspecific Associations/Exclusions. Dwarf waterdogs and Neuse River waterdogs *(N. lewisi)* exhibit a strong dietary overlap and likely compete with each other for food resources in North Carolina in areas where they are syntopic (Braswell and Ashton, 1985). Other associates include lesser sirens *(Siren intermedia)*, many-lined salamanders *(Stereochilus marginatus)*, several species of fishes, and flattened naiads of the dragonfly *Hagenius brevistylus* (Folkerts, 1971).

L. Age/Size at Reproductive Maturity. Dwarf waterdogs reach sexual maturity between 65–70 mm SVL (Hecht, 1958). Folkerts (1971) suggests most individuals become sexually mature during their fifth year.

M. Longevity. A wild-caught animal lived for a little over 5 yr, 8 mo at the Cincinnati Zoo (Snider and Bowler, 1991). If the animal was adult when caught, then Folkerts' estimate of sexual maturity at 5 yr would indicate a longevity in excess of 10 yr.

N. Feeding Behavior. Braswell and Ashton (1985) showed that in sites sympatric with Neuse River waterdogs, adult dwarf waterdogs consumed gastropods (15%), pelecypods, oligochaetes, arachnids, isopods especially (22.2%), plus cladocerans, ostracods, copepods, amphipods, chilopods, and various insect families and orders, especially Trichoptera (22.2%). Prey of adults generally includes other species of salamanders, annelids, crayfish, and insects such as mayflies and chironomids (Brode, 1969; Fedak, 1971; Folkerts, 1971; Meffe and Sheldon, 1987; Gibbons and Semlitsch, 1991). Plant material was also present, although this could have been ingested incidentally (Meffe and Sheldon, 1987). During the breeding season, 54% of adults had empty stomachs, suggesting that some adults either cease feeding or reduce their feeding activity at this time (Meffe and Sheldon, 1987). In March–April many stomachs were empty, suggesting dwarf waterdogs do not feed during the breeding season (Folkerts, 1971). A high percentage of stomachs were empty in mid-October to mid-November, perhaps as a prelude to the breeding season (Meffe and Sheldon, 1987). Folkerts (1971) suggested that feeding may be nocturnal and digestion completed before the daytime collecting. He also suggested that larvae may feed throughout the year. A partially digested, unidentifiable salamander found in one stomach suggests cannibalism (Meffe and Sheldon, 1987).

Prey of juveniles includes annelids (especially earthworms), amphipods, millipedes, and insects such as caddisfly larvae (Folkerts, 1971).

O. Predators. According to Petranka (1998), predators have not been identified but likely include fishes. Unless dwarf waterdogs produce toxic skin secretions, the animals associated with their habitat that are predatory, hence enemies, could include predaceous aquatic insects, crayfish, water snakes, and large salamanders such as sirens, amphiumas, and Neuse River waterdogs. In captivity, Neuse River waterdogs are aggressive towards dwarf waterdogs and will viciously bite them (Petranka, 1998).

P. Anti-Predator Mechanisms. Unknown. Neuse River waterdogs are suspected to produce noxious secretions (Brandon and Huheey, 1985), but this has not been studied in dwarf waterdogs. Escape to hiding places in undercut stream banks or amid debris would seem to be likely ways to avoid predators.

Q. Diseases. Unknown.

R. Parasites. Unknown.

4. Conservation.
Populations likely persist in most areas of the historical range. Recent records suggest that the species occupies areas that probably have long harbored dwarf waterdogs (Tobey, 1985; Roble et al., 1999). Despite stream pollution, they occur widely throughout their range, which also suggests that they occur where earliest records were obtained.

Necturus cf. *beyeri*
LODING'S WATERDOG
Craig Guyer

Editor's note: The nomenclature used here is recommended by Bart et al. (1997) and consistent with the unpublished genetic data of Dr. Guyer.

1. Historical versus Current Distribution.
Loding's waterdogs (*Necturus* cf. *beyeri*; Bart et al., 1997) are found in rivers and streams draining the Gulf Coastal Plains from the Dog River in southwestern Alabama to the Apalachicola River drainage of western Georgia, eastern Alabama, and northern Florida (Viosca, 1937; Hecht, 1958).

2. Historical versus Current Abundance.
This species is still abundant in slow-moving rivers and streams of the Lower

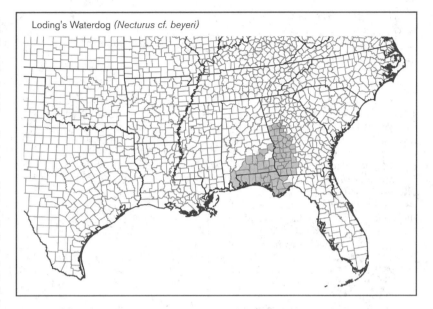

Loding's Waterdog *(Necturus cf. beyeri)*

Coastal Plains where surrounding forests are intact. Their abundance probably is reduced where deforestation increases siltation and decreases seasonal deposition of leaves.

3. Life History Features.
A. Breeding. Reproduction is aquatic.

i. Breeding migrations. As in other members of the genus *Necturus*, mating probably occurs in late autumn to early winter (Shoop, 1965b).

ii. Breeding habitat. Presumably aquatic sites.

B. Eggs.
i. Egg deposition sites. Unknown, but likely to be laid under logs and other cover objects as in Gulf Coast waterdogs (*N. beyeri*; Shoop, 1965b).

ii. Clutch size. Eggs of this species have not been described but they are likely to be laid singly in clusters of 20–40, as in Gulf Coast waterdogs (Shoop, 1965b).

C. Larvae/Metamorphosis.
i. Length of larval stage. Unknown. As with all members of the genus *Necturus*, metamorphosis is gradual and incomplete.

ii. Larval requirements.

a. Food. Unstudied, but larvae are undoubtedly carnivorous and feed on an array of aquatic invertebrates.

b. Cover. Larvae are most easily sampled from submerged piles of leaves in slow-moving streams.

iii. Larval polymorphisms. Unknown and unlikely.

iv. Features of metamorphosis. Unknown.

v. Post-metamorphic migrations. Unknown.

vi. Neoteny. Loding's waterdogs retain gills and remain aquatic as adults.

D. Juvenile Habitat. Juveniles inhabit bottom debris, especially leaf litter, where currents are slow.

E. Adult Habitat. Adults can be found under submerged debris or in submerged piles of leaves.

F. Home Range Size. Unknown, but likely to be small, as in Gulf Coast waterdogs (Shoop and Gunning, 1967).

G. Territories. Unknown.

H. Aestivation/Avoiding Desiccation. Unknown and unlikely.

I. Seasonal Migrations. Unlikely.

J. Torpor (Hibernation). Unlikely.

K. Interspecific Associations/Exclusions. Loding's waterdogs are unlikely to encounter any other species of *Necturus* within their range, except in the drainages surrounding Mobile Bay in Alabama. Here Loding's waterdogs and Gulf Coast waterdogs come in close proximity, and competitive interactions might be expected.

L. Age/Size at Reproductive Maturity. Precise data are lacking. However, this species is likely to be slightly smaller than Gulf Coast waterdogs (Viosca, 1937).

M. Longevity. Unknown.

N. Feeding Behavior. Unknown, but likely to include crustaceans, insects, and small fish.

O. Predators. Predatory fish are likely to be the major predators.

P. Anti-Predator Mechanisms. Unknown.

Q. Diseases. Unknown.

R. Parasites. Unknown.

4. Conservation.
Loding's waterdogs receive no state or federal protection. Keys to maintaining healthy populations are likely to include retention of forested lands along streams to reduce siltation and provide seasonal input of fallen leaves. Additionally, maintaining structural diversity in the form of submerged logs is likely to be important for adult home ranges and nest sites.

Family Rhyacotritonidae

Rhyacotriton cascadae Good
and Wake, 1992
CASCADE TORRENT SALAMANDER

Marc P. Hayes

1. Historical versus Current Distribution.

Cascade torrent salamanders *(Rhyacotriton cascadae)* are restricted to the west slope of the Cascades mountain axis from the west bank of the Skookumchuck River in central Washington (0.6 km [0.4 mi] north of the Thurston–Lewis County line; Wilson and Larsen, 1992; McAllister, 1995) south to the Middle Fork of the Willamette River in central western Oregon (Lane County; J. Applegarth, Bureau of Land Management, personal communication; see also Good et al., 1987; Good and Wake, 1992). Cascade torrent salamanders are known to range in elevation to over 1,219 m (4,000 ft), but areal extent of their geographic range is ambiguous largely because upper- and lower-elevation limits, and how these change with latitude, are poorly understood. A watershed-wide study of the Blue River hydrographic basin near the southern end of their geographic range (Willamette River hydrographic basin, Oregon; Hunter, 1998) revealed Cascade torrent salamanders at 52 of 273 (19%) sites, with the probability of occurrence peaking at an elevation of around 870 m (ca. 2,854 ft). Likelihood of occurrence drops markedly above elevations with heavy snow (J. Applegarth, Bureau of Land Management, personal communication), so the latter point may approximate the elevation above which species occurrence begins to decline. The elevation at which heavy snow begins decreases in elevation with increasing latitude, so the upper elevation limit of Cascade torrent salamanders across their geographic range may parallel that pattern. Further, Hunter's (1998) finding that the maximum size of basins in which Cascade torrent salamanders were detected was 141 ha (348 ac; i.e., relatively small) may indicate the species is infrequent downstream, where gradients are typically lower, thus fewer opportunities exist for appropriate habitat. However, the current distribution pattern may also be influenced by timber harvest, which has been suggested as having a greater influence on lower gradient systems (see "Conservation" below). No attempt has been made to contrast historical versus current distribution. Most data collected on Cascade torrent salamanders are <30 yr old, so an unambiguous assessment of change may not be possible.

2. Historical versus Current Abundance.

No historical abundance data are available, and the few data that provide abundance

Cascade Torrent Salamander *(Rhyacotriton cascadae)*

estimates are relatively recent. Nussbaum and Tait (1977) estimated larval densities of Cascade torrent salamanders at a Columbia River gorge site (Oregon; a small unnamed creek 500 m [ca. 1,640 ft] east of Wahkeena Falls). Density estimates for June–October, which averaged 33.3 larvae/m^2 (3.1 larvae/ft^2) and ranged from 27.6 larvae/m^2 to 41.2 larvae/m^2 (2.6 larvae/ft^2 to 3.8 larvae/ft^2), are among the highest recorded for any species of *Rhyacotriton* (regardless of sampling method) and are also high in comparison to other salamanders (e.g., Spight, 1967c). Nussbaum and Tait (1977) surmised that the absence of vertebrate competitors and predators at this site at least partly explained the high densities. Nijhuis and Kaplan (1998) estimated adult densities at another Columbia River gorge site (Wahkeena Falls) in March–May 1995, which averaged 1.9 adults/m^2 (0.2 adults/ft^2; range = 0.3–8.3 adults/m^2 [<0.1–0.8 adults/ft^2]). At six sites within both the Mt. Hood National Forest and the Andrews Experimental Forest (Oregon), Bury et al. (1991) reported mean densities of 1.40 individuals/m^2 (0.13 individuals /ft^2) and 0.07 individuals/m^2 (<0.01 individuals/ft^2), respectively.

3. Life History Features.

A. Breeding. Reproduction is aquatic.

i. Breeding migrations. Breeding migrations have not been documented.

ii. Breeding habitat. Unknown.

B. Eggs.

i. Egg deposition sites. Nests of Cascade torrent salamanders have not been described. However, nest sites are suspected to have some similarity to those described for the closely related Columbia torrent salamander (*R. kezeri*; Nussbaum, 1969b; Russell et al., 2002). Columbia torrent salamanders place unattached eggs among the coarse substrate spaces of low-velocity flow seeps.

ii. Clutch sizes. Gravid female Cascade torrent salamanders in the Columbia River gorge averaged 8 ova (range = 2–14; Nussbaum and Tait, 1977).

C. Larvae/Metamorphosis.

i. Length of larval stage. Data on Cascade torrent salamanders from the Columbia River gorge indicate a long larval life, estimated between 3–4 yr (Nussbaum and Tait, 1977). This larval interval is long compared with other species of salamanders.

ii. Larval requirements.

a. Food. No data on larval food requirements exist for Cascade torrent salamanders or any other species of *Rhyacotriton*.

b. Cover. No detailed data exist on cover requirements for Cascade torrent salamander larvae, although Nussbaum and Tait (1977) did comment that Cascade torrent salamander larvae were numerous under stones along a narrow stream channel and in the network of fissures in the streambed and bank. Reporting on pooled data that included some information on Cascade torrent salamanders, Bury et al. (1991) indicated that smaller larvae of *Rhyacotriton* frequently occurred under small rocks or in beds of gravel, pebbles, and cobbles. Moreover, data from the closely related southern torrent salamander (*R. variegatus*; Diller and Wallace, 1996; Welsh and Lind, 1996) suggest that stable, slow-flow aquatic microhabitats having unimbedded gravel and cobble (i.e., with open interstices) and reduced levels of fine sediments would be preferred.

iii. *Larval polymorphisms.* No larval polymorphisms have been described for Cascade torrent salamanders or any other species of *Rhyacotriton*.

iv. *Features of metamorphosis.* Metamorphosis of Cascade torrent salamanders in the Columbia River gorge can occur at almost any time, but most individuals appear to metamorphose in late summer to early autumn (Nussbaum and Tait, 1977). The duration of metamorphosis is unknown.

v. *Post-metamorphic migrations.* Post-metamorphic migrations have not been described for Cascade torrent salamanders or any other species of *Rhyacotriton*.

vi. *Neoteny.* Neoteny has not been found in Cascade torrent salamanders (Nussbaum and Tait, 1977), and it is unknown in the genus (Good and Wake, 1992).

D. Juvenile Habitat. Unknown. Based on data for Columbia River gorge Cascade torrent salamanders (Nussbaum and Tait, 1977), little may exist to distinguish the habitat characteristics of juveniles and adults. This comment must be viewed in context of the fact that little effort has been made to distinguish differences in the habitat characteristics of these age groups. Reporting on pooled data that included some information on Cascade torrent salamanders, Bury et al. (1991) found no significant differences either in the shallow-water stream depths or use of rocks by different life stages, but this was a post-metamorphic–larval comparison (juveniles were not partitioned from adults in the post-metamorphic category).

E. Adult Habitat. Except for the general comments by Nussbaum and Tait (1977) that, "Favorite hiding places of adults are the underground water-courses in the rock rubble of stream banks, fissures in stream banks, fissures in stream heads and cracks in wet cliff faces," descriptions of adult habitat are restricted to general accounts (e.g., Leonard et al., 1993; Corkran and Thoms, 1996). Reporting on pooled data that included some information on Cascade torrent salamanders, Bury et al. (1991) noted that *Rhyacotriton* were most often found in riffles, but life stage data were pooled, so how much of this applies to adults is unclear. A few descriptive data exist for the genus. Stebbins and Lowe

(1951) commented that, "The microenvironment of *Rhyacotriton* . . . appears uniform throughout . . . [their geographic] range . . . ." Detailing the microenvironment, they add, ". . . characteristically one of a cold, permanent stream with small water-washed or moss-covered rocks (usually rock rubble) in and along . . . running water . . . seeps and small, trickling tributary streams with rocks of small dimensions . . . ." Stebbins and Lowe (1951) also generalized about the vegetation typical of *Rhyacotriton*-occupied streams, which characterize conditions that provide a cool, wet microenvironment: "Streams harboring *Rhyacotriton* usually have a good leaf canopy, especially during the summer months . . . Abundant understory vegetation, much moss, and a thick leaf mat [characterize] stream banks." They also emphasize the use of rock habitat, stating that *Rhyacotriton* is ". . . less frequently found under moss and wood than . . . rocks." Emphasizing the wet nature of *Rhyacotriton* habitat, Stebbins and Lowe (1951) also noted that it was characterized by low flow velocities: "*Rhyacotriton* usually selects . . . sites where the movement of water tends to be relatively slow. They rest with their vents in shallow water, and one rarely finds an individual [that is] not in contact with free water or at least a saturated substrate." This pattern is likely linked to the fact that *Rhyacotriton* are among the most, if not the most, desiccation-intolerant salamander genus known (Ray, 1958). Desiccation intolerance is likely linked to high dependence on skin surfaces for oxygen uptake (average = 74%) because their lungs are highly reduced (Whitford and Hutchison, 1966b). No data on habitat differences between females and males are available for Cascade torrent salamanders or any other species of *Rhyacotriton*.

F. Home Range Size. Few data are available on home range size per se for Cascade torrent salamanders. Available data suggest that Cascade torrent salamanders are sedentary (Nussbaum and Tait, 1977; Nijhius and Kaplan, 1998), with movements on the scale of a few meters being typical. Using 2-m stream edge increments to distinguish movement, Nussbaum and Tait (1977) found that 134 of 191 (70%) recaptured Cascade torrent salamander larvae made no net movement (i.e., recaptured in the same 2-m interval). Thirty-nine (20%) revealed net movement upstream, and the other 18 (10%) showed net movement downstream, a significant difference. The greatest distance moved, 22 m, was displayed by two larvae, one that moved upstream and the other downstream. Using range length (the distance between the two most widely separated points of capture), Nijhius and Kaplan (1998) found mean range length for Cascade torrent salamander adults captured at least three times during the spring at a

Columbia River gorge site was 2.4 m (7.9 ft; range = 0.2–6.0 m [0.7–19.7 ft]). Nijhius and Kaplan (1998) also used the relatively high percentage of individuals recaptured (36%) in a small portion of available habitat studied to indicate the high degree of sedentation. Further, they found no differences in the magnitude of movements between the age, sex, or size groupings of post-metamorphic Cascade torrent salamanders examined. While the sedentary pattern in these reports is probably real, these data must be interpreted with the knowledge that each of these studies suffers from some degree of bias because recaptures with the intent of identifying movements were made exclusively within highly circumscribed, small areas.

G. Territories. No data are available on territoriality. Available data on Cascade torrent salamanders indicate high densities and restricted movements (Nussbaum and Tait, 1977; Nijhius and Kaplan, 1998; see "Current versus Historical Abundance" and "Home Range Size" above), implying territorial behavior to be unlikely. However, the behavioral data needed to interpret territorial defense are lacking.

H. Aestivation/Avoiding Desiccation. Unknown.

I. Seasonal Migrations. Unknown.

J. Torpor (Hibernation). Unknown.

K. Interspecific Associations/Exclusions. In the Columbia River gorge (Oregon; Nussbaum and Tait, 1977; Nijhius and Kaplan, 1998; personal observations) and probably elsewhere within their range, Cascade torrent salamanders display at least some degree of syntopy with five different amphibian species: coastal tailed frogs (*Ascaphus truei*), Cope's giant salamanders (*Dicamptodon copei*), coastal giant salamanders (*D. tenebrosus*), western red-backed salamanders (*Plethodon vehiculum*), and Dunn's salamanders (*P. dunni*). Patterns of syntopy with each of these taxa have not yet been detailed. Based on studies of other *Rhyacotriton*, giant salamanders are repeatedly mentioned as potential predators that could locally limit *Rhyacotriton* (Stebbins, 1953; Welsh, 1993; Welsh and Lind, 1996). Inverse relationships in abundance between torrent and giant salamander larvae (Welsh and Lind, 1996) and anecdotal observations regarding the relative rarity of syntopy between torrent and giant salamanders (Stebbins, 1953) seem to be the basis of these comments. Recent work revealing that larval southern torrent salamanders are unpalatable to larval coastal giant salamanders (Rundio and Olson, 2001) suggests that the role giant salamanders play in limiting torrent salamanders requires re-evaluation or more thorough investigation. At least one species of *Rhyacotriton* (e.g., southern torrent salamander) is known to move away from injured western red-backed salamanders (Chivers et al., 1997), apparently as a predator-avoidance response. Cascade torrent salamanders may show a similar pattern.

L. Age/Size at Reproductive Maturity. Nussbaum and Tait (1977) estimated that Columbia River Gorge Cascade torrent salamanders require 5.5–6.0 yr to reach reproductive maturity at a minimum size of 41 mm SVL for males, 44 mm for females.

M. Longevity. Unknown. Nussbaum and Tait (1977) imply that a moderately long lifespan (≥ 10 yr) is likely.

N. Feeding Behavior. Unknown. Based on data for other *Rhyacotriton*, Cascade torrent salamanders probably feed in moist forested habitats on invertebrates, notably amphipods, fly larvae, springtails, and stonefly nymphs (Bury and Martin, 1967; Bury, 1970). These taxa occur in the semi-aquatic and aquatic microhabitats where feeding is thought to occur.

O. Predators. No predators of Cascade torrent salamanders have been recorded. Based on studies of other *Rhyacotriton*, giant salamanders are suspected predators (Stebbins, 1953; Nussbaum, 1969b; Welsh, 1993; Welsh and Lind, 1996), but the recent demonstration of the relative unpalatability of southern torrent salamander larvae (Rundio and Olson, 2001) may require re-evaluation of which torrent salamander life stages may be vulnerable to giant salamander predation. Nussbaum et al. (1983) also commented on shrews (Soricidae) rejecting torrent salamanders as unpalatable.

P. Anti-Predator Mechanisms. No anti-predator mechanisms have been recorded for Cascade torrent salamanders. Brodie (1977) described anti-predator postures in Olympic torrent salamanders (*R. olympicus*) *sensu lato* as similar to many *Ambystoma*. The body is coiled, and the tail is elevated or arched and wagged and may be lashed toward a threat. The bright yellow venter contrasting with the drabber dorsal coloration is believed to be aposematic. Partly metamorphosed larvae postured similarly to adults. The recent demonstration that southern torrent salamander larvae are unpalatable (see "Interspecific Associations/Exclusions" above) may agree with an aposematic hypothesis, but unpalatability of post-metamorphic life stages remains to be demonstrated. Based on an experimental study of southern torrent salamanders, avoidance of injured conspecifics and injured western red-backed salamanders (Chivers et al., 1997) may be an anti-predator device also present in Cascade torrent salamanders.

Q. Diseases. Not examined.

R. Parasites. One species of monogenoidean fluke, to date unique to *Rhyacotriton* (Kristsky et al., 1993), has been recorded from Cascade torrent salamanders. However, several other species of digenean flukes (Lynch, 1936; Senger and Macy, 1952; Lehman, 1954; Anderson et al., 1966; Martin, 1966b) and blood parasites (Clark et al. 1969) have been described from other species of *Rhyacotriton* and may occur in Cascade torrent salamanders.

4. Conservation.
Cascade torrent salamanders have state Sensitive status in both the states that encompass their geographic range: Oregon (Marshall et al., 1996) and Washington (K. McAllister, personal communication). In both states, "Sensitive" consists of a watchlist status that lacks legal standing, but is applied to species for which serious concerns exist related to habitat loss in order to increase awareness among resource protection agencies. In the case of both states, Sensitive status was established because concern existed that the rapid rate of conversion of mature and old-growth forests to young stands as a function of timber harvest were limiting habitat quality through increased microhabitat temperatures and sedimentation, and that local extirpation was resulting from these changes (Corn and Bury, 1989a). Concern also existed that headwater streams, seeps, and springs (all habitats presumed to be occupied) lacked adequate protection. Data used to support Sensitive status were an amalgam of information drawn entirely from retrospective studies that addressed related southern torrent salamanders (e.g., Corn and Bury, 1989a; Welsh, 1990, 1993; Welsh and Lind, 1996).

The actual status of Cascade torrent salamanders is unstudied. Lower-gradient, higher-order streams, which may intrinsically provide poor habitat for *Rhyacotriton* (see "Historical versus Current Distribution" above) are often more disturbed as a function of timber harvest (due to greater susceptibility to habitat disturbance because of greater depositional and more limited scour characteristics, a greater upslope area of influence, more harvest rotations, etc.) than higher-gradient, lower-order streams, which may provide better habitat (Diller and Wallace, 1996; Welsh and Lind, 1996; Hunter, 1998). A recent retrospective study of Cascade torrent salamanders focusing on headwater landscapes with stands harvested 0–60 yr ago revealed the highest densities of torrent salamanders associated with stands 21–40 yr old (Steele, 2001). Unfortunately, too few unharvested stands with parallel site criteria were available for comparison. Distinguishing the relative significance of intrinsic habitat limitation from the potential influences of timber harvest will require comparing harvested and unharvested lower- versus higher-gradient sites simultaneously.

In 2000, scientists representing private timber companies, Native American tribes, and state and federal resource agencies assessed the risk of forest management activities to stream-associated vertebrate species. This work, done in preparation for the Washington Forest and Fish Agreement (FFA), concluded that seven species of amphibians (including all three species of torrent salamanders in Washington State) were at high risk of local extirpation from forest management activities. One hoped-for outcome of identifying species at risk and including them in the list of species protected under new forest practice rules was that those species would be studied as part of an innovative, ambitious adaptive management program. The goal of FFA adaptive management is to determine whether new riparian buffer prescriptions designed for headwater streams are effective in protecting resources to which they are linked, including local populations of *Rhyacotriton*. To date, this research has identified the most appropriate sampling methods for landscape detection and relative abundance assessment of Cascade torrent salamanders and the other stream-associated amphibians. These methods will be essential for manipulative studies to test the effectiveness of forest practice rules in protecting *Rhyacotriton*.

Rhyacotriton kezeri Good and Wake, 1992
COLUMBIA TORRENT SALAMANDER

Marc P. Hayes, Timothy Quinn

1. Historical versus Current Distribution.
Columbia torrent salamanders (*Rhyacotriton kezeri*) are restricted to coastal and near-coastal regions of northwestern Oregon and southwestern Washington, from the Little Nestucca River system (Tillamook County, Oregon) in the south to the Chehalis River (Gray's Harbor County, Washington) in the north (Good and Wake, 1992; McAllister, 1995). Across this range, Columbia torrent salamanders occur from near sea level to the highest elevations in the region (i.e., Boisfort Peak, 948 m [3,110 ft], in the Willapa Hills of southwestern Washington, and Saddleback Mountain, 1,001 m [3,283 ft], in the coast ranges of northwestern Oregon). They occur in some upper reaches of the coastal part of the Willamette hydrographic basin (e.g., Upper Grand Ronde system, Oregon [see Good et al., 1987]). However, their distribution across the region and their inland distribution in parts of Washington (especially in the Willapa Hills State Forest area; see McAllister, 1995) are poorly known. In the Willapa Hills, Columbia torrent salamanders were encountered in 53% of 40 near-coastal perennial headwater streams in a managed forest landscape (Wilkins and Peterson, 2000). Wilkins and Peterson (2000) also found that likelihood of occupancy for Columbia torrent salamanders generally increased as channel gradient increased and basin area decreased, so that the likelihood of occupancy generally increased as one moved toward headwaters. Farther inland in the Willapa Hills, Columbia torrent salamanders were detected

in 73% of 70 headwater streams in managed forests (D. Runde, personal communication). A recent study of a large segment of Oregon Columbia torrent salamander range revealed them to be widespread in headwater streams of managed forests (occurring in 58% of such randomly selected streams); this study also showed that Columbia torrent salamanders occurred more often in streams on basalt formations than in streams on marine sediments (Russell et al., 2004). These authors also found that streams with northerly aspects were significantly more likely to have Columbia torrent salamanders than streams with southerly aspects. In a study in the Kilchis River drainage (Oregon), Columbia torrent salamanders were observed at 67% of 33 sites (D. Vesely, Pacific Northwest Research Institute, personal communication). No attempts have been made to contrast historical with current distribution, but the study by Russell et al. (2004), which was conducted over a large segment of the historic Tillamook burn, suggested that this species may be resilient to habitat-altering catastrophic disturbance (at least for this high-precipitation coastal area). The Tillamook burn consisted of a series of stand-replacing events that occurred over an 18-yr period (1933-51) and covered over 121,000 ha (300,000 ac). Still, most data on Columbia torrent salamanders are <30 yr old, so that a range-wide historical assessment may not be possible.

2. Historical versus Current Abundance.

No historical data on abundance are available, but some recent studies provide relative abundance data. In the Willapa Hills, Columbia torrent salamanders made up 21% of the individuals across eight amphibian species and ranked second only to giant salamanders (*Dicamptodon* sp.) in relative abundance (Wilkins and Peterson, 2000). In two other unpublished studies conducted in the Willapa Hills, Columbia torrent salamanders ranked first in abundance among the 9 (D. Runde, Weyerhaeuser Company, personal communication) and 13 (unpublished data) amphibian species recorded; in both cases they represented >60% of individuals of all species. In the Russell et al. (2004) study, Columbia torrent salamander densities ranged from 0–68.3 individuals/m^2; the upper end is the highest recorded for any torrent salamander species. Relative abundance of Columbia torrent salamanders is not always among the highest in amphibian assemblages. In a Kilchis River study, Columbia torrent salamanders ranked fourth in relative abundance among the 10 amphibian species recorded (D. Vesely, personal communication). At least part of the reason for this disparity in relative abundance patterns across studies may depend upon how sampling locations were selected within hydrographic basins, as Russell et al. (2004) indicated that Columbia torrent salamander density increases with proximity to the headwater channel origin.

Two of the three studies in the Willapa Hills found some degree of inverse relationship between relative abundance and gradient. Because these studies were all conducted in landscapes managed for timber harvest, these data are difficult to interpret. The notion that low gradients may intrinsically provide less opportunity for favorable habitat is confounded with the potential for these low-gradient habitats to have been most heavily influenced by timber harvest (see "Conservation" below). Russell et al. (2004) also found that relative abundance of Columbia torrent salamanders was significantly greater in streams with basalt substrates than in streams with marine sedimentary substrates, a finding consistent with Columbia torrent salamander occupancy patterns (Wilkins and Peterson, 2000; see "Historical versus Current Distribution" above). Through the generation of a coarser-grained substrate structure, basalt substrates are thought to provide more suitable habitat than sedimentary substrates (Wilkins and Peterson, 2000), but competence of the substrate, rather than the formation per se, may be the important factor in habitat quality, all else being equal (see Dupuis et al., 2000). Excessive inputs of fine substrates has often been suggested as a factor degrading habitat quality for stream-associated amphibians such as Columbia torrent salamanders (e.g., Corn and Bury, 1989a; Welsh and Lind, 1996; Dupuis et al., 2000).

3. Life History Features.
 A. **Breeding.** Reproduction is aquatic.
 i. Breeding migrations. Breeding migrations have not been documented for this species.
 ii. Breeding habitats. Unknown.
 B. **Eggs.**
 i. Egg deposition sites. Nussbaum (1969b) described two nest sites attributable to Columbia torrent salamanders (the species was not described until Good and Wake [1992]), and more recently, Russell et al. (2002) described three more nests. Four of the nests were from Oregon, the remaining one was from Washington. One of the Nussbaum nests, discovered on 14 December, was found during the excavation of a headwater spring. The other nest Nussbaum described, found on 28 September in a side-slope seep, was revealed by prying loose a large (2.4 m × 1.2 m × 1.2 m [8 ft × 4 ft × 4 ft]) rock block that root action in a seep had separated from the base of a cliff. Both nest sites were in sandstone; had relatively cold (8.3–9.0 °C), slow-flowing water trickling over loose, unattached eggs; had more eggs (16–75) than could have been laid by one female; and lacked an attendant adult (several adults and one larvae were within 60 cm [2 ft] of the 14 December nest, but none were near the eggs, whereas only three young larvae were seen near the

Columbia Torrent Salamander *(Rhyacotriton kezeri)*

28 September nest [Nussbaum 1969b]). Embryos in both nests were in tail-bud stage. Based on these observations, Nussbaum (1969b, 1985) concluded that eggs are not attached to substrate (a condition unique in salamanders), no parental care exists, and communal oviposition may be typical.

The three nests Russell et al. (2002) found broadened the knowledge of nest site variation, as two were from springs at the origin of first-order streams and the third was in a first-order channel, 75 m downstream of the origin. One of the nests found in a spring was under a small boulder (33 cm × 20 cm × 10 cm) underlain by fine sediments and sand, whereas the other two nests were under thick layers of moss (17 cm in the one nest measured) on substrates consisting of a mix of gravels and fine organic sediments. In all three cases, the lithology was an unspecified marine sedimentary formation. Unlike the nests Nussbaum (1969b) described, these nests were discovered between 16–26 July, and the eggs had no evidence of embryological development. Mean egg size from the three Russell et al. (2004) nests ranged from 3.8–4.1 mm in diameter exclusive of jelly capsules; egg diameter data were not obtained from the nests Nussbaum (1969b) described. Further, Russell et al. (2002) reported 1–2 adult female Columbia torrent salamanders in the vicinity of two of the three nests; a distance between adults and eggs was specified only for the two-female nest, in which case both females were within 5 cm of the eggs. Russell et al. (2002) agreed with most interpretations of Nussbaum (1969b, 1985), but considered the question of parental care open. Further, the number of eggs in the nests Russell et al. (2002) described, 7–11, may each represent a single clutch. Overall, these observations suggest that oviposition may be restricted to low-flow headwater habitats, probably because unattached eggs would be at risk from scour at higher flows. That few nest sites that have been found suggests that placement of eggs typically occurs in relatively inaccessible, cryptic locations.

ii. Clutch size. Unknown. If eggs from the Russell et al. (2002) nest descriptions represent single clutches, and no eggs were lost prior to their discovery, clutch size may be 7–11. Based on data for other species of *Rhyacotriton* (Nussbaum and Tait, 1977), fecundity is likely to be low (2–16 eggs).

C. Larvae/Metamorphosis.

i. Length of larval stage. Except for the egg stage, no life history data exist for Columbia torrent salamanders. However, data from closely related species of *Rhyacotriton* (Nussbaum and Tait, 1977) suggest that Columbia torrent salamanders probably remain larvae for a relatively long period (>2 yr).

ii. Larval requirements.
   a. Food. Unknown.
   b. Cover. Unknown. Habitat data from Russell et al. (2004) were not partitioned by life stages. However, Columbia torrent salamander larvae represented 60% of the individuals encountered in surveys, and habitat data generally agree with larval data from closely related species of *Rhyacotriton* (Nussbaum and Tait, 1977; Diller and Wallace, 1996; Welsh and Lind, 1996), suggesting that stable, slow-flow stream microhabitats with loose gravel and cobble, open interstices, and reduced levels of fine sediments are preferred.

iii. Larval polymorphisms. Have not been described and are unlikely.

iv. Features of metamorphosis. Unknown. Data from closely related species of *Rhyacotriton* indicate that metamorphosis can occur at almost any time, but a majority of individuals appear to metamorphose in late summer to early autumn (Nussbaum and Tait, 1977). The duration of metamorphosis is unknown.

v. Post-metamorphic migrations. Have not been described and are unlikely.

vi. Neoteny. Unknown in the genus *Rhyacotriton* (Good and Wake, 1992).

D. Juvenile Habitat. Unknown. Based on related species of *Rhyacotriton* (Nussbaum and Tait, 1997), few data exist to suggest that habitat requirements of adults and juveniles differ. However, almost no effort has been made to distinguish partitioning the habitat requirements of these age groups.

E. Adult Habitat. Except for nest-site habitat characterized by Nussbaum (1969b) and Russell et al. (2002), descriptions of Columbia torrent salamander habitat are limited to landscape and reach scale elements that Russell et al. (2004) discussed (see "Historical versus Current Distribution" and "Historical versus Current Abundance" above). However, some descriptive data encompass the genus. Stebbins and Lowe (1951) stated that "The microenvironment of *Rhyacotriton* appears uniform throughout . . . [their geographic] range . . . ." On microenvironment, they add, ". . . characteristically one of a cold, permanent stream with small water-washed or moss-covered rocks (usually rock rubble) in and along . . . running water . . . seeps and small, trickling tributary streams with rocks of small dimensions are sites [where] numbers of these salamanders may be found." Stebbins and Lowe (1951) also generalized about the vegetation typical of streams where *Rhyacotriton* are found that provide a cool, wet microenvironment: "Streams harboring *Rhyacotriton* usually have a good leaf canopy, especially during the summer months . . . Abundant understory vegetation, much moss, and a thick leaf mat [characterize] stream banks." Emphasizing the use of rock habitat, Stebbins and

Lowe (1951) noted that *Rhyacotriton* are ". . . less frequently found under moss and wood than . . . rocks." In emphasizing the wet nature of *Rhyacotriton* habitat, Stebbins and Lowe (1951) also noted that it was characterized by low flow velocities: "*Rhyacotriton* usually selects . . . sites where the movement of water tends to be relatively slow. They rest with their vents in shallow water, and one rarely finds an individual [that is] not in contact with free water or at least a saturated substrate." This pattern is likely linked to the fact that *Rhyacotriton* are among the most, if not the most, desiccation-intolerant salamander genera known (Ray, 1958). This intolerance is probably linked to a high dependence on skin surfaces for oxygen uptake (on average 74% of uptake is through the skin) because the lungs are highly reduced (Whitford and Hutchison, 1966b). No data on habitat differences between females and males are available for Columbia torrent salamanders or any other species of *Rhyacotriton*.

F. Home Range Size. Unknown. Data from other species of *Rhyacotriton* suggest that individuals are highly sedentary (Nussbaum and Tait, 1977, Welsh and Lind, 1992; Nijhius and Kaplan, 1998), with movements typically limited to a few meters. A highly sedentary pattern may be characteristic of the genus. Yet, all movement studies to date suffer from some degree of bias because recaptures with the intent of delineating home ranges were limited to highly circumscribed, small areas. The latter limitation is likely a function of the high cost associated with thorough searches.

G. Territories. Unknown. Available data on closely related Cascade torrent salamanders *(Rhyacotriton cascadae)* indicate high densities and restricted movements (Nussbaum and Tait, 1977; Nijhius and Kaplan, 1998), implying territorial behavior is unlikely. However, these indirect data must be viewed cautiously because data on behaviors related to territorial defense are lacking.

H. Aestivation/Avoiding Desiccation. Unknown. However, surface activity may be reduced during late summer to early autumn (personal observations), when surface conditions are most dry.

I. Seasonal Migrations. Unknown. Different studies on other species of *Rhyacotriton* appear to suggest conflicting patterns (upstream versus downstream) in larval movements (see Nussbaum and Tait, 1977; Welsh and Lind, 1992) that will require more study for correct interpretation.

J. Torpor (Hibernation). Unknown.

K. Interspecific Associations/Exclusions. Over their range, Columbia torrent salamanders are sympatric with 12 different amphibian species: northwestern salamanders *(Ambystoma gracile)*, long-toed salamanders *(A. macrodactylum)*, Cope's giant salamanders *(Dicamptodon copei)*, coastal giant salamanders *(D. tenebrosus)*, rough-skinned newts *(Taricha granulosa)*, ensatinas *(Ensatina eschscholtzii)*, Van Dyke's salamanders *(Plethodon vandykei)*, western red-backed salamanders *(P. vehiculum)*, Dunn's salamanders *(P. dunni)*, coastal (Pacific) tailed frogs *(Ascaphus truei)*, Pacific treefrogs *(Pseudacris regilla)*, and northern red-legged frogs *(Rana aurora;* McAllister, 1995; Vesely, 1997; D. Vesely, Pacific Northwest Research Institute; D. Runde, Weyerhaeuser Company, personal communication). However, degree of syntopy with each of these taxa has not yet been detailed. Based on studies of other *Rhyacotriton*, giant salamanders are repeatedly mentioned as potential predators that could locally limit *Rhyacotriton* (Stebbins, 1953; Welsh, 1993; Welsh and Lind, 1996). Inverse relationships in abundance between torrent and giant salamander larvae (Welsh and Lind, 1996), and anecdotal observations regarding the relative rarity of syntopy between torrent and giant salamanders (Stebbins, 1953) seem to be the basis of these comments. Recent work revealing that larval southern torrent salamanders are unpalatable to larval giant salamanders (Rundio and Olson, 2001) suggests that the role giant salamanders play in limiting torrent salamanders requires re-evaluation or more thorough investigation. At least one species of *Rhyacotriton* (e.g., southern torrent salamanders *[R. variegatus]*) is known to move away from injured western red-backed salamanders (Chivers et al., 1997) apparently as a predator-avoidance response; Columbia torrent salamanders may show a similar response.

L. Age/Size at Reproductive Maturity. Unknown. Based on data for other *Rhyacotriton* (Nussbaum and Tait, 1977), Columbia torrent salamanders probably have a relatively long interval to reproductive maturity (4 yr) and a relatively small size at reproductive maturity (about 45 mm SVL).

M. Longevity. Unknown. Based on data for other *Rhyacotriton* (Nussbaum and Tait, 1977), a moderately long lifespan (>10 yr) is likely.

N. Feeding Behavior. Unknown. Based on data for other *Rhyacotriton*, Columbia torrent salamanders probably feed on invertebrates dwelling in moist forested habitats, especially amphipods, fly larvae, springtails, and stonefly nymphs (Bury and Martin, 1967; Bury, 1970). These taxa occur in semi-aquatic and aquatic microhabitats where feeding is thought to occur.

O. Predators. Unknown. Based on studies of other *Rhyacotriton*, giant salamanders are suspected predators (Stebbins, 1953; Nussbaum, 1969b; Welsh, 1993; Welsh and Lind, 1996). However, the recent demonstration of the relative unpalatability of southern torrent salamander larvae to giant salamanders (Rundio and Olson, 2001) suggests that more study is

needed to determine which torrent salamander life stages may be vulnerable to giant salamander predation. Nussbaum et al. (1983) also commented on shrews (Soricidae) rejecting torrent salamanders as unpalatable.

P. Anti-Predator Mechanisms. Unknown. Brodie (1977) described anti-predator postures in Olympic torrent salamanders *(R. olympicus) sensu lato* as similar to many mole salamanders *(Ambystoma* sp.). The body is coiled, and the tail is elevated or arched and wagged and may be lashed toward a threat. The bright yellow venter contrasting with the drabber dorsal coloration is believed to be aposematic. Partly metamorphosed larvae posture similarly to adults. The recent demonstration that southern torrent salamander larvae are unpalatable (see "Interspecific Associations/Exclusions" above) may help support the aposematic hypothesis, but unpalatability of post-metamorphic life stages remains to be demonstrated. Based on an experimental study of southern torrent salamanders, avoidance of injured conspecifics and injured western red-backed salamanders (Chivers et al., 1997) may be an antipredator behavior also present in Columbia torrent salamanders.

Q. Diseases. No diseases have been recorded.

R. Parasites. Unknown. However, several species of parasites have been described from other species of *Rhyacotriton*, and may occur in Columbia torrent salamanders. These include one species of monogenoidean fluke, to date unique to *Rhyacotriton* (Kristsky et al., 1993), and several species of digenean flukes (Lynch, 1936; Senger and Macy, 1952; Lehman, 1954; Anderson et al., 1966; Martin, 1966b) and blood parasites (Clark et al., 1969).

4. Conservation.

Columbia torrent salamanders have state Sensitive status in both the states that encompass their geographic range: Oregon (Marshall et al., 1996) and Washington (K. McAllister, personal communication). In both states, "Sensitive" consists of a watchlist status that lacks legal standing, but is applied to species for which there is concern related to habitat loss in order to increase awareness among resource protection agencies. In the case of both states, Sensitive status was established because >97% of the species range was located in young stands (<80 yr old) intensively managed for timber. Much concern for Columbia torrent salamanders was related to the idea that habitat quality would be degraded by increased temperatures and sedimentation following timber harvest. Concern also existed that headwater streams, seeps, and springs (all habitats presumed to be occupied) lacked adequate protection. Data used to support the Sensitive designations were drawn entirely from retrospective studies

that addressed the southern torrent salamander (e.g., Corn and Bury, 1989a; Welsh, 1990).

The actual status of Columbia torrent salamanders is unstudied. Lower-gradient, higher-order streams, which may intrinsically provide poor habitat for *Rhyacotriton* (see "Historical versus Current Distribution" above), are often more disturbed as a function of timber harvest (due to greater susceptibility to habitat disturbance because of greater depositional and more limited scour characteristics, a greater upslope area of influence, more harvest rotations, etc.) than higher-gradient, lower-order streams, which may provide better habitat. Thus, the effect of timber harvest per se on torrent salamanders has been confounded with natural variation in habitat quality (Diller and Wallace, 1996; Welsh and Lind, 1996; Hunter, 1998). Unfortunately, historical harvest patterns have made it difficult to find unharvested stands in low-gradient streams that could serve as experimental units. Distinguishing the relative significance of intrinsic habitat limitation from the potential influences of timber harvest will require comparing harvested and unharvested lower- versus higher-gradient sites. Using surveys conducted on 20 × 40 m plots, Vesely and McComb (2002) showed that torrent salamanders were sensitive to forest practices in riparian areas, and that riparian buffer strips ca. 43 m wide would support a total salamander abundance (including torrent salamanders) similar to that in unlogged forests. Unfortunately, low numbers of torrent salamanders forced Vesely and McComb (2002) to pool the data for Columbia and southern torrent salamanders, making it impossible to determine the response of Columbia torrent salamanders alone.

In 2000, scientists representing private timber companies, Native American tribes, and state and federal resource agencies assessed the risk of forest management activities to stream-associated vertebrate species. This work, done in preparation for the Washington Forest and Fish Agreement (FFA), concluded that seven species of amphibians (including all three species of torrent salamanders in Washington State) were at high risk of local extirpation from forest management. One hoped-for outcome of identifying species at risk and including them in the list of species protected under new forest practice rules was that those species would be studied as part of an innovative, ambitious adaptive management program. The goal of FFA adaptive management is to determine whether new riparian buffer prescriptions designed for headwater streams are effective in protecting resources to which they are linked, including local populations of *Rhyacotriton*. To date, this research has identified the most appropriate sampling methods for landscape detection and relative abundance assessment of Columbia torrent salamanders and the other stream-associated amphibians. These methods will be essential for manipulative studies to test the effectiveness of forest practice rules in protecting *Rhyacotriton*.

Rhyacotriton olympicus (Gaige, 1917)
OLYMPIC TORRENT SALAMANDER

Marc P. Hayes, Lawrence L. C. Jones

1. Historical versus Current Distribution.
Olympic torrent salamanders *(Rhyacotriton olympicus)* are restricted to the region of the Olympic Peninsula (i.e., south to the Chehalis River) in Washington state (Good and Wake, 1992; McAllister, 1995). A notable hiatus is the lack of records for this species in the Black Hills complex near the immediate southern margin of Puget Sound, but north of the Chehalis River (McAllister, 1995), which may reflect the southernmost advance of the last glaciation, which just covered this complex. Limited historical data exist for this species (Slater, 1933, 1955), but no attempts have been made to contrast historical and current distribution. Most data on Olympic torrent salamanders are relatively recent, so only a limited historical assessment may be possible. Recent surveys of Olympic National Park (Bury and Adams, 2000) showed the species to be widespread, occurring in 41% of 168 streams and 47% of

235 seeps surveyed. Few locations were found on the east side of the Park, where conditions are drier and warmer due to the rain shadow of the Olympic Mountains (see also "Conservation" below).

2. Historical versus Current Abundance.
Recent surveys of Olympic National Park (Bury and Adams, 2000) revealed that Olympic torrent salamanders were more abundant in streams with northerly aspects and steep gradients. The latter is a pattern that has been identified in every species of *Rhyacotriton* where it has been studied (e.g., Welsh and Lind, 1996; Russell et al., 2004) and may reflect a process-based greater facility for maintaining suitable habitat in such systems. Although associations with steeper gradients could be confounded with influences from timber harvest that may operate differently under different gradient conditions (see "Conservation" below), such a confound is unlikely for Bury and Adams' (2000) Olympic National Park surveys, because timber harvest has never occurred over most of the area they surveyed. Bury and Adams (2000) also found Olympic torrent salamanders to be less abundant where fine substrates and undercut banks were more frequent. No historic abundance data exist, and future contrasts of potential changes in abundance will either require comparison to Bury and Adams (2000) or establishment of their own baseline for contrast.

3. Life History Features.
   A. Breeding. Reproduction is presumed to be aquatic (see "Egg deposition sites" below).
   i. Breeding migrations. Breeding migrations are undocumented.
   ii. Breeding habitat. Unknown.
   B. Eggs.
   i. Egg deposition sites. No egg deposition sites of this species have been described. Five nests of Columbia torrent salamanders *(R. kezeri)* are known from seeps, springs, or headwater streams with mixed (coarse and fine) substrates; sometimes nests are beneath a layer of moss (Nussbaum, 1969b; Russell et al., 2002). One nest of southern torrent salamanders *(R. variegatus)* was found in the mid-channel of a small, headwater stream (Karraker, 1999). From these observations, low-flow egg deposition sites with similar characteristics are anticipated for Olympic torrent salamanders.
   ii. Clutch size. Unknown, but fecundity is likely to be low because Good and Wake (1992) indicated that gravid female Olympic torrent salamanders have ovarian egg counts averaging eight.
   C. Larvae/Metamorphosis.
   i. Length of larval stage. Unknown. However, data from Cascade torrent salamanders *(R. cascadae)* and southern torrent salamanders (Nussbaum and Tait, 1977) suggest that Olympic torrent salamanders probably have a long larval life (>2 yr).
   ii. Larval requirements.
   a. Food. Unknown.
   b. Cover. Unknown. However, data from other species of *Rhyacotriton* (Nussbaum and Tait, 1977; Diller and Wallace, 1996; Welsh and Lind, 1996) suggest that stable, low-flow volume microhabitats with loose gravel and cobble and open interstices with limited fine sediments may be preferred.
   iii. Larval polymorphisms. Unknown for any species of *Rhyacotriton*.
   iv. Features of metamorphosis. Unknown. Nussbaum and Tait (1977) indicated that Cascade torrent salamanders and southern torrent salamanders can metamorphose at almost any time of the year, but their data indicate most individuals metamorphose in late summer to early autumn. The duration of metamorphosis is unknown.
   v. Post-metamorphic migrations. Have not been described for any species of *Rhyacotriton*.
   vi. Neoteny. Unknown in the genus *Rhyacotriton* (Good and Wake, 1992).
   D. Juvenile Habitat. No data exist on juvenile habitat characteristics. Based on data for related species of *Rhyacotriton* (Nussbaum and Tait, 1977), few differences appear to exist between the habitat characteristics of juveniles and adults. However, almost no effort has been made to distinguish habitat partitioning by these age groups.

Olympic Torrent Salamander *(Rhyacotriton olympicus)*

E. Adult Habitat. Except for the recent habitat analysis of Bury and Adams (2000; see "Current and Historic Distribution" and "Current and Historic Abundance" above), available descriptions of habitat for Olympic torrent salamanders are limited. Leonard et al. (1993) state that "*R. olympicus* are nearly always seen in or very near cold, clear streams, seepages, or waterfalls. Their typical haunt is the splash zone, where a thin film of water runs between or under rocks." Descriptions encompassing the genus add some additional data. Emphasizing the use of rock habitats, Stebbins and Lowe (1951) noted that *Rhyacotriton* are ". . . less frequently found under moss and wood than . . . rocks." In emphasizing the wet nature of *Rhyacotriton* habitat, they also noted that: "*Rhyacotriton* usually selects . . . sites where the movement of water tends to be relatively slow. They rest with their vents in shallow water, and one rarely finds an individual [that is] not in contact with free water or at least a saturated substrate." This pattern is likely linked to the fact that *Rhyacotriton* is among the most, if not the most, desiccation-intolerant salamander genus known (Ray, 1958). Desiccation intolerance is probably linked to a high dependence on skin surfaces for oxygen uptake (average = 74%) because the lungs are highly reduced (Whitford and Hutchison, 1966b). Stebbins and Lowe (1951) also generalized about vegetation typical of streams where *Rhyacotriton* are found. These streams are characteristic of conditions that provide a cool, wet microenvironment: "Streams harboring *Rhyacotriton* usually have a good leaf canopy, especially during the summer . . . Abundant understory vegetation, much moss, and a thick leaf mat [characterize] stream banks."

F. Home Range Size. Unknown. Data from other species of *Rhyacotriton* suggest that the range of movement of individuals is limited (i.e., on a few-meter scale; Nussbaum and Tait, 1977; Welsh and Lind, 1992; Nijhius and Kaplan, 1998). An interpretation of limited movement must be given with the caveat that all studies to date suffer from some degree of bias because recaptures with the intent of identifying movements in a home range context were within highly circumscribed, small areas.

G. Territories. Unknown.

H. Aestivation/Avoiding Desiccation. Olympic torrent salamanders have been encountered when surface active during the summer (Jones and Raphael, 2000), so aestivation may not be typical or necessary in their highly mesic, near hydric habitats.

I. Seasonal Migrations. Unknown.

J. Torpor (Hibernation). Unknown.

K. Interspecific Associations/Exclusions. Olympic torrent salamanders are often associated with coastal tailed frogs (*Ascaphus truei*), Cope's giant salamanders

(*Dicamptodon copei*), Van Dyke's salamanders (*Plethodon vandykei*), and western red-backed salamanders (*P. vehiculum*) in seep and stream habitats (L. Jones, unpublished data; P. Peterson, Stillwater Sciences, personal communication). However, the degree of syntopy with each of these taxa has not yet been detailed. Based on studies of other *Rhyacotriton*, a repeated comment has been that giant salamanders, as potential predators, could locally limit *Rhyacotriton* (Stebbins, 1953; Welsh, 1993; Welsh and Lind, 1996). Inverse relationships in abundance between torrent and giant salamander larvae (Welsh and Lind, 1996) and anecdotal observations regarding the relative rarity of syntopy between torrent and giant salamanders (Stebbins, 1953) seem to be the basis of these comments. Recent work revealing that larval southern torrent salamanders are unpalatable to larval giant salamanders (Rundio and Olson, 2001) will require re-evaluation of the notion of just how giant salamander predation may limit torrent salamanders. At least one species of *Rhyacotriton* (e.g., southern torrent salamander) is known to respond to stimuli from injured western red-backed salamanders (Chivers et al., 1997); a similar pattern may be true for Olympic torrent salamanders.

L. Age/Size at Reproductive Maturity. Unknown. Based on data for other *Rhyacotriton* (Nussbaum and Tait, 1977), a relatively long interval to reproductive maturity (>4 yr) and a relatively small size at reproductive maturity (about 45 mm SVL) are anticipated.

M. Longevity. Unknown. Based on data for other *Rhyacotriton* (Nussbaum and Tait, 1977), a moderately long lifespan (>10 yr) is likely.

N. Feeding Behavior. Unknown. Based on data for other *Rhyacotriton*, Olympic torrent salamanders probably feed on invertebrates dwelling in moist forested habitats, especially amphipods, dipteran larvae, springtails, and stonefly nymphs (Bury and Martin, 1967; Bury, 1970). These taxa occur in the semi-aquatic and aquatic microhabitats where feeding is thought to occur.

O. Predators. Unknown. Based on studies of other *Rhyacotriton*, giant salamanders are suspected predators (Stebbins, 1953; Nussbaum, 1969b; Welsh, 1993; Welsh and Lind, 1996), but the recent demonstration of the relative unpalatability of southern torrent salamander larvae (Rundio and Olson, 2001) may require re-evaluation of which torrent salamander life stages may be vulnerable to giant salamander predation. Nussbaum et al. (1983) also commented on shrews rejecting torrent salamanders as unpalatable.

P. Anti-Predator Mechanisms. Unknown. Brodie (1977) described anti-predator postures in *Rhyacotriton* as being similar to

many *Ambystoma*. In defense, the body is coiled, and the tail is elevated or arched and wagged and may be lashed toward the threat. The bright yellow venter contrasting with the drabber dorsal coloration is believed to be aposematic. Partly metamorphosed larvae posture similarly to adults. The recent demonstration that southern torrent salamander larvae are unpalatable (see "Interspecific Associations/Exclusions" above) may agree with an aposematic hypothesis, but unpalatability of post-metamorphic life stages remains to be demonstrated. Based on experimental study of southern torrent salamanders, avoidance of injured conspecifics and injured western red-backed salamanders (Chivers et al., 1997) may be an anti-predator device also present in Olympic torrent salamanders.

Q. Diseases. Unknown.

R. Parasites. Unknown. However, several species of parasites have been described from other species of *Rhyacotriton* and may occur in Olympic torrent salamanders. These include one species of monogenoidean fluke, to date unique to *Rhyacotriton* (Kristsky et al., 1993), several species of digenean flukes (Lynch, 1936; Senger and Macy, 1952; Lehmann, 1954, 1956; Anderson et al., 1966; Martin, 1966b), and several blood parasites (Clark et al., 1969).

4. Conservation.
Olympic torrent salamanders are state Sensitive in Washington (K. McAllister, personal communication). "Sensitive" is a watchlist status that lacks legal standing, but is applied to species for which concern exists related to habitat loss in order to increase awareness among resource protection agencies. Much concern for the species was related to the idea that habitat quality would be degraded by increased temperatures and sedimentation following timber harvest. Concern also existed that headwater streams, seeps, and springs (all habitats presumed to be occupied) lacked adequate protection. Data used to support the Sensitive status were drawn entirely from retrospective studies that addressed southern torrent salamanders (e.g., Corn and Bury, 1989a; Welsh, 1990).

The actual status of Olympic torrent salamanders is unstudied. Lower-gradient, higher-order streams, which may intrinsically provide poor habitat for *Rhyacotriton* (see "Historical versus Current Distribution" above), are more often disturbed as a function of timber harvest (due to greater susceptibility to habitat disturbance because of greater depositional and more limited scour characteristics, a greater upslope area of influence, more harvest rotations, etc.) than higher-gradient, lower-order streams, which may provide better habitat quality. Thus, the effects of timber harvest per se on torrent salamanders have been confounded with natural variation in habitat quality (Diller and Wallace, 1996; Welsh and Lind, 1996; Hunter, 1998). However, based on Bury and Adams (2000), little question exists that gradient has an influence on habitat in its own right; what is unclear is how much degradation may have occurred in lowland habitats where gradients are lower and the influence of timber harvest is extensive. Distinguishing the relative significance of intrinsic habitat limitation from the potential influences of timber harvest will require comparing harvested and unharvested lower- versus higher-gradient sites simultaneously.

In 2000, scientists representing private timber companies, Native American tribes, and state and federal resource agencies assessed the risk of forest management activities to stream-associated vertebrate species. This work, done in preparation for the Washington Forest and Fish Agreement (FFA), concluded that seven species of amphibians (including all three species of torrent salamander in Washington State) were at high risk of local extirpation from forest management. One hoped-for outcome of identifying species at risk and including them in the list of species protected under new forest practice rules was that those species would be studied as part of an innovative, ambitious adaptive management program. The goal of FFA adaptive management is to determine whether new riparian buffer prescriptions designed for headwater streams are effective in protecting resources to which they are linked, including local populations of *Rhyacotriton*. To date, this research has identified the most appropriate sampling methods for landscape detection and relative abundance assessment of Olympic torrent salamanders and the other stream-associated amphibians. These methods will be essential for manipulative studies to test the effectiveness of forest practice rules in protecting *Rhyacotriton*.

Rhyacotriton variegatus Stebbins
and Lowe, 1951
SOUTHERN TORRENT SALAMANDER
Hartwell H. Welsh, Jr., Nancy E. Karraker

1. Historical versus Current Distribution.
Southern torrent salamanders *(Rhyacotriton variegatus)* are locally distributed in the Pacific Coast ranges of Oregon from the Little Nestucca River southward to Dark Gulch in Mendocino County, California (Good and Wake, 1992; but see Highton, 2000, who suggests that there may be four cryptic species composing a *R. variegatus* species group). Apparently disjunct populations of *Rhyacotriton* in the north Umpqua River drainage of the interior southern Cascade Mountains of Oregon are attributed to southern torrent salamanders (Good and Wake, 1992). However, recent sampling between the Umpqua drainage and the coast ranges of southern Oregon indicates extant populations of southern torrent salamanders occur across this geography (R. Bury, personal communication). There are five specimens known from a disjunct locality about 113 km (70 mi) east of the established range (Stebbins, 1985) in the upper McCloud River drainage in Siskiyou County, California (Jennings and Hayes, 1994a). The distribution of populations within the range of southern torrent salamanders is generally spotty, with occurrences closely linked to headwater habitats. Based on two studies in northern California, Welsh and Lind (1992) found southern torrent salamanders present at 28% and 37%, respectively, of aquatic sites randomly sampled across the range. They found that presence of headwater habitat alone was not a good predictor of salamander presence; only 62.3% and 46.6%, respectively, of sites sampled with headwater microhabitats supported salamanders (Welsh and Lind, 1992). Welsh (1990) reported southern torrent salamanders associated with headwater habitats in late-seral forests througout much of their range in California (see also Welsh and Lind, 1988, 1991, 1996). However, Diller and Wallace (1996) reported no relationship between the presence of southern torrent salamanders and seral stage on coastal redwood commercial timberlands. Possible bioregional differences aside, numerous studies have shown that southern torrent salamanders are negatively impacted by timber harvesting (Corn and Bury, 1989a; Welsh, 1990; Bury et al., 1991; Welsh and Lind, 1996; Welsh et al., 1998; Welsh et al., in press), and thus indicate that their historical distribution may have consisted of more populations across the range than is currently the case. For example, Welsh et al. (1998) reported that southern torrent salamander populations occurred in only 29.0% of headwater habitats in the heavily logged Mattole watershed of southern Humboldt County, California, and 18.9% of headwater habitats on commercial timberlands just to the south in Mendocino County. Both rates of occurrence were significantly lower than the 62.3% and 46.6% random encounter rates reported above (see Welsh and Lind, 1992), or the 76.9% encounter rate found on proximate forest reserve lands in northern Mendocino County. Few studies that directly address this question are available from the Oregon portions of the range (but see Corn and Bury, 1989a). However, a similar trend of region-wide intensive, short-rotation forestry throughout the Oregon Coast ranges has probably severely reduced populations there as well.

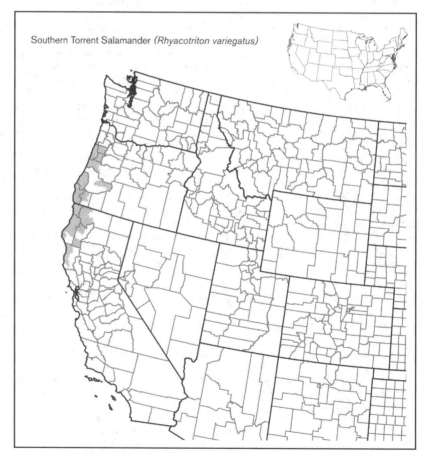

Southern Torrent Salamander *(Rhyacotriton variegatus)*

2. Historical versus Current Abundance.
Historical abundances are unknown; however, evidence from multiple studies on the impacts of timber harvesting (e.g., Corn and Bury, 1989a; Welsh, 1990; Welsh et al., 2000; Welsh et al., in review) indicate that historical abundances were probably higher than at present.

3. Life History Features.
  A. Breeding. Reproduction is aquatic.
  i. Breeding migrations. Undocumented.
  ii. Breeding habitat. Courtship has not been observed in the field, but it is likely that breeding occurs along the margins of low-order streams, springs, and seeps.
  B. Eggs.
  i. Egg deposition sites. Karraker (1999) reported an oviposition site beneath a mid-channel boulder in a first-order stream in Humboldt County, California. Large cobble was the primary substrate in the vicinity of the nest. Another clutch of eggs was found beneath a 42-cm diameter (longest dimension) boulder in a small, coastal stream in Humboldt County, California (G. Hodgson and L. Ollivier, unpublished data). Water temperature at the time of discovery was 10.4 °C, and gravel was the predominant substrate near the oviposition site.
  ii. Clutch sizes. Clutch sizes for the California nest referred to above were 11 eggs (Karraker, 1999) and 8 eggs (G. Hodgson and L. Ollivier, unpublished data). Development of the eggs is slow, >193–220 d (Karraker, 1999). Nussbaum (1969b) reported a hatch time of 210 d for eggs of Columbia torrent salamanders *(R. kezeri)* held at 8 °C in the laboratory.
  C. Larvae/Metamorphosis.
  i. Length of larval stage. Larval development takes 3.5 yr (Nussbaum and Tait, 1977).
  ii. Larval requirements.
  a. Food. Unknown, but probably similar to those of adults (refer to "Feeding Behavior" below).
  b. Cover. The larvae of southern torrent salamanders are morphologically adapted to mountain brook microenvironments (Valentine and Dennis, 1964), which consist primarily of the shallow waters of springs, seeps, and headwater streams with cold (6.5–15 °C), low-velocity flows, over unsorted rock or rock rubble substrates, usually in mesic mature to old-growth forests (Welsh and Lind, 1996, and citations therein).
  iii. Larval polymorphisms. None reported.
  iv. Features of metamorphosis. Smallest transformed individuals ranged from 30.2–38.6 mm SVL (Nussbaum and Tait, 1977). Good and Wake (1992) reported a similar range of 31.0–39.0 mm for museum specimens. The larval period is 3.0–3.5 yr (Nussbaum and Tait, 1977).
  v. Post-metamorphic migrations. Not known; at least some adults appear to remain in proximity to natal streams but

also use adjacent riparian and moist upland habitats.
  vi. Neoteny. Not reported.
  D. Juvenile Habitat. Similar to larvae and adults (see "Larval requirements" above and "Adult Habitat" below).
  E. Adult Habitat. Adults, while occasionally found in adjacent moist riparian vegetation, are usually found in contact with cold (6.5–15 °C) water in springs, seeps, and headwater streams with shallow, slow flows, over unsorted rock or rock rubble substrates, in mesic mature to old-growth forests of the Pacific Northwest (Welsh and Lind, 1996, and citations therein; but see Diller and Wallace, 1996). Stebbins and Lowe (1951) generalized about the vegetation assemblage associated with streams that supported *Rhyacotriton*, noting that it typified riparian conditions that provide a cool, wet microenvironment. This requirement is probably related to the fact that species in the genus *Rhyacotriton* are the most desiccation-intolerant salamanders known (Ray, 1958), a condition that may be linked to a high dependence on skin surfaces for oxygen uptake (average = 74%) as the lungs are highly reduced (Whitford and Hutchison, 1966b). Corn and Bury (1989a), Welsh and Lind (1996), and Welsh and Ollivier (1998) documented the importance of a lack of fine sediments (e.g., sand and fine gravel) which fill in the coarse substrate interstices of the streambed used for cover by both larval and adult salamanders. Diller and Wallace (1996), sampling within commercial timberlands in the coastal redwoods of California, purported to demonstrate a relationship between the presence of southern torrent salamanders and types of parent geology (harder substrate types were more likely to support salamanders) and stream gradient (steeper stream channels were more likely to support salamanders; see also Brode, 1995). Welsh et al. (2000) tested both of these hypotheses using data from across a larger portion of the range of southern torrent salamanders. They found no relationship with gradient or parent geology and concluded that Diller and Wallace's (1996) outcome was the result of their sampling only on harvested commercial timberlands. This limited sampling regime resulted in Diller and Wallace (1996) detecting relationships that were artifacts of the long timber harvesting history in California's coastal redwoods, which has had pronounced negative effects on the amounts and distributions of suitable stream environments for southern torrent salamanders (Welsh et al., 2000).
  F. Home Range Size. Welsh and Lind (1992) marked a total of 188 individuals, recapturing 21% over 3 yr, and found little movement for both larvae and adults (2.2 m/yr and 1 m/yr, respectively). They found that 52%, 32%, and 16% of this

movement was downstream, upstream, or lateral or no net movement, respectively. However, they cautioned that they had no way of determining distances moved by the large portion of animals that either dispersed out of their 12.6 m^2 study area or possibly met other fates such as predation.
  G. Territories. No studies have addressed whether or not they are defended.
  H. Aestivation/Avoiding Desiccation. Except at coastal sites, both larval and adult southern torrent salamanders appear to burrow deeper into streambed substrates in response to warmer water temperatures and lower stream flows associated with summer climate shifts. It is not known whether they are active in the hyporheic zone at interior sites during the summer, but Welsh and Lind (1992) reported a significantly greater weight gain over the winter compared with the summer for both larvae and adults, suggesting a period of inactivity during the hotter and drier summer season.
  I. Seasonal Migration. No studies have addressed whether these salamanders migrate seasonally, but anecdotal evidence indicates that adults may move between headwater and adjacent upland habitats (personal observations).
  J. Torpor (Hibernation). Southern torrent salamanders appear to retreat deeper into stream channel substrates during winter in response to high stream flows, but it is not known whether they remain active in these hyporheic zone refugia. Welsh and Lind (1992) reported over-winter weight gains for both larvae and adults, suggesting that southern torrent salamanders may be active throughout the winter.
  K. Interspecific Associations/Exclusions. Coastal giant salamanders *(Dicamptodon tenebrosus)* are known to feed on southern torrent salamanders, and Stebbins (1953) and Nussbaum (1969b) speculated that the selection of shallow-water microhabitats on the part of southern torrent salamanders was a response to this predation. Welsh (1993), in an analysis of coastal giant salamander habitat associations, reported a weak negative association between these two species, giving credence to this exclusion hypothesis (Stebbins, 1953; Nussbaum, 1969b). A similar analysis of southern torrent salamander associations showed a weak but non-significant (p = 0.09) reciprocal relationship (Welsh and Lind, 1996).
  L. Age/Size at Reproductive Maturity. Nussbaum and Tait (1977) reported that southern torrent salamanders take 4.5–5.0 yr to reach sexual maturity; 3.5 yr in the larval form and 1–1.5 yr after metamorphosis. Adults measured 35–52 mm SVL (n = 49; Welsh and Lind, 1992).
  M. Longevity. No direct measurements are available, but the long maturation time suggests that this is a relatively long-lived species.

N. Feeding Behavior. Adult southern torrent salamanders eat primarily aquatic and semi-aquatic insects, a large portion of which are amphipods and Collembola; they feed mostly on the larval and nymph life stages (Bury and Martin, 1967).

O. Predators. Coastal giant salamanders (see "Interspecific Associations" above) and probably garter snakes (*Thamnophis* sp.) and salmonid fishes (*Oncorhynchus* sp.).

P. Anti-Predator Mechanisms. Stebbins (1953) and Nussbaum (1969b) speculated that the confinement of this species to shallow water flowing through rock rubble may be a response to predation by larval and paedomorphic coastal giant salamanders.

Q. Diseases. None described.

R. Parasites. One species of monogenoidean fluke, previously unique to the genus *Rhyacotriton*, has also been described from Cascade frogs (*Rana cascadae*) from southern Washington (Kritsky et al., 1993). Several digenean flukes have been collected from various species of *Rhyacotriton* (Senger and Macy, 1952; Anderson et al., 1966; Martin, 1966b), and the nematode *Oxyuris dubia* (Lehmann, 1956).

4. Conservation.

Petitions to list southern torrent salamanders under both the Federal Endangered Species Act and that of the State of California, based on both loss of habitat and population declines due to forestry practices, were filed in 1994. Noting that the petition had merit, the California Department of Fish and Game determined the following objectives required addressing: (1) document existing localities; (2) determine populations and habitat status; (3) examine population trends; and (4) determine adequacy of current forest practice rules to protect the species and its habitat. However, following a status report (Brode, 1995) that lacked appropriate statistical analyses of available data, adequate scientific review, and clear and concise recommendations for modification of timber harvest rules to ensure reversal of declines, the California Department of Fish and Game Commission opted not to list the species, promising instead to address problems in its forest practices act that affected torrent salamanders (see Welsh et al., 2000). These problems remain unaddressed to date (Welsh, 2000). The U.S. Fish and Wildlife Service recently issued their "12-month finding" on the federal petition (Federal Register, June 6, 2000) and also concluded that a listing was not warranted. In both the state and federal status reviews, each locality record was considered a population, so the number of extant populations used to make these findings was overestimated. The federal finding reported no results for objectives 2, 3, and 4. Consequently, this finding is based not only on an inflated estimate of

population numbers, but also has no landscape-scale analysis of the species' metapopulation structure, no determination of population trends, and an incomplete analysis of the adequacy of the California Forest Practice Rules to protect southern torrent salamanders (Welsh et al., 2000). This finding does assert that incorrect classification of streams could result in the application of inadequate protection measures to streams potentially containing southern torrent salamanders. However, no assessment is made of the problem of stream misclassification or of the associated impacts to southern torrent salamanders. Because of inadequacies in these findings, and because headwater streams, seeps, and springs receive little protection under the California Forest Practice Rules (Welsh, 2000), it is likely that populations will continue to be extirpated due to timber harvesting and related land-management practices. Such extirpations will further fragment the already disparate metapopulation units of this species (Welsh and Lind, 1992), resulting in their increased isolation and decreased potential for gene flow. This scenario will likely result in these subpopulations disappearing across the landscape, as already appears to be the case in Mendocino County (H.H.W. and colleagues, unpublished data) and the Mattole watershed of Humboldt County (Welsh et al., in press).

Family Salamandridae

Notophthalmus meridionalis (Cope, 1880)
BLACK-SPOTTED NEWT

Kelly J. Irwin, Frank W. Judd

1. Historical versus Current Distribution.
Black-spotted newts (*Notophthalmus meridionalis*) range from northern Veracruz, Mexico, north to southern Texas (Mecham, 1968a). Two subspecies of black-spotted

newts are currently recognized: Texas black-spotted newts (*N. m. meridionalis*) and Mexican black-spotted newts (*N. m. kallerti*). Only Texas black-spotted newts occur in the United States.

In Texas, black-spotted newts inhabit the Coastal Plain of the Tamaulipan Biotic Province, but in Mexico they have been recorded at elevations of 610 and 800 m (Mecham, 1968a). Bishop (1943) mapped the distribution of black-spotted newts from as far north as Waco, Texas (Strecker, 1908c), and as far east as Houston, Texas (Harwood, 1930, 1932). Dixon (2000) provides the most current and accurate distribution map for the species. Black-spotted newts have been recorded from all Texas counties bordering the Gulf Coast, south from Aransas and Refugio counties, and the central portion of the Tamaulipan Province, south from Bexar County. Strecker (1908c, 1922) reported black-spotted newts from Bexar, Falls, and McClennan counties, but these records are now considered questionable or erroneous (B.C. Brown, 1950; Raun and Gehlbach, 1972). Brown (B.C., 1950) suggested that the locality record for Bexar County was labeled as coming from the shipping point rather than the actual collection locality. However, Dixon (2000) mapped the Bexar County record at the northern terminus of the Tamaulipan Biotic Province. The Falls and McClennan County records are undoubtedly erroneous, for they are about 200 km northeast of the Bexar County record and most certainly represent observations of eastern newts (*N. viridescens*). While Strecker (1908b) listed black-spotted newts from Victoria County, both Mecham (1968a) and Dixon (2000) suggest that this record may represent specimens of eastern newts. Dixon (2000) states that the Duval County record was unverified, however a record from McMullen County (Taggart, 1997b) filled the gap between the Bexar

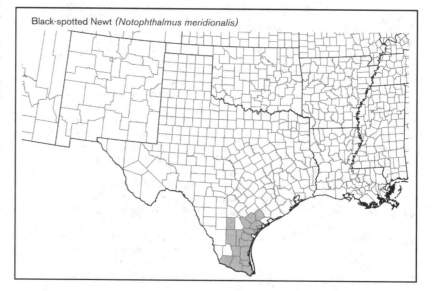

Black-spotted Newt (*Notophthalmus meridionalis*)

and Duval County records, thereby lending credence to earlier records. Taggart (1997b) provided U.S. National Museum (USNM) numbers for specimens from both counties, but did not verify specimen identifications. Boundy (1994b) reported a record from Starr County, substantiating an unverified report of black-spotted newts in Starr County (Irwin, 1993).

2. Historical versus Current Abundance.
Oliver et al. (1980) reported 28 black-spotted newts found beneath fallen (palm?) logs in a marshy clearing on 24 January 1964, at Southmost Ranch, south of Brownsville, Cameron County, Texas. Judd (1985) captured 17 individuals at Laguna Atascosa National Wildlife Refuge, Cameron County, Texas, in March 1983. These newts were seined from an earthen pond on the margin of an old field. Judd (1985) also reported the discovery of three black-spotted newts from an old corral under logs on the outskirts of Brownsville, Cameron County, Texas, on a misty, rainy day in January 1983. Rappole and Klicka (1991) recorded seining 32 newts from a temporary pond on Laguna Atascosa National Wildlife Refuge, Cameron County, Texas, on 25 August 1987. Based on their sample size and pond size, they estimated a population of 533 black-spotted newts and 250 *Siren* sp. (see *Siren texana* account, this volume) at this locality. On the same date, they sampled a pond in northern Tamaulipas, Mexico, estimating a total of 140 *Notophthalmus* and 280 *Siren* at this site. In mid-February 1988, Rappole and Klicka (1991) found 14 newts of various age classes under debris in a junk yard near Vattmannville, Kleberg County, Texas. By digging with a shovel, four more newts were unearthed at this same locality on 29 February 1988.

3. Life History Features.
Mecham (1968a) noted that descriptions of the eggs, early development, and mating behavior are lacking for this species. Indeed, most aspects of the natural history of black-spotted newts are undocumented (Petranka, 1998).

   A. Breeding. Reproduction is aquatic. Mecham (1968a) clarified the description of breeding habits reported by Strecker (1922) as the misidentification of eastern newts (Bishop, 1943; Petranka, 1998). Mecham (1968a,b) reports that mating is tied to rainfall and can occur at any time of year. Rappole and Klicka (1991) provide the following observations on reproduction. Courtship was said to occur throughout the year in captive black-spotted newts, but no specific dates were provided. Courtship was described as similar to that of eastern newts. Females were observed swimming and hovering over deposited spermatophores, but were not observed to grasp spermatophores with the hind feet

as reported in eastern newts. Because captive newts do not feed at temperatures below 10 °C, and little active aquatic life is found in breeding ponds in January and February, Rappole and Klicka (1991) suggest that there is no breeding activity in southern Texas from December–February; they did find reproductive activity among aquatic adults in March and August.

   i. Breeding migrations. No true breeding migrations have been reported for this species (but see "Seasonal Migrations" below).

   ii. Breeding habitat. Mecham (1968a) lists permanent and temporary ponds, roadside ditches, and pools of small streams as habitat for adults, juveniles, and larvae. In Texas, black-spotted newts were found to breed in shallow ephemeral ponds ranging in depth from 0.5–2 m, with firm clay bottoms, and some with rooted macrophytes. Ponds that contained newts had salinities ranging from 0.5–1.0‰. Newts were also found in dried wetland ditches along railroad rights-of-way, which suggests that adults breed in these ditches when they fill with rain water. Newts were not found in water bodies with predatory fish, high salinity, intense cattle usage, or those with agricultural runoff (Rappole and Klicka, 1991).

   B. Eggs. The following information is based on the observations of Rappole and Klicka (1991). Eggs are laid singly, requiring about 3 min to be laid, at intervals of 6–30 min. As an egg was laid it was pinched off, by compressing the cloaca against the hind feet, with soles facing each other. This process was done as the hind feet grasped submerged vegetation and the egg was effectively glued to the surface of the vegetation. Eggs were also laid on other surfaces or in the water column. The female parent or other newts often ate free-floating, unattached eggs. Ova are blue-green in color. Larvae emerged from eggs in 12–14 d.

   i. Egg deposition sites. Eggs are attached to submergent vegetation.

   ii. Clutch size. Rappole and Klicka (1991) did not provide any data on the total number of eggs laid by a single female. References to this species laying 300 eggs should be ascribed to eastern newts until confirmed reports are available.

   C. Larvae/Metamorphosis.
   i. Length of larval stage. Rappole and Klicka (1991) did not report the duration of the larval stage in captive specimens. Larvae were described as being similar to those of eastern newts—aquatic with external gills, smooth skin, green pigmentation, and a laterally compressed tail.

   ii. Larval requirements. Unknown.
   a. Food. Unknown, but larvae are presumably carnivorous as are other members of the genus.
   b. Cover. Unknown.

   iii. Larval polymorphisms. Unknown and unlikely.

   iv. Features of metamorphosis. Additional research is needed—for example to clarify the existence and duration of a true eft stage in this species. Black-spotted newts are considered by some authors to lack a well-defined, dispersive or migratory eft stage (Mecham, 1968a; Petranka, 1998). Based on his seining of newly metamorphosed young and subadults in association with sexually mature adults, Mecham (1968b) raised doubts that a true terrestrial eft stage occurs in black-spotted newts. Conversely, Rappole and Klicka (1991) state that upon metamorphosis the larvae transform into a terrestrial or "eft" stage with cornified skin and round tail, in cross section. Based on research on eastern newts (Gill, 1978; Harris, 1987a), Rappole and Klicka (1991) suggest that both neotenic and eft stages could occur in black-spotted newts. Further, these authors state that the duration of the eft stage could be dependent upon the availability of aquatic breeding sites. They also state that metamorphosis from eft to breeding adult is rapid. Captive terrestrial forms developed first signs of a keel on the tail within 2 d of being placed in an aquarium, and developed fully aquatic, breeding adult characteristics within 2 wk (Rappole and Klicka, 1991). As in other members of the genus, males develop a heavily keeled tail, horny black breeding excrescences on the hind legs, and a distinctly triangular-shaped head with puffy folds of skin. Terrestrial individuals with cloacal swellings contained undeveloped eggs in the ovaries. Rappole and Klicka (1991) propose that it might be possible to sex nonreproductive newts based on head and tail size and shape. These observations suggest that adults revert from a terrestrial phenotype to a breeding aquatic phenotype when conditions are favorable, as is seen in eastern newts and striped newts (*N. perstriatus*), the two other members of this genus.

   v. Post-metamorphic migrations. Unknown. Mecham (1968b) found juveniles and adults under rocks near a recently dried pond in Tamaulipas, Mexico, suggesting that these animals do not disperse great distances from breeding sites.

   vi. Neoteny. Mecham (1968b) collected a reproductively mature female with gill rudiments, suggesting neoteny.

   D. Juvenile Habitat. Mecham (1968b) suggests juveniles may remain aquatic until reproductively mature, unless their pond dries or high temperatures cause them to seek cover on land.

   E. Adult Habitat. In Texas, the presence of black-spotted newts appears to be related to soil type. Deep, poorly drained, clayey sediments (such as the "Tiocano" and "Edroy" clay soils) with slow permeability allow for the formation of ephemeral ponds or wetlands during periods of heavy rain. Most adults have generally been found in the vicinity of, or in, such breeding ponds (Mecham, 1986b; Rappole and Klicka, 1991). Several localities were seasonally dry ditches along railroad rights-of-way or highway borrow pits (Rappole and Klicka, 1991). The presence of intact Tamaulipan thorn forest in clayey soils, with ephemeral wetlands, should be considered optimal adult habitat. With the advent of extensive land clearing for row crop agriculture and rootplowing of native brush for cattle grazing, much of the black-spotted newt's original habitat has been lost. For example, Rappole and Klicka (1991) described a locality as swampy pasture with shrubs and grasses, black loam soil over clay, and two livestock ponds. Many individuals had been found at this site, which was subsequently root plowed, effectively destroying the surrounding terrestrial habitat used by adults and subadults. In addition, Rappole and Klicka (1991) identified several sites in Cameron, Kleberg, and Kenedy counties in Texas that they considered to be metapopulation centers.

F. Home Range Size. Unknown. Mecham (1968a,b) found juveniles and adults under rocks near a recently dried pond, indicating that these animals do not migrate to more suitable, but distant, habitats. Based on several observations (see "Seasonal Migrations" below), newts may disperse greater distances from breeding sites than current information suggests.

G. Territories. Unknown, but has not been observed in this family.

H. Aestivation/Avoiding Desiccation. Black-spotted newts inhabit a region prone to periodic droughts and extreme temperatures. Mecham's (1968a,b) observations indicate that animals under these conditions will seek shelter under cover objects in and near dried breeding sites. Further, Rappole and Klicka (1991) provide evidence of burrowing and use of existing burrows, which may be used to aestivate. In February 1988, four newts were unearthed by digging at a site that had previously produced newts by turning of ground cover in Kleberg County, Texas. In March 1988, these authors dug up a live newt 15 cm below the soil surface along the margins of a known breeding pond in Cameron County, Texas. In the field, newts were found in fissures of dried soil along a drift fence. Newts may also use crab burrows as retreats. Captives utilized crayfish burrows and manmade burrows; they were also observed to burrow in the soil at or below the soil/water interface.

I. Seasonal Migrations. No documented seasonal migrations have been reported. However, Thornton (1977) found one individual crossing a gravel road at night in Cameron County, Texas, and

L. Laack (personal observation) found several newts crossing a road on a rainy night at Laguna Atascosa National Wildlife Refuge, Cameron County, Texas. Taggart (1997b) captured an adult male crossing a paved highway on a rainy night in McClennan County, Texas, at 2140 hr on 31 August 1995. Based on these limited observations, adult migrations to or from breeding sites and dispersal of juveniles are probably made on nights with high humidity or rain.

J. Torpor (Hibernation). Unknown.

K. Interspecific Associations/Exclusions. Judd (1985) recorded the presence of adult *Siren* sp. and Rio Grande leopard frog (*Rana berlandieri*) tadpoles in the same pond as black-spotted newts in Cameron County, Texas. Rappole and Klicka (1991) made several observations of interspecific associations. While searching under railroad ties in January 1988, a newt and a four-lined skink (*Eumeces tetragrammus*) were found in close proximity in Kenedy County, Texas. In early March 1988, a newt was found under railroad ties in a ditch lined with willows in Kleberg County, Texas, where they had previously found a western narrow-mouthed toad (*Gastrophryne olivacea*) and an individual siren (again, see account in this volume for a discussion of the confusion surrounding the identification of south Texas sirens) in late February 1988. Rappole and Klicka (1991) also list barred tiger salamanders (*Ambystoma tigrinum mavortium*) as an associate species in newt breeding ponds, but provide no further details.

The following sympatric species were found on the Audubon Sabal Palm Grove Sanctuary, Cameron County, Texas: Rio Grande chirping frogs (*Eleutherodactylus cystignathoides*), Coastal-Plain toads (*Bufo nebulifer*), Rio Grande leopard frogs, Mexican Smiliscas (*Smilisca baudinii*), four-lined skinks, Texas spotted whiptail lizards (*Cnemidophorus gularis*), Texas spiny lizards (*Sceloporus olivaceus*), green anoles (*Anolis carolinensis*), black-striped snakes (*Coniophanes imperialis*), Texas indigo snakes (*Drymarchon corais*), speckled racers (*Drymobius margaritiferus*), eastern racers (*Coluber constrictor*), rough green snakes (*Opheodrys aestivus*), and Texas coral snakes (*Micrurus tener*; K.J.I., personal observations).

L. Age/Size at Reproductive Maturity. Unknown.

M. Longevity. Unknown.

N. Feeding Behavior. Rappole and Klicka (1991) provide the most extensive information on food and feeding behavior of black-spotted newts. They compared the natural prey items of striped newts, as compiled by Christman and Franz (1973), to prey taxa recorded in black-spotted newt ponds, and found similarity between species of recorded prey taxa with those present in black-spotted newt breeding ponds. Stomach contents of

wild-caught black-spotted newts contained seed shrimp (Ostracoda), small snails (Gastropoda), and unidentified insect larvae and eggs. Although they do not provide any data showing percent composition of diet, they state that ostracods were the primary prey of wild caught individuals. Captive newts readily consumed fairy shrimp (Anostraca), scuds (Amphipoda), odonate larvae (Insecta), eggs of conspecifics, and chunks of flesh of mammals and birds.

Rappole and Klicka (1991) describe two types of feeding behavior in captive specimens: a sit-and-wait approach and stalking. When sitting and waiting, a newt lies motionless, and when a potential prey item is detected, it coils and arches its back, sometimes holding this position for several seconds. Then with extreme rapidity, the newt suddenly uncoils in a snapping motion and attempts to engulf the prey item. When stalking, newts were observed to follow amphipods and then suddenly grab them. During daylight hours, captive newts would remain hidden under terrestrial debris, then forage at night. They observed that captive newts became sluggish at <10°C and did not feed at cooler temperatures.

O. Predators. Rappole and Klicka (1991) found no direct evidence of predation on black-spotted newts, but suggested that turtles or sirens might be possible aquatic predators. They found several newts at one locality that had scars or "bite marks" on the abdomen, neck, and head. Sirens were syntopic at this locality, and as a test, one was placed in an aquarium with a newt. The newt and siren co-existed in the aquarium for 1 wk without any signs of visible interaction. An examination of field-caught siren feces revealed the presence of ostracods, odonates, and snails, but no newt remains.

P. Anti-Predator Mechanisms. As with other members of the genus *Notophthalmus*, black-spotted newts secrete skin toxins that deter or repel potential predators. Investigators found that a skin irritant was present in the skin when touched to the investigator's lips (Rappole and Klicka, 1991). Rappole and Klicka (1991) record the first instance of an unken type of posture in the black-spotted newt. When efts were found under debris or in burrows, they would respond to being touched by immediately contracting into an "S" shape and then flipping onto their backs exposing the bright orange-yellow belly, an aposematic color.

Christman (1959), as cited by Rappole and Klicka (1991), reported production of a soft click sound in eastern newts. Captive black-spotted newts were observed to produce a click vocalization, predominately at night. Black-spotted newts also produced a sound described as a squeal when being captured (Rappole and Klicka, 1991).

Q. Diseases. Unknown.

R. Parasites. Unknown. The descriptions of new species of nematodes (Harwood, 1930, 1932) taken from newts in the vicinity of Houston, Texas, were undoubtedly based on specimens of eastern newts.

4. Conservation.

The Texas Parks and Wildlife Department currently lists black-spotted newts as Endangered, and specimens are protected from collection. Despite this, and the fact that black-spotted newts occur in the United States only in Texas, these animals have no federal listing status. Continued land clearing for agriculture and urban development pose the greatest threats to these salamanders. Over 95% of the original Tamaulipan brushland of the Lower Rio Grande Valley has been destroyed as a result of these activities. In addition, the impacts of exposure to agrichemicals on reproductive success and outright mortality are unknown. The excavation of temporary ponds within existing brushland habitat would create new breeding sites, bolstering maintenance of viable local populations.

Notophthalmus perstriatus (Bishop, 1941[a])
STRIPED NEWT

C. Kenneth Dodd, Jr., D. Bruce Means,
Steve A. Johnson

1. Historical versus Current Distribution.

The range of striped newts (*Notophthalmus perstriatus*) extends from Screven, Jenkins, and Emanuel counties in Georgia, south to Orange, Osceola, and Seminole counties in central Florida, and west to Baker County, Georgia, and Wakulla and Leon counties in Florida (Dodd and LaClaire, 1995; Franz and Smith, 1995; Means and Means, 1998a; Stevenson et al., 1998; Johnson and Dwyer, 2000). Records for Santa Rosa and Glades counties, Florida, are in error (Franz and Smith, 1995). Striped newts appear to occur in two separate regions, one in the Florida Panhandle near Tallahassee (Means and Means, 1998b) and the adjacent Dougherty Plain of southwest Georgia (LaClaire et al., 1995), and a second in the east (populations associated with sand ridges and river terraces on the Atlantic Coastal Plain of southeastern Georgia into peninsular Florida). The hiatus between the Florida groups is roughly 125 km and between the Florida Panhandle and southwest Georgia populations, 100 km. However, it is not known whether these hiatuses represent real biogeographic separations or are artifacts of poor collection and/or habitat loss. Mitochondrial DNA sequences (cytochrome b) support the biogeographic scenario. A neighbor-joining tree revealed distinct western and eastern groups with no sharing of haplotypes between the two regions (Johnson, 2000).

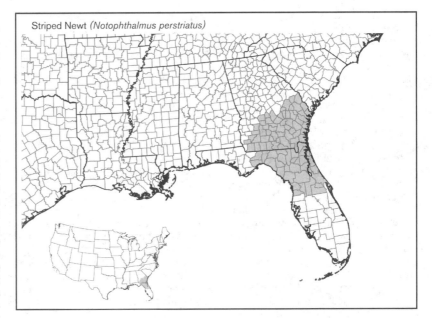

Striped Newt *(Notophthalmus perstriatus)*

In their review, Franz and Smith (1995) recorded 116 historical "observations" of striped newts from 20 Florida counties. They visited 30 of 40 identifiable historical locations and found striped newts at six ponds. They also added 21 new localities. All 27 localities were on public lands. Surveying ponds in the Munson Sand Hills in the Florida Panhandle, Means (2001) found only 20 of 265 (7.5%) ponds containing larvae or adults of striped newts, adding 15 new ponds from Leon County to those surveyed by Franz and Smith (1995), all on public lands. However, only nine of these ponds proved to be breeding ponds with larvae. Also, Means and Means (1998b) surveyed 57 ponds in the Tallahassee Red Hills of panhandle Florida and the Tifton Uplands of adjacent southwestern Georgia, finding one pond on private land containing striped newts. This pond contained striped newts in 1969, but they were not found in 1977, and were probably extirpated because the pond had been deepened and stocked with game-fish. In Georgia, striped newts were reported historically from 11 general areas (treating Trail Ridge as one area). Dodd and LaClaire (1995) found newts at six sites, one of them new. Sites contained from one to a few nearby ponds inhabited by newts. On Trail Ridge, only one pond could be reconfirmed, although many ponds once contained newts in this region (Dodd, 1995c). One additional Georgia location has since been found (Stevenson et al., 1998). Of these seven sites, four are on public land, and one is on a private ecological reserve. At least one Georgia locality (in Ben Hill County) was destroyed in the 1950s. However, surveys in both states were hindered by poor locality data accompanying museum specimens and difficulty in locating

historical sites because of the substantial land changes that have occurred since many of the early specimens were collected. Undoubtedly, many of the historical localities no longer exist.

The historical range of striped newts was probably similar to the current range. Because of extensive habitat modification, however, many populations likely have been lost. It is possible that new populations may be discovered. In that regard, Dodd and LaClaire (1995) suggested areas in Georgia that should be surveyed for this species. Based on habitat potential, Cox and Kautz (2000) listed several public land holdings that might support striped newts in Florida but need to be surveyed.

2. Historical versus Current Abundance.

Striped newts may be extremely abundant at ponds, or ponds may contain only a few individuals. For example, Dodd (1993b) recorded 744 different striped newts in 1987 at a 0.16 ha-pond in north-central Florida, whereas Johnson (1998) marked about 2,400 striped newts at a nearby 0.2 ha-pond during a drift fence study in 1996–97. In contrast, other surveys have recorded <100 striped newts at breeding ponds despite intensive and repeated sampling (Cash, 1994; Means and Means, 1997; Greenberg, 1998; D. Franz, personal communication; C.K.D., S.A.J., unpublished data). The reasons for this disparity are unknown. Both extremes in abundance are found on protected sites, however, and it is likely that considerable variation annually exists from site to site depending on demographic (metapopulation structuring) and environmental conditions. Presumably, similar variation affected striped newt populations historically, although no historical data are available on abundance.

3. Life History Features.

The life history and biology of this species are summarized by Petranka (1998) and Johnson (1998, 2002b). Life history stages include egg, larval, eft, and either transformed or neotenic adults.

A. Breeding. Reproduction is aquatic.

i. Breeding migrations. Striped newts migrate to breeding ponds in fall, winter, and/or spring, depending on weather conditions. Most substantial migrations take place from November–March, although immigration and emigration can take place during nearly any month, except during hot, dry periods (Dodd, 1993b; Means and Means, 1997; Johnson, 1998, 2002). Immigration is correlated with heavy rains that result in pond filling, whereas emigration occurs in response to pond drying (Dodd, 1993b) and metamorphosis (Johnson, 2001, 2002). Dodd and LaClaire (1995) found small larvae in Georgia breeding ponds in May, and Christman and Means (1992), Means et al. (1994), Johnson (1998), and Means (2001) noted the presence of larvae from April–December. Striped newts exhibit a high degree of phenotypic plasticity in the timing of their breeding migrations, which allows individuals to take advantage of variation in seasonal and annual rainfall.

Migrations do not appear to follow corridors or topography, at least in the few studies that have examined travel routes (Dodd and Cade, 1998; Johnson, 2001; D.B.M., unpublished data). At the same time, migration is directionally non-random, reflecting the distribution of suitable habitat surrounding the pond. Patterns of emigration and immigration may vary annually and by sex.

Once at a pond, adults appear to spend several weeks fattening themselves in preparation for breeding; adults entering the pond often are thin and in poor body condition.

ii. Breeding habitat. Breeding occurs in shallow temporary ponds associated with well-drained sands (sandhills ["high pine"], scrubby flatwoods, and scrub communities). In the high forests, ponds may or may not contain uniformly distributed emergent vegetation, although striped newts seem to prefer ponds with substantial amounts of vegetation, particularly grasses (*Panicum* sp.), surrounding the pond margins and *Eleocharis* sp. in the water. Floating mats of *Sphagnum* occur at some sites. Overstory trees (cypress [*Taxodium* sp.] and black gum [*Nyssa* sp.]) may or may not be evenly spaced throughout the pond. Breeding also has been recorded in a few human-created temporary wetlands, such as borrow pits and drainage ditches. Although striped newts are found most often in temporary ponds, they occur in at least one permanent, isolated pond in Ocala National Forest. They do not occur in ponds with predatory fishes.

B. Eggs.

i. Egg deposition sites. Eggs are deposited singly in vegetation, presumably in the shallow water margins of ponds. Although they are deposited one at a time, they may occur in clumps of 2–5 eggs. The female uses her back legs to wrap protective vegetation around the egg.

ii. Clutch size. Unknown.

C. Larvae/Metamorphosis. Larvae hatch at about 8 mm TL (Mecham and Hellman, 1952).

i. Length of larval stage. Larvae have been found from April–December (Christman and Means, 1992; Means et al., 1994; Means, 2001; S.A.J., unpublished data). The length of the larval period probably varies depending on hydroperiod, temperature, and food resources. Dodd (1993b) reported successful metamorphosis at a pond with a hydroperiod of 139 d. At another pond, Johnson (2002) estimated that immature larvae had a larval period of about 6 mo, and mature larvae (i.e., paedotypic individuals; see "Neoteny" below) remained aquatic for about 18 mo. Johnson (2001, 2002) also observed four distinct periods of newt metamorphosis during a 2-yr study at this pond. Two of these metamorphic events occurred during autumn to early winter, and the other two from late spring to summer. The pond held >75 cm of water throughout the 2-yr period. Means (2001) has noted the same biphasic emergence of striped newts in panhandle populations, but sexually mature larvae in his pond remained aquatic for only about 15 mo.

ii. Larval requirements.

a. Food. Unknown. As with adults, larvae probably eat any small invertebrates that they can catch. In captivity, larvae eat mosquito larvae, chironomid larvae, fairy shrimps, clam shrimps, and amphipods (*Hyalella azteca*; D.B.M., S.A.J., unpublished data).

b. Cover. Larvae presumably hide in dense vegetation in shallow water (≤60 cm deep), but nothing is known of their specific cover requirements.

iii. Larval polymorphisms. No larval polymorphisms are known.

iv. Features of metamorphosis. Dodd (1993b) found transformed larvae as small as 18 mm SVL in Florida, although most were 22–25 mm. In a nearby Florida pond, larvae transformed at 20–30 mm (Johnson, 1999).

v. Post-metamorphic migrations. Transformed individuals presumably move to habitats similar to those used by adults. Emigration occurs from summer throughout autumn, depending on the timing of metamorphosis. Emigration occurs both singly and in mass groups. It is not known how long it takes recently transformed individuals to move from breeding ponds to suitable terrestrial habitat. In a Florida Panhandle pond, larvae transformed from slightly <20 mm SVL during pond dry-down, to 20–30 mm for sexually immature larvae, and from 30–42 mm for paedotypic individuals (Means, 2001).

vi. Neoteny. Paedomorphosis has been observed at numerous striped newt breeding ponds. The expression of the paedomorphic phenotype appears to have a substantial genetic component and is likely regulated by a suite of genes (Johnson, 2001). Johnson (unpublished data) found that about 25% of all larvae in one pond became paedomorphic in each of two consecutive years. In a pond in the Ocala National Forest, the vast majority of (if not all) larvae appear to become paedomorphic each year. Although environmental conditions (e.g., pond hydroperiod) certainly affect paedogenesis; food availability, and thus growth rate, did not have a major impact on paedogenesis, of newts from a north Florida pond (Johnson, 2001). Paedomorphic individuals can reproduce at about 1 yr of age, and after reproduction is complete, they initiate metamorphosis and migrate from the breeding pond into the surrounding uplands (Johnson, 2002b). This occurs even if the breeding pond does not dry. If ponds dry, however, observations at several sites show that larviform adults can transform and assume a terrestrial phenotype.

D. Juvenile Habitat. Presumably, juveniles occupy similar terrestrial habitats as adults. According to Petranka (1998) efts generally resemble adults, but unlike adults have a dull orange dorsal background color, rougher skin, and a rounder tail.

E. Adult Habitat. Adults sometimes are found at considerable distances (to > 700 m) from the nearest breeding pond (Dodd, 1996). Adults require shelter from the harsh conditions (heat, cold, and drought) that are present in their environment. Presumably, they stay under cover in subterranean refugia. Occasionally, striped newts are found under logs. The original sandhills vegetation was longleaf pine (*Pinus palustris*) savanna with a rich groundcover of grasses and forbs. A turkey oak (*Quercus laevis*) midstory was suppressed by frequent ground fires in the early lightning season, April–July.

Based on captures at drift fences, Johnson (1999) estimated that 18% of the population of newts at one breeding pond dispersed in excess of 500 m from the pond. Adults require shelter from the harsh environmental conditions (heat, cold, and drought) that are present in their environment. Presumably, they stay under cover in subterranean refugia.

F. Home Range Size. Nothing known.

G. Territories. It is unlikely that adults possess defended terrestrial territories. Adult males vie for females during the aquatic breeding season and attempt to maneuver them in the direction of their spermatophores, but defended territories are unknown.

H. Aestivation/Avoiding Desiccation. Nothing known. However, striped newts are active year-round as long as conditions are appropriate (Dodd, 1993b; Means and Means, 1997; Johnson, 2001, 2002). Dodd (1993b) found that the least activity was in August, one of the hottest months throughout the range of striped newts. However, Johnson (2001, 2002) caught hundreds of efts leaving a pond in August 1998.

I. Seasonal Migrations. After breeding, adults return to terrestrial refugia. The timing of return migrations varies depending on weather and pond conditions. Usually, newts leave the ponds as water levels fall. This occurs from late spring to early summer, depending on rainfall. There is a great degree of variation as to when emigration occurs, however.

J. Torpor (Hibernation). See "Aestivation/Avoiding Desiccation" above.

K. Interspecific Associations/Exclusions. In panhandle Florida and occasionally elsewhere, striped newts are syntopic with eastern newts *(N. viridescens)*, mole salamanders *(Ambystoma talpoideum)*, and dwarf salamanders *(Eurycea quadridigitata;* Means et al., 1994; Means, 1996b; Means and Means, in press). The nature of the interaction between striped newts, eastern newts, and dwarf salamanders is unknown, but predation on striped newts by mole salamanders is highly likely (D.B.M., unpublished data). Striped newts do not breed in ponds inhabited by predatory fish.

L. Age/Size at Reproductive Maturity. The smallest reproductively mature striped newts entering breeding ponds are 28–29 mm SVL. Sexual size dimorphism is not apparent in size at first reproduction, although in general, males are slightly smaller and weigh less than females (Dodd, 1993b). Petranka (1998) suggests that striped newts require 8–24 mo to reach sexual maturity, the same as eastern newts (Johnson, 2001, 2002). Attempts to age striped newts using skeletochronology have proven unsuccessful (G. Zug, personal communication).

M. Longevity. Grogan and P.G. Bystrak (1973b) reported that a striped newt lived at least 11 yr, 2 mo in captivity, but given the large size of the specimen (105 mm TL) and the fact that it was purchased in Maryland (well out of the range of striped newts), we question their species identification. Other *Notophthalmus* live 12–15 yr in the wild (Forester and Likens, 1991). Dodd (1993b) speculated that striped newts lived a long time because of the great variation in reproductive uncertainty due to stochastic environmental conditions.

N. Feeding Behavior. Metamorphosed striped newts are opportunistic feeders taking a wide variety of prey, including frog eggs, worms, snails, fairy shrimp, spiders, and insects (larvae and adults). In Florida, the greatest number of prey items

were benthic dipteran larvae (Christman and Franz, 1973). Christman and Franz (1973) suggested that striped newts were not solely visually oriented predators, but probably used smell to direct their attention toward certain potential food sources (such as frog eggs). These observations were made on aquatic adults in breeding ponds, but nothing is known about food habits during the long terrestrial stage.

O. Predators. Because of toxic skin secretions, probably few predators molest adult striped newts. Larvae undoubtedly fall prey to a host of aquatic predators because of their small size.

P. Anti-Predator Mechanisms. The antipredator mechanisms of striped newts have not been investigated. Other members of the genus *Notophthalmus* have skin glands that produce a potent neurotoxin called tarichatoxin; presumably striped newts do likewise. Toxins may be present in larvae, efts, and aquatic/terrestrial adults to various degrees. Of course, toxins may function not only in defense against predators, but also against external parasites. When molested, adults occasionally assume an "unken" posture, although it is not as intense as in eastern newts.

Q. Diseases. Unknown.

R. Parasites. Unknown.

4. Conservation.
Although striped newts are not protected by federal statutes, the U.S. Fish and Wildlife Service is concerned about their biological status and considers the species as Under Review. Striped newts are listed as Rare in Georgia because of the small number of known localities within the state (Jensen, 1999b). The Florida Natural Areas Inventory considers striped newts as Imperiled in Florida, and the Florida Committee on Rare and Endangered Plants and Animals lists the species as Rare. Although Cox and Kautz (2000) discussed the status and biological requirements of the striped newt in Florida, they are not protected in the state and have no legal protected status. Striped newts have declined substantially throughout their range because of direct habitat loss and habitat degradation (e.g., fire suppression, silvicultural practices, pond drainage, and fish introductions; Dodd and LaClaire, 1995; Franz and Smith, 1995; S.A.J., unpublished data). Presently, they persist at about 15 isolated locations throughout their range, and the majority of these locations are on public property.

Johnson (2001), Dodd and LaClaire (1995), Means (2001), Christman and Means (1992), and Means and Means (in press) provided suggestions for conserving striped newts. In general, the most effective conservation and management plan is one that will closely mimic historical conditions, and therefore facilitate striped newt metapopulation function on a landscape scale (Means and Means, in

press). Breeding ponds must be protected, and a large area of suitable upland core habitat should be protected and managed around the ponds, thus maintaining connectivity among the breeding sites. The best upland habitats are longleaf pine forest with native groundcover. Silvicultural forests of slash, loblolly, and sand pines are not suitable (Means and Means, in press). Prescribed fire is essential to proper management of uplands and wetlands. Furthermore, mechanical disturbance to native ground cover and soils, which is associated with most silvicultural practices, must be avoided. Striped newts should be regularly monitored at sites where they persist and surveys should be conducted to locate additional breeding ponds. Sites on private property should be acquired by state or federal agencies or private conservation organizations, or protected through conservation easements that guarantee permission to manage the sites for striped newts (i.e., conduct prescribed burning).

Notophthalmus viridescens (Rafinesque, 1820)
EASTERN NEWT
Todd W. Hunsinger, Michael J. Lannoo

1. Historical versus Current Distribution.
According to Hurlbert (1969), eastern newts *(Notophthalmus viridescens)* are second only to tiger salamanders *(Ambystoma tigrinum)* among U.S. salamanders in the extent of their distribution. Specifically, eastern newts are found throughout the eastern United States in historically forested areas from northern and central Minnesota, eastern and southern Wisconsin, eastern Iowa, northern and southern Illinois, extreme east-central and southeastern Kansas, and eastern Oklahoma and Texas to the Gulf of Mexico and the Atlantic Ocean. Four subspecies are recognized. Broken-striped newts *(N. v. dorsalis)* are distributed in southeastern North Carolina and northeastern South Carolina. Central newts *(N. v. louisianensis)* have a disjunct distribution. Southern populations are distributed along the Atlantic and Gulf Coastal Plains, except for most of peninsular Florida, and up the Mississippi River drainage into northern Missouri, southern Illinois, and southwestern Indiana. Northern populations are distributed from southeastern Iowa, north around southern and western Lake Superior and surrounding Lake Michigan. Peninsula newts *(N. d. piaropicola)* inhabit the southern 4/5 of peninsular Florida. Red-spotted newts *(N. v. viridescens)* have the largest distribution of the four subspecies and are found along the Appalachian spine from central Indiana and the eastern half of the Lower Peninsula in Michigan to the Atlantic Coast in northern North Carolina and north into and throughout New England.

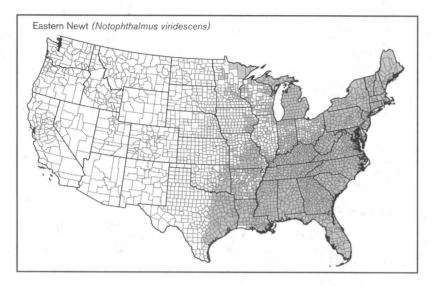

Eastern Newt (*Notophthalmus viridescens*)

The current distribution most likely resembles the historical distribution, with some losses of local demes. Newts have not been reported from Hamilton County, Ohio, in over 70 yr (Smith and Pfingsten, 1989; Davis et al., 1998). However, newts are good colonizers (Fauth and Resetarits, 1991; see also, Brandon and Bremer, 1966; Hurlbert, 1969), occupying suitable habitat modified by beaver *(Castor canadensis)* and created by farm ponds (Petranka, 1998). Newts also benefit from reforestation (Skelly et al., 1999) occurring throughout much of their range.

2. Historical versus Current Abundance.
Little data exist for this far-ranging species. Abundance may vary within and between years (Harris et al., 1988). In southeastern Michigan, with expanding reforestation, eastern newt populations experienced a 25-fold increase in range and colonized five breeding ponds over a 20-yr period (Skelly et al., 1999). Cortwright (1998) reports that southern Indiana populations are widespread and common, and that populations in 36 wetlands showed no upward or downward trends. He also reports that local populations tend to fluctuate more than metapopulations, suggesting that populations connected by dispersive eft stages are more stable than isolated populations. Brown (E. E., 1992) reports eastern newts are rare in the Piedmont of North Carolina.

Because eft movements, feeding, and growth rates are climate dependent (Hurlbert, 1969; Healy, 1973, 1975b), and adults are less tolerant than juveniles to drought conditions (Healy, 1970; Healy, 1974; Morin, 1983a; Gillis and Breuer, 1984), the potential exists for populations to be impacted in the future by climate change, particularly in regions where drought conditions may persist.

3. Life History Features.
Eastern newts have among the most variable life histories of North American amphibians. Most populations have aquatic egg, larval, and adult stages and a terrestrial eft stage. Environmental factors, as well as larval and adult densities, can influence the presence and timing of life history stages between and within populations (Healy, 1970; Morin et al., 1983; Harris, 1987a). Illinois populations have four morphological stages: larvae, terrestrial efts, fully transformed adults, and neotenic adults (Brandon and Bremer, 1966). It is not surprising that Noble (1926, 1929b) and P. H. Pope (1921, 1924, 1928b) engaged in heated exchanges over the ontogeny of efts.

A. Breeding. Reproduction is aquatic.
i. Breeding migrations. Efts migrate from forested terrestrial sites into aquatic habitats and become reproductively mature. Breeding adults that have overwintered on land follow distinct migratory routes, entering the pond at the same point (Hurlbert, 1970b). They will home (Gill, 1979). Migrations are concentrated around rainy days and nights and often follow streams and wet sloughs (Stein, 1938; Hurlbert, 1969). The timing of migration varies, depending on location. Some researchers report a spring migration only, which peaks in late March in Virginia (Gill, 1978; Massey, 1990). Other populations have only fall migrations (Brimley, 1921b; Chadwick, 1944; Healy, 1974). Fall breeding migrations occur from July to early November, including July–August in Massachusetts (Healy, 1975a), September–October in south-central New York (Hurlbert, 1969), and late October in Massachusetts (Stein, 1938). Most fall migrants do not attain reproductive condition until the following spring (Bishop, 1941b). Some New York populations experience fall and spring migrations to the breeding sites (Bishop,

Hurlbert, 1969). Adult densities in breeding ponds peak in February–March in North Carolina (Harris et al., 1988). Eastern newts are reported to be highly philopatric (Gill, 1978; Massey, 1990) but at the same time demonstrate enough flexibility in breeding sites to support metapopulations (Hurlbert, 1969; Gill, 1978; Cortwright, 1998).

Males identify females through chemoreception (Dawley, 1984) and are attracted to larger females that have a greater fecundity (Verrell, 1982, 1985). Males select females using both olfactory and visual cues (Verrell, 1985).

Breeding success is episodic. Semlitsch et al. (1996) report that >15,000 newts metamorphosed from a Carolina bay in 2 years, yet in 9 other years no newts were collected. Factors such as newt density and predator density may influence reproductive success (see Morin, 1983a; Morin et al., 1983).

ii. Breeding habitat. The same as the adult habitats, which include pools, ponds, wetlands, and low-flow areas of streams in forests (see "Adult Habitat" below). Both permanent and semi-permanent bodies of water are used (see Hurlbert, 1969). Male courtship and amplexus of females occurs in shallow water (Pitkin and Tilley, 1982; T. W. H., unpublished observations). Most adults appear to breed annually, although females will skip years (Gill et al., 1983; Gill, 1985). Massey (1990) found that females are capable of sperm storage for upwards of 10 mo.

B. Eggs.
i. Egg deposition sites. Eggs are deposited singly; the female wraps each in a folded leaf of a submerged macrophyte, in a decaying leaf, or in other detritus (Bishop, 1941b; Goin, 1951; Morin et al., 1983). Mature, overwintering adults deposit eggs earlier in the breeding season than do efts returning to the water to breed for the first time (Hurlbert, 1970b). Females in southern populations begin egg deposition in early winter (Goin, 1951). Deposition in a Kansas population occurred in March–April (Ashton, 1977), while dates for northern populations include late April–May (Smallwood, 1928), sometimes late May (Hurlbert, 1970b), and even early July (Gage, 1891; Pope, 1924; Bishop, 1941b; Chadwick, 1944; Worthington, 1969; Ashton, 1977; Harris et al., 1988). Worthington (1969) speculated that extended egg deposition could minimize larval competition with other spring-breeding amphibians.

ii. Clutch size. Each female lays between 200–375 eggs, at a rate of several eggs/day (Bishop, 1941b). Thus, time to complete ovipositing can take many weeks (Bishop, 1941b; Chadwick, 1944; Morin, 1983a). It is unclear whether a female will lay all of her eggs in 1 yr (Bishop, 1941b). Individual eggs have a diameter of about 1.5 mm (Bishop, 1941b; Brandon and Bremer,

1966). Incubation lasts between 20–35 d (Gage, 1891; Bishop, 1941b).

C. Larvae/Metamorphosis.

i. Length of larval stage. The length of the larval stage and size at metamorphosis varies across their range, among water bodies (Hurlbert, 1970b), and among years (Harris et al., 1988). Greater competition between larvae at high densities also affects the timing and size at metamorphosis (Harris, 1987a).

Bishop (1941b) reported that in New York, larvae hatch at about 7.5 mm and metamorphose to the eft stage in 2–3 mo at an average of 36.7 mm TL. Hurlbert (1970b) reported a range of 28–47 mm TL in south-central New York. In a Massachusetts population, the larval period lasts from 2–5 mo, and larvae metamorphose at about 19–21 mm SVL (35–38 mm TL; Healy, 1973, 1974). In Maine, metamorphosis occurs in 3–5 mo at 29–32 mm TL (Pope, 1921). Larvae from Maine raised in the laboratory metamorphosed at 28–34 mm TL (Pope, 1928b). Western North Carolina larvae metamorphose at 34–43 mm TL (Chadwick, 1950), while larvae in the sandhills metamorphose at 24 mm SVL (Harris et al., 1988). Metamorphosis in Illinois occurs at 33–40 mm TL (Brandon and Bremer, 1966). Ashton (1977) found three distinct larval classes in a Kansas pond in late June, including individuals close to metamorphosis. Hurlbert (1970b) noted two size classes of larvae in ponds where females overwinter. Larval mortality rates may be high in some populations (Massey, 1990).

In some populations, the eft stage is abandoned and larvae transform directly into adults. In populations with neotenic adults, metamorphic size ranges from 46–75 mm (Noble, 1929b). Harris et al. (1988) found that some larvae overwinter in the ponds in western North Carolina.

ii. Larval requirements.

a. Food. Newts are carnivorous during all life history stages. Newly hatched larvae feed at night on small invertebrates such as zooplankton; larger larvae include larger invertebrates in their diet and are cannibalistic when confined in the laboratory (Pope, 1924; Walters, 1975; Morin, 1983a; Harris, 1987b; Harris et al., 1988). Cannibalism can occur between different size classes of larvae, but does not occur within size classes (Harris, 1987b). Eastern newts appear to feed on prey in roughly the same proportions to their abundance (Hamilton, 1940; Burton, 1977). Hamilton (1940) concluded that larvae use visual cues in prey selection. Notable prey include protozoans, cladocerans, ostracods, copepods, dipteran larvae, snails, fingernail clams, clams, and mites.

b. Cover. Larval newts are found in water <0.5 m deep (Burton, 1977; see also Forester and Lykens, 1991). During the day, they seek cover under bottom debris (Chadwick, 1950; Ashton, 1977; Morin,

1983a). Larvae exhibit spatial segregation by size class (Harris et al., 1988). This may serve to limit competition or cannibalism with larger conspecifics.

iii. Larval polymorphisms. None reported.

iv. Features of metamorphosis. Newly metamorphosed eastern newts have been observed or collected as follows: July to early November in Massachusetts (Healy, 1974, 1975b) and New York (Bishop, 1941b; Hurlbert, 1970b); July–August in Massachusetts (Smith, 1920; Noble, 1929b); late June in Maryland (Worthington, 1969); mid-August to November in Virginia (Gill, 1978); September in Illinois (Brophy, 1980), coastal North Carolina (Taylor et al., 1988), and western North Carolina (Chadwick, 1950).

Hurlbert (1970b) found that size at metamorphosis is correlated to water temperatures. Due to potential competitive pressures, larvae mature at a larger size in ponds where the adults have emigrated after breeding (Morin, 1983a; Harris, 1987a).

v. Post-metamorphic migrations. Larvae aggregate in one region of the pond just prior to metamorphosis (Hurlbert, 1970b; Ashton, 1977). In populations with efts, newly metamorphosed individuals will migrate away from ponds into upland forests. Migrations occur in waves, usually at night during or following rains (Chadwick, 1950; Hurlbert, 1970b; Healy, 1975b). Efts require about 1 yr to migrate to woodlands 800 m distant (Healy, 1974, 1975a).

vi. Neoteny. Pike (1886) was the first to call attention to neoteny in the newt. Sherwood (1895) noted a scarcity of efts in populations around New York City and Mount Vernon, New York. He suggested that the entire life cycle of newts is aquatic. Noble verified neoteny for populations on Long Island (Noble, 1926) and Cape Cod (Noble, 1929b). Bishop (1943) reported additional populations in Louisiana and Florida. It is now known that neotenic adults occur in populations of all four subspecies, having been reported from Florida, Illinois, Indiana, Louisiana, Massachusetts, New Jersey, New York, North Carolina, and Tennessee (Gage, 1891; Noble, 1926, 1929b; Schmidt and Necker, 1935; Bishop, 1943; Peterson, 1952; Schwartz and Duellman, 1952; Gentry, 1955; Brandon and Bremer, 1966; Healy, 1970; Harris, 1987a).

Neoteny in eastern newts is based on the presence or remnants of external gills, gill slits, and a gular cleft in sexually mature individuals. Newts do not retain as many larval characteristics as neotenic *Ambystoma* (Reilly, 1987).

Many of the populations with neotenic adults occur in sandy, coastal sites, where terrestrial conditions are suboptimal (Noble, 1926; Bishop, 1941b; Brandon and Bremer, 1966). The tendency for neotenic adults to occur in these sites has led to the implication that the terrestrial

eft is selected against (see Petranka, 1998). However, adults will lose their gills and transform if their breeding sites dry (Noble, 1929b; Healy, 1970). Healy (1970) found that the persistence of drought conditions changed a population from nearly all neotenic adults to one of all metamorphosed individuals within a few years. Although environmental factors may influence metamorphosis (Healy, 1970; Harris, 1987a), it has been demonstrated that the extent of neoteny in a population is negatively correlated to larval density (Healy, 1974; Harris, 1987a; Harris et al., 1988). Therefore, metamorphosis may result from resource partitioning between age classes. When compared with other salamander species, in eastern newts, populations with neotenic adults are nearly as genetically variable as populations where adults metamorphose (Petranka, 1998).

D. Juvenile Habitat. Terrestrial efts are usually found in wooded areas (Bishop, 1941b; Evans, 1947; Williams, 1947). Efts are diurnal and nocturnal, moving about on rainy or humid days and nights when the ground is moist (Healy, 1973). Movement peaks during May–June, and again in the autumn (Healy, 1975a). During dry periods, they seek cover in leaf litter or under logs and other objects (Bishop, 1941b; Williams, 1947; Healy, 1974, 1975a). Efts rarely move about when the temperature drops below 10 °C, but will be active above 12 °C (Healy, 1975a). Pough (1974) found that efts select for environmental temperatures of 26–28 °C.

Efts establish home ranges that increase annually (Healy, 1975a); they will home (Gill, 1979). Home ranges averaged 266.9 m^2 in one season and 353 m^2 over two seasons. Individuals have been recaptured within a few centimeters of previous captures in successive years. The average distance from breeding ponds is 800 m.

Efts forage in the forest floor leaf litter, especially during rains. Eft growth, which is temperature and moisture dependent, is concentrated during the summer (Healy, 1973). They feed on invertebrates and have a slight preference for larger prey (MacNamara, 1977). Prey include 58 families from 25 orders of invertebrates. Efts cluster around and under mushrooms in late August to September to feed on dipterans attracted to these fungi (Healy, 1975a).

The eft stage is estimated to last 2–3 yr in New York (Bishop, 1941b); 3–4 yr in Québec (Caetano and Leclair, 1996); 4 yr in the North Carolina mountains (Chadwick, 1944); 3–7 yr in Massachusetts (Healy, 1974); and 4–7 yr, with a mean of 4.4 yr, in a Maryland population (Forester and Likens, 1991). The Québec data suggest that males reach sexual maturity slightly sooner than females (Caetano and Leclair, 1996). Harris et al. (1988) found the eft stage may last only 1 yr. The eft stage may be similarly reduced in southern populations (see Huheey and Brandon,

1974). Chadwick (1944) found that the largest efts exceeded the mean length of aquatic adults in North Carolina.

Where the eft stage is abandoned and larvae transform directly into adults, the juvenile habitat characteristics are similar to larval and adult habitat characteristics. Aquatic larvae grow faster than terrestrial efts, taking a shorter time to metamorphose (Healy, 1973, 1974). Here, juvenile growth occurs throughout the year, with the greatest increment in the spring (Healy, 1973).

E. Adult Habitat. Aquatic adults inhabit pools, ponds, wetlands, sloughs, canals, and quiet areas of streams in upland and bottomland regions (Bishop, 1943; Schwartz and Duellman, 1952; Bellis, 1968; Gates and Thompson, 1982; Petranka, 1998). Adults are found primarily in open, sunny areas. Sites with submergent and emergent vegetation often harbor large populations (Schwartz and Duellman, 1952; George, 1977; Gates and Thompson, 1982). As with efts, adults show a thermal preference. George et al. (1977) found that newts stayed just below the thermocline in mid-summer. During the winter, aquatic adults will sometimes congregate in ice-free areas of ponds where temperatures are at 5–6 °C (Morgan and Grierson, 1932; Pitkin and Tilley, 1982). Both sexes are equally represented.

Adults will leave drying ponds for protected upland sites to avoid desiccation and heat stress (Hurlbert, 1969; Gill, 1978). During this time, adults will mimic eft behavior, hiding under rotten logs and vegetation clumps (Gill, 1978). Adults that remain terrestrial for a long period will develop a more granular skin and reduced tail fin (Noble, 1926; Hurlbert, 1969; Gill, 1978; Massey, 1990).

F. Home Range Size. Petranka (1998) notes that the extent to which adults establish home ranges varies across populations. Bellis (1968) reports aquatic adults staying in the same portion of a pond for weeks (but see Harris, 1981). Reductions in water levels cause adults to move, but following pond-refilling, animals move back to their original sites. Males move more than females. Individuals can identify nearest neighbors (Wise et al., 1993).

G. Territories. Unknown.

H. Aestivation/Avoiding Desiccation. Adults will abandon aquatic sites during dry conditions when water levels drop and water temperatures rise. In North Carolina, newts will leave the ponds after breeding and remain terrestrial until November (Brimley, 1921b). They seek refuge in moist sites beneath logs, rocks, and other detritus (Hurlbert, 1969; Gill, 1978; Massey, 1990).

Efts seek cover in leaf litter and other objects during dry conditions (Bishop, 1941b; Williams, 1947; Healy, 1974, 1975a).

I. Seasonal Migrations. Variable. Efts will leave ponds to forage for 2–7 yr before returning to ponds to breed. Terrestrial adults will return to ponds to breed. The timing of migration varies by gender (Hurlbert, 1969). Males migrate earlier than females in the spring and have a more concentrated migration in the fall. Migrations also vary by latitude and local climate/weather conditions. Hurlbert (1969) found that spring and fall migrations occur during rain episodes when air temperatures exceed 4.4 °C. Newts will migrate out of drying breeding ponds in summer (Hurlbert, 1969). The seasonal point of emigration and immigration is similar at many breeding ponds (Hullbert, 1969). Stein (1938) reported a spectacular fall migration of red efts, consisting of thousands of animals, in Massachusetts. As many as 20 animals were picked up in a single scoop of the hand; 100 animals were collected in 30 min, 1,200 animals were collected from a 5 m² area. Newts use a specialized magnetoreception system to guide homing and migration (Phillips and Borland, 1994).

J. Torpor (Hibernation). Little is known about the hibernacula of newts. Cooper (1956) found newts hibernating in floodplain puddles. An aggregation of newts, lesser sirens (*Siren intermedia*), American bullfrog tadpoles (*Rana catesbeiana*), and a musk turtle (*Sternotherthus odoratus*) was found in 1.5 °C ice-free water in January in Illinois (Cagle and Smith, 1939). During January in a Massachusetts stream, Morgan and Grierson (1932) found newts both singly in tangles of decaying macrophytes and in groups of 20–40 clustered beneath flat rocks. Bothner (1963) excavated two efts, together with a spotted salamander (*A. maculatum*), from the hibernaculum of short-headed garter snakes (*Thamnophis brachystoma*) 43 cm below the surface.

In permanent, deep waterbodies, adult eastern newts may remain aquatic and active throughout the year (Smallwood, 1928; Morgan and Grierson, 1932; George et al., 1977; Pitkin and Tilley, 1982). Although food intake will decrease during the winter, feeding and molting continue, indicating an active metabolism (Morgan and Grierson, 1932; Pitkin and Tilley, 1982).

In populations where the breeding pools are shallow or seasonal, adults migrate out of ponds in summer or autumn (August–September), overwinter on land, and return the following spring to breed (Hurlbert, 1969; Gill, 1978; Massey, 1990).

K. Interspecific Associations/Exclusions. Given both their terrestrial and aquatic life history stages, eastern newts interact with a wide variety of eastern U.S. aquatic and terrestrial amphibians. As well, unlike many North American amphibian species, eastern newts will occupy water bodies containing predaceous fishes (M. J. L., unpublished observations;

Gates and Thompson, 1982; see also Kesler and Munns, 1991).

L. Age/Size at Reproductive Maturity. Size and time to sexual maturity vary depending on life history and location. Onset of sexual maturity is size specific rather than age specific (Healy, 1974; Caetano and Leclair, 1996). Populations with efts take a longer time to reach sexual maturity than populations with neotenic adults (Noble, 1929b; Healy, 1974; Harris, 1987a). Red-spotted newts from western North Carolina are larger than individuals from this subspecies found elsewhere (Bishop, 1943), averaging nearly 10 mm TL larger at the time of sexual maturity than New York specimens (Chadwick, 1944). The smallest sexually mature newts from western North Carolina could have been 85–90 mm TL (Hurlbert, 1969), while the smallest males found in north coastal populations were 51–64 mm (Noble, 1926, 1929b).

Neotenic adults and adults that skip the eft stage mature at 7 mo (Harris, 1987a). On Cape Cod and coastal Massachusetts, adults transforming to the eft stage mature at 2 yr (Harris, 1987a). Females appear to mature later than males and at a larger body size (Caetano and Leclair, 1996).

M. Longevity. As with all eastern newt life history features, longevity varies. Most adults are 3–8 yr old and maximum age in populations varies from 9–15 yr (Petranka, 1998). Forester and Likens (1991) found adults from 4–9 yr old in Maryland. Adults in Québec ranged from 2–13 yr (Caetano and Leclair, 1996). Gill (1978, 1985) reported maximum ages of 15 yr for males and 12 yr for females in Virginia. Female survivorship between years appears to be lower than males (Gill, 1978, 1985; Massey, 1990). This results in a greater number of reproductive seasons for males (Gill, 1978). Morin (1983a) reported high mortality rates during periods of drought accompanied by subfreezing temperatures in North Carolina.

N. Feeding Behavior. Similar to larvae, adults are carnivorous, feeding on any available, palatable prey they can swallow whole. Newts rely on visual and chemical cues to locate food (Martin et al., 1974). Prey include zooplankton, amphipods, mayflies, stoneflies, dipterans, hemipterans, lepidopterans, coleopterans, odonates, oligochaetes, leeches, snails, clams, small fishes, fish eggs, anuran eggs and tadpoles, ambystomatid larvae, conspecific embryos and larvae, and shed skins (Hamilton, 1932; Morgan and Grierson, 1932; Bishop, 1941b; Behre, 1953; Wood and Goodwin, 1954; Ries and Bellis, 1966; Walters, 1975; Burton, 1977; George et al., 1977; Gill, 1978; Morin, 1981, 1983a; Pitkin and Tilley, 1982; Morin et al., 1983; Taylor et al., 1988; Wilbur and Fauth, 1990; Fauth and Resetarits, 1991). Adults exhibit a diel feeding pattern, but most feeding occurs in the early morning and is not influenced by the presence of sunfish (Centrarchidae; Kesler and Munns, 1991). Gut passage times vary from 33 h (25 °C) to > 15 d (5 °C; Jiang and Claussen, 1993). Eastern newts aggregate in vegetation near shorelines but will forage at depths of 9–13 m (George et al., 1977).

Because newts feed opportunistically on seasonally abundant prey, they can influence the relative abundance of zooplankton, insect, and amphibian populations, as well as community structure (Morgan and Grierson, 1932; Morin, 1983a,b; Morin et al., 1983; Fauth and Resetarits, 1991). Morin (1983a) found that the competition with newts leads to decreased size at metamorphosis for tiger salamander (*A. tigrinum*) larvae. Newt predation on the eggs of tiger salamanders. can lead to the exclusion of larvae from breeding ponds (Morin, 1983a). Predation by newts has been documented to influence anuran community composition (Morin, 1983b; Morin et al., 1983; Fauth and Resetarits, 1991), reducing the size of some populations while freeing others from competition.

O. Predators. Many animals are known to feed on eastern newts, despite their anti-predator defenses (see "Anti-Predator Mechanisms" below). In experimental trials, efts and adults were initially rejected by potential predators (e.g., Hurlbert, 1970a). Hurlbert (1970a) found newts to be less acceptable to diurnal terrestrial predators than they were to aquatic and nocturnal terrestrial predators. Adult newts have been found in the stomachs of smallmouth bass (*Micropterus dolomieu*) in New Hampshire (Burton, 1977). Larvae are taken by pumpkinseed sunfish (*Lepomis gibbosus*; Kesler and Munns, 1991). Snakes, such as hognosed snakes (*Heterodon* sp.), will occasionally eat newts (Uhler et al., 1939). Brodie (1968) found eastern garter snakes (*Thamnophis s. sirtalis*) and northern water snakes (*Nerodia s. sipedon*) to have some resistance to newt toxicity, but will not feed consistently on newts. Chelonian predators can include snapping (*Chelydra serpentina*) and painted turtles (*Chrysemys picta*). Hamilton (1932) reported newts in the diet of mudpuppies (*Necturus maculosus*) from New York. Eastern lesser sirens (*Siren intermedia intermedia*; Fauth and Resetaris, 1991), marbled salamanders (*A. opacum*; Walters, 1975), and in controlled studies, tiger salamanders (Morin, 1983a) feed on eggs and larval newts. In laboratory studies American bullfrogs readily consumed newts (Brodie, 1968; Hurlbert, 1970a); however, habitat differences likely limit encounters in the wild (Hurlbert, 1970a). Some eastern newt adults cannibalize embryos (Wood and Goodwin, 1954) and larvae (Gage, 1891; Morgan and Grierson, 1932; Burton, 1977).

Raccoons (*Procyon lotor*) are the main mammalian predator on eastern newts (Hurlbert, 1970a). Shure et al. (1989) found efts that were decapitated and eviscerated by an unspecified predator in North Carolina. Ross (1933) also reported decapitated efts in New York. In Virginia, leeches appear to be a major source of adult mortality (Gill, 1978).

P. Anti-Predator Mechanisms. The skin of newts contains tetrodotoxin, a neurotoxin (Mosher et al., 1964; Wakeley et al., 1966; Brodie, 1968a; Brodie et al., 1974b). Most of this toxin is concentrated on the dorsal surface. The skin of efts is 10 times more toxic than the skin of adults (Brodie, 1968a). This skin toxicity of efts and newts make them unpalatable to many species of crayfish, mammals, birds, fish, reptiles, amphibians, and insects (Hamilton, 1951; Webster, 1960; Brodie, 1968a; Hurlbert, 1970a; Brodie and Brodie, 1980; Brodie and Formanowicz, 1981a; Formanowicz and Brodie, 1982; Kats et al., 1988). In most accounts of predation on adult newts and efts, predators eviscerate the newts after decapitation or puncturing the midventral region, thus avoiding the more toxic dorsal skin (Ross, 1933; Hurlbert, 1970a; Shure et al., 1989). Brook trout (*Salvelinus*) force-fed newts died within hours (Webster, 1960). Brodie (1968a) achieved similar results in trials with American toads (*Bufo americanus*), and many species of reptiles. Common garter snakes (*T. sirtalis*) exhibit mouth gaping and rubbing after predation attempts on newts and efts (Hamilton, 1951; Hurlbert, 1970a). Crayfish (*Cambarus diogenes*) and *Orconectes propinquus*) and beetle larvae (*Dytiscus verticalu*) and *Lethocerus americanus* also exhibited agitated mouth part movements after attempts to prey upon newts at metamorphosis (Formanowicz and Brodie, 1982). Because of the effectiveness of the skin toxins, many animals have a learned avoidance to newts (Brodie and Brodie, 1980; Brodie and Formanowicz, 1982). Pough (1974) speculated that it is a compound other than tetrodotoxin that causes four species of leeches, *Batracobdella phalera*, *Mooreobdella fervida*, *Haemopis marmorata*, and *Helobdella stagnalis*, to avoid parasitizing adult newts.

The bright red color of efts is widely considered to be aposematic—warning off potential predators. The cryptic greenbrown coloration of adults helps them to avoid predation. Eft coloration is such an effective anti-predator mechanism that red salamanders (*Pseudotriton ruber*), mud salamanders (*P. montanus*), and the erythristic morph of the eastern red-backed salamander (*Plethodon cinereus*) represent examples of Batesian mimicry in areas where these species occur with newts (Howard and Brodie, 1970; Brodie and Howard, 1972; Huheey and Brandon, 1974; Brodie and Brodie, 1980). Newt subspecies with bright coloration maintain diurnal activity and have a longer eft stage (Huheey and Brandon, 1974). Cryptic colored central newts (*N. v. louisianensis*) have a shorter eft stage and exhibit more cryptic behavior; this occurs despite the fact that central newts are as toxic as the red subspecies (Brandon et al., 1979a).

The unken posture in newts, first described by Neill (1955), is defined as an upward bending of the head and tail until they nearly touch, and it has been described from observations in the wild (Neill, 1955; Petranka, 1987) and laboratory studies (Brodie and Howard, 1972) of adult newts. This posture exposes the brighter, aposematic coloration of the ventral surface. The unken posture is rarely displayed by efts (Brodie and Howard, 1972). Instead, efts will raise their tail laterally or occasionally in a vertical display when threatened. This may be coupled with an elevation of the hindlimbs (Ducey and Brodie, 1983). Observed behaviors in the wild are similar to laboratory experiments (Brandon et al., 1979b; Ducey and Brodie, 1983).

Newts also exhibit an avoidance response to the tissue of damaged conspecifics (Marvin and Hutchison, 1995; Woody and Mathis, 1997).

Q. Diseases. Unknown.

R. Parasites. The parasites of eastern newt larvae were detailed by Rankin (1937) as follows: Protozoa—*Hexamastix batrachorum*, *Hexamitus intestinalis*, and *Tritrichomonas augusta*; Trematoda—*Plagitura* sp., intestinal wall metacercariae; Nematoda—*Camallanus* sp.

The parasites of eastern newt adults were detailed by Rankin (1937) as follows: Protozoa—*Cryptobia borreli*, *Cytamoeba bacterifera*, *Entamoeba ranarum*, *Euglenamorpha hegneri*, *Eutrichomastix batrachorum*, *Hexamastix batrachorum*, *Hexamitus batrachorum*, *Hexamitus intestinalis*, *Karotomorpha swezi*, *Myxobolus conspicuus*, *Nyctotherus cordiformes*, *Prowazekella longifilis*, and *Tritrichomonas augusta*; Trematoda—*Plagitura* sp., intestinal wall metacercariae.

Adults in Virginia are affected by parasitic leeches (*Batruchobdella picta*) which transmit *Trypanosoma diemyctyli*, a blood endoparasite (Gill, 1978). In response, adults leave the water to rid themselves of leeches.

4. Conservation.
Among U.S. salamanders, eastern newts are second only to tiger salamanders in the extent of their distribution. Their current distribution most likely resembles their historical distribution, with losses of local populations. Eastern newts are good colonizers and will utilize suitable habitat created by beavers and farm pond initiatives (Petranka, 1998). Eastern newts are benefiting from reforestation, which is occurring throughout much of their range (Skelly et al., 1999). Their abundance may vary within and between years (Harris et al., 1988).

Eastern newts are listed as Threatened in Iowa and Kansas, and the potential exists for populations to be impacted in the future by climate change, particularly in regions where drought conditions may persist.

Taricha granulosa (Skilton, 1849)
ROUGH-SKINNED NEWT

Sharyn B. Marks, Darrin Doyle

1. Historical versus Current Distribution.
Rough-skinned newts *(Taricha granulosa)* have the widest distribution among the three species of *Taricha*, ranging from the Coast Range mountains in Santa Cruz County, California, to Admiralty Island, Alaska (Myers, 1942b; Riemer, 1958; Nussbaum and Brodie, 1981; Stebbins, 1985; Petranka, 1998). They extend inland to the east slope of the Cascades in Washington and along the west slope of the Sierra Nevada in northern California. Isolated and extremely small populations exist near Moscow, Idaho, and Thompson Falls, Montana, and may be the result of introductions (Nussbaum et al., 1983; Monello and Wright, 1997). Rough-skinned newts occur from sea level to about 2,800 m in elevation. No recent studies have investigated the distribution of this species, thus, the limits of the current distribution are uncertain.

2. Historical versus Current Abundance.
Rough-skinned newts are one of the most common salamanders in the Pacific Northwest (Nussbaum et al., 1983; Bury et al., 1991; Corn and Bury, 1991; Gilbert and Allwine, 1991). Gomez and Anthony (1996) compared relative abundance of herpetofauna in five forest types in western Oregon and found rough-skinned newts to be the most abundant salamander, representing 86% of salamander captures. Terrestrial adults were encountered more frequently in riparian than upslope habitats and were most abundant associated with deciduous forests. Numerous studies have examined the relationship between forest stand age and amphibian abundance, but results have been equivocal for rough-skinned newts. In the Oregon Coast Range and southern Washington Cascade Range, newts were more abundant in old-growth stands than in mature and young stands, but differences among forest age classes were not statistically significant (Aubry and Hall, 1991; Bury et al., 1991; Corn and Bury, 1991). By contrast, in the Oregon Cascade Range, newts were most abundant in young stands (Gilbert and Allwine, 1991). However, proximity to aquatic breeding sites is probably a more important factor influencing the relative abundance of this species, and probably is a confounding factor in studies examining the effects of forest age on newt abundance (Bury et al.,

Rough-skinned Newt *(Taricha granulosa)*

1991). Estimates of population size, density, and salamander biomass have been made for aquatic rough-skinned newts at Marion Lake, a small (approximately 10-ha) lake in British Columbia. Efford and Mathias (1969) estimated adult population size at 3,965 individuals, average adult density at 300/ha, and maximum adult densities to be as high as 2,700/ha near the edge of the lake. Neish (1971) estimated adult population size at 2,450 individuals and total biomass of aquatic adult newts at approximately 18 kg. Differences in population estimates are due to different methods of newt capture and data analysis and differences in the years during which the studies were conducted. Several reports of the abundance of rough-skinned newts are striking. Farner and Kezer (1953) noted an aggregation of 259 individuals (large larvae, recently metamorphosed animals, and adults), the majority of which were found under a single flat rock 0.9 m^2 (approximately 9 ft^2) in Crater Lake, Klamath County, Oregon. Coates (1970) observed a large aggregation (an oval cluster measuring 2 × 9.2 m) of approximately 5,000 post-reproductive males and females in a channel of water meandering through the lakebed of Clear Lake (Wasco County, Oregon). Both the Crater Lake and Clear Lake aggregations were observed in autumn.

3. Life History Features.
A. Breeding. Reproduction is aquatic.

i. Breeding migrations. Timing of breeding migrations varies with latitude and elevation. Late fall migrations are characteristic of mild-winter areas (Pimentel, 1960), spring migrations characterize low-altitude sites at higher latitudes (Neish, 1971; Oliver and McCurdy, 1974), and summer migrations may be seen at high-elevation sites (Nussbaum et al., 1983). There may also be variation in timing associated with the permanence and depth of bodies of water used for breeding. Pimentel (1952) observed slightly later breeding migrations to temporary ponds relative to nearby permanent ponds, because the former take some time to achieve optimal depth for breeding following the onset of fall rains. Males migrate individually and generally arrive at breeding sites about 1 mo before females, which often migrate in groups (Pimentel, 1960). In general, both sexes participate in breeding migrations. However, at a site on Vancouver Island, British Columbia, only females migrate since males remain in the water throughout the year except for brief terrestrial excursions (Oliver and McCurdy, 1974). Experiments by Pimentel (1960) suggest that olfaction is important for locating breeding sites and that humidity perception and downhill slopes to ponds may be of secondary importance in this regard. Laboratory and field experiments by Landreth and Ferguson (1967a,b) demonstrate that adults locate breeding sites using a sun-compass mechanism to orient themselves in the proper direction. Also see "Seasonal Migrations" below.

ii. Breeding habitat. Mating occurs in quiet water habitats including ponds, lakes, reservoirs, drainage ditches, and slowly flowing sections of streams (Gordon, 1939; Bishop, 1947; Stebbins, 1985). The timing of breeding varies substantially over the range of rough-skinned newts. At low-elevation sites in northern California, mating may occur from late December to June, with peak activity in March and April (Twitty, 1935; Stebbins, 1951). Breeding occurs in winter in western Oregon (Gordon, 1939). At moderate-to-high elevation lakes in northern California and Oregon, mating occurs during summer and autumn (Chandler, 1918; Farner and Kezer, 1953; Garber and Garber, 1978; Marangio, 1978).

B. Eggs.

i. Egg deposition sites. Oviposition usually occurs within a couple of weeks after mating (Storm, 1948) in the same habitats chosen for breeding. An extended oviposition season is typical (Bishop, 1947), with females at a single locality spawning or with eggs in their oviducts from late December to June in northern California (Twitty, 1942). Egg-laying usually occurs in late winter to spring in western Oregon (Chandler, 1918; Gordon, 1939), and occurs during June and July at low elevations in British Columbia (Efford and Mathias, 1969). Spawning probably occurs in autumn in most montane populations in northern California and Oregon (Chandler, 1918; Garber and Garber, 1978; Marangio, 1978). Rough-skinned newts show a stronger preference for spawning in quiet water habitats than any other species in this genus (Twitty, 1942). Females usually deposit eggs singly onto stems and leaves of submerged plants, usually within a few inches of the water's surface (Chandler, 1918; Twitty, 1935). At approximately 1.8 mm in diameter (measured during the morula stage; Twitty, 1935), eggs are smaller than those of other members of the genus (Twitty, 1942). Bishop (1947) reports egg diameters of 1.85–2.0 mm without envelopes or an average of 3.3 mm with envelopes. Eggs appear light tan on the top (animal hemisphere) and cream on the bottom (vegetal hemisphere; Riemer, 1958) and have three envelopes in addition to the vitelline envelope. Oviposition is cyclic, occurring for a few days, followed by a period of non-laying, with this process being repeated at intervals (Oliver and McCurdy, 1974).

ii. Clutch size. No information is available on the total number of eggs laid by an individual female during the breeding season. Embryonic development is complete between 20 and 26 d (Nussbaum et al., 1983), but length of time to complete embryonic development undoubtedly varies considerably with water temperature.

C. Larvae/Metamorphosis.

i. Length of larval stage. Larvae are regarded as pond type, having bushy gills and a large tail fin (Stebbins, 1951). Balancers, a pair of ventrolateral appendages projecting from the side of the head and arising before the forelimbs develop, are always present in hatchlings (Riemer, 1958). The average size of hatchlings is variable, ranging from 7.6–12 mm TL at different localities (Twitty, 1935; Bishop, 1947; Stebbins, 1951; Riemer, 1958). There is also variation in the length of the larval period. At the same locality, larvae may transform in late summer or early fall of the year in which they were hatched (4–5 mo after hatching) or they may overwinter in the water and transform the following summer at larger size (more than a year after hatching; Bishop, 1947). Chandler (1918) and Gordon (1939) report that populations in Oregon may take two summers to complete metamorphosis. Chandler based his conclusion on his observations of two size classes of larvae at a locality in Corvallis, Oregon, suggesting that this may be due to the late breeding time at this locality. However, breeding there occurs from February to mid-July, which is not particularly late relative to other low-elevation populations. Two scenarios seem conceivable to explain these observations. One possibility is that there may be variation within the population, with some larvae metamorphosing after one summer (i.e., offspring of early-breeding females) and others taking two summers (i.e., offspring of late-breeding females). A second possibility is that the two size classes are a direct result of the long breeding season, with the larger larvae a product of early breeding and the smaller ones a product of late breeding. Even the offspring of late-breeding females may have time to metamorphose before the end of fall. At high elevations at Crater Lake, limited data suggest that two seasons of growth are necessary before metamorphosis (Farner and Kezer, 1953).

ii. Larval requirements. In the field, rough-skinned newt larvae were found to be most abundant at water temperatures ranging from 22–26 °C. When exposed to thermal gradients in the lab, larvae chose temperatures ranging from 17–28 °C (Licht and Brown, 1967).

a. Food. Little information is available on the diet of larvae. Pimentel (1952) reports that young larvae initially feed on protozoans, which they scrape off plants, rocks, and other objects in their habitat. Larvae slowly shift to larger food prey until the food-scraping behavior is no longer observed. Larval insects (chironomids and corixids) and small crustaceans (ostracods, copepods, and daphnids) were found in the stomachs of five larvae examined in Oregon (Chandler, 1918). Larger larval rough-skinned newts will eat smaller ones (Pimentel, 1952).

b. Cover. Larvae hide under stones or in vegetation during the day (Pimentel, 1952; Licht and Brown, 1967).

iii. Larval polymorphisms. None reported.

iv. Features of metamorphosis. Timing of metamorphosis varies over the extensive range of this species. Furthermore, even within the same population, larvae apparently either metamorphose in their first summer (around August) or their second summer (see "Length of larval stage" above). It is difficult to compare the size at metamorphosis in different populations because authors variously report snout–vent lengths, snout–pelvis lengths, or total lengths. Larvae at Marion Lake (300 m elevation), British Columbia, metamorphose at 23 mm snout–pelvis length (Efford and Mathias, 1969) or 25–27 mm SVL (Neish, 1970). Twitty (1935) collected larvae ≤75 mm in TL from low-elevation streams in the coast range of northwestern California. Larvae of some populations at high altitudes do not metamorphose until their second summer (June–July) at 70–75 mm TL (Bishop, 1947). Larvae at Crater Lake (1,860 m elevation) attained total lengths as great as 95 mm prior to metamorphosis (Farner and Kezer, 1953). Chandler (1918) reports that 3–4 wk are required to complete metamorphosis based on his observations of larvae in Corvallis, Oregon.

v. Post-metamorphic migrations. After metamorphosis, juveniles exit the water and move onto land, often migrating considerable distances (Chandler, 1918; Twitty, 1955; Efford and Mathias, 1969). Pimentel (1952) suggests that juveniles quickly seek subterranean retreats following emergence and remain there. The length of time they remain on land before returning to the water is in dispute. Chandler

(1918) states that juveniles return to the water the following spring. By contrast, observations by Storm (1948), Pimentel (1952) and Efford and Mathias (1969) indicate that juveniles do not return to water the year after metamorphosis and remain terrestrial until mature. According to Twitty (1955), all species of *Taricha* remain on land until they reach sexual maturity in 3–4 yr.

vi. Neoteny. Adults in some populations retain gills or gill vestiges and remain in permanent bodies of water year-round. Bishop (1947) and Farner and Kezer (1953) described adults retaining varying amounts of external gill tissue at Crater Lake in Oregon; in all cases, these gills had less gill tissue than on an ordinary newt larva. Riemer (1958) reported adults with gill vestiges in San Mateo County, California, and in Latah County, Idaho. Nussbaum and Brodie (1971) described adults bearing gill remnants from Latah County, Idaho, and Kittitas County, Washington. Marangio (1978) found populations of paedomorphic rough-skinned newts at seven high-elevation lakes and ponds in the Cascade Mountains of southern Oregon. Unlike other reports in which the percentage of perennibranchiate individuals was relatively low, in several of these populations the proportion of individuals retaining gills or gill stubs ranged from 87–100%. Gills were in various stages of development, with most individuals possessing three pairs. Some individuals possessed remnants of larval gill arches and gill rakers. Gill rami were present, but fimbriae generally were lacking or few in number, presumably limiting the usefulness of these vestigial gills for respiration. When these animals were held in the laboratory in aged tap water at 20 °C, they resorbed their gill tissue, with or without exposure to exogenous thyroxin, suggesting that the cold water temperatures experienced in nature are important for gill retention. In general, perennibranchiate individuals seem to be associated with winter temperatures below freezing and bodies of water that are deep enough that they do not dry up during the summer (Nussbaum and Brodie, 1971; Marangio, 1978); the San Mateo County specimens are a notable exception to this generalization.

D. Juvenile Habitat. Juveniles are thought to lead a predominantly fossorial existence, and Pimentel (1952) notes that they are found under deeply embedded material in moist situations. However, juveniles are active on the ground surface during wet periods, based on observations that juveniles and adults were captured in terrestrial traps in roughly equal numbers (Twitty et al., 1967b).

E. Adult Habitat. In most populations, adults migrate seasonally between terrestrial and aquatic habitats. Terrestrial adults are found in a variety of habitats such as coniferous forests, redwood forests, oak-woodland, farmlands, and grassland. Terrestrial animals spend much of their time underground in burrows, and also may be found beneath cover objects such as logs, bark, or boards (Bishop, 1947; Stebbins, 1951; Nussbaum et al., 1983). Most adults can be found in underground retreats within 400 km (1/4 m) of ponds used for breeding (Pimentel, 1960). Adults may be found crawling in the open during fall rains (Stebbins, 1951; Pimentel, 1960). Rough-skinned newts are the most aquatic of the three species of *Taricha* (Twitty, 1942). Aquatic habitats include lakes, ponds, roadside ditches, and slow-moving portions of streams (Stebbins, 1951; Nussbaum et al., 1983), but bodies of water lacking surrounding vegetation are avoided (Pimentel, 1960). Rough-skinned newts may be found in bogs of relatively low pH (Pimentel, 1960). The amount of time spent in aquatic versus terrestrial habitats varies with permanence of the aquatic habitat, elevation, locality, and sex. The length of time spent in aquatic habitats generally differs between the sexes, with males spending more time in the water than females (Pimentel, 1960). Furthermore, in northwestern California and western Oregon, adults breeding in permanent lakes and ponds spend more time in aquatic habitats than those occupying temporary bodies of water. At permanent ponds, males spend about 10 mo in the water, whereas females spend about 8 mo. In temporary ponds, males were present for approximately 7 mo, while females were present for only 6 mo (Pimentel, 1960). Studies of lakes in different regions of British Columbia revealed substantial differences between localities with regard to migratory behavior. In southern Vancouver Island, adult males remain mostly aquatic throughout the year, whereas females leave the pond with the onset of fall rains, overwinter on land, and migrate in spring to breeding ponds (Oliver and McCurdy, 1974). This pattern contrasts with the findings of Efford and Mathias (1969) at Marion Lake, a permanent pond on the southwestern British Columbia mainland, in which males and females all leave the water in the fall to overwinter on land. At a few localities in California, Oregon, Washington, and Idaho, permanently aquatic, perennibranchiate individuals of both sexes have been reported, mostly at high-elevation sites (see "Neoteny" above). In California, adults using streams usually remain aquatic all year unless forced to leave when water flow rises during winter floods (Packer, 1961; Twitty, 1942).

F. Home Range Size. It is difficult to apply the concept of home range to this species, because adults in most populations spend part of their life in the water and part on land. To my knowledge, no investigations of home range size have been performed on terrestrial phase newts.

Under certain conditions (during and after the breeding season) aquatic newts form large aggregations, which may serve as a "home range" for the animals inhabiting them, in the sense that displaced individuals rapidly return to the site from which they were captured, traveling ≤ 550 m to do so (Efford and Mathias, 1969; Neish, 1970). However, this philopatry is of a temporary nature, because these concentration centers tend to shift over time. Mark–recapture studies of these aquatic newts indicate that movements of individuals from one area to another is typical; most recaptured individuals showed at least one location shift, and at least one animal was caught in 7 (of a total of 12) areas in the lake over a 21-wk period (Neish, 1970).

G. Territories. Territoriality has not been reported for this species.

H. Aestivation/Avoiding Desiccation. Terrestrial newts spend the summer beneath surface cover objects, in rock crevices, inside decaying logs, and in other animals' burrows (Stebbins, 1985), during which time they are thought to achieve a physiological state of quiescence (Pimentel, 1960). Subterranean retreats offer lower environmental temperatures and higher humidity relative to surface environments. This coupled with a reduced metabolic rate (associated with quiescence) allows for a decreased rate of evaporative water loss through the salamander's skin and lungs. The critical thermal maximum (CTM) for rough-skinned newts is about 36 °C (based on laboratory specimens acclimated at 20 °C; Hutchison, 1961). At the CTM, locomotion becomes disorganized and animals lose their ability to escape from life-threatening situations; from an ecological and evolutionary perspective, the CTM is the lethal temperature.

I. Seasonal Migrations. Based on studies of several low-elevation permanent and temporary ponds in western Oregon and northwestern California, Pimentel (1960) made the following generalizations regarding seasonal migrations in rough-skinned newts. In most populations, individuals breed every other year and there is no evidence that non-reproductive animals migrate to water; these individuals remain on land for approximately 18 mo between breeding periods. Reproductive adults exhibit four basic types of movements: (1) sporadic movements following emergence from subterranean retreat sites; (2) migration to aquatic breeding sites; (3) wandering movements between aquatic and surrounding terrestrial habitats; and (4) post-reproductive migration to subterranean retreats.

i. Sporadic movements. Rainfall is the primary stimulus that initiates emergence from subterranean retreats. After surfacing, adults exhibit sporadic and apparently non-directional movements that are estimated to last from 6–8 wk and may be associated with foraging. Juveniles and non-reproductive adults may be involved in this period of sporadic activity. For reproductive individuals, sporadic movements are followed by a directional migration to breeding sites.

ii. Migration to breeding sites. Rainfall is the triggering mechanism initiating movement towards ponds. However, temperatures must not be too low, or movement will not occur. Newts will not migrate toward water until they have achieved a certain degree of sexual development in preparation for mating. The final approach of animals to ponds is not random but occurs along definite "highways" or preferred entrance routes. These paths may extend for as long as 400 km (0.25 mi), but length varies with locality; distances of ≤ 183 m (200 yd) are probably more typical.

iii. Wandering movements. Once in the water, some individuals exhibit *wandering*, which involves exit from and return to the breeding habitat. The reasons for this behavior are unknown. These movements are usually limited to short distances from the ponds, and the duration of time spent on land is generally brief (only lasting several days), but one male remained out of the pond for as long as 51d. Wandering movements tend to occur in greater frequency during periods of warm rainfall. A similar percentage of males and females engage in wandering movements, but males tend to wander more frequently and for a greater duration of time than females.

iv. Post-reproductive migration. Post-breeding migrations from breeding sites to terrestrial retreats are associated with high water temperature and lowering water level. Both factors are important for animals in temporary ponds, but only the former is relevant for animals in permanent bodies of water. These final exits from the pond occur later in spring and are shorter in duration than the breeding migration to the ponds. Vision and movement toward dark horizons are thought to be important mechanisms for locating underground retreats (Pimentel, 1960).

J. Torpor (Hibernation). The activity of rough-skinned newts is affected by cold temperatures in aquatic and terrestrial habitats. Movement does not occur at air or water temperatures below 5 °C (Pimentel, 1960). Nussbaum and Brodie (1971) reported finding torpid newts at the bottom of an Idaho pond in early spring when water temperatures ranged from −1 °C to +1 °C, suggesting that adults probably overwinter in the pond in a torpid state. By contrast, in the Cascade Range, most adults leave the water after spawning and move to underground retreats to hibernate during winter (Chandler, 1918). At low elevations in Oregon, the behavior is a bit different. Adults emerge from the water in October–November and wander about on land, then curl up in cavities under stumps, logs, or stones in November–December to spend the cold part of the winter, sometimes forming aggregations of > 12 animals. They emerge from their underground retreats on warm days to forage. Chandler (1918) speculates that in the northern portion of their range, rough-skinned newts leave the water to escape being frozen into small pools in their aquatic habitats.

K. Interspecific Associations/Exclusions. Over parts of their range in northwestern California, rough-skinned newts are sympatric with California newts *(T. torosa)* and/or red-bellied newts *(T. rivularis)*, with all three species co-occurring in Sonoma and Mendocino counties. Their interactions are complex, may be correlated with habitat and topography, and vary even over a fairly small geographic area. Twitty (1942) made frequent observations over many years at several different sites in the vicinity of Ukiah, Mendocino County. In this area, the Russian River flows through a narrow valley and receives tributaries from mountains to the east and west. Rough-skinned newts and red-bellied newts were found coexisting in and around mountain brooks flowing from the western side of the valley at certain times of year. Both species entered the water for breeding around the same time, but rough-skinned newts mated and laid their eggs in slower stream sections. Interestingly, California newts were not observed in these streams, even though they contained microhabitats similar to those chosen for spawning by this species in other parts of its range. By contrast, at a nearby site on the eastern side of the valley, rough-skinned newts and California newts were found together in two adjacent artificial reservoirs supplied by a small tributary of the Russian River. Only one adult male red-bellied newt was seen here, even though portions of the stream seemed suitable for this species. At a small spring-fed pool north of Ukiah, only rough-skinned newts were found, even though California newts use these same types of habitats in other parts of their range. This pattern of exclusion of California newts by rough-skinned newts has also been reported at ponds in other parts of their range.

Where their distributions overlap, rough-skinned newts and paedomorphic northwestern salamanders *(Ambystoma gracile)* are frequently syntopic in ponds and lakes. Their interactions have been studied in the Cascade Mountains in Oregon and at Marion Lake, British Columbia. These species are similar in their habitat use and diets (Efford and Mathias, 1969; Neish, 1971; Efford and Tsumura, 1973; Taylor, 1984), so there is potential for competition between them. In the Oregon Cascades, abundance of rough-skinned newts was not related to abundance of northwestern salamanders; however, in lakes with high densities of northwestern salamanders, rough-skinned newts were significantly smaller in length and weight relative to individuals from lakes with low densities of northwestern salamanders. If competition between these two species is occurring, there is neither evidence of resource partitioning nor numerical release when competition is reduced, counter to the predictions of competition theory (Taylor, 1984). In Marion Lake, habitat use by northwestern salamanders and rough-skinned newts was similar, but the distributions were different in that the former species had a widely dispersed, stable distribution, while the latter had a temporally unstable and highly contagious distribution, with individuals frequently forming large aggregations (Neish, 1971). Trout frequently co-occur in lakes with rough-skinned newts and northwestern salamanders, but the abundance of rough-skinned newts is not related to the presence of trout, perhaps not surprisingly since trout do not ordinarily prey on rough-skinned newts (but see "Feeding Behavior" below; Taylor, 1984). Rough-skinned newts also may associate with long-toed salamanders *(A. macrodactylum)*; these two species were found together under rocks and driftwood along the shore of Crater Lake in the Oregon Cascades (Farner and Kezer, 1953) and they have been reported to co-occur in a pond in Benton County, Oregon (Pimentel, 1960). It is likely that they coexist in other parts of their range where they overlap. Stomach content analysis of shoreline Crater Lake specimens reveals that these 2 species have different feeding habits, at least during terrestrial existence (Farner, 1947).

L. Age/Size at Reproductive Maturity. Rough-skinned newts are thought to be reproductively mature at 4–5 yr (Chandler, 1918; Efford and Mathias, 1969). Size at reproductive maturity varies considerably between populations, with adult SVLs ranging from 5.6–8.7 cm (Stebbins, 1985). Males tend to be larger than females (Neish, 1970; Nussbaum et al., 1983; Taylor, 1984). Body size may be linked to mating success; males in amplexus or attempting to gain amplexus were found to be significantly larger than males not engaged in sexual activity (Janzen and Brodie, 1989). During the breeding season, males undergo substantial morphological change, while little change is observed in females. Breeding males develop smooth, turgid, lighter-colored skin; swollen cloacal lips; prominent tail crest; and nuptial excrescences on the underside of hands and feet (Oliver and McCurdy, 1974). In breeding females, the skin is smoother and lighter-colored relative to the terrestrial phase, the cloacal lips appear as a small conical elevation, and the tail crest is not as pronounced as it is in males (Riemer, 1958). The frequency of reproduction varies among populations. Adults breeding at Oak Creek pond and Peavy Arboretum pond near Corvallis, Oregon, reproduce on alternate years (Pimentel, 1960). However, at Marion Lake in British Columbia, adults are thought to reproduce annually (Efford and Mathias, 1969).

M. Longevity. The average longevity is estimated to be 12 yr, based on the size of animals and growth rates calculated from mark–recapture studies at Marion Lake, British Columbia (Efford and Mathias, 1969).

N. Feeding Behavior. Adults feed mostly on soft-bodied, slow-moving prey, though they also may feed on sedentary animals (Chandler, 1918). Direct observation and examination of stomach contents indicate that most feeding takes place at night (Chandler, 1918; Efford and Mathias, 1969; Neish, 1970), but at least in some localities, feeding takes place throughout the day (Kelly, 1951). Adults will approach a prey item slowly and deliberately, and then quickly snap at it (Chandler, 1918). Suction feeding is used for capture of most aquatic prey, though large prey may be grasped with the jaws (Neish, 1970). Visual, tactile, and olfactory cues are important for locating and capturing prey (Chandler, 1918; Neish, 1970). Based on observations made during feeding experiments, Chandler (1918) concluded that conspicuous movement of the prey is important in eliciting feeding behavior, but that individuals may also learn to identify the forms of particular prey items and search specifically for those prey. Olfactory cues are used for locating high concentrations of prey (Neish, 1970), such as hatchling tadpoles. The adult diet consists of a wide range of aquatic and terrestrial invertebrates as well as amphibian eggs and larvae. Invertebrate prey include crustaceans (especially amphipods, copepods, and ostracods), insects, arachnids (spiders and mites), mollusks (gastropods, small freshwater bivalves), annelids (oligochaetes and leeches), and freshwater sponges (Chandler, 1918; Farner, 1947; Evenden, 1948; Packer, 1961; White, 1977; Taylor, 1984). Vertebrate prey consists of amphibian eggs and larvae (Chandler, 1918; Evenden, 1948; Pimentel, 1952; Neish, 1971; Nussbaum et al., 1983; Blaustein et al., 1995b; Rathbun, 1998), including eggs of conspecifics, long-toed salamanders, northwestern salamanders, northern red-legged frogs *(Rana aurora)*, Pacific treefrogs *(Pseudacris regilla)*, and western toads *(Bufo boreas)*, conspecific larvae, and tadpoles of northern red-legged frogs and foothill yellow-legged frogs *(R. boylii)*. Small fishes may be eaten on the rare occasions when they can be caught; upon being collected, a rough-skinned newt regurgitated a 7.5-cm- (3-in-) long rainbow trout (Pimentel, 1952). Algae and plant matter

are found in stomachs (Evenden, 1948; Kelly, 1951), but apparently these are not digested and probably are consumed incidentally with animal prey (Kelly, 1951). Although rough-skinned newts have been considered generalist carnivores, feeding on any soft-bodied prey they encounter and which will fit in their mouths (Pimentel, 1952; White, 1977), there is some evidence for variation between adults regarding food preferences. Chandler (1918) found variation in stomach contents between individuals collected at the same time from the same stream pool; for example, at a couple of localities, most individuals fed on small, shelled mollusks, whereas a few individuals fed almost exclusively on insects. However, in the absence of information on sample sizes or statistical analysis, it is difficult to evaluate the significance of these findings. There are no data concerning the diet of juveniles, but presumably it is similar to that of larvae (see "Food" under "Larval requirements" above) and adults.

O. Predators. As a consequence of being highly toxic, rough-skinned newt adults have few predators. Their only regular predators are common garter snakes *(Thamnophis sirtalis)* and rough-skinned newts themselves. Common garter snakes frequently prey on larval and adult newts (Storm, 1948; Farner and Kezer, 1953; Brodie, 1968b; Gregory, 1978; Nussbaum et al., 1983) and are resistant to the effects of tetrodotoxin (see "Anti-Predator Mechanisms" below). There is a single report of a juvenile Oregon gray garter snake *(T. ordinoides hydrophila)* having eaten a small newt larva (Fitch, 1936). Most snakes do not prey on rough-skinned newts. Nine of ten species of snakes refused to consume newts in laboratory trials (Brodie, 1968b). Rough-skinned newts regularly eat conspecific eggs and larvae (Chandler, 1918; Evenden, 1948; Pimentel, 1952; Nussbaum et al., 1983). There are scattered reports of newts being eaten by fishes, American bullfrogs *(R. catesbeiana)*, and birds, but in most of these cases the individuals that had consumed the newts were found dead, presumably due to the effects of tetrodotoxin. Newts have been found in the stomach of black bass (probably *Micropterus salmoides*), rainbow trout *(Oncorhynchus mykiss)*, and catfish (Siluriformes; Chandler, 1918; Vincent, 1947; Brodie, 1968b). The catfish was found dead with the tail of an adult newt protruding from its mouth (Brodie, 1968b). Apparently, at least in some localities, trout regularly consume newts without lethal consequences. There are several reports of rainbow trout caught with rough-skinned newts in their stomachs (Vincent, 1947; Pimentel, 1952; Twitty, 1966) suggesting that at least some rainbow trout are resistant to ingestion of tetrodotoxin. By contrast, in Marion Lake in British

Columbia, newts were never found in the stomachs of rainbow trout, in spite of detailed investigations of the diets of these fishes (Efford and Mathias, 1969). Newts have been used as bait by black bass fisherman (Twitty, 1966), but whether these fish regularly feed on newts under natural conditions and their degree of resistance to tetrodotoxin is unknown. An American bullfrog was found dead with an adult newt in its stomach, but bullfrogs are nonnative in Oregon, and because these frogs had only recently been introduced into the pond at the time the predation event occurred, this species probably had no previous experience with rough-skinned newts (Brodie, 1968b). Adult newts have been found in the stomachs of dead ducks (Storm, 1948; Nussbaum et al., 1983) and dead pied-billed grebes *(Podilymbus putainodiceps*; McAllister et al., 1997); chickens *(Gallus domesticus)* have been reported dead after consuming newts (Storm, 1948). Humans have occasionally been known to consume newts (under the influence of alcohol and peer pressure) and are highly susceptible to the effects of tetrodotoxin, sometimes with lethal results (Brodie et al., 1974b; Bradley and Klika, 1981).

P. Anti-Predator Mechanisms. Rough-skinned newts possess a variety of anti-predator mechanisms that include body posture, aposematic ventral coloration, and chemical defense. Adults are dark-colored dorsally and bright orange ventrally. The brown or black dorsum is cryptic in many terrestrial and aquatic situations. When threatened by a predator, adults display a characteristic body posture called the *unken reflex*, a rigid U-shaped posture that reveals the bright orange ventral coloration (see Petranka, 1998). During the unken reflex, eyes are closed, limbs extend laterally, the head is raised vertically, the back is depressed, and the tail is raised forward over the body (Stebbins, 1951; Riemer, 1958; Johnson and Brodie, 1975; Brodie, 1977). A release of toxic skin secretions accompanies the defensive posture. The brightly colored venter is a warning to predators that the newt is toxic and unpalatable. The unken reflex is an aposematic cue that elicits avoidance by birds; it has been demonstrated experimentally that the posture plus ventral coloration is a stronger aposematic cue than ventral coloration alone (Johnson and Brodie, 1975).

If the predator is not deterred by the unken reflex, the newt must rely on chemical defense. Adults possess tetrodotoxin (sometimes referred to as tarichatoxin), a potent neurotoxin that is concentrated in the skin, ovaries, muscles, and blood of adults (Wakely et al., 1966, based on studies of California newts). Tetrodotoxin is one of the most toxic non-protein substances known, and it also occurs in pufferfishes and relatives (suborder Tetraodontoidae;

Buchwald et al., 1964; Mosher et al., 1964). The skin from adult rough-skinned newts is several times more toxic than skin from other species of *Taricha* (Brodie et al., 1974b). Tetrodotoxin is also present in the eggs of all species of *Taricha* at levels of toxicity equal to that of the skin of adult California and red-bellied newts (Mosher et al., 1964; Brodie et al., 1974b). Nussbaum et al. (1983) state that tetrodotoxin is present in the skin of larvae, but this conflicts with studies on California newts that conclude that larvae possess little or no toxin (Twitty, 1937). Brodie (1968b) studied the effects of tetrodotoxin on a variety of potential predators and found that all test subjects force-fed adult newt tissue (amphibians, birds, and reptiles) or injected with skin extract (mammals) are susceptible to the toxin, exhibiting symptoms including muscular weakness, loss of righting reflex, convulsions, gasping, gaping, regurgitation, flaccid paralysis, decrease in blood pressure, and continuous heartbeat after cessation of respiration. White mice were killed in <10 min by as little as 0.0002 ml of skin extract taken from the dorsum of rough-skinned newt adults; other mammals and birds are susceptible to similar relative amounts of toxin. Rough-skinned newts were susceptible to tetrodotoxin, but only at high doses. Most snakes were roughly 200 times more resistant than white mice, and garter snakes were 2,000 times more resistant. In contrast to snakes, southern alligator lizards *(Elgaria multicarinata)* were highly susceptible to tetrodotoxin.

Different species of garter snakes on Vancouver Island vary with respect to their susceptibility to tetrodotoxin. When force-fed whole newts, western terrestrial garter snakes *(T. elegans)* and northwestern garter snakes *(T. ordinoides)* showed apparent loss of motor function, though effects were non-lethal and recovery was complete within 3 d. By contrast, most common garter snakes were unaffected, though there was variability between individuals with regard to tetrodotoxin susceptibility (Macartney and Gregory, 1981). Brodie and Brodie (1990) also reported intrapopulational variation in tetrodotoxin resistance in a population of common garter snakes in Oregon that feeds on newts, and they showed that the degree of tetrodotoxin resistance has a genetic basis. There is also interpopulational variation in tetrodotoxin susceptibility. Common garter snakes from a population co-occurring with rough-skinned newts were much more resistant to tetrodotoxin than those from a population outside the range of the newt (Brodie and Brodie, 1990). There is considerable variation in the degree of toxicity of rough-skinned newts in different parts of their range. Skin extracts from Vancouver Island newts were at least 1,000 times less toxic than those of newts from the Willamette Valley of Oregon. The degree of tetrodotoxin resistance

in populations of common garter snakes is roughly correlated with the degree of toxicity of co-occurring populations of rough-skinned newts. Common garter snakes are highly tetrodotoxin-resistant in the Willamette Valley, where newts are highly toxic; whereas on Vancouver Island, common garter snakes have relatively low tetrodotoxin resistance, only slightly higher than conspecifics occurring outside the range of rough-skinned newts (Brodie and Brodie, 1991).

Q. Diseases. None have been reported.

R. Parasites. Adults are parasitized by protozoans, trematodes, nematodes, acanthocephalans, and leeches. Most of the accounts simply report the presence of a particular parasite; only a few studies attempt to correlate the presence of these parasites with the biology of the host.

i. Protozoans (Kingdom Protista). Several flagellates have been reported. *Trypanosoma ambystomae* and *Trypanosoma granulosae* have been found in the blood of rough-skinned newts from Linn County, Oregon, and Sonoma County, California, respectively (Lehmann, 1955, 1959). *Hexamita ovatus*, *Karotomorpha swezi*, and *Tritrichomona augusta* were found in the cloaca in specimens from Marin and Sonoma counties, California (Lehmann, 1960).

ii. Trematodes (Kingdom Animalia, Phylum Platyhelminthes, Class Trematoda). Infestations of the trematode *Ophioxenos microphagus* (= *Megalodiscus microphagus*) have been reported at a few localities in Oregon, Washington, and British Columbia (Macy, 1960; Efford and Tsumura, 1969; Moravec, 1984; Beverly-Burton, 1987). The aquatic snails *Menetus cooperi* and *Gyraulus* sp. serve as intermediate hosts (Macy, 1960), and cercariae emerge from the snails and encyst on the skin of the adult newts. Rough-skinned newts shed their skin in one piece and eat it, thereby ingesting the cysts. Metacercariae emerge in the digestive tract and are found from foregut to cloaca, with the highest concentration in the rectum. Infection occurs mainly in the spring, but continues throughout the aquatic phase of the adult newt's life cycle. This fluke also is found in other species of aquatic amphibians that coexist with rough-skinned newts, but infection rates are much higher in rough-skinned newts, both in terms of the percentage of individuals infected (varying seasonally from 67–100%) and the number of parasites/infected individual (as many as 77 cysts in the gut of one female). However, almost no sexually mature flukes are found in the gut of rough-skinned newts, although large, egg-bearing adults are found in other amphibian species at the same localities, suggesting that rough-skinned newts somehow prevent growth and development of this parasite (Efford and Tsumura, 1969). Sexually mature speci-

mens of another trematode, *Cephalouterina dicamptodonti*, were found in adult rough-skinned newts in British Columbia (Efford and Tsumura, 1969). *Megalodiscus americanus* was found in the rectum of aquatic adults from Benton County, Oregon, and *Brachycoelium salamandrae* was found in the intestines of a terrestrial adult from Marion County, Oregon (Lehmann, 1954).

Ribeiroia ondatrae was recently reported to be present in rough-skinned newts (Johnson et al., 2002). For infected animals, the number of metacercariae/individual ranged from 1–41, with a mean of 12.7. Metacercariae were located subcutaneously, at the base of fore- and hindlimbs, and among the gills or within the ventral head musculature. Aquatic snails in the genus *Planorbella* are first intermediate hosts for *Ribeiroia*, hosting redia and cercaria stages. Although *Ribeiroia* infection has been shown to cause morphological abnormalities in some species of amphibians, its role in causing malformations in rough-skinned newts is somewhat ambiguous. Morphological abnormalities, such as missing limbs and digits, were found in fairly low frequencies (range 4.2–7.4%) both at sites with and without *Ribeiroia*; however, abnormalities significantly exceeded 5% only at sites where *Ribeiroia* was present (Johnson et al., 2002).

iii. Nematodes (Phylum Nematoda) and Acanthocephalans (Phylum Acanthocephala). The nematodes *Megalobatrachonema moraveci* were found in the intestine of adult rough-skinned newts from Vancouver Island; the authors do not indicate if the newts were aquatic or terrestrial when collected. Half of the 12 specimens collected were infected, and the mean intensity of infection was 5.8 worms/host (Richardson and Adamson, 1988). Examination of aquatic and terrestrial adult newts from Benton and Marion counties in Oregon revealed the nematodes *Cosmocercoides dukae* in the intestine and rectum and *Hedruris siredonis* in the stomach (Lehmann, 1954). In the course of stomach content analyses for the purpose of ascertaining diet, parasitic nematodes (*Hedruris* and *Cosmocerca* sp.) and acanthocephalans (*Neoechinorhynchus* sp.) were discovered in the stomach of adult rough-skinned newts (Chandler, 1918; Taylor, 1984).

iv. Leeches (Phylum Annelida, Class Hirudinea). Kelly (1951) found leeches (family Glossiphonidae) in the mouth cavity of rough-skinned newts. In most cases only a single leech was found per animal, but multiple parasites were found in several cases, with one newt infected with 44 leeches. Infection was seasonal, temporary, and widespread, occurring only between May–July; about 90% of the newts examined were infected with leeches during this period.

4. Conservation.

Timber harvesting has the potential to directly impact rough-skinned newt populations, but most studies have been unable to draw strong conclusions regarding the effects of these practices on rough-skinned newts. To my knowledge, there have been no studies on the long-term effects of timber harvest practices on rough-skinned newts, but two studies have attempted to elucidate short-term effects (Cole et al., 1997; Grialou et al., 2000). Conversion of hardwood stands to more profitable Douglas-fir (Pseudotsuga menziesii) stands has been a common practice on federal and private lands in the Oregon Coast Range. For this conversion, hardwoods are typically clearcut, and then the site is burned before planting with Douglas-fir seedlings. During the first few years after planting, herbicides such as glyphosate may be applied to control competing vegetation. Cole et al. (1997) investigated the effects of clearcut logging, broadcast burning, and application of glyphosate on amphibian populations in red alder (Alnus rubra) sites in the Oregon Coast Range. In this study, three treatments were applied at each of three sites: (1) control (uncut); (2) clearcut and burned; and (3) clearcut, burned, and sprayed with glyphosate. Sites were sampled 1 yr before, and 1 and 2 yr after treatment, using pitfall traps. This study did not detect any short-term effects of logging, burning, or glyphosate application on rough-skinned newts; there were no significant differences in capture rates before and after treatments. However, there were relatively low capture rates and high variability in the capture rates between sites, resulting in low statistical power of these comparisons. As a result, there may have been effects that the study design and sampling techniques were unable to detect (Cole et al., 1997). Gomez (1993) demonstrated that newts tend to be less abundant in Douglas-fir stands (regardless of stand age) than in deciduous stands; based on these results Cole et al. (1997) suggest that conversion of red alder stands to Douglas-fir may result in long-term declines in newt populations. Grialou et al. (2000) investigated the short-term effects of clearcutting and thinning on terrestrial salamanders, including rough-skinned newts, in southwestern Washington. Using pitfall traps, they captured rough-skinned newts in both forested areas and clearcuts, but did not capture them in sufficient numbers to conduct statistical tests to determine the response of newts to clearcut harvesting or thinning (Grialou et al., 2000). Both these studies attempted to look at the response of multiple species of amphibians to timber harvest practices, and their sampling techniques and study design may not have been especially well-suited to detecting rough-skinned newts. The life history of this species complicates sampling, since there are restricted periods of time when the animals are terrestrial and above ground, and the number of active individuals is highly variable over time in response to rainfall and temperature patterns. Because pitfall traps sample migratory animals, sampling periods ideally would coincide with peak newt activity periods. Knowledge of local newt migratory pathways would be useful in establishing the placement of pitfall traps.

In the Pacific Northwest, nitrogen fertilizers (e.g., granular urea) are commonly used in commercial timber production to maintain or boost timber production. Urea is commonly applied by helicopters, and potentially could have negative effects on forest amphibians through dermal exposure. Two studies have demonstrated that rough-skinned newts are less sensitive to urea exposure than some other species of forest amphibians. Laboratory experiments demonstrated that adult rough-skinned newts will avoid exposure to urea, yet when newts were exposed to urea for 4 d, at concentrations comparable to recommended application rates for fertilization, no mortality occurred. By contrast, these same urea doses had acute effects on western red-backed salamanders (Plethodon vehiculum) and southern torrent salamanders (Rhyacotriton variegatus), resulting in substantial mortality (60% and 40%, respectively, at the higher of the two urea doses after only 12 hr of exposure; Marco et al., 2001). Hatch et al. (2001) examined the effects of urea exposure on the survival and feeding behavior of four species of newly metamorphosed terrestrial amphibians. Neither mortality nor reduced prey consumption was observed in juvenile rough-skinned newts as a result of a 5-d exposure to urea at an application rate consistent with reported field application rates. Similar results were obtained for long-toed salamanders, but juvenile western toads and Cascade frogs (R. cascadae) exposed to urea suffered substantial mortality and consumed significantly fewer prey items than non-exposed individuals.

Several laboratory studies have investigated potential sublethal effects of UV-B radiation on rough-skinned newts. Belden (2002) demonstrated that the skin of rough-skinned newt larvae darkens in response to short-term exposure (i.e., 5 d) to UV-B; however, whether skin darkening provides protection from UV-B damage remains unclear. A related experiment showed that longer term exposure (i.e., 3 wk) to UV-B led to decreased growth rates in northwestern salamander and long-toed salamander larvae (Belden, 2002). Although Taricha larvae were not investigated in this regard, it is probable that UV-B has similar effects on rough-skinned newts. Blaustein et al. (2000) used rough-skinned newts as a model to examine potential effects of UV-B radiation on amphibian behavior. Adult newts were exposed to low-level doses of UV-B in the laboratory for 14 d, and then tested in the field, where treatment animals were compared to control (i.e., UV-B blocked) animals with regard to (1) orientation towards water and (2) locomotor activity levels. There was no significant difference in orientation between control and treatment animals; both showed significant orientation towards water. However, newts exposed to UV-B were significantly more active than control animals. This increased activity may be a stress response to UV-B exposure, related to an increase in circulating levels of the hormone corticosterone (Blaustein et al., 2000; Belden et al., 2001). Kats et al. (2000) investigated the effects of short-term (i.e., 18 d) exposure to UV-B radiation on the anti-predator behavior of rough-skinned newt larvae. Both UV-B exposed and control (UV-B blocked) larvae responded to chemical cues from conspecific adult predators by increasing the amount of time spent in shelter. UV-B exposed larvae showed a reduced response to conspecific predator cues relative to non-exposed larvae, but these differences were not statistically significant. These studies suggest that UV-B exposure may negatively affect rough-skinned newts, and there is concern that these impacts may be exacerbated by human or natural disturbances (e.g., forest clearcutting, fire) that decrease canopy cover, potentially exposing eggs, larvae, juveniles, and adults to increased doses of ambient UV-B radiation (Blaustein et al., 2000; Kats et al., 2000). Further studies on sublethal effects of UV-B on rough-skinned newts and other amphibians are warranted.

Life history patterns should be considered when developing management or conservation plans for rough-skinned newt populations. Effective management plans must take into account the habitat requirements of all stages of the life cycle, including upland overwintering and foraging sites and migratory pathways between upland sites and aquatic breeding sites, as well as aquatic sites for eggs, larvae, and breeding adults. Information about local migration patterns may be critical to reducing mortality from various anthropogenic disturbances, such as increased vehicular traffic. In addition, the homing behavior of newts (i.e., high degree of site fidelity for breeding) most likely reduces the dispersal of individuals to alternate breeding sites, limits recolonization of sites (Blaustein et al., 1995b), and may increase the likelihood of extinction of local populations in the event of the loss or destruction of aquatic breeding sites.

Rough-skinned newts have been introduced into Idaho, where they are considered an Exotic species (www2.state.id.us).

Taricha rivularis (Twitty, 1935)
RED-BELLIED NEWT

Sharyn B. Marks, Darrin Doyle

1. Historical versus Current Distribution.

Red-bellied newts *(Taricha rivularis)* have the most limited geographical distribution among the three species of *Taricha*. They occur in coastal California north of San Francisco Bay, in Sonoma, Lake, Mendocino, and Humboldt counties, at elevations between 150–450 m. Their distribution is roughly similar to that of the coast redwood *(Sequoia sempervirens)*, but red-bellied newts are not restricted to redwood forests (Myers, 1942b; Stebbins, 1951; Riemer, 1958; Twitty, 1964a). There have been no studies comparing the present distribution with the historical distribution.

Red-bellied Newt *(Taricha rivularis)*

2. Historical versus Current Abundance.

Twitty (1935) described this species as being "very abundantly represented" in mountain streams in Mendocino County. Abundance of red-bellied newts has been quantified only for one locality, Pepperwood Creek (a tributary of the Wheatfield fork of the Gualala River) in Sonoma County. Twitty (1959) remarked that "Pepperwood Creek is a small stream, but it is literally crawling with newts during the breeding season." Using census data from Twitty (1961a, 1966), Hedgecock (1978) estimated that 58,000–60,000

breeding adults occurred in the roughly 2.5-km-long segment of Pepperwood Creek intensively studied by Twitty and collaborators between 1953 and 1961. This creek was chosen in part because of the high abundance of red-bellied newts, so it may not be representative of newt abundance at other localities. No data are available on the current abundance of this species at any locality.

3. Life History Features.

A. Breeding. Reproduction is aquatic.

i. Breeding migrations. Breeding migrations are initiated after a period of terrestrial foraging (see "Seasonal Migrations" below). Beginning in late January, movements become less random, and some individuals begin moving toward the stream; thereafter, the magnitude of the migration increases until it peaks in early

March (Packer, 1960). Red-bellied newts do most of their migrating during cloudy and rainy nights (Grant et al., 1968). In general, rainfall triggers the movement of animals toward the stream, but in heavy amounts it inhibits migration to streams. In the absence of rainfall, the magnitude of the migration is correlated with daily changes in mean evening temperature and minimum relative humidity; when both mean evening temperature and minimum relative humidity increase (relative to values of the previous day), captures increase (Packer, 1960). Adults enter streams

only after the winter floodwaters have receded (Twitty, 1942). Males enter the water a considerable time before females, whose aquatic phase is restricted to a short period during mating and oviposition (Twitty, 1942, 1955; Packer, 1963).

Twitty's (1959, 1961a, 1966) long-term mark–recapture studies at Pepperwood Creek demonstrated that red-bellied newts typically return to the same portion of the home stream for successive breedings, apparently throughout their life. Once in the stream, animals usually do not range >15 m (50 ft) from the original entry point (Packer, 1963). Studies of artificially produced hybrids *(T. rivularis* x *T. torosa* and *T. rivularis* x *T. t. sierrae)* suggest that "imprinting" on the home area takes place sometime early in the life cycle, either during larval life or just after metamorphosis (Twitty, 1961a). This highly developed homing mechanism promotes isolation and genetic divergence among demes in streams (Hedgecock, 1978). Gene flow is likely low between demes within the same stream and extremely low between demes in different streams.

Twitty and colleagues conducted a series of experiments in which they displaced animals from their native stream segments to other portions of the same stream, or to other streams within the watershed (Twitty, 1959, 1961a; Twitty et al., 1964, 1966). These displacements demonstrated the extraordinary homing abilities of red-bellied newts—some are capable of homing to their native stream segment even after displacement distances as great as 8.0 km (5 mi). Displaced animals make the homing journey by moving overland, either parallel to the stream (after intrastream displacement) or directly overland across intervening mountain ranges (after inter-stream displacement; Twitty, 1959; Twitty et al., 1966). Homing is mostly postponed until the animals leave the water at the end of the breeding season and may not really commence until the onset of the fall rains during the next breeding season (Twitty et al., 1964, 1966). In the case of inter-stream displacement, homing may be delayed following residence at the release site for ≥1 yr, but most displaced animals initiate homing the first year following displacement (Twitty et al., 1966). In a few instances, speed of homing has been documented; animals have homed for a distance of about 400 m (0.25 mi) in 1 d, and 4.0 km (2.5 mi) in <1 mo (Twitty et al., 1966).

Of the various senses that might be involved in homing, olfaction is more crucial than sight. Red-bellied newts do most of their migrating during cloudy and rainy nights when celestial cues are not available. Permanently blinded animals, displaced about 0.8 km (0.5 mi) to another part of the stream, are able to find their way back to their home segments with about the same degree of success as sighted animals (Twitty, 1959, 1961a). Experiments in which the olfactory system was damaged surgically or chemically demonstrated that olfaction plays a major role in the ability of red-bellied newts to home to their native stream segments after being displaced (Grant et al., 1968). Olfaction is critical for both initial orientation as well as eventual return to the home site. Grant et al. (1968) proposed that terrestrial odors from vegetation surrounding the home stream segment might provide long-term olfactory cues for homing. Other experiments, which artificially altered the topography of the land, demonstrated that kinesthetic sense is not the mechanism by which migration is oriented (Twitty, 1959, 1961a). Displaced individuals apparently begin their search using oriented (rather than random) movements. Most animals displaced for distances up to 12.9 km (8 mi) oriented their departure from the release site in the homeward direction. However, orientation accuracy decreased with increasing displacement distance (Twitty et al., 1967a).

ii. Breeding habitat. Breeding occurs in mountain streams, where the substratum is clean and rocky (Twitty, 1935; Stebbins, 1951). The initiation of breeding and length of the breeding season depend on weather and stream conditions. Adults enter the water soon after the streams begin to recede from winter floods. The aquatic phase may last from late February to May, though the bulk of breeding occurs in March to early April (Twitty, 1955, 1966; Stebbins, 1985). Heavy rains and consequent flooding may temporarily drive animals from the water, which may extend the breeding season beyond April.

B. Eggs.

i. Egg deposition sites. Oviposition occurs in mountain brooks primarily between mid-March and mid-April, though the oviposition period may span from early March to early May. Eggs are attached in flattened clusters (generally one egg-layer thick) to the undersides of stones, usually in the middle of the stream in swiftly flowing water; occasionally, flattened egg masses are attached to submerged roots along the side of the stream (Twitty, 1935, 1942). In suitable habitats, clutches may occur in great abundance; approximately 70 egg masses were found attached to the underside of a single stone, all within an area of 22 × 30.5 cm (8.5 × 12 in; Twitty, 1935).

ii. Clutch size. Mean clutch size is about 10 eggs (range 6–16 eggs), and mean egg diameter (measured at the blastula stage) is roughly 2.6 mm (range 2.4–2.7 mm; Twitty, 1935, 1964a; Riemer, 1958). Eggs typically are darkly pigmented (dark gray or grayish brown; though completely white eggs are not uncommon) and develop into normally pigmented larvae (Twitty, 1964a). As with other amphibian species, time from oviposition to hatching

varies with environmental temperature. The incubation period of eggs raised in the laboratory at 15 °C ranged from 30–34 d, while eggs raised at 23 °C took only 16–20 d; developmental rates in the field were intermediate between these values (Licht and Brown, 1967). The total length of hatchlings ranges from 10.3–11.0 mm (Riemer, 1958) or 10.5–12.7 mm (mean = 11.8 mm; Twitty, 1964a).

C. Larvae/Metamorphosis.

i. Length of larval stage. The larval stage (from hatching to metamorphosis) lasts about 4–6 mo (Twitty, 1964a,b; Licht and Brown, 1967). Unlike congeneric larvae, pigmentation is distributed uniformly over the dorsum and sides (Twitty, 1935, 1936, 1942, 1964a; Riemer, 1958). Larvae possess a dorsal tail fin that is reduced and does not extend as far anteriorly on the trunk as that of congeners; balancers are either absent or rudimentary in hatchlings (Twitty, 1935, 1936, 1942, 1964a; Riemer, 1958). The more streamlined dorsal tailfin and absence or reduction of balancers have been interpreted as adaptations to a mountain brook habitat (Noble, 1931; Stebbins, 1951; Twitty, 1966). Hindlimb development is considerably advanced at hatching relative to the other species of *Taricha*; at the time of independent feeding, the digits on the forelimb are distinctly forming (Twitty, 1936). This also is probably an adaptation to life in moving water. Petranka (1998) categorized red-bellied newt larvae as "pond-type," but his rationale for this designation is unclear.

ii. Larval requirements.

a. Food. No data are available on larval diet. However, like other larval salamanders (such as rough-skinned newts; *T. granulosa*), red-bellied newt larvae are undoubtedly generalist carnivores, consuming nearly any appropriate prey item that will fit in the mouth.

b. Cover. Stones and vegetation in the water provide retreat sites for larvae during the day. In the field, larvae were most abundant in microhabitats with temperatures between 22–26 °C (Licht and Brown, 1967).

iii. Larval polymorphisms. None reported.

iv. Features of metamorphosis. Metamorphosis occurs during late summer or early fall when larvae reach 45–55 mm in total length (Twitty, 1935, 1936, 1942, 1961a, 1964a; Licht and Brown, 1967). The duration of metamorphosis has not been quantified; however, Twitty (1964a) reported finding abundant larvae at the beginning of September, and only few larvae (all in the process of metamorphosing) <6 wk later in Pepperwood Creek (Sonoma County, California). So, the metamorphic period for that population extends over about 6 wk, but the duration of metamorphosis for individual animals has not been reported. There is no evidence that larvae overwinter before transforming (Riemer, 1958; Twitty, 1964a).

v. Post-metamorphic migrations. Little is known about post-metamorphic migrations except that juveniles leave the stream and go into hiding underground just after metamorphosis (Twitty, 1955, 1961a, 1966; Twitty et al., 1967b; Petranka, 1998).

vi. Neoteny. Not reported.

D. Juvenile Habitat. Juveniles are thought to remain in underground shelters almost continuously from the time of metamorphosis to sexual maturity (estimated to be approximately 5 yr or longer). During a 3-yr study at Pepperwood Creek, few juveniles were observed or captured in land traps during the rainy season when large numbers of adults were evident. Of 8,919 animals captured in land traps, only 1.7% of these were juveniles. This contrasts with the population of rough-skinned newts at Pepperwood Creek, in which adults and juveniles were captured in about equal numbers (Twitty, 1966; Twitty et al., 1967b).

E. Adult Habitat. Adults migrate from terrestrial to aquatic habitats seasonally for breeding. There are no detailed descriptions of terrestrial habitats, and what information is available is somewhat inconsistent between sources. Several sources (Myers, 1942b; Twitty, 1964; Riemer, 1958) state that this species' range is confined to the coast redwood belt, but Riemer (1958) notes that red-bellied newts are not restricted to redwood forests, nor are they particularly abundant in that habitat. However, none of these authors specifically describe the terrestrial habitat for this species. Twitty (1966) comments that California laurel (*Umbellularia californica*) trees are common near his study site at Pepperwood Creek, but no other tree species are mentioned. Petranka (1998) states that red-bellied newts are found predominantly in redwood forests. We have observed terrestrial adults in forest dominated by Douglas-fir (*Pseudotsuga menziesii*), tan oak (*Lithocarpus densiflorus*), and madrone (*Arbutus menziesii*) in southern Humboldt County, and colleagues have seen them within redwood forest in Mendocino County (S. Sillett and J. Spickler, personal communication). Clearly, multiple forest types are used by this species. Adults use terrestrial sites for underground retreats during the dry season (May–October) and for foraging and migration prior to winter breeding. Both Twitty (1966) and Licht and Brown (1967) mentioned that red-bellied newts at their study sites (Pepperwood Creek and Skaggs Springs, respectively, both in Sonoma County) were found on steep, heavily wooded slopes that rise from the south bank of the breeding stream (i.e., north-facing slopes). Packer (1960) noted that at Pepperwood Creek, the banks and north-facing slopes are littered with many fallen trees and branches that provide cover for red-bellied newts and other amphibians.

Aquatic habitats include streams and rivers; red-bellied newts apparently do not use ponds or other standing water habitats for breeding (Riemer, 1958; Stebbins, 1985; Petranka, 1998). Males tend to enter the streams before females and therefore spend more time in the aquatic habitat (Twitty, 1942, 1955; Packer, 1963). Males also tend to breed more frequently than females; males breed usually every 1–2 yr, whereas females usually breed only ≥ 2 yr. Consequently, females may spend several years on land before entering the water again for breeding (Twitty et al., 1964; also see "Age/Size at Reproductive Maturity" below).

F. Home Range Size. Estimates of home range size should take into account the regular movements of individuals between terrestrial and aquatic habitats. Home range size has not been calculated in this way, but there are quantitative data available on aquatic home range size, as well as qualitative data on terrestrial home range size. Mark–recapture studies (e.g., Twitty, 1959, 1961a, 1966) demonstrated that individuals return each breeding year to the same general region of a stream, within roughly 15 m (50 ft) of their original capture point. Once in the water, animals usually do not range >15 m from the original point of entry, although longer jaunts are sometimes taken (Packer, 1963). Based on terrestrial trapping of marked adults, Twitty et al. (1967b) concluded that the terrestrial home range encompasses only a small area of the hillside adjacent to the breeding site. These estimates of home range size are based only on data from a single locality (Pepperwood Creek, Sonoma County) and may not be representative of the species as a whole.

G. Territories. Based on mark–recapture experiments, Twitty (1961a) concluded that adults exhibit territorial behavior in the sense that each member of the population tends to return year after year to the same stream segment for reproduction. However, a demonstration of territoriality must include evidence of the defense of space against intruders. There is no evidence that adults defend their native stream segment from others, so red-bellied newts are probably not truly territorial.

H. Aestivation/Avoiding Desiccation. Adults and juveniles spend the dry season (summer and early autumn) in underground shelters, where temperatures are lower and humidity is higher than at the surface. Packer (1960) remarks that the physiological state of the animals during this period is unknown, so it is uncertain if the adults are aestivating. Gamete formation and even some yolk deposition apparently take place during this time.

I. Seasonal Migrations. Adults emerge from underground retreats in the fall, triggered by rainfall and hormonal activity associated with the reproductive drive (Twitty, 1942, 1959; Packer, 1960).

Typically, it takes several bouts of rainfall before the animals emerge, and as the season progresses, less rainfall is needed to trigger emergence. Not all animals appear at once; emergence is spread out over several months, until the middle of February (Packer, 1960). Salamanders forage on the forest floor before descending to the streams for breeding in the late winter or early spring (Twitty et al., 1964). Beginning in late January, some animals begin to move toward the stream (Packer, 1960). Adults do not enter the water until the streams begin to recede from the winter floods (Twitty, 1942; also see "Breeding migrations" above). If conditions are suitable, animals may begin to enter the water in late February to early March; they may not leave the water until early May. Heavy rainfall, increased stream volume, or increased sediment load, stimulate the animals to temporarily leave the stream (Packer, 1960). Breeding mostly occurs in March–April, provided that heavy rains and flooding do not interrupt the animals' activities too frequently (Twitty, 1966). At the end of the mating season, both sexes usually leave the water abruptly, over a period of only a few days (Riemer, 1958; Twitty, 1959; Packer, 1960). As animals leave the breeding stream, they move uphill, as well as at an angle carrying them in the upstream direction; the reason for the latter behavior is unclear, but may be a carry-over of orientation behavior in the stream, in which animals are oriented upstream (Twitty et al., 1967c). Animals may disperse relatively great distances uphill above the stream, perhaps as far as several hundred yards (Twitty, 1961a), but animals native to one stream confine their migrations to the watershed of that stream (Twitty et al., 1967b). Animals spend the dry summer and early autumn underground and are not seen again until the following fall (Twitty, 1966).

J. Torpor (Hibernation). There are no reports of hibernation in this species. Although freezing temperatures do occur occasionally in coastal northern California, long periods of freezing are unusual. Animals are known to be active above ground from October–February, and in the water from October–May. However, red-bellied newts are not immune to the effects of cold temperatures. Observations of aquatic adults in thermal gradients in the lab indicate that they are immobilized at body temperatures below about 2 °C (Licht and Brown, 1967).

K. Interspecific Associations/Exclusions. Coastal giant salamander (*Dicamptodon tenebrosus*) larvae and small trout are found in mountain brooks occupied by red-bellied newt adults (Twitty, 1935; Bishop, 1947). Where their ranges overlap, red-bellied newts and rough-skinned newts co-occur in mountain brooks. Both species breed in these streams, but the former

tends to breed in faster-moving water whereas the latter prefers slower portions of the stream. There is an overlap in the timing of breeding, although red-bellied newts have a shorter breeding season, entering and leaving the streams earlier than rough-skinned newts (Twitty, 1942). Interspecific amplexus is uncommon between the two species, and existence of hybrids in nature is extremely rare (Davis and Twitty, 1964). Laboratory experiments suggest that species-specific sex attractants released by the females may contribute to reproductive isolation between these species (Davis and Twitty, 1964). Although the geographic ranges of red-bellied newts and California newts (*T. torosa*) overlap in parts of Sonoma and Mendocino counties, these species never occur in the same streams (Twitty, 1942, 1955). In the laboratory, red-bellied newts may be hybridized with any of the other *Taricha* species using artificial cross-fertilization, producing viable and fertile hybrids (Twitty, 1936, 1955, 1959, 1961a); however, natural hybridization is rare between these species (Twitty, 1942, 1955; Hedgecock and Ayala, 1974).

L. Age/Size at Reproductive Maturity. Estimates of age at reproductive maturity range from 4–6 yr (Twitty, 1966) to 6–10 yr (Hedgecock, 1978). Adults range from 5.9–8.1 cm SVL (Stebbins, 1985) or 14–19.5 cm total length (Petranka, 1998). Sexual dimorphism during the breeding season is not as pronounced in red-bellied newts as it is in other species of *Taricha*. The skin of aquatic males is almost entirely smooth, the dorsal fin is not greatly enlarged, and the cloacal lips are not as swollen (Twitty, 1935). There is a difference in breeding frequency between the sexes. Not all individuals breed annually, though annual breeding is more common among males than females. Based on mark–recapture studies, Twitty et al. (1964) concluded that males will commonly, but not always, breed in immediately successive years. About 50–60% of sexually mature males marked one breeding season returned the next; most but not all of the remainder returned in 2 yr time. For females, the interval between breeding seasons varied from one to several years, but no more than about 2–3% of females bred in immediately successive years.

M. Longevity. Nearly 40% of the adult males marked in 1953 were still being caught at Pepperwood Creek in 1964 (Twitty, 1966), meaning that they must have been at least 17 or 18 yr old (assuming it takes about 6 yr to reach sexual maturity). Based on this, Hedgecock (1978) speculated that longevity probably ranges from 20–30 yr.

N. Feeding Behavior. With the onset of winter rains, adults emerge from their burrows to forage in the forest for a short period (Twitty, 1961a). Foraging is restricted to periods of high humidity, late in the day or at night, or during a rainfall (Twitty, 1966). Stomach content analysis of terrestrial and aquatic adults revealed that the diet consists of terrestrial organisms only, predominantly insects. Adults apparently do not feed during their aquatic phase, though they will resume feeding if forced back onto land by winter floods (Packer, 1961). Aquatic adults maintained in the lab also do not feed (Licht and Brown, 1967).

O. Predators. No predators have been specifically reported for red-bellied newts. However, Twitty (1966) reported that at his field station in Sonoma County, garter snakes "brazenly stole" newts from storage containers as he and his assistants looked on. Both red-bellied and rough-skinned newts occurred at the field station, but most of Twitty's work there was on red-bellied newts, so presumably this is the species to which he was referring. He did not identify the garter snakes to species, but most likely they were common garter snakes *(Thamnophis sirtalis)*, which are known to feed on rough-skinned newts (see rough-skinned newt species account for references).

P. Anti-Predator Mechanisms. The anti-predator mechanisms of red-bellied newts are similar to those described for rough-skinned newts: toxicity, aposematic coloration, and defensive posturing. Ovarian eggs and embryos of red-bellied newts contain high levels of tetrodotoxin (Mosher et al., 1964); all three species of *Taricha* show similar levels of toxicity in this regard (Twitty, 1937; Brodie et al., 1974b). The skin of adult red-bellied newts also contains tetrodotoxin. The toxicity of the back skin of adult female red-bellied newts is similar to that of California newts, but adult rough-skinned newts are considerably more toxic. Brodie et al. (1974b) estimated that 1,200–2,500 mice could be killed by the skin of red-bellied newts (as compared to approximately 25,000 mice for rough-skinned newts). Ovarian eggs and adult skin of red-bellied newts have similar toxicity levels. These high levels of tetrodotoxin render newts inedible to nearly all predators (Brodie, 1977). Animals are dark-colored dorsally and bright tomato-red ventrally. The dark dorsum is cryptic against the forest floor. The defensive posture exposes the aposematic coloration of the ventral surface. During the unken reflex, the tail and head are elevated vertically, revealing the bright red underside of the chin and tail; at the same time, toxic skin secretions are released. The defensive display varies somewhat depending on the state of the salamander or with the intensity of stimulation: the low intensity response is a U-shaped posture involving a vertical elevation of head and tail such that both point skyward. With a high intensity response, the head tips further back and the pelvis and hindlimbs are lifted off the substrate such that the

tip of the snout and the base of the tail almost come into contact. The unken reflex display is generally similar in all species of *Taricha*, but red-bellied newts show one striking variation in posture. Instead of elevating the tail, sometimes the pelvis and hindlimbs remain in contact with the ground, while the head, forelimbs, and body are together lifted from the substrate; this variation in posture may be a result of the longer tail of red-bellied newts counterbalancing the body (Brodie, 1977).

Q. Diseases. None reported.

R. Parasites. None reported.

4. Conservation.

The conservation status of red-bellied newts is uncertain, because no ecological studies have been conducted on this species for the last 35 yr or so. However, this species has a limited and somewhat spotty geographic distribution, and human population pressure has intensified considerably over much of its range. Specifically, conversion of native forests and grasslands to vineyards and subdivisions likely poses a serious threat to red-bellied newts. For example, this change in land use has led to the large-scale removal of trees, resulting in the alteration of temperature, sediment load, and physical structure of rivers and streams, such that they are less hospitable to native anadromous salmonid fishes (Giusti and Merenlender, 2002). It is likely that this degradation of aquatic habitat has also negatively impacted aquatic-breeding salamanders, such as red-bellied newts. Removal of trees also affects the microclimate of terrestrial habitats, perhaps rendering them less suitable for terrestrial newts. Finally, increased vehicular traffic associated with housing subdivisions undoubtedly has resulted in increased mortality of terrestrial newts.

Taricha torosa (Rathke, 1833)
CALIFORNIA NEWT

Shawn R. Kuchta

1. Historical versus Current Distribution.
Riemer (1958) conducted the first comprehensive investigation of the distribution and systematics of the genus *Taricha* and recognized 2 allopatric subspecies of *T. torosa* (California newts; Riemer, 1958): *T. t. torosa* (Coast Range newts) and *T. t. sierrae* (Sierra newts). Coast Range newts are distributed from central Mendocino County in northwestern California south through the Coast Ranges to Boulder Creek on the western slope of the peninsular ranges in San Diego County (Stebbins, 1985). Coast Range newts are found from sea level to at least 1,280 m on Mt. Hamilton, Santa Clara County, California (Stebbins, 1959). The southernmost localities in San Diego County compose a

geographic isolate (Stebbins, 1985; Jennings and Hayes, 1994a), and were once recognized as a distinct subspecies (*T. t. klauberi*: Wolterstorff, 1935; Stejneger and Barbour, 1943) or species (*T. klauberi*: Bishop, 1943; Smith and Taylor, 1948). They are genetically distinct (Tan, 1993; Tan and Wake, 1995; Kuchta, 2002) and distinguishable based on morphometric (Riemer, 1958) and osteological (Herre, 1939; Tan, 1993) grounds. However, the initial description was based on pathological animals, and *T. klauberi* was synonymized with *T. torosa* (Myers, 1942b; Twitty, 1942; Stebbins, 1951; Brattstrom and Warren, 1953; see "Parasites" below). Specimens have been reported from northwestern Baja California (Slevin, 1928; Smith and Taylor, 1948), but these records require verification.

Sierra newts occur at elevations below about 2,000 m and range along the western slopes of the Sierra Nevada from Shasta County (Gorman, 1951) south to Kern County (Stebbins, 1985). Many sources report a gap in the distribution between southern Shasta and northern Butte counties, but this may not exist, as Tan (1993) collected specimens in this area (also D. B. Wake, personal communication). Some sources consider Sierra newts to be deserving of specific status (Twitty, 1942; Collins, 1991; Tan, 1993; Kuchta, 2002), but others disagree (Stebbins, 1951; Riemer, 1958; Frost et al., 1992; Montanucci, 1992; Van Devender et al., 1992).

Tan and Wake (1995) outlined the historical biogeography of California newts. Based primarily on mitochondrial DNA evidence, they propose that Coast Range newts and Sierra newts differentiated about 8 million yr ago (mya), when Sierra newts existed in the uplifting central Sierra Nevada, and Coast Range newts inhabited the present day San Diego area. Roughly 5 mya, Coast Range newts expanded their distribution north to Monterey, while Sierra newts spread north and south in the Sierra Nevada. Coastal populations of Coast Range newts invaded the southern Sierra Nevada and differentiated morphologically roughly 2 mya; Tan (1993) and Kuchta (2002) suggest these populations may be sufficiently divergent to warrant species status. Only relatively recently, after the central California inland sea subsided, did Coast Range newts expand north of Monterey to their current distribution. Early workers have suggested, based on differences in larval pigmentation (Twitty, 1942) and preliminary genetic data (Coates, 1967; Hedgecock and Ayala, 1974; Hedgecock, 1976), that Coast Range newts are further divisible into northern and southern "races" located on either side of the Salinas Valley in Monterey County. While a genetic break exists, it is not large relative to other genetic disjunctions in the species (Tan, 1993; Kuchta, 2002).

California Newt (*Taricha torosa*)

2. Historical versus Current Abundance.
Historically, California newts were abundant throughout much of their range, except in the Santa Ynez Mountains of Santa Barbara County, where populations may have always been small (S. Sweet, personal communication, reported in Jennings and Hayes, 1994a). In southern California, suitable habitat is patchy (Jennings and Hayes, 1994a); however, at appropriate sites California newts were historically "common" on the Pacific slope (Klauber, 1928, 1930; Bogert, 1930; Pequegnat, 1945; Dixon, 1967; Brattstrom, 1988).

3. Life History Features.
Some of the best documentation of California newt life history features is by Ritter (1897) and Storer (1925), both of whom worked in the northern part of the range. Unfortunately, both California newts and rough-skinned newts (*T. granulosa*) occur there, and rough-skinned newts were not recognized as a distinct species until the work of Twitty (1935). However, California newts are the more abundant species in this area, and below I cite these authors where I feel the information applies to California newts.
  A. Breeding. Reproduction is aquatic.
  i. Breeding migrations. Coast Range newts migrate to ponds from December–early May, depending on locality, weather conditions, and breeding habitat. Miller and Robbins (1954) state that the migration of Coast Range newts to breeding ponds requires 6–8 wk, but this has not

been rigorously studied. Mating at any one site lasts from 3–12 wk (Miller and Robbins, 1954; De Lisle et al., 1986; Gamradt et al., 1997), and in all situations males arrive before females and remain at the breeding site longer (Ritter, 1897; Smith, 1941; Stebbins, 1951; Miller and Robbins, 1954; Riemer, 1958). In the vicinity of Palo Alto and Berkeley, migrations of pond breeding populations occur from late December to early February (Storer, 1925; Twitty, 1942; Stebbins, 1951; Miller and Robbins, 1954; Riemer, 1958). However, in the same area, stream breeding populations migrate in March and April after spring flooding subsides (Storer, 1925; Twitty, 1942; Stebbins, 1951; Miller and Robbins, 1954; Riemer, 1958). Southern Coast Range newts also migrate in March and April (Storer, 1925; Brame, 1968; Kats et al., 1992), probably because they too breed most commonly in streams. Sierra newts migrate to breeding sites in January and February, and breed from March to early May (Stebbins, 1951; Riemer, 1958).
  Brown (S. C.) and P. S. Brown (1980) investigated water balance in California newts and suggest that they possess some features analogous to terrestrial anurans, such as toads in the family Bufonidae. They discovered that the urinary bladder could store fluids in excess of 50% of the total body weight, and that water could be resorbed from the bladder. They also found that water is pulled from the ventral to the dorsal surface of the body,

probably via capillary action acting on the warty skin of terrestrial individuals (see also Lillywhite and Licht, 1974).
  ii. Breeding habitat. Coast Range newts breed in ponds, reservoirs, and streams (Storer, 1925; Twitty, 1942; Riemer, 1958). Gamradt and Kats (1997) document that stream-breeding newts in southern California commonly lay eggs in deep, slow pools, occasionally in runs, and almost never in riffles. Egg masses are attached to aquatic vegetation, branches, and the outer surfaces of rocks (Ritter, 1897; Storer, 1925; Twitty, 1942). In the central and northern portions of their range, Coast Range newt egg masses are not laid under rocks (Twitty, 1942); but in southern California, egg masses are commonly laid under rocks in quiet stream pools (Gamradt and Kats, 1996, 1997).
  Sierra newts are more stream adapted than Coast Range newts. Relative to Coast Range newts, and analogous to the stream-breeding red-bellied newts (*T. rivularis*), breeding Sierra newt males possess a reduced tail fin, and females lay eggs that are larger and commonly placed on the undersides of rocks in running water (Twitty, 1942). Sierra newts will, however, breed in ditches and other bodies of water with little or no current (Twitty, 1942).
  B. Eggs.
  i. Egg deposition sites. Storer (1925) observed that eggs are most frequently found from 7.5–10 cm deep, and when water levels rise to 30 cm or more the eggs die. Mosher et al. (1964) reports finding eggs from a few centimeters to >1.5 m deep.
  ii. Clutch size. California newts lay eggs in masses ranging from about 7–47 eggs (Ritter, 1897; Storer, 1925; Twitty, 1942; Brame, 1956, 1968; Mosher et al., 1964). Data are sparse, but it seems females lay from 3–6 masses each (Ritter, 1897; Brame, 1968). Miller and Robbins (1954) report for a sample of newts from central coastal California that the average Coast Range newt ovary contains 130–160 ova.
  C. Larvae/Metamorphosis.
  i. Length of larval stage. The larval stage lasts several months, with metamorphic animals leaving the water in summer or fall, depending on conditions. Ritter (1897) observed that the average larval period runs from March–October near Berkeley, California. He estimated the average size at metamorphosis to be 47 mm TL. Bishop (1943) noted that metamorphosis for Coast Range newts begins in early September and continues for some months; he estimates average size at metamorphosis at 50 mm TL, with large individuals measuring 60 mm. Twitty (1935) estimates the size at metamorphosis as approximately 45–55 mm TL. In 1998, at a vernal pool in Sonoma County, I found the average size of 13

metamorphic animals to be 43.8 ± 2.0 mm TL (24.6 ± 1.2 mm SVL). The average weight of 14 metamorphic animals was 0.45 ± 0.1 g. Metamorphosis began on 22 July and ended on 3 August, after the pool dried completely. Metamorphosis in Sierra newts does not seem to have been monitored. Bishop (1943) notes that larvae collected in late August range from 55–62 mm TL.

Kaplan (1985) examined the effects of egg size and food availability on larval development. Larvae from large eggs, when given an abundance of food, grew bigger and metamorphosed sooner than larvae from small eggs. Average time to metamorphosis from the feeding stage ranged from about 75–92 d; size ranged from 23.5–27 mm SVL. Conversely, under food stress, larvae from large eggs, though still metamorphosing at a larger size, took much longer to reach metamorphosis than larvae from small eggs. Average time to metamorphosis from the feeding stage ranged from roughly 105–190 d; size at metamorphosis was 21–23.75 mm SVL. This study suggests, among other things, that larger eggs may not be advantageous when food is limited and there is the threat of habitat desiccation.

Riemer (1958) suggests that only rough-skinned newts are known to overwinter as larvae, yet Storer (1925) collected four California newt larvae in Bailey Canyon, Los Angeles County, in April and May 1909, measuring 52–63 mm that were nearing completion of metamorphosis. Nevertheless, it seems that overwintering by California newt larvae is uncommon.

ii. Larval Requirements.

a. Food. California newt larvae eat small invertebrates, but their diet has not been rigorously studied. Ritter (1897) observed that larvae will eat decomposing organic matter, and perhaps conspecifics. Storer (1925) reports them feeding on mosquito larvae in captivity.

b. Cover. Adult Coast Range newts are cannibalistic (Ritter, 1897; Kats et al., 1992; Hanson et al., 1994; L. Kats, unpublished data, as cited in Elliot et al., 1993). Accordingly, Elliott et al. (1993) showed that lab-reared Coast Range newt larvae (14–17 mm TL) from the Santa Monica Mountains use cover more when provided chemical cues from adult newts. In the same study, chemical cues from odonate naiads and belostomatids (Hemiptera)—both known predators of larvae—did not result in substantially more cover use than controls, although the use of cover increased qualitatively.

In a second study, Kats et al. (1994) found that younger larvae use cover significantly more than older larvae. The behavior does not seem to be size dependent, and the authors suggest that the reduced use of cover is a product of reduced predation, as adult newts typically leave the streams by July.

iii. Larval polymorphisms. Albino larvae have occasionally been found (Riemer, 1958; Wells, 1963; Dyrkacz, 1981).

iv. Features of metamorphosis. Similar to many amphibians, California newt larvae are induced to metamorphose by pond drying (personal observations). At breeding sites with permanent water, the factors influencing metamorphosis remain unstudied. The most detailed description of metamorphosis is by Ritter (1897). The first sign is the development of a dusky color on the dorsum and a tinge of yellow on the venter. Subsequently, the colors deepen rapidly, the tail fin is absorbed from anterior to posterior, and the gills are absorbed. When metamorphic animals leave the water they possess adult coloration (though the larval pigment pattern is visible underneath), the skin is granular, and the gills are reduced to dark stubs. In the lab, this process requires over 2 wk (Ritter, 1897). If newly metamorphosed animals are kept in water, they will drown (Ritter, 1897).

v. Post-metamorphic migrations. California newts leave the aquatic habitat after metamorphosis, and do not return until they breed. Dispersal into the terrestrial landscape—where and how far they travel, the environmental clues employed in dispersal, and the length of the juvenile stage—is in need of study. Trenham (1998) recaptured marked juveniles up to 3,500 m from their natal ponds.

vi. Neoteny. Neoteny is unreported, but in permanent bodies of water California newt adults can remain aquatic year-round, and in this situation males retain their secondary sexual characters (Miller and Robbins, 1954). For example, on land adjacent to the University of California's Hastings Natural History Reserve, in Monterey County, a pond was deepened such that in the summers it no longer dried; the following summer a large population of adults took up permanent residence in the pond (P. Trenham, personal communication).

D. Juvenile Habitat. Unstudied.

E. Adult Habitat. The terrestrial ecology of California newts is incompletely studied, hampering conservation efforts (Jennings and Hayes, 1994a). In central California, Coast Range newts are found in mountainous or rolling woodland and grassland (Riemer, 1958; personal observations). In southern California, Coast Range newts inhabit a drier zone of chaparral, oak woodland, or grassland (Riemer, 1958). Sierra newts are found in mixed Sierran Forest.

F. Home Range Size. Unknown.

G. Territories. Unknown.

H. Aestivation/Avoiding Desiccation. California newts leave the aquatic habitat within a few weeks of breeding, and aestivate terrestrially during the dry summer. Mesic microclimates, such as deep leaf litter and animal burrows, are used as aestivation sites, and some individuals migrate

considerable distances to them (see "Seasonal Migrations" below; Trenham, 1998). Not all aestivation sites are distant from water, however, as Stebbins (1951) reports unearthing 14 aestivating adults 1.5 m from a stream in Los Angeles County. The aestivation site was under a boulder, in a hole partly filled with "slightly damp, coarse sand" (p. 25).

I. Seasonal Migrations. Following metamorphosis, California newts emigrate from the breeding site and spend the next few years growing to sexual maturity (see "Age/Size at Reproductive Maturity" below). When sexually mature, they migrate to a breeding site during winter or spring rains to mate (see "Breeding migrations" above). Following mating, they typically return to the terrestrial environment. Miller and Robbins (1954) found that about half of males and females of breeding size had immature gonads. If California newt breeding resembles red-bellied newt breeding, males typically breed annually while females skip 1, 2, or more years between reproductive events (Twitty, 1966).

California newts can migrate large distances between breeding and aestivation sites. For example, Trenham (1998) recaptured adult newts up to 3,200 m from the breeding pond where they were marked. However, when suitable aestivation sites are nearer, migrations may not be so long (P. Trenham, personal communication; see "Aestivation/Avoiding Desiccation" above). California newts typically return to the same breeding site repeatedly (but see Trenham, 1998). Endler (1970) demonstrated experimentally that California newts exhibit a kinesthetic, or body position, sense in their homing behavior, and observed that migrating newts correct for movements around objects such that a straight course is maintained. Twitty (1966) established the importance of olfaction in the navigation of red-bellied newts, and Landreth and Ferguson (1967a,b) demonstrated the use of celestial cues in the navigation system of rough-skinned newts. None of these experiments have been replicated on California newts.

J. Torpor (Hibernation). Not known to exist.

K. Interspecific Associations/Exclusions. Aside from the recent work of L. Kats and colleagues, little is known about the community ecology of California newts (see "Conservation" below for a discussion of introduced predators). In southern California and the Sierra Nevada, California newts are the only stream-breeding salamander.

L. Age/Size at Reproductive Maturity. Riemer (1958) found the smallest sexually mature animals to be around 50 mm, but there is variation. Both Ritter (1897) and McCurdy (1931) estimate the age of reproductive maturity to be reached after 3 yr; however, neither followed marked animals in the field. Work on experimentally released *rivularis-torosa* and *rivularis-sierrae* hybrids suggests that reproductive maturity requires 7–8 yr or longer (Twitty, 1961b, 1966). Unfortunately, these recapture data were never clearly summarized before Twitty's untimely death, and it is unclear how relevant hybrid data are to California newt sexual maturity.

M. Longevity. The longevity of California newts is unknown, although the animals are certainly long-lived. If red-bellied newts are any indication (Twitty, 1966), many California newts live to be > 20 yr old.

N. Feeding Behavior. Adult California newts feed terrestrially with a highly developed tongue-projection system, including substantial anterior movement of the ceratohyal (Findeis and Bemis, 1990). Conversely, aquatic feeding by adults is gape-and-suck, and further specialization of the feeding apparatus is probably constrained by trade-offs imposed by feeding both terrestrially and aquatically (Findeis and Bemis, 1990).

There has been no systematic study of the diet of terrestrial California newts, but Ritter (1897) reports that newts feed on their own sloughed-off skin, earthworms, insect adults and larvae, sowbugs, small snails and slugs, and other small invertebrates. Adults in the aquatic environment have been observed to feed on coleopterans, lepidopterans, and plecopterans (Hanson et al., 1994), ephemeropterans and the oligochaete *Eisenia rosea* (Kerby and Kats, 1998), and probably on the egg masses of California red-legged frogs (*Rana draytonii*; Rathbun, 1998; newts not identified to species). Furthermore, several workers have documented predation by adults on conspecific egg masses, both during and after oviposition (Ritter, 1897; Pequegnat, 1945; Kaplan and Sherman, 1980; Marshall et at, 1990; Kats et al., 1992), and on conspecific larvae (Ritter, 1897; Kats et al., 1992; L. Kats, unpublished data, as cited in Elliott et al., 1993, and Kats et al., 1994; Hanson et al., 1994). Hanson et al. (1994) found a nestling bird (0.3 g), probably Anna's hummingbird (*Calypte anna*), in the stomach of a small female newt in southern California.

O. Predators. The egg masses, embryos, and adults of California newts possess tetrodotoxin (TTX), a potent neurotoxin, and are thus generally unpalatable (see "Anti-Predator Mechanisms" below). Only common garter snakes (*Thamnophis sirtalis*) have been systematically shown to possess resistance to TTX (Brodie and Brodie, 1970). However, California newt larvae are not poisonous (see "Anti-Predator Mechanisms" below), and individuals in this life history stage may be an important food resource for newborn individuals of some garter snakes (Fitch, 1940, 1941; Fox, 1951), including Federally Endangered San Francisco garter snakes (*T. s. infernalis*; S. Barry, personal communication, as cited by Jennings and Hayes, 1994a). I examined the stomach contents of 23 live Pacific Coast aquatic garter snakes *(T. atratus)* in Sonoma County, California, and found five with one or more California newt larvae in their stomach. Fox (1951) reports finding California newt adults in the stomach of Pacific Coast aquatic garter snakes. Jennings and Cook (1998) found that introduced American bullfrogs (*Rana catesbeiana*) in Sonoma and Riverside counties not only eat larval California newts, but surprisingly also adults. Kats et al. (1998), in Los Angeles County, discovered a ring-necked snake *(Diadophis punctatus)* eating an adult newt, but the snake was lethargic, and in the lab regurgitated the dead newt. Also in southern California, two introduced predators, crayfish and mosquitofish, are causing serious declines (see "Conservation" below) via predation on egg masses and larvae. For a discussion of cannibalism of egg masses, see "Feeding Behavior" above.

P. Anti-Predator Mechanisms. Adults and embryos of California newts contain the potent neurotoxin TTX (Twitty and Johnson, 1934; Twitty, 1937; Wakely et al., 1966; Brodie et al., 1974b). This is the same toxin as found in the ovaries of puffer fish (family Tetraodontidae; Buchwald et al., 1964), as well as in a variety of other taxa, including other amphibians (Daly et al., 1987), echinoderms (Maruyama et al., 1984), cephalopods (Sheumack et al., 1978), and bacteria (Noguchi et al., 1986; Thuesen and Kogure, 1989). It is possible that this neurotoxin is produced by symbiotic gut bacteria (Mosher and Fuhrman, 1984; Noguchi et al., 1986, 1987; Thuesen and Kogure, 1989), but the situation is complex and it is not clear how newts acquire TTX (Mosher and Fuhrman, 1984; Daly et al., 1987). TTX is found in the skin, ovaries and ova, muscle, and blood of adult newts; the liver, viscera, and testes contain only minute amounts (Wakely et al., 1966). Embryos possess TTX throughout their body, but larvae lose their toxicity shortly after the yolk is absorbed (Twitty and Johnson, 1934; Twitty, 1937), suggesting that embryos obtain their poison maternally.

Rough-skinned newts are the most toxic salamanders in the world, and California newts are reported to be about 1/10 as toxic (Brodie et al., 1974b). To put this in perspective, Brodie et al. (1974b) estimated that the skin of adult rough-skinned newts from Benton County, Oregon, could potentially kill 25,000 20-g mice, whereas the skin of adult California newts could kill 1,200–2,500 20-g mice. More recent work has uncovered substantial geographic variation in the toxicity of rough-skinned newts (Hanifin et al., 1999). Variation in California newt toxicity is in need of examination.

California newts have a cryptic brown dorsum and an aposematic yellow-to-orange venter. When harassed, California newts assume a defensive posture called the *unken reflex*, which exposes the ventral aposematic coloration: the belly is pressed down against the substrate, the throat raised up with eyes closed and depressed, the limbs stiffly extended outward, and the tail raised straight over the body (Stebbins, 1951; Riemer, 1958; Johnson and Brodie, 1972, 1975; Brodie, 1977; rough-skinned newts curl their tail tip; Riemer, 1958). Toxicity, aposematic coloration, and defensive posturing form a co-adaptive unit (Johnson and Brodie, 1972, 1975; Brodie, 1977). Naylor (1978) has suggested that the frontosquamosal arch of newts is an anti-predator adaptation designed to add structural support to the skull and protect retracted eyes. California newts produce a repertoire of sounds (Davis and Brattstrom, 1975), and Brodie (1978) suggests that predators may associate some sounds with toxic skin secretions.

Q. Diseases. Unknown. Some animals in San Diego County exhibit extreme wartiness now recognized as pathological (see "Historical versus Current Distribution," above and "Parasites," below).

R. Parasites. Ingles (1936), Lehmann (1954), and Goldberg et al. (1998c) documented the presence of parasites in California newts. Goldberg et al. (1998c) found that in 68 Coast Range newts from Orange and Los Angeles counties, 52 were infected with at least one nematode. Four nematode species were discovered (average number of nematode species/infected newt = 1.3 ± 0.5), and they concluded that the parasite infracommunities of California newts are depauperate relative to other vertebrates. There are no data on the effect of parasite loads on the fitness or ecology of California newts. Some animals in San Diego County exhibiting extreme wartiness were once recognized as a distinct taxon, but it was later determined that the warty phenotype was pathological (see "Historical versus Current Distribution" above).

4. Conservation.

California newt populations are impacted by breeding site degradation (Jennings and Hayes, 1994a), destruction of summer aestivation sites and migration routes (Jennings and Hayes, 1994a; see also Semlitsch, 1998), road kills (Storer, 1925; R. C. Stebbins, personal communication; P. Trenham, personal communication), large-scale commercial exploitation (Jennings and Hayes, 1994a), altered sedimentation dynamics in stream pools resulting from wildfires (Gamradt and Kats, 1997; Kerby and Kats, 1998), and riparian habitat degradation (Faber et al., 1989; Jennings and Hayes, 1994a).

Furthermore, in the Santa Monica Mountains of Los Angeles County, two introduced predators, crayfish (*Procambarus clarkii*) and mosquitofish (*Gambusia affinis*), have caused serious declines and exterminated some populations (Gamradt and Kats, 1996; Gamradt et al., 1997). Crayfish are sold as bait, and mosquitofish are distributed broadly for mosquito control, so both are widespread (Gamradt and Kats, 1996). Also in the Santa Monica Mountains, Anzalone et al. (1998) showed that solar UV-B causes high rates of embryonic mortality in California newt egg masses (see also Blaustein et al., 1998). The genetically distinct San Diego County populations in the Cuyamaca Mountains (see "Historical versus Current Distribution" above) are currently restricted to the Boulder, Ceder, and Conejos Creek systems, where populations persist in small, isolated pockets (from about 15–20 breeding adults; E. Ervin, USGS, personal communication). Exotic predators such as crayfish, green sunfish (*Lepomis cyanellus*), and rainbow trout (*Oncorhynchus mykiss*) occur in the perennial reaches of streams originating in the Cuyamaca Mountains of San Diego County. Because these predators and California newts do not co-occur, they may be excluding newts (E. Ervin, USGS, personal communication). Finally, Fisher and Shaffer (1996) document a decline of California newts in the central valley of California, where these animals are peripherally present.

California newt populations south of the Salinas River in Monterey County are considered by the California Department of Fish and Game to be Species of Special Concern (Jennings and Hayes, 1994a). While this designation carries no legal weight, it is a helpful management tool highlighting concerns about the species' status. Moreover, a large portion of verified populations in San Diego County are extinct (Jennings and Hayes, 1994a), and because of the distinctiveness of the animals in this area (see "Historical versus Current Distribution" above) the status of these populations warrants further investigation.

Family Sirenidae

Pseudobranchus axanthus Netting and Goin, 1942(b)
SOUTHERN DWARF SIREN

Paul E. Moler

1. Historical versus Current Distribution.
Southern dwarf sirens (*Pseudobranchus axanthus*) are restricted to peninsular Florida; their northern boundary includes Alachua, Clay, Duval, Levy, and Putnam counties. Two subspecies, narrow-striped dwarf sirens (*P. a. axanthus*) and Everglades dwarf sirens (*P. a. belli*), are recognized (Moler and Kezer, 1993; Crother et al., 2000). The current distribution of southern dwarf sirens is undoubtedly reduced compared

with their historical distribution, as wetlands in peninsular Florida have been reduced through drainage of surface waters associated with residential, agricultural, and silvicultural development.

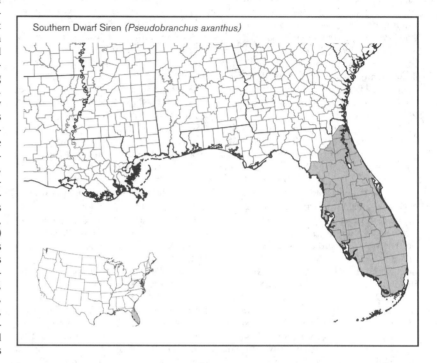

Southern Dwarf Siren (*Pseudobranchus axanthus*)

2. Historical versus Current Abundance.
Southern dwarf sirens are often common in suitable habitat. Current numbers are reduced relative to historical abundance due to the loss of wetland habitat.

3. Life History Features.
A. **Breeding.** Reproduction is aquatic.

i. Breeding migrations. Southern dwarf sirens do not migrate. They breed and permanently reside in the same aquatic habitats.

ii. Breeding habitat. Southern dwarf sirens live and breed in heavily vegetated marshes and shallow lakes.

B. **Eggs.**

i. Egg deposition sites. Eggs are laid singly or in small bunches among aquatic vegetation throughout the spring (Carr, 1940a; Netting and Goin, 1942b). According to Petranka (1998) the oviposition period lasts from early November to March. Newly laid eggs average 3 mm and are surrounded by the vitelline membrane and three jelly envelopes (Noble and Richards, 1932; see also Petranka, 1998).

ii. Clutch size. Unknown.

C. **Larvae.** Newly hatched larvae are between 10–11.5 mm SVL, colored brown dorsally with lighter stripes on the dorsal midline and lateral portions of the body and head (Goin, 1947c; see also Petranka, 1998). They have limb buds (Noble, 1927b). Ashton and Ashton (1988) indicate *Pseudobranchus* larvae make take 2 yr to reach sexual maturity.

D. **Juvenile Habitat.** Same as adult habitat.

E. **Adult Habitat.** Southern dwarf sirens are most abundant in heavily vegetated marshes and shallow lakes (Carr, 1940a). They may be abundant in floating mats of vegetation or in mucky shoreline deposits.

F. **Home Range Size.** Unknown.

G. **Territories.** Unknown.

H. **Aestivation/Avoiding Desiccation.** Southern dwarf sirens burrow into bottom sediments and form a protective cocoon when wetlands dry. Individuals may remain buried for several months until their wetland refills (Freeman, 1958; Etheridge, 1990a).

I. **Seasonal Migrations.** None.

J. **Torpor (Hibernation).** Carr (1940a) reported that southern dwarf sirens "have been found hibernating in deep mud." Nevertheless, although they may become inactive during prolonged periods of cold weather, they may be collected throughout the winter, especially in southern Florida.

K. **Interspecific Associations/Exclusions.** Southern dwarf sirens occur sympatrically, but only occasionally syntopically, with northern dwarf sirens (*P. striatus*) in northern peninsular Florida. In areas of sympatry, northern dwarf sirens are typically found in more acidic habitats than are southern dwarf sirens, but they are known to occur syntopically at a few sites (Moler and Kezer, 1993).

L. **Age/Size at Reproductive Maturity.** Unknown.

M. **Longevity.** Unknown.

N. **Feeding Behavior.** Reported foods include amphipods, chironomid larvae, aquatic oligochaetes, and ostracods

(Harper, 1935; Carr, 1940a; Duellman and Schwartz, 1958; Freeman, 1967).

O. Predators. Southern banded water snakes *(Nerodia fasciata)* have been observed preying on flood-displaced southern dwarf sirens (personal observations). Van Hyning (1932) reported predation by the striped crayfish snake *(Regina alleni)*, and Carr (1940a) reported predation by the mud snake *(Farancia abacura)*. Black swamp snakes *(Seminatrix pygaea)* are also common associates and likely prey on dwarf sirens. Other likely predators include predaceous fish and various species of wading birds, especially where dwarf sirens are concentrated by falling water levels.

P. Anti-Predator Mechanisms. Flight. Southern dwarf sirens are slippery and difficult to hold. Also, when seized, they may produce high-pitched squeaks, but the utility of these sounds as an anti-predator mechanism is unknown.

Q. Diseases. Unknown.

R. Parasites. Unknown.

4. Conservation.
Southern dwarf sirens have a restricted distribution, although they are often common in suitable habitat. The current distribution and abundance of southern dwarf sirens is undoubtedly reduced compared with historical levels.

Pseudobranchus striatus (LeConte, 1824)
NORTHERN DWARF SIREN

Paul E. Moler

1. Historical versus Current Distribution.
Northern dwarf sirens *(Pseudobranchus striatus)* occur in the lower Gulf and Atlantic Coastal Plains from Orangeburg County, South Carolina (Folkerts, 1971), south to central peninsular Florida (Hernando and Volusia counties; Moler and Kezer, 1993), then west to Baker and Lee counties, Georgia (Goin and Crenshaw, 1949), and Walton County, Florida (Moler and Thomas, 1982). Three subspecies are recognized (Moler and Kezer, 1993; Crother et al., 2000): Gulf Hammock dwarf sirens *(P.s. lustricolus)*, slender dwarf sirens *(P.s. spheniscus)*, and broad-striped dwarf sirens *(P.s. striatus)*. The current range of northern dwarf sirens is unchanged from their historical range, although populations have been lost as wetland habitats have been reduced through drainage of surface waters associated with residential, agricultural, and silvicultural development.

2. Historical versus Current Abundance.
Northern dwarf sirens are often common in suitable habitats, but such habitats have been reduced through drainage of surface waters, and current numbers are reduced relative to their historical abundance.

3. Life History Features.
A. Breeding. Reproduction is aquatic.

i. Breeding migrations. Do not migrate. Northern dwarf sirens breed and permanently reside in the same aquatic habitats.

ii. Breeding habitat. Northern dwarf sirens live and breed in cypress *(Taxodium* sp.) or gum *(Nyssa* sp.) ponds and other shallow, acidic, wetlands of the flatwoods.

B. Eggs.

i. Egg deposition sites. Eggs are laid singly or in small bunches among aquatic vegetation (Noble, 1930).

ii. Clutch size. Unknown.

C. Larvae. Not well known. Petranka (1998) notes that larvae are similar to those of southern dwarf sirens *(P. axanthus)*. Ashton and Ashton (1988) indicate *Pseudobranchus* larvae make take 2 yr to reach sexual maturity.

D. Juvenile Habitat. Same as adult habitat.

E. Adult Habitat. Most often associated with cypress or gum ponds and other shallow, acidic wetlands of the flatwoods. Unlike southern dwarf sirens, northern dwarf sirens are not normally found among water hyacinths, which are typically absent from acidic wetlands of the flatwoods. Northern dwarf sirens have been collected from similar floating mats of frog's-bit *(Limnobium spongium;* Moler and Kezer, 1993), but they more typically inhabit decaying bottom vegetation and the soft, mucky soils of pond margins (Le Conte, 1824; Goin and Crenshaw, 1949).

F. Home Range Size. Unknown.

G. Territories. Unknown.

H. Aestivation/Avoiding Desiccation. Northern dwarf sirens burrow into bottom sediments when wetlands dry (Harper, 1935; Goin and Crenshaw, 1949). They will remain buried until their wetland refills.

I. Seasonal Migrations. None.

J. Torpor (Hibernation). Northern dwarf sirens remain buried in mud and bottom debris during cold weather.

K. Interspecific Associations/Exclusions. Northern dwarf sirens occur sympatrically, but only occasionally syntopically, with southern dwarf sirens in northern peninsular Florida. In areas of sympatry, northern dwarf sirens are typically found in more acidic habitats than are southern dwarf sirens, but they are known to occur syntopically at a few sites (Moler and Kezer, 1993).

L. Age/Size at Reproductive Maturity. Unknown. Collections from Okefenokee Swamp, Georgia, suggest that northern dwarf sirens mature in <1 yr (B. Freeman, personal communication). A maximum size of 203 mm was noted by Moler and Mansell (1986).

M. Longevity. Unknown.

N. Feeding Behavior. Unknown, but probably similar to that of southern dwarf sirens.

O. Predators. Unknown. Southern-banded water snakes *(Nerodia fasciata)*, black swamp snakes *(Seminatrix pygaea)*, mud snakes *(Farancia abacura)*, and crayfish snakes *(Regina* sp.) are common associates and likely prey on northern dwarf sirens. Various species of wading birds are likely major predators on *Pseudobranchus*, especially when dwarf sirens are concentrated by falling water levels. Predaceous fish are also probable predators.

P. Anti-Predator Mechanisms. Flight. *Pseudobranchus* are slippery and thus may escape when gripped. When seized, *Pseudobranchus* may produce high-pitched squeaks, but the utility of vocalizations as an anti-predator mechanism is unknown.

Q. Diseases. Unknown.

R. Parasites. Unknown.

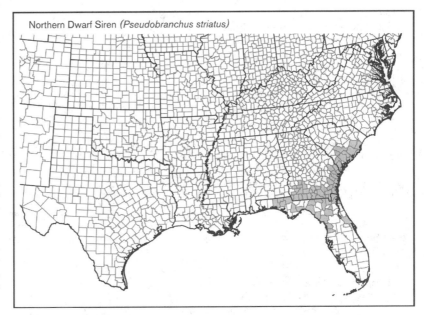
Northern Dwarf Siren *(Pseudobranchus striatus)*

4. Conservation.

The current range of northern dwarf sirens is unchanged from their historical range, although populations have been lost as wetland habitats have been reduced through drainage of surface waters associated with residential, agricultural, and silvicultural development. Northern dwarf sirens are often common in suitable habitats, but such habitats have been reduced through drainage of surface waters, and current numbers are reduced relative to their historical abundance. Northern dwarf sirens are considered Threatened in South Carolina (www.dnr.state.sc.us).

Siren intermedia Barnes, 1826
LESSER SIREN

William T. Leja

1. Historical versus Current Distribution.

Lesser sirens *(Siren intermedia)* inhabit the Atlantic and Gulf Coastal Plain from Virginia to Mexico, and range north in the Mississippi Valley to Illinois, Indiana, and southwestern Michigan (Conant and Collins, 1991). Their range extends south in Mexico to the state of Veracruz (Ramirez-Bautista et al., 1982) and north on the Atlantic Coastal Plain to Caroline County, Virginia (Roble, 1995). Geographical isolates occur in northern Indiana, southwestern Michigan, northeastern North Carolina, and Virginia (Petranka, 1998). Because lesser sirens thrive in pools of fish hatcheries (Hubbs, 1962; Brach, 1995), it is possible that their range has been expanded by fish stocking. Although rotenone, a pesticide used to eliminate undesired fish and amphibian populations, revealed the presence of sirens in Michigan lakes, it could presumably have eliminated them from the state as well (Williams, 1961; Harding, 1997). Several new species within the genus *Siren* will likely be erected following the completion of electrophoretic and karyological studies (P. E. Moler, personal communication). Because the name *Siren intermedia* is assigned to animals throughout most of the current range of the genus, the distribution of this nominal form may decrease as it is divided into newly recognized species.

2. Historical versus Current Abundance.

Throughout their range, lesser sirens occur in scattered populations. With Euro-American settlement, these populations may have become further isolated by flood control programs (Petranka, 1998). The lesser siren populations of south Texas have been threatened by wetland drainage (Bury et al., 1980). In response, the south Texas lesser siren (large form; see *S. texana* account, this volume) has been listed by the State of Texas as Endangered (Campbell, 1995). Within the Great Lakes drainage basin, based on the

Lesser Siren *(Siren intermedia)*

scarcity of recent records, lesser sirens are extremely rare and possibly even extirpated (Harding, 1997).

The former wet prairie of Illinois and Indiana, which covered about 20% of the ground surface in pre-settlement times, likely constituted ideal siren habitat. All the extensive intermittent prairie wetlands in the lower Midwest have been drained to yield prime farmland (Prince, 1997). However, the thousands of farm ponds constructed in the Midwest since the Dust Bowl years may have partially replaced these wetlands as lesser siren habitat.

Methods for capturing sirens have changed over time. Because of the relation of capture methods to qualitative assessments of abundance, they should be reviewed here. Early in this effort, Scroggin and Davis (1956) commented that sirens are highly secretive, and few herpetologists were equipped to capture them. They were successful in capturing them at night by using a seine weighted with a heavy chain. Later, Rossman (1960) noted that lesser sirens were among the most abundant amphibians in Pine Hills Swamp in southern Illinois when electroshocking was used to determine abundance, but that most collectors rarely found them. Electric fish shockers have proven effective at routing sirens from their hiding places during the day (Dundee and Rossman, 1989). Bennett and Taylor (1968) did not discover the presence of sirens in a lake until electrofishing was used, although a variety of netting techniques had been tried earlier. They suggested that seining was probably ineffective because the sirens are buried in the bottom mud during the day. Bennett and Taylor (1968) were also successful using rotenone, but sirens did not appear to be affected until they left their burrows after sunset. In a demonstration of the use of rotenone in fish eradication in a Texas farm pond, Davis and Knapp (1953) collected 209 lesser sirens the next morning. It is noteworthy that

both of the two known Michigan populations were discovered when rotenone was applied to the shallow inlets of lakes for fish management purposes (Harding, 1997). More recently, minnow traps have been successfully used to trap lesser sirens (Gehlbach and Kennedy, 1978; Gibbons and Semlitsch, 1991; Raymond, 1991; Sever et al., 1996).

Lesser sirens can reach high densities in favorable habitats. In an east Texas beaver pond in preferred habitat (shallow water, dense aquatic vegetation, deep sediments) there were 1.3 sirens/m^2; the standing crop was 56.6 g./m^2 (Gehlbach and Kennedy, 1978). Gehlbach and Kennedy (1978) attributed this extreme density in a 5-yr-old aquatic habitat to high fecundity, rapid growth, and successful aestivation during dry periods.

3. Life History Features.

A. Breeding. Reproduction is aquatic.

i. Breeding migrations. Sirens lack an obvious overland dispersal stage in their life cycle (Petranka, 1998), but aquatic migrations to specialized breeding sites are possible.

ii. Breeding habitat. A subset of the adult habitat.

Mating behavior of sirens has not been reported (Zug, 1993; Petranka, 1998). Numerous attempts to breed sirens in captivity have failed (Godley, 1983). However, Brach (1995) stated that "sirens will breed in captivity providing there is sufficient vegetation or debris in which to build their nests." Fertilization is concurrent with oviposition (Sever et al., 1996). In South Carolina, Sever et al. (1996) found that oviposition occurred primarily in February–March, but also possibly in January and April. Oviposition occurred from late December to March in south-central Florida (Godley, 1983). A nest of eggs accompanied by an adult lesser siren was found in early April in Arkansas (Noble and Marshall, 1932).

B. Eggs.

i. Egg deposition sites. Egg masses accompanied by females or adults of undetermined sex have been found at the base of rooted macrophytes, in fibrous mats of water hyacinth roots, and in muddy depressions or plant debris in pond bottoms (Noble and Marshall, 1932; Hubbs, 1962; Godley, 1983).

ii. Clutch size.. The number of eggs produced in a season is highly variable, ranging from about 200–500 or more (Harding, 1997). The incubation period lasts 1.5–2.5 mo, based on the first appearance of small larvae in seasonal samples (Petranka, 1998).

C. Larvae/Metamorphosis. The larvae are about 11 mm long upon hatching. Sexual maturity is reached in about 2 yr after hatching (Pope, 1947; Davis and Knapp, 1953; Harding, 1997; Petranka, 1998).

D. Juvenile Habitat. Juveniles are most active at night and live in burrows or in thick mats of aquatic vegetation, where they forage on small invertebrates (Petranka, 1998). Juvenile sirens feed mostly on zooplankton but also eat larger prey, such as amphipods, craneflies, and lumbriculid worms (Carr, 1940a). Even at night, larval sirens in experimental ponds spend most of the time hidden in leaf litter (Fauth et al., 1990).

E. Adult Habitat. Shallow, warm, quiet water of ponds and sloughs where aquatic vegetation is plentiful (Smith and Minton, 1957). Permanent or semipermanent habitats, including marshes, swamps, farm ponds, ditches, canals, sloughs, and sluggish, vegetation-choked creeks (e.g., Neill, 1949b; see also Petranka, 1998). Temporary floodplain pools and shallow, heavily vegetated sections of ponds with deep sediments provide burrowing sites (Funderburg and Lee, 1967; Gehlbach and Kennedy, 1978). In areas of sympatry, lesser sirens and greater sirens (*S. lacertina*) are partitioned by habitat preferences— lesser sirens tend to inhabit more acidic pH waters, while greater sirens are found in aquatic sites with circumneutral pH (P. E. Moler, personal communication, in McAllister et al., 1994). This appears to be consistent with Carr's (1940) observations in Florida that lesser sirens are characteristic of pine flatwoods ponds and bayheads (a normally acidic environment), while greater sirens are characteristic of sloughs, canals, and drainage ditches, and frequent in lakes. Lesser sirens are restricted to wetlands that hold water for at least 6 mo (Snodgrass et al., 1999).

Low intrinsic metabolic rate, efficient lungs, the ability to withstand prolonged inanition and to undergo facultative anaerobiosis, along with the ability to extract oxygen from waters of low oxygen tension, have allowed sirens to populate warm, shallow, and occasionally hypoxic and hypercarbic waters (Guimond and Hutchison, 1976). Sirens are often abundant in water hyacinth communities, which are, from the viewpoint of gas exchange, one of the most restrictive environments for aquatic organisms (Ultsch, 1976).

F. Home Range Size. In a Texas pond where lesser sirens are abundant, no individual among 60 recaptures moved >12 m from the first capture point. Forty-six (77%) of them traveled fewer than 6 m (Gehlbach and Kennedy, 1978).

G. Territories. Lesser sirens produce trains of pulsed sounds or clicks, often accompanied by head-jerking movements, that may be used in defense of individual space (Gehlbach and Walker, 1970). Social interactions will preclude group formation if sufficient shelter (e.g., deep, soft sediment or debris) is available (Asquith and Altig, 1987).

H. Aestivation/Avoiding Desiccation. Lesser sirens are able to survive drought and the drying of their habitat by retreating into crayfish tunnels to a depth of ≥1 m (Cagle, 1942) or by burrowing into the mud (Harding, 1997). However, the lack of valves on the external nares precludes the possibility of extensive burrowing (Noble, 1929b). If the surrounding mud begins to dry, sirens conserve water by producing a protective cocoon, which covers their bodies except for their mouths (Reno et al., 1972). During aestivation, the gills atrophy and the body shrinks as fat is metabolized; oxygen consumption and heart rate are reduced (Gehlbach et al., 1973). Aestivation can continue for several weeks to >1 yr (Gehlbach et al., 1973; Gibbons and Semlitsch, 1991). In Indiana in June, under the moist soil of a pond bottom that had been drained the previous fall, Blatchley (1899) found numerous sirens at a depth of 8–10 cm while the land was being plowed for row crops. By remaining in the basin of a pond when it dries, sirens are able to prey on the eggs and larvae of even the earliest colonists after the return of water, and are thus likely to have a profound effect on community structure (Fauth and Resetarits, 1991).

I. Seasonal Migrations. Lesser sirens are primarily aquatic, but they have been found on land beneath brush piles and under logs and can move overland on occasion, as they will colonize artificial ponds that have never had a direct connection with natural habitats (Minton, 1972, 2001). Although lesser sirens survive droughts by aestivation in an underground cocoon, they may also migrate to other water bodies. During an unprecedented drought in Louisiana, a lesser siren was collected under oak leaves in flat, mixed woodland about 600 m from the fringe of marshes and cypress swamps (its normal habitat) which border Lake Pontchartrain (Viosca, 1924b). Further, it is possible that lesser sirens forage on land or migrate to other habitats during rains.

Viosca (1924b) found a lesser siren 450 m from the edge of a marsh under a plank in a wet field after heavy rains. Lesser sirens will occasionally leave aquatic habitats in mid-winter: after a January cold snap following a warm December in Arkansas, large numbers of lesser sirens were found frozen within or above the ice (Sugg et al., 1988). Also, after a winter cold snap near St. Louis, Missouri, during which the temperature dropped to -26 °C, hundreds of lesser sirens were found frozen in the ice (Hurter, 1911). Possibly they left the water during the warm rains that often precede winter cold spells. Noble (1929b) commented that the hypertrophy of the Jacobson's organ in sirens may have developed in connection with terrestrial as well as aquatic feeding habits. Lesser sirens have an apparent ability to climb. Gaines (1895) kept a captured lesser siren for 8 mo in a barrel containing some mud and water. On one occasion he placed two laths in the barrel; on the next day he found the animal squirming about on the floor. Scroggin and Davis (1956) found that lesser sirens consumed a large number of terrestrial organisms during rainy periods. They assumed this food was carried in by runoff or by a rising water level, but it is also possible that lesser sirens sometimes feed terrestrially at night during rainy periods. Snodgrass et al. (1999) found lesser sirens in wetlands approximately 0.6 km distant from other aquatic habitats, suggesting that they possess substantial dispersal capabilities.

Lesser sirens demonstrate seasonal activity patterns. In a temporary pond, Raymond (1991) found that adults were considerably more active in fall and winter. He believed that the intensified activity was due either to increased foraging or to reproductive behavior. In contrast, in a permanent pond with deep organic sediments and a high density of lesser sirens, Gehlbach and Kennedy (1978), using the same trapping technique, observed no significant difference between spring and fall catches. In South Carolina, Sever et al. (1996) were able to trap females and males from January–April, but from May–October, males were almost exclusively caught.

Eubanks et al. (2002) found western lesser sirens in Tennessee migrating upstream from one pond to another through a large flooded culvert. They note that the timing of this migration was coincident with the published timing of reproductive activities and suggest these animals were seeking oviposition sites.

J. Torpor (Hibernation). During excavation of a drainage ditch in mid-winter in the dry bed of a slough in southern Illinois, Cockrum (1941) found 60–70 lesser sirens at a depth of 46–102 cm below the ground surface in what appeared to be crayfish burrows. The ground was frozen to a depth of 15 cm and covered with 8 cm of snow. Lesser sirens removed from the ground were active despite near-freezing air temperatures. Cagle and Smith (1939) found what they believed was a hibernating aggregation of 138 lesser sirens in shallow, ice-free water connected to two frozen ponds in early January in southern Illinois. The animals were comparatively inactive and easily secured. Barbour (1971) stated that although lesser sirens remain active most of the winter, individuals spend at least some time in dormancy.

K. Interspecific Associations/Exclusions. In large, temporary ponds of the southeastern United States, lesser sirens and eastern newts *(Notophthalmus viridescens)* are two of the most abundant vertebrate predators. Their larvae are generalized predators that utilize similar types of invertebrate prey, competing as equals in these ephemeral larval environments (Fauth et al., 1990). Until recently, lesser sirens have been overlooked with regard to their role as top predators in pond amphibian communities (Fauth, 1999a). By non-selectively reducing the density of anuran larvae, lesser sirens act to support eastern newts as keystone predators of larval anuran communities in North Carolina ponds (Fauth, 1999a). Eastern newts function as keystone predators by selectively feeding on dominant larval anuran species, thus allowing a greater number of weakly competing anuran species to survive (Fauth and Resetarits, 1991). However, in ostensibly similar ponds in the Francis Marion National Forest in South Carolina, eastern newts and lesser sirens do not function as keystone predators (Fauth, 1999b). Instead, mole salamanders *(Ambystoma talpoideum,* which are not present in the North Carolina ponds) act as top predators (Fauth, 1999b). Lesser sirens affect mole salamanders by limiting their growth and recruitment (Fauth, 1999a).

L. Age/Size at Reproductive Maturity. Lesser sirens become reproductively mature during their second year of life, when males average 18 cm SVL, females 15 cm.

M. Longevity. The record for longevity is held by a specimen at the Rio Grande Zoo in New Mexico that lived for almost 7 yr (Snider and Bowler, 1992).

N. Feeding Behavior. Feeding is primarily nocturnal (Noble and Marshall, 1932). Lesser sirens consume a variety of invertebrate prey, including small crustaceans, insect larvae, snails, and annelid worms (Scroggin and Davis, 1956). Davis and Knapp (1967) found that small crustaceans accounted for up to 87%, and snails and sphaeriid clams accounted for ~10%, of the total number of food items eaten. Altig (1967) concluded that lesser sirens can apparently obtain food by filter feeding—sifting through bottom material and aquatic vegetation. Siren eggs have been noted in the stomachs of adults (Scroggin and Davis, 1956; Collette and Gehlbach, 1961). Cranial anatomy appears

to facilitate feeding within the confines of a burrow (Reilly and Altig, 1996). Lesser sirens will feed on tadpoles (Fauth et al., 1990), including those of the genus *Bufo* (Lefcourt, 1998), and larval salamanders (Fauth and Resitarits, 1991; Fauth, 1999a).

Hurter (1911) stated that lesser sirens feed on worms and minnows. Lesser sirens have been caught by anglers using minnows for bait (Goin, 1957; Scroggin and Davis, 1956), and fish scales have been found in their guts (Altig, 1967). McAllister and McDaniel (1992) believe that third-stage larval anisakid nematodes *(Contracaecum* sp.) infected lesser sirens through ingestion of parasitized fish. However, Pope (1947) questioned Hurter's (1911) statement that lesser sirens feed on minnows. Pope's contention was based on Carr's (1940) observation of the relatively poorly developed prey-capturing skills of greater sirens in Florida. Davis and Knapp (1953) considered lesser sirens to be bottom feeders incapable of active pursuit of larger animals.

O. Predators. Natural predators are poorly documented but undoubtedly include water snakes, fishes, alligators, and wading birds (Petranka, 1998). In Hickman County, Kentucky, lesser sirens make up about 33% of the total food of cottonmouths *(Agkistrodon piscivorus;* Barbour, 1971). Marvel (1972) observed a yellow-bellied water snake *(Nerodia erythrogaster flavigaster)* eating a lesser siren. Predation of lesser sirens by Mississippi green water snakes *(N. cyclopion)* has been reported in Illinois (Garton et al., 1970). Buck (1946) removed a lesser siren from the stomach of a mud snake *(Farancia abacura).* A lesser siren was removed from the stomach of a largemouth bass *(Micropterus salmoides)* in Louisiana (Walker, 1963).

P. Anti-Predator Mechanisms. The nocturnal behavior of lesser sirens is presumed to be an anti-predator behavior that minimizes predation risk from diurnal predators such as fish and wading birds (Petranka, 1998). A lesser siren captured by a water snake in the wild emitted several shrill distress cries (Marvel, 1972). Distress cries of immature frogs captured by water snakes have been suggested to be an anti-predator behavior because they attract large American bullfrogs *(Rana catesbeiana)* and sometimes allow escape of the victim in the ensuing confusion (Smith, 1977). Chemically mediated fright responses of southern leopard frog *(R. sphenocephala)* and southern toad *(Bufo terrestris)* tadpoles to lesser sirens have been examined in the laboratory (Lefcourt, 1996, 1998, respectively).

Q. Diseases. Unknown.

R. Parasites. McAllister et al. (1994) present a summary table of the known helminth fauna of lesser sirens. The following nine helminths have been found in lesser sirens: the trematodes *Allassostomoides louisianensis* (Brooks and Buckner, 1976), *Progorgodera foliata* (Brooks and

Buckner, 1976) and *Diplostomum* sp. (McAllister et al., 1994); the cestoid *Proteocephalus sireni* (Brooks and Buckner, 1976; Brooks, 1978); the acanthocephalans *Fessisentis fessus* (Landewe, 1963; Nickol, 1972; Dunagan and Miller, 1973; Buckner and Nickol, 1979) and *Neoechinorhynchus* sp. (Miller and Dunagan, 1971); and the nematodes *Falcaustra chabaudi* (Dyer, 1973; McAllister et al., 1994), *Capillaria* sp. (McAllister et al., 1994), and *Contracaecum* sp. (McAllister and McDaniel, 1992).

McAllister et al. (1994) comment that there is some degree of host specificity among helminths from different siren taxa, particularly among trematodes of lesser and greater sirens. The two species in the genus *Siren* appear to be partitioned by habitat preferences (P. E. Moler, personal communication, in McAllister et al., 1994); intermediate hosts may be partitioned in the same manner (McAllister et al., 1994).

Cystacanths of the acanthocephalan parasite *Fessisentis fessus* occur in two aquatic isopods, *Asellus forbesi* and *Lirceus lineatus*, in Jackson County, Illinois, where lesser sirens are the usual definitive host; laboratory infections have confirmed the life cycle of the parasite (Buckner and Nickol, 1979). Older lesser sirens harbor more and larger specimens of the acanthocephalan *Fessisentis fessus* than do younger animals (Nickol, 1972).

4. Conservation.
Throughout their range, lesser sirens occur in scattered populations. With Euro-American settlement, these populations became further isolated by flood control programs and wetland drainage (Bury et al., 1980). Within the Great Lakes drainage basin, lesser sirens are possibly extirpated (Harding, 1997). Lesser sirens are considered Threatened in Michigan (Stearns and Lindsley, 1977). The Kentucky Department of Fish and Wildlife Resources lists lesser sirens as Rare and Endangered (Babcock, 1977).

One problem in determining the conservation status of sirens is that they are highly secretive, and methods for capturing them often do not work. Rossman (1960) noted that lesser sirens were among the most abundant amphibians in Pine Hills Swamp in southern Illinois when electroshocking was used to determine abundance, but that most collectors rarely found them.

Acknowledgments. Thanks to Russ Hendricks for providing additional literature sources.

Siren lacertina Linnaeus, 1766
GREATER SIREN

Russ Hendricks

1. Historical versus Current Distribution.
Greater sirens *(Siren lacertina)* are found in the Atlantic and Gulf Coastal Plains from

eastern Virginia (Burch and Wood, 1955) through extreme south Florida, west to southwestern Alabama (Conant and Collins, 1998; Petranka, 1998). A disjunct population occurs in the Rio Grande valley of northern Mexico and southern Texas (Flores-Villela and Brandon, 1992), although Petranka (1998) believes that further genetic characterization is needed to determine the status of sirens in Texas (see *Siren texana* account, this volume). Their historical distribution (Barton, 1808) is most likely similar to current distribution.

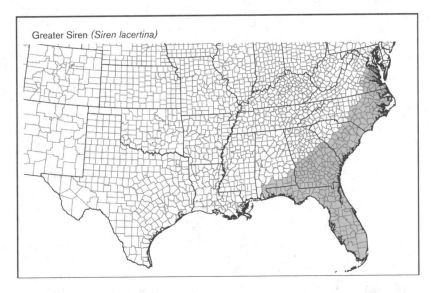

Greater Siren *(Siren lacertina)*

2. Historical versus Current Abundance.
Abundance from the 1960s to the present is not well documented in the literature, except for Florida, southeastern Georgia, and eastern South Carolina (Petranka, 1998) where greater sirens are considered common. Historically, greater sirens have been considered common in the southeastern United States (Barton, 1808; Jobson, 1940; Neill, 1949b), including Florida (Bishop, 1943). In the periphery of their range they are considered Locally Abundant to Rare (Burch and Wood, 1955; Martof, 1956a).

3. Life History Features.
A. Breeding.
i. Breeding migrations. In Gainesville, Florida, mature greater sirens move into shallow water or streams to congregate for breeding in February–March (Ultsch, 1973).
ii. Breeding habitat. In shallow water or streams (Ultsch, 1973).
B. Eggs.
i. Egg deposition sites. In Gainesville, Florida, eggs were found in 15-cm-deep water of a shallow ditch, 1 m from shore in a macrophyte bed predominated by *Myriophyllum* sp. (Goin, 1947c). Little else is known of nest site selection (Petranka, 1998).
ii. Clutch size. Captive females lay eggs singly or in small groups (Ultsch, 1973) that total around 500 eggs (Noble and

Marshall, 1932). Nests may be guarded by females (Ultsch, 1973).
C. Larvae/Metamorphosis.
i. Length of larval stage. Unknown.
ii. Larval requirements.
a. Food. Unknown, although juveniles are reported to feed on small invertebrates (Petranka, 1998).
b. Cover. Young greater sirens are often seen amid water hyacinth roots (Martof, 1973) or other heavy vegetation (Petranka, 1998).
iii. Larval polymorphisms. None reported.
iv. Time to metamorphosis. Difficult to determine. During the first year of life, larval striping is reduced, then lost; juveniles are variously mottled (Duelmann and Schwartz, 1958; see also Petranka, 1998).
v. Post-metamorphic migrations. Do not occur in the usual sense. As greater sirens mature, they are found in increasingly deeper waters (Duelmann and Schwartz, 1958; see also Petranka, 1998).
vi. Neoteny. Called the "most larval" of all salamanders (Noble, 1927b), sirens have external gills and lack hind feet. However, the skin undergoes metamorphosis (Martof, 1974).
D. Juvenile Habitat. Similar to larval and adult characteristics.
E. Adult Habitat. Greater sirens are found in muddy and weed-choked ditches (Funderburg and Lee, 1967), swamps, and ponds (Jobson, 1940; Neill, 1949b), as well as large lakes and streams. No differences have been reported regarding habitats of males versus females.
F. Home Range Size. Unknown.
G. Territories. Unknown.
H. Aestivation/Avoiding Desiccation. When faced with desiccating conditions, greater sirens burrow into the muddy bottom (Freeman, 1958). A cocoon comprised of primarily dead squamous epithelial cells is produced to retard water loss (Etheridge, 1990b), and the gills rapidly

atrophy (Noble, 1927b). A lab-starved specimen from Georgia survived 5.2 yr and lost 86.5% of its original weight before dying. Recovery from aestivation under suitable conditions is likely rapid (Martof, 1969).
I. Seasonal Migrations. No information exists directly regarding seasonal migrations. However, greater sirens in Florida are able to surmount a 46-cm (18-in) high dam with ease (Bishop, 1943). It is reported that greater sirens have the ability to leave water voluntarily (Osterdam, 1769; Barton, 1808; Harlan, 1826b), although this habit has not been noted by more contemporary authors.
J. Torpor (Hibernation). Torpor has not been observed, but feeding activity has been shown to decrease with colder water temperatures in Alabama (Hanlin and Mount, 1978).
K. Interspecific Associations/Exclusions. In Polk County, Florida, greater sirens were found in association with two-toed amphiumas *(Amphiuma means)*, peninsula newts *(Notophthalmus viridescens piarpicola)*, and lesser sirens *(S. intermedia;* Funderburg and Lee, 1967).
L. Age/Size at Reproductive Maturity. Unknown.
M. Longevity. One captive greater siren was reported to live 25 yr (Flower, 1936). Another, housed at the Cincinnati Zoo, lived almost 15 yr (Snider and Bowler, 1992).
N. Feeding Behavior. Primarily nocturnal (Hanlin and Mount, 1978) and carnivorous, although algae found in the digestive tracts of sirens has led some to believe that they may be omnivorous (Dunn, 1924; Ultsch, 1973; Hanlin, 1978). Able to filter feed from bottom debris (Hanlin, 1978) and also feed opportunistically (Hanlin and Mount, 1978). Prey include insects, crustaceans (Duellmann and Schwartz, 1958), gastropods, pelycypods, spiders, mollusks (Hanlin, 1978), crayfish, gastroliths, and small fish (Moler, 1994). Burch and Wood (Ultsch, 1973, 1955) suggest that opportunistic feeding may be accomplished by use of the Jacobsen's organ rather than vision due to the low water clarity that often exists in the habitats of greater sirens.
O. Predators. Information is scant regarding predators, however, greater sirens have been found in the stomachs of red-bellied mud snakes *(Farancia abacura;* Van Hyning, 1932) and American alligators *(Alligator mississippiensis;* Delany and Abercrombie, 1986).
P. Anti-Predator Mechanisms. A variety of noises have been attributed to the greater siren, which may function as a defense mechanism or may be involuntary (Maslin, 1950). Noises include yelping, a sound similar to green treefrogs *(Hyla cinerea)* from a distance (Conant and Collins, 1998), hissing (Barton, 1808), croaking (Ellis, 1767), and a sound like that of young ducks (Osterdam, 1769). Noble and Marshall (1932) reported that

greater sirens can escape disturbance by swimming away rapidly, while Oliver (1955a) states that sirens can produce a painful bite.

Q. Diseases. The only reference to disease is from a captive specimen that died from an infection of *Saprolegnia* sp. fungus (Goin, 1961).

R. Parasites. Platyhelminth parasites, including trematodes, cestodes, and nematodes, have been recorded in Florida, as summarized in McAllister et al. (1994).

4. Conservation.

Greater sirens have been considered common in the more central portions of their range, and from locally abundant to rare in more peripheral sites. In Maryland, one peripheral area, they are listed as Endandered/Extirpated and are afforded legal protection (Levell, 1997). One problem with permanently aquatic salamanders is that populations are difficult to monitor. States that harbor these animals should consider long-term monitoring programs.

"Siren texana" Goin, 1957
RIO GRANDE SIREN(S)

Darrel R. Frost, Michael J. Lannoo

Following Crother et al. (2000, p. 30), we do not include Rio Grande sirens *(Siren texana)* as a separate species from *Siren lacertina*. However, there is a problem regarding the identity of *"Siren texana"* and the taxonomic statuses of the two species of sirens in South Texas, as follows. Flores-Villela and Brandon (1992) found that two species of sirens occur sympatrically in the lower Rio Grande Valley of southern Texas and adjacent northeastern Mexico. One of these they could not separate morphologically from *S. lacertina* (otherwise found only in Florida) and they therefore considered the south Texas population of greater sirens to represent a disjunct population of *S. lacertina*. Flores-Villela and Brandon (1992) also examined the holotype of the south Texas *S. intermedia texana* Goin, 1957, and found that it was not conspecific with the *S. lacertina* population in South Texas, but rather with lesser siren *(S. intermedia)* populations

farther to the north in Texas. Further, they could not distinguish unequivocally *S. i. texana* from *S. i. nettingi* (from farther north), so they placed *S. i. texana* in the synonymy of *S. i. nettingi*. Therefore, as of 1992, two species of sirens were recognized in the lower Rio Grande Valley of south Texas and northeastern Mexico: *S. lacertina* (greater sirens) and *S. intermedia nettingi* (lesser sirens), the latter including the holotype of *S. i. texana*.

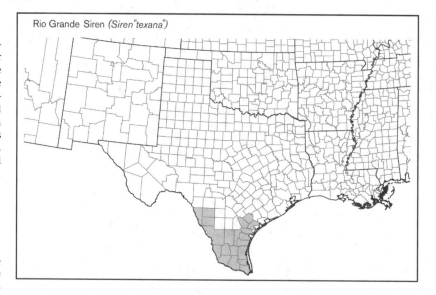

Rio Grande Siren *(Siren "texana")*

Subsequently, Dixon (2000, pp. 51–52) did not accept this arrangement, and instead applied the name *"texana"* (a synonym of lesser sirens, *S. intermedia*, according to Flores-Villela and Brandon, 1992) to greater siren *(S. lacertina)* populations in south Texas. The reasons that he gave for this reversal were that (1) the field guide by Conant and Collins (1998) had failed to recognize Flores-Villela and Brandon's arrangement (note: with no discussion and with no evidence that these authors were aware of the new arrangement; the 1998 treatment by Conant and Collins is identical to their 1991 treatment); (2) Dixon had received a personal communication from P. Moler that unpublished molecular evidence suggested that the greater sirens of south Texas and

Florida were not conspecific (note: which is not evidence that the holotype of *S. i. texana* is conspecific with the south Texas greater sirens, only that the Florida and Texas populations of greater sirens do not represent the same species), and (3) Flores-Villela and Brandon (1992) had found minor differences between the description of *texana* and the holotype itself (note: which is not evidence that *texana* is other than a junior synonym of *S. inter-*

media). So, seemingly on the basis that Flores-Villela and Brandon (1992) *might be* wrong, even though no new evidence that bore on the problem had been presented, Dixon applied the name *texana* to the greater sirens of South Texas, even though this arrangement is not consistent with the published evidence. Unfortunately, because of this action, it is not clear which (or what) naturally occurring population(s) are being protected by the Texas Parks and Wildlife Department under the name *"Siren texana,"* because what evidence has been published suggests that there are two species, not one, of sirens in the lower Rio Grande Valley. Indeed, Texas recently listed "south Texas sirens (large form)" as Threatened (www.tpwd.state.tx.us).

Factors Implicated in Amphibian Population Declines in the United States

DAVID F. BRADFORD

Many species of amphibians have declined substantially in distribution or number of populations in the United States and globally, and a variety of anthropogenic and natural factors have been suggested as causal agents in these declines (e.g., Green, 1997a,b; Lannoo, 1998, 2003; Alford and Richards, 1999; Houlahan et al., 2000, 2001; Semlitsch, 2000; Alford et al., 2001; Lannoo et al., Introduction, Part One this volume). Evidence for the causal action of these agents derives from many types of sources, all of which are important in seeking to understand amphibian population declines. Clear cause-and-effect relationships have been demonstrated in some cases, but the number of such cases is remarkably small given the number of species thought to be declining. More frequently, researchers have reported associations between putative causal agents and population declines. Although the weight of such correlative or circumstantial evidence is strong in many cases, it remains weak in others, usually due to the paucity of research. Nevertheless, such observations can be helpful by providing insight into the relative importance of causal factors and by suggesting which factors warrant the highest priority for further research or management action.

The purpose of this analysis is to evaluate the relative frequency of factors implicated as adversely affecting amphibian populations in the U.S., and to identify taxonomic and geographic patterns for these effects. In this analysis I rely on informed investigators for information about causal factors and thus encompass the gamut of knowledge from well-documented studies to speculations. Such expert opinion may provide the best insight available about the factors responsible for declines. I have used a consistent method of evaluation for each species by reviewing earlier drafts of the accounts written for the U.S. species represented in this book. These species accounts are provided in a standardized format and are written by individuals who have substantial experience with the subject species. To the extent that sufficient information is provided in the accounts, I compile the factors implicated as adversely affecting populations and then assess the status of each of these species with regard to the stability of its distribution and number of populations. I also compare the frequency of adverse factors among the species relative to species status and region of the United States. A secondary purpose of this analysis is to identify taxonomic and geographic

patterns in species status. In particular, I evaluate differences between the western United States and elsewhere, as amphibian declines are often thought to be greatest in the western United States (Hayes and Jennings, 1986; Corn, 2000; Lannoo et al., Introduction, Part One, this volume).

Methods

I reviewed earlier drafts of the species accounts for each amphibian native to the United States (this volume) for information to classify the species by status and to identify factors thought to be affecting its populations. The total number of species accounts available at the time was 267 (91 of 103 anurans and 176 of 185 caudates). The species accounts were written to address a standardized list of topics by authors with extensive experience with the subject species. The status of each species was determined relative to any change in its geographic distribution in the United States, or change in number of sites within this range. Status categories were Increasing, No Change, Net Extirpations, Major Decline, and Not Determined, as defined in Table EP-1. Information for this classification was provided primarily by the sections "Historical versus Current Distribution," "Historical versus Current Abundance" and "Conservation." Two categories (Net Extirpations and Major Decline) were combined for many analyses and are referred to as "adversely affected species."

Although not a specific topic required by the standardized format for the species accounts, most authors provided information on causal factors. A factor was included in the present analysis only if it was implicated by the authors (based on either known or suspected relationships) as affecting the persistence of populations, that is, it was associated with extirpations of whole populations or portions inhabiting a given area, or it was deemed a substantial threat to the persistence of populations. The factor was not included if it was indicated as affecting only individual health or mortality, such as moderate levels of disease or predation. More than one factor was implicated for many species. Factors were categorized according to descriptions in Table EP-2. For the three most commonly implicated factors (i.e., land use change, exotic species, and chemical contamination), the specific land use factors (Table EP-2), exotic taxa, and chemical components or sources were also identified from information presented in the accounts.

Definitions for Species Status with Respect to Change from Historical Geographic Range within the United States, or Number of Sites within This Range

Species Status	Definition
Increase	Recognized net increase in range (excluding introductions beyond historical range)
No Change	No apparent net change in range or number of sites (excluding introductions beyond historical range)
Net Extirpations	At least one population or area-specific extirpation is recognized and losses are not counterbalanced by new populations
Major Decline	Substantial decrease in range or number of sites. Corresponds approximately to Green's (1997a, p. 293) definition of a population decline as "the condition whereby the local loss of populations across the normal range of a species so exceeds the rate at which populations may be established, or reestablished, that there is a definite downward trend in population number."
Not Determined	Information available or provided is insufficient to determine status

NOTE: For many analyses, two categories, Net Extirpations and Major Decline, are collectively referred to as "adversely affected species."

To evaluate the geographic distribution of species by status and implicated factors, species ranges were categorized by region, or combinations of regions, based on species accounts (this volume) and range maps (Stebbins, 1985; Conant and Collins, 1991; and Lannoo et al., this volume). The regions for this analysis include the western United States, defined as the Rocky Mountains, south and west; the eastern United States, defined as the Appalachian Mountains, southern Mississippi and Alabama, and east; and the central United States, which represents the area in between.

Differences in proportions among groups were analyzed using Chi-square tests where sample sizes were sufficient and Fisher exact tests otherwise. In some cases with multiple groups, Tukey-type multiple comparisons among proportions were performed (Zar, 1999).

Results

Sufficient information was provided to classify the status of 74 (81%) of the 91 native anurans considered and 108 (61%) of the 176 native caudates considered with regard to change in their geographic range or number of sites within the range (Table EP-3). Almost half of the 91 anurans were classified as "adversely affected species" (i.e., about 22% as Major Decline and 27% as Net Extirpations), and well over one third of the caudates were also classified this way (i.e., 6% as Major Decline and 32% as Net Extirpations). Approximately one fourth of the species were classified as No Change for both anurans (26%) and caudates (23%). Relatively few species of anurans (5%) and no caudates were classified as increasing (exclusive of introductions beyond their native range). Accounts for four of the five increasing anurans indicated that range increases were associated with anthropogenic changes in land use.

Three anuran families predominate in the United States: Bufonidae (22 native species; 20 in this analysis), Hylidae (28 native species; 26 in this analysis), and Ranidae (28 native extant species; all in this analysis; Fig. EP-1A). The proportion comprising "adversely affected species" (i.e., Major Decline or Net Extirpations) did not differ significantly among the families. In contrast, the proportion of species classified as Major Decline alone differed significantly (χ^2 2 df = 15.46, p < 0.001), with the Ranidae showing by far the greatest frequency of species in this category (46% [13/28] of native

extant species). Among caudates, the Plethodontidae (132 of 139 species were considered) predominate and Ambystomatidae (16 of 16 recognized taxa were considered) rank second in number of species (Fig. EP-1B). The proportion comprising "adversely affected species" differed significantly among the caudate families (χ^2 2 df = 11.34, p < 0.01), with Ambystomidae showing the greatest frequency among the larger families in this category (69% [11/16] taxa). There were too few taxa classified as Major Decline to statistically compare frequencies among caudate families.

Regional differences in the proportion of "adversely affected species" were evident for both anurans (χ^2 4 df = 9.92, p < 0.05) and caudates (χ^2 3 df = 19.26, p < 0.001; Fig. EP-2). For anurans, "adversely affected species" were most prevalent in the western United States, comprising 70.3% (19) of the 27 species restricted to this region. Major Decline species comprised 51.9% (14) of the western species. For caudates, "adversely affected species" were most prevalent in the Central to East region, comprising 61.9% (26) of the 42 species restricted to this region.

As a focus on regional and taxonomic differences in the frequency of "adversely affected species," data are compared for a specious family for both anurans and caudates relative to other families in the western and non-western U.S. (Fig. EP-3). For anurans, this analysis reveals that the greater frequency of "adversely affected species" among ranids (Fig. EP-1A) and western species (Fig. EP-2A) can be attributed to the influence of western ranids alone (Fig. EP-3A). That is, "adversely affected species" were exceptionally frequent among ranids in the western United States (100% [13/13]) in comparison to ranids elsewhere and non-ranids in either the west or non-west (χ^2 4 df = 28.45, p < 0.001; Tukey-type multiple comparison among proportions, p < 0.05), whereas there were no significant differences in frequency of "adversely affected species" among these latter three groups. Moreover, 85% (11/13) of western ranids were classified as Major Decline species. For caudates, the frequency of "adversely affected species" did not differ between plethodontids in the west and non-west, nor between non-plethodontids in the west and non-west (Fig. EP-3B). However, the frequency of "adversely affected species" was significantly greater for non-plethodontids in the non-west (63.3%) than for plethdontids in the non-west (31.0%; χ^2 4 df = 11.79, p < 0.05; Tukey-type multiple comparison among proportions, p < 0.01).

TABLE EP-2

Description of Categories for Factors Affecting Amphibian Populations

Adverse Factor	Specific Land Use Factor	Descriptions Represented in Species Accounts
Land Use		
	Generic habitat change	Habitat loss/destruction/conversion/alteration/modification/degrada-tion/ disturbance/fragmentation (especially wetlands & low lying areas); intensive land use; human development; wetland/swamp/marsh draining; shoreline development; deforestation/forest conversion (to unspecified use); siltation
	Agriculture	Agriculture; cultivated/irrigated agriculture; farming; effects of agriculture on water quality & hydrological regime
	Urban development	Urban/commercial/residential/domestic development/sprawl; effects of urban development on water quality and hydrological regime
	Timber harvest/Silviculture	Timber harvest; logging; silviculture; clear cutting; forestry practices; forest plantation; industrial forests
	Livestock grazing	Livestock grazing; grazing
	Altered fire regime	Fire suppression; disruption of natural fire cycle; burning; altered sedimentation dynamics resulting from wildfires
	Recreational land use or development	Recreation; recreational land use; recreational development
	Road construction/use	Road construction/building; road mortality; vehicular traffic
	Mining	Mining; recreational gold mining effects on streambed habitat (excludes explicit pollution effects, which are categorized as chemical contamination)
	Miscellaneous	Military training; mowing; ski slopes; trails; change of disturbance regime of vegetation
Water Source Manipulation		Impoundments; water storage facilities; water diversions; groundwater pumping; flood control; water flow management practices
Exotic Species		Exotic/introduced species/predators/competitors/defoliators; hybridization with introduced amphibians
Disease		Disease (fungi and other pathogens)
Chemical Contamination		Pollution; acid precipitation/acidification; acid mine drainage; mine water pollution; sewage; heavy metals; (vague descriptions such as "water quality degradation" are not included here)
UV-B Radiation		Increased UV-B; UV-B enhancement of adverse effects of other factors
Collecting/Harvesting		Collecting; harvesting; commercial exploitation; use as fish bait
Other		Extreme weather; displacement by other native amphibians; predation or increased predation by native predators; feral burros; "other unknown factors"

TABLE EP-3

Status of 91 Anuran and 176 Caudate Species Native to the United States with Respect to Change in Geographic Range or Number of Sites within this Range. (Status Categories as Listed in Table EP-1)

Status	Anurans		Caudates	
	Number of Species	Percent of Species	Number of Species	Percent of Species
Increased	5	5.5%	0	0.00%
No Change	24	26.4%	40	22.7%
Net Extirpations	25	27.5%	57	32.4%
Major Decline	20	22.0%	11	6.3%
Status Not Determined	17	18.7%	68	38.6%

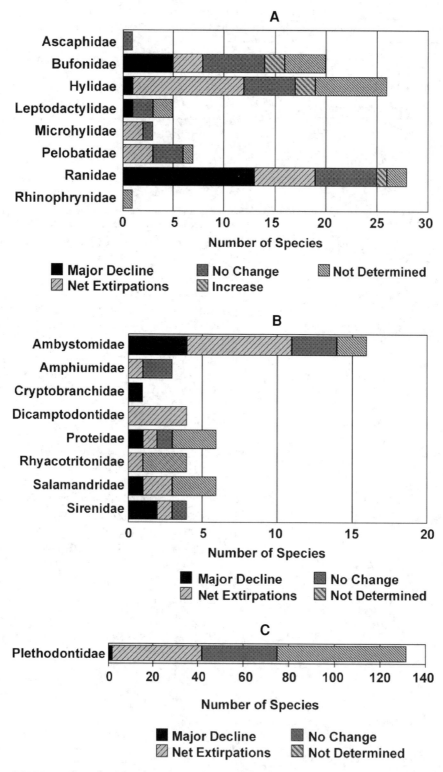

FIGURE EP-1 Status frequency for native species in each family. (A) Anurans; n = 91 species. (B) and (C) Caudates; n = 176 species. Status categories as listed in Table EP-1.

Authors implicated factors affecting the persistence of populations for 58.2% (53/91) of the considered anuran species, and 52.8% (93/176) of the considered caudate species. These values were much higher for "adversely affected species," that is, 84.4% (38/45) of anurans and 88.2% (60/68) of caudates. Among the species with factors implicated, the frequencies for specific factors (e.g., land use, exotic species) did not differ between "adversely affected species" and those species classified in other status categories, except that exotic species were implicated significantly more frequently for "adversely affected" anurans than anurans in other status categories (χ^2 1 df = 4.76, p < 0.05). Because of this general similarity in frequency of factors among status categories, data for adverse factors are presented below without regard to species status.

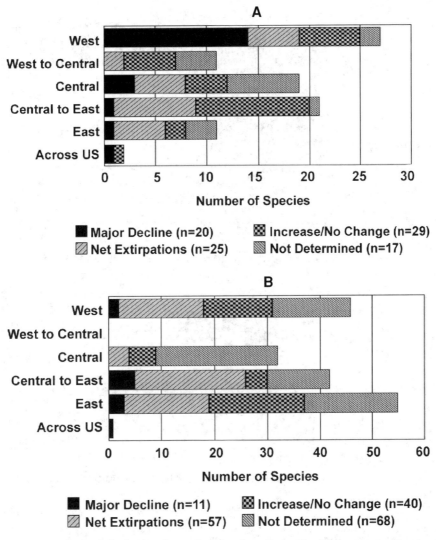

FIGURE EP-2 Regional distribution of status frequency for native species. (A) Anurans; n = 91 species. (B) Caudates; n = 176 species. Status categories as listed in Table1. West, Central, and East refer to the western, central, and eastern United States, exclusively, as defined in Methods. "West to Central" and "Central to East" refer to combinations of two regions.

Of the species with adverse factors implicated, land use was by far the most frequently implicated factor for both anurans (77.4% [41/53]) and caudates (91.4% [85/93]; Fig. EP-4). Exotic species were the secondmost frequently implicated factor for anurans (39.6% [21/53]), and third most for caudates (9.7% [9/93]). Chemical contamination was the third most frequently implicated factor for anurans (18.9% [10/53]), and second most for caudates (17.2% [16/93]). Less frequently implicated factors were disease, water source modification, collecting/harvesting, UV-B radiation, and others.

Among the species for which land use change was implicated, generic habitat change was the most frequently represented, attributed to 43.9% (18/41) of anurans and 52.9 % (45/85) of caudates (Fig. EP-5). The category Generic Habitat Change category was assigned to a species when no specific land use factor (e.g., agriculture) was provided (most cases) and in a few cases when a specific land use factor was implicated but other unspecified land uses or habitat change was also implied. Among the specific land use factors identified for anurans, agriculture and urban development were the most fre-

quently noted, followed by timber harvest/silviculture, livestock grazing, recreational use/development, altered fire regime, road construction/use, mining, and miscellaneous. Among the specific land use factors identified for caudates, timber harvest/silviculture predominated, followed by urban development, agriculture, and road construction/use. Much less frequently implicated were mining, livestock grazing, and altered fire regimes.

Among species for which adverse factors were implicated, exotic species were indicated as a factor more frequently among anurans (40%) than caudates (10%; χ^2 1 df = 16.75, p < 0.001; Fig EP-4). Among species for which exotic species were noted, introduced fishes were identified most frequently for both anurans and caudates (Fig. EP-6). Fish taxa identified were primarily the family Salmonidae, warm-water game fishes in several families, and the mosquitofish (*Gambusia affinis*). Other exotic taxa indicated were American bullfrogs (*Rana catesbeiana*) and other amphibians, including marine toads (*Bufo marinus*) and tiger salamanders (*Ambystoma tigrinum*), crayfish (particularly *Procambarus clarkii* and *Orconectes virilis*), and insects. Insects were in-

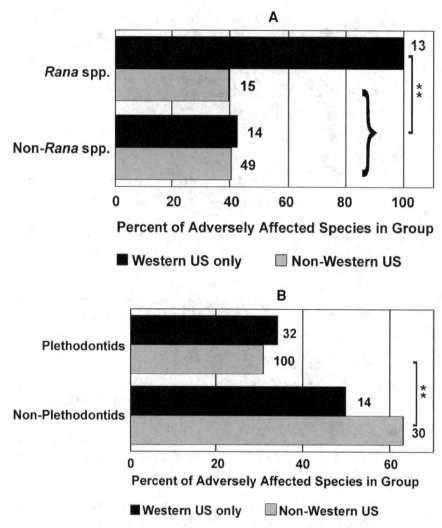

FIGURE EP-3 **(A)** Frequency of *Rana* species (n = 27 species) versus non-*Rana* species (n = 64 species) that are classified as Net Extirpations or Major Decline for the western versus non-western United States. **(B)** Frequency of plethodontid salamanders (n = 132 species) versus non-plethodontids (n = 44 species) that are classified as Net Extirpations or Major Decline for the western versus non-western United States. Non-western United States refers to species whose distribution is partially or totally outside the western United States, as defined in Methods. Numbers adjacent to bars indicate total number of species in each category. Square brackets and asterisks indicate significant differences between groups at 0.01 level (see text).

volved as a result of defoliation of forests by introduced gypsy moths and hemlock wooly adelgids.

Chemical contamination was indicated, in order of frequency, as general pollution, acid precipitation, pesticides/herbicides, mine drainage, and single occurrences of heavy metals and sewage. Specified disease agents identified were primarily chytrid fungus (*Batrachochytrium dendrobatidis*), but included single cases of the fungus *Saprolegnia ferax*, and "other pathogens."

Some of the adverse factors identified differed between the western United States and elsewhere (Fig. EP-7). Conspicuously, among anurans that had adverse factors indicated, exotic species were implicated much more frequently for anurans in the western United States (76.2% [16/21]) than for those in the non-western U.S. (15.6% [5/32]; χ^2 1 df = 16.35, p < 0.001; Fig. EP-7A). Ranid frogs represented most (12 of 14) of the anurans affected by exotic species in the western United States. Also, several factors combined for this analysis (i.e., factors other

than land use, exotic species, and chemical contamination) were implicated significantly more frequently for anurans in the west than for those in the non-west (χ^2 1 df = 6.12, p < 0.05). In contrast, the frequencies for land use change and chemical contamination did not differ significantly between these two regions. For caudates, chemical contamination was significantly more frequently implicated in the non-western United States (23.1% [15/65]) than in the western United States (3.6% [1/28]; χ^2 1 df = 3.947, p < 0.05), whereas land use, exotic species, and all other factors combined did not differ significantly between the two regions (Fig. EP-7B).

Discussion

Several limitations to the above findings should be considered prior to an evaluation of the patterns among species status and adverse factors.

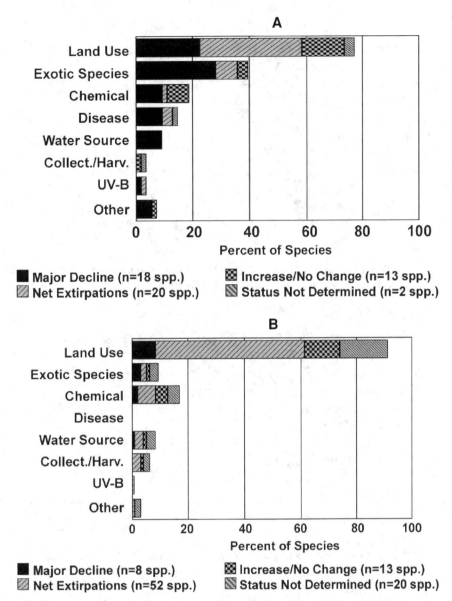

FIGURE EP-4 Frequency of factors implicated (i.e., known or suspected) in affecting persistence of populations. (A) Anurans; n = 53 species. (B) Caudates; n = 93 species. More than one factor may be attributed to a species. Factors as listed in Table EP-2; status categories as listed in Table EP-1.

1. Although the species accounts are all provided in a standardized format, they vary in authorship, extent and nature of data available, level of detail, and degree to which the authors address changes in distribution and known or suspected causes for these changes. In particular, many species of plethodontid salamanders have been described within the past decade and little data are available for many of their biological attributes. Nevertheless, the patterns documented herein are based on 267 species accounts, written by well over 100 authors, such that some general patterns among species are likely to be evident regardless of variation in treatment among the species.

2. The classification of species status (i.e., Major Decline, etc.) is subjective and based solely on the information provided by the authors. Thus, these categorizations are only approximations of reality. For example, some species classified as Net Extirpations may have experienced little change in their overall range or number of populations, whereas some species classified as No Change likely have experienced some population extirpations that were not acknowledged in the species accounts.

3. In most cases, the adverse factors implicated by the authors were based on limited observations. Relatively few studies have firmly established which factors are affecting anuran populations. Nevertheless, for many species expert opinion probably presents the best estimate available for known or suspected adverse factors.

4. Authors noted adverse factors for most species for which status of the species could be determined. Such implications increased from a frequency of 55% for anurans and 38% for caudates for species classified as No Change or Increasing, to 84% and 88%, respectively,

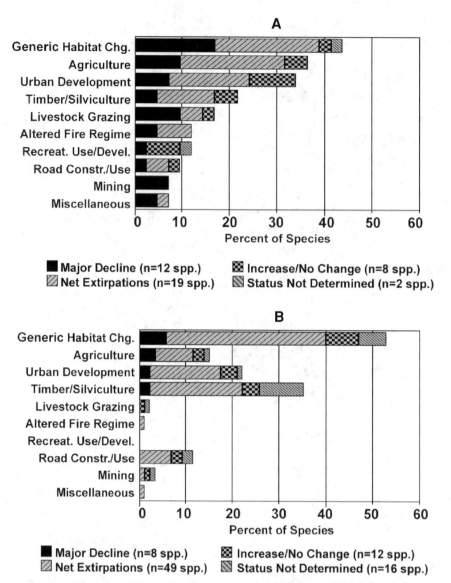

FIGURE EP-5 Frequency of specific land use factors implicated in affecting persistence of populations. (A) Anurans; n = 41 species. (B) Caudates; n = 85 species. More than one factor may be attributed to a species. Factors as listed in Table E-2; status categories as listed in Table EP-1.

for species classified as Major Decline or Net Extirpations. The lower frequencies for the first group are not surprising because this group is comprised of species whose original populations have generally undergone the least change.

5. The above analyses are based on frequency of implication by the authors, with each factor and each species receiving equal weight. For example, if two factors were implicated by an author, the two factors received equal weight even though one may have been viewed as much more important than the other. Thus, the frequency of implication of a factor among species reflects only the commonness of this factor as a problem among species, and does not necessarily reflect the relative importance of the factor within species.

6. Because regional and global factors (e.g., UV-B irradiance, climatic change, long-range pesticide drift) are less readily observable than local factors (e.g., land use,

exotic species, disease), regional and global factors may be underrepresented relative to local factors.

7. Much information has emerged in recent years about some factors, most notably disease, airborne contaminants, and UV-B, that may not be reflected in the earlier versions of the species accounts reviewed.

Several authors have recognized that anuran population declines in the United States appear to have been most pronounced in the west and among ranid frogs (e.g., Hayes and Jennings, 1986; Corn, 2000; Lannoo et al., Introduction, Part One this volume). Although the present study reiterates these two observations based on the frequency of species classified as Major Decline or Net Extirpations, it also suggests that they do not represent general patterns unto themselves. Rather, ranid frogs in the western United States have an exceptionally high frequency of such affected species (i.e., 100% of 13 species restricted to the western United States), whereas non-western

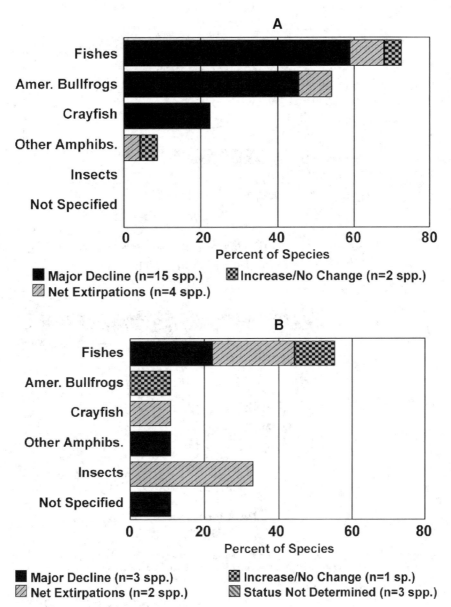

FIGURE EP-6 Frequency of exotic taxa implicated as affecting persistence of populations. (A) Anurans; n = 21 species, (B) Caudates; n = 9 species. More than one taxon may be attributed to a species. Status categories as listed in Table EP-1.

ranids, western non-ranids, and non-western non-ranids do not differ in frequency of affected species (Fig. EP-3). The prevalence of "adversely affected species" among western ranids may be associated with greater exposure to exotic species in the western United States, as implied by the significantly greater frequency of exotic species as factors among adversely affected anurans in the western United States than the non-western United States (Fig. EP-7). Indeed, exotic species were implicated as adverse factors for all 13 western ranids, and these 13 species comprise most of the 17 anurans indicated as affected by exotic species in the western United States. This pattern is not surprising given that western ranids in general evolved in permanent or semi-permanent waters largely in the absence of large aquatic predators and competitors (Hayes and Jennings, 1986). The exotic species implicated, consisting primarily of sport fishes, American bullfrogs, and crayfish, originated mostly from the non-western United States, and are largely restricted

to permanent or semi-permanent waters (Cox, 1999). Another hypothesis for the apparent prevalence of declining species among western ranids is that western aquatic habitats are mostly set in arid landscapes, where degradation and fragmentation of habitat may dramatically reduce dispersal among populations such that recolonization of sites following local extinction is largely precluded (Corn, 2000).

Land use was by far the most frequently implicated factor among both anurans and caudates. The land uses identified would be expected to affect amphibians primarily through habitat loss, alteration, or fragmentation—processes that have long been viewed as a main cause of amphibian population declines worldwide (Wake, 1991). Among the specific land use factors, agriculture, urban development, and timber harvesting/silviculture were the most frequently noted factors for both anurans and caudates, with agriculture and urban development predominating for anurans and timber harvesting/

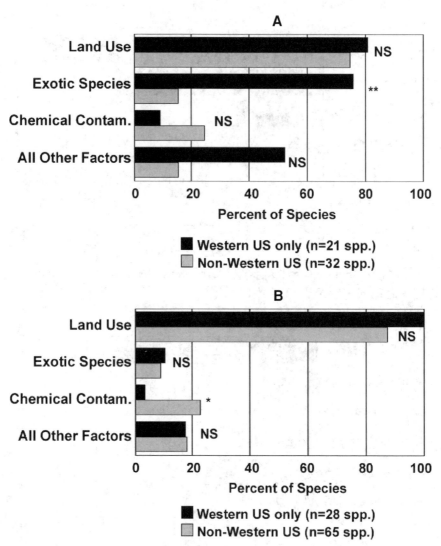

FIGURE EP-7 Frequency of factors implicated as affecting persistence of populations in the western versus non-western United States. (A) Anurans; n = 53 species. (B) Caudates; n = 93 species. NS indicates not significant at 0.05 level; * indicates significance at 0.05 level; ** indicates significance at 0.01 level (χ^2 or Fisher exact test).

silviculture predominating for caudates. Authors of several species accounts commented on the pronounced impact of the conversion of native forest to pine plantations in the southern U.S. on native amphibians, such as gopher frogs *(R. capito)* and several species of *Plethodon* salamanders. Other widespread land uses commonly implicated as affecting amphibian populations were road construction/use, livestock grazing, recreational use/development, altered fire regime, recreational use/development, and mining. In general, these latter factors are less devastating to natural habitats than are agriculture, urban development, and timber harvest/silviculture or operate in smaller patches. Not all land use changes were viewed as detrimental. For several species agriculture and other land uses were beneficial and have apparently led to a range expansion in some cases, such as pig frogs *(R. grylio)* and Woodhouse's toads *(B. woodhousii)*. Also, livestock grazing was viewed as detrimental in some areas, but beneficial in others as a result of the creation of aquatic habitat to support livestock (e.g., stock tanks). Examples are several spadefoot toad species, southwestern ranids, and northwestern salamanders *(A. gracile)*.

Exotic species were the second-most frequently implicated factor for anurans and the third-most for caudates. The frequency was much higher among anurans than caudates perhaps in part because of the relative paucity of pond-breeding salamanders in the western United States, where exotic species are prevalent (see above). The most frequently implicated exotic taxa were fish, followed by American bullfrogs, crayfish, forest-defoliating insects, and other amphibians. Despite the relatively high frequency of exotic species mentioned as an adverse factor, at least among anurans, relatively few studies have conclusively documented the effects of exotic species as agents of amphibian population declines, particularly for bullfrogs, crayfish, and other amphibians (Carey et al., 2003). Many of the affected native amphibians are simultaneously exposed to multiple exotic taxa and habitat degradation (Hayes and Jennings, 1986).

Chemical contamination ranked second among adverse factors for caudates and third for anurans. The frequency of implication may be under-represented in this analysis because this factor was tallied only if it was specifically mentioned; yet some broad land use factors such as agriculture may include a variety of processes such as chemical contamination. The specific types of chemical contamination identified included acid precipitation, pesticides/herbicides, and mine water pollution.

Disease was reported as a factor only for anurans, affecting 15% of the species for which adverse factors were identified. The most commonly identified pathogen was the chytrid fungus, a species not identified until 1999 (Longcore et al., 1999). This disease is likely under-represented among the accounts because most of the accounts were drafted prior to the large increase in knowledge about the extent of chytrid infections since 1999. Water source manipulation was implicated for several anurans and caudates. However, as with chemical contamination, this factor may be under-represented in this analysis because some broad land use factors, such as agriculture or other agents of habitat alteration, may include associated activities such as water source manipulation. Collecting/ harvesting and UV-B were implicated as adverse factors for only a few species.

Summary

Factors affecting amphibian populations in the United States were evaluated using information from species accounts written in a standardized format by numerous authors (this volume). For each species, factors implicated by the authors (i.e., known or suspected) as affecting the persistence of populations were identified. Each species was also classified by status with regards to change in its historical geographic range or number of sites within this range and region of the United States. Information was sufficient to classify the status of 81% of the 91 anuran species native to the United States that were considered, and 61% of the 176 native caudate species considered. Species classified as Major Decline or Net Extirpations (collectively referred to as "adversely affected species") comprised 49% of the anurans and 38% of the caudates. Approximately one fourth of the species were classified as No Change for both anurans (26%) and caudates (23%), and relatively few species were classified as Increase (5% of anurans and 0% of caudates). The frequency of "adversely affected species" was exceptionally high for ranids in the western United States, whereas no differences in frequency were evident among non-western ranids, western non-ranids, and non-western non-ranids. Specific adverse factors were identified for 58% (53) of the 91 anurans considered and 53% (93) of the 176 caudates considered. Of the species with such factors implicated, land use was the most frequently implicated for both anurans (77% of the 53 species) and caudates (91% of the 93 species). Exotic species were the secondmost frequently implicated factor for anurans (40%) and third for caudates (10%). Chemical contamination ranked third for anurans (19%) and second for caudates (17%). Less frequently implicated factors were disease, water source modification, collecting/harvesting, and UV-B radiation. Among the anurans with adverse factors implicated, exotic species were noted significantly more frequently in the western United States (76% of 21 species) than in the other regions (16% of 32 species). Among caudates, chemical contamination was implicated significantly more frequently in the non-western United States (23% of 65 species) than the western United States (4% of 28 species). Among specific land use factors, agriculture, urban development, and timber harvest/silviculture ranked first, second, and third in frequency for anurans and third, second, and first for caudates. Other land use factors were road construction/use, livestock grazing, altered fire regime, recreational use/development, and mining. Among exotic taxa, introduced fishes were implicated most frequently, followed by American bullfrogs, crayfish, defoliating insects, and other amphibians. Specific chemical contaminants included acid precipitation, pesticides/herbicides, and mine water pollution.

Acknowledgments. I am grateful to all species accounts authors, and Chad Cross for assistance with statistical analyses. Notice: The U.S. Environmental Protection Agency (EPA), through its Office of Research and Development, performed the analyses described herein. This chapter has been subjected to EPA's peer and administrative review and has been approved for publication.

CONCLUSION

MICHAEL LANNOO

This work synthesizes and offers direction. In Part One we have attempted to present what we know about the extent and causes of amphibian declines and what we can do about them. In Part Two we present the life history and natural history features needed to manage for amphibians, with a current assessment of their distribution. In assembling the literature for this project, and with a quick look at the species accounts, what immediately is noticeable is that a few species are well known and have a large literature, some species are better known and have a modest literature, and many species are almost unknown. At least part of this problem arises from an increasing tendency for workers to seek research questions from the scientific literature (as proven by the inevitable graduate student lament, "Everything interesting has already been done") rather than from field observations. Therefore, an existing scientific literature creates a future scientific literature and results in a species bias. I would strongly encourage workers to explore species that are not well known and to seek questions from field observations.

A second cause for species bias is that (science being about the knowable and observable) some species are easily found and observable, while others, being usually underground or underwater, are not. Citing the Peterson-Dorcas call recorder (Peterson and Dorcas, 1992, 1994) as an example, I would encourage scientists to explore creative new techniques for observing and monitoring inconvenient animals.

Not since the early 1950s (following the publications of Bishop's [1943] *The Handbook of Salamanders* and Wright and Wright's [1949] *The Handbook of Frogs and Toads*) have we had a comprehensive view of the life histories of North American amphibians. (However, it should be noted that these two classics, while part of the same Handbook Series, differed markedly in style, format, and species treatments.) From that time, we have more than doubled the number of salamander taxa recognized (from 88 to 186) and advanced from 73 to 103 recognized taxa of frogs (indeed, we anticipate the further addition of species as several currently recognized species are split; for an explanation of the paradox between the declining amphibian phenomenon and the continued addition of recognized species, see Hanken, 2000), and we have once again lost our comprehensive consideration. In presenting these standardized accounts, it is my hope that we have re-consolidated the detailed information on U.S. amphibians in a way similar to the information Bent's *Life Histories of North American Birds Series* (published from 1919–1968) provided for birds (unlike amphibian taxonomy, avian taxonomy in the United States has not changed to any appreciable degree in the past 50 years). And, in my view, two fine derivative works (Ehrlich et al's. [1988] The Birder's Handbook and Sibley's [2000] *The Sibley Guide to Birds*) would not have been possible without Bent's original. The fact that Sibley (2000) could consider 810 bird species, while we present here the first detailed, comprehensive consideration of our 289 amphibian taxa, emphasizes the previous lack of centralized knowledge about U.S. amphibians.

The question arises—where do we go from here? First, I encourage people to focus on gaps in the species account information—it should be obvious where we need additional work, and we must begin to fill these gaps. Second, I encourage people to use this information to make informed management decisions so that we may indeed conserve the amphibians of the United States.

LITERATURE CITED

Abbott, C.C. 1882. Notes on the habits of the Savannah cricket frog. American Naturalist 14:707–711.

Abbott, C.G. 1934. Cold-blooded vertebrates. Smithsonian Scientific Series, Volume 8, Washington, D.C.

Abbott, P.L. 1975. On the hydrology of the Edwards Limestone, south-central Texas. Journal of Hydrology 24:251–269.

Abbott, P.L. and C.M. Woodruff Jr. (Eds.). 1986. The Balcones Escarpment: Geology, Hydrology, Ecology and Social Development in Central Texas. Geological Survey of America, San Diego, California.

Abbott, R.T. 1950. Snail invaders. Natural History 59:80–85.

Abdullah, A.R., C.M. Bajet, M.A. Matin, D.D. Nhan and A.H. Suliman. 1997. Ecotoxicology of pesticides in the tropical paddy field ecosystem. Environmental Toxicology and Chemistry 16:59–70.

Abourachid, A. and D.M. Green. 1999. Origins of the frog-kick? Alternate-leg swimming in primitive frogs, families Leiopelmatidae and Ascaphidae. Journal of Herpetology 33:657–663.

Acker, P.M., K.C. Kruse and E.B. Krehbiel. 1986. Aging *Bufo americanus* by skeletochronology. Journal of Herpetology 20:570–574.

Ackerman, K. 1975. Rare and endangered vertebrates of Illinois. Unpublished Technical Report, Bureau of Environmental Science, Illinois Department of Transportation, Springfield, Illinois.

Adams, D.C. 2000. Divergence of trophic morphology and resource use among populations of *Plethodon cinereus* and *P. hoffmani* in Pennsylvania. Pp. 383–394. *In* Bruce, R.C., R.G. Jaeger and L.D. Houck (Eds.), The Biology of Plethodontid Salamanders. Kluwer Academic/Plenum Publishers, New York.

Adams, D.C. and C.K. Beachy. 2001. Historical explanations of phenotypic variation in the plethodontid salamander *Gyrinophilus porphyriticus*. Herpetologica 57:353–364.

Adams, D.R. 1968. Stomach contents of the salamander *Batrachoseps attenuatus* in California. Herpetologica 24:170–172.

Adams, L. 1942. The natural history and classification of the Mount Lyell salamander, *Hydromantes platycephalus*. University of California Publications in Zoology 46:179–204, plates 21–22.

Adams, M.A., D.E. Schindler and R.B. Bury. 2001. Association of amphibians with attenuation of ultraviolet-B radiation in montane ponds. Oecologia 128:519–525.

Adams, M.D. and M.J. Lacki. 1993. Factors affecting amphibian use of road-rut ponds in Daniel Boone National Forest. Transactions of the Kentucky Academy of Science 54:13–16.

Adams, M.J. 1993. Summer nests of the tailed frog *(Ascaphus truei)* from the Oregon Coast Range. Northwestern Naturalist 74:15–18.

Adams, M.J. 1999. Correlated factors in amphibian decline: exotic species and habitat change in western Washington. Journal of Wildlife Management 63:1162–1171.

Adams, M.J. 2000. Pond permanence and the effects of exotic vertebrates on anurans. Ecological Applications 10:559–568.

Adams, M.J. and R.B. Bury. 2002. The endemic headwater stream amphibians of the American Northwest: associations with environmental gradients in a large forested preserve. Global Ecology and Biogeography 11:169–178.

Adams, M.J. and S. Claeson. 1998. Field response of tadpoles to conspecific and heterospecific alarm. Ethology 104:955–961.

Adams, M.J. and A.G. Wilson Jr. 1993. Geographic distribution: *Ascaphus truei* (tailed frog). Herpetological Review 24:64.

Adams, M.J., R.B Bury and S.A. Swarts. 1998. Amphibians of the Fort Lewis Military Reservation, Washington: sampling techniques and community patterns. Northwest Naturalist 79:12–18.

Adams, M.J., S.D. West and L. Kalmbach. 1999. Amphibian and reptile surveys of U.S. Navy lands on the Kitsap and Toandos Peninsulas, Washington. Northwest Naturalist 80:1–7.

Adams, N.G.K. 1967. *Bufo marinus* eaten by *Rattus rattus*. Journal of the North Queensland Naturalists' Club 34:5.

Adams, S.B. and C.A. Frissell. 2001. Thermal habitat use and evidence of seasonal migration by Rocky Mountain tailed frogs, *Ascaphus montanus*, in Montana. Canadian Field-Naturalist 115:251–256.

Adamson, M.L. 1980. *Gyrinicola batrachiensis* (Walton, 1929) n. comb. (Oxyuroidea; Nematoda) from tadpoles in eastern and central Canada. Canadian Journal of Zoology 59:1344–1350.

Adamson, M.L. 1981a. Development and transmission of *Gyrinicola batrachiensis* (Walton, 1929) (Pharyngodonidae: Oxyuroidea). Canadian Journal of Zoology 59:1351–1367.

Adamson, M.L. 1981b. Seasonal changes in populations of *Gyrinicola batrachiensis* (Walton, 1929) in wild tadpoles. Canadian Journal of Zoology 59:1377–1386.

Adamson, M.L. and J.P.M. Richardson. 1991. Seasonal changes in populations of *Chabaudgolvania waldeni* (Quimperiidae; Seuratoidea) in *Ambystoma gracile* (Ambystomatidae; Caudata). Canadian Journal of Zoology 69:163–167.

Adler, K. (Ed.) 1992. Herpetology: Current research on the biology of amphibians and reptiles. Proceedings of the First World Congress of Herpetology. Society of Amphibians and Reptiles, Oxford, Ohio.

Adler, K.K. and D.M. Dennis. 1962. *Plethodon longicrus*, a new salamander (Amphibia: Plethodontidae) from North Carolina. Special Publications of the Ohio Herpetological Society 4:1–14.

Adler, K. and D.H. Taylor. 1981. Toad orientation. Variability of response and its relationship to individuality and environmental parameters. Journal of Comparative Physiology A 144:45–52.

Affa'a, F.-M. and D.H. Lynn. 1994. A review of the classification and distribution of five opalinids from Africa and North America. Canadian Journal of Zoology 72:665–674.

Agassiz, L. 1850. Lake Superior: Its Physical Character, Vegetation and Animals, Compared with those of Other and Similar Regions. Gould, Kendall and Lincoln, Boston, Massachusetts.

Agresti, A. 1996. An Introduction to Categorical Data Analysis. John Wiley and Sons, New York.

Ahlgren, M.O. and S.H. Bowen. 1991. Growth and survival of tadpoles *(Bufo americanus)* fed amorphous detritus derived from natural waters. Hydrobiologia 218:49–51.

Aho, J.M. 1990. Helminth communities of amphibians and reptiles: comparative approaches to understanding patterns and processes. Pp. 157–195. *In* Esch, G.W, A.O. Bush and J.M. Aho (Eds.), Parasite Communities: Patterns and Processes. Chapman and Hall Limited, New York.

Akin, G.C. 1966. Self-inhibiting growth of *Rana pipiens* tadpoles. Physiological Zoology 39:341–356.

Akins, N. and H.W. Wofford. 1999. The effects of s-methoprene on the development of the leopard frog *(Rana pipiens)* tadpole. Journal of the Tennessee Academy of Science 73:107.

Alabama Natural Heritage Program. 1996. Amphibians list. Huntingdon College, Montgomery, Alabama.

Alan Plummer Associates, Incorporated. 2000. Barton Springs pool preliminary algae control plan for city of Austin. Unpublished report, City of Austin, Texas.

Alberta Fish and Wildlife Division. 1991. Alberta's threatened wildlife: northern leopard frog. Alberta Fish and Wildlife Division, Nongame Management Program, Edmonton, Alberta.

Alcala, A.C. 1957. Philippine notes on the ecology of the marine toad. Silliman Journal 4:90–96.

Alcala, A.C. 1996. DAPTF-Philippines: Working Group report 1996. DAPTF Report, Number 1,World Conservation Union (IUCN), Species Survival Commission, Open University, Milton Keynes, United Kingdom.

Alexander, D.G. 1965. An ecological study of the swamp cricket frog, *Pseudacris nigrita feriarum* (Baird) with comparative notes on two other hylids of the Chapel Hill, North Carolina region. Ph.D. dissertation. University of North Carolina, Chapel Hill, North Carolina.

Alexander, M.A. and J.K. Eischeid. 2001. Climate variability in regions of amphibian declines. Conservation Biology 15:930–942.

Alexander, T.R. 1964. Observations on the feeding behavior of *Bufo marinus* (Linne). Herpetologica 20:255–259.

Alexander, W.P. 1927. The Allegheny hellbender and its habitat. Buffalo Society of Natural Science 7:13–18.

Alford, R.A. 1989a. Variation in predator phenology affects predator performance and prey community composition. Ecology 70:206–219.

Alford, R.A. 1989b. Competition between larval *Rana palustris* and *Bufo americanus* is not affected by variation in reproductive phenology. Copeia 1989:993–1000.

Alford, R.A. 1989c. Effects of parentage and competitor phenology on the growth of larval *Hyla chrysoscelis*. Oikos 54:325–330.

Alford, R.A. 1999. Ecology: resource use, competition, and predation. Pp. 240–278. *In* McDiarmid, R.W. and R. Altig (Eds.), Tadpoles: The Biology of Anuran Larvae. University of Chicago Press, Chicago, Illinois.

Alford, R.A. and M.L. Crump. 1982. Habitat partitioning among size classes of larval southern leopard frogs, *Rana utricularia*. Copeia 1982:367–373.

Alford, R.A. and S.J. Richards. 1999. Global amphibian declines: a problem in applied ecology. Annual Review of Ecology and Systematics 30:133–165.

Alford, R.A. and H.M. Wilbur. 1985. Priority effects in experimental pond communities: competition between *Bufo* and *Rana*. Ecology 66:1097–1105.

Alford, R.A., M. Lampo and P. Bayliss. 1995b. The comparative ecology of *Bufo marinus* in Australia and South America. CSIRO *Bufo* Project: An Overview of Research Outcomes, Unpublished report, CSIRO, Australia.

Alford, R.A., M.P. Cohen, M.R. Crossland, M.N. Hearnden and L. Schwarzkopf. 1995a. Population biology of *Bufo marinus* in northern Australia. Proceedings of the Wetland Management in the Wet-Dry Tropics Workshop. Jabiru, Northern Territory, Australia.

Alford, R.A., P.M. Dixon and J.H.K. Pechmann. 2001. Global amphibian population declines. Nature 412:499–500.

Allan, D.M. 1973. Some relationships of vocalization to behavior in the Pacific treefrog, *Hyla regilla*. Herpetologica 29:366–371.

Allard, H.A. 1907. *Bufo fowleri* (Putnam). Science 26:383–384.

Allard, H.A. 1908. *Bufo fowleri* (Putnam) in northern Georgia. Science 28:655–656.

Allard, H.A. 1916. The song of Fowler's toad *(Bufo fowleri)*. Science 44:463–464.

Allen, D. 1937. Some notes on the Amphibia of a waterfowl sanctuary, Kalamazoo County, Michigan. Copeia 1937:190–191.

Allen, E.R. 1938. Notes on Wright's bullfrog, *Rana heckscheri* (Wright). Copeia 1938:50.

Allen, E.R. 1939. Habits of *Rhadinaea flavilata*. Copeia 1939:175.

Allen, E.R. and W.T. Neill. 1949. A new subspecies of salamander (genus *Plethodon*) from Florida and Georgia. Herpetologica 5:112–114.

Allen, E.R. and W.T. Neill. 1956. Effect of marine toad toxins on man. Herpetologica 12:150–151.

Allen, G.M. 1899. Notes on the reptiles and amphibians of Intervale, New Hampshire. Proceedings of the Boston Society of Natural History 29:71.

Allen, J.A. 1868. Catalogue of the reptiles and batrachians found in the vicinity of Springfield, Massachusetts, with notices of all other species known to inhabit the state. Proceedings of the Boston Society of Natural History 12:171–204.

Allen, M.J. 1932. A survey of the amphibians and reptiles of Harrison County, Mississippi. American Museum Novitates, Number 542, American Museum of Natural History, New York.

Allen, M.J. 1933. Report on collection of amphibians and reptiles from Sonora, Mexico, with the description of a new lizard. Occasional Papers of the Museum of Zoology, Number 259, University of Michigan, Ann Arbor, Michigan.

Allmon, W.D. 1994. The value of natural history collections. Curator 37:83–89.

Allran, J.W. and W.H. Karasov. 2000. Effects of atrazine and nitrate on northern leopard frog *(Rana pipiens)* larva exposed in the laboratory from posthatch through metamorphosis. Environmental Toxicology and Chemistry 19:2850–2855.

Allran, J.W. and W.H. Karasov. 2001. Effects of atrazine on embryos, larvae, and adults of anuran amphibians. Environmental Toxicology and Chemistry 20:769–775.

Alm, G. 1952. Year-class fluctuations and span of life in perch. Reports of the Institute for Freshwater Research, Drottningholm 33:17–38.

Altig, R. 1967. Food of *Siren intermedia nettingi* in a spring-fed swamp in southern Illinois. American Midland Naturalist 77:239–241.

Altig, R. 1969. Notes on the ontogeny of the osseous cranium of *Ascaphus truei*. Herpetologica 25:59–62.

Altig, R. 1970. A key to the tadpoles of the continental United States and Canada. Herpetologica 26:180–207.

Altig, R. 1972a. Defensive behavior in *Rana areolata* and *Hyla avivoca*. Quarterly Journal of the Florida Academy of Science 35:212–216.

Altig, R. 1972b. Notes on the larvae and premetamorphic tadpoles of four *Hyla* and three *Rana* with notes on tadpole color patterns. Journal of the Elisha Mitchell Scientific Society 88:113–119.

Altig, R. 1974. Defensive behavior in *Rana areolata* and *Hyla avivoca*. Quarterly Journal of the Florida Academy of Science 35:212–216.

Altig, R. 1981. Status report on the Amargosa toad *(Bufo nelsoni)*. U.S. Fish and Wildlife Service, Boise Area Office, Contract report number 14420-1400-170-00, Boise, Idaho (available from Nevada Natural Heritage Program, Carson City, Nevada).

Altig, R. and E.D. Brodie Jr. 1968. Albinistic and cyanistic frogs from Oregon. Wasmann Journal of Biology 1968:241–242.

Altig, R. and E.D. Brodie Jr. 1971. Foods of *Plethodon larselli, Plethodon dunni,* and *Ensatina eschscholtzii* in the Columbia River Gorge, Multnomah County, Oregon. American Midland Naturalist 85:226–228.

Altig, R. and E.D. Brodie Jr. 1972. Laboratory behavior of *Ascaphus truei* tadpoles. Journal of Herpetology 6:21–24.

Altig, R. and M.T. Christensen. 1981. Behavioral characteristics of the tadpoles of *Rana heckscheri*. Journal of Herpetology 15:151–154.

Altig, R. and K. Dodd Jr. 1987. The status of the Amargosa toad *(Bufo nelsoni)* in the Amargosa River drainage of Nevada. Southwestern Naturalist 32:276–278.

Altig, R. and G.F. Johnson. 1986. Major characteristics of free-living anuran tadpoles. Smithsonian Herpetological Information Service, Number 67, Smithsonian Institution, Washington, D.C.

Altig, R. and J.P. Kelly. 1974. Indices of feeding in anuran tadpoles as indicated by gut characteristics. Herpetologica 30:200–203.

Altig, R. and P.H. Ireland. 1984. A key to salamander larvae and larviform adults of the United States and Canada. Herpetologica 40: 212–218.

Altig, R. and R. Lohoefener. 1983. *Rana areolata*. Pp. 324.1–324.4. Catalogue of American Amphibians and Reptiles. Society for the Study of Amphibians and Reptiles, St. Louis, Missouri.

Altig, R. and R.W. McDiarmid. 1999. Diversity: familial and generic characterizations. Pp. 295–337. *In* McDiarmid, R.W. and R. Altig (Eds.), Tadpoles: The Biology of Anuran Larvae. University of Chicago Press, Chicago, Illinois.

Altig, R. and W.L. Pace 1974. Scanning electron photomicrographs of tadpole labial teeth. Journal of Herpetology 8:247–251.

Altig, R., R.W. McDiarmid, K.A. Nichols and P.C. Ustach. 1998. A key to the anuran tadpoles of the United States and Canada. Contemporary Herpetology Information Series 2:1–53. http://dataserver.calacademy.org/herpetology/herpdocs/chis/1998/2/index.htm.

Altschul, S.F. and D.J. Lipman. 1990. Equal animals. Nature 348:493–94.

Alvarado, R.H. 1967. The significance of grouping on water conservation in *Ambystoma*. Copeia 1967:667–668.

Ameel, D.J. 1938. The morphology and life cycle of *Euryhelmis monorchis* n. sp. (Trematoda) from the mink. Journal of Parasitology 24:219–224.

American Society for Testing and Materials. 1989. Standard guide for conducting acute toxicity tests with fishes, macroinvertebrates, and amphibians. American Society for Testing and Materials, E729-88a, Philadelphia, Pennsylvania.

American Society for Testing and Materials. 1991. Standard guide for conducting the frog embryo teratogenesis assay-*Xenopus* (FETAX). American Society for Testing and Materials, E1439-91, Philadelphia, Pennsylvania.

Amin, O.M. and W.L. Minckley. 1996. Parasites of some fish introduced into an Arizona reservoir, with notes on introductions. Journal of the Helminthological Society of Washington 36:139–200.

Ammon, E.M., C.R. Gourley, K.W. Wilson, D.A. Ross and C.R. Peterson. 2003. Strategies for effective habitat restoration for the Columbia spotted frog *(Rana luteiventris)*: a case study on the Provo River population. Proceedings of the Riparian Habitat and Floodplain Conference, 12–14 March 2001, Sacramento, California.

AmphibiaWeb. 2002. Discussion board, taxonomy issues, status of *Bufo nelsoni*. http://elib.cs.berkeley.edu/aw/discus/index.html.

Anderson, A.M., D.A. Haukos and J.T. Anderson. 1999a. Habitat use by anurans emerging and breeding in playa wetlands. Wildlife Society Bulletin 27:759–769.

Anderson, A.M., D.A. Haukos and J.T. Anderson. 1999b. Diet composition of three anurans from the playa wetlands of northwest Texas. Copeia 1999:515–520.

Anderson, A.R. and J. W. Petranka. 2003. Odonate predator does not affect hatching time or morphology of embryos of two amphibians. Journal of Herpetology 37:65–71.

Anderson, B.D., M.E. Feder and R.J. Full. 1991. Consequences of a gait change during locomotion in toads *(Bufo woodhousii fowleri)*. Journal of Experimental Biology 158:133–148.

Anderson, D.M. 1983. The natural divisions of Ohio. Natural Areas Journal 3:23–33.

Anderson, G.A. 1964. Digenetic trematodes of *Ascaphus truei* in western Oregon. Ph.D. dissertation. Oregon State University, Corvallis, Oregon.

Anderson, G.A. and I. Pratt. 1965. Cercaria and first intermediate host of *Euryhelmis squamula*. Journal of Parasitology 51:13–15.

Anderson, G.A., G.W. Martin and I. Pratt. 1966. The life cycle of the trematode *Cephalouterina dicamptodoni* Senger and Macy, 1953. Journal of Parasitology 52:704–706.

Anderson, G.A., S.C. Schell and I. Pratt. 1965. The life cycle of *Bunoderella metteri* (Allocreadiidae: Bunoderinae), a trematode parasite of *Ascaphus truei*. Journal of Parasitology 51:579–582.

Anderson, J.D. 1961. The courtship behavior of *Ambystoma macrodactylum croceum*. Copeia 1961:132–139.

Anderson, J.D. 1963. Reactions of the western mole to skin secretions of *Ambystoma macrodactylum croceum*. Herpetologica 1963:282–284.

Anderson, J.D. 1965. *Ambystoma annulatum*. Pp. 19.1–19.2. Catalogue of American Amphibians and Reptiles. Society for the Study of Amphibians and Reptiles, St. Louis, Missouri.

Anderson, J.D. 1967a. A comparison of the life histories of coastal and montane populations of *Ambystoma macrodactylum* in California. American Midland Naturalist 77:323–354.

Anderson, J.D. 1967b. *Ambystoma maculatum*. Pp. 51.1–51.4. Catalogue of American Amphibians and Reptiles. Society for the Study of Amphibians and Reptiles, St. Louis, Missouri.

Anderson, J.D. 1968. A comparison of the food habits of *Ambystoma macrodactylum sigillatum*, *Ambystoma macrodactylum croceum*, and *Ambystoma tigrinum californiense*. Herpetologica 24:273–284.

Anderson, J.D. 1969. *Dicamptodon* and *Dicamptodon ensatus*. Pp. 76.1–76.2. Catalogue of American Amphibians and Reptiles. Society for the Study of Amphibians and Reptiles, St. Louis, Missouri.

Anderson, J.D. and R.V. Giacosie. 1967. *Ambystoma laterale* in New Jersey. Herpetologica 23:108–111.

Anderson, J.D. and R.E. Graham. 1967. Vertical migration and stratification of larval *Ambystoma*. Copeia 1967:371–374.

Anderson, J.D. and W.Z. Lidicker Jr. 1963. A contribution of our knowledge of the herpetofauna of the Mexican state of Aguascalientes. Herpetologica 19:40–51.

Anderson, J.D. and P.J. Martino. 1966. The life history of *Eurycea l. longicauda* associated with ponds. American Midland Naturalist 75:257–279.

Anderson, J.D. and P.J. Martino. 1967. Food habits of *Eurycea longicauda*. Herpetologica 23:105–108.

Anderson, J.D. and G.K. Williamson. 1973. The breeding season of *Ambystoma opacum* in the northern and southern parts of its range. Journal of Herpetology 7:320–321.

Anderson, J.D. and G.K. Williamson. 1974. Nocturnal stratification in larvae of the mole salamander, *Ambystoma talpoideum*. Herpetologica 30:28–29.

Anderson, J.D. and G.K. Williamson. 1976. Terrestrial mode of reproduction in *Ambystoma cingulatum*. Herpetologica 32:214–221.

Anderson, J.D., D.D. Hassinger and G.H. Dalrymple. 1971. Natural mortality of eggs and larvae of *Ambystoma t. tigrinum*. Ecology 52: 1107–1112.

Anderson, K. and H.G. Dowling. 1982. Geographic distribution: *Hyla gratiosa* (barking treefrog). Herpetological Review 13:130.

Anderson, K. and P.E. Moler. 1986. Natural hybrids of the pine barrens treefrog *Hyla andersonii* with *H. cinerea* and *H. femoralis* (Anura: Hylidae): morphological and chromosomal evidence. Copeia 1986:70–76.

Anderson, M.E. 1977. Aspects of the ecology of two sympatric species of *Thamnophis* and heavy metal accumulation within the species. Master's thesis. University of Montana, Polson, Montana.

Anderson, P. 1950. A range extension for the ringed salamander, *Ambystoma annulatum* Cope. Herpetologica 6:55.

Anderson, P.D. and S. D'Apollonia. 1978. Aquatic animals. Pp. 187–211. *In* Butler, G.C. (Ed.), Principles of Ecotoxicology. Wiley Publishers, Toronto.

Anderson, P.K. 1942. Amphibians and reptiles of Jackson County, Missouri. Bulletin of the Chicago Academy of Science 6:203–220.

Anderson, P.K. 1951. Albinism in tadpoles of *Microhyla carolinensis*. Herpetologica 7:56.

Anderson, P.K. 1952. Notes on amphibian and reptile populations in a Louisiana pineland area. Ecology 33:274–278.

Anderson, P.K. 1954. Studies in the ecology of the narrow-mouthed toad, *Microhyla carolinensis carolinensis*. Tulane Studies in Zoology 2:15–46.

Anderson, P.K. 1960. Ecology and evolution in island populations of salamanders in the San Francisco Bay region. Ecological Monographs 30:359–385.

Anderson, P.K., E.A. Liner and R.E. Etheridge. 1952. Notes on amphibian and reptile populations in a Louisiana pineland area. Ecology 33:274–278.

Anderson, R.K., D.F. Sorensen, V. Perman, V.A. Dirks, M.M. Snyder and J.E. Bearman. 1971. Selected epizootiologic aspects of bovine leukemia in Minnesota (1961–1965). American Journal of Veterinary Research 32:563–577.

Anderson, R.M. and R.L. May. 1980. Infectious diseases and population cycles of forest insects. Science 210:658–661.

Andrewartha, H.G. and L.C. Birch. 1954. The Distribution and Abundance of Animals. University of Chicago Press, Chicago.

Andrews, J.S. 2000. Vermont herp atlas: Update. Vermont Nongame and Natural Heritage Program, Middlebury, Vermont.

Andrews, K.D., R.L. Lampley, M.A. Gillman, D.T. Corey, S.R. Ballard, M.J. Blasczyk and W.G. Dyer. 1992. Helminths of *Rana catesbeiana* in southern Illinois with a checklist of helminths in bullfrogs of North America. Transactions of the Illinois State Academy of Science 85:147–172.

Angermann, J.E., G.M. Fellers and F. Matsumura. 2002. Polychlorinated biphenyls and toxaphene in Pacific treefrog tadpoles *(Hyla regilla)* from the California Sierra Nevada, USA. Environmental Toxicology and Chemistry 21:2209–2215.

Angle, J.P. 1969. The reproductive cycle of the northern ravine salamander, *Plethodon richmondi richmondi*, in the Valley and Ridge Province of Pennsylvania and Maryland. Journal of the Washington Academy of Science 59:192–202.

Anholt, B.R., D.K. Skelly and E.E. Werner. 1996. Factors modifying antipredator behavior in larval toads. Herpetologica 52:301–313.

Anholt, B.R., E. Werner and D.K. Skelly. 2000. Effect of food and predators on the activity of four larval ranid frogs. Ecology 81:3509–3521.

Ankley, G.T. 1997. Are increases in ultraviolet light a plausible factor contributing to amphibian deformities? NAAMP III Online Paper, http://www.im.nbs.gov/naamp3/papers /deformuv.html.

Ankley, G.T., J.E. Tietge, D.L. DeFoe, K.M. Jensen, G.W. Holcombe, E.J. Durhan and S.A. Diamond. 1998. Effects of ultraviolet light and methoprene on survival and development of *Rana pipiens*. Environmental Toxicology and Chemistry 17:2530–2542.

Ankley, G.T., J.E. Tietge, G.W. Holcombe, D.L. DeFoe, S.A. Diamond, K.M. Jensen and S.J. Degitz. 2000. Effects of laboratory ultraviolet radiation and natural sunlight on survival and development of *Rana pipiens*. Canadian Journal of Zoology 78:1092–1100.

Anonymous (Associated Press). 1977a. Endangered toad may halt building. Dallas Morning News, 24 June.

Anonymous (Associated Press). 1977b. U.S. comes to the defense of a toad. San Francisco Chronicle, 25 June.

Anonymous (Cox News Service). 1990b. Texas to spend $51,000 building tunnels for toads. Fairbanks Daily News-Miner, 25 July.

Anonymous. 1933. Civil minutes, Bastrop County Court, 7 August.

Anonymous. 1934. Civil minutes, Bastrop County Court, 24 May.

Anonymous. 1936. Golf club plans being whipped into shape here. Bastrop Advertiser and County News, 6 August.

Anonymous. 1960. Texas State Parks Board, Bastrop State Park Golf Contract, 2 November.

Anonymous. 1963. Bastrop City Council minutes, 9 December.

Anonymous. 1964. Bastrop City Council minutes, 3 April.

Anonymous. 1966–'68. Texas Parks and Wildlife Department: Planning, design and construction division report for work authorized by Article III, House Bill Number 12, 59th Legislature, FY 1965–'66, 1966–'67, and 1967–'68.

Anonymous. 1966a. Bastrop City Council minutes, 13 January.

Anonymous. 1966b. Bastrop City Council minutes, 24 January.

Anonymous. 1973. Recovery plan for Houston toad, *Bufo houstonensis*. U.S. Fish and Wildlife Service, Washington, D.C.

Anonymous. 1975. Texas Parks and Wildlife Department feasibility study of the construction of an additional nine holes at the Bastrop State Park Golf Course, August.

Anonymous. 1976. Texas Parks and Wildlife Department Commission agenda item, Bastrop State Park Golf Course expansion feasibility study, Exhibit A: History and Operation of the Golf Course, February.

Anonymous. 1978a. Bastrop City Council minutes, 12 June.

Anonymous. 1978b. Defending the toad. Fortune 97:45.

Anonymous. 1980. At the crossroads 1980; A report on California's endangered and rare fish and wildlife. State of California Resources Agency, Fish and Game Commission and Department of Fish and Game, Sacramento, California.

Anonymous. 1987. Rediscovery of the Wyoming toad. Endangered Species Technical Bulletin 12:7.

Anonymous. 1989. Residents rally to protect pines. Bastrop Advertiser and County News, 9 January.

Anonymous. 1990a. 105 trees face the chainsaw. Bastrop Advertiser and County News, 30 June.

Anonymous. 1990c. Bastrop Park golfers frustrated by friends of toad. Bastrop Advertiser, 24 September.

Anonymous. 1993. Postmetamorphic death syndrome. Froglog: Newsletter of the Declining Amphibian Populations Task Force 7:1–2.

Anonymous. 1993a. Transcript of Texas Parks and Wildlife Commission meeting, January.

Anonymous. 1993b. Reverse performance evaluation. Tricks & Trials, July and August.

Anonymous. 1994a. Local park grant project scoring summary, Bastrop State Park golf course expansion. Texas Parks and Wildlife Department internal document, 31 January.

Anonymous. 1994b. Texas Parks and Wildlife Department Commission agenda item Number 1, Exhibit B: Bastrop State Park Lost Pines Golf Course Expansion, January.

Anonymous. 1994c. Houston toad habitat evaluation procedure. U.S. Fish and Wildlife Service, 14 February.

Anonymous. 1994d. Ranchers have little to fear from toad. Smithville Times, 8 September.

Anonymous. 1994e. Environmental assessment for Bastrop State Park golf course expansion. Texas Parks and Wildlife Department. Undated.

Anonymous. 1995a. Transcript from Senate Nominations Committee confirmation hearing for Texas Parks and Wildlife Commissioners, 6 March.

Anonymous. 1995b. National Park Service site inspection report, Bastrop State Park, Bastrop County, Texas; Purpose of visit: to review proposed golf course project and assess its effect on historic district, 13 March.

Anonymous. 1995c. Groups unite to fight golf course. Smithville Times, 16 March.

Anonymous. 1995d. How many golf holes are needed? Smithville Times, 20 April.

Anonymous. 1995e. Finding of no significant impact/decision record; Land and Water Conservation Fund application number 48-01022; Bastrop State Park golf course expansion. National Park Service, Southwest Region, 11 May.

Anonymous. 1995f. Memorandum of agreement between Texas Parks and Wildlife Department, National Park Service, Texas Historical Commission and Advisory Council on Historic Preservation to take into account the effects of the Bastrop State Park golf course expansion on historic properties, 11 May.

Anonymous. 1995g. Park work begins; jail inmates clear trees for nine-hole addition. Bastrop Advertiser, 8 July.

Anonymous. 1995h. Expansion issue fueled. Elgin Courier, 26 July.

Anonymous. 1995i. EarthFirst! halts dozer. Smithville Times, 27 July.

Anonymous. 1995j. Texas Parks and Wildlife Department amendment to Bastrop State Park golf course agreement with Lost Pines Golf Club.

Anonymous. 1995k. Golf course monitored. Smithville Times, 2 November.

Anonymous. 1995l. Ohio sportsman's atlas. Back roads and outdoor recreation. Sportsman's Atlas Company, Lytton, Iowa.

Anonymous. 1996a. Arizona Game and Fish Department herp diversity review, element occurrence notebook. Heritage Data Management System, Phoenix, Arizona.

Anonymous. 1996b. Directory of Ohio's State Nature Preserves. Ohio Department of Natural Resources Division of Natural Areas and Preserves, Columbus, Ohio.

Anonymous. 1996c. Texas property tax guidelines for qualification of agricultural land in wildlife management use. Comptroller of Public Accounts, April.

Anonymous. 1998a. State park honored as national landmark. Bastrop Advertiser, 30 July.

Anonymous. 1998b. Challenge cost-share agreement between USFWS and Lower Colorado River Authority, Bastrop County, Bastrop Board of Realtors, Bastrop County Environmental Network, Bastrop Economic Development Corporation, Champion International. U.S. Fish and Wildlife Service Agreement, Number 1448-20181-98-J609, Bastrop, Texas.

Anonymous. 1998c. County issues bonds for road, bridge repairs. Bastrop Advertiser, 12 November.

Anonymous. 1999. A&M to join toad conservation planning. Bastrop Advertiser, 20 November.

Anonymous. 2000. Bastrop Park expansion approved. Austin American-Statesman, 21 January.

Anonymous. Undated, ca. 1995. Lost Pines Golf Course revenue. Texas Parks and Wildlife Department, internal document.

Anthony, C.D. 1993. Recognition of conspecific odors by *Plethodon caddoensis* and *P. ouachitae*. Copeia 1993:1028–1033.

Anthony, C.D. 1995. Competitive interactions within and between two species of *Plethodon* in the Ouachita Mountains: effects of territoriality and parasitism. Ph.D. dissertation. University of Southwestern Louisiana, Lafayette, Louisiana.

Anthony, C.D. and J.A. Wicknick. 1993a. Geographic distribution: *Plethodon ouachitae* (Rich Mountain salamander). Herpetological Review 24:152.

Anthony, C.D. and J.A. Wicknick. 1993b. Aggressive interactions and chemical communication between adult and juvenile salamanders. Journal of Herpetology 27:261–264.

Anthony, C.D., J.R. Mendelson III and R.R. Simons. 1994. Differential parasitism by sex on plethodontid salamanders and histological evidence for structural damage to the nasolabial groove. American Midland Naturalist 132:302–307.

Anthony, C.D., J.A. Wicknick and R.G. Jaeger. 1997. Social interactions in two sympatric salamanders: effectiveness of a highly aggressive strategy. Behaviour 134:71–88.

Antonelli, A.L., R.A. Nussbaum and S.D. Smith. 1972. Comparative food habits of four species of stream-dwelling vertebrates *(Dicamptodon ensatus, D. copei, Cottus tenuis, Salmo gairdneri)*. Northwest Science 46:277–289.

Anver, M.R. 1992. Amphibian tumors: a comparison of anurans and urodeles. In Vivo 6:435–438.

Anzalone, C.R., L.B. Kats and M.S. Gordon. 1998. Effects of solar UV-B radiation on embryonic development in *Hyla cadavarina*, *Hyla regilla*, and *Taricha torosa*. Conservation Biology 12:646–653.

Applegate, R.D. 1978. Spruce grouse feeds on mink frog. Loon 10:207.

Applegate, R.D. 1990. Natural history notes: *Rana catesbeiana*, *Rana palustris* (bullfrog, pickerel frog). Predation. Herpetological Review 21:90–91.

Araujo, M., A. Goldbloom, K. Nichols, J.S. Seffert and J.A. Deloney. 1990. Texas outdoor recreation plan—Assessment and policy plan. Texas Parks and Wildlife Department, Austin, Texas.

Aresco, M.J. 2001. Natural history notes: *Siren lacertina* (greater siren). Aestivation chamber. Herpetological Review 32:32–33.

Aresco, M.J. 2002. Natural history notes: *Amphiuma means* (two-toed amphiuma). Overland migration. Herpetological Review 33:296–297.

Aresco, M.J. and C. Guyer. 1999. Burrow abandonment by gopher tortoises in slash pine plantations of the Conecuh National Forest. Journal of Wildlife Management 63:26–35.

Aresco, M.J. and R.N. Reed. 1998. Natural history notes: *Rana capito sevosa* (dusky gopher frog). Predation. Herpetological Review 29:40.

Arita, H.T., F. Figueroa, A. Frisch, P. Rodriguez and K. Santos del Prado. 1997. Geographical range size and the conservation of Mexican mammals. Conservation Biology 11:92–100.

Arizona Game and Fish Department, Eds. In preparation. Wildlife of Special Concern in Arizona. Nongame and Endangered Wildlife Program. Arizona Game and Fish Department, Phoenix, Arizona.

Arizona Game and Fish Department. 1988. Threatened native wildlife in Arizona. Arizona Game and Fish Department, Phoenix, Arizona.

Arizona Game and Fish Department. 1996. Wildlife of special concern in Arizona. Arizona Game and Fish Department Publication, Phoenix, Arizona.

Arizona Game and Fish Department. 1999a. Amphibians by county for Arizona. Unpublished report compiled and edited by the Heritage Data Management System, Arizona Game and Fish Department, Phoenix, Arizona.

Arizona Game and Fish Department. 1999b. Amphibians by county. Unpublished report compiled by the Riparian Herpetofauna Database, Arizona Game and Fish Department, Phoenix, Arizona.

Arizona Game and Fish Department. 1999c. Arizona Wildlife Views, Special Edition. Arizona Game and Fish Department, Phoenix, Arizona.

Armentrout, D. and F.L. Rose. 1971. Some physiological responses to anoxia in the Great Plains toad, *Bufo cognatus*. Comparative Biochemistry and Physiology A 38:447–455.

Armstrong, E.P., D.W. Halton, R.C. Tinsley, J. Cable, R.N. Johnston, C.F. Johnston and C. Shaw. 1997. Immunocytochemical evidence for the involvement of an FMRFamide-related peptide in egg production in the flatworm parasite *Polystoma nearticum*. Journal of Comparative Neurology 377:41–48.

Arndt, R.G. 1977. A blue variant of the green frog *Rana clamitans melanota* (Amphibia, Anura, Ranidae) from Delaware. Journal of Herpetology 11:102–103.

Arndt, R.G. and J.F. White. 1988. Geographic distribution: *Hyla gratiosa* (barking treefrog). Herpetological Review 19:16.

Arnold, L.W. 1943. Notes on two species of desert toads. Copeia 1943:128.

Arnold, S.J. 1972. The evolution of courtship behavior in salamanders. Ph.D. dissertation. University of Michigan, Ann Arbor, Michigan.

Arnold, S.J. 1976. Sexual behavior, sexual interference and sexual defense in the salamanders *Ambystoma maculatum*, *Ambystoma tigrinum* and *Plethodon jordani*. Zeitschrift für Tierpsychologie 42:247–300.

Arnold, S.J. 1977. The evolution of courtship behavior in new world salamanders with some comments on old world salamanders. Pp. 141–183. *In* Taylor, D.H. and S.I. Guttman (Eds.), The Reproductive Biology of Amphibians. Plenum Press, New York.

Arnold, S.J. 1982. A quantitative approach to antipredator performance: salamander defense against snake attack. Copeia 1982:247–253.

Arnold, S.J. and T. Halliday. 1986. Life history notes: *Hyla regilla* (Pacific treefrog). Predation. Herpetological Review 17:44.

Arnqvist, G. and D. Wooster. 1995. Meta-analysis: synthesizing research findings in ecology and evolution. TRENDS in Ecology and Evolution 12:236–240.

Aronson, L.R. 1944a. The sexual behavior of Anura. 6. The mating pattern of *Bufo americanus*, *Bufo fowleri*, and *Bufo terrestris*. American Museum Novitates, Number 1250, American Museum of Natural History, New York.

Aronson, L.R. 1944b. Breeding *Xenopus laevis*. American Naturalist 78:131–141.

Ash, A. 1988. Disappearance of salamanders from clearcut plots. Journal of the Elisha Mitchell Scientific Society 104:116–122.

Ash, A.N. 1997. Disappearance and return of plethodontid salamanders to clearcut plots in the southern Blue Ridge Mountains. Conservation Biology 11:983–989.

Ash, A.N. and R.C. Bruce. 1994. Impacts of timber harvesting on salamanders. Conservation Biology 8:300–301.

Ash, A.N., R.C. Bruce, J. Castanet and H. Francillon-Vieillot. 2003. Population parameters of *Plethodon metcalfi* on a 1-year-old clearcut and in a nearby forest in the southern Blue Ridge Mountains. Journal of Herpetology 37:445–452.

Ashley, E.P. and J.T. Robinson. 1996. Road mortality of amphibians, reptiles and other wildlife on the Long Point Causeway, Lake Erie, Ontario. Canadian Field-Naturalist 110:403–412.

Ashton, A.D. and F.C. Rabalais. 1978. Helminth parasites of some anurans of northwestern Ohio. Proceedings of the Helminthological Society of Washington 45:141–142.

Ashton, D.T., A.J. Lind and K.E. Schlock. 1998. Foothill yellow-legged frog *(Rana boylii)* natural history. U.S.D.A. Forest Service, Pacific Southwest Research Station, Redwood Sciences Laboratory, Arcata, California.

Ashton, R.E., Jr. 1975. A study of the movement, home range, and winter behavior of *Desmognathus fuscus* (Rafinesque). Journal of Herpetology 9:85–91.

Ashton, R.E., Jr. 1976. Endangered and threatened amphibians and reptiles in the United States. Edwards, S.R. and G.R. Pisani (Eds.). Herpetological Circular, Number 5, Society for the Study of Amphibians and Reptiles, Athens, Ohio.

Ashton, R.E. 1977. The central newt, *Notophthalmus viridescens louisianensis* (Wolterstorff) in Kansas. Transactions of the Kansas Academy of Science 79:15–19.

Ashton, R.E., Jr. 1985. Field and laboratory observations on microhabitat selection, movements, and home range of *Necturus lewisi* (Brimley). Brimleyana 10:83–106.

Ashton, R.E., Jr. 1990. *Necturus lewisi*. Pp. 456.1–456.2. Catalogue of American Amphibians and Reptiles. Society for the Study of Amphibians and Reptiles, St. Louis, Missouri.

Ashton, R.E., Jr. 1992. Flatwoods salamander. Pp. 39–43. *In* Moler, P.E. (Ed.), Rare and Endangered Biota of Florida. Volume III. Amphibians and Reptiles. University Press of Florida, Gainesville, Florida.

Ashton, R.E., Jr. and P.S. Ashton. 1978. Movements and winter behavior of *Eurycea bislineata bislineata* (Amphibia, Urodela, Plethodontidae). Journal of Herpetology 12:295–298.

Ashton, R.E., Jr. and P.S. Ashton. 1988. Handbook of Reptiles and Amphibians of Florida: Part Three: The Amphibians. Windward Publishing, Miami, Florida.

Ashton, R.E., Jr. and A.L. Braswell. 1979. Nests and larvae of the Neuse River waterdog, *Necturus lewisi* (Brimley) (Amphibia: Proteidae). Brimleyana 1:15–22.

Ashton, R.E., Jr. and J. Peavy. 1986. Black warrior waterdog, *Necturus* sp. Pp. 63–65. *In* Mount, R.H. (Ed.), Vertebrate Animals of Alabama in Need of Special Attention. Alabama Agricultural Experiment Station, Auburn, Alabama.

Ashton, R.E., Jr., S.I. Guttman and P. Buckley. 1973. Notes on the distribution, coloration, and the breeding of the Hudson Bay toad, *Bufo americanus copei*. Journal of Herpetology 7:17–20.

Ashton, T.E. 1966. An annotated check list of order Caudata (Amphibia) of Davidson County, Tennessee. Journal of the Tennessee Academy of Science 41:106–111.

Asquith, A. and R. Altig. 1987. Phototaxis and activity patterns of *Siren intermedia*. Southwestern Naturalist 32:146–148.

Asquith, N.M. 2001. Misdirections in conservation biology. Conservation Biology 15:345–352.

Atrazine Ecological Risk Assessment Panel. 1995. Ecological risk assessment of atrazine in North American surface waters. CIBA Crop Protection, Greensboro, North Carolina.

Aubry, K.B. 2000. Amphibians in managed, second-growth Douglas-fir forests. Journal of Wildlife Management 64:1041–1052.

Aubry, K.B. and P.A. Hall. 1991. Terrestrial amphibian communities in the southern Washington Cascade Range. Pp. 326–338. *In* Ruggiero, L.F., K.B. Aubry, A.B. Carey and M.H. Huff (Tech. Coords.), Wildlife and Vegetation of Unmanaged Douglas-fir Forests. U.S.D.A. Forest Service, General Technical Report, PNW-GTR-285, Pacific Northwest Research Station, Portland, Oregon.

Aubry, K.B., L.L. Jones and P.A. Hall. 1988. Use of woody debris by plethodontid salamanders in Douglas-fir forests in Washington. Pp. 32–37. *In* Szaro, R.C., K.E. Severson and D.R. Patton (Tech. Coords.), Management of Amphibians, Reptiles, and Small Mammals in North America. U.S.D.A. Forest Service, Rocky Mountain Forest and Range Experiment Station, Technical Report, Number RM-166, Fort Collins, Colorado.

Aubry, K.B., C.M. Senger and R.L. Crawford. 1987. Discovery of the Larch Mountain salamander *(Plethodon larselli)* in the central Cascade Range of Washington. Biological Conservation 42:147–152.

Auclair, W. 1961. Monolayer culture of *Rana pipiens* kidney and ecological factors. Pp. 107–113. *In* Duryee, W.R. and L. Warner (Eds.), Proceedings of the Frog Kidney Adenocarcinoma Conference. National Institutes of Health, Bethesda, Maryland.

Auerbach, M.J., E.F. Connor and S. Mopper. 1995. Minor miners and major miners: population dynamics of leaf-mining insects. Pp. 83–110. *In* Cappachino, N. and P. Price (Eds.), Population Dynamics: New Approaches and Synthesis. Academic Press, New York.

Austin, M. and P.C. Heyligers. 1991. Vegetation theory in relation to cost-efficient surveys. Pp. 31–36. *In* Margules, C.R. and M.P. Austin (Eds.), Nature Conservation: Cost Effective Biological Surveys and Data Analysis. Commonwealth Scientific, Industrial, and Research Organization, East Melbourne, Victoria, Australia.

Austin, R.M., Jr. 1993. The reproductive life history of a low altitude population of *Desmognathus quadramaculatus* (Amphibia: Plethodontidae). Master's thesis. Auburn University, Auburn, Alabama.

Austin, R.M., Jr. 2000. Cutaneous microbial flora and antibiosis in *Plethodon ventralis*: inferences for parental care in the Plethodontidae. Pp. 451–462. *In* Bruce, R.C., R.G. Jaeger and L.D. Houck (Eds.), The Biology of Plethodontid Salamanders. Kluwer Academic/Plenum Publishers, New York.

Austin, R.M., Jr. and C.D. Camp. 1992. Larval development of black-bellied salamanders, *Desmognathus quadramaculatus*, in northeastern Georgia. Herpetologica 48:313–317.

Avery, M.I. and R. Leslie. 1990. Birds and Forestry. Poyser, Calton, United Kingdom.

Avila, V.L. and P.G. Frye. 1978. Feeding behavior of the African clawed frog (*Xenopus laevis* Daudin): (Amphibia, Anura, Pipidae): effect of prey type. Journal of Herpetology 12:391–396.

Avise, J.C. and J.L. Hamrick (Eds.). 1996. Conservation Genetics. Chapman and Hall, New York.

Avise J.C. and W.S. Nelson. 1989. Molecular genetic relationships of the extinct dusky seaside sparrow. Science 243:646–648.

Awbrey, F.T. 1963. Homing and home range in *Bufo valliceps*. Texas Journal of Science 15:127–141.

Awbrey, F.T. 1972. "Mating call" of a *Bufo boreas* male. Copeia 1972:579–581.

Awbrey, F.T. 1978. Social interaction among chorusing Pacific treefrogs, *Hyla regilla*. Copeia 1978:208–214.

Axelrod, D.I. and P.H. Raven. 1985. Origins of the Cordilleran flora. Journal of Biogeography 12:21–47.

Axtell, H.H. and R.C. Axtell 1948. *Pseudotriton ruber* in central New York state. Copeia 1948:64.

Axtell, R.W. 1958. Female reaction to the male call in two anurans (Amphibia). Southwestern Naturalist 3:70–76.

Axtell, R.W. 1959. Amphibians and reptiles of the Black Gap National Wildlife Management Area, Brewster County, Texas. Southwestern Naturalist 4:88–109.

Axtell, R.W. and N. Haskell. 1977. An interhiatal population of *Pseudacris streckeri* from Illinois, with an assessment of its postglacial dispersion history. Natural History Miscellanea, Number 202, Chicago Academy of Sciences, Chicago, Illinois.

AZA (American Zoo and Aquarium Association). 1998. Wyoming toad 98 fact sheet. http://www.aza.org/programs/ssp/ssp.cfm

Azevedo-Ramos, C. 1992. Ecology of *Bufo marinus* tadpoles in Central Amazonian savanna: Unpublished Progress Report, Department of Zoology, Belem, Para, Brasil.

Babbitt, K.J. 1995. Natural history notes: *Bufo terrestris* (southern toad). Oophagy. Herpetological Review 26:30.

Babbitt, K.J. and F. Jordan. 1996. Predation on *Bufo terrestris* tadpoles: effects of cover and predator identity. Copeia 1996:485–488.

Babbitt, K.J. and G.W. Tanner. 1997. Effects of cover and predator identity on predation of *Hyla squirella* tadpoles. Journal of Herpetology 31:128–130.

Babbitt, L.H. 1932. Some remarks on Connecticut herpetology. Bulletin of the Boston Society of Natural History 63:23–28.

Babbitt, L.H. 1937. The Amphibia of Connecticut. Bulletin of the State Geological and Natural History Survey of Connecticut 57:1–50.

Babbitt, L.H. and C.H. Babbitt. 1951. A herpetological study of burned-over areas in Dade County, Florida. Copeia 1951:79.

Babcock, J.V. 1977. Endangered plants and animals of Kentucky. Office of Research and Engineering Services, Publication IMMR25-GR4-77, University of Kentucky, Lexington, Kentucky.

Bacchus, S.T., K. Richter and P. Moler. 1993. Geographical distribution: *Xenopus laevis* (African clawed frog). Herpetological Review 24:65.

Bachmann, K. and J.P. Bogart. 1975. Comparative cytochemical measurements in the diploid-tetraploid species pair of hylid frogs *Hyla chrysoscelis* and *Hyla versicolor*. Cytogenetics and Cell Genetics 15:186–194.

Bachmann, M.D. 1984. Defensive behavior of brooding female red-backed salamanders *(Plethodon cinereus)*. Herpetologica 40:436–443.

Bachmann, M.D., R.G. Carlton, J.M. Burkholder and R.G. Wetzel. 1986. Symbiosis between salamander eggs and green algae: microelectrode measurements made inside eggs demonstrate effect of photosynthesis on oxygen concentration. Canadian Journal of Zoology 64:1586–1588.

Badaracco, R. 1962. Amphibians and reptiles of Lassen Volcanic National Park. Lassen Volcanic National Park, Mineral, California.

Bader, R.N. and J.C. Mitchell. 1982. Geographic distribution: *Ambystoma talpoideum* (mole salamander). Herpetological Review 13:23.

Badger, D. and J. Netherton. 1995. Frogs. Voyageur Press, Stillwater, Minnesota.

Bagdonas, K.R. and D. Pettus. 1976. Genetic compatibility in wood frogs (Amphibia, Anura, Ranidae). Journal of Herpetology 10:105–112.

Bahret, R. 1996. Ecology of lake dwelling *Eurycea bislineata* in the Shawangunk Mountains, New York. Journal of Herpetology 30:399–401.

Bailey, J.E. 1992. An ecological study of the Cumberland Plateau salamander, *Plethodon kentucki* Mittleman, in West Virginia. Master's thesis. Marshall University, Huntington, West Virginia.

Bailey, J.E. and T.K. Pauley. 1993. Aspects of the natural history of the Cumberland Plateau salamander, *Plethodon kentucki*, in West Virginia. Association of Southeastern Biologists Bulletin 40:133.

Bailey, J.R. 1937. Notes on plethodont salamanders of the southeastern United States. Occasional Papers of the Museum of Zoology, Number 364, University of Michigan, Ann Arbor, Michigan.

Bailey, M.A. 1989. Migration of *Rana areolata sevosa* and associated winter-breeding amphibians at a temporary pond in the Lower Coastal Plain of Alabama. Master's thesis. Auburn University, Auburn, Alabama.

Bailey, M.A. 1990. Movement of the dusky gopher frog *(Rana areolata sevosa)* at a temporary pond in the lower coastal plain of Alabama. Pp. 27–43. *In* Dodd, C.K., Jr., R.E. Ashton Jr., R. Franz and E. Wester (Eds.), Proceedings of the Eighth Annual Meeting, Gopher Tortoise Council. Florida Museum of Natural History, Gainesville, Florida.

Bailey, M.A. 1991. The dusky gopher frog in Alabama. Journal of the Alabama Academy of Science 62:28–34.

Bailey, M.A. 1992. Black warrior waterdog status survey. Alabama Natural Heritage Program to Alabama Department of Conservation and Natural Resources, Unpublished report, Montgomery, Alabama.

Bailey, M.A. 1994. An investigation of the dusky gopher frog in Alabama. 1994. U.S. Fish and Wildlife Service, Unpublished report, Jackson, Mississippi.

Bailey, M.A. 1995. Performance report, black warrior waterdog survey 1994–95. Alabama Natural Heritage Program, Unpublished report, Alabama Department of Conservation and Natural Resources, Montgomery, Alabama.

Bailey, M.A. and J. Jensen. 1993. Survey for the dusky gopher frog and flatwoods salamander, Conecuh National Forest, Alabama. U.S.D.A. Forest Service, Unpublished report, Andalusia, Alabama.

Bailey, R.M. 1943. Four species new to the Iowa herpetofauna, with notes on their natural histories. Proceedings of the Iowa Academy of Science 50:347–352.

Bailey, R.M. 1944. Iowa's frogs and toads. Iowa Conservationist 3:17–20.

Bailey, R.M. and M.K. Bailey. 1941. The distribution of Iowa toads. Iowa State College Journal of Science 15:169–177.

Bailey, V. 1913. Life zones and crop zones of New Mexico. North American Fauna 35:1–100.

Baird, S.F. 1849 (1850). Descriptions of four new species of salamanders and one new species of scinck. Journal of the Academy of the Natural Sciences of Philadelphia Series 2, 1:292–294.

Baird, S.F. 1850. Revision of the North American tailed batrachia, with description of new genera and species. Journal of the Academy of the Natural Sciences of Philadelphia Series 2, 1:281–292.

Baird, S.F. 1854a. Reptiles of the Boundary. Pp. 1–35. *In* Emory, W.H., United States and Mexican Boundary Survey, Part II, Zoology of the Boundary. Department of the Interior, Washington, D.C.

Baird, S.F. 1854b. Descriptions of new genera and species of North American frogs. Proceedings of the Academy of the Natural Sciences of Philadelphia 7:59–62.

Baird, S.F. 1859. Report upon the reptiles of the route. Pp. 37–45. *In* Explorations and Surveys, R.R. Route from the Mississippi River to the Pacific Ocean, 1853–56. Volume 10, Williamson's Route. Zoological Report, Part 6, Number 4, Washington, D.C.

Baird, S.F. and C. Girard. 1852a. Characteristics of some new reptiles in the Museum of the Smithsonian Institution. Proceedings of the Academy of the Natural Sciences of Philadelphia 6 (2,4,5):68–70, 125–129, 173.

Baird, S.F. and C. Girard. 1852b. Descriptions of new species of reptiles, collected by the U.S. Exploring Expedition, Under the Command of Capt. Charles Wilkes, U.S.N. Proceedings of the Academy of the Natural Sciences of Philadelphia 6:174–177.

Baird, S.F. and C. Girard. 1853a. List of reptiles collected in California by Dr. John L. Le Conte, with descriptions of new species. Proceedings of the Academy of Natural Sciences of Philadelphia 6:301.

Baird, S.F. and C. Girard. 1853b. Communication by Mr. Charles Girard on behalf of Prof. Baird and himself, upon a species of frog and another toad. Proceedings of the Academy of Natural Sciences of Philadelphia 6:378.

Baird, S.F. and C. Girard. 1853c. Descriptions of new species of reptiles collected by the U.S. Exploring Expedition under the command of Capt. Charles Wilkes, U.S.N. Second part. Proceedings of the Academy of the Natural Sciences of Philadelphia 6:378–379.

Baird, T.A. 1983. Influence of social and predatory stimuli in the air-breathing behavior of the African clawed frog, *Xenopus laevis*. Copeia 1983:411–420.

Baker, A.S. 1951. A study of the expression of the burnsi gene in adult *Rana pipiens*. Journal of Experimental Zoology 116:191–229.

Baker, C.L. 1945. The natural history and morphology of Amphiumidae. Report of the Reelfoot Lake Biological Station 9:55–91.

Baker, C.L. 1947. The species of Amphiumae. Journal of the Tennessee Academy of Science 22:9–21.

Baker, C.L., L.C. Baker and M.F. Caldwell. 1947. Observation of copulation in *Amphiuma tridactylum*. Journal of the Tennessee Academy of Science 22:87–88.

Baker, J.K. 1957. *Eurycea troglodytes*: a new blind cave salamander from Texas. Texas Journal of Science 9:328–336.

Baker, J.K. 1961. Distribution of and key to the neotenic *Eurycea* of Texas. Southwestern Naturalist 6:27–32.

Baker, J.M.R. and V. Waights. 1994. The effects of nitrate on tadpoles of the treefrog (*Litoria caerulea*). Herpetological Journal 4:106–108.

Baker, K.N. 1985. Laboratory and field experiments on the responses by two species of woodland salamanders to malathion-treated substrates. Archives of Environmental Contamination and Toxicology 14:685–691.

Baker, L.C. 1937. Mating habits and life history of *Amphiuma tridactylum* Cuvier and effects of pituitary injections. Journal of the Tennessee Academy of Science 12:206–218.

Baker, M.R. 1977. Redescription of *Oswaldocruzia pipiens* (Nematoda, Trichostrongylidae) from amphibians of eastern North America. Canadian Journal of Zoology 55:104–109.

Baker, M.R. 1978a. Morphology and taxonomy of *Rhabdias* sp. (Nematoda: Rhabdiasidae) from reptiles and amphibians of southern Ontario. Canadian Journal of Zoology 56:2127–2141.

Baker, M.R. 1978b. Transmission of *Cosmocercoides dukae* (Nematoda: Cosmocercoidea) to amphibians. Journal of Parasitology 64:765–766.

Baker, M.R. 1979. The free-living and parasitic development of *Rhabdias* sp. (Nematoda: Rhabdiasidae) in amphibians. Canadian Journal of Zoology 57:161–178.

Baker, M.R. 1987. Synopsis of the Nematoda parasitic in amphibians and reptiles. Occasional Papers in Biology, Memorial University of Newfoundland, St. John's, Newfoundland.

Baker, R.H. 1942. The bullfrog, a Texas wildlife resource. Texas Game Fish Oyster Commission Bulletin 23, Austin, Texas.

Bakkegard, K.A. 2001. Activity, foraging, and body size of the Red Hills salamander, *Phaeognathus hubrichti*. Master's thesis. Auburn University, Auburn, Alabama.

Bakkegard, K.A. 2002. Activity patterns of Red Hills salamanders (*Phaeognathus hubrichti*) at their burrow entrances. Copeia 2002:851–856.

Bakker, T.C.M., D. Mazzi and S. Zala. 1997. Parasite-induced changes in behavior and color make *Gammarus pulex* more prone to fish predation. Ecology 78:1098–1104.

Baldauf, R.J. 1947. *Desmognathus f. fuscus* eating eggs of its own species. Copeia 1947:66.

Baldauf, R.J. 1952. Climatic factors influencing the breeding migration of the spotted salamander, *Ambystoma maculatum* (Shaw). Copeia 1952:178–181.

Baldauf, R.J. 1959. Morphological criteria and their use in showing bufonid phylogeny. Journal of Morphology 104:527–560.

Baldauf, R.J. and J.C. Truett. 1964. First record of *Ambystoma talpoideum* (Holbrook) from Texas. Copeia 1964:221.

Baldwin, K.S. and R.A. Stanford. 1987. Life history notes: *Ambystoma tigrinum californiense* (California tiger salamander). Predation. Herpetological Review 18:33.

Baldwin, R.F. and A.J.K. Calhoun. 2002. Natural history notes: *Ambystoma laterale* (blue-spotted salamander) and *Ambystoma maculatum* (spotted salamander). Predation. Herpetological Review 33:44–45.

Baldwin, R.F. and D. Vasconcelos. 2003. Natural history notes: *Ambystoma maculatum* (spotted salamander) and *Rana sylvatica* (wood frog). Habitat. Herpetological Review 34:353–354.

Balinsky, B.I. 1969. The reproductive ecology of amphibians of the Transvaal Highveld. Zoologica Africana 4:37–93.

Balinsky, J.B., E.L. Choritz, C.G.L. Coe and G.S. Van der Schans. 1967. Amino acid metabolism and urea synthesis in naturally aestivating *Xenopus laevis*. Comparative Biochemistry and Physiology 22:59–68.

Ball, R.W. and D.L. Jameson. 1970. Biosystematics of the canyon tree frog, *Hyla cadaverina* Cope (= *Hyla californiae* Gorman). Proceedings of the California Academy of Science, 4th Series 37:363–380.

Ball, S.C. 1936. The distribution and behavior of the spadefoot toad in Connecticut. Transactions of the Connecticut Academy of Arts and Sciences 32:251–379.

Ball, S.C. 1937. Amphibians of Gaspe County, Quebec. Copeia 1937:230.

Ballinger, R.E. 1966. Natural hybridization of the toads *Bufo woodhousei* and *Bufo speciosus*. Copeia 1966:366–368.

Ballinger, R.E. and C.O. McKinney. 1966. Developmental temperature tolerance of certain anuran species. Journal of Experimental Zoology 161:21–28.

Baltar, A.C. 1983. Vertical stratification of some cliff dwelling plethodontid salamanders as related to competitive survival of the Mississippi population of *Aneides aeneus*. Master's thesis. University of Southern Mississippi, Hattiesburg, Mississippi.

Bancroft, G.T., J.S. Godley, D.T. Gross, N.N. Rojas, D.A. Sutphen and R.W. McDiarmid. 1983. Large-scale operations management test of use of the white amur for control of problem plants. The herpetofauna of Lake Conway: species accounts. U.S. Army Engineer Waterways Experiment Station, Aquatic Plant Control Research Program, Miscellaneous Paper A-83-5, Vicksburg, Mississippi.

Banks, B. and T.J.C. Beebee. 1986. A comparison of the fecundities of two species of toad (*Bufo bufo* and *B. calamita*) from different habitat types in Britain. Journal of Zoology, London Series A 208:325–337.

Banks, B. and T.J.C. Beebee. 1988. Reproductive success of natterjack toads *Bufo calamita* in two contrasting environments. Journal of Animal Ecology 57:475–492.

Banks, R.C. and P.F. Springer. 1994. A century of population trends of waterfowl in western North America. Pp. 143–146. *In* Jehl, J.R. and N.K. Johnson (Eds.), A Century of Avifaunal Change in Western North America: Proceedings of an International Symposium at the Centennial Meeting of the Cooper Ornithological Society. Studies in Avian Biology, Number 15, Cooper Ornithological Society, Los Angeles, California.

Banta, A.M. 1914. Sex recognition and the mating behavior of the wood frog, *Rana sylvatica*. Biological Bulletin 26:171–183.

Banta, A.M. and W.L. McAtee. 1906. The life history of the cave salamander, *Spelerpes maculicaudus* (Cope). Proceedings of the U.S. National Museum 30(1443):67–83, plus 3 plates, Washington, D.C.

Banta, B.H. 1961. On the occurrence of *Hyla regilla* in the Lower Colorado River, Clark County, Nevada. Herpetologica 17:106–108.

Banta, B.H. 1965. A distributional checklist of the recent amphibians inhabiting the state of Nevada. Occasional Papers of the Biological Society of Nevada, Number 7, Reno, Nevada.

Banta, B.H. 1966. A six-legged anuran from California. Wasmann Journal of Biology 24:67–69.

Banta, B.H. and G. Carl. 1967. Death-feigning behavior in the eastern gray treefrog *Hyla versicolor*. Herpetologica 23:317–318.

Bantle, J.A., D.J. Fort and B.L. James. 1989. Identification of developmental toxicants using the frog embryo teratogenesis assay-*Xenopus* (FETAX). Hydrobiology 188/189:577–585.

Barach, J.P. 1951. The value of the skin secretions of the spotted salamander. Herpetologica 7:58.

Barbour, J.G. 1973. The eastern hognose snake *Heterodon platyrhinos* near Kingston New York. Engelhardtia 6:11.

Barbour, R.W. 1950. A new subspecies of the salamander *Desmognathus fuscus*. Copeia 1950:277–278.

Barbour, R.W. 1953. The amphibians of Big Black Mountain, Harlan County, Kentucky. Copeia 1953:84–89.

Barbour, R.W. 1957. Observations on the mountain chorus frog *Pseudacris brachyphona* (Cope) in Kentucky. American Midland Naturalist 57:125–128.

Barbour, R.W. 1971. Amphibians and Reptiles of Kentucky. University Press of Kentucky, Lexington, Kentucky.

Barbour, R.W. and L.Y. Lancaster. 1946. Food habits of *Desmognathus fuscus* in Kentucky. Copeia 1946:48–49.

Barbour, R.W. and E.P. Walters. 1941. Notes on the breeding habits of *Pseudacris brachyphona*. Copeia 1941:116.

Barbour, R.W., J.W. Hardin, J.P. Shafer and M.J. Harvey. 1969. Home range, movements, and activity of the dusky salamander, *Desmognathus fuscus*. Copeia 1969:293–297.

Barbour, T. 1920. Herpetological notes from Florida. Copeia 84:55–57.

Barbour, T. 1941. *Gastrophryne carolinensis* in Kentucky. Copeia 1941:262.

Barbour, T. and C.T. Ramsden. 1919. The herpetology of Cuba. Memoirs of the Bulletin of the Museum of Comparative Zoology 47:71–213.

Barden, R.B. and J. Kezer. 1944. The eggs of certain plethodontid salamanders obtained by pituitary gland implantation. Copeia 1944:115–118.

Barica, J. and J.A. Mathias. 1979. Oxygen depletion and winterkill risk in small prairie lakes under extended ice cover. Journal of the Fisheries Research Board Canada 36:980–986.

Baringa, M. 1990. Where have all the froggies gone? Science 247:1033–1034.

Barnes, D.H. 1826. An arrangement of the genera of batrachian animals, with a description of the more remarkable species, including a monograph of the doubtful reptiles. American Journal of Science 11:268–297.

Barnett, H.K. and J.S. Richardson. 2002. Predation risk and competition effects on the life-history characteristics of larval Oregon spotted frog and larval red-legged frog. Oecologia 132:436–444.

Barr, B.R. 1997. Food habits of the American alligator *(Alligator mississippiensis)*, in the southern Everglades. Ph.D. dissertation. University of Miami, Coral Gables, Florida.

Barr, S. (Ed.) 1974. Applications of systematic collections: the environment. Symposium Proceedings, Association of Systematics Collections, Washington, D.C.

Barr, T.C. 1953. Notes on the occurrence of ranid frogs in caves. Copeia 1953:60–61.

Barrentine, C.D. 1991a. Survival of billbugs (*Sphenophorus* sp.) egested by western toads *(Bufo boreas)*. Herpetological Review 22:5.

Barrentine, C.D. 1991b. Food habits of western toads *(Bufo boreas halophilus)* foraging from a residential lawn. Herpetological Review 22:84–87.

Barry, S. 1999. A study of the California red-legged frog *(Rana aurora draytonii)* of Butte County, California. PAR Environmental Services, Technical Report, Number 3, Sacramento, California.

Barry, S.J. and H.B. Shaffer. 1994. The status of the California tiger salamander *(Ambystoma californiense)* at Lagunita: a 50-year update. Journal of Herpetology 28:159–164.

Bart, H.L., Jr. and R.W. Holzenthal. 1985. Feeding ecology of *Necturus beyeri* in Louisiana. Journal of Herpetology 19:402–410.

Bart, H.L., Jr., M.A. Bailey, R.E. Ashton Jr. and P.E. Moler. 1997. Taxonomic and nomenclatural status of the Upper Black Warrior River Waterdog. Journal of Herpetology 31:192–201.

Bart, J., M. Hofschen and B.G. Peterjohn. 1995. Reliability of the breeding bird survey: effects of restricting surveys to roads. Auk 112:758–761.

Barta, J.R. and S.S. Desser. 1984. Blood parasites of amphibians from Algonquin Park, Ontario, Canada. Journal of Wildlife Disease 20:180–189.

Bartelt, P.E. 1998. Natural history notes: *Bufo boreas* (western toad). Mortality. Herpetological Review 29:96.

Bartelt, P.E. 2000. A biophysical analysis of habitat selection in western toads *(Bufo boreas)* in southeastern Idaho. Ph.D. dissertation. Idaho State University, Pocatello, Idaho.

Barthalmus, G.T. 1989. Neuroleptic modulation of oral dyskinesias induced in snakes by *Xenopus* skin mucus. Pharmacology Biochemistry and Behavior 34:95–100.

Barthalmus, G.T. and E.D. Bellis. 1969. Homing in the northern dusky salamander, *Desmognathus fuscus fuscus* (Rafinesque). Copeia 1969:148–153.

Barthalmus, G.T. and E.D. Bellis. 1972. Home range, homing, and the homing mechanism of the salamander, *Desmognathus fuscus*. Copeia 1972:632–642.

Barthalmus, G.T. and I.R. Savidge. 1974. Time: an index of distance as a barrier to salamander homing. Journal of Herpetology 8:251–254.

Barthalmus, G.T. and W.J. Zielinski. 1988. *Xenopus* skin mucus induces oral dyskinesias that promote escape from snakes. Pharmacology Biochemistry and Behavior 30:957–960.

Bartlett, R.D. and P.P. Bartlett. 1999a. A Field Guide to Florida Reptiles and Amphibians. Gulf Publishing Company, Houston, Texas.

Bartlett, R.D. and P.P. Bartlett. 1999b. A Field Guide to Texas Reptiles and Amphibians. Gulf Publishing Company, Houston, Texas.

Barto, W.S. 1999. Predicting potential habitat for the arroyo toad *(Bufo microscaphus californicus)* in San Diego County using a habitat suitability model and digital terrain data. Master's thesis. San Diego State University, San Diego, California.

Barton Springs/Edwards Aquifer Conservation District. 1997. Alternative regional water supply plan for the Barton Springs segment of the Edwards Aquifer. Revised draft. Texas Water Development Board Grant Contract, Number 95-483-079, Austin, Texas.

Barton, B.S. 1808. Some account of the *Siren lacertina*. In a letter from Professor Barton of Philadelphia to Mr. John Schneider of Saxony.

Barton, D. 1995. Helminth parasites of *Bufo marinus* in Australia, Hawaii and South America. Unpublished report. CSIRO *Bufo* Project: An Overview of Research Outcomes, CSIRO, James Cook University, Townsville, Queensland, Australia.

Barton, D. 1997. Introduced animals and their parasites: the cane toad, *Bufo marinus*, in Australia. Australian Journal of Ecology 22:316–324.

Barton, N.H. 2001. Speciation. TRENDS in Ecology and Evolution 16:325.

Bartsch, P. 1944. Observations on *Hyla evittata*. Copeia 1944:187.

Basch, P.F. and R.F. Sturrock. 1969. Life history of *Ribeiroia marini* (Faust and Hoffman, 1934) comb. n. (Trematoda: Cathaemasiidae). Journal of Parasitology 55:1180–1184.

Basey, H.E. 1969. Sierra Nevada amphibians. Sequoia Natural History Association, Three Rivers, California.

Basey, H.E. 1976. Discovering Sierra reptiles and amphibians. Yosemite Natural History Association, Three Rivers, California.

Basey, H.E. and D.A. Sinclear. 1980. Amphibians and reptiles. Pp. 13–74. *In* Verner, J. and A.S. Boss (Tech. Coords.), California Wildlife and Their Habitats: Western Sierra Nevada. U.S.D.A. Forest Service, General Technical Report, PSW-37, Berkeley, California.

Bastien, H. and R.D. Leclair. 1992. Aging wood frogs *(Rana sylvatica)* by skeletochronology. Journal of Herpetology 26:222–225.

Bates, M.F., D.H. de Swardt and S. Louw. 1992. A note on the diet of the yellowbilled egret. Ostrich 63:44.

Batista, W.B. and W.J. Platt. 1997. An old-growth definition for southern mixed hardwood forests. U.S.D.A. Forest Service, Southern Research Station, General Technical Report, SRS-9, Asheville, North Carolina.

Batts, B.S. 1960. Distribution of *Pseudacris nigrita nigrita* and *Pseudacris nigrita feriarum* in the Piedmont and Coastal Plains regions of North Carolina. Herpetologica 16:45–47.

Bauer, M. 1983. Amphibien und die Bewirtschaftung ihrer Laichgewässer. Unpublished thesis. FB Landespflege, Fachhochschule Nürtingen.

Bauer, O.N. 1991. Spread of parasites and diseases of aquatic organisms by acclimatization: a short review. Journal of Fish Biology 39:679–686.

Bauer, O.N. and G.L. Hoffman. 1976. Helminth range extension by translocation of fish. Pp. 163–172. *In* Page, L.A. (Ed.), Wildlife Diseases. Plenum Press, New York.

Baumann, W.L. and M. Huels. 1982. Nests of the two-lined salamander *Eurycea bislineata*. Journal of Herpetology 16:81–83.

Bausmann, G.A. and J.O. Whitaker Jr. 1987. Studies of the habitat and food of sympatric populations of *Plethodon cinereus* (Green) and *Plethodon dorsalis* Cope in south central Parke County, Indiana. Proceedings of the Indiana Academy of Science 97:513–523.

Baxter, G.T. 1947. The amphibians and reptiles of Wyoming. Wyoming Wildlife 11:30–34.

Baxter, G.T. 1952. The relation of temperature to the altitudinal distribution of frogs and toads in southwestern Wyoming. Ph.D. dissertation. University of Michigan, Ann Arbor, Michigan.

Baxter, G.T. and J.S. Meyer. 1982. The status and decline of the Wyoming toad, *Bufo hemiophrys baxteri*. Journal of the Colorado-Wyoming Academy of Science 14:33.

Baxter, G.T. and M.D. Stone. 1985. Amphibians and reptiles of Wyoming. Second edition. Wyoming Game and Fish Department, Cheyenne, Wyoming.

Bayless, L.E. 1966. Comparative ecology of two sympatric species of *Acris* (Anura; Hylidae) with emphasis on interspecific competition. Ph.D. dissertation. Tulane University, New Orleans, Louisiana.

Bayless, L.E. 1969. Ecological divergence and distribution of sympatric *Acris* populations (Anura: Hylidae). Herpetologica 25:181–187.

Beachy, C.K. 1991a. Life history notes: *Ambystoma maculatum* (spotted salamander). Predation. Herpetological Review 22:128.

Beachy, C.K. 1991b. Ecology of larval plethodontid salamanders: size-specific predation, habitat shifts and life history evolution. American Zoologist 31:106A.

Beachy, C.K. 1993a. Differences in variation in egg size for several species of salamanders (Amphibia: Caudata) that use different larval environments. Brimleyana 18:71–81.

Beachy, C.K. 1993b. Guild structure in streamside salamander communities: a test for interactions among larval plethodontid salamanders. Journal of Herpetology 27:465–468.

Beachy, C.K. 1994. Community ecology in streams: effects of two species of predatory salamanders on a prey species of salamander. Herpetologica 50:129–136.

Beachy, C.K. 1995. Effects of larval growth history on metamorphosis in a stream–dwelling salamander (Desmognathus ochrophaeus). Journal of Herpetology 29:375–382.

Beachy, C.K. 1996. Reduced courtship success between parapatric populations of the plethodontid salamander Gyrinophilus porphyriticus. Copeia 1996:199–203.

Beachy, C.K. 1997a. Effect of predatory larval Desmognathus quadramaculatus on growth, survival, and metamorphosis of larval Eurycea wilderae. Copeia 1997:131–137.

Beachy, C.K. 1997b. Courtship behavior in the plethodontid salamander Gyrinophilus porphyriticus. Herpetologica 53:289–296.

Beachy, C.K. and R.C. Bruce. 1992. Lunglessness in plethodontid salamanders is consistent with the hypothesis of a mountain stream origin: a response to Ruben and Boucot. American Naturalist 139:839–847.

Beachy, C.K. and. R.C. Bruce. 1998. Miniaturization in Desmognathus quadramaculatus via accelerated age at maturation. P. 6. In Program and Abstracts of the Fourth Conference on the Biology of Plethodontid Salamanders. Highlands, North Carolina.

Beane, J.C. 1990. Life history notes: Rana palustris (pickerel frog). Predation. Herpetological Review 21:59.

Beane, J.C. 1998. Status of the river frog, Rana heckscheri (Anura: Ranidae), in North Carolina. Brimleyana 25:69–79.

Beane, J.C. and E.L. Hoffman. 1995. Geographic distribution: Rana capito capito (Carolina gopher frog). Herpetological Review 26:153.

Beane, J.C. and E.L. Hoffman. 1997. Geographic distribution: Rana capito capito (Carolina gopher frog). Herpetological Review 28:208.

Beane, J.C. and R.W. Gaul Jr. 1991. Geographic distribution: Ambystoma maculatum (spotted salamander). Herpetological Review 22:133.

Beard, K.H. and J. Baillie. 1998. Natural history notes: Rana catesbeiana (bullfrog). Diet. Herpetological Review 29:40.

Beard, K.H., S. McCullough and A.E. Eschtruth. 2003. Quantitative assessment of habitat preferences for the Puerto Rican terrestrial frog, Eleutherodactylus coqui. Journal of Herpetology 37:10–17.

Beatty, J.J. 1979. Morphological variation in the clouded salamander, Aneides ferreus (Cope) (Amphibia: Caudata: Plethodontidae). Ph.D. dissertation. Oregon State University, Corvallis, Oregon.

Beauregard, N. and R. Leclair. 1988. Multivariate analysis of the summer habitat structure of Rana pipiens Schreber, in Lac Saint Pierre (Quebec, Canada). Pp. 129–143. In Szaro, R.C., K.E. Severson and D.R. Patton (Tech. Coords.), Management of Amphibians, Reptiles, and Small Mammals in North America. U.S.D.A. Forest Service, Rocky Mountain Forest and Range Experiment Station, Technical Report, RM-166, Fort Collins, Colorado.

Beaver, P.C. 1937. Experimental studies with Echinostoma revolutum (Froelich), a fluke from birds and mammals. Illinois Biological Monographs 34:1–96.

Beaver, P.C. 1939. The morphology and life history of Psilostomum ondatrae Price 1931 (Trematoda: Psilostomatidae). Journal of Parasitology 25:383–393.

Beck, C.E. 1997. Effect of changes in resource level on age and size at metamorphosis in Hyla squirella. Oecologia 112:187–192.

Beck, C.W. and J.D. Congdon. 1999. Effects of individual variation in age and size at metamorphosis on growth and survivorship of southern toad (Bufo terrestris) metamorphs. Canadian Journal of Zoology 77:944–951.

Beck, J.M., J. Janovetz and C.R. Peterson. 1998. Amphibians of the Coeur d'Alene basin: a survey of Bureau of Land Management lands. Idaho Bureau of Land Management, Technical Bulletin 98-3, Boise, Idaho.

Bee, M.A., S.A. Perrill and P.C. Owen. 1999. Size assessment in simulated territorial encounters between male green frogs (Rana clamitans). Behavioral Ecology and Sociobiology 45:177–184.

Beebee, T.J.C. 1985. Discriminant analysis of amphibian habitat determinants in southeast England. Amphibia-Reptilia 6:35–43.

Beebee, T.J.C. 1995. Amphibian breeding and climate. Nature 374:219–220.

Beebee, T.J.C. 1996. Ecology and Conservation of Amphibians. Chapman and Hall, London, England.

Behler, J.L. and F.W. King. 1979. The Audubon Society Field Guide to North American Reptiles and Amphibians. Alfred A. Knopf, New York.

Behler, J.L. and F.W. King. 1995. National Audubon Society Field Guide to North American Reptiles and Amphibians. Alfred A. Knopf, New York.

Behler, J.L. and F.W. King. 1998. The Audubon Society Field Guide to North American Reptiles and Amphibians. Alfred A. Knopf, New York.

Behre, E.H. 1953. Food of the salamander Triturus viridescens viridescens. Copeia 1953:60.

Beier, P. 1993. Determining minimum habitat areas and habitat corridors for cougars. Conservation Biology 7:94–108.

Beiswenger, R.E. 1975. Structure and function in aggregations of tadpoles of the American toad, Bufo americanus. Herpetologica 31:222–233.

Beiswenger, R.E. 1977. Diel patterns of aggregative behavior in tadpoles of Bufo americanus in relation to light and temperature. Ecology 58:98–108.

Beiswenger, R.E. 1978. Response of Bufo tadpoles (Amphibia, Anura, Bufonidae) to laboratory gradients of temperature. Journal of Herpetology 12:499–504.

Beiswenger, R.E. 1981. Predation by gray jays on aggregating tadpoles of the boreal toad (Bufo boreas). Copeia 1981:459–460.

Belcher, D.L. 1988. Courtship behavior and spermatophore deposition by the subterranean salamander, Typhlomolge rathbuni (Caudata, Plethodontidae). Southwestern Naturalist 33:124–125.

Belden, L.K. 2002. Sublethal effects of UV-B radiation on larval amphibians. Ph.D. dissertation. Oregon State University, Corvallis.

Belden, L.K. and A.R. Blaustein. 2002a. Exposure of red-legged frog embryos to ambient UV-B radiation in the field negatively affects larval growth and development. Oecologia 130:551–554.

Belden, L.K. and A.R. Blaustein. 2002b. Population differences in sensitivity to UV-B radiation for larval long-toed salamanders. Ecology 83:1586–1590.

Belden, L.K., E.L. Wildy and A.R. Blaustein. 2000. Growth, survival and behaviour of larval long-toed salamanders (Ambystoma macrodactylum) exposed to ambient levels of UV-B radiation. Journal of Zoology London 251:473–479.

Belden, L.K., I.T. Moore, A.C. Hatch and A.R. Blaustein. 2001. The effects of UV-B radiation exposure on activity and circulating corticosterone levels in roughskin newts. Hormones and Behavior 39:313.

Bell, E.L. 1956. Some aspects of the life history of the red salamander, Pseudotriton r. ruber, in Huntington County, PA. Mengel Naturalist 3:10–13.

Bell, G. 1977. The life of the smooth newt, Triturus vulgaris, after metamorphosis. Ecological Monographs 47:279–299.

Bellis, E.D. 1957. The effects of temperature on salientian breeding calls. Copeia 1957:85–89.

Bellis, E.D. 1961a. Growth of the wood frog, Rana sylvatica. Copeia 1961:74–77.

Bellis, E.D. 1961b. Home range and movements of the wood frog in a northern bog. Ecology 46:90–98.

Bellis, E.D. 1962. The influence of humidity on wood frog activity. American Midland Naturalist 68:139–148.

Bellis, E.D. 1968. Summer movement of red-spotted newts in a small pond. Journal of Herpetology 1:68–91.

Bellocq, M.I., K. Kloosterman and S.M. Smith. 2000. The diet of coexisting species of amphibians in Canadian jack pine forests. Herpetological Journal 10:63–68.

Bellows, T.S. 1981. The descriptive properties of some models for density dependence. Journal of Animal Ecology 50:139–156.

Belsky, A.J. and D.M. Blumenthal. 1997. Effects of livestock grazing on stand dynamics and soils in upland forests of the interior West. Conservation Biology 11:315–327.

Belton, J.C. and A. Owczarzak. 1968. Cellular changes associated with the pre-ovulatory deposition and storage of hepatic lipids in the frog Ascaphus truei. Herpetologica 24:113–127.

Beltz, E. 1995. Citations for the original descriptions of North American amphibians and reptiles. Herpetological Circular, Number 24, Society for the Study of Amphibians and Reptiles, St. Louis, Missouri.

Benally, R., K.C. Nishikawa and S.L. Lindstedt. 1996. Adjustment of posture and prey capture kinematics by wild-caught and lab-raised toads (Bufo woodhousii) on platforms of different angles. American Zoologist 36:82.

Beneski, J.T., Jr. 1989. Adaptive significance of tail autotomy in the salamander, Ensatina. Journal of Herpetology 23:322–324.

Beneski, J.T., Jr., J.H. Larsen Jr. and B.T. Miller. 1995. Variation in the feeding kinematics of mole salamanders (Ambystomatidae: Ambystoma). Canadian Journal of Zoology 73:353–366.

Beneski, J.T., Jr., E.J. Zalisko and J.H.J. Larsen. 1986. Demography and migratory patterns of the eastern long-toed salamander, *Ambystoma macrodactylum columbianum*. Copeia 1986:398–408.

Bennett, A.F. and L.D. Houck. 1983. The energetic cost of courtship and aggression in a plethodontid salamander. Ecology 64:979–983.

Bennett, C. and R.J. Taylor. 1968. Notes on the lesser siren, *Siren intermedia* (Urodela). Southwestern Naturalist 13:455–457.

Bennett, G.W. 1971. Management of Lakes and Ponds. Van Nostrand Reinhold Company, New York.

Bennett, P.M. and I.P. Owens. 1997. Variation in extinction risk among birds: chance or eveolutionary predisposition? Proceedings of the Royal Society of London B 264:401–408.

Bennett, R.S., E.E. Klaas, J.R. Coats, M.A. Mayse and E.J. Kolbe. 1983. Fenvalerate residues in nontarget organisms from treated cotton fields. Bulletin of Environmental Toxicology 31:61–65.

Bennett, S.H. and J.B. Nelson. 1991. Distribution and status of Carolina bays in South Carolina. Nongame and Heritage Trust Publications, Number 1, South Carolina Wildlife and Marine Resources Department, Columbia, South Carolina.

Bennett, S.H., J.W. Gibbons and J. Glanville. 1980. Terrestrial activity, abundance and diversity of amphibians in differently managed forest types. American Midland Naturalist 103:412–416.

Bent, A.C. 1919–1968. Life Histories of North American Birds: Series. Smithsonian Institution Press, Washington, D.C.

Bent, A.C. 1926. Life histories of North American Marsh Birds. United States National Museum, Bulletin 135, Smithsonian Institution, Washington, D.C.

Bent, A.C. 1949. Life histories of North American thrushes, kinglets, and their allies. United States National Museum, Bulletin 196, Smithsonian Institution, Washington, D.C.

Berger, L. 1987. Impact of agriculture intensification on Amphibia. Pp. 79–82. *In* van Gelder, J.J., H. Strijbosh and P.J.M. Bergers (Eds.), Proceedings of the Fourth Ordinary General Meeting of the Societas Europaea Herpetologica. Nijmegen, Netherlands.

Berger, L., R. Speare, P. Daszak, D.E. Green, A.A. Cunningham, C.L. Goggin, R. Slocombe, M.A. Ragan, A.D. Hyatt, K.R. McDonald, H.B. Hines, K.R. Lips, G. Marantelli and H. Parkes. 1998. Chytridiomycosis causes amphibian mortality associated with population declines in the rain forests of Australia and Central America. Proceedings of the National Academy of Sciences 95:9031–9036.

Berger-Bishop, L.E. and R.N. Harris. 1996. A study of caudal allometry in the salamander *Hemidactylium scutatum* (Caudata: Plethodontidae). Herpetologica 52:515–525.

Beringer, J. and T.R. Johnson. 1995. Natural history notes: *Rana catesbeiana* (bullfrog). Diet. Herpetological Review 26:98.

Berkhouse, C.S. and J.N. Fries. 1995. Critical thermal maxima of juvenile and adult San Marcos salamanders *(Eurycea nana)*. Southwestern Naturalist 40:430–434.

Bernardo, J. 1994. Experimental analysis of allocation in two divergent, natural salamander populations. American Naturalist 143:14–38.

Bernardo, J. 2000. Early life histories of dusky salamanders, *Desmognathus imitator* and *D. wrighti*, in a headwater seepage in Great Smoky Mountains National Park, USA. Amphibia-Reptilia 21:403–407.

Bernardo, J. 2002. Natural history notes: *Desmognathus carolinensis* (Carolina Mountain dusky salamander) and *Plethodon welleri* (Weller's salamander); *Desmognathus monticola* (seal salamander) and *Desmognathus wrighti* (pigmy salamander); *Gyrinophilus porphyriticus* (spring salamander) and *Desmognathus ocoee* (Ocoee salamander). Intraguild Predation. Herpetological Review 33:121.

Bernardo, J. and S.J. Arnold. 1999. Mass-rearing of plethodontid eggs. Amphibia-Reptilia 20:219–224.

Berrill, M., S. Bertram and B. Pauli. 1997a. Effects of pesticides on amphibian embryos and larvae. Herpetological Conservation 1:233–245.

Berrill, M., S. Bertram and B. Pauli. 1997b. Effects of pesticides on amphibian embryos and tadpoles. Pp. 258–270. *In* Green, D.M. (Ed.), Amphibians in Decline: Canadian Studies of a Global Problem. Herpetological Conservation, Number 1, Society for the Study of Amphibians and Reptiles, St. Louis, Missouri.

Berrill, M., S. Bertram, D. Brigham and V. Campbell. 1992. A comparison of three methods of monitoring frog populations. Occasional Papers of the Canadian Wildlife Service 76:87–93.

Berven, K.A. 1981. Mate choice in the wood frog, *Rana sylvatica*. Evolution 35:707–722.

Berven, K.A. 1982a. The genetic basis of altitudinal variation in the wood frog *Rana sylvatica*. I. An experimental analysis of life-history traits. Evolution 36:962–983.

Berven, K.A. 1982b. The genetic basis of altitudinal variation in the wood frog *Rana sylvatica*. II. An experimental analysis of larval development. Oecologia 52:360–369.

Berven, K.A. 1988. Factors affecting variation in reproductive traits within a population of wood frogs (*Rana sylvatica*). Copeia 1988:605–615.

Berven, K.A. 1990. Factors affecting population fluctuations in larval and adult stages of the wood frog (*Rana sylvatica*). Ecology 71:1599–1608.

Berven, K.A. 1995. Population regulation in the wood frog, *Rana sylvatica*, from three diverse geographic localities. Australian Journal of Ecology 20:385–392.

Berven, K.A. and B.G. Chadra. 1988. The relationship among egg size, density and food level on larval development in the wood frog *(Rana sylvatica)*. Oecologia 75:67–72.

Berven, K.A. and D.E. Gill. 1983. Interpreting geographic variation in life-history traits. American Zoologist 23:85–97.

Berven, K.A. and T.A. Grudzien. 1990. Dispersal in the wood frog *(Rana sylvatica)*: implications for genetic population structure. Evolution 44:2047–2056.

Besharse, J.C. and R.A. Brandon. 1974. Postembryonic eye degeneration in the troglobitic salamander *Typhlotriton spelaeus*. Journal of Morphology 144:381–406.

Besharse, J.C. and R.A. Brandon. 1976. Effects of continuous light and darkness on the eyes of the troglobitic salamander *Typhlotriton spelaeus*. Journal of Morphology 149:527–546.

Besharse, J.C. and J.R. Holsinger. 1977. *Gyrinophilus subterraneus*, a new troglobitic salamander from southern West Virginia. Copeia 1977:624–634.

Bethel, W.M. and J.C. Holmes. 1977. Increased vulnerability of amphipods to predation owing to altered behavior induced by larval acanthocephalans. Canadian Journal of Zoology 55:110–115.

Beurden, E.K. and G.C. Grigg. 1980. An isolated and expanding population of the introduced toad *Bufo marinus* in New South Wales. Australian Wildlife Research 7:305–310.

Beverly-Burton, M. 1987. *Ophioxenos microphagus* (Ingles, 1936) comb. N. (Digenea: Parampistomidae) from ectotherms in western North America with comments on host-parasite relationships. Proceedings of the Helminthological Society of Washington 54:197–199.

Bezy, R.L. and S.R. Goldberg. 1997. Geographic distribution: *Hyla regilla* (Pacific treefrog). Herpetological Review 28:156.

Bezy, R.L., W.C. Sherbrooke and C.H. Lowe. 1966. The rediscovery of *Eleutherodactylus augusti* in Arizona. Herpetologica 22:221–225.

Bider, J.R. 1962. Dynamics and the temporo-spatial relations of a vertebrate community. Ecology 43:634–646.

Biek, R., W.C. Funk, B.A. Maxell and L.S. Mills. 2002. What is missing in amphibian decline research: insights from ecological sensitivity analysis. Conservation Biology 16:728–734.

Bier, C.W. 1985. Geographic distribution: *Aneides aeneus* (green salamander). Herpetological Review 16:60.

Biever, R.C., J.M. Giddings, M. Kiamos, M.F. Annunziato, R. Meyerhoff and K. Racke. 1994. Effects of chlorpyrifos on aquatic mircocosms over a range of off-target spray drift exposures. Pp. 1367–1372. Brighton Crop Protection Conference—Pests and Diseases, 21–24 November, Brighton, United Kingdom.

Bildstein, K.L., D.E. Gawlik, D.P. Ferral, I.L. Brisbin Jr. and G.R. Wein. 1994. Wading bird use of established and newly created reactor cooling reservoirs at the Savannah River site, near Aiken, South Carolina, USA. Hydrobiologia 279/280:71–82.

Billings, J.T. 1973. A partial definition of the fundamental niche of larval *Rana pipiens* (Ranidae, Amphibia). Master's thesis. University of Nebraska, Lincoln, Nebraska.

Bininda-Edmonds, R.R.P., D.P. Vázquez and L.L. Manne. 2000. The calculus of biodiversity: intergrating phylogeny and conservation. TRENDS in Ecology and Evolution 15:92–93.

Biodiversity Legal Foundation. 1994. Amargosa toad *(Bufo nelsoni)*. Petition for a rule to list the Amargosa toad, *Bufo nelsoni*, as "endangered" in the conterminous United States under the Endangered Species Act, 16 U.S.C. Sec 1531 et seq. (1973) as amended. Boulder, Colorado (available from Nevada Natural Heritage Program, Carson City, Nevada).

Birge, W.J., J.A. Black and R.A. Kuehne. 1980. Effects of organic compounds on amphibian reproduction. Water Resources Research Institute Research Report, Number 121, University of Kentucky, Lexington, Kentucky.

Birge, W.J., A.G. Westerman and J.A. Spromberg. 2000. Comparative toxicology and risk assessment of amphibians. Pp. 727–791. *In*

Sparling, D.W., G. Linder and C.A. Bishop (Eds.), Ecotoxicology of Amphibians and Reptiles. Society for Environmental Toxicology and Contaminants (SETAC) Press, Pensacola, Florida.

Birkenholz, D.E. 1963. A study of the life history and ecology of the round-tailed muskrat (*Neofiber alleni* True) in north-central Florida. Ecological Monographs 33:255–280.

Bishop, C.A. 1992. The effects of pesticides on amphibians and the implications for determining causes of declines in amphibian populations. Pp. 67–70. *In* Bishop, C.A. and K.E. Pettit (Eds.), Declines in Canadian Amphibian Populations: Designing a National Monitoring Strategy. Canadian Wildlife Service, Occasional Paper, Number 76, Ottawa, Ontario, Canada.

Bishop, C.A. and K.E. Pettit (Eds.). 1992. Declines in Canadian Amphibian Populations: Designing a National Monitoring Strategy. Canadian Wildlife Service, Occasional Paper, Number 76, Ottawa, Ontario, Canada.

Bishop, C.A., K.E. Pettit, M.E. Gartshore and D.A. Macleod. 1997. Extensive monitoring of anuran populations using call counts and road transects in Ontario (1992 to 1993). Pp. 149–160. *In* Green, D.M. (Ed.), Amphibians in Decline: Canadian Studies of a Global Problem. Herpetological Conservation, Number 1, Society for the Study of Amphibians and Reptiles, St. Louis, Missouri.

Bishop, D.C. 2003. Natural history notes: *Rana okaloosae* (Florida bog frog). Predation. Herpetological Review 34:235.

Bishop, L.A. and T.M. Farrell. 1994. Natural history notes: *Thamnophis sauritus sackenii* (Peninsula ribbon snake). Behavior. Herpetological Review 25:127.

Bishop, S.C. 1919. Notes on the habits and development of the four-toed salamander, *Hemidactylium scutatum* (Schlegel). New York State Museum Bulletin 219:251–282.

Bishop, S.C. 1924. Notes on salamanders. New York State Museum Bulletin 253:87–96.

Bishop, S.C. 1925. The life of the red salamander. Natural History 25:385–389.

Bishop, S.C. 1926. Notes on the habits and development of the mud-puppy *Necturus maculosus* (Rafinesque). New York State Bulletin, Number 268, Albany, New York.

Bishop, S.C. 1927. The Amphibians and Reptiles of Allegheny State Park. New York State Museum, Handbook 3, Albany, New York.

Bishop, S.C. 1928. Notes on some amphibians and reptiles from the southeastern states, with a description of a new salamander from North Carolina. Journal of the Elisha Mitchell Scientific Society 43:153–170.

Bishop, S.C. 1934. Description of a new salamander from Oregon, with notes on related species. Proceedings of the Biological Society of Washington 47:169–171.

Bishop, S.C. 1937. A remarkable new salamander from Oregon. Herpetologica 1:93–95.

Bishop, S.C. 1941a. Notes on salamanders, with descriptions of several new forms. Occasional Papers of the Museum of Zoology, Number 451, University of Michigan, Ann Arbor, Michigan.

Bishop, S.C. 1941b. The Salamanders of New York. New York State Museum Bulletin, Number 324, Albany, New York.

Bishop, S.C. 1943. A Handbook of Salamanders. The Salamanders of the United States, of Canada, and of Lower California. Comstock Publishing Company, Ithaca, New York.

Bishop, S.C. 1944. A new neotenic plethodont salamander, with notes on related species. Copeia 1944:1–5.

Bishop, S.C. 1947. A Handbook of Salamanders. Hafner Publishing Company, New York.

Bishop, S.C. and H.P. Chrisp. 1933. The nests and young of the Allegheny salamander *Desmognathus ochrophaeus* (Cope). Copeia 1933:194–198.

Bishop, S.C. and B.O. Valentine. 1950. A new species of *Desmognathus* from Alabama. Copeia 1950:39–43.

Bishop, S.C. and M.R. Wright. 1937. A new neotenic salamander from Texas. Proceedings of the Biological Society of Washington 50:141–143.

Bitz, A. and R.Thiele. 1992. Bedeutung und Folgewirkungen der Oberflächenentwässerung den Artenschutz, dargestellt am Beispiel rheinhessischer Amphibienpopulationen. Fauna und Flora in Rheinland-Pfalz 6:89–104.

Bizer, J.R. 1978. Growth rates and size at metamorphosis of high elevation populations of *Ambystoma tigrinum*. Oecologia 34:175–184.

Blab, J. 1985. Handlungs-und Forschungsbedarf für den Reptilienschutz. Natur und Landschaft 60:336–339.

Blab, J. 1986. Biologie, Ökologie und Schutz von Amphibien. Third edition. Kilda Verlag, Greven, Germany.

Black, I.H. and K.L. Gosner. 1958. The barking treefrog, *Hyla gratiosa*, in New Jersey. Herpetologica 13:254–255.

Black, J.D. and S.C. Dellinger. 1938. Herpetology of Arkansas. Part two. The amphibians. Occasional Papers of the University of Arkansas Museum, Number 2, Fayetteville, Arkansas.

Black, J.H. 1968. A possible stimulus for the formation of some aggregations in tadpoles of *Scaphiopus bombifrons*. Proceedings of the Oklahoma Academy of Science 1968:13–14.

Black, J.H. 1969a. The frog genus *Rana* in Montana. Northwest Science 43:191–195.

Black, J.H. 1969b. A cave dwelling population of *Ambystoma tigrinum mavortium* in Oklahoma. Journal of Herpetology 3:183–184.

Black, J.H. 1970a. A possible stimulus for the formation of some aggregations in tadpoles of *Scaphiopus bombifrons*. Proceedings of the Oklahoma Academy of Science 49:13–14.

Black, J.H. 1970b. Amphibians of Montana. Montana Wildlife, January, 1970:1–32.

Black, J.H. 1971a. The formation of tadpole holes. Herpetological Review 3:7.

Black, J.H. 1971b. The toad genus *Bufo* in Montana. Northwest Science 45:156–162.

Black, J.H. 1971c. The cave life of Oklahoma. Oklahoma underground. National Speleological Society 4:1–53.

Black, J.H. 1973. A checklist of the cave fauna of Oklahoma: Amphibia. Proceedings of the Oklahoma Academy of Science 53:33–37.

Black, J.H. 1974. Notes on *Plethodon ouachitae* in Oklahoma. Proceedings of the Oklahoma Academy of Science 54:88–89.

Black, J.H. 1975. The formation of "tadpole nests" by anuran larvae. Herpetologica 31:76–79.

Black, J.H. and R.B. Brunson. 1971. Breeding behavior of the boreal toad (*Bufo boreas boreas*) (Baird and Girard) in western Montana. Great Basin Naturalist 31:109–113.

Black, J.H. and W.L. Puckette. 1974. The Ouachita salamander in Oklahoma caves. Oklahoma Underground 6:42–44.

Black, J.H. and G. Sievert. 1989. A Field Guide to Amphibians of Oklahoma. Oklahoma Department of Wildlife Conservation, Oklahoma City, Oklahoma.

Black, J.H., T. Hunkapiller and D. Dawson. 1976. Winter activity in Oklahoma frogs. Bulletin of the Oklahoma Herpetological Society 1:22.

Blackburn, L., P. Nanjappa and M.J. Lannoo. 2000. National atlas of U.S. amphibians. Muncie Center for Medical Education, Indiana University School of Medicine and Ball State University, Muncie, Indiana (see also http:///www.mp2-pwrc.usgs.gov/armiatlas/).

Blair, A. 1939. Records of the salamander *Typhlotriton*. Copeia 1939:108–109.

Blair, A.P. 1941. Variation, isolating mechanisms and hybridization in certain toads. Genetics 26:398–417.

Blair, A.P. 1943. Geographic variation of ventral markings in toads. American Midland Naturalist 29:615–620.

Blair, A.P. 1950. Note on Oklahoma microhylid frogs. Copeia 1950:152.

Blair, A.P. 1951a. Note on the herpetology of the Elk Mountains, Colorado. Copeia 1951:239–240.

Blair, A.P. 1951b. Winter activity in Oklahoma frogs. Copeia 1951:178.

Blair, A.P. 1951c. Note on Oklahoma salamanders. Copeia 1951:178.

Blair, A.P. 1952. Notes on amphibians from Oklahoma and North Dakota. Copeia 1952:114–115.

Blair, A.P. 1955. Distribution, variation, and hybridization in a relict toad (*Bufo microscaphus*) in southwestern Utah. American Museum Novitates, Number 1722, American Museum of Natural History, New York.

Blair, A.P. 1957a. Amphibians. Pp. 211–271. *In* Blair, W.F., A.P. Blair, P. Brodkorb, F.R. Cagle and G.A. Moore (Eds.), Vertebrates of the United States. McGraw-Hill Publishing Company, New York.

Blair, A.P. 1957b. A comparison of living *Plethodon ouachitae* and *P. caddoensis*. Copeia 1957:47–48.

Blair, A.P. 1963. Notes on anuran behavior, especially *Rana catesbeiana*. Herpetologica 19:51.

Blair, A.P. 1967. *Plethodon ouachitae*. Pp. 40.1. Catalogue of American Amphibians and Reptiles. Society for the Study of Amphibians and Reptiles, St. Louis, Missouri.

Blair, A.P. and H.L. Lindsay Jr. 1961. *Hyla avivoca* (Hylidae) in Oklahoma. Southwestern Naturalist 6:202.

Blair, A.P. and H.L. Lindsay Jr. 1965. Color pattern variation and distribution of two large *Plethodon* salamanders endemic to the Ouachita Mountains of Oklahoma and Arkansas. Copeia 1965:95–100.

Blair, J. and R. J. Wassersug. 2000. Variation in the pattern of predator-induced damage to tadpole tails. Copeia 2000:390–401.

Blair, K.B., D. Chiszar and H.M. Smith. 1997. New records for Texas amphibians and reptiles. Herpetological Review 28:99.

Blair, W.F. 1936. A note on the ecology of *Microhyla olivacea*. Copeia 1936:115.

Blair, W.F. 1950. The biotic provinces of Texas. Texas Journal of Science 2:93–117.

Blair, W.F. 1952. Mammals of the Tamaulipan Biotic Province in Texas. Texas Journal of Science 4:230–250.

Blair, W.F. 1953. Growth, dispersal and age at sexual maturity of the Mexican toad (*Bufo valliceps* Wiegmann). Copeia 1953:208–212.

Blair, W.F. 1955a. Mating call and stage of speciation in the *Microhyla olivacea–M. carolinensis* complex. Evolution 9:469–480.

Blair, W.F. 1955b. Size difference as a possible isolation mechanism in *Microhyla*. American Naturalist 89:297–301.

Blair, W.F. 1955c. Differentiation of mating calls in spadefoots, genus *Scaphiopus*. Texas Journal of Science 7:183–188.

Blair, W.F. 1956a. Call difference as an isolation mechanism in southwestern toads (genus *Bufo*). Texas Journal of Science 8:87–106.

Blair, W.F. 1956b. Comparative survival of hybrid toads (*B. woodhousei* x *B. valliceps*) in nature. Copeia 1956:259–260.

Blair, W.F. 1957. Changes in vertebrate populations under conditions of drought. Cold Spring Harbor Symposia on Quantitative Biology 22:273–275.

Blair, W.F. 1958a. Mating call in the speciation of anuran populations. American Naturalist 92:27–51.

Blair, W.F. 1958b. Call difference as an isolation mechanism in Florida species of hylid frogs. Quarterly Journal of the Florida Academy of Science 21:32–48.

Blair, W.F. 1959. Genetic compatibility and species groups in the U.S. toads (genus *Bufo*). Texas Journal of Science 11:427–453.

Blair, W.F. 1960a. A breeding population of the Mexican toad (*Bufo valliceps*) in relation to its environment. Ecology 41:165–174.

Blair, W.F. 1960b. Mating call as evidence of relations in the *Hyla eximia* group. Southwestern Naturalist 5:129–135.

Blair, W.F. 1961a. Calling and spawning seasons in a mixed population of anurans. Ecology 42:99–110.

Blair, W.F. 1961b. Further evidence bearing on intergroup and intragroup genetic compatibility in toads (genus *Bufo*). Texas Journal of Science 13:163–175.

Blair, W.F. 1963a. Intragroup genetic compatibility in the *Bufo americanus* species group of toads. Texas Journal of Science 15:15–34.

Blair, W.F. 1963b. Evolutionary relationships of North American toads of the genus *Bufo*: a progress report. Evolution 17:1–16.

Blair, W.F. 1964a. Isolating mechanisms and interspecies interactions in anuran amphibians. Quarterly Review of Biology 39:334–344.

Blair, W.F. 1964b. Evidence bearing on relationships of the *Bufo boreas* group of toads. Texas Journal of Science 16:181–192.

Blair, W.F. 1965. Amphibian speciation. Pp. 543–556. *In* Wright, H.E. and D.G. Frey (Eds.), The Quaternary of the United States. Princeton University Press, Princeton, New Jersey.

Blair, W.F. 1972a. *Bufo* of North and Central America. Pp. 93–101. *In* Blair, W.F. (Ed.), Evolution in the Genus *Bufo*. University of Texas Press, Austin. Texas.

Blair, W.F. 1972b. Evidence from hybridization. Pp. 196–232. *In* Blair, W.F. (Ed.), Evolution in the Genus *Bufo*. University of Texas Press, Austin. Texas.

Blair, W.F. and D.I. Pettus. 1954. The mating call and its significance in the Colorado River toad (*Bufo alvarius* Girard). Texas Journal of Science 6:72–77.

Blais, D.P. 1996. Movement, home range, and other aspects of the biology of the eastern hellbender (*Cryptobranchus a. alleganiensis*): a radio telemetric study. Master's thesis. State University of New York at Binghamton, Binghamton, New York.

Blais, J.M., D.W. Schindler, D.C.G. Muir, L.E. Kimpe, D.B. Donald and B. Rosenberg. 1998. Accumulation of persistent organochlorine compounds in mountains of western Canada. Nature 395:585–588.

Blakeslee, N. 1999. Smoke and water: Alcoa's coal mine, San Antonio's water, and the heart of central Texas. Texas Observer, 20 August.

Blanchard, C.L. and M. Stromberg. 1987. Acidic precipitation in southeastern Arizona: sulfate, nitrate, and trace-metal deposition. Atmospheric Environment 21:2375–2381.

Blanchard, F.N. 1921. A collection of amphibians and reptiles from northeastern Washington. Copeia 90:5–6.

Blanchard, F.N. 1922. The amphibians and reptiles of western Tennessee. Occasional Papers of the Museum of Zoology, Number 117, University of Michigan, Ann Arbor, Michigan.

Blanchard, F.N. 1923. The life history of the four-toed salamander. American Naturalist 57:262–268.

Blanchard, F.N. 1928a. Topics from the life history and habits of the red-backed salamander in southern Michigan. American Naturalist 62:156–164.

Blanchard, F.N. 1928b. Amphibians and reptiles of the Douglas Lake region in northern Michigan. Copeia 166:42–51.

Blanchard, F.N. 1930. The stimulus to the breeding migration of the spotted salamander, *Ambystoma maculatum* Shaw. American Naturalist 64:154–167.

Blanchard, F.N. 1933a. Spermatophores and the mating season of the salamander *Hemidactylium scutatum* (Schlegel). Copeia 1933:40.

Blanchard, F.N. 1933b. Late autumn collections and hibernating situations of the salamander *Hemidactylium scutatum* (Schlegel) in southern Michigan. Copeia 1933:216.

Blanchard, F.N. 1934a. The relation of the female four-toed salamander to her nest. Copeia 1934:137–138.

Blanchard, F.N. 1934b. The date of egg-laying of the four-toed salamander, *Hemidactylium scutatum* (Schlegel), in southern Michigan. Papers of the Michigan Academy of Sciences, Arts and Letters 19:571–575.

Blanchard, F.N. 1934c. The spring migration of the four-toed salamander, *Hemidactylium scutatum* (Schlegel). Copeia 1934:50.

Blanchard, F.N. 1936. The number of eggs produced and laid by the four-toed salamander, *Hemidactylium scutatum* (Schlegel), in southern Michigan. Papers of the Michigan Academy of Sciences, Arts and Letters 21:567–573.

Blanchard, F.N. and F.C. Blanchard. 1931. Size groups and their characteristics in the salamander *Hemidactylium scutatum* (Schlegel). American Naturalist 65:149–164.

Blaney, R.M. 1971. An annotated checklist and biogeographic analysis of the insular herpetofauna of the Appalachicola region, Florida. Herpetologica 27:406–430.

Blaney, R.M. and P.K. Blaney. 1978. Significance of extreme variation in a cave population of the salamander *Gyrinophilus porphyriticus*. Proceedings of the West Virginia Academy of Science 50:23.

Blaney, R.M. and K. Relyea. 1967. The zigzag salamander, *Plethodon dorsalis* Cope, in southern Alabama. Herpetologica 23:246–247.

Blatchley, W.S. 1892. Notes on the batrachians and reptiles of Vigo County, Indiana. Journal of the Indiana Society of Natural History 124:27.

Blatchley, W.S. 1899. Notes on the batrachians and reptiles of Vigo County, Indiana–II. Indiana Department of Geology and Natural Resources, 24th Annual Report. Department of Geological and Natural Resources, Indianapolis, Indiana.

Blatchley, W.S. 1900 (1901). On a small collection of batrachians, with descriptions of two new species. Pp. 759–763. *In* Indiana Department of Geology and Natural Resources, 25th Annual Report, Indiana Department of Geology and Natural Resources, Bloomington, Indiana.

Blaustein, A.R. 1994. Chicken Little or Nero's fiddle? A perspective on declining amphibian populations. Herpetologica 50:85–97.

Blaustein, A.R. and J.M. Kiesecker. 1997. The effects of ultraviolet-B radiation on amphibians in natural ecosystems. Pp. 175–188. *In* Hader, D.P. (Ed.), The Effects of Ultraviolet Radiation on Aquatic Ecosystems. R.G. Landes Company, Austin, Texas.

Blaustein, A.R. and D.H. Olson. 1991. Amphibian population declines. Science 253:1467.

Blaustein, A.R. and D.B. Wake. 1990. Declining amphibian populations: a global phenomenon? TRENDS in Ecology and Evolution 5:203–204.

Blaustein, A.R. and D.B. Wake. 1995. The puzzle of declining amphibian populations. Scientific American 272:52–57.

Blaustein, A.R., R.K. O'Hara and D.H. Olson. 1984. Kin preference behaviour is present after metamorphosis in *Rana cascadae* frogs. Animal Behaviour 32:445–450.

Blaustein, A.R., D.B. Wake and W.P. Sousa. 1994a. Amphibian declines: judging stability, persistence, and susceptibility of populations to local and global extinctions. Conservation Biology 8:60–71.

Blaustein, A.R., J.J. Beatty, D.H. Olson and R.M. Storm. 1995b. The biology of amphibians and reptiles in old-growth forests in the Pacific Northwest. U.S.D.A. Forest Service, General Technical Report, PNW-GTR-337, Portland, Oregon.

Blaustein, A.R., D.P. Chivers, L.B. Kats and J.M. Kiesecker. 2000. Effects of ultraviolet radiation on locomotion and orientation in roughskin newts (*Taricha granulosa*). Ethology 106:227–234.

Blaustein, A.R., P.D. Hoffman, J.M. Kiesecker and J.B. Hays. 1996. DNA repair activity and resistance to solar UV-B radiation in eggs of the red-legged frog. Conservation Biology 10:1398–1402.

Blaustein, A.R., D.G. Hokit, R.K. O'Hara and R.A. Holt. 1994b. Pathogenic fungus contributes to amphibian losses in the Pacific Northwest. Biological Conservation 67:251–254.

Blaustein, A.R., J.M. Kiesecker, D.P. Chivers and R.G. Anthony. 1997. Ambient UV-B radiation causes deformities in amphibian embryos. Proceedings of the National Academy of Sciences 94:13735–13737.

Blaustein, A.R., B. Edmond, J.M. Kiesecker, J.J. Beatty and D.G. Hokit. 1995a. Ambient ultraviolet radiation causes mortality in salamander eggs. Ecological Applications 5:740–743.

Blaustein, A.R., L.K. Belden, D.H. Olson, D.M. Green, T.L. Root and J.M. Kiesecker. 2001. Amphibian breeding and climate change. Conservation Biology 15:1804–1809.

Blaustein, A.R., P.D. Hoffman, D.G. Hokit, J.M. Kiesecker, S.C. Walls and J.B. Hays. 1994c. UV repair and resistance to solar UV-B in amphibian eggs: a link to population declines? Proceedings of the National Academy of Sciences 91:1791–1795.

Blaustein, A.R., J.M. Kiesecker, D.P. Chivers, D.G. Hokit, A. Marco, L.K. Belden and A. Hatch. 1998. Effects of ultraviolet radiation on amphibians: field experiments. American Zoologist 38:799–812.

Blaustein, A.R., J.B. Hays, P.D. Hoffman, D.P. Chivers, J.M. Kiesecker, W.P. Leonard, A. Marco, D.H. Olson, J.K. Reaser and R.G. Anthony. 1999. DNA repair and resistance to UV-B radiation in western spotted frogs. Ecological Applications 9:1100–1105.

Blazquez, M.C. 1996. Natural history notes: *Nerodia valida* (Pacific water snake). Prey. Herpetological Review 27:83–84.

Bleakney, J.S. 1958. A zoogeographical study of the amphibians and reptiles of eastern Canada. Bulletin of the National Museum of Canada, Number 155, Ottawa, Ontario.

Bleakney, S. 1952. The amphibians and reptiles of Nova Scotia. Canadian Field-Naturalist 66:125–129.

Bleakney, S. 1957. The egg-laying habits of the salamander, *Ambystoma jeffersonianum*. Copeia 1957:141–142.

Bleakney, S. 1958. Cannibalism in *Rana sylvatica* tadpoles, a well known phenomenon. Herpetologica 14:34.

Bleich, K.M. 1978. Preliminary results of a study on *Batrachoseps aridus* in Hidden Palm Canyon, Riverside County, California. Report to California Department of Fish and Game, Sacramento, California.

Blem, C.A., C.A. Ragan and L.S. Scott. 1986. The thermal physiology of two sympatric treefrogs *Hyla cinerea* and *Hyla chrysoscelis* (Anura: Hylidae). Comparative Biochemistry and Physiology A 85:563–570.

Blem, C.R. 1979. *Bufo terrestris*. Pp. 223.1–223.4. Catalogue of American Amphibians and Reptiles. Society for the Study of Amphibians and Reptiles, St. Louis, Missouri.

Blem, C.R. 1992. Lipid reserves and body composition in postreproductive anurans. Comparative Biochemisty and Physiology A 103:653–656.

Blem, C.R. and L.B. Blem. 1989. Tolerance in a Virginia population of the spotted salamander, *Ambystoma maculatum* (Amphibia, Ambystomatidae). Brimleyana 15:37–45.

Blem, C.R. and L.B. Blem. 1991. Cation concentrations and acidity in breeding ponds of the spotted salamander, *Ambystoma maculatum* (Shaw) (Amphibia, Ambystomatidae), in Virginia. Brimleyana 17:67–76.

Blem, C.R., J.W. Steiner and M.A. Miller. 1978. Comparison of jumping abilities of the cricket frogs *Acris gryllus* and *Acris crepitans*. Herpetologica 34:288–291.

Bles, E.J. 1905. The life-history of *Xenopus laevis*, Daud. Transactions of the Royal Society of Edinburgh 41:789–821.

Blihovde, W.B. 1999. Love thy neighbor: gopher frog *(Rana capito)* site fidelity at gopher tortoise burrows. Second Symposium on the Status and Conservation of Florida Turtles, 8–11 October. Eckerd College, St. Petersburg, Florida.

Blihovde, W.B. 2000a. Natural history notes: *Rana capito aesopus* (Florida gopher frog). Ectoparasites. Herpetological Review 31:101.

Blihovde, W.B. 2000b. The territorial behavior of the Florida gopher frog *(Rana capito aesopus)*. Master's thesis. University of Central Florida, Orlando, Florida.

Block, W.M. and M.L. Morrison 1998. Habitat relationships of amphibians and reptiles in California oak woodlands. Journal of Herpetology 32:51–60.

Blouin, M.S. 1989. Life history correlates of a color polymorphism in the ornate chorus frog, *Pseudacris ornata*. Copeia 1989:319–325.

Blouin, M.S. 1990. Evolution of palatability between closely-related treefrogs. Journal of Herpetology 24:309–310.

Blouin, M.S. 1991. Proximate development causes of limb length variation between *Hyla cinerea and Hyla gratiosa* (Anura: Hylidae). Journal of Morphology 209:305–310.

Blouin, M.S. 1992a. Genetic correlations among morphometric traits and rates of growth and differentiation in the green tree frog, *Hyla cinerea*. Evolution 46:735–744.

Blouin, M.S. 1992b. Comparing bivariate reaction norms among species: time and size at metamorphosis in three species of *Hyla* (Anura: Hylidae). Oecologia 90:288–293.

Blouin, M.S. and M.L.G. Loeb. 1991. Effects of environmentally induced development rate variation on head and limb morphology in the green tree frog, *Hyla cinerea*. American Naturalist 138:717–728.

Bloxam, Q.M.C. and S.J. Tonge. 1995. Amphibians: suitable candidates for breeding-release programmes. Biodiversity and Conservation 4:636–644.

Blymyer, M.J. and B.S. McGinnes. 1977. Observations on possible detrimental effects of clearcutting on terrestrial amphibians. Bulletin of the Maryland Herpetological Society 13:79–83.

Boag, D.A. 1986. Dispersal in ponds and snails: potential role of waterfowl. Canadian Journal of Zoology 64:904–909.

Bobka, M.S., R.G. Jaeger and D.C. McNaught. 1981. Temperature dependent assimilation efficiencies of two species of terrestrial salamanders. Copeia 1981:417–421.

Bock, S.F. and J.E. Fauth 1992. Natural history notes: *Pseudotriton* (northern red salamander). Diet. Herpetological Review 23:58.

Boerger, H. 1975. A comparison of the life cycles, reproductive ecologies, and size-weight relationships of *Helisoma anceps*, *H. campanulatum*, and *H. trivolvis* (Gastropoda, Planorbidae). Canadian Journal of Zoology 53:1812–1824.

Bogart, J.P. 1967. Life history and chromosomes of some of the neotenic salamanders of the Edward's Plateau. Master's thesis. University of Texas, Austin, Texas.

Bogart, J.P. 1982. Ploidy and genetic diversity in Ontario salamanders of the *Ambystoma jeffersonianum* complex revealed through electrophoretic examination of larvae. Canadian Journal of Zoology 60:848–855.

Bogart, J.P. 1989. A mechanism for interspecific gene exchange via all-female salamander hybrids. Pp. 170–179. *In* Dawley, R.M. and J.P. Bogart (Eds.), Evolution and Ecology of Unisexual Vertebrates. New York State Museum Bulletin, Number 466, Albany, New York.

Bogart, J.P. and A.P. Jaslow. 1979. Distribution and call parameters of *Hyla chrysoscelis* and *Hyla versicolor* in Michigan. Life Science Contributions of the Royal Ontario Museum 117:1–13.

Bogart, J.P. and M.W. Klemens. 1997. Hybrids and genetic interactions of mole salamanders *(Ambystoma jeffersonianum* and *A. laterale)* (Amphibia: Caudata) in New York and New England. American Museum Novitates, Number 3218, American Museum of Natural History, New York.

Bogart, J.P. and A.O. Wasserman. 1972. Diploid-polyploid cryptic species pairs: a possible clue to evolution by polyploidization in anuran amphibians. Cytogenetics 11:7–24.

Bogart, J.P., R.P. Elinson and L.E. Licht. 1989. Temperature and sperm incorporation in polyploid salamanders. Science 246:1032–1034.

Bogert, C.M. 1930. Annotated list of the amphibians and reptiles of Los Angeles County. Bulletin of the Southern California Academy of Sciences 19:3–14.

Bogert, C.M. 1947. A field study of homing in *Bufo t. terrestris*. American Museum Novitates, Number 1355, American Museum of Natural History, New York.

Bogert, C.M. 1952. Relative abundance, habitats, and normal thermal levels of some Virginian salamanders. Ecology 33:16–30.

Bogert, C.M. 1958. Commentary for recording of "Sounds of North American Frogs." Folkways Records, Number 6116, Folkways Records and Service Corporation, New York.

Bogert, C.M. 1960. The influence of sound on the behaviour of amphibians and reptiles. Pp. 137–320. *In* Lanyon, W.E. and W.N. Tavolga (Eds.). Animal Sounds and Communication. American Institute of Biological Science, Publication Number 7, Washington, D.C.

Bogert, C.M. 1962. Isolation mechanisms in toads of the *Bufo debilis* group in Arizona and western Mexico. American Museum Novitates, Number 2100, American Museum of Natural History, New York.

Bogert, C.M. 1998. Sounds of North American frogs. The biological significance of voice in frogs. Reissue of Folkways 6116 (1958), Track 81, Smithsonian Folkways Recordings, Washington, D.C.

Bogert, C.M. and J.A. Oliver. 1945. A preliminary analysis of the herpetofauna of Sonora. Bulletin of the American Museum of Natural History 83:297–426.

Bohl, E. 1997. Limb deformities of amphibian larvae in Aufsess (Upper Franconia): attempt to determine causes. Munich Contributions to Wastewater Fishery and River Biology 50:160–189.

Böhme, W. 1989. Klimafaktoren und Artenrückgang am Beispiel mitteleuropäischer Eidechsen (Reptilia: Lacertidae). Schriftenreihe für Landschaftspflege und Naturschutz 29:195–202.

Böhmer, J., S. Aniol, A. Bauser-Eckstein, S. Blattner, J. Hildenbrand, S. Mandon, H.P. Straub, K. Zintz and H. Rahmann. 1990. Feldökologische und ultrastrukturelle Aspekte zum Einfluss der Gewässerversauerung auf Amphibien und Insekten im Nordschwarzwald. Verhandlungen der Deutschen Zoologischen Gesellschaft 83:478–479.

Bohnsack, K.K. 1951. Temperature data on the terrestrial hibernation of the green frog, Rana clamitans. Copeia 1951:236–239.

Bohnsack, K.K. 1952. Terrestrial hibernation of the bullfrog, Rana catesbeiana Shaw. Copeia 1952:114.

Bohonak, A.J. 1999. Genetic population structure of the fairy shrimp Branchinecta coloardensis (Anostraca) in the Rocky Mountains of Colorado. Canadian Journal of Zoology 76:2049–2057.

Bohonak, A.J. and H.H. Whiteman. 1999. Dispersal of the fairy shrimp Branchinecta coloradensis (Anostraca): effects of hydroperiod and salamanders. Limnology and Oceanography 44:487–493.

Bolek, M.G. 1997. Seasonal occurrence of Cosmocercoides dukae and prey analysis of the blue-spotted salamander, Ambystoma laterale, in southeastern Wisconsin. Journal of the Helminthological Society of Washington 64:292–295.

Bolek, M.G. and J.R. Coggins. 1998. Helminth parasites of the spotted salamander Ambystoma maculatum and red-backed salamander Plethodon c. cinereus from northwestern Wisconsin. Journal of the Helminthological Society of Washington 65:98–102.

Bolek, M.G. and J.R. Coggins. 2000. Seasonal occurrence and community structure of helminth parasites from the Eastern American toad, Bufo americanus americanus, from southeastern Wisconsin, USA. Comparative Parasitology 67:202–209.

Bonett, R.M. 2002. Analysis of the contact zone between the dusky salamanders Desmognathus fuscus fuscus and Desmognathus fuscus conant (Caudata: Plethodontidae). Copeia 2002:344–355.

Bonett, R.M. and P.T. Chippindale. 2004. Speciation, phylogeography and evolution of life history and morphology in plethodontid salamanders of the Eurycea multiplicata complex. Molecular Ecology 13:1189–1203.

Bonin, J., J.-L. DesGranges, C.A. Bishop, J. Rodrigue, A. Gendron and J.E. Elliott. 1995. Comparative study of contaminants in the mudpuppy (Amphibia) and the common snapping turtle (Reptilia), St. Lawrence River, Canada. Archives of Environmental Contamination and Toxicology 28:184–194.

Bonin, J., M. Ouellet, J. Rodrigue, J. DesGranges, F. Gagne, T. Sharbel and L. Lowcock. 1997a. Measuring the health of frogs in agricultural habitats subjected to pesticides. Herpetological Conservation 1:246–257.

Bonin, J., M. Ouellet, J. Rodrigue, J.L. DesGranges, F. Gagné, T.F. Sharbel and L.A. Lowcock. 1997b. Measuring the health of frogs in agricultural habitats subjected to pesticides. Pp. 258–270. In Green, D.M. (Ed.), Amphibians in Decline: Canadian Studies of a Global Problem. Herpetological Conservation, Number 1, Society for the Study of Amphibians and Reptiles, St. Louis, Missoui.

Bonnaterre, M. l'A. 1789. Tableau Encyclopedique et Methodologie des Trois Regnes de la Nature. Erpetologie. Paris, France.

Boone, M.D. and C.M. Bridges. 1999. The effect of temperature on the potency of carbaryl for survival of tadpoles of the green frog (Rana clamitans). Environmental Toxicology and Chemistry 18:1482–1484.

Boone, M.D. and R.D. Semlitsch. 2001. Interactions of an insecticide with larval density and predation in experimental amphibian communities. Conservation Biology 15:228–238.

Boone, M.D., D.E. Scott and P.N. Niewiarowski. 2002. Effects of hatching time for larval ambystomatid salamanders. Copeia 2002: 511–517.

Boonstra, R. 1994. Population cycles in microtines: the senescence hypothesis. Evolutionary Ecology 8:196–219.

Booth, D.T. and M.E. Feder. 1991. Formation of hypoxic boundary layers and their biological implications in a skin-breathing aquatic salamander, Desmognathus quadramaculatus. Physiological Zoology 64: 1307–1321.

Boray, J. and J.L. Munro. 1998. Economic significance. Pp. 65–77. In Beesley, P.L., G.J.B. Ross and A. Wells (Eds.), Mollusca: The Southern Synthesis. Fauna of Australia. Volume 5. CSIRO Publishing, Melbourne, Victoria, Australia.

Boray, J.C. 1978. The potential impact of exotic Lymnaea sp. on fascioliasis in Australia. Veterinary Parasitology 4:127–141.

Borland, J.R. and R. Rugh. 1943. Evidences of cannibalism in the tadpole of the frog Rana pipiens. American Naturalist 77:282–285.

Bos, D.H. and J.W. Sites Jr. 2001. Phylogeography and conservation genetics of the Columbia spotted frog (Rana luteiventris; Amphibia, Ranidae). Molecular Ecology 10:1499–1513.

Bosakowski, T. 1999. Amphibian macrohabitat associations on a private industrial forest in western Washington. Northwestern Naturalist 80:61–69.

Bosc, L.A.G. 1800. In Daudin, F.M., Histoire Naturelle des Quadrupedes Ovipares. Livraisons Volumes 1 and 2. Marchant et Cie, Paris, France.

Bosc, L.A.G. and F.M. Daudin. 1801. Laraine oculaire, Hyla ocularis, etc. P. 187. In Sonnini, C.S. and P.A. Latreille, Histoire Naturelle des Reptiles. Number 2. Paris, France.

Bosch, K. 1976. Elementare Einführung in die Wahrscheinlichkeitstheorie. Rororo, Hamburg, Germany.

Boschulte, D.S. 1993. Effects of six commonly used herbicides on larval bullfrogs (Rana catesbeiana). Master's thesis. Illinois State University, Normal, Illinois.

Boschulte, D.S. 1995. Effects of six commonly used herbicides on larval bullfrogs (Rana catesbeiana). Beacon 5:7, 10.

Bosma, N.J. 1934. The life history of the trematode Alaria mustelae sp. nov. Transactions of the American Microscopical Society 53:116–152.

Bossert, M., M. Draud and T. Draud. 2003. Natural history notes: Bufo fowleri (Fowler's toad) and Malaclemmys terrapin terrapin (northern diamondback terrapin). Refugia and nesting. Herpetological Review 34:135.

Bothner, R.C. 1963. A hibernaculum of the shortheaded garter snake, Thamnophis brachystoma. Copeia 1963:572–573.

Bothner, R.C. and J.A. Gottlieb. 1991. A study of the New York State populations of the hellbender, Cryptobranchus alleganiensis alleganiensis. Proceedings of the Rochester Academy of Science 17:41–54.

Bouchard, J.L. 1951. The platyhelminths parasitizing some northern Maine Amphibia. Transactions of the American Microscopical Society 70:245–250.

Boulenger, G.A. 1882. Descriptions of a new genus and species of frogs of the family Hylidae. Annals and Magazine of Natural History, Series 5, 10:326–328.

Boulenger, G.A. 1917. Descriptions of new frogs of the genus Rana. Annals of the Magazine Natural History 20:413–418.

Boundy, J. 1994a. Range extensions for Louisiana amphibians and reptiles. Herpetological Review 25:128–129.

Boundy, J. 1994b. County records for Texas amphibians and reptiles. Herpetological Review 25:129.

Boundy, J. 1998. Distributional records for Louisiana amphibians. Herpetological Review 29:251–252.

Boundy, J. 1999. Systematics of the garter snake Thamnophis atratus at the southern end of its range. Proceedings of the California Academy of Sciences 51:311–336.

Boundy, J. 2000. Batrachoseps attenuatus. Pp. 701.1–701.6. Catalogue of American Amphibians and Reptiles. Society for the Study of Amphibians and Reptiles, St. Louis, Missouri.

Boundy, J. and T.G. Balgooyen. 1988. Record lengths for some amphibians and reptiles from the western United States. Herpetological Review 19:26–27.

Boutilier, R.G., P.H. Donohue, G.J. Tattersall and T.G. West. 1997. Hypometabolic homeostasis in overwintering aquatic amphibians. Journal of Experimental Biology 200:387–400.

Bovbjerg, R.V. 1965. Experimental studies on the dispersal of the frog, Rana pipiens. Journal of the Iowa Academy of Science 72:412–418.

Bovbjerg, R.V. 1980. The snails of Lake West Okoboji, twenty years later. Proceedings of the Iowa Academy of Science 87:J-18.

Bovbjerg, R.V. and M.J. Ulmer. 1960. An ecological catalog of the Lake Okoboji gastropods. Proceedings of the Iowa Academy of Science 67:569–577.

Bowen, B.W. and S.A. Karl. 1999. In war, truth is the first casualty. Conservation Biology 13:1013–1016.

Bowen, S.H. 1984. Evidence of a detritus food chain based on consumption of organic precipitates. Bulletin of Marine Science 35:440–448.

Bowers, D.G., D.E. Anderson and N.H. Euliss. 1997. Anurans as indicators of wetland condition in the prairie pothole region of North Dakota. North American Amphibian Monitoring Program III online paper, http://www.nbs.gov/naamp3/papers/2c.html.

Bowers, D.G., D.E. Anderson and N.H. Euliss Jr. 1998. Anurans as indicators of wetland condition in the Prairie Pothole Region of North

Dakota: an environmental monitoring and assessment program pilot project. Pp. 369–378. *In* Lannoo, M.J. (Ed.), Status and Conservation of Midwestern Amphibians. University of Iowa Press, Iowa City, Iowa.

Bowers, J.E. and S.P. McLaughlin. 1995. Flora of the Huachuaca Mountains, Cochise County, Arizona. Pp. 135–143. *In* DeBano, L.F., G.J. Gottfried, R.H. Hamre, C.B. Edminster, P.F. Ffolliott and A. Ortega-Rubio (Eds.), Biodiversity and Management of the Madrean Archipelago: The Sky Islands of Southwestern United States and Northwestern Mexico. Rocky Mountain Forest and Range Experiment Station, Fort Collins, Colorado.

Bowers, J.H. 1966. Food habits of the diamond-backed water snake, *Natrix rhombifera rhombifera*, in Bowie and Red River Counties, Texas. Herpetologica 22:225–229.

Bowker, R.W. and B.K. Sullivan. 1991. Natural history notes: *Bufo punctatus* x *Bufo retiformis* (red-spotted toad, Sonoran green toad). Natural hybridization. Herpetological Review 22:54.

Bowler, J.K. 1977. Longevity of reptiles and amphibians in North American collections as of 1 November, 1975. Herpetological Circular, Number 6, Society for the Study of Amphibians and Reptiles, St. Louis, Missouri.

Bowler, P.A. and T.J. Frest. 1992. The non-native snail fauna of the Middle Snake River, southern Idaho. Proceedings of the Desert and Fishes Council 23:28–42.

Bowles, D.E. (Ed.). 1995. A review of the status of current critical biological and ecological information on the *Eurycea* salamanders located in Travis County, Texas. Texas Parks and Wildlife Department, Austin, Texas.

Boyce, W.M. 1985. The prevalence of *Sebekia mississippiensis* (Pentastomida) in American alligators *(Alligator mississippiensis)* in north Florida and experimental infection of paratenic hosts. Proceedings of the Helminthological Society of Washington 52:278–282.

Boycott, A.E. 1936. The habitats of fresh-water Mollusca in Britain. Journal of Animal Ecology 5:116–186.

Boyer, R. and C.R. Grue. 1995. The need for water quality criteria for frogs. Environmental Health Perspectives 103:352–357.

Boyer, W.D. 1993. Regenerating longleaf pine with natural seeding. Pp. 299–309. *In* Hermann, S.M. (Ed.), The Longleaf Pine Ecosystem: Ecology, Restoration and Management. Proceedings of the Tall Timbers Fire Ecology Conference, Number 18, Tallahassee, Florida.

Brach, V. 1992. Discovery of the Rio Grande chirping frog in Smith County, Texas (Anura: Leptodactylidae). Texas Journal of Science 44:490.

Brach, V. 1995. Those slippery sirens. Tropical Fish Hobbyist 44:164–170.

Bracher, G.A. and J.R. Bider. 1982. Changes in terrestrial animal activity of a forest community after an application of aminocarb matacil. Canadian Journal of Zoology 60:1981–1997.

Bradford, D.F. 1983. Winter kill, oxygen relations, and energy metabolism of a submerged dormant amphibian, *Rana muscosa*. Ecology 64:1171–1183.

Bradford, D.F. 1984. Temperature modulation in a high-elevation amphibian, *Rana muscosa*. Copeia 1984:966–976.

Bradford, D.F. 1989. Allopatric distribution of native frogs and introduced fishes in high Sierra Nevada lakes of California: implication of the negative effect of fish introductions. Copeia 1989:775–778.

Bradford, D.F. 1991. Mass mortality and extinction in a high elevation population of *Rana muscosa*. Journal of Herpetology 25:174–177.

Bradford, D.F. and M.S. Gordon. 1992. Aquatic amphibians in the Sierra Nevada: current status and potential effects of acidic deposition on populations. California Air Resources Board, Sacramento, California.

Bradford, D.F., D.M. Graber and F. Tabatabai. 1993. Isolation of remaining populations of the native frog, *Rana muscosa*, by introduced fishes in Sequoia and Kings Canyon National Parks, California. Conservation Biology 7:882–888.

Bradford, D.F., D.M. Graber and F. Tabatabai. 1994a. Population declines of the native frog, *Rana muscosa*, in Sequoia and Kings Canyon National Parks, California. Southwestern Naturalist 39:323–327.

Bradford, D.F., J.R. Jaeger and R.D. Jennings, 2004. Population status and distribution of a decimated amphibian, the relict leopard frog (*Rana onca*). Southwestern Naturalist 49:218–228.

Bradford, D.F., C. Swanson and M.S. Gordon. 1992. Effects of low pH and aluminum on two declining species of amphibians in the Sierra Nevada, California. Journal of Herpetology 26:369–377.

Bradford, D.F., C. Swanson and M.S. Gordon. 1994c. Effects of low pH and aluminum on amphibians at high elevations in the Sierra Nevada, California. Canadian Journal of Zoology 72:1272–1279.

Bradford, D.F., M.S. Gordon, D.F. Johnson, R.D. Andrews and W.B. Jennings. 1994b. Acidic deposition as an unlikely cause for amphibian population declines in the Sierra Nevada, California. Biological Conservation 69:155–161.

Bradford, D.F., A.C. Neale, M.S. Nash, D.W. Sada and J.R. Jaeger. 2003. Habitat patch occupancy by toads (*Bufo punctatus*) in a naturally fragmented desert landscape. Ecology 84:1012–1023.

Bradley, S.G. and L.J. Klika. 1981. A fatal poisoning from the Oregon rough-skinned newt *(Taricha granulosa)*. Journal of the American Medical Association 246:247.

Brady, M.K. and F. Harper. 1935. A Florida subspecies of *Pseudacris nigrita* (Hylidae). Proceedings of the Biological Society of Washington 48:107–110.

Bragg, A.N. 1937a. A note on the metamorphosis of the tadpoles of *Bufo cognatus*. Copeia 1937:227–228.

Bragg, A.N. 1937b. Observations on *Bufo cognatus* with special reference to the breeding habits and eggs. American Midland Naturalist 18:273–284.

Bragg, A.N. 1940a. Observations on the ecology and natural history of Anura. I. Habits, habitat and breeding of *Bufo cognatus* Say. American Naturalist 74:322–349, 424–438.

Bragg, A.N. 1940b. Observations on the ecology and natural history of Anura. II. Habits, habitat, and breeding of *Bufo woodhousii woodhousii* (Girard) in Oklahoma. American Midland Naturalist 24:306–321.

Bragg, A.N. 1941. New records of frogs and toads for Oklahoma. Copeia 1941:51–52.

Bragg, A.N. 1942. Observations on the ecology and natural history of Anura. X. The breeding habits of *Pseudacris streckeri* Wright and Wright in Oklahoma including a description of the eggs and tadpoles. Wasmann Collector 5:47–62.

Bragg, A.N. 1943a. Observations on the ecology and natural history of Anura. XVI. Life history of *Pseudacris clarkii* (Baird) in Oklahoma. Wasmann Collector 5:129–140.

Bragg, A.N. 1943b. Observations on the ecology and natural history of Anura. XV. The hylids and microhylids of Oklahoma. Great Basin Naturalist 4:62–80.

Bragg, A.N. 1944. The spadefoot toads in Oklahoma with a summary of our knowledge of the group. American Naturalist 78:517–533.

Bragg, A.N. 1945a. Breeding and tadpole behavior of *Scaphiopus hurterii* near Norman Oklahoma in the spring 1945. Wasmann Collector 6:69–78.

Bragg, A.N. 1945b. The spadefoot toads in Oklahoma with a summary of our knowledge of the group. II. American Naturalist 79:52–72.

Bragg, A.N. 1946. Aggregation with cannibalism in tadpoles of *Scaphiopus bombifrons* with some general remarks on the probable evolutionary significance of such phenomena. Herpetologica 3:89–97.

Bragg, A.N. 1947. Tadpole behavior in pools and streams. Proceedings of the Oklahoma Academy of Science 27:59–61.

Bragg, A.N. 1948. Observations on the life history of *Pseudacris triseriata* (Wied) in Oklahoma. Wasmann Collector 7:149–168.

Bragg, A.N. 1949. Observations on the narrow-mouthed salamander. Proceedings of the Oklahoma Academy of Science 29:21–24.

Bragg, A.N. 1950a. The identification of Salientia in Oklahoma. Pp. 9–34. *In* Bragg, A.N., A.O. Weese, H.A. Dundee, H.T. Fisher, A. Richards and C.B. Clard (Eds.), Researches on the Amphibia of Oklahoma. University of Oklahoma Press, Norman, Oklahoma.

Bragg, A.N. 1950b. Salientian range extensions in Oklahoma and a new state record. Pp. 39–44. *In* Bragg, A.N., A.O. Weese, H.A. Dundee, H.T. Fisher, A. Richards and C.B. Clard (Eds.), Researches on the Amphibia of Oklahoma. University of Oklahoma Press, Norman, Oklahoma.

Bragg, A.N. 1950c. Observations on the ecology and natural history of Anura. XVII. Adaptations and distribution in accordance with habitats in Oklahoma. Pp. 59–100. *In* Bragg, A.N., A.O. Weese, H.A. Dundee, H.T. Fisher, A. Richards and C.B. Clard (Eds.), Researches on the Amphibia of Oklahoma. University of Oklahoma Press, Norman, Oklahoma.

Bragg, A.N. 1950d. Frequency of sex calls in some Salientia. Pp. 117–125. *In* Bragg, A.N., A.O. Weese, H.A. Dundee, T.H. Fisher, A. Richards and C.B. Clark (Eds.), Researches on the Amphibia of Oklahoma. University of Oklahoma Press, Norman Oklahoma.

Bragg, A.N. 1950e. Observations on *Microhyla* (Salientia: Microhylidae). Wasmann Journal of Biology 8:113–118.

Bragg, A.N. 1950f. Observations on *Scaphiopus*, 1949 (Salientia: Scaphiopidae). Wasmann Journal of Biology 8:221–228.

Bragg, A.N. 1953. A study of *Rana areolata* in Oklahoma. Wasmann Journal of Biology 11:273–318.

Bragg, A.N. 1955a. The tadpole of *Bufo debilis debilis*. Herpetologica 11:211–212.

Bragg, A.N. 1955b. The Amphibia of Cherokee County, Oklahoma. Herpetologica 11:25–30.

Bragg, A.N. 1956. Dimorphism and cannibalism in tadpoles of *Scaphiopus bombifrons* (Amphibia, Salientia). Southwestern Naturalist 1:105–108.

Bragg, A.N. 1957. Variation in colors and color patterns in tadpoles in Oklahoma. Copeia 1957:36–39.

Bragg, A.N. 1960a. Feeding in the Houston toad. Southwestern Naturalist 5:106.

Bragg, A.N. 1960b. Population fluctuation in the amphibian fauna of Cleveland County, Oklahoma during the past twenty-five years. Southwestern Naturalist 5:165–169.

Bragg, A.N. 1961. The behavior and comparative developmental rates in nature of tadpoles of a spadefoot, a toad, and a frog. Herpetologica 17:80–84.

Bragg, A.N. 1962a. *Saprolegnia* on tadpoles again in Oklahoma. Southwestern Naturalist 7:79–80.

Bragg, A.N. 1962b. Predation on arthropods by spadefoot tadpoles. Herpetologica 18:144.

Bragg, A.N. 1965. Gnomes of the Night. University of Pennsylvania Press, Philadelphia, Pennsylvania.

Bragg, A.N. 1967. Recent studies on the spadefoot toads. Bios 38:75–84.

Bragg, A.N. and W.N. Bragg. 1957. Parasitism of spadefoot tadpoles by *Saprolegnia*. Herpetologica 14:34.

Bragg, A.N. and W.N. Bragg. 1958. Variations in the mouth parts in tadpoles of *Scaphiopus (Spea) bombifrons* Cope (Amphibia: Salientia. Southwestern Naturalist 3:55–69.

Bragg, A.N. and M. Brooks. 1958. Social behavior in juveniles of *Bufo cognatus* Say. Herpetologica 14:141–147.

Bragg, A.N. and O. Sanders. 1951. A new subspecies of the *Bufo woodhousei* group of toads. Wasmann Journal of Biology 9:363–378.

Bragg, A.N. and C.C. Smith. 1942. Observations on the ecology and natural history of Anura. IX. Notes on breeding behavior in Oklahoma. Great Basin Naturalist 3:33–50.

Bragg, A.N. and C.C. Smith. 1943. Observations on the ecology and natural history of Anura. IV. The ecological distribution of toads in Oklahoma. Ecology 24:285–309.

Braid, M.R., C.B. Raymond and W.S. Sanders. 1994. Feeding trials with the dusky gopher frog, *Rana capito sevosa*, in a recirculating water system and other aspects of their culture as part of a "headstarting effort." Journal of the Alabama Academy of Science 65:249–262.

Brame, A.H., Jr. 1956. The number of eggs laid by the California newt. Herpetologica 12:325–326.

Brame, A.H., Jr. 1959. Summer surface activity in certain California salamanders. Bulletin of the Philadelphia Herpetological Society, Number 7, Philadelphia, Pennsylvania.

Brame, A.H., Jr. 1968. The number of egg masses and eggs laid by the California newt, *Taricha torosa*. Journal of Herpetology 2:169–170.

Brame, A.H., Jr. 1970. A new species of *Batrachoseps* (slender salamander) from the desert of southern California. Contributions in Science from the Natural History Museum of Los Angeles County, Number 200, Los Angeles, California.

Brame, A.H., Jr. and R.W. Hansen. 1994. Desert slender salamander. Pp. 248–249. *In* Thelander, C.G. (Ed.), Life on the Edge. Biosystems Books, Santa Cruz, California.

Brame, A.H., Jr. and K.F. Murray. 1968. Three new slender salamanders (*Batrachoseps*) with a discussion of relationships and speciation within the genus. Science Bulletin of the Natural History Museum of Los Angeles County, Number 4, Los Angeles, California.

Brame, A.H., Jr., M.C. Long and A.A. Chiri. 1973. Defensive display of the desert slender salamander, *Batrachoseps aridus*. Herpeton 8:1–3.

Branch, L.C. and R. Altig. 1981. Nocturnal stratification of three species of *Ambystoma* larvae. Copeia 1981:870–873.

Brandon, R.A. 1961. A comparison of the larvae of five northeastern species of *Ambystoma* (Amphibia, Caudata). Copeia 1961:377–383.

Brandon, R.A. 1962. Summary of past investigations on *Typhlotriton spelaeus*, the grotto salamander. Cave Notes 4:9–12.

Brandon, R.A. 1965a. A new race of the neotenic salamander *Gyrinophilus palleucus*. Copeia 1965:346–352.

Brandon, R.A. 1965b. Morphological variation and ecology of the salamander *Phaeognathus hubrichti*. Copeia 1965:67–71.

Brandon, R.A. 1965c. *Typhlotriton, T. nereus,* and *T. spelaeus*. Pp. 20.1–20.2. Catalogue of American Amphibians and Reptiles. Society for the Study of Amphibians and Reptiles, St. Louis, Missouri.

Brandon, R.A. 1966a. Additional localities for *Ambystoma texanum* in Alabama, with comments on site of oviposition. Journal of the Ohio Herpetological Society 5:104–105.

Brandon, R.A. 1966b. Amphibians and reptiles associated with *Phaeognathus hubrichti* habitats. Herpetologica 22:308–310.

Brandon, R.A. 1966c. Systematics of the salamander genus *Gyrinophilus*. Illinois Biological Monographs 53:1–86.

Brandon, R.A. 1966d. A reevaluation of the status of the salamander, *Typhlotriton nereus* Bishop. Copeia 1966:555–561.

Brandon, R.A. 1967a. *Gyrinophilus palleucus*. Pp. 32.1–32.2. Catalogue of American Amphibians and Reptiles. Society for the Study of Amphibians and Reptiles, St. Louis, Missouri.

Brandon, R.A. 1967b. Food and an intestinal parasite of the troglobitic salamander *Gyrinophilus palleucus necturoides*. Herpetologica 23:52–53.

Brandon, R.A. 1967c. *Gyrinophilus porphryriticus*. Pp. 33.1–33.3. Catalogue of American Amphibians and Reptiles. Society for the Study of Amphibians and Reptiles, St. Louis, Missouri.

Brandon, R.A. 1967d. *Haideotriton* and *H. wallacei*. Pp. 39.1–39.2. Catalogue of American Amphibians and Reptiles. Society for the Study of Amphibians and Reptiles, St. Louis, Missouri.

Brandon, R.A. 1970. *Typhlotriton spelaeus*. Pp. 84.1–84.2. Catalogue of American Amphibians and Reptiles. Society for the Study of Amphibians and Reptiles, St. Louis, Missouri.

Brandon, R.A. 1971a. Correlation of seasonal abundance with feeding and reproductive activity in the grotto salamander *(Typhlotriton spelaeus)*. American Midland Naturalist 86:93–100.

Brandon, R.A. 1971b. North American troglobitic salamanders: some aspects of modification in cave habitats, with special reference to *Gyrinophilus palleucus*. National Speleological Society Bulletin 33:1–21.

Brandon, R.A. and N.R. Austin. 1966. Notes on a collection of amphibians and reptiles from Monroe County, Illinois. Transactions of the Illinois Academy of Science 59:296.

Brandon, R.A. and S.R. Ballard. 1998. Status of Illinois chorus frogs in southern Illinois. Pp. 102–112. *In* Lannoo, M.J. (Ed.), Status and Conservation of Midwestern Amphibians. University of Iowa Press, Iowa City, Iowa.

Brandon, R.A. and D.J. Bremer. 1966. Neotenic newts, *Notophthalmus viridescens louisianensis*, in southern Illinois. Herpetologica 22:213–217.

Brandon, R.A. and D.J. Bremer. 1967. Overwintering of larval tiger salamanders in southern Illinois. Herpetologica 23:67–68.

Brandon, R.A. and J.E. Huheey. 1971. Movements and interactions of two species of *Desmognathus* (Amphibia: Plethodontidae). American Midland Naturalist 86:86–92.

Brandon, R.A. and J.E. Huheey. 1975. Diurnal activity, avian predation, and the question of warning coloration and cryptic coloration in salamanders. Herpetologica 31:252–255.

Brandon, R.A. and J.E. Huheey. 1981. Toxicity in the plethodontid salamanders *Pseudotriton ruber* and *Pseudotriton montanus* (Amphibia, Caudata). Toxicon 19:25–31.

Brandon, R.A. and J.E. Huheey. 1985. Salamander skin toxins with special reference to *Necturus lewisi*. Brimleyana 10:75–82.

Brandon, R.A. and E.J. Maruska. 1982. Natural history notes: *Phaeognathus hubrichti* (Red Hills salamander). Reproduction. Herpetological Review 13:46.

Brandon, R.A., G.M. Labanick and J.E. Huheey. 1979a. Learned avoidance of brown efts, *Notophthalmus viridescens louisianensis* (Amphibia, Urodela, Salamandridae), by chickens. Journal of Herpetology 13:171–176.

Brandon, R.A., G.M. Labanick and J.E. Huheey. 1979b. Relative palatability, defensive behavior, and mimetic relationships of red salamanders *(Pseudotriton ruber)*, mud salamanders *(Pseudotriton montanus)*, and red efts *(Notophthalmus viridescens)*. Herpetologica 35:289–303.

Brandt, B.B. 1936a. The frogs and toads of eastern North Carolina. Copeia 1936:215–223.

Brandt, B.B. 1936b. Parasites of certain North Carolina Salientia. Ecological Monographs 6:491–532.

Brandt, B.B. 1952. Albino *Ambystoma maculatum*. Herpetologica 8:3.

Brandt, B.B. 1953. Salientia of Beckley County, Georgia, and vicinity. Herpetologica 9:141–145.

Brandt, B.B. and C.F. Walker. 1933. A new species of *Pseudacris* from the southeastern United States. Occasional Papers of the Museum of Zoology, Number 272, University of Michigan, Ann Arbor, Michigan.

Brandt, R., Mrs. 1947. Transmetamorphic memory in *Ambystoma texanum*. Copeia 1947:171.

Branin, M.L. 1935. Courtship activities and extra-seasonal ovulation in the four-toed salamander, *Hemidactylium scutatum* (Schlegel). Copeia 1935:172–175.

Branning, T. 1979. Frog wars. National Wildlife 17:34–37.

Brannon, M.P. 2000. Niche relationships of two syntopic species of shrews, *Sorex fumeus* and *S. cinereus*, in the southern Appalachian Mountains. Journal of Mammalogy 81:1053–1061.

Branson, B.A. 1975. Claude WHO? another unwanted exotic species. Environmental Journal, June 17–18.

Braswell, A.L. 1985. Geographic distribution: *Ambystoma talpoideum* (mole salamander). Herpetological Review 16:31.

Braswell, A.L. 1988. A survey of the amphibians and reptiles of Nags Head Woods Ecological Preserve. Association of Southeastern Biologists Bulletin 35:199–217.

Braswell, A.L. 1993. Status report on *Rana capito capito* LeConte, the Carolina gopher frog in North Carolina. North Carolina Wildlife Resources Commission, Final report, Raleigh, North Carolina.

Braswell, A.L. and R.E. Ashton Jr. 1985. Distribution, ecology, and feeding habits of *Necturus lewisi* (Brimley). Brimleyana 10:13–35.

Braswell, A.L., H.M. Wilbur and A.S. Weakley. 1990. Geographic distribution: *Ambystoma talpoideum* (mole salamander). Herpetological Review 21:36.

Brattstrom, B.H. 1962a. Homing in the giant toad, *Bufo marinus*. Herpetologica 18:176–180.

Brattstrom, B.H. 1962b. Thermal control of aggregation behavior in tadpoles. Herpetologica 18:38–46.

Brattstrom, B.H. 1963. A preliminary review of the thermal requirements of amphibians. Ecology 44:238–255.

Brattstrom, B.H. 1968. Thermal acclimation in anuran amphibians as a function of latitude and altitude. Comparative Biochemistry and Physiology 24:93–111.

Brattstrom, B.H. 1988. Habitat destruction in California with special reference to *Clemmys marmorata*: a perspective. Pp. 13–24. *In* Lisle, H.F., P.R. Brown, B. Kaufman and B.M. McGurty (Eds.), Proceedings of the Conference on California Herpetology, Special Publication, Number 4, Southwestern Herpetologists Society, Van Nuys, California.

Brattstrom, B.H. and P. Lawrence. 1962. The rate of thermal acclimation in anuran amphibians. Physiological Zoology 35:148–156.

Brattstrom, B.H. and J.W. Warren. 1953. On the validity of *Taricha torosa klauberi* Wolterstorff. Herpetologica 9:180–182.

Brattstrom, B.H. and J.W. Warren. 1955. Observations of the ecology and behavior of the Pacific treefrog, *Hyla regilla*. Copeia 1955:181–191.

Breckenridge, W.J. 1944. Reptiles and Amphibians of Minnesota. University of Minnesota Press, Minneapolis, Minnesota.

Breckenridge, W.J. and J.R. Tester. 1961. Growth, local movements and hibernation of the Manitoba toad, *Bufo hemiophrys*. Ecology 42: 637–646.

Breden, F. 1982. Population structure and ecology of Fowler's toad, *Bufo woodhousei fowleri*, in the Indiana Dunes National Lakeshore. Ph.D. dissertation. University of Chicago, Chicago, Illinois.

Breden, F. 1987. Population structure of Fowler's toad, *Bufo woodhousei fowleri*. Copeia 1987:386–395.

Breden, F. 1988. The natural history and ecology of Fowler's toad, *Bufo woodhousei fowleri* (Amphibia: Bufonidae) in the Indiana Dunes National Lakeshore. Fieldiana Zoology 49:1–16.

Breden, F. and C.H. Kelly. 1982. The effect of conspecific interactions on metamorphosis in *Bufo americanus*. Ecology 63:1682–1689.

Breden, F., A. Lum and R. Wassersug. 1982. Body size and orientation in aggregates of toad tadpoles, *Bufo woodhousei*. Copeia 1982:672–680.

Breder, R.B. 1927. The courtship of the spotted salamander. Bulletin of the New York Zoological Society 30:51–56.

Breitenbach, G.L. 1982. The frequency of joint nesting and solitary brooding in the salamander, *Hemidactylium scutatum*. Journal of Herpetology 16:341–346.

Brekke, D.R., S.D. Hillyard and R.M. Winokur. 1991. Behavior associated with the water absorption response by the toad, *Bufo punctatus*. Copeia 1991:393–401.

Brenner, F.J. 1969. Role of temperature and fat deposition in hibernation and reproduction in two species of frogs. Herpetologica 25:105–113.

Brenowitz, E.A. 1989. Neighbor call amplitude influences aggressive behavior and intermale spacing in choruses of the Pacific treefrog (*Hyla regilla*). Ethology 83:69–79.

Brenowitz, E.A. and G.J. Rose. 1999. Female choice and plasticity of male calling behaviour in the Pacific treefrog. Animal Behaviour 57:1337–1342.

Bretag, A.H., S.R. Dawe and A.G. Moskwa. 1980. Chemically induced myotonia in Amphibia. Nature 286:625–626.

Bretz, J.H., E. Neller, B. Schroeder and H.D. Weaver. 1983. River Cave. Missouri Speleology 23:107–111.

Breuer, P. 1992. Amphibien und Fische-Ergebnisse experimenteller Freilanduntersuchungen. Fauna und Flora in Rheinland-Pfalz 6:117–133.

Breuer, P. and B. Viertel. 1990. Zur Ökologie von Erdkrötenlarven (*Bufo bufo*) und Grasfroschlarven (*Rana temporaria*). I. Die Überlebensrate unter dem Einfluss von Regenbogenforellen (*Oncorhynchus mykiss*). Acta Biologica Benrodis 2:225–244.

Bridges, A.S. and M.E. Dorcas. 2000. Temporal variation in anuran calling behavior: implications for surveys and monitoring programs. Copeia 2000:587–592.

Bridges, C.M. 1997. Tadpole activity and swimming performance affected by sublethal levels of carbaryl. Environmental Toxicology and Chemistry 16:1935–1939.

Bridges, C.M. 1999a. The effects of a chemical stressor on amphibian larvae: individual, population and species level responses. Ph.D. dissertation. University of Missouri, Columbia, Missouri.

Bridges, C.M. 1999b. Effects of a pesticide on tadpole activity and predator avoidance behavior. Journal of Herpetology 33:303–306.

Bridges, C.M. 1999c. Predator-prey interaction between two amphibian species: effects of insecticide exposure. Aquatic Ecology 33:205–211.

Bridges, C.M. 2000. Long-term effects of pesticide exposure at various life stages of the southern leopard frog (*Rana sphenocephala*). Archives of Environmental Toxicology and Chemistry 39:91–96.

Bridges, C.M. and R.D. Semlitsch. 2000. Variation in pesticide tolerance of tadpoles among and within species of Ranidae and patterns of amphibian decline. Conservation Biology 14:1490–1499.

Bridges, C.M. and R.D. Semlitsch. 2001. Genetic variation in insecticide tolerance in a population of southern leopard frogs (*Rana sphenocephala*): implications for amphibian conservation. Copeia 2001: 7–13.

Briggler, J.T., J.E. Johnson and D.D. Rambo. 1999. Demographics of a ringed salamander (*Ambystoma annulatum*) breeding population during migration. 83rd Annual Meeting of the Arkansas Academy of Science, 2 April, (abstract), Russellville, Arkansas.

Briggler, J.T., K.M. Lohraff and G.L. Adams. 2001. Amphibian parasitism by the leech *Desserobdella picta* at a small pasture pond in northwest Arkansas. Journal of Freshwater Ecology. 16:105–111.

Briggs, J.L. 1975. A case of *Bufolucilia elongata* Shannon 1924 (Diptera: Calliphoridae) myiasis in the American toad, *Bufo americanus* Holbrook 1836. Journal of Parasitology 61:412.

Briggs, J.L. 1976. Breeding biology of the Cascade frog, *Rana cascadae*. Herpetological Review 7:75.

Briggs, J.L. 1987. Breeding biology of the Cascade frog, *Rana cascadae*, with comparisons to *R. aurora* and *R. pretiosa*. Copeia 1987:241–245.

Briggs, J.L. and R.M. Storm. 1970. Growth and population structure of the Cascade frog, *Rana cascadae* Slater. Herpetologica 26:283–300.

Briggs, R. and T.J. King. 1952. Transplantation of living nuclei from blastula cells into enucleated frogs' eggs. Proceedings of the National Academy of Sciences 38:455–463.

Bright, C. 1998. Life Out of Bounds: Bioinvasion in a Borderless World. W.W. Norton and Company, New York.

Brimley, C.S. 1896. Batrachia found at Raleigh, North Carolina. American Naturalist 30:501.

Brimley, C.S. 1909. Some notes on the zoology of Lake Ellis, Craven County, North Carolina, with special reference to herpetology. Proceedings of the Biological Society of Washington 22:129–138.

Brimley, C.S. 1910. Records of some reptiles and batrachians from the southeastern United States. Proceedings of the Biological Society of Washington 23:9–18.

Brimley, C.S. 1912. Notes on the salamanders of the North Carolina mountains with descriptions of two new forms. Proceedings of the Biological Society of Washington 25:135–140.

Brimley, C.S. 1920a. Reproduction of the marbled salamander. Copeia 80:25.

Brimley, C.S. 1920b. Notes on *Amphiuma* and *Necturus*. Copeia 77:5–7.

Brimley, C.S. 1921a. Breeding dates of *Ambystoma maculatum* at Raleigh, N.C. Copeia 93:26–27.

Brimley, C.S. 1921b. The life history of the American newt. Copeia 94: 31–32.

Brimley, C.S. 1923. The dwarf salamander at Raleigh, N.C. Copeia 120: 81–83.

Brimley, C.S. 1924. The waterdogs (*Necturus*) of North Carolina. Journal of the Elisha Mitchell Scientific Society 40:166–168.

Brimley, C.S. 1939. Amphibians and reptiles of North Carolina. Carolina Tips, Number 2, Elon College, North Carolina.

Brimley, C.S. 1940. The amphibians and reptiles of North Carolina. Carolina Tips, Number 10, Elon College, North Carolina.

Brimley, C.S. 1944. Amphibians and reptiles of North Carolina. Carolina Tips, Number 7, Elon College, North Carolina.

Brinkhurst, R.O. 1966. Population dynamics of the large pond-skater *Gerris najas* (Hemiptera). Journal of Animal Ecology 35:13–25.

Brinkman, B., T. Ives and S. Fletcher. 2001. Geographic distribution: *Ambystoma maculatum* (spotted salamander). Herpetological Review 32:267.

British Columbia Frogwatch Program. 2002. Northern leopard frog. http://www.gov.bc.ca/frogwatch/

Britson, C.A. and R.E. Kissell Jr. 1996. Effects of food type on developmental characteristics of an ephemeral pond-breeding anuran, *Pseudacris triseriata feriarum*. Herpetologica 52:374–382.

Britson, C.A. and S.T. Threlkeld. 1998a. Abundance, metamorphosis, developmental, and behavioral abnormalities in *Hyla chrysoscelis* tadpoles following exposure to three agrichemicals and methyl mercury in outdoor mesocosms. Bulletin of Environmental Contamnation and Toxicology 61:154–161.

Britson, C.A. and S.T. Threlkeld. 1998b. Agrichemical, methyl mercury, and UV radiation effects on southern leopard frog *(Rana utricularia)* and gray treefrog *(Hyla chrysoscelis)* tadpoles. Abstract at: http://www.mpm.edu/collect/vertzo/herp/Daptf/MWabst.html.

Britton, D. 2000. COSEWIC status report update on the northern cricket frog, *Acris crepitans*, in Canada. Committee on the Status of Endangered Wildlife in Canada, Ottawa, Ontario, Canada.

Britton, J.M. 1986. Habitats and feeding niches of sympatric populations of the salamanders, *Plethodon dorsalis angusticlavius* and *Plethodon glutinosus glutinosus* in northwestern Arkansas. Ph.D. dissertation. University of Arkansas, Fayetteville, Arkansas.

Brocchi, M.P. 1882. Étude des Batraciens de l'Amérique Centrale. Mission scientifique au Mexique et dans l'Amérique Centrale. Recherches Zoologiques, part 3, section 2:1–122.

Brocchi, P. 1877. Sur quelques Batraciens Raniformes et Bufoniformes d l'Amérique Centrale. Bulletin de la Société Philomathique de Paris Série 7, 1:175–197.

Brock, J. and P. Verrell. 1994. Courtship behavior of the seal salamander, *Desmognathus monticola* (Amphibia: Caudata: Plethodontidae). Journal of Herpetology 28:411–415.

Brockelman, W.Y. 1969. An analysis of density effects and predation in *Bufo americanus* tadpoles. Ecology 50:632–644.

Brockelman, W.Y. 1975. Competition, the fitness of offspring, and optimal clutch size. American Naturalist 109:677–699.

Brode, J.M. 1995. Status review of the southern torrent salamander *(Rhyacotriton variegatus)* in California. California Department of Fish and Game, Inland Fisheries Division, Rancho Cordova, California.

Brode, W.E. 1961. Observations on the development of *Desmognathus* eggs under relatively dry conditions. Herpetologica 17:202–203.

Brode, W.E. 1969. A systematic study of the genus *Necturus* Rafinesque. Ph.D. dissertation. University of Southern Mississippi, Hattiesburg, Mississippi.

Brode, W.E. and G. Gunter. 1958. Egg clutches and prehensilism in the slimy salamander. Herpetologica 13:279–280.

Brode, W.E. and G. Gunter. 1959. Peculiar feeding of *Amphiuma* under conditions of forced starvation. Science 130:1758–1759.

Brodie, E.D., Jr. 1968a. Investigations on the skin toxin of the red-spotted newt, *Notophthalmus viridescens viridescens*. American Midland Naturalist 80:276–280.

Brodie, E.D., Jr. 1968b. Investigations on the skin of the adult rough-skinned newt, *Taricha granulosa*. Copeia 1968:307–313.

Brodie, E.D., Jr. 1970. Western salamanders of the genus *Plethodon*: systematics and geographic variation. Herpetologica 26:468–516.

Brodie, E.D., Jr. 1971a. Two more toxic salamanders, *Ambystoma maculatum* and *Cryptobranchus alleganiensis*. Herpetological Review 3:8.

Brodie, E.D. 1971b. *Plethodon stormi*. Pp. 103.1–103.2. Catalogue of American Amphibians and Reptiles. Society for the Study of Amphibians and Reptiles, St. Louis, Missouri.

Brodie, E.D., Jr. 1976. Additional observations on the Batesian mimicry of *Notophthalmus viridescens* efts by *Pseudotriton ruber*. Herpetologica 32:68–70.

Brodie, E.D., Jr. 1977. Salamander antipredator postures. Copeia 1977:523–535.

Brodie, E.D., Jr. 1978. Biting and vocalization as anti-predator mechanisms in terrestrial salamanders. Copeia 1978:127–129.

Brodie, E.D., Jr. 1981. Phenological relationships of model and mimic salamanders. Evolution 35:988–994.

Brodie, E.D., Jr. 1983. Antipredator adaptations of salamanders: evolution and convergence among terrestrial species. Pp. 109–133. *In* Margaris, N.S., M. Arianoutsou-Faraggitaki and R.J. Reiter (Eds.), Plant, Animal and Microbial Adaptations to the Terrestrial Environment. Plenum Publishing Corporation, New York.

Brodie, E.D., Jr. and E.D. Brodie III. 1980. Differential avoidance of mimetic salamanders by free-ranging birds. Science 208:181–182.

Brodie, E.D., Jr. and D.R. Formanowicz Jr. 1981a. Larvae of the predaceous diving beetle *Dytiscus verticalis* acquire an avoidance response to skin secretions of the newt *Notophthalmus viridescens*. Herpetologica 37:172–176.

Brodie, E.D., Jr. and D.R. Formanowicz Jr. 1981b. Palatability and antipredator behavior of the treefrog *Hyla versicolor* to the shrew *Blarina brevicauda*. Journal of Herpetology 15:235–236.

Brodie, E.D., Jr. and D.R. Formanowicz Jr. 1983. Prey size preference of predators: differential vulnerability of larval anurans. Herpetologica 39:67–75.

Brodie, E.D., Jr. and D.R. Formanowicz Jr. 1987. Antipredator mechanisms of larval anurans: protection of palatable individuals. Herpetologica 43:369–373.

Brodie, E.D., Jr. and L.S. Gibson. 1969. Defensive behavior and skin glands of the northwestern salamander, *Ambystoma gracile*. Herpetologica 25:187–194.

Brodie, E.D., Jr. and R.R. Howard. 1972. Behavioral mimicry in the defensive displays of the urodele amphibians *Notophthalmus* viridescens and *Pseudotriton ruber*. BioScience 22:666–667.

Brodie, E.D., Jr. and R.R. Howard. 1973. Experimental study of Batesian mimicry in the salamanders *Plethodon jordani* and *Desmognathus ochrophaeus*. American Naturalist 90:38–46.

Brodie, E.D., Jr., T.G. Dowdey and C.D. Anthony. 1989. Salamander antipredator strategies against snake attack: biting by *Desmognathus*. Herpetologica 45:167–171.

Brodie, E.D., Jr., D.R. Formanowicz Jr. and E.D. Brodie III. 1978. The development of noxiousness of *Bufo americanus* tadpoles to aquatic insect predators. Herpetologica 34:302–306.

Brodie, E.D., Jr., J.L. Hensel and J.A. Johnson. 1974b. Toxicity of the urodele amphibians *Taricha, Notophthalmus, Cynops*, and *Paramesotriton* (Salamandridae). Copeia 1974:506–511.

Brodie, E.D., Jr., J.A. Johnson and C.K. Dodd Jr. 1974a. Immobility as a defensive behavior in salamanders. Herpetologica 30:79–85.

Brodie, E.D., Jr., R.T. Nowak and W.R. Harvey. 1979. The effectiveness of antipredator secretions and behavior of selected salamanders against shrews. Copeia 1979:270–274.

Brodie, E.D., III. 1989. Individual variation in antipredator response of *Ambystoma jeffersonianum* to snake predators. Journal of Herpetology 23:307–309.

Brodie, E.D., III and E.D. Brodie Jr. 1990. Tetrodotoxin resistance in garter snakes: an evolutionary response of predators to dangerous prey. Evolution 44:651–659.

Brodie, E.D., III and E.D. Brodie Jr. 1991. Evolutionary response of predators to dangerous prey: reduction of toxicity of newts and resistance of garter snakes in island populations. Evolution 45:221–224.

Brodkin, M.A., M.P. Simon, A.N. DeSante and K.J. Boyer. 1992. Response of *Rana pipiens* to graded doses of the bacterium *Pseudomonas aeruginosa*. Journal of Herpetology 26:490–495.

Brodman, R. 1993. The effect of acidity on interactions of *Ambystoma* salamander larvae. Journal of Freshwater Ecology 8:209–214.

Brodman, R. 1995. Annual variation in breeding success of two syntopic species of *Ambystoma* salamanders. Journal of Herpetology 29:111–113.

Brodman, R. 1996. Effects of intraguild interactions on fitness and microhabitat use of larva *Ambystoma* salamanders. Copeia 1996: 372–378.

Brodman, R. 1999a. New county records of amphibians and reptiles from northern Indiana. Herpetological Review 30:117–118.

Brodman, R. 1999b. Crowding effects on predation and competition in two species of larval salamanders. Journal of Freshwater Ecology 14:431–447.

Brodman, R. and J.M. Jaskula. 2002. Microhabitat use and activity of five species of salamander larvae. Herpetologica 58:346–354.

Brodman, R. and M. Kilmurry. 1998. Status of amphibians in northwestern Indiana. Pp. 125–136. *In* Lannoo, M.J. (Ed.), Status and Conservation of Midwestern Amphibians. University of Iowa Press, Iowa City, Iowa.

Brodman, R., S. Cortwright and A. Resetar. 2002. Historical changes of reptiles and amphibians of northwestern Indiana fish and wildlife areas. American Midland Naturalist 147:135–144.

Bromfield, L. 1955. From My Experience—The Pleasures and Miseries of Life on a Farm. Harper and Brothers, New York.

Bronmark, C. and P. Edenhamn. 1994. Does the presence of fish affect the distribution of treefrogs *(Hyla arorea)*? Conservation Biology 8:841–845.

Brooks, D.R. 1975. A review of the genus *Allassostomoides* (Trematoda: Paramphistomidae) with a redescription of *Allassostomoides chelydrae*. Journal of Parasitology 61:882–885.

Brooks, D.R. 1976a. Five species of platyhelminths from *Bufo marinus* L. (Anura: Bufonidae) in Columbia with descriptions of *Creptotrema lynchi* sp. n. (Gigenea: Allocreadiidae) and *Glypthelmins robustus* sp. n. (Digenea: Macroderoididae). Journal of Parasitology 62:429–433.

Brooks, D.R. 1976b. Parasites of amphibians of the Great Plains. Part 2. Platyhelminths of amphibians in Nebraska. Bulletin of the University of Nebraska State Museum 10:65–92.

Brooks, D.R. 1978. Systematic status of proteocephalid cestodes from reptiles and amphibians in North America with descriptions of three new species. Proceedings of the Helminthological Society of Washington 45:1–28.

Brooks, D.R. 1979. New records for amphibian trematodes. Proceedings of the Helminthological Society of Washington 46:266–289.

Brooks, D.R. 1984. Platyhelminths. Pp. 247–258. *In* Hoff, G.L, F.L. Frye and E.R. Jacobson (Eds.), Diseases of Amphibians and Reptiles. Plenum Press, New York.

Brooks, D.R. and R.L. Buckner. 1976. Some platyhelminth parasites of sirens (Amphibia: Sirenidae) from North America. Journal of Parasitology 62:906–909.

Brooks, D.R. and N.J. Welch. 1976. Parasites of amphibians of the Great Plains. I. The cercaria of *Cephalogonimus brevicirrus* Ingles, 1932 (Trematoda: Cephalogonimidae). Proceedings of the Helminthological Society of Washington 43:92–93.

Brooks, D.R., R.L. Mayden and D.A. McLennan. 1992. Phylogeny and biodiversity: conserving our evolutionary legacy. TRENDS in Ecology and Evolution 7:55–59.

Brooks, G.R. 1964. An analysis of the food habits of the bullfrog, *Rana catesbeiana*, by body size, sex, month and habitat. Virginia Journal of Science 20:173–186.

Brooks, G.R., Jr. and J.F. Sassaman. 1965. Critical thermal maxima of larval and adult *Eurycea bislineata*. Copeia 1965:251–252.

Brooks, M. 1945. Notes on the amphibians from Bickle's Knob, West Virginia. Copeia 1945:231.

Brooks, M. 1948a. Clasping in the salamanders *Aneides* and *Desmognathus*. Copeia 1948:65.

Brooks, M. 1948b. Notes on the Cheat Mountain salamander. Copeia 1948:239–244.

Broomhall, S.D., W. Osborne and R. Cunningham. 2000. Comparative effects of ambient ultraviolet-B (UV-B) radiation on two sympatric species of Australian frogs. Conservation Biology 14:420–427.

Brophy, T.E. 1980. Food habits of sympatric larval *Ambystoma tigrinum* and *Notophthalmus viridescens*. Journal of Herpetology 14:1–6.

Brophy, T.R. 1995. Natural history, ecology and distribution of *Eurycea cirrigera* in West Virginia. Master's thesis. Marshall University, Huntington, West Virginia.

Brophy, T.R. and T.K. Pauley. in press. Dietary comparison of *Eurycea cirrigera* (southern two-lined salamander) larvae from pond and stream habitats in southern West Virginia. Proceedings of the West Virginia Academy of Science. in press.

Browder, L.W. and J. Davidson. 1964. Spotting variations in the leopard frog. Journal of Heredity 55:234–241.

Brower, J. van Z. and L.P. Brower. 1965. Experimental studies of mimicry. 6. The reaction of toads (*Bufo terrestris*) to honeybees (*Apis mellifera*) and their dronefly mimics (*Eristalis vinetorum*). American Midland Naturalist 96:297–307.

Brown, A., B. Christensen, T.D. Schwaner and M. Nickerson. 2000. Natural history notes: *Bufo microscaphus* (Arizona toad). Reproduction. Herpetological Review 31:168.

Brown, A.L. 1970. The African Clawed Frog *Xenopus laevis*: A Guide for Laboratory Practical Work. Butterworths, London.

Brown, A.P. 1955. Distribution, variation, and hybridization in a relict toad (*Bufo microscaphus*) in southwestern Utah. American Museum Novitates, Number 1722, American Museum of Natural History, New York.

Brown, A.V., W.K. Pierson and K.B. Brown. 1994. Organic carbon resources and the payoff-risk relationship in cave ecosystems. Pp. 67–76. *In* The Second International Conference on Ground Water Ecology. U.S. Environmental Protection Agency and the American Water Resources Association, Washington, D.C.

Brown, B.A., W.W. Cudmore and J.O. Whitaker Jr. 1982. Miscellaneous notes on *Ambystoma texanum* in Vigo County, Indiana. Proceedings of the Indiana Academy of Science 92:473–478.

Brown, B.C. 1942. Notes on *Eurycea neotenes*. Copeia 1942:176.

Brown, B.C. 1950. An Annotated Check List of the Reptiles and Amphibians of Texas. Baylor University Press, Waco, Texas.

Brown, B.C. 1967a. *Eurycea latitans*. Pp. 34.1–34.2. Catalogue of American Amphibians and Reptiles. Society for the Study of Amphibians and Reptiles, St. Louis, Missouri.

Brown, B.C. 1967b. *Eurycea nana*. Pp. 35.1–35.2. Catalogue of American Amphibians and Reptiles. Society for the Study of Amphibians and Reptiles, St. Louis, Missouri.

Brown, B.C. 1967c. *Eurycea neotenes*. Pp. 36.1–36.2. Catalogue of American Amphibians and Reptiles. Society for the Study of Amphibians and Reptiles, St. Louis, Missouri.

Brown, C. 1997. Habitat structure and occupancy patterns of the montane frog, *Rana cascadae*, in the Cascade Range, Oregon, at multiple scales: implications for population dynamics in patchy landscapes. Master's thesis. Oregon State University, Corvallis, Oregon.

Brown, C.W. 1974. Hybridization among the subspecies of the plethodontid salamander *Ensatina eschscholtzii*. University of California Publications in Zoology 98:1–56.

Brown, C.W. and R.C. Stebbins. 1964. Evidence for hybridization between the blotched and unblotched subspecies of the salamander *Ensatina eschscholtzii*. Evolution 18:706–707.

Brown, D. 1994. Transfer of Hamilton's frog, *Leiopelma hamiltoni*, to a newly created habitat on Stephen's Island, New Zealand. New Zealand Journal of Zoology 21:425–430.

Brown, D.R. 1984. Life history notes: *Rana palustris*. Oviposition. Herpetological Review 15:110–111.

Brown, D.S. 1967. A review of the freshwater Mollusca of Natal and their distribution. Annals of the Natal Museum 18:477–494.

Brown, D.S. 1978. Pulmonate molluscs as intermediate hosts for digenetic trematodes. Pp. 287–383. *In* Fretter, A.V. and J. Peake (Eds.), Pulmonates: Systematics, Evolution and Ecology. Volume 2. Academic Press, New York.

Brown, D.S. 1980. Freshwater Snails of Africa and their Medical Importance. Taylor and Francis Limited, London, England.

Brown, E.E. 1979. Some snake food records from the Carolinas. Brimleyana 1:113–124.

Brown, E.E. 1980. Some historical data bearing on the pine barrens treefrog, *Hyla andersoni*, in South Carolina. Brimleyana 3:113–117.

Brown, E.E. 1992. Notes on amphibians and reptiles of the Piedmont of North Carolina. Journal of the Elisha Mitchell Scientific Society 24:14–22.

Brown, H.A. 1967a. High temperature tolerances of the eggs of a desert anuran, *Scaphiopus hammondi*. Copeia 1967:365–370.

Brown, H.A. 1967b. Embryonic temperature adaptations and genetic compatibility in two allopatric populations of the spadefoot toad, *Scaphiopus hammondi*. Evolution 21:742–761.

Brown, H.A. 1969. The heat resistance of some anuran tadpoles (Hylidae and Pelobatidae). Copeia 1969:138–147.

Brown, H.A. 1975a. Temperature and development of the tailed frog, *Ascaphus truei*. Comparative Biochemistry and Physiology A 50: 397–405.

Brown, H.A. 1975b. Reproduction and development of the red-legged frog, *Rana aurora*, in northwestern Washington. Northwest Science 49:241–252.

Brown, H.A. 1976a. The status of California and Arizona populations of the western spadefoot toads (genus *Scaphiopus*). Contributions in Science, Los Angeles County Museum of Natural History, Number 286, Los Angeles, California.

Brown, H.A. 1976b. The time-temperature relation of embryonic development in the northwestern salamander, *Ambystoma gracile*. Canadian Journal of Zoology 54:552–558.

Brown, H.A. 1989a. Developmental anatomy of the tailed frog (*Ascaphus truei*): a primitive frog with large eggs and slow development. Journal of Zoology, London 217:525–537.

Brown, H.A. 1989b. Tadpole development and growth of the Great Basin spadefoot toad, *Scaphiopus intermontanus*, from central Washington. Canadian Field-Naturalist 103:531–534.

Brown, H.A. 1990. Morphological variation and age-class determination in overwintering tadpoles of the tailed frog, *Ascaphus truei*. Journal of Zoology, London 220:171–184.

Brown, J.H. 1989. Patterns, modes, and extents of invasions by vertebrates. Pp. 85–100. *In* Drake, J.A., H.A. Mooney, F. di Castri, R.H. Groves, F.J. Kruger, M. Rejmanek and M. Williamson (Ed.), Biological Invasions: A Global Perspective. John Wiley and Sons, New York.

Brown, K.M. 1982. Resource overlap and competition in pond snails: an experimental analysis. Ecology 63:412–422.

Brown, K.M. 1991. Mollusca: Gastropoda. Pp. 285–314. *In* Thorp, J.H. and A.P. Covich (Eds.), Ecology and Classification of North American Freshwater Invertebrates. Academic Press, San Diego, California.

Brown, K.M. and D.R. DeVries. 1985. Predation and the distribution and abundance of a pulmonate snail. Oecologia 66:93–99.

Brown, K.M., B.K. Leathers and D.J. Minchella. 1988. Trematode prevalence and the population dynamics of freshwater pond snails. American Midland Naturalist 120:289–301.

Brown, L.E. 1967. The significance of natural hybridization in certain aspects of the speciation of some North American toads (genus *Bufo*). Ph.D. dissertation. University of Texas, Austin, Texas.

Brown, L.E. 1969. Natural hybrids between two toad species in Alabama. Quarterly Journal of the Florida Academy of Science 32:285–290.

Brown, L.E. 1970. Interspecies interactions as possible causes of racial size differences in the toads *Bufo americanus* and *Bufo woodhousei*. Texas Journal of Science 21:261–267.

Brown, L.E. 1971a. Natural hybridization and reproductive ecology of two toad species in a disturbed environment. American Midland Naturalist 86:78–85.

Brown, L.E. 1971b. Natural hybridization and trend toward extinction in some relict Texas toad populations. Southwestern Naturalist 16:185–199.

Brown, L.E. 1973a. *Bufo houstonensis*. Pp. 133.1–133.2. Catalogue of American Amphibians and Reptiles. Society for the Study of Amphibians and Reptiles, St. Louis, Missouri.

Brown, L.E. 1973b. Speciation in the *Rana pipiens* complex. American Zoologist 13:73–79.

Brown, L.E. 1974. Behavioral reactions of bullfrogs while attempting to eat toads. Southwestern Naturalist 19:335–337.

Brown, L.E. 1975. The status of the near-extinct Houston toad (*Bufo houstonensis*) with recommendations for its conservation. Herpetological Review 6:37–40.

Brown, L.E. 1976. Letter to C. Kenneth Dodd, 23 December.

Brown, L.E. 1978. Subterranean feeding by the chorus frog *Pseudacris streckeri* (Anura: Hylidae). Herpetologica 34:212–216.

Brown, L.E. 1992. *Rana blairi*. Pp. 536.1–536.6. Catalogue of American Amphibians and Reptiles. Society for the Study of Amphibians and Reptiles, St. Louis, Missouri.

Brown, L.E. 1994. Historical overview of the endangered Houston toad and its interactions with humans. Section 3. Pp. 21–27. *In* Seal, U.S. (Ed.), Houston Toad Population and Habitat Viability Assessment. U.S. Fish and Wildlife Service, and Conservation Breeding Specialist Group, Species Survival Commission, IUCN—The World Conservation Union, Austin, Texas.

Brown, L.E. 1995. The field trip. Beacon 5:1, 12–13.

Brown, L.E. and J.R. Brown. 1972. Mating calls and distributional records of treefrogs of the *Hyla versicolor* complex in Illinois. Journal of Herpetology 6:233–234.

Brown, L.E. and J.R. Brown. 1973. Notes on breeding choruses of two anurans (*Scaphiopus holbrookii, Pseudacris streckeri*) in southern Illinois. Natural History Miscellanea, Number 192, Chicago Academy of Sciences, Chicago, Illinois.

Brown, L.E. and J.E. Cima. 1998. Illinois chorus frog and the Sand Lake dilemma. Pp. 301–311. *In* Lannoo, M.J. (Ed.), Status and Conservation of Midwestern Amphibians. University of Iowa Press, Iowa City, Iowa.

Brown, L.E. and M.A. Ewert. 1971. A natural hybrid between the toads *Bufo hemiophrys* and *Bufo cognatus* in Minnesota. Journal of Herpetology 5:78–82.

Brown, L.E. and M.J. Littlejohn. 1972. Male release call in the *Bufo americanus* group. Pp. 310–323. *In* Blair, W.F. (Ed.), Evolution in the Genus *Bufo*. University of Texas Press, Austin, Texas.

Brown, L.E. and D.B. Means. 1984. Fossorial behavior and ecology of the chorus frog *Pseudacris ornata*. Amphibia-Reptilia 5:261–273.

Brown, L.E. and D. Moll. 1979. The status of the nearly extinct Illinois mud turtle (*Kinosternon flavescens spooneri* Smith 1951) with recommendations for its conservation. Special Publications in Biology and Geology, Number 3. Milwaukee Public Museum, Milwaukee, Wisconsin.

Brown, L.E. and M.A. Morris. 1990. Distribution, habitat, and zoogeography of the plains leopard frog (*Rana blairi*) in Illinois. Illinois Natural History Survey Biological Notes, Number 136, Champaign, Illinois.

Brown, L.E. and J.R. Pierce. 1965. Observations on the breeding behavior of certain anuran amphibians. Texas Journal of Science 17:313–317.

Brown, L.E. and J.R. Pierce. 1967. Male-male interactions and chorusing intensities of the Great Plains toad, *Bufo cognatus*. Copeia 1967:149–154.

Brown, L.E. and G.B. Rose. 1988. Distribution, habitat, and calling season of the Illinois chorus frog (*Pseudacris streckeri illinoensis*) along the lower Illinois River. Illinois Natural History Survey Biological Notes 132:1–13.

Brown, L.E. and R.A. Thomas. 1982. Misconceptions about the endangered Houston toad (*Bufo houstonensis*). Herpetological Review 13:37.

Brown, L.E. and J.H. Thrall. 1974. Similarities in evasive behavior of wolf spiders (Araneae, Lycosidae), American toads (Anura, Bufonidae) and ground beetles (Coleoptera, Carabidae). Great Lakes Entomologist 7:30.

Brown, L.E., H.O. Jackson and J.R. Brown. 1972. Burrowing behavior of the chorus frog, *Pseudacris streckeri*. Herpetologica 28:325–328.

Brown, L.E., M.A. Morris and T.R. Johnson. 1993. Zoogeography of the plains leopard frog (*Rana blairi*). Bulletin of the Chicago Academy of Sciences 15:1–13.

Brown, L.E., W.L. McClure, F.E. Potter Jr., N.J. Scott Jr. and R.A. Thomas. 1984. Recovery plan for the Houston toad (*Bufo houstonensis*). U.S. Fish and Wildlife Service, Department of the Interior, Albuquerque, New Mexico.

Brown, L.R. and P.B. Moyle. 1997. Invading species in the Eel River, California: successes, failures, and relationships with resident species. Environmental Biology of Fishes 49:271–291.

Brown, M. and J.J. Dinsmore. 1986. Implications of marsh size and isolation for marsh bird management. Journal of Wildlife Management 50:392–397.

Brown, M.G. 1942. An adaptation in *Ambystoma opacum* embryos to development on land. American Naturalist 76:222–223.

Brown, R.L. 1974. Diets and habitat preferences of selected anurans in southeast Arkansas. American Midland Naturalist 91:468–473.

Brown, S.C. and P.S. Brown. 1980. Water balance in the California newt, *Taricha torosa*. American Journal of Physiology 238:R113–R118.

Brown, V. 1948. Records of amphibians and of the northern alligator lizard in northwestern California. Copeia 1948:136.

Brown, W.C. and S.C. Bishop. 1948. Eggs of *Desmognathus aeneus*. Copeia 1948:129.

Brown, W.C. and J.R. Slater. 1939. The amphibians and reptiles of the islands of the state of Washington. Occasional Papers of the Department of Biology, College of Puget Sound 4:6–31.

Brown, W.D., C.T. Georgel and J.C. Mitchell. 2003. Natural history notes: *Desmognathus monticola* (seal salamander). Diet/Prey size. Herpetological Review 34:226.

Brownlie, J. 1988. A disjunct population of the blue-spotted salamander, *Ambystoma laterale*, in southwestern Nova Scotia (Canada). Canadian Field-Naturalist 102:263:264.

Bruce, R.C. 1969. Fecundity in primitive plethodontid salamanders. Evolution 23:50–54.

Bruce, R.C. 1970. The larval life of the three–lined salamander, *Eurycea longicauda guttolineata*. Copeia 1970:776–779.

Bruce, R.C. 1971. Life cycle and population structure of the salamander *Stereochilus marginatus* in North Carolina. Copeia 1971:234–246.

Bruce, R.C. 1972a. Variation in the life cycle of the salamander *Gyrinophilus porphyriticus*. Herpetologica 28:230–245.

Bruce, R.C. 1972b. The larval life of the red salamander, *Pseudotriton ruber*. Journal of Herpetology 6:43–51.

Bruce, R.C. 1974. Larval development of the salamanders *Pseudotriton montanus* and *P. ruber*. American Midland Naturalist 92:173–190.

Bruce, R.C. 1975. Reproductive biology of the mud salamander, *Pseudotriton montanus*, in western South Carolina. Copeia 1975:129–137.

Bruce, R.C. 1976. Population structure, life history and evolution of paedogenesis in the salamander *Eurycea neotenes*. Copeia 1976:242–249.

Bruce, R.C. 1977. The pigmy salamander, *Desmognathus wrighti* (Amphibia, Urodela, Plethodontidae), in the Cowee Mountains, North Carolina. Journal of Herpetology 11:246–247.

Bruce, R.C. 1978a. Life-history patterns of the salamander *Gyrinophilus porphyriticus* in the Cowee Mountains, North Carolina. Herpetologica 34:53–64.

Bruce, R.C. 1978b. A comparison of the larval periods of Blue Ridge and Piedmont mud salamanders (*Pseudotriton montanus*). Herpetologica 34:325–332.

Bruce, R.C. 1978c. Reproductive biology of the salamander *Pseudotriton ruber* in the southern Blue Ridge Mountains. Copeia 1978:417–423.

Bruce, R.C. 1979. Evolution of paedogenesis in salamanders of the genus *Gyrinophilus*. Evolution 33:998–1000.

Bruce, R.C. 1980. A model of the larval period of the spring salamander, *Gyrinophilus porphyriticus*, based on size-frequency distributions. Herpetologica 36:78–86.

Bruce, R.C. 1982a. Larval periods and metamorphosis in two species of salamanders of the genus *Eurycea*. Copeia 1982:117–127.

Bruce, R.C. 1982b. Egg-laying, larval periods and metamorphosis of *Eurycea bislineata* and *E. junaluska* at Santeetlah Creek, North Carolina. Copeia 1982:755–762.

Bruce, R.C. 1985a. Larval periods, population structure and the effects of stream drift in larvae of the salamanders *Desmognathus quadramaculatus* and *Leurognathus marmoratus* in a southern Appalachian stream. Copeia 1985:847–854.

Bruce, R.C. 1985b. Larval period and metamorphosis in the salamander *Eurycea bislineata*. Herpetologica 41:19–28.

Bruce, R.C. 1986. Upstream and downstream movements of *Eurycea bislineata* and other salamanders in a southern Appalachian stream. Herpetologica 42:149–155.

Bruce, R.C. 1988a. Life history variation in the salamander *Desmognathus quadramaculatus*. Herpetologica 44:218–227.

Bruce, R.C. 1988b. An ecological life table for the salamander *Eurycea wilderae*. Copeia 1988:15–26.

Bruce, R.C. 1989. Life history of the salamander *Desmognathus monticola*, with a comparison of the larval periods of *D. monticola* and *D. ochrophaeus*. Herpetologica 45:144–155.

Bruce, R.C. 1990. An explanation for differences in body size between two desmognathine salamanders. Copeia 1990:1–9.

Bruce, R.C. 1991. Evolution of ecological diversification in desmognathine salamanders. Herpetological Review 22:44–46.

Bruce, R.C. 1993. Sexual size dimorphism in desmognathine salamanders. Copeia 1993:313–318.

Bruce, R.C. 1995. The use of temporary removal sampling in a study of population dynamics of the salamander *Desmognathus monticola*. Australian Journal of Ecology 20:403–412.

Bruce, R.C. 1996. Life-history perspective of adaptive radiation in desmognathine salamanders. Copeia 1996:783–790.

Bruce, R.C. and N.G. Hairston Sr. 1990. Life-history correlates of body-size differences between two populations of the salamander, *Desmognathus monticola*. Journal of Herpetology 24:124–134.

Bruce, R.C., J. Castanet and H. Francillon-Vieillot. 2002. Skeletochronological analysis of variation in age structure, body size, and life history in three species of desmognathine salamanders. Herpetologica 58:181–193.

Bruce, R.C., Jaeger, R.G. and L.D. Houck. 2000. Preface. Pp. ix–x. *In* Bruce, R.C., R.G. Jaeger and L.D. Houck (Eds.), The Biology of Plethodontid Salamanders. Kluwer Academic Publishers, New York.

Brugger, K.E. 1984. Aspects of reproduction in the squirrel treefrog, *Hyla squirella*. Master's thesis. University of Florida, Gainesville, Florida.

Brune, G. 1981. Springs of Texas. Volume 1. Privately published, Arlington, Texas.

Bruneau, M. and E. Magnin. 1980. Croissance, nutrition, et reproduction des ovaovarons *Rana catesbeiana* Shaw (Amphibia Anura) des Laurentides au nord de Montreal. Canadian Journal of Zoology 58:175–183.

Brunken, G. and T. Meineke. 1984. Amphibien und Reptilien zwischen Harz und Leine. Naturschutz und Landschaftspflege in Niedersachsen, Beiheft 10:1–59.

Brunkow, P.E. and J.P. Collins. 1996. Effects of individual variation in size on growth and development of larval salamanders. Ecology 77:1483–1492.

Bruseth, J.E. and T.K. Perttula. 1995a. Letter to Cynthia Brandimarte, 15 February.

Bruseth, J.E. and T.K. Perttula. 1995b. Letter to Cynthia Brandimarte, 21 February.

Brussock, P.P., III and A.V. Brown. 1982. Selection of breeding ponds by the ringed salamander, *Ambystoma annulatum*. Proceedings of the Arkansas Academy of Science 36:82–83.

Bryan, E.H. 1932. Frogs in Hawaii. Mid-Pacific Magazine 43:61–64.

Bryce, S.A., J.M. Omernik, D.E. Pater, M. Ulmer, J.Schaar, J. Freeouf, R. Johnson, P. Kuck and S.H. Azevedo. 1998. Ecoregions of North Dakota and South Dakota. U.S. Geological Survey, Reston, Virginia.

BSFW (Bureau of Sport Fisheries and Wildlife). 1973. Threatened wildlife of the United States. Bureau of Sport Fisheries and Wildlife, Resource Publication 114, Washington, D.C.

Buchanan, B.W. 1992. Bimodal nocturnal activity pattern of *Hyla squirella*. Journal of Herpetology 27:521–522.

Buchli, G.L. 1969. Distribution, food and life history of tiger salamanders in Devil's Lake, North Dakota. Master's thesis. University of North Dakota, Grand Forks, North Dakota.

Buchwald, H.D., L. Durham, H.G. Fischer, R. Harada, H.S. Mosher, C.Y. Kao and F.A. Fuhrman. 1964. Identity of tarichatoxin and tetrodotoxin. Science 143:474–475.

Buck, D.H. 1946. Food of *Farancia abacura* in Texas. Herpetologica 3:111.

Buckner, R.L. and B.B. Nickol. 1979. Geographic and host-related variation among species of *Fessisentis* (Acanthocephala) and confirmation of the *Fessisentis fessus* life cycle. Journal of Parasitology 65:161–166.

Buhlmann, K.A. 1996. A biological survey of eight caves in Walker and Dade Counties, Georgia. Georgia Department of Natural Resources, Unpublished report, Social Circle, Georgia.

Buhlmann, K.A., J.C. Mitchell and C.A. Pague. 1994. Amphibian and small mammal abundance and diversity in saturated forested wetlands and adjacent uplands of southeastern Virginia. Pp. 1–7. *In* Eckles, S.D., A. Jennings, A. Spingarn and C. Wienhold (Eds.), Proceedings of a Workshop on Saturated Forested Wetlands in the Mid-Atlantic Region: The State of the Science. U.S. Fish and Wildlife Service, Annapolis, Maryland.

Buhlmann, K.A., C.A. Pague, J.C. Mitchell and R.B. Glasgow. 1988. Forestry operations and terrestrial salamanders: techniques in a study of the Cow Knob salamander, *Plethodon punctatus*. Pp. 38–49. *In* Szaro, R.C., K.E. Severson and D.R. Patton (Eds.), Management of Amphibians, Reptiles, and Mammals in North America. U.S.D.A. Forest Service, General Technical Report, RM-166, Rocky Mountain Forest and Range Experiment Station, Fort Collins, Colorado.

Bull, E.L. and B.E. Carter. 1996. Tailed frogs: distribution, ecology, and association with timber harvest in northeastern Oregon. U.S.D.A. Forest Service, PNW-RP-497, Pacific Northwest Research Station, Portland, Oregon.

Bull, E.L., and M.P. Hayes. 2000. Livestock effects on reproduction of the Columbia spotted frog. Journal Range Management 53:291–294.

Bull, E.L. and M.P. Hayes. 2001. Post-breeding season movements of Columbia spotted frogs *(Rana luteiventris)* in northeastern Oregon. Western North American Naturalist 61:119–123.

Bull, E.L. and M.P. Hayes. 2002. Overwintering of Columbia spotted frogs *(Rana luteiventris)* in northeastern Oregon. Northwest Science 76:141–147.

Bull, E.L. and D.B. Marx. 2002. Influence of fish and habitat on amphibian communities in high elevation lakes in northeastern Oregon. Northwest Science 76:240–248.

Bullard, A.J. 1965. Additional records of the treefrog *Hyla andersoni* from the Coastal Plain of North Carolina. Herpetologica 21:154–155.

Bunnell, J.F. and R.A. Zampella. 1999. Acid water anuran pond communities along a regional forest to agro-urban gradient. Copeia 1999: 614–627.

Burch, J.B. 1989. North American Freshwater Snails. Malacological Publications, Hamburg, Michigan.

Burch, P.R. and J.T. Wood. 1955. The salamander *Siren lacertina* feeding on clams and snails. Herpetological Notes 1955:255–256.

Burchett, C.R., Jr. and J.P. Shoemaker. 1990. *Myxidium serotinum* (Myxosporida, Myxidiae) in Kentucky two-lined salamanders. Transactions of the Kentucky Academy of Science 51:189–190.

Bureau of Land Management (B.L.M.). 2002. Amargosa toad. Tonopah Field Station, Tonopah, Nevada, BLM/BM/GI-01/015+9212. (Available at Simandle, E. 2002. The Amargosa Toad. http://www.amargosatoad.org.)

Burgason, B.N. 1999. Four-toed salamander. Pp. 62–65. *In* Hunter, M.L., A.J.K. Calhoun and M. McCollugh, (Eds.), Maine Amphibians and Reptiles. University of Maine Press, Orono, Maine.

Burger, J.W. 1933. A preliminary list of the amphibians of Lebanon County, Pennsylvania, with notes on habits and life history. Copeia 1933:92–94.

Burger, J.W. 1935. *Plethodon cinereus* (Green) in eastern Pennsylvania and New Jersey. American Naturalist 64:578–586.

Burger, W.L. 1950. Novel aspects of the life history of two ambystomas. Journal of the Tennessee Academy of Science 25:252–257.

Burger, W.L. and A.N. Bragg. 1947. Notes on *Bufo boreas* (B. and G.) from the Gothic region of Colorado. Proceedings of the Oklahoma Academy of Science 27:61–65.

Burger, W.L., H.M. Smith and F.E. Potter Jr. 1950. Another neotenic *Eurycea* from the Edwards Plateau. Proceedings of the Biological Society of Washington 63:51–58.

Burgess, R.C., Jr. 1950. Development of spade-foot larvae under laboratory conditions. Copeia 1950:49–51.

Burggren, W.W. 1985. Gas exchange, metabolism, and "ventilation" in gelatinous frog egg masses. Physiological Zoology 58:503–514.

Burggren, W.W., M.E. Feder and A.W. Pinder. 1983. Temperature and the balance between aerial and aquatic respiration in larvae of *Rana berlandieri* and *Rana catesbeiana*. Physiological Zoology 56:263–273.

Burgman, M.A., S. Ferson and H.R. Akçakaya. 1993. Risk Assessment in Conservation Biology. Chapman and Hall, New York.

Burgman, M.A., R. Grimson and S. Ferson. 1995. Inferring threat from scientific collections. Conservation Biology 9:923–928.

Burke, V.C. 1911. Note on *Batrachoseps attenuatus* Esch. American Naturalist 45:413–414.

Burkett, R.D. 1969. An ecological study of the cricket frog, *Acris crepitans*, in northeastern Kansas. Ph.D. dissertation. University of Kansas, Lawrence, Kansas.

Burkett, R.D. 1984. An ecological study of the cricket frog *Acris crepitans*. Pp. 89–103. *In* Seigel, R.A., L.E Hunt, J.L. Knight, L. Malaret and N.L. Zuschlag (Eds.), Vertebrate Ecology and Systematics: A Tribute to Henry S. Fitch. Museum of Natural History, University of Kansas, Lawrence, Kansas.

Burkhart, J.G., J.C. Helgen, D.J. Fort, K. Gallagher, D. Bowers, T.L. Propst, M. Gernes, J. Magner, M.D. Shelby and G. Lucier. 1998. Induction of mortality and malformation in *Xenopus laevis* embryos by water sources associated with field frog deformities. Environmental Health Perspectives 106:841–848.

Burkhart, J.T. 1984. Status of the western green toad *(Bufo debilis insidior)* in Kansas. Kansas Fish and Game Commission, Unpublished report, Contract Number 72, Dodge City, Kansas.

Burkhardt, T., S. Winegarner and R.W. Hansen. 2001. Natural history notes: *Batrachoseps stebbinsi*. Attempted predation. Herpetological Review 32:245.

Burkholder, G. 1998. Natural history notes: *Hyla chrysoscelis* (gray treefrog). Hibernacula. Herpetological Review 29:231.

Burnett, S. 1996. Colonising cane toads cause population declines in some Australian predators: reliable anecdotal information and management implications. Pacific Conservation Biology 3:65–72.

Burns, D.M. 1954. A new subspecies of the salamander *Plethodon vandykei*. Herpetologica 10:83–87.

Burns, D.M. 1962. The taxonomic status of the salamander *Plethodon vandykei larselli*. Copeia 1962:177–181.

Burns, W.C. and I. Pratt. 1953. The life cycle of *Metagonimoides oregonensis* Price (Trematoda: Heterophyidae). Journal of Parasitology 39:60–69.

Burroughs, M. 1999. Making room for the Amargosa toad. Endangered Species Bulletin 24:10–11.

Bursey, C.R. and W.F. DeWolf. 1998. Helminths of the frogs, *Rana catesbeiana*, *Rana clamitans*, and *Rana palustris*, from Coshocton County, Ohio. Ohio Journal of Science 98:28–29.

Bursey, C.R. and S.R. Goldberg. 1998. Helminths of the Canadian toad, *Bufo hemiophrys* (Amphibia: Anura), from Alberta, Canada. Journal of Parasitology 84:617–618.

Bursey, C.R. and S.R. Goldberg. 2001. *Falcaustra lowei* n. sp. and other helminths from the Tarahumara frog, *Rana tarahumarae* (Anura: Ranidae), from Sonora, Mexico. Journal of Parasitology 87:340–344.

Bursey, C.R. and D.R. Schibli. 1995. A comparison of the helminth fauna of two *Plethodon cinereus* populations. Journal of the Helminthological Society of Washington 62:232–236.

Burt, C.E. 1932. Records of amphibians from the eastern and central United States. American Midland Naturalist 13:75–85.

Burt, C.E. 1935. Further records of the ecology and distribution of amphibians and reptiles in the middle west. American Midland Naturalist 16:311–336.

Burt, C.E. 1936. Contributions to the herpetology of Texas I. Frogs of the genus *Pseudacris*. American Midland Naturalist 17:770–775.

Burt, C.E. 1938a. Contributions to Texas herpetology VII. The salamanders. American Midland Naturalist 20:374–380.

Burt, C.E. 1938b. The frogs and toads of the southeastern United States. Transactions of the Kansas Academy of Science 41:331–367.

Burt, C.E. 1942. An aggregation of snakes and salamanders during hibernation. Copeia 1942:262–263.

Burton, S.R. 2001. Amphibian declines in southeast Idaho: using modeling to assess the habitat loss hypothesis *(Bufo boreas, Rana pipiens, Ambystoma tigrinum, Pseudacris maculata)*. Ph. D. dissertation. Idaho State University, Pocatello, Idaho.

Burton, T.M. 1976. An analysis of the feeding ecology of the salamanders (Amphibia: Urodela) of the Hubbard Brook Experimental Forest, New Hampshire. Journal of Herpetology 10:187–204.

Burton, T.M. 1977. Population estimates, feeding habits and nutrient and energy relationships of *Notophthalmus v. viridescens*, in Mirror Lake New Hampshire. Copeia 1977:139–143.

Burton, T.M. and G.E. Likens. 1975a. Energy flow and nutrient cycling in salamander populations in the Hubbard Brook Experimental Forest, New Hampshire. Ecology 56:1068–1080.

Burton, T.M. and G.E. Likens. 1975b. Salamander populations and biomass in the Hubbard Brook Experimental Forest, New Hampshire. Copeia 1975:541–546.

Bury, R.B. 1968. The distribution of *Ascaphus truei* in California. Herpetologica 24:39–46.

Bury, R.B. 1970. Food similarities in the tailed frog, *Ascaphus truei*, and the Olympic salamander, *Rhyacotriton olympicus*. Copeia 1970:170–171.

Bury, R.B. 1972. Small mammals and other prey in the diet of the Pacific giant salamander *(Dicamptodon ensatus)*. American Midland Naturalist 87:524–526.

Bury, R.B. 1973a. The Cascade frog, *Rana cascadae*, in the North Coast Range of California. Northwest Science 47:228–229.

Bury, R.B. 1973b. Western *Plethodon*: systematics and biogeographic relationships of the *elongatus* group (Abstract). Herpetological Information Search Systems News-Journal 1:56–57.

Bury, R.B. 1983. Differences in amphibian populations in logged and old growth redwood forest. Northwest Science 57:167–178.

Bury, R.B. 1998. Evolution and Zoogeography of the Del Norte and Siskiyou Mountain Salamander with Management Recommendations. U.S. Fish and Wildlife Service and Bureau of Land Management, Medford District, Final report, U.S.G.S. Forest and Rangeland Ecosystem Science Center, Corvallis, Oregon.

Bury, R.B. 1999. A historical perspective and critique of the declining amphibian crisis. Wildlife Society Bulletin 27:1064–1068.

Bury, R.B. and M.J. Adams. 1999. Variation in age at metamorphosis across a latitudinal gradient for the tailed frog, *Ascaphus truei*. Herpetologica 55:283–290.

Bury, R.B. and M.J. Adams. 2000. Inventory and monitoring of amphibians in North Cascades and Olympic National Parks, 1995–1998. U.S.G.S. Forest and Rangeland Ecosystem Science Center in cooperation with Olympic National Park, Final report, Corvallis, Oregon.

Bury, R.B. and P.S. Corn. 1987. Evaluation of pitfall trapping in northwestern forests: trap arrays with drift fences. Journal of Wildlife Management 51:112–119.

Bury, R.B. and P.S. Corn. 1988a. Responses of aquatic and streamside amphibians to timber harvest: a review. Pp. 165–181. *In* Raedeke, K.J. (Ed.), Streamside Management: Riparian and Forestry Interactions. Institute of Forest Resources, Contribution 59, University of Washington, Seattle, Washington.

Bury, R.B. and P.S. Corn. 1988b. Douglas-fir forests in the Oregon and Washington Cascades: abundance of terrestrial herpetofauna related to stand age and moisture. Pp. 11–22. *In* Szaro, R.C., K.E. Severson and D. Patton (Tech. Coords.), Management of Amphibians, Reptiles, and Small Mammals in North America. U.S.D.A. Forest Service, Technical Report, RM-166, Rocky Mountain Forest and Range Experiment Station, Fort Collins, Colorado.

Bury, R.B. and P.S. Corn. 1989. Logging in western Oregon: response of headwater habitats and stream amphibians. Forest Ecology and Management 29:39–57.

Bury, R.B. and P.S. Corn. 1991. Sampling methods for amphibians in streams in the Pacific Northwest. U.S.D.A. Forest Service, General Technical Report, PNW-GTR-275, Portland, Oregon.

Bury, R.B. and C.R. Johnson. 1965. Note on the food of *Plethodon elongatus* in California. Herpetologica 21:67–68.

Bury, R.B. and R.A. Luckenbach. 1976. Introduced amphibians and reptiles in California. Biological Conservation 10:1–14.

Bury, R. B. and D.J. Major.1997. Integrated sampling for amphibian communities in montane habitats. Pp. 75–82. *In* Olson, D.H., W. P. Leonard and R. B. Bury (Eds.), Sampling Amphibians in Lentic Habitats. Northwest Fauna, Number 4, Society for Northwestern Vertebrate Biology, Olympia, Washngton.

Bury, R.B. and M. Martin. 1967. The food of the salamander *Rhyacotriton variegatus*. Copeia 1967:487.

Bury, R.B. and M. Martin. 1973. Comparative studies on the distribution and foods of plethodontid salamanders in the redwood region of northern California. Journal of Herpetology 7:331–335.

Bury, R.B. and C.A. Pearl. 1999. Klamath-Siskiyou herpetofauna: biogeographic patterns and conservation strategies. Natural Areas Journal 19:341–350.

Bury, R.B. and S.B. Ruth. 1972. Santa Cruz long-toed salamander: survival in doubt. Herpetological Review 4:20–22.

Bury, R.B. and J.A. Whelan. 1984. Ecology and management of the bullfrog. U.S. Fish and Wildlife Service, Resource Publication Number 155, Washington, D.C.

Bury, R.B., P.S. Corn and K.B. Aubry. 1991a. Regional patterns of terrestrial amphibian communities in Oregon and Washington. Pp. 341–350. *In* Ruggiero, L.F., K.B. Aubry, A.B. Carey and M.H. Huff (Eds.), Wildlife and Vegetation of Unmanaged Douglas-fir Forests. U.S.D.A. Forest Service, General Technical Report, PNW-GTR-285, Pacific Northwest Research Station, Portland, Oregon.

Bury, R.B., C.K. Dodd Jr. and G.M. Fellers. 1980. Conservation of the Amphibia of the United States: A review. United States Fish and Wildlife Service, Resource Publication, Number 134, Washington, D.C.

Bury, R.C., G.M. Fellers and S.B. Ruth. 1969. First records of *Plethodon dunni* in California, and new distributional data on *Ascaphus truei*, *Rhyacotriton olympicus*, and *Hydromantes shastae*. Journal of Herpetology 3:157–161.

Bury, R.B., P. Loafman, D. Rofkar and K.I. Mike. 2001. Clutch sizes and nests of tailed frogs from the Olympic Peninsula, Washington. Northwest Science 75:419–422.

Bury, R.B., P.S. Corn, K.B. Aubry, F.F. Gilbert and L.L.C. Jones. 1991b. Aquatic amphibian communities in Oregon and Washington. *In* Ruggiero, L.F., K.B. Aubry, A.B. Carey and M.H. Huff (Tech. Coords.), Wildlife and Vegetation of Unmanaged Douglas-fir Forests. U.S.D.A. Forest Service, General Technical Report, PNW-GTR-285, Pacific Northwest Research Station, Portland, Oregon.

Bury, R.B., P.S. Corn, C.K. Dodd, W. McDiarmid and N.J. Scott. 1995. Amphibians. Pp. 124–126. *In* Mac, M.J., P.A. Opler, C.E. Puckett Haecker and P.D. Doran (Eds.), Our Living Resources. Volume 1. U.S. Department of the Interior, U.S. Geological Survey, Washington, D.C.

Busby, W.H. and W.R. Brecheisen. 1997. Chorusing phenology and habitat associations of the crawfish frog, *Rana areolata* (Anura: Ranidae), in Kansas. Southwestern Naturalist 42:210–217.

Busby, W.H. and J.R. Parmelee. 1996. Historical changes in a herpetofaunal assemblage in the Flint Hills of Kansas. American Midland Naturalist 135:81–91.

Bush, F.M. 1959. Foods of some Kentucky herptiles. Herpetologica 15:73–77.

Bush, F.M. 1963. Effect of light and temperature on the gross composition of the toad, *Bufo fowleri*. Journal of Experimental Zoology 153:1–13.

Bush, F.M. and E.F. Melnick. 1962. The food of *Bufo woodhousei fowleri*. Herpetologica 18:110–114.

Bussing, W.A. 1987. Peces de las aguas continentales de Costa Rica. Editorial Universidad de Costa Rica, San Jose, Costa Rica.

Bustnes, J.O. and K. Galaktionov. 1999. Anthropogenic influences in the infestation of intertidal gastropods by seabird trematode larvae on the southern Barents Sea Coast. Marine Biology 133:449–453.

Butler, R.W. 1992. Great blue heron. *In* Poole, A., P. Stettenheim and F. Gill (Eds.), The Birds of North America. Number 25. The American Ornithologists' Union, Philadelphia, Academy of Natural Sciences, Washington, D.C.

Butterfield, B.P. 1988. Age structure and reproductive biology of the Illinois chorus frog *(Pseudacris streckeri illinoensis)* from northeastern Arkansas. Master's thesis. Arkansas State University, State University, Arkansas.

Butterfield, B.P., W.E. Meshaka Jr. and C. Guyer. 1997. Nonindigenous amphibians and reptiles. Pp. 123–138. *In* Simberloff, D., D.C. Schmitz and T.C. Brown (Eds.), Strangers in Paradise: Impact and Management of Nonindigenous Species in Florida. Island Press, Washington, D.C.

Butterfield, B.P., W.E. Meshaka and S.E. Trauth. 1989. Fecundity and egg mass size of the Illinois chorus frog, *Pseudacris streckeri illinoiensis* (Hylidae), from northeastern Arkansas. Southwestern Naturalist 34:556–557.

Butts, S.R. and W.C. McComb. 2000. Associations of forest-floor vertebrates with coarse woody debris in managed forests of western Oregon. Journal of Wildlife Management 64:95–104.

Buxton, D.R. 1936. A natural history of the Turkana fauna. Journal of the East Africa and Uganda Natural History Society 13:85–104.

Cable, R.M., R.S. Connor and J.W. Balling. 1960. Scientific survey of Puerto Rico and the Virgin Islands. Digenetic trematodes of Puerto Rican shore birds. New York Academy of Sciences 17:187–255.

Caetano, M.H. and R. Leclair Jr. 1996. Growth and population structure of red-spotted newts (*Notophthalmus viridescens*) in permanent lakes of the Laurentian Shield, Quebec. Copeia 1996:866–874.

Cagle, F.R. 1942. Herpetological fauna of Jackson and Union Counties, Illinois. American Midland Naturalist 28:164–200.

Cagle, F.R. 1948. Observations on a population of the salamander *Amphiuma tridactylum* Cuvier. Ecology 29:479–491.

Cagle, F.R. 1954. Observations on the life history of the salamander *Necturus louisianensis*. Copeia 1954:252–260.

Cagle, F.R. and P.E. Smith. 1939. A winter aggregation of *Siren intermedia* and *Triturus viridescens*. Copeia 1939:232–233.

Cahn, A.R. 1939. The barking frog, *Hyla gratiosa*, in northern Alabama. Copeia 1939:52–53.

Caldwell, J.P. 1982a. Disruptive selection: a tail color polymorphism in *Acris crepitans* in response to differential predation. Canadian Journal of Zoology 60:2818–2828.

Caldwell, J.P. 1982b. *Hyla gratiosa*. Pp. 298.1–298.2. Catalogue of American Amphibians and Reptiles. Society for the Study of Amphibians and Reptiles, St. Louis, Missouri.

Caldwell, J.P. 1986. Selection of egg deposition sites: a seasonal shift in the southern leopard frog, *Rana sphenocephala*. Copeia 1986:249–253.

Caldwell, J.P. 1987. Demography and life history of two species of chorus frogs (Anura: Hylidae) in South Carolina. Copeia 1987:114–127.

Caldwell, J.P., J.H. Thorp and T.O. Jervey. 1980. Predator-prey relationships among larval dragonflies, salamanders, and frogs. Oecologia 46:285–289.

Caldwell, R.S. 1973. Winter congregations of *Plethodon cinereus* in ant mounds, with notes on their food habits. American Midland Naturalist 90:482–485.

Caldwell, R.S. 1975. Observations on the winter activity of the red-backed salamander, *Plethodon cinereus*, in Indiana. Herpetologica 31:21–22.

Caldwell, R.S. and J.E. Copeland. 1992. Status and habitat of the Tennessee cave salamander, *Gyrinophilus palleucus*. Tennessee Wildlife Resources Agency, Final report, Nashville, Tennessee.

Caldwell, R.S. and G.W. Folkerts. 1976. Variation and systematics of the *Desmognathus monticola* complex. Herpetological Review 7:76.

Caldwell, R.S. and W.C. Houtcooper. 1973. Food habits of larval *Eurycea bislineata*. Journal of Herpetology 7:386–388.

Caldwell, R.S. and G.S. Jones. 1973. Winter congregations of *Plethodon cinereus* in ant mounds, with notes on their food habits. American Midland Naturalist 90:482–485.

Calef, G.W. 1973a. Natural mortality of tadpoles in a population of *Rana aurora*. Ecology 54:741–758.

Calef, G.W. 1973b. Spatial distribution and "effective" breeding population of red-legged frogs *(Rana aurora)* in Marion Lake, British Columbia. Canadian Field-Naturalist 87:279–284.

Calef, R.T. 1954. The salamander *Ambystoma tigrinum nebulosum* in southern Arizona. Copeia 1954:223.

California Department of Fish and Game. 1999. Natural Diversity Database: Special Animals. California Department of Fish and Game, Wildlife and Habitat Data Analysis Branch, Natural Diversity Database, Sacramento, California.

California Department of Fish and Game. 2002. California Natural Diversity Database. Habitat Conservation Division, Wildlife and Habitat Data Analysis Branch, Natural Diversity Database, Sacramento, California.

Callaway, D. 1998. The Wyoming toad North American regional studbook. Henry Doorly Zoo, Omaha, Nebraska.

Camiade, E.B. 1946. Letter to city of Bastrop, 31 July.

Camp, C.D. 1986. Distribution and habitat of the southern red-back salamander, *Plethodon serratus* Grobman (Amphibia: Plethodontidae), in Georgia. Georgia Journal of Science 44:136–146.

Camp, C.D. 1988. Aspects of the life history of the southern red-back salamander *Plethodon serratus* Grobman in the southeastern United States. American Midland Naturalist 119:93–100.

Camp, C.D. 1996. Bite scar patterns in the black-bellied salamander, *Desmognathus quadramaculatus*. Journal of Herpetology 30:543–546.

Camp, C.D. 1997a. Natural history notes: *Desmognathus monticola* (seal salamander). Oophagy. Herpetological Review 28:81–82.

Camp, C.D. 1997b. The status of the black-bellied salamander *(Desmognathus quadramaculatus)* as a predator of heterospecific salamanders in Appalachian streams. Journal of Herpetology 31:613–616.

Camp, C.D. 2003a. Natural history notes: *Desmognathus monticola* (seal salamander). Aggression. Herpetological Review 34:227.

Camp, C.D. 2003b. Natural history notes: *Desmognathus ocoee* (ocoee salamander). Mortality. Herpetological Review 34:227.

Camp, C.D. 2004. *Desmognathus folkersi*. p. 782.1–782.3. Catalogue of American Amphibians and Reptiles. Society for the Study of Amphibians and Reptiles, St. Louis, Missouri.

Camp, C.D. and L.L. Bozeman. 1981. Foods of two species of *Plethodon* (Caudata: Plethodontidae) from Georgia and Alabama. Brimleyana 6:163–166.

Camp, C.D. and T.P. Lee. 1996. Intraspecific spacing and interaction within a population of *Desmognathus quadramaculatus*. Copeia 1996:78–84.

Camp, C.D. and D.G. Lovell. 1989. Fishing for "spring lizards": a technique for collecting blackbelly salamanders. Herpetological Review 20:47.

Camp, C.D. and R.D. Payne. 1996. Geographic distribution: *Desmognathus aeneus* (seepage salamander). Herpetological Review 27:86.

Camp, C.D., C.E. Condee and D.G. Lowell. 1990. Oviposition, larval development and metamorphosis in the wood frog, *Rana sylvatica* (Anura: Ranidae) in Georgia. Brimleyana 16:17–21.

Camp, C.D., J.L. Marshall and R.M. Austin Jr. 2000. The evolution of adult body size in black-bellied salamanders (*Desmognathus quadramaculatus*) complex. Canadian Journal of Zoology 78:1712–1722.

Camp, C.D., S.G. Tilley, R.M. Austin Jr. and J.L. Marshall. 2002. A new species of black-bellied salamander (genus *Desmognathus*) from the Appalachian Mountains of Georgia. Herpetologica 58:471–484.

Camp, C.D., J.L Marshall, K.R. Landau, R.M. Austin Jr. and S.G. Tilley. 2000. Sympatric occurrence of the two species of the two-lined salamander (*Eurycea bislineata*) complex. Copeia 2000:572–578.

Camp, C.L. 1915. *Batrachoseps major* and *Bufo cognatus californicus*, new Amphibia from southern California. University of California Publications in Zoology 12:327–334.

Camp, C.L. 1916a. Description of *Bufo canorus*: a new toad from the Yosemite National Park. University of California Publications in Zoology 17:59–62.

Camp, C.L. 1916b. *Spelerpes platycephalus*, a new alpine salamander from the Yosemite National Park. University of California Publications in Zoology 17:11–14.

Camp, C.L. 1917a. An extinct toad from Rancho La Brea. University of California Publication, Bulletin of the Department of Geology 10: 287–292.

Camp, C.L. 1917b. Notes on the systematic status of the toads and frogs of California. University of California Publications in Zoology 17: 115–125.

Camp, J.W. and H.W. Huizinga. 1979. Altered color, behavior and predation susceptibility of the isopod *Asellus intermedius* infected with *Acanthocephalus dirus*. Journal of Parasitology 65:501–510.

Campbell, A. 1999. Declines and disappearances of Australian frogs. Environment Australia, Department of the Environment and Heritage, Canberra, New South Wales, Australia.

Campbell, B. 1931a. *Rana tarahumarae*, a frog new to the United States. Copeia 1931:164.

Campbell, B. 1931b. Notes on *Batrachoseps*. Copeia 1931:131–134.

Campbell, B. 1934. Report on a collection of reptiles and amphibians made in Arizona during the summer of 1933. Occasional Papers of the Museum of Zoology, Number 289, University of Michigan, Ann Arbor, Michigan.

Campbell, C. 1977. Some threatened frogs and toads in Ontario. Pp. 130–131. *In* Mosquin, T. and C. Suchal (Eds.), Proceedings of the Symposium on Canada's Threatened Species and Habitats. Canadian Nature Federation, Special Publication, Number 6, Ottawa, Ontario, Canada.

Campbell, H.W., S.P. Christman and F.G. Thompson. 1980. Geographic distribution: *Stereochilus marginatus* (many-lined salamander). Herpetological Review 11:13.

Campbell, J.A. 1998. Amphibians and Reptiles of Northern Guatemala, the Yucatan, and Belize. University of Oklahoma Press, Norman, Oklahoma.

Campbell, J.A. 1999. Distribution patterns of amphibians in Middle America. Pp. 111–210. *In* Duellman, W.E. (Ed.), Patterns of Distribution in Amphibians. Johns Hopkins Press, Baltimore, Maryland.

Campbell, J.A. and D.R. Frost. 1993. Anguid lizards of the genus *Abronia*: revisionary notes, descriptions of four new species, a phylogenetic analysis and key. Bulletin of the American Museum of Natural History, Number 16, New York.

Campbell, J.A. and J.P. Vannini. 1989. Distribution of amphibians and reptiles in Guatemala and Belize. Western Foundation of Vertebrate Zoology 4:1–21.

Campbell, J.B. 1970a. Food habits of the boreal toad, *Bufo boreas boreas*, in the Colorado Front Range. Journal of Herpetology 4:83–85.

Campbell, J.B. 1970b. Hibernacula of a population of *Bufo boreas boreas* in the Colorado Front Range. Herpetologica 26:278–282.

Campbell, J.B. 1970c. Life history of *Bufo boreas boreas* in the Colorado Front Range. Ph.D. dissertation. University of Colorado, Boulder, Colorado.

Campbell, J.B. 1970d. New elevational records for the boreal toad *Bufo boreas boreas*. Arctic and Alpine Research 2:157–159.

Campbell, J.B. 1976. Environmental controls on boreal toad populations in the San Juan Mountains. Pp. 289–295. *In* Steinhoff, H.W. and J.D. Ives (Eds.), Final Report, San Juan Ecology Project. Colorado State University Publications, Fort Collins, Colorado.

Campbell, J.B. and W.G. Degenhardt. 1971. *Bufo boreas boreas* in New Mexico. Herpetologica 9:157–160.

Campbell, L. 1995. Endangered and threatened animals of Texas. Texas Parks and Wildlife, Austin, Texas.

Campbell, L.A., T.B. Graham, L.P. Thibault and P.A. Stine. 1996. The arroyo toad (*Bufo microscaphus californicus*), ecology, threats, recovery actions, and research needs. National Biological Service, Technical Report, Number NBS/CSC-96-01, California Science Center, Davis, California.

Campbell, P.M. and W.K. Davis. 1968. Vertebrates in stomachs of *Bufo valliceps*. Herpetologica 24:327–328.

Camper, J.D. 1986. Life history notes: *Ambystoma tigrinum tigrinum* (eastern tiger salamander). Feeding. Herpetological Review 17:19.

Camper, J.D. 1988. The status of three uncommon salamanders (Amphibia: Caudata) in Iowa (USA). Journal of the Iowa Academy of Science 95:127–130.

Camper, J.D. 1990. Mode of reproduction in the small-mouthed salamander, *Ambystoma texanum* (Ambystomatidae), in Iowa. Southwestern Naturalist 35:99–100.

Caneris, A. and J. Oliver. 1999. Who will be in Sydney for the 2000 Olympics? Wildlife Australia Autumn 1999.

Cannon, M.S. and J.R. Hostetler. 1976. The anatomy of the parotoid gland in Bufonidae with some histochemical findings. II. *Bufo alvarius*. Journal of Morphology 148:137–160.

Canterbury, R.A. 1991. Ecology of the green salamander, *Aneides aeneus* (Cope and Packard), in West Virginia. Master's thesis. Marshall University, Huntington, West Virginia.

Canterbury, R.A. and T.K. Pauley. 1990. Food habits of *Aneides aeneus* in West Virginia. Proceedings of the West Virginia Academy of Science 62:47–50.

Canterbury, R.A. and T.K. Pauley. 1991. Intra and interspecific competition in the green salamander, *Aneides aeneus*. Association of Southeastern Biologists Bulletin 38:114.

Canterbury, R.A. and T.K. Pauley. 1994. Time and mating and egg deposition of West Virginia populations of the salamander *Aneides aeneus*. Journal of Herpetology 28:431–434.

Cantino, P.D. 2000. Phylogenetic nomenclature: addressing some concerns. Taxon 49:85–93.

Cantino, P.D. and K. de Queiroz. 2000. PhyloCode: a phylogenetic code of biological nomenclature. http://www.ohiou.edu/phylocode/.

Cantino, P.D., H.N. Bryant, K. de Queiroz, M.J. Donoghue, T. Eriksson, D.M. Hillis and M.S.Y. Lee. 1999. Species names in phylogenetic nomenclature. Systematic Biology 48:790–807.

Cappuccino, N. and P.W. Price. 1995. Population Dynamics: New Approaches and Synthesis. Academic Press, New York.

Carey, C. 1976. Thermal physiology and energetics of boreal toads, *Bufo boreas boreas*. Ph.D. dissertation. University of Michigan, Ann Arbor, Michigan.

Carey, C. 1978. Factors affecting body temperatures of toads. Oecologia 35:197–219.

Carey, C. 1979. Aerobic and anaerobic energy expenditure during rest and activity in montane *B.b. boreas* and *Rana pipiens*. Oecologia 39:213–228.

Carey, C. 1987. Status of a breeding population of the western toad, *Bufo boreas boreas* at Lagunitas campground, New Mexico. New Mexico Department of Game and Fish, Unpublished report, Santa Fe, New Mexico.

Carey, C. 1993. Hypotheses concerning the causes of the disappearance of boreal toads from the mountains of Colorado. Conservation Biology 7:355–362.

Carey, C. 1998. Disease and immune function workshop. Froglog: Newsletter of the Declining Amphibian Populations Task Force 30:4.

Carey, C. 2000. Infectious disease and worldwide declines of amphibian populations, with comments on emerging diseases in coral reef organisms and in humans. Environmental Health Perspectives 108, Supplement 1:143–150.

Carey, C. and C.J. Bryant. 1995. Possible interrelations among environmental toxicants, amphibian development, and decline of amphibian populations. Wildlife Development 103:13–17.

Carey, C., N. Cohen and L. Rollins-Smith. 1999. Amphibian declines: an immunological perspective. Developmental and Comprative Immunology 23:459–472.

Carey, C., G.D. Maniero and J.F. Stinn. 1996a. Effects of cold on immune function and susceptibility to bacterial infection in toads (*Bufo marinus*). Pp. 123–129. *In* Geiser, F., A.J. Hulbert and S.C. Nicoll (Eds.), Life in the Cold: Tenth Hibernation Symposium. University of New England Press, Armidale, New South Wales, Australia.

Carey, C., G.D. Maniero, C.W. Harper and G.K. Snyder. 1996b. Measurements of several aspects of immune function in toads (*Bufo marinus*) after exposure to low pH. Pp. 565–577. *In* Stolen, J.S., T.C. Fletcher, C.J. Bayne, C.J. Secombes, J.T. Zelikoff, L.E. Twerdok and D.P. Anderson (Eds.), Modulators of Immune Response: The Evolutionary Trail. SOS Publications, Fair Haven, New Jersey.

Carey, C., D.F. Bradford, J.L. Brunner, J.P. Collins, E.W. Davidson, J.E. Longcore, M. Ouellet, A.P. Pessier and D.M. Schock. 2003. Biotic factors in amphibian population declines. Pp. 153–208. *In* Linder, G., D.W. Sparling and S.K. Krest (Eds.), Multiple Stressors and Declining Amphibian Populations: Evaluating Cause and Effect. Society of Environmental Toxicology and Chemistry (SETAC) Press, Pensacola, Florida.

Carey, C.N., W.R. Heyer, J. Wilkinson, R.A. Alford, J.W. Arntzen, T. Halliday, L. Hungerford, K.R. Lips, E.M. Middleton, S.A. Orchard and A.S. Rand. 2001. Amphibian declines and environmental change: use of remote sensing data to identify environmental correlates. Conservation Biology 15:903–913.

Carey, S.D. 1982. Geographic distribution: *Eleutherodactylus planirostris planirostris* (greenhouse frog). Herpetological Review 13:130.

Carey, S.D. 1984. Geographic distribution: *Amphiuma pholeter* (one-toed amphiuma). Herpetological Review 15:77.

Carey, S.D. 1985. Geographic distribution: *Amphiuma pholeter* (one-toed amphiuma). Herpetological Review 16:31.

Carfioli, H.M., Tiebout III, S.A. Pagano, K.M. Heister and F.C. Lutcher. 2000. Monitoring *Plethodon cinereus* populations: field tests of experimental coverboard designs. Pp. 463–475. *In* Bruce, R.C., R.G. Jaeger and L.D. Houck (Eds.), The Biology of Plethodontid Salamanders. Kluwer Academic/Plenum Publishers, New York.

Cargo, D.G. 1960. Predation of eggs of the spotted salamander, *Ambystoma maculatum*, by the leech, *Macrobdella decora*. Chesapeake Science 1:119–120.

Carl, G.C. 1942. The long-toed salamander on Vancouver Island. Copeia 1942:56.

Carl, G.C. 1943. The amphibians of British Columbia. British Columbia Provincial Museum Handbook, Number 2, Vancouver, British Columbia.

Carl, G.C. 1945. Notes on some frogs and toads of British Columbia. Copeia 1945:52–53.

Carl, G.C. and I.M. Cowan. 1945. Notes on the salamanders of British Columbia. Copeia 1945:43–44.

Carlin, J.L. 1998. Genetic and morphological differentiation between *Eurycea longicauda longicauda* and *E. guttolineata* (Caudata: Plethodontidae). Herpetologica 53:206–217.

Carlson, D.L., L.A. Rollins-Smith and R.G. McKinnell. 1994a. The Lucké herpesvirus genome: its presence in neoplastic and normal kidney tissue. Journal of Comparative Pathology 110:349–355.

Carlson, D.L., W. Sauerbier, L.A. Rollins-Smith and R.G. McKinnell. 1994b. Fate of herpesvirus DNA in embryos and tadpoles cloned from Lucké renal carcinoma nuclei. Journal of Comparative Pathology 111:197–203.

Carlton, J.T. 1992. Introduced marine and estuarine mollusks of North America: an end-of-the-20th-century perspective. Journal of Shellfish Research 11:489–505.

Carmichael, G.J. 1983. Use of formalin to separate tadpoles from largemouth bass fingerlings after harvesting. Progressive Fish-Culturist 45:105–106.

Carpenter, C.C. 1952. Comparative ecology of the common garter snake (*Thamnophis s. sirtalis*), the ribbon snake (*Thamnophis s. sauritis*), and Butler's garter snake (*Thamnophis butleri*) in mixed populations. Ecological Monographs 22:235–258.

Carpenter, C.C. 1953. An ecological survey of the herpetofauna of the Grand Teton–Jackson Hole area of Wyoming. Copeia 1953:170–174.

Carpenter, C.C. 1954. A study of amphibian movement in Jackson Hole Wildlife Park. Copeia 1953:197–200.

Carpenter, C.C. 1955. Aposematic behavior in the salamander *Ambystoma tigrinum melanosticum*. Copeia 1955:311.

Carpenter, C.C. and D.E. Delzell. 1951. Road records as indicators of differential spring migrations of amphibians. Herpetologica 7:63–64.

Carpenter, C.C. and J.C. Gillingham. 1984. Giant centipede (*Scolopendra alternans*) attacks marine toad (*Bufo marinus*). Carribean Journal of Science 20:71–72.

Carpenter, C.C. and J.C. Gillingham. 1987. Water hole fidelity in the marine toad, *Bufo marinus*. Journal of Herpetology 21:158–161.

Carpenter, E. 1915. Ground water in Southeastern Nevada. Water-supply Paper 365. Department of the Interior, U.S. Geological Survey, Government Printing Office, Washington, D.C.

Carr, A.F., Jr. 1939. *Haideotriton wallacei*, a new subterranean salamander from Georgia. Occasional Papers of the Boston Society of Natural History 8:333–336.

Carr, A.F., Jr. 1940a. A Contribution to the Herpetology of Florida. University of Florida Biological Publications Science Series, Volume 3, Number 1, University of Florida Press, Gainesville, Florida.

Carr, A.F., Jr. 1940b. Dates of frog choruses in Florida. Copeia 1940:55.

Carr, A. and C.J. Goin. 1955. Guide to the Reptiles, Amphibians, and Freshwater Fishes of Florida. University of Florida Press, Gainesville, Florida.

Carr, A. and C.J. Goin. 1959. Guide to the Reptiles, Amphibians and Fresh-water Fishes of Florida. University of Florida Press, Gainesville, Florida.

Carr, A.H., R.L. Amborski, D.D. Culley Jr. and G.F. Amborski. 1976. Aerobic bacteria in the intestinal tracts of bullfrogs (*Rana catesbeiana*) maintained at low temperatures. Herpetologica 32:239–244.

Carr, B.A., J.P. Geaghen and D.D. Culley Jr. 1984. Effect on two types of light on survival, growth and ovary development of *Rana catesbeiana*. Aquaculture 40:163–169.

Carr, D.E. and D.H. Taylor. 1985. Experimental evaluation of population interactions among three sympatric species of *Desmognathus*. Journal of Herpetology 19:507–514.

Carr, J.F. and J.K. Hiltunen. 1965. Changes in the bottom fauna of western Lake Erie from 1930–1961. Limnology and Oceanography 10:551–569.

Carr, L.W. and L. Fahrig. 2001. Effect of road traffic on two amphibian species of differing vagility. Conservation Biology 15:1071–1078.

Carr, S.M., A.J. Brothers and A.C. Wilson. 1987. Evolutionary inferences from restriction maps of mitochondrial DNA from nine taxa of *Xenopus* frogs. Evolution 41:176–188.

Carreno, C.A. and R.N. Harris. 1998. Lack of nest defense behavior and attendance patterns in a joint nesting salamander, *Hemidactylium scutatum* (Caudata: Plethodontidae). Copeia 1998:183–189.

Carroll, A., E.L. Blankenship, M.A. Bailey and C. Guyer. 2000. An estimate of maximum local population density of Red Hills salamanders (*Phaeognathus hubrichti*). Amphibia-Reptilia 21:260–263.

Carson, R. 1962. Silent Spring. Houghton Mifflin Company, Boston, Massachusetts.

Carter, H.A. 1934. Georgia records of *Gastrophryne carolinensis* (Holbrook). Copeia 1934:138.

Cartwright, M.E., S.E. Trauth and J.D. Wilhide. 1998. Wood frog (*Rana sylvatica*) use of wildlife ponds in north-central Arkansas. Journal of the Arkansas Academy Science 52:32–34.

Casebere, D.R. and D.H. Taylor. 1976. Population dynamics in breeding populations of the American toad. American Zoologist 16:206.

Casey, T., G. Pritts and B.T. Miller. 1993. Preliminary investigations on the movement of the hellbender (*Cryptobranchus alleganiensis*) in middle Tennessee. Journal of the Tennessee Academy of Science 68:55.

Cash, M.N. and J.P. Bogart. 1978. Cytological differentiation of the diploid-tetraploid species pair of North American treefrogs (Amphibia, Anura, Hylidae). Journal of Herpetology 12:555–558.

Cash, W.B. 1994. Herpetofaunal diversity of a temporary wetland in the southeast Atlantic Coastal Plain. Master's thesis. Georgia Southern University, Statesboro, Georgia.

Casper, G.S. 1996. Geographic distributions of the amphibians and reptiles of Wisconsin—An interim report of the Wisconsin Herpetological Atlas project. Milwaukee Public Museum, Milwaukee, Wisconsin.

Casper, G.S. 1997. Recent amphibian and reptile distribution records for Wisconsin. Herpetological Review 28:214–216.

Casper, G.S. 1998. Review of the status of Wisconsin amphibians. Pp. 199–205. *In* Lannoo, M.J. (Ed.), Status and Conservation of Midwestern Amphibians. University of Iowa Press, Iowa City, Iowa.

Cassells, M. 1970. Another predator of the cane toad. Journal of the North Queensland Naturalists' Club, 37:6.

Castanet, J. 1996. Age estimation in desmognathine salamanders assessed by skeletochronology. Herpetologica 52:160–171.

Castanet, J., H. Francillon-Vieillot and R.C. Bruce. 1996. Age estimation in desmognathine salamanders assessed by skeletochronology. Herpetologica 52:160–171.

Casterlin, M.E. and W.W. Reynolds 1978. Behavioral thermoregulation in *Rana pipiens* tadpoles. Journal of Thermal Biology 3:143–146.

Castillo, L.E., E. de la Cruz and C. Ruepert. 1997. Ecotoxicology and pesticides in tropical aquatic ecosystems of Central America. Environmental Toxicology and Chemistry 16:41–51.

Catalano, P.A. and A.M. White. 1977. New host records for *Haematoloechus complexus* (Seely) Krull, 1933 from *Hyla crucifer* and *Rana sylvatica*. Ohio Journal of Science 77:99.

Catling, P., A. Hertog, R.J. Burt, J.C. Wombey and R.I. Forrester. 1999. The short-term effect of cane toads, *Bufo marinus*, on native fauna in the Gulf Country of the Northern Territory. Wildlife Research 26:161–185.

Caughley, G. 1994. Directions in conservation biology. Journal of Animal Ecology 63:215–244.

Caughley, G. and A. Gunn. 1996. Conservation Biology in Theory and Practice. Blackwell Science, Cambridge, Massachusetts.

Causey, O.R. 1939a. *Aedes* and *Culex* mosquitoes as intermediate hosts of frog filaria *Foleyella* sp. American Journal of Hygiene 29:79–81.

Causey, O.R. 1939b. The development of frog filaria larvae *Foleyella ranae* in *Aedes* and *Culex* mosquitoes. American Journal of Hygiene 29:131–132.

Causey, O.R. 1939c. Development of the larval stages of *Foleyella brachyoptera* in mosquitoes. American Journal of Hygiene 30:69–71.

Ceballos, G., P. Rodriguez and R.A. Medellin. 1998. Assessing conservation priorities in megadiverse Mexico: mammalian diversity, endemicity, and endangerment. Ecological Applications 8:8–17.

Cecil, S.G. and J.J. Just. 1979. Survival rate, population density and development of a naturally occurring anuran larvae *(Rana catesbeiana)*. Copeia 1979:447–453.

Cei, J.M., V. Erspamer and M. Rosenghini. 1968. Taxonomic and evolutionary significance of biogenic amine and polypeptides in amphibian skin. II. Toads of the genera *Bufo* and *Melanophyrniscus*. Systematic Zoology 17:232–245.

Cely, J.E. and J.S. Sorrow Jr. 1983. Distribution, status and habitat of the pine barrens treefrog in South Carolina. South Carolina Wildlife and Marine Resources Department, Study Completion Report, Division of Wildlife and Freshwater Fisheries, Columbia, South Carolina.

Cely, J.E. and J.S. Sorrow Jr. 1986. Distribution and habitat of *Hyla andersonii* in South Carolina. Journal of Herpetology 20:102–104.

Center for North American Herpetology. 2002. http://www.naherpetology.org. March, Lawrence, Kansas.

Chadwick, C.S. 1940. Some notes on the burrows of *Plethodon metcalfi*. Copeia 1940:50.

Chadwick, C.S. 1944. Observations on the life cycle of the common newt in western North Carolina. American Midland Naturalist 32:491–494.

Chadwick, C.S. 1950. Observations on the behavior of the larvae of the common American newt during metamorphosis. American Midland Naturalist 43:392–398.

Chamberlain, T.C. 1897. Studies for students: the method of multiple working hypotheses. Journal of Geology 5:837–848.

Chamberlin, K.M. 1990. Wyoming toad: 1990 field surveys. Fish Division Administrative Report, Contract 5090-30-8301, Wyoming Game and Fish Department, Cheyenne, Wyoming.

Chandler, A.C. 1918. The western newt or water-dog *(Notophthalmus torosus)*, a natural enemy of mosquitos. Oregon Agricultural College Experimental Station Bulletin 152, Corvallis, Oregon.

Chandler, A.C. 1923. Three new trematodes from *Amphiuma means*. Proceedings of the United States National Museum, Number 63, Article 3:1–7, Washington, D.C.

Chandler, A.C. 1942. The morphology and life cycle of a new strigeid, *Fibricola texensis*, parasitic in raccoons. Transactions of the American Microscopical Society 61:156–167.

Chaney, A.H. 1949. The life history of *Desmognathus fuscus auriculatus*. Master's thesis. Tulane University, New Orleans, Louisiana.

Chaney, A.H. 1951. The food habits of the salamander *Amphiuma tridactylum*. Copeia 1951:45–49.

Chaney, A.H. 1958. A comparison of populations of Louisiana and Arkansas populations of *Desmognathus fuscus*. Ph.D. dissertation. Tulane University, New Orleans, Louisiana.

Chang, J.C.W. 1994. Multiple spawning in a female *Rana rugosa*. Japanese Journal of Herpetology, 15:112–115.

Chapel, W.L. 1939. Field notes on *Hyla wrightorum* Taylor. Copeia 1939:225–227.

Chapin, F.S., III, O.E. Sala, I.C. Burke, J.P. Grime, D.U. Hooper, W.K. Laurenroth, A. Lombard, H.A. Mooney, A.R. Mosier, S. Naeem, S.W. Pacala, J. Roy, W.L. Steffen and D. Tilman. 1998. Ecosystem consequences of changing biodiversity. BioScience 48:45–42.

Chapin, P. and S.E. Trauth. 1987. The *Hyla versicolor/chrysoscelis* species complex of gray treefrogs in Arkansas: histological and ultrastructural evidence. Proceedings of the Arkansas Academy of Science 41:20–23.

Charlebois, G. 1977. Carabidae in a fecal pellet from the American toad *Bufo americanus*. Cordulia 3:167.

Chase, J.M. 1998. Size-structured interactions and multiple domains of attraction in pond food webs. Ph.D. dissertation. University of Chicago, Chicago, Illinois.

Chazal, A.C., J.D. Krenz and D.E. Scott. 1996. Relationship of larval density and heterozygosity to growth and survival of juvenile marbled salamanders *(Ambystoma opacum)*. Canadian Journal of Zoology 74:1122–1129.

Cheek, A.O., C.F. Ide, J.E. Bollinger, C.V. Rider and J.A. McLachlan. 1999. Alteration of leopard frog *(Rana pipiens)* metamorphosis by the herbicide acetochlor. Archives of Environmental Contamination and Toxicology 37:70–77.

Chen, C. and M.V. Osuch. 1969. Biosynthesis of bufadienolides-3-hydroxycholanates as precursors in *Bufo marinus* bufadienolides synthesis. Biochemistry and Pharmacology 18:1797–1802.

Chen, G.J. and S.S. Desser. 1989. The Coccidia (Apicomplexa: Eimeriidae) of frogs from Algonquin Park, with descriptions of two new species. Canadian Journal of Zoology 67:1686–1689.

Chen, J., J.F. Franklin and T.A. Spies. 1993. Contrasting microclimates among clearcut, edge, and interior of old-growth Douglass fir forests. Agriculture and Forest Meteorology 63:219–237.

Chen, J., J.F. Franklin and T.A. Spies. 1995. Growing-season microclimatic gradients from clear-cut edges into old-growth Douglass fir forests. Ecological Applications 5:74–86.

Cheng, T.C. 1961. Description, life history and developmental pattern of *Glypthelmins pennsylvaniensis* n. sp. (Trematoda: Brachycoeliidae), new parasite of frogs. Journal of Parasitology 47:469–477.

Chermock, R.L. 1952. A key to the amphibians and reptiles of Alabama. Geological Survey of Alabama, Museum Paper 33, Tuscaloosa, Alabama.

Childs, H.E., Jr. 1953. Selection by predation on albino and normal spadefoot toads. Evolution 7:228–233.

Chippindale, P.T. 1995. Evolution, phylogeny, biogeography, and taxonomy of central Texas spring and cave salamanders, *Eurycea* and *Typhlomolge* (Plethodontidae: Hemidactyliini). Ph.D. dissertation. University of Texas, Austin, Texas.

Chippindale, P.T. 2000. Species boundaries and species diversity in the central Texas hemidactyliine plethodontid salamanders, genus *Eurycea*. Pp. 149–165. *In* Bruce, R.C., R.G. Jaeger and L.D. Houck (Eds.), The Biology of Plethodontid Salamanders. Kluwer Academic/Plenum Publishers, New York.

Chippindale, P.T., D.M. Hillis and A.H. Price. 1990. Molecular studies of Edwards Plateau neotenic salamanders, *Eurycea* and *Typhlomolge*. U.S. Fish and Wildlife Service, Region 2, Section 6 Performance Report, Project E-1-2, Job 3.4, Albuquerque, New Mexico.

Chippindale, P.T., A.H. Price and D.M. Hillis. 1993. A new species of perennibranchiate salamander (*Eurycea*, Plethodontidae) from Austin, Texas. Herpetologica 49:248–259.

Chippindale, P.T., A.H. Price and D.M. Hillis. 1998. Systematic status of the San Marcos salamander, *Eurycea nana* (Caudata: Plethodontidae). Copeia 1998:1046–1049.

Chippindale, P.T., A.H. Price, J.J. Wiens and D.M. Hillis. 2000. Phylogenetic relationships and systematic revision of central Texas hemidactyliine plethodontid salamanders. Herpetological Monographs 14:1–80.

Chitty, D. 1967. The natural selection of self-regulatory behaviour in animal populations. Proceedings of the Ecological Society of Australia 2:51–78.

Chivers, D.P. and J.M. Kiesecker. 1996. Avoidance response of a terrestrial salamander *(Ambystoma macrodactylum)* to chemical alarm cues. Journal of Chemical Ecology 22:1709–1716.

Chivers, D.P., J.M. Kiesecker, E.L. Wildy, M.T. Anderson and A.R. Blaustein. 1997. Chemical alarm signaling in terrestrial salamanders: intra- and interspecific responses. Journal of Herpetology 30:184–191.

Chivers, D.P., J.M. Kiesecker, A. Marco, J. DeVito, M.T. Anderson and A.R. Blaustein. 2001. Predator-induced life history changes in amphibians: egg predation induces hatching. Oikos 92:135–142.

Choi, I.-H., S.H. Lee and R.E. Ricklefs. 1999. Effectiveness and ecological implications of anuran defenses against snake predators. Korean Journal of Biological Science 3:247–252.

Chrapliwy, P.S. and K.L. Williams. 1957. A species of frog new to the fauna of the United States: *Pternohyla fodiens* Boulenger. Natural History Miscellanea, Number 160, Chicago Academy of Sciences, Chicago, Illinois.

Christein, D. and D.H. Taylor. 1978. Population dynamics in breeding aggregations of the American toad, *Bufo americanus* (Amphibia, Anura, Bufonidae). Journal of Herpetology 12:17–24.

Christensen, N.L. 1981. Fire regimes in southeastern ecosystems. Pp. 112–136. *In* Mooney, H.A., T.M. Bonnicksen, N.L. Christensen, J. E. Loatan and W.A. Reiners (Eds.), Proceedings of the Conference on Fire Regimes and Ecosystem Properties. U.S.D.A. Forest Service, General Technical Report, WO-26, Washington, D.C.

Christensen, N.L. 1988. Vegetation of the southeastern Coastal Plain. Pp. 317–363. *In* Barbour, M.G. and W.D. Billings (Eds.), North American Terrestrial Vegetation. Cambridge University Press, New York.

Christian, F.A. and L.L. White. 1973. The genus *Allassostomoides* Stunkard, 1924, with description of *Allassostomoides louisianaensis* n. sp. (Trematoda: Paramphistomidae) from the pig frog, *Rana grylio*, in Louisiana. American Midland Naturalist 90:218–220.

Christian, K.A. 1982. Changes in the food niche during postmetamorphic ontogeny of the frog *Pseudacris triseriata*. Copeia 1982:73–80.

Christiansen, J. and D. Penney. 1973. Anaerobic glycolysis and lactic acid accumulation in cold submerged *Rana pipiens*. Journal of Comparative Physiology 87:237–245.

Christiansen, J.L. 1981. Population trends in Iowa's amphibians and reptiles. Proceedings of the Iowa Academy of Science 88:24–27.

Christiansen, J.L. 1998. Perspectives on Iowa's declining amphibians and reptiles. Proceedings of the Iowa Academy of Science 105:109–114.

Christiansen, J.L. and R.M. Bailey. 1991. The salamanders and frogs of Iowa. Iowa Department of Natural Resources, Nongame Technical Series, Number 3, Des Moines, Iowa.

Christiansen, J.L. and H. Feltman. 2000. A relationship between trematode metacercariae and bullfrog limb abnormalities. Proceedings of the Iowa Academy of Science 107:79–85.

Christiansen, J.L. and C.M. Mabry. 1985. The amphibians and reptiles of Iowa's Loess Hills. Proceedings of the Iowa Academy of Science 92:159–163.

Christiansen, J.L. and C. Van Gorp. 1998. Iowa's declining amphibians. P. 7. *In* Abstracts, Midwest Declining Amphibians Conference, March 20–21. Milwaukee Public Museum, Milwaukee, Wisconsin.

Christman, S.P. 1959. Sound production in newts. Herpetologica 15:13.

Christman, S.P. 1970. *Hyla andersonii* in Florida. Quarterly Journal of the Florida Academy of Science 33:80.

Christman, S.P. 1974. Geographic variation for salt water tolerance in the frog *Rana sphenocephala*. Copeia 1974:773–778.

Christman, S.P. and L.R. Franz. 1973. Feeding habits of the striped newt, *Notophthalmus perstriatus*. Journal of Herpetology 7:133–135.

Christman, S.P. and D.B. Means. 1992. Striped newt *Notophthalmus perstriatus* (Bishop) Pp. 62–65. *In* Moler, P.E. (Ed.), Rare and Endangered Biota of Florida. Volume III. Amphibians and Reptiles. University Press of Florida, Gainesville, Florida.

Christman, S.P., H.W. Campbell, C.R. Smith and H.I. Kochman. 1979a. Geographic distribution: *Rana virgatipes* (carpenter frog). Herpetological Review 10:59.

Christman, S.P., H.W. Campbell, C.R. Smith and H.I. Kockman. 1979b. Geographic distribution: *Stereochilus marginatus* (many-lined salamander). Herpetological Review 10:59.

Churchill, T.A. and K.B. Storey. 1996. Organ metabolism and cryoprotectant synthesis during freezing in spring peepers *Pseudacris crucifer*. Copeia 1996:517–525.

CITIES (Convention on International Trade in Endangered Species of Wild Fauna and Flora). 2000. Amendments to Appendices I and II. Proposal 11.45. Eleventh Meeting of the Conference of the Parties, April 10–20, 2000, Nairobi, Kenya.

City of Austin. 1997. Barton Creek report. Unpublished Report, City of Austin, Texas.

Clanton, W. 1934. An unusual situation in the salamander *Ambystoma jeffersonianum* (Green). Occasional Papers of the Museum of Zoology, Number 290, University of Michigan, Ann Arbor, Michigan.

Clark, D.E., Jr., R. Cantu, D.F. Cowman and D.J. Maxson. 1998. Uptake of arsenic and metals by tadpoles at a historically contaminated Texas site. Ecotoxicology 7:61–67.

Clark, D.R., Jr. 1974. The western ribbon snake *(Thamnophis proximus)*: ecology of a Texas population. Herpetologica 30:372–379.

Clark, G.W., J. Bradford and R. Nussbaum. 1969. Blood parasites of some Pacific Northwest amphibians. Bulletin of the Wildlife Disease Association 5:117–118.

Clark, H.F. and L. Diamond. 1971. Comparative studies on the interaction of benzpyrene with cells derived from poikilothermic and homeothermic vertebrates. Part 2. Effect of temperature on benzpyrene metabolism and cell multiplication. Journal of Cell Physiology 77:385–392.

Clark, K.H. 1998. Grafton Ponds Natural Area Preserve resource management plan. Virginia Department of Conservation and Recreation, Natural Heritage Technical Report, 98-4, Division of Natural Heritage, Richmond, Virginia.

Clark, K.L. 1986a. Distributions of anuran populations in central Ontario relative to habitat acidity. Water, Air, and Soil Pollution 30:727–734.

Clark, K.L. 1986b. Responses of *Ambystoma maculatum* populations in central Ontario to habitat acidity. Canadian Field-Naturalist 100:463–469.

Clark, K.L. and R.J. Hall. 1985. Effects of elevated hydrogen ion and aluminum concentrations on the survival of amphibian embryos and larvae. Canadian Journal of Zoology 63:116–123.

Clark, K.L. and B.D. LaZerte. 1985. A laboratory study of the effects of aluminum and pH on amphibian eggs and tadpoles. Canadian Journal of Fisheries and Aquatic Sciences 42:1544–1551.

Clark, K.L. and B.D. LaZerte. 1987. Intraspecific variation in hydrogen ion and aluminum toxicity in *Bufo americanus* and *Ambystoma maculatum*. Canadian Journal of Fisheries and Aquatic Sciences 44:1622–1628.

Clark, R.J., C.R. Peterson and P.E. Bartelt. 1993. The distribution, relative abundance, and habitat associations of amphibians on the Targhee National Forest. Final report, Targhee National Forest, Victor, Idaho.

Clark, T.W. and J.H. Seebeck (Eds.). 1990. The management and conservation of small populations. Chicago Zoological Society, Brookfield, Illinois.

Clarke, K.C. 1997. Getting Started with Geographic Information Systems. Second edition. Prentice Hall Series in Geographic Information Science. K.C. Clarke, Series Editor. Prentice Hall, Upper Saddle River, New Jersey.

Clarke, K.R. and R.M. Warwick. 1998. A taxonomic distinctness index and its statistical properties. Journal of Applied Ecology 35:523–531.

Clarke, R.D. 1972. The effects of toe clipping on survival in Fowler's toad, *Bufo woodhousei fowleri*. Copeia 1972:182–185.

Clarke, R.D. 1973. The autecology of Fowler's toad *(Bufo woodhousei fowleri)*. Ph.D. dissertation. Yale University, New Haven, Connecticut.

Clarke, R.D. 1974a. Activity and movement patterns in a population of Fowler's toad, *Bufo woodhousei fowleri*. American Midland Naturalist 92:257–274.

Clarke, R.D. 1974b. Postmetamorphic growth rates in a natural population of Fowler's toad, *Bufo woodhousei fowleri*. Canadian Journal of Zoology 52:1489–1498.

Clarke, R.D. 1974c. Food habits of toads, genus *Bufo* (Amphibia, Bufonidae). American Midland Naturalist 91:140–147.

Clarke, R.D. 1977. Postmetamorphic survivorship of Fowler's toad, *Bufo woodhousei fowleri*. Copeia 1977:594–597.

Clarkson, R.B. 1964. Tumult on the Mountains: Lumbering in West Virginia, 1770–1920. McClain Printing Company, Parson, West Virginia.

Clarkson, R.W. and J.C. deVos Jr. 1986. The bullfrog, *Rana catesbeiana* Shaw, in the Lower Colorado River, Arizona–California. Journal of Herpetology 20:42–49.

Clarkson, R.W. and J.C. Rorabaugh. 1989. Status of leopard frogs (*Rana pipiens* complex) in Arizona and southeastern California. Southwestern Naturalist 34:531–538.

Clausnitzer, H.-J. 1987. Gefährdung des Moorfrosches (*Rana arvalis* Nilsson) durch Versauerung der Laichgewässer. Beihefte zur Schriftenreihe für Naturschutz und Landschaftspflege in Niedersachsen 19:131–137.

Claussen, D.L. 1973. The thermal relations of the tailed frog, *Ascaphus truei*, and the Pacific treefrog, *Hyla regilla*. Comparative Biochemistry and Physiology A 44:137–153.

Claussen, D.L. and J.R. Layne Jr. 1983. Growth and survival of juvenile toads, *Bufo woodhousei*, maintained on different diets. Journal of Herpetology 17:107–112.

Clay, W.M., R.B. Case and R. Cunningham. 1955. On the taxonomic status of the slimy salamander, *Plethodon glutinosus* (Green), in southeastern Kentucky. Transactions of the Kentucky Academy of Science 16:57–65.

Clayton, D.R., L.M. Ollivier and H.H. Welsh Jr. 1999. Management recommendations for the Siskiyou Mountain salamander *(Plethodon stormi)*. Interagency Publication of the Regional Ecosystem Office, Portland, Oregon.

Clegg, M. and 55 others. 1995. Science and the Endangered Species Act. National Academy Press, Washington, D.C.

Clemmer, G.H. 1995. Conservation status of *Bufo nelsoni*, the Amargosa toad, in Oasis Valley, Nevada. Bureau of Land Management, Reno, Nevada (available from Nevada Natural Heritage Program, Carson City, Nevada).

Cleveland, W.S. 1979. Robust locally weighted regression and smoothing scatterplots. Journal of the American Statistical Association 74:829–836.

Cliburn, J.W. 1965. A key to the amphibians and reptiles of Mississippi with guides to their study. State Wildlife Museum, Jackson, Mississippi.

Cliburn, J.W. and A.B. Porter. 1986. Comparative climbing abilities of the salamanders *Aneides aeneus* and *Plethodon glutinosus* (Caudata, Plethodontidae). Journal of the Mississippi Academy of Science 31:91–96.

Cliburn, J.W. and A.B. Porter. 1987. Vertical stratification of the salamanders *Aneides aeneus* and *Plethodon glutinosus* (Caudata, Plethodontidae). Journal of the Alabama Academy of Science 58:1–10.

Climate Resources Board. 1979. Carbon dioxide and climate: a scientific assessment. Report of an Adhoc Study Group on Carbon Dioxide and Climate. National Academy of Sciences, Washington, D.C.

Cline, G.R. 1986. Miscellaneous accounts of some herptiles from Ottawa County, Oklahoma. Bulletin of the Oklahoma Herpetological Society 11:32–33.

Cline, G.R. 1990. Evolution in the gray treefrog complex: acoustical morphological, and genetic variation in midwestern populations. Ph.D. dissertation. Oklahoma State University, Stillwater, Oklahoma.

Clovis, J.F. 1979. Tree importance values in West Virginia red spruce forests inhabited by the Cheat Mountain salamander. Proceedings of the West Virginia Academy of Science 51:58–64.

Coates, D. and T.A. Redding-Coates. 1981. Ecological problems associated with irrigation canals in the Sudan with particular reference to the spread of bilharziasis, malaria and aquatic weeds and the ameliorative role of fishes. International Journal of Environmental Studies 16:207–212.

Coates, M. 1967. A comparative study of the serum proteins of the species of *Taricha* and their hybrids. Evolution 21:130–140.

Coates, M. 1970. An unusual aggregation of the newt *Taricha granulosa granulosa*. Copeia 1970:176–178.

Coatney, C.E., Jr. 1982. Home range and nocturnal activity of the Ozark hellbender. Master's thesis. Southwest Missouri State University, Springfield, Missouri.

Cochran, A.G. and P.A. Cochran. 2001. Geographic distribution: *Ambystoma maculatum* (spotted salamander). Herpetological Review 32:267.

Cochran, B.G. and C.D. Cochran. 1989. Geographic distribution: *Ambystoma texanum* (smallmouth salamander). Herpetological Review 20:11.

Cochran, D.M. 1961. Type specimens of Reptiles and Amphibians in the United States National Museum Smithsonian Institution. United States National Museum Bulletin, Number 220, Washington, D.C.

Cochran, D.M. and C.J. Goin. 1970. The New Field Book of Reptiles and Amphibians. G.P. Putnam's Sons, New York.

Cochran, P.A. 1982. Life history notes: *Rana pipiens* (leopard frog). Predation. Herpetological Review 13:45–46.

Cochran, P.A. 1983. Life history notes: *Rana pipiens* (leopard frog). Predation. Herpetological Review 14:18–19.

Cochran, P.A. 1986a. The herpetofauna of the Weaver Dunes, Wabasha County, Minnesota. Prairie Naturalist 18:143–150.

Cochran, P.A. 1986b. Some observations on the herpetofauna of the University of Wisconsin–Madison campus. Bulletin of the Chicago Herpetological Society 21:34–35.

Cochran, P.A. and J.A. Cochran. 2003. Natural history notes: *Pseudacris crucifer* (spring peeper). Predation. Herpetological Review 34:360.

Cochran, P.A. and E.M. Goolish. 1980. Bullfrog eats bird revisited. Bulletin of the Chicago Herpetological Society 15:81.

Cochran, W.G. and G.M. Cox. 1957. Experimental Designs. Second edition. John Wiley and Sons, New York.

Cockran, M.E. 1911. The biology of the red-backed salamander (*Plethodon cinereus erythronotus* Green). Biological Bulletin 20:332–349.

Cockrum, L. 1941. Notes on *Siren intermedia*. Copeia 1941:265.

Coggins, J.R. and R.A. Sajdak. 1982. A survey of helminth parasites in the salamanders and certain anurans from Wisconsin. Proceedings of the Helminthological Society of Washington 49:99–102.

Cohen, M.P. and R.A. Alford. 1993. Growth, survival and activity patterns of recently metamorphosed *Bufo marinus*. Wildlife Research 20:1–13.

Cohen, M.P. and S.E. Williams. 1992. General ecology of the cane toad, *Bufo marinus*, and examination of direct effects on native frog choruses at heathlands, Cape York Peninsula. Cape York Peninsula Scientific Expedition 1992:243–245.

Cohen, N.W. 1952. Comparative rates of dehydration and hydration in some California salamanders. Ecology 33:462–479.

Cohen, N.W. 1975. California anurans and their adaptations. Terra 13:6–13.

Cohen, N.W. and W.E. Howard. 1958. Bullfrog food and growth at the San Joachin Experimental Range, California. Copeia 1958:223–225.

Coker, C.M. 1931. Hermit thrush feeding on salamanders. Auk 48:277.

Colborn, T., F.S. vom Saal and A.M. Soto. 1993. Developmental effects of endocrine-disrupting chemicals in wildlife and humans. Environmental Health Perspectives 101:378–383.

Cole, C.J. 1962. Notes on the distribution and food habits of *Bufo alvarius* at the eastern edge of its range. Herpetologica 18:172–175.

Cole, E.C., W.C. McComb, M. Newton, C.L. Chambers and J.P. Leeming. 1997. Response of amphibians to glyphosate application in the Oregon Coast Range. Journal of Wildlife Management 61:656–664.

Colinvaux, P. 1997. The history of forest on the isthmus from the Ice Age to the Present. Pp. 123–136. *In* Coates, A.G. (Ed.), Central America: A Natural and Cultural History. Yale University Press, New Haven, Connecticut.

Collazo, A. 1990. Development and evolution in the salamander family Plethodontidae. Ph.D. dissertation. University of California, Berkeley, California.

Collazo, A. and S.B. Marks. 1994. Development of *Gyrinophilus porphyriticus*: identification of the ancestral developmental pattern in the salamander family Plethodontidae. Journal of Experimental Zoology 268:239–258.

Collette, B.B. and F.R. Gehlbach. 1961. The salamander *Siren intermedia intermedia* LeConte in North Carolina. Herpetologica 17:203–204.

Colley, S.A., W.H. Keen and R.W. Reed. 1989. Effects of adult presence on behavior and microhabitat use of juveniles of a desmognathine salamander. Copeia 1989:1–7.

Collins, J.P. 1975. A comparative study of the life history strategies in a community of frogs. Ph.D. dissertation. University of Michigan, Ann Arbor, Michigan.

Collins, J.P. 1979. Intrapopulation variation in the body size at metamorphosis and timing of metamorphosis in the bullfrog, *Rana catesbeiana*. Ecology 60:738–749.

Collins, J.P. 1981. Distribution, habitats, and life history variation in the tiger salamander, *Ambystoma tigrinum*, in east-central and southeast Arizona. Copeia 1981:666–675.

Collins, J.P. 1996. A status survey of three species of endangered/sensitive amphibians in Arizona. Arizona Game and Fish Department Heritage Program, Project Report, Number I92014, Phoenix, Arizona.

Collins, J.P. and J.E. Cheek. 1983. Effects of food and density on development of typical and cannibalistic salamander larvae in *Ambystoma tigrinum nebulosum*. American Zoologist 23:77–84.

Collins, J.P. and J.R. Holomuzki. 1984. Intraspecific variation in diet within and between trophic morphs in larval tiger salamanders (*Ambystoma tigrinum nebulosum*). Canadian Journal of Zoology 62:168–174.

Collins, J.P. and M.A. Lewis. 1979. Overwintering tadpoles and breeding season variation in the *Rana pipiens* complex in Arizona. Southwestern Naturalist 24:371–373.

Collins, J.P. and H.M. Wilbur. 1979. Breeding habits and habitats of the amphibians of the Edwin S. George reserve Michigan and notes on the local distribution of fishes. Occasional Papers of the Museum of Zoology, Number 686, University of Michigan, Ann Arbor, Michigan.

Collins, J.P., P.J. Fernandez and M.J. Sredl. 1996. Aging two species of Arizona leopard frogs using skeletochronology. Annual meeting of the Southwestern United States Working Group of the Declining Amphibian Populations Task Force, Tucson, Arizona.

Collins, J.P., T.R. Jones and H.J. Berna. 1988. Conserving genetically distinctive populations: the case of the Huachuca tiger salamander (*Ambystoma tigrinum stebbensi* Lowe). Pp. 45–53. *In* Szaro, R.C., K.E. Severson and D.R. Patton (Tech. Coords.), Management of Amphibians, Reptiles, and Small Mammals in North America. U.S.D.A. Forest Service, Rocky Mountain Forest and Range Experiment Station, Fort Collins, Colorado.

Collins, J.P., J.B. Minton and B.A. Pierce. 1980. *Ambystoma tigrinum*: a multi-species conglomerate? Copeia 1980:938–941.

Collins, J.P., K.E. Zerba and M.J. Sredl. 1993. Shaping intraspecific variation: development, ecology and the evolution of morphology and life history variation in tiger salamanders. Genetica 89:167–183.

Collins, J.T. 1965. A population study of *Ambystoma jeffersonianum*. Journal of the Ohio Herpetological Society 5:61.

Collins, J.T. 1974. Amphibians and reptiles in Kansas. University of Kansas Publications, Museum of Natural History, Public Education Series, Number 1, Lawrence, Kansas.

Collins, J.T. 1979. Geographic distribution: *Gastrophryne olivacea* (Great Plains narrowmouth toad). Herpetological Review 10:101.

Collins, J.T. 1982. Amphibians and reptiles in Kansas. Second edition. University of Kansas Publications, Museum of Natural History, Public Education Series, Number 8, Lawrence, Kansas.

Collins, J.T. 1990. Standard common and current scientific names for North American amphibians and reptiles. Third edition. Herpetological Circular, Number 19, Society for the Study of Amphibians and Reptiles, St. Louis, Missouri.

Collins, J.T. 1991. Viewpoint: a new taxonomic arrangement for some North American amphibians and reptiles. Herpetological Review 22:42–43.

Collins, J.T. 1992. The evolutionary species concept. A reply to Van Devender et al. and Montanucci. Herpetological Review 23:43–46.

Collins, J.T. 1993. Amphibians and reptiles in Kansas. Third edition, revised. Natural History Museum, University of Kansas, Lawrence, Kansas.

Collins, J.T. 1997. Standard common and current scientific names for North American amphibians and reptiles. Fourth edition. Herpetological Circular, Number 25, Society for the Study of Amphibians and Reptiles, St. Louis, Missouri.

Collins, J.T. and S.L. Collins. 1991. Reptiles and amphibians of the Cimarron National Grasslands, Morton County, Kansas. U.S.D.A. Forest Service, Washington, D.C.

Collins, J.T. and T.W. Taggart. 2002. Standard common and current scientific names for North American amphibians, turtles, reptiles, and crocodilians. Fifth edition. Center for North American Herpetology, Lawrence, Kansas.

Collins, J.T., J.E. Huheey, J.L. Knight and H.M. Smith. 1978. Standard common and current scientific names for North American amphibians and reptiles. Herpetological Circular, Number 7, Society for the Study of Amphibians and Reptiles, St. Louis, Missouri.

Collins, J.T., R. Conant, J.E. Huheey, J.L. Knight, E.M. Rundquist and H.M. Smith. 1982. Standard common and current scientific names for North American amphibians and reptiles. Second edition. Herpetological Circular, Number 12, Society for the Study of Amphibians and Reptiles, St. Louis, Missouri.

Comeau, N.M. 1943. Une Ambystome nouvelle. Annales de la Association Canadienne-Francaise pour le Avancement des Sciences 9:124–125.

Comes, P. 1987. Qualitative und quantitative Bestandserfassung von Kreuzkröte (Bufo calamita) und Laubfrosch (Hyla arborea) in der Oberrheinebene zwischen Lörrach und Kehl. Beihefte zu denVeröffentlichungen für Naturschutz und Landschaftspflege in Baden-Württemberg 41:343–378.

Commission for Environmental Cooperation. 1997. Ecological regions of North America: Toward a common perspective. Commission for Environmental Cooperation, Montreal, Quebec, Canada.

Committee on the Status of Endangered Wildlife in Canada. 2002. Canadian species at risk. Canadian Wildlife Service, Environment Canada, Ottawa, Ontario.

Conant, R. 1938. The reptiles of Ohio. American Midland Naturalist 20:1–200.

Conant, R. 1956. Common names for North American amphibians and reptiles. Copeia 1956:172–185.

Conant, R. 1958a. A Field Guide to Reptiles and Amphibians of the United States and Canada east of the 100th Meridian. Houghton Mifflin Company, Boston, Massachusetts.

Conant, R. 1958b. Notes on the herpetology of the Delmarva Peninsula. Copeia 1958:50–52.

Conant, R. 1975. A Field Guide to Reptiles and Amphibians of Eastern and Central North America. Second edition. Houghton Mifflin Company, Boston, Massachusetts.

Conant, R. 1977a. Semiaquatic reptiles and amphibians of the Chihuahuan Desert and their relationships to drainage patterns of the region. Pp. 455–491. In Wauer, R.H. and D.H. Riskind (Eds.), Transactions of the Symposium on the Biological Resources of the Chihuahuan Desert Region United States and Mexico. U.S. Department of the Interior, National Park Service, Transactions and Proceedings Series, Number 3, Washington, D.C.

Conant, R. 1977b. The Florida water snake (Reptilia, Serpentes, Colubridae) established at Brownsville, Texas, with comments on other herpetological introductions in the area. Journal of Herpetology 11:217–220.

Conant, R. 1983. Comment on a frog and lizard newly recorded from central Durango, Mexico. Pp. 399–405. In Rhodin, A.G.J. and K. Miyata (Eds.), Advances in Herpetology and Evolutionary Biology: Essays in Honor of Ernest E. Williams. Museum of Comparative Zoology, Cambridge, Massachusetts.

Conant, R. and J.T. Collins. 1991. A Field Guide to Reptiles and Amphibians: Eastern and Central North America. Third edition. Houghton Mifflin Company, Boston, Massachusetts.

Conant, R. and J.T. Collins. 1998. A Field Guide to Amphibians and Reptiles: Eastern and Central North America. Third edition, expanded. Houghton Mifflin Company, Boston, Massachusetts.

Conaway, C.H. and D.E. Metter. 1967. Skin glands associated with breeding in Microhyla carolinensis. Copeia 1967:672–673.

Congdon, J.D., A.E. Dunham and R.C. Van Loben Sels. 1994. Demographics of common snapping turtles (Chelydra serpentina): implications for conservation and management of long-lived organisms. American Zoologist 34:397–408.

Connell, D.W. and G.J. Miller. 1984. Chemistry and Ecotoxicology of Pollution. John Wiley and Sons, New York.

Connell, J.H. 1983. On the prevalence and relative importance of interspecific competition: evidence from field experiments. American Naturalist 122:661–696.

Conrads, L.M. 1969. Demography and ecology of the fern bank salamander, Eurycea pterophila. Master's thesis. Southwest Texas State University, San Marcos, Texas.

Conroy, M. 1996. Abundance indices. Pp.179–192. In Wilson, D.E., F.R. Cole, J.D. Nichols, R. Rudran and M.S. Foster (Eds.), Measuring and Monitoring Biological Diversity: Standard Methods for Mammals. Smithsonian Institution Press, Washington, D.C.

Converse, K.A., L. Eaton-Poole and J. Mattsson. 1998. Field survey of Midwestern and Northeastern Fish and Wildlife Service lands for the presence of deformed amphibians. P. 8. In Abstracts, Midwest Declining Amphibians Conference, March 20–21. Milwaukee Public Museum, Milwaukee, Wisconsin.

Converse, K.A., J. Mattsson and L. Eaton-Poole. 2000. Field surveys of Midwestern and Northeastern Fish and Wildlife Service lands for the presence of abnormal frogs and toads. Journal of the Iowa Academy of Science 107:160–167.

Cook, D. and M.R. Jennings. 2001. Natural history notes: Rana aurora draytonii (California red-legged frog). Predation. Herpetological Review 32:182–183.

Cook, F.A. 1957. Salamanders of Mississippi. Mississippi Game and Fish Commission Museum, Jackson, Mississippi.

Cook, F.R. 1970. Rare or endangered Canadian amphibians and reptiles. Canadian Field-Naturalist 84:9–16.

Cook, F.R. 1980. Introduction to Canadian amphibians and reptiles. National Museum of Natural Sciences, National Museums of Canada, Ottawa, Ontario, Canada.

Cook, F.R. 1983. An analysis of toads of the Bufo americanus group in a contact zone in central northern North America. National Museum of Natural Sciences Publications in Natural Sciences, Number 3, National Museum of Canada, Ottawa, Ontario.

Cook, F.R. 1984. Introduction to Canadian amphibians and reptiles. National Museum of Natural Sciences, National Museums of Canada, Ottawa, Ontario.

Cook, F.R. and J.C. Cook. 1981. Attempted avian predation by a Canadian toad, Bufo americanus hemiophrys. Canadian Field-Naturalist 95:346–347.

Cook, M.L. and B.C. Brown. 1974. Variation in the genus Desmognathus (Amphibia: Plethodontidae) in the western limits of its range. Journal of Herpetology 8:93–105.

Cook, P. 1997. Water flows towards money: a short course on Texas water law. Beacon 7:1, 10, 12, 14.

Cook, R.P. 1983. Effects of acid precipitation on embryonic mortality of Ambystoma salamanders in the Connecticut Valley of Massachusetts. Biological Conservation 27:77–88.

Cook, R.P. 1989. And the voice of the grey tree frog was heard again in the land ... Park Science 9:6–7.

Cooke, A.S. 1970. The effects of pp'-DDT on tadpoles of the common frog (Rana temporaria). Environmental Pollution 1:57–71.

Coombs, T.L. 1979. Cadmium in aquatic organisms. Pages 93–139. In Webb, M. (Ed.), The Chemistry, Biochemistry, and Biology of Cadmium. Elsevier/North-Holland Biomedical Press, New York.

Cooper, E.L., R.K. Wright, A.E. Klempau and C.T. Smith. 1992. Hibernation alters a frog's immune system. Cryobiology 29:616–631.

Cooper, J.E. 1955. Notes on the amphibians and reptiles of southern Maryland. Maryland Naturalist 23:90–100.

Cooper, J.E. 1956. A Maryland hibernation site for reptiles. Herpetologica 12:238.

Cooper, J.E. 1956. Aquatic hibernation of the red-backed salamander. Herpetologica 1956:165–166.

Cooper, J.E. 1960. The mating antic of the long-tailed salamander. Maryland Naturalist 30:17–18.

Cooper, J.E. 1961. Cave records for the salamander Plethodon r. richmondi Pope, with notes on additional cave-associated species. Herpetologica 17:250–255.

Cooper, J.E. 1962. Stereochilus marginatus in southeastern South Carolina. Copeia 1962:212.

Cooper, J.E. 1968. The salamander *Gyrinophilus palleucus* in Georgia with notes on Alabama and Tennessee populations. Journal of the Alabama Academy of Science 39:182–185.

Cooper, J.E. and R.E. Ashton Jr. 1985. The *Necturus lewisi* study: introduction, selected literature review, and comments on the hydrologic units and their faunas. Brimleyana 10:1–12.

Cooper, J.E. and M.R. Cooper. 1968. Cave-associated herpetozoa II: salamanders of the genus *Gyrinophilus* in Alabama caves. Bulletin of the National Speleological Society 30:19–24.

Cooper, J.G. 1860. Report on explorations and surveys from the Mississippi River to the Pacific Ocean. 36th Congress, First Session, House Executive Document 56, Volume 12 (Book 2, Part 3, Number 4), Washington, D.C.

Cooper, S.D. 1988. The responses of aquatic insects and tadpoles to trout. International Vereinigung für theoretische und angewandte Limnologie, Verhandlungen 23:1858–1861.

Cooper, S.D., T.L. Dudley and N. Hemphill. 1986. The biology of chaparral streams in Southern California. Pp. 139–152. *In* DeVries, J. (Ed.), Proceedings of the Chaparral Ecosystem Research Conference. Report Number 62, California Water Resource Center, Davis, California.

Cope, E.D. 1862. On some new and little known American Anura. Proceedings of the Academy of Natural Sciences of Philadelphia 14:151–159.

Cope, E.D. 1863. On *Trachycephalus*, *Scaphiopus* and other American Batrachia. Proceedings of the Academy of Natural Sciences of Philadelphia 15:43–54.

Cope, E.D. 1865. Third contribution to the herpetology of Tropical America. Proceedings of the Academy of Natural Sciences of Philadelphia 17:185–198.

Cope, E.D. 1866a. On the structures and distribution of the genera of the arciferous Anura. Proceedings of the Academy of the Natural Sciences of Philadelphia, Series 2, 6:67–112.

Cope, E.D. 1866b. Fourth contribution to the herpetology of Tropical America. Proceedings of the Academy of the Natural Sciences of Philadelphia 18:123–132.

Cope, E.D. 1867 (1868). A review of the species of the Ambystomatidae. Proceedings of the Academy of the Natural Sciences of Philadelphia 19:166–211.

Cope, E.D. 1867. On the Reptilia and Batrachia of the Sonoran province of the Nearctic region. Proceedings of the Academy of Natural Science of Philadelphia 18:300–314.

Cope, E.D. 1869. A review of the species of the Plethodontidae and Desmognathidae. Proceedings of the Academy of Natural Sciences of Philadelphia 21:93–118.

Cope, E.D. 1875a. On the Batrachia and Reptilia of Costa Rica. Journal of the Academy of Natural Sciences of Philadelphia (1874–1881) 2:93–154.

Cope, E.D. 1875b. *Rana onca*, sp. nov., Pp. 528–529. *In* Yarrow, H.C. (Ed.), Report upon the collections of batrachians and reptiles made in portions of Nevada, Utah, California, Colorado, New Mexico, and Arizona, during the years 1871, 1872, 1873, 1874. Report upon the U.S. Geographical Explorations and Surveys west of the One Hundredth Meridian 5:509–589.

Cope, E.D. 1875c. Report upon the collections of batrachians and reptiles made in portions of Nevada, Utah, California, Colorado, New Mexico, and Arizona during the years 1871, 1872, 1873, and 1874 by D.H.C. Yarrow. P. 522. *In* Yarrow, D.H.C. (Ed.), U.S. Geographic Survey West of the 100th Meridian. Washington, D.C.

Cope, E.D. 1877. Tenth contribution to the herpetology of tropical America. Proceedings of the American Philosophical Society 17:85–98.

Cope, E.D. 1878a. A new genus of Cystignathidae from Texas. American Naturalist 12:253.

Cope, E.D. 1878b. A Texan cliff frog. American Naturalist 12:186.

Cope, E.D. 1879. Eleventh contribution to the herpetology of tropical America. Proceedings of the American Philosophical Society 18:261–277.

Cope, E.D. 1880. On the zoological position of Texas. Bulletin of the United States National Museum 17:1–51.

Cope, E.D. 1883. Notes on the geographic distribution of Batrachia and Reptilia in western North America. Proceedings of the Academy of the Natural Sciences of Philadelphia 35:10–35.

Cope, E.D. 1886. Synonymic list of the North American species of *Bufo* and *Rana*, with description of some species of Batrachia, from specimens in the National Museum. Proceedings of the American Philosophical Society 23:514–526.

Cope, E.D. 1889. The Batrachia of North America. United States National Museum Bulletin Number 34, Washington, D.C. (reprinted in 1963 by Eric Lundberg, Ashton, Maryland).

Cope, E.D. 1891. A new species of frog from New Jersey. American Naturalist 25:1017–1019.

Cope, E.D. 1894. On the Batrachia and Reptilia of the Plains at latitude 36° 30′. Proceedings of the Academy of the Natural Sciences of Philadelphia 45:386–387.

Cope, E.D. and A.S. Packard Jr. 1881. The fauna of Nickajack Cave. American Naturalist 15:877–882.

Corbett, K.F. 1988. Distribution and status of the sand lizard, *Lacerta agilis agilis*, in Britain. Mertensiella 1:92–100.

Corbit, C.D. 1960. Range extension of *Ascaphus truei* in Idaho. Copeia 1960:240.

Corin, C.W. 1976. The land vertebrates of the Huron Islands, Lake Superior. Jack-pine Warbler 54:138–147.

Corkran, C.C. and C. Thoms. 1996. Amphibians of Oregon, Washington and British Columbia: A Field Identification Guide. Lone Pine Publishing, Edmonton, Alberta.

Corliss, J.O. 1954. The literature on *Tetrahymena*: its history, growth, and recent trends. Protozoology 1:156–169.

Corn, P.S. 1991. Population ecology of the Wyoming toad, *Bufo hemiophrys baxteri*: Interim report (results 1990). U.S. Fish and Wildlife Service, Fort Collins, Colorado.

Corn, P.S. 1992. Population ecology of the Wyoming toad, *Bufo hemiophrys baxteri*: 1991 results. U.S. Fish and Wildlife Service, Fort Collins, Colorado.

Corn, P.S. 1993a. Population ecology of the Wyoming toad, *Bufo hemiophrys baxteri*: 1992 results. U.S. Fish and Wildlife Service, Fort Collins, Colorado.

Corn, P.S. 1993b. Life history notes: *Bufo boreas* (boreal toad). Predation. Herpetological Review 24:57.

Corn, P.S. 1994a. What we know and don't know about amphibian declines in the west. Pp. 59–67. *In* Covington, W.W. and L.F. DeBano (Eds.), Sustainable Ecological Systems: Implementing an Ecological Approach to Land Management. U.S.D.A. Forest Service, General Technical Report, RM-247, Fort Collins, Colorado.

Corn, P.S. 1994b. Straight-line drift fences and pitfall traps. Pp. 109–117. *In* Heyer, W.R., M.A. Donnelly, R.W. McDiarmid, L.-A.C. Hayek and M.S. Foster (Eds.), Measuring and Monitoring Biological Diversity: Standard Methods for Amphibians. Smithsonian Institution Press, Washington, D.C.

Corn, P.S. 1998. Effects of ultraviolet radiation on boreal toads in Colorado. Ecological Applications 8:18–26.

Corn, P.S. 2000. Amphibian declines: a review of some current hypotheses. Pp. 639–672. *In* Sparling, D.W., G. Linder and C.A. Bishop (Eds.), Ecotoxicology of Amphibians and Reptiles. Society for Environmental Toxicology and Contaminants (SETAC) Press, Pensacola, Florida.

Corn, P.S. and R.B. Bury. 1989a. Logging in western Oregon: responses of headwater habitats and stream amphibians. Forest Ecology and Management 29:39–57.

Corn, P.S. and R.B. Bury. 1989b. Small mammals and other prey in the diet of the Pacific giant salamander (*Dicamptodon ensatus*). American Midland Naturalist 87:524–526.

Corn, P.S. and R.B. Bury. 1990. Sampling methods for terrestrial amphibians and reptiles. Pp. 1–34. *In* Carey, A.B. and L.F. Ruggiero (Eds.), Wildlife-Habitat Relationships: Sampling Procedures for Pacific Northwest Vertebrates. U.S.D.A. Forest Service, General Technical Report, PNW-GTR-256, Pacific Northwest Research Station, Portland, Oregon.

Corn, P.S. and R.B. Bury. 1991. Terrestrial amphibian communities in the Oregon Coast Range. Pp. 304–317. *In* Ruggiero, L.F., K.B. Aubry, A.B. Carey and M.H. Huff (Tech. Coords.), Wildlife and Vegetation of Unmanaged Douglas-fir Forests. U.S.D.A. Forest Service, General Technical Report, PNW-GTR-285, Pacific Northwest Research Station, Portland, Oregon.

Corn, P.S. and J.C. Fogleman. 1984. Extinction of montane populations of the northern leopard frog (*Rana pipiens*) in Colorado. Journal of Herpetology 18:147–152.

Corn, P.S. and L.J. Livo. 1989. Leopard frog and wood frog reproduction in Colorado and Wyoming. Northwestern Naturalist 70:1–9.

Corn, P.S. and F.A. Vertucci. 1992. Descriptive risk assessment of the effects of acid deposition on Rocky Mountain amphibians. Journal of Herpetology 26:361–369.

Corn, P.S., M.L. Jennings and E. Muths. 1997. Survey and assessment of amphibian populations in Rocky Mountain National Park. Northwestern Naturalist 78:34–55.

Corn, P.S., E. Muths and W.M. Iko. 2000. A comparison in Colorado of three methods to monitor breeding amphibians. Northwestern Naturalist 81:22–30.

Corn, P.S., W. Stolzenburg and R.B. Bury. 1989. Acid precipitation studies in Colorado and Wyoming: Interim report of surveys of montane amphibians and water chemistry. U.S. Fish and Wildlife Service Biological Report 80 (40.26), Washington, D.C.

Cornejo, D. 1982. Night of the spadefoot toad. Science 82:62–66.

Cornett, J.W. 1981. *Batrachoseps major* (Amphibia: Caudata: Plethodontidae) from the Colorado Desert. Bulletin of the Southern California Academy of Science 80:94–95.

Correa, L.L., M.O.A. Correa, J.F. Vaz, M.I.P.G. Silva, R.M. Silva and M.T. Yamanaka. 1980. Importancia das plantas ornametais dos aquarios como veiculos de propagacao de vetores de *Schistosoma mansoni*. Revista Instituto Adolfo Lutz 40:89–96.

Corrington, J.D. 1929. Herpetology of the Columbia, South Carolina, region. Copeia 172:58–83.

Corse, W.A. and D.E. Metter. 1980. Economics, adult feeding, and larval growth of *Rana catesbeiana* on a fish hatchery. Journal of Herpetology 14:231–238.

Corser, J.D. 2001. Decline of disjunct green salamander, *Aneides aeneus*, populations in southern Appalachians. Biological Conservation 97:119–126.

Cortwright, S.A. 1988. Intraguild predation and competition: an analysis of net growth shifts in larval amphibian prey. Canadian Journal of Zoology 66:1813–1821.

Cortwright, S.A. 1998. Ten to eleven-year population trends of two pond-breeding amphibian species, red-spotted newt and green frogs. Pp 61–71. *In* Lannoo, M.J. (Ed.), Status and Conservation of Midwestern Amphibians. University of Iowa Press, Iowa City, Iowa.

Cortwright, S.A. and C.E. Nelson. 1990. An examination of multiple factors affecting community structure in an aquatic amphibian community. Oecologia 83:123–131.

Cory, L. 1952. Winter kill of *Rana pipiens* in shallow ponds. Herpetologica 8:32.

Cory, L. and J.J. Manion. 1953. Predation on eggs of the woodfrog, *Rana sylvatica*, by leeches. Copeia 1953:66.

Cory, L. and J.J. Manion. 1955. Ecology and hybridization in the genus *Bufo* in the Michigan-Indiana region. Evolution 9:42–51.

Cosineau, M. and K. Rogers. 1991. Observations on sympatric *Rana pipiens*, *Rana blairi* and their hybrids. Journal of Herpetology 25:114–116.

Costanzo, J.P. and R.E. Lee. 1994. Biophysical and physiological responses promoting freeze tolerance in vertebrates. Newsletter of the International Physiological Society 9:252–265.

Costanzo, J.P., M.F. Wright and R.E. Lee Jr. 1992. Freeze tolerance as an overwintering adaptation in Cope's gray treefrog (*Hyla chrysoscelis*). Copeia 1992:565–569.

Couch, J.A. 1966. *Brachycoelium ambystomae* (Trematoda: Brachycoelidae) from *Ambystoma opacum*. Journal of Parisitology 52:46–49.

Counts, C.L., III. 1980. A neural neoplasm in the eastern newt, *Notophthalmus viridescens*. Herpetologica 36:46–50.

Counts, C.L., III and R.W. Taylor. 1977. A xanthoma of indeterminate origin in *Bufo americanus* (Amphibia, Anura, Bufonidae). Journal of Herpetology 11:235–236.

Courchamp, F., T. Clutton-Brock and B. Grenfell. 1999. Inverse density dependence and the Allee effect. TRENDS in Ecology and Evolution 14:405–410.

Courtenay, W.R., Jr. and J.E. Deacon. 1983. Fish introductions in the American southwest: a case history of Rogers Spring, Nevada. Southwestern Naturalist 28:221–224.

Courtois, D., R. Leclair Jr., S. Lacasse and P. Magnan. 1995. Habitats préférentiels d'amphibiens ranidés dans des lacs oligotrophes du Bouclier laurentien, Québec. Canadian Journal of Zoology 73:1744–1753.

Cousineau, M. and K. Rogers 1991. Observations on sympatric *Rana pipiens*, *R. blairi*, and their hybrids in eastern Colorado. Journal of Herpetology 25:114–116.

Couture, M.R. and D.M. Sever. 1979. Developmental mortality of *Ambystoma tigrinum* (Amphibia: Urodela) in northern Indiana. Proceedings of the Indiana Academy of Science 88:173–175.

Covacevich, J. and M. Archer. 1975. The distribution of the cane toad, *Bufo marinus*, in Australia and its effects on indigenous vertebrates. Memoirs of the Queensland Museum 17:305–310.

Cowan, I.M. 1941. Longevity of the red-legged frog. Copeia 1941:48.

Coward, S.J. 1984. An osprey catches a large frog. Oriole 49:13–14.

Cowardin, L.M., V. Carter, F.C. Golet and E.T. LaRoe. 1979. Classification of wetlands and deepwater habitats of the United States. U.S. Fish and Wildlife Service, FWS/OBS 79/31, Washington, D.C.

Cowdry, E.V. 1934. The problem of intranuclear inclusions in virus disease. Archives of Pathology 18:527–542.

Cowles, R.B. and C.M. Bogert. 1936. The herpetology of the Boulder Dam region (Nev., Ariz., Utah). Herpetologica 1:33–42.

Cowman, D.F., D.W. Sparling, G.M. Fellers, J.W. Bickman and T.E. Lacher. 2002. Frogs and pesticides in the Sierra Nevada, CA. Society of Environmental Toxicology and Chemistry (SETAC) 23rd Annual Meeting Abstract, Achieving Global Environmental Quality; Integrating Science and Management, 16–20 November 2002. Salt Lake City, Utah.

Cox, G.W. 1999. Alien species in North America and Hawaii. Island Press, Washington, D.C.

Cox, J.A. and R.S. Kautz. 2000. Habitat conservation needs of rare and imperiled wildlife in Florida. Office of Environmental Services, Florida Fish and Wildlife Conservation Commission, Tallahassee, Florida.

Cracraft, J. 1983a. Species concepts and speciation analysis. Pp. 187–187. *In* Johnston, R.F. (Ed.), Current Ornithology, Volume 1. Plenum Press, New York.

Cracraft, J. 1983b. Species concepts and speciation analysis. Current Ornithology 1:159–187.

Cracraft, J. 1987. Species concepts and the ontology of evolution. Biological Transactions of the Philosophical Society 2:329–346.

Cracraft, J. 1992. The species of the birds-of-paradise (Paradisaeidae): applying the phylogenetic species concept to a complex pattern of diversification. Cladistics 8:1–43.

Cracraft, J. 1997. Species concepts in systematics and conservation biology—an ornithological viewpoint. Pp. 325–339. *In* Claridge, M.F., H.A. Dawah and M.R. Wilson (Eds.), Species. The Units of Biodiversity. Chapman and Hall, London.

Crandall, K.A., O.R.P. Bininda-Evans, G.M. Mace and R.K. Wayne. 2000. Considering evolutionary processes in conservation biology. TRENDS in Ecology and Evolution 15:290–295.

Crawford, J.A. 2000. Investigation of aggressive and satellite behavior in the cricket frog *Acris crepitans*. Master's thesis. Illinois State University, Normal, Illinois.

Crawley, M.J. 1989. Insect herbivores and plant population dynamics. Annual Review of Entomology 34:531–564.

Crawshaw, G.J. 1992. The role of disease in amphibian decline. Pp. 60–62. *In* Bishop, C.A. and K.E. Pettit (Eds.), Declines in Canadian Amphibian Populations: Designing a National Monitoring Strategy. Canadian Wildlife Service, Occasional Paper, Number 76, Ottawa, Ontario, Canada.

Crawshaw, G.J. 1997. Disease in Canadian amphibian populations. Pp. 258–270. *In* Green, D.M. (Ed.), Amphibians in Decline: Canadian Studies of a Global Problem. Herpetological Conservation, Number 1, Society for the Study of Amphibians and Reptiles, St. Louis, Missouri.

Crawshaw, L.I., R.N. Rausch, L.P. Wollmuth and E.J. Bauer. 1992. Seasonal rhythms of development and temperature selection in larval bullfrogs, *Rana catesbeiana* Shaw. Physiological Zoology 65:346–359.

Crayon, J.J. 1997. The African clawed frog (*Xenopus laevis*) on Edwards AFB: Implications for wildlife. Unpublished report to the Environmental Management Division, Edwards Air Force Base, California.

Crayon, J.J. 1998. Natural history notes: *Rana catesbeiana* (bullfrog). Diet. Herpetological Review 29:232.

Crayon, J.J. and R.L. Hothem. 1998. Natural history notes: *Xenopus laevis* (African clawed frog). Predation. Herpetological Review 29:165–166.

Creel, G.C. 1963. Bat as food item of *Rana pipiens*. Texas Journal of Science 15:104–106.

Crenshaw, J.W., Jr. and W.F. Blair. 1959. Relationships in the *Pseudacris nigrita* complex in southwestern Georgia. Copeia 1959:215–222.

Crespi, E.J. 1996. Mountaintops as islands: genetic variability of the pigmy salamander (*Desmognathus wrighti*, family Plethodontidae) in the southern Appalachians. Master's thesis. Wake Forest University, Winston-Salem, North Carolina.

Creusere, F.M. 1971. Range extension of the triploid *Ambystoma platineum*. Journal of Herpetology 5:65–66.

Creusere, F.M. and W.G. Whitford. 1976. Ecological relationships in a desert anuran community. Herpetologica 32:7–18.

Cribb, T.H. 1990. Introduction of a *Brachylaima* species (Digenea: Brachylaimidae) to Australia. International Journal for Parasitology 20:789–796.

Crisafulli, C.M. and C.P. Hawkins. 1998. Ecosystem recovery following a catastrophic disturbance: lessons learned from Mount St. Helens. Pp. 23–26. *In* Mac, M.J., P.A. Opler, C.E. Puckett Haecker and P.D.

Doran (Eds.), Status and Trends of the Nation's Biological Resources. Volume 1. U.S. Department of the Interior, U.S. Geological Survey, Reston, Virginia.

Croes, S.A. and R.E. Thomas. 2000. Freeze tolerance and cryoprotectant synthesis of the Pacific treefrog, *Hyla regilla*. Copeia 2000: 863–868.

Crombie, R.L. 1972. The presence of *Hyla squirella* in the Bahamas. Quarterly Journal of the Florida Academy of Science 35:49–52.

Cromeens, B. 1995a. Federal park officials reject golf course plan. Bastrop Advertiser, 18 February.

Cromeens, B. 1995b. Sansom vows agency will expand golf course. Bastrop Advertiser, 16 March.

Cronin, J.T. and J. Travis. 1986. Size-limited predation on larval *Rana areolata* (Anura: Ranidae) by two species of backswimmer (Insecta: Hemiptera: Notonectidae). Herpetologica 42:171–174.

Crosby, C.R. and S.C. Bishop. 1925. A new genus and two new species of spiders collected by *Bufo quercicus* (Holbrook). Florida Entomologist 9:33–36.

Crossland, M. 1992. The effects of cane toad *(Bufo marinus)* eggs on potential aquatic predators. Cape York Peninsula Scientific Expedition 1992:267–270.

Crossland, M. 1993. In this corner: the cane toad tadpole. Wildlife Australia, Autumn 1993:4–5.

Crossland, M. 1995. Impact of eggs, hatchlings and tadpoles of the introduced cane toad *(Bufo marinus)* on native aquatic fauna. CSIRO *Bufo* Project: An Overview of Research Outcomes, Unpublished report, CSIRO, Australia.

Crossland, M. 1998a. A comparison of cane toad and native tadpoles as predators of native anuran eggs, hatchlings and larvae. Wildlife Research 25:373–381.

Crossland, M. 1998b. Ontogenetic variation in toxicity of tadpoles of the introduced toad *Bufo marinus* to native Australian aquatic invertebrate predators. Herpetologica 54:364–369.

Crossland, M. 1998c. Predation by tadpoles on toxic toad eggs: the effect of tadpole size on predation success and tadpole survival. Journal of Herpetology 32:443–446.

Crossland, M. 2000. Direct and indirect effects of the introduced toad *Bufo marinus* (Anura: Bufonidae) on populations of native anuran larvae in Australia. Ecography 23:283–290.

Crossland, M. and R.A. Alford. 1998. Evaluation of the toxicity of eggs, hatchlings and tadpoles of the introduced toad *Bufo marinus* (Anura: Bufonidae) to native Australian aquatic predators. Australian Journal of Ecology 23:129–137.

Crossland, M. and C. Azevedo-Ramos. 1999. Effects of *Bufo* (Anura: Bufonidae) toxins on tadpoles from native and exotic *Bufo* habitats. Herpetologica 55:192–199.

Crosswhite, E. and M. Wyman. 1920. Notes on an abnormal toad. Journal of Entomology and Zoology, Pomona College 12:78.

Crother, B.I., J. Boundy, J.A. Campbell, K. de Queiroz, R.F. Frost, R. Highton, J.B. Iverson, P.A. Meylan, T.W. Reeder, M.E. Seidel, J.W. Sites Jr. and T.W. Taggart. 2000. Scientific and standard English names of amphibians and reptiles of North America north of Mexico, with comments regarding confidence in our understanding. Herpetological Circular, Number 29, Society for the Study of Amphibians and Reptiles, St. Louis, Missouri.

Crow, J.F. and M. Kimura. 1970. An Introduction to Population Genetic Theory. Harper and Row, New York.

Crowder, W.C., M. Nie and G.R. Ultsch. 1998. Oxygen uptake in bullfrog tadpoles *(Rana catesbeiana)*. Journal of Experimental Zoology 280: 121–134.

Crowell, R.L. 1981. Microhabitat selection and feeding ecology of troglophilic plethodontid salamanders in northwestern Arkansas. Ph.D. dissertation. University of Arkansas, Fayetteville, Arkansas.

Crozier, R.H. 1992. Genetic diversity and the agony of choice. Biological Conservation 61:11–15.

Crozier, R.H. 1997. Preserving the information contents of species: genetic diversity, phylogeny and conservation worth. Annual Review Ecology and Systematics 28:243–268.

Crozier, R.H. and R.M. Kusmierski. 1994. Genetic distances and the setting of conservation priorities. Pp. 227–237. *In* Loeschcke, V., J. Tomiuk and S.K. Jain (Eds.), Conservation Genetics. Birdhauser Verlag, Basel, Switzerland.

Crump, D., M. Berrill, D. Coulson, D. Lean, L. McGillivray and A. Smith. 1999. Sensitivity of amphibian embryos, tadpoles, and larvae to enhanced UV-B radiation in natural pond conditions. Canadian Journal of Zoology 77:1956–1966.

Crump, M.L. 1974. Reproductive strategies in a tropical anuran community. University of Kansas Museum of Natural History, Miscellaneous Publications, Number 61, Lawrence, Kansas.

Crump, M.L. 1979. Intra-population variation in energy parameters of the salamander *Plethodon cinereus*. Oecologia 38:235–247.

Crump, M.L. 1983. Opportunistic cannibalism by amphibian larvae in temporary aquatic environments. American Naturalist 121:281–289.

Crump, M.L. 1984. Intraclutch egg size variability in *Hyla crucifer* (Anura: Hylidae). Copeia 1984:302–308.

Crump, M.L. and N.J. Scott Jr. 1994. Visual encounter surveys. Pp. 84–92. *In* Heyer, W.R., M.A. Donnelly, R.W. McDiarmid, L.-A.C. Hayek and M.S. Foster (Eds.), Measuring and Monitoring Biological Diversity: Standard Methods for Amphibians. Smithsonian Institution Press, Washington, D.C.

Crump, M.L., F.R. Hensley and K.L. Clark. 1992. Apparent decline of the golden toad: underground or extinct? Copeia 1992:413–420.

Crunkilton, R. 1984. Subterranean contamination of Meramec Spring by ammonium nitrate and urea fertilizer and its implication on rare cave biota. From: Proceedings of the 1984 National Cave Management Symposium. Journal of the Missouri Speleological Society 25:151–158.

Csanady, E.R. 1978. The life cycle of the plethodontid salamander *Desmognathus quadramaculatus* (Holbrook) in northeastern Tennessee. Master's thesis. East Tennessee State University, Johnson City, Tennessee.

Csuti, B., S. Polasky, P.H. Williams, R.L. Pressey, J.D. Camm, M. Kershaw, A.R. Kiester, B. Downs, R. Hamilton, M. Huso and K. Sahr. 1997. A comparison of reserve selection algorithms using data on terrestrial vertebrates in Oregon. Biological Conservation 80:83–97.

Culver, D.C. 1973. Feeding behavior of the salamander *Gyrinophilus porphyriticus* in caves. International Journal of Speleology 5:369–377.

Culver, D.C., L.L. Master, M.C. Christman and H.H. Hobbs. 2000. Obligate cave fauna of the 48 contiguous United States. Conservation Biology 14:386–401.

Cunjak, R.A. 1986. Winter habitat of northern leopard frogs, *Rana pipiens*, in a southern Ontario stream. Canadian Journal of Zoology 64:255–257.

Cunningham, A.A., T.E.S. Langton, P.M. Bennett, J.F. Lewin, S.E.N. Drury, R.E. Gough and S.K. MacGregor. 1996. Pathological and microbiological findings from incidents of unusual mortality of the common frog *(Rana temporaria)*. Philosophical Transactions of the Royal Society of London B 351:1539–1557.

Cunningham, J.D. 1954. A case of cannibalism in the toad *Bufo boreas halophilus*. Herpetologica 10:166.

Cunningham, J.D. 1955. Notes on an abnormal *Rana aurora draytonii*. Herpetologica 11:149.

Cunningham, J.D. 1959. Reproduction and food of some California snakes. Herpetologica 15:17–19.

Cunningham, J.D. 1960. Aspects of the ecology of the Pacific slender salamander, *Batrachoseps pacificus*, in southern California. Ecology 41:88–99.

Cunningham, J.D. 1962. Observations on the natural history of the California toad, *Bufo californicus* Camp. Herpetologica 17:255–260.

Cunningham, J.D. 1964. Observations on the ecology of the canyon tree frog, *Hyla californiae*. Herpetologica 20:55–61.

Cunningham, J.D. and D.P. Mullally. 1956. Thermal factors in the ecology of the Pacific treefrog. Herpetologica 12:68–79.

Cunnington, D.C. and R.J. Brooks. 2000. Optimal egg size theory: does predation by fish affect egg size in *Ambystoma maculatum*? Journal of Herpetology 34:46–53.

Cupp, P.V., Jr. 1971. Fall courtship of the green salamander, *Aneides aeneus*. Herpetologica 27:308–310.

Cupp, P.V., Jr. 1980. Territoriality in the green salamander *Aneides aeneus*. Copeia 1980:463–468.

Cupp, P.V., Jr. 1980. Thermal tolerance of five salientian amphibians during development and metamorphosis. Herpetologica 36:234–244.

Cupp, P.V., Jr. 1991. Aspects of the life history and ecology of the green salamander, *Aneides aeneus*, in Kentucky. Journal of the Tennessee Academy of Science 66:171–174.

Cupp, P.V., Jr. 1994. Salamanders avoid chemical cues from predators. Animal Behaviour 48:232–235.

Cupp, P.V. and D.T. Towles. 1983. A new variant of *Plethodon wehrlei* in Kentucky and West Virginia. Transactions of the Kentucky Academy of Science 44:157–158.

Currie, W. and E.D. Bellis. 1969. Home range and movements of the bullfrog, *Rana catesbeiana* Shaw, in an Ontario pond. Copeia 1969: 688–692.

Curtis, B., K.S. Roberts, M. Griffin, S. Bethune, C.J. Hay and H. Kolberg. 1998. Species richness and conservation of Namibian freshwater macro-invertebrates, fish and amphibians. Biodiversity and Conservation 7:447–466.

Cuvier, G.J.L.F. 1827. Sur le genre de reptile batraciens, nomme Amphiuma, et sur une novelle espece de ce genre (Amphiuma tridactylum). Memoirs du Muséum National d' Histoire Naturelle, Paris 14:1–14.

Czopek, J. 1962. Tolerance to submersion in water in amphibians. Acta Biologica Cracoviensis. Series Zoologia 5:241–251.

Czopek, J. 1962. Vascularization of respiratory surfaces in some Caudata. Copeia 1962:576–587.

Dahl, A., M.P. Donovan and T.D. Schwaner. 2000. Egg mass deposited by Arizona toads, Bufo microscaphus, along a narrow canyon stream. Western North American Naturalist 60:456–458.

Dahl, T.E. 1990. Wetland losses in the United States, 1780s to 1980s. U.S. Department of the Interior, Fish and Wildlife Service, Washington, D.C.

Dahl, T.E. and C.E. Johnson. 1991. Status and trends of wetlands in the conterminous United States, mid-1970s to mid-1980s. U.S. Department of the Interior, Fish and Wildlife Service, Washington, D.C.

Dahms, C.W. and B.W. Geils (Tech. eds). 1997. An assessment of forest ecosystem health in the Southwest. U.S.D.A. Forest Service, General Technical Report, RM-GTR-295, Rocky Mountain Forest and Range Experiment Station, Fort Collins, Colorado.

Dale, J.M., B. Freedman and J. Kerekes. 1985a. Acidity and associated water chemistry of amphibian habitats in Nova Scotia Canada. Canadian Journal of Zoology 63:97–105.

Dale, J.M., B. Freedman and J. Kerekes. 1985b. Experimental studies of the effects of acidity and associated water chemistry on amphibians. Proceedings of the Nova Scotia Institute of Science 35:35–54.

Dalrymple, G.H. 1970. Caddisfly larvae feeding upon eggs of Ambystoma t. tigrinum. Herpetologica 26:128–129.

Dalrymple, G.H. 1988. The herpetofauna of Long Pine Key, Everglades National Park, in relation to vegetation and hydrology. Pp. 72–86. In Szaro, R.C., K.E. Stevenson and D.R. Patton (Tech. Coords.), Management of Amphibians, Reptiles, and Small Mammals in North America. U.S.D.A. Forest Service, General Technical Report, RM-166, Flagstaff, Arizona.

Dalrymple, G.H. 1989. The herpetofauna of Long Pine Key, Everglades National Park, in relation to vegetation and hydrology. Pp. 113–114. In Szaro, R. (Ed.), Management of Amphibians, Reptiles and Small Mammals in North America. U.S.D.A. Forest Service, Bulletin Number 29, Washington, D.C.

Dalrymple, G.H. 1990. Habitat suitability index model for the oak toad, Bufo quercicus. Pp. 26–68. In Richter, W. and E. Myers (Eds.), Habitat suitability index models used for the Bird Drive Everglades Basin Special Area Management Plan. Department of Environmental Research Management, Technical Report Number 90–1, Miami, Florida.

Daly, J.W., C.W. Myers and N. Whittaker. 1987. Further classification of skin alkaloids from neotropical poison frogs (Dendrobatidae), with a general survey of toxic/noxious substances in the Amphibia. Toxicon 25:1021–1095.

Danstedt, R.T., Jr. 1975. Local geographic variation in demographic parameters and body size of Desmognathus fuscus (Amphibia: Plethodontidae). Ecology 56:1054–1067.

Danstedt, R.T., Jr. 1979. A demographic comparison of two populations of the dusky salamander (Desmognathus fuscus) in the same physiographic province. Herpetologica 35:164–168.

Danzer, S.R., C.H. Baisan and T.W. Swetnam. 1996. The influence of fire and land-use history on stand dynamics in the Huachuca Mountains of southeastern Arizona. Pp. 265–270. In Ffolliott, P.F., L.F. DeBano, M.B. Baker Jr., G.J. Gottfried, G. Solis-Garaza, C.B. Edminster, D.G. Neary, L.S. Allen and R.H. Hamre (Eds.), Effects of Fire on Madrean Province Ecosystems. Rocky Mountain Forest and Range Experiment Station, Fort Collins, Colorado.

Dapkus, D. 1976. Differential survival involving the burnsi phenotype in the northern leopard frog, Rana pipiens. Herpetologica 32:325–327.

Dapson, R.W. and L. Kaplan. 1975. Biological half-life and distribution of radiocesium in a contaminated population of green treefrogs Hyla cinerea. Oikos 26:39–42.

Dasgupta, S. and M.S. Grewal. 1968. The selective advantage of temperature tolerance among the progeny of frogs with vertebral fusions. Evolution 22:87–92.

Daszak, P., A.A. Cunningham and A.D. Hyatt. 2000. Emerging infectious diseases of wildlife—threats to biodiversity and human health. Science 287:443–449.

Daszak, P., L. Berger, A.A. Cunningham, A.D. Hyatt, D.E. Green and R. Speare. 1999. Emerging infectious diseases and amphibian population declines. Emerging Infectious Diseases 5:735–748.

Daudin, F.M. 1801–1803. Histoire Naturelle, Générale et Particulière des Reptiles; ouvrage faisant suit à l'Histore naturelle générale et particulière, composée par Leclerc de Buffon; et rédigée par C.S. Sonnini, membre de plusieurs sociétés savantes. 8 Volumes. F. Dufart, Paris, France.

Daudin, F.M. 1802. Histoire Naturelle des Rainettes, des Grenouilles, et des Crapauds. Bertrandet, Libraire Levrault, Paris, France.

Daugherty, C.H. 1979. Population ecology and genetics of Ascaphus truei: an examination of gene flow and natural selection. Ph.D. dissertation. University of Montana, Missoula, Montana.

Daugherty, C.H. and A.L. Sheldon. 1982a. Age-determination, growth, and life history of a Montana population of the tailed frog (Ascaphus truei). Herpetologica 38:461–468.

Daugherty, C.H. and A.L. Sheldon. 1982b. Age-specific movement patterns of the frog Ascaphus truei. Herpetologica 38:468–474.

Daugherty, C.H., F.W. Allendorf, W.W. Dunlop and K.L. Knudson. 1983. Systematic implications of geographic patterns of genetic variation in the genus Dicamptodon. Copeia 1983:679–691.

Daugherty, C.H., A. Cree. J.M. Hay and M.B. Thompson. 1990. Neglected taxonomy and continuing extinctions of tuatara (Sphenodon). Nature 347:177–179.

Davic, R.D. 1983. Microgeographic body size variation in Desmognathus fuscus fuscus salamanders from western Pennsylvania. Copeia 1983: 1101–1104.

Davic, R.D. 1991. Ontogenetic shift in the diet of Desmognathus quadramaculatus. Journal of Herpetology 25:108–111.

Davic, R.D. and L.P. Orr. 1987. The relationship between rock density and salamander density in a mountain stream. Herpetologica 43: 357–361.

Davidson, C. 1996. Frog and toad calls of the Rocky Mountains. Library of Natural Sounds, Cornell Laboratory of Ornithology, Ithaca, New York.

Davidson, C. 2000. Spatial patterns of California amphibian declines and ecological limits: Why are frogs disappearing and what does it mean? Ph.D. dissertation. University of California, Davis, California.

Davidson, C., H.B. Shaffer and M.R. Jennings. 2001. Declines of the California red-legged frog: spatial analysis of climate, UV-B, habitat and pesticides hypotheses. Ecological Applications 11:464–479.

Davidson, C., H.B. Shaffer and M.R. Jennings. 2002. Spatial tests of the pesticide drift, habitat destruction, UV-B, and climates-change hypotheses for California amphibian amphibian declines. Conservation Biology 16:1588–1601.

Davidson, J.A. 1956. Notes on the food habits of the slimy salamander Plethodon glutinosus glutinosus. Herpetologica 12:129–131.

Davidson, M. and H. Heatwole. 1960. Late summer oviposition in the salamander, Plethodon cinereus. Herpetologica 16:141–142.

Davis, B., R. Hansen, D. Johns and M. Turner. 2001. Jollyville Plateau water quality and salamander assessment. City of Austin, Austin, Texas.

Davis, J. 1952. Observations on the eggs and larvae of the salamander Batrachoseps pacificus major. Copeia 1952:272–274.

Davis, J.G. and S.A. Menze. 2000. Ohio frog and toad atlas. Ohio Biological Survey Miscellaneous Contributions, Number 6, Columbus, Ohio.

Davis, J.G., P. Krusling and J.W. Ferner. 1998. Status of amphibian populations in Hamilton County, Ohio. Pp. 155–165. In Lannoo, M.J. (Ed.), Status and Conservation of Midwestern Amphibians. University of Iowa Press, Iowa City, Iowa.

Davis, J.I. and K.C. Nixon. 1992. Populations, genetic variation, and the delimitation of phylogenetic species. Systematic Biology 41:421–435.

Davis, J.R. and B.H. Brattstrom. 1975. Sounds produced by the California newts, Taricha torosa. Herpetologica 31:409–412.

Davis, M.S. 1987. Acoustically mediated neighbor recognition in the North American bullfrog, Rana catesbeiana. Behavioral Ecology and Sociobiology 21:185–190.

Davis, M.S. and G.W. Folkerts. 1986. Life history of the wood frog, Rana sylvatica LeConte (Amphibia: Ranidae) in Alabama. Brimleyana 12:29–50.

Davis, T.M. 2002a. Microhabitat use and movements of the wandering salamander, Aneides vagrans, on Vancouver Island, British Columbia, Canada. Journal of Herpetology 36:699–703.

Davis, T.M. 2002b. An ethogram of intraspecific agonistic and display behavior for the wandering salamander, Aneides vagrans. Herpetologica 58:371–382.

Davis, T.M. 2003. Natural history notes: Aneides vagrans (wandering salamander). Reproduction. Herpetological Review 34:133.

Davis, W.B. and J.R. Dixon. 1957. Notes on Mexican amphibians, with a description of a new Microbatrachylus. Herpetologica 13:145–147.

Davis, W.B. and F.T. Knapp. 1953. Notes on the salamander *Siren intermedia*. Copeia 1953:119–121.

Davis, W.C. and V.C. Twitty. 1964. Courtship, behavior and reproductive isolation in the species of *Taricha* (Amphibia, Caudata). Copeia 1964:601–610.

Davis, W.T. 1922. Insects from North Carolina. Journal of the New York Entomological Society 30:74–75.

Davison, A. 1895. A contribution to the anatomy and phylogeny of *Amphiuma means*. Journal of Morphology 11:375–410.

Davison, A.J., W. Sauerbier, A. Dolan, C. Addison and R.G. McKinnell. 1999. Genomic studies of the Lucké tumor herpesvirus (RaHV-1). Journal of Cancer Research and Clinical Oncology 125:232–238.

Davison, J. 1964. A study of spotting patterns in the leopard frog. III. Environmental control of genic expression. Journal of Heredity 55:47–56.

Davison, J. 1966. Chimeric and ex-parabiotic frogs *(Rana pipiens)*: specificity of tolerance. Science 152:1250–1253.

Dawley, E.M. 1984. Identification of sex through odors by male red-spotted newts, *Notophthalmus viridescens*. Herpetologica 40:101–105.

Dawley, E.M. 1986. Behavioral isolating mechanisms in sympatric terrestrial salamanders. Herpetologica 42:156–164.

Dawley, E.M. and R.M. Dawley. 1986. Species discrimination by chemical cues in a unisexual-bisexual complex of salamanders. Journal of Herpetology 20:114–116.

Dawson, B. 1996. Texans love their parks, sometimes way too much. Houston Chronicle, 21 July.

Dawson, J.T. 1982. Kin recognition and schooling in the American toad *(Bufo americanus, B. marinus)*: speed of prey-capture does not determine success. Journal of Comparative Physiology 135:41–50.

Dawson, W.R., J.D. Ligon, J.R. Murphy, J.P. Myers, D. Simberloff and J. Verner. 1987. Report of the scientific advisory panel on the spotted owl. Condor 89:205–229.

Day, J.H. and M.A. Colwell. 1998. Waterbird communities in rice fields subjected to different post-harvest treatments. Colonial Waterbirds 21:185–197.

Dayton, G.H. 2000a. Abiotic and biotic factors affecting nonrandom distributions of Chihuahuan desert anurans. Master's thesis. Texas A&M University, College Station, Texas.

Dayton, G.H. 2000b. Natural history notes: *Gastrophryne olivacea* (narrow-mouthed toad). Vocalization. Herpetological Review 31:40.

Dayton, G.H. and L.A. Fitzgerald. 2001. Competition, predation, and the distribution of four desert anurans. Oecologia 129:430–435.

de Allende, A.L.C. and O. Orías. 1955. Hypophysis and ovulation in the toad *Bufo arenarum* (Hensel). Acta Physiologia Latino Americana 5:57–84.

De Bruyn, L., M. Kazadi and J. Huselmans. 1996. Diet of *Xenopus fraseri* (Anura, Pipidae). Journal of Herpetology 30:82–85.

de Gentile, L., H. Picot, P. Bourdeau, R. Bardet, A. Kerjan, M. Piriou, A. Le Guennic, C. Bayss ade-Defour, D. Chabasse and K.E. Mott. 1996. La dermatite cercarienne en Europe: un probleme de sante publique nouveau? Bulletin of the World Health Organization 74:159–163.

De Lisle, H., G. Cantu, J. Feldner, P. O'Connor, M. Peterson and P. Brown. 1986 (1987). The distribution and present status of the herpetofauna of the Santa Monica Mountains of Los Angeles and Ventura Counties, California. Southwestern Herpetologists Society, Special Publication Number 2, Van Nuys, California.

De Marco, M.N. 1952. Neoteny and the urogenital system in the salamander *Dicamptodon ensatus* (Eschscholtz). Copeia 1952:192–193.

de Queiroz, K. 1995. Phylogenetic approaches to classification and nomenclature and the history of taxonomy. Herpetological Review 26:79–81.

de Queiroz, K. and J. Gauthier. 1992. Phylogenetic taxonomy. Annual Review of Ecology and Systematics 23:449–80.

de Queiroz, K. and J. Gauthier. 1994. Toward a phylogenetic system of biological nomenclature. TRENDS in Ecology and Evolution 9:27–31.

de Rageot, R.H. 1969. Observations regarding three rare amphibians in Surry Co. Virginia Herpetological Society Bulletin 63:3–5.

De Solla, S.R., C.A. Bishop, K.E. Pettit and J.E. Elliott. 2002a. Organochlorine pesticides and polychlorinated biphenyls (PCBs) in eggs of red-legged frogs *(Rana aurora)* and northwestern salamanders *(Ambystoma gracile)* in an agricultural landscape. Chemosphere 46:1027–1032.

De Solla, S.R., K.E. Pettit, C.A. Bishop, K.M. Cheng and J.E. Elliott. 2002b. Effects of agricultural runoff on native amphibians in the lower Fraser River Valley, British Columbia, Canada. Environmental Toxicology and Chemistry 21:353–360.

De Vito, J., D.P. Chivers, J.M. Kiesecker, L.K. Belden and A.R. Blaustein. 1999. Effects of snake predation on aggregation and metamorphosis of Pacific treefrogs *(Hyla regilla)*. Journal of Herpetology 33:504–507.

de Vlaming, V.L. and R.B. Bury. 1970. Thermal selection in tadpoles of the tailed frog, *Ascaphus truei*. Journal of Herpetology 4:179–189.

Dean, R.A. 1966. High temperature tolerances of anuran amphibians. Master's thesis. University of North Dakota, Grand Forks, North Dakota.

Dean, R.J. 1980a. Encounters between bombardier beetles and two species of toads *(Bufo americanus, B. marinus)*: speed of prey-capture does not determine success. Journal of Comparative Physiology 135:41–50.

Dean, R.J. 1980b. Effect of thermal and chemical components of bombardier beetle chemical defense: glossopharyngeal response in two species of toads *(Bufo americanus, B. marinus)*. Journal of Comparative Physiology 135:51–59.

Deban, S.M. 1997. Modulation of prey-capture behavior in the plethodontid salamander *Ensatina eschscholtzii*. Journal of Experimental Biology 200:1951–1964.

Deban, S.M. and K.C. Nishikawa. 1992. The kinematics of prey capture and the mechanism of tongue protraction in the green tree frog *Hyla cinerea*. Journal of Experimental Biology 170:235–256.

Deban, S.M., D.B. Wake and G. Roth. 1987. Salamander with a ballistic tongue. Nature 389:27–28.

DeBenedictis, P.A. 1974. Interspecific competition between tadpoles of *Rana pipiens* and *Rana sylvatica*: an experimental field study. Ecological Monographs 44:129–151.

Deckert, R.F. 1914a. List of Salientia from near Jacksonville, Florida. Copeia 3:7.

Deckert, R.F. 1914b. Further notes on the Salientia of Jacksonville, Fla. Copeia 5:2–4.

Deckert, R.F. 1914c. Salamanders collected in Westchester County, N.Y. Copeia 13:3–4.

Deckert, R.F. 1915. Concluding notes on the Salientia of Jacksonville, Fla. Copeia 20:21–24.

Deckert, R.F. 1916. Note on *Ambystoma opacum*, Grav. Copeia 28:23–24.

Deckert, R.F. 1920. Note on the Florida gopher frog, *Rana aesopus*. Copeia 80:26.

Degenhardt, W.G., C.W. Painter and A.H. Price. 1996. Amphibians and Reptiles of New Mexico. University of New Mexico Press, Albuquerque, New Mexico.

Degos, L., H. Dombret, C. Chomienne, M.T. Daniel, A. Micléa, D. Chastang, S. Castaigne and P. Fenaux. 1995. All-trans retinoic acid as a differentiating agent in the treatment of acute promyelocytic leukemia. Blood 85:2643–2653.

DeGraaf, R.M. and D.D. Rudis. 1981. Forest Habitat for Reptiles and Amphibians of the Northeast. Northeastern Forest Experiment Station and Eastern Region, Forest Service, U.S. Department of Agriculture, Bromall, Pennsylvania.

DeGraaf, R.M. and D.D. Rudis. 1983. Amphibians and Reptiles of New England: Habitats and Natural History. University of Massachusetts Press, Amherst, Massachusetts.

DeGraaf, R.M. and D.D. Rudis. 1990. Herpetofaunal species composition and relative abundance among three New England forest types. Forest Ecology and Management 32:155–165.

DeKay, J.E. 1842. Zoology of New-York, or the New-York Fauna; Comprising Detailed Descriptions of all the Animal Hitherto Observed within the State of New-York, with Brief Notices of Those Occasionally Found near its Borders, and Accompanied by Appropriate Illustrations. Part III. Reptiles and Amphibia. W. and A. White and J. Visscher, Albany, New York.

DeKunder, D. 2000. Long-awaited golf clubhouse could begin construction this year. Bastrop Advertiser, 7 October.

Delany, M.F. and C.L. Ambercrombie. 1986. American alligator food habits in northcentral Florida. Journal of Wildlife Management 50:348–353.

Delis, P.R., H.R. Mushinsky and E.D. McCoy. 1996. Decline of some west-central Florida anuran populations in response to habitat degradation. Biodiversity and Conservation 5:1579–1595.

Delnicki, D. and E. Bolen. 1977. Use of black-bellied tree duck nest sites by other species. Southwest Naturalist 22:275–277.

Delvinquier, B.L.J. and S S. Desser. 1996. Opalinidae (Sarcomastigophora) in North American Amphibia: genus *Opalina* Purkinje and Valentin, 1835. Systematic Parasitology 33:33–51.

Delzell, D.E. 1979. A provisional checklist of amphibians and reptiles in the Dismal Swamp area, with comments on their range of distribution. Pp. 244–260 *In* Kirk, P.W., Jr. (Ed.), The Great Dismal Swamp. University Press of Virginia, Charlottesville, Virginia.

deMaynadier, P.G. and M.L. Hunter. 1995. The relationship between forest management and amphibian ecology: a review of the North American literature. Environmental Reviews 3:230–261.

deMaynadier, P.G. and M.L. Hunter. 1998. Effects of silvicultural edges on the distribution and abundance of amphibians in Maine. Conservation Biology 12:340–352.

deMaynadier, P.G. and M.L. Hunter. 1999. Forest canopy closure and juvenile emigration by pool-breeding amphibians in Maine. Journal of Wildlife Management 63:441–450.

Demlong, M.J. 1997. Head-starting *Rana subaquavocalis* in captivity. Reptiles 5:24–28, 30, 32–33.

Dempster, J.P. and K.H. Lakhani. 1979. A population model for cinnabar moth and its food, ragwort. Journal of Animal Ecology 48:143–164.

Dempster, W.T. 1930. The growth of larvae of *Ambystoma maculatum* under natural conditions. Biological Bulletin 58:182–192.

Den Boer, P.J. 1990. Density limits and survival of local populations in 64 carabid species with different powers of dispersal. Journal of Evolutionary Biology 3:19–48.

Dennis, D.M. 1962. Notes on the nesting habits of *Desmognathus fuscus* (Raf.) in Licking County, Ohio. Journal of the Ohio Herpetological Society 3:28–35.

Denno, R.B. and M.A. Peterson. 1995. Density dependent dispersal and its consequences for population dynamics. Pp. 111–130. *In* Cappucino, N. and P.W. Price (Eds.), Population Dynamics: New Approaches and Synthesis. Academic Press, New York.

deNoyelles, F., W.D. Kettle, C.H. Fromm, M.F. Moffett and S.L. Dewey. 1989. Use of experimental ponds to assess the effects of a pesticide on the aquatic environment. Pp. 41–56. *In* Voshell, J.R. (Ed.), Using Mesocosms to Assess the Aquatic Ecological Risk of Pesticides: Theory and Practice. Entomological Society of America, MPPEAL 75, Lanham, Maryland.

Dent, J.N. and J.S. Kirby-Smith. 1963. Metamorphic physiology and morphology of the cave salamander *Gyrinophilus palleucus*. Copeia 1963:119–130.

Denton, J.S., S.P. Hitchings, T.J.C. Beebee and A. Gent. 1997. A recovery program for the natterjack toad *(Bufo calamita)* in Britain. Conservation Biology 11:1329–1338.

Denver, R.J., N. Mirhadi and M. Phillips. 1998. Adaptive plasticity in amphibian metamorphosis: response of *Scaphiopus hammondii* tadpoles to desiccation. Ecology 76:1859–1872.

Dernell, P.H. 1902. Place-modes of *Acris gryllus* for Madison. Bulletin of the Wisconsin Natural History Society 2:75–82.

Desser, S.S. 1987. *Aegyptianella ranarum* sp. N. (Rickettsiales, Anaplasmataceae): ultrastructure and prevalence in frogs from Ontario. Journal of Wildlife Disease 23:52–59.

Desser, S.S. 1992. Ultrastructural observations on an icosahedral cytoplasmic virus in leukocytes of frogs from Algonquin Park, Ontario. Canadian Journal of Zoology 70:833–836.

Desser, S.S. 2001. The blood parasites of anurans from Costa Rica with reflections on the taxonomy of their trypanosmes. Journal of Parasitology 87:152–160.

Desser, S.S., H. Hong and D.S. Martin. 1995. The life history, ultrastructure, and experimental transmission of *Hepatozoon catesbianae* n. comb., an apicomplexan parasite of the bullfrog, *Rana catesbeiana*, and the mosquito, *Culex territans*, in Algonquin Park, Ontario. Journal of Parasitology 81:212–222.

Desser, S.S., J. Lom and I. Dykova. 1986. Developmental stages of *Sphaerospora ohlmacheri* (Whinery, 1893) n. comb. (Myxozoa: Myxosporea) in the renal tubules of bullfrog tadpoles, *Rana catesbeiana* from Lake of Two Rivers, Algonquin Park, Ontario. Canadian Journal of Zoology 64:2344–2347.

Desser, S.S., M.E. Siddall and J.R. Barta. 1990. Ultrastructural observations on the developmental stages of *Lankesterella minima* (Apicomplexa) in experimentally infected *Rana catesbeiana* tadpoles. Journal of Parasitology 76:97–103.

Dethlefsen, E.S. 1948. A subterranean nest of the Pacific giant salamander, *Dicamptodon ensatus* (Eschscholtz). Wasmann Collector 7:81–84.

Deuchar, E.M. 1975. *Xenopus:* The South African clawed frog. John Wiley and Sons, New York.

Dexter, R.W. 1966. Some herpetological notes and correspondence of Frederick Ward Putnam. Journal of the Ohio Herpetological Society 5:109–114.

Dexter, R.W. 1973. Nomenclatural history and status of Fowler's toad *(Bufo fowleri)*. Smithsonian Herpetological Information Services News Journal 1:155–157.

Di Berardino, M.A. and T.J. King. 1965. Renal adenocarcinomas promoted by crowded conditions in laboratory frogs. Cancer Research 25:1910–1912.

Di Berardino, M.A., T.J. King and R.G. McKinnell. 1963. Chromosome studies of a frog renal adenocarcinoma line carried by serial intraocular transplantation. Journal of the National Cancer Institute 31:769–789.

di Castri, F. 1989. History of biological invasions with special emphasis on the Old World. Pp. 1–30. *In* Drake, J.A., H.A. Mooney, F. di Castri, R.H. Groves, F.J. Kruger, M. Rejmanek and M. Williamson (Eds.), Biological Invasions: A Global Perspective. John Wiley and Sons Limited, New York.

Dial, B.E. 1965. Distributional notes on reptiles and amphibians from northeastern Texas. Southwestern Naturalist 10:143–144.

Diamond, J. 1986. Overview: laboratory experiments, field experiments, and natural experiments. Pp. 3–22. *In* Diamond, J. and T.J. Case (Eds.), Community Ecology. Harper and Rowe, New York.

Diamond, J.M. 1989. The present, past and future of human-caused extinctions. Philosophical Transactions of the Royal Society, London B 325:469–477.

Diana, S.G. and V.R. Beasley. 1998. Amphibian toxicology. Pp. 266–277. *In* Lannoo, M.J. (Ed.), Status and Conservation of Midwestern Amphibians. University of Iowa Press, Iowa City, Iowa.

Dice, L.R. 1939. The Sonoran Biotic Province. Ecology 20:118–129.

Dickerson, M.C. 1906. The Frog Book. Doubleday, Doran, and Company, Garden City, New Jersey.

Dickerson, M.C. 1969. The Frog Book. Dover Publications, New York.

Dickman, C.R. 1991. Habitat selection of *Bufo marinus* in New South Wales and the wet tropics of Queensland. Cane Toad Research Management Committee, CSIRO, ACT, Unpublished report, University of Sydney, New South Wales, Australia.

Dickman, C.R., R.L. Pressey, L. Lim and H.E. Parnaby. 1993. Mammals of particular conservation concern in the Western Division of New South Wales. Biological Conservation 65:219–248.

Dickman, M. 1968. The effect of grazing by tadpoles on the structure of a periphyton community. Ecology 49:1188–1190.

Didiuk, A. 1997. Status of amphibians in Saskatchewan. Pp. 110–116. *In* Green, D.M. (Ed.), Amphibians in Decline: Canadian Studies of a Global Problem. Herpetological Conservation, Number 1, Society for the Study of Amphibians and Reptiles, St. Louis, Missouri.

Diener, E. and J. Marchalonis. 1970. Cellular and humoral aspects of the primary immune response of the toad, *Bufo marinus*. Immunology 18:279–293.

Diener, R.A. 1965. The occurrence of the tadpoles of the green treefrog *Hyla cinerea cinerea*, in Trinity Bay, Texas. British Journal of Herpetology 3:198–199.

Diffendorfer, J.F. 1998. Testing models of source-sink dynamics and balanced dispersal. Oikos 81:417–433.

DiGiovanni, M. and E.D. Brodie Jr. 1981. Efficacy of skin glands in protecting the salamander *Ambystoma opacum* from repeated attacks by the shrew *Blarina brevicauda*. Herpetologica 37:234–237.

Dill, L.M. 1977. "Handedness" in the Pacific treefrog *(Hyla regilla)*. Canadian Journal of Zoology 55:1926–1929.

Diller, J.S. 1907. A salamander-snake fight. Science 26:907–908.

Diller, L.V. and R.L. Wallace. 1994. Distribution and habitat of *Plethodon elongatus* on managed young forests in north coastal California. Journal of Herpetology 28:310–318.

Diller, L.V. and R.L. Wallace. 1996. Distribution and habitat of *Rhyacotriton variegatus* in managed, young growth forests in north coastal California. Journal of Herpetology 30:184–191.

Diller, L.V. and R.L. Wallace. 1999. Distribution and habitat of *Ascaphus truei* in streams on managed, young growth forests in north coastal California. Journal of Herpetology 33:71–79.

DiMauro, D. and M.L. Hunter. 2002. Reproduction of amphibians in natural and anthropogenic temporary pools in managed forests. Forest Science 48:397–406.

Dimmick, W.W., M.J. Ghedotti, M.J. Grose, A.M. Maglia, D.J. Meinhardt and D.S. Penock. 1999. The importance of systematic biology in defining units of conservation. Conservation Biology 13:653–660.

Dimmick, W.W., M.J. Ghedotti, M.J. Grose, A.M. Maglia, D.J. Meinhardt and D.S. Penock. 2001. The evolutionary significant unit and adaptive criteria: a response to Young. Conservation Biology 15:788–790.

Dimmitt, M.A. 1977. Distribution of Couch's spadefoot toad in California (preliminary report). Unpublished report filed with the United States Bureau of Land Management, Riverside District Office, California, Under C-062, 6500, and 1792 Sundesert, May 10, 1977, Riverside, California.

Dimmitt, M.A. and R. Ruibal. 1980a. Environmental correlates of emergence in spadefoot toads (Scaphiopus). Journal of Herpetology 14:21–29.

Dimmitt, M.A. and R. Ruibal. 1980b. Exploitation of food resources by spadefoot toads (Scaphiopus). Copeia 1980:854–862.

Dirig, R. 1978. A large diurnal breeding assemblage of American toads Bufo americanus americanus in the Catskill Mountains New York. Pitch Pine Naturalist 4:1–2.

DiSilvestro, R. 1983. Salamanders. Outdoor Oklahoma 39:38–43.

Ditmars, R.L. 1905. The batrachians of the vicinity of New York City. American Museum Journal 5:192–193.

Dixon, J.R. 1967. Amphibians and Reptiles of Los Angeles County. Los Angeles Museum of Natural History, Los Angeles, California.

Dixon, J.R. 1987. Amphibians and Reptiles of Texas with Keys, Taxonomic Synopses, Bibliography, and Distribution Maps. W.L. Moody Jr., Natural History Series, Number 8, Texas A&M University Press, College Station, Texas.

Dixon, J.R. 1996. Ten year supplement to Texas herpetological county records published in "Amphibians and Reptiles of Texas, 1987." Texas Herpetological Society Special Publication 2:1–64.

Dixon, J.R. 2000. Amphibians and Reptiles of Texas: With Keys, Taxonomic Synopses, Bibliography, and Distribution Maps. W.L. Moody Jr., Natural History Series, Number 25, Texas A&M University Press, College Station, Texas.

Dixon, J.R. and W.R. Heyer. 1968. Anuran succession in a temporary pond in Colima, Mexico. Bulletin of the Southern California Academy of Sciences 67:129–137.

Dizon, A.E., C. Lockyer, W.F. Perrin, D.P. Demaster and J. Sisson. 1992. Rethinking the stock concept: a phylogeographic approach. Conservation Biology 6:24–36.

Dobson, A.P. and J. Foufopoulos. 2001. Emerging infectious pathogens of wildlife. Philosophical Transactions of the Royal Society of London B 356:1001–1012.

Dobson, A.P. and P.J. Hudson. 1992. Regulation and stability of free-living host-parasite system: Trichostrongylus tenuis in red grouse. II. Population models. Journal of Animal Ecology 61:487–498.

Dobson, A.P. and R.M. May. 1986. Patterns of invasions by pathogens and parasites, Pp. 58–76. In Mooney, H.A. and J.A. Drake (Eds.), Ecology of Biological Invasions of North America and Hawaii. Springer-Verlag, New York.

Dobzhansky, T. 1951. Genetics and the Origin of the Species. Columbia University Press, New York.

Dobzhansky, T. and S. Wright. 1941. Genetics of natural populations. V. Relationships between mutation rate and accumulation of lethals in populations of Drosophila pseudoobscura. Genetics 26:23–51.

Dodd, C.K., Jr. 1977a. Immobility in juvenile Bufo woodhousei fowleri. Journal of the Mississippi Academy of Sciences 22:90–94.

Dodd, C.K., Jr. 1977b. Preliminary observations on the reactions of certain salamanders of the genus Ambystoma (Amphibia, Urodela, Ambystomatidae) to a small colubrid snake (Reptilia, Serpentes, Colubridae). Journal of Herpetology 11:222–223.

Dodd, C.K., Jr. 1977c. Letter to Lauren E. Brown, 5 January.

Dodd, C.K., Jr. 1977d. Letter to Lauren E. Brown, 2 December.

Dodd, C.K., Jr. 1978. Endangered and threatened wildlife and plants; determination of critical habitat for the Houston toad. Federal Register 43:4022–4026.

Dodd, C.K., Jr. 1980. Notes on the feeding behavior of the Oklahoma salamander, Eurycea tynerensis (Plethodontidae). Southwestern Naturalist 25:111–113.

Dodd, C.K., Jr. 1982. Correction: Typhlotriton spelaeus not Eurycea tynerensis. Southwestern Naturalist 27:367–368.

Dodd, C.K., Jr. 1988. A re-examination of the status of the Red Hills salamander, Phaeognathus hubrichti. Report to U.S. Fish and Wildlife Service, Jackson, Mississippi.

Dodd, C.K., Jr. 1989. Duration of immobility in salamanders, genus Plethodon (Caudata: Plethodontidae). Herpetologica 45:467–473.

Dodd, C.K., Jr. 1990a. The influence of temperature and body size on duration of immobility in salamanders of the genus Desmognathus. Amphibia-Reptilia 11:401–410.

Dodd, C.K., Jr. 1990b. Line transect estimation of Red Hills salamander burrow density using a Fourier series. Copeia 1990:555–557.

Dodd, C.K., Jr. 1990c. Postures associated with immobile woodland salamanders, genus Plethodon. Florida Science 53:43–49.

Dodd, C.K., Jr. 1991. The status of the Red Hills salamander, Phaeognathus hubrichti, Alabama, USA, 1976–1988. Biological Conservation 55:57–75.

Dodd, C.K., Jr. 1992. Biological diversity of a temporary pond herpetofauna in north Florida sandhills. Biodiversity and Conservation 1:125–142.

Dodd, C.K. 1993a. Strategies for snake conservation. Pp. 362–393. In Seigel, R.A. and J.T. Collins (Eds.), Snakes: Ecology and Behavior. McGraw Hill, New York.

Dodd, C.K., Jr. 1993b. Cost of living in an unpredictable environment: The ecology of striped newts Notophthalmus perstriatus during a prolonged drought. Copeia 1993:605–614.

Dodd, C.K. 1994. The effects of drought on population structure, activity, and orientation of toads (Bufo quercicus and B. terrestris) at a temporary pond. Ethology, Ecology and Evolution 6:331–349.

Dodd, C.K, Jr. 1995a. Reptiles and amphibians in the endangered longleaf pine ecosystem. In Our Living Resources: A Report to the Nation on the Distribution, Abundance, and Health of U.S. Plants, Animals, and Ecosystems. National Biological Service, U.S. Department of the Interior, Washington, D.C.

Dodd, C.K., Jr. 1995b. The ecology of a sandhills population of the eastern narrow-mouthed toad, Gastrophryne carolinensis, during a drought. Bulletin of the Florida Museum of Natural History 38:11–41.

Dodd, C.K., Jr. 1995c. The rare newt of Trail Ridge. Reptile & Amphibian Magazine (July/Aug):36–37, 39–45.

Dodd, C.K., Jr. 1996. Use of terrestrial habitats by amphibians in the sandhill uplands of north-central Florida. Alytes 14:42–52.

Dodd, C.K., Jr. 1997. Imperiled amphibians: a historical perspective. Pp. 165–200. In Benz, G.W. and D.E. Collins (Eds.), Aquatic Fauna in Peril: The Southeastern Perspective. Special Publication Number 1, Southeast Aquatic Research Institute, Lenz Design and Communications, Decatur, Georgia.

Dodd, C.K. 1998. Desmognathus auriculatus at Devil's Millhopper State Geological site, Alachua County, Florida. Florida Scientist 61:38–45.

Dodd, C.K., Jr. and E.D. Brodie Jr. 1976. Observations on the mental hedonic gland-cluster of eastern salamanders of the genus Plethodon. Chesapeake Science 17:129–131.

Dodd, C.K., Jr. and B.S. Cade. 1998. Movement patterns and the conservation of amphibians breeding in small, temporary wetlands. Conservation Biology 12:331–339.

Dodd, C.K. and P.V. Cupp. 1978. The effect of temperature and stage of development on the duration of immobility in selected anurans. British Journal of Herpetology 5:783–788.

Dodd, C.K., Jr. and L.V. LaClaire. 1995. Biogeography and status of the striped newt (Notophthalmus perstriatus) in Georgia, USA. Herpetological Natural History 3:37–46.

Dodd, C.K., Jr. and R.A. Seigel. 1991. Relocation, repatriation and translocation of amphibians and reptiles: are they conservation strategies that work? Herpetologica 47:335–350.

Dodd, C.K., Jr., J.A. Johnson and E.D. Brodie Jr. 1974. Noxious skin secretions of an eastern small Plethodon, P. nettingi hubrichti. Journal of Herpetology 8:89–92.

Dodd, C.K., Jr., G.E. Drewry, R.M. Nowak, J.M. Sheppard and J.D. Williams. 1985. Endangered and threatened wildlife and plants; a review of vertebrate wildlife; notice of review. U.S. Fish and Wildlife Service, Part III, 50 CFR Part 17, Federal Register 50:37958–37967.

Dodson, S.I. 1970. Complementary feeding niches sustained by size-selective predation. Limnology and Oceanography 15:131–137.

Dodson, S.I. 1974. Zooplankton competition and predation: an experimental test of the size-efficiency hypothesis. Ecology 55:605–613.

Dodson, S.I. 1982. Chemical and biological limnology of six west-central Colorado mountain ponds and their susceptibility to acid rain. American Midland Naturalist 107:1733–1739.

Dodson, S.I. and V.E. Dodson. 1971. The diet of Ambystoma tigrinum larvae from western Colorado. Copeia 1971:614–624.

Dole, J.W. 1965a. Spatial relations in natural populations of the leopard frog, Rana pipiens Schreber, in northern Michigan. American Midland Naturalist 74:464–478.

Dole, J.W. 1965b. Summer movements of adult leopard frogs, Rana pipiens Schreber, in northern Michigan. Ecology 46:236–255.

Dole, J.W. 1967. The role of substrate moisture and dew in the water economy of leopard frogs, Rana pipiens. Copeia 1968:141–149.

Dole, J.W. 1968. Homing in leopard frogs, Rana pipiens. Ecology 49:386–399.

Dole, J.W. 1971. Dispersal of recently metamorphosed leopard frogs, *Rana pipiens*. Copeia 1971:221–228.

Dole, J.W. 1972a. Evidence of celestial orientation in newly-metamorphosed *Rana pipiens*. Herpetologica 28:273–276.

Dole, J.W. 1972b. The role of olfaction and audition in the orientation of leopard frogs, *Rana pipiens*. Herpetologica 28:258–260.

Dole, J.W. 1972c. Homing and orientation of displaced toads *Bufo americanus* to their home sites. Copeia 1972:151–158.

Dole, J.W. 1973. Celestial orientation in recently metamorphosed *Bufo americanus*. Herpetologica 29:59–62.

Dole, J.W. 1974. Home range in the canyon treefrog *(Hyla cadaverina)*. Southwestern Naturalist 19:105–119.

Dollfus, R.P. 1950. Trematodes recoltes au Congo Belge par le Professeur Paul Brien. Annals of the Museum Congo Belge, Zoologie Serie 5:1–136.

Donaldson, L.R. 1934. The occurrence of *Ascaphus truei* east of the continental divide. Copeia 1934:184.

Donnelly, M.A. 1994. Amphibian diversity and natural history. Pp. 199–209. *In* McDade, L.A., K.S. Bawa, H.A. Hespenheide and G.S Hartshorn (Eds.), La Selva: Ecology and Natural History of a Neotropical Rainforest. University of Chicago Press, Chicago, Illinois.

Donnelly, M.A. and M.L. Crump. 1998. Potential effects of climate change on two Neotropical amphibian assemblages. Climatic Change 39:541–561.

Donnelly, M.A. and C. Guyer. 1994. Patterns of reproduction and habitat use in an assemblage of Neotropical hylid frogs. Oecologia 98:291–302.

Donovan, L.A. and G.W. Folkerts. 1972. Foods of the seepage salamander, *Desmognathus aeneus* Brown and Bishop. Herpetologica 28:35–37.

Doody, J.S. 1996. Larval growth rate of known age *Ambystoma opacum* in Louisiana under natural conditions. Journal of Herpetology 30:294–297.

Doody, J.S. and J.E. Young. 1995. Temporal variation in reproduction and clutch mortality of leopard frogs *(Rana utricularia)* in south Mississippi. Journal of Herpetology 29:614–616.

Doody, J.S., J.E. Young and G.N. Johnson. 1995. Natural history notes: *Rana capito* (gopher frog). Combat. Herpetological Review 26:202–203.

Doty, T.L. 1978. A study of larval amphibian population dynamics in a Rhode Island vernal pond. Ph.D. dissertation. University of Rhode Island, Kingston, Rhode Island.

Douglas, M.E. 1979. Migration and sexual selection in *Ambystoma jeffersonianum*. Canadian Journal of Zoology 57:2303–2310.

Douglas, M.E. and B.L. Monroe Jr. 1981. A comparative study of topographical orientation in *Ambystoma* (Amphibia: Caudata). Copeia 1981:460–463.

Doving, K.B. 1991. Assessment of animal behaviour as a method to indicate environmental toxicity. Comparative Biochemistry and Physiology C 100:247–252.

Dowdey, T.G. and E.D. Brodie Jr. 1989. Antipredator strategies of salamanders: individual and geographic variation in responses of *Eurycea bislineata* to snakes. Animal Behaviour 38:707–711.

Dowling, H.G. 1956. Geographic relations of Ozarkian amphibians and reptiles. Southwestern Naturalist 1:174–189.

Dowling, H.G. 1957. A review of the amphibians and reptiles of Arkansas. Occasional Papers of the University of Arkansas Museum, Number 3, Fayetteville, Arkansas.

Dowling, H.G. 1993. Viewpoint: a reply to Collins (1991, 1992). Herpetological Review 24:11–13.

Downs, F.L. 1978. Unisexual *Ambystoma* from the Bass Islands of Lake Erie. Occasional Papers of the Museum of Zoology, Number 685, University of Michigan, Ann Arbor, Michigan.

Downs, F.L. 1989a. Family Ambystomatidae. Pp. 87–172. *In* Pfingsten, R.A. and F.L. Downs (Eds.), Salamanders of Ohio. Ohio Biological Survey Bulletin, New Series Volume 7, Number 2, Columbus, Ohio.

Downs, F.L. 1989b. *Ambystoma jeffersonianum* (Green), Jefferson salamander. Pp. 88–101. *In* Pfingsten, R.A. and F.L. Downs (Eds.), Salamanders of Ohio. Ohio Biological Survey Bulletin, New Series, Volume 7, Number 2, Columbus, Ohio.

Downs, F.L. 1989c. *Ambystoma laterale* Hallowell, blue-spotted salamander. Pp. 102–107. *In* Pfingsten, R.A. and F.L. Downs (Eds.), Salamanders of Ohio. Ohio Biological Survey Bulletin, New Series, Volume 7, Number 2, Columbus, Ohio.

Downs, F.L. 1989d. *Ambystoma maculatum* (Shaw), spotted salamander. Pp. 108–125. *In* Pfingsten, R.A. and F.L. Downs (Eds.), Salamanders of Ohio. Ohio Biological Survey Bulletin, New Series, Volume 7, Number 2, Columbus, Ohio.

Downs, F.L. 1989e. *Ambystoma platineum* (Cope), silvery salamander. Pp. 132–139. *In* Pfingsten, R.A. and F.L. Downs (Eds.), Salamanders

of Ohio. Ohio Biological Survey Bulletin, New Series, Volume 7, Number 2, Columbus, Ohio.

Downs, F.L. 1989f. *Ambystoma texanum* (Matthes), smallmouth salamander. Pp. 139–154. *In* Pfingsten, R.A. and F.L. Downs (Eds.), Salamanders of Ohio. Ohio Biological Survey Bulletin, New Series, Volume 7, Number 2, Columbus, Ohio.

Downs, F.L. 1989g. *Ambystoma tigrinum* (Green), tiger salamander. Pp. 155–166. *In* Pfingsten, R.A. and F.L. Downs (Eds.), Salamanders of Ohio. Ohio Biological Survey Bulletin, New Series, Volume 7, Number 2, Columbus, Ohio.

Downs, F.L. 1989h. *Ambystoma tremblayi* Comeau, Tremblay's salamander. Pp. 166–172. *In* Pfingsten, R.A. and F.L. Downs (Eds.), Salamanders of Ohio. Ohio Biological Survey Bulletin, New Series, Volume 7, Number 2, Columbus, Ohio.

Drake, C.J. 1914. The food of *Rana pipiens* Schreber. Ohio Naturalist 14:257–269.

Drewien, R.C. 1968. Ecological relationships of breeding blue-winged teal to prairie potholes. Master's thesis. South Dakota State University, Brookings, South Dakota.

Dreyer, T.F. 1913. The "plathander" (*Xenopus laevis*). Transactions of the Royal Society of South Africa 3:341–355.

Driscoll, D. 1997. Genetic structure, metapopulation processes and evolution influence the conservation strategies for two endangered frog species. Biological Conservation 83:43–54.

Driscoll, D., G. Wardell-Johnson and J.D. Roberts. 1994. Genetic structuring and distribution patterns in rare southwestern Australian frogs: implications for translocation programmes. Pp. 85–90. *In* Serena, M. (Ed.), Reintroduction Biology of Australian and New Zealand Fauna. Surrey Beatty and Sons, Chipping Norton, New South Wales, Australia.

Driscoll, D.A. 1998. Counts of calling males as estimates of population size in the endangered frogs *Geocrinia alba* and *G. vitellina*. Journal of Herpetology 32:475–481.

Driver, E.C. 1936. Observations on *Scaphiopus holbrookii* (Harlan). Copeia 1936:69.

Droege, S. 1990. The North American Breeding Bird Survey. Pp. 1–4. *In* Sauer, J.R. and S. Droege (Eds.), Survey Designs and Statistical Methods for the Estimation of Avian Population Trends. U.S. Fish and Wildlife Service, Biological Report, Number 90, Washington, D.C.

Drost, C.A. and G.M. Fellers. 1996. Collapse of a regional frog fauna in the Yosemite area of the California Sierra Nevada, USA. Conservation Biology 10:414–425.

Drost, C.A. and M.K. Sogge. 1993. Survey of an isolated northern leopard frog population along the Colorado River in Glen Canyon National Recreation Area. National Park Service/Cooperative Park Studies Unit, Northern Arizona University Report, Flagstaff, Arizona.

Drury, R. and W. Gessing Jr. 1940. Additions to the herpetofauna of Kentucky. Herpetologica 2:31–32.

Dubois, A. 1980. Populations, polymorphisme et adaptation: quelques exemples chez lez amphibiens anoures. Pp. 141–158. *In* Barbault, R., P. Blandin and J.A. Meyer (Eds.), Recherches d'Ecologie Theorique, Les Strategies Adaptives. Acutes du Colloque D'Ecologie Theorique Organise a l'Ecole Normale Superieure de Paris les 18, 19, 20 Mai 1978, S.Q. Maloine, Paris.

Dubois, A. 1998. Lists of European species of amphibians and reptiles: will we soon be reaching "stability"? Amphibia-Reptilia 19:1–28.

Dubois, G. and J. Mahon. 1959. Etude de quelques trematodes Nord-Americains suivie díune revision des genres Galactosomum Looss 1899 et Ochetosoma Braun 1901. Bulletin Societe Neuchatelloise Sciences Naturalles 82:191–229.

Dubois, N., D.J. Marcogliese and P. Magnan. 1996. Effects of the introduction of white sucker, *Catostomus commersoni*, on the parasite fauna of brook trout, *Salcelinus fontinalis*. Canadian Journal of Zoology 74:1304–1312.

Ducey, P.K. 1989. Agonistic behavior and biting during intraspecific encounters in *Ambystoma* salamanders. Herpetologica 42:155–160.

Ducey, P.K. and E.D. Brodie Jr. 1983. Salamanders respond selectively to contacts with snakes: survival advantage of alternative antipredator strategies. Copeia 1983:1036–1041.

Ducey, P.K. and J. Dulkiewicz. 1994. Ontogenetic variation in antipredator behavior of the newt *Notophthalmus viridescens*: comparisons of terrestrial adults and efts in field and laboratory tests. Journal of Herpetology 28:530–533.

Ducey, P.K. and J. Heuer. 1991. Effects of food availability on interspecific aggression in salamanders of the genus *Ambystoma*. Canadian Journal of Zoology 69:288–290.

Ducey, P.K. and P. Ritsema. 1988. Intraspecific aggression and responses to marked substrates in *Ambystoma maculatum* (Caudata: Ambystomatidae). Copeia 1988:1008–1013.

Ducey, P.K., K. Schramm and N. Cambry. 1994. Interspecific aggression between the sympatric salamanders, *Ambystoma maculatum* and *Plethodon cinereus*. American Midland Naturalist 131:320–329.

Ducey, R. and E.D. Brodie Jr. 1983. Antipredator adaptations of salamanders: evolution and convergence among terrestrial species. Pp. 109–133. *In* Margaris, N.S., M. Arianoutsou-Faraggitaki and R.J. Reiter (Eds.), Plant, Animal and Microbial Adaptations to the Terrestrial Environment. Plenum Publishing Corporation, New York.

Duellman, W.E. 1948. An *Ambystoma* eats a snake. Herpetologica 4:164.

Duellman, W.E. 1954a. Observations on autumn movements of the salamander *Ambystoma tigrinum tigrinum* in southeastern Michigan. Copeia 1954:156–157.

Duellman, W.E. 1954b. The salamander *Plethodon richmondi* in southwestern Ohio. Copeia 1954:40–45.

Duellman, W.E. 1955. Notes on reptiles and amphibians from Arizona. Occasional Papers of the Museum of Zoology, Number 569, University of Michigan, Ann Arbor, Michigan.

Duellman, W.E. 1961. The amphibians and reptiles of Michoacán, Mexico. University of Kansas Publications, Museum of Natural History, Number 15, Lawrence, Kansas.

Duellman, W.E. 1967. Social organization in the mating calls of some neotropical anurans. American Midland Naturalist 77:156–163.

Duellman, W.E. 1968. *Smilisca baudinii*. Pp. 59.1–59.2. Catalogue of American Amphibians and Reptiles, Society for the Study of Amphibians and Reptiles, St. Louis, Missouri.

Duellman, W.E. 1970. The hylid frogs of Middle America, Volume 2. Monograph of the Museum of Natural History, University of Kansas, Lawrence, Kansas.

Duellman, W.E. 1971. The burrowing toad, *Rhinophrynus dorsalis*, on the Caribbean lowlands of Central America. Herpetologica 27:55–56.

Duellman, W.E. 1990. Herpetofaunas in neotropical rainforests: comparative composition, history and resource use. Pp. 455–508. *In* Gentry, A.H. (Ed.), Four Neotropical Rainforests. Yale University Press, New Haven, Connecticut.

Duellman, W.E. 1993. Amphibian species of the World: Additions and corrections. Museum of Natural History, Special Publication Number 21, University of Kansas, Lawrence, Kansas.

Duellman, W.E. (Ed.). 1999. Patterns of Distribution of Amphibians: A Global Perspective. Johns Hopkins University Press, Baltimore, Maryland, and London, England.

Duellman, W.E. 2001. The hylid frogs of Middle America. Expanded edition. Society for the Study of Amphibians and Reptiles, St. Louis, Missouri.

Duellman, W.E. and A. Schwartz. 1958. Amphibians and reptiles of southern Florida. Bulletin of the Florida State Museum 3:181–324.

Duellman, W.E. and S.S. Sweet. 1999. Distribution patterns of amphibians in the Nearctic Region of North America. Pp. 31–109. *In* Duellman, W.E. (Ed.), Patterns of Distribution of Amphibians. A Global Perspective. Johns Hopkins University Press, Baltimore, Maryland.

Duellman, W.E. and L. Trueb. 1966. Neotropical hylid frogs, genus *Smilisca*. University of Kansas Publications, Museum of Natural History 17:281–375.

Duellman, W.E. and L. Trueb. 1986. Biology of Amphibians. McGraw-Hill Publishing Company, New York.

Duellman, W.E. and J.T. Wood. 1954. Size and growth of the two-lined salamander, *Eurycea bislineata rivicola*. Copeia 1954:92–96.

Dugés, A.1869. Catalogo de animals vertebrados observados en la Republica Mexicana. La Naturaleza 1:137–145.

Dukes, J.S and H.A. Mooney. 1999. Does global change increase the success of biological invaders? TRENDS in Ecology and Evolution 14:135–139.

Dumas, P.C. 1955. Eggs of the salamander *Plethodon dunni* in nature. Copeia 1955:65.

Dumas, P.C. 1956. The ecological relations of sympatry in *Plethodon dunni* and *Plethodon vehiculum*. Ecology 37:484–495.

Dumas, P.C. 1957. Range extension of the salamander *Plethodon vandykei idahoensis*. Copeia 1957:147–148.

Dumas, P.C. 1964. Species-pair allopatry in the genera *Rana* and *Phrynosoma*. Ecology 45:178–181.

Dumas, P.C. 1966. Studies of the *Rana* species complex in the Pacific Northwest. Copeia 1966:60–74.

Duméril A.M.C. and G. Bibron. 1841. Erpetologie general ou histoire naturelle complete des reptiles, Volume 8. Roret, Paris, France.

Duméril, A.-M.-C., G. Bibron and A. Duméril. 1854. Erpetologie generale ou histoire naturelle complete des reptiles, Volume 9, Librairie Enclycopedique de Roret, Paris, France.

Dunagan, T.T. and D.M. Miller. 1973. Some morphological and functional observations on *Fessisentis fessus* Van Cleave (Acanthocephala) from the dwarf salamander, *Siren intermedia* Le Conte. Proceedings of the Helminthological Society of Washington 40:209–216.

Dunbar, J.R. and J.D. Moore. 1979. Correlations of host specificity with host habitat in helminths parasitizing the plethodontids of Washington County, Tennessee. Journal of the Tennessee Academy of Science 54:106–109.

Duncan, M.L. 1967. A study of the reproductive biology of the seal salamanders, *Desmognathus monticola*, in Bedford County, Virginia. Master's thesis. Radford College, Radford, Virginia.

Duncan, R. and R. Highton. 1979. Genetic relationships of the eastern large *Plethodon* of the Ouachita Mountains. Copeia 1979:96–110.

Duncan, R.B. 1998. Bagging the desert slender salamander. Sonoran Herpetologist 11:38–39.

Duncan, R.B. and T.C. Esque. 1986. An ecological study of a slender salamander (*Batrachoseps* species) population at the Guadalupe Creek Study Site of the Santa Rosa Mountains, Riverside County, California. U.S.D.I. Bureau of Land Management, Contract Number CA-950-CT4-11, Indio Resource Area, Riverside, California.

Dundee, H.A. 1947. Note on salamanders collected in Oklahoma. Copeia 1947:117–120.

Dundee, H.A. 1950. Notes on the type locality of *Eurycea multiplicata* (Cope). Herpetologica 6:27–28.

Dundee, H.A. 1957. Partial metamorphosis induced in *Typhlomolge rathbuni*. Copeia 1957:52–53.

Dundee, H.A. 1958. Habitat selection by aquatic plethodontid salamanders of the Ozarks, with studies of their life histories (abstract). Dissertation Abstracts International 19:1480–1481.

Dundee, H.A. 1962. Response of the neotenic salamander *Haideotriton wallacei* to a metamorphic agent. Science 135:1060–1061.

Dundee, H.A. 1965a. *Eurycea multiplicata*. Pp. 21.1–21.2. Catalogue of American Amphibians and Reptiles. Society for the Study of Amphibians and Reptiles, St. Louis, Missouri.

Dundee, H.A. 1965b. *Eurycea tynerensis*. Pp. 22.1–22.2. Catalogue of American Amphibians and Reptiles. Society for the Study of Amphibians and Reptiles, St. Louis, Missouri.

Dundee, H.A. 1971. *Cryptobranchus alleganiensis*. Pp. 101.1–101.4. Catalogue of American Amphibians and Reptiles. Society for the Study of Amphibians and Reptiles, St. Louis, Missouri.

Dundee, H.A. 1998. *Necturus punctatus*. Pp. 663.1–663.5. Catalogue of American Amphibians and Reptiles. Society for the Study of Amphibians and Reptiles, St. Louis, Missouri.

Dundee, H.A. and D.S. Dundee. 1965. Observations on the systematics and ecology of *Cryptobranchus* from the Ozark Plateaus of Missouri and Arkansas. Copeia 1965:369–370.

Dundee, H.A. and E.A. Liner. 1985. Natural history notes: *Phrynohyas venulosa* (veined treefrog). Food. Herpetological Review 16:109.

Dundee, H.A. and D.A. Rossman 1989. The Amphibians and Reptiles of Louisiana. Louisiana State University Press, Baton Rouge, Louisiana.

Dunham, A.E. 1993. Population responses to environmental change: operative environments, physiologically structured models, and population dynamics. Pp. 95–119. *In* Kareiva, P.M., J.G. Kingsolver and R.B. Huey (Eds.), Biotic Interactions and Global Change. Sinauer Associates, Sunderland, Massachusetts.

Dunham, A.E., D.B. Miles and D.N. Reznick. 1988. Life history patterns in squamate reptiles. Pp. 441–522. *In* Gans, C. and R.B. Huey (Eds.), Biology of the Reptilia. Volume 16, Ecology B. Alan R. Liss, New York.

Dunlap, D.G. 1955. Inter- and intraspecific variation in Oregon frogs of the genus *Rana*. American Midland Naturalist 54:314–331.

Dunlap, D.G. 1963. The status of the gray treefrog, *Hyla versicolor*, in South Dakota. Proceedings of the South Dakota Academy of Science 42:136–139.

Dunlap, D.G. 1967. Selected records of amphibians and reptiles from South Dakota. Proceedings of the South Dakota Academy of Science 46:100–106.

Dunlap, D.G. 1977. Wood and western spotted frogs (Amphibian, Anura, Ranidae) in the Big Horn Mountains of Wyoming. Journal of Herpetology 11:85–87.

Dunlap, D.G. and R.M. Storm. 1951. The Cascade frog in Oregon. Copeia 1951:81.

Dunn, E.R. 1916. Two new salamanders of the genus *Desmognathus*. Proceedings of the Biological Society of Washington 29:73–76.

Dunn, E.R. 1917a. Reptile and amphibian collections from the North Carolina mountains, with especial reference to salamanders. Bulletin of the American Museum of Natural History 37:593–634.

Dunn, E.R. 1917b. The breeding habits of *Ambystoma opacum* (Gravenhorst). Copeia 43:40–44.

Dunn, E.R. 1920. Some reptiles and amphibians from Virginia, North Carolina, Tennessee and Alabama. Proceedings of the Biological Society of Washington 33:129–137.

Dunn, E.R. 1923. Mutanda herpetologica. Proceedings of the New England Zoological Club 8:39–40.

Dunn, E.R. 1924. Siren, a herbivorous salamander? Science 59:145.

Dunn, E.R. 1926. Salamanders of the Family Plethodontidae. Smith College 50th Anniversary Publication. Northhampton, Massachusetts.

Dunn, E.R. 1935. The survival value of specific characters. Copeia 1935:85–98.

Dunn, E.R. 1937. The status of *Hyla evittata* Miller. Proceedings of the Biological Society of Washington 50:9–10.

Dunn, E.R. 1940. The races of *Ambystoma tigrinum*. Copeia 1940:154–162.

Dunn, E.R. 1941. Notes on *Dendrobates auratus*. Copeia 1941:88–93.

Dunn, E.R. 1942. An egg cluster of *Aneides ferreus*. Copeia 1942:52.

Dunn, E.R. and A.A. Heinze. 1933. A new salamander from the Ouachita Mountains. Copeia 1933:121–122.

Dunson, W.A. and J. Connell. 1982. Specific inhibition of hatching in amphibian embryos by low pH. Journal of Herpetology 16:314–316.

Dupre, R.K. and J.W. Petranka. 1985. Ontogeny of temperature selection in larval amphibians. Copeia 1985:462–467.

Dupuis, L. 1997. Effects of logging on terrestrial-breeding amphibians on Vancouver Island. Pp. 258–270. *In* Green, D.M. (Ed.), Amphibians in Decline: Canadian Studies of a Global Problem. Herpetological Conservation, Number 1, Society for the Study of Amphibians and Reptiles, St. Louis, Missouri.

Dupuis, L.A. and D. Steventon. 1999. Riparian management and the tailed frog in northern coastal forests. Forest Ecology and Management 124:35–43.

Dupuis, L.A., F.L. Bunnell and P.A. Friele. 2000. Determinants of the tailed frog's range in British Columbia, Canada. Northwest Science 74:109–115.

Dureka, T. 1993. Letter to Andy Sansom, 18 July.

Dureka, T. 1994. Stop–look–listen. Beacon 4:1, 9.

Dureka, T. 1996. TPWD steps on Houston toad in 1995. Beacon 6:6.

Dureka, T. 1998. The Lost Pines under siege. Beacon 8:1, 8–9, 11–12.

Dureka, T. 2001. Alcoa pits neighbors against neighbors. Beacon 11:6.

Dureka, T. and A. Mesrobian. 1999. Showdown in the pines: urban sprawl vs. the Houston toad. Beacon 9:2.

Dureka, T. and A. Mesrobian. 2000. Showdown in the pines: urban sprawl vs. the Houston toad. Available at www.bcen.org/showdown.html.

Durham, F.E. 1956. Amphibians and reptiles of the North Rim, Grand Canyon, Arizona. Herpetologica 12:220–224.

Durham, L. and G.W. Bennet. 1963. Age, growth and homing in the bullfrog. Journal of Wildlife Management 27:107–123.

Durkin, B. 1995. Aging frogs by counting the rings in their bones (yeah—just like with trees!). Unpublished report, Grand Canyon University, Phoenix, Arizona.

DuShane, D.P. and C. Hutchinson. 1944. Differences in size and developmental rate between eastern and midwestern embryos of *Ambystoma maculatum*. Ecology 25:414–424.

Dusi, J. 1949. The natural occurrence of "redleg," *Pseudomonas hydrophila*, in a population of American toads, *Bufo americanus*. Ohio Journal of Science 49:70–71.

Duszynski, D.W. and K.L. Jones. 1973. The occurrence of intradermal mites, *Hannemania sp.* (Acarina: Trombiculidae), in anurans in New Mexico with a histological description of the tissue capsule. International Journal of Parasitology 3:531–538.

Dvornich, K.M., K.R. McAllister and K.B. Aubry. 1997. Amphibians and reptiles of Washington state: location data and predicted distributions. Volume 2. *In* Cassidy, K.M., C.E. Grue and K.M. Dvornich (Eds.). Washington State GAP Analysis—Final Report. Washington Cooperative Fish and Wildlife Unit, University of Washington, Seattle, Washington (http://w3.cqs.washington.edu/~wagap/herps/index.html).

Dye, R.L. 1982. Sandhill cranes prey on amphiumas. Florida Field Naturalist 10:76.

Dyer, W.G. 1973. *Falcaustra chabaudi* sp. n. (Nematoda: Kathlaniidae) from the western lesser siren, *Siren intermedia nettingi* Goin, 1942. Journal of Parasitology 59:994–996.

Dyer, W.G. 1975. Parasitism as an indicator of food sources in a cave-adapted salamander habitat. Bulletin of the Southern California Academy of Sciences 74:72–75.

Dyer, W.G., R.A. Brandon and R.L. Price. 1980. Gastrointestinal helminths in relation to sex and age of *Desmognathus fuscus* (Green 1818) from Illinois. Proceedings of the Helminthological Society of Washington 47:95–99.

Dyrkacz, S. 1974. Rare and/or endangered species of Illinois: reptiles and amphibians. Bulletin of the Chicago Herpetological Society 9:15–16.

Dyrkacz, S. 1981. Recent instances of albinism in North American amphibians and reptiles. Herpetological Circular, Number 11, Society for the Study of Amphibians and Reptiles, St. Louis, Missouri.

Dzubin, A. 1969. Comments on carrying capacity of small ponds for ducks and possible effects of density on mallard production. Pp. 138–160. *In* Saskatoon Wetlands Seminar. Canadian Wildlife Service Report, Number 6, Ottawa, Ontario, Canada.

Eagleson, G.W. 1976. A comparison of the life histories and growth patterns of populations of the salamander *Ambystoma gracile* (Baird) from permanent low altitude and mountain lakes. Canadian Journal of Zoology 54:2098–2111.

Earl, P.R. 1979. Notes on the taxonomy of opalinids (Protozoa), including remarks on continental drift. Transactions of the American Microscopical Society 98:549–557.

Easteal, S. 1981. The history of introductions of *Bufo marinus* (Amphibia: Anura): a natural experiment in evolution. Biological Journal of the Linnean Society 16:93–113.

Easteal, S. 1982. The genetics of introduced populations of the marine toad, *Bufo marinus* (Linnaeus), (Amphibia: Anura); a natural experiment in evolution. Ph.D. dissertation. Griffith University, School of Australian Environmental Studies, Brisbane, Queensland, Australia.

Easteal, S. 1985. The ecological genetics of introduced populations of the giant toad, *Bufo marinus*. III. Geographical patterns of variation. Evolution 39:1065–1075.

Easteal, S. 1986. *Bufo marinus*. Pp. 395.1–395.4. Catalogue of American Amphibians and Reptiles. Society for the Study of Amphibians and Reptiles, St. Louis, Missouri.

Easterla, D.A. 1968. Melanistic spotted salamanders in northeast Arkansas. Herpetologica 24:330–331.

Easterla, D.A. 1971. A breeding concentration of four-toed salamanders, *Hemidactylium scutatum*, in southeastern Missouri. Journal of Herpetology 5:194–195.

Easterla, D.A. 1972. Herpetological records for northwest Missouri. Transactions of the Missouri Academy of Science 6:158–160.

Eaton, T.H., Jr. 1941. Notes on the life history of *Dendrobates auratus*. Copeia 1941:93–95.

Eaton, T.H., Jr. 1953. Salamanders of Pitt Co., North Carolina. Journal of the Elisha Mitchell Scientific Society 69:49–53.

Eaton, T.H., Jr. 1954. *Desmognathus perlapsus* Neill in North Carolina. Herpetologica 10:41–43.

Eaton, T.H., Jr. 1956. Larvae of some Appalachian plethodontid salamanders. Herpetologica 12:303–311.

Eaton, T.H. and G.T. Eaton. 1956. A new locality for the green salamander and wood frog in North Carolina. Herpetologica 12:312.

Echelle, A.A. 1990. In defense of the phylogenetic species concept and the ontological status of hybridogenetic taxa. Herpetologica 46:109–113.

Edgren, R.A. 1949. An autumnal concentration of *Ambystoma jeffersonianum*. Herpetologica 6:137–138.

Edgren, R.A. 1955. The natural history of the hognose snakes, genus *Heterodon*: a review. Herpetologica 11:105–117.

Edgren, R.A. and W.T. Stille. 1948. Checklist of Chicago area amphibians and reptiles. Natural History Miscellanea, Number 26, Chicago Academy of Sciences, Chicago, Illinois.

Edwards, R.J., G. Longley, R. Moss, J. Ward, R. Matthews and B. Stewart. 1989. A classification of Texas aquatic communities with special consideration toward the conservation of endangered and threatened taxa. Texas Journal of Science 41:231–240.

Efford, I.E. and J.A. Mathias. 1969. A comparison of two salamander populations in Marion Lake, British Columbia. Copeia 1969:723–736.

Efford, I.E. and K. Tsumura. 1969. Observations on the biology of the trematode *Megalodiscus microphagus* in amphibians from Marion Lake, British Columbia. American Midland Naturalist 82:197–203.

Efford, I.E. and K. Tsumura. 1973. A comparison of the food of salamanders and fish in Marion Lake, British Columbia. Transactions of the American Fisheries Society 1:33–47.

Ehmann, H. and H. Cogger. 1985. Australia's endangered herpetofauna: a review of criteria and policies. Pp. 435–447. *In* Grigg, G., R. Shine and H. Ehmann (Eds.), Biology of Australasian Frogs and Reptiles. Surrey Beatty and Sons, Sydney, New South Wales, Australia.

Ehrenfeld, J.G. 1983. The effects of changes in land-use on swamps of the New Jersey Pine Barrens. Biological Conservation 25:353–375.

Ehrenfeld, J.G. and J.P. Schneider. 1983. The sensitivity of cedar swamps to the effects of non-point source pollution associated with suburbanization in the New Jersey Pine Barrens. Center for Coastal and Environmental Studies, Rutgers, New Brunswick, New Jersey.

Ehrlich, D. 1979. Predation by bullfrog tadpoles *(Rana catesbeiana)* on eggs and newly hatched larvae of the plains leopard frog *(Rana blairi)*. Bulletin of the Maryland Herpetological Society 15:25–26.

Ehrlich, P.R. 1989. Attributes of invaders and the invading processes: vertebrates. Pp. 315–328. *In* Drake, J.A., H.A. Mooney, F. di Castri, R.H. Groves, F.J. Kruger, M. Rejmanek and M. Williamson (Eds.), Biological Invasions: A Global Perspective. John Wiley and Sons, New York.

Ehrlich, P.R., D.S. Dobkin and D. Wheye. 1988. The Birder's Handbook: A Field Guide to the Natural History of North American Birds. Simon and Schuster, New York.

Einem, G.E. and L.D. Ober. 1956. The seasonal behavior of certain Floridian Salientia. Herpetologica 12:205–212.

Eisenreich, S.J., B.B. Looney and J.D. Thorton. 1981. Airborne organic contaminants in the Great Lakes ecosystem. Environmental Science and Technology 15:30–38.

Eklund, T.J., W.H. McDowell and C.M. Pringle. 1997 Seasonal variation of tropical precipitation chemistry: La Selva, Costa Rica. Atmospheric Environment 31:3903–3910.

El Mofty, M.M. and I.A. Sadek. 1973. The mechanism of action of adrenaline in the induction of sexual reproduction (encystation) in *Opalina sudafricana* parasitic in *Bufo regularis*. International Journal for Parasitology 3:425–431.

Eldredge, N. (Ed.). 1992. Systematics, Ecology, and the Biodiversity Crisis. Columbia University Press, New York.

Eldredge, N. and S.J. Gould. 1972. Punctuated equilibria: an alternative to phyletic gradualism. Pp. 82–115. *In* Schopf, T.J.M (Ed.), Models of Paleobiology. Freeman Cooper Press, San Francisco, California.

Elepfandt, A. 1996. Sensory perception and the lateral line system in the clawed frog, *Xenopus*. Pp. 97–120. *In* Tinsley, R.C. and H.R. Kobel (Eds.), The Biology of *Xenopus*. Clarendon Press, Oxford.

Elgar, M.A. and B.J. Crespi (Eds.). 1992. Cannibalism: Ecology and Evolution Among Diverse Taxa. Oxford University Press, New York.

Elinson, R.P. 1993. Viable triploid hybrids between a large frog, *Rana catesbeiana*, and a small one, *Rana septentrionalis*. Herpetological Review 24:46–47.

Elkan, E. 1960. Some interesting pathological cases in amphibians. Proceedings of the Zoological Society of London 134:275–296.

Elkan, E. and H.H. Reichenbach-Klinke. 1974. Color Atlas of the Diseases of Fishes, Amphibians, and Reptiles. T.F.H. Publications, Limited, Neptune City, New Jersey.

Elliott, S.A., L.B. Kats and J.A. Breeding. 1993. The use of conspecific chemical cues for cannibal avoidance in California newts *(Taricha torosa)*. Ethology 95:186–192.

Ellis, J. 1767. An account of amphibious Bipes. Philosophical Transactions of the Royal Society of London 56:189–192.

Elphick, C.S. and L.W. Oring. 1998. Winter management of Californian rice fields for waterbirds. Journal of Applied Ecology 35:95–108.

Elton, C.S. 1958. The Ecology of Invasions by Animals and Plants. Meuthuen and Company, London.

Ely, C.A. 1944. Development of *Bufo marinus* larvae in dilute sea water. Copeia 1944:256.

Emery, A.R., A.H. Berst and K. Kodaira. 1972. Under-ice observations of wintering sites of leopard frogs. Copeia 1972:123–126.

Emlen, S.T. 1976. Lek organization and mating strategies in the bullfrog. Behavioral Ecology and Sociobiology 1:283–313.

Emlen, S.T. 1977. "Double clutching" and its possible significance in the bullfrog. Copeia 1977:749–751.

Enderson, E.F. 2002. Geographic distribution: *Eleutherodactylus augusti cactorum* (western barking frog). Herpetological Review 33:316.

Enderson, E.F. and R.L. Bezy. 2000. Geographic distribution: *Pternohyla fodiens* (lowland burrowing treefrog). Herpetological Review 31:251–252.

Endler, J. 1970. Kinesthetic orientation in the California newt *(Taricha torosa)*. Behavior 37:15–23.

Enge, K.M. 1997. A standardized protocol for drift-fence surveys. Florida Game and Fresh Water Fish Commission, Technical Report, Number 14, Tallahassee, Florida.

Enge, K.M. 1998. Herpetofaunal drift-fence survey of steephead ravines in the Apalachicola and Ochlockonee river drainages. Florida Game and Freshwater Fish Commission, Tallahassee, Florida.

Enge, K.M. and W.R. Marion. 1986. Effects of clearcutting and site preparation on herpetofauna of a North Florida flatwoods. Forest Ecology and Management 14:177–192.

Enge, K.M. and C.J. Stine. 1987. Encapsulation, translocation, and hatching success of *Ambystoma tigrinum tigrinum* (Green) egg masses. Bulletin of the Maryland Herpetological Society 23:74–83.

Engelhardt, G.P. 1916a. *Ambystoma tigrinum* on Long Island. Copeia 28:20–22.

Engelhardt, G.P. 1916b. *Ambystoma tigrinum* on Long Island II. Records of larvae. Copeia 30:32–35.

Engelhardt, G.P. 1919. The habitat of *Gyrinophilus porphyriticus*. Copeia 69:20–21.

Engle, J.C. and J.C. Munger. 1998. Population structure of spotted frogs in the Owyhee Mountains. Idaho Bureau of Land Management, Technical Bulletin 98-20, Boise, Idaho.

Engles, W.L. 1952. Vertebrate fauna of North Carolina coastal islands II. Shackleford Banks. American Midland Naturalist 47:702–742.

Engstrom, R.T., R.L. Crawford and W.W. Baker. 1984. Breeding bird populations in relation to changing forest structure following fire exclusion: a 15-year study. Wilson Bulletin 96:437–450.

Ensley, B. and R. Cross. 1984. Geographic distribution: *Ambystoma mabeei* (Mabee's salamander). Herpetological Review 15:20.

Epstein, P., B. Sherman, E. Spanger-Siegfried, A. Langston, S. Prasad and B. McJay. 1998. Marine ecosystems: Emerging disease as indicators of change. Health Ecological and Economic Dimenstion (HEED) of Global Change Program, Durham, New Hampshire.

Erdman, S. and D. Cundall. 1984. The feeding apparatus of the salamander *Amphiuma tridactylum*: morphology and behavior. Journal of Morphology 181:175–204.

Ernst, C.H. 1986. Ecology of the turtle, *Sternotherus odoratus*, in southeastern Pennsylvania. Journal of Herpetology 20:341–352.

Ernst, C.H. and R.W. Barbour. 1989. Snakes of Eastern North America. George Mason University Press, Fairfax, Virginia.

Ernst, E.M. 1974. The parasites of the red-backed salamander, *Plethodon cinereus*. Bulletin of the Maryland Herpetological Society 10:108–114.

Erspamer, V., T. Vitali, M. Roseghini and J.M. Cei. 1967. 5-methoxy- and 5-hydroxy-indolealkylamines in the skin of *Bufo alvarius*. Biochemistry and Pharmacology 16:1149–1164.

Ervin, E.E. and R.N. Fisher. 2001. Natural history notes: *Thamnophis hammondii* (two-striped garter snake). Prey. Herpetological Review. 32:265–266.

Ervin, E.L., A.E. Anderson, T.L. Cass and R.E. Murcia. 2001. Natural history notes: *Spea hammondii* (western spadefoot toad). Elevation record. Herpetological Review 32:36.

Ervin, E.L., R.N. Fisher and K. Madden. 2001. Natural history notes: *Hyla cadaverina* (California treefrog). Predation. Herpetological Review 31:234.

Ervin, E.L., S.J. Mullin, M.L. Warburton and R.N. Fisher. 2003 Natural history notes: *Thamnophis hammondii* (two-striped garter snake). Prey. Herpetological Review 34: 74–75.

Erwin, C. 1979. Geographic distribution: *Ambystoma talpoideum* (mole salamander). Herpetological Review 10:23.

Erwin, T.L. 1991. An evolutionary basis for conservation strategies. Science 253:750–52.

Esch, G.W., C.R. Kennedy, A.O. Bush and J.M. Aho. 1988. Patterns of helminth communities in freshwater fish in Great Britain: alternative strategies for colonization. Parasitology 96:519–532.

Eschscholtz, J.F. 1833. Zoologischer Atlas, enthaltend Abbildungen und Beschreibungen neuer Thierarten, wahrend des Flottcapitains von Kotzebue zweiter Reise um die Welt, auf der Russisch-Kaiserlich Kreigsschlupp Predpriaetie in den Jahren 1823–1826. Number 5. Berlin, Germany.

Esco, K. and J. Jensen. 1996. Georgia Herp Atlas: Matrix of Georgia's amphibians and reptiles. Version 1.0. Georgia Department of Natural Resources, Wildlife Resource Division, Nongame Endangered Wildlife Program, Forsyth, Georgia.

Espinoza, F.A., Jr., J.E. Deacon and A. Simmin. 1970. An economic and biostatistical analysis of the bait fish industry in the Lower Colorado River. University of Nevada, Las Vegas, Special Publication, Las Vegas, Nevada.

ESRI (Environmental Systems Research Incorporated). 1999. Getting to Know ArcView GIS: Self-Study Workbook. ESRI Press, Redlands, California.

Esselstyn, J.A. and R.C. Wildman. 1997. Observations of *Juga* in the diet of larval Pacific giant salamanders *(Dicamptodon tenebrosus)*. Northwestern Naturalist 78:70–73.

Estes, R. and J. Tihen. 1964. Lower vertebrates from the Valentine Formation of Nebraska. American Midland Naturalist 72:453–472.

Etges, W.J. 1987. Call site choice in male anurans. Copeia 1987:910–923.

Etheridge, K. 1990a. The energetics of estivating sirenid salamanders (*Siren lacertina* and *Pseudobranchus striatus*). Herpetologica 46:407–414.

Etheridge, K. 1990b. Water balance in estivating sirenid salamanders (*Siren lacertina*). Herpetologica 46:400–406.

Eubanks, J.T., K.W. Pecor and Z.A. Moore. 2002. Natural history notes: *Siren intermedia nettingi* (western lesser siren). Migration and locomotion. Herpetological Review 33:298–299.

Evans, A.L. and D.C. Forester. 1996. Conspecific recognition by *Desmognathus ochrophaeus* using substrate-borne odor cues. Journal of Herpetology 30:447–451.

Evans, A.L., D.C. Forester and B.S. Masters. 1997. Recognition by population and genetic similarity in the mountain dusky salamander, *Desmognathus ochrophaeus* (Amphibia: Plethodontidae). Oecologia 64:413–418.

Evans, C.M. 1993. Amphibian antipredator mechanisms: adhesive strength of skin gland secretions. Master's thesis. University of Texas, Arlington, Texas.

Evans, H.E. 1947. Herpetology of Crystal Lake, Sullivan County, N.Y. Herpetologica 4:19–21.

Evans, M., C. Yáber and J.-M. Hero. 1996. Factors influencing choice breeding site by *Bufo marinus* in its natural habitat. Copeia 1996:904–912.

Evenden, F.G., Jr. 1948. Food habits of *Triturus granulosus* in western Oregon. Copeia 1948:219–220.

Evermann, B.W. 1897. U.S. Fish Commission investigations at Crater Lake. Mazama 1:230–238.

Evermann, B.W. 1920. Lake Maxinkuckee: A physical and biological survey. Indiana Department of Conservation, Publication Number 7, Volume 1, Indianapolis, Indiana.

Eversole, A.G. 1978. Life-cycles, growth and population bioenergetics in the snail *Helisoma trivolvis* (Say). Journal of Molluscan Studies 44:209–222.

Ewert, M.A. 1969. Seasonal movements of the toads *Bufo americanus* and *B. cognatus* in northwestern Minnesota. Ph.D. dissertation. University of Minnesota, Minneapolis, Minnesota.

Eycleshymer, A.C. 1906. The habits of *Necturus maculosus*. American Naturalist 40:123–135.

Faber, P.A., E. Keller, A. Sands and B.M. Massey. 1989. The ecology of riparian habitats of the southern California coastal region: A community profile. U.S. Fish and Wildlife Service, Biological Report 85(7.27), Washington, D.C.

Faeh, S.A., D.K. Nichols and V.R. Beasley. 1998. Infectious diseases of amphibians. Pp. 259–265 *In* Lannoo, M.J. (Ed.), Status and Conservation of Midwestern Amphibians. University of Iowa Press, Iowa City, Iowa.

Fahrig, L. 1997. Relative effects of habitat loss and fragmentation on population extinction. Journal of Wildlife Management 61:603–610.

Fahrig, L., J.H. Pedlar, S.E. Pope, P.D. Taylor and J.F. Wegner. 1995. Effect of road traffic on amphibian density. Biological Conservation 73:177–182.

Fairchild, J.F., T.W. LaPoint, J.L. Zajicek, M.K. Nelso, F.J. Dwyer and P.A. Lovely. 1992. Population-, community- and ecosystem-level responses of aquatic mesocosms to pulsed doses of a pyrethroid insecticide. Environmental Toxicology and Chemistry 11:115–129.

Fairchild, L. 1981. Mate selection and behavioural thermoregulation in Fowler's toads. Science 212:950–951.

Fairchild, L. 1984. Male reproductive tactics in an explosive breeding toad population. American Zoologist 24:407–418.

Faith, D.P. 1992a. Conservation evaluation and phylogenetic diversity. Biological Conservation 61:1–10.

Faith, D.P. 1992b. Systematics and conservation: on predicting feature diversity of subsets of taxa. Cladistics 8:361–373.

Faith, D.P. 1993. Biodiversity and systematics: the use and misuse of divergence information in assessing taxonomic diversity. Pacific Conservation Biology 1:53–57.

Faith, D.P. 1994a. Genetic diversity and taxonomic priorities for conservation. Biological Conservation 68:69–74.

Faith, D.P. 1994b. Phylogenetic diversity: a general framework for the prediction of feature diversity. Pp. 251–68. *In* Forey, P.L., C.J. Humphries and R.I. Vane-Wright (Eds.), Systematics and Conservation Evaluation. Clarendon Press, Oxford, United Kingdom.

Faith, D.P. 1994c. Phylogenetic pattern and the quantification of organismal biodiversity. Philosophical Transactions of the Royal Society of London B 345:45–58.

Faith, D.P. 1996. Conservation priorities and phylogenetic pattern. Conservation Biology 10:1286–1289.

Faith, D.P. 2002. Quantifying biodiversity: a phylogenetic perspective. Conservation Biology 16:248–252.

Faith, D.P. and P.A. Walker. 1993. DIVERSITY: reference and user's guide. CSIRO Division of Wildlife and Ecology, Canberra, New South Wales, Australia.

Faith, D.P. and P.A. Walker. 1995. Environmental diversity: on the best possible use of surrogate data for assessing the relative biodiversity of sets of areas. Biodiversity and Conservation 5:399–415.

Farmer, A.H. and A.H. Parent. 1997. Effects of landscape on shorebird movements at spring migration stopovers. Condor 99:698–707.

Farner, D.S. 1947. Notes on the food habits of the salamanders of Crater Lake, Oregon. Copeia 1947:259–261.

Farner, D.S. and J. Kezer. 1953. Notes on the amphibians and reptiles of Crater Lake National Park. American Midland Naturalist 50:448–462.

Farrar, E.S. and J.D. Hey. 1997. Carnivorous spadefoot (*Spea bombifrons* Cope) tadpoles and fairy shrimp in western Iowa. Journal of the Iowa Academy of Science 104:4–7.

Farrell, M.P. 1971. Effect of temperature and photoperiod acclimations on the water economy of *Hyla crucifer*. Herpetologica 27:41–48.

Farrell, M.P. and J.A. MacMahon. 1969. An eco-physiological study of water economy in eight species of treefrogs (Hylidae). Herpetologica 25:279–294.

Fauth, J.E. 1999a. Interactions between branchiate mole salamanders (*Ambystoma talpoideum*) and lesser sirens *(Siren intermedia)*: asymmetrical competition and intraguild predation. Amphibia-Reptilia 20: 119–132.

Fauth, J.E. 1999b. Identifying potential keystone species from field data—an example from temporary ponds. Ecology Letters 2:36–43.

Fauth, J.E. and W.J. Resetarits Jr. 1991. Interactions between the salamander *Siren intermedia* and the keystone predator *Notophthalmus viridescens*. Ecology 72:827–838.

Fauth, J.E., W.J. Resetarits Jr. and H.M. Wilbur. 1990. Interactions between larval salamanders: a case of competitive equality. Oikos 58:91–99.

Fawcett, D.W. 1956. Electron microscope observations on intracellular virus-like particles associated with cells of the Lucké renal adenocarcinoma. Journal of Biophysical and Biochemical Cytology 2:725–742.

Feaver, P.E. 1971. Breeding pool selection and larval mortality of three California amphibians: *Ambystoma tigrinum californiense* Gray, *Hyla regilla* Baird and Girard, and *Scaphiopus hammondi hammondi* Girard. Master's thesis. Fresno State College, Fresno, California.

Fedak, M.A. 1971. A comparative study of the life histories of *Necturus lewisi* Brimley and *Necturus punctatus* Gibbes (Caudata: Proteidae) in North Carolina. Master's thesis. Duke University, Durham, North Carolina.

Feder, J.H. 1977. Genetic variation and biochemical systematics in western *Bufo*. Master's thesis. University of California, Berkeley, California.

Feder, J.H. 1979. Natural hybridization and genetic divergence between the toads *Bufo boreas* and *Bufo punctatus*. Evolution 33:1089–1097.

Feder, M.E. 1978. Environmental variability and thermal acclimation in Neotropical and temperate zone salamanders. Physiological Zoology 51:7–16.

Feder, M.E. 1983a. Responses of acute aquatic hypoxia in larvae of the frog *Rana berlandieri*. Journal of Experimental Biology 104:79–95.

Feder, M.E. 1983b. The relation of air breathing and locomotion to predation on tadpoles, *Rana berlandieri*, by turtles. Physiological Zoology 56:522–531.

Feder, M.E. 1984. Consequences of aerial respiration for amphibian larvae. Perspectives in Vertebrate Science 3:71–86.

Feder, M.E. 1985. Thermal acclimation of oxygen consumption and cardiorespiratory frequencies in frog larvae. Physiological Zoology 58:303–311.

Feder, M.E. and P.L. Londos. 1984. Hydric constraints upon foraging in a terrestrial salamander, *Desmognathus ochrophaeus* (Amphibia: Plethodontidae). Oecologia 64:413–418.

Feder, M.E. and C.M. Moran. 1985. Effect of water depth on costs of aerial respiration and its alternatives in tadpoles of *Rana pipiens*. Canadian Journal of Zoology 63:643–648.

Feder, M.E. and F.H. Pough. 1975. Temperature selection by the redbacked salamander, *Plethodon c. cinereus* (Green) (Caudata: Plethodontidae). Comparative Biochemistry and Physiology 50:91–98.

Feder, M.E. and S.J. Arnold. 1982. Anaerobic metabolism and behavior during predatory encounters between snakes *(Thamnophis elegans)* and salamanders *(Plethodon jordani)*. Oecologica 53:93–97.

Feder, M.E. and W.W. Burggren (Eds.). 1992. Environmental Physiology of the Amphibians. University of Chicago Press, Chicago, Illinois.

Feder, M.E., J.F. Lynch, H.B. Shaffer and D.B. Wake. 1982. Field body temperatures of tropical and temperate zone salamanders. Smithsonian Herpetological Information Service, Number 52, Smithsonian Institution, Washington, D.C.

Federal Register. 1977. Endangered and threatened wildlife and plants; review of the status of 10 species of amphibians. Federal Register 42(143):39121–39122.

Federal Register. 1982. Endangered and threatened wildlife and plants; review of vertebrate wildlife for listing as endangered or threatened species. Federal Register 47(251):58454–58460.

Federal Register. 1985. Endangered and threatened wildlife and plants; review of vertebrate wildlife. Federal Register 50(181):37958–37967.

Federal Register. 1989. Endangered and threatened wildlife and plants; animal notice of review. Federal Register 54(4):554–579.

Federal Register. 1991. Endangered and threatened wildlife and plants; animal candidate review for listing as endangered or threatened species. Federal Register 56(225):58804–58836.

Federal Register. 1994. Endangered and threatened wildlife and plants; animal candidate review for listing as endangered or threatened species. Federal Register 59(219):58982–59028.

Federal Register. 1995. Endangered and threatened wildlife and plants; notice of finding on a petition to emergency list the Amargosa toad *(Bufo nelsoni)* as endangered. Federal Register 60(56):15280

Federal Register. 1996a. Endangered and threatened wildlife and plants; 12-month finding for a petition to list the Amargosa toad *(Bufo nelsoni)* as endangered. Federal Register 61(42):8018–8019.

Federal Register. 1996b. Endangered and threatened wildlife and plants; review of plant and animal taxa that are candidates for listing as endangered or threatened species. Federal Register 61(40):7596–7613.

Federal Register. 2003. Endangered and threatened wildlife and plants; determination of endangered status for the Sonoma County distinct population segment of the California tiger salamander; final rule. Federal Register 68(53):13497–13520.

Feldmann, R. (Ed.). 1981. Die Amphibien und Reptilien Westfalens. Abhandlungen aus dem Landesmuseum für Naturkunde zu Münster in Westfalen 43:1–161.

Felix, Z.I. 2001. A natural history study of *Desmognathus welteri* in West Virginia. Master's thesis. Marshall University, Huntington, West Virginia.

Felix, Z.I. and T.K. Pauley. 2001. Seasonal, ontogenetic, and diel variation in microhabitat use in three *Desmognathus* salamanders. Association of Southeastern Biologists Bulletin 48:156–157.

Fellers, G.M. 1979a. Aggression, territoriality, and mating behaviour in North American treefrogs. Animal Behaviour 27:107–119.

Fellers, G.M. 1979b. Mate selection in the gray treefrog, *Hyla versicolor*. Copeia 1979:286–290.

Fellers, G.M. 1997. Design of amphibian surveys. Pp. 23–34. *In* Olson, D.H., W.P. Leonard and R.B. Bury (Eds.), Sampling Amphibians in Lentic Habitats. Northwest Fauna Number 4, Society for Northwestern Vertebrate Biology, Olympia, Washington.

Fellers, G.M. and C.A. Drost. 1993. Disappearance of the Cascades frog, *Rana cascadae*, at the southern end of its range, California, USA. Biological Conservation 65:177–181.

Fellers, G.M. and C.A. Drost. 1994. Sampling with artificial cover. Pp. 146–150. *In* Heyer, W.R., M.A. Donnelly, R.W. McDiarmid, L.-C. Hayek and M.S. Foster (Eds.), Measuring and Monitoring Biological Diversity: Standard Methods for Amphibians. Smithsonian Institution Press, Washington, D.C.

Fellers, G.M. and K. Freel. 1995. A standardized protocol for surveying aquatic amphibians. U.S. Department of the Interior, National Biological Service, Technical Report, NPS/WRUC/NRTR-95-01, Washington, D.C.

Fellers, G.M., D.E. Green and J.E. Longcore. 2001. Oral chytridiomycosis in the mountain yellow-legged frog *(Rana muscosa)*. Copeia 2001:945–953.

Fellers, G.M., A.E. Launer, G. Rathbun, S. Bobzien, J. Alvarez, D. Sterner, R.B. Seymour and M. Westphal. 2001. Overwintering tadpoles in the California red-legged frog *(Rana aurora draytonii)*. Herpetological Review 32:156–157.

Fellers, G.M. and L.L. Wood. 2004. Natural history notes: *Rana aurora draytonii* (California red-legged frog). Predation. Herpetological Review 35:163.

Feminella, J.W. and C.P. Hawkins. 1994. Tailed frog tadpoles differentially alter their feeding behavior in response to non-visual cues from four predators. Journal of the North American Benthological Society 13:310–320.

Femmer, S. and D. Metter. 1979. Geographic distribution: *Scaphiopus bombifrons* (plains spadefoot). Herpetological Review 10:23.

Fenolio, D. and M. Ready. 1995. Natural history of the Mexican burrowing frog *Rhinophrynus dorsalis*. Vivarium (Lakeside) 6:18–21.

Fenster, T.L.D. and C.B. Fenster. 1996. Natural history notes: *Plethodon cinereus* (redback salamander). Predation. Herpetological Review 27:194.

Ferguson, D.E. 1952. The distribution of amphibians and reptiles of Wallowa County, Oregon. Herpetologica 8:66–68.

Ferguson, D.E. 1954a. An interesting factor influencing *Bufo boreas* reproduction at high altitudes. Herpetologica 10:199.

Ferguson, D.E. 1954b. An annotated list of the amphibians and reptiles of Union County, Oregon. Herpetologica 10:149–152.

Ferguson, D.E. 1956. Notes on the occurrence of some Oregon salamanders close to the ocean. Copeia 1956:120.

Ferguson, D.E. 1961a. The geographic variation of *Ambystoma macrodactylum* Baird, with the description of two new subspecies. American Midland Naturalist 65:311–338.

Ferguson, D.E. 1961b. The herpetofauna of Tishmingo County, Mississippi, with comments on its zoogeographic affinities. Copeia 1961:391–396.

Ferguson, D.E. 1963. *Ambystoma macrodactylum* (Baird). Pp. 4.1–4.2. Catalogue of American Amphibians. Society for the Study of Amphibians and Reptiles, St. Louis, Missouri.

Ferguson, D.E. and C.C. Gilbert. 1968. Tolerances of three species of anuran amphibians to five chlorinated hydrocarbon insecticides. Journal of the Mississippi Academy of Sciences 13:135–138.

Ferguson, D.E. and J.R. Rhodes. 1958. A new locality for the zigzag salamander in Mississippi. Herpetologica 14:129.

Ferguson, D.E., H.F. Landreth and M.R. Turpinseed. 1965. Astronomical orientation of the southern cricket frog, *Acris gryllus*. Copeia 1965: 58–66.

Ferguson, J.H. and C.H. Lowe. 1969. Evolutionary relationships of the *Bufo punctatus* group. American Midland Naturalist 81:435–446.

Fernandez, J. and G.W. Esch. 1991. The component community structure of larval trematodes in the pulmonate snail *Helisoma anceps*. Journal of Parasitology 77:540–550.

Fernandez, P.J. 1996. A facility for captive propagation of Chiricahua leopard frogs *(Rana chiricahuensis)*. Pp. 7–12. *In* Strimple, P.D. (Ed.), Advances in Herpetoculture, International Herpetological Symposium, Salt Lake City, Utah.

Fernandez, P.J. and J.T. Bagnara. 1991. Effect of background color and low temperature on skin color and circulating a-MSH in two species of leopard frog. General and Comparative Endocrinology 83:132–141.

Fernandez, P.J. and J.T. Bagnara. 1993. Observations on the development of unusual melanization of leopard frog ventral skin. Journal of Morphology 216:9–15.

Fernandez, P.J. and P.C. Rosen. 1996. Effects of the introduced crayfish *Orconectes virilis* on native aquatic herpetofauna in Arizona. Arizona Game and Fish Heritage Program, Project Report Number I94054, Phoenix, Arizona.

Fetkavich, C. and L.J. Livo. 1998. Late-season boreal toad tadpoles. Northwestern Naturalist 79:120–121.

Feyerabend, P. 1980. How to defend society against science. Pp. 55–65. *In* Klemke, E.D., R. Hollinger and A.D. Kline (Eds.), Introductory Readings in the Philosophy of Science. Prometheus Books, Buffalo, New York.

Field, K.J., T.L. Beatty Sr. and T.L. Beatty Jr. 2003. Natural history notes: *Rana subaquavocalis* (Ramsey Canyon leopard frog). Diet. Herpetological Review 34:235.

Field, K.J., M.J. Sredl and M.J. Demlong. 2000. Draft proposal to reestablish Tarahumara frogs *(Rana tarahumarae)* into southcentral Arizona. Arizona Game and Fish Department, Nongame Branch, Phoenix, Arizona.

Figiel, C.R., Jr. and R.D. Semlitsch. 1990. Population variation in survival and metamorphosis of larval salamanders *(Ambystoma maculatum)*. Copeia 1990:818–826.

Figiel, C.R., Jr. and R.D. Semlitsch. 1995. Experimental determination of oviposition site selection in the marbled salamander, *Ambystoma opacum*. Journal of Herpetology 29:452–454.

Finch, M.O. and D.M. Lambert. 1996. Kinship and genetic divergence among populations of tuatara *Sphenodon punctatus* as revealed by minisatellite DNA profiling. Molecular Ecology 5:651–658.

Findeis, E.K. and W.E. Bemis. 1990. Functional morphology of tongue projection in *Taricha torosa* (Urodela: Salamandridae). Zoological Journal of the Linnean Society 99:129–157.

Finkler, M.S., M.T. Sugalski and D.L. Claussen. 2003. Sex-related differences in metabolic rate and locomotor performance in breeding spotted salamanders *(Ambystoma maculatum)*. Copeia 2003:887–893.

Finley, R.B., Jr. 1953. A northern record of *Hyla arenicolor* in western Colorado. Copeia 1953:180.

Finneran, L.C. 1951. Migration to the breeding pond by the spotted salamander. Copeia 1951:81.

Fioramonti, E., R.D. Semlitsch, H.-U. Reyer and K. Fent. 1997. Effects of triphenyltin and pH on the growth and development of *Rana lessonae* and *Rana esculenta* tadpoles. Environmental Toxicology and Chemistry 16:1940–1947.

Firschein, I.L. 1950. A new record of *Spea bombifrons* from northern Mexico and remarks on the status of the *hammondii* group of spadefoot anurans. Herpetologica 6:75–77.

Firschein, I.L. 1951a. Phragmosis and the "unken reflex" in a Mexican hylid frog, *Pternohyla fodiens*. Copeia 1951:74.

Firschein, I.L. 1951b. The range of *Cryptobranchus bishopi* and remarks on the distribution of the genus *Cryptobranchus*. American Midland Naturalist 45:455–459.

Firschein, I.L. and L.S. Miller. 1951. The ringed salamander, *Ambystoma annulatum*, in Oklahoma. Copeia 1951:82.

Fischer, T.D. 1998. Anura of eastern South Dakota: their distribution and characteristics of their wetland habitats, 1997–1998. Master's thesis. South Dakota State University, Brookings, South Dakota.

Fischer, T.D., D.C. Backlund, K.F. Higgins and D.E. Naugle. 1999. A field guide to South Dakota amphibians. South Dakota Agricultural Experiment Station, Bulletin Number 733, South Dakota State University, Brookings, South Dakota.

Fishbeck, D.W. and J.C. Underhill. 1959. A check list of the amphibians and reptiles of South Dakota. Proceedings of the South Dakota Academy of Science 38:107–109.

Fishbeck, D.W. and J.C. Underhill. 1960. Amphibians of eastern South Dakota. Herpetologica 16:131–136.

Fisher, A.K. 1887. *Spelerpes guttolineatus* Holbrook, in the vicinity of Washington, D.C. American Naturalist 21:672.

Fisher, E.M. 1953. A size record for the salamander *Batrachoseps attenuatus attenuatus*. Copeia 1953:62.

Fisher, H.I. 1948. Locality records of Pacific island reptiles and amphibians. Copeia 1948:69.

Fisher, R.N. 1995. Tracing human impacts in natural systems: geckos in the Pacific and amphibians in the Central Valley. Ph.D. dissertation. University of California, Davis, California.

Fisher, R.N. and T.J. Case. 1997. A field guide to the reptiles and amphibians of Coastal Southern California. Department of Biology, University of California, San Diego, California.

Fisher, R.N. and H.B. Shaffer. 1996. The decline of amphibians in California's Great Central Valley. Conservation Biology 10:1387–1397.

Fisher, T.R. 1989. Application and testing of indices of biotic integrity in northern and central Idaho headwater streams. Master's thesis. University of Idaho, Moscow, Idaho.

Fishwild, T.G., R.A. Schemidt, K.M. Jankens, K.A. Berven, G.J. Gamboa and C.M. Richards. 1990. Sibling recognition by larval frogs *(Rana pipiens, R. sylvatica* and *Pseudacris crucifer)*. Journal of Herpetology 24:40–44.

Fitch, H.S. 1936. Amphibians and reptiles of the Rogue River Basin, Oregon. American Midland Naturalist 17:634–652.

Fitch, H.S. 1938. *Rana boylii* in Oregon. Copeia 1938:148.

Fitch, H.S. 1940. A biogeographical study of the ordinoides artenkreis of garter snakes (genus *Thamnophis)*. University of California Publications in Zoology, Number 44, Berkeley, California.

Fitch, H.S. 1941. The feeding habits of California garter snakes. California Department of Fish and Game 27:1–32.

Fitch, H.S. 1956a. A field study of the Kansas ant-eating frog, *Gastrophryne olivacea*. University of Kansas Publication of the Museum of Natural History 8:275–306.

Fitch, H.S. 1956b. Early sexual maturity and longevity under natural conditions in the Great Plains narrow-mouthed frog. Herpetologica 12:281–282.

Fitch, H.S. 1956c. Temperature responses of free-living amphibians and reptiles of northeastern Kansas. Bulletin of the Museum of Natural History University of Kansas 8:417–476.

Fitch, H.S. 1958. Home ranges, territories, and seasonal movements of vertebrates of the Natural History Reservation. University of Kansas Publication of the Museum of Natural History 11:63–326.

Fitch, H.S. 1965. An ecological study of the garter snake *Thamnophis sirtalis*. University of Kansas Publication, Museum of Natural History 15:493–564.

Fitch, H.S. 1975. A demographic study of the ringneck snake *(Diadophis punctatus)* in Kansas. University of Kansas Museum of Natural History, Miscellaneous Publication Number 62, Lawrence, Kansas.

Fite, K.V., A.R. Blaustein, L. Bengston and H.E. Hewitt. 1998. Evidence of retinal light damage in *Rana cascadae*: a declining amphibian species. Copeia 1998:906–914.

Fitzgerald, G.J. and J.R. Bider. 1974a. Evidence of a relationship between age and activity in the toad *Bufo americanus*. Canadian Field-Naturalist 88:499–501.

Fitzgerald, G.J. and J.R. Bider. 1974b. Seasonal activity of the toad *Bufo americanus* in southern Quebec as revealed by a sound transect technique. Canadian Journal of Zoology 52:1–5.

Fitzgerald, K.T., H.M. Smith and L.J. Gillette. 1981. Nomenclature of the diploid species of the diploid-tetraploid *Hyla versicolor* complex. Journal of Herpetology 15:356–360.

Fitzpatrick, J.F. 1986. The pre-Pliocene Tennessee River and its bearing on crawfish distribution (Decapoda: Cambaridae). Brimleyana 12:123–146.

Fitzpatrick, L.C. 1973. Energy allocation in the Allegheny Mountain salamander, *Desmognathus ochrophaeus*. Ecological Monographs 43:43–48.

Flageole, S. and R. Leclair Jr. 1992. Etude demographique d'une population de salamanders *(Ambystoma maculatum)* a l'aide de la methode squeletto-chronologique. Canadian Journal of Zoology 70:740–749.

Flather, C.H. and J.R. Sauer. 1996. Using landscape ecology to test hypotheses about large-scale abundance patterns in migratory birds. Ecology 77:28–35.

Fleming, P.L. 1976. A study of the distribution and ecology of *Rana clamitans* Latreille. Ph.D. dissertation. University Minnesota, Minneapolis, Minnesota.

Fleming, W.J., H. De Chacin, O.H. Pattee and T.G. Lamont. 1982. Parathion accumulation in cricket frogs *Acris crepitans* and its effect on American kestrels *Falco sparverius*. Journal of Toxicology and Environmental Health 10:921–928.

Fletcher, M.D. and B.G. Cochran. 1990. Geographic distribution: *Ambystoma texanum* (smallmouth salamander). Herpetological Review 21:95.

Flint, R.E. 1947. Glacial Geology of the Pleistocene Epoch. John Wiley and Sons, New York.

Flores-Villela, O. 1993. Herpetofauna Mexicana: annotated list of the species of amphibians and reptiles of Mexico, recent taxonomic changes, and new species. Carnegie Museum of Natural History, Special Publication, Number 17, Pittsburgh, Pennsylvania.

Flores-Villela, O. 1993. Herpetofauna of Mexico: distribution and endemism. Pp. 253–280. *In* Ramamoorthy, T.P., R. Bye, A. Lot and J.E. Fa (Eds.), Biological Diversity of Mexico: Origins and Distributions. Oxford University Press, Oxford, United Kingdom.

Flores-Villela, O. and R.A. Brandon. 1992. *Siren lacertina* (Amphibia: Caudata) in northeastern Mexico and southern Texas. Annals of Carnegie Museum 61:289–291.

Flower, S.S. 1936. Further notes on the duration of life in animals—2. Amphibians. Proceedings of the Zoological Society of London 1936:369–394.

Flowers, M.A. and B.M. Graves. 1995. Prey selectivity and size-specific diet changes in *Bufo cognatus* and *B. woodhousii* during early postmetamorphic ontogeny. Journal of Herpetology 29:608–612.

Flowers, M.A. and B.M. Graves. 1997. Juvenile toads avoid chemical cues from snake predators. Animal Behaviour 53:641–646.

Floyd, P.S., Sr., P.S. Floyd Jr. and J.D. Floyd. 1998. Geographic distribution: *Amphiuma pholeter* (one-toed amphiuma). Herpetological Review 29:244.

Floyd, R.B. 1983. Ontogenetic change in the temperature tolerance of larval *Bufo marinus* (Anuran: Bufonidae). Comparative Biochemistry and Physiology A 75:267–271.

Floyd, R.B. 1984. Variation in temperature preference with stage of development of *Bufo marinus* larvae. Journal of Herpetology 18:153–158.

Floyd, R.B. and K.F. Benbow. 1984. Nocturnal activity of a population of cane toads. Koolewong 13:12–14.

Foard, T. and D.L. Auth. 1990. Food habits and gut parasites of the salamander, *Stereochilus marginatus*. Journal of Herpetology 24:428–431.

Foeckler, F. 1990. Charakterisierung und Bewertung von Augewässern des Donauraums Straubing durch Wassermolluskengesellschaften. Berichte der Akademie für Nasturschutz und Landschaftspflege, Beiheft 7:1–154.

Foley, D.H., III. 1994. Short-term response of herpetofauna to timber harvesting in conjunction with streamside-management zones in seasonally flooded bottomland-hardwood forests of southeast Texas. Master's thesis. Texas A&M University, College Station, Texas.

Folkerts, G.W. 1968. The genus *Desmognathus* Baird (Amphibia: Plethodontidae) in Alabama. Ph.D. dissertation. Auburn University, Auburn, Alabama.

Folkerts, G.W. 1971. Notes on South Carolina salamanders. Journal of the Elisha Mitchell Scientific Society 87:206–208.

Folkerts, G.W., M.A. Deyrup and D.C. Sisson. 1993. Arthropods associated with xeric longleaf pine habitats in the southeastern United States: a brief overview. Pp. 159–192. *In* Hermann, S.M. (Ed.), The Longleaf Pine Ecosystem: Ecology, Restoration and Management. Proceedings of the Tall Timbers Fire Ecology Conference, Number 18, Tallahassee, Florida.

Font, W.F. 1997. Improbable colonists: helminth parasites of freshwater fishes on an oceanic island. Micronesia 30:105–115.

Fontenot, C.L., Jr. 1999. Reproductive biology of the aquatic salamander *Amphiuma tridactylum* in Louisiana. Journal of Herpetology 33:100–105.

Fontenot, C.L., Jr. 2003. Natural history notes: *Hyla versicolor* (gray treefrog). Behavior. Herpetological Review 34:358.

Fontenot, C.L., Jr. and L.W. Fontenot. 1989. Life history notes: *Amphiuma tridactylum* (three-toed amphiuma). Feeding. Herpetological Review 20:48.

Forbes, V.E. 1999. Genetics and Ecotoxicology. Taylor and Francis, Philadelphia, Pennsylvania.

Force, E.R. 1925. Notes on reptiles and amphibians of Okmulgee County, Oklahoma. Copeia 141:25–27.

Force, E.R. 1933. The age of the attainment of sexual maturity of the leopard frog *Rana pipiens* in northern Michigan. Copeia 1933:128–133.

Ford, E.B. 1945. Polymorphism. Biological Reviews 20:73–88.

Ford, W.M., B.R. Chapman, M.A. Menzel and R.H. Odum. 2002. Stand age and habitat influences on salamanders in Appalachian cove hardwood forests. Forest Ecology and Management 155:131–141.

Forester, D.C. 1973. Mating call as a reproductive isolating mechanism between *Scaphiopus bombifrons* and *S. hammondi*. Copeia 1973:60–67.

Forester, D.C. 1975. Laboratory evidence for potential gene flow between two species of spadefoot toads, *Scaphiopus bombifrons* and *Scaphiopus hammondi*. Herpetologica 31:282–286.

Forester, D.C. 1977. Comments on the female reproductive cycle and philopatry by *Desmognathus ochrophaeus* (Amphibia, Urodela, Plethodontidae). Journal of Herpetology 11:311–316.

Forester, D.C. 1978. Laboratory encounters between attending *Desmognathus ochrophaeus* (Amphibia, Urodela, Plethodontidae) females and potential predators. Journal of Herpetology 12:537–541.

Forester, D.C. 1979. The adaptiveness of parental care in *Desmognathus ochrophaeus* (Urodela: Plethodontidae). Copeia 1979:332–341.

Forester, D.C. 1981. Parental care in the salamander *Desmognathus ochrophaeus*: female activity pattern and trophic behavior. Journal of Herpetology 15:29–34.

Forester, D.C. 1984. Brooding behavior by the mountain dusky salamander: can the female's presence reduce clutch desiccation? Herpetologica 40:105–109.

Forester, D.C. and C.L. Anders. 2000. Contributions to the life history of the red-backed salamander, *Plethodon cinereus*: preliminary evidence for kin recognition. Pp. 407–416. *In* Bruce, R.C., R.G. Jaeger and L.D. Houck (Eds.), The Biology of Plethodontid Salamanders. Kluwer Academic/Plenum Publishers, New York.

Forester, D.C. and R. Daniel. 1986. Observations on the social behavior of the southern cricket frog, *Acris gryllus*, Anura, Hylidae. Brimleyana 12:5–11.

Forester, D.C. and D.V. Lykens. 1986. Significance of satellite males in a population of spring peepers (*Hyla crucifer*). Copeia 1986:719–724.

Forester, D.C. and D.V. Lykens. 1988. The ability of wood frog eggs to withstand prolonged terrestrial stranding: an empirical study. Canadian Journal of Zoology 66:1733–1735.

Forester, D.C. and D.V. Lykens. 1991. Age structure in a population of red-spotted newts from the Allegheny Plateau of Maryland. Journal of Herpetology 25:373–376.

Forester, D.C. and K.J. Thompson. 1998. Gauntlet behaviour as a male sexual tactic in the American toad (Amphibia, Bufonidae). Behaviour 135:99–119.

Forey, P.L., C.J. Humphries and R.I. Vane-Wright (Eds.). 1994. Systematics and Conservation Evaluation. Clarendon Press, Oxford, United Kingdom.

Forman, R.T.T. 1995. Land Mosiacs: The Ecology of Landscapes and Regions. Cambridge University Press, Cambridge, United Kingdom.

Forman, R.T.T. and M. Godron. 1986. Landscape Ecology. John Wiley and Sons, New York.

Formanowicz, D.R., Jr. and M.S. Bobka. 1989. Predation risk and microhabitat preference: an experimental study of the behavioral responses of prey and predator. American Midland Naturalist 121:379–386.

Formanowicz, D.R., Jr. and E.D. Brodie Jr. 1979. Palatability and antipredator behavior of selected *Rana* to the shrew *Blarina*. American Midland Naturalist 101:456–458.

Formanowicz, D.R., Jr. and E.D. Brodie Jr. 1982. Relative palatabilities of members of a larval amphibian community. Copeia 1982:91–97.

Formanowicz, D.R., Jr. and E.D. Brodie Jr. 1993. Size-mediated predation pressure in a salamander community. Herpetologica 49:265–270.

Formanowicz, D.R., Jr., M.M. Stewart, K. Townsend, F.H. Pough and P.F. Brussard. 1981. Predation by giant crab spiders on the Puerto Rican frog *Eleutherodactylus coqui*. Herpetologica 37:125–129.

Forster, H. 1998. Territorial behavior in the salamander *Plethodon albagula*. Master's thesis. Emporia State University, Emporia, Kansas.

Forstner, J.M., M.R.J. Forstner and J.R. Dixon 1998. Ontogenetic effects on prey selection and food habits of two sympatric east Texas ranids: the southern leopard frog, *Rana sphenocephala*, and the bronze frog, *Rana clamitans clamitans*. Herpetological Review 29:208–211.

Fort, D.J. and E.L. Stover. 1997. Effect of low level copper and pentachlorophenol exposure on various early life stages of *Xenopus laevis*. Pp. 188–203. *In* Bengtson, D.A. and D.S. Henshel (Eds.), Environmental Toxicology and Risk Assessment: Biomarkers and Risk Assessment. Fifth Volume. American Society of Testing and Materials (ASTM), STP 1306, Philadelphia, Pennsylvania.

Fort, D.J., T.L. Propst, E.L. Stover, J.C. Helgen, R.B. Levey, K. Gallagher and J.G. Burkhart. 1999a. Effects of pond water, sediment, and sediment extracts from Minnesota and Vermont, USA, on early development and metamorphosis of *Xenopus*. Environmental Toxicology and Chemistry 18:2305–2315.

Fort, D.J., R.L. Rogers, H.F. Copley, L.A. Bruning, E.L. Stover, J.C. Helgen and J.G. Burkhart. 1999b. Progress toward identifying causes of maldevelopment induced in *Xenopus* by pond water and sediment extracts from Minnesota, USA. Environmental Toxicology and Chemistry 18:2316–2324.

Fortman, J.R. and R. Altig. 1973. Characters of F_1 hybrid tadpoles between six species of *Hyla*. Copeia 1973:411–415.

Fortman, J.R. and R. Altig. 1974. Characters of F_1 hybrid frogs from six species of *Hyla* (Anura: Hylidae). Herpetologica 30:221–234.

Foster, M.S. and R.W. McDiarmid. 1982. Study of aggressive behavior of *Rhinophrynus dorsalis* tadpoles: design and analysis. Herpetologica 38:395–404.

Foster, M.S. and R.W. McDiarmid. 1983. *Rhinophrynus dorsalis* (alma de vaca, sapo borracho, Mexican burrowing toad). Pp. 419–421. *In* Janzen, D.H. (Ed.), Costa Rican Natural History. University of Chicago Press, Chicago, Illinois.

Foster, W.A. 1967. Chorus structure and vocal response in the Pacific tree frog, *Hyla regilla*. Herpetologica 23:100–104.

Foufopoulos, J. and A.R. Ives. 1999. Reptile extinction on land-bridge islands: life history attributes and vulnerability to extinction. American Naturalist 153:1–25.

Fouquette, M.J. and J. Delahoussaye. 1966. Noteworthy herpetological records from Louisiana. Southwestern Naturalist 11:137–139.

Fouquette, M.J., Jr. 1954. Food competition among four sympatric species of garter snakes, genus *Thamnophis*. Texas Journal of Science 5:172–88.

Fouquette, M.J., Jr. 1960. Call structure in frogs of the family Leptodactylidae. Texas Journal of Science 12:201–215.

Fouquette, M.J., Jr. 1968. Remarks on the type specimen of *Bufo alvarius* Girard. Great Basin Naturalist 28:70–72.

Fouquette, M.J., Jr. 1969. *Rhinophrynus dorsalis*. Pp. 78.1–78.2. Catalogue of American Amphibians and Reptiles. Society for the Study of Amphibians and Reptiles, St. Louis, Missouri.

Fouquette, M.J., Jr. 1970. *Bufo alvarius* Girard. Pp. 93.1–93.4. Catalogue of American Amphibians and Reptiles. Society for the Study of Amphibians and Reptiles, St. Louis, Missouri.

Fouquette, M.J., Jr. 1975. Speciation in chorus frogs. I. Reproductive character displacement in the *Pseudacris nigrita* complex. Systematic Zoology 24:16–23.

Fouquette, M.J., Jr. and C. Johnson. 1960. Call discrimination by female frogs of the *Hyla versicolor* complex. Copeia 1960:47–49.

Fouquette, M.J., Jr. and D.A. Rossman. 1963. Noteworthy records of Mexican amphibians and reptiles in the Florida State Museum and the Texas Natural History Collection. Herpetologica 19:185–201.

Fournier, M.A. 1997. Amphibians in the Northwest Territories. Pp. 100–106. *In* Green, D.M. (Ed.), Amphibians in Decline: Canadian Studies of a Global Problem. Herpetological Conservation, Number 1, Society for the Study of Amphibians and Reptiles, St. Louis, Missouri.

Fowler, H.W. and E.R. Dunn. 1917. Notes on salamanders. Proceedings of the Academy of the Natural Sciences of Philadelphia 69:7–28.

Fowler, J.A. 1940. A note on the eggs of *Plethodon glutinosus*. Copeia 1940:133.

Fowler, J.A. 1946. The eggs of *Pseudotriton montanus montanus*. Copeia 1946:105.

Fowler, J.A. 1947. Record for *Aneides aeneus* in Virginia. Copeia 1947:144.

Fowler, J.A. 1951. Preliminary observations on an aggregation of *Plethodon dixi*. Herpetologica 7:147–148.

Fowler, J.A. 1952. The eggs of *Plethodon dixi*. National Speleological Society Bulletin 14:61.

Fowler, J.A. and C.J. Stine. 1953. A new county record for *Microhyla carolinensis carolinensis* in Maryland. Herpetologica 9:167–168.

Fowler, J.A. and R. Sutcliffe. 1952. An additional record for the purple salamander, *Gyrinophilus p. porphyriticus*, from Maine. Copeia 1952: 48–49.

Fox, L.R. 1975. Cannibalism in natural populations. Annual Review of Ecology and Systematics 6:87–106.

Fox, W. 1951. Relationships among the garter snakes of the *Thamnophis elegans* rassenkreis. University of California Publications in Zoology 50:485–530.

Fox, W. 1952. Notes on feeding habits of Pacific coast garter snakes. Herpetologica 8:4–8.

Fox, W., H.C. Dessauer and L.T. Maumus. 1961. Electrophoretic studies of two species of toads and their natural hybrid. Comparative Biochemistry and Physiology 3:52–63.

Frandsen, J.C. and A.W. Grundmann. 1960. The parasites of some amphibians of Utah. Journal of Parasitology 46:678.

Frank, N. and E. Ramus. 1994. State, federal and C.I.T.E.S. Regulations for herpetologists. Reptile and Amphibian Magazine, Pottsville, Pennsylvania.

Frank, N. and E. Ramus. 1995. A Complete Guide to Scientific and Common Names of Reptiles and Amphibians of the World. NG Publishing Company, Pottsville, Pennsylvania.

Frankham, R. 1998. Inbreeding and extinction: island populations. Conservation Biology 12:665–675.

Frankino, W.A. and D.W. Pfennig. 2001. Condition-dependent expression of trophic polyphenism: effects of individual size and competitive ability. Evolutionary Ecology Research 3:939–951.

Franklin, I.R. 1980. Evolutionary change in small populations. Pp. 135–149. *In* Soulé, M.E. and B.A. Wilcox (Eds.), Conservation Biology: An Evolutionary-Ecological Perspective. Sinauer Associates, Sunderland, Massachusetts.

Franz, R. 1964. The eggs of the long-tailed salamander from a Maryland cave. Herpetologica 20:216.

Franz, R. 1970. Egg development of the tailed frog under natural conditions. Bulletin of the Maryland Herpetological Society 6:27–30.

Franz, R. 1986. The Florida gopher frog and Florida pine snake as burrow associates of the gopher tortoise in northern Florida. Pp. 16–20. *In* Jackson, D.R. and R.J. Bryant (Eds.), The Gopher Tortoise and its Community. Proceedings of the Fifth Annual Meeting of the Gopher Tortoise Council, Florida State Museum, Gainesville, Florida.

Franz, R. 1995. An introduction to the amphibians and reptiles of the Katharine Ordway Preserve-Swisher Memorial Sanctuary, Putnam County, Florida. Bulletin of the Florida Museum of Natural History 38:1–10.

Franz, R. and C.J. Chantell. 1978. *Limnaoedus ocularis*. Pp. 209.1–209.2. Catalogue of American Amphibians and Reptiles. Society for the Study of Amphibians and Reptiles, St. Louis, Missouri.

Franz, R. and H. Harris 1965. Mass transformation and movement of the larval long-tailed salamander, *Eurycea longicauda* (Green). Journal of the Ohio Herpetological Society 5:32.

Franz, R. and D.S. Lee. 1970. The ecological and biogeographical distribution of the tailed frog, *Ascaphus truei*, in the Flathead River drainage of northwestern Montana. Bulletin of the Maryland Herpetological Society 6:62–73.

Franz, R. and L.L. Smith. 1995. Distribution and status of the striped newt and Florida gopher frog in peninsular Florida. Florida Fish and Wildlife Conservation Commission, Final report, Tallahassee, Florida.

Franz, R., C.K. Dodd Jr. and C. Jones. 1988. Natural history notes: *Rana areolata aesopus* (Florida gopher frog). Movement. Herpetological Review 19:33.

Fraser, D.F. 1974. Interactions between salamanders of the genus *Plethodon* in the central Appalachians. Ph.D. dissertation. University of Maryland, College Park, Maryland.

Fraser, D.F. 1976a. Coexistence of salamanders in the genus *Plethodon*: a variation of the Santa Rosalia theme. Ecology 57:238–251.

Fraser, D.F. 1976b. Empirical evaluation of the hypothesis of food competition in salamanders of the genus *Plethodon*. Ecology 57:459–471.

Fraser, D.F. 1980. On the environmental control of oocyte maturation in plethodontid salamanders. Oecologia 46:302–307.

Frayer, W.E., D.D. Peters and H.R. Pywell. 1989. Wetlands of the California Central Valley—Status and trends, 1939–1980s. U.S. Fish and Wildlife Service, Unpublished report, Portland, Oregon.

Frazer, N.B. 1992. Sea turtle conservation and halfway technology. Conservation Biology 6:179–184.

Frazer, N.B. 1997. Turtle conservation and halfway technology: what is the problem? Pp. 422–425. *In* Van Abbema, J. (Ed.), Proceedings: Conservation, Restoration, and Management of Tortoises and Turtles. An International Conference. New York Turtle and Tortoise Society, New York.

Freda, J. 1983. Diet of larval *Ambystoma maculatum* in New Jersey. Journal of Herpetology 17:177–179.

Freda, J. 1986. The influence of acidic pond water on amphibians: a review. Water, Air, Soil Pollution 30:439–450.

Freda, J. 1991. The effects of aluminum and other metals on amphibians. Environmental Pollution 71:305–328.

Freda, J. and W.A. Dunson. 1985. The influence of external cation concentration on the hatching of amphibian embryos in water of low pH. Canadian Journal of Zoology 63:2649–2656.

Freda, J. and W.A. Dunson. 1986. Effects of low pH and other chemical variables on the local distribution of amphibians. Copeia 1986: 454–466.

Freda, J. and R.J. Gonzalez. 1986. Daily movements of the treefrog, *Hyla andersonii*. Journal of Herpetology 20:469–471.

Freda, J. and P.J. Morin. 1984. Adult home range of the pine barrens treefrog *(Hyla andersoni)* and the physical, chemical, and ecological characteristics of its preferred breeding ponds. Center for Coastal and Environmental Studies, Division of Pinelands Research. Rutgers, New Brunswick, New Jersey.

Freda, J. and D.H. Taylor. 1992. Behavioral response of amphibian larvae to acidic water. Journal of Herpetology 26:429–433.

Freda, J., B. Gern, D. McLeary and G. Baxter. 1988. 1988 recovery efforts for the endangered Wyoming toad, *Bufo hemiophrys baxteri*. Unpublished report, Cheyenne, Wyoming.

Frederick, D.C. 1998. Letter to Peggy Walicek and Randall Holly, 20 October.

Frederick, D.C. 1999. Land not condemned: wildlife agency sets record straight on plans to protect endangered Houston toads. Bastop Advertiser, 11 September.

Freed, A.N. 1980a. Prey selection and feeding behavior of the green treefrog *(Hyla cinerea)*. Ecology 61:461–465.

Freed, A.N. 1980b. An adaptive advantage of basking behavior in an anuran amphibian. Physiological Zoology 53:433–444.

Freed, A.N. 1982a. A treefrog menu: a selection for an evening's meal. Oecologia 53:20–26.

Freed, A.N. 1982b. To eat or not to eat? That is the question of treefrog prey selection. Dissertation Abstracts International (B) 43(6):1721.

Freed, A.N. 1988. The use of visual cues for prey selection by foraging treefrogs *(Hyla cinerea)*. Herpetologica 44:18–24.

Freed, P.S. and K. Neitman. 1988. Notes on predation on the endangered Houston toad, *Bufo houstonensis*. Texas Journal of Science 40:454–456.

Freeland, W.J. 1985. The need to control cane toads. Search 16:211–215.

Freeland, W.J. 1986. Invasion north: successful conquest by the cane toad. Australian Natural History 22:69–72.

Freeland, W.J. and K.C. Martin. 1985. The rate of range expansion by *Bufo marinus* in Northern Australia, 1980–84. Australian Wildlife Research 12:555–559.

Freeman, J.R. 1958. Burrowing in the salamanders *Pseudobranchus striatus* and *Siren lacertina*. Herpetologica 14:130.

Freeman, J.R. 1967. Feeding behavior of the narrow-striped dwarf siren *Pseudobranchus striatus axanthus*. Herpetologica 23:313–314.

Freeman, S.L. and R.C. Bruce. 2000. Larval period and metamorphosis of the three-lined salamander, *Eurycea guttolineata* (Amphibia: Plethodontidae), in the Chattooga River Watershed. American Midland Naturalist 145:194–200.

Freiburg, R.E. 1951. An ecological study of the narrow-mouthed toad *(Microhyla)* in northeastern Kansas. Transactions of the Kansas Academy of Science 54:374–386.

Freidberg, E.C., G.C. Walker and W. Siede. 1995. DNA Repair and Mutagenesis. ASM Press, Washington, D.C.

French, T.W. 1976. Geographic distribution: *Gyrinophilus porphyriticus* (spring salamander). Herpetological Review 7:122.

French, T.W. and R.H. Mount. 1978. Current status of the Red Hills salamander, *Phaeognathus hubrichti* Highton, and factors affecting its distribution. Journal of the Alabama Academy of Science 49:172–179.

Fried, B., P.L. Pane and A. Reddy. 1997. Experimental infection of *Rana pipiens* tadpoles with *Echinostoma trivolvis* cercariae. Parasitology Research 83:666–669.

Friet, S.C. 1995. Natural history notes: *Plethodon cinereus* (eastern red-backed salamander). Nest behavior. Herpetological Review 26:198–199.

Friet, S.C. and A.W. Pinder. 1990. Hypoxia during natural aquatic hibernation of the bullfrog. American Zoologist. 30:69A.

Frisbie, M.P. and R.L. Wyman. 1992. The effect of soil chemistry on sodium balance in the red-backed salamander: a comparison of two forest types. Journal of Herpetology 26:434–442.

Fritts, T.H., R.D. Jennings and N.J. Scott Jr. 1984. A review of the leopard frogs of New Mexico. New Mexico Department of Game and Fish, Unpublished report, Santa Fe, New Mexico.

Fritz, K. 1987. Die Bedeutung anthropogener Standorte als Lebensraum für die Mauereidechse *(Podarcis muralis)*, dargestellt am Beispiel des südlichen Oberrhein-und des westlichen Hochrheintals. Beihefte zu den Veröffentlichungen für Naturschutz und Landschaftspflege in Baden-Württemberg 41:427–462.

Fritz, K., G. Müller, M. Lehnert and M. Schrenk. 1987. Zur gegenwärtigen Situation der Aspisviper (*Vipera aspis* L.) in Deutschland. Beihefte zu den Veröffentlichungen für Naturschutz und Landschaftspflege in Baden-Württemberg 41:463–472.

Froom, B. 1982. Amphibians of Canada. McClelland and Stewart Limited, Toronto, Ontario.

Frost, C.C. 1993. Four centuries of changing landscape patterns in the longleaf pine ecosystem. Pp. 17–43. *In* Hermann, S.M. (Ed.), The Longleaf Pine Ecosystem: Ecology, Restoration and Management. Proceedings of the Tall Timbers Fire Ecology Conference, Number 18, Tallahassee, Florida.

Frost, D. 1983. Past occurrence of *Acris crepitans* (Hylidae) in Arizona. Southwestern Naturalist 28:105.

Frost, D.R. (Ed.). 1985. Amphibian species of the world: A taxonomic and geographic reference. Association of Systematics Collections, Lawrence, Kansas.

Frost, D.R. 2000a. Anura—frogs. Pp. 6–17. *In* Crother, B.I., J. Boundy, J.A. Campbell, K. de Queiroz, R.F. Frost, R. Highton, J.B. Iverson, P.A. Meylan, T.W. Reeder, M.E. Seidel, J.W. Sites Jr. and T.W. Taggart. Scientific and standard English names of amphibians and reptiles of North America north of Mexico, with comments regarding confidence in our understanding. Herpetological Circular, Number 29, Society for the Study of Amphibians and Reptiles, St. Louis, Missouri.

Frost, D.R. 2000b. *Bufo boreas nelsoni.* Amphibian species of the world. Electronic database version 2.20 (1 September 2000). American Museum of Natural History, New York: http://research.amnh.org/herpetology/amphibia/index.html.

Frost, D.R. 2000c. Amphibian species of the world: An online reference. http://research.amnh.org/herpetology/amphibia/ index.html, electronic database version 2.20. American Museum of Natural History, New York.

Frost, D.R. 2002. *Bufo nelsoni.* Amphibian species of the world. http://research.amnh.org/herpetology/amphibia/index.html, electronic database version 2.21. American Museum of Natural History, New York.

Frost, D.R. and D.M. Hillis. 1990. Species in concept and practice: herpetological applications. Herpetologica 46:87–104.

Frost, D.R., A.G. Kluge and D.M. Hillis. 1992. Species in contemporary herpetology: comments on phylogenetic inference and taxonomy. Herpetological Review 23:46–54.

Frost, J.S. 1982. Functional genetic similarities between geographically separated populations of Mexican leopard frogs (*Rana pipiens* complex). Systematic Zoology 31:57–67.

Frost, J.S. and J.T. Bagnara. 1977. Sympatry between *Rana blairi* and the southern form of leopard frog in southeastern Arizona (Anura: Ranidae). Southwestern Naturalist 22:443–453.

Frost, J.S. and E.W. Martin. 1971. A comparison of distribution and high temperature tolerance in *Bufo americanus* and *Bufo woodhousii fowleri.* Copeia 1971:750–751.

Frost, J.S. and J.E. Platz. 1983. Comparative assessment of modes of reproductive isolation among four species of leopard frogs (*Rana pipiens* complex). Evolution 37:66–78.

Frost, S.W. 1935. The food of *Rana catesbeiana* Shaw. Copeia 1935:15–18.

Fukumoto, J. and S. Herrero. 1998. Observations of the long-toed salamander, *Ambystoma macrodactylum*, in Waterton Lakes National Park, Alberta. Canadian Field-Naturalist 112:579–585.

Fuller, D.D. and A.J. Lind. 1992. Implications of fish habitat improvement structures for other stream vertebrates. Pp. 96–104. *In* Harris, R. and D. Erman (Eds.), Proceedings of the Symposium on Biodiversity of Northwestern California, Santa Rosa, California.

Fulton, M.H. and J.E. Chambers. 1985. The toxic and teratogenic effects of selected organophosphorus compounds on the embryos of three species of amphibians. Toxicology Letters 26:175–180.

Funderburg, J.B., Jr. 1955. The amphibians of New Hanover County, North Carolina. Journal of the Elisha Mitchell Scientific Society 71:19–28.

Funderburg, J.B. and D.S. Lee. 1967. Distribution of the lesser siren, *Siren intermedia* in central Florida. Herpetologica 23:65.

Funk, R.S. 1979. Geographic distribution: *Ambystoma annulatum* (ringed salamander). Herpetological Review 10:101.

Funk, W.C. and W.W. Dunlap. 1999. Colonization of high-elevation lakes by long-toed salamanders (*Ambystoma macrodactylum*) after the extinction of introduced trout populations. Canadian Journal of Zoology 77:1759–1767.

Funk, W.C., D.A. Tallmon and F.W. Allendorf. 1999. Small effective population size in the long-toed salamander. Molecular Ecology 8:1633–1640.

Funkhouser, A. 1976. Observations on pancreas: body weight ratio change during development of bullfrog, *Rana catesbeiana.* Herpetologica 32:370–371.

Futuyma, D.J. 1986. Evolutionary Biology. Sinauer Associates, Sunderland, Massachusetts.

Gabbadon, P.W. and F.A. Chapman. 1996. Use of the lampricide 3-trifluoromethyl-4-nitrophenol (TFM) to control tadpoles in warmwater ornamental fish ponds. Progressive Fish-Culturist 58:23–28.

Gabor, C.R. and R.G. Jaeger. 1999. When salamanders misrepresent threat signals. Copeia 1999:1123–1126.

Gaddy, L.L. and T.L. Kohlstaat. 1987. Recreational impact of the natural vegetation, avifauna, and herpetofauna of four South Carolina Barrier Islands (USA). Natural Areas Journal 7:55–64.

Gadow, H. 1908. Through Southern Mexico. Witherby and Company, London, England.

Gage, S.H. 1891. The life history of the vermilion-spotted newt (*Diemyctylus viridescens* Raf.). American Naturalist 25:1084–1103.

Gaggiotti, O.E. and P.E. Smouse. 1996. Stochastic migration and maintenance of genetic variation in sink populations. American Naturalist 147:919–945.

Gaige, H.T. 1917. Description of a new salamander from Washington. Occasional Papers of the Museum of Zoology, Number 40, University of Michigan, Ann Arbor, Michigan.

Gaige, H.T. 1920. Observations upon the habits of *Ascaphus truei* Stejneger. Occasional Papers of the Museum of Zoology, Number 84, University of Michigan, Ann Arbor, Michigan.

Gaige, H.T. 1931. Notes on *Syrrhophus marnockii* Cope. Copeia 1931:63.

Gaige, H.T. 1932. The status of *Bufo copei.* Copeia 1932:134.

Gaines, A. 1895. Batrachia of Vincennes, Indiana. American Naturalist 29:53–56.

Galatowitsch, S.M. and A.G. van der Valk. 1994. Restoring Prairie Wetlands: An Ecological Approach. Iowa State University Press, Ames, Iowa.

Gallant, A.L., E.F. Binnian, J.M. Omernik and M.B. Shasby. 1995. Ecoregions of Alaska. U.S. Geological Survey, Professional Paper 1567, Washington, D.C.

Gallant, N. and K. Teather. 2001. Differences in size, pigmentation, and fluctuating asymmetry in stressed and nonstressed northern leopard frogs (*Rana pipiens*). Ecoscience 8:430–436.

Gallie, J.A., R.L. Mumme and S.A. Wissinger. 2001. Experience has no effect on the development of chemosensory recognition of predators by tadpoles of the American toad, *Bufo americanus.* Herpetologica. 57:376–383.

Gambs, R.D. and M.J. Littlejohn. 1979. Acoustic behavior of males of the Rio Grande leopard frog (*Rana berlandieri*): an experimental analysis through field playback trials. Copeia 1979:643–650.

Gamradt, S.C. and L.B. Kats. 1996. Effect of introduced crayfish and mosquitofish on California newts. Conservation Biology 10:1155–1162.

Gamradt, S.C. and L.B. Kats. 1997. Impact of chaparral wildfire-induced sedimentation on oviposition of stream-breeding California newts (*Taricha torosa*). Oecologia 110:546–549.

Gamradt, S.C., L.B. Kats and C.B. Anzalone. 1997. Aggression by non-native crayfish deters breeding in California newts. Conservation Biology 11:793–796.

Gans, C. and T. Parsons. 1966. On the origin of the jumping mechanism in frogs. Evolution 20:92–99.

Garber, D.P. and C.E. Garber. 1978. A variant form of *Taricha granulosa* (Amphibia, Urodela, Salamandridae) from northwestern California. Journal of Herpetology 12:59–64.

Garcia, E. 2001. *Bufo boreas*, AmphibiaWeb. http://elib.cs.berkeley.edu/cgi-bin/amphib_query?where-genus=Bufo&where-species=boreas.

García-Paris, M. and S.M. Deban. 1995. A novel antipredator mechanism in salamanders: rolling escape in *Hydromantes platycephalus*. Journal of Herpetology 29:149–151.

Garden, A. 1821. *In* Smith, J.E. 1821. A Selection of the Correspondence of Linnaeus and Other Naturalists. Volume 1. Longman, Hurst, Rees, Orme and Brown, London, England.

Gardiner, D.M. and D.M. Hoppe. 1999. Environmentally induced limb malformations in mink frogs *(Rana septentrionalis)*. Journal of Experimental Zoology 284:207–216.

Gardner, J.D. 1995. Natural history notes: *Pseudacris regilla* (Pacific chorus frog). Reproduction. Herpetological Review 26:32.

Gardner, R.H., B.T. Milne, M.G. Turner and R.V. O'Neill. 1987. Neutral models for the analysis of broad-scale landscape pattern. Landscape Ecology 1:19–28.

Garman, H. 1884. The North American reptiles and batrachians. Bulletin of the Essex Institute 16:1–46.

Garman, H. 1892. A synopsis of the reptiles and amphibians of Illinois. Illinois State Laboratory of Natural History Bulletin 3:215–388.

Garnier, J.H. 1883. The mink or hoosier frog. American Naturalist 17:945–954.

Garrett, C.M. and D.M. Boyer. 1993. Natural history notes: *Bufo marinus* (cane toad). Predation. Herpetological Review 24:148.

Garrett, J.M. and D.A. Barker. 1987. Field Guide to Reptiles and Amphibians of Texas. Texas Monthly Fieldguide Series, Gulf Publishing Company, Houston, Texas.

Garton, E.R., F. Grady and S.D. Carey. 1993. The vertebrate fauna of West Virginia caves. West Virginia Speleological Survey, Bulletin 11, Barrackville, West Virginia.

Garton, J.S. 1972. Courtship of the smallmouth salamander, *Ambystoma texanum*, in southern Illinois. Herpetologica 28:41–45.

Garton, J.S. and R.A. Brandon. 1975. Reproductive ecology of the green treefrog, *Hyla cinerea*, in southern Illinois (Anura: Hylidae). Herpetologica 31:150–161.

Garton, J.S. and H.R. Mushinsky. 1979. Integumentary toxicity and unpalatibility as a defensive mechanism in *Gastrophryne carolinensis*. Canadian Journal of Zoology 57:1965–1973.

Garton, J.S. and B.S. Sill. 1979. The status of the Pine Barrens treefrog, *Hyla andersonii* Baird, in South Carolina. Pp. 131–132. *In* Forsythe, D.M. and W.B. Ezell Jr. (Eds.), Proceedings of the First South Carolina Endangered Species Symposium. South Carolina Wildlife and Marine Resources Department, Columbia, South Carolina.

Garton, J.S., E.W. Harris and R.A. Brandon. 1970. Descriptive and ecological notes on *Natrix cyclopion* in Illinois. Herpetologica 26:454–461.

Gartside, D.F. 1980. Analysis of a hybrid zone between chorus frogs of the *Pseudacris nigrita* complex in the southern United States. Copeia 1980:56–66.

Gartside, D.F. and H.C. Dessauer. 1976. Protein evidence for a narrow zone of hybridization between the chorus frogs *Pseudacris triseriata* and *P. nigrita*. Herpetological Review 7:84.

Gaston, K.J. 1994. Rarity. Chapman and Hall, London, England.

Gaston, K. and W. Kunin. 1997. Rare-common differences: an overview. Pp 12–29. *In* Kunin, W. and K. Gaston (Eds.), The Biology of Rarity. Chapman and Hall, London, England.

Gaston, K.J. and P.H. Williams. 1993. Mapping the world's species—the higher taxon approach. Biodiversity Letters 1:2–8.

Gates, G.O. 1957. A study of the herpetofauna in the vicinity of Wickenburg, Maricopa County, Arizona. Transactions of the Kansas Academy of Science 60:403–418.

Gates, J.E. 1983. The distribution and status of the hellbender (*Cryptobranchus alleganiensis*) in Maryland: I. The distribution and status of *Cryptobranchus alleganiensis* in Maryland. II. Movement patterns of translocated *Cryptobranchus alleganiensis* in a Maryland stream. Maryland Wildlife Administration, Annapolis, Maryland.

Gates, J.E. and E.L. Thompson. 1982. Small pool habitat selection by red-spotted newts in western Maryland. Journal of Herpetology 16:7–15.

Gates, J.E., C.H. Hocutt, J.R. Stauffer Jr. and G.J. Taylor. 1985. The distribution and status of *Cryptobranchus alleganiensis* in Maryland. Herpetological Review 16:17–18.

Gates, W.R. 1988. *Pseudacris nigrita* (LeConte). Pp. 416.1–416.3. Catalogue of American Amphibians and Reptiles. Society for the Study of Amphibians and Reptiles, St. Louis, Missouri.

Gatz, A.J., Jr. 1971. Critical thermal maxima of *Ambystoma maculatum* (Shaw) and *Ambystoma jeffersonianum* (Green) in relation to time of breeding. Herpetologica 27:157–160.

Gatz, A.J., Jr. 1973a. Algal entry into the eggs of *Ambystoma maculatum*. Journal of Herpetology 7:137–138.

Gatz, A.J., Jr. 1973b. Intraspecific variations in critical thermal maxima of *Ambystoma maculatum*. Herpetologica 29:264–288.

Gatz, A.J., Jr. 1981a. Non-random mating by size in American toads, *Bufo americanus*. Animal Behaviour 29:1004–1012.

Gatz, A.J., Jr. 1981b. Size selective mating in *Hyla versicolor* and *Hyla crucifer*. Journal of Herpetology 15:114–116.

Gaudin, A.J. 1964. The tadpole of *Hyla cadaverina* Gorman. Texas Journal of Science 16:80–84.

Gaudin, A.J. 1965. Larval development of the tree frogs *Hyla regilla* and *Hyla californiae*. Herpetologica 21:117–130.

Gaudin, A.J. 1979. *Hyla cadaverina*. Pp. 225.1–225.2. Catalogue of American Amphibians and Reptiles. Society for the Study of Amphibians and Reptiles, St. Louis, Missouri.

Gee, R.W. 1999. Houston toad, we have a problem. Austin American-Statesman, 1 August.

Gehlbach, F.R. 1965. Herpetology of the Zuni Mountains region, northwestern New Mexico. Proceedings of the U.S. National Museum, Number 116 (3505):243–332.

Gehlbach, F.R. 1967a. *Ambystoma tigrinum* (Green). Pp. 52.1–52.4. Catalogue of American Amphibians and Reptiles. Society for the Study of Amphibians and Reptiles, St. Louis, Missouri.

Gehlbach, F.R. 1967b. Evolution of tiger salamanders *(Ambystoma tigrinum)* on the Grand Canyon rims, Arizona. Yearbook of the American Philosophical Society 1967:266–269.

Gehlbach, F.R. 1969. Determination of the relationships of tiger salamander larval populations to different stages of pond succession at the Grand Canyon, Arizona. Yearbook of the American Philosophical Society 1969:299–302.

Gehlbach, F.R. and B.B. Collette. 1959. Distribution and biological notes on the Nebraska herpetofauna. Herpetologica 15:141–143.

Gehlbach, F.R. and S.E. Kennedy. 1978. Population ecology of a highly productive aquatic salamander *Siren intermedia*. Southwestern Naturalist 23:423–430.

Gehlbach, F.R. and B. Walker. 1970. Acoustic behavior of the aquatic salamander, *Siren intermedia*. Bioscience 20:1107–1108.

Gehlbach, F.R., R. Gordon and J.B. Jordan. 1973. Aestivation of the salamander, *Siren intermedia*. American Midland Naturalist 89:455–463.

Gehlbach, F.R., K.A. Arnold, K. Culbertson, D.J. Schmidly, C. Hubbs and R.A. Thomas. 1975. TOES Watch-list of endangered, threatened, and peripheral vertebrates of Texas. Texas Organization of Endangered Species, Publication Number 1, Austin, Texas.

Gelhausen, S., C. Stevens and N. Tuttrup. 1998. Paul and Maria Tuttrup wildlife management plan for 1-d-1-w wildlife use appraisal. Unpublished Report.

Gendron, A.D. in press. Status of the mudpuppy, *Necturus maculosus*, in Canada. Canadian Field Naturalist. in press.

Gentry, G. 1955. An annotated check list of the amphibians and reptiles of Tennessee. Journal of the Tennessee Academy of Science 30:168–176.

Gentry, J.B. and M.H. Smith. 1968. Food habits and burrow associates of *Peromyscus polionotus*. Journal of Mammalogy 49:562–565.

George Washington and Jefferson National Forests, U.S. Fish and Wildlife Service, and Blue Ridge Parkway. 1997. Habitat conservation assessment for the Peaks of Otter salamander *(Plethodon hubrichti)*. George Washington and Jefferson National Forests, Unpublished planning document, Roanoke, Virginia.

George, C.J., C.W. Boylen and R.B. Sheldon. 1977. The presence of the red-spotted newt, *Notophthalmus viridescens rafinesque* (Amphibia, Urodela, Salamandridae) in waters exceeding 12 meters in Lake George, New York. Journal of Herpetology 11:87–90.

George, I.D. 1940. A study of the bullfrog, *Rana catesbeiana* Shaw, at Baton Rouge, Louisiana. Ph.D. dissertation. University of Michigan, Ann Arbor, Michigan.

George, W.O., S.D. Breeding and W.H. Hastings. 1952. Geology and underground resources of Comal County, Texas. U.S. Geological Water Supply, Paper Number 1138, Washington, D.C.

Georgia Natural Heritage Program. 1996. Special concern animals of Georgia. Georgia Department of Natural Resources, Social Circle, Georgia.

Gergits, W.F. 1982. Interference competition and territoriality between the terrestrial salamanders *Plethodon cinereus* and *Plethodon shenandoah*. Master's thesis. State University of New York at Albany, Albany, New York.

Gergits, W.F. and R.G. Jaeger. 1990. Field observations of the behavior of the red-backed salamander, *Plethodon cinereus*: courtship and agonistic interactions. Journal of Herpetology 24:93–95.

Gergus, E.W.A. 1993. Geographic distribution: *Bufo terrestris* (southern toad). Herpetological Review 24:64.

Gergus, E.W.A. 1998. Systematics of the *Bufo microscaphus* complex: allozyme evidence. Herpetologica 54:317–325.

Gergus, E.W.A. 1999. Geographic variation in hylid frogs of southwestern North America: taxonomic and population genetic implications. Ph.D. dissertation. Arizona State University, Tempe, Arizona.

Gergus, E.W.A., K.B. Malmos and B.K. Sullivan. 1999. Natural hybridization among distantly related toads (*Bufo alvarius, Bufo cognatus, Bufo woodhousii*) in central Arizona. Copeia 1999:281–286.

Gergus, E.W.A., B.K. Sullivan and K.B. Malmos. 1997. Call variation in the *Bufo microscaphus* complex: implications for species boundaries and the evolution of mate recognition. Ethology 103:979–989.

Gerhardt, H.C. 1967 Rediscovery of the salamander *Stereochilus marginatus* in Georgia. Copeia 1967:861.

Gerhardt, H.C. 1973. Reproductive interactions between *Hyla crucifer* and *Pseudacris ornata* (Anura: Hylidae). American Midland Naturalist 89:81–88.

Gerhardt, H.C. 1974a. Mating call differences between eastern and western populations of the treefrog *Hyla chrysoscelis*. Copeia 1974: 534–536.

Gerhardt, H.C. 1974b. Behavioral isolation of the treefrogs *Hyla cinerea* and *Hyla andersonii*. American Midland Naturalist 91:424–433.

Gerhardt, H.C. 1975. Sound pressure levels and radiation patterns of the vocalizations of some North American frogs and toads. Journal of Comparative Physiology A 102:1–12.

Gerhardt, H.C., Jr. 1974c. The vocalization of some hybrid treefrogs: acoustic and behavioral analyses. Behavior 49:130–151.

Gerhardt, H.C., S.I. Guttman and A.A. Karlin. 1980. Natural hybrids between *Hyla cinerea* and *Hyla gratiosa*: morphological, vocalization, and electrophoretic analysis. Copeia 1980:577–584.

Gerhardt, H.C., M.B. Ptacek, L. Barnett and K.G. Torke. 1994. Hybridization in the diploid-tetraploid treefrogs *Hyla chrysoscelis* and *Hyla versicolor*. Copeia 1994:51–59.

Getz, L.L. 1958. The winter activities of *Rana clamitans* tadpoles. Copeia 1958:219.

Gibbs, L.R. 1850. On a new species of *Menobranchus*, from South Carolina. P. 169. Third Annual Meeting, Proceedings of the American Association of the Advancement of Science, Charleston, South Carolina.

Gibbons, J.W. 1983. Their Blood Runs Cold. University of Alabama Press, Tuscaloosa, Alabama.

Gibbons, J.W. 2003. Terrestrial habitat: a vital component for herpetofauna of isolated wetlands. Wetlands 23:63–65.

Gibbons, J.W. and D.H. Bennett. 1974. Determination of anuran activity patterns by a drift fence method. Copeia 1974:236–243.

Gibbons, J.W. and J.W. Coker. 1978. Herpetofaunal colonization patterns of Atlantic Coast barrier islands. American Midland Naturalist 99:219–233.

Gibbons, J.W. and S. Nelson Jr. 1968. Observations of the mudpuppy, *Necturus maculosus*, in a Michigan lake. American Midland Naturalist 80:562–564.

Gibbons, J.W. and R.D. Semlitsch. 1981. Terrestrial drift fences with pitfall traps: an effective technique for quantitative sampling of animal populations. Brimleyana 7:1–16.

Gibbons, J.W. and R.D. Semlitsch. 1991. Guide to the Reptiles and Amphibians of the Savannah River Site. University of Georgia Press, Athens, Georgia.

Gibbs, E.L., T.J. Gibbs and P.C. VanDyck. 1966. *Rana pipiens*: health and disease. Laboratory Animal Care 16:142–160.

Gibbs, E.L., G.W. Nance and M.B. Emmons. 1971. The live frog is almost dead. BioScience 21:1027–1034.

Gibbs, J.P. 1993. Importance of small wetlands for the persistence of local populations of wetland-associated animals. Wetlands 13:25–31.

Gibbs, J.P. 1998a. Amphibian movements in response to forest edges, roads, and streambeds in southern New England. Journal of Wildlife Management 62:584–589.

Gibbs, J.P. 1998b. Distribution of woodland amphibians along a forest fragmentation gradient. Landscape Ecology 13:263–268.

Gibbs, J.P., S. Droege and P. Eagle. 1998. Computers in biology: monitoring populations of plants and animals. BioScience 48:935–940.

Gibbs, K.E., T.M. Mingo and D.L. Courtemanch. 1984. Persistence of carbaryl (sevin-4-oil®) in woodland ponds and its effects on pond macroinvertebrates following forest spraying. Canadian Entomology 116:203–213.

Gichuki, C.M. 1987. The diet of the barn owl, *Tyto alba* (Scopoli), in Nairobi, Kenya. Journal of the East Africa Natural History Society and National Museums 75:1–7.

Gilbert, F.F. and R. Allwine. 1991. Terrestrial amphibian communities in the Oregon Cascade Range. Pp. 318–324. *In* Ruggiero, L.F., K.B. Aubry, A.B. Carey and M.H. Huff (Tech. Coords.), Wildlife and Vegetation of Unmanaged Douglas-fir Forests. U.S.D.A. Forest Service, General Technical Report, PNW-GTR-285, Pacific Northwest Research Station, Portland, Oregon.

Gilbert, F., A. Gonzalez and I. Evans-Freke. 1998. Corridors maintain species richness in the fragmented landscapes of a microecosystem. Proceedings of the Royal Society, London B 265:577–582.

Gilbert, M., R. LeClair Jr. and R. Fortin. 1994. Reproduction of the northern leopard frog (*Rana pipiens*) in floodplain habitat in the Richelieu River, P. Quebec, Canada. Journal of Herpetology 28: 465–470.

Gilbert, P.W. 1941. Eggs and nests of *Hemidactylium scutatum* in the Ithaca region. Copeia 1941:47.

Gilbert, P.W. 1942. Observations on the eggs of *Ambystoma maculatum* with special reference to the green algae found within the egg envelopes. Ecology 23:215–227.

Gilbert, P.W. 1944. The alga-egg relationship in *Ambystoma maculatum*, a case of symbiosis. Ecology 25:366–369.

Gilbertson, H. and D.J. Watermolen. 1998. Notes on wrinkled frog (*Rana rugosa*) tadpoles from Hawaii. Bulletin of the Chicago Herpetological Society 33:57–59.

Gilhen, J. 1974. Distribution, natural history, and morphology of the blue-spotted salamanders, *Ambystoma laterale* and *A. tremblayi* in Nova Scotia. Nova Scotia Museum Curatorial Report, Number 22, Halifax, Nova Scotia.

Gilhen, J. 1984. Amphibians and Reptiles of Nova Scotia. Nova Scotia Museum, Halifax, Nova Scotia.

Gill, D.E. 1978. The metapopulation ecology of the red-spotted newt, *Notophthalmus viridescens* (Rafinesque). Ecological Monographs 48: 145–166.

Gill, D.E. 1979. Density dependence and homing behavior in adult red-spotted newts *Notophthalmus viridescens* (Rafinesque). Ecology 60: 800–813.

Gill, D.E. 1985. Interpreting breeding patterns from census data: solution to the Husting dilemma. Ecology 66:344–354.

Gill, D.E., K.A. Berven and B.A. Mock. 1983. The environmental component of evolutionary biology. Pp. 1–36. *In* King, C.E. and P.S. Dawson (Eds.), Population Biology—Retrospect and Prospect. Columbia University Press, New York.

Gillette, J.R., S.E. Kolb, J.A. Smith and R.G. Jaeger. 2000. Pheromonal attractions to particular males by female redback salamanders (*Plethodon cinereus*). Pp. 431–440. *In* Bruce, R.C., R.G. Jaeger and L.D. Houck (Eds.), The Biology of Plethodontid Salamanders. Kluwer Academic/Plenum Publishers, New York.

Gilliam, F.S. and W.J. Platt. 1999. Effects of long-term fire exclusion on tree species composition and stand structure in an old-growth *Pinus palustris* (longleaf pine) forest. Plant Ecology 140:15–26.

Gilliams, J. 1818. Description of two new species of Linnaean *Lacerta*. Journal of the Academy of the Natural Sciences of Philadelphia 1:460–462.

Gilliland, M.G. and P.M. Muzzall. 1999. Helminths infecting froglets of the northern leopard frog (*Rana pipiens*) from Foggy Bottom Marsh, Michigan. Journal of the Helminthological Society of Washington 66:73–77.

Gilliland, M.G. and P.M. Muzzall. 2002. Amphibians, trematodes, and deformities: an overview from southern Michigan. Comparative Parasitology 69:81–85.

Gillis, J.E. 1975. Characterization of a hybridizing complex of leopard frogs. Ph.D. dissertation. Colorado State University, Fort Collins, Colorado.

Gillis, J.E. 1979. Adaptive differences in the water economies of two species of leopard frogs from eastern Colorado. Journal of Herpetology 13: 445–450.

Gillis, J.E. and W.J. Breuer. 1984. A comparison of rates of evaporative water loss and tolerance to dehydration in the red-eft and newt of *Notophthalmus viridescens*. Journal of Herpetology 18:81–82.

Giovannoli, L. 1936. *Scaphiopus holbrookii* in Kentucky. Copeia 1936:69.

Girard, C. 1854. A list of North American bufonids, with diagnosis of new species. Proceedings Academy of Natural Sciences, Philadelphia 7:86–88.

Girard, C. 1855 (1854). Abstract of a report to Lieutenant James McGillis, U.S.N., upon the reptiles collected during the U.S.N. Astronomical Expedition to Chili. Proceedings of the Academy of the Natural Sciences of Philadelphia 7:226–227.

Girard, C. 1859. Reptiles of the Boundary. *In* Baird, S.F. United States and Mexican Boundary Survey. Washington, D.C.

Giuliani, D. 1977. Inventory of habitat and potential habitat for *Batrachoseps* sp. Contract Number CA-101-PH6-805, U.S.D.I. Bureau of Land Management, Bakersfield District Office, Bakersfield, California.

Giuliani, D. 1981. An investigation of pre-selected springs and seepages for additional populations of the desert slender salamander, *Batrachoseps aridus*, in the Santa Rosa Mountains, Riverside County, California. U.S.D.I. Bureau of Land Management, Contract Number CA-060-CT1-2, Indio Resource Area, Riverside, California.

Giuliani, D. 1988. Eastern Sierra Nevada salamander survey. California Department of Fish and Game, Draft report, Contract Number FG 7533, Sacramento, California.

Giuliani, D. 1990. New salamander populations from the eastern Sierra Nevada and Owens Valley region of California with notes on previously known sites. California Department of Fish and Game, Contracts FG7533 and FG8450, Sacramento, California.

Giuliani, D. 1996. Resurvey of eastern Sierra Nevada salamanders. California Department of Fish and Game, Unpublished report, Sacramento, California.

Giusti, G.A. and A.M. Merenlender. 2002. Inconsistent application of environmental laws and policies to California's oak woodlands. Pp. 473–482. *In* Standiford, R.B., D. McCreary and K.L. Purcell (Tech. Coords.), Proceedings of the Fifth Symposium on Oak Woodlands: Oaks in California's Changing Landscape. U.S.D.A. Forest Service, General Technical Report, PSW-GTR-184, Pacific Southwest Research Station, Albany, California.

Given, M.F. 1987. Vocalizations and acoustic interactions of the carpenter frog, *Rana virgatipes*. Herpetologica 43:467–481.

Given, M.F. 1988a. Growth rate and the cost of calling activity in male carpenter frogs, *Rana virgatipes*. Behavioral Ecology and Sociobiology 22:153–160.

Given, M.F. 1988b. Territoriality and aggressive interactions of male carpenter frogs, *Rana virgatipes*. Copeia 1988:411–421.

Given, M.F. 1990. Spatial distribution and vocal interaction in *Rana clamitans* and *R. virgatipes*. Journal of Herpetology 24:377–382.

Given, M.F. 1993. Vocal interactions in *Bufo woodhousii fowleri*. Journal of Herpetology 27:447–452.

Given, M.F. 1996. Intensity modulation of advertisement calls in *Bufo woodhousii fowleri*. Copeia 1996:970–977.

Given, M.F. 1999. Distribution records of *Rana virgatipes* and associated anuran species along Maryland's Eastern Shore. Herpetological Review 30:144–146.

Glahn, J.F., D.S. Reinhold and P. Smith. 1999. Wading bird depredations on channel catfish *Ictalurus punctatus* in northwest Mississippi. Journal of the World Aquaculture Society 30:107–114.

Glandt, D. 1984. Laborexperiment zum Räuber-Beute-Verhältnis zwischen Dreistacheligen Stichlingen, *Gasterosteus aculeatus* L. (Teleostei) und Erdkrötenlarven, *Bufo bufo* (L.) (Amphibia). Zoologischer Anzeiger 213:12–16.

Glandt, D. 1985. Kaulquappen-Fressen durch Goldfische *Carassuis a. auratus* und Rotfedern *Scardinius erythrophthalmus*. Salamandra 21: 180–185.

Glandt, D. and R. Podloucky (Eds.). 1987. Der Moorfrosch. Metelener Artenschutzsymposium. Naturschutz und Landschaftspflege in Niedersachsen, Beiheft 19:1–161.

Glaser, H.S.R. 1970. The distribution of amphibians and reptiles in Riverside County, California. Natural History Series, Number 1, Riverside Museum Press, Riverside, California.

Glass, B.P. 1951. Age at maturity of neotenic *Ambystoma t. mavortium* Baird. American Midland Naturalist 46:391–393.

Glaw, F. and J. Kohler. 1998. Amphibian species diversity exceeds that of mammals. Herpetological Review 29:11–12.

Gleason, R.L. and T.H. Craig. 1979. Food habits of the burrowing owl in southeastern Idaho. Great Basin Naturalist 39:274–276.

Gleaves, K., M. Kuruc and P. Montanio. 1992. The meaning of "species" under the Endangered Species Act. Public Land Law Review 13:25–50.

Glennemeier, K.A. 2001. Roles of corticosterone in the development and physiological ecology of *Rana pipiens* tadpoles and the disruption of this endocrine system by organochlorine contamination. Ph. D dissertation. University of Michigan, Ann Arbor, Michigan.

Glennemeier, K.A. and L.J. Benoche. 2002. Impact of organochlorine contamination on amphibian populations in southwestern Michigan. Journal of Herpetology 36:233–244.

Glennemeier, K.A. and R.J. Denver. 2001. Sublethal effects of chronic exposure to an organochlorine compound on northern leopard frog (*Rana pipiens*) tadpoles. Environmental Toxicology 16:287–297.

Glennemeier, K.A. and R.J. Denver. 2002. Role of corticoids in mediating the response of *Rana pipiens* tadpoles to intraspecific competition. Journal of Experimental Zoology 292:32–40.

Global Biodiversity Assessment. 1996. United Nations Environment Programme, New York.

Glorioso, J.C., R.L. Amborski, G.F. Amborski and D.D. Culley Jr. 1974. Microbiological studies on septicemic bullfrogs (*Rana catesbeiana*). American Journal of Veterinary Research 35:1241–1245.

Gloyd, H.K. 1928. The amphibians and reptiles of Franklin County, Kansas. Transactions of the Kansas Academy of Science 31:115–141.

Gloyd, H.K. and R. Conant. 1990. Snakes of the *Agkistrodon* complex: a monographic review. Contributions to Herpetology, Number 6, Society for the Study of Amphibians and Reptiles, St. Louis, Missouri.

Gnaedinger, L.M. and C.A. Reed. 1948. Contribution to the natural history of the plethodontid salamander *Ensatina eschscholtzii*. Copeia 1948:187–196.

Goater, T.M. 2000. The leech, *Oligobdella biannulata* (Glossiphoniidae) on desmognathine salamanders: potential for trypanosome transmission? American Midland Naturalist 144:434–438.

Goater, T.M., G.W. Esch and A.O. Bush. 1987. Helminth parasites of sympatric salamanders: Ecological concepts at infracommunity, component and compound community levels. American Midland Naturalist 118:289–300.

Goater, T.M., M. Mulvey and G.W. Esch. 1990. Electrophoretic differentiation of two *Halipegus* (Trematoda: Hemiuridae) congeners in an amphibian population. Journal of Parasitology 76:431–434.

Godley, J.S. 1983. Observations on the courtship, nests and young of *Siren intermedia* in southern Florida. American Midland Naturalist 110:215–219.

Godley, J.S. 1992. Gopher frog. Pp. 15–19. *In* Moler, P.E. (Ed.), Rare and Endangered Biota of Florida, Volume 3, Amphibians and Reptiles. Second edition. University Press of Florida, Gainesville, Florida.

Godown, M.E. and A.T. Peterson. 2000. Preliminary distributional analysis of U.S. endangered bird species. Biodiversity and Conservation 14:1313–1322.

Godwin, G.J. and S.M. Roble. 1983. Mating success in male treefrogs, *Hyla chrysoscelis* (Anura: Hylidae). Herpetologica 39:141–146.

Goebel, A. 1996a. Systematics and conservation of bufonids in North America and in the *Bufo boreas* species group. Ph.D. dissertation. University of Colorado, Boulder, Colorado.

Goebel, A. 1996b. Phylogenetic diversity and priorities for conservation: examples from toads (genus *Bufo*) Pp. 556–561. AZA Regional Conference Proceedings. Denver Zoological Gardens, Denver, Colorado.

Goebel, A.M., J.M. Donnelly and M.E. Atz. 1998. PCR primers and amplification methods for 12S ribosomal DNA, the control region, cytochrome oxidase I and cytochrome b in bufonids and other frogs, and an overview of PCR primers which have amplified DNA in amphibians successfully. Molecular Phylogenetics and Evolution 11: 163–199.

Goettl, J.P. and the Boreal Toad Recovery Team. 1997. Boreal toad (*Bufo boreas boreas*) (Southern Rocky Mountain population) recovery plan. Colorado Division of Wildlife, Unpublished report, Denver, Colorado.

Goin, C.J. 1938. A large chorus of *Hyla gratiosa*. Copeia 1938:48.

Goin, C.J. 1939. Notes on *Pseudotriton ruber vioscai* Bishop. Copeia 1939:231.

Goin, C.J. 1943. The lower vertebrate fauna of the water hyacinth community in northern Florida. Proceedings of the Florida Academy of Sciences 6:143–154.

Goin, C.J. 1944. *Eleutherodactylus ricordii* at Jacksonville, Florida. Copeia 1944:192.

Goin, C.J. 1947a. Studies on the life history of *Eleutherodactylus ricordii planirostris* (Cope) in Florida with special reference to the local distribution of an allelomorphic color pattern. University of Florida Studies, Biological Science Series 4, Gainesville, Florida.

Goin, C.J. 1947b. A note on the food of *Heterodon simus*. Copeia 1947:275.

Goin, C.J. 1947c. Notes on the eggs and early larvae of three Florida salamanders. Natural History Miscellanea, Number 10, Chicago Academy of Sciences, Chicago, Illinois.

Goin, C.J. 1951. Notes on the eggs and early larvae of three more Florida salamanders. Annals of the Carnegie Museum 32:253–263.

Goin, C.J. 1957. Description of a new salamander of the genus *Siren* from the Rio Grande. Herpetologica 13:37–42.

Goin, C.J. 1958. Comments upon the origin of the herpetofauna of Florida. Quarterly Journal of the Florida Academy of Science 21:61–70.

Goin, C.J. 1961. The growth and size of *Siren lacertina*. Copeia 1961:139.

Goin, C.J. and J.W. Crenshaw Jr. 1949. Description of a new race of the salamander *Pseudobranchus striatus* (Le Conte). Annals of the Carnegie Museum 31:277–280.

Goin, C.J. and O. Goin. 1953. Temporal variations in a small community of amphibians and reptiles. Ecology 34:406–408.

Goin, C.J. and O. Goin. 1957. Remarks on the behavior of the squirrel treefrog, *Hyla squirella*. Annals of the Carnegie Museum 35:27–36.

Goin, C.J. and O.B. Goin. 1962. Introduction to Herpetology. W.H. Freeman and Company, San Francisco, California.

Goin, C.J. and M.G. Netting. 1940. A new gopher frog from the Gulf Coast, with comments upon the *Rana areolata* group. Annals of the Carnegie Museum 38:137–168.

Goin, C.J. and L.H. Ogren. 1956. Parasitic copepods (Argulidae) on amphibians. Journal of Parasitology 42:154.

Goin, O.B. 1955. World Outside My Door. Macmillan Press, New York.

Goin, O.B. 1958. A comparison of the non-breeding habits of two treefrogs, *Hyla squirella* and *Hyla cinerea*. Quarterly Journal of the Florida Academy of Science 21:49–60.

Goldberg, C.S., K.J. Field, and M.J. Sredl. 2004b. Mitochondrial DNA sequences do not support species status of the Ramsey Canyon leopard frog. Journal of Herpetology 38:313–319.

Goldberg, C.S. and C.R. Schwalbe. 2004a. Considerations for monitoring a rare anuran *(Eleutherodactylus augusti)*. Southwestern Naturalist 49:442–448.

Goldberg, C.S. and C.R. Schwalbe. 2004b. Habitat and spatial population structure of barking frogs *(Eleutherodactylus augusti)* in southern Arizona. Journal of Herpetology 38:26–33.

Goldberg, C.S., B.K. Sullivan, J.H. Malone and C.R. Schwalbe. 2004a. Phylogeography of barking frogs *(Eleutherodactylus augusti)* in Arizona, New Mexico, and Texas. Herpetologica 60:312–320.

Goldberg, S.R. and C.R. Bursey. 2001. Helminths of the California treefrog, *Hyla cadaverina* (Hylidae), from southern California. Bulletin of the Southern California Academy of Science 100:117–122.

Goldberg, S.R. and C.R. Bursey. 2002. Helminth parasites of seven anuran species from northwestern Mexico. Western North American Naturalist 62:160–169.

Goldberg, S.R., C.R. Bursey and H. Cheam. 1998a. Nematodes of the Great Plains narrow-mouthed toad, *Gastrophryne olivacea* (Microhylidae), from southern Arizona. Journal of the Helminthological Society of Washington 65:102–104.

Goldberg, S.R., C.R. Bursey and H. Cheam. 1998b. Helminths of two native frog species *(Rana chiricahuensis, Rana yavapaiensis)* and one introduced frog species *(Rana catesbeiana)* (Ranidae) from Arizona. Journal of Parasitology 84:175–177.

Goldberg, S.R., C.R. Bursey and H. Cheam. 1998c. Composition and structure of helminth communities of the salamanders, *Aneides lugubris, Batrachoseps nigriventris, Ensatina eschscholtzii* (Plethodontidae), and *Taricha torosa* (Salamandridae) from California. Journal of Parasitology 84:248–251.

Goldberg, S.R., C.R. Bursey and H. Cheam. 2000. Helminths of the Channel Islands slender salamander, *Batrachoseps pacificus pacificus* (Caudata: Plethodontidae) from California. Bulletin of the Southern California Academy of Science 99:55–57.

Goldberg, S.R., C.R. Bursey and G. Galindo. 1999. Helminths of the lowland burrowing treefrog, *Pternohyla fodiens* (Hylidae), from southern Arizona. Great Basin Naturalist 59:195–197.

Goldberg, S.R., C.R. Bursey and I. Ramos. 1995. The component parasite community of three sympatric toad species, *Bufo cognatus, Bufo debilis* (Bufonidae), and *Spea multiplicata* (Pelobatidae) from New Mexico. Journal of the Helminthological Society Washington 62:57–61.

Goldberg, S.R., C.R. Bursey, R.G. McKinnell and I.S. Tan. 2001. Helminths of northern leopard frogs, *Rana pipiens* (Ranidae) from North Dakota and South Dakota. Western North American Naturalist 61:248–251.

Goldberg, S.R., C.R. Bursey, B.K. Sullivan and Q.A. Truong. 1996a. Helminths of the Sonoran green toad, *Bufo retiformis* (Bufonidae) from southern Arizona. Journal of the Helminthological Society of Washington 63:120–122.

Goldberg, S.R., C.R. Bursey, E.W.A. Gergus, B.K. Sullivan and Q.A. Truong. 1996b. Helminths from three treefrogs, *Hyla arenicolor, Hyla wrightorum*, and *Pseudacris triseriata* (Hylidae) from Arizona. Journal of Parasitology 82:833–835.

Goldberg, S.R., C.R. Bursey, K.B. Malmos, B.K. Sullivan and H. Cheam. 1996c. Helminths of the southwestern toad *Bufo microscaphus*, Woodhouse's toad *Bufo woodhousii*, and their hybrids from central Arizona, USA. Great Basin Naturalist 56:369–374.

Golden, D.R., G.R. Smith and J.E. Rettig. 2001. Effects of age and group size on habitat selection and activity level in *Rana pipiens* tadpoles. Herpetological Journal 11:69–73.

Goldstein, P.Z., R. DeSalle, G. Amato and A.P. Volger. 2000. Conservation genetics at the species boundary. Conservation Biology 14:120–131.

Gollob, T. 1978. Bullfrogs preying on cedar waxwings. Herpetological Review 9:47–48.

Gomez, D.M. 1993. Small mammal and herpetofaunal abundance in riparian and upslope areas of five forest conditions. Master's thesis. Oregon State University, Corvallis, Oregon.

Gomez, D.M. and R.G. Anthony. 1996. Amphibian and reptile abundance in riparian and upslope areas of five forest types in western Oregon. Northwest Science 70:109–119.

Good, D.A. 1989. Hybridization and cryptic species in *Dicamptodon* (Caudata: Dicamptodontidae). Evolution 43:728–744.

Good, D.A. and D.B. Wake. 1992. Geographic variation and speciation in the torrent salamanders of the genus *Rhyacotriton* (Caudata: Rhyacotritonidae). University of California Publications in Zoology, Number 126, University of California, Berkeley, California.

Good, D.A., G.Z. Wurst and D.B. Wake. 1987. Patterns of geographic variation in allozymes of the Olympic salamander, *Rhyacotriton olympicus* (Caudata: Dicamptodontidae). Fieldiana, Zoology, New Series, Number 32, Field Museum of Natural History, Chicago, Illinois.

Goodacre, W.A. 1947. The giant toad *(Bufo marinus)*: an enemy of bees. Agricultural Gazette of New South Wales 58:374–375.

Goodchild, C.G. 1948. Additional observations on the bionomics and life history of *Gorgodera amplicava* Looss, 1899 (Trematoda: Gorgoderidae). Journal of Parasitology 34:407–427.

Goodchild, C.G. 1953. A subcutaneous, cyst-parasite of bullfrogs: *Histocystidium ranae*, N.G., N. sp. Journal of Parasitology 39:395–405.

Goodman, J.D. 1958. Anuran records from the Upper Coastal Plain of western and central Georgia. Herpetologica 14:123–124.

Goodman, J.D. 1989. *Langeronia brenesi* n. sp. (Trematoda: Lecithodendriidae) in the mountain yellow-legged frog *Rana muscosa* from southern California. Transactions of the American Microscopical Society 108:387–393.

Goodman, R.H., Jr., S.K. Watanabe, K.P. Condon, M.P. Pires and M.S. Benton. 1998. Geographic distribution: *Batrachoseps gabrieli* (San Gabriel Mountain slender salamander). Herpetological Review 29:171.

Goodman, S.M. and S.M. Lanyon. 1994. Scientific collecting. Conservation Biology 8:314–315.

Goodsell, J.A. and L.B. Kats. 1999. Effect of introduced mosquito fish on Pacific treefrogs and the role of alternative prey. Conservation Biology 13:921–924.

Goodyear, C.P. 1971. Y-axis orientation of the oak toad, *Bufo quercicus*. Herpetologica 27:320–323.

Goodyear, C.P. and R. Altig. 1971. Orientation of bullfrogs (*Rana catesbeiana)* during metamorphosis. Copeia 1971:362–364.

Gordon, K. 1939. The Amphibia and Reptilia of Oregon. Oregon State Monographs, Studies in Zoology, Number 1, Oregon State University, Corvallis, Oregon.

Gordon, R. 1997. Letter to Bruce Babbitt, 3 February.

Gordon, R.E. 1952. A contribution to the life history and ecology of the plethodontid salamander *Aneides aeneus* (Cope and Packard). American Midland Naturalist 47:666–701.

Gordon, R.E. 1955. Additional remarks on albinism in *Microhyla carolinensis*. Herpetologica 11:240.

Gordon, R.E. 1960. Young of the salamander, *Plethodon jordani melaventris*. Copeia 1960:26–29.

Gordon, R.E. 1961. The movement of displaced green salamanders. Ecology 42:200–202.

Gordon, R.E. 1966. Some observations on the biology of *Pseudotriton ruber schencki*. Journal of the Ohio Herpetological Society 5: 163–164.

Gordon, R.E. 1967. *Aneides aeneus*. Pp. 30.1–30.2. Catalogue of American Amphibians and Reptiles. Society for the Study of Amphibians and Reptiles, St. Louis, Missouri.

Gordon, R.E. 1968. Terrestrial activity of the spotted salamander, *Ambystoma maculatum*. Copeia 1968:879–880.

Gordon, R.E. and R.L. Smith. 1949. Notes on the life history of the salamander *Aneides aeneus*. Copeia 1949:173–175.

Gordon, R.E., J.A. MacMahon and D.B. Wake. 1962. Relative abundance, microhabitat and behavior of some southern Appalachian salamanders. Zoologica 47:9–14.

Gore, J.A. 1983. The distribution of desmognathine larvae (Amphibia: Plethodontidae) in coal surface impacted streams of the Cumberland Plateau, USA. Journal of Freshwater Ecology 2:13–23.

Gorham, S.W. 1970. The amphibians and reptiles of New Brunswick. Publications of the New Brunswick Museum, Monograph Series, Number 6, St. John, New Brunswick.

Gorham, S.W. 1974. Checklist of World amphibians up to January 1, 1970. New Brunswick Museum, St. John, New Brunswick.

Gorman, J. 1951. Northward range extension of the salamander *Triturus sierrae*. Copeia 1951:78.

Gorman, J. 1954. A new species of salamander from central California. Herpetologica 10:153–158.

Gorman, J. 1956. Reproduction in plethodont salamanders of the genus *Hydromantes*. Herpetologica 12:249–259.

Gorman, J. 1964. *Hydromantes brunus, Hydromantes platycephalus*, and *Hydromantes shastae*. Pp. 11.1–11.2. Catalogue of American Amphibians and Reptiles. Society for the Study of Amphibians and Reptiles, St. Louis, Missouri.

Gorman, J. 1988. The effects of the evolution and ecology of *Hydromantes* on their conservation. Pp. 39–42. *In* De Lisle, H.F., P.R. Brown, B. Kaufman and B.M. McGurty (Eds.), Proceedings of the Conference on California Herpetology. Southwestern Herpetologists Society, Special Publication, Number 4, Van Nuys, California.

Gorman, J. and C.L. Camp. 1953. A new cave species of salamander of the genus *Hydromantes* from California, with notes on habits and habitat. Copeia 1953:39–43.

Gorman, W.L. 1986. Patterns of color polymorphism in the cricket frog, *Acris crepitans*, in Kansas. Copeia 1986:995–999.

Gosdin, W.M. 1975. Texas Parks and Wildlife Department office memorandum: Bastrop State Park expansion and development meeting in 1966.

Gosner, K.L. 1959. Systematic variations in tadpole teeth with notes on food. Herpetologica 15:203–210.

Gosner, K.L. 1960. A simplified table for staging anuran embryos and larvae with notes on identification. Herpetologica 16:183–190.

Gosner, K.L. and I.H. Black. 1955. The effects of temperature and moisture on the reproductive cycle of *Scaphiopus h. holbrookii*. American Midland Naturalist 54:192–203.

Gosner, K.L. and I.H. Black. 1956. Notes on amphibians from the upper Coastal Plain of North Carolina. Journal of the Elisha Mitchell Scientific Society 72:40–47.

Gosner, K.L. and I.H. Black. 1957a. Larval development in New Jersey Hylidae. Copeia 1957:31–36.

Gosner, K.L. and I.H. Black. 1957b. The effects of acidity on the development and hatching of New Jersey frogs. Ecology 38:256–262.

Gosner, K.L. and I.H. Black. 1958a. Notes on larval toads in the eastern United States with special reference to natural hybridization. Herpetologica 14:133–140.

Gosner, K.L. and I.H. Black. 1958b. Observations on the life history of Brimley's chorus frog. Herpetologica 13:249–254.

Gosner, K.L. and I.H. Black. 1967. *Hyla andersonii*. Pp. 54.1–54.2. Catalogue of American Amphibians and Reptiles. Society for the Study of Amphibians and Reptiles, St. Louis, Missouri.

Gosner, K.L. and I.H. Black. 1968. *Rana virgatipes* Cope. Pp. 67.1–67.2. Catalogue of American Amphibians and Reptiles. Society for the Study of Amphibians and Reptiles, St. Louis, Missouri.

Gosner, K.L. and D.A. Rossman. 1959. Observations of the reproductive cycle of the swamp chorus frog, *Pseudacris nigrita*. Copeia 1959: 263–266.

Gosner, K.L. and D.A. Rossman. 1960. Eggs and larval development of the treefrogs *Hyla crucifer* and *Hyla ocularis*. Herpetologica 16:225–232.

Gossling, J., W.J. Loesche, L.D. Ottoni and G.W. Nace. 1980. Passage of material through the gut of hibernating *Rana pipiens* (Amphibia, Anura, Ranidae). Journal of Herpetology 14:407–409.

Gotte, S.W. and C.H. Ernst. 1987. Geographic distribution: *Hyla squirella* (squirrel treefrog). Herpetological Review 18:17.

Gottschalk, J.S. 1970. Conservation of endangered species and other fish or wildlife. United States list of endangered native fish and wildlife. Federal Register 35:16047–16048.

Gould, F. and A. Massey. 1984. Cucurbitacins and predation of the spotted cucumber beetle, *Diabrotica undecimpunctata howardi*. Entomologia Experimentale et Applicata 36:273–278.

Gould, S.J. 1977. Ontogeny and Phylogeny, Belknap Press, Cambridge, Massachusetts.

Goulden, C.E. and L.L. Hornig. 1980. Population oscillations and energy reserves in planktonic cladocera and their consequences to competition. Proceedings of the National Academy of Sciences 77: 1716–1720.

Govindarajulu, P. 2000. Estimation of daily survival rates of *Hyla regilla* tadpoles in the wild. Northwestern Naturalist 81:74.

Goyer, R.A. 1996. Toxic effect of metals. Pp. 696–698. *In* Casarett and Doull's Toxicology: The Basic Science of Poisons. McGraw-Hill Publishing Company, New York.

Gradwell, N. 1971. *Ascaphus* tadpole: experiments on the suction and gill irrigation mechanisms. Canadian Journal of Zoology 49:307–332.

Grady, J.M. and J.M. Quattro. 1999. Using character concordance to define taxonomic and conservation units. Conservation Biology 13: 1004–1007.

Graening, G.O. 2000. Ecosystem dynamics in an Ozark cave. Ph.D. dissertation. University of Arkansas, Fayetteville, Arkansas.

Graening, G.O. and A.V. Brown. 2000. Status survey of aquatic cave fauna in Arkansas. Final report to the Arkansas Game and Fish Commission. Arkansas Water Resource Center, Publication Number MSC-286, University of Arkansas, Fayetteville, Arkansas.

Graening, G.O., M. Slay and A.V. Brown. 2001. Subterranean biodiversity in the Ozark Plateaus of Arkansas. Subterranean Biodiversity Project, Department of Biological Sciences. University of Arkansas, Fayetteville, Arkansas.

Graf, W. 1949. Observations on the salamander *Dicamptodon*. Copeia 1949:79–80.

Graham, K.L. and G.L. Powell. 1999. Status of the long-toed salamander (*Ambystoma macrodactylum*) in Alberta. Alberta Environmental Protection, Fisheries and Wildlife Management Division, and Alberta Conservation Association, Wildlife Status Report, Number 22, Edmonton, Alberta, Canada.

Graham, R.E. 1971. Environmental effects on deme structure, dynamics, and breeding strategy of *Ambystoma opacum* (Amphibia: Ambystomatidae), with an hypothesis on the probable origin of the marbled salamander lifestyle. Ph.D. dissertation. Rutgers University, New Brunswick, New Jersey.

Graham, T.E. 1981. Geographic distribution: *Gyrinophilus porphyriticus porphyriticus* (northern spring salamander). Herpetological Review 12:64.

Graham, T.E. and L. Stevens 1982. Geographic distribution: *Gyrinophilus porphyriticus porphyriticus* (northern spring salamander). Herpetological Review 13:130.

Grant, B.W., K.L. Brown, G.W. Ferguson and J.W. Gibbons. 1994. Changes in amphibian biodiversity associated with 25 years of pine forest regeneration: implications for biodiversity management. Pp. 354–367. *In* Majumdar, S.K., F.J. Brenner, J.E. Lovich, J.F. Shalles, and E.W. Miller (Eds.), Biological Diversity: Problems and Challenges. Pennsylvania Academy of Science, Easton, Pennsylvania.

Grant, B.W., A.D. Tucker, J.E. Lovich, A.M. Mills, P.M. Dixon and J.W. Gibbons. 1992. The use of coverboards in estimating patterns of reptile and amphibian biodiversity. Pp. 379–403. *In* McCullough, D.R. and R.H. Barrett (Eds.), Wildlife 2001. Elsevier Science Publishers, London, England.

Grant, C. 1931. Notes on *Bufo marinus*. Copeia 1931:62–63.

Grant, C. 1936. Herpetological notes from northern Indiana. Proceedings of the Indiana Academy of Science 44:244–246.

Grant, C. 1958. Irruption of young *Batrachoseps attenuatus*. Copeia 1958:222.

Grant, D., O. Anderson and V. Twitty. 1968. Homing orientation by olfaction in newts *(Taricha rivularis)*. Science 160:1354–1356.

Grant, K.P. and L.E. Licht. 1993. Acid tolerance of anuran embryos and larvae from central Ontario. Journal of Herpetology 27:1–6.

Grant, K.P. and L.E. Licht. 1995. Effects of ultraviolet radiation on life-history states of anurans from Ontario, Canada. Canadian Journal of Zoology 73:2292–2301.

Grant, M.P. 1931. Diagnostic stages of metamorphosis in *Amblystoma jeffersonianum* and *Amblystoma opacum*. Anatomical Record 51:1–15.

Grant, W.C., Jr. 1955. Territorialism in two species of salamanders. Science 121:137–138.

Gravenhorst, J.L.C. 1807. Vergleichende des Linneischen und einiger neuern Uebersicht Zoologischen Systeme. H. Dieterich, Gottingen, Germany.

Graves, B.M. and V.S. Quinn. 2000. Temporal persistence of alarm pheromones in skin secretions of the salamander, *Plethodon cinereus*. Journal of Herpetology 34:287–291.

Graves, B.M., C.H. Summers and K.L. Olmstead. 1993. Sensory mediation of aggregation among postmetamorphic *Bufo cognatus*. Journal of Herpetology 27:315–319.

Gray, I.E. 1941. Amphibians and reptiles of the Duke Forest and vicinity. American Midland Naturalist 26:652–658.

Gray, J.E. 1850. Catalog of the Species of Amphibians in the collection of the British Museum, Part 2, Batrachia Gradientia. Pp. 1–72. London, England.

Gray, J.E. 1853. On a new species of salamander from California. Proceedings of the Zoological Society of London 21:11.

Gray, R.H. 1971a. Fall activity and overwintering of the cricket frog *(Acris crepitans)* in central Illinois. Copeia 1971:748–750.

Gray, R.H. 1971b. Ecological studies on color polymorphism in the cricket frog, *Acris crepitans* Baird. Ph.D. dissertation. Illinois State University, Normal, Illinois.

Gray, R.H. 1972. Metachrosis of the vertebral stripe in the cricket frog, *Acris crepitans*. American Midland Naturalist 87:549–551.

Gray, R.H. 1977. Lack of physiological differentiation in three color morphs of the cricket frog *(Acris crepitans)* in Illinois. Transactions of the Illinois State Academy of Science 70:73–79.

Gray, R.H. 1978. Nondifferential predation susceptibility and behavioral selection in three color morphs of Illinois cricket frogs, *Acris crepitans*. Transactions of the Illinois State Academy of Science 71:356–360.

Gray, R.H. 1983. Seasonal, annual and geographic variation in color morph frequencies of the cricket frog, *Acris crepitans*, in Illinois. Copeia 1983:300–311.

Gray, R.H. 1984. Effective breeding size and the adaptive significance of color polymorphism in the cricket frog *(Acris crepitans)* in Illinois, USA. Amphibia-Reptilia 5:101–107.

Gray, R.H. 1995. An unusual color pattern variant in cricket frogs *(Acris crepitans)* from southern Illinois. Transactions of the Illinois State Academy of Science 88:137–138.

Gray, R.H. 2000a. Morphological abnormalities in Illinois cricket frogs, *Acris crepitans*, 1968–1971. Pp. 92–95. *In* Kaiser, H., G.S. Casper and N.P. Bernstein (Eds.), Investigating Amphibian Declines: Proceedings of the 1998 Declining Amphibians Conference, Journal of the Iowa Academy of Science, Volume 107, Cedar Falls, Iowa.

Gray, R.H. 2000b. Historic occurrence of malformations in the cricket frog *Acris crepitans* in Illinois. Transactions of the Illinois State Academy of Science 93:279–284.

Graybeal, A. 1993. The phylogenetic utility of cytochrome b: lessons from bufonid frogs. Molecular Phylogenetics and Evolution 2:256–269.

Graybeal, A. 1997. Phylogenetic relationships of bufonid frogs and tests of alternate macroevolutionary hypotheses characterizing their radiation. Zoological Journal of the Linnean Society 119:297–338.

Green, D. and D. Thomas 1989. Geographic distribution: *Rana c. clamitans* (bronze frog). Herpetological Review 20:11–12.

Green, D.E. and K.A. Converse. 2002. Diagnostic services case report. U.S.G.S. National Wildlife Health Center, Madison, Wisconsin.

Green, D.E. and C. Kagarise Sherman. 2001. Diagnostic histological findings in Yosemite toads *(Bufo canorus)* from a die-off in the 1970s. Journal of Herpetology 35:92–103.

Green, D.E. and R.A. Sohn. 2002. Diagnostic services case report. U.S.G.S. National Wildlife Health Center, Madison, Wisconsin.

Green, D.M. 1979. Treefrog toe pads: comparative surface morphology using scanning electron microscopy. Canadian Journal of Zoology 57:2033–2046.

Green, D.M. 1980. Size difference in adhesive toe-pad cells of treefrogs of the diploid-polyploid *Hyla versicolor* complex. Journal of Herpetology 14:15–19.

Green, D.M. 1982. Mating call characteristics of hybrid toads *(Bufo americanus × B. fowleri)* at Long Point, Ontario. Canadian Journal of Zoology 60:3293–3297.

Green, D.M. 1983. Allozyme variation through a clinal hybrid zone between the toads *Bufo americanus* and *B. hemiophrys* in southeastern Manitoba. Herpetologica 39:28–40.

Green, D.M. 1984. Sympatric hybridization and allozyme variation in the toads *Bufo americanus* and *B. fowleri* in southern Ontario. Copeia 1984:18–26.

Green, D.M. 1985. Natural hybrids between the frogs *Rana cascadae* and *Rana pretiosa* (Anura: Ranidae). Herpetologica 41:262–267.

Green, D.M. 1989. Fowler's toads, *(Bufo woodhousii fowleri)* in Canada: biology and population status. Canadian Field-Naturalist 103:486–496.

Green, D.M. 1992. Fowler's toads, *Bufo woodhousii fowleri*, at Long Point, Ontario: changing abundance and implications for conservation. Pp. 37–45. *In* Bishop, C.A. and K.E. Pettit (Eds.), Declines in Canadian Amphibian Populations: Designing a National Monitoring Strategy. Canadian Wildlife Service, Occasional Publications, Number 76, Environment Canada, Ottawa, Ontario, Canada.

Green, D.M. 1996. The bounds of species. Hybridization in the *Bufo americanus* group of North American toads. Israel Journal of Zoology 42:95–109.

Green, D.M. 1997a. Temporal variation in abundance and age structure in Fowler's toads, *Bufo fowleri*, at Long Point, Ontario. Pp. 45–56. *In* Green, D.M. (Ed.), Amphibians in Decline: Canadian Studies of a Global Problem. Herpetological Conservation, Number 1, Society for the Study of Amphibians and Reptiles, St. Louis, Missouri.

Green, D.M. (Ed.). 1997b. Amphibians in Decline: Canadian Studies of a Global Problem. Herpetological Conservation, Number 1, Society for the Study of Amphibians and Reptiles, St. Louis, Missouri.

Green, D.M. 1997c. Perspectives on amphibian population declines: defining the problem and searching for answers. Pp. 291–308. *In* Green, D.M. (Ed.), Amphibians in Decline: Canadian Studies of a Global Problem. Herpetological Conservation, Number 1, Society for the Study of Amphibians and Reptiles, St. Louis, Missouri.

Green, D.M. and R.W. Campbell. 1984. The Amphibians of British Columbia. British Columbia Provincial Museum, Handbook 45, Victoria, British Columbia.

Green, D.M. and C. Pustowka. 1997. Correlated morphological and allozyme variation in the hybridizing toads *Bufo americanus* and *Bufo hemiophrys*. Herpetologica 53:218–228.

Green, D.M., T.F. Sharbel, J. Kearsley and H. Kaiser. 1996. Postglacial range fluctuation, genetic subdivision and speciation in the western North American spotted frog complex, *Rana pretiosa*. Evolution 50:374–390.

Green, D.M., H. Kaiser, T.F. Sharbel, J. Kearsley and K.R. McAllister. 1997. Cryptic species of spotted frogs, *Rana pretiosa* complex, in western North America. Copeia 1997:1–8.

Green, G.A., R.E. Fitzner, R.G. Anthony and L.E. Rogers. 1993. Comparative diets of burrowing owls in Oregon and Washington. Northwest Scientist 67:88–93.

Green, H.T. 1925. The egg-laying of the purple salamander. Copeia 141:32.

Green, J. 1818. Descriptions of several species of North American Amphibia, accompanied with observations. Journal of the Academy of the Natural Sciences of Philadelphia 1:348–358.

Green, J. 1825. Description of a new species of salamander. Journal of the Academy of the Natural Sciences of Philadelphia 5:116–118.

Green, J. 1827. An account of some new species of salamanders. Contributions of the Maclurean Lyceum 1:3–7.

Green, J. 1830 (1831). Descriptions of two new species of salamander. Journal of the Academy of the Natural Sciences of Philadelphia 6:253–254.

Green, N.B. 1934. *Cryptobranchus alleganiensis* in West Virginia. Proceedings of the West Virginia Academy of Science 7:28–30.

Green, N.B. 1935. Further notes on the food habits of *Cryptobranchus alleganiensis*. Proceedings of the West Virginia Academy of Science 9:36.

Green, N.B. 1938a. The breeding habits of *Pseudacris brachyphona* (Cope) with a description of the eggs and tadpole. Copeia 1938:79–82.

Green, N.B. 1938b. A new salamander, *Plethodon nettingi*, from West Virginia. Annals of the Carnegie Museum 27:295–299.

Green, N.B. 1948. The spade-foot toad, *Scaphiopus h. holbrookii*, breeding in southern Ohio. Copeia 1948:65.

Green, N.B. 1952. A study of the life history of *Pseudacris brachyphona* (Cope) in West Virginia with special reference to behavior and growth of marked individuals. Ph.D. dissertation. Ohio State University, Columbus, Ohio.

Green, N.B. 1955. The ambystomatid salamanders in West Virginia. Proceedings of the West Virginia Academy of Science 27:16–18.

Green, N.B. 1964. Postmetamorphic growth in the mountain chorus frog, *Pseudacris brachyphona* Cope. Proceedings of the West Virginia Academy of Science 36:34–38.

Green, N.B. and P. Brant. 1966. Salamanders found in West Virginia caves. Proceedings of the West Virginia Academy of Science 38:42–45.

Green, N.B. and T.K. Pauley. 1987. Amphibians and Reptiles in West Virginia. University of Pittsburgh Press, Pittsburgh, Pennsylvania.

Green, N.B., P. Brant Jr. and B. Dowler. 1967. *Eurycea lucifuga* in West Virginia: its distribution, ecology, and life history. Proceedings of the West Virginia Academy of Science 41:142–144.

Greenberg, C.H. 1998. Isolated pond use by amphibians in a hardwood-invaded versus savannah-like longleaf pine-wiregrass sandhills upland matrix (1994–1997). U.S.D.A. Forest Service, Bent Creek Experimental Forest, Progress report, Asheville, North Carolina.

Greenberg, L.A. and D.A. Holtzman. 1987. Microhabitat utilization, feeding periodicity, home range and population size of the banded sculpin, *Cottus carolinae*. Copeia 1987:19–25.

Greene, B.D., J.R. Dixon, J.M. Mueller, M.J. Whiting and O.W. Thornton Jr. 1994. Feeding ecology of the Concho water snake, *Nerodia harteri paucimaculata*. Journal of Herpetology 28:165–172.

Greene, H. and J. Losos. 1988. Systematics, natural history, and conservation. BioScience 38:458–462.

Greene, H.W. and C.A. Luke. 1996. Amphibian and reptile diversity in the East Mojave Desert. Pp. 53–58. *In* Luke, C., J. André, and M. Herring (Eds.), Proceedings of the East Mojave Desert Symposium, 7–8 November, 1992. University of California, Riverside, Technical Report, Number 10, Natural History Museum of Los Angeles County, Los Angeles, California.

Greenwalt, L.A. 1975. Endangered species; determination of critical habitat. Federal Register 40:21499–21501.

Greenwalt, L.A. and J.W. Gehringer. 1975. Endangered and threatened species; notice on critical habitat areas. Federal Register 40:17764–17765.

Greenwell, M., V. Beasley and L.E. Brown. 1996. Cricket frog research: the mysterious decline of the cricket frog. Aquaticus 26:48–54.

Gregory, P.T. 1978. Feeding habits and dietary overlap of three species of garter snake *(Thamnophis)* on Vancouver Island. Canadian Journal of Zoology 56:1967–1974.

Gregory, P.T. 1979. Predator avoidance behavior of the red-legged frog *(Rana aurora)*. Herpetologica 35:175–184.

Greig, J.C. 1979. Principles of genetic conservation in relation to wildlife management in Southern Africa. South African Journal of Wildlife Research 9:57–78.

Grenard, S. 1994. Medical herpetology. Reptile and Amphibian Magazine, NG Publishing, Pottsville, Pennsylvania.

Greuter, K. and M.R.J. Forstner. 2003. Natural history notes: *Bufo houstonensis* (Houston toad). Growth. Herpetological Review 34:355–356.

Grialou, J.A., S.D. West and R.N. Wilkins. 2000. The effects of forest clearcut harvesting and thinning on terrestrial salamanders. Journal of Wildlife Management 64:105–113.

Griesenbeck, J.A. Undated, ca. 1976. Letter to Jack Stone.

Griesenbeck, J.A. 1984. Written statement recalling 24 January 1966 meeting with Gosdin and Watson, 30 July.

Griesenbeck, J.A. and A. Patton. 1966. Statement transmitting deed for 1,134 acres from city of Bastrop to the Parks and Wildlife Department of the state of Texas, 25 January.

Griffin, J.W. 1988. The archeology of Everglades National Park: a synthesis. National Park Service, Contract CX 5000-5-0049, Southeast Archeological Center, Tallahassee, Florida.

Griffin, P.C. 1999. *Bufo californicus*, arroyo toad movement patterns and habitat preferences. Master's thesis. University of California, San Diego, California.

Griffis, M.R. and R.G. Jaeger. 1998. Competition leads to an extinction-prone species of salamander: interspecific territoriality in a metapopulation. Ecology 79:2494–2502.

Griffith, G., J. Omernik and S. Azevedo. 1998. Ecoregions of Tennessee. U.S. Geological Survey, Reston, Virginia.

Grigg, G. 2000. Cane toads. Nature Australia, Winter 2000:33–41.

Grimké, S. and R.G. Jaeger. 1998. Tadpole bullies: examining mechanisms of competition in a community of larval anurans. Canadian Journal of Zoology 76:144–153.

Grinnell, J. and C.L. Camp. 1917. A distributional list of the amphibians and reptiles of California. University of California Publications in Zoology 17:127–208.

Grinnell, J. and T.I. Storer. 1924. Animal Life in the Yosemite: An Account of the Mammals, Birds, Reptiles, and Amphibians in a Cross-section of the Sierra Nevada. University of California Press, Berkeley, California.

Grinnell, J., J. Dixon and J.M. Linsdale. 1930. Vertebrate natural history of a section of northern California through the Lassen Peak region. University of California Publications in Zoology, Number 35, Berkeley, California.

Grismer, L.L. 1982. A new population of slender salamander *(Batrachoceps)* from northern Baja California, Mexico. Newsletter of the San Diego Herpetological Society, Number 4, San Diego California.

Grismer, L.L. 1994. The origin and evolution of the peninsular herpetofauna of Baja California, Mexico. Herpetological Natural History 2:51–106.

Grismer, L.L. 2002. Amphibians and Reptiles of Baja California, Including Its Pacific Islands and the Islands in the Sea of Cortés. University of California Press, Berkeley, California.

Grizzell, R.A. 1949. The hibernation site of three snakes and a salamander. Copeia 1949:231–232.

Groan, W.L., Jr. and P.G. Bistroic. 1973. Early breeding activity of *Rana sphenocephala* and *Bufo woodhousei fowleri* in Maryland. Bulletin of Maryland Herpetological Society 9:106.

Grobman, A. 1957. The thyroid gland of *Typhlomolge rathbuni*. Copeia 1957:41–42.

Grobman, A.B. 1941. Variation of the salamander *Pseudotriton ruber*. Copeia 1941:179.

Grobman, A.B. 1944. The distribution of the salamanders of the genus *Plethodon* in eastern United States and Canada. Annals of the New York Academy of Sciences 45:261–316.

Grobman, A.B. 1949. Some recent collections of *Plethodon* from Virginia with the description of a new form. Proceedings of the Biological Society of Washington 63:135–142.

Grogan, W.L., Jr. and D. Bystrak. 1973b. Longevity and size records for the newts, *Notophthalmus perstriatus* and *Notophthalmus v. viridescens*. Smithsonian Herpetological Information Search System News-Journal 1:54.

Grogan, W.L., Jr. and P.G. Bystrak. 1973a. Early breeding activity of *Rana sphenocephala* and *Bufo woodhousei fowleri* in Maryland. Bulletin of the Maryland Herpetological Society 9:106.

Gromko, M.H., F.S. Mason and S.J. Smith-Gill. 1973. Analysis of the crowding effect in *Rana pipiens* tadpoles. Journal of Experimental Zoology 186:63–72.

Gross, S.K. 1982. The influence of habitat variation on the morphology and physiology of *Plethodon wehrlei* (Fowler and Dunn) in West Virginia. Master's thesis. Marshall University, Huntington, West Virginia.

Grossenbacher, K. 1974. Die Amphibien der Umgebung Berns. Mitteilungen der naturforschenden Gesellschaft in Bern 31:3–23.

Grossenbacher, K. 1981. Amphibien und Verkehr. Publication 1. Koordinationsstelle für Amphibien—und Reptilienschutz in der Schweiz, Bern, Switzerland.

Grote, J. 1995a. Amendment to Biological Opinion for Texas Parks and Wildlife's proposed expansion of the Lost Pines Golf Course in Bastrop State Park. 10 February, U.S. Fish and Wildlife Service, Washington, D.C.

Grote, J. 1995b. Letter to Dwight Williford, 26 October.

Grote, J. 1996. Letter to Andy Sansom, 20 February.

Grover, M.C. 1998. Influence of cover and moisture on abundances of the terrestrial salamanders *Plethodon cinereus* and *Plethodon glutinosus*. Journal of Herpetology 32:489–497.

Grover, M.C. 2000. Determinants of salamander distributions along moisture gradients. Copeia 2000:156–168.

Groves, C.R. and C.R. Peterson. 1992. Population trends of Idaho amphibians as determined by mail questionnaire. Idaho Department of Fish and Game, Final report, Boise, Idaho.

Groves, C.R., E.F. Cassirer, D.L. Genter and J.D. Reichel. 1996. Element stewardship abstract: Coeur d'Alene salamander *(Plethodon idahoensis)*. Natural Areas Journal 16:238–247.

Groves, J.D. 1980. Mass predation on a population of the American toad, *Bufo americanus*. American Midland Naturalist 103:202–203.

Grubb, J.C. 1970. Orientation in post-reproductive Mexican toads, *Bufo valliceps*. Copeia 1970:674–680.

Grubb, J.C. 1973a. Olfactory orientation in breeding Mexican toads, *Bufo valliceps*. Copeia 1973:490–497.

Grubb, J.C. 1973b. Orientation in newly metamorphosed Mexican toads, *Bufo valliceps*. Herpetologica 29:95–100.

Grubb, J.C. 1973c. Olfactory orientation in *Bufo woodhousei fowleri*, *Pseudacris clarki* and *Pseudacris streckeri*. Animal Behaviour 21:726–732.

Gruber, E.R. and R.V. Stirling. 1972. Observations on the burrowing habits of the tiger salamander *(Ambystoma tigrinum)*. Herpetological Review 4:85–89.

Gruia-Gray, J. and S.S. Desser. 1992. Cytopathologic observations and epizootiology of frog erythrocytic virus in bullfrogs *(Rana catesbeiana)*. Journal of Wildlife Disease 28:34–41.

Gruia-Gray, J., M. Petric and S. Desser. 1989. Ultrastructural, biochemical and biophysical properties of an erythrocytic virus of frogs from Ontario, Canada. Journal of Wildlife Diseases 25:487–506.

Grumbine, R.E. 1990. Viable populations, reserve size, and federal lands management: a critique. Conservation Biology 4:127–134.

Grzimek, B. 1974. Animal Life Encyclopedia, Volume 5 (Fishes II, Amphibians). Van Nostrand Reinhold Company, New York.

Guarisco, H. 1985. Opportunistic scavenging by the bullfrog, *Rana catesbeiana* (Amphibia, Anura, Ranidae). Transactions of the Kansas Academy of Science 88:38–39.

Guerry, A.D. and M.L. Hunter. 2002. Amphibian distributions in a landscape of forests and agriculture: an examination of landscape composition and configuration. Conservation Biology 16:745–754.

Guillen-Hernandez, S., G. Salgado-Maldonado and R. Lamothe-Argumedo. 2000. Digeneans (Plathelminthes: Trematoda) of seven species of anurans from Los Tuxtlas, Veracruz, Mexico. Studies on Neotropical Fauna and Environment 35:10–13.

Guimond, R.W. and V.H. Hutchison. 1976. Gas exchange of the giant salamanders of North America. Pp. 313–338. *In* Hughes, G.M. (Ed.), Respiration of Amphibious Vertebrates. Academic Press, London.

Gulve, P.S. 1994. Distribution and extinction patterns within a northern metapopulation of the pool frog, *Rana lessonae*. Ecology 75:1357–1367.

Gunderson, J. 1995. The rusty crayfish: a nasty invader; biology, identification, and impacts. Minnesota Sea Grant, Duluth, Minnesota.

Gunter, G. and W.E. Brode. 1964. *Necturus* in the state of Mississippi, with notes on adjacent areas. Herpetologica 20:114–126.

Gunzburger, M.S. 1999. Diet of the Red Hills salamander *Phaeognathus hubrichti*. Copeia 1999:523–525.

Gunzburger, M.S. 2003. Evaluation of the hatching trigger and larval ecology of the salamander *Amphiuma means*. Herpetologica 59:459–468.

Gunzburger, M. and C. Guyer. 1998. Longevity and abandonment of burrows used by the Red Hills salamander *(Phaeognathus hubrichti)*. Journal of Herpetology 32:620–623.

Gurevitch, J. and L.V. Hedges. 1993. Meta-analysis: combining the results of independent experiments. Pp. 378–398. *In* Scheiner, S.M. and J. Gurevitch (Eds.), Design and Analysis of Ecological Experiments. Chapman and Hall, New York.

Gurevitch, J., L.L. Morrow, A. Wallace and J.S. Walsh. 1992. A meta-analysis of competition in field experiments. American Naturalist 140:539–572.

Gustafson, E.J. and G.R. Parker. 1992. Relationships between landcover proportion and indices of landscape spatial pattern. Landscape Ecology 7:101–110.

Gustafson, M.P. 1994. Size-specific interactions among larvae of the plethodontid salamanders *Gyrinophilus porphyriticus* and *Eurycea cirrigera*. Journal of Herpetology 28:470–476.

Guttman, D., J.E. Bramble and O.J. Sexton. 1991. Observations on the breeding immigration of wood frogs *Rana sylvatica* reintroduced in east-central Missouri. American Midland Naturalist 125:269–274.

Guttman, S.I. 1969. Blood protein variation in the *Bufo americanus* species group of toads: evolutionary relationships. Copeia 1969: 243–249.

Guttman, S.I. 1975. Genetic variation in the genus *Bufo*. Part 2: isozymes in northern allopatric populations of the American toad *Bufo americanus*. Pp. 679–697. *In* Markert, C.L. (Ed.), Isozymes IV. Genetics and Evolution. Academic Press, New York.

Guttman, S.I. 1989. *Eurycea longicauda* (Green), longtail salamander. Pp. 204–209. *In* Pfingsten, R.A. and F.L. Downs (Eds.), Salamanders of Ohio. Ohio Biological Survey Bulletin, New Series, Number 7, Columbus, Ohio.

Guttman, S.I. and A.A. Karlin. 1986. Hybridization of cryptic species of two-lined salamanders (*Eurycea bislineata* complex). Copeia 1986: 96–108.

Guttman, S.I., A.A. Karlin and G.M. Labanick. 1978. A biochemical and morphological analysis of the relationship between *Plethodon longicrus* and *Plethodon yonahlossee* (Amphibia, Urodela, Plethodontidae). Journal of Herpetology 12:445–454.

Guttman, S.I., L.A. Weigt, P.E. Moler, R.E. Ashton Jr., B.W. Mansell and J. Peavy. 1990. An electrophoretic analysis of *Necturus* from the southeastern United States. Journal of Herpetology 24:163–175.

Guyer, C. 1997. A status survey of the black warrior waterdog *(Necturus alabamensis)*: Final report. Alabama Department of Conservation and Natural Resources, Unpublished report, Montgomery, Alabama.

Guyer, C. 1998. Historical affinities and population biology of the black warrior waterdog (*Necturus* sp.): Final report. Alabama Department of Conservation and Natural Resources, Montgomery, Alabama.

Guyer, C. and M.A. Bailey. 1993. Amphibians and reptiles of longleaf pine communities. Proceedings of the Tall Timbers Fire Ecology Conference 18:139–158.

Guyer, C. and M. Durflinger. 1999. A demographic study of the black warrior waterdog *(Necturus alabamensis)*: Final report. Alabama Department of Conservation and Natural Resources, Montgomery, Alabama.

Haber, V.R. 1926. The food of the Carolina tree frog, *Hyla cinerea* Schneider. Journal of Comparative Psychology 6:214–216.

Haddad, C.F.B. and J.P. Pombal Jr. 1998. Redescription of *Physalaemus spiniger* (Anura: Leptodactylidae) and description of two new reproductive modes. Journal of Herpetology 32:557–565.

Hadfield, S. 1966. Observations on body temperature and activity in the toad *Bufo woodhousei fowleri*. Copeia 1966:581–582.

Hadorn, E. 1961. Developmental Genetics and Lethal Factors. Methuen and Company, Limited, London, England.

Haertel, J.D. and R.M. Storm. 1970. Experimental hybridization between *Rana pretiosa* and *Rana cascadae*. Herpetologica 26:436–446.

Hager, H.A. 1998. Area-sensitivity of reptiles and amphibians: are there indicator species for habitat fragmentation? Ecoscience 5:139–147.

Hahn, D.E. 1960. Collecting notes on central Korean reptiles and amphibians. Journal of the Ohio Herpetological Society 2:16–24.

Haig, S.M., D.W. Mehlman and L.W. Oring. 1998. Avian movements and wetland connectivity in landscape conservation. Conservation Biology 12:749–758.

Hairston, N.G. 1949. The local distribution and ecology of the plethodontid salamanders of the southern Appalachians. Ecological Monographs 19:47–73.

Hairston, N.G. 1950. Intergradation in Appalachian salamanders of the genus *Plethodon*. Copeia 1950:262–273.

Hairston, N.G. 1951. Interspecific competition and its probable influence upon the vertical distribution of Appalachian salamanders of the genus *Plethodon*. Ecology 32:266–274.

Hairston, N.G. 1973. Ecology, selection and systematics. Breviora 414: 1–21.

Hairston, N.G. 1980a. Species packing in the salamander genus *Desmognathus*: what are the interspecific interactions involved? American Naturalist 115:354–366.

Hairston, N.G. 1980b. The experimental test of an analysis of field distributions: competition in terrestrial salamanders. Ecology 61: 817–826.

Hairston, N.G. 1980c. Evolution under interspecific competition: field experiments on terrestrial salamanders. Evolution 34:409–420.

Hairston, N.G. 1981. An experimental test of a guild: salamander competition. Ecology 62:65–72.

Hairston, N.G. 1983. Growth, survival and reproduction of *Plethodon jordani*: trade offs between selective pressures. Copeia 1983: 1024–1035.

Hairston, N.G. 1996. Predation and competition in salamander communities. Pp. 161–189. *In* Cody, M.L. and J.A. Smallwood (Eds.), Long-term Studies of Vertebrate Communities. Academic Press, San Diego, California.

Hairston, N.G., Sr. 1983. Growth, survival and reproduction of *Plethodon jordani*: trade-offs between selective pressures. Copeia 1983:1024–1035.

Hairston, N.G., Sr. 1986. Species packing in *Desmognathus* salamanders: experimental demonstration of predation and competition. American Naturalist 127:266–291.

Hairston, N.G., Sr. 1987. Community Ecology and Salamander Guilds. Cambridge University Press, New York.

Hairston, N.G. and C.H. Pope. 1948. Geographic variation and speciation in Appalachian salamanders (*Plethodon jordani* group). Evolution 2:266–278.

Hairston, N.G., Sr. and R.H. Wiley. 1993. No decline in salamander (Amphibia: Caudata) populations: a twenty-year study in the southern Appalachians. Brimleyana 18:59–64.

Hairston, N.G., Sr., R.H. Wiley, C.K. Smith and K.A. Kneidel. 1992. The dynamics of two hybrid zones in Appalachian salamanders of the genus *Plethodon*. Evolution 46:930–938.

Hale, S.F. 2001. The status of the Tarahumara frog in Sonora, Mexico, based on a resurvey of selected localities and search for additional populations. Report to the U.S. Fish and Wildlife Service, Phoenix, Arizona.

Hale, S.F. and J.L. Jarchow. 1988. The status of the Tarahumara frog *(Rana tarahumarae)* in the United States and Mexico: Part II. Report to the Arizona Game and Fish Department, Phoenix, Arizona, and the Office of Endangered Species, U.S. Fish and Wildlife Service, Albuquerque, New Mexico.

Hale, S.F. and C.J. May. 1983. Status report for *Rana tarahumarae* Boulenger. Report to the Office of Endangered Species, U.S. Fish and Wildlife Service, Albuquerque, New Mexico.

Hale, S.F., F. Retes and T.R. Van Devender. 1977. New populations of *Rana tarahumarae* (Tarahumara frog) in Arizona. Journal of the Arizona Academy of Science 11:134–135.

Hale, S.F., G.M. Ferguson, P.A. Holm and E.B. Wirt. 1998. Re-survey of selected Tarahumara frog *(Rana tarahumara)* localities in northern Sonora, Mexico, in May 1998. Report to the Arizona Zoological Society and the Tarahumara Frog Conservation Team, Phoenix, Arizona.

Hale, S.F., C.R. Schwalbe, J.L. Jarchow, C.J. May, C.H. Lowe and T.B. Johnson. 1995. Disappearance of the Tarahumara frog. Pp. 138–140. *In* LaRoe, E.T., G.S. Farris, C.E. Puckett, P.D. Doran and M.J. Mac (Eds.), Our Living Resources. A Report to the Nation on the Distribution, Abundance, and Health of U.S. Plants, Animals, and Ecosystems. U.S. Department of the Interior, National Biological Service, Washington, D.C.

Hale, S.M. and C. Guyer. 2000. Effects of temperature and depth on burrows of the Red Hills salamander *(Phaeognathus hubrichti)*. Herpetological Natural History 7:87–90.

Hall, B.G. 2001. Phylogenetic Trees Made Easy. Sinauer Associates, Sunderland, Massachusetts.

Hall, C.D. 2002. Geographic distribution: *Ambystoma maculatum* (spotted salamander). Herpetological Review 33:315.

Hall, H.H. and H.M. Smith. 1947. Selected records of reptiles and amphibians from southeastern Kansas. Transactions of the Kansas Academy of Science 49:447–454.

Hall, J.A. 1993. Post-embryonic ontogeny and larval behavior of the spadefoot toad, *Scaphiopus intermontanus* (Anura: Pelobatidae). Ph.D. dissertation. Washington State University, Pullman, Washington.

Hall, J.A. 1998. *Scaphiopus intermontanus.* Pp. 650.1–650.17. Catalogue of American Amphibians and Reptiles. Society for the Study of Amphibians and Reptiles, St. Louis, Missouri.

Hall, J.A. and J.H. Larsen Jr. 1998. Postembryonic ontogeny of the spadefoot toad, *Scaphiopus intermontanus* (Anura: Pelobatidae): Skeletal morphology. Journal of Morphology 238:179–244.

Hall, J.A., J.H. Larsen Jr. and R.E. Fitzner. 1997. Postembryonic ontogeny of the spadefoot toad *Scaphiopus intermontanus* (Anura: Pelobatidae): external morphology. Herpetological Monographs 11:124–178.

Hall, J.A., J.H. Larsen Jr. and R.E. Fitzner. 2002. Morphology of the prometamorphic larva of the spadefoot toad, *Scaphiopus intermontanus* (Anura: Pelobatidae), with an emphasis on the lateral line system and mouthparts. Journal of Morphology 252:114–130.

Hall, J.D., M.L. Murphy and R.S. Aho. 1978. An improved design for assessing impacts of watershed practices on small streams. Verhandlungen der Internationalen Vereinigung fur Theoretische und Angewandte Limnologie 20:1359–1365.

Hall, R.J. 1976. Summer foods of the salamander, *Plethodon wehrlei* (Amphibia, Urodela, Plethodontidae). Journal of Herpetology 10:129–131.

Hall, R.J. 1977. A population analysis of two species of streamside salamanders, genus *Desmognathus*. Herpetologica 33:109–113.

Hall, R.J. 1994. Herpetofaunal diversity of the Four Holes Swamp, South Carolina. U.S. Department of the Interior, National Biological Survey, Resource Publication 198, Washington, D.C.

Hall, R.J. and P.F.P. Henry. 1992. Assessing the effects of pesticides on amphibians and reptiles: status and needs. Herpetological Journal 2:65–71.

Hall, R.J. and E. Kolbe. 1980. Bioconcentration of organophosphorus pesticides to hazardous levels by amphibians. Journal of Toxicology and Environmental Health 6:853–860.

Hall, R.J. and D.P. Stafford. 1972. Studies in the life history of Wehrle's salamander, *Plethodon wehrlei*. Herpetologica 28:300–309.

Hall, R.J. and D.M. Swineford. 1979. Uptake of methoxychlor from food and water by the American toad *(Bufo americanus).* Bulletin of Environmental Contamination and Toxicology 23:335–337.

Hall, R.J. and D.M. Swineford. 1981. Acute toxicities of toxaphene and endrin to larvae of 7 species of amphibians. Toxicology Letters 8:331–336.

Hall, R.J., R.E. Ashton Jr. and R.M. Prouty. 1985. Pesticide and PCB residues in the Neuse River waterdog, *Necturus lewisi*. Brimleyana 10:107–109.

Halliday, T. 2000. Do frogs make good canaries? Biologist 47:143–146.

Halliday, T.R. and W.R. Heyer. 1997. The case of the vanishing frogs. Technology Review 100:53–63.

Halliday, T.R. and P.A. Verrell 1988. Body size and age in amphibians and reptiles. Journal of Herpetology 22:253–265.

Hallowell, E. 1849. Description of a new species of salamander from upper California. Proceedings of the Academy of the Natural Sciences of Philadelphia 4:126.

Hallowell, E. 1852. One new genus and three new species of reptiles inhabiting North America. Proceedings of the Academy of the Natural Sciences of Philadelphia 6:206–209.

Hallowell, E. 1856 (1857). Notice of a collection of reptiles from Kansas and Nebraska, presented to the Academy of Natural Sciences by Dr. Hammond. Proceedings of the Academy of Natural Sciences of Philadelphia 8:238–253.

Hallowell, E. 1856. Descriptions of two new species of urodeles from Georgia. Proceedings of the Academy of the Natural Sciences of Philadelphia 8:130–131.

Hallowell, E. 1858. On the caducibranchiate urodele batrachians. Journal of the Academy of Natural Sciences of Philadelphia 3:337–366.

Hallowell, E. 1860. Report upon the Reptilia of the North Pacific Exploring Expedition, under command of Capt. John Rogers, U.S.N. Proceedings of the Philadelphia Academy of Natural Sciences 12: 480–510.

Halls, L.K. 1955. Grass production under dense longleaf-slash pine canopies. Southeastern Forest Experiment Station, Resarch notes, U.S.D.A. Forest Service, Asheville, North Carolina.

Hamilton, A.L. 1973. Some taxonomic aspects of certain paedogenetic *Eurycea* of the Blanco River Drainage System in Hays and Blanco Counties, Texas. Master's thesis. Southwest Texas State University, San Marcos, Texas.

Hamilton, G.B., C.R. Peterson and W.A. Wall. 1998. Distribution and habitat relationships of amphibians on the Potlatch Corporation operating area in northern Idaho. Potlatch Corporation, Unpublished report, Lewiston, Idaho.

Hamilton, I.M., J.L. Skilnick, H. Troughton, A.P. Russell and G.L. Powell. 1998. Status of the Canadian toad *(Bufo hemiophrys)* in Alberta. Alberta Environmental Protection, Wildlife Management Division, and the Alberta Conservation Association, Wildlife Status Report, Number 12, Edmonton, Alberta.

Hamilton, R. 1948. The egg-laying process in the tiger salamander. Copeia 1948:212–213.

Hamilton, S.D. 1995. Biological opinion for proposed action to expand the Lost Pines Golf Course in Bastrop State Park, located in Bastrop County, Texas. U.S. Fish and Wildlife Service, Austin, Texas.

Hamilton, W.J., Jr. 1932. The food and feeding habits of some eastern salamanders. Copeia 1932:83–86.

Hamilton, W.J., Jr. 1934. The rate of growth of the toad *(Bufo americanus americanus* Holbrook) under natural conditions. Copeia 1934: 88–90.

Hamilton, W.J. 1940. The feeding habit of larval newts with reference to availability and predilection of food items. Ecology 21:351–356.

Hamilton, W.J., Jr. 1943. Winter habits of the dusky salamander, in central New York. Copeia 1943:192.

Hamilton, W.J., Jr. 1946. Summer habitat of the yellow-barred tiger salamander. Copeia 1946:51.

Hamilton, W.J., Jr. 1948. The food and feeding behavior of the green frog, *Rana clamitans* Latreille, in New York state. Copeia 1948: 203–207.

Hamilton, W.J., Jr. 1950. Notes on the food of the congo eel, *Amphiuma*. Natural History Miscellanea, Number 62, Chicago Academy of Sciences, Chicago, Illinois.

Hamilton, W.J. 1951. The food and feeding of the garter snake in New York State. American Midland Naturalist 46:385–390.

Hamilton, W.J. 1954. The economic status of the toad. Herpetologica 10:37–40.

Hamilton, W.J. 1955. Notes on the ecology of the oak toad in Florida. Herpetologica 11:205–210.

Hamilton, W.J. and J.A. Pollack. 1955. The food of some crotalid snakes from Fort Benning, Georgia. Natural History Miscellanea, Number 140, Chicago Academy of Sciences, Chicago, Illinois.

Hamilton, W.J. and J.A. Pollack. 1956. The food of some colubrid snakes from Fort Benning, Georgia. Ecology 37:519–526.

Hammen, C.S. and V.H. Hutchison. 1962. Carbon dioxide assimilation in the symbiosis of the salamander, *Ambystoma maculatum* and the alga *Oophila amblystomatis*. Life Sciences 10:527–532.

Hammerson, G.A. 1980. Geographic distribution: *Gastrophryne olivacea* (Great Plains narrowmouth toad). Herpetological Review 11:13–14.

Hammerson, G.A. 1982a. Amphibians and Reptiles in Colorado. Colorado Division of Wildlife Publication, DOW-M-I-27-82, Denver, Colorado.

Hammerson, G.A. 1982b. Bullfrog eliminating leopard frogs in Colorado? Herpetological Review 13:115–116.

Hammerson, G.A. 1986. Amphibians and reptiles in Colorado. Colorado Division of Wildlife, Denver, Colorado.

Hammerson, G.A. 1992. Field surveys of amphibians in the mountains of Colorado. Colorado Division of Wildlife, Unpublished report, Denver, Colorado.

Hammerson, G.A. 1999. Amphibians and reptiles in Colorado. Second edition. Colorado Division of Wildlife Publication, University Press of Colorado, Niwot, Colorado.

Hammerson, G.A. and L.J. Livo. 1999. Conservation status of the northern cricket frog (Acris crepitans) in Colorado and adjacent areas in the northwestern edge of its range. Herpetological Review 30: 78–80.

Hammett, J.E. 1992. The shapes of adaptation: historical ecology of anthropogenic landscapes in the Southeastern United States. Landscape Ecology 7:121–135.

Hammond, P. 1992. Species inventory. Pp. 17–39. In Broombridge, B. (Ed.), Global Biodiversity, Status of the Earth's Living Resources. Chapman and Hall, London.

Hammond, P. 1995. Magnitude and distribution of biodiversity: the current magnitude of biodiversity. Pp. 113–138. In Heywood, V. (Ed.), Global Biodiversity Assessment. Cambridge University Press, Cambridge, United Kingdom.

Hamning, V.K., H.L. Yanites and N.L. Peterson. 2000. Characterization of adhesive and neurotoxic components in skin granular gland secretions of Ambystoma tigrinum. Copeia 2000:856–859.

Hanemann, W.M. 1989. Information and the concept of option value. Journal of Environmental Econonomics 16:23–37.

Hanifin, C.T., M. Yotsu-Yamashita, E.D. Brodie III and E.D. Brodie Jr. 1999. Toxicity of dangerous prey: variation of tetrodotoxin levels within and among populations of the newt Taricha granulosa. Journal of Chemical Ecology 25:2161–2175.

Hanken, J. 1999. Why are there so many new amphibian species when amphibians are declining? TRENDS in Ecology and Evolution 14:7–8.

Hanlin, H.G. 1978. Food habits of the greater siren, Siren lacertina, in an Alabama coastal plain pond. Copeia 1978:358–360.

Hanlin, H.G., J.J. Beatty and S.W. Hanlin. 1979. A nest site of the western red-backed salamander Plethodon vehiculum (Cooper). Journal of Herpetology 13:214–215.

Hanlin, H.G., F.D. Martin, L.D. Wike and S.H. Bennett. 2000. Terrestrial activity, abundance and species richness of amphibians in managed forests in South Carolina. American Midland Naturalist 143:70–83.

Hanna, G.D. 1966. Introduced mollusks of western North America. Occasional Papers of the California Academy of Sciences, Number 48, San Francisco, California.

Hansen, K.L. 1957. Movements, area of activity, and growth of Rana heckscheri. Copeia 1957:274–277.

Hansen, K.L. 1958. Breeding pattern of the eastern spadefoot toad. Herpetologica 14:57–67.

Hansen, R., D. Chamberlain and M. Lechner. 1998. Final environmental assessment/habitat conservation plan for issuance of a section 10(a)(1)(B) permit for incidental take of the Barton Springs salamander (Eurycea sosorum) for the operation and maintenance of Barton Springs Pool and adjacent springs. City of Austin and U.S. Fish and Wildlife Service, Austin, Texas.

Hansen, R.W. 1980. The Kern Plateau herpetofauna: a preliminary survey. U.S.D.A. Forest Service, Unpublished report, Sequoia National Forest, Porterville, California.

Hansen, R.W. 1988. Kern Canyon slender salamander (Batrachoseps simatus). California Department of Transportation, Biological Survey Report, District 6, Fresno, California.

Hansen, R.W. 1990. Life history notes: Hydromantes platycephalus (Mount Lyell salamander). Toxicity. Herpetological Review 25:62.

Hansen, R.W. 1997. Salamander diversity in the Kern Valley region. Kern River Research Center Fieldnotes 6:1, 3.

Hansen, R.W. and R. Stafford. 1994a. Kern Canyon slender salamander. Pp. 252–253. In Thelander, C.G. (Ed.), Life on the Edge. Biosystems Books, Santa Cruz, California.

Hansen, R.W. and R. Stafford. 1994b. Tehachapi slender salamander. Pp. 254–255. In Thelander, C.G. (Ed.), Life on the Edge, Biosystems Books, Santa Cruz, California.

Hanski, I. and M. Gilpin. 1991. Metapopulation dynamics: brief history and conceptual domain. Biological Journal of the Linnaean Society 42:3–16.

Hanski, I.A. and M.E. Gilpin (Eds.). 1997. Metapopulation Biology: Ecology, Genetics, and Evolution. Academic Press, San Diego, California.

Hanski, I., P. Turchin, E. Korpimaki and H. Henttonen. 1993. Population oscillations of boreal rodents: regulation by mustelid predators leads to chaos. Nature 364:232–235.

Hanson, J.A. and J.L. Vial. 1956. Defensive behavior and effects of toxins in Bufo alvarius. Herpetologica 12:141–149.

Hanson, K., J. Snyder and L. Kats. 1994. Natural history notes: Taricha torosa (California newt). Diet. Herpetological Review 25:62.

Hardin, E.D. and D.L. White. 1989. Rare vascular plant taxa associated with wiregrass (Aristida stricta) in the southeastern United States. Natural Areas Journal 9:234–245.

Hardin, E.L. and J. Janovy. 1988. Population dynamics of Distoichometra bufonis (Cestoda: Nematotaeniidae) in Bufo woodhousii. Journal of Parasitology 74:360–365.

Hardin, J.W., J.P. Schafer and R.W. Barbour. 1969. Observations on the activity of a seal salamander, Desmognathus monticola. Herpetologica 25:150–151.

Harding, J.H. 1997. Amphibians and Reptiles of the Great Lakes Region. University of Michigan Press, Ann Arbor, Michigan.

Harding, J.H. and J.A. Holman. 1992. Michigan Frogs, Toads, and Salamanders. A Field Guide and Pocket Reference. Michigan State University Museum Press, East Lansing, Michigan.

Hardy, J.D., Jr. 1952. A concentration of juvenile spotted salamanders Ambystoma maculatum (Shaw). Copeia 1952:181.

Hardy, J.D., Jr. 1953. Notes on the distribution of Microhyla carolinensis in southern Maryland. Herpetologica 8:162–166.

Hardy, J.D., Jr. 1964. A new frog, Rana palustris mansuetii, subsp. Nov. from the Atlantic Coastal Plain. Chesapeake Scientist 5:91–100.

Hardy, J.D., Jr. 1969a. A summary of recent studies on the salamander, Ambystoma mabeei. Natural Resources Institute, Unpublished report, Reference Number 69–30, University of Maryland, Solomons, Maryland.

Hardy, J.D. 1969b. Reproductive activity, growth, and movements of Ambystoma mabeei Bishop in North Carolina. Bulletin of the Maryland Herpetological Society 5:65–76.

Hardy, J.D., Jr. 1972. Amphibians of the Chesapeake Bay region. Chesapeake Science 13:123–128.

Hardy, J.D., Jr. and J.D. Anderson. 1970. Ambystoma mabeei. Pp. 81.1–81.2. Catalogue of American Amphibians and Reptiles. Society for the Study of Amphibians and Reptiles, St. Louis, Missouri.

Hardy, J.D. and J.H. Gillespie. 1976. Hybridization between Rana pipiens and Rana palustris in a modified natural environment. Bulletin of the Maryland Herpetological Society 12:41–53.

Hardy, L.M. and M.C. Lucas. 1991. A crystalline protein is responsible for dimorphic egg jellies in the spotted salamander, Ambystoma maculatum (Shaw) (Caudata: Ambystomatidae). Comparative Biochemistry and Physiology A 100:653–660.

Hardy, L.M. and R.W. McDiarmid. 1969. The amphibians and reptiles of Sinaloa, Mexico. University of Kansas Publications, Museum of Natural History 18:39–252.

Hardy, L.M. and L.R. Raymond. 1980. The breeding migrations of the mole salamander, Ambystoma talpoideum, in Louisiana. Journal of Herpetology 14:327–335.

Hardy, L.M. and L.R. Raymond. 1991. Observations on the activity of the pickerel frog, Rana palustris (Anura: Ranidae), in northern Louisiana. Journal of Herpetology 25:220–222.

Harestad, A.S. 1985. Life history notes: Scaphiopus intermontanus (Great Basin spadefoot toad). Mortality. Herpetological Review 16:24.

Harfenist, A., T. Power, K.L. Clark and D.B. Peakall. 1989. A review and evaluation of the amphibian toxicological literature. Canadian Wildlife Service, Technical Report, Series Number 61, Ottawa, Ontario, Canada.

Hargis, C.D., J.A. Bissonette and J.L. David. 1998. The behavior of landscape metrics commonly used in the study of habitat fragmentation. Landscape Ecology 13:167–186.

Hargitt, C.W. 1892. On some habits of Amphiuma means. Science 20:159.

Harker, D.F., Jr., M.E. Medley, W.C. Houtcooper and A. Phillippi. 1980. Kentucky Natural Areas Plan: Appendix A. Kentucky Nature Preserves Commission, Frankfort, Kentucky.

Harlan, R. 1825a. Further observations on the Amphiuma means. Annals of the Lyceum of Natural History of New York 1:269–270.

Harlan, R. 1825b. Description of a variety of the Coluber fulvius, Linn., a new species of Scincus, and two new species of Salamandra. Journal of the Academy of Natural Sciences of Philadelphia 5:154–158.

Harlan, R. 1826a. Notice of a new species of salamander (inhabiting Pennsylvania). American Journal of Science and Arts 10:286–287.

Harlan, R. 1826b. Genera of North American Reptilia, and a synopsis of the species. Journal of the Academy of Natural Sciences of Philadelphia 5:317–372.

Harlan, R. 1835. Medical and Physical Researches, Philadelphia. Lydia R. Bailey, Philadelphia.

Harman, W. J. and A.R. Lawler. 1975. *Dero (Allodero) hylidae*, an oligochaete symbiont in hylid frogs in Mississippi. Transactions of the American Microscopical Society 94:38–42.

Harman, W.N. 1974. Snails (Mollusca: Gastropoda). Pp. 275–312. *In* Hart, C.W., Jr. and S.L.H. Fuller (Eds.), Pollution Ecology of Freshwater Invertebrates. Academic Press, New York.

Harman, W.N. and J.L. Forney. 1970. Fifty years of change in the molluscan fauna of Oneida Lake, New York. Limnology and Oceanography 15:454–460.

Harper, C.A. and D.C. Guynn Jr. 1999. Factors affecting salamander density and distribution within four forest types in the Southern Appalachian Mountains. Forest Ecology and Management 114:245–252.

Harper, F. 1928. Voices of New England toads. Bulletin of the Boston Society of Natural History 46:3–9.

Harper, F. 1931. A dweller in the piney woods. Science Monthly 32:176–181.

Harper, F. 1932. A voice from the pines. Natural History Magazine 32:280–288.

Harper, F. 1935. Records of amphibians in the southeastern states. American Midland Naturalist 16:275–310.

Harper, F. 1937. A season with Holbrook's chorus frog *(Pseudacris ornata)*. American Midland Naturalist 18:260–272.

Harper, F. 1939. Distribution, taxonomy, nomenclature, and habits of the little treefrog *(Hyla ocularis)*. American Midland Naturalist 22:134–149.

Harper, F. 1955. A new chorus frog *(Pseudacris)* from the eastern United States. Natural History Miscellanea, Number 150, Chicago Academy of Sciences, Chicago, Illinois.

Harper, S. 1994. Golf course plan has many hopping mad. Houston Post, 6 February.

Harris, A.J. 1975. Seasonal activity and microhabitat utilization in *Hyla cadaverina* (Anura: Hylidae). Herpetologica 31:236–239.

Harris, H.S., Jr. 1975. Distributional survey (Amphibia/Reptilia): Maryland and the District of Columbia. Bulletin of the Maryland Herpetological Society 11:73–167.

Harris, H.S., Jr. and D.J. Lyons. 1968. The first record of the green salamander *Aneides aeneus* (Cope and Packard) in Maryland. Journal of Herpetology 1:106–107.

Harris, J.P., Jr. 1959a. The natural history of *Necturus*: I. Habitats and habits. Field and Laboratory 27:11–20.

Harris, J.P., Jr. 1959b. The natural history of *Necturus*: III. Food and feeding. Field and Laboratory 27:105–111.

Harris, J.P., Jr. 1959c. The natural history of *Necturus*: IV. Reproduction. Journal of the Graduate Research Center 28:69–81.

Harris, M.L., L. Chora, C.A. Bishop and J.P. Bogart. 2000. Species- and age-related differences in susceptibility to pesticide exposure for two amphibians, *Rana pipiens*, and *Bufo americanus*. Bulletin of Environmental Contamination and Toxicology 64:263–270.

Harris, P.M. 1995. Are autecologically similar species also functionally similar? A test in pond communities. Ecology 76:544–552.

Harris, R.N. 1980. The consequences of within-year timing of breeding in *Ambystoma maculatum*. Copeia 1980:719–722.

Harris, R.N. 1981. Intrapond homing behavior in *Notophthalmus viridescens*. Journal of Herpetology 15:355–356.

Harris, R.N. 1984. Transplant experiments with *Ambystoma* larvae. Copeia 1984:161–169.

Harris, R.N. 1987a. Density-dependent paedomorphosis in the salamander *Notophthalmus viridescens dorsalis*. Ecology 68:705–712.

Harris, R.N. 1987b. An experimental study of population regulation in the salamander, *Notophthalmus viridescens dorsalis* (Urodela: Salamandridae). Oecologia 71:280–285.

Harris, R.N. 1989. Ontogenetic changes in size and shape of the facultatively paedomorphic salamander *Notophthalmus viridescens dorsalis*. Copeia 1989:35–42.

Harris, R.N. and D.E. Gill. 1980. Communal nesting, brooding behavior, and embryonic survival of the four-toed salamander *Hemidactylium scutatum*. Herpetologica 36:141–144.

Harris, R.N., R.A. Alford and H.M. Wilbur. 1988. Density and phenology of *Notophthalmus viridescens dorsalis* in a natural pond. Herpetologica 44:234–242.

Harris, R.N. and P.M. Ludwig. 2004. Resource level and reproductive frequency in female four-toed salamanders, *Hemidactylium scutatum*. Ecology 85:1585–1590.

Harris, R.N., T.J. Vess, J.I. Hammond and C.J. Lindermuth. 2003. Context-dependent kin discrimination in larval four-toed salamanders *Hemidactylium scutatum* (Caudata: Plethodontidae). Herpetologica 59:164–177.

Harris, R.N., W.W. Hames, I.T. Knight, C.A. Carreno and T.J. Vess. 1995. An experimental analysis of joint nesting in the salamander *Hemidactylium scutatum* (Caudata: Plethodontidae): the effects of population density. Animal Behaviour 50:1309–1316.

Harrison, J.R. 1963. Variation in pigmy species of the genus *Desmognathus* (Urodela: Plethodontidae). Ph.D. dissertation. University of Notre Dame, South Bend, Indiana.

Harrison, J.R. 1967. Observations on the life history, ecology, and distribution of *Desmognathus aeneus aeneus* Brown and Bishop. American Midland Naturalist 77:356–370.

Harrison, J.R. 1973. Observations of the life history and ecology of *Eurycea quadridigitata* (Holbrook). Herpetological Information Search Systems News-Journal 1:57–58.

Harrison, J.R. 1992. *Desmognathus aeneus*. Pp. 534:1–534.4. Catalogue of American Amphibians and Reptiles. Society for the Study of Amphibians and Reptiles, St. Louis, Missouri.

Harrison, J.R., III and S.I. Guttman. 2003. A new species of *Eurycea* (Caudata: Plethodontidae) from North and South Carolina. Southeastern Naturalist 2:159–178.

Harrison, S. and N. Cappuccino. 1995. Using density-manipulation experiments to study population regulation. Pp. 131–147. *In* Cappuccino, N. and P.W. Price (Eds), Population Dynamics: New Approaches and Synthesis. Academic Press, New York.

Harte, J. and E. Hoffman. 1989. Possible effects of acidic deposition on a Rocky Mountain population of the tiger salamander *Ambystoma tigrinum*. Conservation Biology 3:149–158.

Harte, J. and E. Hoffman. 1994. Acidification and salamander recruitment. Bioscience 44:126.

Hartman, M.R. 1994. Avian use of restored and natural wetlands in north-central Indiana. Master's thesis. Purdue University, West Lafayette, Indiana.

Hartman, M.T. 1906. Food habits of Kansas lizards and batrachians. Transactions of the Kansas Academy of Science 20:225–229.

Harvey, T. 1997. Fore! Lost Pines opens nine new holes. Bastrop Advertiser, 14 August.

Harwood, P.D. 1930. A new species of *Oxysomatium* (Nematoda) with some remarks on the genera *Oxysomatium* and *Aplectana* and observations on the life history. Journal of Parasitology 17:61–73.

Harwood, P.D. 1932. The helminths parasitic in the Amphibia and Reptilia of Houston, Texas, and vicinity. Proceedings of the U.S. National Museum, Number 81, Article 17, Washington, D.C.

Hasegawa, H. and M. Otsuru. 1977. Life cycle of a frog nematode, *Spinitectus ranae* Morishita, 1926 (Cystidicolidae). Japanese Journal of Parasitology 26:336–344.

Hasegawa, H. and M. Otsuru. 1978. Notes on the life cycle of *Spiroxys japonica* Morishita, 1926 (Nematoda: Gnathostomatidae). Japanese Journal of Parasitology 27:113–122.

Hasegawa, H. and M. Otsuru. 1979. Life history of an amphibian nematode, *Hedrurus ijimai* Morishita, 1926 (Hedruridae). Japanese Journal of Parasitology 28:89–97.

Hasegawa, Y., H. Ueda and M. Sumida. 1999. Clinal geographic variation in the advertisement call of the wrinkled frog, *Rana rugosa*. Herpetologica 55:318–324.

Hassell, M.P. 1986. Detecting density dependence. TRENDS in Ecology and Evolution 1:90–93.

Hassinger, D.D. and J.D. Anderson. 1970. The effect of lunar eclipse on nocturnal stratification of larval *Ambystoma opacum*. Copeia 1970:178–179.

Hassinger, D.D., J.D. Anderson and G.H. Dalrymple. 1970. The early life history and ecology of *Ambystoma tigrinum* and *Ambystoma opacum* in New Jersey. American Midland Naturalist 84:474–495.

Hassler, W.G. 1932. New locality records of two salamanders and a snake in Cattaraugus County, New York. Copeia 1932:94–96.

Hastings, F.L., R.A. Werner, P.J. Shea and E.H. Holston. 1998. Persistence of carbaryl within boreal, temperate and Mediterranean ecosystems. Journal of Economic Entomology 91:665–670.

Hatch, A. and G.A. Burton Jr. 1998. Effects of photoinduced toxicity of fluoranthene on amphibian embryos and larvae. Environmental Toxicology and Chemistry 17:1777–1785.

Hatch, A.C., L.K. Belden, E. Scheessele and A.R. Blaustein. 2001. Juvenile amphibians do not avoid potentially lethal levels of urea on soil substrate. Environmental Toxicology and Chemistry 20:2328–2335.

Hathaway, E.S. 1928. Quantitative study of the changes produced by acclimatization on the tolerance of high temperatures by fishes and amphibians. Bulletin of the United States Bureau of Fisheries 43:1 69–192.

Haurwitz, R.K.M. 2000. How politics ran over the toad. Austin American-Statesman, 16 May.

Hausfater, G., H.C. Gerhardt and G.M. Klump. 1990. Parasites and mate choice in gray treefrogs, *Hyla versicolor*. American Zoologist 30:299–311.

Hauwert, N.M., D.A. Johns and T.J. Aley. 1998. Preliminary report on groundwater tracing studies within the Barton Creek and Williamson Creek watersheds, Barton Springs/Edwards Aquifer. Barton Springs/Edwards Aquifer Conservation District and City of Austin Watershed Protection Department, Austin, Texas.

Haverkos, H.W. 1996. Is Kaposi's sarcoma caused by a new herpesvirus? Biomedicine and Pharmacotherapy 50:318–319.

Hawkins, C.P., L.J. Gottschalk and S.S. Brown. 1988. Densities and habitat of tailed frog tadpoles in small streams near Mt. St. Helens following the 1980 eruption. Journal of the North American Benthological Society 7:246–252.

Hawkins, C.P., M.L. Murphy, N.H. Anderson and M.A. Wilzbach. 1983. Density of fish and salamanders in relation to riparian canopy and physical habitat in streams of the northwestern United States. Canadian Journal of Fisheries and Aquatic Sciences 40:1173–1185.

Hay, O.P. 1888. Observations on *Amphiuma* and its young. American Naturalist 22:315–321.

Hay, O.P. 1892. The batrachians and reptiles of the state of Indiana. Annual Report of the Indiana Department of Geology and Natural Resources 17:412–602.

Hay, R. 1998a. Blanchard's cricket frog (*Acris crepitans blanchardi*) in Wisconsin: a vanishing species. P. 10. *In* Abstracts, Midwest Declining Amphibians Conference, March 20–21. Milwaukee Public Museum, Milwaukee, Wisconsin.

Hay, R. 1998b. Blanchard's cricket frogs in Wisconsin: a status report. Pp. 79–82. *In* Lannoo, M.J. (Ed.), Status and Conservation of Midwestern Amphibians. University of Iowa Press, Iowa City, Iowa.

Hay, W.P. 1902. A list of the Batrachia and Reptilia of the District of Columbia and vicinity. Proceedings of the Biological Society of Washington 15:128.

Hayek, L.C. 1994. Research design for quantitative amphibian studies. Pp. 21–39. *In* Heyer, W.R., M.A. Donnelly, R.W. McDiarmid, L.C. Hayek and M.S. Foster (Eds.), Measuring and Monitoring Biological Diversity: Standard Methods For Amphibians. Smithsonian Institution Press, Washington, D.C.

Hayek, L.C. and R.W. McDiarmid. 1994. Geographic information systems. Pp. 166–171. *In* W.R. Heyer, M.A. Donnelly, R.W. McDiarmid, L.C. Hayek and M.S. Foster (Eds.), Measuring and Monitoring Biological Diversity: Standard Methods for Amphibians. Smithsonian Institution Press, Washington, D.C.

Hayes, F.E. 1989. Antipredator behavior of recently metamorphosed toads (*Bufo a. americanus*) during encounters with garter sakes (*Thamnophis s. sirtalis*). Copeia 1989:1011–1015.

Hayes, M.C. and M.R. Jennings. 1989. Habitat correlates of distribution of the California red-legged frog (*Rana aurora draytonii*) and the foothill yellow-legged frogs (*Rana boylii*): implications for management. Pp. 144–158. *In* Szaro, R.E., K.E. Severson and D.R. Patton (Tech. Coords.), Management of Amphibians, Reptiles, and Small Mammals of North America, Proceedings of the symposium on 19–21 July 1998. U.S.D.A. Forest Service, General Technical Report, RM-166, Flagstaff, Arizona.

Hayes, M.P. 1994. The Oregon spotted frog (*Rana pretiosa*) in western Oregon. Oregon Department of Fish and Wildlife, Technical Report, Number 94-1-01, Portland, Oregon.

Hayes, M.P. 1996. Assessment of the sensitive amphibian and reptile fauna of the proposed Squaw Flat Research Natural Area. Umpqua National Forest under subcontract to Resources Northwest Incorporated, Final report, Kirkland, Washington.

Hayes, M.P. 1997. Status of the Oregon spotted frog (*Rana pretiosa*) in the Deschutes Basin and selected other systems in Oregon and northeastern California with a rangewide synopsis of the species' status. Nature Conservancy under contract to the U.S. Fish and Wildlife Service with assistance from the Oregon Department of Fish and Wildlife, Final report, Portland, Oregon.

Hayes, M.P. and F.S. Cliff. 1982. A checklist of the herpetofauna of Butte County, the Butte Sink, and Sutter Buttes, California. Herpetological Review 13:85–87.

Hayes, M.P. and C.B. Hayes. 2003. Natural history notes: *Rana aurora aurora* (northern red-legged frog). Juvenile growth; male size at maturity. Herpetological Review 34:233–234.

Hayes, M.P. and M.R. Jennings. 1986. Decline of ranid frog species in western North America: are bullfrogs (*Rana catesbeiana*) responsible? Journal of Herpetology 20:490–509.

Hayes, M.P. and M.R. Jennings. 1988. Habitat correlates of distribution of the California red-legged frog (*Rana aurora draytonii*) and the foothill yellow-legged frog (*Rana boylii*): implications for management. Pp. 144–158. *In* Szaro, R.C., K.E. Severson and D.R. Patton (Tech. Coords.), Management of Amphibians, Reptiles and Small Mammals in North America. U.S.D.A. Forest Service, Rocky Mountain Forest and Range Experiment Station, General Technical Report RM-166, Fort Collins, Colorado.

Hayes, M.P. and P.N. Lahanas. 1987. Nesting of the aquatic salamander *Amphiuma means*. Florida Scientist 50 (Supplement 1):16 (Abstract).

Hayes, M.P. and M.M. Miyamoto. 1984. Biochemical, behavioral and body size differences between the red-legged frogs, *Rana aurora aurora* and *Rana aurora draytonii*. Copeia 1984:1018–1022.

Hayes, M.P. and F.C. Schaffner. 1986. Life history notes: *Rana catesbeiana* (bullfrog). Predation. Herpetological Review 17:44–45.

Hayes, M.P. and M.R. Tennant. 1985. Diet and feeding behavior of the California red-legged frog, *Rana aurora draytonii* (Ranidae). Southwestern Naturalist 30:601–605.

Hayes, M.P. and J. Warner. 1985. Life history notes: *Rana catesbeiana* (bullfrog). Food. Herpetological Review 16:109.

Hayes, M.P., C.A. Pearl and C.J. Rombough. 2002. Natural history notes: *Rana aurora aurora* (northern red-legged frog). Movement. Herpetological Review 32:35–36.

Hayes, M.P., J.D. Engler, S. Van Leuven, D.C. Friesz, T. Quinn and D.J. Pierce. 2001. Overwintering of the Oregon spotted frog (*Rana pretiosa*) at Conboy Lake National Wildlife Refuge, Klickitat County, Washington, 2000–2001. Washington Department of Transportation, Final report, Olympia, Washington.

Hayes, M.P., J.D. Engler, R.D. Haycock, D.H. Knopp, W.P. Leonard, K.R. McAllister and L.L. Todd. 1997. Status of the Oregon spotted frog (*Rana pretiosa*) across its geographic range. Oregon Chapter of The Wildlife Society, Corvallis, Oregon.

Hayes, T., K. Haston, M. Tsui, A. Hoang, C. Haeffle and A. Vonk. 2002b. Atrazine-induced hermaphroditism at 0.1 ppb in American leopard frogs (*Rana pipiens*): laboratory and field evidence. Environmental Health Perspectives doi:10.1289/ehp.5932.

Hayes, T., K. Haston, M. Tsui, A. Hoang, C. Haeffle and A. Vonk. 2002c. Feminization of male frogs in the wild. Water-borne herbicide threatens amphibian populations in parts of the United States. Nature 419:895–896.

Hayes, T.B., A. Collins, M. Lee, M. Mendoza, N. Noriegs, A.A. Stuart and A. Vonk. 2002a. Hermaphroditic, demasculinized frogs after exposure to the herbicide atrazine at low ecologically relevant doses. Proceedings of the National Academy of Sciences. 99:5476–5480.

Hayes-Odum, L.A. 1990. Observations on reproduction and embryonic development in *Syrrhophus cystignathoides campi* (Anura: Leptodactylidae). Southwestern Naturalist 35:358–361.

Hays, J.B., A.R. Blaustein, J.M. Kiesecker, P.D. Hoffman, I. Pandelova, D. Coyle and T. Richardson. 1996. Developmental responses of amphibians to solar and artificial UV-B sources: a comparative study. Photochemistry and Photobiology 64:449–456.

Healy, W.R. 1970. Reduction of neoteny in Massachusetts populations of *Notophthalmus viridescens*. Copeia 1970:578–581.

Healy, W.R. 1973. Life history variation and growth of juvenile *Notophthalmus viridescens* from Massachusetts. Copeia 1973:641–647.

Healy, W.R. 1974. Population consequences of alternative life histories in *Notophthalmus v. viridescens*. Copeia 1974:221–229.

Healy, W.R. 1975a. Terrestrial activity and home range in efts of *Notophthalmus viridescens*. American Midland Naturalist 93:131–138.

Healy, W.R. 1975b. Breeding and post larval migrations of the red-spotted newt, *Notophthalmus viridescens*, in Massachusetts. Ecology 56:673–680.

Hearnden, M. 1991. The reproductive and larval ecology of *Bufo marinus* (Anura: Bufonidae). Ph.D. dissertation. James Cook University, Townsville, Queensland, Australia.

Heath, A.G. 1975. Behavioral thermoregulation in high altitude tiger salamanders, *Ambystoma tigrinum*. Herpetologica 31:84–93.

Heath, D.R., D.A. Saugey and G.A. Heidt. 1986. Abandoned mine fauna of the Ouachita Mountains, Arkansas: vertebrate taxa. Proceedings of the Arkansas Academy of Science 40:33–36.

Heatwole, H. 1960. Burrowing ability and behavioral responses to desiccation of the salamander, *Plethodon cinereus*. Ecology 41:661–668.

Heatwole, H. 1961a. Habitat selection and activity of the wood frog, *Rana sylvatica* LeConte. American Midland Naturalist 66:301–313.

Heatwole, H. 1961b. Rates of desiccation and rehydration of eggs in a terrestrial salamander, *Plethodon cinereus*. Copeia 1961:110–112.

Heatwole, H. 1962. Environmental factors influencing local distribution and activity of the salamander, *Plethodon cinereus*. Ecology 43:460–472.

Heatwole, H. and A. Heatwole. 1968. Motivational aspects of feeding behavior in toads. Copeia 1968:692–698.

Heatwole, H. and K. Lim. 1961. Relation of substrate moisture to absorption and loss of water by the salamander, *Plethodon cinereus*. Ecology 41:661–668.

Heatwole, H. and R.C. Newby. 1972. Interaction of internal rhythm and loss of body water in influencing activity levels of amphibians. Herpetologica 28:156–162.

Heatwole, H. and F.H. Test. 1961. Cannibalism in the salamander, *Plethodon cinereus*. Herpetologica 17:143.

Heatwole, H., S.B. de Austin and R. Herrero. 1968. Heat tolerances of tadpoles of two species of tropical anurans. Comparative Biochemistry and Physiology 27:807–815.

Heatwole, H., N. Poran and P. King. 1999. Ontogenetic changes in the resistance of bullfrogs *(Rana catesbeiana)* to the venom of copperheads *(Agkistrodon contortrix contortrix)* and cottonmouths *(Agkistrodon piscivorus piscivorus)*. Copeia 1999:808–814.

Hebard, W.B. and R.B. Brunson. 1963. Hind limb anomalies of a western Montana population of the Pacific tree frog, *Hyla regilla* Baird and Girard. Copeia 1963:570–572.

Hebb, E.A. 1971. Site preparation decreases game food plants in Florida sandhills. Journal of Wildlife Management 35:155–162.

Hecht, M.K. 1958. A synopsis of the mudpuppies of eastern North America. Proceedings of the Staten Island Institute of Arts and Sciences 21:4–38.

Hecht, M.K. and B.L. Matalas. 1946. A review of middle North American toads of the genus *Microhyla*. American Museum Novitates, Number 1315, American Museum of Natural History, New York.

Hecker, L., D.M. Madison, R.W. Dapson and V. Holzherr. 2003. Presence of modified serous glands in the caudal integument of the red-backed salamander *(Plethodon cinereus)*. Journal of Herpetology 37:732–726.

Heckscher, C.M. 1995. Distribution and habitat associations of the eastern mud salamander, *Pseudotriton montanus montanus*, on the Delmarva Peninsula. Maryland Naturalist 39:11–14.

Hecnar, S.J. 1995. Acute and chronic toxicity of ammonium nitrate fertilizer to amphibians from southern Ontario. Environmental Toxicology and Chemistry 14:2131–2137.

Hecnar, S.J. 1997. Amphibian pond communities in southwestern Ontario. Pp. 1–15. *In* Green, D.M. (Ed.), Amphibians in Decline: Canadian Studies of a Global Problem. Herpetological Conservation, Number 1, Society for the Study of Amphibians and Reptiles, St. Louis, Missouri.

Hecnar, S.J. and D.R. Hecnar. 1999. Natural history notes: *Pseudacris triseriata* (western chorus frog). Reproduction. Herpetological Review 30:38.

Hecnar, S.J. and R.T. M'Closkey. 1996. Regional dynamics and the status of amphibians. Ecology 77:2091–2097.

Hecnar, S.J. and R.T. M'Closkey. 1997. Changes in the composition of a ranid frog community following bullfrog extinction. American Midland Naturalist 137:145–150.

Hedeen, S.E. 1967. Feeding behavior of the great blue heron in Itasca State Park, Minnesota. Loon 39:116–120.

Hedeen, S.E. 1970. Summer food habits of *Rana septentrionalis* and sympatric *Rana pipiens* in Minnesota. Herpetological Review 2:6.

Hedeen, S.E. 1971a. Growth of the tadpoles of the mink frog, *Rana septentrionalis*. Herpetologica 27:160–165.

Hedeen, S.E. 1971b. Body temperatures of the mink frog, *Rana septentrionalis* Baird. Journal of Herpetology 5:211–212.

Hedeen, S.E. 1972a. Escape behavior and causes of death of the mink frog, *Rana septentrionalis*. Herpetologica 28:261–262.

Hedeen, S.E. 1972b. Food and feeding behavior of the mink frog, *Rana septentrionalis* Baird, in Minnesota. American Midland Naturalist 88:291–300.

Hedeen, S.E. 1972c. Postmetamorphic growth and reproduction of the mink frog, *Rana septentrionalis* Baird. Copeia 1972:169–175.

Hedeen, S.E. 1977. *Rana septentrionalis*. Pp. 202:1–202.2. Catalogue of American Amphibians and Reptiles. Society for the Study of Amphibians and Reptiles, St. Louis, Missouri.

Hedeen, S.E. 1986. The southern geographic limit of the mink frog, *Rana septentrionalis*. Copeia 1986:239–244.

Hedgecock, D. 1976. Genetic variation in two widespread species of salamanders, *Taricha granulosa* and *Taricha torosa*. Biochemical Genetics 14:561–576.

Hedgecock, D. 1978. Population subdivision and genetic divergence in the red-bellied newt, *Taricha rivularis*. Evolution 32:271–286.

Hedgecock, D. and F.J. Ayala. 1974. Evolutionary divergence in the genus *Taricha* (Salamandridae). Copeia 1974:738–747.

Hedges, S.B. 1989. Evolution and biogeography of West Indian frogs of the genus *Eleutherodactylus*: slow evolving loci and the major groups. Pp. 305–370. *In* Woods, C.A. (Ed.), Biogeography of the West Indies: Past, Present and Future. Sandhill Crane Press, Gainesville, Florida.

Hedges, S.B. 1993. Global amphibian declines: a perspective from the Caribbean. Biodiversity and Conservation 2:290–303.

Hedges, S.B. 1996. The origin of West Indian amphibians and reptiles. Pp. 95–128. *In* Powell, R. and R.W. Henderson (Eds.), Contributions to West Indian Herpetology: A Tribute to Albert Schwartz. Contributions to Herpetology, Volume 12, Society for the Study of Amphibians and Reptiles, Ithaca, New York.

Hedges, S.B. 1999. Distributional patterns of amphibians in the West Indies. Pp. 211–254. *In* Duellman, W.E. (Ed.), Patterns of Distribution in Amphibians. Johns Hopkins Press, Baltimore, Maryland.

Hefner, J.M. and J.D. Brown. 1985. Wetland trends in the southeastern United States. Wetlands 4:1–11.

Heinen, J.T. 1985. Cryptic behavior in juvenile toads. Journal of Herpetology 19:524–527.

Heinen, J.T. 1993a. Aggregations of newly metamorphosed *Bufo americanus*: tests of two hypotheses. Canadian Journal of Zoology 71: 334–338.

Heinen, J.T. 1993b. Substrate choice and predation risk in newly metamorphosed American toads *(Bufo americanus)*: an experimental analysis. American Midland Naturalist 130:184–192.

Heinen, J.T. 1994. Antipredator behavior of newly metamorphosed American toads *(Bufo a. americanus)*, and mechanisms of hunting by eastern garter snakes *(Thamnophis s. sirtalis)*. Herpetologica 50:137–145.

Heinen, J.T. 1995. Predator cues and prey responses: a test using eastern garter snakes *(Thamnophis s. sirtalis)* and American toads *(Bufo a. americanus)*. Copeia 1995:738–741.

Heinen, J.T. and G. Hammond. 1997. Antipredator behaviors of newly metamorphosed green frogs *(Rana clamitans)* and leopard frogs *(Rana pipiens)* in encounters with eastern garter snakes *(Thamnophis sirtalis)*. American Midland Naturalist 137:136–144.

Heinrich, J.H. 1996. August 1995 Oasis Valley Area surveys for the Amargosa toad, *Bufo nelsoni*. Nevada Division of Wildlife, Las Vegas, Nevada (available from Nevada Natural Heritage Program, Carson City, Nevada).

Heinrich, M.L. 1985. Life history notes: *Pseudacris triseriata triseriata* (western chorus frog). Reproduction. Herpetological Review 16:24.

Heinzel, S.J. and C.R. Rossell Jr. 1995. A new record of the mole salamander, *Ambystoma talpoideum*, in Buncombe County, North Carolina. Journal of the Elisha Mitchell Scientific Society 111:130–131.

Helfer, J.R. 1949. Two new cestodes from salamanders. Transactions of the American Microscopical Society 67:359–364.

Helfert, S. 1996. Withdrawal of proposed rule to list the Barton Springs salamander as endangered. Federal Register 61:46608–46616.

Helfrich, L.A., D.J. Orth and R.J. Neves. 1997. Freshwater fish farming in Virginia: selecting the right fish to raise. Virginia Cooperative Extension Service, Publication Number 420-101, Blacksburg, Virginia.

Helgen, J. 1996. The frogs of Granite Falls: frogs as biological indicators. Pp. 55–57. *In* Moriarty, J.J. and D. Jones (Eds.), Minnesota's Amphibians and Reptiles: Their Conservation and Status, Proceedings of a Symposium. Serpent's Tale Press, Excelsior, Minnesota.

Helgen, J., R.G. McKinnell and M.C. Gernes. 1998. Investigation of malformed northern leopard frogs in Minnesota. Pp. 288–297. *In* Lannoo, M.J. (Ed.), Status and Conservation of Midwestern Amphibians. University of Iowa Press, Iowa City, Iowa.

Helgen, J.C., M.C. Gernes, S.M. Kersten, J.W. Chirhart, J.T. Canfield, D. Bowers, R.G. McKinnel and D.M. Hoppe. 2000. Field investigation of malformed frogs in Minnesota 1993–1997. Pp. 96–112. *In* Kaiser, H., G.S. Casper and N. Bernstein (Eds.), Investigating Amphibian Declines: Proceedings of the 1998 Midwest Declining Amphibians Conference. Iowa Academy of Science, Volume 107, Cedar Falls, Iowa.

Helgen, J.C., M.C. Gernes, S.M. Kersten, J.W. Chirhart, J.T. Canfield, D. Bowers, J.Hafernan, R.G. McKinnell and D.M. Hoppe. 2000. Field investigations of malformed frogs in Minnesota 1993–97. Journal of the Iowa Academy of Science 107:96–112.

Hellman, R.E. 1953. A comparative study of the eggs and tadpoles of *Hyla phaeocrypta* and *Hyla versicolor* in Florida. Publications of Ross Allen's Reptile Institute 1:61–74.

Helms, D.R. 1967. Use of formalin for selective control of tadpoles in the presence of fishes. Progressive Fish-Culturist 29:43–47.

Hemesath, L.M. 1998. Iowa's frog and toad survey, 1991–1994. Pp. 206–216. *In* Lannoo, M.J. (Ed.), Status and Conservation of Midwestern Amphibians. University of Iowa Press, Iowa City, Iowa.

Hemmer, H. and J.A. Alcover (Eds.). 1984. Història Biològica del Ferreret. Editorial Moll, Palma de Mallorca, Mallorca.

Hemphill, D.V. 1952. The vertebrate fauna of the boreal areas of the southern Yolla Bolly Mountains, California. Ph.D. dissertation. Oregon State College, Corvallis, Oregon.

Hemphill, N. and S.D. Cooper. 1984. Differences in the community structure of stream pools containing or lacking trout. International Vereinigung für theoretische und angewandte Limnologie, Verhandlungen 22:1858–1861.

Henderson, G.G. 1961. Reproductive potential of *Microhyla olivacea*. Texas Journal of Science 13:355–357.

Hendricks, F.S. 1973. Intestinal contents of *Rana pipiens* Schreber (Ranidae) larvae. Southwestern Naturalist 18:99–101.

Hendricks, L.J. and J. Kezer. 1958. An unusual population of a blind cave salamander and its fluctuation during one year. Herpetologica 14:41–43.

Hendrickson, H.T. 1974. Geographic distribution: *Hyla squirella* (squirrel treefrog). Herpetological Review 5:107.

Hendrickson, J.R. 1954. Ecology and systematics of salamanders of the genus *Batrachoseps*. University of California Publications in Zoology, Number 54, Berkeley, California.

Henle, K. 1995. Biodiversity, people, and a set of important connected questions. Pp. 162–174. *In* Saunders, D., J. Craig and L. Mattiske (Eds.), Nature Conservation 4: The Role of Networks. Surrey Beatty and Sons, Sydney, New South Wales, Australia.

Henle, K. and K. Rimpp. 1993. Überleben von Amphibien und Reptilien in Metapopulationen-Ergebnisse einer 26-jährigen Erfassung. Verhandlungen der Gesellschaft für Ökologie 22:215–220.

Henle, K. and K. Rimpp. 1994. Ergebnisse einer 26-jährigen Erfassung der Herpetofauna in der Umgebung von Rutesheim und Renningen, Kreis Böblingen, Baden-Württemberg. Jahreshefte der Gesellschaft für Naturkunde in Württemberg 150:193–221.

Henle, K. and B. Streit. 1990. Kritische Bemerkungen zum Artenrückgang bei Amphibien und Reptilien und dessen Ursachen. Natur und Landschaft 65:347–361.

Henrard, J.B. 1968. On the occurrence of *Helisoma anceps* (Menke) in Italy. Basteria 32:2–3.

Henrich, T.W. 1968. Morphological evidence of secondary intergradation between *Bufo hemiophrys* Cope and *Bufo americanus* Holbrook in eastern South Dakota. Herpetologica 24:1–13.

Henry, W.V. and V.C. Twitty. 1940. Contributions to the life histories of *Dicamptodon ensatus* and *Ambystoma gracile*. Copeia 1940:247–250.

Hensel, J.L., Jr. and E.D. Brodie Jr. 1976. An experimental study of aposomatic coloration in the salamander *Plethodon jordani*. Copeia 1976:59–65.

Hensher, D. and L.W. Johnson. 1981. Applied Discrete Choice Modelling. Croom Helm, London, England.

Hensley, M. 1959. Albinism in North American amphibians and reptiles. Pp. 135–139. *In* Publications of the Museum of Michigan State University, Series 1, East Lansing, Michigan.

Hensley, M. 1964. The tiger salamander in northern Michigan. Herpetologica 20:203–204.

Hepher, B. and Y. Pruginin. 1981. Commercial Fish Farming. With Special Reference to Fish Culture in Israel. John Wiley and Sons, New York.

Herbeck, L.A. and D.R. Larsen. 1999. Plethodontid salamander response to silvicultural practices in Missouri Ozark forests. Conservation Biology 13:623–632.

Heringhi, H.L. 1969. An ecological survey of the herpetofauna of Alamos, Sonora, Mexico. Master's thesis. Arizona State University, Tempe, Arizona.

Herkert, J.R. 1992. Endangered and threatened species of Illinois: Status and distribution. Volume 2 – Animals. Illinois Endangered Species Protection Board, Springfield, Illinois.

Herkovits, J. and C.S. Pérez-Coll. 1991. Antagonism and synergism between lead and zinc in amphibian larvae. Environmental Pollution 69:217–221.

Hero, J.-M. 1990. An illustrated key to aquatic tadpoles occurring in the Central Amazon rainforest, Manaus, Amazonas, Brasil. Amazoniana 11:201–262.

Hero, J.-M. 1992. CSIRO *Bufo* Project Venezuela: Larval ecology report. Cane Toad Research Management Committee, CSIRO, ACT, Griffith University, Brisbane, Queensland, Australia.

Hero, J.-M. 1994. Larval ecology of *B. marinus* in the Townsville region. Cane Toad Research Management Committee, CSIRO ACT, Unpublished report, Griffith University, Brisbane, Queensland, Australia.

Hero, J.M., Williams, S.E. & Magnusson, W.E. in press. Ecological traits of declining amphibians in upland areas of eastern Australia. Journal of Zoology.

Herre, W. 1939. Studien an asiatischen und nordamerikanischen Salamandriden. Abhandlungen und Berichte aus dem Museum für Naturkunde und Vorgeschichte, Magdeburg 7:79–98.

Herreid, C.F. and S. Kinney. 1966. Survival of Alaskan wood frog (*Rana sylvatica*) larvae. Ecology 47:1039–1041.

Herreid, C.F., II and S. Kinney. 1967. Temperature and development of the wood frog, *Rana sylvatica*, in Alaska. Ecology 48:579–590.

Herreid, C.F., II. 1963. Range extension for *Bufo boreas*. Herpetologica 19:218–219.

Herring, K. and P. Verrell. 1996. Sexual incompatibility and geographic variation in mate recognition systems: tests in the salamander *Desmognathus ochrophaeus*. Animal Behaviour 52:279–287.

Herrington, R.E. 1985. The ecology, reproductive biology and management of the Larch Mountain salamander, (*Plethodon larselli* Burns) with comparisons of two other sympatric plethodons. Ph.D. dissertation. Washington State University, Pullman, Washington.

Herrington, R.E. and J.H. Larsen Jr. 1985. Current status, habitat requirements and management of the Larch Mountain salamander. Biological Conservation 34:169–179.

Herrington, R.E. and J.H. Larsen Jr. 1987. Reproductive biology of the Larch Mountain salamander (*Plethodon larselli*). Journal of Herpetology 21:48–56.

Herrington, R.W. 1988. Talus use by amphibians and reptiles in the Pacific Northwest. Pp. 216–221. *In* Szaro, R.C., K.E. Severson and D.R. Patton (Tech. Coords.), Management of Amphibians, Reptiles, and Small Mammals in North America. U.S.D.A. Forest Service, Rocky Mountain Forest and Range Experiment Station, Technical Report RM-166, Fort Collins, Colorado.

Hess, M.J. 1998. Letter to Andy Sansom, Robert Stanton and Bruce Babbitt, 23 January.

Hess, Z.J. and R.N. Harris. 2000. Eggs of *Hemidactylium scutatum* (Caudata: Plethodontidae) are unpalatable to insect predators. Copeia 2000:597–600.

Heusser, H. 1968. Die Lebensweise der Erdkröte *Bufo bufo* (L.); Wanderungen und Sommerquartiere. Revue Suisse Zoologie 75:927–982.

Heusser, H. 1970. Ansiedlung, Ortstreue und Populationsdynamik des Grasfrosches (*Rana temporaria*) in einem Gartenweiher. Salamandra 6:80–87.

Hewitt, J. and J.H. Power. 1913. A list of South African Lacertilia, Ophidia, and Batrachia in the McGregor Museum, Kimberley; with field-notes on various species. Transactions of the Royal Society of South Africa 3:147–176.

Hewitt, O.H. 1950. The bullfrog as a predator on ducklings. Journal of Wildlife Management 14:244.

Hews, D.K. and A.R. Blaustein. 1985. An investigation of the alarm response of *Bufo boreas* and *Rana cascadae* tadpoles. Behavioral and Neural Biology 43:47–57.

Hey, D. 1949. A report on the culture of the South African clawed frog at the Jonkershoek inland fish hatchery. Transactions of the Royal Society of South Africa 32:45–54.

Hey, J. 2001. The mind of the species problem. TRENDS in Ecology and Evolution 16:326–329.

Hey, J.D. and E.S. Farrar. 1997. Characterizing tadpole morphs of the plains spadefoot toad, *Spea bombifrons*, from western Iowa. Paper presented at the 77th Annual Meeting of the American Society of Ichthyologists and Herpetologists, Seattle, Washington.

Heyer, W.R. 1969. The adaptive ecology of the species groups of the genus *Leptodactylus* (Amphibia: Leptodactylidae). Evolution 23:421–428.

Heyer, W.R. 1971. *Leptodactylus labialis*. Pp. 104:1–104.3. Catalogue of American Amphibians and Reptiles. Society for the Study of Amphibians and Reptiles, St. Louis, Missouri.

Heyer, W.R. 1978. Systematics of the *fuscus* group of the frog genus *Leptodactylus* (Amphibia, Leptodactylidae). Natural History Museum of Los Angeles County Science Bulletin 29:1–85.

Heyer, W.R. 1979. Annual variation in larval amphibian populations within a temperate pond. Journal of the Washington Academy of Science 69:65–74.

Heyer, W.R. 2002. *Leptodactylus fragilis*, the valid name for the Middle American and northern South American white-lipped frog (Amphibia:

Leptodactylidae). Proceedings of the Biological Society of Washington 115:321–322.

Heyer, W.R., A.S. Rand, C.A. Goncalvez da Cruz and O.L. Peixoto. 1988. Decimations, extinctions, and colonizations of frog populations in southeast Brazil and their evolutionary implications. Biotropica 20:230–235.

Heyer, W.R., M.A. Donnelly, R.W. McDiarmid, L.-A.C. Hayek and M.S. Foster (Eds.). 1994. Measuring and Monitoring Biological Diversity: Standard Methods for Amphibians. Smithsonian Institution Press, Washington, D.C.

Heywood, I., S. Cornelius and S. Carver. 1998. An Introduction to Geographical Information Systems. Prentice Hall Series in Geographic Information Science. Clarke, K.C., Series Editor. Prentice Hall, Upper Saddle River, New Jersey.

Higginbotham, A.C. 1939. Studies on amphibian activity. I. Preliminary report on the rhythmic activity of *Bufo americanus americanus* and *Bufo fowleri* Hinckley. Ecology 20:58–70.

Higgins, C. and C. Sheard. 1926. Effects of ultraviolet radiation on the early development of *Rana pipiens*. Journal of Experimental Zoology 46:333–343.

Highton, R. 1956. The life history of the slimy salamander, *Plethodon glutinosus*, in Florida. Copeia 1956:75–93.

Highton, R. 1959. The inheritance of the color phases of *Plethodon cinereus*. Copeia 1959:33–37.

Highton, R. 1961. A new genus of lungless salamander from the Coastal Plain of Alabama. Copeia 1961:65–68.

Highton, R. 1962a. Revision of North American salamanders of the genus *Plethodon*. Bulletin of the Florida State Museum 6:235–367.

Highton, R. 1962b. Geographic variation in the life history of the slimy salamander. Copeia 1962:597–613.

Highton, R. 1970. Evolutionary interactions between species of North American salamanders of the genus *Plethodon*. Part I. Genetic and ecological relationships of *Plethodon jordani* and *P. glutinosus* in the southern Appalachian Mountains. Evolutionary Biology 4:211–241.

Highton, R. 1971 (1972). Distributional interactions among eastern North American salamanders of the genus *Plethodon*. Pp. 139–188. *In* Holt, P.C. (Ed.), The Distributional History of the Biota of the Southern Appalachians, Part III. Vertebrates. Research Division Monograph, Virginia Polytechnic Institute and State University, Blacksburg, Virginia.

Highton, R. 1972. Distributional interactions among eastern North American salamanders of the genus *Plethodon*. Pp. 139–188. *In* Holt, P.C. (Ed.), The Distributional History of the Biota of the Southern Appalachians. Part III: Vertebrates. Research Division Monograph, Virginia Polytechnic Institute and State University, Blacksburg, Virginia.

Highton, R. 1977. Comparison of microgeographic variation in morphological and electrophoretic traits. Evolutionary Biology 10:397–436.

Highton, R. 1979. A new cryptic species of salamander of the genus *Plethodon* from the southeastern United States (Amphibia: Plethodontidae). Brimleyana 1:31–36.

Highton, R. 1984. A new species of woodland salamander of the *Plethodon glutinosus* group from the southern Appalachian Mountains. Brimleyana 9:1–20.

Highton, R. 1985. The width of the contact zone between *Plethodon dorsalis* and *P. websteri* in Jefferson County, Alabama. Journal of Herpetology 19:544–546.

Highton, R. 1986a. *Plethodon aureolus*. P. 381.1. Catalogue of American Amphibians and Reptiles. Society for the Study of Amphibians and Reptiles, St. Louis, Missouri.

Highton, R. 1986b. *Plethodon fourchensis*. P. 391.1. Catalogue of American Amphibians and Reptiles. Society for the Study of Amphibians and Reptiles, St. Louis, Missouri.

Highton, R. 1986c. *Plethodon hoffmani*. Pp. 392.1–392.2. Catalogue of American Amphibians and Reptiles. Society for the Study of Amphibians and Reptiles, St. Louis, Missouri.

Highton, R. 1986d. *Plethodon hubrichti* Thurow. Pp. 393.1–393.2. Catalogue of American Amphibians and Reptiles. Society for the Study of Amphibians and Reptiles, St. Louis, Missouri.

Highton, R. 1986e. *Plethodon kentucki* Mittleman. Pp. 382.1–382.2. Catalogue of American Amphibians and Reptiles. Society for the Study of Amphibians and Reptiles, St. Louis, Missouri.

Highton, R. 1986f. *Plethodon nettingi* Green. Pp. 383.1–383.2. Catalogue of American Amphibians and Reptiles. Society for the Study of Amphibians and Reptiles, St. Louis, Missouri.

Highton, R. 1987a. *Plethodon teyahalee*. Pp. 401.1–401.2. Catalogue of American Amphibians and Reptiles. Society for the Study of Amphibians and Reptiles, St. Louis, Missouri.

Highton, R. 1987b. *Plethodon wehrlei* Fowler and Dunn. Pp. 402.1–402.3. Catalogue of American Amphibians and Reptiles. Society for the Study of Amphibians and Reptiles, St. Louis, Missouri.

Highton, R. 1988a. *Plethodon punctatus* Highton. Pp. 414.1–414.2. Catalogue of American Amphibians and Reptiles. Society for the Study of Amphibians and Reptiles, St. Louis, Missouri.

Highton, R. 1988b. *Plethodon shenandoah*. Pp. 413.1–413.2. Catalogue of American Amphibians and Reptiles. Society for the Study of Amphibians and Reptiles, St. Louis, Missouri.

Highton, R. 1989. Geographic protein variation. Pp. 1–78. *In* Highton, R., G.C. Maha and L.R. Maxson (Eds.), Biochemical Evolution in the Slimy Salamanders of the *Plethodon glutinosus* Complex in the Eastern United States. Illinois Biological Monographs, Number 57, University of Illinois Press, Urbana, Illinois.

Highton, R. 1990. Taxonomic treatment of genetically differentiated populations. Herpetologica 46:114–121.

Highton, R. 1995a. Pigeon Mountain salamander survey. Georgia Department of Natural Resources, Unpublished report, Social Circle, Georgia.

Highton, R. 1995b. Speciation in eastern North American salamanders of the genus *Plethodon*. Annual Review of Ecology and Systematics 26:579–600.

Highton, R. 1997. Geographic protein variation and speciation in the *Plethodon dorsalis* complex. Herpetologica 53:345–356.

Highton, R. 1998. Is *Ensatina eschscholtzii* a ring-species? Herpetologica 54:254–278.

Highton, R. 1999a. Geographic protein variation and speciation in the salamanders of the *Plethodon cinereus* group with the description of two new species. Herpetologica 55:43–90.

Highton, R. 1999b. Hybridization in the contact zone between *Plethodon richmondi* and *Plethodon electromorphus* in northern Kentucky. Herpetologica 55:91–105.

Highton, R. 2000. Detecting cryptic species using allozyme data. Pp. 215–241. *In* Bruce, R.C., R.G. Jaeger and L.D. Houck (Eds.), The Biology of Plethodontid Salamanders. Kluwer Academic/Plenum Publishing, New York.

Highton, R. 2004. A new species of woodland salamander of the *Plethodon cinereus* group from the Blue Ridge Mountains of Virginia. Jeffersoniana 14:1–22.

Highton, R. and A. Larson. 1979. The genetic relationships of the salamanders of the genus *Plethodon*. Systematic Zoology 28:579–599.

Highton, R. and A.H. Brame. 1965. *Plethodon stormi* species nov. Amphibia: Urodela: Plethodontidae. Pilot Register of Zoology, Card 20, Cornell University, Ithaca, New York.

Highton, R. and A.B. Grobman. 1956. Two new salamanders of the genus *Plethodon* from the southeastern United States. Herpetologica 12:185–188.

Highton, R. and J.R. MacGregor. 1983a. *Plethodon kentucki* Mittleman: a valid species of Cumberland Plateau woodland salamander. Herpetologica 39:189–200.

Highton, R. and R.B. Peabody. 2000. Geographic protein variation and speciation in salamanders of the *Plethodon jordani* and *Plethodon glutinosus* complexes in the southern Appalachian Mountains with the description of four new species. Pp. 31–93. *In* Bruce, R.C., R.G. Jaeger and L. Houck (Eds.), The Biology of Plethodontid Salamanders. Kluwer Academic/Plenum Publishers, New York.

Highton, R. and T. Savage. 1961. Functions of the brooding behavior in the female red-backed salamander, *Plethodon cinereus*. Copeia 1961: 95–98.

Highton, R. and T.P. Webster. 1976. Geographic protein variation and divergence in populations of the salamander *Plethodon cinereus*. Evolution 30:33–45.

Highton, R. and R.D. Worthington. 1967. A new salamander of the genus *Plethodon* from Virginia. Copeia 1967:617–626.

Highton, R., G.C. Maha and L.R. Maxson. 1989. Biochemical evolution in the slimy salamanders of the *Plethodon glutinosus* complex in the eastern United States. University of Illinois Biological Monograph, Number 57, Champaign, Illinois.

Hildebrand, S.F. 1938. A new catalogue of the fresh-water fishes of Panama. Pp. 219–359. Field Museum of Natural History, Zoological Series, Number 22, Chicago, Illinois.

Hildenbrandt, H., C. Bender, V. Grimm and K. Henle. 1995. Ein individuenbasiertes Modell zur Beurteilung der Überlebenschancen kleiner Populationen der Mauereidechse (*Podarcis muralis*). Verhandlungen der Gesellschaft für Ökologie 24:207–214.

Hileman, K.S. and E.D. Brodie Jr. 1994. Survival strategies of the salamander *Desmognathus ochrophaeus*: interaction of predator-avoidance and anti-predator mechanisms. Animal Behaviour 47:1–6.

Hill, I.R. 1954. The taxonomic status of the mid-Gulf Coast *Amphiuma*. Tulane Studies in Zoology 1:191–215.

Hill, J. 2000. *Rana heckscheri*. Animal Diversity Web, University of Michigan Museum of Zoology website at: http://animaldiversity.ummz.umich.edu/accounts/rana/r._heckscheri$narrative.html.

Hill, J.E. and P.G. Parnell. 1996. Adiaspiromycosis in bullfrogs *(Rana catesbeiana)*. Journal of Veterinary Diagnostic Investigation 8:496–497.

Hille, S. and J. Thiollay. 2000. The imminent ex1tinction of the kites *Milvus milvus fasciicauda* and *Milvus m. migrans* on the Cape Verde Islands. Bird Conservation International 10:361–369.

Hillis, D.M. 1977. Sex ratio, mortality rate, and breeding stimulus in a Maryland population of *Ambystoma maculatum*. Bulletin of the Maryland Herpetological Society 13:84–91.

Hillis, D.M. 1981. Premating isolating mechanisms among three species of the *Rana pipiens* complex in Texas and southern Oklahoma. Copeia 1981:312–319.

Hillis, D.M. 1982. Morphological differentiation and adaptation of the larvae of *Rana berlandieri* and *Rana sphenocephala* (*Rana pipiens* complex) in sympatry. Copeia 1982:168–174.

Hillis, D.M. 1988. Systematics of the *Rana pipiens* complex: puzzle and paradigm. Annual Review of Ecology and Systematics 19:39–63.

Hillis, D.M. and R. Miller. 1976. An instance of overwintering of larval *Ambystoma maculatum* in Maryland. Bulletin of the Maryland Herpetological Society 12:65–66.

Hillis, D.M., J.T. Collins and J.P. Bogart. 1987. Distribution of diploid and tetraploid species of gray treefrogs (*Hyla chrysoscelis* and *Hyla versicolor*) in Kansas. American Midland Naturalist 117:214–217.

Hillis, D.M, J.S. Frost and D.A. Wright. 1983. Phylogeny and biogeography of the *Rana pipiens* complex: a biochemical evaluation. Systematic Zoology 32:132–143.

Hillis, D.M., A.M. Hillis and R.F. Martin. 1984. Reproductive ecology and hybridization of the endangered Houston toad *(Bufo houstonensis)*. Journal of Herpetology 18:56–72.

Hillis, D.M., C. Moritz and B.K. Mable (Eds.). 1996. Molecular Systematics. Second edition. Sinauer Associates, Sunderland, Massachusetts.

Hillis, D.M., D.A. Chamberlain, T.P. Wilcox and P.T. Chippindale. 2001. A new species of subterranean blind salamander from Austin, Texas, and a systematic revision of central Texas paodomorphic salamanders. Herpetologica 57:266–280.

Hillis, R.E. and E.D. Bellis. 1971. Some aspects of the ecology of the hellbender, *Cryptobranchus a. alleganiensis*, in a Pennsylvania stream. Journal of Herpetology 5:121–126.

Hillman, S.S. 1980. Physiological correlates of differential dehydration tolerance in anuran amphibians. Copeia 1980:125–129.

Hilton, W.A. 1945. Distribution of the genus *Batrachoseps*, especially on the coastal islands of southern California. Bulletin of the Southern California Academy of Science 44:101–129.

Hilton, W.A. 1947. Lateral line sense organs in salamanders. Bulletin of the Southern California Academy of Science 46:97–110.

Hilton, W.A. 1948. Salamander notes from the Northwest. Herpetologica 4:120.

Himes, J.G. and T.W. Bryan. 1998. Geographical distribution: *Bufo americanus charlesmithi* (dwarf American toad). Herpetological Review 29:246.

Hinckley, A.D. 1962. Diet of the giant toad, *Bufo marinus* (L.), in Fiji. Herpetologica 18:253–259.

Hinckley, M.H. 1882. On some differences in the mouth-structure of tadpoles of the anurous batrachians found in Milton, Massachusetts. Proceedings of the Boston Society of Natural History 21:307–314.

Hinderstein, B. 1968. Some noteworthy salamander records from Long Island. Engelhardtia 1:25–27.

Hine, R.L., B.L. Les and B.F. Hellmich. 1981. Leopard frog populations and mortality in Wisconsin, 1974–1976. Wisconsin Department of Natural Resources, Technical Bulletin Number 122, Madison, Wisconsin.

Hinshaw, S. 1992. Northern leopard frog, *Rana pipiens*. Pp. 77–81. *In* Hunter, M.L., J. Albright and J. Arbuckle (Eds.), The Amphibians and Reptiles of Maine. Maine Agricultural Experiment Station, Bulletin 838, Orono, Maine.

Hinshaw, S.H. and B.K. Sullivan. 1990. Predation on *Hyla versicolor* and *Pseudacris crucifer* during reproduction. Herpetologica 24:196–197.

Hirai, T. and M. Matsui. 2000. Myrmecophagy in a ranid frog *Rana rugosa*: specialization or weak avoidance to ant eating? Zoological Science 17:459–466.

Hirai, T. and M. Matsui. 2001. Food partitioning between two syntopic ranid frogs, *Rana nigromaculata* and *R. rugosa*. Herpetological Journal 11:109–115.

Hird, D.W., S.L. Diesch, R.G. McKinnell, E. Gorham, F.B. Martin, C.A. Meadows and M. Gasiorowski. 1983. *Enterobacteriaceae* and *Aeromanas hydrophila* in Minnesota frogs and tadpoles *(Rana pipiens)*. Applied and Environmental Microbiology 46:1423–1425.

Hirschfield, C.J. 1962. Maximum length records for six plethodontid salamanders. Journal of Ohio Herpetological Society 3:52–53.

Hirschfeld, C.J. and J.T. Collins. 1963. Range extension for three amphibians in north-central Kentucky. Copeia 1963:438–439.

Hitchings, S.P. and T.J.C. Beebee. 1997. Genetic substructuring as a result of barriers to gene flow in urban *Rana temporaria* (common frog) populations—implications for biodiversity conservation. Heredity 79:117–127.

Hitchings, S.P. and T.J.C. Beebee. 1998. Loss of genetic diversity and fitness in common toad *(Bufo bufo)* populations isolated by inimical habitat. Journal of Evolutionary Biology 11:269–283.

Hobbs, K.L. 1932. An extension of the range of *Hyla regilla* (Baird and Girard) into Arizona. Copeia 1932:104.

Hoberg, T.D. and C.T. Gause. 1989. Reptiles and amphibians of North Dakota. North Dakota Herpetological Society, Grand Forks, North Dakota.

Hoberg, T.D. and C.T. Gause. 1992. Reptiles and amphibians of North Dakota. North Dakota Outdoors 55:7–18.

Hobson, J.A., C.J. Goin and O.B. Goin. 1967. Sleep behavior of frogs. Quarterly Journal of the Florida Academy of Science 30:184–186.

Hodge, R.P. 1976. Amphibians and Reptiles in Alaska, the Yukon and Northwest Territories. Alaska Northwest Publishing Company, Anchorage, Alaska.

Hoff, J.G. 1977. A Massachusetts hibernation site of the red-backed salamander, *Plethodon cinereus*. Herpetological Review 8:33.

Hoff, K. vS. 1994. Status of the Amargosa toad 1993. Prepared for the Nevada Biodiversity Initiative. *In* Nevada Division of Wildlife. 2000. Conservation agreement for the Amargosa toad *(Bufo nelsoni)* and co-occurring sensitive species in Oasis Valley, Nye County, Nevada (September, 2000).

Hoff, K. vS. 1996. Natural history studies of the Amargosa toad *(Bufo nelsoni)*, 1995. Final report of activities. Contract number 95-33 to Nevada Division of Wildlife, Reno, Nevada (available from Nevada Natural Heritage Program, Carson City, Nevada).

Hoff, K. and R.J. Wassersug. 1986. The kinematics of swimming in larvae of the clawed frog, *Xenopus laevis*. Journal of Experimental Biology 122:1–12.

Hoff, K.S., M.J. Lannoo and R.J. Wassersug. 1985. Kinematics of midwater prey capture by *Ambystoma* (Caudata: Ambystomatidae). Copeia 1985:247–251.

Hoff, K.vS., A.R. Blaustein, R.W. McDiarmid and R. Altig. 1999. Behavior: interactions and their consequences. Pp. 125–239. *In* McDiarmid, R.W. and R. Altig (Eds.), Tadpoles: The Biology of Anuran Larvae. University of Chicago Press, Chicago, Illinois.

Hoffman, A.A. and M.W. Blows. 1993. Evolutionary genetics and climate change: will animals adapt to global warming? Pp. 165–178. *In* Kareiva, P.M., J.G. Kingsolver and R.B. Huey (Eds.), Biotic Interactions and Global Change. Sinauer Associates, Sunderland, Massachusetts.

Hoffman, E.A. and M.S. Blouin. 2004. Evolutionary history of the northern leopard frog: reconstruction of phylogeny, phylogeography, and historial changes in population demography from mitochondrial DNA. Evolution 58:145–159.

Hoffman, G.L. 1955. Notes on the life cycle of *Fibricola cratera* (Trematoda: Strigeida). Journal of Parasitology 41:327.

Hoffman, G.L. 1999. Parasites of North American Freshwater Fishes. Comstock Publishing, Ithaca, New York.

Hoffman, R.L. 1946. The voice of *Hyla versicolor* in Virginia. Herpetologica 3:141–142.

Hoffman, R.L. 1953. *Plethodon welleri* Walker in Tennessee. Journal of the Tennessee Academy of Science 28:86–87.

Hoffman, R.L. 1955. On the occurrence of two species of hylid frogs in Virginia. Herpetologica 11:30–32.

Hoffman, R.L. 1967. Distributional records for three species of *Plethodon* in Virginia. Radford Review, Faculty Research Issue 21:201–214.

Hoffman, R.L. 1980. *Pseudacris brachyphona*. Pp. 234.1–234.2. Catalogue of American Amphibians and Reptiles. Society for the Study of Amphibians and Reptiles, St. Louis, Missouri.

Hoffman, R.L. 1981. On the occurrence of *Pseudacris brachyphona* (Cope) in Virginia. Catesbeiana 1:9–13.

Hoffman, R.L. 1983. *Pseudacris brimleyi*. Pp. 311.1–311.2. Catalogue of American Amphibians and Reptiles. Society for the Study of Amphibians and Reptiles, St. Louis, Missouri.

Hoffman, R.L. 1988. *Hyla femoralis*. Pp. 436.1–436.3. Catalogue of American Amphibians and Reptiles. Society for the Study of Amphibians and Reptiles, St. Louis, Missouri.

Hoffman, R.L. 1990. *Rana utricularia* (southern leopard frog): VA: Pittsylvania Co. VA. Hy 880, ca.0.4 km S of int. with VA 863 at Berry Hill. Catesbeiana 10:44–45.

Hoffman, R.L. and L. Hubricht. 1954. Distributional records of two species of *Plethodon* in the southern Appalachians. Herpetologica 10:191–193.

Hoffman, R.L. and H.I. Kleinpeter. 1948. A collection of salamanders from Mount Rogers, Virginia. Journal of the Washington Academy of Sciences 38:106–108.

Hoffman, R.L. and G.L. Larson. 1999. Natural history notes: *Ambystoma gracile* (northwestern salamander). Predation and cannibalism. Herpetological Review 30:159.

Hoffman, R.L. and J.C. Mitchell. 1996. Records of anurans from Greensville County, Virginia. Banisteria 8:29–36.

Hoffman, R.L. and D.S. Pilliod. 1999. The ecological effects of fish stocking on amphibian populations in high-mountain wilderness lakes. U.S.G.S.-B.R.D. Forest and Rangeland Ecosystem Science Center, Final report, Corvallis, Oregon.

Hoffman, R.L., G.L. Larson and B.J. Brokes. 2003. Habitat segregation of *Ambystoma gracile* and *Ambystoma macrodactylum* in mountain ponds and lakes, Mount Ranier National Park, Washington, USA. Journal of Herpetology 37:24–34.

Hoffpauir, C. and E.O. Morrison. 1966. *Rhabdias ranae* from *Bufo valliceps*. Southwestern Naturalist 11:302.

Hokit, D.G, S.C. Walls and A.R. Blaustein. 1996. Context-dependent kin discrimination in larvae of the marbled salamander, *Ambystoma opacum*. Animal Behaviour 52:17–31.

Holbrook, J.E. 1836. North American Herpetology. Volume 1. J. Dobson and Son, Philadelphia, Pennsylvania.

Holbrook, J.E. 1838a. North American Herpetology. Edition 1, Volume 2. J. Dobson and Son, Philadelphia, Pennsylvania.

Holbrook, J.E. 1838b. North American Herpetology, Edition 1, Volume 3. J. Dobson and Son, Philadelphia, Pennsylvania.

Holbrook, J.E. 1840. North American Herpetology. Edition 1, Volume 4. J. Dobson and Son, Philadelphia, Pennsylvania.

Holbrook, J.E. 1842. North American Herpetology, Volume 5. J. Dobson and Son, Philadelphia, Pennsylvania.

Holden, C.P. and K.B. Storey. 1997. Second messenger and cAMP-dependent protein kinase responses to dehydration and anoxia stresses in frogs. Journal of Comparative Physiology B 167: 305–312.

Holland, D.C. and R.H. Goodman Jr. 1998. A Guide to the Amphibians and Reptiles of MCB Camp Pendleton, San Diego County, California. Report to AC/S Environmental Security, Resource Management Division, Camp Pendleton, California.

Holland, D.C. and N.R. Sisk. 2000. Habitat use and population demographics of the arroyo toad *(Bufo californicus)* on MCB Camp Pendleton, San Diego County, California: Final report for 1998–1999. AC/S Environmental Security, United States Marine Corps, Camp Pendleton, California.

Holland, D.C., M.P. Hayes and E. McMillan. 1990. Late summer movement and mass mortality in the California tiger salamander *(Ambystoma californiense)*. Southwestern Naturalist 35:217–220.

Holland, R.F. 1978. The geographic and edaphic distribution of vernal pools in the Great Central Valley, California. California Native Plant Society, Special Publication, Number 4, Sacramento, California.

Hollenbeck, R.R. 1976. Movements within a population of *Rana pretiosa pretiosa* Baird and Girard in south central Montana. Journal of the Colorado-Wyoming Academy of Science 8:72–73.

Hollingsworth, B. and K. Roberts. 2001. *Pseudacris regilla*, Pacific treefrog. San Diego Natural History Museum Field Guide, http://www.sdnhm.org/fieldguide/herps/hyla-reg.html.

Holloway, W.R. and R.W. Dapson. 1971. Histochemistry of the integumentary secretions of the narrow-mouth toad, *Gastrophryne carolinensis*. Copeia 1971:351–353.

Holm, P.A. and C.H. Lowe. 1995. Status and conservation of sensitive herpetofauna in the Madrean riparian habitat of Scotia Canyon, Huachuca Mountains, Arizona. Arizona Game and Fish Department, Technical Report, Phoenix, Arizona.

Holman, J.A. 1955. Fall and winter food of *Plethodon dorsalis* in Johnson County, Indiana. Copeia 1955:143.

Holman, J.A. 1957. Bullfrog predation on the eastern spadefoot, *Scaphiopus holbrooki*. Copeia 1957:229.

Holman, J.A. 1965a. A polymorphic deme of *Hyla eximia* Baird from Durango, Mexico. Journal of the Ohio Herpetological Society 5:34.

Holman, J.A. 1965b. A late Pleistocene herpetofauna from Missouri. Transactions of the Illinois Academy of Science 58:190–194.

Holman, J.A. 1975. Herpetofauna of the WaKeeney local fauna (Lower Pliocene: Clarendonian) of Trego County, Kansas. Pp. 49–66. University of Michigan Papers on Paleontology, Number 12, Ann Arbor, Michigan.

Holman, J.A. 1995. Pleistocene Amphibians and Reptiles in North America. Oxford University Press, New York.

Holman, J.A. 1998. Amphibian recolonization of midwestern states in the postglacial Pleistocene. Pp. 9–15. *In* Lannoo, M.J. (Ed.), Status and Conservation of Midwest Amphibians. University of Iowa Press, Iowa City, Iowa.

Holomuzki, J.R. 1980. Synchronous foraging and dietary overlap of three species of plethodontid salamanders. Herpetologica 36: 109–115.

Holomuzki, J.R. 1982. Homing behavior of *Desmognathus ochrophaeus* along a stream. Journal of Herpetology 16:307–309.

Holomuzki, J.R. 1985a. Diet of larval *Dytiscus daurius* (Coleoptera: Dytiscidae) in east-central Arizona. Pan-Pacific Entomology 61:229.

Holomuzki, J.R. 1985b. Life history aspects of the predaceous diving beetle, *Dytiscus daurius* (Gebler), in Arizona. Southwestern Naturalist 30:485–490.

Holomuzki, J.R. 1986a. Predator avoidance and diel patterns of microhabitat use by larval tiger salamanders. Ecology 67:737–748.

Holomuzki, J.R. 1986b. Intraspecific predation and habitat use by tiger salamanders *(Ambystoma tigrinum nebulosum)*. Journal of Herpetology 20:439–441.

Holomuzki, J.R. 1986c. Effect of microhabitat on fitness components of larval tiger salamanders, *Ambystoma tigrinum nebulosum*. Oecologia 71:142–148.

Holomuzki, J.R. 1989a. Predation risk and macroalgae use by the stream-dwelling salamander *Ambystoma texanum*. Copeia 1989:22–28.

Holomuzki, J.R. 1989b. Salamander predation and vertical distributions of zooplankton. Freshwater Biology 21:461–472.

Holomuzki, J.R. 1991. Macrohabitat effects on egg deposition and larval growth, survival, and instream dispersal in *Ambystoma barbouri*. Copeia 1991:687–694.

Holomuzki, J.R. 1995. Oviposition sites and fish-deterrent mechanisms of two stream anurans. Copeia 1995:607–613.

Holomuzki, J.R. 1997. Habitat-specific life-histories and foraging by stream-dwelling American toads. Herpetologica 53:445–453.

Holomuzki, J.R. and J.P. Collins. 1983. Diel movement of larvae of the tiger salamander, *Ambystoma tigrinum nebulosum*. Journal of Herpetology 17:276–278.

Holomuzki, J.R. and J.P. Collins. 1987. Trophic dynamics of a top predator, *Ambystoma tigrinum nebulosum* (Caudata: Ambystomatidae), in a lentic community. Copeia 1987:949–957.

Holomuzki, J.R., J.P. Collins and P.E. Brunkow. 1994. Trophic control of fishless ponds by tiger salamander larvae. Oikos 71:55–64.

Holsinger, J.R. 1982. A preliminary report on the cave fauna of Burnsville Cove, Virginia. Bulletin of the National Speleological Society 44:98–101.

Holycross, A.T., B.G. Fodorko and O. Fourie. 1999. Natural history notes: *Bufo alvarius* (Colorado River toad). Habitat. Herpetological Review 20:90.

Hölzinger, J. 1987. Die in Baden-Würtemberg gefährdeten Lurche (Amphibia) und Kriechtiere (Reptilia) Rote Liste. (Second Draft, 31 December 1984.) Beihefte zu den Veröffentlichungen für Naturschutz und Landschaftspflege in Baden-Württemberg 41:157–164.

Holzwart, J. and K.D. Hall. 1984. The absence of glycerol in the hibernating American toad *(Bufo americanus)*. Bios 55:31–36.

Hom, C.L. 1987. Reproductive ecology of female dusky salamanders, *Desmognathus fuscus* (Plethodontidae), in the southern Appalachians. Copeia 1987:768–777.

Hom, C.L. 1988. Cover object choice by female dusky salamanders, *Desmognathus fuscus*. Journal of Herpetology 22:247–249.

Honegger, R.E. 1970. Houston toad *Bufo houstonensis* Sanders 1953. Red Data Book, Volume 3, Amphibia and Reptilia. International Union for Conservation of Nature and Natural Resources, Survival Service Commission, Morges, Switzerland.

Honegger, R.E. 1981. Threatened Amphibians and Reptiles in Europe. Handbuch der Reptilien und Amphibien Europas, Supplementary Volume. Akademische Verlagsgesellschaft, Wiesbaden.

Hood, H.H. 1934. A note on the red-backed salamander at Rochester, New York. Copeia 1934:141–142.

Hooge, P.N. and B. Eichenlaub. 1997. Animal movement extension to ArcView. Version 1.1. Alaska Biological Science Center, U.S. Geological Survey, Anchorage, Alaska.

Hoopes, I. 1930. *Bufo* in New England. Bulletin of the Boston Society of Natural History 57:13–20.

Hoopes, I. 1938. Do you know the mink frog? New England Naturalist 1:4–6.

Hopey, M.E. and J.W. Petranka. 1994. Restriction of wood frogs to fish-free habitats: how important is adult choice? Copeia 1994: 1023–1025.

Hopkins, S.H. 1933. Note on the life history of *Clinostomum marginatum* (Trematoda). Transactions of the American Microscopical Society 52:147–149.

Hopkins, W.A., J. Congdon and J.K. Ray. 2000. Incidence and impact of axial malformations in larval bullfrogs *(Rana catesbeiana)* developing in sites polluted by a coal-burning power plant. Environmental Toxicology and Chemistry 19:862–868.

Hoppe, D. 1996. Unnatural history note: Minnesota deformed frog study. Minnesota Herpetological Society Newsletter 16:9.

Hoppe, D. and E. Mottl. 1997. Anuran species differences in malformation frequency and severity in Minnesota. P. 90. *In* Abstracts, 18th Annual Meeting, November 16–20, San Francisco, California, Society of Toxicology and Chemistry (SETAC), Pensacola, Florida.

Hoppe, D.M. 1996. Historical observations and recent species diversity of deformed anurans in Minnesota. *In* Third annual meeting of the North American Amphibian Monitoring Program. www.im.nbs.gov/naamp3/naamp3.html.

Hoppe, D.M. 2000. History of Minnesota frog abnormalities: do recent findings represent a new phenomenon? Pp. 86–89. *In* Kaiser, H., G.S. Casper and N. Bernstein (Eds.), Investigating Amphibian Declines: Proceedings of the 1998 Midwest Declining Amphibians Conference. Iowa Academy of Science, Volume 107, Cedar Falls, Iowa.

Hoppe, D.M. and R.G. McKinnell. 1991. Minnesota's mutant leopard frogs. Minnesota Volunteer 1991:56–63.

Horne, E.A. 1988. Aggressive behavior of female red-backed salamanders. Herpetologica 44:203–209.

Horne, E.A. and R.G. Jaeger. 1988. Territorial pheromones of female red-backed salamanders. Ethology 78:143–152.

Horne, M.T. and W.A. Dunson. 1995. Effects of low pH, metals, and water hardness on larval amphibians. Archives of Environmental Contamination and Toxicology 29:500–505.

Horner, S., A. Reister and D.L. Carlson. 2000. Cytogenic analysis of deformed frogs. Journal of the Minnesota Academy of Science 64:35–36.

Hosmer, D.W., Jr. and S. Lemeshow. 1989. Applied Logistic Regression. John Wiley and Sons, New York.

Hossack, B.R. 1998. Landscape influences on headwater streams and stream amphibians in northern Idaho. Master's thesis. University of Idaho, Moscow, Idaho.

Hossack, B.R. 2002. Natural history notes: *Ambystoma macrodactylum krausei* (northern long-toed salamander). Vocalization. Herpetological Review 33:121.

Houck, L.D. 1977a. Reproductive biology of a Neotropical salamander, *Bolitoglossa rostrata*. Copeia 1977:70–82.

Houck, L.D. 1977b. Life history patterns and reproductive biology of Neotropical salamanders. Pp. 43–72. *In* Taylor, D.H. and S.I. Guttman (Eds.), The Reproductive Biology of Amphibians. Plenum Press, New York.

Houck, L.D. 1980. Courtship behavior in the plethodontid salamander, *Desmognathus wrighti*. American Zoologist 20:825 (Abstract).

Houck, M.A. and E.D. Bellis. 1972. Comparative tolerance to desiccation in the salamanders *Desmognathus f. fuscus* and *Desmognathus o. ochrophaeus*. Journal of Herpetology 6:209–215.

Houck, L.D., S.G. Tilley and S.J. Arnold. 1985. Sperm competition in a plethodontid salamander: preliminary results. Journal of Herpetology 19:423–425.

Houck, L.D., A.M. Bell, N.L. Reagan-Wallin and R.C. Feldhoff. 1998. Effects of experimental delivery of male courtship pheromones on the timing of courtship in a terrestrial salamander, *Plethodon jordani* (Caudata: Plethodontidae). Copeia 1998:214–219.

Houck, L.D., M.T. Mendonca, T.K. Lynch and D.E. Scott. 1996. Courtship behavior and plasma levels of androgens and corticosterone in male marbled salamanders, *Ambystoma opacum* (Ambystomatidae). General and Comparative Endocrinology 104:243–252.

Houck, W.J. 1969. Albino *Aneides ferreus*. Herpetologica 25:54.

Houlahan, J.E., C.S. Findlay, A.H. Meyer, S.L. Kuzmin and B.R. Schmidt. 2001. Reply to Alford et al. 2001. Nature 412:500.

Houlahan, J.E., C.S. Findlay, B.R. Schmidt, A.H. Meyer and S.L. Kuzmin. 2000. Quantitative evidence for global amphibian population declines. Nature 404:752–755.

Hovey, T.E. and D.R. Bergen. 2003. Natural history notes: *Rana catesbeiana* (bullfrog). Predation. Herpetological Review 34:360–361.

Hovingh, P. 1993. Aquatic habitats, life history observations, and zoogeographic considerations of the spotted frog *(Rana pretiosa)* in Tule Valley, Utah. Great Basin Naturalist 53:168–179.

Hovingh, P. 1997. Amphibians in the eastern Great Basin (Nevada and Utah, USA): a geographical study with paleozoological models and conservation implications. Herpetological Natural History 5:97–134.

Hovingh, P., B. Benton and D. Bornholdt. 1985. Aquatic parameters and life history observations of the Great Basin spadefoot toad in Utah. Great Basin Naturalist 45:22–30.

Howard, J.H. and R.L. Wallace. 1981. Microecological variation of electrophoretic loci in populations of *Ambystoma macrodactylum columbianum* (Caudata: Ambystomatidae). Copeia 1989:466–471.

Howard, J.H. and R.L. Wallace. 1985. Life history characteristics of populations of the long-toed salamander *(Ambystoma macrodactylum)* from different altitudes. American Midland Naturalist 113:361–373.

Howard, J.H., L.W. Seeb and R. Wallace. 1993. Genetic variation and population divergence in the *Plethodon vandykei* species group (Caudata: Plethodontidae). Herpetologica 49:238–247.

Howard, J.H., R.L. Wallace and J.H. Larsen Jr. 1983. Genetic variation and population divergence in the Larch Mountain Salamander *(Plethodon larselli)*. Herpetologica 39:41–46.

Howard, P.C. 1991. Nature conservation in Uganda's tropical forest reserves. World Conservation Union, Gland, Switzerland.

Howard, R.D. 1978a. The evolution of mating strategies in bullfrogs, *Rana catesbeiana*. Evolution 32:850–871.

Howard, R.D. 1978b. The influence of male-defended oviposition sites on early embryo mortality in bullfrogs. Ecology 59:789–798.

Howard, R.D. 1979. Estimating reproductive success in natural populations. American Naturalist 114:221–231.

Howard, R.D. 1980. Mating behavior and mating success in wood frogs, *Rana sylvatica*. Animal Behavior 28:705–716.

Howard, R.D. 1981. Sexual dimorphism in bullfrogs. Ecology 62:303–310.

Howard, R.D. 1988. Sexual selection on male body size and mating behaviour in American toads, *Bufo americanus*. Animal Behaviour 36:1796–1808.

Howard, R.D. and A.G. Kluge. 1985. Proximate mechanisms of sexual selection in wood frogs. Evolution 39:260–277.

Howard, R.D. and J.G. Palmer. 1995. Female choice in *Bufo americanus*: effects of dominant frequency and call order. Copeia 1995:212–217.

Howard, R.D. and J.R. Young. 1998. Individual variation in male vocal traits and female mating preferences in *Bufo americanus*. Animal Behaviour 55:1165–1179.

Howard, R.R. 1971. Avoidance learning of spotted salamanders, *Ambystoma maculatum*, by domestic chickens. American Zoologist 11:637.

Howard, R.R. and E.D. Brodie Jr. 1970. A mimetic relationship in salamanders: *Notophthalmus viridescens* and *Pseudotriton ruber*. American Zoologist 10:475.

Howard, R.R. and E.D. Brodie Jr. 1973. A Batesian mimetic complex in salamanders: responses to avian predators. Herpetologica 29:33–41.

Howard, W.E. 1950. Birds as bullfrog food. Copeia 1950:152.

Howe, G.E., R. Gillis and R.C. Mowbray. 1998. Effects of chemical synergy and larval stage on the toxicity of atrazine and alachlor to amphibian larvae. Environmental Toxicology and Chemistry 17:519–525.

Howell, T. and A. Hawkins. 1954. Variation in topotypes of the salamander *Plethodon jordani melaventris*. Copeia 1954:32–36.

Howell, T. and V. Switzer. 1953. Integrades of the two-lined salamander, *Eurycea bislineata*, in Georgia. Herpetologica 9:152.

Howland, J.M., M.J. Sredl and J.E. Wallace. 1997. Validation of visual encounter surveys. Pp. 21–36. *In* Sredl, M.J. (Ed.), Ranid Frog Conservation and Management. Nongame and Endangered Wildlife Program, Technical Report, Number 121, Arizona Game and Fish Department, Phoenix, Arizona.

Howley, P., D. Ganem and E. Kieff. 1997. DNA viruses. Pp. 168–184. *In* DeVita, V.T., Jr., S. Hellman and S.A. Rosenberg (Eds.), Cancer Principles and Practice of Oncology. Lippincott-Raven, Philadelphia, Pennsylvania.

Hoy, M.D. 1994. Depredations by herons and egrets at bait fish farms in Arkansas. Aquaculture Magazine 20:52–56.

Hranitz, J.M., F.C. Hill, R.G. Sagar and T.S. Klinger. 1989. Abundance size and sex ratios of *Bufo woodhousii fowleri* Hinkley (Chordata: Amphibia) on Assateague Island, Virginia and the adjacent mainland. American Zoologist 29:169a.

Hranitz, J.M., T.S. Klinger, F.C. Hill, R.G. Sagar, T. Mencken and J. Carr. 1993. Morphometric variation between *Bufo woodhousii fowleri* Hinckley (Anura: Bufonidae) on Assateague Island, Virginia and the adjacent mainland. Brimleyana 19:65–75.

Huang, C. and A. Sih. 1990. Experimental studies of behaviorally mediated indirect interactions through a shared predator. Ecology 71: 1515–1522.

Huang, C. and A. Sih. 1991a. An experimental study on the effects of salamander larvae on isopods in stream pools. Freshwater Biology 25: 451–459.

Huang, C. and A. Sih. 1991b. Experimental studies on direct and indirect interactions in a three trophic-level stream system. Oecologia 85: 530–536.

Hubbard, J.P., H.C. Conway, H. Campbell, G. Schmitt and M.D. Hatch. 1979. Handbook of species endangered in New Mexico. New Mexico Department of Game and Fish, Santa Fe, New Mexico.

Hubbard, M.E. 1903. Correlated protective devices in some California salamanders. University of California Publications in Zoology 1: 157–170.

Hubbs, C. 1962. Effects of a hurricane on the fish fauna of a coastal pool and drainage ditch. Texas Journal of Science 14:289–296.

Hubbs, C.E. and N.E. Armstrong. 1961. Minimum developmental temperature tolerance of two anurans, *Scaphiopus couchi* and *Microhyla olivacea*. Texas Journal of Science 13:358–362.

Hubbs, C. and F.D. Martin. 1967. *Bufo valliceps* breeding in artificial pools. Southwestern Naturalist 12:105–106.

Hubbs, C., T. Wright and O. Cuellar. 1963. Developmental temperature tolerance of central Texas populations of two anuran amphibians *Bufo valliceps* and *Pseudacris streckeri*. Southwestern Naturalist 8:142–149.

Hubbs, C.L. 1918. *Bufo fowleri* in Michigan, Indiana and Illinois. Copeia 55:40–43.

Huber, W. 1993. Ecotoxicological relevance of atrazine in aquatic systems. Environmental Toxicology and Chemistry 12:1865–1881.

Hudson, G.E. 1942. The amphibians and reptiles of Nebraska. Nebraska Conservation Bulletin, Number 24, University of Nebraska, Lincoln, Nebraska.

Hudson, G.E. 1972 (reprinted). The amphibians and reptiles of Nebraska. Nebraska Conservation Bulletin, Number 24, University of Nebraska, Lincoln, Nebraska.

Hudson, M.S. 1983. Waterfowl production on three age-classes of stock ponds in Montana. Journal of Wildlife Management 447:112–117.

Hudson, P.J., A.P. Dobson and D. Newborn. 1998. Prevention of population cycles by parasite removal. Science 282:2256–2258.

Hudson, R.G. 1950. A possible hibernating site for the spring peeper. Copeia 1950:59.

Hudson, R.G. 1954. An annotated list of the reptiles and amphibians of the Unami Valley, Pennsylvania. Herpetologica 10:67–72.

Hudson, R.G. 1955. Observations on the larvae of the salamander *Eurycea bislineata bislineata*. Herpetologica 11:202–204.

Hudson, R.G. 1956. The leopard frog *Rana pipiens sphenocephala* in southeastern Pennsylvania. Herpetologica 12:182–183.

Huggins, D.G. 1971. *Scaphiopus bombifrons* Cope, a species new to Iowa. Journal of Herpetology 5:216.

Hughes, J.B., G.C. Daily and P.R. Ehrlich. 1997. Population diversity: its extent and extinction. Science 278:689–692.

Hughes, M., R. Petersen and R.M. Duffield. 1999. Natural history notes: *Plethodon cinereus* (red-backed salamander). Habitat. Herpetological Review 30:160.

Hughes, N. 1965. Comparison of frontoparietal bones of *Scaphiopus bombifrons* and *S. hammondi* as evidence of interspecific hybridization. Herpetologica 21:196–201.

Hughes, R.C. 1928. Studies on the trematode family Stirgeidae (Holostomidae). Number VII. *Tetracotyle pipientis* Faust. Transactions of the American Microscopical Society 47:42–53.

Huheey, J.E. 1959. Notes on habits of some plethodontid salamanders. Herpetologica 15:144.

Huheey, J.E. 1960. Mimicry in the color patterns of certain Appalachian salamanders. Journal of the Elisha Mitchell Scientific Society 76:246–251.

Huheey, J.E. 1964. Use of burrows by the black-bellied salamander. Journal of the Ohio Herpetological Society 4:105.

Huheey, J.E. 1966. The desmognathine salamanders of the Great Smoky Mountains National Park. Journal of the Ohio Herpetological Society 5:63–72.

Huheey, J.E. 1980. Studies in warning coloration and mimicry VIII. Further evidence for a frequency-dependent model of predation. Journal of Herpetology 14:223–230.

Huheey, J.E. and R.A. Brandon. 1961. Further notes on mimicry in salamanders. Herpetologica 17:63–64.

Huheey, J.E. and R.A. Brandon. 1973. Rockface populations of the mountain salamander, *Desmognathus ochrophaeus*, in North Carolina. Ecological Monographs 43:59–77.

Huheey, J.E. and R.A. Brandon. 1974. Studies in warning coloration and mimicry. VI. Comments on the warning coloration of red efts and their presumed mimicry by red salamanders. Herpetologica 30:149–155.

Huheey, J.E. and A. Stupka. 1967. Amphibians and Reptiles of the Great Smoky Mountains National Park. University of Tennessee Press, Knoxville, Tennessee.

Huizinga, H.W. and M.J. Nadakavukaren. 1997. Cellular responses of goldfish, *Carassius auratus* (L.), to metacercariae of *Ribeiroia marini* (Faust and Hoffman, 1934). Journal of Fish Diseases 20:401–408.

Hulse, A.C. 1978. *Bufo retifomis*. Pp. 207.1–207.2. Catalogue of American Amphibians and Reptiles. Society for the Study of Amphibians and Reptiles, St. Louis, Missouri.

Hulse, A.C. and K.L. Hulse. 1992. New county records for amphibians and reptiles from Pennsylvania. Herpetological Review 23:62–64.

Hume, D. 1748. Philosophical Essays Concerning Human Understanding. [Reprinted 1955. An Inquiry Concerning Human Understanding. Liberal Arts Press, New York.].

Humphries, C.J., R.I. Vane-Wright and P.H. Williams. 1991. Biodiversity reserves: setting new priorities for the conservation of wildlife. Parks 2:34–38.

Humphries, C.J., P.H. Williams and R.I. Vane-Wright. 1995. Measuring biodiversity value for conservation. Annual Review of Ecology and Systematics 26:93–111.

Humphries, W.J. 1999. Ecology and population demography of the hellbender, *Cryptobranchus alleganiensis*, in West Virginia. Master's thesis. Marshall University, Huntington, West Virginia.

Humphries, W.J. and T.K. Pauley. 2000. Seasonal changes in nocturnal activity of the hellbender, *Cryptobranchus alleganiensis*, in West Virginia. Journal of Herpetology 34:604–607.

Hunsaker, D., II and P. Breese. 1967. Herpetofauna of the Hawaiian Islands. Pacific Science 21:423–428.

Hunsaker, D. and F.E. Potter Jr. 1960. "Red leg" in a natural population of amphibians. Herpetologica 16:285–286.

Hunt, R.H. 1980. Toad sanctuary in a tarantula burrow. Natural History 89:48–53.

Hunter, B.R., D.L. Carlson, E.D. Seppanen, P.S. Killian, B.K. McKinnell and R.G. McKinnell. 1989. Are renal carcinomas increasing in *Rana pipiens* after a decade of reduced prevalence? American Midland Naturalist 122:307–312.

Hunter, G.W. and W. Hunter. 1934. Studies on fish and bird parasites. Supplement. Pp. 230–234. *In* New York State Conservation Department, 23rd Annual Report, 1933, Albany, New York.

Hunter, M.G. 1998. Watershed-level patterns among stream amphibians in the Blue River watershed, west-central Cascades of Oregon. Master's thesis. Oregon State University, Corvallis, Oregon.

Hunter, M.L., Jr., J. Albright and J. Arbuckle (Eds.). 1992. The amphibians and reptiles of Maine. Bulletin of the Maine Agricultural Experiment Station, Number 838, Orono, Maine.

Hunter, M.L., Jr., A.J.K. Calhoun and M. McCollough (Eds.). 1999. Maine Amphibians and Reptiles. University of Maine Press, Orono, Maine.

Huntington, C., T. Stuhlman and D.J. Cullen. 1993. *Plethodon sequoyah*. Pp. 557.1–557.2. Catalogue of American Amphibians and Reptiles. Society for the Study of Amphibians and Reptiles, St. Louis, Missouri.

Hupf, T.H. 1977. Natural histories of two species of leopard frogs, *Rana blairi* and *Rana pipiens*, in a zone of sympatry in northeastern Nebraska. Master's thesis. University of Nebraska, Lincoln, Nebraska.

Hurlbert, S.H. 1969. The breeding migrations and interhabitat wandering of the vermilion-spotted newt, *Notophthalmus viridescens* (Rafinesque). Ecological Monographs 39:465–488.

Hurlbert, S.H. 1970a. Predator responses to the vermilion-spotted newt *(Notophthalmus viridescens)*. Journal of Herpetology 4:47–55.

Hurlbert, S.H. 1970b. The post-larval migration of the red-spotted newt *Notophthalmus viridescens* (Rafinesque). Copeia 1970:515–528.

Hurlbert, S.H. 1984. Pseudoreplication and the design of ecological field experiments. Ecological Mongraphs 54:187–211.

Hurter, J., Sr. 1911. Herpetology of Missouri. Transactions of the Academy of Science of St. Louis 20:59–274.

Husak, J.F. 1998. Geographic distribution: *Eleutherodactylus marnockii* (cliff chirping frog). Herpetological Review 29:48.

Husting, E.L. 1965. Survival and breeding structure in a population of *Ambystoma maculatum*. Copeia 1965:352–362.

Hutcherson, J.E., C.L. Peterson and R.F. Wilkinson. 1989. Reproductive and larval biology of *Ambystoma annulatum*. Journal of Herpetology 23:181–183.

Hutchison, V.H. 1956. Notes of the plethodontid salamanders, *Eurycea lucifuga* (Rafinesque) and *Eurycea longicauda longicauda* Green. Occasional Papers National Speleological Society, Number 3, Huntsville, Alabama.

Hutchison, V.H. 1958. The distribution and ecology of the cave salamander, *Eurycea lucifuga*. Ecological Monographs 28:1–20.

Hutchison, V.H. 1961. Critical thermal maxima in salamanders. Physiological Zoology 34:92–125.

Hutchison, V.H. 1966. *Eurycea lucifuga*. Pp. 24.1–24.2. Catalogue of American Amphibians and Reptiles. Society for the Study of Amphibians and Reptiles, St. Louis, Missouri.

Hutchison, V.H. and R.K. Dupré. 1992. Thermoregulation. Pp. 206–249. *In* Feder, M.E. and W.W. Burggren (Eds.), Environmental Physiology of the Amphibians. University of Chicago Press, Chicago, Illinois.

Hutchison, V.H. and M.R. Ferrance. 1970. Thermal tolerances of *Rana pipiens* acclimated to daily temperature cycles. Herpetologica 26:1–8.

Hutchison, V.H. and C.S. Hammen. 1958. Oxygen utilization in the symbiosis of embryos of the salamander, *Ambystoma maculatum*, and the alga, *Oophila ambylstomatis*. Biological Bulletin 155:438–489.

Hutchison, V.H. and E.S. Hazard III. 1984. Erythrocytic organic phosphates: diel and seasonal cycles in the frog, *Rana berlandieri*. Comparative Biochemistry and Physiology A 79:533–538.

Hutchison, V.H., W.G. Whitford and M. Kohl. 1968. Relation of body size and surface area to gas exchange in anurans. Physiological Zoology 41:65–85.

Hyatt, A., C. Musso and G. Lopez. 1995. The isolation, identification and characterisation of viruses that might serve as bio-control agents against *B. marinus* in Australia. CSIRO *Bufo* Project: An Overview of Research Outcomes, CSIRO, Unpublished report, Australian Animal Health Laboratory, CSIRO, Geelong, Australia.

Hyde, E.J. 2000. Assessing the diversity and abundance of salamanders in Great Smoky Mountains National Park. Master's thesis. North Carolina State University, Raleigh, North Carolina.

Ideker, J. 1976. Tadpole thermoregulatory behavior facilitates grackle predation. Texas Journal of Science 27:244–245.

Ideker, J. 1979. Adult *Cybister fimbriolatus* are predaceous (Coleoptera: Dytiscidae). Coleopterists Bulletin 33:41–44.

Ildos, A.S. and N. Ancona. 1994. Analysis of amphibian habitat preferences in a farmland area (Po Plain, northern Italy). Amphibia-Reptilia 15:307–316.

Inger, R. and H. Marx. 1961. The food of amphibians. Exploration du Parc National de l'Upemba. Mission G. F. De Witte, Fascicule 64:3–85.

Inger, R.F. 1962. On the terrestrial origin of frogs. Copeia 1962:835–836.

Inger, R.F. 1994. Microhabitat description. Pp. 60–65. *In* Heyer,W.R., M.A. Donnelly, R.W. McDiarmid, L.C. Hayek and M.S. Foster (Eds.), Measuring and Monitoring Biological Diversity: Standard Methods for Amphibians. Smithsonian Institution Press, Washington, D.C.

Ingermann, R.L., D.C. Bencic and V.P. Eroschenko. 1997. Methoxychlor alters hatching and larval startle response in the salamander *Ambystoma macrodactylum*. Bulletin of Environmental Contamination and Toxicology 59:815–821.

Ingersol, C.A., R.F. Wilkinson, C.L. Peterson and R.H. Ingersol. 1991. Histology of the reproductive organs of *Cryptobranchus alleganiensis* (Caudata: Cryptobranchidae) in Missouri. Southwestern Naturalist 36:60–66.

Ingle, D. and D. McKinley. 1978. Effects of stimulus configuration on elicited prey catching by the marine toad (*Bufo marinus*). Animal Behaviour 26:885–891.

Ingles, L.G. 1932a. Four new species of *Heamatoloechus* (Trematoda) from California. University of California Publications in Zoology 37:189–201.

Ingles, L.G. 1932b. *Cephalogonimus brevicirrus*, a new species of trematode from the intestine of *Rana aurora* from California. University of California Publications in Zoology 37:203–210.

Ingles, L.G. 1933a. The specificity of frog flukes. Science 78:168.

Ingles, L.G. 1933b. Studies on the structure and life-history of *Zeugorchis syntomentera* Sumwalt a trematode from the snake *Thamnophis ordinoides* from California. University of California Publications in Zoology 39:163–176.

Ingles, L.G. 1933c. Studies on the structure and life-history of *Ostiolum oxyorchis* (Ingles) from the California red-legged frog *Rana aurora draytoni*. University of California Publications in Zoology 39:135–162.

Ingles, L.G. 1935. Notes on the development of a heterophyid trematode. Transactions of the American Microscopical Society 54:19–21.

Ingles, L.G. 1936. Worm parasites of California Amphibia. Transactions of the American Microscopical Society 55:73–92.

Ingram, G.J. and K.R. McDonald. 1993. An update on the decline of Queensland's frogs. Pp. 297–303. *In* Lunney, D. and D. Ayers (Eds.), Herpetology in Australia: A Diverse Discipline. Surrey Beatty and Sons, Sydney, New South Wales, Australia.

Ingram, W.M. and E.C. Raney. 1943. Additional studies on the movement of tagged bullfrogs, *Rana catesbeiana* Shaw. American Midland Naturalist 29:239–241.

Institute of Medicine (IOM), Board of International Health. 1997. America's Vital Interest in Global Health. National Academy Press, Washington, D.C.

Integrated Taxonomic Information System (IT IS) Database. 2002. http://www.itis.usda.gov/itis_query.html.

International Association of Microbiological Societies. 1992. International Code of Nomenclature of Bacteria, and Statutes of the International Committee on Systematic Bacteriology, and Statutes of the Bacteriology and Applied Microbiology Section of the International Union of Microbiological Societies: Bacteriological Code, 1990 revision. American Society of Microbiology, Washington, D.C.

International Botanical Congress. 2000. International Code of Botanical Nomenclature. Königstein: Koeltz Scientific Books. July–August 1999. Edition adopted by the 16th International Botanical Congress, St. Louis, Missouri.

International Commission on Zoological Nomenclature. 1999. International Code of Zoological Nomenclature. 4th edition. International Trust for Zoological Nomenclature, London, England.

Ireland, P.H. 1973. Overwintering of larval spotted salamanders, *Ambystoma maculatum* (Caudata) in Arkansas. Southwestern Naturalist 17:435–437.

Ireland, P.H. 1974. Reproduction and larval development of the dark-sided salamander, *Eurycea longicauda melanopleura*. Herpetologica 30:338–343.

Ireland, P.H. 1976. Reproduction and larval development of the gray-bellied salamander *Eurycea multiplicata grisegaster*. Herpetologica 32:233–238.

Ireland, P.H. 1979. *Eurycea longicauda*. Pp. 221.1–221.4. Catalogue of American Amphibians and Reptiles. Society for the Study of Amphibians and Reptiles. St. Louis, Missouri.

Ireland, P.H. 1989. Larval survivorship in two populations of *Ambystoma maculatum*. Journal of Herpetology 23:209–215.

Ireland, P.H. 1991. Separate effects of acid-derived anions and cations on growth of larval salamanders of *Ambystoma maculatum*. Copeia 1991:132–137.

Irwin, J.T., J.P. Costanzo and R.E. Lee. 1999. Terrestrial overwintering in the northern cricket frog, *Acris crepitans*. Canadian Journal of Zoology 77:1240–1246.

Irwin, K.J. 1993. A preliminary survey, with management recommendations on the herpetofauna of the Lower Rio Grande Valley National Wildlife Refuge. U.S. Fish Wildlife Service, Final report, Alamo, Texas.

Issacs, J.S. 1971. Temporal stability of vertebral stripe color in a cricket frog population. Copeia 1971:551–552.

IUCN (World Conservation Union). 1980. World Conservation Strategy. Gland, Switzerland.

Iverson, J.B. 1992. Global correlates of species richenss in turtles. Herpetological Journal 2:77–81.

Iverson, J.B. and C.R. Etchberger. 1989. The distributions of the turtles in Florida. Florida Scientist 52:119–144.

Jackman, T.R. 1993. Evolutionary and historical analyses within and among members of the salamander tribe Plethodontini (Amphibia: Plethodontidae). Ph.D. dissertation. University of California at Berkeley, Berkeley, California.

Jackman, T.R. 1999 (1998). Molecular and historical evidence for the introduction of clouded salamanders (genus *Aneides*) to Vancouver Island, British Columbia, Canada, from California. Canadian Journal of Zoology 76:1570–1580.

Jackman, T.R. and D.B. Wake. 1994. Evolutionary and historical analysis of protein variation in the blotched forms of salamanders of the *Ensatina* complex (Amphibia: Plethodontidae). Evolution 43:876–897.

Jackman, T.R., G. Applebaum and D.B. Wake. 1997. Phylogenetic relationships of bolitoglossine salamanders: a demonstration of the effects of combining morphological and molecular data sets. Molecular Biology and Evolution 14:883–891.

Jackson, D.R. and E.G. Milstrey. 1989. The fauna of gopher tortoise burrows. Pp. 86–98. *In* Diemer, J.E., D.R. Jackson, J.L. Landers, J.N. Layne and D.A. Wood (Eds.), Gopher Tortoise Relocation Symposium Proceedings. Florida Game and Fresh Water Fish Commission, Nongame Wildlife Program, Technical Report, Number 5, Tallahassee, Florida.

Jackson, H.H.T. 1914. The land vertebrates of Ridgeway Bog, Wisconsin: their ecological succession and source of ingression. Bulletin of the Wisconsin Natural History Society 11:4–54.

Jackson, L.E. 1989. Mountain Treasures at Risk: The Future of the Southern Appalachian National Forests. Wilderness Society, Washington, D.C.

Jackson, M. 1992. Distribution and habitat characteristics of ranid frogs in the El Paso area. Froglog: Newsletter of the Declining Amphibian Populations Task Force 4:2–3.

Jackson, M.E. and R.D. Semlitsch. 1993. Paedomorphosis in the salamander *Ambystoma talpoideum*: effects of a fish predator. Ecology 74:342–350.

Jackson, M.E., D.E. Scott and R.A. Estes. 1989. Determinants of nest success in the marbled salamander *(Ambystoma opacum)*. Canadian Journal of Zoology 67:2277–2281.

Jackson, S.D. 1996. Underpass systems for amphibians. Pp. 1–4. *In* Evink, G.L., P. Garrett, D. Zeigler and J. Berry (Eds.), Proceedings of the Transportation Related Wildlife Mortality Seminar. Florida Department of Transportation, FL-ER-58-9, Tallahassee, Florida.

Jackson, S.D. 1999. Overview of transportation related wildlife problems. Pp. 1–4. *In* Evink, G.L., P. Garrett and D. Zeigler (Eds.), Proceedings of the Third International Conference on Wildlife Ecology and Transportation. Florida Department of Transportation, FL-ER-73-99, Tallahassee, Florida.

Jackson, S.D. and T.F. Tyning. 1989. Effectiveness of drift fences and tunnels for moving spotted salamanders *Ambystoma maculatum* under roads. Pp. 93–99. *In* Lanston, T.E.S. (Ed.), Amphibians and Roads. ACO (Ahlmann Company) Polymer Products, Limited, Shefford, United Kingdom.

Jacobs, A.J. and D.H. Taylor. 1992. Chemical communication between *Desmognathus quadrimaculatus* and *Desmognathus monticola*. Journal of Herpetology 26:93–95.

Jacobs, D.L. 1950. *Pseudacris nigrita triseriata* on the north shore of Lake Superior. Copeia 1950:154.

Jacobs, J.F. 1987. A preliminary investigation of geographic variation and systematics of the two-lined salamander, *Eurycea bislineata* (Green). Herpetologica 43:423–446.

Jacobs, J.F. and W.R. Heyer. 1994. Collecting tissue for biochemical analysis. Pp. 299–301. *In* Heyer, W.R., M.A. Donnelly, R.W. McDiarmid, L.-A.C. Hayek and M.S. Foster (Eds.), Measuring and Monitoring Biological Diversity. Smithsonian Institution Press, Washington, D.C.

Jacobson, N.L. 1989. Breeding dynamics of the Houston toad. Southwestern Naturalist 34:374–380.

Jacot, A.P. 1936. Spruce litter reduction. Canadian Entomologist 69:31–32.

Jacqmotte, N.J. 1992. The home range and population biology of the Oregon salamander, *Ensatina eschscholtzii oregonensis*, in the Reed College Canyon. Undergraduate thesis. Reed College, Portland, Oregon.

Jaeger, J.R., B.R. Riddle, R.D. Jennings and D.F. Bradford. 2001. Rediscovering *Rana onca*: evidence for phylogenetically distinct leopard frogs from the border region of Nevada, Utah, and Arizona. Copeia 2001:339–354.

Jaeger, R.G. 1970. Potential extinction through competition between two species of terrestrial salamanders. Evolution 24:632–642.

Jaeger, R.G. 1971a. Moisture as a factor influencing the distributions of two species of terrestrial salamanders. Oecologia 6:191–207.

Jaeger, R.G. 1971b. Competitive exclusion as a factor influencing the distributions of two species of terrestrial salamanders. Ecology 52:632–637.

Jaeger, R.G. 1972. Food as a limited resource in competition between two species of terrestrial salamanders. Ecology 53:535–546.

Jaeger, R.G. 1974a. Interference or exploitation? A second look at competition between salamanders. Journal of Herpetology 8:191–194.

Jaeger, R.G. 1974b. Competitive exclusion: comments on survival and extinction of species. BioScience 24:33–39.

Jaeger, R.G. 1978. Plant climbing by salamanders: periodic availability of plant-dwelling prey. Copeia 1978:686–791.

Jaeger, R.G. 1979. Seasonal spatial distributions of the terrestrial salamander *Plethodon cinereus*. Herpetologica 35:90–93.

Jaeger, R.G. 1980a. Density-dependent and density-independent causes of extinction of a salamander population. Evolution 34:617–621.

Jaeger, R.G. 1980b. Microhabitats of a terrestrial forest salamander. Copeia 1980:265–268.

Jaeger, R.G. 1981a. Birds as inefficient predators on terrestrial salamanders. American Naturalist 117:835–837.

Jaeger, R.G. 1981b. Diet diversity and clutch size of aquatic and terrestrial salamanders. Oecologia 48:190–193.

Jaeger, R.G. 1984. Agonistic behavior of the red-backed salamander. Copeia 1984:309–314.

Jaeger, R.G. 1988. A comparison of territorial and non-territorial behaviour in two species of salamander. Animal Behaviour 36:307–310.

Jaeger, R.G. 1994. Transect sampling. Pp. 103–107. *In* Heyer, W.R., M.A. Donnelly, R.W. McDiarmid, L.-C. Hayek and M.S. Foster (Eds.), Measuring and Monitoring Biological Diversity: Standard Methods for Amphibians. Smithsonian Institution Press, Washington, D.C.

Jaeger, R.G. and D.C. Forester. 1993. Social behavior of plethodontid salamanders. Herpetologica 49:163–175.

Jaeger, R.G. and W.F. Gergits. 1979. Intra- and interspecific communication in salamanders through chemical signals on the substrate. Animal Behaviour 27:150–156.

Jaeger, R.G. and S.C. Walls. 1989. On salamander guilds and ecological methodology. Herpetologica 45:111–119.

Jaeger, R.G., D. Kalvarsky and N. Shimizu. 1982. Territorial behaviour of the red-backed salamander: expulsion of intruders. Animal Behaviour 30:490–496.

Jaeger, R.G., D. Fortune, G. Hill, A. Palen and G. Risher. 1993. Salamander homing behavior and territorial pheromones: alternative hypotheses. Journal of Herpetology 27:236–239.

James, H.A. 1969. Studies on the genus *Mesocestoides* Cestoda Cyclophyllidea. Dissertation Abstracts B, Science and Engineering 29:3541-B.

James, J.D. 1998. Status of the Great Plains toad *(Bufo cognatus)* in Alberta. Alberta Environmental Protection, Wildlife Management Division, and Alberta Conservation Association, Wildlife Status Report, Number 14, Edmonton, Alberta.

James, M.T. and T.P. Maslin. 1947. Notes on myiasis of the toad *Bufo boreas boreas*. Journal of the National Academy of Sciences, Washington D.C. 37:366–368.

James, P. 1966. The Mexican burrowing toad, *Rhinophrynus dorsalis*, an addition to the vertebrate fauna of the United States. Texas Journal of Science 18:272–276.

Jameson, D.L. 1950a. The breeding and development of Strecker's chorus frog in central Texas. Copeia 1950:61.

Jameson, D.L. 1950b. The development of *Eleutherodactylus latrans*. Copeia 1950:44–46.

Jameson, D.L. 1952. Local variation, population structure, home range, environmental relations, and breeding behavior in natural populations of the cliff chirping frog, *Syrrhophus marnocki* Cope. Ph.D. dissertation. University of Texas, Austin, Texas.

Jameson, D.L. 1954. Social patterns in the leptodactylid frogs *Syrrhophus* and *Eleutherodactylus*. Copeia 1954:36–38.

Jameson, D.L. 1955. The population dynamics of the cliff frog, *Syrrhophus marnocki*. American Midland Naturalist 54:342–381.

Jameson, D.L. 1956a. Survival of some central Texas frogs under natural conditions. Copeia 1956:55–57.

Jameson, D.L. 1956b. Growth, dispersal and survival of the Pacific tree frog. Copeia 1956:25–29.

Jameson, D.L. 1957. Population structure and homing responses in the Pacific tree frog. Copeia 1957:221–228.

Jameson, D.L. and A.G. Flurry. 1949. Reptiles and amphibians of the Sierra Vieja Range of Southwestern Texas. Texas Journal of Science 1:54–79.

Jameson, D.L. and S. Pequegnat. 1971. Estimation of relative viability and fecundity of color polymorphisms in anurans. Evolution 25:180–194.

Jameson, E.W., Jr. 1944. Food of the red-backed salamander. Copeia 1944:145–147.

Jameson, E.W., Jr. 1947. Food of the western cricket frog. Copeia 1947:212.

Jamieson, D.H. and S.E. Trauth. 1996. Dietary diversity and overlap between two subspecies of spadefoot toads (*Scaphiopus holbrookii holbrookii* and *S. h. hurterii*) in Arkansas. Proceedings of the Arkansas Academy of Science 50:75–78.

Jamieson, D.H., S.E. Trauth and C.T. McAllister. 1993. Food habits of male bird-voiced treefrogs, *Hyla avivoca* (Anura: Hylidae), in Arkansas. Texas Journal of Science 45:45–49.

Jancovich, J.K., E.W. Davidson, J.F. Morado, B.L. Jacobs and J.P. Collins. 1997. Isolation of a lethal virus from the endangered tiger salamander *Ambystoma tigrinum stebbinsi*. Diseases of Aquatic Organisms 31:161–167.

Janzen, D.H. 1994. Priorities in tropical biology. TRENDS in Ecology and Evolution 9:365–368.

Janzen, F.J. and E.D. Brodie Jr. 1989. Tall tales and sexy males: sexual behavior of rough-skinned newts *(Taricha granulosa)* in a natural breeding pond. Copeia 1989:1068–1071.

Jarhrsdoerfer, S.E. and D.M. Leslie Jr. 1988. Tamaulipan brushland of the lower Rio Grande valley of south Texas: description, human impacts, and management options. U.S. Fish and Wildlife Service, Biological Report, Number 88(36), Washington, D.C.

Jaskula, J.M. and R. Brodman. 2000. Density-dependent effects on microhabitat selection and activity of two species of larval *Ambystoma* salamanders. Pp. 146–150. *In* Kaiser, H., G.S. Casper and N.P. Bernstein (Eds.), Investigating Amphibian Declines: Proceedings of the 1998 Declining Amphibians Conference. Journal of the Iowa Academy of Science, Volume 107, Cedar Falls, Iowa.

Jaslow, A.P. and R.C. Vogt. 1977. Identification and distribution of *Hyla versicolor* and *Hyla chrysoscelis* in Wisconsin. Herpetologica 33:201–205.

Jay, J.M. and W.J. Pohley. 1981. *Dermosporidium penneri* sp. n. from the skin of the American toad, *Bufo americanus* (Amphibia: Bufonidae). Journal of Parasitology 67:108–110.

Jenkens, C.L. K. McGarigal and L.R. Gamble. 2002. A comparison of aquatic surveying techniques used to sample *Ambystoma opacum* larvae. Herpetological Review 33:33–35.

Jenni, D.A. 1969. A study of the ecology of four species of herons during the breeding season at Lake Alice, Alachua County, Florida. Ecological Monographs 39:245–270.

Jennings, D.T., H.S. Crawford Jr. and M.L. Hunter Jr. 1991. Predation by amphibians and small mammals on the spruce budworm (Lepidoptera: Tortricidae). Great Lakes Entomologist 24:69–74.

Jennings, M. and A. Anderson. 1997. The Wyoming toad. Endangered Species Bulletin 22:16–17.

Jennings, M. and M. Hayes. 1994. Amphibian and reptile species of special concern in California. California Department of Fish and Game, Sacramento, California.

Jennings, M., R. Beiswinger, S. Corn, M. Parker, A Pessier, B. Spencer and P.S. Miller (Eds.). 2001. Population and habitat viability assessment for the Wyoming toad *(Bufo baxteri)*. Final Workshop Report. World Conservation Union (IUCN)/Species Survival Commission (SSC) Conservation Breeding Specialist Group, Apple Valley, Minnesota.

Jennings, M.R. 1982. Geographic distribution: *Batrachoseps attenuatus* (California slender salamander). Herpetological Review 13:130.

Jennings, M.R. 1987a. Annotated check list of the amphibians and reptiles of California. Southwest Herpetologists Society, Special Publication, Number 3, Second edition, Van Nuys, California.

Jennings, M.R. 1987b. Life history notes: *Rana catesbeiana* (bullfrog). Feeding. Herpetological Review 18:33.

Jennings, M.R. 1988a. *Rana onca*. Pp. 417.1–417.2. Catalogue of American Amphibians and Reptiles. Society for the Study of Amphibians and Reptiles, St. Louis, Missouri.

Jennings, M.R. 1988b. Natural history and decline of native ranid frogs in California. Pp. 61–72. *In* De Lisle, H.F., P.R. Brown, B. Kaufman and B.M. McGurty (Eds.), Proceedings of the Conference on California Herpetology. Southwestern Herpetologists Society, Special Publication, Number 4, Van Nuys, California.

Jennings, M.R. 1994. Use of unverified museum databases for land management decisions: the case of native California frogs. Annual Meeting of American Society of Ichthyologists and Herpetologists, Seattle, Washington.

Jennings, M.R. 1995. Native ranid frogs in California. Pp. 131–134. *In* LaRoe, E.T., G.S. Farris, C.E. Puckett, P.D. Doran and M.J. Mac (Eds.), Our Living Resources: A Report to the Nation on the Distribution, Abundance, and Health of U.S. Plants, Animals, and Ecosystems. U.S. Department of the Interior, National Biological Service, Washington, D.C.

Jennings, M.R. 1996a. Natural history notes: *Ambystoma californiense* (California tiger salamander). Burrowing ability. Herpetological Review 27:194.

Jennings, M.R. 1996b. Status of amphibians. Pp. 921–944. *In* Sierra Nevada Ecosystem Project: Final Report to Congress, Volume II, Assessments and scientific basis for management options. University of California, Davis, Centers for Water and Wildland Resources, Davis California.

Jennings, M.R. and D. Cook. 1998. Natural history notes: *Taricha torosa torosa* (Coast Range newt). Predation. Herpetological Review 29:230.

Jennings, M.R. and M.P. Hayes. 1985. Pre-1900 overharvest of the California red-legged frog *(Rana aurora draytonii)*: the inducement for bullfrog *(Rana catesbeiana)* introduction. Herpetologica 41:94–103.

Jennings, M.R. and M.P. Hayes. 1989. Final report of the status of the California red-legged frog *(Rana aurora draytonii)* in the Pescadero Marsh Natural Preserve. Report for the California Department of Parks and Recreation, Contract Number 4-823-9018, Sacramento, California.

Jennings, M.R. and M.P. Hayes. 1994a. Amphibian and reptile species of special concern in California. California Department of Fish and Game, Inland Fisheries Division, Final report, Rancho Cordova, California.

Jennings, M.R. and M.P. Hayes. 1994b. Decline of native ranid frogs in the desert southwest. Pp. 183–211. *In* Brown, P.R. and J.W. Wright (Eds.), Herpetology of the North American Deserts, Special Publication, Number 5, Southwestern Herpetologists Society, Van Nuys, California.

Jennings, R.D. 1987. The status of *Rana berlandieri*, the Rio Grande leopard frog, and *Rana yavapaiensis*, the lowland leopard frog, in New Mexico. New Mexico Department of Game and Fish, Final report, Santa Fe, New Mexico.

Jennings, R.D. 1988. Ecological studies of the Chiricahua leopard frog, *Rana chiricahuensis*, in New Mexico. Report to Share with Wildlife, New Mexico Department of Game and Fish, Santa Fe, New Mexico.

Jennings, R.D. 1995a. Investigations of recently viable leopard frog populations in New Mexico: *Rana chiricahuensis* and *Rana yavapaiensis*. Report to Endangered Species Program, New Mexico Department of Game and Fish, Santa Fe, New Mexico.

Jennings, R.D. 1995b. Survey for leopard frogs along the Virgin River and adjacent areas. Harry Reid Center for Environmental Studies, University of Nevada-Las Vegas, Las Vegas, Nevada.

Jennings, R.D. and N.J. Scott Jr. 1991. Global amphibian population declines: insights from leopard frogs in New Mexico. Report to Endangered Species Program/Share with Wildlife, New Mexico Department of Game and Fish, Santa Fe, New Mexico.

Jennings, R.D. and N.J. Scott Jr. 1993. Ecologically correlated morphological variation in tadpoles of the leopard frog, *Rana chiricahuensis*. Journal of Herpetology 27:285–293.

Jennings, R.J., B.R. Riddle and D.F. Bradford. 1995. Rediscovery of *Rana onca*, the relict leopard frog, in southern Nevada with comments on the systematic relationships of some southwestern leopard frogs *(Rana pipiens* complex) and the status of populations along the Virgin River. Harry Reid Center for Environmental Studies, University of Nevada at Las Vegas, Las Vegas, Nevada.

Jennings, W.B., D.F. Bradford and D.F. Johnson. 1992. Dependence of the garter snake *Thamnophis elegans* on amphibians in the Sierra Nevada of California. Journal of Herpetology 26:503–505.

Jense, G.K. and R.L. Linder. 1970. Food habits of badgers in eastern South Dakota. Proceedings of the South Dakota Academy of Science 49:37–41.

Jensen, J.B. 1991. The distribution of the Pine Barrens treefrog, *Hyla andersonii*, in Conecuh National Forest, Alabama. Auburn University, Unpublished report, Auburn, Alabama.

Jensen, J.B. 1994. Geographical distribution: *Pseudacris ocularis* (little grass frog). Herpetological Review 25:161.

Jensen, J.B. 1995. Monitoring dusky gopher frog breeding success and management effectiveness, Conecuh National Forest, Alabama. U.S.D.A. Forest Service, Unpublished report, Andalusia, Alabama.

Jensen, J.B. 1996. Natural history notes: *Bufo terrestris* (southern toad). Egg toxicity. Herpetological Review 27:138–139.

Jensen, J.B. 1998. Geographic distribution: *Ambystoma talpoideum* (mole salamander). Herpetological Review 29:244.

Jensen, J.B. 1999a. Pigeon Mountain salamander. P. 104. *In* Protected animals of Georgia. Georgia Department of Natural Resources, Social Circle, Georgia.

Jensen, J.B. 1999b. Striped newt *(Notophthalmus perstriatus)*. Pp. 102–103. *In* Johnson, T.W., J.C. Ozier, J.L. Bohannon, J.B. Jensen and C. Skelton (Eds.), Protected animals of Georgia. Georgia Department of Natural Resources, Social Circle, Georgia.

Jensen, J.B. 2000. Natural history notes: *Rana capito* (gopher frog). Predation. Herpetological Review 31:42.

Jensen, J.B. 2003. Natural history notes: *Ambystoma tigrinum* (tiger salamander). Predation. Herpetological Review 34:132–133.

Jensen, J.B. and L.V. LaClaire. 1995. Geographic distribution: *Rana capito*. Herpetological Review 26:106.

Jensen, J.B. and C. Waters. 1999. The "spring lizard" bait industry in the state of Georgia, USA. Herpetological Review 30:20–21.

Jensen, J.B. and M.R. Whiles. 2000. Diets of sympatric *Plethodon petraeus* and *Plethodon glutinosus*. Journal of the Elisha Mitchell Scientific Society 116:245–250.

Jensen, J.B., C.D. Camp and J.L. Marshall. 2002. Ecology and life history of the Pigeon Mountain salamander. Southeastern Naturalist 1:3–16.

Jensen, J.B., J.G. Palis and M.A. Bailey. 1995. Natural history notes: *Rana capito sevosa* (dusky gopher frog). Submerged vocalization. Herpetological Review 26:98.

Jenssen, T.A. 1967. Food habits of the green frog, *Rana clamitans*, before and during metamorphosis. Copeia 1967:214–218.

Jenssen, T.A. and W.D. Klimstra. 1966. Food habits of the green frog, *Rana clamitans*, in southern Illinois. American Midland Naturalist 76:169–218.

Jenssen, T.A. and W.B. Preston 1968. Behavioral responses of the male green frog, *Rana clamitans*, to its recorded call. Herpetologica 24:181–182.

Jewell, R.D. 1991. Life history, ecology, and morphology of the ravine salamander, *Plethodon richmondi* in northern West Virginia. Master's thesis. Marshall University, Huntington, West Virginia.

Jewell, R.D. and T.K. Pauley. 1995. Notes on the reproductive biology of the salamander *Plethodon richmondi* (Netting and Mittleman) in West Virginia. Herpetological Natural History 3:91–93.

Jewitt, S.G. 1936. Notes on the amphibians of the Portland, Oregon area. Copeia 1936:71–72.

Jia, X., J.A. Richards, W. Gessner (Ed.) and E.E. Ricken (Ed.). 1999. Remote Sensing Digital Image Analysis: An Introduction. Third edition. Springer-Verlag, Berlin, Germany.

Jiang, S. and D.L. Claussen. 1993. The effects of temperature on food passage time through the digestive tract in *Notophthalmus viridescens*. Journal of Herpetology 27:414–419.

Jilek, R. and R. Wolff. 1978. Occurrence of *Spinitectus gracilis* Ward and Magath 1916 (Nematoda: Spiruroidea) in the toad (*Bufo woodhousii fowleri*) in Illinois. Journal of Parasitology 64:619.

Jobson, H.G.M. 1940. Reptiles and amphibians from Georgetown County, South Carolina. Herpetologica 2:39–43.

Jockusch, E.L. 1996. Evolutionary studies in *Batrachoseps* and other plethodontid salamanders: correlated character evolution, molecular phylogenetics, and reaction norm evolution. Ph.D. dissertation. University of California, Berkeley, California.

Jockusch, E.L. 1997a. An evolutionary correlate of genome size change in plethodontid salamanders. Proceedings of the Royal Society of London B 264:597–604.

Jockusch, E.L. 1997b. Geographic variation and phenotypic plasticity of number of trunk vertebrae in slender salamanders, *Batrachoseps* (Caudata: Plethodontidae). Evolution 51:1966–1982.

Jockusch, E.L. 2001. *Batrachoseps campi*. Pp. 722.1–722.2. Catalogue of American Amphibians and Reptiles. Society for the Study of Amphibians and Reptiles, St. Louis, Missouri.

Jockusch, E.L. and M.J. Mahoney. 1997. Communal oviposition and lack of parental care in *Batrachoseps nigriventris* (Caudata: Plethodontidae) with a discussion of the evolution of breeding behavior in plethodontid salamanders. Copeia 1997:697–705.

Jockusch, E.L. and D.B. Wake. 2002. Falling apart and merging: diversification of slender salamanders (Plethodontidae: *Batrachoseps*) in the American West. Biological Journal of the Linnean Society 76:361–391.

Jockusch, E.L., D.B. Wake and K.P. Yanev. 1998. New species of slender salamanders, *Batrachoseps* (Amphibia: Plethodontidae), from the Sierra Nevada of California. Natural History Museum, Los Angeles County, Contributions in Science, Number 472, Los Angeles, California.

Jockusch, E.L., K.P. Yanev and D.B. Wake. 2001. Molecular phylogenetic analysis of slender salamanders, genus Batrachoseps (Amphibia: Plethodontidae), from central coastal California with descriptions of four new species. Herpetological Monographs 15:54–99.

Jofre, M.B. and W.H. Karasov. 1999. Direct effect of ammonia on three species of North American anuran amphibians. Environmental Toxicology and Chemistry 18:1806–1812.

Joglar, R.L. 1998. Los coqíes de Puerto Rico su historia natural y conservación. Editorial de la Universidad de Puerto Rico, San Juan, Puerto Rico.

Joglar, R.L. and P.A. Burrowes. 1996. Declining amphibian populations in Puerto Rico. Pp. 371–380. In Powell, R. and R.W. Henderson (Eds.), Contributions to West Indian Herpetology. A Tribute to Albert Schwartz. Contributions to Herpetology, Volume 12, Society for the Study of Amphibians and Reptiles, St. Louis, Missouri.

John, K.R. and D. Fenster. 1975. The effects of partitions on the growth rates of crowded *Rana pipiens* larvae. American Midland Naturalist 93:123–130.

John-Alder, H.B. and P.J. Morin. 1990. Effects of larval density on jumping ability and stamina in newly metamorphosed *Bufo woodhousii fowleri*. Copeia 1990:856–860.

Johnson, B.K. and J.L. Christiansen. 1976. The food and food habits of Blanchard's cricket frog, *Acris crepitans blanchardi* (Amphibia, Anura, Hylidae) in Iowa. Journal of Herpetology 10:63–74.

Johnson, C. 1959. Genetic incompatibility in the call races of *Hyla versicolor* LeConte in Texas. Copeia 1959:327–335.

Johnson, C. 1961. Cryptic speciation in the *Hyla versicolor* complex. Ph.D. dissertation. University of Texas, Austin, Texas.

Johnson, C. 1963. Additional evidence of sterility between call-types in the *Hyla versicolor* complex. Copeia 1963:139–143.

Johnson, C. 1966. Species recognition in the *Hyla versicolor* complex. Texas Journal of Science 18:361–364.

Johnson, C.R. 1972. Thermal relations and daily variation in the thermal tolerance in *Bufo marinus*. Journal of Herpetology 6:35–38.

Johnson, C.R. and R.B. Bury. 1965. Food of the Pacific treefrog, *Hyla regilla* Baird and Girard, in northern California. Herpetologica 21:56–58.

Johnson, C.R. and C.B. Schreck. 1969. Food and feeding of larval *Dicamptodon ensatus* from California. American Midland Naturalist 81:280–281.

Johnson, D.H. and R.D. Batie. 1996. Surveys of calling amphibians in North Dakota. Northern Prairie Wildlife Research Center, Jamestown, North Dakota. http://www.npwc.usgs.gov/resource/distr/herps/amsurvey/amsurvey/htm.

Johnson, D.H., M.D. Bryant and A.H. Miller. 1948. Vertebrate animals of the Providence Mountains area of California. University of California Publications in Zoology 48:221–376.

Johnson, D.H., S.C. Fowle and J.A. Jundt. 2000. The North American Reporting Center for Amphibian Malformations. Pp. 123–127. *In* Kaiser, H., G.S. Casper and N.P. Bernstein (Eds.), Investigating Amphibian Declines: Proceedings of the 1998 Declining Amphibians Conference. Journal of the Iowa Academy of Science, Volume 107, Cedar Falls, Iowa.

Johnson, J. 1976. Memorandum to Robert A. Thomas, 10 December.

Johnson, J.A. and E.D. Brodie Jr. 1972. Aposematic function of the defensive behavior in the salamander *Taricha granulosa*. American Zoologist 12:647–648.

Johnson, J.A. and E.D. Brodie Jr. 1975. The selective advantage of the defensive posture of the newt *Taricha granulosa*. American Midland Naturalist 93:139–148.

Johnson, J.E. and A.S. Goldberg. 1975. Movement of larval two lined salamanders *(Eurycea bislineata)* in the Mill River, Massachusetts. Copeia 1975:588–589.

Johnson, L.M. 1991. Growth and development of larval northern cricket frogs, *Acris crepitans*, in relation to phytoplankton abundance. Freshwater Biology 25:51–60.

Johnson, M.J. 1984. The distribution and habitat of the pickerel frog in Wisconsin. Master's thesis. University of Wisconsin, Stevens Point, Wisconsin.

Johnson, P.T.J., K.B. Lunde, E.G. Ritchie and A.E. Launer. 1999. The effect of trematode infection on amphibian limb development and survivorship. Science 284:802–804.

Johnson, P.T.J., K.B. Lunde, R.W. Haight, J. Bowerman and A.R. Blaustein. 2001a. *Ribeiroia ondatrae* (Trematoda: Digena) infection induces severe limb malformations in western toads *(Bufo boreas)*. Canadian Journal of Zoology 79:370–379.

Johnson, P.T.J., K.B. Lunde, E.G. Ritchie, J.K. Reaser and A.E. Launer. 2001b. Morphological abnormality patterns in a California amphibian community. Herpetologica 57:336–352.

Johnson, P.T.J., K.B. Lunde, E.M. Thurman, E.G. Ritchie, S.N. Wray, D.R. Sutherland, J.M. Kapfer, T.J. Frest, J. Bowerman and A.R. Blaustein. 2002. Parasite *(Ribeiroia ondatrae)* infection linked to amphibian malformations in the western United States. Ecological Monographs 72:151–168.

Johnson, R.N., D.G. Young and J.F. Butler. 1993. Trypanosome transmission by *Corethrella wirthi* (Diptera: Chaoboridae) to the green treefrog, *Hyla cinerea* (Anura: Hylidae). Journal of Medical Entomology 30:918–921.

Johnson, R.R. and K.F. Higgins. 1997. Wetland resources of eastern South Dakota. Department of Wildlife and Fisheries Sciences, South Dakota State University, Brookings, South Dakota.

Johnson, R.R., D.E. Naugle, M.E. Estey and K.F. Higgins. 1996. Characteristics of eastern South Dakota wetland basins and implications of changes in jurisdictional wetland definitions. Transactions of the North American Wildlife and Natural Resources Conference 61:127–136.

Johnson, S.A. 1998. Natural history and reproductive ecology of the striped newt *(Notophthalmus perstriatus)* in north Florida. U.S. Fish and Wildlife Service, Endangered Species Office, Final report, Jackson, Mississippi.

Johnson, S.A. 2000. Conservation genetics of the striped newt *(Notophthalmus perstriatus)*. Final Report to Florida Fish and Wildlife Conservation Commission, Tallahassee, Florida.

Johnson, S.A. 2001. Life history, ecology, and conservation genetics of the striped newt *(Notophthalmus perstriatus)*. Ph.D. dissertation. University of Florida, Gainesville, Florida.

Johnson, S.A. 2002b. Life history of the striped newt at a north-central Florida breeding pond. Southeastern Naturalist 1:381–402.

Johnson, S.A. and N. Dwyer. 2000. Geographic distribution: *Notophthalmus perstriatus* (striped newt). Herpetological Review 31:249.

Johnson, S.R. 2003. Natural history notes: *Acris crepitans blanchardi* (Blanchard's cricket frog). Behavior. Herpetological Review 34:355.

Johnson, T.R. 1977. The Amphibians of Missouri. University of Kansas, University of Kansas Museum of Natural History Publication, Educational Series 6, Lawrence, Kansas.

Johnson, T.R. 1981. Observations on hellbender, *Cryptobranchus alleganiensis alleganiensis*, eggs and early larvae. Transactions of the Missouri Academy of Science 15:248.

Johnson, T.R. 1987. The Amphibians and Reptiles of Missouri. Missouri Department of Conservation, Jefferson City, Missouri.

Johnson, T.R. 1992. The Amphibians and Reptiles of Missouri. Missouri Department of Conservation, Jefferson City, Missouri.

Johnson, T.R. 1997. The Amphibians and Reptiles of Missouri. Missouri Department of Conservation, Jefferson City, Missouri.

Johnson, T.R. 1998a. Amphibian malformations in Missouri. P. 12. *In* Abstracts, Midwest Declining Amphibians Conference, March 20–21. Milwaukee Public Museum, Milwaukee, Wisconsin.

Johnson, T.R. 1998b. Missouri toad and frog calling survey: the first year. Pp. 357–359. *In* Lannoo, M.J. (Ed.), Status and Conservation of Midwestern Amphibians. University of Iowa Press, Iowa City, Iowa.

Johnson, T.R. 2000. The Amphibians and Reptiles of Missouri. Second edition. Missouri Department of Conservation, Jefferson City, Missouri.

Johnson, W.W. 1961. A life history of *Plethodon wehrlei* in the high plateaus. Ph.D. dissertation. St. Bonaventure University, St. Bonaventure, New York.

Johnston, B. 1998. Terrestrial Pacific giant salamanders *(Dicamptodon tenebrosus* Good)—natural history and their response to forest practices. Master's thesis. University of British Columbia, Vancouver, British Columbia.

Johnston, R.F. and G.A. Schad. 1959. Natural history of the salamander, *Aneides hardii*. University of Kansas Publications, Museum of Natural History 10:573–585.

Jones, D., E. Simandle, C.R. Tracy and B. Hobbs. 2003. Natural history notes: *Bufo nelsoni* (Amargosa toad). Predation. Herpetological Review 34:229.

Jones, F.L. and J.F. Cahlan. 1975. Water. A history of Las Vegas. Volume 1. Las Vegas Water District, Las Vegas, Nevada.

Jones, J.M. 1973. Effects of thirty years of hybridization on the toads *Bufo americanus* and *Bufo woodhousei fowleri* at Bloomington, Indiana. Evolution 27:435–448.

Jones, K.B. 1990. Habitat use and predatory behavior of *Thamnophis cyrtopsis* (Serpentes: Colubridae) in a seasonally variable aquatic environment. Southwestern Naturalist 35:115–122.

Jones, K.B., L. Porzer Kepner and W.G. Kepner. 1983. Anurans of Vekol Valley, central Arizona. Southwestern Naturalist 28:469–470.

Jones, L. 1994. The handling of dangerous reptiles and amphibians for U.S. Fish and Wildlife Service LE personnel, U.S. Government Printing Office, Washington, D.C.

Jones, L.L.C. 1984. Life history notes: *Aneides flavipunctatus flavipunctatus* (speckled black salamander). Behavior. Herpetological Review 15:17.

Jones, L.L.C. 1989. Life history notes: *Plethodon vandykei* (Van Dyke's salamander). Herpetological Review 20:48.

Jones, L.L.C. and J.B. Atkinson. 1989. New records of salamanders from Long Island, Washington. Northwestern Naturalist 70:40–42.

Jones, L.L.C. and K.B. Aubry. 1984. Geographic distribution: *Dicamptodon copei* (Cope's giant salamander). Herpetological Review 15:114.

Jones, L.L.C. and K.B. Aubry. 1985. Natural history notes: *Ensatina eschscholtzii oregonensis* (Oregon ensatina). Reproduction. Herpetological Review 16:26.

Jones, L.L.C. and R.B. Bury. 1983. Larch Mountain salamander surveys. U.S. Fish and Wildlife Service, Denver Wildlife Research Center, Unpublished report, Fort Collins, Colorado.

Jones, L.L.C. and P.S. Corn. 1989. Third specimen of a metamorphosed Cope's giant salamander *(Dicamptodon copei)*. Northwestern Naturalist 70:37–38.

Jones, L.L.C. and M.G. Raphael. 1998. Natural history notes: *Ascaphus truei* (tailed frog). Predation. Herpetological Review 29:39.

Jones, L.L.C. and M.G. Raphael. 2000. Diel patterns of surface activity and microhabitat use by stream-dwelling amphibians in the Olympic Peninsula. Northwestern Naturalist 81:78 [abstract].

Jones, L.L.C., R.B. Bury and P.S. Corn. 1990. Field observation of the development of a clutch of Pacific giant salamander *(Dicamptodon tenebrosus)* eggs. Northwestern Naturalist 71:93–94.

Jones, M.S. (Ed.). 2000. Boreal toad research progress report 1999. Colorado Division of Wildlife, Denver, Colorado.

Jones, M.S. 1999. Henderson/Urad boreal toad studies. Pp. 1–93. *In* Jones, M.S. (Ed.), Boreal Toad Research Progress Report, 1998. Colorado Division of Wildlife, Denver, Colorado.

Jones, M.S. and J.P. Goettl. 1998. Henderson/Urad boreal toad studies. Pp. 21–82. Colorado Division of Wildlife Boreal Toad Research Progress Report, 1995–1997. Denver, Colorado.

Jones, M.S., J.P. Goettl and L.J. Livo. 1999. Natural history notes: *Bufo boreas* (boreal toad). Predation. Herpetological Review 30:91.

Jones, R.L. 1981. Distribution and ecology of the seepage salamander, *Desmognathus aeneus* Brown and Bishop (Amphibia: Plethodontidae) in Tennessee. Brimleyana 7:95–100.

Jones, R.L. 1986. Reproductive biology of *Desmognathus fuscus* and *Desmognathus santeetlah* in the Unicoi Mountains. Herpetologica 42:323–334.

Jones, R.M. 1980. Metabolic consequences of accelerated urea synthesis during seasonal dormancy of spadefoot toads, *Scaphiopus couchi* and *Scaphiopus multiplicatus*. Journal of Experimental Zoology 212:255–267.

Jones, R.M. 1982. Urea synthesis and osmotic stress in the terrestrial anurans *Bufo woodhousei* and *Hyla cadaverina*. Comparative Biochemistry and Physiology A 71:293–297.

Jones, T.M. 1980. A reevaluation of the salamander *Eurycea aquatica* Rose and Bush (Amphibia: Plethodontidae). Master's thesis. Auburn University, Auburn, Alabama.

Jones, T.R. and J.P. Collins. 1992. Analysis of a hybrid zone between subspecies of the tiger salamander *(Ambystoma tigrinum)* in central New Mexico, USA. Journal of Evolutionary Biology 5:375–402.

Jones, T.R., D.K. Skelly and E.E. Werner. 1993. Life history notes: *Ambystoma tigrinum tigrinum* (eastern tiger salamander). Developmental polymorphism. Herpetological Review 24:147–148.

Jones, T.R., J.P. Collins, T.D. Kocher and J.B. Mitton. 1988. Systematic status and distribution of *Ambystoma tigrinum stebbensi* Lowe (Amphibia: Caudata). Copeia 1988:621–635.

Jones, T.R., E.J. Routman, D.J. Begun and J.P. Collins. 1995. Ancestry of an isolated subspecies of salamander, *Ambystoma tigrinum stebbensi* Lowe: the evolutionary significance of hybridization. Molecular Phylogenetics and Evolution 4:194–202.

Jordan, D.S. 1888. A Manual of the Vertebrate Animals of the Northern United States; Including the District North and East of the Ozark Mountains, South of the Laurentian Hills, North of the Southern Boundary of Virginia, and East of the Southern Boundary of Virginia, and East of the Missouri River, Inclusive of Marine Species. Fifth edition, A.C. McClurg, Chicago.

Jordan, H.E. and E.E. Byrd. 1967. The developmental cycle of *Brachycoelium mesorchium* Byrd, 1937 (Trematoda: Brachycoeliinae). Transactions of the American Microscopical Society 86:67.

Jordan, J.R., Jr. 1975. Observations on the natural history and ecology of the Red Hills salamander, *Phaeognathus hubrichti* Highton (Caudata: Plethodontidae). Master's thesis. Auburn University, Auburn, Alabama.

Jordan, O.R., W.W. Byrd and D.E. Ferguson. 1968. Sun-compass orientation in *Rana pipiens*. Herpetologica 20:271–169.

Jordan, P., J.D. Christie and G.O. Unrau. 1980. Schistosomiasis transmission with particular reference to possible ecological and biological methods of control. Acta Tropica 37:95–135.

Jørgensen, C.B. 1983. Pattern of growth in a temperate zone anuran *(Bufo viridis* Laur.). Journal of Experimental Zoology 227:433–439.

Jørgensen, C.B. 1992. Growth and reproduction. Pp. 439–466. *In* Feder, M.E. and W.W. Burggren (Eds.), Environmental Physiology of the Amphibians, University of Chicago Press, Chicago, Illinois.

Jørgensen, C.B., L.O. Larsen and B. Lofts. 1979. Annual cycle of fat bodies and gonads in the toad *Bufo bufo bufo* (L.), compared with cycles in other temperate zone anurans. Biologiske Skrifter Danske Videnskabernes Selskab 22:1–37.

Joy, J.E. and C.A. Bunten. 1997. *Cosmocercoides variabilis* (Nematoda, Cosmocercoidea) populations in the eastern American toad, *Bufo a. americanus* (Salienta, Bufonidae), from western West Virginia. Journal of the Helminthological Society of Washington 64:102–105.

Joy, J.E. and B.T. Dowell. 1994. *Glypthelmins pennsylvaniensis* (Trematoda: Digenea) in the spring peeper, *Pseudacris c. crucifer* (Anura: Hylidae), from southwestern West Virginia. Journal of the Helminthological Society of Washington 61:227–229.

Joy, J.E., T.K. Pauley and M.L. Little. 1993. Prevalence and intensity of *Thelandros magnavulvaris* and *Omeia papillocauda* (Nematoda) in two species of desmognathine salamanders from West Virginia. Journal of the Helminthological Society of Washington 60:93–95.

Judd, F.W. 1985. Status of *Siren intermedia texana, Notophthalmus meridionalis,* and *Crotaphytus reticulatus.* U.S. Fish Wildlife Service, Final report, Cooperative Agreement Number 14-16-0002-81-923, Modification Number 1, Washington, D.C.

Judd, W.W. 1957. The food of the Jefferson's salamander, *Ambystoma jeffersonianum,* in Rondeau Park, Ontario. Ecology 38:77–81.

Judge, K.A., S.J. Swanson and R.J. Brooks. 2000. Natural history notes: *Rana catesbeiana* (bullfrog). Female vocalization. Herpetological Review 31:236–237.

Judge, K.A., J.C. MacDonald, S.J. Swanson and R.J. Brooks. 1999. Abundance, mortality and age distribution of three species of ranids in Algonquin Provincial Park, Ontario: are amphibians declining in undisturbed habitat? Abstract in Canadian Amphibian and Reptile Network, 1999 Conference, Quebec, Canada.

Judy, L. 1969. Meteorological factors controlling the emergence of the eastern spadefoot toad, *Scaphiopus holbrookii holbrookii.* Master's thesis. Marshall University, Huntington, West Virginia.

Jung, R.E. 1993. Blanchard's cricket frogs *(Acris crepitans blanchardi)* in southwest Wisconsin. Transactions of the Wisconsin Academy of Sciences, Arts and Letters 81:79–87.

Jung, R.E. and C.H. Jagoe. 1995. Effects of low pH and aluminum on body size, swimming performance, and susceptibility to predation of green tree frog *(Hyla cinerea)* tadpoles. Canadian Journal of Zoology 73:2171–2183.

Jung, R.E., S. Droege and J.R. Sauer. 1997. DISPro amphibian project: standardized monitoring methods for amphibians in national parks and associations in time and space between amphibian abundance and environmental stressors. Project Proposal from Patuxent Wildlife Research Center to Environmental Protection Agency, Laurel, Maryland.

Jung, R.E., S.W. Droege and J.R. Sauer. 1999. PRIMENet amphibian project at Shenandoah and Big Bend National Parks. U.S.G.S. Patuxent Wildlife Research Center, http://www.pwrc.usgs.gov/resshow/droege4rs/salprime.htm.

Jung, R.E., S.M. Claeson, J.E. Wallace and W.C. Welbourn Jr. 2001. Natural history notes: *Eleutherodactylus guttilatus* (spotted chirping frog), *Bufo punctatus* (red-spotted toad), *Hyla arenicolor* (canyon tree frog), and *Rana berlandieri* (Rio Grande leopard frog). Mite infestation. Herpetological Review 32:33–34.

Jung, R.E., W.L. Ward, C.O. King and L.A. Weir. 2000. Natural history notes: *Plethodon cinereus* (redback salamander). Predation. Herpetological Review 31:98–99.

Justus, J.T., M. Sandomir, T. Urquhart and B.O. Ewan. 1977. Developmental rates of two species of toads from the desert southwest. Copeia 1977:592–594.

Juterbock, J.E. 1978. Sexual dimorphism and maturity characteristics of three species of *Desmognathus* (Amphibia, Urodela, Plethodontidae). Journal of Herpetology 12:217–230.

Juterbock, J.E. 1984. Evidence for the recognition of specific status for *Desmognathus welteri.* Journal of Herpetology 18:240–255.

Juterbock, J.E. 1986. The nesting behavior of the dusky salamander, *Desmognathus fuscus.* I. Nesting phenology. Herpetologica 42:457–471.

Juterbock, J.E. 1987. The nesting behavior of the dusky salamander, *Desmognathus fuscus.* II. Nest site tenacity and disturbance. Herpetologica 43:361–368.

Juterbock, J.E. 1989. *Aneides* (Dope and Packard), green salamander. Pp. 190–195. *In* Pfingsten, R.A. and F.L. Downs (Eds.), The Salamanders of Ohio. Ohio State University, Columbus, Ohio.

Juterbock, J.E. 1998. Population sizes of two endangered Ohio plethodontid salamanders, green salamanders and cave salamanders. Pp. 49–54. *In* Lannoo, M.J. (Ed.), Status and Conservation of Midwestern Amphibians. University of Iowa Press, Iowa City, Iowa.

Kagarise Sherman, C. 1980. A comparison of the natural history and mating system of two anurans: Yosemite toads *(Bufo canorus)* and black toads *(Bufo exsul).* Ph.D. dissertation. University of Michigan, Ann Arbor, Michigan.

Kagarise Sherman, C. and M.L. Morton. 1984. The toad that stays on its toes. Natural History 93:72–78.

Kagarise-Sherman, C. and M.L. Morton. 1993. Population declines of Yosemite toads in the eastern Sierra Nevada of California. Journal of Herpetology 27:186–198.

Kahl, M.P. 1966. A contribution to the ecology and reproductive biology of the Marabou stork *(Leptoptilos crumeniferus)* in East Africa. Journal of Zoology 148:289–311.

Kahl, M.P. 1967. Observations on the behavior of the hamerkop *Scopus umbretta* in Uganda. Ibis 109:25–32.

Kaiser, H., G.S. Casper and N.P. Bernstein (Eds.). 2000. Investigating amphibian declines: Proceedings of the 1998 Declining Amphibians Conference. Journal of the Iowa Academy of Science, Volume 107, Cedar Falls, Iowa.

Kaiser, J. 1997. Deformed frogs leap into spotlight at health workshop. Science 278:2051–2052.

Kaiser, J. 1999. Are pathogens felling frogs? Science 284:728–733.

Kalb, H.J. and G.R. Zug. 1990. Age estimates for a population of American toads, *Bufo americanus* (Salientia: Bufonidae), in northern Virginia. Brimleyana 16:79–86.

Kalela, O. 1962. On the fluctuations in the numbers of arctic and boreal small rodents as a problem of production biology. Annales Academiae Scientarium Fennicae 66:1–38.

Kane, A.S. and D.L. Johnson. 1989. Use of TFM (3-trifluoromethyl-4-nitrophenol) to selectively control frog larvae in fish production ponds. Progressive Fish-Culturist 51:207–213.

Kao, K. and M. Danilchik. 1991. Generation of body plan phenotypes in early embryogenesis. Methods in Cell Biology 36:271–284.

Kaplan, D.L. 1977. Exploitative competition in salamanders: test of a hypothesis. Copeia 1977:234–238.

Kaplan, H.M. and J.G. Overpeck. 1964. Toxicity of halogenated hydrocarbon insecticides for the frog, *Rana pipiens.* Herpetologica 20: 163–169.

Kaplan, R.H. 1980a. The implications of ovum size variability for offspring fitness and clutch size within several populations of salamanders *(Ambystoma).* Evolution 34:51–64.

Kaplan, R.H. 1980b. Ontogenetic energetics in *Ambystoma.* Physiological Ecology 53:43–56.

Kaplan, R.H. 1985. Maternal influences on offspring development in the California newt, *Taricha torosa.* Copeia 1985:1028–1035.

Kaplan, R.H. and M.L. Crump. 1978. The non-cost of brooding in *Ambystoma opacum.* Copeia 1978:99–103.

Kaplan, R.H. and S.N. Salthe. 1979. The allometry of reproduction: an empirical view in salamanders. American Naturalist 113:671–689.

Kaplan, R.H. and P.W. Sherman. 1980. Intraspecific oophagy in California newts. Journal of Herpetology 14:183–185.

Kareiva, P. 1989. Renewing the dialogue between theory and experiments in population ecology. Pp. 68–88. *In* Roughgarden, J., R.M. May and S.A. Levin (Eds.), Perspectives in Ecological Theory. Princeton University Press, Princeton, New Jersey.

Karl, S.A. and B.W. Bowen. 1999. Evolutionary significant units versus geopolitical taxonomy: molecular systematics of an endangered sea turtle (genus *Chelonia*). Conservation Biology 13:990–999.

Karlin, A.A. and S.I. Guttman. 1986. Systematics and geographic isozyme variation in the plethodontid salamander *Desmognathus fuscus* Rafinesque. Herpetologica 42:283–301.

Karlin, A.A. and D.B. Means. 1994. Genetic variation in the aquatic salamander genus *Amphiuma.* American Midland Naturalist 132:1–9.

Karlin, A.A. and R.A. Pfingsten. 1989. *Desmognathus fuscus* (Rafinesque), dusky salamander. Pp. 174–180. *In* Pfingsten, R.A. and F.L. Downs (Eds.). Salamanders of Ohio. Ohio Biological Survey Bulletin, New Series Volume 7, Number 2, Columbus, Ohio.

Karlin, A.A., S.I. Guttman and D.B. Means. 1993. Population structure in the Ouachita Mountain Dusky Salamander, *Desmognathus brimleyorum* (Caudata: Plethodontidae). Southwestern Naturalist 38:36–42.

Karlin, A.A., D.B. Means, S.I. Guttman and D. Lambright. 1982. Systematics and the status of *Hyla andersonii* (Anura: Hylidae) in Florida. Copeia 1982:175–178.

Karlstrom, E.L. 1962. The toad genus *Bufo* in the Sierra Nevada of California. Ecology and Systematic Relationships. University of California Publications in Zoology, Number 62, Berkeley, California.

Karlstrom, E.L. 1973. *Bufo canorus.* Pp. 132.1–132.2. Catalogue of American Amphibians and Reptiles. Society for the Study of Amphibians and Reptiles, St. Louis, Missouri.

Karlstrom, E.L. 1986. Amphibian recovery in the North Fork Toutle River debris avalanche area of Mt. St. Helens. Pp. 334–344. *In* Keller, S.A.C. (Ed.), Mount St. Helens: Five Years Later. Eastern Washington University Press, Cheney, Washington.

Karlstrom, E.L. and R.L. Livezey. 1955. The eggs and larvae of the Yosemite toad *Bufo canorus* Camp. Herpetologica 11:221–227.

Karns, D.R. 1992. Effects of acidic bog habitats on amphibian reproduction in a northern Minnesota peatland. Journal of Herpetology 26:401–412.

Karp, A. P.G. Isaac and D.S. Ingram. 1998. Molecular Tools for Screening Biodiversity. Chapman and Hall, London.

Karraker, N.E. 1999. Natural history notes: *Rhyacotriton variegatus* (southern torrent salamander). Nest site. Herpetological Review 30:160–161.

Karraker, N.E. 2001. Natural history notes: *Ascaphus truei* (tailed frog). Predation. Herpetological Review 32:100.

Karraker, N.E. and G.S. Beyersdorf. 1997. A tailed frog *(Ascaphus truei)* nest site in northwestern California. Northwestern Naturalist 78:110–111.

Kats, L.B. 1986. Natural history notes: *Nerodia sipedon* (northern watersnake). Feeding. Herpetological Review 17:61–62.

Kats, L.B. 1988. The detection of certain predators via olfaction by small-mouthed salamander larvae *(Ambystoma texanum)*. Behavioral and Neural Biology 50:126–131.

Kats, L.B. and A. Sih. 1992. Oviposition site selection and avoidance of fish by streamside salamanders *(Ambystoma barbouri)*. Copeia 1992:468–473.

Kats, L.B., J.A. Breeding and K.M. Hanson. 1994. Ontogenetic changes in California newts *(Taricha torosa)* in response to chemical cues from conspecific predators. Journal of the North American Benthological Society 13:321–325.

Kats, L.B., S.A. Elliott and J. Currens. 1992. Intraspecific oophagy in stream-breeding California newts *(Taricha torosa)*. Herpetological Review 23:7–8.

Kats, L.B., M.C. Linton and R.A. Linton. 1988. Geographic distribution: *Ambystoma maculatum* (spotted salamander). Herpetological Review 19:59–60.

Kats, L.B., J.W. Petranka and A. Sih. 1988. Antipredator defenses and persistence of amphibian larvae with fishes. Ecology 69:1865–1870.

Kats, L.B., J.M. Kiesecker, D.P. Chivers and A.R. Blaustein. 2000. Effects of UV-B radiation on anti-predator behavior in three species of amphibians. Ethology 106:921–931.

Kats, L.B., J.A. Goodsell, N. Matthews, K. Bahn and A. Blaustein. 1998. Natural history notes: *Taricha torosa* (California newt). Predation. Herpetological Review 29:230.

Kauffeld, C.F. 1957. Snakes and Snake Hunting. Hanover House, Garden City, New York.

Kaul, R.B. and S.B. Rolfsmeier. 1993. Native vegetation of Nebraska (Map 1:1,000,000). Conservation and Survey Department, Institute of Agriculture and Natural Resources, University of Nebraska–Lincoln, Lincoln, Nebraska.

Kautz, R. 1993. Trends in Florida wildlife habitat 1936–1987. Florida Scientist 56:7–24.

Kay, P.A. 1982. A perspective on Great Basin paleoclimates. Society for American Archeology 2:76–81.

Kay, W.K. 1989. Movements and homing in the canyon treefrog *(Hyla cadaverina)*. Southwestern Naturalist 34:293–295.

Kaye, J. 1995. Protester locks himself to tractor to halt park clearing. Bastrop Advertiser, 27 July.

Keen, W.H. 1975. Breeding and larval development of three species of *Ambystoma* in central Kentucky (Amphibia: Urodela). Herpetologica 31:18–21.

Keen, W.H. 1979. Feeding and activity patterns in the salamander *Desmognathus ochrophaeus* (Amphibia, Urodela, Plethodontidae). Journal of Herpetology 13:461–567.

Keen, W.H. 1982. Habitat selection and interspecific competition in two species of plethodontid salamanders. Ecology 63:94–102.

Keen, W.H. 1985. Habitat selection by two streamside plethodontid salamanders. Oecologica (Berlin) 66:437–442.

Keen, W.H. and L.P. Orr. 1980. Reproductive cycle, growth, and maturation of northern female *Desmognathus ochrophaeus*. Journal of Herpetology 14:7–10.

Keen, W.H. and R.W. Reed. 1985. Territorial defense of space and feeding sites by a plethodontid salamander. Animal Behaviour 33:1119–1123.

Keen, W.H. and E.E. Schroeder. 1975. Temperature selection and tolerance in three species of *Ambystoma* larvae. Copeia 1975:523–530.

Keen, W.H. and S. Sharp. 1984. Responses of a plethodontid salamander to conspecific and congeneric intruders. Animal Behaviour 32:58–65.

Keen, W.H., J. Travis and J. Juilianna. 1984. Larval growth in three sympatric *Ambystoma* salamander species: species differences and the effects of temperature. Canadian Journal of Zoology 62:1043–1047.

Keiser, E.D. and P.J. Conzelmann. 1969. The red-backed salamander, *Plethodon cinereus* (Green) in Louisiana. Journal of Herpetology 3:189–191.

Kelleher, K.E. and J.R. Tester. 1969. Homing and survival in the Manitoba toad, *Bufo hemiophrys* in Minnesota. Ecology 50:1040–1048.

Keller, C.M. and J.T. Scallan. 1999. Potential roadside biases due to habitat changes along breeding bird survey routes. Condor 101:50–57.

Kellner, A. and D.M. Green. 1995. Age structure and age of maturity in Fowler's toads, *Bufo woodhousii fowleri*, at their northern range limit. Journal of Herpetology 29:417–421.

Kellogg, R. 1932. Mexican tailless amphibians in the United States National Museum. Bulletin of the U.S. National Museum, Number 160, Washington, D.C.

Kelly, R.W. 1951. A preliminary ecological survey of a temporary pond in the Douglas fir forest association with emphasis on the food and feeding habits of the Oregon newt, *Triturus granulosus granulosus*. Master's thesis. University of Oregon, Eugene, Oregon.

Kelsey, K.A. 1995. Responses of headwater stream amphibians to forest practices in western Washington. Ph.D. dissertation. University of Washington, Seattle, Washington.

Kelso, J. 1995. Amphibian amenities show county's fondness for toads. Austin American-Statesman, 16 February.

Kelson, J., B. Beasley, E. Wind and K. Ovaska. 1999. Geographic distribution: *Ensatina eschscholtzii* (ensatina). Herpetological Review 30:171.

Kennedy, C.R. 1993. Introductions, spread and colonization of new localities by fish helminth and crustacean parasites in the British Isles: a perspective and appraisal. Journal of Fish Biology 43:287–301.

Kennedy, D. 1998. Environmental quality and regional conflict. Carnegie Commission on Preventing Deadly Conflict, Carnegie Corporation, New York.

Kennedy, J.P. 1961. Spawning season and experimental hybridization of the Houston toad, *Bufo houstonensis*. Herpetologica 17:239–245.

Kennedy, J.P. 1964. Experimental hybridization of the green treefrog *Hyla cinerea* Schneider (Hylidae). Zoologica 49:211–218.

Kennedy, M.J. 1980. Geographical variation in some representatives of *Haematoloechus looss*, 1899 (Trematoda: Haematoloechidae) from Canada and the U.S. Canadian Journal of Zoology 58:1151–1167.

Kennedy, M.J. 1981. A revision of the species of the genus *Haematoloechus* Looss, 1899 (Trematoda: Haematoloechidae) from Canada and the United States. Canadian Journal of Zoology 59:1836–1846.

Kennicott, R. 1855. Catalogue of animals observed in Cook County, Illinois. Transactions of the Illinois State Agricultural Society 1:591–593.

Kerby, J.L. and L.B. Kats. 1998. Modified interactions between salamander life stages caused by wildfire-induced sedimentation. Ecology 79:740–745.

Kern, W.H., Jr. 1986a. The range of the hellbender, *Cryptobranchus alleganiensis alleganiensis*, in Indiana. Proceedings of the Indiana Academy of Science 95:520–521.

Kern, W.H., Jr. 1986b. Reproduction of the hellbender, *Cryptobranchus alleganiensis*, in Indiana. Proceedings of the Indiana Academy of Science 95:521.

Kern, W.H., Jr. 1986c. Size class distribution of the hellbender, *Cryptobranchus alleganiensis*, in Indiana and its implications for their life history. Proceedings of the Indiana Academy of Science 95:521.

Kerster, H.W. 1964. Neighborhood size in the rusty lizard, *Sceloporus olivaceus*. Evolution 18:445–467.

Kesler, D.H. and W.R. Munns Jr. 1991. Diel feeding by adult red-spotted newts in the presence and absence of sunfish. Journal of Freshwater Ecology 6:267–273.

Kessel, E.L. and B.B. Kessel. 1943a. The rate of growth of the young larvae of the Pacific giant salamander, *Dicamptodon ensatus* (Eschscholtz). Wasmann Collector 5:108–111.

Kessel, E.L. and B.B. Kessel. 1943b. The rate of growth of older larvae of the Pacific giant salamander, *Dicamptodon ensatus* (Eschscholtz). Wasmann Collector 5:141–142.

Kessel, E.L. and B.B. Kessel. 1944. Metamorphosis of the Pacific giant salamander, *Dicamptodon ensatus* (Eschscholtz). Wasmann Collector 6:38–48.

Kezer, J. 1952b. The eggs of *Typhlotriton spelaeus* Stejneger, obtained by pituitary gland implementation. National Speleological Society 14:58–59.

Kezer, J. 1952a. Thyroxin-induced metamorphosis of the neotenic salamanders *Eurycea tynerensis* and *Eurycea neotenes*. Copeia 1952:234–237.

Kezer, J. and D.S. Farner. 1955. Life history patterns of the salamander *Ambystoma macrodactylum* in the high Cascade Mountains of southern Oregon. Copeia 1955:127–131.

Khan, R.A. and J. Thulin. 1991. Influence of pollution on parasites of aquatic animals. Pp. 201–238. *In* Baker, J.R. and R. Muller (Eds.), Advances in Parasitology. Volume 30. Academic Press, Ltd., San Diego, California.

Khonsue, W., M. Matsui, T. Hirai and Y. Misawa. 2001. Age determination of wrinkled frog, *Rana rugosa* with special reference to high variation in postmetamorphic body size (Amphibia: Ranidae). Zoological Science 18:605–612.

Kiesecker, J.M. 1996. pH-mediated predator-prey interaction between *Ambystoma tigrinum* and *Pseudacris triseriata*. Ecological Applications 6:1325–1331.

Kiesecker, J.M. 2002. Synergism between trematode infection and pesticide exposure: a link to amphibian limb deformities in nature? Proceeding of the National Academy of Sciences 99:9900–9904.

Kiesecker, J.M. and A.R. Blaustein. 1995. Synergism between UV-B radiation and a pathogen magnifies amphibian embryo mortality in nature. Proceedings of the National Academy of Sciences 92:11049–11052.

Kiesecker, J.M. and A.R. Blaustein. 1997a. Population differences in responses of red-legged frogs *(Rana aurora)* to introduced bullfrogs. Ecology 78:1752–1760.

Kiesecker, J.M. and A.R. Blaustein. 1997b. Influences of egg laying behavior on pathogenic infection of amphibian eggs. Conservation Biology 11:214–220.

Kiesecker, J.M. and A.R. Blaustein. 1998. Effects of introduced bullfrogs and small mouth bass on microhabitat use, growth, and survival of native red-legged frogs *(Rana aurora)*. Conservation Biology 12:776–787.

Kiesecker, J.M. and A.R. Blaustein. 1999. Pathogen reverses competition between larval amphibians. Ecology 80:2442–2448.

Kiesecker, J.M. and D.K. Skelly. 2000. Choice of oviposition site by gray treefrogs, *Hyla versicolor*: the role of potential parasitic infection. Ecology 81:2939–2943.

Kiesecker, J.M. and D.K. Skelly. 2001. Interactions of disease and pond drying on the growth, development, and survival of the gray treefrog *(Hyla versicolor)*. Ecology 82:1956–1963.

Kiesecker, J.M., A.R. Blaustein and L.K. Belden. 2001a. Complex causes of amphibian population declines. Nature 410:681–684.

Kiesecker, J.M., A.R. Blaustein and C.L. Miller. 2001b. Potential mechanisms underlying the displacement of native red-legged frogs by introduced bullfrogs. Ecology 82:1964–1970.

Kiesecker, J.M., A.R. Blaustein and C.L. Miller. 2001c. Transfer of a pathogen from fish to amphibians. Conservation Biology 15:1064–1070.

Kiesecker, J.M., D.P. Chivers, M. Anderson and A.R. Blaustein. 2002. Effect of predator diet on life-history shifts of red-legged frogs, *Rana aurora*. Journal of Chemical Ecology 28:1007–1015.

Kiesecker, J.M., D.P. Chivers, A. Marco, C. Quilchano, M.T. Anderson and A.R. Blaustein. 1999. Identification of a disturbance signal in larval red-legged frogs, *Rana aurora*. Animal Behaviour 57:1295–1300.

Kiester, A.R. 1971. Species density of North American amphibians and reptiles. Systematic Zoology 20:127–137.

Kiester, A.R., J.M. Scott, B. Csuti, R.F. Noss, B. Butterfield, K. Sahr and D. White. 1996. Conservation prioritization using GAP data. Conservation Biology 10:1332–1342.

Kiffney, P.M. and J.S. Richardson. 2001. Interactions among nutrients, periphyton, and invertebrate and vertebrate *(Ascaphus truei)* grazers in experimental channels. Copeia 2001:422–429.

Kilby, J.D. 1945. A biological analysis of the food and feeding habits of two frogs, *Hyla cinerea cinerea* and *Rana pipiens sphenocephala*. Quarterly Journal of the Florida Academy of Science 8:71–104.

Killebrew, F.C., K.B. Blair, D. Chiszar and H.M. Smith. 1996. New records for amphibians and reptiles from Texas. Herpetological Review 27:90–91.

Killebrew, F.C., K.B. Blair, H.M. Smith and D. Chiszar. 1995. Geographic distribution: *Bufo speciosus* (Texas toad). Herpetological Review 26:152.

Kilpatrick, S.L. 1997. Natural history of the four-toed salamander, *Hemidactylium scutatum*, in West Virginia. Master's thesis. Marshall University, Huntington, West Virginia.

Kim, B., T.G. Smith and S.S. Dresser. 1998. Life history and host specificity of *Hepatozoon clamitae* (Apicomplexa: Adeleorina) and ITS-1 nucleotide sequence variation of *Hepatozoon* species of frogs and mosquitoes from Ontario. Journal of Parasitology 84:789–797.

Kim, C.-H. 1983. The infection status of *Sparganum* and *Gnathostoma* in frogs of southern part of Korea. Korean Journal of Parasitology 21:83–86.

Kimberling, D.N., T.A. Grudzien and S.M. Shuster. 1996b. A comparison of allozyme and RAPD variation among northern leopard frog *(Rana pipiens)* populations in the Southwest. Report to the Arizona Game and Fish Department, Phoenix, Arizona.

Kimberling, D.N., A.R. Ferreira, S.M. Shuster and P. Keim. 1996a. RAPD marker estimation of genetic structure among isolated northern leopard frog populations in the southwestern USA. Molecular Ecology 5:521–529.

King, C. 1984. Immigrant Killers. Introduced Predators and the Conservation of Birds in New Zealand. Oxford University Press, Auckland, New Zealand.

King C.E. and P.S. Dawson. 1973. Population biology and the *Tribolium* model. Evolutionary Biology 5:133–227.

King, F.W. 1932. Herpetological records and notes from the vicinity of Tucson, Arizona, July and August, 1930. Copeia 1932:175–177.

King, O.M. 1960. Observations on Oklahoma toads. Southwestern Naturalist 5:102–103.

King, T.J. and M.A. Di Berardino. 1965. Transplantation of nuclei from the frog renal adenocarcinoma. I. Development of tumor nuclear transplant embryos. Annals of the New York Academy of Sciences 126:115–126.

King, T.J. and R.G. McKinnell. 1960. An attempt to determine the developmental potentialities of the cancer cell nucleus by means of transplantation. Pp. 591–617. *In* Cell Physiology of Neoplasia. University of Texas Press, Austin, Texas.

King, W. 1935. Ecological observations on *Ambystoma opacum*. Ohio Journal of Science 35:4–15.

King, W. 1936. A new salamander *(Desmognathus)* from the southern Appalachians. Herpetologica 1:57–60.

King, W. 1939. A survey of the herpetology of Great Smoky Mountains National Park (Tennessee). American Midland Naturalist 21:531–582.

King, W. and T. Krakauer. 1966. The exotic herpetofauna of southeastern Florida. Quarterly Journal of the Florida Academy of Science 29:149–154.

Kingsland, S.E. 1985. Modeling Nature. University of Chicago Press, Chicago, Illinois.

Kingsley, F.S. 1880. An abnormal foot in *Ambystoma*. American Naturalist 14:594.

Kinsella, J.M. and D.J. Forrester. 1999. Parasitic helminths of the common loon, *Gavia immer*, on its wintering grounds in Florida. Journal of the Helminthological Society of Washington 66:1–6.

Kinsella, J.M., R.A. Cole, D.J. Forrester and C.L. Roderick. 1996. Helminth parasites of the Osprey, *Pandion haliaetus*, in North America. Journal of the Helminthological Society of Washington 63:262–265.

Kirk, J.J. 1988. Western spotted frog *(Rana pretiosa)* mortality following forest spraying of DDT. Herpetological Review 19:51–53.

Kirk, J.J. 1991. *Batrachoseps wrighti*. Pp. 506.1–506.3. Catalogue of American Amphibians and Reptiles. Society for the Study of Amphibians and Reptiles, St. Louis. Missouri.

Kirk, J.J. and R.B. Forbes 1991. Geographic distribution: *Batrachoseps wrighti* (Oregon slender salamander). Herpetological Review 22:22–23.

Kirkland, G.L., Jr., H.W. Snoddy and T.L. Amsler. 1996. Impact of fire on small mammals and amphibians in a central Appalachian deciduous forest. American Midland Naturalist 135:253–260.

Kiviat, E. 1982. Geographic distribution: *Rana grylio* (pig frog). Herpetological Review 13:51.

Kiviat, E. and J. Stapleton. 1983. Life history notes: *Bufo americanus* (American toad). Estuarine habitat. Herpetological Review 14:46.

Klassen, M.A. 1998. Observations on the breeding and development of the plains spadefoot, *Spea bombifrons*, in southern Alberta. Canadian Field-Naturalist 112:387–392.

Klauber, L.M. 1928. A list of the amphibians and reptiles of San Diego County, California. Bulletin of the Zoological Society of San Diego 4:1–8.

Klauber, L.M. 1930. A list of the amphibians and reptiles of San Diego County, California. Bulletin of the Zoological Society of San Diego 5:1–8.

Kleeberger, S.R. 1984. A test of competition in two sympatric populations of desmognathine salamanders. Ecology 65:1846–1856.

Kleeberger, S.R. 1985. Influence of intraspecific density and cover on the home range of a plethodontid salamander. Oecologia 66:404–410.

Kleeberger, S.R. and J.K. Werner. 1982. Home range and homing behavior of *Plethodon cinereus* in northern Michigan. Copeia 1982:409–415.

Kleeberger, S.R. and J.K. Werner. 1983. Post-breeding migration and summer movement of *Ambystoma maculatum*. Journal of Herpetology 17:176–177.

Klein, H.G. 1960. Population estimate of the red-backed salamander. Herpetologica 16:52–54.

Klemens, M.W. 1993. Amphibians and reptiles of Connecticut and adjacent regions. State Geological and Natural History Survey of Connecticut, Bulletin Number 112, Hartford, Connecticut.

Klemt, W.B., T.R. Knowles, G.R. Elder and T.W. Sieh. 1979. Groundwater resources and model applications for the Edwards (Balcones Fault Zone) Aquifer in the San Antonio Region, Texas. Texas Department of Water Resources, Report Number 239, Austin, Texas.

Klewen, R. 1988. Die Verbreitung und Ökologie der Wasserfrösche in Nordrhein-Westfalen und ihre Bestandssituation im Ballungsraum Duisburg/Oberhausen. Jahrbuch für Feldherpetologie 1:73–96.

Klimstra, W.D. 1950. Narrow-mouthed toad taken in Iowa (Microhyla c. carolinensis). Copeia 1950:60.

Klimstra, W.D. and C.W. Myers. 1965. Foods of the toad Bufo woodhousei fowleri Hinckley. Transactions of the Illinois State Academy of Science 58:11–26.

Kline, J. 1998. Monitoring amphibians in created and restored wetlands. Pp. 360–368. In Lannoo, M.J. (Ed.), Status and Conservation of Midwestern Amphibians. University of Iowa Press, Iowa City, Iowa.

Kloss, G.R. 1974. Rhabdias (Nematoda, Rhabditoidea) from the marinus group of Bufo. A study of sibling species. Arquivos de Zoologia 25:61–120.

Kluge, A.G. 1983. Type-specimens of amphibians in the University of Michigan Museum of Zoology. Miscellaneous Publications, Museum of Zoology, Number 106, University of Michigan, Ann Arbor, Michigan.

Kluger, M.J. 1977. Fever in the frog Hyla cinerea. Journal of Thermal Biology 2:79–81.

Knapp, R.A. 1996. Non-native trout in the natural lakes of the Sierra Nevada: an analysis of their distribution and impacts on native aquatic biota. Pp. 363–390. In Sierra Nevada Ecosystem Project. Final Report to Congress. Volume III. Center for Water and Wildland Resources, University of California, Davis, California. Available online at ceres.ca.gov/snep.

Knapp, R.A. and K.R. Matthews. 2000. Non-native fish introductions and the decline of the mountain yellow-legged frog from within protected areas. Conservation Biology 14:428–438.

Knepton, J.C., Jr. 1954. A note on the burrowing habits of the salamander Amphiuma means means. Copeia 1954:68.

Knopf, G.N. 1962. Paedogenesis and metamorphic variation in Ambystoma tigrinum mavortium. Southwestern Naturalist 7:75–76.

Knowlton, G.F. 1944. Some insect food of Rana pipiens. Copeia 1944:119.

Knox, C.B. 1999. Blue-spotted salamander, Ambystoma laterale. Pp. 37–43. In Hunter, M.L., A.J.K. Calhoun and M. McCollough (Eds.), Maine Amphibians and Reptiles. University of Maine Press, Orono, Maine.

Knudsen, J.W. 1960. The courtship and egg mass of Ambystoma gracile and Ambystoma macrodactylum. Copeia 1960:44–46.

Knutson, M.G., J.R. Sauer, D.A. Olsen, M.J. Mossman, L.M. Hemesath and M.J. Lannoo. 1999. Effects of landscape composition and wetland fragmentation on frog and toad abundance and species richness in Iowa and Wisconsin, USA. Conservation Biology 13:1437–1446.

Kobel, H.R., C. Loumont and R.C. Tinsley. 1996. The extant species. Pp. 9–33. In Tinsley, R.C. and H.R. Kobel (Eds.), The Biology of Xenopus. Clarendon Press, Oxford.

Koch, E.D. and C.R. Peterson. 1995. Amphibians and Reptiles of Yellowstone and Grand Teton National Parks. University of Utah Press, Salt Lake City, Utah.

Kocher, T.D. and R.D. Sage. 1986. Further genetic analysis of a hybrid zone between leopard frogs (Rana pipiens complex) in central Texas. Evolution 40:21–33.

Koenings, C.A., C.K. Smith, E.A. Domingue and J.W. Petranka. 2000. Natural history notes: Desmognathus imitator (imitator salamander). Reproduction. Herpetological Review 31:38–39.

Kokko, H. and T. Ebenhard. 1996. Measuring the strength of demographic stochasticity. Journal of Theoretical Biology 183:169–178.

Koller, R.L. and A.J. Gaudin. 1977. An analysis of helminth infections in Bufo boreas (Amphibia: Bufonidae) and Hyla regilla (Amphibia: Hylidae) in southern California. Southwestern Naturalist 21:503–509.

Kolozsvary, M.B. and R.K. Swihart. 1999. Habitat fragmentation and the distribution of amphibians: patch and landscape correlates in farmland. Canadian Journal of Zoology 77:1288–1299.

Komarek, E.V. 1974. Effects of fire on temperate forests and related ecosystems: southeastern United States. Pp. 251–277. In Kozlowski, T.T. and C.E. Ahlgren (Eds.), Fire and Ecosystems. Academic Press, New York.

Komoroski, M.J. 1996. A comparative analysis of amphibian egg lipids. Master's thesis. University of Georgia, Athens, Georgia.

Koonz, W.H. 1992. Amphibians in Manitoba. Pp. 19–20. In Bishop, C.A. and E. Pettit (Eds.), Declines in Canadian Amphibian Populations: Designing a National Monitoring Strategy. Canadian Wildlife Service, Occasional Paper, Number 76, Ottawa, Ontario, Canada.

Kopij, G. 1996. Breeding and feeding ecology of the reed cormorant Phalacrocorax africanus in the Free State, South Africa. Acta Ornithologica 31:89–99.

Kopij, G. 1998. Diet of whitebreasted cormorant Phalacrocorax carbo nestlings in the southeastern Free State, South Africa. South African Journal of Wildlife Research 28:100–102.

Kordges, T. 1988. Zur Wasserfroschproblematik in Ballungsräumen—eine Essener Fallstudie. Jahrbuch für Feldherpetologie 1:97–104.

Korky, J.K. 1978. Differentiation of the larvae of members of the Rana pipiens complex in Nebraska. Copeia 1978:455–559.

Korky, J.K. and R.G. Webb. 1991. Geographic variation in larvae of mountain treefrogs of the Hyla eximia group (Anura: Hylidae). Bulletin of the New Jersey Academy of Science 36:7–12.

Korschgen, L.J. and T.S. Baskett. 1963. Foods of impoundment- and stream-dwelling bullfrogs in Missouri. Herpetologica 19:89–99.

Kosak, K.H. 1999. Genetic variation and sexual isolation across a contact zone between montane and lowland taxa of the Eurycea bislineata species complex (Amphibia: Caudata: Plethodontidae). Master's thesis. Clemson University, Clemson, South Carolina.

Koskela, P. and S. Pasanen. 1975. The reproductive biology of the female common frog, Rana temporaria. L. Aquilo Serie Zoologia 16:1–12.

Kotcher, E. 1941. Studies on the development of frog filariae. American Journal of Hygiene 34:36–65.

Kozlowska, M. 1971. Differences in the reproductive biology of mountain and lowland common frogs, Rana temporaria. L. Acta Biologica Cracoviensia. Series Zoologia 14:17–32.

Kragh, K. 1984. Written statement recalling 24 January 1966 meeting with Gosdin and Watson, 6 August.

Krajewski, C. 1991. Phylogeny and diversity. Science 254:918–919.

Krajewski, C. 1994. Phylogenetic measures of biodiversity: a comparison and critique. Biological Conservation 69:33–39.

Krakauer, T. 1968. The ecology of the neotropical toad, Bufo marinus, in south Florida. Herpetologica 24:214–221.

Krakauer, T. 1970. Tolerance limits of the toad, Bufo marinus, in south Florida. Comparative Biochemistry and Physiology 33:15–26.

Kramek, W.C. 1972. Food of the frog Rana septentrionalis in New York. Copeia 1972:390–392.

Kramek, W.C. 1976. Feeding behavior of Rana septentrionalis (Amphibia, Anura, Ranidae). Journal of Herpetology 10:251–252.

Kramer, D.C. 1973. Movements of western chorus frogs Pseudacris triseriata triseriata tagged with Co[60]. Journal of Herpetology 7:231–235.

Kramer, D.C. 1974. Home range of the western chorus frog Pseudacris triseriata triseriata. Journal of Herpetology 8:245–246.

Kramer, M. 1998. Back to nature: architects, environmentalists find common ground when it comes to course design. Gulf Coast Golfer, 15:8–9, 23–24.

Kramer, P., N. Reichenbach, M. Hayslett and P. Sattler. 1993. Population dynamics and conservation of the Peaks of Otter salamander, Plethodon hubrichti. Journal of Herpetology 27:431–435.

Kramer, P.A. 1996. An analysis of habitat utilization and feeding ecology of Plethodon richmondi and Plethodon cinereus in northern West Virginia. Master's thesis. Marshall University, Huntington, West Virginia.

Krapu, G.L. 1974. Foods of breeding pintails in North Dakota. Journal of Wildlife Management 38:408–417.

Krapu, G.L. 1996. Effects of a legal drain clean-out on wetlands and waterbirds: a recent case history. Wetlands 16:150–162.

Kraus, F. 1985a. A new unisexual salamander from Ohio. Occasional Papers Museum of Zoology, University of Michigan, Number 700, Ann Arbor, Michigan.

Kraus, F. 1985b. Unisexual salamander lineages in northwestern Ohio and southeastern Michigan: a study of the consequences of hybridization. Copeia 1985:309–324.

Kraus, F. 1989. Constraints on the evolutionary history of the unisexual salamanders of the Ambystoma laterale-texanum complex as revealed by mitochondrial DNA analysis. Pp. 218–227. In Dawley, R.M. and J.P. Bogart (Eds.), The Evolution and Ecology of Unisexual Vertebrates. New York State Museum Bulletin, Number 466, Albany, New York.

Kraus, F. 1995. The conservation status of unisexual vertebrate populations. Conservation Biology 9:956–959.

Kraus, F. and J.W. Petranka. 1989. A new sibling species of *Ambystoma* from the Ohio River drainage. Copeia 1989:94–110.

Kraus, F. and G.W. Schuett. 1982. A herpetofaunal survey of the coastal zone of northwest Ohio. Kirtlandia 35–38:21–54.

Kraus, F., E.W. Campbell, A. Allison and T. Pratt. 1999. *Eleutherodactylus* frog introductions to Hawaii. Herpetological Review 30:21–25.

Kraus, F., P.K. Ducey, P. Moler and M.M. Miyamato. 1991. Two new tri-parental unisexual *Ambystoma* from Ohio and Michigan. Herpetologica 47:229–239.

Krebs, C.J. 1992. Population regulation revisited. Ecology 73:714–715.

Krebs, C.J. and J.H. Meyers. 1974. Population cycles in small mammals. Advances in Ecological Research 8:2667–399.

Krebs, J.R. and N.B. Davies. 1981. An Introduction to Behavioural Ecology. Sinauer Associates, Sunderland, Massachusetts.

Krecker, F.H. 1916. *Filaria cingula* parasitic in the skin of *Cryptobranchus alleganiensis*. Journal of Parasitology 2:74–79.

Krenz, J.D. 1995. Fitness traits related to genetic heterozygosity in natural and experimental populations of the marbled salamander, *Ambystoma opacum*. Ph.D. dissertation. University of Georgia, Athens, Georgia.

Krenz, J.D. and D.E. Scott. 1994. Terrestrial courtship affects mating locations in *Ambystoma opacum*. Herpetologica 50:46–50.

Krenz, J.D. and D.M. Sever. 1995. Mating and oviposition in paedomorphic *Ambystoma talpoideum* precedes the arrival of terrestrial males. Herpetologica 51:387–393.

Kricher, J. 1997. A Neotropical Companion. Second edition. Princeton University Press, Princeton, New Jersey.

Kristsky, D.C., E.P. Hoberg and K.B. Aubry. 1993. *Lagarocotyle salamandrae* n. gen., n. sp. (Monogenoidea, Polyoonchoinea, Lagarocotylidea n. ord.) from the cloaca of *Rhyacotriton cascadae* Good and Wake (Caudata, Rhyacotritonidae) in Washington state. Journal of Parasitology 79:322–330.

Krull, W.H. 1931. Life history studies on two frog lung flukes, *Pneumonoeces medioplexus* and *Pneumobites parviplexus*. Transactions of the American Microscopical Society 50:215–277.

Krull, W.H. 1932. Studies on the life history of *Pneumobites longiplexus* (Stafford). Zoologischer Anzeiger 99:231–239.

Krull, W.H. and H.F. Price. 1932. Studies on the life history of *Diplodiscus temperatus* Stafford from the frog. Occasional Papers of the Museum of Zoology, Number 237, University of Michigan, Ann Arbor, Michigan.

Krupa, J.J. 1986a. Multiple clutch production in the Great Plains toad. Prairie Naturalist 18:151–152.

Krupa, J.J. 1986b. Anuran breeding dates in central Oklahoma. Bulletin of the Oklahoma Herpetological Society 11:10–13.

Krupa, J.J. 1986c. Distribution in Oklahoma of the bird-voiced treefrog *(Hyla avivoca)*. Proceedings of the Oklahoma Academy of Sciences 66:37–38.

Krupa, J.J. 1988. Fertilization efficiency in the Great Plains toad *(Bufo cognatus)*. Copeia 1988:800–802.

Krupa, J.J. 1989. Alternative mating tactics in the Great Plains toad. Animal Behavior 37:1035–1043.

Krupa, J.J. 1990. *Bufo cognatus*. Pp. 457.1–457.8. Catalogue of American Amphibians and Reptiles. Society for the Study of Amphibians and Reptiles, St. Louis, Missouri.

Krupa, J.J. 1994. Breeding biology of the Great Plains toad in Oklahoma. Journal of Herpetology 28:217–224.

Krupa, J.J. 1995. Natural history notes: *Bufo woodhousii* (Woodhouse's toad). Fecundity. Herpetological Review 26:142–144.

Kruse, K.C. 1981. Mating success, fertilization potential, and male body size in the American toad *(Bufo americanus)*. Herpetologica 40: 228–233.

Kruse, K.C. 1982. Male-male competition in the American toad, *Bufo americanus*. Biologist 64:78–82.

Kruse, K.C. 1983. Optimal foraging by predaceous diving beetle larvae on toad tadpoles. Oecologia 58:383–388.

Kruse, K.C. and M.G. Francis. 1977. A predation deterrent in larvae of the bullfrog, *Rana catesbeiana*. Transactions of the American Fisheries Society 106:248–252.

Kruse, K.C. and B.M. Stone. 1984. Largemouth bass *(Micropterus salmoides)* learn to avoid feeding on toad *(Bufo)* tadpoles. Animal Behaviour 32:1035–1039.

Krzysik, A.J. 1979. Resource allocation, coexistence, and the niche structure of a streambank salamander community. Ecological Monographs 49:173–194.

Krzysik, A.J. 1980a. Microhabitat selection and brooding phenology of *Desmognathus fuscus fuscus* in western Pennsylvania. Journal of Herpetology 14:291–292.

Krzysik, A.J. 1980b. Trophic aspects of brooding behavior in *Desmognathus fuscus fuscus*. Journal of Herpetology 14:426–428.

Krzysik, A.J. 1998a. Ecological design and analysis: principles and issues in environmental monitoring. Pp. 385–403. *In* Lannoo, M.J. (Ed.), Status and Conservation of Midwestern Amphibians. University of Iowa Press, Iowa City, Iowa.

Krzysik, A.J. 1998b. Amphibians, ecosystems, and landscapes. Pp. 31–41. *In* Lannoo, M.J. (Ed.), Status and Conservation of Midwestern Amphibians. University of Iowa Press, Iowa City, Iowa.

Krzysik, A.J. and E.B. Miller. 1979. Substrate selection by three species of desmognathine salamanders from southwestern Pennsylvania: an experimental approach. Annals of the Carnegie Museum 48:111–117.

Kuchta, S.R. 2002. Systematics, speciation, and mimicry in *Taricha* (Caudata; Salamandridae). Ph.D. dissertation. University of California, Berkeley, California.

Kucken, D.J., J.S. Davis and J.W. Petranka. 1994. Anakeesta Stream acidification and metal contamination: effects on a salamander community. Journal of Environmental Quality 23:1311–1317.

Kudo, R.R. 1943. Further observations on the protozoan *Myxidium serotinum* inhabiting the gall bladder of North American Salientia. Journal of Morphology 72:263–271.

Kuhl, J. 1997. Memorandum to Lisa O'Donnell (USFWS) and Dorinda Scott (TPWD), 24 March.

Kuhn, J. 1994. Lebensgeschichte und Demographie von Erdkrötenweibchen *Bufo bufo bufo* (L.) Zeitschrift für Feldherpetologie 1:3–87.

Kuhn, T.S. 1962. The Structure of Scientific Revolutions. University of Chicago Press, Chicago.

Kuhn, W. and M. Kleyer. 1996. Mapping and assessing habitat models on the landscape level. Pp. 356–362. *In* Settele, J., C. Margules, P. Poschlod and K. Henle (Eds.), Species Survival in Fragmented Landscapes. Kluwer Academic Publishers, Dordrecht, Holland.

Kull, J. 1997. Memorandum to Lisa O'Donnell (U.S.F.W.S.) and Dorinda Scott (T.P.W.D.), 24 March.

Kuntz, R.E. 1941. The metazoan parasites of some Oklahoma Anura. Proceedings of the Oklahoma Academy of Science 21:33–34.

Kunzmann, M.R. and W. Halvorson. 1998. Arizona GAP analysis: amphibians. U.S.G.S. Cooperative Park Studies Unit, University of Arizona, Tucson, Arizona.

Kuperman, B.I., V.E. Matey, R.N. Fisher and E.L. Ervin. 2000. Parasite assemblage in a wild population of the African clawed frog, *Xenopus laevis*, introduced to Southern California. Pp. 67–68. 2000 Joint Meeting of the American Society of Parasitologists and the Society of Protozoologists. June 24–28th, 2000, San Juan, Puerto Rico.

Kupferberg, S.J. 1994. Exotic larval bullfrogs *(Rana catesbeiana)* as prey for native garter snakes: functional and conservation implications. Herpetological Review 25:95–97.

Kupferberg, S.J. 1996a. Hydrologic and geomorphic factors affecting conservation of a river-breeding frog *(Rana boylii)*. Ecological Applications 6:1322–1344.

Kupferberg, S.J. 1996b. The ecology of native tadpoles *(Rana boylii* and *Hyla regilla)* and the impacts of invading bullfrogs *(Rana catesbeiana)* in a northern California river. Ph.D. dissertation. University of California, Berkeley, California.

Kupferberg, S.J. 1997a. Bullfrog *(Rana catesbeiana)* invasion of a California river: the role of larval competition. Ecology 78:1736–1751.

Kupferberg, S.J. 1997b. The role of larval diet in anuran metamorphosis. American Zoologist 37:146–159.

Kupferberg, S.J. 1998. Predator mediated patch use by tadpoles *(Hyla regilla)*: risk balancing or consequence of motionlessness? Journal of Herpetology 32:84–92.

Kupferberg, S.J., J.C. Marks and M.E. Power. 1994. Effects of variation in natural algal and detrital diets on larval anuran *(Hyla regilla)* life-history traits. Copeia 1994:446–457.

Kuris, A.M. 1997. Host behavior modification: an evolutionary perspective. Pp. 293–315. *In* Beckage, N.E. (Ed.), Parasites and Pathogens: Effects on Host Hormones and Behavior. Chapman and Hall, New York.

Kuss, B.D. 1986. A comparison of habitat utilization in the salamanders of the *Plethodon ouachitae* complex and *Plethodon glutinosus*. Ph.D. dissertation. University of Arkansas, Fayetteville, Arkansas.

Kutka, F.J. 1994. Low pH effects on swimming activity of *Ambystoma* salamander larvae. Environmental Toxicology and Chemistry 13:1821–1824.

Kuyt, E. 1991. A communal overwintering site for the Canadian toad, *Bufo americanus hemiophrys*, in the Northwest Territories. Canadian Field Naturalist 105:119–121.

Kuzmin, Y., V.V. Tkach and S.D. Snyder. 2001. *Rhabdias ambystomae* sp. n. (Nematoda: Rhabdiasidae) from the North American spotted salamander *Ambystoma maculatum* (Amphibia: Ambystomatidae). Comparative Parasitology 68:228–235.

Kwet, A. 1996. Zu den natürlichen Feinden des Laichs von Froschlurchen. Salamandra 32:31–44.

La Marca, E. and H.P. Reinthaler. 1991. Population changes in *Atelopus* species of the Cordillera de Merida, Venezuela. Herpetological Review 22:125–128.

Laan, R. and B. Verboom. 1990. Effects of pool size and isolation on amphibian communities. Biological Conservation 54:251–262.

Labanick, G.M. 1976a. Growth rates of recently transformed *Bufo woodhousei fowleri*. Copeia 1976:824–826.

Labanick, G.M. 1976b. Prey availability, consumption and selection in the cricket frog *Acris crepitans*: Amphibia, Anura, Hylidae. Journal of Herpetology 10:293–298.

Labanick, G.M. 1984. Anti-predator effectiveness of autotomized tails of the salamander *Desmognathus ochrophaeus*. Herpetologica 40:110–118.

Labanick, G.M. and R.A. Schleuter. 1976. Diets of sympatric *Acris crepitans* and juvenile *Bufo woodhousei fowleri* in western Indiana, USA. Proceedings of the Indiana Academy of Science 80:460.

Lacher, T.E., Jr. and M.I. Goldstein. 1997. Tropical ecotoxicology: status and needs. Environmental Toxicology and Chemistry 16:100–111.

LaClaire, L.V. 2001. Endangered and threatened wildlife and plants; final rule to list the Mississippi gopher frog distinct population segment of dusky gopher frog as endangered. Federal Register 66:62993–63001.

LaClaire, L.V. and R. Franz. 1991. Importance of isolated wetlands in upland landscapes. Pp. 9–15. *In* Kelly, M. (Ed.), Proceedings of the Second Annual Meeting, Florida Lake Management Society, Winterhaven, Florida.

LaClaire, L.V., R.N. Smith, S. Smith and J. Palis. 1995. Geographic distribution: *Notophthalmus perstriatus*. Herpetolological Review 26:103–104.

Lacy, R.C. 1993. VORTEX: a computer simulation model for population viability analysis. Wildlife Research 20:45–65.

Lacy, R.C., K.A. Hughes and T.J. Kreeger. 1993. VORTEX: A stochastic simulation of the extinction process. Users manual for Version 6. Second edition. IUCN Species Survival Commission, Captive Breeding Specialist Group, Apple Valley, Minnesota.

Lacy, R.C., K.A. Hughes and P.S. Miller. 1995. VORTEX: A stochastic simulation of the extinction process. Users manual for Version 7. IUCN Species Survival Commission, Conservation Breeding Specialist Group, Apple Valley, Minnesota.

Ladd, D.E. 1947. Size of adult *Eurycea l. longicauda* (Green). Herpetologica 4:2.

Laerm, J. and A.S. Hopkins Jr. 1997. Status of possible disjunct populations of the southern toad, *Bufo terrestris*, in the Piedmont and Blue Ridge of Georgia. Herpetological Review 28:162–163.

Lafferty, K.D. and A.K. Morris. 1996. Altered behavior of parasitized killifish increases host susceptibility to predation by bird final hosts. Ecology 77:1390–1397.

Lafferty, K.D. and C.J. Page. 1997. Predation on the endangered tidewater goby, *Eucyclogobius newberryi*, by the introduced African clawed frog, *Xenopus laevis*, with notes on the frog's parasites. Copeia 1997:589–592.

LaGrange, T.G. and J.J. Dinsmore. 1989. Habitat use by mallards during spring migration through central Iowa. Journal of Wildlife Management 53:1076–1081.

Laird, S. 1995. Taxpayers deserve respect. Smithville Times, 16 February.

Lakatos, I. 1970. Falsification and the methodology of scientific research programmes. Pp. 189–195. *In* Lakatos, I. and A. Musgrave (Eds.), Criticism and the Growth of Knowledge. Cambridge University Press, Cambridge, United Kingdom.

Lamb, T. 1984. The influence of sex and breeding conditions on microhabitat selection and diet in the pig frog, *Rana grylio*. American Midland Naturalist 111:311–318.

Lamb, T. 1987. Call site selection in a hybrid population of treefrogs. Animal Behaviour 35:1140–1144.

Lamb, T. and J.C. Avise. 1986. Directional introgression of mitochondrial DNA in a hybrid population of treefrogs: the influence of mating behavior. Proceedings of the National Academy of Sciences 83:2526–2530.

Lamb, T. and J.C. Avise. 1987. Morphological variability in genetically defined categories of anuran hybrids. Evolution 41:157–165.

Lamb, T., R.W. Gaul Jr., M.L. Tripp, J.M. Horton and B.W. Grant. 1998. A herpetofaunal inventory of the lower Roanoke River floodplain. Journal of the Elisha Mitchell Scientific Society 114:43–55.

Lamb, T., J.M. Novak and D.L. Mahoney. 1990. Morphological asymmetry and interspecific hybridization: a case study using hylid frogs. Journal of Evolutionary Biology 3:295–309.

Lamb, T., B.K. Sullivan and K.B. Malmos. 2000. Mitochondrial gene markers for the hybridizing toads *Bufo microscaphus* and *Bufo woodhousii* in Arizona. Copeia 2000:234–237.

Lambert, D.M. and H.G. Spencer (Eds.). 1995. Speciation and the Recognition Concept. Johns Hopkins University Press, Baltimore, Maryland.

Lamoureux, V.S. and D.M. Madison. 1999. Overwintering habitats of radio-implanted green frogs, *Rana clamitans*. Journal of Herpetology 33:430–435.

Lampo, M. 1995. The identification of ticks that infect *Bufo marinus* in South America and their impact on toad populations. CSIRO *Bufo* Project: An overview of research outcomes, Unpublished report, CSIRO, Australia, Centro de Ecologia, Venezuela.

Lampo, M. and P. Bayliss. 1996. Density estimates of cane toads from native populations based on mark–recapture data. Wildlife Research 23:305–315.

Lampo, M. and G.A. De Leo. 1998. The invasion ecology of the toad *Bufo marinus*: From South America to Australia. Ecological Applications 8:388–396.

Lancaster, D.L. and R.G. Jaeger. 1995. Rules of engagement for adult salamanders in territorial conflicts with heterospecific juveniles. Behavioral Ecology and Sociobiology 37:25–29.

Lancaster, D.L. and S.E. Wise. 1996. Differential response by the ringneck snake, *Diadophis punctatus*, to odors of tail-autotomizing prey. Herpetologica 52:98–108.

Lance, S.L. and K.D. Wells. 1993. Are spring peeper satellite males physiologically inferior to calling males? Copeia 1993:1162–1166.

Lancia, R.A., J.D. Nichols and K.H. Pollock. 1994. Estimating the number of animals in wildlife populations. Pp. 215–253. *In* Bookout, T.A. (Ed.), Research and Management Techniques for Wildlife and Habitats. Wildlife Society, Bethesda, Maryland.

Lande, R. 1988. Genetics and demography in biological conservation. Science 241:1455–1460.

Lande, R. 1993. Risks of population extinction from demographic and environmental stochasticity and random catastrophes. American Naturalist 142:911–927.

Lande, R. and G.F. Barrowclough. 1987. Effective population size, genetic variation, and their use in population management. Pp. 87–123. *In* Soulé, M.E. (Ed.), Viable Populations. Cambridge University Press, New York.

Lande, R.L., S. Engen and B.-E. Saether. 1995. Optimal harvesting of fluctuating populations with a risk of extinction. American Naturalist 145:728–745.

Landé, S.P. and S.I. Guttman. 1973. The effects of copper sulfate on the growth and mortality rate of *Rana pipiens* tadpoles. Herpetologica 29:22–27.

Landewe, J.E. 1963. Helminth and arthropod parasites of salamanders from southern Illinois. Master's thesis. Southern Illinois University, Carbondale, Illinois.

Landreth, H.F. and M.T. Christensen. 1971. Orientation of the plains spadefoot toad, *Scaphiopus bombifrons*, to solar cues. Herpetologica 27:454–461.

Landreth, H.F. and D.E. Ferguson. 1966. Evidence of sun-compass orientation in the chorus frog, *Pseudacris triseriata*. Herpetologica 22:106–112.

Landreth, H.F. and D.E. Ferguson. 1967a. Newt orientation by sun-compass. Nature 215:516–518.

Landreth, H.F. and D.E. Ferguson. 1967b. Newts: sun-compass orientation. Science 158:1459–1461.

Landreth, H.F. and D.E. Ferguson. 1968. The sun compass of Fowler's toad, *Bufo woodhousei fowleri*. Behaviour 30:27–43.

Lane, M.A. 2001. The homeless specimens: handling relinquished natural history collections. Museum News (American Association of Museums), January/February 2001:61–63, 82–83.

Lang, B.Z. 1968. The life cycle of *Cephalogonimus americanus* Stafford, 1902 (Trematoda: Cephalogonimidae). Journal of Parasitology 54:945–949.

Lang, B.Z. and L.N. Gleason. 1967. Life cycle of *Metagonimoides oregonensis* Price, 1931 (Trematoda: Heterophyidae) in North Carolina. Journal of Parasitology 53:93.

Lang, C. 1995. Size-fecundity relationships among stream-breeding hylid frogs. Herpetological Natural History 3:193–197.

Lang, C. and R.G. Jaeger. 2000. Defense of territories by male-female pairs in the red-backed salamander. Copeia 2000:169–177.

Lang, J.W. 1972. Geographic distribution: *Ambystoma maculatum*. Herpetological Review 4:170.

Lannoo, M.J. 1986. Vision is not necessary for size-selective zooplanktivory in aquatic salamanders. Canadian Journal of Zoology 44:1071–1078.

Lannoo, M.J. 1987. Neuromast topography in urodele amphibians. Journal of Morphology 191:247–263.

Lannoo, M.J. 1996. Okoboji Wetlands: A Lesson in Natural History. University of Iowa Press, Iowa City, Iowa.

Lannoo, M.J. 1998a. Amphibian conservation and wetland management in the upper midwest: a catch-22 for the cricket frog? Pp. 330–339. *In* Lannoo, M.J. (Ed.), Status and Conservation of Midwestern Amphibians. University of Iowa Press, Iowa City, Iowa.

Lannoo, M.J. (Ed.) 1998b. Status and Conservation of Midwestern Amphibians. University of Iowa Press. Iowa City, Iowa.

Lannoo, M.J. 1998c. Conclusion. Pp. 429–431. *In* Lannoo, M.J. (Ed.), Status and Conservation of Midwestern Amphibians. University of Iowa Press, Iowa City, Iowa.

Lannoo, M.J. 1998d. Introduction. Pp. xi–xviii. *In* Lannoo, M.J. (Ed.), Status and Conservation of Midwest Amphibians. University of Iowa Press, Iowa City, Iowa.

Lannoo, M.J. 2000. Conclusions drawn from the malformity and disease session, Midwest Declining Amphibians Conference, 1998. Pp. 212–216. *In* Kaiser, H., G.S. Casper and N.P. Bernstein (Eds.), Investigating Amphibian Declines: Proceedings of the 1998 Declining Amphibians Conference. Journal of the Iowa Academy of Science, Volume 107, Cedar Falls, Iowa.

Lannoo, M.J. and M.D. Bachmann. 1984a. Aspects of cannibalistic morphs in a population of *Ambystoma t. tigrinum* larvae. American Midland Naturalist 112:103–109.

Lannoo, M.J. and M.D. Bachmann. 1984b. On flotation and air breathing in *Ambystoma tigrinum* larvae: stimuli for and the relationship between these behaviors. Canadian Journal of Zoology 62:15–18.

Lannoo, M.J., L. Lowcock and J. Bogart. 1989. Sibling cannibalism in non-cannibal morph *Ambystoma tigrinum* larvae results in high growth rates and early metamorphosis. Canadian Journal of Zoology 67:1911–1914.

Lannoo, M.J., D.S. Townsend and R.J. Wassersug. 1987. Larval life in the leaves: arboreal tadpoles types, with special attention to the morphology, ecology, and behavior of the oophagous *Osteopilus brunneus* (Hylidae) larva. Fieldiana Zoology New Series Number 38, Field Museum of Natural History, Chicago, Illinois.

Lannoo, M.J., J.A. Holman, G.S. Casper and E. Johnson. 1998. Mummification following winterkill of adult green frogs (Ranidae: *Rana clamitans*). Herpetological Review 29:82–84.

Lannoo, M.J., K. Lang, T. Waltz and G.S. Phillips. 1994. An altered amphibian assemblage: Dickinson County, Iowa, 70 years after Frank Blanchard's survey. American Midland Naturalist 131:311–319.

Lannoo, M.J., D.R. Sutherland, P. Jones, D. Rosenberry, R.W. Klaver, D.M. Hoppe, P.T.J. Johnson, K.B. Lunde, C. Facemire and J.M. Kapfer. 2003. Multiple causes for the malformed frog phenomenon. Pp. 233–262. *In* Linder, G., S. Krest, D. Sparling and E. Little (Eds.), Multiple Stressor Effects in Relation to Declining Amphibian Populations. American Society for Testing Materials International, West Conshoshocken, Pensylvannia.

Lantz, L.A. 1930. Notes on the breeding habits and larval development of *Ambystoma opacum*. Annals of the Magazine of Natural History 5:322–325.

LaPointe, J. 1953. Noteworthy amphibian records from Indiana Dunes State Park. Copeia 1953:129.

Lardie, R.L. and B.P. Glass. 1975. Geographic distribution: *Ambystoma texanum* (small-mouthed salamander). Herpetological Review 6:115.

Larochelle, A. 1974. The American toad as champion carabid beetle collector. Pan-Pacific Entomologist 50:203–204.

Larochelle, A. 1975a. Trente especes de carabidae dans des boulettes fecales de crapaud americain. Cordulia 1:116.

Larochelle, A. 1975b. Le crapaud americain, predateur de coleopteres carabiques. Carabologia 4:151–152.

Larochelle, A. 1976. New data on carabid beetles as prey of the American toad. Cordulia 2:130.

Larochelle, A. 1977a. New data on Carabidae as prey of the American toad. Cordulia 3:38.

Larochelle, A. 1977b. Des boulettes fecales de crapaud *(Bufo)* contenant des carabidae. Fabreries 3:36.

Larochelle, A. 1977c. The Carabidae Coleoptera as prey of the American toad *Bufo americanus*. Cordulia 3:147–152.

Larson, A. 1980. Paedomorphosis in relation to rates of morphological and molecular evolution in the salamander *Aneides flavipunctatus* (Amphibia, Plethodontidae). Evolution 34:1–17.

Larson, A. and P. Chippindale. 1993. Molecular approaches to the evolutionary biology of plethodontid salamanders. Herpetologica 49:204–215.

Larson, A. and R. Highton. 1978. Geographic protein variation and divergence in the salamanders of the *Plethodon welleri* group (Amphibia: Plethodontidae). Systematic Zoology 27:431–448.

Larson, A., D.B. Wake and K.P. Yanev. 1984. Measuring gene flow among populations having high levels of genetic fragmentation. Genetics 106:293–308.

Larson, D.D. 1998. Tiger salamander life history in relation to agriculture in the northern Great Plains: a hypothesis. Pp. 325–329. *In* Lannoo, M.J. (Ed.), Status and Conservation of Midwestern Amphibians. University of Iowa Press, Iowa City, Iowa.

Larson, D.W. 1968. The occurrence of neotenic salamanders, *Ambystoma tigrinum diaboli* Dunn, in Devils Lake, North Dakota. Copeia 1968:620–621.

Larson, K., W. Duffy, E. Johnson and M.J. Lannoo. 1999. "Paedocannibal" morph barred tiger salamanders *(Ambystoma tigrinum mavortium)* from east central South Dakota with comments on the origin of cannibal morph eastern tiger salamanders *(A. t. tigrinum)* from northwest Iowa. American Midland Naturalist 141:124–139.

Larvor, B. 1998. Lakatos: An Introduction. Routledge, New York.

Latham, R. 1968. Notes on the eating of May beetles by a Fowler's toad. Engelhardtia 1:29.

Latham, R. 1970. The diet of shrikes on Long Island. Engelhardtia 3:29.

Latham, R. 1971a. Notes on the Fowler's toad at Orient, Long Island. Engelhardtia 4:57–58.

Latham, R. 1971b. The Fowler's toad on Gardiner's Island. Engelhardtia 4:54.

Latham, R. 1971c. Note on the food of the American bittern. Engelhardtia 4:50.

Latreille, P.A. 1801. *In* Sonnini, C.S. and P.A. Latreille, (Ar. X) Histoire naturelle des reptiles, avec figures dessinees d' après nature. Ches Deterville, Paris, France.

Laurance, W.F. 1996. Catastrophic declines of Australian rainforest frogs: is unusual weather responsible? Biological Conservation 77:203–212.

Laurance, W.F., K.R. McDonald and R. Speare. 1996. Epidemic disease and the catastrophic decline of Australian rain forest frogs. Conservation Biology 10:406–413.

Laurance, W.F., K.R. McDonald and R. Speare. 1997. In defense of the epidemic disease hypothesis. Conservation Biology 11:1030–1034.

Laurin, G. and D.M. Green. 1990. Spring emergence and male breeding behaviour of Fowler's toads, *(Bufo woodhousei fowleri)*, at Long Point, Ontario. Canadian Field-Naturalist 104:429–434.

Lauzon, R.D. and P. Balagus. 1998. New records from the northern range of the plains spadefoot, *Spea bombifrons*, in Alberta. Canadian Field-Naturalist 112:506–509.

Lavery, H.J. 1969. Collisions between aircraft and birds at Townsville, Queensland. Queensland Journal of Agricultural and Animal Sciences 26:447–455.

Lawler, K. and J.-M. Hero. 1997. Palatability of *Bufo marinus* tadpoles to a predatory fish decreases with development. Wildlife Research 24:327–334.

Lawler, S.P. 1989. Behavioral responses to predators and predation risk in four species of larval anurans. Animal Behaviour 38:1039–1047.

Lawler, S.P., D. Dritz, T. Strange and M. Holyoak. 1999. Effects of introduced mosquitofish and bullfrogs on the threatened California red-legged frog. Conservation Biology 13:613–622.

Lawton, J.H., J.R. Prendergast and B.C. Eversham. 1994. The numbers and spatial distributions of species: analyses of British data. Pp. 177–195. *In* Forey, P.L., C.J. Humphries and R.I. Vane-Wright (Eds.), Systematics and Conservation Evaluation. Clarendon Press, Oxford, United Kingdom.

Layne, J.R. 1992. Postfreeze survival and muscle function in the leopard frog *(Rana pipiens)* and the wood frog *(Rana sylvatica)*. Journal of Thermal Biology 17:121–124.

Layne, J.R. and M.A. Romano. 1985. Critical thermal minima of *Hyla chrysoscelis, H. cinerea, H. gratiosa* and natural hybrids (*H. cinerea* x *H. gratiosa*). Herpetologica 41:216–221.

Layne, J.R., Jr. 1991. External ice triggers freezing in freeze-tolerant frogs at temperatures above their supercooling point. Journal of Herpetology 25:129–130.

Layne, J.R., M. Romano and S.I. Guttman. 1989. Responses to desiccation of the treefrogs *Hyla cinerea* and *H. gratiosa* and their natural hybrids. American Midland Naturalist 121:61–67.

Lazell, J.D., Jr. 1976. This Broken Archipelago. Cape Cod and the Islands, Amphibians and Reptiles. Demeter Press, New York.

Lazell, J.D., Jr. 1989. Wildlife of the Florida Keys: A Natural History. Island Press, Washington, D.C.

Lazell, J. 1992. Taxonomic tyranny and the exoteric. Herpetological Review 23:14.

Lazell, J. 1994. Recognition characters and juxtaposition of Florida and Mississippi slimy salamanders (*Plethodon glutinosus* complex). Florida Scientist 57:129–140.

Lazell, J. 1995. Natural history notes: *Plethodon albagula* (western slimy salamander). Foot anomalies. Herpetological Review 26:198.

Lazell, J. 1998. New salamander of the genus *Plethodon* from Mississippi. Copeia 1998:967–970.

Lazell, J.D., Jr. and R.A. Brandon. 1962. A new stygian salamander from the southern Cumberland Plateau. Copeia 1962:300–306.

Lazell, J. and T. Mann. 1991. Geographic distribution: *Bufo americanus charlesmithi* (dwarf American toad). Herpetological Review 22:62, 64.

Lazell, J.D., Jr. and C. Raithel. 1986. Geographic distribution: *Gyrinophilus porphyriticus porphyriticus* (northern spring salamander). Herpetological Review 17:26.

Le Cren, E.D. 1955. Year to year variation in year-class strength of *Perca fluviatilis*. Verhandlungen Internationale Vereingung Limnologie 12:187–192.

Le Cren, E.D. 1965. Some factors regulating the size of populations of freshwater fish. Verhandlungen Internationale Vereingung Limnologie 13:88–105.

Lea, W.A. 1998. Letter to N.D. Hooks, 9 March.

Lean, D.R.S. 1998. Influence of UV-B radiation on aquatic ecosystems. Pp. 1–20. *In* Little, E.E., A.J. DeLonay and B.M. Greenburg (Eds.), Environmental Toxicology and Risk Assessment: Seventh volume. American Society for Testing and Materials, STP 1333, West Conshoshocken, Pennsylvania.

Learm, J., T.C. Carter, M.A. Menzel, T.S. McCay, J.L. Boone, W.M. Ford, L.T. Lepardo, D.M. Krishon, G. Balkcom, N.L. Vad der Maath and M.L. Harris. 1999. Amphibians, reptiles, and mammals of Sapelo Island, Georgia. Journal of the Elisha Mitchell Scientific Society 115:104–126.

Leary, C.J. 2001a. Evidence of convergent character displacement in release vocalizations of *Bufo fowleri* and *Bufo terrestris* (Anura; Bufonidae). Animal Behaviour 61:431–438.

Leary, C.J. 2001b. Investigating opposing patterns of character displacement in release and advertisement vocalizations of *Bufo fowleri* and *Bufo americanus* (Anura; Bufonidae). Canadian Journal of Zoology 79:1577–1585.

Leclair, R. and J. Castanet. 1987. A skeletochronological assessment of age and growth in the frog *Rana pipiens* Schreber (Amphibia, Anura) from southwestern Quebec. Copeia 1987:361–369.

Leclair, R., Jr. and J.P. Bourassa. 1981. Observation et analyse de la larves de Dipteres chironomides, dans la region de Trois-Rivieres (Quebec). Canadian Journal of Zoology 59:1339–1343.

Leclair, R., Jr. and G. Laurin. 1996. Growth and body size in populations of mink frogs *Rana septentrionalis* from two latitudes. Ecography 19:296–304.

Leclair, R., Jr. and L. Vallieres. 1981. Regimes alimentaires de *Bufo americanus* (Holbrook) et *Rana sylvatica* Leconte (Amphibia: Anura) novellement metamorphoses. Naturaliste Canadienne 108:325–329.

Leclair, R., Jr., Y. Alarie and J.P. Bourassa. 1986. Prey choice in larval *Dytiscus harrisii* Kirby and *D. verticalis* Say (Coleoptera, Dytiscidae). Entomologica Basiliensia 11:337–342.

LeClaire, L.V. No date. Status review of Ozark hellbender (*Cryptobranchus bishopi*). U.S. Fish and Wildlife Service, Jackson, Mississippi.

LeConte, J. [E.] 1824. Description of a new species of *Siren*, with some observations on animals of a similar nature. Annals of the Lyceum of Natural History of New York 1:52–58.

LeConte, J. 1825. Remarks on the American species of the genera *Hyla* and *Rana*. Annals of the Lyceum of Natural History of New York 1:278–282.

LeConte, J. 1855. Descriptive catalog of the Ranina of the United States. Proceedings of the Academy of the Natural Sciences of Philadelphia 7:423–431.

LeConte, J. 1856. Description of a new species of *Hyla* from Georgia. Proceedings of the Academy of the Natural Sciences of Philadelphia 8:146.

Ledwith, T. 1996. The effects of buffer strip width on air temperature and relative humidity in a stream side riparian zone. WMC Networker, Summer:6–7.

Lee, D.S. 1968a. Observations on hybrid *Hyla gratiosa* x *cinerea* in central Florida. Bulletin of the Maryland Herpetological Society 4:76–78.

Lee, D.S. 1968b. Herpetofauna associated with central Florida mammals. Herpetologica 24:83–84.

Lee, D.S. 1969a. Floridian herpetofauna associated with cabbage palms. Herpetologica 25:70–71.

Lee, D.S. 1969b. Notes on the feeding behavior of cave-dwelling bullfrogs. Herpetologica 25:211–212.

Lee, D.S. 1969c. Observations on the feeding habits of the Congo eel. Florida Naturalist 42:95.

Lee, D.S. 1969d. A food study of the salamander *Haideotriton wallacei* Carr. Herpetologica 25:175–177.

Lee, D.S. 1973. Seasonal breeding distributions for selected Maryland and Delaware amphibians. Bulletin of the Maryland Herpetological Society 9:101–104.

Lee, D.S. and L.R. Franz. 1973. Red leg in a subterranean population of pickerel frogs. Bulletin of the Maryland Herpetological Society 9:111.

Lee, D.S. and L.R. Franz. 1974. Comments on the feeding behavior of larval tiger salamanders, *Ambystoma tigrinum*. Bulletin of the Maryland Herpetological Society 10:105–107.

Lee, D.S. and A.W. Norden. 1973. A food study of the green salamander, *Aneides aeneus*. Journal of Herpetology 7:53–54.

Lee, J.C. 1996. The Amphibians and Reptiles of the Yucatan Peninsula. Comstock Publishing Associates, Cornell University Press, Ithaca, New York.

Lee, J.C. 2000. A Field Guide to the Amphibians and Reptiles of the Maya World: The Lowlands of Mexico, Northern Guatemala, and Belize. Cornell University Press, Ithaca, New York.

Lee, R.E. and J.P. Costanzo. 1998. Biological ice nucleation and ice distribution in cold-hardy ectothermic animals. Annual Review of Physiology 60:55–72.

Lee, Y.M. 1998. Recent surveys of Blanchard's cricket frogs (*Acris crepitans blanchardi*) in southern Michigan. P. 14. *In* Abstracts, Midwest Declining Amphibians Conference, March 20–21. Milwaukee Public Museum, Milwaukee, Wisconsin.

Lefcort, H. 1996. Adaptive, chemically mediated fright response in tadpoles of the southern leopard frog, *Rana utricularia*. Copeia 1996:455–459.

Lefcort, H. 1998. Chemically mediated fright response in southern toad (*Bufo terrestris*) tadpoles. Copeia 1998:445–450.

Lefcort, H. and A.R. Blaustein. 1995. Disease, predator avoidance, and vulnerability to predation in tadpoles. Oikos 74:469–474.

Lefcort, H. and S.M. Eiger. 1993. Antipredatory behaviour of feverish tadpoles: implications for pathogen transmission. Behaviour 126:13–27.

Lefcort, H., K.A. Hancock, K.M. Maur and D.C. Rostal. 1977. The effects of used motor oil, silt, and the water mold *Saprolegnia parasitica* on the growth and survival of mole salamanders (genus *Ambystoma*). Archives of Environmental Contamination and Toxicology 32:383–388.

Lefcort, H., R.A. Meguire, L.H. Wilson and W.F. Ettinger. 1998. Heavy metals alter the survival, growth, metamorphosis, and antipredatory behavior of Columbia spotted frog (*Rana luteiventris*) tadpoles. Archives of Environmental Contamination and Toxicology 35: 447–456.

Leff, L.G. and M.D. Bachmann. 1986. Ontogenetic changes in predatory behavior of larval tiger salamanders (*Ambystoma tigrinum*). Canadian Journal of Zoology 64:1337–1344.

Leftwich, K.N. and P.D. Lilly. 1992. The effects of duration of exposure to acidic conditions on survival of *Bufo americanus* embryos. Journal of Herpetology 26:70–71.

Leggett, M. 1994. Trees, toads the key issues in Lost Pines golf project. Austin American-Statesman, 26 August.

Leggett, M. 1998. Auditors uncover deficit in parks: multimillion-dollar yearly drain in state system could force big changes in operations. Austin American-Statesman, 3 October.

LeGrand, H.E., Jr. and S.P. Hall. 1995. Natural Heritage Program list of the rare animal species of North Carolina. North Carolina Natural Heritage Program, Division of Parks and Recreation, N.C. Department of Environment, Health and Natural Resources, Raleigh, North Carolina.

LeGrand, H.E., Jr. and S.P. Hall. 1999. Natural Heritage Program list of the rare animal species of North Carolina. North Carolina Natural Heritage Program, Division of Parks and Recreation, North Carolina Department of Environment and Natural Resources, Raleigh, North Carolina.

Lehman, N., A. Eisenhawer, K. Hansen, D.L. Mech, R.O. Peterson, J.P. Gogan and R.K. Wayne. 1991. Introgression of coyote mitochondrial DNA into sympatric North American gray wolf populations. Evolution 45:104–119.

Lehmann, D.L. 1954. Some helminths of West Coast urodeles. Journal of Parasitology 40:231.

Lehmann, D.L. 1955. A new host record for *Trypanosoma ambystomae* with the description of an additional morphological type of the parasite. Journal of Parasitology 41:552.

Lehmann, D.L. 1956. Some helminths of Oregon urodeles. Journal of Parasitology 42:25.

Lehmann, D.L. 1959. *Trypanosoma granulosae* n. sp. from the newt, *Taricha granulosa twittyi*. Journal of Protozoology 6:167–169.

Lehmann, D.L. 1960. Some parasites of central California amphibians. Journal of Parasitology 46:10.

Lehmann, D.L. 1964. Intestinal parasites of northwestern amphibians. Yearbook of the American Philosophy Society 1965:284–285.

Lehmann, D.L. 1966. Two blood parasites of Peruvian Amphibia. Journal of Parasitology 52:613.

Lehtinen, R.M., S.M. Galatowitsch and J.R. Tester. 1999. Consequences of habitat loss and fragmentation for wetland amphibian assemblages. Wetlands 19:1–12.

Leigh, W.H. 1946. Experimental studies on the life cycle of *Glypthelmins quieta* (Stafford, 1900), a trematode of frogs. American Midland Naturalist 35:460–483.

Leighton, F.A. 1995. Pathogens and disease. Pp. 509–518. *In* Hoffman, D.J., B.A. Rattner, G.A. Burton Jr. and J. Cairns Jr. (Eds.), Handbook of Ecotoxicology. Lewis Publishers, Boca Raton, Florida.

Leips, J. and J. Travis. 1994. Metamorphic response to changing food levels in two species of hylid frogs. Ecology 75:1345–1356.

Leja, W.T. 1998. Aquatic habitats in the Midwest: waiting for amphibian conservation initiatives. Pp. 345–353. *In* Lannoo, M.J. (Ed.), Status and Conservation of Midwestern Amphibians. University of Iowa Press, Iowa City, Iowa.

Lemmel, G. 1977. Die Lurche und Kriechtiere Niedersachsens. Naturschutz und Landschaftspflege in Niedersachsen 5:1–76.

Lenaker, R.P. 1972. *Xenopus* in Orange County. Unpublished report, California State Polytechnic University, San Luis Obispo, California.

LeNoir, J.S., L.L. McConnell, G.M. Fellers, T.M. Cahill and James N. Seiber. 1999. Summertime transport of current-use pesticides from California's Central Valley to the Sierra Nevada Mountain Range, USA. Environmental Toxicology and Chemistry 18:2715–2722.

Leonard, M., Vatnick, I., M. Brodkin and M. Simon. 1999. The effect of cold exposure on bacterial content of the gut and spleen of adult northern leopard frogs *(Rana pipiens)*. American Zoologist 39:76A.

Leonard, W.P. 1996. Natural history notes: *Plethodon vehiculum* (western red-backed salamander). Habitat. Herpetological Review 27:195.

Leonard, W.P. and D.M. Darda. 1995. Natural history notes: *Ambystoma tigrinum* (tiger salamander). Reproduction. Herpetological Review 26:29–30.

Leonard, W.P. and O.R. Klaus. 1994. Western long-toed salamander demographics and oviposition in a small vernal wetland of the Puget Sound lowlands. Northwest Science 68:135.

Leonard, W.P. and R.C. Stebbins. 1999. Observations of antipredator tactics of the sharp-tailed snake *(Contia tenuis)*. Northwestern Naturalist 80:74–77.

Leonard, W.P., L. Hallock and K.R. McAllister. 1997. Natural history notes: *Rana pretiosa* (Oregon spotted frog). Behavior and reproduction. Herpetological Review 28:86.

Leonard, W.P., K.R. McAllister and R.C. Friesz. 1999. Survey and assessment of northern leopard frog *(Rana pipiens)* populations in Washington State. Northwest Naturalist 80:51–60.

Leonard, W.P., L.L. Todd and M.A. Leonard. 1998. Geographic distribution: *Dicamptodon copei* (Cope's giant salamander). Herpetological Review 29:244.

Leonard, W.P., H.A. Brown, L.L.C. Jones, K.R. McAllister and R.M. Storm. 1993. Amphibians of Washington and Oregon. Seattle Audubon Society, Trailside Series. Seattle Audubon Society, Seattle, Washington.

Lepage, M., R. Courtois, C. Daigle and S. Matte. 1997. Surveying calling anurans in Quebec using volunteers. Pp. 128–140. *In* Green, D.M. (Ed.), Amphibians in Decline: Canadian Studies of a Global Problem. Herpetological Conservation, Number 1, Society for the Study of Amphibians and Reptiles, St. Louis, Missouri.

Lepage, M., J. DesGranges, J. Rodrigue and M. Ouellet. 1997. Anuran species richness in agricultural landscapes of Quebec: foreseeing long-term results of road call surveys. Pp. 141–148. *In* Green, D.M. (Ed.), Amphibians in Decline: Canadian Studies of a Global Problem. Herpetological Conservation, Number 1, Society for the Study of Amphibians and Reptiles, St. Louis, Missouri.

Leslie, P.H. and R.M. Ranson. 1940. The mortality, fertility and rate of natural increase of the vole *(Microtus agrestis)* as observed in the laboratory. Journal of Animal Ecology 9:27–52.

Levell, J. 1995. A Field Guide to Reptiles and the Law. Serpent's Tale Press, Excelsior, Minnesota.

Levell, J.P. 1997. A Field Guide to Reptiles and the Law. Revised Second edition. Serpent's Tale, Lanesboro, Minnesota.

Levey, R.B. 2000. Investigation of abnormal northern leopard frogs in Vermont. Pp. A-90. *In* Lean, D. and J. Ridal (Eds.), Proceedings of the International Association for Great Lakes Research, Annual Conference, Cornwall, Ontario.

Levins, R. 1969. Some demographic and genetic consequences of environmental heterogeneity for biological control. Bulletin of the Entomological Society of America 15:237–240.

Levins, R. 1970. Extinction. Pp. 77–107. *In* Gerstenhaber, M. (Ed.), Some Mathematical Problems in Biology. American Mathematical Society, Providence, Rhode Island.

Lewis, D.L., G.T. Baxter, K.M. Johnson and M.D. Stone. 1985. Possible extinction of the Wyoming toad, *Bufo hemiophrys baxteri*. Journal of Herpetology 19:166–168.

Lewis, W.M. 1962. Stomach contents of bullfrogs *(Rana catesbeiana)* taken from a minnow hatchery. Transactions of the Illinois State Academy of Science 55:80–83.

Licht, L.E. 1967a. Growth inhibition in crowded tadpoles: intraspecific and interspecific effects. Ecology 48:736–745.

Licht, L.E. 1967b. Death following possible ingestion of toad eggs. Toxicon 5:141–142.

Licht, L.E. 1967c. The initial appearance of the parotoid gland in three species of toads (genus *Bufo*). Herpetologica 23:115–118.

Licht, L.E. 1968. Unpalatability and toxicity of toad eggs. Herpetologica 24:93–98.

Licht, L.E. 1969a. Palatability of *Rana* and *Hyla* eggs. American Midland Naturalist 82:296–298.

Licht, L.E. 1969b. Comparative breeding behavior of the red-legged frog *(Rana aurora aurora)* and the western spotted frog *(Rana pretiosa pretiosa)* in southwestern British Columbia. Canadian Journal of Zoology 47:1287–1299.

Licht, L.E. 1969c. Observations on the courtship behavior of *Ambystoma gracile*. Herpetologica 25:49–52.

Licht, L.E. 1971. Breeding habits and embryonic thermal requirements of the frogs, *Rana aurora aurora* and *Rana pretiosa pretiosa*, in the Pacific Northwest. Ecology 52:116–124.

Licht, L.E. 1974. Survival of embryos, tadpoles and adult frogs, *Rana aurora aurora* and *Rana pretiosa pretiosa*, sympatric in southwestern British Columbia. Canadian Journal of Zoology 52:613–627.

Licht, L.E. 1975a. Comparative life history features of the western spotted frog, *Rana pretiosa*, from low- and high-elevation populations. Canadian Journal of Zoology 53:1254–1257.

Licht, L.E. 1975b. Growth and food of larval *Ambystoma gracile* from a lowland population in southwestern British Columbia. Canadian Journal of Zoology 53:1716–1722.

Licht, L.E. 1976. Sexual selection in toads *(Bufo americanus)*. Canadian Journal of Zoology 54:1277–1284.

Licht, L.E. 1986a. Food and feeding behavior of sympatric red-legged frogs, *Rana aurora*, and spotted frogs, *Rana pretiosa*, in southwestern British Columbia. Canadian Field-Naturalist 100:22–31.

Licht, L.E. 1986b. Comparative escape behavior of sympatric *Rana aurora* and *Rana pretiosa*. American Midland Naturalist 115:239–247.

Licht, L.E. 1989. Reproductive parameters of unisexual *Ambystoma* on Pelee Island, Ontario. Pp. 209–217. *In* Dawley, R.M. and J.P. Bogart (Eds.), The Evolution and Ecology of Unisexual Vertebrates. New York State Museum Bulletin, Number 466, Albany, New York.

Licht, L.E. 1991. Habitat selection of *Rana pipiens* and *Rana sylvatica* during exposure to warm and cold temperatures. American Midland Naturalist 125:259–268.

Licht, L.E. 1992. The effect of food level on growth rate and frequency of metamorphosis and paedomorphosis in *Ambystoma gracile*. Canadian Journal of Zoology 70:87–93.

Licht, L.E. and J.P. Bogart. 1990. Courtship behavior of *Ambystoma texanum* on Pelee Island, Ontario. Journal of Herpetology 24:450–452.

Licht, L.E. and B. Low. 1968. Cardiac response of snakes after ingestion of toad parotoid venom. Copeia 1968:547–551.

Licht, L.E. and D.M. Sever. 1991. Cloacal anatomy of metamorphosed and neotenic salamanders. Canadian Journal of Zoology 69:2230–2233.

Licht, P. and A.G. Brown. 1967. Behavioral thermoregulation and its role in the ecology of the red-bellied newt, *Taricha rivularis*. Ecology 48:598–611.

Ligas, F.J. 1960. The Everglades bullfrog life history and management. Florida Game and Fresh Water Fish Commission, Tallahassee, Florida.

Lillywhite, H.B. 1970. Behavioral temperature regulation in the bullfrog, *Rana catesbeiana*. Copeia 1970:158–168.

Lillywhite, H.B. 1971. Temperature selection by the bullfrog, *Rana catesbeiana*. Comparative Biochemistry and Physiology A 40:213–227.

Lillywhite, H.B. 1974a. Comments on postmetamorphic aggregate of *Bufo boreas*. Copeia 1974:984–986.

Lillywhite, H.B. 1974b. How frogs regulate their body temperature. Environment Southwest 465:3–6.

Lillywhite, H.B. and P. Licht. 1974. Movement of water over toad skin: functional role of epidermal sculpturing. Copeia 1974:165–170.

Lillywhite, H.B. and P. Licht. 1975. A comparative study of integumentary mucous secretions in amphibians. Comparative Biochemistry and Physiology A 51:937–941.

Lillywhite, H.B. and R.J. Wassersug. 1974. Comments on a postmetamorphic aggregate of *Bufo boreas*. Copeia 1974:984–986.

Lillywhite, H.B., P. Licht and P. Chelgren. 1973. The role of behavioral thermoregulation in the growth energetics of the toad, *Bufo boreas*. Ecology 54:375–383.

Limbaugh, B.A. and E.P. Volpe. 1957. Early development of the Gulf Coast toad, *Bufo valliceps* Wiegmann. American Museum Novitates, Number 1842, American Museum of Natural History, New York.

Linam, L.A. and J. Grote 1995. Letter to Lisa O'Donnell, 1 May.

Lind, A.J. 1990. Ontogenetic changes in the foraging behavior, habitat use, and food habits of the western aquatic garter snake, *Thamnophis couchii* at Hurdygurdy Creek, Del Norte Co., California. Master's thesis. Humboldt State University, Arcata, California.

Lind, A.J. and H.H. Welsh Jr. 1990. Predation by *Thamnophis couchii* on *Dicamptodon ensatus*. Journal of Herpetology 24:104–106.

Lind, A.J. and H.H. Welsh Jr. 1994. Ontogenetic changes in foraging behaviour and habitat use by the Oregon garter snake, *Thamnophis atratus hydrophilus*. Animal Behaviour 48:1261–1273.

Lind, A.J., J.B. Bettaso and S.M. Yarnell. 2003. Natural history notes: *Rana boylii* (foothill yellow-legged frog) and *Rana catesbeiana* (bullfrog). Reproductive behavior. Herpetological Review 34:234–235.

Lind, A.J., H.H. Welsh Jr. and R.A. Wilson. 1996. The effects of a dam on breeding habitat and egg survival of the foothill yellow-legged frog (*Rana boylii*) in northwestern California. Herpetological Review 27:62–67.

Lindeman, P.V. 1993. Food of the Coeur d'Alene salamander (*Plethodon idahoensis*) at Elk Creek Falls, Idaho. Northwestern Naturalist 74:58–59.

Lindenmayer, D.B. and H.P. Possingham. 1995. The risk of extinction: ranking management options for Leadbeatter's possum using population vulnerability analysis. Center for Resources and Environmental Studies, Australian National University, Canberra, New South Wales, Australia.

Linder, G. and B. Grillitsch. 2000. Ecotoxicology of metals. Pp. 325–459. *In* Sparling, D.W., G. Linder and C.A. Bishop (Eds.), Ecotoxicology of Amphibians and Reptiles. Society for Environmental Toxicology and Contaminants (SETAC) Press, Pensacola, Florida.

Linder, H.P. 1995. Setting conservation priorities: the importance of endemism and phylogeny in the southern African orchid genus *Herschelia*. Conservation Biology 9:585–595.

Lindsay, H.L. 1954. The narrow-mouthed toads *Microhyla olivacea* and *M. carolinensis* in northeastern Oklahoma. Copeia 1954:158.

Liner, E.A. 1954. The herpetofauna of Lafayette, Terrebonne, and Vermillion parishes, Louisiana. Proceedings of the Louisiana Academy of Science 17:65–85.

Liner, E.A. 1955. A herpetological consideration of the Bayou Tortue Region of LaFayette Parish, Louisiana. Proceedings of the Louisiana Academy of Sciences 18:39–42.

Ling, R.L. and J.K. Werner. 1988. Mortality in *Ambystoma maculatum* larvae due to *Tetrahymena* infection. Herpetological Review 19:26–27.

Ling, R.L., J.P. VanAmburg and J.K. Werner. 1986. Pond acidity and its relationship to larval development of *Ambystoma maculatum* and *Rana sylvatica* in Upper Michigan. Journal of Herpetology 20:230–236.

Link, C.E. 1998. Amphibian colonization and use of constructed ponds on Maryland's eastern shore. Pp. 14–15. *In* Abstracts, Midwest Declining Amphibians Conference, March 20–21. Milwaukee Public Museum, Milwaukee, Wisconsin.

Linnaeus, C. 1737 [1983]. The Critica Botanica. Translated by A. Hort. The Ray Society, London, England.

Linnaeus, C. 1758. Systema Naturae per Regna Tria Naturae, Secundum Classes, Ordines, Genera, Species cum Characteribus, Differentiis, Synonymis, Locis. Tenth edition, Volume 1, L. Salvius, Stockholm, Sweden.

Linnaeus, C. 1766. Systema Naturae. Twelfth edition, Volume 1, Part 2 (Addenda) (not paginated), L. Salvius, Stockholm, Sweden.

Linsdale, J.M. 1927. Amphibians and reptiles of Doniphan County, Kansas. Copeia 164:75–81.

Linsdale, J.M. 1932. Amphibians and reptiles from Lower California. University of California Publications in Zoology 36:345–386.

Linsdale, J.M. 1933a. Records of *Ascaphus truei* in Idaho. Copeia 1933:223.

Linsdale, J.M. 1933b. A specimen of *Rana tarahumarae* from New Mexico. Copeia 1933:222.

Linsdale, J.M. 1938. Environmental responses of vertebrates in the Great Basin. American Midland Naturalist 19:1–206.

Linsdale, J.M. 1940. Amphibians and reptiles in Nevada. Proceedings of the American Academy of Arts and Sciences 73:197–257.

Linzey, D.W. 1967. Food of the leopard frog, *Rana pipiens pipiens* in central New York. Herpetologica 23:11–17.

Lipps, G. and R.A. Odum. 2001. The struggle to save the Wyoming toad: a case history in captive propagation and reintroduction. International Herpetological Symposium, Detroit, Michigan.

Lipps, G.J., Jr. 2000. A survey of the distribution of Blanchard's cricket frog, *Acris creptans blanchardi*, in northern Ohio. Report to the Toledo Zoological Society and Ohio Biological Survey, Toledo and Columbus, Ohio.

Lips, K.R. 1991. Vertebrates associated with tortoise (*Gopherus polyphemus*) burrows in four habitats in south-central Florida. Journal of Herpetology 25:477–481.

Lips, K.R. 1998. Decline of a tropical montane amphibian fauna. Conservation Biology 12:106–117.

Lips, K.R. 1999. Mass mortality and population declines of anurans at an upland site in western Panama. Conservation Biology 13:117–125.

Liss, W.J. and G.L. Larson. 1991. Ecological effects of stocked trout on North Cascades naturally fishless lakes. Park Science 11:22–23.

Liss, W.J., G.L. Larson, E. Deimling, L. Ganio, R. Gresswell, R. Hoffman, M. Kiss, G. Lomnicky, C.D. McIntire, R. Truitt and T. Tyler. 1995. Ecological effects of stocked trout in naturally fishless high mountain lakes, North Cascades. National Park Service, Pacific Northwest Region, Technical Report NPS/PNROSU/NRTR-95-03, Seattle, Washington.

Little, E.E., R. Hurtubise and L. Cleveland. 1998. Photoenhanced toxicity of diluent to the frog, *Rana sphenocephala*. California Department of Fish and Game, Final report, Sacramento, California.

Little, E.L. 1940. Amphibians and reptiles of the Roosevelt Reservoir area, Arizona. Copeia 1940:260–265.

Little, E.L. and J.G. Keller. 1937. Amphibians and reptiles of the Jornada Experimental Range, New Mexico. Copeia 1937:216–222.

Little, M.L. 1983. The zoogeography of the *Hyla versicolor* complex in the central Appalachians, including physiological and morphological analyses. Ph.D. dissertation. University of Louisville, Louisville, Kentucky.

Little, M.L., B.L. Monroe Jr. and J.E. Wiley. 1989. The distribution of the *Hyla versicolor* complex in the Northern Appalachian Highlands. Journal of Herpetology 23:299–303.

Littlejohn, M.J. 1961. Artificial hybridization between some hylid frogs of the United States. Texas Journal of Science 13:176–184.

Littlejohn, M.J. 1971. A reappraisal of mating call differentiation in *Hyla cadavarina* (= *Hyla californiae*) and *Hyla regilla*. Evolution 25:98–102.

Liu, D., K. Thomson and W.M.J. Strachan. 1981. Biodegradation of carbaryl in simulated aquatic environment. Bulletin of Environmental Contamination and Toxicology 27:412–417.

Livezey, R.L. 1950. The eggs of *Acris gryllus crepitans* Baird. Herpetologica 6:139–140.

Livezey, R.L. 1952. Some observations on *Pseudacris nigrita triseriata* (Wied) in Texas. American Midland Naturalist 47:372–381.

Livezey, R.L. 1953. Late breeding of *Hyla regilla* Baird and Girard. Herpetologica 9:73.

Livezey, R.L. 1959. The egg mass and larvae of *Plethodon elongatus* Van Denburgh. Herpetologica 15:41–42.

Livezey, R.L. 1960. Description of the eggs of *Bufo boreas exsul*. Herpetologica 16:48.

Livezey, R.L. 1961. Food of adult and juvenile *Bufo boreas exsul*. Herpetologica 17:266–268.

Livezey, R.L. and H.M. Johnson. 1948. *Rana grylio* in Texas. Herpetologica 4:164.

Livezey, R.L. and A.H. Wright. 1945. Descriptions of four salientian eggs. American Midland Naturalist 34:701–706.

Livezey, R.L. and A.H. Wright. 1947. A synoptic key to the salientian eggs of the United States. American Midland Naturalist 37:179–222.

Livingston, P.G., C.C. Spencer and B.L. Stuart. 1995. Geographic distribution: *Desmognathus aeneus*. Herpetological Review 26:207.

Livo, L.J. 1977. An addition to the herpetofauna of Colorado. Colorado Herpetologist 3:1.

Livo, L.J. 1998. Investigations of boreal toad *(Bufo boreas)* tadpole ecology. Pp. 115–146. *In* Loeffler, C. (Ed.), Colorado Division of Wildlife, Boreal Toad Research Progress Report, 1995–1997. Denver, Colorado.

Livo, L.J. 1999. The role of predation in the early life history of *Bufo boreas* in Colorado. Ph.D. dissertation. University of Colorado, Boulder, Colorado.

Livo, L.J. and D. Yackley. 1997. Comparison of current with historical elevational range in the boreal toad, *Bufo boreas*. Herpetological Review 28:143–144.

Livo, L.J., H.M. Smith and D. Chiszar. 1999. County record tables for Colorado amphibians and reptiles as of January, 1999. Colorado Division of Wildlife, Denver, Colorado.

Lizana, M. and E.M. Pedraza. 1998. The effects of UV-B radiation on toad mortality in mountainous areas of central Spain. Conservation Biology 12:703–707.

Llewellyn, R.L. and C.R. Peterson. 1998. Distribution, relative abundance, and habitat associations of amphibians and reptiles on Craig Mountain, Idaho. Idaho Bureau of Land Management, Technical Bulletin 98-15, Boise, Idaho.

Loafman, P. and L. Jones. 1996. Natural history notes: *Dicamptodon copei* (Cope's giant salamander). Metamorphosis and predation. Herpetological Review 27:136.

Lobdell, R.N. 1936. Field and laboratory studies upon insect pests of south Florida, with particular reference to methods of control. Pp. 123–124. *In* Annual Report of the Agricultural Experiment Station for 1936. University of Florida, Gainesville, Florida.

Lockley, T.C. 1990. Predation on the green treefrog by the star-bellied orb weaver, *Acanthepeira stellata* (Araneae, Araneidae). Journal of Arachnology 18:359.

Lodato, M.J. 1974. Geographic distribution: *Rana heckscheri* (river frog). Herpetological Review 5:20.

Löderbusch, W. 1987. Die Amphibien im Kreis Tübingen. Beihefte zu den Veröffentlichungen für Naturschutz und Landschaftspflege in Baden-Württemberg 41:279–311.

Lodge, D.M., K.M. Brown, S.P. Klosiewski, R.A. Stein, A.P. Covich, B.K. Leathers and C. Bronmark. 1987. Distribution of freshwater snails: spatial scale and the relative importance of physiocochemical and biotic factors. American Malacological Bulletin 51:73–84.

Lodge, D.M., R.A. Stein, K.M. Brown, A.P. Covich, C. Bronmark, J.E. Garvey and S.P. Klosiewski. 1998. Predicting impact of freshwater exotic species on native biodiversity: challenges in spatial scaling. Australian Journal of Ecology 23:53–67.

Lodwick, L. 1974. Texas Parks and Wildlife Department office memorandum: Bastrop State Park—environmental assessment of the proposed golf course addition, 4 August.

Loeb, M.L.G., J.P. Collins and T.J. Maret. 1994. The role of prey in controlling expression of a trophic polymorphism in *Ambystoma tigrinum nebulosum*. Functional Ecology 8:151–158.

Loeffler, C. (Ed.). 1999. Boreal Toad Recovery Team report on the status and conservation of the boreal toad *(Bufo boreas boreas)* in the southern Rocky Mountains. Colorado Division of Wildlife, Unpublished report, Denver, Colorado.

Loennberg, E. 1894. Notes on reptiles and batrachians collected in Florida in 1892 and 1893. Proceedings of the United States National Museum 19:253–254.

Loftus, W.F. and R. Herndon. 1984. Reestablishment of the coqui, *Eleutherodactylus coqui* Thomas, in southern Florida. Herpetological Review 15:23.

Logier, E.B.S. 1932. Some account of the amphibians and reptiles of British Columbia. Transactions of the Royal Canadian Institute 18:311–336.

Loh, J. (Ed.). 2000. Living Planet Report 2000. WWF International, Gland, Switzerland.

Lohman, K. 2002. Annual variation in the density of stream tadpoles in a northern Idaho (USA) watershed. Verhandlungen der Internationalen Vereinigung für Theoretische und Angewandte Limnologie 28:1–5.

Lohoefener, R. and R.L. Jones. 1991. Status survey of *Plethodon fourchensis* and *P. caddoensis*. U.S. Fish and Wildlife Service, Endangered Species Office, Status Survey Report, Jackson, Mississippi.

Lombard, R.E. and D.B. Wake. 1977. Tongue evolution in the lungless salamanders, family Plethodontidae. II. Functional and evolutionary diversity. Journal of Morphology 153:39–79.

Lonard, R.I., J.H. Everitt and F.W. Judd. 1991. Woody plants of the Lower Rio Grande Valley, Texas. Texas Memorial Museum, Miscellaneous Publications, Number 7, University of Texas, Austin, Texas.

Long, A. 1984. Written statement recalling 24 January 1966 meeting with Gosdin and Watson, 30 July.

Long, C.A. 1964. The badger as a natural enemy of *Ambystoma tigrinum* and *Bufo boreas*. Herpetologica 20:144.

Long, C.A. 1982. Rare gigantic toads, *Bufo americanus*, from Lake Michigan isles. University of Wisconsin Museum of Natural History, Reports on the Fauna and Flora of Wisconsin 18:16–19.

Long, D.R. 1987a. A comparison of energy substrates and reproductive patterns of two anurans *Acris crepitans* and *Bufo woodhousei*. Comparative Biochemistry and Physiology A 87:81–92.

Long, D.R. 1987b. Reproductive and lipid patterns of a semiarid-adapted anuran, *Bufo cognatus*. Texas Journal of Science 39:11–13.

Long, D.R. 1989. Energetics and reproduction in female *Scaphiopus multiplicatus* from western Texas. Journal of Herpetology 23:176–179.

Long, L.E., L.S. Saylor and M.E. Soulé. 1995. A pH/UV-B synergism in amphibians. Conservation Biology 9:1301–1303.

Long, M.L. 1970. Food habits of *Rana muscosa* (Anura: Ranidae). Herpeton, Journal of the Southwestern Herpetologists Society 5:1–8.

Long, R.K., Jr. 1999. Could toad protection cost landowner rights? Smithville Times, 2 September.

Longbine, T.W., T.W. Reulbach and T.K. Pauley. 1991. Range of *Ambystoma texanum* and *Ambystoma barbouri* in West Virginia. West Virginia Academy of Science 63:25.

Longcore, J.E., A.P. Pessier and D.K. Nichols. 1999. *Batrachochytrium dendrobatidis*, gen. et sp. nov., a chytrid pathogenic to amphibians. Mycologia 91:219–227.

Longley, G. 1978. Status of *Typhlomolge* (= *Eurycea*) *rathbuni*, the Texas blind salamander. U.S. Fish and Wildlife Service Region 2, Endangered Species Report (2), Albuquerque, New Mexico.

Loomis, R.B. 1956. The chigger mites of Kansas (Acarina, Trombiculidae). University of Kansas Science Bulletin 37:1195–1443.

Loomis, R.B. 1965. The yellow-legged frog, *Rana boylii*, from the Sierra San Pedro Martir, Baja California Norte, Mexico. Herpetologica 21:78–80.

Loomis, R.B. and O.L. Webb. 1951. *Eurycea multiplicata* collected at the restricted type locality. Herpetologica 7:141–142.

Looney, J. 1973. Anticline and featherhead-cooler. Oklahoma Underground 6:19–24.

Looney, N. and B. Puckett. 1970. The Stansberry-January system. Oklahoma Underground 3:6–11.

Lopez, T.J. and L.R. Maxson. 1990. Natural history notes: *Rana catesbeiana* (bullfrog). Polymely. Herpetological Review 21:90.

Loraine, R.K. 1984. Life history notes: *Hyla crucifer crucifer* (northern spring peeper). Reproduction. Herpetological Review 15:16–17.

Lord, R.D., Jr. and W.B. Davis. 1956. A taxonomic study of the relationship between *Pseudacris nigrita triseriata* Wied and *Pseudacris clarki* Baird. Herpetologica 12:115–120.

Loredo, I. and D. Van Vuren. 1996. Reproductive ecology of a population of the California tiger salamander. Copeia 1996:895–901.

Loredo, I., D. Van Vuren and M.L. Morrison. 1996. Habitat use and migration behavior of the California tiger salamander. Journal of Herpetology 30:282–285.

Loring, S.J., W.P. MacKay and W.G. Whitford. 1988. Ecology of small desert playas. Pp. 89–113. *In* Thames, J.L. and C.D. Ziebell (Eds.), Small Water Impoundments in Semi-arid Regions. University of New Mexico Press, Albuquerque, New Mexico.

Loske, R. and P. Rinsche. 1985. Die Amphibien und Reptilien des Kreises Soest. Arbeitsgemeinschaft Biologischer Umweltschutz, Bad-Sassendorf-Lohne.

Losos, E. 1993. The future of the U.S. Endangered Species Act. TRENDS in Ecology and Evolution 8:332–336.

Lotshaw, D.P. 1977. Temperature adaptation and effects of thermal acclimation in *Rana sylvatica* and *Rana catesbeiana*. Comparative Biochemistry and Physiology A 56:287–294.

Lotter, F. 1978. Reproductive ecology of the salamander *Plethodon cinereus* (Amphibia, Urodela, Plethodontidae) in Connecticut. Journal of Herpetology 12:231–236.

Lotter, F. and N.J. Scott. 1977. Correlation between climate and distribution of the color morphs of the salamander *Plethodon cinereus*. Copeia 1977:681–690.

Loveridge, A. 1936. Scientific results of an expedition to rain forest regions in eastern Africa. VII. Amphibians. Bulletin of the Museum of Comparative Zoology (Harvard) 79:369–430.

Loveridge, A. 1942. Scientific results of a fourth expedition to forested areas in East and Central Africa. V. Amphibians. Bulletin of the Museum of Comparative Zoology (Harvard) 91:377–436.

Loveridge, A. 1953. Zoological results of a fifth expedition to East Africa. IV. Amphibians from Nyasaland and Tete. Bulletin of the Museum of Comparative Zoology (Harvard) 110:325–406.

Loveridge, A. 1959. Notes on the present herpetofauna of Ascension Island. Copeia 1959:69–70.

Lovich, J.E. and W.D. Fisher. 1988. Geographic distribution: *Ambystoma maculatum* (spotted salamander). Herpetological Review 19:17.

Lowcock, L.A. 1994. Biotype, genomotype, and genotype: variable effects of polyploidy and hybridity on ecological partitioning in a bisexual-unisexual community of salamanders. Canadian Journal of Zoology 72:104–117.

Lowcock, L.A. and J.P. Bogart. 1989. Electrophoretic evidence for multiple origins for triploid forms in the *Ambystoma laterale-jeffersonianum* complex. Canadian Journal of Zoology 67:350–356.

Lowcock, L.A., H. Griffith and R.W. Murphy. 1991. The *Ambystoma laterale-jeffersonianum* complex in central Ontario: ploidy structure, sex ratio, and breeding dynamics in a bisexual-unisexual community. Copeia 1991:87–105.

Lowcock, L.A., H. Griffith and R.W. Murphy. 1992. Size in relation to sex, hybridity, ploidy, and breeding dynamics in central Ontario populations of the *Ambystoma laterale-jeffersonianum* complex. Journal of Herpetology 26:46–53.

Lowcock, L.A., L.E. Licht and J.P. Bogart. 1987. Nomenclature in hybrid complexes of *Ambystoma* (Urodela: Ambystomatidae): no case for the erection of hybrid species. Systematic Zoology 36:328–336.

Lowe, C.H., Jr. 1950. The systematic status of the salamanders *Plethodon hardii*, with a discussion of biogeographical problems in *Aneides*. Copeia 1950:92–99.

Lowe, C.H., Jr. 1954. A new salamander (genus *Ambystoma*) from Arizona. Proceedings of the Biological Society of Washington 67:243–246.

Lowe, C.H., Jr. 1964. The amphibians and reptiles of Arizona. Pp. 153–174. *In* Lowe, C.H., Jr. (Ed.), The Vertebrates of Arizona. University of Arizona Press, Tucson, Arizona.

Lowe, C.H., Jr. and R.G. Zweifel. 1951. Sympatric populations of *Batrachoseps attenuatus* and *Batrachoseps pacificus* in southern California. Bulletin of the Southern California Academy of Science 50:128–135.

Lowe, J. and N. Nieto. 2003. Natural history notes: *Plethodon elongatus* (Del Norte salamander). Female-female aggression. Herpetological Review 34:354–355.

Lowery, G.R. 1966. Production and food of cutthroat trout in three Oregon coastal streams. Journal of Wildlife Management 30:754–767.

Luce, B., B. Oakleaf, A. Cerovski, L. Hunter and J. Priday. 1997. Atlas of birds, mammals, reptiles, and amphibians in Wyoming. Wyoming Fish and Game Department, Wildlife Division, Biological Services Section, Lander, Wyoming.

Lucké, B. 1934a. A neoplastic disease of the kidney of the frog, *Rana pipiens*. II. On the occurrence of metastasis. American Journal of Cancer 22:326–334.

Lucké, B. 1934b. A neoplastic disease of the kidney of the frog, *Rana pipiens*. American Journal of Cancer 20:352–379.

Lucké, B. 1938a. Carcinoma in the leopard frog: its probable causation by a virus. Journal of Experimental Medicine 68:457–468.

Lucké, B. 1938b. Carcinoma of the kidney of the leopard frog: the occurrence and significance of metastasis. American Journal of Cancer 34:15–30.

Lucké, B. 1952. Carcinoma of the leopard frog: a virus tumor. Annals of the New York Academy of Sciences 54:1093–1109.

Lucké, B. and H. Schlumberger. 1949. Induction of metastasis of frog carcinoma by increase of environmental temperature. Journal of Experimental Medicine 89:269–278.

Lucker, J.T. 1931. A new genus and a new species of trematode of the family Plagiorchidae. Proceedings of the U.S. National Museum, Number 79, Washington, D.C.

Ludwig, D.R., M. Redmer, R. Domazlicky, S. Kobal and B. Conklin. 1992. Current status of amphibians and reptiles in DuPage County, Illinois. Transactions of the Illinois State Academy of Science 85:187–199.

Luepschen, L.K. 1981. Life history notes: *Bufo punctatus* (red-spotted toad). Larval coloration. Herpetological Review 12:79.

Lumsden, R.D. and J.A. Zischke. 1963. Studies on the trematodes of Louisiana birds. Zeitschrift fur Parasitenkunde 22:316–366.

Lun, Z.R. and S.S. Desser. 1996. Analysis of isolates within species of anuran trypanosomes using random amplified polymorphic DNA. Parasitology Research 82:22–27.

Lundelius, E.L., Jr. 1967. Late-Pleistocene and Holocene faunal history of central Texas. Pp. 287–319. *In* Martin, P.S. and H.E. Wright Jr. (Eds.), Pleistocene Extinctions—The Search for a Cause. Volume 6, Proceedings of the VII Congress of the International Association for Quaternary Research, Yale University Press, New Haven, Connecticut.

Lust, J.M., D.L. Carlson, R. Kowles, L. Rollins-Smith, J.W. Williams III and R.G. McKinnell. 1991. Allografts of tumor nuclear transplantation embryos: differentiation competence. Proceedings of the National Academy of Sciences 88:6883–6887.

Lutterschmidt, W.I. and M.L. Thies. 1999. Geographic distribution: *Syrrhophus cystignathoides* (Rio Grande chirping frog). Herpetological Review 30:51.

Lutterschmidt, W.I., G.A. Marvin and V.H. Hutchison. 1994. Alarm response by a plethodontid salamander *(Desmognathus ochrophaeus)*: conspecific and heterospecific "Schreckstoff." Journal of Chemical Ecology 20:2751–2760.

Lutz, B. 1971. Venomous toads and frogs. Pp. 423–473. *In* Bucherl, W. and E.E. Buckley (Eds.), Venomous Animals and their Venoms: Volume II. Venomous Vertebrates. Academic Press, New York.

Lykens, D.V. and D.C. Forester. 1987. Age structure in the spring peeper: so males advertise longevity? Herpetologica 43:216–233.

Lynch, J.D. 1964. Two additional predators of the spadefoot toad, *Scaphiopus holbrookii* (Harlan). Journal of the Ohio Herpetological Society 4:79.

Lynch, J.D. 1965. The Pleistocene amphibians of Pit II, Arredondo, Florida. Copeia 1965:72–77.

Lynch, J.D. 1970. A taxonomic revision of the leptodactylid frog genus *Syrrhophus* Cope. University of Kansas Publications of the Museum of Natural History, Number 20, Lawrence, Kansas.

Lynch, J.D. 1971. Evolutionary relationships, osteology, and zoogeography of leptodactylid frogs. University of Kansas Museum of Natural History, Miscellaneous Publication, Number 53, Lawrence, Kansas.

Lynch, J.D. 1978. The distribution of leopard frogs (*Rana blairi* and *Rana pipiens*) (Amphibia, Anura, Ranidae) in Nebraska. Journal of Herpetology 12:157–162.

Lynch, J.D. 1985. Annotated checklist of the amphibians and reptiles of Nebraska. Transactions of the Nebraska Academy of Sciences 13: 33–57.

Lynch, J.D. and T. Grant. 1998. Dying frogs in western Colombia: catastrophe or trivial observation? Revista de la Academia Colombiana de Ciencias Exactas, Fisicas, y Naturales 22:149–152.

Lynch, J.E. 1936. *Phyllodistomum singulare n. sp.*, a trematode from the urinary bladder of *Dicamptodon ensatus* (Eschscholtz), with notes on related species. Journal of Parasitology 22:42–47.

Lynch, J.E., Jr. 1984. Reproductive ecology of *Plethodon idahoensis*. Master's thesis. University of Idaho, Moscow, Idaho.

Lynch, J.E., Jr. 1987. Field observations of courtship behavior in Rocky Mountain populations of Van Dyke's salamander, *Plethodon vandykei*, with a description of its spermatophore. Journal of Herpetology 21:337–340.

Lynch, J.F. 1974. *Aneides flavipunctatus*. Pp. 158.1–158.2. Catalogue of American Amphibians and Reptiles. Society for the Study of Amphibians and Reptiles, St. Louis, Missouri.

Lynch, J.F. 1981. Patterns of ontogenetic and geographic variation in the black salamander, *Aneides flavipunctatus* (Caudata: Plethodontidae). Smithsonian Contributions to Zoology, Number 324, Washington, D.C.

Lynch, J.F. 1985. The feeding ecology of *Aneides flavipunctatus* and sympatric plethodontid salamanders in northwestern California. Journal of Herpetology 19:328–352.

Lynch, J.F. and D.B. Wake. 1974. *Aneides lugubris*. Pp. 159.1–159.2. Catalogue of American Amphibians and Reptiles. Society for the Study of Amphibians and Reptiles, St. Louis, Missouri.

Lynch, M. 1984. Destabilizing hybridization, general-purpose genotypes and geographic parthenogenesis. Quarterly Review of Biology 59:257–290.

Lynn, W.G. and J.N. Dent. 1941. Notes on *Plethodon cinereus* and *Hemidactylium scutatum* on Cape Cod. Copeia 1941:113–114.

Mabberley, D.J. 1992. Tropical Rain Forest Ecology. Second edition. Chapman and Hall, New York.

Mable, B.K. and L. Rye. 1992. Developmental abnormalities in triploid hybrids between tetraploid and diploid tree frogs (genus *Hyla*). Canadian Journal of Zoology 70:2072–2076.

Mabry, C.M. and J.L. Christiansen. 1991. The activity and breeding cycle of *Scaphiopus bombifrons* in Iowa. Journal of Herpetology 25:116–119.

MacArthur, R.H. and E.O. Wilson. 1967. The Theory of Island Biogeography. Princeton University Press, Princeton, New Jersey.

Macartney, J.M. and P.T. Gregory. 1981. Differential susceptibility of sympatric garter snakes species to amphibian skin secretions. American Midland Naturalist 106:271–281.

MacCormack, Z. 1995. State tries again for Bastrop golf grant. Austin American-Statesman, 7 March.

MacCulloch, R.D. and J.R. Bider. 1975. Phenology, migrations, circadian rhythm and the effect of precipitation on the activity of *Eurycea b. bislineata* in Quebec. Herpetologica 31:433–439.

Mace, T.F. and R.C. Anderson. 1975. Development of the giant kidney worm, *Dioctophyma renale* (Goeze, 1782) (Nematoda: Dioctophymatoidea). Canadian Journal of Zoology 53:1552–1568.

Macey, J.R. 1986. The biogeography of a herpetofaunal transition between the Great Basin and Mojave Deserts. Pp. 119–128. *In* Hall, C.A., Jr. and D.J. Young (Eds.), Natural History of the White-Inyo Range, Eastern California and Western Nevada and High Altitude Physiology. University of California, White Mountain Research Station Symposium, Bishop, California.

Macey, J.R. and T.J. Papenfuss. 1991a. Amphibians. Pp. 277–290. *In* Hall, C.A., Jr. (Ed.), Natural History of the White-Inyo Range, Eastern California. University of California Press, Berkeley, California.

Macey, J.R. and T.J. Papenfuss. 1991b. Reptiles. Pp. 291–360. *In* Hall, C.A., Jr. (Ed.), Natural History of the White-Inyo Range, Eastern California. University of California Press, Berkeley, California.

Macey, J.R., J. Stasburg, J. Brisson, V.T. Vrendenburg, M. Jennings and A. Larson. 2001. Molecular phylogenetics of western North American frogs of the *Rana boylii* species group. Molecular Phylogenetics and Evolution 18:131–143.

Macgregor, H.C. and T.M. Uzzell Jr. 1964. Gynogenesis in salamanders related to *Ambystoma jeffersonianum*. Science 143:1043–1045.

Machovina, B.L. 1994. Ecology and life history of the salamander *Amphiuma means* in Everglades National Park. Master's thesis. Florida International University, Miami, Florida.

Maciolek, J.A. 1983a. Status report: Amargosa toad. National Fisheries Research Center, U.S. Fish and Wildlife Service, Seattle, Washington (available from Nevada Natural Heritage Program, Carson City, Nevada).

Maciolek, J.A. 1983b. Further observations on the distribution and abundance of the Amargosa toad. National Fisheries Research Center, U.S. Fish and Wildlife Service, Seattle, Washington (available from Nevada Natural Heritage Program, Carson City, Nevada).

MacKay, W.P., S.J. Loring, T.M. Frost and W.G. Whitford. 1990. Population dynamics of a desert playa community in the Chihuahuan Desert. Southwestern Naturalist 35:393–402.

MacMahon, J.A. 1964. Additional observations on the courtship of Metcalf's salamander, *Plethodon jordani* (*metcalfi* phase). Herpetologica 20:67–69.

MacNamara, M.C. 1977. Food habits of terrestrial adult migrants and immature red efts of the red-spotted newt *Notophthalmus viridescens*. Herpetologica 33:127–132.

MacTeague, L. and P.T. Northern. 1993. Underwater vocalization by the foothill yellow-legged frog *(Rana boylii)*. Transactions of the Western Section of the Wildlife Society 29:1–7.

Macy, R.W. 1960. On the life cycle of *Megalodiscus microphagus* Ingles (Trematoda: Paramphistomatidae). Journal of Parasitology 46:662.

Maddison, D.R. (Coord./Ed.) 1998. The Tree of Life. A multi-authored, distributed internet project containing information about phylogeny and biodiversity. http://phylogeny.arizona.edu/tree/phylogeny/ html

Madej, R.F. 1998. Discovery of green salamanders in Indiana and a distributional survey. Pp. 55–60. *In* Lannoo, M.J. (Ed.), Status and Conservation of Midwestern Amphibians. University of Iowa Press, Iowa City, Iowa.

Madison, D.M. 1969. Homing behaviour of the red-cheeked salamander, *Plethodon jordani*. Animal Behaviour 17:25–39.

Madison, D.M. 1972. A mechanism of homing orientation in salamanders involving chemical cues. Pp. 485–498. *In* Galler, S.R., K. Schmidt-Koenig, G.J. Jacobs and R.E. Belleville (Eds.), Animal Orientation and Navigation. NASA Special Publication, Number 262, Washington, D.C.

Madison, D.M. 1997. The emigration of radio-implanted spotted salamanders, *Ambystoma maculatum*. Journal of Herpetology 31:542–551.

Madison, D.M. and C.R. Shoop. 1970. Homing behavior, orientation, and home range of salamanders tagged with tantalum-182. Science 168:1484–1487.

Madison, D.M., K. Wareing, J.C. Maerz and V.S. Lamoureux. 1999. Oviposition behavior in the red-backed salamander *(Plethodon cinereus)*: implications of suspending a clutch. Herpetological Review 30:209–210.

Madsen, H and F. Frandsen. 1989. The spread of freshwater snails including those of medical and veterinary importance. Acta Tropica 45:139–146.

Maeda, N. and M. Matsui. 1989. Frogs and Toads of Japan. Bun-ichi Sogo Shuppan Company, Limited, Tokyo, Japan.

Maerz, J.C. and J.M. Karuzas. 2003. Natural history notes: *Plethodon cinereus* (eastern red-backed salamander). Cannibalism. Herpetological Review 34:354.

Maerz, J.C. and D.M. Madison. 2000. Environmental variation and territorial behavior in a terrestrial salamander. Pp. 395–406. *In* Bruce, R.C., R.G. Jaeger and L.D. Houck (Eds.), The Biology of Plethodontid Salamanders. Kluwer Academic/Plenum Publishers, New York.

Maglia, A.M. 1996. Ontogeny and feeding ecology of the red-backed salamander, *Plethodon cinereus*. Copeia 1996:576–586.

Maglia, A.M. 1998. Phylogenetic relationships of extant pelobatid frogs (Anura: Pelobatoidea): Evidence from adult morphology. Natural History Museum, Number 10, University of Kansas, Lawrence, Kansas.

Maglia, A.M. 1999. The adult skeleton of *Spea multiplicata* and a comparison of the osteology of the pelobatid frogs (Anura, Pelobatidae). Alytes 16:148–164.

Maha, G.C., L.R. Maxson and R. Highton. 1983. Immunological evidence for the validity of *Plethodon kentucki*. Journal of Herpetology 17:398–400.

Mahaney, P.A. 1994. Effects of freshwater petroleum contamination on amphibian hatching and metamorphosis. Environmental Toxicology and Chemistry 13:259–265.

Mahler, B.J. and F.L. Lynch. 1999. Muddy waters: temporal variation in sediment discharging from a karst spring. Journal of Hydrology 214:165–178.

Mahler, B.J., F.L. Lynch and P.C. Bennett. 1999. Mobile sediment in an urbanizing karst aquifer: implications for contaminant transport. Environmental Geology 39:25–38.

Mahon, R. and K. Aiken. 1977. The establishment of the North American bullfrog, *Rana catesbeiana* (Amphibia, Anura, Ranidae) in Jamaica. Journal of Herpetology 11:197–199.

Mahoney, M.J. 2001. Molecular systematics of *Plethodon* and *Aneides* (Caudata: Plethodontidae: Plethodontini): phylogenetic analysis of an old and rapid radiation. Molecular Phylogenetics and Evolution 18:174–188.

Mahoney, M. 2004. Molecular systematics and phylogeography of the *Plethodon elongatus* species group: combining phylogenetic and population genetic methods to investigate species history. Molecular Ecology 13:149–166.

Mahony, M.J. 1993. The status of frogs in the Watagan Mountains area of the Central Coast of New South Wales. Pp. 257–264. *In* Lunney, D. and D. Ayers (Eds.), Herpetology in Australia: A Diverse Discipline. Surrey Beatty and Sons, Sydney, New South Wales, Australia.

Mahoney, M.J. 2004. Molecular systematics and phylogeography of the *Plethodon elongatus* group: combining phylogenetic and population genetic methods to investigate species history. Molecular Ecology 13:149–166.

Mahrdt, C.R. 1975. The occurrence of *Ensatina eschscholtzii eschscholtzii* in Baja California, Mexico. Journal of Herpetology 9:240–242.

Mahrdt, C.R. and B.H. Banta. 1997. Natural history notes: *Aneides lugubris* (arboreal salamander). Predation. Herpetological Review 28:81.

Main, A.R. 1961. *Crinia insignifera* (Moore) (Anura: Leptodactylidae) on Rottnest Island. Journal of the Royal Society of Western Australia 44:10–13.

Main, A.R. 1965. Frogs of Southern Western Australia. Western Australia Naturalists Club Handbook, 8, Perth, Western Australia.

Maiorana, V.C. 1976. Size and environmental predictability for salamanders. Evolution 30:599–613.

Maiorana, V.C. 1977a. Observations of salamanders (Amphibia, Urodela, Plethodontidae) dying in the field. Journal of Herpetology 11:1–5.

Maiorana, V.C. 1977b. Tail autotomy, functional conflicts and their resolution by a salamander. Nature 265:533–535.

Maiorana, V.C. 1978a. Difference in diet as an epiphenomenon: space regulates salamanders. Canadian Journal of Zoology 56:1017–1025.

Maiorana, V.C. 1978b. Behavior of an unobservable species: diet selection by a salamander. Copeia 1978:664–672.

Maisonneuve, C. and S. Rioux. 2001. Importance of riparian habitats for small mammal and herpetofaunal communities in agricultural landscapes of southern Quebec. Agriculture, Ecosystems and Environment 83:165–175.

Maksymovitch, E. and P.A. Verrell. 1992. The courtship behavior of the Santeetlah dusky salamander, *Desmognathus santeetlah* Tilley (Amphibia: Caudata: Plethodontidae). Ethology 90:236–246.

Malek, E.A. 1977. Natural infection of the snail *Biomphalaia obstructa* in Louisiana with *Ribeiroia ondatrae* and *Chinoparyphium flexum*, with notes on the genus *Psilostomum*. Tulane Studies in Zoology 19:131–136.

Malewitz, T.D. 1956. Intestinal parasitism of some mid-western salamanders. American Midland Naturalist 55:434–436.

Mallory, M.L. and R. Lariviere. 1998. Wood duck, *Aix sponsa*, eats mink frogs, *Rana septentrionalis*. Canadian Field-Naturalist 112:714–715.

Malmberg, G.T. 1965. Available water supply of the Las Vegas ground water basin Nevada. Department of the Interior, U.S. Geological Survey, Water-supply Paper, Number 1780, Government Printing Office, Washington, D.C.

Malmos, K.B. 1992. Morphology, mating calls, and allozymes of *Bufo a. americanus* and *Bufo w. woodhousii* from a hybrid zone in eastern Nebraska and western Iowa. Master's thesis. University of Nebraska at Omaha, Omaha, Nebraska.

Malmos, K.B., R. Reed and B. Starrett. 1995. Hybridization between *Bufo woodhousii* and *Bufo punctatus* from the Grand Canyon region of Arizona. Great Basin Naturalist 55:368–371.

Malmos, K.B., B.K. Sullivan and T. Lamb. 2001. Calling behavior and directional hybridization between two toads (*Bufo microscaphus* and *B. woodhousii*) in Arizona. Evolution 55:626–630.

Malone, J.H. 1998. Geographic distribution: *Syrrhophus marnockii* (cliff chirping frog). Herpetological Review 29:247.

Malone, J.H. 1999. Natural history notes: *Bufo speciosus* (Texas toad). Diet. Herpetological Review 30:222–223.

Maly, E.J. and M.P. Maly. 1974. Dietary differences between two co-occurring calanoid copepod species. Oecologia 17:325–333.

Maly, E.J., S. Scheonholtz and M.T. Arts. 1980. The influence of flat-worm predation on zooplankton inhabiting small ponds. Hydrobiologica 76:233–240.

Maness, J.D. and V.H. Hutchison. 1980. Acute adjustment of thermal tolerance in vertebrate ectotherms following exposure to critical thermal maxima. Journal of Thermal Biology 5:225–233.

Mangano, F.T., T. Fukuzawa, W.C. Johnson and J.T. Bagnara. 1992. Intrinsic pigment cell stimulating activity in the skin of the leopard frog, *Rana pipiens*. Journal of Experimental Zoology 263:112–118.

Maniero, G.D. and C. Carey. 1997. Changes in selected aspects of immune function in the leopard frog, *Rana pipiens*, associated with exposure to cold. Journal of Comparative Physiology B 167:256–263.

Manion, J.J. and L. Cory. 1952. Winter kill of *Rana pipiens* in shallow ponds. Herpetologica 8:32.

Manis, M.L. and D.L. Claussen. 1986. Environmental and genetic influences on the thermal physiology of *Rana sylvatica*. Journal of Thermal Biology 11:31–36.

Manlow, S.W. 1994. Geographic distribution: *Ascaphus truei* (tailed frog). Herpetological Review 25:31.

Manly, B.F.J. 1977. The determination of key factors from life table data. Oecologia 31:111–117.

Mansueti, R. 1941. Sounds produced by the slimy salamander. Copeia 1941:266–267.

Manville, R.H. 1957. Amphibians and reptiles of Glacier National Park, Montana. Copeia 1957:308–309.

Mao, J., D.E. Green, G.M. Fellers and V.G. Chincar. 1999. Molecular characterization of iridoviruses isolated from sympatric amphibians and fish. Virus Research 63:45–52.

Marangio, M.S. 1975. Phototaxis in larvae and adults of the marbled salamander, *Ambystoma opacum*. Journal of Herpetology 9:293–297.

Marangio, M.S. 1978. The occurrence of neotenic rough-skinned newts (*Taricha granulosa*) in montane lakes of southern Oregon. Northwest Science 52:343–350.

Marangio, M.S. and J.D. Anderson. 1977. Soil moisture preference and water relations of the marbled salamander, *Ambystoma opacum* (Amphibia, Urodela, Ambystomatidae). Journal of Herpetology 11:169–176.

Marchand, P.J. 1991. Life in the Cold: An Introduction to Winter Ecology. Second edition. University Press of New England, Hanover, New Hampshire.

Marchisin, A. and J.D. Anderson. 1978. Strategies employed by frogs and toads (Amphibia, Anura) to avoid predation by snakes (Reptilia, Serpentes). Journal of Herpetology 12:151–155.

Marco, A. and A.R. Blaustein. 1999. The effects of nitrite on behavior and metamorphosis in Cascades frogs (*Rana cascadae*). Environmental Toxicology and Chemistry 18:946–949.

Marco, A., C. Quilchano and A.R. Blaustein. 1999. Sensitivity to nitrate and nitrite in pond-breeding amphibians from the Pacific Northwest, USA. Environmental Toxicology and Chemistry 18:2836–2839.

Marco, A., D. Cash, L.K. Belden and A.R. Blaustein. 2001. Sensitivity to urea fertilization in three amphibian species. Archives of Environmental Contamination and Toxicology 40:406–409.

Marcum, C. 1994. Ecology and natural history of four plethodontid species in the Fernow Experimental Forest, Tucker County, West Virginia. Master's thesis. Marshall University, Huntington, West Virginia.

Mardht, C.R. and F.T. Knefler. 1972. Pet or pest? The African clawed frog. Environment Southwest 446:2–5.

Mardht, C.R. and F.T. Knefler. 1973. The clawed frog—again. Environment Southwest 450:1–3.

Mareel, M., R. Bruneel, K. Tweedell, R. McKinnell and D. Tarin. 1985. Temperature-dependence of PNKT4B frog carcinoma cell invasion *in vitro*. Pp. 335–338. *In* Hellmann, K. and S.A. Eccles (Eds.), Treatment of Metastasis: Problems and Prospects. Taylor and Francis, London.

Maret, T.J. and J.P. Collins. 1994. Individual responses to population size structure: the role of size variation in controlling expression of a trophic polymorphism. Oecologia 100:279–285.

Marinkelle, C.J. and N.J. Williams. 1964. The toad, *Bufo marinus*, as a potential mechanical vector of eggs of *Ascaris lumbricoides*. Journal of Parasitology 50:427–428.

Marks, S.B. 2000. Skull development in two plethodontid salamanders (genus *Desmognathus*) with different life histories. Pp. 261–276. *In* Bruce, R.C., R.G. Jaeger and L.D. Houck (Eds.), The Biology of Plethodontid Salamanders. Kluwer Academic/Plenum Publishers, New York.

Marks, S.B. and A. Collazo. 1998. Direct development in *Desmognathus aeneus* (Caudata: Plethodontidae): a staging table. Copeia 1998:637–648.

Marler, C.A., J. Chu and W. Wilczynski. 1995. Arginine vasotoxin injection increases probability of calling in cricket frogs, but causes call changes characteristic of less aggressive males. Hormones and Behavior 29:554–570.

Marlow, R.W., J.M. Brode and D.B. Wake. 1979. A new salamander, genus *Batrachoseps*, from the Inyo Mountains of California, with a discussion of relationships in the genus. Contributions in Science, Natural History Museum of Los Angeles County, Number 308, Los Angeles, California.

Marnell, L.F. 1977. Herpetofauna of Glacier National Park. Northwestern Naturalist 78:17–33.

Marshall, C.J., L.S. Doyle and R.H. Kaplan. 1990. Intraspecific and sex-specific oophagy in a salamander and a frog: reproductive convergence of *Taricha torosa* and *Bombina orientalis*. Herpetologica 46:395–399.

Marshall, J.L. 1996. Natural history notes: *Eurycea cirrigera* (southern-two lined salamander). Nest site. Herpetological Review 27:75–76.

Marshall, J.L. 1997. Natural history notes: *Plethodon mississippi* (Mississippi slimy salamander). Vocalization. Herpetological Review 28:145.

Marshall, J.L. 1999. The life-history traits of *Eurycea guttolineata* (Caudata, Plethodontidae), with implications for life-history evolution. Alytes 16:97–110.

Marshall, J.L. and C.D. Camp. 1995. Aspects of the feeding ecology of the little grass frog, *Pseudacris ocularis* (Anura: Hylidae). Brimleyana 22:1–7.

Marti, E. and S.G. Fisher. 1998. Factors controlling algal growth in the ponds at Ramsey Canyon Preserve. The Nature Conservancy contract report, Tucson, Arizona.

Martin, A.P., G.J.P. Naylor and S.R. Palumbi. 1992. Rates of mitochondrial DNA evolution in sharks are slow compared with mammals. Nature 357:153–155.

Martin, D.L., R.G. Jaeger and C.P. Labat. 1986. Territoriality in an *Ambystoma* salamander? Support for the null hypothesis. Copeia 1986:725–730.

Martin, D.S. and S.S. Desser. 1990. A light and electron microscopic study of *Trypanosoma fallisi* n. sp. in toads (*Bufo americanus*) from Algonquin Park, Ontario. Journal of Protozoology 37:199–206.

Martin, D.S. and S.S. Desser. 1991a. Infectivity of cultured *Trypanosoma fallisi* Kinetoplastida to various anuran species and its evolutionary implications. Journal of Parasitology 77:498–500.

Martin, D.S. and S.S. Desser. 1991b. Development of *Trypanosoma fallisi* in the leech, *Desserobdella picta*, in toads (*Bufo americanus*), and in vitro. A light and electron microscopic study. Parasitology Research 77:18–26.

Martin, D.S., S.S. Desser and H. Hong. 1992. Allozyme comparison of three *Trypanosoma* sp. (Kinetoplastida Trypanosomatidae) of toads and frogs by starch-gel electrophoresis. Journal of Parasitology 78:317–322.

Martin, D.S., A.D.G. Wright, J.R. Barta and S.S. Desser. 2002. Phylogenetic position of the giant anuran trypanosomes *Trypanosoma chattoni, Trypanosoma fallisi, Trypanosoma mega, Trypanosoma neveulemairei*, and *Trypanosoma ranarum* inferred from 18S rRNA gene sequences. Journal of Parasitology 88:566–571.

Martin, G.W. 1966a. A progenetic trematode (Lecithodendriidae) life cycle involving *Rana aurora*. Ph.D. dissertation. Oregon State University, Corvallis, Oregon.

Martin, G. 1966b. *Caudouterina rhyacotritoni* gen. n. et sp. n. (Trematoda: Digenea) from the Olympic salamander. Journal of Parasitology 52:935–938.

Martin, J.B, N.B. Witherspoon and M.H.A. Keenleyside. 1974. Analysis of feeding behavior in the newt *Notophthalmus viridescens*. Canadian Journal of Zoology 52:277–281.

Martin, P.S. 1958. A biogeography of reptiles and amphibians in the Gomez Farias Region, Tamaulipas, Mexico. Miscellaneous Publications of the Museum of Zoology, Number 101, University of Michigan Ann Arbor, Michigan.

Martin, R.F. 1972. Evidence from osteology. Pp. 279–309. *In* Blair, W.F. (Ed.), Evolution in the Genus *Bufo*. University of Texas Press, Austin Texas.

Martin, R.F. 1973. Osteology of North American *Bufo*: the *americanus, cognatus*, and *boreas* species groups. Herpetologica 29:375–387.

Martof, B.S. 1952. Early transformation of the green frog, *Rana clamitans* Latreille. Copeia 1952:115–116.

Martof, B.S. 1953a. Home range and movements of the green frog, *Rana clamitans*. Ecology 34:529–543.

Martof, B.S. 1953b. The "spring lizard" industry, a factor in salamander distribution and genetics. Ecology 34:436–437.

Martof, B.S. 1955. Observations on the life history and ecology of the amphibians of the Athens area, Georgia. Copeia 1955:166–170.

Martof, B.S. 1956a. Amphibians and Reptiles of Georgia. University of Georgia Press, Athens, Georgia.

Martof, B.S. 1956b. Growth and development of the green frog, *Rana clamitans*, under natural conditions. American Midland Naturalist 55:101–117.

Martof, B.S. 1960. Autumnal breeding of *Hyla crucifer*. Copeia 1960:58–59.

Martof, B.S. 1962a. The behavior of Fowler's toad under various conditions of light and temperature. Physiological Zoology 35:38–46.

Martof, B.S. 1962b. Some aspects of the life history and ecology of the salamander *Leurognathus*. American Midland Naturalist 67:1–35.

Martof, B.S. 1963. Some observations on the herpetofauna of Sapelo Island, Georgia. Herpetologica 19:70–72.

Martof, B.S. 1968. *Ambystoma cingulatum*. Pp. 57.1–57.2. Catalogue of American Amphibians and Reptiles. Society for the Study of Amphibians and Reptiles, St. Louis, Missouri.

Martof, B.S. 1969. Prolonged inanition in *Siren lacertina*. Copeia 1969:285–289.

Martof, B.S. 1970. *Rana sylvatica*. Pp. 86.1–86.4. Catalogue of American Amphibians and Reptiles. Society for the Study of Amphibians and Reptiles, St. Louis, Missouri.

Martof, B.S. 1973. *Siren lacertina*. Pp. 128.1–128.2. Catalogue of American Amphibians and Reptiles. Society for the Study of Amphibians and Reptiles, St. Louis, Missouri.

Martof, B.S. 1974. *Siren*. Pp. 152:1–152.2. Catalogue of American Amphibians and Reptiles. Society for the Study of Amphibians and Reptiles, St. Louis, Missouri.

Martof, B.S. 1975a. *Hyla squirella*. Pp. 168.1–168.2. Catalogue of American Amphibians and Reptiles. Society for the Study of Amphibians and Reptiles, St. Louis, Missouri.

Martof, B.S. 1975b. *Pseudotriton montanus*. Pp. 166.1–166.2. Catalogue of American Amphibians and Reptiles. Society for the Study of Amphibians and Reptiles, St. Louis, Missouri.

Martof, B.S. and R.L. Humphries. 1959. Geographic variation in the wood frog *Rana sylvatica*. American Midland Naturalist 61:350–389.

Martof, B.S. and F.L. Rose. 1963. Geographic variation in southern populations of *Desmognathus ochrophaeus*. American Midland Naturalist 69:376–425.

Martof, B.S. and D.C. Scott. 1957. The food of the salamander *Leurognathus*. Ecology 38:494–501.

Martof, B.S. and E.F. Thompson Jr. 1958. Reproductive behaviour of the chorus frog, *Pseudacris nigrita*. Behaviour 13:243–258.

Martof, B.S., W.M. Palmer, J.R. Bailey and J.R. Harrison III. 1980. Amphibians and Reptiles of the Carolinas and Virginia. University of North Carolina Press, Chapel Hill, North Carolina.

Maruyama, J.T. Noguchi, J.K. Jean, T. Harada and K. Hashimoto. 1984. Occurrence of tetrodotoxin in the starfish *Astropecten latespinosus*. Experimentia 40:1395.

Marvel, B. 1972. A feeding observation on the yellow-bellied water snake, *Natrix erythrogaster flavigaster*. Bulletin of the Maryland Herpetological Society 8:52.

Marvin, G.A. 1990. Behavioral interactions between the salamanders *Plethodon kentucki* and *Plethodon glutinosus*: evidence for interspecific interference competition. Master's thesis. Eastern Kentucky University, Richmond, Kentucky.

Marvin, G.A. 1996. Life history and population characteristics of the salamander *Plethodon kentucki* with a review of *Plethodon* life histories. American Midland Naturalist 136:385–400.

Marvin, G.A. 1998. Territorial behavior of the plethodontid salamander *Plethodon kentucki*: Influence of habitat structure and population density. Oecologia 114:113–144.

Marvin, G.A. 2003. Aquatic and terrestrial locomotor performance in a semiaquatic plethodontid salamander (*Pseudotriton ruber*): influence of acute temperature, thermal acclimation, and body size. Copeia 2003:704–713.

Marvin, G.A. and V.H. Hutchison. 1995. Avoidance response by adult newts (*Cynops pyrrhogaster* and *Notophthalmus viridescens*) to chemical alarm cues. Behavior 132:95–105.

Marvin, G.A. and V.H. Hutchison. 1996. Courtship behavior of the Cumberland Plateau woodland salamander, *Plethodon kentucki* (Amphibia: Plethodontidae), with a review of courtship in the genus *Plethodon*. Ethology 102:285–303.

Maser, C. and J.M. Trappe. 1984. The seen and unseen world of the fallen tree. U.S.D.A. Forest Service, Pacific Northwest Forest and Range Experiment Station, General Technical Report, PNW-164, Washington, D.C.

Maskell, A.J., J. H. Waddle and K.G. Rice. 2003. Natural history notes: *Osteopilus septentrionalis* (Cuban treefrog). Diet. Herpetological Review 34:113.

Maslin, T.P. 1950. The production of sound in caudate Amphibia. Pp. 29–45. University of Colorado Studies in Biology, Number 1, Boulder, Colorado.

Maslin, T.P. 1963a. Notes on a collection of herpetozoa from the Yucatan Peninsula of Mexico. University of Colorado Studies, Series in Biology, Number 9, Boulder, Colorado.

Maslin, T.P. 1963b. Notes on some anuran tadpoles from Yucatan, Mexico. Herpetologica 19:122–128.

Maslin, T.P., Jr. 1939. Egg-laying of the slender salamander (*Batrachoseps attenuatus*). Copeia 1939:209–212.

Massey, A. 1990. Notes on the reproductive ecology of red-spotted newts (*Notophthalmus viridescens*). Journal of Herpetology 24:106–107.

Masta, S.E., B.K. Sullivan, T. Lamb and E.J. Routman. 2002. Molecular systematics, hybridization, and phylogeography of the *Bufo americanus* complex in eastern North America. Molecular Phylogenetics and Evolution 24:302–314.

Masters, B.S. and D.C. Forester. 1995. Kin recognition in a brooding salamander. Proceedings of the Royal Society of London B 261:43–48.

Mathews, K.R., R.A. Knapp and K.L. Pope. 2002. Garter snake distributions in high-elevation aquatic ecosystems: is there a link with declining amphibian populations and nonnative trout introductions? Journal of Herpetology 36:16–22.

Mathews, R.C. and A.C. Echternacht. 1984. Herpetofauna of the spruce-fir ecosystem in the southern Appalachian Mountain regions, with emphasis on the Great Smoky Mountains National Park. Pp. 155–167. *In* White, P.S. (Ed.), The Southern Appalachian Spruce-fir Ecosystem: Its Biology and Threats. National Park Service Research/Resources Management Report SER-71, Great Smoky Mountains National Park, Gatlinburg, Tennessee.

Mathews, R.C., Jr. 1982. Predator stoneflies: role in freshwater stream communities. Journal of the Tennessee Academy of Science 57:82–83.

Mathews, R.C., Jr. and E.L. Morgan. 1982. Toxicity of Anakeesta leachates to shovel-nosed salamanders, Great Smoky Mountains National Park. Journal of Environmental Quality 11:102–106.

Mathews, T.C. 1971. Genetic changes in a population of boreal chorus frogs (*Pseudacris triseriata*) polymorphic for color. American Midland Naturalist 85:208–221.

Mathis, A. 1989. Do seasonal spatial distributions in a terrestrial salamander reflect reproductive behavior or territoriality? Copeia 1989:788–791.

Mathis, A. 1991. Territories of male and female terrestrial salamanders: costs, benefits, and intersexual spatial associations. Oecologia 86:433–440.

Matson, T.O. 1988. The *Hyla chrysoscelis-Hyla versicolor* complex in Ohio. Ph.D. dissertation. Kent State University, Kent, Ohio.

Matson, T.O. 1990. Estimation of numbers for a riverine *Necturus* population before and after TFM lampricide exposure. Kirtlandia 45:33–38.

Matson, T.O. 1998. Evidence for home ranges in mudpuppies and implications for impacts due to episodic applications of the lampricide TFM. Pp. 278–287. *In* Lannoo, M.J. (Ed.), Status and Conservation of Midwestern Amphibians. University of Iowa Press, Iowa City, Iowa.

Matson, T.O. 1999. An introduction to the natural history of the frogs and toads of Ohio. http://www.cmnh.org/research/vertzoo/ frogs.html.

Matthes, B. 1855. Die hemibatrachier im Allgemeinen und die hemibatrachier von Nord-Amerika im Speciellen. Allgemeine Deutsche Naturhistorische Zeitung, N.S. Volume 1:249–280.

Matthews, K. and K. Pope. 1999. Radiotracking mountain yellow-legged frogs. Journal of Herpetology 33:615–624.

Matthews, T. and D. Pettus. 1966. Color inheritance in *Pseudacris triseriata*. Herpetologica 22:269–275.

Matustik, D. 1990. State offers aid where rubber meets the toad. Austin American-Statesman, 25 July.

Mauger, D. 1988. Observations on calling behavior of bullfrogs in relation to male mating strategy. Bulletin of the Chicago Herpetological Society 23:57–59.

Maughan, O.E., P. Laumeyer, R.L. Wallace and M.G. Wickham. 1980. Distribution of the tailed frog, *Ascaphus truei* Stejneger, in several drainages in northcentral Idaho. Herpetological Review 11:15–16.

Maughan, O.E., M.G. Wickham, P. Laumeyer and R.L. Wallace. 1976. Records of the Pacific giant salamander, *Dicamptodon ensatus*, (Amphibia, Urodela, Ambystomatidae) from the Rocky Mountains in Idaho. Journal of Herpetology 10:249–251.

Maunder, J.E. 1997. Amphibians of Newfoundland and Labrador: status changes since 1983. Pp. 93–99. *In* Green, D.M. (Ed.), Amphibians in Decline: Canadian Studies of a Global Problem. Herpetological Conservation, Number 1, Society for the Study of Amphibians and Reptiles, St. Louis, Missouri.

Maurer, E.F. and A. Sih. 1996. Ephemeral habitats and variation in behavior and life history: comparisons of sibling salamander species. Oikos 76:337–349.

Maxell, B.A. 1999. Herpetology in Montana: a history, species checklist, dot distribution maps, museum records, and indexed bibliography. Master's thesis. University of Montana, Missoula, Montana.

Maxell, B.A. 2000. Management of Montana's amphibians: a review of risk factors to population viability and accounts on the identification, distribution, taxonomy, habitat use, natural history, and the status and conservation of individual species. Report to U.S.F.S. Region 1, Order Number 43-0343-0-0224, University of Montana, Wildlife Biology Program, Missoula, Montana.

Maxell, B.A., J.K. Werner, P. Hendricks and D.L. Flath. 2003. Herpetology in Montana: A History, Status Summary, Checklists, Dichotomous Keys, Accounts for Native, Potentially Native, and Exotic Species, and Indexed Bibliography. Society for Northwestern Vertebrate Biology, Northwest Fauna, Number 5, Olympia, Washington.

Maxon, R.D., K.D. Moberg and L.R. Maxon. 1987. Letter to the editor: data from hybrid species supporting the MC'F assortment–exclusion hypothesis. Molecular Biology and Evolution 4:70–73.

Maxson, L.R. and A.C. Wilson. 1974. Convergent morphological evolution detected by studying proteins of treefrogs in the *Hyla eximia* species group. Science 185:66–68.

Maxson, L.R., E. Pepper and R.D. Maxson. 1977. Immunological resolution of a diploid-tetraploid species complex of treefrogs. Science 197:1012–1013.

Maxson, L.R., A.R. Song and R. Lopata. 1981. Phylogenetic relationships among North American toads, genus *Bufo*. Biochemical and Systematic Ecology 9:347–350.

May, R.M. 1976. Models for single populations. Pp. 4–25. *In* May, R.M. (Ed.), Theoretical Ecology: Principles and Applications. W.B. Saunders, Philadelphia, Pennsylvania.

May, R.M. 1981. Patterns in multispecies communities. Pp. 197–227. *In* May, R.M. (Ed.), Theoretical Ecology. Sinauer Associates, Sunderland, Massachusetts.

May, R.M. 1990. Taxonomy as destiny. Nature 347:129–130.

Mayden, R.L. 1997. A hierarchy of species concepts: the denouncement in the saga of the species problem. Pp. 381–424. *In* Claridge, M.F., H.A. Dawah and M.R. Wilson (Eds.), Species: The Units of Biodiversity. Chapman and Hall, London.

Mayden, R.L. and R.M. Wood. 1995. Systematics, species concepts, and the ESU in biodiversity and conservation biology. American Fisheries Society Symposium 17:58–113.

Mayer, F.L. and M.R. Ellersieck. 1986. Manual of acute toxicity: interpretation and data base for 410 chemicals and 66 species of freshwater animals. U.S. Fish and Wildlife Service, Resource Publication, Number 160, Washington, D.C.

Mayhew, W.W. 1965. Adaptations of the amphibian *Scaphiopus couchi* to desert conditions. American Midland Naturalist 74:95–109.

Maynard, E.A. 1934. The aquatic migration of the toad, *Bufo americanus* Le Conte. Copeia 1934:174–177.

Mayr, E. 1942. Systematics and the Origin of Species. Columbia University Press, New York.

Mayr, E. 1969. Principles of Systematic Zoology. Harvard University Press, Cambridge, Massachusetts.

Mayr, E. 1982. The Growth of Biological Thought. Harvard University Press, Cambridge, Massachusetts.

Mayr, E. 1996. What is a species and what is not? Philosophy of Science 63:262–277.

Mazerolle, M.J. 2001. Amphibian activity, movement patterns, and body size in fragmented peat bogs. Journal of Herpetology 35:13–20.

McAlister, W. 1954. Natural history notes on the barking frog. Herpetologica 10:197–199.

McAlister, W.H. 1963. A post-breeding concentration of the spring peeper. Herpetologica 19:293.

McAllister, C.T. 1987. Protozoan and metazoan parasites of Strecker's chorus frog, *Pseudacris streckeri streckeri* (Anura: Hylidae), from north-central Texas. Proceedings of the Helminthological Society of Washington 54:271–274.

McAllister, C.T. 1991. Protozoan, helminth, and arthropod parasites of the spotted chorus frog, *Pseudacris clarkii* (Anura: Hylidae), from north-central Texas (USA). Journal of the Helminthological Society of Washington 58:51–56.

McAllister, C.T. and D.B. Conn. 1990. Occurrence of tetrathydia of *Mesocestoides* sp. (Cestoidea: Cyclophyllidea) in North American anurans (Amphibia). Journal of Wildlife Diseases 26:540–543.

McAllister, C.T. and P.S. Freed. 1992. Larval *Abbreviata* sp. (Spirurida: Physalopteridae) in introduced Rio Grande chirping frogs, *Syrrhophus cystignathoides campi* (Anura: Leptodactylidae), from Houston, Texas. Texas Journal of Science 44:359–361.

McAllister, C.T. and V.R. McDaniel. 1992. Occurrence of larval *Contracaecum* sp. (Ascaridida: Anisakidae) in Rio Grande lesser sirens, *Siren intermedia texana* (Amphibia: Caudata), from south Texas. Journal of the Helminthological Society of Washington 59:239–240.

McAllister, C.T. and S.P. Tabor. 1985. Life history notes: *Gastrophryne olivacea* (Great Plains narrowmouth toad). Coexistence. Herpetological Review 16:109.

McAllister, C.T. and S.E. Trauth. 1993. Geographic distribution: *Bufo speciosus* (Texas toad). Herpetological Review 24:153.

McAllister, C.T. and S.E. Trauth. 1995. New host records for *Myxidium serotinum* (Protozoa: Myxosporea) from North American amphibians. Journal of Parasitology 81:485–488.

McAllister, C.T. and S.E. Trauth. 1996a. Ultrastructure of *Cepidietta virginiensis* (Protista: Haptophryidae) from the gall bladder of the pickerel frog, *Rana palustris*, in Arkansas. Proceedings of the Arkansas Academy of Science 50:133–136.

McAllister, C.T. and S.E. Trauth. 1996b. Food habits of paedomorphic mole salamanders, *Ambystoma talpoideum* (Caudata: Ambystomatidae), from northeastern Arkansas. Southwestern Naturalist 41: 62–64.

McAllister, C.T. and S.J. Upton. 1987a. Parasites of the Great Plains narrowmouth toad *(Gastrophryne olivacea)* from northern Texas. Journal of Wildlife Diseases 23:686–688.

McAllister, C.T. and S.J. Upton. 1987b. Endoparasites of the small-mouthed salamander, *Ambystoma texanum* (Caudata: Ambystomatidae) from Dallas County, Texas. Proceedings of the Helminthological Society of Washington 54:258–261.

McAllister, C.T., S.E. Trauth and C.R. Bursey. 1995b. Parasites of the pickerel frog, *Rana palustris* (Anura: Ranidae), from the southern part of its range. Southwestern Naturalist 40:111–116.

McAllister, C.T., S.E. Trauth and C.R. Bursey. 1995d. Metazoan parasites of the graybelly salamander, *Eurycea multiplicata griseogaster* (Caudata: Plethodontidae), from Arkansas. Journal of the Helminthological Society of Washington 62:70–73.

McAllister, C.T., S.E. Trauth and B.G. Cochran. 1995c. Endoparasites of the ringed salamander, *Ambystoma annulatum* (Caudata: Ambystomatidae), from Arkansas. Southwestern Naturalist 40:327–330.

McAllister, C.T., S.E. Trauth and B.L.J. Delvinquier. 1995a. Ultrastructural observations on *Myxidium serotinum* (Protozoa: Myxosporea) from *Bufo speciosus* (Anura: Bufonidae), in Texas. Journal of the Helminthological Society of Washington 62:229–232.

McAllister, C.T., S.E. Trauth and L.W. Hinck. 1991. *Sphyranura euryceae* (Monogenea) on *Eurycea* sp. (Amphibia:Caudata), from northcentral Arkansas. Journal of the Helminthological Society of Washington 58:137–140.

McAllister, C.T., S.J. Upton and D.B. Conn. 1989. A comparative study of endoparasites in three species of sympatric *Bufo* (Anura: Bufonidae), from Texas. Proceedings of the Helminthological Society of Washington 56:162–167.

McAllister, C.T., S.J. Upton and S.E. Trauth. 1993b. Endoparasites of western slimy salamanders, *Plethodon albagula* (Caudata: Plethodontidae), from Arkansas. Journal of the Helminthological Society of Washington 60:124–126.

McAllister, C.T., S.E. Trauth, S.J. Upton and D.H. Jamieson. 1993a. Endoparasites of the bird-voiced treefrog, *Hyla avivoca* (Anura: Hylidae), from Arkansas. Journal of the Helminthological Society of Washington 60:140–143.

McAllister, C.T., S.J. Upton, S.E. Trauth and C.R. Bursey. 1995e. Parasites of wood frogs, *Rana sylvatica* (Ranidae), from Arkansas, with a description of a new species of *Eimeria* (Apicomplexa: Eimeriidae). Journal of the Helminthological Society of Washington 62: 143–149.

McAllister, C.T., C.R. Bursey, S.J. Upton, S.E. Trauth and B. Conn. 1995f. Parasites of *Desmognathus brimleyorum* (Caudata: Plethodontidae) from the Ouachita Mountains of Arkansas and Oklahoma. Journal of the Helminthological Society of Washington 62:150–156.

McAllister, C.T., S.R. Goldberg, S.E. Trauth, C.R. Bursey, H.J. Holshuh and B.G. Cochran. 1994. Helminths of the western lesser siren, *Siren intermedia nettingi* (Caudata: Sirenidae), from Arkansas. Journal of the Helminthological Society of Washington 61:234–238.

McAllister, K.R. 1995. Distribution of amphibians and reptiles in Washington state. Northwest Fauna 3:81–112.

McAllister, K.R. and W.P. Leonard. 1997. Washington State status report of the Oregon spotted frog. Washington Department of Fish and Wildlife. Olympia, Washington.

McAllister, K.R., W.P. Leonard and R.M. Storm. 1993. Spotted frog *(Rana pretiosa)* surveys in the Puget Trough of Washington, 1989–1991. Northwestern Naturalist 74:10–15.

McAllister, K.R., W.P. Leonard, D.W. Hayes, and R.C. Friesz. 1999. Washington state status report for the northern leopard frog. Washington Department of Fish and Wildlife, Management Program, Olympia, Washington.

McAllister, K.R., J. Skriletz, B. Hall and M.M. Garner. 1997. Natural history notes. *Taricha granulosa* (roughskin newt). Toxicity. Herpetological Review 28:82.

McAlpine, D.F. 1997a. Historical evidence does not suggest New Brunswick amphibians have declined. Pp. 117–127. *In* Green, D.M. (Ed.), Amphibians in Decline. Canadian Studies of a Global Problem. Herpetological Conservation, Number 1, Society for the Study of Amphibians and Reptiles, St. Louis, Missouri.

McAlpine, D.F. 1997b. A simple transect technique for estimating abundance of aquatic ranid frogs. Pp. 180–184. *In* Green, D.M. (Ed.), Amphibians in Decline. Canadian Studies of a Global Problem. Herpetological Conservation, Number 1, Society for the Study of Amphibians and Reptiles, St. Louis, Missouri.

McAlpine, D.F. 1997c. Helminth communities in bullfrogs *(Rana catesbeiana)*, green frogs *(Rana clamitans)*, and leopard frogs *(Rana pipiens)* from New Brunswick, Canada. Canadian Journal of Zoology 75:1883–1890.

McAlpine, D.F. and M.D.B. Burt. 1998. Helminths of bullfrogs, *Rana catesbeiana*, green frogs, *R. clamitans*, and leopard frogs, *R. pipiens* in New Brunswick. Canadian Field-Naturalist 112:50–68.

McAlpine, D.F. and T.G. Dilworth. 1989. Microhabitat and prey size among three species of *Rana* (Anura: Ranidae) sympatric in eastern Canada. Canadian Journal of Zoology 67:2244–2252.

McAlpine, D.F., N.M. Burgess and D.G. Busby. 1998. Densities of mink frogs, *Rana septentrionalis*, in New Brunswick forest ponds sprayed with the insecticide fenitrothion. Bulletin of Environmental Contamination and Toxicology 60:30–36.

McAlpine, D.F., T.J. Flecher, S.W. Gorham and I.T. Gorham. 1991. Distribution and habitat of the tetraploid gray treefrog, *Hyla versicolor*, in New Brunswick and eastern Maine. Canadian Field-Naturalist 105:526–596.

McAtee, W.L. 1921. Homing and other habits of the bull-frog. Copeia 96:39–40.

McAtee, W.L. 1933. Notes on the banded salamander (*Ambystoma opacum*). Copeia 1933:218–219.

McAuley, D. (Ed.). 1993. To save the toad: an exchange. Bastrop Advertiser, 26 August.

McAuley, D. 1990. The editor's uneasy chair: toad power. Bastrop Advertiser and County News, 2 August.

McAuley, D. 1994a. City to huddle on park land suit possibility. Bastrop Advertiser, 21 May.

McAuley, D. 1994b. Golf plan for state park gets okay by board. Bastrop Advertiser, 27 August.

McAuley, D. 1998a. International focus on local toads, frogs. Bastrop Advertiser, 17 January.

McAuley, D. 1998b. Planning effort aims to save toads. Bastrop Advertiser, 17 October.

McAuley, D. 1999a. Houston toad decline poses building issues. Bastrop Advertiser, 17 July.

McAuley, D. 1999b. LCRA drops toad planning effort. Bastrop Advertiser, 11 September.

McAuley, D. 1999c. Area officials getting serious about toads. Bastrop Advertiser, 28 October.

McAuliffe, J.R. 1978. Biological survey and management of sport-hunted bullfrog populations in Nebraska. Nebraska Game and Parks Commission, Lincoln, Nebraska.

McCallum, H. and A. Dobson. 1995. Detecting disease and parasite threats to endangered species and ecosystems. TRENDS in Ecology and Evolution 10:190–194.

McCallum, M.L. and S.E. Trauth. 2001. Are tadpoles of the Illinois chorus frog *(Pseudacris streckeri illinoensis)* cannibalistic? Transactions of the Illinois State Academy of Science 94:171–178.

McCallum, M.L. and S.E. Trauth. 2003. Natural history notes: *Acris crepitans* (northern cricket frog). Communal hibernacula. Herpetological Review 34:228.

McClanahan, L.L., Jr. 1967. Adaptations of the spadefoot toad, *Scaphiopus couchii*, to desert environments. Comparative Biochemistry and Physiology 20:73–99.

McCleary, E.C. 1989. Taxonomic status of a desmognathine salamander in West Virginia. Master's thesis. Kent State University, Kent, Ohio.

McClure, K.A. 1996. Ecology of *Pseudacris brachyphona*: a second look. Master's thesis. Marshall University, Huntington, West Virginia.

McClure, S. 1988. Letter to Robert M. Saunders, 13 April.

McCoid, M.J. 1985. An observation of reproductive behavior in a wild population of African clawed frogs, *Xenopus laevis*, in California. California Fish and Game 71:245–250.

McCoid, M.J. and T.H. Fritts. 1980a. Notes on the diet of feral population of *Xenopus laevis* (Pipidae) in California. Southwestern Naturalist 25:272–275.

McCoid, M.J. and T.H. Fritts. 1980b. Observations of feral populations of *Xenopus laevis* (Pipidae) in Southern California. Bulletin of the Southern California Academy of Sciences 79:82–86.

McCoid, M.J. and T.H. Fritts. 1989. Growth and fatbody cycles in feral populations of the African clawed frog, *Xenopus laevis* (Pipidae), in California [USA] with comments on reproduction. Southwestern Naturalist 34:499–505.

McCoid, M.J. and T.H. Fritts. 1993. Speculations on colonizing success of the African clawed frog *Xenopus laevis* Pipidae in California. South African Journal of Zoology 28:59–61.

McCoid, M.J. and T.H. Fritts. 1995. Female reproductive potential and winter growth of African clawed frogs (Pipidae: *Xenopus laevis*) in California. California Fish and Game 81:39–42.

McCoid, M.J., G.K. Pregill and R.M. Sullivan. 1993. Possible decline of *Xenopus* populations in Southern California. Herpetological Review 24:29–30.

McCollough, M.C.B. 1999. Conservation of Maine's amphibians and reptiles. Pp. 15–24. *In* Hunter, M.L., A.J.K. Calhoun and M. McCollough (Eds.), Maine Amphibians and Reptiles. University of Maine Press, Orono, Maine.

McCollum, S.A. and J. Van Buskirk. 1996. Costs and benefits of predation induced polymorphism in the gray treefrog *Hyla chrysoscelis*. Evolution 50:583–593.

McComb, W.C. and R.E. Noble. 1981. Herpetofaunal use of natural tree cavities and nest boxes. Wildlife Society Bulletin 9:261–267.

McComb, W.C., K. McGarigal and R.G. Anthony. 1993. Small mammal and amphibian abundance in streamside and upslope habitats of mature Douglas-fir stands, western Oregon. Northwest Science 67:7–15.

McCormack, J.C. 1965. Observations on the perch populations of Ullswater. Journal of Animal Ecology 34:463–478.

McCoy, C.J. 1982. Amphibians and reptiles in Pennsylvania. Carnegie Museum of Natural History, Special Publications, Number 6, Pittsburgh, Pennsylvania.

McCoy, C.J. 1992. Rediscovery of the mud salamander (*Pseudotriton montanus*, Amphibia, Plethodontidae) in Pennsylvania, with restriction of the type locality. Journal of the Pennsylvania Academy of Science 66:92–93.

McCoy, E.D. 1994. "Amphibian decline:" a scientific dilemma in more ways than one. Herpetologica 50:98–103.

McCrady, E. 1954. A new species of *Gyrinophilus* (Plethodontidae) from Tennessee caves. Copeia 1954:200–206.

McCranie, J.R. and G. Köhler. 2000. Notes on the type series of *Bufo valliceps* Wiegmann, with the designation of a lectotype. Southwestern Naturalist 45:71–74.

McCranie, J.R. and L.D. Wilson. 1987. The biogeography of the herpetofauna of the pine-oak woodlands of the Sierra Madre Occidental of Mexico. Milwaukee Public Museum Contributions in Biology and Geology, Number 72, Milwaukee, Wisconsin.

McCranie, J.R. and L.D. Wilson. 2002. The amphibians of Honduras. Contributions to Herpetology, Number 19, Society for the Study of Amphibians and Reptiles, St. Louis, Missouri.

McCurdy, H.M. 1931. Development of the sex organs in *Triturus torosus*. American Journal of Anatomy 47:367–403.

McDade, L.A., K.S. Bawa, H.A. Hespenheide and G.S. Hartshorn (Eds.). 1994. La Selva-Ecology and Natural History of a Neotropical Rainforest. University of Chicago Press, Chicago, Illinois.

McDaniel, V.R. 1975. Geographic distribution: *Ambystoma annulatum* (ringed salamander). Herpetological Review 6:115.

McDaniel, V.R. and J.E. Gardner. 1977. Cave fauna of Arkansas: vertebrate taxa. Proceedings of the Arkansas Academy of Science 31:68–71.

McDaniel, V.R. and D.A. Saugey. 1977. Geographic distribution: *Ambystoma annulatum* (ringed salamander). Herpetological Review 8:38.

McDiarmid, R.W. 1994. Amphibian diversity and natural history: an overview. Pp. 5–15. *In* Heyer, W.R., M.A. Donnelly, R.W. McDiarmid, L.C. Hayek and M.S. Foster (Eds.), Measuring and Monitoring Biological Diversity: Standard Methods for Amphibians. Smithsonian Institution Press, Washington, D.C.

McDiarmid, R.W. and K. Adler. 1974. Notes on territorial and vocal behavior of neotropical frogs of the genus *Centrolenella*. Herpetologica 30:75–78.

McDiarmid, R.W. and R. Altig. 1999. Tadpoles: The Biology of Anuran Larvae. University of Chicago Press, Chicago, Illinois.

McDiarmid, R.W. and M.S. Foster. 1975. Unusual sites for two Neotropical tadpoles. Journal of Herpetology 9:264–265.

McDiarmid, R.W. and M.S. Foster. 1987. Cocoon formation in another hylid frog, *Smilisca baudinii*. Journal of Herpetology 21:352–355.

McDiarmid, R.W., G.T. Bancroft and J.S. Godley. 1983. Large-scale operations management test of use of the white amur for control of problem plants. Reports 2 and 3: First and second year poststocking results. Volume 5. The herpetofauna of Lake Conway: Community analysis. Aquatic Plant Control Research Program, Technical Report A-78-2, U.S. Army Engineer Waterways Experiment Station, Vicksburg, Mississippi.

McDonald, M. and R.F. Marsh. 1995. Amphibian inventory of the Jarbridge Resource Area. Boise District. U.S. Bureau of Land Management, Technical Bulletin 95-4, Boise, Idaho.

McGarigal, K. and B.J. Marks. 1995. FRAGSTATS: spatial pattern analysis program for quantifying landscape structure. U.S.D.A. Forest Service, General Technical Report, PNW-GTR-351, Pacific Northwest Research Station, Portland, Oregon.

McGavin, M. 1978. Recognition of conspecific odors by the salamander *Plethodon cinereus*. Copeia 1978:356–358.

McGehee, R.T., R. Reams and M.E. Brown. 2001. Natural history notes: *Bufo valliceps* (Gulf Coast toad). Diet. Herpetological Review 32:101–102.

McGeoch, D.J. and A.J. Davison. 1999. The descent of human herpesvirus 8. Seminars in Cancer Biology 9:201–209.

McGill, M. and D.C. Brindley. 1978. The ultrastructural localization of glucose-6-phosphatase in the parotoid gland of *Bufo alvarius*. Copeia 1978:715–718.

McGinnis, S.M. 1984. Freshwater Fishes of California. California History Guide, Number 49, University of California Press, Berkeley and Los Angeles, California.

McGown, L.S., M.T. Dixon and L.K. Ammerman. 1994. Geographic distribution: *Syrrhophus cystignathoides* (Rio Grande chirping frog). Herpetological Review 25:32.

McGraw, R.L. 1998. Timber harvest effects on metamorphosed and larval long-toed salamanders (*Ambystoma macrodactylum*). Master's thesis. University of Montana, Missoula, Montana.

McGregor, J.H. and W.R. Teska. 1989. Olfaction as an orientation mechanism in migrating *Ambystoma maculatum*. Copeia 1989:779–781.

McKeever, S. 1977. Observations of *Corethrella* feeding on tree frogs (*Hyla*). Mosquito News 37:522–523.

McKenzie, D.S. 1970. Aspects of the autecology of the plethodontid salamander, *Aneides ferreus* (Cope). Dissertation Abstracts International 30B:5299.

McKenzie, D.S. and R.M. Storm. 1970. Patterns of habitat selection in the clouded salamander, *Aneides ferreus* (Cope). Herpetologica 26:450–454.

McKenzie, D.S. and R.M. Storm. 1971. Ontogenetic color patterns of the clouded salamander, *Aneides ferreus* (Cope). Herpetologica 27:142–147.

McKeown, S. 1996. A Field Guide to Reptiles and Amphibians in the Hawaiian Islands. Diamond Head Publishing, Los Osos, California.

McKinnell, R.G. 1962. Interspecific nuclear transplantation in frogs. Journal of Heredity 53:199–207.

McKinnell, R.G. 1964. Expression of the kandiyohi gene in triploid frogs produced by nuclear transplantation. Genetics 49:895–903.

McKinnell, R.G. 1965. Incidence and histology of renal tumors of leopard frogs from the north-central states. Annals of the New York Academy of Sciences 126:85–98.

McKinnell, R.G. 1967. Evidence for seasonal variation in incidence of renal adenocarcinoma in *Rana pipiens*. Proceedings of the Minnesota Academy of Sciences 34:173–175.

McKinnell, R.G. 1973. The Lucké frog kidney tumor and its herpesvirus. American Zoologist 13:97–114.

McKinnell, R.G. 1989. Neoplastic cells: modulation of the differentiated state. Pp. 199–236. *In* Di Berardino, M.A. and L. Etkin (Eds.), Genomic Adaptability in Cell Specialization. Plenum Publishing Company, New York.

McKinnell, R.G. and D.L. Carlson. 1997. The Lucké renal adenocarcinoma, an anuran neoplasm: studies at the interface of developmental biology, pathology, and virology. Journal of Cellular Physiology 173:115–118.

McKinnell, R.G. and W.P. Cunningham. 1982. Herpesviruses in metastatic Lucké renal adenocarcinoma. Differentiation 22:41–46.

McKinnell, R.G. and D.C. Dapkus. 1973. The distribution of burnsi and kandiyohi frogs in Minnesota and contiguous states. American Zoologist 13:81–84.

McKinnell, R.G. and M.A. Di Berardino. 1999. The biology of cloning: history and rationale. BioScience 49:875–885.

McKinnell, R.G. and D.P. Duplantier. 1970. Are there renal adenocarcinoma-free populations of leopard frogs? Cancer Research 30:2730–2735.

McKinnell, R.G. and V.L. Ellis. 1972a. Herpesviruses in tumors of postspawning *Rana pipiens*. Cancer Research 32:1152–1159.

McKinnell, R.G. and V.L. Ellis. 1972b. Epidemiology of the frog renal tumour and the significance of tumour nuclear transplantation to a viral aetiology of the tumour—a review. Pp. 183–197. *In* Biggs, P.M., G. De-Thé and L.N. Payne (Eds), Oncogenesis and Herpesviruses. International Agency for Research on Cancer, Lyon, France.

McKinnell, R.G. and J.C. John. 1995. An unexpectedly high prevalence of spontaneous renal carcinoma found in *Rana pipiens* obtained from northern Vermont, USA. Pp. 279–281. *In* Zwart, P. and G. Matz (Eds), Comptes Rendu Cinquième international de Pathologie des Reptiles et des Amphibians. NRG Repro Facility BV's-Hertogenbosch, Pays-Bas, France.

McKinnell, R.G. and B.K. McKinnell. 1967. An extension of the ranges of the burnsi and kandiyohi variants of *Rana pipiens*. Journal of the Minnesota Academy of Sciences 34:176.

McKinnell, R.G. and B.K. McKinnell. 1968. Seasonal fluctuation of frog renal adenocarcinoma prevalence in natural populations. Cancer Research 28:440–444.

McKinnell, R.G. and D. Tarin. 1984. Temperature-dependent metastasis of the Lucké renal carcinoma and its significance for studies on mechanisms of metastasis. Cancer Metastasis Reviews 3:373–386.

McKinnell, R.G. and J. Zambernard. 1968. Virus particles in renal tumors from spring *Rana pipiens* of known geographic origin. Cancer Research 28:684–688.

McKinnell, R.G., B.A. Deggins and D.D. Labat. 1969. Transplantation of pluripotential nuclei from triploid frog tumors. Science 165:394–396.

McKinnell, R.G., L.M. Steven Jr. and D.D. Labat. 1976. Frog renal tumors are composed of stroma, vascular elements and epithelial cells: what type nucleus programs for tadpoles with the cloning procedure? Pp. 319–330. *In* Müller-Bérat, N., C. Rosenfeld, D. Tarin and D. Viza (Eds.), Progress in Differentiation Research. North Holland Publishing Company, Amsterdam.

McKinnell, R.G., D.L. Carlson, C.G. Christ and J.C. John. 1995. Detection of Lucké tumor herpesvirus DNA sequences in normal and neoplastic tissue obtained from the northern leopard frog, *Rana pipiens*. Pp. 13–16. *In* Zwart, P. and G. Matz (Eds.), Comptes Rendus Cinquième Colloque international de Pathologie des Reptiles et des Amphibiens. NRG Repro Facility BV's-Hertogenbosch, Pays-Bas, France.

McKinnell, R.G., V. Ellis, D.C. Dapkus and L.M. Steven Jr. 1972. Early replication of herpesvirus in naturally occurring frog tumors. Cancer Research 32:1729–1733.

McKinnell, R.G., E. Gorham, F. Martin and J. Schaad IV. 1979. A major reduction in the prevalence of the Lucké renal adenocarcinoma associated with greatly diminished frog populations. Journal of the National Cancer Institute 63:821–824.

McKinnell, R.G., G. De Bruyne, M.M. Mareel, D. Tarin and K.S. Tweedell. 1984. Cytoplasmic microtubules of normal and tumor cells of the leopard frog: temperature effects. Differentiation 26:231–234.

McKinnell, R.G., M.M. Mareel, E.A. Bruneel, E.D. Seppanen and P.R. Mekala. 1986. Invasion in vitro by explants of Lucké renal carcinomas cocultured with normal tissue is temperature dependent. Clinical and Experimental Metastasis 4:237–243.

McKinney, M.L. 1999. High rates of extinction and threat in poorly studied taxa. Conservation Biology 13:1273–1281.

McKitrick, M.C. and R.M. Zink. 1988. Species concepts in ornithology. Condor 90:1–14.

McKnight, M.L., C.K. Dodd Jr. and C.M. Spolsky. 1991. Protein and mitochondrial DNA variation in the salamander *Phaeognathus hubrichti*. Herpetologica 47:440–447.

McLeary, D.A. 1989. Wyoming toad: 1989 field surveys. Wyoming Game and Fish Department, Fish Division Administrative Report, Project 5089-30-8301, Cheyenne, Wyoming.

McLeod, D.S. 1999. A re-survey of amphibian populations in Nebraska after twenty years: a test for declines. Master's thesis. University of Nebraska, Lincoln, Nebraska.

McLister, J.D. and W. Lamond. 1991. A wolf spider, *Bogna helluo*, feeding on a blue-spotted salamander, *Ambystoma laterale*. Canadian Field-Naturalist 105:574–575.

McMillan, J.E. and R.D. Semlitsch. 1980. Prey of the dwarf salamander, *Eurycea quadridigitata*, in South Carolina. Journal of Herpetology 14:424–426.

McMullen, J.E. and R.L. Roudabush. 1936. A new species of trematode, *Cercorchis cryptobranchi*, from *Cryptobranchus alleganiensis*. Journal of Parasitology 22:516–517.

McNeil, C.W. 1948. Abstracts of papers delivered before the annual meeting of the Northwest Scientific Association 27 and 28 December 1948. Northwest Science 23:11.

McPeak, R.H. 2000. Amphibians and Reptiles of Baja California. Sea Challengers. Monterey, California.

McWilliams, S.R. 1992. Courtship behavior of the small-mouthed salamander *(Ambystoma texanum)*: the effects of conspecific males on male mating tactics. Behaviour 121:1–19.

McWilliams, S.R. and M.D. Bachmann. 1989a. Foraging ecology and prey preference of pond-form larval small-mouthed salamanders, *Ambystoma texanum*. Copeia 1989:948–961.

McWilliams, S.R. and M.D. Bachmann. 1989b. Predatory behavior of larval small-mouthed salamanders *(Ambystoma texanum)*. Herpetologica 45:459–467.

McWilliams, W.H., J.R. Mills and W.G. Burkman. 1993. The state of the nation's forestland. National Woodland Magazine 16:8–10, 13.

Meacham, W.R. 1962. Factors affecting secondary intergradation between two allopatric populations in the *Bufo woodhousei* complex. American Midland Naturalist 67:282–304.

Mead, A.R. 1961. The Giant African Snail: A Problem in Economic Malacology. University of Chicago Press, Chicago, Illinois.

Mead, J., T.R. Van Devender, K.L. Cole and D.B. Wake. 1985. Late Pleistocene vertebrates from a packrat midden in the south-central Sierra Nevada, California. Current Research in the Pleistocene 2:107–108.

Mead, L.S., R.S. Nauman, D. Clayton, D.H. Olson and M.E. Pfrender. 2004. Molecular systematics, morphological variation, and phylogeography of the Siskiyou Mountains salamander. Herpetologica 60: in press.

Mead, L.S. and S.G. Tilley. 2000. Ethological isolation and variation in allozymes and dorsolateral pattern between parapatric forms in the *Desmognathus ochrophaeus* complex. Pp. 181–198. *In* Bruce, R.C., R.G. Jaeger and L.D. Houck (Eds.), The Biology of Plethodontid Salamanders. Kluwer Academic/Plenum Publishers, New York.

Mead, L.S., S.G. Tilley and L.A. Katz. 2002. Genetic structure of the Blue Ridge dusky salamander *(Desmognathus orestes)*: inferences from allozymes, mitochondrial DNA and behavior. Evolution 55:2287–2302.

Means, D.B. 1972. Notes on the breeding biology of *Ambystoma cingulatum* (Cope) (Amphibia: Urodela: Ambystomatidae). Association of Southeastern Biologists Bulletin 19:84.

Means, D.B. 1974. The status of *Desmognathus brimleyorum* Stejneger and an analysis of the genus *Desmognathus* (Amphibia: Urodela) in Florida. Bulletin of the Florida State Museum, Biological Science, Number 18, Gainesville, Florida.

Means, D.B. 1975a. Competitive exclusion along a habitat gradient between two species of salamanders *(Desmognathus)* in western Florida. Journal of Biogeography 2:253–263.

Means, D.B. 1975b. Evolutionary ecology studies on salamanders of the genus *Desmognathus*. Part I: Competitive exclusion along a habitat gradient between two species of salamanders *(Desmognathus)* in western Florida; Part II: Life history, growth and body size variation in populations of a streamside salamander *(Desmognathus brimleyorum)* on adjacent mountains. Ph.D. dissertation. Florida State University, Tallahassee, Florida.

Means, D.B. 1977. Aspects of the significance to terrestrial vertebrates of the Apalachicola River drainage basin, Florida. Pp. 37–67. *In* Livingston, R.J. and E.A. Joyce (Eds.), Proceedings of the Conference on the Apalachicola Drainage System, Florida Marine Research Publication Number 26, Tallahassee, Florida.

Means, D.B. 1978. Endangered pine barrens treefrog *(Hyla andersonii* Baird). Pp. 3–4. *In* McDiarmid, R. (Ed.), Rare and Endangered Biota of Florida, Volume 3, Amphibians and Reptiles. University of Florida Press, Gainesville, Florida.

Means, D.B. 1983. The enigmatic pine barrens treefrog. Florida Wildlife 37:16–19.

Means, D.B. 1986a. Threatened pine barrens treefrog, *Hyla andersonii* Baird. Pp. 29–30. *In* Mount, R.H. (Ed.), Vertebrate Animals of Alabama in Need of Special Attention. Alabama Agricultural Experimental Station, Auburn University, Auburn, Alabama.

Means, D.B. 1986b. Threatened dusky gopher frog. Pp. 30–31. *In* Mount, R.H. (Ed.), Vertebrate Animals of Alabama in Need of Special Attention. Alabama Agricultural Experiment Station, Auburn University, Auburn, Alabama.

Means, D.B. 1986c. Poorly known one-toed amphiuma. Pp. 56–57. *In* Mount, R.H. (Ed.), Vertebrate Animals of Alabama in Need of Special Attention, Alabama Agricultural Experiment Station, Auburn University, Auburn, Alabama.

Means, D.B. 1986d. Poorly known southern dusky salamander, *Desmognathus auriculatus* (Holbrook). Pp. 58–59. *In* Mount, R.H. (Ed.), Vertebrate Animals of Alabama in Need of Special Attention. Alabama Agricultural Experimental Station, Auburn University, Auburn Alabama.

Means, D.B. 1991. Florida's steepheads: unique canyonlands. Florida Wildlife 45:25–28.

Means, D.B. 1992a. Rare pine barrens treefrog. Pp. 20–25. *In* Moler, P.E. (Ed.), Rare and Endangered Biota of Florida. Volume III. Amphibians and Reptiles. University Press of Florida, Gainesville, Florida.

Means, D.B. 1992b. Rare one-toed amphiuma. Pp. 34–38 *In* Moler, P.E. (Ed.), Rare and Endangered Biota of Florida, Volume III. Amphibians and Reptiles. University Press of Florida, Gainesville, Florida.

Means, D.B. 1992c. Rare Georgia blind salamander, *Haideotriton wallacei* Carr. Pp. 49–53. *In* Moler, P.E. (Ed.), Rare and Endangered Biota of Florida. Volume III. Amphibians and Reptiles. University Press of Florida, Gainesville, Florida.

Means, D.B. 1992d. Four-toed salamanders. Pp. 54–57. *In* Moler, P.E. (Ed.), Rare and Endangered Biota of Florida, Volume III. Amphibians and Reptiles. University Press of Florida, Gainesville, Florida.

Means, D.B. 1993. *Desmognathus apalachicolae*. Pp. 556.1–556.2. Catalogue of American Amphibians and Reptiles. Society for the Study of Amphibians and Reptiles, St. Louis, Missouri.

Means, D.B. 1996a. *Amphiuma pholeter*. Neill. Pp. 622.1–622.2. Catalogue of American Amphibians and Reptiles. Society for the Study of Amphibians and Reptiles, St. Louis, Missouri.

Means, D.B. 1996b. A preliminary consideration of highway impacts on herpetofauna inhabiting small isolated wetlands in the southeastern U.S. Coastal Plain. Pp. 1–11. *In* Evink, G.L., P. Garrett, D. Zeigler and J. Berry (Eds.), Trends in Addressing Transportation Related Wildlife

Mortality. Proceedings of the Transportation Related Wildlife Mortality Seminar, 30 April–2 May, 1996, Orlando, Florida. Florida Department of Transportation, Tallahassee, Florida.

Means, D.B. 1996c. Longleaf pine forest, going, going . . . Pp. 210–229. *In* Davis, M.B. (Ed). Eastern Old-Growth Forest: Prospects for Rediscovery and Recovery. Island Press, Washington, D.C.

Means, D.B. 1996d. Longleaf pine forest: importance to biodiversity. Pp. 12–14. *In* Kush, J.S. (Ed.), Longleaf Pine: A Regional Perspective of Challenges and Opportunities. Proceedings of the First Longleaf Alliance Conference, Mobile, Alabama.

Means, D.B. 1997. Wiregrass restoration: probable shading effects in a slash pine plantation. Restoration and Management Notes 15:52–55.

Means, D.B. 1998. Salamanders of Eglin Air Force Base: Towards a long-term monitoring program. U.S.A.F. Contract Number F0865197M5602, Final report, Tallahassee, Florida.

Means, D.B. 1999a. *Desmognathus auriculatus*. Catalogue of the American Amphibians and Reptiles Pp. 681.1–681.6. Society for the Study of Amphibians and Reptiles, St. Louis, Missouri.

Means, D.B. 1999b. *Desmognathus brimleyorum*. Pp. 682.1–682.4. Catalogue of American Amphibians and Reptiles. Society for the Study of Amphibians and Reptiles, St. Louis, Missouri.

Means, D.B. 2000. Southeastern U.S. Coastal Plain habitats of the Plethodontidae: the importance of relief, ravines, and seepage. Pp. 287–302. *In* Bruce, R.C., R.J. Jaeger and L.D. Houck (Eds.), The Biology of Plethodontid Salamanders. Plenum Publishing Corporation, New York.

Means, D.B. 2001. Reducing impacts on rare vertebrates that require small isolated water bodies along U.S. Highway 319. Final Report to the Florida Department of Transportation, Tallahassee, Florida.

Means, D.B. submitted. Amphibian declines in the southeastern U.S.: dusky salamanders. Biological Conservation.

Means, D.B. and A.A. Karlin. 1989. A new species of *Desmognathus* from the eastern Gulf Coastal Plain. Herpetologica 45:37–46.

Means, D.B. and C.J. Longden. 1970. Observations on the occurrence of *Desmognathus monticola* in Florida. Herpetologica 26:396–399.

Means, D.B. and C.J. Longden. 1976. Aspects of the biology and zoogeography of the pine barrens treefrog, *Hyla andersonii*, in northern Florida. Herpetologica 32:117–130.

Means, D.B. and R.C. Means. 1997. Use of a temporary pond by amphibians and reptiles in the Munson Sandhills of the Apalachicola National Forest with special emphasis on the striped newt and gopher frog, Year 2: September 1996–September 1997. U.S. Forest Service, Final report, Tallahassee, Florida.

Means, D.B. and R.C. Means. 1998a. Distribution of the striped newt (*Notophthalmus perstriatus*) and gopher frog (*Rana capito*) in the Munson Sand Hills of the Florida panhandle. U.S. Fish and Wildlife Service, Final report, Jackson, Mississippi.

Means, D.B. and R.C. Means. 1998b. Red Hills Survey for breeding pond habitat of the flatwoods salamander (*Ambystoma cingulatum*), gopher frog (*Rana capito*), and striped newt (*Notophthalmus perstriatus*) in the Tallahassee Red Hills of Leon, Gadsden, and Jefferson counties, Florida, and the Tifton Uplands of Thomas and Grady counties, Georgia. U.S. Fish and Wildlife Service, Final report, Jackson, Mississippi.

Means, D.B. and R.C. Means. in press. Effects of sand pine silviculture on pond-breeding amphibians in the Woodville Karst Plain of north Florida. Pp. xxx–xxx. *In* Meshaka, W.E. and K.J. Babbitt (Eds.), Status and Conservation of Florida Amphibians and Reptiles. Krieger Publishing, Malabar, Florida.

Means, D.B. and P.E. Moler. 1979. The pine barrens treefrog: Fire, seepage bogs, and management implications. Pp. 77–83. *In* Odum, R.R. and L. Landers (Eds.), Proceedings of the Rare and Endangered Wildlife Symposium. Georgia Department of Natural Resources, Technical Bulletin, WL-4, Game and Fish Division, Social Circle, Georgia.

Means, D.B., T.E. Ostertag and D. Printiss. 1994. Florida populations of the striped newt, *Notophthalmus perstriatus*, west of the Suwannee River. Contributions to life history, ecology, and distribution. I. Report to U.S. Fish and Wildlife Service, Jackson, Mississippi.

Means, D.B., J.G. Palis and M. Baggett. 1996. Effects of slash pine silviculture on a Florida population of flatwoods salamander. Conservation Biology 10:426–437.

Means, M.L. 1990. Population dynamics and movement of Ozark cavefish in Logan Cave NWR, Benton County, Arkansas with additional baseline water quality information. Master's thesis. Arkansas Technical University, Russellville, Arkansas.

Means, M.L. and J.E. Johnson. 1995. Movement of threatened Ozark cavefish in Logan Cave National Wildlife Refuge, Arkansas. Southwestern Naturalist 40:308–313.

Means, R.C. and D.B. Means. 2000. Effects of sand pine silviculture on pond-breeding amphibians. Annual Meeting of American Society of Ichthyologists and Herpetologists. La Paz, Baja California Sur, Mexico.

Measey, G.J. 1998a. Diet of feral *Xenopus laevis* (Daudin) in South Wales, UK. Journal of Zoology (London) 246:287–298.

Measey, G.J. 1998b. Terrestrial prey capture in *Xenopus laevis*. Copeia 1998:787–791.

Measey, G.J. and R.C. Tinsley. 1998. Feral *Xenopus laevis* in South Wales. Herpetological Journal 8:23–27.

Mecham, J.M. 1958. Isolating mechanisms in anuran amphibians. Pp. 24–61. *In* Blair, W.F. (Ed.), Vertebrate Speciation. University of Texas Press, Austin, Texas.

Mecham, J.S. 1954. Geographic variation in the green frog, *Rana clamitans* Latreille. Texas Journal of Science 6:1–24.

Mecham, J.S. 1958. Some Pleistocene amphibians and reptiles from Friesenhahn Cave, Texas. Southwestern Naturalist 3:17–27.

Mecham, J.S. 1960a. Natural hybridization between the treefrogs *Hyla versicolor* and *Hyla avivoca*. Journal of the Elisha Mitchell Society 66:64–67.

Mecham, J.S. 1960b. Introgressive hybridization between two southeastern treefrogs. Evolution 14:455–457.

Mecham, J.S. 1964. Ecological and genetic relationships of the two cricket frogs, genus *Acris*, in Alabama. Herpetologica 20:84–91.

Mecham, J.S. 1965. Genetic relationships and reproductive isolation in southeastern frogs of the genera *Pseudacris* and *Hyla*. American Midland Naturalist 74:269–308.

Mecham, J.S. 1968a. *Notophthalmus meridionalis*. Pp. 74.1–74.2. Catalogue of American Amphibians and Reptiles. Society for the Study of Amphibians and Reptiles, St. Louis, Missouri.

Mecham, J.S. 1968b. On the relationships between *Notophthalmus meridionalis* and *Notophthalmus kallerti*. Journal of Herpetology 2:121–127.

Mecham, J.S. 1968c. Evidence of reproductive isolation between two populations of the frog, *Rana pipiens*, in Arizona. Southwestern Naturalist 13:35–44.

Mecham, J.S. 1971. Vocalizations of the leopard frog, *Rana pipiens*, and three related Mexican species. Copeia 1971:505–516.

Mecham, J.S. and R.E. Hellman. 1952. Notes on the larvae of two Florida salamanders. Quarterly Journal of the Florida Academy of Science 15:127–133.

Mecham, J.S., M.J. Littlejohn, L.E. Brown, R.S. Oldham and J.R. Brown. 1973. A new species of leopard frog (*Rana pipiens* complex) from the plains of the central United States. Occasional Papers, Museum, Number 18, Texas Technical University, Lubbock, Texas.

Meeks, D.E. and J.W. Nagel. 1973. Reproduction and development of the wood frog, *Rana sylvatica*, in eastern Tennessee. Herpetologica 29:188–191.

Meents, J.K. 1987. Distribution of the Sacramento Mountain salamander (*Aneides hardii*) in the Capitan Mountains, New Mexico. New Mexico Department of Game and Fish, Final report, Santa Fe, New Mexico.

Meffe, G.K. and C.R. Carroll. 1997. Principles of Conservation Biology. Second edition. Sinauer Associates, Sunderland, Massachusetts.

Meffe, G.K. and A.L. Sheldon. 1987. Habitat use by dwarf waterdogs (*Necturus punctatus*) in South Carolina streams, with life history notes. Herpetologica 43:490–496.

Melvin, S.M. and S.M. Roble. 1990. The Massachusetts Wetlands Protection Act: protecting rare wildlife habitat. Pp. 224–227. *In* Mitchell, R.S., C.J. Sheviak and D.J. Leopold (Eds.), Ecosystem Management: Rare Species and Significant Habitats. New York State Museum Bulletin, Number 471, Albany, New York.

Mendelson, J.R., III. 1997a. A new species of *Bufo* (Anura: Bufonidae) from the Pacific Highlands of Guatemala and southern Mexico, with comments on the status of *Bufo valliceps macrocristatus*. Herpetologica 53:14–30.

Mendelson, J.R., III. 1997b. A new species of toad (Anura: Bufonidae) from Oaxaca, Mexico, with comments on the status of *Bufo cavifrons* and *Bufo cristatus*. Herpetologica 53:268–286.

Mendelson, J.R., III. 1998. Geographic variation in *Bufo valliceps* (Anura: Bufonidae), a widespread toad in the United States and Middle America. Natural History Museum, Scientific Papers, Number 9, University of Kansas, Lawrence, Kansas.

Mendelson, J.R., III. 2001. A review of the Guatemalan toad *Bufo ibarrai* (Anura: Bufonidae), with distributional and taxonomic comments of *Bufo valliceps* and *Bufo coccifer*. Pp. 21–30. *In* Johnson, J.D., R. G. Webb and O. Flores-Villela (Eds.), Mesoamerican Herpetology: Systematics, Natural History, and Conservation. University of Texas at El Paso, El Paso, Texas.

MEOTC (Massachusetts Executive Office of Transportation and Construction). 1998. Massachusetts Transportation Facts. Boston, Massachusetts.

Merchant, H. 1972. Estimated population size and home range of the salamanders *Plethodon jordani* and *Plethodon glutinosus*. Journal of the Washington Academy of Science 62:248–257.

Merovich, C.E. and J.H. Howard. 2000. Amphibian use of constructed ponds on Maryland's eastern shore. Journal of the Iowa Academy of Science 107:151–159.

Merrell, D.J. 1965. The distribution of the dominant burnsi gene in the leopard frog, *Rana pipiens*. Evolution 19:69–85.

Merrell, D.J. 1968. A comparison of the estimated and the "effective size" of the breeding populations of the leopard frog, *Rana pipiens*. Evolution 22:274–283.

Merrell, D.J. 1969. Natural selection in a leopard frog population. Journal of the Minnesota Academy of Science 35:86–89.

Merrell, D.J. 1970. Migration and gene dispersal in *Rana pipiens*. American Zoologist 10:47–52.

Merrell, D.J. 1977. Life history of the leopard frog, *Rana pipiens*, in Minnesota. Bell Museum of Natural History, Occasional Paper Number 15, Minneapolis, Minnesota.

Merrell, D.J. and C.F. Rodell. 1968. Seasonal selection in the leopard frog, *Rana pipiens*. Evolution 22:284–288.

Mertz, D.B. 1972. The *Tribolium* model and the mathematics of population growth. Annual Review of Ecology and Systematics 3:51–78.

Mertz, D.B. and J.R. Robertson. 1970. Some developmental consequences of handling, egg-eating, and population density for flour beetle larvae. Ecology 51:989–998.

Meshaka, W.E., Jr. 1993. Hurricane Andrew and the colonization of five invading species in southern Florida. Florida Science 56:193–201.

Meshaka, W.E., Jr. 1994. Ecological correlates of successful colonization in the Cuban treefrog, *Osteopilus septentrionalis* (Anura: Hylidae). Ph.D. dissertation. Florida International University, Miami, Florida.

Meshaka, W.E., Jr. 1996a. Vagility and the Florida distribution of the Cuban treefrog (*Osteopilus septentrionalis*). Herpetological Review 27:37–40.

Meshaka, W.E., Jr. 1996b. Retreat use by the Cuban treefrog (*Osteopilus septentrionalis*): implications for successful colonization in Florida. Journal of Herpetology 30:443–445.

Meshaka, W.E., Jr. 1996c. Occurrence of the parasite *Skrjabinoptera scelopori* (Caballero Rodriguez, 1971) in the Cuban treefrog (*Osteopilus septentrionalis*): mainland and island comparisons. Pp. 271–276. *In* Powell, R. and R.W. Henderson (Eds.), Contributions to West Indian Herpetology: A Tribute to Albert Schwartz. Contributions to Herpetology, Volume 12, Society for the Study of Amphibians and Reptiles, St. Louis, Missouri.

Meshaka, W.E., Jr. 2001. The Cuban Treefrog in Florida: Life History of a Successful Colonizing Species. University Press of Florida, Gainesville, Florida.

Meshaka, W.E., Jr. and J.N. Layne. 2002. Herpetofauna of a long-unburned sandhill habitat in south-central Florida. Florida Scientist 65:35–50.

Meshaka, W.E., Jr. and Layne. 2005. Habitat relationships and seasonal activity of the greenhouse frog, *Eleutherodactylus planirostris*, in southern Florida. Florida Scientist 68:35–43.

Meshaka, W.E., Jr. and P. McLarty. 1988. Geographic distribution: *Ambystoma talpoideum* (mole salamander). Herpetological Review 19:17.

Meshaka, W.E., Jr. and S.E. Trauth. 1995. Reproductive cycle of the Ozark zigzag salamander, *Plethodon dorsalis angusticlavius* (Caudata, Plethodontidae), in north central Arkansas. Alytes 12:175–182.

Meshaka, W.E., Jr., B.P. Butterfield and R.L. Cox. 1989. Geographic distribution: *Ambystoma talpoideum* (mole salamander). Herpetological Review 20:11.

Meshaka, W.E., Jr., B.P. Butterfield and J.B. Hauge. 2004. Exotic Amphibians and Reptiles of Florida. Krieger Publishing Company, Melbourne, Florida.

Meshaka, W.E., Jr., W.F. Loftus and T. Steiner. 2000. The herpetofauna of Everglades National Park. Florida Scientist 63:84–103.

Mesrobian, A. 1994a. TPWD can't see the forest for the tees. Beacon 5:1, 6–7.

Mesrobian, A. 1994b. Turf wars in toadland. Beacon 5:8–9.

Mesrobian, A. 1995. Letter to Roger Kennedy and Jerry Rogers, 8 July.

Mesrobian, A. 1996. Drought stymies endangered Houston toad population distribution survey. Beacon 6:7.

Mesrobian, A. 1997. ToadWatch '97: the Houston toad population distribution study. Beacon 7:4.

Mesrobian, A. 1999. While saving endangered species, humans may be saved. Bastrop Advertiser, 11 September.

Messere, M. and P.K. Ducey. 1998. Forest floor distribution of northern redback salamanders, *Plethodon cinereus*, in relation to canopy gaps: first year following selective logging. Forest Ecology and Management 107:319–324.

Metcalf, A.L. and R. Smartt. 1972. Records of introduced mollusks: New Mexico and Western Texas. Nautilus 85:144–145.

Metcalf, M.M. 1923. The opalinid ciliate infusorians. Smithsonian Institution, United States National Museum Bulletin, Number 120, Washington, D.C.

Metcalf, M.M. 1928. The bell-toads and their opalinid parasites. American Naturalist 62:5–21.

Metcalf, Z.P. 1921. The food capacity of the toad. Copeia 100:81.

Meteyer, C.A., K.I. Loeffler, J.F. Fallon, K.A. Converse, E. Green, J.C. Helgen, S. Kersten, R. Levey, L. Eaton-Poole and J.G. Burkhart. 2000. Hind limb malformations in free-living northern leopard frogs *(Rana pipiens)* from Maine, Minnesota, and Vermont suggest multiple etiologies. Teratology 62:151–171.

Metter, D.E. 1963. Stomach contents of Idaho larval *Dicamptodon*. Copeia 1963:435–436.

Metter, D.E. 1964a. A morphological and ecological comparison of two populations of the tailed frog, *Ascaphus truei* Stejneger. Copeia 1964:181–195.

Metter, D.E. 1964b. On breeding and sperm retention in *Ascaphus*. Copeia 1964:710–711.

Metter, D.E. 1967. Variation in the ribbed frog *Ascaphus truei* Stejneger. Copeia 1967:634–649.

Metter, D.E. 1968. The influence of floods on population structure of *Ascaphus truei* Stejneger. Journal of Herpetology 1:105–106.

Metter, D.E. and R.J. Pauken. 1969. An analysis of the reduction of gene flow in *Ascaphus truei* in the northwest U.S. since the Pleistocene. Copeia 1969:301–307.

Metter, D.E., W.R. Morris and D.A. Kangas. 1970. Great Plains anurans in central Missouri. Copeia 1970:780–781.

Mettrick, D.F. 1963. A revision of the genus *Ribeiroia* Travassos, 1939, with some observations on the family Cathaemasiidae Fuhrmann, 1928, including the erection of a new sub-family Reeselliinae. Revue of Zoology and Botany of Africa 67:137–162.

Metts, B. 2001. Natural history notes: *Ambystoma maculatum* (spotted salamander). Reproduction. Herpetological Review 32:98.

Metzger, N. and R.N. Zare. 1999. Interdisciplinary research: from belief to reality. Science 283:642–643.

Meyer, J. and D. Mikesic. 1998. The status of amphibians in the Chuska Mountains of the Navajo Nation (New Mexico and Arizona): a preliminary assessment. P. 8. *In* Abstracts of the Fourth Annual Meeting of the Southwestern United States Working Group of the Declining Amphibian Populations Task Force, Phoenix, Arizona.

Meyer, J.R. and C.F. Foster. 1996. A Guide to the Frogs and Toads of Belize. Krieger Publishing Company: Malabar, Florida.

Meyer, J.R. and L.D. Wilson. 1971. A distributional checklist of the amphibians of Honduras. Los Angeles County Museum Natural History Contributions to Science, Number 218, Los Angeles, California.

Meyer-Rochow, V. and M. Asashima. 1988. Naturally occuring morphological abnormalities in wild populations of the Japanese newt *Cynops pyrrhogaster* (Salamandridae: Urodela: Amphibia). Zoological Anzeiger 221:70–80.

Meylan, P. 1995. Cladistics: a response to Pritchard. Herpetological Review 26:83–85.

Michaud, T.C. 1962. Call discrimination by females of the chorus frogs, *Pseudacris clarki* and *Pseudacris nigrita*. Copeia 1962:213–215.

Michl, H. and E. Kaiser. 1963. Chemie und biochemie der amphibiengifte. Toxin 1:175–228.

Middleton, E.M., J.R. Herman, E.A. Celarier, J.W. Wilkinson, C. Carey and R.J. Rusin. 2001. Evaluating ultraviolet radiation exposure with satellite data at sites of amphibian declines in Central and South America. Conservation Biology 15:914–929.

Miera, V. and M.J. Sredl. 2000. Range expansion of Rio Grande leopard frogs in central Arizona. Nongame and Endangered Wildlife Program, Arizona Game and Fish Department, Technical Report, Number 171, Phoenix, Arizona.

Mierzwa, K.S. 1989. Legislative update and conservation notes. Bulletin of the Chicago Herpetological Society 24:120.

Mierzwa, K.S. 1998a. Status of northeastern Illinois amphibians. Pp. 115–124. *In* Lannoo, M.J. (Ed.), Status and Conservation of Midwest Amphibians. University of Iowa Press, Iowa City, Iowa.

Mierzwa, K.S. 1998b. Amphibian habitat in the midwestern United States. Pp. 16–23. *In* Lannoo, M.J. (Ed.), Status and Conservation of Midwest Amphibians. University of Iowa Press, Iowa City, Iowa.

Mierzwa, K.S. 1998c. Wetland mitigation and amphibians: preliminary observations at a southwestern Illinois bottomland hardwood forest restoration site. Pp. 15–16. *In* Abstracts, Midwest Declining Amphibians Conference, March 20–21. Milwaukee Public Museum, Milwaukee, Wisconsin.

Mifflin, M.D. and M.M. Wheat. 1979. Pluvial lakes and estimated pluvial climates of Nevada. Nevada Bureau of Mines and Geology Bulletin, Number 94, Mackay School of Mines, University of Nevada, Reno, Nevada.

Mikaelian, I., M. Ouellet, B. Pauli, J. Rodrigue, J.C. Harshbarger and D.M. Green. 2000. *Ichthyophonus*-like infection in wild amphibians from Québec, Canada. Diseases of Aquatic Organisms 40:195–201.

Miles, D.B. 1993. Population differentiation in locomotor performance and the potential response of a terrestrial organism to global environmental change. American Zoology 34:422–436.

Milius, S. 1998. Fatal skin fungus found in U.S. frogs. Science News 154:7–8.

Miller, A.H. and R.C. Stebbins. 1964. The Lives of Desert Animals in Joshua Tree National Monument. University of California Press, Berkeley, California.

Miller, B.K. 1993. Wildlife trends in Indiana. Hoosier Farmland Notes 1:2.

Miller, B.T. 1990. Geographic distribution: *Pseudotriton montanus diastictus* (midland mud salamander). Herpetological Review 21:95.

Miller, B.T. and D.L. Campbell. 1996. Geographic distribution: *Rana capito*. Herpetological Review 27:86–87.

Miller, B.T., J.L. Miller, D.L. Campbell and P. Wyatt. 1998. Natural history notes: *Plethodon dorsalis* (zigzag salamander). Reproduction. Herpetological Review 29:38–39.

Miller, C.E. 1968. Frogs with five legs. Carolina Tips 31:1.

Miller, D.A. and E.M. Hallerman. 1994. Isozyme survey of two-lined salamander *(Eurycea bislineata)* populations in Maryland and Virginia. Bulletin of the Maryland Herpetological Society 30:78–97.

Miller, D.M. and T.T. Dunagan. 1971. Studies on the rostellar hooks of *Macracanthorhynchus hirudinaceus* (Acanthocephala) from swine. Transactions of the American Microscopical Society 90:329–335.

Miller, J.D. 1978. Observations on the diets of *Rana pretiosa*, *Rana pipiens*, and *Bufo boreas* from western Montana. Northwest Science 52:243–249.

Miller, L. 1944. Notes on the eggs and larvae of *Aneides lugubris*. Copeia 1944:224–230.

Miller, M.R. and M.E. Robbins. 1954. The reproductive cycle in *Taricha torosa (Triturus torosus)*. Journal of Experimental Zoology 125:415–446.

Miller, N. 1909a. The American toad *(Bufo lentiginosus americanus* LeConte). A study in dynamic biology. American Naturalist 43:641–668.

Miller, N. 1909b. The American toad *(Bufo lentiginosus americanus* LeConte). II. A study in dynamic biology. American Naturalist 43:730–745.

Miller, R. 1979. Miscellaneous distributional records for Maryland amphibians and reptiles. Bulletin of the Maryland Herpetological Society 15:56–58.

Miller, R., D. Threloff, S. Keeney, D. Becker, G. Knowles, S. Reid and A. Bentivoglio. 2000. California State Agency Report. Abstract from 32nd Annual Desert Fishes Council Meeting, November 16–19, 2000, Furnace Creek Ranch, Death Valley, California.

Miller, S. 1991. Letter to Andrew Sansom, 22 January.

Miller, S. 1992a. Letter to Andrew Sansom, 27 February.

Miller, S. 1992b. The tinkling of a small bell. Beacon 2:1, 4–6.

Miller, S. 1992c. BCEN takes position on golf course expansion. Beacon 3:14.

Miller, S. 1993. Toads at risk: golf course expansion shifts to the fast track. Beacon 3:12.

Miller, S. and BCEN Golf Course Committee. 1992. Bastrop County Environmental Network position paper: proposed Lost Pines Golf Course expansion, 2 August.

Miller, T.J. 1985. Husbandry and breeding of the Puerto Rican toad *(Peltophryne lemur)* with comments on its natural history. Zoo Biology 4:281–286.

Miller, W. deW. and J. Chaplin. 1910. The toads of the northeastern United States. Science 32:315–317.

Mills, G.R. 1996. A study on the life history and seasonal foraging habits of the salamander *Desmognathus quadramaculatus* Holbrook, in West Virginia. Master's thesis. Marshall University, Huntington, West Virginia.

Milne, B.T. 1991. Lessons from applying fractal models to landscape patterns. Pp. 199–235. *In* Turner, M.G. and R.H. Gardner (Eds.), Quantitative Methods in Landscape Ecology. Springer-Verlag, New York.

Milner, J.W. 1874. Report on the fishes of the Great Lakes: The result of inquires in 1871 and 1872. Appendix A. Report of the U.S. Fish Commission 1872:1–75. Washington, D.C.

Milstead, W.W. 1960. Relict species of the Chihuahuan Desert. Southwestern Naturalist 5:75–88.

Milstead, W.W., J.S. Mecham and H. McClintock. 1950. The amphibians and reptiles of the Stockton Plateau in northern Terrell County, Texas. Texas Journal of Science 2:543–562.

Milstead, W.W., A.S. Rand and M. Stewart. 1974. Polymorphism in cricket frogs: an hypothesis. Evolution 28:489–491.

Milstrey, E.G. 1984. Ticks and invertebrate commensals in gopher tortoise burrows: implications and importance. Pp. 4–15. *In* Jackson, D.R. and R.J. Bryant (Eds.), The Gopher Tortoise and its Community. Proceedings of the Fifth Annual Meeting of the Gopher Tortoise Council, Gainesville, Florida.

Minter, S. 1979. Geographic distribution: *Ambystoma annulatum* (ringed salamander). Herpetological Review 10:23.

Minton, S.A. 1954. Salamanders of the *Ambystoma jeffersonianum* complex in Indiana. Herpetologica 10:173–179.

Minton, S.A. 1998. Observations on Indiana amphibian populations: a forty-five-year overview. Pp. 217–220. *In* Lannoo, M.J. (Ed.), Status and Conservation of Midwestern Amphibians. University of Iowa Press, Iowa City, Iowa.

Minton, S.A., Jr. 1954. Salamanders of the *Ambystoma jeffersonianum* complex in Indiana. Herpetologica 10:173–179.

Minton, S.A., Jr. 1959. Observations on amphibians and reptiles of the Big Bend region of Texas. Southwestern Naturalist 3:28–54.

Minton, S.A., Jr. 1972. Amphibians and Reptiles of Indiana. Monograph Number 3, Indiana Academy of Science, Indianapolis, Indiana.

Minton, S.A., Jr. 2001. Amphibians and Reptiles of Indiana. Second edition. Indiana Academy of Science, Indianapolis, Indiana.

Minton, S.A., Jr., J.C. List and M.J. Lodato. 1982. Recent records and status of amphibians and reptiles in Indiana. Proceedings of the Indiana Academy of Science 92:489–498.

Mishler, B.D. 1995. Plant systematics and conservation: science and society. Madrono 42:103–113.

Mishler, B.D. and M.J. Donoghue. 1982. Species concepts: a case for pluralism. Systematic Zoology 31:491–503.

Mitchell, D., A. Jones and J.-M. Hero. 1995. Predation on the cane toad *(Bufo marinus)* by the black kite *(Milvus migrans)*. Queensland Museum Memoirs 38:512.

Mitchell, J.C. 1986. Life history patterns in a central Virginia frog community. Virginia Journal of Science 37:262–271.

Mitchell, J.C. 1990. Life history notes: *Pseudacris feriarum* (upland chorus frog). Predation. Herpetological Review 21:89–90.

Mitchell, J.C. 1991. Amphibians and reptiles. Pp. 411–423. *In* Terwilliger, K. (Coord.), Virginia's Endangered Species. McDonald and Woodward Publishing Company, Blacksburg, Virginia.

Mitchell, J.C. 1994a. The Reptiles of Virginia. Smithsonian Institution Press, Washington, D.C.

Mitchell, J.C. 1994b. Habitat conservation assessment for the Cow Knob salamander *(Plethodon punctatus)* in the George Washington National Forest. U.S. Fish and Wildlife Service, Annapolis, Maryland, and George Washington National Forest, Harrisonburg, Virginia.

Mitchell, J.C. 1996. Distribution and ecology of the Cow Knob salamander *(Plethodon punctatus)* on Shenandoah Mountain. Unpublished report to the George Washington National Forest and the Jefferson National Forest, Roanoke, Virginia.

Mitchell, J.C. 1998. Amphibian decline in the mid-Atlantic region: Monitoring and management of a sensitive resource. Final Report. Legacy Resource Management Program, U.S. Department of Defense, Washington, D.C.

Mitchell, J.C. and J.M. Anderson. 1994. Amphibians and reptiles of Assateague and Chincoteague Islands. Virginia Museum of Natural History, Special Publication, Number 2, Martinsville, Virginia.

Mitchell, J.C. and D.E. Green. 2003. Natural history notes: *Hyla gratiosa* (barking treefrog). Intestinal hernia. Herpetological Review 34:230–231.

Mitchell, J.C. and S.B. Hedges. 1980. *Ambystoma mabeei* Bishop (Caudata: Ambystomatidae): an addition to the salamander fauna of Virginia. Brimleyana 3:119–121.

Mitchell, J.C. and R.K. Reay. 1999. Atlas of amphibians and reptiles in Virginia. Special Publication Number 1, Wildlife Diversity Division,

Virginia Department of Game and Inland Fisheries, Richmond, Virginia.

Mitchell, J.C. and J.A. Wicknick. 1996. Ecology of the Peaks of Otter salamander *(Plethodon hubrichti)*: population responses to timber harvesting practices. Unpublished report to the George Washington and Jefferson National Forests, Roanoke, Virginia.

Mitchell, J.C. and W.S. Woolcott. 1985. Observations of the microdistribution, diet, and predator-prey size relationships in the salamander *Plethodon cinereus* from the Virginia Piedmont. Virginia Journal of Science 36:281–288.

Mitchell, J.C., K.A. Buhlmann and R.L. Hoffman. 1996a. Predation of marbled salamander *(Ambystoma opacum* [Gravenhorst]) eggs by the milliped *Uroblaniulus jerseyi* (Causey). Banisteria 8:55–56.

Mitchell, J.C., S.Y. Erdle and J.F. Pagels. 1998. Notes on the distribution and ecology of some amphibians and reptiles in southeastern Virginia. Banisteria 11:41–46.

Mitchell, J.C., J.A. Wicknick and C.D. Anthony. 1996b. Effects of timber harvesting practices on Peaks of Otter salamander *(Plethodon hubrichti)* populations. Amphibian and Reptile Conservation 1:15–19.

Mitchell, J.C., T.K. Pauley, D.I. Withers, S.M. Roble, B.T. Miller, A.L. Braswell, P.V. Cupp Jr. and C.S. Hobson. 1999. Conservation status of the southern Appalachian herpetofauna. Virginia Journal of Science 50:13–35.

Mitchell, R.W. and J.R. Reddell. 1965. *Eurycea tridentifera*, a new species of troglobitic salamander from Texas and a reclassification of *Typhlomolge rathbuni*. Texas Journal of Science 17:12–27.

Mitchell, R.W. and R.E. Smith. 1972. Some aspects of the osteology and evolution of the neotenic spring and cave salamanders (*Eurycea*, Plethodontidae) of central Texas. Texas Journal of Science 23:343–362.

Mitchell, S.L. and G.L. Miller. 1991. Intermale spacing and calling site characteristics in a southern Mississippi chorus of *Hyla cinerea*. Copeia 1991:521–524.

Mittleman, M.B. 1946. Nomenclatural notes on two southeastern frogs. Herpetologica 3:57–60.

Mittleman, M.B. 1947. American Caudata. I. Geographic variation in *Manculus quadridigitatus*. Herpetologica 3:209–224.

Mittleman, M.B. 1948. American Caudata. II. Geographic variation in *Ambystoma macrodactylum*. Herpetologica 4:81–95.

Mittleman, M.B. 1950. Miscellaneous notes on some amphibians and reptiles from the southeastern United States. Herpetologica 6:20–24.

Mittleman, M.B. 1951. America Caudata. VII. Two new salamanders of the genus *Plethodon*. Herpetologica 7:105–112.

Mittleman, M.B. 1966. *Eurycea bislineata*. Pp. 45.1–45.4. Catalogue of American Amphibians and Reptiles. Society for the Study of Amphibians and Reptiles, St. Louis, Missouri.

Mittleman, M.B. 1967. *Manculus* and *M. quadridigitatus*. Pp. 44.1–44.2. Catalogue of American Amphibians and Reptiles. Society for the Study of Amphibians and Reptiles, St. Louis, Missouri.

Mittleman, M.B. and G.S. Myers. 1949. Geographic variation in the ribbed frog, *Ascaphus truei*. Proceedings of the Biological Society of Washington 62:57–68.

Modzelewski, E. and D.D. Culley Jr. 1974. Occurrence of the nematode *Eustronglides werichii* in lab reared *Rana catesbeiana*. Copeia 1974:1000–1001.

Mohr, C.E. 1931. Observations on the early breeding habits of *Ambystoma jeffersonianum* in central Pennsylvania. Copeia 1931:102–104.

Mohr, C.E. 1943. The eggs of the long-tailed salamander, *Eurycea longicauda longicauda* (Green). Proceedings of the Pennsylvania Academy of Science 17:86.

Mohr, C.E. 1944. A remarkable salamander migration. Proceedings of the Pennsylvania Academy of Science 18:51–54.

Mohr, C.E. 1950. Ozark cave life. National Speleological Association Bulletin 12:3–11.

Mohr, C.E. 1952. The eggs of the zig-zag salamander, *Plethodon cinereus dorsalis*. National Speleological Association Bulletin 14:59–60.

Moir, W.H. and H.M. Smith. 1970. Occurrence of an American salamander, *Aneides hardyi* (Taylor), in tundra habitat. Arctic and Alpine Research 2:155–156.

Moler, P.E. 1980. The Florida population of the pine barrens treefrog *(Hyla andersonii)*, a status review. Unpublished report to the Florida Game and Fresh Water Fish Commission, Tallahassee, Florida.

Moler, P.E. 1981. Notes on *Hyla andersonii* in Florida and Alabama. Journal of Herpetology 15:441–444.

Moler, P.E. 1982. Geographic distribution: *Limnaoedus ocularis* (little grass frog). Herpetological Review 13:131.

Moler, P.E. 1985. A new species of frog (Ranidae: *Rana*) from northwestern Florida. Copeia 1985:379–383.

Moler, P.E. 1992a. Rare and Endangered Biota of Florida. Volume III. Amphibians and Reptiles. University Press of Florida, Gainesville, Florida.

Moler, P.E. 1992b. Rare Florida bog frog *Rana okaloosae* Moler. Pp. 30–33. *In* Moler, P.E. (Ed.), Rare and Endangered Biota of Florida. Volume III. Amphibians and Reptiles. University Press of Florida, Gainesville, Florida.

Moler, P.E. 1993. *Rana okaloosae*. Pp. 561.1–561.3. Catalogue of American Amphibians and Reptiles. Society for the Study of Amphibians and Reptiles, St. Louis, Missouri.

Moler, P.E. 1994. Natural history notes: *Siren lacertina* (greater siren). Diet. Herpetological Review 25:62.

Moler, P.E. and R. Franz. 1987. Wildlife values of small, isolated wetlands in the southeastern coastal plain. Pp. 234–241. *In* Odum, R.R., K.A. Riddleberger and J.C. Ozier (Eds.), Proceedings of the Third Annual Nongame and Endangered Wildlife Symposium, Georgia Department of Natural Resources, Atlanta, Georgia.

Moler, P.E. and J. Kezer. 1993. Karyology and systematics of the salamander genus *Pseudobranchus* (Sirenidae). Copeia 1993:39–47.

Moler, P.E. and B.W. Mansell. 1986. Life history notes: *Pseudobranchus striatus striatus* (broad-striped dwarf siren). Maximum size. Herpetological Review 17:45.

Moler, P.E. and B.W. Mansell. 1987. Geographic distribution: *Stereochilus marginatus* (many-lined salamander). Herpetological Review 18:56.

Moler, P.E. and K. Thomas. 1982. Geographic distribution: *Pseudobranchus striatus* (dwarf siren). Herpetological Review 13:130.

Möller, H. 1987. Pollution and parasitism in the aquatic environment. International Journal of Parasitology 17:353–361.

Monello, R.J. and R.G. Wright. 1997. Geographic distribution: *Taricha granulosa* (roughskin newt). Herpetological Review 28:155.

Monello, R.J. and R.G. Wright. 1999. Amphibian habitat preferences among artificial ponds in the Palouse Region of Northern Idaho. Journal of Herpetology 33:298–303.

Monello, R.J. and R.G. Wright. 2001. Predation by goldfish *(Carassius auratus)* on eggs and larvae of the eastern long-toed salamander *(Ambystoma macrodactylum columbianum)*. Journal of Herpetology 35:350–353.

Monroe, B.L., Jr. and R.W. Giannini. 1977. Distribution of the barking treefrog in Kentucky. Transactions of the Kentucky Academy of Science 38:143–144.

Monsen, K.J. and M.S. Blouin. 2003. Genetic structure in a montane ranid frog: restricted gene flow and nuclear-mitochondrial discordance. Molecular Ecology 12:3275–3286.

Monson, P.D., D.J. Call, D.A. Cox, K. Liber and G.T. Ankley. 1999. Photoinduced toxicity of fluoranthene to northern leopard frogs *(Rana pipiens)*. Environmental Toxicology and Chemistry 18:308–312.

Montague, J.R. and J.W. Poinski. 1978. Note on the brooding behavior in *Desmognathus fuscus fuscus* (Raf.) (Amphibia, Urodela, Plethodontidae) in Columbiana County, Ohio. Journal of Herpetology 12:104.

Montani, M. and T.K. Pauley. 1992. Status of *Eurycea bislineata* (Green) in West Virginia. Proceedings of the West Virginia Academy of Science 64:31–32.

Montanucci, R.R. 1992. Commentary on a proposed taxonomic arrangement for some North American amphibians and reptiles. Herpetological Review 23:9–10.

Montanucci, R.R. and L.A. Wilson. 1980. Determine status, distribution, and breeding habitat association of the pine barrens treefrog, *Hyla andersonii*. Clemson University, Contract Number 18-409, Supplement Number 29, Clemson, South Carolina.

Moody, S.M. 1986. Geographic distribution: *Scaphiopus holbrookii holbrookii* (eastern spadefoot). Herpetological Review 17:91.

Moore, G.A. and R.C. Hughes. 1941. A new plethodontid salamander from Oklahoma. Copeia 1941:139–142.

Moore, G.A. and R.C. Hughes. 1943a. *Acanthocephalus van-cleavi*, a new echinorynchid worm, from a salamander. American Midland Naturalist 29:724–729.

Moore, G.A. and R.C. Hughes. 1943b. *Sphyranura euryceae*, a polystomatid monogenean fluke from *Eurycea tynerensis*. Transactions of the American Microbiology Society 42:286–292.

Moore, J. 1983. Responses of an avian predator and its isopod prey to an acanthocephalan parasite. Ecology 64:1000–10015.

Moore, J. 1984a. Parasites that change the behavior of their host. Scientific Amercian 250:108–115.

Moore, J. 1984b. Altered behavioral responses in intermediate hosts—an acanthocephalan parasite strategy. American Naturalist 123: 572–577.

Moore, J. and N.J. Gotelli. 1990. A phylogenetic perspective on the evolution of altered host behaviors: a critical look at the manipulation hypothesis. Pp. 193–229. *In* Barnard, C.J. and J.M. Behnke (Eds.), Parasitism and Host Behavior. Taylor and Francis, London.

Moore, J. and B. Moore. 1939. Notes on the Salientia of the Gaspé Peninsula. Copeia 1939:104.

Moore, J.A. 1939. Temperature tolerance and rates of development in the eggs of Amphibia. Ecology 20:459–478.

Moore, J.A. 1940. Adaptive differences in the egg membranes of frogs. American Naturalist 74:89–93.

Moore, J.A. 1942. An embryological and genetical study of *Rana burnsi* Weed. Genetics 27:408–416.

Moore, J.A. 1944. Geographic variation in *Rana pipiens* Schreber of eastern North America. Bulletin of the American Museum of Natural History 82:345–370.

Moore, J.A. 1949a. Geographic variation of adaptive characters in *Rana pipiens* Schreber. Evolution 3:1–24.

Moore, J.A. 1949b. Patterns of evolution in the genus *Rana*. Pp. 315–338. *In* Jepsen, G.L., E. Mayr and G.G. Simpson (Eds.), Genetics, Paleontology, and Evolution. Princeton University Press, Princeton, New Jersey.

Moore, J.A. 1952. An analytical study of the geographic distribution of *Rana septentrionalis*. American Naturalist 86:5–22.

Moore, J.E. and E.H. Strickland. 1954. Notes on the food of three species of Alberta amphibians. American Midland Naturalist 52:221–224.

Moore, J.E. and E.H. Strickland. 1955. Further notes on the food of Alberta amphibians. American Midland Naturalist 54:253–256.

Moore, M.K. and V.R. Townsend Jr. 2003. Intraspecific variation in cranial ossification in the tailed frog, *Ascaphus truei*. Journal of Herpetology 37:714–717.

Moore, N.J. and A.G. Matson. 1997. Natural history notes: *Ambystoma texanum* (smallmouth salamander). Reproduction. Herpetological Review 28:199.

Moore, R.D. 1929. *Canis latrans lestes* Merriam feeding on tadpoles and frogs. Journal of Mammalogy 10:255.

Moore, R.D., B. Newton and A. Sih. 1996. Delayed hatching as a response of streamside salamander eggs to chemical cues from predatory sunfish. Oikos 77:331–335.

Moore, R.H. 1976. Reproductive habits and growth of *Bufo speciosus* on Mustang Island, Texas, with notes on the ecology and reproduction of other anurans. Texas Journal of Science 17:173–178.

Moore, W.S. 1995. Inferring phylogenies from mtDNA variation: mitochondrial-gene trees versus nuclear-gene trees. Evolution 49: 718–726.

Moore, W.S. 1997. Mitochondrial-gene trees versus nuclear-gene trees, a reply to Hoelzer. Evolution 51:627–629.

Morafka, D.J. 1976. Biogeographical implications of pattern variation in the salamander *Aneides lugubris*. Copeia 1976:580–586.

Morafka, D.J. 1977. A Biogeographical Analysis of the Chihuahuan Desert Through its Herpetofauna. Dr. W. Junk Publishers, The Hague, Netherlands.

Moravec, F. 1984. Some helminth parasites from amphibians of Vancouver Island, B.C., Western Canada. Vestnik Ceskoslovenske Zoologicke spolecnosti (English translation: Memoirs of the Czech Zoological Society) 48:107–114.

Morek, D.M. 1972. An organ culture study of frog renal tumor and its effects on normal frog kidney *in vitro*. Ph.D. dissertation. University of Notre Dame, South Bend, Indiana.

Morell, V. 1999. Are pathogens felling frogs? Science 284:728–731.

Moreno, G. 1989. Behavioral and physiological differentiation between the color morphs of the salamander, *Plethodon cinereus*. Journal of Herpetology 23:335–341.

Morescalchi, A. 1975. Chromosome evolution in the caudate Amphibia. Evolutionary Biology 8:339–387.

Morey, S.R. 1990. Microhabitat selection and predation in the Pacific treefrog, *Pseudacris regilla*. Journal of Herpetology 24:292–296.

Morey, S.R. 1994. Age and size at metamorphosis in spadefoot toads: a comparative study of adaptation to uncertain environments. Ph.D. dissertation. University of California, Riverside, California.

Morey, S.R. 1998. Pool duration influences age and body mass at metamorphosis in the western spadefoot toad: implications for vernal pool conservation. Pp. 86–91. *In* Witham, C.W., E.T. Bauder, D. Belk, W.R. Ferren and R. Orunduff (Eds.), Ecology, Conservation, and Management of Vernal Pool Ecosystems—Proceedings from a 1996 Conference. California Native Plant Society, Sacramento, California.

Morey, S.R. and D.A. Guinn. 1992. Activity patterns, food habits, and changing abundance in a community of vernal pool amphibians. Pp. 149–158. *In* Williams, D.F., S. Byrne and T.A. Rado (Eds.), Endangered and Sensitive Species of the San Joaquin Valley, California: Their Biology, Management, and Conservation. California Energy Commission, and The Wildlife Society, Western Section, Sacramento, California.

Morey, S.R. and D.N. Janes. 1994. Variation in larval habitat duration influences metamorphosis in *Scaphiopus couchii*. Pp. 159–165. *In* Brown, P.R. and J.W. Wright (Eds.), Herpetology of the North American Deserts. Special Publication Number 5, Southwest Herpetologists Society, Van Nuys, California.

Morey, S. and D. Reznick. 2000. A comparative analysis of plasticity in larval development in three species of spadefoot toads. Ecology 81:1736–1749.

Morey, S. and D. Reznick. 2001. Effects of larval density on postmetamorphic spadefoot toads *(Spea hammondii)*. Ecology 82:510–522.

Morgan, A.H. and M.C. Grierson. 1932. Winter habits and yearly food consumption of adult spotted newts, *Triturus viridescens*. Ecology 13:54–62.

Morgan, L.A. and W.A. Buttemer. 1996. Predation by the non-native fish *Gambusia holbrooki* on small *Litoria aurea* and *L. dentata* tadpoles. Australian Zoology 30:143–149.

Mori, A. 1989. Behavioral responses to an "unpalatable" prey, *Rana rugosa* (Anura: Amphibia), by newborn Japanese striped snakes, *Elaphe quadrivirgata*. Pp. 459–471. *In* Matsui, M., T. Hikida and R.C. Goris (Eds.), Current Herpetology in East Asia. Herpetological Society of Japan, Kyoto, Japan.

Moriarty, E.C. and D.C. Cannatella. 2004. Phylogenetic relationships of the North American chorus frogs (*Pseudacris*: Hylidae). Molecular Phylogenetics and Evolution 30:409–420.

Moriarty, E.C., S.L. Collins and J.T. Collins. 2000. Geographic distribution: *Gastrophryne olivacea* (Great Plains toad). Herpetological Review 31:50.

Moriarty, J.J. 1998. Status of amphibians in Minnesota. Pp. 166–168. *In* Lannoo, M.J. (Ed.), Status and Conservation of Midwestern Amphibians. University of Iowa Press, Iowa City, Iowa.

Moriarty, J.J. and D. Jones. 1997. Minnesota's Amphibians and Reptiles: Their Conservation and Status. Serpent's Tale Natural History Book Distributors, Lanesboro, Minnesota.

Morin, P.J. 1981. Predatory salamanders reverse the outcome of competition among three species of anuran tadpoles. Science 212: 1284–1286.

Morin, P.J. 1983a. Competitive and predatory interactions in natural and experimental populations of *Notophthalmus viridescens dorsalis* and *Ambystoma tigrinum*. Copeia 1983:628–639.

Morin, P.J. 1983b. Predation, competition, and the composition of larval anuran guilds. Ecological Monographs 53:119–138.

Morin, P.J. 1985. Predation intensity, prey survival and injury frequency in an amphibian predator-prey interaction. Copeia 1985:638–644.

Morin, P.J. 1986. Interactions between intraspecific competition and predation in an amphibian predator-prey system. Ecology 67: 713–720.

Morin, P.J. 1987. Predation, breeding asynchrony, and the outcome of competition among treefrog tadpoles. Ecology 68:675–683.

Morin, P.J. 1995. Functional redundancy, non-additive interactions, and supply-side dynamics in experimental pond communities. Ecology 76:133–149.

Morin, P.J., H.M. Wilbur and R.N. Harris. 1983. Salamander predation and the structure of experimental communities: responses of *Notophthalmus* and microcrustacea. Ecology 64:1430–1436.

Morishita, K. 1926. Studies on some nematode parasites of frogs and toads in Japan, with notes on their distribution and frequency. Journal of the Faculty of Science, Imperial University of Tokyo 1:1–32.

Moritz, C. 1994. Defining evolutionarily significant units for conservation. TRENDS in Ecology and Evolution 9:373–375.

Moritz, C. 1995. Uses of molecular phylogenies for conservation. Philosophical Transactions of the Royal Society of London B 349:113–118.

Moritz, C. 1999. Conservation units and translocations: strategies for conserving evolutionary processes. Hereditas 130:217–228.

Moritz, C. and D.P. Faith. 1998. Comparative phylogeography and the identification of genetically divergent areas for conservation. Molecular Ecology 7:419–429.

Moritz, C., C.J. Schneider and D.B. Wake. 1992. Evolutionary relationships within the *Ensatina eschscholtzii* complex confirm the ring species interpretation. Systematic Biology 41:273–291.

Morlan, R.E. and J.V. Matthews Jr. 1992. Range extension for the plains spadefoot, *Scaphiopus bombifrons*, inferred from owl pellets found near Outlook, Saskatchewan. Canadian Field-Naturalist 106: 311–315.

Morowitz, H.J. 1991. Balancing species preservation and economic considerations. Science 253:752–754.

Morris, M.A. 1981. Taxonomic status, reproductive biology, and larval life history of two unisexual forms of *Ambystoma* from Vermilion County, Illinois. Master's thesis. Southern Illinois University, Carbondale, Illinois.

Morris, M.A. 1985. A hybrid *Ambystoma platineum* x *A. tigrinum* from Indiana. Herpetologica 41:267–271.

Morris, M.A. and R.A. Brandon. 1984. Gynogenesis and hybridization between *Ambystoma platineum* and *Ambystoma texanum* in Illinois. Copeia 1984:324–337.

Morris, M.A. and S.M. Meyer. 1980. Longevity records for captive *Bufo americanus* and *Uta stansburiana*. Bulletin of the Chicago Herpetological Society 15:79–80.

Morris, P.A. 1944. They Hop and Crawl. Jaques Cattell Press, Lancaster, Pennsylvania.

Morris, R.L. and W.W. Tanner. 1969. The ecology of the western spotted frog, *Rana pretiosa pretiosa* Baird and Girard. A life history study. Great Basin Naturalist 29:45–81.

Morse, M. 1904. Batrachians and reptiles of Ohio. Proceedings of the Ohio State Academy of Science 4:94–141.

Morton, M.L. and N. Allan. 1990. Effects of snowpack and age on reproductive schedules and testosterone levels in male white-crowned sparrows in a montane environment. Pp. 235–249. *In* Wada, M. (Ed.), Endocrinology of Birds: Molecular to Behavioral. Japan Scientific Society Press, Tokyo, Japan.

Mosher, H.S. and F.A. Fuhrman. 1984. Occurrence and origin of tetrodotoxin. P. 333. *In* Ragelis, E.P. (Ed.), Seafood Toxins. American Chemical Society Symposium Series, Number 262, American Chemical Society, Washington, D.C.

Mosher, H.S., F.A. Fuhrman, H.D. Buchwald and H.G. Fischer. 1964. Tarichatoxin-tetrodotoxin: a potent neurotoxin. Science 144: 1100–1110.

Mosimann, J.E. and G.B. Rabb. 1948. The salamander *Ambystoma mabeei* in South Carolina. Copeia 1948:304.

Moss, B. 2000. Biodiversity in fresh waters—an issue of species preservation or system functioning? Environmental Conservation 27:1–4.

Mossman, M.J. and R.L. Hine. 1984. The Wisconsin frog and toad survey: establishing a long-term monitoring program. Wisconsin Department of Natural Resources, Endangered Resources Report, Number 9, Madison, Wisconsin.

Mossman, M.J., L.M. Hartman, R. Hay, J.R. Sauer and B.J. Dhuey. 1998. Monitoring long-term trends in Wisconsin frog and toad populations. Pp. 169–198. *In* Lannoo, M.J. (Ed.), Status and Conservation of Midwestern Amphibians. University of Iowa Press, Iowa City, Iowa.

Most, A. 1984. *Bufo alvarius*: The Psychedelic Toad of the Sonoran Desert. Privately printed, Venom Press, Gila, Arizona.

Moulis, R.A. 1995a. A survey of suitable habitat in southwestern Georgia for the gopher frog (*Rana capito* ssp.), striped newt (*Notophthalmus perstriatus*), and flatwoods salamander (*Ambystoma cingulatum*). Report to the U.S. Fish and Wildlife Service, Jackson, Mississippi.

Moulis, R.A. 1995b. A survey of eastern Georgia for gopher frogs (*Rana capito* ssp.), with notes on the occurrence of *Notophthalmus perstriatus* and *Ambystoma cingulatum*. U.S. Fish and Wildlife Service, Unpublished report, Jackson, Mississippi.

Moulton, C.A., W.J. Fleming and B.R. Nerney. 1997. The use of PVC pipes to capture hylid frogs. Herpetological Review 27:186–187.

Moulton, J.M. 1954a. A late August breeding of *Hyla cinerea* in Florida. Herpetologica 10:171.

Moulton, J.M. 1954b. Notes on the natural history, collection and maintenance of the salamander *Ambystoma maculatum*. Copeia 1954: 64–65.

Mount, R.H. 1975. The Reptiles and Amphibians of Alabama. Agricultural Experiment Station, Auburn University Press, Auburn, Alabama.

Mount, R.H. 1980. Distribution and status of the Pine Barrens treefrog, *Hyla andersonii*, in Alabama. Unpublished Report to U.S. Fish and Wildlife Service, Auburn, Alabama.

Mount, R.H. 1981. The status of the flattened musk turtle, *Sternotherus minor depressus*, Tinkle and Webb. U.S. Fish and Wildlife Service, Unpublished report, Atlanta, Georgia.

Mount, R.H. 1990. The status of Alabama amphibians and reptiles—an update. Journal of the Alabama Academy of Science 61:117–122.

Moyle, P.B. 1973. Effects of introduced bullfrogs, *Rana catesbeiana*, on the native frogs of the San Joaquin Valley, California. Copeia 1973: 18–22.

Moyle, P.B. 1976. Inland Fishes of California. University of California Press, Berkeley, California.

Moyle, P.B. 2000. Inland Fishes of California. University of California Press, Berkeley, California.

Muchlinsky, A.E. 1985. The energetic cost of the fever response in three species of ectothermic vertebrates. Comparative Biochemistry and Physiology A 81:577–579.

Mückenhausen, E. 1985. Bodenkunde. DLG Verlag, Frankfurt.

Mueller, A.J. 1985. Vertebrate use of nontidal wetlands on Galveston Island, Texas. Texas Journal of Science 37:215–225.

Mueller, N.S. 1980. Mallards capture and eat American toads. Wilson Bulletin 92:523–524.

Mulaik, S. 1937. Notes on *Leptodactylus labialis* (Cope). Copeia 1937: 72–73.

Mulaik, S. 1945. New mites in the family Caeculidae. Bulletin of the University of Utah, Volume 35, Number 17, Biological Series, Volume 8, Number 6, Salt Lake City, Utah.

Mulaik, S. and D. Sollberger. 1938. Notes on the eggs and habits of *Hypopachus cuneus*. Copeia 1938:90.

Mulcahy, D.G. and J.R. Mendelson III. 2000. Phylogeography and speciation of the morphologically variable, widespread species *Bufo valliceps*, based on molecular evidence from mtDNA. Molecular Phylogenetics and Evolution 17:173–189.

Mulcahy, D.G., M.R. Cummer, J.R. Mendelson, B.L. Williams and P.C. Ustach. 2002. Status and distribution of two species of *Bufo* in the northeastern Bonneville Basin of Idaho and Utah. Herpetological Review 33:287–289.

Mulcare, D.J. 1965. The problem of toxicity in *Rana palustris*. Transactions of the Indiana Academy of Science 75:319–324.

Mullally, D.P. 1952. Habits and minimum temperatures of the toad *Bufo boreas halophilus*. Copeia 1952:274–276.

Mullally, D.P. 1953. Observations on the ecology of the toad *Bufo canorus*. Copeia 1953:182–183.

Mullally, D.P. and J.D. Cunningham. 1956. Ecological relations of *Rana muscosa* at high elevations in the Sierra Nevada. Herpetologica 12: 189–198.

Munger, J.C., A. Ames and B. Barnett. 1997a. 1996 Survey for Columbia spotted frogs in the Owyhee Mountains of southwestern Idaho. Idaho Bureau of Land Management, Technical Bulletin 97-13, Boise, Idaho.

Munger, J.C., B.R. Barnett and A. Ames. 1997b. 1996 Sawtooth Wilderness Amphibian Survey. U.S.D.A. Sawtooth National Forest and Boise State University, Challenge Cost Share Report, Boise, Idaho.

Munger, J.C., M. Gerber, M. Carroll, K. Madrid and C. Peterson. 1996. Status and habitat associations of the spotted frog *Rana pretiosa* in Southwestern Idaho. Idaho Bureau of Land Management, Technical Bulletin 96-1, Boise, Idaho.

Munsey, L.D. 1972. Salinity tolerance of the African pipid frog, *Xenopus laevis*. Copeia 1972:584–586.

Murdoch, W.W. 1994. Population regulation in theory and practi[?] Ecology 75:271–287.

Murdock, N.A. 1994. Rare and endangered plants and animals of s[?] ern Appalachian wetlands. Water, Air and Soil Pollution 77:38[?]

Murdock, N.A. and A.L. Braswell. 1985. Geographic distributi[?] *bystoma talpoideum* (mole salamander). Herpetological Revie[?]

Murphy, C.G. 1994a. Determinants of chorus tenure in barkin[?] (*Hyla gratiosa*). Behavioral Ecology and Sociobiology 34:2[?]

Murphy, C.G. 1994b. Chorus tenure of male barking treefr[?] *tiosa*. Animal Behaviour 48:763–777.

Murphy, C.G., S.T. Emlen and P.W. Sherman. 1993. Repr[?] gies of the treefrog *Hyla gratiosa*: implications fc[?] Florida Game and Freshwater Fish Commission, [?] Program, Final report, Gainesville, Florida.

Murphy, D., D. Wilcove, R. Noss, J. Harte, C. Safi[?] Root, V. Sher, L. Kaufman, M. Bean and S. Pimn[?] rization of the Endangered Species Act. Conserv[?]

Murphy, J.C. 1979. *Pseudacris streckeri*, a small frog[?] history. Bulletin of the Chicago Herpetological S[?]

Murphy, M.L. and J.D. Hall. 1981. Varied effects of[?] predators and their habitat in small streams of[?] tains, Oregon. Canadian Journal of Fisheries and[?] 137–145.

Murphy, M.L., C.P. Hawkins and N.H. Anderson. 198[?] modification and accumulated sediment on str[?] Transactions of the American Fisheries Society 11[?]

Murphy, R.C. 1918. The food of *Plethodon cinereus*. C[?]

Murphy, R.W. and R.C. Drewes. 1976. Comments on the occurrence of *Smilisca baudinii* (Duméril and Bibron) (Amphibia: Hylidae) in Bexar County, Texas. Texas Journal of Science 27:406–407.

Murphy, T.D. 1961. Predation on eggs of the salamander, *Ambystoma maculatum*, by caddisfly larvae. Copeia 1961:495–496.

Murphy, T.D. 1962. A study of two breeding populations of the salamanders *Ambystoma maculatum* and *A. opacum*. Journal of the Elisha Mitchell Scientific Society 78:102.

Murphy, T.D. 1965. High incidence of two parasitic infestations and two morphological abnormalities in a population of the frog, *Rana palustris* LeConte. American Midland Naturalist 74:233–239.

Murphy, T.D. 1968. *Hyla versicolor* uses bird house for shelter. Herpetologica 24:78–79.

Murray, I. and C.W. Painter. 2003. Geographic distribution: *Eleutherodactylus augusti*. Herpetological Review 34:161.

Murray, J.A. 1908. The zoological distribution of cancer. Imperial Cancer Research Fund, Third Scientific Report:41–60.

Musgrave, M.E. and D.M. Cochran. 1930. *Bufo alvarius*, a poisonous toad. Copeia 173:96–99.

Mushinsky, H.R. 1975. Selection of substrate pH by salamanders. American Midland Naturalist 93:440–443.

Muths, E. 2003. Home range and movements of boreal toads in undisturbed habitat. Copeia 2003:160–165.

Muths, E. and P.S. Corn. 1997. Basking by adult boreal toads (*Bufo boreas boreas*) during the breeding season. Journal of Herpetology 31:426–428.

Muths, E. and P.S. Corn. 2000. Boreal toad. Pp. 60–65. *In* Reading, R.P. and B. Miller (Eds.), Endangered Animals, a Reference Guide to Conflicting Issues. Greenwood Press, Westport, Connecticut.

Muths, E., T.L. Johnson and P.S. Corn. 2001. Experimental translocation of boreal toad (*Bufo boreas*) embryos, toadlets, and adults in Rocky Mountain National Park. Southwestern Naturalist 46:107–113.

Muths, E., P.S. Corn, A.P. Pessier and D.E. Green. 2003. Evidence for disease-related amphibian decline in Colorado. Biological Conservation. 110:357–365.

Muzzall, P.M. 1990. Endoparasites of the red-backed salamander, *Plethodon c. cinereus*, from southwestern Michigan. Journal of the Helminthological Society of Washington 57:165–167.

Muzzall, P.M. and C.R. Peebles. 1991. Helminths of the wood frog, *Rana sylvatica*, and spring peeper, *Pseudacris c. crucifer*, from southern Michigan. Journal of the Helminthological Society of Washington 58:263–265.

Muzzall, P.M. and D.B. Schinderle. 1992. Helminths of the salamanders *Ambystoma tigrinum tigrinum* and *Ambystoma laterale* (Caudata: Ambystomatidae) from southern Michigan. Journal of the Helminthological Society of Washington 59:201–205.

Muzzall, P.M., C.R. Peebles and T.M. Burton. 1997. Endoparasites of plethodontid salamanders from Paradise Brook, New Hampshire. Journal of Parasitology 83:1193–1195.

Myers, C.W. 1958. Notes on the eggs and larvae of *Eurycea lucifuga* Rafinesque. Quarterly Journal of the Florida Academy of Science 21:125–130.

Myers, C.W. and M.A. Donnelly. 1996. A new herpetofauna from Cerro Yavi, Venezuela: first results of the Robert G. Goelet American Museum-Terramar Expedition to the Northwest Tepuis. American Museum Novitates, Number 3172, American Museum of Natural History, New York.

Myers, G.S. 1924. Amphibians and reptiles from Wilmington, N.C. Copeia 131:59–62.

Myers, G.S. 1927. The differential characters of *Bufo americanus* and *Bufo fowleri*. Copeia 163:50–53.

Myers, G.S. 1930a. The status of the southern California toad, *Bufo californicus* Camp. Proceedings of the Biological Society of Washington 43:73–78.

Myers, G.S. 1930b. Notes on some amphibians in western North America. Proceedings of the Biological Society of Washington 43:55–64.

Myers, G.S. 1931a. *Ascaphus truei* in Humboldt County, California, with a note on the habits of the tadpole. Copeia 1931:56–57.

Myers, G.S. 1931b. The original descriptions of *Bufo fowleri* and *Bufo americanus*. Copeia 1931:94–96.

Myers, G.S. 1942a. The black toad of Seep Springs Valley, Inyo County, California. Occasional Papers of the Museum of Zoology, Number 460, University of Michigan, Ann Arbor, Michigan.

Myers, G.S. 1942b. Notes on Pacific coast *Triturus*. Copeia 1942:77–82.

Myers, G.S. 1943. Notes on *Rhyacotriton olympicus* and *Ascaphus truei* in Humboldt County, California. Copeia 1943:125–126.

Myers, G.S. and T.P. Maslin Jr. 1948. The California plethodont salamander, *Aneides flavipunctatus* (Strauch), with a description of a new

subspecies and notes on other western *Aneides*. Proceedings of the Biological Society of Washington 61:127–138.

Myers, J.H. 1988. Can a general hypothesis explain population cycles in forest Lepidoptera. Advances in Ecological Research 18:179–242.

Myers, J.H. and L.D. Rothman. 1995. Field experiments to study regulation of fluctuating populations. Pp. 229–250. *In* Cappuccino, N. and P.W. Price (Eds.), Population Dynamics: New Approaches and Synthesis. Academic Press, New York.

Myers, R.L. 1990. Scrub and high pine. Pp. 150–193. *In* Myers, R. and J. Ewel (Eds.), Ecosystems of Florida. University of Central Florida Press, Orlando, Florida.

Nace, G.W. and C.M. Richards. 1969. Development of biologically defined strains of amphibians. Pp. 409–418. *In* Mizell, M. (Ed.), Biology of Amphibian Tumors. Springer-Verlag, New York.

Naegele, R.F., A. Granoff and R.W. Darlington. 1974. The presence of the Lucké herpesvirus genome in induced tadpole tumors and its oncogenicity: Koch-Henle postulates fulfilled. Proceedings of the National Academy of Sciences 71:830–834.

Nagel, J.W. 1977. Life history of the red-backed salamander, *Plethodon cinereus*, in northeastern Tennessee. Herpetologica 33:13–18.

Nagel, J.W. 1979. Life history of the ravine salamander (*Plethodon richmondi*) in northeastern Tennessee. Herpetologica 35:38–43.

Nagl, A.M. and R. Hofer. 1997. Effects of ultraviolet radiation on early larval stages of the alpine newt, *Triturus alpestris*, under natural and laboratory conditions. Oecologia 110:514–519.

Najarian, H.H. 1955. Trematodes parasitic in the Salientia in the vicinity of Ann Arbor, Michigan. American Midland Naturalist 53:195–197.

NARCAM. 1997. Northern Prairie Wildlife Research Center. North American Reporting Center for Amphibian Malformations. Jamestown, North Dakota: Northern Prairie Wildlife Research Center Home Page. http://www.npwrc.usgs.gov/narcam (Version 02 Mar 00).

NARCAM. 1999. Northern Prairie Wildlife Research Center. North American Reporting Center for Amphibian Malformations. Northern Prairie Wildlife Research Center Home Page. http://www.npwrc.usgs.gov/narcam/ (Version 24Nov99), Jamestown, North Dakota.

NASA. 1999. Global temperature trends:1998 global surface temperature smashes record. NASA Goddard Institute for Space Studies, 13 January, 1999, http://www.gis.nasa.gov/research/observe/surftemp/.

Nassi, H. 1978. Donnees sur le cycle biologique de *Ribeiroia marini guadeloupensis* n. ssp., trematode sterilisant *Biomphalaria glabrata* en Guadeloupe. Acta Tropica 35:41–56.

National Academy of Sciences. 1974. Amphibians: Guidelines for the breeding, care and management of lab animals. National Academy of Sciences, Washington, D.C.

National Wetlands Inventory. 1993. National wetlands inventory (maps). U.S. Fish and Wildlife Service, Washington, D.C.

Nature Conservancy. 2002. Landmark conservation agreement meets success of broad community partnership. http://nature.org/wherewework/northamerica/states/nevada/news/news455.html.

Naughton, G.P., C.B. Henderson, K.R. Foresman and R.L. McGraw II. 2000. Long-toed salamanders in harvested and intact Douglas-fir forests of western Montana. Ecological Applications 10:1681–1689.

Naugle, D.E., K.F. Higgins, M.E. Estey, R.R. Johnson and S.M. Nusser. 2000. Local and landscape factors influencing black tern habitat suitability. Journal of Wildlife Management 64:253–260.

Naugle, D.E., K.F. Higgins, S.M. Nusser and W.C. Johnson. 1999. Scale-dependent habitat use in three species of prairie wetland birds. Landscape Ecology 14:267–276.

Naugle, D.E., R.R. Johnson, M.E. Estey and K.F. Higgins. 2001. A landscape approach to conserving wetland bird habitat in the Prairie Pothole Region of eastern South Dakota. Wetlands 21:1–17.

Nauman, R.S. and Y. Dettlaff. 1999. Natural history notes: *Rana cascadae* (Cascade frog). Predation. Herpetological Review 30:93.

Naylor, B.G. 1978. The frontosquamosal arch of newts as a defence against predators. Canadian Journal of Zoology 56:2211–2216.

Nebeker, A.V. and G.S. Schuytema. 2000. Effects of ammonium sulfate on growth of larval northwestern salamanders, red-legged and Pacific treefrog tadpoles, and juvenile fathead minnows. Bulletin of Environmental Contamination and Toxicology 64:271–278.

Nebeker, A.V., G.S. Schuytema, W.L. Griffis and A. Cataldo. 1998. Impact of guthion on survival and growth of the frog *Pseudacris regilla* and the salamanders *Ambystoma gracile* and *Ambystoma maculatum*. Archives of Environmental Contamination and Toxicology 35:48–51.

Neck, R.W. 1980a. Geographic distribution: *Gastrophryne olivacea* (Great Plains narrowmouth toad). Herpetological Review 11:36.

Neck, R.W. 1980b. Geographic distribution: *Ambystoma texanum* (smallmouth salamander). Herpetological Review 11:36.

Nee, S. and R.M. May. 1997. Extinction and the loss of evolutionary history. Science 278:692–694.

Needham, J.G. 1924. Observations on the life of the ponds at the head of Laguna Canyon. Journal of Entomology and Zoology 16:1–12.

Needham, P.R. and E.H. Vestal. 1938. Notes on growth of golden trout *(Salmo aguabonita)* in two High Sierra lakes. California Fish and Game 24:273–279.

Neigel, J.E. and J.C. Avise. 1986. Phylogenetic relationships of mitochondrial DNA under various demographic models of speciation. Pp. 513–534. *In* Nevo, E. and S. Karlin, (Eds.), Evolutionary Processes and Theory. Academic Press, London.

Neill, W.T. 1947a. A collection of amphibians from Georgia. Copeia 1947:271–272.

Neill, W.T. 1947b. *Rana grylio* in South Carolina. Copeia 1947:206.

Neill, W.T. 1948a. A new subspecies of tree-frog from Georgia and South Carolina. Herpetologica 4:175–179.

Neill, W.T. 1948b. Hibernation of amphibians and reptiles in Richmond County, Georgia. Herpetologica 4:107–114.

Neill, W.T. 1948c. An unusual variant of *Plethodon glutinosus* in South Carolina. Copeia 1948:247–251.

Neill, W.T. 1949a. Hybrid toads in Georgia. Herpetologica 5:30–32.

Neill, W.T. 1949b. Juveniles of *Siren lacertina* and *S. i. intermedia.* Herpetologica 5:19–20.

Neill, W.T. 1950a. Reptiles and amphibians in urban areas of Georgia. Herpetologica 6:113–116.

Neill, W.T. 1950b. A new species of salamander, genus *Desmognathus,* from Georgia. Pp. 1–6. Publication of the Research Division, Ross Allen's Reptile Institute, Number 1, Silver Springs, Florida.

Neill, W.T. 1951a. A bromeliad herpetofauna in Florida. Ecology 32:140–143.

Neill, W.T. 1951b. Notes on the role of crawfishes in the ecology of reptiles, amphibians, and fishes. Ecology 32:764–766.

Neill, W.T. 1951c. A new subspecies of dusky salamander, genus *Desmognathus,* from south-central Florida. Pp. 25–38. Publication of the Research Division, Ross Allen's Reptile Institute, Number 1, Silver Springs, Florida.

Neill, W.T. 1952. Burrowing habits of *Hyla gratiosa.* Copeia 1952:196.

Neill, W.T. 1954. Taxonomy, nomenclature, and distribution of southeastern cricket frogs, genus *Acris.* American Midland Naturalist 43:152–156.

Neill, W.T. 1955. Posture of chilled newts *(Diemyctylus viridescens louisianensis)* Copeia 1955:61.

Neill, W.T. 1957a. Objections to wholesale revision of type localities. Copeia 1957:140–141.

Neill, W.T. 1957b. Homing by a squirrel treefrog, *Hyla squirella* Latreille. Herpetologica 13:217–218.

Neill, W.T. 1957c. Notes on metamorphic and breeding aggregations of the eastern spadefoot, *Scaphiopus holbrookii* (Harlan). Herpetologica 13:185–187.

Neill, W.T. 1957d. The status of *Rana capito stertens* Schwartz and Harrison. Herpetologica 13:47–52.

Neill, W.T. 1957e. Distributional notes on Georgia amphibians, and some corrections. Copeia 1957:43–47.

Neill, W.T. 1958a. The occurrence of amphibians and reptiles in saltwater areas, and a bibliography. Bulletin of Marine Science of the Gulf and Caribbean 8:1–97.

Neill, W.T. 1958b. The varied calls of the barking treefrog, *Hyla gratiosa* LeConte. Copeia 1958:44–46.

Neill, W.T., Jr. 1963. Notes on the Alabama waterdog, *Necturus alabamensis* Viosca. Herpetologica 19:166–174.

Neill, W.T. 1964a. Frogs introduced on islands. Quarterly Journal of the Florida Academy of Sciences 27:127–130.

Neill, W.T. 1964b. A new species of salamander, genus *Amphiuma,* from Florida. Herpetologica 20:62–66.

Neill, W.T. 1968. Predation on *Bufo valliceps* tadpoles by the predaceous diving beetle *Acilius semisulcatus.* Bulletin of the Ecological Society of America 49:169.

Neill, W.T. 1971. The Last of the Ruling Reptiles: Alligators, Crocodiles, and their Kin. Columbia University Press, New York.

Neill, W.T. and E.R. Allen. 1956. Secondarily ingested food items in snakes. Herpetologica 12:172–174.

Neill, W.T. and J.C. Grubb. 1971. Arboreal habits of *Bufo valliceps* in central Texas. Copeia 1971:347–348.

Neill, W.T. and F.L. Rose. 1949. Nests and eggs of the southern dusky salamander, *Desmognathus auriculatus.* Copeia 1949:234.

Neilson, D. 1997a. State Representative Cook meets with local water council to discuss problems. Smithville Times, 21 August.

Neilson, D. 1997b. New wildlife exemption is explored. Smithville Times, 18 September.

Neilson, D. 1998a. Hostility shown towards logging firm. Smithville Times, 29 January.

Neilson, D. 1998b. Officials continue logging investigation. Smithville Times, 19 February.

Neilson, D. 1998c. Concerns over deforestation are real. Smithville Times, 26 February.

Neilson, D. 1998d. County water council passes resolution. Smithville Times, 26 February.

Neilson, D. 1998e. Saving the Houston toad. Smithville Times, 3 December.

Neish, I.C. 1970. A comparative analysis of the feeding behavior of two salamander populations in Marion Lake, B.C. Ph.D. dissertation. University of British Columbia, Vancouver, British Columbia.

Neish, I.C. 1971. Comparison of size, structure, and distributional patterns of two salamander populations in Marion Lake, British Columbia. Journal of the Fisheries Research Board of Canada 28:49–58.

Nelson, C.E. 1972a. *Gastrophryne olivacea.* Pp. 122.1–122.4. Catalogue of American Amphibians and Reptiles. Society for the Study of Amphibians and Reptiles, St. Louis, Missouri.

Nelson, C.E. 1972b. Systematic studies of the North American microhylid genus *Gastrophryne.* Journal of Herpetology 6:111–137.

Nelson, C.E. 1972c. *Gastrophryne carolinensis.* Pp. 120.1–120.4. Catalogue of American Amphibians and Reptiles. Society for the Study of Amphibians and Reptiles, St. Louis. Missouri.

Nelson, C.E. 1973. Mating calls of the Microhylinae: descriptions and phylogenetic and ecological considerations. Herpetologica 29:163–176.

Nelson, C.E. 1974. Further studies on the systematics of *Hypopachus* (Anura: Microhylidae). Herpetologica 30:250–274.

Nelson, C.E. 1980. What determines the species composition of larval amphibian pond communities in south-central Indiana, USA? Proceedings of the Indiana Academy of Science 89:149.

Nelson, C.E. and M.A. Nickerson. 1966. Notes on some Mexican and Central American amphibians and reptiles. Southwestern Naturalist 11:128–131.

Nelson, J.K. 1999. Monitoring frog and toad populations using vocalization surveys at Rocky Flat Environmental Technology Site, Colorado. Journal of the Colorado-Wyoming Academy of Sciences 31:11.

Nelson, J.M. 1993. Population size, distribution, and life history of *Eurycea nana* in the San Marcos River. Master's thesis. Southwest Texas State University, San Marcos, Texas.

Netting, G.N. 1930. Further distinctions between *Bufo americanus* and *Bufo fowleri.* Papers of the Michigan Academy of Sciences, Arts and Letters 11:437–443.

Netting, M.B. 1933. The amphibians of West Virginia, Part 1: salamanders. West Virginia Wild Life 11:5–6, 15.

Netting, M.B. 1936a. Wehrle's salamander, *Plethodon wehrlei* Fowler and Dunn, in Pennsylvania. Proceedings of the Pennsylvania Academy of Science 10:89–93.

Netting, M.B. 1936b. Wehrle's salamander, *Plethodon wehrlei* Fowler and Dunn, in West Virginia. Proceedings of the West Virginia Academy of Science 10:28–30.

Netting, M.G. 1939. The ravine salamander, *Plethodon richmondi* Netting and Mittleman, in Pennsylvania. Proceedings of the Pennsylvania Academy of Science 13:50–51.

Netting, M.G. and C.J. Goin. 1942a. Additional notes on *Rana sevosa.* Copeia 1942:259.

Netting, M.G. and C.J. Goin. 1942b. Descriptions of two new salamanders from peninsular Florida. Annals of the Carnegie Museum 29:175–196.

Netting, M.G. and C.J. Goin. 1946. The correct names for some toads from the eastern United States. Copeia 1946:107.

Netting, M.G. and M.B. Mittleman. 1938. Description of *Plethodon richmondi,* a new salamander from West Virginia and Ohio. Annals of the Carnegie Museum 27:287–393.

Netting, M.G. and N.D. Richmond. 1932. The green salamander, *Aneides aeneus,* in northern West Virginia. Copeia 1932:101–102.

Netting, M.G., N.B. Green and N.D. Richmond. 1946. The occurrence of Wehrle's salamander, *Plethodon wehrlei* Fowler and Dunn, in Virginia. Proceedings of the Biological Society of Washington 59:157–160.

Nevada Division of Wildlife. 2000. Conservation agreement for the Amargosa toad *(Bufo nelsoni)* and co-occurring sensitive species in Oasis Valley, Nye County, Nevada (September 2000). Available at Simandle, E. 2002. The Amargosa Toad. http://www.amargosatoad.org.

Nevada Natural Heritage Program. 1999. Amphibians found in Nevada: state amphibian list by county. Nevada Natural Heritage Program Database, Carson City, Nevada.

Nevo, E. 1973. Adaptive color polymorphism in cricket frogs. Evolution 27:353–367.

New Jersey Natural Heritage Program. 1995. Database: special vertebrate animals of New Jersey. November, 1995. Division of Parks and Forestry, New Jersey Department of Environmental Protection, Office of Natural Lands Management, Trenton, New Jersey.

New Mexico Department of Game and Fish. 2000. Threatened and endangered species of New Mexico. Biennial review and recommendations. New Mexico Department of Game and Fish, Conservation Service Division, Santa Fe, New Mexico.

Newland, R.E., J.W. Steely, S. Begely and E. Carr. 1997. National Historic Landmark nomination for Bastrop State Park. National Register of Historic Places, National Park Service, Washington, D.C.

Newman, M.C. and P.M. Dixon 1996. Ecologically meaningful estimates of lethal effects in individuals. *In* Newman, M.C. and C.H. Jagoe (Eds.), Ecotoxicology: A Hierarchical Treatment. Lewis Publishers, New York.

Newman, R.A. 1987. Effects of density and predation on *Scaphiopus couchii* tadpoles in desert ponds. Oecologia 71:301–307.

Newman, R.A. 1988. Adaptive plasticity in development of *Scaphiopus couchii* tadpoles in desert ponds. Evolution 42:774–783.

Newman, R.A. 1989. Developmental plasticity of *Scaphiopus couchii* tadpoles in an unpredictable environment. Ecology 70:1775–1787.

Newman, R.A. 1994. Effects of changing density and food level on metamorphosis of a desert amphibian, *Scaphiopus couchii*. Ecology 75:1085–1096.

Newman, R.A. 1998. Ecological constraints on amphibian metamorphosis: interactions of temperature and larval density with responses to changing food level. Oecologia 115:9–16.

Newman, R.A. 1999. Body size and diet of recently metamorphosed spadefoot toads *(Scaphiopus couchii)*. Herpetologica 55:507–515.

Newman, R.A. and A.E. Dunham. 1994. Size at metamorphosis and water loss in a desert anuran *(Scaphiopus couchii)*. Copeia 1994:372–381.

Newman, W.B. 1954a. *Gyrinophilus porphyriticus duryi* (Weller) in Virginia. Herpetologica 10:44.

Newman, W.B. 1954b. A new plethodontid salamander from southwestern Virginia. Herpetologica 10:9–14.

Newmark, W.D. 1987. A land-bridge island perspective on mammalian extinctions in western North American parks. Nature 325:430–432.

Newsom, I.E. and E.N. Stout. 1933. Proventriculitis in chickens due to flukes. Veterinary Medicine 28:462–463.

Ng, M.Y. and H.M. Wilbur. 1995. The cost of brooding in *Plethodon cinereus*. Herpetologica 51:1–8.

Nicholls, J.C., Jr. 1949. A new salamander of the genus *Desmognathus* from east Tennessee. Journal of the Tennessee Academy of Science 24:127–129.

Nichols, D.K., A.P. Pessier and J.E. Longcore. 1998. Cutaneous chytridiomycosis in amphibians: an emerging disease? Pp. 269–271. *In* Proceedings of the American Association of Zoo Veterinarians/American Association of Wildlife Veterinarians, Media, Pennsylvania.

Nichols, R. 2001. Gene trees and species trees are not the same. TRENDS in Ecology and Evolution 16:358–364.

Nickerson, M.A. and S.F. Celino. 2003. Natural history notes: *Rana capito* (gopher frog). Drought shelter. Herpetological Review 34:137–138.

Nickerson, M.A. and C.E. Mays 1968. *Bufo retiformis* Sanders and Smith from the Santa Rosa Valley, Pinal County, Arizona. Journal of Herpetology 1:103.

Nickerson, M.A. and C.E. Mays. 1973a. The hellbenders: North American "giant salamanders." Milwaukee Public Museum Publications in Biology and Geology, Number 1, Milwaukee, Wisconsin.

Nickerson, M.A. and C.E. Mays. 1973b. A study of the Ozark hellbender, *Cryptobranchus alleganiensis bishopi*. Ecology 54:1164–1165.

Nickerson, M.A. and M.D. Tohulka. 1986. The nests and nest site selection by Ozark hellbenders, *Cryptobranchus alleganiensis bishopi* Grobman. Transactions of the Kansas Academy of Science 89:66–69.

Nickerson, M.A., R.E. Ashton Jr. and A.L. Braswell. 1983. Lampreys in the diet of hellbender *Cryptobranchus alleganiensis* (Daudin), and the Neuse River waterdog *Necturus lewisi* (Brimley). Herpetological Review 14:10.

Nickerson, M.A., G.J. Jesmok and J.B. Baier. 1979. A serological investigation of three western populations of dusky salamanders, genus *Desmognathus*. Milwaukee Public Museum Publications in Biology and Geology, Number 24, Milwaukee, Wisconsin.

Nickol, B.B. 1972. *Fessisentis*, a genus of acanthocephalans parasitic in North American poikilotherms. Journal of Parasitology 58: 282–289.

Nie, M., J.D. Crim and G.R. Ultsch. 1999. Dissolved oxygen, temperature, and habitat selection by bullfrog *(Rana catesbeiana)* tadpoles. Copeia 1999:153–162.

Nielson, M., K. Lohman and J. Sullivan. 2001. Phylogeography of the tailed frog *(Ascaphus truei)*: implications for the biogeography of the Pacific Northwest. Evolution 55:147–160.

Nieuwkoop, P.D. and J. Faber (Eds.). 1994. Normal Table of *Xenopus laevis* (Daudin). Garland Publishing, New York and London.

Nigrelli, R.F. 1954. Some longevity records for vertebrates. Transactions of the New York Academy of Science 16:296–299.

Nijhuis, M.J. and R.H. Kaplan. 1998. Movement patterns and life history characteristics in a population of the Cascade torrent salamander *(Rhyacotriton cascadae)* in the Columbia River Gorge, Oregon. Journal of Herpetology 32:301–304.

Nishikawa, K.C. 1990. Intraspecific spatial relationships of two species of terrestrial salamanders. Copeia 1990:418–426.

Nishikawa, K.C. and D.C. Cannatella. 1991. Kinematics of prey capture in the tailed frog *Ascaphus truei* (Anura: Ascaphidae). Zoological Journal of the Linnean Society 103:289–307.

Nishikawa, K.C. and P.M. Service. 1988. A fluorescent marking technique for individual recognition of terrestrial salamanders. Journal of Herpetology 22:351–353.

Nixdorf, W.L., D.H. Taylor and L.G. Isaacson. 1997. Use of bullfrog tadpoles *(Rana catesbeiana)* to examine the mechanisms of lead neurotoxicity. American Zoologist 37:363–368.

Nixon, K.C. and Q.D. Wheeler. 1990. An amplification of the phylogenetic species concept. Cladistics 6:211–223.

Nixon, K.C. and Q.D. Wheeler. 1992. Measures of phylogenetic diversity. Pp. 217–234. *In* Novacek, M.J. and Q.D. Wheeler (Eds.), Extinction and Phylogeny. Columbia University Press, New York.

NOAA (National Oceanographic and Atmosperic Administration). 1997. Decadal-to-centennial climate change: what we know—what we don't. http://www.ncdc.noaa.gov/ol/climate/.

Noble, G.K. 1924. Contributions to the herpetology of the Belgian Congo based on the collection of the American Museum Congo Expedition, 1909–1915. Part III. Amphibia. Bulletin of the American Museum of Natural History 49:147–347.

Noble, G.K. 1925. An outline of the relation of ontogeny to phylogeny within the Amphibia. II. American Museum Novitates, Number 166, American Museum of Natural History, New York.

Noble, G.K. 1926. The Long Island newt: a contribution to the life history of *Triturus viridescens*. American Museum Novitates, Number 228, American Museum of Natural History, New York.

Noble, G.K. 1927a. The plethodontid salamanders; some aspects of their evolution. American Museum Novitates, Number 249, American Museum of Natural History, New York.

Noble, G.K. 1927b. The value of life history data in the study of the evolution of the Amphibia. Annals of the New York Academy of Sciences 30:31–128.

Noble, G.K. 1927c. Distributional list of the reptiles and amphibians of the New York City region. American Museum of Natural History Guide, Leaflet Series Number 69, American Museum of Natural History, New York.

Noble, G.K. 1929a. The relation of courtship to the secondary sexual characters of the two-lined salamander, *Eurycea bislineata* (Green). American Museum Novitates, Number 362, American Museum of Natural History, New York.

Noble, G.K. 1929b. Further observations on the life-history of the newt, *Triturus viridescens*. American Museum Novitates, Number 348, American Museum of Natural History, New York.

Noble, G.K. 1930. The eggs of *Pseudobranchus*. Copeia 1930:52.

Noble, G.K. 1931. The Biology of the Amphibia. McGraw-Hill Publishing Company, New York.

Noble, G.K. 1954. The Biology of the Amphibia. Dover Publications, New York.

Noble, G.K. and L.A. Aronson. 1942. The sexual behavior of Anura: I. The normal mating pattern of *Rana pipiens*. Bulletin of the American Museum of Natural History 80:127–142.

Noble, G.K. and M.K. Brady. 1930. The courtship of the plethodontid salamanders. Copeia 1930:52–54.

Noble, G.K. and M.K. Brady. 1933. Observations on the life history of the marbled salamander, *Ambystoma opacum* Gravenhorst. Zoologica 11:89–133.

Noble, G.K. and G. Evans. 1932. Observations and experiments on the life history of the salamander, *Desmognathus fuscus fuscus* (Rafinesque). American Museum Novitates, Number 533, American Museum of Natural History, New York.

Noble, G.K. and W.G. Hassler. 1936. Three Salientia of geographic interest from southern Maryland. Copeia 1936:63–64.

Noble, G.K. and B.C. Marshall. 1929. The breeding habits of two salamanders. American Museum Novitates, Number 347, American Museum of Natural History, New York.

Noble, G.K. and B.C. Marshall. 1932. The validity of *Siren intermedia* LeConte, with observations on its life history. American Museum Novitates, Number 532, American Museum of Natural History, New York.

Noble, G.K. and R.C. Noble. 1923. The Anderson tree frog (*Hyla andersonii* Baird), observations on its life history. Zoologica 2:416–455.

Noble, G.K. and S.H. Pope. 1928. The effect of light on the eyes, pigmentation, and behavior of the cave salamander, *Typhlotriton*. Anatomical Record 5:21.

Noble, G.K. and P.G. Putnam. 1931. Observations on the life history of *Ascaphus truei* Stejneger. Copeia 1931:97–101.

Noble, G.K. and J.A. Richards. 1932. Experiments on the egg-laying of salamanders. American Museum Novitates, Number 513, American Museum of Natural History, New York.

Noel, J.M., W.J. Platt and E.B. Moser. 1998. Structural characteristics of old- and second-growth stands of longleaf pine *(Pinus palustris)* in the Gulf Coastal region of the USA. Conservation Biology 12:533–548.

Noeske, T.A. and M.A. Nickerson. 1979. Diel activity rhythms in the hellbender, *Cryptobranchus alleganiensis* (Caudata: Cryptobranchidae). Copeia 1979:92–95.

Noguchi, T., D.F. Hwang, O. Arakawa, H. Sugita, Y. Deguchi, Y. Shida and K. Hashimoto. 1987. *Vibrio alginolyticus*, a tetrodotoxin-producing bacterium in the intestines of the fish *Fugu vermicularis vermicularis*. Marine Biology 94:625–630.

Noguchi, T., J. Jeon, O. Arakawa, H. Sugita, Y. Deguchi, Y. Shida and K. Hashimoto. 1986. Occurrence of tetrodotoxin and anhydrotetrodotoxin in *Vibrio* sp. isolated from the intestines of a xanthid crab, *Atergatis floridus*. Journal of Biochemistry 99:311–314.

Noland, R. and G.R. Ultsch. 1981. The role of temperature and dissolved oxygen in microhabitat selection by the tadpoles of a frog *(Rana pipiens)* and a toad *(Bufo terrestris)*. Copeia 1981:645–652.

Nöllert, A. and C. Nöllert. 1992. Die Amphibien Europas. Bestimmung, Gefährdung und Schutz. Franckh-Kosmos Verlag, Stuttgart, Germany.

Norden, A.W. and J.D. Groves. 1974. An *Aneides aeneus* nest in West Virginia. Bulletin of the Maryland Herpetological Society 10:79–80.

Norman, B.R. 1988a. Geographic distribution: *Bufo boreas boreas* (boreal toad). Herpetological Review 19:16.

Norman, B.R. 1988b. Life history notes: *Plethodon vehiculum* (western redbacked salamander). Predation. Herpetological Review 19:34.

Norman, B.R. and M. Swartwood. 1991. Life history notes: *Plethodon vehiculum* (western red-backed salamander). Reproduction. Herpetological Review 22:55.

Norman, W.W. 1900. Remarks on the San Marcos salamander, *Typhlomolge rathbuni* Stejneger. American Naturalist 34:179–183.

Norris, D.O. 1989. Seasonal changes in diet of paedogenetic tiger salamanders *(Ambystoma tigrinum mavortium)*. Journal of Herpetology 23:87–89.

Norris L.A., H.W. Lorz and S.V. Gregory. 1983. Influence of forest and rangeland management on anadromous fish habitat in western North America. Chapter 9: Forest chemicals. U.S.D.A. Forest Service, Pacific Northwest Forest and Range Experiment Station, General Technical Report, PNW-149, Portland, Oregon.

North American Reporting Center for Amphibian Malformations (NARCAM). 1997. Northern Prairie Wildlife Research Center Home Page: http://www.npwrc.usgs.gov/narcam (version 02Mar00). Jamestown, North Dakota.

Northcote, T.G. 1992. Eutrophication and pollution problems. Pp. 551–561. *In* Dejoux, C. and A. Iltis (Eds.), Lake Titicaca: A Synthesis of Limnological Knowledge. Monographiae Biologicae, Volume 68, Kluwer Academic Publishers, Dordrecht, Holland.

Noss, R.F. 1987. From plant communities to landscapes in conservation inventories: a look at The Nature Conservancy (USA). Biological Conservation 41:11–37.

Noss, R.F. 1991. From endangered species to biodiversity. Pp. 227–246. *In* Kohm, K.A. (Ed.), Balancing on the Brink of Extinction. Island Press, Washington, D.C.

Novotny, E.S. 1976. Effects of thermal and water stress on heart rate in *Bufo americanus* and *Bufo marinus*. Herpetological Review 7:93.

Nowak, R.M. 1992. The red wolf is not a hybrid. Conservation Biology 6:593–595.

Nussbaum, R.A. 1969a. Nests and eggs of the Pacific giant salamander, *Dicamptodon ensatus* (Eschscholtz). Herpetologica 25:257–262.

Nussbaum, R.A. 1969b. A nest site of the Olympic salamander, *Rhyacotriton olympicus* (Gaige). Herpetologica 25:277–278.

Nussbaum, R.A. 1970. *Dicamptodon copei*, n. sp., from the Pacific Northwest, USA (Amphibian: Caudata: Ambystomatidae). Copeia 1970:506–514.

Nussbaum, R.A. 1976. Geographic variation and systematics of salamanders of the genus *Dicamptodon* Strauch (Ambystomatidae). Miscellaneous Publications of the Museum of Zoology, Number 149, University of Michigan, Ann Arbor, Michigan.

Nussbaum, R.A. 1983. *Dicamptodon copei*. Pp. 334.1–334.2. Catalogue of American Amphibians and Reptiles. Society for the Study of Amphibians and Reptiles, St. Louis, Missouri.

Nussbaum, R.A. 1985. The evolution of parental care in salamanders. Miscellaneous Publications of the Museum of Zoology, Number 169, University of Michigan, Ann Arbor, Michigan.

Nussbaum, R.A. 1987. Parental care and egg size in salamanders: an examination of the safe harbor hypothesis. Researches on Population Ecology 29:27–44.

Nussbaum, R.A. and E.D. Brodie Jr. 1971. The taxonomic status of the rough-skinned newt, *Taricha granulosa* (Skilton) in the Rocky Mountains. Herpetologica 27:260–270.

Nussbaum, R.A. and E.D. Brodie Jr. 1981. *Taricha granulosa*. Pp. 272.1–272.4. Catalogue of American Amphibians and Reptiles. Society for the Study of Amphibians and Reptiles, St. Louis, Missouri.

Nussbaum, R.A. and G.W. Clothier. 1973. Population structure, growth, and size of larval *Dicamptodon ensatus* (Eschscholtz). Northwest Science 47:218–227.

Nussbaum, R.A. and C. Maser. 1969. Observations of *Sorex palustris* preying on *Dicamptodon ensatus*. Murrelet 50:23–24.

Nussbaum, R.A. and C.K. Tait. 1977. Aspects of the life history and ecology of the Olympic salamander, *Rhyacotriton olympicus* (Gaige). American Midland Naturalist 98:176–199.

Nussbaum, R.A., E.D. Brodie Jr. and R.M. Storm. 1983. Amphibians and Reptiles of the Pacific Northwest. University Press of Idaho, Moscow, Idaho.

Nyman, S. 1986. Mass mortality in larval *Rana sylvatica* attributable to the bacterium, *Aeromonas hydrophilia*. Journal of Herpetology 20:196–201.

Nyman, S. 1987. Life history notes: *Ambystoma maculatum* (spotted salamander). Reproduction. Herpetological Review 18:14–15.

Nyman, S. 1991. Ecological aspects of syntopic larvae of *Ambystoma maculatum* and the *Ambystoma laterale-jeffersonianum* complex in New Jersey. Journal of Herpetology 25:505–509.

Nyman, S., M.J. Ryan and J.D. Anderson. 1988. The distribution of the *Ambystoma jeffersonianum* complex in New Jersey. Journal of Herpetology 22:224–228.

Nyman, S., R.F. Wilkinson and J.E. Hutcherson. 1993. Cannibalism and size relations in a cohort of larval ringed salamanders *(Ambystoma annulatum)*. Journal of Herpetology 27:78–84.

O'Brien, S.J. and E. Mayr. 1991. Bureaucratic mischief: recognizing endangered species and subspecies. Science 251:1187–1188.

O'Brien, S.J., M.E. Roelke, N. Yuhki, K.W. Richards, W.E. Johnson, W.L. Franklin, A.E. Anderson, O.L. Bass Jr., R.C. Belden and J.S. Martenson. 1990. Genetic introgression within the Florida panther *Felis concolor coryi*. National Geographic Research 6:485–494.

O'Conner, H. 1976. Letter to Regional Director, Region 2, U.S. Fish and Wildlife Service, 7 December.

O'Connor, M.P. and C.R. Tracy. 1992. Thermoregulation by juvenile toads of *Bufo woodhousei* in the field and in the laboratory. Copeia 1992:865–876.

O'Connor, T. 1984. A note on the diet of nestling blackheaded herons. Ostrich 55:221–222.

O'Donnell, D.J. 1937. Natural history of the ambystomid salamanders of Illinois. American Midland Naturalist 18:1063–1071.

O'Donnell, L. 1994. Proposal to list the Barton Springs salamander as endangered. Federal Register 59:7968–7978.

O'Donnell, L. 1997. Final rule to list the Barton Springs salamander as endangered. Federal Register 62:23377–23392.

O'Donnell, L. 1998. 90-day finding for a petition to list the robust blind salamander, widemouth blindcat, and toothless blindcat. Federal Register 63:48166–48167.

O'Hara, R.K. 1981. Habitat selection behavior in three species of anuran larvae: Environmental cues, ontogeny and adaptive significance. Ph.D. dissertation. Oregon State University, Corvallis, Oregon.

O'Hara, R.K. and A.R. Blaustein. 1985. *Rana cascadae* tadpoles aggregate with siblings: an experimental study. Oecologia 67:44–51.

O'Neill, J.H. 1996. The golden ghetto: The psychology of affluence. Hazelden Foundation, Center City, Minnesota.

O'Neill, R.V., J.R. Krummel, R.H. Gardner, G. Sugihara, B. Jackson, D.L. DeAngelis, B.T. Milne, M.G. Turner, B. Zygmunt, S.W. Christensen, V.H. Dale and R.L. Graham. 1988. Indices of landscape pattern. Landscape Ecology 1:153–162.

O'Roke, E.C. 1926. The amphibians of South Dakota. Proceedings of the South Dakota Academy of Science 9:13–15.

Odlaug, T.O. 1954. Parasites of Ohio Amphibia. Ohio Journal of Science 54:126–128.

Ogilvie, D.J., R.G. McKinnell and D. Tarin. 1984. Temperature-dependent elaboration of collagenase by the renal adenocarcinoma of the leopard frog, *Rana pipiens*. Cancer Research 44:3438–3441.

Ogren, R.E. 1953. A contribution to the life cycle of *Cosmocercoides* in snails (Nematoda: Cosmocercidae). Transactions of the American Microscopical Society 72:87–91.

Okada, Y. 1938. The oecological studies of the frogs with special reference to their feeding habits. Journal of the Imperial Agriculture Experiment Station 3:275–350.

Okada, Y. 1966. Fauna Japonica: Anura (Amphibia). Biogeographical Society of Japan, Tokyo, Japan.

Okafor, J.I., D. Testrake, H.R. Mushinsky and B.G. Yangio. 1984. A *Basidiobolus* sp. and its association with reptiles and amphibians in southern Florida. Sabouraudia 22:47–51.

O'Laughlin, B.E. and R.N. Harris. 2000. Models of metamorphic timing: an experimental evaluation with the pond-dwelling salamander *Hemidactylium scutatum* (Caudata: Plethodontidae). Oecologia 124:343–350.

Oldfield, B. and J.J. Moriarty. 1994. Amphibians and Reptiles Native to Minnesota. University of Minnesota Press, Minneapolis, Minnesota.

Oldham, M.J. 1990. Ontario herpetofaunal summary. Pp. 195–205. *In* Allen, G.M., P.F.J. Eagles and S.D. Price (Eds.), Conserving Carolinian Canada—Conservation Biology in the Deciduous Forest Region. University of Waterloo Press, Waterloo, Ontario, Canada.

Oldham, M.J. 1992. Declines in Blanchard's cricket frog in Ontario. Pp. 30–31. *In* Bishop, C.A. and K.E. Pettit (Eds.), Declines in Canadian Amphibian Populations: Designing a National Monitoring Strategy. Canadian Wildlife Service, Occasional Paper, Number 76, Ottawa, Ontario, Canada.

Oldham, M.J. and C.A. Campbell. 1990. The status of Blanchard's cricket frog, *Acris crepitans blanchardi*, in Canada. Committee on the Status of Endangered Wildlife in Canada, Ottawa, Ontario, Canada.

Oldham, M.J. and W.F. Weller. 1992. Ontario herpetofaunal summary: compiling information on the distribution and life history of amphibians and reptiles of Ontario. Pp. 21–22. *In* Bishop, C.A. and K.E. Pettit (Eds.), Declines in Canadian Amphibian Populations: Designing a National Monitoring Strategy. Canadian Wildlife Service, Occasional Paper, Number 76, Ottawa, Ontario, Canada.

Oldham, R.S. 1969. Initiation of breeding behavior in the American toad *Bufo americanus*. Canadian Journal of Zoology 47:1083–1086.

Oldham, R.S. 1974. Mate attraction by vocalization in members of the *Rana pipiens* complex. Copeia 1974:982–984.

Oldham, R.S. and H.C. Gerhardt. 1975. Behavioral isolating mechanisms of the treefrogs *Hyla cinerea* and *H. gratiosa*. Copeia 1975:223–231.

Olding, P. 1994. Ecological investigation of the cane toad, *Bufo marinus*, in North Queensland, with emphasis of their invasive potential. Jesus College, Oxford, Great Britain.

Oliver, G.A. 1997 (revised 1998). Inventory of sensitive species and ecosystems in Utah. Inventory of sensitive vertebrate and invertebrate species: A progress report. Utah Division of Wildlife Resources, Salt Lake City, Utah.

Oliver, G.V. 1967. Food habits of the white-throated slimy salamander in central Texas. Transactions of the Oklahoma Junior Academy of Science 1967:500–503.

Oliver, G.V., A.H. Chaney, L. Miller and S. Parker. 1980. Vertebrates of (the) boscaje de la palma (area of) extreme southern Texas. Unpublished Report to the Lower Rio Grande Valley Wildlife Refuge, U.S. Fish and Wildlife Service, Alamo, Texas.

Oliver, J.A. 1949. The peripatetic toad. Natural History 58:30–33.

Oliver, J.A. 1955a. The Natural History of North American Amphibians and Reptiles. D. Van Nostrand Company, Princeton, New Jersey.

Oliver, J.A. 1955b. Giant toads of Florida. Quarterly Journal of the Florida Academy of Science 21:207–211.

Oliver, J.A. and C.E. Shaw. 1953. The amphibians and reptiles of the Hawaiian islands. Zoologica 38:65–95.

Oliver, M.G. and H.M. McCurdy. 1974. Migration, overwintering, and reproductive patterns of *Taricha granulosa* on southern Vancouver Island. Canadian Journal of Zoology 52:541–545.

Olivier, L. 1940. Life history studies on strigeid trematodes of the Douglas Lake region, Michigan. Journal of Parasitology 26:447–477.

Ollivier, L.M. and H.H. Welsh Jr. 1999. Survey protocol for the Del Norte salamander *(Plethodon elongatus)*. Pp. 163–200. *In* Olson, D.H. (Ed.), Survey Protocols for Amphibians under the Survey and Management Provision of the Northwest Forest Plan. U.S. Department of the Interior, Bureau of Land Management, BLM/OR/WA/PT–00/033+1792, Washington, D.C.

Ollivier, L.M. and H.H. Welsh Jr. and D.R. Clayton. 2001. Habitat correlates of the Siskiyou Mountains salamander *Plethodon stormi* (Caudata: Plethodontidae); with comments on the species' range. U.S.D.A. Forest Service, Final report, Pacific Southwest Research Station, Redwood Sciences Laboratory, Arcata, California. Available at: www.rsl.psw.fs.fed.us/pubs/wild90s.html.

Olmstead, R.G. 1995. Species concepts and plesiomorphic species. Systematic Botany 20:623–630.

Olsen, O.W. 1937. Description and life history of the trematode *Haplometrana utahensis* sp. nov. (Plagiorchiidae) from *Rana pretiosa*. Journal of Parasitology 23:13–25.

Olsen, O.W. 1938. *Aplectana gigantica* (Cosmocercidae), a new species of nematode from *Rana pretiosa*. Transactions of the American Microscopical Society 57:200–203.

Olson, D.H. 1989. Predation on breeding western toads *(Bufo boreas)*. Copeia 1989:391–397.

Olson, D.H. 1992. Ecological susceptibility of amphibians to population declines. Pp. 55–62. *In* Harris, R.R. and D.E. Erman (Tech. Coords.), and Kerner, H.M. (Ed.), Biodiversity of Northwestern California. Proceedings of a Symposium 28–30 October 1991, Santa Rosa, California. University of California Wildland Resource Center Report, Number 29, Davis, California.

Olson, D.H. 2001. Ecology and management of montane amphibians of the U.S. Pacific Northwest. Biota 2:51–74.

Olson, D.H., W.P. Leonard and R.B. Bury. 1997. Sampling Amphibians in Lentic Habitats. Society for Northwestern Vertebrate Biology, Northwest Fauna, Number 4, Olympia, Washington.

Olson, R.E. 1956. The amphibians and reptiles of Winnebago County, Illinois. Copeia 1956:188–191.

Olson, R.E. 1959. Notes on some Texas herptiles. Herpetologica 15:48.

Olson, R.E. 1984. Geographic distribution: *Hyla chrysoscelis* (Cope's gray treefrog). Herpetological Review 15:76.

Olson, R.E. 1987. Minnesota herpetological records. Bulletin of the Maryland Herpetolological Society 23:101–104.

Omernik, J.M. 1987. Ecoregions of the conterminous United States. Annals of the Association of American Geographers 77:118–125.

Omernik, J.M. 1994. Ecoregions: a spatial framework for environmental management. Pp. 49–62. *In* Davis, W.S. and T.P. Simon (Eds.), Biological Assessment and Criteria. Lewis Publishers, Boca Raton, Louisiana.

Omland, K.E. 1997. Correlated rates of molecular and morphological evolution. Evolution 51:1381–1391.

Oplinger, C.S. 1966. Sex ratio, reproductive cycles, and time of ovulation in *Hyla crucifer crucifer* Weid. Herpetologica 22:276–283.

Oplinger, C.S. 1967. Food habits and feeding activity of recently transformed and adult *Hyla crucifer crucifer* Weid. Herpetologica 23:209–217.

Orchard, S.A. 1992. Amphibian population declines in British Columbia. Pp. 14–16. *In* Bishop, C.A. and K.E. Pettit (Eds.), Declines in Canadian Amphibian Populations: Designing a National Monitoring Strategy. Canadian Wildlife Service, Occasional Paper, Number 76, Ottawa, Ontario, Canada.

Oregon Natural Heritage Program. 1995. Rare, threatened and endangered plants and animals of Oregon. Oregon Natural Heritage Program, Portland, Oregon.

Organ, J.A. 1958. Courtship and spermatophores of *Plethodon jordani metcalfi*. Copeia 1958:251–259.

Organ, J.A. 1960a. The courtship and spermatophore of the salamander *Plethodon glutinosus*. Copeia 1960:34–40.

Organ, J.A. 1960b. Studies on the life history of the salamander, *Plethodon welleri*. Copeia 1960:287–297.

Organ, J.A. 1961a. Studies of the local distribution, life history, and population dynamics of the salamander genus *Desmognathus* in Virginia. Ecological Monographs 31:189–220.

Organ, J.A. 1961b. Life history of the pigmy salamander, *Desmognathus wrighti*, in Virginia. American Midland Naturalist 66:384–390.

Organ, J.A. 1961c. The eggs and young of the spring salamander, *Pseudotriton porphyriticus*. Herpetologica 17:53–56.

Organ, J.A. 1968. Time of courtship activity of the slimy salamander, *Plethodon glutinosus*, in New Jersey. Herpetologica 24:84–85.

Organ, J.A. 1993. Report on salamander density and diversity in Whitetop Creek Cove and Big Branch Cove, Whitetop Mt., Virginia: A comparison of samples from summer and fall 1957–58 to summer and fall 1991–92. Mount Rogers National Recreational Area, Unpublished report, Marion, Virginia.

Organ, J.A. and D.J. Organ. 1968. Courtship behavior of the red salamander *Pseudotriton ruber*. Copeia 1968:217–223.

Orr, B.K., W.W. Murdoch and J.R. Bence. 1990. Population regulation, convergence and cannibalism in *Notonecta*. Ecology 71:68–82.

Orr, L., J. Neumann, E. Vogt and A. Collier. 1998. Status of northern leopard frogs in northeastern Ohio. Pp. 91–93. *In* Lannoo, M.J. (Ed.), Status and Conservation of Midwestern Amphibians. University of Iowa Press, Iowa City, Iowa.

Orr, L.P. 1962, The possibility of mimicry existing between two species of southern Appalachian salamanders. Ph.D. dissertation. University of Tennessee, Knoxville, Tennessee.

Orr, L.P. 1989. *Desmognathus ochrophaeus*. (Cope), mountain dusky salamander. Pp. 181–189. *In* Pfingsten, R.A. and F.L. Downs (Eds.), Salamanders of Ohio. Ohio Biological Survey Bulletin Number 7, New Series, Columbus, Ohio.

Orr, L.P. and W.T. Maple. 1978. Competition avoidance mechanisms in salamander larvae of the genus *Desmognathus*. Copeia 1978:679–685.

Orser, P.N. and D.J. Shure. 1972. Effects of urbanization on the salamander *Desmognathus fuscus fuscus*. Ecology 53:1148–1154.

Orser, P.N. and D.J. Shure. 1975. Population cycles and activity patterns of the dusky salamander, *Desmognathus fuscus fuscus*. American Midland Naturalist 93:403–410.

Orton, G.L. 1946. Larval development of the eastern narrow-mouthed frog, *Microhyla carolinensis* (Holbrook), in Louisiana. Annals of the Carnegie Museum 30:241–248.

Orton, G.L. 1947. Notes on some hylid tadpoles in Louisiana. Annals of the Carnegie Museum 30:363–383.

Orton, G.L. 1953. The systematics of vertebrate larvae. Systematic Zoology 2:63–75.

Osborne, W. 1989. Distribution, relative abundance and conservation status of corroboree frogs, *Pseudophryne corroboree* Moore (Anura: Myobatrachidae). Australian Wildlife Research 16:537–547.

Osborne, W. 1990. Declining frog populations and extinctions in the Canberra region. Bogong 4:4–7.

O'Siggins, K. 1995. Monitoring palustrine amphibian populations on Salmon National Forest. U.S.D.A. Forest Service, Salmon National Forest, Salmon, Idaho.

Osterdam, A. 1769. *Siren lacertina*. Amoenitates Academicae 7:311–325.

Ouellet, M. 2000. Amphibian deformities: current state of knowledge. Pp. 617–661. *In* Sparling, D.W., G. Linder and C.A. Bishop (Eds.), Ecotoxicology of Amphibians and Reptiles. Society for Environmental Toxicology and Contaminants (SETAC) Press, Pensacola, Florida.

Ouellet, M., J. Bonin, J. Rodrigue, J.L. DesGranges and S. Lair. 1997a. Hindlimb deformities (ectromelia, ectrodactyly) in free-living anurans from agricultural habitats. Journal of Wildlife Diseases 33:95–104.

Ouellet, M., J. Rodrigue, J. Bonin, S. Lair and D.M. Green. 1997b. Developmental abnormalities in free-living anurans from agricultural habitats. NAAMP III online paper, http://www.nbs.gov/naamp3/papers/58df.html.

Outcalt, K.W. and R.M. Sheffield. 1996. The longleaf pine forest: trends and current conditions. U.S.D.A. Forest Service, Resource Bulletin, SRS-9, Southern Research Station, Asheville, North Carolina.

Ovaska, K. 1987. Seasonal changes in agonistic behaviour of the western red-backed salamander, *Plethodon vehiculum*. Animal Behaviour 35:67–74.

Ovaska, K. 1988a. Recognition of conspecific odors by the western red-backed salamander, *Plethodon vehiculum*. Canadian Journal of Zoology 66:1293–1296.

Ovaska, K. 1988b. Spacing and movements of the salamander *Plethodon vehiculum*. Herpetologica 44:377–386.

Ovaska, K. 1997. Vulnerability of amphibians in Canada to global warming and increased ultraviolet radiation. Pp. 206–225. *In* Green, D.M. (Ed.), Amphibians in Decline: Canadian Studies of a Global Problem. Herpetological Conservation, Number 1, Society for the Study of Amphibians and Reptiles, St. Louis, Missouri.

Ovaska, K. and T.M. Davis. 1992. Faecal pellets as burrow markers: intra- and interspecific odour recognition by western plethodontid salamanders. Animal Behaviour 43:931–939.

Ovaska, K. and P.T. Gregory. 1989. Population structure, growth, and reproduction in a Vancouver Island population of the salamander *Plethodon vehiculum*. Herpetologica 45:133–143.

Ovaska, K. and M.A. Smith. 1988. Predatory behavior of two species of ground beetles (Coleoptera: Carabidae) towards juvenile salamanders *(Plethodon vehiculum)*. Canadian Journal of Zoology 66:599–604.

Ovaska, K., T.M. Davis and I. Novales Flamarique. 1997. Hatching success and larval survival of the frogs *Hyla regilla* and *Rana aurora* under ambient and artificially enhanced solar ultraviolet radiation. Canadian Journal of Zoology 75:1081–1088.

Over, W.H. 1943. Amphibians and reptiles of South Dakota. Natural History Studies Number VI, Revised Second edition. University of South Dakota, Vermillion, South Dakota.

Overton, F. 1915. Annual occurrence of spade-foot toads. Copeia 20:17.

Owen, R.D. and S.A. Johnson. 1997. Natural history notes: *Pseudacris ocularis* (little grass frog). Predation. Herpetological Review 28:200.

Owens, D.W. 1941. *Ambystoma talpoideum* in Oklahoma. Copeia 1941:183–184.

Owens, I.P.F. and P.M. Bennett. 2000. Quantifying biodiversity: a phenotypic perspective. Conservation Biology 14:1014–1022.

Oyamada, T., T. Hirata, M. Hara, N. Kudo, T. Oyamada, H. Yoshikawa, T. Yoshikawa, N. Suzuki. 1998. Spontaneous larval *Gnathostoma nipponicum* infection in frogs. Journal of Veterinary Medical Science 60:1029–1031.

Pace, A.E. 1974. Systematic and biological studies of the leopard frogs (*Rana pipiens* complex) of the United States. Miscellaneous Publication of the Museum of Zoology, Number 148, University of Michigan, Ann Arbor, Michigan.

Packard, G.C. 1971. Inconsistency in application of the biological species concept to disjunct populations of anurans in southeastern Wyoming and north-central Colorado. Journal of Herpetology 5:191–193.

Packard, G.C., J.K. Tucker and L.D. Lohmiller. 1998. Distribution of Strecker's chorus frogs *(Pseudacris streckeri)* in relation to their tolerance for freezing. Journal of Herpetology 32:437–440.

Packer, W.C. 1960. Bioclimatic influences on the breeding migration of *Taricha rivularis*. Ecology 41:509–517.

Packer, W.C. 1961. Feeding behavior in adult *Taricha*. Copeia 1961:351–352.

Packer, W.C. 1963. Observations on the breeding migration of *Taricha rivularis*. Copeia 1963:378–382.

Paetkau, D. 1999. Using genetics to identify intraspecific conservation units: a critique of current methods. Conservation Biology 13:1507–1509.

Pague, C.A. 1984. Notes on the local distribution of *Desmognathus wrighti* and *Plethodon welleri* in Virginia. Catesbeiana 4:10–11.

Pague, C.A. 1989. King of the mountain. Virginia Wildlife 50:8–13.

Pague, C.A. 1991. Pigmy salamander. Pp. 433–435. *In* Terwilliger, K. (Coord.), Virginia's Endangered Species. McDonald and Woodward Publishing Company, Blacksburg, Virginia.

Pague, C.A. and K.A. Buhlmann. 1992. Fieldnotes: *Rana virgatipes* (carpenter frog). Catesbeiana 12:9.

Pague, C.A. and J.C. Mitchell. 1987. The status of amphibians in Virginia. Virginia Journal of Science 38:305–318.

Pague, C.A. and J.C. Mitchell. 1991a. Mabee's salamander. Pp. 427–429. *In* Terwilliger, K. (Coord.), Virginia's Endangered Species, McDonald and Woodward Publishing Company, Blacksburg, Virginia.

Pague, C.A. and J.C. Mitchell. 1991b. Peaks of Otter salamander. Pp. 436–437 *In* Terwilliger, K. (Coord.), Virginia's Endangered Species, McDonald and Woodward Publishing Company, Blacksburg, Virginia.

Paine, F.L., J.D. Miller, G. Crawshaw, B. Johnson, R. Lacy, C.F. Smith III and P.J. Tolson. 1989. Status of the Puerto Rican crested toad. International Zoo Yearbook 28:53–58.

Painter, C.W. 1985. Herpetology of the Gila and San Francisco river drainages of southwestern New Mexico. Unpublished Report to the New Mexico Department of Game and Fish, Santa Fe, New Mexico.

Painter, C.W. 1986. Geographic distribution: *Hyla eximia* (mountain treefrog). Herpetological Review 17:26.

Painter, C.W. 2000. Status of endemic New Mexico salamanders. New Mexico Department of Game and Fish Federal Aid Report to U.S. Fish and Wildlife Service, Albuquerque, New Mexico.

Painter, C.W. and R.D. Burkett. 1991. Geographic distribution: *Pseudacris clarkii* (spotted chorus frog). Herpetological Review 22:64.

Painter, C.W., N.J. Scott Jr. and M.J. Altenbach. 1999. Natural history notes: *Thamnophis elegans vagrans* (wandering garter snake). Diet. Herpetological Review 30:48.

Palazzo, T.L. 1994. Bagging salamanders. Pp. 250–251. *In* Thelander, C.G. (Ed.), Life on the Edge. Biosystems Books, Santa Cruz, California.

Palen, W.J., D.E. Schindler, M.J. Adams, C.A. Pearl, R.B. Bury and S.A. Diamond. 2002. Optical characteristics of natural waters protect amphibians from UV-B in the U.S. Pacific Northwest. Ecology 83: 2951–2957.

Palis, J.G. 1995a. Species stewardship summary for *Rana capito*. Pp. 77–89. *In* Jordan, R.S., K.S. Wheaton and W.M. Wheiler (Eds.), Integrated Endangered Species Management Recommendations for Army Installations in the Southeastern United States: Assessment of Army-wide Guidelines for the Red-cockaded Woodpecker on Associated Endangered, Threatened, and Candidate Species. Nature Conservancy, Southeast Regional Office, Chapel Hill, North Carolina.

Palis, J.G. 1995b. Larval growth, development, and metamorphosis of *Ambystoma cingulatum* on the Gulf Coastal Plain of Florida. Florida Scientist 58:352–358.

Palis, J.G. 1996a. Element stewardship abstract. Flatwoods salamander (*Ambystoma cingulatum* Cope). Natural Areas Journal 16:49–54.

Palis, J.G. 1996b. Natural history notes: *Ambystoma opacum* (marbled salamander). Communal nesting. Herpetological Review 27:134.

Palis, J.G. 1997a. Breeding migration of *Ambystoma cingulatum* in Florida. Journal of Herpetology 31:71–78.

Palis, J.G. 1997b. Distribution, habitat, and status of the flatwoods salamander (*Ambystoma cingulatum*) in Florida, USA. Herpetological Natural History 5:53–65.

Palis, J.G. 1998. Breeding biology of the gopher frog, *Rana capito*, in western Florida. Journal of Herpetology 32:217–223.

Palis, J.G. 2000a. Natural history notes: *Scaphiopus holbrookii* (eastern spadefoot). Predation. Herpetological Review 31:42–43.

Palis, J.G. 2000b. Natural history notes: *Rana sphenocephala* (southern leopard frog). Subterranean vocalization. Herpetological Review 31:42.

Palis, J.G. and J.B. Jensen. 1995. Distribution and breeding biology of the flatwoods salamander (*Ambystoma cingulatum*) and gopher frog (*Rana capito*) on Eglin Air Force Base, Florida. Final Report. Florida Natural Areas Inventory, Tallahassee, Florida.

Palmer, W.M. and A.L. Braswell. 1995. Reptiles of North Carolina. University of North Carolina Press (Chapel Hill), for the North Carolina State Museum of Natural Sciences, Chapel Hill, North Carolina.

Palmer, W.M. and D.E. Whitehead. 1961. Herpetological collections and observations in Hyde and Tyrrell Counties, North Carolina. Journal of the Elisha Mitchell Scientific Society 77:280–289.

Pamilo, P. and M. Nei. 1988. Relationships between gene trees and species trees. Molecular Biology and Evolution 5:568–583.

Panek, F.M. 1978. A developmental study of *Ambystoma jeffersonianum* and *A. platineum* (Amphibia, Urodele, Ambystomatidae). Journal of Herpetology 12:265–266.

Panik, H.R. and S. Barrett. 1994. Distribution of amphibians and reptiles along the Truckee River system. Northwestern Science 68:197–204.

Panitz, E. 1967. *Thelandros salamandrae* (Oxyuroidea) Schad 1960, in *Plethodon neomexicanus* from the Jemez Mountains, New Mexico. Canadian Journal of Zoology 45:1296–1297.

Papenfuss, T.J. 1986. Amphibian and reptile diversity along elevational transects in the White-Inyo Range. Pp. 129–136. *In* Hall, C.A., Jr. and D.J. Young (Eds.), Natural History of the White-Inyo Range, Eastern California and Western Nevada and High Altitude Physiology: University of California, White Mountain Research Station Symposium, Bishop, California.

Papenfuss, T.J. and P. Brouha. 1979. Shasta salamander *Hydromantes shastae*. Shasta-Trinity National Forest comprehensive species management plan and a species status report. U.S.D.A. Forest Service, Unpublished report, Redding, California.

Papenfuss, T.J. and J.R. Macey. 1986. A review of the population status of the Inyo Mountains salamander. U.S. Fish and Wildlife Service, Order Number 10188-5671-5, Endangered Species Office, Sacramento, California.

Paperna, I. and R. Lainson. 1995. *Alloglugea bufonis* nov. gen., nov. sp. (Microsporea: Glugeidae), a microsporidian of *Bufo marinus* tadpoles and metamorphosing toads (Amphibia: Anura) from Amazonian Brazil. Diseases of Aquatic Organisms 23:7–16.

Parham, J.F., C.K. Dodd Jr. and G.R. Zug. 1996. Skeletochonological age estimates for the Red Hills salamander, *Phaeognathus hubrichti*. Journal of Herpetology 30:401–404.

Park, J.M., J.-E. Jung and B.J. Lee. 1994. Antimicrobial peptides from the skin of a Korean frog, *Rana rugosa*. Biochemistry and Biophysics Research Commununications 205:948–954.

Park, O., J.A. Lockett and D.J. Myers. 1931. Studies in nocturnal ecology with special reference to climax forest. Ecology 12:709–727.

Parker, E.D. and M. Niklasson. 1995. Desiccation resistance among clones in the invading parthenogenetic cockroach, *Pycnoscelus surinamensis*: a search for the general-purpose genotype. Journal of Evolutionary Biology 8:331–337.

Parker, F., S.L. Robbins and A. Loveridge. 1947. Breeding, rearing and care of the South African clawed frog *(Xenopus laevis)*. American Naturalist 81:38–49.

Parker, G.E. 1967. The influence of temperature and thyroxin on oxygen consumption in *Rana pipiens* tadpoles. Copeia 1967:610–616.

Parker, J. 1998. The Wyoming toad. University of Wyoming website: http://uwadmnweb.uwyo.edu.

Parker, J. 2000. Habitat use and movements of the Wyoming toad, *Bufo baxteri*: a study of wild juvenile, adult, and released captive-raised toads. Master's thesis. University of Wyoming, Laramie, Wyoming.

Parker, K.M., R.J. Sheffer and P.W. Hedrick. 1999. Molecular variation and evolutionary significant units in the endangered Gila topminnow. Conservation Biology 13:109–116.

Parker, M.L. and M.L. Goldstein. 2004. Diet of the Rio Grande leopard frog (*Rana berlandieri*) in Texas. Journal of Herpetology 38:127–130.

Parker, M.S. 1991. Relationship between cover availability and larval Pacific giant salamander density. Journal of Herpetology 25:355–357.

Parker, M.S. 1993a. Predation by Pacific giant salamander larvae on juvenile steelhead trout. Northwestern Naturalist 74:77–81.

Parker, M.S. 1993b. Size-selective predation on benthic macroinvertebrates by stream-dwelling salamander larvae. Archiv für Hydrobiologie 128:385–400.

Parker, M.S. 1994. Feeding ecology of stream-dwelling Pacific giant salamander larvae *(Dicamptodon tenebrosus)*. Copeia 1994:705–718.

Parker, M.V. 1937. Some amphibans and reptiles from Reelfoot Lake. Journal of the Tennessee Academy of Science 12:60–86.

Parker, M.V. 1939. The amphibians and reptiles of Reelfoot Lake and vicinity, with a key for the separation of species and subspecies. Journal of the Tennessee Academy of Sciences 14:72–101.

Parker, M.V. 1951. Notes on the bird-voiced treefrog, *Hyla phaeocrypta*. Journal of the Tennessee Academy of Science 26:208–213.

Parmelee, J.R. 1993. Microhabitat segregation and spatial relationships among four species of mole salamander (genus *Ambystoma*). Occasional Papers of the Museum of Natural History, Number 160, University of Kansas, Lawrence, Kansas.

Parmesan, C. 1996. Climate and species' range. Nature 383:765–766.

Parmley, D. 1982. Food items of roadrunners from Palo Pinto County, north central Texas. Texas Journal of Science 34:94–95.

Parmley, D. 1988a. Additional Pleistocene amphibians and reptiles from the Seymour Formation, Texas. Journal of Herpetology 22:82–87.

Parmley, D. 1988b. Middle Holocene herpetofauna of Klein Cave, Kerr County, Texas. Southwestern Naturalist 33:378–382.

Parmley, D. and J.D. Tyler. 1978. Herpetological prey at a burrowing owl hole. Bulletin of the Oklahoma Herpetological Society 3:13.

Parra-Olea, G., M. Garcia-Paris and D.B. Wake. 1999. Status of some populations of Mexican salamanders (Amphibia: Plethodontidae). Revista de Biologia Tropical 47:217–223.

Parris, K.M. and M.A. McCarthy. 2001. Identifying effects of toe clipping on anuran return rates: the importance of statistical power. Amphibia-Reptilia 22:275–289.

Parris, M.J. 1998. Terrestrial burrowing ecology of newly metamorphosed frogs (*Rana pipiens* complex). Canadian Journal of Zoology 76:2124–2129.

Parris, M.J. and R.D. Semlitsch. 1998. Asymmetric competition in larval amphibian communities: conservation implications for the northern crawfish frog, *Rana areolata circulosa*. Oecologia 116:219–226.

Pasanen, S. and J. Sorjonen. 1994. Partial terrestrial wintering in a northern common frog population (*Rana temporaria* L.). Annales Zoologici Fennici 31:275–278.

Patch, C.L. 1922. Some amphibians and reptiles from British Columbia. Copeia 1922:74–79.

Paten, P.W.C., R.S. Egan, J.E. Osenkowski, C.J. Raithel and R.T. Brooks. 2003. Natural history notes: *Rana sylvatica* (wood frog). Breeding behavior. Herpetological Review 34:236–237.

Pater, D.E., S.A. Bryce, T.D. Thorson, J. Kagan, C. Chappell, J.M. Omernik, S.H. Azevedo and A.J. Woods. 1998. Ecoregions of western Washington and Oregon. U.S. Geological Survey, Reston, Virginia.

Paterson, H.E.H. 1985. The recognition concept of species. Pp. 21–30. In Vrba, E.S. (Ed.), Species and Speciation. Transvaal Museum Monograph, Number 4, Pretoria, South Africa.

Patla, D.A. 1997. Changes in a population of spotted frogs in Yellowstone National Park between 1953 and 1995: the effects of habitat modification. Master's thesis. Idaho State University, Pocatello, Idaho.

Patla, D.A. and C.R. Peterson. 1994. The effects of habitat modification on a spotted frog population in Yellowstone National Park. University of Wyoming/National Park Service Research Center, Final report, University of Wyoming, Cheyenne, Wyoming.

Patla, D.A. and C.R. Peterson. 1999. Are amphibians declining in Yellowstone National Park? Yellowstone Science 7:2–11.

Patla, D.A., C.R. Peterson and R. Van Kirk. 2000. Distribution and status of amphibians in watersheds of the Greater Yellowstone ecosystem. P. 235. In Abstracts of the 14th Annual Meeting of the Society for Conservation Biology, University of Montana, Missoula, Montana.

Patterson, K.K. 1978. Life history aspects of paedogenic populations of the mole salamander, Ambystoma talpoideum. Copeia 1978:649–655.

Pauley, B.A. 1998. The use of emergent rocks and refugia for the Cheat Mountain salamander, Plethodon nettingi Green. Master's thesis. Marshall University, Huntington, West Virginia.

Pauley, B.A. and T.K. Pauley. 1990. Competitive interactions between two sympatric Plethodon salamanders. Proceedings of the West Virginia Academy of Science 62:19.

Pauley, T.K. 1978a. Food types and distribution as a Plethodon habitat partitioning factor. Bulletin of the Maryland Herpetological Society 14:79–82.

Pauley, T.K. 1978b. Plants as indicators of occurrence of two sympatric Plethodon species. Bulletin of the Maryland Herpetological Society 14:29–35.

Pauley, T.K. 1978c. Moisture as a factor regulating habitat partitioning between two sympatric Plethodon (Amphibia, Urodela, Plethodontidae) species. Journal of Herpetology 12:491–493.

Pauley, T.K. 1978d. Temperature and insolation as factors regulating the partitioning of habitats of two sympatric Plethodon species. Proceedings of the West Virginia Academy of Science 50:77–84.

Pauley, T.K. 1980a. Field notes on the distribution of terrestrial amphibians and reptiles of the West Virginia mountains above 970 meters. Proceedings of the West Virginia Academy of Science 52:84–92.

Pauley, T.K. 1980b. The ecological status of the Cheat Mountain salamander (Plethodon nettingi). U.S. Forest Service, Monongahela National Forest, Unpublished report, Elkins, West Virginia.

Pauley, T.K. 1981. The range and distribution of the Cheat Mountain salamander, Plethodon nettingi. Proceedings of the West Virginia Academy of Science 53:31–35.

Pauley, T.K. 1986. Additional notes on the range and distribution of the Cheat Mountain salamander, Plethodon nettingi. Proceedings of the West Virginia Academy of Science 58:56–57.

Pauley, T.K. 1987. Range of the Cheat Mountain salamander. Herpetological Review 18:39.

Pauley, T.K. 1991a. Report on studies conducted on Cheat Mountain salamander populations. Report to the West Virginia Division of Natural Resources, Elkins, West Virginia.

Pauley, T.K. 1991b. Cheat Mountain salamander (Plethodon nettingi) Recovery Plan. U.S. Fish and Wildlife Service, Northeast Region, Newton Corner, Massachusetts.

Pauley, T.K. 1993a. Amphibians and reptiles in the upland forest. Pp. 179–196. In Stephenson, S.L. (Ed.), Upland Forests of West Virginia. McClain Printing Company, Parsons, West Virginia.

Pauley, T.K. 1993b. Report of Upland Vertebrates in the New River Gorge National River. Volumes I–III. Report to the United States Department of the Interior-National Park Service, Glen Jean, West Virginia.

Pauley, T.K. 1994. Impact of habitat disturbances on amphibians in West Virginia. Herpetologists League and Society for the Study of Amphibians and Reptiles Annual Meeting, Athens, Georgia (Abstract).

Pauley, T.K. 1995a. Terrestrial salamanders. Pp. 42–52. In Reardon, R.C. (Handbook Coord.), Effects of Diflubenzuron on Non-Target Organisms in Broadleaf Forested Watersheds in the Northeast. U.S.D.A. Forest Service, Report FHM-NC-05095, Morgantown, West Virginia.

Pauley, T.K. 1995b. Surveys for Plethodon punctatus in the George Washington National Forest, West Virginia. Unpublished report to the U.S. Fish and Wildlife Service, Elkins, West Virginia.

Pauley, T.K. 1995c. Aquatic salamanders. Pp. 14–22. In Reardon, R.C. Handbook Coord.), Effects of diflubenzuron on non-target organisms in broadleaf forested watersheds in the northeast. U.S. Department of Agriculture, National Center of Forest Health Management, FHM-NC-05-95, Washington, D.C.

Pauley, T.K. 1997. Study of a relocated population of Plethodon nettingi Green. Association of Southeastern Biologists Bulletin 44:134.

Pauley, T.K. 1998. Surveys for Plethodon punctatus in areas outside the George Washington National Forest, West Virginia. Unpublished report to the U.S. Fish and Wildlife Service, Elkins, West Virginia.

Pauley, T.K. and J.W. Barron. 1995. Natural history and ecology of anurans. Pp. 158–171. In Evans, D.K. and H.H. Allen (Eds.), Mitigated Wetland Restoration: Environmental effects at Green Bottom Wildlife Management Area, West Virginia. U.S. Army Corps of Engineers, Wetlands Research Program, Technical Report, WRP-RE-10, Waterways Experiment Station, Washington, D.C.

Pauley, T.K. and W.H. England. 1969. Time of mating and egg deposition in the salamander, Plethodon wehrlei Fowler and Dunn, in West Virginia. Proceedings of the West Virginia Academy of Science 41:155–160.

Pauley, T.K. and E.C. Keller Jr. 1993. Relationships of burrow size, soil temperature, and body size and temperature in the salamanders Plethodon wehrlei and Plethodon cinereus. Proceedings of the West Virginia Academy of Science 65:2–8.

Pawar, K.R., H.V. Ghate and M. Katdare. 1983. Effect of malathion on embryonic development of the frog Microhyla ornata (Dumeril and Bibron). Bulletin of Environmental Contamination and Toxicology 31:170–176.

Pawling, R.O. 1939. The amphibians and reptiles of Union County, Pennsylvania. Herpetologica 1:165–170.

Payne, R.T. 1966. Food web complexity and species diversity. American Naturalist 100:65–78.

Peacock, R.L. and R.A. Nussbaum. 1973. Reproductive biology and population structure of the western red-backed salamander, Plethodon vehiculum. Journal of Herpetology 7:215–224.

Peacor, S.D. and E.E. Werner. 1997. Trait-mediated indirect interactions in a simple aquatic food web. Ecology 78:1146–1156.

Pearl, C.A. 1999. The Oregon spotted frog (Rana pretiosa) in the Three Sisters Wilderness Area/Willamette National Forest: 1998 summary of findings. Report Prepared for U.S. Fish and Wildlife Service, Portland, Oregon.

Pearl, C.A. 2003. Natural history notes: Ambystoma gracile (northwestern salamander). Egg predation. Herpetological Review 34:352–353.

Pearl, C.A., M.J. Adams, R.B. Bury and B. McCreary. 2004. Asymmetrical effects of introduced bullfrogs (Rana catesbeiana) on native ranid frogs in Oregon, USA. Copeia 2004:11–20.

Pearl, C.A., M.J. Adams, N. Leuthold and R.B. Bury. 2005. Amphibian occurrence and aquatic invaders in a changing landscape: implications for wetland mitigation in the Willamette Valley, Oregon, USA. Wetlands 25: in press.

Pearl, C.A. and M.P. Hayes. 2002. Predation by Oregon spotted frogs (Rana pretiosa) on western toads (Bufo boreas) in Oregon. American Midland Naturalist 147:145–152.

Pearl, C.A., D.A. Major and R.B. Bury. 2002. Natural history notes: Ascaphus truei (tailed frog). Albinism. Herpetological Review 33:123.

Pearman, P.B. 1995. Effects of pond size and consequent predator density on two species of tadpoles. Oecologia 102:1–8.

Pearman, P.B. 2001. Conservation value of independently evolving units: sacred cow or testable hypotheses? Conservation Biology 15:780–783.

Pearse, A.S. 1911. Concerning the development of frog tadpoles in sea water. Philippine Journal of Science 6:219–220.

Pearse, A.S. 1921. Habits of the Necturus, an enemy of food fishes. U.S. Bureau of Fisheries, Economic Circular, Number 49, Washington, D.C.

Pearse, B. 1979. A population and home range study of Bufo marinus. Honours thesis. Griffith University, Brisbane, Australia.

Pearson, H.A., R.R. Lohoefener and J.L. Wolfe. 1987. Amphibians and reptiles on longleaf-slash pine forests in southern Mississippi. Pp. 157–165. In Pearson, H.A., F.E. Smeins and R.E. Thill (Eds.), Ecological, Physical, and Socioeconomic Relationships Within Southern National Forests. Proceedings of the Southern Evaluation Project Workshop. U.S.D.A. Forest Service, Southern Forest Experiment Station, New Orleans, Louisiana.

Pearson, J.C. 1956. Studies on the life cycles and morphology of the larval stages of Alaria arisaemoides Augustine and Uribe, 1927, and Alaria canis La Rue and Fallis. Canadian Journal of Zoology 34:295–387.

Pearson, P.G. 1955. Population ecology of the spadefoot toad, *Scaphiopus h. holbrooki* (Harlan). Ecological Monographs 25:233–267.

Pearson, P.G. 1957. Further notes on the population ecology of the spadefoot toad. Ecology 38:580–586.

Pearson, S.M. 1993. The spatial extent and relative influence of landscape-level factors on wintering bird populations. Landscape Ecology 8:3–18.

Pechmann, J.H.K. 1994. Population regulation in complex life cycles: aquatic and terrestrial density-dependence in pond-breeding amphibians. Ph.D. dissertation. Duke University, Durham, North Carolina.

Pechmann, J.H.K. 1995. Use of large field enclosures to study the terrestrial ecology of pond-breeding amphibians. Herpetologica 51:434–450.

Pechmann, J.H.K. and R.D. Semlitsch. 1986. Diel activity patterns in the breeding migrations of winter-breeding anurans. Canadian Journal of Zoology 64:1116–1120.

Pechmann, J.H.K. and D.B. Wake. 1997. Declines and disappearances of amphibian populations. Pp. 135–137. *In* Meffe, G.K. and C.R. Carroll (Eds.), Principles of Conservation Biology. Second edition. Sinauer Associates, Sunderland, Massachusetts.

Pechmann, J.H.K. and H. Wilbur. 1994. Putting declining amphibian populations in perspective: natural fluctuations and human impacts. Herpetologica 50:65–84.

Pechmann, J.H.K., R.A. Estes, D.E. Scott and J.W. Gibbons. 2001. Amphibian colonization and use of ponds created for trial mitigation of wetland loss. Wetlands 21:93–111.

Pechmann, J.H.K., D.E. Scott, J.W. Gibbons and R.D. Semlitsch. 1989. Influence of wetland hydroperiod on diversity and abundance of metamorphosing juvenile amphibians. Wetlands Ecology and Management 1:3–11.

Pechmann, J.H.K., D.E. Scott, R.D. Semlitsch, J.P. Caldwell, L.J. Vitt and J.W. Gibbons. 1991. Declining amphibian populations: the problem of separating human impacts from natural fluctuations. Science 253:892–895.

Peck, S.B. 1973. Feeding efficiency in the cave salamander, *Haideotriton wallacei*. International Journal of Speleology 5:15–19.

Peck, S.B. 1974. The food of the salamanders *Eurycea lucifuga* and *Plethodon glutinosus* in caves. National Speleological Society Bulletin 36:7–10.

Peck, S.B. and B.L. Richardson. 1976. Feeding ecology of the salamander *Eurycea lucifuga* in the entrance of twilight and dark zones of caves. Annales de Speleologie 31:175–182.

Peckham, R.S. and C.F. Dineen. 1954. Spring migrations of salamanders. Proceedings of the Indiana Academy of Science 64:278–280.

Pedersen, S.C. 1993. Skull growth in cannibalistic tiger salamanders, *Ambystoma tigrinum*. Southwestern Naturalist 38:316–324.

Peele, P.L. 1992. Behavioral interactions among desmognathine salamanders: *Desmognathus ochrophaeus*, *D. aeneus*, and *D. wrighti*. Master's thesis. Western Carolina University, Cullowhee, North Carolina.

Peet, R.K. and D.J. Allard. 1993. Longleaf pine vegetation of the southern Atlantic and eastern Gulf Coast regions: a preliminary classification. Pp. 45–81. *In* Hermann, S.M. (Ed.), The Longleaf Pine Ecosystem: Ecology, Restoration and Management. Proceedings of the Tall Timbers Fire Ecology Conference, Number 18, Tallahassee, Florida.

Pehek, E.L. 1995. Competition, pH, and the ecology of larval *Hyla andersonii*. Ecology 76:1786–1793.

Pelgen, J.L. 1951. A *Rana catesbeiana* with six functional legs. Herpetologica 7:138–139.

Pemberton, C.E. 1949. Longevity of the tropical American toad, *Bufo marinus* L. Science 110:512.

Pendergast, J.R., R.M. Quinn, J.H. Lawton, B.C. Eversham and D.W. Gibbons. 1993. Rare species, the coincidence of diversity hotspots and conservation strategies. Nature 365:335–337.

Penn, G.H., Jr. 1943. Herpetological notes from Cameron Parish, Louisiana. Copeia 1943:58–59.

Pennock, D.S. and W.W. Dimmick. 1997. Critique of the evolutionary significant unit as a definition for "distinct population segments" under the U.S. Endangered Species Act. Conservation Biology 11:611–619.

Pennock, D.S. and W.W. Dimmick. 2000. Distinct population segments and congressional intent: reply to Waples. Conservation Biology 14:567–569.

Pentecost, E.D. and R.C. Vogt. 1976. Amphibians and reptiles of the Lake Michigan drainage basin. Environmental Status of the Lake Michigan Region. Argonne National Laboratory, Volume 16, Argonne, Illinois.

Pequegnat, W.E. 1945. A report on the biota of the Santa Ana Mountains. Journal of Entomology and Zoology 37:1–7.

Perez Vigueras, I. 1940. Notas sobre algunas especies nuevas de trematodes y sobre otras poco conocidas. Publications and Reviews of the University of Habana 5:217–242.

Pérez-Ponce de Leon, G., V. León-Règuagnon, L. García-Prieto, U. Razo-Mendivil and A. Sánchez-Alvarez. 2000. Digenean fauna of amphibians from central Mexico: Nearctic and Neotropical influences. Comparative Parasitology 67:92–106.

Perrill, S.A. 1984. Male mating behavior in *Hyla regilla*. Copeia 1984:727–732.

Perrill, S.A. and R.E. Daniel. 1983. Multiple egg clutches in *Hyla regilla*, *H. cinerea*, and *H. gratiosa*. Copeia 1983:513–516.

Pessier, A.P., D.K. Nichols, J.E. Longcore and M.S. Fuller. 1999. Cutaneous chytridiomycosis in poison dart frogs (*Dendrobates* sp.) and White's tree frogs (*Litoria caerulea*). Journal of Veterinary Diagnostic Investigation 11:194–199.

Peterjohn, B.G., J.R. Sauer and C.S. Robbins. 1995. Population trends from the North American Breeding Bird Survey. Pp. 3–39. *In* Martin, T.E. and D.M. Finch (Eds.), Ecology and Management of Neotropical Migratory Birds: A Synthesis and Review of Critical Issues. Oxford University Press, Oxford, United Kingdom.

Peters, E.L. and C.J. McCoy. 1978. Geographic distribution: *Bufo alvarius* (Colorado River toad). Herpetological Review 9:107.

Peters, J.A. 1946. Records of certain North American salamanders. Copeia 1946:106.

Peters, J.A. 1968. Endangered—Houston toad *Bufo houstonensis* (Sanders) [sic]. Sheet RA-10. *In* Rare and Endangered Fish and Wildlife of the United States. Bureau of Sport Fisheries and Wildlife, Resource Publication, Number 34, U.S. Department of the Interior, Washington, D.C.

Peters, R.H. 1991. A Critique for Ecology. Cambridge University Press, Cambridge, United Kingdom.

Peters, R.L. 1988. Effects of global warming on species and habitats: an overview. Endangered Species 5:1–8.

Peters, W. 1882. Vorlegung dreier neuen Batrachier. Sitzungsberichte Gessell. Naturforsch. Freunde Berlin 10:145–148.

Petersen, J.A. and A.R. Blaustein. 1992. Relative palatabilies of anuran larvae to natural aquatic insect predators. Copeia 1992:577–584.

Peterson, A.T. 1998. New species and new species limits in birds. Auk 115:555–558.

Peterson, A.T. and A.G. Navarro-Sigüenza. 1999. Alternate species concepts as bases for determining priority conservation areas. Conservation Biology 13:427–431.

Peterson, C.L. 1985. Comparative demography of four populations of the hellbender, *Cryptobranchus alleganiensis*, in the Ozarks. Ph.D. dissertation. University of Missouri, Columbia, Missouri.

Peterson, C.L. 1987. Movement and catchability of the hellbender, *Cryptobranchus alleganiensis*. Journal of Herpetology 21:197–204.

Peterson, C.L. 1988. Breeding activities of the hellbender in Missouri. Herpetological Review 19:28–29.

Peterson, C.L. 1989. Seasonal food habits of *Cryptobranchus alleganiensis* (Caudata: Cryptobranchidae). Southwestern Naturalist 34:438–441.

Peterson, C.L. and R.F. Wilkinson. 1996. Home range size of the hellbender (*Cryptobranchus alleganiensis*) in Missouri. Herpetological Review 27:126–127.

Peterson, C.L., C.A. Ingersol and R.F. Wilkinson. 1989. Winter breeding of *Cryptobranchus alleganiensis bishopi* in Arkansas. Copeia 1989:1031–1035.

Peterson, C.L., D.E. Metter and B.T. Miller. 1988. Demography of the hellbender *Cryptobranchus alleganiensis* in the Ozarks. American Midland Naturalist 119:291–303.

Peterson, C.L., M.S. Topping, R.F. Wilkinson Jr. and C.A. Taber. 1985. An examination of long term growth of *Cryptobranchus alleganiensis* predicted by linear regression methods. Copeia 1985:492–496.

Peterson, C.L., R.F. Wilkinson, D. Moll and T. Holder. 1991. Premetamorphic survival of *Ambystoma annulatum*. Herpetologica 47:96–100.

Peterson, C.L., R.F. Wilkinson, D. Moll and T. Holder. 1992. Estimating the number of female *Ambystoma annulatum* (Caudata: Ambystomatidae) based on oviposition. Southwestern Naturalist 37:425–426.

Peterson, C.L., R.F. Wilkinson Jr., M.S. Topping and D.E. Metter. 1983. Age and growth of the Ozark hellbender (*Cryptobranchus alleganiensis bishopi*). Copeia 1983:225–231.

Peterson, C.R. 1997. Checklist for amphibian survey reports. Pp. 113–122. *In* Olson, D.H., W.P. Leonard and R.B. Bury (Eds.), Sampling Amphibians in Lentic Habitats. Northwest Fauna, Number 4, Society for Northwestern Vertebrate Biology, Olympia, Washington.

Peterson, C.R. and M.E. Dorcas. 1992. The use of automated data-acquisition techniques in monitoring amphibians and reptile populations. Pp. 369–378. *In* McCullough, D.R. and R.H. Barrett (Eds.), Wildlife 2001: Populations. Elsevier Science Publishers, Essex, England.

Peterson, C.R. and M.E. Dorcas. 1994. Automated data-acquisition. Pp. 47–57. *In* Heyer, W.R., M.A. Donnelly, R.W. McDiarmid, L.-A.C. Hayek and M.S. Foster (Eds.), Measuring and Monitoring Biological Diversity: Standard Methods for Amphibians. Smithsonian Institution Press, Washington, D.C.

Peterson, H.G., C. Boutin, P.A. Martin, K.E. Freemark, N.J. Ruecker and M.J. Moody. 1994. Aquatic phyto-toxicity of 23 pesticides applied at expected environmental concentrations. Aquatic Toxicology 28:275–292.

Peterson, H.W. 1952. A new salamander from the Everglades of southern Florida. Herpetologica 8:102–106.

Peterson, H.W., R. Garrett and J.P. Lantz. 1952. The mating period of the giant treefrog *Hyla dominicensis*. Herpetologica 8:63.

Peterson, J.R. and A.R. Blaustein. 1991. Unpalatability in anuran larvae as a defense against natural salamander predators. Ethology, Ecology and Evolution 3:63–72.

Peterson, M.G., J.R. Gillette, R. Franks and R.G. Jaeger. 2000. Alternative life styles in a terrestrial salamander: do females preferentially associate with each other? Pp. 417–429. *In* Bruce, R.C., R.G. Jaeger and L.D. Houck (Eds.), The Biology of Plethodontid Salamanders. Kluwer Academic/Plenum Publishers, New York.

Peterson, P.J. and B.J. Alloway. 1979. Cadmium in soils and vegetation. Pp. 45–92. *In* Webb, M. (Ed.), The Chemistry, Biochemistry, and Biology of Cadmium. Elsevier/North-Holland Biomedical Press, New York.

Peterson, R.L. 1950. Amphibians and reptiles of Brazos County, Texas. American Midland Naturalist 43:157–164.

Petit, R.J., A. El Mousadik and O. Pons. 1998. Identifying populations for conservation on the basis of genetic markers. Conservation Biology 12:844–855.

Petranka, J.W. 1979. The effects of severe winter weather on *Plethodon dorsalis* and *Plethodon richmondi* populations in central Kentucky. Journal of Herpetology 13:369–371.

Petranka, J.W. 1982a. Courtship behaviour of the small-mouthed salamander *(Ambystoma texanum)* in central Kentucky. Herpetologica 38:333–336.

Petranka, J.W. 1982b. Geographic variation in the mode of reproduction and larval characteristics of the small-mouthed salamander *(Ambystoma texanum)* in the east-central United States. Herpetologica 38:475–485.

Petranka, J.W. 1983. Fish predation: a factor affecting the spatial distribution of a stream-breeding salamander. Copeia 1983:624–628.

Petranka, J.W. 1984a. Breeding migrations, breeding season, clutch size and oviposition of stream-breeding *Ambystoma texanum*. Journal of Herpetology 18:106–112.

Petranka, J.W. 1984b. Ontogeny of the diet and feeding behavior of *Eurycea bislineata* larvae. Journal of Herpetology 18:48–55.

Petranka, J.W. 1984c. Sources of interpopulational variation in growth responses of larval salamanders. Ecology 65:1857–1865.

Petranka, J.W. 1984d. Incubation, larval growth, and embryonic and larval survivorship of smallmouth salamanders *(Ambystoma texanum)* in streams. Copeia 1984:862–868.

Petranka, J.W. 1987. Natural history notes: *Notophthalmus viridescens dorsalis* (broken–striped newt). Behavior. Herpetological Review 18: 72–73.

Petranka, J.W. 1989a. Response of toad tadpoles to conflicting chemical stimuli: predator avoidance versus 'optimal' foraging. Herpetologica 45:283–292.

Petranka, J.W. 1989b. Chemical interference competition in tadpoles: does it occur outside laboratory aquaria? Copeia 1989:921–930.

Petranka, J.W. 1989c. Density-dependent growth and survival of larval *Ambystoma*: evidence from whole-pond manipulations. Ecology 70:1752–1767.

Petranka, J.W. 1990. Observations on nest site selection, nest desertion, and embryonic survival in marbled salamanders. Journal of Herpetology 24:229–234.

Petranka, J.W. 1994. Response to impact of timber harvesting on salamanders. Conservation Biology 8:302–304.

Petranka, J.W. 1998. Salamanders of the United States and Canada. Smithsonian Institution Press, Washington, D.C.

Petranka, J.W. and L. Hayes. 1998. Chemically mediated avoidance of a predatory odonate *(Anax junius)* by American toad *(Bufo americanus)* and wood frog *(Rana sylvatica)* tadpoles. Behavioral Ecology and Sociobiology 42:263–271.

Petranka, J.W. and C.A. Kennedy. 1999. Pond tadpoles with generalized morphology; is it time to reconsider their functional roles in aquatic communities? Oecologia 120:621–631.

Petranka, J.W. and S.M. Murray. 2001. Effectiveness of removal sampling for determining salamander density and biomass: a case study in an Appalachian streamside community. Journal of Herpetology 35: 36–44.

Petranka, J.W. and J.G. Petranka. 1980. Selected aspects of the larval ecology of the marbled salamander in the southern portion of its range. American Midland Naturalist 104:352–363.

Petranka, J.W. and J.G. Petranka. 1981a. On the evolution of nest-site selection in the marbled salamander, *Ambystoma opacum*. Copeia 1981:387–391.

Petranka, J.W. and J.G. Petranka. 1981b. Notes on the nesting biology of the marbled salamander, *Ambystoma opacum*, in the southern portion of its range. Journal of the Alabama Academy of Science 52:20–24.

Petranka, J.W. and D.A.G. Thomas. 1995. Explosive breeding reduces egg and tadpole cannibalism in the wood frog, *Rana sylvatica*. Animal Behavior 50:731–739.

Petranka, J.W. and A. Sih. 1986. Environmental instability, competition, and density–dependent growth and survivorship of a stream-dwelling salamander. Ecology 67:729–736.

Petranka, J.W., M.E. Eldridge and K.E. Haley. 1993. Effects of timber harvesting on southern Appalachian salamanders. Conservation Biology 7:363–370.

Petranka, J.W., J.J. Just and E.C. Crawford. 1982. Hatching of amphibian embryos: the physiological trigger. Science 217:257–259.

Petranka, J.W., L.B. Kats and A. Sih. 1987b. Predator-prey interactions among fish and larval amphibians: use of chemical cues to detect predatory fish. Animal Behaviour 35:420–425.

Petranka, J.W., A.W. Rushlow and M.E. Hopey. 1998. Predation by tadpoles of *Rana sylvatica* on embryos of *Ambystoma maculatum*: implications of ecological role reversals by *Rana* (predator) and *Ambystoma* (prey). Herpetologica 54:1–13.

Petranka, J.W., M.E. Brannon, M.E. Hopey and C.K. Smith. 1994. Effects of timber harvesting on low elevation populations of southern Appalacian salamanders. Forest Ecology and Management 67:135–147.

Petranka, J.W., A. Sih, L.B. Kats and J.R. Holomuzki. 1987a. Stream drift, size-selective predation and the evolution of ovum size in an amphibian. Oecologia 71:624–630.

Petranka, J.W., M.E. Hopey, B.T. Jennings, S.D. Baird and J. Boone. 1994. Breeding habitat segregation of wood frogs and American toads: the role of interspecific tadpole predation and adult choice. Copeia 1994:691–697.

Pettus, D. and G.M. Angleton. 1967. Comparative reproductive biology of montane and piedmont chorus frogs. Evolution 21:500–507.

Petzing, J.E. and C.A. Phillips. 1998a. Geographic distribution: *Hyla squirella* (squirrel treefrog). Herpetological Review 29:107.

Petzing, J.E. and C.A. Phillips. 1998b. Geographic distribution: *Ambystoma texanum* (smallmouth salamander). Herpetological Review 29:105.

Petzing, J.E., C.A. Phillips, M.J. Dreslik, A.K. Wilson, M. Redmer, T.G. Anton, D. Mauger and M.J. Blanford. 1998. New amphibian and reptile records in Illinois. Herpetological Review 29:179–182.

Pfennig, D.W. 1990. The adaptive significance of an environmentally-cued developmental switch in an anuran tadpole. Oecologia 85: 101–107.

Pfennig, D.W. and J.P. Collins. 1993. Kinship affects morphogenesis in cannibalistic salamanders. Nature 262:836–838.

Pfennig, D.W. and W.A. Frankino. 1997. Kin mediated morphogenesis in facultatively cannibalistic tadpoles. Evolution 51:1993–1999.

Pfennig, D.W. and P.J. Murphy. 2002. How fluctuating competition and phenotypic plasticity mediate species divergence. Evolution 56: 1217–1228.

Pfennig, D.W., M.L.G. Loeb and J.P. Collins. 1991. Pathogens as a factor limiting the spread of cannibalism in tiger salamanders. Oecologia 88:161–166.

Pfennig, D.W., A. Mabry and D. Orange. 1991. Environmental causes of correlations between age and size at metamorphosis in *Scaphiopus multiplicatus*. Ecology 72:2240–2248.

Pfennig, D.W., P.W. Sherman and J.P. Collins. 1994. Kin recognition and cannibalism in polyphenic salamanders. Behavioral Ecology 5: 225–232.

Pfennig, K.S. 2000. Female spadefoot toads compromise on mate quality to ensure conspecific matings. Behavioral Ecology 11:220–227.

Pfennig, K.S. and R.C. Tinsley. 2002. Different mate preferences by parasitized and unparasitized females potentially reduces sexual selection. Journal of Evolutionary Biology 15:399–406.

Pfingsten, R.A. 1966. The reproduction of *Desmognathus ochrophaeus* in Ohio. Master's thesis. Kent State University, Kent, Ohio.

Pfingsten, R.A. 1989a. The status and distribution of the hellbender, *Cryptobranchus alleganiensis*, in Ohio. Ohio Journal of Science 89:3.

Pfingsten, R.A. 1989b. *Plethodon cinereus* (Green), redback salamander. Pp. 229–242. *In* Pfingsten, R.A. and F.L. Downs (Eds.), Salamanders of Ohio. Bulletin of the Ohio Biological Society 7(2), College of Biological Sciences, Ohio State University, Columbus, Ohio.

Pfingsten, R.A. 1989c. *Plethodon richmondi* (Netting and Mittleman), ravine salamander. Pp. 253–260. *In* Pfingsten, R.A. and F.L. Downs (Eds.), Salamanders of Ohio. Bulletin of the Ohio Biological Society 7(2), College of Biological Sciences, Ohio State University, Columbus, Ohio.

Pfingsten, R.A. 1989d. Genus *Plethodon*. Pp. 229–264. *In* Pfingsten, R.A. and F.L. Downs (Eds.), Salamanders of Ohio. Ohio Biological Survey Bulletin Number 7, New Series, Columbus, Ohio.

Pfingsten, R.A. 1989e. *Plethodon wehrlei* Fowler and Dunn, Wehrle's salamander. Pp. 261–264. *In* Pfingsten, R.A. and F.L. Downs (Eds.), Salamanders of Ohio. Bulletin of the Ohio Biological Survey, Volume 7, Number 2, Columbus, Ohio.

Pfingsten, R.A. 1989f. *Pseudotriton montanus* Baird, mud salamander. Pp. 265–269. *In* Pfingsten, R.A. and F.L. Downs (Eds.), Salamanders of Ohio. Ohio Biological Survey Bulletin, New Series 7(2), Columbus, Ohio.

Pfingsten, R.A. 1989g. *Pseudotriton ruber* (Latreille), red salamander. Pp. 269–275. *In* Pfingsten, R.A. and F.L. Downs (Eds.), Salamanders of Ohio. Ohio Biological Survey Bulletin, New Series 7(2), Columbus, Ohio.

Pfingsten, R.A. 1998. Distribution of Ohio amphibians. Pp. 221–258. *In* Lannoo, M.J. (Ed.), Status and Conservation of Midwest Amphibians. University of Iowa Press, Iowa City, Iowa.

Pfingsten, R.A. and F.L. Downs. 1989. Salamanders of Ohio. Ohio Biological Survey Bulletin, New Series, Volume 7, Number 2, Columbus, Ohio.

Pfingsten, R.A. and A.M. White. 1989. *Necturus maculosus* (Rafinesque), mudpuppy. Pp. 72–78. *In* Pfingsten, R.A. and F.L. Downs (Eds.), Salamanders of Ohio. Ohio Biological Survey. Volume 7, Columbus, Ohio.

Phelps, J.P. and R.A. Lancia. 1995. Effects of a clearcut on the herpetofauna of a South Carolina bottomland swamp. Brimleyana 22:31–45.

Philibosian, R., R. Ruibal, V.H. Shoemaker and L.L. McClanahan. 1974. Nesting behavior and early larval life of the frog *Leptodactylus bufonius*. Herpetologica 30:381–386.

Phillips, C.A. 1991a. Geographic distribution: *Ambystoma maculatum* (spotted salamander). Herpetological Review 22:22.

Phillips, C.A. 1991b. Geographic distribution: *Ambystoma maculatum* (spotted salamander). Herpetological Review 22:133.

Phillips, C.A. 1992. Variation in metamorphosis in spotted salamanders *Ambystoma maculatum* from eastern Missouri. American Midland Naturalist 128:276–280.

Phillips, C.A. 1994. Geographic distribution of mitochondrial DNA variants and the historical biogeography of the spotted salamander, *Ambystoma maculatum*. Evolution 48:597–607.

Phillips, C.A. and O.J. Sexton. 1989. Orientation and sexual differences during breeding migrations of the spotted salamander, *Ambystoma maculatum*. Copeia 1989:17–22.

Phillips, C.A., R.A. Brandon and E.O. Moll. 1999. Field guide to amphibians and reptiles of Illinois. Illinois Natural History Survey, Manual 8, Champaign, Illinois.

Phillips, C.A., G. Suau and A.R. Templeton. 2000. Effects of Holocene climate fluctuation on mitochondrial DNA variation in the ringed salamander, *Ambystoma annulatum*. Copeia 2000:542–545.

Phillips, C.A., M.J., Dreslik. J.R. Johnson and J.E. Petzing. 2001. Application of population estimation to pond breeding salamanders. Transactions of the Illinois State Academy of Science 94:111–118.

Phillips, C.A., J.R. Johnson, M.J. Dreslik and J.E. Petzing. 2002. Effects of hydroperiod on recruitment of mole salamanders (genus *Ambystoma*) at a temporary pond in Vermilion County, Illinois. Transactions of the Illinois State Academy of Science 95:131–139.

Phillips, C.A., T. Uzzell, C.M. Spolsky, J.M. Serb, R.E. Szafoni and T.R. Pollowy. 1997. Persistent high levels of tetraploidy in salamanders of the *Ambystoma jeffersonianum* complex. Journal of Herpetology 31:530–535.

Phillips, J.B. and C. Borland. 1994. Use of a specialized magnetoreception system for homing by the eastern red-spotted newt *Notophthalmus viridescens*. Journal of Experimental Biology 188:275–291.

Phillips, K. 1990. Where have all the frogs and toads gone? Bioscience 40:422–424.

Phillips, K. 1994. Tracking the Vanishing Frogs: An Ecological Mystery. St. Martin's Press, New York.

Phillips, K.M. 1995. *Rana capito capito*, the Carolina gopher frog, in southeast Georgia: reproduction, early growth, adult movement patterns, and tadpole fright response. Master's thesis. Georgia Southern University, Statesboro, Georgia.

Phillips, P.C. and M.J. Wade. 1990. Life history notes: *Rana sylvatica* (wood frog). Reproductive mortality. Herpetological Review 21:59.

Phisalix, M. 1922. Animaux venimeux et venins. Masson et Cie., Paris, France.

Pickens, A.L. 1927a. Intermediate between *Bufo fowleri* and *B. americanus*. Copeia 162:25–26.

Pickens, A.L. 1927b. Amphibians of upper South Carolina. Copeia 165:106–110.

Picker, M.D. 1985. Hybridization and habitat selection in *Xenopus gilli* and *Xenopus laevis* in the south-western Cape Province. Copeia 1985:574–580.

Picker, M.D. 1994. Adaptive cannibalism in *Xenopus* (Anura: Pipidae). Bulletin of the Ecological Society of America 75:180–181.

Pickett, S.T.A. and J.N. Thompson. 1978. Patch dynamics and the design of nature reserves. Biological Conservation 13:27–37.

Pickwell, G. 1947. Amphibians and Reptiles of the Pacific States. Dover Publications, New York.

Pickwell, G. 1948. Amphibians and Reptiles of the Pacific States. Stanford University Press, Stanford, California.

Pielou, E.C. 1967. The use of information theory in the study of the diversity of biological populations. Proceedings of the Fifth Berkeley Symposium on Mathematics and Statistical Probability 4: 163–77.

Pierce, B.A. 1985. Acid tolerance in amphibians. Bioscience 35:239–243.

Pierce, B.A. and J.M. Harvey. 1987. Geographic variation in acid tolerance of Connecticut wood frogs. Copeia 1987:94–103.

Pierce, B.A. and J. Montgomery. 1989. Effects of short-term acidification on growth rates of tadpoles. Journal of Herpetology 23:97–102.

Pierce, B.A. and P.H. Whitehurst. 1990. *Pseudacris clarkii*. Pp. 458.1–458.3. Catalogue of American Amphibians and Reptiles. Society for the Study of Amphibians and Reptiles, St. Louis, Missouri.

Pierce, B.A., J.B. Mitton and F.L. Rose. 1981. Allozyme variation among large, small and cannibal morphs of the tiger salamander inhabiting the Llano Estacado of west Texas. Copeia 1981:590–595.

Pierce, B.A., J.B. Mitton, L. Jacobson and F.L. Rose. 1983. Head shape and size in cannibal and noncannibal larvae of the tiger salamander from west Texas. Copeia 1983:1006–1012.

Pierce, G.B. and W.C. Speers. 1988. Tumors as caricatures of the process of tissue renewal: prospects for therapy by directing differentiation. Cancer Research 48:1996–2004.

Pierce, J.R. 1975. Genetic compatibility of *Hyla arenicolor* with other species in the family Hylidae. Texas Journal of Science 16:431–441.

Pierce, J.R. 1976. Distribution of two mating call types of the plains spadefoot, *Scaphiopus bombifrons*, in southwestern United States. Southwestern Naturalist 20:578–582.

Pierce, V.A. and J.B. Mitton. 1980. Patterns of allozyme variation in *Ambystoma tigrinum mavortium* and *A. t. nebulosum*. Copeia 1980:594–605.

Piersol, W.H. 1910. The habits and larval state of *Plethodon cinereus erythronotus*. Transactions of the Royal Canadian Institute 8:469–492.

Piersol, W.H. 1914. The egg-laying habits of *Plethodon cinereus*. Transactions of the Royal Canadian Institute 10:121–126.

Pike, N.M. 1886. Some notes on the life-history of the common newt. American Naturalist 20:17–25.

Pilliod, D.S. 1999. Natural history notes: *Rana luteiventris* (Columbia spotted frog). Cannibalism. Herpetological Review 30:93.

Pilliod, D.S. 2001. Ecology and conservation of high elevation amphibian populations in historically fishless watersheds with introduced trout. Ph.D. dissertation. Department of Biological Sciences, Idaho State University, Pocatello, Idaho.

Pilliod, D.S. 2002. Clark's nutcracker (*Nucifraga columbiana*) predation on tadpoles of the Columbia spotted frog (*Rana luteiventris*). Northwestern Naturalist 83:59–61.

Pilliod, D.S. and C.R. Peterson. 2000. Evaluating effects of fish stocking on amphibian populations in wilderness lakes. Pp. 328–335. *In* Cole, D.N., S.F. McCool, W.T. Borrie and J. O'Loughlin (Eds.), Proceedings:

Wilderness Science in a Time of Change. Volume 5. U.S.D.A. Forest Service, Rocky Mountain Research Station, RMRS-P-15-VOL-5, Ogden, Utah.

Pilliod, D.S. and C.R. Peterson. 2001. Local and landscape effects of introduced trout on amphibian populations in historically fishless watersheds. Ecosystems 4:322–333.

Pilliod, D.S., C.R. Peterson and P.I. Ritson. 2003. Seasonal migration of Columbia spotted frogs *(Rana luteiventris)* among complementary resources in a high mountain basin. Canadian Journal of Zoology 80:1849–1862.

Pimentel, R.A. 1952. Studies on the biology of *Triturus granulosus* Skilton. Ph.D. dissertation. Oregon State College, Corvallis, Oregon.

Pimentel, R.A. 1960. Inter- and intrahabitat movements of the rough-skinned newt, *Taricha torosa granulosa* (Skilton). American Midland Naturalist 63:470–496.

Pimm, S. and T. Brooks. 2000. The sixth extinction: how large, where, and when? Pp. 46–62. *In* Raven, P. (Ed.), Nature and Human Society. National Academy Press, Washington, D.C.

Pimm, S.L. 1991. The Balance of Nature: Ecological Issues in the Conservation of Species and Communities. University of Chicago Press, Chicago, Illinois.

Pimm, S.L., G.J. Russell, J.L. Gittelman and T.M. Brooks. 1995. The future of biodiversity. Science 269:347–350.

Pinder, A.W. 1985. Respiratory physiology of two amphibians, *Rana pipiens* and *Rana catesbeiana*. Ph.D. dissertation. University of Massachusetts, Amherst, Massachusetts.

Pinder, A.W. and S.C. Friet. 1994. Oxygen transport in egg masses of the amphibians *Rana sylvatica* and *Ambystoma maculatum*: convection, diffusion and oxygen production by algae. Journal of Experimental Biology 197:17–30.

Pinder, A.W., K.B. Storey and G.R. Ultsch. 1992. Estivation and hibernation. Pp. 250–274. *In* Feder, M.E. and W.W. Burggren (Eds.), Environmental Physiology of the Amphibians. University of Chicago Press, Chicago, Illinois.

Pineda, R., O. Andrade, S. Paramo, L. Trejo, J. Almeyda, D. Osoria and G. Perez-Ponce de Leon. 1985. Estudio del Control Sanitario de la Piscifactoria Benito Juarez y en los vasos de las Presas de Malpaso y La Angostura, Chiapas. *Memoria*. Universidad Juarez Autonoma de Tabasco, Sepesca, Mexico.

Pisani, G.R. 1973. A guide to preservation techniques for amphibians and reptiles. Herpetological Circular, Number 1, Society for the Study of Amphibians and Reptiles, St. Louis, Missouri.

Pitkin, R.B. and S.G. Tilley. 1982. An unusual aggregation of adult *Notophthalmus viridescens*. Copeia 1982:185–186.

Placyk, J.S., Jr., L. Torretti and B.M. Graves. 2000. Natural history notes: *Plethodon cinereus* (red-backed salamander). Intraspecific aggregation. Herpetological Review 31:167.

Platin, T.J. 1994. Wetland amphibians *Ambystoma gracile* and *Rana aurora* as bioindicators of stress associated with watershed urbanization. Master's thesis. University of Washington, Seattle, Washington.

Platt, D.R., J.T. Collins and R.E. Ashton. 1974. Rare, endangered and extirpated species in Kansas. II. Amphibians and reptiles. Transactions of the Kansas Academy of Science 76:185–192.

Platt, D.R., F.B. Cross, D. Distler, O.S. Fent, E.R. Hall, M. Termain, J. Walstrom and J. Zimmerman. 1973. Rare, endangered and extirpated species in Kansas. II. Amphibians and reptiles. Transactions of the Kansas Academy of Science 76:185–192.

Platt, J.R. 1964. Strong inference. Science 146:347–353.

Platt, S.G. and L.W. Fontenot. 1993. Bullfrog *(Rana catesbeiana)* predation on Gulf Coast toads *(Bufo valliceps)* in Louisiana. Bulletin of the Chicago Herpetological Society 28:189–190.

Platt, W.J. 1999. Southeastern pine savannas. Pp. 23–38. *In* Anderson, R.C., J.S. Fralish and J.M. Baskin (Eds.), Savannas, Barrens, and Rock Outcrop Plant Communities of North America. Cambridge University Press, New York.

Platt, W.J., G.W. Evans and M.M. Davis. 1988. Effects of fire season on flowering of forbs and shrubs in longleaf pine forests. Oecologia 76:353–363.

Platz, J.E. 1972. Sympatric interaction between two forms of leopard frog *(Rana pipiens* complex) in Texas. Copeia 1972:232–240.

Platz, J.E. 1981. Suture zone dynamics: Texas populations of *Rana berlandieri* and *Rana blairi*. Copeia 1981:733–734.

Platz, J.E. 1984. Status report for *Rana onca*. Unpublished report submitted to Office of Endangered Species, U.S. Fish and Wildlife Service, Albuquerque, New Mexico.

Platz, J.E. 1988. *Rana yavapaiensis*. Pp. 418.1–418.2. Catalogue of American Amphibians and Reptiles. Society for the Study of Amphibians and Reptiles, St. Louis, Missouri.

Platz, J.E. 1989. Speciation within the chorus frog *Pseudacris triseriata*: morphometric and mating call analyses of the boreal and western subspecies. Copeia 1989:704–712.

Platz, J.E. 1991. *Rana berlandieri*. Pp. 508.1–508.4. Catalogue of American Amphibians and Reptiles. Society for the Study of Amphibians and Reptiles, St. Louis, Missouri.

Platz, J.E. 1993. *Rana subaquavocalis*, a remarkable new species of leopard frog *(Rana pipiens* complex) from southeastern Arizona that calls under water. Journal of Herpetology 27:154–162.

Platz, J.E. 1996. *Rana subaquavocalis*: Conservation assessment/ Conservation strategy. U.S. Fish and Wildlife contract report, Phoenix, Arizona.

Platz, J.E. 1997. Status survey of the Ramsey Canyon leopard frog, *Rana subaquavocalis*. Report to Arizona Game and Fish Department, Phoenix, Arizona.

Platz, J.E. and D.C. Forester. 1988. Geographic variation in mating call among the four subspecies of the chorus frog: *Pseudacris triseriata*. Copeia 1988:1062–1066.

Platz, J.E. and J.S. Frost. 1984. *Rana yavapaiensis*, a new species of leopard frog *(Rana pipiens* complex). Copeia 1984:940–948.

Platz, J.E. and T. Grudzien. 1993. Geographic distribution and genetic variability assessment of the Ramsey Canyon leopard frog. Department of Biology, Creighton University, Omaha, Nebraska.

Platz, J.E. and T.A. Grudzien. 2003. Limited genetic heterozygosity and status of two populations of the Ramsey Canyon leopard frog: *Rana subaquavocalis*. Journal of Herpetology 37:758–761.

Platz, J.E. and A. Lathrop 1993. Body size and age assessment among advertising male chorus frogs. Journal of Herpetology 27:109–111.

Platz, J.E. and J.S. Mecham. 1979. *Rana chiricahuensis*, a new species of leopard frog *(Rana pipiens* complex) from Arizona. Copeia 1979: 383–390.

Platz, J.E. and A.L. Platz. 1973. *Rana pipiens* complex: hemoglobin phenotypes of sympatric and allopatric populations in Arizona. Science 179:1334–1336.

Platz, J.E., R.W. Clarkson, J.C. Rorabaugh and D.M. Hillis. 1990. *Rana berlandieri*: recently introduced populations in Arizona and southeastern California. Copeia 1990:324–333.

Platz, J.E., A. Lathrop, L. Hofbauer and M. Vradenburg. 1997. Age distribution and longevity in the Ramsey Canyon leopard frog, *Rana subaquavocalis*. Journal of Herpetology 31:552–557.

Pleijel, F. 1999. Phylogenetic taxonomy, a farewell to species, and a revision of Heteropodarke (Hesionidae, Polychaeta, Annelida). Systematic Biology 48:755–789.

Plotkin, M. and R. Atkinson. 1979. Geographic distribution: *Eleutherodactylus planirostris planirostris* (greenhouse frog). Herpetological Review 10:59.

Plotnick, R.E., R.H. Gardner and R.V. O'Neill. 1993. Lacunarity indices as measures of landscape texture. Landscape Ecology 8:201–211.

Plummer, M.V. 1977. Observations on breeding migrations of *Ambystoma texanum*. Herpetological Review 8:79–80.

Plummer, M.V. 1982. The status of the Caddo Mountain and Fourche Mountain salamanders *(Plethodon caddoensis, P. fourchensis)* in Arkansas. Report to the Arkansas Natural Heritage Commission, Little Rock, Arkansas.

Plummer, M.V. and G. Turnipseed. 1982. Geographic distribution: *Scaphiopus bombifrons* (plains spadefoot). Herpetological Review 13:80.

Podloucky, R. 1988. Zur Situation der Zauneidechse *Lacerta agilis* Linnaeus, 1758 in Niedersachsen—Verbreitung, Gefährdung und Schutz. Mertensiella 1:146–166.

Poethke, H.J., A. Seitz and C. Wissel. 1996. Species survival and metapopulations: conservation implications from ecological theory. Pp. 81–92. *In* Settete, J., C.R. Margules, P. Poschlod and K. Henle (Eds.), Species Survival in Fragmented Landscapes. Kluwer, Amsterdam.

Pohlmann, K.F., D.J. Campagna, J.B. Chapman and S. Earman. 1998. Investigation of the origin of springs in the Lake Mead National Recreation Area. Water Resources Center, Desert Research Institute, Publication Number 41161, University and Community College System of Nevada, Las Vegas, Nevada.

Pointier, J.P., W. Lobato Paraense and V. Mazille. 1993. Introduction and spreading of *Biomphalaria straminea* (Dunker, 1848) (Mollusca: Pulmonata: Planorbidae) in Guadeloupe, French West Indies. Memorias Instituto Oswaldo Cruz (Rio de Janiero) 88:449–455.

Polis, G.A. 1981. The evolution and dynamics of intraspecific predation. Annual Review of Ecology and Systematics 12:225–251.

Polis, G.A. 1988. Exploitation competition and the evolution of interference, cannibalism, and intraguild predation in age/size-structured populations. Pp. 185–202. *In* Ebenman, B. and L. Persson (Eds.), Size-structured Populations. Springer-Verlag, New York.

Pollack, E.D. 1971. A simple method for the removal of protozoan parasites from *Rana pipiens* larvae. Copeia 1971:557.

Pollio, C.A. 1993. Interactions of adult and larval salamanders (*Desmognathus quadramaculatus* and *Eurycea cirrigera*) in Keeney Creek, Fayette County, West Virginia. Master's thesis. West Virginia Graduate College, South Charleston, West Virginia.

Pollio, C.A. 2000. Natural history notes: *Eurycea cirrigera* (southern two-lined salamander). Larval period/size-class determination. Herpetological Review 31:166–167.

Pomerance, R., J.K. Reaser and P.O. Thomas. 1999. Coral bleaching, coral mortality, and global climate change. Report to the U.S. Coral Reef Task Force. U.S. Department of State, Washington, D.C. http://www.state.gov/www/global/global_issues/coral_reefs/990305_coralreef_rpt.html

Pomeroy, L.V. 1981. Developmental polymorphism in the tadpoles of the spadefoot toad *Scaphiopus multiplicatus*. Ph.D. dissertation. University of California, Riverside, California.

Poole, R.W. 1974. An Introduction to Quantitative Ecology. McGraw-Hill Publishing Company, New York.

Pope, C.H. 1924. Notes on North Carolina salamanders, with especial reference to the egg-laying habitats of *Leurognathus* and *Desmognathus*. American Museum Novitates, Number 153, American Museum of Natural History, New York.

Pope, C.H. 1928. Some plethodontid salamanders from North Carolina and Kentucky with the description of a new race of *Leurognathus*. American Museum Novitates, Number 306, American Museum of Natural History, New York.

Pope, C.H. 1944. Amphibians and Reptiles of the Chicago Area. Chicago Natural History Museum Press, Chicago, Illinois.

Pope, C.H. 1947. Amphibians and Reptiles of the Chicago Area. Chicago Natural History Museum Press, Chicago, Illinois.

Pope, C.H. 1950. A statistical and ecological study of the salamander, *Plethodon yonahlossee*. Bulletin of the Chicago Academy of Science 9:79–106.

Pope, C.H. 1964a. Amphibians and Reptiles of the Chicago Area. Chicago Natural History Museum Press, Chicago, Illinois.

Pope, C.H. 1964b. *Plethodon caddoensis*. Pp. 14.1. Catalogue of American Amphibians and Reptiles. Society for the Study of Amphibians and Reptiles, St. Louis, Missouri.

Pope, C.H. and J.A. Fowler. 1949. A new species of salamander (*Plethodon*) from southwestern Virginia. Natural History Miscellanea, Number 47, Chicago Academy of Sciences, Chicago, Illinois.

Pope, C.H. and N.G. Hairston. 1947. The distribution of *Leurognathus* a southern Appalachian genus of salamanders. Fieldiana Zoology 31:155–162.

Pope, C.H. and S.H. Pope. 1949. Notes on growth and reproduction of the slimy salamander *Plethodon glutinosus*. Fieldiana Zoology 31: 251–261.

Pope, C.H. and S.H. Pope. 1951. A study of the salamander *Plethodon ouachitae* and the description of an allied form. Bulletin of the Chicago Academy of Science 9:129–152.

Pope, K.L. 1999. Natural history notes: *Rana muscosa* (mountain yellow-legged frog). Diet. Herpetological Review 30:163–164.

Pope, P.H. 1919. Some notes on the amphibians of Houston, Texas. Copeia 74:93–98.

Pope, P.H. 1921. Some doubtful points in the life history of *Notophthalmus viridescens*. Copeia 91:14–15.

Pope, P.H. 1924. The life history of the common water newt (*Notophthalmus viridescens*) together with observations on the sense of smell. Annals of the Carnegie Museum 15:305–362.

Pope, P.H. 1928a. The longevity of *Ambystoma maculatum* in captivity. Copeia 169:99–100.

Pope, P.H. 1928b. The life-history of *Triturus viridescens*—some further notes. Copeia 168:61–73.

Pope, P.H. 1937. Notes of the longevity of an *Ambystoma maculatum* in captivity. Copeia 1937:140–141.

Pope, S.E., L. Fahrig and H.G. Merriam. 2000. Landscape complementation and metapopulation effects on leopard frog populations. Ecology 81:2498–2508.

Popper, K.R. 1959. The Logic of Scientific Discovery. Basic Books, New York.

Porter, K.R. 1962. Mating calls and noteworthy collections of some Mexican amphibians. Herpetologica 18:165–171.

Porter, K.R. 1968. Evolutionary status of a relict population of *Bufo hemiophrys* Cope. Evolution 22:583–594.

Porter, K.R. 1969a. Evolutionary status of the Rocky Mountain population of wood frogs. Evolution 23:163–170.

Porter, K.R. 1969b. Description of *Rana maslini*, a new species of wood frog. Herpetologica 25:212–215.

Porter, K.R. 1970. *Bufo valliceps*. Pp. 94.1–94.4. Catalogue of American Amphibians and Reptiles. Society for the Study of Amphibians and Reptiles, St. Louis, Missouri.

Porter, K.R. and D.E. Hakanson. 1976. Toxicity of mine drainage to embryonic and larval boreal toads (Bufonidae: *Bufo boreas*). Copeia 1976:327–331.

Portnoy, J.W. 1990. Breeding biology of the spotted salamander *Ambystoma maculatum* (Shaw) in acidic, temporary ponds at Cape Cod, USA. Biological Conservation 53:61–75.

Potter, F.E., Jr. 1963. Gross morphological variation in the genus *Typhlomolge* with a description of a new species. Master's thesis. University of Texas, Austin, Texas.

Potter, F.E. and E.L. Rabb. 1960. Thyroxin induced metamorphosis in a neotenic salamander, *Eurycea nana* Bishop. Zoological Series, Number 1, Texas A&M, College Station, Texas.

Potter, F.E., Jr. and S.S. Sweet. 1981. Generic boundaries in Texas cave salamanders and a redescription of *Typhlomolge robusta* (Amphibia: Plethodontidae). Copeia 1981:64–75.

Pough, F.H. 1971. Leech-repellent property of eastern red-spotted newts, *Notophthalmus viridescens*. Science 174:144–146.

Pough, F.H. 1974. Natural daily temperature acclimation of eastern red efts, *Notophthalmus v. viridescens* (Rafinesque) (Amphibia: Caudata). Comparative Biochemistry and Physiology 47:71–78.

Pough, F.H. 1976. Acid precipitation and embryonic mortality in spotted salamanders, *Ambystoma maculatum*. Science 192:68–70.

Pough, F.H. and R.E. Wilson. 1970. Natural daily temperature stress, dehydration, and acclimation in juvenile *Ambystoma maculatum* (Shaw) (Amphibia: Caudata). Physiological Zoology 43:194–205.

Pough, F.H. and R.E. Wilson. 1977. Acid precipitation and reproductive success of *Ambystoma* salamanders. Water, Air and Soil Pollution 7:307–316.

Pough, H.F., E.M. Smith, D.H. Rhodes and A. Collazo. 1987. The abundance of salamanders in forest stands with different histories of disturbance. Forest Ecology and Management 20:1–9.

Pough, F.H., R.M. Andrews, J.E. Cadle, M.L. Crump, A.H. Savitzky and K.D. Wells. 1998. Herpetology. Prentice Hall, Upper Saddle River, New Jersey.

Poulin, R. 1995. Phylogeny, ecology, and the richness of parasite communities in vertebrates. Ecological Monographs 65:283–302.

Poulin, R. 1998. Evolution and phylogeny of behavioral manipulation of insect hosts by parasites. Parasitology 116:S3–S11.

Pounds, J.A. and M.L. Crump. 1994. Amphibian declines and climate disturbance: the case of the golden toad and the harlequin frog. Conservation Biology 8:72–85.

Pounds, J,A., M.P.L. Fogden and J.H. Campbell. 1999. Biological response to climate change on a tropical mountain. Nature 398:611–615.

Pounds, J.A., M.P. Fogden, J.M. Savage and G.C. Gorman. 1997. Tests of null models for amphibian declines on a tropical mountain. Conservation Biology 11:1307–1322.

Powders, V.N. 1967. Altitudinal distribution of the astomatous ciliate *Cepedietta michiganensis* (Woodhead) in a new host, *Plethodon jordani* Blatchley. Transactions of the American Microscopical Society 86:336–338.

Powders, V.N. 1970. Altitudinal distribution of the protozoan *Cepedietta michiganensis* in the salamanders *Plethodon glutinosus* and *Plethodon jordani* in eastern Tennessee. American Midland Naturalist 83:393–402.

Powders, V.N. 1973. Cannibalism by the slimy salamander, *Plethodon glutinosus* in eastern Tennessee. Journal of Herpetology 7:139–140.

Powders, V.N. and R. Cate. 1980. Food of the dwarf salamander, *Eurycea quadridigitata*, in Georgia. Journal of Herpetology 14:81–82.

Powders, V.N. and W.L. Tietjen. 1974. The comparative food habits of sympatric and allopatric salamanders, *Plethodon glutinosus* and *Plethodon jordani* in eastern Tennessee and adjacent areas. Herpetologica 30:167–175.

Powell, G.L., A.P. Russell, J.D. James, S.J. Nelson and S.M. Watson. 1997. Population biology of the long-toed salamander, *Ambystoma macrodactylum* in the front range of Alberta. Pp. 37–44. *In* Green, D.M. (Ed.), Amphibians in Decline: Canadian Studies of a Global Problem.

Society for the Study of Amphibians and Reptiles, Herpetological Conservation, Number 1, St. Louis, Missouri.

Powell, R., J.T. Collins and E.D. Hooper Jr. 1998. A Key to Amphibians and Reptiles of the Continental United States and Canada. Kansas University Press, Lawrence, Kansas.

Powell, R., K.P. Bromeier, N.A. Laposha, J.S. Parmerlee and B. Miller. 1982. Maximum sizes of amphibians and reptiles from Missouri. Transactions of the Missouri Academy of Science 16:99–106.

Powell, R., T.R. Johnson and D.D. Smith. 1995. New records of amphibians and reptiles in Missouri 1995. Missouri Herpetological Association Newsletter 8:9–12.

Powell, R., T.R. Johnson and D.D. Smith. 1996. New records of amphibians and reptiles in Missouri 1996. Missouri Herpetological Association Newsletter 9:9–14.

Powell, R.L. and C.S. Lieb. 2003. Natural history notes: Hyla arenicolor (canyon tree frog). Toxic skin secretions. Herpetological Review 34:230.

Power, M.E., D. Tilman, J.A. Estes, B.A. Menge, W.J. Bond, L.S. Mills, G. Daily, J.C. Castilla, J. Lubchenco and R.T. Paine. 1996. Challenges in the quest for keystones. BioScience 46:609–620.

Powers, A.L. and B.H. Banta. 1972. Recent amphibians of Nate Harrison Grade, Palomar Mountain, San Diego County, California. Bulletin of the Maryland Herpetological Society 8:96–99.

Powers, J.H. 1903. The causes of acceleration and retardation in the metamorphosis of Ambystoma tigrinum: a preliminary report. American Naturalist 37:385–410.

Powers, J.H. 1907. Morphological variation and its causes in Ambystoma tigrinum. University of Nebraska Studies 7:197–274.

Pratt, I. and J.E. McCauley. 1961 Trematodes of the Pacific Northwest: An Annotated Catalog. Oregon State Monographs. Oregon State University Press, Corvallis, Oregon.

Preest, M.R., D.G. Brust and M.L. Wygoda. 1992. Cutaneous water loss and the effects of temperature and hydration state on aerobic metabolism of canyon treefrogs, Hyla arenicolor. Herpetologica 48:210–219.

Prentice, M.A. 1983. Displacement of Biomphalaria glabrata by the snail Thiara granifera in field habitats in St. Lucia, West Indies. Annals of Tropical Medicine and Parasitology 77:51–59.

Pressey, R.L., H.P. Possingham and C.R. Margules. 1996. Optimality in reserve selection algorithms: when does it matter and how much? Biological Conservation 76:259–267.

Preston, W.B. 1982. Amphibians and reptiles of Manitoba. Manitoba Museum of Man and Nature, Winnipeg, Manitoba, Canada.

Preston, W.B. and D.R.M. Hatch. 1986. The plains spadefoot, Scaphiopus bombifrons, in Manitoba. Canadian Field-Naturalist 100: 123–125.

Price, A.H. 1990. Final report: status survey of the Houston toad (Bufo houstonensis) along State Highway 21, Bastrop County, Texas. Unpublished report submitted to U.S. Fish and Wildlife Service, Albuquerque, New Mexico.

Price, A.H. 1992. Houston toad (Bufo houstonensis) status survey. Texas Parks and Wildlife Department, Performance Report, Project E-1-3, Job Number 8.0, Austin, Texas.

Price, A.H. 2003. The Houston toad in Bastrop State Park 1990–2002: A narrative. Wildlife Diversity Branch, Open-file Report 03-0401, Texas Parks and Wildlife Department, Austin Texas.

Price, A.H. and B.K. Sullivan. 1988. Bufo microscaphus. Pp. 415.1–415.3. Catalogue of American Amphibians and Reptiles. Society for the Study of Amphibians and Reptiles, St. Louis, Missouri.

Price, A.H., P.T. Chippindale and D.M. Hillis. 1995. A status report on the threats facing populations of perennibranchiate hemidactyliine plethodontid salamanders of the genus Eurycea north of the Colorado River in Texas. Texas Parks and Wildlife Department, Final Section 6 Report, Part III, Austin, Texas.

Price, E.W. 1931. Four new species of trematode worms from the muskrat, Ondatra zibethica with a key to the trematode parasites of the muskrat. Proceedings of the United States National Museum 79:1–13.

Price, E.W. 1938. A redescription of Clinostomum intermedius Lamont (Trematoda: Clinostomidae) with a key to the species of the genus. Proceedings of the Helminthological Society of Washington 5: 11–13.

Price, E.W. 1942. A new trematode of the family Psilostomidae from the lesser scaup duck, Marila affinis. Proceedings of the Helminthological Society of Washington 9:30–31.

Price, J. and D. Price. 1991. Geographic distribution: Gastrophryne olivacea. Herpetological Review 22:133.

Price, J., S. Droege and A. Price. 1995. The Summer Atlas of North American Birds. Academic Press, San Diego, California.

Price, R.L. and T. St. John. 1980. Helminth parasites of the small-mouthed salamander, Ambystoma texanum Matthes, 1855 from Williamson County, Illinois. Proceedings of the Helminthological Society of Washington 47:273–274.

Primack, R.B. 1993. Essentials of Conservation Biology. Sinauer Associates, Sunderland, Massachusetts.

Prince, H. 1997. Wetlands of the American Midwest. University of Chicago Press, Chicago, Illinois.

Pringle, C.M. 2000. Riverine conservation in tropical versus temperate latitudes. Pp. 367–378. In Boon, P.J., B. Davies and G.E. Petts (Eds.), Global Perspective On River Conservation: Science, Policy, and Practice. John Wiley and Sons, New York.

Prinsloo, J.F., H.J. Schoonbee and J.G. Nxiweni. 1981. Some observations on biological and other control measures of the African clawed frog, Xenopus laevis (Daudin) (Pipidae, Amphibia) in fish ponds in Transkei. Water South Africa 7:88–96.

Printiss, D.J. and D.L. Hipes. 1999. Rare amphibian and reptile survey of Eglin Air Force Base, Florida. Florida Natural Areas Inventory, Final report, Tallahassee, Florida.

Pritchard, P.C.H. 1994. Cladism: the great delusion. Herpetological Review 25:103–110.

Pritchard, P.C.H. 1999. Status of the black turtle. Conservation Biology 13:1000–1003.

Promislow, D.E.L. 1987. Courtship behavior of a plethodontid salamander, Desmognathus aeneus. Journal of Herpetology 21:298–306.

Propper, C.R., S.D. Hillyard and W.E. Johnson. 1995. Central angiotensin-II induces thirst-related responses in an amphibian. Hormones and Behavior 29:74–84.

Prosser, D.T. 1911. Habits of Ambystoma tigrinum at Tolland, Colorado. University of Colorado Studies 8:257–263.

Prudhoe, S. and R.A. Bray. 1982. Platyhelminth Parasites of the Amphibia. British Museum (Natural History), London and Oxford University Press, Oxford, United Kingdom.

Ptacek, M.B. 1992. Calling sites used by male gray treefrogs, Hyla versicolor and Hyla chrysoscelis, in sympatry and allopatry in Missouri. Herpetologica 48:373–382.

Ptacek, M.B. 1996. Interspecific similarity in life-history traits in sympatric populations of gray treefrogs, Hyla chrysoscelis and Hyla versicolor. Herpetologica 52:323–332.

Ptacek, M.B., H.C. Gerhardt and R.D. Sage. 1994. Speciation by polyploidy in treefrogs: multiple origins of the tetraploid, Hyla versicolor. Evolution 48:898–908.

Puckette, B.G. 1962. Ophiophagy in Hyla versicolor. Herpetologica 18:143.

Pudney, J., J.A. Canick and G.V. Callard. 1985. The testis and reproduction in male Necturus, with emphasis on N. lewisi (Brimley). Brimleyana 10:53–74.

Pulliam, H.R. 1988. Sources, sinks, and population regulation. American Naturalist 132:652–661.

Punzo, F. 1991a. Feeding ecology of spadefooted toads (Scaphiopus couchi and Spea multiplicata) in western Texas. Herpetological Review 22:79–80.

Punzo, F. 1991b. Group learning in tadpoles of Rana heckscheri (Anura: Ranidae). Journal of Herpetology 25:214–217.

Punzo, F. 1992a. Dietary overlap and activity patterns in sympatric populations of Scaphiopus holbrooki (Pelobatidae) and Bufo terrestris (Bufonidae). Florida Scientist 55:38–44.

Punzo, F. 1992b. Socially facilitated behavior in tadpoles of Rana catesbeiana and Rana heckscheri (Anura: Ranidae). Journal of Herpetology 26:219–222.

Punzo, F. 1993. Ovarian effects of a sublethal concentration of mercuric chloride in the river frog, Rana heckscheri (Anura: Ranidae). Bulletin of Environmental Contamination and Toxicology 50:385–391.

Punzo, F. 1995. An analysis of feeding in the oak toad, Bufo quercicus (Holbrook) (Anura: Bufonidae). Florida Scientist 58:16–20.

Punzo, F. 1997. The effects of azadirachtin on larval stages of the oak toad, Bufo quercicus (Holbrook) (Anura: Bufonidae). Florida Scientist 60:158–165.

Purvis, A., P.-M. Agapow, J.L. Gittleman and G.M. Mace. 2000. Nonrandom extinction and the loss of evolutionary history. Science 287: 328–330.

Puttlitz, M.H., D.P. Chivers, J.M. Keisecker and A.R. Blaustein. 1999. Threat-sensitive predator avoidance by larval Pacific treefrogs (Amphibia, Hylidae). Ethology 105:449–456.

Pyburn, W.F. 1956. Population structure and inheritance and distribution of vertebral stripe color in local populations of the cricket frog, Acris gryllus. Ph.D. dissertation. University of Texas, Austin, Texas.

Pyburn, W.F. 1958. Size and movements of a local population of cricket frogs *(Acris crepitans)*. Texas Journal of Science 10:325–342.

Pyburn, W.F. 1960. Hybridization between *Hyla versicolor* and *H. femoralis*. Copeia 1960:55–56.

Pyburn, W.F. 1961a. The inheritance and distribution of vertebral stripe color in the cricket frog. Pp. 235–261. *In* Blair, W.F. (Ed.), Vertebrate Speciation. University of Texas Press, Austin, Texas.

Pyburn, W.F. 1961b. Inheritance of green vertebral stripe in *Acris crepitans*. Southwest Naturalist 6:164:167.

Pyburn, W.F. and J.P. Kennedy. 1961. Hybridization in U.S. treefrogs in the genus *Hyla*. Proceedings of the Biological Society of Washington 74:157–160.

Pyle, J.C. 1964. A population study of the Ozark blind cave salamander, *Typhlotriton spelaeus*. Master's thesis. Central Missouri State College, Warrensburg, Missouri.

Pylka, J.M. and R.D. Warren. 1958. A population of *Haideotriton* in Florida. Copeia 1958:334–336.

Quicke, D.L.J. 1993. Principles and Techniques of Contemporary Taxonomy. Blackie Academic and Professional, Glasgow, United Kingdom.

Quinn, H.R. and G. Mengden. 1984. Reproduction and growth of *Bufo houstonensis* (Bufonidae). Southwestern Naturalist 29:189–195.

Quinn, T., J. Gallie and D.P. Volen. 2001. Amphibian occurrences in artificial and natural wetlands of the Teanaway and Lower Sauk River Drainages of Kittitas County, Washington. Northwest Science 75: 84–89.

Quinn, T.W., G.F. Shields and A.C. Wilson. 1991. Affinities of the Hawaiian Goose based on two types of mitochondrial DNA data. Auk 108:585–593.

Rabalais, F.C. 1970. Trematodes from some Caudata in Louisiana. American Midland Naturalist 84:265–267.

Rabb, G.B. 1956. Some observations on the salamander *Stereochilus marginatum*. Copeia 1956:119.

Rabb, G.B. 1966. *Stereochilus* and *S. marginatus*. Pp. 25.1–25.2. Catalogue of American Amphibians and Reptiles. Society for the Study of Amphibians and Reptiles, St. Louis, Missouri.

Rabinowe, J.H., J.T. Serra, M.P. Hayes and T. Quinn. 2002. Natural history notes: *Rana aurora aurora* (northern red-legged frog). Diet. Herpetological Review 33:128.

Rabor, D.S. 1952. Preliminary notes on the giant toad *Bufo marinus* (Linn.) in the Philippine Islands. Copeia 1952:281–282.

Radke, M.F. 1998. Ecology of the barking frog *(Eleutherodactylus augusti)* in Caves County, New Mexico. Report to New Mexico Department of Game and Fish, Share with Wildlife, Albuquerque, New Mexico.

Rafferty, K.A., Jr. 1967. The biology of spontaneous renal carcinoma of the frog. Pp. 301–315. *In* King, J.S. (Ed.), Renal Neoplasia. Little, Brown and Company, Boston, Massachusetts.

Rafferty, K.A., Jr. and N.S. Rafferty. 1961. High incidence of transmissible kidney tumors in uninoculated frogs maintained in a laboratory. Science 133:702–703.

Rafinesque, C.S. 1818. American Monthly Magazine Critical Reviews 4:41.

Rafinesque, C.S. 1820. Annals of nature or annual synopsis of new genera and species of animals, plants, etc. discovered in North America. III Class. Erpetia: The Reptiles, Number 1, Thomas Smith, Printer, Lexington, Kentucky.

Rafinesque, C.S. 1822. On two new salamanders of Kentucky. Kentucky Gazette, N.S. Volume 1(9):3, Lexington, Kentucky.

Raimondo, S.M., C.L. Rowe and J.D. Congdon. 1998. Exposure to coal ash impacts swimming performance and predator avoidance in larval bullfrogs *(Rana catesbeiana)*. Journal of Herpetology 32:289–292.

Ralin, D.B. 1968. Ecological and reproductive differentiation in the cryptic species of the *Hyla versicolor* complex (Hylidae). Southwestern Naturalist 13:283–299.

Ralin, D.B. 1977a. Evolutionary aspects of mating call variation in a diploid-tetraploid species complex of treefrogs (Anura). Evolution 31:721–736.

Ralin, D.B. 1977b. Hybridization of *Hyla cinerea* of the United States and *Hyla arborea savignyi* (Amphibia, Anura, Hylidae) of Israel. Journal of Herpetology 11:105–106.

Ralin, D.B. 1981. Ecophysiological adaptation in a diploid-tetraploid complex of treefrogs (Hylidae). Comparative Biochemistry and Physiology A 68:175–179.

Ralin, D.B. and J.S. Rogers. 1972. Aspects of tolerance to desiccation in *Acris crepitans* and *Pseudacris streckeri*. Copeia 1972:519–525.

Ralin, D.B. and J.S. Rogers. 1979. A morphological analysis of a North American diploid-tetraploid complex of treefrogs (Amphibia, Anura, Hylidae). Journal of Herpetology 13:261–269.

Ralin, D.B. and R.K. Selander. 1979. Evolutionary genetics of treefrogs of the genus *Hyla*. Evolution 33:595–608.

Ralin, D.B., M.A. Romano and C.W. Kilpatrick. 1983. The tetraploid treefrog *Hyla versicolor*: evidence for a single origin from the diploid *H. chrysoscelis*. Herpetologica 39:212–225.

Ralls, K. 1997. On becoming a conservation biologist: autobiography and advise. *In* Clemmons, J.R. and R. Buchholz (Eds.), Behavioral Approaches to Conservation in the Wild. Cambridge University Press, Cambridge, United Kingdom.

Ralph, P.H. 1938. *Cercaria concavocorpa* Sizemore becomes *Tetrapapillatrema*, a new telorchiid-like genus of Plagiorchioidea Dollfus. Transactions of the American Microscopical Society 57: 376–382.

Ramer, J.D., T.H. Jenssen and C.J. Hurst. 1983. Size-related variation in the advertisement call of *Rana clamitans* (Anura: Ranidae), and its effect on conspecific males. Copeia 1983:141–155.

Ramirez, R.S., Jr. 2000. Arroyo toad *(Bufo californicus)* radio telemetry study, Little Rock Creek, Los Angeles County, California. U.S.D.A. Forest Service, Arcadia, California.

Ramirez-Bautista, A., O. Flores-Villela and G. Casas-Andreu. 1982. New herpetological state records for Mexico. Bulletin of the Maryland Herpetological Society 18:167–169.

Ramos Ramos, P. 1995. Some trematodes of vertebrates from the Miguel Aleman dam in Temascal, Oaxaca, Mexico. Anales del Instituto de Biologia. Serie Zoologica 66:241–246.

Ramotnik, C.A. 1986. Status report, *Plethodon neomexicanus* Stebbins and Riemer, Jemez Mountains salamander. U.S. Fish and Wildlife Service, Ft. Collins, Colorado.

Ramotnik, C.A. 1988. Habitat requirements and movements of Jemez Mountains salamanders, *Plethodon neomexicanus*. Master's thesis. Colorado State University, Fort Collins, Colorado.

Ramotnik, C.A. 1997. Conservation assessment of the Sacramento Mountain salamander. U.S.D.A. Forest Service, General Technical Report, RM-GTR-293, Fort Collins, Colorado.

Ramotnik, C.A. and N.J. Scott Jr. 1988. Habitat requirements of New Mexico's endangered salamanders. Pp. 54–63. *In* Szaro, R.C., K.E. Severson and D.R. Patton (Eds.), Management of Amphibians, Reptiles, and Small Mammals in North America. U.S.D.A. Forest Service, Technical Report, RM-166, Rocky Mountain Forest and Range Experimental Station, Fort Collins, Colorado.

Ramsey, L.W. and J.W. Forsyth. 1950. Breeding dates for *Ambystoma texanum*. Herpetologica 6:70.

Ramus, E. (Ed.). 1998. 1998–1999 directory: The herpetology sourcebook. Reptile and Amphibian Magazine, Pottsville, Pennsylvania.

Rand, A.S. 1950. Leopard frogs in caves in winter. Copeia 1950:324.

Rand, A.S. 1952. Jumping ability of certain anurans with notes on endurance. Copeia 1952:15–20.

Rand, A.S. and P.J. Rand. 1966. The relation of size and distance jumped in *Bufo marinus*. Herpetologica 22:206–209.

Rand, G.M., P.G. Wells and L.S. McCarty. 1995. Introduction to aquatic toxicology. *In* Rand, G.M. (Ed.), Fundamentals of Aquatic Toxicology: Effects, Environmental Fate, and Risk Assessment. Taylor and Francis, Washington, D.C.

Raney, E.C. 1940. Summer movements of the bullfrog, *Rana catesbeiana* Shaw, as determined by the jaw-tag method. American Midland Naturalist 23:733–745.

Raney, E.C. and W.M. Ingram. 1941. Growth of tagged frogs (*Rana catesbeiana* Shaw and *Rana clamitans* Daudin) under natural conditions. American Midland Naturalist 26:201–206.

Rankin, J.S. 1937. An ecological study of parasites of some North Carolina salamanders. Ecological Monographs 7:169–269.

Rankin, J.S., Jr. 1938. Studies on the trematode genus *Brachycoelium* Duj. I. Variation in specific characters with reference to the validity of the described species. Transactions of the American Microscopical Society 57:358–375.

Rankin, J.S. 1939. The life cycle of the frog bladder fluke, *Gorgoderina attenuata* Stafford, 1902 (Trematoda: Gorgoderidae). American Midland Naturalist 21:476–488.

Rankin, J.S. 1944. A review of the trematode genus *Glypthelmins* Stafford, 1905, with an account of the life cycle of *G. quieta* (Stafford, 1900) Stafford, 1905. Transactions of the American Microscopical Society 63:30–43.

Rankin, J.S., Jr. 1945. An ecological study of the helminth parasites of amphibians and reptiles of western Massachusetts and vicinity. Journal of Parasitology 31:142–150.

Ranta, E. 1992. Gregariousness versus solitude: another look at parasite faunal richness in Canadian freshwater fishes. Oecologia 89: 150–152.

Rapaport, R.A., N.R. Urban, P.D. Capel, J.E. Baker, B.B. Looney, S.J. Eisenreich and E. Gorham. 1985. New DDT inputs to North America: atmospheric deposition. Chemosphere 14:1167–1173.

Raphael, M.G. 1988. Long-term trends in abundance of amphibians, reptiles, and mammals in Douglas-fir forests of northwestern California. Pp. 23–31. *In* Szaro, R.C., K.E. Severson and D. Patton (Tech. Coords.), Management of Amphibians, Reptiles, and Small Mammals in North America. U.S.D.A. Forest Service, Technical Report, RM-166, Rocky Mountain Forest and Range Experiment Station, Fort Collins, Colorado.

Rappole, J.H. and J. Klicka. 1991. Status of the black-spotted newt *(Notophthalmus meridionalis)* in Texas and Mexico. U.S. Fish and Wildlife Service, Lower Rio Grande Valley Wildlife Refuge, Unpublished report, Alamo, Texas.

Rassner, G. 1991. Of toads and birdies. Beacon 1:1, 5, 8.

Rassner, G. 1994. Song of the open road. Beacon 4:3, 8.

Rathbun, G.B. 1998. Natural history notes: *Rana aurora draytonii* (California red-legged frog). Egg predation. Herpetological Review 29:165.

Rathbun, G.B. and T.G. Murphey. 1996. Evaluation of a radio-belt for ranid frogs. Herpetological Review 27:187–189.

Rathke, M.H. 1833. *In* Eschscholtz, F. 1833. Zoologischer Atlas. Part 5. Reimer, Berlin, Germany.

Raun, G.G. 1959. Terrestrial and aquatic vertebrates of a moist, relict area in central Texas. Texas Journal of Science 11:158–171.

Raun, G.G. and F.R. Gehlbach. 1972. Amphibians and reptiles in Texas. Dallas Museum of Natural History Bulletin, Number 2, Dallas, Texas.

Ray, G.C. 1958. The vital limits and rates of desiccation in salamanders. Ecology 39:75–83.

Raymond, L.R. 1991. Seasonal activity of *Siren intermedia* in northwestern Louisiana (Amphibia: Sirenidae). Southwestern Naturalist 36:144–147.

Raymond, L.R. and L.M. Hardy. 1990. Demography of a population of *Ambystoma talpoideum* (Caudata: Ambystomatidae) in northwestern Louisiana. Herpetologica 46:371–382.

Raymond, L.R. and L.M. Hardy. 1991. Effects of a clearcut on a population of the mole salamander, *Ambystoma talpoideum*, in an adjacent unaltered forest. Journal of Herpetology 25:509–512.

Read, J.L. and M.J. Tyler. 1994. Natural levels of abnormalities in the trilling frog (*Neobactrachus centralis*) at the Olympic Dam Mine. Bulletin of Environmental Contamination and Toxicology 53:25–31.

Reading, C.J. 1998. The effect of winter temperatures on the timing of breeding activity in the common toad *Bufo bufo*. Oecologia 117:469–475.

Reagan, D.P. 1972. Ecology and distribution of the Jemez Mountains salamander, *Plethodon neomexicanus*. Copeia 1972:486–492.

Real, L.A. 1996. Disease ecology. Ecology 77:989.

Reaser, J.K. 1996a. The elucidation of amphibian declines. Amphibian and Reptile Conservation 1:4–9.

Reaser, J.K. 1996b. Natural history notes: *Rana pretiosa* (spotted frog). Vagility. Herpetological Review 27:196.

Reaser, J.K. 1997a. Amphibian declines: conservation science and adaptive management. Ph.D. dissertation. Stanford University, Stanford, California.

Reaser, J.K. 1997b. Natural history notes: *Batrachoseps attenuatus*. Predation. Herpetological Review 28:81.

Reaser, J.K. 2000a. Amphibian declines: an issue overview. U.S. Taskforce on Amphibian Declines and Deformities, Washington, D.C.

Reaser, J.K. 2000b. Demographic analyses of the Columbia spotted frog (*Rana luteiventris*): case study in spatio-temporal variation. Canadian Journal of Zoology 78:1158–1511.

Reaser, J.K. 2003. Occurrence of the California red-legged frog *(Rana aurora draytonii)* in Nevada, USA. Western North American Naturalist 63:400–401.

Reaser, J.K. and R.E. Dexter. 1996a. Natural history notes: *Rana pretiosa.* (spotted frog). Predation. Herpetological Review 27:75.

Reaser, J.K. and R.E. Dexter. 1996b. Natural history notes: *Rana pretiosa.* (spotted frog). Toe clipping effects. Herpetological Review 27:195–196.

Reaser, J.K. and C. Galindo-Leal. 1999. La Desaparacion de las Rasas: Mesoamerica y el Caribe. Vanishing Frogs: Mesoamerica and the Caribbean. U.S. Federal Taskforce on Amphibian Declines and Defomities (TADD) and IUCN Declining Amphibian Population Task Force (DAPTF), Washington, D.C.

Reaser, J.K. and P.T. Johnson. 1997. Amphibian abnormalities—a review. Froglog: Newsletter of the Declining Amphibian Populations Task Force 24:2–4.

Reaser, J.K., R. Pomerance and P.O. Thomas. 2000. Coral bleaching and global climate change: scientific findings and policy recommendations. Conservation Biology 14:1500–1511.

Reddell, J.R. 1970. A checklist of the cave fauna of Texas. VI. Additional records of Vertebrata. Texas Journal of Science 22:139–158.

Redman, W.H. and R.L. Jones. 1985. Geographic distribution: *Plethodon wehrlei* (Wehrle's salamander). Herpetological Review 16:31.

Redmer, M. 1995. Natural history notes: *Ambystoma texanum* (smallmouth salamander). Maximum size. Herpetological Review 26:29.

Redmer, M. 1998a. Natural history notes: *Hyla avivoca* (bird-voiced treefrog). Amplexus and oviposition. Herpetological Review 29:230–231.

Redmer, M. 1998b. Status and distribution of two uncommon frogs, pickerel frogs and wood frogs in Illinois. Pp. 83–90. *In* Lannoo, M.J. (Ed.), Status and Conservation of Midwestern Amphibians. University of Iowa Press, Iowa City, Iowa.

Redmer, M. 2000. Demographic and reproductive characteristics of a southern Illinois population of the crayfish frog, *Rana areolata*. Pp. 128–133. *In* Kaiser, H., G.S. Casper and N.P. Bernstein (Eds.), Investigating Amphibian Declines: Proceedings of the 1998 Declining Amphibians Conference. Journal of the Iowa Academy of Science, Volume 107, Cedar Falls, Iowa.

Redmer, M. 2002. Natural history of the wood frog (*Rana sylvatica*) in the Shawnee National Forest, southern Illinois. Illinois Natural History Survey Bulletin 36:163–194.

Redmer, M. and S.R. Ballard. 1995. Recent distribution records for amphibians and reptiles in Illinois. Herpetological Review 26:49–53.

Redmer, M. and S.J. Karsen. 1990. Winter activity of the eastern spadefoot (*Scaphiopus holbrookii*) in southern Missouri. Bulletin of the Chicago Herpetological Society 25:88.

Redmer, M. and G. Kruse. 1998. Updates to the list of Illinois' endangered and threatened amphibians and reptiles. Bulletin of the Chicago Herpetological Society 33:244–245.

Redmer, M. and K.S. Mierzwa. 1994. A review of the distribution and zoogeography of the pickerel frog, *Rana palustris*, in northern Illinois. Bulletin of the Chicago Herpetological Society. 29:21–30.

Redmer, M., L.E. Brown and R.A. Brandon. 1999a. Natural history of the bird-voiced treefrog (*Hyla avivoca*) and green treefrog (*Hyla cinerea*) in southern Illinois. Bulletin of the Illinois Natural History Survey 36:37–66.

Redmer, M., D.L. Jamieson and S.E. Trauth. 1999b. Notes on the diet of female bird-voiced treefrogs (*Hyla avivoca*) in Illinois. Transactions of the Illinois State Academy of Science 92:271–275.

Redmer, M., J.B. Camerer, J.K. Tucker and J. Capps. 1995. Geographic distribution: *Rana palustris* (pickerel frog), Illinois. Herpetological Review 26:42.

Redmond, W.H. 1980. Notes on the distribution and ecology of the Black Mountain dusky salamander *Desmognathus welteri* Barbour (Amphibia: Plethodontidae) in Tennessee. Brimleyana 4:123–131.

Redmond, W.H. and A.F. Scott. 1996. Atlas of Amphibians in Tennessee. The Center for Field Biology, Miscellaneous Publication, Number 12, Austin Peay State University, Clarksville, Tennessee.

Redmond, W.H., A.F. Scott and D. Roberts. 1982. Comments on the distribution of *Ambystoma talpoideum* (Holbrook) in Tennessee. Herpetological Review 13:83–85.

Reed, C.A. and R. Borowsky. 1967. The "world's largest toad" and other herpetological specimens from southern Suriname. Studies on the Fauna of Suriname and other Guyana 50:159–171.

Reed, H.D. and A.H. Wright. 1909. The vertebrates of the Cayuga Lake Basin, N.Y. Proceedings of the American Philosophical Society 48:370–459.

Reeder, A.L., G.L. Foley, D.K. Nichols, L.G. Hansen, B. Wikoff, S. Faeh, J. Eisold, M.B. Wheeler, R. Warner, J.E. Murphy and V.R. Beasely. 1998. Forms and prevalence of intersexuality and effects of environmental contaminants on sexuality in cricket frogs (*Acris crepitans*). Environmental Health Perspectives 106:261–266.

Rees, W.J. 1965. The aerial dispersal of Mollusca. Proceedings of the Malacological Society of London 36:269–282.

Reese, A.M. 1933. The fauna of West Virginia caves. Proceedings of the West Virginia Academy of Science 7:39–53.

Reese, R.W. 1969. The taxonomy and ecology of the tiger salamander (*Ambystoma tigrinum*) of Colorado. Ph.D. dissertation. University of Colorado, Boulder, Colorado.

Regester, K.J. 2000a. *Plethodon richmondi*. Pp. 707.1–707.3. Catalogue of American Amphibians and Reptiles. Society for the Study of Amphibians and Reptiles, St. Louis, Missouri.

Regester, K.J. 2000b. *Plethodon electromorphus*. Pp. 706.1–706.3. Catalogue of American Amphibians and Reptiles. Society for the Study of Amphibians and Reptiles, St. Louis, Missouri.

Regosin, J.V., B.S. Windmiller and J.M. Reed. 2003. Terrestrial habitat use and winter densities of the wood frog (Rana sylvatica). Journal of Herpetology 37:390–394.

Reh, W. and A. Seitz. 1989. Untersuchungen zum Einfluß der Landnutzung auf die genetische Struktur von Populationen des Grasfrosches (Rana temporaria L.). Verhandlung der Gesellschaft für Ökologie 18:793–796.

Reh, W. and A. Seitz. 1990. The influence of land use on the genetic structure of populations of the common frog (Rana temporaria). Biological Conservation 54:239–249.

Reiber, R.J. 1941. Nematodes of Amphibia and Reptilia. I. Reelfoot Lake, Tennessee. Journal of the Tennessee Academy of Sciences 5:92–99.

Reichel, J.D. 1996. Preliminary amphibian and reptile survey of the Helena National Forest: 1995. U.S. Department of Agriculture, Helena National Forest, Helena, Montana.

Reichel, J.D. and D.L. Flath. 1995. Identification of Montana's amphibians and reptiles. Montana Outdoors 26:15–34.

Reichenbach-Klinke, H. and E. Elkan. 1965. The Principal Diseases of Lower Vertebrates. Academic Press, New York.

Reigle, N.J. 1967. The occurrence of Necturus in the deeper waters of Green Bay. Herpetologica 23:232–233.

Reilly, S.M. 1987. Ontogeny of the hyobranchial apparatus in the salamanders Ambystoma talpodium (Ambystomatidae) and Notophthalmus viridescens (Salamandridae): the ecological morphology of two neotenic strategies. Journal of Morphology 191:205–214.

Reilly, S.M. and R. Altig. 1996. Cranial ontogeny in Siren intermedia (Caudata: Sirenidae): paedomorphic, metamorphic, and novel patterns of heterochrony. Copeia 1996:29–41.

Reilly, S.M., G.V. Lauder and J.P. Collins. 1992. Performance consequences of a trophic polymorphism: feeding behavior in typical and cannibal phenotypes of Ambystoma tigrinum. Copeia 1992:672–679.

Reimchen, T.E. 1991. Introduction and dispersal of the Pacific treefrog, Hyla regilla, on the Queen Charlotte Islands, British Columbia. Canadian Field-Naturalist 105:288–290.

Reimer, W.J. 1958. Giant toads of Florida. Quarterly Journal of the Florida Academy of Science 21:207–211.

Reinbold, S.L. 1979. Habitat comparisons of two sympatric salamander species of the genus Plethodon (Amphibia, Caudata, Plethodontidae). Journal of Herpetology 13:504–506.

Reinert, H.K. 1991. Translocation as a conservation strategy for amphibians and reptiles: some comments, concerns, and observations. Herpetologica 47:357–363.

Reis, K.M. and E.D. Bellis. 1966. Spring food habits of the red-spotted newt in Pennsylvania. Herpetologica 22:152–155.

Reister, A., S. Horner and D. Carlson. 1998. Chromosomal analysis of deformed frogs. Proceedings of the South Dakota Academy of Science 77:95.

Relyea, R.A. 2000. Trait-mediated indirect effects in larval anurans: reversing competitive outcomes with the threat of predation. Ecology 81:2278–2289.

Relyea, R.A. 2001. Morphological and behavioral plasticity of larval anurans in response to different predators. Ecology 82:523–540.

Relyea, R.A. and E.E. Werner. 1999. Quantifying the relation between predator-induced behavior and growth performance in larval anurans. Ecology 80:2117–2124.

Relyea, R.A. and E.E. Werner. 2000. Morphological plasticity of four anurans distributed along an environmental gradient. Copeia 2000:178–190.

Renaud, M. 1977. Polymorphic and polytypic variation in the Arizona treefrog (Hyla wrightorum). Ph.D. dissertation. Arizona State University, Tempe, Arizona.

Renken, R.B., W.K. Gram, D.K. Fantz, S.C. Richter, T.J. Miller, K.B. Ricke, B. Russell and X. Wang. 2004. Effects of forest management on amphibians and reptiles in Missouri Ozark Forests. Conservation Biology 18:174–188.

Reno, H.W., F.R. Gehlbach and R.A. Turner. 1972. Skin and aestivational cocoon of the aquatic amphibian, Siren intermedia Le Conte. Copeia 1972:625–631.

Repenning, R.W. and R.F. Labisky. 1985. Effects of even-age timber management on bird communities of the longleaf pine forest in northern Florida. Journal of Wildlife Management 49:1088–1098.

Reques, R. and M. Tejedo. 1995. Negative correlation between length of larval period and metamorphic size in natural populations of natterjack toads (Bufo calamita). Journal of Herpetology 29:311–314.

Resetar, A. 1998. Locating historical information on amphibian populations. Pp. 379–384. In Lannoo, M.J. (Ed.), Status and Conservation of Midwestern Amphibians. University of Iowa Press, Iowa City, Iowa.

Resetarits, W.J., Jr. 1986. Ecology of cave use by the frog, Rana palustris. American Midland Naturalist 116:256–266.

Resetarits, W.J., Jr. 1991. Ecological interactions among predators in experimental stream communities. Ecology 72:1782–1793.

Resetarits, W.J., Jr. 1995. Competitive asymmetry and coexistence in size-structured populations of brook trout and spring salamanders. Oikos 73:188–198.

Resetarits, W.J., Jr. 1998. Differential vulnerability of Hyla chrysoscelis eggs and hatchlings to larval insect predators. Journal of Herpetology 32:440–443.

Resetarits, W.J. and R.D. Aldridge. 1988. Reproductive biology of a cave-associated population of the frog Rana palustris. Canadian Journal of Zoology 66:329–333.

Resetarits, W.J., Jr. and H.M. Wilbur. 1989. Choice of oviposition site by Hyla chrysoscelis: role of predators and competitors. Ecology 70:220–228.

Reynolds, A.E. 1959. Observations on various stages of topotypic Plethodon jordani. Herpetologica 15:183–192.

Reynolds, T.D. and T.D. Stephens. 1984. Multiple ectopic limbs in a wild population of Hyla regilla. Great Basin Naturalist 44:166–169.

Richards, C.M. 1958. The inhibition of growth in crowded Rana pipiens tadpoles. Physiological Zoology 31:138–151.

Richards, S.J., K.R. McDonald and R.A. Alford. 1993. Declines in populations of Australia's endemic tropical rainforest frogs. Pacific Conservation Biology 1:66–77.

Richardson, C.J. 1983. Pocosins: vanishing wastelands or valuable wetlands? Bioscience 33:626–633.

Richardson, C.J. 1991. Pocosins: an ecological perspective. Wetlands 11 (Special Issue):335–354.

Richardson, J.P.M. and M.L. Adamson. 1988. A new Kathlaniidae (Cosmocercoidea; Nematoda), Megalobatrachonema (Chabaudgolvania) moraveci sp. n. from the intestine of the rough-skinned newt, Taricha granulosa. Proceedings of the Helminthological Society of Washington 55:155–159.

Richman, J.B. 1973. A range extension for the Tehachapi slender salamander, Batrachoceps stebbinsi [sic]. Herpetological Information Search Systems News-Journal 1:97.

Richmond, N.D. 1945. Nesting of the two-lined salamander on the Coastal Plain. Copeia 1945:170.

Richmond, N.D. 1947. Life history of Scaphiopus holbrookii (Harlan). Part I: larval development and behavior. Ecology 28:53–66.

Richmond, N.D. 1952. First record of the green salamander in Pennsylvania, and other range extensions in Pennsylvania, Virginia and West Virginia. Annals of the Carnegie Museum 32:313–318.

Richmond, N.D. 1964. The green frog (Rana clamitans melanota) developing in one season. Herpetologica 20:132.

Richter, K.O. and A.L. Azous. 1995. Amphibian occurrence and wetland characteristics in the Puget Sound Basin. Wetlands 15:305–312.

Richter, K.O. and A.L. Azous. 2000. Amphibian distribution, abundance, and habitat use. Pp. 143–166. In Azous, A.L. and R.R. Horner (Eds.), Wetlands and Urbanization: Implications for the Future. CRC Press LLC., Boca Raton, Florida.

Richter, S.C. 1998. The demography and reproductive biology of gopher frogs, Rana capito, in Mississippi. Master's thesis. Southeastern Louisiana University, Hammond, Louisiana.

Richter, S.C. 2000. Larval caddisfly predation on the eggs and embryos of Rana capito and Rana sphenocephala. Journal of Herpetology 34:590–593.

Richter, S.C. and R.A. Seigel. 1997. Demography and reproduction of gopher frogs in Mississippi. Unpublished 1997 report to the U.S. Fish and Wildlife Service, Jackson, Mississippi.

Richter, S.C. and R.A. Seigel. 1998. Demography and reproduction of gopher frogs in Mississippi. Unpublished 1998 report to the U.S. Fish and Wildlife Service, Jackson, Mississippi.

Richter, S.C. and R.A. Seigel. 2002. Annual variation in the population ecology of the endangered gopher frog, Rana sevosa Goin and Netting. Copeia 2002:962–972.

Richter, S.C., J.E. Young, G.N. Johnson and R.A. Seigel. 2003. Stochastic variation in reproductive success of a rare frog, Rana sevosa: implications for conservation and for monitoring amphibian populations. Biological Conservation 111:171–177.

Richter, S.C., J.E. Young, R.A. Seigel and G.N. Johnson. 2001. Postbreeding movements of the dark gopher frog, Rana sevosa Goin and Netting: implications for conservation and management. Journal of Herpetology 35:316–321.

Ricker, W.E. and E.B.S. Logier. 1935. Notes on the occurrence of the ribbed toad (*Ascaphus truei* Stegneger) in Canada. Copeia 1935:46.

Ridgway, R. 1889. The Ornithology of Illinois. Descriptive Catalogue. Volume 1, Natural History Survey of Illinois, State Laboratory of Natural History (1913 reprint). Pantagraph Printing and Stationery Company, Bloomington, Illinois.

Riemer, W.J. 1955. Comments on the distribution of certain Mexican toads. Herpetologica 11:17–23.

Riemer, W.J. 1958. Variation and systematic relationships within the salamander genus *Taricha*. University of California Publications in Zoology 56:301–390.

Riesecrer, J.M., M.T. Anderson, D.P. Chivers, E.L. Wildy, J. Devito, A. Marco, A.R. Blaustein, J.J. Beatty and R.M. Storm. 1996. Natural history notes: *Plethodon dunni* (Dunn's salamander). Cannibalism. Herpetological Review 27:194.

Riffel, M. and M. Braun. 1987. Zwischenbericht zur Amphibienfauna im Raum Bruchsal. Beihefte zu den Veröffentlichungen für Naturschutz und Landschaftspflege in Baden-Württemberg 41:313–320.

Riggin, G.T. 1956. A note on *Ribeiroia ondatrae* (Price, 1931) in Puerto Rico. Proceedings of the Helminthological Society of Washington 23:28–29.

Riley, S.P.D., H.B. Shaffer, S.R. Voss and B.M. Fitzpatrick. 2003. Hybridization between a rare, native tiger salamander (*Ambystoma californiense*) and its introduced congener. Ecological Applications 13:1263–1275.

Rimbach, D.N. 1968. Low Water Bridge Cave. Missouri Speleology 10:122–124.

Rimpp, K. 1992. Amphibien und Reptilien in Schönbuch und Gäu. Pp. 155–178. *In* Schubert, W. (Ed.), Die Teirwelt in Schönbuch und Gäu. Die Wirbeltiere und ihr Schutz. Natur-Rems-Murr Verlag, Remshalden-Buoch.

Rimpp, K. and G. Hermann. 1987. Die Amphibien des Landkreises Böblingen. Jahrbuch für Feldherpetologie 1:3–17.

Rios-López, N. and R.L. Joglar. 1999. Geographic distribution: *Rana grylio* (pig frog). Herpetological Review 30:230–231.

Ripert, C.L. and C.P. Raccurt. 1987. The impact of small dams on parasitic diseases in Cameroon. Parasitology Today 3:287–289.

Rising, J.D. and F.W. Schueler. 1980. Screech owl eats fish and salamander in winter. Wilson Bulletin 92:250–251.

Rist, L., R.D. Semlitsch, H. Hotz and H.-U. Reyer. 1997. Feeding behavior, food consumption, and growth efficiency of hemiclonal and parental tadpoles of the *Rana esculenta* complex. Functional Ecology 11:735–742.

Ritchie, S.A. 1982. The green treefrog (*Hyla cinerea*) as a predator of mosquitoes in Florida. Mosquito News 42:619.

Ritke, M.E. and J.G. Babb. 1991. Behavior of the gray treefrog (*Hyla chrysoscelis*) during the breeding season. Herpetological Review 22:5–8.

Ritke, M.E. and C.A. Lessman. 1994. Longitudinal study of ovarian dynamics in female gray treefrogs (*Hyla chrysoscelis*). Copeia 1994:1014–1022.

Ritke, M.E., J.G. Babb and M.K. Ritke. 1990. Life history of the gray treefrog (*Hyla chrysoscelis*) in western Tennessee. Herpetologica 24:135–141.

Ritke, M.E., J.G. Babb and M.K. Ritke. 1991a. Annual growth rates of adult gray treefrogs (*Hyla chrysoscelis*). Journal of Herpetology 25:382–385.

Ritke, M.E., J.G. Babb and M.K. Ritke. 1991b. Breeding-site specificity in the gray treefrog (*Hyla chrysoscelis*). Journal of Herpetology 25:123–125.

Ritland, K., L.A. Dupuis, F.L. Bunnell, W.L.Y. Hung and J.E. Carlson. 2000. Phylogeography of the tailed frog (*Ascaphus truei*) in British Columbia. Canadian Journal of Zoology 78:1749–1758.

Ritter, W.E. 1897. The life-history and habits of the Pacific Coast newt (*Diemyctylus torosus* Esch.). Proceedings of the National Academy of Sciences, Series 3, 1:73–114.

Ritter, W.E. 1903. Further notes on the habits of *Autodax lugubris*. American Naturalist 37:883–886.

Ritter, W.E. and L. Miller. 1899. A contribution to the life history of *Autodax lugubris*, a Californian salamander. American Naturalist 33:691–704.

Rittschof, D. 1975. Some aspects of the natural history and ecology of the leopard frog, *Rana pipiens*. Ph.D. dissertation. University of Michigan, Ann Arbor, Michigan.

Rivero, J.A. 1978. Los Anfibios Y Reptiles De Puerto Rico. University of Puerto Rico, San Juan, Puerto Rico.

Robbins, C.S. 1981. Bird activity levels related to weather. Studies in Avian Biology, Number 6:301–310.

Robbins, C.S., D. Bystrak and P.H. Geissler. 1986. The Breeding Bird Survey: its first fifteen years, 1965–1979. U.S. Fish and Wildlife Service, Resource Publication, Number 157, Washington, D.C.

Robbins, C.S., D.K. Dawson and B.A. Dowell. 1989. Habitat area requirements of breeding forest birds of the Middle Atlantic States. Wildlife Monograph, Number 103, Wildlife Society, Bethesda, Maryland.

Roberts, D.T., D.M. Schleser and T.L. Jordan. 1995. Notes on the captive husbandry and reproduction of the Texas salamander *Eurycea neotenes* at the Dallas Aquarium. Herpetological Review 26:23–25.

Roberts, K.A. and R.B. Page. 2003. Natural history notes: *Hyla cinerea* (green tree frog). Reproduction. Herpetological Review 34:136.

Roberts, L.S. and J. Janovy Jr. 1996. Foundations of Parasitology. Fifth edition. William C. Brown Publishers, Dubuque, Iowa.

Roberts, L.S. and J. Janovy Jr. 2000. Gerald D. Schmidt and Larry S. Roberts' Foundations of Parasitology. Sixth edition. McGraw-Hill Publishing Company, Dubuque, Iowa.

Roberts, M.E. 1963. Studies on the transmissibility and cytology of the renal carcinoma of *Rana pipiens*. Cancer Research 23:1709–1714.

Roberts, W. 1981. What happened to the leopard frogs? Alberta Naturalist 1:1–4.

Roberts, W. 1992. Declines in amphibian populations in Alberta. Pp. 14–16. *In* Bishop, C.A. and K.E. Pettit (Eds.), Declines in Canadian Amphibian Populations: Designing A National Monitoring Strategy. Canadian Wildlife Service, Occasional Paper, Number 76, Ottawa, Ontario, Canada.

Roberts, W. and V. Lewin. 1979. Habitat utilization and population densities of the amphibians of northeastern Alberta. Canadian Field-Naturalist 93:144–154.

Robertson, D.G. and R.D. Slack. 1995. Landscape change and its effects on the wintering range of a lesser snow goose (*Chen caerulescens caerulescens*) population: a review. Biological Conservation 71:179–185.

Robertson, W. B. and E.L. Tyson. 1950. Herpetological notes from eastern North Carolina. Journal of the Elisha Mitchell Scientific Society 66:130–147.

Robinson, T.S. 1957. Notes on the development of a brood of Mississippi kites in Barber County, Kansas. Transactions of the Kansas Academy of Science 60:174–180.

Robinson, T.S. and K.T. Reichard. 1965. Notes on the breeding biology of the midland mud salamander, *Pseudotriton montanus diastictus*. Journal of the Ohio Herpetological Society 5:29.

Robison, H.W. and S. Winters. 1978. Geographic distribution: *Ambystoma talpoideum* (mole salamander). Herpetological Review 9:21.

Roble, S.M. 1979. Dispersal movements and plant associations of juvenile gray treefrogs, *Hyla versicolor* LeConte. Transactions of the Kansas Academy of Science 82:235–245.

Roble, S.M. 1985. Observations on satellite males in *Hyla chrysoscelis*, *Hyla picta*, and *Pseudacris triseriata*. Journal of Herpetology 19:432–436.

Roble, S.M. 1995. Geographic distribution: *Siren intermedia intermedia* (eastern lesser siren). Herpetological Review 26:150–151.

Roble, S.M. 1998. A zoological inventory of the Grafton Ponds sinkhole complex, York County, Virginia. Virginia Department of Conservation and Recreation, Division of Natural Heritage (Richmond), Natural Heritage Technical Report 98–3, U.S. Environmental Protection Agency, Philadelphia, Pennsylvania.

Roble, S.M. and C.S. Hobson. 1995. Records of amphibians and reptiles from the Clinch Ranger District, Jefferson National Forest. Catesbeiana 15:3–14.

Roble, S.M., D.J. Stevenson and C.S. Hobson. 1999. Distribution of the dwarf waterdog (*Necturus punctatus*) in Virginia, with comments on collecting techniques. Banisteria 14:39–44.

Rogers, A.M. 1999. Ecology and natural history of *Rana clamitans melanota* in West Virginia. Master's thesis. Marshall University, Huntington, West Virginia.

Rogers, C.P. 1996. Natural history notes: *Rana catesbeiana* (bullfrog). Predation. Herpetological Review 27:19.

Rogers, J.S. 1973a. Biochemical and morphological analysis of potential introgression between *Bufo cognatus* and *B. speciosus*. American Midland Naturalist 90:127–142.

Rogers, J.S. 1973b. Protein polymorphism, genetic heterozygosity and divergence in the toads *Bufo cognatus* and *B. speciosus*. Copeia 1973:322–330.

Rogers, Je. 1995a. Letter to Andrew Sansom, 16 February.

Rogers, Je. 1995b. E-mail message, acquired through Freedom of Information Act, to Ed Shellenberger, Neil Mangum, Joe Sovick, Cecilia Matic, Linda McClelland, Ethan Carr, Kitty L. Roberts, Joan Anzelmo, Destry Jarvis, Anne Badgley, Pat Tiller and Kate Stevenson, 22 February.

Rogers, Je. 1995c. E-mail message, acquired through Freedom of Information Act, to Roger Kennedy, 28 March.

Rogers, Je. 1995d. E-mail message, acquired through Freedom of Information Act, to Destry Jarvis, 17 April.

Rogers, Jo. 1994. Draft biological opinion for proposed action to expand the Lost Pines Golf Course in Bastrop State Park, located in Bastrop County, Texas. U.S. Fish and Wildlife Service, Austin, Texas.

Rogers, K. 1999. Toads to get own pad. Las Vegas Review Journal, Sunday 23 May, 1999, http://www.lvrj.com/lvrj_home/1999/May-23-Sun-1999/news/11175557.html.

Rohlf, D.J. 1994. There's something fishy going on here: a critique of the National Marine Fisheries Service's definition of species under the Endangered Species Act. Environmental Law 24:617–671.

Rojas, M. 1992. The species problem and conservation: what are we protecting? Conservation Biology 6:170–178.

Rollins, L.A. and R.G. McKinnell. 1980. The influence of glucocorticoids on survival and growth of allografted tumors in the anterior eye chamber of leopard frogs. Developmental and Comparative Immunology 4:283–294.

Rombough, C.J., D.J. Jordan and C.A. Pearl. 2003. Natural history notes: *Rana cascadae* (Cascade frog). Cannibalism. Herpetological Review 34:138.

Romspert, A.P. 1976. Osmoregulation of the African clawed frog, *Xenopus laevis*, in hypersaline media. Comparative Biochemistry and Physiology A 54:207–210.

Rorabaugh, J.C. and J. Humphrey. 2002. The Tarahumara frog: return of a native. Endangered Species Bulletin 27:24–26.

Rorabaugh, J.C. and M.J. Sredl. 2002. Continued invasion by an introduced frog *(Rana berlandieri)* in Arizona, California, and Mexico: implications and opportunities for management. Pp. 112–113. *In* Halvorson, W.L. and G.S. Gebow (Eds.), Fourth Conference on Research and Resource Management in the Southwestern Deserts. Extended abstracts. Tucson, Arizona.

Rorabaugh, J.C., J.M. Howland and R.D. Babb. 2004. Distribution and habitat use of the Pacific treefrog *(Pseudacris regilla)* on the lower Colorado River and in Arizona. Southwestern Naturalist 49:94–99.

Rorabaugh, J.C., M.J. Sredl, V. Miera and C.A. Drost. 2002. Continued invasion of an introduced frog *(Rana berlandieri)*: southwestern Arizona, southeastern California, and Rio Colorado, Mexico. Southwestern Naturalist 47:12–20.

Rose, F.L. 1966a. Reproductive potential of *Amphiuma means*. Copeia 1966:598–599.

Rose, F.L. 1966b. Weight change during starvation in *Amphiuma means*. Herpetologica 22:312–313.

Rose, F.L. 1966c. Homing to nests by the salamander *Desmognathus auriculatus*. Copeia 1966:251–253.

Rose, F.L. 1967. Seasonal changes in lipid levels of the salamanders *Amphiuma means*. Copeia 1967:662–666.

Rose, F.L. 1971. *Eurycea aquatica*. Pp. 116.1–116.2. Catalogue of American Amphibians and Reptiles. Society for the Study of Amphibians and Reptiles, St. Louis, Missouri.

Rose, F.L. and D. Armentrout. 1976. Adaptive strategies of *Ambystoma tigrinum* Green inhabiting the Llano Estacado of west Texas. Journal of Animal Ecology 45:713–729.

Rose, F.L. and F.M. Bush. 1963. A new species of *Eurycea* (Amphibia: Caudata) from the southeastern United States. Tulane Studies in Zoology 10:121–128.

Rose, F.L. and J.L. Dobie. 1963. *Desmognathus monticola* in the coastal plain of Alabama. Copeia 1963:564–565.

Rose, F.L., D. Armentrout and P. Roper. 1971. Physiological responses of paedogenic *Ambystoma tigrinum* to acute anoxia. Herpetologica 27:101–107.

Rose, S.M. and F.C. Rose. 1952. Tumor agent transformation in amphibia. Cancer Research 12:1–12.

Rose, W. 1950. The Reptiles and Amphibians of Southern Africa. Maskew Miller, Limited, Cape Town, South Africa.

Rosen, M. and R.E. Lemon. 1974. The vocal behavior of spring peepers, *Hyla crucifer*. Copeia 1974:940–950.

Rosen, P.C. and C.H. Lowe. 1996. Ecology of the amphibians and reptiles at Organ Pipe Cactus National Monument, Arizona. National Park Service, Technical Report, Number 53, United States Department of the Interior, Tucson, Arizona.

Rosen, P.C. and C.R. Schwalbe. 1998. Using managed waters for conservation of threatened frogs. Pp. 180–195. *In* Feller, J.M. and D.S. Strouse (Eds.), Environmental, Economic, and Legal Issues Related to Rangeland Water Developments. Center for the Study of Law, Science and Technology, Arizona State University, Tempe, Arizona.

Rosen, P.C., S.S. Sartorius, C.R. Schwalbe, P.A. Holm and C.H. Lowe. 1996. Draft annotated checklist of the amphibians and reptiles of the Sulphur Springs Valley, Cochise County, Arizona. Final Report, Part 1. Submitted to Arizona Game and Fish Department, Phoenix, Arizona.

Rosen, P.C., C.R. Schwalbe, D.A.J. Parizek, P.A. Holm and C.H. Lowe. 1995. Introduced aquatic vertebrates in the Chiricahua region: effects on declining ranid frogs. Pp. 251–261. *In* DeBano, L.F., G.J. Gottfried, R.H. Hamre, C.B. Edminster, P.F. Ffolliott and A. Ortega-Rubio (Eds.), Biodiversity and Management of the Madrean Archipelago: The Sky Islands of Southwestern United States and Northwestern Mexico. Rocky Mountain Forest and Range Experiment Station, Fort Collins, Colorado.

Rosen, R. and R. Manis. 1976. Trematodes of Arkansas amphibians. Journal of Parasitology 62:833–834.

Rosenberg, E.A. and B.A. Pierce. 1995. Effect of initial mass on growth and mortality at low pH in tadpoles of *Pseudacris clarkii* and *Bufo valliceps*. Journal of Herpetology 29:181–185.

Rosenshield, M.L., M.B. Jofre and W.H. Karasov. 1999. Effects of polychlorinated biphenyl 126 on green frog *(Rana clamitans)* and leopard frog *(Rana pipiens)* hatching success, development, and metamorphosis. Environmental Toxicology and Chemistry 18:2478–2486.

Rosenthal, G.M. 1957. The role of moisture and temperature in the local distribution of the plethodontid salamander *Aneides lugubris*. University of California Publications in Zoology 54:371–420.

Ross, B.A., J.R. Tester and W.J. Breckenridge. 1968. Ecology of mima-type mounds in northwestern Minnesota. Ecology 49:172–177.

Ross, D.A. and B.J. Richardson. 1995. Natural history notes: *Rana pretiosa* (spotted frog). Basking behavior. Herpetological Review 26:203.

Ross, D.A, J.K. Reaser, P. Kleeman and D.L. Drake. 1999. Natural history notes: *Rana luteiventris*. (Columbia spotted frog). Mortality and site fidelity. Herpetological Review 30:163.

Ross, D.A, D.L. Shirley, P.A. White and L.D. Lentsch. 1993. Distribution of the spotted frog along the Wasatch Front in Utah, 1991–1992. Utah Division of Wildlife, Resources Publication, Number 93-4, Salt Lake City, Utah.

Ross, D.A, M.C. Stranger, K. McDonald, D.L. Shirley, P.A. White and L.D. Lentsch. 1994. Distribution, habitat use, and relative abundance indices of spotted frogs in the West Desert, Utah, 1993. Utah Division of Wildlife, Resources Publication, Number 93-15, Salt Lake City, Utah.

Ross, D.E., T.C. Esque, R.A. Fridell and P. Hovingh. 1995. Historical distribution, current status, and a range extension of *Bufo boreas* in Utah. Herpetological Review 26:187–189.

Ross, M.R. 1933. A preliminary biological survey of the Ringwood Wild Flower Preserve. Ph.D. dissertation. Cornell University, Ithaca, New York, USA.

Rossi, J.V. 1983. The use of olfactory cues by *Bufo marinus*. Journal of Herpetology 17:72–73.

Rossman, D.A. 1958. A new race of *Desmognathus fuscus* from the south-central United States. Herpetologica 14:158–160.

Rossman, D.A. 1960. Herpetofaunal survey of the Pine Hills area of southern Illinois. Quarterly Journal of the Florida Academy of Science 22:207–225.

Rossman, D.A. 1965a. Rediscovery of the tiger salamander, *Ambystoma tigrinum*, in Louisiana. Proceedings of the Louisiana Academy of Science 27:17–20.

Rossman, D.A. 1965b. The Blue Ridge two-lined salamander, *Eurycea bislineata wilderae*, in southern Alabama. Herpetologica 20:287–288.

Rossman, D.A., N.B. Ford and R.A. Siegel. 1996. The Garter Snakes: Evolution and Ecology. University of Oklahoma Press, Norman, Oklahoma.

Roth, A.H. and J.F. Jackson. 1987. The effect of pool size on recruitment of predatory insects and on mortality in a larval anuran. Herpetologica 43:224–232.

Rothermel, B. 2003. Movement behavior, migratory success, and demography of juvenile amphibians in a fragmented landscape. Ph.D dissertation. University of Missouri, Columbia, Missouri.

Roudebush, R.E. 1988. A behavioral assay for acid sensitivity in two desmognathine species of salamanders. Herpetologica 44:392–395.

Roudebush, R.E. and D.H. Taylor. 1987a. Behavioral interactions between two desmognathine salamander species: importance of competition and predation. Ecology 68:1453–1458.

Roudebush, R.E. and D.H. Taylor. 1987b. Chemical communication between two desmognathine salamanders. Copeia 1987:744–748.

Rous, P. 1967. The challenge to man of the neoplastic cell. Cancer Research 27:1919–1924.

Rouse, J.D., C.A. Bishop and J. Struger. 1999. Nitrogen pollution: an assessment of its threat to amphibian survival. Environmental Health Perspectives 107:799–803.

Roushdy, M.Z. and M. El-Eniani. 1981. A natural population of *Heliosoma duryi* in the River Nile in Egypt. Egyptian Journal of Bilharziasis 8:87–89.

Routman, E. 1993. Population structure and genetic diversity of metamorphic and paedomorphic populations of the tiger salamander, *Ambystoma tigrinum*. Journal of Evolutionary Biology 6:329–357.

Rowe, C.L. and W.A. Dunson. 1993. Relationships among biotic parameters and breeding effort by three amphibians in temporary wetlands of central Pennsylvania. Wetlands 13:237–246.

Rowe, C.L. and W.A. Dunson. 1994. The value of simulated pond communities in mesocosms for studies of amphibian ecology and ecotoxicology. Journal of Herpetology 28:346–356.

Rowe, C.L. and W.A. Dunson. 1995. Impacts of hydroperiod on growth and survival of larval amphibians in temporary ponds of central Pennsylvania. Oecologia 102:397–403.

Rowe, C.L. and J. Freda. 2000. Effects of acidification on amphibians at multiple levels of biological organization. Pp. 545–571. *In* Sparling, D.W., G. Linder and C.A. Bishop (Eds.), Ecotoxicology of Amphibians and Reptiles. Society for Environmental Toxicology and Contaminants (SETAC) Press, Pensacola, Florida.

Rowe, C.L., O.M. Kinney and J.D. Congdon. 1998a. Oral deformities in tadpoles of the bullfrog *(Rana catesbeiana)* caused by conditions in a polluted habitat. Copeia 1998:244–246.

Rowe, C.L., W.J. Sadinski and W.A. Dunson. 1994. Predation on larval and embryonic amphibians by acid-tolerant caddisfly larvae *(Ptilostomis postica)*. Journal of Herpetology 28:357–364.

Rowe, C.L., O.M. Kinney, R.D. Nagle and J.D. Congdon. 1998b. Elevated maintenance costs in an anuran *(Rana catesbeiana)* exposed to a mixture of trace elements during the embryonic and early larval periods. Physiological Zoology 71:27–35.

Rowell, G.A. (Ed.). 1999. A status report on the threats facing populations of perennibranchiate hemidactyliine plethodontid salamanders of the genus *Eurycea* north of the Colorado River in Texas. U.S. Fish and Wildlife Service, Region 2, Final Section 6 Report, Part III, Grant E-1-4, Project 3.4, Section 6 Performance Report, Project E-1-2, Job 3.4, Albuquerque, New Mexico.

Rowe-Rowe, D.T. 1977. Food ecology of otters in Natal, South Africa. Oikos 28:210–219.

Roy, M.S., E. Geffen, D. Smith and R.K. Wayne. 1996. Molecular genetics of pre-1940 red wolves. Conservation Biology 10:1413–1424.

Roy, M.S., E. Geffen, D. Smith, E.A. Ostrander and R.K. Wayne. 1994. Patterns of differentiation and hybridization in North American wolflike canids, revealed by analysis of microsatellite loci. Molecular Biology and Evolution 11:553–570.

Royama, T. 1992. Analytical Population Dynamics. Chapman and Hall, London, England.

Royder, S. 1995a. Letter to Edwin Shellenberger, 31 January.

Royder, S. 1995b. Letter to Claudia Nissley, Advisory Council on Historic Preservation, 13 February.

Royder, S. 1995c. Letter to Roger Kennedy, 13 July.

Royder, S. 1995d. Letter to Roger Kennedy, 28 September.

Royder, S. 1995e. Letter to Nancy Kauffman, 24 October.

Royder, S. 1998. Letter to William Seawell, 29 January.

Rubenstein, N.M. 1969. A study of the salamanders of Mount Cheaha, Cleburne County, Alabama. Journal of Herpetology 3:33–47.

Rubin, D. 1965. Amphibians and reptiles of Vigo County, Indiana. Master's thesis. Indiana State University, Terre Haute, Indiana.

Rubin, D. 1969. Food habits of *Plethodon longicrus* Adler and Dennis. Herpetologica 25:102–105.

Rubin, D. 1971. *Desmognathus aeneus* and *D. wrighti* on Wayah Bald. Journal of Herpetology 5:66.

Rubin, J.A. and A. Boucot. 1989. The origin of the lungless salamanders (Amphibia: Plethodontidae). American Naturalist 134:161–169.

Rubin, J.A., N.L. Reagan, P.A. Verrell and A.J. Boucot. 1993. Plethodontid salamander origins: a response to Beachy and Bruce. American Naturalist 142:1038–1051.

Rubinoff, D. 1996. Natural history notes: *Aneides lugubris* (arboreal salamander). Predation. Herpetological Review 27:135.

Rudolph, D.C. 1978. Aspects of the larval ecology of five plethodontid salamanders of the western Ozarks. American Midland Naturalist 100:141–159.

Rudolph, D.C. 1980. Competition among five species of cave associated salamanders (Family Plethodontidae). Ph.D. dissertation. Texas Tech University, Lubbock, Texas.

Ruffner, B.M. 1933. Practical frog raising. Southern Frog Farms, Jennings, Louisiana.

Ruggiero, L.F., L.L.C. Jones and K.B. Aubry. 1991. Plant and animal habitat associations in Douglas-fir forests of the Pacific Northwest: an overview. Pp. 447–462. *In* Ruggiero, L.F., K.B. Aubry, A.B. Carey and M.H. Huff (Tech. Coords.), Wildlife and Vegetation of Unmanaged Douglas-fir Forests. U.S.D.A. Forest Service, General Technical Report, PNW-GTR-285, Portland, Oregon.

Ruibal, R. 1959. The ecology of a brackish water population of *Rana pipiens*. Copeia 1959:189–195.

Ruibal, R. 1962. The ecology and genetics of a desert population of *Rana pipiens*. Copeia 1962:189–195.

Ruibal, R. and S. Hillman. 1981. Cocoon structure and function in the burrowing hylid frog, *Pternohyla fodiens*. Journal of Herpetology 15:403–408.

Ruibal, R., L. Tevis Jr. and V. Roig. 1969. The terrestrial ecology of the spadefoot toad *Scaphiopus hammondii*. Copeia 1969:571–584.

Ruiz Garcia, F.N. 1987. Anfibios de Cuba. Editorial Gente Nueva, Palacio del Segundo Cabo, O'Reilley Number 4, Habana Vieja, La Habana.

Rundquist, E.M. 1979. The status of *Bufo debilis* and *Opheodrys vernalis* in Kansas. Transactions of the Kansas Academy of Science 82:67–70.

Russell, A.P. and A.M. Bauer. 1993. The Amphibians and Reptiles of Alberta. University of Calgary Press and University of Alberta Press, Calgary, Alberta.

Russell, A.P. and G.L. Powell and D.R. Hall. 1996. Growth and age of Alberta long-toed salamanders (*Ambystoma macrodactylum krausei*): a comparison of two methods of estimation. Canadian Journal of Zoology 74:397–412.

Russell, G.J., T.M. Brooks, M.M. McKinney and C.G. Anderson. 1997. Present and future taxonomic selectivity in bird and mammal extinctions. Conservation Biology 12:1365–1376.

Russell, K.R. and R.L. Wallace. 1992. Occurrence of *Halipegus occidualis* (Digenea: Derogenidae) and other trematodes in *Rana pretiosa* (Anura: Ranidae) from Idaho, USA. Transactions of the American Microscopic Society 111:122–127.

Russell, K.R., T.J. Mabee and M.B. Cole. in press. Distribution and habitat of Columbia torrent salamanders at multiple scales in managed forests of northwestern Oregon. Journal of Wildlife Management 68:in press.

Russell, K.R., A.A. Gonyaw, J.D. Strom, K.E. Diemer and K.C. Murk. 2002. Three new nests of the Columbia torrent salamander, *Rhyacotriton kezeri*, in Oregon with observations of nesting behavior. Northwestern Naturalist 83:19–22.

Russell, R.W. and J.D. Anderson. 1956. A disjunct population of the long-nosed salamander from the coast of California. Herpetologica 12:137–140.

Russell, W.H. 1976. Distribution of troglobitic salamanders in the San Marcos area, Hays Co., Texas. Texas Association for Biological Investigation of Troglobitic *Eurycea*, Report Number 7601, Austin, Texas.

Russell, W.H. 1993. The Buttercup Creek Karst, Travis and Williamson counties, Texas: geology, biology, and land development. Report to the University Speleological Society, University of Texas, Austin, Texas.

Russell-Hunter, W.D. 1978. Ecology of Freshwater Pulmonates. Pp. 335–383. *In* Fretter, A.V. and J. Peake (Eds.), Pulmonates: Systematics, Evolution and Ecology. Volume 2. Academic Press, New York.

Ruth, B.C., W.A Dunson, C.L. Rowe and S.B. Blair. 1993. A molecular and functional evaluation of the egg mass color polymorphism of the spotted salamander, *Ambystoma maculatum*. Journal of Herpetology 27:306–314.

Ruth, F.S. 1961. Seven-legged bullfrog, *Rana catesbeiana*. Turtox News 39:232.

Ruth, S.B. 1974. The current status of the Santa Cruz long-toed salamander—an endangered animal. Herpetological Review 5:27–28.

Ruth, S.B. 1988. The life history and current status of the Santa Cruz long-toed salamander *Ambystoma macrodactylum croceum*. Pp. 89–110. *In* DeLisle, H.F., P.R. Brown, B. Kaufman and B. McGurty (Eds.), Proceedings of the Conference of California Herpetology, Special Publication, Number 4, Southwestern Herpetologists Society, Van Nuys, California.

Ruthven, A.G. 1907. A collection of reptiles and amphibians from southern New Mexico and Arizona. Bulletin of the American Museum of Natural History 23:483–604.

Ruthven, A.G. 1912. Description of a new salamander from Iowa. Proceedings of the U.S. National Museum 41:517–519.

Ruthven, A.G. 1917. On the occurrence of *Bufo fowleri* in Michigan. Occasional Papers of the Museum of Zoology, Number 47, University of Michigan, Ann Arbor, Michigan.

Ruthven, A.G. 1919. The amphibians of the University of Michigan—Walker Expedition to British Guiana. Occasional Papers of the Museum of Zoology, Number 69, University of Michigan, Ann Arbor, Michigan.

Ruthven, A.G. and H.T. Gaige. 1915. The reptiles and amphibians collected in northeastern Nevada by the Walker-Newcomb Expedition of the University of Michigan. Occasional Papers of the Museum of Zoology, Number 8, University of Michigan, Ann Arbor, Michigan.

Ruwaldt, J.J., L.D. Flake and J.M. Gates. 1979. Waterfowl pair use of natural and man-made wetlands in South Dakota. Journal of Wildlife Management 43:375–383.

Ruxton, G.D., W.S. Gurney and A.M. deRoos. 1992. Interference and generation cycles. Theoretical Population Biology 442:235–253.

Ryan, M.J. 1980. The reproductive behavior of the bullfrog *(Rana catesbeiana)*. Copeia 1980:108–114.

Ryan, M.J., K.M. Warkentin, B.E. McClelland and W. Wilczynski. 1995. Fluctuating asymmetries and advertisement call variation in the cricket frog, *Acris crepitans*. Behavioral Ecology 6:124–131.

Ryan, P.G. 1992. Fiscal shrike feeding by plunge diving. Ostrich 63:42.

Ryan, R.A. 1953. Growth rates of some ranids under natural conditions. Copeia 1953:73–80.

Ryan, T.J. 1997. Larva of *Eurycea junaluska* (Amphibia: Caudata: Plethodontidae), with comments on distribution. Copeia 1997:210–215.

Ryan, T.J. 1998a. Larval life history and abundance of a rare salamander, *Eurycea junaluska*. Journal of Herpetology 32:10–17.

Ryan, T.J. 1998b. Natural history notes: *Eurycea junaluska* (Junaluska salamander). Morphology. Herpetological Review 29:163.

Ryan, T.J. and R.C. Bruce. 2000. Life history evolution and adaptive radiation of hemidactyliine salamanders. Pp. 303–326. *In* Bruce, R.C., R.G. Jaeger and L.D. Houck (Eds.), The Biology of Plethodontid Salamanders. Kluwer Academic/Plenum Publishers, New York.

Ryder, O.A. 1986. Species conservation and systematics. The dilemma of subspecies. TRENDS in Ecology and Evolution 1:9–10.

Rye, L.A., W.J. Cook and J.P Bogart. 1997. The value of monitoring genetic diversity: distribution of ambystomatid salamander lineages in Ontario. Pp. 87–92. *In* Green D.M. (Ed.), Amphibians in Decline: Canadian Studies of a Global Problem. Herpetological Conservation, Number 1, Society for the Study of Amphibians and Reptiles, St. Louis, Missouri.

Ryser, J. 1986. Alterstruktur, Geschlechtsverhältnis und Dynamik einer Grasfrosch-Population (*Rana temporaria* L.) aus der Schweiz. Zoologischer Anzeiger 217:234–251.

Ryser, J. 1988. Clutch parameters in a Swiss population of *Rana temporaria*. Herpetological Journal 1:310–311.

Sabath, M.D., W.C. Boughton and S. Easteal. 1981. Expansion of the range of the introduced toad *Bufo marinus* in Australia from 1935 to 1974. Copeia 1981:676–680.

Saber, P.A. and W.A. Dunson. 1978. Toxicity of bog water to embryonic and larval amphibians. Journal of Experimental Zoology 204:33–42.

Saccheri, I., M. Kuussaari, M. Kankare, P. Vikman, W. Fortelius and I. Hanski. 1998. Inbreeding and extinction in a butterfly metapopulation. Nature 392:491–494.

Sachs, L. 1982. Applied Statistics. Springer Verlag, New York, Heidelberg and Berlin.

Sachs, L. 1993. Regulators of normal development and tumor suppression. International Journal of Developmental Biology 37:51–59.

Sadinski, W.J. and W.A. Dunson. 1992. A multilevel study of effects of low pH on amphibians of temporary ponds. Journal of Herpetology 26:413–422.

Saetersdal, M., J.M. Line and H.J.B. Birks. 1993. How to maximize biological diversity in nature reserve selections: vascular plants and breeding birds in deciduous woodlands of western Norway. Biological Conservation 66:131–138.

Sage, R.D. and R.K. Selander. 1979. Hybridization between species of the *Rana pipiens* complex in central Texas. Evolution 33:1069–1088.

Sagoff, M. 1985. Fact and value in environmental science. Environmental Ethics 7:99–116.

Sagor, E.S., M. Ouellet, E. Barten and D.M. Green. 1998. Skeletochronology and geographic variation in age structure in the wood frog, *Rana sylvatica*. Journal of Herpetology 32:469–474.

Sait, S.M., M. Begon and D.J. Thompson. 1994. Long-term population dynamics of the Indian meal moth *Plodia interpunctella* and its granulosis virus. Journal Animal Ecology 63:861–870.

Salt, G.W. 1952. The belltoad, *Ascaphus truei*, in Mendocino County, California. Copeia 1952:193–194.

Saltarelli, W.A. 1977. Parasites of *Eurycea lucifuga* (Rafinesque), *E. longicauda melanopleura* (Green), and *E. multiplicata* (Cope) (Amphibia: Plethodontidae) from northwestern Arkansas. Master's thesis. University of Arkansas, Fayetteville, Arkansas.

Salthe, S.N. 1963. The egg capsules in the Amphibia. Journal of Morphology 113:161–171.

Salthe, S.N. 1969. Reproductive modes and the number and sizes of ova in urodeles. American Midland Naturalist 81:467–490.

Salthe, S.N. 1973a. *Amphiuma means*. Pp. 148.1–148.2. Catalogue of American Amphibians and Reptiles. Society for the Study of Amphibians and Reptiles, St. Louis, Missouri.

Salthe, S.N. 1973b. *Amphiuma tridactylum*. Pp. 149.1–149.3. Catalogue of American Amphibians and Reptiles. Society for the Study of Amphibians and Reptiles, St. Louis, Missouri.

Salthe, S.N. and J.S. Mecham. 1974. Reproductive and courtship patterns. Pp. 309–421. *In* Lofts, B. (Ed.), Physiology of the Amphibia. Volume 2. Academic Press, New York.

Salwasser, H., C. Schonewald-Cox and R. Baker. 1987. The role of interagency cooperation in managing for viable populations. Pp. 147–173. *In* Soulé, M.E. (Ed.), Viable Populations. Cambridge University Press, New York.

Samollow, P.B. 1980. Selective mortality and reproduction in a natural population of *Bufo boreas*. Evolution 34:18–39.

Sampson, F.B. and F.L. Knopf (Eds.). 1996. Ecosystem Management: Selected Readings. Springer-Verlag, New York.

Sanders, H.O. 1970. Pesticide toxicities to tadpoles of the western chorus frog *Pseudacris triseriata* and Fowler's toad *Bufo woodhousii fowleri*. Copeia 1970:246–251.

Sanders, O. 1953. A new species of toad, with a discussion of morphology of the bufonid skull. Herpetologica 9:25–47.

Sanders, O. 1961. Indications for the hybrid origin of *Bufo terrestris* Bonnaterre. Herpetologica 17:145–156.

Sanders, O. 1973. A new leopard frog (*Rana berlandieri brownorum*) from southern Mexico. Journal of Herpetology 7:87–92.

Sanders, O. 1978. *Bufo woodhousei* in central Texas. Bulletin of the Maryland Herpetological Society 14:55–66.

Sanders, O. 1986. The heritage of *Bufo woodhousei* Girard in Texas (Salientia: Bufonidae). Occasional Papers of the Strecker Museum, Number 1, Baylor University, Waco, Texas.

Sanders, O.E. 1987. Evolutionary hybridization and speciation in North American indigenous bufonids. Privately published, Dallas, Texas.

Sanders, O. and H.M. Smith. 1951. Geographic variation in toads of the *debilis* group of *Bufo*. Field and Laboratory 19:141–160.

Sanders, O. and H.M. Smith. 1971. Skin tags and ventral melanism in the Rio Grande leopard frog. Journal of Herpetology 51:31–38.

Sanders, R.M. 1950. A herpetological survey of Ventura County, California. Master's thesis. Stanford University, Stanford, California.

Sangster, G. 2000. Taxonomic stability and avian extinctions. Conservation Biology 14:579–581.

Sansom, A. 1993. Letter to Tom Dureka, 12 August.

Sansom, A. 1995. Letter to Sam Hamilton, 9 February, 1995.

Santiago, J. 1999. Influences of relative humidity, soil and air temperatures, and lunar phase on occurrence of terrestrial plethodontid salamanders at high elevation sites. Master's thesis. Marshall University, Huntington, West Virginia.

Santos-Barrera, G. 1994. Opalinid protozoans of anurans from Los Tuxtlas, Veracruz, Mexico. Anales del Instituo de Biología, Universidad Nacional Autonoma de Mexico (Seria Zoología) 65:191–193.

Sarkar, S. 1996. Ecological theory and anuran declines. BioScience 46:199–207.

Sartorius, S.S. and R.C. Rosen. 2000. Breeding phenology of the lowland leopard frog *(Rana yavapaiensis)*: implications for conservation and ecology. Southwestern Naturalist 45:267–273.

Sattler, P. and N. Reichenbach. 1998. The effects of timbering on *Plethodon hubrichti*: short-term effects. Journal of Herpetology 32:399–404.

Sattler, P.W. 1980. Genetic relationships among selected species of North American *Scaphiopus*. Copeia 1980:605–610.

Sattler, P.W. 1985. Introgressive hybridization between the spadefoot toads *Scaphiopus bombifrons* and *S. multiplicatus* (Salentia: Pelobatidae). Copeia 1985:324–332.

Sauer, J.R., B.G. Peterjohn and W.A. Link. 1994. Observer differences in the North American Breeding Bird Survey. Auk 111:50–62.

Sauer, J.R., S. Schwartz and B. Hoover. 1996. The Christmas Bird Count home page. Version 95.1. Patuxent Wildlife Research Center, Laurel, Maryland. http://www.mbr.nbs.gov/bbs/cbc.html.

Sauerbier, W., L.A. Rollins-Smith, D.L. Carlson, C.S. Williams, J.W. Williams III and R.G. McKinnell. 1995. Sizing of the Lucké tumor herpesvirus genome by field inversion electrophoresis and restriction analysis. Herpetopathologia 2:137–143.

Saugey, D.A., G.A. Heidt and D.R. Heath. 1985. Summer use of abandoned mines by the Caddo Mountain salamander, *Plethodon caddoensis* (Plethodontidae) in Arkansas. Southwestern Naturalist 30:318–319.

Saugey, D.A., D.G. Saugey and N.H. Douglas. 1986. Geographic distribution: *Ambystoma texanum* (smallmouth salamander). Herpetological Review 17:49.

Saumure, R.A. 1993. Life history notes: *Rana catesbeiana* (bullfrog). Predation. Herpetological Review 24:30–31.

Saumure, R.A. and J.S. Doody. 1998. Natural history notes: *Amphiuma tridactylum* (three-toed amphiuma). Ectoparasites. Herpetological Review 29:163.

Saunders, D.A. 1996. Habitat networks in the fragmented landscape of the Western Australian wheatbelt: preliminary results, involvement with landcare groups, and experience in implementation. Pp. 69–80. *In* Settele, J., C. Margules, P. Poschlod and K. Henle (Eds.), Species Survival in Fragmented Landscapes. Kluwer Academic Publishers, Dordrecht, Holland.

Saunders, D.A., R.J. Hobbs and P.R. Ehrlich. 1993. Nature Conservation 3: The Reconstruction of Fragmanted Ecosystems. Surrey Beatty and Sons, Sydney, New South Wales, Australia.

Saunders, D.A., R.J. Hobbs and C.R. Margules. 1991. Biological consequences of ecosystem fragmentation: a review. Conservation Biology 5:18–32.

Saunders, R. 1994. Letter to Ann Mesrobian and all petition signatories, 12 October.

Savage, J.M. 1954. A revision of the toads of the *Bufo debilis* complex. Texas Journal of Science 6:83–112.

Savage, J.M. 1960a. Geographic variation in the tadpole of the toad, *Bufo marinus*. Copeia 1960:233–236.

Savage, J.M. 1960b. A preliminary biosystematic analysis of toads of the *Bufo boreas* group in Nevada and California. Yearbook of the American Philosophical Society 1959:251–254.

Savage, J.M. 1967. The systematics of the toad *Bufo politus* Cope, 1862. Copeia 1967:225–226.

Savage, J.M. 1982. The enigma of the Central American herpetofauna: dispersals or vicariance? Annals of the Missouri Botanical Garden 69:464–547.

Savage, J.M. and F.W. Schuierer. 1961. The eggs of toads of the *Bufo boreas* group, with descriptions of the eggs of *Bufo exsul* and *Bufo nelsoni*. Bulletin of the Southern California Academy of Science 60:93–99.

Savage, J.M. and J. Villa. 1986. Herpetofauna of Costa Rica. Contributions to Herpetology, Number 3, Society for the Study of Amphibians and Reptiles, St. Louis, Missouri.

Savage, R.M. 1963. A speculation on the pallid tadpoles of *Xenopus laevis*. British Journal of Herpetology 3:74–76.

Sawyer, R.T. 1972. North American freshwater leeches, exclusive of the Piscicolidae, with a key to all species. Illinois Biological Monographs, Number 46, University of Illinois, Urbana, Illinois.

Sawyer, R.T. and R.H. Shelley. 1976. New records and species of leeches (Annelida: Hirudinea) from North and South Carolina. Journal of Natural History 10:65–97.

Say, T. 1823. *In* James, E. Account of an expedition from Pittsburgh to the Rocky Mountains, performed in the years 1819 and '20, by order of the Honorable J.C. Calhoun, Secretary of War, under the command of Major Stephen H. Long. H.C. Carey and I. Lea. Volume 2. Philadelphia, Pennsylvania.

Sayler, A. 1966. The reproductive ecology of the red-backed salamander, *Plethodon cinereus*, in Maryland. Copeia 1966:183–193.

Scarlett, H. 1977a. Toads may cut FHA's backing of area growth. Houston Post, 21 June.

Scarlett, H. 1977b. Expert has plan—toad compromise possible. Houston Post, 22 June.

Scarlett, H. 1977c. Home builders to help save toad, group's president says. Houston Post, 23 June.

Schaaf, R.T., Jr. 1969. Toxicity of *Rana palustris*. Journal of Herpetology 3:198 (abstract).

Schaaf, R.T. and J.S. Garton. 1970. Raccoon predation on the American toad *Bufo americanus*. Herpetologica 26:334–335.

Schaaf, R.T. and P.W. Smith. 1970. Geographic variation in the pickerel frog. Herpetologica 26:240–254.

Schaaf, R.T. and P.W. Smith. 1971. *Rana palustris* LeConte. Pp. 117.1–117.3. Catalogue of American Amphibians and Reptiles. Society for the Study of Amphibians and Reptiles, St. Louis, Missouri.

Schad, G.A. 1960. The genus *Thelandros* (Nematoda: Oxyuroidea) in North American salamanders, including a description of *Thelandros salamandrae* n. sp. Canadian Journal of Zoology 38:115–120.

Schardien, B.J. and J.A. Jackson. 1982. Kildeers feeding on frogs. Wilson Bulletin 94:85–87.

Schaub, D.L. and J.H. Larsen Jr. 1978. The reproductive ecology of the Pacific treefrog (*Hyla regilla*). Herpetologica 34:409–416.

Schell, S.C. 1962. Development of the sporocyst generations of *Glypthelmins quieta* (Stafford 1900) (Trematoda: Plagiorchioidea), a parasite of frogs. Journal of Parasitology 48:387–394.

Schell, S.C. 1964. *Bunoderella metteri* gen. and sp. n. (Trematoda: Allocreadiidae) and other trematode parasites of *Ascaphus truei* Stejneger. Journal of Parasitology 50:652–655.

Schell, S.C. 1965. The life history of *Haematoloechus breviplexus* Stafford, 1902 (Trematoda: Haplometridae McMullen, 1937), with emphasis on the development of the sporocysts. Journal of Parasitology 51:587–593.

Schell, S.C. 1985. Handbook of Trematodes of North America. University Press of Idaho, Moscow, Idaho.

Scherff-Norris, K.L. 1999. Final report: Experimental reintroduction of boreal toads, *Bufo boreas boreas*. Colorado Division of Wildlife, Denver, Colorado.

Scheuer, J.H. 1993. Biodiversity: beyond Noah's Ark. Conservation Biology 7:206–207.

Schindler, D.W., P.J. Curtis, B.R. Parker and M.P. Stainton. 1996. Consequences of climate warming and lake acidification for UV-B penetration in North American boreal lakes. Nature 379:705–708.

Schlefer, E.K., M.A. Romano, S.I. Guttman and S.B. Ruth. 1986. Effects of twenty years of hybridization in a disturbed habitat on *Hyla cinerea* and *Hyla gratiosa*. Journal of Herpetology 20:210–221.

Schlichter, L.C. 1981. Low pH affects the fertilization and development of *Rana pipiens* eggs. Canadian Journal of Zoology 59:1693–1699.

Schlüpmann, M. 1981. Grasfrosch-*Rana t. temporaria* Linnaeus, 1758. Abhandlungen aus dem Landesmuseum für Naturkunde zu Münster in Westfalen 43:103–112.

Schlüpmann, M. 1982. Bestand, Lebensraum und Lebensweise der Erdkröte (*Bufo bufo*) im Hohenlimburger Raum (MTB: 4611). Beobachtungen bis 1980. Natur und Heimat 42:65–81.

Schlupp, I. and R. Podloucky. 1994. Changes in breeding site fidelity: a combined study of conservation and behaviour in the common toad *Bufo bufo*. Biological Conservation 69:285–291.

Schmid, W.D. 1965a. High temperature tolerances of *Bufo hemiophrys* and *Bufo cognatus*. Ecology 46:559–560.

Schmid, W.D. 1965b. Some aspects of the water economies of nine species of amphibians. Ecology 46:261–269.

Schmid, W.D. 1982. Survival of frogs in low temperature. Science 215:697–698.

Schmidt, G.D. 1992. Essentials of Parasitology. W.C. Brown Publishers, Dubuque, Iowa.

Schmidt, K.A. and B. Fried. 1997. Prevalence of larval trematodes in *Helisoma trivolvis* (Gastropoda) from a farm pond in Northampton County, Pennsylvania with special emphasis on *Echinostoma trivolvis* (Trematoda) cercariae. Journal of the Helminthological Society of Washington 64:157–159.

Schmidt, K.P. 1953. A Checklist of North American Amphibians and Reptiles. Sixth edition. American Society of Ichthyologists and Herpetologists. University of Chicago Press, Chicago, Illinois.

Schmidt, K.P. and W.L. Necker. 1935. Amphibians and reptiles of the Chicago region. Bulletin Chicago Academy of Science 5:57–77.

Schmidt, K.P. and T.F. Smith. 1944. Amphibians and reptiles of the Big Bend region of Texas. Field Museum of Natural History, Zoological Series 29:75–96.

Schmidt, R.S. 1971. Neural mechanisms of mating call orientation in female toads (*Bufo americanus*). Copeia 1971:545–548.

Schneider, J.G. 1792. Beschreibung und Abbildung einer neuen Art von Wasserschildkrote. Schriften Gesellschaft Naturforschender Freunde Berlin 10:259–283.

Schneider, S.H and T.L. Root. 1998. Climate change. Pp. 89–105, 108–116. *In* Mac, M.J., P.A. Opler, C.E. Puckett Haecker, P.D. Doran (Eds.), Status and Trends of the Nation's Biological Resources. Volume 1. U.S. Department of the Interior, U.S. Geological Survey, Reston, Virginia.

Schoener, T.W. 1972. Mathematical ecology and its place among the sciences. Science 178:389–391.

Schoener, T.W. 1983. Field experiments on interspecific competition. American Naturalist 122:240–285.

Schoener, T.W. 1985. Some comments on Connell's and my reviews of field experiments on interspecific competition. American Naturalist 125:730–740.

Schoener, T.W. 1986. Overview: kinds of ecological communities—ecology becomes pluralistic. Pp. 467–479. In Diamond, J. and T.J. Case (Eds.), Community Ecology. Harper and Row, New York.

Schoener, T.W. and D.A. Spiller. 1992. Is extinction related to temporal variability in population size? An empirical answer for orb spiders. American Naturalist 139:1176–1207.

Schoenherr, A.A. 1976. The herpetofauna of the San Gabriel Mountains, Los Angeles County, California. Special Publication of the Southwestern Herpetologists Society, Van Nuys, California.

Schoenherr, A.A. 1992. A Natural History of California. University of California Press, Berkeley, California.

Schoenherr, A.A., C.R. Feldmeth and M.J. Emerson. 1999. Natural History of the Islands of California. University of California Press, Berkeley, California.

Schonberger, C.F. 1944. Food of salamanders in the northwestern United States. Copeia 1944:257.

Schonberger, C.F. 1945. Food of some amphibians and reptiles of Oregon and Washington. Copeia 1945:120–121.

Schonewald-Cox, C.M. 1983. Conclusions: guidelines to management: a beginning attempt. Pp. 141–445. In Schonewald-Cox, C.M., S.M. Chambers, B. MacBryde and W.L. Thomas (Eds.), Genetics and Conservation: A Reference for Managing Wild Plant and Animal Populations. Benjamin/Cummings, Menlo Park, California.

Schoonbee, H.J., V.S. Nakani and J.F. Prinsloo. 1979. The use of cattle manure and supplementary feeding in growth studies of the Chinese silver carp in Transkei. South African Journal of Science 75:489–495.

Schoonbee, H.J., J.F. Prinsloo and J.G. Nxiweni. 1992. Observations on the feeding habits of larvae, juvenile and adult stages of the African clawed frog Xenopus laevis in impoundments in Transkei. Water South Africa 18:227–236.

Schotthoefer, A.M., R.A. Cole and V.R. Beasley. 2003. Relationship of tadpole stage to location of echinostome cercariae encystment and the consequences for tadpole survival. Journal of Parasitology 89:475–482.

Schramm, M. 1986. Control of Xenopus laevis (Amphibia: Pipidae) in fish ponds with observations on its threat to fish fry and fingerlings. Water South Africa 13:53–56.

Schreber, H.1782. Der Naturforscher. Volume 18. Johann Jacob Gebaur, Halle.

Schreiber, R.K. and J.R. Newman. 1988. Acid precipitation effects in forest habitats: implications for wildlife. Conservation Biology 2:249–259.

Schroeder, E.E. 1968a. Aggressive behavior in Rana clamitans. Journal of Herpetology 1:95–96.

Schroeder, E.E. 1968b. Movements of subadult green frogs, Rana clamitans. Journal of Herpetology 1:119.

Schroeder, E.E. and T.S. Baskett. 1968. Age estimation, growth rates, and population structure in Missouri bullfrogs. Copeia 1968:583–592.

Schueler, F.W. 1975. Geographic variation in the size of Rana septentrionalis in Quebec, Ontario, and Manitoba. Journal of Herpetology 9:177–185.

Schueler, F.W. 1987. Natural history notes: Rana septentrionalis (mink frog). Terrestrial activity. Herpetological Review 18:72.

Schueler, F.W. 1999. Rapport sur la situation de la rainette faux-grillon de l'ouest (Pseudacris triseriata) au Québec. Canadian Field-Naturalist 113:699.

Schuette, B. 1980. Geographic distribution: Ambystoma maculatum (spotted salamander). Herpetological Review 11:114.

Schuierer, F.W. 1961. Remarks on the natural history of Bufo exsul Myers, the endemic toad of Deep Springs Valley, Inyo County, California. Herpetologica 17:260–265.

Schuierer, F.W. 1962. Notes on two populations of Bufo exsul Myers and a commentary on speculation within the Bufo boreas Group. Herpetologica 18:262–267.

Schuierer, F.W. 1972. The current status of the endangered species Bufo exsul Myers, Deep Springs Valley, Inyo County, California. Herpetological Review 4:81.

Schulz, T.F. 1998. Kaposi's sarcoma-associated herpesvirus (human herpesvirus 8). Journal of General Virology 79:1573–1591.

Schuytema, G.S. and A.V. Nebeker. 1998. Comparative toxicity of diuron on survival and growth of Pacific treefrog, bullfrog, red-legged frog, and African clawed frog embryos and tadpoles. Archives of Environmental Contamination and Toxicology 34:370–376.

Schuytema, G.S. and A.V. Nebeker. 1999a. Comparative toxicity of ammonium and nitrate compounds to Pacific treefrog and African clawed frog tadpoles. Environmental Toxicology and Chemistry 18:2251–2257.

Schuytema, G.S. and A.V. Nebeker. 1999b. Effects of ammonium nitrate, sodium nitrate, and urea on red-legged frogs, Pacific treefrogs, and African clawed frogs. Bulletin of Environmental Contamination and Toxicology 63:357–364.

Schwalbe, C.R. and P.C. Rosen. 1988. Preliminary report on effect of bullfrogs on wetland herpetofauna in southeastern Arizona. Pp. 166–173. In Szaro, R.C., K.E. Severson and D.R. Patton (Eds.), Proceedings of the Symposium on Management of Amphibians, Reptiles and Small Mammals in North America. U.S.D.A. Forest Service, General Technical Report, RM-166, Fort Collins, Colorado.

Schwalbe, C.R. and P.C. Rosen. 1999. Bullfrogs–the dinner guests we're sorry we invited. Sonorensis 19:8–10.

Schwalbe, C., B. Alberti and M. Gilbert. 1997. The limestone troll. Bajada 5:1.

Schwaner, T.D. and J.M. Anderson. 1991. Geographic distribution: Ambystoma maculatum (spotted salamander). Herpetological Review 22:62.

Schwaner, T.D. and R.H. Mount. 1970. Notes on the distribution, habits, and ecology of the Red Hills salamander Phaeognathus hubrichti Highton. Copeia 1970:571–573.

Schwaner, T.D., D.R. Hadley, K.R. Jenkins, M.P. Donovan and B. Al Tait. 1998. Population dynamics and life history studies of toads: short-and long-range needs for understanding amphibian populations in southern Utah. Pp. 197–202. In Hill, L.M. (Ed.), Learning from the land: Grand Staircase-Escalante National Monument Science Symposium Proceedings. U.S. Department of the Interior, Bureau of Land Management, BLM/UT/GI–98006+1220, Salt Lake City, Utah.

Schwartz, A. 1952. Hyla septentrionalis Duméril and Bibron on the Florida mainland. Copeia 1952:117–118.

Schwartz, A. 1955. The chorus frog Pseudacris brachyphona in North Carolina. Copeia 1955:138.

Schwartz, A. 1957a. Chorus frogs (Pseudacris nigrita LeConte) in South Carolina. American Museum Novitates, Number 1838, American Museum of Natural History, New York.

Schwartz, A. 1957b. Albinism in the salamander Amphiuma means. Herpetologica 13:75–76.

Schwartz, A. 1974. Eleutherodactylus planirostris (Cope). Pp. 154.1–154.4. Catalogue of American Amphibians and Reptiles. Society for the Study of Amphibians and Reptiles, St. Louis, Missouri.

Schwartz, A. and W.E. Duellman. 1952. The taxonomic status of the newts, Diemictylus viridescens, of peninsular Florida. Bulletin of the Chicago Academy of Sciences 9:219–227.

Schwartz, A. and R. Etheridge. 1954. New and additional herpetological records from the North Carolina Coastal Plain. Herpetologica 10:167–171.

Schwartz, A. and R. W. Henderson. 1985. A guide to the identification of the amphibians and reptiles of the West Indies exclusive of Hispaniola. Milwaukee Public Museum, Milwaukee, Wisconsin.

Schwartz, A. and R.W. Henderson. 1991. Amphibians and Reptiles of the West Indies Descriptions, Distributions, and Natural History. University of Florida Press, Gainesville, Florida.

Schwartz, A. and R. Thomas. 1975. A check list of West Indian amphibians and reptiles. Carnegie Museum of Natural History, Special Publication, Number 1, Pittsburgh, Pennsylvania.

Schwartz, J.J. 1989. Graded aggressive calls of the spring peeper, Pseudacris crucifer. Herpetologica 45:172–181.

Schwartz, J.J., M.A. Bee and S.D. Tanner. 2000. A behavioral and neurobiological study of the responses of gray treefrogs, Hyla versicolor, to the calls of a predator, Rana catesbeiana. Herpetologica 56:27–37.

Schwartz, J.S. 1976. A biological study of Cathedral Cave Crawford County, Missouri. Journal of the Missouri Speleological Society 16:1–21.

Scott, A.F. and J. Koons. 1993a. Geographic distribution: Hyla cinerea (green treefrog). Herpetological Review 24:64.

Scott, A.F. and J. Koons. 1993b. Geographic distribution: Hyla cinerea (green treefrog). Herpetological Review 24:64.

Scott, A.F. and R.M. Johnson. 1972. Geographic distribution: *Ambystoma texanum*. Herpetological Review 4:93.

Scott, A.F., S. Sutton and H. Sutton. 2001. Geographic distribution: *Ambystoma maculatum* (spotted salamander). Herpetological Review 32:267.

Scott, D.E. 1990. Effects of larval density in *Ambystoma opacum*: an experiment in large-scale field enclosures. Ecology 71:296–306.

Scott, D.E. 1993. Timing in reproduction of paedomorphic and metamorphic *Ambystoma talpoideum*. American Midland Naturalist 129:397–402.

Scott, D.E. 1994. The effect of larval density on adult demographic traits in *Ambystoma opacum*. Ecology 75:1383–1396.

Scott, D.E. 1999. An amphibian's eye-view of wetlands. Savannah River Ecology Laboratory Outreach Publication, Number 3, Aiken, South Carolina.

Scott, D.E. and M.R. Fore. 1995. The effect of food limitation on lipid levels, growth, and reproduction in the marbled salamander, *Ambystoma opacum*. Herpetologica 51:462–471.

Scott, J.M., B. Csuti and K.A. Smith. 1990. Playing Noah while paying the devil. Bulletin of the Ecological Society of America 71:156–159.

Scott, J.M., T. Tear and F.W. Davis. 1996. Gap Analysis: A Landscape Approach to Biodiversity Planning. American Society of Photogrannetry and Remote Sensing, Bethesda, Maryland.

Scott, J.M., F. Davis, B. Csuti, R. Noss, B. Butterfield, C. Groves, H. Anderson, S. Caicco, F. D'Erchia, T.C. Edwards, J. Ulliman and R.G. Wright. 1993. Gap analysis: a geographical approach to protection of biological diversity. Wildlife Monographs 123:1–41.

Scott, N.J., Jr. 1990. Studies of the biology of the Sacramento Mountain salamander, *Aneides hardii*. Annual report submitted to the U.S. Forest Service, Alamogordo, New Mexico.

Scott, N.J., Jr. 1992. Ranid frog survey of the Gray Ranch with recommendations for management of frog habitats, August 1990–September 1991. Report to Gray Ranch, Animas, New Mexico.

Scott, N.J., Jr. 1993. Postmetamorphic death syndrome. Froglog: Newsletter of the Declining Amphibian Populations Task Force 7:1–2.

Scott, N.J., Jr. and R.D. Jennings. 1985. The tadpoles of five species of New Mexican leopard frogs. Museum of Southwestern Biology, Occasional Papers, Number 3, University of New Mexico, Albuquerque, New Mexico.

Scott, N.J., Jr. and C.A. Ramotnik. 1992. Does the Sacramento Mountain salamander require old-growth forests? Pp. 170–178. *In* Kaufman, M.R., W.H. Moir and B.L. Bassett (Eds.), Old-Growth Forests in the Southwest and Rocky Mountain Regions: Proceedings of a workshop. U.S.D.A. Forest Service, General Technical Report, RM-213, Portal, Arizona.

Scott, N.J., C.A. Ramotnik, M.J. Altenbach and B.E. Smith. 1987. Distribution and ecological requirements of endemic salamanders in relation to forestry management: Summary of 1987 activities. Part 1: Lincoln National Forest. U.S.D.A. Forest Service, Final report, Albuquerque, New Mexico.

Scribner, K.T., J.W. Arntzen and T. Burke. 1997. Effective number of breeding adults in *Bufo bufo* estimated from age-specific variation at mini-satellite loci. Molecular Ecology 6:701–712.

Scroggin, J.B. and W.B. Davis. 1956. Food habits of the Texas dwarf siren. Herpetologica 12:231–237.

Scudday, J.F. 1996. Documentation of the amphibians and reptiles of Big Bend Ranch State Park with notes on their biogeographical and ecological affinities and their present status. Texas Parks and Wildlife Department, Unpublished report, Austin, Texas.

Seabloom, R.W., R.D. Crawford and M.G. McKenna. 1978. Vertebrates of southwestern North Dakota: Amphibians, Reptiles, Birds, Mammals. Institute for Ecological Studies, University of North Dakota, Grand Forks, North Dakota.

Seal, U.S. (Ed.). 1994. Population and habitat viability assessment. Houston toad, *Bufo houstonensis*. Report of workshop conducted by Conservation Breeding Specialist Group, Species Survival Commission, IUCN—The World Conservation Union, in partial fulfillment of U.S.F.W.S. Contract Number 94–172, Austin, Texas.

Seale, D.B. 1980. Influence of amphibian larvae on primary production, nutrient flux, and competition in a pond ecosystem. Ecology 61:1531–1550.

Seale, D.B. 1982a. Obligate and facultative suspension feeding in anuran larvae: feeding regulation in *Xenopus* and *Rana*. Biology Bulletin 162:214–231.

Seale, D.B. 1982b. Physical factors influencing oviposition by the wood frog, *Rana sylvatica*, in Pennsylvania. Copeia 1982:627–635.

Seale, D.B. and N. Beckvar. 1980. The comparative ability of anuran larvae (genera: *Hyla, Bufo* and *Rana*) to ingest suspended blue-green algae. Copeia 1980:495–503.

Seawell, W. 1998. Letter to Scott Royder, 10 February.

Seburn, C.N.L. 1992. The status of amphibian populations in Saskatchewan. Pp. 17–18. *In* Bishop, C.A. and K.E. Pettit (Eds.), Declines in Canadian Amphibian Populations: Designing a National Monitoring Strategy. Canadian Wildlife Service, Occasional Paper, Number 76, Ottawa, Ontario, Canada.

Seburn, C.N.L., D.C. Seburn and C.A. Paszkowski. 1997. Northern leopard frog *(Rana pipiens)* dispersal in relation to habitat. Pp. 64–72. *In* Green, D.M. (Ed.), Amphibians in Decline: Canadian Studies of a Global Problem. Herpetological Conservation, Number 1, Society for the Study of Amphibians and Reptiles, St. Louis, Missouri.

Secor, S.M. 1988. Perch sites of calling male bird-voiced treefrogs, *Hyla avivoca*, in Oklahoma. Proceedings of the Oklahoma Academy of Sciences 68:71–73.

Secretario de Desarrollo Social. 1994. NORMA Oficial Mexicana NOM-059-ECOL-1994, que determina las especies y subespecies de flore y fauna silvestres terrestres y acuaticas en peligro de extincion, amenasadas, raras y las sujetas a proteccion especial, y que establece especificaciones para su proteccion. Diaro Oficial, Lunes 16 de Mayo de 1994.

Seeman, S.E. 1996. *Desmognathus welteri* and *Desmognathus quadramaculatus*: a distributional survey and microhabitat comparison in southern West Virginia. Master's thesis. Kent State University, Kent, Ohio.

Seibert, H.C. and R.A. Brandon. 1960. The salamanders of southeastern Ohio. Ohio Journal of Science 60:291–303.

Seifert, W. 1978. Geographic distribution: *Bufo americanus charlesmithi* (dwarf American toad). Herpetological Review 9:61.

Seigel, R.A. and C.K. Dodd Jr. 2000. Manipulation of turtle populations for conservation: half-way technologies or viable options? Pp. 218–238. *In* Klemens, M. (Ed.), Turtle Conservation. Smithsonian Institution Press, Washington, D.C.

Seigel, R.A. and C.K. Kennedy. 2000. Demography and reproduction of gopher frogs in Mississippi. U.S. Fish and Wildlife Service, Unpublished report, Jackson, Mississippi.

Seim, S.G. and M.J. Sredl. 1994. Lowland leopard frog micro-habitat use in Arizona. Annual Meeting of the Arizona and New Mexico Chapters of the Wildlife Society, Sierra Vista, Arizona.

Sekerak, C.M., G.W. Tanner and J.G. Palis. 1996. Ecology of flatwoods salamander larvae in breeding ponds in Apalachicola National Forest. Proceedings of the Annual Conference of the Southeastern Association of Fish and Wildlife Agencies 50:321–330.

Selby, M.F., S.C. Winkel and J.W. Petranka. 1996. Geographic uniformity in agonistic behaviors of Jordan's salamander. Herpetologica 52:108–115.

Semlitsch, R.D. 1980a. Growth and metamorphosis of larval dwarf salamander *(Eurycea quadridigitata)*. Herpetologica 36:6–16.

Semlitsch, R.D. 1980b. Geographic and local variation in population parameters of the slimy salamander *Plethodon glutinosus*. Herpetologica 36:6–16.

Semlitsch, R.D. 1981. Terrestrial activity and summer home range of the mole salamander *(Ambystoma talpoideum)*. Canadian Journal of Zoology 59:315–322.

Semlitsch, R.D. 1983a. Structure and dynamics of two breeding populations of the eastern tiger salamander, *Ambystoma tigrinum*. Copeia 1983:608–616.

Semlitsch, R.D. 1983b. Terrestrial movements of an eastern tiger salamander, *Ambystoma tigrinum*. Herpetological Review 14:112–113.

Semlitsch, R.D. 1983c. Burrowing ability and behavior of salamanders of the genus *Ambystoma*. Canadian Journal of Zoology 61:616–620.

Semlitsch, R.D. 1983d. Growth and metamorphosis of larval red salamanders *(Pseudotriton ruber)* on the coastal plain of South Carolina. Herpetologica 39:48–52.

Semlitsch, R.D. 1985a. Analysis of climatic factors influencing migrations of the salamander *Ambystoma talpoideum*. Copeia 1985:477–489.

Semlitsch, R.D. 1985b. Reproductive strategy of a facultatively paedomorphic salamander *Ambystoma talpoideum*. Oecologia 65:305–313.

Semlitsch, R.D. 1987. Relationship of pond drying to the reproductive success of the salamander *Ambystoma talpoideum*. Copeia 1987:61–69.

Semlitsch, R.D. 1988. Allotopic distribution of two salamanders: effects of fish predation and competitive interactions. Copeia 1988:290–298.

Semlitsch, R.D. 1990. Effects of body size, sibship, and tail injury on the susceptibility of tadpoles to dragonfly predation. Canadian Journal of Zoology 68:1027–1030.

Semlitsch, R.D. 1993. Effects of different predators on the survival and development of tadpoles from the hybridogenetic *Rana esculenta* complex. Oikos 67:40–46.

Semlitsch, R.D. 1998. Biological delineation of terrestrial buffer zones for pond breeding salamanders. Conservation Biology 12:1113–1119.

Semlitsch, R.D. 2000a. Size does matter: the value of small isolated wetlands. National Wetlands Newsletter, January–February 200:5–6,13.

Semlitsch, R.D. 2000b. Principles for management of aquatic-breeding amphibians. Journal of Wildlife Management 64:615–631.

Semlitsch, R.D. and J.R. Bodie. 1998. Are small, isolated wetlands expendable? Conservation Biology 12:1129–1133.

Semlitsch, R.D. and J.P. Caldwell. 1982. Effects of density on growth, metamorphosis, and survivorship of tadpoles of *Scaphiopus holbrooki*. Ecology 63:905–911.

Semlitsch, R.D. and J.W. Gibbons. 1985. Phenotypic variation in metamorphosis and paedomorphosis in the salamander *Ambystoma talpoideum*. Ecology 66:1123–1130.

Semlitsch, R.D. and J.W. Gibbons. 1988. Fish predation in size-structured populations of tree frog tadpoles. Oecologia 75:321–326.

Semlitsch, R.D. and J.W. Gibbons. 1990. Effects of egg size on success of larval salamanders in complex aquatic environments. Ecology 71:1789–1795.

Semlitsch, R.D. and J.B. Jensen. 2001. Core habitat, not buffer zone. National Wetlands Newsletter, July–August 2001:5–6,11.

Semlitsch, R.D. and M.A. McMillian. 1980. Breeding migrations, population size structure, and reproduction of the dwarf salamander, *Eurycea quadridigitata*, in South Carolina. Brimleyana 3:97–105.

Semlitsch, R.D. and J.H.K. Pechmann. 1985. Diel pattern of migratory activity for several species of pond-breeding salamanders. Copeia 1985:86–91.

Semlitsch, R.D. and S.C. Walls. 1990. Geographic variation in the egg-laying strategy of the mole salamander, *Ambystoma talpoideum*. Herpetological Review 21:14–15.

Semlitsch, R.D. and S.C. Walls. 1993. Competition in two species of larval salamanders: a test of geographic variation in competitive ability. Copeia 1993:587–595.

Semlitsch, R.D. and C.A. West. 1983. Aspects of the life history and ecology of Webster's salamander, *Plethodon websteri*. Copeia 1983:339–346.

Semlitsch, R.D. and H.M. Wilbur. 1988. Effects of pond drying time on metamorphosis and survival in the salamander *Ambystoma talpoideum*. Copeia 1988:978–983.

Semlitsch, R.D., J.W. Gibbons and T.D. Tuberville. 1995. Timing of reproduction and metamorphosis in the Carolina gopher frog *(Rana capito capito)* in South Carolina. Journal of Herpetology 29:612–614.

Semlitsch, R.D., R.N. Harris and H.M. Wilbur. 1990. Paedomorphosis in *Ambystoma talpoideum*: maintenance of population variation and alternative life-history pathways. Evolution 44:1604–1613.

Semlitsch, R.D., D.E. Scott and J.H.K. Pechmann. 1988. Time and size at metamorphosis related to adult fitness in *Ambystoma talpoideum*. Ecology 69:184–192.

Semlitsch, R.D., D.E. Scott, J.H.K. Pechmann and J.W. Gibbons. 1993. Phenotypic variation in the arrival time of breeding salamanders: individual repeatability and environmental influence. Journal of Animal Ecology 62:334–340.

Semlitsch, R.D., D.E. Scott, J.H.K. Pechmann and J.W. Gibbons. 1996. Structure and dynamics of an amphibian community: evidence from a 16-year study of a natural pond. Pp. 217–248. *In* Cody, M.L. and J.A. Smallwood (Eds.), Long-term Studies of Vertebrate Communities. Academic Press, San Diego, California.

Semlitsch, R.D., M. Foglia, A. Mueller, I. Steiner, E. Fioramonti and K. Fent. 1995. Short-term exposure to triphenyltin affects the swimming and feeding behavior of tadpoles. Environmental Toxicology and Chemistry 14:1419–1423.

Semonsen, V.J. 1998. Natural history notes: *Ambystoma californiense* (California tiger salamander). Survey technique. Herpetological Review 29:96.

Senger, C.M. and R.M. Macy. 1952. Helminths of northwest mammals. Part III. The description of *Euryhelmis pacificus* n. sp. and notes on its life cycle. Journal of Parasitology 38:481–486.

Sepulveda, M.S., M.G. Spalding, J.M. Kinsella and D.J. Forrester. 1999. Parasites of the great egret *(Ardea albus)* in Florida and a review of the helminths reported for the species. Journal of the Helminthological Society of Washington 66:7–13.

Servage, D.L. 1979. Biosystematics of the *Ambystoma jeffersonianum* complex in Ontario. Master's thesis. University of Guelph, Guelph, Ontario, Canada.

Sessions, S.K. 1982. Cytogenetics of diploid and triploid salamanders of the *Ambystoma jeffersonianum* complex. Chromosoma 84:599–621.

Sessions, S.K. and J. Kezer. 1987. Cytogenetic evolution in the plethodontid salamander genus *Aneides*. Chromosoma 95:17–30.

Sessions, S.K. and S.B. Ruth. 1990. Explanations for naturally occurring supernumery limbs in amphibians. Journal of Experimental Zoology 254:38–47.

Sessions, S.K., R.A. Franssen and V.L. Horner. 1999. Morphological clues from multilegged frogs: are retinoids to blame? Science 284:800–802.

SETAC (Society of Environmental Toxicology and Chemistry). 1997. Abstracts, 18th Annual Meeting, November 16–20. San Francisco, California.

Settele, J., K. Henle and C. Bender. 1996. Metapopulationen und Biotopverbund: theorie und Praxis am Beispiel von Schmetterlingen und Reptilien. Zeitschrift für Ökologie und Naturschutz 5:187.206.

Sever, D.M. 1972. Geographic variation and taxonomy of *Eurycea bislineata* (Caudata: Plethodontidae) in the upper Ohio River Valley. Herpetologica 28:314–324.

Sever, D.M. 1975. Morphology and seasonal variation of the mental hedonic glands of the dwarf salamander, *Eurycea quadridigitata* (Holbrook). Herpetologica 31:214–251.

Sever, D.M. 1976. Identity of an enigmatic *Eurycea* (Urodela: Plethodontidae) from the Great Smoky Mountains of Tennessee. Herpetological Review 7:98.

Sever, D.M. 1978a. Male cloacal glands of *Plethodon cinereus* and *Plethodon dorsalis*. Herpetologica 34:1–20.

Sever, D.M. 1978b. Female cloacal anatomy of *Plethodon cinereus* and *Plethodon dorsalis*. Journal of Herpetology 12:397–406.

Sever, D.M. 1979. Male secondary sexual characters of the *Eurycea bislineata* (Amphibia, Urodela, Plethodontidae) complex in the southern Appalachian Mountains. Journal of Herpetology 13:245–253.

Sever, D.M. 1983a. Observations on the distribution and reproduction of the salamander *Eurycea junaluska* in Tennessee. Journal of the Tennessee Academy of Science 58:48–50.

Sever, D.M. 1983b. *Eurycea junaluska*. Pp. 321.1–321.2. Catalogue of American Amphibians and Reptiles. Society for the Study of Amphibians and Reptiles, St. Louis. Missouri.

Sever, D.M. 1984. The discovery of *Eurycea junaluska*. Bulletin of the Chicago Herpetological Society 19:75–84.

Sever, D.M. 1985. Sexually dimorphic glands of *Eurycea nana*, *Eurycea neotenes* and *Typhlomolge rathbuni* (Amphibia: Plethodontidae). Herpetologica 41:71–84.

Sever, D.M. 1989. Comments on the taxonomy and morphology of two-lined salamanders of the *Eurycea bislineata* complex. Bulletin of the Chicago Herpetological Society 24:70–74.

Sever, D.M. 1999a. *Eurycea bislineata*. Pp. 683.1–683.5. Catalogue of American Amphibians and Reptiles. Society for the Study of Amphibians and Reptiles, St. Louis, Missouri.

Sever, D.M. 1999b. *Eurycea cirrigera*. Pp. 684.1–684.6. Catalogue of American Amphibians and Reptiles. Society for the Study of Amphibians and Reptiles, St. Louis, Missouri.

Sever, D.M. 1999c. *Eurycea wilderae*. Pp. 685.1–685.4. Catalogue of American Amphibians and Reptiles. Society for the Study of Amphibians and Reptiles, St. Louis, Missouri.

Sever, D.M. and H.L. Bart Jr. 1996. Ultrastructure of the spermathecae of *Necturus beyeri* (Amphibia: Proteidae) in relation to its breeding season. Copeia 1996:927–937.

Sever, D.M. and C.F. Dineen. 1978. Reproductive ecology of the tiger salamander, *Ambystoma tigrinum*, in northern Indiana. Proceedings of the Indiana Academy of Science 87:189–203.

Sever, D.M. and N.M. Kloepfer. 1993. Spermathecal cytology of *Ambystoma opacum* (Amphibia: Ambystomatidae) and the phylogeny of sperm storage in female salamanders. Journal of Morphology 217:115–127.

Sever, D.M., H.A. Dundee and C.D. Sullivan. 1976. A new *Eurycea* (Amphibia: Plethodontidae) from southwestern North Carolina. Herpetologica 32:26–29.

Sever, D.M., L.C. Rania and J.D. Krenz. 1996. Reproduction of the salamander *Siren intermedia* Le Conte with especial reference to oviductal anatomy and mode of fertilization. Journal of Morphology 227:335–348.

Sever, D.M., J.D. Krenz, K.M. Johnson and L.C. Rania. 1995. Morphology and evolutionary implications of the annual cycle of secretion and sperm storage in spermathecae of the salamander *Ambystoma opacum* (Amphibia: Ambystomatidae). Journal of Morphology 223:35–46.

Sever, D.M., E.C. Moriarty, L.C. Rania and W.C. Hamlett. 2001. Sperm storage in the oviduct of the internal fertilizing frog *Ascaphus truei*. Journal of Morphology 248:1–21.

Sexton, O.J. and J.R. Bizer. 1978. Life history patterns of *Ambystoma tigrinum* in montane Colorado. American Midland Naturalist 99:101–118.

Sexton, O.J. and C. Phillips. 1986. A qualitative study of fish-amphibian interactions in 3 Missouri ponds. Transactions of the Missouri Academy of Science 20:25–35.

Sexton, O.J., C. Phillips and J.E. Bramble. 1990. The effects of temperature and precipitation on the breeding migration of the spotted salamander *(Ambystoma maculatum)*. Copeia 1990:781–787.

Sexton, O.J., C.A. Phillips and E. Routman. 1994. The response of naive breeding adults of the spotted salamander to fish. Behavior 130:113–121.

Sexton, O.J., J. Bizer, D.C. Gayou, P. Freiling and M. Moutseous. 1986. Field studies of breeding spotted salamanders, *Ambystoma maculatum*, in eastern Missouri, USA. Milwaukee Public Museum, Contributions in Biology and Geology, Number 67, Milwaukee, Wisconsin.

Sexton, O.J., C.A. Phillips, T.J. Bergman, E.B. Wattenberg and R.E. Preston. 1998. Abandon not hope: status of repatriated populations of spotted salamanders and wood frogs at the Tyson Research Center, St. Louis County, Missouri. Pp. 340–344. *In* Lannoo, M.J. (Ed.), Status and Conservation of Midwestern Amphibians. University of Iowa Press, Iowa City, Iowa.

Seyle, C.W., Jr. 1985a. Geographic distribution: *Ambystoma talpoideum* (mole salamander). Herpetological Review 16:59–60.

Seyle, C.W., Jr. 1985b. Life history notes: *Amphiuma means* (two-toed amphiuma). Reproduction. Herpetological Review 16:51–53.

Seyle, C.W., Jr. 1985c. Geographic distribution: *Necturus punctatus* (dwarf waterdog). Herpetological Review 16:60.

Seyle, C.W. 1994. Distribution and status of the gopher frog *(Rana capito)* and flatwoods salamander *(Ambystoma cingulatum)* in Georgia. U.S. Fish and Wildlife Service, Unpublished report, Jackson, Mississippi.

Seyle, C.W., Jr. and S.E. Trauth. 1982. Life history notes: *Pseudacris ornata* (ornate chorus frog). Reproduction. Herpetological Review 13:45.

Seymour, R.S. 1972. Behavioral thermoregulation by juvenile green toads, *Bufo debilis*. Copeia 1972:572–575.

Seymour, R.S. 1973. Gas exchange of spadefoot toads beneath the ground. Copeia 1973:452–460.

Shaffer, C.L. 1990. Nature Reserves: Island Theory and Conservation Practice. Smithsonian Institution Press, Washington, D.C.

Shaffer, H.B. and J.E. Juterbock. 1994. Night driving. Pp. 163–171. *In* Heyer, W.R., M.A. Donnelly, R.W. McDiarmid, L-A.C. Hayek and M.S. Foster (Eds.), Measuring and Monitoring Biological Diversity: Standard Methods for Amphibians. Smithsonian Institution Press, Washington, D.C.

Shaffer, H.B. and G.V. Lauder. 1985. Patterns of variation in aquatic ambystomatid salamanders: kinematics of the feeding mechanism. Evolution 39:83–92.

Shaffer, H.B. and M.L. McKnight. 1996. The polytypic species revisited: differentiation and molecular phylogenetics of the tiger salamander *Ambystoma tigrinum* (Amphibia: Caudata) complex. Evolution 50:417–433.

Shaffer, H.B., C.C. Austin and R.B. Huey. 1991. The consequences of metamorphosis on salamander *(Ambystoma)* locomotor performance. Physiological Zoology 64:212–231.

Shaffer, H.B., G.M. Fellers, A. Magee and S.R. Voss. 2000. The genetics of amphibian declines: population substructure and molecular differentiation in the Yosemite toad *Bufo canorus* (Anura, Bufonidae) based on the single-strand conformational polymorphism analysis (SSCP) and mitochondrial DNA sequence data. Molecular Ecology 9:245–257.

Shaffer, H.B., G.M. Fellers, S.R. Voss, J.C. Oliver and G.B. Pauley. 2004. Species boundaries, phylogeography, and conservation genetics of the red-legged frog *(Rana aurora/draytonii)* complex. Molecular Ecology 13:2667–2677.

Shaffer, L.L. 1991. Pennsylvania amphibians and reptiles. Pennsylvania Fish Commission, Harrisburg, Pennsylvania.

Shaffer, M.L. 1983. Determining minimum population sizes for the grizzly bear. International Conference on Bear Research and Management 5:133–139.

Shaffer, M.L. 1987. Minimum population sizes for species conservation. Pp. 69–86. *In* Soulé, M.E. (Ed.), Viable Populations for Conservation. Cambridge University Press, New York.

Shannon, F.A. 1949. A western subspecies of *Bufo woodhousii* hitherto erroneously associated with *Bufo compactilis*. Bulletin of the Chicago Academy of Science, Number 16, Chicago, Illinois.

Shannon, F.A. 1953. *Scaphiopus bombifrons*, a state record for Arizona. Herpetologica 9:127–128.

Shannon, F.A. 1957. Further comments upon the western range of *Scaphiopus bombifrons*. Copeia 1957:145–146.

Shannon, F.A. 1959. The reptiles and amphibians of Korea. Herpetologica 12:22–49.

Shannon, F.A. and C.H. Lowe. 1955. A new subspecies of *Bufo woodhousei* from the inland Southwest. Herpetologica 11:185–190.

Shapin S. 1996. The Scientific Revolution. University of Chicago Press, Chicago, Illinois.

Shapovalov, L. 1937. Third record of *Ascaphus truei* in California. Copeia 1937:234.

Sharbel, T.F. and J. Bonin. 1992. Northernmost record of *Desmognathus ochrophaeus*: biochemical identification in the Chateauguay River drainage basin, Quebec. Journal of Herpetology 26:505–508.

Sharbel, T.F., J. Bonin, L.A. Lowcock and D.M. Green. 1995. Partial genetic compatibility and unidirectional hybridization in syntopic populations of the salamanders *Desmognathus fuscus* and *D. ochrophaeus*. Copeia 1995:466–469.

Sharitz, R.R. 2003. Carolina bays: hydrologically variable depression wetlands of the southeastern Atlantic Coastal Plain. Wetlands 23: 550–562.

Shaw, G. 1802. General Zoology or Systematic Natural History. Volume 3. Amphibians. G. Kearsley, London, England.

Shay, R. 1973. Oregon's rare or endangered wildlife. Oregon State Game Commission Bulletin 29:3–8.

Shealy, R.M. 1975. Factors influencing activity in the salamanders *Desmognathus ochrophaeus* and *D. monticola* (Plethodontidae). Herpetologica 31:94–102.

Shelford, V.E. 1913. Animal Communities of Temperate America. University of Chicago Press, Chicago, Illinois.

Shepard, D.B. and H.M. Burdett. 2000. Geographic distribution: *Pseudacris clarkii* (spotted chorus frog). Herpetological Review 31:50.

Sheppard, R.F. 1977. The ecology and home range movements of *Ambystoma macrodactylum krausei* (Amphibia: Urodela). Master's thesis. University of Calgary, Calgary, Alberta, Canada.

Sherman, E. 1980. Cardiovascular responses of the toad *Bufo marinus* to thermal stress and water deprivation. Comparative Biochemistry and Physiology A 24:47–54.

Sherman, E. and S.G. Stadlen. 1986. The effect of dehydration on rehydration and metabolic rate in a lunged and lungless salamander. Comparative Biochemistry and Physiology A 85:483–487.

Sherwood, W.L. 1895. The salamanders found in the vicinity of New York City, with notes upon extra-limital or allied species. Proceedings of the Linnaean Society of New York 7:21–37.

Sheumack, D.D., M.E.H. Howden, I. Spence and R.J. Quinn. 1978. Maculotoxin: a toxin from the venom glands of the octopus *Hapalachlaena maculosa* identified as tetrodotoxin. Science 199:188.

Shields, J.D. 1987. Pathology and mortality of the lung fluke *Haematoloechus longiplexus* (Trematoda) in *Rana catesbeiana*. Journal of Parasitology 73:1005–1013.

Shields, S.E., D.J. Ogilvie, R.G. McKinnell and D. Tarin. 1984. Degradation of basement membrane collagens by metalloproteases released by human, murine, and amphibian tumours. Journal of Pathology 143:193–197.

Shinn, E.A. and J.W. Dole. 1978. Evidence for a role for olfactory cues in the feeding response of leopard frogs, *Rana pipiens*. Herpetologica 34:167–172.

Shirose, L.J. and R.J. Brooks. 1995a. Growth rate and age at maturity in syntopic populations of *Rana clamitans* and *Rana septentrionalis* in central Ontario. Canadian Journal of Zoology 73:1468–1473.

Shirose, L.J. and R.J. Brooks. 1995b. Age structure, mortality, and longevity in syntopic populations of three species of ranid frogs in central Ontario. Canadian Journal of Zoology 73:1878–1886.

Shirose, L.J. and R.J. Brooks. 1997. Fluctuations in abundance and age structure in three species of frogs (Anura: Ranidae) in Algonquin Park, Canada, from 1985 to 1993. Pp. 16–26. *In* Green, D.M. (Ed.), Amphibians in Decline. Canadian Studies of a Global Problem. Herpetological Conservation, Number 1, Society for the Study of Amphibians and Reptiles, St. Louis, Missouri.

Shirose, L.J., C.A. Bishop, D.M. Green, C.J. MacDonald, R.J. Brooks and N.J. Helferty. 1995. Validation study of a calling amphibian survey in Ontario. Proceedings of the Second North American Amphibian Monitoring Program Conference, Burlington, Ontario.

Shirose, L.J., C.A. Bishop, D.M. Green, C.J. MacDonald, R.J. Brooks and N.J. Helferty. 1997. Validation tests of an amphibian call count survey technique in Ontario, Canada. Herpetologica 53:312–320.

Shively, J.N., J.G. Songer, S. Prchal, M.S. Keasey III and C.O. Thoen. 1981. *Mycobacterium marinum* infection in Bufonidae. Journal of Wildlife Diseases 17:3–7.

Shoemaker, V.H., L. McClanahan and R. Ruibal. 1969. Seasonal changes in body fluids in a field population of spadefoot toads. Copeia 1969:585–591.

Shonewald-Cox, C.M., S.M. Chambers, B. MacBryde and L. Thomas. 1983. Genetics and Conservation. A Reference for Managing Wild Animal and Plant Populations. Benjamin/Cummings Publishing Company, Menlo Park, California.

Shoop, C.R. 1960. The breeding habits of the mole salamander, *Ambystoma talpoideum* (Holbrook), in southeastern Louisiana. Tulane Studies in Zoology 8:65–82.

Shoop, C.R. 1965a. Orientation of *Ambystoma maculatum* movements to and from breeding ponds. Science 149:558–559.

Shoop, C.R. 1965b. Aspects of reproduction in Louisiana *Necturus* populations. American Midland Naturalist 74:357–367.

Shoop, C.R. 1967. Relation of migration and breeding activities to time of ovulation in *Ambystoma maculatum*. Herpetologica 23: 319–321.

Shoop, C.R. 1968. Migratory orientation of *Ambystoma maculatum*: movements near breeding ponds and displacements of migrating individuals. Biological Bulletin 135:230–238.

Shoop, C.R. 1974. Yearly variation in larval survival of *Ambystoma maculatum*. Ecology 55:440–444.

Shoop, C.R. and T.L. Doty. 1972. Migratory orientation by marbled salamanders (*Ambystoma opacum*) near a breeding area. Behavioral Biology 7:131–136.

Shoop, C.R. and G.E. Gunning. 1967. Seasonal activity and movements of *Necturus* in Louisiana. Copeia 1967:732–737.

Shoop, W.L. and K.C. Corkum. 1987. Maternal transmission by *Alaria marcianae* (Trematoda) and the concept of amphiparatenesis. Journal of Parasitology 73:110–115.

Shoop, W.L., W.F. Font and P.F. Malatesta. 1990. Transmammary transmission of mesocercariae of *Alaria marcianae* (Trematoda) in experimentally infected primates. Journal of Parasitology 76: 869–873.

Shrader-Frechette, K.R. and E.D. McCoy. 1993. Method in Ecology: Strategies for Conservation. Cambridge University Press, Cambridge, United Kingdom.

Shrader-Frechette, K.R. and E.D. McCoy. 1994. What ecology can do for environmental management. Journal of Environmental Management 41:293–307.

Shure, D.J., L.A. Wilson and C. Hochwender. 1989. Predation on aposematic efts of *Notophthalmus viridescens*. Journal of Herpetology 23:437–439.

Sibbing, J.M. 1995. Ohio wetlands. National Audubon Society, Great Lakes Regional Office, Columbus, Ohio.

Sibley, D.A. 2000. The Sibley Guide to Birds. Alfred A. Knopf, New York.

Siddall, M.E. and S.S. Desser. 1992. Alternative leech vectors for frog and turtle trypanosomes. Journal of Parasitology 78:562–563.

Siegfried, W.R. 1975. On the nest of the hamerkop. Ostrich 46:267.

Sievert, G.R. 1986. An investigation of the distribution and population status of the Rich Mountain salamander (*Plethodon ouachitae*) in Oklahoma. Final Report. Oklahoma Department of Wildlife Conservation, Oklahoma City, Oklahoma.

Sievert, L.M. 1991. Thermoregulatory behaviour in the toads *Bufo marinus* and *Bufo cognatus*. Journal of Thermal Biology 16:309–312.

Sih, A. 1992. Integrative approaches to the study of predation: general thoughts and a case study on sunfish and salamander larvae. Annales Zoologici Fennici 29:183–198.

Sih, A. and L.B. Kats. 1991. Effects of refuge availability on the responses of salamander larvae to chemical cues from predatory green sunfish. Animal Behaviour 42:330–332.

Sih, A. and L.B. Kats. 1994. Age, experience and the response of streamside salamander hatchlings to chemical cues from predatory sunfish. Ethology 96:253–259.

Sih, A. and E. Maurer. 1992. Effects of cryptic oviposition on egg survival for stream-breeding streamside salamanders. Journal of Herpetology 26:114–116.

Sih, A. and R.D. Moore. 1993. Delayed hatching of salamander eggs in response to enhanced larval predation risk. American Naturalist 142: 947–960.

Sih, A. and J.W. Petranka. 1988. Optimal diets: simultaneous search and handling of multiple-prey loads by salamander larvae. Behavioral Ecology and Sociobiology 23:335–339.

Sih, A., L.B. Kats and R.D. Moore. 1992. Effects of predatory sunfish on the density, drift, and refuge use of stream salamander larvae. Ecology 73:1418–1430.

Silverstone, P.A. 1975. A revision of the poison-arrow frogs of the genus *Dendrobates* Wagler. Natural History Museum of Los Angeles County, Science Bulletin, Number 27, Los Angeles, California.

Simandle, E. 2002. The Amargosa Toad. http://www.amargosatoad.org.

Simberloff, D.S. 1983. Competition theory, hypothesis testing and other community ecology buzzwords. American Naturalist 122:626–635.

Simberloff, D.S. 1988. The contribution of population and community biology to conservation science. Annual Review of Ecology and Systematics 19:473–511.

Simberloff, D.S. and L.G. Abele. 1976. Island biogeography theory and conservation practice. Science 191:285–286.

Simmons, D. 1975. The evolutionary ecology of *Gyrinophilus palleucus*. Master's thesis. University of Florida, Gainesville, Florida.

Simmons, D. 1976. A naturally metamorphosed *Gyrinophilus palleucus* (Amphibia, Urodela, Plethodontidae). Journal of Herpetology 10: 255–257.

Simmons, H.M. 2000. Habitat use in winter by northern leopard frogs, *Rana pipiens*, in Washington state: a comparison to breeding locations. Department of Biological Sciences, Central Washington University, Ellensburg, Washington.

Simmons, J.E. 1987. Herpetological collecting and collections management. Herpetological Circular, Number 16, Society for the Study of Amphibians and Reptiles, St. Louis, Missouri.

Simmons, R. and J.D. Hardy Jr. 1959. The river-swamp frog, *Rana heckscheri* Wright, in North Carolina. Herpetologica 15:36–37.

Simon, T.P., J.O. Whitaker Jr., J.S. Castrale and S.A. Minton. 1992. Checklist of the vertebrates of Indiana. Proceedings of the Indiana Academy of Science 101:95–126.

Simons, L.H. 1998. Natural history notes: *Rana cascadae* (Cascades frog). Predation. Herpetological Review 29:232.

Simons, L.H. and M. Simons. 1998. Geographic distribution: *Ascaphus truei* (tailed frog). Herpetological Review 29:106.

Simons, R.R., R.G. Jaeger and B.E. Felgenhauer. 1997. Competitor assessment and area defense by terrestrial salamanders. Copeia 1997: 70–76.

Simons, T.R. and E. Johnson. 1999. Assessing the diversity and habitat associations of salamanders in Great Smoky Mountains National Park. National Park Service, 1998 Annual report, Gatlinburg, Tennessee.

Simovich, M.A. 1994. The dynamics of spadefoot toads (*Spea multiplicata* and *S. bombifrons*) hybridization system. Pp. 167–182. *In* Brown, P.R. and J.W. Wright (Eds.), Herpetology of the North American Deserts. Southwest Herpetologists Society, Special Publication, Number 5, Van Nuys, California.

Simovich, M.A., C.A. Sassaman and A. Chovnick. 1991. Post-mating selection of hybrid toads (*Scaphiopus multiplicatus* and *Scaphiopus bombifrons*). Proceedings of the San Diego Society of Natural History 5:1–6.

Simpson, G.G. 1961. Principles of Animal Taxonomy. Columbia University Press, New York.

Simpson, N.R. 1984. Written statement recalling 24 January 1966, meeting with Gosdin and Watson, 31 July.

Simpson, N.S. and R.G. McKinnell. 1964. The burnsi gene as a nuclear marker for transplantation experiments in frogs. Journal of Cell Biology 23:371–375.

Sinclair, R.M. 1950. Notes on some salamanders from Tennessee. Herpetologica 6:49–51.

Singer, F.J., W.T. Swank and E.C.C. Clebsch. 1982. Some ecosystem responses to European wild boar rooting in a deciduous forest. National Park Service, Southeast Regional Office, Uplands Field Research Laboratory, Research/Resource Management Report, Number 54, Gatlinburg, Tennessee.

Sinsch, R. 1991. Cold acclimation in frogs (*Rana*): microhabitat choice, osmoregulation, and hydromineral balance. Comparative Biochemistry and Physiology A 98:469–477.

Sinsch, U. 1991. Mini-review: the orientation behaviour of amphibians. Herpetological Journal 1:541–544.

Sinsch, U. 1997. Postmetamorphic dispersal and recruitment of first breeders in a *Bufo calamita* metapopulation. Oecologia 112:42–47.

Sites, J.W., Jr. 1978. The foraging strategy of the dusky salamander, *Desmognathus fuscus* (Amphibia: Urodela: Plethodontidae): an empirical approach to predation theory. Journal of Herpetology 12:373–383.

Sizemore, P.D. 1936. *Cercaria concavocorpa* n. sp. Transactions of the American Microscopical Society 55:483–486.

Sjoberg, J. 2000. Amargosa toad monitoring (draft). Nevada Division of Wildlife, Las Vegas, Nevada.

Sjoberg, J. 2001. NDOW Amargosa toad monitoring report, 2001 (draft). Nevada Division of Wildlife, Las Vegas, Nevada.

Sjögren, P. 1991. Extinction and isolation gradients in metapopulations: the case of the pool frog *(Rana lessonae)*. Biological Journal of the Linnean Society 42:135–147.

Skagen, S.K. and F.L. Knopf. 1994. Residency patterns of migrating sandpipers at a midcontinental stopover. Condor 96:949–958.

Skelly, D.K. 1992. Field evidence for a cost of behavioral antipredator response in a larval amphibian. Ecology 73:704–708.

Skelly, D.K. 1996. Pond drying, predators, and the distribution of *Pseudacris* tadpoles. Copeia 1996:599–605.

Skelly, D.K. 1997. Tadpole communities: pond permanence and predation are powerful forces shaping the structure of tadpole communities. American Scientist 85:36–45.

Skelly, D.K. and E.E. Werner. 1990. Behavioral and life-historical responses of larval American toads to an odonate predator. Ecology 71:2313–2322.

Skelly, D.K., E.E. Werner and S.A. Cortwright. 1999. Long-term distributional dynamics of a Michigan amphibian assemblage. Ecology 80: 2326–2337.

Skilton, A.J. 1849. Description of two reptiles from Oregon. American Journal of Science and the Arts Series 2, 7:202.

Skrjabin, K.I., N.P. Shikhobalova and E.A. Lagodovskaya. 1961. *Oxyurata* of animals and man. Part 2. *In* Skrjabin, K.I. (Ed.), Essentials of Nematology. Volume 10. Translated from Russian by Israel Program for Scientific Translations, Jerusalem (1974).

Slater, J.R. 1931. The mating of *Ascaphus truei* Stejneger. Copeia 1931: 62–63.

Slater, J.R. 1933. Notes on Washington salamanders. Copeia 1933:44.

Slater, J.R. 1936a. Notes on *Ambystoma gracile* Baird and *Ambystoma macrodactylum* Baird. Copeia 1936:234–236.

Slater, J.R. 1936b. Amphibians of Mt. Rainier National Park. Mount Rainier National Park Nature Notes 1936:1–137.

Slater, J.R. 1939. Description and life history of a new *Rana* from Washington. Herpetologica 1:145–149.

Slater, J.R. 1955. Distribution of Washington amphibians. College of Puget Sound, Occasional Papers, 16:122–154.

Slater, J.R. and W.C. Brown. 1941. Island records of amphibians and reptiles for Washington. Department of Biology, College of Puget Sound, Occasional Papers 13:74–77.

Slater, J.R. and J.W. Slipp. 1940. A new species of *Plethodon* from northern Idaho. Department of Biology, College of Puget Sound, Occasional Papers 8:38–43.

Slatkin, M. 1981. Estimating levels of gene flow in natural populations. Genetics 99:323–335.

Slevin, J.R. 1928. The amphibians of Western North America. An account of the species known to inhabit California, Alaska, British Columbia, Washington, Oregon, Idaho, Utah, Nevada, Arizona, Sonora and Lower California. Occasional Papers of the California Academy of Sciences, Number 16, San Francisco, California.

Slevin, J.R. 1931. Range extensions of certain western species of reptiles and amphibians. Copeia 1931:140–141.

Slobodkin, L.B. 1986. On the susceptibility of different species to extinction: elementary instructions for owners of the world. Pp. 226–242. *In* Norton, B.G. (Ed.), The Preservation of Species. Princeton University Press, Princeton, New Jersey.

Slonim, A.R. 1986. Acute toxicity of some hydrazine compounds to salamander larvae, *Ambystoma* sp. Bulletin of Environmental Contamination and Toxicology 37:739–746.

Slonim, A.R. and E.E. Ray. 1975. Acute toxicity of beryllium sulfate to salamander larvae (*Ambystoma* sp.). Bulletin of Environmental Contamination and Toxicology 13:307–312.

Smallwood, W.M. 1905. Adrenal tumors in the kidney of the frog. Anatomischer Anzeiger 26:652–658.

Smallwood, W.M. 1928. Notes on the food of some Onondaga Urodela. Copeia 169:89–98.

Smith, A.K. 1977. Attraction of bullfrogs (Amphibia, Anura, Ranidae) to distress calls of immature frogs. Journal of Herpetology 11: 234–235.

Smith, B.G. 1907. The life history and habits of *Cryptobranchus alleganiensis*. Biological Bulletin 13:5–39.

Smith, B.G. 1911a. Notes on the natural history of *Ambystoma jeffersonianum, A. punctatum* and *A. tigrinum*. Bulletin of the Wisconsin Natural History Society 9:14–27.

Smith, B.G. 1911b. The nests and larvae of *Necturus*. Biological Bulletin 20:191–200.

Smith, B.G. 1912a. The embryology of *Cryptobranchus alleghaniensis*, including comparisons with some other vertebrates. I. Introduction: the history of the egg before cleavage. Journal of Morphology 23:61–157.

Smith, B.G. 1912b. The embryology of *Cryptobranchus alleghaniensis*, including comparisons with some other vertebrates. II. General embryonic and larval development, with special reference to external features. Journal of Morphology 23:455–579.

Smith, C.C. 1960. Notes on the salamanders of Arkansas. 1. Life history of a neotenic stream-dwelling form. Proceedings of the Arkansas Academy of Science 13:66–74.

Smith, C.C. and A.N. Bragg. 1949. Observations on the ecology and natural history of Anura, VII. Food and feeding habits of the common species of toads in Oklahoma. Ecology 30:333–349.

Smith, C.C. and S.D. Fretwell. 1974. The optimal balance between size and number of offspring. American Naturalist 108:499–506.

Smith, C.J. and R.A. Pfingsten. 1989. *Notophthalmus viridescens* (Rafinesque), eastern newt. Pp. 79–87. *In* Pfingsten, R.A. and Downs (eds.), Salamanders of Ohio. Bulletin of the Ohio Biological Survey, Volume 7, Number 2, Columbus, Ohio.

Smith, C.K. 1983. Notes on breeding period, incubation period, and egg masses of *Ambystoma jeffersonianum* (Green) (Amphibia: Caudata) from the southern limits of its range. Brimleyana 9:135–140.

Smith, C.K. 1988. Variation in body size and competition within larval populations of the salamander *Ambystoma opacum*. Ph.D. dissertation. University of North Carolina, Chapel Hill, North Carolina.

Smith, C.K. 1990. Effects of variation in body size on intraspecific competition among larval salamanders. Ecology 71:1777–1788.

Smith, C.K. and J.W. Petranka. 1987. Prey size distributions and size-specific foraging success of *Ambystoma* larvae. Oecologia 71: 239–244.

Smith, C.K., J.W. Petranka and R. Barwick. 1996. Life history notes: *Desmognathus welteri* (Black Mountain salamander). Reproduction. Herpetological Review 27:136.

Smith, D.C. 1987. Adult recruitment in chorus frogs: effects of size and date at metamorphosis. Ecology 68:344–350.

Smith, D.D. 1985. Life history notes: *Ambystoma tigrinum* (tiger salamander). Behavior. Herpetological Review 16:77.

Smith, D.D. and R. Powell. 1983. Life history notes: *Acris crepitans blanchardi* (Blanchard's cricket frog). Anomalies. Herpetological Review 14:118–119.

Smith, D.G., C.R. Wilson and H.H. Frost. 1972. The biology of the American kestral in central Utah. Southwestern Naturalist 17:73–83.

Smith, E.M. and F.H. Pough. 1994. Intergeneric aggression in salamanders. Journal of Herpetology 28:41–45.

Smith, E.P. and G. van Belle. 1984. Non-parametric estimation of species richness. Biometrics 40:119–129.

Smith, G.C. 1976. Ecological energetics of three species of ectothermic vertebrates. Ecology 57:252–264.

Smith, G.R. 1999. Microhabitat preferences of bullfrog tadpoles *(Rana catesbeiana)* of different ages. Transactions of the Nebraska Academy of Sciences 25:73–76.

Smith, G.R. and L.D. Johnson. 1999. New Missouri amphibian and reptile distribution records from a catalogued college teaching collection. Herpetological Review 30:58–59.

Smith, G.R., A. Todd, J.E. Rettig and F. Nelson. 2003. Microhabitat selection by northern cricket frogs *(Acris crepitans)* along a west-central Missouri creek: field and experimental observations Journal of Herpetology 37:383–385.

Smith, G.R., J.E. Rettig, G.G. Mittlebach, J.L. Valiulis and S.R. Schaack. 1999. The effects of fish on assemblages of amphibians in ponds: a field experiment. Freshwater Biology 41:829–837.

Smith, H.A. and B. Herrington. 1989. Sub-surface feeding abilities of three northwestern plethodontid salamanders. Georgia Journal of Science 47:8–9.

Smith, H.M. 1932. *Ascaphus truei* Stejneger in Montana. Copeia 1932: 100.

Smith, H.M. 1934. The amphibians of Kansas. American Midland Naturalist 15:377–528.

Smith, H.M. 1950. Handbook of amphibians and reptiles of Kansas. University of Kansas Museum of Natural History, Miscellaneous Publication, Number 2, Lawrence, Kansas.

Smith, H.M. 1963. New distributional records of amphibians and reptiles from South Dakota and Wyoming. Herpetologica 19:147–148.

Smith, H.M. and S. Barlowe. 1978. Amphibians of North America. A Guide to Field Identification. Golden Press, New York.

Smith, H.M. and H.K. Buechner. 1947. The influence of the Balcones Escarpment on the distribution of amphibians and reptiles in Texas. Bulletin of the Chicago Academy of Sciences 8:1–16.

Smith, H.M. and P.S. Chrapliwy. 1958. New and noteworthy Mexican herptiles from the Lidicker Collection. Herpetologica 13:267–271.

Smith, H.M. and A.J. Kohler. 1977. A survey of herpetological introductions in the United States and Canada. Transactions of the Kansas Academy of Science 80:1–24.

Smith, H.M. and F.E. Potter Jr. 1946. A third neotenic salamander of the genus *Eurycea* from Texas. Herpetologica 3:105–109.

Smith, H.M. and E.H. Taylor. 1948. An annotated checklist and key to the Amphibia of Mexico. U.S. National Museum Bulletin, Number 194, Washington, D.C.

Smith, H.M., D. Chiszar and R. Montanucci. 1997. Subspecies and classification. Herpetological Review 28:13–17.

Smith, H.M., C.W. Nixon and P.E. Smith. 1947. Notes on the tadpoles and natural history of *Rana areolata circulosa*. Bulletin of the Ecological Society 28:73.

Smith, H.M., C.W. Nixon and P.E. Smith. 1948. A partial description of the tadpole of *Rana areolata circulosa* and notes on the natural history of the race. American Midland Naturalist 39:608–614.

Smith, H.M., D. Chiszar, J.T. Collins and F. van Breukelen. 1998. The taxonomic status of the Wyoming toad, *Bufo baxteri* Porter. Contemporary Herpetology 1, http://dataserver.calacademy.org/herpetology/herpdocs/ch/1998/1/index.htm.

Smith, L. 1920. Some notes on *Notophthalmus viridescens*. Copeia 80:22–24.

Smith, L. 1921. A note on the eggs of *Ambystoma maculatum*. Copeia 97:41.

Smith, L.L. and R. Franz. 1994. Use of Florida round-tailed muskrat houses by amphibians and reptiles. Florida Field Naturalist 22:69–74.

Smith, L.L., R. Franz and C.K. Dodd Jr. 1993. Additions to the herpetofauna of Edgemont Key, Hillsborough County, Florida. Florida Scientist 56:231–234.

Smith, P.B. and M.C. Michener. 1962. An adult albino *Ambystoma*. Herpetologica 18:67–68.

Smith, P.W. 1947. The reptiles and amphibians of eastern central Illinois. Chicago Academy of Science Bulletin 8:21–40.

Smith, P.W. 1948a. A cestode infestation in *Typhlotriton*. Herpetologica 4:152.

Smith, P.W. 1948b. Food habits of cave dwelling amphibians. Herpetologica 4:205–208.

Smith, P.W. 1950. A third locality for the ringed salamander, *Ambystoma annulatum*. Copeia 1950:228.

Smith, P.W. 1956. The status, correct name, and geographic range of the boreal chorus frog. Proceedings of the Biological Society of Washington 69:169–176.

Smith, P.W. 1961. The amphibians and reptiles of Illinois. Bulletin of the Illinois Natural History Survey, Number 28, Urbana, Illinois.

Smith, P.W. 1966a. *Hyla avivoca*. Pp. 28.1–28.2. Catalogue of American Amphibians and Reptiles. Society for the Study of Amphibians and Reptiles, St. Louis, Missouri.

Smith, P.W. 1966b. *Pseudacris streckeri*. Pp. 27.1–27.2. Catalogue of American Amphibians and Reptiles. Society for the Study of Amphibians and Reptiles, St. Louis, Missouri.

Smith, P.W. and J.C. List. 1955. Notes on Mississippi amphibians and reptiles. American Midland Naturalist 53:115–125.

Smith, P.W. and S.A. Minton Jr. 1957. A distributional summary of the herpetofauna of Indiana and Illinois. American Midland Naturalist 58:341–351.

Smith, P.W. and D.M. Smith. 1952. The relationships of the chorus frogs, *Pseudacris nigrita feriarum* and *Pseudacris n. triseriata*. American Midland Naturalist 48:165–180.

Smith, R.E. 1941. Mating behavior in *Triturus torosus* and related newts. Copeia 1941:255–262.

Smith, R.J. 1967. Ancylid snails as intermediate hosts of *Megalodiscus temperatus* and other digenetic trematodes. Journal of Parasitology 53:287–291.

Smith, S.D. and A.L. Braswell. 1994. Preliminary investigation of acidity in ephemeral wetlands and the relationship to amphibian usage in North Carolina. Report to the Nongame and Endangered Wildlife Program, North Carolina Wildlife Resources Commission, Raleigh, North Carolina.

Smith, S.N. 1997. Geographic distribution: *Ascaphus truei* (tailed frog). Herpetological Review 28:47.

Smith, S.N. 1998. Geographic distribution: *Plethodon serratus* (southern redback salamander). Herpetological Review 29:246.

Smith, T.B. and R.K. Wayne. 1996. Molecular Genetic Approaches in Conservation. Oxford University Press, Oxford, United Kingdom.

Smith, T.G., S.H. Kopko and S.S. Desser. 1996. Life cycles, morphological characteristics, and host specificity of *Hepatozoon* species infecting eastern garter snakes from Ontario. Canadian Journal of Zoology 74:1850–1856.

Smith, T.G., B. Kim, H. Hong and S.S. Desser. 2000. Intraerythrocytic development of species of *Hepatozoon* infecting ranid frogs: evidence for convergence of life cycle characteristics among apicomplexans. Journal of Parasitology 83:451–458.

Smith, W.H. 1887. Report on the amphibians and reptiles of Ohio. Ohio Geological Survey 4:633–734.

Smith-Gill, S.J. 1974. Morphogenesis of the dorsal pigmentary pattern in wild-type and mutant *Rana pipiens*. Developmental Biology 37:153–170.

Smith-Gill, S.J. and K.A. Berven. 1979. Predicting amphibian metamorphosis. American Naturalist 113:563–585.

Smith-Gill, S.J. and D.E. Gill. 1978. Curvilinearities in the competition equations: an experiment with ranid tadpoles. American Naturalist 112:557–570.

Smits, A.W. 1984. Activity patterns and thermal biology of the toad *Bufo boreas halophilus*. Copeia 1984:689–696.

Smits, A.W. and D.L. Crawford. 1984. Emergence of toads to activity: a statistical analysis of contributing cues. Copeia 1984:696–701.

Smyers, S.D., M.J. Rubbo and R.G. Jaeger. 2001. Interactions between juvenile ambystomatid salamanders in a laboratory experiment. Copeia 2001:1017–1025.

Smyers, S.D, M.J. Rubbo, V.R. Townsend Jr. and C.C. Swart. 2002. Intra- and interspecific characterizations of burrow use and defense by juvenile ambystomatid salamanders. Herpetologica 58:422–429.

Smyth, J.D. and M.M. Smyth. 1980. Frogs as Host-parasite Systems. I. Macmillan Press, Hong Kong, China.

Snider, A. and K. Bowler. 1992. Longevity of reptiles and amphibians in North American Collections. Second edition. Herpetological Circular, Number 21, Society for the Study of Amphibians and Reptiles, St. Louis, Missouri.

Snieder, R. 2000. The tube worm turns. Science 406:939.

Snodgrass, J.W., J.W. Ackerman, A.L. Bryan Jr. and J. Burger. 1999. Influence of hydroperiod, isolation, and heterospecifics on the distribution of aquatic salamanders (*Siren* and *Amphiuma*) among depression wetlands. Copeia 1999:107–113.

Snyder, D.H. 1971. The function of brooding behavior in the plethodontid salamander, *Aneides aeneus*: a field study. Ph.D. dissertation. University of Notre Dame, South Bend, Indiana.

Snyder, D.H. 1991. The green salamander (*Aneides aeneus*) in Tennessee and Kentucky, with comments on the Carolinas' Blue Ridge populations. Journal of Tennessee Academy of Science 66:165–169.

Snyder, G.K. and G.A. Hammerson. 1993. Interrelationships between water economy and thermoregulation in the canyon tree-frog *Hyla arenicolor*. Journal of Arid Environments 25:321–329.

Snyder, J.O. 1923. Eggs of *Batrachoseps attenuatus*. Copeia 1923:86–88.

Snyder, R.C. 1956. Comparative features of the life histories of *Ambystoma gracile* (Baird) from populations at low and high altitudes. Copeia 1956:41–50.

Snyder, W.E. and S.G. Platt. 1997. Anuran records from the Piedmont of South Carolina, USA. Herpetological Review 28:53.

Snyder, W.F. and D.L. Jameson. 1965. Multivariable geographic variation of mating calls in populations of the Pacific tree frog (*Hyla regilla*). Copeia 1965:129–142.

Society for the Study of Amphibians and Reptiles Monetary Value of Amphibians Subcommittee. 1989. Monetary value of U.S. Amphibians. Herpetological Review 20(2S):1–4.

Sogandares-Bernal, F. and H. Grenier. 1971. Life cycles and host-specificity of the plagiorchiid trematodes *Ochetosoma kansensis* (Crow, 1913) and *O. laterotrema* (Byrd and Denton, 1938). Journal of Parasitology 57:297.

Soiseth, C.R. 1992. The pH and acid neutralizing capacity of ponds containing *Pseudacris regilla* larvae in an alpine basin of the Sierra Nevada. California Fish and Game 78:11–19.

Sokal, R.R. and F.J. Rohlf. 1995. Biometry. Third edition. W.H. Freeman and Company, New York.

Soledad Sepulveda, M., M.G. Spalding, J.M. Kinsella and D.J. Forrester. 1996. Parasitic helminths of the little blue heron, *Egretta caerulea*, in southern Florida. Journal of the Helminthological Society of Washington 63:136–140.

Soledad Sepulveda, M., M.G. Spalding, J.M. Kinsella and D.J. Forrester. 1999. Parasites of the great egret (*Ardea albus*) in Florida and a review of helminths reported for the species. Journal of the Helminthological Society of Washington 66:7–13.

Solomon, K.R., D.B. Baker, R.P. Richards, K.R. Dixon, S.J. Klaine, T.W. LaPoint, R.J. Kendall, C.P. Weisskopf, J.M. Giddings, J.P. Giesy, L.W. Hall and W.M. Williams. 1996. Ecological risk assessment of atrazine in North American surface waters. Environmental Toxicology and Chemistry 15:31–76.

Soltis, P. and M.A. Gitzendanner. 1999. Molecular systematics and the conservation of rare species. Conservation Biology 13:471–483.

Somers, A.B. 1990. Geographic distribution: *Ambystoma talpoideum* (mole salamander). Herpetological Review 21:95.

Sonin, M.D. 1968. Filariata of animals and man and diseases caused by them. Part 2. Diplotriaenoidea. *In* Skrjabin, K.I. (Ed.), Essentials of Nematology. Volume 31. Translated (1975) from Russian by Israel Program for Scientific Translations, Jerusalem.

Sonnini, C.S. and P.A. Latreille. 1801. Histoire naturelle des reptiles. Volume 2. Paris, France.

Souder, W. 2000. A Plague of Frogs: The Horrifying True Story. Hyperion Press, New York.

Soulé, M.E. (Ed.). 1987. Viable Populations. Cambridge University Press, New York.

Soulé, M.E. and B.A. Wilcox. 1980. Conservation Biology: An Evolutionary-Ecological Perspective. Sinauer Associates, Sunderland, Massachusetts.

Sousa, W.P. and E.D. Grosholz. 1991. The influence of habitat structure on the transmission of parasites. Pp. 300–324. *In* Bell, S.S., E.D. McCoy and H.R. Mushinsky (Eds.), Habitat Structure: The Physical Arrangement of Objects in Space. Chapman and Hall, London.

Southerland, M.T. 1986a. Behavioral interactions among four species of the salamander genus *Desmognathus*. Ecology 67:175–181.

Southerland, M.T. 1986b. Coexistence of three congeneric salamanders: the importance of habitat and body size. Ecology 67:721–728.

Southerland, M.T. 1986c. Organization in desmognathine salamander communities: the roles of habitat and biotic interactions. Dissertation Abstracts International, Part B, Sciences and Engineering 46:137.

Southerland, M.T. 1986d. The effects of variation in streamside habitats on the composition of mountain salamander communities. Copeia 1986:731–741.

Southerland, M.T. 1986e. Behavioral niche expansion in *Desmognathus fuscus* (Amphibia: Caudata: Plethodontidae). Copeia 1986:235–237.

Southerland, M.T. 1986f. Life history notes: *Leurognathus marmoratus* (shovelnose salamander). Climbing. Herpetological Review 17:45.

Southgate, V.R. 1997. Schistosomiasis in the Senegal River Basin: before and after the construction of the dams at Diama, Senegal and Manatali, Mali and future prospects. Journal of Helminthology 71:125–132.

Sovick, J. 1995. E-mail message, acquired through Freedom of Information Act, to Jerry Rogers, 17 April.

Sower, S.A., K.L. Reed and K.J. Babbitt. 2000. Limb malformations and abnormal sex hormone concentrations in frogs. Environmental Health Perspectives 108:1085–1090.

Spalding, R.F. and D.D. Snow. 1989. Stream levels of agrichemicals during a spring discharge event. Chemosphere 19:1129–1140.

Spalding, R.F., D.D. Snow, D.A. Cassada and M.E. Burbach. 1994. Study of pesticide occurrence in two closely spaced lakes in Northeastern Nebraska. Journal of Environmental Quality 23:571–578.

Sparks, T.C. 1996. Effects of predation risk on population variation in adult size in a stream-dwelling isopod. Oecologia 106:85–92.

Sparling, D.W. 1998. Field evidence for linking Altosid applications with increased amphibian deformities in southern leopard frogs. Abstract at: http://www.mpm.edu/collect/vertzo/herp/Daptf/MWabst.html.

Sparling, D.W. 2000. Ecotoxicology of organic compounds to amphibians. Pp. 461–494. *In* Sparling, D.W., G. Linder and C.A. Bishop (Eds.), Ecotoxicology of Amphibians and Reptiles. Society for Environmental Toxicology and Contaminants (SETAC) Press, Pensacola, Florida.

Sparling, D.W. and T.P. Lowe. 1996. Metal concentrations of tadpoles in experimental ponds. Environmental Pollution 91:149–159.

Sparling, D.W., C.A. Bishop and G. Linder. 2000. The current status of amphibian and reptile ecotoxicology research. Pp. 1–13. *In* Sparling, D.W., G. Linder and C.A. Bishop (Eds.), Ecotoxicology of Amphibians and Reptiles. Society for Environmental Toxicology and Contaminants (SETAC) Press, Pensacola, Florida.

Sparling, D.W., G.M. Fellers and L.L. McConnell. 2001. Pesticides and amphibian declines in California, USA. Environmental Toxicology and Chemistry 20:1591–1595.

Sparling, D.W., G. Linder and C.A. Bishop (Eds.). 2000. Ecotoxicology of Amphibians and Reptiles. Society of Environmental Toxicology and Chemistry (SETAC), SETAC Press, Pensacola, Florida.

Sparling, D.W., T.P. Lowe, D. Day and K. Dolan. 1995. Responses of amphibian populations to water and soil factors in experimentally treated aquatic macrocosms. Archives Environmental Contamination and Toxicology 29:455–461.

Speare, R. 1990. A review of the diseases of the cane toad, *Bufo marinus*, with comments on biological control. Australian Wildlife Research 17:387–410.

Spencer, B. 1999. The Wyoming toad SSP. Endangered Species Bulletin 24:18–19.

Spencer, J. 1995. Protest at Bastrop State Park results in arrest of Austin man. Daily Texan, 25 July.

Spicer, J.I. and K.J. Gaston. 1999. Physiological Diversity and its Ecological Implications. Blackwell Science, Oxford, United Kingdom.

Spight, T.M. 1967a. The water economy of salamanders: water uptake after dehydration. Comparative Biochemistry and Physiology 20:767–771.

Spight, T.M. 1967b. The water economy of salamanders: exchange of water with the soil. Biological Bulletin 132:126–132.

Spight, T.M. 1967c. Population structure and biomass production by a stream salamander. American Midland Naturalist 78:437–447.

Spight, T.M. 1968. The water economy of salamanders: evaporative water loss. Physiological Zoology 41:195–203.

Spolsky, C., C.A. Phillips and T. Uzzell. 1992. Gynogenetic reproduction in hybrid mole salamanders (genus *Ambystoma*). Evolution 46:1935–1944.

Spotila, J.R. 1972. Role of temperature and water in the ecology of lungless salamanders. Ecological Monographs 42:95–125.

Spotila, J.R. 1976. Courtship behavior of the ringed salamander (*Ambystoma annulatum*): observations in the field. Southwestern Naturalist 21:412–413.

Spotila, J.R. and R.J. Beumer. 1970. The breeding habits of the ringed salamander, *Ambystoma annulatum* (Cope), in northwestern Arkansas. American Midland Naturalist 84:77–89.

Spotila, J.R. and P.H. Ireland. 1970. Notes on the eggs of the gray-bellied salamander, *Eurycea multiplicata griseogaster*. Southwestern Naturalist 14:366–368.

Springer, S. 1938. On the size of *Rana sphenocephala*. Copeia 1938:49.

Sprules, W.G. 1972. Effects of size-selective predation and food competition on high altitude zooplankton communities. Ecology 53:375–386.

Sprules, W.G. 1974a. The adaptive significance of paedogenesis in North American species of *Ambystoma* (Amphibia: Caudata): an hypothesis. Canadian Journal of Zoology 52:393–400.

Sprules, W.G. 1974b. Environmental factors and the incidence of neoteny in *Ambystoma gracile* (Baird) (Amphibia: Caudata). Canadian Journal of Zoology 52:1545–1552.

Sredl, M.J. 1996. From rags to riches: evolution of ranid frog conservation and management in Arizona. Annual meeting of The Wildlife Society, Western Section, Sparks, Nevada.

Sredl, M.J. and D. Caldwell. 2000a. Update on chytrid surveys of frogs from Arizona. Website at: http://www.jcu.edu.au/school/phtm/PHTM/frogs/sredl)).htm.

Sredl, M.J. and D. Caldwell. 2000b. Wintertime populations surveys—call for volunteers. Sonoran Herpetologist 13:1.

Sredl, M.J. and J.P. Collins. 1992. The interaction of predation, competition, and habitat complexity in structuring an amphibian community. Copeia 1992:607–614.

Sredl, M.J. and J.M. Howland. 1995. Conservation and management of Madrean populations of the Chiricahua leopard frog, *Rana chiricahuensis*. Pp. 379–385. *In* DeBano, L.F., G.J. Gottfried, R.H. Hamre, C.B. Edminster, P.F. Ffolliott and A. Ortega-Rubio (Eds.), Biodiversity and management of the Madrean Archipelago: The Sky Islands of southwestern United States and northwestern Mexico. Rocky Mountain Forest and Range Experiment Station, Fort Collins, Colorado.

Sredl, M.J. and L.S. Saylor. 1998. Conservation and Management Zones and the role of earthen cattle tanks in conserving Arizona leopard frogs on large landscapes. Pp. 211–225. *In* Feller, J.M. and D.S. Strouse (Eds.), Environmental, Economic, and Legal Issues Related to Rangeland Water Developments. Center for the Study of Law, Science and Technology, Arizona State University, Tempe, Arizona.

Sredl, M.J., E.P. Collins and J.M. Howland. 1997. Mark–recapture studies of Arizona leopard frogs. Pp. 1–35. *In* Sredl, M.J. (Ed.), Ranid Frog

Conservation and Management. Nongame and Endangered Wildlife Program, Technical Report, Number 121, Arizona Game and Fish Department, Phoenix, Arizona.

Sredl, M.J., P.C. Rosen and G.A. Bradley. 2000. Chytrid fungus widespread in Arizona. Annual Meeting of the American Society of Ichthyologists and Herpetologists, La Paz, Mexico.

Sredl, M.J., J.M. Howland, J.E. Wallace and L.S. Saylor. 1997. Status and distribution of Arizona's native ranid frogs. Pp. 45–101. In Sredl, M.J. (Ed.), Ranid Frog Conservation and Management. Nongame and Endangered Wildlife Program, Technical Report, Number 121, Arizona Game and Fish Department, Phoenix, Arizona.

St. Amant, J.A. 1975. Exotic visitor becomes permanent resident. Terra, Los Angeles County Museum of Natural History Quarterly 13:22–23.

St. Amant, J.A. and F.G. Hoover. 1969. Addition of Misgurnus anguillicaudatus (Cantor) to the California fauna. California Fish and Game 55:330–331.

St. Amant, J.A., F.G. Hoover and G.R. Stewart. 1973. African clawed frog, Xenopus laevis (Daudin), established in California. California Fish and Game 59:151–153.

St. Cloud, S.F. 1966. Observation by J. James at Tinaroo Creek, February, 1966. Journal of the North Queensland Naturalists' Club 34:6.

St. John, A.D. 1985. The herpetology of the Owyhee River drainage of Malheur County, Oregon. Nongame Wildlife Program, Technical Report, Number 85-5-03, Oregon Department of Fish and Wildlife, Portland, Oregon.

St. John, A.D. 1987. The herpetology of the Willamette Valley, Oregon. Oregon Department of Fish and Wildlife, Nongame Wildlife Program, Technical Report, Number 87-3-01, Portland, Oregon.

Stabler, R.M. and T. Chen. 1936. Observations on an Endamoeba parasitizing opalinid ciliates. Biological Bulletin 70:56–71.

Stacey, P.B. and M. Taper. 1992. Environmental variation and the persistence of small populations. Ecological Applications 2:18–29.

Stackpole, C.W. 1969. Herpes-type virus of the frog renal adenocarcinoma. I. Virus development in tumor transplants maintained at low temperature. Journal of Virology 4:75–93.

Stafford, J. 1905. Trematodes from Canadian vertebrates. Zoologischer Anzeiger 28:681–694.

Stafford, J. 1971. Heron populations of England and Wales 1928–1970. Bird Study 18:218–221.

Stallard, R.F. 2001. Possible environmental factors underlying amphibian decline in eastern Puerto Rico: analysis of U.S. government data archives. Conservation Biology 15:943–953.

Standaert, W.F. 1967. Growth, maturation, and population ecology of the carpenter frog (Rana virgatipes, Cope). Ph.D. dissertation. Rutgers University, New Brunswick, New Jersey.

Stangel, P.W. 1983. Least sandpiper predation on Bufo americanus and Ambystoma maculatum larvae. Herpetological Review 14:112.

Stangel, P.W. 1988. Premetamorphic survival of the salamander Ambystoma maculatum, in eastern Massachusetts. Journal of Herpetology 22:345–347.

Stangier, U. 1988. Kleingewässerrückgang im westlichen Münsterland und heutige potentielle Vernetzung der Amphibienpopulationen. Jahrbuch für Feldherpetologie, Beiheft 1:117–127.

Starnes, S.M., C.A. Kennedy and J.W. Petranka. 2000. Sensitivity of embryos of southern Appalachian amphibians to ambient solar UV-B radiation. Conservation Biology 14:277–282.

Starrett, P. 1960. Descriptions of tadpoles of Middle American frogs. Miscellaneous Publications of the Museum of Zoology, Number 110, University of Michigan, Ann Arbor, Michigan.

Staub, N. 1986. A status survey of the Sacramento Mountain salamander, Aneides hardii, with an assessment of the impact of logging on salamander abundance. Final report submitted to the U.S. Fish and Wildlife Service, Endangered Species Office, Albuquerque, New Mexico.

Staub, N. 1989. The evolution of sexual dimorphism in the salamander genus Aneides (Amphibia: Plethodontidae). Ph.D. dissertation. University of California, Berkeley, California.

Staub, N.L. 1993. Intraspecific agonistic behavior of the salamander Aneides flavipunctatus (Amphibia: Plethodontidae) with comparisons to other plethodontid species. Herpetologica 49:271–282.

Staub, N.L. 1995. Natural history notes: Plethodon idahoensis (Coeur d'Alene salamander). Predation. Herpetological Review 26:199.

Staub, N.L. and R. Anderson. 2001. A novel behavior for urodeles: micturition in the plethodontid salamander Aneides lugubris. Herpetological Review 32:158–159.

Staub, N.L., C.W. Brown and D.B. Wake. 1995. Patterns of growth and movements in a population of Ensatina eschscholtzii platensis (Cau-

data: Plethodontidae) in the Sierra Nevada, California. Journal of Herpetology 29:593–599.

Stauffer J.R., M.E. Arnegard, M. Cetron, J.J. Sullivan, L.A. Chitsulo, G.F. Turner, A. Chiotha and K.R. McKaye. 1997. Controlling vectors and hosts of parasitic diseases using fishes. A case history of schistosomiasis in Lake Malawi. Bioscience 47:41–48.

Stauffer, J.R., Jr., J.E. Gates and W.L. Goodfellow. 1983. Preferred temperature of two sympatric Ambystoma larvae: a proximate factor in niche segregation? Copeia 1983:1001–1005.

Stearns, B.P. and S.C. Stearns. 1999. Watching, From the Edge of Extinction. Yale University Press, New Haven, Connecticut.

Stearns, F. and D. Lindsey. 1977. Environmental status of the Lake Michigan Region. Volume 11. Natural areas of the Lake Michigan Drainage Basin and endangered or threatened plant and animal species. Argonne National Laboratory, Report ANL/ES-40, Argonne, Illinois.

Stearns, S.C. 1992. The Evolution of Life Histories. Oxford University Press, Oxford, United Kingdom.

Stebbins, R.C. 1947. Tail and foot action in the locomotion of Hydromantes platycephalus. Copeia 1947:1–5.

Stebbins, R.C. 1949a. Observations on the laying, development, and hatching of the eggs of Batrachoseps wrighti. Copeia 1949:161–168.

Stebbins, R.C. 1949b. Speciation in salamanders of the plethodontid genus Ensatina. University of California Publications in Zoology 48:377–526.

Stebbins, R.C. 1951. Amphibians of Western North America. University of California Press, Berkeley, California.

Stebbins, R.C. 1953. Southern occurrence of the Olympic salamander, Rhyacotriton olympicus. Herpetologica 11:238–239.

Stebbins, R.C. 1954a. Amphibians and Reptiles of Western North America. McGraw-Hill Publishing Company, New York.

Stebbins, R.C. 1954b. Natural history of the salamanders of the plethodontid genus Ensatina. University of California Publications in Zoology 54:47–124.

Stebbins, R.C. 1957. Intraspecific sympatry in the lungless salamander Ensatina eschscholtzii. Evolution 11:265–270.

Stebbins, R.C. 1959. Reptiles and Amphibians of the San Francisco Bay Region. University of California Press, Berkeley, California.

Stebbins, R.C. 1962. Amphibians of Western North America. University of California Press, Berkeley, California.

Stebbins, R.C. 1966. A Field Guide to Western Reptiles and Amphibians. Houghton Mifflin Company, Boston, Massachusetts.

Stebbins, R.C. 1972. California Amphibians and Reptiles. University of California Press, Berkeley, California.

Stebbins, R.C. 1985. A Field Guide to Western Reptiles and Amphibians. Second edition. Houghton Mifflin Company, Boston, Massachusetts.

Stebbins, R.C. and N.W. Cohen. 1995. A Natural History of Amphibians. Princeton University Press, Princeton, New Jersey.

Stebbins, R.C. and C.H. Lowe Jr. 1951. Subspecific differentiation in the Olympic salamander, Rhyacotriton olympicus. University of California Publications in Zoology 50:465–484.

Stebbins, R.C. and W.J. Reimer. 1950. A new species of plethodontid salamander from the Jemez Mountains of New Mexico. Copeia 1950:73–80.

Stebbins, R.C. and H.C. Reynolds. 1947. Southern extension of the range of the Del Norte salamander in California. Herpetologica 4: 41–42.

Steele, C.A. 2001. Effect of forest age on density of stream-breeding amphibians in Skamania County, Washington. Master's thesis. Utah State University, Logan, Utah.

Steele, C.A., E.D. Brodie Jr. and J.G. MacCracken. 2003. Natural history notes: Dicamptodon copeia (Cope's giant salamander). Reproduction. Herpetological Review 34:227–228.

Steele, C.W., S. Strickler-Shaw and D.H. Taylor 1991. Failure of Bufo americanus tadpoles to avoid lead-enriched water. Journal of Herpetology 25:241–243.

Steele, C.W., S. Strickler-Shaw and D.H. Taylor. 1999. Effects of sublethal lead exposure on the behaviors of green frog (Rana clamitans), bullfrog (Rana catesbeiana) and American toad (Bufo americanus) tadpoles. Marine and Freshwater Behavior Physiology 32:1–16.

Stein, J. 1996. Amargosa toad (Bufo nelsoni) summer survey report—1996. Nevada Division of Wildlife, Las Vegas, Nevada (available from Nevada Natural Heritage Program, Carson City, Nevada).

Stein, J. 1999. Amargosa toad (Bufo nelsoni). Population monitoring and habitat evaluation of the Amargosa toad in Oasis Valley, Nevada. 1998 report of activities. Nevada Division of Wildlife, Las Vegas, Nevada (available from Nevada Natural Heritage Program, Carson City, Nevada).

Stein, J., B. Hobbs and G.A. Wasley. 2000. Population monitoring of the Amargosa toad (Bufo nelsoni) and habitat evaluation in Oasis Valley, Nevada. Nevada Division of Wildlife, Las Vegas. Cited in Nevada Division of Wildlife. 2000. Conservation agreement for the Amargosa toad (Bufo nelsoni) and co-occurring sensitive species in Oasis Valley, Nye County, Nevada (September 2000).

Stein, K.F. 1938. Migration of Triturus viridescens. Copeia 1938:86–88.

Stein, R.J. 1989. Mantis preys on salamander. Bulletin of the Maryland Herpetological Society 25:60–61.

Steiner, G. 1924. Some nemas from the alimentary tract of the Carolina green treefrog (Hyla carolinensis Pennant). Journal of Parasitology 11:1–32.

Steinwascher, K. 1978a. The effect of coprophagy on the growth of Rana catesbeiana tadpoles. Copeia 1978:130–134.

Steinwascher, K. 1978b. Interference and exploitative competition among tadpoles of Rana utricularia. Ecology 57:1289–1296.

Steinwascher, K. and J. Travis. 1983. Influence of food quality and quantity on early larval growth in two anurans. Copeia 1983:238–242.

Stejneger, L.H. 1893 (1892). Preliminary description of a new genus and species of blind cave salamander from North America. Proceedings of the United States National Museum 15:115–117.

Stejneger, L.H. 1893. Annotated list of the reptiles and batrachians collected by the Death Valley Expedition in 1891, with descriptions of new species. North American Fauna 7:159–228.

Stejneger, L.H. 1894 (1895). A new salamander from Arkansas with notes on Ambystoma annulatum. Proceedings of the U.S. National Museum 17:597–599.

Stejneger, L.H. 1896. Description of a new genus and species of blind tailed batrachians from the subterranean waters of Texas. Proceedings of the U.S. National Museum 18:619–621.

Stejneger, L.H. 1896. Description of a new genus and species of blind Texan cave salamander, Typhlomolge rathbuni. Proceedings of the U.S. National Museum, Volume 18, Washington, D.C.

Stejneger, L.H. 1899. Description of a new genus and species of discoglossid toad from North America. Proceedings of the United States National Museum 21:899–901.

Stejneger, L.H. 1901. A new species of bullfrog from Florida and the Gulf Coast. Proceedings of the United States National Museum 24:211–215.

Stejneger, L.H. 1906. A new salamander from North California. Proceedings of the United States National Museum 30:559–562.

Stejneger, L.H. 1907. Herpetology of Japan and adjacent territory. Bulletin of the United States National Museum, Number 58, Washington, D.C.

Stejneger, L.H. 1915. A new species of tailless batrachian from North America. Proceedings of the Biological Society of Washington 28:131–132.

Stejneger, L.H. and T. Barbour. 1917. A Check List of North American Amphibians and Reptiles. Harvard University Press, Cambridge, Massachusetts.

Stejneger, L.H. and T. Barbour. 1923. A Check List of North American Amphibians and Reptiles. Second edition. Harvard University Press, Cambridge, Massachusetts.

Stejneger, L.H. and T. Barbour. 1933. A Check List of North American Amphibians and Reptiles. Third edition. Harvard University Press, Cambridge, Massachusetts.

Stejneger, L.H. and T. Barbour. 1939. A Check List of North American Amphibians and Reptiles. Fourth edition. Harvard University Press, Cambridge, Massachusetts.

Stejneger, L.H. and T. Barbour. 1943. A Check List of North American Amphibians and Reptiles. Fifth edition. Bulletin of the Museum of Comparative Zoology, Number 93, Harvard University Press, Cambridge, Massachusetts.

Stelmock, J.J. and A.S. Harestad. 1979. Food habits and life history of the clouded salamander (Aneides ferreus) on northern Vancouver Island, British Columbia. Syesis 12:71–75.

Stenhouse, S.L. 1984. Coexistence of the salamanders Ambystoma maculatum and Ambystoma opacum: predation and competition. Ph.D. dissertation. University of North Carolina at Chapel Hill, North Carolina.

Stenhouse, S.L. 1985a. Migratory orientation and homing in Ambystoma maculatum and Ambystoma opacum. Copeia 1985:631–637.

Stenhouse, S.L. 1985b. Interdemic variation in predation on salamander larvae. Ecology 66:1706–1717.

Stenhouse, S.L. 1987. Embryo mortality and recruitment of juveniles of Ambystoma maculatum and Ambystoma opacum in North Carolina. Herpetologica 43:496–501.

Stenhouse, S.L., N.G. Hairston and A.E. Cobey. 1983. Predation and competition in Ambystoma larvae: field and laboratory experiments. Journal of Herpetology 17:210–220.

Stenseth, N.T. 1993. The Biology of Lemmings. Academic Press, New York.

Stephens, P.A. and W.J. Sutherland. 1999. Consequences of the Allee effect for behavior, ecology and conservation. TRENDS in Ecology and Evolution 14:401–405.

Stephenson, J.R. and G.M. Calcarone. 1999. Southern California mountains and foothills assessment: habitat and species conservation issues. U.S.D.A. Forest Service, Pacific Southwest Research Station, General Technical Report, GTR-PSW-172, Albany, California.

Stern, S.L. and C.F. Mueller. 1972. Diurnal variation in the oxygen consumption of Plethodon cinereus. American Midland Naturalist 88:502–506.

Stevens, M.J. 1987. Hydrochemische Untersuchungen an einigen Laichplätzen der Echten Wassermolche (Gattung Triturus Rafinesque, 1815) im Kreis Viersen (Caudata: Salamandridae). Salamandra 23:166–172.

Stevenson, D.J. and A.E. Davis Jr. 1995. A summary of rare herpetofaunal species surveys at Fort Stewart, Georgia. Nature Conservancy, Fort Stewart Inventory, Unpublished report, Savannah, Georgia.

Stevenson, D.J., J.B. Jensen and K.R. Tassin. 1998. Geographic distribution: Notophthalmus perstriatus (striped newt). Herpetological Review 29:245–246.

Stevenson, H.M. 1967. Additional specimens of Amphiuma pholeter from Florida. Herpetologica 23:134.

Stevenson, H.M. 1976. Vertebrates of Florida. University Press of Florida, Gainesville, Florida.

Stewart, G.D. 1953. Notes on a collection of amphibians from central Korea. Herpetologica 9:146–148.

Stewart, G.D. and E.D. Bellis. 1970. Dispersion patterns of salamanders along a brook. Copeia 1970:86–89.

Stewart, G.R., M.R. Jennings and R.H. Goodman Jr. in press. Sensitive species of snakes, frogs, and salamanders in southern California conifer forest areas: status and management. In (Eds.) Proceedings of the Symposium on Planning for Biodiversity: Bringing Research and Management Together. U.S.G.S. Western Ecological Research Center and the U.S.D.A. Forest Service, Pacific Southwest Research Station, Arcata, California.

Stewart, J.E. 1991. Introductions as factors in diseases of fish and aquatic invertebrates. Canadian Journal of Fish and Aquatic Science 45:110–117.

Stewart, M.McB. and L.F. Biuso. 1982. A bibliography of the green frog, Rana clamitans Latreille, 1801–1981. Smithsonian Herpetological Information Service, Number 56, Smithsonian Institution, Washington, D.C.

Stewart, M.M. 1956. The separate effects of food and temperature differences on development of marbled salamander larvae. Journal of the Elisha Mitchell Scientific Society 72:47–56.

Stewart, M.M. 1968. Population dynamics of Eurycea bislineata in New York. Journal of Herpetology 2:176–177 (Abstract).

Stewart, M.M. 1974. Parallel pattern polymorphism in the genus Phrynobatrachus (Amphibia: Ranidae). Copeia 1974:823–832.

Stewart, M.M. 1979. The role of introduced species in a Jamaican frog community. Pp. 113–146. In Wolda, H. (Ed.), Proceedings of the IVth Symposium of Tropical Ecology, Panama City, Panama.

Stewart, M.M. 1985. Arboreal habitat use and parachuting by a subtropical forest frog. Journal of Herpetology 19:391–401.

Stewart, M.M. 1995. Climate driven population fluctuations in rainforest frogs. Journal of Herpetology 29:437–446.

Stewart, M.M. and F.H. Pough. 1983. Population density of tropical forest frogs: relation to retreat sites. Science 221:570–572.

Stewart, M.M. and A.S. Rand. 1991. Vocalizations and the defense of retreat sites by male and female frogs, Eleutherodactylus coqui. Copeia 1991:1013–1024.

Stewart, M.M. and J. Rossi. 1981. The Albany Pine Bush: A northern outpost for southern species of amphibians and reptiles in New York. American Midland Naturalist 106:282–292.

Stewart, M.M. and P. Sandison. 1972. Comparative food habits of sympatric mink frogs, bullfrogs, and green frogs. Journal of Herpetology 6:241–244.

Stewart, M.M. and L.L. Woolbright. 1996. Amphibians. Pp. 273–320. In Reagan, D.P. and R.B. Waide (Eds.), The Food Web of a Tropical Rain Forest. University of Chicago Press, Chicago, Illinois.

Stewart, R.E. and H.A. Kantrud. 1971. Classification of natural ponds and lakes in the glaciated prairie region. U.S. Fish and Wildlife Service, Resource Publication, Number 92, Washington, D.C.

Stiassny, M.L.J. 1992. Phylogenetic analysis and the role of systematics in the biodiversity crisis. Pp. 109–120. *In* Eldredge, N. (Ed.), Systematics, Ecology and the Biodiversity Crisis. Columbia University Press, New York.

Stiassny, M.L.J. and M.C.C. de Pinna. 1994. Basal taxa and the role of cladistic patterns in the evaluation of conservation priorities: a view from freshwater. Pp. 81–100. *In* Forey, P.L., C.J. Humphries and R.I. Vane-Wright (Eds.), Systematics and Conservation Evaluation. Clarendon Press, Oxford, United Kingdom.

Stickley, A.R., J.F. Glahn, J.O. King and D.T. King. 1995. Impact of great blue heron depredations on channel catfish farms. Journal of the World Aquaculture Society 26:194–199.

Stiling, P. 1988. Density-dependent process and key factors in insect populations. Journal of Animal Ecology 57:581–594.

Still, C.J., P.N. Fosters and S.H. Schneider. 1999. Simulating the effects of climate change on tropical montane cloud forests. Nature 398: 608–610.

Stille, W.T. 1952. The nocturnal amphibian fauna of the southern Lake Michigan beach. Ecology 33:149–162.

Stille, W.T. 1954. Eggs of the salamander *Ambystoma jeffersonianum* in the Chicago area. Copeia 1954:300.

Stille, W.T. 1958. The water absorption response of an anuran. Copeia 1958:217–218.

Stine, C.J., Jr., J.A. Fowler and R.S. Simmons. 1954. Occurrence of the eastern tiger salamander, *Ambystoma tigrinum tigrinum* (Green) in Maryland, with notes on its life history. Annals of the Carnegie Museum 33:145–148.

Stinner, J., N. Zarlinga and S. Orcutt. 1994. Overwintering behavior of adult bullfrogs, *Rana catesbeiana*, in northeastern Ohio. Ohio Journal of Science 94:8–13.

Stoddard, H.L. 1935. Use of controlled fire in southeastern upland game management. Journal of Forestry 33:346–351.

Stohlgren, T.J., T.N. Chase, R.A. Pielke, T.G. Kittel and J.S. Baron. 1998. Evidence that local land use practices influence regional climate, vegetation, and stream flow patterns in adjacent natural areas. Global Change Biology 4:495–505.

Stokes, D.E. 1997. Pasteur's Quadrant. Brookings Institution Press, Washington, D.C.

Stokes, W.L. 1986. Geology of Utah. Utah Geological Mineral Survey and the Utah Museum of Natural History, Utah Museum of Natural History Publication, Number 6, Salt Lake City, Utah.

Stolarek, J. 1997. County's water planning needs discussed. Bastrop Advertiser, 16 August.

Stone, L.S. 1964a. The structure and visual function of the eye of larval and adult cave salamanders *Typhlotriton spelaeus*. Journal of Experimental Zoology 156:201–218.

Stone, L.S. 1964b. Return of vision in transplanted larval eyes of cave salamanders. Journal of Experimental Zoology 156:219–228.

Stone, M.D. 1991. Wyoming Toad Recovery Plan. U.S. Fish and Wildlife Service, Denver, Colorado.

Stone, W. 1932. Terrestrial activity of spade-foot toads. Copeia 1932: 35–36.

Stone, W.B. and R.D. Manwell. 1969. Toxoplasmosis in cold blooded hosts. Journal of Protozoology 16:99–102.

Storer, T.I. 1922. The eastern bullfrog in California. California Fish and Game 8:219–224.

Storer, T.I. 1925. A synopsis of the Amphibia of California. University of California Publications in Zoology, Number 27, Berkeley, California.

Storey, K.B. 1984. Freeze tolerance in the frog, *Rana sylvatica*. Experientia 40:1261–1262.

Storey, K.B. and J.M. Storey. 1986. Freeze tolerance and intolerance as strategies of winter survival in terrestrially-hibernating amphibians. Comparative Biochemistry and Physiology A 83:613–617.

Storey, K.B. and J.M. Storey. 1987. Persistence of freeze tolerance in terrestrially hibernating frogs after spring emergence. Copeia 1987:720–726.

Storez, R.A. 1969. Observations on the courtship of *Ambystoma laterale*. Journal of Herpetology 3:87–95.

Storfer, A. 1999a. Gene flow and local adaptation in a sunfish-salamander system. Behavioral Ecology and Sociobiology 46:273–279.

Storfer, A. 1999b. Gene flow and population subdivision in the streamside salamander, *Ambystoma barbouri*. Copeia 1999:174–178.

Storfer, A. and A. Sih. 1998. Gene flow and ineffective antipredator behavior in a stream-breeding salamander. Evolution 52:558–565.

Storfer, A., J. Cross, V. Rush and J. Caruso. 1999. Adaptive coloration and gene flow as a constraint to local adaptation in the streamside salamander, *Ambystoma barbouri*. Evolution 53:889–898.

Storfer, S. 1996. Quantitative genetics; a promising approach for the assessment of genetic variation in endangered species. TRENDS in Ecology and Evolution 11:343–348.

Storm, R.M. 1947. Eggs and young of *Aneides ferreus*. Herpetologica 4:60–62.

Storm, R.M. 1948. The herpetology of Benton County, Oregon. Ph.D. dissertation. Oregon State College, Corvallis, Oregon.

Storm, R.M. 1955. Northern and southern range limits of Dunn's salamander, *Plethodon dunni*. Copeia 1955:64–65.

Storm, R.M. 1960. Notes on the breeding biology of the red-legged frog (*Rana aurora aurora*). Herpetologica 16:251–259.

Storm, R.M. 1966. Endangered plants and animals of Oregon: II. Amphibians and reptiles. Agricultural Experiment Station. Oregon State University, Special Report 206, Corvallis, Oregon.

Storm, R.M. 1986. Current status of Oregon amphibians and reptiles–a brief review. Oregon Nongame Wildlife Management Plan, Appendix 8, Oregon Department of Fish and Wildlife, Portland, Oregon.

Storm, R.M. and A.R. Aller. 1947. Food habits of *Aneides ferreus*. Herpetologica 4:59–60.

Storm, R.M. and E.D. Brodie Jr. 1970a. *Plethodon dunni*. Pp. 81.1–81.2. Catalogue of American Amphibians and Reptiles. Society for the Study of Amphibians and Reptiles, St. Louis, Missouri.

Storm, R.M. and E.D. Brodie Jr. 1970b. *Plethodon vehiculum*. Pp. 82.1–82.2. Catalogue of American Amphibians and Reptiles. Society for the Study of Amphibians and Reptiles, St. Louis, Missouri.

Storm, R.M. and R.A. Pimentel. 1954. A method for studying amphibian breeding populations. Herpetologica 10:161–166.

Stout, B.M., III, K.K. Stout and C.W. Stihler. 1992. Predation by the caddisfly *Banksiola dossuaria* on egg masses of the spotted salamander *Ambystoma maculatum*. American Midland Naturalist 127: 368–372.

Strauch, A. 1870. Revision der salamandriden Gattungen. Memoires de l'Academy imp. Sciences, Series 7, Number 16, St. Petersburg, Russia.

Straughan, I.R. 1966. The natural history of the "cane toad" in Queensland. Australian Natural History 15:230–232.

Straughan, I.R. 1975. An analysis of the mechanisms of mating call discrimination in the frogs *Hyla regilla* and *H. cadaverina*. Copeia 1975:415–424.

Strecker, J.K. 1902. Reptiles and batrachians on McLennan County, Texas. Proceedings and Transactions of the Texas Academy of Science 4, Part 2:95–101.

Strecker, J.K. 1908a. Notes on the habits of two Arkansas salamanders and a list of batrachians and reptiles collected at Hot Springs. Proceedings of the Biological Society of Washington 21:85–89.

Strecker, J.K. 1908b. The reptiles and batrachians of Victoria and Refugio Counties, Texas. Proceedings of the Biological Society of Washington 21:47–52.

Strecker, J.K. 1908c. A preliminary annotated list of the Batrachia of Texas. Proceedings of the Biological Society of Washington 21:53–62.

Strecker, J.K. 1910. Description of a new solitary spadefoot (*Scaphiopus hurterii*) from Texas, with other herpetological notes. Proceedings of the Biological Society of Washington 23:115–122.

Strecker, J.K. 1922. An annotated catalogue of the amphibians and reptiles of Bexar County, Texas. Science Society of San Antonio 4:1–31.

Strecker, J.K. 1926. Chapters from the life-histories of Texas reptiles and amphibians. Part I. Contributions of the Baylor University Museum, Number 8, Waco, Texas.

Strecker, J.K. 1927. Observations on the food habits of Texas amphibians and reptiles. Copeia 162:6–9.

Strecker, J.K. 1933. Collecting at Helotes, Bexar County, Texas. Copeia 1933:77–79.

Strecker, J.K. and W.J. Williams. 1928. Field notes on the herpetology of Bowie County, Texas. Contributions of the Baylor University Museum, Number 17, Waco, Texas.

Strotthatte-Moormann, M. and D. Forman. 1992. Gullys als verheerende Kleinteirfallen in einem Aachener Parkgelände. Beiträge zur Erforschung der Dortmunder Herpetofauna 18:177–191.

Strüssmann, C., M.B.R.D. Vale and M.H. Meneghini. 1984. Diet and foraging mode of *Bufo marinus* and *Leptodactylus ocellatus*. Journal of Herpetology 18:138–146.

Stuart, J.N. 1992. Status survey of the spotted chorus frog *(Pseudacris clarkii)* in New Mexico. Unpublished report to the New Mexico Department of Game and Fish, Santa Fe, New Mexico.

Stuart, J.N. 1995. Natural history notes: *Rana catesbeiana* (bullfrog). Diet. Herpetological Review 26:33.

Stuart, J.N. and C.W. Painter. 1993. Life history notes: *Rana catesbeiana* (bullfrog). Cannibalism. Herpetological Review 24:103.

Stuart, J.N. and C.W. Painter. 1994. A review of the distribution and status of the boreal toad, *Bufo boreas boreas*, in New Mexico. Bulletin of the Chicago Herpetological Society 29:113–116.

Stuart, J.N. and C.W. Painter. 1996. Natural history notes on the Great Plains narrowmouth toad, *Gastrophryne olivacea*, in New Mexico. Bulletin of the Chicago Herpetological Society 31:44–47.

Stuart, L.C. 1934. A contribution to a knowledge of the herpetological fauna of El Petén, Guatemala. Occasional Papers of the Museum of Zoology, Number 292, University of Michigan, Ann Arbor, Michigan.

Stuart, L.C. 1935. A contribution to a knowledge of the herpetology of a portion of the savanna region of central El Petén, Guatemala. Miscellaneous Publications of the Museum of Zoology, Number 29, University of Michigan, Ann Arbor, Michigan.

Stuart, L.C. 1948. The amphibians and reptiles of Alta Verapaz, Guatemala. Miscellaneous Publications of the Museum of Zoology, Number 69, University of Michigan, Ann Arbor, Michigan.

Stuart, L.C. 1951. The distributional implications of temperature tolerances and hemoglobin values in the toads *Bufo marinus* (Linnaeus) and *Bufo bocourti* Brocchi. Copeia 1951:220–229.

Stuart, L.C. 1961. Some observations on the natural history of tadpoles of *Rhinophrynus dorsalis* Duméril and Bibron. Herpetologica 17:73–79.

Stubbe, M., D. Heidecke, D. Dolch, J. Teubner, R. Labes, H. Ansorge, H. Mau and D. Blanke. 1993. Monitoring Fischotter 1985–1991. Tiere im Konflikt 1:11–59.

Stumpel, A.H.P. 1992. Successful reproduction of introduced bullfrogs *Rana catesbeiana* in northwestern Europe: a potential threat to indigenous amphibians. Biological Conservation 60:61–62.

Stumpel, A.H.P. 1993. The terrestrial habitat of *Hyla arborea*. Pp. 47–54. *In* Stumpel, A.H.P. and U. Tester (Eds.), Ecology and Conservation of the European Tree Frog. DLO Institute for Forestry and Nature Research, Wageningen, Netherlands.

Stupka, A. 1953. Some notes relating to the mortality of screech owls in Great Smoky Mountains National Park. Migrant 24:3–5.

Sturdivant, H.P. 1949. The sperm cycle in *Amphiuma*. Journal of the Tennessee Academy of Science 24:4–11.

Suchomel, J. 1998a. Pine tree loggers pull out under fire. Bastrop Advertiser, 22 January.

Suchomel, J. 1998b. Pine loggers raising concern. Bastrop Advertiser, 24 January.

Suchomel, J. 1998c. Logger labels BCEN charge unfair to him. Bastrop Advertiser, 14 February.

Sugalski, M.T. and D.L. Claussen. 1997. Preference for soil moisture, soil pH, and light intensity by the salamander, *Plethodon cinereus*. Journal of Herpetology 31:245–250.

Sugg, D.W., A.A. Karlin, C.R. Preston and D.R. Heath. 1988. Morphological variation in a population of the salamander, *Siren intermedia nettingi*. Journal of Herpetology 22:243–247.

Sullivan, B.K. 1982a. Male mating behaviour in the Great Plains toad *(Bufo cognatus)*. Animal Behaviour 30:939–940.

Sullivan, B.K. 1982b. Sexual selection in Woodhouse's toad *(Bufo woodhousei)*. I. Chorus organization. Animal Behaviour 30:680–686.

Sullivan, B.K. 1983. Sexual selection in the Great Plains toad *(Bufo cognatus* Say). Behaviour 84:258–264.

Sullivan, B.K. 1984. Advertisement call variation and observations on breeding behavior of *Bufo debilis* and *B. punctatus*. Journal of Herpetology 18:406–411.

Sullivan, B.K. 1986a. Advertisement call variation in the Arizona treefrog *Hyla wrightorum* Taylor, 1938. Great Basin Naturalist 46:378–381.

Sullivan, B.K. 1986b. Hybridization between the toads *Bufo microscaphus* and *Bufo woodhousei* in Arizona: morphological variation. Journal of Herpetology 20:11–21.

Sullivan, B.K. 1989a. Desert environments and the structure of anuran mating systems. Journal of Arid Environments 17:175–183.

Sullivan, B.K. 1989b. Mating system variation in Woodhouse's toad *(Bufo woodhousii)*. Ethology 83:60–68.

Sullivan, B.K. 1990. Natural hybrid between the Great Plains toad *(Bufo cognatus)* and the red-spotted toad *(Bufo punctatus)* from central Arizona. Great Basin Naturalist 50:371–372.

Sullivan, B.K. 1992a. Sexual selection and calling behavior in the American toad *Bufo americanus*. Copeia 1992:1–7.

Sullivan, B.K. 1992b. Calling behavior in the southwestern toad *(Bufo microscaphus)*. Herpetologica 48:383–389.

Sullivan, B.K. 1993. Distribution of the southwestern toad *(Bufo microscaphus)* in Arizona. Great Basin Naturalist 53:402–406.

Sullivan, B.K. 1995. Temporal stability in hybridization between *Bufo microscaphus* and *Bufo woodhousii* (Anura: Bufonidae): behavior and morphology. Journal of Evolutionary Biology 8:233–247.

Sullivan, B.K. and P.J. Fernandez. 1999. Breeding activity, estimated age-structure, and growth in Sonoran Desert anurans. Herpetologica 55:334–343.

Sullivan, B.K. and S.H. Hinshaw. 1990. Variation in advertisement calls and male calling behavior in the spring peeper *(Pseudacris crucifer)*. Copeia 1990:1146–1150.

Sullivan, B.K. and T. Lamb. 1988. Hybridization between the toads *Bufo microscaphus* and *Bufo woodhousei* in Arizona: variation in release calls and allozymes. Herpetologica 44:325–333.

Sullivan, B.K. and K.B. Malmos. 1994. Call variation in the Colorado River toad *(Bufo alvarius)*: behavioral and phylogenetic implications. Herpetologica 50:146–156.

Sullivan, B.K. and E.A. Sullivan. 1985. Variation in advertisement calls and male mating success of *Scaphiopus bombifrons*, *S. couchi*, and *S. multiplicatus* (Pelobatidae). Southwestern Naturalist 30:349–355.

Sullivan, B.K. and W.E. Wagner Jr. 1988. Variation in advertisement and release calls, and social influences on the calling behavior in the Gulf Coast toad *(Bufo valliceps)*. Copeia 1988:1014–1020.

Sullivan, B.K., K.B. Malmos and M.F. Given. 1996a. Systematics of the *Bufo woodhousii* complex (Anura: Bufonidae); advertisement call variation. Copeia 1996:274–280.

Sullivan, B.K., R.W. Bowker, K.B. Malmos and E.W.A. Gergus. 1994. Arizona distribution of three Sonoran Desert anurans: *Bufo retiformis*, *Gastrophryne olivacea*, and *Pternohyla fodiens*. Arizona Game and Fish, Contract report, Phoenix, Arizona.

Sullivan, B.K., R.W. Bowker, K.B. Malmos and E.W.A. Gergus. 1996b. Arizona distribution of three Sonoran Desert anurans: *Bufo retiformis*, *Gastrophryne olivacea*, and *Pternohyla fodiens*. Great Basin Naturalist 56:38–47.

Sullivan, B.K., K.B. Malmos, E.W.A. Gergus and R.W. Bowker. 2000. Evolutionary implications of advertisement call variation in *Bufo debilis*, *B. punctatus*, and *B. retiformis*. Journal of Herpetology 34:368–374.

Summers, K. 1989. Sexual selection and intra-female competition in the green poison dart frogs, *Dendrobates auratus*. Animal Behaviour 37:797–805.

Surface, H.A. 1906. The serpents of Pennsylvania. Bulletin of the Division of Zoology, Pennsylvania State Department of Agriculture 4:113–208.

Surface, H.A. 1913. First report of the economic features of the amphibians of Pennsylvania. Pennsylvania Zoology Bulletin, Division of Zoology, Pennsylvania Department of Agriculture 3:68–152.

Sutherst, R.W., R.B. Floyd and G.F. Maywald. 1995. The potential geographical distribution of the cane toad, *Bufo marinus* L., in Australia. Conservation Biology 9:294–299.

Sutton, K.B. and K.N. Paige. 1980. Geographic distribution: *Ambystoma talpoideum* (mole salamander). Herpetological Review 11:13.

Suzuki, S., Y. Ohe, T. Okubo, T. Kakegawa and K. Tatemoto. 1995. Isolation and characterization of novel antimicrobial peptides, rugosins A, B and C, from the skin of the frog, *Rana rugosa*. Biochemistry and Biophysics Research Commununications 212:249–254.

Svihla, A. 1933. Extension of the ranges of some Washington Amphibia. Copeia 1933:39.

Svihla, A. 1936. *Rana rugosa* (Schlegel): Notes of the life history of this interesting frog. Mid-Pacific Magazine 49:124–125.

Svihla, A. 1953. Diurnal retreats of the spadefoot toad *Scaphiopus hammondi*. Copeia 1953:186.

Svihla, A. and R.D. Svihla. 1933a. Notes on *Ascaphus truei* in Kittitas County, Washington. Copeia 1933:37–38.

Svihla, A. and R.D. Svihla. 1933b. Amphibians and reptiles of Whitman County, Washington. Copeia 1933:125–128.

Swanson, D.L. and B.M. Graves. 1995. Supercooling and freeze intolerance in overwintering juvenile spadefoot toads *(Scaphiopus bombifrons)*. Journal of Herpetology 29:280–285.

Swanson, D.L., B.D. Graves and K.L. Koster. 1996. Freezing tolerance/intolerance and cryoprotection synthesis in terrestrially overwintering anurans in the Great Plains, USA. Journal of Comparative Physiology B166:110–119.

Swanson, P.L. 1948. Notes on the amphibians and reptiles of Venango County, Pennsylvania. American Midland Naturalist 40:362–371.

Sward, S. 1990. Secretary Lujan and the squirrels: Interior Chief calls Endangered Species Act "too tough." 12 May, San Francisco Chronicle, San Francisco, California.

Swaringen, K. 1996. Conservation spotlight: Wyoming toad. Endangered Species Update 13:9.

Sweeney, R.C.H. 1961. Snakes of Nyasaland. Nyasaland Society and Nyasaland Government, Blantyre and Zomba.

Sweet, S.S. 1977a. *Eurycea tridentifera*. Pp. 199.1–199.2. Catalogue of American Amphibians and Reptiles. Society for the Study of Amphibians and Reptiles, St. Louis, Missouri.

Sweet, S.S. 1977a. Natural metamorphosis in *Eurycea neotenes*, and the generic allocation of the Texas *Eurycea* (Amphibia: Plethodontidae). Herpetologica 33:364–375.

Sweet, S.S. 1977b. *Eurycea tridentifera*. Pp. 199.1–199.2. Catalogue of American Amphibians and Reptiles. Society for the Study of Amphibians and Reptiles, St. Louis, Missouri.

Sweet, S.S. 1977b. Natural metamorphosis in *Eurycea neotenes*, and the generic allocation of the Texas *Eurycea* (Amphibia: Plethodontidae). Herpetologica 33:364–375.

Sweet, S.S. 1978a. The evolutionary development of the Texas *Eurycea* (Amphibia: Plethodontidae). Ph.D. dissertation. University of California, Berkeley, California.

Sweet, S.S. 1978b. On the status of *Eurycea pterophila* (Amphibia: Plethodontidae). Herpetologica 34:101–108.

Sweet, S.S. 1982. A distributional analysis of epigean populations of *Eurycea neotenes* in central Texas, with comments on the origin of troglobitic populations. Herpetologica 38:430–444.

Sweet, S.S. 1984. Secondary contact and hybridization in the Texas cave salamanders *Eurycea neotenes* and *E. tridentifera*. Copeia 1984: 428–441.

Sweet, S.S. 1992. Initial report on the ecology and status of the arroyo toad *(Bufo microscaphus californicus)* on the Los Padres National Forest of southern California, with management recommendations. U.S.D.A. Forest Service, Contract report, Goleta, California.

Sweet, S.S. 1993. Second report on the biology and status of the arroyo toad *(Bufo microscaphus californicus)* on the Los Padres National Forest of southern California. U.S. Department of Agriculture, Forest Service, Goleta, California.

Sweetman, H.L. 1944. Food habits and molting of the common tree frog. American Midland Naturalist 32:499–501.

Swift, C.C., J.L. Nelson, C. Maslow and C. Stein. 1989. Biology and distribution of the tidewater goby *Eucyclogobius newberryi* (Pisces: Gobiidae) of California. Contributions to Science 404:1–19.

Swofford, D.L. 1993. PAUP: Phylogenetic analysis using parsimony, version 3.1. Illinois Natural History Survey, Champaign, Ilinois.

Sype, W.E. 1975. Breeding habits, embryonic thermal requirements and embryonic and larval development of the Cascade frog, *Rana cascadae* Slater. Ph.D. dissertation. Oregon State University, Corvallis, Oregon.

Systematics Agenda 2000. 1994a. Systematics Agenda 2000: Charting the Biosphere. Society of Systematic Biologists, American Society of Plant Taxonomists, Willi Hennig Society, Association of Systematics Collections, New York.

Systematics Agenda 2000. 1994b. Systematics Agenda 2000: Charting the Biosphere. Technical Report. Society of Systematic Biologists, American Society of Plant Taxonomists, Willi Hennig Society, Association of Systematics Collections, New York.

Szafoni, R.E., C.A. Phillips and M. Redmer. 1999. Translocations of amphibian species outside their native range: a comment on Thurow (1994, 1997). Transactions of the Illinois State Academy of Science 92:277–283.

Taber, C.A., R.F. Wilkinson Jr. and M.S. Topping. 1975. Age and growth of hellbenders in the Niangua River, Missouri. Copeia 1975: 633–639.

Taft, S.J., K. Suchow and M. Van Horn. 1993. Helminths from some Minnesota and Wisconsin Raptors. Journal of the Helminthological Society of Washington 60:260–263.

Taggart, T.W. 1997a. Status of *Bufo debilis* (Anura: Bufonidae) in Kansas. Kansas Herpetological Society Newsletter 109:7–12.

Taggart, T.W. 1997b. Geographic distribution: *Notophthalmus meridionalis* (black-spotted newt). Herpetological Review 28:47.

Taigen, T.L. 1983. Activity metabolism of anuran amphibians: implications for the origin of endothermy. American Naturalist 121: 94–109.

Taigen, T.L., S.B. Emerson and F.H. Pough. 1982. Ecological correlates of anuran exercise physiology. Oecologia 52:49–56.

Takats, L. and J. Willis. 2000. A northern leopard frog *(Rana pipiens)* observed in central Alberta: parkland management staff keep an eye open for endangered amphibians. Alberta Naturalist 30:52–53.

Talbot, S.B. 1933. Life history studies of trematodes of the subfamily Reniferinae. Parasitology 25:518–545.

Talentino, K.A. and E. Landre. 1991. Comparative development of two species of sympatric *Ambystoma* salamanders. Journal of Freshwater Ecology 6:395–401.

Tallmon, D.T., W.C. Funk, F.W. Allendorf and W.W. Dunlap. 2000. Genetic differentiation of long-toed salamander (*Ambystoma macrodactylum*) populations. Copeia 2000:27–35.

Tamsitt, J.R. 1962. Notes on a population of the Manitoba toad *(Bufo hemiophrys)* in the Delta Marsh Region of Lake Manitoba, Canada. Ecology 43:147–150.

Tan, A.M. 1993. Systematics, phylogeny and biogeography of the northwest American newts of the genus *Taricha* (Caudata: Salamandridae). Ph.D. dissertation. University of California, Berkeley, California.

Tan, A.M. and D.B. Wake. 1995. MtDNA phylogeography of the California newt, *Taricha torosa* (Caudata, Salamandridae). Molecular Phylogenetics and Evolution 4:383–394.

Tanner, V.M. 1929. A distributional list of the amphibians and reptiles of Utah, No. 3. Copeia 171:46–52.

Tanner, V.M. 1931. A synoptical study of Utah Amphibia. Transactions of the Utah Academy of Science 8:159–198.

Tanner, V.M. 1939. A study of the genus *Scaphiopus*, the spade-foot toads. Great Basin Naturalist 1:3–20.

Tanner, W.W. 1950. Notes on the habits of *Microhyla carolinensis olivacea* (Hallowell). Herpetologica 6:47–48.

Tanner, W.W. 1953. Notes on the life history of *Plethopsis wrighti* Bishop. Herpetologica 9:139–140.

Tanner, W.W. 1989a. Amphibians of western Chihuahua. Great Basin Naturalist 49:38–70.

Tanner, W.W. 1989b. Status of *Spea stagnalis* Cope (1875), *Spea intermontanus* Cope (1889), and a systematic review of *Spea hammondii* Baird (1839) (Amphibia: Anura). Great Basin Naturalist 49: 503–510.

Tanner, W.W., D.L. Fisher and T.J. Willis. 1971. Notes on the life history of *Ambystoma tigrinum nebulosum* Hallowell in Utah. Great Basin Naturalist 31:213–222.

Tardell, J.H., R.C. Yates and D.H. Schiller. 1981. New records and habitat observations of *Hyla andersonii* Baird (Anura: Hylidae) in Chesterfield and Marlboro Counties, South Carolina. Brimleyana 6: 153–158.

Tarkhnishvili, D.N. 1997. The status of amphibian species in Georgia (C.I.S.). DAPTF Report, Number 2, World Conservation Union (IUCN), Species Survival Commission, Open University, Milton Keynes, United Kingdom.

Tattersall, G.J. and P.A. Wright. 1996. The effects of ambient pH on nitrogen excretion in early life stages of the American toad *(Bufo americanus)*. Comparative Biochemistry and Physiology 113:369–374.

Taub, F.B. 1961. The distribution of red-backed salamanders, *Plethodon c. cinereus*, within the soil. Ecology 42:681–698.

Taubert, D.B., P.W. Shetley, D.P. Philipp and T. Harrison. 1982. Breeding biology and distribution of the Illinois chorus frog *(Pseudacris streckeri illinoensis)* in Illinois. Illinois Department of Conservation, Unpublished report, Springfield, Illinois.

Taylor, A.D. 1990. Metapopulations, dispersal, and predator-prey dynamics: an overview. Ecology 71:429–233.

Taylor, A.D. 1998. Environmental variability and the parasitoid-host metapopulation models. Theoretical Population Biology 52:98–107.

Taylor, B.E. and D.E. Scott. 1997. Effects of larval density dependence on population dynamics of *Ambystoma opacum*. Herpetologica 53: 132–145.

Taylor, B.E., R.A. Estes, J.H.K. Pechmann and R.D. Semlitsch. 1988. Trophic relations in a temporary pond: larval salamanders and their microinvertebrate prey. Canadian Journal of Zoology 66:2191–2198.

Taylor, B.L. and A.E. Dizon. 1999. First policy then science: why a management unit based solely on genetic criteria cannot work. Molecular Ecology 8:S11–S16.

Taylor, C.L., R.F. Wilkinson and C.L. Peterson. 1990. Reproductive patterns of five plethodontid salamanders from the Ouachita Mountains. Southwestern Naturalist 35:468–472.

Taylor, D.W. 1981 Freshwater mollusks of California: a distributional checklist. California Fish and Game 67:140–163.

Taylor, E.H. 1929. List of reptiles and batrachians of Morton County, Kansas, reporting species new to the state fauna. University of Kansas Science Bulletin 19:63–65.

Taylor, E.H. 1936. Notes on herpetological fauna of the Mexican state of Sinaloa. University of Kansas Science Bulletin 24:505–537.

Taylor, E.H. 1938a. Frogs of the *Hyla eximia* group in Mexico, with descriptions of two new species. University of Kansas Science Bulletin 25:421–445.

Taylor, E.H. 1938b. New species of Mexican tailless Amphibia. University of Kansas Science Bulletin 25:385–399.

Taylor, E.H. 1941. A new plethodont salamander from New Mexico. Proceedings of the Biological Society of Washington 54:77–80.

Taylor, J. 1983a. Size-specific associations of larval and neotenic northwestern salamanders, *Ambystoma gracile*. Journal of Herpetology 17:203–209.

Taylor, J. 1983b. Orientation and flight behavior of the neotenic salamanders *(Ambystoma gracile)* in Oregon. American Midland Naturalist 109:40–49.

Taylor, J. 1984. Comparative evidence for salamander competition between the salamanders *Ambystoma gracile* and *Taricha granulosa*. Copeia 1984:672–683.

Taylor, S.K., E.S. Williams and K.W. Mills. 1999. Mortality of captive Canadian toads from *Basidiobolus ranarum* mycotic dermatitus. Journal of Wildlife Diseases 35:64–69.

Taylor, S.K., E.S. Williams, E.T. Thorne, K.W. Mills, D.I. Withers and A.C. Pier. 1999. Causes of mortality of the Wyoming toad. Journal of Wildlife Disease 35:49–57.

Teberg, E.K. 1963. An extension into Montana of the known range of the salamander *Plethodon vandykei idahoensis*. Herpetologica 19:287.

Teberg, E.K. 1965. Range extensions of the salamander *Plethodon vandykei idahoensis*. Copeia 1965:244.

Technical Advisory Panel. 1990. Technical factors in Edwards Aquifer use and management. Prepared for Special Committee on the Edwards Aquifer, Texas State Legislature, Austin, Texas.

Tejedo, M. 1992. Effect of body size and timing of reproduction on reproductive success in female natterjack toads *(Bufo calamita)*. Journal of Zoology, London 228:545–555.

Tejedo, M. and R. Reques. 1994. Does larval growth history determine timing of metamorphosis in anurans? A field experiment. Herpetologica 50:113–118.

Telford, S.R. 1952. A herpetological survey in the vicinity of Lake Shipp, Polk County, Florida. Quarterly Journal of the Florida Academy of Sciences 15:175–185.

Temminck, C.J. and H. Schlegel. 1838. *In* Von Siebold, P.F. (Ed.), Fauna Japonica sive Descriptio animalium, quae in itinere per Japonianum, jussu et auspiciis superiorum, qui summum in India Batava Imperium tenent, suscepto, annis 1823–1830 colleget, notis observationibus et adumbrationibus illustratis. Volume 3, (Chelonia, Ophidia, Sauria, Batrachia), J.G. Lalau, Leiden, Germany.

Templeton, A.R. 1989. The meaning of species and speciation: a genetic perspective. Pp. 3–27. *In* Otte, D. and J.A. Endler (Eds.), Speciation and Its Consequences. Sinauer Associates, Sunderland, Massachusetts.

Templeton, A.R., E. Routman and C.A. Phillips. 1995. Separating population structure from population history: a cladistic analysis of the geographical distribution of mitochondrial DNA haplotypes in the tiger salamander, *Ambystoma tigrinum*. Genetics 140: 767–782.

Tennessee Wildlife Resources Agency. 1994. Tennessee Wildlife Resources Commission Proclamation, Endangered or Threatened Species, Proceedings Number 94–17, Nashville, Tennessee.

Tennessen, J.A. and K.R. Zamudio. 2003. Early-male precedence, multiple paternity, and sperm storage in an amphibian aggregate breeder. Molecular Ecology 12:1567–1576.

Teska, W.R. 1990. Geographic distribution: *Ambystoma talpoideum* (mole salamander). Herpetological Review 21:36.

Test, F.H. 1955. Seasonal differences in populations of the red-backed salamander in southeastern Michigan. Papers of the Michigan Academy Sciences, Arts and Letters 40:137–153.

Test, F.H. 1958. Butler's garter snake eats amphibian. Copeia 1958: 151–152.

Test, F.H. and B.A. Bingham. 1948. Censuses of a population of the red-backed salamander *(Plethodon cinereus)*. American Midland Naturalist 39:362–372.

Test, F.H. and H. Heatwole. 1962. Nesting sites of the red-backed salamander, *Plethodon cinereus*, in Michigan. Copeia 1962:206–207.

Test, F.H. and R.G. McCann. 1976. Foraging behavior of *Bufo americanus* tadpoles in response to high densities of micro-organisms. Copeia 1976:576–578.

Tester, J.R. and W.J. Breckenridge. 1964a. Winter behavior patterns of the Manitoba toad, *Bufo hemiophrys*, in northwestern Minnesota. Annales Academiae Scientiarum Fennicae, Series A. IV Biologica 71: 423–431.

Tester, J.R. and W.J. Breckenridge. 1964b. Population dynamics of the Manitoba toad, *Bufo hemiophrys*, in northwestern Minnesota. Ecology 45:592–601.

Tester, J.R., M.A. Ewert and D.B. Siniff. 1970. Effects of ionizing radiation on natural and laboratory populations of Manitoba toads, *Bufo hemiophrys*. Radiation Research 44:379–389.

Tester, J.R., A. Parker and D.B. Siniff. 1965. Experimental studies on habitat preference and thermoregulation of *Bufo americanus*, *B. hemiophrys*, and *B. cognatus*. Journal of the Minnesota Academy of Science 33:27–32.

Tester, U. 1990. Artenschützerisch relevante Aspekte zur Ökologie des Laubfroschs *(Hyla arborea* L.). B. Schlattmann, Basel, Switzerland.

Tevis, L. 1966. Unsuccessful breeding by desert toads *(Bufo punctatus)* at the limit of their ecological tolerance. Ecology 47:766–775.

Theron, J., L.C. Hoffman and J.F. Prinsloo. 1992. Aspects of the chemical control of *Xenopus laevis* tadpoles under laboratory conditions. South African Journal of Wildlife Research 22:110–113.

Thibaudeaux, D.G. and R. Altig. 1988. Sequence of ontogenic development and atrophy of the oral apparatus of six anuran tadpoles. Journal of Morphology 197:63–69.

Thielcke, G., C.P. Herrn, C.P. Hutter and R.L. Schreiber. 1983. Rettet die Frösche. Pro natur, Stuttgart, Germany.

Thiesmeier, B., O. Jäger and U. Fritz. 1994. Erfolgreiche reproduktion des ochsenfrosches *(Rana catesbeiana)* im nördlichen landkreis Böblingen (Baden-Württemberg). Zeitschrift für Feldherpetologie 1:169–176.

Thomas, L.A. and J. Allen. 1997. Natural history notes: *Bufo houstonensis* (Houston toad). Behavior. Herpetological Review 28:40–41.

Thomas, L.J. 1939. Life cycle of a fluke, *Halipegus eccentricus*, n. sp. found in the ears of frogs. Journal of Parasitology 25:207–221.

Thomas, R. 1966. New species of Antillean *Eleutherodactylus*. Quarterly Journal of the Florida Academy of Sciences 28:375–391.

Thomas, R.A. 1976a. Houston toad critical habitat. Recommendation prepared for the U.S. Fish and Wildlife Service, 3 November.

Thomas, R.A. 1976b. Letter to James Johnson, U.S. Fish and Wildlife Service, 16 December.

Thomas, R.A. 1997. Bob Thomas' perspective on critical habitat designation for the Houston toad. Summary prepared for L.E. Brown, 4 December.

Thomas, R.A., S.A. Nadler and W.L. Jagers. 1984. Helminth parasites of the endangered Houston toad, *Bufo houstonensis* Sanders, 1953 (Amphibia, Bufonidae). Journal of Parasitology 70:1012–1013.

Thompson, C. 1915. Notes on the habits of *Rana areolata* Baird and Girard. Occasional Papers of the Museum of Zoology, Number 9, University of Michigan, Ann Arbor, Michigan.

Thompson, D.B. and T.R. Jones. 1992. The occurrence of paedomorphic cave-dwelling tiger salamanders in central New Mexico. Pp. 3–6. *In* Belski, D. (Ed.), GYPKAP Report 2 1988–1991. Southwestern Region National Speleological Society, Adobe Press, Albuquerque, New Mexico.

Thompson, E.L. and J.E. Gates. 1982. Breeding pool segregation by the mole salamanders, *Ambystoma jeffersonianum* and *A. maculatum*, in a region of sympatry. Oikos 38:273–279.

Thompson, E.L., J.E. Gates and G.J. Taylor. 1980. Distribution and breeding habitat selection of the Jefferson salamander, *Ambystoma jeffersonianum*, in Maryland. Journal of Herpetology 14:113–120.

Thompson, H.B. 1913. Description of a new subspecies of *Rana pretiosa* from Nevada. Proceedings of the Biological Society of Washington 26:53–56.

Thompson, M. 2001. An unusually adept ambystomatid, the long toed salamander, coping at northern extremes. Boreal Dip Net 5:8–10.

Thompson, S. 1976. Collections of South Dakota Amphibians and Reptiles. South Dakota Department of Game, Fish and Parks, Unpublished report, Pierre, South Dakota.

Thompson, W.L., G.C. White and C. Gowan. 1998. Monitoring Vertebrate Populations. Academic Press, San Diego, California.

Thornton, O.W. 1977. The impact of man upon herpetological communities in the lower Rio Grande Valley, Texas. Master's thesis. Texas A&M University, College Station, Texas.

Thornton, W.A. 1955. Interspecific hybridization in *Bufo woodhousei* and *Bufo valliceps*. Evolution 9:455–468.

Thornton, W.A. 1960. Population dynamics in *Bufo woodhousei* and *Bufo valliceps*. Texas Journal of Science 12:176–200.

Thorpe, J.P. 1982. The molecular clock hypothesis: biochemical evolution, genetic differentiation and systematics. Annual Review of Ecology and Systematics 13:139–168.

Thrall, J. 1971. Excavation of pits by juvenile *Rana catesbeiana*. Copeia 1971:751–752.

Thuesen, E.V. and K. Kogure. 1989. Bacterial production of tetrodotoxin in four species of Chaetognatha. Biological Bulletin 176:191–194.

Thul, J.E., D.J. Forrester and C.L. Abercrombie. 1985. Ecology of parasite helmniths of wood ducks, *Aix sponsa*, in the Atlantic flyway. Proceedings of the Helminthological Society of Washington 52:297–310.

Thurn, V., A.P. Block and M. Hennecke. 1984. Amphibien und Reptilien im Rems-Murr-Kreis. Jahreshefte der Gesellschaft für Naturkunde in Württemberg 139:161–193.

Thurow, G.R. 1954. A range extension of the salamander *Gyrinophilus porphyriticus porphyriticus*. Copeia 1954:221–222.

Thurow, G.R. 1955. Taxonomic and ecological studies on the zig-zag salamander *(Plethodon dorsalis)* and the red-backed salamander *(Plethodon cinereus)*. Ph.D. dissertation. Indiana University, Bloomington, Indiana.

Thurow, G.R. 1956. A new subspecies of *Plethodon welleri*, with notes on other members of the genus. American Midland Naturalist 55:343–356.

Thurow, G.R. 1957. A new *Plethodon* from Virginia. Herpetologica 13:59–66.

Thurow, G.R. 1963. Taxonomic and ecological notes on the salamander, *Plethodon welleri*. University of Kansas Science Bulletin 44:87–108.

Thurow, G.R. 1964. *Plethodon welleri*. Pp. 12.1–12.2. Catalogue of American Amphibians and Reptiles. Society for the Study of Amphibians and Reptiles, St. Louis, Missouri.

Thurow, G.R. 1966. *Plethodon dorsalis*. Pp. 29.1–29.3. Catalog of American Amphibians and Reptiles. Society for the Study of Amphibians and Reptiles, St. Louis, Missouri.

Thurow, G.R. 1976. Aggression and competition in eastern *Plethodon* (Amphibia, Urodela, Plethodontidae). Journal of Herpetology 10:277–291.

Thurow, G.R. 1994. Experimental return of wood frogs to west-central Illinois. Transactions of the Illinois State Academy of Science 87:83–97.

Thurow, G.R. 1997. Ecological lessons from two-lined salamander translocations. Transactions of the Illinois State Academy of Science 90:79–88.

Thurow, G.R. 1999. New *Plethodon shenandoah* localities and their significance. Bulletin of the Chicago Herpetological Society 34:269–273.

Thurston, J.P. 1967. The morphology and life-cycle of *Cephalochlamys namaquensis* (Cohn, 1906) (Cestoda: Pseudophyllidea) from *Xenopus muelleri* and *X. laevis*. Parasitology 57:187–200.

Tietge, J.E., S.A. Diamond, G.T. Ankley, D.L. DeFoe, G.W. Holcombe, K.M. Jensen, S.J. Degitz, G.E. Elonen and E. Hammer. 2001. Ambient solar UV radiation causes mortality in larvae of three species of *Rana* under controlled exposure conditions. Photochemistry and Photobiology 74:261–268.

Tihen, J.A. 1937. Additional distributional records of amphibians and reptiles in Kansas counties. Transactions of the Kansas Academy of Science 40:401–409.

Tihen, J.A. 1952. *Rana grylio* from the Pleistocene of Florida. Herpetologica 8:107.

Tihen, J.A. 1962. Osteological observations of New World *Bufo*. American Midland Naturalist 67:157–183.

Tihen, J.A. and C.J. Chantell. 1963. Urodele remains from the Valentine Formation of Nebraska. Copeia 1963:505–510.

Tilley, S.G. 1968. Size-fecundity relationships and their evolutionary implications in five desmognathine salamanders. Evolution 22:806–816.

Tilley, S.G. 1969. Variation in the dorsal pattern of *Desmognathus ochrophaeus* at Mt. Mitchell, North Carolina, and elsewhere in the southern Appalachian Mountains. Copeia 1969:161–175.

Tilley, S.G. 1972. Aspects of parental care and embryonic development in *Desmognathus ochrophaeus*. Copeia 1972:532–540.

Tilley, S.G. 1973a. Life histories and natural selection in populations of the salamander *Desmognathus ochrophaeus*. Ecology 54:3–17.

Tilley, S.G. 1973b. Observations on the larval period and female reproductive ecology of *Desmognathus ochrophaeus* (Amphibian: Plethodontidae) in western North Carolina. American Midland Naturalist 89:394–407.

Tilley, S.G. 1974. Structures and dynamics of populations of the salamander *Desmognathus ochrophaeus* Cope in different habitats. Ecology 55:808–817.

Tilley, S.G. 1977. Studies of life histories and reproduction in North American plethodontid salamanders. Pp. 1–41. *In* Taylor, D.H. and S.I. Guttman (Eds.), The Reproductive Biology of Amphibians. Plenum Press, New York.

Tilley, S.G. 1980. Life histories and comparative demography of two salamander populations. Copeia 1980:806–821.

Tilley, S.G. 1981. A new species of *Desmognathus* (Amphibia: Caudata: Plethodontidae) from the southern Appalachian Mountains. Occasional Papers of the Museum of Zoology, Number 695, University of Michigan, Ann Arbor, Michigan.

Tilley, S.G. 1985. *Desmognathus imitator*. Pp. 359.1–359.2. Catalogue of American Amphibians and Reptiles. Society for the Study of Amphibians and Reptiles, St. Louis, Missouri.

Tilley, S.G. 1988. Hybridization between two species of *Desmognathus* (Amphibia: Caudata: Plethodontidae) in the Great Smoky Mountains. Herpetological Monographs 2:27–39.

Tilley, S.G. 1997. Patterns of genetic differentiation in Appalachian desmognathine salamanders. Journal of Heredity 88:305–315.

Tilley, S.G. 2000a. The systematics of *Desmognathus imitator*. Pp. 121–148. *In* Bruce, R.C., R.G. Jaeger and L.D. Houck (Eds.), The Biology of Plethodontid Salamanders. Kluwer Academic/Plenum Publishers, New York.

Tilley, S.G. 2000b. *Desmognathus santeetlah*. Pp. 703.1–703.3. Catalogue of American Amphibians and Reptiles. Society for the Study of Amphibians and Reptiles, St. Louis, Missouri.

Tilley, S.G. and J. Bernardo. 1993. Life history evolution in plethodontid salamanders. Herpetologica 49:154–163.

Tilley, S.G. and J.R. Harrison. 1969. Notes on the distribution of the pigmy salamander, *Desmognathus wrighti* King. Herpetologica 25:178–180.

Tilley, S.G. and J.S. Hausman. 1976. Allozymic variation and multiple inseminations in populations of the salamander *Desmognathus ochrophaeus*. Copeia 1976:734–741.

Tilley, S.G. and M.J. Mahoney. 1996. Patterns of genetic differentiation in salamanders of the *Desmognathus ochrophaeus* complex (Amphibia: Plethodontidae). Herpetological Monographs 10:1–42.

Tilley, S.G. and D.W. Tinkle. 1968. A reinterpretation of the reproductive cycle and demography of the salamander *Desmognathus ochrophaeus*. Copeia 1968:299–303.

Tilley, S.G., B.L. Lundrigan and L.P. Brower. 1982. Erythrism and mimicry in the salamander *Plethodon cinereus*. Herpetologica 38:409–417.

Tilley, S.G., P.A. Verrell and S.J. Arnold. 1990. Correspondence between sexual isolation and allozyme differentiation: a test in the salamander *Desmognathus ochrophaeus*. Proceedings of the National Academy of Sciences 87:2715–2719.

Tilley, S.G., R.B. Merritt, B. Wu and R. Highton. 1978. Genetic differentiation in salamanders of the *Desmognathus ochrophaeus* complex (Plethodontidae). Evolution 32:93–111.

Tiner, R.W. 1984. Wetlands of the United States: current status and trends. U.S. Department of the Interior, Fish and Wildlife Service, National Wetland Inventory, Washington, D.C.

Ting, H.-P. 1951. Duration of the tadpole stage of the green frog, *Rana clamitans*. Copeia 1951:82.

Tinker, S.W. 1938. Animals of Hawaii. Nippu Jiji Company, Honolulu, Hawaii.

Tinkle, D.W. 1952. Notes on the salamander, *Eurycea longicauda guttolineata* in Florida. Field and Laboratory 29:105–108.

Tinkle, D.W. 1959. Observations of reptiles and amphibians in a Louisiana swamp. American Midland Naturalist 62:189–205.

Tinkle, D.W. 1965. Population structure and effective size of a lizard population. Evolution 19:569–573.

Tinkle, D.W. and G.N. Knopf. 1964. Biologically significant distribution records of amphibians and reptiles of northwest Texas. Herpetologica 20:42–47.

Tinsley, R.C. 1973. Studies on the ecology and systematics of a new species of clawed toad, the genus *Xenopus*, from western Uganda. Journal of Zoology (London) 169:1–27.

Tinsley, R.C. 1995. Parasitic disease in amphibians: control by the regulation of worm burdens. Parasitology 111:S153–S178.

Tinsley, R.C. 1996. Parasites of *Xenopus*. Pp. 233–261. *In* Tinsley, R.C. and H.R. Kobel (Eds.), The Biology of *Xenopus*. Clarendon Press, Oxford, United Kingdom.

Tinsley, R.C. 1997. Parasitology and ecology of *Scaphiopus*. Southwestern Research Station News, 2–3.

Tinsley, R.C. and C.M. Earle. 1983. Invasion of vertebrate lungs by the polystomatid monogeneans *Pseudodiplorchis americanus* and *Neodiplorchis scaphiopodis*. Parasitology 86:501–517.

Tinsley, R.C. and M.J. McCoid. 1996. Feral populations of *Xenopus* outside Africa. Pp. 81–94. *In* Tinsley, R.C. and H.R. Kobel (Eds.), The Biology of *Xenopus*. Clarendon Press, Oxford, United Kingdom.

Tinsley, R.C. and K. Tocque. 1995. The population dynamics of a desert anuran, *Scaphiopus couchii*. Australian Journal of Ecology 20:376–384.

Tinsley, R.C., H.R. Kobel and M. Fischberg. 1979. The biology and systematics of a new species of *Xenopus* (Anura: Pipidae) from the highlands of Central Africa. Journal of Zoology (London) 188:69–102.

Tinsley, R.C., C. Loumont and H.R. Kobel. 1996. Geographical distribution and ecology. Pp. 35–59. *In* Tinsley, R.C. and H.R. Kobel (Eds.), The Biology of *Xenopus*. Clarendon Press, Oxford, United Kingdom.

Titus, T.A. 1990. Genetic variation in two subspecies of *Ambystoma gracile* (Caudata: Ambystomatidae). Herpetologica 24:107–111.

Titus, T.A. and A. Larson. 1996. Molecular phylogenetics of desmognathine salamanders (Caudata: Plethodontidae): a reevaluation of evolution in ecology, life history, and morphology. Systematic Biology 45:451–472.

Tkadlec, E. and J. Zejda. 1998. Small rodent population fluctuations: the effects of age structure and seasonality. Evolutionary Ecology 12:191–210.

Tobey, F.J. 1985. Virginia amphibians and reptiles: a distributional survey. Virginia Herpetological Society, Purcellville, Virginia.

Tocque, K. 1993. The relationship between parasite burden and host resources in the desert toad (*Scaphiopus couchii*), under natural environmental conditions. Journal of Animal Ecology 62:683–693.

Todd, M. 1995. Golf course protester arrested for locking himself to bulldozer. Austin American-Statesman, 25 July.

Todd, M. 1996. Obstacles slow golf course expansion. Austin American-Statesman, 5 November.

Toft, C.A. 1981. Feeding ecology of Panamanian litter anurans: patterns of diet and foraging mode. Journal of Herpetology 15:139–144.

Toft, C.A. 1995. Evolution of diet specialization in poison-dart frogs (Dendrobatidae). Herpetologica 51:202–216.

Tome, M.E. and F.H. Pough. 1982. Responses of amphibians to acid precipitation. Pp. 245–254. *In* Haines, T.A. and R.E. Johnson (Eds.), Acid Rain/Fisheries: Proceedings of an International Symposium on Acid Precipitation and Fisheries Impacts in Northeastern North America. American Fisheries Society, Bethesda, Maryland.

Tomson, O.H. and D.E. Ferguson. 1972. Y-axis orientation in larvae and juveniles of three species of *Ambystoma*. Herpetologica 28:6–9.

Toner, G.C. and N.D. St. Remy. 1941. Amphibians of eastern Ontario. Copeia 1941:10–13.

Topping, M.S. and C.A. Ingersol. 1981. Fecundity in the hellbender, *Cryptobranchus alleganiensis*. Copeia 1981:873–876.

Topping, M.S. and C.L. Peterson. 1985. Movement in the hellbender, *Cryptobranchus alleganiensis*. Transactions of the Missouri Academy of Science 19:121.

Tordoff, W., III. 1980. Report of study of limestone salamander on the Merced River. U.S. Bureau of Land Management, Folsom District Office, Contract Number CA-040-CTO-09, Folsom, California.

Townsend, D.S. 1989. The consequences of microhabitat choice for male reproductive success in a tropical frog (*Eleutherodactylus coqui*). Herpetologica 45:451–458.

Townsend, D.S. and M.M. Stewart. 1985. Direct development in *Eleutherodactylus coqui* (Anura:Leptodactylidae): a staging table. Copeia 1985:423–436.

Townsend, D.S. and M.M. Stewart. 1986. Courtship and mating behavior of a Puerto Rican frog, *Eleutherodactylus coqui*. Herpetologica 42:165–170.

Townsend, D.S., M.M. Stewart and F.H. Pough. 1984. Male parental care and its adaptive significance in a neotropical frog. Animal Behaviour 32:421–431.

Townsend, D.S., M.M. Stewart, F.H. Pough and P.F. Brussard. 1981. Internal fertilization in an oviparous frog. Science 212:469–471.

Tracy, C.R. 1971. Evidence for the use of celestial cues by dispersing immature California toads (*Bufo boreas*). Copeia 1971:145–147.

Tracy, C.R. 1973. Observations on social behavior in immature California toads (*Bufo boreas*) during feeding. Copeia 1973:342–345.

Tracy, C.R. and J.W. Dole. 1969. Orientation of displaced California toads, *Bufo boreas*, to their breeding sites. Copeia 1969:693–700.

Trapido, H. and R.T. Clausen. 1940. The larvae of *Eurycea bislineata major*. Copeia 1940:244–246.

Trapp, M.M. 1956. Range and natural history of the ringed salamander, *Ambystoma annulatum* Cope (Ambystomatidae). Southwestern Naturalist 1:78–82.

Trapp, M.M. 1959. Studies of the life history of *Ambystoma annulatum* (Cope). Master's thesis. University of Arkansas, Fayetteville, Arkansas.

Trauth, S.E. 1980. Geographic distribution: *Ambystoma annulatum* (ringed salamander). Herpetological Review 11:114.

Trauth, S.E. 1983. Reproductive biology and spermathecal anatomy of the dwarf salamander (*Eurycea quadridigitata*) in Alabama. Herpetologica 39:9–15.

Trauth, S.E. 1984. Spermathecal anatomy and the onset of mating in the slimy salamander (*Plethodon glutinosus*) in Alabama. Herpetologica 40:314–321.

Trauth, S.E. 1988. Egg clutches of the Ouachita dusky salamander, *Desmognathus brimleyorum* (Caudata: Plethodontidae), collected in Arkansas during a summer drought. Southwestern Naturalist 33:234–236.

Trauth, S.E. 1992. Distributional survey of the bird-voiced treefrog, *Hyla avivoca* (Anura: Hylidae), in Arkansas. Proceedings of the Arkansas Academy of Science 46:80–82.

Trauth, S.E. 2000. Winter breeding as a common occurrence in the ringed salamander, *Ambystoma annulatum* (Caudata: Ambystomatidae), in the Ozark National Forest of northcentral Arkansas. Journal of the Arkansas Academy of Science 54:157–158.

Trauth, S.E. and C.T. McAllister. 1983. Geographic distribution: *Rana clamitans melanota* (green frog). Herpetological Review 14:83.

Trauth, S.E. and C.T. McAllister. 1995. Vertebrate prey of selected Arkansas snakes. Proceedings of the Arkansas Academy of Science 49:188–192.

Trauth, S.E. and J.W. Robinette. 1990. Notes on distribution, mating activity, and reproduction in the bird-voiced treefrog, *Hyla avivoca*, in Arkansas. Bulletin of the Chicago Herpetological Society 25:218–219.

Trauth, S.E., B.P. Butterfield and W.E. Meshaka Jr. 1989a. Geographic distribution: *Scaphiopus bombifrons* (plains spadefoot). Herpetological Review 20:12.

Trauth, S.E., M.E. Cartwright and W.E. Meshaka. 1989b. Reproduction in the wood frog, *Rana sylvatica* (Anura: Ranidae), from Arkansas. Proceedings of the Arkansas Academy of Science 43:105–108.

Trauth, S.E., M.E. Cartwright and W.E. Meshaka. 1989c. Winter breeding in the ringed salamander, *Ambystoma annulatum* (Caudata: Ambystomatidae), from Arkansas. Southwestern Naturalist 34:145–146.

Trauth, S.E., M.L. McCallum and M.E. Cartwright. 2000. Breeding mortality in the wood frog, *Rana sylvatica* (Anura: Ranidae), from north-central Arkansas. Journal of the Arkansas Academy of Science 54:154–156.

Trauth, S.E., W.E. Meshaka Jr. and R.L. Cox. 1999. Post-metamorphic growth and reproduction in the eastern narrowmouth toad (*Gastrophryne carolinensis*) from northeastern Arkansas. Journal of the Arkansas Academy of Science 53:120–124.

Trauth, E.S., H.W. Robison and M.V. Plummer. 2004. The Amphibians and Reptiles of Arkansas. University of Arkansas Press, Fayetteville, Arkansas.

Trauth, S.E., D.M. Sever and R.D Semlitsch. 1994. Cloacal anatomy of paedomorphic female *Ambystoma talpoideum* (Caudata: Ambystomatidae), with comments on intermorph mating and sperm storage. Canadian Journal of Zoology 72:2147–2157.

Trauth, S.E., J.D. Wilhide and P. Daniel. 1992a. Status of the Ozark hellbender, *Cryptobranchus bishopi* (Urodela: Cryptobranchidae), in the Spring River, Fulton County, Arkansas. Proceedings of the Arkansas Academy of Science 46:83–86.

Trauth, S.E., J.D. Wilhide and P. Daniel. 1992b. Geographic distribution: *Cryptobranchus bishopi* (Ozark hellbender). Herpetological Review 23:121.

Trauth, S.E., J.D. Wilhide and P. Daniel. 1993b. The Ozark hellbender, *Cryptobranchus bishopi*, in Arkansas: distributional survey for 1992. Bulletin of the Chicago Herpetological Society 28:81–85.

Trauth, S.E., M.E. Cartwright, J.D. Wilhide and D.H. Jamieson. 1995a. A review of the distribution and life history of the wood frog, *Rana sylvatica* (Anura: Ranidae), in north-central Arkansas. Bulletin of the Chicago Herpetological Society 30:46–51.

Trauth, S.E., R.L. Cox Jr., J.D. Wilhide and H.J. Worley. 1995b. Egg mass characteristics of terrestrial morphs of the mole salamander, *Ambystoma talpoideum* (Caudata: Ambystomatidae), from northeastern Arkansas and clutch comparisons with other *Ambystoma* species. Proceedings of the Arkansas Academy of Science 49:193–196.

Trauth, S.E., M.L. McCallum, R.R. Jordan and D.A. Saugey. 2003. Brooding postures and nest site fidelity in the western slimy salamander, *Plethodon albagula* (Caudata: Plethodontidae), from an abandoned mine shaft in Arkansas. Herpetological Natural History 9:141–149.

Trauth, S.E., B.G. Cochran, D.A. Saugey, W. Posey and W.A. Stone. 1993a. Distribution of the mole salamander, *Ambystoma talpoideum* (Caudata: Ambystomatidae), in Arkansas with notes on paedomorphic populations. Proceedings of the Arkansas Academy of Science 47:154–156.

Trauth, S.E., R.L. Cox, B.P. Butterfield, D.A. Saugey and W.E. Meshaka. 1990. Reproductive phenophases and clutch characteristics of selected Arkansas amphibians. Proceedings of the Arkansas Academy of Science 44:107–113.

Travassos, L. 1939. Un novo trematodeo parasito da garcas: *Ribeiroia insignis* n.g., n.sp. Boletim Biologico (N.S.) 4:301–304.

Travis, J. 1980. Phenotypic variation and the outcome of interspecific competition in hylid tadpoles. Evolution 34:40–50.

Travis, J. 1980a. Genetic variation for larval specific growth rate in the frog *Hyla gratiosa*. Growth 44:167–181.

Travis, J. 1980b. Phenotypic variation and the outcome of interspecific competition in hylid tadpoles. Evolution 34:40–50.

Travis, J. 1981. A key to the tadpoles of North Carolina. Brimleyana 6:119–127.

Travis, J. 1983a. Variation in growth and survival of *Hyla gratiosa* larvae in experimental enclosures. Copeia 1983:232–237.

Travis, J. 1983b. Variation in development patterns of larval anurans in temporary ponds. I. Persistent variation within a *Hyla gratiosa* population. Evolution 37:496–512.

Travis, J. 1984. Anuran size at metamorphosis: experimental test of a model based on intraspecific competition. Ecology 65:1155–1160.

Travis, J. and D.J. Futuyma. 1993. Global change: lessons from and for evolutionary biology. Pp. 251–261. *In* Kareiva, P.M., J.G. Kingsolver and R.B. Huey (Eds.), Biotic Interactions and Global Change. Sinauer Associates, Sunderland, Massachusetts.

Travis, J., W.H. Keen and J. Julianna. 1985. The role of relative body size in a predator-prey relationship between dragonfly naiads and larval anurans. Oikos 45:59–65.

Travis, T.J. and R.C. Bruce. 2000. Life history evolution and adaptive radiation of hemidactyliine salamanders. Pp. 303–326. *In* Bruce, R.C., R.G. Jaeger and L.D. Houck (Eds.), The Biology of Plethodontid Salamanders. Kluwer Academic/Plenum Publishers, New York.

Treanor, R.R. 1975. Management of the bullfrog *(Rana catesbeiana)* resource in California. California Department of Fish and Game, Inland Fish Administrative Report 75-1, Sacramento, California.

Treanor, R.R. and S.J. Nichola. 1972. A preliminary report of the commercial and sporting utilization of the bullfrog, *Rana catesbeiana* Shaw in California. California Department of Fish and Game, Inland Fish Administrative Report 72-4, Sacramento, California.

TreeBase. A database of phylogenetic information. http://www.herbaria.harvard.edu/treebase.

Trenham, P.C. 1998. Demography, migration, and metapopulation structure of pond breeding salamanders. Ph.D. dissertation. University of California, Davis, California.

Trenham, P.C. 2001. Terrestrial habitat use by adult *Ambystoma californiense*. Journal of Herpetology 35:343–346.

Trenham, P.C., W.D. Koenig and H.B. Shaffer. 2001. Spatially autocorrelated demography and interpond dispersal in the California tiger salamander, *Ambystoma californiense*. Ecology 82:3519–3530.

Trenham, P.C., H.B. Shaffer, W.D. Koenig and M.R. Stromberg. 2000. Life history and demographic variation in the California tiger salamander *(Ambystoma californiense)*. Copeia 2000:365–377.

Trombulak, S.C. and C.A. Frissell. 2000. Review of ecological effects of roads on terrestrial and aquatic communities. Conservation Biology 14:18–30.

Trowbridge, A.H. 1937. New records of Amphibia for Oklahoma. Copeia 1937:71–72.

Trowbridge, A.H. and M.S. Trowbridge. 1937. Notes on the cleavage rate of *Scaphiopus bombifrons* Cope, with additional remarks on certain aspects of its life history. American Naturalist 71:460–480.

Trueb, L. 1969. *Pternohyla, P. dentata, P. fodiens.* Pp. 77.1–77.4. Catalogue of American Amphibians and Reptiles. Society for the Study of Amphibians and Reptiles, St. Louis, Missouri.

Trueb, L. and C. Gans. 1983. Feeding specializations of the Mexican burrowing toad, *Rhinophrynus dorsalis* (Anura: Rhinophrynidae). Journal of Zoology (London) 199:189–208.

Truitt, J.O. 1964. Observations on the defensive attitude of a southern toad *(Bufo terrestris)*. British Journal of Herpetology 3:167.

Tschudi, J.J. 1838. Classification der batrachien, mit bercksichtigung der fossilen thiere dieser abteilung der reptilien. Memoires de la Societe des Sciences Naturelles Neuchatel 2:1–100.

Tucker, J.K. 1994. A comment on bullfrog *(Rana catesbeiana)* predation on Gulf Coast toads *(Bufo valliceps)* as reported by Platt and Fontenot. Bulletin of the Chicago Herpetological Society 29:73.

Tucker, J.K. 1995. Early post-transformational growth in the Illinois chorus frog *(Pseudacris streckeri illinoensis)*. Journal of Herpetology 29:314–316.

Tucker, J.K. 1997a. Fecundity in the Illinois chorus frog *(Pseudacris streckeri illinoensis)* from Madison County, Illinois. Transactions of the Illinois State Academy of Science 90:167–170.

Tucker, J.K. 1997b. Description of newly transformed froglets of the Illinois chorus frog *(Pseudacris streckeri illinoensis)*. Transactions of the Illinois State Academy of Science 90:161–166.

Tucker, J.K. 1997c. Food habits of the fossorial frog *Pseudacris streckeri illinoensis*. Herpetological Natural History 5:83–87.

Tucker, J.K. 1998. Status of Illinois chorus frogs in Madison County, Illinois. Pp. 94–101, *In* Lannoo, M.J. (Ed.), Status and Conservation of Midwestern Amphibians. University of Iowa Press, Iowa City, Iowa.

Tucker, J.K. 2000a. Natural history notes: *Pseudacris streckeri illinoensis* (Illinois chorus frog). Frost injuries. Herpetological Review 31:41–42.

Tucker, J.K. 2000b. Growth and survivorship in the Illinois chorus frog *(Pseudacris streckeri illinoensis)*. Transactions of the Illinois State Academy of Science 93:63–68.

Tucker, J.K. and D.P. Philipp. 1993. Population status of the Illinois chorus frog *(Pseudacris streckeri illinoensis)* in Madison County, Illinois, with emphasis on the new Poag Road/FAP 413 interchange and FAP 413 wetland mitigation site. Illinois Department of Transportation, Technical report, Springfield, Illinois.

Tucker, J.K. and M.E. Sullivan. 1975. Unsuccessful attempts by bullfrogs to eat toads. Transactions of the Illinois State Academy of Science 68:167.

Tucker, J.K., J.B. Camerer and J.B. Hatcher. 1995. Natural history notes: *Pseudacris streckeri illinoensis* (Illinois chorus frog). Burrows. Herpetological Review 26:32–33.

Tucker, R.B. 1998. Ecology and natural history of the Cow Knob salamander, *Plethodon punctatus*, in West Virginia. Master's thesis. Marshall University, Huntington, West Virginia.

Tuggle, B.N. and S.K. Schmeling. 1982. Parasites of the bald eagle *(Haliateetus leucocephalus)* of North America. Journal of Wildlife Diseases 18:501–506.

Tugwell, N. and K. Schwartz. 1991. Geographic distribution: *Scaphiopus holbrookii hurterii* (Hurter's spadefoot). Herpetological Review 22:134.

Tumlison, R. and G.R. Cline. 1997. Further notes on the habitat of the Oklahoma salamander, *Eurycea tynerensis*. Proceedings of the Oklahoma Academy of Sciences 77:103–106.

Tumlison, R., G.R. Cline and P. Zwank. 1990a. Morphological discrimination between the Oklahoma salamander *(Eurycea tynerensis)* and the graybelly salamander *(Eurycea multiplicata griseogaster)*. Copeia 1990:242–246.

Tumlison, R., G.R. Cline and P. Zwank. 1990b. Prey selection in the Oklahoma salamander *(Eurycea tynerensis)*. Journal of Herpetology 24:222–225.

Tumlison, R., G.R. Cline and P. Zwank. 1990c. Surface habitat associations of the Oklahoma salamander *(Eurycea tynerensis)*. Herpetologica 46:169–175.

Tupa, D.D. and W.K. Davis. 1976. Population dynamics of the San Marcos salamander, *Eurycea nana* Bishop. Texas Journal of Science 27:179–195.

Turchin, P. 1995. Population regulation: old arguments and a new synthesis. Pp. 19–37. *In* Cappuccino, N. and P.W. Price (Eds.), Population Dynamics: New Approaches and Synthesis. Academic Press, New York.

Turchin, P., A.D. Taylor and J.D. Reeve. 1999. Dynamical role of predators in population cycles of a forest insect. Science 285:1068–1071.

Turgeon, D.D., J.F. Quinn Jr., A.E. Bogan, E.V. Coan, F.G. Hochberg, W.G. Lyons, P.M. Mikkelsen, R.J. Neves, C.F.E. Roper, G. Rosenberg, B. Roth, A. Scheltema, F. G. Thompson, M. Vecchione and J.D. Williams. 1998. Common and scientific names of aquatic invertebrates from the United States and Canada: Mollusks. Second edition. Committee on Scientific and Vernacular Names of Mollusks of the Council of Systematic Malacologists and the American Malacological Union, Bethesda, Maryland.

Turner, F.B. 1958a. Life history of the western spotted frog in Yellowstone National Park. Herpetologica 14:96–100.

Turner, F.B. 1958b. Some parasites of the western spotted frog in Yellowstone National Park. Journal of Parasitology 44:182.

Turner, F.B. 1959a. Some features of the ecology of *Bufo punctatus* in Death Valley, California. Ecology 40:175–181.

Turner, F.B. 1959b. An analysis of the feeding habits of *Rana p. pretiosa* in Yellowstone National Park, Wyoming. American Midland Naturalist 61:403–413.

Turner, F.B. 1960a. Postmetamorphic growth in anurans. American Midland Naturalist 64:327–338.

Turner, F.B. 1960b. Population structure and dynamics of the western spotted frog, *Rana p. pretiosa* Baird and Girard, in Yellowstone Park, Wyoming. Ecological Monographs 30:251–278.

Turner, F.B. 1962a. The demography of frogs and toads. Quarterly Review of Biology 37:303–314.

Turner, F.B. 1962b. An analysis of geographic variation and distribution of *Rana pretiosa*. Year Book. American Philosophical Society 1962: 325–328.

Turner, F.B. and P.C. Dumas. 1972. *Rana pretiosa*. Pp. 119:1–119.4. Catalogue of American Amphibians and Reptiles. Society for the Study of Amphibians and Reptiles, St. Louis, Missouri.

Turner, F.B. and R.H. Wauer. 1963. A survey of the herpetofauna of the Death Valley area. Great Basin Naturalist 23:119–128.

Turner, M.C. 1997. The ecology, natural history, and distribution of *Desmognathus quadramaculatus* in West Virginia. Master's thesis. Marshall University, Huntington, West Virginia.

Turner, M.C. and T.K. Pauley. 1992. An ecological study of *Desmognathus quadramaculatus* (Holbrook) in Keeney's Creek, Fayette County, West Virginia. Proceedings of the West Virginia Academy of Science 64:35–36.

Turner, M.G. 1989. Landscape ecology: the effect of pattern on process. Annual Review of Ecology and Systematics 20:171–197.

Turner, M.G. and R.H. Gardner (Eds.). 1991. Quantitative Methods in Landscape Ecology. Springer-Verlag, New York.

Turnipseed, G. and R. Altig. 1975. Population density and age structure of three species of hylid tadpoles. Journal of Herpetology 9:287–291.

Turnipseed, G. and R. Altig. 1991. Geographic distribution: *Ambystoma annulatum* (ringed salamander). Herpetological Review 22:133.

Turtle, S. 2000. Embryonic survivorship of the spotted salamander (*Ambystoma maculatum*) in roadside and woodland vernal pools in southeastern New Hampshire. Journal of Herpetology 34:60–67.

Twedt, B. 1993. A comparative ecology of *Rana aurora* Baird and Girard and *Rana catesbeiana* Shaw at Freshwater Lagoon, Humboldt County, California. Master's thesis. Humboldt State University, Arcata, California.

Tweedell, K.S. 1967. Induced oncogenesis in developing frog kidney cells. Cancer Research 27:2042–2052.

Twitty, V.C. 1935. Two new species of *Triturus* from California. Copeia 1935:73–80.

Twitty, V.C. 1936. Correlated genetic and embryological experiments on *Triturus*. I. and II. Journal of Experimental Zoology 74:239–302.

Twitty, V.C. 1937. Experiments on the phenomenon of paralysis produced by a toxin occurring in *Triturus* embryos. Journal of Experimental Zoology 76:67–104.

Twitty, V.C. 1941. Data on the life history of *Ambystoma tigrinum californiense* Gray. Copeia 1941:1–4.

Twitty, V.C. 1942. The species of California *Triturus*. Copeia 1942:65–76.

Twitty, V.C. 1955. Field experiments on the biology and genetic relationships of the California species of *Triturus*. Journal of Experimental Zoology 129:129–148.

Twitty, V.C. 1959. Migration and speciation in newts. Science 130:1735–1743.

Twitty, V.C. 1961a. Experiments on homing behavior and speciation in *Taricha*. Pp. 415–459. *In* Blair, W.F. (Ed.), Vertebrate Speciation. University of Texas Press, Austin, Texas.

Twitty, V.C. 1961b. Second-generation hybrids of the species of *Taricha*. Proceedings of the National Academy of Sciences 47:1461–1486.

Twitty, V.C. 1964a. *Taricha rivularis*. Pp. 9.1–9.2. Catalogue of American Amphibians and Reptiles. American Society of Ichthyologists and Herpetologists, St. Louis, Missouri.

Twitty, V.C. 1964b. Fertility of *Taricha* species-hybrids and viability of their offspring. Proceedings of the National Academy of Sciences 52:156–161.

Twitty, V.C. 1966. Of Scientists and Salamanders. W.H. Freeman and Company, San Francisco, California.

Twitty, V.C. and H.H. Johnson. 1934. Motor inhibition in *Amblystoma* produced by *Triturus* transplants. Science 80:78–79.

Twitty, V.C., D. Grant and O. Anderson. 1964. Long distance homing in the newt *Taricha rivularis*. Proceedings of the National Academy of Sciences 52:51–58.

Twitty, V.C., D. Grant and O. Anderson. 1966. Course and timing of the homing migration in the newt *Taricha rivularis*. Proceedings of the National Academy of Sciences 56:864–871.

Twitty, V.C., D. Grant and O. Anderson. 1967a. Initial homeward orientation after long distance displacement in the newt *Taricha rivularis*. Proceedings of the National Academy of Sciences 57:342–348.

Twitty, V.C., D. Grant and O. Anderson. 1967b. Home range in relation to homing in the newt *Taricha rivularis* (Amphibia: Caudata). Copeia 1967:649–653.

Twitty, V.C., D. Grant and O. Anderson. 1967c. Amphibian orientation: an unexpected observation. Science 155:352–353.

Tyler, J.D. and H.N. Buscher. 1980. Notes on a population of larval *Ambystoma tigrinum* (Ambystomatidae) from Cimarron County, Oklahoma. Southwestern Naturalist 25:391–395.

Tyler, M.J. 1975. The cane toad, *Bufo marinus*. An historical account and modern assessment. Ph.D. dissertation. University of Adelaide, Adelaide, South Australia, Australia.

Tyler, M.J. 1989. Australian Frogs. Viking O'Neil, Penguin Books Australia Limited, South Yarra, Victoria, Australia.

Tyler, M.J. 1991. Declining amphibian populations—a global phenomenon? An Australian perspective. Alytes 9:43–50.

Tyler, M.J. 1998. Australian Frogs: A Natural History. Cornell University Press, Ithaca, New York.

Tyler, T.J., W.J. Liss, R.L. Hoffman and L.M. Ganio. 1998b. Experimental analysis of trout effects on survival, growth, and habitat use of two species of ambystomatid salamanders. Journal of Herpetology 32:345–349.

Tyler, T.J., C.D. McIntire, B. Samora, R.L. Hoffman and G.L. Larson. 2002. Inventory of aquatic breeding amphibians, Mount Rainier National Park, 1994–1999. National Park Service, Final report, U.S. Forest and Rangeland Ecosystem Science Center, Corvallis, Oregon.

Tyler, T.J., W. Liss, L.M. Ganio, G.L. Larson, R. Hoffman, E. Deimling and G. Lomnicky. 1998a. Interaction between introduced trout and larval salamanders (*Ambystoma macrodactylum*) in high-elevation lakes. Conservation Biology 12:94–105.

Tyning, T.F. 1990. A Guide to Amphibians and Reptiles. Little, Brown and Company, Boston, Massachusetts.

U.S. Congress, Office of Technology Assessment, Harmful Non-Indigenous Species in the United States. 1993. U.S. Government Printing Office, OTA-F-565, Washington, D.C.

U.S. Department of the Interior. 1991. Endangered and threatened wildlife and plants; animal notice of review. Federal Register, 50 CFR Part 17, 56:58804–58836.

U.S. Environmental Protection Agency. 1975. Methods for acute toxicity tests with fish, macroinvertebrates, and amphibians. Committee on Methods of Acute Toxicity Tests with Aquatic Organisms, National Environmental Research Center, EPA-660/3-75-009, Corvalis, Oregon.

U.S. Environmental Protection Agency. 1999. Draft map of Level III ecoregions of the continental United States. National Health and Environmental Effects Research Laboratory, Corvallis, Oregon.

U.S. Environmental Protection Agency. 2000a. Level III ecoregions of the continental United States (revision of Omernik, 1987), Map M-1. U.S. Environmental Protection Agency, National Health and Environmental Effects Research Laboratory, Corvallis, Oregon.

U.S. Environmental Protection Agency. 2000b. Draft level III and IV ecoregions of the southeastern states. National Health and Environmental Effects Research Laboratory, Corvallis, Oregon.

U.S. Fish and Wildlife Service. 1976. Determination that the Red Hills salamander is a threatened species. Federal Register 41:53032–53034.

U.S. Fish and Wildlife Service. 1977. Endangered and threatened wildlife and plants; proposed determination of critical habitat for the Houston toad. Federal Register 42:27009–27011.

U.S. Fish and Wildlife Service. 1982. Desert slender salamander recovery plan. U.S. Fish and Wildlife Service, Portland, Oregon.

U.S. Fish and Wildlife Service. 1989a. Determination of threatened status for the Cheat Mountain salamander and endangered status for the Shenandoah salamander. Federal Register 54:34464–34468.

U.S. Fish and Wildlife Service. 1989b. Endangered and threatened wildlife and plants: proposal to determine threatened status for the Cheat Mountain salamander and endangered status for the Shenandoah salamander. Federal Register, 50 CFR Part 17, Volume 53(188):37814–37818.

U.S. Fish and Wildlife Service. 1989c. Endangered and threatened wildlife and plants; animal notice of review. Federal Register 54: 554–558.

U.S. Fish and Wildlife Service. 1990. Flattened musk turtle *(Sternotherus depressus)* recovery plan. USFWS, Jackson, Mississippi.

U.S. Fish and Wildlife Service. 1991. Endangered and threatened wildlife and plants; animal candidate review for listing as endangered or threatened species, proposed rule. Federal Register 56: 58804–58836.

U.S. Fish and Wildlife Service. 1993. Endangered and threatened wildlife and plants; finding on petition to list the spotted frog. Federal Register 58:38553.

U.S. Fish and Wildlife Service. 1994a. Endangered and threatened wildlife and plants; determination of endangered status for the arroyo southwestern toad. Federal Register 59:64859–64866.

U.S. Fish and Wildlife Service. 1994b. Endangered and threatened wildlife and plants; 12-month petition finding for the California tiger salamander. Federal Register 59:18353–18354.

U.S. Fish and Wildlife Service. 1994c. Endangered and threatened wildlife and plants; animal candidate review for listing as endangered or threatened species; proposed rule. Federal Register 59(219): 58982–59028.

U.S. Fish and Wildlife Service. 1994d. Shenandoah salamander *(Plethodon shenandoah)* recovery plan. U.S. Fish and Wildlife Service, Hadley, Massachusetts.

U.S. Fish and Wildlife Service. 1995a. Endangered and threatened wildlife and plants; 12-month finding for a petition to list the Southern Rocky Mountain population of the boreal toad as endangered. Federal Register 60:15281–15283.

U.S. Fish and Wildlife Service. 1995b. San Marcos and Comal Springs and associated aquatic ecosystems (revised) recovery plan. U.S.Fish and Wildlife Service Region 2, Albuquerque, New Mexico.

U.S. Fish and Wildlife Service. 1995c. Administrative 12-month finding on a not warranted petition to list the Amargosa toad under the Endangered Species Act. Memorandum to Director, Fish and Wildlife Service from Acting Regional Director, Region 1, dated 26 October, 1995.

U.S. Fish and Wildlife Service. 1996a. Endangered and threatened wildlife and plants; review of plant and animal taxa that are candidates for listing as endangered or threatened species. Federal Register 61(40):7596–7613.

U.S. Fish and Wildlife Service. 1996b. San Marcos and Comal Springs and associated aquatic ecosystems (revised) recovery plan. United States Fish and Wildlife Service Region 2 Office, Albuquerque, New Mexico.

U.S. Fish and Wildlife Service. 1996c. Endangered and threatened wildlife and plants: determination of threatened status for the California red-legged frog. Federal Register 61(101):25813-25833.

U.S. Fish and Wildlife Service. 1997a. Endangered and threatened species; review of plant and animal taxa; proposed rule. [Federal Register: September 19, 1997 (Volume 62, Number 182)] 50 CFR Part 17, Department of the Interior, Washington, D.C.

U.S. Fish and Wildlife Service. 1997b. Endangered and threatened wildlife and plants; notice of reclassification of ten candidate species. Federal Register 62:49191–49193.

U.S. Fish and Wildlife Service. 1997c. Endangered and threatened wildlife and plants; proposed rule to list the flatwoods salamander as threatened. Federal Register 62(241):65787–65794.

U.S. Fish and Wildlife Service. 1998a. Final environmental assessment/habitat conservation plan for issuance of a section 10(a)(1)(B) permit for incidental take of the Barton Springs salamander *(Eurycea sosorum)* for the operation and maintenance of Barton Springs Pool and adjacent springs, Austin, Texas.

U.S. Fish and Wildlife Service. 1998b. Headwaters Forest Project Final Environmental Impact Statement/ Environmental Impact Review and Pacific Lumber Company Habitat Conservation Plan. U.S. Fish and Wildlife Service, Arcata, California.

U.S. Fish and Wildlife Service. 1999a. Arroyo southwestern toad *(Bufo microscaphus californicus)* recovery plan. U.S. Department of the Interior, Fish and Wildlife Service, Portland, Oregon.

U.S. Fish and Wildlife Service. 1999b. Agency Draft. California red-legged frog *(Rana aurora draytonii)* Recovery Plan, Portland, Oregon.

U.S. Fish and Wildlife Service. 1999c. Santa Cruz long-toed salamander *(Ambystoma macrodactylum croceum)* draft revised recovery plan. U.S. Fish and Wildlife Service, Portland, Oregon.

U.S. Fish and Wildlife Service. 2000a. Review of endangered and threatened wildlife and plants; proposal to list the Chiricahua leopard frog as threatened with a special rule. Federal Register 65:37343–37357.

U.S. Fish and Wildlife Service. 2000b. Endangered and threatened wildlife and plants; emergency rule to list the Santa Barbara County distinct population of the California tiger salamander as endangered. Federal Register 65:3095–3109.

U.S. Fish and Wildlife Service. 2001a. Tarahumara frog conservation program. http://arizonaes.fws.gov/T-frog3.htm.

U.S. Fish and Wildlife Service. 2001b. Changes in list of species in appendices to the Convention on International Trade in Endangered Species of Wild Fauna and Flora. Federal Register 66(97): 27601–27615.

U.S. Fish and Wildlife Service. 2002a. Endangered and threatened wildlife and plants; listing of the Chiricahua leopard frog *(Rana chiricahuensis)*. Federal Register 67(114):40790–40810.

U.S. Fish and Wildlife Service. 2002b. Endangered and threatened wildlife and plants; determination of endangered status for the Southern California distinct vertebrate segment of the mountain yellow-legged frog *(Rana muscosa)*. Federal Register 67(127): 44382– 44392.

U.S. Fish and Wildlife Service. 2002c. Recovery plan for the California red-legged frog *(Rana aurora draytonii)*. U.S. Fish and Wildlife Service, Portland, Oregon.

U.S. Fish and Wildlife Service. 2002d. Status review for the Columbia spotted frog *(Rana luteiventris)* on the Wasatch Front, Utah. U.S. Department of the Interior, U.S. Fish and Wildlife Service, Region 6, Denver, Colorado.

U.S. Forest Service. 1988. Endangered Species Act of 1973 as Amended Through the 100th Congress. U.S. Department of the Interior, Washington, D.C.

U.S. General Accounting Office. 1991. Rangeland management: BLM's hot desert grazing program merits reconsideration. Washington, D.C.

U.S. Senate and House of Representatives. 1973. Endangered Species Act of 1973. Public Law 93–205.

Ubelaker, J.E., D.W. Duszynski and D.L. Beaver. 1967. Occurrence of the trematode, *Glypthelmins pennsylvaniensis* Cheng, 1961, in chorus frogs, *Pseudacris triseriata*, in Colorado. Bulletin of the Wildlife Disease Association 3:177.

Uchida, A., H. Itagaki and H. Inoue. 1980. Studies on the amphibian helminths in Japan. VII. *Rhacophotrema itagakii* n. g. et n. sp, and *Opisthioglyphe japonicus* n. sp. (Digenea: Omphalometridae) from frogs. Japanese Journal of Parasitology 29:109–113.

Uhlenhuth, E. 1919. Observations on the distribution of the blind Texan cave salamander, *Typhlomolge rathbuni*. Copeia 69:26–27.

Uhlenhuth, E. 1921. Observations on the distribution and habitats of the blind Texan cave salamander, *Typhlomolge rathbuni*. Biological Bulletin 15:73–104.

Uhler, F.M., C. Cottam and T.E. Clarke. 1939. Food of snakes of the George Washington National Forest, Virginia. Pp. 605–622. *In* Transactions of the Fourth North American Wildlife Conference, Washington, D.C.

Ulmer, M.J. 1970. Studies on the helminth fauna of Iowa. I. Trematodes of amphibians. American Midland Naturalist 83:38–64.

Ulmer, M.J. and H.A. James. 1976. Studies on the helminth fauna of Iowa. II. Cestodes of amphibians. Proceedings of the Helminthological Society of Washington 43:191–200.

Ultsch, G.R. 1973. Observations on the life history of *Siren lacertina*. Herpetologica 29:304–305.

Ultsch, G.R. 1976. Eco-physiological studies of some metabolic and respiratory adaptations of sirenid salamanders. Pp. 287–311. *In* Hughes, G.M. (Ed.), Respiration of Amphibious Vertebrates. Academic Press, London.

Ultsch, G.R. and S.J. Arceneaux. 1988. Gill loss in larval *Amphiuma tridactylum*. Journal of Herpetology 22:347–348.

Ultsch, G.R., D.F. Bradford and J. Freda. 1999. Physiology: coping with the environment. Pp. 189–214. *In* McDiarmid, R.W. and R. Altig (Eds.), Tadpoles: The Biology of Anuran Larvae. University of Chicago Press, Chicago, Illinois.

Ultsch, G.R., T.E. Graham and C.E. Crocker. 2000. An aggregation of overwintering leopard frogs, *Rana pipiens*, and common map turtles, *Graptemys geographica*, in northern Vermont. Canadian Field-Naturalist 114:314–315.

Umber, R.W. and L.D. Harris. 1974. Effects of intensive forestry on succession and wildlife in Florida sandhills. Proceedings of the Annual Conference of Southeastern Association of Game and Fish Commissioners 28:686–693.

Underhill, D.K. 1966. An incidence of spontaneous caudal scoliosis in tadpoles of *Rana pipiens* Schreber. Copeia 1966:582–583.

Underhill, D.K. 1968a. Heritability of dorsal spot number and snout-vent length in *Rana pipiens*. Journal of Heredity 59:235–240.

Underhill, D.K. 1968b. Albino eggs and larvae of *Ambystoma texanum* in central Illinois. Herpetologica 24:266.

Underhill, J.C. 1958. Notes on the toads of eastern South Dakota. Copeia 1958:149–151.

Underhill, J.C. 1961. Intraspecific variation in the Dakota toad, *Bufo hemiophrys*, from northeastern South Dakota. Herpetologica 17:220–227.

Upchurch, D.J. 1971. Geographic distribution: *Ambystoma maculatum*. Herpetological Review 3:89.

Upton, S.J. and C.T. McAllister. 1988. The Coccidia (Apicomplexa: Eimeriidae) of Anura, with descriptions of four new species. Canadian Journal of Zoology 66:1822–1830.

Upton, S.J., C.T. McAllister and S.E. Trauth. 1993. The Coccidia (Apicomplexa: Eimeriidae) of Caudata (Amphibia), with descriptions of two new species from North America. Canadian Journal of Zoology 71:2410–2418.

URS Company. 1977. Las Vegas Wash interim report Number 2, Water Quality Series: Clark County 208 Water Quality Management Plan, Clark County, Nevada. U.R.S. Company, Las Vegas, Nevada.

Usher, M.B. (Ed.). 1986. Wildlife Conservation Evaluation. Chapman and Hall, London, England.

Utah Division of Wildlife Resources. 1991. Distribution and status of the western spotted frog in central Utah. Prepared for the U.S. Fish and Wildlife Service, Salt Lake City, Utah.

Uzendoski, U.V. and P.A. Verrell. 1993. Sexual incompatibility and mate-recognition systems: a study of two species of sympatric salamanders (Plethodontidae). Animal Behaviour 46:267–278.

Uzzell, T. 1963. Natural triploidy in salamanders related to *Ambystoma jeffersonianum*. Science 139:113–115.

Uzzell, T. 1967a. *Ambystoma platineum*. Pp. 49.1–49.2. Catalogue of American Amphibians and Reptiles. Society for the Study of Amphibians and Reptiles, St. Louis, Missouri.

Uzzell, T. 1967b. *Ambystoma tremblayi*. Pp. 50.1–50.2. Catalogue of American Amphibians and Reptiles. Society for the Study of Amphibians and Reptiles, St. Louis, Missouri.

Uzzell, T. 1969. Notes on spermatophore production by salamanders of the *Ambystoma jeffersonianum* complex. Copeia 1969:602–612.

Uzzell, T.M. 1952. Additional notes on the range of *Rana grylio* Stejneger in South Carolina. Copeia 1952:118.

Uzzell, T.M. 1963. Natural triploidy in salamanders related to *Ambystoma jeffersonianum*. Science 139:113–115.

Uzzell, T.M. 1964. Relations of the diploid and triploid species of the *Ambystoma jeffersonianum* complex (Amphibia: Caudata). Copeia 1964:257–300.

Uzzell, T.M. 1967a. *Ambystoma jeffersonianum*. Pp. 47.1–47.2. Catalogue of American Amphibians and Reptiles. Society for the Study of Amphibians and Reptiles, St. Louis, Missouri.

Uzzell, T.M. 1967b. *Ambystoma laterale*. Pp. 48.1–48.2. Catalogue of American Amphibians and Reptiles. Society for the Study of Amphibians and Reptiles, St. Louis, Missouri.

Uzzell, T.M. 1969. Notes on the spermatophore production by salamanders of the *Ambystoma jeffersonianum* complex. Copeia 1969: 602–612.

Uzzell, T.M. 1969. Unisexual species of salamanders. Discovery 4: 99–108.

Uzzell, T.M. and S.M. Goldblatt. 1967. Serum proteins of salamanders of the *Ambystoma jeffersoniamum* complex and the origin of the triploid species of this group. Evolution 21:345–354.

Uzzell, T.M., Jr. 1963. Natural triploidy in salamanders related to *Ambystoma jeffersonianum*. Science 139:113–115.

Uzzell, T.M., Jr. 1964. Relations of the diploid and triploid species of the *Ambystoma jeffersonianum* complex (Amphibia, Caudata). Copeia 1964:257–300.

Valentine, B.D. 1961. Variation and distribution of *Desmognathus ocoee* Nicholls (Amphibia: Plethodontidae). Copeia 1961:315–322.

Valentine, B.D. 1963a. The salamander genus *Desmognathus* in Mississippi. Copeia 1963:130–139.

Valentine, B.D. 1963b. The plethodontid salamander *Phaeognathus*: collecting techniques and habits. Journal of the Ohio Herpetological Society 4:49–54.

Valentine, B.D. 1963c. Notes on the early life history of the Alabama salamander, *Desmognathus aeneus chermocki* Bishop and Valentine. American Midland Naturalist 69:182–188.

Valentine, B.D. 1964. *Desmognathus ocoee*. Pp. 7.1–7.2. Catalogue of American Amphibians and Reptiles. Society for the Study of Amphibians and Reptiles, St. Louis, Missouri.

Valentine, B.D. 1974. *Desmognathus quadramaculatus*. Pp. 153.1–153.4. Catalogue of American Amphibians and Reptiles. Society for the Study of Amphibians and Reptiles, St. Louis, Missouri.

Valentine, B.D. and D.M. Dennis. 1964. A comparison of the gill-arch system and fins of three genera of larval salamanders, *Rhyacotriton*, *Gyrinophilus*, and *Ambystoma*. Copeia 1964:196–201.

Valerio, C.E. 1971. Ability of some tropical tadpoles to survive without water. Copeia 1971:364–375.

Valett, B.B. and D.L. Jameson. 1961. The embryology of *Eleutherodactylus augusti latrans*. Copeia 1961:103–109.

Valtonen, E.T., J.C. Holmes and M. Koskivaara. 1997. Eutrophication, pollution, and fragmentation: effects on parasite communities in roach *(Rutilus rutilus)* and perch *(Perca flutviatilis)* in four lakes in central Finland. Canadian Journal of Fisheries and Aquatic Sciences 54:572–585.

Van Buskirk, J. 1988. Interactive effects of dragonfly predation in experimental pond communities. Ecology 69:857–867.

Van Buskirk, J. 1992. Competition, cannibalism, and size class dominance in a dragonfly. Oikos 65:455–464.

Van Buskirk, J. 2002. A comparative test of the adaptive plasticity hypothesis: relationships between habitat and phenotype in anuran larvae. American Naturalist 160:87–102.

Van Buskirk, J. and S.A. McCollum. 1999. Plasticity and selection explain variation in tadpole phenotype between ponds with different predator communities. Oikos 85:31–39.

Van Buskirk, J. and S.A. McCollum. 2000a. Functional mechanisms of an inducible defense in tadpoles: morphology and behaviour influence mortality and risk from predation. Journal of Evolutionary Biology 13:336–347.

Van Buskirk, J. and S.A. McCollum. 2000b. Influence of tail shape on tadpole swimming performance. Journal of Experimental Biology 203:2149–2158.

Van Buskirk, J. and R.A. Relyea. 1998. Natural selection for phenotypic plasticity: predator-induced morphological responses in tadpoles. Biological Journal of the Linnean Society 65:301–328.

Van Buskirk, J. and D.C. Smith. 1991. Density-dependent population regulation in a salamander. Ecology 72:1747–1756.

Van Damme, R., D. Bauwens and R.F. Verheyen. 1986. Selected body temperatures in the lizard *Lacerta vivipara*: variation within and between populations. Journal of Thermal Biology 11:219–222.

Van Damme, R., D. Bauwens and R.F. Verheyen. 1990. Evolutionary rigidity of thermal physiology: the case of the cool temperature lizard *Lacerta vivipara*. Oikos 57:61–67.

Van Damme, R., D. Bauwens and R.F. Verheyen. 1991. The thermal dependence of feeding behaviour, food consumption and gut-passage time in the lizard *Lacerta vivipara* Jacquin. Functional Ecology 5:507–517.

Van Denburgh, J. 1895. Notes on the habits and distribution of *Autodax iëcnus*. Proceedings of the California Academy of Science 5:776–778.

Van Denburgh, J. 1906. Description of a new species of the genus *Plethodon (Plethodon vandykei)* from Mt. Rainier, Washington. Proceedings of the California Academy of Sciences 4:61–63.

Van Denburgh, J. 1916. Four species of salamanders new to the state of California, with a description of *Plethodon elongatus*, a new species, and notes on other salamanders. Proceedings of the California Academy of Science, 4th Series, 6:215–221.

Van Denburgh, J. 1924. Notes on the herpetology of New Mexico, with a list of species known from that state. California Academy of Science, Proceedings Fourth Series, Volume XIII(12):189–230.

Van Denburgh, J. and J.R. Slevin. 1914. Reptiles and amphibians of the islands of the West Coast of North America. Proceedings of the California Academy of Sciences 4:129–152.

Van Denburgh, J. and J.R. Slevin. 1921. A list of the amphibians and reptiles of Nevada, with notes on the species in the collection of the Academy. Proceedings of the California Academy of Sciences XI(2): 27–38.

van der Leun, J.C., X. Tang and M. Tevini (Eds.). 1998. Environmental effects of ozone depletion: 1998 assessment. Photochemistry and Photobiology 46:1–108.

Van Devender, T.R. 1969. A record of albinism in the canyon tree frog, *Hyla arenicolor* Cope. Herpetologica 25:68–69.

Van Devender, T.R. and C.H. Lowe. 1977. Amphibians and reptiles of Yepomera, Chihuahua, Mexico. Journal of Herpetology 11:41–50.

Van Devender, T.R., C.H. Lowe, H.K. McCrystal and H.E. Lawler. 1992. Viewpoint: reconsider suggested systematic arrangements for some North American amphibians and reptiles. Herpetological Review 23:10–14.

Van Devender, W. 1998. Vertebrate collections at smaller schools: good or bad? Part 1. Association of Southeastern Biologists Bulletin 45:188–191.

van Dijk, D.E. 1977. Habitats and dispersal of Southern African Anura. Zoologica Africana 12:169–181.

Van Hyning, O.C. 1932. Food of some Florida snakes. Copeia 1932:37.

Van Hyning, O.C. 1933. Batrachia and Reptilia of Alachua County, Florida. Copeia 1933:3–7.

Van Kirk, R., L. Benjamin and D. Patla. 2000. Riparian area assessment and amphibian status in the watersheds of the Greater Yellowstone Ecosystem. Greater Yellowstone Coalition, Project completion report, Bozeman, Montana.

Van Valen, L. 1974. A natural model for the origin of some higher taxa. Journal of Herpetology 8:109–121.

Van Valen, L. 1976. Ecological species, multispecies, and oaks. Taxonomy 25:233–239.

Van Wagner, T.J. 1996. Selected life-history and ecological aspects of a population of foothill yellow-legged frogs (Rana boylii) from Clear Creek, Nevada Co., CA. Master's thesis. Department of Biological Sciences, California State Chico, Chico, California.

Van Winkle, K. 1922. Extension of the range of Ascaphus truei Stejneger. Copeia 102:4–6.

Van Zandt, P.A. and S. Mopper. 1998. A meta-analysis of adaptive deme formation in phytophagous insect populations. American Naturalist 152:595–604.

Vanderburgh, D.J. and R.C. Anderson. 1987a. The relationship between nematodes of the genus Cosmocercoides Wilkie, 1930 (Nematoda: Cosmocercoidea) in toads (Bufo americanus) and slugs (Deroceras laeve). Canadian Journal of Zoology 65:1650–1661.

Vanderburgh, D.J. and R.C. Anderson. 1987b. Preliminary observations on seasonal changes in prevalence and intensity of Cosmocercoides variabilis (Nematoda: Cosmocercoidea) in Bufo americanus (Amphibia). Canadian Journal of Zoology 65:1666–1667.

Vane-Wright, R.I., C.J. Humphries and P.H. Williams. 1991. What to protect? Systematics and the agony of choice. Biological Conservation 55:235–54.

Vane-Wright, R.I., C.R. Smith and I.J. Kitchling. 1994. Systematic assessment of taxic diversity by summation. Pp. 309–326. In Forey, P.L., C.J. Humphries and R.I. Vane-Wright (Eds.), Systematics and Conservation Evaluation. Clarndon Press, Oxford, United Kingdom.

VanKirk, E.A. 1980. Report on population of Bufo hemiophrys on the Laramie plains, Albany County, Wyoming. Wyoming Natural Heritage Program, Nature Conservancy, Cheyenne, Wyoming.

VanNorman, D.E. and A.F. Scott. 1987. The distribution and breeding habitat of the barking treefrog, Hyla gratiosa LeConte, in south-central Kentucky and north-central Tennessee. Journal of the Tennessee Academy of Science 62:7–11.

Varhegyi, G., S.M. Mavroidis, B.M. Walton, C.A. Conaway and A.R. Gibson. 1998. Amphibian surveys in the Cuyahoga Valley National Recreation Area. Pp. 137–154. In Lannoo, M.J. (Ed.), Status and Conservation of Midwestern Amphibians. University of Iowa Press, Iowa City, Iowa.

Varley, G.C. and G.R. Gradwell. 1960. Key factors in population studies. Journal of Animal Ecology 29:399–401.

Varley, G.C. and G.R. Gradwell. 1970. Recent advances in insect population dynamics. Annual Review of Entomology 15:1–24.

Varley, G.C., B.R. Gradwell and M.P. Hassell. 1973. Insect Population Ecology. Blackwell Scientific Publications, Oxford, United Kingdom.

Vatnick, I., M.A. Brodkin, M.P. Simon, B.W. Grant, C.R. Conte, M. Gleave, R. Myers and M.M. Sadoff. 1999. The effects of exposure to mild acidic conditions on adult frogs (Rana pipiens and Rana clamitans): mortality rates and pH preferences. Journal of Herpetology 33:370–374.

Veldman, R.B. 1997. Trend toward extinction of the cricket frog (Acris crepitans): prediction of declines using field data and the Vortex simulation program. Master's thesis. Illinois State University, Normal, Illinois.

Veloso, A. and J. Navarro. 1988. Systematic list and geographic distribution of amphibians and reptiles from Chile. Museo Regionale di Scienze Naturali Bollettino (Turin) 6:481–540.

Vences, M., M.W. Penuel-Matthews, D.R. Vieites and R. Altig. 2003. Natural history notes: Rana temporaria (common frog) and Bufo fowleri (Fowler's toad). Protozoan infestation. Herpetological Review 34: 237–238.

Veni, G. 1988. The caves of Bexar County. Second edition. Speleological Monograph, Texas Memorial Museum, University of Texas, Austin, Texas.

Veni, G. and Associates. 1987. Valdina Farms Sinkhole: Hydrogeologic and biologic evaluation. Edwards Underground Water District, Unpublished report, San Antonio, Texas.

Verma, N. and B.A. Pierce. 1994. Body mass, developmental stage, and interspecific differences in acid tolerance of larval anurans. Texas Journal of Science 46:319–327.

Vernberg, F.J. 1953. Hibernation studies of two species of salamanders, Plethodon cinereus cinereus and Eurycea bislineata bislineata. Ecology 34:55–62.

Vernberg, F.J. 1955. Correlation of physiological and behavioral indexes of activity in the study of Plethodon cinereus (Green) and Plethodon glutinosus (Green). American Midland Naturalist 54:382–393.

Verrell, P.A. 1982. Male newts prefer large females as mates. Animal Behavior 30:1254–1255.

Verrell, P.A. 1985. Male mate choice for large, fecund females in the red-spotted newt, Notophthalmus viridescens: how is size assessed? Herpetologica 41:382–386.

Verrell, P.A. 1989. Male mate choice for fecund females in a plethodontid salamander. Animal Behaviour 38:1086–1088.

Verrell, P.A. 1990a. Tests for sexual isolation among sympatric salamanders of the genus Desmognathus. Amphibia-Reptilia 11:147–153.

Verrell, P.A. 1990b. Frequency of interspecific mating in salamanders of the plethodontid genus Desmognathus: different experimental designs may yield different results. Journal of Zoology (London) 221:441–451.

Verrell, P.A. 1990c. Sexual compatibility among plethodontid salamanders: tests between Desmognathus apalachicolae, and D. ochrophaeus and D. fuscus. Herpetologica 46:415–422.

Verrell, P.A. 1994a. Courtship behaviour of the salamander Desmognathus imitator (Amphibia: Caudata: Plethodontidae). Amphibia-Reptilia 15:135–142.

Verrell, P.A. 1994b. Is decreased frequency of mating among conspecifics a cost of sympatry in salamanders? Evolution 48:921–925.

Verrell, P.A. 1995. The courtship behavior of the spotted dusky salamander, Desmognathus fuscus conanti (Amphibia: Caudata: Plethodontidae). Journal of the Zoological Society (London) 235:515–523.

Verrell, P. 1997. Courtship behaviour of the Ouachita dusky salamander, Desmognathus brimleyorum, and a comparison with other desmognathine salamanders. Journal of Zoology 243:21–27.

Verrell, P. 1999. Bracketing the extremes: courtship behaviour of the smallest-and largest-bodied species in the salamander genus Desmognathus (Plethodontidae: Desmognathinae). Journal of Zoology 247: 105–111.

Verrell, P. 2000. Methoxychlor increases susceptibility to predation in the salamander Ambystoma macrodactylum. Bulletin of Environmental Contamination and Toxicology 64:85–92.

Verrell, P.A. and S.J. Arnold. 1989. Behavioral observations of sexual isolation among allopatric populations of the mountain dusky salamander, Desmognathus ochrophaeus. Evolution 43:745–755.

Verrell, P.A. and A. Donovan. 1991. Male-male aggression in the plethodontid salamander Desmognathus ochrophaeus. Evolution 43:745–755.

Verrell, P. and M. Mabry. 2000. The courtship of plethodontid salamanders; form, function, and phylogeny. Pp. 371–380. In Bruce, R.C., R.G. Jaeger and L.D. Houck (Eds.), The Biology of Plethodontid Salamanders. Kluwer Academic/Plenum Publishers. New York.

Verrell, P. and J. Pelton. 1996. The sexual strategy of the central long-toed salamander, Ambystoma macrodactylum columbianum, in southeastern Washington. Journal of Zoology 240:37–50.

Verrell, P.A. and S.G. Tilley. 1992. Population differentiation in plethodontid salamanders: divergence of allozymes and sexual compatibility among populations of Desmognathus imitator and Desmognathus ochrophaeus (Caudata: Plethodontidae). Zoological Journal of the Linnean Society 104:67–80.

Vertucci, F. and S. Corn. 1994. Reply: acidification and salamander recruitment. Bioscience 44:126–127.

Vertucci, F.A. and P.S. Corn. 1996. Evaluation of episodic acidification and amphibian declines in the Rocky Mountains. Ecological Applications 6:449–457.

Vesely, D.G. 1997. Terrestrial amphibian abundance and species richness in headwater riparian buffer strips, Oregon Coast Range. Master's thesis. Oregon State University, Corvallis, Oregon.

Vesely, D.G. and W.C. McComb. 2002. Salamander abundance and amphibian species richness in riparian buffer strips in the Oregon Coast Range. Forest Science 48:291–297.

Vial, J.L. 1968. The ecology of the tropical salamander, Bolitoglossa subpalmata, in Costa Rica. Revista de Biologia Tropical 15:13–115.

Vial, J.L. and F.B. Preib. 1966. Antibiotic assay of dermal secretions from the salamander *Plethodon cinereus* (Green). Herpetologica 22: 284–287.

Vial, J.L. and L. Saylor. 1993. The status of amphibian populations: a compilation and analysis. Declining Amphibian Populations Task Force Working Document Number 1, World Conservation Union (IUCN), Species Survival Commission, Open University, Milton Keynes, United Kingdom.

Vickers, C.R., L.D. Harris and B.F. Swindel. 1985. Changes in herpetofauna resulting from ditching of cypress ponds on Coastal Plains flatwoods. Forest Ecology and Management 11:17–29.

Villa, J. 1972. Anfibios de Nicaragua. Introducción a su Sistemática, Vida y Costumbres. Instituto Geografico Nacional & Banco Central de Nicaragua. Managua, Nicaragua.

Villa, J. 1984. Geographic distribution: *Rhinophrynus dorsalis* (burrowing toad; sapo borracho). Herpetological Review 15:52.

Vincent, W.S. 1947. A checklist of amphibians and reptiles of Crater Lake National Park. Crater Lake National Park Nature Notes 13:19–22.

Vinklarek, D. 1993. Officials meet on toad. Smithville Times, 30 December.

Vinklarek, D. 1994. County makes toad plan; regional approach taken to development. Smithville Times, 17 February.

Viosca, P., Jr. 1923. An ecological study of the cold-blooded vertebrates of southeastern Louisiana. Copeia 115:35–44.

Viosca, P., Jr. 1924a. Observations of the life history of *Ambystoma opacum*. Copeia 134:86–88.

Viosca, P., Jr. 1924b. A terrestrial form of *Siren lacertina*. Copeia 136:102–104.

Viosca, P., Jr. 1928. A new species of *Hyla* from Louisiana. Proceedings of the Biological Society of Washington 41:89–92.

Viosca, P., Jr. 1937. A tentative revision of the genus *Necturus* with descriptions of three new species from the southern Gulf drainage area. Copeia 1937:120–138.

Viosca, P., Jr. 1944. Distribution of certain cold-blooded animals in Louisiana in relationship to the geology and physiography of the state. Proceedings of the Louisiana Academy of Sciences 8:47–62.

Visalli, D. and W.P. Leonard. 1994. Geographic distribution: *Ascaphus truei* (tailed frog). Herpetological Review 25:31.

Vitousek, P.M., C.M. D'Antonio, L.L. Loope and R. Westbrooks. 1996. Our mobile society is redistributing the species on the earth at a pace that challenges ecosystems, threatens human health, and strains economies. American Scientist 84:468–478.

Vitt, L.J. 1992. Lizard mimics millipede. National Geographic Research and Exploration 8:76–95.

Vitt, L.J. and R.D. Ohmart. 1978. Herpetofauna of the lower Colorado River: Davis Dam to the Mexican border. Proceedings of the Western Foundation of Vertebrate Zoology 2:35–72.

Vogler, A.P. and R. DeSalle. 1994. Diagnosing units of conservation management. Conservation Biology 8:354–363.

Vogt, R.C. 1981. Natural history of amphibians and reptiles in Wisconsin. Milwaukee Public Museum, Milwaukee, Wisconsin.

Voice, P., Jr. 1923. An ecological study of the cold blooded vertebrates of southeastern Louisiana. Copeia 115:35–44.

Voice, P., Jr. 1931. Principles of bullfrog (*Rana catesbeiana*) culture. Transactions of the American Fisheries Society 60:262–269.

Voice, P., Jr. 1938. Notes on the winter frogs of Alabama. Copeia 1938:201.

Volpe, E.P. 1952. Physiological evidence for natural hybridization of *Bufo americanus* and *Bufo fowleri*. Evolution 6:393–406.

Volpe, E.P. 1955a. A taxo-genetic analysis of the status of *Rana kandiyohi* Weed. Systematic Zoology 4:75–82.

Volpe, E.P. 1955b. Intensity of reproductive isolation between sympatric and allopatric populations of *Bufo americanus* and *Bufo fowleri*. American Naturalist 89:303–317.

Volpe, E.P. 1956. Reciprocal mis-matings between *Hyla squirella* and *Microhyla carolinensis*. Copeia 1956:261–262.

Volpe, E.P. 1957a. Embryonic temperature tolerance and rate of development in *Bufo valliceps*. Physiological Zoology 30:164–176.

Volpe, E.P. 1957b. Embryonic temperature adaptations in highland *Rana pipiens*. American Naturalist 9:303–310.

Volpe, E.P. 1957c. The early development of *Rana capito sevosa*. Tulane Studies in Zoology 5:207–225.

Volpe, E.P. 1959. Experimental and natural hybridization between *Bufo terrestris* and *Bufo fowleri*. American Midland Naturalist 61:295–312.

Volpe, E.P. 1960. Evolutionary consequences of hybrid sterility and vigor in toads. Evolution 14:181–193.

Volpe, E.P. 1961. Variable expressivity of a mutant gene in leopard frog. Science 134:102–104.

Volpe, E.P. 1963. Interplay of mutant and wild-type pigment cells in chimeric leopard frogs. Developmental Biology 8:205–221.

Volpe, E.P. 1980. The Amphibian Embryo in Transplantation Immunity. Karger, Basel, Switzerland.

Volpe, E.P. and J.L. Dobie. 1959. The larva of the oak toad, *Bufo quercicus* Holbrook. Tulane Studies in Zoology 7:145–152.

Volpe, E.P. and B.M. Gebhardt. 1966. Evidence from cultured leucocytes of blood cell chimerism in ex-parabiotic frogs. Science 154:1197-1199.

Volpe, E.P. and R.G. McKinnell. 1966. Successful tissue transplantation in frogs produced by nuclear transfer. Journal of Heredity 57:167–174.

Volpe, E.P., M.A. Wilkens and J.L. Dobie. 1961. Embryonic and larval development of *Hyla avivoca*. Copeia 1961:340–349.

von Bloeker, J.C., Jr. 1942. Amphibians and reptiles of the dunes. Bulletin of the Southern California Academy of Science 41:29–38.

Von Volkenberg, H.L. 1935. Biological control of an insect pest by a toad. Science 82:278–279.

Vonesh, J.R. and O. De la Cruz. 2002. Complex life cycles and density dependence: assessing the contribution of egg mortality to amphibian declines. Oecologia 133:325–333.

Voris, H.K. and J.P. Bacon Jr. 1966. Differential predation on tadpoles. Copeia 1966:594–598.

Voris, H.K. and R.F. Inger. 1995. Frog abundance along streams in Bornean forest. Conservation Biology 9:679–683.

Vos, C.C. and J.P. Chardon. 1998. Effects of habitat fragmentation and road density on the distribution pattern of the moor frog, *Rana arvilis*. Journal of Applied Ecology 35:44–56.

Vos, C.C. and A.H.P. Stumpel. 1995. Comparison of habitat-isolation parameters in relation to fragmented distribution patterns in the tree frog *(Hyla arborea)*. Landscape Ecology 11:203–214.

Voss, S.R. 1993a. Effect of temperature on body size, developmental stage, and timing of hatching in *Ambystoma maculatum*. Journal of Herpetology 27:329–333.

Voss, S.R. 1993b. Relationship between stream order and length of larval period in the salamander *Eurycea wilderae*. Copeia 1993:736–742.

Voss, S.R., D.G. Smith, C.K. Beachy and D.G. Heckel. 1997. Allozyme variation in neighboring isolated populations of the plethodontid salamander *Leurognathus marmoratus*. Journal of Herpetology 29: 493–497.

Voss, W.J. 1961. Rate of larval development and metamorphosis of the spadefoot toad, *Scaphiopus bombifrons*. Southwestern Naturalist 6: 168–174.

Vredenburg, V.T. 2000. Natural history notes: *Rana muscosa* (mountain yellow-legged frog). Egg predation. Herpetological Review 31: 170–171.

Vucetich, J.A. and T.A. Waite. 1998. Number of censuses required for demographic estimation of effective population size. Conservation Biology 12:1023–1030.

Vucetich, J.A., T.A. Waite and L. Nunney. 1997. Fluctuating population size and the ratio of effective to census population size. Evolution 51:2017–2021.

Wacasey, J.W. 1961. An ecological study of two sympatric species of salamanders, *Ambystoma maculatum* and *Ambystoma jeffersonianum*, in southern Michigan. Ph.D. dissertation. Michigan State University, East Lansing, Michigan.

Wager, V.A. 1965. The Frogs of South Africa. Purnell and Sons, Cape Town, South Africa.

Wagner, G. 1997. Status of the northern leopard frog (*Rana pipiens*) in Alberta. Wildlife Status Report, Number 9, Alberta Environmental Protection, Wildlife Management Division, Edmonton, Alberta.

Wagner, W.E., Jr. 1986. Tadpoles and pollen: observations on the feeding behavior of *Hyla regilla* larvae. Copeia 1986:802–804.

Wagner, W.E., Jr. 1989a. Fighting assessment and frequency alteration in Blanchard's cricket frog. Behavioral Ecology and Sociobiology 25:429–436.

Wagner, W.E., Jr. 1989b. Social correlates of variation in male calling behavior in Blanchard's cricket frog, *Acris crepitans blanchardi*. Ethology 82:27–45.

Wagner, W.E., Jr. 1989c. Graded aggressive signals to Blanchard's cricket frog vocal responses to opponent proximity and size. Animal Behaviour 38:1025–1038.

Wagner, W.E., Jr. 1992. Deceptive or honest signaling of fighting ability? A test of alternate hypotheses for the function of changes in call dominant frequency by male cricket frogs. Animal Behaviour 44:449–462.

Wagner, W.E., Jr. and B.K. Sullivan. 1992. Chorus organization in the Gulf Coast toad (*Bufo valliceps*): male and female behavior and the opportunity for sexual selection. Copeia 1992:647–658.

Wahbe, T.R. and F.L. Bunnell. 2001. Preliminary observations on movements of tailed frog tadpoles (Ascaphus truei) in streams through harvested and natural forests. Northwest Science 75:77–83.

Wahlenberg, W.G. 1946. Longleaf pine: Its use, ecology, regeneration, protection, growth, and management. Charles Lathrop Pack Forestry Foundation, Washington, D.C.

Waitz, J.A. 1961. Parasites of Idaho amphibians. Journal of Parasitology 47:89.

Wake, D. and T. Jackman. 1999 (1998). Description of a new species of plethodontid salamander from California. Appendix. In Jackman, T.R. 1999 (1998). Molecular and historical evidence for the introduction of clouded salamanders (genus Aneides) to Vancouver Island, British Columbia, Canada, from California. Canadian Journal of Zoology 76:1579–1580.

Wake, D.B. 1961. The distribution of the Sinoloa narrow-mouthed toad, Gastrophryne mazatlanensis. Bulletin of the Southern California Academy of Sciences 60:88–92.

Wake, D.B. 1965. Aneides ferreus. Pp. 16.1–16.2. Catalogue of American Amphibians and Reptiles. Society for the Study of Amphibians and Reptiles, St. Louis, Missouri.

Wake, D.B. 1966. Comparative osteology and evolution of the lungless salamanders, family Plethodontidae. Memoirs of the Southern California Academy of Sciences, Number 4, Los Angeles, California.

Wake, D.B. 1987. Adaptive radiation of salamanders in Middle American cloud forests. Annals of the Missouri Botanical Garden 74:242–264.

Wake, D.B. 1991. Declining amphibian populations. Science 253:860.

Wake, D.B. 1993. Phylogenetic and taxonomic issues relating to salamanders of the family Plethodontidae. Herpetologica 49:229–237.

Wake, D.B. 1996. A new species of Batrachoseps (Amphibia: Plethodontidae) from the San Gabriel Mountains, southern California. Contributions in Science from the Natural History Museum of Los Angeles County, Number 463, Los Angeles, California.

Wake, D.B. 1997. Incipient species formation in salamanders of the Ensatina complex. Proceedings of the National Academy of Sciences 94:7761–7767.

Wake, D.B. 1998. Action on amphibians. TRENDS in Ecology and Evolution 13:379–380.

Wake, D.B. and J. Castanet. 1995. A skeletochronological study of growth and age in relation to adult size in Batrachoseps attenuatus. Journal of Herpetology 29:60–65.

Wake, D.B. and I.G. Dresner. 1967. Functional morphology and evolution of tail autotomy in salamanders. Journal of Morphology 122:265–306.

Wake, D.B. and E.L. Jockusch. 2000. Detecting species borders using diverse data sets: examples from plethodontid salamanders in California. Pp. 95–119. In Bruce, R.C., R.G. Jaeger and L.D. Houck (Eds.), The Biology of Plethodontid Salamanders, Kluwer Academic/Plenum Publishers, New York.

Wake, D.B. and H.J. Morowitz. 1991. Declining amphibian populations—a global phenomenon? Findings and recommendations. Report to Board on Biology, National Research Council, Workshop on Declining Amphibian Populations, Irvine, California. Reprinted in Alytes 9:33–42.

Wake, D.B. and C.J. Schneider. 1998. Taxonomy of the plethodontid salamander genus Ensatina. Herpetologica 54:279–298.

Wake, D.B. and J. Woods. 1968. The cave salamander, Eurycea lucifuga, in Mississippi. Herpetologica 24:89.

Wake, D.B. and K.P. Yanev. 1986. Geographic variation in allozymes in a "ring species," the plethodontid salamander Ensatina eschscholtzii of western North America. Evolution 40:702–715.

Wake, D.B., L.R. Maxson and G.Z. Wurst. 1978. Genetic differentiation, albumin evolution, and their biogeographic implications in plethodontid salamanders of California and southern Europe. Evolution 32:529–539.

Wake, D.B., G. Roth and M. Wake. 1983. On the problem of stasis in organismal evolution. Journal of Theoretical Biology 101:211–224.

Wake, T.A., D.B. Wake and M.H. Wake. 1983. The ossification sequence of Aneides lugubris, with comments on heterochrony. Journal of Herpetology 17:10–22.

Wake, D.B., K.P. Yanev and C.W. Brown. 1986. Intraspecific sympatry in a "ring species," the plethodontid salamander Ensatina eschscholtzii, in southern California. Evolution 40:866–868.

Wake, D.B., K.P. Yanev and M.M. Freelow. 1989. Sympatry and hybridization in a "ring species": the plethodontid salamander Ensatina eschscholtzii. Pp. 134–157. In Otte, D. and J.A. Endler (Eds.), Speciation and its consequences. Sinauer, Sunderland, Massachusetts.

Wake, D.B., K.P. Yanev and R.W. Hansen. 2002. New species of slender salamander, genus Batrachoseps, from the southern Sierra Nevada of California. Copeia 2002:1016–1028.

Wakeley, J.F., G.J. Fuhrman, F.A. Fuhrman, H.G. Fischer and H.S. Mosher. 1966. The occurrence of tetrodotoxin (tarichatoxin) in Amphibia and the distribution of the toxin in the organs of newts (Taricha). Toxicon 3:195–203.

Waldick, R. 1997. Effects of forestry practices on amphibian populations in eastern North America. Pp. 191–205. In Green, D.M. (Ed.), Amphibians in Decline: Canadian Studies on a Global Problem. Herpetological Conservation, Number 1, Society for the Study of Amphibians and Reptiles, St. Louis, Missouri.

Waldman, B. 1981. Sibling recognition in toad tadpoles: the role of experience. Zeitschrift für Tierpsychologie 56:341–358.

Waldman, B. 1982a. Sibling association among schooling toad tadpoles: field evidence and implications. Animal Behaviour 30:700–713.

Waldman, B. 1982b. Adaptive significance of communal oviposition in wood frogs (Rana sylvatica). Behavioral Ecology and Sociobiology 10:169–174.

Waldman, B. 1985a. Sibling recognition in toad tadpoles: are kinship labels transferred among individuals? Zeitschrift für Tierpsychologie 68:41–57.

Waldman, B. 1985b. Olfactory basis of kin recognition in toad tadpoles. Journal of Comparative Physiology A 156:565–577.

Waldman, B. 1986. Preference for unfamiliar siblings over familiar non-siblings in American toad (Bufo americanus) tadpoles. Animal Behaviour 34:48–53.

Waldman, B. and K. Adler. 1979. Toad tadpoles associate preferentially with siblings. Nature 282:611–613.

Waldman, B. and M.J. Ryan. 1983. Thermal advantages of communal egg mass deposition in wood frogs (Rana sylvatica). Journal of Herpetology 17:169–174.

Waldron, J.L. 2000. Ecology and sympatric relations of crevice salamanders in Randolph County, West Virginia. Master's thesis. Marshall University, Huntington, West Virginia.

Waldron, J.L., T.K. Pauley, Z.I. Felix, W.J. Humphries and A.J. Longenecker. in press. The herpetofauna of the Bluestone National Wild and Scenic River. Proceedings of the West Virginia Academy of Science. in press.

Walker, C.F. 1931. Description of a new salamander from North Carolina. Proceedings of the Junior Society of Natural Science 2:48–51.

Walker, C.F. 1934. Plethodon welleri at White Top Mountain, Virginia. Copeia 1934:190.

Walker, C.F. 1946. The amphibians of Ohio. Part I. Frogs and toads. Ohio State Museum of Science Bulletin, Volume 1, Number 3, Columbus, Ohio.

Walker, C.F. and W. Goodpaster. 1941. The green salamander, Aneides aeneus, in Ohio. Copeia 1941:178.

Walker, J. 1993. Rare vascular plant taxa associated with the longleaf pine ecosystems: patterns in taxonomy and ecology. Proceedings of the Tall Timbers Fire Ecology Conference 18:105–125.

Walker, J.M. 1963. Amphibians and reptiles of Jackson Parish, Louisiana. Proceedings of the Louisiana Academy of Science 26:91–101.

Walker, K.F. 1998. Molluscs of inland waters. Pp. 59–65. In Beesley, P.L., G.J.B. Ross and A. Wells (Eds.), Mollusca: The Southern Synthesis. Fauna of Australia, Volume 5. CSIRO Publishing, Melbourne, Victoria, Australia.

Walker, P.A. and D.P. Faith. 1995. Diversity-PD: procedures for conservation evaluation based on phylogenetic diversity. Biodiversity Letters 2:132–139.

Walker, R.F. and W.G. Whitford. 1970. Soil water absorption capabilities in selected species of anurans. Herpetologica 26:411–418.

Wall, D., H. Mooney, G. Adams, G. Boxshall, A. Dobson, T. Nakashizuka, J. Seyani, C. Samper and J. Sarukhán. 2001. An international biodiversity observation year (IBOY). TRENDS in Ecology and Evolution 16:52–54.

Wallace, J.M. 1975. Biochemical genetics of Eurycea bislineata and Eurycea aquatica (Amphibia: Plethodontidae) in Davidson County, Tennessee. Master's thesis. Austin Peay State University, Clarksville, Tennessee.

Wallace, J.T. 1969. A study of Plethodon richmondi from Mason County, Kentucky, with notes on its distribution within the state. Transactions of the Kentucky Academy of Science 30:38–44.

Wallace, J.T. and R.W. Barbour. 1957. Observations on the eggs and young of Plethodon richmondi. Copeia 1957:48.

Wallace, R.L. and L.V. Diller. 1998. Length of the larval cycle of *Ascaphus truei* in coastal streams of the redwood region, northern California. Journal of Herpetology 32:404–409.

Walley, H. 1991. Geographic distribution: *Rana palustris* (pickerel frog). Herpetological Review 22:24.

Walls, S.C. 1990. Interference competition in postmetamorphic salamanders: interspecific differences in aggression by co-existing species. Ecology 71:307–314.

Walls, S.C. 1991. Ontogenetic shifts in the recognition of siblings and neighbors by juvenile salamanders. Animal Behaviour 42: 423–434.

Walls, S.C. 1995. Differential vulnerability to predation and refuge use in competing larval salamanders. Oecologia 101:86–93.

Walls, S.C. 1996. Differences in foraging behaviour explain interspecific growth inhibition in competing salamanders. Animal Behaviour 52:1157–1162.

Walls, S.C. and R. Altig. 1986. Female reproductive biology and larval life history of *Ambystoma* salamanders: a comparison of egg size, hatchling size, and larval growth. Herpetologica 42:334–345.

Walls, S.C. and A.R. Blaustein. 1994. Does kinship influence density dependence in a larval salamander? Oikos 71:459–468.

Walls, S.C. and A.R. Blaustein. 1995. Larval marbled salamanders, *Ambystoma opacum*, eat their kin. Animal Behaviour 50:537–545.

Walls, S.C. and R.G. Jaeger. 1987. Aggression and exploitation as mechanisms of competition in larval salamanders. Canadian Journal of Zoology 65:2938–2944.

Walls, S.C. and R.E. Roudebush. 1991. Reduced aggression towards siblings as evidence of kin recognition in cannibalistic salamanders. American Naturalist 138:1027–1038.

Walls, S.C. and R.D. Semlitsch. 1991. Visual and movement displays function as agonistic behavior in larval salamanders. Copeia 1991: 936–942.

Walls, S.C. and M.G. Williams. 2001. The effect of community composition on persistence of prey with their predators in an assemblage of pond-breeding amphibians. Oecologia 128:134–141.

Walls, S.C., S.S. Belanger and A.R. Blaustein. 1993b. Morphological variation in a larval salamander: dietary induction of plasticity in head shape. Oecologia 96:162–168.

Walls, S.C., J.J. Beatty, B.N. Tissot, D.G. Hokit and A.R. Blaustein. 1993a. Morphological variation and cannibalism in a larval salamander (*Ambystoma macrodactylum columbianum*). Canadian Journal Zoology 71:1543–1551.

Walsh, R. 1998. An extension of the known range of the long-toed salamander *Ambystoma macrodactylum*, in Alberta. Canadian Field-Naturalist 112:331–333.

Walters, B. 1975. Studies of interspecific predation within an amphibian community. Journal of Herpetology 9:267–279.

Walters, V. 1955. Records of the spadefoot toad *Scaphiopus bombifrons*, from Arizona. Copeia 1955:252–253.

Walters, W.D., Jr. 1997. The heart of the cornbelt: An illustrated history of corn farming in McLean County. McLean County Historical Society, Bloomington, Illinois.

Walton, A.C. 1929. Studies on some nematodes of North American frogs I. Journal of Parasitology 15:227–240.

Walton, A.C. 1931. Note on some larval nematodes found in frogs. Journal of Parasitology 17:228–229.

Walton, A.C. 1935. The Nematoda as parasites of Amphibia II. Journal of Parsitology 21:27–50.

Walton, A.C. 1938. The trematodes as parasites of Amphibia. Lists of hosts. Contributions of the Biological Laboratory of Knox College 62:1–24.

Walton, A.C. 1940. Notes on amphibian parasites. Proceedings of the Helminthological Society of Washington 7:897–910.

Walton, A.C. 1942. The parasites of the Ambystomoidea (Amphibia: Caudata). Journal of Parasitology 28:29 (Abstract number 62).

Walton, A.C. 1946. Protozoan parasites of the Bufonidae (Amphibia). Illinois Academy of Science Transactions 39:143–147.

Walton, A.C. 1947. Parasites of the Ranidae (Amphibia). American Society of Zoologists 99:684–685.

Walton, A.C. 1949. Parasites of the Ranidae (Amphibia). Anatomical Record 105:628.

Walton, A.C. 1963. Amphibia as hosts of protozoan parasites. II. Journal of Parasitology 49:31–32.

Walton, M. 1988. Relationships among metabolic, locomotory, and field measures of organismal performance in the Fowler's toad (*Bufo woodhousei fowleri*). Physiological Zoology 61:107–118.

Walton, M. and B.D. Anderson. 1988. The aerobic cost of saltatory locomotion in the Fowler's toad (*Bufo woodhousei fowleri*). Journal of Experimental Biology 136:273–288.

Waples, R.S. 1991. Pacific salmon, *Oncorhynchus* sp., and the definition of "species" under the Endangered Species Act. Marine Fisheries Review 53:11–22.

Waples, R.S. 1995. Evolutionary significant units and the conservation of biological diversity under the Endangered Species Act. Pp. 8–27. *In* Nielsen, J.L. (Ed.), Evolution and the Aquatic Ecosystem: Defining Unique Units in Population Conservation. Symposium 17. American Fisheries Society, Bethesda, Maryland.

Waples, R.S. 1998. Evolutionarily significant units, distinct population segments, and the Endangered Species Act: reply to Pennock and Dimmick. Conservation Biology 12:718–721.

Ward, D. and O.J. Sexton. 1981. Anti-predator role of salamander egg membranes. Copeia 1981:724–726.

Ware, S., C. Frost and P.D. Doerr. 1993. Southern mixed hardwood forest: the former longleaf pine forest. Pp. 447–493. *In* Martin, W.H., S.G. Boyce and A.C. Echternacht (Eds.), Biodiversity of the Southeastern United States. John Wiley and Sons, New York.

Warfel, H.E. 1937. A new locality record for *Gyrinophilus porphyriticus* (Green) in Massachusetts. Copeia 1937:74.

Warkentin, K.M. 1992a. Effects of temperature and illumination on feeding rates of green frog tadpoles (*Rana clamitans*). Copeia 1992:725–730.

Warkentin, K.M. 1992b. Microhabitat use and feeding rate variation in green frog tadpoles (*Rana clamitans*). Copeia 1992:731–740.

Warner, J. 1971. The distribution of *Plethodon glutinosus* (Green) in central Louisiana with a taxonomic comparison to neighboring populations. Journal of Herpetology 5:115–119.

Warner, S.C., J. Travis and W.A. Dunson. 1993. Effect of pH variation on interspecific competition between two species of hylid tadpoles. Ecology 74:183–194.

Warren, G.L. 1997. Nonindigenous freshwater invertebrates. Pp. 101–108. *In* Simberloff, D., D.C. Schmitz and T.C. Brown (Eds.), Strangers in Paradise. Island Press, Washington, D.C.

Warwick, R.M. and K.R. Clarke. 1998. Taxonomic distinctness and environmental assessment. Journal of Applied Ecology 35:532–543.

Washburn, F.L. 1899. A peculiar toad. American Naturalist 33:139–141.

Washington Department of Wildlife. 1993. Status of the Larch Mountain salamander (*Plethodon larselli*) in Washington. Washington Department of Wildlife, Unpublished report, Olympia, Washington.

Wasserman, A.O. 1957. Factors affecting interbreeding in sympatric species of spadefoots (*Scaphiopus*). Evolution 11:320–338.

Wasserman, A.O. 1958. Relationships of allopatric populations of spadefoots (genus *Scaphiopus*). Evolution 12:311–318.

Wasserman, A.O. 1968. *Scaphiopus holbrokii*. Pp. 70.1–70.4. Catalogue of American Amphibians and Reptiles. Society for the Study of Amphibians and Reptiles, St. Louis, Missouri.

Wasserman, A.O. 1970. Polyploidy in the common gray tree toad *Hyla versicolor* LeConte. Science 167:385–386.

Wassersug, R.J. 1973. Aspects of social behavior in anuran larvae. Pp. 273–297. *In* Vial, J.L. (Ed.), Evolutionary Biology of the Anurans. University of Missouri Press, Columbia, Missouri.

Wassersug, R.J. 1975. The adaptive significance of the tadpole stage with comments on the maintenance of complex life cycles in anurans. American Zoologist 15:405–417.

Wassersug, R.J. 1976. Internal oral features in *Hyla regilla* (Anura: Hylidae) larvae: an ontogenic study. Occasional Papers of the Museum of Natural History, Number 49, University of Kansas, Lawrence, Kansas.

Wassersug, R.J. 1996. The biology of *Xenopus* tadpoles. Pp. 195–211. *In* Tinsley, R.C. and H.R. Kobel (Eds.), The Biology of *Xenopus*. Clarendon Press, Oxford, United Kingdom.

Wassersug, R.J. and M.E. Feder. 1983. The effects of aquatic oxygen concentration, body size, and respiratory behaviour on the stamina of obligate aquatic (*Bufo americanus*) and facultative air-breathing (*Xenopus laevis* and *Rana berlandieri*) anuran larvae. Journal of Experimental Biology 105:173–190.

Wassersug, R.J. and C.M. Hessler. 1971. Tadpole behaviour: aggregation in larval *Xenopus laevis*. Animal Behaviour 19:386–389.

Wassersug, R.J. and E.A. Seibert. 1975. Behavioral responses of amphibian larvae to variation in dissolved oxygen. Copeia 1975: 86–103.

Wassersug, R.J. and D.G. Sperry. 1977. The relationship of locomotion to differential predation on *Pseudacris triseriata* (Anura: Hylidae). Ecology 58:830–839.

Wassersug, R.J., R.J. Lum and M.J. Potel. 1981. An analysis of school structure for tadpoles (Anura: Amphibia). Behavioral Ecology and Sociobiology 9:15–22.

Watermolen, D.J. 1995. A key to the eggs of Wisconsin's amphibians. Wisconsin Department of Natural Resources, Research Report Number 165, Madison, Wisconsin.

Watermolen, D.J. 1996. Natural history notes: *Plethodon cinereus* (redback salamander). Brooding behavior. Herpetological Review 27:136–137.

Waters, D.L. 1992. Geographic distribution: *Pseudacris regilla* (Pacific treefrog). Herpetological Review 23:24–25.

Waters, J.R., C.J. Zabel, K.S. McKelvey and H. H. Welsh Jr. 2001. Vegetation patterns and abundances of amphibians and small mammals along small streams in a northwestern California watershed. Northwest Science 75:37–52.

Watkins, T.B. 1996. Predator-mediated selection on burst swimming performance in tadpoles of the Pacific treefrog, *Pseudacris regilla*. Physiological Zoology 69:154–167.

Watkins-Colwell, K.A. and G.J. Watkins-Colwell. 1998. Differential survivorship of larval leopard frogs *(Rana pipiens)* exposed to temporary and permanent reduced pH conditions. Bulletin of the Maryland Herpetological Society 34:64–75.

Watney, G.M.S. 1941. Notes on the life history of *Ambystoma gracile* Baird. Copeia 1941:14–17.

Watson, J.W., K.R. McAllister and D.J. Pierce. 2003. Home ranges, movements, and habitat selection of Oregon spotted frogs *(Rana pretiosa)*. Journal of Herpetology 37:292–300.

Watson, J.W., K.R. McAllister, D.J. Pierce and A. Alvarado. 1998. Movements, habitat selection, and population characteristics of a remnant population of Oregon spotted frogs *(Rana pretiosa)* in Thurston County, Washington. Washington Department of Fish and Wildlife, Olympia, Washington.

Watson, S. 1997. Food level effects on metamorphic timing in the long-toed salamander, *Ambystoma macrodactylum krausei*. Master's thesis. University of Calgary, Edmonton, Alberta.

Waye, H.L. and C.H. Shewchuk. 1995. Natural history notes: *Scaphiopus intermontanus* (Great Basin spadefoot). Production of odor. Herpetological Review 26:98–99.

Wayne, R.K. and S.M. Jenks. 1991. Mitochondrial DNA analysis implying extensive hybridization of the endangered red wolf *Canis rufus*. Nature 351:565–568.

Weatherby, C.A. 1982. Introgression between the American toad, *Bufo americanus*, and the southern toad, *B. terrestris*, in Alabama. Ph.D. dissertation. Auburn University, Auburn, Alabama.

Weaver, H.D. 1987. The commercial caves of Camden County, Missouri. Journal of the Missouri Speleological Society 27:91–132.

Webb, R.G. 1960. Notes on some amphibians and reptiles from northern Mexico. Transactions of the Kansas Academy of Sciences 63:289.

Webb, R.G. 1963. The larva of the casque-headed frog *Pternohyla fodiens* Boulenger. Texas Journal of Science 15:89–97.

Webb, R.G. 1965. Observations on breeding habits of the squirrel treefrog Bosc in Daudin. American Midland Naturalist 74:500–501.

Webb, R.G. 1969. Survival adaptations of tiger salamanders *(Ambystoma tigrinum)* in the Chihuahuan Desert. Pp. 256–276. *In* Hoff, C.C. and M.L. Riedull (Eds.), Physiological Systems in Semiarid Environments. University of New Mexico Press, Albuquerque, New Mexico.

Webb, R.G. 1971. Egg deposition of the Mexican smilisca, *Smilisca baudinii*. Journal of Herpetology 5:185–187.

Webb, R.G. 1984. Herpetogeography in the Mazatlán-Durango region of the Sierra Madre Occidental, Mexico. Pp. 217–241. *In* Seigel, R.A., L.E. Hunt, J.L. Knight, L. Malaret and N.L. Zuschlag. Vertebrate Ecology and Systematics—A Tribute to Henry S. Fitch. Museum of Natural History, University of Kansas, Lawrence, Kansas.

Webb, R.G. and J.K. Korky. 1977. Variation in tadpoles of frogs of the *Rana tarahumarae* group in western Mexico (Anura: Ranidae). Herpetologica 33:73–82.

Webb, R.G. and W.L. Rouche. 1971. Life history aspects of the tiger salamander *(Ambystoma tigrinum mavortium)* in the Chihuahuan desert. Great Basin Naturalist 31:193–212.

Webb, R.G., Jr., J.K. Jones Jr. and G.W. Byers. 1962. Some reptiles and amphibians from Korea. University of Kansas Publications, Museum of Natural History 15:149–173.

Webber, L.A. and D.C. Cochran. 1984. Laboratory observations on some freshwater vertebrates and secerial saline fishes exposed to a monomolecular organic surface film (ISA-20E). Mosquito News 44:68–69.

Weber, A., B. Bailey and W.L. Bailey. 1993. Ohio State Parks Guidebook. Glovebox Guidebook Publishing Company, Clarkston, Michigan.

Weber, J.A. 1928. Herpetological observations in the Adirondack Mountains, New York. Copeia 169:106–112.

Weber, J.A. 1944. Observations on the life history of *Amphiuma means*. Copeia 1944:61–62.

Webster, D.A. 1960. Toxicity of the spotted newt, *Notophthalmus viridescens*, to trout. Copeia 1960:74–75.

Weeber, R.C. and M. Vallianatos. 2000. The Marsh Monitoring Program 1995–1999: Monitoring Great Lakes wetlands and their amphibian and bird inhabitants. Published by Bird Studies Canada in cooperation with Environment Canada, Nepean, Ontario, Canada, and the U.S. Environmental Protection Agency, Washington, D.C.

Weed Sciences Society of America. 1983. Herbicide handbook of the Weed Science Society of America. Fifth edition. Champaign, Illinois.

Weed, A.C. 1922. New frogs from Minnesota. Proceedings of the Biological Society of Washington 35:107–110.

Wegmann, D.L. 1997. Natural history, ecology and potential environmental stress of *Ambystoma maculatum* (LeConte) and *Rana sylvatica* (Shaw) at Dolly Sods, West Virginia. Master's thesis. Marshall University, Huntington, West Virginia.

Weichert, C.K. 1945. Seasonal variation in the mental gland and reproductive organs of the male *Eurycea bislineata*. Copeia 1945: 78–84.

Weigmann, D.L. and R. Altig. 1975. Anaerobic glycolosis in two larval amphibians. Journal of Herpetology 4:355–357.

Weigmann, D.L., M. Hakkila, K. Whitmore and R.A. Cole. 1980. Survey of Sacramento Mountain salamander habitat of the Cloudcroft and Mayhill districts of the Lincoln National Forest. Final report submitted to the U.S. Forest Service, Lincoln National Forest, Alamogordo, New Mexico.

Weil, A.T. and W. Davis. 1994. *Bufo alvarius*: a potent hallucinogen of animal origin. Journal of Ethnopharmacology 41:1–8.

Weins, J.A. 1970. Effects of early experience on substrate pattern selection in *Rana aurora* tadpoles. Copeia 1970:543–548.

Weis, J.S. 1975. The effect of DDT on tail regeneration in *Rana pipiens* and *R. catesbeiana* tadpoles. Copeia 1975:765–767.

Weisbrod, B.A. 1964. Collective-consumption services of individualized-consumption goods. Quarterly Journal of Economics 78:471–477.

Weitzel, N.H. and H.R. Panik. 1993. Long-term fluctuations of an isolated population of the Pacific chorus frog *(Pseudacris regilla)* in Northwestern Nevada. Great Basin Naturalist 53:379–384.

Weitzman, M.L. 1992a. On diversity. Quarterly Journal of Economics 107:363–406.

Weitzman, M.L. 1992b. Diversity functions. Harvard Institute of Economic Research, Discussion paper Number 1610, Harvard University, Cambridge, Massachusetts.

Weitzman, M.L. 1993. What to preserve? An application of diversity theory to crane conservation. Quarterly Journal of Economics 108: 157–183.

Welbourn, W.C., Jr. and R.B. Loomis. 1975. *Hannemania* (Acarina: Trombiculidae) and their anuran hosts at Fourtynine Palms Oasis, Joshua Tree National Monument, California. Bulletin of the Southern California Academy of Sciences 74:15–19.

Weller, W.F. and D.M. Green. 1997. Checklist and current status of Canadian amphibians. Pp. 309–328. *In* Green, D.M. (Ed.), Amphibians in Decline: Canadian Studies of a Global Problem. Herpetological Conservation, Number 1, Society for the Study of Amphibians and Reptiles, St. Louis, Missouri.

Weller, W.H. 1930. Records of some reptiles and amphibians from Chimney Rock Camp, Chimney Rock, N.C., and vicinity. Proceedings of the Junior Society of Natural Sciences of Cincinnati 1:51–54.

Weller, W.H. 1931. A preliminary list of the salamanders of the Great Smoky Mts. of North Carolina and Tennessee. Proceedings of the Junior Society of Natural Sciences of Cincinnati 2:21–32.

Wells, C.S. and R.N. Harris. 2001. Activity level and the tradeoff between growth and survival in the salamanders *Ambystoma jeffersonianum* and *Hemidactylium scutatum*. Herpetologica 57:116–127.

Wells, K.D. 1976. Multiple egg clutches in the green frog *(Rana clamitans)*. Herpetologica 32:85–87.

Wells, K.D. 1977a. The social behavior of anuran amphibians. Animal Behaviour 25:666–693.

Wells, K.D. 1977b. Territoriality and male mating success in the green frog *(Rana clamitans)*. Ecology 58:750–762.

Wells, K.D. 1978a. Courtship and paternal behavior in a Panamanian poison-arrow frog *(Dendrobates auratus)*. Herpetologica 34: 148–155.

Wells, K.D. 1978b. Territoriality in the green frog *(Rana clamitans):* vo- calizations and agonistic behaviour. Animal Behaviour 26:1051–1063.

Wells, K.D. 1980. Spatial associations among individuals in a population of slimy salamanders *(Plethodon glutinosus).* Herpetologica 36:271–275.

Wells, K.D. and T.L. Taigen. 1984. Reproductive behavior and capacities of male American toads *(Bufo americanus):* is behavior constrained by physiology? Herpetologica 40:292–298.

Wells, K.D. and T.L. Taigen. 1986. The effect of social interactions on calling energetics in the gray treefrog *Hyla versicolor.* Behavioral Ecol- ogy and Sociobiology 19:9–18.

Wells, K.D. and R.A. Wells. 1976. Patterns of movement in a population of the slimy salamander, *Plethodon glutinosus,* with observations on aggregations. Herpetologica 32:156–162.

Wells, M.M. 1963. An incidence of albinism in *Taricha torosa.* Herpeto- logica 19:291.

Wells, P.H. and W. Gordon. 1958. Brooding slimy salamanders, *Pletho- don glutinosus glutinosus* (Green). National Speleological Society Bul- letin 20:23–24.

Welsh, H.H., Jr. 1985. Geographic distribution: *Ascaphus truei* (tailed frog). Herpetological Review 16:59.

Welsh, H.H., Jr. 1986. Life history notes: *Dicamptodon ensatus* (Pacific giant salamander). Behavior. Herpetological Review 17:19.

Welsh, H.H., Jr. 1990. Relictual amphibians and old-growth forests. Conservation Biology 4:309–319.

Welsh, H.H., Jr. 1993. A hierarchical analysis of the niche relationships of four amphibians of forested habitats in northwestern California. Ph.D. dissertation. University of California, Berkeley, California.

Welsh, H.H., Jr. 2000. California forest management and aquatic/ riparian ecosystems in the redwoods. Pp. 255–261. *In* Noss, R. (Ed.), The Redwood Forest: History, Ecology, and Management of the Coast Redwoods. Island Press, Covelo, California.

Welsh, H.H., Jr. and A.J. Lind. 1988. Old growth forests and the distri- bution of the terrestrial herpetofauna. Pp. 439–455. *In* Szaro, R.C., K.E. Severson and D.R. Patton (Eds.), Management of Amphibians, Reptiles and Mammals in North America. U.S.D.A. Forest Service, Rocky Mountain Forest and Range Experimental Station, General Technical Report, RM-166, Fort Collins, Colorado.

Welsh, H.H., Jr. and A.J. Lind. 1991. The structure of the herpetofaunal assemblage in the Douglas fir/hardwood forests of northwestern California and southwestern Oregon. Pp. 394–413. *In* Ruggiero, L.F., K.B. Aubry, A.B. Carey and M.H. Huff (Tech. Coords.), Wildlife and Vegetation of Unmanaged Douglas-fir Forests. U.S.D.A. Forest Ser- vice, General Technical Report, PNW-GTR-285, Pacific Northwest Research Station, Portland, Oregon.

Welsh, H.H., Jr. and A.J. Lind. 1992. Population ecology of two relictual salamanders from the Klamath Mountains of northwestern California. Pp. 419–437. *In* McCullough, D.R. and R.H. Barrett (Eds.), Wildlife 2001: Populations. Elsevier Applied Science, London.

Welsh, H.H., Jr. and A.J. Lind. 1995. Habitat correlates of the Del Norte salamander, *Plethodon elongatus* (Caudata: Plethodontidae), in north- western California. Journal of Herpetology 29:198–210.

Welsh, H.H., Jr. and A.J. Lind. 1996. Habitat correlates of the southern torrent salamander, *Rhyacotriton variegatus* (Caudata: Rhyacotritonidae), in northwestern California. Journal of Herpetology 30:385–398.

Welsh, H.H., Jr. and A.J. Lind. 2002. Multiscale habitat relationships of stream amphibians in the Klamath-Siskiyou region of California and Oregon. Journal of Wildlife Management 66:581–602.

Welsh, H.H., Jr. and L.M. Ollivier. 1998. Stream amphibians as indicators of ecosystem stress: a case study from California's redwoods. Ecologi- cal Applications 8:1118–1132.

Welsh, H.H., Jr. and R.J. Reynolds. 1986. Life history notes: *Ascaphus truei* (tailed frog). Behavior. Herpetological Review 17:19.

Welsh, H.H., Jr. and R.A. Wilson. 1995. Life history notes: *Aneides fer- reus* (clouded salamander). Reproduction. Herpetological Review 26: 196–197.

Welsh, H.H., Jr., G.R. Hodgson and A.J. Lind. in review. An analysis of the ecogeographic relationships of the herpetofauna of a northern California watershed: linking species patterns to landscape processes. Ecography **xx:xxx–xxx.**

Welsh, H.H., Jr., T.D. Roelofs and C.A. Frissell. 2000. Aquatic ecosystems of the redwood region. Pp. 165–199. *In* Noss, R. (Ed.), The Redwood Forest: History, Ecology, and Management of the Coast Redwoods. Is- land Press, Covelo, California.

Welsh, H.H., Jr., A.J. Lind, L.M. Ollivier, G.R. Hodgson and N.E. Kar- raker. 1998. Comments on the PALCO HCP/SYP and EIS/EIR with re- gard to the maintenance of riparian, aquatic, and late seral ecosystems and their associated amphibian and reptile species. U.S.D.A. Forest Service, Pacific Southwest Research Station, Unpub- lished report, Redwood Sciences Laboratory, Arcata, California.

Welter, W.A. and R.W. Barbour. 1940. Additions to the herpetofauna of northeastern Kentucky. Copeia 1940:132–133.

Welter, W.A. and K. Carr. 1939. Amphibians and reptiles of northeastern Kentucky. Copeia 1939:128–130.

Wendelken, P.W. 1968. Differential predation and behavior as factors in the maintenance of the vertebral stripe polymorphism in *Acris crepi- tans.* Master's thesis. University of Texas, Austin, Texas.

Wente, W.H. and M.J. Adams. 2002. Amphibian research and moni- toring initiative regional project: status of amphibian populations at historical sites in Oregon and Nevada. Report published by U.S.G.S. Forest and Rangeland Ecosystems Science Center, Corvallis, Oregon.

Werler, J.C. and J. McCallion. 1951. Notes on a collection of reptiles and amphibians from Princess Anne County, Virginia. American Midland Naturalist 45:245–252.

Werner, E.E. 1986. Amphibian metamorphosis: growth rate, predation risk, and the optimal size at transformation. American Naturalist 128:319–341.

Werner, E.E. 1991. Nonlethal effects of a predator on competitive inter- actions between two anuran larvae. Ecology 72:1709–1720.

Werner, E.E. 1992. Competitive interactions between wood frog and northern leopard frog larvae: the influence of size and activity. Copeia 1992:26–35.

Werner, E.E. and K.S. Glennemeier. 1999. Influence of forest canopy cover on the breeding pond distributions of several amphibian species. Copeia 1999:1–12.

Werner, E.E. and M.A. McPeek. 1994. Direct and indirect effects of pred- ators on two anuran species along an environmental gradient. Ecol- ogy 75:1368–1382.

Werner, E.E., G.A. Wellborn and M.A. McPeek. 1995. Diet composition in postmetamorphic bullfrogs and green frogs: implications for in- terspecific predation and competition. Journal of Herpetology 29: 600–607.

Werner, J.K. 1969. Temperature-photoperiod effects on spermatogenesis in the salamander *Plethodon cinereus.* Copeia 1969:592–602.

Werner, J.K. 1971. Notes on the reproductive cycle of *Plethodon cinereus* in Michigan. Copeia 1971:161–162.

Werner, J.K. and M.B. McCune. 1979. Seasonal changes in anuran populations in a northern Michigan pond. Journal of Herpetology 13:101–104.

Werner, J.K. and J.D. Reichel. 1994. Amphibian and reptile survey of the Kootenai National Forest: 1994. U.S. Department of Agriculture For- est Service, Helena National Forest, Helena, Montana.

Werner, J.K. and K. Walewski. 1976. Amphibian trypanosomes from the McCormick Forest Michigan. Journal of Parasitology 62:20–25.

Werner, J.K., J.S. Davis and K.S. Slaght. 1988. Trypanosomes of *Bufo americanus* from northern Michigan. Journal of Wildlife Diseases 24:647–649.

Werner, J.K., T. Plummer and J. Weaselhead. 1998. Amphibians and rep- tiles of the Flathead Indian Reservation. Intermountain Journal of Science 4:33–49.

Wernz, J.G. 1969. Spring mating of *Ascaphus.* Journal of Herpetology 3:167–169.

Wernz, J.G. and R.M. Storm. 1969. Pre-hatching stages of the tailed frog, *Ascaphus truei* Stejneger. Herpetologica 25:86–93.

Werschkul, D.F. and M.T. Christensen. 1977. Differential predation by *Lepomis macrochirus* on the eggs and tadpoles of *Rana.* Herpetologica 33:237–241.

Wershler, C. and W. Smith. 1992. Status of the Great Plains Toad in Alberta—1990. World Wildlife Fund Canada (Prairie for Tomorrow)/ Alberta Forestry Lands and Wildlife, Edmonton, Alberta.

West, L.B. 1960. The nature and growth inhibitory material from crowded *Rana pipiens* tadpoles. Physiological Zoology 33:232–239.

Wetzel, E.J. and G.W. Esch. 1996. Seasonal population dynamics of *Halipegus occidualis* and *Halipegus eccentricus* (Diagenea: Hemiuridae) in their amphibian host, *Rana clamitans.* Journal of Parasitology 82:414–422.

Weyerhaeuser. 1999. Forest learning center at Mount St. Helens. http://www.weyerhaeuser.com/sthelens/mtsthelens/wildlife.htm.

Weygoldt, P. 1989. Changes in the composition of mountain stream frog communities in the Atlantic mountains of Brazil: frogs as indicators of environmental deteriorations? Studies of Neotropical Fauna and Environment 243:249–255.

Weylan, P.R., C.N. Caviedes and M.E. Quesada. 1996. Interannual variability of monthly precipitation in Costa Rica. Journal of Climate 9:2606–2613.

Wharton, C.H. 1978. The natural environments of Georgia. Georgia Department of Natural Resources, Atlanta, Georgia.

Wheeler, C.A., H.H. Welsh Jr. and L.L. Heise. 2003. Natural history notes: *Rana boylii* (foothill yellow-legged frog). Oviposition. Herpetological Review 34:234.

Wheeler, G.C. and J. Wheeler. 1966. The Amphibians and Reptiles of North Dakota. University of North Dakota Press, Grand Forks, North Dakota.

Wheeler, Q.D. 1995. Systematics, the scientific basis for inventories of biodiversity. Biodiversity Conservation 4:476–489.

Wheeler, Q.D. and R. Meier (Eds.). 2000. Species Concepts and Phylogenetic Theory: A Debate. Columbia University Press, New York.

Wheeler, T.A., M. Roberts, M. Beverley-Burton and D.G. Sutton. 1989. *Brachylaimia apoplania* n. sp. (Digenea: Brachylaimidae) from the Polynesian rat, *Rattus exulans* (Rodentia: Muridae), in New Zealand: origins and zoogeography. Journal of Parasitology 75:680–684.

Whitaker, J.O., Jr. 1961. Habitat and food of mousetrapped young *Rana pipiens* and *Rana clamitans*. Herpetologica 17:173–179.

Whitaker, J.O., Jr. 1971. A study of the western chorus frog, *Pseudacris triseriata*, in Vigo County, Indiana. Journal of Herpetology 5:127–150.

Whitaker, J.O., Jr. and D.C. Rubin. 1971. Food habits of *Plethodon jordani metcalfi* and *Plethodon jordani shermani* from North Carolina. Herpetologica 27:81–86.

Whitaker, J.O., Jr., W.W. Cudmore and B.B. Brown. 1982. Foods of larval, subadult, and adult smallmouth salamanders, *Ambystoma texanum*, from Vigo County, Indiana. Indiana Academy of Science 90:461–464.

Whitaker, J.O., Jr., D. Rubin and J.R. Munsee. 1977. Observations on food habits of four species of spadefoot toads, genus *Scaphiopus*. Herpetologica 33:468–475.

Whitaker, J.O., S.P. Cross, J.M. Slovlin and C. Maser. 1983. Food habits of the spotted frog *(Rana pretiosa)* from managed sites in Grant County, Oregon. Northwest Science 57:147–154.

Whitaker, J.O., Jr., C. Maser, R.M. Storm and J.J. Beatty. 1986. Food habits of clouded salamanders *(Aneides ferreus)* in Curry County, Oregon (Amphibia: Caudata: Plethodontidae). Great Basin Naturalist 46:228–240.

White, L.D., L.D. Harris, J.E. Johnston and D.G. Milchunas. 1975. Impact of site preparation on flatwoods wildlife habitat. Proceedings of the Southeastern Game and Fish Commission Conference 29:347–353.

White, M. and J.A. Kolb. 1974. A preliminary study of *Thamnophis* near Sagehen Creek, California. Copeia 1974:126–136.

White, R.L. 1977. Prey selection by the rough skinned newt *(Taricha granulosa)* in two pond types. Northwest Science 51:114–118.

Whitehurst, P.H. and B.A. Pierce. 1991. The relationship between allozyme variation and life-history traits of the spotted chorus frog, *Pseudacris clarkii*. Copeia 1991:1032–1039.

Whiteman, H.H. 1994a. Evolution of facultative paedomorphosis in salamanders. Quarterly Review of Biology 69:205–221.

Whiteman, H.H. 1994b. Evolution of alternative life histories in the tiger salamander, *Ambystoma tigrinum nebulosum*. Ph.D. dissertation. Purdue University, Lafayette, Indiana.

Whiteman, H.H. 1994c. Alternative evolutionary mechanisms of facultative paedomorphosis in salamanders. Quarterly Review of Biology 69:205–221.

Whiteman, H.H. 1997. Maintenance of polymorphism promoted by sex-specific fitness payoffs. Evolution 51:2039–2044.

Whiteman, H.H. and R.D. Howard. 1998. Conserving alternative amphibian phenotypes: is there anybody out there? Pp. 317–324. *In* Lannoo, M.J. (Ed.), Status and Conservation of Midwestern Amphibians. University of Iowa Press, Iowa City, Iowa.

Whiteman, H.H. and S.A. Wissinger. 1991. Differences in the antipredator behavior of three plethodontid salamanders to snake attack. Journal of Herpetology 25:352–355.

Whiteman, H.H., J.J. Gutrich and R.S. Moorman. 1999. Courtship behavior in a polymorphic population of the tiger salamander, *Ambystoma tigrinum nebulosum*. Journal of Herpetology 33:348–351.

Whiteman, H.H., R.D. Howard and K.A. Whitten. 1995. Effects of pH on embryo tolerance and adult behavior in the tiger salamander, *Ambystoma tigrinum tigrinum*. Canadian Journal of Zoology 73:1529–1537.

Whiteman, H.H., S.A. Wissinger and A.J. Bohonak. 1994. Seasonal movement patterns in a subalpine population of the tiger salamander, *Ambystoma tigrinum nebulosum*. Canadian Journal of Zoology 72:1780–1787.

Whiteman, H.H., S.A. Wissinger and W.S. Brown. 1996. Growth and foraging consequences of facultative paedomorphosis in the tiger salamander, *Ambystoma tigrinum nebulosum*. Evolutionary Ecology 10:429–422.

Whiteman, H.H., R.D. Howard, X. Spray and J. McGrady-Steed. 1998. Facultative paedomorphosis in an Indiana population of the eastern tiger salamander, *Ambystoma tigrinum tigrinum*. Herpetological Review 29:141–143.

Whitford, W.G. 1968. Physiological responses to temperature and desiccation in the endemic New Mexico plethodontids, *Plethodon neomexicanus* and *Aneides hardii*. Copeia 1968:247–251.

Whitford, W.G. and V.H. Hutchison. 1965. Body size and metabolic rate in salamanders. Physiological Zoology 40:127–133.

Whitford, W.G. and V.H. Hutchison. 1966a. Effect of photoperiod on pulmonary and cutaneous respiration in the spotted salamander, *Ambystoma maculatum*. Copeia 1966:53–58.

Whitford, W.G. and V.H. Hutchison. 1966b. Cutaneous and pulmonary gas exchange in ambystomatid salamanders. Copeia 1966:573–577.

Whitford, W.G. and V.H. Hutchison. 1967. Body size and metabolic rate in salamanders. Physiological Zoology 40:127–133.

Whitford, W.G. and J. Ludwig. 1976. The biota of the Baca geothermal site. Whitford Ecological Consultants, 2712 Ridgeway Court, Las Cruces, New Mexico.

Whitford, W.G. and M. Massey. 1970. Responses of a population of *Ambystoma tigrinum* to thermal and oxygen gradients. Herpetologica 26:372–376.

Whitford, W.G. and A. Vinegar. 1966. Homing, survivorship, and overwintering of larvae in spotted salamanders *Ambystoma maculatum*. Copeia 1966:515–519.

Whiting, M.J. and A.H. Price. 1994. Not the last picture show: new collection records from Paris, Texas (and other places). Herpetological Review 25:130.

Whitney, C.L. 1980. The role of the "encounter" call in spacing of Pacific treefrogs, *Hyla regilla*. Canadian Journal of Zoology 58:75–78.

Whitney, C.L. 1981. The monophasic call of *Hyla regilla* (Anura: Hylidae). Copeia 1981:230–233.

Whitney, C.L. and J.R. Krebs. 1975. Spacing and calling in Pacific treefrogs, *Hyla regilla*. Canadian Journal of Zoology 35:1519–1527.

Wicknick, J.A. 1995. Interspecific competition and territoriality between a widespread species of salamander and a species with limited range. Ph.D. dissertation. University of Southwestern Louisiana, Lafayette, Louisiana.

Widener, D. 1998. Short doesn't mean easy at Lost Pines Golf Club. Gulf Coast Golfer 15:18–19.

Wiedenheft, W.D. 1983. Life history and secondary production of tiger salamanders *(Ambystoma tigrinum)* in prairie pothole lakes. Master's thesis. North Dakota State University, Fargo, North Dakota.

Wied-Neuwied, M. Prinze zu. 1838. Reise in das innere Nord-Amerika in den Jahren 1832 bis 1834. Volume 1. J. Hoelscher, Coblenz.

Wiegmann, A. F.A. 1833. Herpetologische Beitrage. Isis von Oken 26:651–662.

Wiens, J.J. 1989. Ontogeny of the skeleton of *Spea bombifrons* (Anura: Pelobatidae). Journal of Morphology 202:29–51.

Wiens, J.J. and T.A. Titus. 1991. A phylogenetic analysis of *Spea* (Anura: Pelobatidae). Herpetologica 47:21–28.

Wiens, J.J., P.T. Chippindale and D.M. Hillis. 2003. When are phylogenetic analyses misled by convergence? A case study in Texas cave salamanders. Systematic Biology 52:501–514.

Wiest, J.A., Jr. 1982. Anuran succession at temporary ponds in a post oak-savanna region of Texas. Pp. 39–47. *In* Scott, N.J. (Ed.), Herpetological Communities. U.S. Fish and Wildlife Service Wildlife Research Report 13, Washington, D.C.

Wiewandt, T.A., C.H. Lowe and M.W. Larson. 1972. Occurrence of *Hypopachus variolosus* (Cope) in the short-tree forest of southern Sonora, Mexico. Herpetologica 28:162–164.

Wiggins, D.A. 1992. Foraging success of leopard frogs *(Rana pipiens)*. Journal of Herpetology 26:87–88.

Wigley, T.B., S.W. Sweeney and J.M. Sweeney. 1999. Final Report: Southeastern Coastal Plain amphibian survey. National Council for Air and Stream Improvement, Clemson, South Carolina.

Wiken, E.B. 1986. Terrestrial ecozones of Canada. Ecological Land Classification Series, Number 19, Environment Canada, Ottawa, Ontario, Canada.

Wilber, C.G. 1954. Toxicity of Sarin in bullfrogs. Science 120:322.

Wilbur, H.M. 1971. The ecological relationship of the salamander *Ambystoma laterale* to its all-female, gynogenetic associate. Evolution 25:168–179.

Wilbur, H.M. 1972. Competition, predation, and the structure of the *Ambystoma-Rana sylvatica* community. Ecology 53:3–21.

Wilbur, H.M. 1977a. Density-dependent aspects of growth and metamorphosis in *Bufo americanus*. Ecology 58:196–200.

Wilbur, H.M. 1977b. Interactions of food level and populations density in *Rana sylvatica*. Ecology 58:206–209.

Wilbur, H.M. 1977c. Propagule size, number, and dispersion pattern in *Ambystoma* and *Asclepias*. American Naturalist 111:43–68.

Wilbur, H.M. 1980. Complex life cycles. Annual Review of Ecology and Systematics 11:67–93.

Wilbur, H.M. 1982. Competition between tadpoles of *Hyla femoralis* and *Hyla gratiosa* in laboratory experiments. Ecology 63:278–282.

Wilbur, H.M. 1984. Complex life cycles and community organization in amphibians. Pp. 195–224. *In* Price, P.W., C.N. Slobodchikoff and W.S. Gaud (Eds.), A New Ecology: Novel Approaches to Interactive Systems. John Wiley and Sons, New York.

Wilbur, H.M. 1987. Regulation of structure in complex ecosystems: experimental temporary pond communities. Ecology 68:1437–1452.

Wilbur, H.M. 1990. Coping with chaos: toads in ephemeral ponds. TRENDS in Ecology and Evolution 5:37.

Wilbur, H.M. 1997. Experimental ecology of food webs: complex systems in temporary ponds. Ecology 78:2279–2302.

Wilbur, H.M. and R.A. Alford. 1985. Priority effects in experimental pond communities: responses of *Hyla* to *Bufo* and *Rana*. Ecology 66:1106–1114.

Wilbur, H.M. and J.P. Collins. 1973. Ecological aspects of amphibian metamorphosis. Science 182:1305–1314.

Wilbur, H.M. and J.E. Fauth. 1990. Experimental aquatic food webs: interactions between two predators and two prey. American Naturalist 135:176–204.

Wilbur, H.M. and P.J. Morin. 1988. Life history evolution in turtles. Pp. 387–439. *In* Gans, C. and R.B. Huey (Eds.), Biology of the Reptilia. Volume 16, Ecology B. Alan R. Liss, New York.

Wilbur, H.M., D.I. Rubenstein and L. Fairchild. 1978. Sexual selection in toads: the roles of female choice and male body size. Evolution 32:264–270.

Wilder, H.H. 1899. *Desmognathus fuscus* and *Spelerpes bilineatus* (Green). American Naturalist 33:231–246.

Wilder, I.W. 1913. The life history of *Desmognathus fusca*. Biological Bulletin 24:251–342.

Wilder, I.W. 1924a. The developmental history of *Eurycea bislineata* in western Mass. Copeia 133:77–80.

Wilder, I.W. 1924b. The relation of growth to metamorphosis in *Eurycea bislineata* (Green). Journal of Experimental Zoology 40:1–112.

Wilder, I.W. and E.R. Dunn. 1920. The correlation of lunglessness in salamanders with a mountain brook habitat. Copeia 84:63–68.

Wildy, E.L., A.R. Blaustein and H.R. Hepburn. 2001. Learned recognition of intraspecific predators in larval long-toed salamanders *Ambystoma macrodactylum*. Ethology 107:479–493.

Wildy, E.L., D.P. Chivers and A.R. Blaustein. 1999. Shifts in life-history traits as a response to cannibalism in larval long-toed salamanders (*Ambystoma macrodactylum*). Journal of Chemical Ecology 25:2337–2346.

Wildy, E.L., D.P. Chivers, J.M. Kiesecker and A.R. Blaustein. 1998. Cannibalism enhances growth in larval long-toed salamanders, (*Ambystoma macrodactylum*). Journal of Herpetology 32:286–289.

Wiley, E.O. 1978. The evolutionary species concept reconsidered. Systematic Zoology 27:17–26.

Wiley, E.O. 1981. Phylogenetics. The Theory and Practice of Phylogenetic Systematics. Wiley-Liss, New York.

Wiley, J.E. and M.L. Little. 2000. Replication patterns of the diploid-tetraploid treefrogs *Hyla chrysoscelis* and *Hyla versicolor*. Cytogenetics and Cell Genetics 88:11–14.

Wilkins, R.N. and N.P. Peterson. 2000. Factors related to amphibian occurrence and abundance in headwater streams draining second-growth Douglas-fir forests in southwestern Washington. Forest Ecology and Management 139:79–91.

Wilkinson, C., O. Linden, H. Cesar, G. Hodgson, J. Rubens and A.E. Strong. 1999. Ecological and socioeconomic impacts of 1988 coral mortality in the Indian Ocean: an ENSO impact and a warning of future change? Ambio 28:198–196.

Wilkinson, J.W. (Ed.). 1997. 1996–1997 collected DAPTF Working Group reports. DAPTF Report, Number 3, World Conservation Union (IUCN), Species Survival Commission, Open University, Milton Keynes, United Kingdom.

Wilkinson, R.F., C.L. Peterson, D. Moll and T. Holder. 1993. Reproductive biology of *Plethodon dorsalis* in northwestern Arkansas. Journal of Herpetology 27:85–87.

Wilks, B.J. 1963. Some aspects of the ecology and population dynamics of the pocket gopher (*Geomys bursarius*) in southern Texas. Texas Journal of Science 15:241–283.

Williams, A.A. 1980. Fluctuations in a population of the cave salamander, *Eurycea lucifuga*. National Speleological Society Bulletin 42: 49–52.

Williams, A.A. 1998. Geographic distribution: *Pseudacris crucifer* (spring peeper). Herpetological Review 29:108.

Williams, D.D. 1978. Helminth parasites of some anurans of northwestern Ohio. Proceedings of the Helminthological Society of Washington 45:141–142.

Williams, D.D. and S.J. Taft. 1980. Helminths of anurans from NW Wisconsin. Proceedings of the Helminthological Society of Washington 47:278.

Williams, E.C. 1947. The terrestrial form of the newt, *Triturus viridescens*, in the Chicago region. Natural History Miscellanea, Number 5, Chicago Academy of Sciences, Chicago, Illinois.

Williams, E.E., R. Highton and D.M. Cooper. 1968. Breakdown of polymorphism of the red-backed salamander on Long Island. Evolution 22:76–86.

Williams, J.E. 1961. The western lesser siren in Michigan. Copeia 1961:355.

Williams, J.W., III, D.L. Carlson, C.S. Williams, R.L. Gadson, L. Rollins-Smith and R.G. McKinnell. 1993a. Cytogenetic analysis of triploid renal carcinoma in *Rana pipiens*. Cytogenetics and Cell Genetics 64:18–22.

Williams, J.W., III, K.S. Tweedell, D. Sterling, N. Marshall, C.G. Christ, D.L. Carlson and R.G. McKinnell. 1996a. Oncogenic herpesvirus DNA absence in kidney cell lines established from the northern leopard frog, *Rana pipiens*. Diseases of Aquatic Organisms 27:1–4.

Williams, K.L. and R.E. Gordon. 1961. Natural dispersal of the salamander *Aneides aeneus*. Copeia 1961:353.

Williams, P.H. 1993. Measuring more of biodiversity for choosing conservation areas using taxonomic relatedness. Pp. 194–227. *In* Moon, T.Y. (Ed.), International Symposium on Biodiversity and Conservation. Korea University, Seoul, Korea.

Williams, P.H. 1996. Worldmap: software and help document 4.2. Distributed privately. Available from http://www.nhm.ac.uk/science/projects/worldmap.

Williams, P.H. and K.J. Gaston. 1994. Measuring more of biodiversity: can higher-taxon richness predict wholesale species richness? Biological Conservation 67:211–217.

Williams, P.H. and C.J. Humphries. 1994. Biological diversity, taxonomic relatedness, and endemism in conservation. Pp. 269–287. *In* Forey, P.L., C.J. Humphries and R.I. Vane-Wright (Eds.), Systematics and Conservation Evaluation. Clarendon Press, Oxford, United Kingdom.

Williams, P.H., K.J. Gaston and C.J. Humphries. 1994. Do conservationists and molecular biologists value differences between organisms in the same way? Biodiversity Letters 2:67–78.

Williams, P.H., C.J. Humphries and R.I. Vane-Wright. 1991. Measuring biodiversity: taxonomic relatedness for conservation priorities. Australian Systematic Botany 4:665–669.

Williams, P.H., R.I. Vane-Wright and C.J. Humphries. 1993b. Measuring biodiversity for choosing conservation areas. Pp. 309–328. *In* LaSalle, J. and I.D. Gauld (Eds.), Hymenoptera and Biodiversity. CAB International, Wallingford, United Kingdom.

Williams, P.H., D. Gibbons, C. Margules, A. Rebelo, C. Humphries and R. Pressey. 1996b. A comparison of richness hotspots, rarity hotspots, and complementary areas for conserving diversity of British birds. Conservation Biology 10:155–174.

Williams, P.K. 1973. Seasonal movements and population dynamics of four sympatric mole salamanders, genus *Ambystoma*. Ph.D. dissertation. Indiana University, Bloomington, Indiana.

Williams, R.D., J.E. Gates, C.H. Hocutt and G.T. Taylor. 1981. The hellbender: a nongame species in need of management. Wildlife Society Bulletin 9:94–100.

Williams, R.W. 1960. Observations on the life history of *Rhabdias sphaerocephala* Goodey, 1924 from *Bufo marinus* L., in the Bermuda Islands. Journal of Helminthology 34:93–98.

Williams, S.E. and J.-M. Hero. 1998. Rainforest frogs of the Australian wet tropics: guild classification and the ecological similarity of declining species. Proceedings of the Royal Society of London B 265:597–602.

Williams, S.R. 1972a. The Jemez Mountains salamander, *Plethodon neomexicanus*. Pp. 118–127. *In* Symposium on rare and endangered wildlife of the southwestern United States. New Mexico Department of Game and Fish, Albuquerque, New Mexico.

Williams, S.R. 1972b. Reproduction and ecology of the Jemez Mountains salamander, *Plethodon neomexicanus*. Master's thesis. University of New Mexico, Albuquerque.

Williams, S.R. 1973. *Plethodon neomexicanus*. Pp. 131.1–131.2. Catalogue of American Amphibians and Reptiles. Society for the Study of Amphibians and Reptiles, St. Louis, Missouri.

Williams, S.R. 1976. Comparative ecology and reproduction of the endemic New Mexico plethodontid salamanders, *Plethodon neomexicanus* and *Aneides hardii* (Amphibia, Urodela, Plethodontidae). Ph.D. dissertation. University of New Mexico, Albuquerque, New Mexico.

Williams, S.R. 1978. Comparative reproduction of the endemic New Mexico plethodontid salamanders, *Plethodon neomexicanus* and *Aneides hardii*. Journal of Herpetology 12:471–476.

Williams, T. 2000. False forests. Mother Jones May/June 2000:72–79.

Williams, T.A. and J.H. Larsen Jr. 1986. New function for the granular skin glands of the eastern long-toed salamander, *Ambystoma macrodactylum columbianum*. Journal of Experimental Zoology 239:329–333.

Williamson, G.K. and R.A. Moulis. 1979. Distribution of Georgia amphibians and reptiles in the Savannah Science Museum Collection. Savannah Science Museum, Special Publication, Number 1, Savannah, Georgia.

Williamson, G.K. and R.A. Moulis. 1994. Distribution of amphibians and reptiles in Georgia. Savannah Science Museum, Special Publication, Number 3, Volumes 1 and 2, Savannah, Georgia.

Williamson, I. 1995. Impacts of *B. marinus* on native anurans on the Darling Downs. CSIRO *Bufo* Project: An Overview of Research Outcomes. CSIRO, Unpublished report, Queensland University of Technology, Brisbane, Australia.

Williford, D. 1995a. Letter to Jana Grote, 3 October.

Williford, D. 1995b. Letter to Jana Grote, 8 December.

Williford, D. 1996. Letter to Steve Helfert, 16 January.

Willis, L.D. and A.V. Brown. 1985. Distribution and habitat requirements of the Ozark cavefish, *Amblyopsis rosae*. American Midland Naturalist 114:311–317.

Willis, Y.L., D.L. Moyle and T.S. Baskett. 1956. Emergence, breeding, hibernation, movements and transformation of the bullfrog, *Rana catesbeiana*, in Missouri. Copeia 1956:30–41.

Wilmott, G.B. 1933. The salamanders of Staten Island, N.Y., in 1931. Staten Island Institute of Arts and Science 6:161–164.

Wilson, A.G. and J.H. Larsen Jr. 1992. Geographic distribution: *Rhyacotriton olympicus* (Olympic salamander). Herpetological Review 23:85.

Wilson, A.G., Jr. and J.H. Larsen Jr. 1988. Activity and diet in seepage-dwelling Coeur d'Alene salamanders *(Plethodon vandykei idahoensis)*. Northwest Science 62:211–217.

Wilson, A.G., Jr. and P. Ohanjanian. 2002. *Plethodon idahoensis*. Pp. 741.1–741.4. Catalogue of American Amphibians and Reptiles. Society for the Study of Amphibians and Reptiles, St. Louis, Missouri.

Wilson, A.G. and E.M. Simon. 1985. Life history notes: *Plethodon vandykei idahoensis* (Coeur d' Alene salamander). Predation. Herpetological Review 16:111.

Wilson, D.J. and H. Lefcort. 1993. The effect of predator diet on the alarm response of red-legged frog, *Rana aurora*, tadpoles. Animal Behaviour 46:1017–1019.

Wilson, E.O. 1992. The Diversity of Life. W.W. Norton and Company, New York.

Wilson, E.O. 1994. Biophilia. Harvard University Press, Cambridge, Massachusetts.

Wilson, F.H. 1940. The life cycle of *Amphiuma* in the vicinity of New Orleans based on a study of the gonads and gonoducts (abstract). Anatomical Record 78 (Supplement):104.

Wilson, F.H. 1942. The cycle of egg and sperm production in *Amphiuma tridactylum* Cuvier (abstract). Anatomical Record 84 (Supplement):532.

Wilson, J. 1975a. Letter to Will W. Cox, 2 July.

Wilson, J. 1975b. Letter to Will W. Cox, 26 August.

Wilson, L.A. 1995. Land manager's guide to the amphibians and reptiles of the South. Nature Conservancy, Southeastern Region, Chapel Hill, North Carolina.

Wilson, L.D. 1966. Occurrence of the slimy salamander, *Plethodon glutinosus*, west of the Mississippi in Louisiana. Proceedings of the Louisiana Academy of Science 32:38–40.

Wilson, L.D. and J.R. McCranie. 1998. Amphibian population decline in a Honduran National Park. Froglog: Newsletter of the Declining Amphibian Populations Task Force 25:1–2.

Wilson, L.D. and L. Porras. 1983. The ecological impact of man on the south Florida herpetofauna. University of Kansas Museum of Natural History, Special Publication, Number 9, Lawrence, Kansas.

Wilson, L.D., J.R. McCranie, and M.R. Espinal, 2001. The ecogeography of the Honduran herpetofauna and the design of biotic reserves, pp. 109–158. In J. D. Johnson, R.G. Webb, and O. Flores-Villela (eds.) Mesoamerican Herpetology: Systematics, Zoogeography, and Conservation. Centennial Museum, University of Texas El Paso, Special Publication No. 1.

Wilson, R.E. 1976. An ecological study of *Ambystoma maculatum* and *Ambystoma jeffersonianum*. Ph.D. dissertation. Cornell University, Ithaca, New York.

Wilson, R.L. 1970. *Dicamptodon ensatus* feeding on a microtine. Journal of Herpetology 4:93.

Wiltenmuth, E.B. 1996. Agonistic and sensory behaviour of the salamander *Ensatina eschscholtzii* during asymmetrical contests. Animal Behaviour 52:841–850.

Wiltenmuth, E.B. 1997a. Agonistic behavior and use of cover by stream-dwelling larval salamanders *(Eurycea wilderae)*. American Midland Naturalist 21:531–582.

Wiltenmuth, E.B. 1997b. Effects of dehydration on behavior and habitat choice in three species of salamanders (genus *Plethodon*). Ph.D. dissertation. Northern Arizona University, Flagstaff, Arizona.

Wiltenmuth, E.B. and K.C. Nishikawa 1998. Geographical variation in agonistic behaviour in a ring species of salamander, *Ensatina eschscholtzii*. Animal Behaviour 55:1595–1606.

Windmiller, B.S. 1996. The pond, the forest, and the city: spotted salamander ecology and conservation in a human-dominated landscape. Ph.D. dissertation. Tufts University, Medford, Massachusetts.

Winker, K. 1996. The crumbling infrastructure of biodiversity: the avian example. Conservation Biology 10:703–707.

Winn, B., J.B. Jensen and S. Johnson. 1999. Geographic distribution: *Eleutherodactylus planirostris* (greenhouse frog). Herpetological Review 30:49.

Winokur, R.M. and S. Hillyard. 1992. Pelvic cutaneous musculature in toads of the genus *Bufo*. Copeia 1992:760–769.

Winter, D.A., W.M. Zawada and A.A. Johnson. 1986. Comparison of the symbiotic fauna of the family Plethodontidae in the Ouachita Mountains of western Arkansas. Proceedings of the Arkansas Academy of Science 40:82–85.

Winter, T.C. 1988. Conceptual framework for assessment of cumulative impacts on the hydrology of non-tidal wetlands. Environmental Management 12:605–620.

Wise, S.E., K.S. Siex, K.M. Brown and R.G. Jaeger. 1993. Recognition influences social interactions in red-spotted newts. Journal of Herpetology 27:149–154.

Wisely, S.M. and R.T. Golightly. 2003. Behavioral and ecological adaptations to water economy in two plethodontid salamanders, *Ensatina eschscholtzii* and *Batrachoseps attenuatus*. Journal of Herpetology 37:659–665.

Wissel, C. and T. Stephan. 1994. Bewertung des aussterberisikos und das minimum-viable-populations-konzept. Zeitschrift für Ökologie und Naturschutz 3:155–159.

Wissinger, S.A. and H.H. Whiteman. 1992. Fluctuation in a Rocky Mountain population of salamanders: anthropogenic acidification or natural variation? Journal of Herpetology 26:377–391.

Wissinger, S.A., A.J. Bohonak, H.H. Whiteman and W.S. Brown. 1999a. Subalpine wetlands in Colorado: habitat permanence, salamander predation, and invertebrate communities. Pp. 757–790. *In* Batzer, D.P., R.B. Rader and S.A. Wissinger (Eds.), Invertebrates in Freshwater Wetlands. John Wiley and Sons, New York.

Wissinger, S.A., G.B. Sparks, G.L. Rouse, W.S. Brown and H. Steltzer. 1996. Intraguild predation and cannibalism among larvae of detritivorous caddisflies in subalpine wetlands. Ecology 77: 2421–2430.

Wissinger, S.A., H.H. Whiteman, G.B. Sparks, G.L. Rouse and W.S. Brown. 1999b. Foraging tradeoffs along a predator-permanence gradient in subalpine wetlands. Ecology 80:2102–2116.

Withers, D. 1992. The Wyoming toad (*Bufo hemiophrys baxteri*): an analysis of habitat use and life history. Master's thesis. University of Wyoming, Laramie, Wyoming.

Withers, D.I. 1996. Tennessee Natural Heritage Program rare vertebrates list. Tennessee Department of Environment and Conservation, Division of Natural Heritage, Nashville, Tennessee.

Witting, L. and V. Loeschcke. 1995. The optimization of biodiversity conservation. Biological Conservation 71:205–207.

Wittliff, J.L. 1964. Venom constituents of *Bufo fowleri, Bufo valliceps* and their natural hybrids analysed by electrophoresis and chromography. Pp. 457–464. *In* Leone, C.A. (Ed.), Taxonomic Biochemistry and Serology. Ronald Press, New York.

Woiwood, I.P. and I. Hanski. 1992. Patterns of density dependence in moths and aphids. Journal of Animal Ecology 61:619–629.

Wojnowski, D. 2000. Hurricane Floyd's effect on the nesting success of the marbled salamander (*Ambystoma opacum*) at Fall's Lake, North Carolina. Journal of the Elisha Mitchell Scientific Society 116: 171–175.

Wolf, K.R. 1994. Untersuchungen zur Biologie der Erdkröte *Bufo bufo* L. unter besonderer Berücksichtigung des Einflusses von Migrationshindernissen auf das Wanderverhalten und die Entwicklung von vier Erdkrötenpopulationen im Stadtgebiet von Osnabrück. Mellen University Press, Hemmoor and Lewiston, Germany.

Wollmuth, L.P., L.I. Crawshaw, R.B. Forbes and D.A. Grahn. 1987. Temperature selection during development in a montane anuran species, *Rana cascadae*. Physiological Zoology 60:472–480.

Wolterstorff, W. 1935. Ueber eine eigentümliche Form des kalifornischen Wassermolches, *Taricha torosa* (Rathke). Blätter für Aquarien- und Terrarienkunde 46:178–184.

Wolton, A.C. 1941. Notes on some helminths from California Amphibia. Transactions of the American Microscopical Society 60: 53–57.

Woo, P.T.K. and J.P. Bogart. 1984. *Trypanosoma* sp. (Protozoa: Kinetoplastida) in *Hyla* (Anura) from eastern North America, with notes on their distribution and prevalences. Canadian Journal of Zoology 62:820–824.

Woo, P.T.K. and J.P. Bogart. 1986. Trypanosome infection in salamanders (Order: Caudata) from eastern North America with notes on the biology of *Trypanosoma ogwali* in *Ambystoma maculatum*. Canadian Journal of Zoology 64:121–127.

Wood, F.E., G. Carey and H.R. Rageot. 1955. The nesting and ovarian eggs of the dusky salamander, *Desmognathus f. fuscus* Raf. in southeastern Virginia. Virginia Journal of Science 1955:149–153.

Wood, J.T. 1945a. *Plethodon richmondi* in Greene County, Ohio. Copeia 1945:49.

Wood, J.T. 1945b. Ovarian eggs in *Plethodon richmondi*. Herpetologica 2:206–210.

Wood, J.T. 1947a. Description of juvenile *Plethodon glutinosus shermani* Stejneger. Herpetologica 3:188.

Wood, J.T. 1947b. Juveniles of *Plethodon jordani* Blatchley. Herpetologica 3:185–188.

Wood, J.T. 1948. *Microhyla c. carolinensis* in an ant nest. Herpetologica 4:226.

Wood, J.T. 1949. *Eurycea bislineata wilderae* Dunn. Herpetologica 5:61–62.

Wood, J.T. 1953a. The nesting of the two-lined salamander, *Eurycea bislineata*, on the Virginia Coastal Plain. Natural History Miscellanea, Number 122, Chicago Academy of Sciences, Chicago, Illinois.

Wood, J.T. 1953b. Observations on the complements of ova and nesting of the four-toed salamander in Virginia. American Naturalist 87:77–86.

Wood, J.T. 1955. The nesting of the four-toed salamander, *Hemidactylium scutatum* (Schlegel), in Virginia. American Midland Naturalist 53:381–389.

Wood, J.T. and R.F. Clarke. 1955. The dusky salamander: oophagy in nesting sites. Herpetologica 11:150–151.

Wood, J.T. and W.E. Duellman. 1951. Ovarian egg complements in the salamander *Eurycea bislineata rivicola* Mittleman. Copeia 1951:181.

Wood, J.T. and M.E. Fitzmaurice. 1948. Eggs, larvae, and attending females of *Desmognathus f. fuscus* in southwestern Ohio and southeastern Indiana. American Midland Naturalist 39:93–95.

Wood, J.T. and O.K. Goodwin. 1954. Observations on the abundance, food, and feeding behavior of the newt, *Notophthalmus viridescens viridescens* (Rafinesque) in Virginia. Journal of the Elisha Mitchell Science Society 70:27–30.

Wood, J.T. and H.N. McCutcheon. 1954. Ovarian egg complements and nests of the two-lined salamander, *Eurycea b. bislineata* x *cirrigera*, from southeastern Virginia. American Midland Naturalist 52:433–436.

Wood, J.T. and H.R. Rageot. 1955. The eggs of the slimy salamander in Isle of Wright County, Virginia. Virginia Journal of Science 6:85–87.

Wood, J.T. and H.R. Rageot. 1963. The nesting of the many-lined salamander in the Dismal Swamp. Virginia Journal of Science 14:121–125.

Wood, J.T. and F.E. Wood. 1955. Notes on the nests and nesting of the Carolina mountain dusky salamander in Tennessee and Virginia. Journal of the Tennessee Academy of Science 39:36–39.

Wood, K.V., J.D. Nichols, H.F. Percival and J.E. Hines 1998. Size-sex variation in survival rates and abundance of pig frogs, *Rana grylio*, in northern Florida wetlands. Journal of Herpetology 32:527–535.

Wood, T.S. 1977. Food habits of *Bufo canorus*. Master's thesis. Occidental College, Los Angeles, California.

Wood, W.F. 1934. Notes on the salamander, *Plethodon elongatus*. Copeia 1934:191.

Wood, W.F. 1935. Encounters with the western spadefoot, *Scaphiopus hammondii*, with a note on a few albino larvae. Copeia 1935:100–102.

Wood, W.F. 1936. *Aneides flavipunctatus* in burnt-over areas. Copeia 1936:171.

Wood, W.F. 1939. Amphibian records from northwestern California. Copeia 1939:110.

Woodbury, A.M. 1952. Amphibians and reptiles of the Great Salt Lake Valley. Herpetologica 8:42–50.

Woodruff, C.M., Jr. and P.L. Abbott. 1986. Stream piracy and evolution of the Edwards Aquifer along the Balcones Escarpment, central Texas. Pp. 77–89. *In* Abbott, P.L. and C.M. Woodruff Jr. (Eds.), The Balcones Escarpment: Geology, Hydrology, Ecology and Social Development in Central Texas. Geological Survey of America, San Diego, California.

Woodruff, D.S., M. Mulvey and M.W. Yipp. 1985. The continued introduction of intermediate host snails of *Schistosoma mansoni* into Hong Kong. Bulletin of the World Health Organization 63:621:622.

Woods, A.J., J.M. Omernik, D.D. Brown. 1996. Level III and IV ecoregions of Pennsylvania and the Blue Ridge Mountains, the Ridge and Valley, and the Central Appalachians of Virginia, West Virginia, and Maryland. U.S. Environmental Protection Agency, EPA/600R-96/077, National Health and Environmental Effects Research Laboratory, Corvallis, Oregon.

Woods, A.J., J.M. Omernik, J.A. Nesser, J. Shelden and S.H. Azevedo. 1999. Ecoregions of Montana. U.S. Geological Survey, Reston, Virginia.

Woods, A.J., J.M. Omernik, C.S. Brockman, T.D. Gerber, W.D. Hosteter and S.H. Azevedo. 1998. Ecoregions of Indiana and Ohio. U.S. Geological Survey, Reston, Virginia.

Woods, J.E. 1969. The ecology and natural history of a Mississippi population of *Aneides aeneus* and associated salamanders. Ph.D. dissertation. University of Southern Mississippi, Hattiesburg, Mississippi.

Woodward, B.D. 1982a. Sexual selection and nonrandom mating patterns in desert anurans (*Bufo woodhousei, Scaphiopus couchi, S. multiplicatus* and *S. bombifrons*). Copeia 1982:351–355.

Woodward, B.D. 1982b. Tadpole competition in a desert anuran community. Oecologia 54:96–100.

Woodward, B.D. 1982c. Local intraspecific variation in clutch parameters in the spotted salamander (*Ambystoma maculatum*). Copeia 1982: 157–160.

Woodward, B.D. 1983a. Tadpole size and predation in the Chihuahuan Desert. Southwestern Naturalist 29:232–233.

Woodward, B.D. 1983b. Predator-prey interactions and breeding-pond use by temporary-pond species in a desert anuran community. Ecology 64:1549–1555.

Woodward, B.D. 1984. Operational sex ratios and sex biased mortality in *Scaphiopus* (Pelobatidae). Southwestern Naturalist 29:232–233.

Woodward, B.D. 1987a. Clutch parameters and pond use in some Chihuahuan Desert anurans. Southwestern Naturalist 32:13–19.

Woodward, B.D. 1987b. Intra- and interspecific variation in spadefoot toad (*Scaphiopus*) clutch parameters. Southwestern Naturalist 32: 127–131.

Woodward, B.D. 1987c. Interactions between Woodhouse's toad tadpoles (*Bufo woodhousii*) of mixed sizes. Copeia 1987:380–385.

Woodward, B.D. and S. Mitchell. 1985. The distribution of *Bufo boreas* in New Mexico. Unpublished report, New Mexico Department of Game and Fish, Santa Fe, New Mexico.

Woodward, B.D. and S. Mitchell. 1990. Predation on frogs in breeding choruses. Southwestern Naturalist 35:449–450.

Woodward, B.D. and S.L. Mitchell. 1991. The community ecology of desert anurans. Pp. 223–248. *In* Polis, G.A. (Ed.), The Ecology of Desert Communities. University of Arizona Press, Tucson, Arizona.

Woody, D.R. and A. Mathis. 1997. Avoidance of areas labeled with chemical stimuli from damaged conspecifics by adult newts, *Notophthalmus viridescens*, in a natural habitat. Journal of Herpetology 31:316–318.

Woolbright, L.L. 1985. Patterns of nocturnal movement and calling by the tropical frog, *Eleutherodactylus coqui*. Herpetologica 41:1–9.

Woolbright, L.L. 1989. Sexual dimorphism in *Eleutherodactylus coqui*: selection pressures and growth rates. Herpetologica 45:68–74.

Woolbright, L.L. 1991. The impact of Hurricane Hugo on forest frogs in Puerto Rico. Biotropica 23:462–467.

Woolbright, L.L. and M.M. Stewart. 1987. Foraging success of the tropical frog *Eleutherodactylus coqui*: the cost of calling. Copeia 1987:69–75.

Wooley, H.P. 1971. The ecology and distribution of *Eurycea longicauda melanopleura* and *Eurycea lucifuga* in western Missouri. Herpetological Review 3:13.

Worrest, R.C. and D.J. Kimeldorf. 1975. Photoreactivation of potentially lethal, UV-induced damage to boreal toad *(Bufo boreas boreas)* tadpoles. Life Sciences 17:1545–1550.

Worrest, R.C. and D.J. Kimeldorf. 1976. Distortions in amphibian development induced by ultraviolet-B enhancement (290–315 nm) of a simulated solar spectrum. Photochemistry and Photobiology. 24: 377–382.

Worthing, P. 1993. Endangered and threatened wildlife and plants; finding on petition to list the spotted frog. Federal Register 58: 38553.

Worthington, R.D. 1968. Observations on the relative sizes of three species of salamander larvae in a Maryland pond. Herpetologica 24: 242–246.

Worthington, R.D. 1969. Additional observations on sympatric species of salamander larvae in a Maryland pond. Herpetologica 25:227–229.

Worthington, R.D. 1974. High incidence of anomalies in a natural population of spotted salamanders, *Ambystoma maculatum*. Herpetologica 30:216–220.

Worthington, S. and E.B. Worthington. 1933. Inland Waters of Africa. Macmillan Press, London.

Worthylake, K.M. and P. Hovingh. 1989. Mass mortality of salamanders *(Ambystoma tigrinum)* by bacteria *(Acinetobacter)* in an oligotrophic seepage mountain lake. Great Basin Naturalist 49:364–372.

Worton, B.J. 1989. Kernel methods for estimating the utilization distribution in home-range studies. Ecology 70:164–168.

Wright, A.H. 1908. Notes on the breeding habits of *Ambystoma punctatum*. Biological Bulletin 14:286.

Wright, A.H. 1914. North American Anura: Life histories of the Anura of Ithaca, New York. Publication 197, Carnegie Institute, Washington, D.C.

Wright, A.H. 1918. Notes on Muhlenberg's turtle. Copeia 52:7.

Wright, A.H. 1923. The Salientia of the Okefinokee Swamp, Georgia. Copeia 115:34.

Wright, A.H. 1924. A new bullfrog *(Rana heckscheri)* from Georgia and Florida. Proceedings of the Biological Society of Washington 37: 141–152.

Wright, A.H. 1926. The vertebrate life of the Okefinokee Swamp in relation to the Atlantic Coastal Plain. Ecology 7:77–95.

Wright, A.H. 1929. Synopsis and description of North American tadpoles. Proceedings of the United States National Museum 74:1–70.

Wright, A.H. 1932. Life Histories of the Frogs of the Okefinokee Swamp, Georgia. North American Salientia (Anura) Number 2. Macmillan Press, New York.

Wright, A.H. and A.A. Allen. 1909. The early breeding habits of *Ambystoma punctatum*. American Naturalist 43:687–692.

Wright, A.H. and J.M. Haber. 1922. The carnivorous habits of the purple salamander. Copeia 105:31–32.

Wright, A.H. and H. Trapido. 1940. *Pseudotriton montanus montanus* in West Virginia. Copeia 1940:133.

Wright, A.H. and A.A. Wright. 1924. A key to the eggs of the Salientia east of the Mississippi River. American Naturalist 58:375–381.

Wright, A.H. and A.A. Wright. 1933. Handbook of Frogs and Toads. Comstock Press, Ithaca, New York.

Wright, A.H. and A.A. Wright. 1949. Handbook of Frogs and Toads of the United States and Canada. Third edition. Comstock Publishing Associates, Ithaca, New York.

Wright, A.H. and A.A. Wright. 1995. Handbook of Frogs and Toads. Third edition. Comstock Publishing Associates, Ithaca, New York.

Wright, A.N. and K.R. Zamudio. 2002. Color pattern asymmetry as a correlate of habitat disturbance in spotted salamanders *(Ambystoma maculatum)*. Journal of Herpetology 36:129–133.

Wright, H.P. and G.S. Myers. 1927. *Rana areolata* at Bloomington, Indiana. Copeia 159:173–175.

Wright, J.W. 1966. Predation on the Colorado River toad, *Bufo alvarius*. Herpetologica 22:127–128.

Wright, S. 1931. Evolution in Mendelian populations. Genetics 16:97–159.

Wright, S. 1948. On the roles of directional and random changes in gene frequency in the genetics of populations. Evolution 2: 279–294.

Wrobel, D.J., W.F. Gergits and R.G. Jaeger. 1980. An experimental study of interference competition among terrestrial salamanders. Ecology 61:1034–1039.

Wu, C.-I. and W.-H., Li. 1985. Evidence for higher rates of nucleotide substitution in rodents than in man. Proceedings of the National Academy of Sciences 82:1741–1745.

Wygoda, M. 1989a. A comparative study of heating rates in arboreal and nonarboreal frogs. Journal of Herpetology 23:141–145.

Wygoda, M.L. 1984. Low cutaneous evaporative water loss in arboreal frogs. Physiological Zoology 57:329–337.

Wygoda, M.L. 1988a. Adaptive control of water loss resistance in an arboreal frog. Herpetologica 44:251–257.

Wygoda, M.L. 1988b. A comparative study of time constants of cooling in green treefrogs *(Hyla cinerea)* and southern leopard frogs *(Rana sphenocephala)*. Herpetologica 44:261–265.

Wygoda, M.L. 1989b. Body temperature in arboreal and nonarboreal anuran amphibians: differential effects of a small change in water vapor density. Journal of Thermal Biology 14:239–242.

Wygoda, M.L. and R.H. Garman. 1993. Ontogeny of resistance to water loss in green tree frogs, *Hyla cinerea*. Herpetologica 49:365–374.

Wygoda, M.L. and A.A. Williams. 1991. Body temperature in free-ranging green tree frogs *(Hyla cinerea)*: a comparison with "typical" frogs. Herpetologica 47:328–335.

Wyman, R.L. 1971. The courtship behavior of the small-mouthed salamander, *Ambystoma texanum*. Herpetologica 27:491–498.

Wyman, R.L. 1988a. Salamanders, soil acidification and forest ecology. Newsletter of the Edmund Niles Huyck Preserve, Winter 1987–1988, Rensselaerville, New York.

Wyman, R.L. 1988b. Soil acidity and moisture and the distribution of amphibians in five forests of southcentral New York. Copeia 1988: 394–399.

Wyman, R.L. 1990. What's happening to the amphibians? Conservation Biology 4:350–352.

Wyman, R.L. and D.S. Hawksley-Lescault. 1987. Soil acidity affects distribution, behavior, and physiology of the salamander *Plethodon cinereus*. Ecology 68:1819–1827.

Wyman, R.L. and J.H. Thrall. 1972. Sound production by the spotted salamander, *Ambystoma maculatum*. Herpetologica 28:210–212.

Wynn, A.H. 1991. Shenandoah salamander. Pp. 439–442. *In* Terwilliger, K. (Coordinator), Virginia's Endangered Species. McDonald and Woodward Publishing Company, Blacksburg, Virginia.

Wynn, A.H., R. Highton and J.F. Jacobs. 1988. A new species of rock-crevice dwelling *Plethodon* from Pigeon Mountain, Georgia. Herpetologica 44:135–143.

Yaffee, S.L. 1982. Prohibitive Policy—Implementing the Federal Endangered Species Act. MIT Press, Cambridge, Massachusetts.

Yaffee, S.L., A.F. Phillips, I.C. Frantz, P.W. Hardy, S.M. Maleki and B.E. Thorpe. 1996. Ecosystem Management in the United States. Island Press, Washington, D.C.

Yamaguti, S. 1935. Studies on the helminth fauna of Japan. Part 10. Amphibian nematodes. Japanese Journal of Zoology 6:387–392.

Yamaguti, S. 1961. Nematodes of vertebrates. Pp. 1–679. *In* Systema Helminthum. Volume 3, Part 1. Interscience Publishing, New York.

Yamaguti, S. 1971. Synopsis of Digenetic Trematodes of Vertebrates. Volume 1. Keigaku Publishing Company, Tokyo, Japan.

Yamamura, K. 1999. Key-factor/key-stage analysis for life table data. Ecology 80:533–537.

Yan, N.D., W. Keller, N.M. Scully, D.R.S. Lean and P.J. Dillon. 1996. Increased UV-B penetration in a lake owing to drought-induced acidification. Nature 381:141–143.

Yanev, K.P. 1978. Evolutionary studies of the plethodontid salamander genus *Batrachoseps*. Ph.D. dissertation. University of California, Berkeley, California.

Yanev, K.P. 1980. Biogeography and distribution of three parapatric salamander species in coastal and borderland California. Pp. 531–550. *In* Power, D.M. (Ed.), The California Islands: Proceedings of a Multidisciplinary Symposium. Santa Barbara Museum of Natural History, Santa Barbara, California.

Yanev, K.P. and D.B. Wake. 1981. Genic differentiation in a relict desert salamander, *Batrachoseps campi*. Herpetologica 37:16–28.

Yang, X.B. and D.O. TeBeest. 1992. Green treefrogs as vectors of *Colletotrichum gloeosporioides*. Plant Disease 76:1266–1269.

Yang, X.B., D.O. TeBeest and E.L. Moore. 1992. Biological vectors for dispersal of *Colletotrichium gloeosporioides*. Proceeding of the Arkansas Academy of Sciences 46:96–99.

Yeatman, H.C. and H.B. Miller. 1985. A naturally metamorphosed *Gyrinophilus palleucus* from the type-locality. Journal of Herpetology 19:304–306.

Yeo, J.J. and C.R. Peterson. 1998. Amphibian and reptile distribution and habitat relationships in the Lost River Mountains and Challis-Lemhi Resource Areas. Idaho Bureau of Land Management, Technical Bulletin 98-10, Boise, Idaho.

Yoffe, E. 1992. Silence of the frogs. New York Times Magazine 6:36–38, 64–76.

Yom, S. 1990. It's no use telling these jumpy lovers to look before you leap. Wall Street Journal, 26 July.

Young, B.E., K.R. Lips, J.K. Reaser, R. Ibáñez, A.W. Salas, J.R. Cedeño, L.A. Coloma, S. Ron, E. LaMarca, J.R. Meyer, A. Muñoz, F. Bolaños, G. Chavez and D. Romo. 2001. Population declines and priorities for amphibian conservation in Latin America. Conservation Biology 15:1213–1223.

Young, F.N. and J.R. Zimmerman. 1956. Variations in temperature in small aquatic situations. Ecology 37:609–611.

Young, J.E. and B.I. Crother. 2001. Allozyme evidence for the separation of *Rana areolata* and *Rana capito* and for the resurrection of *Rana sevosa*. Copeia 2001:382–388.

Young, K.A. 2001. Defining units of conservation for intraspecific biodiversity: reply to Dimmick et al. Conservation Biology 15:784–787.

Zacuto, B.J. 1975. The status of the African clawed frog (*Xenopus laevis*) in Agua Dulce and Soledad Canyons. California Department of Fish and Game, Unpublished report, Sacramento, California.

Zaffaroni, N.P., E. Arias and T. Zavanella. 1992. Natural variation in the limb skeletal pattern of the crested newt, *Triturus carnifex* (Amphibia: Salamandridae). Journal of Morphology 213:265–273.

Zaga, A., E.E. Little, C.F. Rabeni and M.R. Ellersieck. 1998. Photoenhanced toxicity of a carbamate insecticide to early life stage anuran amphibians. Environmental Toxicology and Chemistry 17: 2543–2553.

Zaldivar-Riverson, A., V. Leon-Regagnon and A. Nieto-Montes de Oca. 2004. Phylogeny of Mexican coastal leopard frogs of the *Rana berlandieri* group based on mtDNA sequences. Molecular Phylogenetics and Evolution 30:38–49.

Zambernard, J. and R.G. McKinnell. 1969. "Virus-free" renal tumors obtained from pre-hibernating leopard frogs of known geographic origin. Cancer Research 29:653–657.

Zambernard, J., A.E. Vater and R.G. McKinnell. 1966. The fine structure of nuclear and cytoplasmic inclusions in primary renal tumors of mutant leopard frogs. Cancer Research 26:1688–1700.

Zampella, R.A. and J.F. Bunnell. 2000. The distribution of anurans in two river systems of a Coastal Plain watershed. Journal of Herpetology 34:210–221.

Zamudio, K.R. and W.K. Savage. 2003. Historical isolation, range expansion, and secondary contact of two highly divergent mitochondrial lineages in spotted salamanders (*Ambystoma maculatum*). Evolution 57:1631–1652.

Zappalorti, R.T. 1998. Ecology and habitat use of tiger salamander (*Ambystoma tigrinum*) in the coastal Pine Barrens of southern New Jersey. Herpetological Associates, Unpublished report, Beachwood, New Jersey.

Zappler, L. 1994. Watch out for harnessed toads. Smithville Times, 10 March.

Zar, J.H. 1996. Biostatistical Analysis. Third edition. Prentice Hall, Upper Saddle River, New Jersey.

Zar, J.H. 1999. Biostatistical analysis. Fourth edition. Prentice Hall, Upper Saddle River, New Jersey.

Zaret, T.M. 1980. Predation and Freshwater Communities. Yale University Press, New Haven, Connecticut.

Zell, G.A. 1986. The clawed frog: an exotic from Africa invades Virginia. Virginia Wildlife, February:28–29.

Zelmer, D.A., E.J. Wetzel and G.W. Esch. 1999. The role of habitat in structuring *Halipegus occidualis* metapopulations in the green frog. Journal of Parasitology 85:19–24.

Zenisek, C.J. 1963. A study of the natural history and ecology of the leopard frog, *Rana pipiens* Schraber. Ph.D. dissertation. Ohio State University, Columbus, Ohio.

Zerba, K.E. and J.P. Collins. 1992. Spatial heterogeneity and individual variation in diet of an aquatic top predator. Ecology 73: 268–279.

Zhang, Y. and O.A. Ryder. 1995. Different rates of mitochondrial DNA sequence evolution in Kirk's dik-dik (*Madoqua kirkii*) populations. Molecular Phylogenetics and Evolution 4:291–297.

Zhao, E.-M. and K. Adler. 1993. Herpetology of China. Contributions to Herpetology 10:1–522.

Ziehmer, B. and T. Johnson. 1992. Status of the Ozark hellbender in Missouri. Missouri Department of Conservation, Natural History Division, Unpublished report, Jefferson City, Missouri.

Zielinski, W.J. and G.T. Barthalmus. 1989. African clawed frog skin compounds: antipredatory effects on African and North American water snakes. Animal Behaviour 38:1083–1086.

Zimmerman, B.L. 1994. Auto strip transects. Pp. 92–97. *In* Heyer, W.R., M.A. Donnelly, R.W. McDiarmid, L-A.C. Hayek and M.S. Foster (Eds.), Measuring and Monitoring Biological Diversity: Standard Methods for Amphibians. Smithsonian Institution Press, Washington, D.C.

Zimskind, P.D. and R.M. Schisgall. 1955. Photorecovery from ultraviolet induced pigmentation changes in anuran larvae. Journal of Cellular and Comparative Physiology 45:167–175.

Zink, R.M. and H.W. Kale. 1995. Conservation genetics of the extinct dusky seaside sparrow *Ammodramus maritimus nigrescens*. Biological Conservation 74:69–71.

Zink, R.M., G.F. Barrowclough, J.L. Atwood and R.C. Blackwell-Rago. 2000. Genetics, taxonomy, and conservation of the threatened California gnatcatcher. Conservation Biology 14:1394–1405.

Zug, G.R. 1985. Anuran locomotion fatigue and jumping performance. Herpetologica 41:188–194.

Zug, G.R. 1993. Herpetology. Academic Press, San Diego, California.

Zug, G.R. and P.B. Zug. 1979. The marine toad, *Bufo marinus*: a natural history resume of native populations. Smithsonian Contributions in Zoology, Number 284, Smithsonian Institution, Washington, D.C.

Zug, G.R., E. Lindgren and J.R. Pippet. 1975. Distribution and ecology of the marine toad, *Bufo marinus*, in Papua New Guinea. Pacific Science 29: 31–50.

Zuiderwijk, A., G. Smit and H. van den Bogert. 1993. Die Anlage künstlicher Eiablageplätze: eine einfache Möglichkeit zum Schutz der Ringelnatter (*Natrix natrix* L., 1758). Mertensiella 3:227–234.

Zweifel, R.G. 1949. Comparison of food habits of *Ensatina eschscholtzii* and *Aneides lugubris*. Copeia 1949:285–287.

Zweifel, R.G. 1955. Ecology, distribution, and systematics of frogs of the *Rana boylii* group. University of California Publications in Zoology 54:207–292.

Zweifel, R.G. 1956a. A survey of the frogs of the *augusti* group, genus *Eleutherodactylus*. American Museum Novitates, Number 1813, American Museum of Natural History, New York.

Zweifel, R.G. 1956b. Two pelobatid frogs from the Tertiary of North America and their relationships to fossil and recent forms. American Museum Novitates, Number 1762, American Museum of Natural History, New York.

Zweifel, R.G. 1958. Results of the Puritan-American Museum of Natural History Expedition to western Mexico. 2. Notes on reptiles and amphibians from the Pacific Coastal islands of Baja California. American Museum Novitates, Number 1895, American Museum of Natural History, New York.

Zweifel, R.G. 1961. Larval development of the treefrogs *Hyla arenicolor* and *Hyla wrightorum*. American Museum Novitates, Number 2056, American Museum of Natural History, New York.

Zweifel, R.G. 1967. *Eleutherodactylus augusti*. Pp. 41.1–41.4. Catalogue of American Amphibians and Reptiles. Society for the Study of Amphibians and Reptiles, St. Louis, Missouri.

Zweifel, R.G. 1968a. Effects of temperature, body size and hybridization on mating call of toads, *Bufo a. americanus* and *Bufo woodhousei fowleri*. Copeia 1968:269–285.

Zweifel, R.G. 1968b. Reproductive biology of anurans of the arid southwest, with adaptation of embryos to temperature. Bulletin of the Museum of Natural History 140:1–64.

Zweifel, R.G. 1968c. *Rana tarahumarae*. Pp. 66.1–66.2. Catalogue of American Amphibians and Reptiles. Society for the Study of Amphibians and Reptiles, St. Louis, Missouri.

Zweifel, R.G. 1970. Descriptive notes on larvae of toads of the *debilis* group, genus *Bufo*. American Museum Novitates, Number 2407, American Museum of Natural History, New York.

Zweifel, R.G. 1977. Upper thermal tolerances of anuran embryos in relation to stage of development and breeding habits. American Museum Novitates, Number 2617, American Museum of Natural History, New York.

Zweifel, R.G. 1989. Calling by the frog, *Rana sylvatica* outside the breeding season. Journal of Herpetology 23:185–186.

Zwinger, A. 1986. John Xántus: The Fort Tejon Letters 1857–1859. University of Arizona Press, Tucson, Arizona.

INDEX